Steroids in the Laboratory and Clinical Practice

Steroids in the Laboratory and Clinical Practice

John William Honour
Institute for Women's Health,
University College London,
London, United Kingdom

ELSEVIER

Elsevier
Radarweg 29, PO Box 211, 1000 AE Amsterdam, Netherlands
The Boulevard, Langford Lane, Kidlington, Oxford OX5 1GB, United Kingdom
50 Hampshire Street, 5th Floor, Cambridge, MA 02139, United States

Notices
Knowledge and best practice in this field are constantly changing. As new research and experience broaden
our understanding, changes in research methods, professional practices, or medical treatment may become
necessary.

Practitioners and researchers must always rely on their own experience and knowledge in evaluating and using
any information, methods, compounds, or experiments described herein. In using such information or
methods they should be mindful of their own safety and the safety of others, including parties for whom they
have a professional responsibility.

To the fullest extent of the law, neither the Publisher nor the authors, contributors, or editors, assume any
liability for any injury and/or damage to persons or property as a matter of products liability, negligence or
otherwise, or from any use or operation of any methods, products, instructions, or ideas contained in the
material herein.

ISBN: 978-0-12-818124-9

For information on all Elsevier publications
visit our website at https://www.elsevier.com/books-and-journals

Publisher: Stacy Masucci
Acquisitions Editor: Patricia M. Osborn
Editorial Project Manager: Samantha Allard
Production Project Manager: Selvaraj Raviraj
Cover Designer: Miles Hitchen

Typeset by STRAIVE, India

Working together
to grow libraries in
developing countries

www.elsevier.com • www.bookaid.org

I have dedicated the book to Cedric Shackleton (my mentor, colleague, and friend) for the introduction to a fascinating career. Finally I owe my deepest thanks to my family for their love and support. My warmest gratitude to Pauline (my wife of 55 years) particularly for her tolerance during the hours I was at my computer whilst writing this book during a pandemic. My son James, a commercial air line pilot will now understand the difficulties when father is away from home as was often the case during my career. I doubt if he will ever need to refer to the book except to boost his paramedic knowledge.

Contents

Part III
Steroids in clinical practice

About the Author

John William Honour, PhD FRCPath, received his bachelor's degree in applied biochemistry from Bath University in 1972 where he conducted his first research in steroids (with Professor M.W. Hardisty) by studying changes in enzyme activity of lamprey testes as they migrated from salt to fresh water to reproduce and die. After graduation Dr. Honour started his professional life as Senior Research Officer at the MRC Clinical Research Centre (CRC) in Harrow, England, where he completed his PhD in biochemistry from the University of London in 1978 supervised by Cedric Shackleton and Professor Vivian James. His education in the application of mass spectrometry for steroids and continued interests were due to Cedric. The PhD thesis was titled "Application of gas chromatography and mass spectrometry to the investigation of human mineralocorticosteroid metabolism." Dr. Honour had worked with pediatrician Dr. Bernard Valman to study the role of aldosterone in salt loss in the urine of preterm infants that gave rise to his first steroid publication. The urine steroid profile of a patient with 17-hydroxylase deficiency was shown to reflect the enterohepatic circulation of corticosterone and bacterial metabolism of steroids in the intestine, as previously described in rats with corticosterone as the principal glucocorticoid. This started another area of future interest to the author.

Cedric was among the first to use capillary column gas chromatography and at the CRC they also had a good mass spectrometry facility. The main mass spectrometer was a magnetic sector Varian MAT 731 Mattauch-Herzog double focusing instrument that filled a room of 4×7 m. A Becker 409 gas chromatograph was modified for capillary columns held in place at both ends with heat shrink PTFE tubing and had automated injection with dropping sample vials cut from hematocrit tubes. GC-MS was in 1972, really new, and remains a powerful tool with better resolution than high-pressure liquid chromatography popularized later. The success of HPLC came many years later when spray techniques at the interface with a mass spectrometer enabled coupling of the two elements. LC-MS does not always require time-consuming steps of derivative formation.

Dr. Honour also collaborated with Dr. David Grant who moved from the CRC to the children's hospital at Great Ormond St, London (GOSH). He then provided steroid assays for clinicians at GOSH including Mike Dillon. Professor David Kirk was a collaborator of the author. He ran the Medical Research Council Steroid Reference Collection that was a great source of small amounts of pure steroids. Upon the sad death of David, the collection passed to the Sigma-Aldrich Collection of Rare Chemicals, but there appears to be no catalogue of the contents and access sadly seems to no longer exist.

Cedric was previously a postdoctoral fellow at the Karolinska Institute and was already into the use of solid-phase extractions (SPE) of steroids using Amberlite XAD-2 resin that replaced toxic solvents that had been used previously. The training Dr. Honour had from Cedric proved valuable for life, and he remained a collaborator and friend. Pauline Honour (Dr. Honour's wife) was a bacteriologist with a veterinary background, which was valuable to her when she was the first to see the anaerobe, *Clostridium difficile*, in biological samples from patients with irritable bowel syndrome. At a meeting of the Anaerobe Discussion Group in London, Dr. Honour set up a collaboration with Victor Bokkenheuser, and his name is encountered frequently in publications with Dr. Honour and Cedric.

Dr. Honour was awarded the Postdoctoral Fellowship of the American Heart Foundation with Edward Biglieri at University of California, San Francisco. He explored further there in 1978–79 the role of bacterial steroid metabolism in the intestine having recognized the similarity of corticosterone metabolism in patients with 17-hydroxlase deficiency to that in rats. Reactions such as 21-dehydroxlation had earlier been shown by Jan Sjovall and colleagues at the Karolinska Institute to be absent in germ-free rats. On returning to the CRC, Dr. Honour was able to conduct experiments that showed experimental hypertension in rats could be prevented by treatment with antibiotics. This was the first time that bacterial steroid metabolism was linked to hypertension, predating claims made in 2015 when a new term, the microbiome, was popularized. In 1980–82, with Ursula and Detlev Ganten in the laboratory of Professor Paul Vecsei in Heidelberg, antibiotics were also shown to prevent hypertension in stroke-prone rats. The changes in the intestinal bacterial flora were assumed to affect the nature of the steroids reabsorbed from the intestine to reduce the impact on blood pressure. Peter Borriello at the CRC started work to characterize the intestinal flora of rats, but the work at that time was laborious and inconclusive, with most of the organisms being anaerobic and acutely sensitive to air. Work in collaboration with Pauline and Peter Borriello would have taken a different journey if 16S RNA sequencing had been available then. At that time bacteriology was limited to culture in liquid and on plates, crude biochemical tests making gas and pH changes, microscopy, and antibiotic sensitivity. After 30 years, the gut microbiome was acknowledged to contribute to many pathological processes. The number of bacteria in the gut is realized to exceed the number of cells in the human body and much research on the impact of the microbiome on homeostasis and pathology will follow. The technology for bacterial identification had advanced considerably with 16S RNA sequencing being the means to identification of the organisms in biological materials.

In 1983 Dr. Honour accepted a senior lecturer position at the Middlesex Hospital Medical School in London, England. The post had responsibility for steroid endocrinology in the Department of Clinical Chemistry (Head Arthur Miller) then based in the Courtauld Institute (directed by Professor Peter Campbell) and allowed research and development in the Cobbold Laboratories, which at the time was directed by Professor (later Sir) John Nabarro. He collaborated in research for the early use of in vitro fertilization (Professor Howard Jacobs), pediatric endocrinology (Professor Charles Brooks), and hypertension (Professor JDH [Willie] Slater) in the illustrious company of Professors Roger Ekins (who should have had a Nobel Prize for hormone-binding assays akin to the immunoassay developed by Rosalyn Yallow and Sol Berson), Jeffrey O'Riordan, Debra Donike, and Ivan Roitt were part of the endocrine establishment at the Middlesex. Aldosterone had been discovered at the Middlesex Hospital in 1953 by Jim and Sylvia Tait. Earlier work by E.C. Dodds between 1922 and 1952 had looked at many endocrine problems including the use of stilboesterol to treat certain cancers. Dr. Honour recalls a framed set of capon feathers in the Courtauld Institute that showed how Dodds demonstrated the feminization of chickens from the administration of a reduced phenanthrene (an early endocrine disruptor). The feathers and a watercolor of the frame were on the wall of Professor Campbell's office were a daily reminder of this early work.

Professor Charles Brook had a Carlo-Erba Fractovap gas chromatograph previously used by Chris Kelnar, but having established use in my laboratory Dr. Honour realized the oven was lined with asbestos which in 1983 was recognized as a hazard so Dr. Honour was given funds by the Courtauld Institute to purchase a Packard Model 437A. Funding was granted from the Middlesex Hospital Special Trustees to

buy a reconditioned VG 70S mass spectrometer, another monster taking up 3×5 m floor space. SPE had progressed and SepPak C18 cartridges were in use. Polar steroids such as 1- and 6-hydroxy-cortisol metabolites were extractable for the first time.

The Middlesex Hospital was closed in 1988, and Dr. Honour continued his steroid research in University College London Hospital and took more responsibility for clinical liaison and Clinical Chemistry department management. Research into the molecular genetics of steroidogenic enzymes started in 1988 with the appointment of Dr. Gillian Rumsby. His publication record and grant supported funds for research led to a promotion to Reader in Steroid Endocrinology in 1991, which became an honorary contract when moving to Consultant Clinical Scientist. In 1991 Dr. Honour was awarded Fellow of the Royal College of Pathology on the basis of his published works.

Dr. Honour at the time of his retirement in 2011 had 45 years of experience in analyzing steroids mainly by gas chromatography and mass spectrometry and applying these results to the investigation of patients with endocrine disorders. From 1990 the laboratory was equipped with bench top GC-MS systems (HP-MSD, Shimadzu 2010 and 5050 models). His work has included the diagnosis of inherited disorders of blood pressure, sex determination, infertility, osteoporosis, and salt balance in the body. Dr. Honour, as shown above, has long had an interest in the ability of bacteria to change steroids in the body and has used bacteria in test-tubes to achieve some changes in steroid structure that could not be easily achieved with chemicals (15-hydroxylation with *Streptomyces roseochromogenes*).

Several research projects have examined the safety of inhaled steroids with respect to effects on the hypothalamic-pituitary-adrenal axis. In children taking inhaled corticosteroids, growth was suppressed. Blood samples were taken at regular intervals throughout the night to see whether growth hormone secretion was affected. Using these samples, the pattern of cortisol release during the night was studied too. Cortisol was released at 90-minute intervals; the peaks increased in amplitude with the last and highest peak being around 0800h. ICS were found to suppress the pattern of cortisol; the number of peaks lost related to the dose of ICS. This work revealed several important facts about the HPA axis. The diurnal rhythm of cortisol is really an ultradian rhythm. Further studies showed that not all subjects peaked at 0800h; the variation was from 0600 to 1000h. By merging data from several subjects, a wide diurnal pattern was seen with low concentrations (<150 nmol/L) at midnight and a range of 200–800 nmol/L because the pulsatile patterns of individuals were out of phase. A collaboration with Peter

Hindmarsh at GOSH has led to the use of an insulin pump to deliver hydrocortisone subcutaneously in a more physiological manner. Originally tested in patients with congenital adrenal hyperplasia, it is now being used by patients with Addison disease and hypopituitarism.

In collaboration with Willie Slater, we pursued experiments on low-dose ACTH tests (1 µg instead of 250 µg). In initial pulsing at 1-h intervals, cortisol concentrations did not return to baseline before the next incremental dose. Research Fellows in 1990–98 with Professor Brook looked to refine the test. A dose of 1 µg per meter squared of body surface gave a peak comparable with that achieved at 30 min from 250 µg. When precursors were measured, the peaks were earlier than 30 min. Although we reported this in 1998, the protocols in the literature have not refined the test to include samples earlier than 30 min, and the literature is very confusing.

In 1999, Dr. Honour became Consultant Clinical Scientist in Clinical Biochemistry at UCLH and Head of the specialist service for steroid endocrinology. The steroid unit had clinical pathology accreditation and was designated a supraregional assay (SAS) center for plasma aldosterone, plasma renin activity, 17-hydroxyprogesterone, and urinary steroid profiles. In addition to reporting quantitative results, the service offered interpretation of the results and advice for further investigations. This is possible from extensive research and experience in the investigations of disorders of the adrenal cortex and gonads. Comprehensive reference ranges were established and validated for all of the tests. The steroid laboratory maintains strong links with the clinical endocrinologists at UCL and the Hospital for Children at Great Ormond Street. Research in the unit covered the validation and use of mass spectrometry in steroid analysis for disorders of sex determination, adrenal diseases, polycystic ovary syndrome, safety of corticosteroids in treatment of asthma, links between birthweight and cardiovascular risk, and the enterohepatic circulation of steroids.

Dr. Honour was successful with many research grant applications with funding to develop a sensitive estradiol immunoassay, exploit the application of mass spectrometry for steroid analysis, study peripheral 21-hydroxylase, adrenal function in HIV, children with asthma, children in relation to birthweight (Avon Longitudinal Study Parent and Child—ALSPAC), and osteoporosis from a vegetarian diet. In April 1999, with collaborator Dr. Jonathan Fry, an international symposium "Neurosteroids" was held at UCL bringing together many "stars" in the field at that time. An application in 2002 for funds from a government scheme entitled "New and Emerging Analytical Technology" (NEAT) to purchase an LC-MS/MS system was declined because it was not "new"!

Dr. Honour traveled widely to scientific conferences across Europe, the United States, and Australia. This afforded the opportunity for networking with other steroid "anoraks," an important element in recognition and exchange of results and ideas. He presented 177 papers at scientific conferences.

Dr. Honour has more than 170 refereed publications and 19 book chapters covering a range of analytical, genetic, and clinical aspects of steroids. He was scientific adviser to UKNEQAS for immunoassays, steroid accuracy, and pediatric investigations. Dr. Honour served on the Editorial Board of Clinical Endocrinology. He served on two working parties for the European Society for Paediatric Endocrinology with the Lawson Wilkins Pediatric Endocrine Society that produced guidelines on congenital adrenal hyperplasia and newborn screening. Dr. Honour was Treasurer of the British Society for Paediatric Endocrinology and Diabetes (BSPED) for 12 years, introducing a sponsorship that took reserves from £200 to £200,000. An EQA scheme for urinary steroids was operated from UCL globally with the collaboration of the Dutch Foundation for Quality in Clinical Chemistry (SKML). Dr. Honour has maintained an interest in the detection of abuse of anabolic steroids in sport and has acted as an expert witness in many cases of drug abuse in sport. Dr. Honour was Associate Editor of the *Annals of Clinical Biochemistry* (2010–2020) and is a member of the Independent Review panel for UK Anti-doping. Dr. Honour is Honorary Senior Research Associate at the Institute for Women's Health at University College London.

This book is somewhat a summary of the broad experience of steroid endocrinology over 45 years. The author takes full responsibility for errors and omissions. The book was partly written in the year 2020/2021 when the world was struck with a pandemic through a SARS virus (loosely named COVID-19). The author like many in the United Kingdom of advancing years and health problems was placed by the UK government into lockdown, which meant not being able to leave the house for nearly 24 months. It was a peculiar time with all except key workers asked to work from home where possible and only leave the house for essential shopping and a limited period of exercise. Everyone had to keep a 2 m area around them away from people. Similar constraints were applied locally across the world. The virus caused a severe respiratory tract infection with fever and loss of taste and smell. Deaths around the world were huge (tens of thousands) with the United Kingdom particularly affected because of the high density of population as one contributory factor. This disease emphasized the vulnerability of human to zoonotic diseases. The lockdown enabled the author to focus on writing the book for a period when some struggled with purpose of life when confined to the house.

The author is grateful to the many clinical scientists, biomedical scientists, post-graduate students and clinical research fellows that worked in the steroid laboratory contributing to the success of the unit. Collaboration with clinicians at the Middlesex Hospital, UCLH and GOSH and across the country was also very important.

The author acknowledges the support of Professor Peter Hindmarsh, Hospital for Children, Great Ormond Street, London, and Professor Jonathan Fry, Physiology, UCL, London England for advice, criticism, and support on drafts of the book. The author also thanks the Elsevier staff, especially Tari Broderick, Samantha Allard, Stalin Viswanathan, Praveen Sachidanandam, and Selvaraj Raviraj, for bringing to print this *magnus opum*.

Foreword

Steroids are chemicals in the body that contribute to the endocrine system, which with nerves and blood vessels are the communication systems in the body. A hormone is a chemical produced in one part of the body (gland) and transported in the blood to target sites. The name hormone comes from a Greek word meaning to "arouse." Hormones differ from chemicals produced elsewhere that are sent to target sites directly through ducts (tubes); examples are the gall bladder and part of the pancreas (digestive juice). The pancreas secretes a digestive juice into the duodenum (exocrine function) but also has an endocrine function with the hormone insulin secreted into the blood. Steroid hormones in humans are produced from cholesterol in the adrenal glands, the gonads (testes and ovaries), and during pregnancy in the placenta. These glands are part of the endocrine system that includes the pituitary gland in the brain, thyroid and parathyroid glands in the neck, and pancreas in the abdomen.

The endocrine system is a network of glands that secrete hormones into the blood. Hormones can be:

- steroids,
- short chains of amino acids and molecular weight less than around 1000 Da (oxytocin),
- proteins with a single chain of amino acids (adrenocorticotrophin),
- proteins with be more than one chain of amino acids (insulin),
- amine derivatives (catecholamines), or
- tyrosine derivatives (thyroxine).

Endocrine glands are sometimes called ductless glands. Steroids are also produced in nonendocrine tissues like heart, liver, and adipose tissues. This is called the intracrine system or intracrinology. Some hormones are secreted from the nervous system (pituitary).

Chemical signaling is the key to the endocrine system, complementing the nervous system that uses electrical signals. Most hormones travel in the blood to target cells whose function or behavior they regulate. Steroids in the blood are largely bound to specific proteins; the free hormones are the effector molecules at the targets. Steroid hormones are important for the control of growth, development of systems, the control of glucose and electrolytes (homeostasis), metabolism (chemical changes in the body), and reproduction. In the target cells, hormones first react with proteins called receptors that are for steroids in the cell cytoplasm, whereas peptide hormones have receptors in the cell membrane. Steroids are transcription factors that alter gene expression. Steroids are slower in their action than the protein hormones that operate through membrane receptors and second messengers such as cyclic AMP, inositol triphosphate, and calcium or membrane-bound kinases.

Cortisol and aldosterone are important products of the adrenal cortex affecting glucose and salt metabolism, respectively, and thus called glucocorticoid and mineralocorticoid, respectively. The major secretion of the adrenal cortex in adult life is dehydroepiandrosterone sulfate (DHEAS). This is a weak male sex hormone (androgen) and an antiglucocorticoid. The precise role of this steroid is largely unknown although there is considerable interest in this steroid as an antiaging hormone. During fetal life, DHEAS is produced by a fetal zone of the adrenal cortex. The purpose of the DHEAS is not known, but the activity of the fetal zone continues into the newborn period and has to be considered when measuring the concentrations of other steroids.

Testosterone and estradiol are known as the sex steroid hormones. Testosterone is synthesized by the testes in the male and is responsible for male secondary sexual characteristics (beard growth, muscle development, deepening of voice, and growth and function of prostate and epididymis). Male hormones are androgens. Estradiol, which is secreted by the ovaries, varies widely in concentration in plasma throughout the adult female menstrual cycle and influences changes in the reproductive tract. Estradiol is responsible for female secondary sex characteristics (breast development, ovarian follicle development, and uterine enlargement). Steroids with action like estradiol are estrogens produced in vast amounts during the second and third trimesters of pregnancy. This estrogen is made in the placenta from weak adrenal androgens produced by the adrenal cortex of the developing fetus. Progesterone is a product of the ovary during the second half of the menstrual cycle and continued secretion of progesterone after implantation of a fertilized egg is essential to maintain pregnancy. Related steroids are progestogens.

Steroids are made on demand then transported around the body by the blood mainly bound to proteins. Steroids are also produced in peripheral tissues through autocrine, paracrine (locally), and intracrine (within cells) mechanisms. Steroids were first recognized because of their biological actions and functional assays were necessary to assist the search for, and chemical identification of, steroids in large amounts of tissues (kilogram quantities) and body fluids. The body controls sugar, salt, sex, and homeostasis through the effects of steroid hormones. Hormones such as thyroxine are stored in the glands where they are produced (thyroid glands) and secreted on demand.

The apparent chemical similarity between the steroids in pictorial form can make it difficult for medical and biology students to grasp the important relationships between structure and function. All steroids are based on the structure of cholesterol from whence they are made. The numbers given to the carbon atoms are used when systematically naming steroids. The synthesis of steroids from cholesterol was largely explored by chemical and isotopic studies, but the application of molecular biology has grown and in some instances extended our understanding of normal routes of steroid production and action as well as enabled the explanation of disease processes. Many issues are not fully understood. A pathway for steroid production has survived for many years, but new "nonclassical" and "backdoor" paths have been proposed in recent years.

The pituitary gland at the base of the brain is only about the size of a pea, but among several functions it is an important regulator of steroid production through protein hormones sent via the blood stream to the steroid producing glands. Adrenocorticotrophic hormone (ACTH), luteinizing hormone (LH), and follicle-stimulating hormone (FSH) are involved with steroid production. The pituitary gland also secretes prolactin, growth hormone, thyroid-stimulating hormone, antidiuretic hormone (vasopressin), and oxytocin.

The adrenal glands have a separate regulatory system for a steroid called aldosterone, which acts in the kidney to regulate blood concentrations of sodium potassium and hydrogen. A series of proteins initiated by renin from the kidney and angiotensin in the blood stimulate aldosterone production. Although the renin angiotensin system has been studied for many years, this axis is in the process of expanding to many more proteins through new knowledge.

To measure concentrations of steroids in biological samples, immunoassay (IA) became the mainstay of research and clinical investigations from the 1970s. Early methods often included purification by chromatography, but with improvements in the recognition of steroids by the antibodies there were recognized limitations in the specificity of the methods when steroid concentrations were determined directly. As new methods based on detection with mass spectrometry are introduced, they will slowly replace IA methods. The assays on automated platforms will continue because of the convenience of results in parallel with other analytes. Laboratory staff need some training particularly in unfamiliar basic laboratory practices and handling complex technology.

The mass spectrometer is the detector after a chromatographic separation of the steroids. High-pressure liquid chromatography has advantages over GC because the preparation of thermostable derivatives is avoided, which speeds up the analysis. LC-MS/MS now enables several steroids to be quantified in one analysis with high specificity. Clinicians have got the feeling that these methods are "gold standard," but that is not necessarily the case unless the benefits and disadvantages are weighed up. The laboratory will be asked to provide the concentrations of many steroids in body fluids or tissues for clinicians with patients in basal circumstances and during tests to stimulate or suppress steroid production. There are many factors that influence steroids during life and the laboratory needs to provide appropriate reference ranges for the tests. Gender and age are minimal factors. The simultaneous determination of several steroids in a blood sample is known as a profile. Urinary steroid profiling with gas chromatography coupled with mass spectrometry has provided a powerful diagnostic technique in reference laboratories for decades and is an introduction to the expanding field of metabolomics.

Rare steroid disorders can in many cases be traced to abnormalities of known genes, but this area is expanding all the time and in some cases the underlying cause of a steroid related abnormality in a patient is an opportunity to seek new mechanisms of disease. Genetic defects of electron transport, mitochondrial function, regulatory hormones, and steroid receptors are found as well as defects of steroidogenic enzymes. The disorders diminish life quality and pose an economic burden to society. Fertility, hypertension, aging, cancers, and neuronal disorders are targets for treatment. Diet and environmental factors add to the complexities in studying steroids. A laboratory, within a network, may be needed to help with specific gene defects. This book will not attempt to be up to date with all the methods for defining the genetic causes of steroid disorders because of the rapid changes outside the remit of the author to cover the biochemistry of steroid disorders. Proteomic and metabolomics has yet to find new biomarkers to match the genomic data beyond the recognized steroid pathways. Regardless of the assay technology, the laboratory work in partnership with clinicians and geneticists to be sure that the clinicians are made familiar with the technology used and its limitations since they bear responsibility for the interpretation of the results for the patient. A dialogue between the laboratory and clinics is therefore essential. In addition to genetic disorders around steroid production and action, several autoimmune diseases also contribute to disorders. Genetic findings are increasingly used to inform clinical management of many disorders including cancers. The genes involved in the adrenal and gonadal disorders have also been shown to play a role in the pathogenesis of sporadic adrenocortical tumors. The clinicians will need to call upon imaging (X-ray, ultrasound, CT, and MRI) and clinical assessment of genitalia, secondary sex characteristics, bone development, and growth of the patients.

Obesity, hypertension, cancer, and infertility are important health issues that a steroid endocrinologist will encounter. Children change in their body proportions and steroid test results need to take this into account particularly during the process of puberty as they achieve potential for reproduction. From the outset, a newborn child may be born of uncertain gender making it difficult for new parents to respond to their first question: "Is it a boy or a girl?" Laboratory tests are needed with urgency at this time so that a gender can be given to the baby.

Steroids are prescribed as medicines for a number of different situations, and the term steroid is confusing for lay people. Patients may be prescribed steroids for different reasons, and they encounter one or more of the following problems. The effects of steroids on sugar, salt, and sex cover most of the actions of these compounds. **Cortisol** or **hydrocortisone** is well known to patients with disorders of cortisol production, autoimmune adrenal insufficiency due to Addison's disease, or hypopituitarism because cortisol is needed when production fails. The detail of steroid replacement is a real challenge to replicate the natural rhythms. Some patients with adrenal insufficiency will also take **fludrocortisone** to stop the loss of salt in the urine. This replaces the natural steroid called **aldosterone,** which was discovered in 1953, at the Middlesex Hospital in London. Aldosterone is essential for life. Many children these days seem to be given steroids to treat asthma and respiratory conditions. **Fluticasone** and **Becotide** are commonly used as **anti-inflammatory** steroids. These are potent steroids applied locally to the airways. An array of synthetic steroids with a range of actions has been developed by the pharmaceutical industry, which can be used in the treatment of a number of ailments, conditions, and diseases. Currently, only females take steroids for oral **contraceptives**, but a male treatment is being pursued. The only real use for testosterone beyond replacement therapy is to stimulate red blood cells and muscle growth in a small number of patients, but we are all aware that steroids such as 19-nortestosterone are used in sport to enhance performance and elsewhere to enhance physique. These anabolic steroids, at high doses, encourage muscle development but have many side effects such as acne, unwanted hair, shrunken testes, clitoromegaly, infertility, baldness, and breast development. These steroids are called **androgenic-anabolic steroids** (AAS) because they have two actions. Androgens stimulate secondary sex characteristics of men, whereas these steroids also stimulate muscle growth, which is the anabolic action. Sportspersons want anabolic effects without the androgenic actions, but no steroid has been produced with anabolic effects alone. AAS are also taken to enhance muscle as a cosmetic feature. Steroids are abused, notably in sport, so methods of steroid analysis now include the detection of synthetic steroids to deter illegal use. Steroids can increase blood pressure that increases mortality. Many pharmaceutical products are targeted at the renin-angiotensin-aldosterone system to lower blood pressure into the normal range.

The industrial manufacture of steroids relied for many years on reactions with microorganism, both fungi and bacteria initially starting with bile but later from plant and soya sources. In recent years, genes for enzymic reactions have been inserted into the DNA of yeast and other organisms such that today many of the earlier processes have been refined. Although the action of bacteria in the human intestine on steroids excreted in bile has been known for many years, the identification of the organisms required long winded and thus expensive experiments conducted anaerobically and aerobically to characterize the biochemical properties of the isolated cells. These days microorganisms are identified by their RNA and fast methods of identification are now available. Studies of the microbiome are very recent and the literature in this area will multiply in the next decade or so.

Steroids bind to specific cytoplasmic proteins called receptors that after a series of intracellular events and translocation to the nucleus stimulate or suppress protein transcription. A number of genetic disorders affecting steroid action have been ascribed to defects in these receptors. Steroids also have faster nongenomic actions that as yet are not completely understood. Transporters of steroids has been an exciting area for research. The isolation of a mitochondrial membrane transporter (called StAR for steroidogenic acute regulatory protein) was possibly the beginning of research on intracellular and intercellular communication of steroids encompassing steroid hormones and the biosynthetic intermediates. How steroids move between mitochondria and the endoplasmic reticulum and move out of endocrine cells and into target organs is unknown. Dehydroepiandrosterone sulfate (DHEAS) is an important steroid without a clear function. A whole research program in Germany is looking at the role of DHEAS in reproductive processes. A group of organic anion transporters has been identified.

This book acknowledges the separate difficulties faced by laboratorians and clinicians when investigating patients, so the book endeavors to explain steroid production, laboratory measurements, physiological mechanisms, and clinical significance. Early biochemical assessment, particularly of adrenal function, is important to provide adequate clinical diagnosis to prevent life-threatening adrenal crises. Studies with animals in vivo, animal organs, cells, subcellular fractions, and reconstituted cells have been important to our understanding of many endocrine processes.

Comparative endocrinology continues to challenge and define our understanding of new processes. The book, however, focuses on humans. Genetic testing complements the biochemical findings and a balance between the two has to be struck for diagnosis of a clinical problem. The preference will be dictated by speed, cost, and fashion. The book should be a comprehensive source for understanding steroid hormones and the disorders of steroid production, action metabolism, and pharmacology.

Endocrinology is a dynamic, evolving subject, so the book can only reflect knowledge to the year 2018. The scientist or clinician should always be on the lookout for the unusual, and they will be surprised by little in endocrinology. Each chapter ends with suggestions for further reading with particular consideration of the latest issues in genetics. Steroids have fascinated the author of this book for more than 40 years, and the book thus reflects the diverse areas of interest, publications, and lectures during that time.

Part I

Chemistry and biochemistry of steroids

Steroids when drawn out on paper look like chicken wire until the **chemistry** is mastered, then the subject is easier to comprehend. The basis of steroids is the four-membered ring combination. Three rings of six carbon atoms in a phenanthrene arrangement with a saturated cyclopentane. The numbering of the carbon atoms need to be understood. The flow is anticlockwise around the A and B rings (C-1 to C-10), then clockwise through the C ring (C-11 to C-14) and anticlockwise in the D ring (C-15 to C-17). For the novice, two methyl groups (C-19 and C-18 at C-10 and C-13, respectively) tend to be forgotten when drawing the structure. The 19 carbon structure is common to the male androgens. The female oestrogens have no C-19 methyl group at C-10 and the A ring is phenolic. The corticosteroids have a 2 carbon side chain at C-17 making a total of 21 carbon atoms. The student must learn the structure of cholesterol that has 27 carbon atoms with a hydroxyl group at C-3 and a double bond at C-5. There are issues around the three-dimensional arrangement of steroids with an upper beta face and a lower alpha face. The C-3 hydroxyl group in cholesterol is thus a 3β-hydroxyl. The first step in steroid hormone synthesis is side-chain shortening with loss of 6 carbon atoms. The steroid hormones are created from cholesterol, and the C_{21} product is pregnenolone with a carbonyl at C-20.

Understanding the route of **steroid hormone synthesis** is fundamental to appreciating the structure-function relationships of steroid hormones. A number of enzymes located in the mitochondria and endoplasmic reticulum are involved. Through electron transport, water molecules become the basis for hydroxyl groups to be added to the steroid hydrocarbon at specific sites C-21, C-11, and C-17. Cytochromes are used to catalyze reactions. Apart from the oestrogens, all active steroid hormones have a 3-keto, delta-4 structure. A hydroxyl group at C-17 in a 21 carbon corticosteroid distinguishes glucocorticoids (sugar steroids) from mineralocorticoids (salt steroids).

The adrenal cortex, gonads, and placenta are the major **sources** of steroid hormones that are secreted into the blood circulation in a controlled manner. These organs change in structure and function during life, and the clinical investigator needs to be aware of the impact on endocrine tests. Periods of fetal, neonatal, adrenarche, puberty, and menopause are important events in life with distinct characteristics. **Steroids in blood and tissues** are moved by the action of transporters. Steroids in blood are largely (>95%) bound to nonspecific proteins (albumin) and specific binding globulins. Cholesterol is moved with lipoproteins into the cells that produce the steroid hormones, then steroids intermediate to the final products are moved between subcellular compartments within the glands producing the steroids. The chapter on **regulation** is about how the body maintains the concentrations of steroids in the circulation within limits or in response to external stimuli or homeostatic changes. Glucocorticoid and sex steroid synthesis is controlled by demand in processes involving the hypothalamus and pituitary glands at the base of the brain akin to the domestic heating system in the home (a negative feedback). The brain is like the thermostat, and the adrenal or gonads are boilers. The mineralocorticoids are controlled by a cascade of proteins from the kidney that senses changes in blood flow, blood pressure, and sodium concentration in the kidney to stimulate renin secretion (Note: renin (pronounced re-nin) not rennin (pronounced renn in), which is an enzyme in ruminant animals).

For a long time, steroids were thought to be recognized at nuclear target sites to stimulate or repress transcription of specific proteins that are effectors of hormone action. Not all actions of steroids fit this model in terms of speed of action. Rapid effects can be explained to some extent by activation of protein kinases and ion channels. The physiological effects are apparent when investigating clinical problems with high and low levels of steroids. In addition to endocrine effects, steroid scan operates in paracrine and intracrine modes.

The breakdown and clearance of steroids were originally confined to liver metabolism and renal clearance. To this can be added excretion in bile, metabolism in the intestine, and reabsorption to the circulation. Steroids are found in other biological samples. The actions of certain enzymes in tissues around the body have led to our understanding of local metabolism.

Chapter 1.1

Chemistry of steroids

1.1.1 Cholesterol and it's derivatives

1.1.1.1 Introduction

Cortisone was the first steroid to be recognized as a drug to treat inflammation in rheumatoid arthritis. There is now demand for steroids as oral contraceptives and in the treatment of asthma among many steroid drugs. Tons of steroids are created in the pharmaceutical industry, and from some predictions, the global financial market could reach near $5000 million by 2021. This area has advanced over the years and chemical reactions with plant and animal extracts have been replaced by the use of biological processes.

The low concentrations of steroids in body fluids can only be detected by sensitive tests. The first chemical tests involved reactions of reagents with structural features of steroids to give colored products. For quantitative assays, the reaction of steroids with antibodies became the basis for immunoassay in 1969 and has been the main test for steroids since. Steroids needed to be chemically modified into a larger molecule that is antigenic or into a competitive reagent in the immunoassay. Radioimmunoassay was the first method replaced by other detection agents (chemiluminescent for example). The lack of specificity in immunoassay has led to the introduction of mass spectrometric (MS) tests. A stable isotope-labeled steroid is now used in a MS dilution analysis.

The steroid hormones are derived from cholesterol which is a constituent of membranes probably in every cell in animals. Cholesterol is so named after isolation of a crystalline compound from gall stones (*chole* means bile, *stereos* stands for solid). Steroids and sterols are members of a class of naturally occurring organic compounds called terpenes (Hillier and Lathe, 2019) which are widely distributed in nature. The terpenoids are the result of linking isoprene units (C_5H_8) "head to tail" to form chains, as in rubber, then be arranged to form rings, as in cholesterol (Fig. 1.1.1). Steroids are tri-terpenoids. The biosynthesis of cholesterol and the production of the steroid hormones will be covered in more detail in the "Biosynthesis of steroids" Chapter 1.3. The hydrocarbon nucleus of cholesterol and steroids is comprised of three saturated six-membered (hexagonal) rings (perhydrophenanthrene, making up rings A, B and C) and a five-membered ring (pentagonal)

(cyclopentane, ring D) (Fig. 1.1.1). This is the basis for steroid nomenclature.

Seventeen of the carbon atoms in cholesterol are numbered in a sequence defined by the International Union of Pure and Applied Chemistry (IUPAC-IUB, Moss, 1993). Two methyl groups are sited at the junctions of the A and B, and C and D rings (carbons 19 and 18, respectively). Cholesterol has an 8-carbon side chain (carbons numbered 20–27) and as a C_{27} carbon unit the hydrocarbon structure is of CHOLESTANE. Cholesterol has a double bond at C5 and a hydroxyl group at C3 and is strictly speaking called 3β-hydroxy-cholest-5-ene.

Steroid hormones have a skeleton derived from cholesterol differing by one or more bond scissions or ring expansions or contractions. The steroid hormones which derive from cholesterol have 21 carbon atoms (pregnanes), 19 carbons (androstanes) and 18 carbon atoms (estranes). Vitamin D is a hormone from cholesterol via cholecalciferol which is formed in the skin by the action of sunlight to open

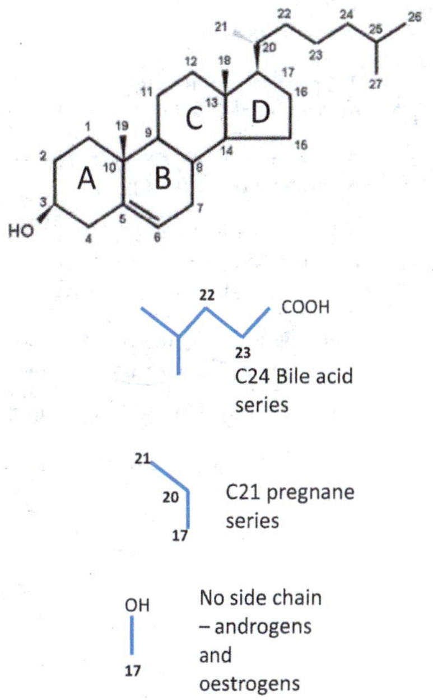

FIG. 1.1.1 Cholesterol synthesis from linear isoprene units to cyclic squalene, numbering of the carbon atoms and substituents of steroid hormones and outline structure of the related bile acids and vitamin D. (*Author original.*)

Steroids in the Laboratory and Clinical Practice. https://doi.org/10.1016/B978-0-12-818124-9.00020-6

the B-ring of 7-dehydrocholesterol by scission of the C9 to C10 bond. The bile acids have 24 carbon atoms (cholane), also formed from metabolism of cholesterol (see Fig. 1.1.1).

1.1.2   Functional substitution

A number of chemical prefixes and suffixes are used when referring to functional groups added to the steroid base which can be estr-, androst- and pregn- for C_{18}, C_{19} and C_{21} parent steroid molecules (see Table 1.1.1). Compounds are distinguished by orientation of the H atom or a substituent at C5; substituent at C10 (CH_3 or 19-nor); substituents at C17 (presence and length of the side chain); number and position of double bonds; number and position of substituents; type of substituent on the steroid skeleton.

- A double bond is shortened to ene and a C_{21} with a double bond becomes a pregnene and more specifically pregn-4-ene to indicate the double bond is between carbons 4 and 5. If the double bond cannot be accurately described by the single number then the carbons linked are referred to as in estra-1,3,5 (10)-triene.
- Hydroxyl groups can be numbered from the carbon number which is substituted (3β-hydroxy for example)

or the suffix -ol can be used (cholest-5-en-3β-ol). Both of those bracketed names when describing cholesterol.
- Carbonyl groups add the suffix -one or the prefix -keto as in 4-pregnen-3,20 dione for progesterone or 11-keto-testosterone for 4-androst-3,11,dione-17β-ol.
- For an aldehyde function, the suffix -al is used as in 4-pregn-11β,17α,21-triol-3,20-dione-18-al for aldosterone.
- C_{21} steroids can have a diol, triol, glycerol, dihydroxy-acetone or glyceraldehyde side chain through C-17, C-20 and C-21 according to substitution with hydroxyl and carbonyl groups. Reactions with these groupings were used in early chemical determinations of steroids.
- Carboxylic acid derivatives of corticosteroids have been described with the suffix -oic acid.

Steroid hormones are given systematic chemical names, trivial names (Table 1.1.2) and abbreviations, according to the alphabetical sequence of isolation by the chemists

TABLE 1.1.1 Steroid substituents.

Prefix	Steroid modification
cis	cis indicates functional groups on the same side of the carbon chain
Cyclo	With a bond between two carbons that are normally not united
Dehydro	Removal of two hydrogen atoms to leave a double bond between two carbon atoms
Deoxy	Removal of atomic oxygen—removal of hydroxyl with hydrogen
Dihydro	Addition of two hydrogen atoms to saturate a double bond between two carbon atoms or an oxygen from ketone to hydroxyl
epi	Differing in configuration of a carbon atom bonded to two other carbon atoms
Hexahydro	Addition of six hydrogen atoms to a steroid
homo	One additional carbon atom
iso	Differing in configuration of a carbon atom bonded to three other carbon atoms
nor	Removal of a carbon, such as the C19 methyl group from testosterone
seco	With a carbon-carbon bond broken
tetrahydro	Addition of four hydrogen atoms to a steroid
trans	Trans indicates functional groups on opposite sides of the carbon chain

TABLE 1.1.2 Trivial and systematic nomenclature for the steroid hormones.

Cholesterol $C_{27}H_{46}O$ 386	5-Cholesten-3β-ol
Pregnenolone $C_{21}H_{32}O_2$ 316	5-Pregnene-3β-ol-20-one
Progesterone	4-Pregnene-3,20-dione
Dehydroepiandrosterone	3β-Hydroxy-5-androsten-17-one
17-Hydroxypregnenolone	3β,17α-Dihydroxy-5-pregnen-20-one
17-Hydroxyprogesterone	17α-Hydroxy-4-pregnen-3,20-dione
11-Deoxycorticosterone (DOC)[a]	21-Hydroxy-4-pregnen-3,20-dione
Corticosterone (B)[b]	11β,21-Dihydroxy-4-pregnen-3,20-dione
Aldosterone	11β,21-Dihydroxy-4-pregnen-3,20-dione-18-al
11-Deoxycortisol (S)[b]	17α,21-Dihydroxy-4-pregnen-3,20-dione
21-Deoxycortisol	11β,17α-Dihydroxy-4-pregnen-3,20-dione
Cortisol	11β,17α,21-Trihydroxy-4-pregnen-3,20-dione
Androstenedione	Androsten-4-en-3,17-dione
Testosterone	17β-Hydroxy-4-androsten-3-one
Estrone	3-Hydroxy-estra-1,3,5 (10) trien-17-one
Estradiol	Estra-1,3,5 (10)-trien-3,17β-diol

[a]Trivial name.
[b]Kendall/Reichstein.

TABLE 1.1.3 Steroids by letters (Kendall, Reichstein, Wintersteiner).

Steraloids		Kendall	Reichstein	Wintersteiner
Q3690	A	11-Dehydrocorticosterone	5α-Pregnan-3β,11β,17α,20β,21-pentol	5α-Pregnan-3β,11β,17α,20β,21-pentol
Q1550	B	Corticosterone	H	
	C	Allo-THF	Allo-THF	D
	D	Adrenosterone	5α-Pregnan-3β,17α,21-triol-11,20-dione	A 11-Dehydrocorticosterone
Q2500	E	Cortisone	4-Pregnene-11β,17α,20β,21-tetrol-3-one	F Cortisol
		Epi E 20β-dihydrocortisol		
Q3880	F	Cortisol	M	
	Fa		Cortisone	
Q5200	G	5α-Pregnan-3β,17α,21-triol-11,20-dione	D Adrenosterone—Androst-4-ene-3,11,17-trione	B Corticosterone
Q2230	H	5α-Pregnan-3β,21-diol-11, 20-dione	Corticosterone	
	J	5α-Pregnan-3β,17α,20β-triol	J 5α-Pregnan-3β,17β,20β-triol	
	K	5α-Pregnan-3β,17β,20β,21-tetrol	K 5α-Pregnan-3β,17α,20β,21-tetrol	
	L		5α-Pregnan-3β,17α-diol-20-one	5α-Pregnan-3β,17α,21-triol-11,20-dione
	M		Cortisol	
	O		5α-Pregnan-3β,17α,20α-triol	
	P	P	5α-Pregnan-3β,17α,21-triol-20-one	
	Q	Dihydro-DOC	21-Hydroxy-4-pregnene-3,20-dione	
	R	R	5α-Pregnan-3β,11β,21-triol-20-one	
	S	11-Deoxycortisol	11-Deoxycortisol	
	T		20β-Dihydro-11-dehydrocorticosterone	
	U		20β-Dihydro cortisone	
	V		5α-Pregnan-3β,11β,17α,21-tetrol-20-one	

Cortexone is sometimes used for 11-deoxycorticosterone.
Uro is used for tetrahydro- cortisol and cortisone metabolites in urine—urocortisol, urocortisone.

Kendall and Reichstein (Table 1.1.3). The naming of metabolites (breakdown products) and pharmaceuticals will be dealt with later in the chapter on metabolism.

1.1.3 Three-dimensional structure

The steroid molecules have three-dimensional structures. The six membered A, B and C rings in the steroid hormones each take the conformation (shape) of a chair, whereas the 5 membered D-ring is like an open envelope (Fig. 1.1.2A). The angular methyl groups project above the plane of the steroid molecule which is called the β-configuration while below the molecule is the α-side.

Two hydrogen atoms are attached to many of the carbon atoms. Functional groups can be described in the axial (perpendicular to the ring) or equatorial plane (in the general plane of the ring) (Fig. 1.1.2B and C). By convention the axial atoms are drawn as a solid line if in the β configuration and dotted lines if in the α configuration below the plane. The term epi- is sometimes used for a β-substitution. Other structures that affect the names of steroids are in Table 1.1.1. Reduction of the double bond at C4–C5 can give a 5β or 5α dihydrosteroid. A hydrogen at C-5 can be either on the same side of the steroid as the C-19 methyl group (cis configuration or 5β) or on opposite face (trans or 5α). When the double bond at C4–5 is reduced, the 5β-hydrogen changes the A ring configuration to the shape of a boat, whereas the 5α steroid is in a chair shape. The A-ring of estrogens is phenolic and is flattened (not shown). The 5α reduced steroids are in the allo series. There are three ring fusions in steroids and in humans the junctions are mainly trans-trans-trans (Fig. 1.1.2D). The cis-trans-trans arrangement is found in plant glycosides.

The above rules for the steroid nucleus do not apply to the side chain since the carbon atoms are not in the same plane as the ring system. Fieser and Fieser based a system on the relationship to C-20 when viewed from the front. The other carbon atoms are viewed as the eye passes along the chain looking to the right (α) and left (β) of substituents at each junction. A more basic system by Ingold correctly describes the three-dimensional position of each substituent on any carbon in all molecules. The symbol R is used for right (from Latin rectus) and S for left (sinister). Progesterone with 20α-ol is 20S, whereas 20α,21-diol is 20R and 17α,20α,21 triol is 20S (Fig. 1.1.2E).

1.1.4 Steroids by function

Free steroids hormones favor a lipid environment (lipophilic) rather than an aqueous phase (hydrophobic) which is exploited when chemists extract steroids from biological samples. This property affects transport in the blood and movement to compartments of the cells. Water solubility of steroid hormones range from around 6 μg/mL for pregnenolone and 17-hydroxyprogesterone to 320 μg/mL for cortisol. Steroids are cleared from the body as metabolites after conjugation with glucuronic acid or sulfuric acid which increases their water solubility. Steroids are usually dissolved in organic solvents for use in the laboratory. The highest purity grades of steroid reagents should be used. Soda-lime glass containers should be avoided to avoid degradation of the steroids by the glass (Burstein, 1976). Plastics are often used these days, but when solvents have been in contact with the plastic then concentrated for further analysis of the extracts, there can be interference problems especially with steroid binding to antibodies in immunoassay methods (McManus and Sharifi, 2020). Solvents should be purchased in glass not plastic containers. Albumin and gelatin, which enhance the solubility of steroids in aqueous media, are useful additives to buffers when redissolving the steroids from a dried extract.

Steroids were originally isolated and characterized from large amounts of endocrine tissues from animals through bioassays. Three main groups led by Reichstein, Kendall and Wintersteiner, were involved in the characterization of steroids, the results of their work are summarized in Table 1.1.3. In those early days (1929–53), it was not known that steroid endocrine glands do not store the steroids. Plant sterols such as diosgenin, tigogenin, solasodine and hecogenin (Fernandes et al., 2003) have been used commercially for partial synthesis of steroids (Fig. 1.1.3).

Throughout this book steroids will be referenced by function to those affecting sugar, salt and sex (Fig. 1.1.4) and other responses. Corticosteroids are C_{21} steroids produced by the outer regions of the adrenal cortex, the zona glomerulosa and zona fasciculata (see Chapter 1.2 on "Sources of steroid hormones").

The active steroids from these zones are then aldosterone and cortisol (in humans, corticosterone in rats and other species), respectively. One distinguishing feature of those steroids is that aldosterone (and corticosterone) does not have a 17-hydroxyl group. Aldosterone influences salt metabolism and is a **mineralocorticoid**. Among the biological activities of cortisol is the role in glucose (sugar) metabolism which is called **glucocorticoid** action. Sex steroids are produced in the zona reticularis of the adrenal cortex and in the gonads and the placenta. Progesterone is a C_{21} steroid produced by the ovaries and placenta. Androgens are C_{19} steroids by the adrenal zona reticularis, testes and ovaries. The estrogens are produced in the ovaries and placenta and have 18 carbons with an aromatic A-ring, whereas most other steroid hormones have a 3-keto-4-ene configuration. Different activities relate to changes in the position of functional groups.

1.1.4.1 Corticosteroids

Cortisol affects glucose metabolism in a manner that opposes the actions of insulin. Cortisol promotes deposition of glycogen in the liver and breakdown of body proteins (catabolism). Cortisol is the major **glucocorticoid** in man. It is a 21 carbon steroid (pregn) with the characteristic 3-keto-4-ene structure of active steroids (Fig. 1.1.4). Hydroxyl groups are at carbons 11, 17 and 21. There is a further carbonyl group at C20. The dihydroxyacetone side chain of cortisol was the target for chemical methods of cortisol determination. Corticosterone (cortisol without the C-17 hydroxyl group) is the major glucocorticoid in many animals. Aldosterone influences sodium homeostasis so it is therefore called a **mineralocorticoid** and differs in structure from cortisol in the absence of the hydroxyl at C-17 and possession of an aldehyde at C-18.

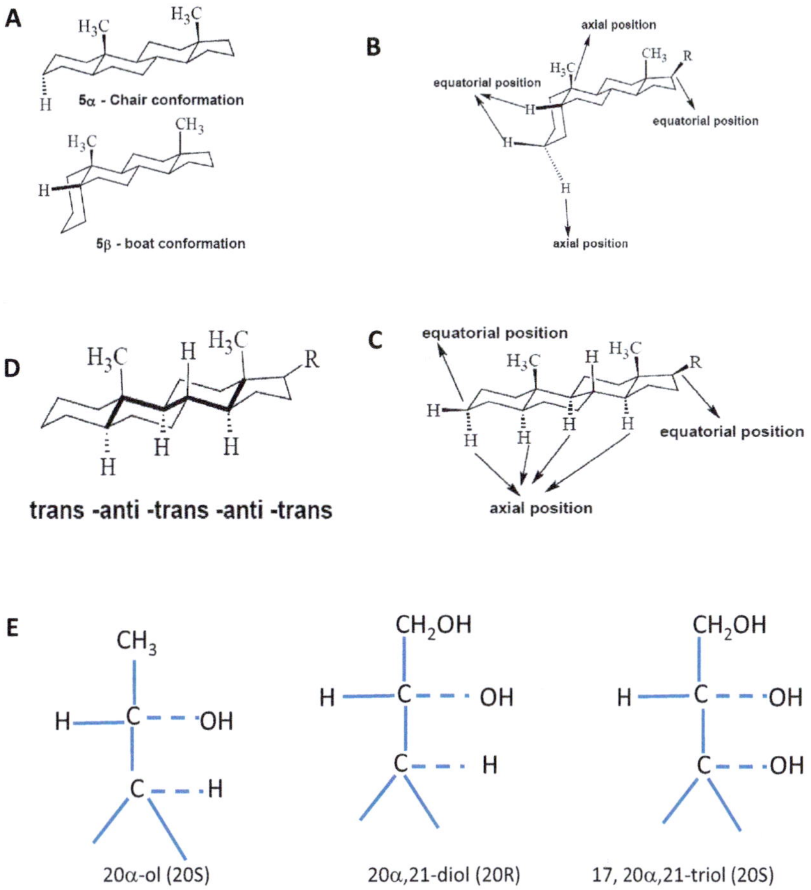

A
H3C
H3C
H
5α - Chair conformation
CH3
H3C
H
5β - boat conformation
B
axial position
CH3
R
H3C
equatorial position
H
H
equatorial position
H
axial position
D
H3C
H
H3C
R
H
H
H
trans -anti -trans -anti -trans
C
equatorial position
H3C
H
H3C
R
H
H
H
H
H
equatorial position
axial position
E
CH3
H
C
OH
C
H
20α-ol (20S)
CH2OH
H
C
OH
C
H
20α,21-diol (20R)
CH2OH
H
C
OH
C
OH
17, 20α,21-triol (20S)

FIG. 1.1.2 Conformation (A) and configuration (B, C, D) and nomenclature of steroids. *(Author original.)*

HO
H
H
H
O
O
Diosgenin
HO
H
H
H
H
O
O
Tigogenin
O
HO
H
H
H
O
O
Hecogenin
HO
H
H
H
O
N
H
Solasodine

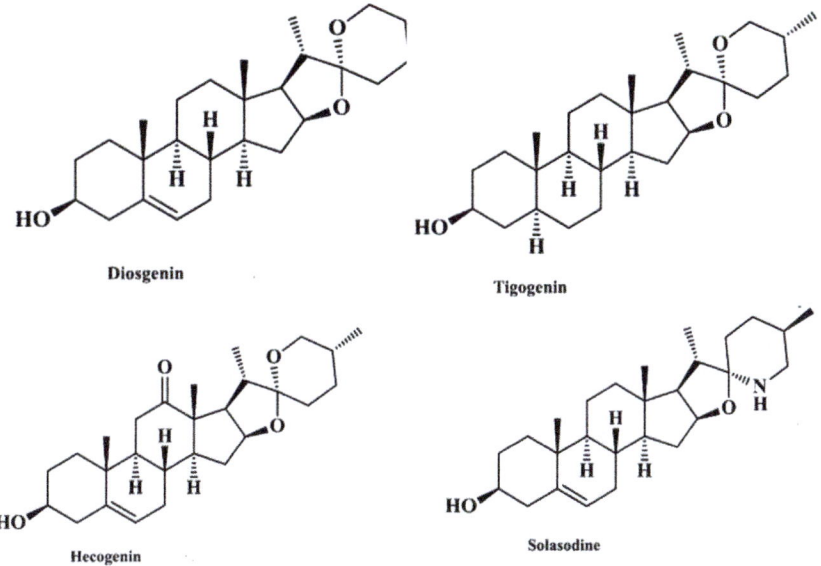

FIG. 1.1.3 Steroid sapogenins. *(Modified from Al-Jasem Y, Khan M, Taha A, Thiemann T. Preparation of steroidal hormones with an emphasis on transformations of phytosterols and cholesterol—a review. Mediterr J Chem 2014;3(2):796–830. Fig. 3 p. 801.)*

Sex steroid Androgen C19	Sex steroid Oestrogen C18	Sex steroid Progestogen C21	Corticosteroid Glucocorticoid C21	Corticosteroid Mineralocorticoid C21
Testosterone	Oestradiol	Progesterone	Cortisol	Aldosterone

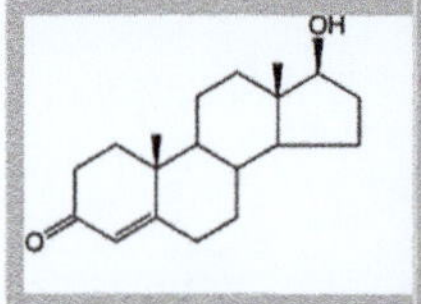

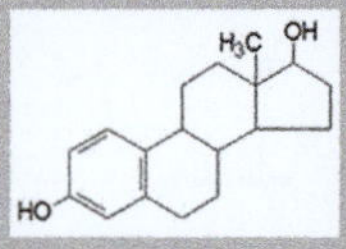

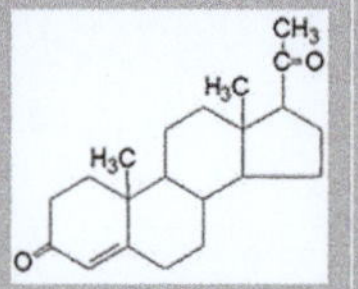

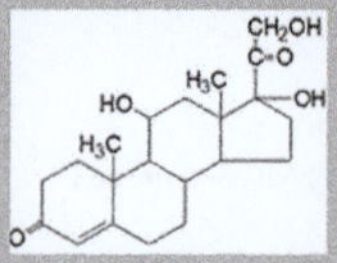

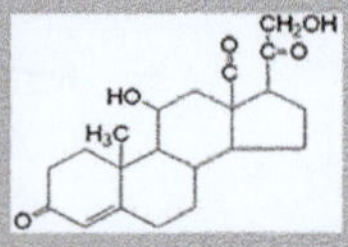

FIG. 1.1.4 Examples of the major steroid hormones according to carbon number (18, 19 or 21) and biological function. *(Author original.)*

1.1.4.2 Sex steroids

The sex steroids influence the sexual characteristics of male or female. **Androgens** are the 19 carbon steroids for male features and **estrogens** are 18 carbon steroids with no C-19 methyl group that affect the sexual characteristics of females (Fig. 1.1.4). Testosterone and estradiol are the main sex steroids in males and females respectively. Dehydroepiandrosterone (DHEA) as a C3-sulfate conjugate (DHEAS) is the most abundant weak androgen in circulation. Androgens also encourage muscle development (anabolic effect) and are thus frequently implicated in issues of doping in sport to enhance performance. The C_{21} steroids without an hydroxyl group at C-21 are **progestins** (short for *promote gestation*) of which progesterone is the main steroid in circulation that prepares the uterus for implantation of a fertilized ovum in each menstrual cycle. If pregnancy is established progesterone has a supportive role. Initially progesterone is secreted by a corpus luteum that develops in the ovary when the dominant follicle releases the egg in the menstrual cycle. The corpus luteum persists if the egg is fertilized. Later in pregnancy, the placenta takes over the production of progesterone to continue to support the pregnancy.

1.1.4.3 Neurosteroids

A number of 3β-hydroxy-5-ene steroids in the brain have been called "neurosteroids" because (Corpéchot et al., 1983; Robel and Baulieu, 1985; Le Goascogne et al., 1987) they were synthesized within nervous tissue. They may have a role in anxiety, depression, premenstrual syndrome (now called premenstrual dysphoric disorder) and development of disorders in the central nervous system such as Alzheimer's and Parkinson's disease. The neurosteroids include **allo-pregnanolone** (Fig. 1.1.5).

The precise identities of the neurosteroids has cast some doubt on their forms in brain and possibility of artifact formation during analytical procedures (Liere et al., 2004; Ebner et al., 2006; Dury et al., 2016).

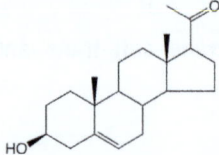

Pregnenolone (above)	3β-hydroxypregn-5-en-20-one
Epi-pregnanolone	3β-hydroxy-5β-pregnan-20-one
Allopregnanolone	3α-hydroxy-5α–pregnan-20-one
Pregnanolone	3α–hydroxy-5β-pregnan-20-one
Epiallopregnanolone	3β-hydroxy-5α-pregnan-20-one
Androsterone	3α-hydroxy-5α-androstan-20-one

FIG. 1.1.5 Structures of neurosteroids. *(Modified from Dury AY, Ke Y, Labrie F. Precise and accurate assay of pregnenolone and five other neurosteroids in monkey brain tissue by LC-MS/MS. Steroids 2016;113:64–70. Fig. 1 p. 65.)*

1.1.4.4 Progestogens

The primary target for progesterone is the uterine endometrium upon which it exerts a strong secretory effect provided the tissue has been previously under the influence of estrogen. Progesterone also has a secretory effect on the vaginal epithelium and endocervix. Progesterone is primarily the hormone of pregnancy and is essential for the establishment and continuation of pregnancy. Progestogens (Fig. 1.1.6) are synthetic relatives of 17-hydroxyprogesterone that have higher progestogenic activity than progesterone. Derivatives of 19-nortestosterone (Norethisterone and Dienogest) and 19-norprogesterone (Nomegestrol acetate) also have progestational activity.

Several progestagens (Gestodene, medroxyprogesterone acetate [MPA] and norethisterone) are used in contraceptives in combination with an estrogen and at higher doses these are used in the treatments of cancers. Mefipristone RU486 behaves as an antiprogestagen, sensitizes the myometrium and softens and dilates the cervix. The drug RU486 is used in combination with the prostaglandin

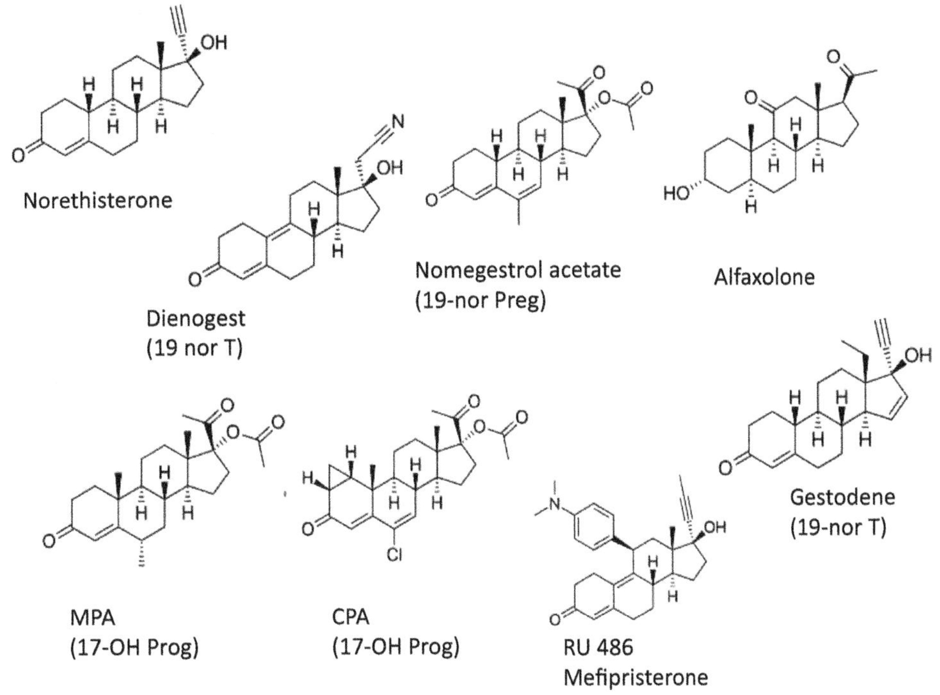

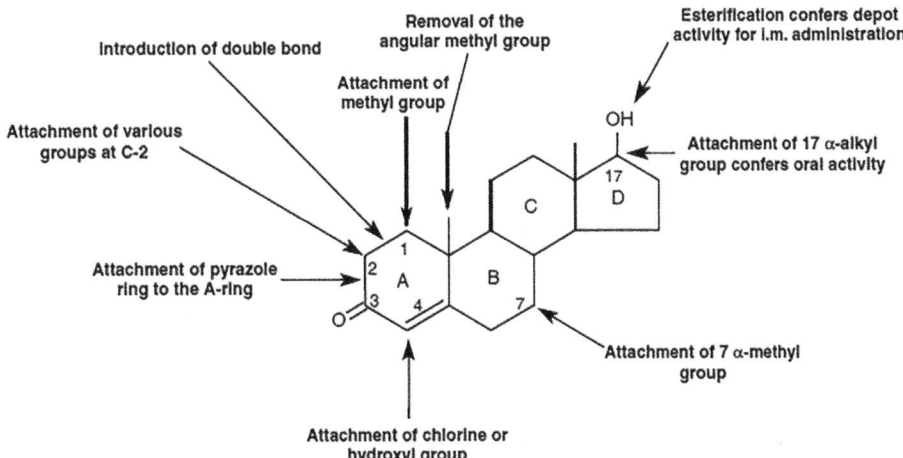

FIG. 1.1.6 Structures of synthetic progestogens. *(Author original.)*

Norethisterone

Dienogest
(19 nor T)

Nomegestrol acetate
(19-nor Preg)

Alfaxolone

MPA
(17-OH Prog)

CPA
(17-OH Prog)

RU 486
Mefipristerone

Gestodene
(19-nor T)

Introduction of double bond

Removal of the
angular methyl group

Esterification confers depot
activity for i.m. administration

Attachment of
methyl group

Attachment of various
groups at C-2

Attachment of 17 α-alkyl
group confers oral activity

Attachment of pyrazole
ring to the A-ring

Attachment of 7 α-methyl
group

Attachment of chlorine or
hydroxyl group

FIG. 1.1.7 Structural modifications to the A- and B-rings of testosterone that may increase anabolic activity of the related steroids; substitution at C-17 confers oral or depot activity. *(Kicman AT. Pharmacology of anabolic steroids. Br J Pharmacol 2008;154(3):502–21. Fig 2 p 505.)*

(gemeprost) for the termination of pregnancy. Some progestogens have antiandrogenic properties (cyproterone acetate) or anesthetic properties (Alphaxolone).

1.1.4.5 Androgenic anabolic steroids

A pure anabolic steroid affecting muscle without affecting male sexual characteristics would have many clinical advantages and drug companies have sought such a compound by modifying the testosterone molecule. A new group of selective androgen receptor modulators (SARMs) are available (see later Chapter 1.8). No compound with sole anabolic activity and minimal androgenic activity has been found so the term AAS is now the term used for steroids with both activities. There are many modifications to the steroid molecule that affect activity, mode of delivery (tablet or injection for example), metabolism and clearance (Fig. 1.1.7).

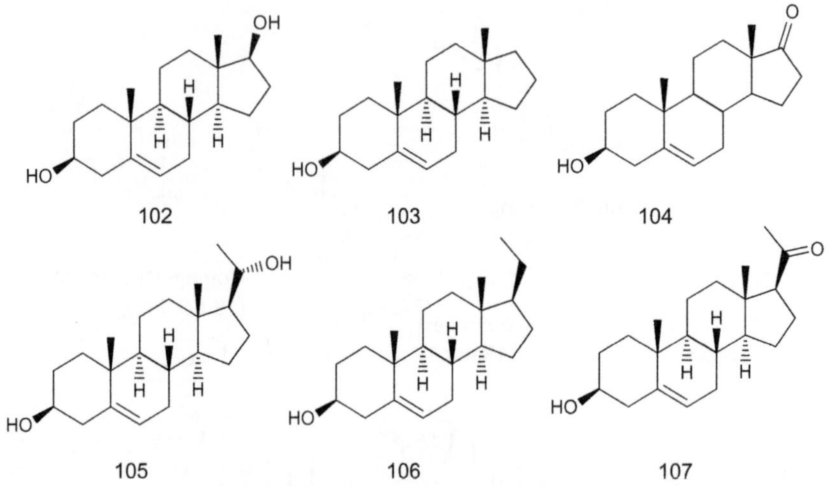

FIG. 1.1.8 Reaction products from the auto-oxidation of cholesterol in air are androst-5-ene-3β,17β-diol (**102**), androst-5-en-3-β-ol (**103**), 3β-hydroxyandrost-5-en-17-one (**104**), pregn-5-ene-3β,20α-diol (**105**), pregn-5-en-3β-ol (**106**) and to 3β-hydroxypregn-5-en-20-one (**7**). *(From Al-Jasem Y, Khan M, Taha A, Thiemann T. Preparation of steroidal hormones with an emphasis on transformations of phytosterols and cholesterol—a review. Mediterr J Chem 2014;3 (2):796–830. Fig. 6 p. 810 (Originally described by van Lier et al. J Org Chem 1970; 35:2627–2632).)*

1.1.5 Chemistry of steroids

A reputable supplier of steroids, Steraloids, state that steroids should be stored in a closed container away from direct sunlight, excessive heat and moisture. Steroids do not need to be stored in the fridge or freezer. The steroids are stable for years but to confirm the products have not changed visually inspect for discoloration, run a melting point check and a thin layer analysis every 3–5 years. Cholesterol in air can oxidize into steroids of the androstane and the pregnane series (Al-Jasem et al., 2014) (Fig. 1.1.8). The oxidation probably proceeds via the 20α- and 25-hydroperoxy derivatives, which can be detected when cholesterol is reacted with air at elevated temperatures.

In general, as stated above, steroids are stable molecules but some need particular respect because of interactions of functional groups across the molecules in acid and alkali conditions. The ketol and dihydroxyacetone side chains of steroids resemble the terminal groups of the ketol sugars, it is therefore not surprising that the corticosteroid side chain of C_{21} steroids is susceptible to degradation by dilute alkali with formation of etienic acids and 20-hydroxy acids. This degradation is retarded when nitrogen replaces oxygen (Monder, 1968).

Microchemical reactions have been published in books (Fieser and Fieser, 1959; Bush, 1961; Djerassi, 1963; Lednicer, 2011). A one-step procedure to convert Δ5 steroids to their corresponding a-ketols was recently published (Salvador et al., 2006). A dictionary of steroids (Hill et al., 1991) is an important reference work, now dated and still expensive. Entries present structure, physical properties, medicinal uses and sources of steroids from literature reviewed until 1990. An atlas of steroid spectra (Neubert and Ropke, 1965) includes IR spectra of 900 compounds, 41 UV absorption curves and 95 NMR spectra.

1.1.5.1 Aldosterone

The structure of aldosterone is near unique among the steroids secreted by the adrenal gland in having a C-18 aldehyde group. The close proximity of the C-18-aldehyde with the C-11 oxygen and with the α-ketol side chain accounts for a number of changes in conformation of aldosterone under normal circumstances and during the procedures to isolate, identify and measure aldosterone (Lantos et al., 1987). The structure of aldosterone in solution is considered to be an equilibrium mixture of the 11,18-hemiacetal (Fig. 1.1.9 #1a) and of the bicyclic 11,18:18,20-diepoxy (Fig. 1.1.9 #1b) forms.

When crystalline there is no evidence from nuclear magnetic resonance spectroscopy for the former C-20 ketone compound. An 17-iso aldosterone (Fig. 1.1.9 #2) can be formed under alkaline conditions. This isomerization can occur on storage to a significant extent particularly if soda glass vessels or ampoules are used. The 17-iso aldosterone is also formed during HPLC separation of aldosterone with both normal and reversed phase columns. The 11,18:18,20 bicyclic form of aldosterone can be produced during HPLC separation of the hormone. Mass spectrometry of reference 17-iso aldosterone, 18,21 aldosterone diacetate and of 18,21 aldosterone bis-trimethylsilyl ether clearly confirmed the exclusive formation of 17-iso aldosterone. The use of methanol for dissolution of aldosterone should be avoided since impurities in the solvent affect the stability of labeled steroid. Wherever possible ethanol containing 0.1% triethylamine should be used (Roy et al., 1976).

Dilute alkali (0.03 M potassium carbonate and 0.07 M sodium hydroxide in aqueous 92% ethanol) on aldosterone at room temperature leads to a rearrangement to 11,18:18,21-diepoxy-20,21-dihydroxy-pregn-4-en-3-one (apo-aldosterone) (Kirk and Miller, 1982) (Fig. 1.1.9 #5). The presence of the dissolved oxygen caused simultaneous

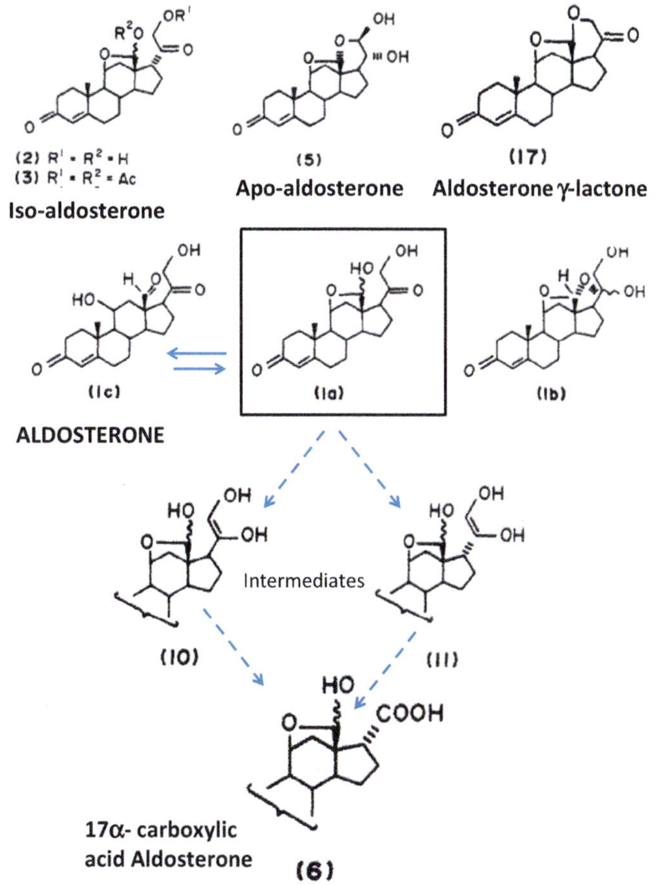

FIG. 1.1.9 Aldosterone in acid and alkaline conditions. *(Modified from Kirk DN, Miller BW. 18-Substituted steroids—9. Studies on the stability of aldosterone in dilute alkali. J Steroid Biochem 1982;16(2):269–76. Fig. 1 p. 273.)*

degradation to 17-iso aldosterone and alkaline hydrolysis to a 17α-carboxylic acid (#6). X-ray crystallographic studies showed that two isomers of 11,18:18,21-diepoxy-20,21-dihydroxy-pregn-4-en-3-one are formed during the reaction of aldosterone with alkali. They are the 18R,20S,21S and 18R,20S,21R (Fig. 1.1.9 #10,#11) forms. The conformation of the A, B, C rings are similar in the various forms of aldosterone. The 18-acetal-21-hemiketal isomers however, fixes the 21-hydroxyl group. This could influence receptor binding and also affect the antibody recognition of aldosterone. Reactions with acids are utilized for the detection of aldosterone and in a number of quantitative methods for the hormone.

In solution, therefore aldosterone is sensitive to changes in pH (both acid and alkaline) with and without heating and to changing oxygen concentration and salt. A discussion of these changes is highly relevant to this book because many manipulations of aldosterone in sample collection and processing or in assay development may have affected the nature of the steroid molecule. For example, during the chemical synthesis of hapten complexes of aldosterone (by which antibodies are raised for immunoassay techniques) and of ligands of aldosterone containing radioactive

atoms. Furthermore a widely used, convenient assay for an important aldosterone metabolite (aldosterone 18-glucuronide in urine) involves acid hydrolysis of the conjugate prior to a radioimmunoassay of the liberated free aldosterone. Acid hydrolysis affords a small amount of aldosterone gamma lactone (Fig. 1.1.9 #17) as well as an acetal 11α,18;18,21-diepoxy-pregn-4-ene-3,20-dione (Fig. 1.1.9 #3). In the collection of urine samples for the estimation of aldosterone, most workers take no precautions to buffer pH to maintain neutral conditions. In recent years, high-performance liquid chromatography (HPLC) has become a popular separation technique for steroids in extracts of biological material. Although the columns used in these methods offer superior resolution, the bonded phases which are not capped, and thus have residual active sites, may have chemical effects on steroids particularly aldosterone and its precursors.

1.1.5.2 The 18-hydroxy steroids

18-Hydroxycorticosterone (18-hydroxy B) and 18-hydroxy-DOC are of special interest because of their potential role as

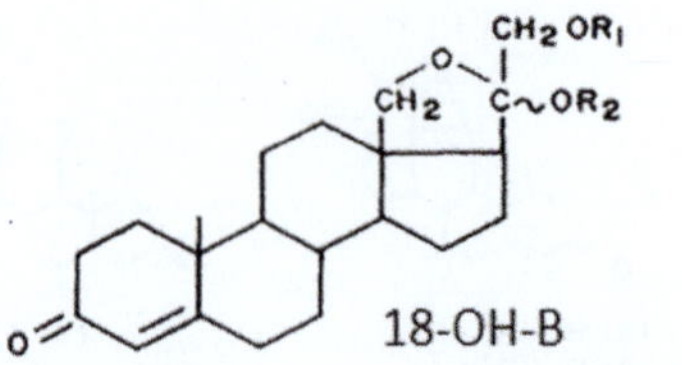

FIG. 1.1.10 Chemistry of 18-hydroxy precursors to aldosterone. 18-Hydroxy-11-deoxycorticosterone (18-OH-DOC) has been observed to exist in two interconvertible forms of markedly different chromatographic mobility. The more polar form is the cyclic hemiketal (1) and the less polar is a mixed ketal at C-20 derived by reaction with an alcoholic solvent (iV). This reaction is catalyzed by traces of acidic impurities present in most commercial sources of reagent grade methanol or ethanol, but can be abolished by removal or neutralization of these impurities. *(Modified from Roy AK, Ramirez LC, Ulick S. Structure and mechanism of formation of the two forms of 18-hydroxy-11-deoxycorticosterone. J Steroid Biochem 1976;7: 81–87. Fig. 1 p. 82.)*

	R1	R2
I	H	H
II	CH$_3$CO	H
III	H	CH$_3$
IV	H	CH$_2$CH$_3$
V	CH$_3$CO	CH$_3$
VI	CH$_3$CO	CH$_2$CH$_3$
VII	CH$_3$CO	CH$_3$CO

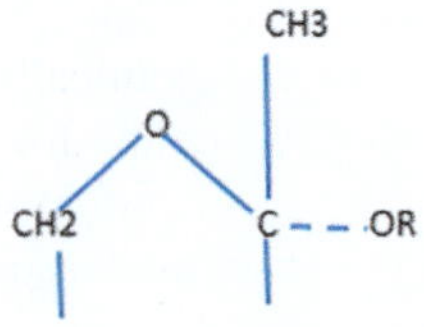

precursors of aldosterone. 18-Hydroxy-B is, however, a poor substrate for aldosterone synthesis in vitro. These compounds present considerable difficulties to the analyst because of the lability of these steroids due to proximity of the 18-hydroxyl group with oxygen functions at C-11, C-17 and C-20 (Usa et al., 1979; Roy et al., 1976; Damasco and Lantos, 1975; Dominguez, 1965). A variety of products may occur as a consequence of spontaneous formation of cyclic ketals or hemiketals (Aragones et al., 1978). The number of possible products increases with the number of oxygen groups and with exposure of the steroids to acid, base, alcohol, etc. (Fig. 1.1.10).

In the 1950s and 1960s, some of the 18-hydroxy steroids were structurally characterized by their formation of acetates, their ability to reduce tetrazolium salts and by their chromatographic properties. Infrared spectra of 18-hydroxy-B and 18-hydroxy-DOC do not show absorption peaks at $1700\,cm^{-1}$ characteristic of unconjugated carbonyl groups. The C-18 to C-20 oxygen bridge has characteristic absorption in the spectrum. The NMR spectrum of 18-hydroxy-DOC has no signal corresponding to the carbonyl in the alpha-ketol side chain.

In organic solvents, 18-hydroxylated steroids form, spontaneously and reversibly, compounds of lower polarity. Thus for 18-hydroxy-DOC the free 18-hydroxyl group and the 20-18-hemiketal forms are possible structures with two diastereoisomers of the 20-18-hemiketal. The two products of 18-hydroxy-DOC were a less polar form (L) and a more polar form (M) (Aragones et al., 1978). The former is probably the alkyl ketal product of the M-form but a dimer of the L-form, compared with the constitutional monomer of the L-form, was thought likely. The alkyl ketal derivative may form with traces of acid in alcoholic solutions. On acetylating 18-hydroxy-DOC which has been left standing in ethanol, two products were characterized by mass spectrometry—an 18-hydroxy-DOC-monoacetate and the less polar 20-ethyl ketal,21-acetate. Corresponding methoxy compounds were found in methanolic solutions. The reactions are faster in anhydrous solvents and in more dilute solutions of the steroid. The reactions are prevented by redistilling solvent from sodium hydroxide pellets or more conveniently by adding 0.1% triethylamine to all steroid solutions though this has not been published in recent papers.

1.1.6 Industrial production of steroids

The first achievement in producing steroids for pharmaceutical use was the preparation of corticosteroids from cholic acids in ox bile. Later, the degradation of the side chain of sapogenins was found to be a cheap source of precursors for steroid synthesis. The synthesis of cortisone from deoxycholic acid was a multistep chemical process (31 steps) characterized by low mass yields (0.16%) and high economic costs ($200/g in 1949). Other early extractions are summarized in Table 1.1.4.

TABLE 1.1.4 Early isolation of steroids.

1931	15,000 L male urine	15 mg androsterone
	4 tons sows ovaries	12 mg estradiol
	625 kg sows ovaries	20 mg progesterone
	Hundreds of gallons pregnant human urine	Estrone (Theelin)
	100 kg bulls testes	10 mg testosterone
	100 kg adrenals (20,000 cows)	75 mg E; 50 mg cortisol
	615 kg beef bile—deoxycholic acid	1 mg cortisone
1953	500 kg beef adrenals	21 mg electrocortin

The yields were far too low for the demands of the pharmaceutical industry. The combination of chemistry and biology was required in the processing of sapogenins extracted from plants. A 16,17-epoxy group is then inserted chemically and 16-epoxyprogesterone was the starting point for the synthesis of many steroids—prednisone, prednisolone, hydrocortisone, dexamethasone and beclomethasone for examples (Fig. 1.1.11A–D, respectively).

Sapogenins have been progressively replaced by several natural sterols that can also be biotransformed into steroidal derivatives with properties similar to certain sex hormones (Fernandes et al., 2003). The use of enzymatic reactions started just after World War II. Since then biotransformations and enzymatic processes joined chemistry to produce corticosteroids in the pharmaceutical industry starting from stigmasterol, diosgenin, hecogenin, solasodine, sitosterol and campesterol (Fig. 1.1.12) extracted from plant roots, e.g., yams.

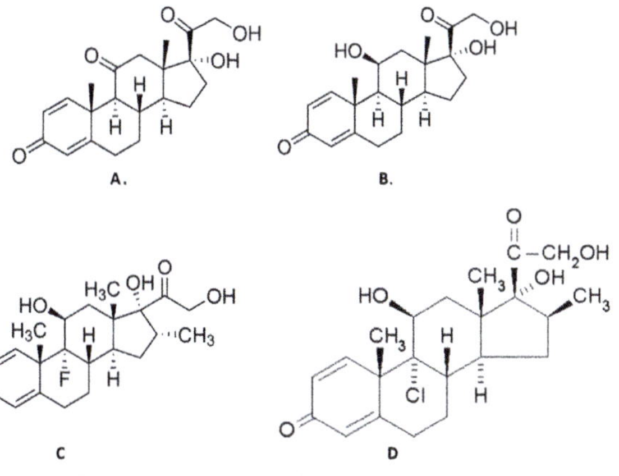

The intermediate 16-dehydropregnenolone was key from the first chemical stages. In 1950, Murray and Petersen were able to introduce a hydroxyl group in position 11 alpha by fermentation of progesterone with a mold of the genus *Rhizopus*. Several other microorganisms were then used to replace multiple chemical steps. Hydroxylations and side chain cleavage were important requirements. The potential of microbial steroid biotransformation was used for several decades since they offered a number of advantages over chemical synthesis:

(i) regio- and/or stereospecific functionalization of molecules at positions not always available with chemical agents,
(ii) multiple consecutive reactions carried out in a single operation step and
(iii) more ecofriendly processes (i.e., mild reaction conditions and aqueous media).

Other chemical steps have been replaced by microbial bioconversions in steroid synthesis processes in the last decades, leading to more competitive and robust industrial processes. For example, the steroid hormone testosterone is chemically synthesized from the steroidal intermediate 4-androstene-3,17-dione, which is previously obtained from natural sterols by microbial biotransformation. In recent years, new bioprocesses have been also designed by recombinant DNA technology approaches, that open up new opportunities for the construction of more robust and versatile microbial cell factories (MCF) for the production of steroids *à la carte*.

Several types of phytosterols are by-products of other industries (e.g., from soybean, pine, paper industry wastes), used generally as industrial feedstock instead of cholesterol (obtained from animal fats and oils) due to the exhaustive quality controls required for the use of any type of animal basic precursor. The successful implementation of metabolic engineering approaches would not have been possible

FIG. 1.1.11 Structures of synthetic glucocorticosteroids (A) prednisone, (B) prednisolone, (C) dexamethasone (9α-fluoro-16α-methyl prednisolone), (D) beclomethasone (9α-chloro-16β-methyl prednisolone). *(Author original.)*

FIG. 1.1.12 Steroids from raw materials. *(Modified from Fernandes P, Cruz A, Angelova B, Pinheiro HM, Cabral JM. Microbial conversion of steroid compounds: recent developments. Enzyme Microbial Technol 2003;12:688–705. Fig. 2 p. 691.)*

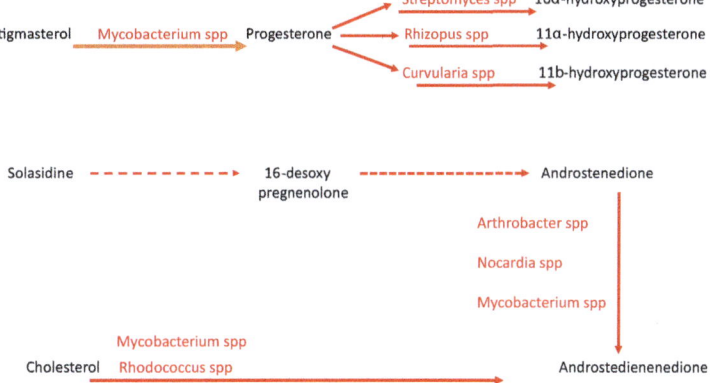

without the numerous studies of catabolic pathways of steroids developed in various model of actinobacteria (e.g., *Mycobacterium smegmatis*), the development of new molecular biology tools for genetic manipulation of these nonmodel bacteria and the sequencing and annotation of their genomes.

De novo biosynthesis of progesterone and hydrocortisone from simple carbon sources (e.g., galactose, ethanol) was successfully achieved in recombinant strains of *Saccharomyces cerevisiae*, by engineering the endogenous sterol biosynthesis pathway to generate a cholesterol-like molecule that served as a precursor to a multienzymatic heterologous route mimicking human steroid biosynthesis (Szczebara et al., 2003; Woodward et al., 1952). Most of the limitations observed in current industrial bioprocesses are directly related to the intrinsic properties of steroidal molecules (e.g., low solubility in aqueous media, high cell toxicity). With the aim of overcoming these limitations, different technological approaches such as micronization or emulsification with surfactants of the steroidal substrates, it is possible to assemble an artificial biosynthetic route of sterols with a partially interrupted cholesterol catabolic pathway to synthesize steroidal intermediates of interest.

The bioconversion of lanosterol into steroidal intermediates of interest by *Mycobacterium* sp. NRRL B-3805 has been described. Although in recent years the optimization of several bioprocesses has been addressed through recombinant DNA technology approaches, these techniques have hardly not been applied for steroid synthesis. This fact could in part be explained by the intrinsic difficulties above and the difficult genetic manipulation and bad reputation of the MCFs commonly used in these bioprocesses (i.e., mycobacterial species). The main challenge to become competitive with the current industrial chemical processes is to design more robust bioprocesses with higher substrate conversion yields and product selectivity. To achieve all these challenges, it will be necessary to construct new MCFs based on the implementation of synthetic biology and systems biology approaches.

Now that steroids are determined by mass based methods, the carbon 13 content isotope of plant derived steroids is different to the steroids produced in humans in vivo. This is exploited when detecting abuse of endogenous anabolic steroids in sport. The urinary steroids are isolated before gas chromatographic separation. The steroids are then converted to carbon dioxide by combustion then the C^{13} to C^{12} ratio is determined of metabolites.

In recent years, high-performance liquid chromatography (HPLC) has become a popular separation technique for steroids in extracts of biological material. Although the columns used in these methods offer superior resolution, the bonded phases which are not capped, and thus have residual active sites, are not without chemical effects on steroids, particularly aldosterone and its precursors.

1.1.7 Chemical reactions with steroids

Acetylation, benzoylation, hydrazone and ether formation reactions were used extensively in the early years of characterization of steroids usually combined with chromatography of the products. Ketone reduction can be performed with sodium borohydride or with additional cerium chloride heptahydrate for α,β-unsaturated ketones like testosterone. Reduction of C-17 ketones gives primarily 17β-hydroxyl steroids as the sole diastereoisomer. Reduction of C-3 or C-20 ketones gives diastereoisomers. 3-Keto reduction gives about 10 times more of the 3β hydroxy isomer. C-20 reduction of pregnenolone favors the 20R diastereoisomer by 6 to 1 ratio.

Mild oxidations of secondary hydroxyl groups to ketone groups and oxidation of side chains to give 17-keto steroids are achieved with a number of reagents, e.g., dilute chromic acid, periodate, sodium bismuthate and manganese. Some of the reactions were incorporated into quantitative assays by combining with formation of colored products, e.g., 17-ketosteroids to 2,4-dinitrophenylhydazones (Bartos and Pesez, 1979). Other reactions include hydroxylation of double bonds, dehydration, reductive elimination with zinc. The above reactions are rarely used in the laboratory today. A phenomenon of epimerization of methylated steroids has been observed in the analysis of androgenic anabolic steroids (Schänzer et al., 1992) (Fig. 1.1.13).

1.1.8 Commercial sources of steroids

In addition to the common suppliers of chemicals, there are a number of specialist manufacturers and suppliers for steroids.

These include:

- Steraloids P.O. Box 689, Newport, Rhode Island 02840, United States.
- RESEARCH PLUS INC, P.O. Box 712, Farmingdale, NJ 07727.
- National Measurement Institute, Sydney, Australia.
- Ceriliant Corporation | 811 Paloma Drive, Suite A | Round Rock, Texas 78665.

1.1.9 Steroid metabolites

The steroid hormones are inactivated mainly by hepatic reduction of the 3-keto-4-ene group. Water solubility of steroids is desirable for urinary excretion. This is achieved by hydroxylation of the steroids and conjugation, usually as glucuronide or sulfate esters. Steroids in human and animal excreta get into soil, groundwater, wastewater and sewage treatment plant effluents. The removal of steroids is now an environmental issue. The steroids can get broken down by the actions of bacteria, fungi and algae.

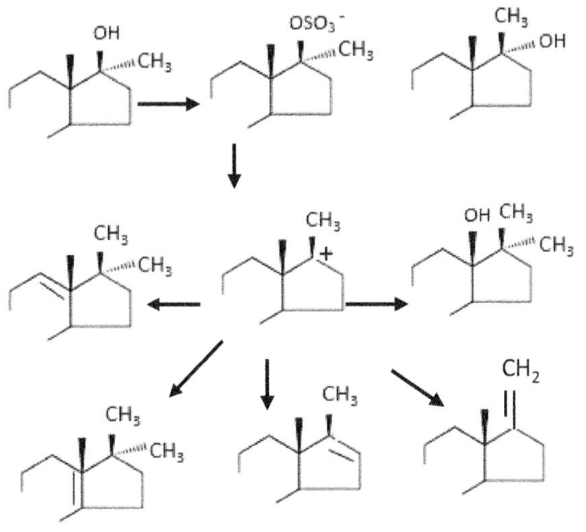

FIG. 1.1.13 Reaction scheme for 17-epimerization. (1) 17β-Hydroxy-17α-methyl steroid; (2) 17,17-dimethyl-18-nor-13-ene; (3) 17,17-dimethyl-18-norandrost-12-ene; (4) 16-ene; (5) 17-methylene; (6) 13-hydroxy-17,17-dimethyl rearrangement product; (7) 17α-hydroxy-17β-methyl epimer. *(From Schänzer W, Opfermann G, Donike M. 17-Epimerization of 17 alpha-methyl anabolic steroids in humans: metabolism and synthesis of 17 alpha-hydroxy-17 beta-methyl steroids. Steroids 1992;57(11):537–50. Fig. 2 p. 542.)*

The products may have influences aquatic organisms in a hormonal, negative or endocrine disrupting manner. The increasing environmental and public health risk requires novel ways to eliminate some compounds from the environment. Incomplete removal of estrogenic compounds will lead to residual estrogenic activity in domestic water supplies.

Syntheses of many steroids for reference purposes have been achieved with chemical reactions or by the use of enzyme activities in microorganisms (Fernandes et al., 2003; Donova and Egorova, 2012; Herráiz, 2017; Fernández-Cabezón et al., 2018; Batth et al., 2020). Multiple hydroxylases are encountered among steroids in urine, the newborn infant is particularly interesting in this regard. Hydroxylation at C1, C6, C15 and C18 is found among the urinary metabolites of adrenal steroids (Taylor et al., 1978; Joannou, 1981; Kraan et al., 1993; Christakoudi et al., 2010, 2012a,b, 2013) Tentative identification of many such steroids in urine of newborns with genetic defects of steroid 21-hydroxylase was achieved using gas chromatography retention time and mass spectral shifts with GC-tandem mass spectrometry. During solvolysis procedures 6-hydroxy-3-keto 4-ene is unstable (Kornel and Motohashi, 1965) and cholesterol appeared to be a precursor for pregnenolone and DHEA obtained from the fraction assumed to contain sulfated steroids (Liere et al., 2009). An estrogen that contains 1,2-dihydroxybenzene in its structure is called a catechol estrogen, both 2-hydroxyestradiol and 4-hydroxyestradiol are examples. Methoxylated estrogen metabolites are called guaicol estrogens. Steroids in horses are 7-dehydro-estrogens and for estrone this gives the metabolite called Equilin.

1.1.9.1 Steroid conjugates

The reduced metabolites are made more water soluble by conjugation usually at the C3 or C21of the corticosteroids or C17 of the sex steroids. The glucuronides and sulfates are the commonest forms removed from the body in urine, bile and feces. Mono-, di- and mixed conjugates are encountered. A number of other conjugates have been reported including cysteine, acetylcysteine, acetylglucosaminides, glutathione and fatty acid esters (e.g., stearate; Appendix 1.1.1). Sapogenins and cardiac glycosides are related compounds based on the steroid nucleus. In the terpenes, conjugation with a much wider range of chemicals is found, e.g., sugars such as arabinose, galactose, glucose, rhamnose and xylose.

1.1.10 Synthesis of steroid conjugates

With the interest in the analysis of metabolites in biological materials (metabolomics) there is a need for reference preparations of steroid conjugates as well as the free steroids. Steroids can be conjugated at one or more sites usually with glucuronic or sulfuric acids singly, doubly or mixed groups (Pranata et al., 2019) (Fig. 1.1.14). Some examples of the preparation of steroid conjugates are in Appendix 1.1.2.

1.1.10.1 Mono-glucuronides

Glucuronide conjugates of steroids have been prepared chemically by condensation of steroids with tri-*O*-acetyl-1-bromo-1-deoxy-a-D-glucosiduronate. Yields were less than 40%. Glucuronylation is today better performed

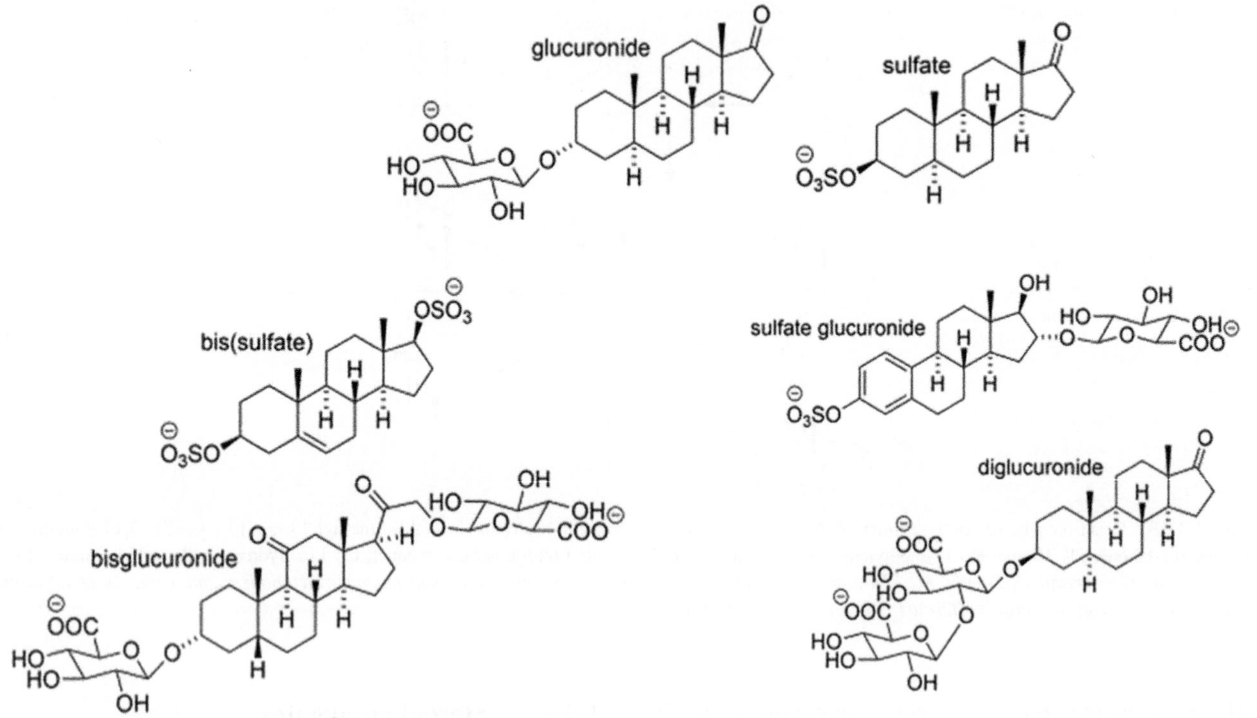

FIG. 1.1.14 Examples of doubly conjugated steroid metabolites. *(Modified from Pranata A, Fitzgerald CC, Khymenets O, Westley E, Anderson NJ, Ma P, Pozo OJ, McLeod MD. Synthesis of steroid bisglucuronide and sulfate glucuronide reference materials: unearthing neglected treasures of steroid metabolism. Steroids 2019;143:25–40. Fig. 2 p. 28.)*

enzymatically with *Escherichia coli* glucuronyl synthase as catalyst (Ma et al., 2014) using α-ᴅ-glucuronyl fluoride as the glucuronide donor (Fig. 1.1.15).

This has been applied to hydroxylated keto-steroids with various structures and stereochemistries and successfully produced a range of steroid monoglucuronides with 5%–90% conversion. Direct glucuronylation of 5α-androstane-3β,17α-diol afforded 5α-androstane-3β,17α-diol 3-glucuronides as the sole conjugated product (Doué et al., 2015; Badoud et al., 2013; Fabregat et al., 2013a,b).

1.1.10.2 Bis-glucuronides

When the *E. coli* glucuronyl synthase reaction is applied to estradiol there is a mixture of estradiol bisglucuronide, estradiol 3-glucuronide and estradiol 17-glucuronide. Steroid bisglucuronides were prepared chemically by Mattox and colleagues in the 1980's but these days glucuronide bis-conjugates (such as 5α-androstane-3β,17β-diol) are prepared in a single glucuronylation reaction of the steroid diols using an excess of α-ᴅ-glucuronyl fluoride donor with the glucuronyl synthase enzyme (Pranata et al., 2019) (Fig. 1.1.16).

Steroid diols gave >98% conversion to conjugated steroid mixtures although the 3α alcohol did not react. A weak anion solid phase extraction (WAX SPE) purification was performed to remove the unreacted steroid diols (see

Chapter 2.3 Steroid Purification for more on solid phase extraction SPE). The more polar compound (steroid bisglucuronide) was eluted with lower concentrations of methanol in water and the less polar compound in the mixture (steroid diol mono-glucuronides) were subsequently eluted with 100% methanol. Many steroid bisglucuronides could be eluted selectively with 15%–25% v/v methanol in water. An even lower methanol concentration was required to selectively elute estradiol bisglucuronide (10% v/v methanol in water) and the less polar compounds based on pregnene skeletons that needed 50% and 40% v/v methanol in water, respectively.

1.1.10.3 Sulfates

A range of methods have been developed to prepare steroid sulfates (Waller and McLeod, 2014; Mitamura et al., 2014; Okihara et al., 2010; Al-Horani and Desai, 2010), including the reaction of the parent steroid with sulfate salts and acetic anhydride, chlorosulfonic acid, amine complexes of sulfur trioxide, sulfuric acid and carbodiimides, sulfamic acid or more recently by novel sulfuryl imidazolium salts. These reactions however, while effective in affording the desired sulfate compounds, generally require significant chemical expertise and may also require harsh or hazardous conditions, specialized reagents or complicated purification methods. These factors make small-scale synthesis of

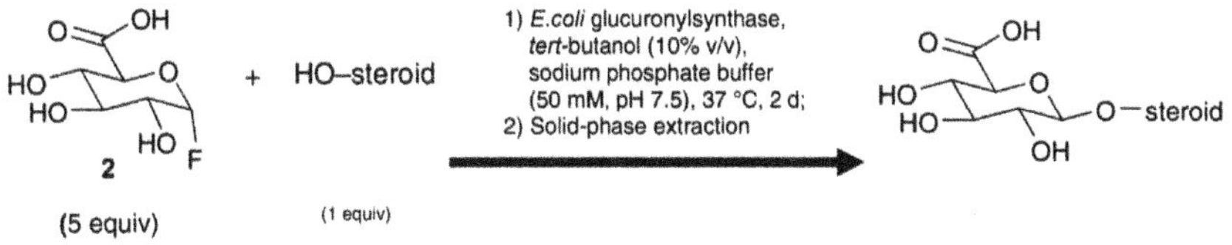

FIG. 1.1.15 The glucuronylsynthase protocol. *(Modified from Ma P, Kanizaj N, Chan SA, Ollis DL, McLeod MD. The* Escherichia coli *glucuronylsynthase promoted synthesis of steroid glucuronides: improved practicality and broader scope. Org Biomol Chem 2014;12(32):6208–14. Scheme 2 p. 6209.)*

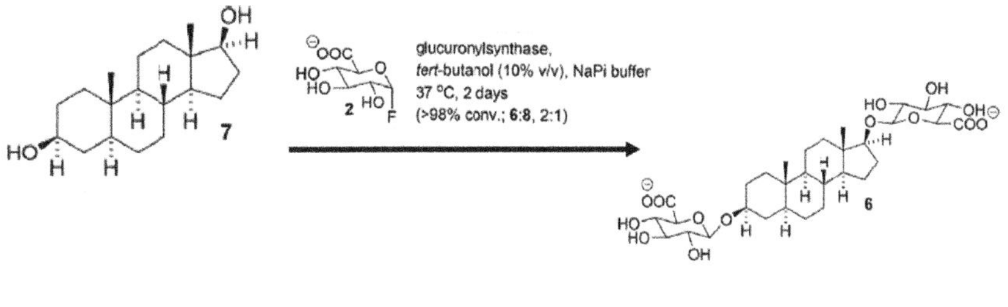

FIG. 1.1.16 One-step synthesis of 5α-androstane-3β,17β-diol bisglucuronide promoted by the enzyme *E. coli* glucuronyl synthase. *(From Pranata A, Fitzgerald CC, Khymenets O, Westley E, Anderson NJ, Ma P, Pozo OJ, McLeod MD. Synthesis of steroid bisglucuronide and sulfate glucuronide reference materials: unearthing neglected treasures of steroid metabolism. Steroids 2019;143:25–40.)*

steroid sulfates for analytical purposes a challenging undertaking. Simple synthetic routes to steroid sulfates would facilitate the identification of metabolites and assist in the development of methods targeting these analytes.

A method suitable for use by analytical laboratories takes advantage of a rapid purification by solid-phase extraction (SPE) with potential in chemical synthesis (Waller and McLeod, 2014) (Fig. 1.1.17).

The application of sulfur trioxide amine complexes appeared to offer the greatest utility due to their commercial availability, ease of handling, reasonable stability to residual moisture and mild reaction conditions as opposed to competing methods. Sulfur trioxide pyridine complex could be weighed in the laboratory without special precautions. A solution of sulfur trioxide pyridine complex in DMF (100 mg/mL) was used for the sulfation reactions and when stored in a sealed vial at 4°C maintained activity for 2 weeks. In contrast to typical steroid sulfation reactions which use pyridine as the reaction solvent, DMF and 1,4-dioxane are used instead to maintain compatibility with the SPE protocol and to reduce toxicity and odor concerns. Pregnenolone and DHEA have been sulfated using sulfur trioxide in chloroform. Testosterone (1 mg) could be reliably converted to testosterone 17-sulfate with >98%

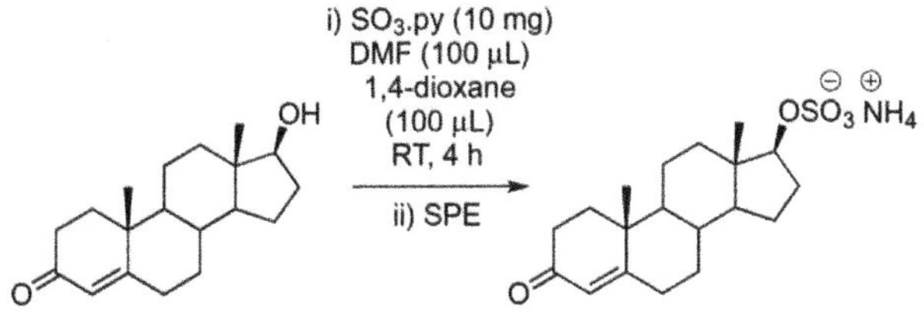

conversion. On a larger scale (10 mg) synthesis of testosterone 17-sulfate, the isolated yield (94%) showed reasonable concordance with this high conversion. These conditions enabled the synthesis of a wide range of secondary alcohol-derived steroid sulfates. Those results highlight the power of this approach for the small-scale synthesis of steroid sulfate compounds for analytical purposes.

Sulfation of xenobiotics is known to occur in the liver, and the enzymes required for the donor 3′-phosphoadenosine-5′-phosphosulfate (PAPS) production are present in liver preparations. In vitro technologies have been applied that typically make use of enzymatic products from liver tissue. Homogenized liver is centrifuged at 9000g for 20 min to isolate a supernatant commonly referred to as the S9 fraction. This fraction includes disrupted membranes of the endoplasmic reticulum (microsomes) and the soluble components of the cytosol. The S9 fraction can be fractionated further by ultracentrifugation at 100,000g to isolate the microsomal fraction (pellet) from the cytosolic fraction (supernatant) and all three preparations are commercially available. From the perspective of steroid metabolism: microsomes are a concentrated source of cytochrome P450, flavin mono-oxygenase and uridine 5′-diphospho-glucuronosyltransferase (UGT) enzymes; liver cytosol contains aldehyde oxidase and sulfotransferase (SULT) enzymes and S9 fraction contains all these components.

Cofactors must be added to these liver extracts for in vitro metabolism: glucuronylation by S9 fraction or microsomes with uridine 5′-diphosphoglucuronic acid (UDPGA) or sulfation (sometimes called sulfonation) by S9 fraction or cytosol with (PAPS) (Venkatachalam, 2003). However, the PAPS cofactor required for sulfation is prohibitively expensive (around US$1600 for 25 mg of PAPS, Sigma-Aldrich 2019 price) and chemically unstable, and as a result in vitro studies are limited. There have been some reports detailing biological synthesis of steroid sulfates using in vitro systems, but these have

i) SO₃.py (10 mg)
DMF (100 μL)
1,4-dioxane (100 μL)
RT, 4 h
ii) SPE

testosterone (1 mg) → **1a** (>98% conversion)

FIG. 1.1.17 Small-scale synthesis and solid phase purification (SPE) of steroid sulfates. *(Modified from Waller CC, McLeod MD. A simple method for the small scale synthesis and solid-phase extraction purification of steroid sulfates. Steroids 2014;92:74–80. Scheme 1 p. 27.)*

not been widely adopted by laboratories. The limitations associated with PAPS can be overcome by in situ synthesis. An in vitro approach for the preparation of sulfate metabolites has been described that employs a series of six bacterial enzymes derived from *Rhizobium meliloti* and *E. coli* for the in situ generation of PAPS. Many of the bacteria are not be readily available to laboratories. The approach involves ATP-sulfurylase to catalyze the sulfation of ATP to generate adenosine-5′-phosphosulfate (APS). This compound is subsequently phosphorylated by APS-kinase to generate PAPS and adenosine diphosphate (ADP). In animal cells, these two enzymes are expressed as a bifunctional protein molecule named PAPS synthase (PAPSS).

Following PAPS synthesis, a sulfotransferase (SULT) can then catalyze the sulfation of a target molecule hydroxyl group. In addition to the sulfated metabolite, 3′-phosphoadenosine-5′-phosphate (PAP) is released and is subsequently dephosphorylated and re-phosphorylated in several enzyme catalyzed steps to afford ATP. Overall this protocol has been able to generated a number of sulfate compounds in high yield, but may be difficult to implement due to the requirement for the bacterial enzymes.

The PAPSS isoform PAPSS2b localizes in the cytoplasm and is therefore available for PAPS generation in the liver S9 fraction or cytosol. ATP (around US$50 for 1000 mg, Sigma Aldrich, 38 Castle Hill, Australia in 2019) and sodium sulfate has been used as inexpensive precursors for the generation of PAPS in situ for in vitro metabolism targeting sulfate metabolites. Optimized conditions have been used to generate synthetic anabolic steroids as sulfate conjugates (Weththasinghe et al., 2018).

1.1.10.4 Mixed sulfate—Glucuronide steroids

In addition to the limited lists of steroid conjugates from the above suppliers of steroids, many publications have addressed the problem of steroid supply, the authors may be willing to provide or sell the products. Chemical and biological methods of preparation have been used (Pranata et al., 2019; Hintikka et al., 2008; Jäntti et al., 2007). The purity of the compounds will need to be verified according to the qualitative or quantitative requirements. Steroids conjugated with cysteine and glutathione have been found as well as sulfates and glucuronides. Mixed conjugates are also found. A selected list of references is presented in Appendix 1.1.1 at the end of the chapter.

Specialist commercial suppliers include:

- Steraloids, United States,
- NMI, Australia and
- LGC standards, Milano, Italy.

1.1.11 Reagents for immunoassay

Immunoassay for quantitative analysis of steroids became popular from 1973 and is still used for rapid results although the specificity has been challenged and the technique is being phased out of use. The procedure (to be described in more detail later in Chapter 2.4) requires a preparation of steroid antibodies and a steroid label. The steroid label is mixed with the sample containing the steroid and the antibody such that the antibody is around 50% saturated with steroid. The label and the unlabeled steroid compete for antibody binding sites so that as more unlabeled steroid is in the sample the less will be the binding of the label. A calibration curve is prepared with known concentrations of the steroid plotted against the bound or free label. Samples with unknown concentrations of the steroid are then included in the assay and results compared with the calibration.

1.1.11.1 Preparation of antisera for steroid immunoassays

For many years, the quantitative measurement of steroids relied upon immunoassay techniques which are described in more detail in the Chapter 2.4 on analysis. The essential component of these assays is steroid antibodies. Several editions of the Immunoassay Handbook have been published with thorough reviews of all aspects of the technique (Wild, 2013). In order to generate antibodies, an animal such as a rabbit is injected with an immunogen. Molecules less than 2000 kDa molecular weight are not immunogenic. Steroids are therefore conjugated as haptens to an immunogenic carrier protein such as albumin, bovine γ-globulin, bovine thyroglobulin or keyhole limpet hemocyanin. The steroid component is the desired antigen. The specificity of the resulting antisera to the antigen however will critically depend on the nature of its conjugation to the immunogenic protein (Bauminger et al., 1974).

Carboxy derivatives of steroids are the basis for the preparation of many of the haptens after amide linkage between the protein and the carboxy derivative of the steroid. The specificity of binding of hapten to antibody is determined largely by the chemical groups of the hapten far from the point of attachment of the original steroid derivative to the protein immunogen. The linkage is generally a peptide bound between the carboxyl group on the steroid derivative and a free amino group on the protein (e.g., side chain of lysine residues). Chemical bridges such as steroid *O*-carboxymethyl oximes or hemisuccinates are universally used to facilitate coupling to the protein via the reactive carboxyl group. For example, to raise antibodies to estradiol a conjugate such as the 6-(carboxymethyl)-oxime with bovine serum albumin has been used and estradiol 6-(*O*-carboxymethyl)-oximino-(2)-^{125}I iodohistamine is

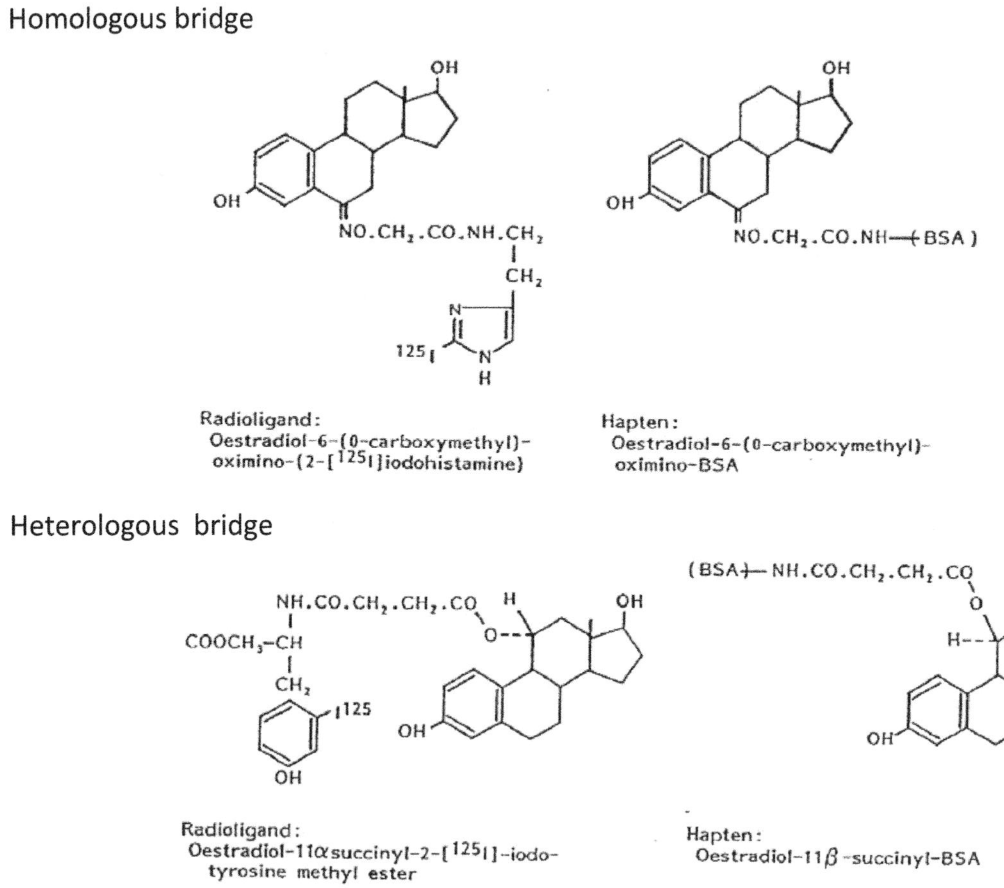

Homologous bridge

Radioligand:
Oestradiol-6-(0-carboxymethyl)-
oximino-(2-[^{125}I]iodohistamine)

Hapten:
Oestradiol-6-(0-carboxymethyl)-
oximino-BSA

Heterologous bridge

Radioligand:
Oestradiol-11α succinyl-2-[^{125}I]-iodo-
tyrosine methyl ester

Hapten:
Oestradiol-11β -succinyl-BSA

FIG. 1.1.18 Haptens and ligands for estradiol radioimmunoassay. The bridges were homologous or heterologous. *(Honour and Holownia, unpublished.)*

used as the radioligand. Examples are found in Fig. 1.1.18 and this will be discussed later in the Analysis chapter (Chapter 2.4).

A comprehensive guide to derivatives for steroid assays can be found in the review by Pratt (1978). The hapten is mixed with killed mycobacteria or pertussis vaccine in oil as adjuvant before injecting to sheep, goats or rabbits. The characterization of antibodies is discussed later. In recent years, monoclonal antibodies have been produced. A monoclonal antibody was able to discriminate testosterone and dihydrotestosterone (Kohen et al., 1982). Assays are available for steroid conjugates (Kohen et al., 1980a,b; Barnard et al., 1981).

1.1.11.2 Synthesis of radioisotope-labeled steroids

There are now few commercial suppliers of radioactive steroids or the compounds that might be suitable starting materials for in-house developments. Tritium and carbon-14 labels were the most common but needed liquid scintillation counters for detection of the weak beta radiation. Substituted steroids (e.g., histamine, tyrosine, methyl ester

and tyramine) were labeled with ^{131}I and were used in immunoassays with gamma counters. The half-lives of iodine radioisotopes are short, so the materials have to be replaced regularly. The demand for these compounds has fallen as radioimmunoassays and autoradiography have been replaced with alternative labels in binding assays. Mass spectrometry and ^{13}C and ^{1}H nuclear magnetic resonance has been used to characterize products in some studies. Estrogens have been labeled with C14 (Wang et al., 2014; Lan et al., 2019). Radioactive steroids are still useful for metabolic, pharmacokinetic and environmental studies. Tritium continues to play a central role in drug discovery but this will become less important as metabolomic studies expand. Commercial radioactive-labeled aldosterone is often impure (Brien and Slater, 1967) because aldosterone labeled with tritium or carbon 14 is destroyed on storage even if it is kept under ideal conditions, i.e., in a benzene:ethanol (95:5 v/v) solution in a deep freeze. The products of destruction have to be removed before use and this is achieved by HPLC. Purified solvents should be used when dissolving steroid because impurities (e.g., peroxides in ethyl acetate) may concentrate when solvents are evaporated (Marques et al., 2015; Valleix et al., 2006).

Where derivatives of steroids can be used, such as in quantitative methods on gas chromatography, then radioactivity in the reagents can be used in preparation of steroid derivatives as of the internal standards. Labeled Girard reagent and labeled silylating reagent for the formation of radioactive-labeled trimethylsilyl ether and other derivatives are available. The reaction should be performed in parallel with the use of unlabeled reagent.

Commercial sources of radioisotope-labeled steroids and reagents are

- Perkin Elmer https://www.perkinelmer.com/category/steroids,
- American Radiolabeled Chemicals (Saint Louis, MO, United States) https://www.arcincusa.com/ and
- CEA, Gif-sur-Yvette, France (Euriso-Top, CEA-Saclay, France) https://www.eurisotop.com/.

1.1.11.3 Labeled steroids for immunoassay detection (manual and platform assays)

The earliest immunoassays used radioactive labels to mark steroids in the reaction with antibodies. Concern over exposure of staff to radioactivity and regulations for disposal of radioactive material stimulated the development of alternative labels. There was also a drive for greater sensitivity in steroids assays since immunoassays could achieve steroid detection above background of zeptomoles (10^{-21}). Steroid immunoassays have changed format over 60 years from labor intensive tests in tubes to incorporation into the format of automated platform assays. Immunoassays will continue for a while but in many laboratories methods based on mass spectrometry are replacements because much greater specificity can be achieved. Immunoassay methods on analytical platforms will continue because automated sample processing is in line with the other analytes for clinical practice and there is no need for skilled personnel. Mass spectrometric methods with gas or liquid chromatography for separating steroids from an extract of a biological fluid enables the quantification of several steroids in one analysis. For these reasons, the reader is advised to seek detailed information on immunoassays from the literature at the end of the Quantitative Analysis chapter (Chapter 2.4).

An enzyme can be used as the label, in some cases utilizing its catalytic properties to generate colored, fluorescent or luminescent compounds from a neutral substrate. The reagents are more stable than radioisotopes. The enzymes most commonly used are horseradish peroxidase (HRP) and alkaline phosphatase (AP). Steroid is conjugated to lysine groups in HRP and amino groups in AP. HRP is an oxidoreductase that catalyzes a number of hydrogen donors to reduce hydrogen peroxide and generate colored, fluorescent or luminescent products. AP catalyzes hydrolysis of phosphate esters of primary alcohols, phenols and amines.

Automated immunoassays can be in direct competitive binding or immunometric protocols. In some systems, the antibody is coated to magnetic particles, the second antibody is labeled with the tracer. Free and bound components are then separated by magnetic separation of the micron-sized paramagnetic particle solid phase reagent. A bound chemiluminescent tracer is measured by a luminometer after the tracer is oxidized yielding a flash of light in a box where light is completely eliminated that is concentration dependent. For example, an isoluminol derivative is attached covalently to a carboxy group on a steroid. A hemisuccinate of 11α-hydroxyprogesterone is converted to an activated N-succinyl ester then reacted with the isoluminol compound (Fig. 1.1.19A). Chemiluminescence with an acridinium ester tracer for signal generation has been shown to be at least as sensitive as radioimmunoassay (Kohen et al., 1979a,b, 1980a,b, 1981; Pazzagli et al., 1981a,b,c,d; Fig. 1.1.19B).

Steroids can also be biotinylated, to compete with steroid in the sample for binding sites of antibody coated onto a solid phase of magnetic particle (Dressendörfer et al., 1992). Streptavidin with an enzyme tracer is used to bind to the bound biotin. After a wash step, the substrate for the enzyme is added to generate a signal. Alkaline phosphatase, HRP, fluorescein and rhodamine have been used in the streptavidin conjugates. Patients taking biotin supplements have been found to have abnormal immunoassay test results when streptavidin-biotin assays are used.

1.1.11.4 Other labeled steroids

Fluorescent-labeled steroids are used for visualization of steroids in tissues and cellular compartments to understand steroid trafficking (Králová et al., 2018; Fig. 1.1.20). Two steroids P1 and P2 are substituted with pyridine and then attached to fluorophores P1 is abiraterone. Fluorophores are attached on P1 via the 3-hydroxyl group to give FP-1 to FP4 or on pyridyl group for probes F5-F8. Six further probes have been prepared (not shown) In FP9 and FP-10, fluorophores are substituted at both sites. Fluorophores F11 to F14 were prepared from P2. FP-5 was the most promising probe to track cholesterol trafficking.

1.1.12 Reagents for analysis of steroids with mass spectrometry

The concentrations of steroids in biological fluids has over the past 50 years largely been by forms of immunoassay. The reaction of steroids in the sample is compared with the concentration effects in a calibration curve. The lack

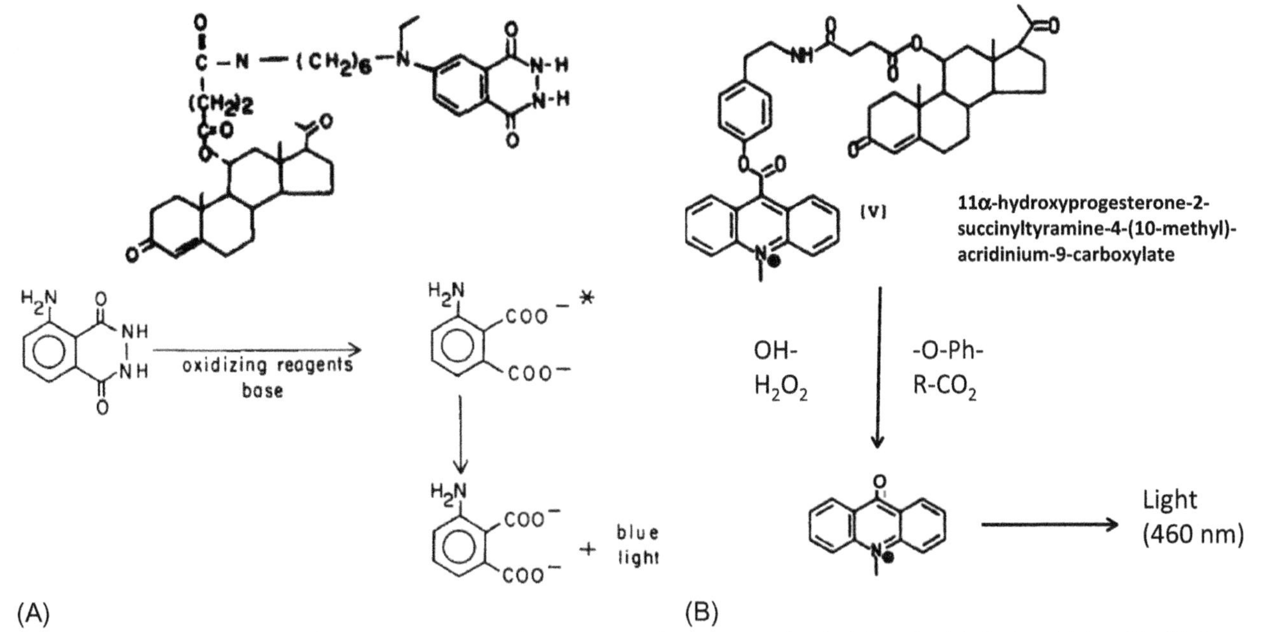

FIG. 1.1.19 (A) Acridinium ester of progesterone. (B) Luminol conjugate of progesterone. Luminol (3-aminophthalhydrazide) is oxidized in base to yield chemiluminescence corresponding to the fluorescence of the aminophthalate product. ((A) Modified from Klinger W, Wiegang G, Knuppen R. Chemiluminescent labels for steroid immunoassays. J Steroid Biochem 1987;27:41–45. Fig. 5 p. 43. (B) Modified from Kohen F, Kim JB, Lindner HR, Collins WP. Development of a solid-phase chemiluminescence immunoassay for plasma progesterone. Steroids 1981;38(1):73–88. Fig. 1 p. 76.)

of specificity in these assays has led to the increased adoption of mass spectrometric techniques. The quantitation of steroids by mass spectrometry is often essentially an isotope dilution analysis when a steroid containing a stable isotope with mass higher than the native steroid is used. The nature and quality of the reagents is important in whatever method is used. The traceability of the reagents now has to be described in publications based on the analysis of steroids.

An important question for calibration curves is what matrix is used to be comparable with the biological sample. Tests are subject to interferences and commercial materials are available.

1.1.12.1 Synthesis of stable isotope-labeled (SIL) steroids

1.1.12.1.1 Internal standards in quantitative analysis

Mass spectrometric methods for quantitative steroid assays use stable isotope-labeled (SIL) internal standards. For many steroids labeled with stable isotopes, there are commercial sources prepared in much the same way as radio-isotopes, usually exchange reactions. The sites of isotope incorporation are not necessarily appropriate for modern uses. The demand for SIL-steroids has risen as methods

for steroid analysis based on mass spectrometry have been demanded with concerns about the specificity of immuno-assays. Some of these are now available commercially. The ^{13}C and deuterium SIL are by far the commonest in use.

The ^{13}C-Labeled steroids have major advantages for metabolism studies in vivo because of the stability of the label and the avoidance of isotope effects to the action of enzymes. The synthesis of these isotopes however can be synthetically rather complex, laborious and expensive. Ideally when used as internal standards the isotope enrichment should exceed 95%.

Deuterium labeling is relatively inexpensive and available at high isotopic purity in a larger number of chemical reagents. The incorporation of deuterium in the steroid molecule is generally less difficult than C^{13}-labeling techniques. Two early reviews of the synthesis of deuterium-labeled steroid hormones and their application to quantitative and metabolism studies were published by Johnson et al. in 1981 and Wudy in 1990 (Johnson et al., 1981; Wudy, 1990). Tom Baillie prepared deuterium-labeled steroid sulfate (Baillie et al., 1975a,b). Since then, a relatively small variety of practical synthetic routes leading to the major deuterium-labeled steroid hormones have been published. Appendix 1.1.3 summarizes synthetic procedures for the preparation of deuterium-labeled and C^{13}-labeled biologic steroid hormones suitable for

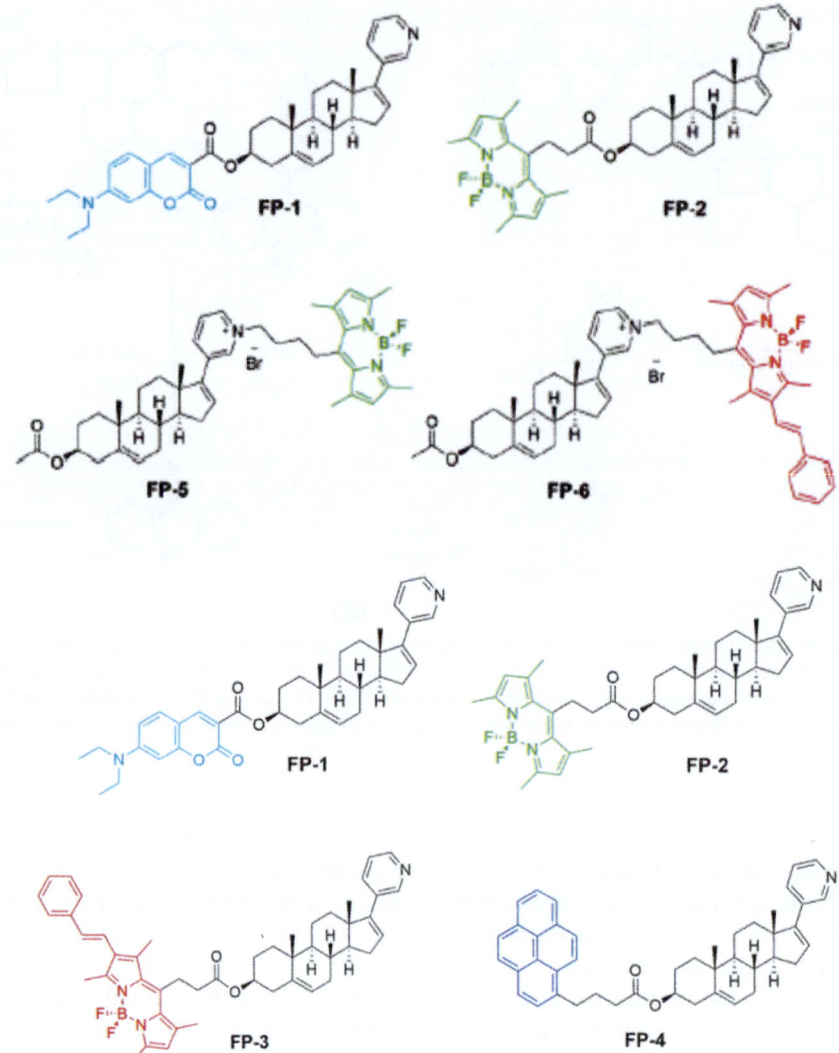

FIG. 1.1.20 Examples of sterol probes for monitoring sterol trafficking. *(Modified from Králová J, Jurášek M, Krčová L, Dolenský B, Novotný I, Dušek M, Rottnerová Z, Kahle M, Drašar P, Bartůněk P, Král V. Heterocyclic sterol probes for live monitoring of sterol trafficking and lysosomal storage disorders. Sci Rep 2018;8(1):14428. Fig. 1 p. 2.)*

application to studies using isotope dilution mass spectrometry (IDMS) since 1990.

Papers dealing with the deuterium labeling of steroid molecules for internal standards are summarized in Appendix 1.1.3; however, this does not include compounds for the elucidation of stereochemistry which were largely the earlier work from the laboratory of Carl Djerassi. For a general summary of deuteration techniques in the steroid field, the reader is referred to the comprehensive review by Tokes and Throop (1972).

The major drawback in the use of deuterium-labeled steroid hormones is their susceptibility to loss of the label either due to chemical exchange, biological reactions or chemical processes during the analytic procedure. As a rule of thumb, the easier a deuterium label is introduced, the easier it can be lost. For example, deuterium labels adjacent to carbonyl groups are easily introduced by exchange reactions but are susceptible to back exchange unless the carbonyl group is removed or modified. Therefore, isotopic labels have to be incorporated at positions that are chemically stable and are not subject to biological attack.

Hydrogen isotopes such as deuterium have many properties of ideal tracer nuclides. Deuterium enrichment of compounds can be detected with very high sensitivity, by conventional mass spectrometry. In recent years, the rapid development of high-resolution mass spectrometry has led to a significant increase in deuterium labeling applications, because there is no need for chemical derivative formation as with GC-MS. The popularity of the hydrogen isotopes in the life sciences stems from their ability to allow for the direct incorporation of a unique detection signal into a target molecule without changing its chemical structure, physical properties or biological activity. Consequently, hydrogen isotopes enable the detection and quantification of drug related material or the discovery of new biological pathways in experimental animals or humans. Deuterium is a stable isotope and can thus be handled under standard laboratory conditions without special permissions, handling licenses or radiation safety measures.

The methods described in the literature for deuterium labeling of steroids are often multistep processes with relatively poor overall yields and sometime requiring expensive starting reagents or materials. In some methods, the starting material may have required additional multistep preparation. The incorporation of deuterium into an organic molecule can be achieved by two principal routes, namely either by direct hydrogen isotope exchange (HIE) or by a conventional multistep synthesis. Depending on the complexity of the chemistry, the chemical structure of the target molecule and the labeling positions, a classical synthesis approach, starting from appropriate commercially available labeled precursors, can be very time- and resource-consuming. Therefore, methods for the fast and convenient late-stage introduction of deuterium or tritium into organic molecules have been extensively investigated in recent years (Furuta et al., 1999).

Commercial stable isotopes of steroids are often synthesized in the same way as radioactive steroids and the positions for substitution tend to be the same. Deuterium is by far the commonest stable label as immunoassays are being replaced with GC or LC coupled with MS and MS/MS in isotope dilution methods. The deuterium is often introduced by exchange reactions, it is important that the label does not back exchange to hydrogen during the application. The number of atoms of deuterium introduced needs to be considered carefully in the context of detection by mass spectrometry so that there is no interference from steroids of the same mass of the parent molecule and in some cases of fragments.

The Isosciences Company (King of Prussia, PA, United States) described the spectrum of commercial d8-aldosterone (2, 2, 4, 6, 6, 17α, 21, 21-d8) where the molecular ion of the d8 isotopomer had 60.7% enrichment and isotopomers at M+7 and M+6 meant that the incorporation of deuterium was actually 7.7d/molecule. The deuterium labels were at base-sensitive positions. Loss of deuterium is possible during the analytical procedure depending on pH and solvents used. The analyst must verify the integrity of the steroid as an internal standard and this may be unanswered. That requires analysis of the full range m/z of analyte and internal standard and that is not usually done. Isosciences now produce a (9α,11α,12,12)-d4-aldosterone with labels at completely nonexchangeable positions and where d4 is 96.9% enrichment.

Steroids are a special problem in this regard where dihydro and tetrahydro reduced metabolites for the steroid hormones will be present in the biological samples. Steroids with three or more deuterium atoms substituted are thus used but the extra deuterium atoms shift the retention times of the labeled steroid in chromatography so that they may cease to act as carriers for the analyte. This will be considered in more detail in Chapter 2.4 steroid analysis.

Carbon 13 substitution is less commonly used mainly because the synthesis is more difficult so the compounds are more expensive. However the label is more stable when in the core of the steroid molecule (Zomer and Stavenuiter, 1990). Carbons at 1, 2, 4 and 19 replaced with ^{13}C are commonly generated. The ability to measure isotope distribution at natural abundance with high accuracy and high precision has increased the application of GC-combustion-isotope ratio mass spectrometry in recent years particularly in doping control. The ratio of carbon 13 isotope to carbon 12 isotope is determined. Some specialist steroid syntheses are summarized in Table 1.1.5.

Where labeled steroids are used as internal standards for isotope dilution mass spectrometric quantitative assays, very small amounts are used.

This creates problems in the storage of such compounds in dilute solution format. Some products are supplied in

TABLE 1.1.5 Publications with synthesis of stable isotope-labeled steroids.

Steroid	Labeling	Purity	Senior author	Year	Publication
6-Hydroxy cortisol	1,1,19,19,19-^{2}H$_5$	90	Furuta T	2003	Steroids 68 (7–8):693–703
6-Hydroxy cortisol	1,2,4,19-^{13}C$_4$ 1,1,19,19,19-^{2}H$_5$	92	Furuta T		
6-Hydroxy cortisone	1,1,19,19,19-^{2}H$_5$		Furuta T		
6-Hydroxy testosterone	1α,16,16,17-^{2}H$_4$		Furuta T		
Cortisol	9,11,12,12-d$_4$		Shibasaki H	1992	Steroids 57 (1) 13–17
Cortisol	1,1,9,11,12,12,19,19,19-d$_9$		Shibasaki H		
Cortisol	1,2,4,19-^{13}C$_4$		Furuta T	2000	Steroids 65(4) 180–189
Cortisone	1,2,4,19-^{13}C$_4$		Furuta T		

Continued

TABLE 1.1.5 Publications with synthesis of stable isotope-labeled steroids—cont'd

Steroid	Labeling	Purity	Senior author	Year	Publication
Cortisol	$1,2,4,19\text{-}^{13}C_4$ $1,1,19,19,19\text{-}^2H_5$		Furuta T		
Cortisone	$1,2,4,19\text{-}^{13}C_4$ $1,1,19,19,19\text{-}^2H_5$		Furuta T		
Tetrahydrocortisol	$1,2,3,4,5\text{-}d_5$	86	Furuta T	1999	Steroids 64 (12) 805–810
Allo-tetrahydrocortisol	$1,2,3,4,5\text{-}d_5$	75	Furuta T		
Tetrahydrocortisone	$1,2,3,4,5\text{-}d_5$	82	Furuta T		
Testosterone; cortisol; progesterone; estradiol	$3\text{-}^{13}C$; $4\text{-}^{13}C$; $3,4\text{-}^{13}C_2$; $1,2,3,4\text{-}^{13}C_4$		Zomer G	1990	Steroids 55 (10) 440–442
Androstenedione			Johnson DW	1985	J Steroid Biochem 22 (3) 349–353
Testosterone					
Androstenedione	$19\text{-}d_3$, $19\text{-}d_2$ and $19\text{-}d_1$		Dyer RL	1979	Steroids 33 (6) 617–624
Estradiol	$2,4\text{-}^2H_2$		Murphy RC	1974	Steroids 24 (3) 343–350
Estrone	$2,4\text{-}^2H_2$				

liquid form, others are solids. The relative insolubility of steroids in water will affect the mode of delivery of the steroid for any study in vivo.

1.1.12.1.2 Clinical uses

In whole body production rate (PR) studies, up to 12 mg per subject have been used. The procedure for PR is not without error because of the perturbation of the steady state. The labeled steroids are safe to use in vivo and have been for studies even in pregnant women. The metabolism and production rates of 3β-hydroxy-5α-pregnan-20-one sulfate and the 3- and 3,20-sulfates of 5α-pregnane-3β,20α-diol in pregnant women were studied (Baillie et al., 1976). These steroids labeled with deuterium in the 3α,11,11- or 3α,11,11,20β positions were injected intravenously, and the deuterium content of 13 steroids in the mono- and disulfate fractions from blood samples drawn at different times after the injection was determined by gas chromatography-mass spectrometry. Production rates and metabolic clearance rates were studied using two-pool models. Hydroxylation at C16 and 20-oxidoreduction was demonstrated.

In my laboratory, labeled progesterone ([11,11,12, 12-$^2H_4$] progesterone; [19-$^2H_3$] progesterone and [18-$^2H_3$]

progesterone) were synthesized in order to investigate the possible extra-adrenal 21-hydroxylase (Kirk et al., 1990a, b). Despite approval of the proposal by the Ethical Committee, we were unable to get insurance cover so failed to complete the study in patients and this may be a problem for others. There are problems with these studies because of the distribution of the steroid in pools in the body. In pregnancy, there is a complex relationship between mother, fetus and placenta which in theory can be overcome by using different labels in each pool.

The pharmacokinetic behavior of cortisol was studied after a trace amount (5 mg) of stable isotopically labeled cortisol [1,1,19,19,19 $^2H_5$ cortisol] was administered orally (Kasuya et al., 1995). Cortisol production rates were determined using 1,2,3,4-^{13}C-cortisol (Chapman et al., 1987) and 9,12,12-^2H-cortisol in normal subjects and patients with congenital adrenal hyperplasia (Chapman et al., 1991). Cortisol 9,12,12-$^2H3$ was used in several studies (Brandon et al., 1999; Klopfenstein et al., 2011, for examples). The interconversion of cortisol with cortisone by the enzymes 11β-hydroxysteroid dehydrogenase with time has been studied by Bryan Walker and Ruth Andrew in a number of studies (Stirrat et al., 2018; Cobice et al., 2018). D4-labeled cortisol was used to study the transfer and metabolism of cortisol by the isolated perfused human placenta.

Along with improved quantitative determinations of steroid concentrations in biological fluids, steroids in tissues can now be detected by mass spectrometry and the data compared with chemically stained tissue in adjacent histology section to localize sites of activity (Cobice et al., 2018). Some publications have addressed specific needs outside the availability from commercial suppliers. The uncertainty of risk in experimental studies in humans has limited the application of this powerful technique.

Different clinical-experimental approaches have been devised to study the metabolism of cholesterol and particularly the synthesis of bile acids, its main catabolic products. Most evidence in humans has derived from studies utilizing the administration of labeled sterols; these have several advantages over in vitro assay of enzyme activity and expression, requiring an invasive procedure such as a liver biopsy, or the determination of fecal sterols, which is cumbersome and not commonly available. Stable isotope studies (Bertolotti et al., 2012) have overcome radioactivity exposure. Isotope enrichment studies during tracer infusion has enabled steps to characterize changes in the degradation of cholesterol via the "classical" and the "alternative" pathways of bile acid synthesis. Evidence brought by tracer studies in vivo provides an exceptional tool for the investigation of sterol metabolism and integrate the studies in vitro on human tissue.

1.1.12.2 Stable-labeled steroid conjugates

Glucuronylation of some steroids has been performed using $^{13}C_6$-α-D-glucuronyl fluoride to label the conjugate group. ^{18}O has also been incorporated with the glucuronic acid (Stachulski and Meng, 2013).

1.1.12.3 Alternatives to SIL-steroids

Nonnaturally occurring isomers of steroids can be used. For example, 3β,5α-tetrahydroaldosterone can be the internal standard to quantify 3α,5β-tetrahydroaldosterone in urine (Honour and Shackleton, 1977). The two isomers separate by GC so the carrier effect will be lost (see later). Another indirect approach is to prepare derivatives using stable isotope-labeled derivatizing reagent for example trimethylsilyl ether or Girard reagent.

1.1.12.4 Commercial sources of SIL-steroids include

- *Cambridge Isotope Laboratories, Inc.*, 3 Highwood Drive, Tewksbury, MA 01876. https://www.isotope.com/.
- *Santa Cruz Biotechnology, Inc.*, 10410 Finnell Street, Dallas, Texas 75220, United States. https://www.scbt.com/.
- *Steraloids*, P.O. Box 689, Newport, RI 02840, United States. sales@steraloids.com.
- *National Measurement Institute*, GPO Box 2013, Canberra ACT 2601. General enquiries: info@measurement.gov.au.
- *Toronto Research Chemicals*, 20 Martin Ross Avenue, North York, M3J 2K8 Ontario, Canada. www.trc-canada.com.
- *C/D/N Sainte-Claire*, C/D/N Isotopes Inc., 88 Leacock Street, Pointe-Claire, Quebec, Canada H9R 1H1. https://cdnisotopes.com.
- *Isosciences*, 340 Mathers Road, Ambler, PA 19002, United States. https://isosciences.com/.
- *Orphachem*, Rue Michel Renaud, Saint-Beuzire 63360, France.
- *Cerilliant*, Corporation 811 Paloma Drive, Suite A | Round Rock, Texas 78665. https://www.cerilliant.com/.
- *CEA-Labeled Compounds*, Gif sur yvette, France. www.cea.fr.
- *Qmx Laboratories*, Bolford Street, Thaxted, Essex, CM6 2PY United Kingdom. https://www.qmx.com.

Commercial sources of deuterium-labeled reagents include:

- *Sigma-Aldrich for d9-trimethylchlorosilane.*
- *Avanti Polar Lipids*, 700 Industrial Park Drive, Alabaster, Alabama 35007-9105. https://avantilipids.com/contact-us. *Avantilipids for DMABA-d4-NHS ester and d5-Girard reagent.*

1.1.12.5 Kits for quantitative MS measurement of steroid panels

Many methods for analysis of steroids by isotope dilution mass spectrometry have been developed within laboratories and are so called "home brew" methods. The validation of these methods is time consuming and expensive so there is a need for "kits" such as those in clinical laboratories for immunoassays. The following are examples:

- Chromsystems Instruments & Chemicals GmbH, Am Haag 12, 82166 Gräfelfing/Munich. https://www.chromsystems.com.
- Thermo-Fisher, San Jose, CA, United States. www.thermofisher.com.
- AbsoluteIDQstero17 Kit (Biocrates, 6020 Innsbruck, Austria). https://www.biocrates.com/.
- SteroIDQ Kit—Biocrates Life Sciences AG, 6020 Innsbruck Austria. https://www.biocrates.com/.
- CHS MS MS steroid kit from Perkin Elmer. www.perkinelmer.com/ContactUs.
- RECIPE Chemicals + Instruments GmbH, Dessauerstraße 3, 80992 München. www.info@recipe.de.

- EUREKA LAB DIVISION Srl. via Enrico Fermi, 25 60033 Chiaravalle, Ancona, Italy. Tel +39 071 74.50.790. Fax +39 071 74.96.579. Email: info@eurekaone.com.
- Cambridge Isotope Laboratories Inc. (www.isotope. com) supply a mixture of 10 reference stable isotope-labeled steroids suitable for diagnosis of congenital adrenal hyperplasia (NSK-S-CAH).

1.1.12.6 Commercial sources of biological matrices for the preparation of calibrants

A negative urine control (Surine) is available from Sigma-Aldrich and Cerilliant.

Charcoal stripped serum can be prepared in the laboratory or purchased from a number of suppliers.

A simple artificial urine can be made to the formulation in Table 1.1.6.

TABLE 1.1.6 Peptone and yeast extract were obtained from Unipath, Basingstoke and all other reagents were of Analar quality obtained from Merck, Poole, UK.

Compound	Quantity (g)	Concentration (mmol/L)
Peptone L37	1	
Yeast extract	0.005	
Lactic acid	0.1	1.1
Citric acid	0.4	2
Sodium bicarbonate	2.1	25
Urea	10	170
Uric acid	0.07	0.4
Creatinine	0.8	7
Calcium chloride 2H$_2$O	0.37	2.5
Sodium chloride	5.2	90
Iron II sulphate 7H$_2$O	0.0012	0.005
Magnesium sulphate 7H$_2$O	0.49	2
Sodium sulphate 10H$_2$O	3.2	10
Potassium hydrogen phosphate	0.95	7
Dipotassium hydrogen phosphate	1.2	7
Ammonium chloride	1.3	25
Distilled water	To 1 L	

Interference testing reagents are available from Molecular Depot, 7915 Silverton Ave #313, San Diego, CA 92126, United States. Email: info@moleculardepot.com

- K2010001 contains ascorbic and six endogenous substances to use in assay routine interference testing.
- Interference Test Kit includes concentrated stock solutions of Ascorbic Acid (176 mg/mL), Free Bilirubin (20 mg/mL), Conjugated Bilirubin (20 mg/mL), Human Hemoglobin (200 mg/mL), Human Serum Proteins (20 g/dL) and a Triglycerides mix (1000 mg/mL).
- Interference Test Kit conveniently includes the diluents to be used when testing each of the interfering substances.
- S2010001 Rheumatoid Factor (catalog #S2010002) consists of human plasma containing an extremely high level of RF (>1900 IU/mL). This product can be used to test the RF tolerance of immunoassays. It can be only diluted with the Rheumatoid Factor Diluent (catalog #S2010002).
- C2010001 Triglyceride mix for interference testing. A mixture of five saturated triglycerides that can be used for interference studies, assaying or chromatography. Mixture consists of 20% (m/m) Triacetin (C2:0), 20% (m/m) Tributyrin (C4:0), 20% (m/m) Tricaproin (C6:0), 20% (m/m) Tricaprylin (C8:0) and 20% (m/m) Tricaprin (C10:0). Available in neat solution of 1000 mg.

1.1.13 Summary

The chemistry of steroid hormones is based on a defined nomenclature around the nucleus of 18, 19 and 21 carbon atoms, all derived from cholesterol with 27 carbon atoms. An understanding of steroid chemistry is needed in healthcare in order to deal not only with the complexities of steroid biosynthesis and metabolism but also the diagnostic methods in the laboratory. Steroid determinations have moved from colorimetric tests, through immunoassay to chromatography coupled with mass spectrometry, each with needs for standards and labeled internal standards. The reagents for quantitative analysis of steroids can be commercial or "home-brew." The pharmaceutical markets for steroids are immense and methods to meet world demand will continue to be developed and refined. This book is not designed to cover all of the chemistry needed in the pharmaceutical industry. In the 1950s, the production of steroid hormones required up to nearly 60 steps mainly chemical. The world demand necessitated improvements. All sterol modifications to cortisol, androgens and progestagens are now achieved by microbiological processes. Total synthesis of steroids is now possible and steps in that process can be modified when introducing ^{13}C label into steroids to produce stable isotope-labeled steroids for mass spectrometric quantification of steroids.

Appendix 1.1.1. Conjugates other than glucuronides and sulfates

1. L.J. Meng, W.J. Griffiths, J. Sjövall, The identification of novel steroid *N*-acetylglucosaminides in the urine of pregnant women. J Steroid Biochem Mol Biol. 58 (1996) 585–98.
2. M. Nakagomi, K. Yamada, Y. Matsuki, H. Kurihara, E. Suzuki, Enzyme immunoassay for the measurement of 17alpha-estradiol 17-*N*-acetylglucosaminide in rabbit urine. Steroids. 64 (1999) 301–7.
3. R. Ramanathan, K. Cao, E. Cavalieri, M.L. Gross, Mass spectrometric methods for distinguishing structural isomers of glutathione conjugates of estrone and estradiol. J Am Soc Mass Spectrom. 9 (1998) 612–9.
4. M. Nakagomi, E. Suzuki, Quantitation of catechol estrogens and their N-acetylcysteine conjugates in urine of rats and hamsters. Chem Res Toxicol. 13 (2000) 1208–13.
5. H.J. Jung, W.Y. Lee, B.C. Chung, M.H. Choi, Mass spectrometric profiling of saturated fatty acid esters of steroids separated by high-temperature gas chromatography. J Chromatogr A. 1216 (2009) 1463–8.
6. F. Wang, A. Koskela, E. Hämäläinen, U. Turpeinen, H. -Savolainen-Peltonen, T.S. Mikkola, V. Vihma, H. Adlercreutz, M.J. Tikkanen, Quantitative determination of dehydroepiandrosterone fatty acyl esters in human female adipose tissue and serum using mass spectrometric methods. J Steroid Biochem Mol Biol. 124 (2011) 93–8.
7. C. Gómez, O.J. Pozo, A. Fabregat, J. Marcos, K. Deventer, P. Van Eenoo, J. Segura, R. Ventura. Detection and characterization of urinary metabolites of boldione by LC-MS/MS. Part I: Phase I metabolites excreted free, as glucuronide and sulfate conjugates, and released after alkaline treatment of the urine, Drug Test Anal. 4 (2012) 775–85.
8. O.J. Pozo, C. Gómez, J. Marcos, J. Segura, R. Ventura. Detection and characterization of urinary metabolites of boldione by LC–MS/MS. Part II: Conjugates with cysteine and *N*-acetylcysteine, Drug Test Anal. 4 (2012) 786–97.
9. A. Fabregat, A. Kotronoulas, J. Marcos, J. Joglar, I. Alfonso, J. Segura, R. Ventura, O.J. Pozo. Detection, synthesis and characterization of metabolites of steroid hormones conjugated with cysteine, Steroids. 78 (2013) 327–36.
10. A. Fabregat, J. Marcos, L. Garrostas, J. Segura, O.J. Pozo, R. Ventura, Evaluation of urinary excretion of androgens conjugated to cysteine in human pregnancy by mass spectrometry. J Steroid Biochem Mol Biol. 139 (2014) 192–200.
11. A. Fabregat, L. Garrostas, J. Segura, O.J. Pozo, R. Ventura. Evaluation of urinary excretion of androgens conjugated to cysteine in human pregnancy by mass spectrometry, J Steroid Biochem Mol Biol. 139 (2014) 192–200.
12. J. Marcos, M. Pol, A. Fabregat, R. Ventura, N. Renau, F.A. Hanzu, G. Casals, S. Marfà, B. Barceló, A. Barceló, J. Robles, J. Segura, O.J. Pozo. Urinary cysteinyl progestogens: Occurrence and origin. J Steroid Biochem Mol Biol. 152 (2015) 53–61.
13. A. Kotronoulas, A. Gomez-Gomez, J. Segura, R. Ventura, J. Joglar, O.J. Pozo. Evaluation of two glucuronides resistant to enzymatic hydrolysis as markers of testosterone oral administration. J Steroid Biochem Mol Biol. 165 (2017) 212–218.

Appendix 1.1.2. Synthesis of steroid conjugates

1. Mattox VR, Nelson AN, Vrieze WD, Jardine I. Synthesis of mono- and diglucosiduronates of metabolites of deoxycorticosterone and corticosterone and analysis by a new mass spectrometric technique. Steroids. 1983 Oct;42(4):349–64.
2. Shackleton CH, Mattox VR, Honour JW. Analysis of intact steroid conjugates by secondary ion mass spectrometry (including FABMS) and by gas chromatography. J Steroid Biochem. 1983 Jul;19(1A):209–17.
3. Mattox VR, Goodrich JE, Nelson AN. Chemical synthesis of glucuronidated metabolites of cortisol. J Steroid Biochem. 1983 Feb;18(2):153–9.
4. Fenselau C, Johnson LP. Analysis of intact glucuronides by mass spectrometry and gas chromatography–mass spectrometry. A review. Drug Metab Dispos. 1980 Jul-Aug;8(4):274–83. Review.
5. Cantrall EW, McGrath MG, Bernstein S. Steroid conjugates. II. The synthesis of a sulfoglucuronide derivative of 17-beta-estradiol. Steroids. 1966 Dec;8(6):967–75.
6. Kaspersen FM, Van Boeckel CA. A review of the methods of chemical synthesis of sulfate and glucuronide conjugates. Xenobiotica. 1987 Dec;17(12):1451–71.Review.
7. H. Hosoda, H. Yokohama, T. Nambara, Synthesis of cortol 3-glucuronides and cortolone 3-glucuronides. Chem Pharm Bull (Tokyo). 32 (1984) 4023–8.
8. H. Hosoda, K. Osanai, I. Fukasawa, T. Nambara, Chemical conversion of corticosteroids to 3 alpha,5 alpha-tetrahydro derivatives. Synthesis of allotetrahydrocortisol glucuronides and allotetrahydrocortisone glucuronides. Chem Pharm Bull (Tokyo). 38 (1990) 1949–52.
9. H. Hosoda, K. Osanai, T. Nambara, Chemical conversion of corticosteroids to 3alpha,5 alpha-tetrahydro derivatives. Synthesis of 5 alpha-cortol 3-glucuronides and 5 alpha-cortolone 3-glucuronides. Chem Pharm Bull (Tokyo). 39 (1991) 3283–6.
10. R. Okihara, K. Mitamura, M. Hasegawa, M. Mori, A. Muto, G. Kakiyama, S. Ogawa, T. Iida, M. Shimada, N. Mano, S. Ikegawa. Potential corticoid metabolites: chemical synthesis of 3- and 21-monosulfates and their double-conjugates of tetrahydrocorticosteroids in the 5alpha- and 5beta-series. Chem Pharm Bull (Tokyo). 58 (2010) 344–53.
11. A.V. Stachulski, J.R. Harding, J.C. Lindon, J.L. Maggs, B.K. Park, I.D. Wilson. Acyl glucuronides: biological activity, chemical reactivity, and chemical synthesis, J Med Chem. 49 (2006) 6931–45.

12. R.A. Al-Horani, U.R. Desai. Chemical Sulfation of Small Molecules - Advances and Challenges. Tetrahedron. 66 (2010) 2907–2918.

13. C.C. Waller, M.D. McLeod. A simple method for the small scale synthesis and solid-phase extraction purification of steroid sulfates, Steroids. 92 (2014) 74–80.

14. S.E. Jäntti, A Kiriazis, R.R. Reinilä, R.K. Kostiainen, R.A. Ketola. Enzyme-assisted synthesis and characterization of glucuronide conjugates of neuroactive steroids. Steroids. 72 (2007) 287–96.

15. P. Ma, N. Kanizaj, S.A. Chan, D.L. Ollis, M.D. McLeod. The *Escherichia coli* glucuronylsynthase promoted synthesis of steroid glucuronides: improved practicality and broader scope. Org Biomol Chem. 12 (2014) 6208–14.

16. A.V. Stachulski, J.R. Harding, J.C. Lindon, J.L. Maggs, B.K. Park, I.D. Wilson. Acyl glucuronides: biological activity, chemical reactivity, and chemical synthesis, J Med Chem. 49 (2006) 6931–45.

17. F. Qin, Y.Y. Zhao, M.B. Sawyer, X.K. Li. Hydrophilic interaction liquid chromatography-tandem mass spectrometry determination of estrogen conjugates in human urine, Anal Chem. 80 (2008) 3404–11.

18. M. Doué, G. Dervilly-Pinel, K. Pouponneau, F. Monteau, B. Le Bizec. Analysis of glucuronide and sulfate steroids in urine by ultra-high-performance supercritical-fluid chromatography hyphenated tandem mass spectrometry, Anal Bioanal Chem. 407 (2015) 4473–84.

19. F. Badoud, E. Grata, J. Boccard, D. Guillarme, J.L. Veuthey, S. Rudaz, M. augy, Quantification of glucuronidated and sulfated steroids in human urine by ultra-high pressure liquid chromatography quadrupole time-of-flight mass spectrometry. Anal Bioanal Chem. 400 (2013) 503–16.

20. A. Fabregat, O.J. Pozo, J. Marcos, J. Segura, R. Ventura, Use of LC–MS/MS for the open detection of steroid metabolites conjugated with glucuronic acid. Anal Chem. 85 (2013) 5005–14.

21. A. Fabregat, A. Kotronoulas, J. Marcos, J. Joglar, I. Alfonso, J. Segura, R. Ventura, O.J. Pozo. Detection, synthesis and characterization of metabolites of steroid hormones conjugated with cysteine, Steroids. 78 (2013) 327–36.

22. M. Nakagomi, E. Suzuki, Quantitation of catechol estrogens and their *N*-acetylcysteine conjugates in urine of rats and hamsters. Chem Res Toxicol. 13 (2000) 1208–13.

23. O.J. Pozo, P. Van Eenoo, W. Van Thuyne, K. Deventer, F.T. Delbeke. Direct quantification of steroid glucuronides in human urine by liquid chromatography-electrospray tandem mass spectrometry. J Chromatogr A. 1183 (2008) 108–18.

24. A. Fabregat, O.J. Pozo, J. Marcos, J. Segura, R. Ventura. Use of LC–MS/MS for the open detection of steroid metabolites conjugated with glucuronic acid. Anal Chem. 85 (2013) 5005–14

25. H. Zhang, J. Henion. Quantitative and qualitative determination of estrogen sulfates in human urine by liquid chromatography/tandem mass spectrometry using 96-well technology. Anal Chem. 71 (1999) 3955–64.

26. K. Mitamura, R. Satoh née Okihara, M. Kamibayashi, K. Sato, T. Iida T, S. Ikegawa, Simultaneous determination of 18 tetrahydrocorticosteroid sulfates in human urine by liquid chromatography/electrospray ionization-tandem mass spectrometry. Steroids. 85 (2014) 18–29.

27. A. Fabregat, L. Garrostas, J. Segura, O.J. Pozo, R. Ventura. Evaluation of urinary excretion of androgens conjugated to cysteine in human pregnancy by mass spectrometry, J Steroid Biochem Mol Biol. 139 (2014) 192–200.

28. J. Marcos, M. Pol, A. Fabregat, R. Ventura, N. Renau, F.A. Hanzu, G. Casals, S. Marfà, B. Barceló, A. Barceló, J. Robles, J. Segura, O.J. Pozo. Urinary cysteinyl progestogens: Occurrence and origin. J Steroid Biochem Mol Biol. 152 (2015) 53–61.

29. M. Nakagomi, K. Yamada, Y. Matsuki, H. Kurihara, E. Suzuki, Enzyme immunoassay for the measurement of 17alpha-estradiol 17-*N*-acetylglucosaminide in rabbit urine. Steroids. 64 (1999) 301–7.

30. F. Wang, A. Koskela, E. Hämäläinen, U. Turpeinen, H. Savolainen-Peltonen, T.S. Mikkola, V. Vihma, H. Adlercreutz, M.J. Tikkanen, Quantitative determination of dehydroepiandrosterone fatty acyl esters in human female adipose tissue and serum using mass spectrometric methods. J Steroid Biochem Mol Biol. 124 (2011) 93–8.

31. L.J. Meng, W.J. Griffiths, J. Sjövall, The identification of novel steroid *N*-acetylglucosaminides in the urine of pregnant women. J Steroid Biochem Mol Biol. 58 (1996) 585–98.

32. R. Ramanathan, K. Cao, E. Cavalieri, M.L. Gross, Mass spectrometric methods for distinguishing structural isomers of glutathione conjugates of estrone and estradiol. J Am Soc Mass Spectrom. 9 (1998) 612–9.

33. A. Fabregat, J. Marcos, L. Garrostas, J. Segura, O.J. Pozo, R. Ventura, Evaluation of urinary excretion of androgens conjugated to cysteine in human pregnancy by mass spectrometry. J Steroid Biochem Mol Biol. 139 (2014) 192–200.

34. H.J. Jung, W.Y. Lee, B.C. Chung, M.H. Choi, Mass spectrometric profiling of saturated fatty acid esters of steroids separated by high-temperature gas chromatography. J Chromatogr A. 1216 (2009) 1463–8.

Appendix 1.1.3. Preparation of stable isotope-labeled steroids as internal standards in quantitative analysis by chromatography coupled with mass spectrometry

1. S. Ikegawa, K. Nagae, T. Mabuchi, R. Okihara, M. Hasegawa, T. Minematsu, T. Iida, K. Mitamura, Synthesis of 3- and 21-monosulfates of [2,2,3β,4,4-^{2}H$_5$]-tetrahydrocorticosteroids in the 5β-series as internal standards for mass spectrometry. Steroids. 76 (2011) 1232–40.

2. Sanaullah, L.D. Bowers. Facile synthesis of [16,16,17-2H3]-testosterone, −epitestosterone and their glucuronides and sulfates. J Steroid Biochem Mol Biol. 58 (1996) 225–34.

3. T. Furuta, N. Eguchi, H. Shibasaki, Y. Kasuya. Simultaneous determination of endogenous and 13C-labelled cortisols and cortisones in human plasma by stable isotope dilution mass spectrometry. J Chromatogr B Biomed Sci Appl. 738 (2000) 119–27.

4. T. Furuta, N. Eguchi, A. Yokokawa, H. Shibasaki, Y. Kasuya, Synthesis of multi-labeled cortisols and cortisones with (2)H and (13)C for study of cortisol metabolism in humans. Steroids. 65 (2000) 180–9.

5. Murphy RC. Facile synthesis of deuterated estrogens. Steroids. 1974;24:343–50.

6. Dyer RL, Harrow TA. Synthesis of C-19 deuterium labelled steroids. Steroids.1979;33(6):617–24.

7. Johnson DW, McEvoy M, Seamark RF, Cox LW, Phillipou G. Deuterium labelled steroid hormones: tracers for the measurement of androgen plasma clearance rates in women. J Steroid Biochem. 1985;22:349–53.

8. Zomer G, Stavenuiter JF. Synthesis of 13C-labeled steroid hormones. Steroids. 1990;55(10):440–2. Review.

9. Furuta T, Kusano K, Kasuya Y. Simultaneous measurements of endogenous and deuterium-labelled tracer variants of androstenedione and testosterone by capillary gas chromatography–mass spectrometry. J Chromatogr. 1990;525(1):15–23.

10. Shibasaki H, Furuta T, Kasuya Y. Stable isotope dilution mass spectrometry for the simultaneous determination of cortisol, cortisone, prednisolone and prednisone in plasma. J Chromatogr. 1992;579(2):193–202.

11. Furuta T, Namekawa T, Shibasaki H, Kasuya Y. Synthesis of deuterium-labeled tetrahydrocortisol and tetrahydrocortisone for study of cortisol metabolism in humans. Steroids. 1999;64(12):805.

12. Furuta T, Suzuki A, Matsuzawa M, Shibasaki H, Kasuya Y. Syntheses of stable isotope-labeled 6 beta-hydroxycortisol, 6 beta-hydroxycortisone, and 6 beta-hydroxytestosterone. Steroids. 2003;68(7–8):693–703.

13. Suzuki A, Shibasaki H, Kasuya Y, Furuta T. Simultaneous determination of endogenous and stable isotope-labelled 6beta-hydroxycortisols in human urine by stable isotope dilution mass spectrometry. J Chromatogr B Analyt Technol Biomed Life Sci. 2003;794(2):373–80.

14. Numazawa M, Handa W. Reduction of 1,4-dien-3-one steroids with LiAl$_2$H$_4$ or NaB$_2$H$_4$: stereospecific deuterium-labeling at the c-1alpha position of a 4-en-3-one steroid. Chem Pharm Bull (Tokyo). 2006;54 (4):554–6.

15. Sulima A, Prisinzano TE, Spande T, Deschamps JR, Whittaker N, Hochberg Z, Jacobson AE, Rice KC. A concise method for the preparation of deuterium-labeled cortisone: synthesis of [6,7-2H]cortisone. Steroids. 2005;70(11):763–9.

16. Shinohara Y, Baba S. Stable isotope methodology in the pharmacokinetic studies of androgenic steroids in humans. Steroids. 1990;55(4):170–6. Review.

17. Wudy SA. Synthetic procedures for the preparation of deuterium-labeled analogs of naturally occurring steroids. Steroids. 1990;55(10):463–71. Review.

18. Minagawa K, Kasuya Y, Baba S, Knapp G, Skelly JP. Identification and quantification of 6 beta-hydroxydexamethasone as a major urinary metabolite of dexamethasone in man. Steroids. 1986 Feb-Mar;47(2–3):175–88.

19. Goad LJ, Breen MA, Rendell NB, Rose ME, Duncan JN, Wade AP. Synthesis of deuterium labeled cholesterol and steroids and their use for metabolic studies. Lipids. 1982;17(12):982–91.

20. Knuppen R, Haupt O, Hoppen HO. Synthesis of [^{2}H$_8$] estradiol, [^{2}H$_7$] estrone, [^{2}H$_6$] 2-hydroxyestrone and [^{2}H$_6$] 4-hydroxyestrone as internal standards for selected ion monitoring. Steroids. 1982;39(6):667–73.

21. Johnson DW, Phillipou G, Seamark RF. Deuterium labelled steroid hormones: syntheses and application in quantitation and endocrinology. J Steroid Biochem. 1981;14(8):793–800. Review.

References

Al-Horani RA, Desai UR. Chemical sulfation of small molecules—advances and challenges. Tetrahedron 2010;66(16):2907–18.

Al-Jasem Y, Khan M, Taha A, Thiemann T. Preparation of steroidal hormones with an emphasis on transformations of phytosterols and cholesterol—a review. Mediterr J Chem 2014;3(2):796–830.

Aragones A, Gros EG, Lantos CP, Locascio GA. Less polar forms and derivatives of 18 hydroxy-corticosterone. J Steroid Biochem 1978;9 (2):175–80.

Badoud F, Boccard J, Schweizer C, Pralong F, Saugy M, Baume N. Profiling of steroid metabolites after transdermal and oral administration of testosterone by ultra-high pressure liquid chromatography coupled to quadrupole time-of-flight mass spectrometry. J Steroid Biochem Mol Biol 2013;138:222–35.

Baillie TA, Sjövall J, Herz JE. Synthesis of specifically deuterium-labelled pregnanolone and pregnanediol sulphates for metabolic studies in humans. Steroids 1975a;26(4):438–57.

Baillie TA, Eriksson H, Herz JE, Sjövall J. Specific deuterium labelling and computerized gas chromatography—mass spectrometry in studies on the metabolism in vivo of a steroid sulphate in the rat. Eur J Biochem 1975b;55(1):157–65.

Baillie TA, Anderson RA, Sjövall K, Sjövall J. Identification and quantitation of 16alpha-hydroxy C21 steroid sulphates in plasma from pregnant women. J Steroid Biochem 1976;7(3):203–9.

Barnard G, Collins WP, Kohen F, Lindner HR. The measurement of urinary estriol-16 alpha-glucuronide by a solid-phase chemiluminescence immunoassay. J Steroid Biochem 1981;14(9):941–8.

Bartos J, Pesez M. IUPAC analytical chemistry division. Colorimetric and fluorometric determination of steroids. Pure Appl Chem 1979;51:2157–69.

Batth R, Nicolle C, Cuciurean IS, Simonsen HT. Biosynthesis and industrial production of androsteroids. Plants (Basel) 2020;9(9):1144.

Bauminger S, Kohen F, Lindner HR, Weinstein A. Antiserum to 5alpha-dihydrotestosterone: production, characterization and use in radioimmunoassay. Steroids 1974;24(4):477–88.

Bertolotti M, Crosignani A, Del Puppo M. The use of stable and radioactive sterol tracers as a tool to investigate cholesterol degradation to bile acids in humans in vivo. Molecules 2012;17(2):1939–68.

Brandon DD, Isabelle LM, Samuels MH, Kendall JW, Loriaux DL. Cortisol production rate measurement by stable isotope dilution using gas chromatography-negative ion chemical ionization mass spectrometry. Steroids 1999;64(6):372–8.

Brien TG, Slater JD. Isomerization of radioactive aldosterone at C-17: enzymic purification of the 17-alpha- and beta-isomers. J Endocrinol 1967;38(2):197–8.

Burstein S. Decomposition of 11-deoxycorticosterone and corticosterone in soda-lime glass. Steroids 1976;27(4):493–6.

Bush IE. The chromatography of steroids. Pergamon Press; 1961.

Chapman TE, Kraan GP, Drayer NM, Nagel GT, Wolthers BG. Determination of the urinary cortisol production rate using (1,2,3,4-13C) cortisol. Isotope dilution analyses at very small enrichments. Biomed Environ Mass Spectrom 1987;14(2):73–82.

Chapman TE, Kraan GP, Nagel GT, Wolthers BG, Drayer NM. Measurement of the cortisol production rate in two sisters with 17 alpha-hydroxylase deficiency using [1,2,3,4-13C]cortisol and isotope dilution mass spectrometry. J Steroid Biochem Mol Biol 1991;38:489–96.

Christakoudi S, Cowan DA, Taylor NF. Steroids excreted in urine by neonates with 21-hydroxylase deficiency: characterization, using GC-MS and GC-MS/MS, of the D-ring and side chain structure of pregnanes and pregnenes. Steroids 2010;75(1):34–52.

Christakoudi S, Cowan DA, Taylor NF. Steroids excreted in urine by neonates with 21-hydroxylase deficiency. 2. Characterization, using GC-MS and GC-MS/MS, of pregnanes and pregnenes with an oxo-group on the A- or B-ring. Steroids 2012a;77(5):382–93.

Christakoudi S, Cowan DA, Taylor NF. Steroids excreted in urine by neonates with 21-hydroxylase deficiency. 3. Characterization, using GC-MS and GC-MS/MS, of androstanes and androstenes. Steroids 2012b;77(13):1487–501.

Christakoudi S, Cowan DA, Taylor NF. Steroids excreted in urine by neonates with 21-hydroxylase deficiency. 4. Characterization, using GC-MS and GC-MS/MS, of 11oxo-pregnanes and 11oxo-pregnenes. Steroids 2013;78(5):468–75.

Cobice DF, Livingstone DEW, McBride A, MacKay CL, Walker BR, Webster SP, et al. Quantification of 11β-hydroxysteroid dehydrogenase 1 kinetics and pharmacodynamic effects of inhibitors in brain using mass spectrometry imaging and stable-isotope tracers in mice. Biochem Pharmacol 2018;148:88–99.

Corpéchot C, Synguelakis M, Talha S, Axelson M, Sjövall J, Vihko R, et al. Pregnenolone and its sulfate ester in the rat brain. Brain Res 1983;270 (1):119–25.

Damasco MC, Lantos CP. The existence of two interconvertible forms of 18-hydroxycorticosterone: is one of them an active precursor of aldosterone? J Steroid Biochem 1975;6(1):69–74.

Djerassi C. Steroid reactions. An outline for organic chemists. San Francisco: Holden-Day Inc; 1963.

Dominguez OV. The presence of two interconvertible forms of 18-hydroxy-11-deoxycorticosterone. Steroids 1965;Suppl. 2:29–49.

Donova MV, Egorova OV. Microbial steroid transformations: current state and prospects. Appl Microbiol Biotechnol 2012;94(6):1423–47.

Doué M, Dervilly-Pinel G, Pouponneau K, Monteau F, Le Bizec B. Analysis of glucuronide and sulfate steroids in urine by ultra-high-performance supercritical-fluid chromatography hyphenated tandem mass spectrometry. Anal Bioanal Chem 2015;407(15):4473–84.

Dressendörfer RA, Kirschbaum C, Rohde W, Stahl F, Strasburger CJ. Synthesis of a cortisol-biotin conjugate and evaluation as a tracer in an immunoassay for salivary cortisol measurement. J Steroid Biochem Mol Biol 1992;43(7):683–92.

Dury AY, Ke Y, Labrie F. Precise and accurate assay of pregnenolone and five other neurosteroids in monkey brain tissue by LC-MS/MS. Steroids 2016;113:64–70.

Ebner MJ, Corol DI, Havlíková H, Honour JW, Fry JP. Identification of neuroactive steroids and their precursors and metabolites in adult male rat brain. Endocrinology 2006;147(1):179–90.

Fabregat A, Kotronoulas A, Marcos J, Joglar J, Alfonso I, Segura J, et al. Detection, synthesis and characterization of metabolites of steroid hormones conjugated with cysteine. Steroids 2013a;78(3):327–36.

Fabregat A, Pozo OJ, Marcos J, Segura J, Ventura R. Use of LC-MS/MS for the open detection of steroid metabolites conjugated with glucuronic acid. Anal Chem 2013b;85(10):5005–14.

Fernandes P, Cruz A, Angelova B, Pinheiro HM, Cabral JM. Microbial conversion of steroid compounds: recent developments. Enzyme Microbial Technol 2003;12:688–705.

Fernández-Cabezón L, Galán B, García JL. New insights on steroid biotechnology. Front Microbiol 2018;9:958.

Fieser LF, Fieser M. Steroids. Reinhold Publishing; 1959.

Furuta T, Namekawa T, Shibasaki H, Kasuya Y. Synthesis of deuterium-labeled tetrahydrocortisol and tetrahydrocortisone for study of cortisol metabolism in humans. Steroids 1999;64(12):805–11.

Herráiz I. Chemical pathways of corticosteroids, industrial synthesis from sapogenins. Methods Mol Biol 2017;1645:15–27.

Hill RA, Kirk DN, Makin HLJ, Murphy GM. Dictionary of steroids. Chemical data, structures and bibliographies. London: Chapman & Hall; 1991, ISBN:0-412-27060-9.

Hillier SG, Lathe R. Terpenes, hormones and life: isoprene rule revisited. J Endocrinol 2019;242(2):R9–R22.

Hintikka L, Kuuranne T, Aitio O, Thevis M, Schänzer W, Kostiainen R. Enzyme-assisted synthesis and structure characterization of glucuronide conjugates of eleven anabolic steroid metabolites. Steroids 2008;73(3):257–65.

Honour JW, Shackleton CH. Mass spectrometric analysis of tetrahydroaldosterone. J Steroid Biochem 1977;8:299–305.

Jäntti SE, Kiriazis A, Reinilä RR, Kostiainen RK, Ketola RA. Enzyme-assisted synthesis and characterization of glucuronide conjugates of neuroactive steroids. Steroids 2007;72(3):287–96.

Joannou GE. Identification of 15 beta-hydroxylated C21 steroids in the neonatal period: the role of 3 alpha,15 beta,17 alpha-trihydroxy-5 beta-pregnan-20-one in the perinatal diagnosis of congenital adrenal hyperplasia (CAH) due to a 21-hydroxylase deficiency. J Steroid Biochem 1981;14(9):901–12.

Johnson DW, Philipou G, Seamark RF. Deuterium labeled steroid hormones: syntheses and applications in quantitation and endocrinology. J Steroid Biochem 1981;14:793–800.

Kasuya Y, Iwano M, Shibasaki H, Furuta T. Pharmacokinetic studies of cortisol after oral administration of deuterium-labelled cortisol to a normal human subject. Rapid Commun Mass Spectrom 1995. Spec No:S29-34.

Kirk DN, Miller BW. 18-Substituted steroids—9. Studies on the stability of aldosterone in dilute alkali. J Steroid Biochem 1982;16(2):269–76.

Kirk DN, Smith CZ, Honour JW. Synthesis of [11,11,12,12-^{2}H$_4$] progesterone for mass spectral investigations of peripheral metabolism. Steroids 1990a;55(5):222–7.

Kirk DN, Smith CZ, Varley MJ, Honour JW. Synthesis of [19-^{2}H$_3$] progesterone and [18-^{2}H$_3$] progesterone. J Chem Soc Perkin Trans 1990b;1:2745–8.

Klopfenstein BJ, Purnell JQ, Brandon DD, Isabelle LM, DeBarber AE. Determination of cortisol production rates with contemporary liquid chromatography-mass spectrometry to measure cortisol-d(3) dilution after infusion of deuterated tracer. Clin Biochem 2011;44:430–4.

Kohen F, Pazzagli M, Kim JB, Lindner HR, Boguslaski RC. An assay procedure for plasma progesterone based on antibody-enhanced chemiluminescence. FEBS Lett 1979a;104(1):201–5.

Kohen F, Hollander Z, Boguslaski RC. Non-radioisotopic homogeneous steroid immunoassays. J Steroid Biochem 1979b;11(1A):161–7.

Kohen F, Pazzagli M, Kim JB, Lindner HR. An immunoassay for plasma cortisol based on chemiluminescence. Steroids 1980a;36(4):421–37.

Kohen F, Kim JB, Barnard G, Lindner HR. An assay for urinary estriol-16 alpha-glucuronide based on antibody-enhanced chemiluminescence. Steroids 1980b;36(4):405–19.

Kohen F, Kim JB, Lindner HR, Collins WP. Development of a solid-phase chemiluminescence immunoassay for plasma progesterone. Steroids 1981;38(1):73–88.

Kohen F, Lichter S, Eshhar Z, Lindner HR. Preparation of monoclonal antibodies able to discriminate between testosterone and 5 alpha-dihydrotestosterone. Steroids 1982;39(4):453–9.

Kornel L, Motohashi K. Conversion of 6-hydroxy-delta-4-3-ketosteroids to steroid 5-alpha-ane-3,6-diones and delta-4-ene-3,6-diones during routine solvolysis procedures. Steroids 1965;6(1):9–30.

Kraan GP, Wolthers BG, van der Molen JC, Nagel GT, Drayer NM, Joannou GE. New identified 15 beta-hydroxylated 21-deoxypregnanes in congenital adrenal hyperplasia due to 21-hydroxylase deficiency. J Steroid Biochem Mol Biol 1993;45(5):421–34.

Králová J, Jurášek M, Krčová L, Dolenský B, Novotný I, Dušek M, et al. Heterocyclic sterol probes for live monitoring of sterol trafficking and lysosomal storage disorders. Sci Rep 2018;8(1):14428.

Lan X, Wang T, Ewald F, Chen Z, Cui K, Schäffer A, et al. (14)C-Labelling of the natural steroid estrogens 17α-estradiol, 17β-estradiol, and estrone. J Hazard Mater 2019;5(375):26–32.

Lantos CP, Damasco MC, Aragonés A, Ceballos NR, Burton G, Cozza EN. Versatile steroid molecules at the end of the aldosterone pathway. J Steroid Biochem 1987;27(4–6):791–800.

Le Goascogne C, Robel P, Gouézou M, Sananès N, Baulieu EE, Waterman M. Neurosteroids: cytochrome P-450scc in rat brain. Science 1987;237 (4819):1212–5.

Lednicer D. Steroid chemistry at a glance. Chichester, UK: John Wiley; 2011, ISBN:9780470660850.

Liere P, Pianos A, Eychenne B, Cambourg A, Liu S, Griffiths W, et al. Novel lipoidal derivatives of pregnenolone and dehydroepiandrosterone and absence of their sulfated counterparts in rodent brain. J Lipid Res 2004;45(12):2287–302.

Liere P, Pianos A, Eychenne B, Cambourg A, Bodin K, Griffiths W, et al. Analysis of pregnenolone and dehydroepiandrosterone in rodent brain: cholesterol autoxidation is the key. J Lipid Res 2009;50(12):2430–44.

Ma P, Kanizaj N, Chan SA, Ollis DL, McLeod MD. The *Escherichia coli* glucuronylsynthase promoted synthesis of steroid glucuronides: improved practicality and broader scope. Org Biomol Chem 2014;12 (32):6208–14.

Marques R, Helmy R, Waterhouse D. Enhancing radiolytic stability upon concentration of tritium-labeled pharmaceuticals utilizing centrifugal evaporation. J Labelled Comp Radiopharm 2015;58(6):261–3.

McManus JM, Sharifi N. Structure-dependent retention of steroid hormones by common laboratory materials. J Steroid Biochem Mol Biol 2020;198, 105572.

Mitamura K, Satoh née Okihara R, Kamibayashi M, Sato K, Iida T, Ikegawa S. Simultaneous determination of 18 tetrahydrocorticosteroid sulfates in human urine by liquid chromatography/electrospray ionization-tandem mass spectrometry. Steroids 2014;85:18–29.

Monder C. Stability of corticosteroids in aqueous solutions. Endocrinology 1968;82(2):318–26.

Moss G. IUPAC-IUB Joint Commission on Biochemical Nomenclature (JCBN). The nomenclature of steroids. Recommendations 1989. Eur J Biochem. 1989;186(3):429–58. Erratum in: Eur J Biochem 1993;213(1):2.

Neubert W, Ropke H. Atlas of steroid spectra. Heidelberg: Springer-Verlag; 1965.

Okihara R, Mitamura K, Hasegawa M, Mori M, Muto A, Kakiyama G, et al. Potential corticoid metabolites: chemical synthesis of 3- and 21-monosulfates and their double-conjugates of tetrahydrocorticosteroids in the 5alpha- and 5beta-series. Chem Pharm Bull (Tokyo) 2010;58 (3):344–53.

Pazzagli M, Kim JB, Messeri G, Kohen F, Bolelli GF, Tommasi A, et al. Luminescent immunoassay (LIA) of cortisol—1. Synthesis and evaluation of two chemiluminescent labels of cortisol. J Steroid Biochem 1981a;14(10):1005–12.

Pazzagli M, Kim JB, Messeri G, Kohen F, Bolelli GF, Tommasi A, et al. Luminescent immunoassay (LIA) of cortisol—2. Development and validation of the immunoassay monitored by chemiluminescence. J Steroid Biochem 1981b;14(11):1181–7.

Pazzagli M, Kim JB, Messeri G, Martinazzo G, Kohen F, Franceschetti F, et al. Luminescent immunoassay (LIA) for progesterone in a heterogeneous system. Clin Chim Acta 1981c;115(3):287–96.

Pazzagli M, Kim JB, Messeri G, Martinazzo G, Kohen F, Franceschetti F, et al. Evaluation of different progesterone-isoluminol conjugates for chemiluminescence immunoassay. Clin Chim Acta 1981d;115(3): 277–86.

Pranata A, Fitzgerald CC, Khymenets O, Westley E, Anderson NJ, Ma P, et al. Synthesis of steroid bisglucuronide and sulfate glucuronide reference materials: unearthing neglected treasures of steroid metabolism. Steroids 2019;143:25–40.

Pratt JJ. Steroid immunoassay in clinical chemistry. Clin Chem 1978;24 (11):1869–90.

Robel P, Baulieu EE. Neuro-steroids: 3β-hydroxy-Δ(5)-derivatives in the rodent brain. Neurochem Int 1985;7(6):953–8.

Roy AK, Ramirez LC, Ulick S. Structure and mechanism of formation of the two forms of 18-hydroxy-11-deoxycorticosterone. J Steroid Biochem 1976;7(2):81–7.

Salvador JA, Moreira VM, Hanson JR, Carvalho RA. One-pot, high yield synthesis of alpha-ketols from delta 5-steroids. Steroids 2006;71(3):266–72.

Schänzer W, Opfermann G, Donike M. 17-Epimerization of 17 alpha-methyl anabolic steroids in humans: metabolism and synthesis of 17 alpha-hydroxy-17 beta-methyl steroids. Steroids 1992;57(11):537–50.

Stachulski AV, Meng X. Glucuronides from metabolites to medicines: a survey of the in vivo generation, chemical synthesis and properties of glucuronides. Nat Prod Rep 2013;30(6):806–48.

Stirrat LI, Sengers BG, Norman JE, Homer NZM, Andrew R, Lewis RM, et al. Transfer and metabolism of cortisol by the isolated perfused human placenta. J Clin Endocrinol Metab 2018;103(2):640–8.

Szczebara FM, Chandelier C, Villeret C, Masurel A, Bourot S, Duport C, et al. Total biosynthesis of hydrocortisone from a simple carbon source in yeast. Nat Biotechnol 2003;21(2):143–9.

Taylor NF, Curnow DH, Shackleton CH. Analysis of glucocorticoid metabolites in the neonatal period: catabolism of cortisone acetate by an infant with 21-hydroxylase deficiency. Clin Chim Acta 1978;85(3):219–29.

Tokes L, Throop LJ. Introduction of deuterium into the steroid system. In: Fried J, Edwards JA, editors. Organic reactions in steroid chemistry, vol. 1. New York: Van Nostrand Reinhold Company; 1972. p. 145–221.

Usa T, Ganguly A, Weinberger MH. M and L forms of 18-hydroxy-11-deoxycorticosterone and 18-hydroxycorticosterone: factors influencing conversion, stability and immunological properties. J Steroid Biochem 1979;10(5):557–62.

Valleix A, Carrat S, Caussignac C, Léonce E, Tchapla A. Secondary isotope effects in liquid chromatography behaviour of ²H and ³H labelled solutes and solvents. J Chromatogr A 2006;1116(1–2):109–26.

Venkatachalam KV. Human 3′-phosphoadenosine 5′-phosphosulfate (PAPS) synthase: biochemistry, molecular biology and genetic deficiency. IUBMB Life 2003;55(1):1–11 [review].

Waller CC, McLeod MD. A simple method for the small scale synthesis and solid-phase extraction purification of steroid sulfates. Steroids 2014;92:74–80.

Wang J, Wang Y, Wang T, Cui K, Wang L, Ji R. Synthesis and characterization of 14C-labelled sulfate conjugates of steroid oestrogens. J Labelled Comp Radiopharm 2014;57(7):470–6.

Weththasinghe SA, Waller CC, Fam HL, Stevenson BJ, Cawley AT, McLeod MD. Replacing PAPS: in vitro phase II sulfation of steroids with the liver S9 fraction employing ATP and sodium sulfate. Drug Test Anal 2018;10(2):330–9.

Wild D. The immunoassay handbook. 4th ed. Elsevier Ltd; 2013, ISBN:978-0080970370.

Woodward RB, Sondheimer F, Taub D, Heusler K, McLamore WM. The total synthesis of steroids. J Am Chem Soc 1952;74:4223–51.

Wudy SA. Synthetic procedures for the preparation of deuterium-labeled analogs of naturally occurring steroids. Steroids 1990;55(10):463–71.

Zomer G, Stavenuiter JF. Synthesis of ¹³C-labeled steroid hormones. Steroids 1990;55(10):440–2.

Chapter 1.2

Sources of steroid hormones

1.2.1 Introduction

The steroids were first isolated from large amounts of biological materials such as bovine adrenal glands, bulls' testes, sow ovaries, beef bile, urine in human pregnancy and postmenopausal human urine and the urine of horses. Steroids are now known to be produced in the adrenal glands, male and female reproductive tracts and the placenta, each in a controlled manner, and the steroid hormones are not stored as are the hormones of some other endocrine glands. The structures of the steroid hormones are based on cholesterol which is in the diet and taken up from the intestine. Cholesterol is also derived by de novo synthesis from acetyl coenzyme A (see Chapter 1.3: Biosynthesis). Cholesterol is important for the maintenance of the structure, permeability and fluidity of cell membranes but also in signaling pathways and synthesis of steroid hormones, bile acids and vitamin D. Cholesterol from the diet is transformed into low density lipoprotein cholesterol (LDL) that is transported into steroidogenic cells. The adrenal and gonadal glands synthesizes steroid hormones from cholesterol. The glands develop in the first trimester of the fetus in pregnancy. Mullerian ducts are lost in the male but form the fallopian tubes and reproductive tract of the female. Wolffian ducts are lost in the female but become the vas deferens and seminal vesicle in the male to carry sperm from the testis. The steroidogenic glands have different activities during life in the male and female with notable changes at puberty. An understanding of these changes is important to the nature of investigations in clinical situations. During pregnancy, the adrenal cortex and the placenta interact in what is called a fetoplacental unit.

1.2.2 Dietary cholesterol

Cholesterol in the diet is an important source of cholesterol in the body and is the substrate for steroids, bile salts and vitamin D production. Cholesterol is esterified in the jejunum by addition of fatty acids at C3. Cholesterol esters are hydrolyzed and combined with bile salts in the duodenum to form minute droplets called micelles which are then absorbed from the jejunum (Afonso et al., 2018). Over 40 noncholesterol sterols are also present in the human diet from plants, fungi and yeast. Plants convert squalene to stigmasterol, sitosterol, campesterol, ergosterol, etc., while shellfish produce fucosterol (Kapourchali et al., 2016). Phytosterols, are often consumed in nuts, seeds and vegetable oils. Most individuals absorb nearly 50% of dietary cholesterol but less than 5% of the dietary plant sterols reach the circulation. Plasma concentrations of phytosterols are normally less than 0.5% that of cholesterol. Free cholesterol is delivered to intestinal cells by a specific transporter called Niemann-Pick protein (NPC1L1) (Yu et al., 2019). The NPC1L1 also takes up plant stanols but these are returned back to the lumen of the jejunum through the actions of two ATP-binding cassettes (ABC) (ABCG8 and ABCG5) transporters (Fig. 1.2.1).

In enterocytes of the jejunum, free cholesterol with triacylglycerols are assembled by microsomal triacylglycerol transfer proteins (MTP) into chylomicrons that are transported to the mesenteric lymphatics. Triglycerides are glycerol esterified at each of the three hydroxyls with a mixture of fatty acids (Fig. 1.2.2), typically 18 carbons in stearic acid and 16 carbons in palmitic acid. Phospholipids are similar to triglycerides but with one fatty acid residue replaced by phosphate and a nitrogenous base.

Chylomicrons are the largest of a family of globular particles called lipoproteins that have a nonpolar core of triglyceride and cholesteryl esters (CE) surrounded by a surface layer of phospholipids, cholesterol and apolipoproteins (Fig. 1.2.3). Analytically, lipoproteins can be separated by ultracentrifugation. Minus their associated lipid, the apoproteins are classified into structurally related subgroups A, B, C and E. The chylomicrons get apolipoprotein B48, which is a specific apoB, from intestinal cells and the particles acquire surface apoC1, C2, C3, A and E proteins in the liver and periphery.

Chylomicrons access the plasma from the thoracic lymph duct into the subclavian vein and are rapidly processed by lipoprotein lipase on the luminal surface of endothelial cells with progressive removal of triacylglycerols to yield chylomicron remnants (Julve et al., 2016). ApoC particles are shed and apoE proteins are gained. When the chylomicron remnants reach the liver, they are 90% triglyceride. Cholesterol may account for 50% of the lipid mass

Steroids in the Laboratory and Clinical Practice. https://doi.org/10.1016/B978-0-12-818124-9.00009-7

FIG. 1.2.1 Absorption of cholesterol from gastrointestinal tract. Scheme of cholesterol absorption from the gastrointestinal tract. *BCG8/5*, ATP binding cassette Family G member 5 and 8; *CE*, cholesterol ester; *FC*, free cholesterol; *MTP*, microsomal transport protein; *NPC1li*, Nieman Pick C1 like intestinal sterol transporter; *ST*, sterol transporter; *TAGs*, triacylglycerides; chylomicron—triglyceride ▇ cholesterol ▇ phospholipid ▇ apoB48 ▇. *(Author original.)*

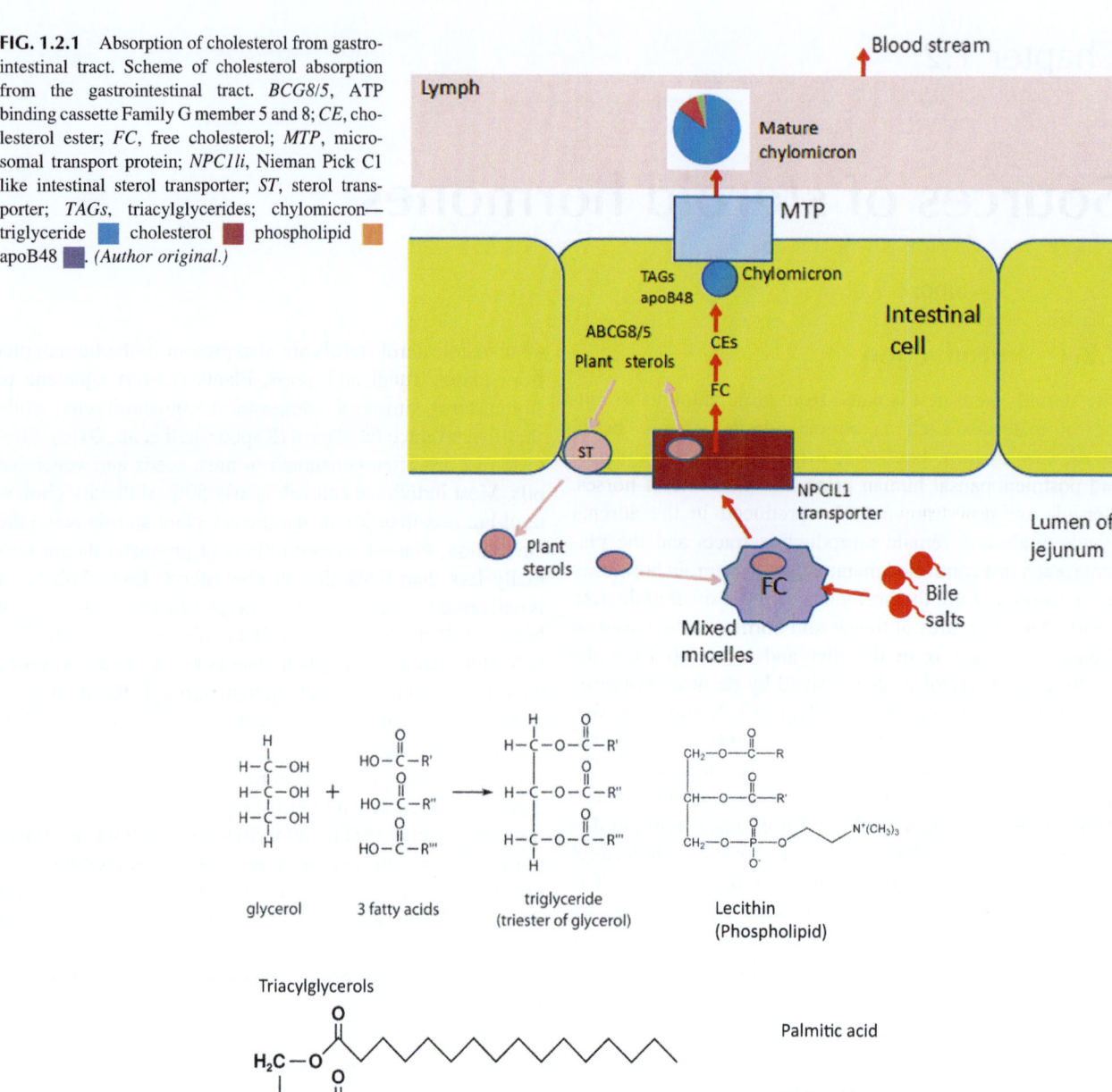

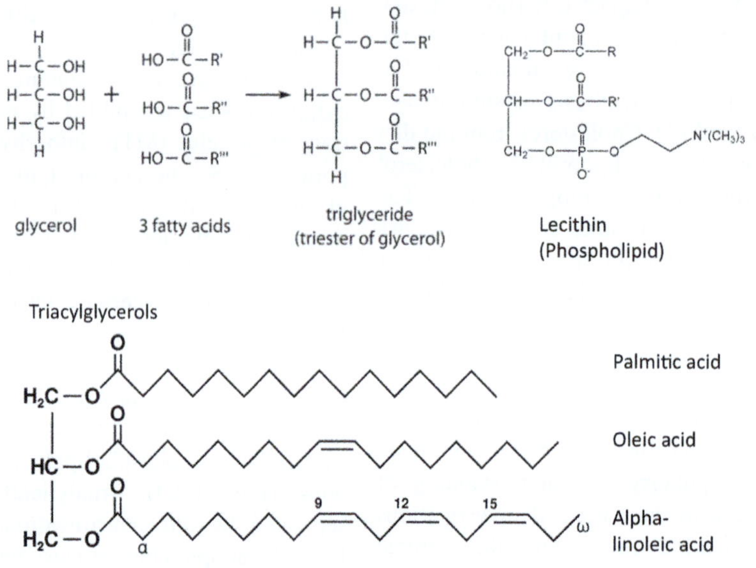

FIG. 1.2.2 Structures of fatty acids, glycerol, phospholipids and triacylglycerols. *(Author original.)*

and apoE is the most significant ligand that reacts with LDL receptor (LDLR) and a remnant receptor (LDLR-related protein LRP). Triglycerides are removed from chylomicrons by the action of lipoprotein lipase in the liver, releasing free fatty acids. Very low density lipoprotein (VLDL) is formed in the liver after loss of triglycerides from the chylomicrons. Cholesterol, phospholipids, apo E and apoC are released at the same time and transferred to

HDL. VLDL is converted to intermediate density lipoprotein (IDL). There is continuous change in composition of lipoprotein with increasing losses of TAG's by exchange of lipid components between the various lipoproteins (see Table 1.2.1).

CE are transferred to IDL by cholesterol ester transfer protein (CETP). Lipoprotein a (LPa) is a larger and more dense relative of LDL of unknown function. IDL has one

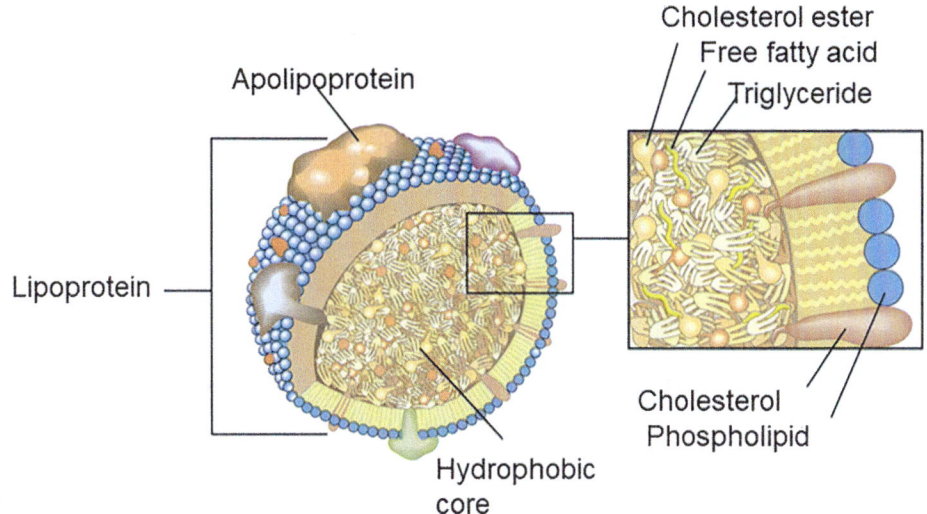

FIG. 1.2.3 Chylomicrons. A core of cholesteryl ester and triglyceride is surrounded by phospholipids, cholesterol and apoproteins. *(Author original.)*

TABLE 1.2.1 Properties of lipoproteins (ranges of diameter and mean percentages).

	Diameter (nm)	CE	FC	TAGs	PL	Apoprotein
Chylomicrons	80–500	2	1	90	4	1 C,B-48,E,A
VLDL	30–80	13	7	65	10	5 B-100,C,E
IDL	25–30	28	8	30	20	15 B-100,E
LDL	19–25	40	8	10	20	15 B-100
Lp(a)	0–25	33	9	4	22	32 a, B-100
HDL$_2$	8–11	18	5	5	25	55 A,C,E
HDL$_3$	6–9	14	3	5	30	50 A,C,E

molecule of apoA for each apoB molecule. Triglyceride is removed from IDL by hepatic triglyceride lipase thus converting to LDL which now comprises mainly CE esters and apoB-100. Phospholipid transfer protein (PLTP) promotes the interchange of phospholipid molecules between lipoprotein particles especially HDL. Lecithin cholesterol acyl transfer protein (LCAT) catalyzes the transfer of fatty acids from phospholipids to the 3-hydroxyl group of cholesterol and form cholesterol esters.

LDL is the principal CE carriers in the bloodstream. LDL can pass junctions between capillary and endothelial cells and attach to LDL receptors on cell membranes of steroidogenic cells through recognition of apoB-100 (Kraemer, 2007) (Fig. 1.2.4). The receptors cluster in a clathrin-coated pit. The receptor-LDL complex moves inwards and form vesicles within the cytoplasm. Endosomes are the result of fusion of several vesicles. In the endosome, LDL is released and the LDLR is recycled back to the membrane. Free cholesterol is released after internalization and lysosomal lipase enzyme action. Unesterified cholesterol is also derived by synthesis from acetyl CoA via the rate limiting HMGCoA reductase enzyme that is inhibited by cholesterol.

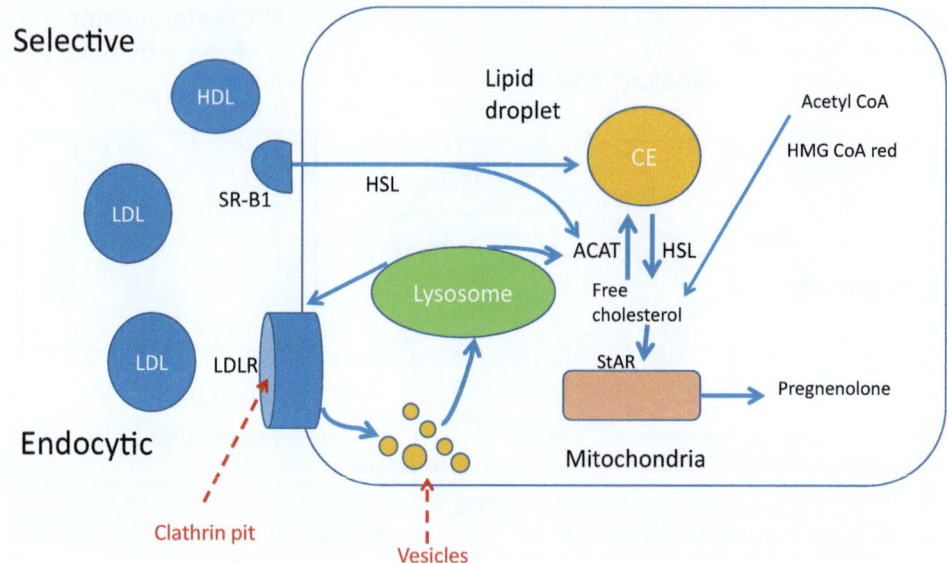

FIG. 1.2.4   Cholesterol ester uptake. Cholesteryl esters are taken up by endocytosis on LDL receptors (LDLR), whereas LDL and HDL are "selectively" taken up by scavenger receptor class B, type 1 (SR-BI). Cholesteryl esters are hydrolyzed to free cholesterol in lysosomes, whereas selectively delivered cholesteryl esters are hydrolyzed by hormone-sensitive lipase (HSL). Free cholesterol is also formed from the hydrolysis of cholesteryl esters (CE) stored in lipid droplets through the actions of HSL or by synthesis from acetyl CoA via hydroxymethylglutaryl coenzyme A reductase (HMG CoA Red). Free cholesterol can be esterified for storage in lipid droplets by acyl CoA:cholesterol acyltransferase (ACAT) or transported into mitochondria via steroidogenic acute regulatory protein (StAR) for metabolism by cholesterol side chain cleavage enzyme (CYP11A1) to produce pregnenolone from cholesterol. *(Modified from Kraemer FB. Adrenal cholesterol utilization. Mol Cell Endocrinol 2007;265–266:42–5. Fig. 1 p. 43.)*

High density lipoproteins (HDL) are assembled from hepatic and tissue lipids and apolipoprotein A. HDL particles bind to the scavenger receptor, class B type 1 (SR-B1) on the steroidogenic cell and deliver CE into the cell without internalization. CE can be stored in droplets then hydrolyzed by hormone sensitive lipase (HSL) which is regulated by the trophic hormones. The cholesterol can move to mitochondria for steroidogenesis or can be re-esterified in the ER by acylCoA:cholesterol acyltransferase (ACAT) and be transported to cholesterol ester rich droplets for storage. The vast majority of lipoprotein derived cholesterol for steroid synthesis is obtained through this route (Buitenwerf et al., 2017a,b).

The above description of cholesterol uptake from the diet is a very simplified account of a complex process. A number of proteins are associated with lipid droplets, including structural proteins, lipid synthesizing enzymes, lipases and a number of proteins involved in vesicular transport (Ikonen, 2018; Elustondo et al., 2017; Shen et al., 2016). Soluble NSF attachment protein receptor (SNARE) proteins, certain synaptosomal-associated proteins (SNAP 25) and syntaxin-5 and syntaxin-17 (Iaea et al., 2020; Kraemer, 2007), are important components in the trafficking of cholesterol to the outer mitochondrial membrane (OMM) for steroidogenesis and the precise regulation of these steps in individual steroidogenic cells are unclear (Luo et al., 2020). The lipid droplets may be capable of steroidogenesis but the in vitro data here need to be treated with caution in case of contamination of cell fractions. Our understanding of the precise routes for cholesterol trafficking requires characterization of further binding proteins.

1.2.3   Steroidogenic glands

The adrenal cortex and gonads are sources of steroid hormones from cholesterol formed by de novo synthesis as well as dietary sources. These glands, as well as the kidneys, originate in the embryo from the urogenital ridge which lies between the coelomic epithelium and the aorta. Specification of the kidney and adrenal occurs first. The adrenal glands develop at 3–8 weeks in gestation but continues to change throughout fetal life and beyond in neonate, child and adult (Ishimoto and Jaffe, 2011; Malendowicz, 2010). The outer portion of the adrenal, the adrenal cortex, undergoes much change in structure, mass and function during the life of a human being before and after birth from a single fetal zone before addition of a definitive zone that becomes the adrenal cortex in the adult. The adrenal cortex zones are involved in the synthesis of mineralocorticoids (outer zona glomerulosa), glucocorticoids (zona fasciculata) and after 6 years, androgenic steroids (zona reticularis) and these will be considered in both normal and in later chapters in pathological conditions. In addition to biosynthetic enzymes, a number of genes, epigenetics and proteins have been found to contribute to adrenal diseases so these will be discussed in Part III of the book.

The gonads arise in the urogenital ridge at around 6 weeks postconception. Primary sex determination is determined by the presence of a Y chromosome and differentiation of Sertoli cells in the testis and granulosa cells in the ovaries. The male sex is directed by the presence of an Sry gene (for sex determining region of the Y chromosome). The process is becoming clearer with many factors known to be involved with an ever increasing list of genes, epigenetics and cofactors involved. Many of these have come to light from genetic studies of patients and these will be addressed in the relevant clinical chapters. This knowledge can still not provide an explanation for many adrenal and DSD cases (differences in sexual development; defects in SD). Adjacent to each gonad are two primitive ducts that give rise to the male and female reproductive tracts. If a male gonad is formed the Sertoli cells produce anti-Mullerian hormone which enables the Wolffian ducts

to develop into the epididymis, vas deferens and seminal vesicles through the actions of testosterone from the Leydig cells. In the female, the Mullerian ducts develop into fallopian tubes, uterus and vagina; the wolffian ducts regress.

1.2.3.1 Adrenal cortex

Adrenal cortex morphogenesis starts at weeks 3–4 of intrauterine life with appearance in the newly formed body cavity of local thickening of coelomic epithelium, called the adrenal anlage. Subsequently, cells of the anlage migrate to the cranial end of the mesonephros (weeks 4–6 of intrauterine development) forming the adrenal placode (adrenal blastema) (Fig. 1.2.5). Cells originating from the intermediate mesoderm also migrate to the adrenal placode, thus forming the adrenogonadal primordium (AGP).

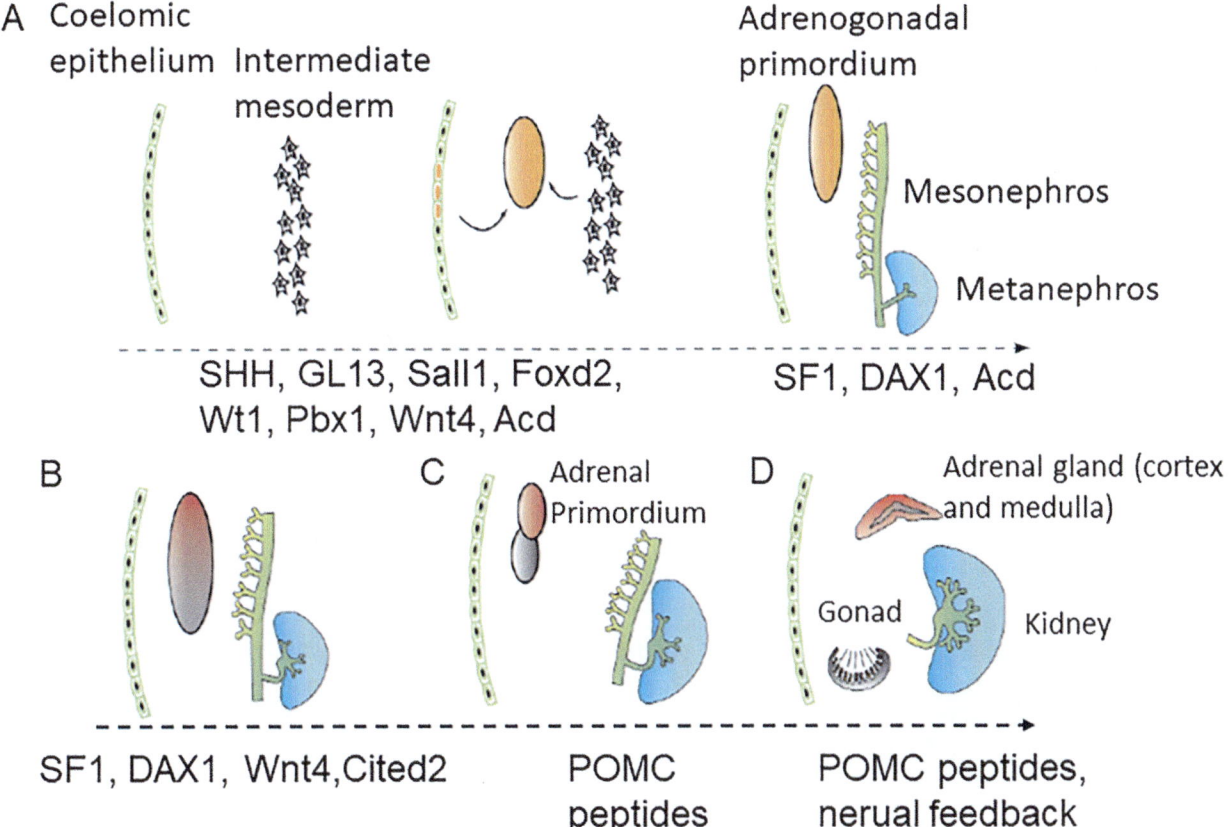

FIG. 1.2.5 Scheme of factors involved in the development of the adrenogonadal primordium from the urogenital ridge to the adult adrenal gland. Within the urogenital ridge, the adrenogonadal primordium is formed by cells from the coelomic epithelium and/or intermediate mesoderm (A). Cells of the intermediate mesoderm also give rise to renal precursors (mesonephros) and the definitive kidney (metanephros). Within the adrenogonadal primordium, the adrenal primordium becomes evident (B) and further differentiates into adrenocortical steroidogenic cells (C). The adult adrenal gland is localized cranial to the kidney (D). In parallel to the depiction of adrenocortical development, factors are annotated which have been implicated in these steps. Abbreviations: *Sall1*, Sal-like 1; *Wt1*, Wilms tumor 1. *(From Malendowicz LK. 100th anniversary of the discovery of the human adrenal fetal zone by Stella Starkel and Lesław Węgrzynowski: how far have we come? Folia Histochem Cytobiol 2010;48(4):491–506. Fig. 2 p. 493.)*

A number of genes known to influence the process are indicated although the precise manner of their action is unknown so cannot yet be discussed. This Sonic hedgehog gene (SHH) is a signaling factor in many processes.

At this stage of development, expression of **SF1 (steroidogenic factor 1 gene)** takes place in cells destined to become the steroidogenic cells of the adrenals and gonads. The adrenal cortex originates from the cephalic part of the AGP, while steroidogenic cells of the gonads originate from its caudal portion.

At 9 weeks of development, the primitive or fetal adrenal cortex is surrounded by a second wave of mesothelial derived cells which eventually becomes the cortex of the adult gland (Ross and Louw, 2015). The differentiation of the outer neocortex and the inner fetal cortex is seen histochemically at approximately 6–8 weeks of gestation (Fig. 1.2.6). The inner, fetal zone, occupies approximately 85% of the total volume of the gland at this stage.

Around the developing adrenals, specialized mesenchymal cells, migrating from the area of Bowman's capsule, form the adrenal capsule (weeks 8–9 of intrauterine development). At this stage of development, neural crest-derived cells migrate into the gland and they will differentiate into chromaffin cells of the adrenal medulla. During fetal life,

these chromaffin cells only secrete noradrenaline. Numerous islands of small, neural-crest derived cells are dispersed among the steroidogenic cells, especially within the fetal zone.

At this stage of development, steroidogenic cells are organized into two distinct components. Centrally localized cells, with abundant, eosinophilic cytoplasm, form the large adrenal fetal zone, while small steroidogenic cells with basophilic cytoplasm, form a very narrow definitive cortex only in the vicinity of adrenal capsule (Kempná and Flück, 2008).

Around the 10th week of fetal development, mesenchymal cells, which surround the fetal cortex, differentiate into fibroblasts and lay down the collagenous capsule of the gland. Blood and nerve supply also starts to develop during this stage. As the adrenal continues to enlarge, mainly due to expansion of the fetal zone, it assumes an extended, flattened shape, which allows growth without any further increase in total cortical thickness. The latter is thought to be important with respect to the blood supply—the venous end of the capillary bed is close to the arterial supply. In the adult, cortisol produced in the adrenal cortex passes in the blood through to the adrenal medulla where it ensures adrenaline production.

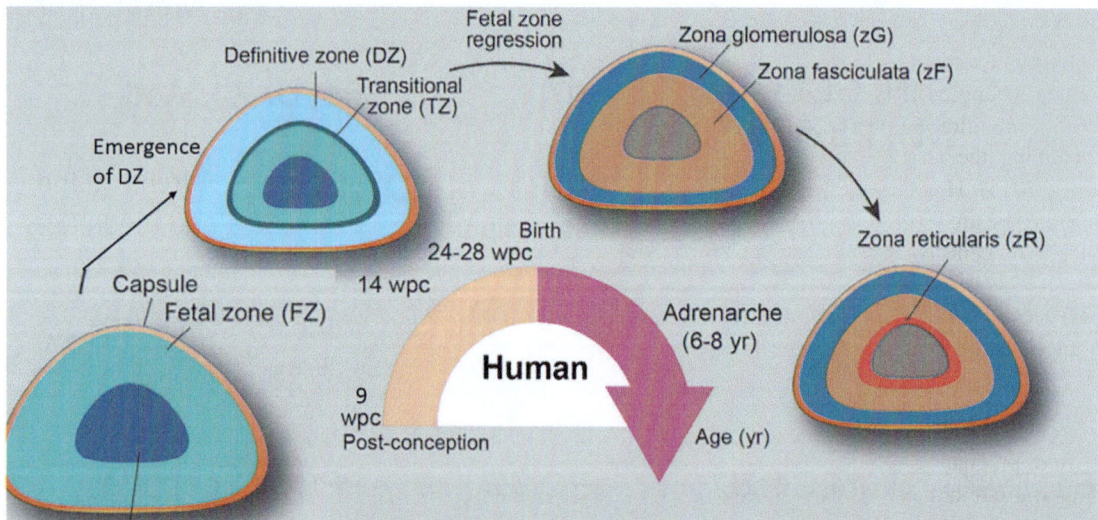

FIG. 1.2.6   Adrenal development. The human adrenal cortex, kidneys and gonads originate from the urogenital ridge. In the embryonic stage, following the growth of adrenogonadal primordium (AGP) on both sides, the adrenal progenitor population on the medial side of the AGP and the gonadal progenitor population on the lateral side of the AGP separate to form the adrenal primordium (AP) and the gonadal primordium (GP), respectively. From 6 weeks postconception (wpc), the neural crest cells that later become the adrenal medulla invade the AP and the mesenchymal cells that become the capsule encapsulate them to establish the fetal adrenal gland. The fetal adrenal cortex separates from the other anlagen and at around 9 weeks of gestation divides into a fetal zone and a definitive zone. The enlarged fetal zone (FZ) is gradually replaced by the outer definitive zone (DZ). After birth, the FZ regresses through apoptosis and the adrenal cortex starts the zonation of the DZ into the zona glomerulosa (zG) and zona fasciculate (zF). The fetal zone begins to involute after birth. By 6 months of age the definitive zone has become two zones, the outer zona glomerulosa and inner zona fasciculata, among the three major cortical zones, the zona reticularis (zR) is the last to develop. At around 6–8 years old, a period known as adrenarche, the zR is formed in the cortical-medullary boundary of the adrenal cortex. The production of adrenal androgens is clearly observed from this stage onwards. *(From Kim J-H, Choi MH. Embryonic development and adult regeneration of the adrenal gland. Endocrinol Metab 2020;35:765–73. Fig. 1 p. 766.)*

1.2.3.1.1 Cell differentiation and adrenal growth

During both embryonic and postnatal growth, most cell division takes place in the outer adrenal cortex, the zona glomerulosa and zona fasciculata or in the larger definitive zone in the fetal adrenal cortex (Kempná and Flück, 2008). The actual mechanism of restricting cell division to the outer region of the cortex is not known, but is probably related to the pattern of blood supply and transcription factors. Once the adrenal cortex has reached its mature size, the rate of cell division in the outer cortex decreases to that required to balance the rate of loss of cells due to cell death (apoptosis) at the cortico-medullary junction. Initially it was assumed that each zone of the adrenal cortex was self maintaining, i.e., that cells in each zone were derived solely from other cells in that zone, as there were functional differences recognized between the zones—a concept known as the "zonal theory." The observed distribution of mitosis in the adrenal cortex however, makes this concept unlikely. The zona glomerulosa exceeds its rate of production of cells for self maintenance, while the rate of cell division in the zona reticularis is insufficient to maintain the zone. In addition to the high rate of cell death in the zona reticularis, yellow-brown deposits of age pigment (lipofuscin) are found in this zone, probably the result of lipid peroxidation.

A number of observations support the concept that cells in the inner zona glomerulosa are pushed inwards into the cortex by the pressure of cell division and become fasciculata cells and further in become reticularis cells (Dumontet et al., 2018). When young, the adrenocyte secretes aldosterone, after leaving the glomerulosa it produces corticosteroids and on reaching the reticularis it produces sex hormones (Xing et al., 2015). Half the cells are thought to die on the way, while the rest are finally eliminated in the reticular zone. This concept of centripetal migration of adrenocytes was named the "escalator" or cell migration theory and for some time the term of "the streaming adrenal cortex" had been proposed (Zajicek et al., 1986).

1.2.3.1.2 Adrenal function in Neonatal life

By the term of pregnancy, the fetal adrenal steroidogenesis is proceeding at a rate higher than at any other stage of development, at least five times that of the adult gland (Pasqualini, 2005). The gland is capable of responding to signals mediated by pituitary ACTH or the renin angiotensin system. Plasma ACTH concentrations after vaginal birth are very much higher than with Caesarian section, suggesting that the pituitary at term can respond to stress, the rise presumably mediated by increased secretion of CRF. During the newborn period, placental steroids such as progesterone and estradiol are rapidly cleared from the neonatal circulation, sulphated conjugates are cleared more slowly. There is a striking improvement in adrenal 3β-hydroxysteroid dehydrogenase activity and a corresponding increase in the ability of the neonate to secrete steroids such as cortisol and aldosterone. The fetal adrenal zone regresses by apoptosis over 6 months after birth. Analysis of steroids in urine during infancy shows that progesterone metabolites disappear by the end of the first week (Dhayat et al., 2015). A study in children up to 2 years of age has shown that 10 times more androstenedione is found in the adrenals than in the testes. After the first 6 months, until puberty the adrenals are the main source of testosterone. The inner fetal cortex involutes shortly after birth with the first signs being about 5 days before delivery with the appearance of vacuoles in the fetal zone cells (Ross and Louw, 2015). The fetal cortex has normally regressed totally and been replaced by adult cortex by the age of 12 months. Due to the loss of the fetal cortex the gland decreases in weight over the first weeks after birth and does not regain its original birth weight until at least the end of the third year of life. By 6 years of life all the zones of the cortex are well established.

1.2.3.1.3 Adrenarche

As the child develops, starting at around 5–7 years of age until the middle of the second decade of life, there is a progressive rise in the circulating concentrations of the adrenal C19 steroids DHEA and DHEAS. A 10–20-fold rise in plasma levels of these steroids are seen in this period while cortisol remains constant. These changes occur in parallel with progressive development of the zona reticularis and are preceded by the loss of the fetal zone. The rise of adrenal androgen production during childhood is known as adrenarche and is responsible for the development of pubic and axillary hair that precedes normal sexual maturation (Auchus, 2011). Adrenarche occurs independently of puberty during childhood and may be a trigger for the latter. These changes have been calculated from the ratio of metabolites with substrate and product relationship. 17,20 desmolase and 17α-hydroxylase activities increased during childhood while there is a decline in 3β-hydroxysteroid dehydrogenase activity and a fall in 11β-hydroxylase activity in the adrenal cortex. The overall effect is a reduced potential for cortisol synthesis. This accounts for the increased ACTH secretion to maintain normal cortisol levels. The overall effect of these changes is a rise in adrenal androgen secretion. 11β-hydroxylase activity is inhibited by adrenal androgen. It is possible that this production is variable among individuals due to small changes in gene structure (for example, point mutations) and this may account for hypertension due to greater deoxycorticosterone production. The ability to measure concentrations of more steroids using LC-MS/MS techniques has shown 11-ketotestosterone to be the dominant androgen in adrenarche (Rege et al., 2018).

The importance of DHEA/DHEAS in initiating the early physical signs of sexuality in children is different in the two sexes with serum DHEA concentrations in girls being double that seen in boys when compared according to stage of pubic hair growth. Recent work in dogs progressing in adrenarche demonstrated that the morphological and functional development of the zona reticularis may be subject to dopaminergic control and therefore could represent an important step in the initiation of adrenarche. Studies in animals are restricted to species near humans that also have a fetal adrenal zone (chimpanzee for example). Mice have an X-zone that has similar properties to the human fetal adrenal but is not an ideal taxonomic model (Huang and Kang, 2019). DHEAS is crucial in brain maturation on entering puberty (Cumberland et al., 2021).

1.2.3.1.4 Cytology

The adult human adrenal glands, which together weigh less than 15 g, have a pyramidal shape within fat above each kidney. In the adult adrenal, three cortical zones are structurally distinct (Fig. 1.2.7). The outermost zona glomerulosa, which occupies 5%–10% of the cortex, is composed of closely packed, ill-defined clusters of cells.

These zona glomerulosa cells contain little rough endoplasmic reticulum (RER), but abundant smooth endoplasmic reticulum (SER) (Seccia et al., 2018). There is a relatively high nuclear to cytoplasmic ratio and some lipid droplets are present. The zona fasciculata on the other hand

has an abundance of lipid droplets, containing cholesterol esters and ascorbic acid. This zone occupies 75% of the cortex and the lipid droplets give the gland its yellow color. The fasciculata cells are larger than cells in the zona glomerulosa and are in long cords arranged radially with respect to the medulla. The cords are separated by straight cortical capillaries. In the fasciculata cells, there is again an abundance of SER.

The innermost cortical zone—the zona reticularis—is composed of a network of short cords with interdigitating capillaries. There is less extensive SER in the cells and fewer lipid droplets, but the cells are seen to contain more lysosomes and larger lipofuscin granules, which increase in number with age. Cell contacts between cortical cells in all the zones involve desmosomes. Large and numerous gap junctions are found, functionally coupling the cells in the two inner zones.

New research has revealed that adrenocortical cells are markedly influenced by their microenvironment both in physiological and pathophysiological conditions. The cell types which interact with adrenal steroidogenic cells especially include mast cells, macrophages, lymphocytes and neurons through local release of various bioactive signals (reviewed by Lopez et al., 2021). These cell-to-cell communication processes can be involved in the pathogenesis of adrenal disorders, particularly adrenal tumors and thus likely to be relevant targets for future pharmacological treatments of steroid excess syndromes.

1.2.3.1.5 Vasculature and innervation

The adrenal glands, although only 0.02% of total body weight in an adult, may receive 0.14% of the cardiac output. The blood supply for each gland is supplied from a circle of different arteries (Ross and Louw, 2015). The superior suprarenal from inferior phrenic artery, the medial suprarenal from the aorta, and the inferior suprarenal from the renal artery (Fig. 1.2.8).

Smaller vessels from these main trunks pierce the adrenal capsule and break in capillary plexi. Three types of vessels are found: capsular, cortical and medullary. The cortical cells descend from the capsular plexus and form the capillary bed which supplies the cortical parenchyma. Straight capillaries between the fasciculata cells join in the zona reticularis and empty into the medullary vascular bed where cortisol induces enzymes in catecholamine synthesis. The adrenal medullary venules collect the blood which then empties to a central vein. Blood from the left adrenal drains into the left renal vein and from the right adrenal into the inferior vena cava. Studies of micrographs of rat adrenal glands collected under conditions of stress or ACTH administration have shown that every cortical cell is adjacent to a blood vessel.

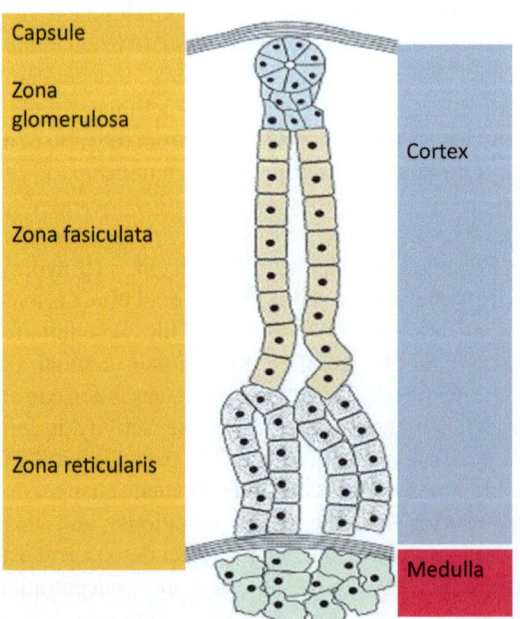

FIG. 1.2.7 Section through the adrenal gland showing three outer zones of the adrenal cortex and the inner medulla. *(Author original.)*

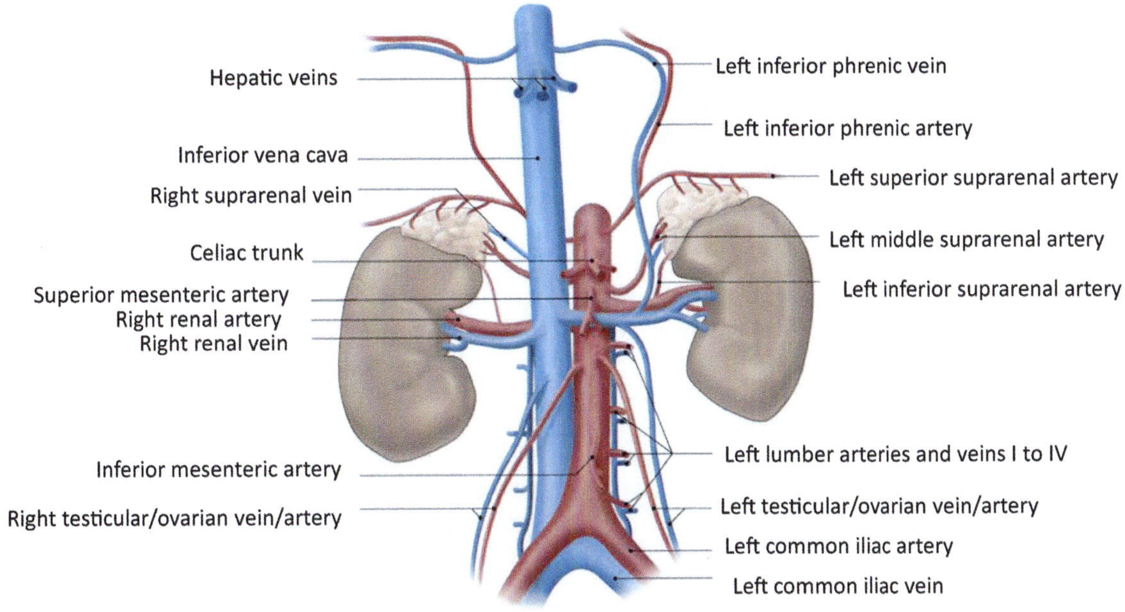

FIG. 1.2.8 The arterial supply, venous drainage of the adrenal glands located above the kidneys. *(From Ross IL, Louw GJ. Embryological and molecular development of the adrenal glands. Clin Anat 2015;28(2):235–42. Fig. 1 p. 236.)*

Some cortical blood can leave the organ via the alar veins when the other small veins are contracted (and hence obstructed) by longitudinal muscle columns into an emissary vein. As blood vessels are constricted there is a build-up of blood at first in the zona reticularis and inner zona fasiculata which eventually moves outwards through the cortex. The increase in blood space and temporary slowing down of the passage of blood enables adrenocorticotrophic hormone (ACTH) to come in contact with the cells with greater ease. This explains changes in the morphology of the cortex from the zona reticularis outwards in response to stress with, for example, lipid droplet depletion.

The innervation of the gland is derived from the splanchnic nerves which arise from the lateral horn preganglionic sympathetic neurons at spinal cord levels T8–T11. Some of these fibers innervate the blood vessel walls of the gland, as mentioned above, while others enter the medulla and end in cholinergic synapses. The cells of the adrenal cortex are thought not to have any secretomotor innervation.

The adrenal cortex is capable of dramatic regenerative growth when required (Dumontet et al., 2018). A functional, zoned adrenal cortex can regenerate from fragments of adrenocortical tissue when transplanted elsewhere in the body. This was common surgical practice years ago, on adrenalectomy for steroid dependent cancers, to maintain some adrenal function which had to be monitored. Nowadays medical steroid replacement is used. The pattern of adrenal regrowth is similar to that occurring naturally during normal development, with the formation of a morphologically disorganized cell mass followed by a period of growth with mitosis in the outer region of the regenerating tissue and finally re-establishment of functional zones.

ACTH is the major hormonal regulator of adrenocortical growth. Excess pituitary secretion of ACTH in diseases such as Cushing syndrome, or the various forms of congenital adrenal hyperplasia (CAH) are associated with a large adrenal gland and conversely, diseases of ACTH deficiency with a failure of adrenal development or adrenal atrophy, depending on when the ACTH deficiency arises. The removal of one adrenal gland stimulates compensatory growth of the other. The mechanism for regulation of adrenocortical growth is still being elucidated. Some polypeptide hormones (e.g., fibroblast growth factor) are thought to be direct mitogens of the adrenal gland (Guasti et al., 2013), whereas ACTH is probably an indirect mitogen (as it directly inhibits replication) acting to increase the delivery of growth factors to adrenocortical tissue by, for example, effects on the adrenal vasculature. ACTH will act at the adrenal cortex to increase cell size rather than cell number or DNA content. Blocking ACTH secretion from the pituitary results in a decrease in adrenocortical growth including DNA synthesis and cell division. This is later followed by adrenal atrophy with the loss of protein, DNA and RNA content.

1.2.3.2 Gonads

1.2.3.2.1 Structures and functions of the gonads

The reader needs some appreciation of the processes involved in assembling the steroidogenic gonads with blood supplies and nervous inputs in normal development, pathology and tumor initiation and progression. Many genes produce factors for the movement of cells to create the gonads, to determine the sex of the fetus and direct differentiation of the internal reproductive tracts (Kim et al., 2017). The numbers of genes involved keeps changing as new genes are recognized and a literature search is needed to keep up-to-date on this topic. The names and abbreviations for the proteins arise from different origins. The transcription factors Emx2, Lin1, Wt-1 and Wnt4 direct the formation of the AdP. Wt-1 upregulates Sf1 in the AdP. Other combinations of transcription factors dictate the formation of the kidneys, gonads and adrenals. This will be discussed in clinical sections which is where the involvement of many such factors has been first encountered.

At the time of fertilization of the oocyte, until 8 weeks of gestation, the sex of the fetus is not established. The development of the reproductive tract in the period of 8–16 weeks gestation leads to the distinction between the structures and functions of the male or female gonads (Fig. 1.2.9).

Sex determination and differentiation is a sequential process that involves successively:

- the establishment of chromosomal or genetic sex in the fertilized egg near the time of conception,
- the determination of gonadal sex according to the genetic sex and
- the differentiation of the genital characteristics.

Sex determination is concerned with the control of the primary or gonadal sex. The differentiation of the primitive gonads into testes or ovaries in utero is genetically determined in humans (Morohashi et al., 2013; Ungewitter and Yao, 2013; Biason-Lauber, 2010). Both male and female embryos possess indifferent, common factors that direct female sex differentiation *unless* there is active interference by factors which direct male sex differentiation. The protein SRY is the first switch for activation of the bipotential gonad to a testis through upregulation of SOX9 (Vining et al., 2021). An ovary differentiates unless the indifferent fetal gonad is diverted by factors produced from genes on the Y-chromosome. The development of the female genital tract occurs independently of male hormones. The sexual dimorphism that results from sex differentiation in humans is mediated by the fetal testis and its secretions (testosterone and Mullerian inhibitory hormone—AMH; see Section 1.2.3.4). More genes are being identified all the time in an active research area, these can be common or specific to the gender development.

Sex differentiation is concerned with all the events that follow after gonadal organogenesis. These processes are regulated by at least 30 specific genes located on sex chromosomes and autosomes that act through a variety of mechanisms including:

- (i) organizing factors,
- (ii) sex steroids,
- (iii) peptides and
- (iv) specific tissue receptors.

During fetal development, the gonads are an important source of androgens, with DHT being more important than T. After birth, there is a 6-month period in the neonate called the mini-puberty where gonadotrophins and testosterone are raised for up to 6 months, and then the gonads remain quiescent for 5–8 years until before adolescence when they are activated by hormones from the anterior pituitary (gonadotrophins). Hormones secreted by the gonads at this time cause the appearance of sex specific secondary

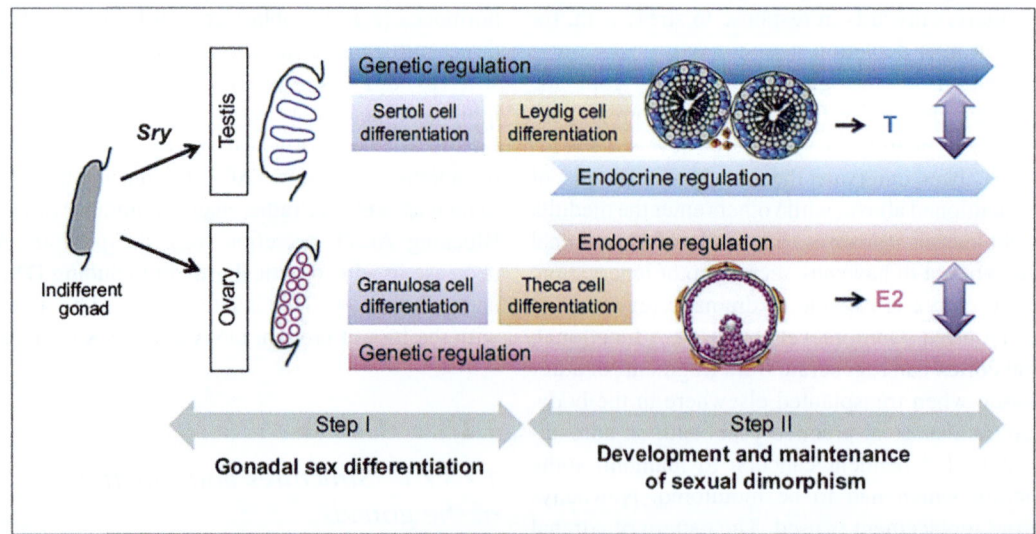

FIG. 1.2.9 Diagrammatic illustration of differentiation of indifferent gonads into testes and ovary. Step I is the gonadal sex differentiation triggered by SRY. Sertoli and Leydig cells differentiate in the testes, whereas granulosa and theca cells differentiate in the ovary. In step II, the cells produce the steroids for sexual dimorphism. The testis produces testosterone that acts through the androgen receptor (AR) and the ovary produces estradiol that binds to the estrogen receptor (ER). The activities of the steroidogenic cells have endocrine regulatory systems to maintain male and female functions. *(From Morohashi K, Baba T, Tanaka M. Steroid hormones and the development of reproductive organs. Sex Dev 2013;7(1–3):61–79. Fig. 5 p. 73.)*

characteristics of the adult male and female and the onset of sexual cycles in the female (this period is puberty). In both sexes, the gonads have the dual functions to produce germ cells and secrete sex hormones:

- androgens are the steroid sex hormones that are masculinizing in their action—testosterone is the principal testicular steroid,
- estrogens are steroids that are feminizing—estradiol is the main estrogen produced by the ovary and
- progestins are produced by the ovary and serve to prepare the adult uterus for pregnancy.

1.2.3.2.2 Chromosomal sex

Sex is determined genetically by two chromosomes called the sex chromosomes to distinguish them from the other somatic chromosomes called autosomes. In humans, the sex chromosomes are called the X and Y. The Y chromosome contains a gene that codes for the production of the testes (Sry—Sex determining region of Y-chromosome). Male cells with a diploid number of chromosomes contain an X and a Y chromosome whereas female cells contain two X chromosomes. As a consequence of meiosis in gametogenesis each normal ovum contains one X chromosome, but half the normal male sperm contain a Y chromosome. When a sperm containing a Y chromosome fertilizes an ovum, an XY pattern results and the zygote develops into a genetic male. When

fertilization occurs with an X-containing sperm, an XX pattern and a genetic female result.

1.2.3.2.3 Development of the gonads

On each side of the embryo, a primitive gonad arises from the genital ridge. Each gonad develops with a cortex and a medulla. Until the sixth week of fetal development, these structures are identical in both sexes (Mäkelä et al., 2019; Zhao and Yao, 2019). In genetic males, the gonadal medulla then develops into a testis and the cortex regresses.

Testicular development is dependent on products of the Y-chromosome (Nef et al., 2019; Sobel et al., 2004). Sertoli cells differentiate, then Leydig cells appear and produce testosterone (T) and other factors are secreted. Current knowledge on the genetic programs and epigenetic mechanisms that regulate gonadal sex determination is based largely on experiments in mice supplemented with facts derived by studies of humans with disorders of sexual development (Stévant and Nef, 2019). An increase in Sry expression upregulates fibroblast growth factor 9 (Fgf9) gene expression to increase the synthesis of prostaglandin D2.

1.2.3.2.4 Embryology of the genitalia

In the seventh week of gestation, the embryo has two primordial genital ducts. The indifferent gonad in the genital ridge consists of an outer cortex and inner medulla (Fig. 1.2.10). Under the influence of testis determining

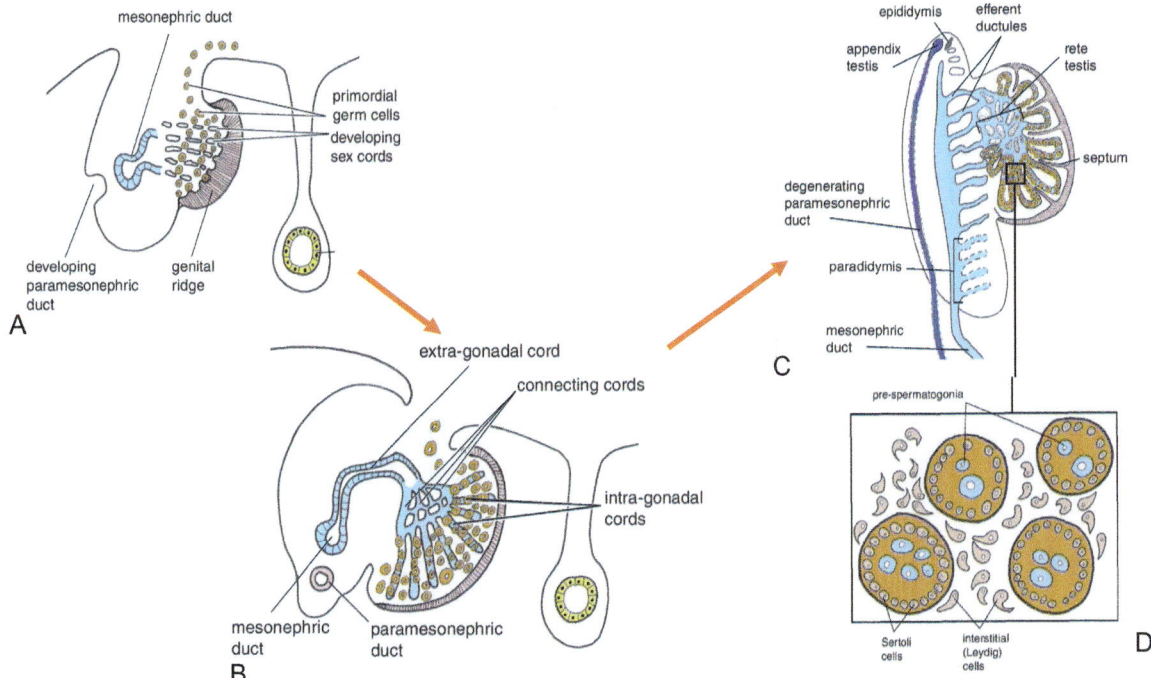

FIG. 1.2.10 Diagram illustrating the differentiation of indifferent gonads into testes (A) indifferent gonads from a 6-week-old fetus; (B) at 7 weeks showing testes developing under the influence of high concentrations of AMH; (C) testis at 20 weeks showing the rete testis and seminiferous tubules derived from medullary cords (D) contents of a seminiferous tubule from a 20-week fetus. *(Modified from https://veteriankey.com/male-and-female-reproductive-systems/ Figs. 21.1, 21.2, 21.3.)*

factor the testis begins to form with medullary cords inside a tunica albuginea. By 20 weeks there is a central network of tubes called the rete testis and seminiferous tubule derived from the medullary cords. Spermatogonia can be seen in the space between the seminiferous tubules.

In the normal male fetus, the Wolffian duct system on each side develops into the epididymis and vas deferens (Eid and Biason-Lauber, 2016; Wilhelm and Koopman, 2006). In the normal female fetus, the Mullerian duct system develops into oviducts and a uterus. The external genitalia are similarly bipotential at 8 weeks. Thereafter the urogenital slit closes and the male genitalia form, or alternatively, it remains open and female external genitalia form.

When there are functional testes in the embryo, male internal and external genitalia develop (Fig. 1.2.11).

The fetal testes secrete two hormones which influence male sex development:

- the steroid hormone testosterone and
- a peptide called Mullerian inhibitory substance/factor (MIS/MIF) or anti-Mullerian hormone (AMH).

Testosterone, which needs to be reduced with hydrogen to a more active form called 5α-dihydrotestosterone (DHT), acts

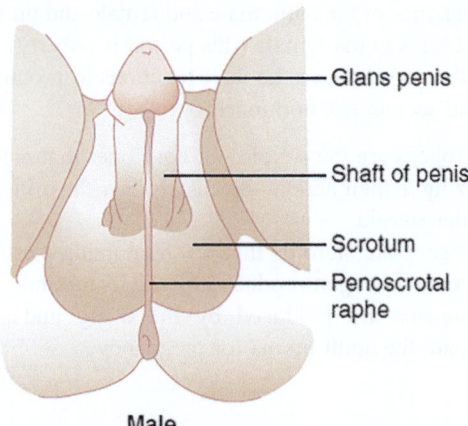

Male

FIG. 1.2.12 Differentiation of the indifferent stage into male external genitalia. *(Modified from Strauss J, Barbieri R, editors. Yen & Jaffe's reproductive endocrinology. 8th ed. Philadelphia, PA; Elsevier; 2019. Fig. 16.3 p. 368.)*

to induce the formation of the male external genitalia. There is an increase in steroidogenesis from gestation week 8 (Savchuk et al., 2019). Deletion of the 5-alpha reductase prevents the development of the male phenotype (Imperato-McGinley et al., 2002; Andersson et al., 1991). The genital folds respond to DHT, to elongate and fuse to form the shaft of the penis; the genital swellings fuse to form the scrotum and the genital tubercle becomes the glans penis (Fig. 1.2.12).

MIF acts alone to assist regression of the Mullerian ducts. MIF and testosterone are probably needed to develop the Wolffian ducts into epididymis, vas deferens and seminal vesicles on each side; the urogenital sinus gives rise to the prostate and prostatic urethra. MIF is glycoprotein product of fetal testicular Sertoli cells (Sobel et al., 2004). The time course of male sexual development is summarized in Fig. 1.2.13.

1.2.3.2.5 Anti-Mullerian hormone (AMH), Mullerian inhibitory factor (MIF) and Mullerian inhibitory substance (MIS)

AMH is a member of the transforming growth factor beta subfamily (TGF-β). AMH is a 140 kDa protein consisting of two identical glycoprotein subunits linked by disulfide bonds coded by a gene on chromosome 19. The AMH gene is 275 base pairs long in 5 exons and codes for pre-pro-AMH of 560 amino acids. A signal peptide of 25 amino acids is cleaved from the N-terminus of pre-pro-AMH followed by a proteolytic cleavage by pro-protein convertases to release the N-terminal AMH and C-terminal AMH dimers that are 110 kDa and 25 kDa, respectively. The AMH receptor gene has a total length of 8.7 kbp from 11 exons carried by chromosome 12 (13q.12). Exons 1–3 code for the extracellular domain of the receptor, exon #4 codes

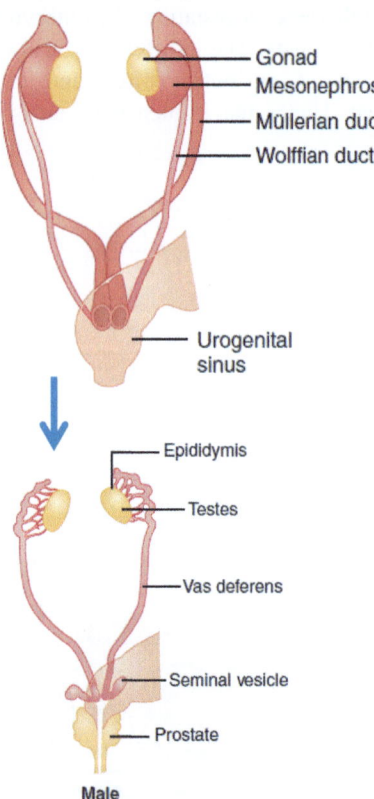

Male

FIG. 1.2.11 Differentiation of the male genital tracts from Wolffian ducts. *(Based on Sobel V, Zhu YS, Imperato-McGinley J. Fetal hormones and sexual differentiation. Obstet Gynecol Clin N Am 2004;31 (4):837–56, x–xi. Fig. 2 p. 844, Redrawn.)*

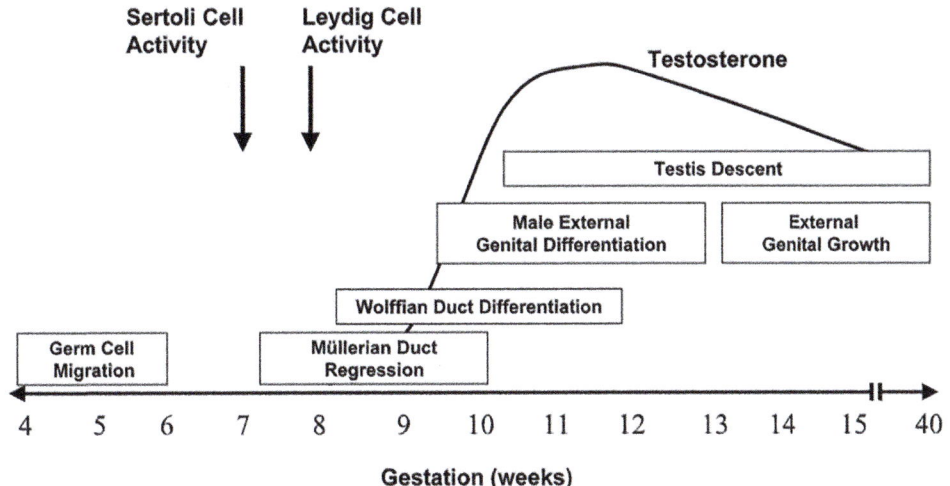

FIG. 1.2.13 Time course of male sex determination and differentiation. *(Modified from Waugh A, Grant A, editors. Ross & Wilson anatomy and physiology in health and illness. 13th ed. Philadelphia, PA: Elsevier; 2018. Fig. 18.13 p. 498.)*

for most of the transmembrane domain and exons 5–11 code the intracellular serine/threonine kinase domain. AMH is a new assay in pediatric investigations though only measured in specialist centers for DSD's, whereas measurements of AMH are more often measured in adult females being studied for ovarian reserve so this will be discussed further in the chapter on regulation of steroids. There are many issues over sample collection and storage as well as the quality of the assays that until recently had no reference preparation of AMH for assay standardization.

1.2.3.3 The male reproductive tract

The male reproductive tract (Figs. 1.2.14 and 1.2.15) comprises several structures with discrete activities:

- a pair of testes that produce spermatozoa,
- the male sex accessory organs that receive, store and transport sperm,
- the sex accessory glands that add substances to the contents of the ducts and
- the external genitalia.

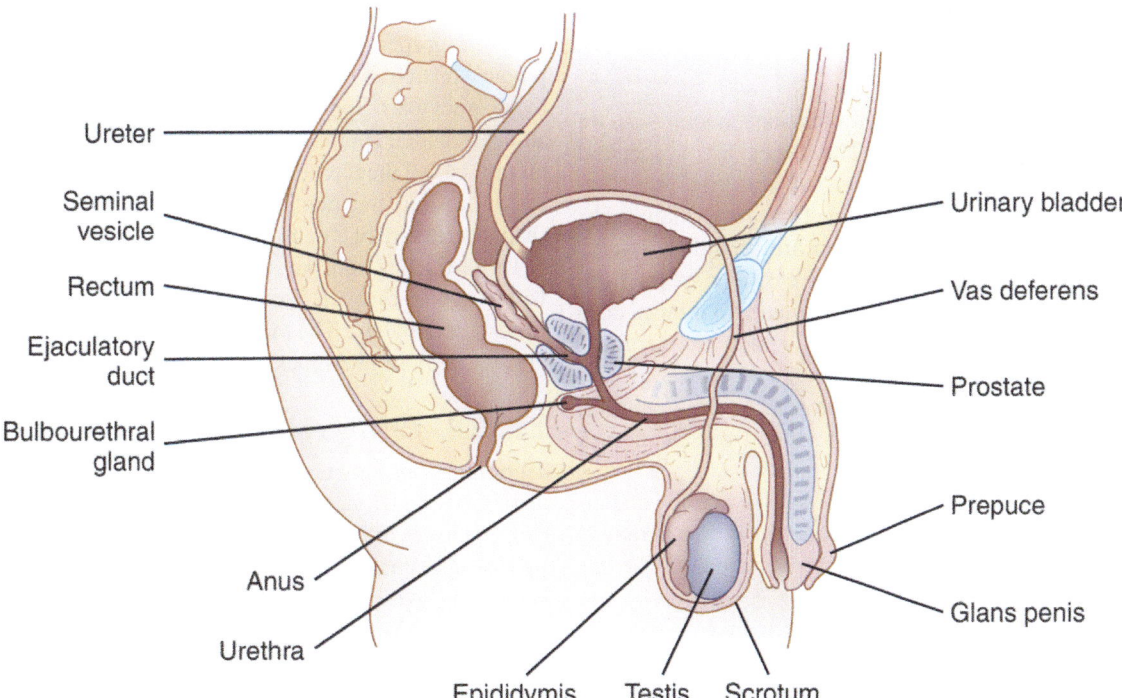

FIG. 1.2.14 Schematic representation of male pelvic region showing reproductive organs. *(Modified from Norman AW, Henry HL. Hormones. 3rd ed. Academic Press; 2015, Fig. 12.1 p. 256.)*

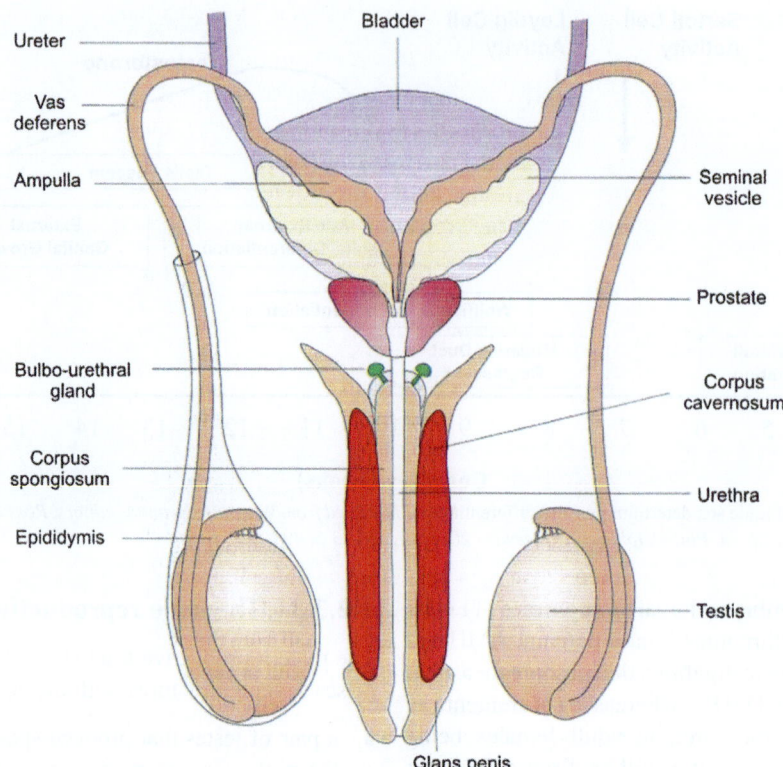

FIG. 1.2.15 Schematic representation of a posterior view of the male pelvic region showing male sex accessory glands and ducts. *(From Pocock G, Richards CD, Richards DA. Human physiology. 5th ed. Oxford University Press; 2017. Fig 48.1 p 766.)*

There are also secondary sex characteristics that distinguish male from females these include facial hair, musculature, deeper voice and height.

1.2.3.3.1 Testis

An adult testis is oval with dimensions about 4 cm in length and 2.5 cm in width. The testes are suspended outside the body in the scrotum. A testis is linked to the body through the inguinal canal by a spermatic cord that contains a testicular nerve, an artery and two veins and the vas deferens that is the tube carrying sperm from the gland. The testis is divided into about 250 lobules containing coiled seminiferous tubules within the seminiferous epithelium. The wall is lined by germ cells and Sertoli cells. The space outside the seminiferous tubules, the interstitial space, is occupied by blood vessels, nerves and Leydig cells.

The testis has two functions that are the actions of separate compartments. Sperm are produced in the pyramid-shaped Sertoli cells in the seminiferous tubules whereas sex steroids are synthesized in Leydig cells (Svechnikov and Söder, 2008). Germ cells (spermatogonia) next to the basement membrane divide initially by mitosis to start the process of spermatogenesis with diploid cells (46 chromosomes). A primary spermatocyte is formed by meiosis which is haploid (23 chromosomes). On a second meiosis, four secondary spermatocytes are formed, which completes spermatogenesis. Follicle stimulating hormone (FSH) stimulates mitosis of spermatogonia and is necessary for transformation of spermatids to spermatozoa. The heads of maturing spermatids embed in the Sertoli cells which also produce testicular fluid and remove degenerating germ cells by phagocytosis. Sperm are released from the Sertoli cells in an event called spermiation and these pass to a network of tubules called the rete testis. They are carried along the vasa efferentia into the ductus epididymis which is a comma shaped structure of 2 up to 4 cm in length that lies on the side of the testis then continues to the vas deferens. The tubules within the epididymis secrete substances that help sperm survive and mature. The tubules are lined with cilia to move the sperm along the tubes.

1.2.3.3.2 Leydig cells

Tight junctions between Sertoli cells prevent free exchange of steroids and water soluble compounds from the Leydig cells (Inoue et al., 2018; Tremblay, 2015). This blood-testis barrier operates in both directions to prevent damage to sperm and avoid leakage of Sertoli products to the circulation where an immune response would be stimulated. Sertoli cells secrete androgen binding protein into the tubular lumen to concentrate the sex steroids and support spermatogenesis.

Testosterone and dihydrotestosterone in the adult male are produced in the Leydig cells on stimulation by luteinizing hormone (LH). Near 98% of the testosterone in circulation is bound to proteins—sex hormone binding globulin (SHBG) and some is bound to albumin. In contrast, fetal Leydig cells produce androstenedione and need Sertoli cells to synthesize testosterone (Wen et al., 2016).

1.2.3.3.3 Vas deferens

This 45 cm long tube in the adult passes through the inguinal wall into the pelvic cavity. The vas deferens makes a loop around the top of the bladder to join the urethra which carries urine from the bladder. The extended end of the vas deferens is a reservoir for the sperm and leads to an ejaculatory duct prior to the urethra. The walls of the vas deferens have smooth muscle which contacts in waves to propel sperm into the ejaculate.

1.2.3.3.4 Accessory glands

Seminal vesicles join the ejaculatory duct at the base of the bladder. The paired pouch like glands secretes seminal plasma to be mixed with sperm into seminal fluid, a viscous alkaline fluid with high fructose content. Prostate glands also secrete alkaline fluid into the seminal fluid. The prostate is a doughnut shaped organ below the bladder about the size of a walnut. The next accessory glands are the Cowper's or bulbourethral glands which are pea size on the side of the urethra secreting mucus into the ejaculate.

1.2.3.3.5 Urethra

The urethra is the tube from the bladder to the external opening of the penis or external meatus. The urethra takes urine and sperm out of the pelvic cavity though the inguinal canal.

1.2.3.3.6 Male external genitalia

The penis and scrotum are in the groin (Liu et al., 2018; Baskin et al., 2018). The testes and some of the accessory ducts are in the scrotum which is a pouch divided into two. The skin of the scrotum is wrinkled (rugose) with a ridge in the midline called the raphe. The scrotum helps to keep the testes away from the body and maintain a lower temperature that favors spermatogenesis. Testes move from the pelvic cavity into the scrotum usually before birth. The penis has an acorn-shaped head, the glans penis, to the shaft. Skin from the shaft covers the glans as the foreskin or prepuce which is partially removed at circumcision. The shaft comprises three cylindrical masses of spongy tissue filled with blood sinuses. The cylinders on each side are the corpora cavernosa and the central cylinder is the corpora spongiosum. The shaft fills with blood on erection of the penis taking the length from a range of 8 to 11 cm to a range of 12 to 24 cm.

1.2.3.4 The female reproductive tract

Before the seventh week of gestation the female genitalia are similar to the male. In the absence of a TDF, the cortex of the indifferent gonad forms cortical cords (Fig. 1.2.16).

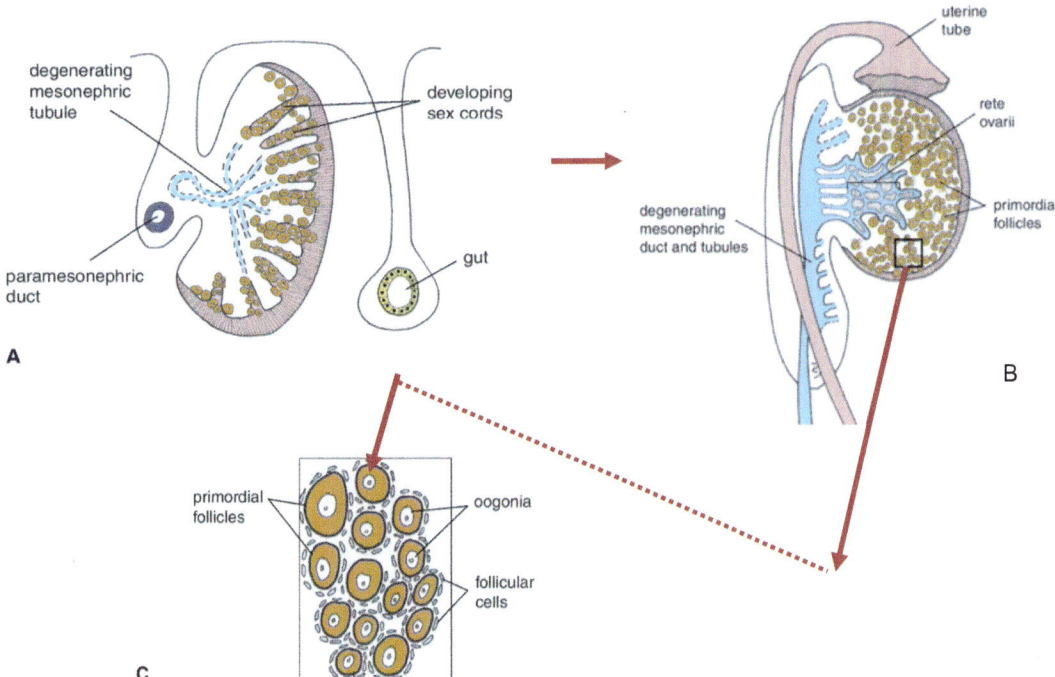

FIG. 1.2.16 Cross-section (A) and ventral view (B) of differentiation of the ovary from the undifferentiated gonad, showing the formation of primordial follicles and the uterine tube. (C) Primordial follicles. *(Modified from https://veteriankey.com/male-and-female-reproductive-systems/ Figs. 21.2, 21.4.)*

The primordial germ cells concentrate in the cortical cords and the medulla degenerates (Lamothe et al., 2020; Hummitzsch et al., 2015). Differentiation of ovaries occurs a few weeks later in the female than the testes in the male.

The remnants of the medullary cords persist as *rete ovari* in the developing ovary. The enlarging cortex, which contains oogonia derived from primordial germ cells, forms the ovary. The oogonia undergo mitotic division over three months. Most of the oogonia degenerate, but the survivors enter meiosis and become oocytes which are surrounded by a single layer of granulosa cells.

The main players in female gonadal differentiation are WNT4 (Wingless type MMTV integration site) and R-spondin1 linked to the β-catenin pathway and the forkhead box L2 (FOXL2) transcription factor. WTN4 is necessary to maintain the development of female germ cells (Lamothe et al., 2020). WTN4 and RSPO1 genes are located on chromosome 1 at p.35 and p34, respectively. FOXL2 is located at chromosome 3.q.22.3 and is expressed early in ovary development. In genetic females, the cortex develops into an ovary and the medulla regresses. Granulosa cells then theca cells differentiate. Together these cells produce estradiol (E2).

The Mullerian ducts develop into female reproductive system comprising oviducts, uterus, vagina and external genitalia (clitoris and urethra) linking to the paired ovaries (Fig. 1.2.17).

The external genitalia include the vaginal orifice, the labia majora and labia minora, the clitoris and vestibule (Fig. 1.2.18). The labia are skin folds at the boundary of the vagina.

The vaginal exit to the exterior (vaginal introitus) is bordered by two layers of fleshy folds, the labia majora and labia minora which have differentiated from the urogenital folds. The clitoris differentiates from the genital tubercle and sits at the upper junction of the labia above the urethral orifice.

The shaft of the clitoris is, like the partially homologous penis, largely filled with two spongy cylinders of tissue, the corpora cavernosa. The clitoris is partially covered with a clitoral prepuce homologous to the glans of the penis. The clitoris has many nerves with pressure and temperature receptors.

1.2.3.4.1 Pelvic region

The female reproductive tract includes a pair of ovaries with oviducts leading to the uterus, cervix, vagina and external genitalia (Fig. 1.2.19).

1.2.3.4.2 Uterus

The uterus develops from the Mullerian ducts to an inverted pear-shaped organ in the pelvic cavity located above the bladder and in front of the rectum (Fig. 1.2.20). The uterus

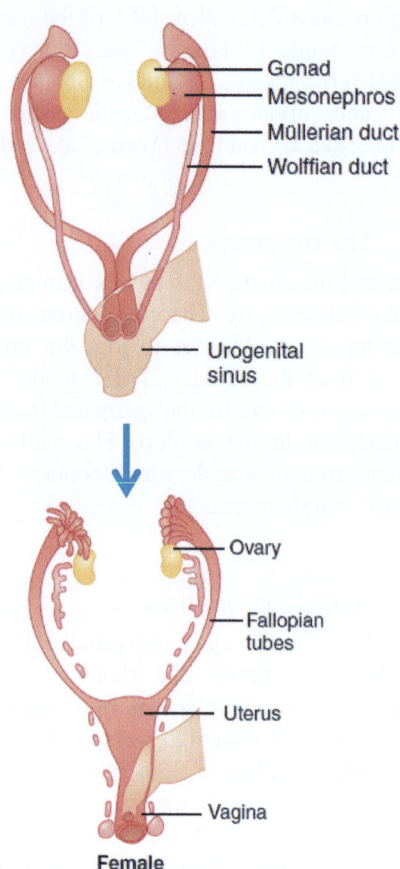

FIG. 1.2.17 Differentiation of the female genital tract from Mullerian ducts. *(From Sobel V, Zhu YS, Imperato-McGinley J. Fetal hormones and sexual differentiation. Obstet Gynecol Clin N Am 2004;31 (4):837–56, x–xi. Fig. 2 p. 844, Redrawn.)*

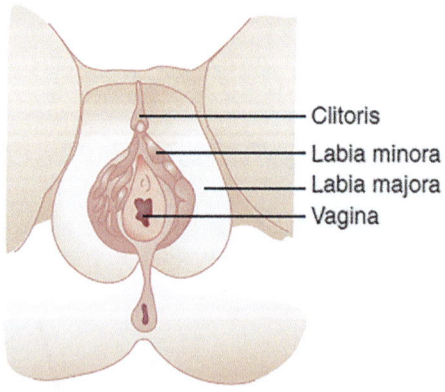

FIG. 1.2.18 Differentiation of the indifferent stage into female external genitalia. *(Modified from Strauss J, Barbieri R, editors. Yen & Jaffe's reproductive endocrinology. 8th ed. Philadelphia, PA; Elsevier; 2019. Fig. 16.3 p. 368.)*

is attached by ligaments to the pelvic wall and the tail bone. Two oviducts enter the dome or fundus of the uterus and the base leads to the cervix that connects to the vagina.

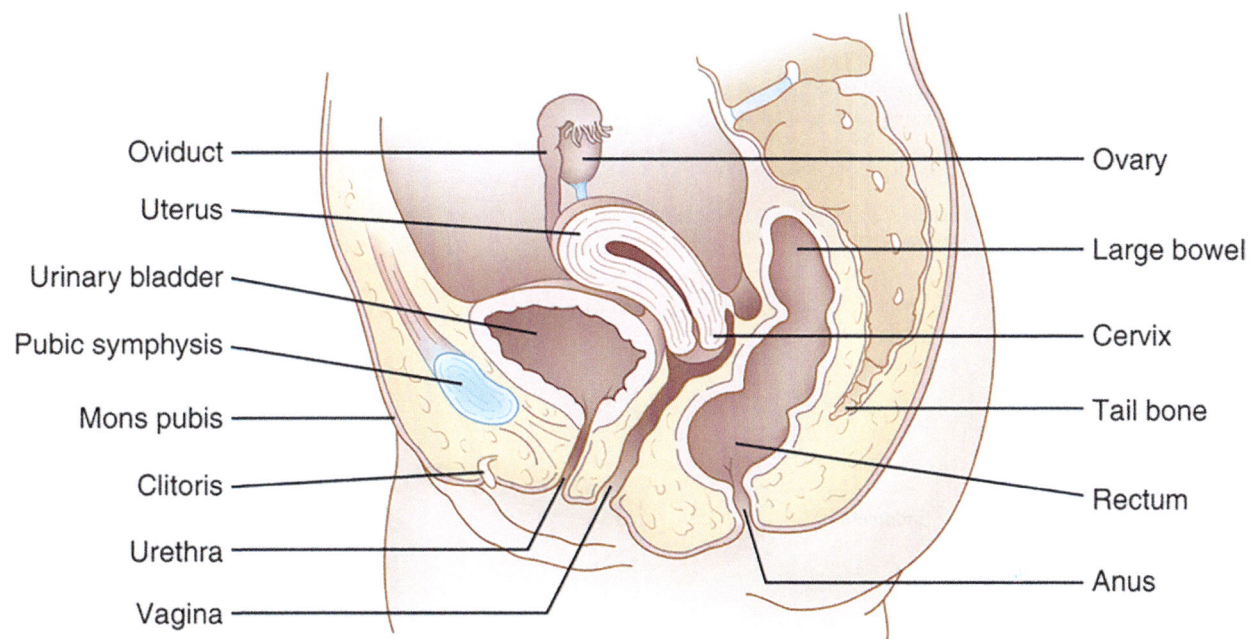

FIG. 1.2.19 Anatomy of the female reproductive system. A midsagittal view of the organs and structures in the pelvic region. *(Modified from Norman AW, Henry HL. Hormones. 3rd ed. Academic Press; 2015, Fig. 13.1 p. 276.)*

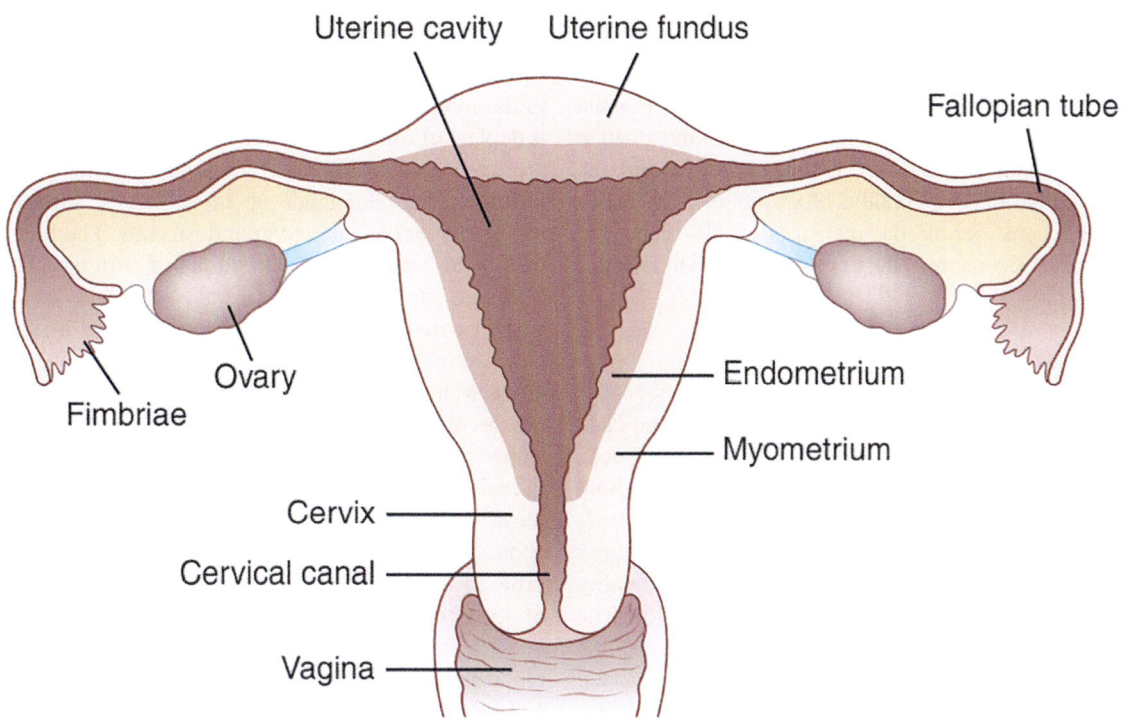

FIG. 1.2.20 Internal female reproductive organs. Midfrontal section. *(Modified from Norman AW, Henry HL. Hormones. 3rd ed. Academic Press; 2015, Fig. 13.2 p. 277.)*

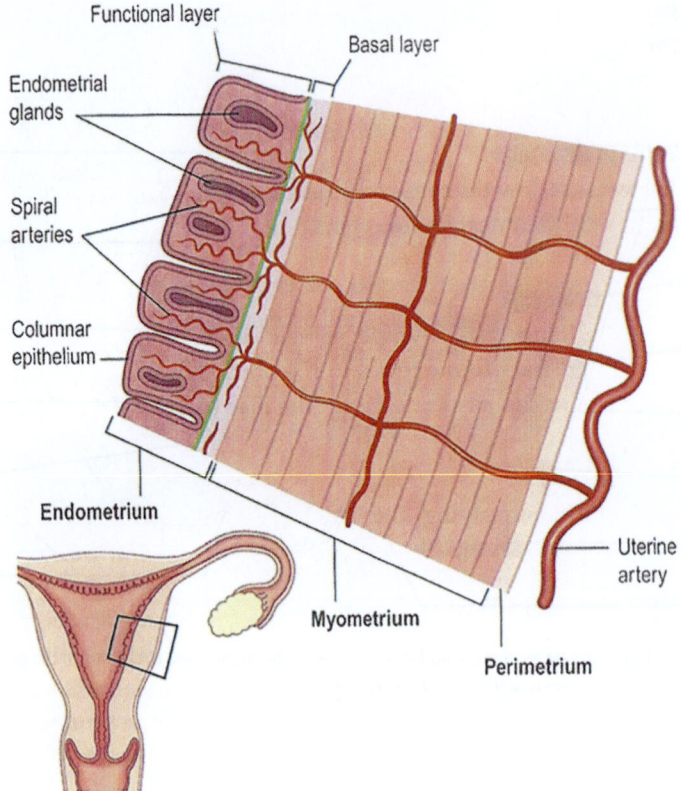

FIG. 1.2.21 Structure of uterine wall. *(Modified from Waugh A, Grant A, editors. Ross & Wilson anatomy and physiology in health and illness. 13th ed. Philadelphia, PA: Elsevier; 2018. Fig. 18.5 p. 491.)*

The wall of the uterus has three layers that undergo changes during the menstrual cycle. A thin outer membrane, the perimetrium, a thick layer of muscle, the myometrium and lining the uterine cavity is the endometrium (Cunha et al., 2018). The endometrium is divided into the *stratum functionalis* which is shed during menstruation, and the inner stratum basalis that contains blood vessels. The endometrium thickens during the menstrual cycle under the influence of estrogens from the developing ovarian follicles (Fig. 1.2.21).

1.2.3.4.3 Ovary

The ovaries of the adult are 2–4 cm paired organs on each side of the upper pelvic cavity at the ends of the oviducts. The ovary is a complex structure that changes throughout a female's lifetime in three distinct stages Prenatal development of ovaries and germ cells up until arrest of follicular growth in utero, changes at puberty and adult functionality of ovarian cycles and ovulation (Rojas et al., 2015). The numbers of follicles change dramatically during life (Shah et al., 2018). The maximum number of follicles is found in the fetus mid-gestation thereafter the numbers decrease by atresia. At birth, there are estimated to be 500,000 follicles which declines to about 80,000 at puberty and about 30,000 at 35 years of age. At any time in the reproductive years, most follicles degenerate in the periphery of the ovary by the age of 30 the pool of follicles is about 12% of the initial numbers and by 50 years at the menopause the ovary is deplete of follicles (Hale et al., 2014). Primordial follicles are seen under the microscope as structures less than 50 μm in diameter made up of an oocyte surrounded by a single layer of flattened granulosa cells. The few follicles that grow under the direction of follicle-stimulating hormone (FSH) become primary follicles with cube-shaped cells about 100 μm in diameter and a highly vascular connective tissue, theca cells, with a membrane covering. Many of the estimated 160,000 primary follicles do not develop beyond this stage and are lost by apoptosis. In some primary follicles, the granulosa cells divide by mitosis and secrete fluid (antral fluid) between the granulosa cells. Approximately 6000 secondary follicles are about 200 μm in diameter. About 100 secondary follicles grow to tertiary follicles that have a granulosa membrane of three or four cell layers and the theca that has two layers. The inner theca contains glandular cells and small blood vessels and the theca external has connective tissue and large blood vessels.

In a section under the microscope, there is a thin sheet of tissue on the outer surface (surface epithelium). The picture

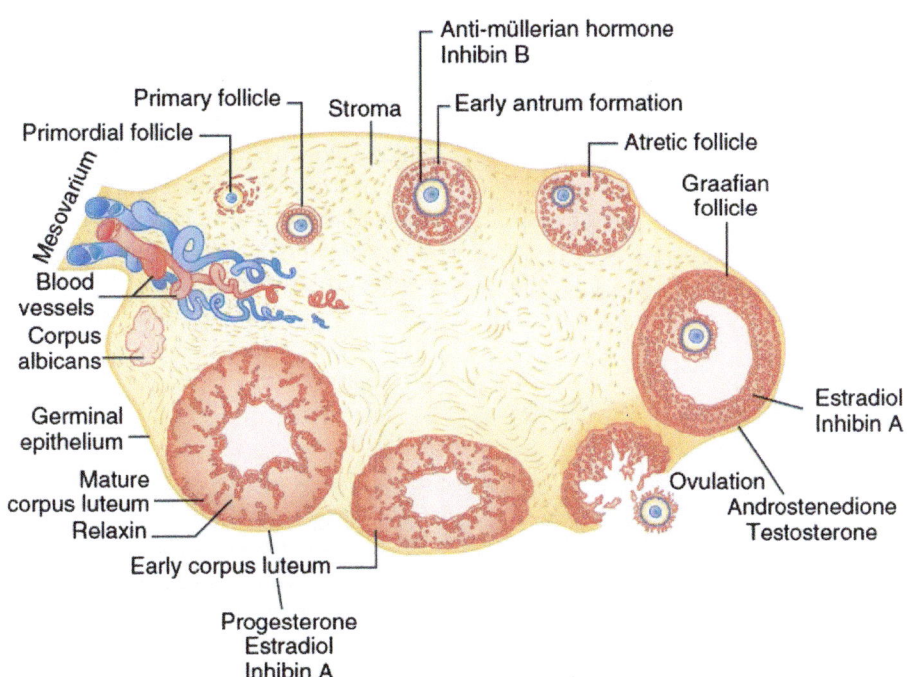

FIG. 1.2.22 Stages in the development of an ovarian follicle. Primordial follicles consists of an immature ovum surrounded by a single layer of granulosa cells under a basement membrane. The oocytes mature and granulosa cells proliferate. Of the many follicles that develop the majority become atretic. At the secondary follicle stage, a thecal cell layer develops. A fluid filled antrum appears which in one follicle will have a mature oocyte. The follicle has a highly differentiated wall consisting of theca externa, theca interna and granulosa cells. Section through the ovary at different stages of the menstrual cycle depicting the component parts (clockwise from primordial follicle) (top left). Following ovulation the remaining follicle differentiates into lutein cells of the corpus luteum which in the absence of pregnancy degenerates after 2 weeks to the corpus albicans. *(Modified from Strauss J, Barbieri R, editors. Yen & Jaffe's reproductive endocrinology. 8th ed. Philadelphia, PA; Elsevier; 2019. Fig. 8.1 p. 168.)*

changes over 1 month and this progression is depicted in Fig. 1.2.22. Underneath this tissue there is a tissue framework, the ovarian stroma which is divided into cortex and medulla containing the germ cells (oocytes) within the membrane (zona pellucida) of the ovarian follicles.

One Graafian follicle in each monthly menstrual cycle is large enough to be seen as a blister, or stigma, on the surface of the ovary. This follicle develops from a primordial follicle to a Graafian follicle when the oocyte is surrounded by a ring of granulosa cells (corona radiata) suspended by a stalk of granulosa cells (cumulus granulosa) in antral fluid (clockwise from top left of figure). Graafian follicles grow to about 15–25 mm in diameter under the influence of follicle stimulating hormone (Lew, 2019). The ovum is a haploid ootid with 23 chromosomes having undergone two meiotic divisions since the primary oocyte stage. The ovarian follicles produce and secrete estrogens through the actions of enzymes in the theca and granulosa cells. The theca interna is capable of androgenic steroid hormone production and the granulosa cells have aromatase activity and synthesize estrogens when the ovarian follicle has reached the late secondary or early tertiary stage (Jamnongjit and Hammes, 2006). Follicle growth and steroid hormone secretion are controlled by the pituitary hormones LH and FSH. The stigma ruptures and the ovum is expelled with follicular fluid (ovulation) following a surge of

luteinizing hormone (LH) which stimulates progesterone production by the granulosa cells.

After ovulation the follicle wall is like a collapsed sac with a blood clot from torn blood vessels on the surface so it is called the corpus hemorrhagicum. The luteal cells divide and invade the antral cavity to form a yellow corpus luteum. Blood vessels from the thecal layer penetrate the luteal mass. Luteal cells secrete progesterone through stimulation by LH. The corpus luteum grows up to 20 mm in diameter and degenerates over 15 or so days until menstruation starts. The hormonal changes in the menstrual cycle are detailed in Chapter 2.5. Inhibins and gonadotrophin surge attenuating factor (GnSAF) have roles in the menstrual cycle (Messinis et al., 2014).

1.2.3.4.4 Vagina

The vagina is a 10 cm long tube that increases in length and diameter during sexual arousal. The vagina acts as the passage for menstrual blood, accommodating the penis during intercourse and as part of the birth canal at the end of pregnancy. Above the vaginal introitus is the urethral orifice where urine passes from the body. To the sides of the urethral orifice are the vestibular glands which are the equivalent of the prostate glands in the male.

1.2.3.4.5 Pregnancy

A mature follicle in the ovary ruptures to release an egg into the fallopian tube. The empty follicle will become a corpus luteum which persists if pregnancy ensues. The ovum will undergo meiosis but if fertilized by sperm the pronuclei will fuse to a full chromosome complement (Fig. 1.2.23). The chromosomes replicate and divide and the fertilized egg or zygote undergoes cleavage. Successive cleavages, as the zygote moves along the tube, produces a morula (with

a solid ball of 16–32 cells) then with further increase in size an early blastocyst with 70–100 cells.

The wall of the blastocyst is one cell thick except in one area where it is three or four cells thick where the embryo will develop. The cells of the blastocyst differentiate into two types of cells, trophoblast, that develop into a fetal part of the **placenta** and an inner cell mass that will become the embryo (embryonic disc) (Fig. 1.2.24A). The blastocyst is a single cell layer around a fluid filled cavity, the blastocoel

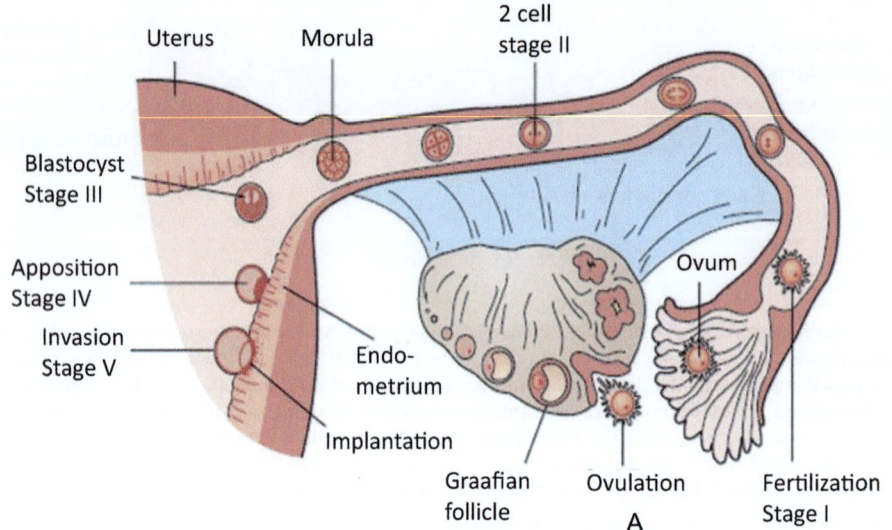

FIG. 1.2.23 Diagram of ovulation. The stigma ruptures and the ovum is expelled. The ovulated ovum in second meiotic division is penetrated by sperm. After fertilization and further cell division the blastocyst is formed and implanted into the endometrium. *(Modified from Strauss J, Barbieri R, editors. Yen & Jaffe's reproductive endocrinology. 8th ed. Philadelphia, PA; Elsevier; 2019. Fig. 9.14 p. 225.)*

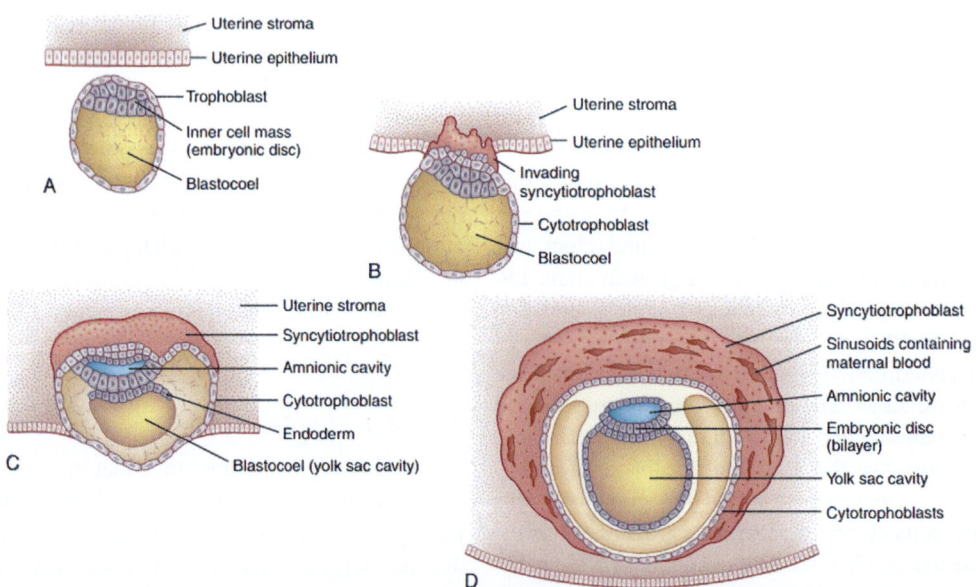

FIG. 1.2.24 (A) Implantation of the human embryo (blastocyst formation); (B) the trophoblast has penetrated the epithelium and is beginning to invade the stroma: (C) the blastocyst penetrates further into the stroma and the amniotic cavity has appeared: (D) the uterine tissue has grown over the implantation site and blood sinusoids have appeared in the syncytiotrophoblast. *(Modified from Norman AW, Henry HL, editors. Hormones. 3rd ed. Academic Press; 2015, Fig. 14.3 p. 300)*

that will become the yolk sac. The fluid filled blastocyst begins to implant into the uterine wall on days 6–7. By this time the uterine lining will have responded to increased concentrations of progesterone from the corpus luteum and be responsive to the fertilized egg.

During the early stages of implantation, the trophoblast differentiates into syncytiotrophoblast (Fig. 1.2.24B) that secretes proteases to breakdown the uterine epithelium and invades the stroma. The cells lose their cell membranes and have linked cytoplasm. Soon after implantation the cell mass within the cytotrophoblast differentiates into two layers called the ectoderm and endoderm then a third middle layer (Fig. 1.2.24C). As the blastocyst move further into the stroma an amniotic cavity appears then the uterine tissue grows over the point of implantation (Fig. 1.2.24D). Irregular, maternal blood vessels appear in the syncytiotrophoblast from which the maternal-fetal blood circulation is established.

At this stage, the embryo is an inner cell mass at the end of a ring of trophoblast cells in a blastocyst. The trophoblast, within a zona pellucida, differentiates into an inner cytotrophoblast and an inner syncytiotrophoblast. Finger projections called chorionic villi extend from the syncytiotrophoblast into the vascular uterine stroma to initiate formation of the placenta. The placenta forms after the human embryo invades the uterine wall at around 7–10 days after conception. Enzymes are secreted to dissolve the blood vessel walls and the mother's blood bathes the villi. A thin membrane separates the embryo blood in the villi from the mother's blood that is in the space surrounding the villi. Oxygen, glucose and other small compounds can diffuse through the wall of the chorionic villi. The cells of the uterine stroma grow over the implantation site (deciduoma response) to surround the blastocyst and complete implantation. Some of the cells from the placenta develop into the outer layer of membrane (chorion) around the blastocyst. An inner layer of membranes (amnion) develop from other cells that form the amniotic sac now containing the embryo and a yolk sac.

The amniotic sac fills with fluid and expands to envelop the embryo in amniotic fluid (Fig. 1.2.25). The inner cell mass, suspended from the amnion by a stalk, differentiates into two layers of cells called the ectoderm and endoderm and a layer (mesoderm) appears between these. This is now 3 weeks since fertilization and the fifth week of pregnancy. The heart and major blood vessels have appeared. The embryo now develops within the amniotic sac against the wall of the uterus. The embryo elongates and the brain and spinal cord begin to develop, a process which continue to develop through pregnancy. Many organs have developed by 10 weeks after fertilization, the skeleton is forming and the embryo is now a fetus filling the uterus. The placenta grows and it extends villi into the wall of the uterus.

1.2.3.4.6 *Fetal steroidogenesis*

In pregnancy, a close metabolic interrelationship exists between the fetal adrenal, the fetal liver, the placenta and the maternal circulation. Progesterone, estrogens, androgens and cortisol are needed to establish and maintain a pregnancy. These steroids are derived from cholesterol delivered to or made in the placenta (Chatuphonprasert et al., 2018). This has in the past been interpreted as evidence for a fetoplacental steroidogenic unit (Fig. 1.2.26).

According to this concept the fetal adrenal is intrinsically deficient in 3β-hydroxysteroid dehydrogenase and therefore entirely dependent on the supply of placental progesterone for the synthesis of cortisol. Since the placenta is deficient in the ACTH dependent enzymes 17-hydroxylase and 17,20-desmolase, the major purpose of the fetal adrenal is the provision of dehydroepiandrosterone sulfate (DHEAS) as essential precursor for placental estrogen biosynthesis.

Cortisol production is carefully regulated by the fetus and serum cortisol concentrations depend on four variables:

(i)   pituitary ACTH secretion,
(ii) inhibition of adrenal 3β-hydroxysteroid dehydrogenase by placental and maternal steroids, particularly the estrogens,
(iii) rapid placental clearance of cortisol by conversion to cortisone and
(iv) placental transport of maternal cortisol.

Studies where adrenal tissue is cocultured with pituitary or placental tissue have shown that the inhibition of 3β-hydroxysteroid dehydrogenase is due to material of placental origin. Although fetal low density lipoprotein (LDL) cholesterol appears to be the principal substrate, placental progesterone may be utilized to some degree to circumvent the relative deficiency of 3β-hydroxysteroid dehydrogenase. Up to 30% of fetal adrenal steroidogenesis may be derived from de novo cholesterol biosynthesis in the adrenal itself. Placental progesterone is used as a substrate for glucocorticoid and mineralocorticoid production by the fetal adrenal.

In the human fetus, ACTH appears in the pituitary by 5 weeks gestation. Since ACTH does not cross the placenta, fetal plasma concentrations depend entirely on the integrity of the fetal hypothalamic pituitary unit. As early as 8–12 week's gestation negative feedback regulation of fetal ACTH secretion can be demonstrated. A fall in cord cortisol and DHEAS levels and maternal estrogen excretion is seen following administration of glucocorticoids to the mother. This is the basis of controversial treatment of the fetus affected with congenital adrenal hyperplasia so as to prevent early virilization (see later). ACTH administered to the fetus usually increases total cortisol production and maternal estrogen excretion. The occasional lack of an acute response

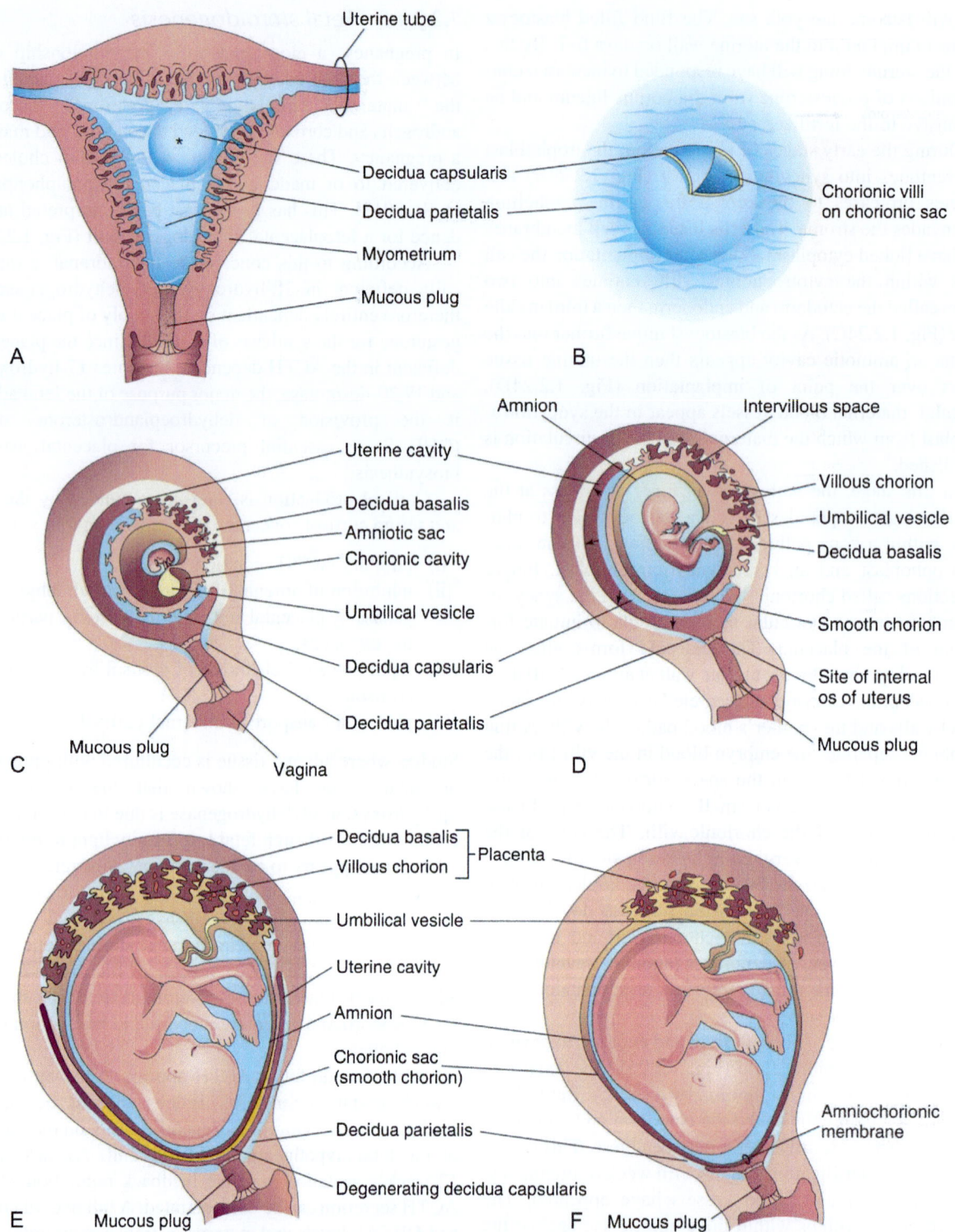

FIG. 1.2.25 Development of the placenta and fetal membranes. (A) Frontal section of the uterus showing elevation of the decidua capsularis by the expanding chorionic sac of a 4-week-old embryo implanted in the endometrium on the posterior wall (asterisk). (B) Enlarged drawing of the implantation site. The chorionic villi were exposed by cutting an opening in the decidua capsularis. (C–F) Sagittal sections of the gravid (pregnant) uterus from weeks 5 to 22 showing the changing relations of the fetal membranes to the decidua. In (F), the amnion and chorion are fused with each other and the decidua parietalis, thereby obliterating the uterine cavity. Note in (D) to (F) that the chorionic villi persist only where the chorion is associated with the decidua basalis. *(From Moore KL, Persaud TV, Torchia MG. The developing human—clinically oriented embryology. 10th ed. Elsevier; 2016. Fig. 71 p. 108.)*

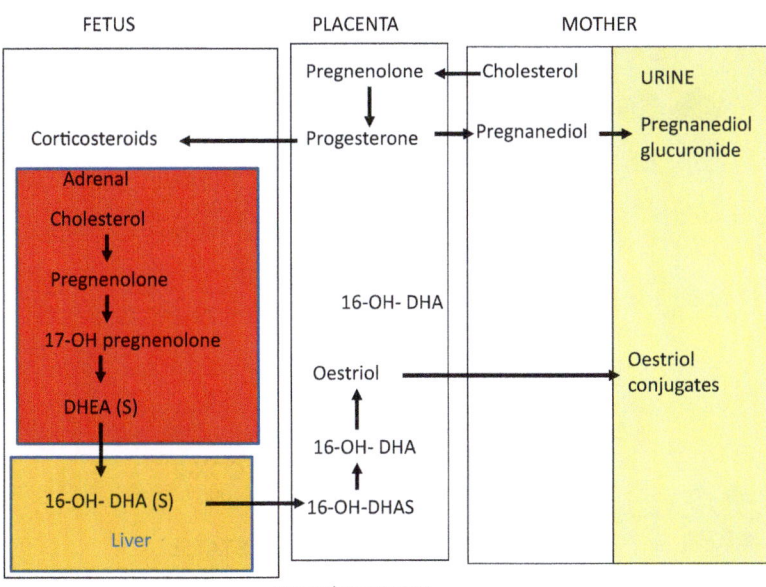

FETUS   PLACENTA   MOTHER

FIG. 1.2.26 Estrogen biosynthesis in the human feto-placental unit. A number of biochemical steps in the synthesis of the steroid hormone estriol, including 16a-hydroxylation, aromatization and reduction of important C19 steroid precursors such as androstenedione and DHEA, are reliant on both the fetus and the placenta. In this process, the fetus and the placenta work as a unit whereby the placenta synthesizes progesterone from maternal cholesterol and pregnenolone, and the fetal adrenal cortex is responsible for the conversion of this progesterone to DHEA. This may then be converted to DHEA-S which undergoes 16a-hydroxylation in the fetal liver, and subsequent aromatization in the placenta through 16a-hydroxy-androstenedione to estriol. *(Author original.)*

may indicate that the adrenal is already maximally stimulated. Experiments of this nature have been done mainly on early gestation fetuses prior to termination. In anencephalic (brainless) fetuses, the adrenal glands are also small as little or no corticotrophin releasing factor (CRF) is released from the grossly malformed or absent hypothalamus. ACTH concentrations in such fetuses are markedly reduced and as a result adrenal steroidogenesis shown by, for example, serum DHEAS levels or maternal estriol excretion is negligible.

The fetal cortex is capable of steroid production at an early stage of gestation and glucocorticoids produced from placental progesterone by the adrenal are involved in a number of important processes including:

(i) production of surfactant from type II cells in the alveoli of the lung. A deficiency of surfactant results in respiratory distress syndrome in the newborn,
(ii) development of hypothalamic function and of the pituitary thyroid axis,
(iii) sequential changes in placental structure and in the ionic composition of amniotic fluid,
(iv) induction of thymic involution and
(v) initiation of endocrine changes for parturition.

At the end of fetal life, the adrenal gland can be as much as twenty times bigger relative to other organs. The adrenal gland size is comparable to that of the adult adrenal gland. Zonation of the cortex begins late in fetal life, but only the zona glomerulosa and zona fasciculata are present at birth. The third zone—the zona reticularis—is not obvious until the third or fourth year of postnatal life.

1.2.3.4.7 Parturition

The initiation of parturition is not fully understood yet. Most work has been performed in the sheep. The sequence of events and the nature of signals that lead to the onset of parturition in the ewe have not been confirmed in the human. In sheep, there is an increase in fetal cortisol production. This acts on the **placenta** to reduce progesterone secretion and to increase estrogen formation. The reduction in progesterone is again at least partially controlled by the inhibition of 3β-hydroxysteroid dehydrogenase. In the last third of human pregnancy, the progesterone:estrogen ratio progressively decreases. Near term, estrogen production can be 1000 times that of a nonpregnant woman. The rise in cortisol production provides the essential signal for parturition in several species. In man, this relationship is not as clear although anencephaly (absence of the major part of the brain) is frequently associated with extended pregnancy due to absence of HPA axis. The intra-amniotic administration of cortisol may initiate labor in women with prolonged pregnancy. Just before and during labor there is also an increase in production of certain prostaglandins and these are thought to be critical in the initiation of labor.

During pregnancy, increased progesterone concentrations acting through PR can maintain myometrial quiescence via several mechanisms (Fig. 1.2.27), including:

- interaction with NF-κB p65;
- recruitment of corepressors to block p65 transcriptional activation;
- transcriptional induction of the NF-κB inhibitor, IκBα and/or the MAPK inhibitor, MKP-1/DUSP1 and

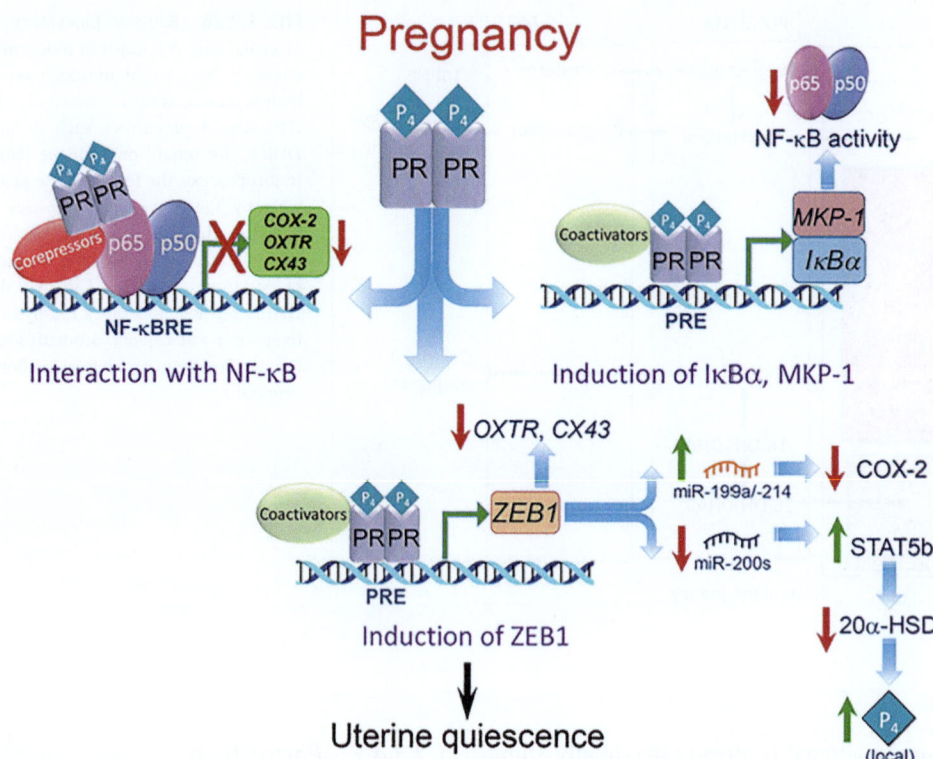

FIG. 1.2.27   Mechanisms for progesterone (P4)/and progesterone receptors (PR) maintenance of myometrial quiescence throughout most of pregnancy (left panel) and for the initiation of term and preterm parturition (right panel). During pregnancy, increased P4 acting through PR can maintain myometrial quiescence via several mechanisms, including: interaction with NF-κB p65 and recruitment of corepressors to block p65 transcriptional activation; transcriptional induction of the NF-κB inhibitor, IkBa and/or the MAPK inhibitor, MKP-1/DUSP1 and transcriptional induction of ZEB1, which suppresses transcription of myometrial contractile genes. *(From Mendelson CR, Gao L, Montalbano AP. Multifactorial regulation of myometrial contractility during pregnancy and parturition. Front Endocrinol (Lausanne) 2019;10:714. Fig. 4 p. 9.)*

- transcriptional induction of ZEB1, which suppresses transcription of myometrial genes for contractile proteins.

The initiation of labor at term follows increased secretion of surfactant components from the fetal lung, increased uterine stretch and increased circulating estradiol in the normal situation (Gao et al., 2015) (Fig. 1.2.28). In a preterm delivery, an infection with chorioamnionitis is accompanied by activation of macrophages possibly of fetal origin. These macrophages migrate to the maternal uterus with release of proinflammatory cytokines. This results in an increase in myometrial NF-κB and ERα transcriptional activity and upregulation of contractile and proinflammatory genes, as well as a decline in progesterone function, which further accelerates the inflammatory process leading to labor.

1.2.4   Steroidogenesis in peripheral tissues

1.2.4.1   Skin

Skin and the subcutaneous adipose tissue is a protective layer to the body that has up to seven layers of ectodermal tissue guarding muscles, bones, ligaments and internal organs.

Over the body, hair grows in follicles through the top two layers of skin, the outer epidermis above the dermis (Fig. 1.2.29). At the base of the hair is a dermal papilla. The root sheath is composed of an external and internal root sheath. In the hair bulb, cells divide and grow to build the hair shaft. Blood vessels nourish the cells in the hair bulb, and deliver hormones that modify hair growth and structure at different times of life. In the middle of the hair follicle, a bulge has stem cells that regenerate hair follicles, sebaceous glands and epidermis. The bulge also has arrector pili, the muscle tissue that makes hair stand on the ends when goosebumps are experienced during stress. Sebaceous glands produce sebum.

Skin can be classified as an intracrine steroidogenic tissue because it expresses the enzyme, CYP11A1 (also known as cytochrome P450scc) which initiates steroid synthesis by converting cholesterol to pregnenolone, as in other steroidogenic tissues. The activity of CYP11A1 in the skin has been confirmed with mitochondria prepared from immortalized keratinocytes which convert cholesterol to pregnenolone, albeit at a lower rate than placental mitochondria. CYP11A1 activity has

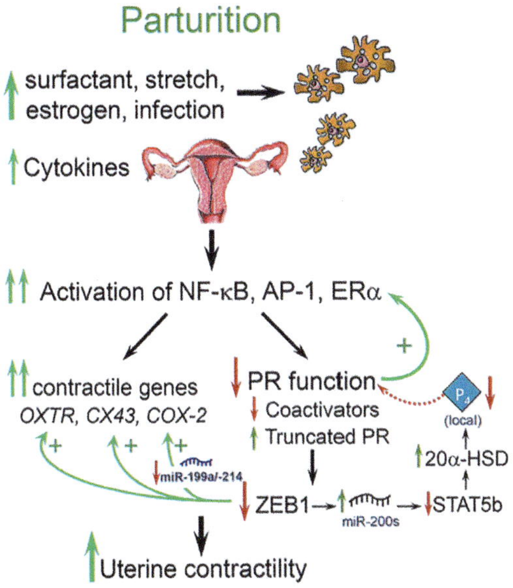

Parturition

↑ surfactant, stretch, estrogen, infection

↑ Cytokines

↑↑ Activation of NF-κB, AP-1, ERα

↑↑ contractile genes
OXTR, CX43, COX-2

↓ PR function
↓ Coactivators
↑ Truncated PR

↓ miR-199a/-214

↓ ZEB1 → ↑ miR-200s → ↓ STAT5b

P₄ (local)

↑ 20α-HSD

↑ Uterine contractility

FIG. 1.2.28 The initiation of labor at term (by increased secretion of surfactant components from the fetal lung, increased uterine stretch and increased circulating E2) and preterm (by infection with chorioamnionitis) is accompanied by activation of macrophages that may be of fetal origin, their migration to the maternal uterus and release of proinflammatory cytokines. This results in an increase in myometrial NF-κB and ERa transcriptional activity and upregulation of contractile and proinflammatory genes, as well as a decline in PR function, which further accelerates the inflammatory process leading to labor. *(From Mendelson CR, Gao L, Montalbano AP. Multifactorial regulation of myometrial contractility during pregnancy and parturition. Front Endocrinol (Lausanne) 2019;10:714. Fig. 2 p. 3.)*

also been demonstrated in cultured skin cells from squamous cell carcinoma, melanoma and keratinocytes. Immortalized sebaceous gland cells convert 22R-hydroxycholesterol to pregnenolone and 17α-hydroxypregnenolone. CYP11A1 expression has been detected at the mRNA or protein level in sebocytes, keratinocytes, melanocytes, squamous cell carcinoma, dermal fibroblasts and melanoma cells (Fig. 1.2.29B).

As well as expressing the genes (CYP11A1, StAR protein and MLN64) required to initiate steroidogenesis, the skin expresses CYP17A1, CYP21A2, CYP11B1 and 3βHSD with the potential to synthesize glucocorticoids. The production of deoxycorticosterone, corticosterone, cortisol and 18-hydroxydeoxycorticosterone in skin samples has been documented; assuming the pathways for the synthesis of these steroids follows the usual paths of other endocrine glands.

The skin can also synthesize testosterone from circulating DHEA sulfate and thus does not require CYP17A1 or CYP11A1. Sulfatase to produce free DHEA, 3βHSD1 and 17βHSD type 5 are all expressed in skin. Testosterone synthesis has been demonstrated from endogenous cholesterol in skin. Human skin also expresses CYP19A1 (aromatase) and thus can synthesize estrogens with conversion of androstenedione to estrone being demonstrated by a human skin culture system. The physiological importance of excess adipose tissue in the etiology of type 2 diabetes and cardiovascular disease is of interest in the context of the obesity epidemic in affluent societies (Blouin et al., 2006).

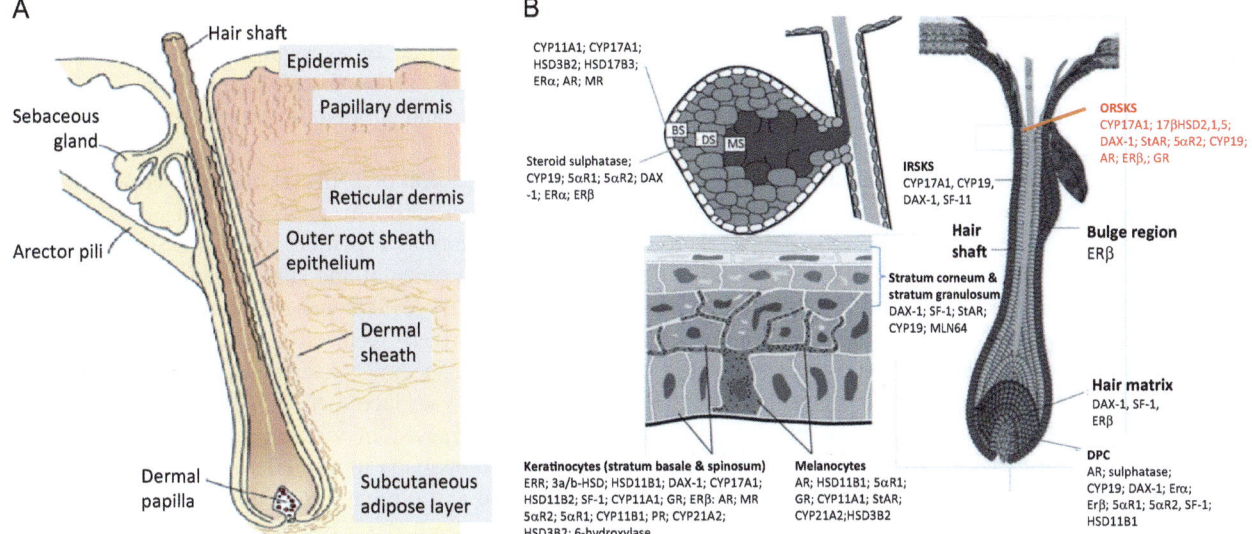

FIG. 1.2.29 (A) Diagram of a human hair follicle and surrounding skin. The outer root sheath merges with the epidermis, and the dermal sheath connects with the papillary layer of the dermis. (B) Compartmental expression of enzymes and cofactors involved in cutaneous steroidogenesis. *AR*, androgen receptor; *BS*, basal sebocytes; *DAX-1*, dosage-sensitive sex-reversal-adrenal hypoplasia congenital critical region on the X-chromosome, gene 1; *DPCs*, dermal papilla cells; *DS*, differentiating sebocytes; *ER/*, estrogen receptor or; *ERR*, estrogen-related receptor; *GR*, glucocorticoid receptor; *HF*, hair follicle; *IRSKS*, inner root sheath keratinocytes; *MR*, mineralocorticoid receptor; *MS*, mature sebocytes; *ORSKs*, outer root sheath keratinocytes; *PR*, progesterone receptor; *SF-1*, steroidogenic factor 1; *SREBP-1*, sterol response binding protein-1. *((A) From Jahoda CA, Reynolds AJ. Hair follicle dermal sheath cells: unsung participants in wound healing. Lancet 2001;358(9291):1445–8. Fig. 1 p. 1446; (B) From Slominski A, Zbytek B, Nikolakis G, Manna PR, Skobowiat C, Zmijewski M, Li W, Janjetovic Z, Postlethwaite A, Zouboulis CC, Tuckey RC. Steroidogenesis in the skin: implications for local immune functions. J Steroid Biochem Mol Biol 2013;137:107–23. Fig. 3 p. 113.)*

The pilosebaceous unit, including hair, has all the requirements to synthesize cholesterol de novo and utilize sex steroid precursors for transformation to more potent sex hormones

1.2.4.2 Adipose tissue

Adipose tissues are involved in the storage of triglycerides but also in synthesis and release of numerous protein hormones, growth factors and cytokines known as "adipokines." Adipose tissues are important sites for steroid hormone synthesis and metabolism. White adipose tissue can be under the skin, around the limbs, muscles and blood vessels, around vital organs (heart and intestines). Brown adipose tissue can be on the back, above the shoulder blades and the kidneys. Thus, estrogens in men and in postmenopausal women are principally produced by adipose tissue from circulating steroid precursors. Sex steroids have general metabolic roles in adipose tissue in different parts of the body regulating lipolysis and adipose tissue deposition. Subcutaneous adipose tissue is closely associated with the skin, which also synthesizes and metabolizes steroids, and the action of environmental factors initiated in the upper layers of the skin may extend down into the subcutaneous layer of fat. Visceral fat is located in the abdomen in the omentum and mesentery and surrounds organs such as the kidneys and the heart. Visceral fat accounts for approximately 10%–20% of the body adipose tissue. Obesity is associated with common diseases such as type 2 diabetes and cardiovascular diseases as well as certain types of cancer. Adipose tissue can have endocrine, paracrine/autocrine or intracrine systems regulatory roles (Fig. 1.2.30).

Adipose tissue is undoubtedly a major reservoir for steroid hormones in the human body. Plasma dehydroepiandrosterone (DHEA), DHEA sulfate (DHEAS), androstenedione and testosterone can be taken up and transformed to active estrogens or androgens in adipose tissue by various steroid-converting enzymes (Li et al., 2015). Steroid hormone synthesis in adipose tissue may affect circulating steroid concentrations, but the metabolic effects are considered to be mainly local. DHEA and androstenedione in obese men can be around 17 ng/g and 6 ng/g, respectively, similar to pre- and postmenopausal women, and greater than in lean men. The concentration of estrone in subcutaneous and visceral adipose tissue in both sexes, around 0.5 ng/g, is lower than the precursors DHEA and androstenedione. In pre- and postmenopausal women, progesterone, dehydroepiandrosterone, androstenedione and estrone concentrations are all 5–20 times higher in subcutaneous and visceral adipose tissue than in serum. Concentrations of 17β-estradiol [0.02 ng/g (70 fmol/g)] in subcutaneous adipose tissue after the menopause are several times less than in serum. Adipose cortisol and cortisone concentrations in women (15 ng/g and 3 ng/g, respectively) on the other hand are 10%–20% of those

found in the serum. In obese men, there is a positive adipose tissue/plasma gradient for estrone and estradiol.

Gender-, age- and depot-specific adipose steroidogenic potential will lead to large variation in the identity and amount of the final steroid products. The outcome of steroid biosynthesis in adipose tissue will depend on the relative expression or activity of steroidogenic enzymes. The mRNA expression of StAR shows a circadian rhythmic pattern in subcutaneous adipose tissue explants from obese women, with its level peaking 12 h after initiation of the culture. No studies so far have reported the gene expression of CYP11A1 in adipose tissue from men, and no protein expression, enzyme activity or intracellular location of adipose CYP11A1 has been determined in adipose tissue either from humans or animals. CYP11A1 expression may vary between preadipocytes, differentiated adipocytes and mature adipocytes.

HSD3B1 and HSD3B2 genes have been detected in adipose tissue from women, with slightly higher concentrations of both isoforms in subcutaneous fat. 11βHSD1 is highly expressed in human adipose tissue, particularly in visceral fat, and contribute to metabolic disease. The activity of 11βHSD1, regenerating active cortisol in stromal cells from human omental adipose tissue, is higher than that derived from subcutaneous abdominal adipose tissue. The higher production of cortisol in visceral fat may influence in regional adiposity. The expression of CYP11B1 has not yet been confirmed in human adipose tissue. Aromatase is expressed in undifferentiated adipose fibroblasts and its activities occur in human adipose tissue. In women after the menopause and in men, adipose tissue is the major source of estrogens. Aromatase activity is positively related to body weight in pre- and postmenopausal women and aromatase expression increases with advancing age.

Gene expression of 17βHSD has been shown in human adipose tissue. mRNA expression of the androgen-activating isoforms, 17βHSD3 and 17βHSD5, is detected in both human omental and subcutaneous adipose tissues. The activity of 17βHSD3 was measureable in cultured preadipocytes derived from abdominal subcutaneous and omental sites. The mRNA expression ratio of 17βHSD3 to aromatase was decreased in abdominal subcutaneous tissue and increased in visceral adipose tissue of obese women in comparison to normal weight women

The gene and protein expression of STS and sulfatase activity have been described in subcutaneous abdominal adipose tissue from both men and women. The mRNA expression of STS gene appears to be significantly higher in subcutaneous compared with visceral adipose tissue, in obese men as well as in obese and lean women. In obese men, STS expression in subcutaneous adipose tissue was positively related to estradiol concentrations in the tissue. Extra-adrenal and extra-gonadal steroidogenesis most

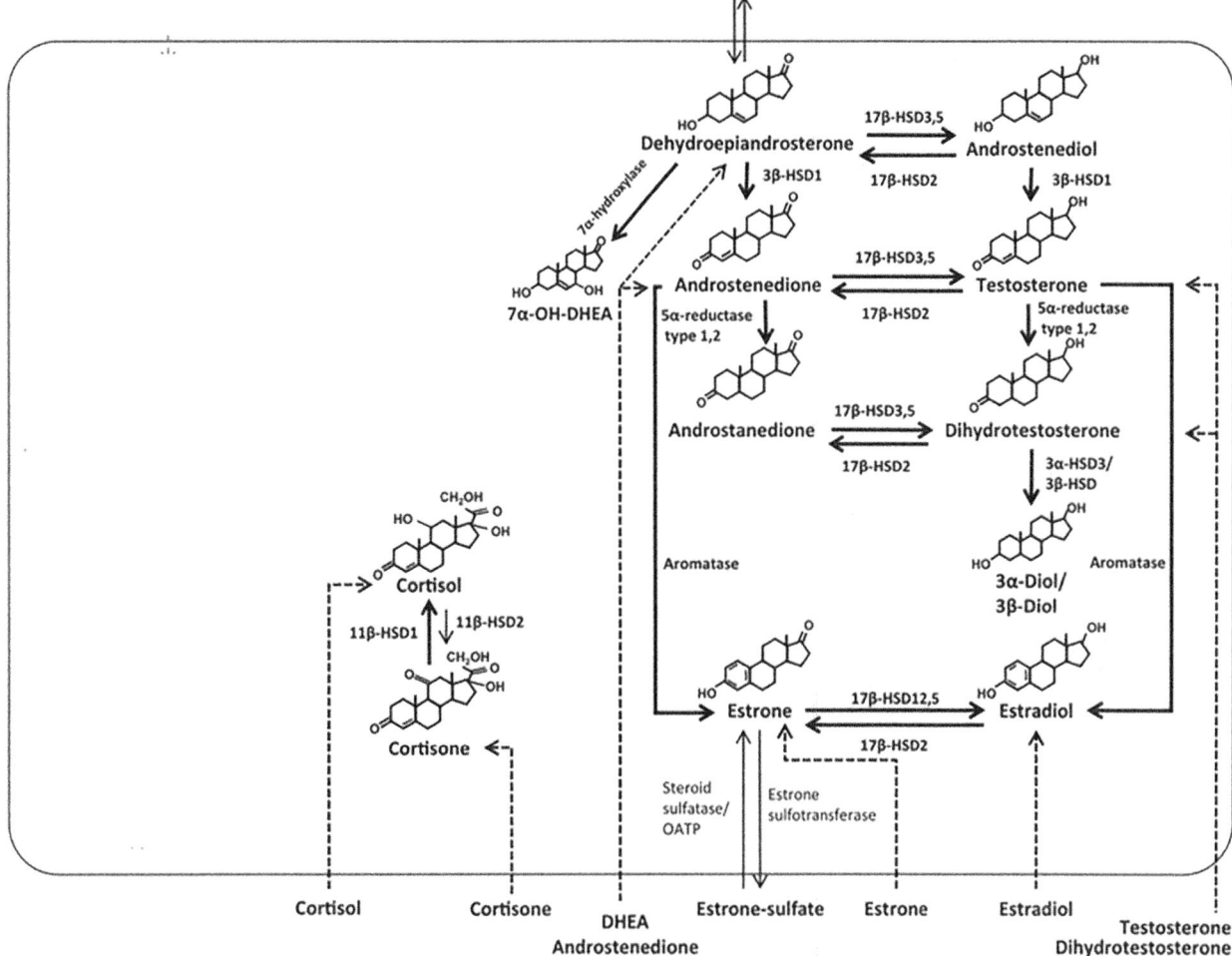

FIG. 1.2.30 Pathways of steroid hormone metabolism in adipose tissue. *Gray arrows* and steroids indicate putative pathways requiring confirmation. *Black arrows* and steroids indicate confirmed pathways. *11/18-hydroxylase*, aldosterone synthase; *11β-HSD1*, 11β-hydroxysteroid dehydrogenase type 1; *11β-HSD2*, 11β-hydroxysteroid dehydrogenase type 2; *17β-HSD12*, 17β-hydroxysteroid dehydrogenase type 12; *17β-HSD2*, 17β-hydroxysteroid dehydrogenase type 2; *17β-HSD3*, 17β-hydroxysteroid dehydrogenase type 3; *17β-HSD5 (AKR1C3)*, 17β-hydroxysteroid dehydrogenase type 5 or aldo-ketoreductase 1C type 3; *3α-HSD3 (AKR1C2)*, 3α-hydroxysteroid dehydrogenase or aldo-ketoreductase 1C2; *3β-HSD*, 3β-hydroxysteroid dehydrogenase; *3β-HSD1*, 3β-hydroxysteroid dehydrogenase type 1; *OATP*, transporter from the organic anion transport polypeptide family; *P450SCC*, side-chain cleavage enzyme; *SRD5A1-2*, steroid 5α-reductase type 1–2; *StAR*, steroidogenic acute regulatory protein. *: progesterone can also be transformed by 20α-hydroxysteroid dehydrogenase (20α-HSD or AKR1C1) into 20α-hydroxyprogesterone and a mixture of 5α-, 5β-, 20α- and 3α/β-reduced metabolites through activity of other enzymes. *(From Tchernof A, Mansour MF, Pelletier M, Boulet MM, Nadeau M, Luu-The V. Updated survey of the steroid-converting enzymes in human adipose tissues. J Steroid Biochem Mol Biol 2015;147:56–69. Fig. 2 p. 64.)*

likely regulates local homeostasis in a paracrine or autocrine manner. Whether de novo synthesis of steroids from cholesterol in adipose tissue is physiologically relevant in humans remains to be established (Hannen et al., 2011).

1.2.5 Summary

Steroid hormones are produced mainly in the adrenal cortex and gonads but also in skin and adipose tissues and recently appreciated in the nervous system. Cholesterol is the source of steroids and is derived by absorption from the diet and by

de novo synthesis. The adrenal cortex and gonads as well as kidneys develop from the urogenital ridge of the fetus. Mullerian and Wolffian ducts. Hormones from the primitive testes determine the development of the male reproductive tract from the Wolffian ducts and in the absence of testosterone, dihydrotestosterone and anti-Mullerian hormone the Mullerian ducts develop to the uterus and fallopian tubes. The proteins from many genes are now known to be involved but the processes remain to be fully elucidated.

Sex determination is a complex developmental process by which the bipotential gonads develop as either testes or ovaries. Despite several decades of research revealing the delicate balance of male and female genes mediating sex determination, the majority of DSD (differences in or disorders of sexual development) cases related to gonadal development and differentiation remain undiagnosed at the genetic level. The success rate of identifying causal variants through Exome sequencing in patients with DSD has not identified causal variants in comparison to other genetic disorders. Pathogenic variants and may be localized in intergenic regions, perhaps regulating the balance of expression of key sex-determining genes. Enhancers, silencers and insulators are attractive candidates, but the lack of knowledge regarding the locations of these crucial gene-regulatory elements limits the ability to study the mechanisms that control gene expression. In the near future, through the use of more sensitive genome-wide approaches, a precise characterization of the dynamics of gene expression during the process of sex determination will be available for each of the major cell-lineages of the gonad, both in the mouse model and human. The functionalities of the steroidogenic organs change throughout life, and the clinician and laboratory need to understand these issues when investigating patients with disorders of sugar metabolism, homeostasis and sex development.

The changes in the adrenal cortex and gonads during life need to be understood when evaluating the role of the gland in normal and pathological circumstances. The purpose of a fetal adrenal cortex during pregnancy remains enigmatic but the presence of steroids from this adrenal zone (especially DHEAS, pregnenolone sulfate and 17-hydroxypregnenolone sulfate) in the circulation during the neonatal period causes a number of problems in the analysis of other steroid hormones. The execution and interpretation of laboratory tests need special considerations particularly in childhood with consideration of reference ranges for age, development and in some cases body size.

References

Afonso MS, Machado RM, Lavrador MS, Quintao ECR, Moore KJ, Lottenberg AM. Molecular pathways underlying cholesterol homeostasis. Nutrients 2018;10(6). pii: E760.

Andersson S, Berman DM, Jenkins EP, Russell DW. Deletion of steroid 5 alpha-reductase 2 gene in male pseudohermaphroditism. Nature 1991;354(6349):159–61.

Auchus RJ. The physiology and biochemistry of adrenarche. Endocr Dev 2011;20:20–7.

Baskin L, Shen J, Sinclair A, Cao M, Liu X, Liu G, et al. Development of the human penis and clitoris. Differentiation 2018;103:74–85.

Biason-Lauber A. Control of sex development. Best Pract Res Clin Endocrinol Metab 2010;24(2):163–86.

Blouin K, Richard C, Brochu G, Hould FS, Lebel S, Marceau S, et al. Androgen inactivation and steroid-converting enzyme expression in abdominal adipose tissue in men. J Endocrinol 2006;191(3):637–49.

Buitenwerf E, Kerstens MN, Links TP, Kema IP, Dullaart RPF. High-density lipoproteins and adrenal steroidogenesis: a population-based study. J Clin Lipidol 2017a;11(2):469–76.

Buitenwerf E, Dullaart RPF, Muller Kobold AC, Links TP, Sluiter WJ, Connelly MA, et al. Cholesterol delivery to the adrenal glands estimated by adrenal venous sampling: an in vivo model to determine the contribution of circulating lipoproteins to steroidogenesis in humans. J Clin Lipidol 2017b;11(3):733–8.

Chatuphonprasert W, Jarukamjorn K, Ellinger I. Physiology and pathophysiology of steroid biosynthesis, transport and metabolism in the human placenta. Front Pharmacol 2018;9:1027.

Cumberland AL, Hirst JJ, Badoer E, Wudy SA, Greaves RF, Zacharin M, et al. The enigma of the adrenarche: identifying the early life mechanisms and possible role in postnatal brain development. Int J Mol Sci 2021;22(9):4296.

Cunha GR, Robboy SJ, Kurita T, Isaacson D, Shen J, Cao M, et al. Development of the human female reproductive tract. Differentiation 2018;103:46–65.

Dhayat NA, Frey AC, Frey BM, d'Uscio CH, Vogt B, Rousson V, et al. Estimation of reference curves for the urinary steroid metabolome in the first year of life in healthy children: Tracing the complexity of human postnatal steroidogenesis. J Steroid Biochem Mol Biol 2015;154:226–36. Erratum in: J Steroid Biochem Mol Biol. 2018;183:238.

Dumontet T, Sahut-Barnola I, Septier A, Montanier N, Plotton I, Roucher-Boulez F, et al. PKA signaling drives reticularis differentiation and sexually dimorphic adrenal cortex renewal. JCI Insight 2018;3(2), e98394.

Eid W, Biason-Lauber A. Why boys will be boys and girls will be girls: Human sex development and its defects. Birth Defects Res C Embryo Today 2016;108(4):365–79.

Elustondo P, Martin LA, Karten B. Mitochondrial cholesterol import. Biochim Biophys Acta Mol Cell Biol Lipids 2017;1862(1):90–101.

Gao L, Rabbitt EH, Condon JC, Renthal NE, Johnston JM, Mitsche MA, et al. Steroid receptor coactivators 1 and 2 mediate fetal-to-maternal signaling that initiates parturition. J Clin Invest 2015;125(7):2808–24.

Guasti L, Candy Sze WC, McKay T, Grose R, King PJ. FGF signalling through Fgfr2 isoform IIIb regulates adrenal cortex development. Mol Cell Endocrinol 2013;371(1–2):182–8.

Hale GE, Robertson DM, Burger HG. The perimenopausal woman: endocrinology and management. J Steroid Biochem Mol Biol 2014;142:121–31.

Hannen RF, Michael AE, Jaulim A, Bhogal R, Burrin JM, Philpott MP. Steroid synthesis by primary human keratinocytes; implications for skin disease. Biochem Biophys Res Commun 2011;404(1):62–7.

Huang CC, Kang Y. The transient cortical zone in the adrenal gland: the mystery of the adrenal X-zone. J Endocrinol 2019;241(1):R51–63.

Hummitzsch K, Anderson RA, Wilhelm D, Wu J, Telfer EE, Russell DL, et al. Stem cells, progenitor cells, and lineage decisions in the ovary. Endocr Rev 2015;36(1):65–91.

Iaea DB, Spahr ZR, Singh RK, Chan RB, Zhou B, Bareja R, et al. Stable reduction of STARD4 alters cholesterol regulation and lipid homeostasis. Biochim Biophys Acta Mol Cell Biol Lipids 2020;1865(4):158609. https://doi.org/10.1016/j.bbalip.2020.158609. Epub 2020 Jan 7 31917335. PMC6996790.

Ikonen E. Mechanisms of cellular cholesterol compartmentalization: recent insights. Curr Opin Cell Biol 2018;53:77–83.

Imperato-McGinley J. 5alpha-reductase-2 deficiency and complete androgen insensitivity: lessons from nature. Adv Exp Med Biol 2002;511:121–31.

Inoue M, Baba T, Morohashi KI. Recent progress in understanding the mechanisms of Leydig cell differentiation. Mol Cell Endocrinol 2018;468:39–46.

Ishimoto H, Jaffe RB. Development and function of the human fetal adrenal cortex: a key component in the feto-placental unit. Endocr Rev 2011;32 (3):317–55.

Jamnongjit M, Hammes SR. Ovarian steroids: the good, the bad, and the signals that raise them. Cell Cycle 2006;5(11):1178–83.

Julve J, Martín-Campos JM, Escolà-Gil JC, Blanco-Vaca F. Chylomicrons: advances in biology, pathology, laboratory testing, and therapeutics. Clin Chim Acta 2016;455:134–48.

Kapourchali FR, Surendiran G, Goulet A, Moghadasian MH. The role of dietary cholesterol in lipoprotein metabolism and related metabolic abnormalities: a mini-review. Crit Rev Food Sci Nutr 2016;56 (14):2408–15.

Kempná P, Flück CE. Adrenal gland development and defects. Best Pract Res Clin Endocrinol Metab 2008;22(1):77–93.

Kim JH, Kang E, Heo SH, Kim GH, Jang JH, Cho EH, et al. Diagnostic yield of targeted gene panel sequencing to identify the genetic etiology of disorders of sex development. Mol Cell Endocrinol 2017;444:19–25.

Kraemer FB. Adrenal cholesterol utilization. Mol Cell Endocrinol 2007;265-266:42–5.

Lamothe S, Bernard V, Christin-Maitre S. Gonad differentiation toward ovary. Ann Endocrinol (Paris) 2020;81(2-3):83–8.

Lew R. Natural history of ovarian function including assessment of ovarian reserve and premature ovarian failure. Best Pract Res Clin Obstet Gynaecol 2019;55:2–13.

Li J, Papadopoulos V, Vihma V. Steroid biosynthesis in adipose tissue. Steroids 2015;103:89–104.

Liu X, Liu G, Shen J, Yue A, Isaacson D, Sinclair A, et al. Human glans and preputial development. Differentiation 2018;103:86–99.

Lopez AG, Duparc C, Wils J, Naccache A, Castanet M, Lefebvre H, et al. Steroidogenic cell microenvironment and adrenal function in physiological and pathophysiological conditions. Mol Cell Endocrinol 2021;535, 111377.

Luo J, Yang H, Song BL. Mechanisms and regulation of cholesterol homeostasis. Nat Rev Mol Cell Biol 2020;21(4):225–45.

Mäkelä JA, Koskenniemi JJ, Virtanen HE, Toppari J. Testis Development. Endocr Rev 2019;40(4):857–905.

Malendowicz LK. 100th anniversary of the discovery of the human adrenal fetal zone by Stella Starkel and Lesław Węgrzynowski: how far have we come? Folia Histochem Cytobiol 2010;48(4):491–506.

Messinis IE, Messini CI, Dafopoulos K. Novel aspects of the endocrinology of the menstrual cycle. Reprod BioMed Online 2014;28(6):714–22.

Morohashi K, Baba T, Tanaka M. Steroid hormones and the development of reproductive organs. Sex Dev 2013;7(1–3):61–79.

Nef S, Stévant I, Greenfield A. Characterizing the bipotential mammalian gonad. Curr Top Dev Biol 2019;134:167–94.

Pasqualini JR. Enzymes involved in the formation and transformation of steroid hormones in the fetal and placental compartments. J Steroid Biochem Mol Biol 2005;97(5):401–15.

Rege J, Turcu AF, Kasa-Vubu JZ, Lerario AM, Auchus GC, Auchus RJ, et al. 11-Ketotestosterone is the dominant circulating bioactive androgen during normal and premature adrenarche. J Clin Endocrinol Metab 2018;103(12):4589–98.

Rojas J, Chávez-Castillo M, Olivar LC, Calvo M, Mejías J, Rojas M, et al. Physiologic course of female reproductive function: a molecular look into the prologue of life. J Pregnancy 2015;2015, 715735.

Ross IL, Louw GJ. Embryological and molecular development of the adrenal glands. Clin Anat 2015;28(2):235–42.

Savchuk I, Morvan ML, Antignac JP, Kurek M, Le Bizec B, Söder O, et al. Ontogenesis of human fetal testicular steroidogenesis at early gestational age. Steroids 2019;141:96–103.

Seccia TM, Caroccia B, Gomez-Sanchez EP, Gomez-Sanchez CE, Rossi GP. The biology of normal zona glomerulosa and aldosterone-producing adenoma: pathological implications. Endocr Rev 2018;39 (6):1029–56.

Shah JS, Sabouni R, Cayton Vaught KC, Owen CM, Albertini DF, Segars JH. Biomechanics and mechanical signaling in the ovary: a systematic review. J Assist Reprod Genet 2018;35(7):1135–48.

Shen WJ, Azhar S, Kraemer FB. Lipid droplets and steroidogenic cells. Exp Cell Res 2016;340(2):209–14.

Sobel V, Zhu YS, Imperato-McGinley J. Fetal hormones and sexual differentiation. Obstet Gynecol Clin N Am 2004;31(4):837–56. x–xi.

Stévant I, Nef S. Genetic control of gonadal sex determination and development. Trends Genet 2019;35(5):346–58.

Svechnikov K, Söder O. Ontogeny of gonadal sex steroids. Best Pract Res Clin Endocrinol Metab 2008;22(1):95–106.

Tremblay JJ. Molecular regulation of steroidogenesis in endocrine Leydig cells. Steroids 2015;103:3–10.

Ungewitter EK, Yao HH. How to make a gonad: cellular mechanisms governing formation of the testes and ovaries. Sex Dev 2013;7 (1–3):7–20.

Vining B, Ming Z, Bagheri-Fam S, Harley V. Diverse regulation but conserved function: SOX9 in vertebrate sex determination. Genes (Basel) 2021;12(4):486.

Wen Q, Cheng CY, Liu YX. Development, function and fate of fetal Leydig cells. Semin Cell Dev Biol 2016;59:89–98.

Wilhelm D, Koopman P. The makings of maleness: towards an integrated view of male sexual development. Nat Rev Genet 2006;7 (8):620–31.

Xing Y, Lerario AM, Rainey W, Hammer GD. Development of adrenal cortex zonation. Endocrinol Metab Clin N Am 2015;44(2):243–74.

Yu XH, Zhang DW, Zheng XL, Tang CK. Cholesterol transport system: an integrated cholesterol transport model involved in atherosclerosis. Prog Lipid Res 2019;73:65–91.

Zajicek G, Ariel I, Arber N. The streaming adrenal cortex: direct evidence of centripetal migration of adrenocytes by estimation of cell turnover rate. J Endocrinol 1986;111(3):477–82.

Zhao F, Yao HH. A tale of two tracts: history, current advances, and future directions of research on sexual differentiation of reproductive tracts. Biol Reprod 2019;101(3):602–16.

Chapter 1.3

Steroid biosynthesis

1.3.1 Introduction

Cholesterol can be synthesized in the steroidal cells or derived from **cholesterol esters** in plasma **low-density** or **high-density lipoproteins** (LDL/HDL) that import dietary cholesterol. In steroidogenic cells, cholesterol and the esters traffic between lipid droplet stores, plasma membrane and cell compartments (Deng et al., 2019). Cholesterol needs to be moved into mitochondria to initiate steroidogenesis and this is a rate-limiting step. **Steroidogenic acute regulatory protein (StAR)** (Stocco et al., 2017) and **translocator protein** (Papadopoulos et al., 2018) are among current candidate proteins. The enzymes for steroidogenesis are distributed in the endoplasmic reticulum and mitochondria. The movement of steroids between compartments in the cells that produce these hormones is poorly understood.

Steroid synthesis from cholesterol involves **cytochrome P450** and **dehydrogenase enzymes**. The side chain of the 27 carbon cholesterol is reduced in length to the 21 carbon pregnenolone in mitochondria, which is converted to progesterone and deoxycorticosterone in the endoplasmic reticulum.

In the **adrenal zona fasciculata**, 17-hydroxyprogesterone and 11-deoxycortisol are intermediates to cortisol synthesis. In the outer **zona glomerulosa**, corticosterone is an intermediate to aldosterone. The adrenal steroid cortisol is responsible for regulating carbohydrate metabolism (glucocorticoid) and stress management while deoxycorticosterone and aldosterone affect salt balance (mineralocorticoids). In the adult, a new zona reticularis appears between the adrenal cortex and the inner medulla, dehydroepiandrosterone (DHEA) sulfate (DHEAS) is secreted from the zona reticularis and becomes the steroid at highest concentration in blood (2–10 μmol/L).

Testicular androgenic steroids are essential for reproductive function and male secondary characteristics while ovarian estrogens and progestins maintain female secondary sex characteristics and reproductive function. An additional class of steroids is produced in the brain and is called neurosteroids.

Steroids are produced on demand and not stored in endocrine glands like other hormones. Pathways for steroid production have been known for many years but new routes have been found, recently these include a nonclassic path to dihydrotestosterone. In pregnancy, the fetus and placenta work as a unit to produce steroids. Steroids can be sulfated by the fetus before secretion and the sulfate group can be removed by sulfatase and this is applicable before or after uptake by other target cells. The transport of steroids and steroid sulfates into cells is now thought to be an active process involving carrier or transport proteins. This will be addressed later.

1.3.2 De novo synthesis of cholesterol

Steroids are a class of organic compounds biologically derived from six isopentyl pyrophosphate units. Acetate units (2C) are converted to isoprenoid units (5C) which are condensed to form the squalene unit (30C). The pathway for de novo synthesis of cholesterol is divided into two parts divided by squalene which is a linear compound then lanosterol that is a cyclic compound (Miziorko, 2011).

The sequence for the first part toward squalene (Fig. 1.3.1) is thus:

- two molecules of acetyl-CoA converted to acetoacetyl-CoA in cytosol by β-ketothiolase.
- acetyl-CoA is added by HMG-CoA synthase to give 3-hydroxy-3-methylglutaryl-CoA (HMG-CoA) a six carbon compound.
- HMG-CoA is reduced to mevalonate by a reductase (HMGCR). The electrons are donated by two molecules of NADPH.
- Mevalonate kinase catalyzes two ATP dependent steps to form diphosphomevalonate.
- Diphosphomevalonate decarboxylase then phosphorylates the 3-hydroxyl group and catalyzes decarboxylation by releasing carbon dioxide to produce isopentylpyrophosphate (five carbon atoms).
- Isopentylpyrophosphate (IPP) can isomerize to dimethyallyl pyrophosphate (DMAPP).

Three enzymatic steps then take IPP and DMAPP, each with 5 carbon atoms, via geranyl pyrophophate (10 carbons) and farnesyl pyrophosphate (15 carbons) to squalene (30 carbon atoms)

Steroids in the Laboratory and Clinical Practice. https://doi.org/10.1016/B978-0-12-818124-9.00013-9

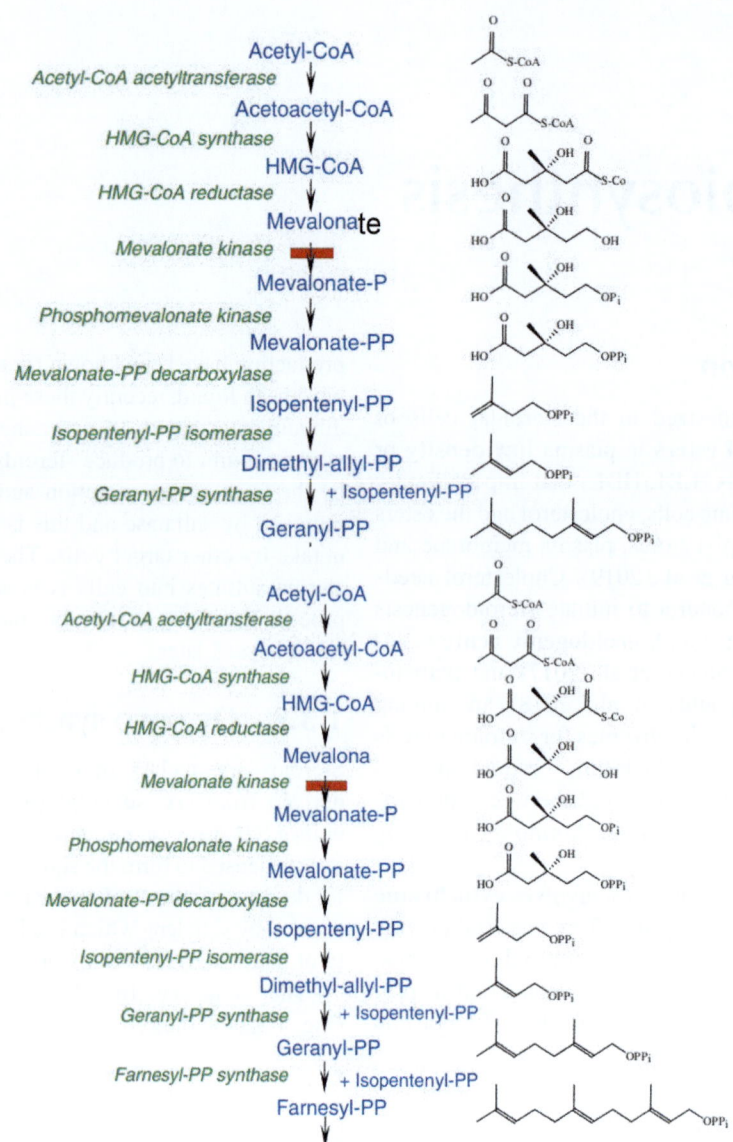

FIG. 1.3.1 Isoprenoid biosynthesis. The first step of steroid synthesis takes acetyl CoA to 3-isopentenyl pyrophosphate (IPP) and dimethylallyl pyrophosphate reacts (DMPP) thence to squalene synthesis. Dimethylallyl pyrophosphate (DMPP) reacts with 3-isopentenyl pyrophosphate (IPP) to form geranyl pyrophosphate (GPP). *(Modified from Waterham HR. Defects of cholesterol biosynthesis. FEBS Lett 2006;580(23):5442–9. Fig. 1 p. 5443.)*

- Isopentylpyrophosphate and dimethyallyl pyrophosphate condense to form geranylpyrophosphate (Fig. 1.3.1).
- Geranylpyrophosphate reacts with a second molecule of isopentylpyrophosphate to produce farnesyl pyrophosphate (Fig. 1.3.2A).
- Squalene synthase joins two molecules of farnesyl pyrophosphate to produce squalene (Fig. 1.3.2B).
- Squalene is cyclized to the triterpene lanosterol after formation of 2,3-epoxysqualene. Ring closures lead to lanosterol formation (Fig. 1.3.3).

Six enzymes are involved in the production of cholesterol from lanosterol (Fig. 1.3.4):

- sterol Δ14 reductase,
- 4,4-demethylase,
- Δ8,7-isomerase,
- sterol C5-desaturase,
- sterol Δ7-reductase and
- sterol Δ 24,25-reductase.

The synthesis of cholesterol from acetate via squalene and lanosterol is summarized in Fig. 1.3.5.

Some of the detail of cholesterol synthesis has become clearer from studies of the fate of radioactive precursors and of patients with disorders of genes for enzymes in the pathway considered later (see Part III).

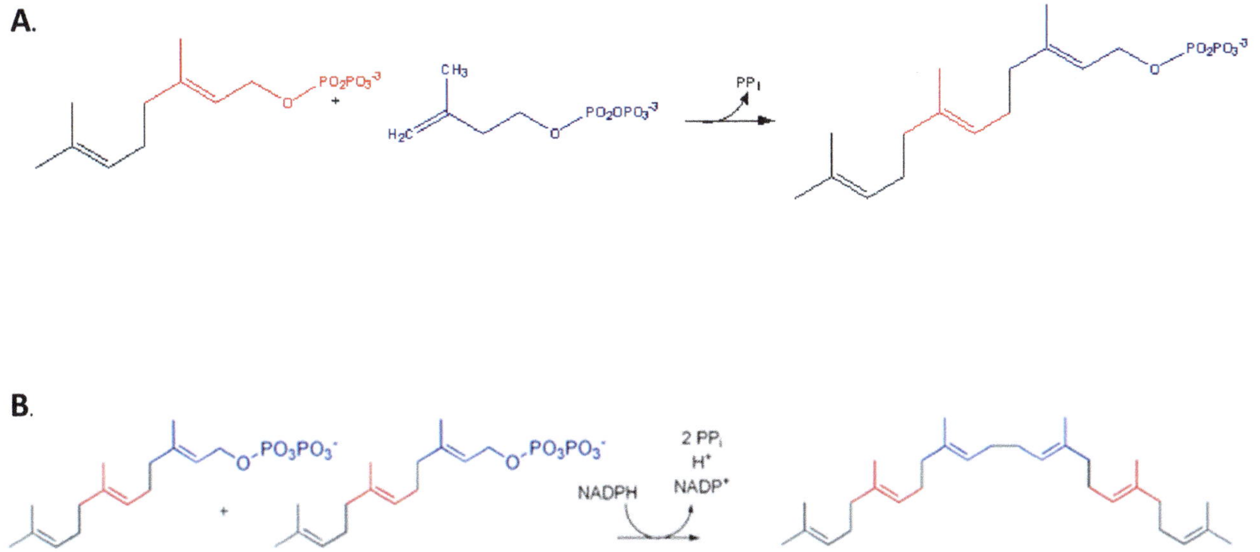

A.

B.

FIG. 1.3.2 (A) Geranyl pyrophosphate then reacts with another molecule of 3-isopentenyl pyrophosphate to form farnesyl pyrophosphate (FPP). (B) Two molecules of farnesyl pyrophosphate condense with reduction by NADPH to form squalene by squalene synthase.

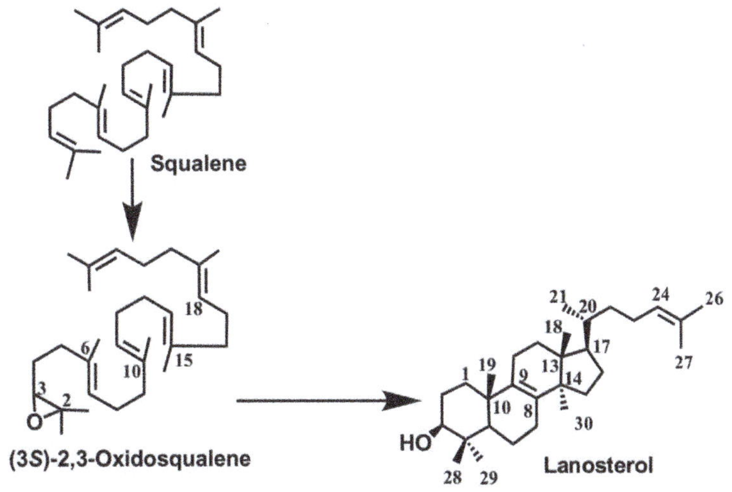

FIG. 1.3.3 Cyclisation of squalene to lanosterol. *(Modified from Wu TK, Chang CH, Liu YT, Wang TT. Saccharomyces cerevisiae oxidosqualene-lanosterol cyclase: a chemistry-biology interdisciplinary study of the protein's structure-function-reaction mechanism relationships. Chem Rec 2008;8(5):302–25. Scheme 1 p. 305.)*

1.3.3 Cholesterol uptake and trafficking

Cholesterol from the diet is a source for steroid hormone synthesis in addition to the de novo synthesis just described. **Cholesterol esters** in the blood are delivered to steroidogenic cells by **lipoproteins LDL** and **HDL** then converted to **free cholesterol** by acid lipase within lysosomes or hormone sensitive lipase, respectively (Fig. 1.3.6).

The free cholesterol, along with de novo synthesized cholesterol, traffics to the endoplasmic reticulum where it is esterified by acyl-CoA cholesterol acyl transferase (ACAT) and stored in cytoplasmic lipid droplets or transferred to the plasma membrane or outer mitochondrial membrane (Kraemer, 2007). The lipid droplets are rapidly depleted of cholesterol on stimulation of steroidogenesis by adrenocorticotrophic hormone (ACTH) (Shen et al., 2016). Hormone sensitive lipase (HPL) will liberate free cholesterol. There is intracellular trafficking of cholesterol between membranes and by cholesterol binding proteins. Lipid droplets are rapidly moved to be in close proximity to mitochondria probably involving the soluble N-ethylmaleimide sensitive factor (NSF), attachment protein receptor (SNARE) proteins and other factors (e.g., syntaxin) (Deng et al., 2019) but this is not yet entirely clear. In general, data are derived in subcellular reconstitution assays which may not represent the reality of cells in vivo.

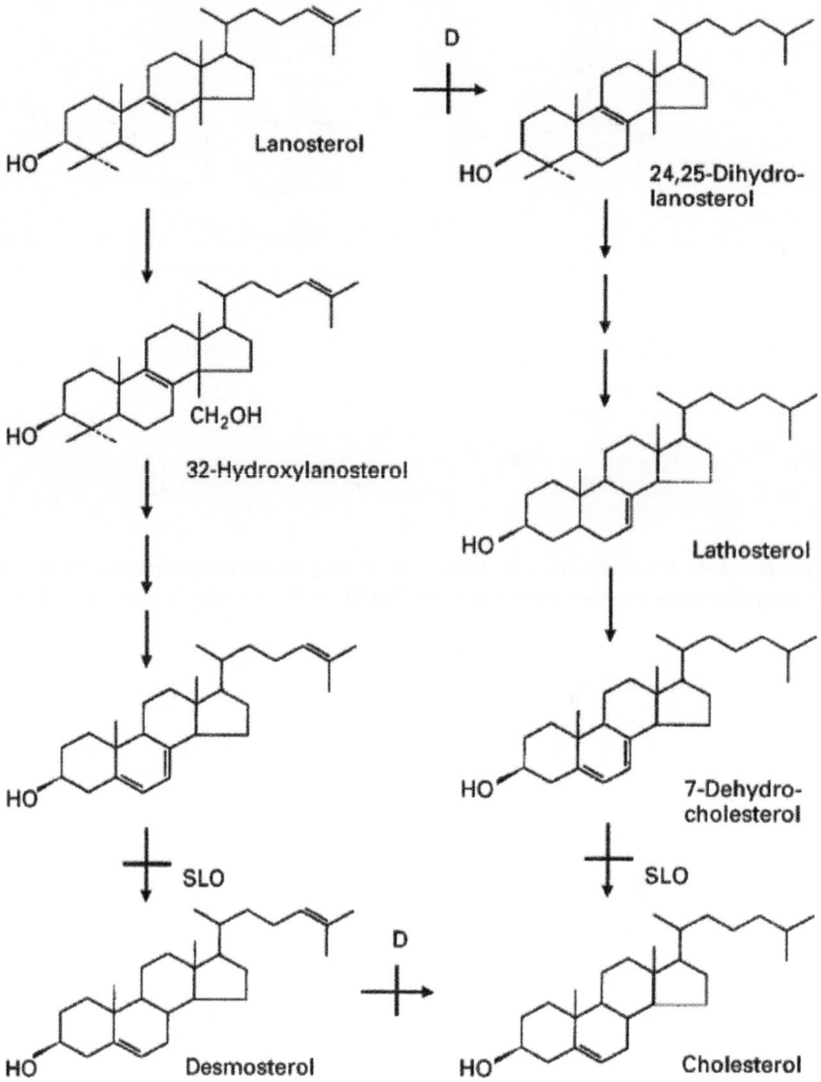

FIG. 1.3.4 Final steps of cholesterol synthesis from lanosterol. Inborn errors of enzymes can disrupt cholesterol synthesis and lead to desmosterolosis (D) and Smith-Lemli-Opitz syndrome (SLO). *(Modified from Clayton PT. Disorders of cholesterol biosynthesis. Arch Dis Child 1998;78(2):185–9. Fig 2 p. 187.)*

1.3.3.1 Steroidogenic acute regulatory protein

Cholesterol needs to be transported into mitochondria to initiate the synthesis of steroid hormones. **StAR** protein facilitates the transfer of cholesterol from the outer to the inner mitochondrial membrane. Several related proteins called **StAR related lipid transfer proteins (START)** (Clark, 2020) are probably involved in intracellular cholesterol transfer. Within the START family there are a number of related **START domain proteins (StARD)** (Iaea et al., 2020) of which StARD4 and 5 are likely involved in cholesterol transfer. The mechanisms of cholesterol transfer in adrenal cells are not well understood and no further debate is attempted in this review. Further discussion

can be found in a comprehensive review on the initial steps of steroid synthesis (Miller, 2017a,b). Some clinical disorders (Smith-Lemli-Opitz syndrome, Wolman disease, Nieman-Pick C syndrome [NPC]) have given clues for the involvement of other cholesterol transfer steps like NPC1 and NPC2, SNARE (soluble NSF attachment protein receptor) and oxysterol binding protein. Other proteins may be involved in cholesterol movement into the mitochondria. The benzodiazepine receptor (translocator protein TSPO) (Denora and Natile, 2017) and its ligand, the ATP binding cassette subfamily D (ABCD1) and sterol carrier protein 2 (SCP2) (Burgardt et al., 2017) may act as a docking site for StAR but more research is required to delineate all the players in cholesterol transport.

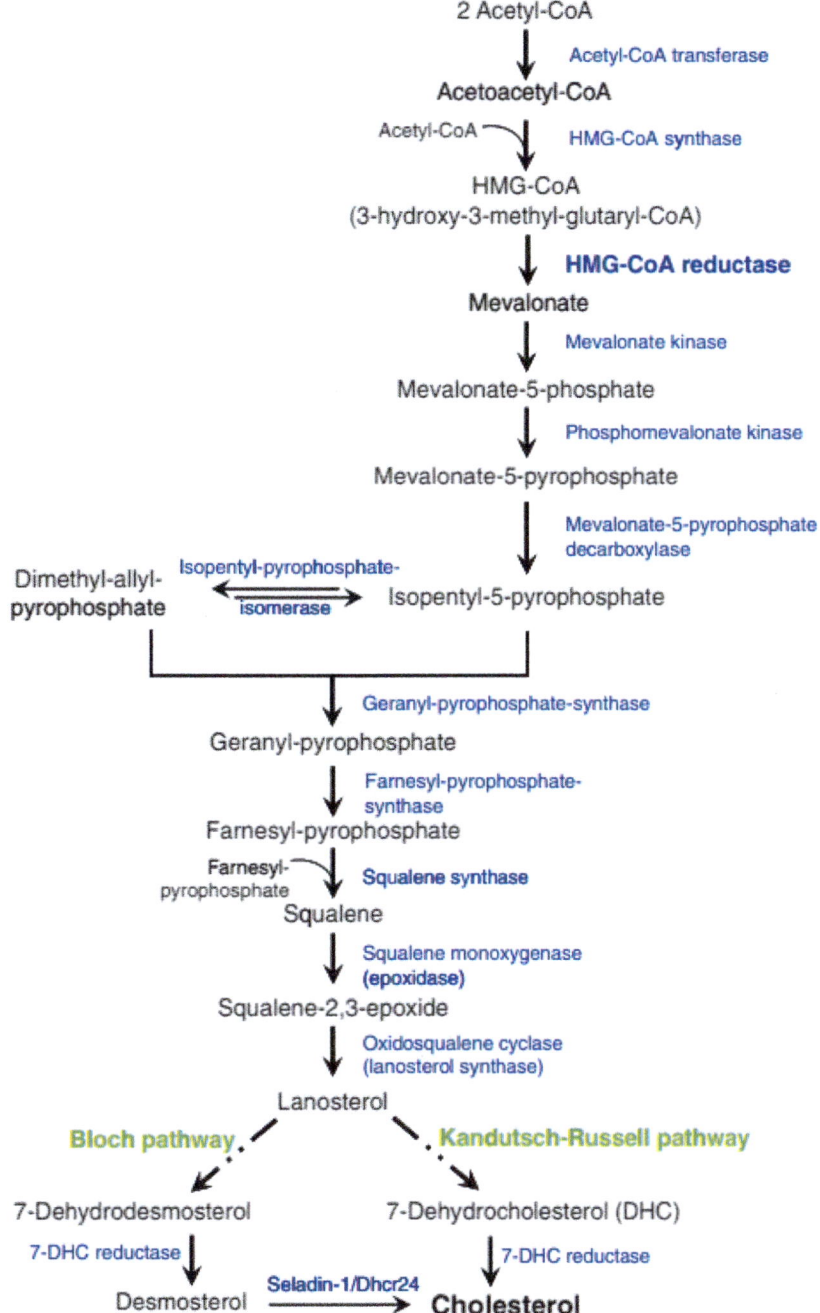

2 Acetyl-CoA

Acetyl-CoA transferase

Acetoacetyl-CoA

Acetyl-CoA ——

HMG-CoA synthase

HMG-CoA
(3-hydroxy-3-methyl-glutaryl-CoA)

HMG-CoA reductase

Mevalonate

Mevalonate kinase

Mevalonate-5-phosphate

Phosphomevalonate kinase

Mevalonate-5-pyrophosphate

Mevalonate-5-pyrophosphate
decarboxylase

Dimethyl-allyl-
pyrophosphate

Isopentyl-pyrophosphate-
isomerase

Isopentyl-5-pyrophosphate

Geranyl-pyrophosphate-synthase

Geranyl-pyrophosphate

Farnesyl-pyrophosphate-
synthase

Farnesyl-pyrophosphate

Farnesyl-
pyrophosphate

Squalene synthase

Squalene

Squalene monoxygenase
(epoxidase)

Squalene-2,3-epoxide

Oxidosqualene cyclase
(lanosterol synthase)

Lanosterol

Bloch pathway **Kandutsch-Russell pathway**

7-Dehydrodesmosterol

7-Dehydrocholesterol (DHC)

7-DHC reductase

7-DHC reductase

Desmosterol ——Seladin-1/Dhcr24——> **Cholesterol**

FIG. 1.3.5 Schematic representation of the main steps involved in cholesterol biosynthesis from precursor acetyl-CoA. The rate-limiting enzyme of the pathway is HMG-CoA reductase which catalyzes the synthesis of mevalonate from HMG-CoA. The post-lanosterol steps of cholesterol biosynthesis have been divided into the Bloch and Kandutsch–Russell pathways, which share the same enzymatic steps but differ by the stage at which the C24 double bond is reduced. *(Modified from Maulik M, Westaway D, Jhamandas JH, Kar S. Role of cholesterol in APP metabolism and its significance in Alzheimer's disease pathogenesis. Mol Neurobiol 2013;47(1):37–63. Fig. 2 p. 42.)*

1.3.4 Steroid-converting enzymes

The steroid hormones are synthesized from cholesterol mainly by a series of cytochrome P450 hydroxylation reactions and steps to add hydrogen (reduce) to double bonds and carbonyl groups (Miller and Auchus, 2011). Sulfation and desulfation of steroids are increasingly becoming of importance though further research is needed. Five classes of enzyme are thus involved in overall steroid synthesis. These include:

- six cytochrome P450 enzymes (CYP),
- short-chain dehydrogenase/reductase (SDR) enzymes,
- aldo/keto reductase (AKR) enzymes,

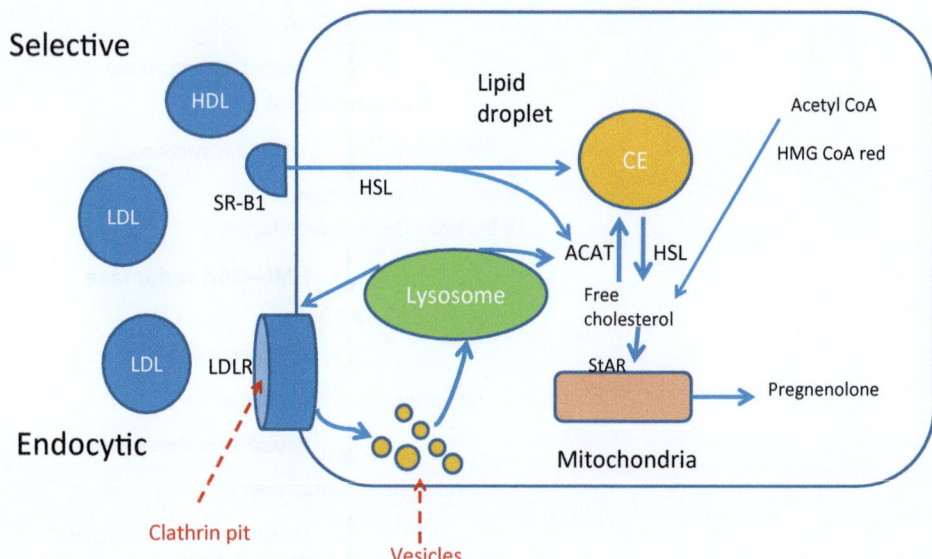

FIG. 1.3.6 Cholesterol uptake for steroidogenesis. A fuller account of this process can be found in earlier Chapter 1.2.2. *(Modified from Kraemer FB. Adrenal cholesterol utilization. Mol Cell Endocrinol 2007;265–266:42–5. Fig. 1 p. 43.)*

- sulfotransferases (SULTs) and
- sulfatases (STS).

The structures of the enzymes are known either from crystallography or in silico modeling. The catalytic functions and active site topography of the enzymes are becoming better understood although there are still unanswered questions. Many of these enzymes are targets for drug design, for example, in the treatment of hormone dependent cancers. In this regard, structures of enzyme-inhibitor complexes have revealed the molecular basis of inhibition and assisted drug discovery and further optimization of these inhibitors. The steroid binding pockets of the enzymes are lined with a series of hydrophobic residues to accommodate the apolar steroid nucleus, whereas the two hydrophilic ends of the steroid are often anchored to membranes by polar residues. The activity of an enzyme can require and be affected by important cofactors.

1.3.4.1 P450 enzymes

Six of the enzymes in the steroid synthetic pathways to cortisol, aldosterone and the sex steroids are members of the cytochrome P450 superfamily of hemoproteins which catalyze mono-oxygenation of lipophilic and exogenous compounds (Guengerich, 2018). All of the CYP enzymes exhibit a characteristic shift in light absorbance from 420 to 450 nm upon reduction with carbon monoxide. Cytochrome P450's all reduce atmospheric oxygen with electrons from NADPH via one or more electron transport intermediates (Fig. 1.3.7).

The cytochrome P450 enzymes all have a common nomenclature and a standardized nomenclature based on the abbreviation CYP. There is some confusion when reading earlier literature because over time different terms have been used. The hydroxylase enzymes have a nomenclature for the site of hydroxylation in addition to being identified as P450 class enzymes. There are around 60 identified CYP genes in the human genome. The italicized root symbol *CYP* represents the gene for a cytochrome P450, followed by the number of the family and a letter denoting the subfamily. Pseudogenes are identified by "P" after the gene number and enzymes by nonitalicized capitals, for example, 21-hydroxylase is identified as CYP21A1P or CYP21A, respectively. The mitochondrial enzymes, cholesterol side chain cleavage enzyme and 11-hydroxylase, are members of the same family, but are in different subfamilies and their respective genes are denoted CYP11A and CYP11B1, respectively. The steroid hydroxylase genes are selectively expressed in different tissues, discrete zones of those tissues and different subcellular locations. For example, C-11 hydroxylases, CYP11B1 and CYP11B2, are confined to the adrenal gland, whereas 17-hydroxylase (CYP17) is found in adrenal zona fasciculata and the zona reticularis of testis and the ovarian thecal cells but is not found in the adrenal zona glomerulosa nor ovarian granulosa cells. The mechanism of tissue specific expression is unclear but the genes share a common sequence which binds a protein found only in steroidogenic cells, the Ad-4 binding protein.

The **mitochondrial** enzymes are the terminal oxidases in an electron transport system in which **adrenodoxin reductase** accepts hydrogen atoms from NADPH (Fig. 1.3.7A). Adrenodoxin, a soluble iron-sulfur protein, also called ferredoxin, acts as a nonspecific electron shuttle transferring reducing equivalents from adrenodoxin reductase, a flavoprotein, to several mitochondrial P450 enzymes.

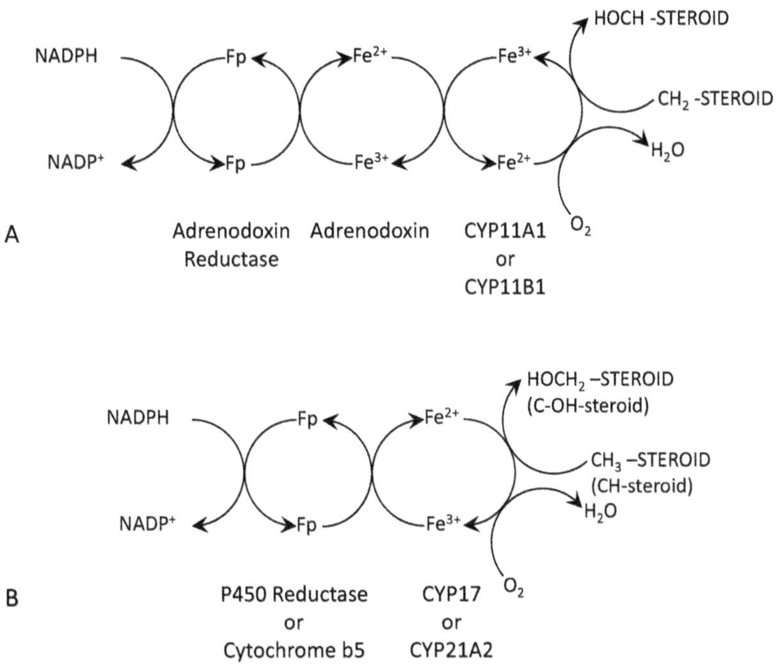

FIG. 1.3.7 Electron transport. (A) Mitochondria and (B) endoplasmic reticulum.

The genes for adrenodoxin reductase and adreno-doxin have been mapped to chromosomes 17q and 11q, respectively.

The **microsomal** P450's obtain electrons from NADPH P450 reductase and cytochrome $b5$ (Fig. 1.3.7B). The catalytic cycle includes two transfers of an electron from reduced nicotinamide adenine dinucleotide phosphate (NADPH) via **cytochrome P450 oxidoreductase (POR)**, followed by binding of the substrate and oxygen to form an active heme-iron complex with oxygen and finally to water release and substrate oxidation.

NNT Nicotinamide nucleotide transdehydrogenase is an antioxidant defense gene on chromosome 5. Under most circumstances NNT uses energy from the mitochondrial proton gradient to generate NADPH which reduces thioredoxin and glutathione and other processes where a supply of reductive power is required to protect the mitochondria against oxidative stress (Meimaridou et al., 2018). NNT is located on the inner mitochondria membrane and provides NADPH for CYP11B1 and CP11B2 via the flavoprotein ferrodoxin reductase. The NNT gene is located at chromosome 5p12. It consists of 22 exons and codes for a 1086 amino acid protein. NNT is a highly conserved enzyme integrated in the inner mitochondrial membrane. It comprises three domains: two mitochondrial matrix domains and one transmembrane domain (including 14 a-helices) that spans the mitochondrial inner membrane.

Thioredoxin reductase is another essential enzyme in maintenance of cellular redox homeostasis. The enzyme is kept in a reduced state by the action of the mitochondrial

selenoprotein TXNRD2 that transfers electrons from NADPH in an intricate mechanism (Biterova et al., 2005). The action of the protein together with the glutathione system is to eliminate superoxides which are generated by steroidogenesis.

The electron transport intermediates are not specific for steroidogenic enzymes and defects involving manganese superoxide dismutase (MnSOD), glutathione reductase (GR), peroxiredoxin (PRDX3), **thioredoxin reductase (TXNRD2)** and **nicotinamide nucleotide transhydrogenase (NNT)** have been shown to be lethal. Some steroidogenic defects are now associated with genetic defects of the electron transport chain and the finding of apparent multiple defects in adrenal steroidogenesis has been explained by a lesion in the flavoprotein.

The steroidogenic cytochrome P450 enzymes share common features (Werck-Reichhart and Feyereisen, 2000). Near the carboxy terminus is a heme-binding region that contains a conserved cysteine residue that serves as the fifth coordinating ligand of the heme iron. The amino terminus is characterized in a number of microsomal cytochrome P450s by a region of hydrophobic amino acids that form a membrane anchoring domain.

P450 enzymes are heavily involved in steroid biosynthesis and metabolism and are responsible for steps in the formation of both estrogens and androgens. P450scc (CYP11A1) catalyzes a multiple-step reaction, with sequential hydroxylation at carbons 20 and 22 followed by C—C bond cleavage. This enzyme catalyzed reaction converts cholesterol to pregnenolone, the first step in steroid hormone biosynthesis. P450c17 (CYP17A1) is responsible for the conversion of pregnenolone (a C_{21} steroid), the precursor molecule for all active sex steroid hormones, to DHEA (a C_{19} steroid). Aromatase (CYP19A1) plays a central role in conversion of androgens to estrogens. These three P450 enzymes are membrane-bound and display a canonical fold, like other P450 proteins, harboring an iron-containing heme group.

The chronic response to ACTH in the adrenal cortex directs transcription of the genes encoding the steroidogenic enzymes. Each steroidogenic gene utilizes unique cAMP-responsive sequences (CRS) found in the promoters of each gene, which bind a diverse array of transcription factors. Moreover, once specific transcription factors are bound to the promoters of the steroidogenic genes, increased gene expression requires posttranslational modification (phosphorylation/dephosphorylation) of the transcription factors and binding of co-activator proteins.

1.3.4.1.1   NADPH regeneration

The first and last steps of cortisol production occur in the mitochondria and require a constant supply of reductant. The enzyme **P450 oxidoreductase (POR)** contains 680 amino acids and is encoded by the POR gene on chromosome 7. POR transfers electrons from reduced nicotinamide adenine dinucleotide phosphate to microsomal P450 enzymes, which are important in steroidogenesis (e.g., CYP17A1, CYP19A1, CYP21A2) (Burkhard et al., 2017; Pandey and Flück, 2013). Mitochondrial NADPH is regenerated by thioredoxin and glutathione pathways through NNT and is the major mitochondrial enzymatic source of NADPH contributing 45% of the total NADPH supply (Chortis et al., 2018; Meimaridou et al., 2018). It exists as a dimer and spans the inner mitochondrial membrane modulating hydrogen ion movement and supplying the high concentrations of NADPH required for the detoxification of ROS by glutathione and thioredoxin pathways (Fig. 1.3.8).

Under normal physiological conditions NNT uses energy from the mitochondrial protein gradient to generate high concentrations of NADPH. This is required for many processes in the cell including the supply of reductive oxidative power to a network of antioxidant enzymes. The physiological roles of NNT are only gradually being revealed by studies in mutant mice and humans with defects in steroidogenesis.

1.3.4.1.2   Cofactors

Cytochrome b5

Cyt $b5$ is a small (15.4 kDa) hemoprotein taking part in many biochemical reactions including fatty acid metabolism and cholesterol biosynthesis. The action of cyt $b5$ with P450 17A1 (CYP17) is critical to the biosynthesis of

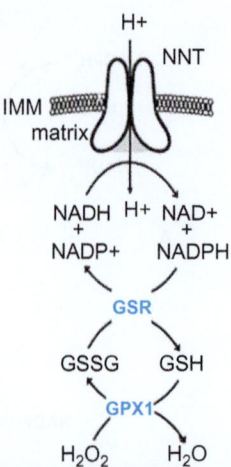

FIG. 1.3.8   Nicotinamide nucleotide transhydrogenase (NNT) pumps protons across the inner mitochondrial membrane (IMM). The detoxification of reactive oxygen species (ROS) depends upon maintenance of a high GSH/GSSG ratio. *(From Meimaridou E, Goldsworthy M, Chortis V, Fragouli E, Foster PA, Arlt W, Cox R, Metherell LA. NNT is a key regulator of adrenal redox homeostasis and steroidogenesis in male mice. J Endocrinol 2018;236(1):13–28. Fig. 2 p. 15.)*

androgen precursors where, and without which, the steroid pathway branches to glucocorticoid based steroids (Duggal et al., 2018; Bhatt et al., 2017). In mammals, cyt $b5$ is present in three different isoforms;

- a soluble cytosolic form (soluble cyt $b5$, 98 amino acids) which is predominantly found in erythrocytes,
- an endoplasmic reticular membrane bound form (i.e., microsomal cyt $b5$, 134 amino acids),
- an outer mitochondrial membrane bound form (mitochondrial cyt $b5$, 146 amino acids).

The soluble cyt $b5$ has key roles in several important biochemical reactions across different tissue types, including adrenal glands, gonads, liver, kidneys, spleen, lungs, brain and adipose tissue. In general, the soluble cytosolic cyt $b5$ is formed by four alpha-helices that create a hydrophobic cleft that incorporates the heme moiety with its edge exposed to the cytosol. This structure is similar to the membrane bound cyt $b5$ albeit with an additional transmembrane domain. The microsomal cyt $b5$ form is a cylindrical-shaped, membrane-bound protein comprised of 20% glutamate or aspartate residues and hence is overall acidic. Microsomal cyt $b5$ consists of three distinct domains.

- (i) an N-terminal, cytosolic, heme-containing soluble domain (70 amino acids),
- (ii) a linker or hinge region (15 amino acids) and
- (iii) a C-terminal, hydrophobic, membrane-binding domain (40 amino acids).

The high structural integrity of cyt $b5$ is maintained by the interactions between pairs of aromatic amino acid residues with the heme cofactor. The heme iron completes the coordination with two highly conserved histidine residues and further stabilized by two highly conserved phenylalanine residues, to create so called "face-to-face p-stacking and edge-to-face T-stacking" for interactions with other proteins.

1.3.4.2 Hydroxysteroid dehydrogenases

Reactions with double bonds and carbonyl groups involve several human isoforms, some of which are preferential oxidases, whereas others are principally reductases. They fall into **short chain dehydrogenases (SDR)** and **aldo-keto reductases (AKR)**. The oxidative and reductive reactions are regulated by the presence of excess of a suitable cofactor. The reductive enzymes utilize NAD(P)H as the cofactor and the oxidative enzymes utilize NAD(P)$^+$. The isoforms acting at C17 share only 20%–30% sequence identity, and yet their secondary and tertiary structures are remarkably similar. These enzymes use NAD(P)(H) as cofactors. The reaction catalyzed depends on cofactor preference and cellular localization. SDR and AKR members share conservation in their catalytic mechanisms despite differences in protein folding.

1.3.4.2.1 Aldo-keto reductase A KR superfamily

The **AKR** family is usually monomeric, soluble enzymes operating in the reductive direction with NADPH as co-factor in vivo. The protein structure forms a $(\alpha/\beta)8$-barrel fold. The structures of the **AKR** family enzymes 17β-HSD5 (AKR1C3), 3α-HSD3 (AKR1C2) and steroid 5β-reductase (AKR1D1) have been elucidated (Penning et al., 2019). AKR1C family enzymes display substrate promiscuity, due to the larger steroid binding pocket in these enzymes and the nature of C-19 steroid substrates. The alternative binding modes of steroid molecules present in the same active site pocket explains the broad specificity of these enzymes. These steroid-converting enzymes have a critical role in a number of diseases, especially in hormone-dependent cancers, and have long been targets of drug design.

The 17β-HSD-V, initially cloned as a 3β-hydroxysteroid dehydrogenase, primarily catalyzes the conversion of Δ4 androstenedione to testosterone but also has 20β-HSD activity for progesterone. 17β-HSD-V is expressed in most tissues, including liver, kidney, blood vessels, testis, prostate, adrenal, bone and ovary. 17β-HSD-V is the enzyme responsible for the last step in testosterone production in the ovary.

1.3.4.2.2 Short chain dehydrogenases SDR superfamily

Short-chain dehydrogenases/reductases (SDR) comprise a large family of functionally heterogeneous proteins that participate in the metabolism of steroids, prostaglandins, retinoids, aliphatic alcohols and xenobiotics. Members of the SDR superfamily are found in the cytoplasm, mitochondria, nuclei, peroxisomes and endoplasmic reticulum. Some of the SDR enzymes act on the same endogenous substrates but exhibit different subcellular localization, cofactor specificity, substrate affinity and tissue distribution. Due to the cellular ratios of NAD$^+$/NADH and NADP$^+$/NADPH, the SDR enzymes that prefer NAD$^+$ as a cofactor function in the oxidative direction in vivo, whereas those that prefer NADP$^+$ function in the reductive direction. The **SDR** family of enzymes are usually membrane bound, operating in the oxidizing direction with NAD as the preferred cofactor in vivo. The protein contains a fold for NAD(H) binding. Representatives of these acting on steroids include the SDR family enzyme 17β-HSD1, whose structure was the first-determined for any human steroid-converting enzyme.

Type I 17βHSD is expressed in the ovary, where it catalyzes the last step in estradiol synthesis from estrone, and is expressed in the placenta where it contributes to the production of estriol. 17βHSD-I, encoded by a gene on chromosome 17q21 near the BRCA locus, is also expressed in endometrium, breast, testis, adipose tissue, skin, liver and prostate. 17βHSD-I is a homodimer that uses NADPH as its cofactor and converts estrone to estradiol.

Type 2 17β-HSD-II oxidizes estradiol to estrone and testosterone to Δ4-androstenedione with equal efficiency using NAD$^+$ as a cofactor. This enzyme can also oxidize C-20 substrates (20α-HSD activity), converting 20α-hydroxyprogesterone to progesterone. 17β-HSD-II, encoded on chromosome 16q24, shares only 20% amino acid sequence identity with 17β-HSD-I. 17β-HSD-II is found in the placenta, breast, liver, small intestine, prostate, secretory endometrium, kidney and ovary.

17β-HSD-III is an androgenic enzyme that uses NADPH as a cofactor and reduces androstenedione to testosterone and DHEA to androstenediol. The 17β-HSD-III gene on chromosome 9q22 is expressed primarily in the testis but also in adipose tissue, but it is not expressed in the ovary.

1.3.4.3 Sulfotransferases

Sulfonation reactions in all organisms require the conversion of sulfate to the universal sulfonate (SO$_3$ K) donor, phosphoadenosine phosphosulfate (PAPS). In the cytosol, PAPS is generated by the enzyme, PAPS synthetase, that sulfurylates ATP to form APS followed by phosphorylation to form PAPS (Fig. 1.3.9). The sulfonate group from PAPS is then transferred to the target substrate via sulfotransferase enzymes, which can be grouped into two classes:

(i) cytosolic sulfotransferases that sulfonate steroids, neurotransmitters, bile acids and xenobiotics and

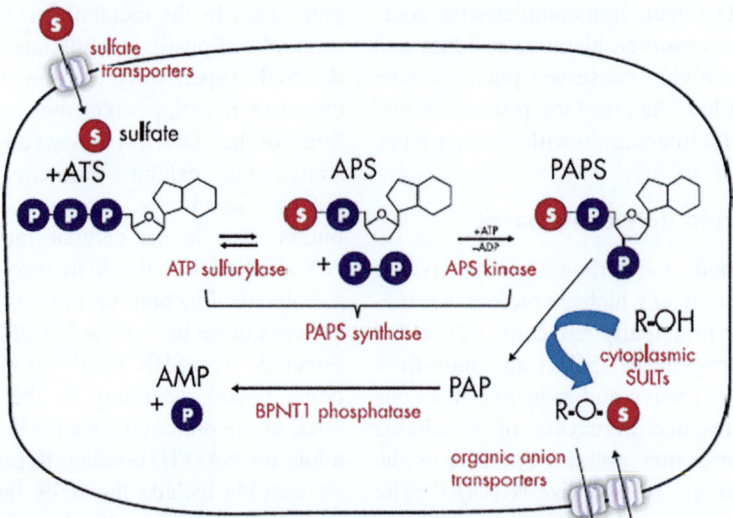

FIG. 1.3.9 Steroid sulfation with sulfotransferase and phosphoadenosine phosphosulfate (PAPS).

(ii) Golgi-located sulfotransferases that have proteoglycan and lipid substrates and rely on PAPS transporters (PAPST1 and PAPST2) to mediate the translocation of PAPS from the cytosol into the Golgi (Venkatachalam, 2003).

The human DHEA-sulfotransferase (*SULT2A1*) gene is localized to chromosome region 19q13.3 with three variable 3′ nontranslated regions. SULT2A1 is a 285 amino acid protein with a calculated molecular mass of 33,765 Da. The structures of human DHEA sulfo-transferase and human estrogen sulfotransferase (SULT1E1) have been determined in the presence of steroid ligand and/or PAPS. The cofactor binding site and the catalytic residues are highly conserved in this family, and their main difference lies in the region/loops governing substrate specificity (Mueller et al., 2015).

Steroid sulfonation was initially thought to produce only excretable steroids, due to their increased polarity and inability to bind to steroid receptors. In human blood, dehydroepiandrosterone sulfate (DHEAS) is present in concentrations up to 100-fold higher than unconjugated DHEA and the half-life of DHEAS is estimated to be 20 times higher than that of DHEA. The levels of DHEAS are age-dependent, reaching their highest concentrations during the 20s or early 30s, to later fall continuously. In a similar way, pregnenolone sulfate (PregS) levels decline with age. These findings suggest that the sulfonated steroids act as a pool of steroids involved in physiological processes such as reproduction, immunity and possibly cognition. Some sulfated steroids like DHEAS and PregS are neuromodulators. DHEAS exhibits other physiological activities, including enhancement of enzymatic side-chain cleavage of cholesterol. Sulfated steroids cannot passively cross cell membranes like their unconjugated counterparts due to their increased hydrophilicity. Their uptake is mediated by specific transporters in organs such as the human testis or the brain.

1.3.4.4 Sulfatases

Once sulfated steroids have entered the cell, they can be desulfonated in order to regain physiological activity as unconjugated steroids. Some of them can be transformed directly into other sulfated steroids without cleavage of their sulfo groups at positions 3β- or 17β, as demonstrated in vitro and in vivo. These reactions are catalyzed by the same enzymes required for steroid biosynthesis of unconjugated steroids. DHEAS, for example, was obtained from cholesterol sulfate (CS), but the physiological importance of this contribution has not been characterized.

The gene for the human STS is located on the distal short arm of the X-chromosome and maps to Xp22.3-Xpter. The gene consists of 10 exons and spans 146 kb, with the intron sizes ranging from 102 bp up to 35 kb. It encodes a protein of 583 amino acids, with a signal peptide and four potential glycosylation sites of which at least two are used, at asparagine residues 47 and 259. The steroid sulfatase crystal structure contains long hydrophobic transmembrane helices and a globular head harboring the catalytic site and the Ca^{2+} cofactor. The sulfate group is covalently linked to a hydroxyformylglycine residue resulting from posttranslational modification. Steroid sulfatase (STS), catalyzes the cleavage of the sulfo group in sulfonated steroids such as DHEAS.

Conversion of PregS into its hydroxylated form, 17α-hydroxy-pregnenolone sulfate (17OHPregS) has been recently confirmed in a reconstituted in vitro system as well as in a human cell line with identification by liquid chromatography-tandem mass spectrometry (LC-MS/MS)

(Neunzig et al., 2014). Other examples are the conversion of DHEAS into androstenediol-3β-sulfate (AnDiolS) in human testes, the aromatization of testosterone sulfate into estradiol-17-sulfate in human placental microsomes or the in vitro hydroxylation of CS at position C-27. All of the above information indicates a potential role of sulfated steroids in human physiology.

1.3.5 Cholesterol side chain cleavage enzyme (CYP11A)

On the mitochondrial inner membrane, a cytochrome P450 for side chain cleavage enzyme (scc) is located. The conversion of **cholesterol to pregnenolone** (Fig. 1.3.10) involves:

- 22-hydroxylation of cholesterol,
- 20-hydroxylation of 22R-hydroxycholesterol and
- cleavage of the C20 to C22 bond with release of isocaproaldehyde and pregnenolone.

Cytochrome P450 side chain cleavage (scc) (desmolase) requires electrons from a flavoprotein called adrenodoxin reductase, an iron sulfur protein called ferredoxin and NADPH (Bose et al., 2020; Chien et al., 2017). The expression of scc requires the action of the zinc-finger transcription factor, **steroidogenic factor 1 (SF1)**. The single gene, *CYP11A*, encoding this enzyme maps to chromosome 15q23-q24. The mRNA is approximately 1850 bp in length, excluding the poly A tail, coding for a protein of 521 amino acids. Expression has been shown in adrenals, testes and placenta. A fetus with mutation in the CYP11A1 gene is at high risk for being miscarried or delivered prematurely, as P450scc is necessary for steroidogenesis in the placenta. Progesterone suppresses uterine contractility and thus maintains the pregnancy.

1.3.6 Biosynthesis of steroid hormones

Apart from estrogens all active steroid hormones have 3-keto-4-ene structure created by the action of HSD3B enzymes on pregnenolone and 17-hydroxypregnenolone followed by a number of hydroxylation steps. Cortisol synthesis from pregnenolone involves four steps including

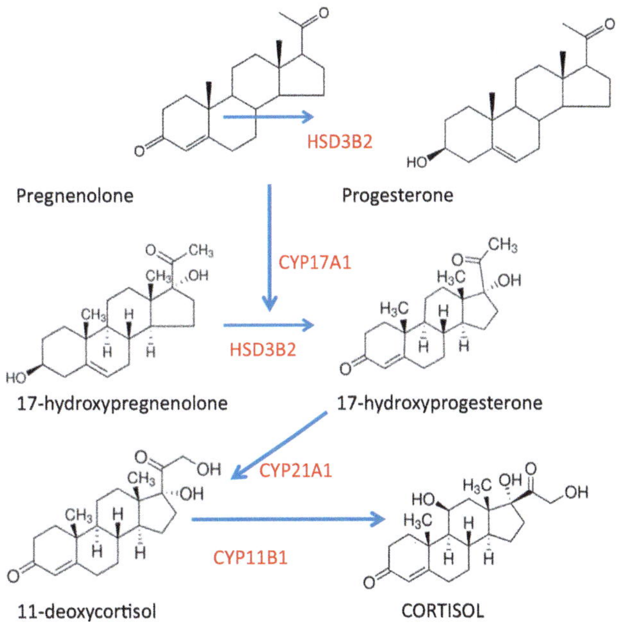

FIG. 1.3.11 Overview of cortisol synthesis. *(Author original.)*

hydroxylation at C17 (Fig. 1.3.11), aldosterone synthesis is in three steps without 17-hydroxylation but with formation of an aldehyde group at C18. Further side chain carbon-carbon cleavage of pregnenolone leads to the formation of C19 steroids (androgens) and removal of the C19 methyl group is involved in the synthesis of C18 steroids (estrogens).

1.3.6.1 3β-Hydroxysteroid dehydrogenase/4–5 isomerase

Two active genes for 3β-HSD and three pseudogenes have been described. The type I (HSD3B1) and II (HSD3B2) genes map to chromosome 1p11-p13 (32,33) and are 93% homologous in the coding regions. Each gene has four exons, the first one and part of the second exon are untranslated and is contained within a 7.8 kb fragment (Miller, 2002). The 1676 bp HSD3B2 mRNA is the predominant species detectable in the human adrenal cortex and is also present in the gonads. In both organs, it encodes a 41 kDa protein of 371 amino acids, one amino acid less than the

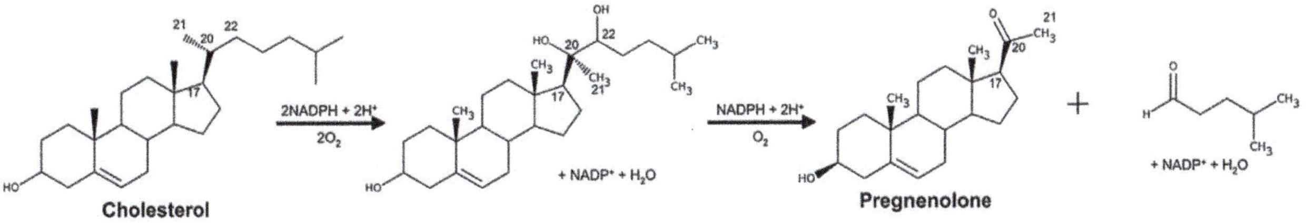

FIG. 1.3.10 Reaction catalyzed by P450scc (CYP11A1). Cholesterol is first converted to the intermediate 20,22-dihydroxycholesterol before the C20-C22 bond is cleaved to obtain pregnenolone and isocaproic aldehyde. *(Modified from Manenda MS, Hamel CJ, Masselot-Joubert L, Picard ME, Shi R. Androgen-metabolizing enzymes: a structural perspective. J Steroid Biochem Mol Biol 2016;161:54–72. Fig. 2 p. 57.)*

type 1 enzyme. These enzymes not only transform the hydroxyl group at C3 to a ketone but also isomerize the double bond from Δ5 to Δ4 positions. These changes create the structural requirements for active steroid hormones. The type I enzyme with identical intron/exon organization, is found in placenta, breast and other extraglandular tissues, such as skin. Mutations have been identified only in the 3β-HSD-II gene.

Pregnenolone, is converted to progesterone, the first biologically important steroid hormone in the sex steroid pathway by HSD3B2. The hydroxyl group at C3 is converted to a ketone group and the double bond at C5 in the B ring (Δ5 steroids) is isomerized to C4 in the A ring (Δ4 steroids). The single enzyme can also convert 17α-hydroxypregnenolone to 17α-hydroxyprogesterone (17-OHP), dehydroepiandrosterone (DHEA) to androstenedione, as well as androstenediol to testosterone, all with the same catalytic efficiency. This enzyme is essential for the synthesis of sex steroids and corticoids.

1.3.6.2 Cortisol biosynthesis

The intermediates in cortisol synthesis from pregnenolone are 17-hydroxypregnenolone, 17-hydroxyprogesterone and 11 deoxycortisol. Pregnenolone in the adrenal zona fasciculata is the principal starting point for cortisol synthesis. The CYP450 17-hydroxylase in the endoplasmic reticulum (CYP17A1) acts to convert pregnenolone to 17-hydroxypregnenolone and progesterone to 17-hydroxyprogesterone. CYP17A1 has a second activity to produce androgens by side chain cleavage of 17-hydroxypregnenolone, this will be addressed later (see Section 1.3.6.4). 17-Hydroxylation is the division between mineralocorticoid production from cortisol and sex steroids. CYP17 is expressed in adrenals and gonads but is not expressed in the adrenal zona glomerulosa or ovarian granulosa cells. The lyase activity is greatest in the gonads. The enzyme can also catalyze 16-hydroxylation of progesterone.

In the endoplasmic reticulum of the zona fasciculata cells, pregnenolone is converted to progesterone. The HSD3B2 is also involved in converting 17-hydroxypregnenolone to 17-hydroxyprogesterone which by either route is converted by CYP21 to 11-deoxycortisol, which precedes cortisol formation by 11-hydroxylation.

Microsomal CYP21A2 (or CYP21B) catalyzes the C-21 hydroxylation of 17-hydroxyprogesterone to 11-deoxycortisol. This enzyme receives electrons from NADPH via cytochrome P450 (POR). In evolution, the CYP21 gene has become duplicated (CYP21A1P or CYP21A) although mutations in the second gene prevent transcription of active protein. Exchange between the two genes is common which is the basis for some cases of 21-hydroxylase deficiency. The final step in cortisol

biosynthesis is catalyzed by the mitochondrial enzyme CYP11B1 which also has 18-hydroxylase activity.

1.3.6.2.1 Steroid 21-hydroxylase

The CYP21A2 gene (previously called CYP21 or CYP21B, GeneID 1589) is located in the HLA class III region in the major histocompatibility (MHC) locus on the short arm of chromosome 6 (band 6p21.3) together with a highly homologous pseudogene, CYP21A1P (previously called CYP21P or CYP21A) (Fig. 1.3.12). In evolution, the gene became duplicated (CYP21A1P or CYP21A) and mutations in the second gene prevent transcription of active protein. Exchange between the two genes is common and is the basis for some cases of 21-hydroxylase deficiency. These 2 genes share about 98% homology in their 10 exons and about 96% in the introns, but CYP21A1P is inactive because of several deleterious mutations. Both genes are arranged in tandem repeat with the C4A and C4B genes encoding the fourth component of complement (Pignatelli et al., 2019).

The C4/CYP21 unit is flanked by a telomeric serine/threonine kinase (RP) gene and a centromeric tenascin (TNX) gene, forming what is referred to as RCCX modules (RP-C4-CYP21-TNX) (Miller and Merke, 2018). RP1 encodes a putative nuclear protein similar to DNA helicase. RP2 is a truncated, nonfunctional form of RP1. Most haplotypes have a bimodular form composed of two sets of four genes arranged in tandem as follows: RP1-C4A-CYP21A1P-TNXA-RP2-C4B-CYP21A2-TNXB (Fig. 1.3.12). The CYP21A2 receives electrons from NADPH via the **flavoprotein cytochrome P450 oxidoreductase (POR)**.

TNXB encodes an extracellular matrix protein, tenascin X, which overlaps the CYP21A2 gene on the opposite strand, and TNXA is a truncated copy of TNXB overlapping CYP21A1P, also on the opposite DNA strand.

1.3.6.2.2 17-Hydroxylase/17,20-lyase

The gene for *CYP17*, has been mapped to chromosome 10q24-25 and contains 8 exons spanning 6.6kb. The 1.7kb mRNA, which is expressed in both adrenal and testis, encodes an enzyme of 508 amino acids and a molecular weight of approximately 57,000kDa. CYP17 catalyzes not only 17-hydroxylation but also side chain cleavage (lyase) at carbon 17 of the steroid molecule and is essential for both glucocorticoid and sex steroid production. A number of factors determine whether a steroid molecule will remain on the single active site of CYP17 and undergo 17,20 bond scission following 17-hydroxylation including the availability of reducing equivalents, the amount of P450 reductase and cytochrome *b*5 available in addition to the membrane lipid environment (Storbeck et al., 2015; Miller and Tee, 2015). The enzyme is expressed in adrenals and gonads but is not expressed in the adrenal zona

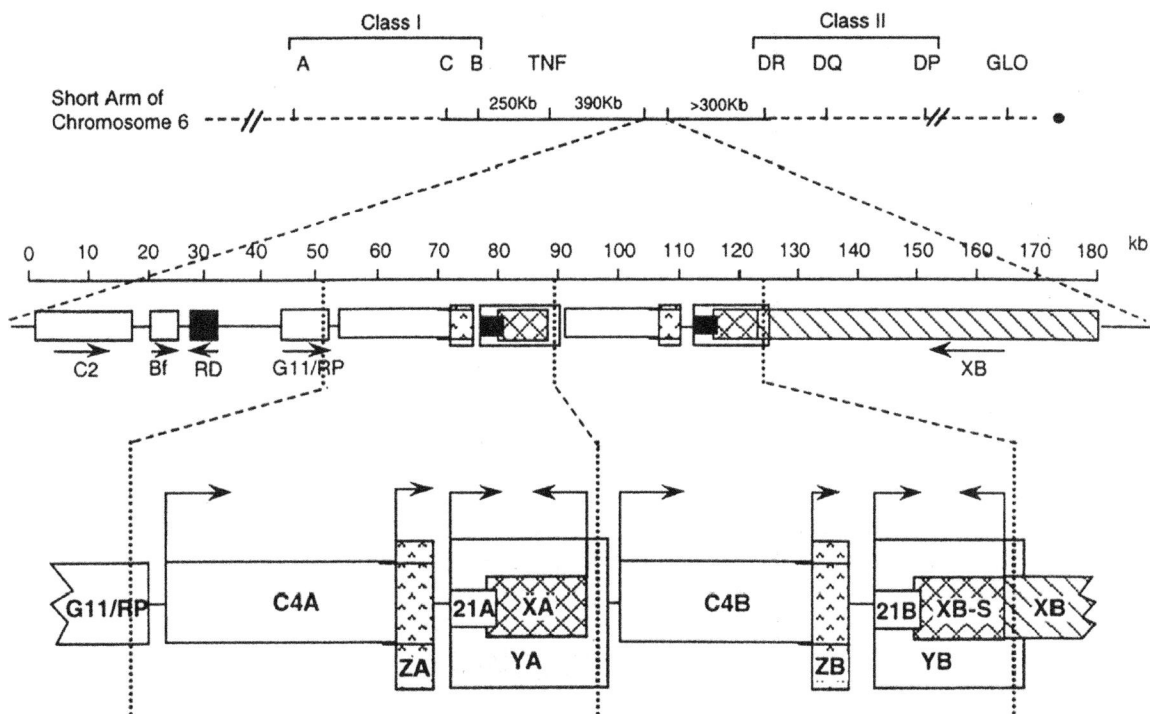

FIG. 1.3.12 21-Hydroxylase genes. (Top) Scheme for the cytochrome P450 *CYP21* gene locus on chromosome 6 including HLA Class I and II regions. (Middle) The enlarged segment 6p21.33. Transcriptional orientation (indicated by arrows) showing C2 (complement factor C2 gene), Bf (properdin factor Bf) G11RP (serine threonine kinase 10—STK 19) reading toward the centromere and XB (the TNXB gene for tenascin) and RD (now called NEFLE negative elongation factor subunit E) reading away from the centromere. (Bottom) The duplicated C4A and C4B genes for complement C4; 21A the CYP21A1P pseudogene; 21B the active CYP21A2 gene. Also indicated are XA, YA and YB that lack open reading frames, XB-S and adrenal specific form of tenascin of unknown function. ZA and ZB are essential in CYP21A1P and CYP21A1 promoters. *(From Miller WL, Merke DP. Tenascin-X, congenital adrenal hyperplasia, and the CAH-X syndrome. Horm Res Paediatr 2018;89(5):352–361. Fig. 1 p. 12.)*

glomerulosa or ovarian granulosa cells. The preferred substrate for formation of C_{19} steroids by CYP17 appears to be 17-hydroxypregnenolone, the human enzyme exhibiting extremely low activity toward progesterone and 17-hydroxyprogesterone. The enzyme can also catalyze 16-hydroxylation of progesterone.

No major deletions have been observed at this locus but a number of smaller deletions and duplications in addition to point mutations have been described in patients with CAH. The cysteine residue involved in heme binding is located at codon 441. A number of mutations have been located downstream of this site in the carboxy terminal region indicating that this region, although not believed to play a role in the active site, is important for the function of the enzyme.

1.3.6.2.3 11β-Hydroxylase/aldosterone synthase

The final steps in cortisol biosynthesis is catalyzed by one of two related 11β-hydroxylase enzymes, CYP11B1 with transcription prompted by ACTH whereas aldosterone synthase (CYP11B2) has similar functions regulated by angiotensin II (Mornet et al., 1989). The proteins are encoded by two highly homologous genes approximately 40 kb apart on chromosome 8q22 (Schiffer et al., 2015). Angiotensin II increases transcription of both CYP11B2 and CYP11B1 in cultured glomerulosa cells.

The CYP11B1 gene has nine exons and shares 93% base sequence homology with CYP11B2 in coding regions although only 48% in the 5′ promoter region. Multiple transcripts of CYP11B1 of 4.2, 3.6 and 2.2 kb are found whereas the mRNA of CYP11B2 is 3.1 kb, possibly the result of multiple polyadenylation signals as the two proteins are of identical size, having 503 amino acids including a 24 residue mitochondrial signal peptide. Transcripts of CYP11B1 are restricted to the zona fasciculata while CYP11B2 is localized predominantly in the zona glomerulosa with a trace amount in the zona fasciculata, although contamination in tissue experiments in vitro could not be excluded.

In view of the duplicated gene arrangement, it might be expected that gene conversion, the common source of mutation in CYP21, occurs at the CYP11B1 and B2 loci. While such events have been described in the CYP11B1 and CYP11B2 genes they are not the direct cause of pathological mutations, possibly because both genes encode active proteins with similar activities and such sequence changes may have minimal effect.

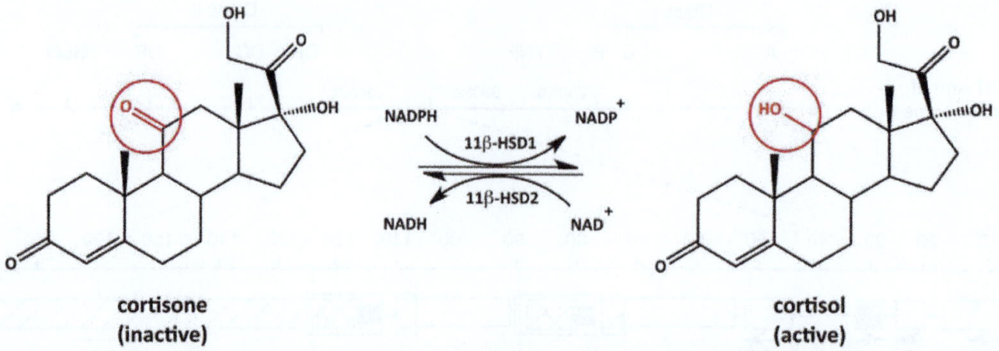

FIG. 1.3.13 Regulation of cortisone and cortisol interconversion by 11β-HSDs. 11β-HSD1 predominantly functions as a reductase by catalyzing the interconversion of inactive cortisone to receptor-active cortisol in vivo. The reverse reaction is mediated by the unidirectional NAD-dependent 11β-HSD2. *(Redrawn from Loerz C, Maser E. The cortisol-activating enzyme 11β-hydroxysteroid dehydrogenase type 1 in skeletal muscle in the pathogenesis of the metabolic syndrome. J Steroid Biochem Mol Biol 2017;174:65–71. Fig. 1 p. 66.)*

The CYP11B1 promoter region contains a single CRE (cAMP response element) at −78 and two putative AP-1 elements. Constructs containing up to the −1093 nucleotide from the transcription start site of CYP11B1 has strong promoter activity in the presence of dibutyryl cAMP which was lost on deletion to the −105 nucleotide. In contrast, dibutyryl cAMP had no effect on CYP11B2 constructs in spite of there being three putative CREs and two putative AP-2 binding sites in the region. Following deletion of the CYP11B2 promoter from −2015 to −64 bp from the initiation start site some increase in basal transcription is suggestive of the presence of a negative regulatory element in the 5′ flanking region of CYP11B2. The CYP11B2 gene has a putative CRE at nucleotide 396.

1.3.6.2.4 11-Hydroxysteroid dehydrogenases

Two 11β-hydroxysteroid dehydrogenases are particularly important for cortisol activity (Fig. 1.3.13). The HSD11B2 gene is located on chromosome 16q22.1 and is composed of 5 exons that encode a 405 amino acid protein. The HSD11B2 gene is expressed primarily in aldosterone-responsive tissues, such as those of the distal tubules of the nephrons of the kidneys and in salivary glands. In these tissues, the HSD11B2 enzyme is responsible for inactivating cortisol in order to prevent inappropriate activation of the mineralocorticoid receptor (MR).

Although the normal receptor for cortisol is the glucocorticoid receptor (GR), the MR has nearly identical affinities for aldosterone (the mineralocorticoid) and cortisol. Since cortisol is present at 1000 times the concentration of aldosterone, inactivation of cortisol is essential to the effects of aldosterone. Mutations in the HSD11B2 gene are associated with a condition of apparent mineralocorticoid excess which induces hypertension due to the lack of ability to inactivate cortisol. As opposed to the use of $NADP^+$ by the HSD11B1 enzyme in the direction of cortisol and corticosterone oxidation, the HSD11B2 enzyme utilizes

NAD^+ as its cofactor. The expression of the HSD11B2 gene is also found in cells that do not also express the MR such as the placenta where the function of the HSD11B2 enzyme is to protect those cells from the growth-inhibiting effects of cortisone. Licorice and carbenoxolone inhibit HSD11B2 causing a state of apparent mineralocorticoid excess. The active component of licorice is glycyrrhetinic acid and related compounds are called GALFS (Morris et al, 2007)

The primary function of the HSD11B1 enzyme is to reduce the 11-oxo groups (11-oxoreductase activity) in cortisone and 11-dehydrocorticosterone which generates the active glucocorticoids, cortisol and corticosterone, respectively. This enzyme can also in reverse inactivate (dehydrogenase activity) cortisol and corticosterone by the reactions converting cortisol to cortisone and corticosterone to 11-dehydrocorticosterone. The primary determinant of the oxo-reductase activity of HSD11B1 is the ratio of NADPH to $NADP^+$. Of clinical significance mutations in the H6PD gene are associated with glucocorticoid deficiency. The enzyme encoded by the HSD11B2 gene possesses only the 11β-dehydrogenase activity and prefers NAD presence.

1.3.6.3 Aldosterone synthesis

In the outer zona glomerulosa of the adrenal cortex the microsomal P450 enzyme (CYP21A2 or CYP21B) catalyzes the C-21 hydroxylation of progesterone to deoxycorticosterone (DOC). This enzyme receives electrons from NADPH via a flavoprotein called cytochrome P450 (POR). The actions of a mitochondrial enzyme are required to complete the synthesis of aldosterone. CYP11B1 is active in the zona glomerulosa and the fasciculata to introduce a 11β-hydroxyl group to DOC or 11-deoxycortisol, respectively, yielding corticosterone or cortisol. In the zona glomerulosa, only CYP11B2 (aldosterone synthase) catalyzes 11 hydroxylation, 18-hydroxylation and oxidation to a

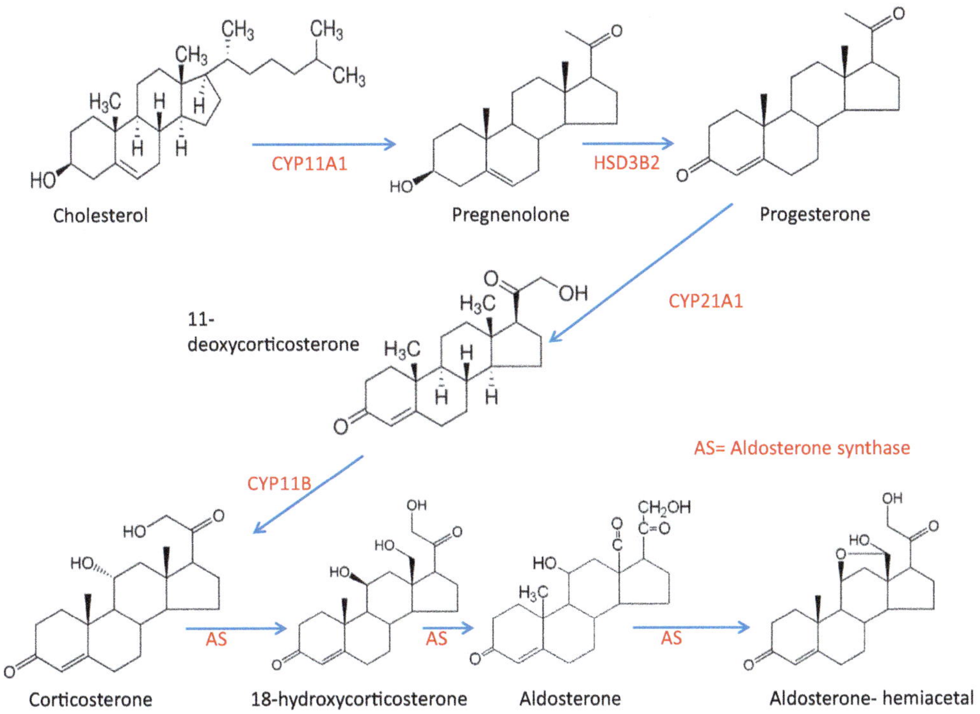

FIG. 1.3.14 Overview of aldosterone synthesis. *(Author original.)*

C-18 aldehyde (corticosterone methyl oxidase). Both enzymes use ferredoxin, ferredoxin reductase to transfer electrons from NADPH. SF1 and other transcription factors are needed although the precise mechanisms are uncertain. This is the late rate-limiting step of aldosterone synthesis.

The final steps from deoxycorticosterone to aldosterone are thus achieved with one mitochondrial enzyme, aldosterone synthase (Fig. 1.3.14) (Reddish and Guengerich, 2019).

CYP11B2 can carry out 11-hydroxylation and the subsequent 18-methyl oxidation reaction to produce aldosterone in the zona glomerulosa. The ability to synthesize aldosterone lies in the 3′ end of the gene sequence from exon 5 of CYP11B2 as determined by expression of hybrid constructs composed of the 5′ end of CYP11B1 and the 3′ end of CYP11B2 in Cos-1 cells. Constructs containing up to exon 4 of CYP11B1 all convert deoxycorticosterone to corticosterone, 18-hydroxycorticosterone and aldosterone. In contrast, in cells transfected with constructs containing exon 5 and up of CYP11B1, no aldosterone synthesis was detected. These findings have been reinforced by the description of cases of glucocorticoid responsive hyperaldosteronism in which a chimeric gene created by nonhomologous recombination, has the 5′ end of the CYP11B1 abutted to the 3′ end of CYP11B2 (Dluhy and Lifton, 1995) (Fig. 1.3.15). Thus the 11-hydroxylase enzyme acquires aldosterone synthesizing properties under the control of a promoter responsive to ACTH.

1.3.6.4 Sex steroid formation

The synthesis of the sex steroids requires removal of the side chain through the action of 17–20 lyase which is a second function of CYP17A1. A number of factors determine whether a steroid molecule will undergo 17,20 bond scission following 17-hydroxylation including the availability of reducing equivalents, the amount of P450 reductase and cytochrome $b5$ available in addition to the membrane lipid environment. The active human androgens are C_{19} steroids with the 3-keto-4-ene configuration of the A/B rings, and an hydroxyl group at C17 on the D-ring. The synthesis from cholesterol to pregnenolone and 17-hydroxypregnenolone is essentially the same for all steroid hormones. Estrogens are C_{18} steroids with a phenolic A-ring (Fig. 1.3.16).

1.3.6.4.1 Cytochrome P450 17A1 and side-chain cleavage

The first step in sex steroid production involves side chain cleavage of pregnenolone by CYP17A1 involves the 17-hydroxylase and 17,20-lyase reaction of the same enzyme to produce dehydroepiandrosterone (DHEA). Human P450c17 catalyzes the 17α-hydroxylation of Δ5 pregnenolone and Δ4 progesterone with equal efficiency, but poorly catalyzes the 17,20-lyase conversion of 17-OHP to Δ4 androstenedione. Most sex steroid synthesis therefore proceeds through DHEA and little through 17-OHP. The 17α-hydroxylase reaction is 20- to 25-fold more efficient

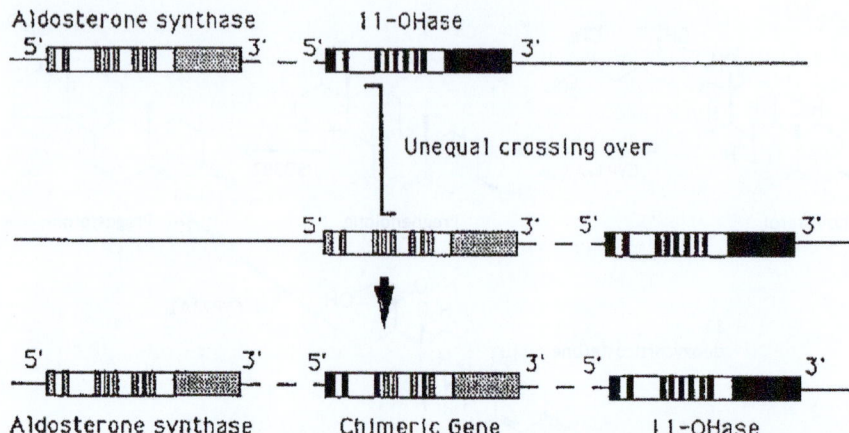

FIG. 1.3.15 Arrangement of the two 11-hydroxylase genes, transcribed from left to right, on chromosome 8q code for aldosterone synthase (11-hydroxylase and 18-hydroxylase and aldehyde synthesis) (CYP11B2) and 11-hydroxylase (CYP11B1) for cortisol synthesis. The locations of exons are depicted in each gene by *black blocks* in CYP11B1 and *stippled* in CYP11B2. CYP11B2 is preceded by an angiotensin regulatory region whereas CYP11B1 follows an ACTH responsive region. A chimeric gene has been found due to misalignment of chromosomes and unequal crossing over. *(From Dluhy RG, Lifton RP. Glucocorticoid-remediable aldosteronism (GRA): diagnosis, variability of phenotype and regulation of potassium homeostasis. Steroids 1995;60(1):48–51. Fig. 1 p. 49.)*

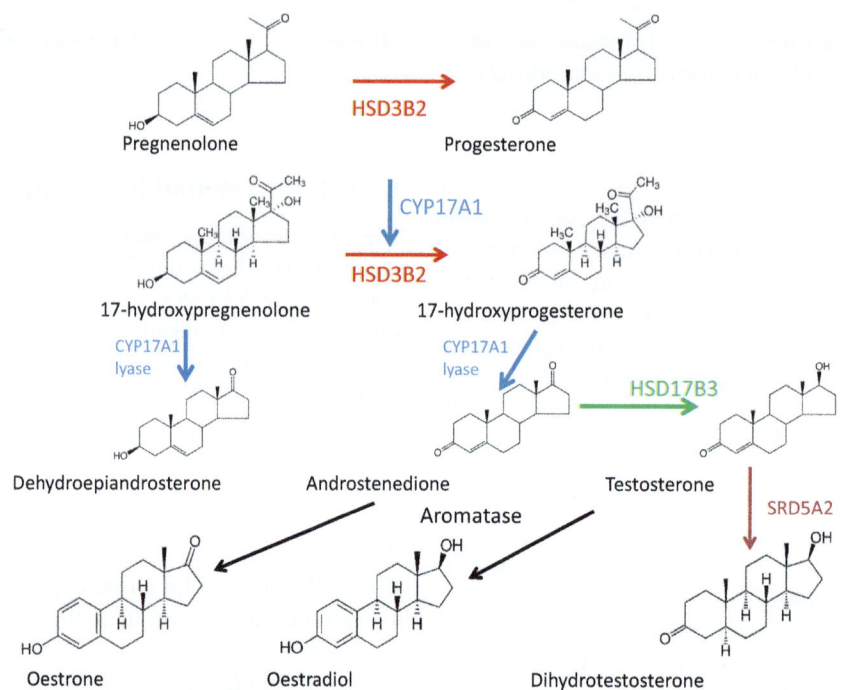

FIG. 1.3.16 Overview of androgens and estrogens synthesis. *(Author original.)*

(higher Vmax/Km) than the 17,20-lyase reaction. The major factor regulating the 17,20-lyase reaction and thus the production of all androgens and estrogens, is electron transfer.

Like all microsomal P450 enzymes, P450c17 receives electrons from NADPH via the membrane-bound POR and mediates catalysis via the iron atom in the P450 heme ring. POR receives two electrons from NADPH and transfers them one at a time to the P450 CYP17.

The 17,20-lyase activity of P450c17 is increased when the molar ratio of POR to P450c17 is higher or by factors that increase the affinity of POR for P450c17. Two post-translational mechanisms, the presence of cytochrome *b*5 and the serine phosphorylation of P450c17, facilitate the interaction of P450c17 with POR to optimize electron transfer (Miller and Tee, 2015; Tee and Miller, 2013).

The adrenal zona reticularis, the site of production of adrenal androgen precursors, contains abundant cytochrome *b*5, whereas the glomerulosa and fasciculata adrenal zones have almost no activity. Cytochrome *b*5 acts as an allosteric factor to promote the interaction of POR and P450c17, favoring 17,20 lyase activity. This has been confirmed from studies of genetic material from patients with loss-of-function mutations in CYP17 and scc. The activity of 17,20 lyase is facilitated by the phosphorylation of serine and/or threonine residues on P450c17 in experimental studies but this has not been confirmed in humans. CYP17A1 is a target for the treatment of prostate and breast cancers that proliferate in response to sex hormones.

1.3.6.4.2 Estrogen synthesis

Estrogen production from androgens is catalyzed by a single aromatase enzyme CYP19 in a series of complex reactions. Estrone is produced from androstenedione by concerted steps of carbonyl reduction, aromatization and formyl removal, aromatase switches traditional male hormones (androgens) into female hormones (estrogens) (Fig. 1.3.17). These steps each require O_2, NADPH and POR. CYP19 in peripheral tissues, especially fat, can convert substantial portions of circulating androstenedione and testosterone in women to estrone and estradiol (Santen et al., 2009).

The aromatase enzyme is encoded by a single gene on chromosome 15q21.1 and expressed in the endoplasmic reticulum. The gene uses several different alternative transcriptional start sites and first exons driven by different upstream promoter sequences, permitting the same protein to be expressed under different control in different cell types.

1.3.6.4.3 Estrogen production in ovary

The mechanism for estradiol production by a dominant follicle is loosely called a "two cell two gonadotrophin" system (Hillier et al., 1994). LH stimulates androstenedione production by the theca interstitial cells. Androstenedione diffuses into the follicular fluid where high concentrations are found. FSH stimulation of the aromatase in granulosa cells leads to synthesis of estrone which is reduced to estradiol (Fig. 1.3.18).

1.3.6.4.4 Progesterone production
Progesterone formation

In the endoplasmic reticulum of steroidogenic cells, pregnenolone is converted to progesterone with oxidation of the 3β-hydroxyl to a ketone and switch in the double bond in the B-ring to the A-ring by HSD3B2. NAD is the cofactor for this enzyme. Pregnenolone and progesterone are branch points in the synthesis of aldosterone, cortisol and sex steroids. After ovulation, progesterone is the principal secretion from the ovarian corpus luteum. If a pregnancy ensues the placenta is the source of progesterone.

1.3.6.4.5 Activation of androgens and estrogens
Reduction of C17 ketone

A C-17 hydroxyl group on a C_{19} steroid is important for androgen action. The 17β-hydroxysteroid dehydrogenases interconvert 17-ketosteroids and the corresponding

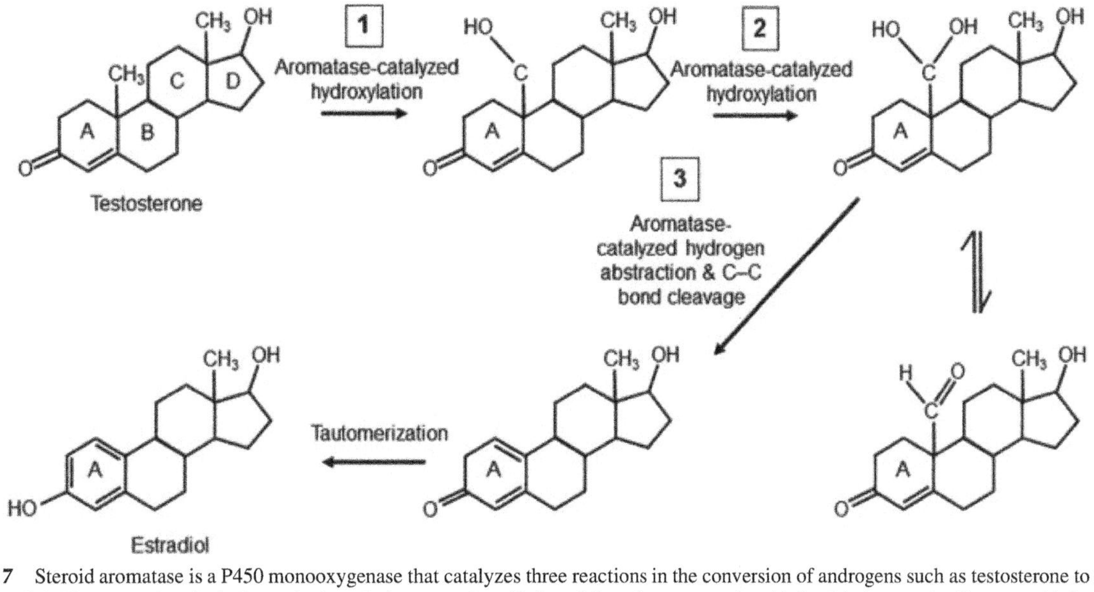

FIG. 1.3.17 Steroid aromatase is a P450 monooxygenase that catalyzes three reactions in the conversion of androgens such as testosterone to estrogens such as estradiol. These reactions include two hydroxylations at carbon 19 (1 and 2) and an uncommon third oxidation step leading to multiple molecular rearrangements, resulting in carbon 19 elimination (3), reduction of the A-ring keto group to a hydroxyl, and A-ring aromatization. *(Modified from Spiering MJ. On the trail of steroid aromatase: the work of Kenneth J. Ryan. J Biol Chem 2019;294(28):10743–5. Fig. 1 p. 10744.)*

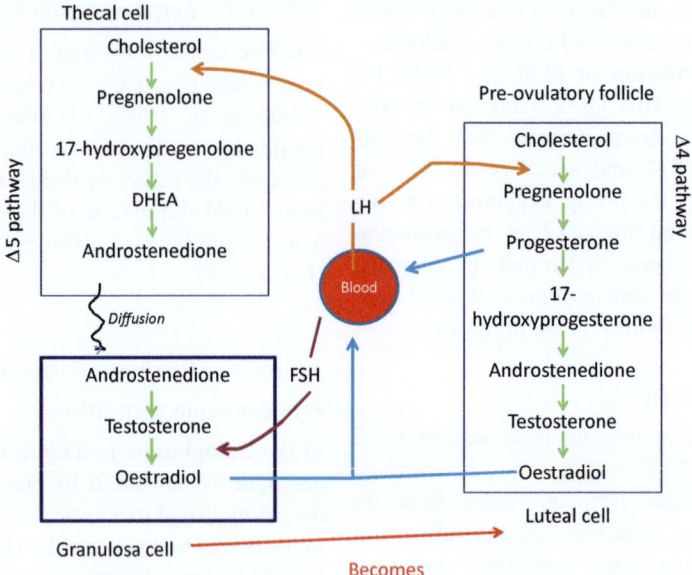

FIG. 1.3.18 Two compartments for steroid synthesis in the ovary.

17-hydroxysteroids and therefore control the synthesis and metabolism of sex steroids.

1.3.6.4.6 Androgen production in testis

Luteinizing hormone (LH) from pituitary gland regulates testosterone production in testicular Leydig cells via the LH receptor, which then promotes the 3′,5′-cyclic adenosine monophosphate (cAMP)/protein kinase A (PKA) signaling pathway by modulating G proteins. Intracellular cAMP promotes the cAMP/PKA pathway, which is mediated by the activities of adenylate cyclases (ACs) and phosphodiesterases (PDEs). The elevation of cAMP regulates the activation of PKA to stimulate signaling in downstream steroidogenic proteins. Steroidogenic acute regulatory protein (StAR) regulates cholesterol transfer within the mitochondria in a rate-limiting step to initiate steroidogenesis in Leydig cells. CYP11A catalyzes a cascade of steroidogenic reactions that firstly converts cholesterol to pregnenolone, the metabolic intermediate in the biosynthesis of most of the steroid hormones including testosterone. CYP17 then hydroxylates pregnenolone at C17 to form 17-hydroxpregnenolone, before cleavage of the 17–20 carbon bond (17,20 lyase) to leave a carbonyl group at C17 of the C_{19} steroid DHEA. The further actions of HSD3B2 and HSD17B3 are required for testosterone synthesis. In the periphery, testosterone is reduced to dihydrotestosterone by 5-alpha reductase (SRD5A1). This is called the **classic pathway for dihydrotestosterone synthesis**. A route involving pregnenolone to progesterone, 17-hydroxyprogesterone and androstenedione is not favored in the testis. Testosterone production is low until puberty

with the exception of a 6 month period after birth called the "minipuberty."

Dihydrotestosterone (DHT) is the 5-alpha-reduced metabolite of testosterone (T) that is produced principally in target organs such as prostate, skin and liver. Intracellular DHT is a more potent androgenic agonist than T, and its presence is necessary for the full prostate organ development and function of and other tissues in the reproductive tract. DHT is the most potent endogenous androgen based on four critical aspects of its binding to the androgen receptor AR:

1. DHT has a relative binding affinity for the AR that is roughly four times that of T.
2. The rate of dissociation from the AR is about three times slower than T
3. The binding of DHT to the AR transforms the AR to its DNA-binding state.
4. DHT upregulates AR synthesis and reduces AR turnover.

The molecular structure of DHT favors tight linkage to the steroid binding site on SHBG. Compared with T, DHT has roughly a fivefold greater binding affinity to SHBG.

DHT is one of four principal androgens in humans and is synthesized primarily via the irreversible action of microsomal steroid 5α-reductase SRD5A (both types I and II). In the periphery, T is converted directly to DHT by 5α-reductase SRD5A1. Hair follicles also use this route. In a further synthetic pathway to DHT, namely, the 5α-androstanedione pathway, 5α-androstanedione is converted by HSD17-B3 to DHT. Synthesis can also occur from precursors other than testosterone, but these pathways, although potentially important in tissues such as the prostate, were thought to be minor.

1.3.6.4.7 Promiscuity in hydroxysteroid dehydrogenases (HSD)

17β-Hydroxysteroid dehydrogenase 3 (17β-HSD3) is a microsomal enzyme with 310 amino acids. It is expressed predominantly in testes and catalyzes the conversion of Δ4-androstenedione (AD) to testosterone using NADPH as a cofactor. The encoding gene, HSD17B3, is mapped to human chromosome 9q22 and contains 11 exons. Some steroids can bind to a particular HSD with two alternative orientations. In one instance, they bind the C17 carbon at the active site but in other cases the HSD can bind the steroid with the C3 carbon oriented at the active site. The pseudo-symmetric feature of C19-steroids makes it possible for some enzymes to bind a steroid hormone in both the normal and reverse orientations. The alternative binding of some steroid hormones has been confirmed by crystallographic and biochemical studies.

The crystal structure of human 17β-HSD1 complex with testosterone, places the orientation of testosterone in this complex in the reverse manner to the previously known orientation for estradiol in the enzyme, due to rotation around the steroid central vertical axis. This theme is continued in the AKR family, but extends beyond the pseudosymmetry of C19 steroids in the narrow binding site of 17β-HSD1. DHT binds to human 3α-HSD3 (AKR1C2) in two orientations, which can simultaneously lead to the oxidized product or the reduced product in two monomers from one crystal asymmetric unit, thus enzyme activity is observed at both the C17 and C3 reaction sites. When DHT is bound in a normal orientation (C3 pointing toward the catalytic site) the 3α-HSD3 shows 3α activity and reduces DHT to 5α-androstane-3α,17β-diol. When it bound DHT in the reverse orientation, after a rotation so that the reactive hydroxyl group (C17) points toward the catalytic site, then 3α-HSD3 can oxidizes DHT to A-dione (5α-androstane-3,17-dione).

The binding pocket of 3α-HSD3 has only one amino acid (Val54) different from the binding pocket of human 20α-hydroxysteroid dehydrogenase (human 20α-HSD; AKR1C1). When this amino acid was mutated to mimic the human 20α-HSD binding pocket (Val54Leu), the alternative binding was lost in human 3α-HSD3/V54L and only an orientation that supported 20α-HSD activity was observed. Testosterone can bind to human type 5 17β-hydroxysteroid dehydrogenase (human 17β-HSD5; AKR1C3) in two orientations—either its O3 or O17 can penetrate the active site. The promiscuity of the AKR1C1, AKR1C2, AKR1C3 and AKR1C4 enzymes toward their steroid substrates has also been described (Penning et al., 2015).

One determinant of steroid orientation in the HSD binding pocket may be the oxidation state of the cofactor and/or the microenvironment. With an oxidative cofactor like NADP$^+$, 17β-HSD1 prefers to oxidize the C17 position of DHT, whereas with the reductive cofactor NADPH it prefers to reduce the C3 position of DHT to yield 3β-diol. That is, when 17β-HSD1 cooperates with an oxidative cofactor/microenvironment, it prefers to bind DHT in a normal C17-orientation, whereas when cooperating with a reductive cofactor/microenvironment 17β-HSD1 prefers to bind DHT in a reverse orientation. NADPH is the predominant reducing cofactor however, in the cell and will likely dictate preference for the orientation of steroid binding that favors reduction. The structures of the reduced and oxidized cofactors can assume a boat or chair conformation whereas the oxidized cofactors will be planar.

Progesterone (Prog) can also bind to human 3α-HSD3 in two orientations and with a rotation about the steroid C3–C17 long axis. In both orientations, the C20 oxygen atom is in proximity to the catalytic tetrad, however, the alpha and beta faces of the steroid are flipped. In both orientations, the steroid C20 ketone groups as well as the distances between the C20 atoms and the C4 position of the nicotinamide ring are incompatible with hydride transfer and represent steroid binding modes before or after the 20α-hydroxysteroid dehydrogenase reaction.

1.3.6.4.8 Backdoor path to DHT

DHT can be synthesized from substrates other than T, for example, from 17α-hydroxypregnenolone and 17α-hydroxyprogesterone in what is termed the "backdoor" pathway and from 5α-androstane-3α,17β-diol via the intracrine reverse synthesis pathway (Reisch et al., 2019; Miller and Auchus, 2019). This was first described in marsupials, rodents and spotted hyaena, The steroidal conversions in the backdoor pathway involve a 5α-reductase, 17,20 lyase, a 17β-hydroxysteroid dehydrogenase and both reductive and oxidative 3α-hydroxysteroid dehydrogenases (3αHSD) (see Fig. 1.3.19).

Studying diseases resulting from mutations in the genes encoding steroidogenic enzymes and other factors has contributed to new understanding of human processes. The potential for the backdoor DHT pathway in humans became likely as an explanation for poor masculinization of patients with 3α-hydroxysteroid dehydrogenase deficiency. The index family had combined partial disorders of two AKR1C enzymes and mutations genes for AKR1C2 and AKR1C4, whereas another patient had 46,XY DSD only with disordered AKR1C2. This unique, newly described DSD supports the idea that the backdoor pathway is essential for normal male sexual development, thus revising former views of the mechanisms of normal and pathological masculinization (Flück et al., 2011).

Alternative routes to DHT

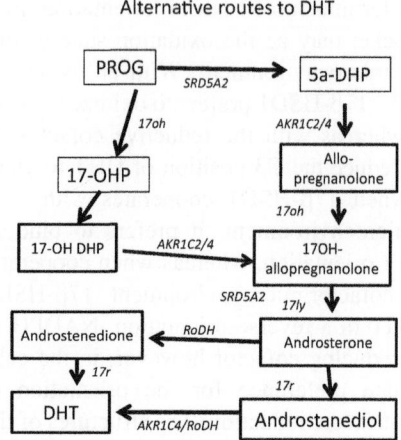

FIG. 1.3.19 Backdoor path to DHT.

The precise enzymes involved in the backdoor path to DHT are not certain. AKR1C2 appears to participate in the backdoor pathway to DHT in the human fetal testis, but not in the adult testis, and its mutation appears to cause incomplete male genital development. Retinol dehydrogenase RoDH (17βHSD6) is detected in the fetal testes where it may catalyze the oxidative 3αHSD reaction needed to complete the backdoor pathway to DHT, but this identification remains uncertain. However, 5α-androstane-3α,17β-diol was identified as a major oxidative product in the prostate (Bauman et al., 2006). The identification of the major oxidative 3alpha-hydroxysteroid dehydrogenase in human prostate that converts 5alpha-androstane-3alpha,17beta-diol to 5alpha-dihydrotestosterone is taken as evidence for a potential therapeutic target for androgen-dependent disease. Adult adrenals express abundant AKR1C2, so that the backdoor pathway to DHT production can be justified in situations where intra-adrenal 17OHP concentrations are high, such as in 21-hydroxylase deficiency. AKR1C2 is not expressed in the fetal ovary, consistent with the fact that androgens are not crucial in female sex development, whereas its mRNA is readily detectable in the adult gonad (Marti et al., 2017).

1.3.6.4.9 Glucocorticoid synthesis in extra-adrenal tissues (liver, adipose tissue, and skeletal muscle)

There is mounting evidence for extra-adrenal synthesis of steroids in lymphoid organs, intestine, skin, brain and possibly heart. Local enzyme activity can be inferred from measurements of steroid concentrations in arterial blood compared with the tissue, even after adrenalectomy. The case is strengthened by demonstrations with immunohistochemistry of enzyme expression and with molecular biology techniques (PCR) for the presence of mRNA for steroidogenic enzymes (Ahmed et al., 2019; Taves et al., 2011;

Noti et al., 2009). A summary of the distribution in tissues of the proteins involved in the initial rate-controlling steps of steroidogenesis can be found in article by Slominski et al. (2020). The list includes CYP11A1 and StAR, and downstream steroidogenic enzymes reported to be in these tissues including immune cells.

Cortisol can be generated from cortisone in liver, adipose tissue and skeletal muscle thorough a pathway within the endoplasmic reticulum (ER). The HSD11B1 responds to changes in nutrient levels in hepatocytes, adipocytes and the pancreas cells. States of overfeeding and prolonged fasting/starvation result in the activation of endoplasmic reticulum stress response pathways. The overfeeding induced increases in intracellular glucocorticoids plays an important role in the pathology of obesity, type II diabetes, the metabolic syndrome, hypertension and depression. The critical participants are glucose-6-phosphate and glucose-6-phosphate dehydrogenase activity, which exists in two forms. The H form of glucose-6-phosphate dehydrogenase activity is identified as hexose-6-phosphate dehydrogenase and resides within the ER and the sarcoplasmic reticulum (SR). The G form of glucose-6-phosphate dehydrogenase is a cytoplasmic enzyme with a critical role in the oxidative reactions of the pentose phosphate pathway and also as glucose 1-dehydrogenase (Fig. 1.3.20).

In the ER, hexose-6-phosphate dehydrogenase converts glucose-6-phosphate and NADP$^+$ to 6-phosphogluconate and NADPH in a single step, whereas in the cytosol this process requires two enzymes. In addition to glucose-6-phosphate, H6PD can metabolize other hexose-6-phosphates, glucose-6-sulfate and glucose. The ER- and SR-localized NADPH maintain redox homeostasis within these organelles. NADPH produced by ER-localized hexose-6-phosphate dehydrogenase is important for reducing equivalents reductases in the ER. In steroid synthesis, within adipose tissue and skeletal muscle, the intracellular concentration of glucose-6-phosphate is a direct function of the blood levels of both glucose and insulin. In these tissues, insulin action, through its receptor, results in mobilization of GLUT4 transporters to the plasma membrane leading to enhanced glucose uptake from the blood. In the liver, the uptake of glucose, via the GLUT2 transporter, is solely dependent on glucose levels in the blood. Glucose uptake by the liver only occurs, to a significant extent, during the postfeeding period because the Km of GLUT2 for glucose is high (on the order of 15mM). As the levels of glucose-6-phosphate rise in these tissues there is a concomitant increase in the activities of the ER-localized proteins that are involved in intracellular glucocorticoid activation. These ER-localized proteins include the H6PD and HSD11B1 encoded proteins, as well as the ER membrane-localized glucose-6-phosphate transporter encoded by the G6PT1 gene. The increase in G6PT1,

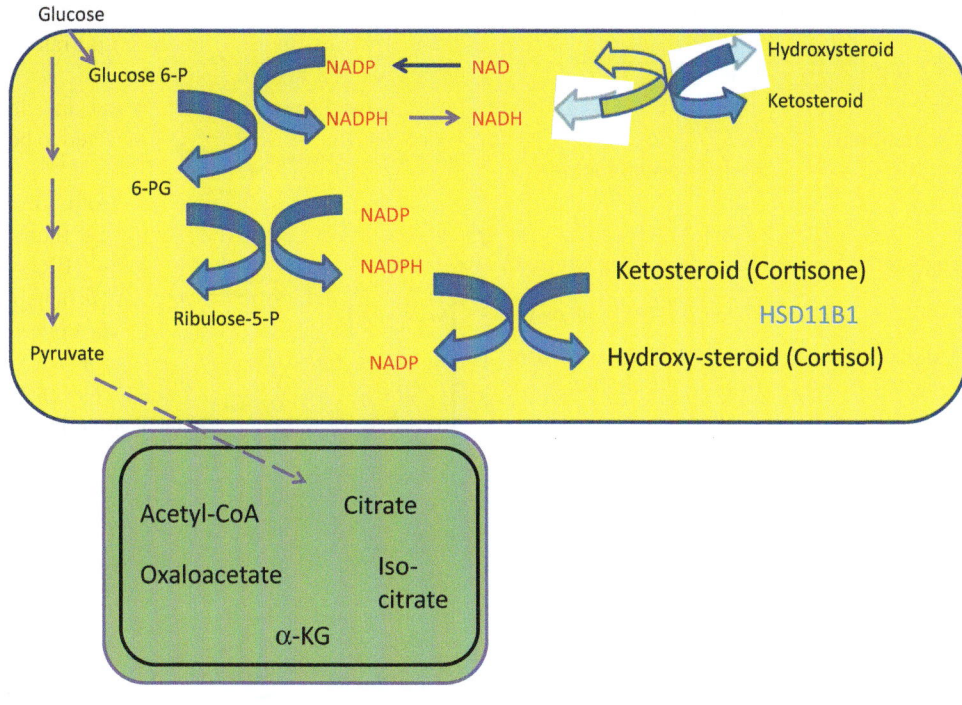

FIG. 1.3.20 Electron transport and links with glucose metabolism.

H6PD and HSD11B1 activities leads to increased conversion of inactive glucocorticoids to their active forms (cortisol and corticosterone).

Although glucose consumption directly results in increased intracellular glucocorticoid activation, the consumption of fatty acids will indirectly activate this pathway. Free fatty acids are known to interfere with glucose oxidation via the glucose-fatty acid cycle. The oxidation of fatty acids leads to increased mitochondrial NADH levels which impair the movement of carbon through the TCA cycle resulting in citrate transport to the cytosol which in turn leads to inhibition of the 6-phosphofructo-1-kinase (PFK1) activity of glycolysis. The increased mitochondrial NADH level also inhibits the pyruvate dehydrogenase complex (PDHc) reaction further impairing the oxidation of glucose. Thus, one of the effects of overeating, either carbohydrate or lipids, particularly within adipose tissue and skeletal muscle, is enhanced glucocorticoid activation.

When the H6PD gene was knocked out in mice the pathology that resulted included fasting hypoglycemia, low hepatic glycogen content, increased sensitivity to insulin and decreased negative feedback on the hypothalamic-pituitary-adrenal axis. Those results strongly implicate an important role for the triad of G6PT1, H6PD and HSD11B1 in the metabolic modifications that result in response to feeding. Excess nutrient intake, either in the form of carbohydrate or lipid, can result in increased

intracellular glucocorticoid activation, especially in adipose tissue and skeletal muscle, both of which are critical, insulin-responsive tissues. Glucocorticoids have been known for quite a while to induce a state of insulin resistance in both adipose tissue and skeletal muscle. Glucocorticoids interfere with insulin signaling in these tissues resulting in impaired GLUT4 mobilization to the plasma membrane, impaired glucose oxidation and impaired glycogen synthesis. Within visceral adipose tissue glucocorticoids stimulate preadipocyte differentiation and triglyceride synthesis. In the liver, glucocorticoids stimulate gluconeogenesis with leads to an exacerbation of the hyperglycemia that is the result of insulin resistance in skeletal muscle and adipose tissue. All of these metabolic disturbances contribute to the development of the metabolic syndrome and the onset of type 2 diabetes.

1.3.6.4.10 Intracrine synthesis

A tissue specific, local production of steroids, without significant release into the circulation constitutes an intracrine mechanism for activation of steroids (Fig. 1.3.21).

In this regard, DHEAS is a precursor for active androgens, the importance of which is seen when lost at the menopause and local enhancement in the etiology of pathologies including cancers of the breast and endometrium. The increasing use of more sensitive steroid assays has enabled studies to provide evidence for changes in tissue

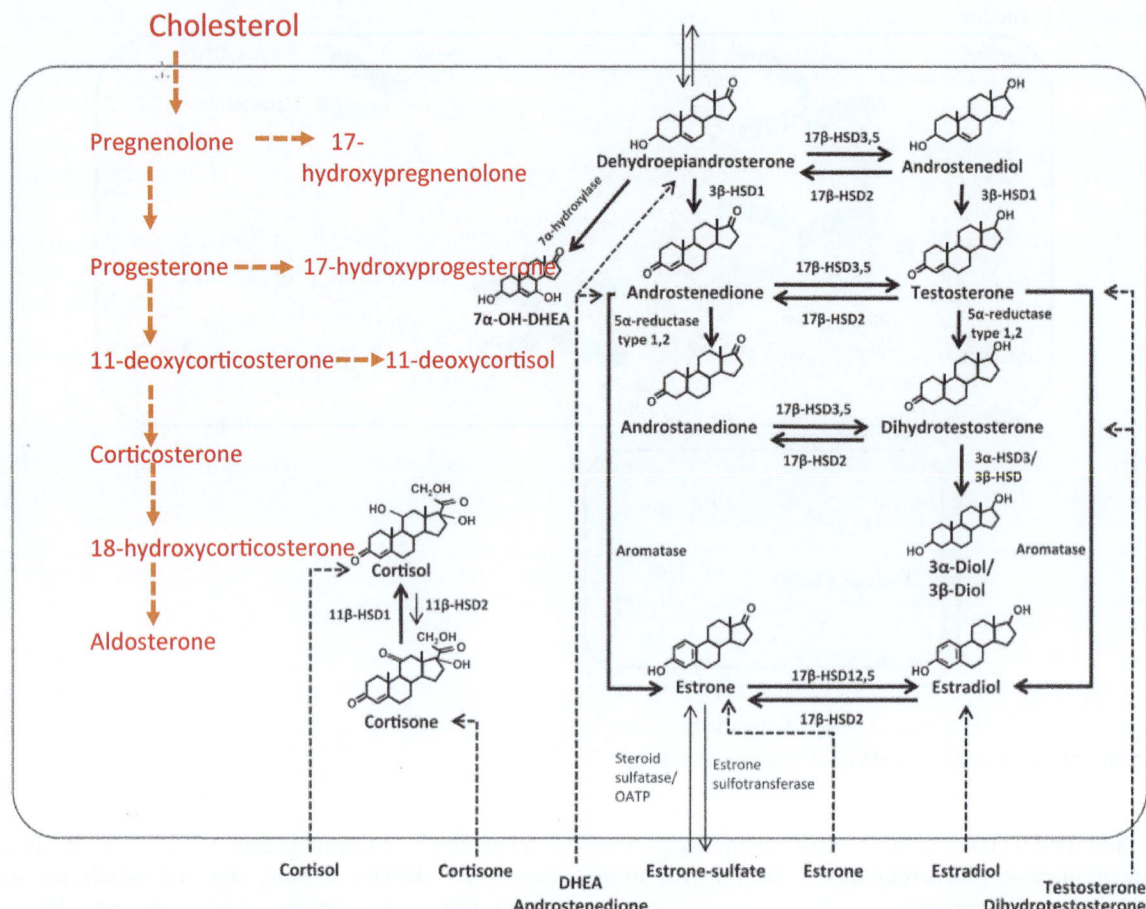

FIG. 1.3.21 Steroid hormone synthesis in the skin (intracrine mode). Pathways of steroid hormone metabolism in adipose tissue. *Black arrows* are confirmed pathways *Red arrows* are putative reactions. *(From Tchernof A, Mansour MF, Pelletier M, Boulet MM, Nadeau M, Luu-The V. Updated survey of the steroid-converting enzymes in human adipose tissues. J Steroid Biochem Mol Biol 2015;147:56–69. Fig. 2 p. 64.)*

specific concentrations of steroids that did not parallel those in blood. The expression of isoenzymes of steroid dehydrogenases and reductases also differs between tissues (Tchernof et al., 2015).

1.3.6.4.11 Peripheral 21-hydroxylase

Steroid 21-hydroxylase was found many years ago in kidney tissue of the human adult and fetus creating the potential for the in situ formation of deoxycorticosterone (DOC) from circulating progesterone in a tissue site of DOC action (Gomes et al., 2009). In addition, the extra-adrenal conversion of plasma progesterone to DOC is a major source of plasma DOC during the mid-luteal phase of the ovarian cycle and during pregnancy. Extra-adrenal 21-hydroxylation appears to be catalyzed by other microsomal P450s such as CYP2C9, CYP2C19 and CYP3A4. Gene expressions of CYP21, CYP3A4, CYP2C9 and CYP2C19 were all detected in human adipose tissue (subcutaneous and visceral). However, 21-hydroxylase activity of these enzymes has not been demonstrated in fat cells. Incubation

of radiolabeled progesterone in primary adipocytes, primary preadipocytes and differentiated adipocytes from the subcutaneous and omental fat of obese women identified 20α-hydroxyprogesterone as the main metabolic product of progesterone, together with a smaller amount of other metabolites including 5α/β-pregnanedione, 5α/β-pregnane-20α-ol-3-one, 5α/β-pregnane-α/β-ol-20-one, 5α/β-pregnane-3 α/β-20-ol-20-one and 5α/β-pregnane-3 α/β-20α-diol. These metabolites are most likely formed through the activity of 20αHSD, 3αHSD3 and 17βHSD5, rather than via 21-hydroxylase (Zhang et al., 2009). The influence of a local presence of CYP21 in adipose tissue warrants future studies.

1.3.6.4.12 Neurosteroids

The CNS synthesizes cholesterol via desmosterol (the Bloch pathway) or lathosterol (the Kandutsch-Russel pathway) in astrocytes and neurons, respectively. In 1981, Etienne-Emile Baulieu and colleagues reported the levels of dehydroepiandrosterone (DHEA) sulfate in the brain of both gonadectomized and adrenalectamized adult male rats and suggested a possible endogenous synthesis of DHEA sulfate in the brain, independent of the secretions of adrenals and testes. The term "neurosteroids" was then coined to refer to the steroids produced by the nervous system. Studies in my laboratory with Professor Jonathan Fry and Martin Ebner failed to confirm the presence of DHEAS in rat brain (Ebner et al., 2006). The methodology critically separated steroids according to their conjugation and detected steroids as methyloxime-trimethylsilyl ethers and heptafluorobutyrate derivatives. After extraction from adult male rat brain, steroids were fractionated to free steroids and separately their sulfate esters and then converted them to thermostable gas chromatography analysis and selected ion monitoring mass spectrometry. In the free steroid fraction, corticosterone, 3α,5α-tetrahydrodeoxycorticosterone, testosterone and dehydroepiandrosterone were found in the absence of detectable precursors usually found in endocrine glands, indicating peripheral sources and/or alternative synthetic pathways in brain. Conversely, the potent neuroactive steroid 3α,5α-tetrahydroprogesterone (allopregnanolone) was found in the presence of its precursors pregnenolone, progesterone and 5α-dihydroprogesterone. Furthermore, the presence of 3β-, 11β-, 17α- and 20α-hydroxylated metabolites of 3α,5α-tetrahydroprogesterone implicated possible inactivation pathways for this steroid. The 20α-reduced metabolites could also be found for pregnenolone, progesterone and 5α-dihydroprogesterone, introducing a possible regulatory diversion from the production of 3α,5α-tetrahydroprogesterone (Fig. 1.3.22). In the steroid sulfate fraction, dehydroepiandrosterone sulfate was identified but not pregnenolone sulfate. Subsequent studies using radioactive tracers showed both DHEAS and PregS to be

extensively desulfated as they entered the adult male rat brain (Qaiser et al., 2017).

The presence of pregnenolone, progesterone and reduced metabolites were confirmed by Liere working in the Baulieu department (Liere et al., 2004, 2009) and explained the original measurement of DHEAS by immunoassay to be an artifact of the isolation procedure (Fig. 1.3.23).

Using immunohistochemistry the presence of StAR and translocator protein (TSPO) has been demonstrated in various brain tissue slices, so cholesterol can be transported into mitochondria. Oligodendrocytes can convert mevalonate to pregnenolone. Sulfated steroids have been detected directly by using LC-based separation instead of gas-phase separation. High performance liquid chromatography-electrospray ionization tandem mass spectrometry (HPLC-ESI-MS/MS) and capillary column HPLC-nanoelectrospray ionization-MS/MS, have failed to detect sulfates of PREG or of DHEA in rodent brain (Lionetto et al., 2017; Dury and Labrie, 2016; Higashi et al., 2016).

Besides the detection of steroid in brain, a second criterion for a neurosteroid is to demonstrate the enzymes required in the brain for steroid interconversion. Expression of the enzyme P450 side chain cleavage (P450scc) in the brain is evidence for this enzyme in the first step of steroidogenesis, converting cholesterol in pregnenolone (PREG) (Chiang et al., 2011). Such conversion was observed in oligodendrocytes, demonstrating the existence of neurosteroids. P450scc mRNA has been shown in different brain regions.

The nervous system is a source as well as a target of steroid hormones although the mechanism of action through receptors has not been confirmed. Neurosteroids probably act directly to modulate plasma membrane ion channels

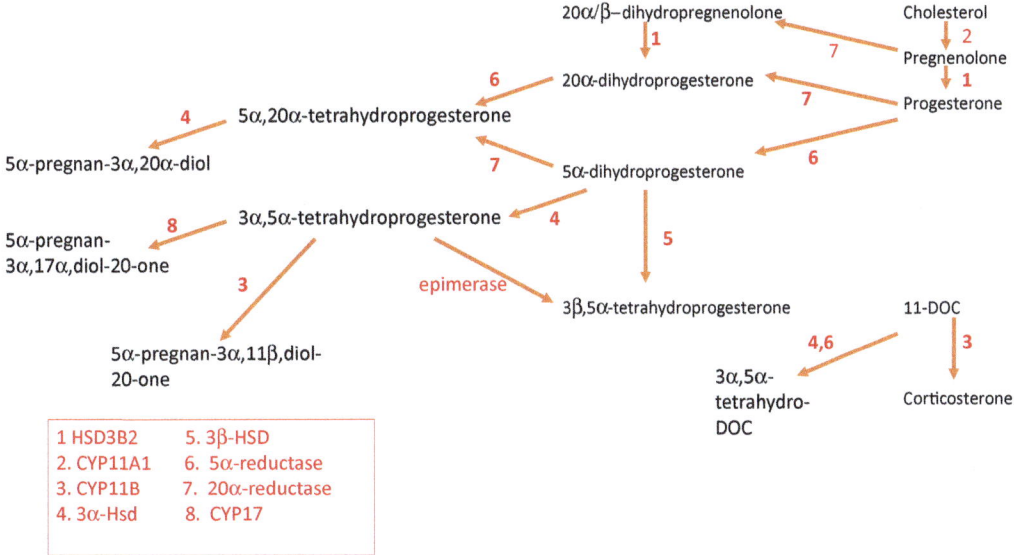

FIG. 1.3.22 Steroids as MO-TMS derivatives identified by GC_MS in rat brain. *(Redrawn from Ebner MJ, Corol DI, Havlíková H, Honour JW, Fry JP. Identification of neuroactive steroids and their precursors and metabolites in adult male rat brain. Endocrinology 2006;147(1):179–90. Fig. 4 p. 188.)*

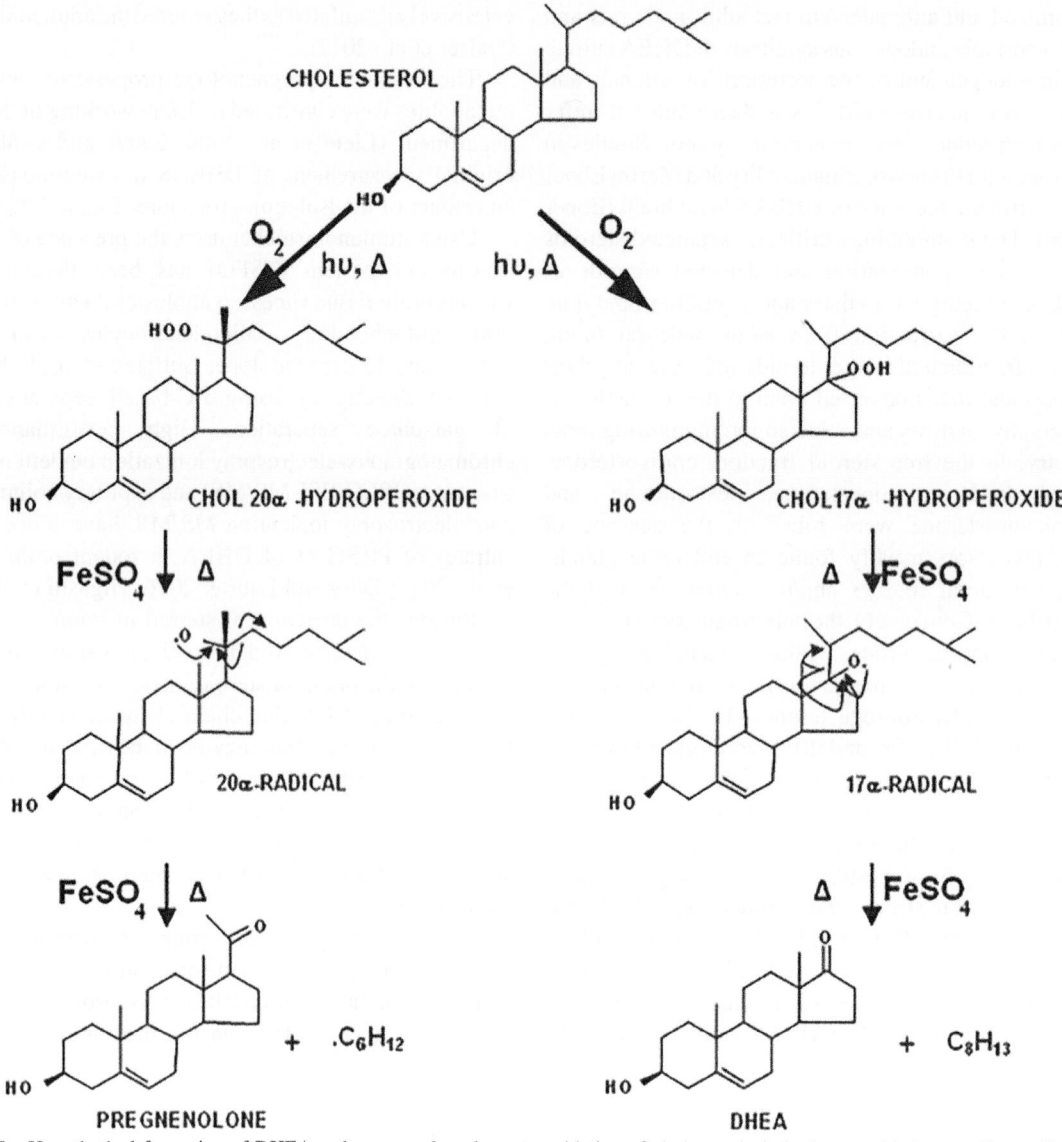

FIG. 1.3.23 Hypothetical formation of DHEA and pregnenolone by auto-oxidation of cholesterol via hydroperoxide intermediates. *(From Liere P, Pianos A, Eychenne B, Cambourg A, Bodin K, Griffiths W, Schumacher M, Baulieu EE, Sjövall J. Analysis of pregnenolone and dehydroepiandrosterone in rodent brain: cholesterol autoxidation is the key. J Lipid Res 2009;50(12):2430–44. Fig. 9 p. 2441.)*

(Lloyd-Evans and Waller-Evans, 2020). The mitochondrial cytochrome P450scc (CYP11A1) which catalyzes the de novo synthesis of pregnenolone (PREG) from cholesterol, is expressed throughout the rodent brain and has been detected in most cell types (glia, neurons, astrocytes). By immunohistochemistry 3-hydroxysteroid dehydrogenase (3β-HSD), is also distributed throughout the brain and spinal cord. mRNA of the CYP17 enzyme has also been detected in the adult rat brain hippocampus and spinal cord and neurons. Astrocytes in culture have been shown to convert PREG to DHEA. The HSD3B2 isoform is also largely expressed in the human brain and spinal cord.

In mice, the mRNA of the PREG sulfotransferase SULT2B1a is most prominent in brain and spinal cord,

suggesting that PREGS could be formed within the rodent CNS. Ebner et al. did not find PregS in rat brain. During development, SULT2B1a is expressed as early as embryonic day 8.5 (E 8.5) in mice and throughout embryonic life. In rats, SULT2B1a mRNA is expressed in both brain and testis. Neither of the SULT2B1 isoforms has been detected in human brain. SULT2A1, SULT2B1a or SULT2B1b expression has not been confirmed in human temporal lobe. A SULT-like cDNA has been cloned from human, rat and mouse brain (Liyou et al., 2003). The enzyme has been named SULT4A1, but is currently considered as an orphan member of the SULT family for which no activity or functions have so far been identified.

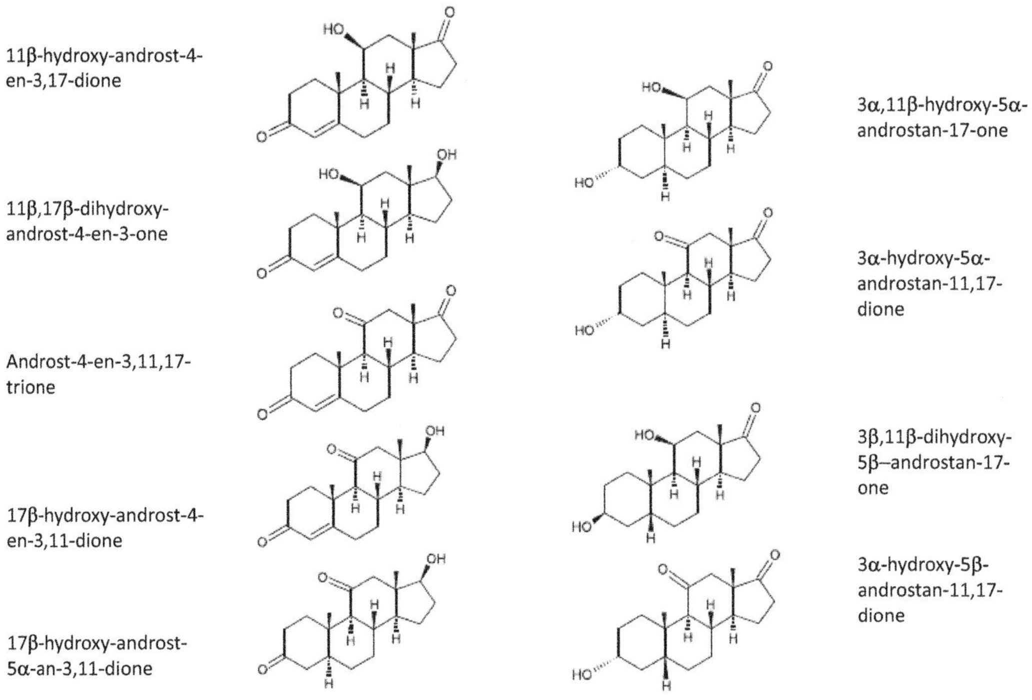

11β-hydroxy-androst-4-
en-3,17-dione

11β,17β-dihydroxy-
androst-4-en-3-one

Androst-4-en-3,11,17-
trione

17β-hydroxy-androst-4-
en-3,11-dione

17β-hydroxy-androst-
5α-an-3,11-dione

3α,11β-hydroxy-5α-
androstan-17-one

3α-hydroxy-5α-
androstan-11,17-
dione

3β,11β-dihydroxy-
5β—androstan-17-
one

3α-hydroxy-5β-
androstan-11,17-
dione

FIG. 1.3.24 11-Hydroxy and 11-keto androgens found in plasma and urine.

The steroids allo-pregnanolone and tetrahydrodeoxycorticosterone have been found in brain and act to inhibit activity of the gamma-aminobutyric acid receptor (GABA$_A$R) or excite N-methyl-D-aspartate receptor (NMDA) (Guennoun, 2020). An isoform of 5α-reductase (SRD5A1) has been detected in the brain as protein by immunohistochemistry and analysis of mRNA. Activity has been demonstrated with testosterone and progesterone. Estradiol synthesis has been demonstrated in brain to act through rapid, nongenomic actions on membrane receptors that constitute "nonclassical" mechanisms of estrogens regulating signaling cascades. Behavior and physiology including aggression and under active investigation (Balthazart et al., 2018). For 50–70 years, steroids have been known to have anticonvulsant (Aird and Gordon, 1951) and anesthetic properties (Green et al., 1978). Much further research is needed to enhance our understanding of neurosteroids. The evidence to date may be subject to misinterpretation, as was the case with pregnenolone sulfate. Lack of specificity in steroid and protein determinations by immunological tests has been found, which hampers our understanding of neurosteroid production and function in normal and disease situations. Most of the data is confined to studies in birds, rats and mice.

1.3.6.4.13 11-Hydroxylated androgens

In the past decade, there has been excitement over 11-oxygenated androgens (Bloem et al., 2013)

(Fig. 1.3.24). Like the investigations with neurosteroids, methods have been developed to quantify the steroids and proof for the action of enzymes and the steroid products been sought.

An important fact to consider is the biological activity of 11-oxygenated androgens. For many years, it was known that 11-hydroxyandrostenedione was inactive, in cock comb assays, for example. Recent work has shown binding of 11-keto testosterone and 11-keto-DHT to the androgen receptor is comparable to DHT (Gent et al., 2019). The question arises if a side chain cleavage of cortisol or an 11-hydroxylation of androstenedione is responsible for 11-hydroxyandrostenedione.

1.3.6.4.14 Pregnancy

At the time of conception, the corpus luteum in the ovary is maintained to keep progesterone production during the first trimester. The human placenta forms at the interface of maternal and fetal circulation. It participates in biosynthesis and metabolism of steroids as well as their regulated exchange between maternal and fetal compartment. The placenta expresses large amounts of aromatase, protecting the fetus from maternal androgens and may represent a signaling mechanism disposal of fetal C-19 steroids. Although the placenta produces huge amounts of estriol this hormone is not needed for normal pregnancy, as shown by the normal development, labor and parturition of fetuses that have genetic lesions of aromatase or sulfatase that

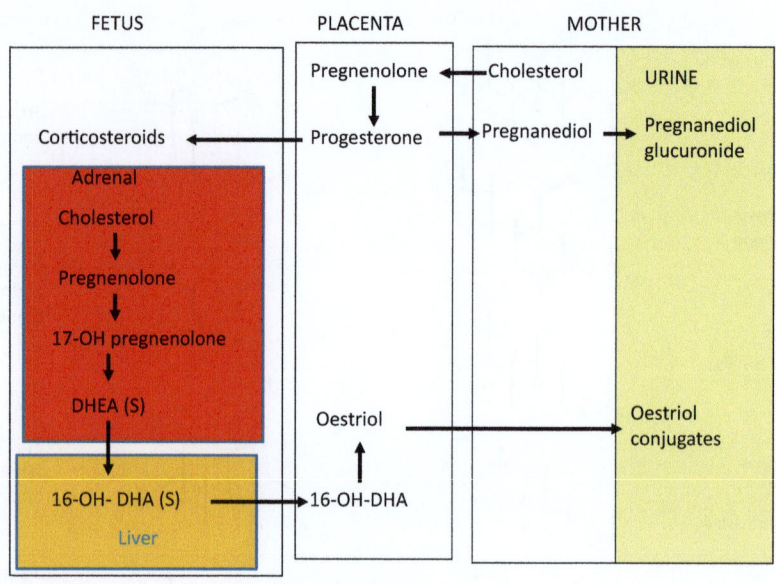

FIG. 1.3.25 Steroidogenesis in the feto-placental unit. *(Author original.)*

prevent estriol production (Noyola-Martínez et al., 2019; Chatuphonprasert et al., 2018).

Cholesterol is transported from mother to offspring involving lipoprotein receptors such as low-density lipo-protein receptor (LDLR) and scavenger receptor class B type I (SRB1) as well as ATP-binding cassette (ABC)-transporters, ABCA1 and ABCG1. Cholesterol is also a pre-cursor for placental progesterone and estrogen synthesis. Hormone synthesis is predominantly performed by members of the cytochrome P-450 (CYP) enzyme family including CYP11A1 or CYP19A1 and hydroxysteroid dehydrogenases (HSDs) such as 3β-HSD and 17β-HSD (Fig. 1.3.25). Placental estrogen synthesis requires delivery of sulfate-conjugated precursor molecules from fetal and maternal serum. Placental uptake of these precursors is mediated by members of the solute carrier (SLC) family including sodium-dependent organic anion transporter (SOAT) organic anion transporter 4 (OAT4) and organic anion transporting polypeptide 2B1 (OATP2B1). Maternal-fetal glucocorticoid transport has to be tightly reg-ulated in order to ensure healthy fetal growth and devel-opment. For that purpose, the placenta expresses the enzymes 11-HSD 1 and 2 as well as the transporter ABCB1.

16-Hydroxy-DHEA is the major product of the fetal adrenal gland and this is sulfated in the fetal liver before conversion to estriol by the placenta. The sulfate is removed by sulfatase then aromatase acts to create estrone then estradiol in the placenta. Concentrations of estriol in the maternal blood or urine increase from the second trime-ster and used to be used as an index for fetal viability. This is now superseded by ultrasound examination of the abdomen.

1.3.7 Summary

Steroid biosynthesis occurs principally in the adrenal cortex, ovary, testis, placenta and central nervous system. The uptake and movement of cholesterol is unclear but several participants have been defined and other candidates have yet to find a clear action point. The enzymes and cofactors involved in the synthesis of cortisol, aldosterone and sex steroids have been characterized and the genes for their production have been located on chromosomes, cloned and sequenced in normal and pathological states. Our understanding of neurosteroids needs clarification. Studying diseases resulting from mutations in the genes encoding steroidogenic enzymes and other factors has con-tributed to new understanding of human processes, which will be discussed in Part III of the book.

1.3.8 Further reading

Many other references were read during the preparation of this chapter and some not cited may be useful to the reader during a literature search. These have been grouped according to a theme of study.

1.3.8.1 Cholesterol and StAR

Bose et al. (2008), Horenkamp et al. (2018), Larsen et al. (2020), Lucki et al. (2012), Rodriguez-Agudo et al. (2019), Simpson et al. (1979), and Strauss et al. (2003).

1.3.8.2 Steroidogenic enzymes

Agrawal and Miller (2013), Brignac-Hubner et al. (2016), Gupta et al. (2003), Miller (2008), Miller and Bose

(2011), Miller and Auchus (2011), Miller (2013a), Petrunak et al. (2014), Tee et al. (2008, 2011), Wang et al. (2010, 2019), and Yoo et al. (2019).

1.3.8.3 11-Oxygenated androgens

Bloem et al. (2013), Chapman et al. (2013), and Gent et al. (2019).

1.3.8.4 Tissues/organs

Hillier et al. (1994), Hong et al. (2019), Labrie (2019), Lerner et al. (2020), Li et al (2015), Liu et al. (2003), Savchuk et al. (2019), Schiffer et al. (2018), Schumacher et al. (2008), Spitschak and Hoeflich (2018), Stocco (2012), Terao and Katayama (2016), Thiboutot et al. (2003), Zhu et al. (2019), and Zielinska et al. (2017).

1.3.8.5 Redox

Iyanagi (2019), Miller (2005), Penning (1999), Pinto and Cooper (2014), Tsachaki and Odermatt (2019), and Yazawa et al. (2020).

1.3.8.6 Sulphates

Klymiuk et al. (2018) and von Figura et al. (1998).

1.3.8.7 Mitochondria

Miller (2013b) and Papadopoulos and Miller (2013).

1.3.8.8 Genetics

Pignatelli et al. (2019) and Speiser et al. (2018).

1.3.8.9 Analytical

Dury et al. (2016).

References

Agarwal AK, Auchus RJ. Minireview: cellular redox state regulates hydroxysteroid dehydrogenase activity and intracellular hormone potency. Endocrinology 2005;146(6):2531–8.

Agrawal V, Miller WL. P450 oxidoreductase: genotyping, expression, purification of recombinant protein, and activity assessments of wild-type and mutant protein. Methods Mol Biol 2013;987:225–37.

Ahmed A, Schmidt C, Brunner T. Extra-adrenal glucocorticoid synthesis in the intestinal mucosa: between immune homeostasis and immune escape. Front Immunol 2019;10:1438.

Aird RB, Gordon GS. Anticonvulsive properties of desoxycorticosterone. J Am Med Assoc 1951;145(10):715–9.

Balthazart J, Choleris E, Remage-Healey L. Steroids and the brain: 50 years of research, conceptual shifts and the ascent of non-classical and membrane-initiated actions. Horm Behav 2018;99:1–8.

Bauman DR, Steckelbroeck S, Williams MV, Peehl DM, Penning TM. Identification of the major oxidative 3alpha-hydroxysteroid dehydrogenase in human prostate that converts 5alpha-androstane-3alpha,17beta-diol to 5alpha-dihydrotestosterone: a potential therapeutic target for androgen-dependent disease. Mol Endocrinol 2006;20(2):444–58.

Bhatt MR, Khatri Y, Rodgers RJ, Martin LL. Role of cytochrome b5 in the modulation of the enzymatic activities of cytochrome P450 17α-hydroxylase/17,20-lyase (P450 17A1). J Steroid Biochem Mol Biol 2017;170:2–18.

Biterova EI, Turanov AA, Gladyshev VN, Barycki JJ. Crystal structures of oxidized and reduced mitochondrial thioredoxin reductase provide molecular details of the reaction mechanism. Proc Natl Acad Sci USA 2005;102(42):15018–23.

Bloem LM, Storbeck KH, Schloms L, Swart AC. 11β-hydroxyandrostenedione returns to the steroid arena: biosynthesis, metabolism and function. Molecules 2013;18(11):13228–44.

Bose M, Whittal RM, Miller WL, Bose HS. Steroidogenic activity of StAR requires contact with mitochondrial VDAC1 and phosphate carrier protein. J Biol Chem 2008;283(14):8837–45.

Bose HS, Marshall B, Debnath DK, Perry EW, Whittal RM. Electron transport chain complex ii regulates steroid metabolism. iScience 2020;23(7):101295.

Brignac-Huber LM, Park JW, Reed JR, Backes WL. Cytochrome P450 organization and function are modulated by endoplasmic reticulum phospholipid heterogeneity. Drug Metab Dispos 2016;44(12):1859–66.

Burgardt NI, Gianotti AR, Ferreyra RG, Ermácora MR. A structural appraisal of sterol carrier protein 2. Biochim Biophys Acta Proteins Proteom 2017;1865(5):565–77.

Burkhard FZ, Parween S, Udhane SS, Flück CE, Pandey AV. P450 oxidoreductase deficiency: analysis of mutations and polymorphisms. J Steroid Biochem Mol Biol 2017;165(Pt A):38–50.

Chapman K, Holmes M, Seckl J. 11β-Hydroxysteroid dehydrogenases: intracellular gate-keepers of tissue glucocorticoid action. Physiol Rev 2013;93(3):1139–206.

Chatuphonprasert W, Jarukamjorn K, Ellinger I. Physiology and pathophysiology of steroid biosynthesis, transport and metabolism in the human placenta. Front Pharmacol 2018;9:1027.

Chiang YF, Lin HT, Hu JW, Tai YC, Lin YC, Hu MC. Differential regulation of the human CYP11A1 promoter in mouse brain and adrenals. J Cell Physiol 2011;226(8):1998–2005.

Chien Y, Rosal K, Chung BC. Function of CYP11A1 in the mitochondria. Mol Cell Endocrinol 2017;441:55–61.

Chortis V, Taylor AE, Doig CL, Walsh MD, Meimaridou E, Jenkinson C, et al. Nicotinamide nucleotide transhydrogenase as a novel treatment target in adrenocortical carcinoma. Endocrinology 2018;159(8):2836–49.

Clark BJ. The START-domain proteins in intracellular lipid transport and beyond. Mol Cell Endocrinol 2020;504, 110704.

Deng B, Shen WJ, Dong D, Azhar S, Kraemer FB. Plasma membrane cholesterol trafficking in steroidogenesis. FASEB J 2019;33(1):1389–400.

Denora N, Natile G. An updated view of translocator protein (TSPO). Int J Mol Sci 2017;18(12):2640.

Dluhy RG, Lifton RP. Glucocorticoid-remediable aldosteronism (GRA): diagnosis, variability of phenotype and regulation of potassium homeostasis. Steroids 1995;60(1):48–51.

Duggal R, Denisov IG, Sligar SG. Cytochrome b_5 enhances androgen synthesis by rapidly reducing the CYP17A1 oxy-complex in the lyase step. FEBS Lett 2018;592(13):2282–8.

Dury AY, Ke Y, Labrie F. Precise and accurate assay of pregnenolone and five other neurosteroids in monkey brain tissue by LC-MS/MS. Steroids 2016;113:64–70.

Ebner MJ, Corol DI, Havlíková H, Honour JW, Fry JP. Identification of neuroactive steroids and their precursors and metabolites in adult male rat brain. Endocrinology 2006;147(1):179–90.

von Figura K, Schmidt B, Selmer T, Dierks T. A novel protein modification generating an aldehyde group in sulfatases: its role in catalysis and disease. Bioessays 1998;20(6):505–10.

Flück CE, Meyer-Böni M, Pandey AV, Kempná P, Miller WL, Schoenle EJ, et al. Why boys will be boys: two pathways of fetal testicular androgen biosynthesis are needed for male sexual differentiation. Am J Hum Genet 2011;89(2):201–18.

Gent R, du Toit T, Bloem LM, Swart AC. The 11β-hydroxysteroid dehydrogenase isoforms: pivotal catalytic activities yield potent C11-oxy C_{19} steroids with 11βHSD2 favouring 11-ketotestosterone, 11-ketoandrostenedione and 11-ketoprogesterone biosynthesis. J Steroid Biochem Mol Biol 2019;189:116–26.

Gomes LG, Huang N, Agrawal V, Mendonça BB, Bachega TA, Miller WL. Extraadrenal 21-hydroxylation by CYP2C19 and CYP3A4: effect on 21-hydroxylase deficiency. J Clin Endocrinol Metab 2009;94(1):89–95.

Green CJ, Halsey MJ, Precious S, Wardley-Smith B. Alphaxolone-alphadolone anaesthesia in laboratory animals. Lab Anim 1978;12(2):85–9.

Guengerich FP. Mechanisms of cytochrome P450-catalyzed oxidations. ACS Catal 2018;8(12):10964–76.

Guennoun R. Progesterone in the Brain: Hormone. Neurosteroid and Neuroprotectant Int J Mol Sci 2020;21(15):5271.

Gupta MK, Guryev OL, Auchus RJ. 5Alpha-reduced C21 steroids are substrates for human cytochrome P450c17. Arch Biochem Biophys 2003;418(2):151–60.

Higashi T, Aiba N, Tanaka T, Yoshizawa K, Ogawa S. Methods for differential and quantitative analyses of brain neurosteroid levels by LC/MS/MS with ESI-enhancing and isotope-coded derivatization. J Pharm Biomed Anal 2016;117:155–62.

Hillier SG, Whitelaw PF, Smyth CD. Follicular oestrogen synthesis: the 'two-cell, two-gonadotrophin' model revisited. Mol Cell Endocrinol 1994;100(1–2):51–4.

Hong SH, Kim SC, Park MN, Jeong JS, Yang SY, Lee YJ, et al. Expression of steroidogenic enzymes in human placenta according to the gestational age. Mol Med Rep 2019;19(5):3903–11.

Horenkamp FA, Valverde DP, Nunnari J, Reinisch KM. Molecular basis for sterol transport by StART-like lipid transfer domains. EMBO J 2018;37(6), e98002.

Iaea DB, Spahr ZR, Singh RK, Chan RB, Zhou B, Bareja R, et al. Stable reduction of STARD4 alters cholesterol regulation and lipid homeostasis. Biochim Biophys Acta Mol Cell Biol Lipids 2020;1865(4), 158609.

Iyanagi T. Molecular mechanism of metabolic NAD(P)H-dependent electron-transfer systems: The role of redox cofactors. Biochim Biophys Acta Bioenerg 2019;1860(3):233–58.

Klymiuk MC, Neunzig J, Bernhardt R, Sánchez-Guijo A, Hartmann MF, Wudy SA, et al. Efficiency of the sulfate pathway in comparison to the Δ4- and Δ5-pathway of steroidogenesis in the porcine testis. J Steroid Biochem Mol Biol 2018;179:64–72.

Kraemer FB. Adrenal cholesterol utilization. Mol Cell Endocrinol 2007;265–266:42–5.

Labrie F. Intracrinology and menopause: the science describing the cell-specific intracellular formation of estrogens and androgens from DHEA and their strictly local action and inactivation in peripheral tissues. Menopause 2019;26(2):220–4.

Larsen MC, Lee J, Jorgensen JS, Jefcoate CR. STARD1 functions in mitochondrial cholesterol metabolism and nascent HDL formation. Gene expression and molecular mRNA imaging show novel splicing and a 1:1 mitochondrial association. Front Endocrinol 2020;11, 559674.

Lerner A, Owens LA, Coates M, Simpson C, Poole G, Velupillai J, et al. Differential expression of genes involved in steroidogenesis pathway in human oocytes obtained from patients with polycystic ovaries. J Reprod Immunol 2020;142, 103191.

Li J, Papadopoulos V, Vihma V. Steroid biosynthesis in adipose tissue. Steroids 2015;103:89–104.

Liere P, Pianos A, Eychenne B, Cambourg A, Liu S, Griffiths W, et al. Novel lipoidal derivatives of pregnenolone and dehydroepiandrosterone and absence of their sulfated counterparts in rodent brain. J Lipid Res 2004;45(12):2287–302.

Liere P, Pianos A, Eychenne B, Cambourg A, Bodin K, Griffiths W, et al. Analysis of pregnenolone and dehydroepiandrosterone in rodent brain: cholesterol autoxidation is the key. J Lipid Res 2009;50(12):2430–44.

Lionetto L, De Andrés F, Capi M, Curto M, Sabato D, Simmaco M, et al. LC-MS/MS simultaneous analysis of allopregnanolone, epiallopregnanolone, pregnanolone, dehydroepiandrosterone and dehydroepiandrosterone 3-sulfate in human plasma. Bioanalysis 2017;9(6):527–39.

Liu S, Sjövall J, Griffiths WJ. Neurosteroids in rat brain: extraction, isolation, and analysis by nanoscale liquid chromatography-electrospray mass spectrometry. Anal Chem 2003;75(21):5835–46.

Liyou NE, Buller KM, Tresillian MJ, Elvin CM, Scott HL, Dodd PR, et al. Localization of a brain sulfotransferase, SULT4A1, in the human and rat brain: an immunohistochemical study. J Histochem Cytochem 2003;51(12):1655–64.

Lloyd-Evans E, Waller-Evans H. Biosynthesis and signalling functions of central and peripheral nervous system neurosteroids in health and disease. Essays Biochem 2020;64(3):591–606.

Lucki NC, Li D, Sewer MB. Sphingosine-1-phosphate rapidly increases cortisol biosynthesis and the expression of genes involved in cholesterol uptake and transport in H295R adrenocortical cells. Mol Cell Endocrinol 2012;348(1):165–75.

Marti N, Galván JA, Pandey AV, Trippel M, Tapia C, Müller M, et al. Genes and proteins of the alternative steroid backdoor pathway for dihydrotestosterone synthesis are expressed in the human ovary and seem enhanced in the polycystic ovary syndrome. Mol Cell Endocrinol 2017;5(441):116–23.

Meimaridou E, Goldsworthy M, Chortis V, Fragouli E, Foster PA, Arlt W, et al. NNT is a key regulator of adrenal redox homeostasis and steroidogenesis in male mice. J Endocrinol 2018;236(1):13–28.

Miller WL. Androgen biosynthesis from cholesterol to DHEA. Mol Cell Endocrinol 2002;198(1–2):7–14.

Miller WL. Minireview: regulation of steroidogenesis by electron transfer. Endocrinology 2005;146(6):2544–50.

Miller WL. Steroidogenic enzymes. Endocr Dev 2008;13:1–18.

Miller WL. A brief history of adrenal research: steroidogenesis—the soul of the adrenal. Mol Cell Endocrinol 2013a;371(1–2):5–14.

Miller WL. Steroid hormone synthesis in mitochondria. Mol Cell Endocrinol 2013b;379(1–2):62–73.

Miller WL. Disorders in the initial steps of steroid hormone synthesis. J Steroid Biochem Mol Biol 2017a;165(Pt A):18–37.

Miller WL. Steroidogenesis: unanswered questions. Trends Endocrinol Metab 2017b;28(11):771–93.

Miller WL, Auchus RJ. The molecular biology, biochemistry, and physiology of human steroidogenesis and its disorders. Endocr Rev 2011;32(1):81–151.

Miller WL, Auchus RJ. The "backdoor pathway" of androgen synthesis in human male sexual development. PLoS Biol 2019;17(4), e3000198.

Miller WL, Bose HS. Early steps in steroidogenesis: intracellular cholesterol trafficking. J Lipid Res 2011;52(12):2111–35.

Miller WL, Merke DP. Tenascin-X, congenital adrenal hyperplasia, and the CAH-X syndrome. Horm Res Paediatr 2018;89(5):352–61.

Miller WL, Tee MK. The post-translational regulation of 17,20 lyase activity. Mol Cell Endocrinol 2015;408:99–106.

Miziorko HM. Enzymes of the mevalonate pathway of isoprenoid biosynthesis. Arch Biochem Biophys 2011;505(2):131–43.

Mornet E, Dupont J, Vitek A, White PC. Characterization of two genes encoding human steroid 11 beta-hydroxylase (P-450(11) beta). J Biol Chem 1989;264(35):20961–7. 2592361.

Morris DJ, Latif SA, Hardy MP, Brem AS. Endogenous inhibitors (GALFs) of 11beta-hydroxysteroid dehydrogenase isoforms 1 and 2: derivatives of adrenally produced corticosterone and cortisol. J Steroid Biochem Mol Biol 2007;104(3–5):161–8.

Mueller JW, Gilligan LC, Idkowiak J, Arlt W, Foster PA. The regulation of steroid action by sulfation and desulfation. Endocr Rev 2015;36(5):526–63.

Neunzig J, Sánchez-Guijo A, Mosa A, Hartmann MF, Geyer J, Wudy SA, et al. A steroidogenic pathway for sulfonated steroids: the metabolism of pregnenolone sulfate. J Steroid Biochem Mol Biol 2014;144(Pt B):324–33.

Noti M, Sidler D, Brunner T. Extra-adrenal glucocorticoid synthesis in the intestinal epithelium: more than a drop in the ocean? Semin Immunopathol 2009;31(2):237–48.

Noyola-Martínez N, Halhali A, Barrera D. Steroid hormones and pregnancy. Gynecol Endocrinol 2019;35(5):376–84.

Pandey AV, Flück CE. NADPH P450 oxidoreductase: structure, function, and pathology of diseases. Pharmacol Ther 2013;138(2):229–54.

Papadopoulos V, Miller WL. Role of mitochondria in steroidogenesis. Best Pract Res Clin Endocrinol Metab 2012;26(6):771–90.

Papadopoulos V, Fan J, Zirkin B. Translocator protein (18 kDa): an update on its function in steroidogenesis. J Neuroendocrinol 2018;30(2). https://doi.org/10.1111/jne.12500.

Penning TM. Molecular determinants of steroid recognition and catalysis in aldo-keto reductases. Lessons from 3alpha-hydroxysteroid dehydrogenase. J Steroid Biochem Mol Biol 1999;69(1–6):211–25.

Penning TM, Chen M, Jin Y. Promiscuity and diversity in 3-ketosteroid reductases. J Steroid Biochem Mol Biol 2015;151:93–101.

Penning TM, Wangtrakuldee P, Auchus RJ. Structural and functional biology of aldo-keto reductase steroid-transforming enzymes. Endocr Rev 2019;40(2):447–75.

Petrunak EM, DeVore NM, Porubsky PR, Scott EE. Structures of human steroidogenic cytochrome P450 17A1 with substrates. J Biol Chem 2014;289(47):32952–64.

Pignatelli D, Carvalho BL, Palmeiro A, Barros A, Guerreiro SG, Macut D. The complexities in genotyping of congenital adrenal hyperplasia: 21-hydroxylase deficiency. Front Endocrinol 2019;10:432.

Pinto JT, Cooper AJ. From cholesterogenesis to steroidogenesis: role of riboflavin and flavoenzymes in the biosynthesis of vitamin D. Adv Nutr 2014;5(2):144–63.

Qaiser MZ, Dolman DEM, Begley DJ, Abbott NJ, Cazacu-Davidescu M, Corol DI, et al. Uptake and metabolism of sulphated steroids by the blood-brain barrier in the adult male rat. J Neurochem 2017;142(5):672–85.

Reddish MJ, Guengerich FP. Human cytochrome P450 11B2 produces aldosterone by a processive mechanism due to the lactol form of the intermediate 18-hydroxycorticosterone. J Biol Chem 2019;294(35):12975–91.

Reisch N, Taylor AE, Nogueira EF, Asby DJ, Dhir V, Berry A, et al. Alternative pathway androgen biosynthesis and human fetal female virilization. Proc Natl Acad Sci U S A 2019;116(44):22294–9.

Rodriguez-Agudo D, Malacrida L, Kakiyama G, Sparrer T, Fortes C, Maceyka M, et al. StarD5: an ER stress protein regulates plasma membrane and intracellular cholesterol homeostasis. J Lipid Res 2019;60(6):1087–98.

Santen RJ, Brodie H, Simpson ER, Siiteri PK, Brodie A. History of aromatase: saga of an important biological mediator and therapeutic target. Endocr Rev 2009;30(4):343–75.

Savchuk I, Morvan ML, Antignac JP, Kurek M, Le Bizec B, Söder O, et al. Ontogenesis of human fetal testicular steroidogenesis at early gestational age. Steroids 2019;141:96–103.

Schiffer L, Anderko S, Hannemann F, Eiden-Plach A, Bernhardt R. The CYP11B subfamily. J Steroid Biochem Mol Biol 2015;151:38–51.

Schiffer L, Arlt W, Storbeck KH. Intracrine androgen biosynthesis, metabolism and action revisited. Mol Cell Endocrinol 2018;465:4–26.

Schumacher M, Liere P, Akwa Y, Rajkowski K, Griffiths W, Bodin K, et al. Pregnenolone sulfate in the brain: a controversial neurosteroid. Neurochem Int 2008;52(4–5):522–40.

Shen WJ, Azhar S, Kraemer FB. Lipid droplets and steroidogenic cells. Exp Cell Res 2016;340(2):209–14.

Simpson ER, Carr BR, Parker Jr CR, Milewich L, Porter JC, MacDonald PC. The role of serum lipoproteins in steroidogenesis by the human fetal adrenal cortex. J Clin Endocrinol Metab 1979;49(1):146–8.

Slominski RM, Tuckey RC, Manna PR, Jetten AM, Postlethwaite A, Raman C, et al. Extra-adrenal glucocorticoid biosynthesis: implications for autoimmune and inflammatory disorders. Genes Immun 2020;21(3):150–68.

Speiser PW, Arlt W, Auchus RJ, Baskin LS, Conway GS, Merke DP, et al. Congenital adrenal hyperplasia due to steroid 21-hydroxylase deficiency: an Endocrine Society Clinical Practice Guideline. J Clin Endocrinol Metab 2018;103(11):4043–88.

Spitschak M, Hoeflich A. Potential functions of IGFBP-2 for ovarian folliculogenesis and steroidogenesis. Front Endocrinol 2018;9:119.

Stocco C. Tissue physiology and pathology of aromatase. Steroids 2012;77(1–2):27–35.

Stocco DM, Zhao AH, Tu LN, Morohaku K, Selvaraj V. A brief history of the search for the protein(s) involved in the acute regulation of steroidogenesis. Mol Cell Endocrinol 2017;441:7–16.

Storbeck KH, Swart AC, Fox CL, Swart P. Cytochrome b5 modulates multiple reactions in steroidogenesis. J Steroid Biochem Mol Biol 2015;151:66–73.

Strauss III JF, Kishida T, Christenson LK, Fujimoto T, Hiroi H. START domain proteins and the intracellular trafficking of cholesterol in steroidogenic cells. Mol Cell Endocrinol 2003;202(1–2):59–65.

Taves MD, Gomez-Sanchez CE, Soma KK. Extra-adrenal glucocorticoids and mineralocorticoids: evidence for local synthesis, regulation, and function. Am J Physiol Endocrinol Metab 2011;301(1):E11–24.

Tchernof A, Mansour MF, Pelletier M, Boulet MM, Nadeau M, Luu-The V. Updated survey of the steroid-converting enzymes in human adipose tissues. J Steroid Biochem Mol Biol 2015;147:56–69.

Tee MK, Miller WL. Phosphorylation of human cytochrome P450c17 by p38α selectively increases 17,20 lyase activity and androgen biosynthesis. J Biol Chem 2013;288(33):23903–13.

Tee MK, Dong Q, Miller WL. Pathways leading to phosphorylation of p450c17 and to the posttranslational regulation of androgen biosynthesis. Endocrinology 2008;149(5):2667–77.

Tee MK, Huang N, Damm I, Miller WL. Transcriptional regulation of the human P450 oxidoreductase gene: hormonal regulation and influence of promoter polymorphisms. Mol Endocrinol 2011;25(5):715–31.

Terao M, Katayama I. Local cortisol/corticosterone activation in skin physiology and pathology. J Dermatol Sci 2016;84(1):11–6.

Thiboutot D, Jabara S, McAllister JM, Sivarajah A, Gilliland K, Cong Z, et al. Human skin is a steroidogenic tissue: steroidogenic enzymes and cofactors are expressed in epidermis, normal sebocytes, and an immortalized sebocyte cell line (SEB-1). J Invest Dermatol 2003;120(6):905–14.

Tsachaki M, Odermatt A. Subcellular localization and membrane topology of 17β-hydroxysteroid dehydrogenases. Mol Cell Endocrinol 2019;489:98–106.

Venkatachalam KV. Human 3′-phosphoadenosine 5′-phosphosulfate (PAPS) synthase: biochemistry, molecular biology and genetic deficiency. IUBMB Life 2003;55(1):1–11.

Wang YH, Tee MK, Miller WL. Human cytochrome p450c17: single step purification and phosphorylation of serine 258 by protein kinase A. Endocrinology 2010;151(4):1677–84.

Wang Y, Li H, Zhu Q, Li X, Lin Z, Ge RS. The cross talk of adrenal and Leydig cell steroids in Leydig cells. J Steroid Biochem Mol Biol 2019;192, 105386.

Werck-Reichhart D, Feyereisen R. Cytochromes P450: a success story. Genome Biol 2000;1(6). REVIEWS3003.

Yazawa T, Imamichi Y, Uwada J, Sekiguchi T, Mikami D, Kitano T, et al. Evaluation of 17β-hydroxysteroid dehydrogenase activity using androgen receptor-mediated transactivation. J Steroid Biochem Mol Biol 2020;196, 105493.

Yoo SE, Yi M, Kim WY, Cho SA, Lee SS, Lee SJ, et al. Influences of cytochrome b5 expression and its genetic variant on the activity of CYP2C9, CYP2C19 and CYP3A4. Drug Metab Pharmacokinet 2019;34(3):201–8.

Zhang HJ, Yang J, Zhang MN, Zhang W, Liu JM, Wang WQ, et al. Variations in the promoter of CYP21A2 gene identified in a Chinese patient with simple virilizing form of 21-hydroxylase deficiency. Clin Endocrinol (Oxf) 2009;70(2):201–7.

Zhu P, Wang W, Zuo R, Sun K. Mechanisms for establishment of the placental glucocorticoid barrier, a guard for life. Cell Mol Life Sci 2019;76(1):13–26.

Zielinska AE, Fletcher RS, Sherlock M, Doig CL, Lavery GG. Cellular and genetic models of H6PDH and 11β-HSD1 function in skeletal muscle. Cell Biochem Funct 2017;35(5):269–77.

Chapter 1.4

Regulation of steroid production

1.4.1 Introduction

Steroid hormones are not stored in the endocrine glands, they are produced on demand in a controlled manner. Each of the classes of steroids have their own protein regulators. The receptors for these regulators are located in cell membranes and link to cyclic AMP generation or intracellular calcium concentrations.

1.4.2 Adrenal cortex

The adrenal cortex is the source of two important hormones, aldosterone and cortisol that are essential for survival. Cortisol is an important hormone with effects on many tissues in the body. It plays a major role in metabolism by promoting protein breakdown, in muscle and connective tissues, and the release of glycerol and fatty acids from adipose tissue. Cortisol provides the substrates necessary for gluconeogenesis in the liver. Cortisol and glucocorticoids act also as antiinflammatory or immunosuppressant agents. Aldosterone is responsible for promoting sodium reabsorption and potassium excretion particularly in the kidney but also in the colon, sweat glands and salivary glands. These effects are called mineralocorticoid actions. The electrolytes are part of the internal environment that needs control in the process of homeostasis. During fetal life and adulthood, dehydroepiandrosterone sulfate is produced in the largest quantities of all the adrenal steroids. From 6 months of age to around 5–8 years, there is little DHEAS production until a new are (Zona reticularis) in the adrenal cortex appears inward of the zona fasciculata which again restores DHEAS production as the dominant steroid. The actual regulator of DHEAS synthesis is unknown. The role of this steroid is largely unknown but research is currently active focusing on DHEAS as a precursor for androgens after transport into cells and desulfation.

1.4.2.1 Regulation of cortisol production by adrenocorticotrophic hormone (ACTH)

The zona fasciculata of the adrenal cortex is directed to cortisol synthesis when stimulated by adrenocorticotrophic hormone (ACTH) from corticotrophs of the anterior pituitary gland at the base of the brain. Low cortisol concentrations in blood are sensed by the hypothalamus and signal, through corticotrophic releasing factor (CRF), the pituitary to secrete ACTH. This constitutes the hypothalamic-pituitary-adrenal axis or HPA for short (Miller, 2018) (Fig. 1.4.1).

The secretion of ACTH from the pituitary is pulsatile, under negative feedback control (Oster et al., 2017). Pulses of increasing amplitude at 90 min intervals start very early in the morning (around 12.30–02.00 h) with highest peak around 0800 h but this can vary from around 0600 to 1000 h. In a population, there is a circadian rhythm pattern with highest circulating cortisol concentrations around 0800 h (200–800 nmol/L) and lowest around midnight (<150 nmol/L) (Fig. 1.4.2). The range of results at any time point reflect the fact that individuals may be around a peak or trough in the pulsatile pattern or anywhere between.

ACTH release is under the influence of a number of other hormones (corticotrophin releasing hormone, vasopressin) and factors (cytokines, catecholamines) secreted in the body. Physiological activities such as eating and stress (Berger et al., 2019; Nicolaides et al., 2017) can also affect the release of ACTH. The HPA system is known to vary with aging and there are gender differences. The corticotroph cells of the pituitary have high content of the ACTH precursor, pro-opiomelanocortin (POMC) (Cawley et al., 2016) which has high glycoprotein content in the N-terminal glycopeptide that histologically reacts positively with periodic acid-Schiff staining. POMC is 266 amino acids long and is sequentially cleaved by prohormone convertases (PC1 and PC2) to release a N-terminal POMC fragment (N-POC) and β-lipotrophin (β-LPH) (Fig. 1.4.3). The N-POC is further cleaved to three products β-melanocyte stimulating hormone (β-MSH), ACTH and joining peptide.

β-LPH is further cleaved to γ-LPH and β-endorphin. ACTH can be cleaved to α-MSH and corticotrophin like intermediate peptide (CLIP). α-MSH induces skin pigmentation, is involved in appetite regulation and the immune response. β-Endorphin exerts potent analgesic effects through opiate receptors. ACTH is a polypeptide of 39 amino acids of which the 12 N-terminal amino acids are critical in the adrenal response for cortisol synthesis. ACTH induces steroidogenesis firstly by activating adrenal cortical

Steroids in the Laboratory and Clinical Practice. https://doi.org/10.1016/B978-0-12-818124-9.00017-6

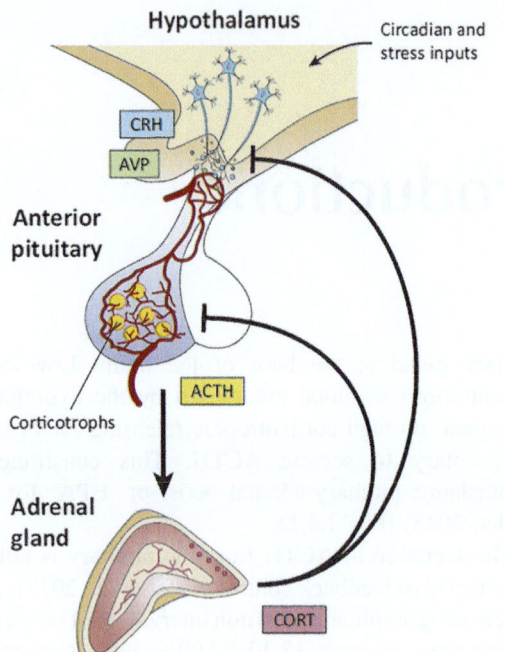

FIG. 1.4.1 HPA axis. Corticotrophs in the anterior lobe of the pituitary, are stimulated by corticotrophin releasing factor (CRF) from the hypothalamus in response to stress, to release adrenocorticotrophin (ACTH) which in turn stimulates cortisol synthesis by the adrenal cortex. Cortisol then inhibits ACTH production in a negative feedback manner. ACTH is released from a precursor protein pro-opiomelanocortin (POMC). *(From Zavala E, Wedgwood KCA, Voliotis M, Tabak J, Spiga F, Lightman SL, Tsaneva-Atanasova K. Mathematical modelling of endocrine systems. Trends Endocrinol Metab 2019;30(4):244–57. Open access Fig. 2 p. 249.)*

cell surface receptors via adenyl cyclase to regulate enzyme transcription.

POMC is encoded by a single-copy gene on chromosome 2p23.3 over 8 kb with a $5'$ promoter and three exons coding for the hydrophobic signal peptide, the 18 amino-acids of the N-terminal glycopeptide and the 833 bp exon 3. POMC transcription is positively regulated by cortico-trophin releasing hormone (CRH). CRH acts via its G-protein coupled receptor to activate adenylate cyclase, increase intracellular cAMP and stimulate protein kinase-A (Dedic et al., 2018). cAMP and Ca^{2+} are important in the intracellular signaling of corticotrophs. POMC mRNA transcription in corticotrophs is negatively regulated by glucocorticoids.

1.4.2.1.1 ACTH assays

The introduction of immunoassay methodology enabled for the first time the measurement of the low concentrations of ACTH in normal plasma replacing the earlier bioassays that needed large plasma sample volumes. For the first immuno-assays, an antibody was directed to a single epitope such as the ACTH N-terminal region (within ACTH 5–18). Those assays were subject to interferences from ACTH fragments and different results were obtained from N- and C-terminal

ACTH-directed antisera. Assays based on single antibodies were thus replaced. The two-site **immunometric assays** (IRMA) now in use rely on two antibodies binding two different epitopes of the ACTH peptide (Findling et al., 1990). Specificity in relation to ACTH fragments is improved considerably with IRMA. The development of IRMA improved sensitivity enabling measurements of low normal ACTH concentrations without the need to extract large volumes of plasma. In general, antibodies have been raised to N and C-terminal regions of ACTH in order to ensure that ACTH 1–39 is detected rather than fragments derived from ACTH, such as α-MSH and CLIP. An IRMA employs excess antibody in excess reagent or noncompetitive assay formats which overcame the problem of requiring a high-affinity antibody needed for an RIA. Loss of specificity may arise from hemolyzed blood samples, which can interfere in the assay. Erroneous results of suppressed endogenous ACTH concentrations following synthetic $ACTH_{1-24}$ infusion have been reported when measured by IRMA, attributed to binding of synthetic $ACTH_{1-24}$ to the N-terminal capture antibody but not the C-terminal antibody without forming a detectable sandwich complex. Therefore, $ACTH_{1-24}$ negatively interferes with $ACTH_{1-39}$ binding when antibody concentrations become limited. Comprehensive data on negative interference is not provided in the kit inserts produced by manufacturers. Erroneously low "ACTH" concentrations may therefore be reported in conditions where ACTH precursors are high, such as in the ectopic ACTH syndrome, Nelson's syndrome and patients with large invasive pituitary macro-adenomas. It is important to know the degree of cross-reactivity of precursors, although few of the commercially available assays have this information. Ideally information on assay recognition of αMSH and CLIP, pro-ACTH and POMC as a minimum should be provided.

High concentrations of ACTH precursors can also cause a "hook effect" in some assays, resulting in a falsely low ACTH concentration. At very high analyte concentrations, capture and labeled antibodies become limiting, leading to competition between analyte and analyte-labeled antibody for free solid-phase capture antibodies. The problem can be overcome by the sequential addition of reagents using excess capture and labeled antibody or by assaying samples at two dilutions.

The quantification of ACTH is now routinely performed in clinical laboratories, with nonradioisotopic IRMA methods. There is no generally accepted human international reference preparation for ACTH. The assays depend on a limited selection of human purified and synthetic preparations. There are significant differences in IRMA performance between the synthetic preparations and purified biological standards. Materials from The National Institute of Biological Standards and Control (NIBSC) and the National Hormone and Peptide Program, United States

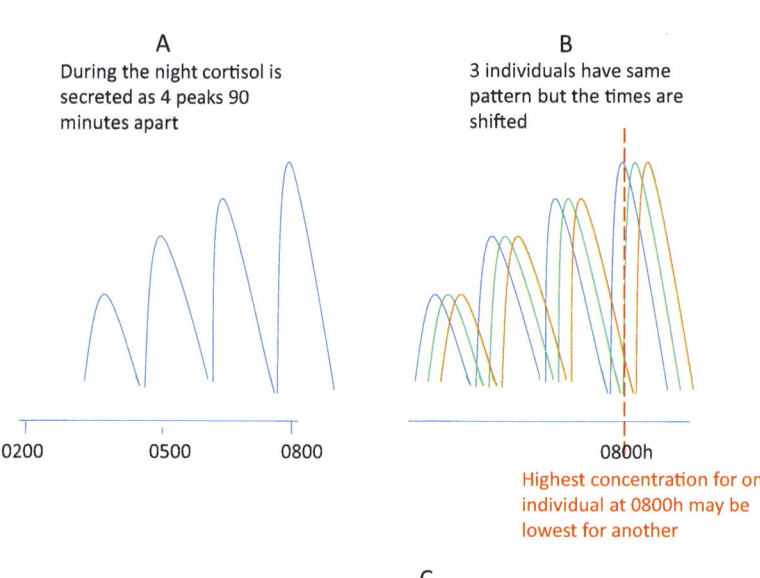

A

During the night cortisol is secreted as 4 peaks 90 minutes apart

B

3 individuals have same pattern but the times are shifted

0200   0500   0800

0800h

Highest concentration for one individual at 0800h may be lowest for another

FIG. **1.4.2** Ultradian rhythm of cortisol (A) Schematic of cortisol pulses in night of one subject, increasing amplitude and 90 min intervals. (B) Cortisol pulses in night for several individuals not in phase. At any one time point, one individual may be at a peak concentration whereas another may be at the trough and others may be between. (C) The range of cortisol concentrations in a group increases throughout the night. *(Author original.)*

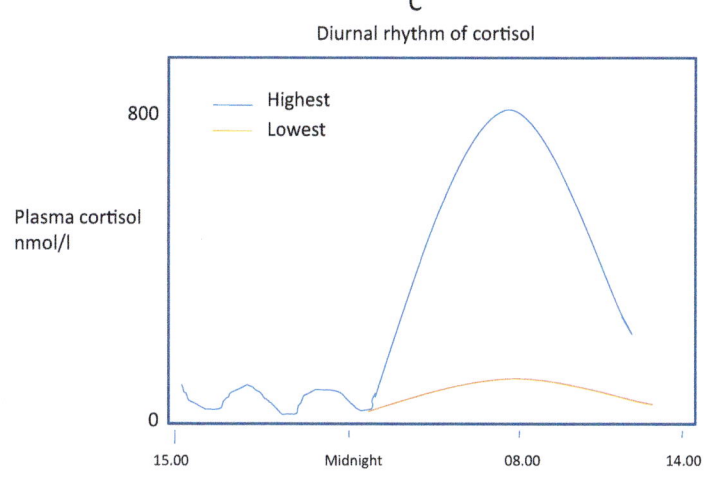

C

Diurnal rhythm of cortisol

— Highest
— Lowest

800

Plasma cortisol nmol/l

0

15.00   Midnight   08.00   14.00

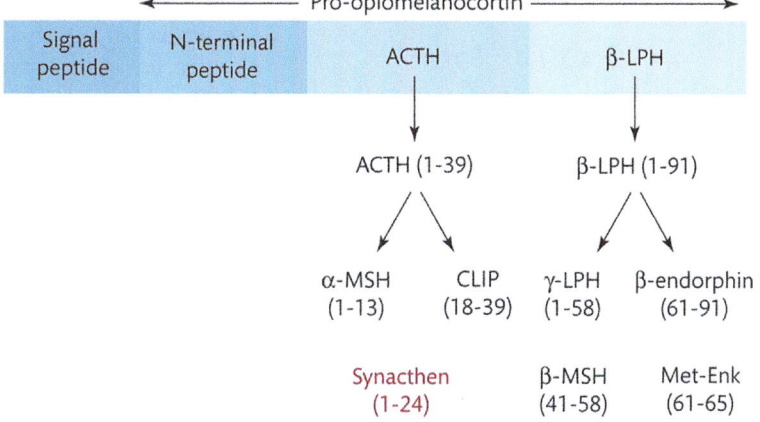

Pro-opiomelanocortin

FIG. 1.4.3 Schematic diagram of POMC that encodes multiple peptides through cleavage by prohormone convertases. *ACTH*, adrenocorticotrophic hormone; *CLIP*, corticotrophin like intermediate peptide; *JP*, joining peptide; *MSH*, melanocyte stimulating hormone; *RSP*, regulated secretory protein. *(Author original.)*

(NHPP) have been assigned a biological potency (in terms of international units, IU) after calibration against the third international standard for ACTH (of porcine origin, known to contain impurities). The World Health Organization (WHO) establishes international standards (IS) and reference preparations (IRP) for substances of biological or synthetic origin that cannot be adequately characterized by chemical means alone. An IS has an IU of activity assigned, generally on the basis of an international collaborative study. The IU values assigned to the NIBSC (12.4 IU per 50 mg) and NHPP (4.7 IU per 50 mg) preparations differ significantly such that the NIBSC standard is approximately three-fold higher. The immunoreactive potencies of the two standards in the ACTH IRMA are similar. Until an IRP for human ACTH is available it is essential that each laboratory measuring ACTH defines the standard used in its assay system and derives a normal reference range for the assay based on this preparation.

1.4.2.1.2 Corticotropin releasing hormone (CRH)

CRH is a 41 amino-acid neuropeptide derived from a 196-amino acid prohormone. CRH is probably involved in behavioral, autonomic and hormonal stress responses. On histology of the pituitary and hypothalamus, CRH immunoreactivity is mainly found in the paraventricular (PV) nuclei of the hypothalamus, often co-localized with arginine vasopressin.

CRH binds to a seven-transmembrane domain receptor, which is coupled to adenylate cyclase via $G_{s\alpha}$, stimulating cAMP synthesis and PK-A activity (Dedic et al., 2018). Two CRH receptor (CRH-R) genes have been identified so far in humans. **CRH-R1** mediates the action of CRH at the corticotroph by binding to both CRH and urocortin I. **CRH-R2** also binds to both CRH and urocortin, but with over 20-fold higher affinity for urocortin. CRH secretion is regulated by neurotransmitters and cytokines that include acetylcholine, noradrenaline, histamine, serotonin, interleukin-1β and tumor necrosis factor (TNF) that increase hypothalamic CRH expression. Gamma-aminobutyric acid (GABA) is inhibitory. CRH stimulates POMC transcription and ACTH biogenesis, CRH also stimulates the release of ACTH, in a biphasic response with the fast release of stored ACTH, and a slower and sustained release of newly synthesized ACTH.

1.4.2.1.3 Rhythms

If blood samples are taken at 10 min intervals, ACTH concentrations fluctuate, reflecting secretion in about 40 pulses over 24 h (Fig. 1.4.4). There is then for each pulse, a 15–20 min delay in the secretion of cortisol. Activity is greatest between 0200 and 1000 h with peaks at 90 min intervals of increasing amplitude (Lightman et al., 2020; Oster et al., 2017). The pattern is reproduced between individuals but not synchronized, giving the widely published diurnal pattern of plasma cortisol concentrations highest at 0800 h and lowest at midnight, such that at 0800 h plasma cortisol concentrations in individuals vary from 200 to 800 nmol/L and at midnight cortisol is less than 200 nmol/L.

The circadian rhythm is mediated via a master oscillator in the supra-chiasmatic nucleus (SCN). Arginine vasopressin (AVP) may therefore be more important than CRH. The adrenal rhythm of cortisol secretion persists after

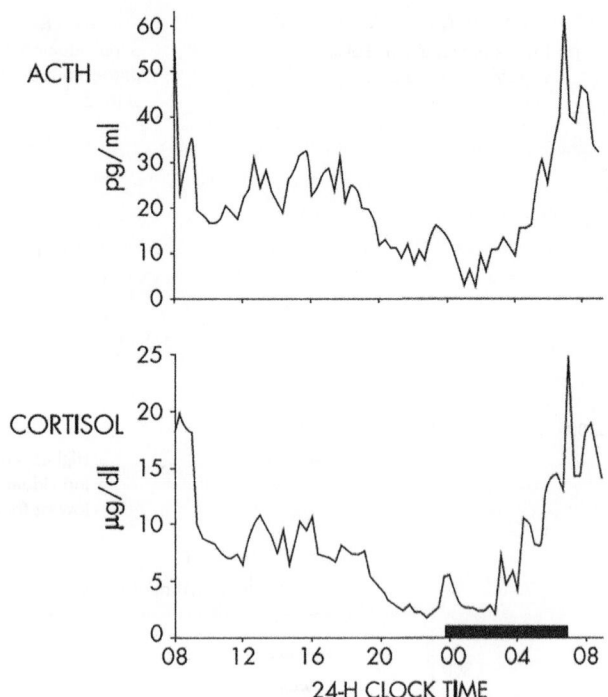

FIG. 1.4.4 Plasma ACTH and cortisol concentrations over 24 h. Representative 24 h profiles in healthy male of ACTH and cortisol at 15 min intervals. During sleep (*black bar*), the sampling catheter was extended for remote sampling. *(From Oster H, Challet E, Ott V, Arvat E, de Kloet ER, Dijk DJ, Lightman S, Vgontzas A, Van Cauter E. The functional and clinical significance of the 24-hour rhythm of circulating glucocorticoids. Endocr Rev 2017;38(1):3–45. Open access Fig. 5 p. 11.)*

hypophysectomy, and light pulses can induce glucocorticoid secretion independent of ACTH secretion. This HPA axis-independent pathway is mediated by the sympathetic nervous system innervation of the adrenals. The adrenal glands also possess an independent circadian oscillator.

1.4.2.1.4 Stress

Stress is the body's reaction to harmful situations, real or perceived. When threatened, the body initiates chemical reactions that prevent injury (Berger et al., 2019; Nicolaides et al., 2017). During the initial stage, there is an immediate increase in catecholamines via activation of the sympathetic preganglionic neurons in the spinal cord, which in turn stimulates adrenal medulla production of catecholamines via splanchnic nerve innervation. The catecholamines released will also collectively affect peripheral effector organs into the classical fight-or-flight response. During the stress response, the heart rate increases, breathing quickens, muscles tighten and blood pressure rises. Adrenalin release from the adrenal medulla is the first response and activation of the HPA follows. Stress increases the release of ACTH mainly by increasing the transcription of CRH and AVP. Stress is mimicked in humans by induction of hypoglycemia during the insulin tolerance test, and by venepuncture. The magnitude of the cortisol rise correlates positively with the severity of the insult. In animals, experimental stress paradigms include hemorrhage, oxidative stress, intraperitoneal hypertonic saline, restraint/immobilization, foot shock and forced swimming. During acute stress, an immediate activation of the autonomic nervous system takes place, followed by a delayed response via the HPA axis-mediated release of glucocorticoids. The delayed response of stress involves activation of the HPA axis, leading to an increase in glucocorticoid level, which in turn can terminate the effects of the sympathetic response together with the reflex parasympathetic activation.

1.4.2.1.5 HPA axis

Glucocorticoid feedback occurs at the pituitary, at the hypothalamus, and most importantly, at the hippocampus, which contains the highest concentration of glucocorticoid receptors in the central nervous system. This feedback is mediated through inhibition of CRH and AVP synthesis and release in the PVN, inhibition of POMC transcription and inhibition of ACTH release induced by the fall in CRH and AVP.

Plasma cortisol concentration is well known to rise after eating. Carbohydrates lead to equal stimulation of two mechanisms:

(i) by direct stimulation of the HPA axis and
(ii) regeneration of cortisol from cortisone by stimulation of 11-hydroxysteroid dehydrogenase type 1 (11βHSD1).

Fat and protein lead to greater stimulation of the HPA axis compared with 11βHSD1. Gut hormones (incretins) are the most likely candidates that stimulate the HPA axis after eating. Among these peptides released in response to food, glucagon-like peptide-1_{7-36} (GLP-1_{7-36}) has been shown to stimulate cortisol and ACTH secretion, possibly through a direct effect on the hypothalamus and/or pituitary (Stimson et al., 2014). Gastric inhibitory peptide (GIP), has been proven to stimulate cortisol secretion only in Cushing's syndrome where the bilateral adrenal hyperplastic glands have ectopic GIP receptors. Ghrelin has been shown to increase cortisol secretion when given in infusion, but ghrelin is suppressed after eating, making it an unlikely mediator of the postprandial cortisol response.

1.4.2.1.6 ACTH receptor

ACTH exerts its effects on the adrenals via a highly selective receptor, **MC2**, a member of the melanocortin (MC) receptor superfamily of type 1 G protein-coupled receptors which are seven transmembrane proteins (Novoselova et al., 2018, 2019). The family includes five members, each having characteristic size, tissue distribution and biological significance. The MC system and its

receptors affect many physiological processes including skin pigmentation, glucocorticoid production, food intake and energy balance. The MC2 receptor is a 297 amino acid, G-protein coupled receptor. In humans, it maps to chromosome 18p11.2. The MC2 receptor is localized in all three zones of the adrenal cortex. The zona fasciculata has many more binding sites per cell of higher affinity than the other zones. Activation of the MC2 receptor initiates a cascade of events at several steps in adrenal cortisol production. **The MC2 receptor effects** includes the cAMP-dependent transcription factors **CREM** (cAMP responsive element modulator) and **CREB** (cAMP responsive element binding protein) that result in transcriptional activation of steroidogenic enzymes, cell proliferation and differentiation. Activation of the MC2 receptor leads to **stimulation of Fos and Jun transcription**. The Fos gene family consists of four members, c-Fos, FosB, Fra1 and Fra2, while the Jun family consists of three members, c-Jun, JunB and JunD. These proteins form hetero- or homo-dimers inducing transcription through binding to AP1-binding sites. Activation of AP1-dependent transcriptions leads to the production of several pro-mitotic proteins while its inhibition results in a blockade of cell cycle of the G1 to S phase transition.

Binding of ACTH to the MC2 receptor triggers the G-proteins to increase cAMP formation and activate PKA and PKC (Fig. 1.4.5). Several intermediate molecules are involved including kinases and transcription factors that orchestrate the ACTH actions on adrenal cells through increased expression of StAR and hormone sensitive lipase.

ACTH bound to the MC2 receptor activates MAP Kinases (ERK1 and ERK2) that are important in ACTH-triggered mitogenic effects. In normal adrenal cortical cells, MC2 signals lead to activation of the Stress Activated Protein Kinase (SAPK) JNK (c-Jun NH$_2$-terminal kinase).

Multiple transcription initiation sites are likely in human adrenal cortical cells. The MC2 receptor gene has one untranslated exon (exon one), an 18 kb intron, and the coding exon (exon two). An alternate exon 1 (exon1f) is transcribed in adipose tissue but not in the adrenals with a different promoter region from that in the adrenal, conferring tissue specificity. The MC2 receptor promoter contains binding sites for several transcription factors, that are nuclear proteins modifying the expression of genes by binding to specific DNA sequences usually located upstream of gene promoters. Phosphorylation of a transcription factor results in its activation and modulation of the transcriptional activity of a promoter containing response elements for the specific factor.

The MC2 receptor depends, for its trafficking to the cell surface, on a small, single transmembrane domain protein the **MC2 accessory protein (MRAP)** (Clark and Chan, 2019). Studies of certain patients with apparent ACTH resistance and other in vitro studies have shown that the MRAP protein is necessary not only for the trafficking of

FIG. 1.4.5 ACTH binds to its G-protein coupled receptor MCR2 resulting in a Gs protein activation that activates adenyl cyclase resulting in enhanced cAMP production. The binding of cAMP to protein kinase A (PKA) allows the release of subunits that catalyze phosphorylation of several targets including the transcription factor CREB (cAMP response element binding protein) to stimulate cAMP-dependent gene transcription. cAMP degradation by phosphodiesterases act as negative regulators of the pathway. Steroidogenesis is activated by increased expression of steroidogenic acute regulatory (StAR) and hormone sensitive lipase. MCR2 is trafficked to the membrane by melanocortin 2 receptor-associated protein (MRAP). *(Modified from Bonnet-Serrano F, Bertherat J. Genetics of tumors of the adrenal cortex. Endocr Relat Cancer. 2018;25(3):R131–R152, Fig. 1, R 134.)*

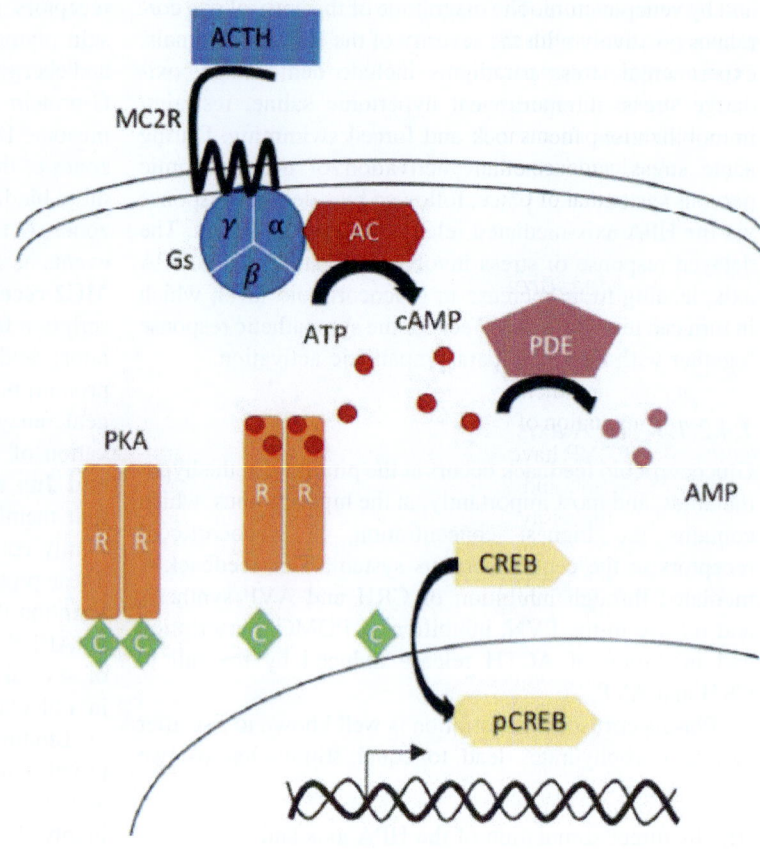

the receptor, but also for conformational changes in MC2 necessary for the binding of the ACTH ligand. The MRAP gene is mapped in human chromosome 21. Two isoforms have been identified, each conferring a different affinity of the MC2 receptor toward ACTH, thus explaining the observed two subpopulations of MC2 receptor in terms of affinity toward the ACTH (Rouault et al., 2017). The precise mechanisms in ACTH regulation of adrenal steroidogenesis are still unclear.

1.4.2.1.7 Arginine-vasopressin (AVP)

AVP was the first hypothalamic factor that was recognized to possess ACTH-releasing properties even before the characterization of CRH. The neurohypophyseal nona-peptide AVP has an important role in the regulation of ACTH secretion in addition to the antidiuretic and pressor actions. AVP is both a major secretagogue of ACTH and a potentiator of the ACTH-releasing activity of corticotropin-releasing hormone (CRH). The pituitary vasopressin receptor has been designated as V3 (or V1b) receptor, and is distinct from two other receptors (V1 and V3).

ACTH and cortisol responses to the administration of AVP (usually lysine vasopressin, LVP), have been documented in healthy subjects and patients with pituitary disorders. LVP and AVP stimulate the secretion of ACTH and cortisol in patients with Cushings disease (CD) and have a synergistic effect with CRH. LVP is not now used in the diagnostic workup of ACTH-dependent Cushings syndrome (CS) as CRH has replaced it with better discrimination between CD and ectopic ACTH syndrome and with less side effects compared with LVP (nausea, abdominal pain and flushing) (see Chapter 2.7—Endocrine tests). AVP is synthesized from precursor called preprovasopressin, which is broken down into three peptides: AVP, neurophysin II and copeptin. The latter is the 39 amino acids, C-terminal part of pro-AVP and has been easier to measure in blood samples because of its small size and simple structure compared with neurophysin which has seven disulfide bridges. Mean concentrations of copeptin decline from 8 to 4 pmol/- between 20 and 45 years of age.

1.4.2.1.8 Desmopressin (1-deamino-S-D-arginine vasopressin, DDAVP)

A synthetic analog of AVP, DDAVP is now used in the investigation of the HPA axis and more specifically for the workup of CS. In general, administration of desmopressin is well tolerated with only minimal and transient side effects, such as small increases in blood pressure and heart rate, slight head heaviness, nausea, flushing and cold sensation. Water intoxication may be a risk, particularly in children and in patients with congestive heart failure on a high fluid intake so care must be taken to advise patients to restrict fluids on the day of the test. Overall, the incidence of adverse reactions is considered low, comparable to that of CRH and much less than those of LVP.

1.4.2.1.9 Measurement of AVP and copeptin

AVP has important osmo-regulatory and fluid regulatory functions, and its biochemical detection has implications for diagnosis and therapy of disorders related to osmoregulation and regulation of fluid levels. Competitive radioimmunoassays for AVP have been available since the 1970s, and several extraction and assay procedures, that are critical to success with the assay, have been reported, including competitive ELISAs. There are pitfalls with immunoassays for AVP (Leng and Sabatier, 2016) and in general with ELISA for small peptides (Smith et al., 1993) The combination of high sensitivity, small sample volume and acceptable turn-around time that is required for routine clinical use, together with an appropriate physiological detection limit, has proved challenging, and assays for AVP that meet these criteria are not yet commercially available. AVP assays may require more than 1 mL of plasma or serum and generally take 48 h to produce a result. Preanalytical factors can complicate the detection of AVP in the circulation. Extraction or concentration of plasma AVP is generally required to bring the sample within the range of detection of the assay. A large amount of circulating AVP is bound to platelets expressing vasopressin V1 receptors. Incomplete removal of platelets from plasma or long-term storage of unprocessed blood leads to falsely elevated or varying measurements of free plasma AVP. The short in vivo half-life of AVP (30 min) is a challenge to detection although AVP in plasma may be stable for 48 h at 4°C. A sensitive method for AVP using LC-MS/MS was published (Tsukazaki et al., 2016) with results in 22 samples that correlated well with immunoassay.

Copeptin assay kits are available commercially for automated and manual methods and used as an alternative to vasopressin measurements. An immunoluminometric assay and a version for the Brahms KRYPTOR automated platform have similar performance characteristics (Sailer et al., 2021). Copeptin assays can be performed with as little as 50 µL of sample, with no extraction step, and give a result in under 2.5 h (Sailer et al., 2021). Copeptin assays can detect the peptide in the plasma of healthy volunteers, regardless of osmolality. Copeptin is highly stable ex vivo, and samples can be stored for ≥7 days at room temperature or ≥14 days at 4°C (Balanescu et al., 2011). Assays for copeptin give variable results in normal subjects and in some cases poor discrimination, for example, in patients with subarachnoid hemorrhage.

1.4.2.2 Regulation of aldosterone production

Many factors are involved in the regulation of aldosterone production (summarized in Fig. 1.4.6). Aldosterone synthesis is mainly controlled by changes of the membrane potential in the adrenal cortex and of calcium homeostasis (MacKenzie et al., 2019; Bollag, 2014). The activities and expression of StAR and CYP11B2 are stimulated at higher calcium concentrations in the mitochondria. The cytosolic concentrations of adrenal glomerulosa cells is about 100–200 nmol, whereas the extracellular calcium concentration is 1–2 mmol.

The sodium/calcium exchanger (NCX) and plasma calcium ATPases (PMCA) export calcium to maintain the resting low calcium concentration (Yang et al., 2018). Calcium influx is controlled by the store operated channel (SOC) and two voltage activated calcium channels (CaV) that are activated by depolarization of the cell membrane. The membrane potential and cytosolic calcium concentrations depend on the actions of potassium channels (TASK or KCN), calcium channels (CaV), sodium/potassium exchangers (ATP1A1), sodium/calcium exchanger (NCX), sodium/hydrogen exchanger (NHE) and plasma membrane calcium ATPases (PMCA). T type CaV is partially open at resting potential but the L-type CaV is closed at resting potential and opened at higher levels of membrane depolarization. The specific contributions of these pathways to calcium signaling in glomerulosa cells is largely unknown. More detail can be found in the 2014 review by Bollag. Recurrent somatic mutations in Primary Aldosteronism affect potassium and calcium channels (Seccia et al., 2018). This is discussed further in Chapter 3.3.1.

1.4.2.2.1 Angiotensin II

After binding to the angiotensin II type 1 (AT1) receptor subtype in cells of zona glomerulosa, Ang II activates phosphoinositide-specific phospholipase C, which hydrolyzes phosphatidylinositol 4,5-bisphosphate, to generate the secondary messengers inositol 1,4,5-trisphosphate (IP3) and diacylglycerol. IP3 triggers aldosterone secretion by eliciting a transient increase in cytosolic calcium (Ca^{2+}) concentration and activating Ca^{2+}/calmodulin-dependent protein kinases (CaMKs). Diacylglycerol stimulates protein kinase C, which regulates StAR levels via cAMP response element binding (CREB) and activator protein-1 mitochondrial production of NADPH, activated transcription factors CREB, NURR1 and NGF1B, activation of StAR by phosphorylation and expression of CYP11B2 through cAMP-dependent response elements in the 5' region of *CYP11B2* gene. A low sodium diet leads to upregulation of AT1R levels which affects a physiological response to increase aldosterone production. Activation of the renin-angiotensin-aldosterone system by low Na^+ intake induces expression of CYP11B2 without affecting CYP11B1,

Regulation of aldosterone secretion

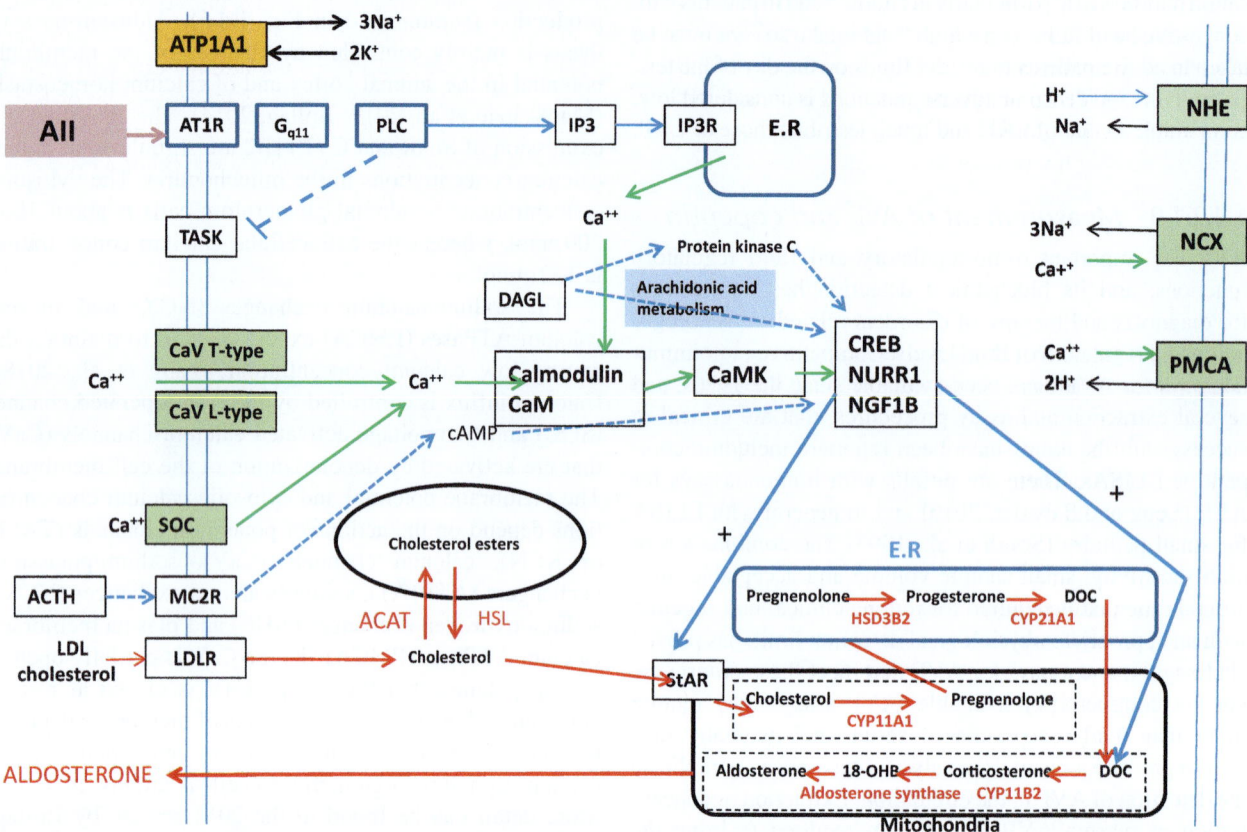

FIG. 1.4.6 Regulation of aldosterone synthesis by angiotensin II through calmodulin (Cam) and calcium signaling pathways. Angiotensin II binds to AT1R, a G-protein coupled through Gq/11 to phospholipase C which hydrolyzes phosphatidylinositol-bisphosphate (PIP2) to yield inositol-triphosphate (IP3) and diacylglycerol (DAG). IP3 binds to receptors on the endoplasmic reticulum to release calcium from intracellular stores. The increase in calcium concentrations in the cytosol activates calmodulin-dependent protein kinases (CamK) which phosphorylates substrates controlling steroidogenic response in mitochondria. Elevated extracellular potassium depolarizes the glomerulosa cell membrane by decreasing the electrochemical gradient for potassium efflux. The small change in membrane potential activates T-type calcium voltage channel (CaV T) and SOC to increase calcium influx and activate CamK. The activities of NCX and PMCA can lower cytosolic levels of calcium ACTH binds to melanocortin receptors (MCR2), a G-protein coupled to adenyl-cyclase. Increased cAMP levels stimulate protein kinase A and activates CamK. PKA and CamK phosphorylate factors to induce StAR and CYP11B2 expression. *ACT*, acyl-CoA cholesterol transferase; *ACTH*, adrenocorticotrophic hormone; *AII*, angiotensin II; *AS*, aldosterone synthase; *AT1R*, angiotensin II receptor type 1; *ATPA1*, sodium/potassium ATPase; *CAM*, calmodulin; *CAMK*, calmodulin kinase; *cAMP*, cyclic AMP; *CaV-L*, voltage-dependent calcium channel type L; *CaV-T*, voltage-dependent Ca channel type T; *CREB*, cAMP responsive element binding protein 3; *CYP11A1*, 11β-hydroxylase type 1; *CYP11B2*, 11β-hydroxylase type 2; *CYP21A1*, 21-hydroxylase; *DAGL*, diacylglycerol; *DOC*, 11-deoxycorticosterone; *E.R*, endoplasmic reticulum; $G_{q/11}$, guanine protein subunit alpha 11; *HSD3B2*, 3β-hydroxysteroid dehydrogenase and Δ5-Δ4-isomerase; *HSL*, hormone sensitive lipase; *IP1*, inositol triphosphate; *IP3R*, inositol triphosphate receptor; *LDLR*, low density lipoprotein receptor; *MC2R*, melanocortin 2 receptor; *NCX*, sodium/calcium exchanger; *NGF1B*, nerve growth factor 1B like receptor; *NHE*, sodium/hydrogen exchanger; *NURR1*, nuclear receptor subfamily 4 member 2; *PLC*, phospholipase C; *PMCA*, plasma membrane calcium ATPase; *SOC*, calcium release-activated calcium channel protein; *StAR*, steroidogenic acute regulatory protein; *TASK*, potassium two pore domain channel subfamily K member 3 (KCN3). (*Author original.*)

probably because of the greater expression of AT1 receptors in zona glomerulosa than in zona fasciculata cells.

Ang II regulates both Ca^{2+} influx into the zona glomerulosa cell via voltage-transient T- and long-lasting L-type Ca^{2+} channels and intracytoplasmic Ca^{2+} movement, thus increasing cytosolic Ca^{2+} concentrations, leading to activation of CaMK II complexes. CaMK II complexes mediate

the transcription of nuclear receptor–related 1 protein (also known as NR4A2), activating transcription factors 1 and 2 and CREB, which initiate transcription of CYP11B2. Thus, activation of the AT1 receptor stimulates both the early and late regulatory steps of aldosterone synthesis, which include phosphorylation of StAR and expression of the steroidogenic enzymes, respectively.

The major role for Ca^{2+} in the regulation of aldosterone biosynthesis is unambiguously supported by the blunted Ang II–induced aldosterone secretion after inhibition of Ca^{2+} influx with Ca^{2+} channel antagonists. Chronic Ang II stimulation maintains aldosterone synthesis not only through increased CYP11B2 transcription but also by increasing cholesterol uptake by glomerulosa cells from high-, low- and very low-density lipoproteins and increasing StAR protein, necessary for cholesterol transport into the mitochondria.

Angiotensin II depolarizes the cell membrane and inhibits the TASK potassium channels that maintain resting potassium concentrations. AII also inhibits the NA/K ATPase (ATP1A1). Depolarization of the membrane also activates voltage-dependent calcium channels (CaV).

Angiotensin II can also bind to angiotensin II receptor type 2 (AT2R). This protein has 34% sequence homology to AT1R and is expressed widely in fetal tissues but then restricted in the adult to brain, adrenal, heart, kidney, myometrium, skin and ovary. Expression has been increased in pathological conditions such as vascular injury, myocardial infarction, renal failure and brain ischemia. The mechanisms of action are not clear but Gi protein coupling, activation of serine/threonine phosphatase PP2A, phosphotyrosine phosphatase, phospholipase, release of bradykinin and nitric oxide have been reported. Cyclic GMP mediates actions of nitric oxide leading to vasodilatation and natriuresis.

1.4.2.2.2 Potassium

In addition to the renin-angiotensin system which is the major regulatory mechanism, serum potassium concentrations and to a lesser extent adrenocorticotrophic hormone (ACTH) provide transient control mechanisms. The plasma potassium concentrations also influence aldosterone synthesis, a small increase in extracellular K^+, when potassium efflux is prevented from efflux through the G-protein activated inward rectifier potassium channel 4 (GIRK4), causes membrane depolarization increasing intracellular calcium concentrations which increases expression of genes for aldosterone synthesis. Conversely hypokalemia inhibits aldosterone synthesis. Potassium acts through several signaling steps (TASK, ATP1A1).

Extracellular K^+ is a potent regulator of aldosterone synthesis. K^+ channels, mostly the leak K^+ channels TASK-1, TASK-2 and TASK-3 (coded by Kcnk3, KcnK5 and Kcnk9 genes), the tandem of p domains in a weakly inward rectifying K^+ channel (TWIK)-related acid sensitive K^+ channel and the G protein inward rectifying potassium channel Kir3.4 (KCNJ5), maintain adrenocortical cells hyperpolarized under resting conditions. The two-pore domain channels TASK-1 and TASK-2 are highly expressed in the normal human adrenal cortex and generate background K^+ currents, blunting aldosterone production. Inhibition of TASK channels by Ang II results in depolarization, and

mice with deletions of TASK-1, TASK-3 or both genes have hyperaldosteronism. Increases in extracellular K^+ depolarize the plasma membrane and activate the voltage-dependent calcium channels, leading to Ca^{2+} influx and the signaling mechanisms described above. Inhibition of Ca^{2+} influx abolishes K^+-stimulated aldosterone secretion. A role of K^+ in zona glomerulosa cell growth is also suggested by the increased thickness of zona glomerulosa found in rats with a chronically high K^+ intake.

1.4.2.2.3 ACTH

ACTH binds to the melanocortin receptor (MC2R) in the cells of the zona glomerulosa to stimulate cyclic AMP production via the heterotrimeric G-protein G_8 which activates protein kinase A and cAMP-dependent protein kinase which increases hormone sensitive lipase as well as increasing StAR and aldosterone synthase gene expression via NURR1 and NGF1B. There is a small and rapid response of aldosterone to physiological levels of ACTH (as synacthen) that requires blood sampling at 0, 10, 20 and 30 min to record (Honour et al., 2008). A normal protocol of samples at 0, 30 and 60 min would miss the response.

1.4.2.2.4 Calcium homeostasis

Aldosterone secretion is influenced by calcium homeostasis. The plasma ionized calcium concentration is regulated to within a very narrow range (1.3–1.5 mmol/L). This is achieved by both the parafollicular cells of the thyroid gland and the parathyroid glands sensing the concentration of calcium ions in the blood flowing through them. When the calcium concentration rises the thyroid gland increases the secretion of calcitonin into the blood. At the same time, the parathyroid glands reduce their rate of parathyroid hormone (PTH) secretion into the blood. The resulting high concentrations of calcitonin in the blood stimulate the skeleton to remove calcium from the blood plasma, and deposit it as bone. The reduced concentrations of PTH inhibit removal of calcium from the skeleton. The low levels of PTH have several other effects including increased loss of calcium in the urine and reduced, loss of phosphate ions via that route. Phosphate ions will therefore be retained in the plasma where they form insoluble salts with calcium ions, thereby removing them from the ionized calcium pool in the blood. The low levels of PTH also inhibit the formation of calcitriol (1,25-dihydroxyvitamin D_3) from cholecalciferol (vitamin D_3) by the kidneys. The reduction in the blood calcitriol concentration acts slowly on the epithelial enterocytes of the duodenum inhibiting calcium absorption from the intestinal contents. The low calcitriol levels also act on bone causing the osteoclasts to release fewer calcium ions into the blood plasma.

When the plasma ionized calcium level is low the opposite happens. Calcitonin secretion is inhibited and

PTH secretion is stimulated, resulting in calcium being removed from bone to correct the plasma calcium level. The high plasma PTH levels inhibit calcium loss in the urine and stimulates the excretion of phosphate ions by that route. The kidneys secrete calcitriol, which enhances calcium absorption from the intestinal contents into the blood, by stimulating the production of calbindin. The PTH stimulated production of calcitriol also causes calcium to be released from bone into the blood, by the release of RANKL (a cytokine) from the osteoblasts which increases the bone resorptive activity by the osteoclasts although these are relatively slow processes. So regulation of the plasma ionized calcium level primarily involves rapid movements of calcium into or out of the skeleton in the short term. Longer term regulation is achieved by controlling the amount of calcium absorbed from the gut or lost via the feces. Studies in humans have shown that vitamin D decreases activity of the R-A-A system by suppressing expression of the renin gene via a cis-DNA element in renin production. This is independent of calcium metabolism.

1.4.2.2.5 The renin-angiotensin-aldosterone system (RAAS)

In the juxtaglomerular apparatus of the kidney, low plasma sodium concentrations or low renal blood flow stimulates release into the circulation of renin from prorenin. Renin, a proteolytic enzyme, then acts on angiotensinogen to free a 10 amino acid angiotensin I (once considered inert).

Angiotensinogen is synthesized in the liver. A further two amino acids are cleaved, by angiotensin converting enzyme (ACE) at several sites including the lung, from AI to produce angiotensin II (AII). The AII has two actions, firstly as the most potent vasoconstrictor but secondly at the adrenal cortex it binds to the angiotensin receptor (AT1R) to stimulate aldosterone secretion (Balakumar and Jagadeesh, 2014).

Renin is synthesized as the inactive pro-renin which then loses a pro-segment that normally masks the active site and prevents access of the renin substrate, angiotensinogen. The renal juxtaglomerular cells are the only known site of production for renin. A number of extra-renal tissues, including the adrenal glands, ovary, testis, placenta and retina, produce pro-renin. The importance of these extra-renal sites to pro-renin production is clarified by the plasma pro-renin concentrations in anephric individuals, which are around half the values in normal individuals. Pro-renin concentrations in plasma are 10-fold higher than for renin. Pro-renin exists in two different conformations. In nearly all samples of plasma, the prosegment of pro-renin masks the active site and the protein is inactive. In around 2% of plasma pro-renin, the prosegment does not mask the active site and is accessible to angiotensinogen. The open form of prorenin is enzymatically active (renin forms) (Fig. 1.4.7).

Activation of pro-renin occurs in blood samples within 8h at room temperature. If plasma is cooled to between −5°C and 4°C the pro-segment can unfold and be cleaved by plasma proteases. After 12h at 0°C, approximately 5% of prorenin is activated. Incubation of plasma at 22°C for

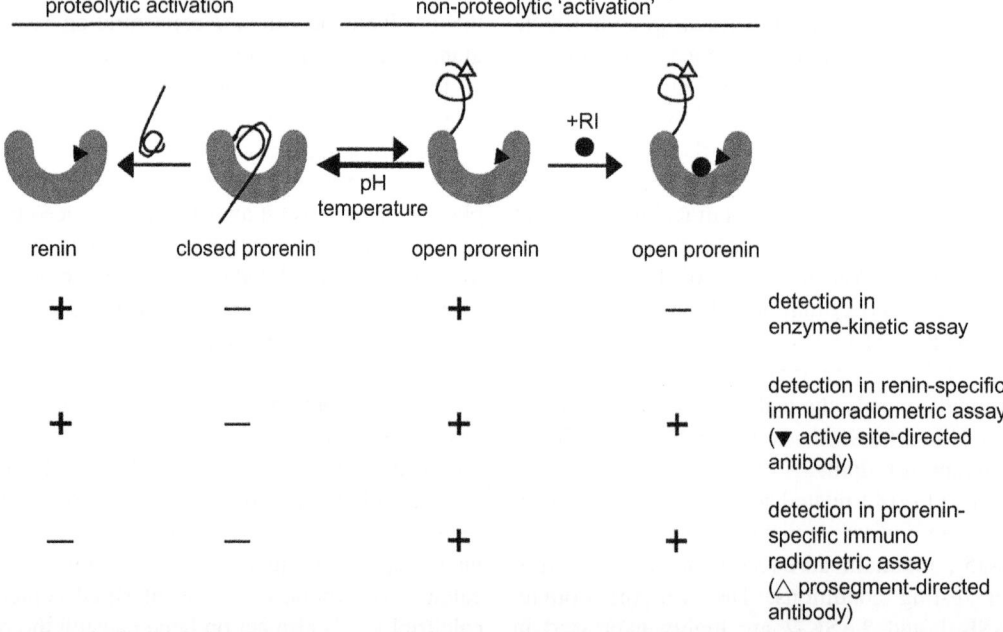

FIG. 1.4.7 Renin is synthesized as the inactive pro-renin which exists in two different conformations. In plasma, the prosegment of prorenin prosegment masks the active site and more than 98% of the protein is inactive. In around 2% of plasma prorenin, the prosegment does not mask the active site and is accessible to angiotensinogen. *(Modified from Danser AH, Deinum J. Renin, prorenin and the putative (pro)renin receptor. Hypertension 2005;46(5): 1069–76. Fig. 3 p. 1073.)*

24h during a renin immunoassay increases conversion of the plasma prorenin to renin. The renin immunoassay may be more susceptible to unfolding of the pro-renin pro-segment than the activity assay. The prorenin may refold during the incubation at 37°C in the activity assay, whereas refolding is less likely to occur because the antibodies rapidly lock the pro-renin in an open conformation.

1.4.2.2.6 The renin-angiotensin II-aldosterone system and angiotensin II fragments

In recent years, our understanding of the RAAS has been extended with knowledge of the generation of angiotensin peptides downstream of AI (Fig. 1.4.8). The [A(1–8)] is a product of ACE action and then A(1–7) through the angiotensin converting enzyme type 2 (ACE2). This enzyme can also convert AI [A(1–10)] to A(1–9).

The A(1–7) binds to a mas receptor (MrgD-R) that leads to inhibition of mitogen activated protein kinase (MAPK), stimulation of cellular phosphatase, inhibition of cyclooxygenase 2 (COX2) and facilitation of nitric oxide release. A(1–7) is implicated in the pathogenesis of cardiovascular diseases.

A(1–8) acts at the AT2R. A peptide, A(1–12) has also been discovered as a carboxyl-terminally extended AI with valine and isoleucine. A(1–12) is present in plasma and tissues including aorta and kidney. A(1–12) is converted to AII by chymase in the heart. The conversion of AI to A(1–12) has been proven in rat aorta experiments. AII can also be converted to AIII [A(2–8)] and AIV [A(3–8)] by aminopeptidases A (APA) and N (APN), respectively. AIV can be further cleaved to A(5–8) by endopeptidase and A(5–7) by carboxypeptidase (CPP). AIV inhibits the

insulin receptor aminopeptidase (IRAP). The AIV/IRAP axis induces the release of nitric oxide and promotes antiinflammatory and antifibrotic effects. Dipeptidyl peptidase III (DPP III) cleaves two amino acids from the N-terminus of AIV and A(3–7) to give A(5–8) and A(5–7) oligopeptides. Angiotensin A is a newly characterized peptide related to AII with Ala instead of Asp in the first position of the sequence generated by aspartate decarboxylase. Alamandine is A(1–7) with alanine instead of aspartate in the first amino acid position that binds to a novel G-protein coupled receptor Mas (MrgD-R). Angioprotectin is like AII with Pro and Glu instead of Asp and Arg. Angioprotectin has high affinity for the mas receptor and has vasorelaxing properties. There is evidence for an intra-renal RASS as a local feed-forward system for augmentation of intra-renal generation or action of RASS that plays a role in pathogenesis of hypertension and renal disease. There are several components such as PRR, Wnt/b-catenin signaling and PEG2/EP4 as positive elements and Klotho, VDR and LXR as negative components. These discoveries open potential for the development of new drugs.

The RAAS is continually being revised and at August 2020 there were five axes identified:

- the classical angiotensinogen-renin-ACE-angiotensin II-AT1R-aldosterone axis
- prorenin receptor (PRR)-mitogen activated protein (MAP)-kinase pathway
- the AIII-APN-AIV-IRAP-AT4 receptor axis
- AII-APA-AIII–AT2R-NO-cGMP axis where ATR2 is the angiotensin II receptor type 2; NO is nitric oxide and cGMP is cyclic GMP
- AI-AII-ACE2/ang(1–7)-mas receptor axis

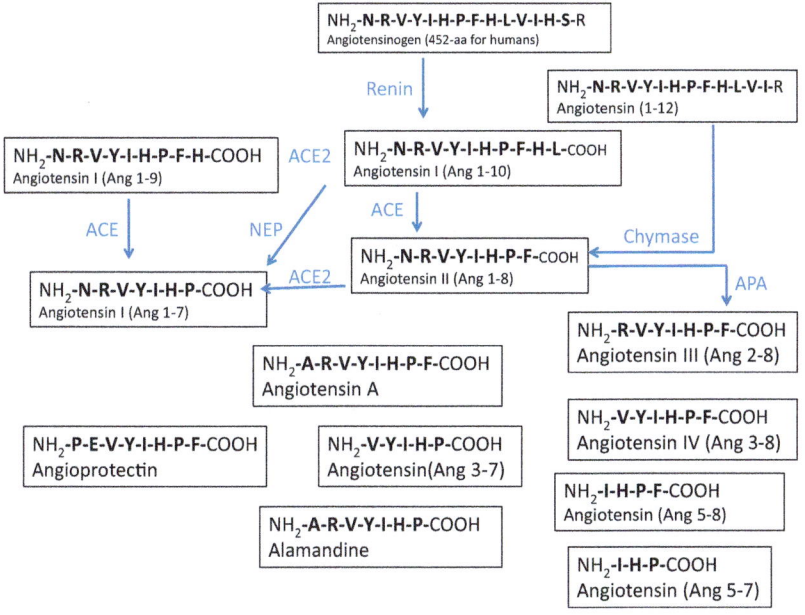

FIG. 1.4.8 New RAAS. Current view of angiotensin metabolism by ACE angiotensin converting enzyme; *APA*, aminopeptidase A; *APN*, aminopeptidase N; *DPP*, dipeptidyl aminopeptidase; *MLDAD*, mononuclear leucocyte derived aspartate decarboxylase; *NEP*, neutral endopeptidase; *NPN*, neprolysin; *PCP*, prolyl carboxypeptidase; *PEP*, prolyl endopeptidase; *TOP*, thimet oligopeptidase. (*Author original*).

The first three axes are powerful vasopressor systems, whereas the others are vasodepressor and cardiorenal protective axes. The effects have been described as Devil/Angel or Yang/Yin of vasoactive systems. The action of angiotensin II through AT2R is independent of aldosterone.

The idea that AII is the only active peptide of the RAAS is outdated because angiotensinogen is the source of several peptides with new biological actions in adrenal, cardiac, renal and brain tissues and our understanding of their roles in pathophysiological states is incomplete.

1.4.2.2.7 Renin assays
Measurement of plasma renin activity (PRA)

PRA refers to the production of Ang I by the enzymatic activity of plasma renin acting on endogenous plasma angiotensinogen. The amount of Ang I produced by a constant amount of renin is linearly related to the angiotensinogen concentration. The PRA is dependent on the angiotensinogen concentration and renin concentration and to interpret the PRA, the laboratory and the clinician needs to be aware of conditions that affect the renin concentration and that affect the substrate concentration. Plasma angiotensinogen concentrations are increased during pregnancy, glucocorticoid excess and estrogen administration and are decreased when angiotensinogen production is reduced in liver disease. Renin concentrations may be sufficiently increased to decrease angiotensinogen concentrations during sodium deficiency and during therapy with angiotensin-converting enzyme inhibitors and with type 1 Ang II receptor blockers. High renin concentrations may cleave a large proportion of plasma angiotensinogen, and low angiotensinogen concentrations occur when high renin concentrations are due to decreased hepatic angiotensinogen production in adrenal insufficiency and heart failure.

In the classic method of PRA measurement, inhibitors of angiotensinase and angiotensin-converting enzyme are added to plasma to prevent degradation of Ang I and its conversion to Ang II during incubation at 37°C. The rate of Ang I production in the PRA assay depends on the pH and the extent of plasma dilution in the assay, and differences in methodology between laboratories make comparisons difficult. The duration of incubation can be adjusted according to the renin concentration and thus the rate of Ang I production. The incubation time can be extended for several hours to allow sufficient Ang I formation to allow precise measurement of low PRA values.

A widely applicable mass spectrometric method for PRA has been described (Owen et al., 2014) and in separate laboratories automated direct renin has been combined with aldosterone by LC-MS/MS (Juutilainen et al., 2014).

Plasma renin concentration (PRC) measured by activity assay

To avoid the influence of variation in plasma angiotensinogen concentration, exogenous angiotensinogen is added to the plasma before incubation. Plasma from nephrectomized sheep is suitable. This addition permits a much higher rate of Ang I production. An advantage of the use of exogenous angiotensinogen is that the PRC can be calibrated against a renin calibrator, such as the International Reference Preparation of human renin, and the concentration can be expressed in international units per liter. The antibody-capture method of activity assay can be used to measure ac-PRC.

Measurement of renin, by immunoassay

The renin immunoassay is a sandwich assay, because the protein being measured is "sandwiched" between two different antibodies: a capture antibody that is immobilized on a bead or other surface, and a detection antibody that may be radioactive or have a chemiluminescent label. The capture antibody binds to a region of the renin molecule that is distant from the active site and the prosegment. The detection antibody for the renin immunoassay binds to a region adjacent to the active site that is exposed when the prosegment is absent (as for renin) or in an open conformation (as for open prorenin) but that is masked when the prosegment is in the closed conformation (Fig. 1.4.7). Sandwich immunoassays offer particular advantages with respect to specificity and limit of detection.

Renin and aldosterone can be measured together on automated immunochemiluminescence analysers (e.g., IDS-i-sys). Immunoassays are being replaced with LC-MS/MS methods for angiotensinogen (Dahabiyeh et al., 2019), the AI generated using immuno-MALDI (Li et al., 2017; Popp et al., 2015; Chappell et al., 2012). PRA and PRC assays give different information, not only for patients who are undergoing renin inhibitor therapy but also for patients who are not receiving such therapy, whereas the PRC assay measures only immunoreactive renin (and open prorenin), Ang I production by the PRA assay is influenced by both renin (and open prorenin) and angiotensinogen concentrations.

1.4.2.2.8 Angiotensin assays

The routine quantification of angiotensin II is performed using antibody based fluorescence assays. However, commercially available monoclonal antiangiotensin II-antibodies are characterized by high cross reactivity because of sequence homologies of different angiotensin peptides. These methodological difficulties may lead to the extremely divergent angiotensin II levels reported in different studies. These vary in healthy subjects in the range of 3–85 pmol. Mass spectrometry-based methods, using stable isotope labeled peptides, allow absolute quantification in low femtomole range and are especially useful when no analyte-free matrix is available. A highly selective, sensitive, accurate and precise method for absolute quantification of endogenous angiotensin II levels in human plasma has been published based on combination of the immuno-affinity purification and mass spectrometric detection by using stable isotope (^{13}C- and ^{15}N-) labeled angiotensin II as an internal standard (Schulz et al., 2014). Angiotensin II, ^{13}C and ^{15}N labeled is available from ClearPoint (AS-64805, AnaSpec Inc., Fremont, California, United States).

Analysis of ANG peptides has been performed using HPLC combined with radioimmunoassay. Liquid chromatography-mass spectrometry (LC-MS) features high selectively and high sensitivity. LC-MS is a useful tool for biological analysis and is widely used for identification and quantification of proteins and peptides. Usually, LC-MS is used to provide initial identification based on molecular weight, LC-MS/MS provide further confirmation via structural specific fragmentation (He et al., 2020). The combination of LC-MS and LC-MS/MS analysis allows sensitive and unambiguous analysis of peptides in complex sample matrices. The collection and storage of samples and the addition of enzyme inhibitors is not fully resolved for the conversions beyond AII formation. Further research is required.

A new concept has been measurement of angiotensin II generation in an in vivo incubation without enzyme inhibitors called **equilibrium angiotensin** (van Rooyen et al., 2016; Guo et al., 2020; Bernstone et al., 2021) and commercial versions of the tests at https://www.attoquant.com/aaa/#AAA.

1.4.2.2.9 Angiotensin receptors

The activation of the angiotensin receptor by ATII inhibits the activity of the ATP1A1 Na/K ATPase and decreases the K permeability of the KCNJ5 channel. Polarization of the plasma membrane opens the voltage gated calcium channel and flow of calcium into the cell increases cytosolic concentrations leading to activation of the calcium-sensitive protein kinase (CAMK), promoting phosphorylation of transcription factors CREB, NURR1, NGF1B and ATF1 which promote transcription of aldosterone synthase (CYP11B2).

Specific cell surface receptors for at least three distinct angiotensin peptides produce distinct cellular signals that regulate system-wide physiological response to RAS. The two well characterized receptors are angiotensin type 1 receptor (AT1 receptor) and type 2 receptor (AT2 receptor). They respond to the octapeptide hormone angiotensin II. The oncogene product MAS is a putative receptor for Ang (1–7). While these are G-protein coupled receptors

(GPCRs), the in vivo angiotensin IV binding sites may be type 2 transmembrane proteins. These four receptors together regulate cardiovascular, hemodynamic, neurological, renal and endothelial functions; as well as cell proliferation, survival, matrix-cell interactions and inflammation. Angiotensin receptors are important therapeutic targets for several diseases. Researchers and pharmaceutical companies are thus focusing on drugs targeting AT1 receptor rather than the AT2 receptor, MAS and AngIV binding sites. AT1 receptor blockers are the cornerstone of current treatment for hypertension, heart failure, renal failure and many types of vascular diseases including atherosclerosis, aortic aneurism and Marfan syndrome.

Most of the known effects of Ang II are mediated through the AT1 receptor, e.g., vasoconstriction, aldosterone and vasopressin release, salt and water retention and sympathetic activation without neglecting the important autocrine and paracrine effects of Ang II on cell proliferation and migration and on extracellular matrix formation. The function of the AT2 receptor has become unraveled over the last few years through sophisticated approaches including gene transfection and deletion. The AT2 receptor is now thought to counterbalance the effect of the AT1 receptor in vitro as well as in vivo. There is an inactivation of MAPK, antiproliferation, promotion of apoptosis, differentiation and regeneration, opening of delayed-rectifier K1 channels and closing of T-type Ca21 channels. The re-expression of the AT2 receptor in various diseases suggests a role of this receptor in pathophysiology.

Recently a new angiotensin receptor type, AT4, has been discovered and characterized that preferentially binds Ang II(3–8), a fragment of angiotensin II, now referred to as Ang IV (Chai et al., 2004). This receptor site is prominent among brain structures concerned with cognitive processing, motor and sensory functions. Specifically, high densities of AT4 sites have been localized in neocortex, hippocampus and amygdala. Major motor structures with high levels of AT4 receptors include basal ganglia, vestibular and reticular nuclei of the hindbrain, motor trigeminal nucleus, cerebellum and ventral horn of the spinal cord. Significant sensory structures include thalamus, lateral olfactory tract and primary sensory neocortex. Peripheral tissues that reveal heavy distributions of AT4 sites are kidney, bladder, heart, spleen, prostate, adrenals and colon. The AT4 receptor appears to be involved in memory acquisition and recall. Like the AT2 receptor, it may also oppose the effect of the AT1 receptor as it regulates renal blood flow, inhibits tubular sodium reabsorption and affects cardiac hypertrophy.

1.4.2.3 Regulation of DHEAS

Little is known of the regulation of DHEAS production. ACTH can stimulate DHEAS. Plasma DHEA-S concentrations however do not exhibit a circadian rhythm because of the much longer half-life of this sulfated steroid. There is indirect evidence that other pituitary hormones may be involved (Parker, 1991). A direct effect of PRL on adrenal steroidogenesis has been suggested from studies of clinical disorders characterized by hyperprolactinemia. PRL receptors have been demonstrated in the adrenal glands of patients with hyperprolactinemia and in women with PRL-secreting tumors there is a correlation between PRL concentration and DHEA-S. The adrenal androgens can then be suppressed with bromocriptine or dopamine. LH-HCG receptors have been demonstrated in zona reticularis and fasciculata. The receptor bearing cells were also positive for steroidogenic enzymes, indicating that the receptors could be coupled to DHEAS secretion. PRL does not seem to contribute to the age-related changes in DHEAS. Women with chronic anovulation have high DHEAS and LH with normal ACTH concentrations. When adrenal cells were examined by immunocytochemistry an LH/hCG receptor was found in zona reticularis cells. Side chain cleavage enzyme was detected but the response of the cells to LH was not tested.

The production of growth hormone is higher at puberty Tanner stage 5 than at stages 2–4. There is also a decline in GH after the second decade of life. These findings are similar to changes in DHEAS before puberty. A number of other endocrine signals have been proposed as co-regulators of adrenal androgen secretion. The prominent compounds are estrogen, epidermal growth factor, angiotensin, lipotropin and β-endorphin.

The effects of estrogen have been studied in adrenarche when high DHEAS did not correlate with prepubertal concentrations of estrogens. In gonadal dysgenesis, the gonadotrophins did not correlate with DHAS concentrations and when patients with gonadal dysgenesis were treated with estrogens there was no increase in DHEAS.

High DHEAS concentrations are found in salt wasting syndromes and in some cases correlate with high angiotensin II. The production of epidermal growth factor increases in childhood with a further rise until the second decade of life before a decline in older subjects. This pattern is again similar to DHEAS. Further research is needed to understand the regulation and role of DHEAS.

1.4.3　Regulation of the reproductive organs

The male reproductive tract is functional for life from puberty but in the female the ovaries are controlled in a monthly cycle that terminates when the ovaries are depleted of oocytes by the fifth decade of life when the menopause starts and menstruation ceases, whereas the male continues to be fertile. As with the regulation of the adrenal cortex, the reproductive hormones are regulated in a hypothalamic-pituitary-gonadal axis (HPG) (Fig. 1.4.9).

1.4.3.1　Gonadotrophins and gonadotrophin receptors

The pituitary gonadotropin hormones LH and FSH, as well as the placentally derived LH paralog hCG, are all large, heterodimeric glycoproteins that consist of α and β subunits (Fig. 1.4.10). Along with the pituitary-derived TSH, this group of hormones is termed the glycoprotein hormones (Anderson et al., 2018; Caltabiano et al., 2008). The glycoprotein hormone heterodimers are composed of a 92-amino acid (aa) α-subunit encoded by a single gene common to all three hormones and a hormone-specific β-subunit (LH: 121 aa; hCG: 145 aa; FSH: 111 aa and TSH: 112 aa). LH and hCG differ only by a C-terminal, 24-aa peptide [carboxyterminal peptide (CTP)] extension in hCG, that confers an increased half-life to hCG.

The α and β subunits include 10 and 12 cysteine residues, respectively. The intra subunit tertiary structure is governed by disulfide bonds. A cysteine loop of the β-subunit encircles loop two of the a-subunit. The α and β subunits contribute to receptor binding. The mature gonadotrophins are heavily glycosylated. The common α subunit has two glycosylation sites while the β subunits of LH and FSH are glycosylated at N30 or N7 and N24.

The secretion of LH and FSH are sexually dimorphic and in females the circulating hormone levels vary throughout the menstrual cycle. GnRH is released in a pulsatile manner, each pulse results in a concomitant pulse of LH secretion. Pulses of FSH follow GnRH. High frequency pulses of GnRH favor LH secretion, whereas low frequency pulses promote FSH secretion (Cui et al., 2018).

The glycoprotein hormone receptor subfamily of the G-protein coupled receptors have the prototypical seven transmembrane domains with three intracellular loops and three extracellular loops. The C-terminus is intracellular while the N-terminal is extra-cellular. Ligand activation promotes GTP displacement and activated Gα subunits activate adenyl cyclase to release messenger cAMP. Inositol phosphate calcium ions, MAP kinase and phosphatidylinositol kinase signaling effects have also been reported (Fig. 1.4.11). The gonadotrophin receptors have very large extracellular domains followed by several leucine reach repeats which binds the gonadotrophin.

1.4.3.1.1　Gonadotrophin assays

Assay methods for LH Luteinizing hormone is usually by immunoassay. All current commercial LH immunoassays are based on the sandwich principle using two monoclonal or in some cases one monoclonal capture antibody and a polyclonal detector antibody. The choice of reagents is critical because some antibodies do not detect a fairly common variant of LH that has two point mutations in the gene for LHβ. Some monoclonal antibodies do not recognize

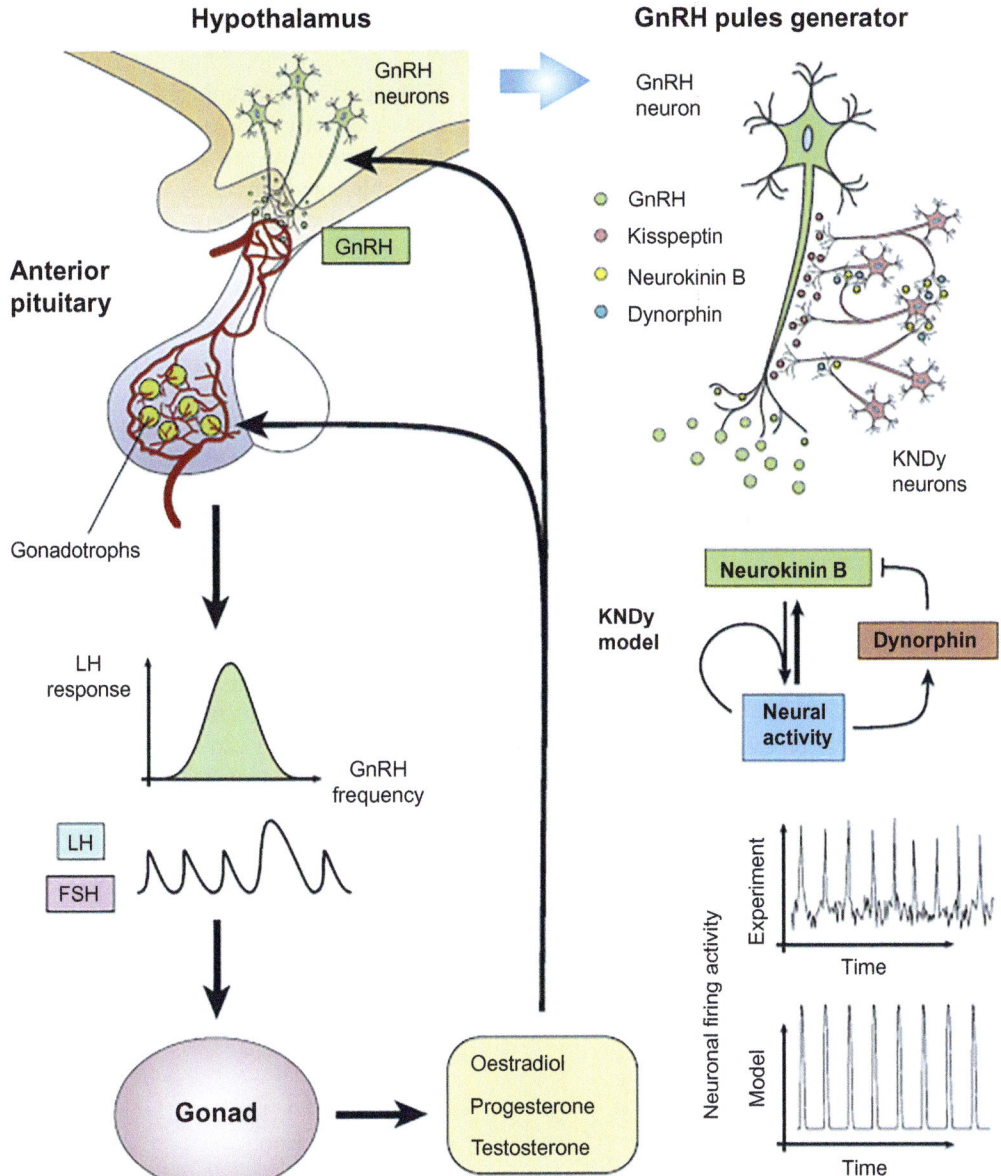

FIG. 1.4.9 Hypothalamic-pituitary-gonadal axis in male and female. Gonadotrophin releasing hormones (GnRH) secreted by GnRH neurons in the hypothalamus stimulate the release of the gonadotrophic hormones luteinizing hormone (LH) and follicle stimulating hormone (FSH) from the pituitary. Gonadotrophins act on the gonads, initiating processes involved in gametogenesis and ovulation and triggering the release of sex steroids (estradiol, testosterone and progesterone) that feedback on the pituitary gland to modulate GnRH and LH/FSH secretion dynamics. Hypothalamic neurons co-expressing kisspeptin, neurokinin and dynorphin control the pulsatile dynamics of GnRH secretion. *(From Zavala E, Wedgwood KCA, Voliotis M, Tabak J, Spiga F, Lightman SL, Tsaneva-Atanasova K. Mathematical modelling of endocrine systems. Trends Endocrinol Metab 2019;30(4):244–57. Open access Fig. 3 p. 252.)*

the LH variant while other antibodies underestimate it. This explains the considerable variation in LH results obtained in earlier used assays. Antibodies detecting wild-type and variant LH equally are available however the data from quality assessment programs indicate there are two-fold differences in the results obtained by some LH assays. Like hCG, LH is excreted into urine at concentrations similar to those in serum. A substantial part of LH is degraded into LHbcf (LH beta core fragment) during passage through

the kidneys. This fragment is detected by some but not by other immunoassays. In menstruating women, the peak of LH immunoreactivity in urine often appears 2–3 days after the LH peak in serum and this peak is partly caused by delayed excretion of LHbcf (O'Connor et al., 1998).

Immunoassays are widely used for clinical determination of FSH for diagnosis and in physiological studies because they are rapid, readily available, relatively cheap and sensitive. Immunoassays are generally considered to

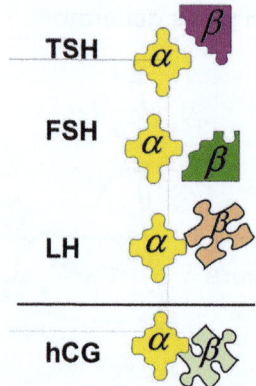

FIG. 1.4.10 Glycoprotein hormones (GPHs) are the most complex molecules with hormonal activity. They include three pituitary hormones, the gonadotropins follicle-stimulating hormone (FSH; follitropin) and luteinizing hormone (LH; lutropin) as well as thyroid-stimulating hormone (TSH; thyrotropin). Chorionic gonadotropin (CG) is secreted by the placenta. Each GPH consists of two different glycoprotein subunits, called α and β, that are noncovalently associated. The α-subunits of all GPHs are encoded by a same and unique gene that is expressed in pituitary gonadotroph and thyrotroph cells. The β-subunits are different and specific for each hormone. *(Author original.)*

be more precise than in vivo bioassays although they are assumed to give no information about biological activity. Early immunoassays for FSH were based on the RIA format, which gave robust assays that are now rarely used. Most commercially available assays are now based on sandwiches of monoclonal or monoclonal-polyclonal antibodies with a variety of detection modes and are generally more sensitive and precise than one-site assays. Assays of apparently high specificity are commercially available but there is variability between the results of different assays for

gonadotropins. The reasons for such variations have been ascribed to differences in:

- calibration of different assay kits,
- cross-reactivity between gonadotropin subunits,
- nonlinearity between kit standard and internationally available standards,
- differential recognition of different samples obtained, for example, at different stages of the menstrual cycle and
- varying dose-response characteristics between different assay systems.

The precise details of the methods of calibration and sources and composition of kit and in-house standards are frequently unknown to the user, which makes evaluation of the standardization of calibration and hence comparisons between different immunoassay kits difficult. When an unusual result is found with a patient sample it can be difficult to work out the source of the problem so analysis on a different assay platform may be needed.

The nine main antigenic epitopes of the subunits of FSH have been identified; five on the α-subunit, two on the β- and two dependent on the conformation of the dimer. Although the α-subunits of the glycoprotein hormones have the same amino acid sequences in each of the four hormones, some of the identified epitopes on the FSH α-subunit are different between the different glycoprotein hormone heterodimers (Casarini and Crépieux, 2019). Assay of variable epitopes in a hormone-specific fashion may enable further observations on how structural features of hormones change with different physiological conditions. Antibody responses to epitopes in the region covered by amino acid residues 33–53 of the FSH β-subunit have been found to both enhance and neutralize FSH action.

FIG. 1.4.11 Following GPCR activation by endogenous neurotransmitters both the Gα subunit and the Gβγ complex functionally dissociate from the receptor and go on to stimulate or inhibit a range of intracellular effectors such as cyclic AMP. The Gα activation process involves a loss of bound GDP (inactive form) in exchange for GTP (active form). Signaling is terminated by the hydrolysis of the bound GTP back to GDP by the intrinsic GTPase activity of the Gα subunit.

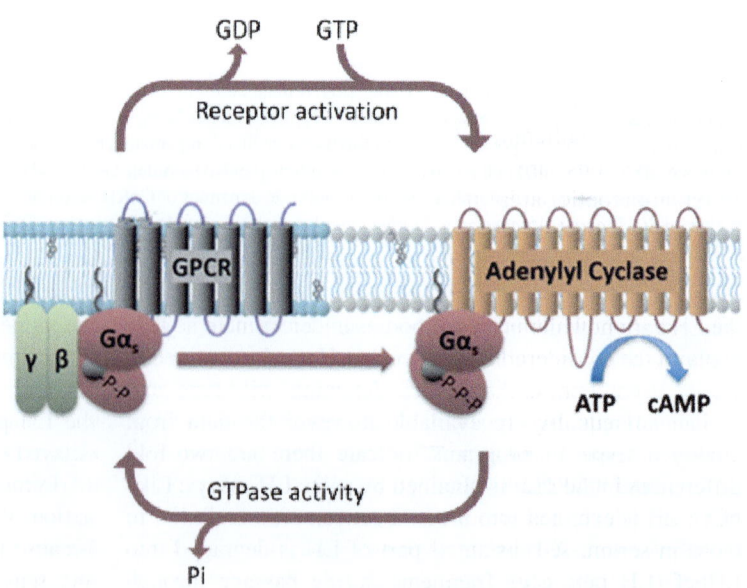

In the diagnosis of clinical conditions, the level of biologically active FSH in the circulation is of importance. However, diagnostic tests are generally made by immunoassay that may recognize FSH molecules that are not biologically active. Conversely, some highly specific assays may discriminate against some forms of FSH. There are reports of genetic variants of FSH that have been associated with delayed sexual development and infertility but these are rare and result in drastic alterations to the FSH molecule. The particular selectivity of any one assay system is seldom known, and it is therefore not possible to define FSH in terms of immunoreactivity. Since FSH levels and isoform composition change drastically throughout life and through menstrual cycles, clinical determinations usually require some additional clinical data such as stage of cycle or repeat measurements to distinguish between different diagnostic possibilities. A combination of quantitative and qualitative assays would be a major advance in clinical utility of gonadotropin determinations. Other complications of the gonadotropin system that have been implicated in disorders of reproduction and that may affect clinical diagnoses based on FSH determinations include mutations in the FSH receptor.

1.4.3.1.2 Control of gonadotrophin secretion

Gonadotropin-releasing hormone (GnRH), also known as luteinizing-hormone (LH) releasing hormone (LHRH), has been known as the master molecule of reproduction in a wide range of vertebrate species. GnRH is a tropic peptide hormone responsible for the release of gonadotropins, LH and follicle stimulating hormone (FSH), from the anterior pituitary, and the released gonadotropins in turn stimulate gamete production and concomitant steroid hormone release. In addition, GnRH neurons send their fibers throughout the brain and play an important role in the control of reproductive behavior. GnRH thus controls different aspects of reproduction and plays a central role in the reproductive axis (Zavala et al., 2019).

The identity of GnRH was first clarified by Nobel Laureates, Roger Guillemin and Andrew V. Schally, in 1977. GnRH1 neurons are generally present in the ventral forebrain-preoptic area (POA)-basal hypothalamus and send neuronal fibers to the median eminence (ME) in mammals which represents its primary role in hypophysiotropic function to stimulate the release of gonadotropins. The extrahypothalamic GnRH1 fibers influence sexual behavior or nonreproductive functions. GnRH2 neurons are exclusively present in the midbrain of most vertebrates.

In mammals, two types of GnRH receptors have been identified (type I and type II) along with their ligands GnRH1 and GnRH2 (Maggi et al., 2016). The type II receptor, which has a specific receptor for GnRH2, is probably expressed as a nonfunctional truncated form because of frame-shift, deletion or internal stop codon.

In the absence of the type II GnRH receptor, the expression of the type I GnRH receptor compensates because it can bind with both GnRH1 and GnRH2. GnRH receptor is expressed in several regions including the hippocampus, amygdala and oculomotor pathways in the brain.

GnRH1 neurons are regulated by many stimulatory and inhibitory factors including gonadal steroids, neurotransmitters, neuropeptides and neurosteroids. Neurotransmitters such as gamma-aminobutyric acid (GABA), noradrenaline, dopamine and β-endorphin are also known to directly act on GnRH1 neurons to alter their activities. GABA-ergic neurons have been shown to form synapses on GnRH1 neurons and GABA stimulates secretion of GnRH1 through activation of GABA receptors expressed on GnRH1 neurons. Noradrenaline also stimulates GnRH1 release through a1 receptors located on GnRH1 neurons. Immunohistochemical studies suggest direct action of β-endorphin and dopamine on GnRH neurons. The GnRH neuronal regulation by environmental factors however seems to be mediated mainly by steroids and other hormonal signals. The regulatory mechanisms of GnRH neurons are important for the understanding of animal reproduction where environmental conditions, such as temperature, food availability, energy storage and social status need to be controlled.

Estrogen and progesterone secreted from the ovary are the regulators of the synthesis and release of GnRH1 and gonadotropins. The changes in plasma estrogen concentrations show a clear relationship with GnRH1 surges during human menstrual cycle (Moenter et al., 2019). Thus the steroid feedback is considered to play an important role for the rhythmic secretion of GnRH1 observed during the estrous cycle. The mechanism of steroid feedback to GnRH1 neurons is still debated. The GnRH1 promoter has been shown to have functional estrogen response elements in mammals, suggesting direct action of estrogen on GnRH1 gene expression. In addition, expression of both subtype of estrogen receptors, ERa and ERb, has been observed in rat GnRH1 neurons. The failure to observe mRNAs and proteins for ERa, nor-progesterone and androgen receptors in GnRH neurons, led to the general hypothesis that steroids regulate GnRH1 neurons transsynaptically or via glial interactions.

A recently found neuropeptide family, RFamide peptides, could provide the missing parts of knowledge of the control of GnRH neurons. The RFamide peptides, possessing an Arg-Phe-NH_2 motif at the C-terminus, have been well characterized in all groups of vertebrates. The first RFamide peptide was isolated in 1977 as a cardio-excitatory peptide in the Venus clam. Several RFamide peptides have since been identified in mammals and other species, and shown to be involved in a broad range of biological activities, such as nociceptive stimuli, arterial blood pressure regulation, food intake control and modulation of pituitary hormone secretion.

1.4.3.1.3 GnIH

Gonadotropin-Inhibitory Hormone (GnIH) family, which includes GnIH (mammals, birds) is characterized by a C-terminal amino acid sequence amide (LPXRFa) motif (X = L or Q). cDNAs encoding the precursors of GnIH peptides have been identified in mammals (Tsutsui and Ubuka, 2021). The precursor proteins encompass two to four putative/mature peptides with the LPXRFa motif, which are variously named: RFamide-related peptide (RFRP)-1, RFRP-2 RFRP-3 in mammals. There are further types in birds, frogs, amphibians and fish.

Mammalian RFRP-3 is referred to as GnIH because of it is known to inhibit gonadotropin secretion. Mammalian RFRP-1 is considerably similar to RFRP-3, but RFRP-2 does not possess the same RFamide motif as RFRP-1 and RFRP-3. The receptor for GnIH is a member of G protein coupled receptors expressed in the pituitary and brain areas including hypothalamus.

In mammals, GnIH neurons have been located in the periventricular nucleus of the hypothalamus, dorsomedial hypothalamic nucleus and ventromedial hypothalamic nucleus but fibers of GnIH neurons spread across the brain. GnIH acts directly at the level of the pituitary to inhibit LH secretion a role conserved in evolution. The presence of GnIH fiber projections and receptor expression in the hypothalamus suggests that GnIH also acts centrally to suppress the reproductive system through direct inhibitory action on GnRH1 neurons. The inhibitory role of GnIH on GnRH release is also supported by morphological evidences.

1.4.3.1.4 Kisspeptin

The newest member of RFamide peptides, kisspeptin, is among a family of peptides encoded by KISS1 gene, which includes metastin (kisspeptin-54) and kisspeptin-10, with the ability to activate a G protein-coupled receptor GPR54 (also known as KISS1 receptor) (Skorupskaite et al., 2014). All biologically active products of KISS1 found in the human placenta derive from a common precursor of 145 aa in length and share a common RFamide C-terminus. The C-terminal 10-aa sequence (YNW/LNSFGLRY/F) is highly conserved during evolution, whereas the complete sequences have low homology. The shortest endogenous 10-aa kisspeptin exerts equal receptor binding activity to the other longer endogenous fragments. A C-terminal free form of kisspeptin-54 has 10,000-fold less activity compared with kisspeptin-10. These observations, indicate that the core 10-aa region with C-terminal amidation is important for receptor binding.

1.4.3.1.5 Kisspeptin receptors (GPR54)

Kisspeptin receptor (Kiss1r) was first cloned in the rat in 1999 as an orphan GPCR called GPR54. Subsequently, the ortholog of Kiss1r has been cloned in humans and named AXOR12 or hOT7T175. In the human brain, expression of KISS1R mRNA has been demonstrated by RT-PCR in the thalamus, caudate nucleus, substantia nigra, hippocampus, cerebellum, corpus callous and amygdala. GPR54 was reported in 2003 to play an important role in reproduction. Kisspeptin/GPR54 system is crucial to the normal initiation of puberty in mammals mediated by GnRH neurons. Kiss1 peptide has in fact been shown to directly stimulate GnRH in mammalian species. Several genetic studies have revealed that mutations and deletions of Gpr54 causes idiopathic hypogonadotropic hypogonadism in humans (Franssen and Tena-Sempere, 2018).

The expression of Kiss1 and GPR54 in the mammalian pituitary suggests that kisspeptin has a local action. Kisspeptin stimulates the release of FSH and LH in vitro at the level of the pituitary in the rat, bovine and porcine. Furthermore the expression of kiss1 and kiss2 in the pituitary varies during different reproductive stages (Uenoyama et al., 2018). Further studies are required to elucidate the precise role of kisspeptin in the pituitary.

1.4.3.1.6 Analysis of GnRH and GnIH

Gonadotropin-inhibitory hormones (GnIH) and gonadotropin-releasing hormones (GnRH) are neuropeptides essential for the regulation of reproduction in all vertebrate animals examined. Determination of neuropeptides in the biological sample is highly challenging due to their complex matrix and weak stability. The wide variety of peptides or protein degradation products often interferes with the determination of the target peptide. A specific ultrahigh performance liquid chromatography-tandem mass spectrometry method for simultaneous determination of nine critical neuropeptides in biological samples has been reported (Bussy et al., 2015). A separation method by ultra-performance liquid chromatography coupled to a multiple reaction monitoring (MRM) by tandem mass spectrometry allows the selective determination of the neuropeptides in brain and plasma matrices after solid-phase extraction. Specific MSMS transitions were optimized using MRM of multiple charged peptides generated by electrospray ionization in positive mode. The resulting analytical method was fully validated for stability, recovery, matrix effect and intra- and interday accuracy and precision in brain and plasma. The optimized method has limit of quantification ranging from 0.1 to 0.75 ng/mL. With slight modification, this method can be applied to other biological samples.

1.4.3.1.7 Gonadotrophin surge attenuating factor (GnSAF)

In vivo and in vitro experiments indicate that GnSAF is a nonsteroidal ovarian substance that plays an important role during the normal menstrual cycle. The bioactivity of this factor is high in the early- to midfollicular phase, decreasing gradually thereafter until the midcycle. This factor antagonizes the sensitizing effect of estradiol on the pituitary response to GnRH during the greater part of the follicular phase (Messinis et al., 2018). With its decreasing activity in the late follicular phase, GnSAF facilitates the full expression of the positive feedback mechanism and the occurrence of a normal midcycle LH surge. It is suggested that GnSAF is the "missing link" between the ovaries and the hypothalamo-pituitary system, which maintains the pituitary in a state of low responsiveness to GnRH pulses in the early- to midfollicular phase, limiting thus the release of LH to low but adequate amounts for normal folliculogenesis. This factor appears to affect the pulsatile aspect of LH release that is controlled centrally by GnRH and modulated by the ovarian steroids. GnSAF therefore provides a negative action in the context of the positive feedback loop.

1.4.3.2 Testis

The testes produce testosterone in the Leydig cells and sperm in Sertoli cells and the testes are regulated by hormones from the hypothalamus and pituitary. The testicular artery enters the testis on its posterior surface, sending a network of branches that run deep to the tunica albuginea before entering the substance of the testis. The venous drainage passes posteriorly and emerges at the upper pole of the testis as a plexus of veins termed the pampiniform plexus.

GnRH from the hypothalamus stimulates the pituitary to secrete luteinizing hormone (LH) and follicle stimulating hormone (FSH) into the blood stream. LH stimulates Leydig cells to produce testosterone. Testosterone acts by a negative feedback to suppress GnRH. The Sertoli cells also produce inhibin B that also limits GnRH secretion. Testosterone and FSH acts on the Sertoli cells that produce sperm. Spermatogenesis is dependent on androgen-**secretion** by the Leydig cells; androgens stimulate and maintain germ cell development throughout life. Spermatogenesis relies on many intrinsic and extrinsic factors. Testicular testosterone levels are higher than those required for the initiation and maintenance of spermatogenesis. Androgen action on receptors within Leydig cells, peritubular myoid cells and Sertoli cells is essential for normal steroidogenesis and spermatogenesis. While testosterone is essential for spermatogenesis, exogenous testosterone administration resulting in even slightly raised serum levels suppresses gonadotropin **secretion**. While FSH is not essential for spermatogenesis, optimal spermatogenesis requires the combined actions of both androgen and FSH, with both hormones having independent, co-operative and synergistic effects to promote maximal sperm output (Campo et al., 2019). FSH is secreted in multiple forms with oligosaccharides that endow glycoforms with different activities,

The testis produces retinoic acid (RA) from circulating retinol and RA is a major driver of spermatogenesis. A pulse of retinoic acid acts directly on spermatogonia to stimulate the pathway to meiosis. Sertoli cells have an internal clock that appears to be set by retinoic acid, although the germ cells themselves influence the clock.

Germ cells particularly in the testis express noncoding RNAs, including microRNAs, small interfering RNAs, piRNAs and long noncoding RNAs (Azhar et al., 2020). These noncoding RNAs are required for the transcriptional program during meiosis and spermiogenesis. Epigenetic processes such as DNA methylation and histone modifications regulate chromatin structure and modulate gene transcription and silencing. The epigenome in sperm can be altered by diet, lifestyle and exposure to environmental factors so can influence both fertility and the health of the next generation.

Leydig cells have the capacity to synthesize cholesterol from acetate or to take up cholesterol from lipoproteins. Leydig cells have abundant smooth endoplasmic reticulum and mitochondria with tubular cristae that are typical of steroidogenic cells. The enzymes required for steroidogenesis are located in the mitochondria and in endoplasmic reticulum requiring. Intermediates in testosterone synthesis move between these organelles to achieve successful androgen production. Leydig cells also produce the peptide hormone, insulin-like factor 3 (INSL3), insulin, IGF1 and IGF2 family. INSL3 acts via its receptor, RXFP2 on meiotic and postmeiotic germ cells and on the Leydig cells themselves. Leydig cell differentiation and cell number contribute to increasing testicular volume at the onset of puberty or treatment for hypogonadism.

Testosterone is the major androgen secreted by the Leydig cells in the intertubular spaces of the testis. A normal male produces approximately 7 mg of testosterone daily but also produces lesser amounts of weaker androgens such as androstenedione and dehydroepiandrosterone. The more potent androgen dihydrotestosterone is produced by the testis in smaller amounts through the actions of the enzyme 5α reductase. The testis also contributes approximately 25% of the total daily production of 17β-estradiol through the local action of the enzyme aromatase.

LH enhances the transcription of genes that encode a range of enzymes in the steroidogenic pathway. Continued LH stimulation results in Leydig cell hypertrophy and hyperplasia. The testosterone production of the human testis declines in aging men due to a reduction in the response LH pulses.

Leydig cell development and function is critically dependent on other testicular cell types including Sertoli-, germ-, macrophages and peritubular myoid cells. Sertoli cells recruit and maintain their progenitors and regulate steroidogenesis. LH targets the Leydig cells to stimulate androgen biosynthesis, and the androgens act on receptors within the seminiferous epithelium to stimulate and support spermatogenesis. FSH targets receptors in the Sertoli cells. Androgens and FSH have independent effects on Sertoli cells but also act co-operatively and synergistically to initiate and maintain normal spermatogenesis. Androgens act on androgen receptors (AR) in the testis to support normal spermatogenic function. Androgen action on Sertoli cells is needed for meiosis and spermiogenesis. Androgens are necessary for the formation of tight junctions between Sertoli cells which contribute to the blood-testis-barrier and they drive the expression and translation of many genes expressed in the Sertoli cells. The process of spermatogenesis, taking more than 60 days, involves an incredibly complex program whereby the transcription and translation of thousands of genes is precisely constrained as the germ cell proceeds through proliferation, meiosis and spermiogenesis.

1.4.3.2.1 The hypothalamic-pituitary-testis axis

The pituitary secretion of FSH and LH by the gonadotrophs is also controlled by the feedback inhibition that occurs via the steroids, testosterone and estradiol. The secretion of FSH and LH is also regulated by protein inhibitors, inhibin, secreted by the gonads and follistatin, produced locally within the pituitary by the follicular-stellate cells. Hypothalamic secretion of GnRH in turn stimulates FSH and LH to act on the testis. The mechanism for the differential regulation of FSH and LH secretion is unknown. Kisspeptins, a family of neuropeptides localized to the arcuate nucleus of the brain are upstream regulators of GnRH secretion. GnIH acts both upstream of GnRH and possibly at the gonads as an autocrine/paracrine regulator of steroidogenesis. Follistatin inhibits FSH secretion by blocking the actions of the activins A and B produced by the pituitary gland.

Testosterone acts at the hypothalamus by decreasing GnRH pulse frequency without a change in pulse amplitude, whereas estradiol appears to predominantly act at the pituitary where it decreases LH pulse amplitude without changing pulse frequency. Testosterone and estradiol suppress FSH in the male but inhibin B has a specific negative feedback inhibition on FSH secretion at the pituitary level. There are two circulating forms of inhibin dimer, consisting of an alpha inhibin subunit associated with a βA or βB subunit to form inhibin A or inhibin B and that RIAs relying on use of antibodies to the inhibin alpha subunit did not reflect both forms (see Fig. 1.4.12). The alpha subunit is produced as a pro-alphaN-alphaC precursor that could combine with the β subunits to produce several different forms of inhibin, both inhibin A and inhibin B have the capacity in vitro to inhibit FSH secretion by pituitary cells in culture. In contrast, activins A, B and AB all have the capacity to stimulate FSH secretion by pituitary cells. Follistatin, has the capacity to suppress FSH secretion

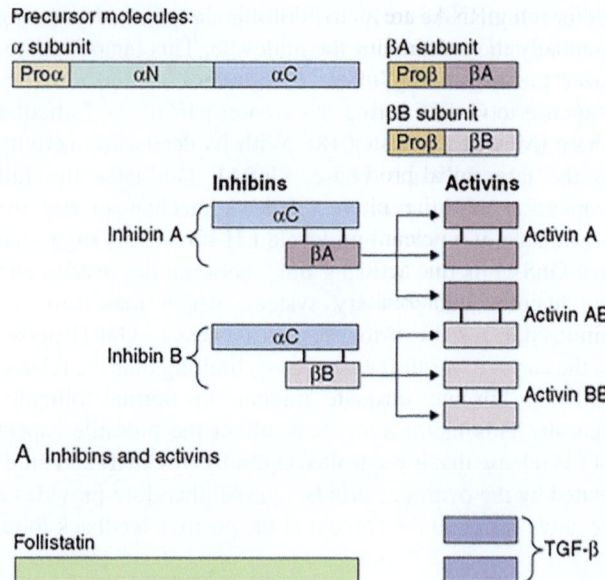

FIG. 1.4.12 Activins, inhibins, follistatin, and TGFβ. A schematic representation of the formation of different dimeric proteins from three basic subunits, α, βA, and βB. Inhibins are heterodimers of α and β subunits (α–βA, α–βB); activins are homodimers of the β subunits (βAβA, βAβB, βBβB). Links between the units represent disulfide bridges. *(From Makanji Y, Zhu J, Mishra R, Holmquist C, Wong WP, Schwartz NB, Mayo KE, Woodruff TK. Inhibin at 90: from discovery to clinical application, a historical review. Endocr Rev 2014;35(5):747–94. Fig. 1 p. 750.)*

specifically by pituitary cells by neutralizing the actions of activin and thus suppressing FSH secretion.

In the male, inhibin is produced by the Sertoli cell and is secreted into the basement membrane of the seminiferous tubule and into the lumen. The concentrations of inhibin B in males are inversely related to the levels of FSH. FSH predominantly stimulates inhibin α subunit production and does not alter the β subunit message. The testis predominantly secretes inhibin rather than activin. A subunit of inhibin can be produced by Leydig cells and increased LH levels result in the release of α subunit products into the circulation. In men, gonadotrophin suppression by testosterone reduces circulating inhibin B and α subunit, suggesting that secretion is not fully gonadotropin-dependent. Although Sertoli cells, Leydig cells and peritubular myoid cells can produce activin, castration does not decrease circulating activin A levels. Activin acts on the pituitary but it also acts within the testis to stimulate spermatogonial mitosis, Sertoli cell mitosis during testis development and possibly acts directly on germ cells.

The gonadal feedback signal on FSH secretion is through inhibin B. Activin and follistatin can exert a paracrine role directly in the pituitary gland. The α and β subunit mRNAs are present in gonadotrophs in the pituitary gland. These substances exert a local action on FSH secretion.

Follistatin mRNAs are also present in a number of different pituitary cell types including the folliculo-stellate cells. This local production of follistatin also has the capacity to regulate the actions of activin.

GnRH and the sex steroids estradiol and testosterone can modulate the local production of α, βA, βB and follistatin mRNAs within the pituitary but these interactions are complex with no clear answer to the relative roles of paracrine and endocrine actions of these glycoprotein hormones. The actions of inhibin are predominantly exerted through secretion from the testis into the peripheral circulation, whereas the actions of activins and follistatin on FSH secretion occur through paracrine actions at the pituitary gland.

1.4.3.2.2 Inhibin assays

The first reliable and reproducible enzyme linked immunoassays for inhibin were based on antibodies raised to intact dimeric inhibin but subsequently shown to recognize the alpha subunit of inhibin, or specifically the N-terminus of the alpha C subunit of inhibin. While inhibin was recognized as a dimer of an alpha subunit, coupled with a β subunit, there were inconsistencies between inhibin and FSH levels in women and men. The first data in the menstrual cycle using an RIA showed that inhibin levels were present throughout the menstrual cycle, increased only in the very late follicular and were high throughout the luteal phase of the cycle. The early RIAs could not distinguish between nonbiologically active free alpha inhibin and biologically active inhibin A or inhibin B. In addition, in most female species it appeared that both inhibin A and B were produced by the ovary and RIA's were of limited use for diagnostic purpose.

There is minimal production of inhibin A by early developing follicles at the start of the menstrual cycle and inhibin A is secreted first in any quantity by the dominant preovulatory follicle(s) and then by the corpus luteum. ELISA's were developed specific for inhibin A, inhibin B and the pro-alpha C forms of inhibin. Using the ELISA specific for inhibin A very low levels were found early in the follicular phase, that increased in the late follicular phase of the menstrual cycle and were further increased in the luteal phase, correlating with the RIA data. In early pregnancy, hCG stimulated a further increase in inhibin A secretion by the corpus luteum and the placenta then maintains inhibin secretion throughout pregnancy. These changes in inhibin A matched fairly closely those with the RIAs which detected dimerized or free alpha inhibin subunits.

Results from a specific ELSA for inhibin B showed that levels increased as soon as follicle development was stimulated by FSH in the early follicular phase of the menstrual cycle and then declined toward the stages of preovulatory follicle selection when inhibin A secretion increased.

Furthermore, while inhibin A had been undetectable in men, inhibin B was shown to be the inhibin secreted by men. There is species variability to be aware of when reading the literature.

Circulating INHB, named for its ability to inhibit FSH production, peaks shortly after birth, then decreases and remains low but measurable until puberty when it rises again, first as a consequence of FSH stimulation and then as a result of the combined regulation by FSH and the ongoing spermatogenesis.

The inhibin A and B ELISAs are sandwich assays in which the monoclonal antibodies E4 (anti-A-subunit) and C5 (anti-B-subunit) are utilized as capture antibodies and alkaline phosphatase conjugated R1 monoclonal antibodies are used as the detection antibody. R1 was raised against a synthetic peptide corresponding to AA1–32 of the mature inhibin A-subunit. The inhibin A-specific antibody, E4, was raised against AA82–114 of the mature A-subunit, and the inhibin B-specific antibody, C5, was raised against AA82–114 of the mature B-subunit. The E4 antibody also serves as the capture and detection antibody in the activin A ELISA, although the C5 antibody could not be utilized in the same manner for an activin B ELISA. To improve the sensitivity of the inhibin A and B ELISAs, samples are subjected to hydrogen peroxide treatment that oxidizes the Met residues in the consensus sequence Met-Ser-Met on the A- and B-subunits. In those assays, the serum samples required sodium dodecyl sulfate treatment and boiling to improve sensitivity to 7 pg/mL. Later the 46A/F antibody raised specifically against an as-yet uncharacterized region of the mature human B-subunit was utilized in activin B and inhibin B ELISAs. Serum samples in those assays do not require hydrogen peroxide, sodium dodecyl sulfate or heat treatment of samples and have improved sensitivity to 4 pg/mL. The total inhibin ELISA was primarily developed to detect all inhibin forms, including the free α-subunit, in order to monitor various ovarian cancers. This ELISA replaced a two-site total inhibin C immunofluorometric assay that utilized polyclonal antibodies (no. 41 was detection antibody and 128 was capture antibody) raised against a human C-subunit fusion protein. In contrast, the broadly specific total inhibin ELISA utilizes monoclonal antibodies raised against the inhibin C-subunit and alkaline phosphatase fused-R1 (as detection antibodies) and a combination of PO#14 and PO#23 (as capture antibodies). The total inhibin ELISA has improved specificity and sensitivity in detecting the various molecular mass forms of inhibins in postmenopausal women and women with various types of ovarian cancers. The set of available inhibin assays has become an essential tool in the understanding of the functional roles of inhibin A and inhibin B in various physiological and disease processes.

The inhibin B assays from OBI and DSL have been highly cited for the measurement of Inhibin B in plasma and serum. Both the assays have provided significant clinical information related to various normal and disease states. The OBI and DSL Inhibin B assays use the monoclonal antibodies C5 and R1 developed by Groome and co-workers. Both C5 and R1 were raised to the synthetic peptides made from the βB- and α-subunits of inhibin, respectively. Both assays require a methionine oxidation step(in the β-subunit) with hydrogen peroxide to allow the C5 antibody to recognize its epitope. The OBI assay also uses SDS detergent and heat pretreatment of the sample. From an immunoassay view point, the need for sample oxidation creates unnecessary complexity. For example, the boiling step in the OBI assay may lead to the sample becoming gelatinized in the presence of SDS. The oxidation of different recombinant Inhibin B calibrator preparations with respect to the serum/plasma specimens may vary. Groome and co-workers successfully raised a monoclonal antibody (46A/F) against activin B that does not require the oxidation of the βB-subunit for binding. The second generation Inhibin B assay uses this 46A/F and the biotinylated R1 monoclonal antibodies as capture and detection, respectively. The second generation assay uses a very simple procedure that is extremely robust and specific, with no sample pretreatment requirement.

Challenges can arise when different methods produce different results for the same analyte. In an effort to minimize the variation between the measurements of Inhibin B methods, the second generation Inhibin B assay was standardized to the highly cited OBI assay. The calibrators used in the second generation assay (recombinant Inhibin B in FBS) are traceable to the WHO Inhibin B reference reagent (96/784) using a factor of 2.47. The WHO preparation is composed of a mixture of inhibin forms immunopurified from human follicular fluid. The "Factor" with respect to WHO will vary between methods for the same recombinant calibrator materials due to the unavailability of the pure WHO preparation.

Serum inhibins A and B are well-recognized ovarian tumor markers (Robertson et al., 2004). Sex-cord tumors, particularly granulosa cell tumors, almost always have increased inhibins, especially inhibin B, but normal CA125. Ovarian epithelial tumors, including endometrioid, serous and clear cell and undifferentiated tumors have elevated CA125 but usually normal inhibins. Ovarian mucinous tumors frequently have raised CA125 and/or inhibins. In investigating virilized females with increased androgens, the presence of raised inhibin A and particularly B would be supportive of an ovarian tumor. Although absent in investigative algorithms, inhibins should be included in the investigative protocols of virilized females with hyperandrogenemia, since they may be of diagnostic value, particularly in postmenopausal females where they should be undetectable. Inhibins may also have a role in detecting recurrence of androgen-secreting ovarian

tumors, since they may recur even though they are very rarely malignant.

1.4.3.2.3 Anti-Mullerian hormone

AMH is an important hormone for sex differentiation of the male and measurements of serum AMH concentrations are useful in patients with disorders of sexual development. AMH, a dimeric glycoprotein, is a member of the transforming growth factor β family of growth and differentiation factors (see earlier Chapter 1.2). During normal male fetal sexual development, testicular production of AMH, secreted by Sertoli cells from the 7th week of gestation, plays a vital role in male sexual differentiation. It is responsible for the involution/regression of the female Müllerian duct while testosterone from Leydig cells stimulates differentiation of the male Wolffian ducts, the urogenital sinus and the external genitalia. AMH concentrations in boys are 500–2000 pmol/L and remain very high (300–1300 pmol/L) until puberty when they drop significantly following the increased testosterone production that downregulates AMH, probably via activation of the Sertoli cell androgen receptor. In pediatric practice, AMH measurement can be useful in distinguishing cryptorchidism from anorchidism and in persistent Müllerian duct syndrome, hypogonadotropic hypogonadism and as a marker of testicular function in Klinefelter syndrome (XXY).

There are many issues around the assays for AMH which are discussed later (see Section 1.4.3.3.1).

1.4.3.3 Ovary

During each monthly cycle, the follicular phase is timed from the first day of menses until ovulation. Lower temperatures on a basal body temperature chart, and the development of ovarian follicles, characterize this phase. Folliculogenesis begins during the last few days of the preceding menstrual cycle until the release of the mature follicle at ovulation.

Declining steroid production by the corpus luteum and the dramatic fall of inhibin A allows for follicle stimulating hormone (FSH) to rise during the last few days of the menstrual cycle (Fig.1.4.13). Development of the dominant follicle can be described in three stages:

1. recruitment,
2. selection and
3. dominance.

There is an increase in GnRH pulsatile secretion secondary to a decline in both estradiol and progesterone levels.

The elevation in FSH leads to recruitment of a cohort of ovarian follicles in each ovary, one of which is destined to ovulate during the next menstrual cycle. The recruitment stage takes place during days 1 through 4 of the menstrual

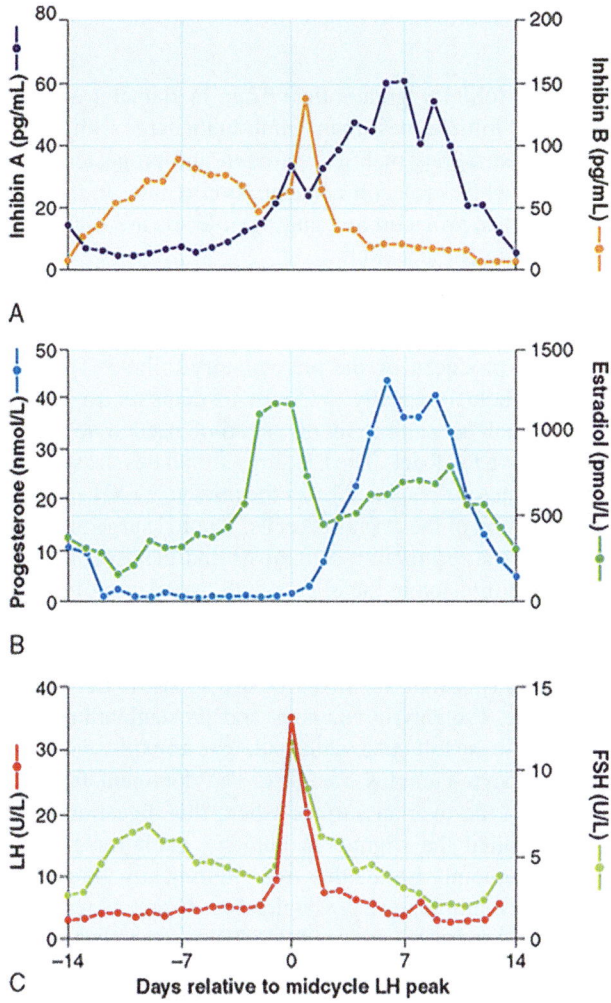

A

B

C

FIG. 1.4.13 Hormones and menstrual cycle. Plasma hormone concentrations during the female menstrual cycle. Top—Inhibin A and B; Middle—progesterone and estradiol; Bottom—LH and FSH.

cycle. Once menses ensues, FSH levels begin to decline due to the negative feedback of estrogen and the negative effects of inhibin B produced by the developing follicle. FSH activates the aromatase enzyme in granulosa cells. A decline in FSH levels leads to the production of higher androgen concentrations within follicles adjacent to the growing dominant follicle. Between cycle days 5 and 7, selection of a follicle takes place whereby only one follicle is selected from the cohort of recruited follicles to ovulate, and the remaining follicles will undergo atresia. The granulosa cells of the growing follicle also secrete a variety of peptides that may play an autocrine/paracrine role to inhibit the development of the adjacent follicles. **Anti-Müllerian hormone (AMH)**, a product of granulosa cells, is believed to play a role in the selection of the dominant follicle. By cycle day 8, one follicle exerts its dominance by promoting its own growth and suppressing the maturation of the other ovarian follicles thus becoming the dominant follicle.

During the follicular phase, serum estradiol levels rise in parallel to the growth of follicle size as well as to the increasing number of granulosa cells. FSH receptors exist exclusively on the granulosa cell membranes. Increasing FSH levels during the late luteal phase leads to an increase in the number of FSH receptors and ultimately to an increase in estradiol by granulosa cells.

The rise in estradiol secretion appears to increase the total number of estradiol receptors on the granulosa cells. In the presence of estradiol, FSH stimulates the formation of LH receptors on granulosa cells that secrete small quantities of progesterone and 17-hydroxyprogesterone (17-OHP) that may exert a positive feedback on the estrogen-primed pituitary to augment LH release. FSH also stimulates and 3β-hydroxysteroid dehydrogenase (HSD3B2). Table 1.4.1 shows the production rates of sex steroids during the

TABLE 1.4.1 Production rates of steroids in the menstrual cycle.

Sex steroid	Early follicular phase	Preovulatory	Midluteal
Progesterone (mg)	1.0	4.0	25.0
17-Hydroxyprogesterone (mg)	0.5	4.0	4.0
Dehydroepiandrosterone (mg)	7.0	7.0	7.0
Androstenedione (mg)	2.6	4.7	3.4
Testosterone (mg)	144	171	126
Estrone (mg)	50	350	250
Estradiol (mg)	36	380	250

follicular phase, luteal phase and at the time of ovulation. LH receptors are located on theca cells during all stages of the menstrual cycle. LH stimulates androstenedione production and to a lesser extent testosterone production in theca cells. Androstenedione is transported to the granulosa cells where it is aromatized to estrone and estradiol by 17-β-hydroxysteroid dehydrogenase type I. This is known as the two-cell, two-gonadotropin hypothesis of regulation of estrogen synthesis in the human ovary.

In the ovary, the primordial follicles are surrounded by a single layer of granulosa cells and are arrested in the diplotene stage of the first meiotic division. After puberty, each primordial follicle enlarges and develops into a preantral follicle. The preantral follicle is now surrounded by several layers of granulosa cells as well as by theca cells. The preantral follicle has acquired FSH receptors. The preantral follicle then develops a cavity and is now known as an antral follicle. Finally, it becomes a preovulatory follicle on its way toward ovulation. Due to the presence of 5α-reductase, preantral and early antral follicles produce more androstenedione and testosterone in relation to estrogens. 5α-reductase is the enzyme responsible for converting testosterone to dihydrotestosterone (DHT). Once testosterone has been 5α-reduced, DHT cannot be aromatized. However, the dominant follicle is able to secrete large quantities of estradiol, due to high levels of aromatase (CYP19). This shift from an androgenic to an estrogenic follicular microenvironment may play an important role in selection of the dominant follicle from the large number of other follicles that will become atretic. Development of the follicle to the preantral stage is gonadotropin independent, but follicular growth beyond this point requires gonadotropin interaction.

FSH is elevated during the early follicular phase and then begins to decline until ovulation. In contrast, LH is low during the early follicular phase and begins to rise by the mid-follicular phase due to the positive feedback from the rising estrogen levels, that need to be greater than 200 pg/mL for approximately 50 h in duration. The frequency and amplitude of the gonadotrophin pulses vary according to the phase of the menstrual cycle. During the early follicular phase, LH secretion occurs at a pulse frequency of 60–90 min with relatively constant pulse amplitude. During the late follicular phase prior to ovulation, the pulse frequency increases and the amplitude may begin to increase. The LH pulse amplitude begins to increase after ovulation takes place.

There are numerous substances found in follicular fluid, such as steroids, pituitary hormones, plasma proteins, proteoglycans and nonsteroidal ovarian factors, that contribute to the microenvironment of the ovary and regulate steroidogenesis in granulosa cells. Growth factors such as insulin-like growth factor 1 and 2 (IGF1, IGF2) and epidermal growth factor (EGF) play important roles in oocyte development and maturation. The concentration of ovarian steroids is much higher in follicular fluid in comparison to plasma concentrations. There are two populations of antral follicles:

1. large follicles, greater than 8 mm in diameter and.
2. small follicles, less than 8 mm. In the large follicles, the concentrations of FSH, estrogen and progesterone are high while prolactin concentration is low. In the small follicles, prolactin and androgen levels are higher than in large antral follicles.

Ovulation occurs approximately 10–12 h after the LH peak (Fig. 1.4.13). The LH surge is initiated by a dramatic rise of estradiol produced by the preovulatory follicle. The dominant follicle is usually >15 mm in diameter to produce the critical concentration of estradiol needed to initiate the positive feedback. The LH surge stimulates the synthesis of progesterone responsible for the midcycle FSH surge and luteinization of the granulosa cells. The LH surge is responsible for the resumption of meiosis and the completion of reduction division in the oocyte with the release of the first polar body.

Prostaglandins and proteolytic enzymes, such as collagenase and plasmin, are increased in response to LH and progesterone. Proteolytic enzymes and prostaglandins digest collagen in the follicular wall, leading to an explosive release of the oocyte-cumulus complex. The dominant follicle is closest to the ovarian surface where this digestion occurs and is called the stigma. In humans, ovulation probably occurs randomly from either ovary during any given cycle. The concentrations of prostaglandins E and F series and hydroxyeicosatetraenoic acid (HETE) reach a peak level in follicular fluid just prior to ovulation. Prostaglandins may stimulate proteolytic enzymes while HETEs may stimulate angiogenesis and hyperemia. Infertility patients are advised to avoid taking prostaglandin synthetase inhibitors, as well as cyclo-oxygenase (COX) inhibitors, around the time of ovulation to avoid a luteinized, unruptured follicle.

Estradiol levels fall dramatically immediately prior to the LH peak due to LH downregulation of its own receptor or because of direct inhibition of estradiol synthesis by progesterone. Progesterone is also responsible for stimulating the midcycle rise in FSH. Elevated FSH levels at this time are thought to free the oocyte from follicular attachments, stimulate plasminogen activator and increase granulosa cell LH receptors. The decline in LH postovulation may be due to the loss of the positive feedback effect of estrogen, due to the increasing inhibitory feedback effect of progesterone, or due to a depletion of LH content of the pituitary from downregulation of GnRH receptors.

The luteal phase is usually 14 days long after ovulation. The remaining granulosa cells continue to enlarge, become vacuolated in appearance, and begin to accumulate a yellow pigment called lutein. The luteinized granulosa cells combine with the newly formed theca-lutein cells and surrounding stroma in the ovary to become what is known as the corpus luteum. The corpus luteum predominantly secretes progesterone, to prepare the estrogen primed endometrium for implantation of the fertilized ovum. The basal lamina dissolves and capillaries invade into the granulosa layer of cells in response angiogenic factors released by the granulosa and thecal cells. Around 8 days after ovulation, approximately around the time of expected implantation, peak vascularization is achieved. This time also corresponds to peak serum levels of progesterone and estradiol. The central cavity of the corpus luteum may also accumulate blood and become a hemorrhagic corpus luteum. The life span of the corpus luteum depends upon continued LH support. Corpus luteum function declines by the end of the luteal phase unless human chorionic gonadotropin is produced by a pregnancy. If pregnancy does not occur, the corpus luteum undergoes luteolysis under the influence of estradiol and prostaglandins and forms a scar tissue called the corpus albicans.

Estrogen levels rise and fall twice during the menstrual cycle. Levels rise during the mid-follicular phase and then drop precipitously after ovulation. This is followed by a secondary rise in estrogen levels during the mid-luteal phase with a decrease at the end of the menstrual cycle. The secondary rise in estradiol parallels the rise of serum progesterone and 17α-hydroxyprogesterone levels.

The mechanism by which the corpus luteum regulates steroid secretion is not completely understood. Regulation may be determined in part by LH secretory pattern and LH receptors or variations in the levels of 3β-HSD, CYP17, CYP19 or side chain cleavage enzymes. The number of granulosa cells formed during the follicular phase and the amount of LDL cholesterol may also play a role in steroid regulation by the corpus luteum. The luteal cell population consists of large and the small cells. Small cells are thought to have been derived from thecal cells while the large cells are from granulosa cells. The large cells are more active in steroidogenesis and are influenced by various autocrine/paracrine factors such as inhibin, relaxin and oxytocin. The secretion of progesterone and estradiol during the luteal phase is episodic and correlates with pulses of LH (Fig. 1.4.13). The frequency and amplitude of LH secretion during the follicular phase regulates subsequent luteal phase function and is consistent with the regulatory role of LH during the luteal phase. Reduced levels of FSH during the follicular phase can lead to a shortened luteal phase and the development of a smaller corpus lutea.

The corpus luteum function begins to decline around 10 days after ovulation through unknown mechanisms. Estrogen is believed to play a role in the luteolysis of the corpus luteum. Prostaglandin F2α inhibits steroidogenesis and stimulates the release of a growth factor, tumor necrosis factor alpha (TNFα), that induces cell apoptosis. Oxytocin and vasopressin exert their luteotropic effects via an autocrine/paracrine mechanism. Matrix metalloproteinases also appear to play a role in luteolysis. Androgens, glucocorticoids and pituitary hormones, excluding LH and FSH, undergo only minimal fluctuation in the menstrual cycle. Due to extra-adrenal 21-hydroxylation of progesterone, plasma levels of deoxycorticosterone are increased during the luteal phase.

1.4.3.3.1 Anti-Mullerian hormone

AMH may be detected in female subjects from fetal stage until the menopause. The levels of the hormone in blood are an order of magnitude lower than males. AMH participates in the initial selection of follicles for further developmental from preantral and small antral follicles. AMH is secreted by granulosa cells in antral follicles and can be measured in follicular fluid and blood (Bedenk et al., 2020). The development and growth of the dominant follicle is dependent on FSH stimulation. Requests for AMH measurements have increased in number for infertile couples as well as increasing of women at an advanced reproductive age trying to conceive evoke. The test is used as a prediction of their ovarian function. The levels of AMH reflect the number of preantral follicles and thus are a marker of oocyte pool. Plasma levels of AMH correlate with the number of mature follicles (AFC = antral follicle count), as assessed by transvaginal ultrasonography. The determination of plasma AMH in women of fertile age enables assessment of the extent of ovarian reserve more specifically than determination of FSH with steroid hormones and inhibin. AMH is not involved in feed-back mechanisms of the HPG axis. AMH levels are therefore, almost independent on the phase of the menstrual cycle, and usually a single measurement is sufficient.

AMH is measured by enzymatic immunoassay methods. The first diagnostic methods for AMH were benchtop or semiautomatic ELISAs released by commercial manufacturers. First generation kits were supplied by Diagnostic Systems Laboratories (DSL) and Immunotech (IOT), which gave quantitatively different results. Beckman Coulter then amalgamated these two methodologies into a second generation kit, which utilizes the AMH antibodies from the DSL kit and the AMH calibrators from the IOT kit. After initial problems, attributed to interference from complement in the sample, a predilution step was introduced into the protocol resulting in more reliable results. The Beckman Coulter Gen II AMH ELISA an anti-AMH capture antibody and a secondary horseradish peroxidase conjugated antibody, allowing for detection with tetramethylbenzidine (TMB) substrate, which can be measured at 450 nm. The measured absorbance is directly related to the AMH concentration in the original sample. This assay has a limit of detection (LOD) of 0.57 pmol/L and a limit of quantitation (LOQ) of 1.14 pmol/L (at 20% total imprecision). Calibrators are quantified in terms of mass, i.e., ng/mL but serum values are normally reported as concentrations, i.e., pmol/L ($1\,ng/mL = 7.14\,pmol/L$).

A more sensitive ELISA kit, used mainly for research, is also available from Ansh laboratories. The LOD and LOQ values for this assay are 0.16 pmol/L and 0.43 pmol/L, respectively. This method utilizes the same assay technology as the Gen II but uses recombinant human AMH protein for the calibrators rather than the bovine derived calibrators used in the Gen II kit. These two ELISAs not surprisingly give quantitatively different values.

Autoanalyzer methods for the detection of AMH have been available since 2014. These are derived from the Gen II assay but have been adapted to autoanalyzer signal detection methods thus enabling a higher throughput and real time, rather than batch, analysis. The two major autoanalyzer methods in use in the United Kingdom are the Beckman Access assay and the Roche Elecsys AMH assays, both of which use the same antibodies as the Beckman Coulter Gen II ELISA. Beckman Access AMH assay which takes 39 min quote LOD 0.02 pmol/L and LOQ 0.036 pmol/L (at 20% total imprecision) whereas The Roche Elecsys assay has a sample run time of 18 min with LOD 0.071 pmol/L and LOQ 0.21 pmol/L. The latter has been assessed for comparability between five laboratories (Anckaert et al., 2016).

A reference reagent based on recombinant AMH is under evaluation (Li et al., 2021; Ferguson et al., 2018, 2020). In parallel, measurement by isotope dilution mass spectrometry (Whiting et al., 2018) improvements in assay quality are anticipated. There are at least 20 immunoassay methods and platforms in recent validation of assay performance (Ferguson et al., 2020).

1.4.3.3.2 Menstruation

In the absence of a pregnancy, steroid hormone levels begin to fall due to declining corpus luteum function. Progesterone withdrawal results in increased coiling and constriction of the spiral arterioles. This eventually results in tissue ischemia due to decreased blood flow to the superficial endometrial layers, the spongiosa and compacta. The endometrium releases prostaglandins that cause contractions of the uterine smooth muscle and sloughing of the degraded endometrial tissue. The use of prostaglandin synthetase inhibitors decreases the amount of menstrual bleeding and can be used as therapy in women with excessive menstrual bleeding or menorrhagia. Menstrual fluid is composed of desquamated endometrial tissue, red blood cells, inflammatory exudates and proteolytic enzymes. Within 2 days after the start of menstruation and while endometrial shedding is still occurring, estrogen produced by the growing follicles starts to stimulate the regeneration of the surface endometrial epithelium. The estrogen secreted by the growing ovarian follicles, causes prolonged vasoconstriction enabling the formation of a clot

over the denuded endometrial vessels. Also, the regeneration and remodeling of the uterine connective tissue is regulated in part by the matrix metalloproteinase (MMP) system. The average duration of menstrual flow can be from as little as 2 days up to 8 days. The average amount of menstrual blood loss is 30 mL and greater than 80 mL is considered abnormal.

1.4.4 Redox regulation of steroidogenesis

The mitochondria have a central role in regulating steroidogenesis firstly in supplying cholesterol and then as source of hydrogen transfer to the cytochromes for steroid hormone synthesis.

1.4.4.1 Nicotinamide nucleotide transhydrogenase (NNT)

The first and last steps of cortisol production occur in the mitochondria and require a constant supply of reductant NADPH. This NADPH is regenerated from NADP by a few pathways including the thioredoxin and glutathione pathways, which are ultimately enabled by NNT (Meimaridou et al., 2018). Several selenoproteins, including the thioredoxin reductases and glutathione peroxidases, contribute significantly to redox regulation (Ren et al., 2017). The glutathione and thioredoxin systems maintain reduced peroxiredoxin 3 (PRDX3) (Fig. 1.4.14), which is integral for redox regulation within the adrenal.

NNT provides the thioredoxin and glutathione systems with high concentrations of NADPH required for this process (Meimaridou et al., 2018). NNT is the major mitochondrial enzymatic source of NADPH contributing 45% of the total NADPH supply. It exists as a dimer and spans the inner mitochondrial membrane modulating H^+ movement and supplying the high concentrations of NADPH required for the detoxification of ROS by glutathione and thioredoxin pathways. The significance of NNT is only gradually being revealed by the study of a C57BL/6J mouse that has a spontaneous mutation in NNT (an in-frame 5-exon deletion), resulting in the truncation of the message and absence of the protein.

Detoxification of free radicals in the mitochondria. NNT encodes a protein, integral to the inner mitochondrial membrane, which under normal physiological conditions uses energy from the mitochondrial proton gradient to generate high concentrations of NADPH. This is required for many processes in the cell including the supply of reductive power to a network of antioxidant enzymes, specifically the glutathione (GSH/GSSG) and thioredoxin (Trx(SH)2/TrxS2) systems, to allow the detoxification of H_2O_2. Manganese superoxide dismutase (MnSOD) converts O_2^- into

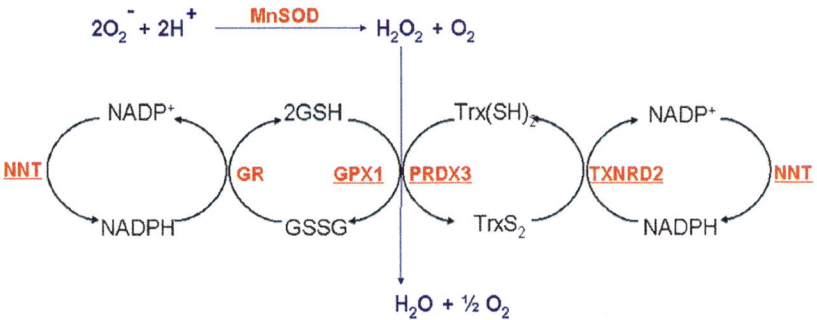

FIG. 1.4.14 NNT (nicotinamide nucleotide transhydrogenase) protects mitochondria from oxidative stress. NNT uses energy from the mitochondrial proton gradient to generate high concentrations of NADPH that is used for many processes in the cell. A number of free radicals such as H_2O_2 are detoxified, specifically by glutathione (GSH/GGSG) and thioredoxin ($Trx(SH)_2/TrxS_2$) reductases. Oxygen is converted to H_2O_2 by manganese superoxide dismutase to protect reactive oxygen sensitive proteins from damage. The H_2O_2 is removed by glutathione peroxidases (e.g., GPX1) or peroxiredoxins (e.g., PRDX3) using glutathione (GSH) $Trx(SH)_2$ as co-factors. *(Modified from Meimaridou E, et al. NNT is a key regulator of adrenal redox homeostasis and steroidogenesis in male mice. J Endocrinol. 2018;236(1):13–28. Fig, 2, p. 15.)*

H_2O_2 and protects ROS-sensitive proteins from oxidative damage. H_2O_2 is then removed by glutathione peroxidases (e.g., GPX1) or peroxiredoxins (e.g., PRDX3) using GSH and $Trx(SH)2$ as co-factors. GSH and $Trx(SH)2$ can be regenerated by glutathione reductase (GR) and thioredoxin reductase-2 (TXNRD2), respectively, using the reducing power from NADPH. Without NNT, the production of NADPH is compromised, causing the mitochondria to become more sensitive to oxidative stress.

H295R adrenal tumor cells, where NNT has been stably knocked down, undergo oxidative stress as demonstrated by low glutathione levels and increased mitochondrial superoxide production. Similar defects in energy metabolism due to NNT ablation have also been demonstrated in other mouse tissues (heart, liver, pancreas) emphasizing the importance of NNT for cellular bioenergetics. The mechanism by which loss of NNT causes adrenal-specific pathology is unclear.

1.4.4.2 Thioredoxin reductase

TXNRD2, one of three thioredoxin reductases, is mitochondria specific and exists as an antiparallel homodimer. The N- and C-terminal redox active centers of the two subunits functionally interact and transfer electrons from NADPH/H to thioredoxin 2 (TXN2) and other substrates. TXNRD2 catalyzes the reduction of the active disulfide of thioredoxin 2 and other substrates. As a selenoprotein, it requires insertion of a selenocysteine residue for catalytic activity. PRDX3 is reported to be the most important H_2O_2-eliminating enzyme in the mitochondria of the adrenal cortex with hyperoxidation of PRDX3 resulting in diminished steroidogenesis. During H_2O_2 elimination, two reduced PRDX3 subunits are converted to an oxidized disulfide-linked dimer that is reduced again by TXN2. PRDX3, together with mitochondrial-specific TXN2 and TXNRD2, thus provide a primary line of defense against

H_2O_2 produced by the mitochondrial respiratory chain in the adrenal gland.

A glutaredoxin 2 has recently been identified as another electron donor for PRDX3. Glutaredoxin 2 (GLRX2) itself is reduced by TXNRD2 as well as GSH, and the mitochondrial thioredoxin and GSH systems function in parallel to protect against oxidative stress (Fig. 1.4.15). In an in vitro knockdown adrenocortical model, the glutathione system is unable to fully compensate for the TXNRD2 deficiency leading to increased mitochondrial superoxide production.

Oxidative stress impedes steroidogenesis, and steroidogenesis itself induces oxidative stress as a result of electron leak throughout the steroidogenic pathway. Approximately 40% of the total electron flow from NAPDH directed at reactive oxygen species production during steroidogenesis is required for CYP11B1 within the mitochondria, at the final step of cortisol production. This may explain the particular susceptibility of the zona fasciculata to oxidative stress, and hence, individuals with TXNRD2 and NNT mutations primarily develop glucocorticoid deficiency.

1.4.4.3 Sphingosine 1 phosphate

In adrenal and gonadal steroidogenesis, several studies have reported various roles for sphingolipids as structural components of cell membranes and key signaling mediators of many cellular processes. As a bioactive lipid mediator, S1P regulates a broad array of physiological functions, including cell proliferation and survival and protection against ceramide-mediated apoptosis (Spiegel and Milstien, 2002; Cuvillier et al., 1996). S1P has also been shown to stimulate cortisol production in zona fasciculata bovine adrenal cells in a protein kinase C (PKC)- and Ca^{2+}-dependent manner (Rábano et al., 2003) and promote aldosterone secretion in bovine glomerulosa cells via the phosphoinositide 3-kinase (PI3K)/Akt and mitogen-activated protein kinase (MAPK/ERK) pathways (Brizuela et al., 2007) (Fig. 1.4.16). S1P also mediates

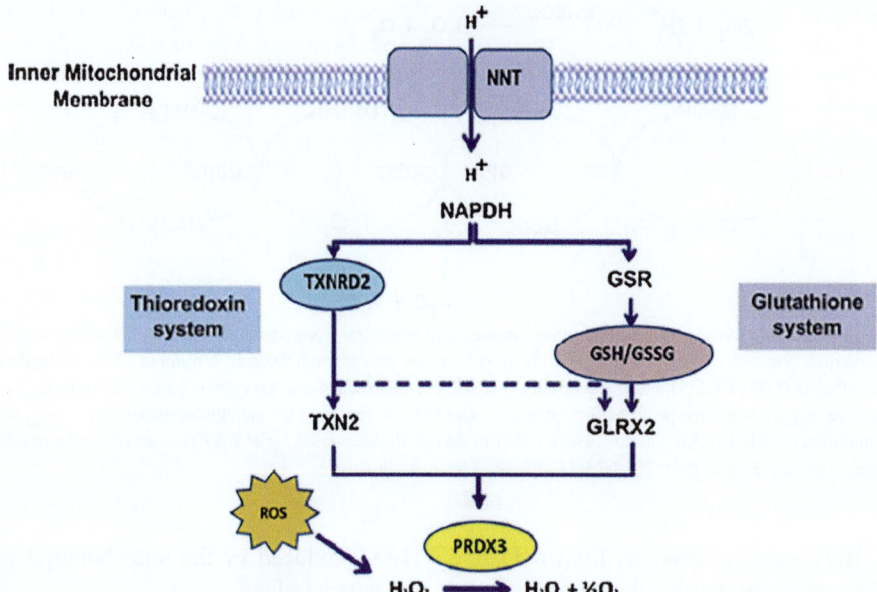

FIG. 1.4.15 The thioredoxin and glutathione redox homeostatic pathways. NADPH is the electron donor which drives the thioredoxin and glutathione pathways; NNT being the primary regenerator of NADPH levels. TXNRD2 reduces both TXN2 and GLRX2 (glutathione reductase and GSH play a further role in reduction of GLRX2) which subsequently reduces PRDX3. PRDX3, a highly conserved thioredoxin-dependent peroxide reductase protects the cellular environment from oxidative stress due to its antioxidant properties. Abbreviations: *GLRX2*, glutaredoxin 2; *GSH/GSSG*, glutathione/oxidized glutathione; *GSR*, glutathione reductase; *NNT*, nicotinamide nucleotide transhydrogenase; *PRDX3*, peroxiredoxin 3; *ROS*, reactive oxygen species; *TXN2*, thioredoxin 2; *TXNRD2*, thioredoxin reductase 2. *(From Maharaj A, Maudhoo A, Chan LF, Novoselova T, Prasad R, Metherell LA, Guasti L. Isolated glucocorticoid deficiency: genetic causes and animal models. J Steroid Biochem Mol Biol 2019;189:73–80. Fig. 2 p. 77.)*

cAMP-dependent cortisol secretion in H295R human adrenocortical cells (Lucki et al., 2012) by promoting the SREBP-1-dependent transcription of CYP17. S1P not only functions as an intracellular messenger but also exerts many of its effects through cell surface G-protein coupled receptors (Alvarez et al., 2007).

The mechanism of S1P export from cells is not completely understood, however, studies have provided evidence for the involvement of the ATP-binding cassette (ABC) family of transporters in this process (Takabe et al., 2010). Five S1P receptors (S1PR1–5) have been identified.

ACTH rapidly stimulates sphingolipid metabolism in H295R cells (Ozbay et al., 2004). ACTH and the cAMP analog Bt2cAMP decrease cellular amounts of sphingomyelin, ceramide and sphingosine, while simultaneously increasing the secretion of S1P. The S1P produced stimulates cortisol secretion from H295R cells by promoting the maturation and binding of sterol regulatory element binding protein 1 (SREBP1) to the CYP17 promoter, thereby inducing gene transcription. These findings implicate S1P as a paracrine mediator of ACTH-dependent CYP17 transcription. S1P rapidly increases cortisol biosynthesis and the mRNA expression of multiple genes involved in the acute phase of steroid hormone production including StAR, TSPO, SR-BI and LDLR.

S1P mediates steroid hormone biosynthesis by activating various intracellular cascades. S1P-mediated cortisol secretion in bovine adrenocortical cells is dependent on

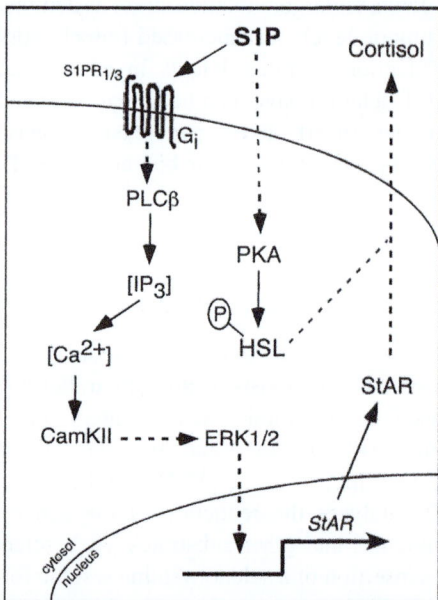

FIG. 1.4.16 Induction of StAR gene expression by sphingosine signaling. Sphingosine-1-phosphate (S1P) binds to S1PR receptors and activates phospholipase C (PLC), thereby increasing IP$_3$ and cytosolic Ca^{2+} that in turn activates CamKII (calmodulin-dependent kinase) and phosphorylation of ERK1/2 (extracellular signal regulated kinases). Activation of the S1P pathway culminates in an increase in cortisol secretion. *(From Lucki NC, Li D, Sewer MB. Sphingosine-1-phosphate rapidly increases cortisol biosynthesis and the expression of genes involved in cholesterol uptake and transport in H295R adrenocortical cells. Mol Cell Endocrinol 2012;348(1):165–75. Fig. 7 p. 173.)*

PKC and intracellular Ca^{2+} in a pertussis toxin-sensitive manner. A similar mechanism of action for S1P in human H295R cells was demonstrated, although PLC activation was also required. S1P induces the phosphorylation of HSL at Ser563, providing evidence that S1P regulates steroidogenesis by acting via multiple mechanisms. Given that ACTH/cAMP promotes the secretion of S1P from H295R cells, it is plausible that adrenocortical cells might utilize an ACTH/S1P feed-forward mechanism to facilitate rapid steroidogenic output and fine-tune sustained hormone production (Lucki et al., 2012).

1.4.5 Other regulators

The above description of regulation of steroidogenesis has failed to define the basis of adrenal and gonadal insufficiency in rare cases, some additional factors have come to light mainly through the application of molecular genetic tests. The factors described above can be classified as primary regulators. Some other factors are implicated though the mechanisms of action of so-called secondary regulators are unclear. This is an area of active research particularly in the context of drug discovery and tumor development. The detail is beyond the scope of this book and is taking endocrinology into extremely complicated mechanisms requiring an ever expanding glossary, new databases and interpretative tools. Only two factors will be considered in this chapter. Elsewhere epigenetics is another area where gene and protein modulation contributes to regulatory mechanisms (see Chapter 1.6—Recognition of steroids). DNA methylation and histone modification are the main mechanisms. The synthesis of steroids is also affected by circulating microRNA species (Azhar et al., 2020) (also briefly discussed in Chapter 1.6)

1.4.5.1 NRF 1

Granulosa cells (GCs) constitute the vast majority of follicular cells in mammalian ovary and functions in steroid synthesis during follicular development. Sex steroid hormones progesterone (P$_4$) and estradiol (E$_2$) participate in regulating ovarian function and act as antioxidant protecting follicles from oxidative stress and atresia. In mammals, more than 99% of follicles will undergo atresia during follicular growth and development and only one or a few ovarian follicles will ovulate. Follicular atresia is normally considered to be a hormonally controlled, apoptotic process, regulated by endocrine and paracrine changes among which, granulosa cell (GC) apoptosis is considered as a main mechanism via several signaling pathways (Asselin et al., 2000). Some factors known to regulate GC apoptosis in the goat include reproductive hormones, cytokines and apoptotic related factors, while the intracellular regulation of GC apoptosis is unclear (Matsuda et al., 2012).

Mitochondria play central roles in the initiation of apoptosis triggered by intrinsic death signals. Mitochondrial dysfunction in GCs results in follicular atresia (Glister et al., 2014). **Nuclear respiratory factor 1 (*NRF1*)**, as a major transcription factor, is involved in mitochondrial biogenesis, signal transduction and protein synthesis (Okoh et al., 2015; Mattingly et al., 2008). Goat GCs have a reduced *NRF1* expression in atretic follicles compared with healthy follicles (Zhang et al., 2015). The regulation of *NRF1* in human GC apoptosis and follicular atresia remains to be elucidated. In addition, E$_2$-induced DNA synthesis in MCF-7 breast cancer cells depends on mitochondrial oxidant signaling to *NRF1*, knockdown of *NRF1* blocks E$_2$ stimulation of mitochondrial biogenesis and activity. These studies indicate that *NRF1* may be involved in the regulation of steroidogenesis, oxidative stress and cell apoptosis.

Overall, aberrant expression of *NRF1* could induce mitochondrial dysfunction and disturb the cellular redox balance, which lead to disturbance of steroid hormone synthesis, and trigger LGC apoptosis through the mitochondria-dependent pathway. These findings will be helpful for understanding the role of *NRF1* in human ovarian follicular development and atresia (Zhang et al., 2017).

1.4.5.2 LGR4

WNT signaling has been demonstrated to have essential functions in the ovary but because WNT signaling is very complicated, the role of specific components in the ovary, especially in CL, is still not clear. The **leucine-rich repeat-containing GPCR 4** (LGR4) is one of the receptors for **R-spondins**, which augment the WNT signaling pathway. LGR4, also called GPR48, contains 17 leucine-rich repeats in its ectodomain and belongs to the evolutionarily conserved LGR subgroup of membrane GPCRs, which includes FSH receptor, LHCGR, TSH receptor and relaxin receptor (LGR7 and LGR8). Using genetically modified mouse models, Lgr4 is widely expressed and plays important roles in both embryonic and postnatal development of multiple tissues and organs, including the eye, kidney, intestine, bone, male reproductive tract, uterus and others.

LGR4 is a newly identified receptor for R-spondin1–4, which amplify both the canonical and noncanonical WNT signaling to regulate DNA synthesis and gap junction assembly in GCs through catenin B11 (CTNNB1) (classical pathway for WNt path). The function of CTNNB1 in granulosa-lutein (GL) cells is obscure. WNT signaling is critical to promoting CL maturation and progesterone production. CTNNB1 promotes preovulatory follicular development and represses LH-induced ovulation and luteinization, so it is possible that the activity of WNT/CTNNB1 would be suppressed during a critical time frame of ovulation and reactivated to promote luteinization. WNT signaling in GL cells is partially mediated through LGR4 signaling although all four R-spondins have the activity to potentiate

LGR4-mediated WNT signaling in vitro. The precise endogenous ligand for LGR4 in GL cells needs to be found.

EGFR-ERK signaling also plays critical roles in oocyte maturation, ovulation and luteinization. Disruption of the ERK pathway, specifically in the mouse, GCs resulted in the failure of oocyte resumption of meiosis and ovulation (Pan et al., 2014). Overall the expression of LGR4 was specifically induced in GCs during luteinization. After binding with R-spondins (RSPO2, for example), LGR4 augments WNT/CTNNB1 activity, which induces Mmp9 mRNA transcription. MMP9 releases the active form of HB-EGF, which further activates EGFR-ERK (epidermal growth factor receptor activated extracellular-signal regulated kinase) signaling for CL maturation and progesterone secretion in GL cells (Fig. 1.4.17).

Lgr4 deficiency downregulated the WNT-EGFR-ERK signaling, which led to the impairment of the differentiation of GL cells and CL maturation. Because luteal-phase insufficiency is one of the major causes of female infertility, identifying the essential roles of LGR4 in these patients represents an advance that has implications for diagnosis and as a potential drug target for luteal-phase insufficiency.

1.4.6 Microbiota and regulation of steroids

Since the start of this century the impact of the gut microbiota, and the bacteria on other mucosal surfaces, on health has developed as an exciting area for research. Forty trillion microbial cells live on and inside the human body.

In the terminal ileum, and colon concentrations of micro-organisms can reach 10^{11} per gram. The gut microbiome is dominated by Firmicutes, that includes Lactobacillus, Clostridium and Enterococcus species and Bacteroides. Firmicutes are mostly gram positive bacteria, whereas Bacteroidetes are gram negative. Many other phyla are present including Actinobacteria (e.g., Bifidobacteria, Proteobacteria (e.g., *Escherichia coli*), Fusobacteria, Verrucomicrobia and Cyanobacteria).

The impact of the gut microbiome on circulating plasma steroid concentrations has been known for a long time in man (Eriksson et al., 1969; Eriksson, 1971a,b; Adlercreutz et al., 1979) and rats reared normally or germfree (Eriksson et al., 1969; Eriksson, 1979b). The intestinal bacteria sit within the enterohepatic circulation of steroids (Eriksson, 1971a,b; Adlercreutz et al., 1979). Steroid concentrations in bile (Cronholm et al., 1972), the bacteria (Ervin et al., 2019) and feces (Eriksson et al., 1970) and the circulation (Shin et al., 2019; Flores et al., 2012) highlight differences through bacterial metabolism The gut flora has been associated with developmental diseases, obesity, intestinal illnesses, inflammatory conditions, cancers and psychiatric problems (Jackson et al., 2018; Morris and Brem, 2019). At the present time, bacteria can be easily characterized by 16S RNA analysis and this has opened the potential to determine the effects of the microbiome on health and to study if manipulation of the gut microflora can be used therapeutically. The gut flora of infants has been shown to be different depending on route of delivery, a vaginal delivery (Al'Abri et al., 2021; Reyman et al., 2019) leads to a microbiome of the infant

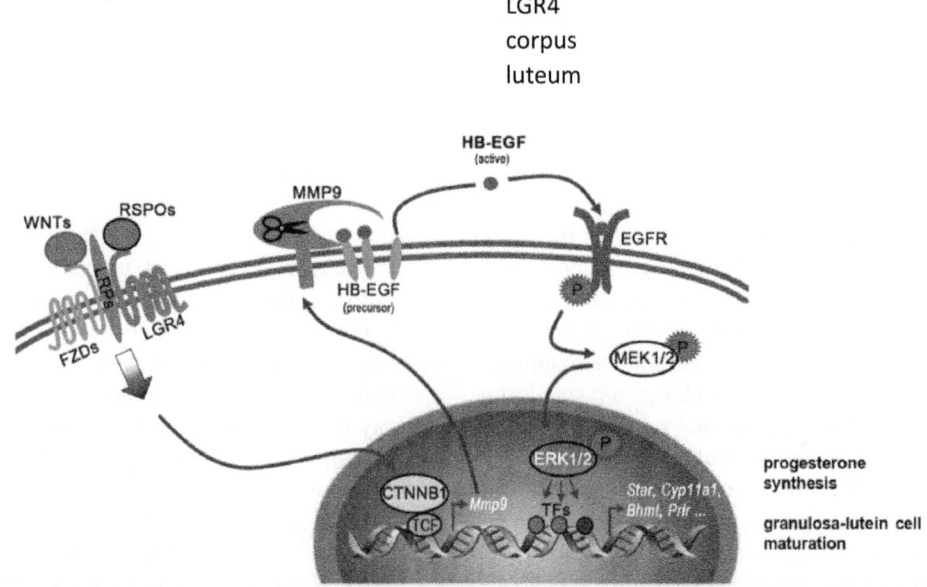

FIG. 1.4.17 Schematic diagram of LGR-4 (**leucine-rich repeat-containing G-protein coupled receptor 4**) mediated WNT signaling. LGR-4 is the receptor for R-spondins that potentiates the canonical Wnt signaling pathway. Upon binding to R-spondins, LGR-4 associates with phosphorylated LRP6 and frizzled receptors that are activated by extracellular Wnt receptors, triggering the canonical Wnt signaling pathway to increase expression of target genes in the EGFR-ERK pathway through activation of MMP9 (matrix metallopeptidase 9). *(From Pan H, Cui H, Liu S, Qian Y, Wu H, Li L, Guan Y, Guan X, Zhang L, Fan HY, Ma Y, Li R, Liu M, Li D. Lgr4 gene regulates corpus luteum maturation through modulation of the WNT-mediated EGFR-ERK signaling pathway. Endocrinology 2014;155(9):3624–37. Fig. 7 p. 3634.)*

nearer the gut flora of the mother than if by cesarean section. Changes in life will depend on gender, diet, exposure to antibiotics and endocrine disrupting chemicals which considerably influence standardization of experiments with human subjects. Dysbiosis is a term used to reflect change in the microbial flora. The actions of bacterial enzymes include glucuronidases, dehydrogenases, dehydroxylations and side chain cleavage which are described in more detail in Chapter 1.8—Breakdown and clearance of steroids. Unwanted pregnancy when taking oral contraceptives and antibiotics that disrupted the EHC of steroids is a long known consequence of estrogen dysbiosis (Adlercreutz et al., 1984) and other effects will be characterized with the passage of time and research on this axis. The impact of the bacterial flora on hypertension has been studied in rats (Morris and Brem, 2019; Honour, 2015) and man including the production of glycyrrhetinic acid like factors (GALFs) (Morris and Ridlon, 2017) and hypertension have also been described in Chapter 1.7.

1.4.7 Summary

Steroid synthesis is a highly regulated system with ever changing components of the orchestra. The pituitary gland is still an important conductor but there are further elements at, or from the steroidogenic organs, where receptor mediated mechanisms locally, or acting back to the pituitary, are subject to much variation. The pituitary peptides act through cell membrane receptors and cyclic AMP production. Mineralocorticoid synthesis is determined by a kidney activated system and angiotensin production. Just as steroid assays will be seen to have improved over 70 years, the measurement of regulatory protein concentrations in plasma have depended on specificity and accuracy of analysis. Standardization of protein assays is a real challenge. New proteins such as melatonin, leptin, kisspeptin and neurokinin B are being included in endocrine investigations and others will undoubtedly follow. Epigenetic regulation of genes through methylation and acetylation can result in heritable changes in production of the final protein product of the gene. In recent years, microRNAs (single-stranded, short, noncoding sequences of RNA) have been found to regulate the expression of target genes by blocking translation or promoting degradation. Our understanding of these areas will develop in the course of time. There is much left to be learned about the regulation of steroid synthesis with the role of the microbiome being a new contribution.

REFERENCES

Adlercreutz H, Martin F, Järvenpää P, Fotsis T. Steroid absorption and enterohepatic recycling. Contraception 1979;20(3):201–23.

Adlercreutz H, Pulkkinen MO, Hämäläinen EK, Korpela JT. Studies on the role of intestinal bacteria in metabolism of synthetic and natural steroid hormones. J Steroid Biochem 1984;20(1):217–29.

Al'Abri IS, Durmusoglu D, Crook N. What *E. coli* knows about your 1-year-old infant: antibiotic use, lifestyle, birth mode, and siblings. Cell Host Microbe 2021;29(6):854–5.

Alvarez SE, Milstien S, Spiegel S. Autocrine and paracrine roles of sphingosine-1-phosphate. Trends Endocrinol Metab 2007;18(8):300–7.

Anckaert E, Öktem M, Thies A, Cohen-Bacrie M, Daan NM, Schiettecatte J, et al. Multicenter analytical performance evaluation of a fully automated anti-Müllerian hormone assay and reference interval determination. Clin Biochem 2016;49(3):260–7. Erratum in: Clin Biochem. 2020;76:47–8.

Anderson RC, Newton CL, Anderson RA, Millar RP. Gonadotropins and their analogs: current and potential clinical applications. Endocr Rev 2018;39(6):911–37.

Asselin E, Xiao CW, Wang YF, Tsang BK. Mammalian follicular development and atresia: role of apoptosis. Biol Signals Recept 2000; 9(2):87–95.

Azhar S, Dong D, Shen WJ, Hu Z, Kraemer FB. The role of miRNAs in regulating adrenal and gonadal steroidogenesis. J Mol Endocrinol 2020;64(1):R21–43.

Balakumar P, Jagadeesh G. A century old renin-angiotensin system still grows with endless possibilities: AT1 receptor signaling cascades in cardiovascular physiopathology. Cell Signal 2014;26(10):2147–60.

Balanescu S, Kopp P, Gaskill MB, Morgenthaler NG, Schindler C, Rutishauser J. Correlation of plasma copeptin and vasopressin concentrations in hypo-, iso-, and hyperosmolar states. J Clin Endocrinol Metab 2011;96(4):1046–52.

Bedenk J, Vrtačnik-Bokal E, Virant-Klun I. The role of anti-Müllerian hormone (AMH) in ovarian disease and infertility. J Assist Reprod Genet 2020;37(1):89–100.

Berger I, Werdermann M, Bornstein SR, Steenblock C. The adrenal gland in stress—adaptation on a cellular level. J Steroid Biochem Mol Biol 2019;190:198–206.

Bernstone L, Adaway JE, Keevil BG. An LC-MS/MS assay for analysis of equilibrium angiotensin II in human serum. Ann Clin Biochem 2021;58(5):422–33.

Bollag WB. Regulation of aldosterone synthesis and secretion. Compr Physiol 2014;4:1017–55.

Brizuela L, Rábano M, Gangoiti P, Narbona N, Macarulla JM, Trueba M, et al. Sphingosine-1-phosphate stimulates aldosterone secretion through a mechanism involving the PI3K/PKB and MEK/ERK 1/2 pathways. J Lipid Res 2007;48(10):2264–74.

Bussy U, Wang H, Chung-Davidson YW, Li W. Simultaneous determination of gonadotropin-inhibitory and gonadotropin-releasing hormones using ultra-high performance liquid chromatography electrospray ionization tandem mass spectrometry. Anal Bioanal Chem 2015;407(2):497–507.

Caltabiano G, Campillo M, De Leener A, Smits G, Vassart G, Costagliola S, et al. The specificity of binding of glycoprotein hormones to their receptors. Cell Mol Life Sci 2008;65(16):2484–92.

Campo S, Andreone L, Ambao V, Urrutia M, Calandra RS, Rulli SB. Hormonal regulation of follicle-stimulating hormone glycosylation in males. Front Endocrinol 2019;10:17.

Casarini L, Crépieux P. Molecular mechanisms of action of FSH. Front Endocrinol 2019;10:305 [review].

Cawley NX, Li Z, Loh YP. 60 YEARS OF POMC: biosynthesis, trafficking, and secretion of pro-opiomelanocortin-derived peptides. J Mol Endocrinol 2016;56(4):T77–97.

Chai SY, Fernando R, Peck G, Ye SY, Mendelsohn FA, Jenkins TA, et al. The angiotensin IV/AT4 receptor. Cell Mol Life Sci 2004;61 (21):2728–37. https://doi.org/10.1007/s00018-004-4246-1. 15549174.

Chappell DL, McAvoy T, Weiss B, Weiner R, Laterza OF. Development and validation of an ultra-sensitive method for the measurement of plasma renin activity in human plasma via LC-MS/MS. Bioanalysis 2012;4(23):2843–50.

Clark AJL, Chan L. Stability and turnover of the ACTH receptor complex. Front Endocrinol 2019;10:491.

Cronholm T, Eriksson H, Gustafsson JA. Excretion of steroid hormone metabolites in bile of male rats. Steroids 1972;19(4):455–70.

Cui H, Zhao G, Wen J, Tong W. Follicle-stimulating hormone promotes the transformation of cholesterol to estrogen in mouse adipose tissue. Biochem Biophys Res Commun 2018;495(3):2331–7.

Cuvillier O, Pirianov G, Kleuser B, Vanek PG, Coso OA, Gutkind S, et al. Suppression of ceramide-mediated programmed cell death by sphingosine-1-phosphate. Nature 1996;381(6585):800–3.

Dahabiyeh LA, Tooth D, Carrell RW, Read RJ, Yan Y, Pipkin FB, et al. Measurement of the total angiotensinogen and its reduced and oxidised forms in human plasma using targeted LC-MS/MS. Anal Bioanal Chem 2019;411(2):427–37.

Dedic N, Chen A, Deussing JM. The CRF family of neuropeptides and their receptors—mediators of the central stress response. Curr Mol Pharmacol 2018;11(1):4–31.

Eriksson H. Absorption and enterohepatic circulation of neutral steroids in the rat. Eur J Biochem 1971a;19(3):416–23.

Eriksson H. Steroids in germfree and conventional rats. Metabolites of (4-14C)pregnenolone and (4-14C)corticosterone in urine and faeces from male rats. Eur J Biochem 1971b;18(1):86–93.

Eriksson H, Gustafsson JA, Sjövall J. Steroids in germfree and conventional rats. 21-dehydroxylation by intestinal microorganisms. Eur J Biochem 1969;9(4):550–4.

Eriksson H, Gustafsson JA, Sjövall J. Excretion of steroid hormones in adults. C19 and C21 steroids in faeces from pregnant women. Eur J Biochem 1970;12(3):520–6.

Ervin SM, Li H, Lim L, Roberts LR, Liang X, Mani S, et al. Gut microbial β-glucuronidases reactivate estrogens as components of the estrobolome that reactivate estrogens. J Biol Chem 2019;294(49):18586–99.

Ferguson JM, Pépin D, Duru C, Matejtschuk P, Donahoe PK, Burns CJ. Towards international standardization of immunoassays for Müllerian inhibiting substance/anti-Müllerian hormone. Reprod Biomed Online 2018;37(5):631–40.

Ferguson J, Hockley J, Rigsby P, Burns C. Establishment of a WHO reference reagent for anti-Mullerian hormone. Reprod Biol Endocrinol 2020;18(1):86.

Findling JW, Engeland WC, Raff H. The use of immunoradiometric assay for the measurement of ACTH in human plasma. Trends Endocrinol Metab 1990;1(6):283–7.

Flores R, Shi J, Fuhrman B, Xu X, Veenstra TD, Gail MH, et al. Fecal microbial determinants of fecal and systemic estrogens and estrogen metabolites: a cross-sectional study. J Transl Med 2012;10:253.

Franssen D, Tena-Sempere M. The kisspeptin receptor: a key G-protein-coupled receptor in the control of the reproductive axis. Best Pract Res Clin Endocrinol Metab 2018;32(2):107–23.

Glister C, Hatzirodos N, Hummitzsch K, Knight PG, Rodgers RJ. The global effect of follicle-stimulating hormone and tumour necrosis factor α on gene expression in cultured bovine ovarian granulosa cells. BMC Genomics 2014;15:72.

Guo Z, Poglitsch M, McWhinney BC, Ungerer JPJ, Ahmed AH, Gordon RD, et al. Measurement of equilibrium angiotensin II in the diagnosis of primary aldosteronism. Clin Chem 2020;66(3):483–92.

He C, Hu S, Zhou W. Development of a novel nanoflow liquid chromatography-parallel reaction monitoring mass spectrometry-based method for quantification of angiotensin peptides in HUVEC cultures. PeerJ 2020;8:e9941.

Honour JW. Historical perspective: gut dysbiosis and hypertension. Physiol Genomics 2015;47(10):443–6.

Honour JW, Bridges NA, Conway-Phillips E, Hindmarsh PC. Plasma aldosterone response to the low-dose adrenocorticotrophin (ACTH 1-24) stimulation test. Clin Endocrinol (Oxf) 2008;68(2):299–303.

Jackson MA, Verdi S, Maxan ME, Shin CM, Zierer J, Bowyer RCE, et al. Gut microbiota associations with common diseases and prescription medications in a population-based cohort. Nat Commun 2018;9(1):2655.

Juutilainen A, Savolainen K, Romppanen J, Turpeinen U, Hämäläinen E, Kemppainen J, et al. Combination of LC-MS/MS aldosterone and automated direct renin in screening for primary aldosteronism. Clin Chim Acta 2014;433:209–15.

Leng G, Sabatier N. Measuring oxytocin and vasopressin: bioassays, immunoassays and random numbers. J Neuroendocrinol 2016;28(10). https://doi.org/10.1111/jne.12413.

Li H, Popp R, Frohlich B, Chen MX, Borchers CH. Peptide and protein quantification using automated Immuno-MALDI (iMALDI). J Vis Exp 2017;126:55933.

Li HWR, Robertson DM, Burns C, Ledger WL. Challenges in measuring AMH in the clinical setting. Front Endocrinol 2021;12, 691432.

Lightman SL, Birnie MT, Conway-Campbell BL. Dynamics of ACTH and cortisol secretion and implications for disease. Endocr Rev 2020;41(3):470–90.

Lucki NC, Li D, Sewer MB. Sphingosine-1-phosphate rapidly increases cortisol biosynthesis and the expression of genes involved in cholesterol uptake and transport in H295R adrenocortical cells. Mol Cell Endocrinol 2012;348(1):165–75.

MacKenzie SM, van Kralingen JC, Davies E. Regulation of aldosterone secretion. Vitam Horm 2019;109:241–63.

Maggi R, Cariboni AM, Marelli MM, Moretti RM, Andrè V, Marzagalli M, et al. GnRH and GnRH receptors in the pathophysiology of the human female reproductive system. Hum Reprod Update 2016;22(3):358–81.

Matsuda F, Inoue N, Manabe N, Ohkura S. Follicular growth and atresia in mammalian ovaries: regulation by survival and death of granulosa cells. J Reprod Dev 2012;58(1):44–50.

Mattingly KA, Ivanova MM, Riggs KA, Wickramasinghe NS, Barch MJ, Klinge CM. Estradiol stimulates transcription of nuclear respiratory factor-1 and increases mitochondrial biogenesis. Mol Endocrinol 2008;22(3):609–22.

Meimaridou E, Goldsworthy M, Chortis V, Fragouli E, Foster PA, Arlt W, et al. NNT is a key regulator of adrenal redox homeostasis and steroidogenesis in male mice. J Endocrinol 2018;236(1):13–28.

Messinis IE, Messini CI, Anifandis G, Garas A, Daponte A. Gonadotropin surge-attenuating factor: a nonsteroidal ovarian hormone controlling GnRH-induced LH secretion in the normal menstrual cycle. Vitam Horm 2018;107:263–86.

Miller WL. The hypothalamic pituitary adrenal axis: a brief history. Horm Res Paediatr 2018;89:212–23.

Moenter SM, Silveira MA, Wang L, Adams C. Central aspects of systemic oestradiol negative- and positive-feedback on the reproductive neuroendocrine system. J Neuroendocrinol 2019; e12724.

Morris DJ, Brem AS. Role of gut metabolism of adrenal corticosteroids and hypertension: clues gut-cleansing antibiotics give us. Physiol Genomics 2019;51(3):83–9.

Morris DJ, Ridlon JM. Glucocorticoids and gut bacteria: "the GALF hypothesis" in the metagenomic era. Steroids 2017;125:1–13.

Nicolaides NC, Charmandari E, Kino T, Chrousos GP. Stress-related and circadian secretion and target tissue actions of glucocorticoids: impact on health. Front Endocrinol 2017;28(8):70.

Novoselova TV, Chan LF, Clark AJL. Pathophysiology of melanocortin receptors and their accessory proteins. Best Pract Res Clin Endocrinol Metab 2018;32(2):93–106.

Novoselova T, King P, Guasti L, Metherell LA, Clark AJL, Chan LF. ACTH signalling and adrenal development: lessons from mouse models. Endocr Connect 2019. pii: EC-19-0190.R1.

Okoh VO, Garba NA, Penney RB, Das J, Deoraj A, Singh KP, et al. Redox signalling to nuclear regulatory proteins by reactive oxygen species contributes to oestrogen-induced growth of breast cancer cells. Br J Cancer 2015;112(10):1687–702.

O'Connor JF, Kovalevskaya G, Birken S, Schlatterer JP, Schechter D, McMahon DJ, et al. The expression of the urinary forms of human luteinizing hormone beta fragment in various populations as assessed by a specific immunoradiometric assay. Hum Reprod 1998;13(4):826–35.

Oster H, Challet E, Ott V, Arvat E, de Kloet ER, Dijk DJ, et al. The functional and clinical significance of the 24-hour rhythm of circulating glucocorticoids. Endocr Rev 2017;38(1):3–45.

Owen LJ, Adaway J, Morris K, Lockhart S, Keevil BG. A widely applicable plasma renin activity assay by LC-MS/MS with offline solid phase extraction. Ann Clin Biochem 2014;51(Pt 3):409–11.

Ozbay T, Merrill Jr AH, Sewer MB. ACTH regulates steroidogenic gene expression and cortisol biosynthesis in the human adrenal cortex via sphingolipid metabolism. Endocr Res 2004;30(4):787–94. https://doi.org/10.1081/erc-200044040. 15666826.

Pan H, Cui H, Liu S, Qian Y, Wu H, Li L, et al. Lgr4 gene regulates corpus luteum maturation through modulation of the WNT-mediated -EGFR-ERK signaling pathway. Endocrinology 2014;155(9):3624–37.

Parker LN. Control of adrenal androgen secretion. Endocrinol Metab Clin North Am 1991;20(2):401–21.

Popp R, Malmström D, Chambers AG, Lin D, Camenzind AG, van der Gugten JG, et al. An automated assay for the clinical measurement of plasma renin activity by immuno-MALDI (iMALDI). Biochim Biophys Acta 2015;1854(6):547–58.

Rábano M, Peña A, Brizuela L, Marino A, Macarulla JM, Trueba M, et al. Sphingosine-1-phosphate stimulates cortisol secretion. FEBS Lett 2003;535(1–3):101–5.

Ren X, Zou L, Zhang X, Branco V, Wang J, Carvalho C, et al. Redox signaling mediated by thioredoxin and glutathione systems in the central nervous system. Antioxid Redox Signal 2017;27(13):989–1010.

Reyman M, van Houten MA, van Baarle D, Bosch AATM, Man WH, Chu MLJN, et al. Impact of delivery mode-associated gut microbiota dynamics on health in the first year of life. Nat Commun 2019;10(1):4997.

Robertson DM, Burger HG, Fuller PJ. Inhibin/activin and ovarian cancer. Endocr Relat Cancer 2004;11(1):35–49.

Rouault AAJ, Srinivasan DK, Yin TC, Lee AA, Sebag JA. Melanocortin receptor accessory proteins (MRAPs): functions in the melanocortin system and beyond. Biochim Biophys Acta Mol Basis Dis 2017;1863(10 Pt. A):2462–7.

Sailer CO, Refardt J, Blum CA, Schnyder I, Molina-Tijeras JA, Fenske W, et al. Validity of different copeptin assays in the differential diagnosis of the polyuria-polydipsia syndrome. Sci Rep 2021;11(1):10104.

Schulz A, Jankowski J, Zidek W, Jankowski V. Absolute quantification of endogenous angiotensin II levels in human plasma using ESI-LC-MS/MS. Clin Proteomics 2014;11(1):37.

Seccia TM, Caroccia B, Gomez-Sanchez EP, Gomez-Sanchez CE, Rossi GP. The biology of normal zona glomerulosa and aldosterone-producing adenoma: pathological implications. Endocr Rev 2018;39(6):1029–56.

Shin JH, Park YH, Sim M, Kim SA, Joung H, Shin DM. Serum level of sex steroid hormone is associated with diversity and profiles of human gut microbiome. Res Microbiol 2019;170(4–5):192–201.

Skorupskaite K, George JT, Anderson RA. The kisspeptin-GnRH pathway in human reproductive health and disease. Hum Reprod Update 2014;20(4):485–500.

Smith SC, McIntosh N, James K. Pitfalls in the use of ELISA to screen for monoclonal antibodies raised against small peptides. J Immunol Methods 1993;158(2):151–60.

Spiegel S, Milstien S. Sphingosine 1-phosphate, a key cell signaling molecule. J Biol Chem 2002;277(29):25851–4.

Stimson RH, Mohd-Shukri NA, Bolton JL, Andrew R, Reynolds RM, Walker BR. The postprandial rise in plasma cortisol in men is mediated by macronutrient-specific stimulation of adrenal and extra-adrenal cortisol production. J Clin Endocrinol Metab 2014;99(1):160–8.

Takabe K, Kim RH, Allegood JC, Mitra P, Ramachandran S, Nagahashi M, et al. Estradiol induces export of sphingosine 1-phosphate from breast cancer cells via ABCC1 and ABCG2. J Biol Chem 2010;285(14):10477–86.

Tsukazaki Y, Senda N, Kubo K, Yamada S, Kugoh H, Kazuki Y, et al. Development of a high-sensitivity quantitation method for arginine vasopressin by high-performance liquid chromatography tandem mass spectrometry, and comparison with quantitative values by radioimmunoassay. Anal Sci 2016;32(2):153–9.

Tsutsui K, Ubuka T. Gonadotropin-inhibitory hormone (GnIH): a new key neurohormone controlling reproductive physiology and behavior. Front Neuroendocrinol 2021;61, 100900.

Uenoyama Y, Inoue N, Maeda KI, Tsukamura H. The roles of kisspeptin in the mechanism underlying reproductive functions in mammals. J Reprod Dev 2018;64(6):469–76.

van Rooyen JM, Poglitsch M, Huisman HW, Mels C, Kruger R, Malan L, et al. Quantification of systemic renin-angiotensin system peptides of hypertensive black and white African men established from the RAS-fingerprint®. J Renin Angiotensin Aldosterone Syst 2016;17(4). pii: 1470320316669880.

Whiting G, Ferguson J, Fang M, Pepin D, Donahoe P, Matejtschuk P, et al. Quantification of Müllerian inhibiting substance/anti-Müllerian hormone polypeptide by isotope dilution mass spectrometry. Anal Biochem 2018;560:50–5.

Yang T, He M, Hu C. Regulation of aldosterone production by ion channels: from basal secretion to primary aldosteronism. Biochim Biophys Acta Mol Basis Dis 2018;1864(3):871–81.

Zavala E, Wedgwood KCA, Voliotis M, Tabak J, Spiga F, Lightman SL, et al. Mathematical modelling of endocrine systems. Trends Endocrinol Metab 2019;30(4):244–57.

Zhang G, Wan Y, Zhang Y, Lan S, Jia R, Wang Z, et al. Expression of mitochondria-associated genes (PPARGC1A, NRF-1, BCL-2 and BAX) in follicular development and atresia of goat ovaries. Reprod Domest Anim 2015;50(3):465–73.

Zhang GM, Deng MT, Lei ZH, Wan YJ, Nie HT, Wang ZY, et al. Effects of *NRF* on steroidogenesis and apoptosis in goat luteinized granulosa cells. Reproduction 2017;154(2):111–22.

Chapter 1.5

Steroids in blood and tissues

1.5.1 Introduction

For years it was assumed steroids diffuse across cell membranes but in recent years the intracellular transport and mechanisms for steroids crossing membranes have been recognized for movement and efflux of cholesterol through steroid synthesis and delivery of steroid hormones to target organs for physiological effects. Steroids as hormones are, by definition, transported from the endocrine glands to target sites in the blood where the majority is bound to proteins. The steroid hormones have relatively specific-binding globulins in the circulation thus cortisol is bound to cortisol-binding globulin (CBG) and for testosterone, estradiol and progesterone there is sex hormone-binding globulin (SHBG). Albumin in blood is a nonspecific steroid-binding protein.

1.5.2 Cholesterol transport

Cholesterol is needed in the body as an essential component of cell membranes as well as a source of sterols and steroids. Cholesterol is absorbed from the intestine, by the action of several transporters that prevent excess uptake, or is synthesized in the body. Low density lipoprotein (LDL) is the main carrier for steroids in humans. Most of the cholesterol derives in the small intestine through a receptor mediated endocytotic uptake of **low density lipoprotein (LDL)** into vesicles that fuse with lysosomes. Cholesterol is insoluble in water and is transported in the blood mainly as cholesterol esters bound to lipoproteins. LDL is 35%–45% cholesteryl ester, 5%–10% free cholesterol, 5%–12% triacylglycerol, 20%–25% phospholipids and 20%–25% protein. The predominant apoprotein of LDL is apoB. Cholesterol contained within HDL can be delivered to cells following binding of the lipoproteins to the **scavenger receptor, class B Type 1 (SR-B1)** (Shen et al., 2018a,b) on the cell surface with transfer of cholesterol esters across cell membrane to the interior without endocytosis of the intact lipoprotein particle. Intracellular cholesterol trafficking employs two mechanisms: "vesicular transport," where membranes fuse with other membranes to deliver cholesterol from one compartment to another, or "nonvesicular transport," which involves cholesterol-binding proteins. The cholesterol can be esterified by acyl-CoA cholesterol acyl transferase (ACAT) (Chang et al., 2009) and stored in lipid droplets (Fig. 1.5.1).

The **cholesterol esters are hydrolyzed by hormone sensitive lipase** before delivery of cholesterol to the mitochondria. In the liver, LDL is removed from the blood by hepatic receptor **LDLR** which binds the apoB protein. Cholesterol is excreted from the body to maintain homeostasis.

1.5.2.1 Cholesterol uptake

The mechanisms of cholesterol transfer in adrenal cells are not well understood. Much of the evidence for cholesterol transport mechanisms come from cell organelle reconstitution assay systems and cell knockdown experiments which may create artificial data. Further discussion can be found in a comprehensive reviews on the initial steps of steroid synthesis (Miller, 2017a,b). Some clinical disorders (Smith-Lemli-Opitz syndrome, Wolman disease, **Nieman-Pick C** syndrome [NPC]) have given clues for the involvement of other cholesterol transfer steps like NPC1 and NPC2 (**soluble NSF attachment protein receptor**) (SNARE) (Kraemer et al., 2017) and **oxysterol-binding protein** (Pfisterer et al., 2016). These conditions will be discussed later in clinical sections (Part III).

Other proteins may be involved in cholesterol movement into the mitochondria (Midzak and Papadopoulos, 2014). The benzodiazepine receptor (now called **translocator protein TSPO**) (Papadopoulos et al., 2018; Selvaraj et al., 2015) and its ligand, **the ATP-binding cassette subfamily D (ABCD1)** and the **sterol carrier protein 2 (SCP2)** (Shen et al., 2018a,b) may act as a docking site for **Steroidogenic acute regulatory protein (StAR)** (Tugaeva and Sluchanko, 2019; Clark, 2020) but more research is required to delineate all the players in cholesterol transport.

1.5.2.2 Steroidogenic acute regulatory protein (StAR)

There is intracellular trafficking of cholesterol between touching membranes and by movement with cholesterol-binding proteins. Steroidogenic acute regulatory protein (StAR) facilitates the transfer of cholesterol from the outer to the inner mitochondrial membrane (Stocco et al., 2017).

Steroids in the Laboratory and Clinical Practice. https://doi.org/10.1016/B978-0-12-818124-9.00011-5

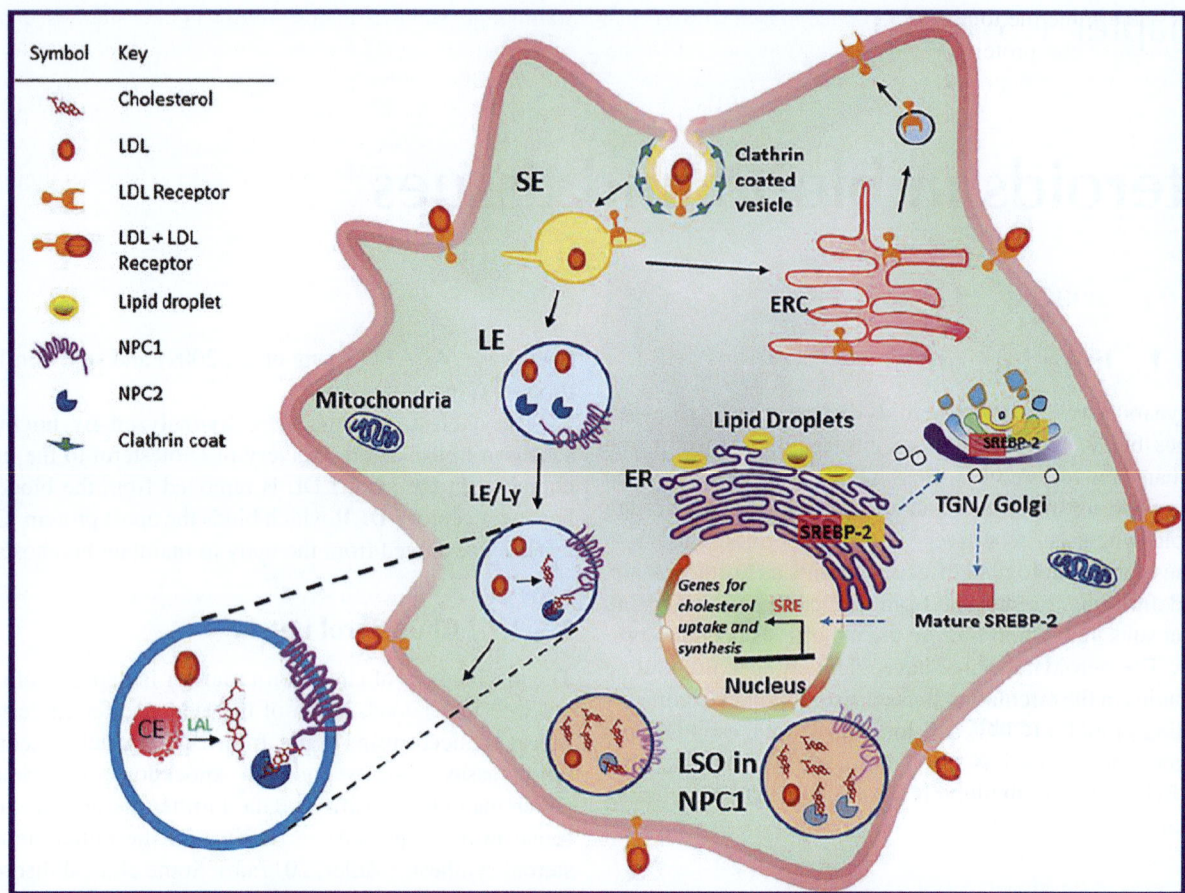

FIG. 1.5.1 Schematic diagram of intracellular sterol trafficking. Cholesterol is synthesized in the endoplasmic reticulum (ER). If cells are depleted of cholesterol the SREBP2-SCAP complex migrates to the Golgi where SREBP2 undergoes proteolytic cleavage. The mature SREBP2 translocates to the nucleus where it activates the sterol response element (SRE) responsible for regulating genes involved in cholesterol uptake and synthesis. Cells also obtain cholesterol by receptor mediated endocytosis of low density lipoproteins (LDL) (bottom, left) which are delivered to late endosomes (LE) or lysosomes (Ly) where the cholesterol esters in LDL are hydrolyzed by lysosomal acid lipase (LAL). Free cholesterol is distributed among various organelles by a combination of vesicular and nonvesicular pathways. Excess cholesterol is esterified in the ER by acylCoA:cholesterol acyl transferase (ACAT) and stored in lipid droplets. *(From Maxfield FR, Iaea DB, Pipalia NH. Role of STARD4 and NPC1 in intracellular sterol transport. Biochem Cell Biol 2016;94(6):499–506. Fig. 1 p. 500.)*

Several related proteins called START (StAR related lipid transfer proteins) are likely involved in intracellular cholesterol transfer. Within the START family there are a number of **START domain proteins (StARD)** (Soccio et al., 2002) of which StARD4 and 5 are likely involved in cholesterol transfer (Iaea et al., 2020; Maxfield et al., 2016).

1.5.3 Steroids in blood

Only very small amounts of the steroid hormones in the blood exist in the unbound state. These are thought to be the biologically active components (Henley et al., 2016). In the case of cortisol, from 5% to 10% of the hormone is free in plasma, whereas for testosterone and estradiol it is between 1% and 3%, respectively. Steroids bind to **albumin** as well as to specific-binding proteins. The concentrations of **Sex hormone-binding globulin (SHBG)** and **cortisol-binding globulin (CBG)** influence the interpretation of sex steroids in clinical investigations (Hammond, 2016).

Dialysis studies revealed that with increasing hormone concentration in plasma there was increasing unbound hormone. Note this starts from physiological levels of cortisol and extends through the pathological levels found in cortisol excess states. Because the steroids in blood are largely protein bound the low concentrations of the free steroids are difficult to determine. When steroids are removed from the binding proteins the combined bound and free steroids can be measured as the total steroid concentrations which are normal practice though in many commercial methods the mechanism is not declared. The free steroid may be the biologically active steroid although there is a debate about the uptake of the steroid bound to proteins.

1.5.3.1 Total steroids

The binding properties of CBG and SHBG were first studied in 1981 (Dunn et al., 1981) and later for SHBG compared with other proteins (Hong et al., 2015). In general, steroid assays determine the total steroid concentrations (after displacement of the protein bound steroids) are added to the free hormone component is serum/plasma. The assay technology is described in detail in Chapters 2.3 and 2.4 of the book and the concentrations of the total serum steroids are in Chapter 2.5 according to age, gender and a number of clinical scenarios.

1.5.3.2 SHBG

The first clear evidence for sex hormone binding came from plasma electrophoretic studies which showed that radioactive estradiol added to the sample was bound to a β-globulin distinct from albumin and CBG (Sandberg et al., 1957). Purification of SHBG proved difficult but was achieved using affinity chromatography, gel filtration and ion-exchange chromatography. SHBG has a molecular weight of around 52,000 Da. It is readily denatured by heating to 60°C; steroid-binding affinity is higher at low temperatures. In terms of steroid structure, a planar A-ring and a 17β-hydroxyl group are necessary for binding. Among the androgens there is an apparent relationship between biological activity and ability to bind to SHBG. A small proportion of the total testosterone in blood is bound to cortisol-binding globulin (CBG); however, the main testosterone fraction is bound to sex hormone-binding globulin (SHBG). There are changes in SHBG concentrations in the menstrual cycle (Jia et al., 1992), with use of oral contraceptives (Hammond et al., 1984) and in pregnancy. The concentration of SHBG is 15% higher in the luteal phase than the follicular phase.

1.5.3.2.1 Measurement of SHBG

Three methods have been used to quantify SHBG concentrations in fluids.

1. Equilibrium dialysis of diluted plasma
2. Electrophoresis
3. Competitive adsorption

With the availability of tritium labeled steroids the binding can be measured quantitatively. The steroid can be added to plasma and after equilibration protein precipitated with ammonium sulfate added to 50% saturation. Radioactive dihydrotestosterone is the preferred ligand for the assay because of its higher affinity for SHBG than T and because DHT does not bind to CBG which is also precipitated. Radioimmunoassay (RIA) and immunoradiometric assays (IRMA) (see Chapter 2.3) are now used to directly measure SHBG and CBG. Free androgen index is a widely used index of excess free androgen in the blood. In general, measurement does not add a great deal to the diagnosis or management of patients with sex steroid problems. Bioavailable T can be measured after precipitation with

ammonium sulfate, or non-SHBG bound steroids can be derived from the difference between the total testosterone and SHBG antibody bound testosterone (Keevil and Adaway, 2019; Raverot et al., 2010) (see Chapter—Endocrine tests Chapter 2.7). Methods for free testosterone were reviewed by Goldman et al. (2017) highlighting considerable variability in results between laboratories.

1.5.3.3 Cortisol-binding globulin (CBG)

Cortisol as well as corticosterone, 17-hydroxyprogesterone, progesterone and 21-deoxycortisol but not aldosterone is bound to CBG. The liver produces CBG and SHBG. Concentrations are reduced in case of liver disease and nephrosis and are elevated by estrogens. Like SHBG, steroid-binding affinity is higher at low temperatures (Henley et al, 2016; Vogeser and Briegel, 2007) (Fig. 1.5.2).

Cortisol, like testosterone and estradiol, is bound to albumin with low binding affinity but the extents of binding are important because of the very high capacity of the albumin concentration in blood. The specifically bound steroids are thought not to be biologically available but the albumin bound steroid may be dissociated in certain tissues.

1.5.3.4 Physiological changes in binding protein concentrations

The binding T to SHBG is of high affinity. In men, SHBG is nearly saturated since the molar concentration of SHBG in adult human male plasma is only marginally greater than the molar concentration of T. In female plasma, the SHBG concentration is 2-fold higher than males but the T concentration is 10-fold lower than in men therefore most of the binding sites are unoccupied. SHBG concentrations are high in young children and fall during puberty. Women have

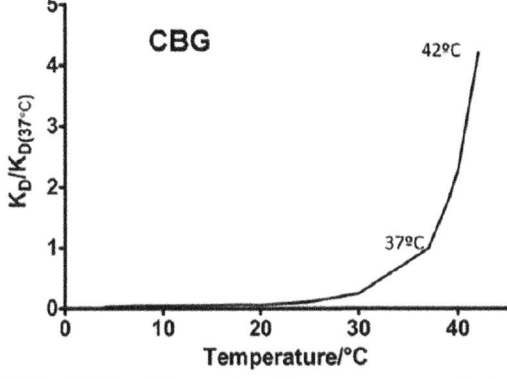

FIG. 1.5.2 CBG and temperature. Temperature responsive hormone release. Plot of change in ratio of binding affinities ($K_D/K_{D(37°C)}$) for CBG with temperature. There will be a near fivefold increase in free cortisol with a rise in body temperature to 42°C. *(From Henley D, Lightman S, Carrell R. Cortisol and CBG—getting cortisol to the right place at the right time. Pharmacol Ther 2016;166:128–35. Fig. 2 p. 131.)*

higher levels of SHBG than men and rise during estrogen treatment and particularly during pregnancy. SHBG in women with menstrual cycles needs to be judged in relation to the timing of the cycle (Fig. 1.5.3).

SHBG is here plotted with time in relation to the mid-cycle increase in circulating estradiol.

SHBG correlates inversely with body mass index and in general there is low SHBG concentration and a high free androgen index with obesity. This factor should be considered if SHBG is to be interpreted usefully since many patients under investigation have polycystic ovaries and obesity which inevitably go with low SHBG. There is limited evidence that a low calorie diet can lead to a significant reduction in body weight and in SHBG concentrations.

SHBG production may be related to insulin or insulin-like growth factor. During this diet, a reduction in the hyperinsulinemia, often seen in obese patients with PCO, can be recorded and also a fall in IGF-1.

CBG concentrations show similar relationship between males and females as were seen with SHBG (Hammond et al., 1984). They are also influenced by estrogens and pregnancy. Pregnancy has a definite influence on CBG concentrations (Fig. 1.5.4) due to its increased synthesis under high progesterone and estrogen exposure. The biological half-lives of the binding proteins are around 5 days.

The concentration of free cortisol in serum shows a good correlation with the total cortisol concentration (see Fig. 1.5.5). CBG concentration is not often measured in clinical investigations. An important consequence of the free cortisol in blood is when filtered in the kidney to be excreted in urine; the measurement of cortisol in urine is a good screening test of patients with Cushing's syndrome (Keevil and Adaway, 2019). Since cortisol production is normally below 6–9 mg per day it is difficult to accurately

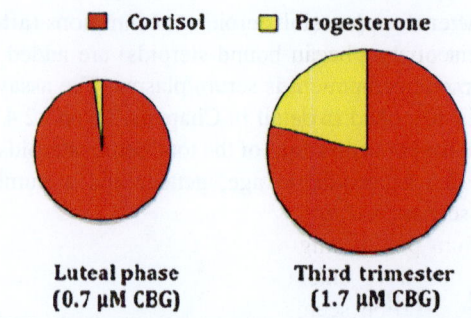

FIG. 1.5.4   Proportional occupation of CBG in serum during the luteal phase of the menstrual cycle vs the third trimester of pregnancy as estimated computationally from data of serum CBG, cortisol and progesterone levels (Dunn et al., 1981). The percentage of free steroid was determined by centrifugal ultrafiltration dialysis. *(From Hammond GL. Plasma steroid-binding proteins: primary gatekeepers of steroid hormone action. J Endocrinol 2016;230(1):R13–25. Fig. 1 p. R15.)*

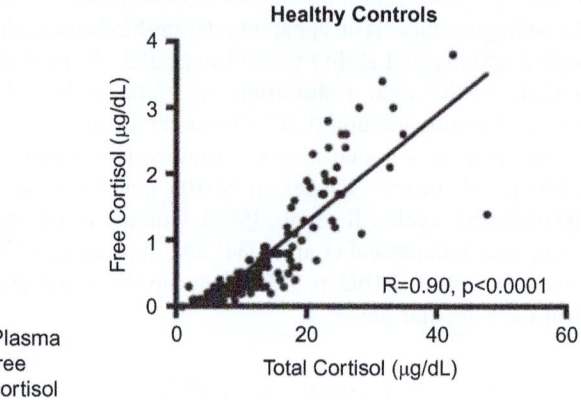

Plasma free cortisol

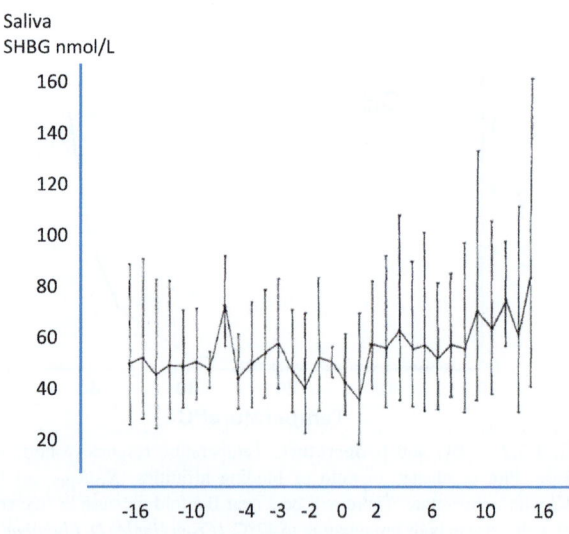

Saliva SHBG nmol/L

Menstrual cycle days relative to ovulation

FIG. 1.5.3   SHBG menstrual cycle. *(Author original.)*

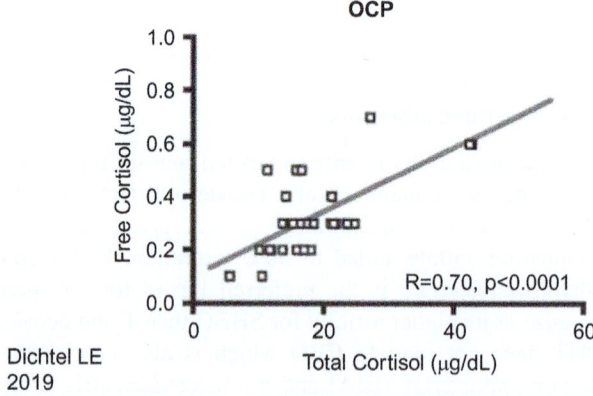

Dichtel LE 2019

FIG. 1.5.5   Correlation between plasma free cortisol and total cortisol in healthy controls and women talking oral contraceptives (OCP). *(From Dichtel LE, Schorr M, Loures de Assis C, Rao EM, Sims JK, Corey KE, Kohli P, Sluss PM, McPhaul MJ, Miller KK. Plasma free cortisol in states of normal and altered binding globulins: implications for adrenal insufficiency diagnosis. J Clin Endocrinol Metab 2019;104(10):4827–4836. Fig. 4 p. 4833.)*

assess adrenal suppression on the basis of reduction in urine free cortisol output because of imprecision of measurements at low concentrations. Furthermore a point is reached at which CBG is not saturated.

The binding of many steroids to CBG and SHBG were summarized many years ago (Dunn et al., 1981), methodology will have changed over the years (filtration, dialysis, equilibrium partitioning and capillary electrophoresis) but applications have changed less so. A method based on affinity extraction seems to be faster than some methods in use (Zheng et al., 2015).

1.5.3.5 Effects of binding proteins on steroid hormone assays

In order to measure the total steroid hormone concentration in blood plasma, it is necessary either to extract all steroid into an organic solvent or provide some means of displacing steroid from the binding proteins. This can be achieved in a number of ways.

- heat denaturation of the protein,
- pH effects on tertiary structure and
- use of a steroid with stronger affinity for the binding protein than the analyte.

1.5.3.6 Free steroid measurements

Free steroids can be determined after **ultra-filtration of plasma**. Assays for the free steroid concentrations in plasma are time consuming and not widely available but an indication of free hormone concentrations can be obtained by measurements of steroid concentrations in saliva or urine. Like the kidney, the salivary glands filter free hormones from plasma and actively excrete certain **free steroids into saliva** which thus offers a noninvasive means of quantifying the presumed biologically active hormone levels (Fig. 1.5.6). This has been used for monitoring ovulation at home by collections of samples without coming to hospital. There is a good correlation between free cortisol in saliva and plasma total cortisol concentration above the physiological normal level. However, cortisol can be inactivated to cortisone in salivary glands through the oxidase action of an enzyme (11β-hydroxysteroid dehydrogenase type 2). Once CBG is saturated, additional cortisol will be in the free fraction in blood and hence in saliva and urine as seen earlier. Cortisol and cortisone can be measured by LC-MS/MS methods (Kalaria et al., 2020; Bäcklund et al., 2020). Repeated saliva sample collection has been used to examine the effect of treatment on adrenal steroid production in patients with congenital adrenal hyperplasia. 17-Hydroxyprogesterone is measured as the indicator of adrenal cortical function. It must be remembered that most adrenal activity takes place during the night when it is not possible to collect saliva without waking the subject.

1.5.3.7 Defects in binding protein production

Patients have been found to have inactive SHBG (Wu and Hammond, 2014) or CBG (Simard et al., 2015; Dichtel et al., 2019) because of gene mutations and polymorphisms (Lin et al., 2012) that create inactive proteins. This will be discussed in clinical chapters (see Part III).

1.5.4 Steroid transport in target cells

Steroids were thought to enter cells by diffusion but the importance of the uptake transporters and exporters in determining steroid disposition are increasingly appreciated (Neuman and Bashirullah, 2018). This is a relatively new area of active research and knowledge is very incomplete. Steroid conjugates (sulfates and glucuronides) are of particular interest. Local deconjugation influences the extent of further metabolism.

1.5.4.1 Multidrug resistance-associated proteins (MRP2 and MRP3)

Steroid uptake in the liver and metabolite excretion is accomplished by transport proteins located at the basolateral and canalicular membrane of hepatocytes. The transport systems in the plasma membrane are complemented by intracellular enzymes for phase 1 and phase 2 metabolic reactions. Phase 1 reactions largely take steroid hormones with 3-keto-delta-4 structure to A-ring saturated metabolites and hydroxylated steroids through the actions of cytochromes. Phase 2 reactions involve glucuronide and sulfate conjugate formation. More water soluble metabolites are formed that can be excreted in bile or returned to the vasculature for renal excretion. Elimination of steroid conjugates into bile is accomplished by **multidrug-resistant-associated protein 2 (MRP2)** (Järvinen et al., 2018) which is a member of the **ATP-binding cassette (ABC)** transporter family which **utilizes ATP** to actively transport compounds across biological membranes. A related **breast cancer resistance protein (BRCP)** (now known as ABCG2) may be responsible for transport of estrogen glucuronides and sulfates. Some members of the ABC transporter family function as efflux transporters and are able to move sulfated steroids. Efflux transporters in epithelial cells may act in the **blood brain barrier (BBB)** and the **blood testis barrier (BTB)**. Sulfatase in the testis can cleave the sulfate from DHEAS to DHEA which contributes to overall sex steroid hormone synthesis. The active sites of sulfatase are located in the lumen of the

FIG. 1.5.6 Correlation between saliva cortisol and serum cortisol. *(Redrawn from Mezzullo M., Fanelli F., Fazzini A., Gambineri A., Vicennati V., Di Dalmazi G., et al. Validation of an LC-MS/MS salivary assay for glucocorticoid status assessment: evaluation of the diurnal fluctuation of cortisol and cortisone and of their association within and between serum and saliva. J Steroid Biochem Mol Biol 2016;163:103–12. Fig. 4 p. 110.)*

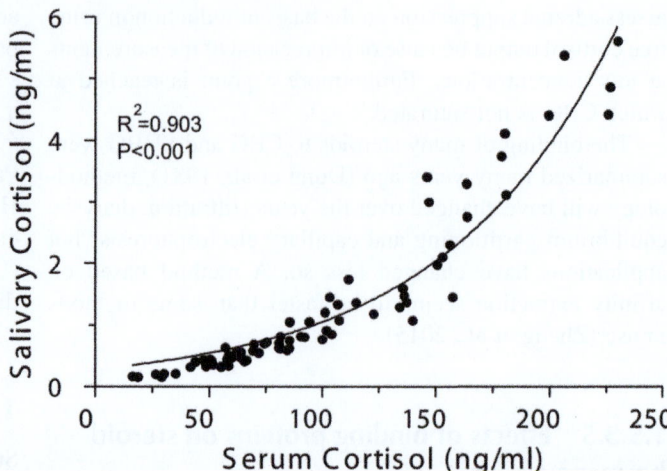

endoplasmic reticulum so DHEAS has to cross the ER membrane to reach their site of hydrolysis. The liberated DHEA very likely needs a transporter to move it back to the blood or to sites for activation to androstenedione and testosterone. Sulfated steroids therefore can no longer be considered just as metabolites for excretion but also as steroid hormone precursor molecules.

1.5.4.2   Organic anion transport proteins (OATP)

Sodium independent transporters are members of the **organic anion transporting peptides (OATP)** with 12 transmembrane domains (TMDs) and an extracellular N-terminus and C-terminus. The **solute-linked carrier** protein **SLC10** was found to primarily **transport sulfated steroids** and can be classed within the **sodium dependent organic ion transporter family (SOAT)** (Claro da Silva et al., 2013). SLC10 was originally described as bile acid influx transporters. SLC10 has a nine-transmembrane structure with an outward facing N-terminus and an inward facing C terminus (Fig. 1.5.7).

Many OATPs are now known to transport steroid hormones and their conjugates (Li et al., 2020). Uptake and efflux carriers (SOAT, OATP6A1) and efflux transporters (MRP2 and 3) have been detected in human testis. **Sulfated steroids** can be taken up by **OATP2B1** (McFeely et al., 2019) into **Sertoli cells** where active sex steroids are produced, after sulfatase action, that pass to germ cells as free steroids Sulfated steroids are transported out of Sertoli cells by **efflux transporters (MRP1, MRP4)** into the intercellular space above the BTB (Hartmann et al., 2016). From there they can be taken up by germ cells (SOAT, OATP2B1) and reactivated by sulfotransferase activity. Two sulfatase pathways therefore likely exist in humans due to localization in Sertoli cells and germ cells (Fig. 1.5.8). DHEAS uptake is now recognized in breast cancer cells.

For intracellular action, sulfated steroid hormones (red triangle plus "S") have to be taken up by specific uptake carriers (blue, e.g., SLC10A6, OATPs) into the cell. There, the sulfate residue is cleaved by the activity of STS. Afterwards, unconjugated steroid hormones are able to bind to nuclear steroid receptors (AR, ERs), translocate into the nucleus and bind to hormone responsive elements (HREs). By this, gene expression may be activated or repressed. This is considered as the genomic pathway. Within the cell,

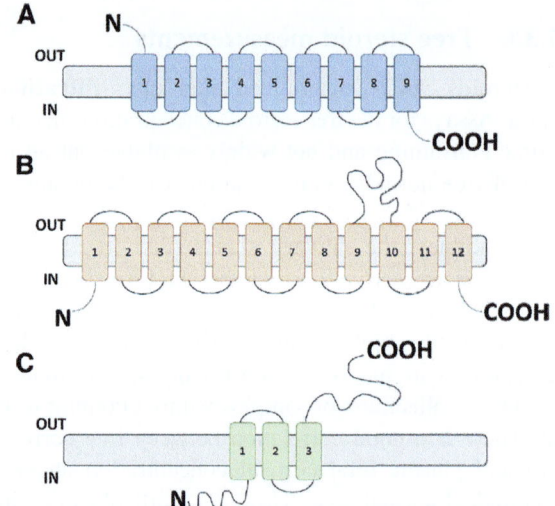

FIG. 1.5.7   Scheme of the membrane topologies of the Solute carriers NTCP and SOAT (A), for OATPs (B) and OSCP (C). (A) Members of the SLC10 family consists of nine TMDs with an outward facing N-terminus and an inward facing C-terminus (e.g., NTCP and SOAT). (B) Members of the OATP family reveal 12 TMDs with N- and C-terminus both facing to the intracellular space. Notice the prominent extracellular loop between TMD 9 and 10. (C) Organic solute carrier protein (OSCP1) exhibits only three TMDs, an inward facing N-terminus and an outward facing C terminus. Other groups stated OSCP1 to be a soluble and cytoplasmically located protein. *(From Fietz D. Transporter for sulfated steroid hormones in the testis—expression pattern, biological significance and implications for fertility in men and rodents. J Steroid Biochem Mol Biol 2018;179:8–19. Fig. 2 p. 9.)*

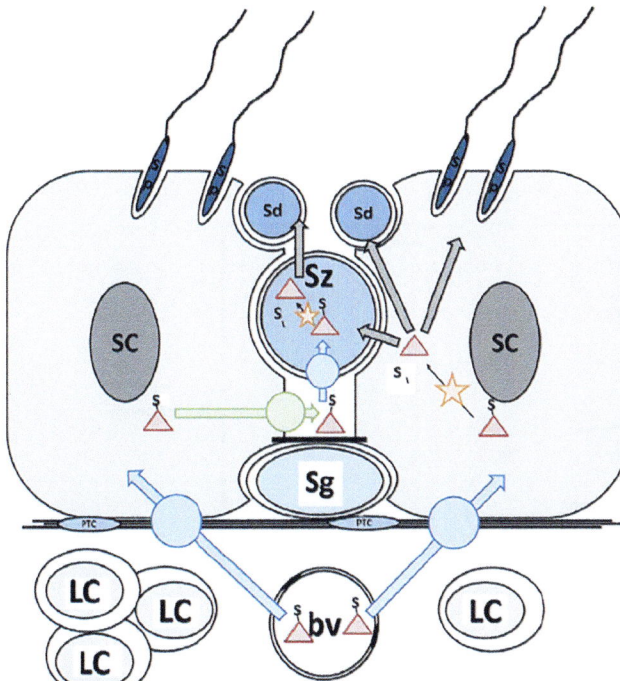

FIG. 1.5.8 Action of sulfatase in testis. Sulfated steroids (*red triangle* plus "S") can be taken up by specific uptake carrier (*blue*, e.g., OATP2B1, OATP3A1) into Sertoli cells. From there, two possible sulfatase pathways are likely in germ cells (left) and Sertoli cells (right). Left: sulfated steroids are transported out of the Sertoli cells by specific efflux transporters (*green*, e.g., MRP1, MRP4) into the intercellular space above the BTB. From there, they can be taken up by germ cells (*blue*, e.g., SOAT, OATP2B1, OATP3A1, OATP6A1, OSCP1) and re-activated by STS activity (*orange star*). As free steroid hormones (*red triangle*), they are now able to pass through cell membranes by diffusion (*gray arrow*). Right: In Sertoli cells, sulfated steroids may not only be transported but can also be reactivated by STS activity and pass from Sertoli cells to germ cells as free steroids by diffusion. *bv*, blood vessels; *LC*, Leydig cells; *PTC*, peritubular cells; *SC*, Sertoli cells; *Sd*, round spermatids; *Sg*, spermatogonia; *Sp*, elongated spermatids; *Sz*, spermatocytes. *(From Fietz D. Transporter for sulfated steroid hormones in the testis—expression pattern, biological significance and implications for fertility in men and rodents. J Steroid Biochem Mol Biol 2018;179:8–19. Fig. 4 p. 16.)*

unconjugated steroid hormones can be inactivated by adding a sulfate residue (catalyzed by sulfotransferase SULT, purple star) (Fig. 1.5.9). As conjugated steroid hormones have longer half-lives, they are considered as a local supply. They may also leave the target cell by specific efflux transporters (ATP-binding cassette transporters, ABCs). Sulfated steroid hormones can also bind to membrane-associated G-protein coupled receptors (GPCRs, e.g., GPER, Gnα11) and activate a nongenomic pathway.

Activating of MAP kinase (MAPK) cascade via Ras/RAF and ERK leads to phosphorylation of the transcription factors CREB/ATF which act on CREB inducible genes (Molina and Adjei, 2006). This pathway has already been shown to be initiated by DHEAS binding. A second possible nongenomic pathway of steroid hormone action is the depolarization of cell membranes leading to a Ca^{2+} influx. By binding of steroid hormones to membrane receptors (probably GPCRs), phospholipase C (PLC) is activated and cleaves phosphatidylinositol 4,5-bisphosphate (PIP2) into inositol 1,4,5-trisphosphate (IP3) and diacyl glycerol (DAG). DAG inhibits K^+ ATPase channels, which leads to a Ca^{2+} influx. Furthermore, IP3 binds to IP3 receptors on the endoplasmic reticulum which leads to a Ca^{2+} influx.

1.5.4.3 Cytoplasmic-nuclear transfer of steroid receptors

Numerous cellular processes, including steroid receptors, navigate cytoplasmic compartments to the nucleus other than by simple diffusion. Proteins called **importin** and **exportin** are implicated, for example in nuclear translocation of **G-protein coupled receptors** (Bhosle et al., 2019). The cytoplasm and nucleus are separated by a double membrane nuclear envelope bearing nuclear pores. The molecular details and functioning of transport pathways through membranes are becoming apparent but are challenging because of the numbers already found with undoubtedly more to follow and beyond the scope of this book. The reader should explore this through current literature beyond 2020. Two reviews might start this search (Chang and Hsia, 2021; Shah et al., 2019).

1.5.5 Transporters affected by steroids

The studies of human disease have provided much information about physiological processes that involve steroid transporters induced by steroids. This will be discussed further in clinical sections of the book (see Part III).

1.5.5.1 Epithelial sodium transport channel (ENaC)

In the distal convoluted tubules, aldosterone binds to the mineralocorticoid receptor (MR) resulting in a complex that in the nucleus initiates the transcription of several proteins including **serum and glucocorticoid regulated kinase 1 (SGK1)** that was found to modulate the surface expression and function of several classes of ion channels that regulate sodium, potassium, calcium, magnesium and chloride (Valinsky et al., 2018).

The **amiloride sensitive epithelial sodium transport channel (ENaC)** (Zaika et al., 2013; Mamenko et al., 2012), that belongs to the **degenerin** family of ion channels, is implicated in a variety of physiological functions. The channel is composed of three homologous units called α-, β- and γ-NaC (Fig. 1.5.10). Each subunit has two transmembrane domains (M1 and M2) yielding a protein with a large extracellular loop and short cytoplasmic N- and C-terminal

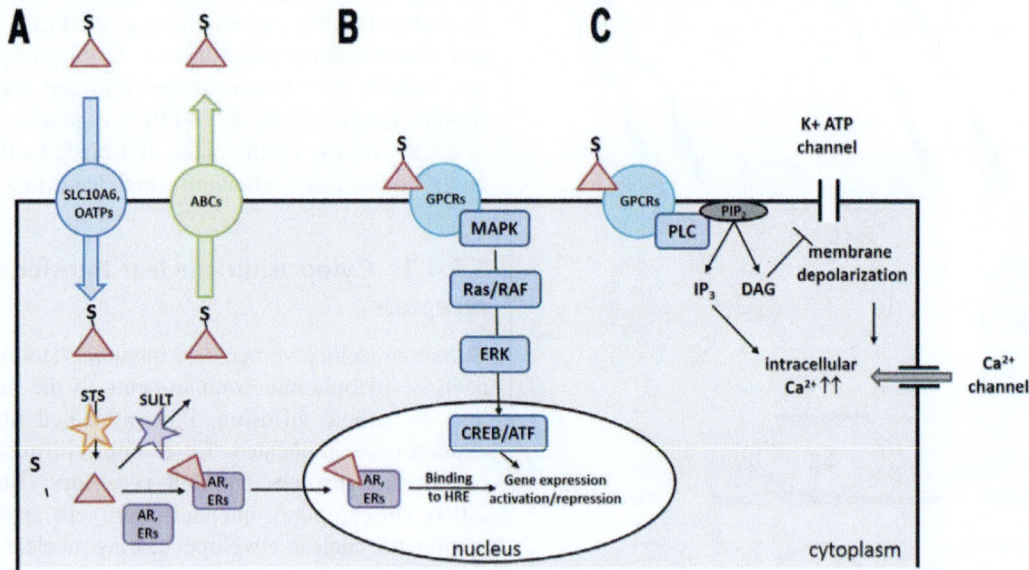

The sulphatase pathway and signalling pathways for sulphated steroid hormones

FIG. 1.5.9 **Schematic drawing of the sulfatase pathway and signaling pathways for sulfated steroid hormones.** (A) For intracellular action, sulfated steroid hormones (*red triangle* plus "S") have to be taken up by specific uptake carriers (*blue*, e.g., SLC10A6, OATPs) into the cell. There, the sulfate residue is cleaved by the activity of STS (*orange star*). Afterwards, unconjugated steroid hormones (*red triangle*) are able to bind to nuclear steroid receptors (AR, ERs), translocate into the nucleus and bind to hormone responsive elements (HREs). By this, gene expression may be activated or repressed. This is considered as the genomic pathway. Within the cell, unconjugated steroid hormones can be inactivated by adding a sulfate residue (catalyzed by sulfotransferase (SULT), *purple star*). As conjugated steroid hormones have longer half-lives, they are considered as a local supply. They may also leave the target cell by specific efflux transporters (*green*, ATP-binding cassette transporters, ABCs). (B) Sulfated steroid hormones can also bind to membrane-associated G-protein coupled receptors (GPCRs, e.g., GPER, Gnα11) and activate a nongenomic pathway. Activating of MAP kinase (MAPK) cascade via Ras/RAF and ERK leads to phosphorylation of the transcription factors CREB/ATF which act on CREB inducible genes. (C) A second possible nongenomic pathway of steroid hormone action is the depolarization of cell membranes leading to a Ca^{2+} influx. By binding of steroid hormones to membrane receptors (probably GPCRs), phospholipase C (PLC) is activated and cleaves phosphatidylinositol 4,5-bisphosphate (PIP2) into inositol 1,4,5-trisphosphate (IP3) and diacyl glycerol (DAG). DAG inhibits K^+ ATPase channels, which leads to a Ca^{2+} influx. Furthermore, IP3 binds to IP3 receptors on the endoplasmic reticulum which leads to a Ca^{2+} influx. *(From Fietz D. Transporter for sulfated steroid hormones in the testis—expression pattern, biological significance and implications for fertility in men and rodents. J Steroid Biochem Mol Biol 2018;179:8–19. Fig. 1 p. 9.)*

domains. A disulfide bridge across cysteine rich domains between M1 and M2 is critical for folding and channel expression at the membrane.

Aldosterone activates sodium uptake by ENaC directly through **SGK1** and indirectly through the ubiquitin ligase **neural precursor cell expressed developmentally down-regulated protein (Nedd4-2)** (Fig. 1.5.11) which when nonphosphorylated causes ubiquitination and internalization of ENaC from the plasma membrane.

SGK1 phosphorylates Nedd4-2 which reduces the affinity of Nedd4-2 for ENaC. The E3 ubiquitin ligase, Nedd4-2, regulates ENaC in principal cells by binding to ENaC through motifs in the carboxy terminus in α-, β- and γ-ENaC, regulating the expression levels at the plasma membrane through internalization, polyubiquitination and degradation.

Aldosterone after binding to the MR regulates the ubiquitin ligase **Nedd4-2** that mediates the MR-dependent regulation of **pendrin** in β-intercalated cells where it functions as a Cl^-/HCO_3^- exchanger. Angiotensin II and aldosterone

upregulates pendrin. **Serine/threonine like kinase 1 (ULK1)** robustly phosphorylates MR. As an upstream regulator of ULK1, **Mammalian Target of Rapamycin (mTOR)** is known to inhibit ULK1 by phosphorylating the ULK1. Stimulation with AngII results in increased phosphorylation of the **ribosomal protein S6K (S6 kinase)** which is a known substrate of mTOR complex1, confirming that AngII activates mTOR. Therefore, mTOR and ULK1 regulate MR activity in intercalated cells (Fig. 1.5.12), by controlling pendrin activity. The extent to which pendrin modulates blood pressure remains to be determined (Wall et al., 2020).

The thick ascending loop (TAL) is responsible for reabsorption of around 30% of filtered sodium. The apically located Na-K-2Cl transporter (NKCC2) is the pacemaker of NaCl reabsorption in the TAL and mediates uptake of sodium, potassium and chloride across the apical membrane of TAL cell using the chemical gradient of sodium generated by the basolateral Na-K ATPase (Mount, 2014) (Fig. 1.5.13).

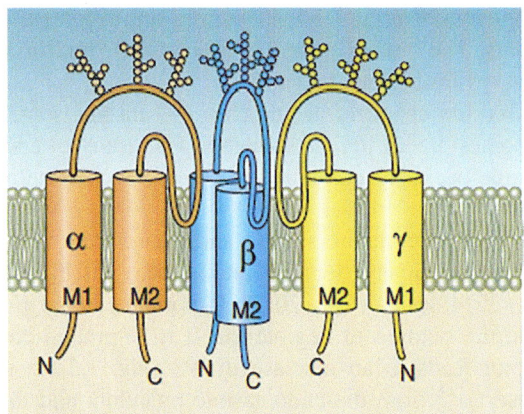

FIG. 1.5.10 Schematic heteromeric structure of ENaC. ENaC is made of three homologous subunits (α, β, γ) sharing around 30%–40% identity at the protein level. Each subunit has short cytoplasmic NH₂ and COOH termini and two transmembrane domains (M1 and M2) with a very large (60 kDa) extracellular loop characterized by two cysteine-rich domains (CRD) and 6–12 glycosylation sites. The extracellular domain includes about 70% of the sequence of amino acids of an ENaC subunit. *(From Rossier BC, Baker ME, Studer RA. Epithelial sodium transport and its control by aldosterone: the story of our internal environment revisited. Physiol Rev 2015;95(1):297–340. Fig. 13 p. 324.)*

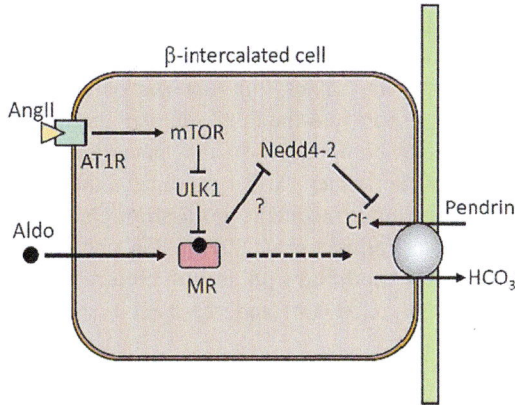

FIG. 1.5.12 Mechanism of pendrin regulation by ANG II and aldosterone. mTOR and ULK1. *(From Yamazaki O, Ishizawa K, Hirohama D, Fujita T, Shibata S. Electrolyte transport in the renal collecting duct and its regulation by the renin-angiotensin-aldosterone system. Clin Sci (Lond) 2019;133(1):75–82. Fig. 2 p. 78.)*

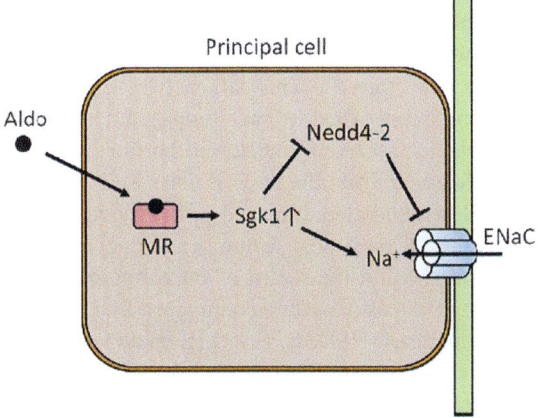

FIG. 1.5.11 Mechanism of ENaC regulation by aldosterone. *(From Yamazaki O, Ishizawa K, Hirohama D, Fujita T, Shibata S. Electrolyte transport in the renal collecting duct and its regulation by the renin-angiotensin-aldosterone system. Clin Sci (Lond) 2019;133(1):75–82. Fig. 1 p. 76.)*

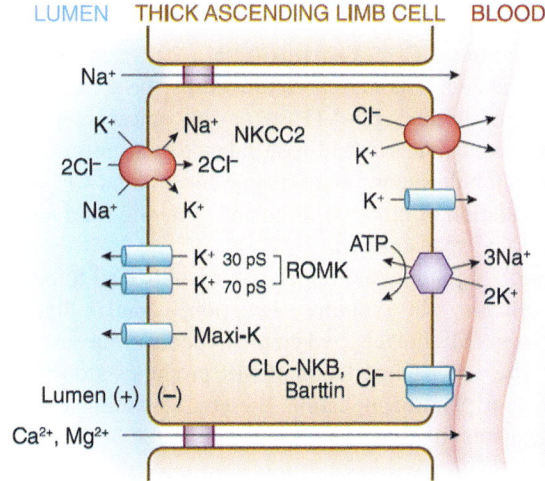

FIG. 1.5.13 Transepithelial Na1-Cl2 transport pathway in the TAL. The three subtypes of the apical K^+ channels are a 30-picosiemen (pS) channel, a 70 pS channel and a calcium activated maxi K^+ channel. *Barttin,* Cl2 channel subunit; *CLC-NKB,* human Cl2 channel; *KCC4,* K1-Cl2 cotransporter-4; *Maxi-K,* calcium-activated maxi K1 channel (also known as the BK channel); *NKCC2,* Na1-K1-2Cl2 cotransporter-2; *ROMK,* renal outer medullary K1 channel. *(From Mount DB. Thick ascending limb of the loop of Henle. Clin J Am Soc Nephrol 2014;9(11):1974–86. Fig. 2 p. 1975.)*

The cells of the medullary TAL, cortical TAL and macula densa share the same basic transport mechanisms. Na^+, K^+ and Cl^- are cotransported across the apical membrane by NKCC2, an electroneutral Na^+-K^+-$2Cl^-$ cotransporter that is exquisitely sensitive to furosemide, a "loop" diuretic known for four decades to inhibit transepithelial Cl^- transport by the TAL. This transporter generally requires the simultaneous presence of all three ions such that the transport of Na^+ and Cl^- across the epithelium is mutually codependent and dependent on the luminal presence of K^+.

Chloride carried into the TAL cell exits mainly through ClC-Kb. The function and cell surface expression of ClC-Kb and the related ClC-Ka are facilitated by their essential B-subunit called Barttin. Potassium entering the cell via NKCC2 is recycled into the lumen via ROMK. A tight co-ordination of the activities of NKCC2, ROMK and ClC-Kb is needed to ensure proper TAL function (Mount, 2014).

Extracellular potassium homeostasis, maintained by the regulation of renal potassium excretion, is dependent on the activity of weakly inward rectifying potassium channels (SK)1 that are expressed on the apical membrane of epithelial cells in the distal nephron. The ROMK (Kir 1.1 or KCNJ1) gene codes for these Kir channels which are thought to be the major, but not exclusive, route for potassium transport into the tubule lumen and constitute a final regulated component of the potassium secretory machinery of the kidney. SGK1 modulates the expression levels of renal outer medullary **potassium** channels in the distal nephron (Valinsky et al., 2018). SGK1 phosphorylates the **renal outer medullary potassium channel (ROMK1)** (Palmada et al., 2003a,b) and the Kir1.1 subtype (Yoo et al., 2003).

1.5.5.2   Apical ATPase and pendrin

In cortical collecting ducts, about 30% of the cells are intercalated cells of two types α and β. The α intercalated cells excrete proton via apical H^+-ATPase, whereas β-intercalated cells express the Cl^-/HCO_3 exchanger, **pendrin** and H^+-ATPase at the apical and basolateral membranes, respectively, leading to secretion of HCO_3^- and reabsorption of Cl^- (Yamazaki et al., 2019). Non-α, non-β intercalated cells are characterized by the expression of both pendrin and H^+-ATPase at the apical membrane. Pendrin contributes to acid-base balance by secreting HCO_3^- into the tubular lumen in exchange for luminal Cl^- and to the regulation of blood pressure and systemic fluid balance via the Cl^- reabsorption.

AngII signaling and mineralocorticoids lead to pendrin upregulation. In volume depletion, AngII signaling decreases MR levels, whereas hyperkalemia increases MR. The dephosphorylation of MR promotes aldosterone-dependent pendrin induction, thereby activating Cl^- flux mechanisms involving intercalated cells.

1.5.5.3   Chloride channels (CLC)

Chloride is the most important anion in the human body for homeostasis. Serum chloride is around 120 mmol, whereas intercellular concentrations are 5–10 mmol with sodium and potassium as counterions. Chloride transfer across membranes generates electrical currents and therefore influences excitability of neurons, muscles and endocrine cells. Chloride channels can exist as anion channels and anion/proton exchangers. The ClC proteins are 18x-transmembrane helices formed from two subunits with separate active centers that transport anions independently. Each subunit has a carboxy terminal domain that forms intramolecular dimeric complexes. Four ClC channels (ClC 1and 2, ClC-Ka, Kb) have been characterized in plasma membranes with links to steroid disorders (Jentsch and Pusch, 2018). The role of CLC-Ka and Kb was discovered through investigations of a genetic disease

causing hyponatremia, hypovolemia and hypotension in patients. The mutant gene has been called **Barttin** and is discussed further in Chapter 3.3.1.

Chloride channels in branches 3–7 in endosomes and lysosomes are $2Cl/H^+$ exchangers. ClC transporters have a similar backbone structure and operate in chlorine/proton cycles.

Most tissues express CLC and transporters that meet diverse functions (Jentsch and Pusch, 2018; Stölting et al., 2014; Zaika et al., 2013). In normal adrenal glomerulosa, the binding of angiotensin II to G-protein coupled receptor leads to an increase in IP3 (Fig. 1.5.14) which release Ca^{2+} from the endoplasmic reticulum and inhibits potassium channels which leads to depolarization of the membrane and opening of calcium channels and further increase in cytosolic calcium concentration.

1.5.6   Steroids in peripheral tissues (intracrinology)

The steroid hormones in some tissues will bind to specific receptors to activate transcription of genes that implement the functions of the steroids. Steroid hormones can also act in the nonclassical manner. This is the focus of the chapter on Steroid recognition (Chapter 1.6). Within some tissues, steroids can undergo local conversion to more active/inactive steroids. This sits between synthesis and metabolism and is called **intracrinology**. Local conversion often serves to inactivate the steroid hormones to metabolites for clearance from the body is discussed in the chapter on Breakdown and clearance and primarily covers steroids in the liver and intestines. A number of enzyme catalyzed reactions are within the realm of intracrinology including sulfatase hydrolysis of sulfate conjugates (SULT), dehydrogenase/reductases (HSD3B, HSD11B enzymes, HSD17B5, SRD5), and aromatase. The transfer of sulfated steroids into cells before desulfation by sulfatases is a new area of active research.

This area is becoming incredibly complicated. Genetic disorders in humans and genetically modified mice (knock out; mutations leading to activation or overexpression of the protein) have contributed significantly to our knowledge base but the outcomes of these studies are made difficult to interpret by a number of issues that include:

- difficulties in measuring low concentrations of steroid hormones and metabolites,
- local steroid metabolism (e.g., testosterone/dihydrotestosterone; estrone/estradiol),
- differences in responses to low, normal (physiological) and high concentrations (pharmacological) of steroids,
- co-factor provision,
- tissue content (adipose tissue, stem cell and glandular tissue),

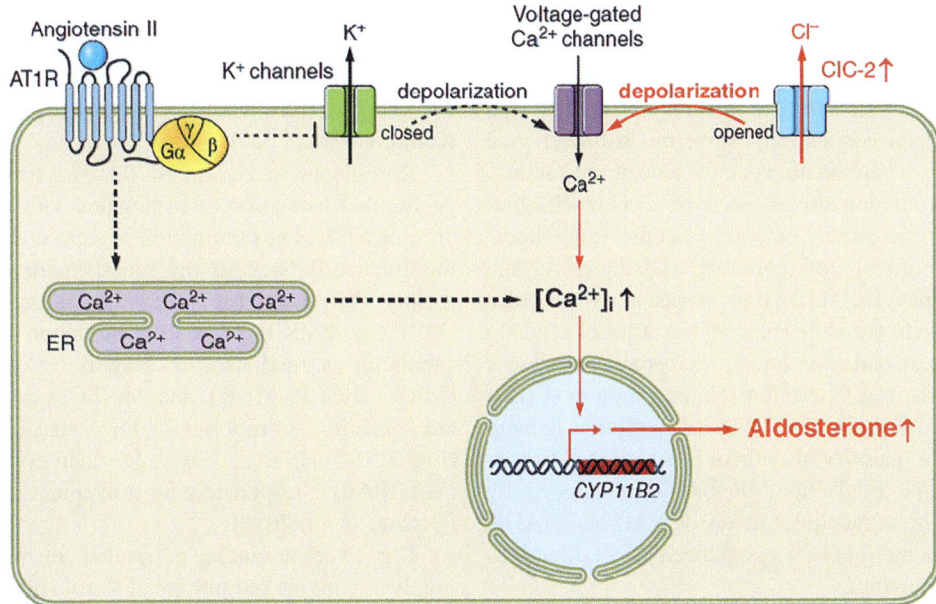

FIG. 1.5.14 In normal adrenal glomerulosa cells, binding of angiotensin II to its G protein-coupled receptor leads to an increase in IP3 which releases Ca^{2+} from the ER, as well as to an inhibition of plasma membrane K^+ channels that largely determine the resting potential of these cells. This inhibition leads to a depolarization, which opens voltage-dependent Ca^{2+} channels, leading to a further increase in cytosolic Ca^{2+}. A rise in extracellular K^+ concentration can also directly depolarize glomerulosa cells. *(From Jentsch TJ, Pusch M. CLC chloride channels and transporters: structure, function, physiology, and disease. Physiol Rev 2018;98(3):1493–1590. Fig. 7 p. 1519.)*

- response through classical receptor channel or nonclassical manner,
- receptor isoforms,
- interpretation of immunohistochemical evidence with poor-quality antibodies,
- differences in interpretation of results according to sex,
- disparities in results between human and animal models (man vs mouse) and
- cross-talk with receptors (Gr vs PR; AR vs ER).

1.5.6.1 Kidney

The kidney has a major role in homeostasis by controlling movement of electrolytes in the kidney. The juxtaglomerular apparatus is a site near the glomerulus that senses plasma sodium concentration in the tubule and releases **renin** into the circulation when sodium concentrations fall. Renin is the first in a series of chemical messengers that make up the renin-angiotensin-aldosterone system. Sodium chloride reabsorption from the distal convoluted tubules is controlled by the action of **aldosterone**. The kidney also controls water loss from the renal tubules through the action of **antidiuretic hormone** secreted by the posterior pituitary gland when there is a rise in plasma osmolality. ADH acts on vasopressin (V2) receptors to control water movement in **aquaporin** channels (Su et al., 2020).

The kidney is a target site for mineralocorticosteroids. Aldosterone acts in a classical manner by binding at the mineralocorticoid receptors in the late distal convoluted tubules (DCT), the connecting tubule (CNT) and the collecting duct (CD) to regulate circulating sodium concentrations by inducing many genes notably including activity and transcription of the **epithelial sodium transport channel (ENaC)** (Zaika et al., 2013). Aldosterone after binding to the MR regulates Nedd4-2. Na^+ reabsorbed through ENaC is transported to the extracellular fluid via the Na^+/K^+ ATPase to increase its activity (Yamazaki et al., 2019).

The kidney maintains systemic fluid homeostasis mainly through reabsorption of sodium chloride. Up to 10% of filtered sodium is reabsorbed in the DCT and CNT and then about 1% in the CD. A sodium chloride transporter (NCC) is active early in the DCT and increases with sodium concentration and phosphorylation by a complex system of kinases (Brown et al., 2021; Wu et al., 2019). The **Na^+/H^+ exchanger** in proximal tubules, the **$Na^+ K^+ 2Cl^-$ co-transporter** in the thick ascending loop of Henle and the **Na^+-Cl^- co-transporter** (NCC) act in tandem with the effects of aldosterone. **Pendrin** is another protein expressed in the kidney (Wall et al., 2020; Yamazaki et al., 2019) and has been localized to the cortical collecting duct where it is involved in bicarbonate secretion. The handling of sodium and potassium in the kidney was thoroughly reviewed (Rossi et al., 2020a,b) with useful diagrams of electrolyte movements in sections of the DCT, CNT and CD. Much of our understanding of these channels arose from studies of the genetic disorders of the proteins. A number of disorders of hypertension and salt wasting have been described with the associated gene mutations. The pathology associated with these channels will be discussed later (see Chapter 3.3.1 and 3.3.2, respectively) (see Rossi et al., 2020a,b and Downie et al., 2021 for comprehensive reviews).

Cortisol and progesterone can also bind to the MR and since blood cortisol concentrations are much higher than aldosterone there would be no place for aldosterone action. 11β-hydroxysteroid dehydrogenase type 2 (11β-HSD or HSD11B2) is a microsomal enzyme responsible for inter-conversion of cortisol and cortisone (Diederich et al., 2000). In the kidney, HSD11B2 is expressed in renal tubules to locally inactivate cortisol. The enzyme requires NAD to convert cortisol to cortisone and in the process produces NADH that binds to a C-terminal binding protein (CtBP) that acts as a co-repressor of DNA transcription (Zhang et al., 2002). The majority of cortisol is inactivated to cortisone, the residue binds the MR but is inactivated by NADH-CtBP. The intracellular levels of NAD and NADH are regulated in a number of ways that can affect directionality of certain enzymes.

The physiological significance of HSD11B2 has become apparent from clinical studies of an enzyme deficient state called **apparent mineralocorticoid excess (or AME)** (Fan et al., 2020). This is a rare cause of hypertension occurring principally in childhood (see Chapter 3.3.1 on Hypertension).

The kidney is the main site for steroid excretion from the body. Free steroids are filtered from the blood at the renal corpuscle and pass into the ultrafiltrate. Free steroids pass along the renal tubules into the urine. Cortisol can bind to the MR in kidney but this action is prevented by the inactivation to cortisone by HSD11B. Free cortisol in the serum is filtered in the kidney glomerulus and is not reabsorbed in the tubules and excreted in the urine. Cortisol in urine is thus a marker of free cortisol. Steroid glucuronides are cleared in urine by renal MRP transporters (Li et al., 2019).

1.5.6.2 Liver

The liver is an important site for detoxification of various endogenous and exogenous substances and the breakdown of blood. The organ also synthesizes proteins and biochemicals that are necessary for digestion and growth. Bile is produced in the liver which passes through the bile duct to the intestine. It is an important site for glycogen storage (see Section 1.7.1.1) and cholesterol regulation (see Section 1.2.1). GCH increase synthesis of PEPCK and G-6-phophatase and inactivate phosphorylase. GCH have a permissive role promoting optimal ability of glucagon, epinephrine and cyclic-AMP in gluconeogenesis. Many substances produced in the liver return to the circulation.

The liver is the primary organ for the metabolic degradation of steroids and xenobiotics (chemicals foreign, e.g., drugs) in to the body. Multiple processes are involved acting at several sites. The uptake of steroids (and xenobiotics) in hepatocytes is mediated by organic anion transporting polypeptides (OATPs) and by organic anion and cation transporters (OATs and OCTs) starting with the action of transmembrane transport proteins (Jetter and Kullak-Ublick, 2020).

Steroids are inactivated in the liver by saturation of the A-ring and formation of conjugates with sulfuric and glucuronic acids. The elimination of steroid glucuronides from the liver cell back to the bloodstream is accomplished mainly by multidrug resistance-associated protein 3 (MRP3) and MRP4, while the elimination toward the biliary canaliculi is mediated by several different transporters (MRP2, BCRP, MDR1 and MATE1). MRP3 and MRP2 are efflux transporters for steroid glucuronides (Fig. 1.5.15) (Li et al., 2019; Jarvinen et al., 2018), SOAT (SLC10A6) is responsible for movement of steroid sulfates (Grosser et al., 2018).

The liver produces a number of proteins that are involved with the physiology of steroid hormones. Albumin and binding globulins are transporters of steroid hormones in the circulation and deliver free hormone to target tissues. The liver is the site of **angiotensinogen** production which is the substrate for renin in the renin-aldosterone axis for sodium homeostasis.

1.5.6.3 Female organs

In female reproductive tissues, **ATP-binding cassette transporters (ABC transporters)** have various, poorly-understood, roles in steroidogenesis, fertilization, implantation, nutrient transport and immune responses. ATP transporters are trans-membrane proteins that modulate transfer of substrates, including steroids, cholesterol and cytokines from cytosol to extracellular space. ATP hydrolysis is behind the movement of substrates. P-glycoprotein (P-gp) is a member of the protein family coded by the ABCB1 gene that in humans acts as drug resistant protein and transporter of steroids in key reproductive processes including uterine function, pregnancy and fetal development. P-gp is responsive to LH and progesterone so may play a role in gametogenesis, granulosa cell differentiation, oogenesis, endometrium changes, placental function, embryo development, fetal membranes, myometrial function and steroidogenesis (Bloise et al., 2016).

1.5.6.3.1 Endometrium and uterus

Steroids induce uterine and endometrial expression of various growth and angiogenic factors. P-gp expression is responsive to hormonal changes in the menstrual cycle and thus may play a role in the implantation process and initiation of pregnancy by promoting efflux of cytokines and prostaglandins (Mendes et al., 2019; Viganò et al., 2003).

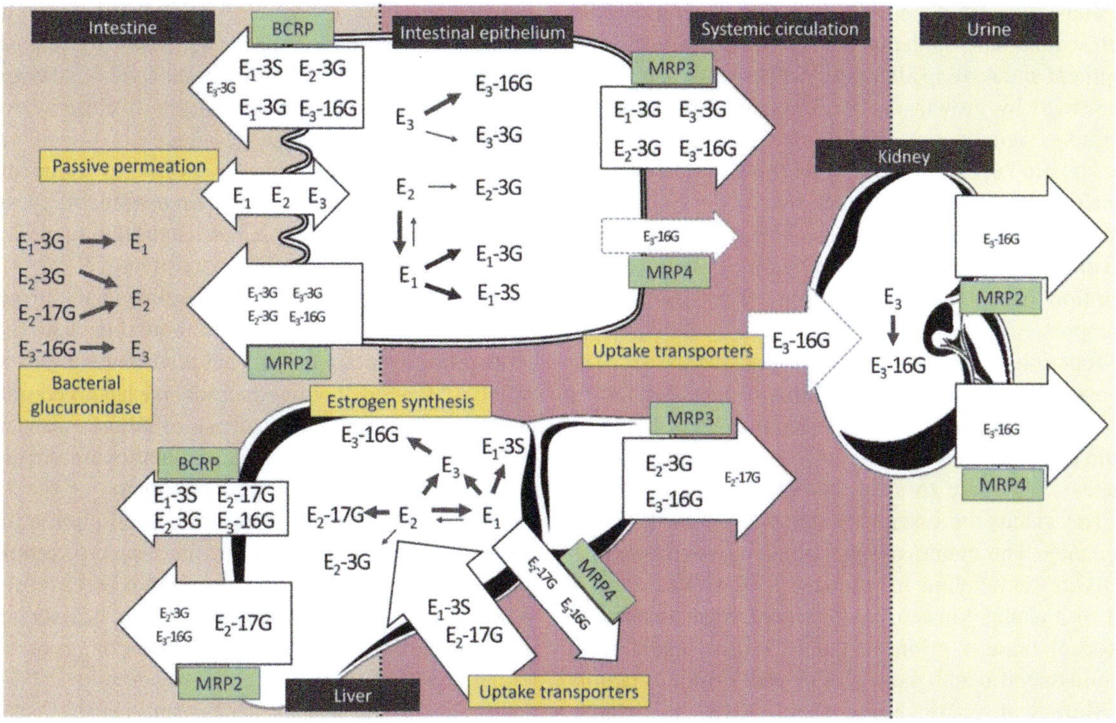

FIG. 1.5.15 **Disposition of human estrogen conjugates**. Transporters are represented as *white arrows*, the width of which indicates whether it is relatively highly or lowly expressed in the tissue and *dashed outlines* stand for speculative transporters or mechanisms that have limited evidence. The liver schema stands for a typical hepatocyte. Conjugated and unconjugated estrogens are represented by abbreviations containing numbers and letters. The abbreviation of a compound denotes the type of estrogen (E1, E2 and E3 standing for estrone, estradiol and estriol, respectively) and the latter part describes the conjugation position and the type of conjugate. Larger compound names indicate higher transport rates of the given substrate by the specific transporter. *Black arrows* inside the cells represent estrogen biotransformation reactions and the relative extent of these reactions. *(From Järvinen E, Deng F, Kidron H, Finel M. Efflux transport of estrogen glucuronides by human MRP2, MRP3, MRP4 and BCRP. J Steroid Biochem Mol Biol 2018;178:99–107. Fig. 6 p. 105.)*

The relative proportions of ER and PR and their interaction determine the expression of specific genes upon steroidal stimulation. Dysregulations of steroid modulated expression is believed to be involved in the pathogenesis of many endometrial diseases. Irregular bleeding induced by steroidal contraception, for example, is thought to involve aberrant endometrial vascular development and expression of angiogenic growth factors. The antiestrogen tamoxifen induces growth factors like vascular endothelial growth factor and adrenomedullin which may be key mediators of endometrial neoplastic effects.

1.5.6.3.2 Mammary glands

The mammary glands transition through distinct stages of life in the human female from development in the fetus when the rudimentary ductal structure is formed (Macias and Hinck, 2012). There is some growth of the mammary glands at puberty and further growth in lactation. Postnatally the development of the normal mammary gland (mammogenesis) is dependent mainly on sex steroids, E2 and P (Berryhill et al., 2016). Ducts and secretory alveoli develop from epithelium. Proteins and growth factors are responsible for

branching of the ducts and lactation. ER mediated signaling is essential for ductal morphogenesis, while PR signaling is critical for lobulo-alveolar development.

Differentiation and development of the mammary gland are influenced by estrogens, glucocorticoids and progesterone. Specific binding sites for each of these classes of steroid hormones have been characterized in vivo and in cell-free preparations of lactating mammary gland. Some studies have demonstrated cross-talk between the ER and PR and the receptors for certain growth factors, suggesting an important functional role for estrogens and P in the lactation cycle in co-operation with these growth factors.

An increasing list of local growth factors has been shown to modulate survival and apoptosis in the mammary gland. A stimulating role in the proliferation and/or differentiation of mammary epithelial cells is suggested for EGF, TGF, AR and IGF. The cytokine TNF may be involved in stimulating survival or cell death depending on the presence or absence of other factors, whereas TGF has been found only to inhibit growth and induce apoptosis.

Tissue parenchyma is the functional unit of an organ and all of the remaining cells within that organ collectively make up the tissue stroma. The stroma includes fibroblasts, endothelial cells, immune cells and nerves. Interactions between stromal and epithelial cells are essential for tissue development and healing after injury. These interactions are also governed by growth factors, inflammatory cytokines and hormone signaling cascades.

The **steroid receptor coactivator (SRC)** family of proteins includes three transcriptional coactivators that facilitate the assembly of multiprotein complexes to induce gene expression in response to activation of many cellular transcription factor signaling cascades. They are ubiquitously expressed and are especially critical for the developmental function of steroid hormone responsive tissues. The SRCs are overexpressed in multiple cancers including breast, ovarian, prostate and endometrial cancers.

In the adult, with each ovarian cycle the lateral buds differentiate to give rise to small alveolar buds (Shyamala, 1997). The glands are essentially quiescent until the onset of pregnancy. The mammary gland undergoes further profound tissue remodeling in response to RANKL (RANK ligand), Jk2 (Janus kinase), Stat5 (**signal transducer and activator of transcription 5**) and Elf5 (a member of an epithelium-specific subclass of the transcription factors ETS) (Shin et al., 2019; Napso et al., 2018); to **progesterone and prolactin during pregnancy**. The signaling pathways involved are beyond the scope of this book but include RANK (**receptor activator of nuclear factor κ B**) (Cordero et al., 2016).

Lactation has been studied extensively in animals because of the commercial interests in milk production. Milk is a complex fluid (Ontsouka and Albrecht, 2014) composed of proteins, sugars, lipids, minerals, vitamins, trace elements and cells with a caloric and nutrient content to support survival of the infant (Napso et al., 2018). The mechanisms that co-ordinate the activities of secretory and synthetic processes of mammary glands are poorly understood (Golan and Assaraf, 2020; Cordero et al., 2016; Sackmann-Sala et al., 2015; McManaman et al., 2006). Triglycerides represent more than 95% of the total lipids in milk; cholesterol represents less than 0.5% of the total milk lipid. Calcium transferred from mother to fetus and neonate is provided by the degradation of the female bone by increased osteoclastic activity, which is regulated by the RANK/RANKL axis. The transport of cholesterol across the epithelial barrier may involve ABC transporters (Ontsouka and Albrecht, 2014; Mani et al., 2010). Oxytocin, synthesized in the hypothalamus, is secreted into the circulation to induce contractions of cells in the mammary gland resulting in milk ejection.

Postmenopausal women are at high risk of developing breast cancer due to estrogen production in peripheral tissues of the body other than ovaries. Aromatase is present in breast tissue, leading to local estrogen production which can be inhibited by a variety of steroidal and nonsteroidal aromatase inhibitors. There are many aromatase inhibitors available in clinical practice like exemestane, formestane, anastrozole, letrozole, fadrozole, vorozole and so forth, but the major challenge in antibreast cancer therapy is the toxicity associated with aromatase inhibitors, especially the steroidal class of drugs. It is, therefore, urgently required to develop novel anticancer drugs having better safety and efficacy for the treatment of breast cancer. Many studies of aromatase inhibitors report benefits in the management of breast cancer.

The incidence of gynecomastia has dramatically increased over the last 20 years, implying that the endogenous or exogenous sex-steroid environment has changed, which is associated with other adverse health consequences in men such as an increased risk of prostate cancer, metabolic syndrome, type 2 diabetes or cardiovascular disorders.

The rise in production of estrogens at the onset of puberty lead to increased epithelial proliferation to form a tree-like pattern of ducts from the nipple to terminal end buds. The younger age at onset of breast development, which has been declining in recent years is associated with increased breast cancer risk independent of age at menarche (Houghton et al., 2018).

1.5.6.3.3 Placenta

The placenta is a temporary organ during pregnancy that controls nutrient, gas and waste exchange between the fetus and the mother. It connects the fetus to the uterine wall by the umbilical cord which has arteries and one vein. The perfusion of intervillous spaces with maternal blood allows the transfer of nutrients and oxygen to the fetus through the umbilical veins. IgG antibodies can pass to the fetus to provide immune protection, the antibodies persist after birth. Deoxygenated fetal blood passes through the arteries to the placenta. Waste products and carbon dioxide of the fetus are passed to the maternal blood by diffusion across the placenta. The fetus has to be protected from the normal immune response of the mother otherwise it would be rejected. The fetus and the placenta have immune tolerance.

The placenta produces estrogens from fetal androgens and itself produces progesterone as well as human chorionic gonadotrophin, placental lactogen, placental growth hormone, relaxin and kisspeptin. At the start of pregnancy, progesterone from the corpus luteum assists implantation of the embryo then maintains the integrity of the uterine wall. Progesterone suppresses contraction of muscles in the uterus. Estrogen production increases in pregnancy to stimulate growth of the uterus and prepares the uterus for labor. The mammary glands enlarge in response to estrogen in preparation for milk production.

Estrogen synthesis is dependent on supply of C_{19} steroid sulfates from the fetus. Uptake of conjugated fetal steroids requires placental expression of appropriate transport systems. The human OAT4 (SLC22A11) is expressed in

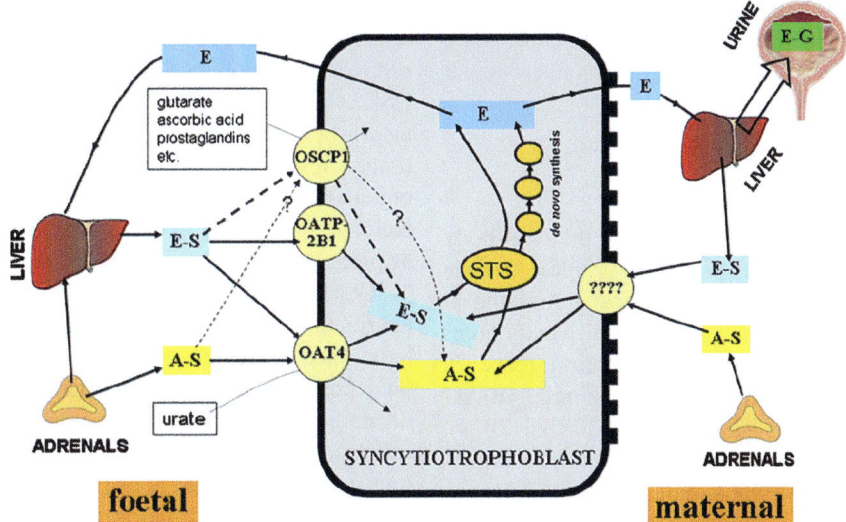

FIG. 1.5.16 **Scheme of placental estrogen metabolism with special focus on transport of sulfoconjugated steroids**. On the fetal side of syncytio-trophoblast OAT4 transports sulfoconjugated estrogens as well as sulfoconjugated C19-steroid precursors for placental de novo synthesis of estrogens. OATP2B1 may be not involved in de novo synthesis of placental estrogens but may contribute to the clearance of estrogen sulfates from placental circulation, only. *In contrast, transporters for the uptake of* maternal steroid sulfates on the apical side of the syncytiotrophoblast are unknown. *A-S*, sulfoconjugated C19-steroids; *E*, estrogens; *E-G*, estrogen glucuronides; *E-S*, estrogen sulfates; *OAT4*, organic anion transporter 4; *OATP2B1*, organic anion transporting polypeptide 2B1; *OSCP1*, organic solute carrier protein 1; *STS*, steroid sulfatase. *(Modified from Ugele B., Bahn A., Rex-Haffner M. Functional differences in steroid sulfate uptake of organic anion transporter 4 (OAT4) and organic anion transporting polypeptide 2B1 (OATP2B1) in human placenta. J Steroid Biochem Mol Biol 2008;111(1–2):1–6. Fig. 3 p. 3.)*

the placenta to play a critical role in the first step toward estrogen synthesis (Fig. 1.5.16). Following the uptake of androgen sulfates the placenta sulfatase liberates and androgens which are substrates for aromatase. The resulting estrogens can be sulfated by sulfotransferase. OAT2B1 is most likely to be the transporter for estrogen sulfates to the maternal circulation (Ugele et al., 2008).

Many ABC transporters are expressed in reproductive tissues (Joshi et al., 2016). The multidrug transporters, p-glycoprotein (P-gp), BCRP (breast cancer-related protein) and MRP1–5, and the cholesterol transporters, ABCA1 and ABCG1, are the most well described. Understanding ABC efflux activity, expression patterns, substrate specificity and regulation of the ABC transporters will provide substantial insight into normal reproductive function and associated pathologies, including infertility and pregnancy complications. It may also hold the key to development of innovative new interventions to treat these pathologies (Staud et al., 2012). Cholesterol from the maternal circulation has to cross two membrane barriers to reach the fetal circulation. Several cell types are involved and mechanisms change with gestational age. The process is not clear (Kallol and Albrecht, 2020).

Glucocorticoids are important in pregnancy and the fetus is protected by placental enzyme activity. HSD11B2 is highly expressed in syncytiotrophoblast of the placenta as well as colon and sweat glands. Cortisol levels are substantially higher in maternal versus cord blood due to placental

HSD11B2 enzyme. HSD11B1 activity is localized predominantly in the chorionic trophoblast (Sun et al., 2002). HSD11B1 insufficiency may enhance cardiometabolic risk in the offspring. Overexposure to endogenous or exogenous glucocorticoids is associated with hypothalamus-pituitary-adrenal axis reprogramming and cardiometabolic dysfunction. The implications of placental HSD11B2 for postnatal metabolic health in humans are still under investigation (Jahnke et al., 2021; Wang et al., 2020).

1.5.6.4 Male sex glands

The blood testis barrier prevents the entry of large or lipophilic molecules from the blood stream (Miller and Cherrington, 2018). ABC efflux transporters prevent accumulation of small or lipophilic compounds that may be harmful to germ cells. The most abundant ABC transporters in the testes include P-gp, BCRP and MRP1 and 4. MRP1 has a role in testicular steroidogenesis and efflux of testosterone from Leydig cells (Klein et al., 2014). Sertoli cells degrade tubular cytoplasmic residues from elongated spermatids and ABC transporters in protecting Sertoli cells from lipotoxicity of this material by efflux from the cells (Selva et al., 2004).

Androgens regulate the development and/or function of many tissues in the human body particularly the male sexual organs, secondary sex characteristics and male behavior. The normal ratio of plasma T to DHT is 10 in males and

2 in females although the male concentrations are 10 times higher than in the adult female (see Chapter 2.5 Steroids in blood).

1.5.6.5   Musculoskeletal system

1.5.6.5.1   Muscles

Skeletal muscles account for about 40% of the total body weight. Androgens act on myofibers to promote muscle growth. The effects of androgens are most noticeable at puberty when boy musculature changes significantly compared with girls. Androgens activate anabolic effects via IGF-1, phosphoinositide-3-kinase-protein kinase (B/Akt (PI3K/Aky)) and follistatin pathways (Seo et al., 2019). GCH promote protein degradation in skeletal muscle to generate gluconeogenic amino acids (e.g., alanine). GCH preserve plasma glucose by inhibiting glucose uptake and utilization in muscle (Kuo et al., 2013). Skeletal muscle stores glycogen High levels of GCH induce loss of muscle and strength.

1.5.6.5.2   Bones

The human skeleton is made up of nearly 300 bones that appear to be stable in composition, albeit subject to growth (Emmanuelle et al., 2021). Bone not only provides structure and support for the body that protects many important organs and interestingly produces red and white blood cells. Bone has a honeycomb-like matrix based on collagen which is mineralized in order to make the tissue hard. Osteoblasts and osteoclasts are continuously involved in the formation and breakdown of bone, respectively. The osteoblasts form a protective layer on the surface of the bone. Bone carries nerves and blood vessels.

Androgens have a role in building the skeleton although there is increasing evidence that part of the action of androgens is actually mediated by estrogens formed by the local action of aromatase (Tyagi et al., 2017; Clarke and Khosla, 2009). The conversion of testosterone to 5-alpha dihydrotestosterone is another important step in androgen action (Thu et al., 2017). Androgen receptors are expressed in chondrocytes in growth plates, osteoblasts and osteocytes GCH increase osteoclast activity and decrease osteoblasts leading to bone resorption and decrease formation leading to osteoporosis (Hardy and Cooper, 2010).

1.5.6.5.3   Skin and adipose tissue

The skin is a very large, complex organ in three layers—epidermis, dermis and subcutaneous layers. Skin acts as a physical and biochemical barrier. The keratinocytes, sebocytes and sebaceous glands express CYP117A1, CYP19A1 and HSD17B enzymes. Hair follicles sit in the dermis. Skin has the potential for steroidogenesis from cholesterol since it expresses CYP11A1 and other enzymes to produce progesterone. Skin is one of several tissues that can locally activate cortisol from cortisone by the action of HSD11B1. Fat in the subcutaneous layer can be either brown or white with different distribution. White adipose tissue (WAT) is mainly subcutaneous or around internal organs (visceral). GCH enhance glucose uptake and oxidation and increase lipolysis in WAT to release glycerol as precursor for gluconeogenesis and fatty acids.

GCH regulate adipose tissue differentiation, function, distribution, insulin sensitivity and in excess lead to pathological states. Patients with GCH excess develop abdominal fat deposition associated with metabolic syndrome (dyslipidemia, insulin resistance and hypertension). Multidrug resistance associated proteins (MRP, ABCC) and SOAT transporters are found to be expressed in skin (Osman-Ponchet et al., 2014) that are involved in uptake of compounds from the epidermal compartment as well as secretion into the blood and sweat glands.

1.5.6.6   Lungs

The importance of steroids in lung development has been known for more than 50 years (Liggins and Howie, 1972). Since then it has become common clinical practice to treat respiratory distress with corticosteroids (Briceño-Pérez et al., 2019). Fetal lungs in humans express mRNA of SULT1E1, StAR, CYP11A1, HSD3B1, HSD117B1, HSD17B2 and 5 as well as steroid receptors. ATP-binding cassettes (ABC transporters) are found in mammalian lung (Fig. 1.5.17). They are involved in cholesterol and phospholipid movement in lung cells (Chai et al., 2017).

In 2020, a global outbreak of COVID-19 infection was a new respiratory disease that started in China in December 2019. Vaccines became available by early 2021. Dexamethasone was found to be a helpful treatment as an antiinflammatory agent that also suppressed cortisol secretion. Inhibition of the MR with spironolactone was also beneficial (Edwards, 2021).

1.5.7   Steroid transport out of cells (efflux)

1.5.7.1   Saliva

The salivary glands consist of a system of blind ending ducts. There are three pairs in humans (parotid, submandibular and sublingual) (Fig. 1.5.18). The parotid gland is composed of serous secretory cells only, whereas submandibular and sublingual glands have serous and mucous cells.

Human salivary glands have a number of membrane transporters that provide efflux and influx functions although the focus of research has been on electrolytes

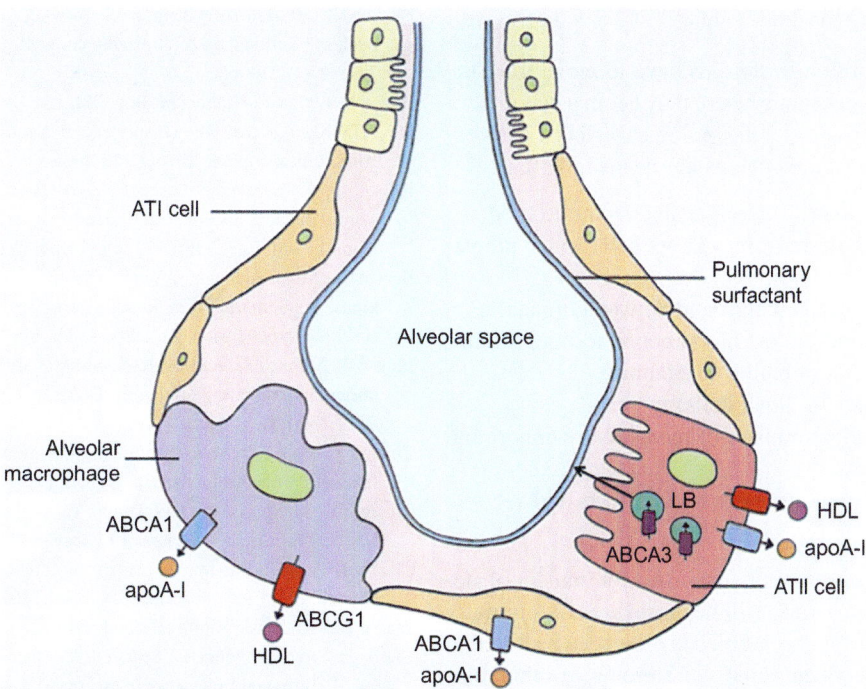

FIG. 1.5.17   ABC transporters expressed in various cell types of the alveolus. Three ABC transporters, namely ABCA1, ABCG1 and ABCA3, are expressed in lung cells present in the alveolus, notably the alveolar epithelial type I and II cells (ATI and ATII, respectively) that line the alveoli or air sacs and alveolar macrophages that are phagocytes of the pulmonary immune system. *LB*, lamellar bodies, where ABCA3 contributes lipids that eventually are secreted as surfactants (represented via the *black arrow*). *(From Chai AB, Ammit AJ, Gelissen IC. Examining the role of ABC lipid transporters in pulmonary lipid homeostasis and inflammation. Respir Res 2017;18(1):41. Fig. 1 p. 3.)*

and water (Sneyd et al., 2021). Aquaporins play a part in salivary gland functions (D'Agostino et al., 2020). There were discrepancies between data from mRNA expression and protein content recently characterized by LC-MS/MS after extraction from membranes (Lapczuk-Romanska et al., 2019; Sun et al., 2008) but more work is needed in this area.

Saliva is a complex mixture of water, proteins and electrolytes, maintained at pH between 6.2 and 7.6 by a bicarbonate buffering system. At the bottom of the ducts, acinar cells filter the blood into the saliva fluid. The composition of saliva is modified in the canals but this does not affect steroids. The concentration of steroid in saliva is not much dependent on quantities of plasma-binding proteins but 11β-hydroxysteroid dehydrogenase type 2 is expressed in salivary glands so cortisol is converted to cortisone.

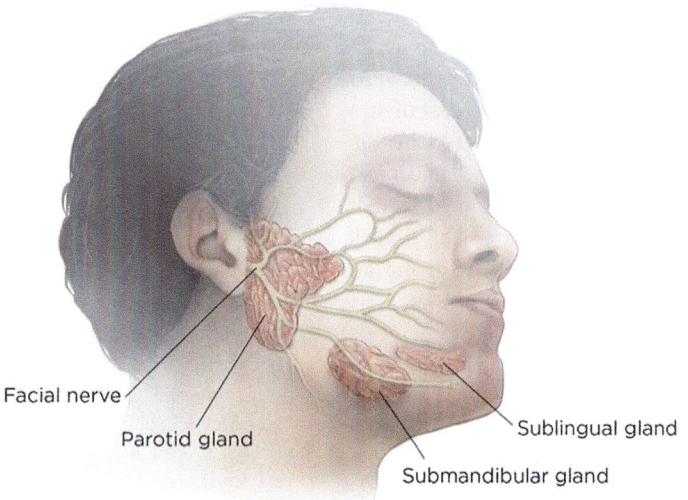

FIG. 1.5.18   Locations of the saliva glands.

1.5.8 Summary

By definition as hormones, steroids have to move from the site of origin to target organs where they act in a genomic or nongenomic route to achieve the biological effects. At every step in this journey, **transporters** are needed, be that

- uptake of cholesterol in the intestine from the food,
- movement of cholesterol into target cells and synthesis of the steroid hormones,
- movement of steroids within and between organelles,
- transport of steroids in the blood bound to lipoproteins,
- binding to plasma globulins or albumin,
- uptake of steroids in target cells,
- cytoplasmic-nuclear transfer of receptors by import and export proteins,
- transcription of new genes or rapid effects and
- movement of water.

In the course of time, further mechanisms for transfer of steroids, electrolytes and water will be discovered. A summary of the concentrations of the steroids in blood can be found in Chapter 2.5. The concentrations of steroids in other biological fluids and tissues are addressed in Chapter 2.6. In Part III of the book, the effects of hyper and hypo steroid states highlight areas where steroids have additional effects on tissues in the body particularly with hormone dependent cancers. Steroid transport, uptake and distribution needs to be understood where administering replacement steroids and peptide hormones with new routes (e.g., nasal, intradermal). The prescription of steroids for insufficiency can be from childhood for immaturity of lung function through many disorders of steroid production and action.

The number of identified interactions between uptake and efflux transporters and phase II conjugates continues to increase. This increased understanding of transporter involvement will further improve our ability of predict drug disposition and possible interindividual variability due to intrinsic (e.g., genotype, disease) and extrinsic factors (e.g., concomitant medication). Many studies provide in vitro and clinical information that indicate the involvement of steroids, their receptors and the transporters reviewed in this chapter, in the development and progression of cancers in the organs and form the basis of how treatments are directed. There is much still to learn of steroids in blood and tissues with the transport of steroids, especially DHEAS being an exciting area for more research.

References

Bäcklund N, Brattsand G, Israelsson M, Ragnarsson O, Burman P, Edén Engström B, et al. Reference intervals of salivary cortisol and cortisone and their diagnostic accuracy in Cushing's syndrome. Eur J Endocrinol 2020;182(6):569–82.

Berryhill GE, Trott JF, Hovey RC. Mammary gland development—it's not just about estrogen. J Dairy Sci 2016;99(1):875–83.

Bhosle VK, Rivera JC, Chemtob S. New insights into mechanisms of nuclear translocation of G-protein coupled receptors. Small GTPases 2019;10(4):254–63.

Bloise E, Ortiga-Carvalho TM, Reis FM, Lye SJ, Gibb W, Matthews SG. ATP-binding cassette transporters in reproduction: a new frontier. Hum Reprod Update 2016;22(2):164–81.

Briceño-Pérez C, Reyna-Villasmil E, Vigil-De-Gracia P. Antenatal corticosteroid therapy: historical and scientific basis to improve preterm birth management. Eur J Obstet Gynecol Reprod Biol 2019;234:32–7.

Brown A, Meor Azlan NF, Wu Z, Zhang J. WNK-SPAK/OSR1-NCC kinase signaling pathway as a novel target for the treatment of salt-sensitive hypertension. Acta Pharmacol Sin 2021;42(4):508–17.

Chai AB, Ammit AJ, Gelissen IC. Examining the role of ABC lipid transporters in pulmonary lipid homeostasis and inflammation. Respir Res 2017;18(1):41.

Chang CC, Hsia KC. More than a zip code: global modulation of cellular function by nuclear localization signals. FEBS J 2021;288(19):5569–85.

Chang TY, Li BL, Chang CC, Urano Y. Acyl-coenzyme A: cholesterol acyltransferases. Am J Physiol Endocrinol Metab 2009;297(1):E1–9.

Clark BJ. The START-domain proteins in intracellular lipid transport and beyond. Mol Cell Endocrinol 2020;504:110704.

Clarke BL, Khosla S. Androgens and bone. Steroids 2009;74(3):296–305.

Claro da Silva T, Polli JE, Swaan PW. The solute carrier family 10 (SLC10): beyond bile acid transport. Mol Aspects Med 2013;34(2–3):252–69.

Cordero A, Pellegrini P, Sanz-Moreno A, Trinidad EM, Serra-Musach J, Deshpande C, et al. Rankl impairs lactogenic differentiation through inhibition of the prolactin/Stat5 pathway at midgestation. Stem Cells 2016;34(4):1027–39.

D'Agostino C, Elkashty OA, Chivasso C, Perret J, Tran SD, Delporte C. Insight into salivary gland aquaporins. Cells 2020;9(6):1547.

Dichtel LE, Schorr M, Loures de Assis C, Rao EM, Sims JK, Corey KE, et al. Plasma free cortisol in states of normal and altered binding globulins: implications for adrenal insufficiency diagnosis. J Clin Endocrinol Metab 2019;104(10):4827–36.

Diederich S, Quinkler M, Burkhardt P, Grossmann C, Bähr V, Oelkers W. 11Beta-hydroxysteroid-dehydrogenase isoforms: tissue distribution and implications for clinical medicine. Eur J Clin Invest 2000;30(Suppl 3):21–7.

Downie ML, Lopez Garcia SC, Kleta R, Bockenhauer D. Inherited tubulopathies of the kidney: insights from genetics. Clin J Am Soc Nephrol 2021;16(4):620–30. Erratum in: Clin J Am Soc Nephrol. 2021;16(7):1100.

Dunn JF, Nisula BC, Rodbard D. Transport of steroid hormones: binding of 21 endogenous steroids to both testosterone-binding globulin and corticosteroid-binding globulin in human plasma. J Clin Endocrinol Metab 1981;53(1):58–68.

Edwards C. New horizons: does mineralocorticoid receptor activation by cortisol cause ATP release and COVID-19 complications? J Clin Endocrinol Metab 2021;106(3):622–35.

Emmanuelle NE, Marie-Cécile V, Florence T, Jean-François A, Françoise L, Coralie F, et al. Critical role of estrogens on bone homeostasis in both male and female: from physiology to medical implications. Int J Mol Sci 2021;22(4):1568.

Fan P, Lu YT, Yang KQ, Zhang D, Liu XY, Tian T, et al. Apparent mineralocorticoid excess caused by novel compound heterozygous mutations in HSD11B2 and characterized by early-onset hypertension and hypokalemia. Endocrine 2020;70(3):607–15.

Golan Y, Assaraf YG. Genetic and physiological factors affecting human milk production and composition. Nutrients 2020;12(5):1500.

Goldman AL, Bhasin S, Wu FCW, Krishna M, Matsumoto AM, Jasuja R. A reappraisal of testosterone's binding in circulation: physiological and clinical implications. Endocr Rev 2017;38(4):302–24.

Grosser G, Bennien J, Sánchez-Guijo A, Bakhaus K, Döring B, Hartmann M, et al. Transport of steroid 3-sulfates and steroid 17-sulfates by the sodium-dependent organic anion transporter SOAT (SLC10A6). J Steroid Biochem Mol Biol 2018;179:20–5.

Hammond GL. Plasma steroid-binding proteins: primary gatekeepers of steroid hormone action. J Endocrinol 2016;230(1):R13–25.

Hammond GL, Langley MS, Robinson PA, Nummi S, Lund L. Serum steroid binding protein concentrations, distribution of progestogens, and bioavailability of testosterone during treatment with contraceptives containing desogestrel or levonorgestrel. Fertil Steril 1984;42 (1):44–51.

Hardy R, Cooper MS. Adrenal gland and bone. Arch Biochem Biophys 2010;503(1):137–45.

Hartmann K, Bennien J, Wapelhorst B, Bakhaus K, Schumacher V, Kliesch S, et al. Current insights into the sulfatase pathway in human testis and cultured Sertoli cells. Histochem Cell Biol 2016;146(6):737–48.

Henley D, Lightman S, Carrell R. Cortisol and CBG—getting cortisol to the right place at the right time. Pharmacol Ther 2016;166:128–35.

Hong H, Branham WS, Ng HW, et al. Human sex hormone-binding globulin binding affinities of 125 structurally diverse chemicals and comparison with their binding to androgen receptor, estrogen receptor, and α-fetoprotein. Toxicol Sci 2015;143(2):333–48.

Houghton LC, Knight JA, De Souza MJ, Goldberg M, White ML, O'Toole K, et al. Comparison of methods to assess onset of breast development in the LEGACY girls study: methodological considerations for studies of breast cancer. Breast Cancer Res 2018;20(1):33.

Iaea DB, Spahr ZR, Singh RK, Chan RB, Zhou B, Bareja R, et al. Stable reduction of STARD4 alters cholesterol regulation and lipid homeostasis. Biochim Biophys Acta, Mol Cell Biol Lipids 2020;1865(4):158609.

Jahnke JR, Terán E, Murgueitio F, Cabrera H, Thompson AL. Maternal stress, placental 11β-hydroxysteroid dehydrogenase type 2, and infant HPA axis development in humans: psychosocial and physiological pathways. Placenta 2021;104:179–87.

Järvinen E, Deng F, Kidron H, Finel M. Efflux transport of estrogen glucuronides by human MRP2, MRP3, MRP4 and BCRP. J Steroid Biochem Mol Biol 2018;178:99–107.

Jentsch TJ, Pusch M. CLC chloride channels and transporters: structure, function, physiology, and disease. Physiol Rev 2018;98(3):1493–590.

Jetter A, Kullak-Ublick GA. Drugs and hepatic transporters: a review. Pharmacol Res 2020;154:104234.

Jia MC, Zhou LY, Ren S, Dong L, Xiao B. Serum SHBG levels during normal menstrual cycle and after insertion of levonorgestrel-releasing IUD. Adv Contracept 1992;8(1):33–40.

Joshi AA, Vaidya SS, St-Pierre MV, Mikheev AM, Desino KE, Nyandege AN, et al. Placental ABC transporters: biological impact and pharmaceutical significance. Pharm Res 2016;33(12):2847–78.

Kalaria T, Agarwal M, Kaur S, Hughes L, Sharrod-Cole H, Chaudhari R, et al. Hypothalamic-pituitary-adrenal axis suppression—the value of salivary cortisol and cortisone in assessing hypothalamic-pituitary-adrenal recovery. Ann Clin Biochem 2020;57(6):456–60.

Kallol S, Albrecht C. Materno-fetal cholesterol transport during pregnancy. Biochem Soc Trans 2020;48(3):775–86.

Klein DM, Wright SH, Cherrington NJ. Localization of multidrug resistance-associated proteins along the blood-testis barrier in rat, macaque, and human testis. Drug Metab Dispos 2014;42(1):89–93.

Kraemer FB, Shen WJ, Azhar S. SNAREs and cholesterol movement for steroidogenesis. Mol Cell Endocrinol 2017;441:17–21.

Kuo T, Harris CA, Wang JC. Metabolic functions of glucocorticoid receptor in skeletal muscle. Mol Cell Endocrinol 2013;380(1–2):79–88.

Keevil BG, Adaway J. Assessment of free testosterone concentration. J Steroid Biochem Mol Biol 2019;190:207–11.

Lapczuk-Romanska J, Busch D, Gieruszczak E, Drozdzik A, Piotrowska K, Kowalczyk R, et al. Membrane transporters in human parotid gland-targeted proteomics approach. Int J Mol Sci 2019;20(19):4825.

Li CY, Basit A, Gupta A, Gáborik Z, Kis E, Prasad B. Major glucuronide metabolites of testosterone are primarily transported by MRP2 and MRP3 in human liver, intestine and kidney. J Steroid Biochem Mol Biol 2019;191:105350.

Li CY, Gupta A, Gáborik Z, Kis E, Prasad B. Organic anion transporting polypeptide-mediated hepatic uptake of glucuronide metabolites of androgens. Mol Pharmacol 2020;98(3):234–42.

Liggins GC, Howie RN. A controlled trial of antepartum glucocorticoid treatment for prevention of the respiratory distress syndrome in premature infants. Pediatrics 1972;50(4):515–25.

Lin HY, Underhill C, Lei JH, Helander-Claesson A, Lee HY, Gardill BR, et al. High frequency of SERPINA6 polymorphisms that reduce plasma corticosteroid-binding globulin activity in Chinese subjects. J Clin Endocrinol Metab 2012;97(4):E678–86.

Macias H, Hinck L. Mammary gland development. Wiley Interdiscip Rev Dev Biol 2012;1(4):533–57.

Mamenko M, Zaika O, Ilatovskaya DV, Staruschenko A, Pochynyuk O. Angiotensin II increases activity of the epithelial Na+ channel (ENaC) in distal nephron additively to aldosterone. J Biol Chem 2012;287(1):660–71.

Mani O, Körner M, Sorensen MT, Sejrsen K, Wotzkow C, Ontsouka CE, et al. Expression, localization, and functional model of cholesterol transporters in lactating and nonlactating mammary tissues of murine, bovine, and human origin. Am J Physiol Regul Integr Comp Physiol 2010;299(2):R642–54.

Maxfield FR, Iaea DB, Pipalia NH. Role of STARD4 and NPC1 in intracellular sterol transport. Biochem Cell Biol 2016;94(6):499–506.

Molina JR, Adjei AA. The Ras/Raf/MAPK pathway. J Thorac Oncol 2006;1(1):7–9.

McFeely SJ, Wu L, Ritchie TK, Unadkat J. Organic anion transporting polypeptide 2B1—more than a glass-full of drug interactions. Pharmacol Ther 2019;196:204–15.

McManaman JL, Reyland ME, Thrower EC. Secretion and fluid transport mechanisms in the mammary gland: comparisons with the exocrine pancreas and the salivary gland. J Mammary Gland Biol Neoplasia 2006;11(3–4):249–68.

Mendes S, Timóteo-Ferreira F, Almeida H, Silva E. New insights into the process of placentation and the role of oxidative uterine microenvironment. Oxid Med Cell Longev 2019;2019:9174521.

Midzak A, Papadopoulos V. Binding domain-driven intracellular trafficking of sterols for synthesis of steroid hormones, bile acids and oxysterols. Traffic 2014;15(9):895–914.

Miller WL. Disorders in the initial steps of steroid hormone synthesis. J Steroid Biochem Mol Biol 2017a;165(Pt A):18–37.

Miller WL. Steroidogenesis: unanswered questions. Trends Endocrinol Metab 2017b;28(11):771–93.

Miller SR, Cherrington NJ. Transepithelial transport across the blood-testis barrier. Reproduction 2018;156(6):R187–94.

Mount DB. Thick ascending limb of the loop of Henle. Clin J Am Soc Nephrol 2014;9(11):1974–86.

Napso T, Yong HEJ, Lopez-Tello J, Sferruzzi-Perri AN. The role of placental hormones in mediating maternal adaptations to support pregnancy and lactation. Front Physiol 2018;9:1091.

Neuman SD, Bashirullah A. Reconsidering the passive diffusion model of steroid hormone cellular entry. Dev Cell 2018;47(3):261–2.

Ontsouka EC, Albrecht C. Cholesterol transport and regulation in the mammary gland. J Mammary Gland Biol Neoplasia 2014;19(1):43–58.

Osman-Ponchet H, Boulai A, Kouidhi M, Sevin K, Alriquet M, Gaborit A, et al. Characterization of ABC transporters in human skin. Drug Metabol Drug Interact 2014;29(2):91–100.

Palmada M, Embark HM, Wyatt AW, Böhmer C, Lang F. Negative charge at the consensus sequence for the serum- and glucocorticoid-inducible kinase, SGK1, determines pH sensitivity of the renal outer medullary K + channel, ROMK1. Biochem Biophys Res Commun 2003a;307 (4):967–72.

Palmada M, Embark HM, Yun C, Böhmer C, Lang F. Molecular requirements for the regulation of the renal outer medullary K(+) channel ROMK1 by the serum- and glucocorticoid-inducible kinase SGK1. Biochem Biophys Res Commun 2003b;311(3):629–34.

Papadopoulos V, Fan J, Zirkin B. Translocator protein (18 kDa): an update on its function in steroidogenesis. J Neuroendocrinol 2018;30(2).

Pfisterer SG, Peränen J, Ikonen E. LDL-cholesterol transport to the endoplasmic reticulum: current concepts. Curr Opin Lipidol 2016;27 (3):282–7.

Raverot V, Lopez J, Grenot C, Pugeat M, Déchaud H. New approach for measurement of non-SHBG-bound testosterone in human plasma. Anal Chim Acta 2010;658(1):87–90.

Rossi GM, Regolisti G, Peyronel F, Fiaccadori E. Recent insights into sodium and potassium handling by the aldosterone-sensitive distal nephron: a review of the relevant physiology. J Nephrol 2020a;33 (3):431–45.

Rossi GM, Regolisti G, Peyronel F, Fiaccadori E. Recent insights into sodium and potassium handling by the aldosterone-sensitive distal nephron: implications on pathophysiology and drug discovery. J Nephrol 2020b;33(3):447–66.

Sackmann-Sala L, Guidotti JE, Goffin V. Minireview: prolactin regulation of adult stem cells. Mol Endocrinol 2015;29(5):667–81.

Sandberg AA, Jr SWR, Antoniades HN. The binding of steroids and steroid conjugates to human plasma proteins. Recent Prog Horm Res 1957;13:209–60 [discussion 260–7].

Selva DM, Hirsch-Reinshagen V, Burgess B, Zhou S, Chan J, McIsaac S, et al. The ATP-binding cassette transporter 1 mediates lipid efflux from Sertoli cells and influences male fertility. J Lipid Res 2004;45 (6):1040–50.

Selvaraj V, Stocco DM, Tu LN. Minireview: translocator protein (TSPO) and steroidogenesis: a reappraisal. Mol Endocrinol 2015;29(4):490–501.

Seo JY, Kim JH, Kong YY. Unraveling the paradoxical action of androgens on muscle stem cells. Mol Cells 2019;42(2):97–103.

Shah P, Chaumet A, Royle SJ, Bard FA. The NAE pathway: autobahn to the nucleus for cell surface receptors. Cell 2019;8(8):915.

Shen WJ, Asthana S, Kraemer FB, Azhar S. Scavenger receptor B type 1: expression, molecular regulation, and cholesterol transport function. J Lipid Res 2018a;59(7):1114–31.

Shen WJ, Azhar S, Kraemer FB. SR-B1: a unique multifunctional receptor for cholesterol influx and efflux. Annu Rev Physiol 2018b;80:95–116.

Shin HY, Hennighausen L, Yoo KH. STAT5-driven enhancers tightly control temporal expression of mammary-specific genes. J Mammary Gland Biol Neoplasia 2019;24(1):61–71.

Shyamala G. Roles of estrogen and progesterone in normal mammary gland development insights from progesterone receptor null mutant mice and in situ localization of receptor. Trends Endocrinol Metab 1997;8 (1):34–9.

Simard M, Hill LA, Lewis JG, Hammond GL. Naturally occurring mutations of human corticosteroid-binding globulin. J Clin Endocrinol Metab 2015;100(1):E129–39.

Sneyd J, Vera-Sigüenza E, Rugis J, Pages N, Yule DI. Calcium dynamics and water transport in salivary acinar cells. Bull Math Biol 2021;83(4):31.

Soccio RE, Adams RM, Romanowski MJ, Sehayek E, Burley SK, Breslow JL. The cholesterol-regulated StarD4 gene encodes a StAR-related lipid transfer protein with two closely related homologues, StarD5 and StarD6. Proc Natl Acad Sci U S A 2002;99(10):6943–8.

Staud F, Cerveny L, Ceckova M. Pharmacotherapy in pregnancy; effect of ABC and SLC transporters on drug transport across the placenta and fetal drug exposure. J Drug Target 2012;20(9):736–63.

Stocco DM, Zhao AH, Tu LN, Morohaku K, Selvaraj V. A brief history of the search for the protein(s) involved in the acute regulation of steroidogenesis. Mol Cell Endocrinol 2017;441:7–16.

Stölting G, Fischer M, Fahlke C. CLC channel function and dysfunction in health and disease. Front Physiol 2014;5:378.

Su W, Cao R, Zhang XY, Guan Y. Aquaporins in the kidney: physiology and pathophysiology. Am J Physiol Renal Physiol 2020;318(1):F193–203.

Sun K, He P, Yang K. Intracrine induction of 11beta-hydroxysteroid dehydrogenase type 1 expression by glucocorticoid potentiates prostaglandin production in the human chorionic trophoblast. Biol Reprod 2002;67(5):1450–5.

Sun QF, Sun QH, Du J, Wang S. Differential gene expression profiles of normal human parotid and submandibular glands. Oral Dis 2008;14 (6):500–9.

Thu HE, Mohamed IN, Hussain Z, Shuid AN. Dihydrotestosterone, a robust promoter of osteoblastic proliferation and differentiation: understanding of time-mannered and dose-dependent control of bone forming cells. Iran J Basic Med Sci 2017;20(8):894–904.

Tugaeva KV, Sluchanko NN. Steroidogenic acute regulatory protein: structure, functioning, and regulation. Biochemistry (Mosc) 2019;84 (Suppl 1):S233–53.

Tyagi V, Scordo M, Yoon RS, Liporace FA, Greene LW. Revisiting the role of testosterone: are we missing something? Rev Urol 2017;19(1):16–24.

Ugele B, Bahn A, Rex-Haffner M. Functional differences in steroid sulfate uptake of organic anion transporter 4 (OAT4) and organic anion transporting polypeptide 2B1 (OATP2B1) in human placenta. J Steroid Biochem Mol Biol 2008;111(1–2):1–6.

Valinsky WC, Touyz RM, Shrier A. Aldosterone, SGK1, and ion channels in the kidney. Clin Sci (Lond) 2018;132(2):173–83.

Viganò P, Mangioni S, Pompei F, Chiodo I. Maternal-conceptus cross talk—a review. Placenta 2003;24(Suppl. B):S56–61.

Vogeser M, Briegel J. Effect of temperature on protein binding of cortisol. Clin Biochem 2007;40(9–10):724–7.

Wall SM, Verlander JW, Romero CA. The renal physiology of pendrin-positive intercalated cells. Physiol Rev 2020;100(3):1119–47.

Wang WS, Guo CM, Sun K. Cortisol regeneration in the fetal membranes, a coincidental or requisite event in human parturition? Front Physiol 2020;11:462.

Wu TS, Hammond GL. Naturally occurring mutants inform SHBG structure and function. Mol Endocrinol 2014;28(7):1026–38.

Wu A, Wolley M, Stowasser M. The interplay of renal potassium and sodium handling in blood pressure regulation: critical role of the WNK-SPAK-NCC pathway. J Hum Hypertens 2019;33(7):508–23.

Yamazaki O, Ishizawa K, Hirohama D, Fujita T, Shibata S. Electrolyte transport in the renal collecting duct and its regulation by the renin-angiotensin-aldosterone system. Clin Sci (Lond) 2019;133 (1):75–82.

Yoo D, Kim BY, Campo C, Nance L, King A, Maouyo D, et al. Cell surface expression of the ROMK (Kir 1.1) channel is regulated by the aldosterone-induced kinase, SGK-1, and protein kinase A. J Biol Chem 2003;278(25):23066–75.

Zaika O, Mamenko M, Staruschenko A, Pochynyuk O. Direct activation of ENaC by angiotensin II: recent advances and new insights. Curr Hypertens Rep 2013;15(1):17–24.

Zhang Q, Piston DW, Goodman RH. Regulation of corepressor function by nuclear NADH. Science 2002;295(5561):1895–7.

Zheng X, Bi C, Brooks M, Hage DS. Analysis of hormone-protein binding in solution by ultrafast affinity extraction: interactions of testosterone with human serum albumin and sex hormone binding globulin. Anal Chem 2015;87(22):11187–94.

Chapter 1.6

Recognition of steroids

1.6.1 Introduction

All the steroid hormones exert their actions by passing through the plasma membrane and **binding as ligands to receptors** within target cells. Steroid receptors are found in the cytosol or the nucleus in the absence of ligand. The inactivated receptor resides as part of a complex consisting of chaperone heat shock proteins (HSPs 90, 70 and 50), immunophilins and other proteins. On ligand binding, the receptor undergoes a conformational change that results in dissociation from this complex and translocation to the nucleus. Within the nucleus, the receptor binds as a dimer to tandem response elements through specific nucleotide sequences in the DNA of responsive genes, identified as **hormone response elements**, HREs. The interaction of steroid-receptor complexes with DNA leads to **altered rates of transcription** of selected genes. This contrasts with the peptide hormone receptors that cross the plasma membrane and bind ligand outside the cell nucleus. When steroids act as transcription factors the response is slow, taking hours or days but rapid responses to steroids are seen and these effects are due to membrane reactions that do not involve the nucleus and protein synthesis. Much of the processes involved are poorly understood at this time.

1.6.2 Nuclear receptors

The human steroid receptors are members of the **superfamily** of **steroid/thyroid/retinoic acid nuclear receptor** transcription factor proteins that influence the transcription rates of numerous target genes in a positive or negative fashion.

1.6.2.1 Steroid recognition

The steroid receptors include **the androgen receptor (AR), the progesterone receptor (PR), the estrogen receptor (ER)**, the thyroid hormone receptor (TR), the vitamin D receptor (VDR), the retinoic acid receptors (RARs), **the mineralocorticoid receptor (MR), and the glucocorticoid receptor (GR)**.

The steroid family of receptors each has distinct functional domains in the protein (Fig. 1.6.1).

- DNA-binding domain (DBD),
- hinge region
- ligand-binding domain (LBD),
- N-terminal domain (NTD) and
- a transcriptional regulatory domain, referred to as the activation function domain (e.g., AF-1).

The NTD in the A/B region is the site of the major posttranslational regulation of receptor function. The NTD or immunogenic domain is encoded by exon 2 and interacts with coactivators. Serine residues are sites for phosphorylation in the NTD of the receptor. The N-terminal A/B region harbors an autonomous activation function. The structure of the NTD of the steroid receptors is between a fully folded state and a structured folded conformation with a tendency to form helical structures.

The central C region, corresponding to the DBD, is involved in DNA binding and receptor dimerization. Exons 3 and 4 code for the amino acids that form the DBD into a compact globular structure with the characteristic motif of **two zinc fingers**, that facilitates the interaction between the receptor and its target DNA sequences in the promoter regions of responsive genes (Fig. 1.6.2).

The DBD is the best conserved region in the members of the receptor superfamily with a high content of basic amino acids and by nine conserved cysteine residues. The folding of the DBD is similar for the AR, GR and ER. Each zinc finger contains one zinc atom which interacts via coordination bonds with four cysteine residues. The two zinc fingers are both C-terminally flanked by an α-helix. The two zinc clusters are structurally and functionally different and are encoded from different exons. The α-helix of the most N-terminal sited zinc cluster interacts directly with nucleotides of the hormone response element (HRE) in the major groove of the DNA. Three amino acid residues at the N-terminus of this α-helix are responsible for the specific recognition of the DNA-sequence of the responsive element. The residues are different at homologous positions in the ER. The P(roximal)-box [Gly; Ser; Val;] are identical in the AR, PR, GR and MR and can therefore recognize the same response element. DNA-sequences flanking the HRE, code for receptor interactions with other proteins and receptor concentrations are important for the hormone and

Steroids in the Laboratory and Clinical Practice. https://doi.org/10.1016/B978-0-12-818124-9.00008-5

Domains receptors

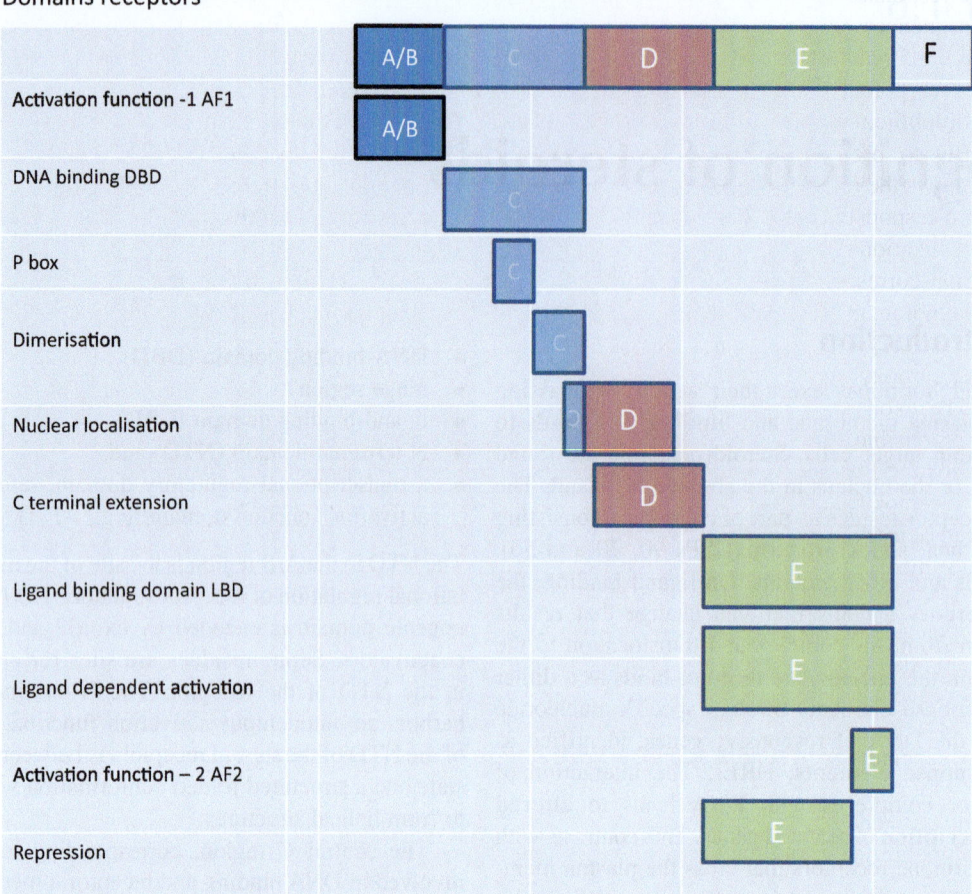

FIG. 1.6.1 General scheme for the domains of steroid receptors. *(Author original.)*

FIG. 1.6.2 The DBD consist of two zinc fingers, between the NBD and the hinge region. Variation in the function of GR can be correlated with the domains. *D*, distal; *P*, proximal. Regions important in GR function are indicated below the protein. *A*, acetylation; *AF*, activation function; *N*, nitrosylation; *NES*, nuclear export signal; *NLS*, nuclear localization signal; *NRS*, nuclear retention signal; *O*, oxidation; *P*, phosphorylation; *S*, sumoylation; *U*, ubiquitination. Zinc fingers. *(Based on Vandevyver S, Dejager L, Libert C. Comprehensive overview of the structure and regulation of the glucocorticoid receptor. Endocr Rev 2014;35(4):671–93. Fig. 3 p. 677.)*

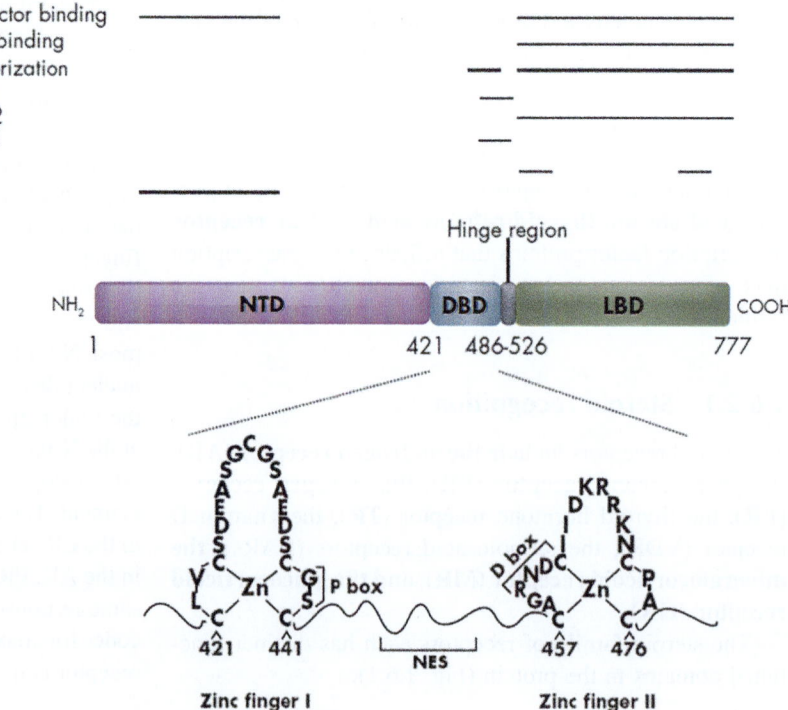

tissue specific responses of the different receptors. Some palindromic sequences (nucleotides sequences that read the same from the 5′-end to the 3′-end) and half-sites have been identified as potential receptor binding sites The D region is a hydrophilic region and it forms a **hinge** between DBD and LBD to provide structural flexibility to the receptor and allows the interaction of the LBD with several different steroid-responsive genes. The hinge region harbors two nuclear localization signals identified as NL1 and NL2.

The E region corresponds to the C-terminal LBD and acts for ligand binding, interaction with heat-shock proteins, dimerization, nuclear targeting, and hormone-dependent activation. The carboxyl-terminal fragment of the receptor is the LBD. The exons 5–9 code for amino acids in the LBD that is responsible for the binding of the receptor to steroids, then for the translocation of ligand-induced activation of the receptor from the cytoplasm to the nucleus, and for the transactivation and interaction of the receptor with coactivator molecules (AF-1 domain). A second activation function (AF2) domain is found in the extreme C-terminal region of the receptors. The AF2 domain, like the AF1 domain, also binds to transcriptional coregulators in a ligand-dependent manner. A shortened receptor can act as a repressor.

The hormone binding domain is a highly conserved region encoded by approximately 250 amino acid residues in the C-terminal end of the molecule. The three-dimensional structure takes the conformation of a nuclear receptor ligand binding domain fold. The ligand binding pocket can accommodate ligands with different structures. On hormone binding, the fold of the ligand binding domain results in a globular structure with an interaction surface for binding of interacting proteins like coactivators (e.g., AF2). Following transcriptional activation or inhibition of the steroid-responsive genes, the receptor dissociates from the ligand. The unliganded receptor remains within the nucleus for a considerable length of time and is then exported to the cytoplasm; both within the nucleus and within the cytoplasm the receptor may be recycled and/or degraded in the proteasome.

1.6.3 Steroid receptors

1.6.3.1 General mode of action

After entering a target cell, the steroid binds to its receptor, in some cases (as will become apparent later in this chapter) after local conversion to a related steroid (for testosterone reduction to 5α-dihydrotestosterone). The receptor protein is in a complex with chaperone proteins (e.g., heat shock proteins) in the cytoplasm, which is collectively called the "**foldosome**" and also has functions beyond the classical role in the cytosol (Weikum et al., 2017; Cano et al., 2013). The chaperone dissociates from the complex as the steroid

binds to the receptor component. Simultaneously there is a conformational change of the receptor protein resulting in a transformation and translocation to the nucleus (Fig. 1.6.3).

On binding specific DNA-sequences in the nucleus (hormone response element—HRE or GRE if referring to glucocorticoid specifically), the receptor dimerizes with a second molecule and the homodimer entity recruits further proteins (e.g., **coactivators**, general transcription factors, RNA-polymerase II) through specific interaction motifs (Skowron et al., 2019; Fuller et al., 2017). The bound receptor stimulates the transcription of target genes by facilitating the formation of a complex with RNA polymerase II and its ancillary components to positively or negatively regulate expression. The outcome is thus transcriptional activation or suppression of specific steroid responsive genes.

Within the nucleus the steroid can act at many transcription regulatory factors and different HREs with different coregulators to activate or repress transcription of particular target genes. The transcription site may be within 100 base pairs or further from the target genes. Multiple genes are affected depending on the tissue involved which partly explains the plasticity seen with steroids in the body overall. Various tissue specific GRα isoforms result from phenomena referred to as **ribosome shunting** and/or **leaky ribosome scanning**. The GR can also affect transcriptional activity via a mechanism referred to as **tethering.** The GR is tethered to chromatin by its interaction with other transcription factors. Tethering of GR has been shown to occur with many transcription factors including CREB, AP-1, NF-κB, GATA1, STAT1, and STAT3 (Timmermans et al., 2019; Ratman et al., 2013). GR as a monomer binds to half sites or reverse half-sites. The GR can further transrepress gene-expression by competing for DNA binding-sites (BS), by sequestrating transactivation factors (TF)s and by competing for cofactors with other TFs. GR might also function as a tetramer, but its function is not known. Two GR monomers have been found to bind on opposite sides of the DNA, called inverted repeat glucocorticoid repeat site (ir-GBS), negating expression of activity.

There is much to be learnt about the recognition of steroids and the processes controlled.

1.6.4 Receptors for adrenal steroids

1.6.4.1 Glucocorticoid receptor

The GR is encoded by the NR3C1 (nuclear receptor subfamily 3 group C member 1) gene. The NR3C1 gene is located on chromosome 5q31.3 and is composed of 8 protein coding exons (exons 2–9) and 13 variants of exon 1 that differ as a result of upstream promoter elements (Timmermans et al., 2019; Vandevyver et al., 2014). The GR in the absence of ligand remains in the cytosol within

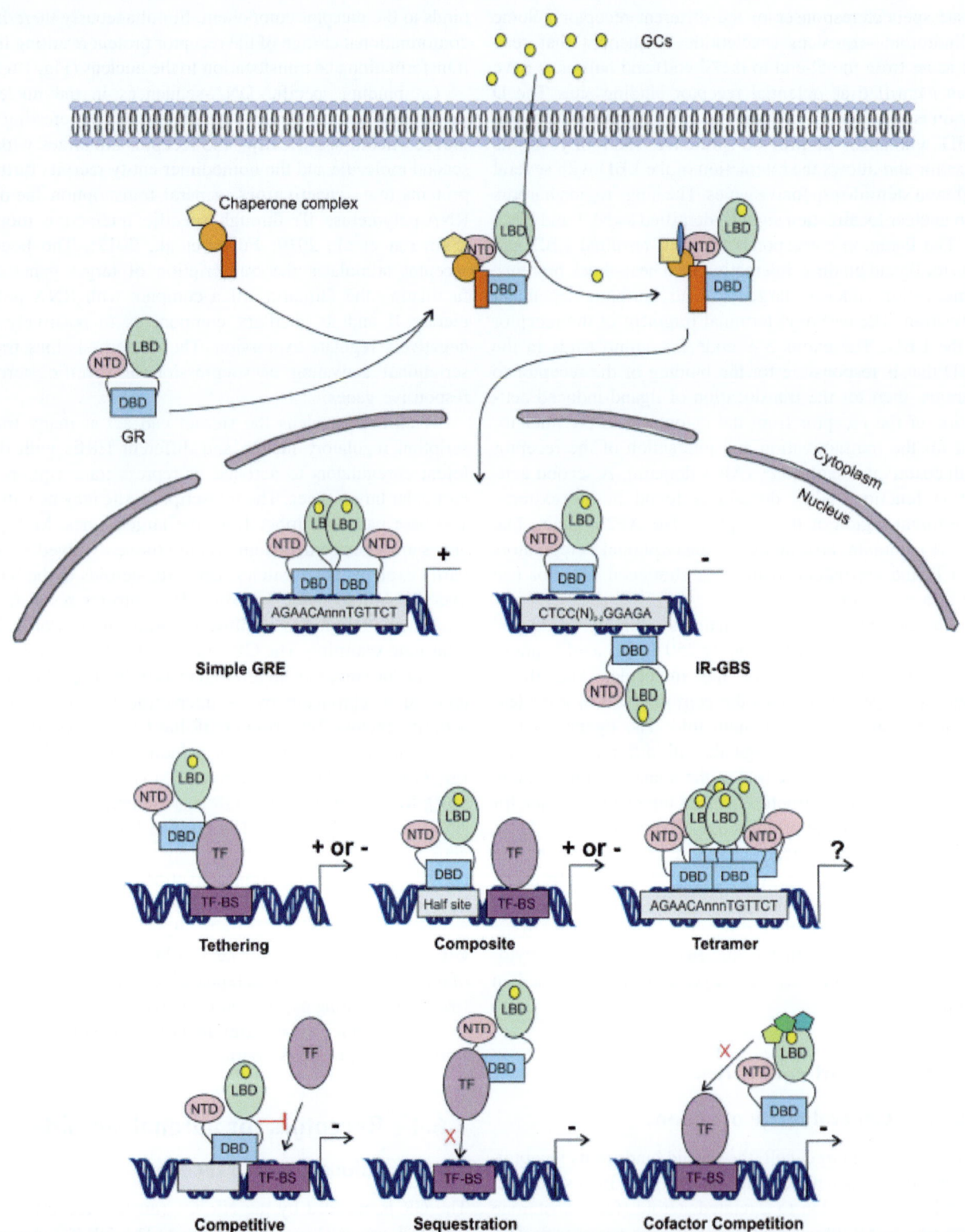

FIG. 1.6.3 Glucocorticoid receptor activation and function. Glucocorticoids (GCs) diffuse through the cell membrane and bind the glucocorticoid receptor (GR) in the cytoplasm. This induces a change in the chaperone complex bound to GR, after which GR translocates to the nucleus to transactivate (+) or transrepress (−) gene transcription as a monomer or a dimer. The GR can transactivate genes by binding to glucocorticoid responsive elements (GRE) as a dimer, or as a monomer bound to other transcription factors (TF) through tethering or by binding to composite-elements. The GR can further transrepress gene-expression by binding to inverted repeat GR-binding sequences (IR-GBS), by **tethering**, by composite-elements, by competing for DNA binding-sites (BS), by sequestering TFs and by competing for cofactors with other TFs. GR might also function as a tetramer of unknown function. *(From Timmermans S, Souffriau J, Libert C. A general introduction to glucocorticoid biology. Front Immunol 2019;10:1545. Fig. 5 p. 7.)*

multiprotein chaperone complexes (called a foldosome) that includes heat-shock protein 90 (hsp90), hsp70, the co-chaperone called p23 and various immunophilins (e.g., FKBP51 and FKBP52), that prevent its degradation and assist in its maturation. On contact with steroid, the GR complex changes conformation leading to release of the ligand-bound receptor and two nuclear localization signals in the GR allowing for rapid transport into the nucleus. Within the nucleus, the receptor binds as a dimer to tandem glucocorticoid-response elements (GREs) in the promoter regions of target genes, to regulate their expression positively or negatively through the formation of the transcription initiation complex with RNA polymerase II and its ancillary components (Louw, 2019) (Fig. 1.6.4).

In the nucleus, the ligand-bound GR interacts with HRE in target genes. This HRE has a general sequence identified as AGAACAnnnTGTTCT (n represents any nucleotide). Examination of numerous glucocorticoid target genes has identified a glucocorticoid response element (GRE) with the sequences GGAACAnnnTGTTCT. The GR binds to the GRE as a homodimer with each of the half sites being bound by one receptor subunit in the homodimer (Timmermans et al., 2019). There are also target genes that are repressed by GR binding to target sequences and these are referred to as is $CTCC(n)_{0-2}GGAGA$ and in these elements each half site binds a monomeric GR.

The interaction of GR with GRE causes additional conformational changes that facilitates the recruitment of, and interaction with, transcriptional coregulators that assist the actions of the GR (Escoter-Torres et al., 2019; Weikum et al., 2017). The SRC (steroid receptor coactivator) proteins are an important family of GR coactivators. The GR binding to the nGRE elements on the other hand results in the recruitment of, and interaction with, transcriptional corepressors such as nuclear receptor corepressor 1 (NCoR1: encoded by the NCOR1 gene) and silencing mediator of retinoic and thyroid receptors (SMRT: encoded by the NCOR2 gene).

Two major isoforms, GRα and GRβ, are the result of alternative splicing in exon 9 (Fig. 1.6.5). When exon 8 is spliced to the 5'-end of exon 9 the resultant protein is the GRα isoform.

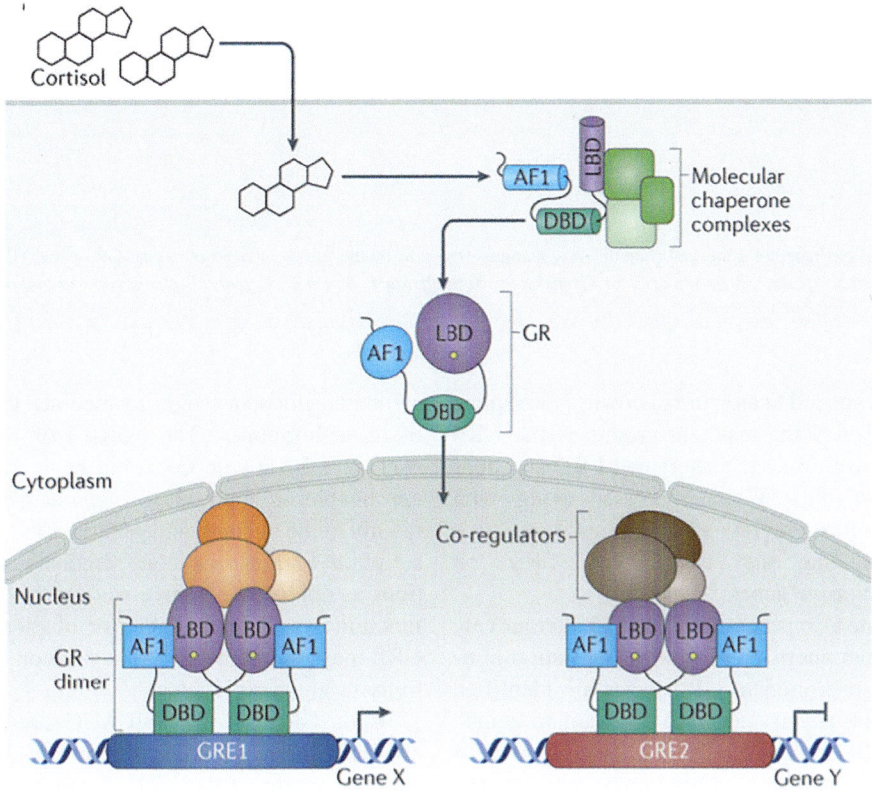

FIG. 1.6.4 Overview of signaling mediated by the natural GR ligand, cortisol. Activating ligand interacts with monomeric GR associated with molecular chaperone-containing complexes such as Hsp90 in the cytosol. This induces local and remote allosteric changes that potentiate nuclear transport and other activities. Within the nucleus, GR readily dimerizes and the nucleus acquires transcription regulatory complexes containing various other transcriptional regulatory factors (TRFs) such as AP_1, NK-B, Smad and transcriptional coregulators at different glucocorticoid response elements (GREs) to activate or repress transcript ion of particular target genes. GRE1 and GRE2 represent distinct GREs within the genome, Gene X and Gene Y represent the genes under the control of GRE1 and GRE2, respectively. *(From Weikum ER, Knuesel MT, Ortlund EA, Yamamoto KR. Glucocorticoid receptor control of transcription: precision and plasticity via allostery. Nat Rev Mol Cell Biol 2017;18(3):159–74. Fig. 1 p. 160.)*

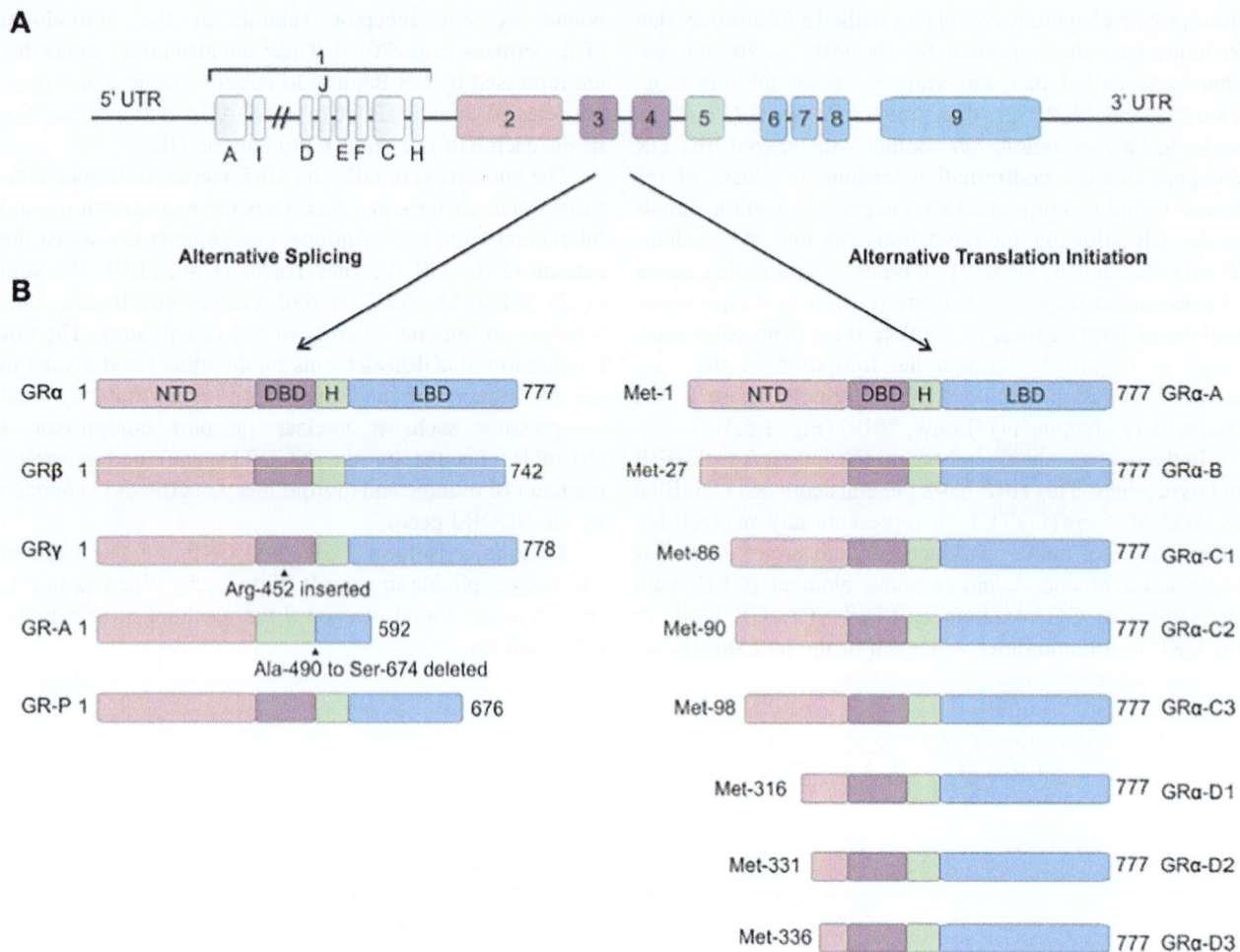

FIG. 1.6.5 Glucocorticoid receptor gene and protein. (A) Genomic structure of the glucocorticoid receptor (GR) gene. (B) Alternative splice and translation-initiation variants of the GR protein. *(From Timmermans S, Souffriau J, Libert C. A general introduction to glucocorticoid biology. Front Immunol 2019;10:1545. Fig. 3 p. 3.)*

When exon 8 is spliced to an internal downstream splice acceptor site in exon 9 the resultant protein is the GRβ isoform. The GRβ isoform has a shortened LBD and does not bind glucocorticoids. When expressed along with GRα the GRβ isoform serves as a dominant negative inhibitor, antagonizing the actions of GRα on glucocorticoid-responsive genes.

The NR3C1 gene also generates at least 15 alternatively spliced mRNAs from alternative in-frame translation initiation codons. The predominant GR species are identified as **GRα** (777 amino acids) and **GRβ** (742 amino acids). The human (h) GR influences the transcription rate of numerous glucocorticoid target genes in a positive or negative fashion. GR isoforms are the result of alternative processing of the primary mRNA in terms of splicing and translation. The isoforms differ in their expression patterns, gene regulatory networks, and other functional aspects. The hGRα is located primarily in the cytoplasm of cells and is the classic glucocorticoid receptor that binds natural and

synthetic glucocorticoids to mediate the genomic actions of these hormones. The hGRβ isoforms share the same NTD and DBD with GRα, but have a **unique LBD**. GRβ are unable to bind known glucocorticoids, are located mainly in the nucleus, and generally fail to activate the transcription of glucocorticoid-responsive genes. GRβ functions as dominant negative isoform of GRα promoters and functions as a natural inhibitor of glucocorticoid actions. GRβ may play a role in the regulation of target cell sensitivity to glucocorticoids.

Eight GRα isoforms (all AUG start codons) are due to alternative translational initiation sites in exon 2 harboring various lengths of the NTD. These various GRα isoforms are identified as GRα-A, -B, -C1, -C2, -C3, -D1, -D2, and -D3. The GRα-C isoforms are expressed at highest levels in the pancreas and colon, the GRα-D isoforms are highest in spleen and lungs. The GRα isoforms also exhibit different transcriptional activities in response to glucocorticoid binding. The GRα-C isoforms enhance the induction of pro-apoptotic genes and are more efficient in the recruitment of transcriptional coactivators. Additional GR isoforms result from other alternative splicing events.

- The GRγ isoform, with the insertion of a single arginine residue (R452) between the two zinc fingers in the DNA-binding domain.
- The GR-A isoform with exons 5 through 7 spliced out such that amino acids 490–674 are deleted.
- The GR-P isoform with exons 8 and 9 spliced out
- The GRδ isoform with exon 2 spliced out such that amino acids 313–338 are absent (this isoform is sometimes expressed as GRΔ313–338).
- The GRδ isoform with several sites for phosphorylation of the receptor deleted and these sites are important for the transactivation potential of the receptor.
- The GR-S1 isoform with retention of intron H between exons 8 and 9, and the splicing out of exons 8 and 9.
- The GR-DL1 isoform is a truncated receptor protein due to a single nucleotide deletion in exon 2.

The hGRα uses AF-1 and AF-2 domains to interact with nuclear receptor coactivators and chromatin-remodeling complexes and to initiate transcription (Fig. 1.6.6).

Several coactivators (P300), cAMP-responsive element-binding protein (CREB)-binding protein (CBP), P160 and the steroid receptor coactivator-1 (SRC-1), SRC-2 and SRC-3 are involved with the transmission of the glucocorticoid signal to the RNA polymerase II (Skowron et al., 2019). Following transcriptional activation or inhibition of the glucocorticoid-responsive genes, the hGRα dissociates from the ligand. The unliganded hGRα remains within the nucleus for a considerable length of time and is then exported to the cytoplasm; the hGR may be recycled and/or degraded in the proteasome.

The expression levels of hGR proteins vary considerably among tissues. The splice and translational hGR isoforms expressed from different promoters appear to form more than 250 different combinations of homo- and heterodimers with varying transcriptional activities. The marked complexity in the transcription/translation of the *NR3C1* gene enables target tissues to differentially respond to circulating glucocorticoid concentrations and accounts for the random nature of the glucocorticoid signaling pathway.

1.6.4.2 Mineralocorticoid receptor

The MR is found in the cytoplasm and nucleus in the apical membrane of epithelial cells of the distal convoluted tubule as well as in cells of other tissues involved with conservation of salt, such as colon, sweat glands, lung and tongue

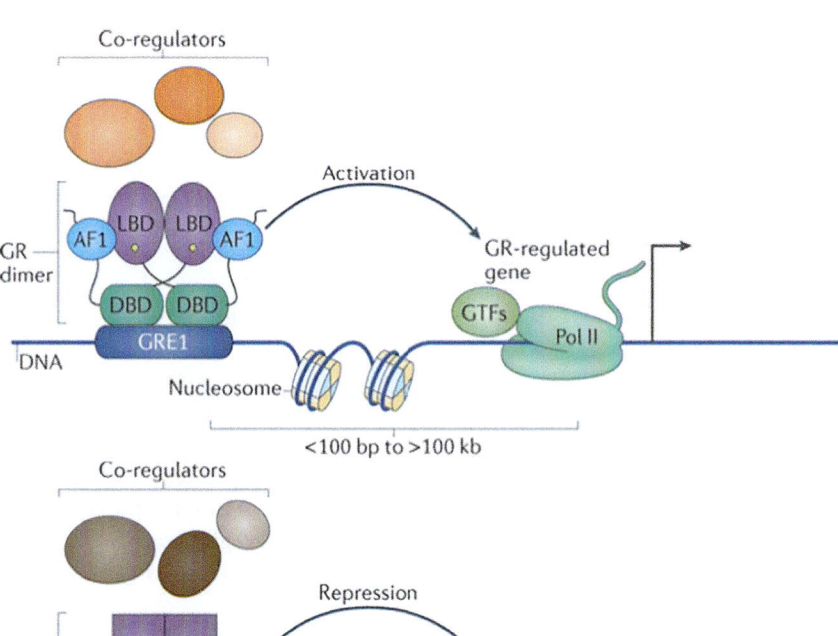

FIG. 1.6.6 Gene regulation. Glucocorticoid receptor (GR)-regulated genes are commonly linked to multiple GR-occupied regions (GORs), usually >10 kb from the transcription start site of the regulated gene, one or more of which may be a functional glucocorticoid response element (GRE) for that gene. GREs may be near (<100 bp) to or far (>100 kb) from their target genes. *(From Weikum ER, Knuesel MT, Ortlund EA, Yamamoto KR. Glucocorticoid receptor control of transcription: precision and plasticity via allostery. Nat Rev Mol Cell Biol 2017;18(3):159–74. Fig. 3 p. 163.)*

where the sodium channel targets are expressed (Fuller et al., 2019). The human MR (hMR) is similar to hGR in the DNA-binding domains (94% homology in the amino acid) and has very similar ligand-binding domains (57%), but divergent N-terminal A/B regions (<15%). The hMR gene is mapped to chromosome 4q31.1-31.2 and hMR cDNA encodes 984 amino acids in the 107-kDa polypeptide. The MR gene consists of 10 exons (Fig. 1.6.7). The MR has two exons 1 (exon 1α and exon 1β), each with an alternative promoter although the finally translated MR protein is the same. Exons 1 are untranslated regions, exons 2 codes for the immunogenic domain (A/B), exons 3 and 4 for the DBD (C), and exons 5–9 for the hinge region (D) and the LBD (E). Expression of the different hMR variants is under the control of two different promoters that contain no obvious TATA element, but multiple GC boxes. Both hMRα and hMRβ mRNAs are expressed at approximately the same level in the mineralocorticoid target tissues.

MRs in its unliganded state is located in the cytoplasm, in a complex containing heat shock proteins 90, 70 and 50. The receptor-ligand complex dissociates from the heat shock proteins, homo- or heterodimerizes on binding with their ligand, and translocates into the nucleus (Fuller et al., 2017). Homodimers or heterodimers of the MR interact with hormone-responsive elements (HRE) and/or other transcription factors in the promoter regions of target genes that affect the subunits of the **epithelial sodium transport channel (ENaC) or** other proteins related to this channel and sodium transport in general, and modulates the transcription rates of these genes. The MR is comprised of distinct functional domains (activation function AF-1a, AF-1b and AF-2) and nuclear localization signals (NLS0, NLS1 and NSL2), as well as one nuclear export signal (NES) (Fig. 1.6.7).

1.6.4.2.1 Role of HSD11B2

Aldosterone, deoxycorticosterone and corticosterone are the principal steroids recognized by the MR. Most glucocorticoids, including cortisol also bind to the mineralocorticoid receptor (MR encoded by the NR3C2 gene) and therefore can exhibit mineralocorticoid-like activities. The action of cortisol on the MR however is protected by local conversion to inactive cortisone by the enzymes HSD11B2 (Fig. 1.6.8).

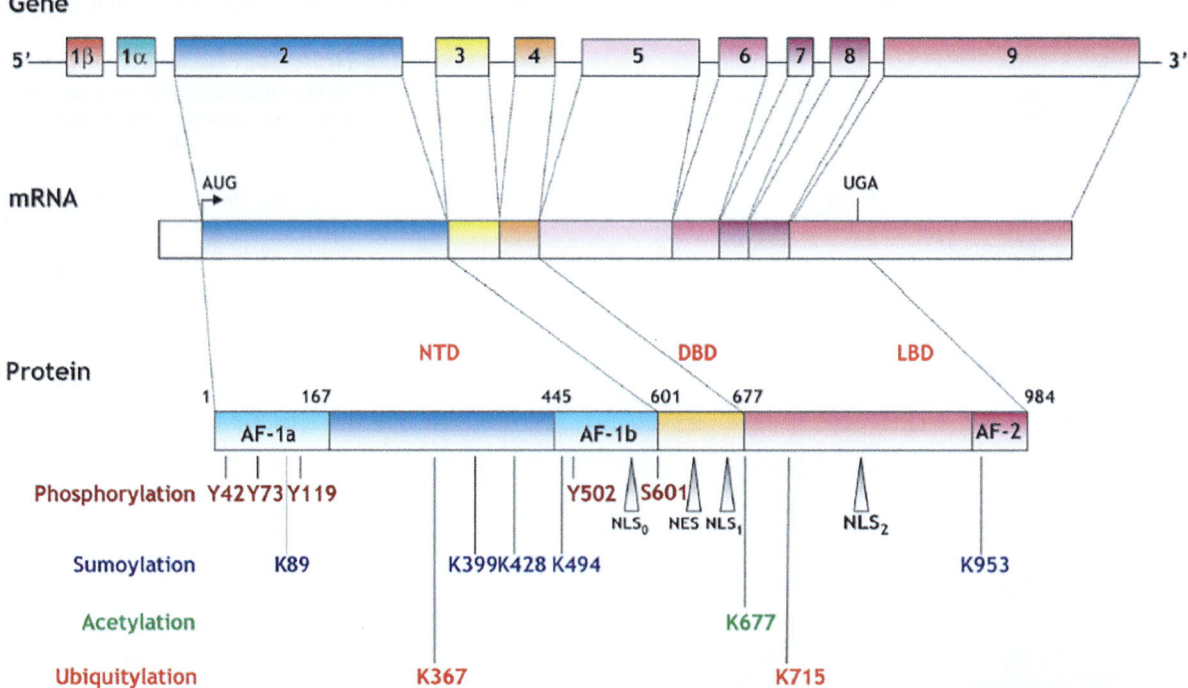

FIG. 1.6.7 Schematic representation of human MR structure. MR gene, mRNA, protein, functional domains and associated posttranslational modifications are depicted. The hMR gene is composed of 10 exons, including 2 untranslated first exons (1α and 1β). The AUG translational initiation start codon is located 2 bp after the beginning of exon 2, while the stop codon is located in exon 9. Multiple mRNA isoforms generated by alternative transcription or splicing events are translated into various protein variants, including those generated by utilization of alternative translation initiation sites (not shown). The receptor is comprised of distinct functional domains (activation function AF-1a, AF-1b and AF-2) and nuclear localization signals (NLS0, NLS1 and NSL2), as well as one nuclear export signal (NES). The positioning of amino acids targeted for phosphorylation, sumoylation, acetylation and ubiquitylation is indicated for the human MR sequence. *(From Viengchareun S, Le Menuet D, Martinerie L, Munier M, Pascual-Le Tallec L, Lombès M. The mineralocorticoid receptor: insights into its molecular and (patho)physiological biology. Nucl Recept Signal 2007;5:e012. Fig. 1 p. 3.)*

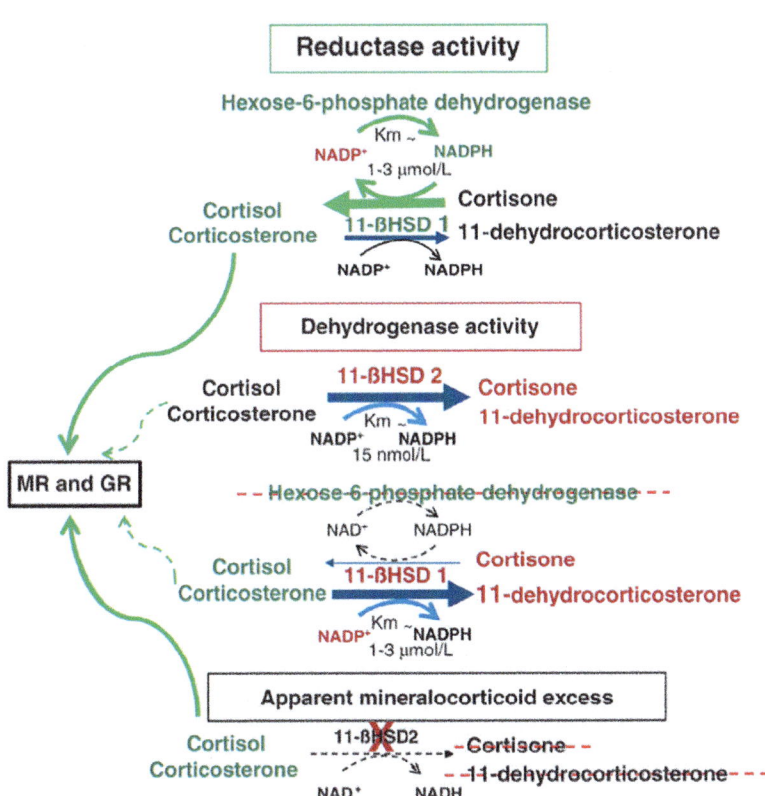

Reductase activity

Hexose-6-phosphate dehydrogenase

Km ~
NADP⁺ → NADPH
1-3 µmol/L

Cortisol
Corticosterone

Cortisone
11-dehydrocorticosterone

11-ßHSD 1

NADP⁺ NADPH

Dehydrogenase activity

Cortisol
Corticosterone

11-ßHSD 2

Cortisone
11-dehydrocorticosterone

Km ~
NADP⁺ NADPH
15 nmol/L

Hexose-6-phosphate dehydrogenase

NAD⁺ NADPH

Cortisol
Corticosterone

11-ßHSD 1

Cortisone
11-dehydrocorticosterone

Km ~
NADP⁺ NADPH
1-3 µmol/L

MR and GR

Apparent mineralocorticoid excess

Cortisol
Corticosterone

11-ßHSD2

Cortisone
11-dehydrocorticosterone

NAD⁺ NADH

FIG. 1.6.8 HSD11B2 at MR target sites converts cortisol to cortisone which does not bind to the MR, thus allowing aldosterone action. *(From Gomez-Sanchez E, Gomez-Sanchez CE. The multifaceted mineralocorticoid receptor. Compr Physiol 2014;4(3):965–94. https://doi.org/10.1002/cphy.c130044. PMID: 24944027; PMCID: PMC4521600. Fig. P.)*

Cortisol circulates at plasma concentrations several orders of magnitude higher than those of aldosterone, and cortisol has a high affinity for the MR. Cortisol would overwhelm the MR in mineralocorticoid target tissues and cause mineralocorticoid excess. A local enzyme, 11β-hydroxysteroid dehydrogenase type 2 (11β-HSD2), however, converts active cortisol to inactive cortisone, and protects the MRs from the effects of cortisol (White, 2018; Bailey, 2017; Krozowski, 1992). 11β-HSD2 catalyzes the interconversion of hormonally active C11-hydroxylated corticosteroids (cortisol in humans or corticosterone in rodents) and their inactive C11-keto metabolites (cortisone in humans or 11-keto androgens in rodents). An isozyme, 11β-HSD1 has been identified, 11β-HSD type 1 (11β-HSD1) which differs in the biological properties and tissue distributions, acting primarily as a reductase.

11β-HSD2 rapidly inactivates glucocorticoids. The human 11β-HSD2 gene encodes 405 amino acids and its molecular weight is approximately 40 kDa. 11β-HSD2 has a hydrophilic N-terminal domain that is thought to anchor the protein into membranes. 11β-HSD2 is localized as a dimer in the nucleus and cytoplasm of cells of the cortical collecting duct and colon. Prednisolone and prednisone are substrates for both 11β-HSD isozymes but dexamethasone is only slightly metabolized by 11β-HSD2. Licorice derivatives, such as glycyrrhizic acid, and the hemisuccinate derivative carbenoxolone are inhibitors of

11β-HSD2 (Dellow et al., 1999). Inhibition of 11β-HSD2 confers mineralocorticoid potency to physiologic concentrations of endogenous glucocorticoids in the kidney and colon. Thus, in normal physiology, 11β-HSD2 protects the MR by converting cortisol to the inactive cortisone and allows aldosterone-selective access to the inherently nonselective MR in mineralocorticoid target tissues.

1.6.5 Sex steroids

1.6.5.1 Androgen receptor

The androgen receptor gene (NR3C4; Nuclear Receptor subfamily 3, group C, gene 4) (AR) gene in humans is coded on the X chromosome. The AR gene is located at Xq11.2 -12 and spans 186,587 kilobases (kb) in total coding for a protein of about 110 kDa. The gene has eight coding exons organized like the genes coding for the other steroid hormone receptors (Fig. 1.6.9). Exon/intron boundaries are highly conserved. Two androgen receptor mRNA species of 8.5 and 11 kb, respectively, have been identified in cell lines and may be represented differently in tissues. In the prostate and in genital skin of humans fibroblasts, the larger mRNA is predominantly expressed (Pihlajamaa et al., 2015).

Testosterone diffuses through the cell membrane into the cytoplasm and binds to the AR, inducing dissociation from

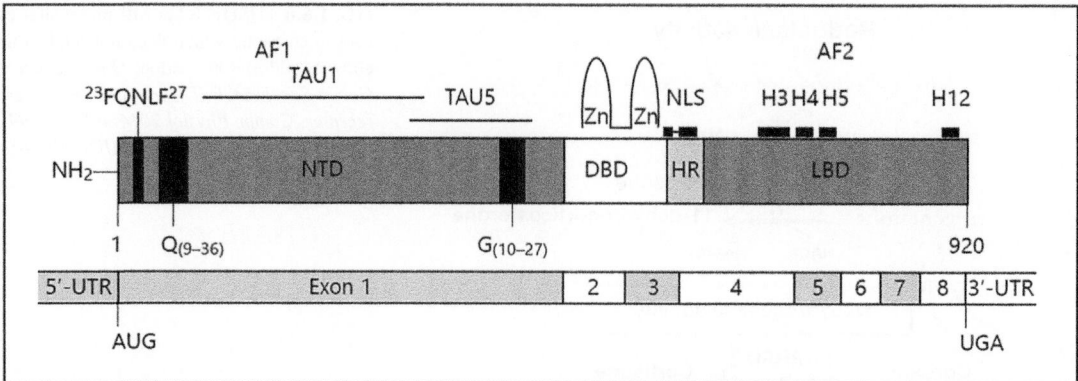

FIG. 1.6.9 Functional domains of the AR and relation to the exon structure of its mRNA. *AUG*, start codon; *G(10–27)*, polymorphic glycine repeat; *H*, helix; *HR*, hinge region; *NLS*, bipartite nuclear localization signal; *Q(9–36)*, polymorphic glutamine repeat; *UGA*, stop codon; *UTR*, untranslated region. *(From Werner R, Holterhus PM. Androgen action. Endocr Dev 2014;27:28–40. https://doi.org/10.1159/000363610. Epub 2014 Sep 9. PMID: 25247642. Fig. 1 p. 29.)*

chaperone proteins. The AR translocates to the nucleus where it binds to androgen response element in genes, inducing transcriptional activation and repression. These classical genomic effects of androgens are slow and occur after several hours because they require transcription and translation of the new proteins.

DHT has a more potent androgenic effect than T because its binding affinity to AR is approximately double that of T. Reduction of T precedes binding of androgen to the AR. There are site specific and developmental changes of the reductases. A backdoor path of DHT synthesis has also been found particularly active in patients with congenital adrenal hyperplasia due to 21-hydroxylase deficiency. Another new discovery involves 11-hydroxylation of androgens to give 11-hydroxytestosterone and 11-hydroxandrostenedione which are oxidized to 11-ket androgens that have stronger affinity to the AR than testosterone (Pretorius et al., 2017).

Binding of T or DHT to the AR is followed by dissociation of chaperone protein complexes (e.g., heat shock proteins) in the cytoplasm, along with a conformational change of the receptor protein resulting in a transformation and a translocation to the nucleus. T and DHT are ligands with different expression of the AR however only one AR cDNA has been identified and cloned (Takayama, 2017; Davey and Grossmann, 2016) (Fig. 1.6.10).

The promoter for the androgen receptor gene drives transcription primarily by the Zn-finger transcription factor Sp1, which binds to GC-boxes upstream of the transcription start site (−46 to −41 bps) and within the 5′UTR (+429 to +442), because TATA and CCAAT elements are lacking. The DBD and LBD of the AR have a high homology with the equivalent domains of the other steroid receptors but the NH$_2$-terminal domain of the AR has low homology with the other steroid receptors. The NH$_2$-terminal domain has a poly-glutamine stretch, encoded by a polymorphic (CAG) nCAA repeat and the length of the repeat has been used

for identification of X-chromosomes for carrier detection in pedigree analyses. In the normal population, between 9 and 38 glutamine residues are observed which may lead to a very mild modulation of AR activity and may be relevant to the development of prostate cancer. Male infertile patients may have longer (CAG)nCAA repeat, with the risk of defective spermatogenesis.

The NCBI reference sequence (NM 000044.3) for AR cDNA is different from the original numbering scheme from the Gen-Bank mRNA sequence M20132.1. The AR protein of the new reference sequence is 920 residues, leading to a +2 shift in amino acid numbering between residues 78 and 449 and to a +1 shift between residues 472 and 919 compared with the previously used standard reference sequence. The +1 shift involves all the amino acid residues in the DNA-binding domain (DBD) and ligand-binding domain (LBD). This affects the amino acid sequence numbering and in order to correctly identify published mutations from earlier publications, some changes are needed:

- the variable polyglutamine tract length is two amino acids longer (23 instead of 21) and
- the variable polyglycine tract length is one shorter (23 instead of 24).

The 539 amino acids in androgen receptor NH$_2$-terminal domain (NTD) codes for the major transcription activation functions. Two activation domains have been identified. Activation function 1 (AF-1) (between residues 103 and 372) which is essential for transcriptional activity of full length AR, and activation function 5 (AF-5) (between residues 362 and 486) which is required for transactivity of a constitutively active AR, that lacks its LBD. The AF-5 region in the receptor NTD interacts with a glutamine rich domain in p160 cofactors. The NTD also binds to the COOH-terminal LBD (N/C interaction). The NTD regions required for the binding of the LBD have been mapped to

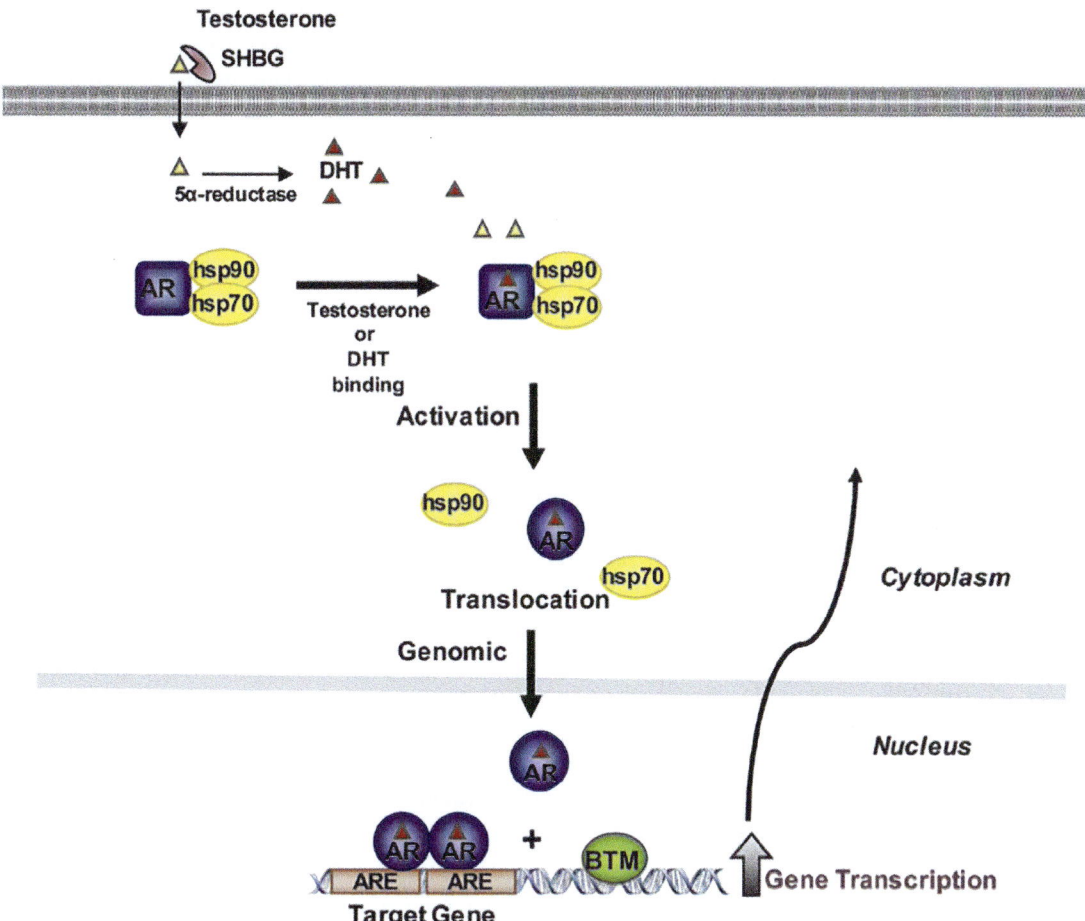

FIG. 1.6.10 Classical mechanisms of androgen signaling in a target cell. In circulation, testosterone is bound to the serum sex hormone-binding globulin (SHBG). Testosterone dissociates from its binding carrier protein and diffuses freely through the cell membrane. Once in the cytoplasm, testosterone can be converted to its active metabolite 5α-dihydrotestosterone (DHT). Testosterone or DHT can directly bind to the androgen receptor (AR) leading to the dissociation of heat shock proteins (hsp) from the inactive receptor. The ligand-bound receptor then exerts its effects by genomic mechanisms involving the activated AR translocation into the nucleus. In the nucleus, the androgen receptor binds as homodimer to specific DNA elements present as enhancers in upstream promoter sequences of androgen target genes. On AR binding, coactivators are recruited and the basal transcription machinery (BTM) (e.g., RNA-polymerase II [RNA-Pol II], TATA box binding protein [TBP], TBP associating factors [TAFs], general transcription factors [GTFs]) is activated. The interactions between AR, coactivators and the BTM results in gene transcription. *(From Cruz-Topete D, Dominic P, Stokes KY. Uncovering sex-specific mechanisms of action of testosterone and redox balance. Redox Biol 2020;31:101490. Fig. 2 p. 5.)*

the first 36 amino acids and the residues 372–495. The androgen dependent interaction of the NTD with the LBD serves to stabilize the AR dimer complex and the steroid receptor complex. The AR N/C interaction is rapid in the cytoplasm on hormone binding and is followed by an N/C interaction in the nucleus for receptor dimerization. The N/C interaction also protects the coactivator binding groove. The N/C interaction is lost allowing cofactor binding. The cysteine rich DBD is composed of two zinc finger motifs which are encoded by exons 2 and 3, respectively. The zinc fingers recognize binding sites in AR response element and stabilizes the DNA binding.

The hinge region encoded by part of exon 4 is important for nuclear localization of the AR and corepressor binding. Acetylation and phosphorylation can occur in the hinge region. The LBD is encoded by about 250 amino acid

residues in the C-terminal end of the molecule. The AR can use different transactivation domains (AF1 and AF5, respectively, in the NTD domain and AF2 in the COOH-terminal domain). The AF2 function in the ligand binding domain is strongly dependent on the presence of nuclear receptor coactivators.

The crystal structures of the human AR in complex with 5α-dihydrotestosterone and with the synthetic ligand methyltrienolone (R1881) have been determined. The three-dimensional structure has typical nuclear receptor ligand binding domain fold. The LBD pocket consists of 18 amino acid residues interacting with the bound steroid, through specific hydrogen-bonds and hydrophobic interactions. The ligand binding pocket can accommodate different ligands. The structural data are being used in developing **selective androgen receptor modulators (SARMs)**.

Crystallographic data on the LBD complexed with androgen predict 11 helices with two antiparallel β-sheets arranged in a so-called helical sandwich pattern. The carboxy-terminal helix 12 is positioned in an orientation to close the ligand binding pocket. On hormone binding, the fold of the LBD results in a globular structure with a site for the binding of coactivators.

AR signaling in Sertoli cells is most important in meiosis during spermatogenesis which is arrested before the first meiotic division (diplotene stage). The functional AR in germ cells is not essential for spermatogenesis and male fertility in mice. Testosterone acts through the AR in the somatic Sertoli cells to induce the production of factors that support the maturation of adjacent germ cells into spermatozoa. AR is the only specific receptor for androgen that has been identified despite dihydrotestosterone being more active.

There is a vast regulatory network of androgens with tissue specific enhancers and transcription factor binding events that are dictated by the local chromatin context. Future work utilizing existing and novel approaches is likely to reveal additional regulatory layers and clinically important features of the AR pathway (Pihlajamaa et al., 2015).

1.6.5.2 Estrogen receptor

Two distinct genes code for ERα and ERβ isoforms are found in humans. ERα is predominantly expressed in mammary glands, pituitary, hypothalamus, ovarian theca cells and reproductive tract, whereas ERβ is primarily expressed in ovarian granulosa cells, lung and prostate. Human ERα consists of 595 amino acids. Human ERα gene has been mapped to chromosome 6q24-q27. Multiple promoter and regulatory regions in the 59-untranslated sequence of the human has been described, but only a single open reading frame appears to exist (Fuentes and Silveyra, 2019; Gustafsson et al., 2019). Twenty different splice variants of ERα have been found in tumor cells. The human ERβ gene is located at chromosome 14q22-q24 and the expressed protein has 530 amino acids (Fig. 1.6.11). The isoforms are distinguished by size, with a shorter N9 terminus in ERβ protein. Human ERβ also has several isoforms due to alternative splicing of exons 8 and 9, or deletion of 1 or more coding exons to give ERβ1, ERβ2, ERβ3, etc. Like most other nuclear receptors, ERs contain a domain with ligand-independent activation function (AF-1) at the N-terminus, a DNA-binding domain (DBD domain) followed by a hinge domain, and a ligand-binding/dimerization domain (LBD) at the C-terminus that contains a ligand-dependent transcription activation domain (AF-2) (Ohnemus et al., 2006).

ERβ may be considered as a dominant-negative regulator of ER modulating transcriptional responses to estrogens. The ratio of ERα vs. β within a cell may determine the cell sensitivity to estrogens and its biological responses to the hormone (Böttner et al., 2014). In early studies of the ER, particularly using immunohistochemical techniques, a number of problems were encountered. To stain the PR in the nucleus, the tissue section was incubated with 0.5% triton which led to loss of PR in the cytoplasm. Another problem came using antibodies against the N-terminus of ERβ expressed in E. coli β. These antibodies worked well in E. coli but not immunochemistry of tissues which was attributed to posttranslational modifications. ERβ splice variants have also been inconsistently detected by C-termini antibodies. In breast cancer tissues, ERβ was assumed to be present because N-terminal antibodies bound to the PR when what was actually present was a splice variant of ERβ (Gustafsson et al., 2019).

1.6.5.3 Progesterone receptor

Human PRs are encoded by a single gene located on chromosome 11 (11q22-q23). Two promoters control expression of PR isoforms to produce two major mRNA transcripts that encode two isoform proteins:

- the full-length PR-B (116 kDa) controlled by the distal PR-B promoter region and initiated from the first AUG translational start codon and
- PR-A (94 kDa) that is controlled by the proximal PR-A promoter region and initiated from the second AUG translational start codon that is 492 bases from the PR-A start codon (Fig. 1.6.12).

Other PR isoforms can be generated by the initiation of translation from further downstream AUG start sites (e.g., PR-C), exon splicing or exon insertions, respectively, but their physiologic significance is unknown (Grimm et al., 2016; Jacobsen and Horwitz, 2012).

Both PR isoforms consist of multiple domains, like the ER protein, such as the AF-1 in the N-terminus, the DBD and the LBD which contains AF-2. The PR-B isoform has an additional 164 amino acids in the N-terminus and an additional activation domain (AF-3) that endows a transactivation function specific to the PR-B protein, and plays an essential role in specifying target genes activated only by PR-B. The PR-B is the functional form in humans, but both isoforms are expressed in mouse uteri. PR-A and PR-B can therefore differentially regulate the expression of targeting genes in response to progesterone, involving different transactivation capabilities in different targeting tissues. The PR undergoes extensive posttranslational modifications that modulate various PR functions including nuclear translocation, dimerization, DNA binding, protein stability, hormone sensitivity, cell cycle progression and transcriptional activity.

1.6.5.4 Determinations of steroid receptors

Receptor assays are based on the principle that the receptor proteins are measured in the form of a complex with

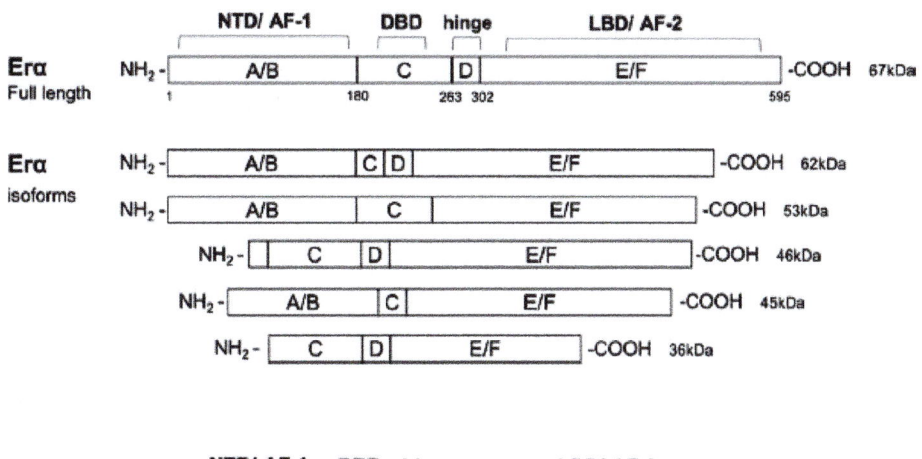

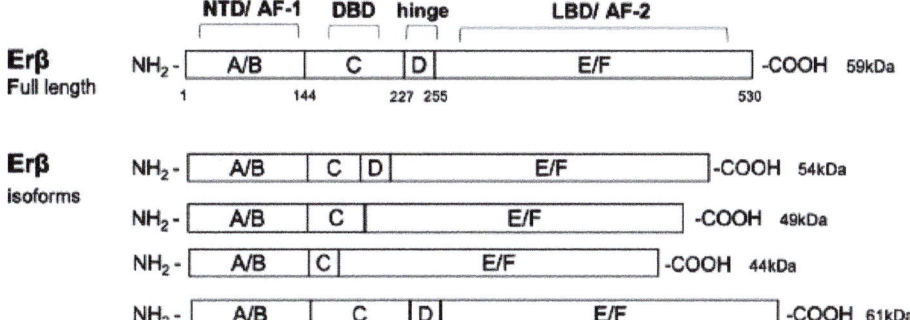

FIG. 1.6.11 ER Structural organization of estrogen receptor α (ERα) (595 aa) and ERβ (530 aa) with domains labeled A–F. Both receptors have six different structural and functional domains. The domain organization of the full-length 595 amino acid ERα (67kDa), and truncated shorter isoforms (62kDa, 53kDa, 46kDa, 45kDa, and 36kDa) resulting from alternative splicing and/or alternate translation start sites are illustrated. The domain organization of the full-length ERβ (59kDa) and truncated shorter isoforms (54kDa, 49kDa, and 44kDa), and elongated isoform (61kDa), resulting from alternative splicing and/or alternate translation start sites are shown. Estrogen binds to ERs which dimerizes and translocates to the nucleus inducing transcriptional changes in estrogen responsive genes. ER can also with other classes of transcription factors (TF) (4), or by regulation of transcription factor phosphorylation (5). *(From Fuentes N, Silveyra P. Estrogen receptor signaling mechanisms. Adv Protein Chem Struct Biol 2019;116:135–70. Fig. 6, p. 146.)*

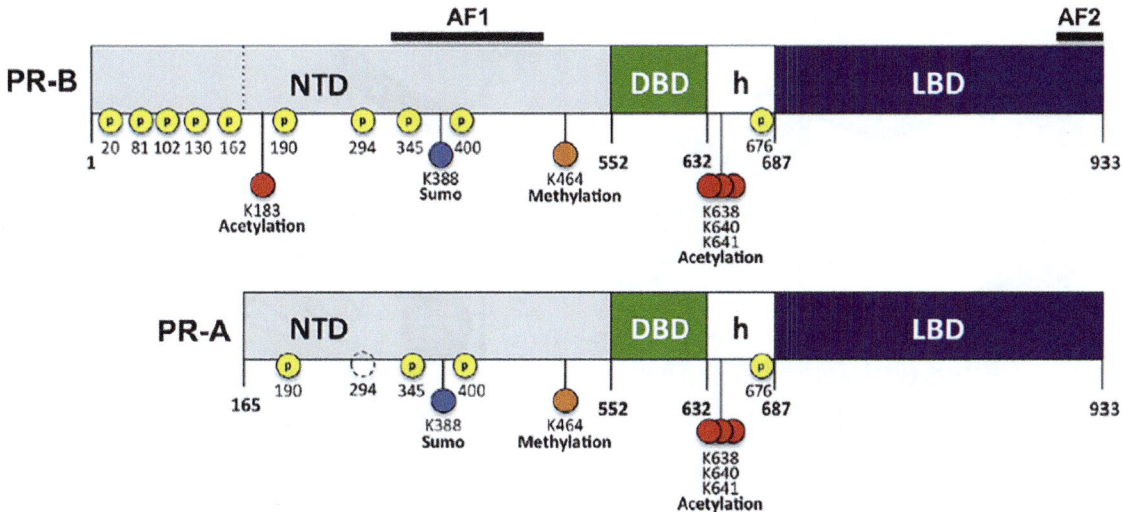

FIG. 1.6.12 Structural domains of human PR and posttranslational modifications. Ligand-binding domain (LBD), DNA-binding domain (DBD), amino-terminal domain (NTD), transcriptional activation or functional domains (AF1, AF2), and phosphorylated serine residues (p). *(From Grimm SL, Hartig SM, Edwards DP. Progesterone receptor signaling mechanisms. J Mol Biol 2016;428(19):3831–49. Fig. 1 p. 3832.)*

radioactive steroid ligands (see Chapter 2.7). A common method is to perform a binding analysis using increasing concentrations of the radioactive ligand so that after equilibrium is reached a saturation curve can be constructed. The binding data are generally analyzed by a Scatchard plot, a mathematical transformation derived from the Law of Mass Action. In a parallel series of incubations, excess of the radioactive ligand is added to determine nonspecific binding. The difference gives the specific binding by the receptor. Sucrose density gradient centrifugation, isoelectric focusing, polyacrylamide gel electrophoresis and agar gel electrophoresis are also used to distinguish specific from nonspecific binding. Posttranslational modifications of receptors, particularly ubiquitinylation and Sumoylation are amenable to mass spectrometric techniques.

1.6.5.5 Posttranslational modifications of receptors

The steroid receptor isoforms are subject to various posttranslational modifications (PTM), such as phosphorylation, ubiquitinylation, methylation, acetylation and SUMOylation. These covalent modifications of GR affect its stability, subcellular localization, transcriptional activity, and interaction with other proteins (Fig. 1.6.13). Ubiquitin is a 8.6-kDa protein of 76 amino acids that affects several intracellular processes and regulates protein degradation (Faus and Haendler, 2006).

Sumoylation is a posttranslational modification of a protein that involves covalent addition of a **Small Ubiquitin-like Modifier** through an isopeptide link between a lysine in the protein and the terminal glycine of the SUMO. Sumoylation is a similar process to ubiquitinylation adding a globular protein of around 100 amino acids, 12 kDa, that affects the protein structure and subcellular localization. **Phosphorylation** of a receptor can occur at serine, threonine and tyrosine residues by specific kinases and can be directly or indirectly linked to activation of hormone binding, altering of nuclear cytoplasmic shuttling, modulation of DNA binding and transcriptional activity. Most phosphorylation sites are in the N-terminal domain. All the reactions are reversible and consequently enzymes that mediate dephosphorylation, deacetylation, deubiquitination, demethylation and de-SUMOylation are also potential regulators of receptor activity. The modifications may affect receptor stability, subcellular localization,

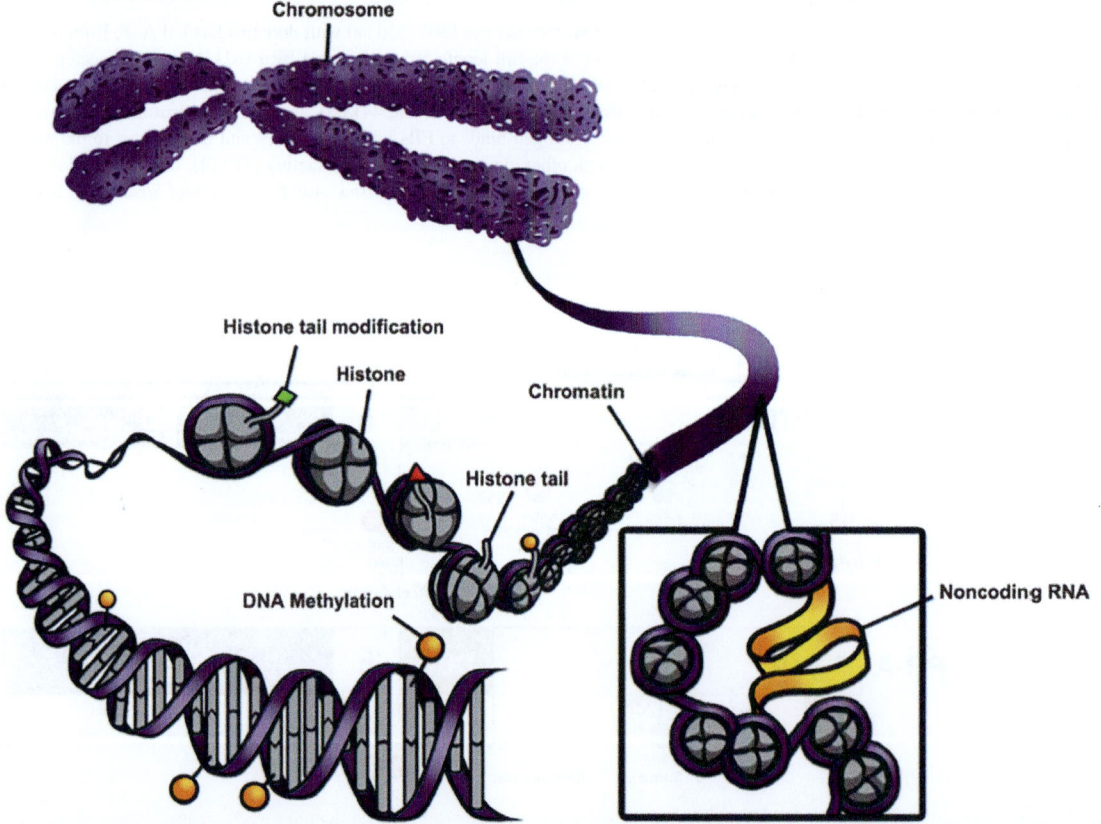

FIG. 1.6.13 An overview of epigenetic mechanisms including chromatin modifications, DNA methylation, and ncRNA interactions. Chromatin is made of DNA, histone proteins, and nucleosomes and regulates gene expression by controlling the access of transcription factors to DNA. DNA methylation creates a physical barrier that generally impedes transcription factor binding. Noncoding RNA interacts with chromatin and modifies domains. *(From Rattan. Biol Reprod 2019;101:635. Fig. 2 p. 638.)*

together with the interaction between the receptor and other proteins. The structures can be determined by mass spectrometric methods (Heap et al., 2017; Kessler et al., 2017).

1.6.6   Coregulators

The mechanisms for steroid action are much more complex than the long-held process where once the ligand was bound, a transformation occurred in the receptor enabling the receptor to move into the nucleus, interact with DNA and switch genes on or off. The ligand-receptor interaction is a dynamic interaction, with the receptor exhibiting considerable plasticity. The resulting signal transduction is determined by interactions with an extensive, cell-specific repertoire of coregulatory molecules. There are ligand-, gene-, and cell-specific steroid responses determined by **coactivator** interactions. **Steroid receptor coactivator-1 (SRC-1)** is one of the hundreds of coregulators that have been described, in addition there are **Corepressors**, that reverse many of metabolic processes (Stashi et al., 2014).

The knowledge of steroid recognition is now applied to the design of drugs that selectively modulate the action of a particular steroid. The drugs can also be degrader or deregulator substances in action. The drugs are called selective (hormone) receptor modulators so **SARM, SERM, SPRM** for androgen, estrogen and progesterone receptors, respectively, or selective (hormone) receptor deregulator such **SERD** for estrogen. The coactivators promote acetylation of histones, coregulators inhibit gene expression by inducing histone deacetylase activity. Receptor/coregulator interactions determine whether for example an ER ligand is agonist (estradiol), a SERM (tamoxifen) or an antagonist (fulvestrant).

1.6.7   Epi-genetics

Genes are located within chromatin, in the nuclei of eukaryotic cells, comprising genomic DNA that is noncovalently polymerized to the histone proteins (Fig. 1.6.14). Expression is a term used for the activation of specific genes for the purpose of transcription and translation into specific protein gene products.

Epigenetic modulation of gene expression may result from environment-gene interactions that alter the structures of histones and genomic DNA mainly by processes of acetylation and methylation (Martinez-Arguelles and Papadopoulos, 2010). Epigenetic regulation can result in heritable changes in production of the final protein product of the gene. Epigenetics refers to the regulation of gene expression via an array of molecular processes comprising **posttranslational** histone **modifications**, methylation of cytosine bases on DNA, the actions of transcription regulators, translational regulation of mRNAs by microRNAs, RNA splicing and RNA editing. Transposable elements, including retrotransposons and DNA transposons, comprise about 40% of the human genome. The data suggest that these elements have a regulatory role in stress and aging in the brain. But more research is needed to clarify the effects of epigenetics on steroid action.

1.6.7.1   Methylation at CpG islands

CpG islands are short sequences of genomic DNA in which the frequency of the linear 5′-CpG-3′ sequence is higher than at other regions of the gene, where "p" indicates the phosphodiester bond that connects cytosine and guanine nucleotides. The promoter CpG sequences in genes inactive in a particular cell or tissue are typically methylated at cytosine (to 5-methyl-cytosine), with consequent suppression of their expression. Methylation of the CpG sequences located at the promoter regions of genes that are essential for general cell functions (housekeeping genes) is not usual. The methylated cytosine at CpG islands may be converted to thymine by spontaneous deamination through a highly inefficient mechanism of mismatch repair, whereas cytosine-to-uracil mutations are efficiently repaired.

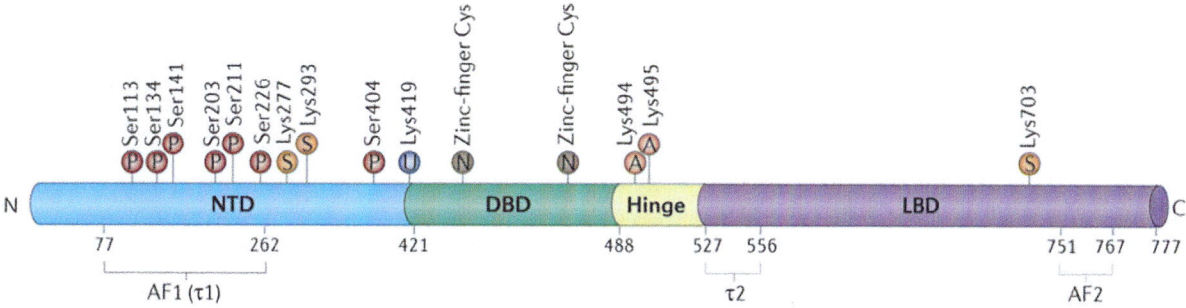

FIG. 1.6.14   Sites of glucocorticoid receptor posttranslational modifications. Major reported modifications, including phosphorylation (P), sumoylation (S), ubiquitylation (U), acetylation (A) and nitrosylation (N), are mapped onto the glucocorticoid receptor domain schematic. *AF,* activation function domain; *DBD,* DNA-binding domain; *LBD,* ligand-binding domain; *NTD,* amino-terminal domain. *(From Weikum ER, Knuesel MT, Ortlund EA, Yamamoto KR. Glucocorticoid receptor control of transcription: precision and plasticity via allostery. Nat Rev Mol Cell Biol 2017;18(3):159–74. Fig. 5 p. 167.)*

Methylated CpG sequences that are converted to TpG sequences are thought to be responsible for the relative scarcity of CpG sequences in currently inactive gene sequences that have been inherited over an evolutionary timescale. Such CpG sequences may be useful for identifying similarities in noncoding regions of genes and predicting their phylogenetic origins.

The methylation pattern of a cell is passed on from one generation of cells to the next generation because of the capacity of the enzyme DNA methyltransferase to bind specifically to the CpG sequences at sites of methylation. Different cell types have specific methylation patterns that reflect their differentiation and specialization. Consequently, changes in methylation patterns imply changes in cell development and functions.

1.6.7.2 Role of histone acetylation

H3 is one of five histones that maintain the structure of chromatin in eukaryotic cells. The N-terminal tail of histone H3 is easily identifiable and subject to posttranslational modification by attaching methyl or acetyl groups to lysine and arginine sites, as well as phosphorylating serine or threonine sites. Hypermethylation of lysine-9 is associated with decreased gene expression (gene silencing), whereas monomethylation is associated with gene activation. Histone acetyltransferase enzymes acetylate histone H3 at different lysine positions, and the site is indicative of different events; for example, acetylation at lysine-14 indicates active transcription of DNA into RNA.

1.6.7.3 Micro RNAs

Micro RNAs (MiRNAs) are endogenous single-stranded small RNAs of ~22 nucleotides in length that posttranscriptionally downregulate gene expression. A single RNA could suppress hundreds of genes, and a single gene could be modulated by multiple miRNAs. Although miRNA regulation of steroidogenesis is an emerging field, extensive literature exists about the important role played by miRNAs in the regulation of many other cellular processes such as inflammatory and immune responses, cell-cycle progression and proliferation, differentiation, tissue remodeling, apoptosis and disease pathology. The limited information that is currently available for three steroidogenic tissues, adrenal, ovary and testis, already suggests that miRNAs are putatively involved in virtually every aspect of the steroidogenic process, including receptor-mediated lipoprotein cholesterol delivery to the cell interior, intracellular cholesterol processing and transport to and within the mitochondria, other steroidogenesis related accessory protein and transcription factors, and steroidogenic enzymes themselves. Table 1.6.1 depicts the current understanding of the roles of specific miRNAs in steroidogenesis.

The large amount of data generated through genome-wide analyses of steroid producing adrenal and gonadal tissues, particularly ovarian granulosa cells, should provide ample opportunity for the identification and functional characterization of miRNAs that will likely lead to a much greater mechanistic understanding of the posttranscriptional regulation of steroidogenesis. Future studies with the

TABLE 1.6.1 Potential targets of miRNAs in steroidogenic pathways.

Tissue/cells	Target	miRNA#	Reference
Adrenal			
Human tissue and H295R cells	StAR	132	Hu et al. (2013)
	CYP11A1	125a-5p	Robertson et al. (2013)
		579	Clayton et al. (2018)
	CYP11B1	10b	Robertson et al. (2013)
		24	Robertson et al. (2013)
		561	Clayton et al. (2018)
	CYP11B2	10b	Robertson et al. (2013)
		24	Robertson et al. (2013)
		125b-5p	Robertson et al. (2013)
		320a-3p	Clayton et al. (2018)
		561	Clayton et al. (2018)
	CYP21	125a-5p	Hu et al. (2013)
	CYP17A	579	Clayton et al. (2018)
	HSD11B1	561	Han et al. (2013)

TABLE 1.6.1 Potential targets of miRNAs in steroidogenic pathways—cont'd

Tissue/cells	Target	miRNA#	Reference
Ovary			
Human granulosa cells	E2F1	320	Sun et al. (2015)
	SF-1	320	Yu and Li (2016)
Human ovarian cumulus granulosa cells	SREBP2	122	Menon et al. (2015)
	LHCGR		Menon et al. (2017)
Human granulosa-like tumor cell line KGN	Foxl2	133b	Dai et al. (2013)
	LHCGR		Toms et al. (2018)
	NURR1	592	Wang et al. (2003)
	SP1	27a-3p	Jiang et al. (2014)
	StAR	150, 763-3p	Zhou et al. (2019)
	SF-1	509-3p	Pan et al. (2016)

Targets are *E2F1*, ESF transcription factor; *FoxL2*, forkhead box L2; *Nurr 1*, Nurr 1 transcription factor; *SP1, specificity protein 1; SREB2*, sterol regulatory element binding protein 2.

introduction and availability of high-throughput, next-generation RNA sequencing tools, as well as technical advances in sequencing methodologies and newer bioinformatics, should enable precise identification of novel, or low abundance miRNAs in steroidogenic tissues and their pathophysiological relevance to steroidogenesis. Studies to determine the expression and role of miRNAs in pathological settings should lead to identification of miRNAs as biomarkers, validated targets and functions that are likely to aid in the clinical management of steroid hormone related diseases and/or reproductive diseases.

1.6.7.3.1 Related receptors

In addition to the nuclear receptors discussed so far, other family members are the liver X receptors (LXRs), farnesoid X receptors (FXRs), the pregnane X receptor (PXR), the estrogen related receptors (ERRβ and ERRγ), the retinoid-related orphan receptor (RORα), and the constitutive androstane receptor (CAR). The RXRs serve as obligatory partners for these other members of the nuclear receptor family including PPARs, LXRs, and FXRs.

The **retinoid X receptors (RXRs)** are a further class of nuclear receptors that represents a second class of transcription factors that bind the retinoid 9-*cis*-retinoic acid. There are three isotypes of the RXRs: RXRα, RXRβ, and RXRγ and each isotype is composed of several isoforms. In the absence of a heterodimeric-binding partner, the RXRs are bound to hormone response elements (HREs) in DNA and are complexed with corepressor proteins that include a histone deacetylase (HDAC) and silencing mediator of retinoid and thyroid hormone receptor (SMRT) or nuclear receptor corepressor 1 (NCoR). RXRα is widely expressed at high in liver, kidney, spleen, placenta, and skin. RXRα null mice are embryonic lethals; RXRβ is important for spermatogenesis and RXRγ has a restricted expression in the brain and muscle. The **RAR**s exhibit highest affinity for **all-*trans*-retinoic acid** (all-*trans*-RA). The RXRs have been shown to enhance the DNA-binding activity of RARs and the thyroid hormone receptors (TRs).

Peroxisome proliferator-activated receptors (PPARs). are also members of the superfamily. There are three family members: PPARα, PPARβ/δ, and PPARγ. Each of these receptors forms a heterodimer with the RXRs. PPARα is the endogenous receptor for polyunsaturated fatty acids that is highly expressed in the liver, skeletal muscle, heart, and kidney. In the liver, it induces peroxisomal fatty acid oxidation during periods of fasting. Expression of PPARα is also seen in macrophage foam cells and vascular endothelium where it is thought to activate antiinflammatory and antiatherogenic effects. The first family member identified was **PPARα** binds to the fibrate class of antihyperlipidemic drugs or peroxisome proliferators. **PPARδ** is expressed in most tissues in promotion of mitochondrial fatty acid oxidation, energy consumption, and thermogenesis. PPARδ serves as the receptor for polyunsaturated fatty acids and VLDLs.

PPARγ is highly expressed in both colonic epithelial cells and adipose tissues. Macrophages and lymphocytes are also sites for PPARγ expression. PPARγ actively participates in regulation of colon inflammation and oxidative stress, mainly through modulation of the expression of key transcription factors and kinases involved in inflammatory signaling cascades such as nuclear factor kappa-light-chain-enhancer of activated B cells (NF-κB), c-Jun, c-Fos, and nuclear factor of activated T cell; inhibition of mucosal production of inflammatory cytokines (i.e., interleukin (IL)-1β and TNF-α) and chemokines; proliferation of inflammatory cells; and expression of some adhesion molecules.

An increasing number of other nuclear proteins have been identified with protein structure homologous with that of nuclear receptors, but without a known ligand. These so-called **"orphan" receptors** are a subfamily of transcription factors acting either in the absence of any ligand or with as yet unknown endogenous ligands.

1.6.7.3.2 Epigenetics of GR

A number of **posttranslational modifications (PTMs)** of the GR are known. **Phosphorylation, ubiquitination, acetylation and sumoylation** of the GR affect receptor stability, subcellular localization, as well as the interaction between GR and other proteins (Fig. 1.6.15) (Holmes Jr et al., 2019; Bartlett et al., 2019; Gray et al., 2017). The ligand-activated hGRα monomer can also modulate gene expression independently of binding to GREs, by interacting with other **transcription factors**, such as **activator protein-1 (AP-1), nuclear factor-κB (NF-κB), p53 and signal transducers and activators of transcription (STATs)**. hGRα may affect signal transduction cascades by influencing their ability to stimulate or inhibit the transcription rates of respective target genes. Suppression of transactivation of other transcription factors through protein-protein interactions may be important in suppression of immune function and inflammation by glucocorticoids by the interaction between GR and NF-kB, AP-1 and STATs.

1.6.7.3.3 Epigenetics of MR

An important action of aldosterone impacts epithelial Na+ channel (ENaC) activity in the distal nephron through multiple mechanisms including epigenetic effects. The positions of amino acids targeted for phosphorylation, sumoylation, acetylation and ubiquitylation are indicated in the scheme throughout the human MR sequence (Viengchareun et al., 2007) (see earlier Fig. 1.6.7). Aldosterone transcriptionally upregulates the ENaCα subunit

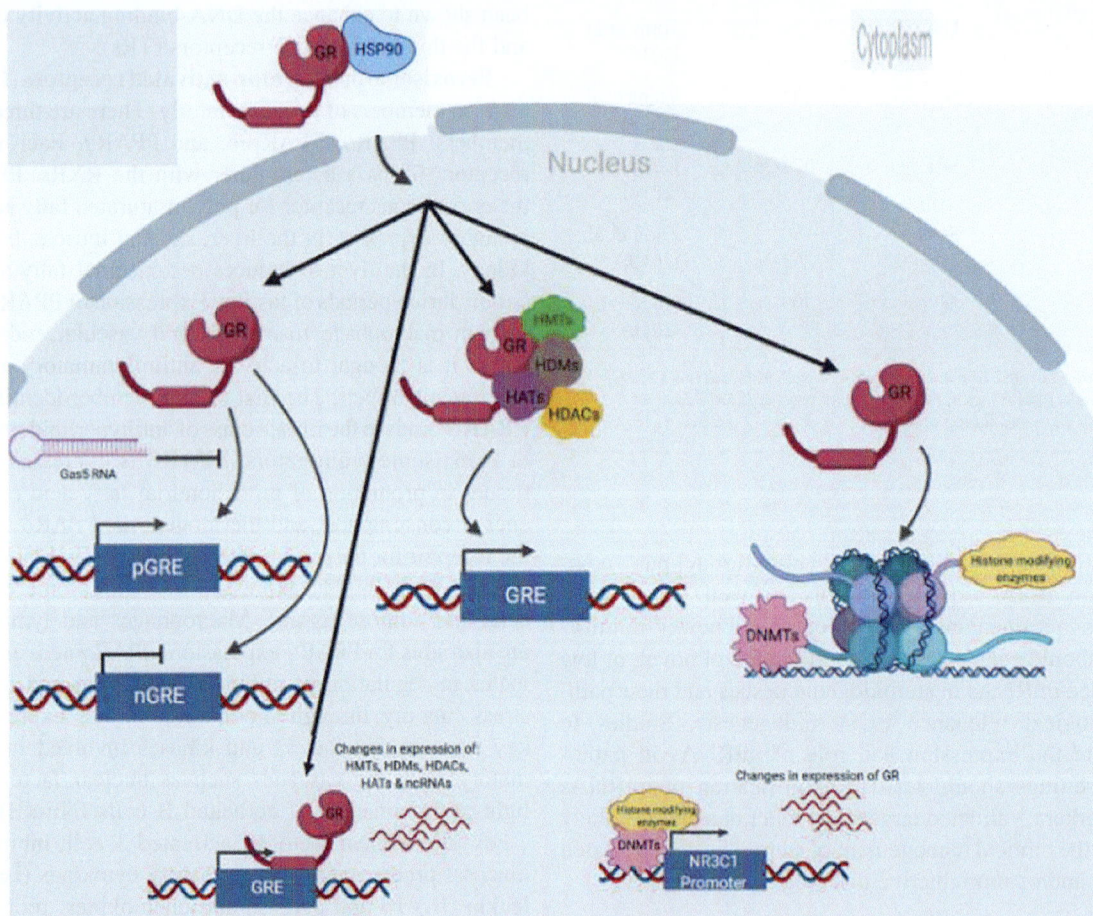

FIG. 1.6.15 Epigenetics of the GR. The GR translocates into the nucleus following ligand activation, regulated in part by histone deacetylase 6 (HDAC6). The glucocorticoid receptor then binds to positive (pGRE) or negative glucocorticoid response elements (nGRE) in the DNA, thereby enhancing or repressing transcription, respectively. Gas5 noncoding RNA (ncRNA) sequesters ligand-bound receptors. The glucocorticoid receptor recruits histone methyltransferases (HMTs), histone demethylases (HDMs), histone acetyltransferases (HATs), and HDACs to promoters. Histone-modifying enzymes and DNA methyltransferases (DNMTs) recruit the glucocorticoid receptor to local chromatin. The glucocorticoid receptor changes the expression of HMTs, HDMs, HDACs, HATs, and ncRNAs. Reciprocally, DNMTs and histone-modifying enzymes act at the NR3C1 promoter to regulate glucocorticoid receptor expression. *(From Bartlett AA, Lapp HE, Hunter RG. Epigenetic mechanisms of the glucocorticoid receptor. Trends Endocrinol Metab 2019;30(11):807–18. Fig. 4 p. 815.)*

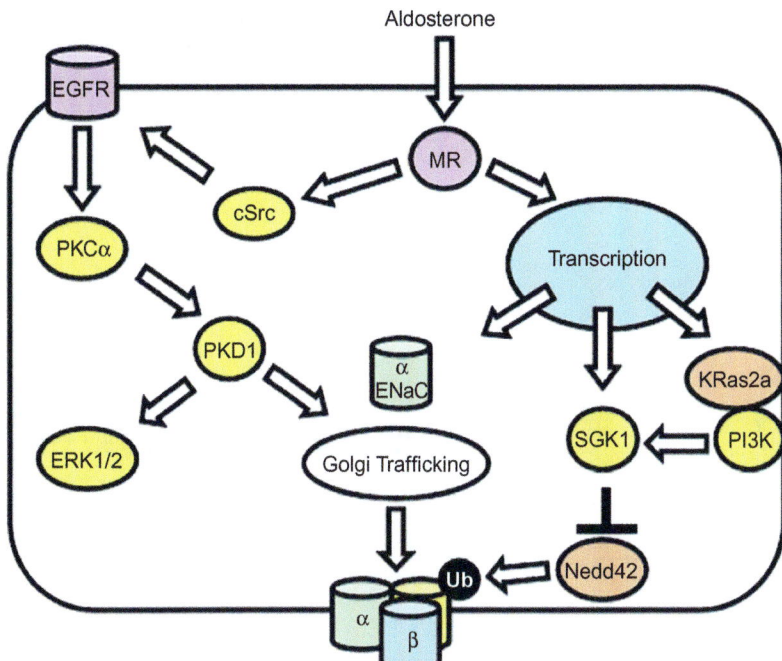

FIG. **1.6.16** Aldosterone-bound mineralocorticoid receptor (MR) upregulates the transcription on ENaC, serum glucocorticoid regulated kinase (SGK)-1 and KRas 2A. Aldosterone bound to MR upregulates the transcription of ENaC activity. KRas 2A enhances phosphatidylinositol 3 kinase (PI3K) signaling to enhance SGK-1 activity. SGK-1 phosphorylates and inactivates Nedd 4–2 ubiquitin ligase, so stabilizing ENaC subunits that are cycling between the cell membrane and a subapical vesicle pool. MR activates protein kinase D1 (PKD1) through an epidermal growth factor receptor (EGFR) trans-activation pathway. PKD1 activation affects the rate of Golgi to membrane trafficking and so the impacts upon the rate of ENaC integration into the apical membrane of renal epithelial cells. *(From Quinn S, Harvey BJ, Thomas W. Rapid aldosterone actions on epithelial sodium channel trafficking and cell proliferation. Steroids 2014;81:43–8. https://doi.org/10.1016/j.steroids.2013.11.005. Epub 2013 Nov 20. PMID: 24269741. Fig. 3 p. 47.)*

and also up regulates serum and glucocorticoid-regulated kinase-1 (SGK1) that indirectly regulates the ubiquitination of ENaC subunits (Fig. 1.6.16). Aldosterone promotes the activation of protein kinase D1 (PKD1) through an epidermal growth factor receptor (EGFR). PKD1 can modify the activity of ENaC and other transporters through effects on sub-cellular trafficking.

The sequence of events from MR binding to gene transcription and protein translation is shown in Fig. 1.6.17. A number of coregulators and repressors in the cytoplasm protect stability of the multiprotein complex and dissociation thereof in the nucleus with MR binding to the hormone response element (HRE). The complex includes two HSP90, other HSP's (40 and 70) and HSP organizing protein (HOP), stabilized by p23. MR binds to the complex with release of HOP then association with FKBP51 which exchanges with FKBP52. Coactivators, such as steroid receptor coregulator 1 (SRC-1), PPAR-γ coactivator-α (PGC1-α), p300/CREB-binding protein (CBP) and eleven-nineteen lysine-rich leukemia (ELL) are sequentially recruited to the MR to allow histone acetylation and target gene transcription to occur.

1.6.7.3.4 Epigenetics of AR

Posttranslational modifications of the AR protein have been found in at least 23 sites. Phosphorylation of the AR can occur at serine, threonine and tyrosine residues by specific kinases. The AR protein can also be modified by methylation, acetylation, ubiquitination and SUMOylation (Vigodner et al., 2020) or phosphorylation (Treviño and

Weigel, 2013; Gioeli and Paschal, 2012) to activate hormone binding, alter nuclear cytoplasmic/nuclear shuttling, DNA binding and transcriptional activity (Gioeli and Paschal, 2012) (Fig. 1.6.18).

Immediately after translation the AR becomes phosphorylated resulting in two isoforms separable by SDS-polyacrylamide gel electrophoresis. The faster migrating 110 kDa nonphosphorylated isoform is converted into a 112 kDa phospho-isoform. Phosphorylation of the AR can occur at serine, threonine and tyrosine residues by specific kinases and can be directly or indirectly linked to activation of hormone binding, altering of nuclear cytoplasmic shuttling, modulation of DNA binding and transcriptional activity. Phosphorylation of the AR can furthermore play a role in the hormone-independent activation of the AR by protein kinases in the MAPK and AKT (protein kinase B) signaling pathways.

1.6.8 Nongenomic actions of steroids

Some actions of steroids are much faster than fit with a genomic mechanism. Acute vasodilatation by estradiol and immediate anesthetic effects of progesterone were recognized in the 1940s as rapid effects of steroids. The nongenomic mechanism is partially characterized by experiments that showed persistence of the effects after preincubation of cells with actinomycin or cycloheximide that inhibited transcription and translation. Further experiments on cells without nuclei still responded to steroid, and isolated mitochondrial membranes also responded to steroids. The concept of the nongenomic effects of steroids was

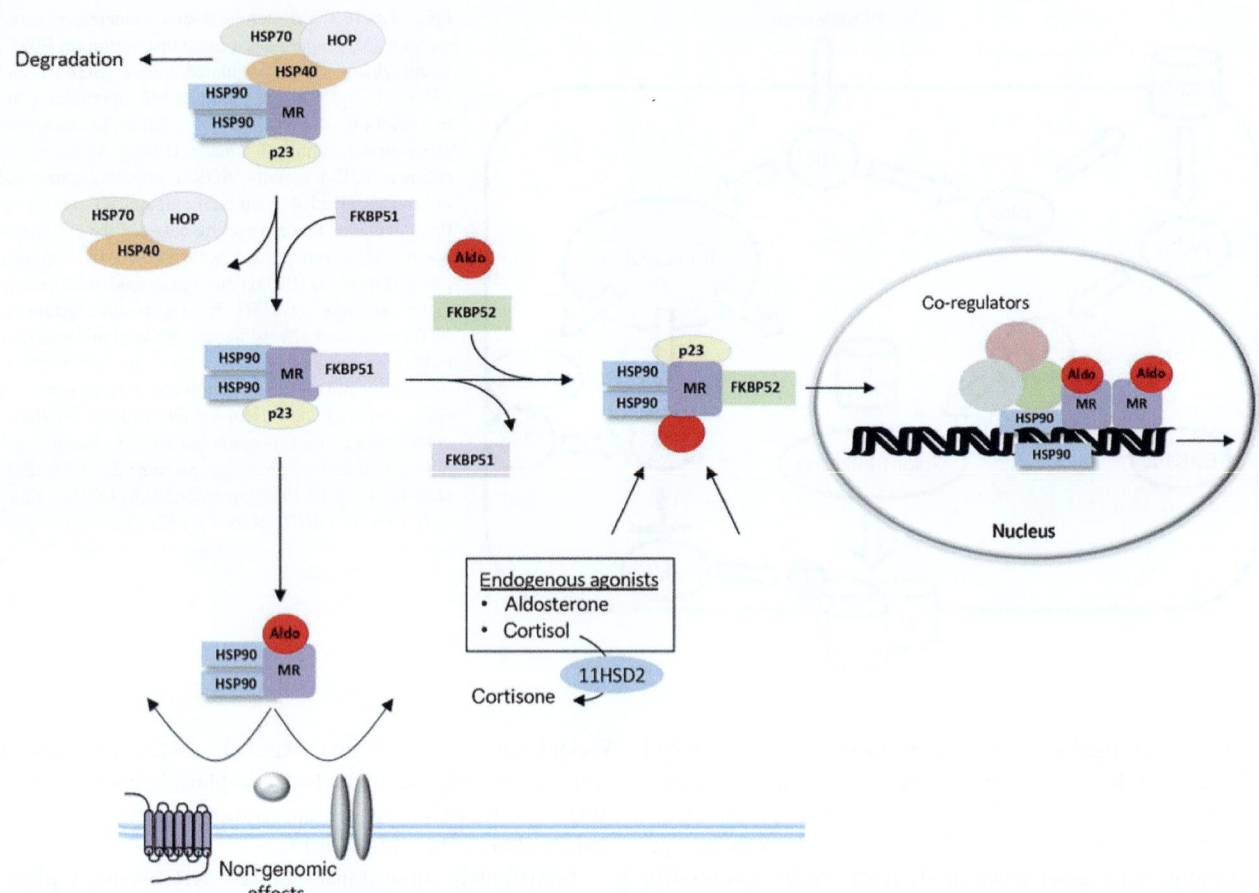

FIG. 1.6.17 Schematic representation of MR signaling upon agonist ligand binding. Key cellular and molecular events regulating MR action start with binding of the ligand to the MR that resides in the cytoplasm in a multiprotein complex. The key chaperone in this complex is heat shock protein (HSP), which holds the MR inactive in the cytoplasm. Other chaperones include protein p23, HSP70-90-organizing protein (HOP) and FK506-binding protein (FKBP). Repressors include silencing mediator of retinoic acid and thyroid hormone receptor (SMRT), nuclear receptor corepressor (NcoR), protein inhibitor of activated STAT 1 (PIAS1) and Gem-associated protein 4 (GEMIN4). When aldosterone binds to the MR there is exchange of FKB51 for FKB52 that mediates trafficking along microtubules to the nucleus. The MR is polyubiquitinated via the ubiquitin-specific protease 2–45 (Usp2–45). Ligand binding induces a conformational change in the LBD of the MR that enables interaction with co-regulatory molecules, and through the action of transportin translocates to the nucleus where it binds as a dimer to a hormone response element (HRE). The target genes for ion channels include the epithelial sodium transport channel (ENaC), large conductance voltage- and Ca2+-regulated K^+ channel (BK); Na^+/H^+ exchanger (NHE) and Na^+, K^+-ATPase (NKA). Target genes for regulators of ion channels include serum- and glucocorticoid regulated kinase 1 (SGK1), kidney-specific with-no-lysine kinase 1 (KS-WNK1), connector enhancer of kinase suppressor of ras 3 (CNKSR3); glucocorticoid-induced leucine zipper (GILZ), kidney-specific with-no-lysine kinase 1 (KS-WNK1), period circadian regulator 1 (PER1) and proviral integration site of Moloney murine leukemia virus 3 kinase (PIM3). Tissue dependent ligand specificity is achieved by a pre-receptor enzyme 11b-hydroxy-steroid dehydrogenase type 2 (HSD11B2). *(From Grossmann C, Almeida-Prieto B, Nolze A, Alvarez de la Rosa D. Structural and molecular determinants of mineralocorticoid receptor signalling. Br J Pharmacol 2021;179(13):3103–18. Fig. 2. p. 3108.)*

controversial for many years. The nongenomic response is distinguished by actions on components of signal transduction pathways and range from changes in activity of adenyl cyclase, MAPKs, and PI3K to increases in intracellular calcium ion concentrations. Further evidence for nonclassical effects of steroids comes from experiments with gene knockouts and when steroid is bound to protein (bovine serum albumin for example) that does not get through the cell membrane. Some of the rapid effects of steroids may be mediated through G-protein coupled receptor (GPCR) signaling pathways. The most clearly defined

actions occur thorough membrane-bound classical steroid receptors and the use of antibodies against different regions of the steroid receptor. There are many examples where steroids regulate cellular processes in a purely cytoplasmic manner. The effect in one particular cell type may not be replicated in other cell types. Humans with genetic defects in the classical steroid hormone pathway and mice carrying targeted mutations of specific residues in the receptors provide important evidence in attempts to understand the physiological significance of rapid effects of steroids. The lack of a single, or a few mechanisms, makes detailed

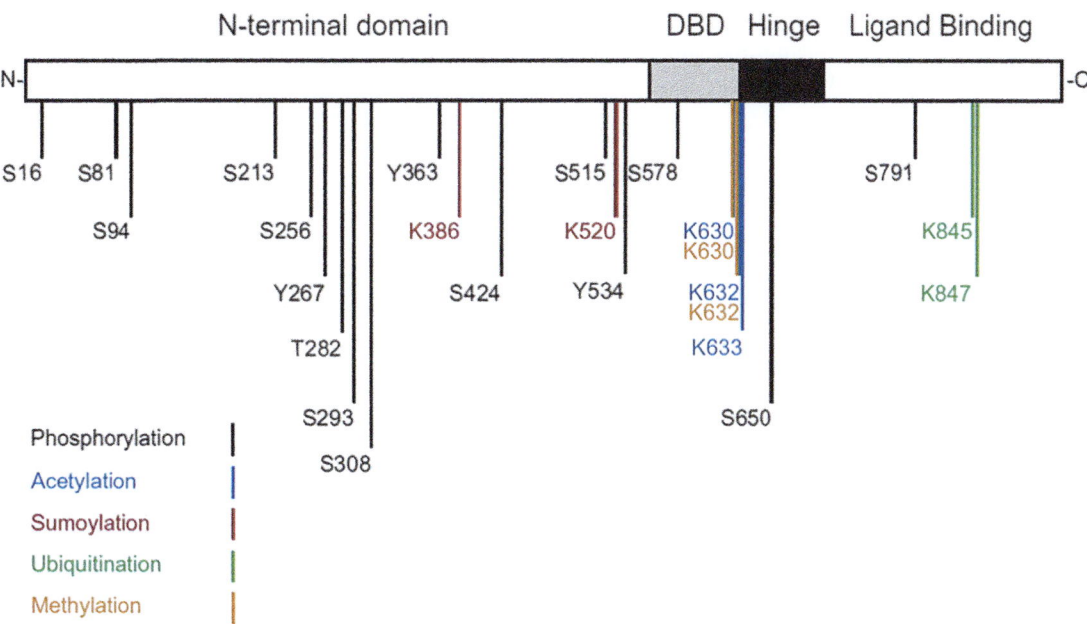

FIG. 1.6.18 Post trans modifications AR. The major domains of the AR are shown along with the known posttranslational modifications. Phosphorylation is indicated in *black*, acetylation in *blue*, SUMOylation in *red*, and ubiquitination in *green*. *(From Gioeli D, Paschal BM. Post-translational modification of the androgen receptor. Mol Cell Endocrinol 2012;352(1–2):70–8. https://doi.org/10.1016/j.mce.2011.07.004. Epub 2011 Jul 24. PMID: 21820033. Fig. 1 p. 71.)*

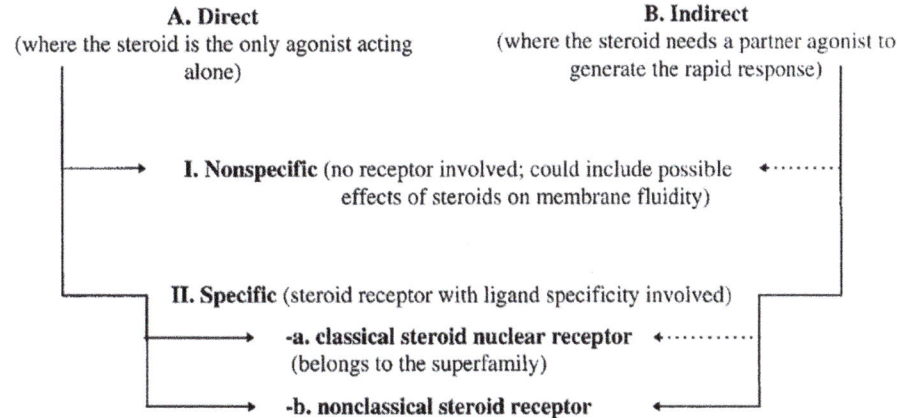

FIG. 1.6.19 Mannheim classification of nongenomic steroid actions. *Dotted arrows* indicate a hypothetical category with no example yet known. Other *arrows* indicate examples for categories. *(From Falkenstein E, Norman AW, Wehling M. Mannheim classification of nongenomically initiated (rapid) steroid action(s). J Clin Endocrinol Metab 2000;85(5):2072–5. Fig. 1 p. 2073.)*

discussion of the topic beyond the scope of the book nevertheless the principles will become important in understanding some disease processes and pharmacologies of targeted drug treatments. The history of estrogen receptor research was reviewed by Balthazart et al. (2018).

In the year 2000, an important meeting in Mannheim proposed a classification of rapid responses according to the route followed by the steroid either directly or indirectly (Falkenstein et al., 2000) (Fig. 1.6.19). In this classification, BII-1 is the classic intracellular steroid receptor. The

mechanisms are somewhat cell specific but can be considered to involve five overlapping distinct mechanisms.

A direct action involving insertion of the steroid into the cell membrane, altering fluidity of the membrane and creation of a vesicle, that makes the membrane permeable (AI). ERs located at the cell membrane, generally in caveolae, initiate cytosolic signaling events that modulate protein phosphorylation and the activation of different protein kinases, that then regulate second-messenger systems such as Ca^{2+}, K^+ and nitric oxide (NO), leading to the activation of tyrosine kinase receptors, epidermal growth factor receptor (EGFR), insulin-like growth factor 1 receptor (IGF1R) and protein kinase B (PKB).

1.6.8.1 Rapid action of glucocorticoids

The glucocorticoids have some signal effects, within seconds or minutes, that do not require hGRα transcriptional activity. The mechanisms may be triggered by membrane-bound GRs (Fig. 1.6.20) but are not fully understood. The activity of kinase signaling pathways are induced, such as the mitogen-activated protein kinase (MAPK) or the phosphatidylinositol 3-kinase (PI$_3$K) cascades, and these are called nongenomic actions. The immediate suppression of CRF/ACTH release from the anterior lobe of pituitary possibly through endocannabinoid release (Di et al., 2003) is an example.

Sites in the D-loop of the mitochondrial genome have functional GREs so glucocorticoids may exert some effects through mitochondrial hGRs. Ligand-activated hGRα has been shown to translocate from the cytoplasm to mitochondrion and influence mitochondrial gene expression (Scheller et al., 2000). Many mitochondrial RNA-processing enzymes are expressed under the control of nuclear hGRα (Morgan et al., 2016). The mitochondrial hGRα is recognized as a potential therapeutic agent because of its involvement in the programmed cell death (apoptosis) of malignant cells. Glucocorticoid might act via a putative membrane receptor and activate theErk1/2 MAPK through the PKC, which might be mediated via cRaf-1 (Qiu et al., 2001).

1.6.8.2 Nongenomic action of aldosterone

Aldosterone was among the first steroid for which a rapid, clinically relevant effect was shown even in humans. After

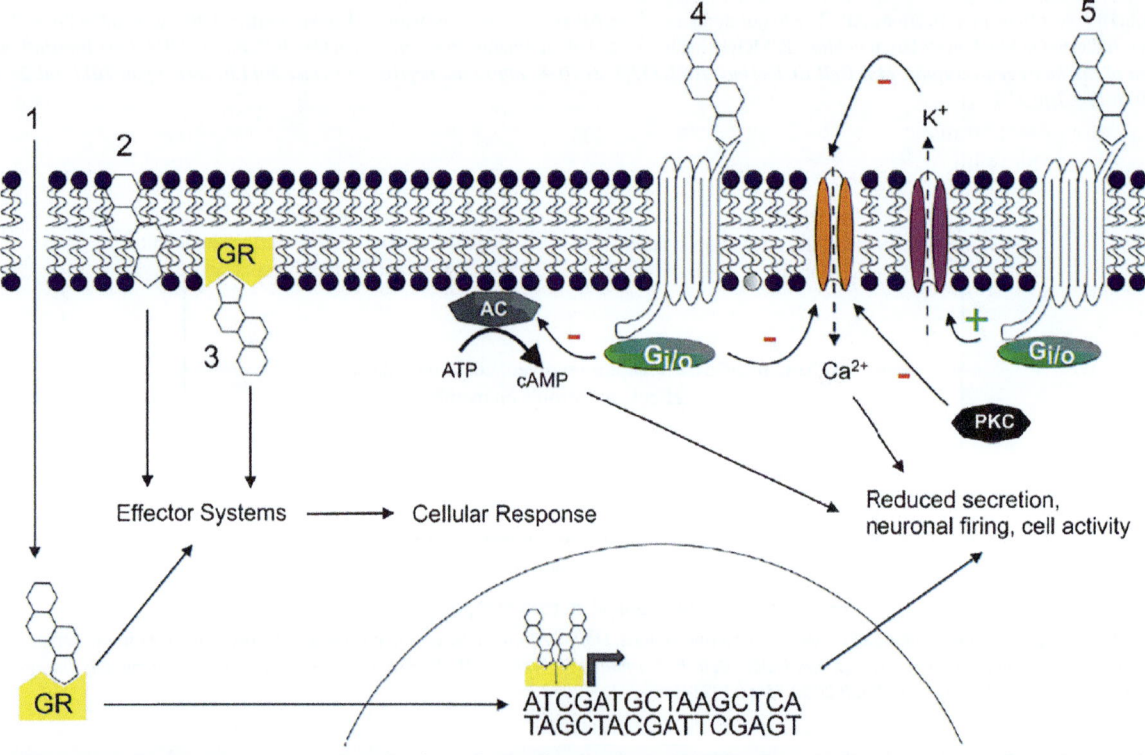

FIG. 1.6.20 Schematic of putative nongenomic mechanisms of action for glucocorticoids (GC). (Route 1) The classical genomic mechanism of steroid action in which the GC traverses the plasma membrane, binds to the intracellular GC receptor (GR), and enters the nucleus where it alters gene transcription leading to a cellular response that is typically slower than rapid, nongenomic GC responses. Putative nongenomic actions: (Route 2) The GC directly influences membrane fluidity and composition, likely through incorporation of the steroid into the lipid bilayer, that leads to regulation of signal-transducers/effector systems, producing the cell response. (Route 3) The GC binds an intracellular (or intracellular-like) GR localized in the cytoplasm or in the plasma membrane leading to regulation of signal-transducers/effector systems producing a cellular response. (Route 4–5) For actions where GCs may rapidly suppress cell responses (for example in pituitary, adrenochromaffin cells, and neurons), the steroid may bind a putative pertussis-toxin sensitive inhibitory G-protein (Gi/o) coupled receptor that leads to reductions in intracellular cAMP (via suppression of adenylyl cyclase (AC)), intracellular Ca^{2+}, or both. Declines in intracellular Ca^{2+} resulting from reduced influx of extracellular Ca^{2+} may result from direct inactivation of L-type voltage-gated Ca^{2+} channels or indirectly through activation of inward rectifying K^+ channels that leads to membrane hyperpolarization. Evidence suggests that GCs may also enhance protein kinase C (PKC) activity which could lead to inactivation of Ca^{2+} channels via phosphorylation. *(From Johnstone WM III, Honeycutt JL, Deck CA, Borski RJ. Nongenomic glucocorticoid effects and their mechanisms of action in vertebrates. Int Rev Cell Mol Biol 2019;346:51–96. Fig. 3 p. 60.)*

over 30 years of research, the nature of the rapid actions of aldosterone is still unclear. There is consistent evidence for effects involving structurally different receptors ("membrane receptors") that are insensitive to classic antagonists and persistent in knockout models (Wehling, 2018; Ong and Young, 2017). Aldosterone impacts on epithelial Na$^+$ channel (ENaC) activity in the distal nephron through multiple mechanisms. Near the plasma membrane, the MR seems to be associated with caveolin and striatin as well as with receptor tyrosine kinases like EGFR, PDGFR and IGF1R and G protein-coupled receptors like AT1 and GPER1, which then mediate nongenomic aldosterone effects. GPER1 has also been named a putative novel MR (Fig. 1.6.21). In the cytosol, rapid MR signaling influences not only ROS homeostasis and the activity of different signaling molecules like small GTPases like RhoA and KRas but also kinases like c-src, PKC, PI3K, MAP kinases (including ERK, JNK and p38) and Ca^{2+} signaling.

GPER1 has not been found to bind aldosterone. GPER1 might therefore be an intermediary signaling enhancer of mineralocorticoid action as shown for epithelial growth factor receptors (Fig. 1.6.21).

Identification of II-b-receptors is highly desirable and essential for clinical translation. Type IIa and b may coexist in the same cell with IIa augmenting early IIb effects. Cloning of IIb receptors has so far been unsuccessful. Some effects of aldosterone on the vascular system are rapid and may thus involve nongenomic mechanisms (Mihailidou et al., 2019; Hermidorff et al., 2017). Nongenomic effects may be MR dependent or MR independent involving other

receptors such as GPR30. Aldosterone, in an MR-dependent pathway, induces rapid activation of JNK, p38 MAPK, and ERK1/2 through the transactivation of EGFR and c-Src (a nonreceptor tyrosine kinase). c-Src induces ROS (reactive oxygen species) production, which activates ROCK and MAPK. Aldosterone, through its interaction with GPR30, activates ERK1/2 nongenomically without intervention of the classical MR (Fuller et al., 2019).

HDACs are important enzymes in epigenetic gene silencing, acting as corepressors of transcription by deacetylating the ε-amino group of histone lysine residues. HDAC facilitates transcriptional activity of the mineralocorticoid receptor (MR) (Fig. 1.6.22).

There are many indications for a functional interaction between aldosterone/MR and GPER1 signaling, especially in cardiovascular cells (endothelial cells (ECs), vascular smooth muscle cells (VSMCs) and cardiomyocytes) as well as in tumor cells. Vascular smooth muscle cells (VSMCs) are one of the best studied models for nongenomic MR signaling pathways (Fig. 1.6.23). Nongenomic aldosterone signaling has been demonstrated to lead to vasoconstriction although the signaling pathways are complex,

Nongenomic aldosterone signaling often seems to rely on classical MR and a crosstalk with membrane-associated signaling cascades, including receptor tyrosine kinase and GPCR pathways. Although many participating signaling components have been identified, the precise spatial and temporal sequence of events has not been elucidated. Also the existence and identity of a possible additional aldosterone receptor is still not clear.

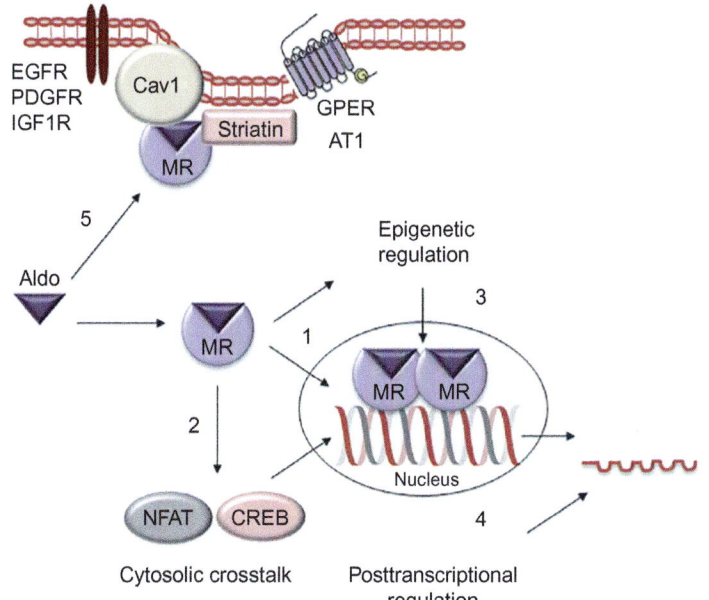

FIG. 1.6.21 Aldosterone can also bind to MR attached to the plasma membrane by scaffolding proteins like Cav1 and striatin. There it may elicit nongenomic effects by interacting with receptors, i.e., receptor tyrosine kinases like EGFR, PDGFR and IGF1R or GPCR like AT1 or GPER1. *(From Ruhs S, Nolze A, Hübschmann R, Grossmann C. 30 Years of the mineralocorticoid receptor: nongenomic effects via the mineralocorticoid receptor. J Endocrinol 2017;234(1):T107–T124. Fig. 1 p. T108.)*

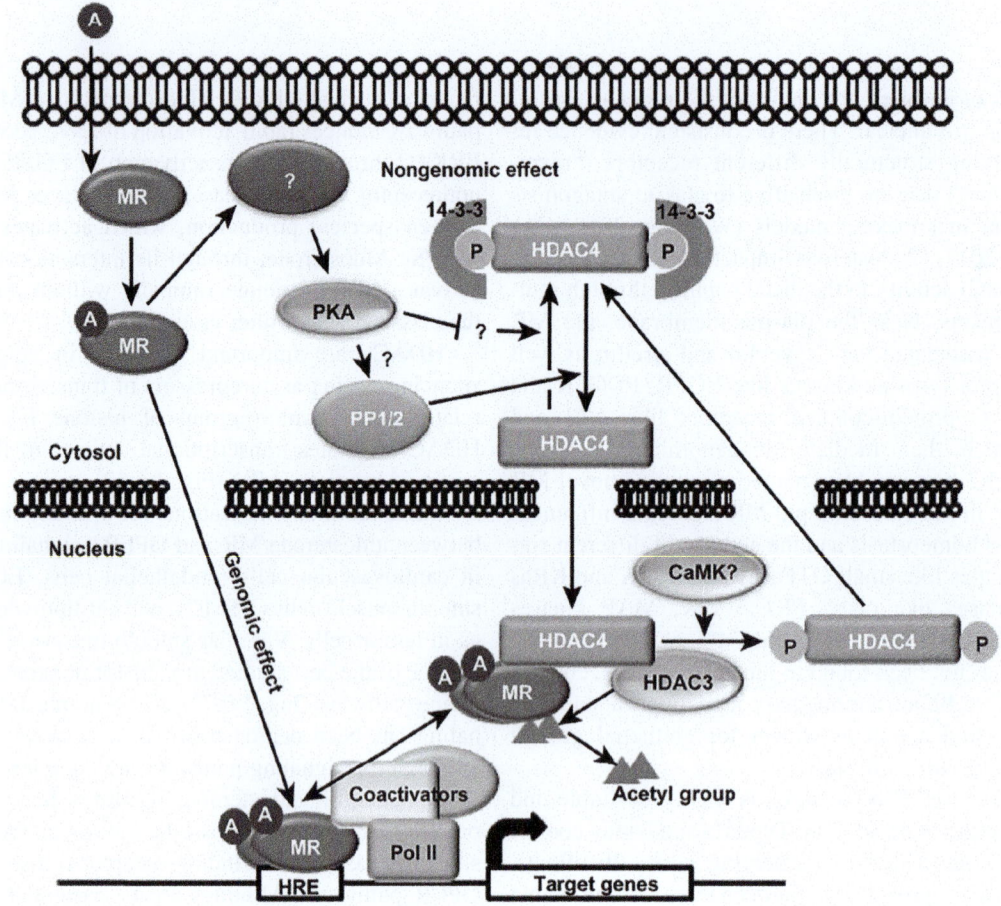

FIG. 1.6.22 Ligand-bound MR translocates from the cytosol to the nucleus where it acts as a transcription factor (genomic effect of the MR). HDAC3 deacetylates MR, which increases MR DNA binding affinity. Interaction between MR and HDAC3 is mediated by HDAC4, which is imported to the nucleus through PKA and PP1 and 2 pathways (nongenomic effect of MR). *(From Lee HA, Song MJ, Seok YM, Kang SH, Kim SY, Kim I. Histone deacetylase 3 and 4 complex stimulates the transcriptional activity of the mineralocorticoid receptor. PLoS One 2015;10(8):e0136801. Fig. 8 p. 13.)*

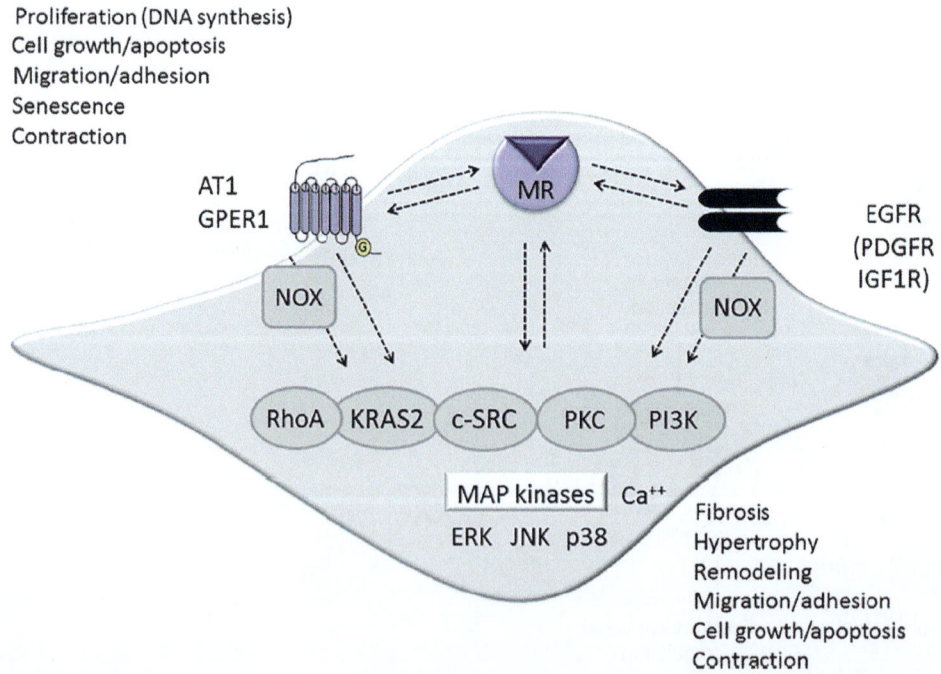

FIG. 1.6.23 Components of rapid aldosterone signaling in VSMCs. As common interaction partners of the MR, GPCR like AT1 and GPER1 as well as receptor tyrosine kinases like EGFR, PDGFR and IGF1R have been identified at the membrane. *(From Ruhs S, Nolze A, Hübschmann R, Grossmann C. 30 Years of the mineralocorticoid receptor: nongenomic effects via the mineralocorticoid receptor. J Endocrinol 2017;234(1):T107–T124. Fig. 3 p. T115.)*

1.6.8.3 Nongenomic actions of androgens

Some effects of androgens, such as the dynamics of tight junctions, are effected too quickly to be explained by genomic actions so these effects have been postulated to involve nongenomic mechanisms which can be mediated by the phosphorylation signaling pathway (Deng et al., 2017) and AR trafficking is mediated by the signaling pathway which mediates spermatogenesis. Akt (Ser473), Erk1/2 (Thr202/Tyr204), and MEK1/2 (Ser217/221), are phosphorylated with 5 min exposure to physiological testosterone (Fig. 1.6.24). The phosphorylation levels of the identified kinase returned to basal levels within another 10-min period. After being phosphorylated, the AR is translocated to the membrane to retain phosphorylation signaling while some ARs are translocated to the nucleus to induce gene transcriptional activation or repression.

The release of mature sperm from Sertoli cells is regulated by Src kinase family members through nongenomic signaling. The activation of kinases in the Sertoli cells is important for maintaining spermatogenesis. The non-AR tyrosine kinase Src is regarded as a major factor in signal transduction. Once activated, Src caused EGFR to activate the kinases RAF, MEK, and Erk. Src will also phosphorylate FAK, β-catenin and N-cadherin proteins that contribute to the formation of adhesion complexes between Sertoli cells and the mature elongated spermatids.

Testosterone concentrations, similar to or lower than those found in the testes, could cause the AR in the cytoplasm to translocate to and associate with the plasma membrane. This association is mediated by the MEK and the Akt phosphorylation signaling pathways, which result in the activation of kinase Src, initiated by testosterone binding to the membrane localized AR (mAR). Testosterone action can thus be through the classical AR and the nonclassical pathways.

Four intriguing candidates for mARs, all of them distinct multifunction proteins, have recently been proposed (Thomas, 2019). Two of the proteins, OXER1 and GPR6A,

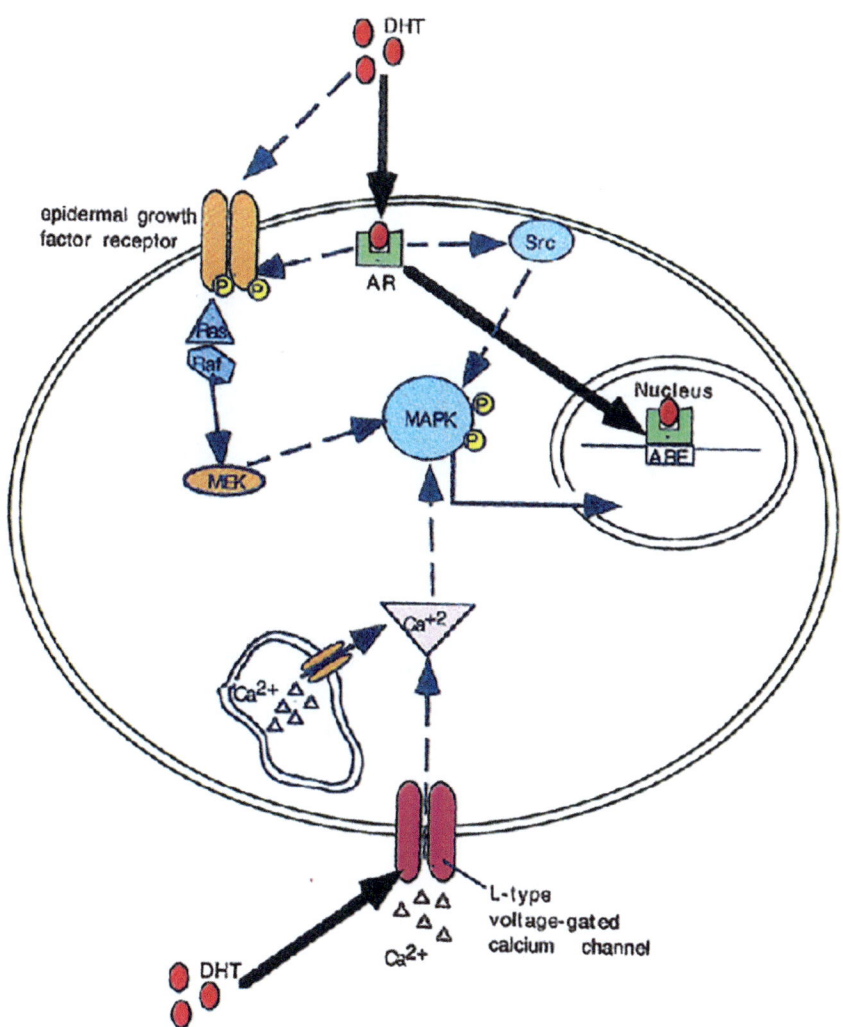

FIG. 1.6.24 A model for the genomic and nongenomic actions of androgens. The biological activity of androgens is thought to occur predominantly through binding to the androgen receptor (AR) that functions as a ligand activated transcription factor (*black arrow* represents genomic androgen action). However, androgens have also been reported to induce the rapid activation of kinase signaling cascades and modulate intracellular calcium levels. These effects are considered to be nongenomic (*blue arrows*) because they are observed to occur too rapidly to involve changes in gene transcription. Such nongenomic effects of androgens may occur through AR functioning in the cytoplasm to induce the MAP kinase signal cascade. *(From Kang HY, Tsai MY, Chang C, Huang KE. Mechanisms and clinical relevance of androgens and androgen receptor actions. Chang Gung Med J 2003;26(6):388–402. PMID: 12956285. Fig. 3 p. 392.)*

are G protein-coupled receptors (GPCRs), whereas the two other proteins, TRPM8 and ZIP9, are ion channels/transporters for Ca2+ and zinc, respectively. Some actions of androgens are misinterpreted because the action of aromatase in the system under investigation produces estrone then estradiol. Nevertheless the actions of androgens can also occur in a nongenomic, rapid and sex-nonspecific way by crosstalk with the Src, Raf-1, Erk-2 pathway.

1.6.8.4 Nongenomic actions of estrogens

The genomic action of estrogen for secondary sex characteristics (breast development, fat deposition) is a slow process but estradiol is reported to induce rapid effects through nongenomic mode in some cell types (Fig. 1.6.25). The action is mediated through signaling pathways like protein kinase C (PLC), extracellular activated kinase (ERK/SRC) phosphatidyl-inositol 3 kinase (PI3K), mitogen activated protein kinase (MPK), Janus kinase signal transducer and activators of transcription (JAK-STAT) (Alexander et al., 2017).

Membrane signaling is reported to be mediated through mERα46 and mERα36 in endothelial cells and breast cancer cells. Estrogenic signaling also occurs via membrane-bound G protein-coupled estrogen receptor 1 (GPER1),

which is localized at the cell membrane and on the endoplasmic reticulum. Estradiol binds to GPER1 and activates multiple cellular effectors, such as JUN amino-terminal kinases (JNKs), mitogen-activated protein kinases (MAPKs), phosphoinositide 3-kinase (PI3K) and other rapid cellular processes. CAV1, caveolin 1. A direct action of estrogen stimulates nitric oxide synthase activity (AII-a) which is inhibited by classic receptor antibodies and Tamoxifen that binds to the ER but insensitive to actinomycin.

Estrogens markedly influence numerous CNS functions, including learning and memory, and neuronal survival (Fig. 1.6.26). The classical ERs, ERa and ERb contribute to some rapid CNS estrogenic responses. However, emerging evidence suggests that the recently identified GPER1 is one of the main estrogen sensitive receptors responsible for the rapid nongenomic actions of estrogen.

Recent localization studies indicate that GPER1 is enriched in the brain and it is in a prime position to regulate synaptic function as it is also highly expressed at synapses. Activation of GPER1 regulates aspects of hippocampal synaptic function, with potent rapid effects reported on dendritic morphology and activity-dependent synaptic plasticity. GPER1 activation can also influence higher brain functions such as cognition, although the precise cellular mechanisms involved in GPER1 regulation of memory

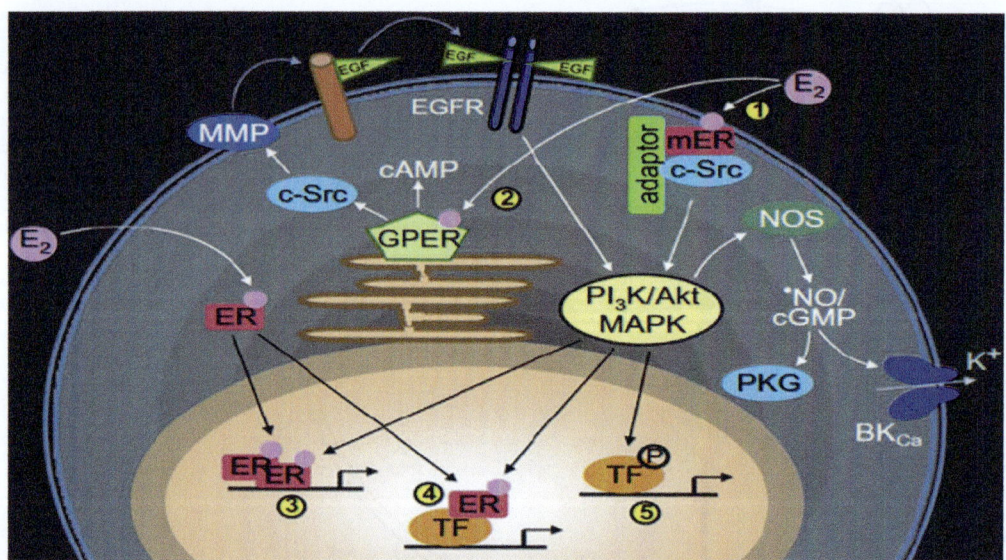

FIG. 1.6.25 Estrogen receptor (ER)-activated pathways involved in rapid estrogen signaling. Natural (endogenous) 17b-estradiol (E2) activates membrane subpopulations of ERa and ERb (mER) that interact with adaptor proteins (adaptor) and signaling molecules such as c-Src (1), thereby modifying downstream ER-induced signaling via PI3K/Akt and MAPK pathways. E2 also binds to G protein-coupled estrogen receptor GPER, which is primarily localized to the endoplasmic reticulum (2). GPER functions through activation of downstream effectors, such as adenylate cyclase (resulting in cAMP production), and c-Src; c-Src, in turn, activates matrix metalloproteinases (MMP), which cleave proheparin-bound-epidermal growth factor receptors (EGFR). EGFR transactivation causes several intracellular responses, including activation of MAPK and PI3K. Once activated, PI3K stimulates NO generation by NO synthase (NOS) in vascular endothelial (eNOS) or smooth muscle cells (nNOS). Membrane-permeable NO, in turn, results in vascular smooth muscle cell guanylate cyclase activation resulting in cGMP production. The NO/cGMP pathway mediates vasodilation and associated signaling events, such as protein kinase G (PKG)-dependent regulation of myosin light chain phosphorylation, and activation of large-conductance calcium-activated potassium (BKCa) channels. *(From Barton M. Position paper: the membrane estrogen receptor GPER—clues and questions. Steroids 2012;77(10):935–42. Fig. 1 p. 936.)*

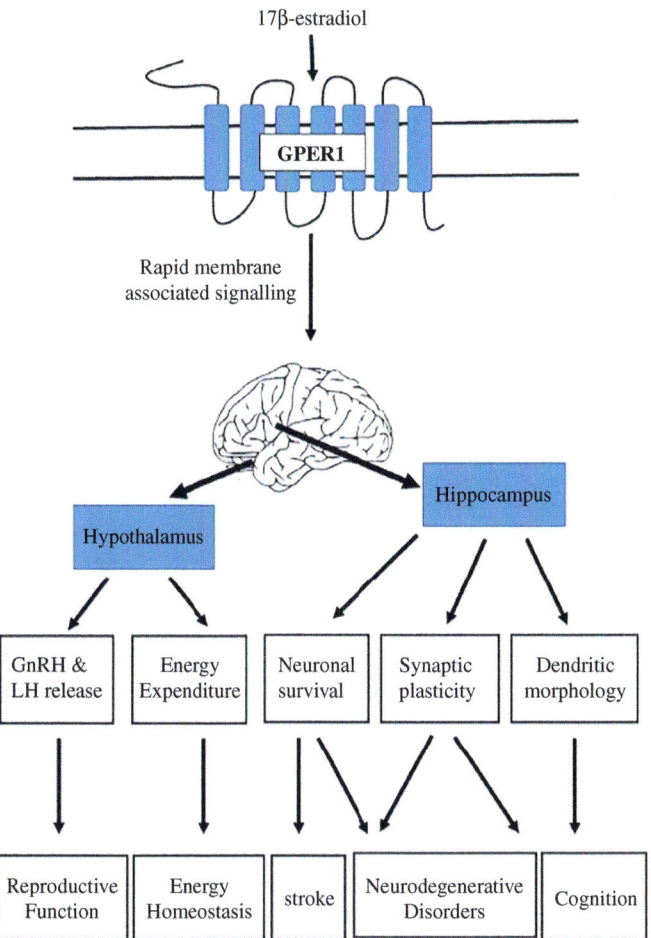

17β-estradiol

GPER1

Rapid membrane
associated signalling

Hypothalamus

Hippocampus

| GnRH & LH release | Energy Expenditure | Neuronal survival | Synaptic plasticity | Dendritic morphology |

| Reproductive Function | Energy Homeostasis | stroke | Neurodegenerative Disorders | Cognition |

FIG. 1.6.26 Schematic representation of the key neuronal functions that are regulated by GPER1. E2 activation of membrane-associated GPER in hippocampal neurons results in the initiation of rapid signaling events and subsequent modulation of key hippocampal and hypothalamic functions. The ability of GPER to regulate hippocampal synaptic function is likely to contribute to GPER-mediated enhancement of hippocampal memory and cognition. Moreover, activation of GPER may be beneficial in neurodegenerative disorders as GPER activation is reported to enhance neuronal survival in cellular models of Parkinson's disease. GPER1 activation is also thought to play a key role in mediating the effects of E2 on key hypothalamic functions, including energy homeostasis and reproductive function. *(From Alexander A, Irving AJ, Harvey J. Emerging roles for the novel estrogen-sensing receptor GPER1 in the CNS. Neuropharmacology. 2017;113(Pt B):652–60. Fig. 2 p. 657.)*

and cognition remained to be established. GPER1 has the ability to protect neurons from a variety of toxic insults. GPER activation promotes neuronal survival in models suggesting a potential role for GPER1 in various neurodegenerative disorders.

1.6.8.5 Nongenomic actions of progesterone

Rapid progesterone responses are mediated by activation of cell membrane receptors (Garg et al., 2017; Boonyaratanakornkit et al., 2001), cytoplasmic PR and receptor independent intracellular signaling cascades (Fig. 1.6.27).

Membrane progesterone receptors have been characterized as having seven trans-membrane domains similar to G protein coupled receptors expressed in ovary, uterus, nervous system and intestine. Activation of G protein coupled membrane receptors in turn activates pathways related to cAMP dependent protein kinases and P13K/Akt and ERK.MAPK. MEK signaling in macrophages may contribute to inflammatory responses (Lei et al., 2019) and regulation of labor.

1.6.8.6 Nongenomic actions of neurosteroids

Behavioral studies have demonstrated rapid changes in estrogen response, following either from a single injection of a high dose of estradiol or from the acute inhibition of aromatase activity. The bioavailability of estrogens in the brain can change according to two complementary mechanisms in time by which estrogens modulate behavior. Estrogens produced locally in the brain could be considered as neuromodulators and perhaps neurotransmitters rather than just neuroactive steroids.

In neural tissue, cholesterol is transported by translocator protein TSPO and converted to pregnenolone in the ER then to progesterone. The side-chain cleavage enzyme is expressed in neural tissues to produce DHEA and T. The action of 5α-reductase creates DHP and DHT which can then be converted by 3α and 3β-hydroxysteroid reductases to tetrahydroprogesterone (allo-pregnanolone and iso-pregnanolone) and androstanediols (5α-androstane- 3α,17β and 3β,17β-diols) (Giatti et al., 2019). Some of the metabolites bind to classic hormone receptors others activate non-classic membrane receptors that include *N*-methyl-ᴅ-aspartate receptors (NMDA), dopamine 1 receptor (D1),

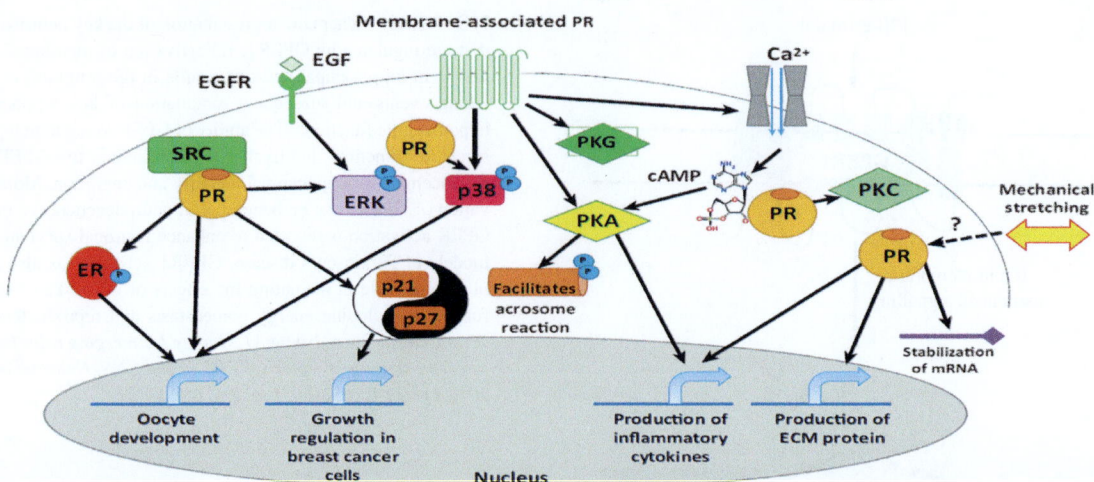

FIG. 1.6.27   Key pathways involved in nonclassical progesterone signaling are illustrated schematically. Nonclassical signaling has been shown to affect several processes including oocyte development, production of inflammatory cytokines, and extracellular matrix (ECM) protein production induced by mechanical stimulation of myocytes. P indicates phosphorylation; the question mark indicates that the mechanism is currently unknown. The nonclassical pathway involves the PR ligand-independent activation by membrane-associated kinases and the activation of multiple G protein-coupled membrane receptors of progesterone (mPRs), which in turn activates pathways related to cAMP-dependent protein kinase A (PKA), Ca2C-dependent protein kinase C (PKC), PI3K/Akt and ERK/MAPK. Abbreviations: *EGF*, epidermal growth factor; *EGFR*, EGF receptor; *ER*, estrogen receptor; *ERK*, extracellular signal-regulated kinase; *p38*, p38 mitogen activated protein kinase; *PKA*, protein kinase A; *PKC*, protein kinase C; *PKG*, protein kinase G; *PR*, progesterone receptor. *(From Garg D, Ng SSM, Baig KM, Driggers P, Segars J. Progesterone-mediated non-classical signaling. Trends Endocrinol Metab 2017;28(9):656–68. Fig. 2 p. 659.)*

nicotine receptors, α-1 adrenergic receptor, the sigma-1 receptor and the l-type Ca^{2+} channel (Fig. 1.6.28). The action of THP and 3α-Adiol on the GABA receptors at synaptic and extra-synaptic sites affects behavior in animals in a positive or negative manner depending on the concentrations of the steroids but the precise mechanisms are unknown (González-Orozco and Camacho-Arroyo, 2019).

1.6.9   Sexual dimorphism

Significant differences are found between the male and female expression of many biological functions beyond sex determination and incidence of disease, which is called sexual dimorphism. This has been a difficult area for investigations because in females the estrogen, androgen and progestogen hormone levels vary so much more than in males making selection of matched subjects for experimentation difficult to control. Testosterone is an important androgen but in tissues the steroid can be converted to estrone through the action of aromatase which is widely expressed. Many pharmacological and other studies have only been performed in males. There are differences found in prevalence, age of onset, severity or disease progression. Sex differences can be attributed to X and Y chromosomes, where females have two X chromosomes and males only one, females have no Y chromosome, whereas males have one Y chromosome. Mitochondria derive from the oocyte at conception, so a critical area for future research. Nearly all genes undergo alternative splicing leading to different

isoforms. Advances in genetic manipulation and metabolomics will enable investigations into the effects of sex on human homeostasis and disorders with more experimentation being needed in females to match studies in males.

Males and females differ in energy consumption and nutritional requirements. Females have enhanced capability of producing antibodies leading to disease phenotypes with autoimmunity more often in females and cancers more in males. There is a plethora of evidence suggesting a sex bias in innate and adaptive immunity that can impact response to infections, vaccination and onset of various diseases there is no consensus on treating diseases based on the sex of the patient. Cardiovascular disease (CVD) is the leading cause of death in men and women. Premenopausal women are protected from CVD relative to age-matched males but this protection is lost after the menopause. These issues are addressed in appropriate clinical disorders in Part III of the book.

1.6.10   Brain

In addition to the effects of steroids in the hypothalamus and pituitary for reproductive function and homeostasis, research has focused on understanding the effects of steroids in the brain and their role in behavior and regulating emotion and cognition. Glucocorticoids are essential for adaptation to stressors (allostasis) and in maladaptation resulting from allostatic load and overload. GR, MR, AR, ER and PR receptors can be detected in the hippocampus and other brain regions. Chronic stress can reshape the

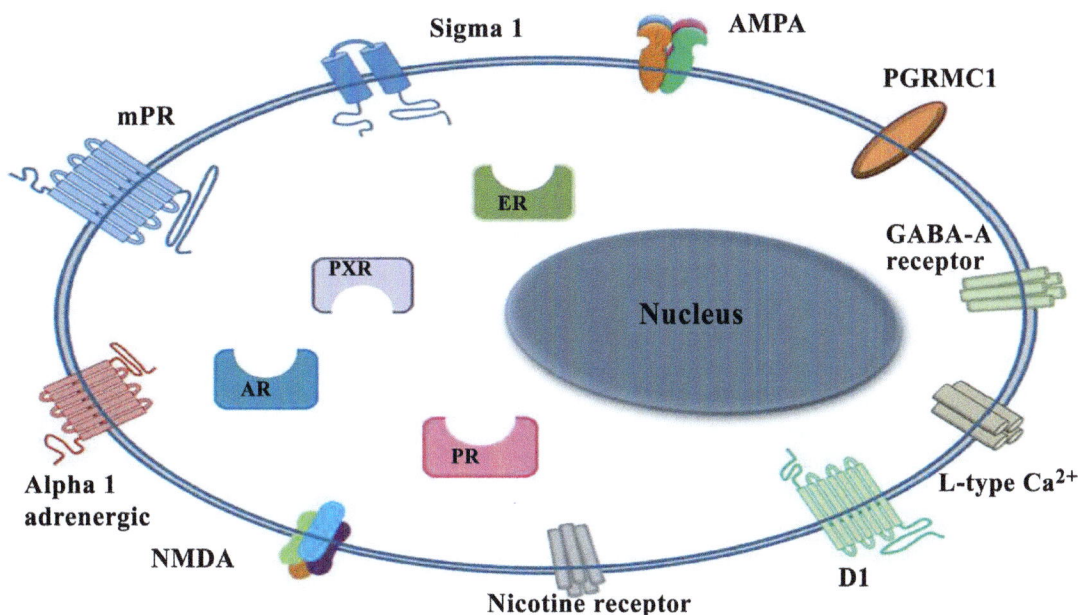

FIG. 1.6.28 Schematic representation of the different mechanisms of action of neuroactive steroids. Neuroactive steroids bind to intracellular receptors such as estrogen receptors (ER), pregnane X receptor (PXR), androgen receptor (AR) and progesterone receptors (PR). The nonclassical action of progesterone includes the modulation of the γ-aminobutyric acid type A (GABA-A) receptors following its conversion into allopregnanolone In addition, neuroactive steroids also regulate neural function acting on membrane molecules like for instance NMDA and AMPA receptors, progestin membrane receptor component 1 (PGRMC1), GABA-A receptor, L-type Ca^{2+} channels, dopamine 1 receptor (D1), nicotine receptor, alpha 1 adrenergic receptor, membrane progesterone receptor (mPR) and sigma 1 receptor. *(From Giatti S, Garcia-Segura LM, Barreto GE, Melcangi RC. Neuroactive steroids, neurosteroidogenesis and sex. Prog Neurobiol 2019;1176:1–17. Fig. 2 p. 3.)*

hypothalamic-pituitary-adrenal axis through epigenetic modification of genes in the hippocampus, hypothalamus and other stress-responsive brain regions. Glucocorticoids exert their effects on the brain through genomic mechanisms that involve both glucocorticoid receptors and mineralocorticoid receptors directly binding to DNA, as well as by nongenomic mechanisms. 11β-hydroxysteroid dehydrogenase activity is important to the effects of cortisol on the MR. Furthermore, glucocorticoids synergize both slowly (genomically) and rapidly (nongenomically) with neurotransmitters, neurotrophic factors, sex hormones and other stress mediators to determine the responses to a stressful environment. These issues will be discussed further in relevant clinical sections.

1.6.11 Summary

The actions of steroids have evolved from the original concept of specific binding of a steroid to a receptor protein prior to a transformation enabling the receptor to move into the nucleus, interact with DNA and regulate expression of genes. A family of nuclear receptors is central to current comprehension of steroid actions. The mechanisms however are now seen to be much more complex and therefore the full actions of steroids are not completely understood (see Balthazart et al., 2018 for history of

estrogen receptor research). The ligand-receptor interaction is a dynamic interaction, with the receptor conformation determined by the ligand. The resulting signal-transduction is determined by interactions with chromatin, domains within the receptor, other transcription factors and an extensive, cell-specific repertoire of coregulatory molecules. Isoforms of receptors are encoded from a single gene, transcribed from different promoters to generate subclasses of receptor mRNAs.

The past decade has witnessed tremendous progress in the understanding of the steroid receptors, their molecular mechanism of action, and implications for physiology and pathophysiology. After the initial cloning of the receptors and the identification of the genes structure and promoters, they are now recognized as major actors in protein-protein interaction networks. The role of transcriptional coregulators and the determinants of steroid selectivity have been elucidated. This chapter describes the structure, molecular mechanism of action and transcriptional regulation mediated by steroid receptors, emphasizing the most recent developments at the cellular and molecular level.

The nuclear route for steroid action occurs over hours to days but a rapid signaling is now accepted that occurs in seconds to minutes. Rapid responses to steroid hormones can be transduced by nuclear receptor remaining in the cytosol and often involves transactivation of membrane receptors such as EGFR. Nongenomic signaling may also be effected by membrane associated receptors such as mER, GPER, intracellular signaling, mechanical signaling and other key membrane receptor effects. Membrane binding sites have been described for all steroids. Aldosterone for example stimulates generation of IP3 with release of calcium from IP3 stores in pulmonary epithelial cells. Some steroids (allo-pregnanolone and THDOC) alter the excitability of neurons via GABAergic action (BII-b). There is significant cross-talk between genomic and nongenomic signaling pathways to affect biological responses in ion transport, metabolism, cell differentiation and proliferation in many tissues. Our understanding of these actions is far from complete and the specific events in the signaling pathways need to be resolved. There are species, sex and cell type differences. As the role of the steroid receptors, in an expanding range of tissues, becomes more fully appreciated (see Chapter 1.7 "Physiological effects of steroids") the complexity and diversity of the interactions of the steroid receptors has, and will continue, to provide insights into disease processes and opportunities for therapeutics particularly in cancer treatments as introduced elsewhere in the book. A limited discussion of the clinical relevance of epigenetics will be encountered in Part III of the book.

1.6.12 Further reading

Some references not cited in the text may also be useful to the reader advancing knowledge of the subject. These include history of GPER (Barton et al., 2018), history of progesterone receptor (Beato et al., 2020), vascular actions of aldosterone (Briet and Schiffrin, 2013), PXR and CAR (Buchman et al., 2018), HSD11B2 (Gent et al., 2019), sex steroids in hair follicles and target tissues (Grymowicz et al., 2020; Wierman 2007), CYP11B1 and 2 (Nusrin et al., 2014), nongenomic effects of steroids (Panettieri et al., 2019) and micro RNA's (Tesfaye et al., 2018; Troppmann et al., 2014; Yin et al., 2014).

References

Alexander A, Irving AJ, Harvey J. Emerging roles for the novel estrogen-sensing receptor GPER1 in the CNS. Neuropharmacology 2017;113 (Pt B):652–60.

Bailey MA. 11β-Hydroxysteroid dehydrogenases and hypertension in the metabolic syndrome. Curr Hypertens Rep 2017;19(12):100.

Balthazart J, Choleris E, Remage-Healey L. Steroids and the brain: 50 years of research, conceptual shifts and the ascent of non-classical and membrane-initiated actions. Horm Behav 2018;99:1–8.

Bartlett AA, Lapp HE, Hunter RG. Epigenetic mechanisms of the glucocorticoid receptor. Trends Endocrinol Metab 2019;30(11):807–18.

Barton M, Filardo EJ, Lolait SJ, Thomas P, Maggiolini M, Prossnitz ER. Twenty years of the G protein-coupled estrogen receptor GPER: historical and personal perspectives. J Steroid Biochem Mol Biol 2018;176:4–15.

Beato M, Wright RHG, Dily FL. 90 Years of progesterone: molecular mechanisms of progesterone receptor action on the breast cancer genome. J Mol Endocrinol 2020;65(1):T65–79.

Boonyaratanakornkit V, Scott MP, Ribon V, Sherman L, Anderson SM, Maller JL, et al. Progesterone receptor contains a proline-rich motif that directly interacts with SH3 domains and activates c-Src family tyrosine kinases. Mol Cell 2001;8(2):269–80.

Böttner M, Thelen P, Jarry H. Estrogen receptor beta: tissue distribution and the e still largely enigmatic physiological function. J Steroid Biochem Mol Biol 2014;139:245–51.

Briet M, Schiffrin EL. Vascular actions of aldosterone. J Vasc Res 2013;50:89–99. 95.

Buchman CD, Chai SC, Chen T. A current structural perspective on PXR and CAR in drug metabolism. Expert Opin Drug Metab Toxicol 2018;14(6):635–47.

Cano LQ, Lavery DN, Bevan CL. Mini-review: foldosome regulation of androgen receptor action in prostate cancer. Mol Cell Endocrinol 2013;369(1–2):52–62.

Clayton SA, Jones SW, Kurowska-Stolarska M, Clark AR. The role of microRNAs in glucocorticoid action. J Biol Chem 2018;293 (6):1865–74.

Dai A, Sun H, Fang T, Zhang Q, Wu S, Jiang Y, et al. MicroRNA-133b stimulates ovarian estradiol synthesis by targeting Foxl2. FEBS Lett 2013;587(15):2474–82.

Davey RA, Grossmann M. Androgen receptor structure, function and biology: from bench to bedside. Clin Biochem Rev 2016;37(1):3–15.

Dellow EL, Unwin RJ, Honour JW. Pontefract cakes can be bad for you: refractory hypertension and liquorice excess. Nephrol Dial Transplant 1999;14(1):218–20.

Deng Q, Zhang Z, Wu Y, Yu WY, Zhang J, Jiang ZM, et al. Non-genomic action of androgens is mediated by rapid phosphorylation and regulation of androgen receptor trafficking. Cell Physiol Biochem 2017;43(1):223–36.

Di S, Malcher-Lopes R, Halmos KC, Tasker JG. Nongenomic glucocorticoid inhibition via endocannabinoid release in the hypothalamus: a fast feedback mechanism. J Neurosci 2003;23(12):4850–7.

Escoter-Torres L, Caratti G, Mechtidou A, Tuckermann J, Uhlenhaut NH, Vettorazzi S. Fighting the fire: mechanisms of inflammatory gene regulation by the glucocorticoid receptor. Front Immunol 2019; 10:1859.

Falkenstein E, Norman AW, Wehling M. Mannheim classification of nongenomically initiated (rapid) steroid action(s). J Clin Endocrinol Metab 2000;85(5):2072–5.

Faus H, Haendler B. Post-translational modifications of steroid receptors. Biomed Pharmacother 2006;60(9):520–8.

Fuentes N, Silveyra P. Estrogen receptor signaling mechanisms. Adv Protein Chem Struct Biol 2019;116:135–70.

Fuller PJ, Yang J, Young MJ. 30 Years of the mineralocorticoid receptor: coregulators as mediators of mineralocorticoid receptor signalling diversity. J Endocrinol 2017;234(1):T23–34.

Fuller PJ, Yang J, Young MJ. Mechanisms of mineralocorticoid receptor signaling. Vitam Horm 2019;109:37–68.

Garg D, Ng SSM, Baig KM, Driggers P, Segars J. Progesterone-mediated non-classical signaling. Trends Endocrinol Metab 2017;28(9):656–68.

Gent R, du Toit T, Bloem LM, Swart AC. The 11β-hydroxysteroid dehydrogenase isoforms: pivotal catalytic activities yield potent C11-oxy C_{19} steroids with 11βHSD2 favouring 11-ketotestosterone, 11-ketoandrostenedione and 11-ketoprogesterone biosynthesis. J Steroid Biochem Mol Biol 2019;189:116–26.

Giatti S, Garcia-Segura LM, Barreto GE, Melcangi RC. Neuroactive steroids, neurosteroidogenesis and sex. Prog Neurobiol 2019;1176:1–17.

Gioeli D, Paschal BM. Post-translational modification of the androgen receptor. Mol Cell Endocrinol 2012;352(1–2):70–8. https://doi.org/10.1016/j.mce.2011.07.004. Epub 2011 Jul 24 21820033.

González-Orozco JC, Camacho-Arroyo I. Progesterone actions during central nervous system development. Front Neurosci 2019;13:503.

Gray JD, Kogan JF, Marrocco J, McEwen BS. Genomic and epigenomic mechanisms of glucocorticoids in the brain. Nat Rev Endocrinol 2017;13(11):661–73.

Grimm SL, Hartig SM, Edwards DP. Progesterone receptor signaling mechanisms. J Mol Biol 2016;428(19):3831–49.

Grymowicz M, Rudnicka E, Podfigurna A, Napierala P, Smolarczyk R, Smolarczyk K, et al. Hormonal effects on hair follicles. Int J Mol Sci 2020;21(15):5342.

Gustafsson JA. Steroids and the scientist. Mol Endocrinol 2005; 19(6):1412–7.

Gustafsson JA, Strom A, Warner M. Update on ERbeta. J Steroid Biochem Mol Biol 2019;191:105312.

Han Y, Staab-Weijnitz CA, Xiong G, Maser E. Identification of microRNAs as a potential novel regulatory mechanism in HSD11B1 expression. J Steroid Biochem Mol Biol 2013;133:129–39.

Heap RE, Gant MS, Lamoliatte F, Peltier J, Trost M. Mass spectrometry techniques for studying the ubiquitin system. Biochem Soc Trans 2017;45(5):1137–48.

Hermidorff MM, de Assis LV, Isoldi MC. Genomic and rapid effects of aldosterone: what we know and do not know thus far. Heart Fail Rev 2017;22(1):65–89.

Holmes Jr L, Shutman E, Chinaka C, Deepika K, Pelaez L, Dabney KW. Aberrant epigenomic modulation of glucocorticoid receptor gene (NR3C1) in early life stress and major depressive disorder correlation: systematic review and quantitative evidence synthesis. Int J Environ Res Public Health 2019;16(21). pii: E4280.

Hu Z, Shen WJ, Cortez Y, Tang X, Liu LF, Kraemer FB, et al. Hormonal regulation of microRNA expression in steroid producing cells of the ovary, testis and adrenal gland. PLoS One 2013;8(10), e78040.

Jacobsen BM, Horwitz KB. Progesterone receptors, their isoforms and progesterone regulated transcription. Mol Cell Endocrinol 2012;357 (1–2):18–29.

Jiang J, Lv X, Fan L, Huang G, Zhan Y, Wang M, et al. MicroRNA-27b suppresses growth and invasion of NSCLC cells by targeting Sp1. Tumour Biol 2014;35(10):10019–23.

Kessler BM, Bursomanno S, McGouran JF, Hickson ID, Liu Y. Biochemical and mass spectrometry-based approaches to profile SUMOylation in human cells. Methods Mol Biol 2017;1491:131–44.

Krozowski Z. 11 Beta-hydroxysteroid dehydrogenase and the short-chain alcohol dehydrogenase (SCAD) superfamily. Mol Cell Endocrinol 1992;84(1–2):C25–31.

Lei Y, Yang Y, Zhao J, Gao H, Chen R, Bai B, et al. P-AKT2/SPK1 (P-SPK1) and P-MEK/P-ERK cell signaling pathways are involved in LPS-induced macrophage migration. Am J Transl Res 2019;11:2725–41.

Louw A. GR dimerization and the impact of GR dimerization on GR protein stability and half-life. Front Immunol 2019;10:1693.

Martinez-Arguelles DB, Papadopoulos V. Epigenetic regulation of the expression of genes involved in steroid hormone biosynthesis and action. Steroids 2010;75(7):467–76.

Menon B, Gulappa T, Menon KM. miR-122 regulates LH receptor expression by activating sterol response element binding protein in rat ovaries. Endocrinology 2015;156(9):3370–80.

Menon B, Gulappa T, Menon KM. Molecular regulation of LHCGR expression by miR-122 during follicle growth in the rat ovary. Mol Cell Endocrinol 2017;442:81–9.

Mihailidou AS, Tzakos AG, Ashton AW. Non-genomic effects of aldosterone. Vitam Horm 2019;109:133–49.

Morgan DJ, Poolman TM, Williamson AJ, Wang Z, Clark NR, Ma'ayan A, et al. Glucocorticoid receptor isoforms direct distinct mitochondrial programs to regulate ATP production. Sci Rep 2016;6:26419.

Nusrin S, Tong SK, Chaturvedi G, Wu RS, Giesy JP, Kong RY. Regulation of CYP11B1 and CYP11B2 steroidogenic genes by hypoxia-inducible miR-10b in H295R cells. Mar Pollut Bull 2014; 85(2):344–51.

Ohnemus U, Uenalan M, Inzunza J, Gustafsson JA, Paus R. The hair follicle as an estrogen target and source. Endocr Rev 2006;27(6): 677–706.

Ong GS, Young MJ. Mineralocorticoid regulation of cell function: the role of rapid signalling and gene transcription pathways. J Mol Endocrinol 2017;58(1):R33–57.

Pan Y, Robertson G, Pedersen L, Lim E, Hernandez-Herrera A, Rowat AC, et al. miR-509-3p is clinically significant and strongly attenuates cellular migration and multi-cellular spheroids in ovarian cancer. Oncotarget 2016;7(18):25930–48.

Panettieri RA, Schaafsma D, Amrani Y, Koziol-White C, Ostrom R, Tliba O. Non-genomic effects of glucocorticoids: an updated view. Trends Pharmacol Sci 2019;40(1):38–49.

Pihlajamaa P, Sahu B, Jänne OA. Determinants of receptor- and tissue-specific actions in androgen signaling. Endocr Rev 2015; 36(4):357–84.

Pretorius E, Arlt W, Storbeck KH. A new dawn for androgens: novel lessons from 11-oxygenated C19 steroids. Mol Cell Endocrinol 2017;441:76–85.

Qiu J, Wang P, Jing Q, Zhang W, Li X, Zhong Y, et al. Rapid activation of ERK1/2 mitogen-activated protein kinase by corticosterone in PC12 cells. Biochem Biophys Res Commun 2001;287(4):1017–24.

Ratman D, Vanden Berghe W, Dejager L, Libert C, Tavernier J, Beck IM, et al. How glucocorticoid receptors modulate the activity of other transcription factors: a scope beyond tethering. Mol Cell Endocrinol 2013;380(1–2):41–54.

Robertson S, MacKenzie SM, Alvarez-Madrazo S, Diver LA, Lin J, Stewart PM, et al. MicroRNA-24 is a novel regulator of aldosterone and cortisol production in the human adrenal cortex. Hypertension 2013;62 (3):572–8.

Scheller K, Sekeris CE, Krohne G, Hock R, Hansen IA, Scheer U. Localization of glucocorticoid hormone receptors in mitochondria of human cells. Eur J Cell Biol 2000;79(5):299–307.

Skowron KJ, Booker K, Cheng C, Creed S, David BP, Lazzara PR, et al. Steroid receptor/coactivator binding inhibitors: an update. Mol Cell Endocrinol 2019;493:110471.

Stashi E, York B, O'Malley BW. Steroid receptor coactivators: servants and masters for control of systems metabolism. Trends Endocrinol Metab 2014;25(7):337–47.

Sun JY, Xiao WZ, Wang F, Wang YQ, Zhu YH, Wu YF, et al. MicroRNA-320 inhibits cell proliferation in glioma by targeting E2F1. Mol Med Rep 2015;12(2):2355–9.

Takayama KI. The biological and clinical advances of androgen receptor function in age-related diseases and cancer [review]. Endocr J 2017;64(10):933–46.

Tesfaye D, Gebremedhn S, Salilew-Wondim D, Hailay T, Hoelker M, Grosse-Brinkhaus C, et al. MicroRNAs: tiny molecules with a significant role in mammalian follicular and oocyte development. Reproduction 2018;155(3):R121–35.

Thomas P. Membrane androgen receptors unrelated to nuclear steroid receptors. Endocrinology 2019;160(4):772–81.

Timmermans S, Souffriau J, Libert C. A general introduction to glucocorticoid biology. Front Immunol 2019;10:1545.

Toms D, Pan B, Li J. Endocrine regulation in the ovary by MicroRNA during the estrous cycle. Front Endocrinol (Lausanne) 2018;8:378.

Treviño LS, Weigel NL. Phosphorylation: a fundamental regulator of steroid receptor action. Trends Endocrinol Metab 2013;24(10):515–24.

Troppmann B, Kossack N, Nordhoff V, Schüring AN, Gromoll J. MicroRNA miR-513a-3p acts as a co-regulator of luteinizing hormone/chorionic gonadotropin receptor gene expression in human granulosa cells. Mol Cell Endocrinol 2014;390(1-2):65–72.

Vandevyver S, Dejager L, Libert C. Comprehensive overview of the structure and regulation of the glucocorticoid receptor. Endocr Rev 2014;35(4):671–93.

Viengchareun S, Le Menuet D, Martinerie L, Munier M, Pascual-Le Tallec L, Lombès M. The mineralocorticoid receptor: insights into its molecular and (patho)physiological biology. Nucl Recept Signal 2007;5, e012.

Vigodner M, Lucas B, Kemeny S, Schwartz T, Levy R. Identification of sumoylated targets in proliferating mouse spermatogonia and human testicular seminomas. Asian J Androl 2020;22(6):569–77.

Wang Z, Benoit G, Liu J, Prasad S, Aarnisalo P, Liu X, et al. Structure and function of Nurr1 identifies a class of ligand-independent nuclear receptors. Nature 2003;423(6939):555–60.

Wehling M. Rapid actions of aldosterone revisited: receptors in the limelight. J Steroid Biochem Mol Biol 2018;176:94–8.

Weikum ER, Knuesel MT, Ortlund EA, Yamamoto KR. Glucocorticoid receptor control of transcription: precision and plasticity via allostery. Nat Rev Mol Cell Biol 2017;18(3):159–74.

White PC. Alterations of cortisol metabolism in human disorders. Horm Res Paediatr 2018;89(5):320–30.

Wierman ME. Sex steroid effects at target tissues: mechanisms of action. Adv Physiol Educ 2007;31(1):26–33.

Yin M, Wang X, Yao G, Lü M, Liang M, Sun Y, et al. Transactivation of microRNA-320 by microRNA-383 regulates granulosa cell functions by targeting E2F1 and SF-1 proteins. J Biol Chem 2014;289(26):18239–57.

Yu X, Li Z. The role of microRNAs in the adrenocortical carcinomas. Tumour Biol 2016;37(2):1515–9.

Zhou R, Miao Y, Li Y, Li X, Xi J, Zhang Z. MicroRNA-150 promote apoptosis of ovine ovarian granulosa cells by targeting STAR gene. Theriogenology 2019;127:66–71.

Chapter 1.7

Physiological effects of steroids

1.7.1 Introduction

The evidence for the effects of steroids in physiology comes from clinical observations, experiments that deplete the sources of steroids, hormone replacement, hormone treatment and recently genomic studies with gene knockout and overactive models. Most of the physiological effects of steroids are through their nuclear receptors and transcribed proteins that were discussed in the previous chapter.

1.7.2 Adrenal corticosteroids

The absence of the effects of adrenal steroids was first recognized in man when Thomas Addison 1855 described patients suffering from chronic fatigue, muscle degeneration and weight loss. Tuberculosis would have been a more common cause of adrenal insufficiency in 1855 than is the case today. It was many years before the autoimmune basis for adrenal insufficiency, and the darkening of the skin in some of those patients was explained. From 1946, steroids from the adrenal cortex were synthesized and used to treat inflammatory diseases such as rheumatoid arthritis. Replacement of cortisol or corticosterone into adrenalectomized animals however did not totally achieve survival and in 1953 a separate hormone was characterized that was responsible for most of the biological activity affecting electrolyte metabolism in the adrenal cortex secretions. Sylvia and James Tait originally described this steroid as electrocortin but it was replaced with the name aldosterone which has persisted.

Cortisol synthesis is controlled from the hypothalamus and pituitary in a negative feedback fashion. In response to low circulating concentrations of cortisol, CRF is secreted from the hypothalamus into the hypophyseal portal system. The pituitary gland secretes ACTH which is detected by the adrenal cortex to secrete cortisol. Cortisol has major effects on metabolism in the liver, muscles and adipose tissues by stimulating gene transcription of enzymes (Fig. 1.7.1). Cortisol inhibits activity in the immune and cardiovascular systems. Cortisol inhibits the synthesis of cytokines that in the absence of cortisol activate pain receptors. High levels of cortisol from long-term stress can increase blood cholesterol, triglycerides, blood sugar and blood pressure that are common risk factors for heart disease.

1.7.2.1 Physiologic effects of glucocorticoids (GCH)

When cortisol is released from the adrenal cortex to the circulation, cortisol is almost entirely bound to CBG or albumin and only 5%–10% of circulating cortisol is free and biologically active. The function of CBG is to transport cortisol in the blood and regulate distribution of the hormone into tissues. The main target for cortisol is the glucocorticoid receptor (GR) and transcription of proteins. Following uptake into tissues cortisol bioavailability is further regulated by two enzymes. The **11β-hydroxysteroid dehydrogenase type 2** (encoded by the **HSD11B2** gene) oxidizes cortisol to its inactive metabolite, cortisone (Fig. 1.7.2) which circumvents the action of cortisol on the mineralocorticoid receptor (MR).

The other enzyme, **11β-hydroxysteroid dehydrogenase 1** (encoded by the **HSD11B1** gene) reduces the 11-oxo group (11-oxoreductase activity) in cortisone and 11-dehydrocorticosterone generating the active glucocorticoids, cortisol and corticosterone, respectively. All of these factors influence the delivery of cortisol to the target cells (Morgan et al., 2013, 2016).

The **glucocorticoid hormones** (GCH) are a class of steroid that modulate the **metabolism of carbohydrates**. The primary human glucocorticoid is cortisol, but corticosterone is the major GCH in many other species. The best known and studied effects of glucocorticoids are on **carbohydrate metabolism** and **immune function** although the precise manner by which GCH regulates glucose homeostasis is still unclear. Physiological increases in the production of GCH is a response to meet the increased energy demands associated with **stress**. GCH are essential for the hepatic contribution to maintaining blood glucose concentrations. GCH exert effects on the liver during periods of stress to ensure adequate release of stored glucose. Genomic and rapid effects of corticosteroids are seen in response to stress (Groeneweg et al., 2011).

Steroids in the Laboratory and Clinical Practice. https://doi.org/10.1016/B978-0-12-818124-9.00015-2

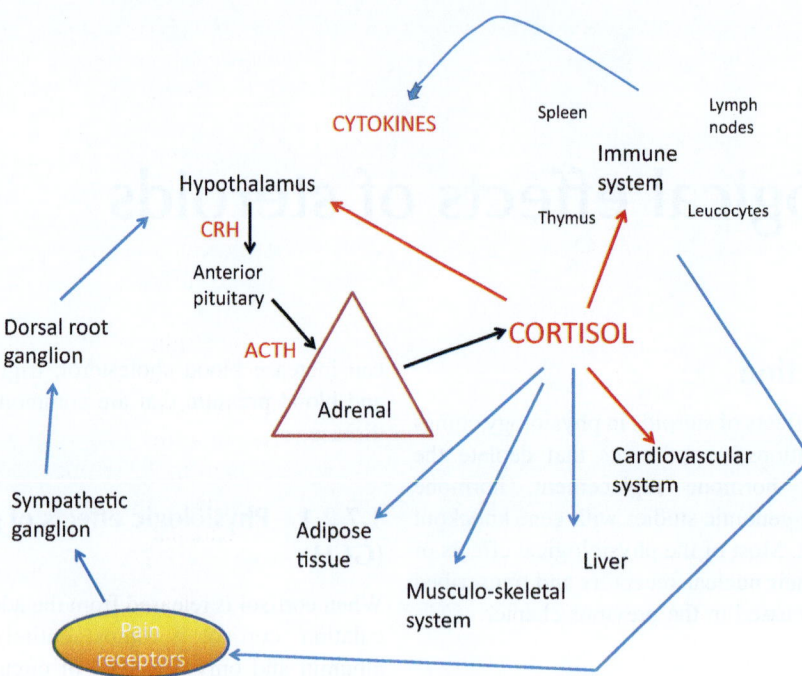

FIG. 1.7.1   Physiological effects of GCH in tissues and immune system. *Red lines* denote inhibition and *black arrows* represent activation paths .

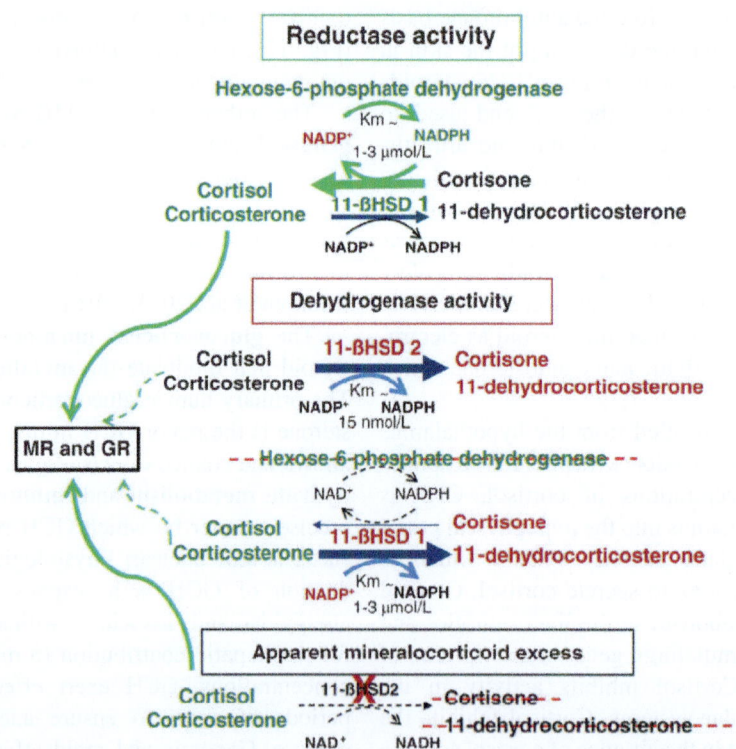

FIG. 1.7.2   Actions of HSD11B enzymes to protect MR from the effects of cortisol. *(From Gomez-Sanchez E, Gomez-Sanchez CE. The multifaceted mineralocorticoid receptor. Compr Physiol 2014;4(3):965–994.)*

The GCH have other physiologically and biochemically significant functions thus GCH also regulate:

- overall energy homeostasis,
- stress responses that affect survival and reproduction, and
- modulate embryonic and postnatal development.

At high concentrations of corticosteroids there are other effects of cortisol on bone (Hachemi et al., 2018; Hardy et al., 2018; Hartmann et al., 2016) and psychology that are considered in more detail in Chapter 3.1.1. Exogenous corticosteroids are problematic because treatment of inflammation or replacement therapy is difficult without exposure to excess hormone.

Cortisol is essential to normal development and tissue maturation of the fetus and overexposure to cortisol in the fetus causes lower birth weight and in later life is associated with cardiovascular risk factors including high blood pressure, insulin resistance as well as mental health and cognitive problems (McGowan and Matthews, 2018; Pervanidou and Chrousos, 2018; Hoffman et al., 2017). Nutritional deprivation early in life is associated with risk for nutrition-related chronic diseases (Barker hypothesis), is based on a number of studies that have found a link between poor nutrition early in life and mental health and cognitive development. The fetus is protected from high maternal GCH by the placental HSD11B2.

1.7.2.1.1 Carbohydrate metabolism

Cortisol acts as an **insulin antagonist** and suppresses the release of insulin, leading to reduced glucose uptake in muscle and adipose tissue but enhanced hepatic glucose synthesis (**gluconeogenesis**) (Alwashih et al., 2017; Petersen et al., 2017). The blood glucose concentrations are further enhanced through the increased **breakdown of skeletal muscle protein** and **adipose tissue triglycerides** (Lee et al., 2018; Kuo et al., 2015; Gathercole et al., 2011) which provides energy and substrates for gluconeogenesis. The precise mechanisms for muscle wasting are still not clear. In the fasted state, cortisol production increases and maintains normal concentrations of glucose in blood through several processes, in consort with the effects of other hormones such as insulin.

GCH *stimulate gluconeogenesis*, especially in the liver and brain (Stanley et al., 2019; Petersen et al., 2017) by enhancing the expression of enzymes involved in the synthesis of glucose from amino acids and lipids and is the best known action of GCH by activation of three steps in gluconeogenesis:

- phosphoenolpyruvate carboxykinase, mPCK1, converts oxaloacetate to phosphoenolpyruvate is stimulated by GCH

- fructose-1,6-biphosphatase PFK1 converts fructose-1,6-bisphosphate to fructose-6-phosphate, and
- glucose-6-phosphatase, converts G6PC glucose-6-phosphate to glucose leading to **reversal of the glycolysis** path from glucose to pyruvate (Fig. 1.7.3) is stimulated by GCH.

Maximal expression of both the PEPCK1 and G6PC genes requires additional factors including the nuclear receptor, PPARα.

Lactate and alanine (Fig. 1.7.3, near bottom, lower left) are converted to pyruvate, which enters the mitochondria and is converted to oxaloacetate (OAA) by PC. OAA exits the mitochondria to form phosphoenolpyruvate (PEP) which feeds into the cytosolic gluconeogenic pathway to glucose-6-phosphate. The final product, glucose, is produced in the endoplasmic reticulum (ER) by the enzyme G6PC. OAA can be converted directly to PEP in the mitochondria by mPCK1. Glycerol can be metabolized to dihydroxyacetone-phosphate (DHAP) which is then converted directly or indirectly through G3P to F1,6BP.

Glycolysis is a central pathway that produces important phosphated precursor metabolites: six-carbon compounds of **glucose-6P** and **fructose-6P** and three-carbon compounds of **glycerone-P, glyceraldehyde-3P, glycerate-3P, phosphoenolpyruvate and pyruvate** as well as generating small amounts of ATP (energy) and NADH (reducing power) (Rooney et al., 1993).

Acetyl-CoA, an important **cholesterol precursor** metabolite, is produced by oxidative **decarboxylation of pyruvate**. In the liver, GCH stimulate glycogen deposition (glycogenesis) by **increasing glycogen synthase** and **inhibiting glycogen phosphorylase** which is a glycogen mobilizing enzyme (glycogenolysis).

Amino acids are mobilized from extrahepatic tissues by high doses of cortisol (Brillon et al., 1995). Tyrosine aminotransferase (TAT) is induced by GCH. TAT catalyzes the transamination of tyrosine with ketoglutarate to form glutamate and hydroxy-PEP that is further metabolized to acetyl CoA via pyruvate, fumarate and acetate that are substrates for gluconeogenesis. All amino acids can be processed via the citric acid cycle for gluconeogenesis (Fig. 1.7.4). Cortisol is known to bring about induction of several aminotransferases in the liver.

Inhibition of protein synthesis. When GCH levels are elevated, such as during pharmacological treatment with corticosteroids or under pathological conditions of hypercortisolaemia, there is resistance to insulin, **inhibition of protein synthesis** and **enhanced proteolysis** within **skeletal muscle** (Darmaun et al., 1988; Simmons et al., 1984) The enhanced proteolysis serves to provide amino acids to the liver for gluconeogenesis. These pathophysiological consequences within skeletal muscle result in muscle weakness and atrophy. The increased rate of protein

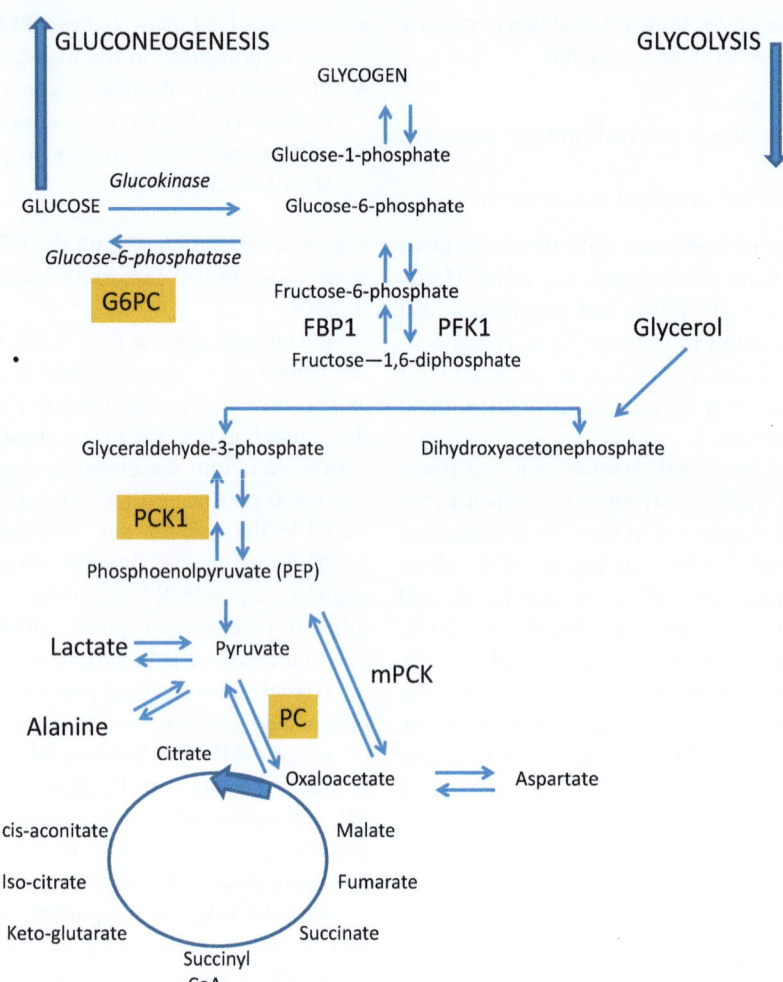

FIG. 1.7.3 Effects of glucocorticosteroids (GCH) on glucose homeostasis in the liver. The *yellow boxes* are enzymes induced by GCH. Abbreviations. *PC*, pyruvate carboxylase; *mPCK1*, mitochondrial PEP carboxykinase; *G6PC*, glucose-6-phosphate glucose 6 phosphatase, catalytic subunit. *(Author original.)*

metabolism leads to increased urinary nitrogen excretion and the induction of urea cycle enzymes. These metabolic changes through GCH inhibit several energy-consuming processes such as digestion, reproduction and other immune responses.

1.7.2.1.2 Adipose tissue

Under normal physiological conditions, GCH promote differentiation of preadipocytes into mature adipocytes. During GCH-induced adipocyte differentiation, the triglyceride synthesis, lipid transport and lipid storage are activated. During periods of fasting, stress or during exercise, catecholamines (adrenaline) and GCH stimulate glycogen breakdown. GCH effects on adipose tissue encompass increased lipogenesis through adipocyte differentiation and increased lipolysis.

Fat breakdown in adipose tissue is stimulated by cortisol. The fatty acids released by lipolysis are used for production of energy in tissues like muscle; **glycerol** is released to provide another substrate for gluconeogenesis. Fatty acids are converted to lipid mediators such as diacylglycerol and ceramides. There is an increase in free fatty acid concentrations, total cholesterol and triglycerides. HDL cholesterol levels fall. GCH induce enzymes in the urea cycle and ketone body synthesis.

In periods of fasting or stress, the **energy required for gluconeogenesis** is derived by the **oxidation of fatty acids from adipose tissue.** GCH regulate several genes encoding lipogenesis and triglyceride synthesis. Acetyl-CoA-carboxylase (ACACA) for the conversion of acetyl CoA to malonyl CoA is the rate controlling enzyme in fatty acid synthesis, and the expression of ACACA is increased by GCH (Wang et al., 2012) GCH also increases the activity of fatty acid synthase (FAS) (Fig. 1.7.5).

GCH activate the expression of genes coding enzymes in triglyceride synthesis in adipocytes namely stearyol CoA

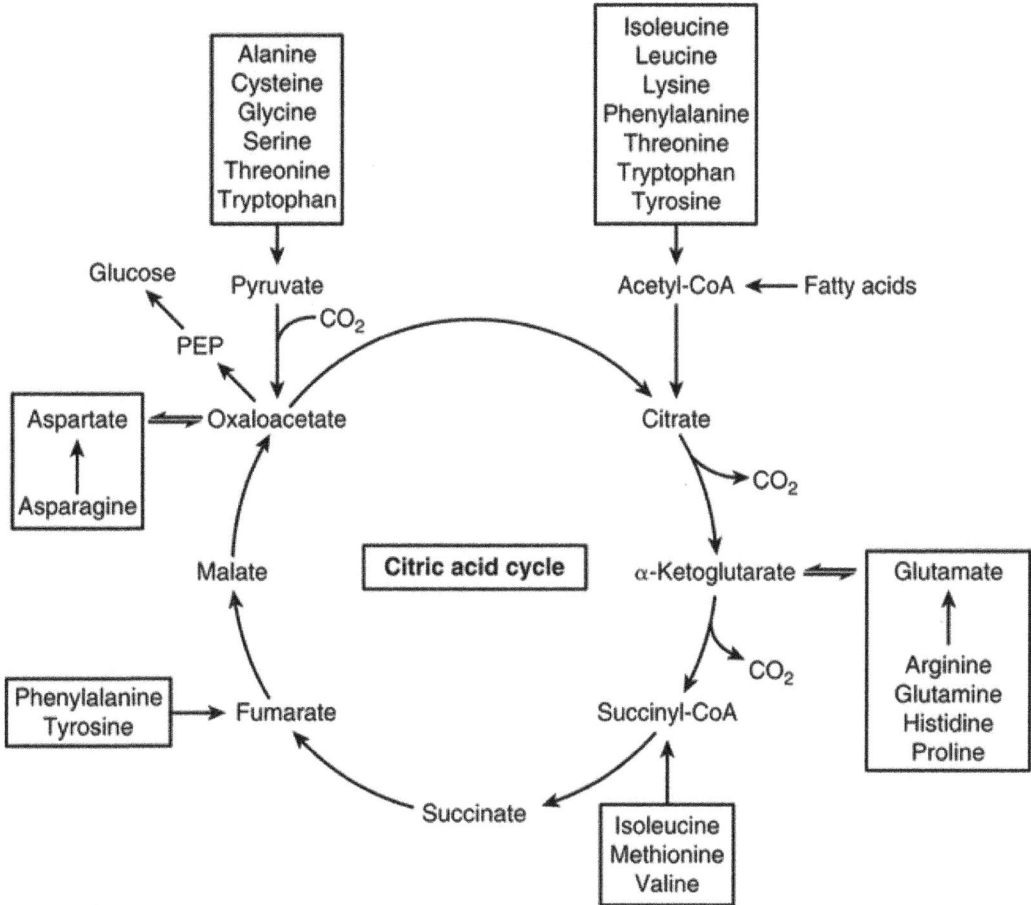

FIG. 1.7.4 Amino acids are metabolized by aminotransferases and the citric acid cycle. *(Author original.)*

desaturase (Scd) then glycerol phosphate acyltransferase (GPT) and lipin (LPIN1) that has phosphatide phosphatase activity. Multiple glucocorticoid response elements (GRE) are involved in activation of enzymes in lipogenesis (Andrews et al., 2002). The effects of GCH on acylglycerolphosphate acyltransferase (Agpat) and diacyglycerl-O-acyltransferase (DGAT) are uncertain. Triacylglycerol is hydrolyzed to free fatty acids by lipoprotein lipase (Lpl). GCH and insulin act together to increase expression of Lpl. Expression of lipin is the only gene in this path to be inhibited by insulin.

The triglyceride synthesis genes turned on by cortisol include several GPAT genes (acylglycerol-3-phosphate acyltransferases) (Yu et al., 2018). GPATs catalyze the first step of synthesis of triacylglycerol (Fig. 1.7.6) which acts as the rate limiting step for de novo acid release from adipose tissue is through expression of the **hormone-sensitive lipase gene** (LIPE or HSL) and the gene encoding **monoglyceride lipase** (MGLL).

GPATs competitively catalyze acyl-CoA and glycerol-3-phosphate (G3P) to produce lysophosphatidic acid (LPA) and protect acyl-CoA from β-oxidation. Then, LPA and acyl-CoA are converted to phosphatidic acid

(PA) by AGPAT. Consequently, PA is dephosphorylated by lipin to diacylglycerols (DAG). DAG and acyl-CoA are catalyzed by DGAT to form triacylglycerol (TAG). Furthermore, TAG synthesizes lipid droplets (LD). The intermediate products (LPA, PA, DAG) are responsible for intercellular signal transduction.

GCH elevate the expression of all genes in the lipolytic pathways in adipocytes. Insulin represses the expression or activity of these enzymes (Fig. 1.7.7). Lipe and Mgll encode enzymes in the lipolytic pathway. Angptl4 is a secreted protein that likely binds to an unknown receptor, to increase cAMP levels in adipocytes, which in turn activates PKA (Lee et al., 2018).

PKA phosphorylates Lipe, which increases its activity and translocation from cytosol to lipid droplet. GC also increases the expression of Pnpla2 and decrease the expression of Pde3b. The mechanisms for these events are unknown. Adipocyte lipolysis is a complex process that is precisely controlled through integration of multiple and diverse hormonal and biochemical signals. Breakdown of this regulation may contribute to the development of obesity and associated pathologies. Many exciting advances have been made recently; however, much remains to be

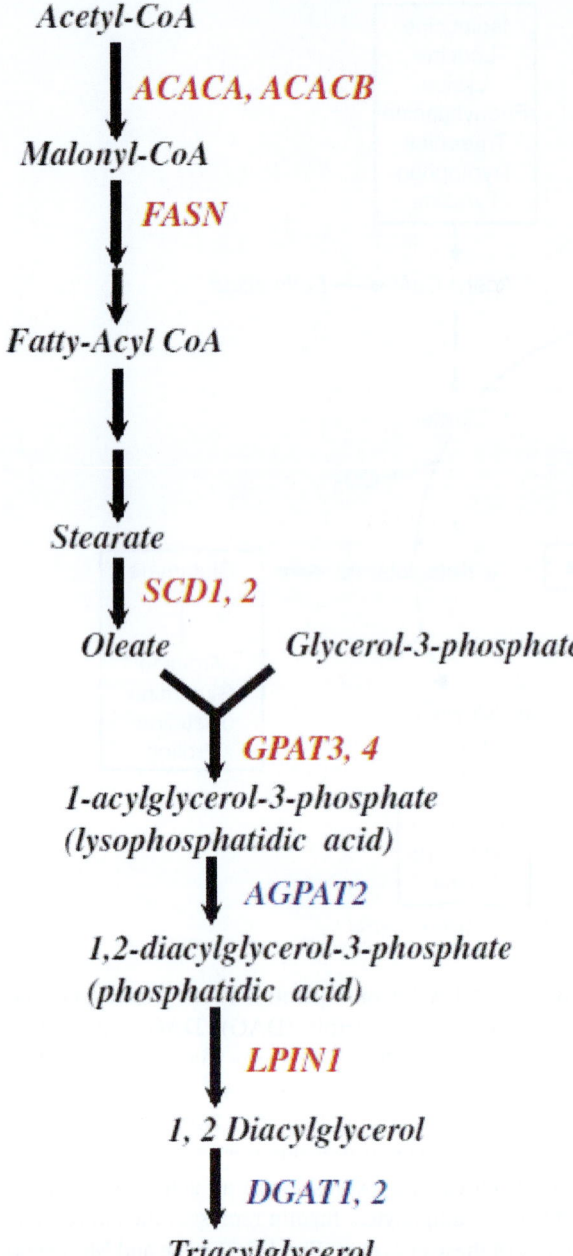

FIG. 1.7.5 GCH regulated genes in lipogenic pathway. Red indicates potential GCH primary target genes, whereas blue indicates genes regulated by glucocorticoids with no glucose binding region (GBR) is identified in or nearby their genomic regions. *(From Wang JC, Gray NE, Kuo T, Harris CA. Regulation of triglyceride metabolism by glucocorticoid receptor. Cell Biosci 2012;2(1):19. Fig. 1 p. 2.)*

elucidated regarding the in vivo functioning and relative contribution of these lipases to overall adipocyte lipolysis.

1.7.2.1.3 Inflammation and immune response

GCH's reduce lymphocyte numbers and increase neutrophil counts while eosinophil count falls. Inhibition of cytokine production from lymphocytes is mediated through inhibition of NF-κB (nuclear factor kappa light chain enhancer of activated B cells). GCH activate many antiinflammatory genes and repress proinflammatory genes that have been activated in inflammation (Escoter-Torres et al., 2019; Cain and Cidlowski, 2017; Barnes, 2011) (Table 1.7.1).

GCH switch on genes encoding β2 adrenergic receptors and the antiinflammatory proteins leukoprotease inhibitor and mitogen activated protein kinase phosphatase-1 (MKP-1) that inhibits the MAPK pathway. GCH act on macrophage production of cytokines such as interleukin-1b which is involved in T-cell activation. GCH also inhibit B-cell proliferation hence antibody production.

GCH **switches off inflammatory genes** that encode **cytokines, chemokines, adhesion molecules, inflammatory enzymes, and receptors** (Fig. 1.7.8). GCH thus have potent antiinflammatory and immunosuppressive properties, particularly when at high concentrations, but is also important in normal immune responses. GCH inhibit wound healing through suppression of proliferation and the inflammatory response.

GC's act through several mechanisms to exert antiinflammatory effects:

(1) nongenomic pathways involving GC receptor mediated direct interactions with second messenger proteins, including MAPK protein JNK, inhibiting the activation of this signaling pathway.
(2) GR-mediated transactivation of key inflammatory genes involves direct binding of both GR dimers and monomers/multimers to GC-response elements (GRE) in the promoter region of target gene.
(3) *trans*-Repressions of proinflammatory genes does not require direct DNA binding of GR but rather tethering of GR monomers to DNA bound proinflammatory transcription factors.

Glucocorticoids and the glucocorticoid receptor head a regulatory network that blocks several inflammatory pathways (Fig. 1.7.9). Glucocorticoids can inhibit prostaglandin production through three independent mechanisms:

1. the induction and activation of annexin I,
2. the induction of MAPK phosphatase 1, and
3. the repression of NF-kB and transcription of cyclooxygenase 2 (COX-2).

1. Annexin I (also called lipocortin-1) is an antiinflammatory protein that physically interacts with and inhibits cytosolic phospholipase A2a (cPLA2a). The activation of cPLA2a occurs with the movement of the enzyme from the cytosol to the perinuclear membrane, where it hydrolyzes phospholipids containing arachidonic acid. Cortisol can induce annexin I, which by inhibiting cPLA2a, blocks the release of arachidonic acid and its

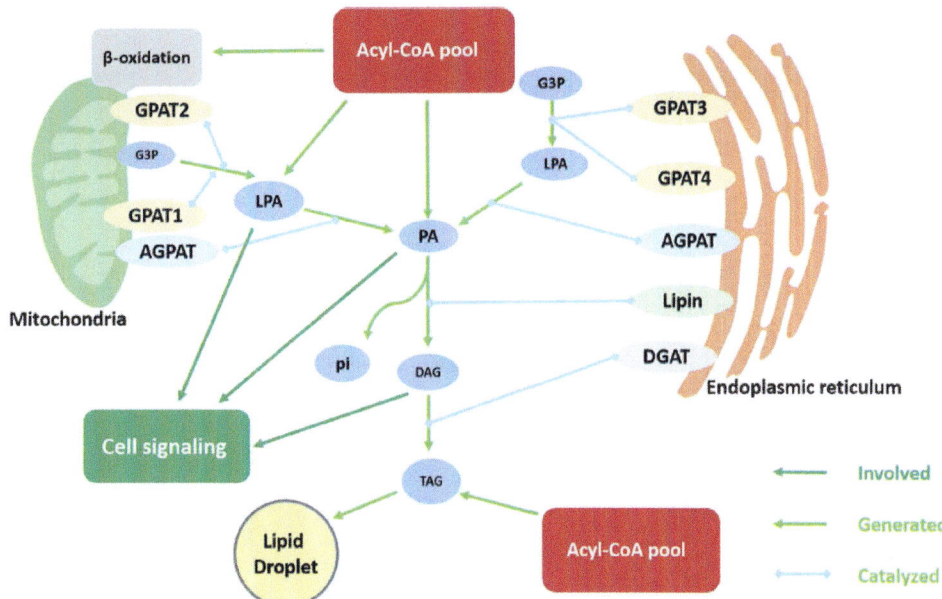

FIG. 1.7.6 GPAT and triacylglycerol synthesis in adipose tissue. Glycerophospholipid pathway. Glycerol-3-phosphate acyltransferase 1 (GPAT1) and GPAT2 are localized in the mitochondrial outer membrane, while GPAT3, GPAT4, phosphatidic acid phosphatase (PAP/lipin) and diacylglycerol acyltransferase (DGAT) are localized in the endoplasmic reticulum (ER) membrane. In addition, 1-acyl glycerol-3-phosphate acyltransferase (AGPAT) is localized in both the mitochondrial outer membrane and the endoplasmic reticulum (ER) membrane. *(From Yu J, Loh K, Song ZY, Yang HQ, Zhang Y, Lin S. Update on glycerol-3-phosphate acyltransferases: the roles in the development of insulin resistance. Nutr Diabetes 2018;8(1):34. Fig. 1 p. 2.)*

Lipolysis

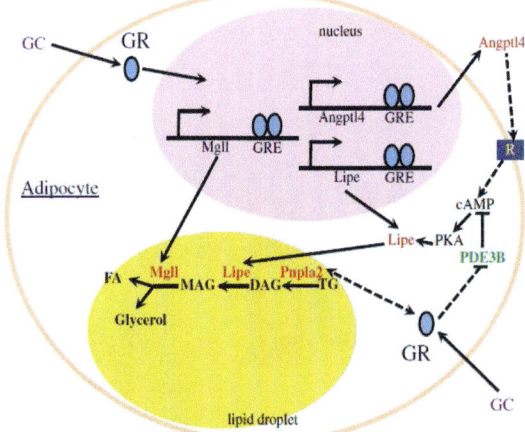

FIG. 1.7.7 Glucocorticoids promote lipolysis in adipocytes. Glucocorticoids (GC) activate the transcription of Lipe, Mgll and Angptl4 gene transcription. *(From Wang JC, Gray NE, Kuo T, Harris CA. Regulation of triglyceride metabolism by glucocorticoid receptor. Cell Biosci 2012; 2(1):19. Fig. 2 p. 4.)*

TABLE 1.7.1 Effects of glucocorticoids on transcription of genes in immune system.

1. *trans*-Activation—Increased transcription
   - β2-adrenoreceptors
   - Antiinflammatory cytokines:
     - Interleukin IL10, IL-12, IL-1
   - Inhibitor of NF-κB (INFκb-α)
   - GCH inducible leucine zipper (GILZ)
   - Lipocortin-1
   - Inhibits MAP kinase pathways:
     - Mitogen activated kinase phosphatase 1 (MKP1)

2. *trans*-Repression (decreased transcription)
   - Adhesion molecules:
     - Intracellular cell adhesion molecule ICAM-1,
     - vascular cell adhesion molecule VCAM-1
   - Chemokines—CCL1, CCL5, CCL11, CXXL8
   - Inflammatory cytokines:
     - Interleukin IL2,3,4,5,6,13,15;
     - tumor necrosis factor TNF-a,
     - Granulocyte macrophage colony stimulatory factor GM-CSF,
     - Stem cell factor (SCF)
     - Thymic stromal lymphopoeitin (TSLP)
   - Inflammatory enzymes
     - inducible nitric oxide synthse (iNOS),
     - inducible cyclo-oxygenase (COX-2)
   - Inflammatory peptides—Endothelin-1
   - Mediator receptors—neurokinin (NK1) bradyknin B2 receptors

subsequent conversion to prostaglandins. There is a strong correlation between basal and corticotropin-stimulated cortisol levels and the expression of annexin I in neutrophils in humans, but the clinical importance of annexin I as an antiinflammatory protein is unknown.

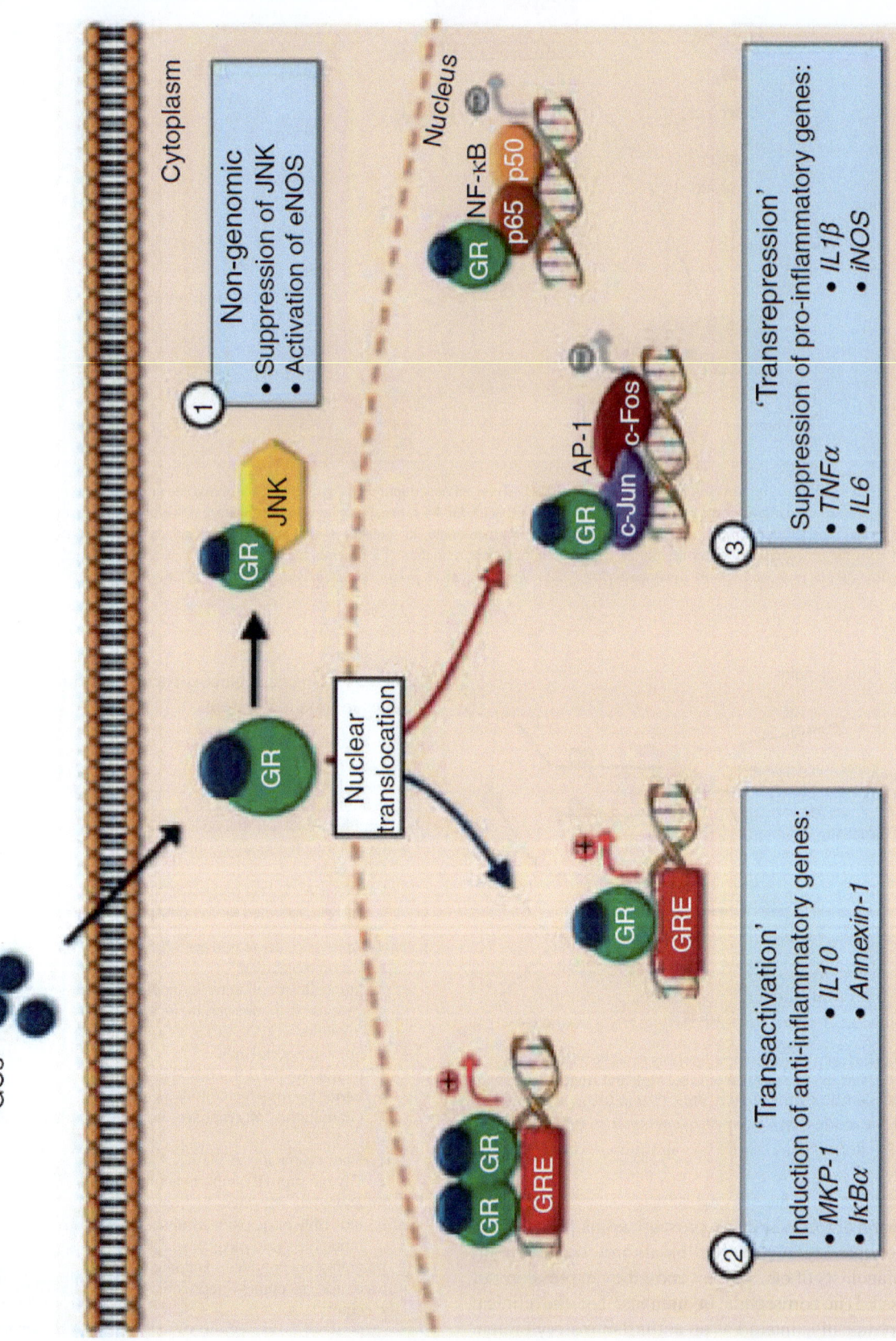

FIG. 1.7.8 Steroids and immune system. GGC effects on inflammatory signaling. *(From Nixon M, Upreti R, Andrew R. 5α-Reduced glucocorticoids: a story of natural selection. J Endocrinol 2012; 212(2):111–27. Fig. 1 p. 112.)*

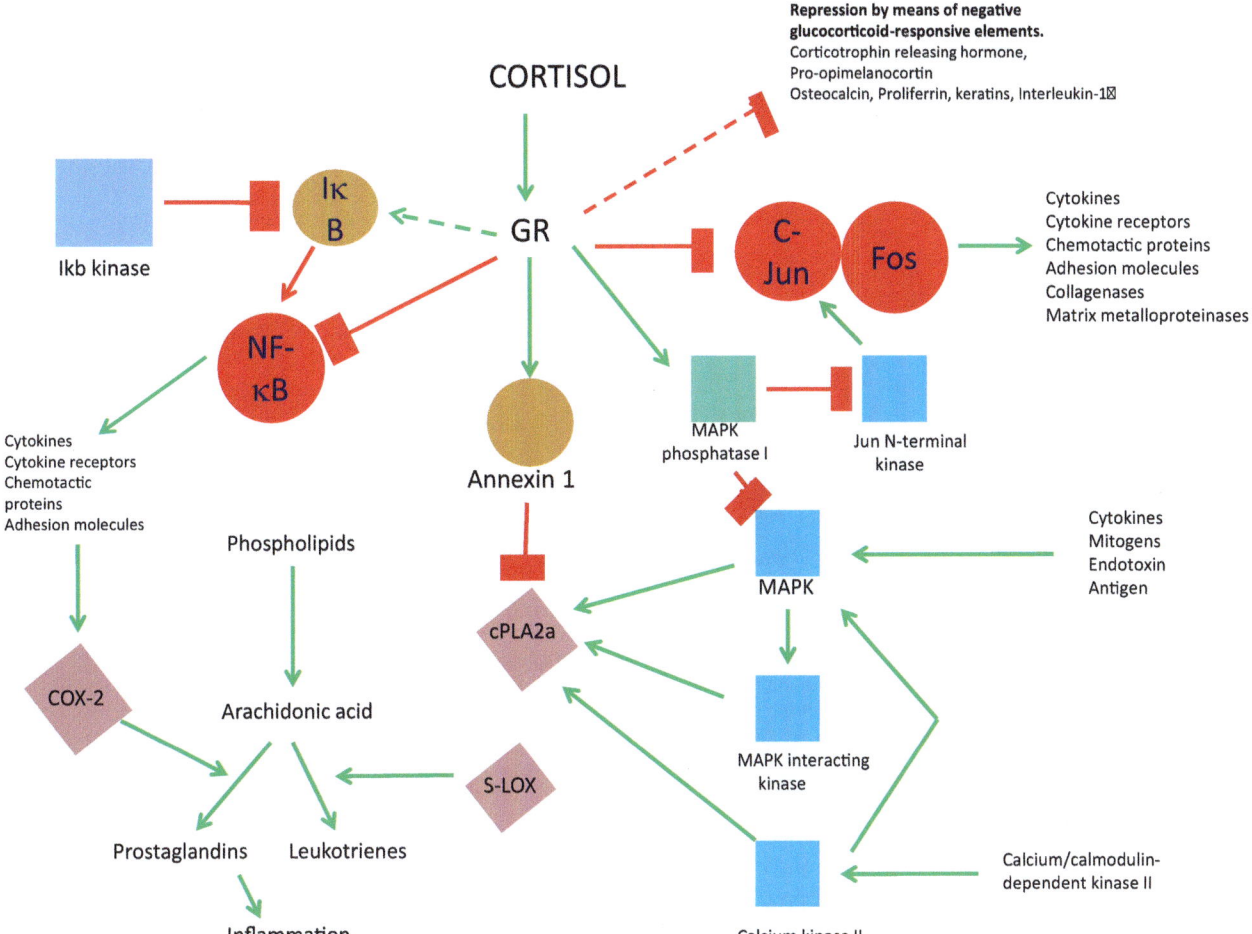

FIG. 1.7.9 Glucocorticoid induced antagonism and inflammation. Inflammatory pathways are characterized by positive feedback loops (i.e., cytokines activate NF-*k*B, which in turn stimulates the synthesis of more cytokines) and by redundancy (i.e., cytokines also activate c-Jun-Fos). The glucocorticoid receptor inhibits these pathways at multiple points by directly blocking the transcription of inflammatory proteins by NF-kB and activator protein 1 and by inducing the expression of antiinflammatory proteins such as IkB, annexin I and MAPK phosphatase I. 5-LOX denotes 5-lipoxygenase and COX-2 cyclo-oxygenase 2 (*red lines* denote inhibition and *green arrows* activation). *(From Rhen T, Cidlowski JA. Antiinflammatory action of glucocorticoids—new mechanisms for old drugs. N Engl J Med 2005;353(16):1711–23. Fig. 4 p. 1718.)*

2. MAPK phosphatase (MPK1) is a second antiinflammatory protein induced by glucocorticoids. MAPK phosphatase 1 also dephosphorylates and inactivates all members of the MAPK family of proteins. Cytokines, bacterial and viral infections and ultraviolet radiation are some of the inflammatory signals that activate MAPK cascades. MAPK phosphatase 1 may also inhibit cPLA2a activity by blocking its phosphorylation by MAPKs and MAPK-interacting kinase.

3. Glucocorticoid-induced antagonism of NF-kB and repression of COX-2 is the third mechanism for the inhibition of prostaglandin synthesis after the induction of the antagonists of cPLA2a, annexin I and MAPK phosphatase 1. The cortisol-glucocorticoid receptor complex also physically interacts with NF-kB to block its

transcriptional activity. In its inactive state, NF-kB is sequestered in the cytoplasm by an inhibitory protein named IkB. TNF-a, interleukin-1, microbial pathogens, viral infections and other inflammatory signals trigger signaling cascades that activate IkB kinases. Phosphorylation of IkB leads to its ubiquitination and degradation by the proteasome, unmasking a nuclear localization signal on NF-kB. In the nucleus, NF-kB binds DNA sequences called NF-kB elements and stimulates the transcription of cytokines, chemokines, cell adhesion molecules, complement factors and receptors for these molecules. NF-kB also induces the transcription of cyclooxygenase 2 (COX-2), an enzyme essential for prostaglandin production. Direct interactions between the glucocorticoid receptor and NF-kB probably

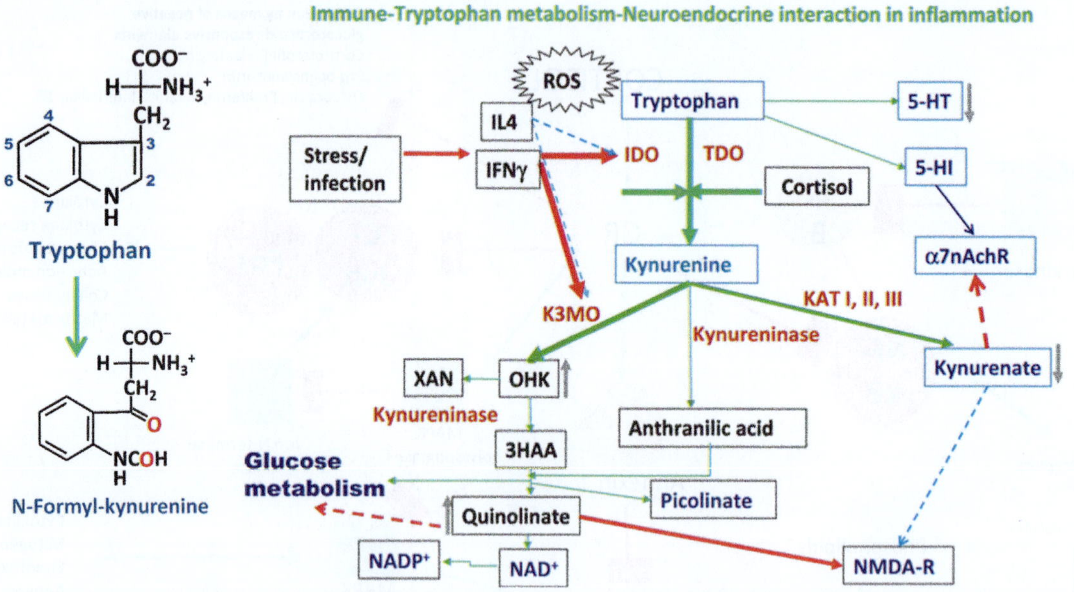

FIG. 1.7.10 Transformation of tryptophan to N-formylkynurenine *(green arrows)*. Tryptophan dioxygenase (TD) is induced by GCH and glucagon; the mechanisms involved are different, although the effects are at least partially additive. GCH cause induction of the new mRNA and protein synthesis, unlike the increase in activity observed in the presence of higher than normal amounts of tryptophan or heme. In response to the administration of the synthetic glucocorticoid, dexamethasone there is increased transcription of the rat liver tryptophan dioxygenase gene, resulting in a 10-fold increase in tryptophan dioxygenase mRNA in the liver. The *red arrows* indicate negative effects and the *blue arrows* indicate positive effect. *IDO*, indole oxygenase; *OHK*, 3-hydroxykynurenine; *HAA*, 3-hydroxyanthranilic acid; *KAT*, kynurenine aminotransferases. *(From Myint A.M. Kynurenines: from the perspective of major psychiatric disorders. FEBS J 2012;279(8): 1375–1385.)*

account for most of the inhibitory effects of glucocorticoids on NF-kB signaling. Despite the analogous nature of glucocorticoid receptor-mediated repression of activator protein 1 and NF-kB, different parts of the surface of the glucocorticoid receptor contact each transcription factor.

4. Ultraviolet light triggers a kinase cascade that phosphorylates and activates Jun N-terminal kinase, which in turn phosphorylates the transcription factor c-Jun. Glucocorticoid-induced MAPK phosphatase 1 dephosphorylates and inactivates Jun N-terminal kinase, thereby inhibiting c-Jun-mediated transcription. Cortisol and the glucocorticoid receptor directly interfere with c-Jun-mediated transcription.

Thus GCH action on the immune system is exerted partly through inhibition of **phospholipase A₂** (PLA₂) activity by reduced annexin and a consequent reduction in the release of arachidonic acid from membrane phospholipids GCH also inhibits NF-κB and cyclooxygenase (COX) required for prostaglandins and thromboxanes and lipo-oxygenase action for leukotriene production (Rhen and Cidlowski, 2005) and inducing antiinflammatory proteins.

Catabolism of the essential amino acid tryptophan is also involved in immunoregulation. **Tryptophan pyrrolase** (TP) is induced by GCH thus catalyzing the oxidation of tryptophan to N-formylkynurenine and further

reactions that result in **formation of nicotinamide adenine dinucleotide (NAD⁺)**. The first step of the pathway (Fig. 1.7.10) can be carried out by any of the three enzymes, indoleamine 2,3-dioxygenase (IDO or IDO1), IDO2 or tryptophan 2,3-dioxygenase (TDO) (Myint, 2012). Quinolinic acid, a downstream kynurenine pathway metabolite, can induce apoptosis in T-cells and also stimulate cytokine production and activation of the innate immune system (Sorgdrager et al., 2019).

There are complex interactions between the immune system and the kynurenine pathway metabolites. Recent studies have shown that kynurenine can regulate the immune system through highly specific mechanisms involving binding to the aryl hydrocarbon receptor. The immune modulating effects of GCH are exploited pharmacologically in the antiinflammatory effects of drugs with variable effects over time. Dexamethasone for example is a long-acting steroid whereas prednisone is an intermediate-acting steroid. The use of these synthetic GCH can lead to severe side effects including hypertension, diabetes, osteoporosis and glaucoma.

The effects of glucocorticoids on peripheral vasculature are tested by skin blanching tests, on the negative feedback of cortisol by testing the dexamethasone suppression of cortisol in saliva and suppression of the immune system by determining interleukin-6 and TNFα production by circulating leukocytes (Ebrecht et al., 2000). Such tests are useful when questioning resistance to cortisol in a patient.

1.7.2.2 Adrenal mineralocorticoid effects (aldosterone)

In the kidney, aldosterone regulates Na reabsorption and K secretion and, therefore, plays an important role in electrolyte and water balance, hence the title mineralocorticoid hormone (MCH) assignment acting through the mineralocorticoid receptor (MR) (Rossier et al., 2015). Aldosterone also has profound effects on acid-base balance (Yamazaki et al., 2019; Valinsky et al., 2019) in a regulated cycle (Fig. 1.7.11). The juxtaglomerular apparatus of the kidney secretes renin in response to a reduction in blood flow and low sodium concentrations.

Renin acts on angiotensinogen in the circulation to release angiotensin I that is cleaved to angiotensin II by the action of angiotensin converting enzymes (ACE). Angiotensin II acts on cells of the adrenal zona glomerulosa to secrete aldosterone. The distal nephron is the principal site of action of aldosterone to stimulate exchange of sodium and potassium through three mechanisms. Aldosterone also acts in sodium reabsorption in sweat glands, stomach, salivary glands and colon. Aldosterone causes the kidney to:

(1) increase sodium reabsorption back into the bloodstream to decrease loss of sodium in the urine,
(2) increase water reabsorption, and
(3) increase renal excretion of potassium.

Aldosterone acts as a classical hormone through nuclear receptors and by nongenomic pathways (Cannavo et al.,

2018). Aldosterone increases the transcription of the genes associated with the basolateral $Na^+/K^{\pm}ATPase$ and the apical epithelial sodium transport channel (ENaC), leading to Na^+ reabsorption (Fig. 1.7.12).

Na^+ transport through ENaC in the distal tubule produces a lumen-negative potential, providing a driving force for Cl^- reabsorption or K^+ excretion through renal outer medullary K^+ channel (ROMK) (Kamel et al., 2018; Palmer and Schnermann, 2015). Aldosterone also activates the transcription of the gene encoding serum kinase 1 (SGK1) that activates phosphorylation of its kinase domain. SGK1 increases ENaC numbers in the luminal membrane of cells in the distal nephron and also increases the ubiquitin ligase (NEDD4-2) (Valinsky et al., 2018). Aldosterone induces the production of ENaC activating protease (prostasin) in the apical membrane that mediates cleavage of the ENaC at specific sites within the extracellular loops of the α and γ units.

Since cortisol also binds to the MR, an important function in the distal nephron is the localization of 11β-hydroxysteroid dehydrogenase (HSD11B2) that oxidizes cortisol to cortisone that does not bind to the MR (see earlier Fig. 1.7.2). Liqourice was known for many years to cause hypertension and the action was assumed to be through the MR until it was found that glycyrrhetinic acid, the active ingredient, inhibited HSD11B2 (Stewart et al., 1987). In the brain, MR \may promote resilience to stress (Kanatsou et al., 2019) although more research is needed.

The retention of sodium by aldosterone leads to an increase in blood pressure. This is usually detected with a

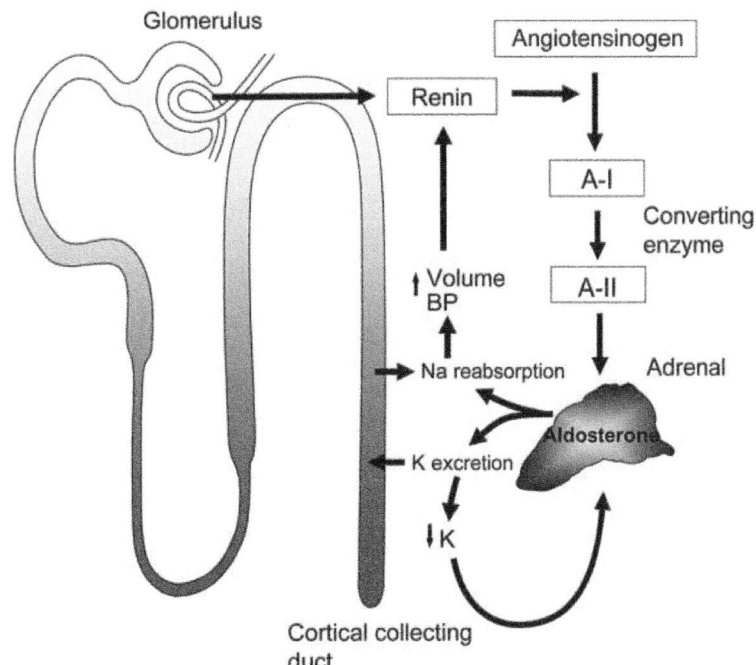

FIG. 1.7.11 Sodium retention and the renin-angiotensin-aldosterone pathway (RAAS).

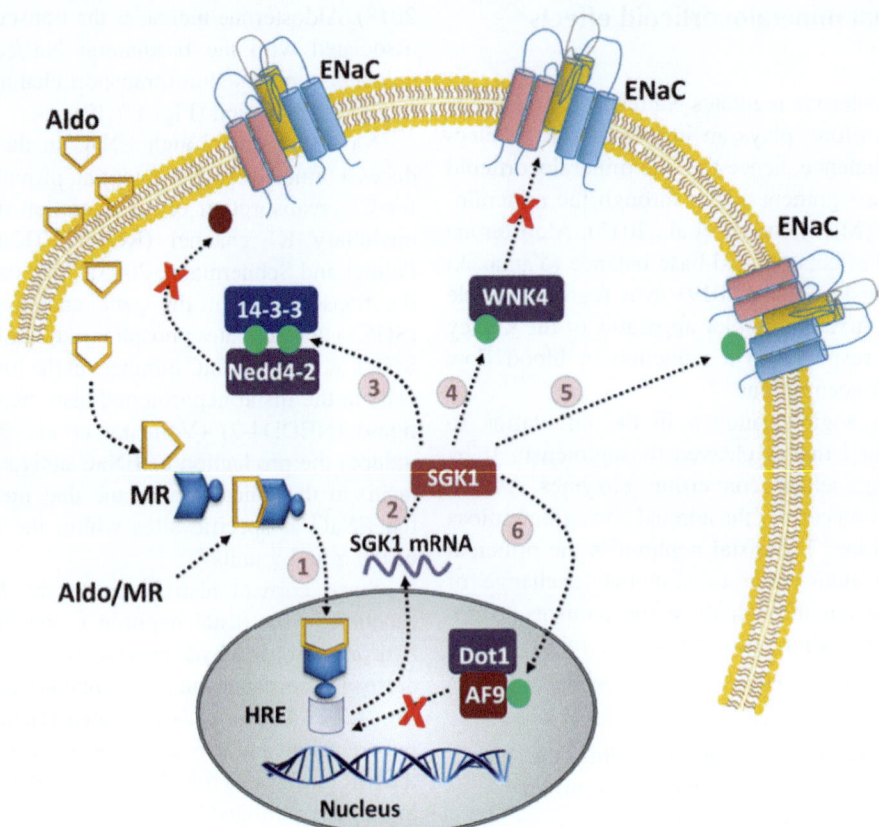

FIG. 1.7.12 Schematic of aldosterone interactions. Aldosterone cross the cell membrane and binds to the MR. The aldo/MR complex translocates to the nucleus, binds to specialized hormone response elements (HRE) and promotes the transcription of aldosterone regulated genes, including SGK1 which translated into protein SGK1. Newly synthesized SGK1 upregulates ENAC activity through several pathways that reduce ENaC ubiquitination through bi-phosphorylation of Nedd4-2, prevent ENaC endocytosis by phosphorylation of WNKK4, recruit silent ENaCs to active ones by direct phosphorylation and inhibit the transcriptional repressor complex Dot1a-AF9 via phosphorylation of AF9. *(From Valinsky WC, Touyz RM, Shrier A. Aldosterone, SGK1, and ion channels in the kidney. Clin Sci (Lond). 2018;132(2):173-183. Fig 1. P 175.)*

sphygmomanometer that includes an inflatable cuff, a measuring unit (mercury manometer) a mechanism for inflating the cuff and a stethoscope to detect 5 Korotkoff sounds indicators of the systolic and diastolic pressures. After inflating the cuff to a point where blood flow is not heard the pressure is slowly released, the first change in sound is at the systolic pressure and the fifth change is the diastolic pressure. Digital meters employ oscillometric measurements and electronic calculations rather than auscultation. It is important to choose the correct cuff for the patient. A cuff 4×8 cm is used for a newborn child, increasing to 6×12 cm and 9×18 cm in children; 12×22 for small adult and 16×42 large adult. Ambulatory monitoring of BP can be achieved by wearing a digital meter for 24 h and recording BP at 15 or 30 min intervals. This overcomes the patient sensitive to direct measurements by a clinician (white coat syndrome).

Blood pressure increases through childhood with boys and girls similar from newborn mean 90/90 to 105/70 at age 13 years when boys have higher ranges (Fig. 1.7.13).

There are further increases in adults over the next 50 years to mean 134/87 mmHg at 65 years.

1.7.3 Sex steroids

The sex steroids are produced in the adrenal cortex, the gonads and the periphery.

1.7.3.1 Adrenal androgens—Dehydroepiandrosterone sulfate (DHEAS)

The function of this adrenal androgen is not well understood apart from the maintenance of secondary sexual characteristics, particularly in the growth and maintenance of pubic hair. The appearance of pubic hair at 5–8 years of age is due to the increased production of DHEAS by the adrenal cortex. This is called the **adrenarche**. There are five stages in the development of pubic hair (Marshall and Tanner, 1969, 1970) (Fig. 1.7.14).

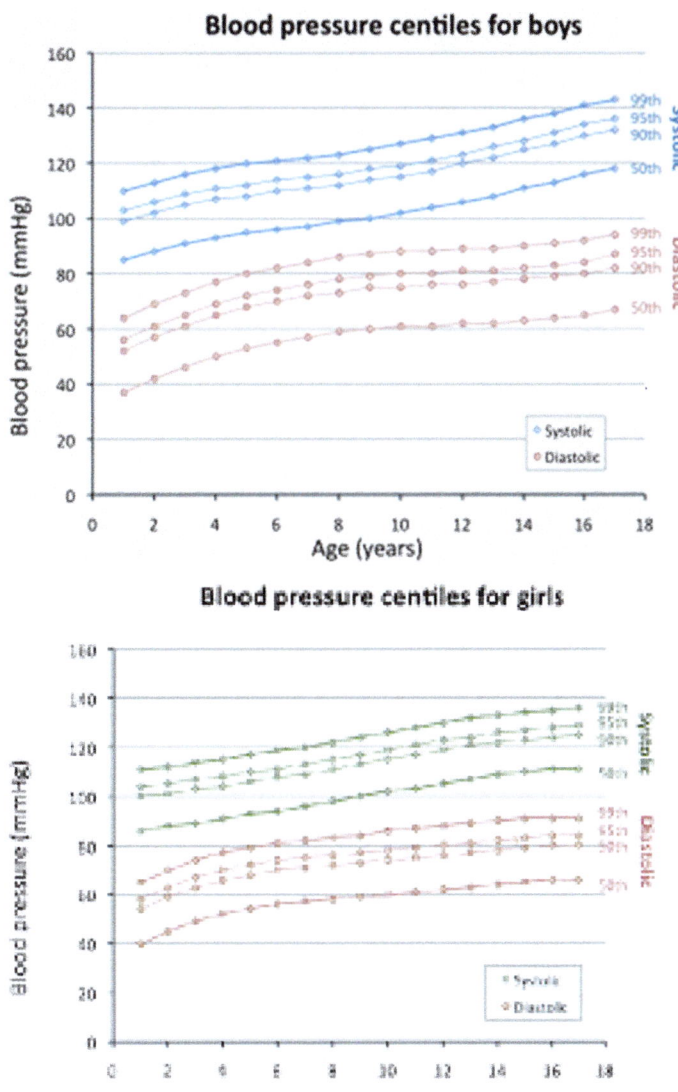

FIG. 1.7.13 Normal systolic blood pressure with age in children.

- Stage 1: Essentially there is no pubic hair
- Stage 2: There is sparse growth of long, pigmented, downy hairs. In girls, there are straight or slightly curved appearing along the female labia and in boys hairs appear at the base of the penis.
- Stage 3: Hair is considerably darker, coarser and more curled. The hair spreads over the junctions of the pubes.
- Stage 4: Hair is no adult in type but the area covered is smaller than in most adults.
- Stage 5: Adult in quantity and type distributed as an inverse triangle of the feminine pattern, spreading to the medial surface of the thighs.

DHEAS, and the androgens in general, may also be involved in the regulation of bone mineral density, muscle mass and may be have beneficial actions against type 2 diabetes and obesity. DHEAS is taken up into cells by organic anion transporter systems. Sulfatase enzymes in the cell then releases free DHEA which can be further metabolized to androstenedione, testosterone and dihydrotestosterone (see Chapter 1.6). These steps may be relevant to testis and ovary as well as peripheral tissues such as adipose tissues. In pregnancy, fetal DHEAS is the precursor for estrogen synthesis. Early growth in childhood was associated, in two large studies, with low birth weight and higher adrenal androgen exposure (Honour et al., 2007; Ong et al., 2004). In adults, androgens may play a role in the menstrual cycle during decidualization to increase endometrial receptivity and hence be an additive treatment for infertility.

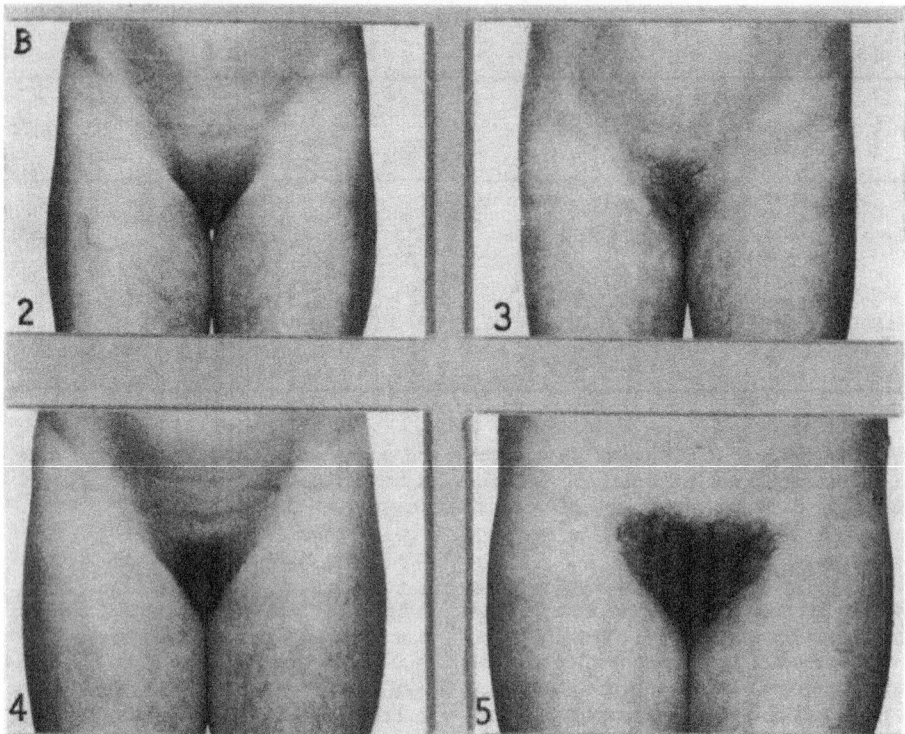

FIG. 1.7.14 Development of pubic hair is mainly through the action of DHEA. Pubic hair ratings from 1 to 5. *(From Marshall WA, Tanner JM. Variations in pattern of pubertal changes in girls. Arch Dis Child. 1969;44(235):291-303. Fig 2 p. 293.)*

1.7.3.2 Gonadal sex steroids

1.7.3.2.1 Estradiol

Estradiol is synthesized in granulosa cells under FSH control. Estradiol has physiological functions in the reproductive system but also acts on tissues around the body such as the vascular and skeletal systems. At puberty, estrogen stimulates enlargement of the uterus, ovaries and vagina. In the adult, female estrogens are essential in maintaining the menstrual cycle. The nonreproductive functions of estrogens are in cognition and energy balance, bone mass maintenance, pancreatic cell function, lipid and glucose homeostasis, wound healing, cardiovascular health and adipose health.

Ovary

The ovary can be visualized with abdominal ultrasound. Follicles can be counted and the development of follicles to a dominant state can be seen. A polycystic ovary has a ring of follicles of 5–7 mm around the periphery of the ovary. There are many clinical features to PCOS including hyperandrogenism.

Uterus

The uterus grows and changes shape in puberty from a tubular to a pear shape (Kelsey et al., 2016). *trans-*Abdominal pelvic ultrasound or MRI is safe technique to determine the size and configuration of the uterus (Hagen et al., 2015; Gilligan et al., 2019). The increase in uterine volume at puberty (Fig. 1.7.15) follows similar paths to other sites of estradiol action.

Endometrium

The endometrium can be visualized with sonography. The endometrium thickness is usually between 0.5 and 5 mm. In a classical 28-day menstrual cycle, ovulation occurs on day 14. Histological changes can be seen throughout the cycle due to the changes in steroid hormones (Fig. 1.7.16).

On cycle day 17, the endometrial glands become more tortuous and dilated. On cycle day 18, the vacuoles in the epithelium decrease in size and are frequently located next to the nuclei. Also, glycogen is now found at the apex of the endometrial cells. By cycle day 19, vacuolation almost completely disappears. On cycle day 21 or 22, the endometrial stroma begins to become edematous. On cycle day 23, stromal cells surrounding the spiral arterioles begin to enlarge and stromal mitoses become apparent. On cycle day 24, predecidual cells appear around the spiral arterioles and stromal mitoses become more apparent. On cycle day 25, the predecidua begins to differentiate under the surface epithelium. On cycle day 27, there is a marked lymphocytic infiltration and the upper endometrial stroma appears as a solid sheet of well-developed decidua-like cells. On cycle day 28, menstruation begins.

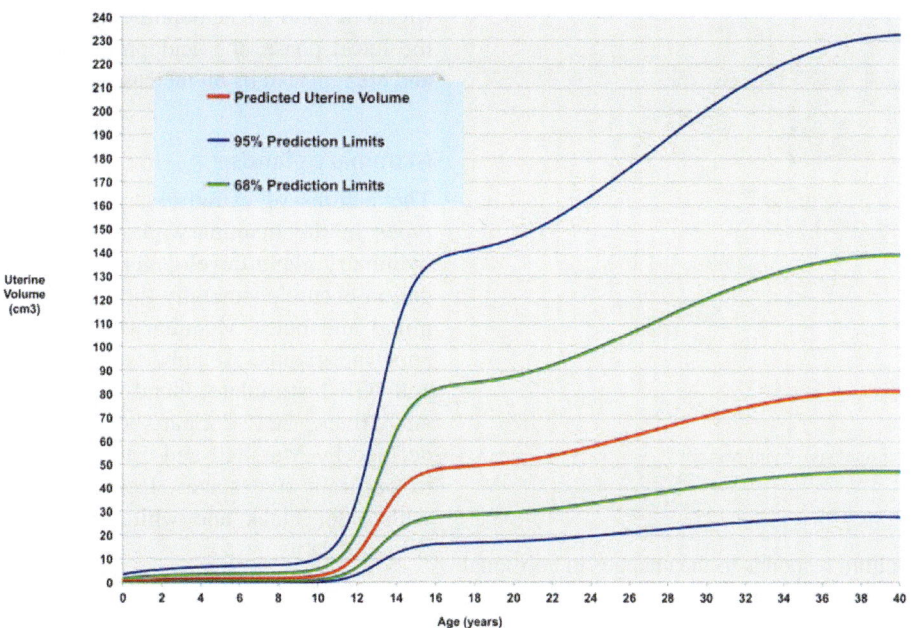

FIG. 1.7.15 Uterine volume. Log-adjusted data and normative model. The observed volumes (*blue points*) are shown with the predicted volume (*red line*), the 95% confidence interval for the predicted volume (*brown lines*) and the 95% prediction limits for the model (*purple lines*). (*From Kelsey TW, Ginbey E, Chowdhury MM, Bath LE, Anderson RA, Wallace WH. A validated normative model for human uterine volume from birth to age 40 years. PLoS ONE 2016;11(6):e0157375.*)

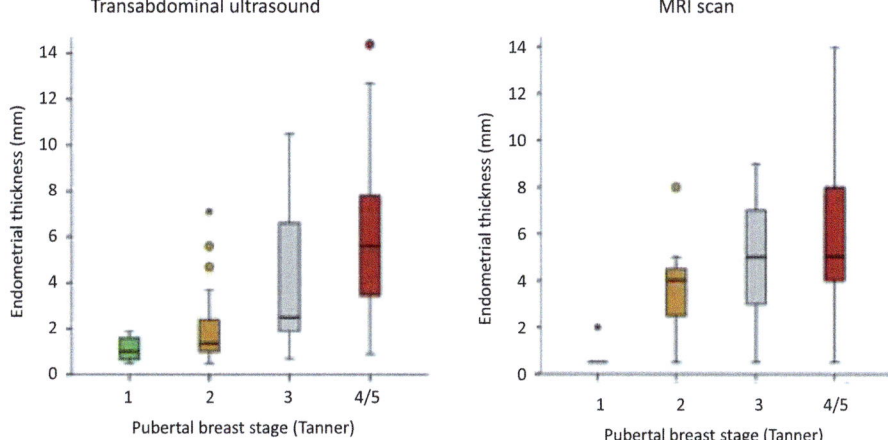

FIG. 1.7.16 Endometrial thickness by ultrasound and MRI according to pubertal breast stages. 25–75th percentile in boxes and 2.5 to 97.5th percentile (hinge). (*From Hagen CP, Mouritsen A, Mieritz MG, Tinggaard J, Wohlfahrt-Veje C, Fallentin E, Brocks V, Sundberg K, Jensen LN, Juul A, Main KM. Uterine volume and endometrial thickness in healthy girls evaluated by ultrasound (3-dimensional) and magnetic resonance imaging. Fertil Steril 2015;104(2):452–9.e2.*)

Cervix

Immediately after menstruation, the cervical mucous is scant and viscous. During the late follicular phase, under the influence of rising estradiol levels, the cervical mucous becomes clear, copious and elastic. A 30 fold increase in the quantity of cervical mucous occurs compared with the early follicular phase. The stretchability or elasticity of the

cervical mucous can be evaluated between two glass slides (Cohen et al., 1952) (Fig. 1.7.17).

When examined under the microscope, the cervical mucous will display a characteristic ferning or palm-leaf arborization appearance. After ovulation, as progesterone levels rise, the cervical mucous once again becomes thick, viscous and opaque and the quantity produced by the endo-cervical cells decreases.

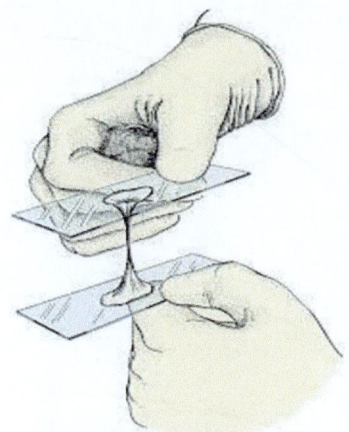

FIG. 1.7.17 Cervical mucous—Spinnbarkeit.

Vagina

The vaginal epithelium responds to the changes in hormonal levels of estrogen and progesterone. During the early follicular phase, vaginal epithelial cells have vesicular nuclei and are basophilic. In the late follicular phase, in response to the rising estradiol levels, the **nuclei of vaginal epithelial cells** become dense and compact and begins to fragment (karyorrhexis) resulting in spheres of dark-staining nuclear chromatin

(pyknotic) and are acidophilic. As progesterone rises during the luteal phase, the acidophilic cells decrease in number and are replaced by an increasing number of leukocytes.

Mammary glands

The actions of estradiol are mediated by specific cell types in the mammary glands. Estradiol is critical for mammary gland development and function. Ductal elongation occurs postnatally through the proliferation of terminal and buds. Mammary tissues arise from modified apocrine glands. At puberty estrogens, growth hormone and IGF-1 stimulate ductal breast development (Macias and Hinck, 2012). Charts of breast development were defined by Marshall and Tanner more than 50 years ago to occur in five stages. Recent images are comparable with older black and white images (Sun et al., 2012) (Fig. 1.7.18)

Stage 1: Preadolescent
Stage 2: Breast bud stage, elevation of breast as a small mound, enlargement of areola diameter
Stage 3: Further enlargement of breast and areola
Stage 4: Formation of a secondary mound above the level of the breast

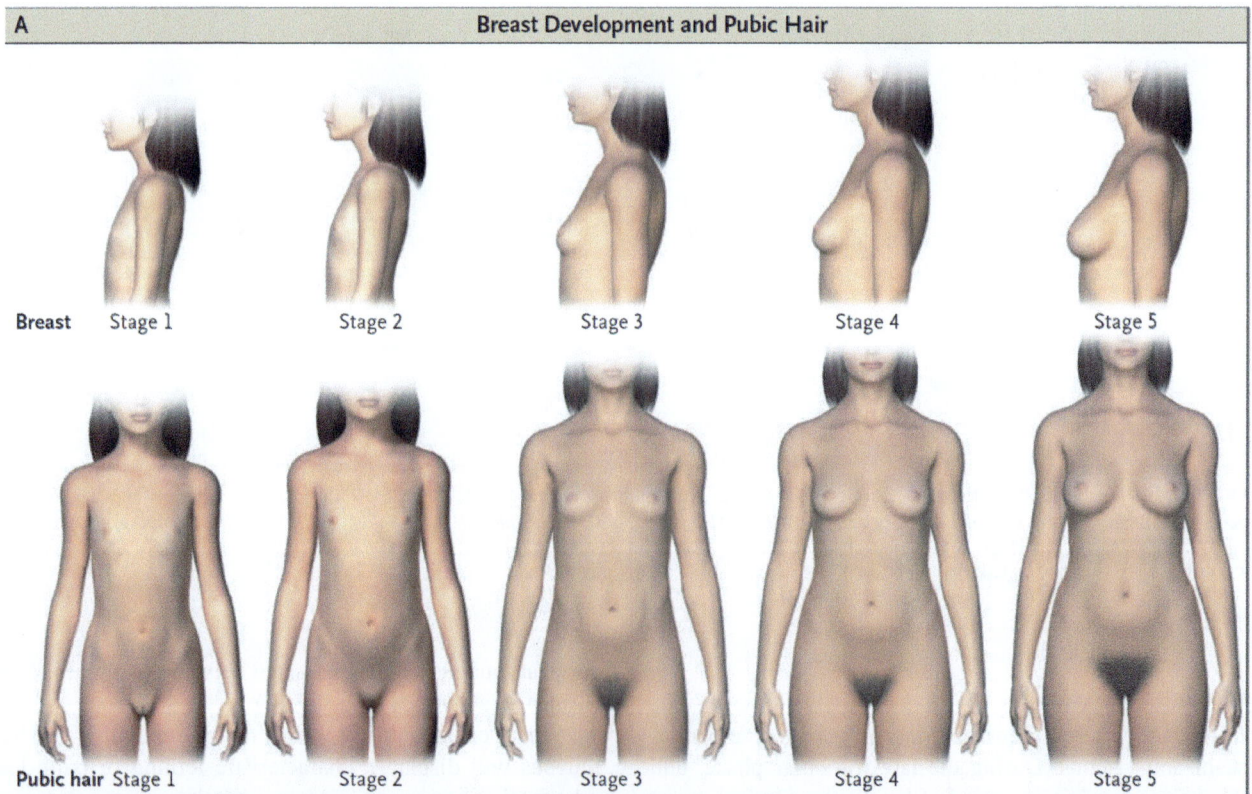

FIG. 1.7.18 Stages of breast development and pubic hair. *(From Sun Y, Tao FB, Su PY; China Puberty Research Collaboration. Self-assessment of pubertal Tanner stage by realistic colour images in representative Chinese obese and non-obese children and adolescents. Acta Paediatr 2012;101 (4):e163–6. Suppl. Fig. 2.)*

Stage 5: Mature stage, projection of papilla only, recession of the areola to the general contour of the breast

Ultrasound is now used for evidential documentation of breast development (Bruserud et al., 2020).

Skeletal system

From birth, the bones change in size and shape with an acceleration of growth rate at puberty (10 years in girls and 12 years in boys). The height, weight and head circumference of a child can be measured and compared with published standards which are gender specific (Fig. 1.7.19).

Growth is mainly in the long bones (femur, tibia, fibula, humerus, and radius) and the phalanges of the fingers and toes. The legs account for nearly half of the total height in the adult. The distance of the knee to the floor (knemometry) has also been charted for normal children. From the growth chart, a final height can be predicted and this relates to the heights of the parents. Girls stop growing around 16 years of age and boys at 18 years. The bones grow near the ends at the epiphyses which slowly calcify.

Girls have maximum height spurt on average at 12 years which precedes menarche 13 years (Fig. 1.7.20). There is rapid breast development through stages 2, 3, and 4 at 11, 12, and 13 years. Pubic hair appears before 11 years of age.

Estradiol is formed from testosterone in granulosa cells by the action of aromatase. Most of the principal physiological events that take place in the developing and mature

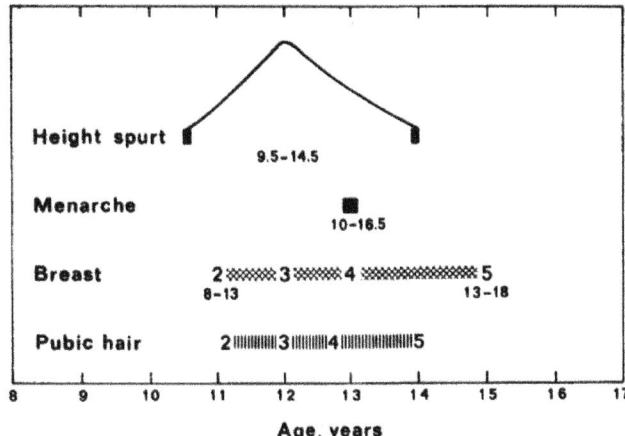

FIG. 1.7.20 Sequence of events with age at puberty for average girls. *(From Marshall WA, Tanner JM. Variations in the pattern of pubertal changes in boys. Arch Dis Child. 1970;45(239):13–23. Fig 8 p. 22.)*

male bone are now considered to be under the control of estrogen (Rochira et al., 2015). Estrogen determines the acceleration of bone elongation at puberty, epiphyseal closure, harmonic skeletal proportions, the achievement of peak bone mass and the maintenance of bone mass (Russell and Grossmann, 2019). Furthermore, it seems to crosstalk with androgen even in the determination of bone size, a more androgen-dependent phenomenon. At puberty, epiphyseal closure and growth arrest occur when a critical number of estrogens are reached.

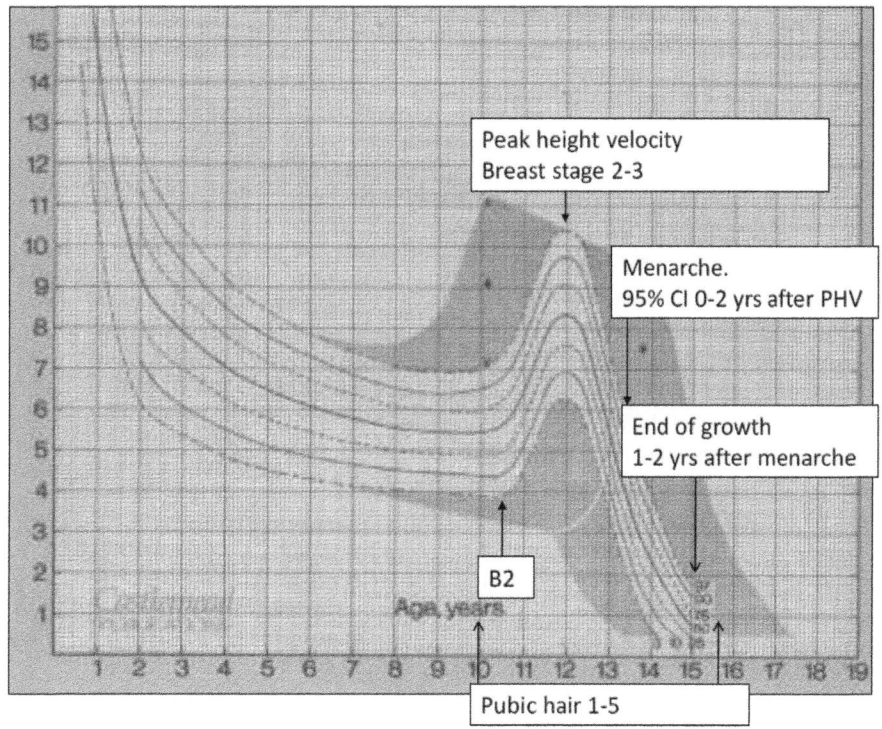

FIG. 1.7.19 Growth chart of girls.

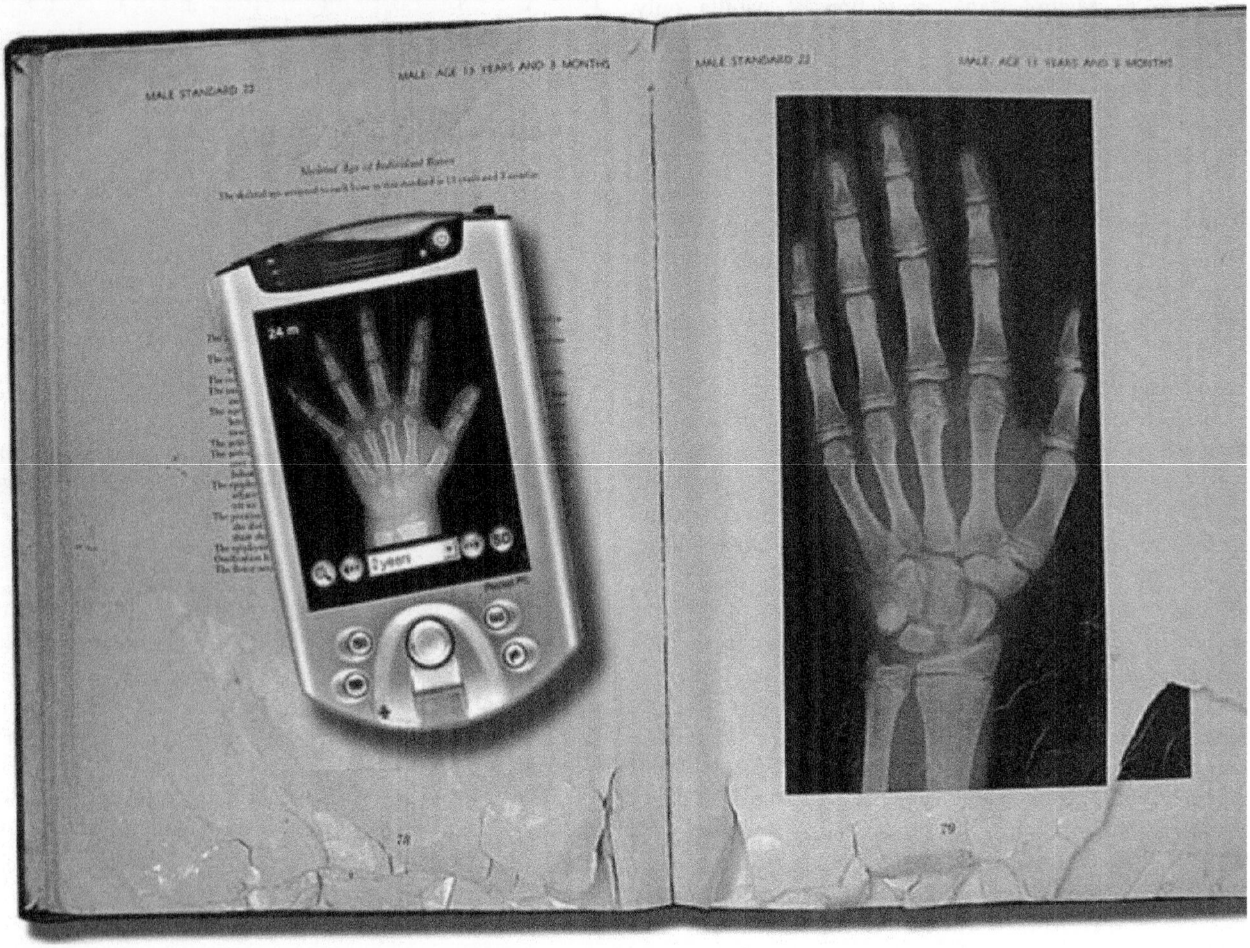

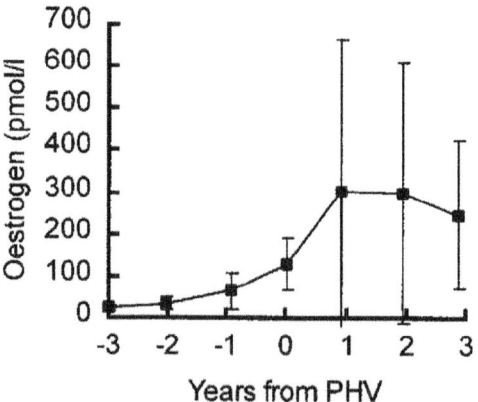

FIG. 1.7.21 Comparison of the traditional Greulich and Pyle atlas used for determination of bone maturity from hand radiographs and the electronic alternative, a digital atlas of "idealized" hand radiographs that can be reviewed on standard hand-held PDAs. *(From Gilsanz V, Ratib O. Hand bone age—a digital atlas of skeletal maturity. Heidelberg: Springer Verlag; 2005. ISBN: 978-3-540-27070-6. Fig. 2.2 Page 7.)*

Estradiol is a factor in growth

In pediatrics, an X-ray of the hand and wrist is taken and compared with images in an atlas created by Greulich and Pyle (1959) (Fig. 1.7.21) and still used today although with the application of artificial intelligence interpretation is now being used to report the results. In childhood, carpal and tarsal bones appear, separated by a layer of cartilage.

Sex steroids influence body strength and this was studied across puberty in groups of boys and girls followed longitudinally. PHV was determined from the peak in alkaline phosphatase. Estradiol in plasma increased with puberty (Fig. 1.7.22) (Round et al., 1999) after PHV.

Estrogens inhibit the functions of osteoblasts, inhibits the effects of parathyroid hormone leading to bone loss through a reduction in the rate of bone resorption. Estrogens are important to bone growth in males as well as females (Kalkwarf et al., 2007). Estradiol acting via the ERα receptor appears to be the principal sex steroid involved

FIG. 1.7.22 Plasma estradiol in relation to age of peak height velocity (PHV). Values are given as mean SD of the values for girls in each year for whom the age of PHV could be clearly identified. *(From Round JM, Jones DA, Honour JW, Nevill AM. Hormonal factors in the development of differences in strength between boys and girls during adolescence: a longitudinal study. Ann Hum Biol 1999;26(1):49–62. Fig. 2 p. 53.)*

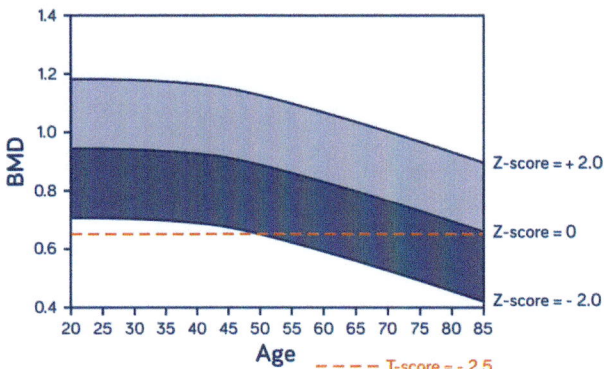

FIG. 1.7.23 Bone mineral density with age (Royal Osteoporosis Society).

in bone growth and maintenance of bone density and microstructure. Bone densitometry by dual energy X-ray absorptiometry (DEXA scan) is used to determine bone density which slowly increases to a peak at 30–40 years of age then declines (Fig. 1.7.23).

In children, interpretation of DEXA data is complicated by the third dimension of growth (Khalatbari et al., 2021).

Central nervous system

Estrogens modulate neuropeptides and neurotransmitters in particular estrogen increases serotonin synthesis thus acting as a serotoninergic agonist to affect mood positively.

Cardiovascular system

Estrogen receptors are found on the smooth muscle cells of coronary arteries and endothelial cells. Estrogen causes short term vasodilation by stimulating nitric oxide and prostacyclin release form endothelial cells. Cardiovascular incidence rates are lower in premenopausal than postmenopausal women.

1.7.3.2.2 Testosterone

Male development in utero is partly dependent on testosterone from the eighth week of gestation as described in Chapter 1.2 along with a Y-chromosome and antimullerian hormone. At puberty, testosterone is important for secondary sex characteristics.

The Leydig cells are responsible for testosterone synthesis in the male under LH control. Some development of Sertoli cells takes place in the mini-puberty period but is more obvious in puberty (Koskenniemi et al., 2017). This accounts for some increase in testicular volume. Inhibin B is secreted by Sertoli cells and AMH is lost due to the high intratesticular testosterone concentrations.

The most prevalent male sex hormone is testosterone, but the reduced form dihydrotestosterone (DHT) is more biologically active and is are necessary for normal

spermatogenesis and development, secondary sex characteristics, increased muscle mass, bone mass, fat distribution, red blood cell production, sexual function, sweat, body hair, and decreased risk of osteoporosis (Imperato-McGinley and Zhu, 2002) (Fig. 1.7.24). Some testosterone is converted to estrogen through the action of aromatase.

The process of fetal sex differentiation involves the establishment of genetic sex, differentiation of the gonads and the development of phenotypic sex (see Chapter 1.2). At around the eighth week of gestation, the testes develop and secrete testosterone. From studies of patients with genetic orders of steroid synthesis, dihydrotestosterone is now known to have more potent effects than testosterone on the development of the male reproductive tract (Fig. 1.7.25). Testosterone is important for development of the vas deferens and seminal vesicles but DHT affects the prostate and growth of the penis.

The size of the testes and penis increase during puberty through 5 stages (Fig. 1.7.26).

Stage 1: Preadolescent (penis stretched length 4 cm, testes 2 mL)
Stage 2: The scrotum and testes (3-4 mL) have enlarged, some change in texture and reddening of the scrotum (9–14 years)
Stage 3: Lengthening (mean 6 cm) and broadening of penis (11–15 years)
Stage 4: Further growth of penis (mean 8 cm), testes (mean 15 mL) and scrotum with development of the glans (11–16 years)
Stage 5: Genitalia adult in size (penis 9 cm long; testes mean 20 mL) (12–17 years)

The pubic hair pattern goes through 5 stages.

Stage 1: Preadolescent. The velus over the pubes is no further developed than that over the abdominal wall, i.e. no pubic hair. Stage 2: Sparse growth of long, slightly pigmented, downy hair, straight or only slightly curled, appearing chiefly at the base of the penis. This stage is difficult to see on photographs, particularly of fair-haired subjects. Stage 3: Considerably darker, coarser and more curled. The hair spreads sparsely over the junction of the pubes. This and subsequent stages were clearly recognizable on the photographs. Stage 4: Hair is now adult in type, but the area covered by it is still considerably smaller than in most adults. There is no spread to the medial surface of the thighs. Stage 5: Adult in quantity and type, distributed as an inverse triangle of the classically feminine pattern. Spread to the medial surface of the thighs but not up the linea alba or elsewhere above the base of the inverse triangle.

Pubic hair is not easily seen in photographs but starts at around 8 years of age in boys and continues through puberty. Androgens act at androgen receptors in hair and skin

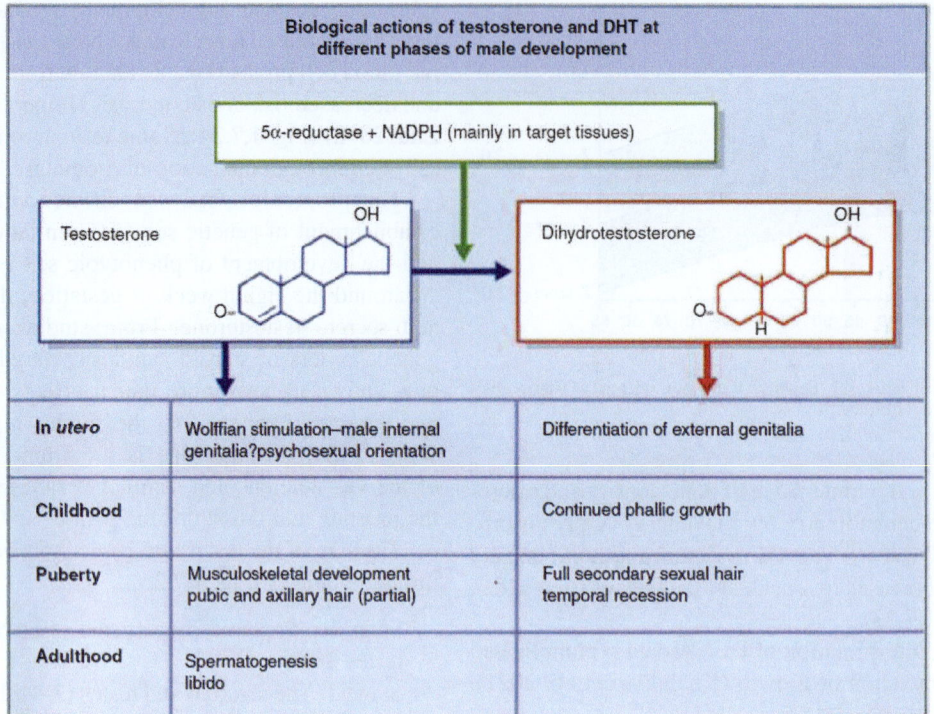

FIG. 1.7.24 Dihydrotestosterone is more active than testosterone. *(From Belchetz PE, Barth JH, Kaufman JM. Biochemical endocrinology of the hypogonadal male. Ann Clin Biochem 2010;47(Pt 6):503–15. Fig. 3 p. 505.)*

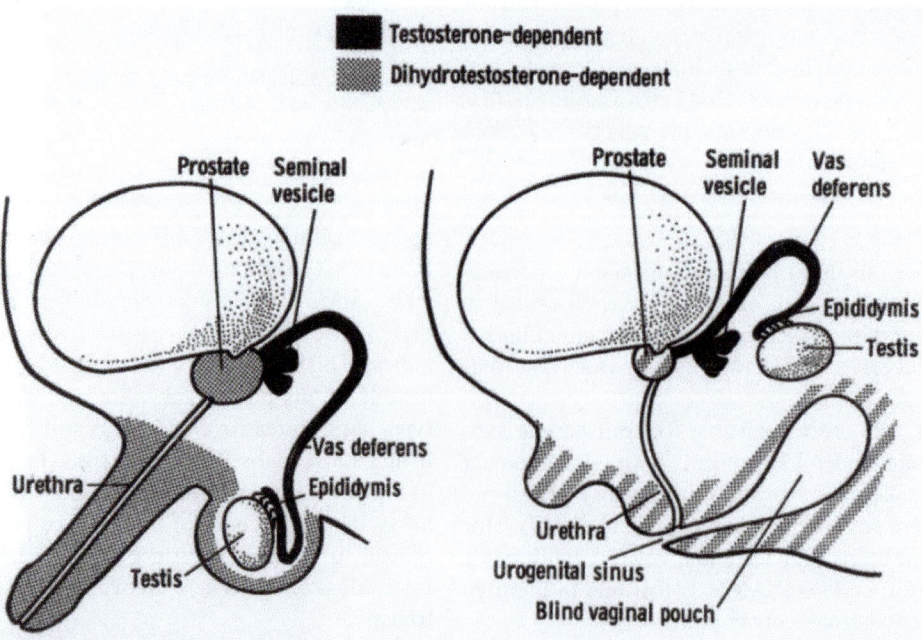

FIG. 1.7.25 Schematic for effects of testosterone and dihydrotestosterone on male development in the fetus. The diagram illustrates the roles of testosterone and DHT in male sexual differentiation in utero. *(From Imperato-McGinley J, Zhu YS. Androgens and male physiology the syndrome of 5alpha-reductase-2 deficiency. Mol Cell Endocrinol 2002;198(1-2):51–9. Fig. 1 p. 53.)*

B | **Genital Development and Pubic Hair**

| Stage 1 | Stage 2 | Stage 3 | Stage 4 | Stage 5 |

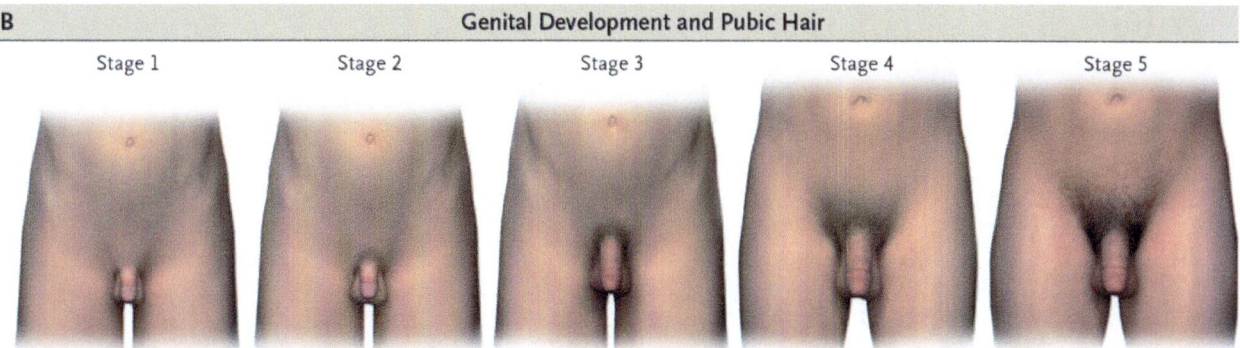

FIG. 1.7.26 Pubic hair and genital development of boys in puberty. Testis growth starts around 10 years of age and continues to 18 years of age, growth of the penis and height continues from 10 to 17 years of age with maximum growth spurt at 14 years of age. *(From Sun Y, Tao FB, Su PY; China Puberty Research Collaboration. Self-assessment of pubertal Tanner stage by realistic colour images in representative Chinese obese and non-obese children and adolescents. Acta Paediatr 2012;101(4):e163–6. Suppl.)*

(Ceruti et al., 2018; Condorelli et al., 2018) leading to hair growth and sebum production. Cells in the skin are capable of steroidogenesis using cholesterol and DHEAS as substrates.

Testicular size can be assessed clinically by comparing the size of the testes in one hand to a string of modeled testes in the other hand (Prader orchidometer) (Fig. 1.7.27). This assessment is subjective and varies between observers.

The size can be reliably measured with ultrasound or MRI scanning (Oehme et al., 2020; Madsen et al., 2020; Ankarberg-Lindgren et al., 2011) (Fig. 1.7.28). This of course produces data that can be filed with patient notes.

Sex steroids influence growth with testosterone being more potent than estradiol (Round et al., 1999) (Fig. 1.7.29). Further measurements in children of quad strength and biceps strength showed significantly greater strength in boys than girls after peak height velocity had been reached (Fig. 1.7.30) that were attributed to testosterone. In boys, the greater biceps strength was also influenced by longer upper arm length (data not shown here).

Peak height velocity boys are 2 years later than seen in girls (Fig. 1.7.31) compared with Fig. 1.7.18.

The body shape of the male also gains body mass in comparison with the female (Fig. 1.7.32) with broadening of the shoulders and hair growth on face and torso is seen.

The HPG axis is operational in girls around 2 years before boys (Fig. 1.7.33). Breast development and adrenarche in girls are earlier than growth of scrotum and testes and pubic hair growth in boys.

Testosterone in puberty produces many of the male secondary sex characteristics such as broad shoulders, square jaw and greater height, which become obvious cues to the male sex. Testosterone and related androgens develop bones structure in the facial bones like the jaw and brow ridge and lengthen the vocal cords. The tissues of the larynx are rich in androgen receptors. T causes the diameter of the larynx to increase and the vocal folds to thicken and lengthen (Zamponi et al., 2021). The male voice has a lower pitch and greater intensity partly due to larger lungs in a broader thorax.

During puberty, androgens surge via the hypothalamic-pituitary-gonadal axis. These high levels of androgens promote muscle growth and increase the myonuclei number per myofiber leading to an increase in muscle mass during puberty. However, after puberty, muscle mass and myonuclei number per myofiber reach a steady state and remain stable during adulthood (Seo et al., 2019).

A number of other measures of androgen activity can be quantified:

- Finger ratio—2D to 4D ratio fingers (Fig. 1.7.34). Exposure of the fetus to androgens is linked to 2D:4D ratios (Manning et al., 1998). This is controversial so that compared with normal children, patients with congenital adrenal hyperplasia (high androgens) (Brown et al., 2002) have an increase in the ratio and

FIG. 1.7.27 Prader orchidometer.

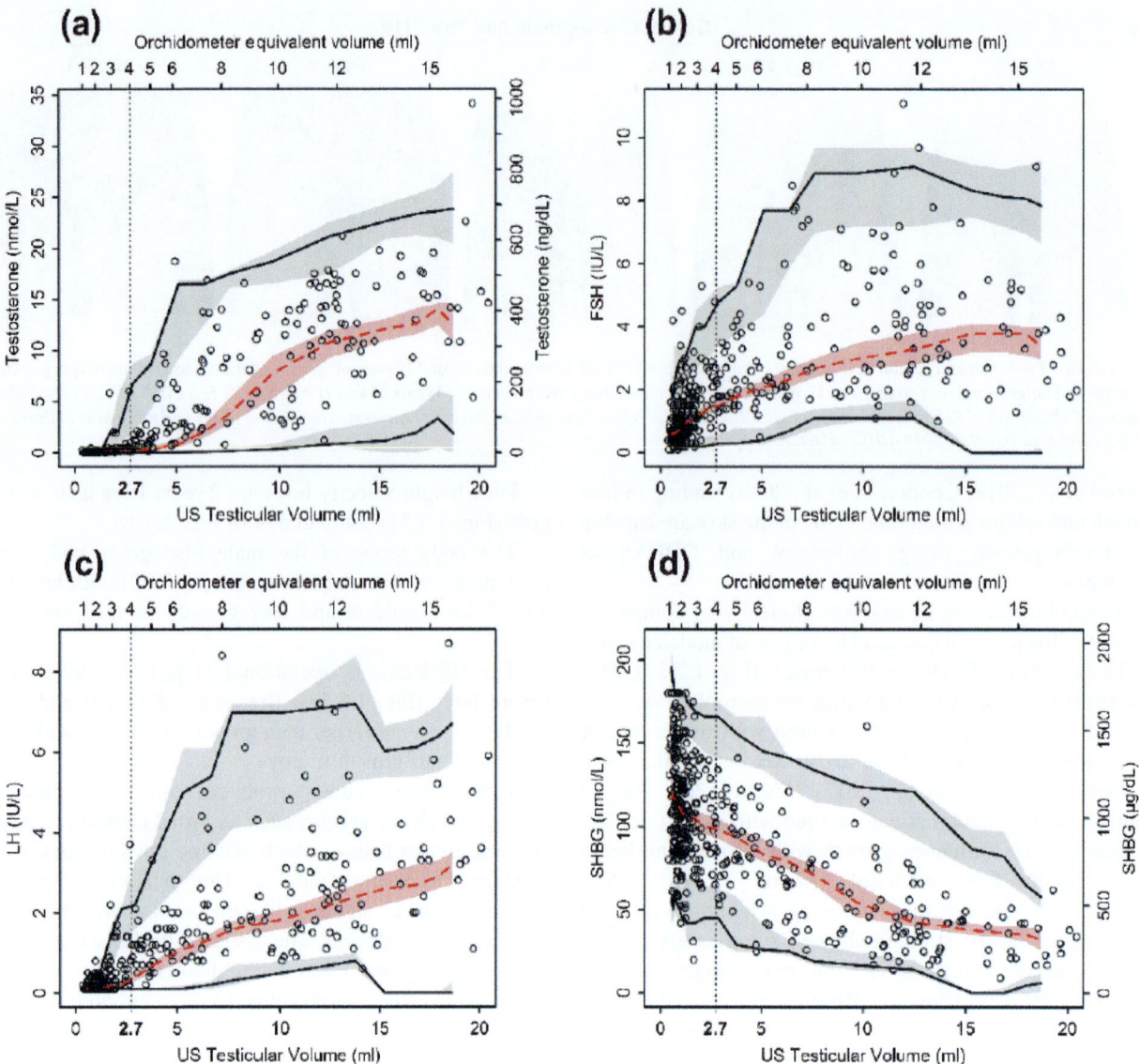

FIG. 1.7.28 LMS-smoothed reference chart of ultrasound (US) measured testicular volume in 514 healthy Norwegian boys, aged 6–16 years. Corresponding equivalent Prader orchidometer volumes are shown on the right axis. *(Data are modified for the skew (L for lambda), the median (M for mu) and the coefficient of variation (S for sigma); Oehme NHB, Roelants M, Saervold Bruserud I, Madsen A, Eide GE, Bjerknes R, Rosendahl K, Juliusson PB. Reference data for testicular volume measured with ultrasound and pubic hair in Norwegian boys are comparable with northern European populations. Acta Paediatr. 2020. https://doi.org/10.1111/apa.15159 [Epub ahead of print]. PMID: 31899821. Fig. 1 p. 3.)*

patients with Klinefelters (low androgens) also have an increase in 2D;4D ratios (Zhang et al., 2020)

Ano-genital distance (Fig. 1.7.35). The AGD relates to androgen exposure in utero (Sathyanarayana et al., 2010; Fowler et al., 2016). Short AGD refers to the distance from the anus to the perineo-scrotal junction (AGDAS) in boys and from the anus to the fourchette (AGDAF) in girls (Fischer et al., 2020). Similarly, long AGD is the distance from the anus to the anterior insertion of the penis (AGDAP) in boys and from the anus to the clitoris (AGDAC) in girls Two studies (so called Cambridge and TIDES (The Infant

Development and Environmental Study) for the study of infant development and the environment) sought to standardize the position for holding the baby. The TIDES method places the infant in a supine position with the lower half of the body exposed and the legs lifted in a froglike posture (with a 60–90 degrees angle from the torso at the hip) and knees pulled back towards to shoulders (Fig. 1.7.36). The Cambridge method places the infant in a supine position with both hips flexed, feet placed on the surface and light pressure exerted onto the thighs.

All measures of AGD were significantly shorter in females than males (Fig. 1.7.37). Further studies are

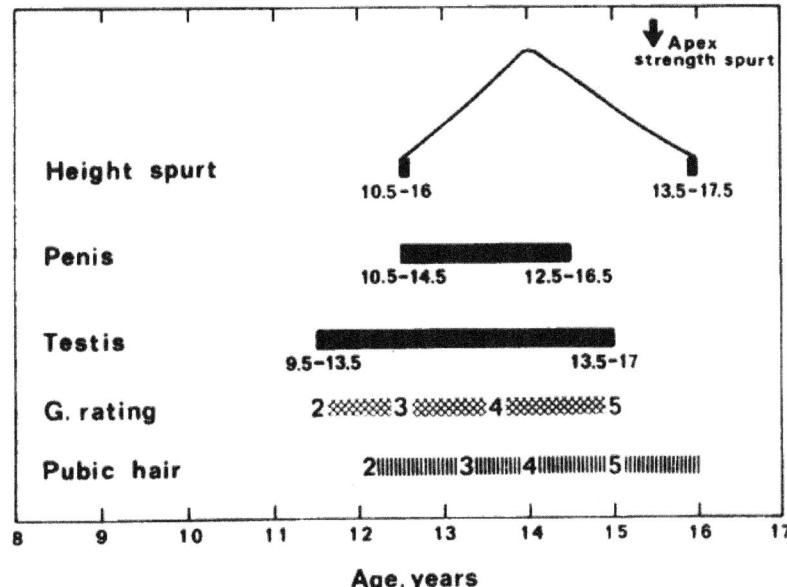

FIG. 1.7.29   Changes in bodies of boys in puberty.

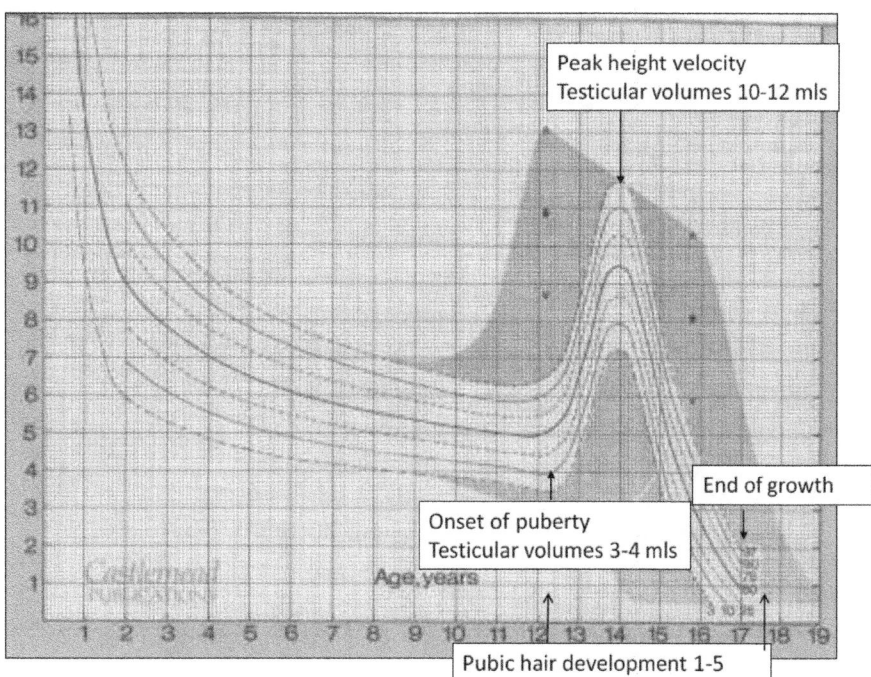

FIG. 1.7.30   Growth chart for boys.

on-going to see if this measure of androgen exposure in utero relates to adult diseases such as obesity and cardio-vascular disease.

- Stretched penile length (Fig. 1.7.38) is another parameter of testosterone action (Jaiswal et al., 2019; Boas et al., 2006). This measure is used when boys with poor sexual development are treated with exogenous steroids to assess likelihood of penis development.

Spermatogenesis

The process of generating sperm in seminiferous tubules, called spermatogenesis, occurs in three phases—mitotic, meiotic and spermiogenesic over 64 days. The process is supported by FSH and testosterone although the concept that high testosterone is required for spermatogenesis has been challenged (Huhtaniemi, 2018) . Within the seminif-erous tubules not all areas of the seminiferous epithelium

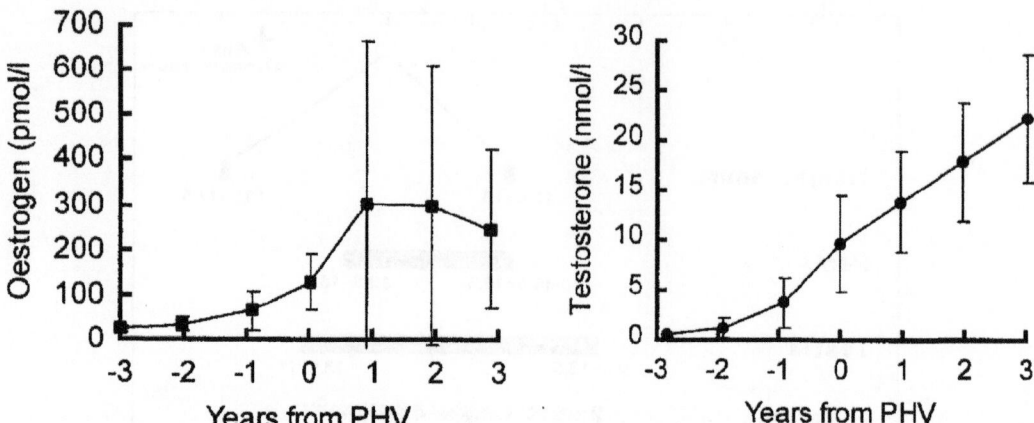

FIG. 1.7.31   Circulating testosterone in relation to PHV. Plasma testosterone in boys plotted against the years before or after the time of peak height velocity. Values are given as mean SD of the values for children in each year for whom the age of PHV could be clearly identified. *(From Round JM, Jones DA, Honour JW, Nevill AM. Hormonal factors in the development of differences in strength between boys and girls during adolescence: a longitudinal study. Ann Hum Biol 1999;26(1):49–62. Fig. 5 p. 55.)*

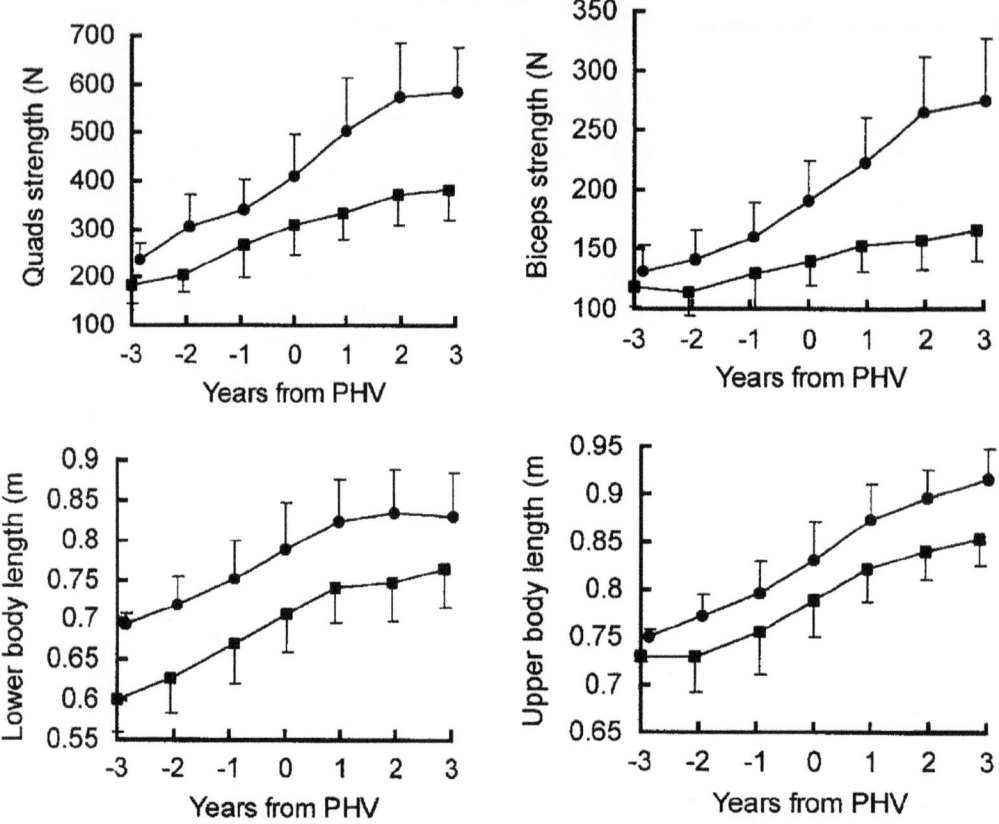

FIG. 1.7.32   Development of strength in boys and girls in relation to PHV. (A) quad strength, (B) biceps strength, (C) lower body length, (D) upper body length. Values are given as mea ± SD of the values for children in each year. *Black circles* are for boys; filled squares for girls. *(Modified from Round JM, Jones DA, Honour JW, Nevill AM. Hormonal factors in the development of differences in strength between boys and girls during adolescence: a longitudinal study. Ann Hum Biol 1999;26(1):49–62. Figa. 3, 4 p. 54; Fig. 7 p. 57.)*

will be in the same stage. The diploid germ cells (spermatogonia) have 46 chromosomes lying next to the basement membrane of seminiferous tubules, multiply by mitosis to renew stem cells. At the initiation of spermatogenesis, some spermatogonia, by FSH stimulated mitosis, differentiate into primary spermatocytes that contain diploid numbers of chromosomes. The primary spermatocytes then undergo two meiotic divisions to form spermatids that have a haploid number ($n = 23$) of chromosomes. Testosterone stimulates the first meiotic division (Smith and Walker, 2014).

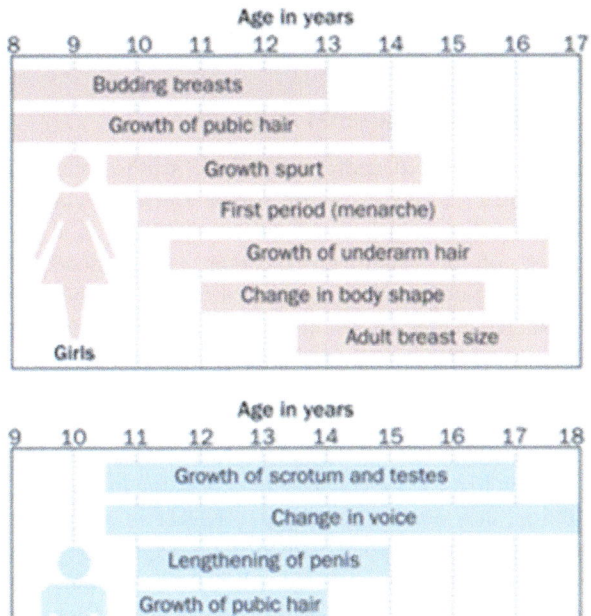

FIG. 1.7.33 Course of development of boys and girls through puberty. Sequence of events with age at puberty for average boys. *(Based on Marshall WA, Tanner JM. Variations in the pattern of pubertal changes in boys. Arch Dis Child. 1970;45(239):13–23.)*

The initial round can take 2 weeks whereas secondary spermatocytes only last 8 h. The final stage of spermatogenesis is the maturation of round to elongated spermatids that then form tailed spermatozoa with haploid complement of chromosomes. FSH is necessary for this transformation. Spermatozoa have a head containing condensed chromosomes with an acrosomal cap containing proteolytic enzymes needed for penetration of an ovum. The more mature spermatids embed in the Sertoli cells to mature by the action of FSH into spermatozoa (spermiogenesis) and later released (spermiation) and moved into the lumen of the seminiferous tubules. The sperm move through the tubules into the rete testis then carried through vasa efferentia into the ductus epididymis then into the vas deferens.

Semen analysis

The World Health Organization (WHO) regards the ejaculate of 1.5–5 mL as normal with 1.5 mL as the lower reference limit. A sperm count is the total number of spermatozoa in the ejaculate. The lower reference limit is 39 million per ejaculate and WHO considers a total sperm count of 15 million to be adequate although the 5th percentile is 19 million of spermatozoa. Sperm motility is the proportion of moving sperms and ideally should be between 40% and 50%. More than 4% of the sperm should have normal morphology with an oval head and an acrosome covering 40%–70% of the head area. A normal spermatozoon has no neck, midpiece, tail abnormalities nor cytoplasmic droplets larger than 50% of the sperm head. WHO specify pH to be in the 7.2–7.8 range. Normal semen

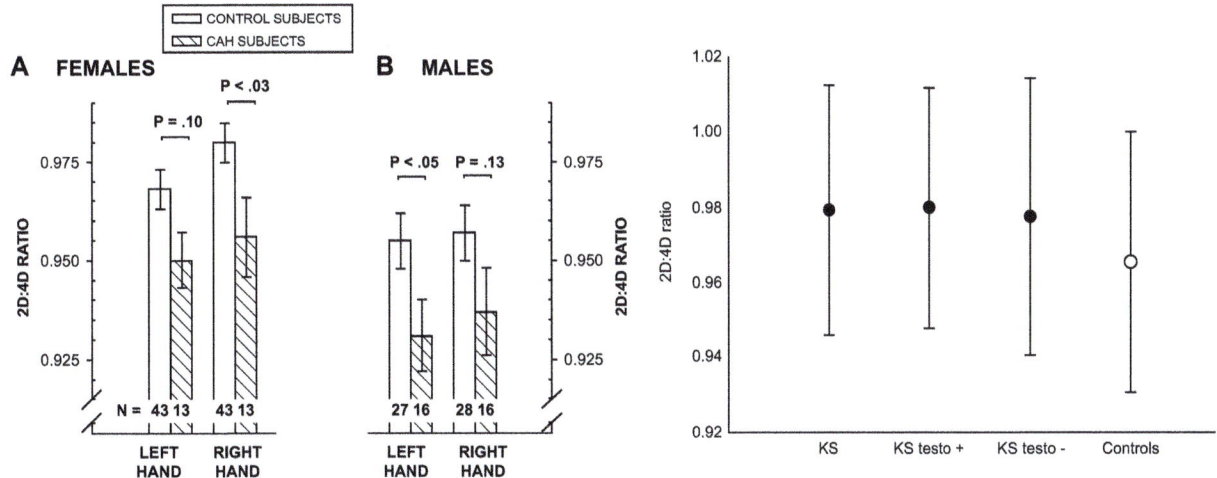

FIG. 1.7.34 Index finger (2D) to ring finger (4D) ratio (2D/4D ratio). (LEFT) (A) Mean finger length ratios of females with CAH and control females for both the left and right hands. Females with CAH have a significantly smaller 2D:4D on the right hand than do control females. (B) Mean finger length ratios of males with CAH and control males for both the left and right hands. (Right) Males with CAH have a significantly smaller 2D:4D on the left hand compared with control males. Error bars represent standard errors of the means; all P values are two-tailed. (RIGHT) 2D:4D ratio in Klinefelter syndrome naive (KS) and testosterone untreated and treated (matched KS) and controls (Mean ± 2SD). *(LEFT: From Brown WM, Hines M, Fane BA, Breedlove SM. Masculinized finger length patterns in human males and females with congenital adrenal hyperplasia. Horm Behav 2002;42(4):380–6. Fig. 1 p. 382. RIGHT: From Chang S, Skakkebæk A, Trolle C, Bojesen A, Hertz JM, Cohen A, Hougaard DM, Wallentin M, Pedersen AD, Østergaard JR, Gravholt CH. Anthropometry in Klinefelter syndrome—multifactorial influences due to CAG length, testosterone treatment and possibly intrauterine hypogonadism. J Clin Endocrinol Metab 2015;100(3):E508–17. Fig 1. P E512.)*

TIDES

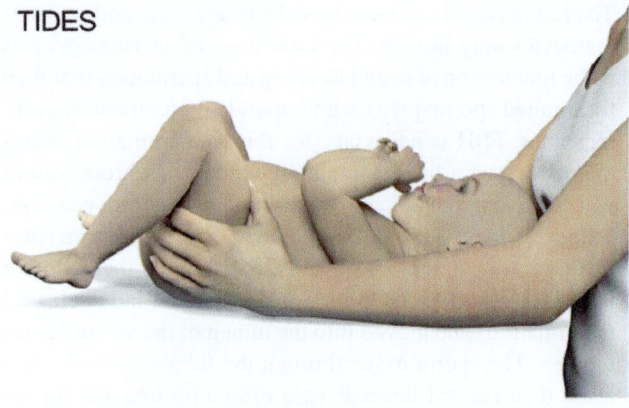

b)

Cambridge

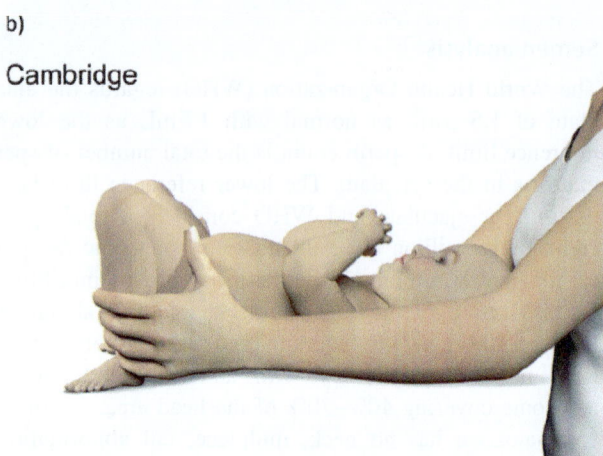

FIG. 1.7.35 Positioning of boy before measurements of ano-genital distance (Cambridge vs TIDES). *(From Fischer MB, Ljubicic ML, Hagen CP, Thankamony A, Ong K, Hughes I, Jensen TK, Main KM, Petersen JH, Busch AS, Upners EN, Sathyanarayana S, Swan SH, Juul A. Anogenital distance in healthy infants: method-, age- and sex-related reference ranges. J Clin Endocrinol Metab 2020;105(9):2996–3004. Fig. 1 p. 3.)*

contains fructose for energy. Normal semen is thick at the time of ejaculation but within 20–30min is more liquid.

1.7.3.3 Other adrenal androgens

The 11-oxygenated androgens can be synthesized by the action of 17,20-lyase on cortisol or 11-hydroxylation of androstenedione. 11-keto testosterone (11KT) is a potent agonist of the androgen receptor (AR, encoded by *NR3C4*). In cell models engineered to express the human AR and a luciferase reporter, the maximum androgenic activity of 11KT approached that of testosterone, whereas that of 11OHT was less active (Campana et al., 2016). The induction of expression of the AR target gene was also similar for both testosterone and 11KT. Similar results have been seen in a COS-1 cell system expressing human AR; in addition, they demonstrated that 11KDHT is equipotent to DHT. All of the steroids activated the expression of AR-regulated genes and promoted cell growth. Further studies are needed to clarify the effects of 11-keto-androstenedione in pregnancy and development. Publications from the Amanda Swart group should be followed.

1.7.3.4 Progesterone

The steroid progesterone has a critical role in reproduction. A progestin is required for follicular growth and ovulation and responsible for ovarian changes associated with the luteal phase of the menstrual cycle (Bull et al., 2019; Taraborrelli, 2015). Progesterone raises the body temperature by about 0.5 degrees C at the time of ovulation (Fig. 1.7.39).

Progesterone converts the endometrium into a secretory type in preparation for implantation of a blastocyst (Chi et al., 2020) and is essential for the maintenance of

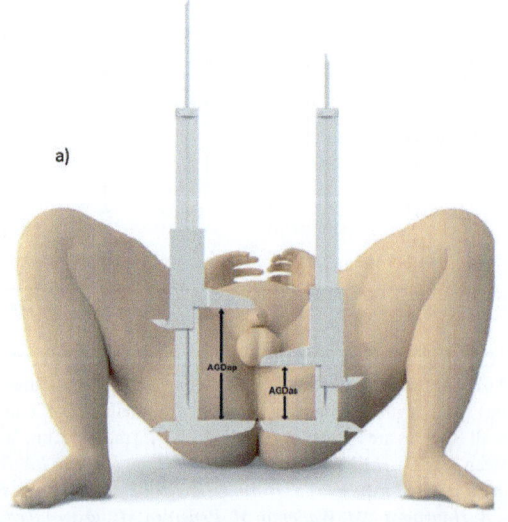

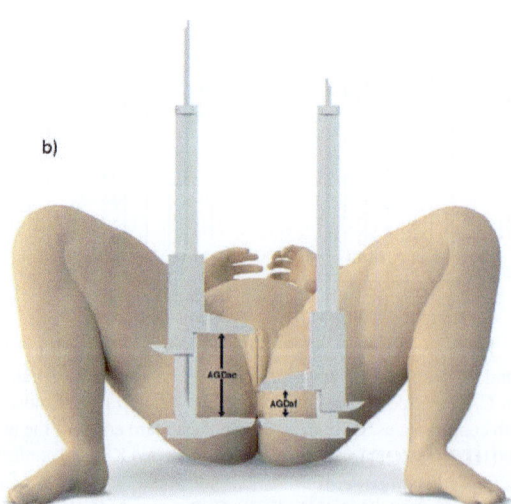

FIG. 1.7.36 Measurement of AGD position. *(From Fischer MB, Ljubicic ML, Hagen CP, Thankamony A, Ong K, Hughes I, Jensen TK, Main KM, Petersen JH, Busch AS, Upners EN, Sathyanarayana S, Swan SH, Juul A. Anogenital distance in healthy infants: method-, age- and sex-related reference ranges. J Clin Endocrinol Metab 2020;105(9):2996–3004. Fig. 2 p. 3.)*

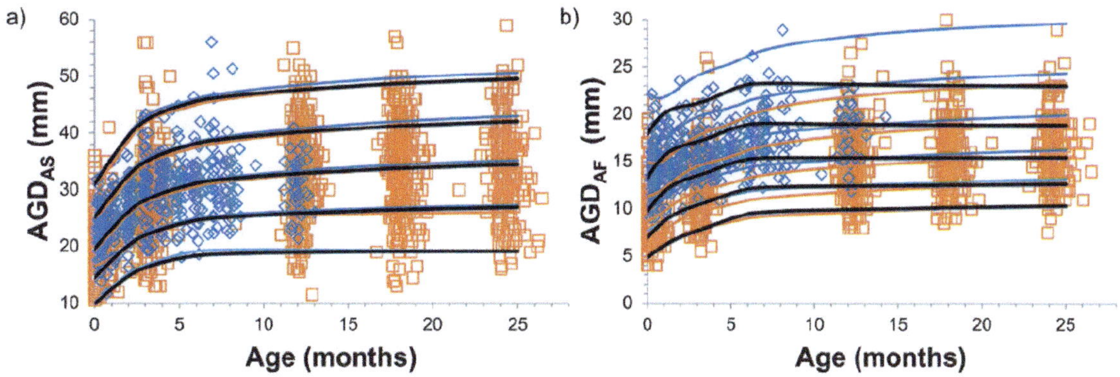

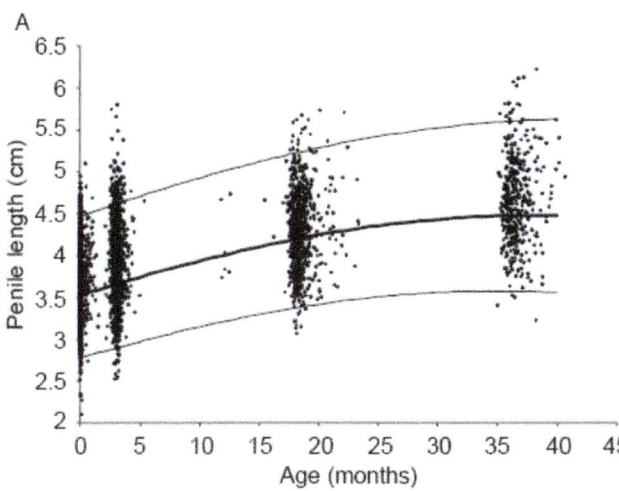

FIG. 1.7.37 AGD measures Ano-genital distance. Illustration of AGD measurements in boys (A) and girls (B), using the Cambridge method. Reference ranges according to age for short AGDas (boys) (A) **and AGDAF (girls)** (B) **by center. Center 2 (orange) and** center 4 (*blue*). Dots indicate individual values, lines indicate +2SD, +1SD, mean, −1SD, −2SD. *Black lines* indicate the combined reference levels for three centers participating in comparison. Abbreviations: *AGDAP*, anogenital distance (ano-penile); *AGDAS*, anogenital distance (ano-scrotal); *AGDAC*, anogenital distance (ano-clitoral); *AGDAF*, anogenital distance (ano-fourchettal). *(From Fischer MB, Ljubicic ML, Hagen CP, Thankamony A, Ong K, Hughes I, Jensen TK, Main KM, Petersen JH, Busch AS, Upners EN, Sathyanarayana S, Swan SH, Juul A. Anogenital distance in healthy infants: method-, age- and sex-related reference ranges. J Clin Endocrinol Metab 2020;105(9):2996–3004. Fig. 3 p. 5.)*

FIG. 1.7.38 Stretched penile length (cm) in 1962 healthy boys versus (A) chronological age (months), (B) length (cm) and (C) weight (kg). Dots represent individual measurements, the lines describe the estimated mean with the 2.5 and 97.5 percentiles. *(From Boas M, Boisen KA, Virtanen HE, Kaleva M, Suomi AM, Schmidt IM, et al. Postnatal penile length and growth rate correlate to serum testosterone levels: a longitudinal study of 1962 normal boys. Eur J Endocrinol. 2006;154(1):125–129. Fig 1 p. 127.)*

early pregnancy (Szekeres-Bartho and Schindler, 2019). Luteinized granulosa cells and luteal cells are responsible for progesterone synthesis under LH control.

Progesterone can have some art in the development of breast cancer (Brisken et al., 2015) (see Clinical Chapter 3.2.1).

1.7.4 Life events

Sex steroids are produced in the male testes in the first 6 months after birth (the mini puberty) under the influence of LH which is then turned off until puberty. There are no obvious physiological changes in the mini puberty though there may be developments in the testes that have later effects. The next event in boys and girls is the adrenarche with an increase in DHEAS production by a new zone of the adrenal cortex. Pubic hair growth in this period is the feature of this event which precedes puberty. After puberty males and females are fertile although only females go through monthly cycles of egg production. When the female ovary is almost depleted of eggs the production of sex steroids declines through the menopause.

1.7.4.1 Puberty

The reproductive organs and gonads reach maturity during puberty which is a physiological process during development when in males and females there are notable changes in the gonads, increased secretion of sex steroids, development of secondary sex characteristics and fertility potential. The mechanisms for the onset of puberty are not fully understood. The timing of puberty is uncertain so standards of development have been produced for hormonal and physical changes. The central feature of puberty is the reactivation of the HPG axis which drives the development and appearance of secondary sex characteristics. The axis is quiescent from the mini-puberty, the mechanism for which is an enigma.

Neurons in the hypothalamus release GnRH which passes in the hypophyseal portal system to the pituitary where gonadotrophes secrete gonadotrophins. Kisspeptin along with NKB (Neurokinin) and Dyn (Dynorphin), collectively neuropeptides regulate GnRH release through NK3R and KOR (kappa-opioid receptor) in kisspeptin neurons. Kisspeptin is a G-protein-coupled receptor derived in the same cells that release GnRH. Kisspeptin neurons mediate the negative feedback effect on GnRH/LH secretion.

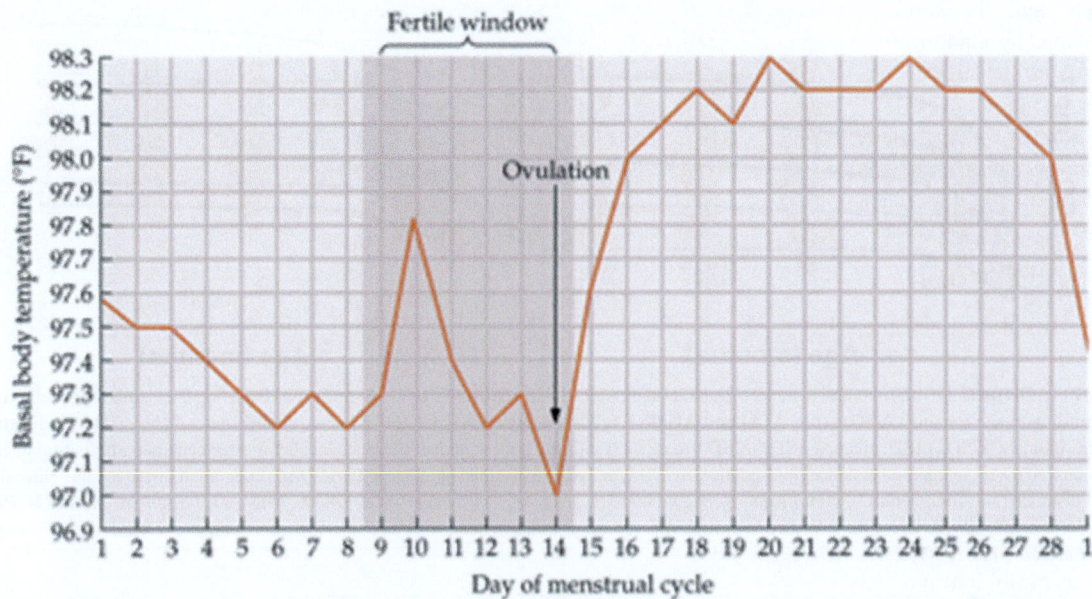

FIG. 1.7.39 Temperature in the menstrual cycle. Typical temperature chart in a biphasic menstrual cycle as seen in the app. Shown here are the fertile/nonfertile days (*red/green*) returned by the algorithm. The fertile window days are darker red. Days with measurements are shown as filled circles. The cycle average temperature (cover line) is the gray horizontal line at 36.37°C. (*From Bull JR, Rowland SP, Scherwitzl EB, Scherwitzl R, Danielsson KG, Harper J. Real-world menstrual cycle characteristics of more than 600,000 menstrual cycles. NPJ Digit Med 2019;2:83. Fig. 6 p. 6.*)

NKB stimulates GnRH, dynorphin mediates the inhibitory feedback of progesterone on GnRH pulses.

Estrogen strongly suppresses *Kiss1* expression by the arcuate nucleus (ARC) via direct and indirect pathways: estrogen-responsive neurons in the preoptic area (POA) mediate estrogen negative feedback action to ensure the prepubertal suppression of ARC *Kiss1* expression (Fig. 1.7.40). During the pubertal transition, the sensitivity to the estrogen negative feedback action on ARC *Kiss1* expression somehow decreases, which results in an increase in *Kiss1*

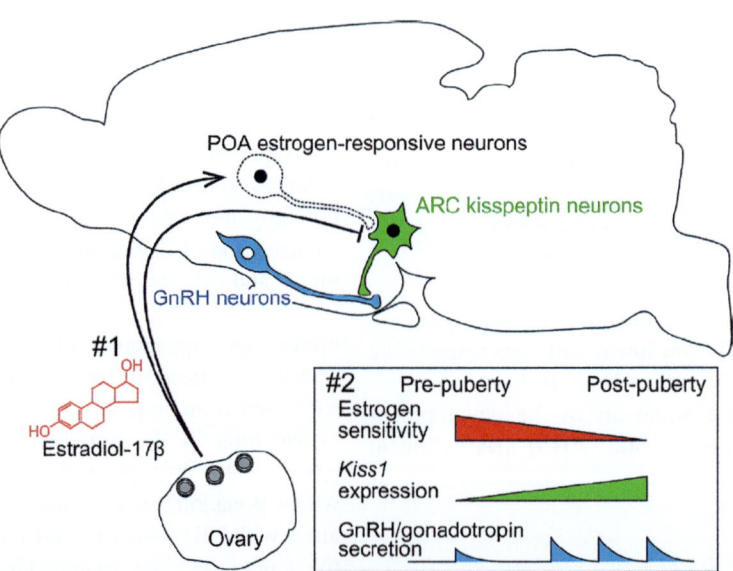

FIG. 1.7.40 Effects of estradiol with kisspeptin in puberty development. (*From Uenoyama Y, Inoue N, Maeda KI, Tsukamura H. The roles of kisspeptin in the mechanism underlying reproductive functions in mammals. J Reprod Dev. 2018;64(6):469–476. Fig 4 p. 474.*)

expression, and the subsequent secretion of kisspeptin triggers GnRH/gonadotropin secretion at pubertal onset (Uenoyama et al., 2018).

There is much to be learnt about the mechanism of puberty. A makorin-ring protein (MKRN3) is a new inhibitory candidate with plasma concentrations that fall at the time of puberty.

1.7.4.2 Adult females

Theca-interstitial cells of the ovarian follicle as well as theca-luteal cells of the corpus luteum are the major site of ovarian androgen production. The androgens are substrates for ovarian estrogen biosynthesis by granulosa cells. Androgens maintain follicular growth through promotion of early stage folliculogenesis and prevention of atresia. Androgens are synthesized in the theca-interstitial cells then are transported into granulosa cells where they are converted to estradiol and other estrogens by aromatase. High circulating concentrations of androgens can affect body smell/odor in either sex. Skin acne is also a feature of hyperandrogenism (see Chapter 3.2.1).

Both steroids act through cytoplasmic receptors and two receptors for each steroid, alpha and beta (ER-α and ER-β; PR-α and PR-β) are now recognized, sometimes they antagonize each other. Isoforms of each receptor exist that have different tissue expression patterns and function which affects gene expression in normal and tumor tissue hence ligand treatment. ER-α is mainly expressed in reproductive tissues, breast, kidney, bone, white adipose tissue and liver. ER-β is present in ovary, central nervous system (CNS), lung, colon, kidney and immune system. PR-α and PR-β are virtually identical in structure except PR-β has an additional 164 amino acids at the N terminus. PR-α and PR-β are expressed equally in human tissues. The classic view of steroids as dimeric transcription factors that bind a hormone response element is now thought to be oversimplified. The phosphorylation, sumoylation of the receptors as well as the interaction of coregulatory proteins, particularly with PR, affects transcriptional response. Estrogens are important in maintenance of lipid and glucose homeostasis. They regulate food intake and energy expenditure by action on the CNS. Estrogens also act to promote synthesis of proteins that maintain peripheral energy homeostasis. Progesterone leads to gene transcription of proteins that regulate functions of the uterus, ovary, mammary gland and brain.

1.7.4.3 Pregnancy

In pregnancy, estrogens stimulate placental angiogenesis and induce gap junction formation in the myometrium. Estradiol activates serine threonine protein kinase (SGK1) that suppresses apoptosis and promotes the antiinflammatory response in decidual stromal cells to sustain pregnancy. Towards term estrogens increase oxytocin and α-adrenergic receptor concentrations, acts to open calcium-dependent ion channels and stimulates the expression of PGE2 and PGF2α receptors in the uterus to induce the labor (Chang and Zhang, 2008) although much needs to be learnt about the process.

The placenta integrates and communicates environmental factors such as nutritional status and maternal stress to the developing fetus by changes in placental structure, oxygen transport, nutrient concentrations and placental hormone secretion (Pasqualini and Chetrite, 2016). Progesterone and GCH have roles in preventing rejection of the fetus through intolerance (Di Renzo et al., 2016). The fetus can respond to environmental cues by adapting to its environment. Total protein in the placenta is involved with the regulation of maternal-fetal tolerance. The activity of the kynurenine pathway increases under inflammatory conditions, and the different metabolites produced by this pathway can exert potent immunoregulatory functions.

Placental syncytiotrophoblasts express the enzyme encoded by *HSD11B2*. This enzyme limits fetal exposure to maternal GCH because the predominant activity of placental HSD11B2 is oxidation of maternal cortisol to the inactive GCH, cortisone (Zhu et al., 2019). However, this enzyme only prevents approximately 80%–90% of maternal cortisol from crossing the placenta. Expression of HSD11B2 increases throughout pregnancy with a rapid decline late in the third trimester. This increased cortisol exposure later in gestation promotes terminal differentiation of many organs in preparation for life outside of the uterus. GCH actions in the fetal lung include thinning of the alveolar septa, maturation of the alveoli and production and release of surfactant proteins and phospholipids. In addition, cortisol promotes the expression of fetal gluconeogenic enzymes.

The fetal hypothalamic-pituitary-adrenal (HPA) axis is particularly susceptible to long-term programming by glucocorticoids; these effects can persist throughout life (Hoffman et al., 2017). Dysfunction of the HPA axis as a result of fetal programming has been associated with impaired brain growth, altered behavior and increased susceptibility to chronic disease (such as metabolic and cardiovascular disease). Moreover, the effects of glucocorticoid-mediated programming are evident in subsequent generations, and transmission of these changes can occur through both maternal and paternal lineages. Glucocorticoid hormones (GCH) have a role in promoting maturation of the fetal lung and production of the surfactant necessary for extrauterine lung function after birth. Adrenal insufficiency of the fetus and infant can lead to death from pulmonary immaturity.

The introduction of ultrasound scanning in pregnancy to monitor development of the fetus has negated may of the steroid tests used in the past for this purpose. Low estrogen production was a useful trigger for investigation of endocrine disorders (see Chapter 3.2.2).

At the end of pregnancy, a fully differentiated mammary gland is required for milk production after stimulation by the newborn infant suckling and release of prolactin (Macias and Hinck, 2012). In the uterus, estradiol regulates the proliferation of the uterine compartments. Actions of estradiol in mammary gland and uterus are mediated through ERalpha.

The placenta is the source of progesterone to maintain late pregnancy. This inhibits the action of prolactin until after delivery when lactation can take place. Progesterone is the main steroid in the maintenance of pregnancy by preventing of, or delaying, the onset of labor. Progesterone has important effects

1. on the immune system that enable tolerance of the fetus,
2. acts on maternal T-lymphocytes to increases the synthesis of antiinflammatory cytokines,
3. favors the decidualization of the endometrium,
4. increases homeobox (HOXA) 10 gene expression,
5. stimulates nitric oxide synthesis in the umbilical vein,
6. acts on trophoblast invasion by inhibiting the activity of matrix-specific metalloproteases,
7. induces relaxation of vascular smooth muscle cells in umbilical cord,
8. decreases contractile activity of myometrium,
9. decreases production of prostaglandin E2,
10. down regulates myometrial gap junction density.

1.7.4.4 Menopause

The cessation of monthly menstruation is called the menopause and occurs typically between the age of 40 and 60 years. The potential for reproduction declines and ceases because the ovaries become depleted of follicles from birth. A transition period in mid-life, for 2–10 years, when menstruation is less regular is called the perimenopause. The menopause is associated with a significant decline in plasma concentrations of sex hormones and consequently an increase in the concentrations of the gonadotrophins and changes in other directly involved hormones such as the inhibins. These changes are superimposed with effects of aging, social and metabolic factors. Daily activity and well-being are influential. Age at the menopause reflects the complex interactions of health and socioeconomic factors including for example ethnicity, diet, education, oral contraceptive use, weight, employment, exposure to endocrine disrupting chemicals, alcohol use, smoking and physical activity (Hale et al., 2014). Some of the biochemical changes at the menopause are attributed to estrogen and or progesterone depletion (Honour, 2018).

In perimenopausal women, an increase in the cyclical early follicular concentrations of FSH is a consistent finding. Concentrations of estrogens and inhibins are lower than seen in earlier years which correlate with the depletion of follicles in the ovaries and the resistance of the ovaries to gonadotrophins with a reduction in estrogen and inhibin production. The menopause transition goes through four phases (Table 1.7.2) from normal ovarian activity (called D), through a stage with prolonged follicular phase and no luteal activity (B). There is a stage with normal follicular activity and insufficient luteal phase (C) and finally the menopause with low estrogens and progesterone (A). This division is based on data from multiple, annual blood tests in 13 women for between 4 and 9 years up to the menopause. Blood samples were taken three times a week for a 4 week period in each year. These changes are summarized in Table 1.7.2. B cycles that became longer may be delayed ovulatory cycles or anovulatory cycles. As the process moves on, FSH concentrations are higher and inhibins lower than in the D follicular phases. Intervals of 24 up to 38 days separate menses in early phase and the interval can be

TABLE 1.7.2 Extract of NICE guideline for menopause.

1.2. Diagnosis of perimenopause and menopause

1.2.1. Diagnose the following without laboratory tests in otherwise healthy women aged over 45 years with menstrual symptoms

- perimenopause based on vasomotor symptoms and irregular periods
- menopause in women who have not had a period for at least 12 months and are not using hormonal contraception
- menopause based on symptoms in women without a uterus

1.2.2. Take into account that it can be difficult to diagnose menopause in women who are taking hormonal treatment for example for the treatment of heavy periods

1.2.3. Do not use the following laboratory and imaging tests to diagnose perimenopause or menopause in women age over 45 years

- anti-Mullerian hormone
- inhibin A
- inhibin B
- estradiol

1.2.4. Do not use a serum follicle stimulating hormone (FSH) test to diagnose menopause in women using estrogen and progestogen contraception or high dose progestogen

1.2.5. Consider using a FSH test to diagnose menopause only

- in women aged 40–45 years with menopausal symptoms including a change in their menstrual cycle
- in women aged under 40 years in whom menopause is suspected

TABLE 1.7.3 Staging of menopause.

Terminology	Early menopausal transition	Late menopausal transition	Early post menopause	Late post menopause
	Perimenopause	Perimenopause 1–3 years	Perimenopause	
Menstrual cycles	Variable length, >7 day difference in cycle length	>60 days amenorrhea		
FSH	High	>25 IU/L	High variable	Very high stable
AMH	Low	Low	Low	Very low
Inhibin B	Low	Low	Low	Very low

greater than 60 days in later phase. FSH concentrations above 25 IU/L are seen in the late transitional period. Once menstruation has ceased FSH is above 60 IU/L with LH raised (10–45 IU/L), but estradiol (<200 pmoL/L), progesterone and inhibins (B < 25; A < 10 ng/L) concentrations are low. Hormone measurements other than FSH during the perimenopause are generally considered to be of little diagnostic value. The transition may take 4 or more years. The current guidelines from the National Institute for Clinical Excellence (NICE) for tests in the menopause are shown in Table.1.7.3.

Hot flushes, night sweats and vaginal dryness are common symptoms of the menopause from estrogen withdrawal. A hot flush is a sensation of heat of unknown cause in the upper body, face and neck lasting for 3-5 min accompanied by higher heart rate and peripheral blood flow with a rise in skin temperature. With estrogen loss there is a reduction in bone density and an increase in cardiovascular risk although the latter may be related more to concentrations of high density lipoprotein cholesterol.

The reference ranges for many diagnostic biochemical tests are influenced by the menopause and these ranges need to be used when interpreting results in clinical investigations for patient management. Changes in methodology need to be borne in mind when the laboratory is interpreting results against reference ranges. The introduction of methods based on chromatography combined with mass spectrometry is having a significant impact on quality of assays for many of the analytes discussed in this review. Reference ranges by the best available technology where possible will be used. The standardization and harmonization of assays is being addressed. Many women now choose to develop their career before bearing children and the health service has had to change services around this by offering new tests to confirm potential for fertility and freeze the embryos for later implantation. Estrogen treatment of postmenopausal women leads to an increase in bone density of hip and spine with a reduction in the incidence of facture.

1.7.5   Nongenomic actions of steroids

Acute vasodilatation by estradiol and testosterone and immediate anesthetic effects of progesterone were recognized in the 1940s as rapid effects of steroids. The nongenomic mechanism was characterized by experiments that showed persistence of the effects after preincubation of cells with actinomycin or cycloheximide that inhibited transcription and translation. Further experiments on cells without nuclei still responded to steroid and conjugation of steroids prevented access to the nucleus. Mitochondrial membranes responded to steroids. The concept of the nongenomic effects of steroids was controversial for many years. Most of the data supporting the rapid effects of steroids are based on animal experiments particularly over behavior. The data is humans is limited and will not be discussed further here.

1.7.6   Other actions

1.7.6.1   Neurosteroids

In the early 1940s, Hans Seyle showed that high doses of **progesterone** can very **rapidly** modulate brain excitability by inducing **anesthesia** in rats. In the 1970s, Karavolas and collaborators showed that progesterone was converted to 5α-dihydroprogesterone and allopregnanolone within the rat hypothalamus and pituitary gland (Karavolas et al., 1979). In 1986, came the demonstration that both allopregnanolone (3α,5α-tetrahydroprogesterone) and 3α,5α-tetrahydrodeoxycorticosterone are natural positive modulators of neuronal GABA-A receptors (Majewska et al., 1986), providing a mechanistic insight into the rapid psychopharmacological actions of progesterone and its metabolites, including anxiolytic, antidepressant, anesthetic, anticonvulsant and analgesic effect.

Reduced metabolites of progesterone, such as 3α,5α-THP (Fig. 1.7.41), are thus considered to play an important physiological role to locally modulate neuronal excitability

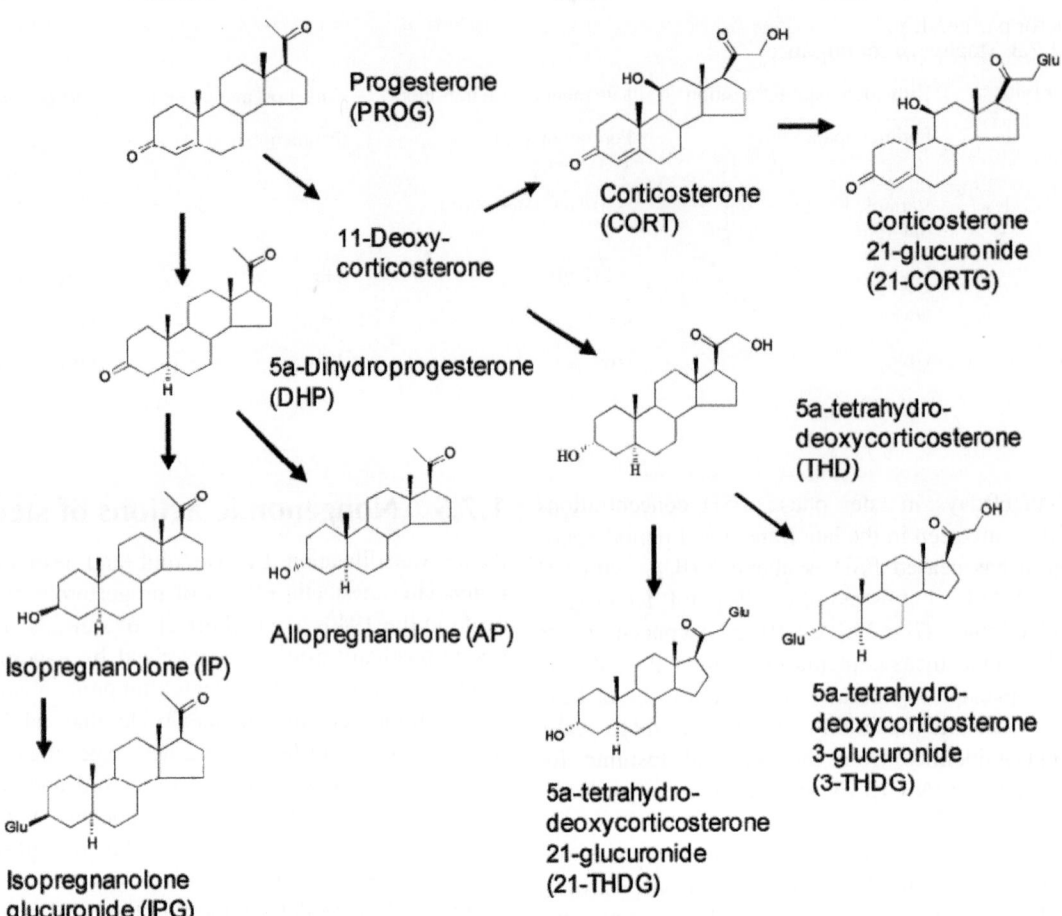

FIG. 1.7.41   Reduced metabolites of progesterone in brain. *(From Jäntti SE, Tammimäki A, Raattamaa H, Piepponen P, Kostiainen R, Ketola RA. Determination of steroids and their intact glucuronide conjugates in mouse brain by capillary liquid chromatography-tandem mass spectrometry. Anal Chem 2010;82(8):3168–75. Fig. 1. p. 3170.)*

by fine-tuning the action of gamma-amino-butyric acid (GABA) at GABA(A) receptors.

Progesterone metabolites are classified as **neuroactive steroids**. In regions of the brain such as the hippocampus, nerve cells exhibit inhibitory GABA-ergic transmission as a phasic component mediated by GABA(A)-R and by a tonic component at extra-synaptic GABA(A)-R.

In pregnancy, there is a marked increase in neuroactive steroid levels in plasma and brain that could contribute to changes in mood, anxiety and other psychiatric conditions.

1.7.6.2   Metabolites and precursors

Although steroid metabolism serves to inactivate steroids, metabolites of aldosterone and other 5-alpha reduced corticosteroids have been found to have antinatriuretic properties.

Liquorice was originally thought to act at the mineralo-corticoid receptor to produce hypertension. This view was challenged with the discovery of the protective effect of HSD11B2 to prevent cortisol binding at the MR (Stewart

et al., 1987). Liquorice blocks the action of HSD11B2. The active ingredient of liquorice is glycerrhitinic acid and the name **GALFs'** for **glycerrhetinic acid like factors** was introduced for compounds with this activity. Some 21-deoxycorticosteroids are potent in this regard and are known to be formed in the intestine by bacterial action so in reviewing the physiological effects of steroids the role of gut bacteria in the enterohepatic circulation of steroids must be considered. Other compounds may act as GALF's. These are going to be of renewed interest because of the interest in the microbiome (Honour, 2015; Morris et al., 2014).

1.7.7   Summary

Clinicians need to understand the physiological effects of steroids and how they are influenced by normal changes with development, age and gender through life events and changes in the circulating concentrations of steroid hormones. In the days of telemedicine, the visual examination of the patient and the observed semiotic changes are still important in the diagnostic evaluation and defining the patient/doctor partnership. The introduction of new technologies particularly for imaging not only improves visualization of changes in the body (ultrasound in pregnancy for viability and development of the fetus) but also provide valuable quantitative data (e.g., testicular volume by ultrasound) that are superior to manual assessments and can be documented from reliable measurements. CT and MRI scans provide detailed information about the skeleton, tissues, the cardiovascular system, blood vessels, the digestive, respiratory and reproductive tracts and defines abnormalities anywhere in the body. Many scans can now be interpreted by the associated computer software using artificial intelligence and large databases. The imaging department can therefore provide quicker information to the clinician. Many physiological actions of steroids are known but there is going to be of renewed interest in GALF's because of the new interest in the microbiome (Honour, 2015; Morris et al., 2014).

References

Alwashih MA, Stimson RH, Andrew R, Walker BR, Watson DG. Acute interaction between hydrocortisone and insulin alters the plasma metabolome in humans. Sci Rep 2017;7(1):11488.

Andrews RC, Herlihy O, Livingstone DE, Andrew R, Walker BR. Abnormal cortisol metabolism and tissue sensitivity to cortisol in patients with glucose intolerance. J Clin Endocrinol Metab 2002; 87(12):5587–93.

Ankarberg-Lindgren C, Westphal O, Dahlgren J. Testicular size development and reproductive hormones in boys and adult males with Noonan syndrome: a longitudinal study. Eur J Endocrinol 2011; 165(1):137–44.

Barnes PJ. Glucocorticosteroids: current and future directions. Br J Pharmacol 2011;163(1):29–43.

Boas M, Boisen KA, Virtanen HE, Kaleva M, Suomi AM, Schmidt IM, et al. Postnatal penile length and growth rate correlate to serum testosterone levels: a longitudinal study of 1962 normal boys. Eur J Endocrinol 2006;154(1):125–9.

Brillon DJ, Zheng B, Campbell RG, Matthews DE. Effect of cortisol on energy expenditure and amino acid metabolism in humans. Am J Physiol 1995;268(3 Pt 1):E501–13.

Brisken C, Hess K, Jeitziner R. Progesterone and overlooked endocrine pathways in breast cancer pathogenesis. Endocrinology 2015;156 (10):3442–50.

Brown WM, Hines M, Fane BA, Breedlove SM. Masculinized finger length patterns in human males and females with congenital adrenal hyperplasia. Horm Behav 2002;42(4):380–6.

Bruserud IS, Roelants M, Oehme NHB, Madsen A, Eide GE, Bjerknes R, et al. References for ultrasound staging of breast maturation, tanner breast staging, pubic hair, and menarche in Norwegian girls. J Clin Endocrinol Metab 2020;105(5). pii: dgaa107.

Bull JR, Rowland SP, Scherwitzl EB, Scherwitzl R, Danielsson KG, Harper J. Real-world menstrual cycle characteristics of more than 600,000 menstrual cycles. NPJ Digit Med 2019;2:83.

Cain DW, Cidlowski JA. Immune regulation by glucocorticoids. Nat Rev Immunol 2017;17(4):233–47.

Campana C, Rege J, Turcu AF, Pezzi V, Gomez-Sanchez CE, Robins DM, et al. Development of a novel cell based androgen screening model. J Steroid Biochem Mol Biol 2016;156:17–22.

Cannavo A, Bencivenga L, Liccardo D, Elia A, Marzano F, Gambino G, et al. Aldosterone and mineralocorticoid receptor system in cardiovascular physiology and pathophysiology. Oxid Med Cell Longev 2018;2018:1204598.

Ceruti JM, Leirós GJ, Balañá ME. Androgens and androgen receptor action in skin and hair follicles. Mol Cell Endocrinol 2018;465:122–33.

Chang K, Zhang L. Review article: steroid hormones and uterine vascular adaptation to pregnancy. Reprod Sci 2008;15(4):336–48.

Chi RA, Wang T, Adams N, Wu SP, Young SL, Spencer TE, et al. Human endometrial transcriptome and progesterone receptor cistrome reveal important pathways and epithelial regulators. J Clin Endocrinol Metab 2020;105(4).

Cohen MR, Stein Sr IF, Kaye BM. Spinnbarkeit: a characteristic of cervical mucus; significance at ovulation time. Fertil Steril 1952;3(3):201–9.

Condorelli RA, Cannarella R, Calogero AE, La Vignera S. Evaluation of testicular function in prepubertal children. Endocrine 2018;62 (2):274–80.

Darmaun D, Matthews DE, Bier DM. Physiological hypercortisolemia increases proteolysis, glutamine, and alanine production. Am J Physiol 1988;255(3 Pt 1):E366–73.

Di Renzo GC, Giardina I, Clerici G, Brillo E, Gerli S. Progesterone in normal and pathological pregnancy. Horm Mol Biol Clin Investig 2016;27(1):35–48.

Ebrecht M, Buske-Kirschbaum A, Hellhammer D, Kern S, Rohleder N, Walker B, et al. Tissue specificity of glucocorticoid sensitivity in healthy adults. J Clin Endocrinol Metab 2000;85(10):3733–9.

Escoter-Torres L, Caratti G, Mechtidou A, Tuckermann J, Uhlenhaut NH, Vettorazzi S. Fighting the fire: mechanisms of inflammatory gene regulation by the glucocorticoid receptor. Front Immunol 2019;10:1859.

Fischer MB, Ljubicic ML, Hagen CP, Thankamony A, Ong K, Hughes I, et al. Anogenital distance in healthy infants: method-, age- and sex-related reference ranges. J Clin Endocrinol Metab 2020;105(9): 2996–3004.

Fowler PA, Filis P, Bhattacharya S, le Bizec B, Antignac JP, Morvan ML, et al. Human anogenital distance: an update on fetal smoke-exposure and integration of the perinatal literature on sex differences. Hum Reprod 2016;31(2):463–72.

Gathercole LL, Morgan SA, Bujalska IJ, Hauton D, Stewart PM, Tomlinson JW. Regulation of lipogenesis by glucocorticoids and insulin in human adipose tissue. PLoS ONE 2011;6(10):e26223.

Gilligan LA, Trout AT, Schuster JG, Schwartz BI, Breech LL, Zhang B, et al. Normative values for ultrasound measurements of the female pelvic organs throughout childhood and adolescence. Pediatr Radiol 2019;49(8):1042–50.

W.W. Greulich, S. Idell Pyle. Radiographic atlas of skeletal development of the hand and wrist. Calif Med 1959, 91(1): 53.

Groeneweg FL, Karst H, de Kloet ER, Joëls M. Rapid non-genomic effects of corticosteroids and their role in the central stress response. J Endocrinol 2011;209(2):153–67.

Hachemi Y, Rapp AE, Picke AK, Weidinger G, Ignatius A, Tuckermann J. Molecular mechanisms of glucocorticoids on skeleton and bone regeneration after fracture. J Mol Endocrinol 2018;61(1):R75–90.

Hagen CP, Mouritsen A, Mieritz MG, Tinggaard J, Wohlfahrt-Veje C, Fallentin E, et al. Uterine volume and endometrial thickness in healthy girls evaluated by ultrasound (3-dimensional) and magnetic resonance imaging. Fertil Steril 2015;104(2):452–9. e2.

Hale GE, Robertson DM, Burger HG. The perimenopausal woman: endocrinology and management. J Steroid Biochem Mol Biol 2014; 142:121–31.

Hardy RS, Zhou H, Seibel MJ, Cooper MS. Glucocorticoids and bone: consequences of endogenous and exogenous excess and replacement therapy. Endocr Rev 2018;39(5):519–48.

Hartmann K, Koenen M, Schauer S, Wittig-Blaich S, Ahmad M, Baschant U, et al. Molecular actions of glucocorticoids in cartilage and bone during health, disease, and steroid therapy. Physiol Rev 2016;96(2): 409–47.

Hoffman DJ, Reynolds RM, Hardy DB. Developmental origins of health and disease: current knowledge and potential mechanisms. Nutr Rev 2017;75(12):951–70.

Honour JW. Historical perspective: gut dysbiosis and hypertension. Physiol Genomics 2015;47(10):443–6.

Honour JW. Biochemistry of the menopause. Ann Clin Biochem 2018; 55(1):18–33.

Honour JW, Jones R, Leary S, Golding J, Ong KK, Dunger DB. Relationships of urinary adrenal steroids at age 8 years with birth weight, postnatal growth, blood pressure, and glucose metabolism. J Clin Endocrinol Metab 2007;92(11):4340–5.

Huhtaniemi I. Mechanisms in endocrinology: hormonal regulation of spermatogenesis: mutant mice challenging old paradigms. Eur J Endocrinol 2018;179(3):R143–50.

Imperato-McGinley J, Zhu YS. Androgens and male physiology the syndrome of 5alpha-reductase-2 deficiency. Mol Cell Endocrinol 2002; 198(1-2):51–9.

Jaiswal VK, Khadilkar V, Khadilkar A, Lohiya N. Stretched penile length and testicular size from birth to 18 years in boys from Western Maharashtra. Indian J Endocrinol Metab 2019;23(1):3–8.

Kalkwarf HJ, Zemel BS, Gilsanz V, Lappe JM, Horlick M, Oberfield S, et al. The bone mineral density in childhood study: bone mineral content and density according to age, sex, and race. J Clin Endocrinol Metab 2007;92(6):2087–99.

Kanatsou S, Joels M, Krugers H. Brain mineralocorticoid receptors and resilience to stress. Vitam Horm 2019;109:341–59.

Kamel KS, Schreiber M, Halperin ML. Renal potassium physiology: integration of the renal response to dietary potassium depletion. Kidney Int 2018;93(1):41–53.

Karavolas HJ, Hodges DR, O'Brien DJ, MacKenzie KM. In vivo uptake of [3H]progesterone and [3H]5 alpha-dihydroprogesterone by rat brain and pituitary and effects of estradiol and time: tissue concentration of progesterone itself or specific metabolites? Endocrinology 1979; 104(5):1418–25.

Kelsey TW, Ginbey E, Chowdhury MM, Bath LE, Anderson RA, Wallace WH. A validated normative model for human uterine volume from birth to age 40 years. PLoS ONE 2016;11(6), e0157375.

Khalatbari H, Binkovitz LA, Parisi MT. Dual-energy X-ray absorptiometry bone densitometry in pediatrics: a practical review and update. Pediatr Radiol 2021;51(1):25–39.

Koskenniemi JJ, Virtanen HE, Toppari J. Testicular growth and development in puberty. Curr Opin Endocrinol Diabetes Obes 2017;24 (3):215–24.

Kuo T, McQueen A, Chen TC, Wang JC. Regulation of glucose homeostasis by glucocorticoids. Adv Exp Med Biol 2015;872: 99–126.

Lee RA, Harris CA, Wang JC. Glucocorticoid receptor and adipocyte biology. Nucl Receptor Res 2018;5, 101373.

Macias H, Hinck L. Mammary gland development. Wiley Interdiscip Rev Dev Biol 2012;1(4):533–57.

Madsen A, Oehme NB, Roelants M, Bruserud IS, Eide GE, Viste K, et al. Testicular ultrasound to stratify hormone references in a cross-sectional Norwegian study of male puberty. J Clin Endocrinol Metab 2020;105(6): dgz094. https://doi.org/10.1210/clinem/dgz094. 31697832.

Majewska MD, Harrison NL, Schwartz RD, Barker JL, Paul SM. Steroid hormone metabolites are barbiturate-like modulators of the GABA receptor. Science 1986;232(4753):1004–7.

Manning JT, Scutt D, Wilson J, Lewis-Jones DI. The ratio of 2nd to 4th digit length: a predictor of sperm numbers and concentrations of testosterone, luteinizing hormone and oestrogen. Hum Reprod 1998;13 (11):3000–4.

Marshall WA, Tanner JM. Variations in pattern of pubertal changes in girls. Arch Dis Child 1969;44(235):291–303.

Marshall WA, Tanner JM. Variations in the pattern of pubertal changes in boys. Arch Dis Child 1970;45(239):13–23.

McGowan PO, Matthews SG. Prenatal stress, glucocorticoids, and developmental programming of the stress response. Endocrinology 2018;159 (1):69–82.

Morgan SA, Gathercole LL, Simonet C, Hassan-Smith ZK, Bujalska I, Guest P, et al. Regulation of lipid metabolism by glucocorticoids and 11β-HSD1 in skeletal muscle. Endocrinology 2013;154(7): 2374–84.

Morgan SA, Hassan-Smith ZK, Lavery GG. Mechanisms in endocrinology: tissue-specific activation of cortisol in Cushing's syndrome. Eur J Endocrinol 2016;175(2):R83–9.

Morris DJ, Latif SA, Brem AS. An alternative explanation of hypertension associated with 17α-hydroxylase deficiency syndrome. Steroids 2014;79:44–8.

Myint AM. Kynurenines: from the perspective of major psychiatric disorders. FEBS J 2012;279(8):1375–85.

Oehme NHB, Roelants M, Saervold Bruserud I, Madsen A, Eide GE, Bjerknes R, et al. Reference data for testicular volume measured with ultrasound and pubic hair in Norwegian boys are comparable with Northern European populations. Acta Paediatr 2020. https://doi.org/ 10.1111/apa.15159 [Epub ahead of print] 31899821.

Ong KK, Potau N, Petry CJ, Jones R, Ness AR, Honour JW, et al. Opposing influences of prenatal and postnatal weight gain on adrenarche in normal boys and girls. J Clin Endocrinol Metab 2004;89(6):2647–51.

Palmer LG, Schnermann J. Integrated control of Na transport along the nephron. Clin J Am Soc Nephrol 2015;10(4):676–87.

Pasqualini JR, Chetrite GS. The formation and transformation of hormones in maternal, placental and fetal compartments: biological implications. Horm Mol Biol Clin Investig 2016;27(1):11–28.

Pervanidou P, Chrousos GP. Early-life stress: from neuroendocrine mechanisms to stress-related disorders. Horm Res Paediatr 2018;89(5): 372–9.

Petersen MC, Vatner DF, Shulman GI. Regulation of hepatic glucose metabolism in health and disease. Nat Rev Endocrinol 2017;13(10): 572–87.

Rhen T, Cidlowski JA. Antiinflammatory action of glucocorticoids—new mechanisms for old drugs. N Engl J Med 2005;353(16):1711–23.

Rochira V, Kara E, Carani C. The endocrine role of estrogens on human male skeleton. Int J Endocrinol 2015;2015, 165215.

Rooney DP, Neely RD, Cullen C, Ennis CN, Sheridan B, Atkinson AB, et al. The effect of cortisol on glucose/glucose-6-phosphate cycle activity and insulin action. J Clin Endocrinol Metab 1993;77(5): 1180–3.

Rossier BC, Baker ME, Studer RA. Epithelial sodium transport and its control by aldosterone: the story of our internal environment revisited. Physiol Rev 2015;95(1):297–340.

Round JM, Jones DA, Honour JW, Nevill AM. Hormonal factors in the development of differences in strength between boys and girls during adolescence: a longitudinal study. Ann Hum Biol 1999;26(1):49–62.

Russell N, Grossmann M. Mechanisms in endocrinology: estradiol as a male hormone. Eur J Endocrinol 2019;181(1):R23–43.

Sathyanarayana S, Beard L, Zhou C, Grady R. Measurement and correlates of ano-genital distance in healthy, newborn infants. Int J Androl 2010;33(2):317–23.

Seo JY, Kim JH, Kong YY. Unraveling the paradoxical action of androgens on muscle stem cells. Mol Cells 2019;42(2):97–103.

Simmons PS, Miles JM, Gerich JE, Haymond MW. Increased proteolysis. An effect of increases in plasma cortisol within the physiologic range. J Clin Invest 1984;73(2):412–20.

Smith LB, Walker WH. The regulation of spermatogenesis by androgens. Semin Cell Dev Biol 2014;30:2–13.

Sorgdrager FJH, Naudé PJW, Kema IP, Nollen EA, Deyn PP. Tryptophan metabolism in inflammaging: from biomarker to therapeutic target. Front Immunol 2019;10:2565.

Stanley S, Moheet A, Seaquist ER. Central mechanisms of glucose sensing and counter regulation in defense of hypoglycemia. Endocr Rev 2019;40(3):768–88.

Stewart PM, Wallace AM, Valentino R, Burt D, Shackleton CH, Edwards CR. Mineralocorticoid activity of liquorice: 11-beta-hydroxysteroid dehydrogenase deficiency comes of age. Lancet 1987;2(8563):821–4.

Sun Y, Tao FB, Su PY, China Puberty Research Collaboration. Self-assessment of pubertal Tanner stage by realistic colour images in representative Chinese obese and non-obese children and adolescents. Acta Paediatr 2012;101(4):e163–6.

Szekeres-Bartho J, Schindler AE. Progestogens and immunology. Best Pract Res Clin Obstet Gynaecol 2019;60:17–23.

Taraborrelli S. Physiology, production and action of progesterone. Acta Obstet Gynecol Scand 2015;94(Suppl. 161):8–16.

Uenoyama Y, Inoue N, Maeda KI, Tsukamura H. The roles of kisspeptin in the mechanism underlying reproductive functions in mammals. J Reprod Dev 2018;64(6):469–76.

Valinsky WC, Touyz RM, Shrier A. Aldosterone, SGK1, and ion channels in the kidney. Clin Sci (Lond) 2018;132(2):173–83.

Valinsky WC, Touyz RM, Shrier A. Aldosterone and ion channels. Vitam Horm 2019;109:105–31.

Wang JC, Gray NE, Kuo T, Harris CA. Regulation of triglyceride metabolism by glucocorticoid receptor. Cell Biosci 2012;2(1):19.

Yamazaki O, Ishizawa K, Hirohama D, Fujita T, Shibata S. Electrolyte transport in the renal collecting duct and its regulation by the renin-angiotensin-aldosterone system. Clin Sci (Lond) 2019;133(1):75–82.

Yu J, Loh K, Song ZY, Yang HQ, Zhang Y, Lin S. Update on glycerol-3-phosphate acyltransferases: the roles in the development of insulin resistance. Nutr Diabetes 2018;8(1):34.

Zamponi V, Mazzilli R, Mazzilli F, Fantini M. Effect of sex hormones on human voice physiology: from childhood to senescence. Hormones (Athens) 2021;20(4):691–6.

Zhang K, Yang X, Zhang M, Wang C, Fang P, Xue M, et al. Revisiting the relationships of 2D:4D with androgen receptor (AR) gene and current testosterone levels: replication study and meta-analyses. J Neurosci Res 2020;98(2):353–70.

Zhu P, Wang W, Zuo R, Sun K. Mechanisms for establishment of the placental glucocorticoid barrier, a guard for life. Cell Mol Life Sci 2019;76(1):13–26.

Chapter 1.8

Breakdown and clearance of steroids

1.8.1 Introduction

Many changes in steroid hormone structure occur after glandular secretion by the adrenal cortex and gonads primarily with the aim of deactivating then removing steroids from the body. Hepatic metabolism is an important part of steroid catabolism and excretion, but peripheral metabolism can inactivate steroids or in some cases be an additional source of active hormones (see Steroids in blood and tissues). In many cases, adrenal steroids especially the androgens serve as precursors for peripheral modification which is particularly important in skin, fat, brain, sex organs and accessory sex organs, placenta, and kidney. Two comprehensive books by Dorfman & Ungar from 1965 and Bornstein from 1970 may still be useful to the serious reader (Dorfman and Ungar, 1965; Bornstein and Solomon, 1970). Reviews by Storbeck et al. (2019) and Schiffer et al. (2019) will be valuable for clinical aspects of steroid metabolism.

Some of the enzyme names will be familiar from steroid biosynthesis, although isoenzymes and different gene promoters will often be involved with variable tissue specificities. Our understanding of glandular and peripheral metabolism has improved as more has been learned about the genes which code for enzymes in steroid metabolism. A number of clinical problems have also provided important clues to the significance of peripheral metabolism and new pathways. These will be discussed in Part III of the book and include some disorders of sex differentiation, hypertension, hirsutism, breast cancer, prostate cancer as well as the effects of diet and antibiotic administration on risk of osteoporosis and effectiveness of contraception. The pharmaceutical industry is now applying this information and directing much attention to the design of inhibitors of peripheral steroid metabolism for clinical use. It is worth emphasizing that one third of all cancers, namely, breast, prostate, ovarian and uterine are hormone sensitive and are important candidates for control based on inhibiting local hormone production.

1.8.2 Hepatic catabolism of steroids

Hormones circulating in plasma are inactivated by metabolism in two phases primarily in the liver. The **phase 1** reactions convert the steroids by enzymatically catalyzed reactions into more polar compounds, inactivate the steroid and partly facilitate elimination from the body. **Phase 2** reactions are to conjugate the phase 1 products with glucuronic or sulfuric acid also in the liver. These steps are important to decrease receptor binding, increase the solubility of steroids prior to renal extraction and excretion in urine and in some cases through bile thence to feces. The uptake of steroids from blood into the hepatocytes has not been thoroughly studied, bearing in mind that most of the steroids are bound to specific proteins or albumin. In view of the quantities of steroid metabolites identified in urine, then metabolism of the free steroids alone in the liver is unlikely. Proteases acting on CBG and SHBG may release bound steroids. Steroids such as DHEAS are likely taken into liver cells through the action of organic anion transporters then desulfated before further metabolism as discussed below.

1.8.2.1 Phase 1

There are three major phase 1 pathways of steroid catabolism in the liver and several further pathways that vary with gender and age.

1. **Reduction of the double bond at C-4** with accompanying **reduction of the 3-keto group** to a secondary alcohol group involves the addition in total of four hydrogen atoms (tetrahydro products). The 5α- and 5β-reductases are involved for the conversion of androst-4-enes to 5-androstanes (dihydro-) followed by 3-keto reduction by 3α-hydroxysteroid dehydrogenase which is an aldo-keto reductase (principally **AKR1C4**) (Penning et al., 2000) that inactivates steroid hormones (Penning et al., 1997). The 5α reductase is a short chain alcohol dehydrogenase (**SDR5A1**) in the

endoplasmic reticulum of hepatic and peripheral cells (Uemura et al., 2008; Thigpen et al., 1992, 1993), whereas the 5β reductase (AKR1D1) is in the liver cytoplasm (Chen and Penning, 2014; Chen et al., 2011). Reduction to 5β products is an integral part of bile acid metabolism and is relatively more active for reduction of androstenedione, progesterone and 17-hydroxyprogesterone (Palermo et al., 2008). Both enzymes have NADPH as a cofactor. Reduction of the "A"-ring with four hydrogen atoms to form a 3α-hydroxy-5β-tetrahydro metabolite is the major fate of cortisol, cortisone (Fig. 1.8.1), aldosterone (Ulick, 1961; Flood et al., 1967), DOC (Calvin and Lieberman, 1962) and 18-hydroxy DOC (Melby et al., 1971, 1972). In contrast in man, the principal metabolites of corticosterone belong to the 3α,5α pregnane series (Migeon et al., 1956).

5-Dihydroaldosterone, and the 3α, 5α and 3β epimers of tetrahydroaldosterone, were characterized following the synthesis of reference steroids for comparison, hormone degradation and infra-red spectroscopy of the steroids, but the urinary excretion of each represented less than 1% of a dose of aldosterone (Kelly et al., 1962, 1963a). Improved syntheses and identification using nuclear magnetic resonance spectroscopy enabled the

characterization of 3α,5β-tetrahydroaldosterone-3-β-glucosiduronic acid after isolation from a 12 h urine collection using ion-exchange chromatography. The recovery was 34% of oral dose (Alvarez et al., 1976) which is in good agreement with other reported figures. 3α,5β-Tetrahydroaldosterone is inactive but the excretion of 3α,5α-tetrahydroaldosterone has not been measured and this would be an interesting experiment in sodium deprivation and certain categories of hypertension.

2. 17β-Hydroxysteroid dehydrogenase

Metabolites of androgens are mainly 17-keto reduced forms due to the actions of HSD17B enzymes. These are members of the short chain dehydrogenase family (Moeller and Adamski, 2009) with different substrate and co-factor requirements (Lin et al., 2013; Adamski and Jakob, 2001). The metabolites can be measured by LC-MS/MS methods (Pozo et al., 2010).

3. Hydroxylation

Cytochromes have important metabolic roles in steroid breakdown and clearance. The C-6 hydroxylation is an important fate for cortisol through the action of the cytochrome P450 enzyme from *CYP3A4* that is induced by certain drugs. This enzyme acts on other 3-keto-Δ4 steroids. **CYP3A4** is the most abundant in

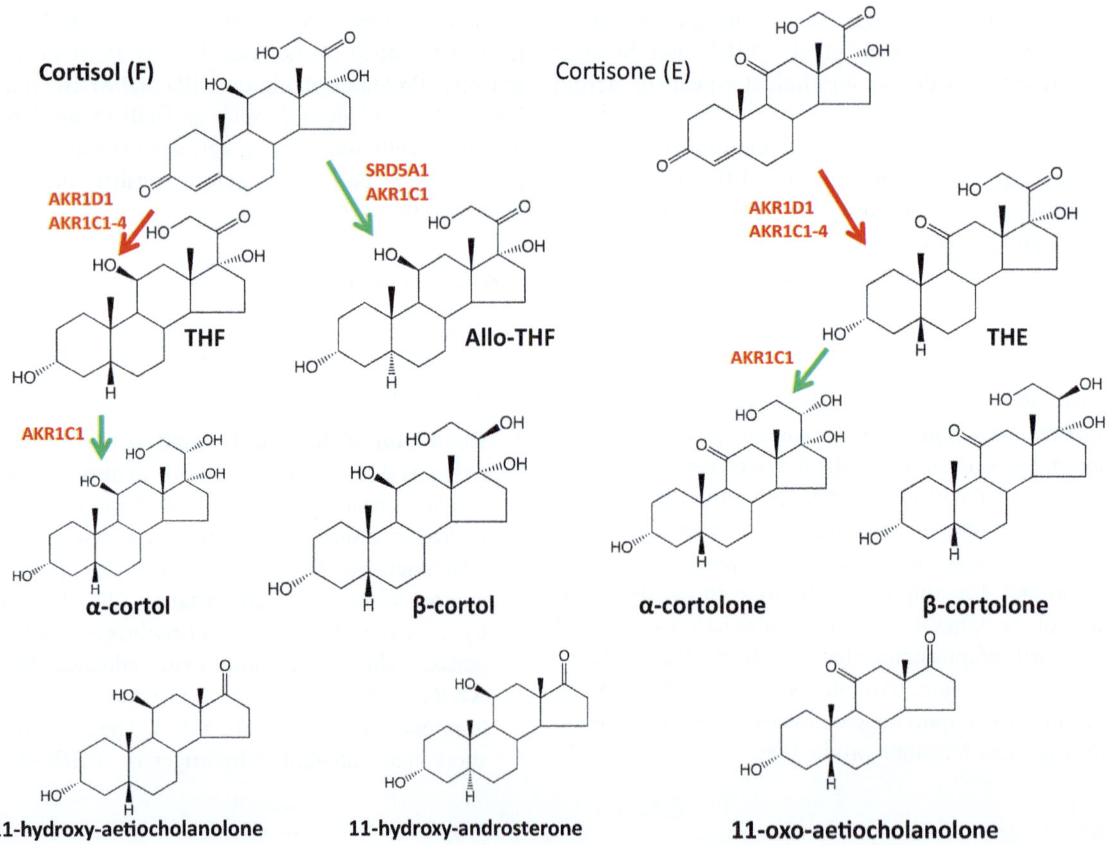

FIG. 1.8.1 Metabolism of cortisol and cortisone by A-ring reduction (tetrahydro compounds), oxidation of C-11 hydroxyl and reduction of side chain. *(Author original.)*

FIG. 1.8.2 Aldosterone metabolites. *(Author original.)*

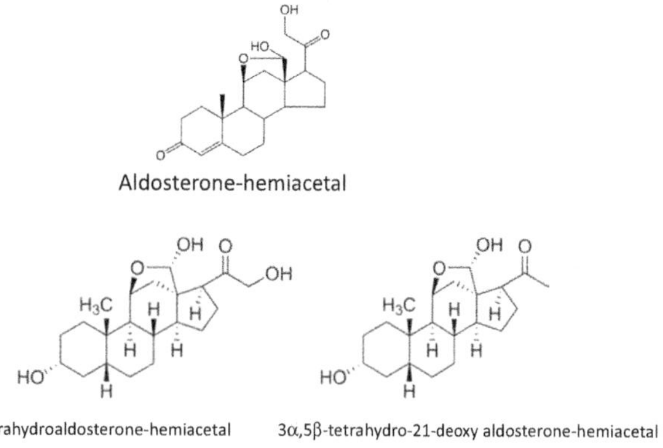

Aldosterone-hemiacetal

3α,5β-tetrahydroaldosterone-hemiacetal   3α,5β-tetrahydro-21-deoxy aldosterone-hemiacetal

the liver but is also expressed in the intestine and is responsible for the metabolism of steroids as well as drugs, prostaglandins and fatty acids. This enzyme is known to simultaneously bind multiple ligands and display atypical enzyme kinetics.

A number of further cytochrome P450 enzymes are involved in steroid hydroxylations (CYP3A5, CYP3A7, CYP2C19) (Niwa et al., 2015; Lee et al., 2003). In pregnancy, C-1 and C-6 hydroxylations of steroids are strong, continuing from the fetus into the perinatal period (Derks and Drayer, 1978; Taylor et al., 1978; Shackleton et al., 1979a,b,c; Kraan et al., 1980, 1993a,b). Urinary 6β-hydroxycorticosterone, 6β-hydroxycortisol (6β-OHcortisol), 6β-hydroxycortisone (6β-OHcortisone) excretion rates and a combination of the two, as well as plasma 4β-hydroxycholesterol (4β-OH-cholesterol) have been reported as predictive markers of CYP3A activity. The CYP3A7 is highly expressed in the fetus, while CYP3A4 is low.

Hydroxylation can also be at carbons **2, 6, 15, 16,** 17, 18 and 21 depending on the steroid substrate (Colombier et al., 1992; Shackleton and Taylor, 1975a,b; Taylor and Shackleton, 1978; Kraan et al., 1993a,b; Reeder and Joannou, 1995; Christakoudi et al., 2010a,b, 2012a,b). The genes for these reactions are **prominent in the newborn** but have yet to be characterized, many drugs can inhibit CYP3A4 function by binding to the active site. Some further steroid transformations are mediated by members of the cytochrome P450 family. Hydroxylation of the aldosterone nucleus has been demonstrated as for other steroids at C2α, 6β, 19 (Latif et al., 1989; Kirk et al., 1989, 1993; Harnik et al., 1985) but not 1, 7, 15, or 16. However, 21-deoxy and 19-nor aldosterone have been identified. Many metabolites have been detected in chromatography of biological samples without characterization (Morris and Tsai, 1981). The enzymes responsible for these reactions are uncertain.

Androgens (such as DHEA) are hydroxylated at C7 and other sites (C1, C15, C16, and C18 particularly in the fetus and newborn infant) (Shackleton and Taylor, 1975a,b; Lacroix et al., 1997; Christakoudi et al., 2013), although the precise cytochromes involved in those reactions is unknown. These steps decline over the first 6 months after birth. **CYP7B1** performs a broad spectrum of functions in diverse tissues. This enzyme functions as a steroid 7α-hydroxylase in the brain using pregnenolone and dehydroepiandrosterone as substrates. In the liver, CYP7B1 also catalyzes bile acid synthesis with both 25- and 27-hydroxycholesterol as substrates. CYP7B1 is involved in the inactivation of the 19-carbon steroid, 5α-androstane-3β,17β-diol that binds to the estrogen receptor an important action in the prostate gland. DHEAS has abundant effects due to conversion to potent downstream metabolites following actions of HSD3B1 alleles (Naelitz and Sharifi, 2020).

Estrogens can be formed by the action of **aromatase** which is the product of *CYP19* which is expressed in extra-gonadal tissues in men and women (Smy and Straseski, 2018; Thomas and Potter, 2013; Stocco, 2012). Estrogens are hydroxylated at C-2 and C-4 (catecholestrogens) (Fig. 1.8.3). In humans, **CYP1A1 and CYP1A2** are the primary enzymes catalyzing the 2-hydroxylation of estradiol and production of 2 and 4 hydroxyestradiol.

Catechol-*O*-methyltransferase (**COMT**) acts on 2-hydroxyestradiol to give 2-methoxyestradiol (Bai et al., 2007). Conjugation of parent estrogens to sulfate and glucuronide moieties; of catechol estrogens to methyl, sulfate, and glucuronide conjugates; and of catechol estrogen quinones to glutathione conjugates all represent potential "detoxification" reactions.

The **CYP2C19** in liver microsomes has been shown to catalyze progesterone conversion to form 21-hydroxyprogesterone as a major product, with 16α- and 17α-hydroxyprogesterone as minor products. **CYP2C9 and 3A4** can also oxidize progesterone to 16α-, 6β-, 2β-hydroxyprogesterone. The physiological significance of these metabolic pathways is unclear.

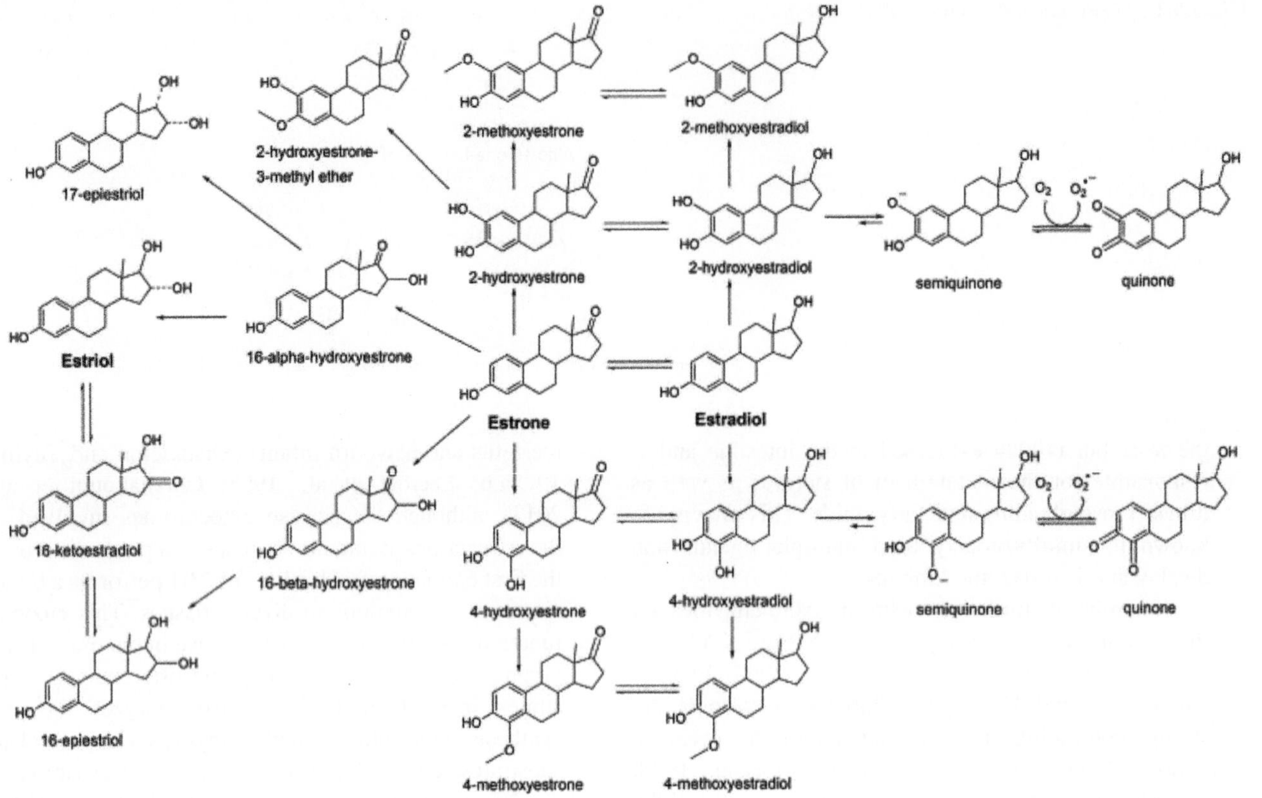

FIG. 1.8.3 Overview of estrogen metabolism by C2 and C4 hydroxylations (to catechol estrogens) and C-16 hydroxylation. The catechol estrogens may be converted to semiquinones then quinones. The catechol estrogens are metabolized by catechol-*O*-methyltransferase into methoxy derivatives. *AKR*, aldo-keto reductase; *COMT*, catechol-*O*-methyl transferase; *CYP*, cytochrome. *(From Smy L, Straseski JA. Measuring estrogens in women, men, and children: recent advances 2012–2017. Clin Biochem 2018;62:11–23. https://doi.org/10.1016/j.clinbiochem.2018.05.014.)*

Much research is still needed to understand the clinical significance of these hydroxylation reactions. Cytochromes are likely subject to much **genetic and epigenetic variation** (e.g., **methylation**) with considerable **ethnic** variation due to polymorphisms within the genes. **CYP variants** are classified as "**loss-of-function** variants" and "**gain-of-function** variants." The loss-of-function variants reduce clearance, and increase plasma concentrations, while gain-of-function variants increase clearance and lower drug concentrations. Loss-of-function polymorphisms in CYP genes frequently affect splicing and expression, as opposed to transcription or protein structure. The gain of function polymorphisms of CYP genes may result from **copy number variants (CNV)** with an increased number of functional gene copies or promoter variants and amino acid variants with increased substrate turnover of CYP genes. Drugs can induce or inhibit the activities of cytochromes. CYP induction is mostly transcriptional, although nontranscriptional mechanisms such as stabilization of mRNA, enzyme stabilization, or inhibition of the protein degradation pathway have also been reported.

The group of Professor David Kirk and others examined a number of microbial transformations to complement chemical techniques for production of steroids for the Steroid Reference Collection funded by the Medical Research Council. With several organisms he achieved hydroxylations at 7α, 7β, 15β as well as some 11α, 12α, 14α, 15β hydroxylated products of progesterone, testosterone and androstenedione (Smith et al., 1988, 1989a,b). The steroids were identified primarily by NMR. The products would have needed extensive purification (Smith et al., 1991).

4. Minor reductases

Steroid reductions are also found at C20 and C11. Hydroxysteroid dehydrogenases interconvert potent and relatively inactive forms of steroids using NADPH/NADP and NADH/NAD co-factors. In most cases, the enzymes function in one direction in vitro but in vivo intracellular metabolism and redox state can alter the HSD activity depending on cofactor abundance (Agarwal and Auchus, 2005). Further research has shown that the steroid binding pocket consists of 14 amino acids located on 5 protein loops, 3 of which are flexible. The aldo-keto reductases demonstrate promiscuity (Penning et al., 2015) in terms of regioselectivity, positional specificity and stereospecificity (Penning et al., 2019). The effects vary according to tissues beyond the liver that metabolize steroids now called intracrinology (Schiffer et al., 2019; Labrie et al., 1997). These processes have considerable effect in certain cancers so are of interest particularly in the search for enzyme inhibitors (Lin et al., 2013).

1. **Reduction of the 20-keto group in C_{21} steroids** to a secondary alcohol group (if after tetrahydro steroid formation gives hexahydro product) by 20α and 20β-reductases. The cortisol metabolites THE and THF are converted to cortols and cortolones, respectively (Morgan et al., 2017). Progesterone is reduced to 20α-hydroxyprogesterone through the 3-ketosteroid reductase actions of **AKR1C1 or AKR1C3** although the latter has relatively higher activity for androstenedione reduction to testosterone conversion in adipose tissue. The hepatic 20-Dihydro reduced metabolites of both corticosterone (20α- and β) and, 21-hydroxy-4-pregnene-3,11, 20-trione (20β only) were identified after ingestion of a large dose of the hormone (Bulaschenko et al., 1960). Pregnane-3α,11β,20,21-tetrols (hexahydrocorticosterone, 5β- and 5α-) have been identified in the urine of a healthy man also following a pharmacological dose of the hormone (Exley, 1965). The 5α/5β ratio was about 2 and the conversion rate was at least 10% in the first 24 h urine collection, making these metabolites as significant as tetrahydro corticosterone. In the absence of suitable reference compounds at that time, the stereochemistry at C-20 was not established. The reduction of C20 ketone to 20α hydroxyl is catalyzed by AKR1C1.

2. **Oxidation of the 11β-hydroxyl group of cortisol** by 11-hydroxysteroid dehydrogenase type 2 (**HSD11B2**) produces cortisone, then related metabolites, with a carbonyl group at C11. **HSD11B1** is a reversible enzyme that primarily acts as a reductase in the periphery leading to reduction of cortisone to cortisol (Terao and Katayama, 2016; Gathercole et al., 2013; Tomlinson et al., 2004). The enzyme co-localizes with hexose-6-phosphate dehydrogenase (H6PDH). The HSD11B1 can also inactivate cortisol to cortisone depending on cofactor affinity, cellular redox state and pH in the tissue. The activity is influenced by gender and body fat distribution and insulin resistance. HSD11B2 is exclusively an oxidase and notably co-localizes in the kidney with the mineralocorticoid receptor (MR). By oxidizing cortisol to cortisone the product is inactive at the MR allowing access to aldosterone which is present at much lower concentrations. Normally, most of the corticosterone metabolites retain the hydroxyl at C-11, although 18-hydroxy 11-dehydro tetrahydrocorticosterone (18-hydroxy-tetrahydroCompound A) is

the principal metabolite of 18-hydroxy corticosterone (Ulick and Vetter, 1962) by the action of HSD11B2. Aldosterone in blood and urine exists primarily as 11–18 hemiacetal, so the C-11 hydroxyl is not normally available for oxidation (Fig. 1.8.2) and in the major metabolites the A-ring is fully reduced.

Although 11-dehydroaldosterone (21-hydroxy-4-pregnene 3,11,20-trione-18-al) was an effective precursor in vitro for aldosterone biosynthesis by rabbit adrenal tissue (Fazekas and Kokai, 1967; Fazekas and Sandor, 1969), no aldosterone metabolites in man with a carbonyl function at C-11 have been positively characterized. Primarily on the basis of its infra-red spectrum, 21-hydroxy-4-pregnene-3,11,20-trion-18-al was identified in urine, but this has never been confirmed. The type 1 11-hydroxysteroid dehydrogenase (HSD11B1) will activate cortisone.

Consuming liquorice is known to cause hypertension (Dellow et al., 1999) and originally thought to act like aldosterone. The active component is glycyrrhizin and this recognized to inhibit HSD11B2 (Stewart et al., 1987) which is an important reaction in the kidney to normally act on cortisol so as to prevent binding to the mineralocorticoid receptor (MR). Other substances are now known to inhibit HSD11B2 and these have become to be characterized with glycyrrhetinic acid like activity (GALF) (Morris and Ridlon, 2017; Morris et al., 2014). For example, 3α,5α-tetrahydroaldosterone possesses GALF activity, potently inhibiting HSD11B2 but not HSD11B1.

The HSD11B enzymes act on substrates other than glucocorticoids including bile acids, 7-keto steroids, neurosteroids and xenobiotics (Gomez-Sanchez and Gomez-Sanchez, 2021). Key functions include regulation of immune cell differentiation, cytokine production and inflammation. Activities of the enzymes are influenced by age and obesity. HSD11B1 can act on 7-hydroxy steroids (Muller et al., 2006).

3. **Oxidative cleavage of the side-chain of C_{21}** steroids (scc) such as pregnenolone and cortisol leads to the production of DHEA and C11 oxygenated androgens, respectively (Miller, 2017), although the latter steroids are now known to also be produced in a synthetic path by 11-hydroxylation and oxidation of androgen.

4. **Cortoic acids.** In 1973 and after, Leon Bradlow and Carl Monder reported the identification of cortoic acids from cortisol, each being 21-oic acids of the cortols and cortolones (Bradlow et al., 1973). They comprised in urine 5%–25% of administered cortisol. The most abundant (8% of the dose) was in essence

cortolonic acid and was not a metabolite of 21-dehydrocortisol (Monder et al., 1975). Even earlier deoxycorticosterone had been shown in vitro to be converted to acidic metabolites (Schneider, 1964).

1.8.2.2 Phase 2

The above reactions serve to inactivate the steroid hormones but the polarity of these metabolites is further increased by conjugation. **Glucuronide** conjugates at C-3 and C-21 are more readily excreted in the urine than sulfate esters which are also formed. 5β-Androstane-3α,17α-diol glucuronide is an example of these conjugates (Fig. 1.8.4) which has been measured as an indicator of androgen excess (Li et al., 2019) (discussed in Chapter 3.2.1).

Glucuronidation reactions are catalyzed by enzymes of the **UDP-glucuronosyltransferase (UGT)** superfamily (now called **glycosyltransferase** because many of the proteins do not preferentially use UDP glucuronic acid), which have been classified into families and subfamilies based on sequence identity. The UGT superfamily is divided into UGT1, UGT2, UGT3 and UGT8 based on a variable number of exons (5, 6, 7 and 5/6, respectively) (Sten et al., 2009a; Knights et al., 2009; Reed, 2004). Eighteen catalytically active human UDP-glucuronosyltransferases (UGTs) are found in liver and kidney. In humans, UGT's 2B7, 2B15 and 2B17 are the main catalysts of glucuronidation of steroids. The human UGT2B genes are clustered at chromosome 4q13-21. A gene deletion of UGT2B17 was associated with a reduced excretion in urine of testosterone that was found to affect the TG to epitestosterone-G (T/e) ratio that was the basis of detecting testosterone abuse in athletes (Wilson III et al., 2004). Korean/Asian athletes had been found not to be failing doping tests for testosterone. There had been a suspicion that the concurrent use of nonsteroidal antiinflammatory drugs was masking the metabolism of testosterone (Sten et al., 2009a,b) but NSAIDs are conjugated by different UGTs. Polymorphisms in UGT2B17 were found in Asians and the T/e ratio was less than 2 and rarely above 6, which is the cut-off above which athletes are suspected of testosterone abuse (Okano et al., 2013). The UGTs are now known to be expressed beyond the liver and into mammary tissues, skin, uterus, ovary,

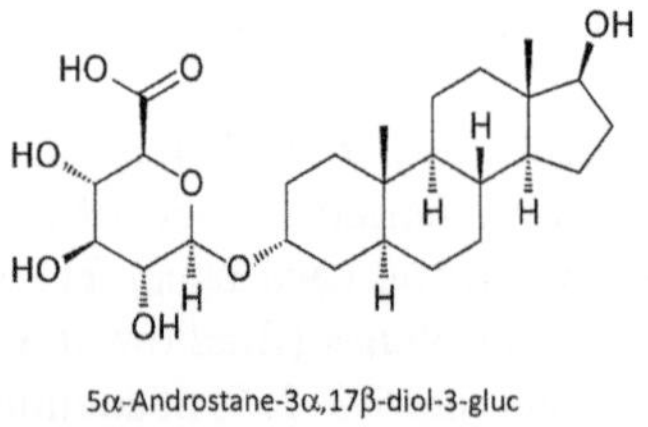

olfactory system and prostate glands. Conjugation may serve to locally inactivate steroids in the tissues such as breast tissue as well as increase hepatic and renal excretion (Reed, 2004; Pearlman et al., 1985).

Sulfation is important in the biosynthesis of steroid hormones and in modulating signaling pathways mediated by steroids and sterols (Mueller et al., 2015). The adrenal secretes high concentrations of androgen sulfates especially in the fetus, newborn infant and particularly DHEAS in the adult. **Sulfation** is one of the major conjugating pathways responsible for the detoxification and subsequent elimination from the body of a host of xenobiotics and endogenous small molecules. These biotransformation reactions are catalyzed by members of the cytosolic **sulfotransferase (SULT)** superfamily (Table 1.8.1) all of which utilize the universal sulfuryl donor **3′-phosphoadenosine 5′-phosphosulfate (PAPS)** in the transfer of SO_3 to acceptor substrates through hydroxyl functions (Mueller et al., 2018). These are soluble, cytosolic proteins that exist as homodimers although some family members are predicted to exist as monomers due to disruption of the dimer formation. The enzymes show degrees of substrate promiscuity.

PAPS is synthesized in a two-step reaction from inorganic sulfate and ATP by enzymes called PAPS synthetases (PAPSS). These are bifunctional enzymes in vertebrates with the two domains catalyzing the sequential ATP sulfurylase and APS kinase reactions required for the synthesis of PAPS. There are differences in tissue distribution (Meloche and Falany, 2001) and in vitro enzyme kinetics, however, more recently it has been demonstrated that there are marked differences in protein stability between the two human PAPSS enzymes that are directly influenced by the intermediate APS. Pregnenolone metabolites as mono- and di-sulfates are important in bile when rats were cannulated (Cronholm et al., 1971) but not when urine was collected in intact rats. This was early work on the enterohepatic circulation of steroids.

There are two SULT2 genes in humans, **SULT2A1 and SULT2B1**, that code for three functional sulfotransferases involved in the sulfation of steroids such as dehydroepiandrosterone (DHEA) and pregnenolone, and sterols including cholesterol and oxysterols (Fuda et al., 2002). The importance of sulfated steroids as circulating precursors for the biosynthesis of receptor-active hormones such as 17β-estradiol, testosterone and dihydrotestosterone are well known. DHEA sulfate (DHEAS) is the major circulating steroid in humans, and rosterone sulfate is a significant metabolite of androsterone in urine (Chang et al., 2004). Maintaining tissue levels of active steroid hormones involves, a delicate interplay between sulfation, desulfation by steroid sulfatase and the transport of steroid sulfates into and out of cells by members of the OATP and MRP transporter families, respectively, in addition to a host of

5α-Androstane-3α,17β-diol-3-gluc

FIG. 1.8.4 Androgen glucuronide metabolites. *(Author original.)*

TABLE 1.8.1 Sulfotransferases and their steroid substrates.

Steroid	SULT	K_m values (µmol)	References
DHEA	SULT2A1	0.8–3.7	Fuda et al. (2002); Gulcan and Duffel (2011), Meloche and Falany (2001), Ekuase et al. (2011), Chang et al. (2001, 2004), Falany et al. (1989), and Nishiyama et al. (2002)
	SULT1E1	0.2	Hempel et al. (2000)
Androsterone	SULT2A1	2.1	Chang et al. (2004)
Pregnenolone	SULT2A1	1.9–4.9	Fuda et al. (2002) and Meloche and Falany (2001)
	SULT2B1a	4.4	Fuda et al. (2002)
Estrone	SULT1E1	0.2	Hempel et al. (2000)
Estradiol	SULT1E1	0.04–0.3	Kester et al. (2002), Zhang et al. (1998), Nishiyama et al. (2002), Hempel et al. (2000), and Sun and Leyh (2010)
	SULT1A1	0.24	Gamage et al. (2005)
Cholesterol	SULT2B1b	1.2	Fuda et al. (2002)

From Mueller JW, Gilligan LC, Idkowiak J, Arlt W, Foster PA. The regulation of steroid action by sulfation and desulfation. Endocr Rev 2015;36(5):526–63.

biosynthetic enzymes (Davies, 2018; Gulcan and Duffel, 2011; Sun and Leyh, 2010; Chang et al., 2004). In addition to their function as a reservoir in the circulation and tissues, some sulfated steroids possess strong biological activity of their own. This is particularly the case for the neurosteroids, which are capable of modulating neurotransmission through their ability to interact with a range of neurotransmitter receptors, including GABA and NMDA, and other signaling pathways. Both DHEAS and pregnenolone sulfate are known to be functional neurosteroids.

The SULT2A1 enzyme sulfates a wide range of endogenous steroids (including DHEA, testosterone, dihydrotestosterone, androsterone and pregnenolone). Most sulfotransferases are prone to substrate inhibition which results when there is the formation of an enzyme-PAP-substrate dead-end complex The SULT2A1 enzyme is widely expressed in human tissues. In the adult liver, it is the second most abundant sulfotransferase.

More than one conjugate group has been added to steroids to give reference preparations of bis-glucuronides, bis-sulfates and combined glucuronic and sulfate derivatives (Pranata et al., 2019). Of the compounds synthesized, pregn-5-ene-3β,20S-diol 3-sulfate 20-glucuronide, 5α-androstane-3β,17β-diol 3-sulfate 17-glucuronide and androst-5-ene-3β,17β-diol 3-sulfate 17-glucuronide were detected in urine samples from males ($n = 11$) and females ($N = 11$) except for three urines out of 8 samples collected during pregnancy.

Conjugation of steroids is not limited to glucuronides and sulfates, other reactions have been observed with **glutathione, cysteine** (Marcos et al., 2015) (Fig. 1.8.5) and

N-acetylglucosamine (Fig. 1.8.6). These have usually been observed with estrogens but have been described also for 5-pregnenediol, progesterone and recently cortisol (Fabregat et al., 2013). Lipoidal derivatives have also been found with pregnenolone, estradiol (Schatz and Hochberg, 1981; Hochberg et al., 1979); **fatty acid acyl esters** of steroids have also been found in adipose tissue (Wang et al., 2013; Hochberg et al., 1991) and mammary gland tissue (Pearlman et al., 1985; Hampel et al., 1978). The significance of those conjugates will need further investigations.

Conjugates are amenable now to direct analysis with LC-MS/MS techniques and this will be an important area for development to clarify the extent of different conjugate patterns (Marcos and Pozo, 2016). The characterization of glucuronides and sulfates was based largely on the successful deconjugation with relatively specific enzymes. This practice led to confusion in the source and identity of steroids in brain. Ion exchange SPE techniques now enable the isolation of steroid conjugate groups. In the LC-MS/MS analysis, the molecular ions of intact conjugates can be seen and fragments due to loss of the conjugate groupings.

1.8.2.3 Metabolism of aldosterone

Aldo is metabolized to dihydro and tetrahydro derivatives that are subsequently glucuronidated in the liver (Knights et al., 2009). Additionally, Aldo as the hemiacetal is glucuronidated directly to form aldosterone 18β-glucuronide (ALDO 18β-G) (Fig. 1.8.2). Tetrahydroaldosterone glucuronide (THA-G) and ALDO 18β-G account for the majority

A

B

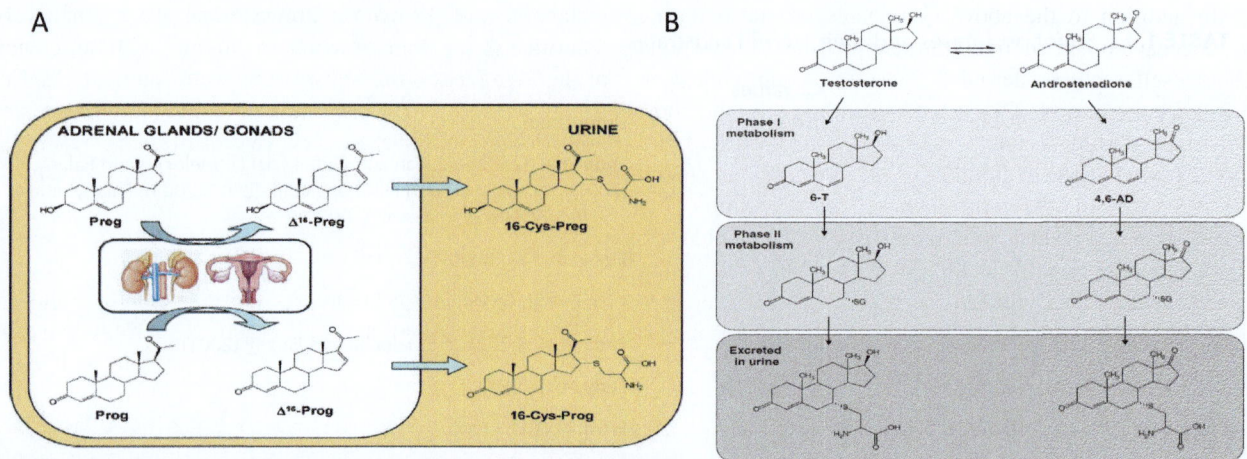

FIG. 1.8.5 Formation of cysteinyl conjugates. *(From Marcos J, Pol M, Fabregat A, Ventura R, Renau N, Hanzu FA, Casals G, Marfà S, Barceló B, Barceló A, Robles J, Segura J, Pozo OJ. Urinary cysteinyl progestogens: occurrence and origin. J Steroid Biochem Mol Biol 2015;152:53–61. Fig. 6 p. 60.)*

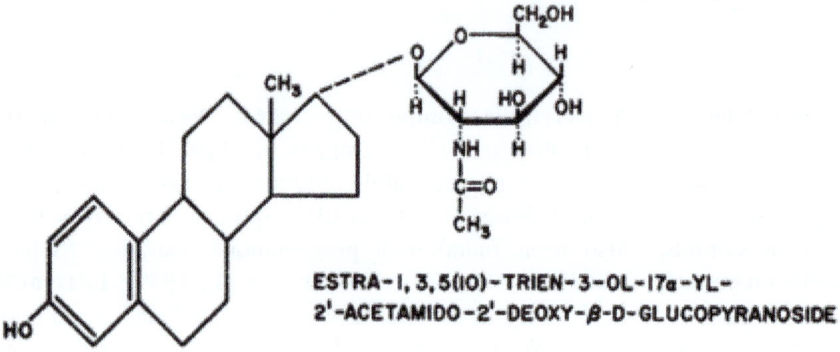

ESTRA-I,3,5(IO)-TRIEN-3-OL-17α-YL-
2'-ACETAMIDO-2'-DEOXY-β-D-GLUCOPYRANOSIDE

FIG. 1.8.6 *N*-acetylglucosamine conjugate of estradiol. *(Modified from Layne DS, Sheth NA, Kirdani RY. The isolation from rabbit urine of a conjugate of 17-alpha-estradiol with N-acetylglucosamine. J Biol Chem. 1964;239: 3221–5. Fig. 3, p. 3224.)*

of daily aldosterone elimination in urine (Ulick, 1961). However, unlike many other steroids, which are exclusively synthesized in the liver, the majority of ALDO 18β-G is thought to be formed in kidney (~80%) with a minor hepatic component.

UGT2B7 and UGT2B4 have been implicated in the glucuronidation of ALDO and THA, respectively. The common involvement of UGT2B7 in the metabolism of ALDO and NSAIDs raises the possibility of a metabolic interaction. David Morris found several hydroxylated metabolites in a "polar" fraction on HPLC analysis (Morris and Tsai, 1981). Hydroxylation at C2, C6 and C19 were found after isolation and synthesis of reference compounds (Kirk et al., 1989, 1993; Latif et al., 1989; Morris et al., 1990).

THAldosterone metabolites are conjugated and excreted mainly as glucuronides at C3, C21 and C18 but also as sulfates at the C3 and C21 positions. The C18 glucuronide is a unique conjugate of aldosterone that occurs in

human urine as a highly polar metabolite from which free aldosterone can be liberated by hydrolysis at pH 1 (Underwood and Tait, 1964). The structure was confirmed, by degradation and synthesis, as the C-18-β-D-glucosiduronate of aldosterone hemiacetal (Carpenter and Mattox, 1976). This metabolite is produced primarily in the kidney although an increased excretion of aldosterone-18-glucosiduronate in the third trimester of pregnancy has been attributed to metabolism in other organs. The synthesis of aldosterone-21-sulfate preceded the detection and measurement of this compound in urine. Approximately 0.7% of aldosterone is excreted as its C-21 sulfate ester (Grose et al., 1973). This was a minor product of injected aldosterone recovered in the sulfate fraction of a 48 h urine collection. 6β-Hydroxycorticosterone was found to be excreted in urine as the free steroid and equally as its C-21 sulfate ester. This was a minor product but may become important if 6β-hydroxylase enzyme in human liver is induced by certain drugs.

In addition to the above conjugates produced during hepatic catabolism of free steroid hormones, the adrenal gland itself secretes steroid C-21 sulfates, particularly in the newborn period. DHEAS is quantitatively the most important steroid secreted by the adrenal cortex except for the period of one to around 6 years of age. A comparison of the fate of ^{3}H-corticosterone sulfate with ^{14}C-corticosterone injected simultaneously indicated that the conjugate was rapidly metabolized by adults to polar urinary metabolites without hydrolysis, whereas most of the free corticosterone was converted to glucuronide conjugated tetrahydro-derivatives (Pasqualini et al., 1967). This metabolism may be important in pregnancy and the newborn period because the fetal adrenal secretes large amounts of corticosterone sulfate (Klein et al., 1973). The sulfated steroids are of immense interest in recent years.

1.8.2.4 Corticoid precursors

Progesterone is reduced by four hydrogens to the trivial named pregnanediol. The steroids deoxycorticosterone, corticosterone and 11-deoxycortisol are metabolized to tetrahydro and hexahydro reduced metabolites. Tetrahydro and hexahydro-metabolites of 17-hydroxyprogesterone (Fig. 1.8.7) are more usually referred to as 17-hydroxypregnanolone and pregnanetriol, respectively.

The major urinary steroid from 17-hydroxypregnenolone is pregnenetriol after addition of 2 hydrogen atoms. A large number of hydroxylated metabolites of 17-hydroxyprogesterone were tentatively identified in the urine of newborn children with deficiency of the 21-hydroxylase. Substituents were found at C2, C6, C15, C16, and C18 (Christakoudi et al., 2013). Those steroids disappeared over the first year of life.

1.8.2.5 Sex steroids

17β-Hydroxysteroids (androgens and estrogens) are subject to enzymatic oxidation by 17β-hydroxysteroid dehydrogenase (AKR1C3) to form the 17-keto steroid which are the main excreted products of C18 and C19 steroids (Manenda et al., 2016; Meier et al., 2009; Mueller et al., 2015). Testosterone is oxidized at C-17 to androstenedione. Ring A reduction first gives 5α- and 5β-androstanedione then principally 3α-reduction to give the trivial named metabolites androsterone and etiocholanolone, respectively. These two steroids along with DHEA are the principal **17-oxosteroids** (Fig. 1.8.8A). Testosterone and dihydrotestosterone give rise to small amounts of androstanediol isomers by A-ring reduction alone. Androstenedione is subject to similar fate (Fig. 1.8.8B).

A number of androgen hydroxylation products have been found at C2, 12, 16, 18, and 19 in newborn infants. Later 5α-androstane-3α,17β-diol glucuronide is a marker for peripheral androgen production and correlates with clinical signs of androgen excess in some cases (discussed in Chapter 3.2.1).

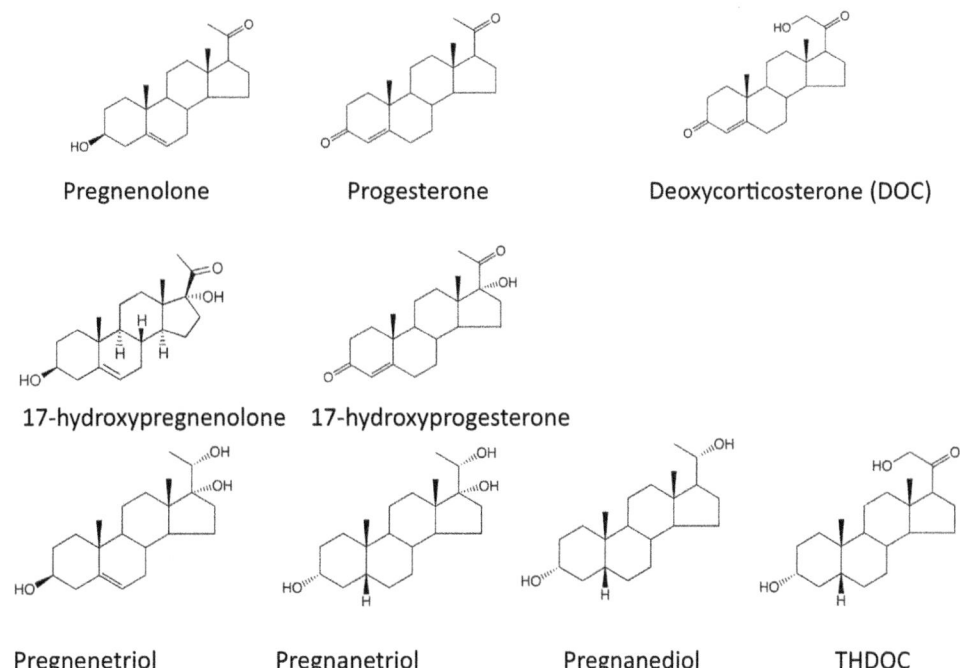

Pregnenolone   Progesterone   Deoxycorticosterone (DOC)

17-hydroxypregnenolone   17-hydroxyprogesterone

Pregnenetriol   Pregnanetriol   Pregnanediol   THDOC

FIG. 1.8.7 Metabolites of pregnenolone, progesterone and 17-hydroxyprogesterone. *(Author original.)*

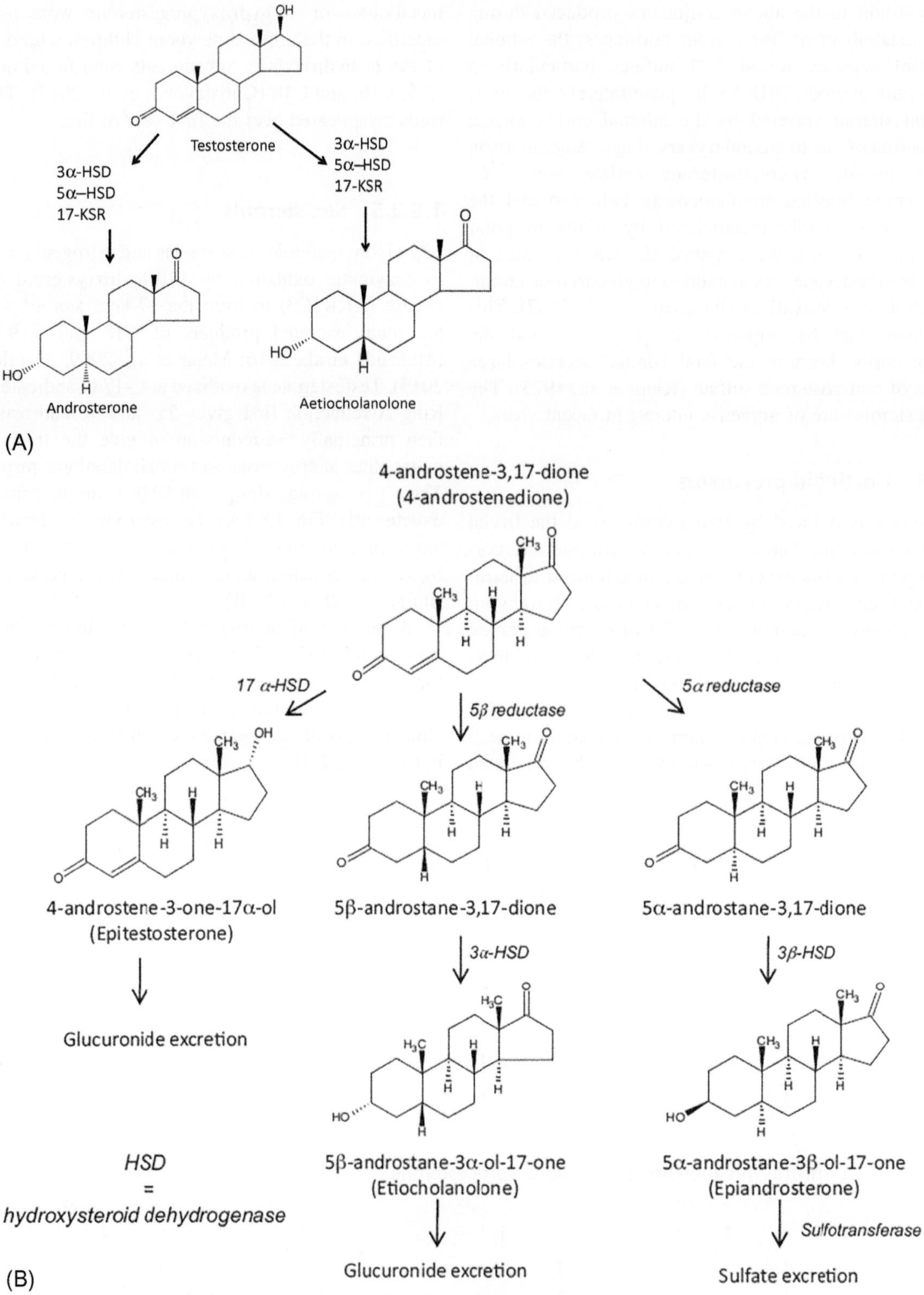

OH
Testosterone
3α-HSD
5α–HSD
17-KSR
3α-HSD
5α–HSD
17-KSR
O
HO
H
Androsterone
O
HO
H
Aetiocholanolone
(A)
4-androstene-3,17-dione
(4-androstenedione)
CH3
CH3
H
H
H
17 α-HSD
5β reductase
5α reductase
4-androstene-3-one-17α-ol
(Epitestosterone)
Glucuronide excretion
5β-androstane-3,17-dione
3α-HSD
5β-androstane-3α-ol-17-one
(Etiocholanolone)
Glucuronide excretion
5α-androstane-3,17-dione
3β-HSD
5α-androstane-3β-ol-17-one
(Epiandrosterone)
Sulfotransferase
Sulfate excretion
HSD
=
hydroxysteroid dehydrogenase
(B)

FIG. 1.8.8 Phase 1 metabolism of androgens (A) testosterone (B) androstenedione. *(Author original.)*

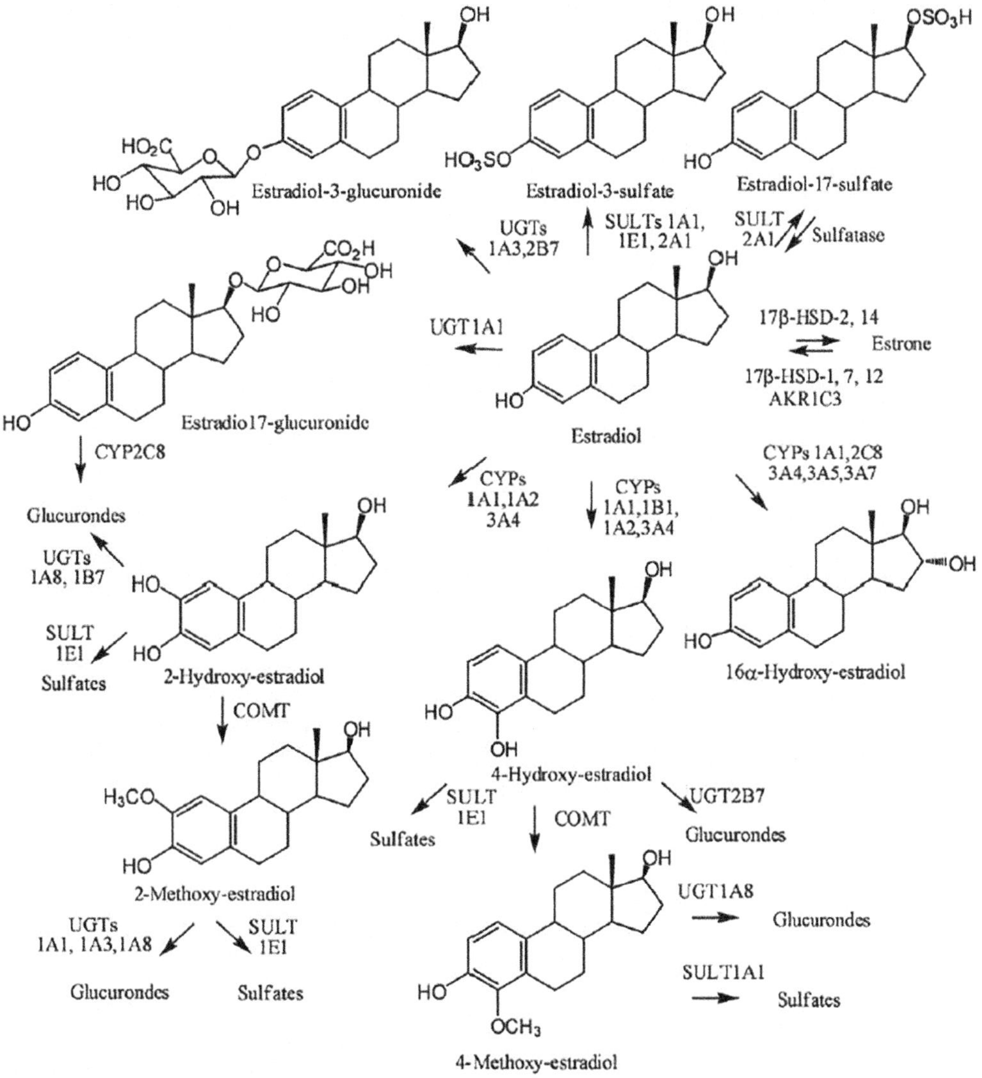

FIG. 1.8.9 Enzymes involved in estrogens metabolism and methylation. *(From Blair IA. Analysis of estrogens in serum and plasma from postmeno-pausal women: past present, and future. Steroids 2010;75(4–5):297–306. Fig. 2 p. 300.)*

Estrogens are metabolized by hydroxylation at the C2, C4, or C16 positions of the steroid rings producing hydroxy derivatives with possible subsequent methylation (methoxy derivatives) by catechol-*O*-methyl transferase (COMT) (Fig. 1.8.9), conjugation to glucuronic and sulfuric acids as well as to glutathione, *N*-acetylglucosamine though the latter two are not well characterized in the human. Studies in animals used measurements by immunoassay for which specificity is an issue that has not yet been addressed.

Past research proposed that estrogen metabolism by C2-hydroxylation pathway, rather than the C16-hydroxylation pathway, would lower breast cancer risk (Meilahn et al., 1998). Additionally, the hydroxy derivatives are considered more carcinogenic than ethoxy derivatives. Hydroxy derivatives can be further oxidized to mutagenic quinone products, while methoxy derivatives divert the metabolic

reaction away from these products. Only since the development of LC-MS/MS methods to detect the low concentrations of the estrogens and their metabolites have researchers been able to study their association with breast cancer.

1.8.3 Transport of steroids for breakdown and renal clearance

The mechanisms for steroid uptake and release by hepatocytes and renal tubular cells have been little addressed. **OATPs**, **SLC** polypeptides, and **MRP** proteins are likely involved but the precise routes of action have not been confirmed. The conjugation of steroids renders them more water soluble.

Receptor mediated endocytosis is involved in clearing steroids and many substances by cellular uptake and degradation or metabolic transformation. Cholesterol is taken up from plasma by LDL receptors and related receptors Megalin and Cubilin, although the exact mechanism of cholesterol trafficking between cellular compartments is a matter for further investigations.

A new orphan solute carrier, SLC22A24, was found in a momentous study using genome-wide association (Yee et al., 2019). Interestingly SLC22A24 correlated with levels of androsterone glucuronide and etiocholanolone-glucuronide. Studies of this nature will be more common in the search for pharmacological targets to regulate steroid levels.

1.8.4 Excretion of steroids in bile

Three major classes of sterols and steroids (with the exception of cortisol) pass in bile through to the gastrointestinal tract to interact with the gut micro-organisms to undergo microbial transformations. These classes are

- cholesterol originating from the diet, liver, intestinal epithelium, and other tissues,
- bile acids synthesized from cholesterol in the liver and excreted via the biliary tract, and
- steroid hormone metabolites synthesized from cholesterol in the adrenal cortex and gonads, cleared by the liver, and excreted via the biliary tract mainly as conjugates.

Between 400 and 1000 mg of cholesterol, 100–500 mg of bile acids, and less than 2 mg of steroids per day pass through the intestine in a healthy nonpregnant human subject on a "mixed western" diet. The cholesterol and the steroid hormones are generally not water-soluble. In bile, they form mixed polymolecular aggregates (micelles) with the highly soluble, ionized bile acid conjugates and other polar lipids such as phospholipids. During the passage through the lower gastrointestinal tract, all three types of steroids interact with the intestinal flora. The efficacy of oral contraceptives is reduced when antibiotics are taken because of loss of steroid from the gut.

The efficacy of oral contraceptives is reduced when antibiotics (particularly rifampicin types) are taken because of loss of steroid from the gut. Bile samples are not easy to obtain. In patients undergoing cholecystectomy, bile can be collected from a T-tube previously inserted at surgery for cholecystitis (removal of stones). The cumulative excretion of activity after corticosterone administration in the first 12 h bile collection reached 20 and 31% of the labeled dose, although no further change was noted (Migeon et al., 1956). A similar proportion of radioactivity was found in bile after intravenous administration of DOC (Casey and MacDonald, 1982). A 16% recovery of a dose of 18-hydroxy-DOC in bile from a patient with incomplete biliary fistula, and estimated that for a normal bile flow

the true excretion could account for 24%–39% of the total dose. In several studies (Melby et al., 1972), among the steroid conjugates excreted in bile tetrahydro DOC was identified as a mono-and-di-sulfate while 5α (and β) tetrahydrocorticosterone and 3α, 21-dihydroxy-5β (and α)-pregnane-11,20-dione were excreted only as monosulfates (Laatikainen, 1970).

1.8.4.1 Bacterial metabolism of steroids in the gut

The intestinal tract of adults contains approximately 1 kg of bacteria, equivalent to 10^{14} organisms in at least 400 distinct species. The organisms are almost exclusively obligate anaerobes. The numbers of bacteria increases from jejunum through the ileum and constitutes the bulk of the intestinal contents in the colon. Thus both dietary and endogenous substances come intimately in contact with this bacterial population in the lower gut. A range of microbial species, both anaerobes and facultative aerobes, can transform the steroids so that the physical and biological properties of the steroids undergo drastic changes. Estrogen metabolism has been studied in vivo and several organisms of the fecal microflora shown to have many activities including reduction of estrone, 16α-hydroxyestrone and 15α-hydroxyestrone (Järvenpää et al., 1980). Steroidal metabolites may often be absorbed and returned to the liver where further metabolism can take place. Bacteria in the intestine have been shown to possess glucuronidase and sulfohydrolase activity so some of the freed metabolites may be reabsorbed into the pool of neutral metabolites (Bouguen et al., 2015). Recycled metabolites can then be partially excreted again in the bile but some return to the blood for renal excretion. Information on these steroid transformations and their biological significance is gleamed through experiments with humans and certain animal models, by culture with feces cultures, isolation of pure bacterial cultures, and crude or purified enzyme systems. Transformations. The clearance of steroids from the liver into and bile and the reabsorption of steroids and return to the liver are called the **enterohepatic circulation (EHC)** (Gorbach and Goldin, 1987; Winter and Bokkenheuser, 1987; Adlercreutz et al., 1979; Eriksson, 1971a,b).

1.8.4.2 Hydrolysis of steroid glucuronides and sulfates by bacteria in the gut

In the normal human colon, the biliary steroids are deconjugated by the action of sulfatases and glucuronidases (Hawksworth et al., 1971). Bacterial sulfatases can hydrolyze steroid sulfates conjugated at the 3α-, 3β-, 17β-, and 21-hydroxyl positions (Huijghebaert et al., 1984). Glucuronidases are present in the intestinal wall as well as the bacteria of the gut. The bacterial enzymes are absent from newborn human feces but active in increasing concentrations with age particularly if delivered vaginally and breast milk fed. In the rat, glucuronidase activity is related to the presence of *Bacteroides*; similar types of *Bacteroides* are also present in human fecal flora (Adlercreutz and Martin, 1980). The hydrolysis of glucuronides or sulfates is very important because bacteria cannot modify the conjugated molecules. *E. lentum*, for example, is incapable of 21-dehydroxylation of DOC-21-sulfate, but does so in the presence of sulfatase.

1.8.4.3   Dehydroxylation of the 21-hydroxyl group

21-Dehydroxylating bacteria are present in stools of normal subjects regardless of the diet. An enzyme is synthesized by *Eubacterium lentum* (Bokkenheuser and Winter, 1980; Winter et al., 1979) and similar organisms of which there are about 10^7/g of human feces. Colonization takes place during the first year of life, and may well take place at birth because the organisms have been isolated from vagina of 25% of pregnant women. 21-Dehydroxylase is a constitutive enzyme with a pH optimum between 6.4 and 6.8. A hydroxyl group at C-20 protects the molecule against the action of the enzyme. The enzyme has no effect on hydroxyl groups at C-11 and C-17, and it acts independently of the configuration of ring A.

1.8.4.4   Reduction of ring A

The saturation of ring **A** in Δ4–3-keto steroids proceeds in two steps. Both reactions can be carried out by enzymes secreted by a single microbial strain but other organisms have the enzyme catalyzing only one of these reductions (Lombardi et al., 1978). When Δ4–3-keto steroids are incubated with intestinal flora there is stereo-specific reduction to the 3α-hydroxy-5β-configuration (Schubert et al., 1967). Complete reduction is performed with *C. paraputrificum* while partial reduction is achieved by *L. leichmanii*, *Bifidobacterium adolescentis*, and *Peptostreptococcus* (Bokkenheuser et al., 1976). Bacteria have been used for the synthesis of rare and expensive reference compounds, e.g., 3α-hydroxy, 5β-derivatives of 18-hydroxylated corticoids (Shackleton et al., 1979b). *C. paraputrificum* incubated with A4–3-keto steroids reduces the double bond before formation of a hydroxyl group at C-3. A methyl function at C-6 together with a 6, 7 double bond increases the resistance of the molecule to bacterial reduction (Cooke and Vallance, 1965). Furthermore, 1,4-dienes and 4,6-dienes are more resistant to *C. paraputrificum* than 4-mono-ene steroids and megestrol acetate, the most active oral progestin, owes its resistance to the 6-methyl, 4–6-diene configuration rendering it refractory to both hepatic and bacterial reduction.

1.8.4.5   Epimerization of 5α-hydroxyl group

Organisms related to *E. lentum* possess a 3α-HSDH responsible for the epimerization of the 3-hydroxyl group from α to β configuration. For example, incubation of pregnanolone with *E. lentum* strains yields pregnanedione, **3β**- and 20-pregnanolone under mild anaerobic conditions (Bokkenheuser et al., 1979a,b).

1.8.4.6   Introduction of a double bond conjugated to a keto group

Nuclear dehydrogenation may be carried out in vitro by some strains of *C. welchii* and *C. paraputrificum* in the presence of a hydrogen acceptor (Goddard et al., 1975). Nuclear dehydrogenation takes place in the gut because of the highly reduced environment. Delta-1 steroids are the major products.

1.8.4.7   Side chain cleavage

The formation of 17-keto steroids from cortisol was observed in patients given hydrocortisone *per rectum* at the Middlesex Hospital in London in 1959 (Wade et al., 1959) and confirmed in Stockholm (Gustafsson et al., 1968a,b; Eriksson, 1971a,b). 17KS production was reduced when antibiotics were taken. Side chain cleavage was also demonstrated in rats raised conventionally, whereas the reaction was not found in germ-free rats. Recently, C19 metabolites, 5ε-androstane-3,11β,17-triol and 5ε-androstane-3α, 11β-diol-17-one, were isolated from incubation of cortisol with human fecal flora and with *Clostridium scindens* (Cerone-McLernon et al., 1981; Doden and Ridlon, 2021; Ridlon et al., 2013). Free-living soil organisms can also remove the side chain of C_{21} steroids in the absence of the 17-hydroxyl group (Dorfman and Ungar, 1965).

1.8.4.8   Reduction of a 20-ketone to an alcohol

Organisms in the fecal flora such as *Bacteroides fragilis* converts DOC to a metabolite tentatively identified as 20,21-**dihydroxy-5β-pregnan-3-one** in 10% yield; but side chain cleavage cannot be produced from related structures such as THDOC and pregnanolone suggesting that the 20-HSDH of *B. fragilis* requires an intact 3-keto-4-ene structure for its action (Bokkenheuser et al., 1975). A different 20-HSDH in a common anaerobe in the intestinal flora, *Bifidobacterium adolescentis* is more efficient than the enzyme elaborated by *B. fragilis* and acts regardless of an unsaturation in the A ring (Winter et al, 1983). The enzyme shows a wide substrate specificity and reduces the 20-keto group to a 20β-hydroxyl group. It is affected by presence hydroxyl groups at C3, C11, C17, or C21. Once formed, 20-hydroxyl compounds are resistant to further bacterial metabolism of the side chain.

1.8.4.9 16α-Dehydroxylation

Eubacteria are also responsible for the conversion of 16α-hydroxy progesterone to 17α-progesterone in a two-step reaction (Bokkenheuser et al , 1981). The first, the removal of the hydroxyl group with formation of Δ^{16}-progesterone, and the second step, reduction of the Δ^{16} bond (Eriksson et al., 1968; Calvin and Lieberman, 1962).

1.8.4.10 Summary of bacterial metabolism of steroids in the intestine

From the above discussion of bacterial steroid metabolism (summarized in Fig. 1.8.10 largely from historical studies), many changes in steroid structure can be accounted. A few research groups were responsible for this work over 30 years to about 1983, the impact of this has been underestimated but with the current interest in the microbiome the process is being investigated for many health problems such as cardiovascular disease (Liuzzo and Galiuto, 2021); arthritis (Costello et al., 2015); cancer (El-Sayed et al., 2021) and bowel disease through dysbiosis (Kastl Jr et al., 2020). Changes in diet and antibiotic prescriptions have been shown to improve insulin resistance (Hwang et al., 2015).

The earlier bacteriological studies were hampered by poor techniques for identification, particularly of the important anaerobes, but bacteria can now be characterized by analysis of 16S RNA. Better methods for steroid analysis are also now available.

1.8.4.10.1 Estrogens

Conjugated estrogens are hydrolyzed by the intestinal microflora; and administration of antibiotics causes a huge

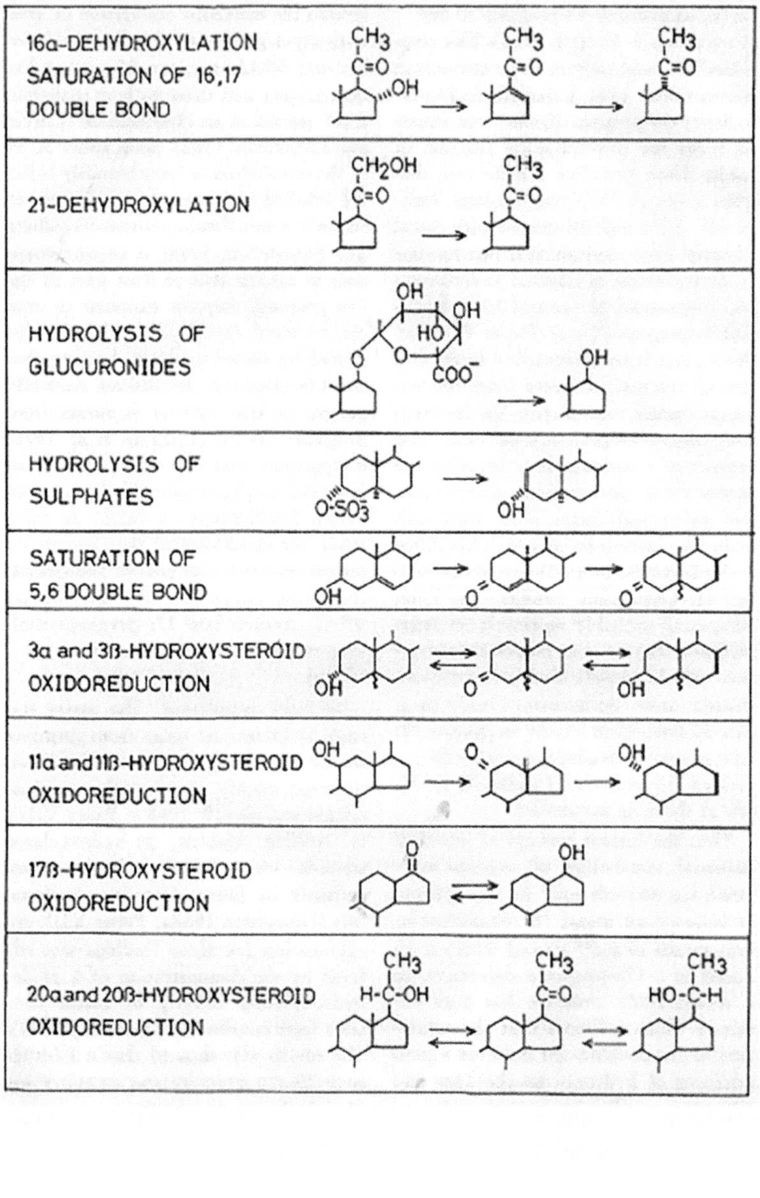

FIG. 1.8.10 Summary of bacterial metabolism of steroids. *(From Eriksson H. Studies on the metabolism of steroid hormones in the rat. Opuscula Medica Suppl XVIII. Thesis; 1971. Fig. 2 p. 13.)*

increase in the excretion of conjugates in the feces (Adlercreutz et al., 1979; Adlercreutz and Martin, 1980). Estrone and 15α-hydroxy-estrone can be reduced in the 17-position and 16-oxo-estradiol is converted to 16-epiestriol (Adlercreutz, 1967). Experiments with fecal flora or pure bacterial cultures are needed to clarify the precise mechanism and site of the transformation.

1.8.4.11 Historical studies of overall steroid metabolism

Many of the above studies were performed between 1965 and 1985 with now outdated technology. The bacteria from intestinal contents were then cultured on agar plates with enriched media grown aerobically and anaerobically. Individual colonies were then further processed in sugars and other media to type metabolism by acid and gas production. The bacteria were observed under the microscope after staining. The analysis of anaerobes was particularly difficult then but has changed dramatically.

Interest in the microbiome has arisen since 2015 with the realization that the bacterial content of the intestine is greater than the number of cells in our bodies. The phenomena will need re-investigation now that new technology based on RNA analysis has reclassified the bacterial species. The ability to examine culture media for metabolites is now much improved using LC-MS/MS. Many associations between disease and the microbiome have been found but the mechanisms are unknown.

Bacterial metabolism of corticosterone in a patient with congenital adrenal hyperplasia due to 17-hydroxylase deficiency leads to 50% of urinary metabolites without the 21-hydroxyl group (Honour et al., 1978). Tetrahydrocorticosterone and related steroids in the liver are excreted in bile as glucuronide and sulfate conjugates (Fig. 1.8.11). Bacteria in the intestine can hydrolyze the conjugates and have other actions on the steroids of which 21-dehydroxylation is probably a prominent reaction. 21-Deoxy steroids are reabsorbed from the intestine to the portal vein for return to the liver, thus completing the enterohepatic circulation.

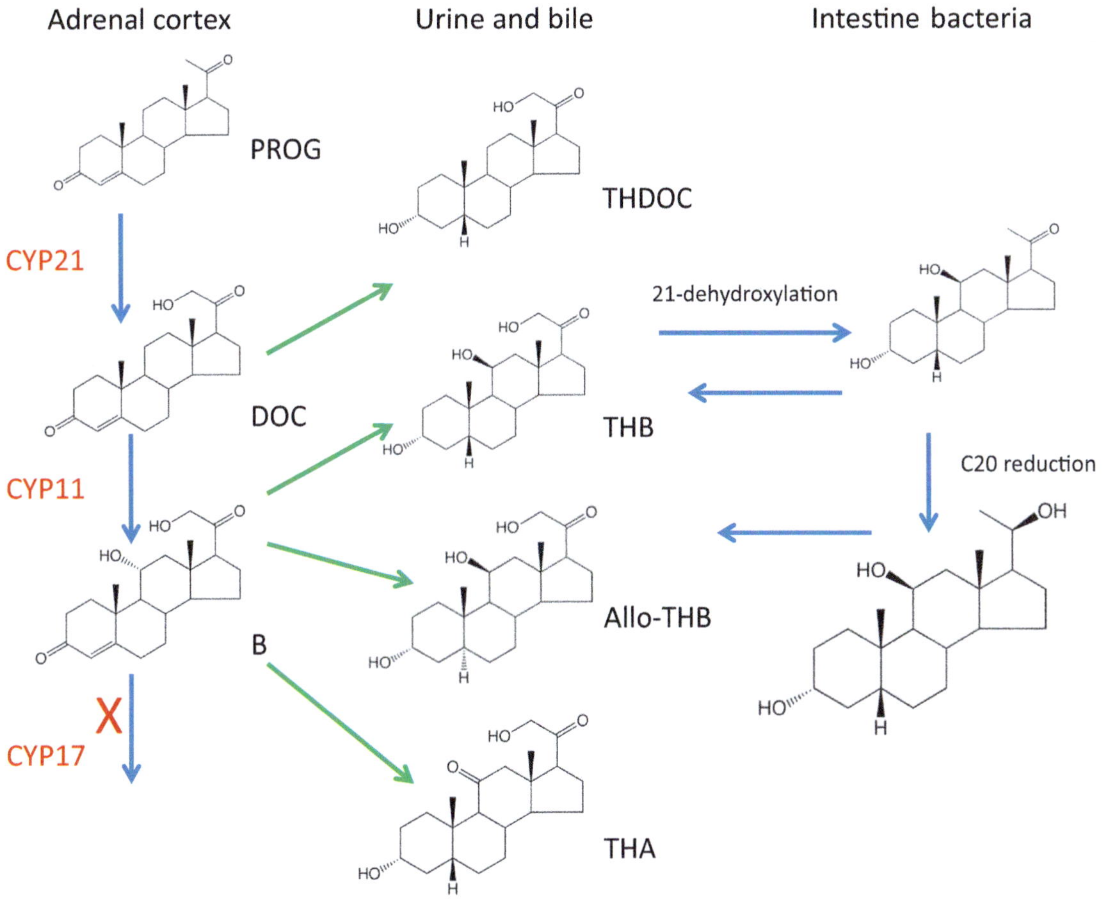

FIG. 1.8.11 Summary of corticosterone metabolism in a patient with 17-hydroxylase deficiency. Metabolism of corticosterone in liver by A-ring reduction, oxidation of C-11 hydroxyl and reduction of side chain. 21-Dehydroxylation takes place in the small intestine by the action of bacterial enzymes. *(Author original.)*

The changes in excretion rates of the 21-deoxy steroids in provocative tests (dexamethasone suppression and ACTH stimulation) were 1 day later than for the hepatic metabolites supporting the time for the enterohepatic circulation (Shackleton et al., 1979a,b,c).

After intravenous injection of radio-labeled aldosterone to a normal subject, 90% of the activity is recovered from the urine (Flood et al., 1967). The recovery of activity after injection of other mineralocorticosteroids is much lower, typical values being 79% for corticosterone (Migeon et al., 1956), about 45% for DOC (Bradlow et al., 1977) and about 50% for 18-hydroxy DOC (Melby et al., 1972). A comparison of the fate of radiolabeled corticosterone with a separate study of cortisol injected intravenously into the same patients and volunteer indicated major difference in the metabolism of these hormones (Migeon et al., 1956). The rate of disappearance from plasma of free corticosterone was faster than for cortisol, yet in a 2 day collection of urine less corticosterone metabolites were excreted than cortisol metabolites.

Two bicyclic acetal metabolites of tetrahydroaldosterone were characterized (Kelly et al., 1962) and formed in each case by reduction of the C-20 carbonyl to a secondary hydroxyl group which then reacts with the 11–18 hemiacetal with the elimination of water (Fig. 1.8.12).

The minor acetal metabolite, representing 1.4% of the oral dose of aldosterone-21-acetate can be regarded as a product of hexahydroaldosterone which itself was not identified. A 21-deoxy metabolite of tetrahydroaldosterone was characterized as the hemiacetal of 3α, 11β-dihydroxy-5β-pregnane-20-one-18-al, and the second acetal (8% of the dose) was a ketal of this metabolite justified by the general observation that 21-deoxy compounds formed by the bacterial metabolism of 21-hydroxy 20-keto precursors (Bokkenheuser et al., 1979a,b) had always been recovered in urine as the C-20 alcohols. The structures were confirmed by synthesis of the two compounds. This suggests that tetrahydroaldosterone may also be excreted in bile and

reabsorbed from the intestine after bacterial 21-dehydroxylation which may be the precursor for a compound M1 described by Kelly in 1963 (Fig. 1.8.13).

No 21-deoxy metabolites of 18 hydroxy DOC or its tetrahydro metabolites have yet been identified in urine or feces.

The neutral corticosteroid metabolites described so far may still represent 70% or less of the total radio activity recovered in urine. However, with improved methods for steroid extraction (e.g., solid phase extraction) it has become possible to recover from aqueous solution some further activity not extracted with organic solvents and, for example, an additional 5%–30% of the total metabolites of cortisol can now be accounted in polar fractions of steroids acids with a 20-oic acid group. An enzyme that oxidizes corticosteroids at position 21 has been purified from human liver (Monder and Bradlow, 1977) obtained at postmortem but the activity with 17-deoxy-C_{21} steroid substrates is considerably lower than for cortisol and evidence in vivo suggests that acidic metabolite formation from DOC may account for only 0.5%–2% of the hormone, although by measuring the transfer of radiolabel from [21-^{3}H]-DOC into body water the oxidation of DOC to an acid may be from 2% to 8%. Some of this activity can probably be attributed to bacterial metabolism of DOC in the intestine since loss of tritium has been demonstrated when [21-^{3}H]-DOC was incubated with human fecal flora. The investigations of steroid acids for biological activity and influence of disease on these metabolic changes have still to be evaluated. The possibility must also be considered that nonsteroidal acidic or neutral metabolites might be excreted after degradation of the steroid nucleus.

1.8.4.11.1 Effects of changes in intestinal bacterial flora on steroid metabolism

Males excrete about 13 mg and females 6.5 mg of neutral steroids in the bile per **24** h, but less than 2 mg arrives in the colon. The steroids are initially conjugated with both

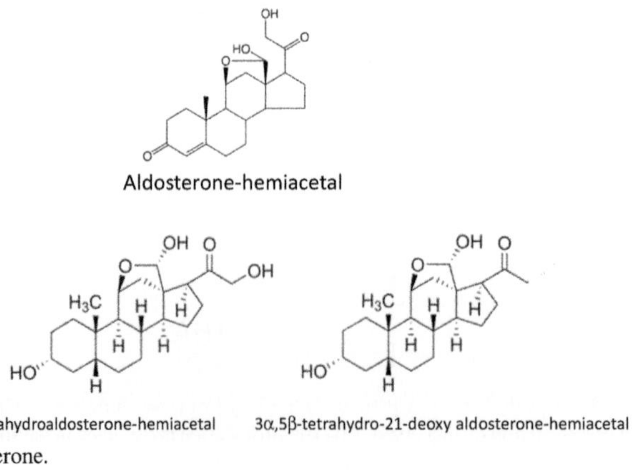

Aldosterone-hemiacetal

3α,5β-tetrahydroaldosterone-hemiacetal   3α,5β-tetrahydro-21-deoxy aldosterone-hemiacetal

FIG. 1.8.12 Bicyclic metabolites of aldosterone.

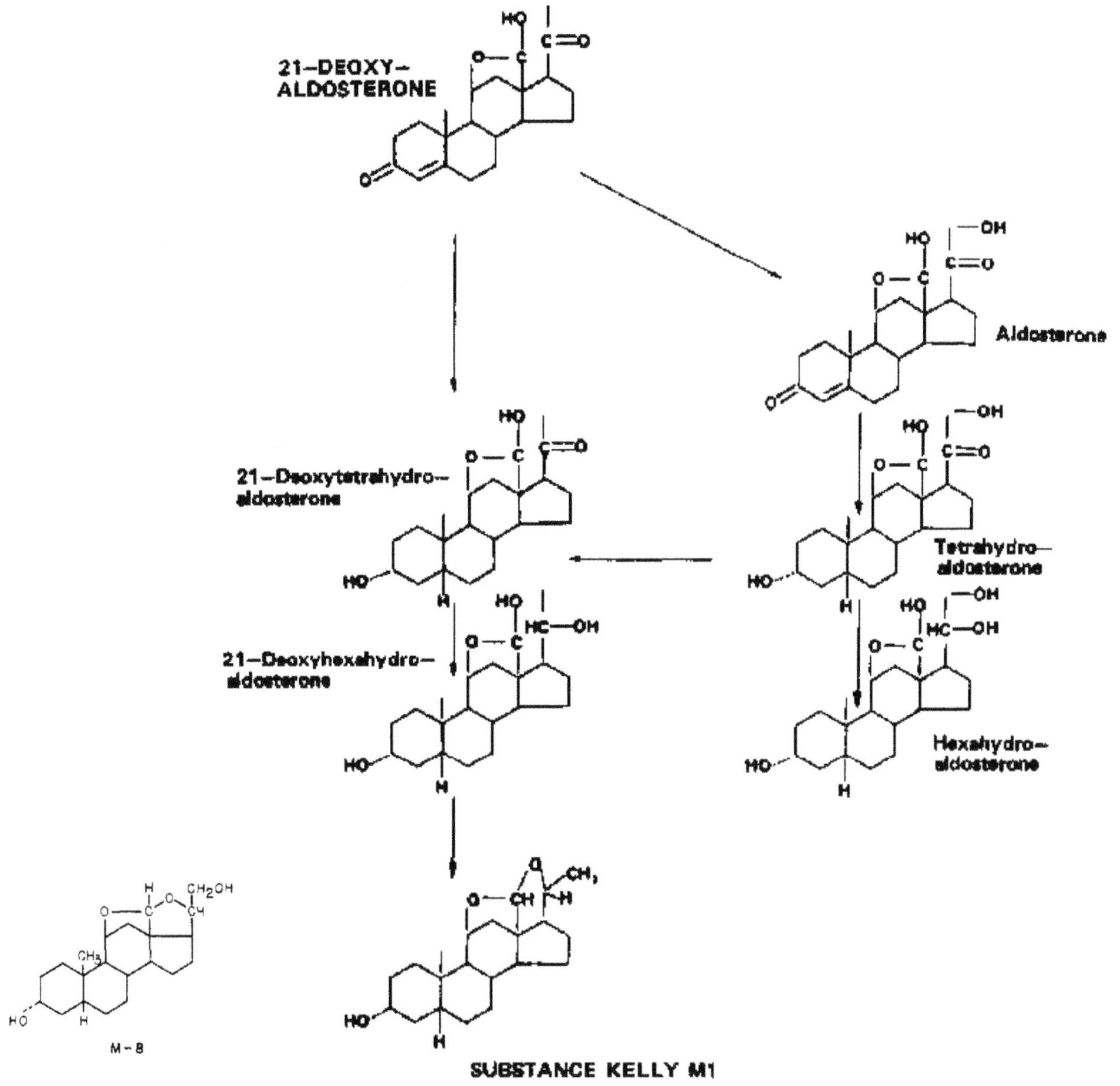

FIG. 1.8.13 Formation of compound Kelly M1 from 21-deoxyaldosterone. *(From Lewicka S, Vecsei P, Bige K, Fisher T, Winter J, Abdelhamid S, Heinrich U, Bokkenheuser VD. Urinary excretion of aldosterone metabolite Kelly-M1 in patients with adrenal dysfunction. J Steroid Biochem 1988;29(3):333–9. Scheme 1 p. 338.)*

sulfuric and glucuronic acids, e.g., estriol is secreted in the bile as the 3-sulfate, 16α-glucuronide, while estrone is conjugated with sulfuric acid only and estradiol with glucuronic acid only. Quantification has only been achieved in patients with bile fistula. Nonpregnant and pregnant women excrete around 15 mg and 3500 mg of estriol, respectively, per 24 h (Adlercreutz, 1967). estrone and estradiol, constitute less than 15% of the biliary phenolic steroids. The bulk of estrogens in the bile are hydroxylated at the 15 or 16 position in the liver. After absorption, the products are transported to the liver. The bulk of the progestins also escape ring **A** reduction in the liver and enters the circulation via

the sinusoids. A small proportion of the compounds are reduced in the liver, despite the protective ethinyl group, conjugated, and excreted in the bile.

Most of the synthetic estrogens taken orally probably also pass unaltered through the liver, but sooner or later circulating estrogen metabolites are taken up by the liver cells, conjugated, and excreted in the bile. Most of the conjugated biliary steroids, normal or synthetic, are deconjugated in the gut, and reabsorbed before or after bacterial alterations although synthetic steroids have a higher rate of fecal excretion than the natural steroids. So, following intravenous administration, 5%–7% of estrone and estradiol, but 30% of ethinyl estradiol is excreted in the feces (Sandberg and Slaunwhite, 1951; Reed et al., 1972). The fecal excretion of norgestrel and norethisterone is even higher (Smith, 1974). The hepatic uptake and conjugation of steroids appear to be abnormal in certain diseases. In patients with Gilbert's disease (a genetic disorder associated with glucuronyl transferase deficiency), for example, the hepatic uptake and conjugation of steroids are decreased. In patients with Dubin-Johnson syndrome (an inherited liver transport defect of conjugated bilirubin), the excretion of conjugated anions is reduced. Cholestasis decreases biliary excretion of estrogen glucuronides. Synthetic steroids also undergo EHC. Pregnancy can be a risk if antibiotics, particularly rifampin, are prescribed when contraception is based on synthetic steroids.

Adlercreutz et al. (1979) showed differences between the steroid content of bile and feces indicating bacterial changes before some reabsorption into the circulation (EHC). The administration of antibiotics that affected the bacterial population of the intestine, made the fecal content more like the composition of bile.

1.8.4.11.2 Germ free rats

A comparison of steroid metabolism in vitro and in vivo using germ free and conventional rats demonstrated extensive metabolism of steroid hormones and their metabolites by intestinal bacteria (Eriksson, 1971a,b). Subsequently, bacteria isolated from the human intestine have been shown in culture to reduce hydroxyl groups in steroids at the 16α- and 21-positions. Since no 21-dehydroxylating enzymes have been reported in human tissues, it must be concluded that the earlier isolation of pregnanediol and 11-oxo-pregnanediol from urine after the administration of DOC and 21-hydroxy-4-pregnene-3,11,20-trione, respectively, was due to bacterial metabolism of biliary steroids which had been reabsorbed from the intestine. 11-Oxo-pregnanolone and allo-pregnanolone (Fig. 1.8.14) may inhibit 11β-hydroxysteroid dehydrogenase type 2 and thus act as glycyrrhetinic acid like compounds (GALFs) as proposed recently by Morris et al. (2014).

The 33% of an intravenous radioactive dose of DOC is recovered from feces but this is not a significant excretory route for other mineralocorticosteroids since they are effectively reabsorbed from the intestine. The predominant isomers of the steroids both remaining in the free steroid fractions of feces and reabsorbed from the intestine have a $3\alpha,5\beta$ or $3\beta,5\alpha$ configuration which indicates that the bacteria also possess reductase and isomerase activities.

Hypertension in rats can be induced by a number of mechanisms including administration of mineralocorticoid steroids and adrenal stimulation with ACTH. In my laboratory in 1981, we were able to prevent hypertension from ACTH or corticosterone by administration of neomycin

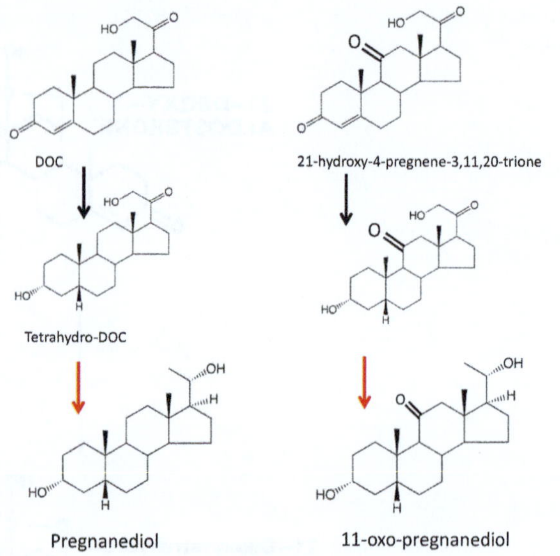

FIG. 1.8.14 Metabolism of tetrahydo-11-deoxycorticosterone. *(Author original.)*

(Fig. 1.8.15A) or vancomycin (Fig. 1.8.15B). We speculated that the reabsorption of steroids that influence the response of blood pressure from the intestine was prevented. The rise in hypertension in spontaneous hypertensive rats of stroke prone substrain was also prevented by neomycin (Fig. 1.8.16).

In the 1980s, the identification of bacteria from intestinal samples was difficult because most of the bacteria in the gut are anaerobes which were difficult to culture before biochemical tests for identification. From 2015, there have been several papers looking at dysbiosis of the gut and hypertension, these studies have been association studies and the mechanism has yet to be elucidated. Bacteria are now identified from 16S RNA. Morris et al. (2014) has likened the possible effects of intestinal steroids on blood pressure glycyrrhetinic acid, a constituent of licorice.

1.8.4.11.3 Effects of vegetarian diet

Estrogens are excreted in high amounts in the bile, in fact following the administration of estriol to women, 23% is excreted in bile (Goebelsmann et al., 1966). The estriol in bile is conjugated with glucuronic acid and sulfuric acid as single or mixed conjugates, of which the major product is the glucuronide at the C-16 position (Goldin et al., 1982). In the intestine, bacterial enzymes hydrolyze the estrogen conjugates to release free estrogen. During reabsorption of the steroid through the intestinal wall, estriol-3-glucuronide is formed (Clifford et al., 1986). This steroid is regarded as a specific intestinal mucosal product. Its concentrations in plasma or, better still, its excretion rate in urine can be used to monitor the influence of the intestine and diet on estrogen metabolism.

FIG. 1.8.15 (A) Systolic blood pressures (mmHg) in rats treated with corticosterone (*triangle*), treated with corticosterone and neomycin (*square*) and injected with diluent only (*circles*). Blood pressures in the morning are represented by open symbols and in the evening with closed symbols. The mean±2 SEM are shown for six male rats. (B) Effects of antibiotics on hypertension induced with corticosterone (4 mg/day). Systolic pressures (mmHg) in rats pretreated with neomycin (*closed squares*) or vancomycin (*diamond*) are compared with normal rats given steroid only (*triangle*). The mean±2 SEM are shown for six rats. The rats had been pretreated with oral antibiotics. (*Author original.*)

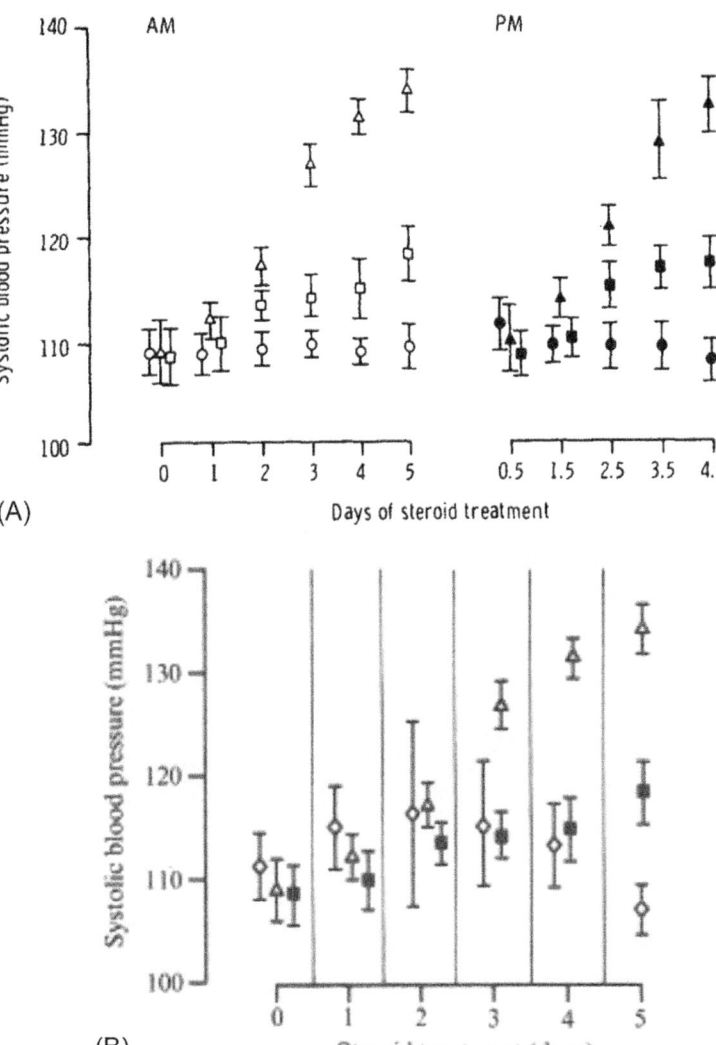

(A)

(B)

Diet can influence gut motility, fecal bulk and pH as well as the bacterial flora. In these ways, diet may affect several controlling factors during the enterohepatic circuit of estrogens and thus influence the availability of sex steroids to the tissues. Such a mechanism has been explored to account for the significant correlation between dietary fat consumption and the increase in breast cancer on a worldwide basis (Clifford et al., 1986). Other dietary components like fiber plays a role in the regulation of the enterohepatic circulation of estrogens (Adlercreutz et al., 1987). That intestinal bacteria play a significant role in estrogen metabolism has been demonstrated in numerous studies that show a change in steroid metabolism when the intestinal microflora is altered by treatment with antibiotics such as oxytetracycline (Hämäläinen et al., 1987; Hämäläinen et al., 1991). An increased calorie intake in adulthood may increase total estrogen production. A vegetarian diet is associated with reduced levels of plasma unbound estrogen and

of prolactin (Goldin et al., 1982). This may partly arise from the greater fecal excretion of estrogen in vegetarians than in nonvegetarians, based upon an interruption of the enterohepatic circulation because of the diet induced change in the intestinal flora. Low estrogen exposure is regarded as protective against breast cancer conversely this could create risk for osteoporosis.

A longitudinal survey of nutrition, blood chemistry and bone mineral density (BMD) of the spine and hip was conducted in free-living vegetarian women, by Jasmine Challis at the Steroid Laboratory of the Middlesex Hospital, later University College London Hospitals. The vegetarians living in South East England were studied at intervals over 2 years. Lifestyle questionnaires and blood samples were analyzed during the study. Ninety healthy premenopausal women vegetarian for a mean of 13.5 years were compared with 35 age-matched healthy premenopausal omnivorous women. The women were of normal weight were

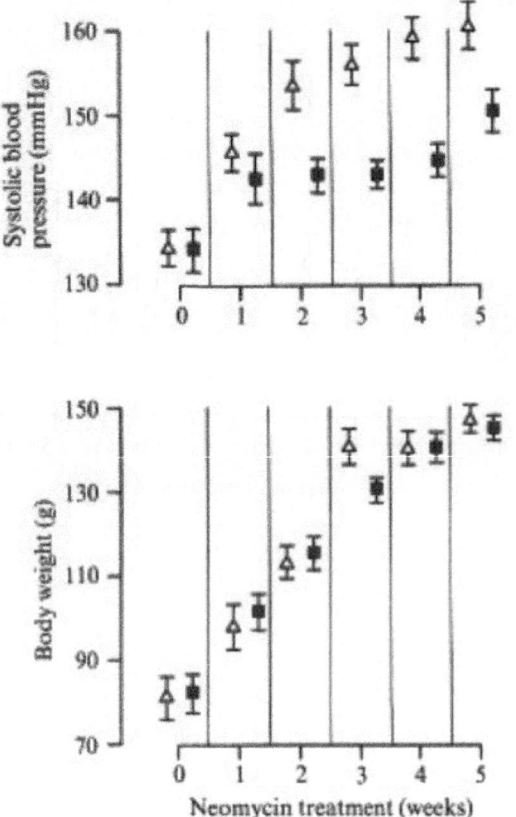

FIG. 1.8.16 Effects of antibiotics on the development of hypertension and body weight of hypertensive rats of stroke-prone substrain. The mean ± SEM are shown for control (*triangle*) and neomycin treated (*closed square*) rats n = 10. *(Author original.)*

nonsmokers and were aged 35–45 years at entry to the study. The numbers of vegans (n = 11), ovovegetarians (n = 1) and lactovegetarians (n = 7) were small compared with the ovo-lactovegetarians (n = 71). Blood tests revealed normal hormonal status in most subjects; one case of hypothyroidism emerged among the vegetarians. Mean BMD in both femoral neck and total hip were significantly lower in vegetarians than in omnivores (P < .05) (Table 1.8.2). Osteopenia, by WHO standards using the Hologic young normal values, was found in the femoral neck of 47% of vegetarians and 20% of the omnivores (P < .01) (Challis et al., unpublished).

Using total hip BMD, 31% of vegetarians showed **osteopenia** compared with 10% of omnivores (P < .05). Femoral neck BMD correlated with weight and duration of vegetarianism but not with oral contraceptive pill usage, number of pregnancies, age at menarche, alcohol intake, or hours of total or weight bearing exercise per week. There was no significant difference in BMD of the spine between omnivores and vegetarians although 17% and 19%, respectively, were osteopenic. So, reduced bone mineral density of the hip was found in vegetarian women and osteopenia correlated with duration of vegetarianism. Larger studies are needed to confirm these findings.

1.8.5 Inhibitors of steroid metabolism

Components in licorice and chemically related carbenoxolone inhibit HSD11B2 and thus lead to a block in cortisol inactivation and a state of apparent mineralocorticoid excess. The active component of licorice is glycyrrhetinic acid (Fig. 1.8.17) and some steroids act as glycyrrhetinic acid like factors (GALFs).

TABLE 1.8.2 Bone density (mean ± SE) of vegetarians and omnivores (*P <.05) (Challis and Honour, unpublished).

	Omnivores (n = 35) first scan	Vegetarians (n = 90) first scan	Om (n = 32) final scan	Veg (n = 77) final scan
	Mean (SE)	Mean (SE)	Mean (SE)	Mean (SE)
BMD spine	1.05 (0.02)	1.06 (0.01)	1.05 (0.02)	1.05 (0.01)
BMD total hip	0.973 (0.02)	0.925 (0.01)*	0.986 (0.02)	0.937 (0.02)*
BMD fem neck	0.860 (0.02)	0.808 (0.01)*	0.858 (0.02)	0.798 (0.01)*
t score spine	0.15 (0.2)	0.10 (0.12)	0.02 (0.2)	0.08 (0.14)
t score total hip	0.05 (0.14)	−0.39 (0.11)*	0.10 (0.17)	−0.29 (0.14)
t score fem neck	−0.24 (0.21)	−0.89 (0.12)*	−0.39 (0.19)	−0.88 (0.15)

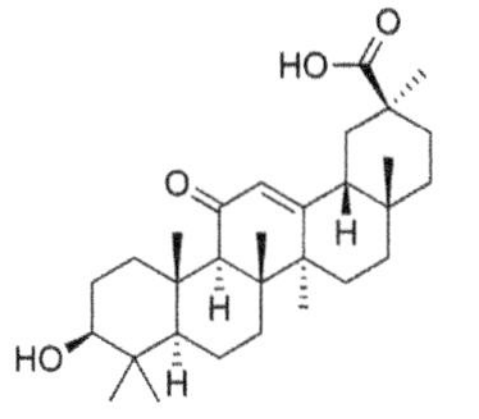

FIG. 1.8.17 Glycyrrhetinic acid is a compound in licorice that inhibits conversion of cortisol to cortisone. Compounds with similar activities are glycyrrhetinic acid like factors (GALFs).

Because aromatase mediates estrogen formation and the estrogens have a role in mammary tumor growth the development of aromatase inhibitors is of particular interest for the control of breast cancer. Inhibition of aromatase can be achieved by a variety of inhibitors both steroidal and nonsteroidal. Recent research efforts have focused on enzyme-activated irreversible inhibitors (suicide substrates), e.g., 4-hydroxyandrostenedione. Inhibitors of HSD17B3 are also used to treat breast and prostate cancers.

1.8.6 Peripheral steroid metabolism

1.8.6.1 Steroid reduction

In the periphery, a number of enzymes serve to activate or inactivate steroid hormones. Testosterone activity is increased by the action of 5α-reductase (SRD5A2) (Blouin et al., 2009). The importance of this enzyme is apparent in male patients with inactive SRD5A2 (see Chapter 3.2.2 Disorders of puberty) who are born with female phenotype but take on a male physique at the time of puberty. This indicates that DHT must be important for expression of male sex characteristics in the fetus but T at puberty is sufficient to develop skeletal muscles.

HSD3B2 is expressed in adrenals and gonads but an HSD3B1 is expressed in skin, breast and placenta. HSD3B1 in placenta is a reversible enzyme usually acting to convert DHEA to androstenedione and oxidize pregnenolone to progesterone. Inhibition of HSD3B1 in pregnancy with 17-hydroxyprogesterone caproate has been tried to prevent spontaneous labor (Manuck, 2017). 3β-Hydroxysteroid dehydrogenase plus 5/4 isomerase has a key role in the synthesis of active steroid hormones early in the biosynthetic pathway. 3β-HSD is found not only in steroidogenic tissues, namely adrenal cortex, ovary, testis and placenta but also in several peripheral tissues including skin, adipose tissue, breast, lung, endometrium, prostate, liver, kidney and brain. Two genes have now been cloned in man. Type 2 gene expresses the adrenal and gonadal enzymes, while type I codes for the peripheral enzyme. The impact of the latter

in peripheral tissues will clearly be dependent on the extent of sulfatase acting to release DHA from DHAS which needs to be better understood.

The short chain alcohol dehydrogenases (SCDH) groups in the periphery include isoforms of 11-hydroxysteroid dehydrogenase (HSD11B1) and 17-ketosteroid reductase. There were 12 human isoforms of 17β-hydroxysteroid dehydrogenases known in 2009 of which HSD17B2 is the major oxidative form. HSD17B5 should really be called AKR1C3. Reduction of androstenedione to testosterone is a notable action in the periphery. Sulfatase and sulfotransferase activities are significant in peripheral tissues (Cornel et al., 2019) and with other enzymes can fine tune tissue activity (Gibson et al., 2018a,b) and there is interest in the roles of sulfated steroids in females (Gibson et al., 2018a,b) and males (Hartmann et al., 2016). At the menopause, adrenal DHEA and DHEAS becomes the major source of testosterone (Labrie et al., 2017; Hetemäki et al., 2017).

1.8.6.2 Aromatase

Estrogen production is catalyzed by the enzyme aromatase which transforms C19 steroids into C18 steroids and generates a phenolic A-ring. The aromatase process involves three hydroxylation steps. The principal sites of estrogen synthesis are the ovarian granulosa cells in premenopausal women, the placenta in pregnancy and the adipose tissue in postmenopausal women but aromatase is also expressed in a number of other cells and tissues including the testis, skin, liver, brain and mammary tissues.

Estriol-3-glucuronide and estriol-16-glucuronide are hepatic and intestinal products of estrogen metabolism. The impact of the enterohepatic circulation of estrogen on overall hormone economy can be accessed from the ratio of these two metabolites measured separately by immunoassay or GC-MS. Estradiol, like testosterone, is oxidized to estrone. Hydroxylation at C-16 can give estriol and 16-hydroxy-estrone, respectively. Estrone is converted to estrone sulfate and estrone glucuronide.

1.8.6.3 Metabolism of dihydrotestosterone

DHT formed in peripheral tissues is extensively metabolized before its metabolites appear in the circulation. Metabolism of DHT to inactive steroids occurs primarily via the initial actions of 3α,17β-hydroxysteroid dehydrogenase (3α-HSD) and 3β,17β-hydroxysteroid dehydrogenase (3β-HSD) in liver, intestine, skin, and androgen-sensitive tissues. Subsequent conjugation by uridine 5′-diphospho-(UDP)-glucuronyltransferase (UGT) is the major pathway for urinary and biliary elimination of DHT metabolites and, locally, is the principal irreversible step to protect tissues from high concentrations of this potent androgen. Of the UGTs, only UGT2 isozymes participate in DHT metabolism. In this regard, UGT2B7, B15, and B17 have remarkable capacities to conjugate androgens and are abundant in androgen-sensitive tissues. Differential expression of UGT2 isozymes has been reported and likely plays a role in tissue DHT concentrations independent of circulating androgen levels, particularly in androgen sensitive tissue. For example, transcripts of UGT2B7, B15, and B17 have been identified in liver, intestine, skin, breast, uterus, and ovary, but adipose tissue expresses only UGT2B15, whereas in prostate, UGT2B15 and B17 are expressed only in luminal and basal cells, respectively. This differential localization combined with other local differences in androgen-metabolizing enzymes provides a finely tuned mechanism for control of intracellular androgen concentrations. Polymorphisms of UGT2B15 (that is highly effective in conjugating DHT and its metabolites) have been identified and are postulated to protect prostate tissue from high DHT concentrations and thus lower prostate cancer risk. Conversely, increased prostate cancer risk had been observed in white but not African American men with UGT2B17 deletion polymorphism.

The glucuronides (G) of testosterone (T), androsterone (A), aetiocholanolone (Ae) and DHT (TG, AG, AetioG and DHTG) are substrates for the multidrug resistance-associated protein transporters, MRP2 and MRP3 expressed in basolateral membranes of epithelial cells in the liver but also intestine and kidney (Li et al., 2019; Järvinen et al., 2018). Androgen glucuronides can be passed to bile or return to the circulation. Biliary metabolites can be deconjugated by bacteria in the intestine or excreted in feces. The importance of steroid transporters is not fully appreciated yet.

1.8.6.4 Desulfation of steroid sulfates

In recent years, the fate of steroid sulfates has begun to be addressed. DHEAS is now accepted to enter cells by the assistance of transporter proteins. In the cells, **aryl sulfatase C (STS)** will hydrolyze the steroid sulfate esters to their unconjugated, often active, forms. DHEAS, estrone sulfate, pregnenolone sulfate and cholesterol sulfate are the main substrates. STS is an endoplasmic reticulum membrane bound enzyme coded by a gene on short arm of the X-chromosome. A precursor protein of 63 kDa is formed that undergoes many posttranslational modifications including reduction to final size of 61 kDa and N-glycosylations. A cysteine residue at C75 which is in the catalytic site is modified to a formylglycine residue that attacks the sulfate group of the steroid sulfates. The formation of formylglycine requires the coenzyme formylglycine generating enzyme (FGE) that is encoded by the **sulfatase**

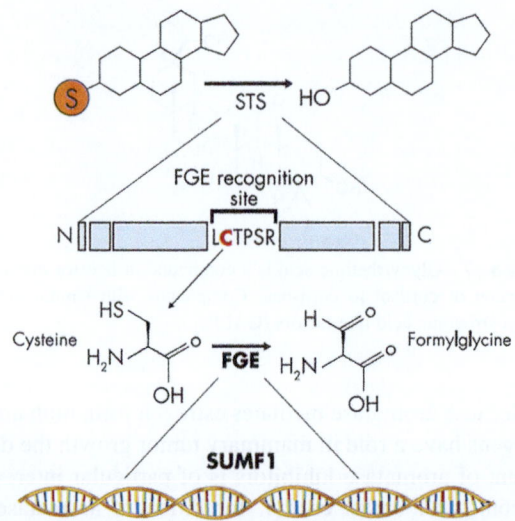

FIG. 1.8.18 Desulfation of steroid sulfates by STS. A cysteine residue, in the FGE recognition site, is modified by SUMF1 to produce a formylglycine. *(From Mueller JW, Gilligan LC, Idkowiak J, Arlt W, Foster PA. The regulation of steroid action by sulfation and desulfation. Endocr Rev 2015;36(5):526–63. Fig. 2 p. 530.)*

modifying factor gene (SUMF1) (Fig. 1.8.18). There is much still to be learnt about the balance between steroid sulfation and the transport, uptake and function of desulfation of steroids conjugates.

1.8.7 Metabolism of specific steroids

From measurements of steroids in blood and urine, ratios of metabolites are used to assess enzyme activity when the metabolites of a substrate and product of one enzyme are calculated.

1.8.7.1 Testosterone/epi-testosterone

Epi-testosterone is secreted by the gonads in man but is not a metabolite of synthetic testosterone. A testosterone/epi-testosterone (T/E) ratio is therefore used to detect abuse by athletes of testosterone drugs. The original description of the test was based on analysis of the glucuronides of T and e but some operating procedures have erred from what is acceptable. Under normal circumstances both these metabolites are minor products of androgen metabolism in the body and this T/E ratio is less than six in man and three in women. The results are different for Asian athletes because of a polymorphism in UDPG (Okano et al., 2013). If testosterone esters have been taken there is extra exogenous testosterone in the urine but less endogenous T and e because of gonadotrophin suppression and the ratio is raised. Mass spectrometry is used for this determination in the steroid profile used in antidoping laboratories and confirmed by carbon isotope ratio determination of the urine metabolites (Donike et al., 1982). The synthetic testosterone derivatives are made in the pharmaceutical industry from plant precursors that have a different composition of ^{13}C and ^{12}C. After gas chromatographic (GC) separation of steroid derivatives, the effluent is passed through a combustion furnace to discharge CO_2 to the mass spectrometer (see Chapter 2.4 Quantitative analysis).

1.8.7.2 Dehydroepiandrosterone to androstenedione

The conversion of DHEA to A4 by 3β-hydroxysteroid dehydrogenase and 5/4 isomerase can occur in the gonads and adrenal cortex, in fact A4 is a true adrenal androgen with the rate of testosterone production being much lower than in the gonads. The mean ratio of DHEA to A4 in serum is 1 or less. Adrenal androgen production is unique to higher primates. After the menopause when the ovaries have ceased sex steroid production, some estrogen production is sustained by the action of steroid metabolizing enzymes in the periphery.

DHEAS is quantitatively the most important circulating steroid in man. In adults, the concentration can be 2–10 μmol (which is 100 times more than cortisol, 500 times more than testosterone in males, more than 1000 times greater than testosterone in females). The uptake of DHEAS by active transport is an important source of DHEA after the action of sulfotransferase. DHEA can be produced by the actions of sulfatases.

The above ratios are useful clinically when investigating patients with possible genetic enzyme deficiencies (see later, Chapter 3.2.2).

1.8.8 Cholesterol to bile acids

The main route for cholesterol removal from the body is by hepatic bile acid synthesis and biliary excretion. About 50% of cholesterol in the body is metabolized in the liver to bile acids which are stored in the gall bladder and excreted into the intestine in the bile. Bile acids are hydroxylated, C24 sterols and a multicarbon side chain. Bile acids can be assigned to three general structural classes according to the length of the side chain and functionality of the terminal polar group hence 27-carbon (C27) bile alcohols, C27 bile acids, and 24-carbon (C24) bile acids, with C24 bile acids being the predominant form in humans. Cholic acid (3α-, 7α-,12α-tri-hydroxycholan-24-oic) and chenodeoxycholic acid (3α-,7α-di-hydroxy-cholan-24-oic acid) are the primary bile acids.

Twelve enzymes are involved in bile acid synthesis located in the cytoplasm (C), microsomes (ER), mitochondria (M) and peroxisomes (P). Microsomal 7α-hydroxylase is probably the rate limiting step (Pandak and Kakiyama, 2019). The major pathway for the synthesis of the bile acids is initiated by hydroxylation of cholesterol at the 7 position through the action of cholesterol 7α-hydroxylase (CYP7A1) which is an ER localized enzyme (Fig. 1.8.19). CYP7A1 is a member of the cytochrome P450 family of metabolic enzymes.

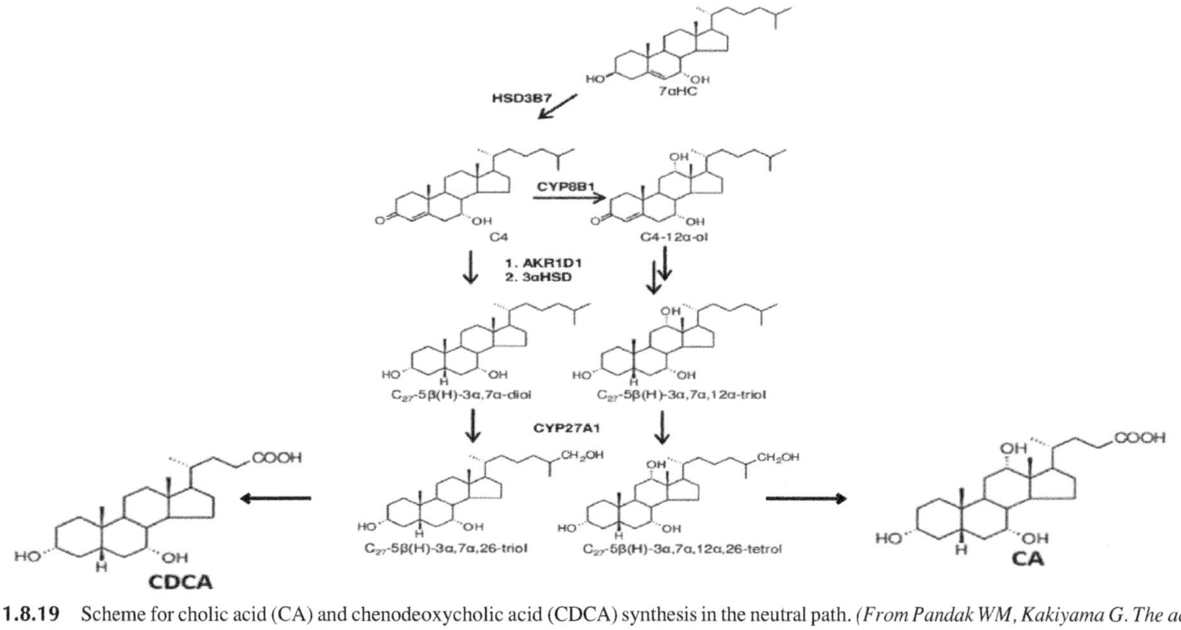

FIG. 1.8.19 Scheme for cholic acid (CA) and chenodeoxycholic acid (CDCA) synthesis in the neutral path. *(From Pandak WM, Kakiyama G. The acidic pathway of bile acid synthesis: not just an alternative pathway. Liver Res 2019;3(2):88–98. Fig. 1 p. 20.)*

The activity of CYP7A1 is inhibited by bile acids and the synthesis of the bile acids is thus under negative feedback control. A microsomal 3β-hydroxy-Δ5-C$_{27}$-steroid dehydrogenase/isomerase produces 7α-hydroxy-4-cholest-3-one, a branch point in bile acid synthesis to chenodeoxycholic acid (CDCA) and the C-12 hydroxylated cholic acid (CA) formed by the activity of sterol 12α-hydroxylase (CYP8B1). The pathway initiated by CYP7A1 is referred to as the "classic" or "neutral" pathway of bile acid synthesis. An alternative pathway that involves hydroxylation of cholesterol at the 27 position by the mitochondrial enzyme sterol 27-hydroxylase (CYP27A1) is referred to as the "acidic" pathway that in humans accounts for around 5% (Fig. 1.8.20). The bile acid intermediates generated via the action of CYP27A1 can be hydroxylated at C24, C25 or C26 and these are subsequently hydroxylated on the 7 position by oxysterol 7α-hydroxylase (CYP7B1).

The hydroxyl group on cholesterol at the 3 position is in the β-orientation and is epimerized to the α-orientation during the synthesis of the bile acids. Epimerization is a two-step conversion with 3β-hydroxyl to a 3-oxo group catalyzed by 3β-hydroxy-Δ^5-C$_{27}$-steroid oxidoreductase (HSD3B7) before the action of HSD3B7.

Cholic aid and chenodeoxycholic acid are released from the CoA products by bacterial action in the intestine. The primary bile acids are secreted into the canalicular lumen after they are conjugated via an amide bond at the terminal carboxyl group to either of the amino acids glycine or taurine. These conjugation reactions yield **glycoconjugates** and **tauroconjugates**, respectively. The conjugates are more amphipathic making them more easily secretable as well as less cytotoxic. Bile is a complex mixture of cholesterol, cholesteryl esters, phospholipids, bile salts, bilirubin, biliverdin and minor amounts of plasma proteins (prealbumin, albumin, haptoglobin, and transferrin), immunoglobulins, electrolytes and water. The conjugated bile acids are the major solutes in human bile. The bile salts taurocholate and glycocholate assist absorption from the small intestine of dietary cholesterol and fatty acids formed by the action of lipase in the secretions of the exocrine pancreas.

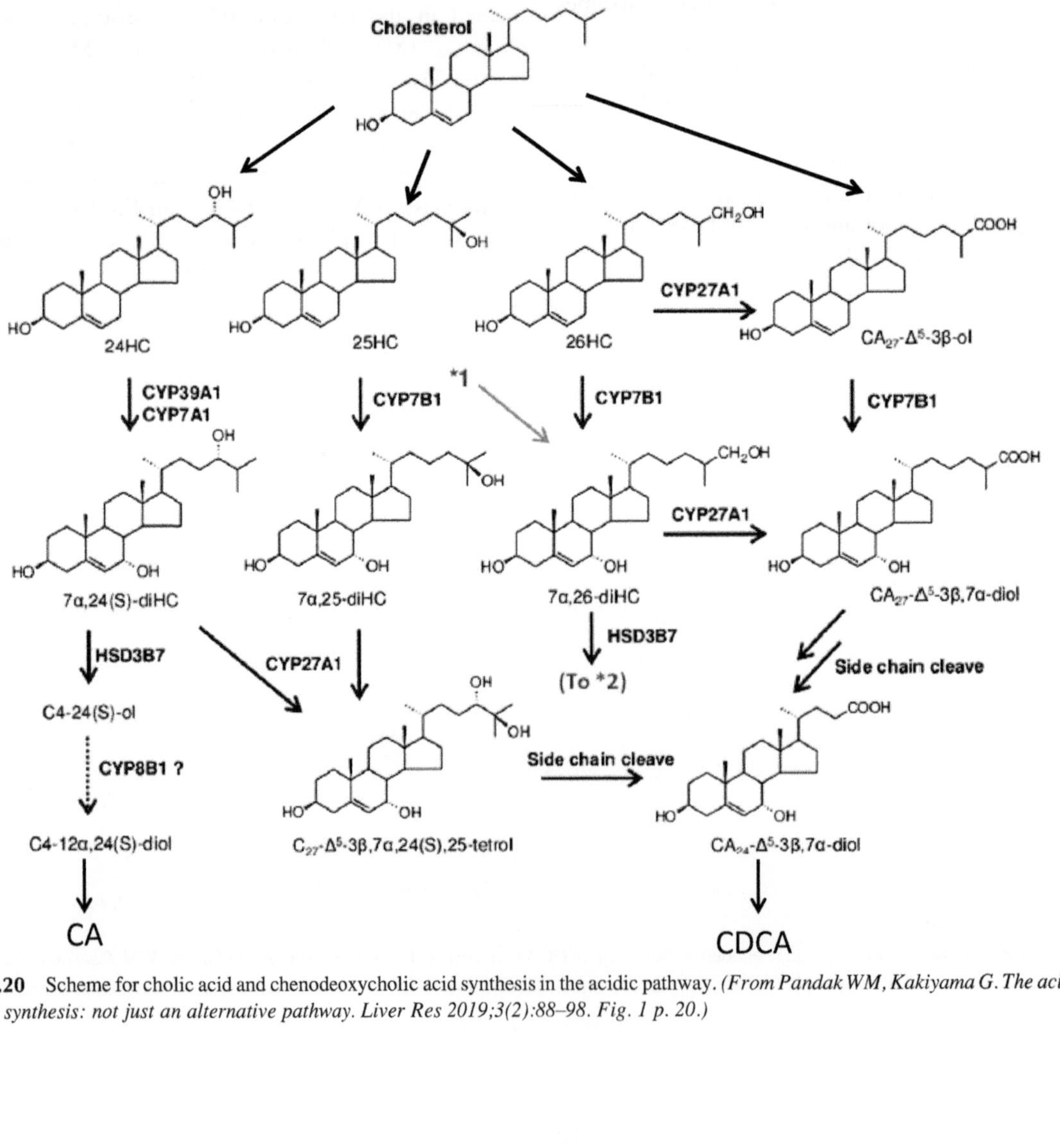

FIG. 1.8.20 Scheme for cholic acid and chenodeoxycholic acid synthesis in the acidic pathway. *(From Pandak WM, Kakiyama G. The acidic pathway of bile acid synthesis: not just an alternative pathway. Liver Res 2019;3(2):88–98. Fig. 1 p. 20.)*

1.8.9 Biliary excretion of cholesterol and enterohepatic circulation

The composition of bile is nearly 70% bile salts, around 20% phospholipids, and 5% cholesterol. Bile also contains electrolytes, minerals, minor levels of proteins, plus bilirubin and biliverdin pigments. The bilirubin and biliverdin are what impart the yellow-green or even orange hue to bile. Bile acids, produced in the liver are excreted through the canalicular membrane into bile ductules. The bile acids are at an alkaline pH in the bile in the gallbladder. Eating causes the gallbladder to contract and to release bile into the common bile duct. The primary bile acids are secreted into the bile by a primary active transport mechanism. The bile, mixed with pancreatic secretions, enters into the proximal small intestine, where bile acids facilitate the enzyme reactivity of lipases and proteases. Bile acids have important physiological functions in the gut. Bile acids are natural detergents that aid intestinal absorption of cholesterol. The proximal portion of the small intestine also contains acidic luminal fluid from the stomach. Bile acids are mixed with water, insoluble fats and phospholipids into micelles. Once bile salts are secreted into the duodenum to emulsify fats, around 95% are reabsorbed into the distal ileum. A small percentage of the bile salts are not reabsorbed and undergo deconjugation by intestinal microbiota before either being absorbed or converted into secondary bile acids (Fig. 1.8.21).

Anaerobic bacteria present in the colon modify the primary bile acids converting them to the secondary bile acids, deoxycholate and ursodeoxycholate (DCA and UDCA, from cholate) and lithocholate (LCA, from chenodeoxycholate) (Bhatti and Khera, 2012; Philipp, 2011). These secondary bile acids are either passively absorbed from the colon or they are excreted in the feces. After excretion to the gastrointestinal tract, a portion of bile acids subject to bacterial metabolism, including deconjugation, oxidation, esterification, desulfation, and epimerization, leading to more than 50 different secondary bile acids in human feces, and numerous species of tertiary bile acids. The initial step to deconjugate the taurine and glycine conjugates is mediated by bile salt hydrolase (BSH). Deconjugation of the polar conjugates makes bile acids more hydrophobic and less dangerous to bacteria. Such increased hydrophobicity reduces re-absorption of bile acids and rather favors excretion in feces. Bacteria can remove the side chain of sterols (Ahmad et al., 1992).

The bile acids are reabsorbed into the portal blood by the terminal ileum. This portal blood returns to the liver and the bile acids are taken up from the sinusoidal blood by a Na^+-dependent co-transporter. Fats, in the form of mixed micelles, are absorbed by the jejunum. In contrast, bile acids are actively absorbed by the ileum. Bile salt reabsorption occurs via the apical sodium-dependent bile transporter (ASBT) present in the brush border membrane of the enterocyte. Ileal bile acid-binding protein (IBABP; aka fatty acid-binding protein subclass 6: FABP6) is involved in the transport of bile salts across the enterocyte cytosol to the basolateral membrane. Once bile salts reach the basolateral membrane they are transported into the blood by the organic solute transporters OSTα/OSTβ. The primary and secondary bile acids and salts are transported back to

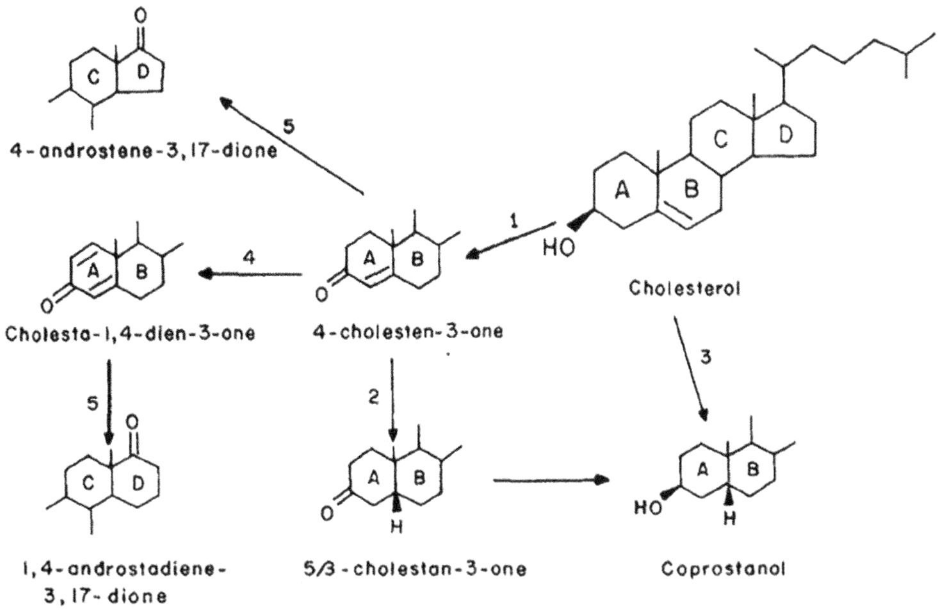

FIG. 1.8.21 Cholesterol metabolism to coprostanol. *(From Macdonald IA, Bokkenheuser VD, Winter J, McLernon AM, Mosbach EH. Degradation of steroids in the human gut. J Lipid Res 1983;24(6):675–700. Fig. 3 p. 682.)*

the liver where most are actively transported into hepatocytes by the sodium (Na^+)-taurocholate co transporting polypeptide (NTCP/SLC10A1) and organic anion transporters (OATP) (such as OAT1B2) for the uptake of bile salts and bile acids, respectively. Most of the bile acids are reconjugated in the liver and re-excreted through the canalicular membrane for another cycle. The recirculation of bile acids through the enterohepatic circulation occurs many times each day.

1.8.9.1 Defects in bile acid synthesis

Defects in bile acid synthesis result in decreased bile formation which causes malabsorption of vitamins and fats. Some of the intermediates that accumulate are toxic. Defects of cholesterol-7α-hydroxylase, 3β-hydroxy-Δ5-C_{27}-steroid dehydrogenase/isomerase, $Δ^{4-3}$ oxosteroid 5β-reductase and sterol 27 hydroxylase have been identified. A newborn harboring a mutation in the CYP7B1 gene presented with severe cholestasis (blockage in bile flow from liver) with cirrhosis and liver dysfunction. Children harboring mutations in the HSD3B7 gene children develop progressive liver disease that is characterized by cholestatic jaundice.

1.8.9.2 Metabolism of cholesterol by bacteria in the intestine

Cholesterol is in the gut from bile adding to any unabsorbed cholesterol from the food. The intestine contains large number of bacteria that act on cholesterol during transit.

1.8.9.2.1 Reduction of cholesterol to coprostanol

Two major pathways have been proposed for the conversion of cholesterol to coprostanol (Fig. 1.8.21). The first involves oxidation of the 3βOH group and isomerization of the double bond to form the intermediate 4-cholesten-3-one which then undergoes nuclear reduction to 5β-cholestan-3-one and further reduction of the ketone to yield coprostanol.

The second pathway is direct reduction of the double bond at C-5. The relative importance of different bacterial species in catalyzing cholesterol to coprostanol is uncertain but most likely is *Eubacteria*.

More than 50% of the total fecal sterol can be present in the form of coprostanol depending on the diet. Coprostanone and lesser amounts of cholestanol are detected in normal feces. The latter saturated sterol arises predominantly from the bacterial reduction of cholestenone. Mixed cultures of human or rat feces readily transform cholesterol to coprostanol and common intestinal bacteria;

Bifidobacterium sp., *Clostridium* sp., and *Bacteroides* spp. produce coprostanol (Crowther et al., 1977). *Eubacterium* spp. have also been isolated from human and rat feces that reduced A5-3β-hydroxysteroids to 5β-saturated derivatives and should be solely responsible for the conversion of cholesterol to coprostanol (Eyssen et al., 1973, 1977; Sadzikowski et al., 1977; Björkhem et al., 1973).

1.8.9.2.2 Side chain cleavage of cholesterol

The gut flora of guinea pig is capable of removing the side chain of cholesterol, whereas in the rat only a 3-carbon unit is removed. Both activities are suppressed by administering antibiotics to the animals, thus emphasizing the role of bacteria in the conversion (Goddard and Hill, 1974). Certain strains of *E. coli* can cleave the side chain of cholesterol and nuclear dehydrogenate ring A leading to 4-cholesten-3-one, androsta-4-en-3,17-dione, cholesta-1,4-dien-3-one and androsta-1,4-dien-3,17-dione (Owen et al., 1978). All four metabolites were produced in aerobic conditions for incubation while only the two latter compounds were obtained under anaerobic conditions. The clinical significance of cholesterol and its metabolites coprostanol and coprostanone in the human intestine in health and disease is still far from clear.

1.8.9.3 Steroid degradation

From an environmental point of view, large amounts of steroids made and used by humans and animals are passed to sewerage and farm sludge. Studies have shown that many bacteria strains can open the steroid ring structure (Olivera and Luengo, 2019) by oxygen dependent opening and subsequent hydrolytic cleavage of rings A and B, degradation of C and D rings occurs by unknown mechanisms (Bergstrand et al., 2016). Pyruvate, acetyl CoA, propionyl CoA and succinyl CoA are final products. Sterols, bile acids and vitamin D are likely to follow similar fates.

1.8.10 Steroids in urine

Urine is the major route for elimination of steroids from the body although measurements of steroids in bile and feces have indicated that steroids are also excreted into and from the intestine. If the urine steroid excretion of radioactive hormone is monitored after injection into the blood some different patterns of excretion with time can be seen depending on peripheral metabolism and the extent of enterohepatic circulation. The mechanisms for transfer of steroids from blood into urine have not been studied. OATS maybe involved in renal tubular cells for steroid conjugates.

1.8.10.1 Measurement of steroid metabolites

Many endocrine investigations today can be conducted by measurements of steroids in blood provided that samples are collected at appropriate times and the relationship with trophic hormones is determined. Analysis of steroids in urine still has some advantages. The daily activity of the adrenal gland, for example, can be measured by analysis of the steroids in a 24 h urine collection. Since much of the activity of this gland takes place at night this time period can be included when it will be less amenable to study by blood sampling. However, the highest concentrations of cortisol metabolites in urine are between 1000 and 1600 h presumably because of the transcription of clock genes. Urine collection is considered easier when wanting to exclude drug abuse by athletes (see Chapter 3.2.1). Abuse of steroids in sport. Large volumes of urine can be collected and concentrated in order to increase the detection limit. General techniques such as gas chromatography can be used to examine all of the steroids in urine rather than set about an analysis based on selection of specific analytes. From the variety of metabolic steps which have been considered above, there may be a number of metabolites arising from each hormone so that interpretation of all the products is quite skilled. Machine learning is now being applied to steroid profile analysis techniques. Another problem is that most of the steroid metabolites are conjugated and most analytical methods for steroids have been easier for the steroid alone so a method is often needed to remove the conjugates before analysis of the free steroid products. Chemical and enzymatic methods are available both of which are rather slow for routine use.

Many urine steroid methods have until recently required analysis of free steroids rather than conjugates and hydrolysis of conjugates is required first. This can be chemical or enzymic. The direct analysis of steroid conjugates is now possible and methods for this will increase because of the time saved from the hydrolysis step. There is a lack of stable isotope labeled steroid conjugates that are needed as internal standards for quantitative analysis. Estrone glucuronide excretion rate is measured in urine as a good indicator of ovarian function in the follicular phase.

1.8.10.2 Individual steroids

1. **Aldosterone-18-glucuronide** is a unique metabolite not hydrolyzed with the enzymes usually used for steroids. Free aldosterone is created by acid hydrolysis of the conjugate in urine after dichloroform extraction of the free steroids. The optimum pH for the hydrolysis reaction is 1.0 at 30°C overnight. The urine extract is mixed with a twofold volume of 0.2 N HCl and pH should be checked and adjusted if necessary. The aldosterone has been measured by RIA, GC-MS of the gamma-lactone of aldosterone and LC-MS/MS. The aldosterone 18-glucuronide excretion in urine shows a useful relationship with urine sodium excretion rate (Fig. 1.8.22). Aldosterone-18-glucuronide excretion rate is from 18 to 85 μg/day.

2. Urine **free cortisol** excretion rate is a very useful indicator of excess cortisol production and is therefore a **screening test for Cushing's syndrome** (Oßwald et al., 2019). Daily output is normally less than 200 μg/day. Unfortunately, it has been widely used when looking for adrenal suppression by inhaled and injected corticosteroids which is not appropriate. Cortisol is not excreted in bile but in vitro cortisol is metabolized by bacteria to androgens (Bokkenheuser et al., 1984; Winter et al., 1984a,b,c; Ridlon et al., 2013). There is some confusion here because "desmolase" is used for 17,20 lyase.

Physicians concerned about **adrenal suppression in asthmatic patients receiving inhaled corticosteroids (ICS)** have recourse to a number of diagnostic procedures. These include the measurement of plasma cortisol at 0800 h, integrated plasma cortisol concentrations over 24 h, urinary 17-hydroxycorticosteroids (17-OHCS), the measurement of saliva or urinary free cortisol (UFC) and the plasma cortisol response to adrenocorticotrophic hormone. Of these options, UFC is the most common used because of the noninvasive nature of sample collection. While determination of UFC concentrations using automated immunoassays may be convenient, some systems may be subject to interferences from compounds produced by the metabolism of oral steroids or ICS. Theoretically, this cross-reactivity could result in falsely elevated results, thereby masking endogenous cortisol suppression. A second drawback is the question of sensitivity, since commercially available immunoassays were originally introduced for the detection of cortisol excess (Cushing's syndrome) and not for the detection of adrenal suppression. Under normal circumstances serum cortisol binding globulin is saturated and binds most of the circulating cortisol. The fraction of "free cortisol" filtered by the kidney and excreted in the urine is thus small and UFC concentrations are inherently low. A small degree of suppression may therefore result in even lower UFC concentrations (<50 nmol/24 h of urine), at which point the precision of many commercial immunoassays may be unacceptable (for diagnostic purposes a precision <10% is recommended).

Free cortisol in the urine (UFC) is frequently measured in clinical research to assess whether inhaled corticosteroids (ICS) cause suppression of the hypothalamic-pituitary-adrenal axis. Thirteen healthy male subjects received single inhaled doses (of molar equivalence) of fluticasone propionate (FP), triamcinolone acetonide (TAA), budesonide (BUD), and placebo

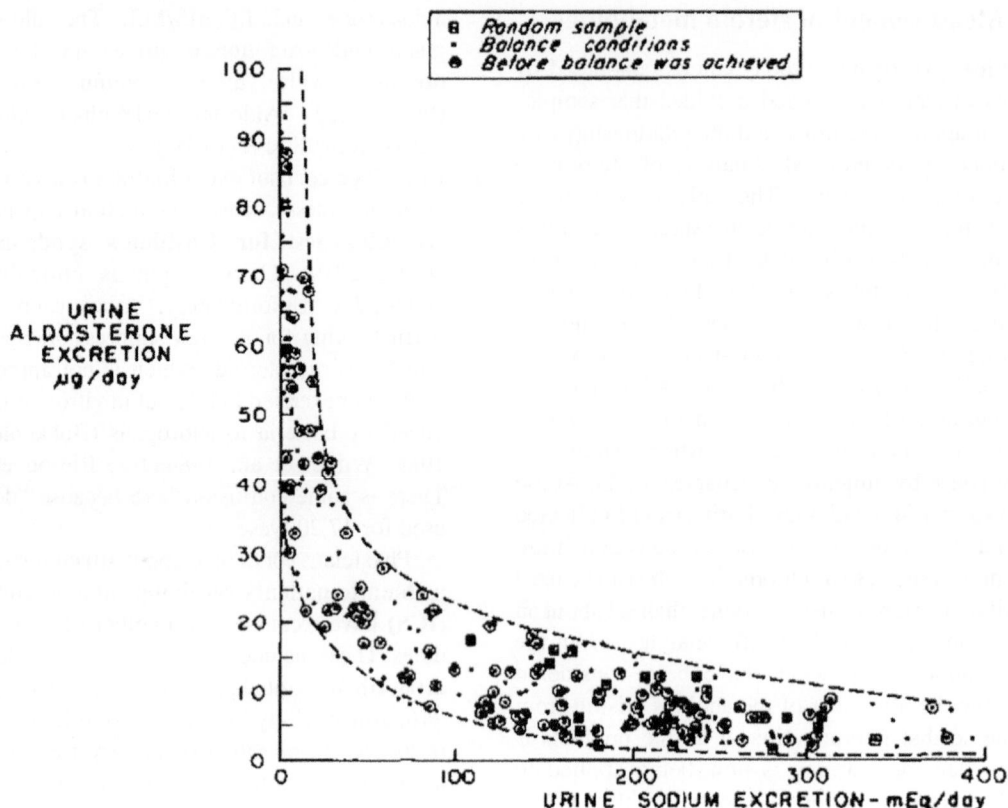

FIG. 1.8.22 Correlation of urine aldosterone-18-glucuronide with sodium excretion. *(From Sealey JE, Bühler FR, Laragh JH, Manning EL, Brunner HR. Aldosterone excretion. Physiological variations in man measured by radioimmunoassay or double-isotope dilution. Circ Res 1972;31(3):367–78. Fig. 3 p. 372.)*

in a single blind, randomized, cross-over study. UFC output was measured using four commercial immuno-assays in samples collected in 12-h aliquots over 24 h (Fink et al., 2002). Cortisol was measured in **Abbott TDX, Bayer ACS 180, DOC Immunlite and ICN Biomedical Corticote assays. For comparison**, the cortisol production rate was separately assessed from the outputs of **cortisol metabolites determined by GC-MS** (Fig. 1.8.23). The UFC in the Abbott TDX assay showed a 100% increase over placebo levels after the administration of BUD. This result is likely to reflect cross-reactivity of BUD metabolites with the anticortisol antibody used in the assay.

The main BUD metabolites formed in man are 6β-hydroxybudesonide and 16α-hydroxyprednisolone, one or both of which may be detected by the assay. The potential problem of exogenous oral corticosteroid and/or cortisol metabolite interference had been recognized previously and immunoassay kit manufacturers recommend extraction procedures to reduce this problem. In spite of pretreatment with dichloromethane, the Abbott assay in particular was subject to interference generated by BUD or its metabolites. Although others have

recognized cross-reactions with immunoassays in patients receiving oral prednisolone, it is not acceptable to adjust the measured cortisol for the presumed cross reactivity. This was the first report showing that a commonly used ICS can perturb assays used in the routine clinical laboratory. The marked interference seen with BUD was not apparent with the other assays used, although it may have been present to a lesser extent. Of the ICS tested BUD is chemically the nearest to prednisolone. Reports in which the effects of BUD on the HPA axis have been evaluated using the Abbott system should be reviewed with caution, although the importance of this n serum or plasma was not evaluated in this study.

The immunoassays detected variable suppression (ranging from 29% to 61% suppression for FP, 30%–62% suppression for TAA, and 25% suppression to 100% stimulation for BUD) (Fig. 1.8.23). Suppression was more pronounced in the first 12 h after TAA and in the second 12 h after FP. Similar suppression was found in each 12-h period after BUD.

UFC estimation based on immunoassays after ICS may be an unreliable surrogate marker of adrenal

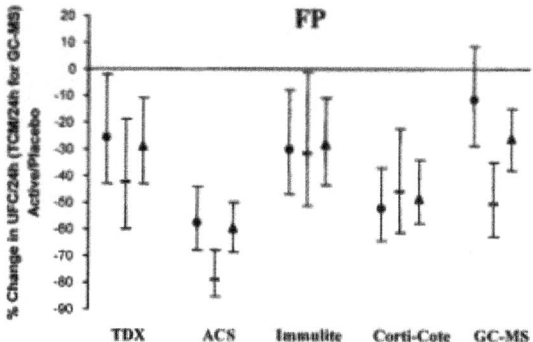

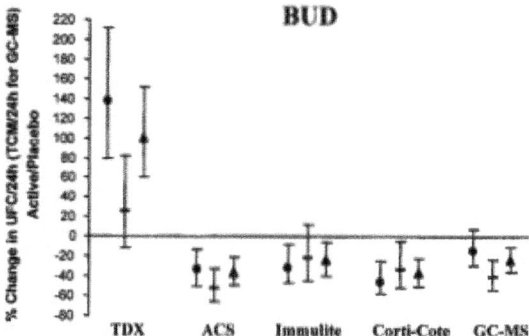

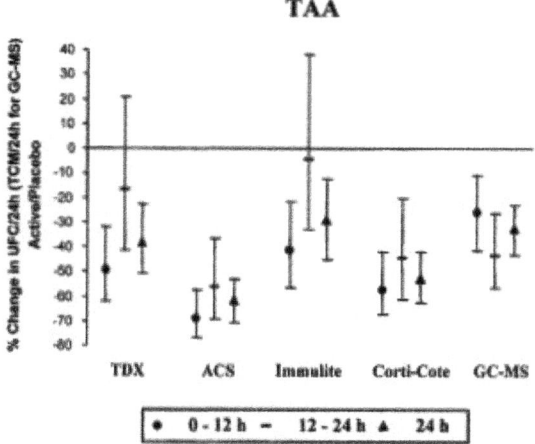

FIG. 1.8.23 Urine free cortisol output as measured by immunoassays (TDX, ACS, Immulite, Corti-Cote) and total cortisol metabolites by GC-MS over 24 h after fluticasone propionate (FP), budesonide (BUD) and triamcinolone acetate (TAA). *(From Fink RS, Pierre LN, Daley-Yates PT, Richards DH, Gibson A, Honour JW. Hypothalamic-pituitary-adrenal axis function after inhaled corticosteroids: unreliability of urinary free cortisol estimation. J Clin Endocrinol Metab 2002;87(10):4541–6. Fig. 2 p. 4545.)*

cortisol suppression following ICS administration is influenced by the analytical method. The same urine sample could yield up to a twofold difference in observed HPA axis suppression when comparing the immunoassays and up to fivefold when compared with the reference GC-MS method (Fig. 1.8.23). Thus, depending on the assay used, entirely different conclusions may be drawn on the potential for, and magnitude of, HPA effects induced by these ICS (Fig. 1.8.24).

Importantly, this variability was only seen in the presence of ICS. Following placebo, no significant difference was detected between the immunoassays. Lack of specificity and poor precision at low concentrations are the likely explanations for this difference. Further, although significant suppression was detected for BUD, FP and TAA by TCM measurement, the reductions observed were generally lower than those seen with the immunoassays. These data may be more representative of the actual HPA axis effects as the aforementioned caveats are not applicable.

The comparison of day time urine collections with overnight samples (Fig. 1.8.23) demonstrates that the timing of urine collection can significantly influence the cortisol suppression detected. The pharmacology and metabolism of the drugs may explain these findings. The interference following BUD in the Abbott TDX system was most prominent in the first 12 h postdose urine collection. This supports the conclusion that the apparent increase in UFC is due to metabolite cross-reactivity and is consistent with the short half-life of BUD and its metabolites.

TAA produced marked suppression in the first 12 h period, however, particularly when using the Abbott TDX or DPC Immulite, but no significant effect was seen in the overnight collection. This is also likely to reflect the 2.5 h half-life of TAA with approximately 97% of the dose being eliminated over the first 12 h collection period. Therefore, an overnight collection following a single morning dose of TAA would be inappropriate when assessing the HPA effects of this ICS. On the other hand, FP has a plasma half-life of 10 h from the MDI and in some assays demonstrated more suppression in the 12–24 collection than the 0–12 h collection. Despite these considerations, 12-h overnight collections are frequently used in clinical medicine and clinical research to assess cortisol production. In view of the number of available ICS and their respective pharmacokinetic profiles, collection periods should be chosen to obtain meaningful information in all cases.

The results have implications beyond patient management. Immunoassays are used extensively in clinical trials to assess and compare the safety of ICS and a huge body of literature has evolved over the past 10 years. When evaluating these reports, the investigator should

suppression. Many of the published studies describing or comparing the safety of different ICS should be re-evaluated, and some should be interpreted with caution.

In addition to the interference seen with cortisol after BUD and analysis on the Abbott TDX, the extent of

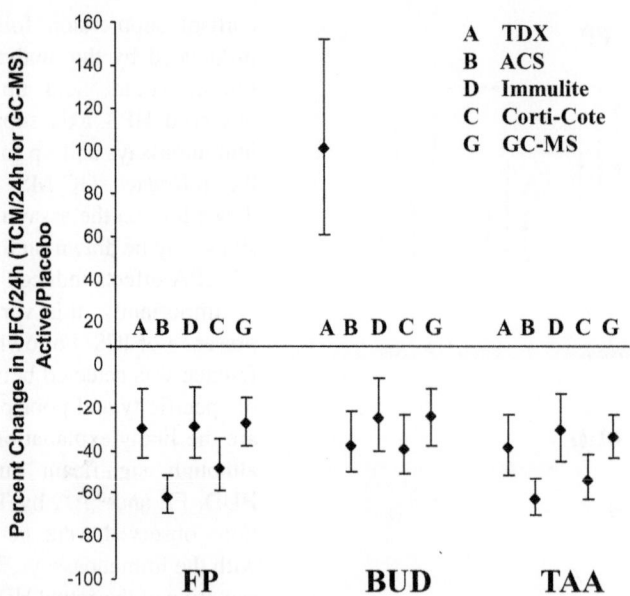

FIG. 1.8.24 Cortisol or TCM excretion rates after FP, Bud and TAA treatment in 0–12 h, 12–24 h and 24 h urine collection periods. Values are percent change with 95% confidence limits. *(From Fink RS, Pierre LN, Daley-Yates PT, Richards DH, Gibson A, Honour JW. Hypothalamic-pituitary-adrenal axis function after inhaled corticosteroids: unreliability of urinary free cortisol estimation. J Clin Endocrinol Metab 2002;87(10):4541–6. Fig. 3 p. 4545.)*

consider whether potential cross-reacting substances were effectively removed from the urine, and which immunoassay was selected for the ICS under study. For example, based on our data, an estimated 56% and 79% reduction in overnight UFC excretion would be observed in a study in which TAA and FP are compared with samples measured on the Bayer ACS. In contrast, a study utilizing 24 h collections and the same analytical platform would demonstrate reductions of 62% and 61%. The reference assay of TCM indicated suppression of 34% and 27% for TAA and FP, respectively, in 24 h collections. When examining UFC values below the reference range (50 nmol/24 h used here and other authors have used similar limits) discrepant results would also be observed. For example, the combination of FP and the Bayer ACS would show UFC suppression in 6/12 subjects. BUD and the Bayer ACS would show suppression in 3/13. However, with the Corti-Cote system, these identical samples would show suppression in 2/12 and 2/13 cases, respectively. Clearly the selected assay significantly influences the outcome and conclusions in such studies. The findings of the study may go some way to explain the discrepancies in exogenous steroid (inhaled and intranasal) HPA axis effects described in the literature. Although unfortunately, the details of assays and sample treatment are often omitted in published papers on this topic.

In summary, UFC results in subjects receiving ICS are subject to variables within the methodology. These variables exert such powerful effects that suppression or indeed normal function can be diagnosed with equal facility in the same individual depending on the assay and collection protocol used. This has critical implications for patient management and also for clinical research where conclusions about ICS safety may need to be re-evaluated because of issues raised in this study. The results of this study demonstrate that the assessment of adrenal suppression by measuring UFC is more problematic than previously realized and is markedly influenced by the assay methods used as well as the protocol for urine collection.

3. **Pregnanediol**, as the main metabolite of progesterone, has been used to monitor activity of a corpus luteum after ovulation in the menstrual cycle and by the placenta in pregnancy but this is today covered by blood measurements of progesterone and ultrasound imaging of the ovaries. From week 12–38 in pregnancy, PD output can increase to 30 mg/day. The assay has been effectively used in women with ovarian failure undergoing oocyte donation (Schneider et al., 1993) to determine the timing of placental support in the pregnancy (Fig. 1.8.25A). Exogenous steroids were given in the first trimester. Estradiol valerate and progesterone were given at increasing doses from 3 days before implantation of an embryo until the hCG test on day 15 after implantation (Fig. 1.8.25A and B).

After confirmation of the pregnancy, estradiol valerate was continued at 8 mg/day and progesterone at 100 mg/day. The steroid support was decreased from week 10. From week 6, the estradiol and pregnanediol excretion in the oocyte transfer women was greater than seen in normal pregnancies.

FIG. 1.8.25 Outputs of steroids in urine in normal pregnancies (A) estriol and (B) pregnanediol. *(From Schneider MA, Davies MC, Honour JW. The timing of placental competence in pregnancy after oocyte donation. Fertil Steril 1993;59 (5):1059–64.)*

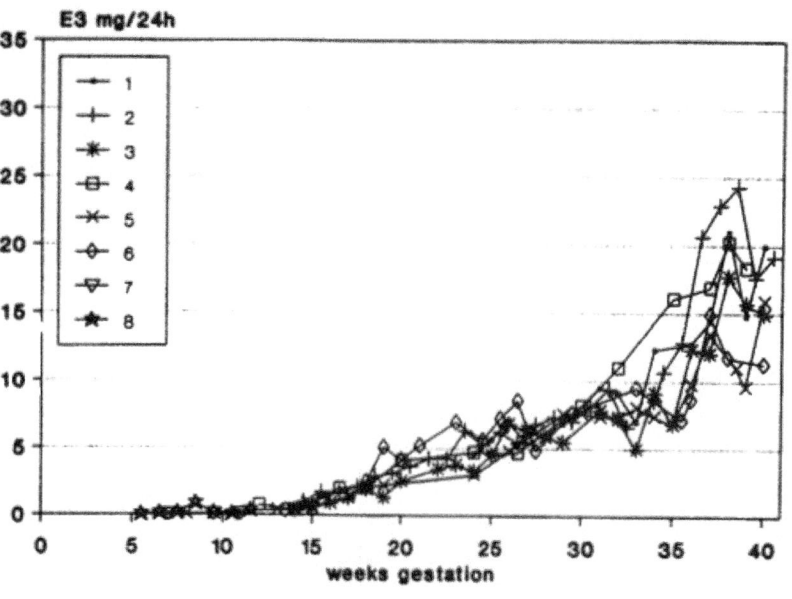

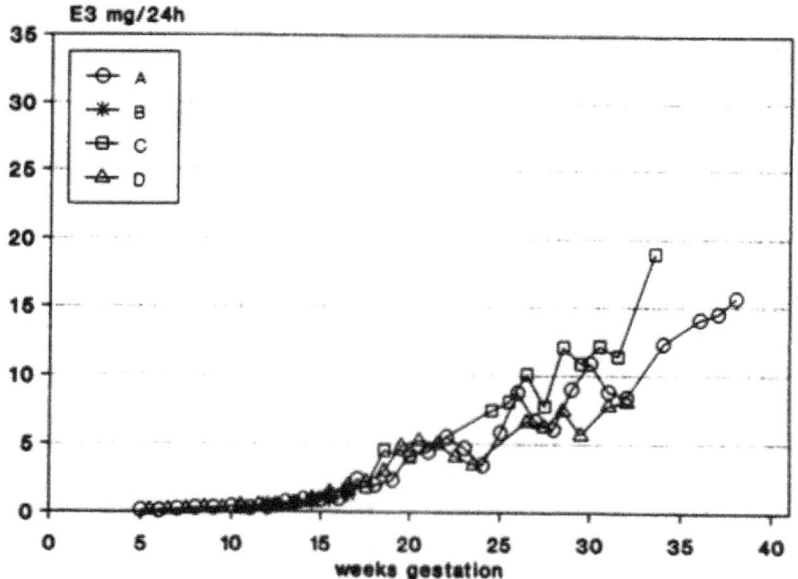

4. **Pregnanetriol** is an important metabolite of 17-hydroxyprogesterone and was used to assist diagnosis of congenital adrenal hyperplasia due to 21-hydroxylase deficiency. Sometimes the assay was used to monitor treatment. The diagnosis of 21-hydroxylase deficiency is needed most often in the investigation of infants with ambiguous genitalia.

5. **Estriol** is excreted in large amounts (up to 25 mg/day) during pregnancy. It is the end result of steroid secretion by the adrenal cortex of the developing fetus (mainly secreting DHEAS), fetal hepatic steroid metabolism (16-hydroxylation) and placental modification of DHEAS and 16-hydroxy-DHEAS (sulfatase and aromatase). These combined activities are called the feto-placental unit. Estriol measurements were used in the past as an index for fetal viability but this is now achieved with ultrasound imaging of fetal movements.

6. **18-Hydroxycortisol** in urine and blood is raised in rare patients with glucocorticoid remediable hyperaldosteronism (GRA; DSH) who have a chimeric gene between CYP11B1 and CYP11B2 that has higher 18-hydroxylase activity for cortisol than for corticosterone that is regulated by ACTH. 18-Hydroxycortisol can be regarded as a chimeric steroid, normal mean excretion is 117 µg/day and 600–2476 µg/day in patients with GRA (see Chapter 3.3.1 Hypertension).

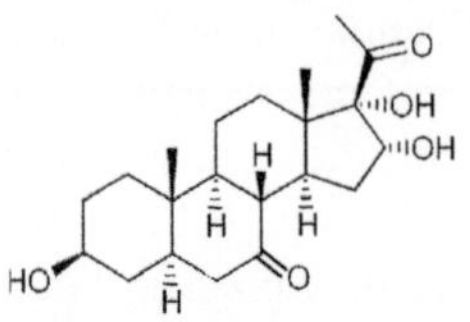

FIG. 1.8.26 Structure of 3β,16α,17α-trihydroxy-5α-pregnane-7,20-dione. This is an early urine marker for CAH due to 21-hydroxylase deficiency. *(Based on Christakoudi S, Cowan DA, Taylor NF. A new marker for early diagnosis of 21-hydroxylase deficiency: 3beta,16alpha,17alpha-trihydroxy-5alpha-pregnane-7,20-dione. J Steroid Biochem Mol Biol. 2010;121(3–5):574–81.)*

7. Christakoudi et al. found **3β,16α,17α-trihydroxy-5α-pregnane-7,20-dione** (Fig. 1.8.26) was raised in the first days of life of a child with congenital adrenal hyperplasia due to 21-hydroxylase deficiency.

This steroid is a metabolite of 17-hydroxyprogesterone and was found to be an early biomarker for the condition. Investigations of ambiguous genitalia are undertaken and the parents cannot initially be told whether they have a boy or girl. An early diagnosis is important to the timing when the clinicians can truly discuss with the anxious parents the impact of the condition on their child.

1.8.10.3 Group methods

The group methods for steroid measurements rely on common structural features and usually a chemical endpoint. With the introduction of specific analytical methods such as immunoassay the group methods are not used in many countries but in the United States they are still used and results are encountered in the literature so the reader should know of their existence.

1. **Oxosteroids** are androgen metabolites with a 17-keto or oxo group. They are detected after reaction with dinitrobenzene—the Zimmerman reaction. An interesting increase in oxosteroid excretion has been reported when hydrocortisone was administered per rectum attributed to bacterial metabolism and absorption of side chain cleaved cortisol.
2. **17-Oxogenic steroids** are the 17-oxosteroids formed after reduction and oxidation of 17-hydroxy-C_{21} steroids. They are a measure of cortisol metabolites. Steroids such as tetrahydrocortisol have a dihydroxyacetone side-chain which on reduction gives a glycerol side chain (which is found in the cortols and cortolones). Periodate oxidation frees the side chain leaving 17-keto steroids which are measured by the Zimmerman reaction. This analysis will also measure 17-hydroxyprogesterone and 11-deoxycortisol metabolites. In the United States, cortisol metabolites are measured by reaction of the dihydroxyacetone side chain of with phenylhydrazine (Porter-Silber chromogens).

1.8.10.4 Urinary steroid profile

Gas chromatography has been used for many years in order to examine in one analysis the range of steroid metabolites in urine. This is a particularly good method for detecting enzyme defects. The analysis will reveal any steroids in excess and reveal if the production of terminal products is inadequate. Gas chromatography can be coupled with mass spectrometry (GC-MS) and the output from this combined technology is a series of mass spectra which are fingerprints for each steroid. Steroids are extracted from urine and conjugates are hydrolyzed enzymatically. The liberated steroids are recovered from the aqueous enzyme solution and steroids derivatized to stabilize them for GC analysis at high temperatures. Carbonyl groups are usually converted to methyloxime derivatives and hydroxyl groups silylated to trimethylsilyl ethers.

A normal profile will show metabolites of cortisol and adrenal androgens. Peaks will also be seen for metabolites of progesterone, 17-hydroxyprogesterone, corticosterone and other intermediates. In patients with defects in steroid production, there will be a number of steroids in excess. Positive identification localizes the metabolic defect. Tumors may produce a range of steroids not all of which are commonly measured in the blood (see later Chapter 3.1.1 Steroid secreting tumors). By characterizing the spectrum of steroids prior to surgery a specific follow-up can be offered for recurrence.

1.8.10.5 Steroid profile in sport

Urine has been the principal sample for analysis from athletes in the antidoping program administered by the World Anti-Doping Authority (WADA). An important target has been the detection of anabolic steroids in urine of athletes who have abused these compounds. All **anabolic agents** fall into three structural categories (1a) exogenous anabolic androgenic steroids; (1b) endogenous androgens; (2) other anabolic agents (clenbuterol and selective androgen receptor modulators). Initially the androgenic steroids were screened by immunoassay techniques but final proof, and now much of the screening, depends on confirmation by mass spectrometry (Schänzer and Thevis, 2017; Parr and Schänzer, 2010). The 19-nor anabolic steroids are recognized because their metabolites have molecular weights 14 mass units less than natural androgen metabolites. 17α-(M)ethyl anabolic steroid metabolites give mass spectra clearly reflecting the modifying group at C-17 (Nováková et al., 2015a,b). The topic is addressed in more detail in Chapter 3.2.1.

Corticosteroids are now forbidden in sports competition (Vernec et al., 2020; Coll et al., 2018; Collomp et al., 2016) when taken in a nonsystemic way since they alleviate pain and allow athletes to compete when injured. The ban is thus over concern for the welfare of the athlete but also the image of drugs in sport. It didn't look good when several swimmers were seen at the side of an Olympic event using corticosteroid inhalers for lung problems. Receptor modulators are also now banned in sport (Thevis and Schänzer, 2018).

In order to analyze a large number of urine samples, the antidoping laboratories from 1976 have extracted the steroids with SPE, rapidly hydrolyze conjugates then prepare TMS derivatives for GC-MS analysis with selected ion monitoring. A large number of AAS and some of their metabolites can be detected (Thevis and Schänzer, 2018; Parr and Schänzer, 2010). The abuse of testosterone itself created problems for the laboratories because it is present naturally in urine particularly from males. Initially the ratio of testosterone to epi-testosterone was used on the grounds that natural T and epi-T production will be reduced when exogenous T suppresses LH secretion (Donike et al., 1982). A further approach was based on the premise that all anabolic steroids are created from plants such as yam which have a $^{12}C/^{13}C$ ratio different to testosterone synthesized from cholesterol in the body. A carbon isotope ratio technique was therefore developed to detect testosterone abuse (de la Torre et al., 2019; Piper et al., 2013; Aguilera et al., 1996). The laboratories have had to continuously exploit developments in technology so analysis of urinary steroids has progressed to LC-MS/MS (Görgens et al., 2016; Gosetti et al., 2013) and LC-HRMS (Ponzetto et al., 2019). In order to complement the analysis of steroids in urine, blood samples are now taken at frequent intervals to trend changes in steroids in blood as well as gonadotrophins, growth hormone, GH releasing factors, erythropoietin in a Biological passport (Faiss et al., 2019; Robinson et al., 2011).

1.8.11 Steroids in feces

Some steroids are excreted in the feces but this has not been much studied in the last 35 years during which time there have been many changes in the analytical techniques for steroid analysis. This is an area that now needs re-investigation because of the interest in the microbiome (Rastelli et al., 2019) where new methods for identification of the organisms on the basis of 16S RNA analysis have been adopted. In addition to determining the numbers of organisms, the overall colonization of intestinal contents is judged by richness and diversity (Flores et al., 2012). Richness is defined as the number of species identified, whereas diversity takes account of the amounts of the species. The microbiome in terms of cell numbers is as large as the numbers of cells in the body tissues. The available data on the steroid composition of feces is based largely on GC-MS analysis and radioimmunoassay of steroids purified by gel and paper chromatography (see Chapter 2.4). Steroids have also been studied in bile

collected during intestinal surgery. The differences in composition of bile and feces reflect the metabolism by the intestinal bacteria or the intestine itself. The estrogens and oral contraceptive steroids have been the most studies mainly because of the high concentrations of estrogens produced during pregnancy or studies of the impact of diet and drugs (particularly antibiotics) on the efficacy of contraception. The effects of the antibiotics ampicillin, oxytetracycline, on steroid metabolism have been studied in normal women and in pregnancy. Administration of antimicrobial agents to women taking oral contraceptives has been reported to lead to contraceptive failure and subsequent pregnancy (Adlercreutz et al., 1984).

1.8.12 Summary

Steroid hormones are inactivated mainly in the liver by metabolism in two phases, essentially Phase One for reduction of double bonds and carbonyl groups of the steroid hormones then Phase two conjugation with glucuronic and sulfuric acids. The steroids can return to the circulation for renal excretion or transported into the bile. In the main 3β-hydroxy-5-ene steroids or saturated steroids with 3α-hydroxy-5α or 5β, reduced structures are found. Bile contains high concentrations of some steroids particularly those conjugated as sulfates. Sulfated steroids are now of interest in a number of areas including reproduction (Geyer et al., 2017) and oncology (Poisson Paré et al., 2009). Sterols are conjugated with glycine and taurine before excretion in bile. Steroids excreted in bile and modified by intestinal bacterial enzymes are also reabsorbed from the intestine and some appear in the urine. Steroid metabolism in the intestine can be attributed to the action of bacteria. Steroids are actively conjugated by glucuronidase in the intestine wall. When the fecal steroid content is compared with that in bile, differences are observed both in the amounts of steroids and also the structure and conjugation of the products. Although in bile the steroids are mostly conjugated, in feces mostly free steroids are present. In feces, the steroids are mainly 3β-hydroxy-5α reduced steroids. This suggests that extensive metabolism has taken place in the intestine. The impact of gut metabolism and enterohepatic circulation of steroids can be monitored in the case of estriol by the relative excretion rates of estriol-16-glucuronide and estriol-3-glucuronide. Furthermore, during the administration of antibiotics the excretion of steroids in feces resembles the steroid composition of bile. Other data in support of this hypothesis comes from animal studies such as a comparison of germ-free rats with equivalent conventionally reared rats. Evidence for this in man comes from a study of corticosterone metabolism in patient with congenital adrenal hyperplasia due to 17-hydroxylase deficiency. In the future, there is much more to be learnt about the fate of steroid conjugates.

Most of the literature on steroid metabolism is quite historical, before 1986, and will need to be re-investigated with modern analytical techniques. The microbiome is an exciting area for research being almost as important as the steroid metabolism in liver and periphery. Facilities for germ-free rats became too expensive to maintain and almost disappeared in the 1970s. Germ free and gnotobiote mice (where the numbers of specific organisms in the microbiome are reduced or adjusted), bacterial/fecal transplants and specific pathogen free species are cheaper options now being investigated (Delannoy-Bruno et al., 2021). The use of radioactive steroids in tracer studies has also declined. The use of stable isotope labeled steroids in human studies has been limited and may still present ethical issues.

References

Much of this literature predates the year 1980 and methodology is at most based on gas chromatography coupled with mass spectrometry. Many studies examine the fate of large doses of a steroid.

Books

Bornstein SS, Solomon S. Chemical and biological aspects of steroid conjugation. New York, USA: Springer-Verlag; 1970. First critical review of steroid conjugates.

Dorfman RI, Ungar F. Metabolism of steroid hormones. London: Academic Press; 1965. Steroids isolated from natural sources; Biosynthesis of steroid hormones; Steroid transformations by micro-organisms; Catabolic reactions of steroids; Enzymes and mechanisms of reactions; A system of steroid metabolism; Relationships between tissue steroids and metabolites in blood; Steroid hormone production rates.

Papers

Adamski J, Jakob FJ. A guide to 17beta-hydroxysteroid dehydrogenases. Mol Cell Endocrinol 2001;171(1–2):1–4.

Adlercreutz H. Biochemical and clinical aspects of enterohepatic circulation of estrogens. Acta Endocrinol 1967;(Suppl. 124):101–40.

Adlercreutz H, Martin F. Biliary excretion and intestinal metabolism of progesterone and estrogens in man. J Steroid Biochem 1980;13(2):231–44.

Adlercreutz H, Martin F, Järvenpää P, Fotsis T. Steroid absorption and enterohepatic recycling. Contraception 1979;20(3):201–23.

Adlercreutz H, Pulkkinen MO, Hämäläinen EK, Korpela JT. Studies on the role of intestinal bacteria in metabolism of synthetic and natural steroid hormones. J Steroid Biochem 1984;20(1):217–29.

Adlercreutz H, Höckerstedt K, Bannwart C, Bloigu S, Hämäläinen E, Fotsis T, et al. Effect of dietary components, including lignans and phytoestrogens, on enterohepatic circulation and liver metabolism of estrogens and on sex hormone binding globulin (SHBG). J Steroid Biochem 1987;27(4–6):1135–44.

Agarwal AK, Auchus RJ. Minireview: cellular redox state regulates hydroxysteroid dehydrogenase activity and intracellular hormone potency. Endocrinology 2005;146(6):2531–8.

Aguilera R, Becchi M, Casabianca H, Hatton CK, Catlin DH, Starcevic B, et al. Improved method of detection of testosterone abuse by gas chromatography/combustion/isotope ratio mass spectrometry analysis of urinary steroids. J Mass Spectrom 1996;31(2):169–76.

Ahmad S, Garg SK, Johri BN. Biotransformation of sterols: selective cleavage of the side chain. Biotechnol Adv 1992;10(1):1–67.

Alvarez MN, Carpenter PC, Mattox VR. Isolation of tetrahydroaldosterone 3beta-glucosiduronic acid from urine. J Steroid Biochem 1976;7(9):661–4.

Bai HW, Shim JY, Yu J, Zhu BT. Biochemical and molecular modeling studies of the O-methylation of various endogenous and exogenous catechol substrates catalyzed by recombinant human soluble and membrane-bound catechol-O-methyltransferases. Chem Res Toxicol 2007;20(10):1409–25.

Bergstrand LH, Cardenas E, Holert J, Van Hamme JD, Mohn WW. Delineation of steroid-degrading microorganisms through comparative genomic analysis. mBio 2016;7(2), e00166. Erratum in: MBio. 2016;7(4). pii: e00865-16.

Bhatti HN, Khera RA. Biological transformations of steroidal compounds: a review. Steroids 2012;77(12):1267–90.

Björkhem I, Gustafsson JA, Wrange O. Microbial transformation of cholesterol into coprostanol. Properties of a 3-oxo-4-steroid-5 beta-reductase. Eur J Biochem 1973;37(1):143–7.

Blouin K, Veilleux A, Luu-The V, Tchernof A. Androgen metabolism in adipose tissue: recent advances. Mol Cell Endocrinol 2009;301(1–2):97–103.

Bokkenheuser VD, Winter J. Biotransformation of steroid hormones by gut bacteria. Am J Clin Nutr 1980;33(11 Suppl):2502–6.

Bokkenheuser VD, Suzuki JB, Polovsky SB, Winter J, Kelly WG. Metabolism of deoxycorticosterone by human fecal flora. Appl Microbiol 1975;30(1):82–90.

Bokkenheuser VD, Winter J, Dehazya P, de Leon O, Kelly WG. Formation and metabolism of tetrahydrodeoxycorticosterone by human fecal flora. J Steroid Biochem 1976;7(10):837–43.

Bokkenheuser VD, Winter J, Honour JW, Shackleton CH. Reduction of aldosterone by anaerobic bacteria: origin of urinary 21-deoxy metabolites in man. J Steroid Biochem 1979a;11(2):1145–9.

Bokkenheuser VD, Winter J, Finegold SM, Sutter VL, Ritchie AE, Moore WE, et al. New markers for *Eubacterium lentum*. Appl Environ Microbiol 1979b;37(5):1001–6.

Bokkenheuser VD, Winter J, Hylemon PB, Ayengar NK, Mosbach EH. Dehydroxylation of 16 alpha-hydroxyprogesterone by fecal flora of man and rat. J Lipid Res 1981;22(1):95–102. PMID: 7217789.

Bokkenheuser VD, Morris GN, Ritchie AE, Holdeman LV, Winter J. Biosynthesis of androgen from cortisol by a species of *Clostridium* recovered from human fecal flora. J Infect Dis 1984;149(4):489–94.

Bouguen G, Dubuquoy L, Desreumaux P, Brunner T, Bertin B. Intestinal steroidogenesis. Steroids 2015;103:64–71.

Bradlow HL, Zumoff B, Monder C, Lee HJ, Hellman L. Isolation and identification of four new carboxylic acid metabolites of cortisol in man. J Clin Endocrinol Metab 1973;37(5):811–8.

Bradlow HL, Monder C, Zumoff B. Studies in the biotransformation of cortisol to cortoic acids in man. III. 21-Oxidation of 4-14C,21-3H-desoxycorticosterone. J Clin Endocrinol Metab 1977;45(5):960–4.

Bulaschenko H, Richardson EM, Dohan FC. Urinary excretion of C-20 reduction products of corticosterone and 11-dehydrocorticosterone. Arch Biochem Biophys 1960;87:81–7.

Calvin HI, Lieberman S. Studies on the metabolism of 16alpha-hydroxyprogesterone in humans; conversion to urinary 17-isopregnanolone. Biochemistry 1962;1:639–45.

Carpenter PC, Mattox VR. Isolation, determination of structure and synthesis of the acid-labile conjugate of aldosterone. Biochem J 1976;157(1):1–14.

Casey ML, MacDonald PC. Metabolism of deoxycorticosterone and deoxycorticosterone sulfate in men and women. J Clin Invest 1982;70 (2):312–9.

Cerone-McLernon AM, Winter J, Mosbach EH, Bokkenheuser VD. Side-chain cleavage of cortisol by fecal flora. Biochim Biophys Acta 1981;666(3):341–7.

Chang HJ, Zhou M, Lin SX. Human dehydroepiandrosterone sulfotransferase: purification and characterization of a recombinant protein. J Steroid Biochem Mol Biol 2001;77(2–3):159–65.

Chang HJ, Shi R, Rehse P, Lin SX. Identifying androsterone (ADT) as a cognate substrate for human dehydroepiandrosterone sulfotransferase (DHEA-ST) important for steroid homeostasis: structure of the enzyme-ADT complex. J Biol Chem 2004;279(4):2689–96.

Chen M, Penning TM. 5β-Reduced steroids and human Δ(4)-3-ketosteroid 5β-reductase (AKR1D1). Steroids 2014;83:17–26.

Chen M, Drury JE, Penning TM. Substrate specificity and inhibitor analyses of human steroid 5β-reductase (AKR1D1). Steroids 2011;76 (5):484–90.

Christakoudi S, Cowan DA, Taylor NF. Steroids excreted in urine by neonates with 21-hydroxylase deficiency: characterization, using GC-MS and GC-MS/MS, of the D-ring and side chain structure of pregnanes and pregnenes. Steroids 2010a;75(1):34–52.

Christakoudi S, Cowan DA, Taylor NF. A new marker for early diagnosis of 21-hydroxylase deficiency: 3beta,16alpha,17alpha-trihydroxy-5alpha-pregnane-7,20-dione. J Steroid Biochem Mol Biol 2010b;121 (3–5):574–81.

Christakoudi S, Cowan DA, Taylor NF. Steroids excreted in urine by neonates with 21-hydroxylase deficiency. 3. Characterization, using GC-MS and GC-MS/MS, of androstanes and androstenes. Steroids 2012a;77(13):1487–501.

Christakoudi S, Cowan DA, Taylor NF. Steroids excreted in urine by neonates with 21-hydroxylase deficiency. 2. Characterization, using GC-MS and GC-MS/MS, of pregnanes and pregnenes with an oxo-group on the A- or B-ring. Steroids 2012b;77(5):382–93.

Christakoudi S, Cowan DA, Christakudis G, Taylor NF. 21-Hydroxylase deficiency in the neonate—trends in steroid anabolism and catabolism during the first weeks of life. J Steroid Biochem Mol Biol 2013;138:334–47.

Clifford CK, Butrum RR, Greenwald P, Yates JW. Clinical trials of low fat diets and breast cancer prevention. Prog Clin Biol Res 1986;222:93–115.

Coll S, Matabosch X, Garrostas L, Perez-Maña C, Ventura R. Effect of glucocorticoid administration on the steroid profile. Drug Test Anal 2018;10(6):947–55.

Collomp K, Arlettaz A, Buisson C, Lecoq AM, Mongongu C. Glucocorticoid administration in athletes: performance, metabolism and detection. Steroids 2016;115:193–202.

Colombier M, Gachancard-Bouya JL, Bègue RJ, Prost M. Identification of 2 alpha-hydroxy-4-pregnene-3,20-dione in human pregnancy urine. J Steroid Biochem Mol Biol 1992;41(2):191–6.

Cooke BA, Vallance DK. Metabolism of megestrol acetate and related progesterone analogues by liver preparations in vitro. Biochem J 1965;97 (3):672–7.

Cornel KMC, Bongers MY, Kruitwagen RPFM, Romano A. Local estrogen metabolism (intracrinology) in endometrial cancer: a systematic review. Mol Cell Endocrinol 2019;489:45–65.

Costello ME, Robinson PC, Benham H, Brown MA. The intestinal microbiome in human disease and how it relates to arthritis and spondyloarthritis. Best Pract Res Clin Rheumatol 2015;29(2):202–12.

Cronholm T, Eriksson H, Gustafsson JA. Excretion of endogenous steroids and metabolites of (4-^{14}C)pregnenolone in bile of female rats. Eur J Biochem 1971;19(3):424–32.

Crowther JS, Drasar BS, Goddard P, Hill MJ. The effect of a chemically defined diet on the faecal flora and a faecal steroid concentration. Gut 1977;14:490–3.

Davies W. Sulfation pathways: the steroid sulfate axis and its relationship to maternal behaviour and mental health. J Mol Endocrinol 2018;61(2): T199–210.

de la Torre X, Jardines D, Curcio D, Colamonici C, Botrè F. Isotope ratio mass spectrometry in antidoping analysis: the use of endogenous reference compounds. Rapid Commun Mass Spectrom 2019;33(6):579–86.

Delannoy-Bruno O, Desai C, Raman AS, Chen RY, Hibberd MC, Cheng J, et al. Evaluating microbiome-directed fibre snacks in gnotobiotic mice and humans. Nature 2021;595(7865):91–5.

Dellow EL, Unwin RJ, Honour JW. Pontefract cakes can be bad for you: refractory hypertension and liquorice excess. Nephrol Dial Transplant 1999;14(1):218–20.

Derks HJ, Drayer NM. Polar corticosteroids in human neonatal urine; synthesis and gas chromatography-mass spectrometry of ring A reduced 6-hydroxylated corticosteroids. Steroids 1978;31(1):9–22.

Doden HL, Ridlon JM. Microbial Hydroxysteroid dehydrogenases: from alpha to omega. Microorganisms 2021;9(3), 469.

Donike M, Barwald KR, Klostermann K, Scchnazer W, Zimmermann J. Nachweis von exogenem Testosterone.Spot: Leitsung und Gesundheit/Kongesbd. Dtsch. Sportatzekongres; 1982. p. 293–8.

Ekuase EJ, Liu Y, Lehmler HJ, Robertson LW, Duffel MW. Structure-activity relationships for hydroxylated polychlorinated biphenyls as inhibitors of the sulfation of dehydroepiandrosterone catalyzed by human hydroxysteroid sulfotransferase SULT2A1. Chem Res Toxicol 2011;24(10):1720–8.

El-Sayed A, Aleya L, Kamel M. Microbiota's role in health and diseases. Environ Sci Pollut Res Int 2021;1–17.

Eriksson H. Steroids in germfree and conventional rats. Metabolites of (4-14C)pregnenolone and (4-14C)corticosterone in urine and faeces from male rats. Eur J Biochem 1971a;18(1):86–93.

Eriksson H. Absorption and enterohepatic circulation of neutral steroids in the rat. Eur J Biochem 1971b;19(3):416–23.

Eriksson H, Gustafsson JA, Sjövall J. Steroids in germfree and conventional rats. 4. Identification and bacterial formation of 17 alpha-pregnane derivatives. Eur J Biochem 1968;6(2):219–26.

Exley D. Urinary excretion of C-20-reduction products of corticosterone. Biochem J 1965;94(1):271–8.

Eyssen HJ, Parmentier GG, Compernolle FC, De Pauw G, Piessens-Denef M. Biohydrogenation of sterols by *Eubacterium* ATCC 21,408—nova species. Eur J Biochem 1973;36(2):411–21.

Eyssen H, Smets L, Parmentier G, Janssen G. Sex-linked differences in bile acid metabolism of germfree rats. Life Sci 1977;21(5):707–12.

Fabregat A, Kotronoulas A, Marcos J, Joglar J, Alfonso I, Segura J, et al. Detection, synthesis and characterization of metabolites of steroid hormones conjugated with cysteine. Steroids 2013;78(3):327–36.

Faiss R, Saugy J, Saugy M. Fighting doping in elite sports: blood for all tests! Front Sports Act Living 2019;1:30.

Falany CN, Vazquez ME, Kalb JM. Purification and characterization of human liver dehydroepiandrosterone sulphotransferase. Biochem J 1989;260(3):641–6.

Fazekas AG, Kokai K. Conversion of tritiated aldosterone into 11-dehydroaldosterone by rabbit adrenal tissue in vitro. Steroids 1967;10(1):71–4.

Fazekas AG, Sandor T. Unusual pathways of aldosterone biosynthesis in the rabbit adrenal. Steroids 1969;14(2):161–77.

Fink RS, Pierre LN, Daley-Yates PT, Richards DH, Gibson A, Honour JW. Hypothalamic-pituitary-adrenal axis function after inhaled corticosteroids: unreliability of urinary free cortisol estimation. J Clin Endocrinol Metab 2002;87(10):4541–6.

Flood C, Pincus G, Tait JF, Tait SA, Willoughby S. A comparison of the metabolism of radioactive 17-isoaldosterone and aldosterone administered intravenously and orally to normal human subjects. J Clin Invest 1967;46(5):717–27.

Flores R, Shi J, Fuhrman B, Xu X, Veenstra TD, Gail MH, et al. Fecal microbial determinants of fecal and systemic estrogens and estrogen metabolites: a cross-sectional study. J Transl Med 2012;10:253.

Fuda H, Lee YC, Shimizu C, Javitt NB, Strott CA. Mutational analysis of human hydroxysteroid sulfotransferase SULT2B1 isoforms reveals that exon 1B of the SULT2B1 gene produces cholesterol sulfotransferase, whereas exon 1A yields pregnenolone sulfotransferase. J Biol Chem 2002;277(39):36161–6.

Gamage NU, Tsvetanov S, Duggleby RG, McManus ME, Martin JL. The structure of human SULT1A1 crystallized with estradiol. An insight into active site plasticity and substrate inhibition with multi-ring substrates. J Biol Chem 2005;280(50):41482–6.

Gathercole LL, Lavery GG, Morgan SA, Cooper MS, Sinclair AJ, Tomlinson JW, et al. 11β-Hydroxysteroid dehydrogenase 1: translational and therapeutic aspects. Endocr Rev 2013;34(4):525–55.

Geyer J, Bakhaus K, Bernhardt R, Blaschka C, Dezhkam Y, Fietz D, et al. The role of sulfated steroid hormones in reproductive processes. J Steroid Biochem Mol Biol 2017;172:207–21.

Gibson DA, Simitsidellis I, Collins F, Saunders PTK. Endometrial intracrinology: oestrogens, androgens and endometrial disorders. Int J Mol Sci 2018a;19(10):3276.

Gibson DA, Foster PA, Simitsidellis I, Critchley HOD, Kelepouri O, Collins F, et al. Sulfation pathways: a role for steroid sulphatase in intracrine regulation of endometrial decidualisation. J Mol Endocrinol 2018b;61(2):M57–65.

Goddard P, Hill MJ. The in vivo metabolism of cholesterol by gut bacteria in the rat and guinea-pig. J Steroid Biochem 1974;5(6):569–72. https://doi.org/10.1016/0022-4731(74)90106-x. PMID: 4427454.

Goddard P, Fernandez F, West B, Hill MJ, Barnes P. The nuclear dehydrogenation of steroids by intestinal bacteria. J Med Microbiol 1975;8(3):429–35.

Goebelsmann U, Cooke I, Wiqvist N, Diczfalusy E. Comparison of the metabolism of oestriol-3-glucosiduronate and oestriol-16-glucosiduronate in pregnant women. Acta Endocrinol 1966;52(1):30–42.

Goldin BR, Adlercreutz H, Gorbach SL, Warram JH, Dwyer JT, Swenson L, et al. Estrogen excretion patterns and plasma levels in vegetarian and omnivorous women. N Engl J Med 1982;307(25):1542–7.

Gomez-Sanchez EP, Gomez-Sanchez CE. 11β-Hydroxysteroid dehydrogenases: a growing multi-tasking family. Mol Cell Endocrinol 2021;15(526), 111210.

Gorbach SL, Goldin BR. Diet and the excretion and enterohepatic cycling of estrogens. Prev Med 1987;16(4):525–31.

Görgens C, Guddat S, Thomas A, Wachsmuth P, Orlovius AK, Sigmund G, et al. Simplifying and expanding analytical capabilities for various classes of doping agents by means of direct urine injection high performance liquid chromatography high resolution/high accuracy mass spectrometry. J Pharm Biomed Anal 2016;131:482–96.

Grose JH, Nowaczynski W, Kuchel O, Genest J. Isolation of aldosterone urinary metabolites, glucuronides and sulfate. J Steroid Biochem 1973;4(6):551–66.

Gulcan HO, Duffel MW. Substrate inhibition in human hydroxysteroid sulfotransferase SULT2A1: studies on the formation of catalytically non-productive enzyme complexes. Arch Biochem Biophys 2011;507(2):232–40.

Gustafsson BE, Gustafsson JA, Sjövall J. Steroids in germfree and conventional rats. 3. Solvolyzable sterol conjugates in germfree and conventional rat faeces. Eur J Biochem 1968a;4(4):574–7.

Gustafsson BE, Gustafsson JA, Sjövall J. Steroids in germfree and conventional rats. 2. Identification of 3 alpha-,16 alpha-dihydroxy-5 alpha-pregnan-20-one and related compounds in faeces from germfree rats. Eur J Biochem 1968b;4(4):568–73.

Hämäläinen E, Korpela JT, Adlercreutz H. Effect of oxytetracycline administration on intestinal metabolism of oestrogens and on plasma sex hormones in healthy men. Gut 1987;28(4):439–45.

Hämäläinen E, Fotsis T, Adlercreutz H. A gas chromatographic method for the determination of neutral steroid profiles in urine, including studies on the effect of oxytetracycline administration on these profiles in men. Clin Chim Acta 1991;199(2):205–20.

Hampel MR, Peng LH, Pearlman MR, Pearlman WH. Acylation of [3H] corticosterone by acini from mammary gland of lactating rats. Localization of the acylated glucocorticoid in the nuclear fraction. J Biol Chem 1978;253(23):8545–53.

Harnik M, Kashman Y, Aharonowitz Y, Morris DJ. Synthesis of 19-hydroxyaldosterone and the 3 beta-hydroxy-5-ene analog of aldosterone, active mineralocorticoids. J Steroid Biochem 1985;23(2):207–18.

Hartmann K, Bennien J, Wapelhorst B, Bakhaus K, Schumacher V, Kliesch S, et al. Current insights into the sulfatase pathway in human testis and cultured Sertoli cells. Histochem Cell Biol 2016;146(6):737–48.

Hawksworth G, Drasar BS, Hill MJ. Intestinal bacteria and the hydrolysis of glycosidic bonds. J Med Microbiol 1971;4(4):451–9.

Hempel N, Barnett AC, Bolton-Grob RM, Liyou NE, McManus ME. Site-directed mutagenesis of the substrate-binding cleft of human estrogen sulfotransferase. Biochem Biophys Res Commun 2000;276(1):224–30.

Hetemäki N, Savolainen-Peltonen H, Tikkanen MJ, Wang F, Paatela H, Hämäläinen E, et al. Estrogen metabolism in abdominal subcutaneous and visceral adipose tissue in postmenopausal women. J Clin Endocrinol Metab 2017;102(12):4588–95.

Hochberg R, Bandy L, Ponticorvo L, Welch M, Lieberman S. Naturally occurring lipoidal derivatives of 3 beta-hydroxy-5-pregnen-20-one; 3 beta,17 alpha-dihydroxy-5-pregnen-20-one and 3 beta-hydroxy-5-androsten-17-one. J Steroid Biochem 1979;11(4):1333–40.

Hochberg RB, Pahuja SL, Zielinski JE, Larner JM. Steroidal fatty acid esters. J Steroid Biochem Mol Biol 1991;40(4–6):577–85.

Honour JW, Tourniaire J, Biglieri EG, Shackleton CHL. Urinary steroid excretion in 17-hydroxylase deficiency. J Steroid Biochem 1978;9:495–506.

Huijghebaert SM, Sim SM, Back DJ, Eyssen HJ. Distribution of estrone sulfatase activity in the intestine of germfree and conventional rats. J Steroid Biochem 1984;20(5):1175–9.

Hwang I, Park YJ, Kim YR, Kim YN, Ka S, Lee HY, et al. Alteration of gut microbiota by vancomycin and bacitracin improves insulin resistance via glucagon-like peptide 1 in diet-induced obesity. FASEB J 2015;29:2397–411.

Järvenpää P, Kosunen T, Fotsis T, Adlercreutz H. In vitro metabolism of estrogens by isolated intestinal micro-organisms and by human faecal microflora. J Steroid Biochem 1980;13(3):345–9.

Järvinen E, Deng F, Kidron H, Finel M. Efflux transport of estrogen glucuronides by human MRP2, MRP3, MRP4 and BCRP. J Steroid Biochem Mol Biol 2018;178:99–107.

Kastl Jr AJ, Terry NA, Wu GD, Albenberg LG. The structure and function of the human small intestinal microbiota: current understanding and future directions. Cell Mol Gastroenterol Hepatol 2020;9(1):33–45.

Kelly WG, Bandi L, Shoolery JN, Lieberman S. Isolation and characterization of aldosterone metabolites from human urine; two metabolites bearing a bicyclic acetal structure. Biochemistry 1962;1:172–81.

Kelly WG, Bandi L, Lieberman S. Isolation and characterization of human urinary metabolites of aldosterone. V. Dihydroaldosterone and 21-deoxytetrahydroaldosterone. Biochemistry 1963a;2:1249–54.

Kester MH, Bulduk S, van Toor H, Tibboel D, Meinl W, Glatt H, et al. Potent inhibition of estrogen sulfotransferase by hydroxylated metabolites of polyhalogenated aromatic hydrocarbons reveals alternative mechanism for estrogenic activity of endocrine disrupters. J Clin Endocrinol Metab 2002;87(3):1142–50.

Kirk DN, Burke PJ, Toms HC, Latif SA, Morris DJ. 18-Substituted steroids. Part 15. 6 beta-hydroxylation of aldosterone by liver. Steroids 1989;54(2):169–84.

Kirk DN, Schröder MH, Latif SA, Souness GW, Morris DJ. 18-Substituted steroids. Part 18. Chemical synthesis and mineralocorticoid activity of 2 alpha- and 2 beta-hydroxyaldosterone. Steroids 1993;58(2):59–63. https://doi.org/10.1016/0039-128x(93)90053-p. 8484185.

Klein GP, Baden M, Giroud CJ. Quantitative measurement and significance of 5 plasma corticosteroids during the perinatal period. J Clin Endocrinol Metab 1973;36(5):944–50.

Knights KM, Winner LK, Elliot DJ, Bowalgaha K, Miners JO. Aldosterone glucuronidation by human liver and kidney microsomes and recombinant UDP-glucuronosyltransferases: inhibition by NSAIDs. Br J Clin Pharmacol 2009;68(3):402–12.

Kraan GP, Derks HJ, Drayer NM. Quantification of polar glucocorticosteroids in the urine of pregnant and nonpregnant women: a comparison with 6 alpha-hydroxylated metabolites of cortisol in neonatal urine and amniotic fluid. J Clin Endocrinol Metab 1980;51(4):754–8.

Kraan GP, van Wee KT, Wolthers BG, van der Molen JC, Nagel GT, Drayer NM, et al. Synthesis and characterization of the 6 alpha- and 6 beta-hydroxylated derivatives of corticosterone, 11-dehydrocorticosterone, and 11-deoxycortisol. Steroids 1993a;58(10):495–503.

Kraan GP, Wolthers BG, van der Molen JC, Nagel GT, Drayer NM, Joannou GE. New identified 15 beta-hydroxylated 21-deoxy-pregnanes in congenital adrenal hyperplasia due to 21-hydroxylase deficiency. J Steroid Biochem Mol Biol 1993b;45(5):421–34.

Laatikainen T. Quantitative studies on the excretion of glucuronide and mono- and disulphate conjugates of neutral steroids in human bile. Ann Clin Res 1970;2(4):338–49.

Labrie F, Bélanger A, Cusan L, Candas B. Physiological changes in dehydroepiandrosterone are not reflected by serum levels of active androgens and estrogens but of their metabolites: intracrinology. J Clin Endocrinol Metab 1997;82(8):2403–9.

Labrie F, Bélanger A, Pelletier G, Martel C, Archer DF, Utian WH. Science of intracrinology in postmenopausal women. Menopause 2017;24(6):702–12.

Lacroix D, Sonnier M, Moncion A, Cheron G, Cresteil T. Expression of CYP3A in the human liver—evidence that the shift between CYP3A7 and CYP3A4 occurs immediately after birth. Eur J Biochem 1997;247(2):625–34.

Latif SA, Morris DJ, Wei L, Kirk DN, Burke PJ, Toms HC, et al. 18-Substituted steroids—Part 17. 2 Alpha-hydroxylated liver metabolites of aldosterone identified by high-field [1H]NMR spectroscopy. J Steroid Biochem 1989;33(6):1119–25.

Lee AJ, Conney AH, Zhu BT. Human cytochrome P450 3A7 has a distinct high catalytic activity for the 16alpha-hydroxylation of estrone but not 17beta-estradiol. Cancer Res 2003;63(19):6532–6.

Li CY, Basit A, Gupta A, Gáborik Z, Kis E, Prasad B. Major glucuronide metabolites of testosterone are primarily transported by MRP2 and MRP3 in human liver, intestine and kidney. J Steroid Biochem Mol Biol 2019;191, 105350.

Lin SX, Poirier D, Adamski J. A challenge for medicinal chemistry by the 17β-hydroxysteroid dehydrogenase superfamily: an integrated biological function and inhibition study. Curr Top Med Chem 2013;13(10):1164–71.

Liuzzo G, Galiuto L. Modulating the gut microbiome with dietary interventions to reduce cardiometabolic disease risk. Eur Heart J 2021;42(22):2152–3. https://doi.org/10.1093/eurheartj/ehab261. 34097727.

Lombardi P, Goldin B, Boutin E, Gorbach SL. Metabolism of androgens and estrogens by human fecal microorganisms. J Steroid Biochem 1978;9(8):795–801.

Manenda MS, Hamel CJ, Masselot-Joubert L, Picard MÈ, Shi R. Androgen-metabolizing enzymes: a structural perspective. J Steroid Biochem Mol Biol 2016;161:54–72.

Manuck TA. 17-Alpha hydroxyprogesterone caproate for preterm birth prevention: where have we been, how did we get here, and where are we going? Semin Perinatol 2017;41(8):461–7.

Marcos J, Pozo OJ. Current LC-MS methods and procedures applied to the identification of new steroid metabolites. J Steroid Biochem Mol Biol 2016;162:41–56.

Marcos J, Pol M, Fabregat A, Ventura R, Renau N, Hanzu FA, et al. Urinary cysteinyl progestogens: occurrence and origin. J Steroid Biochem Mol Biol 2015;152:53–61.

Meier M, Möller G, Adamski J. Perspectives in understanding the role of human 17beta-hydroxysteroid dehydrogenases in health and disease. Ann N Y Acad Sci 2009;1155:15–24.

Meilahn EN, De Stavola B, Allen DS, Fentiman I, Bradlow HL, Sepkovic DW, et al. Do urinary oestrogen metabolites predict breast cancer? Guernsey III cohort follow-up. Br J Cancer 1998;78(9):1250–5.

Melby JC, Dale SL, Wilson TE. 18-Hydroxy-deoxycorticosterone in human hypertension. Circ Res 1971;28(5):143–52. Suppl. 2.

Melby JC, Dale SL, Grekin RJ, Gaunt R, Wilson TE. 18-Hydroxy-11-deoxycorticosterone (18-OH-DOC) secretion in experimental and human hypertension. Recent Prog Horm Res 1972;28:287–351.

Meloche CA, Falany CN. Expression and characterization of the human 3beta-hydroxysteroid sulfotransferases (SULT2B1a and SULT2B1b). J Steroid Biochem Mol Biol 2001;77(4–5):261–9.

Migeon CJ, Paul AC, Samuels LT, Sandberg AA. Metabolism of 4-C14-corticosterone in man. J Clin Endocrinol Metab 1956;16(10):1291–8.

Miller WL. Disorders in the initial steps of steroid hormone synthesis. J Steroid Biochem Mol Biol 2017;165(Pt A):18–37.

Moeller G, Adamski J. Integrated view on 17beta-hydroxysteroid dehydrogenases. Mol Cell Endocrinol 2009;301(1–2):7–19.

Monder C, Bradlow HL. Carboxylic acid metabolites of steroids. J Steroid Biochem 1977;8(8):897–908.

Monder C, Zumoff B, Bradlow HL, Hellman L. Studies in the biotransformation of cortisol to the cortoic acids in man. I. Metabolism of 21-dehydrocortisol J Clin Endocrinol Metab 1975;40(1):86–92.

Morgan RA, Beck KR, Nixon M, Homer NZM, Crawford AA, Melchers D, et al. Carbonyl reductase 1 catalyzes 20β-reduction of glucocorticoids, modulating receptor activation and metabolic complications of obesity. Sci Rep 2017;7(1):10633.

Morris DJ, Ridlon JM. Glucocorticoids and gut bacteria: "The GALF hypothesis" in the metagenomic era. Steroids 2017;125:1–13.

Morris DJ, Tsai R. Chromatographic separation of aldosterone and its metabolites. Adv Chromatogr 1981;19:261–85.

Morris DJ, Latif SA, Conca TJ, Wei LT, Watlington CO, Kirk DN, Toms HC, Shackleton CH. Synthesis of 6 beta-hydroxyaldosterone by A6 (toad kidney) cells in culture. Steroids 1990;55(11):482–7.

Morris DJ, Latif SA, Brem AS. An alternative explanation of hypertension associated with 17α-hydroxylase deficiency syndrome. Steroids 2014;79:44–8.

Mueller JW, Gilligan LC, Idkowiak J, Arlt W, Foster PA. The regulation of steroid action by sulfation and desulfation. Endocr Rev 2015;36(5):526–63.

Mueller JW, Idkowiak J, Gesteira TF, Vallet C, Hardman R, van den Boom J, et al. Human DHEA sulfation requires direct interaction between PAPS synthase 2 and DHEA sulfotransferase SULT2A1. J Biol Chem 2018;293(25):9724–35.

Muller C, Pompon D, Urban P, Morfin R. Inter-conversion of 7alpha- and 7beta-hydroxy-dehydroepiandrosterone by the human 11beta-hydroxysteroid dehydrogenase type 1. J Steroid Biochem Mol Biol 2006;99(4–5):215–22.

Nabarro JD, Moxham A, Walker G, Slater JD. Rectal hydrocortisone. Br Med J 1957;2(5039):272–4.

Naelitz BD, Sharifi N. Through the looking-glass: reevaluating DHEA metabolism through HSD3B1 genetics. Trends Endocrinol Metab 2020;31(9):680–90.

Nishiyama T, Ogura K, Nakano H, Kaku T, Takahashi E, Ohkubo Y, et al. Sulfation of environmental estrogens by cytosolic human sulfotransferases. Drug Metab Pharmacokinet 2002;17(3):221–8.

Niwa T, Murayama N, Imagawa Y, Yamazaki H. Regioselective hydroxylation of steroid hormones by human cytochromes P450. Drug Metab Rev 2015;47(2):89–110.

Nováková L, Grand-Guillaume Perrenoud A, Nicoli R, Saugy M, Veuthey JL, Guillarme D. Ultra high performance supercritical fluid chromatography coupled with tandem mass spectrometry for screening of doping agents. I: Investigation of mobile phase and MS conditions. Anal Chim Acta 2015a;853:637–46.

Nováková L, Rentsch M, Grand-Guillaume Perrenoud A, Nicoli R, Saugy M, Veuthey JL, et al. Ultra high performance supercritical fluid chromatography coupled with tandem mass spectrometry for screening of doping agents. II: Analysis of biological samples. Anal Chim Acta 2015b;853:647–59.

Okano M, Ueda T, Nishitani Y, Kano H, Ikekita A, Kageyama S. UDP-glucuronosyltransferase 2B17 genotyping in Japanese athletes and evaluation of the current sports drug testing for detecting testosterone misuse. Drug Test Anal 2013;5(3):166–81.

Olivera ER, Luengo JM. Steroids as environmental compounds recalcitrant to degradation: genetic mechanisms of bacterial biodegradation pathways. Genes (Basel) 2019;10(7):512.

Oßwald A, Wang R, Beuschlein F, Hartmann MF, Wudy SA, Bidlingmaier M, et al. Performance of LC-MS/MS and immunoassay based 24-h urine free cortisol in the diagnosis of Cushing's syndrome. J Steroid Biochem Mol Biol 2019;190:193–7.

Owen RW, Tenneson ME, Bilton RF, Mason AN. The degradation of cholesterol by Escherichia coli isolated from human faeces [proceedings]. Biochem Soc Trans 1978;6(2):377–9.

Palermo M, Marazzi MG, Hughes BA, Stewart PM, Clayton PT, Shackleton CH. Human Delta4-3-oxosteroid 5beta-reductase (AKR1D1) deficiency and steroid metabolism. Steroids 2008;73(4):417–23.

Pandak WM, Kakiyama G. The acidic pathway of bile acid synthesis: not just an alternative pathway. Liver Res 2019;3(2):88–98.

Parr MK, Schänzer W. Detection of the misuse of steroids in doping control. J Steroid Biochem Mol Biol 2010;121(3–5):528–37.

Pasqualini JR, Wiqvist N, Diczfalusy E. Studies on the metabolism of corticosteroids in the human foeto-placental unit. 1. Metabolism of corticosterone sulphate administered into the umiblical circulation. Acta Endocrinol 1967;56(2):308–20.

Pearlman WH, LaMay EN, Peng LH, Pearlman MR, Hass JR. In vitro metabolism of [3H] corticosterone by mammary glands from lactating rats. Isolation and identification of 21-acyl[3H]corticosterone. J Biol Chem 1985;260(9):5296–301.

Penning TM, Bennett MJ, Smith-Hoog S, Schlegel BP, Jez JM, Lewis M. Structure and function of 3 alpha-hydroxysteroid dehydrogenase. Steroids 1997;62(1):101–11.

Penning TM, Burczynski ME, Jez JM, Hung CF, Lin HK, Ma H, et al. Human 3 alpha-hydroxysteroid dehydrogenase isoforms (AKR1C1-AKR1C4) of the aldo-keto reductase superfamily: functional plasticity and tissue distribution reveals roles in the inactivation and formation of male and female sex hormones. Biochem J 2000;351(Pt 1):67–77.

Penning TM, Chen M, Jin Y. Promiscuity and diversity in 3-ketosteroid reductases. J Steroid Biochem Mol Biol 2015;151:93–101.

Penning TM, Wangtrakuldee P, Auchus RJ. Structural and functional biology of aldo-keto reductase steroid-transforming enzymes. Endocr Rev 2019;40(2):447–75.

Philipp B. Bacterial degradation of bile salts. Appl Microbiol Biotechnol 2011;89(4):903–15.

Piper T, Emery C, Thomas A, Saugy M, Thevis M. Combination of carbon isotope ratio with hydrogen isotope ratio determinations in sports drug testing. Anal Bioanal Chem 2013;405(16):5455–66.

Poisson Paré D, Song D, Luu-The V, Han B, Li S, Liu G, et al. Expression of estrogen sulfotransferase 1E1 and steroid sulfatase in breast cancer: a Immunohistochemical study. Breast Cancer (Auckl) 2009;20(3):9–21.

Ponzetto F, Boccard J, Nicoli R, Kuuranne T, Saugy M, Rudaz S. Steroidomics for highlighting novel serum biomarkers of testosterone doping. Bioanalysis 2019;11(12):1171–87.

Pozo OJ, Marcos J, Ventura R, Fabregat A, Segura J. Testosterone metabolism revisited: discovery of new metabolites. Anal Bioanal Chem 2010;398(4):1759–70.

Pranata A, Fitzgerald CC, Khymenets O, Westley E, Anderson NJ, Ma P, Pozo OJ, McLeod MD. Synthesis of steroid bisglucuronide and sulfate glucuronide reference materials: unearthing neglected treasures of steroid metabolism. Steroids 2019;143:25–40.

Rastelli M, Cani PD, Knauf C. The gut microbiome influences host endocrine functions. Endocr Rev 2019;40(5):1271–84.

Reed MJ. Role of enzymes and tissue-specific actions of steroids. Maturitas 2004;48(Suppl. 1):S18–23.

Reed MJ, Fotherby K, Steele SJ. Metabolism of ethynyloestradiol in man. J Endocrinol 1972;55(2):351–61.

Reeder AY, Joannou GE. 15 beta-hydroxysteroids (part IV). Steroids of the human perinatal period: the synthesis of 3 alpha,15 beta,17 alpha-trihydroxy-5 alpha-pregnan-20-one and its A/B-ring configurational isomers. Steroids 1995;60(12):796–801.

Ridlon JM, Ikegawa S, Alves JM, Zhou B, Kobayashi A, Iida T, et al. *Clostridium scindens*: a human gut microbe with a high potential to convert glucocorticoids into androgens. J Lipid Res 2013;54(9):2437–49.

Robinson N, Saugy M, Vernec A, Pierre-Edouard S. The athlete biological passport: an effective tool in the fight against doping. Clin Chem 2011;57(6):830–2.

Sadzikowski MR, Sperry JF, Wilkins TD. Cholesterol-reducing bacterium from human feces. Appl Environ Microbiol 1977;34(4):355–62. https://doi.org/10.1128/aem.34.4.355-362.1977. PMID: 335969; PMCID: PMC242663.

Sandberg AA, Slaunwhite WR. Studies on phenolic steroids in human subjects II. The metabolic fate and hepatobiliary circulation on C^{14}-estrone and C^{14} estradiol in women. J Clin Invest 1951;36:1266–78.

Schänzer W, Thevis M. Human sports drug testing by mass spectrometry. Mass Spectrom Rev 2017;36(1):16–46.

Schatz F, Hochberg RB. Lipoidal derivative of estradiol: the biosynthesis of a nonpolar estrogen metabolite. Endocrinology 1981;109(3): 697–703.

Schiffer L, Barnard L, Baranowski ES, Gilligan LC, Taylor AE, Arlt W, et al. Human steroid biosynthesis, metabolism and excretion are differentially reflected by serum and urine steroid metabolomes: a comprehensive review. J Steroid Biochem Mol Biol 2019;194, 105439.

Schneider JJ. In vitro conversion of deoxycorticosterone to some acidic metabolites. In: Martini L, Pecile A, editors. Hormonal steroids. New York: Academic Press; 1964. p. 127–35.

Schneider MA, Davies MC, Honour JW. The timing of placental competence in pregnancy after oocyte donation. Fertil Steril 1993;59 (5):1059–64.

Schubert K, Schlegel J, Böhme KH, Hörhold C. Mikrobielle hydrierungs- und dehydrierungs-Reaktionen bei delta 4-3-Ketosteroiden mit einer 6-Hydroxygruppe [Microbial hydrogenation and dehydrogenation of delta 4-3-ketosteroids having a 6-hydroxy group]. Biochim Biophys Acta 1967;144(1):132–8 [in German].

Shackleton CH, Taylor NF. Identification of the androstenetriolones and androstenetetrols present in the urine of infants. J Steroid Biochem 1975a;6(10):1393–9.

Shackleton CH, Taylor NF. Conversion of 3beta, 15beta, 16beta-trihydroxy-5-androsten-17-one to 1,3,5(10)-oestratriene-3,15beta,16beta,17beta-tetrol by placental homogenates. J Steroid Biochem 1975b;6(10):1401–3.

Shackleton CH, Biglieri EG, Roitman E, Honour JW. Metabolism of radiolabeled corticosterone in an adult with the 17 alpha-hydroxylase deficiency syndrome. J Clin Endocrinol Metab 1979a;48(6):976–82.

Shackleton CH, Honour JW, Winter J, Bokkenheuser VD. Urinary metabolites of 18-hydroxylated corticosteroids: microbial preparation of reference compounds. J Steroid Biochem 1979b;11(2):1141–4.

Shackleton CH, Honour JW, Taylor NF. Metabolism of fetal and neonatal adrenal steroids. J Steroid Biochem 1979c;11(1B):523–9.

Smith RL. Biliary excretion and hepatotoxicity of contraceptive steroids. Acta Endocrinol Suppl 1974;185:149–68.

Smith KE, Latif S, Kirk DN, White KA. Microbial transformations of steroids—I. Rare transformations of progesterone by *Apiocrea chrysosperma*. J Steroid Biochem 1988;31(1):83–9.

Smith KE, Latif S, Kirk DN. Microbial transformations of steroids—V. Transformation of progesterone by whole cells and extracts of *Botryosphaerica obtusa*. J Steroid Biochem 1989a;33(5):927–34.

Smith KE, Latif S, Kirk DN, White KA. Microbial transformations of steroids—IV. 6,7-Dehydrogenation; a new class of fungal steroid transformation product. J Steroid Biochem 1989b;33(2):271–6. https://doi.org/10.1016/0022-4731(89)90304-x. 2770300.

Smith KE, Latif SA, Kirk DN. Microbial transformation of steroids—VII. Hydroxylation of progesterone by extracts of *Phycomyces blakesleeanus*. J Steroid Biochem Mol Biol 1991;38(2):249–56.

Smy L, Straseski JA. Measuring estrogens in women, men, and children: recent advances 2012-2017. Clin Biochem 2018;62:11–23.

Sten T, Finel M, Ask B, Rane A, Ekström L. Non-steroidal anti-inflammatory drugs interact with testosterone glucuronidation. Steroids 2009a;74(12):971–7.

Sten T, Kurkela M, Kuuranne T, Leinonen A, Finel M. UDP-glucuronosyltransferases in conjugation of 5alpha- and 5beta-androstane steroids. Drug Metab Dispos 2009b;37(11):2221–7.

Stewart PM, Wallace AM, Valentino R, Burt D, Shackleton CH, Edwards CR. Mineralocorticoid activity of liquorice: 11-beta-hydroxysteroid dehydrogenase deficiency comes of age. Lancet 1987;2(8563):821–4.

Stocco C. Tissue physiology and pathology of aromatase. Steroids 2012;77 (1–2):27–35.

Storbeck KH, Schiffer L, Baranowski ES, Chortis V, Prete A, Barnard L, et al. Steroid metabolome analysis in disorders of adrenal steroid biosynthesis and metabolism. Endocr Rev 2019;40(6):1605–25.

Sun M, Leyh TS. The human estrogen sulfotransferase: a half-site reactive enzyme. Biochemistry 2010;49(23):4779–85.

Taylor NF, Shackleton CH. 15alpha-Hydroxyoestriol and other polar oestrogens in pregnancy monitoring. Ann Clin Biochem 1978;15(1):1–11.

Taylor NF, Curnow DH, Shackleton CH. Analysis of glucocorticoid metabolites in the neonatal period: catabolism of cortisone acetate by an infant with 21-hydroxylase deficiency. Clin Chim Acta 1978;85 (3):219–29.

Terao M, Katayama I. Local cortisol/corticosterone activation in skin physiology and pathology. J Dermatol Sci 2016;84(1):11–6.

Thevis M, Schänzer W. Detection of SARMs in doping control analysis. Mol Cell Endocrinol 2018;464:34–45.

Thigpen AE, Davis DL, Milatovich A, Mendonca BB, Imperato-McGinley J, Griffin JE, Francke U, Wilson JD, Russell DW. Molecular genetics of steroid 5 alpha-reductase 2 deficiency. J Clin Invest 1992;90 (3):799–809.

Thigpen AE, Silver RI, Guileyardo JM, Casey ML, McConnell JD, Russell DW. Tissue distribution and ontogeny of steroid 5 alpha-reductase isozyme expression. J Clin Invest 1993;92(2):903–10.

Thomas MP, Potter BV. The structural biology of oestrogen metabolism. J Steroid Biochem Mol Biol 2013;137:27–49.

Tomlinson JW, Walker EA, Bujalska IJ, Draper N, Lavery GG, Cooper MS, et al. 11Beta-hydroxysteroid dehydrogenase type 1: a tissue-specific regulator of glucocorticoid response. Endocr Rev 2004;25 (5):831–66.

Uemura M, Tamura K, Chung S, Honma S, Okuyama A, Nakamura Y, et al. Novel 5 alpha-steroid reductase (SRD5A3, type-3) is overexpressed in hormone-refractory prostate cancer. Cancer Sci 2008;99 (1):81–6.

Ulick S. Stereospecificity in the metabolism of aldosterone in man. J Biol Chem 1961;236:680–4.

Ulick S, Vetter KK. Identification of two C18 oxygenated corticosteroids isolated from human urine. J Biol Chem 1962;237:3364–74.

Underwood RH, Tait JF. Purification, partial characterization and metabolism of an acid labile conjugate of aldosterone. J Clin Endocrinol Metab 1964;24:1110–24.

Vernec A, Slack A, Harcourt PR, Budgett R, Duclos M, Kinahan A, et al. Glucocorticoids in elite sport: current status, controversies and innovative management strategies-a narrative review. Br J Sports Med 2020;54(1):8–12.

Wade AP, Slater JD, Kellie AE, Holliday ME. Urinary excretion of 17-ketosteroids following rectal infusion of cortisol. J Clin Endocrinol Metab 1959;19(4):444–53.

Wang F, Vihma V, Soronen J, Turpeinen U, Hämäläinen E, Savolainen-Peltonen H, et al. 17β-Estradiol and estradiol fatty acyl esters and estrogen-converting enzyme expression in adipose tissue in obese men and women. J Clin Endocrinol Metab 2013;98(12):4923–31.

Wilson III W, Pardo-Manuel de Villena F, Lyn-Cook BD, Chatterjee PK, Bell TA, Detwiler DA, et al. Characterization of a common deletion polymorphism of the UGT2B17 gene linked to UGT2B15. Genomics 2004;84(4):707–14.

Winter J, Bokkenheuser VD. Bacterial metabolism of natural and synthetic sex hormones undergoing enterohepatic circulation. J Steroid Biochem 1987;27(4–6):1145–9.

Winter J, Bokkenheuser VD, Ponticorvo L. Bacterial metabolism of corticoids with particular reference to the 21-dehydroxylation. J Biol Chem 1979;254(8):2626–9.

Winter J, Cerone-McLernon A, O'Rourke S, Ponticorvo L, Bokkenheuser VD. Formation of 20 beta-dihydrosteroids by anaerobic bacteria. J Steroid Biochem 1983;17(6):661–7.

Winter J, Morris GN, O'Rourke-Locascio S, Bokkenheuser VD, Mosbach EH, Cohen BI, et al. Mode of action of steroid desmolase and reductases synthesized by Clostridium "scindens" (formerly Clostridium strain 19). J Lipid Res 1984a;25(10):1124–31.

Winter J, Shackleton CH, O'Rourke S, Bokkenheuser VD. Bacterial formation of aldosterone metabolites. J Steroid Biochem 1984b;21 (5):563–9.

Winter J, O'Rourke-Locascio S, Bokkenheuser VD, Mosbach EH, Cohen BI. Reduction of 17-keto steroids by anaerobic microorganisms isolated from human fecal flora. Biochim Biophys Acta 1984c;795 (2):208–11.

Yee SW, Stecula A, Chien HC, Zou L, Feofanova EV, van Borselen M, et al. Unraveling the functional role of the orphan solute carrier, SLC22A24 in the transport of steroid conjugates through metabolomic and genome-wide association studies. PLoS Genet 2019;15(9), e1008208.

Zhang H, Varlamova O, Vargas FM, Falany CN, Leyh TS. Sulfuryl transfer: the catalytic mechanism of human estrogen sulfotransferase. J Biol Chem 1998;273(18):10888–92. Erratum in: J Biol Chem 1998;273(27):17296.

Further reading

Abaffy T, Matsunami H. 19-Hydroxy steroids in the aromatase reaction: review on expression and potential functions. J Endocr Soc 2021;5 (7), bvab050.

Adlercreutz H, Järvenpää P. Assay of estrogens in human feces. J Steroid Biochem 1982;17(6):639–45.

Adnan S, Nelson JW, Ajami NJ, Venna VR, Petrosino JF, Bryan Jr RM, et al. Alterations in the gut microbiota can elicit hypertension in rats. Physiol Genomics 2017;49(2):96–104.

Aizawa H, Niimura M, Kon Y. Influence of oral metronidazole on the endocrine milieu and sebum excretion rate. J Dermatol 1992;19(12): 959–63.

Ballatori N, Christian WV, Lee JY, Dawson PA, Soroka CJ, Boyer JL, et al. OSTalpha-OSTbeta: a major basolateral bile acid and steroid transporter in human intestinal, renal, and biliary epithelia. Hepatology 2005;42(6):1270–9.

Bélanger A, Pelletier G, Labrie F, Barbier O, Chouinard S. Inactivation of androgens by UDP-glucuronosyltransferase enzymes in humans. Trends Endocrinol Metab 2003;14(10):473–9.

Birzniece V. Hepatic actions of androgens in the regulation of metabolism. Curr Opin Endocrinol Diabetes Obes 2018;25(3):201–8.

Eriksson H. Steroids in germfree and conventional rats. Unconjugated metabolites of [4-14C]pregnenolone and [4-14C]corticosterone in faeces from female rats. Eur J Biochem 1970;16(2):261–7.

Eriksson H, Gustafsson JA. Excretion of steroid hormones in adults. Steroids in urine from a pregnant woman. Eur J Biochem 1970a;16 (2):268–77.

Eriksson H, Gustafsson JA. Steroids in germfree and conventional rats. Steroids in the mono- and disulphate fractions of faeces from female rats. Eur J Biochem 1970b;16(2):252–60.

Eriksson H, Gustafsson JA. Steroids in germfree and conventional rats. Distribution and excretion of labelled pregnenolone and corticosterone in male and female rats. Eur J Biochem 1970c;15(1):132–9.

Eriksson H, Gustafsson JA. Steroids in germfree and conventional rats. Sulpho- and glucuronohydrolase activities of caecal contents from conventional rats. Eur J Biochem 1970d;13(1):198–202.

Eriksson H, Gustafsson JA. Metabolism of corticosterone in the isolated perfused rat liver. Eur J Biochem 1971a;20(2):231–6.

Eriksson H, Gustafsson JA. Excretion of steroid hormones in adults. Steroids in faeces from adults. Eur J Biochem 1971b;18(1):146–50.

Eriksson H, Gustafsson JA, Sjövall J. Steroids in germfree and conventional rats. 21-dehydroxylation by intestinal microorganisms. Eur J Biochem 1969a;9(4):550–4.

Eriksson H, Gustafsson JA, Sjövall J. Steroids in germfree and conventional rats. Free steroids in faeces from conventional rats. Eur J Biochem 1969b;9(2):286–90.

Eriksson H, Gustafsson JA, Sjövall J. Studies on the structure, biosynthesis and bacterial metabolism of 15-hydroxylated steroids in the female rat. Eur J Biochem 1971;19(3):433–41.

Gong L, Aranibar N, Han YH, Zhang Y, Lecureux L, Bhaskaran V, et al. Characterization of organic anion-transporting polypeptide (Oatp) 1a1 and 1a4 null mice reveals altered transport function and urinary metabolomic profiles. Toxicol Sci 2011;122(2):587–97.

Gosetti F, Mazzucco E, Gennaro MC, Marengo E. Ultra high performance liquid chromatography tandem mass spectrometry determination and profiling of prohibited steroids in human biological matrices. A review. J Chromatogr B Analyt Technol Biomed Life Sci 2013;927:22–36.

Gustafsson JA. Steroids and germfree and conventional rats. 7. Identification of C19 and C21 steroids in faeces from conventional rats. Eur J Biochem 1968;6(2):248–55.

Gustafsson JA, Sjövall J. Steroids in germfree and conventional rats. 6. Identification of 15 alpha- and 21-hydroxylated C21 steroids in faeces from germfree rats. Eur J Biochem 1968a;6(2):236–47.

Gustafsson JA, Sjövall J. Steroids in germfree and conventional rats. 5. Identification of C19 steroids in faeces from germfree rats. Eur J Biochem 1968b;6(2):227–35.

Gustafsson BE, Gustafsson JA, Sjövall J. Intestinal and fecal sterols in germfree and conventional rats. Bile acids and steroids 172. Acta Chem Scand 1966;20(7):1827–35.

Hennebert O, Montes M, Favre-Reguillon A, Chermette H, Ferroud C, Morfin R. Epimerase activity of the human 11beta-hydroxysteroid dehydrogenase type 1 on 7-hydroxylated C19-steroids. J Steroid Biochem Mol Biol 2009;114(1–2):57–63.

Honour JW. Hypertensive effect of steroids produced in the intestine by bacterial metabolism. In: Serono symposium: endocrinology of hypertension, Padova; October 1981. p. 30. Abstract # 13.

Honour JW. The possible involvement of intestinal bacteria in steroidal hypertension. Endocrinology 1982;110:285–7.

Honour JW, Ganten U. In: Effect of neomycin on hypertension in normal and SHR rats. Sixth International Congress on Hormonal Steroids, Jerusalem; September 1982.

Honour JW, Kent J. Dissociation of the hypertensive effect from other physiological responses to steroids in rats. In: Proceedings of the society for endocrinology 164th meeting. London: Society of Endocrinology; 1981. p. 7. Abstract #13.

Honour JW, Millar G, Roitman E, Shackleton CHL. Steroid excretion in man during suppression and stimulation of adrenals in 17-hydroxylase deficiency syndrome. J Clin Endocrinol Metabol 1981;52:1039–42.

Honour JW, Borriello SP, Ganten U, Honour P. Antibiotics attenuate experimental hypertension in rats. J Endocrinol 1985;105:347–50.

Huang A, Roth CL. The link between obesity and puberty: what is new? Curr Opin Pediatr 2021;33(4):449–57.

Iannone M, Botrè F, Martinez-Brito D, Matteucci R, de la Torre X. Development and application of analytical procedures for the GC-MS/MS analysis of the sulfates metabolites of anabolic androgenic steroids: The pivotal role of chemical hydrolysis. J Chromatogr B Analyt Technol Biomed Life Sci 2020;1155, 122280.

Järvinen E, Kidron H, Finel M. Human efflux transport of testosterone, epitestosterone and other androgen glucuronides. J Steroid Biochem Mol Biol 2020;197, 105518.

Kelly WG, Bandi L, Lieberman S. Isolation and characterization of human urinary metabolites of aldosterone. IV. The synthesis and stereochemistry of two bicyclic acetal metabolites. Biochemistry 1963b;2:1243–9.

Li J, Zhao F, Wang Y, Chen J, Tao J, Tian G, et al. Gut microbiota dysbiosis contributes to the development of hypertension. Microbiome 2017;5 (1):14.

Liang Q, Xu W, Hong Q, Xiao C, Yang L, Ma Z, et al. Rapid comparison of metabolites in humans and rats of different sexes using untargeted UPLC-TOFMS and an in-house software platform. Eur J Mass Spectrom (Chichester) 2015;21(6):801–21.

Lloyd-Evans E, Waller-Evans H. Biosynthesis and signalling functions of central and peripheral nervous system neurosteroids in health and disease. Essays Biochem 2020;64(3):591–606.

Macdonald IA, Bokkenheuser VD, Winter J, McLernon AM, Mosbach EH. Degradation of steroids in the human gut. J Lipid Res 1983;24(6): 675–700.

Mackenzie PI, Rodbourne L, Stranks S. Steroid UDP glucuronosyltransferases. J Steroid Biochem Mol Biol 1992;43 (8):1099–105.

Mell B, Jala VR, Mathew AV, Byun J, Waghulde H, Zhang Y, et al. Evidence for a link between gut microbiota and hypertension in the Dahl rat model. Physiol Genomics 2015;47:187–97.

Mobley JE, Headstream JW, Melby J. Primary aldosteronism. Preoperative preparation with spironolactone. JAMA 1962;180:1056–8.

Pluznick JL, Protzko RJ, Gevorgyan H, Peterlin Z, Sipos A, Han J, et al. Olfactory receptor responding to gut microbiota-derived signals plays a role in renin secretion and blood pressure regulation. Proc Natl Acad Sci U S A 2013;110:4410–5.

Rižner TL, Penning TM. Role of aldo-keto reductase family 1 (AKR1) enzymes in human steroid metabolism. Steroids 2014;79:49–63.

Santen RJ, Brodie H, Simpson ER, Siiteri PK, Brodie A. History of aromatase: saga of an important biological mediator and therapeutic target. Endocr Rev 2009;30(4):343–75.

Schiffer L, Arlt W, Storbeck KH. Intracrine androgen biosynthesis, metabolism and action revisited. Mol Cell Endocrinol 2018;465:4–26.

Sealey JE, Bühler FR, Laragh JH, Manning EL, Brunner HR. Aldosterone excretion. Physiological variations in man measured by radioimmunoassay or double-isotope dilution. Circ Res 1972;31(3):367–78.

Theiler-Schwetz V, Zaufel A, Schlager H, Obermayer-Pietsch B, Fickert P, Zollner G. Bile acids and glucocorticoid metabolism in health and disease. Biochim Biophys Acta Mol Basis Dis 2019;1865(1):243–51.

Thevis M, Volmer DA. Mass spectrometric studies on selective androgen receptor modulators (SARMs) using electron ionization and electrospray ionization/collision-induced dissociation. Eur J Mass Spectrom (Chichester) 2018;24(1):145–56.

Vural M, Gilbert B, Üstün I, Caglar S, Finckh A. Mini-review: human microbiome and rheumatic diseases. Front Cell Infect Microbiol 2020;10(10), 491160. https://doi.org/10.3389/fcimb.2020.491160. 33304855. PMC7693548.

Wan R, Kong X, Yang Y, Tao S, Chen Y, Teichmann AT, et al. Role of human 3α-hydroxysteroid dehydrogenase isoforms (AKR1C1-AKR1C3) in the extrahepatic metabolism of the steroidal aromatase inactivator Formestane. J Steroid Biochem Mol Biol 2020;198, 105527.

Weththasinghe SA, Waller CC, Fam HL, Stevenson BJ, Cawley AT, McLeod MD. Replacing PAPS: in vitro phase II sulfation of steroids with the liver S9 fraction employing ATP and sodium sulfate. Drug Test Anal 2018;10(2):330–9.

Yan Q, Gu Y, Li X, Yang W, Jia L, Chen C, et al. Alterations of the gut microbiome in hypertension. Front Cell Infect Microbiol 2017;7:381.

Yang T, Santisteban MM, Rodriguez V, Li E, Ahmari N, Carvajal JM, et al. Gut dysbiosis is linked to hypertension. Hypertension 2015;65 (6):1331–40.

Zanger UM, Schwab M. Cytochrome P450 enzymes in drug metabolism: regulation of gene expression, enzyme activities, and impact of genetic variation. Pharmacol Ther 2013;138(1):103–41.

Zumoff B, Monder C, Bradlow HL. Studies in the biotransformation of cortisol to the cortoic acids in man. II. The central role of tetrahydrocortisol and tetrahydrocortisone as intermediates. J Clin Endocrinol Metab 1977;44(4):647–50.

Part II

Laboratory analysis of steroids

The clinicians may request that **biological samples** are collected and sent to the laboratory so that concentrations of hormones can be determined. Nurses, health care assistants, and phlebotomists may be dedicated to these steps. The steps involved constitute pre**analytical steps before the actual analysis of steroids in the laboratory**. The early stages are crucial to the process in order to avoid the concept of "rubbish in, rubbish out." All staff involved need training for the procedures. Blood is the commonest sample because this takes steroids to the target sites. This is a complex mixture of cells electrolytes, minerals, proteins, and lipids in a fluid (**plasma**). Blood naturally sticks together into clots with exclusion of some fluid (**serum**). Blood spots can be collected onto designated filter paper. Analysis of these **dried blood spots** is used in newborn screening and sometimes as a procedure by the patient at home to monitor a treatment. Anticoagulant chemicals can be included in the blood collection tube when plasma is needed for analysis. Venous blood samples from the arm are the commonest biological samples for steroid analysis. **Urine** is a **biological fluid** that can be analyzed for steroid content. A single sample of urine can be collected in a small container (100 mL), many voids over 24 h can be collected in a large amount (1–2.5 L), it is surprisingly difficult to remember to keep all voids. There are a few other liquids that can be collected under different circumstances—**saliva** collection requires no intervention, but **ovarian follicular fluid, amniotic fluid, etc.** will be collected under anesthesia.

Hair is a biological sample that can be used to determine steroid concentrations as a measure of long-term exposure. **Feces** are rarely collected except in animal studies where blood and other samples are difficult to collect without restraint. The new interest in the microbiome will direct investigations of steroid content of **bile** and **intestinal contents** as well as the intestinal microflora.

On arrival in the laboratory, many steroid tests are still performed by immunoaassay often incorporated into the automated chemistry analysers. These tests will continue to be front-line investigations for some years because of the need to process large numbers of samples quickly, but as we now know at the expense of specificity. The most accurate means of steroid analysis these days involve isotope dilution whereby a stable isotope labeled internal standard is added to the sample to act as a reference point in the analysis because at every step recovery of the analyte from the sample and the added internal standard are the same. A **sample preparation** step is needed, a solvent or solid-phase extraction ideally for many samples at a time (96-well plate for example). The extract will contain many steroids and other chemicals that will need **purification** before the actual analytical step. Gas chromatography and liquid chromatography are used to create a point of time for **analysis** where a purer extract is submitted to the mass spectrometer. Tens of steroids can now be measured in a single extract. High-resolution mass spectrometry is a further level of analysis at high specificity. A cost-benefit analysis is essential in optimizing the instruments needed in the laboratory.

The clinician will need to judge the meaning of the steroid results first by comparison with the most appropriate normal (reference) range for the biological sample then in relation to the trophic hormone concentrations. Age, sex, and development of the patient will have to be considered. Further investigations may be needed including some provocative and genetic tests although the author is not an expert in the latter.

Chapter 2.1

Samples for analysis

2.1.1 Introduction

Many biological materials are used for analysis to answer questions about endocrine functions in normal and pathological situations. The ease of collection of samples and detectability of the steroids therein are important questions. Blood samples in most cases require venipuncture which for some people and animals can be unpleasant even a little painful. Each step in the process of drawing blood samples (phlebotomy) can affect the quality of the specimen and is thus important for preventing laboratory error, patient injury and even death. There are useful published guidelines on best practices for blood sampling (World Health Organization, 2010; CLSI (formerly NCCLS), 2003, 2007; Simundic et al., 2018). The jarring and shaking of blood tubes in transit can cause lysis of red blood cells causing false laboratory results. Errors in completing forms and identifying patients are common, costly and preventable. Patients can suffer bruising at the site of puncture, fainting, nerve damage and hematomas. Phlebotomy is a risk for health workers. A needle stick injury can result in transmission of disease.

Urine is collected with ease and deciliter volumes of aqueous material yields plenty of steroid without the interfering protein and cells found in blood. A saliva sample is not difficult to obtain but the steroid concentrations are very low because the steroids are derived as an ultrafiltrate of blood and are an index of the steroid free in plasma not bound to proteins. Hair can be analyzed as an historical index of steroid exposure. Brain is a difficult matrix from which to extract steroids due to the high lipid content and the material is only available from living animals after sacrifice.

This book is directed at human investigations but animal studies and cultured cells have been important in revealing mechanisms in steroid endocrinology. Two such studies are highlighted here: **(1) the newborn Tammar Wallaby**, which provides interesting access to sexual development in the pouch that in other species would be completed while in utero and **(2) the female spotted hyena**, which has been studied because of unusual development of a fused scrotum and enlarged clitoris that is similar to the male penis of the species. The underlying mechanisms are discussed elsewhere (see Chapter 3.2.1).

In nearly all investigations the nature of the sample at the time of collection must be preserved, hence any factor before the analysis that could affect the steroids in the sample must be verified. Where material is not available from organs or fluids then cultured cells can provide a stable supply of a tissue but care must be taken to prevent changes in the genome. In some cases, the genome can be modified by gene deletion or enrichment techniques in cells or animals. Samples from transgenic animals have answered many questions in steroid endocrinology and some examples will be given. Steroids are widely used as pharmaceutical agents or nonprescription sources. Passage of steroids into the environment then has to be investigated. The measurement of steroids in many cases goes hand in hand with knowledge of the proteins in the regulatory pathways. Samples for these protein measurements require special preanalytical considerations. The stability of samples must be considered and tested for effects of temperature, time of storage and repeated freeze-thaw cycles.

Other sources include analysis of placental tissue, perfusion studies of the placenta, amniotic fluid, composition of milk, sweat and nails which will be considered elsewhere in the book where relevant. The studies include exploration of the unique immunological role of the placenta, the placenta as a barrier to immune rejection, reproductive status, the socioeconomic effects of body dour, long-term exposure to steroids through disease or excessive use and the use of prohibited steroids.

2.1.2 Preanalytical factors

The preanalytical phase encompasses all the procedures before the start of laboratory testing. This phase of the testing process is responsible for the majority of errors that affect the outcome of the analytical procedure (Rizner and Adamski, 2019; Zhou et al., 2017; Carraro et al., 2012). The related procedures involve many sorts of supporting staff without a laboratory background working outside the laboratory setting and without direct supervision by the laboratory staff. Therefore, organization or management of both personnel and procedures that regard, for example, blood specimen collection by venipuncture are of fundamental importance, since the various steps for performing blood collection represent per se sources of laboratory variability. Samples for bio-banking may need different requirements (Kirwan et al., 2018) so a plan and specifications need to be in place at the outset (Ciplea et al., 2018). All items of equipment that come in contact with the sample need to be reviewed for possible steroid retention (McManus and Sharifi, 2019).

2.1.2.1   Test ordering and retesting

Inappropriate test requesting has long been known as a waste of resources and results in inappropriate patient pathways. Inappropriate tests are those that could be avoided with no detriment to patient care, but when considering appropriate utilization of the laboratory, it is important to include tests which were not requested but which would have been clinically relevant at the time of the initial request. The area should therefore ensure that all appropriate tests are performed and inappropriate tests suppressed. There are various mechanisms to ensure appropriate test ordering, both via reducing inappropriate tests and by ensuring the correct tests for a condition are performed. A standard operating procedure must be in place (Tuck et al., 2009).

Minimum retesting intervals can be used to put blocks into place to prevent a test being repeated in a time window that is not clinically relevant. There are also some well-documented processes to achieve the reduction in inappropriate testing, such as reflex testing, where a frontline test is performed and subsequent tests added only when clinically useful in light of the original result.

Reflex testing, as well as preventing some tests from being performed, also ensures that required tests are performed in a timely manner and often without recollecting specimens from the patients. An alternative approach is condition-specific requesting, a growing area where clinicians use information technology systems to click on a condition; preset algorithms then determine what tests are required as a frontline. Further tests are then cascaded within the laboratory, as required.

2.1.2.2   Patient preparation

There are a large number of preanalytical variables that fall under the umbrella of patient preparation and can affect laboratory tests. These include diet, drugs (medication and recreational), herbal remedies, vitamins, physical activity, smoking, alcohol and many more. Patients and medical staff should be provided with the right information for the tests requested. This is something that does not always currently occur.

Each patient should be asked about fasting status, physical activity and therapeutic-drug intake prior to the any sample collection by health care assistant or clinical staff. Briefly, fasting time of 12 h before the majority of blood tests; any exercise shall be avoided 72 h before blood sampling. Information about therapeutic-drug intake should be made available to laboratory personnel. A light meal, such as a breakfast with around 550 kcal can lead to laboratory variability for both clinical chemistry, hematology and coagulation testing. Indeed, if fasting time is avoided—for complete blood count, albumin, bilirubin, phosphate, calcium, magnesium and potassium—more risks than benefits can derive for patients. However, in particular situations, blood specimen drawing is not achievable under optimal conditions.

Evacuated tubes systems are preferred to syringes to perform blood collection. The phlebotomist needs to understand appropriate mixing of blood and additives (i.e., anticoagulants or clot activators) by inverting not shaking of the tubes. Evacuated tubes, holders and needles are collectively known as specimen collection apparatus, while each one has its own technical characteristics devised for specific use. Needles can be purchased from manufacturers for either use in evacuated sample tube systems or in connection with a syringe, either ready for straight use or joined to a so-called "winged infusion set or butterfly needle" system. Needles are standardized by gauge (G) and the larger the gauge number, the smaller the diameter of the bore (Fig. 2.1.1). Routine venipunctures are usually performed with needles ranging from 19 to 25G. The 19–21G needles are used primarily for large antecubital veins; 23G needles for either smaller antecubitals, medium-size forearm, hands or foot veins and 25G or smaller needles are used only for the smallest veins, or in newborns and children. Blood should not be taken from catheters.

2.1.2.3   Patient identification and tube labeling

Patients must be asked, if conscious, to state their name, address, ID number and/or date of birth. Any discrepancy must be recorded and reported as determined by local policy. If the patient is unconscious and emergency testing is required, this practice may not be applicable to safeguard patient's life. If printed labels are not provided, then three methods of ID should be written legibly onto the tubes and information again checked against patient information.

2.1.2.4   Sampling

Preanalytical factors are the greatest source of errors attributable to laboratory diagnostics. The procedures for collection of biological samples should be designed to preserve the integrity of the sample to the original chemical composition. Blood samples are the most widely used.

NEEDLE GAUGES FOR INJECTIONS CHART SIZE

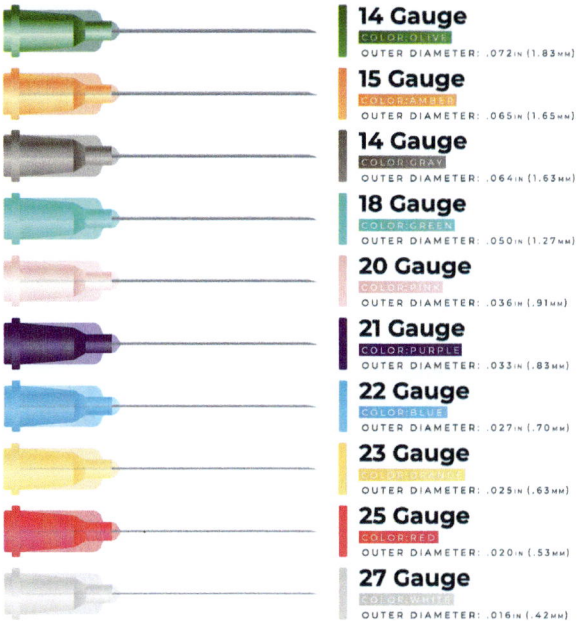

14 Gauge
COLOR:OLIVE
OUTER DIAMETER: .072IN (1.83MM)

15 Gauge
COLOR:AMBER
OUTER DIAMETER: .065IN (1.65MM)

14 Gauge
COLOR:GRAY
OUTER DIAMETER: .064IN (1.63MM)

18 Gauge
COLOR:GREEN
OUTER DIAMETER: .050IN (1.27MM)

20 Gauge
COLOR:PINK
OUTER DIAMETER: .036IN (.91MM)

21 Gauge
COLOR:PURPLE
OUTER DIAMETER: .033IN (.83MM)

22 Gauge
COLOR:BLUE
OUTER DIAMETER: .027IN (.70MM)

23 Gauge
COLOR:ORANGE
OUTER DIAMETER: .025IN (.63MM)

25 Gauge
COLOR:RED
OUTER DIAMETER: .020IN (.53MM)

27 Gauge
COLOR:WHITE
OUTER DIAMETER: .016IN (.42MM)

FIG. 2.1.1 Needles.

Blood sample quality can be affected by hemolysis, insufficient or inappropriate sample, wrong container, undue clotting, contamination with infusion fluids, cross contamination with blood tube additives, inappropriate sample storage and repeated freeze-thaw cycles. Hemolysis of samples is a risk with blood collection. It is not just the release of hemoglobin as red blood cells also discharge structural proteins, lipids, carbohydrates and enzymes. Hemolysis can be due to traumatic venepuncture collection with inappropriate devices (needles too small) or vigorous shaking of the samples. Needles should be 21 gauge for adults (0.82 mm outer diameter and 0.51 inner) and for children 23 gauge needles (0.64 outer and 0.34 inner diameter) should be used. Blood samples for serum should be taken preferably into glass tubes with no additive. The staff taking blood samples have to be trained in relation to written guidelines. The venipuncture site should be cleansed and allowed to dry. Cleaning prevents infection by skin microorganisms. The patient should be asked to just lightly clench his/her hand (never request the patient to "pump"). The clenching of the forearm before venipuncture can modify the concentration of several analytes (i.e., potassium). A tourniquet can be applied, and the venipuncture site and vein chosen. Start the venipuncture and once blood flow begins, request the patient to open his/her hand. To prevent venous stasis and hemolysis, also release and remove the tourniquet. Staff should check the tests against the most appropriate preservative for the sample and select the correct numbers and types of tubes that are needed to cover the different needs for sample

preservation and these are identified by colored caps (Fig. 2.1.2).

The tubes should be filled to the correct level to standardize the concentration of the preservative through the sample, using the correct order of draw. Samples for plasma require an anticoagulant which blocks the fibrinogen activated clotting cascade. Heparin is an antithrombin activator and inactivator of serine proteases. EDTA and citrate chelate calcium ions. EDTA and heparin are in solid form coated onto the blood collection tube. Sodium citrate and potassium/EDTA should not be used if mass spectrometric methods are used for analysis because some analytes react with salts to form sodium and potassium in different combinations of potassium and/or sodium, with chloride and/or formate anions, that may cause ion suppression or enhancement. Vacutainer tubes with hemoguard stoppers are designed with a safety shield for safer tube handling. The rubber stopper is recessed inside the plastic shield, so any drops left by a blood collection needle remain isolated from potential contact. The shield is designed to protect lab personnel from direct contact with blood on the stopper or around the outer rim of the tube, as well as from blood splattering upon opening the tube.

Where several tests are requested that involve the use of different sample tubes, then the order in which the samples should be taken from the patient depend on the tests and nature of coagulation in the following sequence:

- blood culture,
- citrate (clotting studies),
- serum, serum separating tubes (SST),
- heparin, plasma,
- EDTA, plasma and
- glycolysis inhibitors.

Samples with additives should be inverted (not shaken) a few times to ensure mixing of the chemicals throughout the sample. Whatever method is used to separate blood cells from serum or serum the concentrations of steroids in the samples should not be compromised and needs full validation (Snaterse et al., 2021).

2.1.2.5 Transport to laboratory

Pneumatic tube systems are often used, sometimes samples are taken by porters. These staff should be trained to treat samples accordingly.

2.1.2.6 Receipt in laboratory

The clotting times should be standardized (30–60 min) although this can create problems in some areas of the hospital. The samples need centrifugation usually for 10 min at $2000 \times g$ at room temperature.

Most clinical chemistry platform analyzers determine serum indices based on hemolysis, icterus (measure of

Blood Collection Tubes

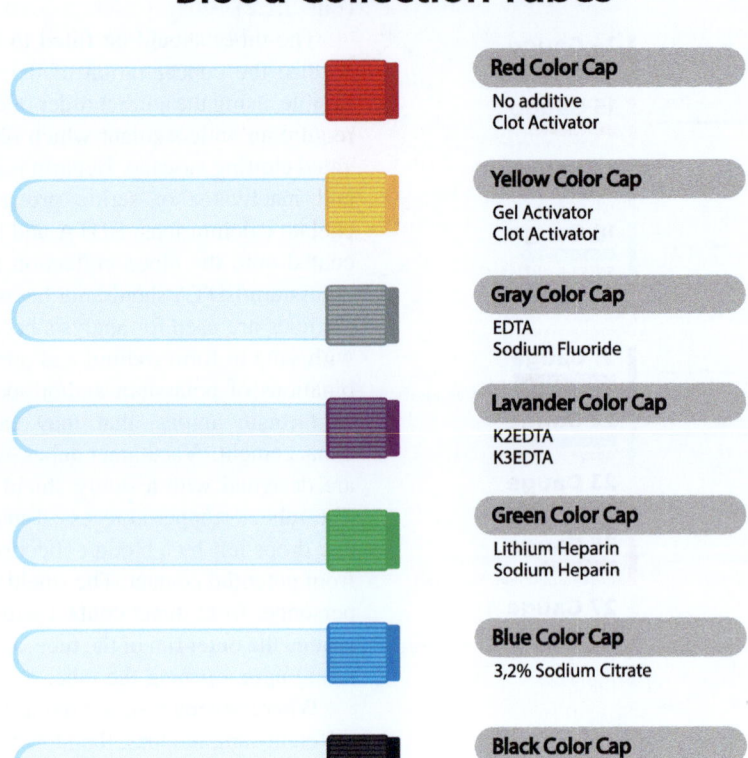

Red Color Cap

No additive
Clot Activator

Yellow Color Cap

Gel Activator
Clot Activator

Gray Color Cap

EDTA
Sodium Fluoride

Lavander Color Cap

K2EDTA
K3EDTA

Green Color Cap

Lithium Heparin
Sodium Heparin

Blue Color Cap

3,2% Sodium Citrate

Black Color Cap

3,8% Sodium Citrate

FIG. 2.1.2 Tubes for blood collection.

jaundice) and lipemia. During sample storage at 4°C, steroids may be released from binding proteins and albumin. Few hospital laboratories have −80°C freezers for long term storage of samples so −20°C freezers will be used. Stability of steroids in stored samples will have to be tested. Thawing and re-freezing is not desirable. It is important to identify and understand all possible factors that may introduce variation into the analysis. Sample collection and handling should be documented and standard operating procedures (SOP) produced and implemented.

Blood spots can be collected on pure cotton fiber paper designed for the purpose. Dried blood spots (DBS) have several advantages. The collection of blood from the finger prick site is less invasive than venepuncture. Filter paper cards were specifically designed for newborn screening of inherited disorders (Guthrie cards), it is important that sufficient blood is applied to fill the marked circles. The sample is easy to transport and store.

2.1.3 Biological samples

2.1.3.1 Blood samples

2.1.3.1.1 Collection of samples

The staff used for blood collection are not always laboratory scientists and are often ancillary staff or nurses. It is therefore important that the sample collection procedures

must be considered and protocols with standard operating procedures must be developed and adhered to by all staff involved with collection and storage of samples (Rizner and Adamski, 2019). If the treatment of samples is substandard, substantial variation or bias may result in analytical results that are then not accurate. Procedures based on validated methodologies should whenever possible be used. Staff training to follow protocols is needed especially when there is no previous scientific education. The samples in clinical laboratories are usually for targeted analysis which includes a particular analyte or set of analytes belonging to a specific class involved in a metabolic pathway. With the introduction of technology for untargeted analysis in metabolomics the consistent collection of samples will become more important (Kirwan et al., 2018; Gika et al., 2014). Here as many metabolites as possible are analyzed without prior selection bias and usually focuses on comparing the metabolome between a patient group and controls. Metabolomics allows the diagnosis of a condition at a less advanced stage, the monitoring of drug toxicity and response to treatment (Hernandes et al., 2017). More than one pathway can be assessed with metabolomics so it is worth considering related paths and displayed for example in a Kyoto Encyclopedia of genes and genomes (KEGG) pathway (https://www.genome.jp/kegg/pathway.html) or the review published by Ronda Greaves for instance (Greaves et al., 2014).

Blood samples for serum and plasma are the commonest samples for clinical investigations, less often tests are performed on saliva, whole blood and dried blood spots. There are a number of guidelines (see References section for this chapter—World Health Organization, 2010; CLSI (formerly NCCLS), 2003, 2007; Simundic et al., 2018). A crucial decision is whether to aim for plasma or serum. Selecting the correct sample matrix is of paramount importance which need to be evaluated for their possible influence on metabolomic studies. Blood samples are not usually taken by the analysts so detailed procedures should be in place. All efforts from the analysts are in vain if sample quality is poor. Plasma is obtained by the addition of the blood sample to a tube containing an anticoagulant. EDTA not only inhibits the coagulation process but also inhibits magnesium dependent enzymes. The sample size taken should be appropriate for the tube so that any preservative (heparin, EDTA) is at the correct concentration. The subject needs to be identified and tubes accurately labeled with the patient demographics from name, date of birth, gender, numbers in the hospital or national system. Blood samples should in general only be taken from intravenous catheters when essential.

Blood is usually taken from an **antecubital vein** after application of a tourniquet. The site is cleaned with a swab of 70% isopropyl alcohol which is left to dry before the sampling. Too much pressure on the swab when cleaning and locating the vein should be avoided. Clenching of the fist and pumping is not necessary. The tourniquet should be released as soon as possible, ideally as soon as the first sample is aspirated. This and other steps should be documented in the safety procedures. Strong aspiration of the blood from a syringe, through a needle, into the tube can cause hemolysis. The date and time that a blood sample is taken should be recorded since there are many reasons for changes in the concentrations of steroids in the samples. These can be age, sex, time of day or night, day in the menstrual cycle.

Samples may be taken into several different tubes and the order of draw is important.

1. Blood culture
2. Blue tube for coagulation (sodium citrate)
3. Red no gel
4. Gold SST (plain tube w/gel and clot activator additive)
5. Green and dark green (heparin, with and without gel)
6. Lavender (EDTA)
7. Pink—blood bank (EDTA)
8. Gray (oxalate/fluoride)

Sample tubes for plasma can be put into iced water to avoid exposure of the samples to room temperature. Lithium from lithium-heparin tubes and sodium salts (fluoride, citrate) have variable interferences and matrix effects can affect ionization in LC-MS/MS methods. Sodium fluoride can inhibit esterases in the circulation and reduce release of testosterone from orally administered testosterone undecanoate. Serum needs to clot at room temperature for 30 min during which time activated platelets can release lipids and proteases into the serum. Platelet numbers in the sample vary from 20,000 to 1,000,000 per microliter. Plasma samples are cleaner than serum which is important when samples get analyzed by mass spectrometry. The tube composition (glass or plastic polymers), additives, protease inhibitors can influence the results and the impact of the tubes on all laboratory tests must be tested beforehand to avoid noise in the analysis. Gentle mixing of the contents of the tubes is required. Breakdown of red blood cells, erythrocytes and leucocytes can release ions, proteins and intracellular enzymes into the plasma or serum. Prolonged venous stasis should be avoided. Samples are usually sent to the laboratory in a pneumatic tube system.

Serum or plasma for steroid measurements should be separated from the blood cells as soon as possible. Hormone concentrations will change with prolonged contact with blood cells or the gels used in barrier tubes. ACTH as measured by RIA may be unstable at room temperature and plasma should be frozen as soon as possible. ACTH is destroyed by repeated freezing and thawing. Protein is absorbed onto glass so all tubes for ACTH should be plastic and kept cold. A stressful venepuncture can cause within 30 min a marked increase in ACTH output. If repeated blood samples are requested then in this situation, to avoid stress and trauma from many needle stabbings, the samples can be collected via a reliable intravenous catheter inserted 90 min before the start of sampling. The tubing can be maintained patent with a heparin-saline solution but this must be taken away into a syringe before any blood is taken, hence as already stated not desirable.

The sample for steroid analysis should be centrifuged at $2000 \times g$ for 10 min at room temperature except when protocol specifies at 4°C and plasma or serum separated within 2 h of arrival in the laboratory. The plasma should be removed without disturbing the buffy coat layer. The clinician will need details of diet, appropriate fasting, recent physical exercise, BMI, dietary supplements, smoking, alcohol intake, liver and kidney functions and drugs but these may not usually be known unless crucial to an investigative study. When necessary 12 h fasting is recommended. Hemolysis should be avoided and this can happen at several points in the sample processing from the phlebotomy use of a tourniquet, rapid syringing of the blood, shaking tubes, centrifugation at high speed. The total duration of time from collection through centrifugation to storage should be minimal. Both plasma and serum are separated from whole blood via centrifugation. Blood sample tubes that contain a separating gel (often called SST tubes) are often used to obtain blood serum for laboratory tests. After centrifugation, the inert acrylic gel at the bottom of the tube normally occupies the middle position between the cells (clot) and the serum, as its density is intermediate between theirs.

The gel then serves as a barrier to diffusion, preventing contamination of the serum with cellular components.

It is important that the collection of blood for the measurement of aldosterone and renin is made under strictly controlled conditions. The patient should have a normal dietary intake of sodium and have been given potassium supplements if hypokalemic. Blood is ideally taken after the patient has been lying down overnight or after every effort to keep the child lying down quietly for 2h before the test. The blood must be collected into heparinized tubes, centrifuged and plasma immediately frozen and stored frozen until the assay. The electrolyte and acid base status should be confirmed. The use of antihypertensive drugs should be recorded because many influence the renin-angiotensin-aldosterone axis.

For some analytes, such as ACTH, all tubes should be cold. Hemolysis, icterus and lipemia can affect the analysis and needs to be negated in the assay validation. Many automated analytical platforms detect and flag up these samples which are called serum indices. Hemolysis can usually be seen by the red coloration of the yellow plasma. Lipemic samples have a white lipid layer on the surface of the plasma. Samples should be kept cold until analysis then frozen. Most clinical laboratories only have $-20°C$ freezers and steroids will remain stable for months at this level. The concentration of steroids are bound to albumin (and other proteins) can change in this time and this needs to be considered in frozen samples. Repeated freeze-thaw cycles are unavoidable and tests need to be performed in the method validation to ensure stability of samples in freeze thaw cycles. During the freezing process of serum, the water will freeze first and become less dense resulting in accumulation of proteins and salts at the bottom of the tube. Precipitation and denaturation of proteins is likely. A cryoprecipitate may form primarily of lipids. The sample should therefore be mixed before and after centrifugation. Concentrations of analytes can vary and must be tested in the method validation.

In some cases, finger prick blood samples can be collected and spotted onto dedicated filter papers such has been used for newborn screening for many years. The sample is taken from the fleshy part of heel of the child after warming the area and cleaning with isopropyl alcohol. An index finger is warmed to increase the blood flow. The skin should be wiped clean with an alcohol swab and the skin is pierced with a lancet to produce a small volume of blood. The first exudate from the puncture should be wiped from the finger with sterile cotton. For valid **dried blood spot (DBS)**, drops of blood must be collected to fill completely the marked area on the card (Ostler et al., 2014; Choi et al., 2019). The blood on the card should be air dried for 3h at room temperature and can then be stored at 4°C for less than 2 years storage and -20 to -80 for long term storage. Spots are usually punched to 3.2 or 6mm discs and steroids eluted with buffer or a solvent. Steroids on the DBS card are remarkably stable for years (Török et al., 2002). DBS are analyzed for many analytes of which 17-hydroxyprogesterone was the first steroid to be tested as a biomarker of the commonest steroid genetic defects causing congenital adrenal hyperplasia. Finger prick samples are used more generally for DBS samples when monitoring steroid treatment. There have been many attempts to overcome issues with hematocrit and homogeneity of the sample including whole blood spot analysis, precut spots, correction of results according to potassium concentration.

Some novel collection devices have been proposed. A volumetric absorption microsampling (VAMS) device has a volumetric porous pad which absorbs by wicking a fixed volume of sample (Heussner et al., 2017; Protti et al., 2019) (Fig. 2.1.3). Up to 10 samplers can be in a plastic holder (Mitra micro sampling device, Neoteryx LLC, Torrance, CA 90501).

A fixed volume sampler in the channel of a card (Hemaxis DB10, Route des Avouillons 6, CH1196, Gland, Switzerland) is used for blood collection that transfers the sample to filter paper. To get repeated blood samples at short time intervals, an automated blood sampling system for rats (www.Culex.net) has been modified for 10min interval blood sampling over 24 use in humans (Henley et al., 2009) (Fig. 2.1.4).

Microdialysis is a bioanalytical sampling technique used to monitor the chemistry of the extracellular space in vivo of tissues. A permeable dialysis membrane is placed in an extremely small scale device so that unbound analytes can be continuously sampled. Microdialysis provides a preview of what goes on in tissues before changes in systemic blood levels. This was first used in endocrinology for glucose measurements but has been tested for free cortisol (Bhake et al., 2020). MicroEye is a microdialysis system that transforms point-of-care blood sampling from "point in time" to automated continuous monitoring of a vast range of potential analytes, for improved outcomes in critical care. It is easy-to-use device inserted using common peripheral venous catheters. It delivers dialysate to an external sensor system; so blood stays in the patient. The device enables real-time data to standalone monitors, bedhead array systems or hospital information systems (Bhake et al., 2020). Monitoring vital blood analytes and drug concentrations in premature neonates may require up to 25 heel sticks a day over several days, or even weeks; so a modification, NanoEye, is small enough to be inserted via a 24G cannula, to form the basis for a rapid point-of-care system for critically ill neonates and children.

Adrenal, testicular and ovarian veins are occasionally used for blood samples to identify a local source. Blood taken from veins draining tumors can also be useful material for tests of function. The samples are taken while monitoring the course of catheters under X-rays. Steroids and in some cases the regulatory proteins are analyzed. In the

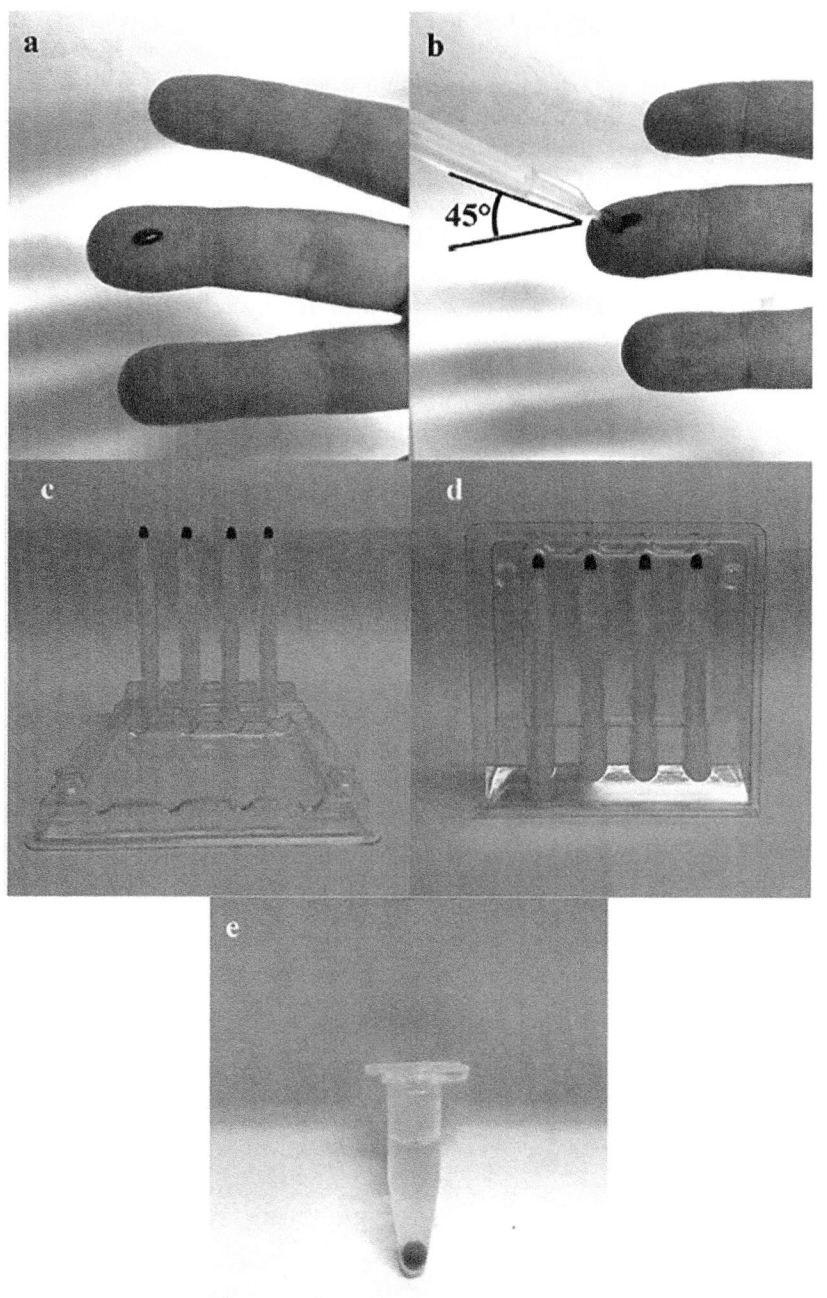

FIG. 2.1.3 Capillary blood VAMS sampling: (A) a blood drop is obtained by finger pricking and then (B) sampled by touching the blood surface with the VAMS device; multiple VAMS blood samples (C) right after sampling and (D) enclosed in the dedicated clamshell; sampled VAMS tip detached from the handle, in a microtube containing extraction solvent (E) *(From Protti M, et al. Tutorial: volumetric absorptive microsampling (VAMS). Anal Chim Acta 2019;1046:32–47. Fig. 3 p. 37.)*

distant past, samples from adrenal veins have been studied in sheep after auto-transplantation of the adrenal gland to the neck for easy access (Arcus et al., 1979) (Fig. 2.1.5).

2.1.3.2 Urine

Urine is a useful biological fluid because sample collection is painless. Any sample container should be tested to be free of any interferents in the analysis and if soda-lime glassware is used that may absorb steroids the glass should be silanized

before use (McManus and Sharifi, 2019). Random or timed samples may be collected. A benefit of analyzing a 24h collection is that the results provide information about daily production of steroids (Côté et al., 2008). A 24h urine collection from an adult can be one to two liters so a large container is needed. For children, there are large variations in the 24h volumes according to age and gender (Table 2.1.1) with a rough guide being 100 mL of urine per year of life.

Larger samples of urine than blood can be analyzed, so the yield of steroids is greater in quantity. There is a lack of

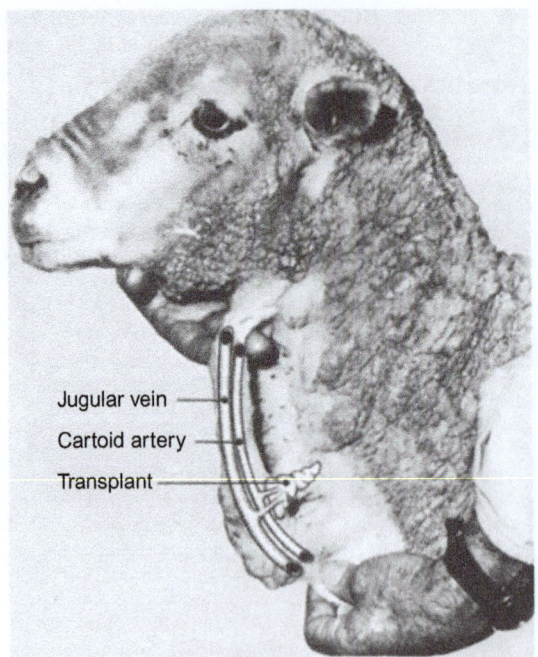

FIG. 2.1.4 General view of endocrine gland autotransplanted to neck of sheep. *(From Arcus AC, Beaven DW, Hart DS, Holland GW. Long-term hormonal secretion from the autotransplanted sheep pancreas. Diabetologia 1979;16(5):325–30. Fig. 1 p. 325.)*

information on stability of steroids in urine so the general rule is to keep the sample cool in the short term or frozen for long term preservation. Samples should ideally be received in the laboratory within hours but are unlikely to be analyzed immediately so refrigeration is recommended (Laparre et al., 2017). Filtration through cellulose filters

with 0.2–0.45 μm pore size is another means of removing bacteria, cells from the urogenital tract and debris from urine samples. Sodium azide is a potential preservative. Metabolomic studies of stored samples need to demonstrate stability of steroids in the samples (Fernandez-Peralbo and de Castro, 2012; Remer et al., 2014; Ryan et al., 2011).

The salt content of urine is high and protein concentrations are normally low, in contrast to plasma. Urine contains phosphate and sulfate ions and saccharo-1:4-lactone that can inhibit enzymes used for hydrolysis of conjugates. The timing of a sample is important because excretion rates of steroids vary during the day and night. The highest concentrations of cortisol metabolites in urine are between 1000 and 1800h, presumably because hepatic metabolizing genes are then best expressed within the 24h period (Jerjes et al., 2006). This is later than the peak concentrations seen in blood in the early morning. Increased outputs of progestagens and androgens have been reported in night-shift compared with day-shift workers. A diet with high soya content will lead to contamination of the steroid fraction with lignan products. Ingestion of citrus juice can upregulate the catabolism and urine clearance of steroids and bias interpretation of results. The use of prescription or over-the-counter medication should be declared and ideally withdrawn for a few days before the sample collection.

A spot urine sample after awakening, or the second sample of urine in the morning have been used for many investigations with recommendations to take the sample mid-stream to avoid particulates collected in the bladder overnight. A random urine from a newborn infant is acceptable when testing for excess metabolites in a child with steroid metabolic defect. If a 24h urine collection

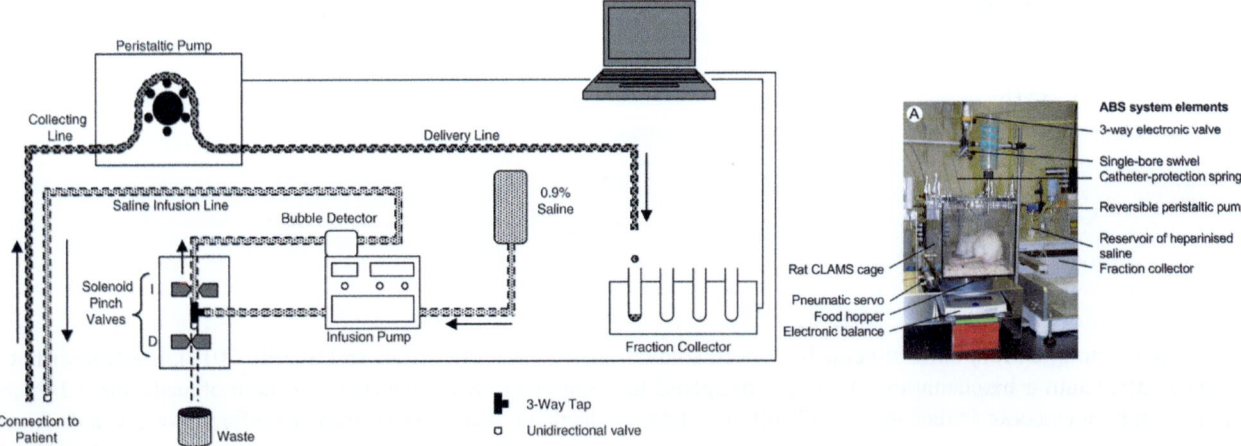

FIG. 2.1.5 Autosampling of blood at intervals over 24 h. (Left) *Light* and *dark* lines represent 0.9% saline and blood, respectively. *Arrows* indicate direction of flow. The divert valve (V) is depicted in the closed state with the infusion valve (I) open such that 0.9% saline is directed toward the participant at this instant. (Right—A) Automated blood sampling system. CLAMS is for comprehensive laboratory monitoring which includes food monitoring. *Left panel: From Henley DE, Leendertz JA, Russell GM, Wood SA, Taheri S, Woltersdorf WW, Lightman SL. Development of an automated blood sampling system for use in humans. J Med Eng Technol 2009;33(3):199–208. Fig 1 page 200. Right panel: Modified from Tilston TW, Brown RD, Wateridge MJ, Arms-Williams B, Walker JJ, Sun Y, Wells T. A novel automated system yields reproducible temporal feeding patterns in laboratory rodents. J Nutr 2019;149(9):1674–84. Suppl Fig 1.*

TABLE 2.1.1 24-h Urine collections children and adolescents (Wudy et al., 2007).

Age years	Boys Mean excretion	Boys 2 × s.d range	Girls Mean excretion	Girls 2 × s.d range
3–4	543	180–910	504	80–930
5–6	485	75–895	596	170–1024
7–8	710	296–1128	664	136–1192
9–10	767	850–1280	706	294–1118
11–12	922	100–1748	772	152–1392
13–14	871	385–1367	1043	248–1837
15–16	1108	264–1952	1070	50–1260
17–18	1204	50–2396	1237	171–2270

has to be in more than one container, the urines of the separate containers should be well mixed together. A quick test of the urine with a dip stick can be used to confirm suitable pH, glucose concentration, protein, specific gravity and nitrite as an indicator of contamination with bacteria or blood. The underlying explanation of blood in the sample may be a bleeding disorder or vaginal contamination during menstruation. The correct procedure for collecting samples is going to have more importance as we move into the metabolomics era and many more metabolites will be analyzed in one sample.

In sports antidoping tests there are special requirements around the chain of custody needed, the actual voiding of the sample will be observed to detect any physical manipulation of the sample that would negate the sample being from the athlete. The sample will be divided into tamper proof sealed bottles. The urine of females can be contaminated with intestinal bacteria from the perineum if hygiene is poor. Degradation of steroids in urine has been documented but so far there is no requirement from the World Anti-Doping Authority to preserve the quality of the sample. Hydrolysis of conjugates and an increase in free testosterone are the result of bacterial action on steroids in the urine. A high concentration of androstanedione in the sample is taken by antidoping laboratories as evidence for bacterial degradation based on a small study. In forensics as well as doping control, the samples are not subjected to preservation techniques so the impact of bacterial degradation is rarely accounted for. In some SOPs for urine collection washing of the introitus of women or the glans penis of men is included to minimize bacterial contamination of the sample. Soaps should be avoided because many contain surfactants that can affect analysis. A preservative is being tested in antidoping tests. This includes sodium azide as a general antibacterial salt, penicillin, streptomycin, amphotericin to prevent growth of a range of bacteria and a mixture of enzyme inhibitors that could affect the concentrations of proteins such as growth hormone and erythropoietin that can be used to enhance performance in sport.

For children and adults, the urine can be collected directly into a container but neonates pose additional challenges. In-dwelling catheters provide for reliable collection of urine, adhesive bags are available and diapers can be processed to yield representative urine samples (Heckmann et al., 2005). Urine on a diaper can be collected from the material by centrifugation in a fruit juice extractor, for example.

The collection of urine for 24 h is difficult to achieve without forgetting to collect all voidings. The determination of creatinine concentration or specific gravity has been used to assess if the sample collection is complete and includes all voidings (Singh et al., 2015). This is reliable between samples from an individual but very poor between individuals as reference ranges are large. Urinary creatinine concentrations increases throughout childhood (Remer et al., 2002) (Fig. 2.1.6) with Tanner stages of sexual development, age and male gender but urine specific gravity is not affected by these factors.

The patient should be given a description of the 24 h urine test emphasizing it is about 24 h urine production and collecting the appropriate sample. The test should start and finish with an empty bladder. A clear instruction is needed. A very practical procedure with minimal disturbance would be:

- Keep the large urine container provided next to a toilet.
- Empty the bladder to the toilet before going to bed and note the time.
- From this time, collect all urine into the large container (including any voids if up in the night).
- At the end of the 24 h, the final void should be collected in the container, thus emptying the bladder.

2.1.3.3 Saliva

Saliva samples are easy to obtain and collection can be repeated many times during the day which is less practical with blood samples unless through a catheter (Bhattarai et al., 2018; Gröschl, 2017). Subjects or research assistants can with minimal training easily collect saliva samples. Saliva is not considered a biohazard unless contaminated with blood. The subject under investigation needs to avoid stress if samples are collected for cortisol analysis. The collection of saliva samples is beneficial in pediatrics. Steroids in saliva reflect the free hormones in blood that are filtered in the salivary glands, concentrations are thus low, for example, cortisol can be around 0.2 nmol/L, but estradiol <5 pg/mL. Only 1% of the DHEAS in blood reaches the saliva. Saliva samples will contain epithelial cells and food

FIG. 2.1.6 Urinary creatinine with age. Twenty four hour urinary creatinine excretion in children aged 3–7 years. (A) Absolute excretion rates; (B) body surface area related excretion rates. *(From Remer T, Neubert A, Maser-Gluth C. Anthropometry-based reference values for 24-h urinary creatinine excretion during growth and their use in endocrine and nutritional research. Am J Clin Nutr 2002;75 (3):561–9. Fig. 1 p. 563.)*

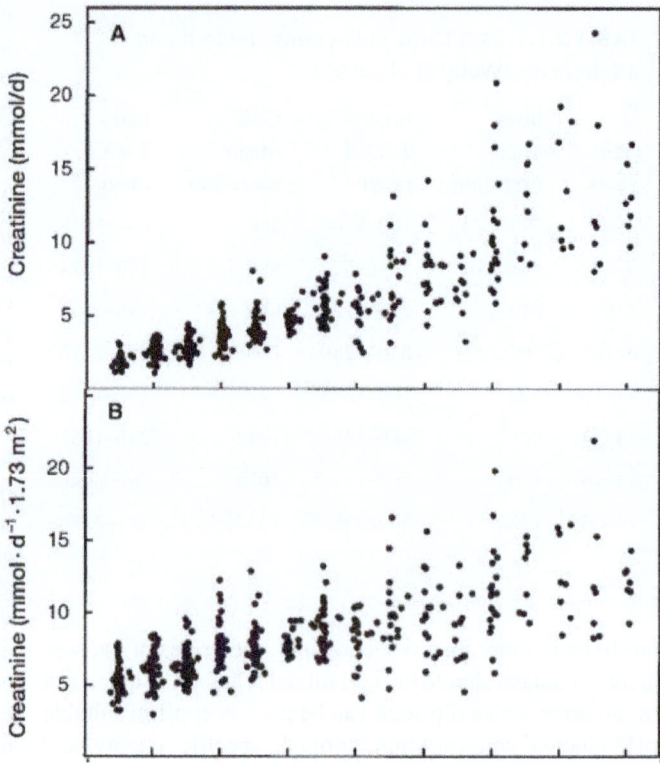

particles. The mouth should be rinsed with water before sample collection. Contamination of the saliva with blood from gum trauma should be avoided, transferrin concentrations in saliva are used as a marker to check the samples. Saliva can be obtained as a passive drool (Fig. 2.1.7) or active spitting into a tube.

Chewing paraffin wax or chewing gum increases the flow of sample. Care must be taken not to locate the swab near to one particular salivary gland. Results for cortisol

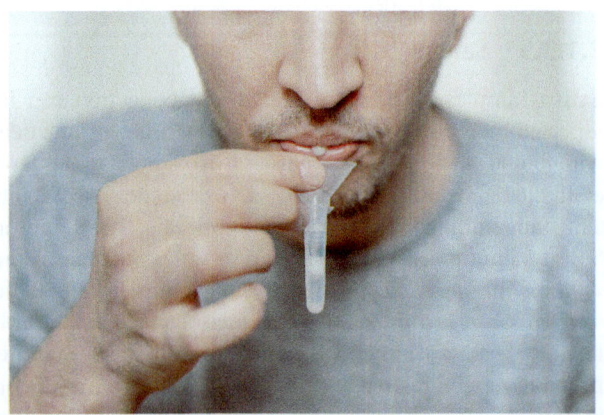

FIG. 2.1.7 Saliva collection by passive drool (A) Place ribbed end of saliva collection aid into a prelabeled collection tube, (B) Pool saliva in the mouth. With head tilted forward gently guide saliva through the collection aid into the tube. Fill to the required level.

can vary between drool and swabs depending on whether the sample reflects the saliva in the mouth from several glands (Fig. 2.1.8).

Parafilm should not be used since it absorbs lipophilic substances and releases contaminants in some assays. Cotton rolls (Salivette, Sarstedt, Intercept, Quantisal) absorb saliva which is released on centrifugation. Saliva can be stored frozen at −20°C long term without affecting significantly the steroid concentrations.

A midnight saliva sample for cortisol is a good marker for the loss of the diurnal rhythm that characterizes Cushing's syndrome. The ovarian cycle can be monitored to differentiate between the luteal and follicular phases. Cortisol is readily oxidized by HSD11B2 in the salivary glands and measurements of cortisone have gained favor (Ponzetto et al., 2020; Titman et al., 2020) and in some laboratories a profile of 10 or more steroids is possible (Gomez-Gomez and Pozo, 2020).

2.1.3.4 Other biological samples

The analysis of tissues usually requires the tissue to be broken up into the smallest particles before extraction of the analyte. The conditions of homogenization (homogenizer, solvent, time, temperature, dilution fold) need to be standardized. If an internal standard is added to the tissue, does this equilibrate to the same extent as the

The Salivary Glands

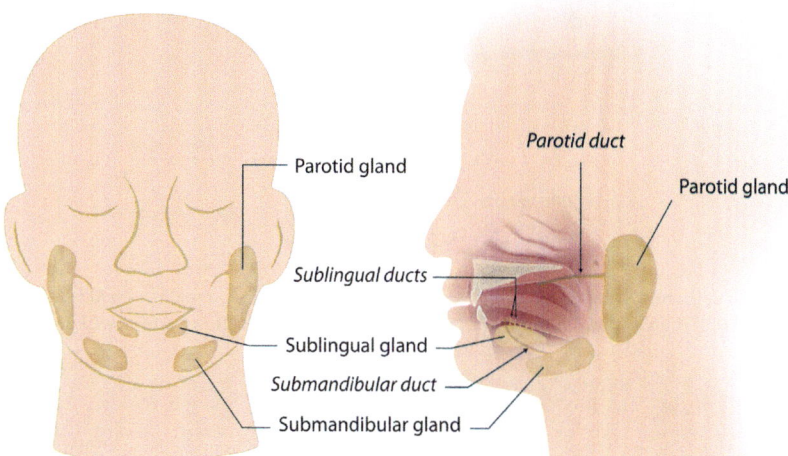

FIG. 2.1.8 Location of salivary glands.

analyte? If a quantitative method of analysis is required, it will be important at the outset to consider the nature of the sample matrix for the calibration so that issues of interference, extractability, recovery, specificity, sensitivity, precision and stability can be validated (see Chapter 2.4 for validation criteria). Does such a suitable matrix exist or is a surrogate to be used? Are samples from children be adequate for investigations in the adult. Is tissue from another organ acceptable? Is the same tissue from another animal species acceptable? Plasma may have to be the nearest matrix for some tissues. Most studies have been validated against a surrogate matrix and the analyst will need to consider how well the criteria for LLOQ, LLOD and absence of matrix effects have been addressed. To some extent, the impact of matrix effects is more likely in a nonspecific analysis (ligand binding assay) than in an LC-MS/MS analysis although interferences from matrix effects such as phospholipids will depend on the ionization mode.

2.1.3.4.1   Hair

Incorporation of steroid in hair occurs via diffusion from capillaries in the follicle into the medulla of the hair shaft. Steroids may also be deposited on the hair shaft from sweat and sebaceous gland secretions. This can be removed by washing the hair before mincing or grinding. Hair grows about 1 cm per month. The concentrations of steroids along the distance from the proximal root hair fibers shows specific changes in relation to time in the past of exposure to steroid. Many studies have measured cortisol in humans (Greff et al., 2019; Stalder and Kirshbaum, 2012) and animals (Yamanashi et al., 2016). Hair samples can be taken

without the stress associated with taking blood or the inconvenience of collecting urine. Sampling the hair at the posterior vortex of the head is the most common site (Wester and van Rossum, 2015) (Fig. 2.1.9).

2.1.3.4.2   Feces

The analysis of feces is gaining interest as a biosample since feces can be easily collected, contain not only the endogenous human metabolites from bile but also intestinal bacteria metabolites thereof and are thus a source to interrelate the microbiome and human metabolome (Palme, 2005). In recent years, the gut microbiome has been linked with many disease processes so sample preservation is a key factor when collected from animals in vivo. The 16S RNA of bacteria is analyzed in order to identify the organisms and this has enabled the identification of hundreds of species including many that could not previously have been detected because of high sensitivity to oxygen and iron concentrations. The human gut contains more bacterial cells than the number of somatic and germ cells in the body. Samples have been lyophilized before extraction with ice cold methanol:water (8:2) in a bath ultrasonicator. Homogenization with dry ice has also been used before extraction. Fecal incontinence is found often in women over 50 years of age. Stool steroid metabolites are under investigation for possible effects on the intestine in irritable bowel and cancer. The steroids in feces end up in sewage sludge and ultimately in rivers that can be used for our water supply.

The collection of feces is a noninvasive method for a biological sample (Loftfield et al., 2016; Deda et al., 2015) and has been particularly useful with animal studies

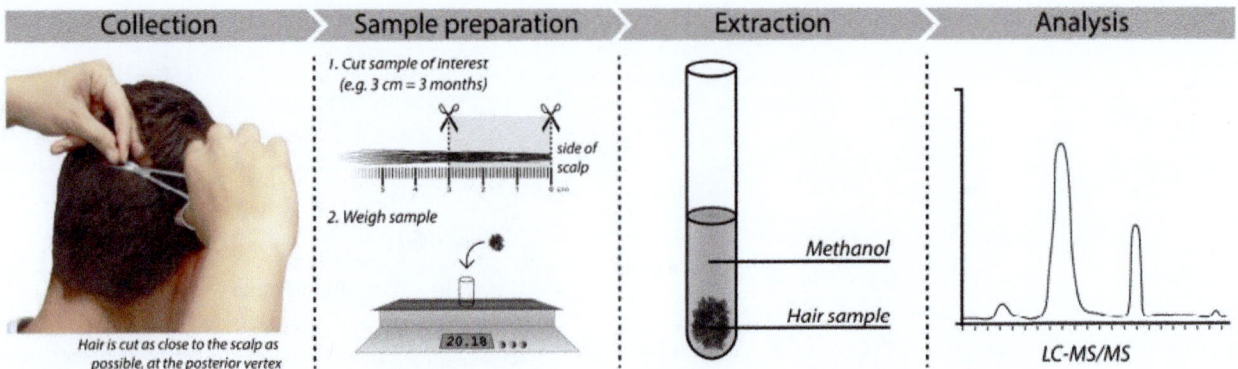

FIG. 2.1.9 Scalp hair analysis. Hair collection, sample processing, and LC/MS analysis. *(Based on van Manen MJG, Wester VL, van Rossum EFC, van den Toorn LM, Dorst KY, de Rijke YB, Wijsenbeek MS. Scalp hair cortisol and testosterone levels in patients with sarcoidosis. PLoS One 2019;14(6): e0215763. Fig 1 P4/15.)*

that would require collection of other samples under anesthesia. In animal studies, the diet can be controlled, this is less practical in humans and the gut microbiome is more diverse.

Monitoring of steroid concentrations in the feces of exotic animal species has provided valuable basic information on the endocrine characteristics of the ovarian cycle and pregnancy in the Indian rhinoceros (*Rhinoceros unicornis*) as well as the African black (*Diceros bicornis*) *and white rhinoceros (Ceratotherium simum)* (Fig. 2.1.10).

There are considerable species differences in steroid metabolism and urinary hormone excretion within the *Rhinocerotidae*. For example, estrone sulfate and estrone glucuronide are the major estrogens in the urine of the Indian and black rhinoceros whereas the white rhinoceros predominantly excretes estradiol glucuronide. Furthermore, whereas pregnanediol glucuronide has been identified as the most abundant urinary progesterone metabolite in the Indian rhinoceros, the white and black rhinoceros mainly excretes 20α-dihydroprogesterone into urine. Fecal progesterone metabolites have been measured to monitor luteal function and pregnancy in both African rhinoceros species.

In the period 1968–74, Hakan Eriksson in the Sjovall laboratory studied steroids in rat feces and cecal contents. He had the fortune of access to germ free rats and was able to delineate the contribution of micro-organisms to the fate of progesterone, pregnenolone and corticosterone (Eriksson, 1971a,b). The fecal steroids were very different to the steroids that entered the bowel through bile, notably with the major proportion having lost sulfate and glucuronide conjugation through the activities of micro-organisms. Eriksson described bacterial 21-dehydroxylation and 15-hydroxylation (Eriksson et al., 1971; Eriksson and Gustafsson, 1971a,b). He confirmed an enterohepatic circulation by feeding rats with bile from rats injected with radiolabeled steroids and collecting bile from the second set of rats.

Herman Adlercreutz studied the steroid contents of human feces and bile in pregnancy and used antibiotics to break the enterohepatic circulation of the estrogens (Adlercreutz and Martin, 1980). Such studies warrant re-examination with improved extractions and analytical techniques. There will be experiments to modify the intestinal flora by taking probiotics, fecal transplant experiments and the effects of bariatric surgery.

2.1.3.4.3   Meconium

The first stool of the newborn infant has been studied to quantify prenatal steroid concentrations in relation to the sex development of the offspring (St Pyrek, 1987; Shackleton et al., 1970). The data were gained more than 50 years ago and should be re-examined with better techniques. GC and GC-MS were used in the early days of those technologies. One more recent study has examined differences in meconium of males and females (Frey et al., 2017) using isotope dilution liquid chromatography with high resolution MS. The standard deviations of steroid concentrations for androgens are very much greater than the means so nonparametric statistics are needed.

2.1.3.5   Other fluids

2.1.3.5.1   Bile

This fluid is usually obtained through a biliary fistula during surgery. Conjugates of progesterone, pregnenolone and corticosterone are found in the bile. Herman Adlercreutz studied estrogens in bile over many years from around 1960 to 1980 (Adlercreutz and Martin, 1980). The content of sulfated steroids is recently proving to be an interesting area of investigation. The results of applying new methodology to a re-examination of steroids in bile is awaited.

FIG. 2.1.10 White and black rhinoceros.

2.1.3.5.2 Ovarian follicular fluid

During ovarian stimulation for IVF, the fluid contents of fol-licles are collected with a needle through the abdomen in a laparoscopic or transvaginal procedure with ultrasonog-raphy (Walters et al., 2019). Steroids and particularly androgens have been determined in the fluid usually during the course of hormone treatment with ovarian hyperstimulation.

2.1.3.5.3 Cerebrospinal fluid (CSF)

A lumbar puncture is required to get a sample of CSF. The first sample is usually contaminated with blood and should be discarded unless insufficient sample is otherwise obtained. The sample when clear should be centrifuged at $2000 \times g$ for 10 min to remove cells. The time from col-lection to storage should be kept to a minimum. An increase in pregnenolone and isopregnanolone with a decrease in dihydroprogesterone and tetrahydroprogesterone has been seen in men with multiple sclerosis (Caruso et al., 2014).

2.1.3.6 Tears

Steroids are present in tears at extremely low concentrations. Ultra-performance liquid chromatography coupled with mass spectrometry (UPLC-MS) and tandem mass spec-trometry (MS/MS) methods enable concurrent detection and estimation of 14 sex steroid and metabolite compounds in human tears (Gibson et al., 2019). Progesterone, androsterone-glucuronide (ADT-G) and 3αDiol-G were suc-cessfully detected in the tear extract and their concentrations in the pooled tear sample were estimated (with 95% confi-dence intervals) to be 0.10 ± 0.03 pg/μL, 30.9 ± 18.3 pg/μL and 9.8 ± 4.3 pg/μL, respectively. Other steroids were below the limits of detection. The investigation of sex steroids in tears and ocular surface tissue will aid understanding of the influence of sex steroids on ocular surface tissues facilitating better targeted treatment for dry eye.

2.1.3.7 Tissues

Samples of ovaries, testis, prostate and liver can be obtained at autopsy or biopsy material can collected under anesthesia. Flash freezing is recommended. Fetal tissues are not easy to obtain but have been used to examine features of sexual development (Savchuk et al., 2017a,b, 2018; Ben Maamar et al., 2017).

The tissue is homogenized or manually crushed in a mortar and pestle with liquid nitrogen. Interpretation of results from analysis of intact tissues, cell suspensions, venous and peripheral blood do not always agree (Vinson et al., 1985) usually because of the mixed populations of cells (Falk et al., 2008). The analytical techniques have changed over 60 years in nature, specificity, sensitivity (Maeda et al., 2013). When mass spectrometry is used care must be taken with ion suppression and suitable matrix for spiking will be critical though not always to obtain (Ho and Gao, 2015). Mass spectrometry can now be applied directly to tissue sections (Cobice et al., 2013).

2.1.3.7.1 Liquid biopsy

The term liquid biopsy has been adopted for the study of tumor cells and tumor-derived genetic materials in body fluids of patients with cancer (Di Nunno et al., 2018; Di Meo et al., 2017). Tumor cells are shed into the circulation such that tumor DNA and micro RNA can be purified from body fluids. Tumor-specific mutations, gene re-arrangements and copy number variants can be analyzed. Liquid biopsies are a noninvasive alternative to tissue biopsies. Liquid biopsies are a promising approach for personalized medicine that will enable the prediction and monitoring of disease and offer evidence for more appropriate therapies (Kim et al., 2015). Samples from peripheral sources can include postop-erative seroma (fluid filled pocket), spinal fluid, lavage washings from various body sites, pap smears and even men-strual blood. Cancer cells are often larger than normal blood cells and physical techniques are used to enrich the sample. The low concentrations of nucleic acid products require next generation sequencing and polymerase chain reaction for amplification.

2.1.4 Cell lines

A number of in vitro systems have been used to study human adrenal function, including tissue slices, cell suspensions from dispersed tissue and primary and secondary monolayer cultures from normal adrenals. Each model has advantages and disadvantages with regard to the acute and chronic studies of adrenal differentiated function. Dispersed cells and primary cultures suffer, however, from the continued need to isolate cells from fresh tissue unless immortalized (Vinson et al., 1985). Cell culture systems do not reach the complexity of tissues or interactions of cell to cell and

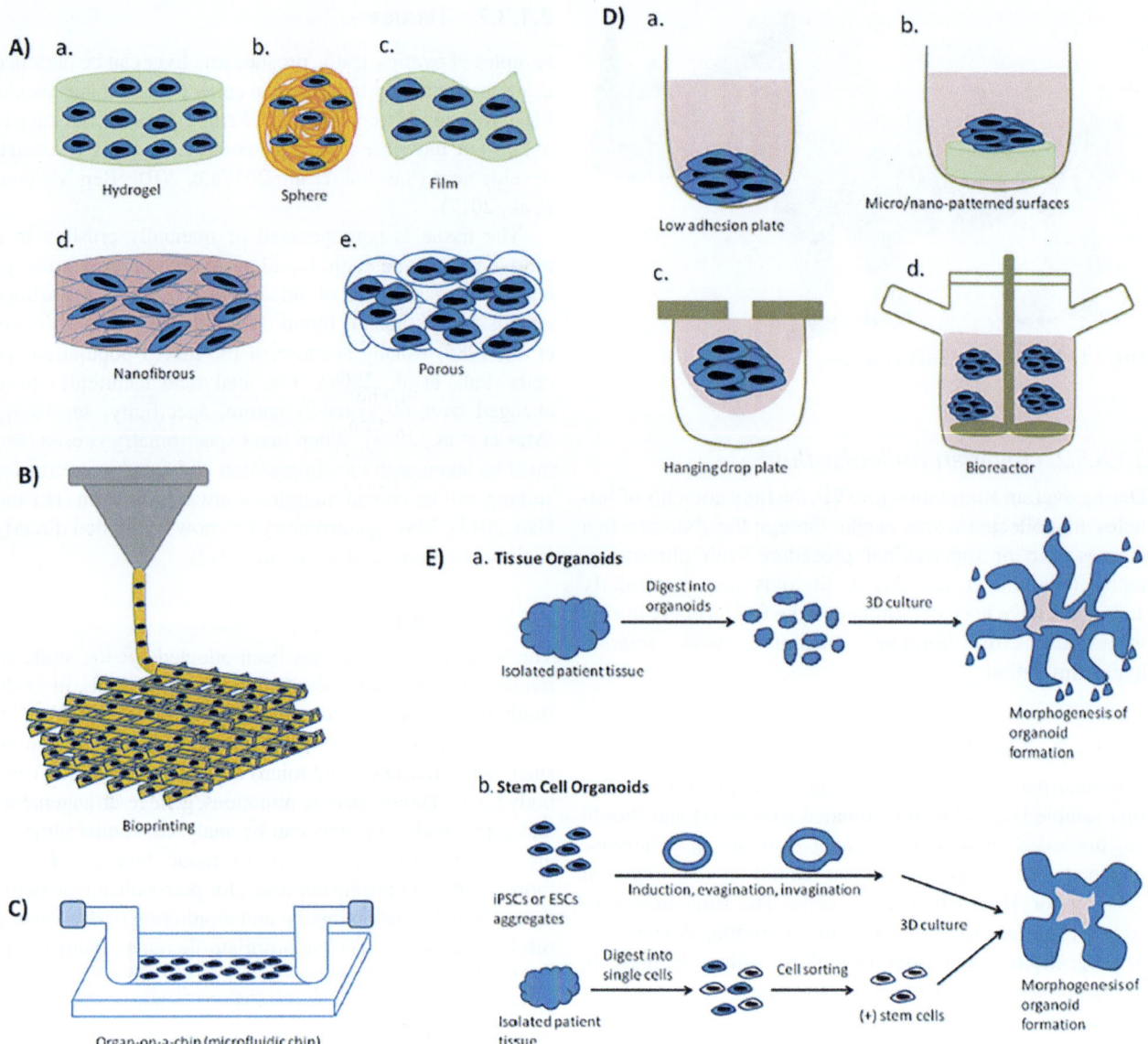

FIG. 2.1.11 3D cell culture systems. (A) Scaffold systems; a. hydrogel, b. sphere, c. film, d. nanofiber, e. porous forms, (B) Bioprinting system, (C) Organ-on-chips system, (D) Spheroid systems; a. low adhesion plate, b. micro/nano-patterned surfaces, c. hanging drop plate, d. bioreactor, (E) Organoid systems; a. tissue organoid, b. stem cell organoids. *(Based on Saglam-Metiner P, Gulce-Iz S, Biray-Avci C. Bioengineering-inspired three-dimensional culture systems: organoids to create tumor microenvironment. Gene 2019;686:203–12.)*

cell to extracellular matrix. three-dimensional culture systems are being developed in scaffolds, 3D printing, organs-on-chips, spheroids and organoids (Saglam-Metiner et al., 2019) (Fig. 2.1.11).

A cell line is a permanently established cell culture that will proliferate indefinitely given appropriate fresh medium and space. Lines differ from cell strains in that they become immortalized. Care must be taken when interpreting the results as cell lines do not always accurately replicate the primary cells.

Cell culture and cell lines have assumed an important role in studying physiological, patho-physiological and the differentiation processes of specific cells. They allow

the examination of stepwise alterations in the structure, biology and genetic makeup of the cell under controlled environments. This is especially valuable for complex tissues composed of various cell types, where examination of individual cells is difficult, if not impossible. In general, animal cells, particularly fibroblasts, can be more successfully cultured than human cells, and human fibroblasts are easier to culture than epithelial cells. Also, different epithelial cells show different responses to culture conditions. Despite advances in culturing techniques, human epithelial cells cannot be maintained in culture for long time periods. The problem is the tendency of human cells to undergo senescence after a certain cell division.

2.1.4.1 Adrenal cortex cell lines

Long term culture of nontumor adrenocortical cells is possible but cells cultured in this manner lose many of the differentiated functions desired for study. The advantages of an adrenocortical cell line include general availability to many laboratories together with the ability to undertake experiments requiring large numbers of cells (Wang and Rainey, 2012). Transfection with the *E6E7* gene of human papilloma virus or with the small and large T antigen of the simian virus (SV) 40 has partially overcome the senescence and has increased cell longevity in vitro but has not led to immortality of the cells. The genetic manipulations limit the use of these cells for molecular biological studies, especially for changes that occur during cell differentiation and transformation. The introduction of foreign genes alters the function of the host's regulatory genes. The quality of the culture medium and the cell preparation technique are very important for the maintenance of human epithelial cells in culture.

A method to prolong the lifespan of human cells is the infection of cells with telomerase, an enzyme that prevents telomere loss by de novo addition. It restores the length of telomeres, which otherwise shorten with each cell proliferation, leading to senescence. Attempts have been made to identify and culture stem cells of specific tissues because these cells can better adjust to the environmental conditions and can give rise to a variety of mature cells under specific environments. These cultures provide the ability to investigate differentiation pathways and provide a tool to test the effects of natural and synthetic substances such as cytokines, growth factors, nutrients and factors in the maturation or death of the cells.

The mechanisms of malignant transformation can be studied in vitro using cell lines treated with a carcinogen or radiation in culture. Genetic changes (e.g., DNA adduct levels, alkylations, mutations) and chromosomal changes can be investigated. Human epithelial cells in culture have not yet been transformed, so the need for animal models still exists. Rodents are much more susceptible to carcinogenicity than humans.

The Y-l mouse adrenal cell line is an adrenal cell line that has been widely used for studying acute and chronic regulation of steroidogenesis, as well as the mechanisms of transcription of steroidogenic enzymes (Rainey et al., 2004). Y-l cells, do not have the ability to produce mineralocorticoids, glucocorticoids or C_{19} steroids. Indeed, even if these cells expressed the normal complement of mouse adrenal steroidogenic enzymes they would still not produce cortisol or C_{19} steroids because P45Oc17 is not expressed in the mouse adrenal.

The NCI-H295 cell line was established in 1980 from an invasive primary adrenocortical carcinoma from a patient who showed symptoms of mineralocorticoid, glucocorticoid and androgen excess. The tumor (a $14 \times 13 \times 11$-cm right adrenal mass) was surgically removed. Because of fibroblast growth, a population of tumor cells which grew as a suspension was used to select and establish the H295 cell line. The initial description and analysis of the steroidogenic properties of the H295 cell line was performed after the cells had been in culture for 7–10 years. More than 30 steroids were detected in the culture medium from H295 cells, of which about 20 were identified. The H295 cells also produced low levels of testosterone and estrogens. The H295 cell line is available through the American Type Culture Collection (ATCC). The human H295 adrenocortical tumor cell line has contributed to the molecular and biochemical characterization of human adrenal cell function. The continued ability of the H295 cells to produce aldosterone provides a model for research on the mechanisms causing certain forms of endocrine hypertension. H295 cell expression of the full complement of human adrenocortical enzymes should allow the definition of the transcriptional and posttranscriptional mechanisms controlling the expression of these enzymes. The continued expression of the mRNA transcripts for the ACTH and angiotensin II AI1 receptors could also allow studies on the mechanisms controlling hormonal responsiveness. Furthermore, the ability of the H295 cell to be manipulated into a cell producing mineralocorticoids, glucocorticoids or C19-steroids provides a valuable model for elucidating the molecular mechanisms which give rise to each of the three distinct zones of the human adrenal cortex.

The most frequently used substrain, cell line H295R, was derived by selecting for cells attached to culture dishes lead to the establishment of the H295R cell line. The H295R cells were selected to grow in a serum substitute called Nu-Serum type I. The H295R cells are available through ATCC. The H295R cell line expresses the required enzymes and transport proteins needed for steroid hormone synthesis, as the parent cell line. The H295R cell line also retains the capability to produce and secrete sex hormones, whereas another substrain, H295A, does not. Compared with the H295A cell line, H295R produce slightly lower amounts of aldosterone. H295R cells respond to cAMP inducers, Ang II and weakly to ACTH. Treatment with Ang II or potassium makes the H295R cells adopt a more "glomerulosa-like" behavior with increased secretion of aldosterone. Forskolin treatment leads to a more "fasciculata-like" behavior of the cells, with increased synthesis of cortisol and adrenal androgens. Forskolin is a specific adenylyl cyclase activator and stimulates steroidogenesis through activation of cAMP and protein kinase A. Forskolin is used to mimic ACTH stimulation, since the H295R cells are only weakly responsive to ACTH. Forskolin-stimulated steroidogenic gene expression in H295R cells mimics mRNA expression levels of human adrenal cells in vivo. Effects on steroidogenic gene expression levels have been reported by real time PCR following treatment with pesticides, natural compounds, pharmaceuticals and different food additives. Hence, the H295R cell line is a promising tool for the assessment of effects of chemicals on steroidogenic gene expression, as well as effects on hormone secretion (Pinto et al., 2018). The H295R cell line has attracted much attention as a potential screening model for toxic effects on sex hormone secretion, since the cell line secretes testosterone and estrogens. The H295R cell line has undergone validation in a program for the Organization for Economic Cooperation and Development (OECD) as a potential tool for screening purposes of the effects on testosterone and estradiol secretion.

The COS-7 cell line was developed in the early 1980s. It is derived from the CV-1 African green monkey kidney fibroblast cell line transformed by a mutant strain of Simian Virus 40 (SV40) that codes for the wild-type T-antigen. This cell line has unique characteristics of fibroblast-like growth and virus susceptibility. These characteristics make COS-7 cells a popular research tool and an excellent choice for DNA plasmid transfection experiments (Nozaki et al., 2018). Many previous studies have reported that COS-7 cells are nonsteroidogenic cells. COS-7 cells have been used for transfection experiments to analyze the functions of steroidogenic genes, steroid receptors and the effects of steroids on functional molecules. The expression of steroidogenic enzymes in the kidney of humans and rodents has been reported. These results suggest that COS-7 cells may metabolize steroids. The mRNAs of P450scc, 3β-HSDs, P450$7\alpha$, P450c17, 17β-HSDs, 5α-reductases, P450c21, P450c11β and P450arom are expressed in COS-7 cells. In addition, the equivalent steroidogenic enzymes were active in cultured COS-7 cells.

Expression of P450c11β protein has been shown by western blotting and immunohistochemistry in normal human kidney. In addition, northern blot analysis has detected 17β-HSD type XI mRNA expression in the human kidney. In male and female rat kidneys, P450scc and 3β-HSDs mRNAs and 3β-HSD protein have been detected. The expression of P450scc in the rat kidney localizes to the cortical distal tubules and is high during the first days of life. Labeled [^{14}C]progesterone is metabolized to 5α-dihydroprogesterone, 11-deoxycorticosterone, 17α-hydroxyprogesterone, androstenedione and testosterone in the kidney tissue, from rats of both sexes. These studies suggest that the kidney has the ability for local steroid production. The steroidogenic enzyme activities of 3β-HSD, P450$7\alpha$, 5α-reductase, P450c21 and P450c11β were low compared with the activity of 17β-HSD in cultured COS-7 cells. Thus, the activities of these enzymes might not influence the production of the steroid hormones by the cells.

2.1.4.2 Ovary cell lines

Primary culture systems of human granulosa cells isolated from ovarian follicles have often been used. Human granulosa cells are obtainable mainly from in vitro fertilization programs when retrieving oocytes and follicular fluid. They are not obtainable in large numbers and do not survive in culture for extended cell generations. Such difficulties have often prevented the performance of the detailed analyses on molecular and cell biological levels. A few human granulosa cell lines have been established. These human granulosa cell lines include:

(1) lines established by the long-term culture of human granulosa cell tumor cell,
(2) human granulosa-lutein cells immortalized with SV40 large T antigen,
(3) a line established by a forced introduction of the human papillomavirus gene to the primary human granulosa cells and
(4) a line established by triplet transfection of primary human granulosa cells with the SV-40 DNA, Ha-*ras* oncogene and a temperature-sensitive (ts) mutant of the p53 tumor suppresser gene.

Although these human cell lines are useful and offer some promise of further discoveries in this field, none express a functional FSH receptor. A new line of human granulosa-like cells is available derived from a tumor specimen enucleated from a patient who showed a local recurrence of a granulosa cell tumor after menopause. This cell line, KGN, is unique and useful, because it maintains most physiological activities, including the expression of functional FSH receptor (Nagao et al., 2019; Tremblay and Sirard, 2017; Nishi et al., 2001).

2.1.4.3 Testis cell lines

Testicular cell lines have usually been derived from spontaneous or experimentally induced tumors (Rahman and Huhtaniemi, 2004). A MSC-1 immortalized Sertoli cell line was isolated from transgenic mice containing Sertoli cells transformed by the small and large T-antigens of the SV40 virus, which were targeted to Sertoli cells using the promoter for Mullerian inhibiting substance. (Note MSC here used for mouse Sertolic ell rather than mesenchymal stem cells more often used). MSC-1 cells were similar to primary Sertoli cells morphologically and expressed many of the same genes as primary Sertoli cells. Follicle-stimulating hormone receptor (FSHr) and Mullerian inhibiting substance were not detected in MSC-1 cells.

Sertoli cells are well known to interact with other cell types in the local environment and therefore these cells are particularly vulnerable to deficiencies of the isolated or enriched culture environment. MA-10, BLTK1 and

TM3cell lines were found to be unsuitable models to investigate interferences by substances of Leydig cell T synthesis (Engeli et al., 2018; Odermatt et al., 2016). Human Leydig cell lines expressing the entire set of steroidogenic enzymes should be developed for high throughput screening of potential endocrine disrupting chemicals.

The above activities were largely studied with immunoassay procedures. There is a need to replace antibody-based steroid quantification methods by mass-spectrometry-based methods(GC-MS/MS, LC-MS/MS) when analyzing samples from biological systems producing several steroid metabolites to improve specificity and sensitivity (Engeli et al., 2018).

2.1.4.4 Prostate cell lines

Many prostate cancer (PCa) tissues express the key steroidogenic enzymes and PCa cells in culture are capable of synthesizing testosterone and DHT from cholesterol (Luu-The et al., 2008). Some cell lines are androgen dependent (LNCaP) others are independent (PC-3, DU145) that enable survival in androgen-depleted media. Protein and mRNA expression for HSD17B3 has been confirmed in a prostate cancer (LNCaP) xenograft mouse model and mRNA in human PCa cells. RNA transcripts for CYP11A1 were found in human PCa tissues, but the protein was not analyzed. Recent reports show that mRNA for these enzymes and other steroidogenic enzymes are expressed weakly and variably in PCa cell lines and mouse xenograft models, but no immunohistochemical (IHC) analysis has been performed for these enzymes, as a group, in localized PCa.

PCa, regardless of disease stage, have an inherent ability to adapt and survive independently of androgens. However, regardless of androgen source, most castrate resistant prostate cancers (CRPC) rely on AR activation by androgen. A new consideration is that some CRPC phenotypes are androgen self-sufficient, through synthesis of androgens from sources other than the testes. Dehydroepiandrosterone sulfate is a candidate. Therapies targeted at steroidogenic enzymes, such as abiraterone acetate against CPY17A1, have proven beneficial for interrupting intratumoral and adrenal steroidogenesis. For patients with PCa, monitoring all steroidogenic metabolites from cholesterol to testosterone may prove beneficial so that individual therapies that disrupt steroidogenesis may be administered.

2.1.4.5 Breast cancer cell lines

The first breast cancer cell line to be established was BT-20 in 1958 and there was a 20 year gap, before breast cancer cell lines became more widespread, including the MD Anderson series. The most commonly used breast cancer cell line in the world, MCF-7 was established in 1973 at the Michigan Cancer Foundation. The popularity of MCF-7 is largely due to its hormone sensitivity through expression of estrogen receptor (ER), making it an ideal model to study hormone response (Comşa et al., 2015; Lee et al., 2015). Relatively few breast cancer cell lines have been established in the more recent past, mainly because of difficulties in culturing homogeneous populations without significant stromal and metastatic contamination and, at least in the United Kingdom, partly due to rigorous ethical regulations surrounding obtaining human tissue for research. Cell lines were derived from either breast primary tumors, pleural effusions or various metastatic sites in individual patients. These cell lines are now widely available through commercial cell banks.

Breast cancer cell classification was based on histological type, tumor grade, lymph node status and the presence of predictive markers such as ER and, more recently, human epidermal growth factor receptor 2 (HER2) (Holliday and Speirs, 2011). DNA microarrays proved this heterogeneity, demonstrating gene expression and IHC expression of ERα, progesterone receptor (PR) and HER2. Breast cancer could be classified into at least five subtypes: luminal A, luminal B, HER2, basal and normal. Cell lines have become contaminated over the years and subject to a high mutational frequency with many uncertainties. Breast cancer lines are now regarded as only crude models for tumors of the same subtype.

2.1.4.6 "Body on a chip"

Cells are now propagated in vitro in three-dimensional cellular cluster systems called spheroids and organoids that better mimic natural tissue existence. By including these organoids in microfluidic devices the interactions of cell types (Leydig and Sertoli cells; fetal adrenal cortex) or organs (liver, testis, muscle, brain) can be studied in a so-called "body on a chip" device (Skardal et al., 2020; Richer et al., 2019). Cell adhesion molecules and extracellular matrix effects have been studied. These systems in combination with metabolomics will enable studies of the complex, integrated actions within and around organs in endocrine research and be cheaper than animal studies. The technical advances in genome editing using CRISP-Cas9 may allow mutations in cells from patients to be corrected for implantation of healthy back to the patient.

2.1.5 Certified reference materials

Public confidence in measurement results is important in many aspects of modern society, including health care, food and environmental protection. Certified Reference Materials (CRMs) allow calibration of instruments, validation of methods and quality control of methods and laboratories based on traceability and comparability of measurement results. EC-JRC BCR certified reference materials (BCR is a registered trademark of the EC) are produced with research funding of the European Commission, DG Research and ERM certified reference materials (ERM is a registered trademark of the EC). These CRMs are produced according to specific Guidelines of the European Commission, which are in agreement with the relevant ISO Guides 34 and 35. The JRC-Geel has been accredited for the production of reference materials since 2004 (accreditation certificate BELAC 268-RM). The products for hormone content are summarized in Table 2.1.2.

Certificates carry a certified value with its uncertainty which is traceable either to a SI unit or an internationally accepted reference. The intended use for each CRM is stated on the certificate. CRMs are stored under controlled conditions which ensure their stability. Monitoring programs have been set up to control CRM stability during the whole shelf-life. At present, JRC-Geel's Reference Materials Unit offers about 700 different CRMs. A complete list of these CRMs, including description and price as well as the CRM catalogue and all certificates and certification reports can be accessed directly via the online catalogue: https://crm.jrc.ec.europa.eu.

The National Measurement Institute in Australia (NMI) has produced a freeze dried urine reference material certified for the concentrations of the endogenous steroids testosterone and epi-testosterone (T and E) to allow laboratories to reliably compare the measurements of those steroids. The T/E ratio gives an initial indication of testosterone abuse in sport. The carbon isotope ratios of these materials will be useful when certified. The certification of this material for other endogenous is awaited. This will be useful in in detecting steroid abuse by individuals who are genetically disposed to have low T/E ratios even when doping.

TABLE 2.1.2 European reference serum.

High progesterone	BCR 348R	26.9 nmol/L (8.5 µg/L)
Progesterone	DA 347	10.13 nmol/L (3.19 ng/ml)
Oestradiol low	BCR 576	0.114 nmol/L
Oestradiol medium	BCR 577	0.69 nmol/L
Oestradiol	BCR 578	1.34 nmol/L
Cortisol unspiked	DA 192	273 nmol/L (98.8 µg/L
Cortisol spiked	DA 193	763 nmol/L (277 µg/L)
Serum cortisol	DA 451	17 samples 83–764 nmol/L

2.1.6 Animals and models for human disease

The human and the mouse are genetically very similar, with many of the disease-related genes nearly identical. The mouse has become the foremost mammalian model for studying human disease and human health. Both the zebrafish and the rat play a role as disease models in the investigation of the genetics of human disease. Each occupies a narrower niche than the mouse for several reasons.

2.1.6.1 Mouse models

The mouse is small and breeds well, making it a low cost experimental resource. Scientists have accumulated much knowledge about mouse physiology, anatomy and its genes, which today can be manipulated with molecular tools. The human and mouse genomes have been sequenced so the usual molecular constitutions are known. The sequences vary between individuals and between the two species. Even two healthy people have millions of small sequence and structure differences that help define who those people are as individuals. Making small tweaks to individual mouse genes or even sequences within a specific gene allows scientists to see what happens as a result (Gurumurthy and Lloyd, 2019; Rosen et al., 2015; Doyle et al., 2012).

Small blood samples (20–40 µL) can be obtained from the tail or saphenous vein after clipping the fur (Hoggatt et al., 2016; Rathkolb et al., 2013a,b; Donovan and Brown, 2013). The skin can be punctured with a needle and blood collected into a capillary tube and spotted onto blotting paper. The procedure can be repeated up to three times without stressing the animal which can be restrained without anesthesia in a suitably sized tube (Falcon tube for example). The local animal ethics committee must approve all such sampling procedures.

The International Mouse Phenotyping Consortium (IMPC) is an international effort by 19 research institutions to identify the function of every protein-coding gene in the mouse genome (Mallon et al., 2012). The entire genome of many species has now been published and whole genome sequencing is becoming relatively quick and cheap to complete. Despite these advancements the function of the majority of genes remains unknown. The IMPC's mission is to fill this knowledge gap and create a comprehensive catalogue of mammalian gene function that is freely available for researchers.

The IMPC is systematically switching off or "knocking out" each of the roughly 20,000 genes that make up the mouse genome. Subsequently, the knockout mice undergo standardized physiological tests (phenotyping tests) in order to infer gene function, before the data is made freely available to the research community on our website (www.mousephenotype.org). IMPC data can be used in a variety of ways, such as to investigate basic biology mechanisms that can lead to new therapeutic targets or to narrow down a suspected list of genes in patients. In the past few years, the IMPC have made major discoveries in parts of the genome that were hitherto unexplored, with new genes discovered that relate to areas such as deafness, diabetes and several rare diseases. The overall aim of the project is not only to understand the function of every gene and provide insights into the genetic basis of diseases that will impact upon clinical diagnosis and management and ultimately prevent, detect, diagnose and treat disease.

There are other "mouse clinics" at European EUMODIC program (www.eumodic.org), the Mouse Genetics Project (MGP) at the Sanger Institute (www.sanger.ac.uk/resources/mouse), KOMP312 at UC Davis (www.kompphenotype.org) and the German Mouse Clinic (GMC; www.mouseclinic.de).

2.1.6.2 Rats

Research involving the rat has for many years lagged behind that of the mouse in terms of developing the techniques for manipulating its genetic systems (Mellon et al., 1995). This, coupled with the expense of producing mutations in the rat, has been the primary reason for it having been used less widely than the mouse for the study of the genetics of disease processes even though a complete genome sequence has been published.

Blood sampling from the rat is usually from the tail or saphenous vein (Parasuraman et al., 2010). A system has been developed for repeated blood sampling and use of this equipment has enabled studies of hormone concentrations at intervals say over a 24h period (see earlier Fig. 2.1.5).

Several inbred rat cell lines have been developed. Many of these have been characterized for diseases such as diabetes and hypertension for which the rat is a particularly trac3 model. Hypertension models are divided according to sex differences in blood pressure and response to treatment, vascular target organ damage in different experimental models (Salt sensitive, renovascular, NO dependency, Angiotensin II dependency), cerebrovascular pathologies (Lerman et al., 2019). Rats are the preferred species for these diseases because their large size is more suitable for the use of the technologies available for the measurement of phenotypes such as blood pressure (Table 2.1.3).

Comparisons between inbred lines have revealed a significant amount of variation in disease phenotypes. Genetic crosses between them show significant phenotypic differences and allow the genetic regions involved to be mapped and ultimately identified. For example, the genetics of hypertension is a major area for study in the rat and a number of genes have been identified that are involved in determining blood pressure.

TABLE 2.1.3 Transgenic rat (R) and mouse (M) in hypertension.

Transgenic technique	Species	Examples
Gene over-expression	M	Mendelian models of hypertension
	M/R	Salt sensitive hypertension
Inducible expression	R	Inducible hypertension and end-organ damage
Bacterial artificial chromosomes	M	Reduced risk of hypertension
Global gene knockout	M	Mendelian models of hypertension
	M/R	Salt sensitive hypertension
	M	Pulmonary hypertension; Renin hypertension
Gene knock-in	M/R	Renin; Renin/angiotensin/aldosterone humanization
Targeted gene knockout	M	Kidney specific HSD11B2 knockout
Conditional knockout	M	Pulmonary hypertension
Zinc finger nuclease	R	Salt sensitive hypertension
CRISPR/Cas 9	R	Humanization and reverse orientation slice acceptor ROSA26
Gene knockdown/anti-microRNA; small interfering RNA	M/R	Angiotensin 1A receptor; reduced hypertension in (a) aldosterone salt treated mice and (b) spontaneously hypertensive rats

(Selection from Lerman LO, Kurtz TW, Touyz RM, Ellison DH, Chade AR, Crowley SD, et al. Animal models of hypertension: a scientific statement from the American Heart Association. Hypertension 2019;73(6):e87–e120.)

Spontaneous hypertensive rat (SHR) was generated from cross between Wistar Kyoto male rats with noticeable elevated blood pressure and females with slight elevation of blood pressure, followed by selective inbreed of the offspring with highest blood pressure. The SHR is used as an experimental model for genetically induced hypertension. Twelve-week-old SHRs are hypertensive, hyperinsulinemic and insulin resistant compared with Wistar-Kyoto rats. Spontaneously hypertensive rats generally do not develop hypercholesterolemia and hyperlipidemia unless they are put on a special diet regimen, such as high-cholesterol or high-fructose high-fat diet. Modification of SHR, known as obese SHR or Koletsky rat, was obtained by crossing a female SHR with a normotensive Sprague-Dawley male. Koletsky rats carry a nonsense mutation in the leptin receptor and possess interesting phenotypes, including obesity at 5 weeks of age, hypertriglyceridemia even with standard diet, hyperinsulinemia with normal blood glucose and severe hypertension at 3 months of age. Koletsky rats have been suggested as a more appropriate animal model for metabolic syndrome (MetS) compared with SHRs.

The POUND mouse (C57BL/6NCrl-*Lepr*^db-lb^/Crl) was established in the last decade as another model fulfilling all the metabolic syndrome (MetS) criteria in a single animal. The animals are fed with Purina Diet ad libitum and show obesity at 1 month of age, hyperinsulinemia and hyperglycemia at 18 weeks of age, increased leptin levels at 17–18 weeks of age, as well as increased cholesterol levels at 14 weeks of age.

2.1.6.2.1 Rat adrenal enucleation

Adrenal enucleation involves slicing open the capsule and extruding the medulla and most of the cortex has been a model used since 1939 to address issues about adrenal zonation. Following enucleation of both adrenal glands in the adult rat; proliferation of the residual cortical tissue (glomerulosal cells) restores a normal appearing cortex within a month (Greep and Deane, 1949; Takeda et al., 1964). The regenerated cortex possesses the usual zona glomerulosa, zona fasciculata and zona reticularis. No evidence has been found for the conversion of capsular fibroblasts to cortical cells. Within 8 days after the operation, ascorbic acid and droplets giving the reactions characteristic of steroid-producing glands reappear in the immediate subcapsular cells and eventually come to occupy all of the cells of the cortex.

One month after enucleation of their adrenals, rats respond in the normal fashion to a prolonged fast, i.e., the blood-sugar level declines temporarily, then returns to 100 mg. per deciliter within 5 days. Adrenalectomized rats will succumb within 4 days in hypoglycemic coma. The daily administration of deoxycorticosterone acetate over a period of a month causes a suppression of secretory activity in the zona glomerulosa of the regenerated gland. The regenerated cortex apparently secretes hormones governing both salt and sugar metabolism, and the normal functional zonation is restored. Because of the regeneration of the cortex, the operation is generally termed demedullation. The regeneration is dependent upon the presence of the pituitary.

2.1.6.3 Animal models of PCOS

Clinical observations have supported a primary role for androgen action in the origins of human PCOS and there is significant experimental evidence pointing to an important role for direct AR-mediated androgen actions in the pathogenesis of PCOS. The generation of animal PCOS models in rodents, notably global and cell specific mouse models (such as the aromatase deficient specific knock out mouse, ARKO), sheep and nonhuman primates, has created opportunities for mechanistic experiments to be undertaken (Walters et al., 2018; Noroozzadeh et al., 2017). The results strongly support a direct role for androgen actions via the AR in the origins and evolution of PCOS. Several androgen-induced PCOS animal models have confirmed that elevated androgen exposure can replicate many PCOS features including AR actions in the brain playing a key role in driving the PCOS phenotype. Several possible treatments to reduce the impact of androgen action or metabolic derangements such as decanoic acid or metformin may offer potential new treatments for PCOS, with scope for further in animal PCOS models prior to clinical testing.

2.1.6.4 Germ free animals

Steroid research with germ free rats was conducted for many years (c. 1968–76) at the Karolinska Institute in Sweden and useful data on the enterohepatic circulation of steroids was obtained. This work predated the recognition of the microbiome from around 2003 as an important area for research into links between the gut bacteria and diseases in man. The human intestine has kilograms of bacteria within. Germ-free animals are expensive to maintain and some centers moved to keeping specific pathogen free (gnotobiote) animals. The use of germfree mice is likely to increase however now that the characterization of the micro-organisms by 16S RNA analysis has improved the analysis of the microbiome. The importance of the microbiome became recognized when cell numbers of microbes in the gut were realized to be comparable with cell numbers in the body.

2.1.6.5 Nonhuman primates

Next to man, other related primates (Fig. 2.1.12) have been studied, although this is controversial. The cost of maintaining animals in captivity is very high and in recent years fewer research centers now have primate facilities.

All primates share fundamental similarities in life history, typically being relatively large-brained, long-lived species that usually produce a single offspring that requires a high degree of parental and relatives' care. Reproductive endocrinology and physiology is broadly similar across primates, monkeys, apes and humans have hemochorial placentas. Studies of nonhuman primates have improved understanding of the evolutionary history of the hominin lineage and the primate order overall.

FIG. 2.1.12 Primates used in research.

The chimpanzee (*Pan troglodytes*) is a species of great ape, closest to man in the Hominidae. The chimpanzee has a bare face, fingers, toes, palms of the hands and soles of the feet. It weighs between 40 and 65 kg and measuring about 63–95 cm. Its gestation period is 8 months. The infant is weaned at about 3 years old. It reaches puberty at the age of 8–10. Its lifespan in the wild can exceed 40 years. Few centers conduct research with chimpanzees.

The next closest relative of man is the gorilla. The Homininae genera, all diverged from a common ancestor about 7 million years ago. Human gene sequences differ only 1.5% on average from the sequences of corresponding gorilla genes. Adult gorilla males are 1.4–1.8 m tall, with an arm span that stretches around 2.5 m. Female gorillas are shorter at 1.25–1.5 m with smaller arm spans. The average weight of 47 wild adult male gorillas is 150 kg. Gorilla are very expensive to keep.

Baboons are primates in the genus *Papio*, one of the 23 genuses of Old World monkeys. Baboons vary in size and weight depending on the species which has determined the species used in captivity for research. The smallest, the Guinea baboon, is 50 cm in length and weighs only 15 kg. Baboons have been studied over many years by Gerald Pepe.

Common marmosets (*Callithrix jacchus*) are in the subfamily Callitrichinae (family Cebidae) and are found primarily in the scrub and swamp habitat of coastal north eastern Brazil. Common marmosets offer unique opportunities as nonhuman primate models. They are relatively small (around 400 g in adulthood, about the size of an adult Sprague-Dawley rat), achieve sexual maturity relatively quickly (16–20 months of age), can produce offspring twice a year, have a typical lifespan in captivity of 6 years. They have been used in studies of reproductive biology and much knowledge of primate pregnancy, development, lactation and parenting has been gained over nearly 30 years of research.

Squirrel monkeys are New World monkeys of the genus *Saimiri*. Squirrel monkeys live in the tropical forests of Central and South America in the canopy layer. Squirrel monkeys grow from 25 to 35 cm in body length with a 35–42 cm tail. Male squirrel monkeys weigh 750–1100 g. Females weigh 500–750 g. Both males and females have long and hairy tails, flat nails and pointed claws. Female squirrel monkeys have pseudo-penises, which they use to display dominance over smaller monkeys, in much the same way that the male squirrel monkeys display their dominance. Squirrel monkeys have very high cortisol production due to defects in action of the glucocorticoid receptor that have yet to be clarified. Ken Setchell and Cedric Shackleton in the Medical Research Council facilities in London studied urine from several of the above species. The studies will be difficult to replicate because few species are now kept in research centers because of very high costs.

2.1.6.6 Zebrafish

There has been a significant increase in the use of zebrafish for the study of disease processes in humans. Around 70% of human genes are found in zebrafish. Moreover, zebrafish have two eyes, a mouth, brain, spinal cord, intestine, pancreas, liver, bile ducts, kidney, esophagus, heart, ear, nose, muscle, blood, bone, cartilage and teeth. In theory, any type of disease that causes changes in these body parts in humans could theoretically be modeled in zebrafish (Clelland and Peng, 2009). The one-cell-stage fertilized eggs can be easily injected with DNA or RNA to permanently modify their genetic makeup in order to generate transgenic or knock-out zebrafish lines.

Zebrafish reproduce easily and quickly and have morphological and physiological similarities to mammals. The use of the species will lead to progress in several aspects of the drug development process, including target identification, disease modeling, lead discovery and toxicology (Tokarz et al., 2013; Hsu et al., 2006). The study of the zebrafish genome is relatively well advanced and near a complete genome sequence. It has been the focus of several major forward genetic screens for a variety of diseases and other phenotypes. Zebrafish models have been developed for several human diseases, including blood disorders, diabetes, muscular dystrophy and neurodegenerative diseases. The transparency of the developing zebrafish embryo has enhanced its usefulness for studying the genetics of development.

One area where much progress has been made is in the study of the genetics of the development of the heart and vascular system. Increased understanding about the genes involved has also contributed to understanding of these processes in vertebrates. Many mutant fish with defects in steroidogenesis have been developed. Blocking the action of CYP11A1 results in defects of embryonic cell migration during epiboly—a process of blastula and gastrula cell movement (Hsu et al., 2009).

Sulfotransferase function is also being studied in zebrafish and a number of zebrafish SULTs have also been shown to be involved in the sulfation of steroid hormones. Zebrafish SULT1 ST2 and SULT1 ST6 exhibited the strongest activities toward estrone and 17b-estradiol, with SULT1 ST6 having a high degree of substrate specificity for estrogens. SULT1 ST6 is likely the counterpart of human SULT1E1 since it has its highest sequence homology to SULT1E1 among the 12 known human SULTs. All members of zebrafish SULT2 and SULT3 family displayed sulfating activities toward hydroxysteroids including pregnenolone and dehydroepiandrosterone (DHEA). Hydroxysteroid-sulfating activity of zebrafish SULT2s is in line with human SULT2 enzymes, SULT2A1 and SULT2B1 (Kurogi et al., 2013). The SULT2 enzymes can displayed preferences for pregnenolone, DHEA and corticosterone, respectively. Knock down/out experiments targeting zebrafish SULT1 ST5 and SULT1 ST6 genes may provide valuable information concerning their physiological involvement. Recent research has shown that two steroid-sulfating SULT3 STs may play differential roles in the metabolism and regulation of steroids during zebrafish development and in adulthood.

Deficiency of ferredoxin 1B has been induced in zebrafish mutants producing embryos in which concentrations of cortisol and androgens were reduced (Oakes et al., 2019). The testes of mutant males lacked seminiferous tubule structure. 21-Hydroxylase deficient mutant zebrafish showed upregulation of the hypothalamic pituitary interrenal axis and interrenal hyperplasia (Eachus et al., 2017).

FIG. 2.1.13 Spotted hyena and Tammar Wallaby.

2.1.6.7 Spotted hyena and tammar wallaby

Some unusual animals (Fig. 2.1.13) have been studied because of their differences in sexual development compared with other animals (Glickman et al., 2005).

Virilization of the urogenital tract of Tammar wallaby is under the control of testicular androgens (Leihy et al., 2004) whereas the female external genitalia of the spotted hyena are noted for extreme masculinization (Cunha et al., 2014). They have provided important evidence for biosynthetic pathways for production of dihydrotestosterone that do not conform to classical paths in sexual differentiation. The studies will be described in more detail in the biosynthesis chapter under the heading "Backdoor path to dihydrotestosterone" and in the chapter on Androgen excess.

2.1.6.8 Brain

Although blood, urine and saliva have been the biological fluids most often examined for the presence of steroids there is increasing research interest in steroids in brain (for so-called neurosteroids). Steroids in the brain have important physiological roles and influence brain development, behavior, cognition, neuroplasticity and neuroinflammation. In addition to steroids that enter the brain from the periphery, steroids are synthesized within the brain itself. High steroid concentrations and studies of steroidogenic enzyme mRNA, protein and activity in the brain, support the brain having all functions as a steroid-synthetic. The high lipid content of this tissue has made difficult the preparation of extracts sufficiently clean for analysis of the steroids. Reports of detection and measurement of steroids in brain have to be treated with some reservations without characterization of the steroids to high analytical standards. DHEAS and pregnenolone sulfate were thought to be present in brain but this was almost certainly artifactual. In some cases, a combination of techniques is needed to ensure purity of the final product. An indirect approach is to measure steroid concentrations in blood leaving the brain (jugular vein) or other sites (carotid, brachial).

Animal brain can be obtained after euthanasia which should be performed without any stress to the animal during the procedure. An overdose of an anesthetic (Isoflurane, halothane, carbon dioxide, chloral hydrate, pentobarbital, ketamine and clonazepam) can be given but experiments will be needed to confirm these drugs have no effect on the steroid concentrations in the brain. Animals can be quickly sacrificed by decapitation or cervical dislocation. If necessary, blood can be removed by transcardial perfusion of saline. Blocks and sections of brain tissue can be cut with a scalpel based on position relative to neuroanatomical landmarks (midline, ventricles). Histological sections of brain can be examined by matrix assisted laser desorption ionization mass spectrometry after formation of Girard P reagents (Guo et al., 2020). This will be used to examine brain for steroidogenesis.

2.1.7 Samples for analysis of regulatory peptides

2.1.7.1 ACTH

ACTH is secreted in a pulsatile manner with a mean pulse frequency for men and women of 18 and 10 peaks per 24 h, respectively, with a mean peak amplitude of 16.8 and 10.3 ng/L (3.7 and 2.3 pmol/L). Depending on the sampling frequency there is a diurnal circadian rhythm with a maximum between 06:00 and 08:00 h. Due to the poor sensitivity of early assays, it was customary to take samples between 08:00 and 10:00 h, near when ACTH concentrations would be at their highest. The peaks of ACTH plasma concentrations are less during the early afternoon. It is advised to collect whole blood into lithium heparin or

EDTA plastic tubes, taking note of the sample requirement of the immunoassay to be used. Glass tubes must be silanized as ACTH sticks to glass. Samples should be kept cool and centrifuged within 1 h of collection, and the plasma transferred into plastic or silanized glass tubes. Grossly hemolyzed samples cannot be used. Multiple freeze-thaw cycles should be avoided; ideally, two separate aliquots (minimum 500 mL) of the sample should be stored at less than $-20°C$. Samples should be transported frozen on dry ice if going to a reference laboratory. ACTH is not measured in urine.

ACTH can be extracted from tissue samples such as pituitary and extra-pituitary tumors. Tissue should be chopped or ground frozen (with dry ice) prior to homogenization in acid (1 M acetic or 0.1 M HCl) with protease inhibitors (e.g. phenylsulfonylfluoride 0.3 mg/mL and pepstatin 10 mg/mL). Following centrifugation, the supernatant is lyophilized. The dried sample is then reconstituted in 0.01 M HCl and diluted in appropriate assay buffer.

ACTH was historically judged on immunoassay data to be unstable (ACTH half-life spiked into thawed plasma is 200–300 min and into whole blood is around 50 min). Consequently, special precautions for the handling of blood specimens are widely advocated and samples are spun immediately and the plasma collected and stored at $-70°C$ until assay. This is the method of choice because certain samples may degrade more rapidly than others. The endogenous bioactive and immunoreactive ACTH are stable for at least 1 h in heparinized blood and 2 h in plasma. Thus, ACTH is relatively stable and it may even be possible to undertake one or two freeze-thaw cycles before samples degrade. If a result is inappropriately low, it is important to assess the storage history.

Ectopic tumors can produce ACTH and are localized by catheter studies usually around the thorax since tumors can be in heart, lungs and elsewhere (see clinical chapters Bilateral inferior petrosal sinus sampling (IPSS) is an invasive procedure in which adrenocorticotropic hormone (ACTH) levels are sampled from the veins that drain the pituitary gland; these levels are then compared with the ACTH levels in the peripheral blood to determine whether a pituitary tumor (as opposed to an ectopic source of ACTH) is responsible for ACTH-dependent Cushing's syndrome. IPSS can also be used to establish on which side of the pituitary gland the tumor is located.

2.1.7.2 Renin

Special precautions are needed in the preanalytical stage to avoid unfolding and cleavage of the prorenin prosegment. Blood samples should be promptly centrifuged and the plasma quickly frozen. The plasma for renin measurement can be safely kept in an unopened EDTA Vacutainer (BD Medical Systems) for up to 24 h at room temperature before centrifugation. Leaving blood at ambient temperature however may cause unfolding of the prorenin prosegment and lead to overestimation of renin activity and concentration. Another reason for promptly centrifuging blood samples is the wide variation in ambient temperature that occurs by season. Blood should be centrifuged within 30 min of collection, preferably within 10 min, and the plasma sample should be rapidly frozen if it is not assayed immediately. Care must be taken to ensure that frozen samples do not thaw during storage or during transport to the laboratory. Frozen plasma should be rapidly thawed before assay and only once. These precautions are especially important for plasma samples from patients with primary aldosteronism, pregnancy or diabetes where renin is low. A new reason for promptly centrifuging samples is that patients are getting renin inhibitor therapy so prorenin conversion needs to be minimized.

2.1.7.3 LH/FSH/hCG

Gonadotropins are much less stable in urine than in serum and care should be taken to avoid loss of immunoreactivity. Addition of glycerol to serum to a concentration of 7% and storage of the samples at −20°C has been shown to retain LH immunoreactivity. The assay used may have recognized LH fragments, which might have formed during storage. LH in serum is stable for several weeks when stored at 4°C with sodium azide (0.02%) to prevent bacterial growth. When stored at −20°C a highly variable loss of immunoreactivity may occur, in some samples more than 90% is lost. The loss depends on storage time, but in some samples it occurs after a single round of freezing and thawing. HCG is somewhat more stable than LH, but freezing at −20°C may cause rapid loss of immunoreactivity. In some samples, the loss of hCG is accompanied by a corresponding increase in hCGβ, indicating dissociation of the subunits, but in some samples all forms of immunoreactivity are reduced. Fresh urine contains rather little hCGβ and the presence of this form in urine is usually a sign of inappropriate storage. The beta core fragment (HCGβcf) is somewhat more stable than hCG, but there is large variation between samples. Like LH, hCG is quite stable when stored at 4°C with 0.02% azide. Nonspecific adsorption of urinary proteins to the walls of certain plastic tubes is a further preanalytical problem.

2.1.8 Nonbiological samples

Steroids may also need to be measured in unexpected matrices. These include:

- **Pharmaceutical tablets** for low levels of impurities. Many of the tests for pharmaceutical grade chemicals are relatively insensitive and impurities may be missed. Impurities can be due to impure starting materials, side

reactions and instability of the formulation. The standards for pharmaceuticals have followed progress in technology from column chromatography, through thin layer and gas chromatography and HPLC (Görög, 2012).
- **Herbal medicines** that may contain pharmaceuticals. Hydrocortisone and synthetic glucocorticosteroids as antiinflammatory agents have been found in order to add these properties to a herbal preparation.

Food and nutritional supplements. The finding of high concentrations of 19-nor testosterone in boar meat became an issue of antidoping for banned anabolic steroids (Hülsemann et al., 2020). There is limited knowledge on the contribution of dietary supplement (Walpurgis et al., 2020). Plant sterols (Kopylov et al., 2021). In sport, nutritional supplements are widely used but have given rise to positive tests in doping control.

2.1.8.1 Water and sewage effluents

Oral contraceptives, the other pharmaceutical steroids and endocrine disruptors can re-enter the food chain via sewerage despite treatment (Zhao et al., 2016). Concentrations of steroids can be very low but sample volume is not limited so extraction is an important step in the analytical process.

2.1.9 Summary

The correct collection of samples and transfer to the laboratory are critical steps before the analysis of the content of the samples. This is often performed by staff that do not have the training of the laboratory staff and therefore education and installation of procedures is essential. Samples are collected at different times into different containers and standardization is critical to achieving useable results that offer answers to questions on the need for the samples. The need for proper sample collection must be understood by all of the staff involved. The European Federation of Clinical Chemistry and Laboratory Medicine have produced a useful checklist toward prevention of reanalytical errors (Lippi et al., 2020). As long as the same blood preparation procedure is used, either plasma or serum will generate similar results for steroids in clinical investigations. There are many constraints with human and animal studies. Research with cells in vitro may not provide answers to the state in vivo. Genetic manipulations of animals are now becoming a common research tool, the extent of this has not been fully addressed here.

2.1.10 Further reading

2.1.10.1 Methodology

Lippi et al. (2020), Thevis et al. (2016), and Thiry et al. (2017).

2.1.10.2   Urine

Wudy et al. (2007).

2.1.10.3   Bile, enterohepatic circulation, faeces

Adlercreutz and Järvenpää (1982), Adlercreutz et al. (1987), Cronholm et al. (1971), Higashi et al. (2016), Järvenpää et al. (1980), and St Pyrek (1987).

2.1.10.4   Saliva

Bessonneau et al. (2015).

2.1.10.5   Tissue samples

Bennett et al. (2012), du Toit et al. (2017), Falk et al. (2008), Gonzalez-Riano et al. (2016), Nikolakis et al. (2016), Slominski et al. (2015), Stalder and Kirschbaum (2012), and Taves et al. (2011).

2.1.10.6   Metabolomics

Hirsch et al. (2012) and Kaur and Dufour (2012).

2.1.10.7   Reproduction

Koskenniemi et al. (2017) and Tremblay and Sirard (2017).

2.1.10.8   Research models

Alves-Lopes and Stukenborg (2018), Hanahan et al. (2007), Poli et al. (2019), Ruiz-Babot et al. (2018), and Walsh et al. (2017).

2.1.10.9   Enzyme inhibitors

Mangelis et al. (2019) and Neunzig et al. (2017).

2.1.10.10   Receptor assays

Cadwallader et al. (2011), Jarque et al. (2016), Keiler et al. (2018), Mertl et al. (2014), and Zierau et al. (2019).

References

Adlercreutz H, Martin F. Biliary excretion and intestinal metabolism of progesterone and estrogens in man. J Steroid Biochem 1980;13(2):231–44.

Adlercreutz H, Järvenpää P. Assay of estrogens in human feces. J Steroid Biochem 1982;17(6):639–45.

Adlercreutz H, Höckerstedt K, Bannwart C, Bloigu S, Hämäläinen E, Fotsis T, et al. Effect of dietary components, including lignans and phytoestrogens, on enterohepatic circulation and liver metabolism of estrogens and on sex hormone binding globulin (SHBG). J Steroid Biochem 1987;27(4–6):1135–44.

Alves-Lopes JP, Stukenborg JB. Testicular organoids: a new model to study the testicular microenvironment in vitro? Hum Reprod Update 2018;24 (2):176–91.

Arcus AC, Beaven DW, Hart DS, Holland GW. Long-term hormonal secretion from the autotransplanted sheep pancreas. Diabetologia 1979;16(5):325–30.

Ben Maamar M, Lesné L, Hennig K, Desdoits-Lethimonier C, Kilcoyne KR, Coiffec I, et al. Ibuprofen results in alterations of human fetal testis development. Sci Rep 2017;7:44184.

Bennett NC, Hooper JD, Lambie D, Lee CS, Yang T, Vesey DA, et al. Evidence for steroidogenic potential in human prostate cell lines and tissues. Am J Pathol 2012;181(3):1078–87.

Bessonneau V, Boyaci E, Maciazek-Jurczyk M, Pawliszyn J. In vivo solid phase microextraction sampling of human saliva for non-invasive and on-site monitoring. Anal Chim Acta 2015;856:35–45.

Bhake R, Russell GM, Kershaw Y, et al. Continuous free cortisol profiles in healthy men. J Clin Endocrinol Metab 2020;105(4):dgz002.

Bhattarai KR, Kim HR, Chae HJ. Compliance with saliva collection protocol in healthy volunteers: strategies for managing risk and errors. Int J Med Sci 2018;15(8):823–31.

Cadwallader AB, Lim CS, Rollins DE, Botrè F. The androgen receptor and its use in biological assays: looking toward effect-based testing and its applications. J Anal Toxicol 2011;35(9):594–607.

Carraro P, Zago T, Plebani M. Exploring the initial steps of the testing process: frequency and nature of pre-preanalytic errors. Clin Chem 2012;58(3):638–42.

Caruso D, Melis M, Fenu G, Giatti S, Romano S, Grimoldi M, et al. Neuroactive steroid levels in plasma and cerebrospinal fluid of male multiple sclerosis patients. J Neurochem 2014;130(4):591–7.

Choi R, Park HD, Oh HJ, Lee K, Song J, Lee SY. Dried blood spot multiplexed steroid profiling using liquid chromatography tandem mass spectrometry in Korean neonates. Ann Lab Med 2019;39 (3):263–70.

Ciplea AM, Laeer S, Burckhardt BB. A feasibility study prior to an international multicentre paediatric study to assess pharmacokinetic/pharmacodynamic sampling and sample preparation procedures, logistics and bioanalysis. Contemp Clin Trials Commun 2018;12:32–9.

Clelland E, Peng C. Endocrine/paracrine control of zebrafish ovarian development. Mol Cell Endocrinol 2009;312(1–2):42–52.

CLSI (formerly NCCLS). Tubes and additives for venous blood specimen collection; approved standard. CLSI document H1-A5. 5th ed. Wayne, PA, USA: Clinical and Laboratory Standards Institute; 2003, ISBN:1-56238-519-4. http://specimencare.com/main.aspx?cat=711&id=3048.

CLSI (formerly NCCLS). Procedures for the collection of diagnostic blood specimens by venipuncture; approved standard. CLSI document H3-A6. 6th ed. Wayne, PA, USA: Clinical and Laboratory Standards Institute; 2007, ISBN:1-56238-650-6.

Cobice DF, Mackay CL, Goodwin RJ, McBride A, Langridge-Smith PR, Webster SP, et al. Mass spectrometry imaging for dissecting steroid intracrinology within target tissues. Anal Chem 2013;85(23):11576–84.

Comşa Ş, Cîmpean AM, Raica M. The story of MCF-7 breast cancer cell line: 40 years of experience in research. Anticancer Res 2015;35 (6):3147–54.

Côté AM, Firoz T, Mattman A, Lam EM, von Dadelszen P, Magee LA. The 24-hour urine collection: gold standard or historical practice? Am J Obstet Gynecol 2008;199(6):625.e1–6.

Cronholm T, Eriksson H, Gustafsson JA. Excretion of endogenous steroids and metabolites of (4-14C)pregnenolone in bile of female rats. Eur J Biochem 1971;19(3):424–32.

Cunha GR, Risbridger G, Wang H, Place NJ, Grumbach M, Cunha TJ, et al. Development of the external genitalia: perspectives from the spotted hyena (Crocuta crocuta). Differentiation 2014;87(1–2):4–22.

Deda O, Gika HG, Wilson ID, Theodoridis GA. An overview of fecal sample preparation for global metabolic profiling. J Pharm Biomed Anal 2015;113:137–50.

Di Meo A, Bartlett J, Cheng Y, Pasic MD, Yousef GM. Liquid biopsy: a step forward towards precision medicine in urologic malignancies. Mol Cancer 2017;16(1):80.

Di Nunno V, Gatto L, Santoni M, Cimadamore A, Lopez-Beltran A, Cheng L, et al. Recent advances in liquid biopsy in patients with castration resistant prostate cancer. Front Oncol 2018;8:397.

Donovan J, Brown P. Care and handling of laboratory mice. Curr Protoc Microbiol 2013;31:A.3N.1–A.3N.18.

Doyle A, McGarry MP, Lee NA, Lee JJ. The construction of transgenic and gene knockout/knockin mouse models of human disease. Transgenic Res 2012;21(2):327–49.

du Toit T, Bloem LM, Quanson JL, Ehlers R, Serafin AM, Swart AC. Profiling adrenal 11β-hydroxyandrostenedione metabolites in prostate cancer cells, tissue and plasma: UPC(2)-MS/MS quantification of 11β-hydroxytestosterone, 11keto-testosterone and 11keto-dihydrotestosterone. J Steroid Biochem Mol Biol 2017; 166:54–67.

Eachus H, Zaucker A, Oakes JA, Griffin A, Weger M, Güran T, et al. Genetic disruption of 21-hydroxylase in zebrafish causes interrenal hyperplasia. Endocrinology 2017;158(12):4165–73.

Engeli RT, Fürstenberger C, Kratschmar DV, Odermatt A. Currently available murine Leydig cell lines can be applied to study early steps of steroidogenesis but not testosterone synthesis. Heliyon 2018;4(2), e00527.

Eriksson H. Absorption and enterohepatic circulation of neutral steroids in the rat. Eur J Biochem 1971a;19(3):416–23.

Eriksson H. Steroids in germfree and conventional rats. Metabolites of (4-14C)pregnenolone and (4-14C)corticosterone in urine and faeces from male rats. Eur J Biochem 1971b;18(1):86–93.

Eriksson H, Gustafsson JA. Metabolism of corticosterone in the isolated perfused rat liver. Eur J Biochem 1971a;20(2):231–6.

Eriksson H, Gustafsson JA. Excretion of steroid hormones in adults. Steroids in faeces from adults. Eur J Biochem 1971b;18(1):146–50.

Eriksson H, Gustafsson JA, Sjövall J. Studies on the structure, biosynthesis and bacterial metabolism of 15-hydroxylated steroids in the female rat. Eur J Biochem 1971;19(3):433–41.

Falk RT, Gentzschein E, Stanczyk FZ, Brinton LA, Garcia-Closas M, Ioffe OB, et al. Measurement of sex steroid hormones in breast adipocytes: methods and implications. Cancer Epidemiol Biomarkers Prev 2008;17(8):1891–5.

Fernandez-Peralbo MA, de Castro MD. Preparation of urine samples prior to targeted or untargeted metabolomics mass spectrometry analysis. Trends Anal Chem 2012;41:75–85.

Frey AJ, Park BY, Schriver ER, Feldman DR, Parry S, Croen LA, et al. Differences in testosterone and its precursors by sex of the offspring in meconium. J Steroid Biochem Mol Biol 2017;167: 78–85.

Gibson EJ, Bucknall MP, Golebiowski B, Stapleton F. Comparative limitations and benefits of liquid chromatography—mass spectrometry techniques for analysis of sex steroids in tears. Exp Eye Res 2019;179:168–78.

Gika HG, Theodoridis GA, Plumb RS, Wilson ID. Current practice of liquid chromatography-mass spectrometry in metabolomics and metabonomics. J Pharm Biomed Anal 2014;87:12–25.

Glickman SE, Short RV, Renfree MB. Sexual differentiation in three unconventional mammals: spotted hyenas, elephants and tammar wallabies. Horm Behav 2005;48(4):403–17.

Gomez-Gomez A, Pozo OJ. Determination of steroid profile in hair by liquid chromatography tandem mass spectrometry. J Chromatogr A 2020;1624, 461179.

Gonzalez-Riano C, Garcia A, Barbas C. Metabolomics studies in brain tissue: a review. J Pharm Biomed Anal 2016;130:141–68.

Görög S. The paradigm shifting role of chromatographic methods in pharmaceutical analysis. J Pharm Biomed Anal 2012;69:2–8.

Greaves RF, Jevalikar G, Hewitt JK, Zacharin MR. A guide to understanding the steroid pathway: new insights and diagnostic implications. Clin Biochem 2014;47(15):5–15.

Greep RO, Deane HW. Histological, cytochemical and physiological observations on the regeneration of the rat's adrenal gland following enucleation. Endocrinology 1949;45(1):42–56.

Greff MJE, Levine JM, Abuzgaia AM, Elzagallaai AA, Rieder MJ, van Uum SHM. Hair cortisol analysis: an update on methodological considerations and clinical applications. Clin Biochem 2019;63:1–9.

Gröschl M. Saliva: a reliable sample matrix in bioanalytics. Bioanalysis 2017;9(8):655–68.

Guo S, Tang W, Hu Y, Chen Y, Gordon A, Li B, et al. Enhancement of on-tissue chemical derivatization by laser-assisted tissue transfer for MALDI MS imaging. Anal Chem 2020;92(1):1431–8.

Gurumurthy CB, Lloyd KCK. Generating mouse models for biomedical research: technological advances. Dis Model Mech 2019;12(1). pii: dmm029462.

Hanahan D, Wagner EF, Palmiter RD. The origins of oncomice: a history of the first transgenic mice genetically engineered to develop cancer. Genes Dev 2007;21(18):2258–70.

Heckmann M, Hartmann MF, Kampschulte B, Gack H, Bödeker RH, Gortner L, et al. Assessing cortisol production in preterm infants: do not dispose of the nappies. Pediatr Res 2005;57(3):412–8.

Henley DE, Leendertz JA, Russell GM, Wood SA, Taheri S, Woltersdorf WW, et al. Development of an automated blood sampling system for use in humans. J Med Eng Technol 2009;33(3):199–208.

Hernandes VV, Barbas C, Dudzik D. A review of blood sample handling and pre-processing for metabolomics studies. Electrophoresis 2017;38(18):2232–41.

Heussner K, Rauh M, Cordasic N, Menendez-Castro C, Huebner H, Ruebner M, et al. Adhesive blood microsampling systems for steroid measurement via LC-MS/MS in the rat. Steroids 2017; 120:1–6.

Higashi T, Yamagata K, Kato Y, Ogawa Y, Takano K, Nakaaze Y, et al. Methods for determination of fingernail steroids by LC/MS/MS and differences in their contents between right and left hands. Steroids 2016;109:60–5.

Hirsch A, Hahn D, Kempná P, Hofer G, Nuoffer JM, Mullis PE, et al. Metformin inhibits human androgen production by regulating steroidogenic enzymes HSD3B2 and CYP17A1 and complex I activity of the respiratory chain. Endocrinology 2012;153(9): 4354–66.

Ho S, Gao H. Surrogate matrix: opportunities and challenges for tissue sample analysis. Bioanalysis 2015;7(18):2419–33.

Hoggatt J, Hoggatt AF, Tate TA, Fortman J, Pelus LM. Bleeding the laboratory mouse: not all methods are equal. Exp Hematol 2016;44(2):132–137.e1.

Holliday DL, Speirs V. Choosing the right cell line for breast cancer research. Breast Cancer Res 2011;13(4):215.

Hsu HJ, Hsu NC, Hu MC, Chung BC. Steroidogenesis in zebrafish and mouse models. Mol Cell Endocrinol 2006;248(1–2):160–3.

Hsu HJ, Lin JC, Chung BC. Zebrafish cyp11a1 and hsd3b genes: structure, expression and steroidogenic development during embryogenesis. Mol Cell Endocrinol 2009;312(1–2):31–4.

Hülsemann F, Fußhöller G, Lehn C, Thevis M. Excretion of 19-norandrosterone after consumption of boar meat. Drug Test Anal 2020;12(11–12):1581–6.

Jarque S, Bittner M, Hilscherová K. Freeze-drying as suitable method to achieve ready-to-use yeast biosensors for androgenic and estrogenic compounds. Chemosphere 2016;148:204–10.

Järvenpää P, Kosunen T, Fotsis T, Adlercreutz H. In vitro metabolism of estrogens by isolated intestinal micro-organisms and by human faecal microflora. J Steroid Biochem 1980;13(3):345–9.

Jerjes WK, Cleare AJ, Peters TJ, Taylor NF. Circadian rhythm of urinary steroid metabolites. Ann Clin Biochem 2006;43(Pt 4):287–94.

Kaur G, Dufour JM. Cell lines: valuable tools or useless artifacts. Spermatogenesis 2012;2(1):1–5.

Keiler AM, Zierau O, Wolf S, Diel P, Schänzer W, Vollmer G, et al. Androgen- and estrogen-receptor mediated activities of 4-hydroxytestosterone, 4-hydroxyandrostenedione and their human metabolites in yeast based assays. Toxicol Lett 2018;292:39–45.

Kim J, Abdulwahab S, Choi K, Lafrenière NM, Mudrik JM, Gomaa H, et al. A microfluidic technique for quantification of steroids in core needle biopsies. Anal Chem 2015;87(9):4688–95.

Kirwan JA, Brennan L, Broadhurst D, Fiehn O, Cascante M, Dunn WB, et al. Preanalytical processing and biobanking procedures of biological samples for metabolomics research: a white paper, community perspective (for "Precision Medicine and Pharmacometabolomics Task Group"-The Metabolomics Society initiative). Clin Chem 2018;64(8):1158–82.

Kopylov AT, Malsagova KA, Stepanov AA, Kaysheva AL. Diversity of plant sterols metabolism: the impact on human health, sport, and accumulation of contaminating sterols. Nutrients 2021;13(5):1623.

Koskenniemi JJ, Virtanen HE, Toppari J. Testicular growth and development in puberty. Curr Opin Endocrinol Diabetes Obes 2017;24(3):215–24.

Kurogi K, Liu TA, Sakakibara Y, Suiko M, Liu MC. The use of zebrafish as a model system for investigating the role of the SULTs in the metabolism of endogenous compounds and xenobiotics. Drug Metab Rev 2013;45(4):431–40.

Laparre J, Kaabia Z, Mooney M, Buckley T, Sherry M, Le Bizec B, et al. Impact of storage conditions on the urinary metabolomics fingerprint. Anal Chim Acta 2017;951:99–107.

Lee AV, Oesterreich S, Davidson NE. MCF-7 cells—changing the course of breast cancer research and care for 45 years. J Natl Cancer Inst 2015;107(7):djv073.

Leihy MW, Shaw G, Wilson JD, Renfree MB. Penile development is initiated in the tammar wallaby pouch young during the period when 5alpha-androstane-3alpha,17beta-diol is secreted by the testes. Endocrinology 2004;145(7):3346–52.

Lerman LO, Kurtz TW, Touyz RM, Ellison DH, Chade AR, Crowley SD, et al. Animal models of hypertension: a scientific statement from the American Heart Association. Hypertension 2019;73(6):e87–e120.

Lippi G, von Meyer A, Cadamuro J, Simundic AM. European Federation of Clinical Chemistry and Laboratory Medicine (EFLM) working Group for Preanalytical Phase (WG-PRE). PREDICT: a checklist for preventing preanalytical diagnostic errors in clinical trials. Clin Chem Lab Med 2020;58(4):518–26.

Loftfield E, Vogtmann E, Sampson JN, Moore SC, Nelson H, Knight R, et al. Comparison of collection methods for fecal samples for discovery metabolomics in epidemiologic studies. Cancer Epidemiol Biomarkers Prev 2016;25(11):1483–90.

Luu-The V, Bélanger A, Labrie F. Androgen biosynthetic pathways in the human prostate. Best Pract Res Clin Endocrinol Metab 2008;22(2):207–21.

Maeda N, Tanaka E, Suzuki T, Okumura K, Nomura S, Miyasho T, et al. Accurate determination of tissue steroid hormones, precursors and conjugates in adult male rat. J Biochem 2013;153(1):63–71.

Mallon AM, Iyer V, Melvin D, Morgan H, Parkinson H, Brown SD, et al. Accessing data from the international mouse phenotyping consortium: state of the art and future plans. Mamm Genome 2012;23(9–10):641–52.

Mangelis A, Jühlen R, Dieterich P, Peitzsch M, Lenders JWM, Hahner S, et al. A steady state system for in vitro evaluation of steroidogenic pathway dynamics: application for CYP11B1, CYP11B2 and CYP17 inhibitors. J Steroid Biochem Mol Biol 2019;188:38–47.

McManus JM, Sharifi N. Structure-dependent retention of steroid hormones by common laboratory materials. J Steroid Biochem Mol Biol 2019;198:105572. https://doi.org/10.1016/j.jsbmb.2019.105572 [Epub ahead of print] 31883923.

Mellon SH, Compagnone N, Sander M, Cover C, Ganten D, Djavidani B. Rodent models for studying steroids and hypertension: from fetal development to cells in culture. Steroids 1995;60(1):59–64.

Mertl J, Kirchnawy C, Osorio V, Grininger A, Richter A, Bergmair J, et al. Characterization of estrogen and androgen activity of food contact materials by different in vitro bioassays (YES, YAS, ERα and AR CALUX) and chromatographic analysis (GC-MS, HPLC-MS). PLoS One 2014;9(7), e100952.

Nagao S, Iwata N, Soejima Y, Takiguchi T, Aokage T, Kozato Y, et al. Interaction of ovarian steroidogenesis and clock gene expression modulated by bone morphogenetic protein-7 in human granulosa cells. Endocr J 2019;66(2):157–64.

Neunzig J, Milhim M, Schiffer L, Khatri Y, Zapp J, Sánchez-Guijo A, et al. The steroid metabolite 16(β)-OH-androstenedione generated by CYP21A2 serves as a substrate for CYP19A1. J Steroid Biochem Mol Biol 2017;167:182–91.

Nikolakis G, Stratakis CA, Kanaki T, Slominski A, Zouboulis CC. Skin steroidogenesis in health and disease. Rev Endocr Metab Disord 2016;17(3):247–58.

Nishi Y, Yanase T, Mu Y, Oba K, Ichino I, Saito M, et al. Establishment and characterization of a steroidogenic human granulosa-like tumor cell line, KGN, that expresses functional follicle-stimulating hormone receptor. Endocrinology 2001;142(1):437–45.

Noroozzadeh M, Behboudi-Gandevani S, Zadeh-Vakili A, Ramezani Tehrani F. Hormone-induced rat model of polycystic ovary syndrome: a systematic review. Life Sci 2017;191:259–72.

Nozaki M, Haraguchi S, Miyazaki T, Shigeta D, Kano N, Lei XF, et al. Expression of steroidogenic enzymes and metabolism of steroids in COS-7 cells known as non-steroidogenic cells. Sci Rep 2018;8(1):2167.

Oakes JA, Li N, Wistow BRC, Griffin A, Barnard L, Storbeck KH, et al. Ferredoxin 1b deficiency leads to testis disorganization, impaired spermatogenesis and feminization in zebrafish. Endocrinology 2019. https://doi.org/10.1210/en.2019-00068. pii: en.2019-00068. [Epub ahead of print] 31/08/2019.

Odermatt A, Strajhar P, Engeli RT. Disruption of steroidogenesis: cell models for mechanistic investigations and as screening tools. J Steroid Biochem Mol Biol 2016;158:9–21.

Ostler MW, Porter JH, Buxton OM. Dried blood spot collection of health biomarkers to maximize participation in population studies. J Vis Exp 2014;(83), e50973.

Palme R. Measuring fecal steroids: guidelines for practical application. Ann N Y Acad Sci 2005;1046:75–80.

Parasuraman S, Raveendran R, Kesavan R. Blood sample collection in small laboratory animals. J Pharmacol Pharmacother 2010;1(2):87–93.

Pinto CL, Markey K, Dix D, Browne P. Identification of candidate reference chemicals for in vitro steroidogenesis assays. Toxicol In Vitro 2018;47:103–19.

Poli G, Sarchielli E, Guasti D, Benvenuti S, Ballerini L, Mazzanti B, et al. Human fetal adrenal cells retain age-related stem- and endocrine-differentiation potential in culture. FASEB J 2019;33:2263–77.

Ponzetto F, Settanni F, Parasiliti-Caprino M, Rumbolo F, Nonnato A, Ricciardo M, et al. Reference ranges of late-night salivary cortisol and cortisone measured by LC-MS/MS and accuracy for the diagnosis of Cushing's syndrome. J Endocrinol Invest 2020;43(12):1797–806.

Protti M, Mandrioli R, Mercolini L. Tutorial: volumetric absorptive microsampling (VAMS). Anal Chim Acta 2019;1046:32–47.

Rahman NA, Huhtaniemi IT. Testicular cell lines. Mol Cell Endocrinol 2004;228(1–2):53–65.

Rainey WE, Saner K, Schimmer BP. Adrenocortical cell lines. Mol Cell Endocrinol 2004;228(1–2):23–38.

Rathkolb B, Fuchs H, Gailus-Durner V, Aigner B, Wolf E, Hrabě de Angelis M. Blood collection from mice and hematological analyses on mouse blood. Curr Protoc Mouse Biol 2013a;3(2):101–19.

Rathkolb B, Hans W, Prehn C, Fuchs H, Gailus-Durner V, Aigner B, et al. Clinical chemistry and other laboratory tests on mouse plasma or serum. Curr Protoc Mouse Biol 2013b;3(2):69–100.

Remer T, Neubert A, Maser-Gluth C. Anthropometry-based reference values for 24-h urinary creatinine excretion during growth and their use in endocrine and nutritional research. Am J Clin Nutr 2002;75(3):561–9.

Remer T, Montenegro-Bethancourt G, Shi L. Long-term urine biobanking: storage stability of clinical chemical parameters under moderate freezing conditions without use of preservatives. Clin Biochem 2014;47(18):307–11.

Richer G, Baert Y, Goossens E. In-vitro spermatogenesis through testis modelling: toward the generation of testicular organoids. Andrology 2019. https://doi.org/10.1111/andr.12741 [published online ahead of print, 2019].

Rizner TL, Adamski J. Paramount importance of sample quality in preclinical and clinical research-need for standard operating procedures (SOPs). J Steroid Biochem Mol Biol 2019;186:1–3.

Rosen B, Schick J, Wurst W. Beyond knockouts: the international knockout mouse consortium delivers modular and evolving tools for investigating mammalian genes. Mamm Genome 2015;26(9–10):456–66.

Ruiz-Babot G, Balyura M, Hadjidemetriou I, Ajodha SJ, Taylor DR, Ghataore L, et al. Modeling congenital adrenal hyperplasia and testing interventions for adrenal insufficiency using donor-specific reprogrammed cells. Cell Rep 2018;22(5):1236–49.

Ryan D, Robards K, Prenzler PD, Kendall M. Recent and potential developments in the analysis of urine: a review. Anal Chim Acta 2011;684 (1–2):8–20.

Saglam-Metiner P, Gulce-Iz S, Biray-Avci C. Bioengineering-inspired three-dimensional culture systems: organoids to create tumor microenvironment. Gene 2019;686:203–12.

Savchuk I, Morvan ML, Antignac JP, Gemzell-Danielsson K, Le Bizec B, Söder O, et al. Androgenic potential of human fetal adrenals at the end of the first trimester. Endocr Connect 2017a;6(6):348–59.

Savchuk I, Morvan ML, Søeborg T, Antignac JP, Gemzell-Danielsson K, Le Bizec B, et al. Resveratrol inhibits steroidogenesis in human fetal adrenocortical cells at the end of first trimester. Mol Nutr Food Res 2017b;61(2).

Savchuk I, Morvan ML, Antignac JP, Gemzell-Danielsson K, Le Bizec B, Söder O, et al. The human genital tubercle is steroidogenic organ at early pregnancy. Mol Cell Endocrinol 2018;477:148–55.

Shackleton CH, Gustafsson JA, Sjövall J. The identification of epimeric 5-androstene-3, 16, 17-triols in plasma from the umbilical cord and in meconium, feces and urine from infants. Steroids 1970;15(1):131–7.

Simundic AM, Bölenius K, Cadamuro J, Church S, Cornes MP, van Dongen-Lases EC, et al, Working Group for Preanalytical Phase (WG-PRE), of the European Federation of Clinical Chemistry and Laboratory Medicine (EFLM) and Latin American working Group for Preanalytical Phase (WG-PRE-LATAM) of the Latin America Confederation of Clinical Biochemistry (COLABIOCLI). Joint EFLM-COLABIOCLI recommendation for venous blood sampling. Clin Chem Lab Med 2018;56(12):2015–38.

Singh GK, Balzer BW, Desai R, Jimenez M, Steinbeck KS, Handelsman DJ. Requirement for specific gravity and creatinine adjustments for urinary steroids and luteinizing hormone concentrations in adolescents. Ann Clin Biochem 2015;52(Pt 6):665–71.

Skardal A, Aleman J, Forsythe S, et al. Drug compound screening in single and integrated multi-organoid body-on-a-chip systems. Biofabrication 2020;12(2), 025017.

Slominski AT, Manna PR, Tuckey RC. On the role of skin in the regulation of local and systemic steroidogenic activities. Steroids 2015;103:72–88.

Snaterse G, van Dessel LF, Taylor AE, Visser JA, Arlt W, Lolkema MP, et al. Validation of circulating steroid hormone measurements across different matrices by liquid chromatography-tandem mass spectrometry. Steroids 2021;167, 108800.

Stalder T, Kirschbaum C. Analysis of cortisol in hair—state of the art and future directions. Brain Behav Immun 2012;26(7):1019–29.

St Pyrek J. Constituents of human meconium—III. Identification of 3,20-dihydroxypregnan-21-oic acids. Steroids 1987;49(4–5):313–33.

Takeda R, Morimoto S, Miyabo S, Murakami M. A study of adrenal regeneration hypertension: effect of the delayed adrenal enucleation on the development of hypertension and salt appetite. Endocrinol Jpn 1964;11:19–24.

Taves MD, Ma C, Heimovics SA, Saldanha CJ, Soma KK. Measurement of steroid concentrations in brain tissue: methodological considerations. Front Endocrinol (Lausanne) 2011;2:39.

Thevis M, Geyer H, Tretzel L, Schänzer W. Sports drug testing using complementary matrices: advantages and limitations. J Pharm Biomed Anal 2016;130:220–30.

Thiry J, Evrard B, Nys G, Fillet M, Kok MGM. Sampling only ten microliters of whole blood for the quantification of poorly soluble drugs: itraconazole as case study. J Chromatogr A 2017;1479:161–8.

Titman A, Price V, Hawcutt D, Chesters C, Ali M, Cacace G, et al. Salivary cortisol, cortisone and serum cortisol concentrations are related to age and body mass index in healthy children and young people. Clin Endocrinol (Oxf) 2020;93(5):572–8.

Tokarz J, Möller G, de Angelis MH, Adamski J. Zebrafish and steroids: what do we know and what do we need to know? J Steroid Biochem Mol Biol 2013;137:165–73.

Török D, Mühl A, Votava F, Heinze G, Sólyom J, Crone J, et al. Stability of 17alpha-hydroxyprogesterone in dried blood spots after autoclaving and prolonged storage. Clin Chem 2002;48(2):370–2.

Tremblay PG, Sirard MA. Transcriptomic analysis of gene cascades involved in protein kinase A and C signaling in the KGN line of human ovarian granulosa tumor cells†. Biol Reprod 2017;96(4):855–65.

Tuck MK, Chan DW, Chia D, Godwin AK, Grizzle WE, Krueger KE, et al. Standard operating procedures for serum and plasma collection: early detection research network consensus statement standard operating procedure integration working group. J Proteome Res 2009;8(1): 113–7.

Vinson GP, Hinson JP, Raven PW. The relationship between tissue preparation and function; methods for the study of control of aldosterone secretion: a review. Cell Biochem Funct 1985;3(4):235–53.

Walpurgis K, Thomas A, Geyer H, Mareck U, Thevis M. Dietary supplement and food contaminations and their implications for doping controls. Foods 2020;9(8):1012.

Walsh NC, Kenney LL, Jangalwe S, Aryee KE, Greiner DL, Brehm MA, et al. Humanized mouse models of clinical disease. Annu Rev Pathol 2017;12:187–215.

Walters KA, Bertoldo MJ, Handelsman DJ. Evidence from animal models on the pathogenesis of PCOS. Best Pract Res Clin Endocrinol Metab 2018;32(3):271–81.

Walters KA, Eid S, Edwards MC, Thuis-Watson R, Desai R, Bowman M, et al. Steroid profiles by liquid chromatography-mass spectrometry of matched serum and single dominant ovarian follicular fluid from women undergoing IVF. Reprod Biomed Online 2019;38(1): 30–7.

Wang T, Rainey WE. Human adrenocortical carcinoma cell lines. Mol Cell Endocrinol 2012;351(1):58–65.

Wester VL, van Rossum EF. Clinical applications of cortisol measurements in hair. Eur J Endocrinol 2015;173(4):M1–10.

World Health Organization. Guidelines on drawing blood: best practices in phlebotomy; 2010, ISBN:978 92 4 159922 1.

Wudy SA, Hartmann MF, Remer T. Sexual dimorphism in cortisol secretion starts after age 10 in healthy children: urinary cortisol metabolite excretion rates during growth. Am J Physiol Endocrinol Metab 2007;293(4):E970–6.

Yamanashi Y, Teramoto M, Morimura N, Hirata S, Suzuki J, Hayashi M, et al. Analysis of hair cortisol levels in captive chimpanzees: effect of various methods on cortisol stability and variability. MethodsX 2016;3:110–7.

Zhao YG, Zhang Y, Zhan PP, Chen XH, Pan SD, Jin MC. Fast determination of 24 steroid hormones in river water using magnetic dispersive solid phase extraction followed by liquid chromatography-tandem mass spectrometry. Environ Sci Pollut Res Int 2016;23(2): 1529–39.

Zhou Z, Chen Y, He J, Xu J, Zhang R, Mao Y, et al. Systematic evaluation of serum and plasma collection on the endogenous metabolome. Bioanalysis 2017;9(3):239–50.

Zierau O, Kolodziejczyk A, Vollmer G, Machalz D, Wolber G, Thieme D, et al. Comparison of the three SARMs RAD-140, GLPG0492 and GSK-2881078 in two different in vitro bioassays, and in an in silico androgen receptor binding assay. J Steroid Biochem Mol Biol 2019;189:81–6.

Chapter 2.2

Steroid determination—Sample preparation

2.2.1 Introduction

Steroid hormones are produced from cholesterol primarily in the adrenal glands, gonads (testis and ovary), placenta and brain and secondarily in other tissues (mammary, skin, etc.). A steroid has the cyclopentanoperhydrophenanthrene nucleus and in physico-chemical terms is a lipophilic molecule with poor solubility in water (so hydrophobic and non-polar). Modifications of the nucleus or on the side-chain by the addition of hydroxyl or keto groups makes the molecule more hydrophilic (water soluble or lipophobic, polar). The structures of the steroids include molecules of 18, 19 or 21 carbon atoms at least two oxygen functions with zero to three carbonyl groups and zero to five hydroxyl groups (see Chemistry Chapter 1.1).

Steroid hormones are carried in the bloodstream bound to plasma proteins with specific and nonspecific properties. The specific steroid-binding globulins have high affinity but low capacity (e.g., cortisol binding globulin, sex hormone binding globulin) whereas albumin has low affinity but high capacity. In general, the concentrations of steroids in blood serum are determined as the total, after the bound steroid is displaced from binding proteins and combined with the free steroid. The concentrations of the free steroids are not usually determined but methods are available after equilibrium dialysis. Accurate measurements of the low levels of sex steroids in normal children and in women after the menopause is critical and requires understanding of normal development when seeking the diagnosis of steroid disorders. For estradiol, the first generation immunoassays had limits of quantitation (LOQ) of 20 pg/mL and so only covered the range for premenopausal women with normal menstrual cycles. Newer LC-MS/MS methods have LOQ of 10 pg/mL which is not suitable for all prepubertal children. Second generation IA were slightly better with LOQ 6–10 pg/mL. With high specimen volume MS methods can have LOQ to 1 pg/mL. The method of analysis must be considered when using laboratory data in these circumstances both on terms of accuracy and specificity.

The metabolism of endogenous androgens, as well as exogenous androgens used in the context of doping in sport, was elegantly presented by Schänzer in 1996 with detail that is largely applicable to all steroids. Phase 1 metabolism is mainly through enzymes which catalyze the chemical reduction of the steroids (see Metabolism Chapter 1.8). Hormonal steroids with 3-keto-4-ene structure are reduced to 3α-hydroxy metabolites and can also be 5α- or 5β-reduced to dihydro steroids, hence 3α-hydroxy androstane or pregnane steroids. There are further reductions of ketones at C-20 (α and β isomers) and at C-11. Phase 2 metabolism involves conjugation of steroids to assist clearance from the body. Glucuronides and sulfates are the main metabolites. So, there are many more steroids in urine than the hormones in blood.

The analysis of steroids of biological importance to humans requires methods for the determination of the typically very low concentrations of steroids found in body fluids and sometimes in biological tissues or formed during experimental studies in vitro or in vivo. To put the quantities in some context, it is worth looking at how much steroid is involved so (Fig. 2.2.1) a teaspoon amount of testosterone dissolved in an Olympic size swimming pool of water gives a concentration of testosterone seen in adult females with hyperandrogenism.

This book primarily addresses studies in humans but much research is conducted in animals. There are a number of additional factors to consider then which have been addressed in a recent review called "A users guide to HPA axis research" (Spencer and Deak, 2017). That review encourages the researcher, using rats, mice and other species, to consider many issues from sourcing of animals, to transport, housing, acclimatization to handling and limitations of samples in volume and number.

The excretion rates of steroids in urine from children were first determined by paper chromatography and colorimetric visualization of the steroids. Chemical fractionation was needed to separate the androgens from the corticosteroids before paper chromatography in by-gone years. The chemical methods for the steroid group determination such as 17-oxosteroids,

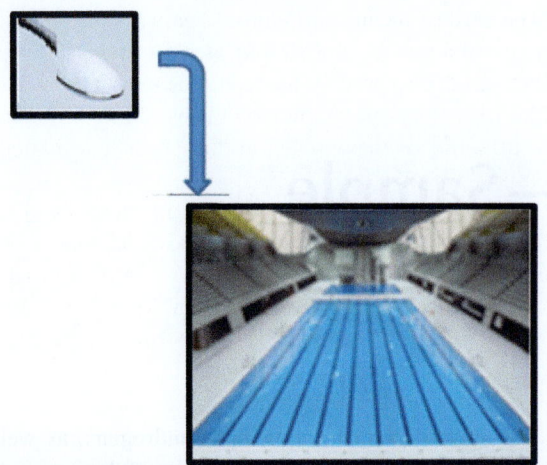

FIG. 2.2.1 Steroids in the blood serum are at low concentrations for example achieved by dissolving a teaspoon of testosterone (c. 3 g) in an Olympic size swimming pool of water (2,500,000 L). This would produce a solution at 4 nmol/L which would be a low concentration in an adult male (typically 10–30 nmol/L) and high in an adult female (normal range 0.5–2.5 nmol/L). *(I am grateful to Prof. J.P. Fry for this picture from his inaugural lecture.)*

ketogenic steroids and 17-hydroxycorticosteroids are not now used globally because the determination of individual steroids and profiles has greater diagnostic power (Nozaki, 2001). Gas chromatography (GC) of steroids was pioneered by the Horning's (Horning et al., 1969) and this has been the mainstay of urinary steroid profiling (USP) until recently. A GC was first coupled with a mass spectrometer (MS) by Ryhage in 1964 with a system (separator) to reduce carrier gas flow to the MS. The introduction of capillary columns overcame this interface problem and the column is now linked directly into the MS. USP with GC-MS has most commonly in clinical laboratories used methyloxime-trimethysilyl ether (MO-TMS) derivatives. Quantitation can employ the peak height or area in total ion chromatograms or specific selected ion monitoring (SIM). The use of SIM increases the sensitivity over the total ion scanning mode.

The determination of individual free steroid concentrations in urine after HPLC separation was achieved with immunoassay (Schöneshöfer and Weber, 1983) of fractions collected of the eluates from the HPLC. This was clearly only a research tool, too lengthy for routine hospital laboratory use. Fluorescent detection of cortisol, cortisone and tetrahydro-metabolites in biological fluids required formation of derivatives such as 9-anthroyl-nitrile before HPLC.

Steroid analysis has advanced considerably in recent years. Since around 1970 it has been possible to measure by immunoassay a steroid in a single sample extract of almost any biological sample with or without a chromatographic separation. Rosalyn Yallow had received the Nobel prize for developing the immunoassay technique which was first introduced for the measurement of insulin.

Concentrations of steroids are especially low in saliva and were not detectable in some immunoassays although Salimetrics is a company that pioneered commercial kits for steroids in saliva. The poor specificity of immunoassay has driven the move to using chromatography combined with mass spectrometry for steroid measurements.

Steroid metabolites formed by hepatic catabolism generally become more hydrophilic by reduction or hydroxylation and esterification (conjugation) with glucuronic or sulfuric acid. Traditionally, conjugates were hydrolyzed before analysis of the liberated steroid, but analysis of intact conjugates is now possible. The analysis of steroid metabolites in urine has a strong place in clinical investigations and steroids such as dehydroepiandrosterone sulfate are being recognized with important activity. The availability of reference steroids, particularly stable isotope labeled steroids, is a limitation to the accurate quantitative analysis of steroid conjugates. Steroids are all, to varying degrees, soluble in organic solvents and can thus be extracted from aqueous media by a solvent or solvent mixture of suitable polarity.

Steroids constitute the major group of drugs abused in sport. The antidoping laboratories have to try to keep ahead of the athletes in detecting abuse, so new methods, instruments and techniques are adopted early by the laboratories and the developments are worth watching for potential application elsewhere such as in clinical work. Steroid residues are now a recognized group of environmental contaminants, poorly removed from conventional waste water treatment. The analysis of sewage, sludge, soils and sediments is important but is not covered in this book.

The dynamics of steroid production in the human body is such that no single measurement is likely to have any particular meaning. The variations already seen include:

- time of day,
- day of month in women of reproductive age,
- age,
- dependence upon trophic hormones,
- sex,
- pregnancy,
- levels of binding proteins and
- rate of clearance from the body.

The clinician may not have total diagnostic confidence in results from a single, base-line sample so a number of provocative tests are used to stimulate or suppress components of an endocrine axis under likely circumstances. These will be addressed later in the book (Chapter 2.7). Hospital laboratories operate under pressure to meet the demands and reduce diagnosis and treatment times. Since the results of the analysis directly affect the clinical decision making process, the quality of the tests must satisfy the rules of regulatory bodies. Automation of key steps in the sample journey is important. Early methods of steroid analysis were directed to quantitative determination of steroids with a common chemical structure (e.g., dihydroxyacetone side chain) then with the introduction of immunoassay to one steroid. The simultaneous measurement of several steroids is now recognized to be more informative in many clinical situations. The technology has advanced in the past 40 years and this chapter describes these developments such that some methods are now rarely used. The reader will be able to chart these changes to the betterment of current methodology. Immunoassay enabled the separate measurement of many steroids and analysis on automated platforms will continue because of the convenience within the work-flow of hospital laboratories but the specificity of immunoassay has been challenged. Manufacturers of the automated platforms have addressed this by searching for more specific antibodies and showing comparability of the results with those obtained by MS methods. Chromatography has improved the ability to separate the steroid components of a sample, resolution was enhanced from relatively poor separation with paper, thin-layer and column chromatography when gas and high pressure liquid chromatography became available. A mass spectrometer as the detection system affords a more specific analysis with greater sensitivity.

The stability of steroids, interferences from related steroids, the ability to measure several steroids in a single analysis, quality control and potential for automation of stages in the analysis are recurring issues addressed in this chapter.

2.2.2 Analysis of steroids in serum and plasma

In any measurement of steroid content of biological materials, the specificity of the final quantitative determination and the requirements of the assay must be taken into consideration. The first question to consider, for a routine test in clinical laboratories for diagnostic purposes, is whether the steroid can be measured directly in the medium without pretreatment, thus avoiding the need for extraction. Immunoassay tests are often now performed with serum, or plasma or urine processed without any extraction or purification of the steroids. In some cases, a nonextraction assay may provide a semiquantitative estimate of the concentration of the analyte that is of value to clinicians. The result may not be accurate, but may be precise and reproducible, and still have value (Makin and Gower, 2010). High specificity, quantitative procedures are possible with minimum prepurification. It is very important to ensure that the omission of purification steps do not compromise the final measurement and ensure that the method is only applied for the situations for which it was developed. Methods developed for one situation should not be applied uncritically to other situations in which they may not be valid.

The earliest methods of steroid analysis were based on chemical reaction of the steroid at a functional group in order to produce a product that could be measured by a colorimetric or fluorometric method (Table 2.2.1). The chemistry of some of these reactions has been considered in Chemistry Chapter 1.1.

These methods were not specific for a steroid and were subject to many interferences. Since these methods are now rarely used they will be reviewed only briefly so that the reader is at least aware of the terminology. The simpler methods depend on the reducing properties of the α-ketol side-chain at C-17. Other more elaborate methods involve liberation of formaldehyde from the side-chain by periodic acid, or a color reaction with phenylhydrazine in the presence of sulfuric acid. Oxidation of the side-chain with sodium bismuthate had the great advantage that it could be done without previous hydrolysis of the conjugates. The Zimmermann reaction is a chemical reaction between an alkaline solution of *meta*-dinitrobenzene and an active methylene group (carbon-16) of 17-ketosteroids; it is then used as the basis of the 17-ketogenic steroid assay test after

TABLE 2.2.1 Colorimetric and fluorimetric assays for steroids.

Reaction name	Function detected	Determination
Zimmerman	17-Ketosteroid reaction with dinitrobenzene	Red color, colorimetry
	3-Keto-pregnene-d4 17,20,21-triol with periodate	Colorimetry
Schryver	Dihydroxyacetone side chain 3,20 keto, 17,21 diol with periodate ferricyanide	Red color, colorimetry
	Dihydroxyacetone side chain. 3,20 keto, 17,21 diol with bismuthate, periodate and Zimmerman	Red color, colorimetry
Norymberski	21-Deoxy-17,20 diols with periodate and Zimmerman	Red color, colorimetry
Porter-Silber	3,20 Keto, 17,21 diol with phenylhydrazine in acid	Yellow color, colorimetry
Mattingly	11-Hydroxycorticosteroids in acid	Fluorescence
Kober (Brown, 1955)	Estrogens with acid and phenol/hydroquinone	Pink color, colorimetry later fluorescence

There have been many modifications for manual assays and early automated methods.

removal of the side chain. Some of those tests were automated in bubble flow systems which are now obsolete.

There are several steps in an analytical procedure (Fig. 2.2.2).

1. extraction steroids from the sample into an organic solvent,
2. further **sample preparation** may include hydrolysis of conjugates as well as **derivative** formation,
3. purification of steroids in the extract usually by **chromatography** (see Chapter 2.3) and
4. quantitative determination (Chapter 2.4) which was mainly by immunoassay but now more commonly by **mass spectrometry**. The **endpoints** can be single steroids, groups of steroids (panels) and profiles.

The analyst needs to be trained in a number of classical skills:

1. Understanding analytical standard (reference material) and traceability
   - Certified reference materials (CRM)
   - Surrogate internal standard
   - Volumetric internal standards

2. Measurement and dispensing
   - the analytical balance and sources of uncertainty in weighing,
   - standard volumetric flasks,
   - glass and micropipettes,
   - loop injectors and
   - syringes.
3. Preparation of solutions for calibration
   - the surrogate matrix,
   - QC samples and
   - Calibration with internal standard.
4. Chromatography
   - stationary phases,
   - mobile phase,
   - sample injection,
   - separation and
   - interfaces to mass spectrometers.
5. Mass analyzers in quantitative analysis

Although blood, urine and saliva have been the biological fluids most often examined for the presence of steroids there is increasing interest in steroids in tissues including brain (so-called neurosteroids). Tissues when frozen can be

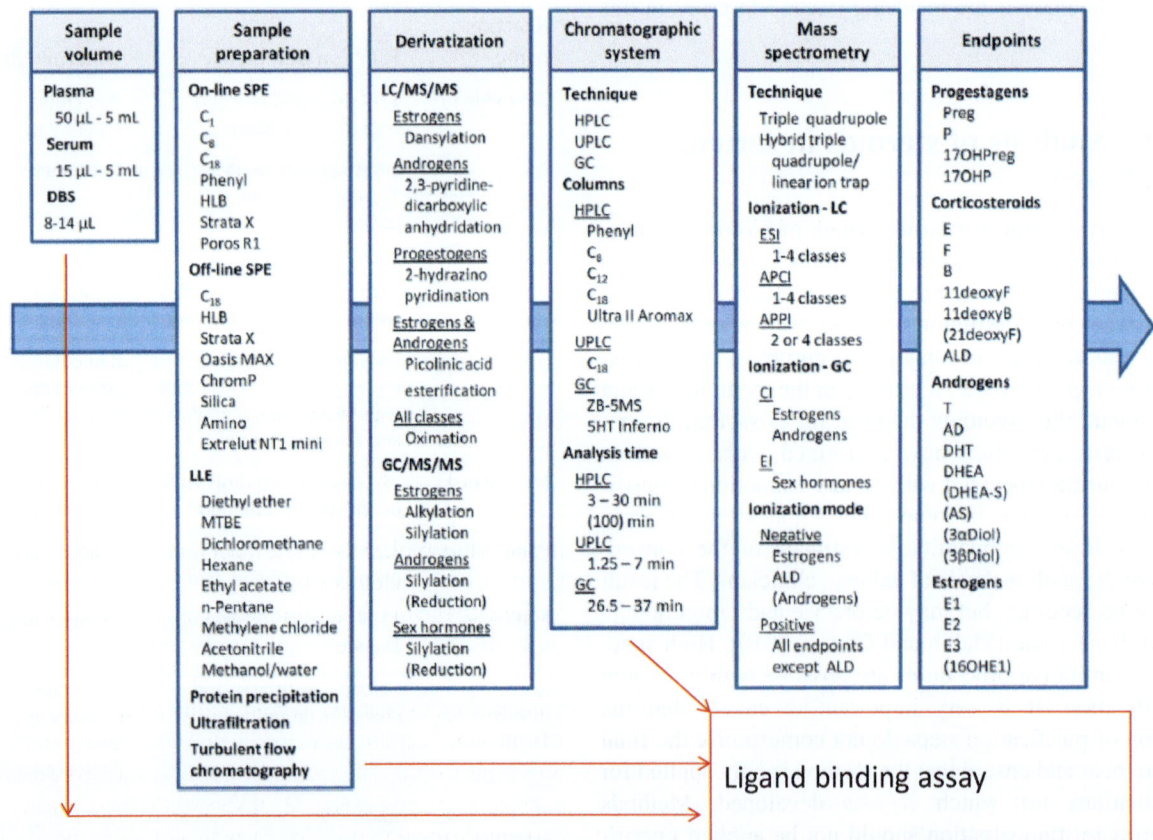

FIG. 2.2.2 Overview of the steps in steroid analysis.

pulverized in a mortar and pestle to a fine powder from which steroids can be extracted into solvents. The vessel can be placed in an ultrasonic bath or an ultrasound probe can be used to further break up tissues. The high lipid content of the tissue has not only made difficult the preparation of extracts sufficiently clean for analysis of the steroids but there are reports of detection and measurement of steroids in brain such as DHEAS and pregnenolone sulfate that were almost certainly artefactual and have to be treated with some reservation without characterization of the steroids to high analytical standards (Liere et al., 2009). In some cases, a combination of techniques is needed to ensure purity of the final product. The steroids in urine are mainly conjugates with glucuronic and sulfuric acid and in the early days the steroids were liberated before any further analysis.

Purification prior to the quantitative determination inevitably affects recovery of the analyte and the more extensive the purification, the greater the losses sustained. Ideally there must therefore in any quantitative method be some means of assessing losses through the extraction and purification procedures. In methods using saturation analysis (competitive binding assays, immunoassays or receptor assays), a radio-labeled analyte is added to the sample matrix in such small quantities that it does not significantly contribute to mass in the final concentration. Tritium ^{3}H-labeled steroids of high specific activity are available that meet this requirement. Fewer steroids are available labeled with ^{14}carbon than tritium and detection required the addition of liquid scintillation fluid. By switching to ^{125}I-radiolabels a gamma counter could be used directly. The cost of disposing radioactive materials led to a further change in labels. Steroid assays using saturation analysis these days use nonisotopic assays such as chemiluminescent and fluorescent labels.

An internal standard or radioactive (or stable isotope labeled) recovery marker which is added to the matrix at the start of the analysis also meets the requirement to assess recovery. There are several factors which must be considered in the selection of internal standards which is of course constrained by the choice of the quantitative procedure to be used. The internal standard after addition to the matrix must be:

- distributed in the same way as the analyte through all elements of the sample in the biological matrix (e.g., bound to any protein) and
- indistinguishable from the steroid analyte during the process of extraction and purification but must be recognizable at the final quantitation stage.

The usual practice is to add the internal standard to the sample then incubate the mixture for a period of time at 37°C, while gently shaking the fluid. This would not be appropriate for a sample such as brain that contains steroid metabolizing enzymes. Steroids are hydrophobic compounds and do not dissolve significantly in aqueous media and are thus added to the matrix dissolved in ethanol, methanol or other polar solvent which is miscible with water. The volume of such solvent should be as small as possible in comparison to the volume of fluid used for analysis to prevent denaturation of any protein. In some instances, the steroid is added to a glass container before the solvent evaporated off and the biological fluid for analysis added. Steroids when added in this way can be adsorbed to the surface of the glass and may not dissolve completely in the aqueous matrix. It is good practice to inactivate the glass by prior silylation through treatment with dimethyldichlorosilane (1%, v/v in cyclohexane). It is important to wash the glassware after to remove acid released during the silanization reaction. Some steroids can be destroyed when evaporated to dryness on soda glass surfaces. The glass should not have been cleaned by treatment with chromic acid as used to be common practice. Steroids that are present in very low concentrations can become susceptible to loss once extracted from the matrix by adsorption to glass surfaces, destruction by metal surfaces and oxidation. A quantity of internal standard larger than the endogenous concentration can prevent or minimize such losses, by acting as a carrier. For GC-MS and LC-MS/MS methods, this process is the so-called **isotope dilution mass spectrometry (IDMS)**. For isotopic internal standards hydrogen, carbon and oxygen atoms in the analyte are replaced with deuterium (^{2}H) or ^{13}C or rarely ^{18}O. If steroids are labeled with two or more deuterium atoms, then the greater the incorporation the larger the LC separation of the two molecules due to differences in physico-chemical properties between the analyte and the stationary phase of the column packing.

The ideal internal standard should have deuterium or carbon-13 labeled forms of the analyte and such standards should contain at least two but preferably three extra stable labeled atoms. If less than three deuterium atoms are incorporated then natural isotopes of the target analyte may contribute to the intensity of the molecular ion of the internal standard and distort the standard curve. The endogenous steroids have two natural isotopes at $M+1$ and $M+2$ and there may be interference between the $M+2$ of the analyte and the M^+ of the IS if twice labeled with deuterium. The internal standard should not itself have $M-1$ or $M-2$ isotopes through incomplete labeling. Carbon 13 is replaced in the hydrocarbon nucleus of the steroid and this does not give changes in the LC separation of the IS and the analyte. Multiple labeling of the steroids (steroids with up to 8 deuterium atoms are being used) and can lead to chromatographic separation from the natural steroid, reducing the carrier effect. In making internal standards deuterium is usually introduced into steroid molecules by acid-catalyzed deuterium exchange. Care must be taken therefore during any extraction or prepurification to avoid acid conditions, which can lead to loss of deuterium by back-exchange.

A **chemical analog** of the analyte can also be used as internal standard, for example Δ^1-testosterone or 19-nor-testosterone has been used in the measurement of testosterone by GC/MS or LC-MS/MS, although the latter can occur naturally. These compounds have physico-chemical properties very similar to testosterone, but can be separated from testosterone in the chromatographic step.

In some research assays an alternative to addition of the standard to the fluid under analysis is to add the analyte to a second sample and process each side by side, the **parallel recovery** being assessed by the difference between the two results. Such procedures are not economical in a clinical setting. The use of two different internal standards is an exceptional possibility. The analyte is quantified by relation to each separately. The analyte concentration should be the same irrespective of which standard is used. If this is not found it suggests that the standard which gives the lower result is not completely resolved from a contaminating peak.

2.2.3 Sample preparation

2.2.3.1 No sample preparation (direct assays)

The first decision in many assay methods for steroids is whether the analysis can be performed on the native sample or an extraction step is needed to get the steroid from the biological matrix to a chemical matrix. In the clinical context "free" or unbound plasma/serum concentrations of steroids may become of value, in the same way that thyroid tests for free T3 and free T4 have been adopted. Measurements of steroids in urine and saliva, which are ultrafiltrates, are indirect methods of assessing free steroid in the circulation although a number of problems have been identified in salivary testosterone measurements. For many steroid assays, today serum or plasma is used directly in platform methods with immunoassay. The total steroid content is determined when the free steroid and the protein bound steroids are released into the sample matrix before the analysis. Heat or protein precipitation will denature the proteins and release bound steroids. Alternatively, a steroid or other reagent with stronger binding to the proteins can be added to the sample. This technique is used in many methods although if using a commercial kit, the actual steroid used is not disclosed. In fact, commercial assays keep confidential most of the detail of the methods, which makes difficult any trouble-shooting problems. Danazol, and 8-anilino-naphthalene sulfonic acid (ANSA) are general agents for displacing steroids from binding proteins. Extremes of pH are also effective for steroid displacement (pH 4 or pH 10). Heat denaturation and enzyme digestion of proteins have also been used. Dihydrotestosterone or a synthetic steroid like mesterolone, is used for more specific displacement of testosterone and estradiol from SHBG. Cortisol and danazol will displace progesterone from cortisol binding globulin. Typical concentrations of these steroid displacing agents are 200–400 ng per tube (micromolar) which is adequate even when CBG concentrations are raised above normal. The synthetic steroids and cortisol, when used, need to have low cross-reaction with the antisera used in the assay.

2.2.3.2 Ultrafiltration/equilibrium dialysis (free steroids)

For actual separation of the free steroid fraction in serum some sort of equilibrium dialysis or ultrafiltration are still the methods of choice today, usually combined with sensitive GC-MS as the concentrations involved are at least an order of magnitude lower than total concentrations (Faix, 2013; Södergård et al., 1982). Assays which measure the serum concentrations of free steroid hormone need to be extremely sensitive and are not generally available. Total steroid concentrations in serum are very much easier to measure but the steroids have to be displaced from binding proteins. Ultrafiltration is the best method for isolating **free steroids** from plasma and many devices are commercially available.

An indirect measurement of free testosterone (FT) is accomplished by adding ^{3}H-T to the sample and after equilibrium has been attained measuring free ^{3}H-T. The amount of total testosterone (TT) is determined in a separate assay of the same plasma. The fraction of free ^{3}H-T is multiplied by the amount of TT. The assay is compromised by the purity of the ^{3}H-T and the accuracy of the TT measurement (Keevil and Adaway, 2019; Ray et al., 2012; Kirchhoff et al., 2011).

The isolation of free steroids does not fit in with hospital laboratory work-flow patterns. The alternatives are to determine the binding protein concentrations and calculate a free steroid component of the measured total steroid concentration using the Law of Mass Action. The calculation depends on the measurement of TT, total SHBG and total albumin and the use of equilibrium dissociation constants (Kd) for the binding of SHBG with T and albumin with T. The **free androgen index (FAI)** is the ratio of T/SHBG with no units. The correlation of FAI and FT is less good at low concentrations because T measurement is imprecise. **Bioavailable-T** is measured by adding ^{3}H-T to serum and precipitating ^{3}H-T with ammonium sulfate. The fraction of ^{3}H-T not precipitated is used to calculate Bio-T.

For testosterone, the simple equation will be:

$$TT = FT + Albumin - bound\ T + SHBG - bound\ T$$

The calculation needs some assumptions:

- Binding constant for albumin is 3.6×10^4
- Binding constant for SHBG is 10^9.

The calculations can be tedious. For worked examples, see Vermeulen et al. (1999).

Similar tests can be used for measurements of free estradiol and cortisol (Ray et al., 2012; Kirchhoff et al., 2011). There are technical and fundamental limitations of all measures and there remains a degree of arbitrariness. The tests can be improved by using stable isotope labeled steroids with mass spectrometric measurement of the isotope dilution (Ray et al., 2012; Kirchhoff et al., 2011), and equilibrium dialysis can be automated. The measurements of steroids in saliva and urine are indices of free hormones in the blood because the fluids are in essence ultrafiltrates of the serum.

2.2.3.3 Extraction

Most analytical methods for steroids can be improved by some form of extraction and/or prepurification before quantitation. Over the past six decades the methods employed have changed considerably, particularly as the toxicities of reagents were considered. The use of solvents such as benzene and toluene will no longer be tolerated. Liquid and solid means have been employed. The extraction step may serve as a partial purification of the sample.

The ideal extraction solvent must do two things, it must firstly extract the steroid of interest quantitatively and leave behind in the aqueous medium other steroids and nonspecific interfering substances and secondly it must totally disrupt the binding of the steroid to protein. For the quantitative extraction of neutral unconjugated steroids from aqueous media such as urine, bile, plasma/serum, saliva, in vitro incubation media, etc., the addition of an equal volume or excess of polar solvent such as ethanol, methanol, acetone or methyl cyanide to the aqueous phase is very effective. In practice such solvent extractions will also extract a number of other steroids of similar polarity and thus similar structure to the analyte which may well interfere in the final quantitation.

It may then be necessary to include purification steps prior to the quantitative analysis. Some steroids are incorporated into the lipoproteins and it may be necessary to add ammonium sulfate or other chemicals to disrupt the lipoprotein structure prior to extraction. A 0.5 M ammonium acetate (pH 5.5) solution can be used to dissociate steroids from binding proteins.

2.2.3.3.1 Protein precipitation

Sample preparation techniques including protein precipitation are usually deemed necessary to reduce blockages in subsequent steps and "noise" in the analytical stage, which in certain maneuvers are called matrix effects. Protein precipitation works well for steroids that are in high concentration in the fluid such as DHEAS and cortisol. The following protein precipitant solutions are typically used:

- aqueous ammonium sulfate (saturated at room temperature),
- aqueous aluminum chloride (5%, w/v),
- aqueous m-phosphoric acid (5%, w/v),
- aqueous trichloroacetic acid (TCA) (10%, w/v), (1:1),
- aqueous zinc sulfate heptahydrate (10%, w/v),
- 0.5 N sodium hydroxide,
- acetonitrile,
- ethanol and
- methanol.

The precipitant is typically added to plasma in volume ratios of 2:1. Solutions are vortexed for 20 s, left to stand for 20 min (acid precipitated solutions are refrigerated at 4°C) and centrifuged for 10 min at 3000g (3000 rpm if rotor arm length: 12 cm). Zinc sulfate, acetonitrile and trichloroacetic acid are most effective. The three precipitants consistently remove in excess of 90% of plasma proteins (Polson et al., 2003). There are times when acetonitrile was in short supply; methanol is recommended as the best substitute for extraction and chromatography particularly if phospholipid interference is a problem.

The different protein precipitation techniques (organic solvent, acid, salt and metal ion) have different modes of action. Protein solubility results from polar interactions with the aqueous solvent, ionic interactions with salts and repulsive electrostatic forces between like charged molecules. At the isoelectric point (pI), there is no net charge on a protein, and consequently the protein has minimum solubility in aqueous solvent. Above the pI, a protein has a net negative charge while below its pI, it has a net positive charge. Precipitants exert specific effects on proteins to facilitate their precipitation from solution. Organic solvent precipitants lower the dielectric constant of the plasma protein solution, which increases the attraction between charged molecules and facilitates electrostatic protein interactions. The organic solvent also displaces the ordered water molecules around the hydrophobic regions on the protein surface. Hydrophobic interactions between proteins are minimized as a result of the surrounding organic solvent, while electrostatic interactions become predominant and lead to protein aggregation. Acidic reagents form insoluble salts with the positively charged amino acid groups of the protein molecules at pH's below their pI. Proteins are precipitated from solutions with high salt concentrations as the salt ions become hydrated and the available water molecules decrease, drawing the water away from the protein hydrophobic surface regions which in turn results in aggregation of protein molecules via protein-protein hydrophobic interactions. The binding of positively charged metal ions reduces protein solubility by changing its isoelectric point (pI). Metal ions are in competition with solution protons for the coordination sites on the exposed amino acids. The stronger binding metal ions displace the protons from the binding sites, resulting in a lowering of the solution pH. The combination of changing the protein's p*I* and lowering the pH generally succeeds in precipitation of proteins. For an LC-MS/MS method, a combination of protein precipitation and solid phase extraction may be needed to reduce matrix effects (Yuan et al., 2020).

2.2.3.3.2 Solvent extraction

Steroid hormones circulate bound to albumin and/or specific binding globulins and it is generally accepted that the physiologically active steroid is the unbound portion which comprises around 10% or less of the total. Steroids may also need to be measured in tissues. If an internal standard (IS) is being used in the analysis this should be added to the sample before the extraction step. If the IS needs to be dissolved in ethanol then when it is added to the sample, especially serum, there is only a low concentration of aqueous ethanol so that protein is not precipitated before equilibration has been achieved. Solvent extraction, with or without the use of protein denaturing agents and/or binding displacers disrupts the binding and thus extracts the total steroid.

There is no single procedure which efficiently and selectively removes all the steroids from the medium in which they are found because of the occurrence of the range of polarities of steroids from estrogens, to the corticosteroids. A single secreted steroid may be metabolized to several metabolites of differing degrees of polarity and may also be conjugated with polar acids to assist in excretion. The reverse may happen, and steroid polarity may be decreased by esterification of hydroxyl groups with fatty acids. For example, cholesterol esters are important constituents of the lipoproteins. Esters of estrogens have also been described and the process of de-esterification, saponification by incubation with alkali, usually KOH, is widely used as a first step in the extraction of a number of steroids from food and other material.

The wide variety of steroids and the metabolism, conjugation and/or esterification which they can undergo is not the only problem facing the analyst. Choice of method must also take account of the sample medium selected for analysis. The physical characteristics of the sample matrix are clearly of importance and methods adopted for the analysis of steroids in human urine will not be appropriate for example, in the analysis of a steroid in sunflower oil or metabolites of corticosteroids in serum or plasma.

Organic solvents are used for liquid-liquid extractions. Health and safety now limits the use and disposal of toxic and/or flammable solvents. There can be problems with emulsion formation on the surface or interface of the liquids. Dichloromethane, diethyl ether, ethyl acetate, chloroform and methyl tertiary butyl ether (MTBE) have been used for extraction of the steroids. Salts and phospholipids do not partition into the organic solvent. Progesterone may be extracted from plasma or serum with hexane or petroleum ether (40–60°C boiling point). Testosterone and estradiol are usually extracted with diethyl ether and cortisol with dichloromethane. Careful choice of solvents can provide a considerable degree of selectivity. A solvent extraction will not discriminate between steroids of similar polarity although it is possible to achieve quite simple separations by judicious choice of solvent. There are numerous examples of such solvent selectivity in the literature, the use of isopentane to separate etiocholanolone (96% extracted) from 11β-hydroxyetiocholanolone (2.5% extracted) (Few, 1968) and the separation of estradiol and estrone (>95% extracted) from estriol (4% extracted) using benzene:light petroleum (1:1, v/v) (Brown, 1955). Such steps are rarely used now. A method of solvent extraction by ion pairing of testosterone and epi-testosterone glucuronides from 3 mL of urine using 1-butyl-3-methylimidazolium chloride (a hydrophilic ionic liquid) (0.2 g) and dipotassium hydrogen phosphate(3.4 g) (He et al., 2005) has been replaced with solid phase extraction (Zhan et al., 2011). The efficiency of extraction of steroid sulfates using ion-pairing (Cawley et al., 2005) was evaluated with a number of reagents but this is now replaced with SPE technology.

Solvents of high quality should be used but may need to be purified before use if high blanks are found in the detection system whether that be immunoassay or mass spectrometry. Fresh ether should be used, even when this solvent is supplied with inhibitors of peroxide formation. A wash with 5% (w/v) ferrous sulfate will remove peroxides. Where solvent extraction is used to recover steroids from serum, the extraction efficiency should exceed 95%. This can be checked by adding radioactive steroid to samples and counting the recovered radioactivity. A liquid-liquid extraction often results in emulsion formation requiring a centrifugation step to break, which is a time-consuming step.

Solvent extraction generally precludes the use of glassware and silicone grease though this is less often used today. Considerable care must be taken with all plastics since the occurrence of plasticizers (phthalates) from the manufacturing process may get into in extracts and interfere in the final detection particularly if mass spectrometry is used (see Chapter 2.4). When processing a sample extract dried into a plastic tube a common practice is to add a small amount of polar solvent (ethanol, methanol, etc.), which is totally miscible with water, to dissolve the steroid prior to addition of the aqueous medium or alternatively that the standard is added directly to the aqueous medium in a suitable solvent in sufficiently small volumes not to disrupt any binding or to denature any enzyme. Glassware may also present other hazards and it is not advisable to evaporate solvents and leave small quantities of steroids in the dry state for long periods.

Removal of solvents

Solvents used for a liquid-liquid extraction, such as ether, can be removed after separation of the phases by immersion of the extraction tubes in a freezing mixture (e.g., ethanol or acetone in solid carbon dioxide) so freezing the lower aqueous layer before decanting the upper solvent layer into the assay tube. For more dense solvents such as chloroform, aliquots of the solvent phase can be carefully pipetted after vacuum aspiration of the majority of the aqueous layer. Considerable practice is needed in order to maintain precision in an assay which includes these procedures. Positive displacement (i.e., capillary) pipettes are useful when transferring organic solvents between tubes.

A rotary evaporator can be used to remove the solvent by evaporation under vacuum. The water bath should be kept below 50°C. This is limited to one or a few samples at a time and therefore of more use in a research laboratory. After the extraction, **solvents can also be evaporated** in a stream of air or oxygen-free nitrogen at temperatures less than 60°C. The flow of gas should be regulated to minimize spluttering or splashing out of the sample. The number of samples will depend on the number of gas outlets. The procedure is laborious and time consuming and needs to be performed in a fume cupboard. Various instruments are available, however, to hold 100 tubes and which have heaters, operate with a vacuum and with a vortex or centrifugal action to speed up solvent removal.

Freeze drying is another technique for solvent removal from multiple samples in a single run. The product can be rehydrated very quickly and if the product is at the bottom of the tube, then it is possible to dissolve it in a smaller volume of solvent. However, the recovered sample must be checked, because the residue on concentrating some frozen liquid can be spread over a large area and reconstitution in the solvent may be difficult, particularly if only a small volume is used, which can affect the performance of the assay.

Steroids share their essential nonpolar nature with a large number of other lipids and methods which effectively extract a wide range of steroids are likely also to extract a large amount of nonsteroidal lipid material. Such lipid material must not be ignored since it can well interfere in subsequent separation or quantitation procedures. Lipid can be removed from extracts by partitioning with organic solvents. This is an important step when isolating steroids from brain tissue, Ebner et al. (2006) used isooctane. When determining estradiol in plasma, a solvent extraction with ethyl acetate:hexane (40/60, v/v) will contain many nonpolar compounds such as fatty acids and phospholipids so after this extraction a further wash with a basic aqueous solution (0.2 M ammonium carbonate buffer, pH 8.0) will remove polar compounds (note pK_A of estradiol is 10.5) but not extract estrogens.

The immunoassay of 17-hydroxyprogesterone is used in the early diagnosis of congenital adrenal hyperplasia and a falsely high result could have severe consequences. Steroids from the fetal adrenal zone were suspected of interfering in the assay. The interference was demonstrated to be due to 5-en-3β-ol steroids conjugated with sulfuric acid, particularly 17-hydroxypregnenolone sulfate, formed in the fetal adrenal cortex and present at high concentrations in neonatal plasma (Wong et al., 1992). An extraction of the plasma with a relatively nonpolar solvent mixture, isopropanol:hexane (3:97, v/v) recovered the 17-hydroxyprogesterone, leaving behind in the aqueous matrix the interfering steroid sulfates.

2.2.3.3.3 Supported-liquid extraction

For small samples (up to 10 mL), liquid extraction can be achieved with a modified diatomaceous earth such as Celite packaged in syringe-like devices or 96 well plates. The technique is described as **supported liquid extraction** with partition between the aqueous layer on the particles and the eluting solvent. The total procedure consists of adding the sample, which percolates through the material under gravity, waiting for 5 min then eluting with a solvent. Pressure may need to be applied if flow is too slow. Steroids usually elute best with dichloromethane but others (for example ethyl acetate, 50/50 hexane:dichloromethane) should be tested and conditions optimized. The protocol can be optimized according to the nature of the steroid, drug or analyte.

There are a number of formats (syringe, pipette, plate) available commercially (Fig. 2.2.3). SLE is marketed by Biotage as Isolute SLE+ or Chem Elut S (a synthetic sorbent) and by Agilent.

There are new forms of LLE such as single-drop microextraction (SDME) (Jeannot and Cantwell, 1997) and hollow fiber liquid phase (HF-LPME) (Luo et al., 2016) but these do not achieve sufficiently high preconcentration of the analytes in the extract from small samples of urine and they have not been adopted for steroid analysis of urine.

2.2.3.3.4 Sephadex extraction systems

Sephadex is a trademark for cross-linked dextran gel used for gel filtration. In 1959 when launched the name was derived from **se**paration **Pha**rmacia **dex**tran. The media are composed of macroscopic beads synthetically derived from the polysaccharide dextran. Bead sizes fall in discrete ranges between 20 and 300 μm. The dextran chains are cross-linked to give a three-dimensional network having functional ionic groups attached by ether linkages to glucose units of the polysaccharide chains. Sephadex is used to separate molecules by molecular weight. Sephadex is a faster alternative to dialysis without dilution. Sephadex is also used for buffer exchange and the removal of small

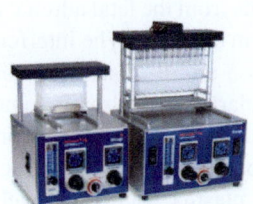

Positive pressure manifolds SPE

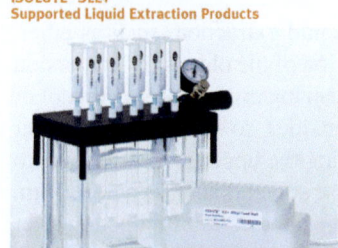

ISOLUTE® SLE+
Supported Liquid Extraction Products

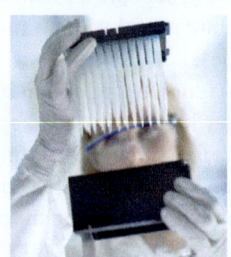

FIG. 2.2.3 The SLE products can be held in manifolds in negative and positive pressure.

molecules during the preparation of large biomolecules. Sephadex can be chemically modified to produce anion and cation exchangers.

Sephadex was designed for use with hydrophilic compounds in polar aqueous solvents but hydroxyalkoxy-modified Sephadex was developed and can be used for lipophilic materials. **Sephadex LH-20** is a beaded, cross linked dextran with 2-hydroxyethyl groups bound in ether link to hydroxyl groups of the dextran chain (Murphy, 1971) (Fig. 2.2.4). The Sjovall group pioneered the use of further **modified Sephadex** columns to be **lipophilic** in nature which in recent years have been have applied to the analysis of steroids in fluids, tissues including brain. This will be discussed in more detail in Chapter 2.3 on Purification of samples.

Lipidex is hydroxyalkoxypropyl-Sephadex is used for gel filtration of steroids and is used as a means of purifying steroid trimethylsilyl ethers formed by incubation with trimethylsilylimidazole (see GC-MS section), the excess reagent being retained in the Lipidex, while the steroid derivative is eluted with hexane. A new simpler method involves addition of water to the derivative then extracting the derivatives back into cyclohexane.

Extraction procedures can be extended for a steroid of interest in a matrix by using immobilized antibodies in **immunoaffinity** chromatography (IAC) (Su et al., 2005). The antibody is chemically attached to Sepharose or Sephacryl and kits are available to carry out this procedure. The immobilized antibody is then packed into a column and the matrix percolated through it. After washing, the analyte(s) can be released from the antibody and thus eluted from the column by altering the salt concentration of the mobile phase. Highly specific antibodies obviously provide a selective extraction but in some cases a less selective, broad spectrum extraction can be achieved by using an antibody with broader specificity. The application of IAC has preceded determinations by GC-MS, or prior to HPLC-ToF (Time of flight) MS in the analysis of estrogens in biological samples. The multitarget IAC for estrone, estradiol and estriol in urine prior to separation by **micellar electrokinetic chromatography** (MIKC) is another example but not commonly used. Adrenal steroids have also been determined by microfluidic chip using MIKC (Shen et al., 2009).

2.2.3.3.5 Solid phase extractions (SPE)

Extraction of steroids from aqueous media has been carried out using neutral polystyrene-divinylbenzene resins (Amberlite XAD-2) (Bradlow, 1977) and ion-exchange resins (DEAE-Sephadex) (Fotsis et al., 1981) packed in columns but is now achieved with small cartridges (e.g., Waters Oasis SPE) which work well for the extraction of polar steroids. Dowex AG resins have been used in conjunction with adsorption cartridges for the purification of catechol and guaiacol estrogens. Over the last 20–25 years the most important step forward in steroid analysis has been the advent of **solid phase extraction (SPE)**, particularly the availability of microparticulate silica that is coated with octadecasilane (C18) packed into syringes or cartridges (Shackleton and Whitney, 1980). There are a wide variety of SPE available for use in the extraction of steroids that fall into two distinct groups.

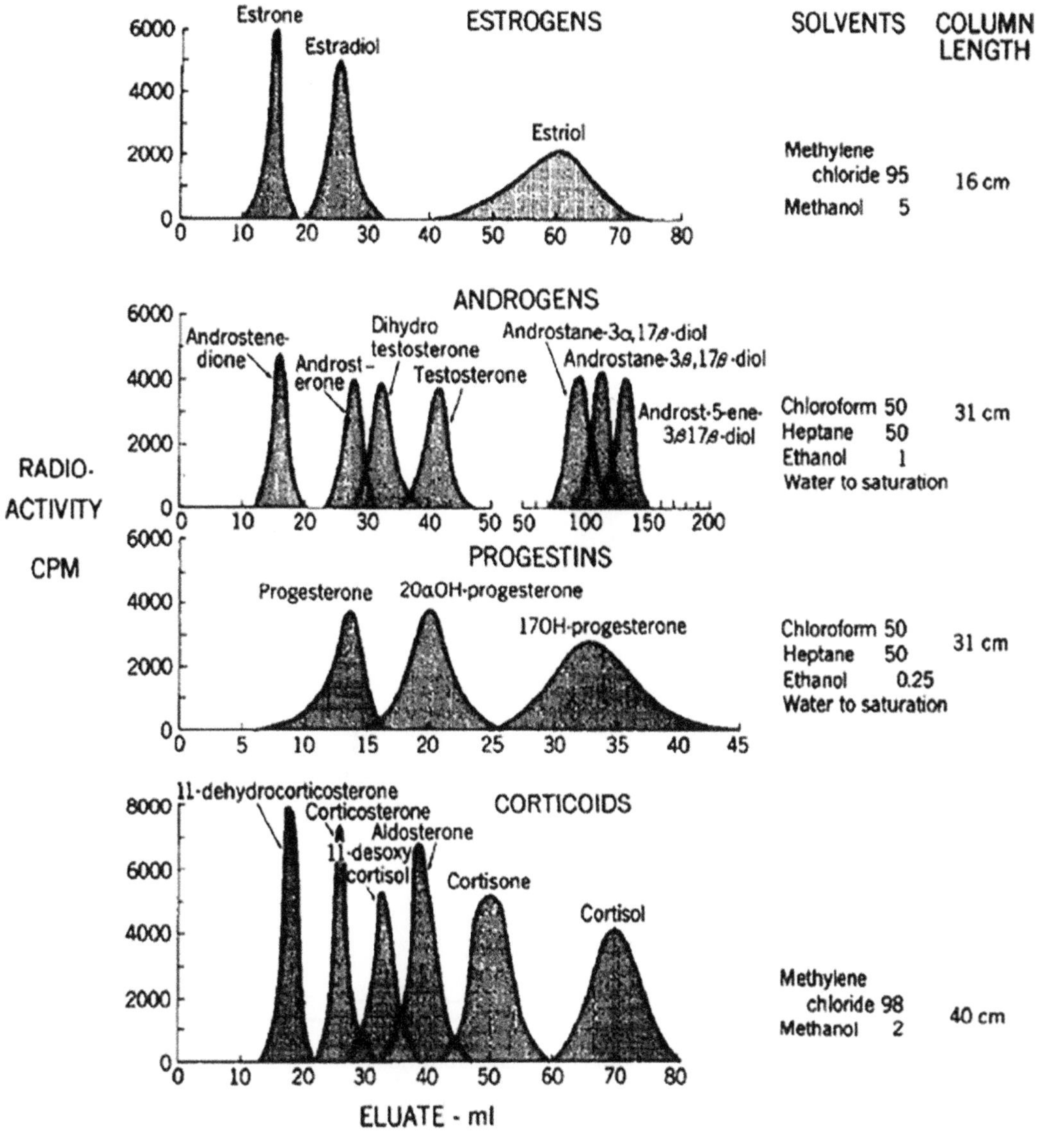

FIG. 2.2.4 Separations of steroids with Sephadex LH-20. The columns vary in length and are eluted with different solvents. Radioactive steroids were used to monitor elution from the columns and thus calibrate the column for extracts of biological samples. *(From Murphy BE. "Sephadex" column chromatography as an adjunct to competitive protein binding assays of steroids. Nat New Biol 1971;232(27):21–4. Fig. 1 p. 22.)*

Firstly, systems based on **Celite** treated commercially in unspecified ways to inactivate the material, presumably by the addition of carbon chains, before being sieved into different size ranges (Schöneshöfer, 1977). The material is then packed into syringes, cartridges of various sizes and shapes made from a variety of different plastics and 96 well plates (Fig. 2.2.5).

These sorbents of varying sizes can cope with differing sample loads. A 96 well plate format is also available. These SPE systems are based upon a varying mixture of adsorption and partition. SPE needs a water flush before and after an ethanol priming. The aqueous medium is poured onto the material, which takes up the water and the steroids are then eluted with organic solvents. This process would appear to be a simple liquid-liquid partition chromatography process similar to the celite partition column chromatography of the past. It is possible to inadvertently exceed the binding capacity of the column and if aqueous material passes through the column, steroids of interest will also pass through still dissolved in the aqueous matrix. Tox Elut (Varian), a similar type of system using a more granular absorption material designed for the analysis of drugs of abuse, has a dye incorporated into it, which indicates how far down the column the added aqueous medium has reached. SPE preparation systems can be incorporated on-line with HPLC column and solvent flow switching (Márta et al., 2018).

FIG. 2.2.5 Formats for solid phase extraction for single and multiple samples. Cartridges can also be used in a test tube which is the centrifuged to draw sample or eluent through the packing. *(Illustration by Supelco.)*

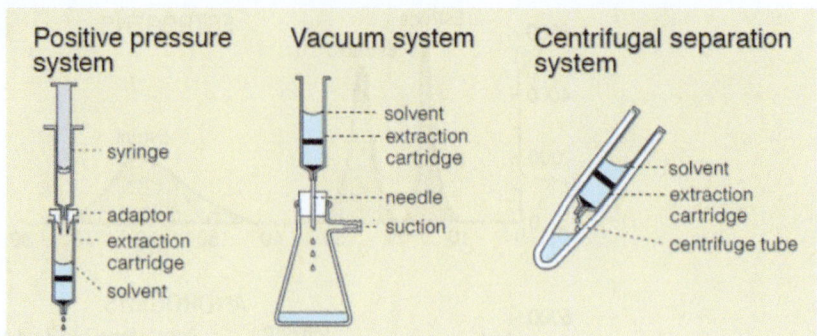

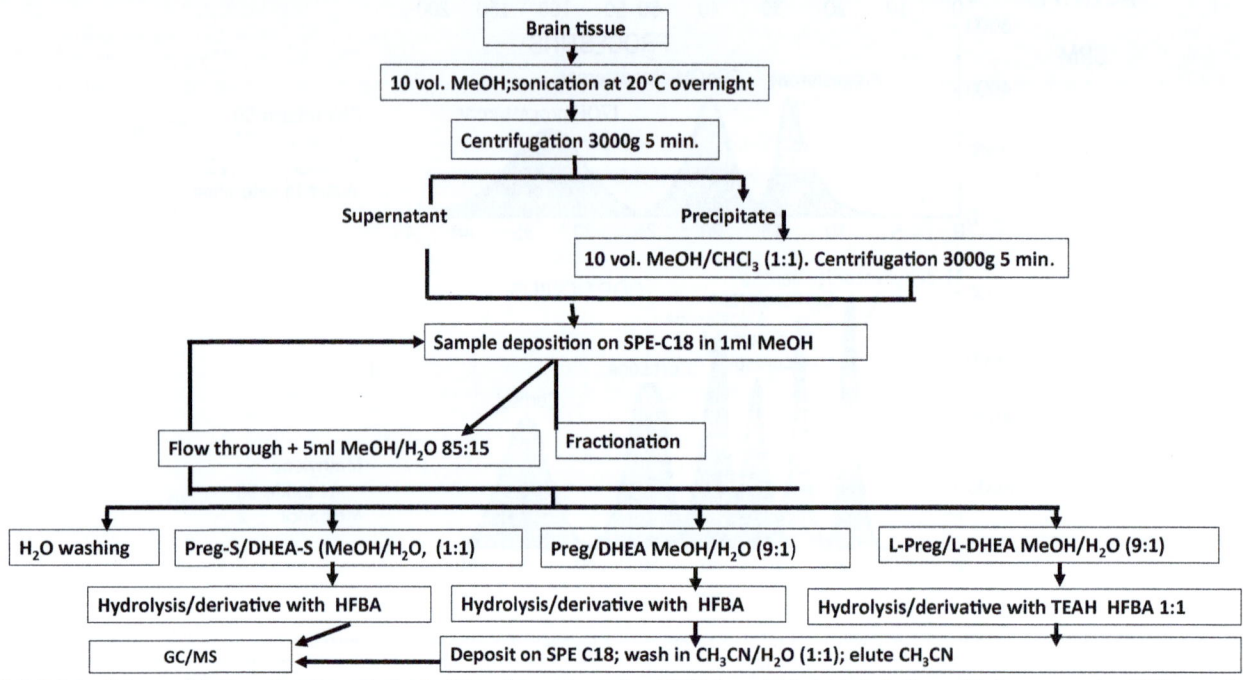

FIG. 2.2.6 Scheme of the procedure for analysis of sulphated, unconjugated and lipidal steroids in brain including solid phase extraction (SPE) recycling and fractionation procedure. Heptafluorobutyrate derivatives were separated by GC-MS. *(Redrawn from Liere P, Pianos A, Eychenne B, Cambourg A, Liu S, Griffiths W, Schumacher M, Sjövall J, Baulieu EE. Novel lipoidal derivatives of pregnenolone and dehydroepiandrosterone and absence of their sulfated counterparts in rodent brain. J Lipid Res. 2004;45(12):2287–302. Fig.1. p. 2289).*

SPE was used when separating steroids from the methanolic and ethanolic extracts of human and rat brains (Liere et al., 2009) (Fig. 2.2.6).

There has been controversy on the source and conjugate status of steroids in brain following the characterization of "neurosteroids" by Etienne Baulieu. The original methodology using solvent extraction to separate free from conjugated steroids has been criticized for giving an incomplete separation and thereby an overestimate of the conjugates pregnenolone sulfate and DHEAS in brain. This has been resolved by careful handling of extracts at every step in the analytical process and using charged SPE (e.g., Oasis MAX) for clear separation of free and conjugated steroids.

The **second type** of SPE material is mainly based upon **microparticulate silica**, either used directly or modified in a variety of different ways. The SPE sorbents range in nature from nonpolar, polar and ion exchange sorbents which are available today from a number of commercial suppliers (Table 2.2.2). The move of formats from cartridges to plates has enabled some automation of the process.

TABLE 2.2.2 Commercial sample prep techniques.

Method	Thermo	Biotage	Waters	Phenomenex	Supelco
Protein precipitation (PP)	Pierce PP plates	Isolute PPT	Sirocco PP	Impact plate	96 well PP filter plate
SLE	Hypersep SLE	Isolute SLE		Novum SLE	
SPE	Hypersep	Isolute SPE	Oasis µElution	Strata plates	Discovery SPE plates

Most steroid extractions are carried out using **reverse-phase** methodology where the silica has been modified by linking the silanol groups to hydrocarbons of varying chain length, the most popular being octadecane, forming **octadecasilyl silica** (ODS or C18), although other chain lengths have been used (e.g., C8, C2) for extraction of steroids. These cartridges do not tolerate drying out. Depending on upon the treatment used to form the ODS material, a proportion of silanol groups on the silica may still be unchanged. Further treatment, with a silylating material to produce trimethylsilyl groups to remove these extraneous groups, known as "end-capping" can be carried out. The presence of untreated silanol groups leads to a significant adsorption as well as a reverse-phase partition and deliberately produced nonfully end-capped material (Bond-Elut C18-OH) has been put to good use by combining the extraction and subsequent separation of steroid metabolites on the same cartridge by changing solvents so-called "phase-switching." A wash step with alkali similar to that used for organic solvent extracts can be used. A similar purification can be achieved with amino (NH_2) columns in series with C-18 extraction columns. These silica-based sorbent columns too have finite capacity but are now available, as mentioned above, in different sizes or alternatively the packing material can be supplied alone and appropriate amounts made up in columns of suitable sizes for the particular application. There was a problem with these cartridges if the material dried out, trapping the steroids between the hydrocarbon chains.

The Oasis **HLB (Hydrophilic-Lipophilic-Balance)** Sorbent is a macroporous copolymer made from a balanced ratio of two monomers, the lipophilic divinylbenzene and the hydrophilic N-vinylpyrrolidone which is water-wettable. Oasis cartridges produced by the Waters, comparable products by Phenomenex and other companies are available in five or more sorbent chemistries which are designed to meet just about all of the sample preparation needs (Fig. 2.2.7). It is claimed that these cartridges tolerate drying-out but the user should confirm this is the case

for the procedure adopted in the method. These cartridges have been used during the analysis of corticosteroids (Zheng et al., 2015) and aldosterone (Hinchliffe et al., 2013).

Other SPE systems

There are **several new techniques** that have not yet been widely used for steroids (Nováková, 2013) (Fig. 2.2.8). **Stir-bar** sorptive extraction (Kawaguchi et al., 2006; Liu et al., 2012) and **Graphitized carbon** have been used for the extraction of estrogens and their conjugates from urine, serum, milk and amniotic fluid (Caccamo et al., 1988). Steroids form inclusion compounds with **cyclodextrin** (Manaf et al., 2018) and this has proved valuable to extract steroids prior to GC-MS. SPE has been used to remove lipids from brain extracts. A **Hollow fiber LPME (HF-LPME)** has some unique advantages besides high sensitivity, fast equilibration, good reliability and good stability. At equilibrium, the ratio of the concentrations inside and outside the hollow fiber membrane is related to partition coefficient (log P) of the specific compound. For example, for a compound with $\log P = 4$, this would mean that, at equilibrium, the concentration inside the hollow fiber membrane would be 10,000-fold that outside the hollow fiber. Another important advantage is the filtration effect, which can prevent large molecules such as proteins from entering the inside of the hollow fiber. HF-LPME is relatively cheap and there is no carryover because the fibers are normally discharged after extraction. Progesterone has been extracted from human serum and analyzed by GC-MS after sample cleanup with HF-LPME (Kawaguchi and Takatsu, 2009) with a limit of detection (LOD) of $0.5\,\text{ng}\,\text{mL}^{-1}$ and a limit of quantitation (LOQ) of $2.0\,\text{ng}\,\text{mL}^{-1}$. HF-LPME is a very promising technology, but has its limitations. The lack of standard devices and procedures makes it hard to develop repeatable methods and procedures. Secondly, the throughput of current HF-LPME does not meet the needs of hospital laboratory use. Thirdly there is no report comparing application of throughput HF-LPME and any other throughput pretreatment methods.

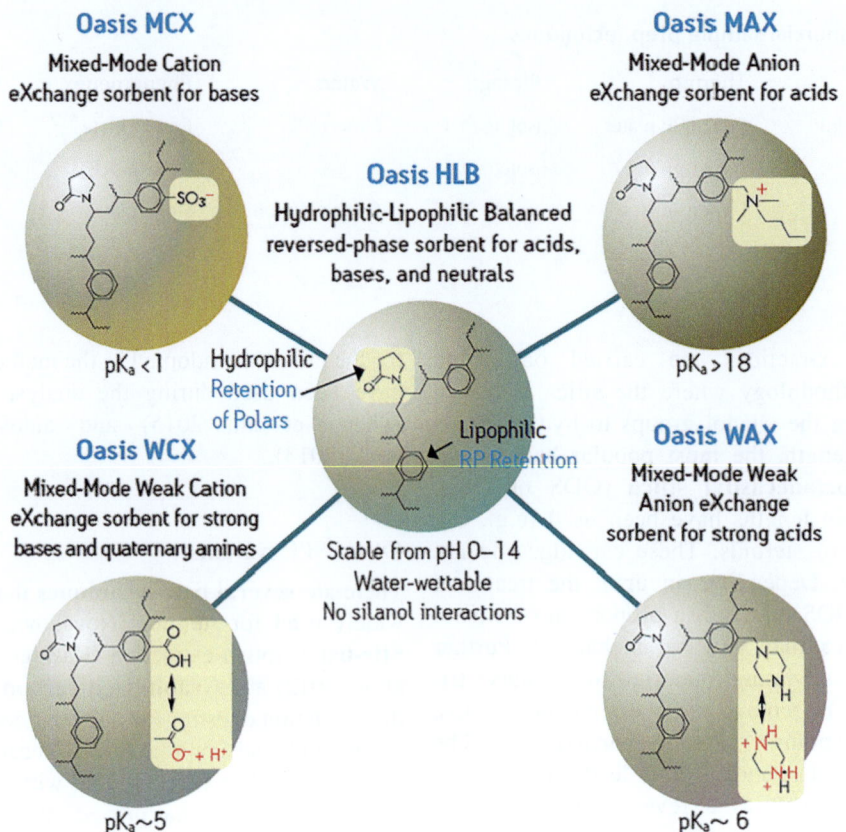

FIG. 2.2.7 Oasis ion exchange sorbents (Waters Corp).

Molecularly imprinted polymers (MIPs) are synthetic multifunctional materials with high affinity and specificity for target molecules (synthetic antibodies) (Whitcombe et al., 2014; Prieto et al., 2011; Alexander et al., 2006). These materials are synthesized by the polymerization of template molecules and functional monomers in the presence of cross-linker monomers (Fig. 2.2.9). After polymerization, the templates are removed from the complex and then "tailor-made" binding sites are formed for templates rebinding.

Owing to their excellent physical/chemical stability and recognition properties, MIPs have been suggested to be a promising class of materials for sample preparation but rarely used since first described for corticosteroids in 1996 by Olof Ramstrom et al. Magnetic MIPs have been developed for the separation of hydrocortisone and dexamethasone from cosmetic samples. The introduction of magnetic nanoparticles into MIPs field has expanded the application of MIPs in sample preparation but there have been few applications in clinical work because faster procedures that are amenable to automation are needed. Likewise, refinements of the technique such as **reversible addition-**fragmentation chain transfer (RAFT)** polymers have attracted little clinical attention despite potential for diverse monomers. RAFT is a versatile polymerization process offering predetermined molecular weight selection. **Restricted access material (RAM)** coupled on-line to LC-MS has also been described (Ye et al., 2012) but with few clinical applications to warrant further discussion. With the increased sensitivity of mass spectrometers sample size can be reduced and there are now a number of **micro-SPE systems** (Table 2.2.3) for this.

Microextraction by packed sorbent (MEPS)

Miniaturization of SPE (MEPS) is a more recent technique which can be produced in the laboratory. A MEPS can be made in a syringe barrel or disposable pipette tip in dispersive extraction technology (DPX). This uses low volumes of sample and organic solvents that is quick and can be automated but there is a high cost and reduced number of commercial tips available (Fig. 2.2.10). A small amount (1–4 mg) of sorbent is packed inside a cartridge directly placed in a pipette tip or syringe (100–250 μL) between the barrel and the needle (Vlčková et al., 2017).

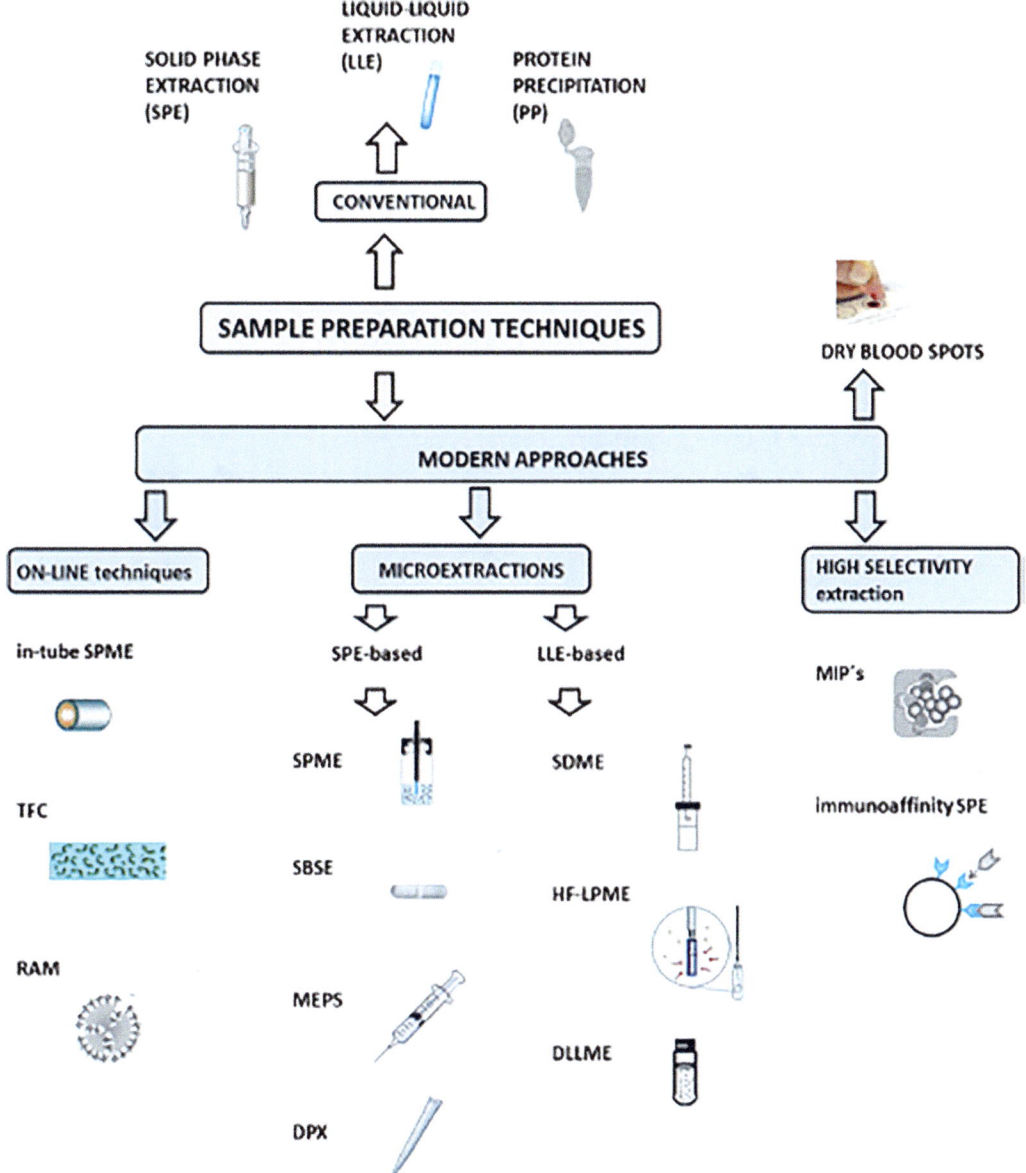

FIG. 2.2.8 Overview of microsample preparation techniques. *DLLME*, dispersive liquid-liquid microextraction; *DPX*, disposable pipettes extraction; *HF-LPME*, hollow fiber liquid phase microextraction; *MEPS*, micro-extraction by packed sorbent; *MPIs*, molecularly imprinted polymers; *RAM*, restricted access material; *SBSE*, stir-bar sorptive extraction; *SDME*, single drop microextraction; *SPME*, solid phase micro-extraction. As the sensitivity of steroid detection increased then sample size could be reduced and extractor size could also be reduced. *(From Nováková L. Challenges in the development of bioanalytical liquid chromatography-mass spectrometry method with emphasis on fast analysis. J Chromatogr A 2013;1292:25–37. Fig. 2 p. 28.)*

Different types of sorbents are commercially available such as reversed (C_{18}, C_8 and C_2), normal (silica) or ion exchange (SCX) stationary phases. The sorbent can be used several times with an adapted washing and reconditioning to avoid carry-over and to keep the adsorption power of the phase. Because of miniaturization, extraction time, sample size and solvent volumes are considerably reduced and the elution extract is directly compatible with an on-line injection in LC, GC or (of less capacity) capillary electrophoresis (CE).

In a typical application the C_{18} MEPS sorbent is first conditioned with 100 μL MeOH and 100 μL H_2O. The sample is applied three times with 100 μL each time before washing the phase with 100 μL H_2O and 80 μL hexane (Prieto et al., 2011). The "steroid" fraction is eluted with 2× 90 μL of a MeOH/ethyl acetate mixture (30:70, v/v).

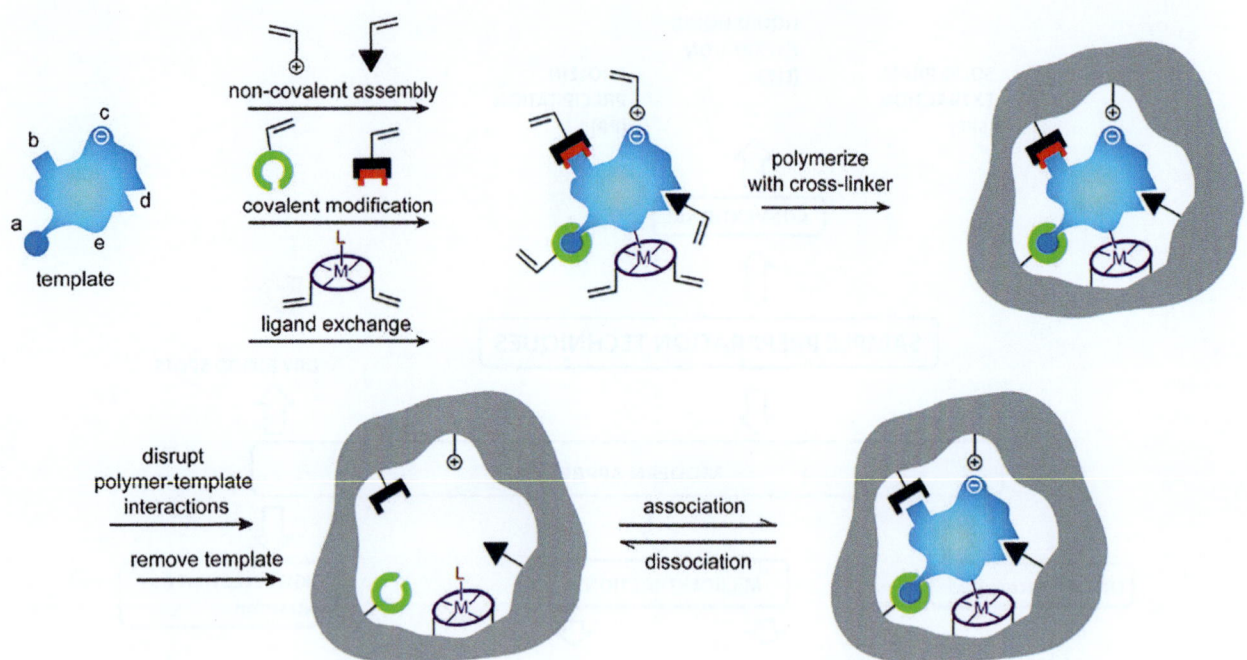

FIG. 2.2.9 Schematic for MIP preparation. The formation of reversible interactions between the template and polymer may involve reversible covalent bonds, covalently attached polymer binding groups, electrostatic interactions, hydrophobic or van der Walls interactions or co-ordination with a metal center. *(From Whitcombe MJ, Kirsch N, Nicholls IA. Molecular imprinting science and technology: a survey of the literature for the years 2004-2011. J Mol Recognit 2014;27(6):297–401. Fig. 2 p. 298.)*

TABLE 2.2.3 Micro-SPE in pipette tips.

Supplier	Name of tip	Type of sorbent	Chemistry	Tip volume (μL)
Merck	ZipTips	Silica particle	C18, C8, SCX	20, 200
Thermo	HyperSepSPE tips	Silica	C18, C4	10
SunChrom	NuTips	Monolith	C18,C8, C4	10, 200
GLScience	MonoTips	Monolith	C18	10, 200
Agilent	OMIX	Monolith	C18, C4	2, 10, 100
Velocity	SiliaPrep	Silica		
Interchim	UptiTip		C18, HILIC, SAC	10, 200

Finally, the sorbent is rinsed successively with 2× 90 μL of the elution mixture and 2× 100 μL water to avoid any carryover phenomenon. The extracts are evaporated under N_2, and trimethylsilylation with 20 μL of MSTFA/DTE/TMIS (1000:5:5, v/m/v) mixture was performed at 60°C during 50 min before injection.

Some sample preparation techniques are called QuEChERS for Quick, Easy, CHeap, Rugged, Safe but these have only been applied in meat, food and water analysis. Automated liquid sampling handling systems are now being used to improve analytical efficiency. These can be bench top and free-standing instruments (Fig. 2.2.11). SPE can be incorporated into such procedures and has been tested for example in urine steroid analysis in antidoping methods, foods and pharmaceuticals but not in clinical laboratories.

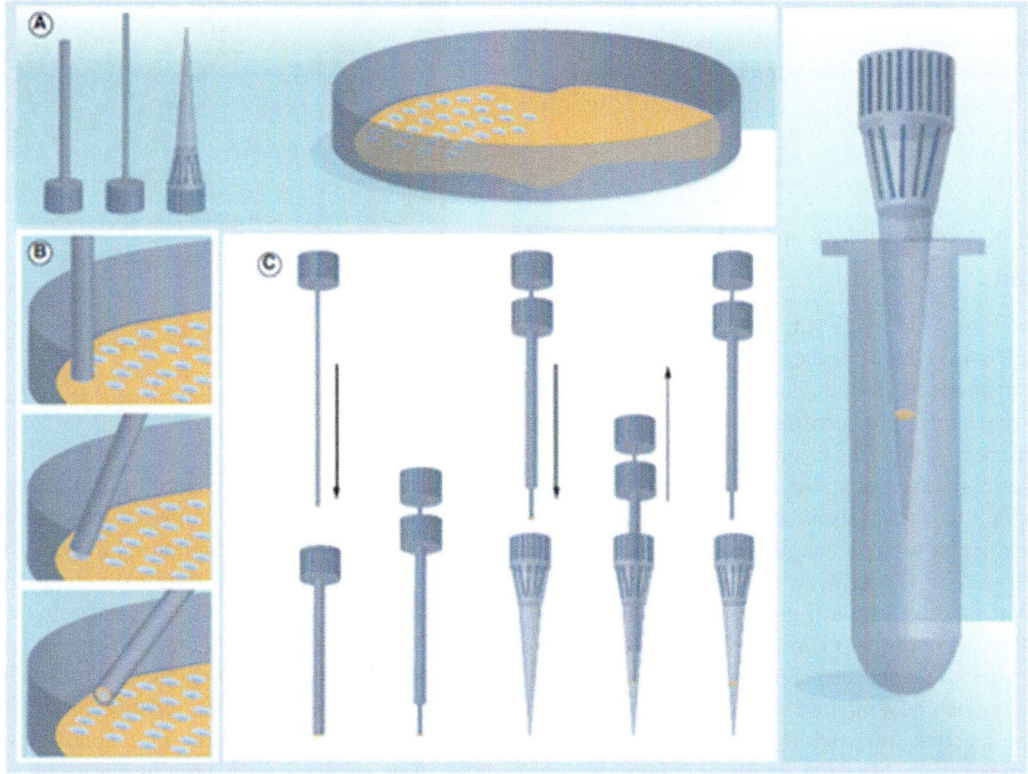

FIG. 2.2.10 The preparation of homemade micro-SPE tips. (A) Necessary tools; (B) the process of the sorbent cutting; (C) the process of the sorbent insertion into the tip and (D) the arrangement for μ-SPE-PT performed by the centrifugation approach. *μ-SPE-PT*, micro-SPE in pipette tips. *(From Vlčková H, Pilařová V, Novák O, Solich P, Nováková L. Micro-SPE in pipette tips as a tool for analysis of small-molecule drugs in serum. Bioanalysis 2017;9(11):887–901.)*

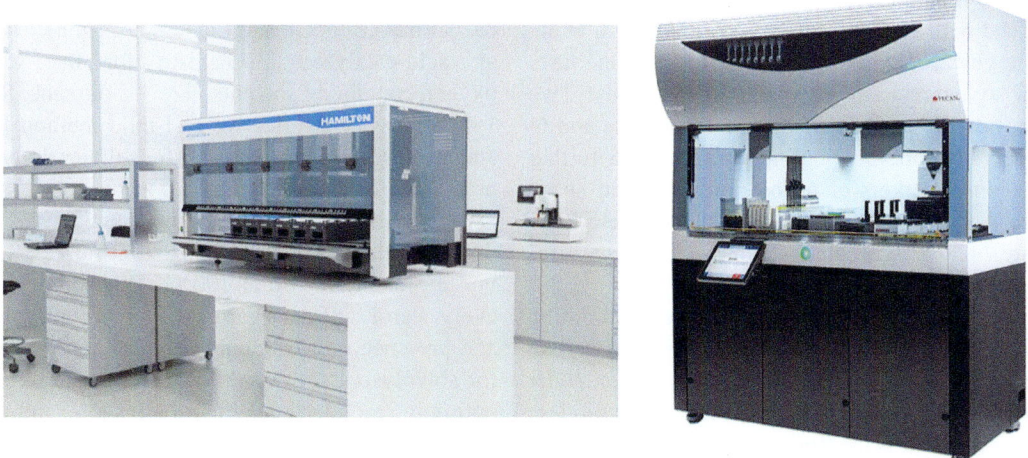

FIG. 2.2.11 Modern benchtop and free-standing liquid sample handling for large numbers of samples.

2.2.4 Extraction of steroids from urine

Urine is a complex medium which contains a wide variety of metabolic products of the secreted steroid hormones (androgens, estrogens, corticosteroids, pregnanediols, cholesterol, etc.) which may be unconjugated or conjugated with glucuronic or sulfuric acids. A comprehensive analysis of urine presents therefore a considerable analytical challenge. The analysis of urine for clinical diagnostic purposes is gaining in popularity particularly the use of steroid profiles by capillary GC/MS or LC-MS/MS because it provides very valuable information and has now moved from research laboratories and now falls in line with the acceptance of these methods in clinical laboratories. SPE is by far the commonest method for extraction of steroids from urine, occasionally a LLE is used, chloroform is used to extract cortisol from urine.

2.2.4.1 Liberation of free steroids from conjugation

Conjugation also increases the water solubility and excretion in urine. New research is showing that the in vivo conjugation and deconjugation of active steroid hormones are mechanisms of controlling the activity of the hormone in specific tissues. In primates, plasma DHEAS circulating at high concentrations is now considered to be a prohormone. In the analysis of steroids it has been common practice to remove conjugate groups and analyze the liberated steroids. **Hydrolysis** of **steroid glucuronides** is usually carried out using a β-glucuronidase preparation although effective hydrolysis can also be achieved chemically with periodic acid that removes the glucuronide residue, leaving behind a mixture of free steroid and steroid formate. The reaction times with the enzyme can be from 15 min to more than 24 h. For enzymic hydrolysis, the optimum conditions of temperature and pH need to be established. The enzyme from abalone (*Haliotis* spp.) entrails for example is best used at pH 5.2 and 42°C. Higher temperatures and faster reactions are possible with *E. coli* glucuronidase. Commercial β-glucuronidase preparations are crude and some contain cholesterol oxidase activity, this activity can be reduced with sodium ascorbate (Christakoudi et al., 2008). Some glucuronide metabolites of testosterone have been found to be resistant to enzyme hydrolysis. Two metabolites (6-hydroxy androsterone-3-glucuronide and 6-hydroxyetiocholanolone-3-glucuronide) may, with further research, become markers of testosterone abuse in sport. Urine contains numerous inhibitors for glucuronidase some of which are of low molecular weight such as saccharo-1,4-lactone, gluconic acid and saccharic acid.

Hydrolysis of **steroid sulfates** is usually carried out by acid solvolysis in ethyl acetate. Sulfatase preparations are available from a variety of sources (*Patella vulgata*, *Helix pomatia*, etc.) although most of them have β-glucuronidase activity. *Helix pomatia* is the Roman snail (Fig. 2.2.12) and the enzyme is harvested from the intestines. *Helix ampullaria* is another source (Shibasaki et al., 2001).

The glucuronidase from *E. coli* does not have sulfatase activity. A sulfatase preparation from *Pseudomonas aeruginosa* has no glucuronidase activity (Stevenson et al., 2015). Some problems have been encountered, for example the recovery of DHEA from urine treated with *Helix pomatia* preparations decreases dramatically as the concentration of enzyme increases. In preparations containing both

FIG. 2.2.12 *Helix pomatia* snail. The intestinal juice is rich source of glucuronidase and sulfatase for hydrolysis of steroid conjugates.

glucuronidase and sulfatase activity, sulfatase activity can be inhibited by incubating in pH 4.5 phosphate.

An alternative to the enzymatic hydrolysis is the chemical hydrolysis or solvolysis. Traditionally acidic hydrolysis at elevated temperatures was used for deconjugation of sterol sulfates. The drastic conditions that are required for this hydrolysis however, including high amounts of mineral acid and refluxing, can lead to degradation or transformation of some sterols. Solvolysis under mild conditions is preferred and can be achieved by extracting the sterol sulfates from an acidified (with sulfuric acid) aqueous sample with ethyl acetate and storing this moist organic phase for 24 h at 39°C or with trimethylchlorosilane in methanol (methanolysis). The ability of oxygen-containing solvents, especially ethers, to cleave sterol sulfates in presence of minor amounts of water and acid was investigated in 1958 by Burstein and Lieberman (Burstein and Lieberman, 1958). They proposed an acid-catalyzed mechanism for the solvolysis in oxygen containing solvents, like 1,4-dioxane. Solvolysis in 1,4-dioxane is an effective and mild method for both 3α- and 3β-sterol sulfates as well as for sulfates derived from saturated and unsaturated sterols.

A modification of the method published for sterols in insects (Hutchins and Kaplanis, 1969) uses 1% acetic acid in 1,4-dioxane for 6 h at 100°C. This solvolysis works for every sterol sulfate regardless of the configuration at C3 and presence of a Δ5-double bond. The best reaction time for solvolysis is 3 h for a full range of sterol sulfates. The optimum reaction times for solvolysis for individual sterol sulfates can vary and must be optimized. Some anabolic steroid metabolites have been found to be released from conjugates by alkaline hydrolysis but more research is needed on this topic.

Methanolysis, a strong acid-catalyzed sulfonic acid cleavage mechanism utilizing acetyl chloride in methanol, can be used in steroid analysis to simultaneously cleave both sulfate and glucuronide conjugated metabolites (Viljanto et al., 2018). Strong acidic conditions cause the protonation of the oxygen attached to C-17 and the nucleophilic attack by methanol, causing cleavage of sulfate and glucuronide moieties. There is also a risk of partial dehydration occurring (formation of double bonds) in the steroid due to the harsh acidic environment. Even so, due to good recoveries and ease of use, methanolysis is widely used particularly in doping-control laboratories as a convenient alternative to the combination of solvolysis and enzyme hydrolysis. Care should be taken when selecting certain types of deuterium labeled internal standards to be used in procedures employing conditions that promote keto-enol tautomerism such as methanolysis. Steroids that are deuterium labeled on an α-carbon adjacent to a ketone functional group(s), such as within D_9-progesterone, are more likely to be subject to hydrogen exchange, while other labeled analytes are unlikely to pose any problems. When methanolysis is carried out the deuterium ions of D_9-progesterone are exchanged with hydrogen ions. This is caused by acid-catalyzed enol tautomer formation. In keto-enol tautomerism, the double bond is rapidly moving between the keto and enol forms. The equilibrium usually favors the keto tautomer, but this can be shifted by acidic or alkaline conditions. Multiple deuterium atoms in the same carbon will also be exchanged which would have adverse effects on analytical results.

Hydrolysis can be speeded up by placing the reaction tube in an ultrasonic bath (Alvarez-Sánchez et al., 2009). Conditions will have to be optimized by the user. Most studies have not controlled the efficiency of the deconjugation process. Conjugates of methylumbelliferone have been proposed as standards for this evaluation (Grignon et al., 2017). The use of 2,4,16,16-d_4 estradiol-sulfate from Toronto Research Chemicals (Toronto, Canada) and 2,2,4,4-d_4-androsterone-d4-glucuronide from NMI (National Measurement Institute, West Lindfield, NSW 2070, Australia) has been described and others may be available as standards (Wang et al., 2021) for the control of treatment steps.

2.2.4.2 Separation of conjugates

Steroid conjugation is the final step in the metabolism of these hormones, leading to their deactivation and elimination from the body. However, the concept has been considered that certain types of steroid conjugates are biologically active compounds, and that the conjugation serves, in some instances, a purpose other than merely rendering the hormone more water soluble, to facilitate its excretion. In particular, the sulfate conjugates of steroids are not only synthesized in the steroid-producing glands but also in various "peripheral" tissues that possess the ability to form sulfate conjugates. The importance of this finding is becoming clearer. To gain more insight into the biological role of steroid conjugates, there have been many attempts to isolate conjugate "fractions." A systematic study was first made by Ludwig Kornel firstly with paper chromatography (Kornel and Hill, 1961; Kornel et al., 1969) then high voltage electrophoresis (Kornel, 1964). The methodology would be totally impractical today and the detail will not be described here.

Some of the steroid conjugates are available commercially as reference preparations (see Chemistry Chapter 1.1). In order to study the intact cortisol conjugates the synthesis was required of reference steroids in glucuronide and sulfate conjugated forms at C-3 and C-21 positions. Stable isotope labeled steroid sulfate conjugates have also been prepared. Chemical methods and enzymatic methods have been used for conjugate synthesis. Purity and traceability of all steroids will be an issue if these compounds are used in a method needing accreditation. An internal standard of 9,12,12,21,21-d_5-THE-3-glucuronide was used as the reference point for the quantitative analysis of 12 THF and THE glucuronide related steroids (Ikegawa et al., 2009). Recovery experiments of added steroid conjugates have been satisfactory. Carbon 13 labeled steroids conjugates have also been prepared.

The steroid glucuronide and sulfate conjugates and sometimes even the polyhydroxylated C_{21} steroids are very polar and not always effectively extracted even by the polar solvents. The addition of ammonium sulfate at saturation concentrations to urine and acidification to pH2 with 10% (v/v) HCl prior to extraction with ether:isopropanol mixture (3:1, v/v) overcomes this problem and the majority of steroid glucuronides and sulfates are effectively extracted by this procedure. The steroid ester conjugates are present in plasma (especially C19 glucuronides and sulfates and C21 sulfates) at lower concentrations than in urine and are extracted with varying efficiency into polar organic solvents. It is possible by careful choice of solvents for extraction, to provide a considerable degree of selectivity for one steroid of interest, although such selectivity is seldom exclusive and there are many examples of the use of this type of selection. Isopentane, heptane, cyclohexane, dichloromethane, chloroform, ether and ethyl acetate are solvents of increasing polarity that are used for extraction of steroids and are immiscible with water. In contrast acetone, propanol, ethanol, methanol and pyridine (again in order of polarity) are miscible with water.

Size exclusion or gel filtration chromatography based on Sephadex, a cross linked dextran, is still used for the separation of steroid groups, such as sulfates, glucuronides, etc. (see Urinary steroid profile analysis by GC-MS). Since 1968 Sephadex LH20 columns have been used for the fractionation of steroids dissolved in salt saturated methanol:chloroform (1:1, v/v) to a 4g LH20 column (Setchell and Shackleton, 1973). Glucuronides are eluted with 30mL of the same solvent. A further elution with methanol releases the sulfates.

Ion-exchange columns have been applied for steroid separation. The group of Herman Adlercreutz isolated estrogen conjugate groups using ion exchange chromatography. Sephadex was modified to produce, for example, diethylaminoethyl (**DEAE**)-**substituted Sephadex** which acts as an ion-exchange column but still with size exclusion properties (Fotsis et al., 1981). After application of the sample to a column containing DEAP-LH-20 (0.6 g) in the acetate form in 72% ethanol, pressure is applied to the column to produce a flow rate of approximately 25 mL/h, and the neutral steroids are eluted from the column with 25 mL of 72% ethanol. The solvent is then changed to 0.25 M formic acid in 72% ethanol (25 mL) and the glucuronide conjugates are recovered. Steroid monosulfates are eluted with 25 mL of 0.3 M acetic acid-potassium acetate, pH 6.3, in 72% ethanol. Finally, steroid disulfates are recovered by elution with 25 mL of 0.5 M potassium acetate solution in 727; ethanol adjusted to pH 10.0 with potassium hydroxide (Fig. 2.2.13). These procedures are too time consuming for routine clinical laboratory use.

In some studies, free steroids are extracted into ether before hydrolysis of the conjugates. Urinary steroids can be fractionated into free, glucuronide and sulfate fractions, after extraction using Sep Pak C_{18} cartridges, using columns of DEAE-Sephadex A25. The use of triethylaminohydroxypropyl (**TEAP**)-**LH-20** for an extensive and efficient fractionation of steroid groups (Fig. 2.2.14) prior to GC-MS has been described (Axelson and Sahlberg, 1983; Meng and Sjövall, 1997). The steroids were eluted sequentially with methanol:chloroform (2:1, v/v), 72% aqueous methanol

saturated with CO_2, 0.4 M formic acid in 72% aqueous methanol, 0.3 M acetate buffer pH 6.3 in 72% methanol and finally 0.5 M potassium acetate-potassium hydroxide in 72% aqueous methanol pH 10. The method would not be practical for routine use. There are earlier papers from Sjovall's laboratory.

Other examples of the use of modified Sephadex for the separation of steroids have been described but not widely adopted. Commercial preparations are expensive.

2.2.4.2.1 Ion exchange cartridges

Free and conjugated steroids are now separated with **mixed mode anion exchange SPE** cartridges. WAX, MAX and SAX cartridges have been compared. A weak anion exchanger bonded on a styrene divinylbenzene polymer (**MM4 cartridge from Interchim, France**), a strong anion exchanger bonded on a HLB phase (**MAX** cartridge) and a strong anion exchanger on a functionalized silica (**SAX**) have been used (Anizan et al., 2010). The starting conditions and elution pattern are summarized in Table 2.2.4.

Conjugates can be separated after the urine steroid extracts are loaded into SPE-WAX cartridges that contain a mixed mode polymeric/weak anion exchange resin conditioned with acetate buffer (2 m, pH 5.2). After washing with water, the glucuronide steroids are eluted with methanol/10% formic acid in water (95:5, v/v). The cartridge is then washed with methanol/5% ammonium hydroxide in water

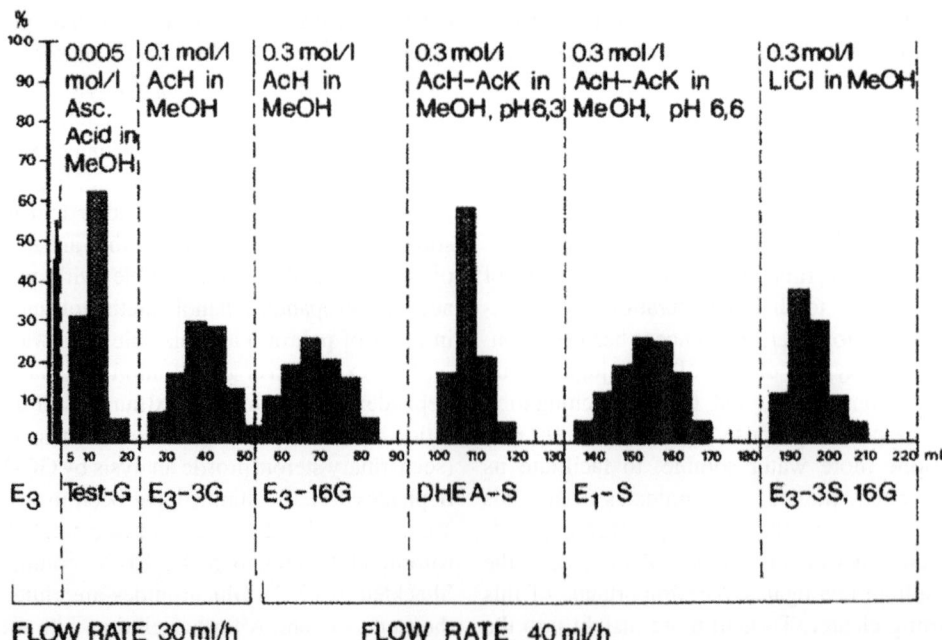

FIG. 2.2.13 Ion exchange separation of steroid conjugates. Large volumes of solvent were needed and the process was too slow for a routine hospital test today. *(From Fotsis T, Adlercreutz H, Järvenpää P, Setchell KD, Axelson M, Sjövall J. Group separation of steroid conjugates by DEAE-Sephadex anion exchange chromatography. J Steroid Biochem 1981;14(5):457–63. Fig. 1 p. 460.)*

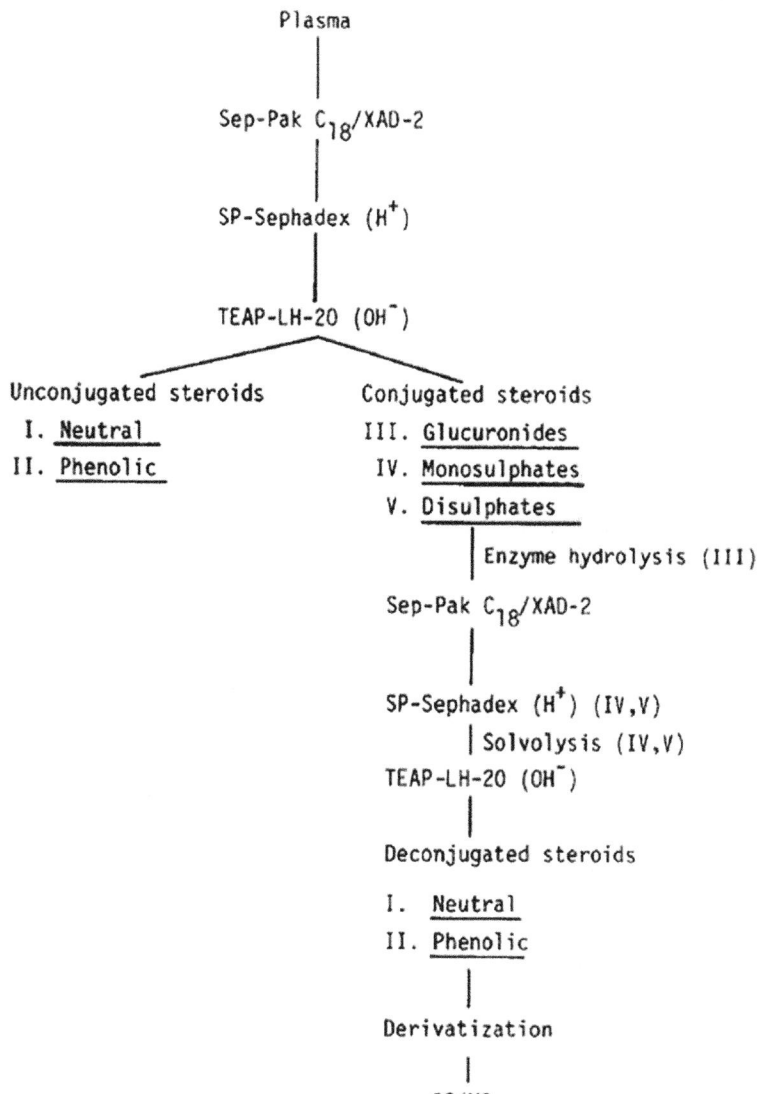

FIG. 2.2.14 Group separation plasma steroids with ion-exchange Sephadex LH-20. *(Redrawn from Axelson M, Sahlberg BL. Group separation and gas chromatography–mass spectrometry of conjugated steroids in plasma. J Steroid Biochem 1983;18(3):313–21. Fig. 1 p. 114.)*

TABLE 2.2.4 Conditions for conditioning, loading, washing and elution SPE.

	MM4	C18/HLB	MAX	SAX
Conditioning	6 mL MeOH	6 mL MeOH	6 mL MeOH	6 mL MeOH
Conditioning	6 mL H_2O	6 mL H_2O	6 mL H_2O	6 mL H_2O
Loading	2 mL	2 mL	2 mL	2 mL
Washing	5 mL H_2O	5 mL H_2O	5 mL H_2O	5 mL H_2O
Washing 2		5 mL cyclohexane/ethyl acetate 20:80	5 mL MeOH	5 mL MeOH/H_2O 25/75 (v/v)
Elution 1	13 mL MeOH, 2% NH_4OH	5 mL ethyl acetate/ MeOH 70:30	5 mL MeOH, 5% formic acid	5 mL MeOH/2% formic acid
Elution 2			5 mL MeOH, 10% NH_4OH	5 mL MeOH, 10% NH_4OH

Redrawn from Anizan S, Di Nardo D, Bichon E, Monteau F, Cesbron N, Antignac JP, Le Bizec B. Targeted phase II metabolites profiling as new screening strategy to investigate natural steroid abuse in animal breeding. Anal Chim Acta 2011;700(1–2):105–13.

(20:80, v/v) before elution of the sulfates with methanol/5% ammonium hydroxide in water (20:80, v/v). The recoveries of the glucuronide compounds with MM4 and of the sulfate compounds with MAX appeared very low. Indeed, the MM4 did not sufficiently retain the glucuronide compounds which were eluted during the washing step (not shown). On the other hand, the MAX cartridge presented too strong ionic affinities regarding the sulfate compounds to allow their elution. Both these cartridges were then rejected. C_{18} and HLB materials have given equivalent recovery yields, which were found a little bit higher compared with the ones obtained with the SAX cartridge. However, in this last case, a mixture of MeOH/formic acid (98:2, v/v) specifically allowed the elution of glucuronic forms (recoveries >13%) while a mixture of MeOH/ammonium hydroxide 32% (90:10, v/v) permitted to elute the sulfate forms (recoveries >55%). Purification based on a SAX cartridge has given the best results. Splitting the extract in two fractions increases the reliability and sensitivity of the UPLC-MS/MS analyses, respectively, by decreasing the matrix effect and increasing the MRM transition dwell time or the precursor ion scan duration. The main drawback was a longer analysis duration (including two acquisition times per sample) but this one, though less than 17 min. Concerning the solvent modifier, formic acid was preferred to the other one assessed, such as ammonium acetate (pH 4.2 or 6.8). Indeed, the separation was found slightly better with formic acid, but the main reason was the higher sensitivity (five times higher) probably due to a better ionization.

At University College London, postgraduate student Martin Ebner working with Jonathan Fry, used MAX cartridges to separate free and conjugated steroids from extract of rat brains. The cartridge was primed with 15 mL ethyl acetate, 10 mL 0.25 M ammonium acetate pH 7, 15 mL 96% aqueous ethanol then washed with 3 mL water. After loading the sample extract the cartridge was washed with 5 mL ethanol at 20% (v/v) ammonium acetate (20 mM, pH 7 v/v). Free steroids were eluted with 4 mL ethyl acetate. Glucuronides were eluted with 20 mL ethanol at 60% (v/v) in formate/pyridine buffer (20 mM, pH 3). Finally, the steroid sulfates were eluted with acidified ethyl acetate but this solvolyzed the sulfates and is replaced with 30 mL of 60% aqueous ethanol containing ammonium carbonate 1% (v/v).

Cholesterol is an integral part of the structure of the plasma lipoproteins and these lipoproteins may form part of the structure of cells and tissues. Autoxidation of cholesterol during sample processing constitutes a potential source of error not only in the analysis of oxysterols but also in the analysis of C_{19} and C_{21} steroids with a 3β-hydroxy-5-ene structure (Liere et al., 2009), because there is the potential these could be formed as extraction artifacts (Fig. 2.2.15).

In light of this, the analytical results of many studies of sulfated, free and fatty acid esters of PREG and DHEA in brain tissue need to be re-evaluated. Recent data confirms that in rat brain, free PREG is endogenous, Ebner et al. (2006) could detect DHEAS by GC-MS. In other methods, levels of DHEAS and PREGS are below the detection limits of the currently used methods present method in contrast to early results when neurosteroids were first identified. Subsequent work showed that DHEAS and PregS are rapidly desulfated as they enter the brain from the plasma (Qaiser et al., 2017).

Ion exchange systems are in use for conjugate separations. The sulfonic acid (MCX) and quaternary amine (MAX) derivatives of Oasis HLB provide dual modes of retention, enabling greater cleanup selectivity and sensitivity for both acid and/or basic compounds and they tolerate the sorbent running dry. Oasis WCX (weak cation exchanger) and WAX (weak anion exchanger) are derivatives specifically designed to offer the same benefits and features as HLB with the ability to retain and release strong acids (e.g., sulfonates) and bases (e.g., quaternary amines) (Lavén et al., 2009). All of the five patented Oasis chemistries are available in several device formats (e.g., cartridges, 96-well plate and μElution plate) for purpose. These cartridges have been used by Ebner et al. (2006) in the processing of steroids from brain extracts and the method works for steroid conjugates in urine. Microparticulate silica coated with ODS and attached to quaternary amine (Bond-Elut SAX) have been used for the fractionation of steroids and bile acids into conjugate groups prior to further analysis.

The plastics in which the SPE material is packed (including cartridge, syringe, frits, etc.) can react with solvents and material may be eluted from the plastic which is innocuous if used in one method but devastating when the eluate is used in another. Small SPE columns can be inserted into HPLC systems. Before being applied to the column, any steroid-protein binding must be disrupted and in the case of the silica-C_{18} extraction procedures this is usually carried out by the use of acetonitrile or methanol which is added in equal amounts before a vortex mix and centrifugation to create a protein plug in the bottom of the tube. The acetonitrile:water mix is then applied to the SPE cartridge and after washing with various concentrations of water in methanol to elute polar material the steroids can be eluted with methanol. Sometimes the protein plug is re-suspended in acetonitrile and combined with the extract applied to the Silica C_{18} SPE system. This procedure is very effective usually leading to near quantitative recovery.

2.2.5 Other sample matrices

2.2.5.1 Processing hair samples

Steroids exist in hair in an intercellular space between the cuticle and cortical cells. The hair lipids include squalene, wax esters, triglycerides, free fatty acids, ceramides and cholesterol sulfate. Steroids can be extracted from hair using

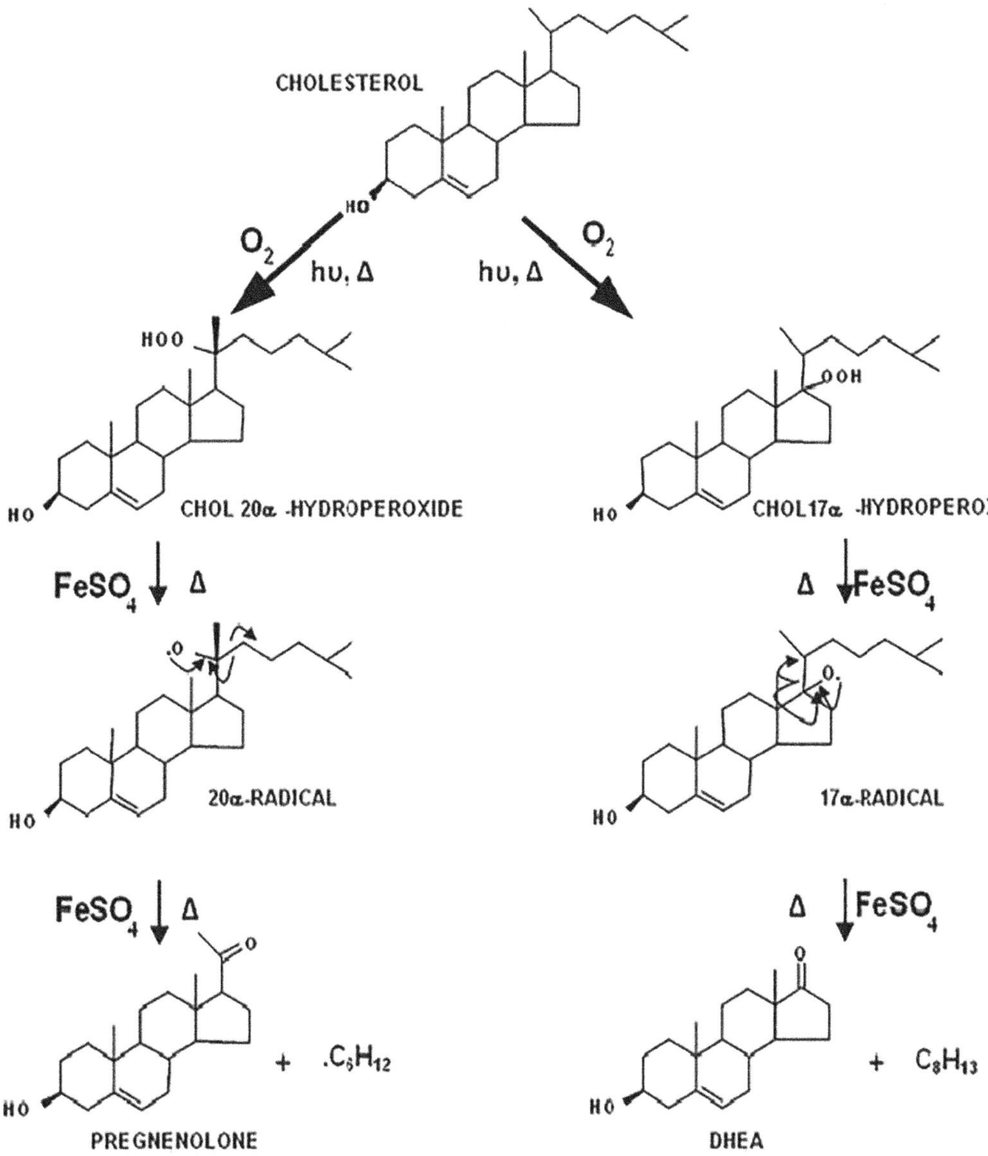

FIG. 2.2.15 Auto-oxidation of cholesterol. *(From Liere P, Pianos A, Eychenne B, Cambourg A, Bodin K, Griffiths W, Schumacher M, Baulieu EE, Sjövall J. Analysis of pregnenolone and dehydroepiandrosterone in rodent brain: cholesterol autoxidation is the key. J Lipid Res 2009;50(12):2430–44.)*

solvent mixtures such as chloroform with methanol. The analysis of steroids in hair as a means of detecting ingestion of anabolic steroids has been described (Voegel et al., 2020; Greff et al., 2019; Gao et al., 2016; Cooper et al., 2012; Kintz, 2003; Wennig, 2000). Fatty acids can be easily removed from a nonwater miscible solvent extract by an alkaline wash but this may also remove acidic steroids such as estrogens or bile acids. Saponification, apart from liberating steroids from their esters, also has the added advantage of hydrolyzing many neutral lipids which can then be removed by alkali or acid washes. Despite these precautions, significant amounts of nonspecific, nonsteroidal, neutral lipid material may be present in supposedly clean steroid extracts and this may not always be appreciated

since this interfering material usually does not have any recognizable characteristic which immediately betrays its presence. While in most instances, such material can be ignored, there are situations where it can assume importance, for example in immunoassays where steroids extracted from biological matrices may not always behave in the same way as standard steroids used for the standard curve. When mass spectrometry is used, lipids can have matrix effects on the analysis. Fingernails are modified hair and are now used to determine steroid content (Binz et al., 2018). The material is harder than hair so needs harsher methods to break down the nails before extraction. There are religious reasons where nails are acceptable samples when hair is not appropriate.

2.2.5.2 Tissue samples

Tissues can be fresh or frozen or freeze dried. Tissues are usually chopped/homogenized/blended/ground using a number of manual and instrument devices before the extraction step. As with blood and urine samples the internal standard mix for the analysis will be added early in this procedure. The extraction process can be assisted by ultrasonic techniques. Such extraction systems take up steroid conjugates with varying degrees of efficiency when used in combination with acid or alkaline pH and added salts. The use of chloroform:methanol mixtures have the added advantage that they also disrupt steroid protein binding by denaturing the protein. The protein precipitate can then be removed by centrifugation, leaving a solvent:water mixture which can be further extracted by the addition of suitable solvent such as chloroform which is immiscible with water. This causes the formation of two layers, the bottom of which contains the unconjugated steroids which can be removed. Acetonitrile was widely used as an extraction solvent for steroid metabolites in a similar fashion and the aqueous extract, after removal of the protein by centrifugation, was used for further processing and solid-phase purification (e.g., using Sep-Pak silica cartridges) if necessary.

2.2.6 Summary of sample preparation

Many techniques are available for sample preparation that in general avoid the use of large volumes of solvents and over time, considerably reduced sample sizes and costs of reagents. Liquid-liquid extraction is a cheap option but solid extraction systems are now used more often. The description of some early methods of sample preparation has been included since they can still be useful depending on circumstances of the experiment.

Reverse-phase extraction using C_{18} (ODS-) cartridges is today by far the most popular material for SPE of steroids. Small ODS-coated cartridges has also been used prior to HPLC as a guard column. Diluted samples can be injected into the column, washed and then eluted onto the analytical HPLC column with methanol. Ion exchange systems will be increasingly used for conjugated steroids. Several automated techniques for sample preparation are now available.

Whatever method is used for the extraction of the steroids, some form of purification before quantitation may be necessary. Purification can firstly remove potentially interfering compounds of similar structure which can be present in higher concentrations than the analyte itself. High performance liquid chromatography (HPLC) and gas chromatography (GLC) provide high resolution and thus achieve specificity which is further improved when these systems are coupled to a mass spectrometer (MS). Capillary electrophoresis has also been used but has not proven to be used in clinical assays.

2.2.7 Further reading

Some references were excluded from citation in chapter text but may interest the reader.

2.2.7.1 Preanalytical issues

Kirwan et al. (2018), Ma and Chowdhury (2011), Pizzato et al. (2017), and Sharma et al. (2019).

2.2.7.2 Steroid profiles

Ankarberg-Lindgren et al. (2020), Fiet et al. (1980), Greaves et al. (2014), Honour et al. (2018), Saracino et al. (2014), Shackleton (2012), Taylor (2013), Voegel et al. (2018), and Yuan et al. (2020).

2.2.7.3 Extraction techniques

Aufartová et al. (2011), Derks and Drayer (1978), Egawa et al. (2005), Giton et al. (2009), He et al. (2005), Hennion (1999), Jeannot and Cantwell (1997), Kawaguchi et al. (2009), Kole et al. (2011), Kumazawa et al. (2010), Luque-Córdoba and Priego-Capote (2021), Olędzka et al. (2013), O'Shannessy and Renwick (1983), Ramautar and de Jong (2014), Romstrom et al. (1996), and Wei and Mizaikoff (2007).

2.2.7.4 Chromatography

Casals et al. (2014), Chen et al. (2009), Cimpoiu et al. (2006), Kornel and Hill (1961), Luque-Córdoba et al. (2020), Miękus et al. (2017), Morineau et al. (1997), Nováková and Vlcková (2009), Poole and Dias (2000), Setchell and Shackleton (1973), Snyder et al. (2010), and Volin (2001).

2.2.7.5 Conjugate processing

Gagné et al. (2008), Gomes et al. (2009), Iannone et al. (2020), Jänne et al. (1969), Messeri et al. (1984), and Setchell et al. (1976).

2.2.7.6 Method comparison

Gomez et al. (2014) and Pujos et al. (2005).

References

Alexander C, Andersson HS, Andersson LI, Ansell RJ, Kirsch N, Nicholls IA, et al. Molecular imprinting science and technology: a survey of the literature for the years up to and including 2003. J Mol Recognit 2006;19(2):106–80.

Alvarez-Sánchez B, Priego-Capote F, Luque de Castro MD. Ultrasound-enhanced enzymatic hydrolysis of conjugated female steroids as pretreatment for their analysis by LC-MS/MS in urine. Analyst 2009;134(7):1416–22.

Anizan S, Bichon E, Monteau F, Cesbron N, Antignac JP, Le Bizec B. A new reliable sample preparation for high throughput focused steroid profiling by gas chromatography-mass spectrometry. J Chromatogr A 2010;1217(43):6652–60.

Ankarberg-Lindgren C, Andersson MX, Dahlgren J. Determination of estrone sulfate, testosterone, androstenedione, DHEAS, cortisol, cortisone, and 17α-hydroxyprogesterone by LC-MS/MS in children and adolescents. Scand J Clin Lab Invest 2020;80(8):672–80.

Aufartová J, Mahugo-Santana C, Sosa-Ferrera Z, Santana-Rodríguez JJ, Nováková L, Solich P. Determination of steroid hormones in biological and samples using green microextraction techniques: an overview. Anal Chim Acta 2011;704(1–2):33–46.

Axelson M, Sahlberg BH. Group separation and gas chromatography-mass spectrometry of conjugated steroids in plasma. J Steroid Biochem 1983;18:313–21.

Binz TM, Gaehler F, Voegel CD, Hofmann M, Baumgartner MR, Kraemer T. Systematic investigations of endogenous cortisol and cortisone in nails by LC-MS/MS and correlation to hair. Anal Bioanal Chem 2018;410(20):4895–903.

Bradlow HL. Modified technique for th elution of polar steroid conjugates from Amberlite-XAD-2. Steroids 1977;30(4):581–2.

Brown JB. A chemical method for the determination of oestriol, oestrone and oestradiol in human urine. Biochem J 1955;60(2):185–93.

Burstein S, Lieberman S. Hydrolysis of ketosteroid hydrogen sulfates by solvolysis procedures. J Biol Chem 1958;233(2):331–5.

Caccamo F, Carfagnini G, Di Corcia A, Samperi R. Measurement of urinary estriol glucuronides during the menstrual cycle by high-performance liquid chromatography. J Chromatogr 1988;434(1):61–70.

Casals G, Marcos J, Pozo OJ, Alcaraz J, Martínez de Osaba MJ, Jiménez W. Microwave-assisted derivatization: application to steroid profiling by gas chromatography/mass spectrometry. J Chromatogr B Analyt Technol Biomed Life Sci 2014;960:8–13.

Cawley AT, Kazlauskas R, Trout GJ, George AV. Determination of urinary steroid sulfate metabolites using ion paired extraction. J Chromatogr B Analyt Technol Biomed Life Sci 2005;825(1):1–10.

Chen HX, Deng QP, Zhang LW, Zhang XX. Quantification of testosterone and epitestosterone in biological samples by capillary electrophoresis with immunoaffinity extraction. Talanta 2009;78 (2):464–70.

Christakoudi S, Cowan DA, Taylor NF. Sodium ascorbate improves yield of urinary steroids during hydrolysis with *Helix pomatia* juice. Steroids 2008;73:309–19.

Cimpoiu C, Hosu A, Hodisan S. Analysis of some steroids by thin-layer chromatography using optimum mobile phases. J Pharm Biomed Anal 2006;41(2):633–7.

Cooper GA, Kronstrand R, Kintz P, Society of Hair Testing. Society of Hair Testing guidelines for drug testing in hair. Forensic Sci Int 2012;218 (1–3):20–4.

Derks HJ, Drayer NM. Improved methods for isolating cortisol metabolites from neonatal urine. Clin Chem 1978;24(7):1158–62.

Ebner MJ, Corol DI, Havlíková H, Honour JW, Fry JP. Identification of neuroactive steroids and their precursors and metabolites in adult male rat brain. Endocrinology 2006;147(1):179–90.

Egawa Y, Shimura Y, Nowatari Y, Aiba D, Juni K. Preparation of molecularly imprinted cyclodextrin microspheres. Int J Pharm 2005;293(1–2):165–70.

Faix JD. Principles and pitfalls of free hormone measurements. Best Pract Res Clin Endocrinol Metab 2013;27(5):631–45.

Few JD. A simple method for the separate estimation of 11-deoxy and 11-oxygenated 17-hydroxycortico-steroids in human urine. J Endocrinol 1968;41(2):213–22.

Fiet J, Gourmel B, Villette JM, Brerault JL, Julien R, Cathelineau G, et al. Simultaneous radioimmunoassay of androstenedione, dehydroepiandrosterone and 11-beta-hydroxyandrostenedione in plasma. Horm Res 1980;13(3):133–49.

Fotsis T, Adlercreutz H, Järvenpää P, Setchell KD, Axelson M, Sjövall J. Group separation of steroid conjugates by DEAE-Sephadex anion exchange chromatography. J Steroid Biochem 1981;14(5):457–63.

Gagné S, Laterreur J, Mahrouche L, Sørensen D, Gauthier JY, Truong VL, et al. Selective isolation of in vitro phase II conjugates using a lipophilic anionic exchange solid phase extraction method. J Chromatogr B Analyt Technol Biomed Life Sci 2008;863(2):242–8.

Gao W, Kirschbaum C, Grass J, Stalder T. LC-MS based analysis of endogenous steroid hormones in human hair. J Steroid Biochem Mol Biol 2016;162:92–9.

Giton F, Guéchot J, Fiet J. New reusable Celite/ethylene glycol cartridges for selective chromatography of steroids before immunoassay. Clin Biochem 2009;42(16–17):1735–8.

Gomes L, Meredith W, Snape CE, Sephton MA. Analysis of conjugated steroid androgens: deconjugation, derivatisation and associated issues. J Pharm Biomed Anal 2009;49:1133–40.

Gomez C, Fabregat A, Pozo OJ, Marcos J, Segura J, Ventura R. Analytical strategies based on mass spectrometric techniques for the study of steroid metabolism. Trends Anal Chem 2014;53:106–16.

Greaves RF, Jevalikar G, Hewitt JK, Zacharin MR. A guide to understanding the steroid pathway: new insights and diagnostic implications. Clin Biochem 2014;47(15):5–15.

Greff MJE, Levine JM, Abuzgaia AM, Elzagallaai AA, Rieder MJ, van Uum SHM. Hair cortisol analysis: an update on methodological considerations and clinical applications. Clin Biochem 2019;63:1–9.

Grignon C, Dupuis A, Albouy-Llaty M, Condylis M, Barrier L, Carato P, et al. Validation of a probe for assessing deconjugation of glucuronide and sulfate phase II metabolites assayed through LC-MS/MS in biological matrices. J Chromatogr B Analyt Technol Biomed Life Sci 2017;1061-1062:72–8.

He C, Li S, Liu H, Li K, Liu F. Extraction of testosterone and epitestosterone in human urine using aqueous two-phase systems of ionic liquid and salt. J Chromatogr A 2005;1082(2):143–9.

Hennion MC. Solid-phase extraction: method development, sorbents, and coupling with liquid chromatography. J Chromatogr A 1999;856:3–54 [review].

Hinchliffe E, Carter S, Owen LJ, Keevil BG. Quantitation of aldosterone in human plasma by ultra high performance liquid chromatography tandem mass spectrometry. J Chromatogr B Analyt Technol Biomed Life Sci 2013;913–914:19–23.

Honour JW, Conway E, Hodkinson R, Lam F. The evolution of methods for urinary steroid metabolomics in clinical investigations particularly in childhood. J Steroid Biochem Mol Biol 2018;181:28–51.

Horning MG, Chambaz EM, Brooks CJ, Moss AM, Boucher EA, Horning EC, et al. Characterization and estimation of urinary steroids of the newborn human by gas-phase analytical methods. Anal Biochem 1969;31(1):512–31.

Hutchins RF, Kaplanis JN. Sterol sulfates in an insect. Steroids 1969;13 (5):605–14.

Iannone M, Botrè F, Martinez-Brito D, Matteucci R, de la Torre X. Development and application of analytical procedures for the GC-MS/MS analysis of the sulfates metabolites of anabolic androgenic steroids: the pivotal role of chemical hydrolysis. J Chromatogr B Analyt Technol Biomed Life Sci 2020;1155, 122280.

Ikegawa S, Hasegawa M, Okihara R, Shimidzu C, Chiba H, Iida T, et al. Simultaneous determination of twelve tetrahydrocorticosteroid glucuronides in human urine by liquid chromatography/electrospray ionization-linear ion trap mass spectrometry. Anal Chem 2009;81 (24):10124–35.

Jänne O, Vihko R, Sjövall J, Sjövall K. Determination of steroid mono- and disulfates in human plasma. Clin Chim Acta 1969;23:405–12.

Jeannot MA, Cantwell FF. Solvent microextraction as a speciation tool: determination of free progesterone in a protein solution. Anal Chem 1997;69(15):2935–40.

Kawaguchi M, Takatsu A. Miniaturized hollow fiber assisted liquid-phase microextraction and gas chromatography-mass spectrometry for the measurement of progesterone in human serum. J Chromatogr B Analyt Technol Biomed Life Sci 2009;877(3):343–6.

Kawaguchi M, Ito R, Saito K, Nakazawa H. Novel stir bar sorptive extraction methods for environmental and biomedical analysis. J Pharm Biomed Anal 2006;40(3):500–8.

Kawaguchi M, Fujii S, Itoh N, Ito R, Nakazawa H, Takatsu A. Development of vial wall sorptive extraction and its application to determination of progesterone in human serum. J Chromatogr A 2009;1216(44):7553–7.

Keevil BG, Adaway J. Assessment of free testosterone concentration. J Steroid Biochem Mol Biol 2019;190:207–11.

Kintz P. Testing for anabolic steroids in hair: a review. Leg Med (Tokyo) 2003;5(Suppl 1):S29–33.

Kirchhoff F, Briegel J, Vogeser M. Quantification of free serum cortisol based on equilibrium dialysis and isotope dilution-liquid chromatography-tandem mass spectrometry. Clin Biochem 2011;44 (10–11):894–9.

Kirwan JA, Brennan L, Broadhurst D, Fiehn O, Cascante M, Dunn WB, et al. Preanalytical processing and biobanking procedures of biological samples for metabolomics research: a white paper, community perspective (for "Precision Medicine and Pharmacometabolomics Task Group"-The Metabolomics Society Initiative). Clin Chem 2018;64 (8):1158–82.

Kole PL, Venkatesh G, Kotecha J, Sheshala R. Recent advances in sample preparation techniques for effective bioanalytical methods. Biomed Chromatogr 2011;25:199–217.

Kornel L. Studies on steroid conjugates. 3. Separation of conjugated steroids by means of high voltage paper electrophoresis. J Clin Endocrinol Metab 1964;24:956–64.

Kornel L, Hill Jr SR. Paper chromatographic pattern of endogenous urinary corticosteroids in normal subjects. Metabolism 1961; 10:18–26.

Kornel L, Starnes WR, Hill A. Studies on steroid conjugates. VI. Quantitative paper chromatography of urinary corticosteroids in essential hypertension. J Clin Endocrinol Metab 1969;29(12):1608–17.

Kumazawa T, Hasegawa C, Lee X-O, Sato K. New and unique methods of solid phase extraction for use before instrumental analysis of xenobiotics in human specimens. Forensic Toxicol 2010;28:61–8.

Lavén M, Alsberg T, Yu Y, Adolfsson-Erici M, Sun H. Serial mixed-mode cation- and anion-exchange solid-phase extraction for separation of basic, neutral and acidic pharmaceuticals in wastewater and analysis

by high-performance liquid chromatography-quadrupole time-of-flight mass spectrometry. J Chromatogr A 2009;1216(1):49–62.

Liere P, Pianos A, Eychenne B, Cambourg A, Bodin K, Griffiths W, et al. Analysis of pregnenolone and dehydroepiandrosterone in rodent brain: cholesterol autoxidation is the key. J Lipid Res 2009;50(12): 2430–44.

Liu W, Zhang L, Fan L, Lin Z, Cai Y, Wei Z, et al. An improved hollow fiber solvent-stir bar microextraction for the preconcentration of anabolic steroids in biological matrix with determination by gas chromatography-mass spectrometry. J Chromatogr A 2012;1233:1–7.

Luo G, Li Y, Bao JJ. Development and application of a high-throughput sample cleanup process based on 96-well plate for simultaneous determination of 16 steroids in biological matrices using liquid chromatography-triple quadrupole mass spectrometry. Anal Bioanal Chem 2016;408(4):1137–49.

Luque-Córdoba D, Priego-Capote F. Fully automated method for quantitative determination of steroids in serum: an approach to evaluate steroidogenesis. Talanta 2021;224, 121923.

Luque-Córdoba D, López-Bascón MA, Priego-Capote F. Development of a quantitative method for determination of steroids in human plasma by gas chromatography-negative chemical ionization-tandem mass spectrometry. Talanta 2020;220, 121415.

Ma S, Chowdhury SK. Analytical strategies for assessment of human metabolites in preclinical safety testing. Anal Chem 2011;83 (13):5028–36.

Manaf NA, Saad B, Mohamed MH, Wilson LD, Latiff AA. Cyclodextrin based polymer sorbents for micro-solid phase extraction followed by liquid chromatography tandem mass spectrometry in determination of endogenous steroids. J Chromatogr A 2018;1543:23–33.

Márta Z, Bobály B, Fekete J, Magda B, Imre T, Mészáros KV, et al. Simultaneous determination of thirteen different steroid hormones using micro UHPLC-MS/MS with on-line SPE system. J Pharm Biomed Anal 2018;150:258–67.

Meng LJ, Sjövall J. Method for combined analysis of profiles of conjugated progesterone metabolites and bile acids in serum and urine of pregnant women. J Chromatogr B Biomed Sci Appl 1997;688(1):11–26.

Messeri G, Cugnetto G, Moneti G, Serio M. Helix pomatia induced conversion of some 3 beta-hydroxysteroids. J Steroid Biochem 1984;20:793–6.

Miękus N, Konieczna L, Kowiański P, Moryś J, Bączek T. HILIC-MS rat brain analysis, a new approach for the study of ischemic attack. Transl Neurosci 2017;8:70–5.

Morineau G, Gosling J, Patricot MC, Soliman H, Boudou P, al Halnak A, et al. Convenient chromatographic prepurification step before measurement of urinary cortisol by radioimmunoassay. Clin Chem 1997;43(5):786–93.

Murphy BE. "Sephadex" column chromatography as an adjunct to competitive protein binding assays of steroids. Nat New Biol 1971;232 (27):21–4. https://doi.org/10.1038/newbio232021a0. 5284423.

Nováková L. Challenges in the development of bioanalytical liquid chromatography-mass spectrometry method with emphasis on fast analysis. J Chromatogr A 2013;1292:25–37.

Nováková L, Vlcková H. A review of current trends and advances in modern bio-analytical methods: chromatography and sample preparation. Anal Chim Acta 2009;656(1–2):8–35.

O'Shannessy DJ, Renwick AG. Extraction and separation of androstenedione from products of aromatase assays on micro-columns of magnesium oxide. J Chromatogr 1983;278(1):151–5.

Olędzka I, Kowalski P, Dziomba S, Szmudanowski P, Bączek T. Optimization of a pre-MEKC separation SPE procedure for steroid molecules in human urine samples. Molecules 2013;18(11):14013–32.

Pizzato EC, Filonzi M, Rosa HSD, de Bairros AV. Pretreatment of different biological matrices for exogenous testosterone analysis: a review. Toxicol Mech Methods 2017;27(9):641–56.

Polson C, Sarkar P, Incledon B, Raguvaran V, Grant R. Optimization of protein precipitation based upon effectiveness of protein removal and ionization effect in liquid chromatography-tandem mass spectrometry. J Chromatogr B Analyt Technol Biomed Life Sci 2003;785(2):263–75.

Poole CF, Dias NC. Practitioner's guide to method development in thin-layer chromatography. J Chromatogr A 2000;892(1–2):123–42.

Prieto A, Vallejo A, Zuloaga O, Paschke A, Sellergen B, Schillinger E, et al. Selective determination of estrogenic compounds in water by microextraction by packed sorbents and a molecularly imprinted polymer coupled with large volume injection-in-port-derivatization gas chromatography-mass spectrometry. Anal Chim Acta 2011;703 (1):41–51.

Pujos E, Flament-Waton MM, Paisse O, Grenier-Loustalot MF. Comparison of the analysis of corticosteroids using different techniques. Anal Bioanal Chem 2005;381(1):244–54.

Qaiser MZ, Dolman DEM, Begley DJ, Abbott NJ, Cazacu-Davidescu M, Corol DI, et al. Uptake and metabolism of sulphated steroids by the blood-brain barrier in the adult male rat. J Neurochem 2017;142 (5):672–85.

Ramautar R, de Jong GJ. Recent developments in liquid-phase separation techniques for metabolomics. Bioanalysis 2014;6(7):1011–26.

Ray JA, Kushnir MM, Bunker A, Rockwood AL, Meikle AW. Direct measurement of free estradiol in human serum by equilibrium dialysis-liquid chromatography-tandem mass spectrometry and reference intervals of free estradiol in women. Clin Chim Acta 2012;413(11–12):1008–14.

Romstrom O, Ye L, Mosbach K. Artificial antibodies to corticosteroids prepared by molecular imprinting. Chem Biol 1996;3(6):471–7.

Saracino MA, Iacono C, Somaini L, Gerra G, Ghedini N, Raggi MA. Multimatrix assay of cortisol, cortisone and corticosterone using a combined MEPS-HPLC procedure. J Pharm Biomed Anal 2014;88:643–8.

Schänzer W. Metabolism of anabolic androgenic steroids. Clin Chem 1996;42:1001–20.

Schöneshöfer M. Simultaneous determination of eight adrenal steroids in human serum by radioimmunoassay. J Steroid Biochem 1977;8 (9):995–1009.

Schöneshöfer M, Weber B. Specific estimation of fifteen unconjugated, non-metabolized steroid hormones in human urine. J Steroid Biochem 1983;18(1):65–73.

Setchell KD, Shackleton CH. The group separation of plasma and urinary steroids by column chromatography on sephadex LH-20. Clin Chim Acta 1973;47(3):381–8.

Setchell KD, Almé B, Axelson M, Sjövall J. The multicomponent analysis of conjugates of neutral steroids in urine by lipophilic ion exchange chromatography and computerised gas chromatography-mass spectrometry. J Steroid Biochem 1976;7:615–29.

Shackleton CH. Role of a disordered steroid metabolome in the elucidation of sterol and steroid biosynthesis. Lipids 2012;47(1):1–12.

Shackleton CH, Whitney JO. Use of Sep-Pak cartridges for urinary steroid extraction; evaluation of the method prior to gas chromatographic analysis. Clin Chim Acta 1980;107:581–2.

Sharma MK, Dhakne P, Nn S, Reddy PA, Sengupta P. Paradigm shift in the arena of sample preparation and bioanalytical approaches involving liquid chromatography mass spectroscopic technique. Anal Sci 2019;35(10):1069–82.

Shen S, Li Y, Wakida S, Takeda S. Determination of adrenal steroids by microfluidic chip using micellar electrokinetic chromatography. Environ Monit Assess 2009;153(1–4):201–8.

Shibasaki H, Tanabe C, Furuta T, Kasuya Y. Hydrolysis of conjugated steroids by the combined use of beta-glucuronidase preparations from *Helix pomatia* and *ampullaria*: determination of urinary cortisol and its metabolites. Steroids 2001;66:795–801.

Snyder LR, Kirkland JJ, Dolan JW. Introduction to modern liquid chromatography. Wiley; 2010, ISBN:978 0 470 16754 0.

Södergård R, Bäckström T, Shanbhag V, Carstensen H. Calculation of free and bound fractions of testosterone and estradiol-17 beta to human plasma proteins at body temperature. J Steroid Biochem 1982;16 (6):801–10.

Stevenson BJ, Waller CC, Ma P, et al. *Pseudomonas aeruginosa* arylsulfatase: a purified enzyme for the mild hydrolysis of steroid sulfates. Drug Test Anal 2015;7(10):903–11.

Su P, Zhang XX, Chang WB. Development and application of a multitarget immunoaffinity column for the selective extraction of natural estrogens from pregnant women's urine samples by capillary electrophoresis. J Chromatogr B Analyt Technol Biomed Life Sci 2005;816(1–2):7–14.

Taylor NF. Urinary steroid profiling. Methods Mol Biol 2013;1065:259–76.

Vermeulen A, Verdonck L, Kaufman JM. A critical evaluation of simple methods for the estimation of free testosterone in serum. J Clin Endocrinol Metab 1999;84(10):3666–72.

Viljanto M, Pita CH, Scarth J, Walker CJ, Kicman AT, Parkin MC. Important considerations for the utilisation of methanolysis in steroid analysis. Drug Test Anal 2018;10(9):1469–73.

Vlčková H, Pilařová V, Novák O, Solich P, Nováková L. Micro-SPE in pipette tips as a tool for analysis of small-molecule drugs in serum. Bioanalysis 2017;9(11):887–901.

Voegel CD, La Marca-Ghaemmaghami P, Ehlert U, Baumgartner MR, Kraemer T, Binz TM. Steroid profiling in nails using liquid chromatography-tandem mass spectrometry. Steroids 2018;140:144–50.

Voegel CD, Hofmann M, Kraemer T, Baumgartner MR, Binz TM. Endogenous steroid hormones in hair: investigations on different hair types, pigmentation effects and correlation to nails. Steroids 2020;154, 108547.

Volin P. Analysis of steroidal lipids by gas and liquid chromatography. J Chromatogr A 2001;935(1–2):125–40.

Wang R, Hartmann MF, Wudy SA. Targeted LC-MS/MS analysis of steroid glucuronides in human urine. J Steroid Biochem Mol Biol 2021;205, 105774.

Wei S, Mizaikoff B. Binding site characteristics of 17beta-estradiol imprinted polymers. Biosens Bioelectron 2007;23(2):201–9.

Wennig R. Potential problems with the interpretation of hair analysis results. Forensic Sci Int 2000;107(1–3):5–12.

Whitcombe MJ, Kirsch N, Nicholls IA. Molecular imprinting science and technology: a survey of the literature for the years 2004–2011. J Mol Recognit 2014;27(6):297–401.

Wong T, Shackleton CH, Covey TR, Ellis G. Identification of the steroids in neonatal plasma that interfere with 17 alpha-hydroxyprogesterone radioimmunoassays. Clin Chem 1992;38(9):1830–7.

Ye L, Wang Q, Xu J, Shi ZG, Xu L. Restricted-access nanoparticles for magnetic solid-phase extraction of steroid hormones from environmental and biological samples. J Chromatogr A 2012; 1244:46–54.

Yuan TF, Le J, Wang ST, Li Y. An LC/MS/MS method for analyzing the steroid metabolome with high accuracy and from small serum samples. J Lipid Res 2020;61(4):580–6.

Zhan Y, Musteata FM, Basset FA, Pawliszyn J. Determination of free and deconjugated testosterone and epitestosterone in urine using SPME and LC-MS/MS. Bioanalysis 2011;3(1):23–30.

Zheng L, Luo X, Zhu L, Xie W, Liu S, Cheng Z. Simultaneous determination of cortisol, cortisone, 6β-hydroxycortisol and 6β-hydroxycortisone by HPLC. J Chromatogr Sci 2015;53(4):451–5.

Books

Makin HLJ, Gower DB. Steroid analysis. 2nd ed. Springer; 2010, ISBN:978 1 4020 9774 4.

Reviews

Nozaki O. Steroid analysis for medical diagnosis. J Chromatogr A 2001;935(1–2):267–78.

Spencer RL, Deak T. A users guide to HPA axis research. Physiol Behav 2017;2017(178):43–65.

Chapter 2.3

Steroid determination—Purification of extracts

2.3.1 Introduction

After the extraction of the steroids from a biological sample, some form of purification before a quantitative determination may be necessary. Purification can first remove potentially interfering compounds of similar structure which can be present in higher concentrations than the analyte itself. High-performance liquid chromatography (HPLC) and gas chromatography (GLC) are now the major methods that provide high resolution and thus achieve specificity which is further improved when these systems are coupled to an immunoassay system or a mass spectrometer (MS). Capillary electrophoresis has limited use. The steroid metabolites, most of which are conjugated, are of interest in addition to the plasma steroid hormones. The removal of the conjugate groups by hydrolysis was discussed in the previous chapter. This has been the usual practice but there are now methods capable of the analysis of the intact conjugates so the free, glucuronide and sulfate groups need to be separated or distinguished. Cedric Shackleton wrote an excellent review covering publications from 1986 to 2018 and the whole topic is covered in the Makin and Gower book of 2010. In clinical laboratories where rapid methods are needed, many of the methods described in this chapter will not be needed except possibly during development of a method when getting to understand more about the matrix of the samples under investigation.

2.3.2 Chromatography

Steroids in a mixture can be separated by chromatography by distributing its components between two phases. The **stationary phase** remains fixed in place, while the **mobile phase** carries the components of the mixture through the medium being used. The stationary phase acts as a constraint on many of the components in a mixture, slowing them down to move slower in the mobile phase. The stationary phase requires a **support** which can be particles of silica, synthetic beads or the internal walls of a very fine tube. The movement of the components in the mobile phase is controlled by the significance of the interactions between the mobile and/or stationary phases. Because of the differences in factors such as the solubility of certain components in the mobile phase and the strength of their affinities for the stationary phase, some components will move faster than others, thus facilitating the separation of the components within that mixture. The mobile phase can be a liquid or a gas and the power of separation ranges from crude liquid-solid chromatography in a **column**, **paper** sheet or **thin layer** plate to **high-pressure liquid chromatography** and **gas chromatography** with immense resolution. In gas chromatography, the stationary phase is coated on particles that fill a tube or coated on the wall itself. For analysis by gas chromatography, the steroids have to be chemically modified to lower the temperatures at which they are volatile and remain stable at the high temperatures used. Capillary column gas chromatography achieves much higher resolution than liquid chromatography with particles in a tube as the stationary phase. The reader should consult textbooks on the theory of chromatography for more detail.

2.3.2.1 Column chromatography

The stationary phase material coated onto particles packed into columns in this technique enables separation of the steroids on the basis of solubility in the mobile phase and interaction with the stationary phase. The column is usually glass but can be inert plastic, of any diameter and/or length. A column can be filled with the adsorbent material and steroids can be separated by selective adsorption to this material, being eluted from the column by solvents of increasing polarity passing through the column. Florisil (magnesium silicate), aluminum oxide, and silica have been used in a glass or inert plastic column. The basis of the separation can be:

- partition chromatography either by normal or reverse phase chromatography. Normal phase chromatography is based on a polar stationary phase and nonpolar mobile phase. Reverse phase chromatography separates compounds with hydrophobic moieties and
- absorption chromatography.

Steroids in the Laboratory and Clinical Practice. https://doi.org/10.1016/B978-0-12-818124-9.00002-4

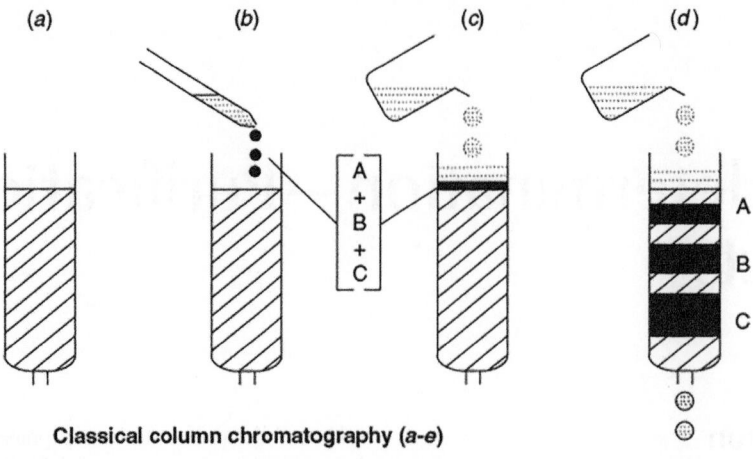

Classical column chromatography (a-e)

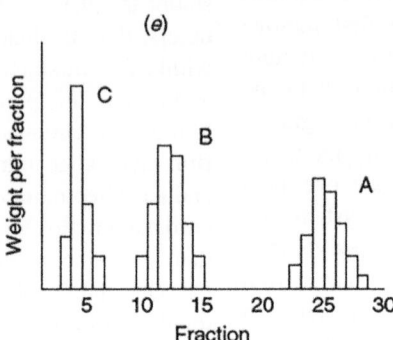

FIG. 2.3.1 Schematic of classical column chromatography.

The steroid of interest is separated in partition chromatography by its relative solubilities in mobile phase (usually an organic solvent) and the stationary phase (usually a hydrophilic water-based material) (Fig. 2.3.1).

Microparticulate silica gives excellent separations and replaced previous column separations using alumina or ordinary silica. Use of microparticulate silica can provide rapid and simple purification and if cartridges are used they may, after washing, be re-used many times and can thus be very cheap. Large numbers of samples have to be processed in clinical laboratories so any kind of column chromatography can be cumbersome and time consuming since columns are used by their very nature in sequential manner These adsorptive materials contain varying degrees of water, so the separation process still involves partition to a small degree.

The **stationary phase** is mixed with an inert support, often **Celite** (trade name for a form of finely powdered silica from a diatomaceous earth which has been washed and sieved), and the mixture is carefully packed into a column. The **mobile phase is** allowed to pass through by gravity usually from a reservoir attached to the top of the column. The stationary phase after drying can be modified with ethylene glycol or similar liquid in order to increase absorption/

adsorption to the stationary phase. The mobile phase can remain the same throughout the separation (isocratic elution) or it can be varied (gradient elution). The gradient can be achieved by precolumn mixing from two or more separate reservoirs and or using two pumps that change the flow rates of the separate solvents. In an early paper by Siiteri (1975), a 43 cm × 4.5 cm column was eluted in four steps (Fig. 2.3.2). The first consists of three 10 mL washes with 100 mL of isooctane. Step 2 is a linear gradient from mixing 200 mL of isooctane and 200 mL isooctane: ethyl acetate (7:3). More polar metabolites are eluted using a linear gradient from 200 mL of isooctane:ethyl acetate (7:3) and 200 mL ethyl acetate. Finally to elute very polar steroids like cortisol and estriol pure ethyl acetate is passed through the column. Three milliliter fractions were collected so a large multitube device was required.

Those columns do not approach the resolution which is achieved now using high-performance liquid chromatography (HPLC), although it was possible to measure, by separate immunoassays, up to eight steroids in small volumes of plasma (Anderson et al., 1976) (Fig. 2.3.3).

The column techniques are cheap, relatively simple to set up and are capable of dealing with large quantities of material but only practical in research studies. Both

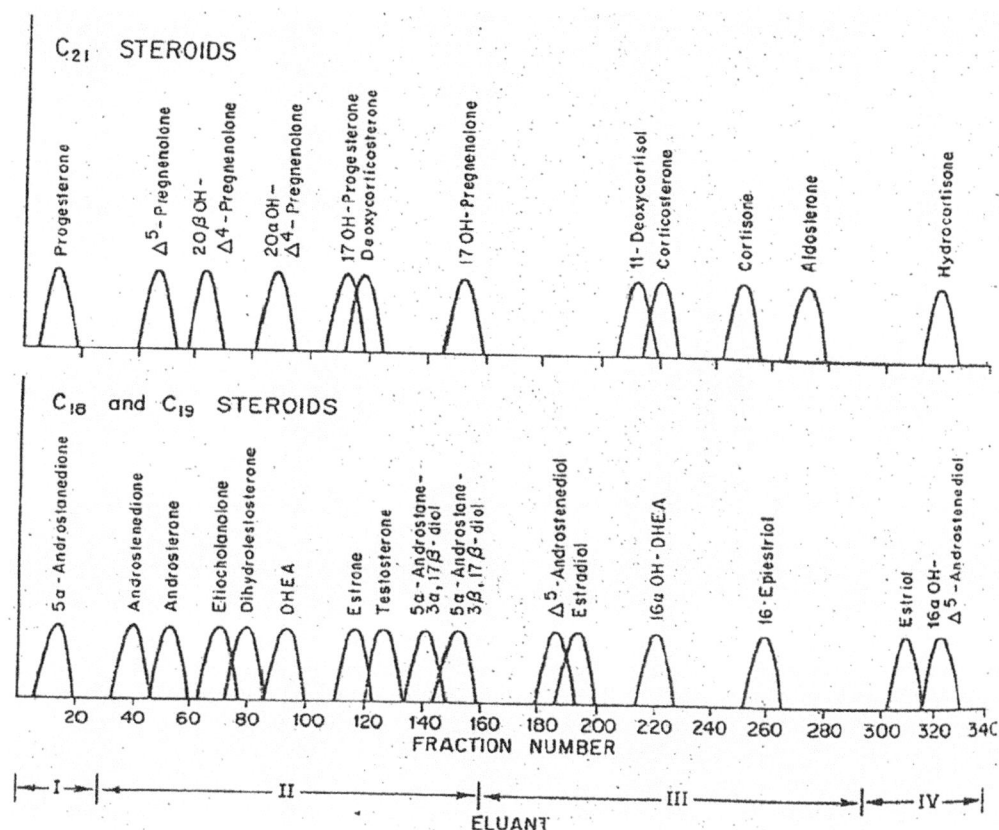

FIG. 2.3.2 Chromatography of steroids on a 120 g column Celite mixed with ethylene glycol. Samples are applied in 1 mL ethylene glycol and 0.5 mL isooctane. Eluant 1—130 mL isooctane; Eluant 2—a linear gradient of 200 mL of isooctane and 200 mL isooctane:ethylacetate 7:3; Eluant 3—a linear gradient of 200 mL of isooctane:ethyl acetate (7:3) and 200 mL ethylacetate; Eluant 4—100 mL ethyl acetate. Fractions of 3 mL were collected. *(From Siiteri PK. A universal chromatographic system for the separation of steroid hormones and their metabolites. Methods Enzymol 1975;36:485–89. Fig. 1 p. 486.)*

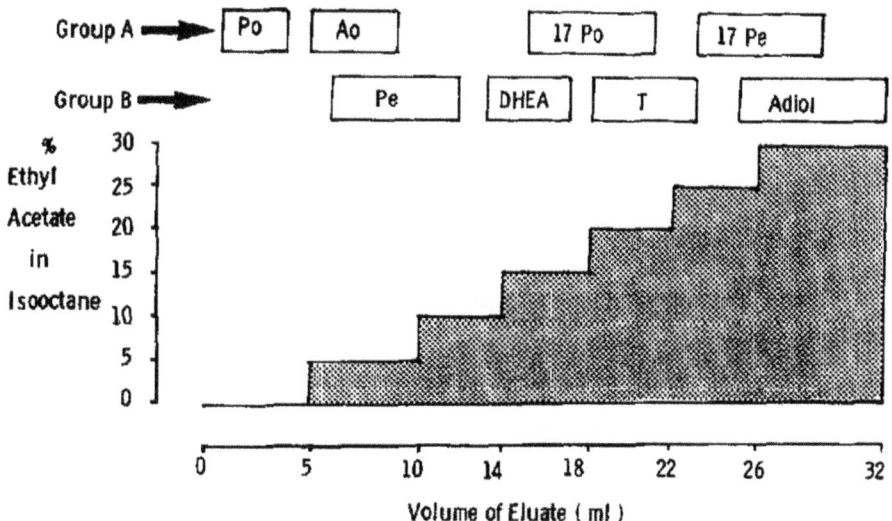

FIG. 2.3.3 Group fractionation of eight steroids. Steroids were extracted from plasma into 3 vols diethyl ether. The ether layer was dried and extract dissolved in 1 mL isooctane (I-O). The column was eluted with 5 mL I-O then increasing concentrations of ethyl acetate in I-O at 5% (5 mL), 10% (4 mL), 15% (4 mL), 20% (4 mL), 25% (4 mL), and 30% (6 mL). *Adiol*, androstenediol; *Ao*, androstenedione; *Pe*, pregnenolone; *Po*, progesterone; *17Pe*, 17-hydroxy-pregnenolone; *17Po*, 17-hydroxyprogesterone. *(From Anderson DC, Hopper BR, Lasley BL, Yen SS. A simple method for the assay of eight steroids in small volumes of plasma. Steroids 1976;28(2):179–96. Fig. 1 p. 182.)*

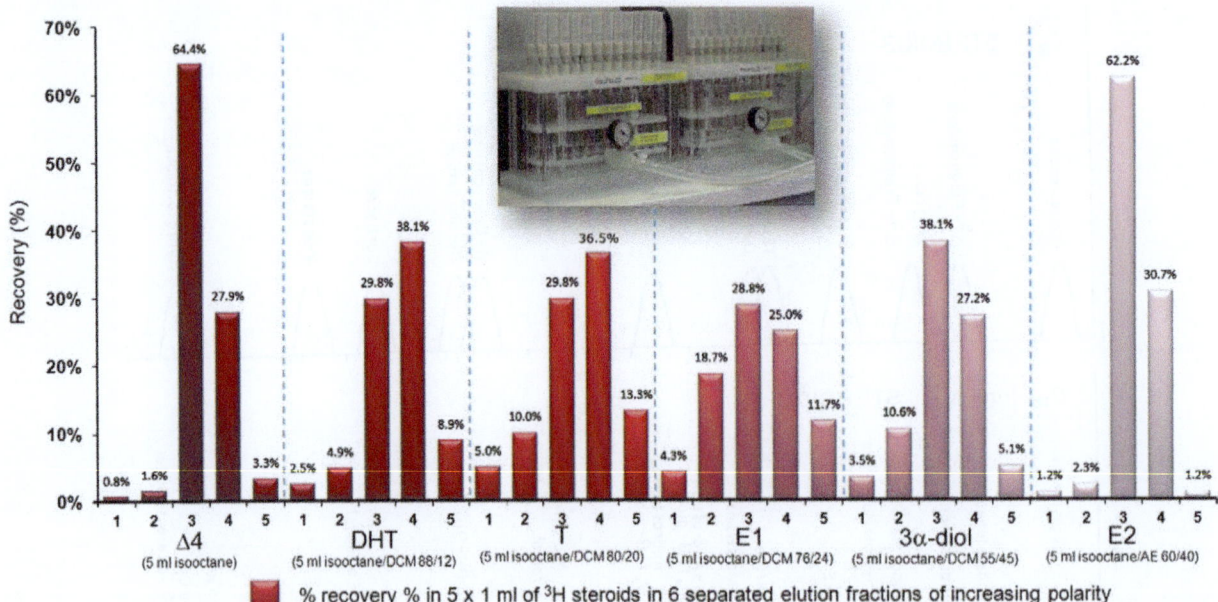

FIG. 2.3.4 Separation of steroids with Celite mini-columns in device for multiple cartridges. Columns contain 700 mg of Celite coated with ethylene glycol. Steroid fractions were eluted sequentially with isooctane, four fractions with I-O and increasing proportion of dichloromethane and finally with I-O and ethyl acetate (60:40). Column material was regenerated with methanol before reuse. *(From Giton F, Guechot J, Fiet J. New resuable Celite/ethylene glycol cartridges for selective chromatography of steroids before immunoassay. Clin Biochem 2009;42:1735–8. Fig. 1 p. 1737.)*

straight-phase and reverse-phase partition columns are available but rarely used today. The purity of the stationary phase should be checked before use, there are even batch to batch variations from the same supplier. Automated techniques were developed for up to six steroids (Giton et al., 2009) (Fig. 2.3.4).

Sephadex LH-20 is a hydroxypropylated, cross-linked dextran that swells in polar solvents. LH-20 is prepared from Sephadex G25. Chromatographic separation of organic compounds can be achieved on the gel by molecular sieving, reversible solute gel interactions or partition between solvent mixtures. Benzene and methanol were originally suitable solvent systems. Heptane, isooctane, hexane and cyclohexane were the basic solvents used for chromatography of various neutral steroids. These systems required the addition of small amounts of benzene, ethyl acetate or chloroform together with methanol. In some solvent mixtures, the gel floats in the column. Ken Setchell (Setchell and Shackleton, 1973) developed a method to separate steroids on a 6 g column of Sephadex LH-20 where the gel did not float. The column was eluted with 90 mL of cyclohexane: ethanol in 80:20 proportions, fractions are collected for analysis by immunoassay or mass spectrometry (Fig. 2.3.5). The steroids from plasma elute in fraction 1—progesterone, fraction 2—17-hydroxyprogesterone, 11-deoxycorticosterone, fraction 3—corticosterone, 11-deoxycortisol, 11-dehydrocorticosterone, fraction 4—cortisone, fraction 5—cortisol with increasing solvent volumes from 12 to 22 then 37, 55, 67 and 93 mL of solvent,

respectively. This takes several hours and is not compatible with rapid sample processing in clinical laboratories. The eluates from the columns did not affect the immunoassay process, unlike eluates from alumina and some silica columns.

2.3.2.1.1 Ion exchange

The ion-exchange DEAE-Sephadex was used to separate estrogen conjugates but this has not been widely used. The column was eluted with a sodium chloride in water gradient but was a lengthy procedure and only useful in a research setting for preparative separation of sulfate and glucuronide conjugates. The aqueous extracts needed further extraction back to organic solvents. The group led by Jan Sjovall refined Sephadex LH-20 in order to create gels that worked with a range of organic solvents and later had ion exchange properties. Lipidex was prepared from LH-20 by hydroxyalkylation at two concentrations (Lipidex 1000 with 10% of gel weight and Lipidex 5000 with 50% of gel weight). A number of anion exchange derivatives were prepared from chloro- and bromohydroxypropyl intermediates (Sjövall and Axelson, 1984). By using mixtures of water, alcohols and hydrocarbons both reverse and normal phase systems were created.

The gels can be made in the laboratory or purchased and are reusable. Lipophilic anion exchange derivatives of Sephadex have been used for group separation of neutral steroids, estrogens and estrogen glucuronides. DEAE-Sephadex

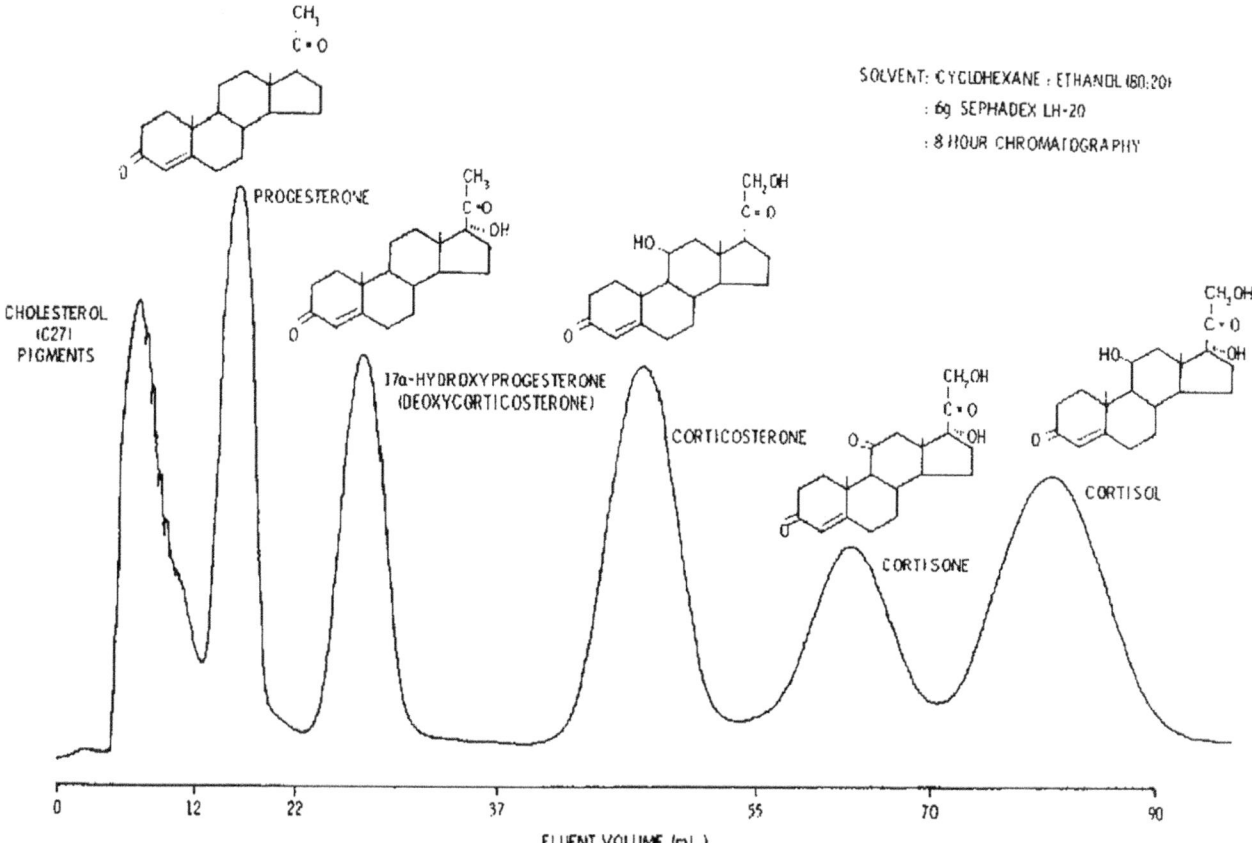

FIG. 2.3.5 Separation of steroids with Sephadex LH-20. A 6 g column of Sephadex LH-20 was eluted with cyclohexane:ethanol (80/20). *(From Setchell KD, Shackleton CH. The group separation of plasma and urinary steroids by column chromatography on sephadex LH-20. Clin Chim Acta 1973;47 (3):381–8. Fig. 3 p. 384.)*

A25 was used (Fotsis et al., 1981) to separate groups of estrogen conjugates. TEAP-LH-20 is stronger anion exchanger than DEAP-LH-20. TEAP-LH-20 was used to separate conjugate groups of steroids (Fig. 2.3.6) (Axelson and Sahlberg, 1983; Setchell et al., 1976). A cation exchanger (SP-Sephadex) is used to remove basic contaminants or to remove counter-ions from ion-pair extractions.

Neutral steroids are eluted in the first 4 mL rinse of 72% methanol. All further eluents are made in 72% methanol. Unconjugated phenolic steroids are eluted with 8 mL 72% methanol saturated with carbon dioxide. Monoglucuronides are eluted with 10 mL of 0.4 M formic, the monosulfates with 10 mL of 0.3 M acetic acid-potassium acetate pH 6.5, and the disulfates with 15 mL of potassium acetate-potassium hydroxide, 0.5 M in acetate apparent pH of 10.0. The whole process took more than 2 days so faster methods have been developed.

Ion exchange SPE is now being used more often than column techniques to purify extracts of plasma, urine and brain into conjugate groups. Ebner et al. (2006) homogenized rat brain and, after removal of lipid, steroids were extracted on a reverse phase Oasis HLB cartridge. Extracts were dissolved in 3.75 vol of 20% ethanol in phosphate buffer and passed through a 60 mg reverse phase Oasis mixed anion exchange (MAX) cartridge. After a wash with 5 mL of 20% ethanol in ammonium acetate buffer, the free steroids were eluted 4 mL ethyl acetate. Steroid glucuronide could be eluted with 20 mL of 60% ethanol in formate/pyridine buffer. Finally after a wash with 2 mL dry ethyl acetate the steroid sulfates were eluted in 15 mL of 50 mm benzene sulfonic acid in ethyl acetate saturated with 2 M H_2SO_4. Pranata et al. (2019) used WAX cartridges to purify steroid bisglucuronides and sulfate/glucuronide conjugates they had synthesized.

There will be further use of ion exchange SPE because of simplicity in use and because of the interest in conjugates in physiology and pathophysiology. There is however a shortage of reference preparations of the conjugates in native form and with stable isotope labels.

2.3.2.2 Paper chromatography

Paper chromatographic systems, widely used in the 1950s and 1960s, are seldom used today and a literature search will only occasionally unearth the use of a paper

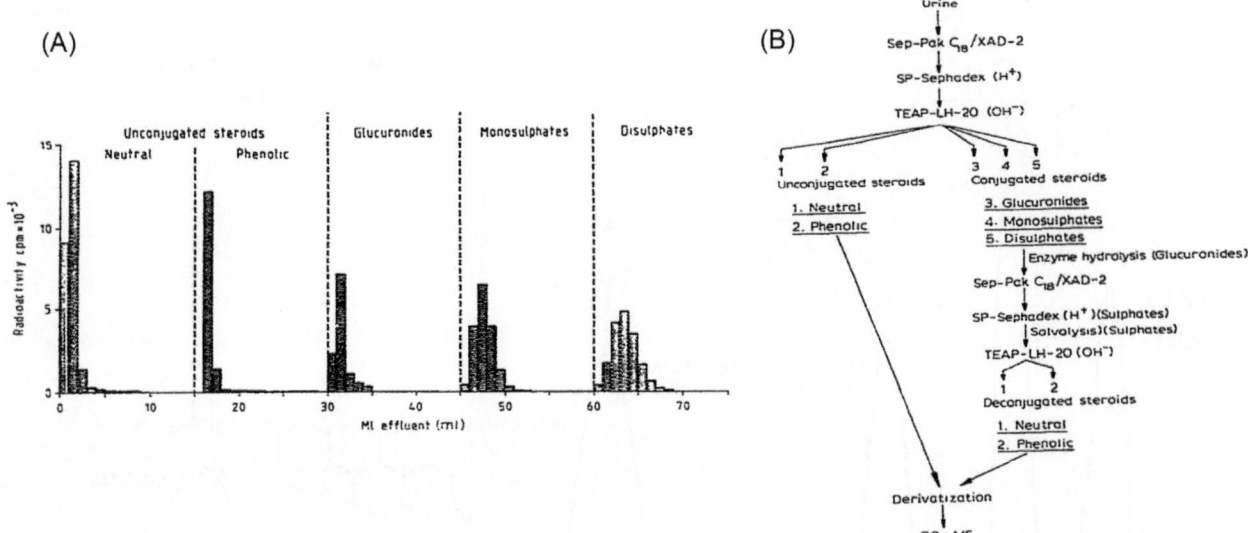

FIG. 2.3.6 Lipophilic ion exchange separation of steroid conjugates. (A) Group separation on TEAP-LH-20 (40 × 4 mm, OH-form in 72% v/v aqueous methanol) of radioactively labeled neutral steroids (cortisol, progesterone), phenolic steroids (estrone, estradiol, estriol), steroid glucuronides (testosterone and estriol glucuronides), steroid monosulfates (dehydroepiandrosterone sulfate) and steroid disulfates (5-androstene-3β,17β-diol disulfate) added to an extract of urine. (B) Steroids in conjugate groups were hydrolyzed and free steroids were recovered with Sep Pak cartridges or Amberlite XAD-2 columns. Extracts were purified with SP-Sephadex (a strong cation exchanger) and repeat TEAP-LH-20. *(From Axelson M, Sahlberg BL, Sjovall J. Analysis of profiles of conjugated steroids in urine by ion-exchange separation and gas chromatography-mass spectrometry. J Chromatogr 1981;224:355–70.)*

chromatographic separation. Radiolabeled steroids can be purified by paper chromatography prior to use and may sometimes be used for the purification of a steroid prior to quantitation. Paper chromatography, using solvent systems of methanol and water developed by Ivor Bush, is time consuming and significant amounts of potentially interfering nonspecific material is usually eluted from the paper together with the steroid of interest and closely related steroids (Fig. 2.3.7), giving rise to high blank values in immunoassays. Careful washing of the paper prior to chromatography can often reduce the blank values but the use of paper chromatographic systems for steroid separation today will not compete with GC and LC-MS/MS systems.

Comparison of steroid results by direct analysis and methods that incorporate chromatography need to be carefully assessed for recovery and accuracy. Large glass tanks [typically around 45 (w), 15 (d) and 60 cm (h)] were used in temperature-controlled rooms. Solvents such as benzene, toluene, and petroleum ether are not desirable these days. The mobile phase is used in the descending or ascending manner.

An early example is the analysis of pregnanediol in urine after enzyme hydrolysis of conjugates and extraction into benzene (Eberlein and Bongiovanni, 1958). The paper was developed with an ascending mixture of isooctane: methanol and water. Pregnanediol was separated from pregnanetriol and 3α-steroids that might be expected in urine but not resolved from the 3β-steroids that were included in the analysis but would not be present in urine.

The determination of allo-tetrahydrocortisol (allo-THF) in urine was a real challenge in 1958 and is only described here so the reader can appreciate the power of such methods. Steroids were extracted from urine and hydrolyzed to release conjugates. A preliminary fractionation was performed on paper developed with light petroleum:benzene: methanol:water (Bush and Willoughby, 1957). The resolution is low (Fig. 2.3.8A) and detection of the steroids required several methods. The steroids in fraction 3 that reacted with blue tetrazolium were eluted for a second chromatography developed with benzene:methanol:water.

Urine was collected from a subject after ACTH stimulation. After extraction and hydrolysis, steroids were separated with paper chromatography (Fig. 2.3.8B Lane B).

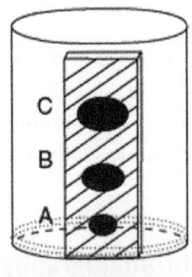

Paper or thin-layer chromatography

FIG. 2.3.7 Schematic for thin layer or paper chromatography.

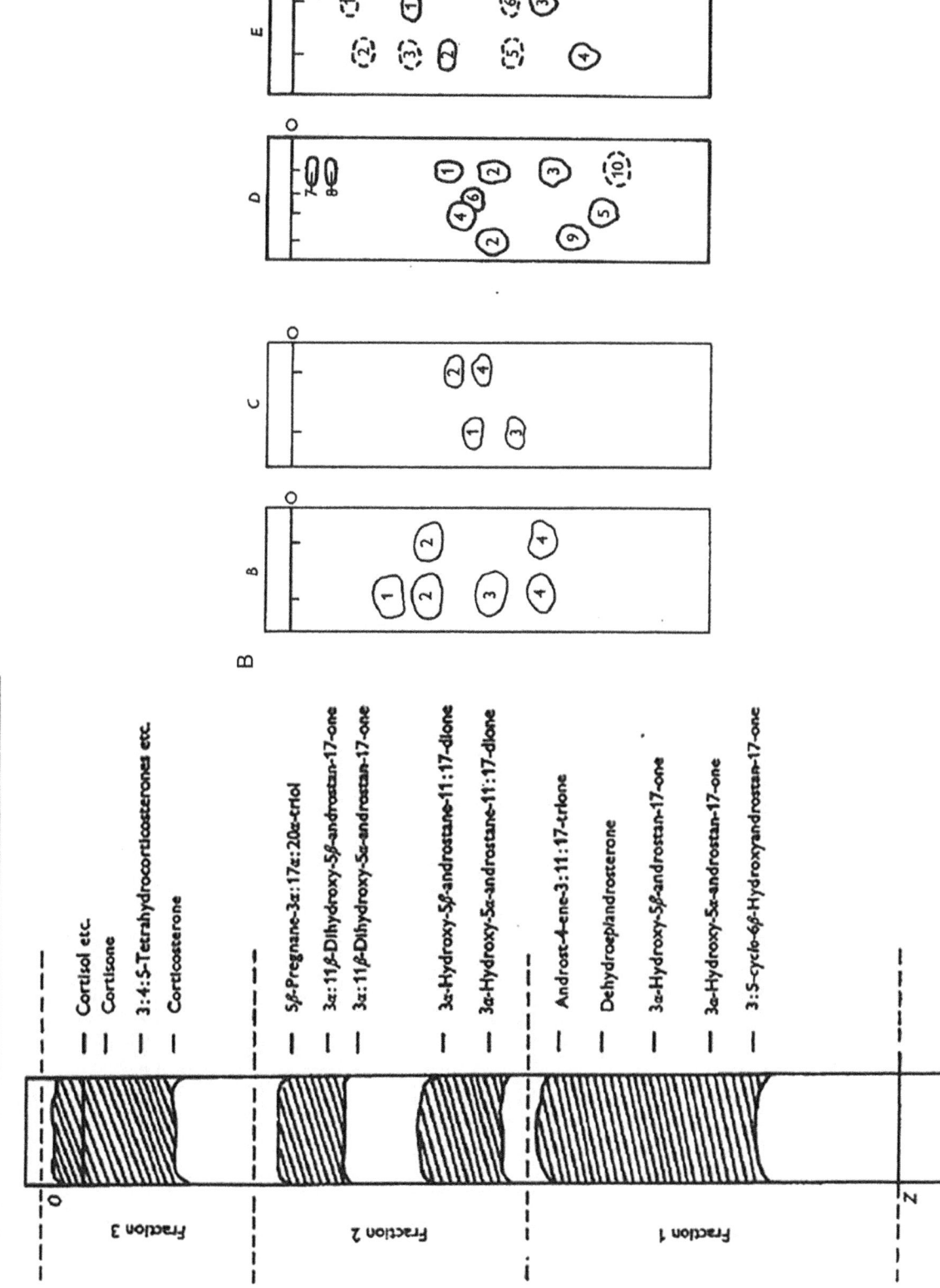

FIG. 2.3.8 (A) Preliminary fractionation of urine extracts en route to determine excretion of allo-THF in urine. Whatman no. 3MM paper was equilibrated for 3h and run for 1.75h in light petroleum–benzene–methanol–water (66:33:80:20 by vol.) (System 2). The positions of the main urinary neutral steroids are shown (right) and the fractions taken for subsequent quantitative chromatography (left). The paper is cut at the *dotted lines* for elution of the fractions. O is origin of the sample; Z the solvent front; OZ measures 40cm at 21–24°C. The *shaded zones* are those obtained when a urine extract is examined with alkaline m-dinitrobenzene (Fractions 1 and 2) or blue tetrazolium (Fraction 3). (B) Paper chromatography of fraction 3 developed with B. (1) Tetrahydrocortisol (THF); (2) allo-tetrahydrocortisol (aTHF); (3) tetrahydrocortisone (THE); (4) allo-tetrahydrocortisone (aTHE). Mobile phase benzene:methanol:water (2:1:1) (System 1). Blue tetrazolium detection. (C) Diacetates of THF; aTHF; THE and aTHE. Mobile phase light petroleum:benzene:methanol:water (66:33:80:20) (System 2). (D) Bismuthate oxidation products—(1) Dehydroepiandrosterone; (2) 3α-hydroxy-5β-androstan-17-one; (3) 3α-hydroxy-5α-androstan-17-one; (4) 3β-hydroxy-5α-androstan-17-one; (5) 3β-hydroxy-5β-androstan-17-one; (6) 3α-hydroxy-5β-androstan-9(11)-17-one; (7) 3α,11β-dihydroxy-5β-androstan-17-one; (8) 3α-hydroxy-5β-androstan-11,17-dione; (9) 6β-hydroxy-3;5-cycloandrostan-17-one; (10) 3α-acetoxy-11β-hydroxy-5α-androstan-17-one. Mobile phase System 2. Mobile phase for acetates of those products was light petroleum:methanol:water (25:24:1) (System 3). M-dinitrobenzene detection. (E) Chromate oxidation products. (1) 3α,11β-Dihydroxy-5β-androstan-17-one; (2) 3α,11β-dihydroxy-5α-androstan-17-one; (3) 3α-hydroxy-5α-androstan-11,17-dione; (4) 3α-hydroxy-5α-androstan-3,11,17-trione; (5) 5α-androstan-3,11,17-trione; (6) 5β-androstan-3,11,17-trione. M-dinitrobenzene detection. Mobile phase was System 3. ((A) *From Bush IE, Willoughby M. The excretion of allo tetrahydrocortisol in human urine. Biochem J 1957;67(4):689–700. Fig. 4 p. 695.); (B–E) Bush IE, Willoughby M. The excretion of allo tetrahydrocortisol in human urine. Biochem J 1957;67(4):689–700. Fig. 2 p. 692.; (B–E) Bush IE, Willoughby M. The excretion of allo tetrahydrocortisol in human urine. Biochem J 1957;67(4):689–700. Fig. 4 p. 695.)*

The extract was acetylated and products separated on paper chromatography (Fig. 2.3.8B Lane C). The extract was subject to bismuthate oxidation and products separated (Fig. 2.3.8B Lane D). The bismuthate products were subject to chromic acid oxidation and paper chromatography (Fig. 2.3.8B Lane E). At each step, steroids were analyzed with paper chromatography and Bush solvent systems. Reducing steroids were detected with blue tetrazolium, 17-oxosteroids detected with n-dinitrobenzene. A glycerol or 17,20-dihydroxy side chain were detected by Porter-Silber method with phenylhydrazine. In addition, the NaOH fluorescence reaction described by Bush for α:β-unsaturated 3-ketosteroids and the long wave ultraviolet (Raymaster) lamp for fluorescent areas were utilized. The two examples above illustrate how it took many hours to perform over several days to complete the experiments. This would certainly not be acceptable today but almost never now used.

Steroid conjugates were separated by paper chromatography (Kornel et al., 1975 and several earlier papers) including modifications incorporating high voltage paper electrophoresis (Kornel, 1964). Using **two-dimensional TLC** total of 37 steroids were separated and many detected by UV absorption or reaction with antimony trichloride (Brown et al., 1988).

2.3.2.3 Thin layer chromatography (TLC)

Steroid separation by TLC is still widely carried out to display conversions of steroids in enzymic reactions such as gene expression studies. TLC plates are now usually purchased as the material coated onto aluminum foil, which is lighter than the original glass plates, is unaffected by solvents used for elution and the areas of interest can be removed by cutting out with scissors rather than scraping off the adsorbent material from the glass. The adsorptive materials used are usually silica gel although aluminum oxide has been used for the separation of C19 androgens. TLC has the advantage that 8–12 samples can be processed in a single chromatographic run. Binary and tertiary solvent mixtures are used in ascending mode to effect separation, so chloroform:methanol; ethyl acetate:chloroform, cyclohexane:ethyl acetate to ethyl acetate:chloroform:water to cyclohexane:toluene:methanol:acetic acid have been used. The development time is quicker than paper and column chromatography that can take many hours. TLC gives sharper spots of the steroid than with paper chromatography which is prone to streaks. The difficulty with TLC is the need to identify the areas on the plate which correspond to the steroid of interest. Steroids which are UV absorbing (for example delta-4-ones absorb at 240 nm) can be visualized with a UV lamp and the adsorptive material can incorporate a fluorescent compound (e.g., F254) which is

quenched by UV absorbance properties of the steroids of interest. Steroids can also be detected by placing the TLC in the tank with iodine vapor. Spraying the plate with sulfuric acid gives colored or fluorescent zones which after charring give black on white chromatograms (Heftmann et al., 1966). For steroids that do not absorb in the UV, a narrow side strip of the TLC plate can be removed and the steroids located in that strip. The strip can be sprayed and/or heated with a variety of reagents which may or may not be specific for particular types of steroids (17-ketosteroids, 17-hydroxycorticoids, etc.). The position of standards which have been run together with the samples of interest can be located. Radioactive steroids can be identified by placing the glass plate in contact with the X-ray film or photostimulated luminescence film from Fuji (https://www.fujifilm.com/products/medical/technologies/focused_phosphor_technology/) producing an autoradiogram. Several oxidized products of cortisol were identified in urine in this way after intravenous administration of ^{14}C cortisol (Ito, 1968) (Fig. 2.3.9). The aluminum foil plate can be cut with scissors into 1 cm bands for subsequent liquid scintillation counting.

The instability of cortisol in aqueous solutions was addressed by separation of the material with TLC (Monder, 1968) (Fig. 2.3.10). 21-Dehydroxycortisol was a product. The 11-hydroxylated androgens have attracted much interest since 2015 particularly by the Amanda Swart group (du Toit et al., 2018; Bloem et al., 2015; Swart and Storbeck, 2015).

The Carl Monder paper in 1968 has important messages for the analyst today on scrupulous attention to cleanliness of reagents and glassware. Oxidation of cortisol to 21-dehydrocortisol is demonstrated in carbonate more than phosphate buffers.

Cortisol in plasma was determined by TLC using isonicotinic acid hydrazide to visualize the steroids (Fenske, 2008). Microparticulate silica has enabled the introduction of so called "high-performance" TLC which is claimed to have increased resolving power but in practice the differences are small. Reverse-phase systems for TLC have been described but are no longer widely used. Many examples of the use of thin-layer chromatography for steroid separation in the period from 1995 to 2008 are summarized in Makin book published in 2010.

Using **two-dimensional TLC**, total of 37 steroids were separated and many detected by UV absorption or reaction with antimony trichloride (Brown et al., 1988). Two-dimensional TLC improves separation of steroids which can be useful when assessing activities of enzymes in tissues such as subcutaneous adipocytes (Hannen et al., 2011) (Fig. 2.3.11). The tissue was incubated with ^{14}C progesterone before ether extraction. After application of the concentrated extract to the silica gel 60 TLC plate, two

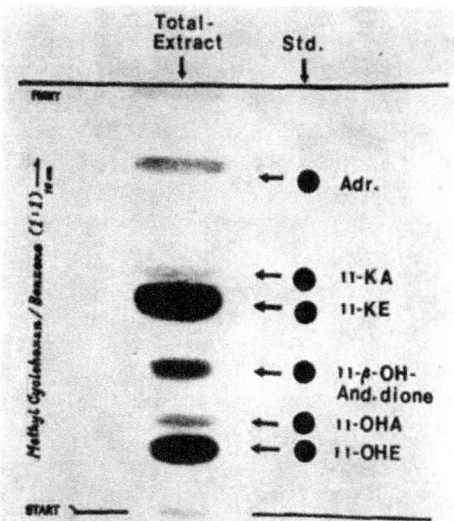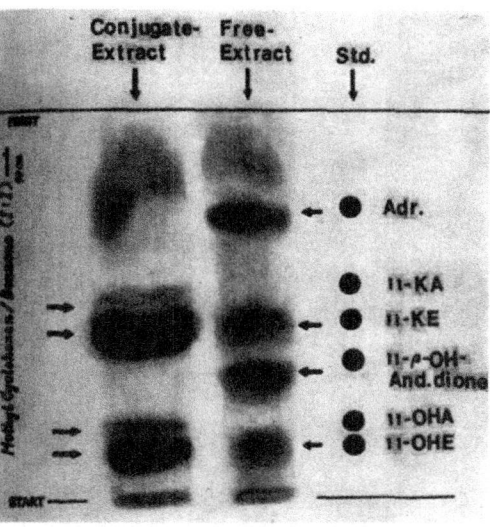

FIG. 2.3.9 Autoradiograms of TLC (left) of the oxidized products from a total extract from normal urine (right) the oxidized products obtained from the free and conjugated extracts from normal urine. *A*, adrenosterone; *11KE*, 11-ketoetiocholanolone; *11-KA*, 11-keto androsterone; *11β-OHAnd dione*, 11-hydroxyandrostenedione; *11-OHA*, 11β-hydroxyandrosterone; *11-OHE*, 11β-hydroxyetiocholanolone. *(From Ito Y. C-20-reduced metabolites of cortisol in normal human urine. Endocrinol Jpn 1968;15(1):13–9. Fig. 1 p. 16.)*

BUFFER	9.1 CO_3		9.1 CO_3		9.1 CO_3		9.1 CO_3		9.1 CO_3		9.1 CO_3		7.2 PO_4	
INCUBATION TIME (hrs)	0		0.25		1		2		3		24		24	
DISTRIBUTION ^{3}H		%		%		%		%		%		%		%
						2.8		1.3		1.6		9.3		0.5
⊚ =21-dehydrocortisol				0.8				4.6		4.9		2.4		0.8
						1.4		4.0		6.0		4.9		2.7
⊘ =Cortisol												33.9		
		100		99.2		95.8		90.1		87.5		47.7		95.5
														0.5
	x		x		x		x		x		1.6	x	

FIG. 2.3.10 Chromatographic evidence for the instability of cortisol with conversion of cortisol to 21-dehydrocortisol. Cortisol was stored in carbonate buffer (CO_3) at pH 9.1 or 9.3 for up to 24 h or in phosphate buffer (PO_4) at pH 7.2 for 24 h. TLC plate was developed using chloroform-ethanol (25:1). Steroids visualized under UV lamp. *(From Monder C. Stability of corticosteroids in aqueous solutions. Endocrinology 1968;82(2):318–26. Fig. 2 p. 320; Fig. 3 p. 321.)*

successive migrations were performed with chloroform: ether (10:3). The plate was turned through 90 degrees then developed in two migrations of hexane: ethyl acetate (5:2). The study showed that mature adipocytes efficiently generate progesterone metabolites.

Although TLC is still commonly used for steroid separations, very little research and development has been carried out in the last 30 years. TLC is still widely used to demonstrate purity of pharmaceuticals and to demonstrate the reactivity of tissues and transfected genes. Recent work (published by Qaiser et al. in 2017) demonstrated uptake of DHEAS and more rapidly pregnenolone sulfate (Fig. 2.3.12) into rat brain. Free steroids were liberated by sulfatase activity. Pregnenolone was not further metabolized but DHEA underwent 17-hydroxylation to form androstenediol.

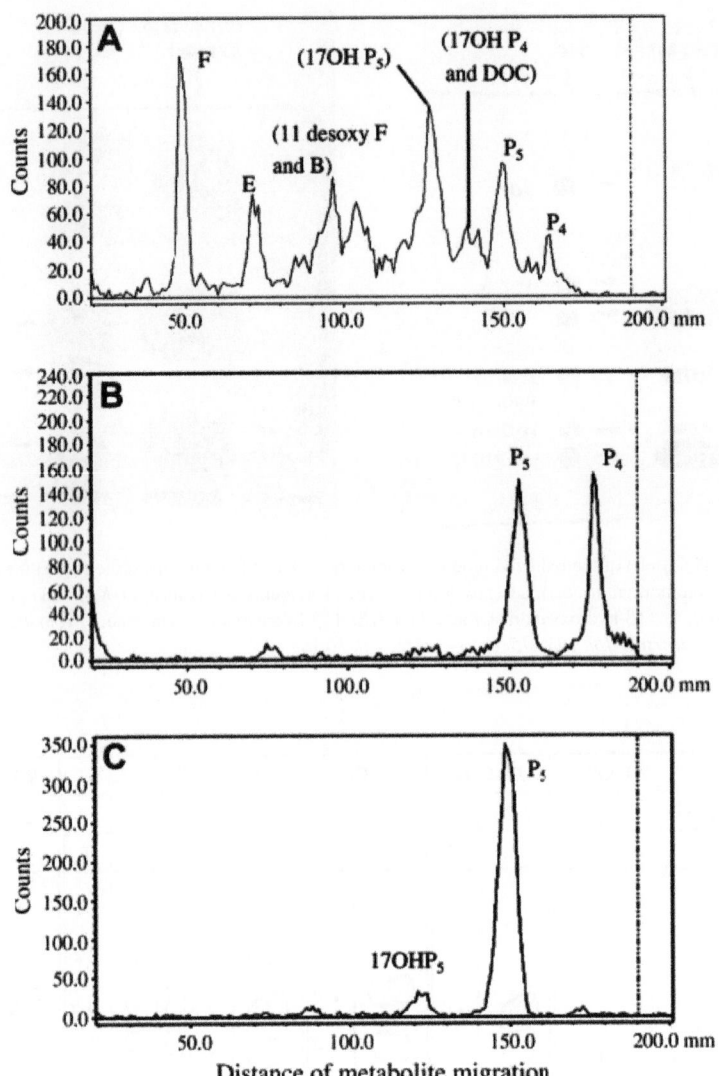

FIG. 2.3.11 Keratinocyte metabolism of 7-³H-pregnenolone to cortisol. Steroids were extracted from culture media and resolved by TLC with chloroform:95% ethanol v/v mobile phase (Lane A). Migration of labeled steroids was monitored using an AR2000 Bioscan radiochromatogram scanner (Lablogic, Sheffield, United Kingdom). Lanes B and C are reference steroids. *B*, corticosterone; *DOC*, 11-deoxycorticosterone; *E*, cortisone; *F*, cortisol; *11-desoxy-F*, 11-desoxycortisol; *17OHP4*, 17a-hydroxyprogesterone; *17OHP5*, 17a-hydroxypregnenolone; *P4*, progesterone; *P5*, pregnenolone. *(From Hannen RF, Michael AE, Jaulim A, Bhogal R, Burrin JM, Philpott MP. Steroid synthesis by primary human keratinocytes; implications for skin disease. Biochem Biophys Res Commun 2011;404(1):62–7. Fig. 2 p. 65.)*

A net efflux of sulfated steroid was demonstrated, supporting transport mechanisms through unknown routes not blocked by inhibitors of ATP-binding cassette transporter B1 (GF120918 or MK571) and multidrug resistance associated proteins, respectively. Overall, there was an intake of liberated free steroid.

Pharmaceutical impurities of less than 0.5% are possible which may be undetectable (false negative) with some methods of quantitation. The data from TLC experiments may thus need to be supplemented by more sensitive tests such as GC, GC-MS, HPLC, and LC-MS. For example, a

TLC plate can now be scanned with MALDI-MS (matrix assisted laser desorption) after derivatization of steroids with pyridine and 3-bromopropionyl chloride in a composite mixture containing 1,8,9-anthracenetriol, mixed with graphite dispersed in glycerol (Esparza et al., 2018) (Fig. 2.3.13).

A review article covered the analysis of steroids in biological samples from animals and plants as well as pharmaceuticals (Bhawani et al., 2010) shows that the technique is not completely negated. Some tips on choice of mobile phase can be found in another review (Cimpoiu et al., 2006).

(a)

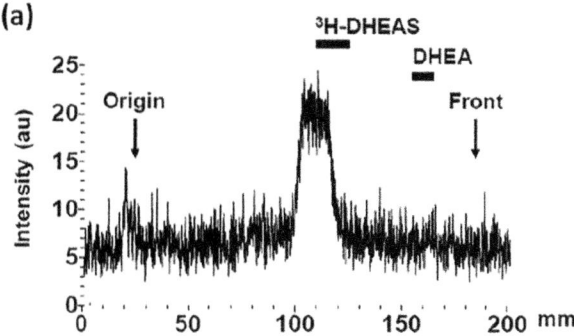

(b)

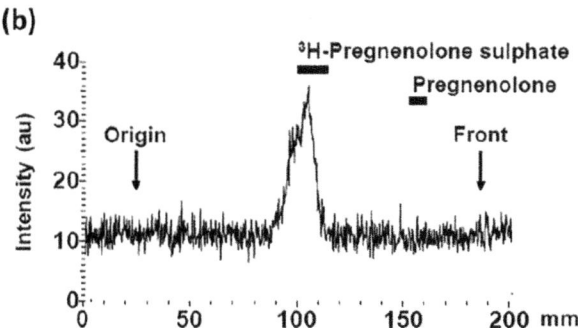

FIG. 2.3.12   Stability of DHEAS and preg sulfate in isolation of steroids from brain. The TLC plate was developed with ethyl acetate:ethanol: ammonia (5:5:1). Tritiated steroids were detected by placing TLC plate in contact with imaging film (BAS-TR202 = 40S, Raytek, Sheffield, United Kingdom) and subsequent visualization in a Typhoon 9410 variable mode phosphorimager. *(From Qaiser MZ, Dolman DEM, Begley DJ, et al. Uptake and metabolism of sulphated steroids by the blood-brain barrier in the adult male rat. J Neurochem 2017;142(5):672–85. Fig. S2 p. 5.)*

2.3.2.4   Gas-liquid chromatography (GLC or GC)

The GLC procedure is a partition system where the steroid solute is in the vapor phase. GLC of steroids has to be carried out at high temperatures, usually in excess of 200°C. The vaporized steroid, once introduced into the GLC column by an injection system, is carried through the system by an inert gas (commonly helium, argon, or nitrogen) to the detector (Fig. 2.3.14).

Hydrogen, because it is less dense than helium or nitrogen, gives improved separation and can be created in laboratory by electrolysis of water in a hydrogen generator and is therefore less expensive than the purchase of cylinders of gas. With hydrogen there are safety issues to be considered. Analysis time is shortened and chromatography improved (Impens et al., 2001).

Separation was originally carried out using packed columns (stationary phase coated onto Celite particles in column 3–4 m long and 2–5 mm bore) but today capillary columns of glass or fused silica are used almost exclusively. Fused silica capillary columns of 15–30 m length are almost exclusively used for GC applications.

Their wide acceptance is due not only to improved resolution but to the ease with which they can be interfaced with MS. The carrier gas flow rates are sufficiently low (0.5–2 mL/min) that the column is passed through a simple seal in the ion source housing and terminated close to the ionization region. In GC, the mobile phase is an inert gas such as helium or hydrogen and the stationary phase is the thin layer of the viscous liquid coated or chemically bonded on the walls of the capillary column. Steroids vaporize and dissolve in the gas phase on injection to the column. The steroids partition with the column coating and the gas. Steroids are resolved through their relative affinities in the stationary phase. A temperature gradient is applied to speed up the chromatography and improve the desired separation. The user needs to appreciate the work-up procedures for sample preparations and interpretation of the fragmentation of derivatives (trimethylsilyl ether and methyloxime-trimethylsilyl ethers) of steroids and sterols. Reviews from the Shackleton laboratories have detailed the use of GC and GC-MS for clinical steroid analysis and have detailed descriptions of the optimal extraction, hydrolysis, and derivatization methods for urinary steroid analysis. High-resolution GC-MS-SIM (selected ion monitoring) and MS-MS techniques can also be used for sensitive steroid analysis and for tentative identification in complex mixtures. Christakoudi published four papers (Christakoudi et al., 2010a,b, 2012a,b, 2013a,b), in which 230 pregnene/pregnane steroids then 76 androgens were tentatively characterized in the urine of newborn infants with congenital adrenal hyperplasia. Martin Hill has identified 95 glucuronides and sulfates in pregnancy urine (Hill et al., 2019). GC with mass spectrometry is clearly a powerful technique for steroid analysis.

2.3.2.4.1   Column technology

The stationary phase used to be coated onto particles of fused silica packed into a long glass column (typically 5 mm × 3 M) but is now coated on the wall of a capillary tube (0.1–0.3 mm × 30 M), and even better separation is achieved through the greater number of theoretical plates with capillary column compared with packed columns. The steroids of interest which have been converted to thermostable compounds are separated by their relative solubility in the stationary phase. There are a wide variety of stationary phases and precoated columns available commercially. The capillary columns are usually wall-coated open tubular (WCOT) with dimethylpolysiloxane OV-1 or 101 stationary phases coated on the inside of columns of diameter of approximately 0.2 mm, although megabore columns (ID 0.5–0.75 mm) with greater capacity, but

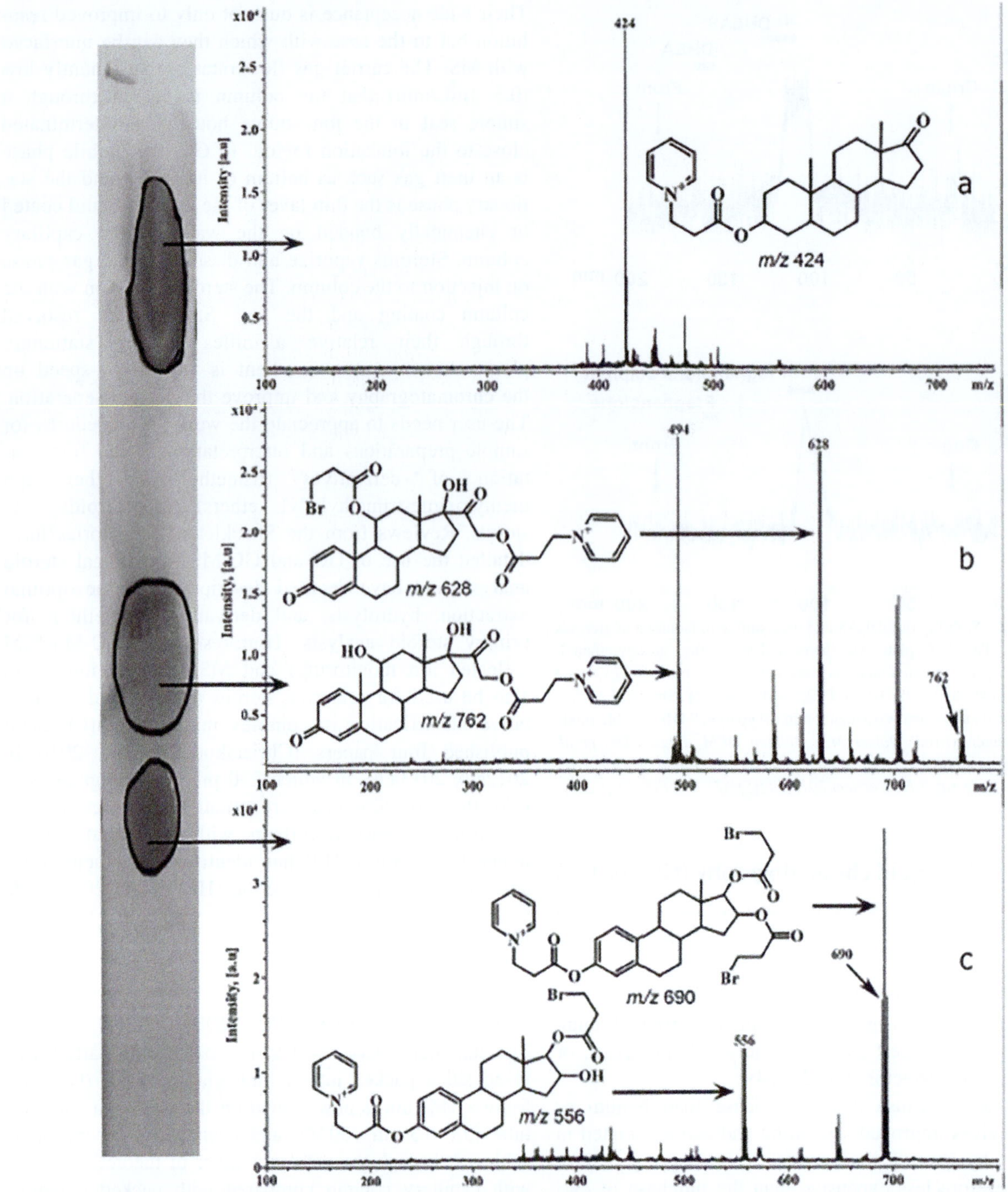

FIG. 2.3.13 MALDI mass spectra of steroid alcohols derivatized on TLC plate: (a) trans-androsterone; (b) prednisolone; (c) estriol. In (b) and (c), the positions of bromopropionyl and pyridiniumpropionyl groups may vary. The plate was developed in ethyl acetate:acetonitrile (5:1). *(From Esparza C, Borisov RS, Polovkov NY, Zaikin VG. Post-chromatographic fixed-charge derivatization for the analysis of hydroxyl-containing compounds by a combination of thin-layer chromatography and matrix-assisted laser desorption/ionization mass spectrometry. J Chromatogr A 2018;1560:97–103. Fig. 2 p. 100.)*

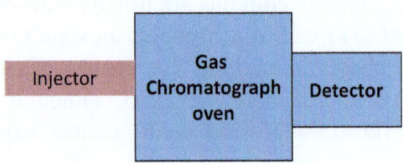

FIG. 2.3.14 Schematic of a gas chromatography system. The capillary column in the oven between injector and detector. *(Author original.)*

reduced resolution, are also available. Several suppliers offer GC columns of similar nature (Table 2.3.1). Chemical substitution of other groups onto the siloxane polymer gives rise to further stationary phases which may have selective characteristics. At high temperatures, some of the coating can be eluted from the column which partly contributes to

TABLE 2.3.1 Examples of capillary GLC columns and suppliers.

Supplier	Column
Agilent J & W	ZB-1
Restek	Rex-1
SGE	BP1; BP5
Sigma-Aldrich	SBP-1
Supelco	SE30
Macherey-Nagel	Optima 1

the rise in baseline, so called column "bleed," but this is less with capillary columns than packed columns and has been largely overcome by the use of chemically bonded stationary phases.

The thickness of the coating (from around $0.1\,\mu m$ upwards) affects the capacity of the column. The stationary phase can now be chemically bonded to the wall of the columns. The process of chemical bonding can affect retention times of the steroids. Short (5 m) and micro bore columns are becoming fashionable for fast chromatography but the GC needs an injector and flow control that withstands very high pressures if narrow columns are used. Short columns can enable faster separation while narrow bore columns can be easily overloaded with sample. Steroid conjugates as methyl ether, methyloxy-trimethylsilyl ether derivatives have been separated with such columns (Shackleton et al., 1983). The upper temperature limit of columns is normally around 300°C but for sustained periods

to 400°C can be reached with some new columns without significant column degradation. When components are not completely resolved on one column it may be necessary to compare the content of peaks in separate columns.

2.3.2.4.2 Sample injection

Injection systems can be for solid or liquid extracts. The most generally applicable injection system for steroid derivatives is **splitless** injection with **cold-trapping** (Fig. 2.3.15).

This system allows a large injection volume ($\sim 2\,mL$) to be used. This vaporizes in a removable glass insert in the heated injection port and the heavier (less volatile) components condense in the initial part of the column which is maintained at near ambient temperature (cold-trapping). In injection systems where solubilized material goes directly into the column (split injection, on-column injection), the first few centimeters of the column can become contaminated over several weeks but a length can be cut off periodically. This results in a gradual shortening of column length which does not significantly affect resolution but retention times of expected components change markedly.

The column temperature is usually programmed to increase. The solutes re-vaporize when they get to their volatilization temperature and are duly separated. This system keeps the column relatively clean since most of the involatile material is deposited in the injector glass insert. There are refinements to injection systems with pulsed and pressure features. Injection liners come in many shapes to affect maximum delivery of the sample depending on volatility, sample volume, and sensitivity. *On-column* injection from a syringe results in heavy column contamination.

The split / splitless injector

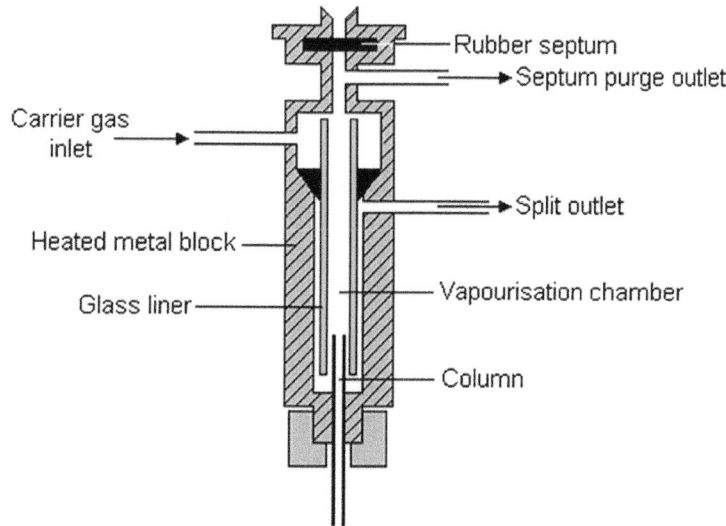

FIG. 2.3.15 Schematic of the split/splitless injector for the GC. *(Author original.)*

2.3.2.4.3 Derivative formation

GLC requires the steroid solute to be present in the vapor phase and capable of analysis without destruction at the high temperatures required for this. Androgens and estrogens can be analyzed by GLC without derivatization since they are stable at the oven temperatures required but have long retention times and occasionally have been shown to be dehydrated. However, the presence of hydroxyl groups impairs the chromatographic resolution due to adsorption during chromatography. C_{21} steroids with the 17-hydroxycorticosteroid side chain undergo side chain cleavage giving 17-oxo steroids and to prevent this break up chemical derivatization is usually carried out prior to chromatography. The majority of methods for steroids using GC (GC-MS) for identification and measurements of steroids and metabolites with low detection limits require previous formation of derivatives. John Halket has published a number of excellent reviews on this topic. The choice of derivative which is formed will depend upon the method of quantitation as well as the need for good chromatography. The range of derivatives include:

- Trimethylsilyl ether and other alkylsilyl ethers (such as tertiary-butyldimethylsilyl ether) (Gaskell and Brooks, 1976) are suitable for hydroxyl groups. A number of silylating reagents are available (Fig. 2.3.16). Dimethyldichlorosilane, trimethylchlorosilane, and bis-trimethylsilylacetamide (BSA) or bis-trimethyltrifluoroacetamide (BSTFA) will form trimethylsilyl derivatives on most of the steroid hydroxyls. N,O-bis(trimethylsilyl)acetamide (BSA) or N,O-bis(trimethylsilyl)trifluoroacetamide (BSTFA) alone will only form such derivatives on nonsterically hindered groups. In the Shackleton method for urinary steroids silylation is achieved with trimethylsilylimidazole (TSIM) in pyridine at 100°C (Shackleton and Honour, 1976).

- O-Methyloxime to protect ketones. Syn- and antiproducts that separate in the GC are formed, so two peaks have to be determined (Horning et al., 1968; Shackleton and Honour, 1976).

- Enol TMS ethers are formed on ketones with N-methyl-N-trimethylsilylfluoroacetamide-trimethylsilyliodosilane (MSTFA-TMSI). This derivative is used in doping control for the detection of anabolic steroids and for corticosteroids because a single reaction saves time (Saugy et al., 2000).

- Pentafluorophenyldimethylsilyl-TMS ether (PTDMS, flophemesyl) derivatives for electron capture detection or give intense molecular ions in MS that are attractive for GC-MS analysis (Choi and Chung, 1999; Choi et al., 2002).

- Heptafluorobutyrates have been used for quantitative analysis of several of the plasma steroids (Wudy, 2000, 2001, 2002; Exley and Chamberlain, 1967).

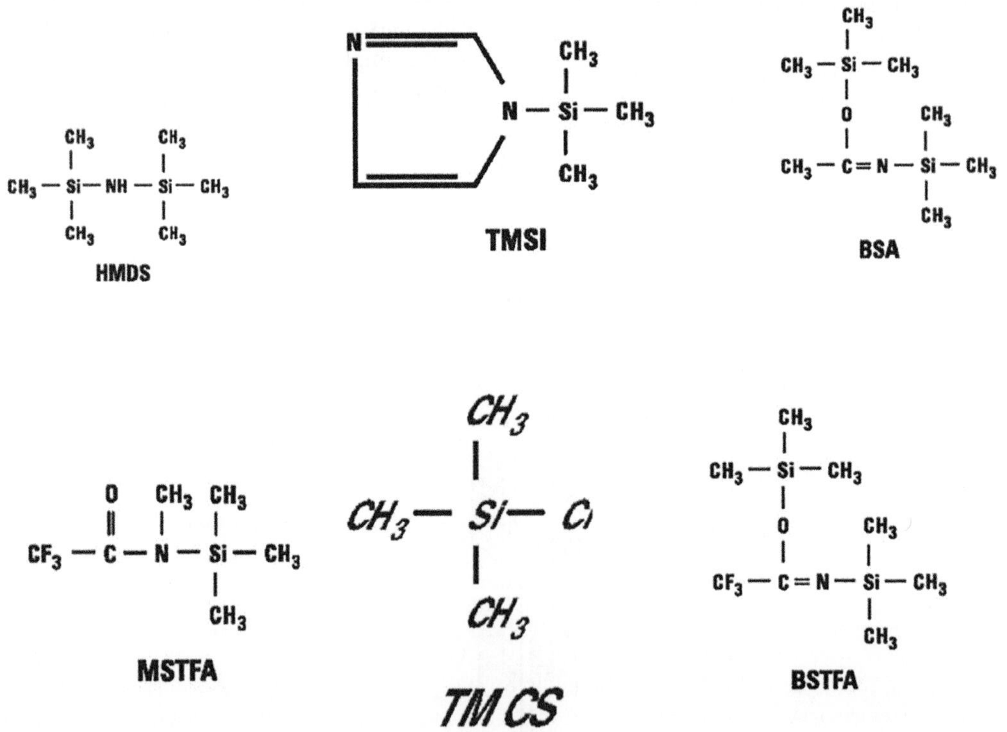

FIG. 2.3.16 Silylation reagents. *BSA*, N,O-bis(trimethylsilyl)acetamide; *BSTFA*, N,O-bis (trimethylsilyl)trifluoroacetamide; *HMDS*, hexamethyldisilazane; *MSTFA*, N-methyl-N-trimethylsilyltrifluoroacetamide; *TMCS*, trimethylchlorosilane; *TMSI*, trimethylsilylimidazole.

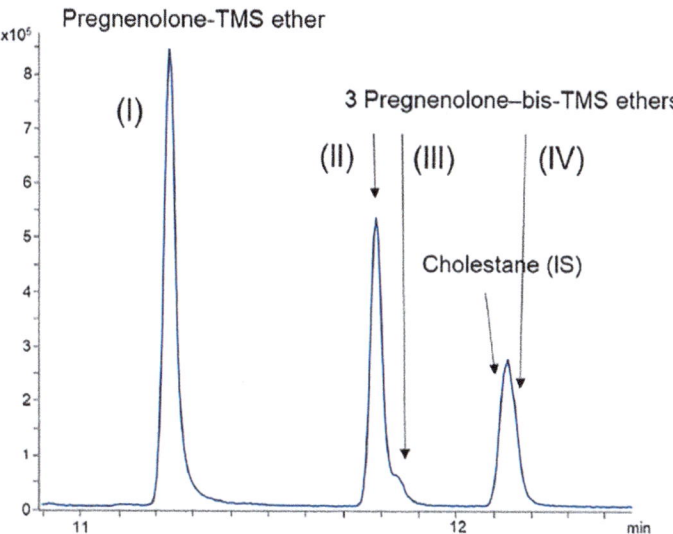

FIG. 2.3.17 GC trace for TMS derivatives of pregnenolone and cholestane (internal standard) after reaction with MSTFA/TSIM (9:1). *(From Junker J, Chong I, Kamp F, Steiner H, Giera M, Müller C, Bracher F. Comparison of strategies for the determination of sterol sulfates via GC-MS leading to a novel deconjugation-derivatization protocol. Molecules 2019;24(13). pii: E2353. Fig. 3 p. 6.)*

- Bismethylenedioxy-pentafluoropropionate derivatives have been used rarely in analysis of cortisol metabolites (Furuta et al., 1998, 2000).
- Trimethylsilylimidazole (TSIM) will react with hydroxyl groups even the sterically hindered groups such as the 17α-hydroxyl in the corticosteroids. The reagent is not volatile and for many years the derivative was cleaned up on a small Lipidex column. A water wash is satisfactory, despite the claims that derivatives are anhydrous conditions.
- Formates (Makin, 1970) and acetates (van de Calseyde et al., 1970; Brooks, 1965) have also been used of hydroxyl groups.
- The carboxyl group of steroid acids must also be derivatized prior to GLC and usually methyl esters are formed (Shackleton et al., 1980, 1983). For the reaction, trimethylsilyldiazomethane is now preferred to diazomethane. The acid of glucuronide can also be methylated before oxime and trimethylsilylation to stabilize the intact steroid glucuronide (Shackleton, 1981).
- Tertiary butyldimethylsilyl ethers are commonly used for quantitative GC-MS as these derivatives give mass spectra without extensive fragmentation (Gaskell and Brooks, 1976).
- Vicinal hydroxyls can be derivatized to give cyclic alkyl boronate esters and the formation of these derivatives with unknown steroids gives an indication of the structure of the steroid (Brooks and Harvey, 1969). Mixed esters can also be used, such as formation of cyclic boronate esters across vicinal hydroxyl groups and subsequent trimethylsilyl ether formation on the remaining hydroxyl groups.

Pregnenolone when derivatized with MSTFA:TSIM (9:1) can form one main product for pregnenolone with only one TMS ether and three artifacts (Junker et al., 2019) (Fig. 2.3.17) (see Quantitation Chapter 2.4 for detail of the products).

In sports doping laboratories, TMS derivatives are used because a single derivative reaction makes for a faster method for analysis. In the reaction, carbonyl groups are converted to enol-TMS derivatives. Clinical laboratories convert the carbonyl groups to oxime derivatives (usually methyl oximes). The preparation of derivatives has been a time-consuming step and a deterrent to the use of GC techniques. The process can now be speeded up using microwave assisted derivatization (Casals et al., 2014).

2.3.2.4.4 GLC detectors

Steroids separated by GLC are detected by one of three main methods, flame ionization (FID), electron capture (ECD), and nitrogen (ND) detectors or coupled to a mass spectrometer.

2.3.2.4.5 Flame ionization

The commonest detector for steroids is the **flame ionization detection (FID)** which responds to all steroids with varying response (Shackleton and Honour, 1976). In order to use GC-FID for quantitation of steroids, therefore, it is necessary to set up a standard curve or establish a response factor for the steroid of interest. In the FID, hydrogen and air are mixed with the carrier gas and the mixture is burnt at the tip of a jet (Fig. 2.3.18).

The Flame Ionisation Detector

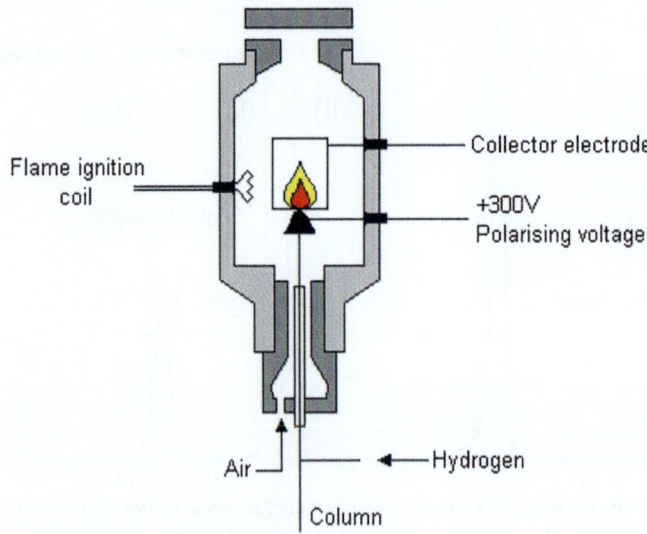

Flame ignition coil

Collector electrode

+300V
Polarising voltage

Air — Hydrogen

Column

FIG. 2.3.18 Schematic of flame ionization detector. *(Author original.)*

A current is applied to the tip of a fine jet and a collector electrode is sited above to complete an electrical circuit. Since cylinders or supplies of compressed gases (hydrogen and air to combust the steroids) are needed, the lifetime cost of a GC-FID system may exceed that of a GC-MS, so may not be a good choice. The GC retention time is the only measure of specificity unless the steroid forms two or more derivatives. The use of capillary gas chromatography for the analysis of urinary steroid profiles is a very good example of the valuable use of GLC-FID using trimethylsilyl ether-*O*-methyloxime derivatives and when such systems are also linked to an MS, they can be of immense value.

2.3.2.4.6 Nitrogen detector

Compounds containing nitrogen atoms can be detected with **a nitrogen-phosphorus detector** (NPD) but as most steroids commonly encountered do not contain nitrogen, to use this detection system requires formation of nitrogen containing derivatives such as methyloximes. This detection system can be quite useful for the selective measurement and detection of steroids containing oxo groups after oxime derivative formation (Sear et al., 1980).

2.3.2.4.7 Electron capture

The third method of detection involves the use of **electron capture** which is extremely sensitive and detects steroids containing halogen groups. The electron capture detector is used for detecting steroids as halogenated derivatives in the output of the gas chromatograph. Gas flow is increased with a makeup gas usually nitrogen because it has low

excitation energy which makes it easy to remove an electron from the nitrogen. The ECD has a radioactive electron (beta particle) emitter, typically a metal foil holding 10 mCi (370 MBq) of the radionuclide ^{63}Ni. The electrons from the electron emitter collide with the molecules of the makeup gas, resulting in many more free electrons. The electrons are accelerated toward a positively charged anode, generating a small current. There is a background signal from the nitrogen. As the derivatized steroids in the sample reach the detector in the carrier gas, electron-absorbing analyte molecules capture electrons and thereby reduce the current between the collector anode and the cathode. Over a wide range of concentrations the rate of electron capture is proportional to the analyte concentration.

Halogenated silyl ether derivatives have been used in conjunction with an electron capture detector for the measurement of DHA in plasma after formation of iodomethyl-dimethylsilyl ethers. Aldosterone, DOC, 18-hydroxy-DOC, corticosterone, cortisol, and 11-deoxycortisol were measured in plasma by GC with ECD (Mason and Fraser, 1975). Tetrahydroaldosterone in urine was determined by GC and ECD (Nicolis et al., 1968). Reports of methods using GLC with electron capture are becoming increasingly rare as these detection systems are difficult to use and are susceptible to detector contamination. Pentafluorobenzyl and pentafluoropropionate derivatives have been used as electron capturing derivatives for estrogens by GC-NI-CI. GLC-EC, although extremely sensitive, is also very difficult to use quantitatively and it has largely been replaced by interfacing the **GLC with a mass spectrometer** which when operated correctly can be as sensitive as electron capture and more selective.

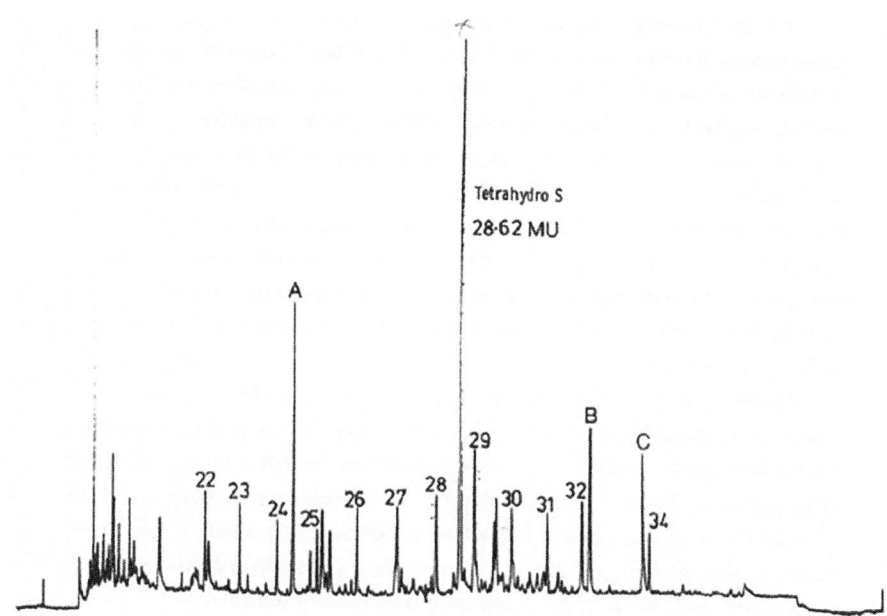

FIG. 2.3.19 Elution of tetrahydro-11-deoxycortisol (substance S) in relation to a series of alkanes. The steroid has a retention of 28.62 methylene units. GC was performed on a 30 m OV-101 capillary column with helium carrier gas. A temperature programme from 180°C to 280°C was used. *(Author original.)*

2.3.2.4.8 Identification

The retention time of a steroid in a GC analysis is important data toward identification. The actual time is not sufficient and relative retention times are used because they are reproducible. A number of systems are used. A retention relative to linear hydrocarbons has been very effective. For the range of steroids in biological samples as TMS and MO-TMS derivatives, the hydrocarbons C_{20} to C_{44} are needed and retention relative to the hydrocarbon is calculated as **methylene units**. An example is shown in Fig. 2.3.19.

2.3.2.4.9 Gas chromatography with detector for beta-emitting radiolabeled steroids (RGC)

RGC consists of three processes, separation of radioactive components by a GC column, mass detection and radioactivity detection (Baba, 1989). Detection is usually performed by flame ionization detection (FID), electron-capture detection (ECD), or mass spectrometry (MS). The instruments for general detection (e.g., FID) and radioactivity detection (RAD) are usually arranged in parallel. The effluents from a GC column are split in a suitable ratio, the minor portion passes through the FID, and the major is transferred for RAD, although with tritiated compounds with a high specific radioactivity, the splitting ratio should be reversed. Methods for the detection of radioactivity in the effluents from a GC column fall into two categories, namely, "continuous methods," in which the gamma radioactivity is monitored continuously, and "discontinuous

methods," in which the β-isotope (e.g., tritium) labeled components in the effluents are trapped as they emerge from the column for separate radioassay by scintillation counting. Continuous methods have the advantage that the RAD trace is produced simultaneously with the mass detector trace. The major criticism of continuous methods is that they are inherently less sensitive than discontinuous methods because of the relatively short counting time. Although the discontinuous methods are laborious and time-consuming, they have higher sensitivity Moreover, trapping techniques can be simple and inexpensive, and a liquid scintillation counter can be used for the radioactivity determination. In most situations, however, continuous methods for RGC are used and hence trapping techniques are less frequently utilized.

Most RGC systems use gas flow proportional counters. RGC can be classified into two types, those with and those without decomposition of the effluent The techniques in which the effluents are directly transferred for RAD have not been generally accepted, however, because the radioactive components often condense on the inside wall of the transfer line and detector, resulting in high background counts. A decomposition step was almost invariably carried out by a combustion technique. Fig. 2.3.20 illustrates the typical arrangement of an RGC system which has been most commonly used.

The oxidation-reduction (or combustion) tube is a quartz tube packed with copper oxide and iron filings, placed in an electric furnace The inner volumes of most combustion tubes used in RGC are several milliliters. The temperature

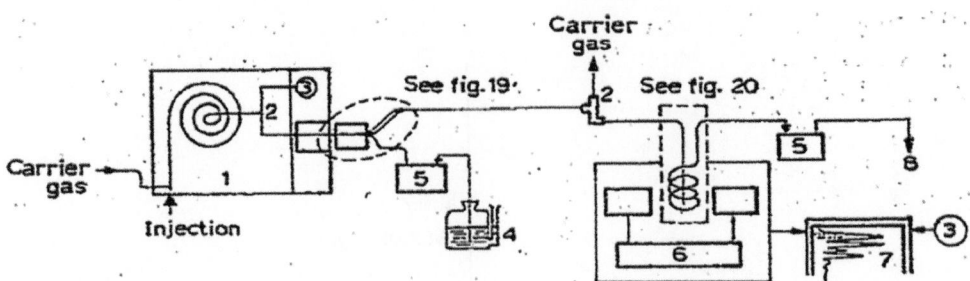

FIG. 2.3.20 Scheme of radio-gas chromatography with continuous detection of radioactivity by liquid scintillation. 1=oven of gas chromatograph; 2=gas splitter; 3=flame ionization detector; 4=container with liquid scintillation liquid; 5=peristaltic pump; 6=liquid scintillation spectrometer; 7=two pen recorder; 8=fraction collector. *(From Matucha M, Smolková E. Gas chromatography of 3H- and 14C-labelled compounds. J Chromatogr 1976;127(3):163–201. Fig. 2, p. 168.)*

of the inside of the furnace is between 700°C and 800°C. The extent of deterioration of the copper oxide can be followed by its color change (from black to red). Labeled compounds in the effluents are oxidized to carbon dioxide and water. If ^3H is present, the tritiated water produced is then reduced to ^3H. The effluent containing ^{14}CO, and/or ^3H, is mixed with counting gas (methane, propane, or carbon dioxide) and then passed through a gas flow proportional counter. The counter tube is a metal tube with an inner volume usually between 10 and 30 mL. The optimum voltage applied to the counter tube can be found by use of external radiation sources. The flowrates of the carrier and counting gases had to be strictly regulated, because the detection sensitivity is severely influenced by the composition and flow-rate of gas passing through RAD. The radioactivity in RGC can be recorded by analog means with a rate meter or digital recording of the radioactivity per certain time period (sampling time) as resulting in a histogram. The latter is more convenient for the quantitation of radioactivity and is more accurate. Most of the RGC systems use analog recording because digital recording suffers from severe statistical fluctuations of counts owing to a limited counting time, especially with samples of low radioactivity.

There are ethical and practical reasons for this analysis being rare today. Radioactive isotopes can now be replaced with stable isotopes which pose no hazard in use and stable isotopes can be detected with mass spectrometry which is becoming more commonly used and will continue to be adopted in hospital laboratories.

2.3.2.5 High-performance liquid chromatography (HPLC or LC)

The HPLC technique remains important for steroids because:

(i) high temperatures are not required,
(ii) material can be recovered from the column eluates for further analytical procedures,
(iii) the resolution achieved by HPLC is superior to TLC and paper chromatography, and
(iv) HPLC offers the potential and versatility for separation of intact conjugates.

An HPLC system (Fig. 2.3.21) has reservoirs for solvents, since often more than one solvent is mixed then pumped through the column to the detector. The sample is injected at the start of the column.

In some systems, a chemical reactor is placed at either end of the column to produce products that increase the sensitivity at the detection system. HPLC coupled with tandem mass spectrometry seeks to use the flexibility of the detector to obviate the need for extensive cleanup of samples.

2.3.2.5.1 Columns

The separation of steroids with HPLC can be affected by absorption, partition, ion-exchange, reversed-phase (RP), and reversed-phase ion-pair chromatography. High-performance silica and alumina columns give excellent separation of steroids. RP columns eluted with polar binary solvent mixtures, usually methanol or acetonitrile with water, are now used widely. In fact, there are times when acetonitrile is in short supply, methanol is the best substitute. RP columns using microparticulate silica coated with C_{18}, C_8, C_2, and phenyl materials (examples are listed in Table 2.3.2) have been used. Careful selection of the stationary phase from the range of commercially available products can enable a system to be devised with high selectivity.

Silica packings to which are bonded octadecyl or diol groups are most popular for general use. Supports differ in particle size, porosity, and levels of residual accessible silanol groups. Synthetic polymers may be more inert than silica. Supports have variable and often incomplete coverage of residual silanol groups ("uncapped") which affects separation, peak shape and recovery. Some packings with about 5% of uncapped silanol groups are chemically reactive with steroids due to intramolecular hydrogen bonding. This leaves the phase acidic and may explain

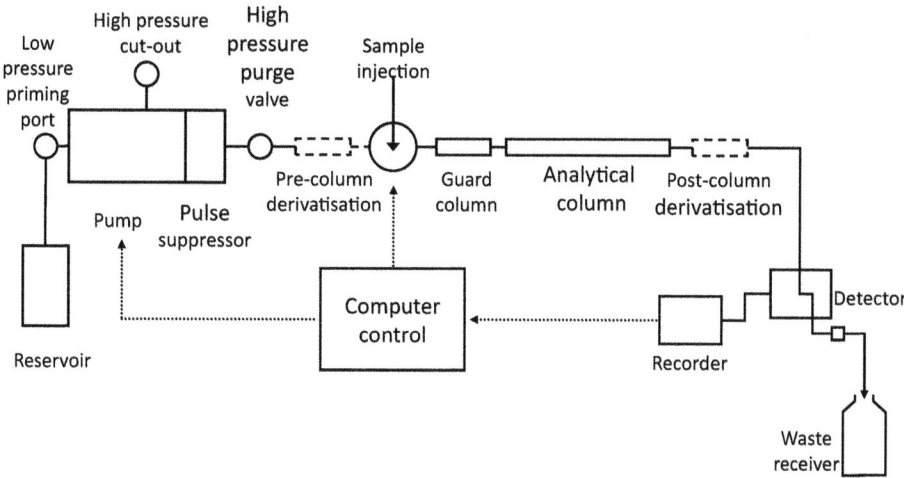

FIG. 2.3.21 Schematic HPLC system. *(Author original.)*

TABLE 2.3.2 Selection of HPLC columns and sample preparations in analysis of steroids from saliva, serum and tissues.

Steroids	Column	Sample	Sample prep	Ref
F	C8	Serum	PPT	Owen et al.(2013)
E1, E2	C8	Plasma	SPE	Zhao et al. (2014)
T, free T	C12	Serum	SPE	Salameh et al. (2014)
F, B, DOC, A4, 17OHP, T, 21-OHP, E	C18	Serum	LLE	Dury et al. (2015)
16 x steroids	C18	Matrices		Luo et al. (2016)
AndroG, aetioG, ADiolG	C18	Serum		Ke et al. (2015)
Aldo, B, F, E, DOC, DHEA, DHEAS, 17-OHP, Prog, T	Biphenyl	Serum	PPT	Lindner et al. (2017)
Estrogens, androgens, progestagens, glucocorticoids, mineralocorticoids	Biphenyl	Serum, plasma endometrium	Toluene	Häkkinen et al. (2018)
Estrogens, dansylated	Phenyl hexyl	Serum	Ethyl acetate	Szarka et al. (2013)
Aldosterone	C18	Serum	Dichloromethane/ether	Turpeinen et al. (2008)
Catechol estrogens	C18 PFP	Plasma		Denver et al. (2019)
18 x tetrahydrocorticoids	TSKgel ODS	Urine	WAX	Mitamura et al. (2014)
	RP18	Plasma, serum, urine	SPE	Gaudl et al. (2016)
15 steroids	BEH C18	Serum	PPT	Travers et al. (2017)

Steroid abbreviations: *17OHP*, 17-hydroxyprogesterone; *21-OHP*, 21-hydroxyprogesterone; *A4*, androstenedione; *ADiol*, androstanediol; *aetio*, aetiocholanone; *Andro*, androsterone; *B*, corticosterone; *DOC*, 11-deoxyB; *E1*, estrone; *E2*, estradiol; *F*, cortisol; *G*, glucuronide; *Prog*, progesterone; *T*, testosterone.
Technology abbreviations: *LLE*, liquid-liquid extraction; *ODS*, octadecylsilyl (C18); *PFP*, pentafluorophenyl; *PPT*, protein precipitation; *SPE*, solid phase extraction; *WAX*, weak anion exchanger.

the instability of certain steroids in such systems. Aldosterone and 18-hydroxylated steroids are susceptible to a number of reactions on certain columns which can influence the quality of the HPLC result (O'Hare et al., 1980). Acid, such as may be found on uncapped HPLC supports, can lead to ring closure of such steroids with a bridge of C-18 to C-20 or C-21. In the presence of methanol, this may lead to formation of methyl ethyl ketals. Other products, dimers and isomers are possible leading to the production of a number of peaks in the HPLC analysis of a single compound. These products can have retention times spread throughout a solvent gradient elution of a RP column. This may be disastrous in interpretation of a metabolic study unless products are characterized by other means. Some supports are not recommended for aldosterone and related steroids, e.g., 18-hydroxycorticosterone. The extent to which a packing is not covered (end-capped) can be determined by a methyl red absorption test.

In recent years, there has been a move to narrower bore columns with smaller particle size. Methods for a single steroid determination are now rare with panels of 10–20 steroids now the norm. Run times have shortened from 30 min to less than 5 min. This makes for sharp peaks where shape is more difficult to interpret and the near co-elution of isobaric components less easy to see. The chromatography depends largely on partition so that selectivity will vary to some extent with the carbon chain length and the nature of the mobile phase. A risk with RP packings is that very nonpolar material will accumulate on the columns and decrease separation. This can be prevented to some extent by the use of a guard column (30–70 mm in length) containing the pellicular equivalent of the analytical column. Guard columns are cheap and can be dry-packed. The first few millimeters of packing from the analytical column can also be replaced at intervals.

RP columns with 60,000–80,000 theoretical plates per meter are common. These offer excellent resolution and sharp peaks permit detection by UV absorption of around 1 ng of steroid injected onto the column. Typically, columns are 100–300 mm in length and around 4–5 mm internal diameter. Cyano and amino phases have been used to effect separation of corticosteroids and for estrogens. HPLC of polar estrogens has been achieved on ion-exchange columns. Micro bore columns (<2 mm i.d.) may permit increased sensitivity by narrowing the elution peak but depending on the volume of sample and the total mass of material in the extract there may be a loss of peak shape and resolution.

The complete separation of naturally occurring mixtures of steroid hormones (Fig. 2.3.22) poses problems due to the wide range of polarities and the tendency for steroids of similar polarity derived from different metabolic pathways to elute in clusters.

The physical characteristics of many packings have been studied with various solvent gradients. There seems to be no easy means to identify the most suitable packing for a particular separation. Selective differences cannot be firmly attributed to alkyl chain length or to shape of the packing. Immobilized cyclodextrins, macrocyclic polymers of glucose, which have been used for the extraction of steroids because of their ability to form inclusion complexes, have also been introduced as stationary phases (cyclobond) for steroid chromatography and may have advantages to offer but that may depend on carbon load. Carbon-coated zirconia was compared with porous graphitic carbon stationary phase for separation of equine conjugated estrogens giving separations superior to C_{18} and alkyl bonded silica phases (Reepmeyer et al., 2005). In a similar fashion graphitized carbon, which has been used for crude steroid fractionation may also have use as a stationary phase for steroid HPLC (Andreolini et al., 1987), since it is micro-crystalline and contains no unreacted silanol groups such as those on silica-based materials and thus may be considered to be a suitable inert material for RP chromatography. Biphenyl columns are useful for separation of isobaric steroids.

2.3.2.5.2 Mobile phases

Chromatographic systems suitable for HPLC of steroids are based upon or can be tested with TLC. Useful separations of steroids can be achieved using isocratic chromatography on silica gel with binary solvents. The separation of a range of steroids is best achieved with gradient elution. Additional pH, ion-pair, and modifier effects can be incorporated. Retention times are reproducible between runs provided that the column is equilibrated to the starting solvent mixture. Methanol:water gradients affect the separation of the major adrenal steroids. Dioxane is a better choice for the separation of polar adrenal steroids and acetonitrile is preferred for resolving testicular steroids. Peak shape, resolution, and reproducibility can be improved by maintaining the column at a fixed temperature above ambient, e.g., at 45–60°C that may need optimizing for the required separation. At these temperatures, the eluent viscosity is reduced. If working at ambient temperature, it is advisable to have a room with well controlled temperature to achieve reproducible retention times or use a jacketed column with temperature control. Temperature gradients have been tested with C_8 and C_{18} columns. The difficulties in choosing the appropriate column packing for a particular separation have been eased to some extent by using three and four solvents in a mobile phase system. Systematic, statistical procedures for solvent optimization have been developed. Column packings may not be consistent and chromatographic conditions may have to be adjusted.

An HPLC separation of a complex mixture containing 14 androgenic anabolic steroids (natural and synthetic)

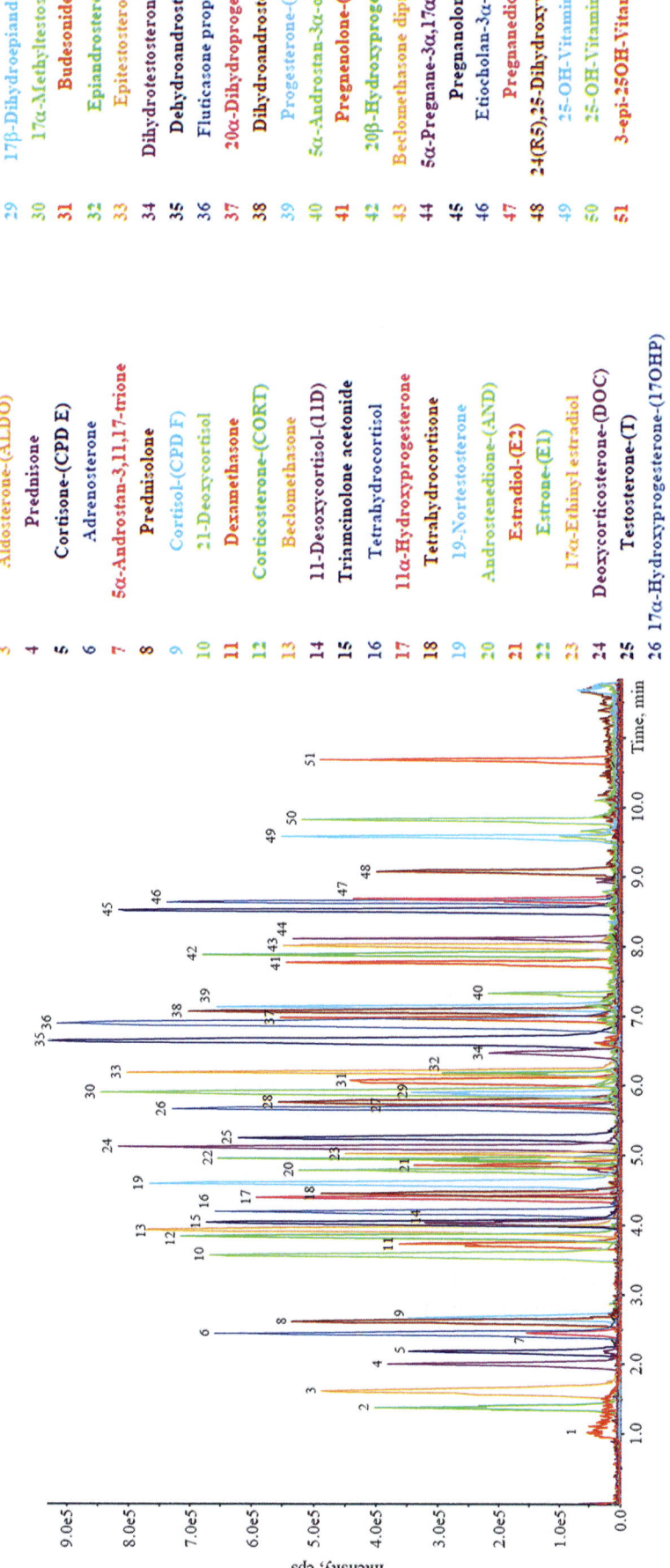

1	Dehydroepiandrosterone Sulfate		29	17β-Dihydroepiandrosterone
2	Estriol		30	17α-Methyltestosterone
3	Aldosterone-(ALDO)		31	Budesonide
4	Prednisone		32	Epiandrosterone
5	Cortisone-(CPD E)		33	Epitestosterone
6	Adrenosterone		34	Dihydrotestosterone-(DHT)
7	5α-Androstan-3,11,17-trione		35	Dehydroandrosterone
8	Prednisolone		36	Fluticasone propionate
9	Cortisol-(CPD F)		37	20α-Dihydroprogesterone
10	21-Deoxycortisol		38	Dihydroandrosterone
11	Dexamethasone		39	Progesterone-(PG)
12	Corticosterone-(CORT)		40	5α-Androstan-3α-ol-17-one
13	Beclomethasone		41	Pregnenolone-(PN)
14	11-Desoxycortisol-(11D)		42	20β-Hydroxyprogesterone
15	Triamcinolone acetonide		43	Beclomethasone dipropionate
16	Tetrahydrocortisol		44	5α-Pregnane-3α,17α,20α-triol
17	11α-Hydroxyprogesterone		45	Pregnanolone
18	Tetrahydrocortisone		46	Etiocholan-3α-diol
19	19-Nortestosterone		47	Pregnanediol
20	Androstenedione-(AND)		48	24(R,S),25-Dihydroxyvitamin D3
21	Estradiol-(E2)		49	25-OH-Vitamin D3
22	Estrone-(E1)		50	25-OH-Vitamin D2
23	17α-Ethinyl estradiol		51	3-epi-25OH-Vitamin D3
24	Deoxycorticosterone-(DOC)			
25	Testosterone-(T)			
26	17α-Hydroxyprogesterone-(17OHP)			
27	Dehydroepiandrosterone-(DHEA)			
28	17α-Hydroxypregnenolone-(17OPN)			
29	17β-Dihydroepiandrosterone			

FIG. 2.3.22 HLC-MS of steroids. Discrete transitions (>2 per analyte) were monitored for each of 51 steroids. HPLC column was a Zorbax XDBC18, 100 × 3 mm eluted with gradient with solvent A: 5:95 MeOH:H₂O; solvent B: MeOH—40%B initially held for 10s, then ramped to 100%B over 13.37 min, hold 100%B 30s, step to 40%B 90s. There was an APCI interface, and the MS was operated in positive ion mode. Optimized per transition Dwell time 5 ms. *(Data kindly provided by Russell Grant.)*

has been used for anabolic steroid screening purposes. The optimization of the method assessed the use of binary, ternary and quaternary mobile phases containing acetonitrile, methanol, or tetrahydrofuran as organic modifiers (Izquierdo-Hornillos and Gonzalo-Lumbreras, 2003). The optimum separation was achieved by using a water–acetonitrile (55:45, v:v) mobile phase and a Hypersil ODS (250 mm × 4.6 mm) 5 μm column (30°C) in about 38 min, allowing the separation of all 14 compounds tested (when danazol was excluded, 13 of the 14 were separated in 23 min). Calibration graphs were obtained using bolasterone, methyltestosterone, and canrenone as internal standards. Detection limits were in the range 0.012–0.11 μg/mL.

For separation of catechol estrogens phosphate was incorporated into the mobile phase (Shimada et al., 1979). With these systems the buffer anion and pH exert significant effects on the separation. Salts used in the eluting solvent may in the long-term corrode the steel of the columns and tubing. Estrogens can be effectively separated when silver nitrate is included in the mobile phase to give 2 g of silver nitrate with 60 mL methanol, 40 mL water at 0.55 mL/min (Tscherne and Capitano, 1977). To prevent metallic silver building up on the column, a water: methanol (50:50, v/v) mobile phase is used each evening to flush excess silver nitrate from the system. Even so a small build-up can occur which requires a rinse with dilute nitric acid or replacement of the tubing when backpressure rises.

2.3.2.5.3 Ion-pair chromatography (IPC)

The mobile phase is supplemented with an ion-pairing reagent often large ionic molecules having a charge opposite to the analyte of interest as well as a substantial hydrophobic region that allows interaction with the stationary phase, plus associated counter-ions (Andreolini et al., 1987). In total, IPC results in different retention of analytes to facilitate separation. In general, ion-pair reagents have only been used for analysis of synthetic corticosteroids (dexamethasone phosphate, budesonide epimers, 17-oxosteroids prelabeled with dansylhydrazine). Excellent separations of estrogens have been described with RP-18 packings by eluting with a gradient of acetonitrile/ methanol and phosphate buffer containing cetyltrimethylammonium bromide (Andreolini et al., 1987). Gradient elution is usually necessary to elute a series of steroids and bile acids after extraction from biological fluids. Gradient elution reduces analysis times and depending upon the gradient shape can optimize separation and improve peak symmetry. Nonlinear, stepped, and linear gradients have been used, largely dictated by the available facilities for programming the pumps. Flow and temperature programming can also be used. The technique may become useful for analysis of steroid conjugates.

2.3.2.5.4 Effects of cyclodextrins and micellar chromatography

Addition of cyclodextrins to the mobile phase has been claimed to improve the HPLC separation of a variety of steroids, from C_{21} corticosteroids and estrogens mainly from Shimada's laboratory although it is not entirely clear how this improvement is achieved. There are a variety of different cyclodextrins and not all confer the same improvement in resolution. It is often necessary to try a number of the polymers of different ring size and binding affinity of the steroids before discovering the best for the particular separation. Addition of other compounds to the mobile phase has also provided advantages such as the use of micellar chromatography which can be achieved by the addition of sodium dodecyl sulfate (SDS). Use of micellar chromatography and a two-column system has allowed the measurement of urinary free cortisol by direct injection of urine onto the first column, washing off protein and elution of the concentrated analyte onto the second analytical column (Nozaki et al., 1991, 1992). Cetyltrimethylammonium bromide was used to improve separation of betamethasone and dexamethasone (Peña-García-Brioles et al., 2004).

2.3.2.5.5 Column switching

The availability of switching valves which can be operated automatically has, over the last 10 years, enabled the use of multiple columns, automatically switching selected peaks from one column to another—so-called "column switching" or "heart cutting." This has been described for the measurement of aldosterone in human serum and plasma, triamcinolone in urine (Schöneshöfer et al., 1985) and later for the measurement of urinary free cortisol, and for 20-reduced metabolites of cortisol and cortisone in urine (Schöneshöfer et al., 1986). These column switching techniques can be of considerable use as a means of automating assays involving HPLC and in addition offer a convenient method of measuring steroids present in trace amounts, by concentrating the analyte on the first column, "heart cutting" onto the second column, and finally by back flushing, the final analytical column is presented with a tight band of concentrated analyte, without nonspecific material from the matrix, which has been largely removed by the first column.

2.3.2.5.6 Hydrophilic interaction liquid chromatography (HILIC)

An HILIC stationary phase has a thin layer of water covering the stationary phase particles. HILIC can be described as a hydrophilic interaction (hence HILIC) partitioning process or as a liquid/liquid extraction system between an immobilized water-rich layer on the surface of the

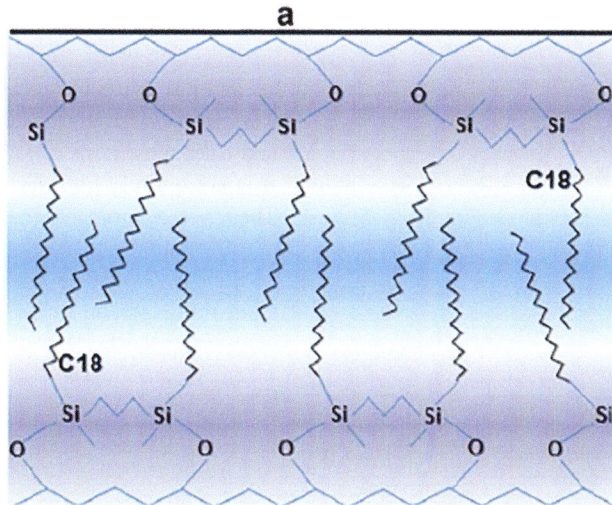

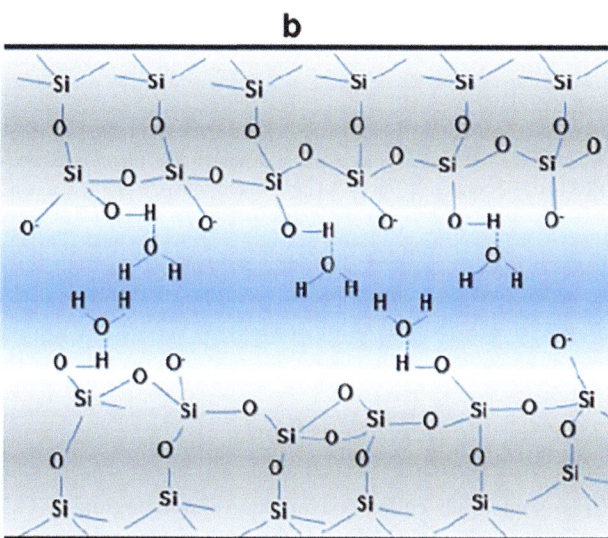

FIG. 2.3.23 Schematic of typical reverse phase HPLC column (A) and HILIC column (B). *(From Tang DQ, Zou L, Yin XX, Ong CN. HILIC-MS for metabolomics: an attractive and complementary approach to RPLC-MS. Mass Spectrom Rev 2016;35(5):574–600. Fig. 1 p. 576.)*

stationary phase and a hydrophobic environment that is created by the highly organic mobile phase (Tang et al., 2016) (Fig. 2.3.23).

In addition to the partitioning effects, weak electrostatic or, depending on the charge state of the analytes, even ionic interactions between the analytes and charged residues in the chemically bonded stationary phase contribute to the main retention mechanism in HILIC. Polar compounds can partition in and out of the water. Polar compounds have stronger interactions and can be retained by hydrogen bonding and electrostatic interactions. Ion exchange is a secondary interaction as many polar stationary phases are mildly acidic or basic.

HILIC mobile phases are mixtures of aqueous buffer systems such as acetonitrile or methanol and organic additives. HILIC uses a high-organic, low-aqueous mobile phase. HILIC offers several advantages, such as enhanced sensitivity in ESI-MS, higher speed due to the lower viscosity of the mobile phase compared with that of standard RP and direct injection of the solid-phase extract. HILIC is a promising approach for the separation of highly polar, hydrophilic, and charged compounds. The eluents are compatible with coupling to MS. HILIC stationary phases consist of polar-modified silica gels or polymers (diol-, amino-, or zwitterionic residues). Compared with RPLC, HILIC ensures considerable retention of compounds, which show less or no retention on RP materials. In contrast to RPLC, polar and hydrophilic analytes are stronger retained on HILIC phases because of predominant solubilization in the immobilized water layer. During gradient elution, an increasing water content results in an enhanced polarity of the mobile phase and, therefore, in a reduced difference in polarity between the water-rich layer on the stationary phase and the actual hydrophobic mobile phase, which leads to an enhanced solubilization of the analytes in the mobile phase and to elution from the column.

The determination of estrogen conjugates in urine samples requires separation in order to eliminate matrix interference. To improve sensitivity, hydrophilic interaction liquid chromatography (HILIC) separation was examined as an alternative strategy to the commonly used RPLC separation. In one case, samples were analyzed by HILIC after a simple SPE using an OASIS HLB cartridge without tedious and time-consuming pretreatment (Qin et al., 2008) (Fig. 2.3.24).

The HILIC column was a TSKgel Amide-80 (2.0 mm × 150 mm, 5 μm, 80 Å; Tosoh Bioscience, Montgomeryville, PA), with carbamoyl groups attached onto the surface of silica through an aliphatic carbon chain. The HILIC eluent had a high-organic/low-aqueous mobile phase against the polar stationary phase to retain and separate polar compounds. The optimized mobile phase consisted of acetonitrile/aqueous ammonium acetate (5 mM, pH 6.80) (85/15, v/v). Paired ions from each conjugate with the higher intensity were established and used for quantification, but both pairs of ions were used for confirmation. Compound-dependent parameters for the LC-MS interface are optimized by direct infusion of individual standards (100 ng/mL) using a syringe pump. These included compound-dependent parameters: declustering potential (DP), entrance potential (EP), collision energy (CE), and collision cell exit potential (CXP).

The HILIC-MS/MS method achieved higher sensitivity than an RPLC-MS/MS method in the determination of estrogen conjugates in human urine. Estrogen conjugates in the urine of a pregnant woman and two breast cancer

MRM Conditions Used for HILIC–MS/MS Analysis of the Estrogen Conjugates and the Internal Standard (ESI, Negative Mode)

analytes	precursor ion (*m/z*)	product ion (*m/z*)	DP	EP	CE	CXP
E1-3S	349	269[a] / 145	−60	−10	−54	−13
E2-3S	351	271[a] / 80	−65	−10	−50	−13
E3-3S	367	287[a] / 80	−65	−10	−50	−15
E1-3G	445	269[a] / 113	−75	−10	−50	−15
E2-3G	447	271[a] / 113	−75	−10	−50	−15
E3-16G	463	287[a] / 113	−75	−10	−50	−15
E3-3G	463	287[a] / 113	−75	−10	−50	−15
IS	467	289[a] / 291	−80	−10	−50	−15

[a] The more abundant product ion was used for quantitative analysis.

Declustering potential (DP), entrance potential (EP), collision energy (CE), and collision cell exit potential (CXP).

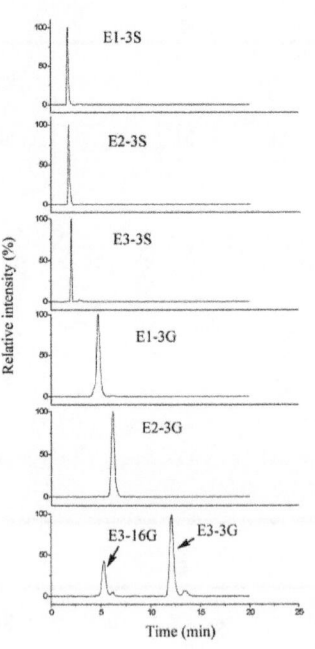

FIG. 2.3.24 Estrogen conjugates in reverse phase HILIC separation of estrogens in river water after column switching part of RP-HPLC eluates. Mobile phase A was acetonitrile/aqueous ammonium acetate (5 mM, pH 6.8) (5/95 v/v); and B was ACN/aqueous ammonium acetate (5 mM, pH 6.8) (75:25 v/v) HILIC was eluted with gradient elution of A to B. HILIC was performed on a ZIC-*p*HILIC column (10 mm × 2.1 mm) eluted with A/B gradient mixture. HPLC was conducted on a Luna C18 column (100 mm × 2 mm; 3 μm packing) and step gradient elution (20%, 30%, 80% B). Fraction 2 containing estrogen conjugates were switched to HILIC. A hybrid quadrupole/linear ion trap (AP Sciex) with Turbo ion spray source was use in negative ion mode. Precursor and product ions (multiple reaction monitoring) are recorded in the chromatograms. *(From Qin F, Zhao YY, Sawyer MB, Li XF. Column-switching reversed phase-hydrophilic interaction liquid chromatography/tandem mass spectrometry method for determination of free estrogens and their conjugates in river water. Anal Chim Acta 2008a;627(1):91–8. Fig. 1 p. 95.)*

patients were studied. The study demonstrated unique profiles and changes in the estrogen conjugates during different stages of pregnancy and in breast cancer patients before and after treatment. HILIC enabled several advantages, such as enhanced sensitivity in ESI-MS, higher speed due to the lower viscosity of the mobile phase compared with that of standard RP, and direct injection of the solid.

The retention and separation of the glucuronides on the column exhibited typical HILIC characteristics. Decreasing the acetonitrile content from 90% to 80%, or increasing the salt concentration from 1 to 10 mM increased significantly the retention and resolution of the estrogen glucuronides. Varying the pH of the mobile phase from 3.0 to 6.8 increased slightly the retention of the four glucuronides, which may be due to increased hydrophilicity of the conjugates at higher pH (pK_a of the glucuronide conjugate was 3.68). Changing the mobile phase conditions described above had little effect on the retention behavior of the three sulfate conjugates on the HILIC column-phase extraction (SPE) elute onto the separation column without reconstitution. The low number of publications shows that from 1998 to 2018 the use of HILIC has increased but it is not very popular.

2.3.2.5.7 Supercritical fluid chromatography (SFC)

SFC, sometimes called convergence chromatography (CC), utilizes compressed CO_2 in the chromatographic mobile phase at pressures between 100 and 400 times atmospheric pressure which makes the supercritical CO_2. SFC is a convergence of ultrahigh-pressure gas chromatography with classical liquid chromatography. SFC can potentially offer better selectivity and shorter analysis time than HPLC due to the low viscosity and high diffusivity of supercritical CO_2 (Nováková et al., 2015; Parr et al., 2018) (Fig. 2.3.25). The CO_2 is often mixed with liquid co-solvents like methanol, ethanol, isopropanol, and acetonitrile to modulate chromatographic retention. The LC conditions will have to be optimized but the technique is going to be useful for metabolomic studies (West et al., 2017; de Kock et al., 2018; Desfontaine et al., 2018a). It is difficult to predict to what extent this will be part of clinical investigations.

CO_2 is the preferred gas due to its low critical parameters, noninflammability and, nontoxicity. Full compatibility of SFC with mass spectrometry detection makes it an attractive analytical platform for metabolomics. The

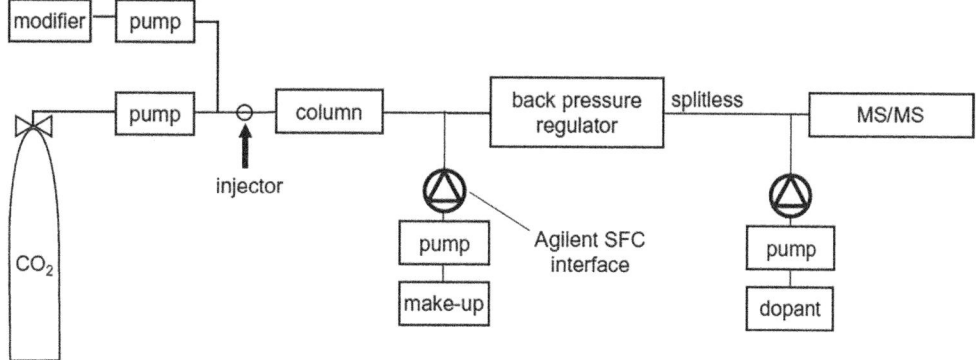

FIG. 2.3.25 Configuration of SFC-MS/MS. *(From Parr MK, Wüst B, Teubel J, Joseph JF. Splitless hyphenation of SFC with MS by APCI, APPI, and ESI exemplified by steroids as model compounds. J Chromatogr B Analyt Technol Biomed Life Sci 2018;1091:67–78. Fig. 2 p. 70.)*

configuration of SFC-MS/MS involves four pumps, the chromatographic separation of which will need to be optimized. The operator will also need to couple the analytical column to the MS in addition to maintaining the operation of the MS. This is a tall order and few analysts will master the technique. There has been a push for commercialization of the complete package of equipment. SFC has been successfully used with multiple ionization sources (Guillarme et al., 2018). The reader will find a useful tutorial published by Novakova et al. in 2014.

One of the major advantages of SFC is that it can simultaneously analyze both polar and nonpolar compounds. Polar stationary phases are best suited for polar compounds retention in SFC. A variety of stationary phases, including bare silica, aminopropyl-, and cyanopropyl-bonded, have been tested and show adequate retention for a range of polar compounds. The silica phases are the most acidic, strongly interacting with basic solutes while the nitrogen-containing phases, like amine or triazole-type, are the most basic and strongly interact with acidic solutes. The neutral phases of amide and alcohol type display an intermediate behavior. Broader acceptance of SFC for the analysis of steroids is needed.

The 2.6–2.7 μm SP phases provide excellent kinetic performance in SFC at a lower pressure than columns packed with sub-2 μm particles. Selecting the proper column for the analysis of a particular metabolite or metabolite class remains challenging. This is especially difficult for metabolomics applications where it is important to separate many compound classes and achieve optimal separation of individual compounds within each compound class. Many column chemistries that have been developed for HPLC and UHPLC analysis have not been extensively tested in SFC with large enough diversity of compound classes.

In addition to optimizing chromatographic conditions, it is also critical to optimize mass spectrometry detection parameters to achieve maximal sensitivity in SFC-MS analysis. The mobile phase composition and the flow rate of the make-up solvent for MS detection are needed for a wide range of compounds. As with any LC-MS/MS system there are issues with matrix effects (Desfontaine et al., 2018b). Transfer of the existing LCMS methods to SFC-MS may be more challenging than expected.

Separations with UHPLC and UHPSFC (Storbeck et al., 2018) are compared in Fig. 2.3.26.

In UHPLC chromatography, hydrophobic interactions and electrostatic repulsion are involved whereas in UHPSFC hydrogen bonding and dipole-dipole interactions affect chromatography. Separations of steroids panels (Quanson et al., 2016) and conjugates (Doué et al., 2015) have been described.

SFC-MS will not fully replace other separation techniques in general us, but it will definitely become a major analytical platform in metabolomics to complement other analytical techniques like NMR, GC-MS and LC-MS. By using methanol/carbon dioxide linear gradients as a mobile phase faster separations and higher efficiencies than conventional GC can be achieved. This nonpolar solvent system has a low backpressure, allowing higher flow rates than LC due to the viscosity of the mobile phase being more similar to gas rather than liquid. The implementation of SFC has remained very limited in the clinical laboratory. Proponents of supercritical fluid chromatography suggest that this separation method harnesses the advantage of GC separation to LC analysis.

2.3.2.5.8 Sample injection onto LC systems

The injection system for HPLC is usually a loop into which the sample is loaded and eluted onto the column by diverting solvent flow. Extracts are usually dissolved in the mobile phase. The addition of a suitable macromolecular matrix, e.g., polyethylene glycol to the extracting solvent prior to evaporation improves the recovery of steroids suggesting that some steroids dissolve poorly in the mobile phase alone. Injectors which encompass rubber septa should be avoided.

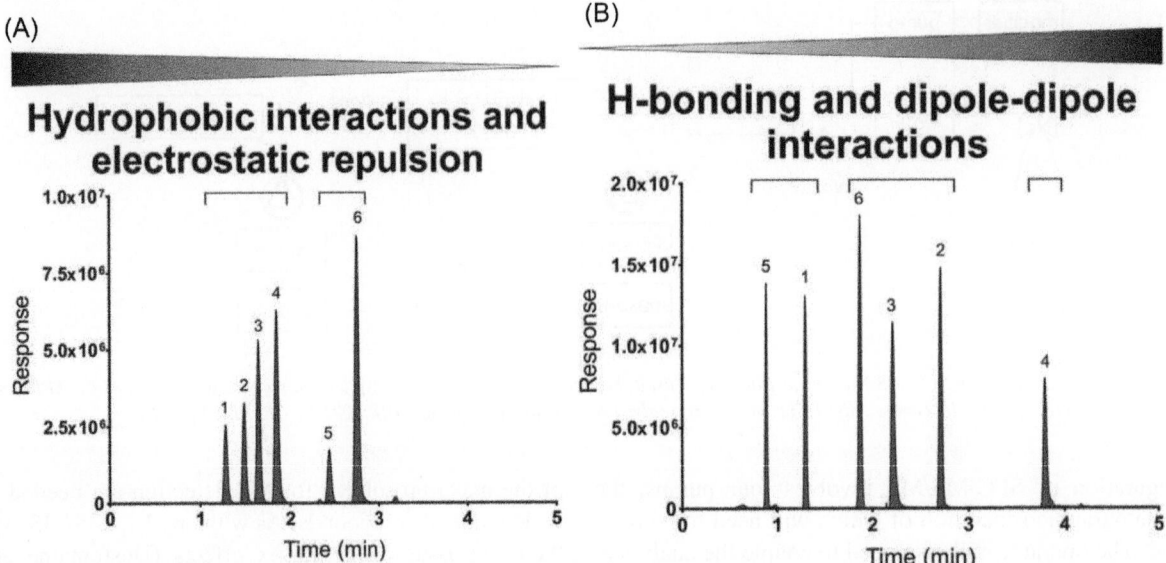

FIG. 2.3.26 (A) UHPLC. Steroid metabolites were separated with a UPLC high strength silica (HSS) column (2.1 mm × 50 mm; 1.8 μm packing). Mobile phase was (A) 0.1% formic acid and (B) methanol. A linear gradient from 55% to 75% was used. Column temperature was 50°C. A Xevo TQ triple quad MS was used in MRM mode with ESI probe in positive mode. (B) UHPSFC. Steroid metabolites separated with Acquity UPC² system with an Acquity UPC² BEH 2EP column (3 mm × 100 mm; 1.7 μm particle size) (Quanson et al., 2016; Storbeck et al., 2018) eluted with liquid CO_2 modified with methanol. A linear gradient from 2% to 9.5% methanol was used. Column temperature and automated back pressure regulator were at 60°C. A Xevo TQ triple quad MS was used in MRM mode with ESI probe in positive mode. Steroids: 11-ketoandrostenedione (1), 11-ketotestosterone (2), 11β-hydroxyandrostenedione (3), 11β-hydroxytestosterone (4), androstenedione (5) and testosterone (6). *((A) From Storbeck KH, Gilligan L, Jenkinson C, et al. The utility of ultra-high performance supercritical fluid chromatography-tandem mass spectrometry (UHPSFC-MS/MS) for clinically relevant steroid analysis. J Chromatogr B Analyt Technol Biomed Life Sci 2018;1085:36–41. Fig. 2 p. 38; (B) From Storbeck KH, Gilligan L, Jenkinson C, et al. The utility of ultra-high performance supercritical fluid chromatography-tandem mass spectrometry (UHPSFC-MS/MS) for clinically relevant steroid analysis. J Chromatogr B Analyt Technol Biomed Life Sci 2018;1085:36–41. Fig. 3 p. 38.)*

At the high instrument sensitivities often used for the analysis of steroid hormones, such procedures may lead to the production of spurious and irreproducible peaks may occur in the chromatogram. These may reflect the action of injected solvents on the septum. Injection valves are, therefore, preferred. As column dimensions have been reduced the sample can be between 5 and 20 μL. The sample size, solvent and flow rate need to be optimized for the analyte. If sample size is high, the peak detected can be distorted from the desirable symmetry and co-elution of analytes may be missed. With careful optimization it may be possible to increase the sample size and gain sensitivity with a two to fivefold gain in the LOQ.

2.3.2.5.9 Detectors for HPLC

Steroid hormones with unsaturated A-ring can be detected by UV absorption (peak 240 nm) or photo-diode array detector but sensitivity is low. Estrogens can be detected at 10–100 pg/mL through peak absorption of the phenolic A-ring at 280 nm. Although most steroid metabolites are virtually without ultraviolet absorption, which is the most useful of current detectors, some further metabolites of steroids can be detected with a refractive index or an electrochemical detector or by the use of **pre- or postcolumn**

reaction with compounds which cause enhanced UV absorbance, or fluorescence. A mass spectrometer now offers the best option for steroid detection because the quality of the determination can be demonstrated.

2.3.3 Summary

In the past 60 years, the increased power of chromatography has enabled better separation of steroids and hence purity of the product. The analysis times have also reduced from hours to minutes. The move to couple the chromatography with mass spectrometry has led to improved specificity in the analysis.

2.3.4 Further reading

The following references were not cited in the chapter but may direct the reader to further research.

2.3.4.1 Steroid analysis review

Griffiths et al. (2006), Makin et al. (2010), Shackleton (1986, 2008, 2012), Sjövall and Axelson (1979, 1982), Smy and Straseski (2018), and Zhang and Rockwood (2015).

2.3.4.2 Biological samples

Gröschl (2017), Hennig et al. (2018), Higashi et al. (2016), Kim et al. (2015a,b), Li et al. (2018), Ostler et al. (2014), Qaiser et al. (2017), Siiterii et al. (1982), Wagner et al. (2016), and Wang et al. (2019a,b).

2.3.4.3 Extraction

Abdel-Khalik et al. (2013), Andersson and Sjövall (1983), Cawley et al. (2005), Enderle et al. (2016), Hennion (1999), Manaf et al. (2018), Sjövall et al. (1983), Uusijärvi et al. (1989), van der Berg et al. (2020), Vanluchene and Vandekerckhove (1988), and Zarzycki et al. (2009).

2.3.4.4 Chromatography

Bennett and Heftmann (1962), Brooks and Keates (1969), Bush (1952), Bush and Crowshaw (1965), Carr et al. (1971), Ellingboe et al. (1968, 1970), Fenske (2006), Heikkinen et al. (1983), Hobkirk et al. (1969), Joustra et al. (1967), Kulikov and Galat (2009), Li et al. (1995), Lisboa (1970), Miękus et al. (2017), Murphy and D'Aux (1975), Randazzo et al. (2016), Ruiz-Ángel et al. (2009), Salentijn et al. (2018), Watson and Bartosik (1971), and Zaffaroni et al. (1950).

2.3.4.5 Conjugate analysis

Alvarez-Sánchez et al. (2009), Christakoudi et al. (2008), Esquivel et al. (2017), Fotsis (1987), Fotsis and Adlercreutz (1987), Gagné et al. (2008), Galuska et al. (2013), Kuehnbaum and Britz-McKibbin (2011), Ma et al. (2014), Mattox et al. (1975a,b), Meng et al. (1996), Mitamura et al. (2014), Plager (1966), Stevenson et al. (2015), and Strahm et al. (2009).

2.3.4.6 GC-MS

Ahmadkhaniha et al. (2010), Anderson et al. (1974), Anizan et al. (2020), Axelson and Sjövall (1977), Doué et al. (2015), Halket and Zaikin (2003), Halket and Zaikin (2004, 2005, 2006), Honour et al. (1982), Lien et al. (2012), Little (1999), Magnisali et al. (2008), Matsuzaki et al. (2004), Moon et al. (2009, 2010), Moros et al. (2017), Rossi et al. (1994), Shackleton et al. (1979), Tetsuo et al. (1980), Teubel et al. (2018), Wudy et al. (1999, 2000, 2001, 2002), and Zaikin and Halket (2005, 2006).

2.3.4.7 Isotope ratio GC-MS

Brailsford et al. (2012) and Casilli et al. (2016).

2.3.4.8 LC-MS

Bae et al. (2016), Boelen et al. (2016), Boggs et al. (2016), Bonfiglio et al. (1999), Cao et al. (2018), Carvalho et al. (2008), Ceglarek et al. (2009a,b), Dai et al. (2012), Domínguez-Romero et al. (2014), du Toit et al. (2017, 2020), Dury et al. (2015), Fahlbusch et al. (2015), Fanelli et al. (2011), Fiet et al. (2017), Frey et al. (2016), Gao et al. (2016), Gaudl et al. (2016), Guo et al. (2004), Havlíková et al. (2013), Higashi and Ogawa (2016), Janzen et al. (2008), Karvaly et al. (2018), Keller et al. (2008), Khan et al. (2006), Kulle et al. (2011, 2013), Kushnir et al. (2006), Liu et al. (2003), Matysik and Liebisch (2017), Methlie et al. (2013), Napoli et al. (2010), Nováková et al. (2014), Rauh (2010), Rauh et al. (2006), Ray et al. (2014), Rossi et al. (2010), Søeborg et al. (2017), Stone and Fitzgerald (2018), Suhr et al. (2016), Vallano et al. (2005), Wang et al. (2016), Wudy and Choi (2016), Xu et al. (2006), and Yamashita et al. (2007).

2.3.4.9 Bile acid analysis

Almé et al. (1977).

2.3.4.10 High resolution MS

Abushareeda et al. (2017).

References

Abdel-Khalik J, Björklund E, Hansen M. Development of a solid phase extraction method for the simultaneous determination of steroid hormones in H295R cell line using liquid chromatography-tandem mass spectrometry. J Chromatogr B Analyt Technol Biomed Life Sci 2013;935:61–9.

Abushareeda W, Lyris E, Kraiem S, Wahaibi AA, Alyazidi S, Dbes N, et al. Gas chromatographic quadrupole time-of-flight full scan high resolution mass spectrometric screening of human urine in antidoping analysis. J Chromatogr B Analyt Technol Biomed Life Sci 2017;1063:74–83.

Ahmadkhaniha R, Shafiee A, Rastkari N, Khoshayand MR, Kobarfard F. Quantification of endogenous steroids in human urine by gas chromatography mass spectrometry using a surrogate analyte approach. J Chromatogr B Analyt Technol Biomed Life Sci 2010;878(11–12):845–52.

Almé B, Bremmelgaard A, Sjövall J, Thomassen P. Analysis of metabolic profiles of bile acids in urine using a lipophilic anion exchanger and computerized gas-liquid chromatorgaphy-mass spectrometry. J Lipid Res 1977;18(3):339–62.

Alvarez-Sánchez B, Priego-Capote F, Luque de Castro MD. Ultrasound-enhanced enzymatic hydrolysis of conjugated female steroids as pretreatment for their analysis by LC-MS/MS in urine. Analyst 2009;134(7):1416–22.

Anderson RA, Defaye G, Madani C, Chambaz EM, Brooks CJ. Lipophilic gel and gas-phase analysis of steroid hormones application to the human newborn. J Chromatogr 1974;99:485–94.

Anderson DC, Hopper BR, Lasley BL, Yen SS. A simple method for the assay of eight steroids in small volumes of plasma. Steroids 1976;28 (2):179–96.

Andersson SH, Sjövall J. A method combining solvent and gel extraction for isolation and preliminary purification of steroids in tissues. Anal Biochem 1983;134(2):309–12.

Andreolini F, Borra C, Caccamo F, Di Corcia A, Samperi R. Estrogen conjugates in late-pregnancy fluids: extraction and group separation by a graphitized carbon black cartridge and quantification by high-performance liquid chromatography. Anal Chem 1987;59(13):1720–5.

Anizan S, Bichon E, Monteau F, Cesbron N, Antignac JP, Le Bizec B. A new reliable sample preparation for high throughput focused steroid profiling by gas chromatography-mass spectrometry. J Chromatogr A 2010;1217 (43):6652–60.

Axelson M, Sahlberg BL. Group separation and gas chromatography—mass spectrometry of conjugated steroids in plasma. J Steroid Biochem 1983;18(3):313–21.

Axelson M, Sjövall J. Analysis of unconjugated steroids in plasma by liquid-gel chromatography and glass capillary gas chromatography mass spectrometry. J Steroid Biochem 1977;8(6):683–92.

Baba S. Radio-gas chromatography. J Chromatogr 1989;492:137–65.

Bae YJ, Gaudl A, Jaeger S, Stadelmann S, Hiemisch A, Kiess W, et al. Immunoassay or LC-MS/MS for the measurement of salivary cortisol in children? Clin Chem Lab Med 2016;54(5):811–22.

Bennett RD, Heftmann E. Thin-layer chromatography of corticosteroids. J Chromatogr 1962;9:348–52.

Bhawani SA, Sulaiman O, Hashim R, Mohamad MN. Thin-layer chromatographc analysis of steroids. A review. Trop J Pharm Res 2010;9:310–3.

Bloem LM, Storbeck KH, Swart P, du Toit T, Schloms L, Swart AC. Advances in the analytical methodologies: profiling steroids in familiar pathways-challenging dogmas. J Steroid Biochem Mol Biol 2015;153:80–92.

Boelen A, Ruiter AF, Claahsen-van der Grinten HL, Endert E, Ackermans MT. Determination of a steroid profile in heel prick blood using LC-MS/MS. Bioanalysis 2016;8(5):375–84.

Boggs AS, Bowden JA, Galligan TM, Guillette Jr LJ, Kucklick JR. Development of a multi-class steroid hormone screening method using liquid chromatography/tandem mass spectrometry (LC-MS/MS). Anal Bioanal Chem 2016;408(15):4179–90.

Bonfiglio R, King RC, Olah TV, Merkle K. The effects of sample preparation methods on the variability of the electrospray ionization response for model drug compounds. Rapid Commun Mass Spectrom 1999;13(12):1175–85.

Brailsford AD, Gavrilović I, Ansell RJ, Cowan DA, Kicman AT. Two-dimensional gas chromatography with heart-cutting for isotope ratio mass spectrometry analysis of steroids in doping control. Drug Test Anal 2012;4(12):962–9.

Brooks CJ. Studies of acetylated corticosteroids and related 20-oxopregnane derivatives by gas liquid chromatography. Anal Chem 1965;37:636–41.

Brooks CJ, Harvey DJ. Comparison of various alkylboronic acids for the characterization of corticosteroids by gas-liquid chromatography-mass spectrometry. Biochem J 1969;114(1):15P.

Brooks CJ, Keates RA. Gel filtration in lipophilic solvents using hydroxyalkoxypropyl derivatives of Sephadex. J Chromatogr 1969;44 (3):509–21.

Brown JW, Carballeira A, Fishman LM. Rapid two-dimensional thin-layer chromatographic system for the separation of multiple steroids of secretory and neuroendocrine interest. J Chromatogr 1988;439(2): 441–7.

Bush IE. Methods of paper chromatography of steroids applicable to the study of steroids in mammalian blood and tissues. Biochem J 1952;50(3):370–8.

Bush IE, Crowshaw K. A technique for rapid paper chromatography. J Chromatogr 1965;19(1):114–29.

Bush IE, Willoughby M. The excretion of allo tetrahydrocortisol in human urine. Biochem J 1957;67(4):689–700.

Cao BY, Gong CX, Wu D, Liang XJ, Li WJ, Liu M, et al. Liquid chromatography-tandem mass spectrometry-based characterization of steroid hormone profiles in healthy 6 to 14-year-old male children. Chin Med J (Engl) 2018;131(7):862–6.

Carr BR, Mikhail G, Flickinger GL. Column chromatography of steroids on Sephadex LH-20. J Clin Endocrinol Metab 1971;33(2):358–60.

Carvalho VM, Nakamura OH, Vieira JG. Simultaneous quantitation of seven endogenous C-21 adrenal steroids by liquid chromatography tandem mass spectrometry in human serum. J Chromatogr B Analyt Technol Biomed Life Sci 2008;872(1–2):154–61.

Casals G, Marcos J, Pozo OJ, Alcaraz J, Martínez de Osaba MJ, Jiménez W. Microwave-assisted derivatization: application to steroid profiling by gas chromatography/mass spectrometry. J Chromatogr B Analyt Technol Biomed Life Sci 2014;960:8–13.

Casilli A, Piper T, de Oliveira FA, Padilha MC, Pereira HM, Thevis M, et al. Optimization of an online heart-cutting multidimensional gas chromatography clean-up step for isotopic ratio mass spectrometry and simultaneous quadrupole mass spectrometry measurements of endogenous anabolic steroid in urine. Drug Test Anal 2016;8(11–12):1204–11.

Cawley AT, Kazlauskas R, Trout GJ, George AV. Determination of urinary steroid sulfate metabolites using ion paired extraction. J Chromatogr B Analyt Technol Biomed Life Sci 2005;825(1):1–10.

Ceglarek U, Kortz L, Leichtle A, Fiedler GM, Kratzsch J, Thiery J. Rapid quantification of steroid patterns in human serum by on-line solid phase extraction combined with liquid chromatography-triple quadrupole linear ion trap mass spectrometry. Clin Chim Acta 2009a;401 (1–2):114–8.

Ceglarek U, Leichtle A, Brügel M, Kortz L, Brauer R, Bresler K, et al. Challenges and developments in tandem mass spectrometry based clinical metabolomics. Mol Cell Endocrinol 2009b;301(1–2):266–71.

Choi MH, Chung BC. GC-MS determination of steroids related to androgen biosynthesis in human hair with pentafluorophenyldimethylsilyl-trimethylsilyl derivatisation. Analyst 1999;124(9):1297–300.

Choi MH, Hahm JR, Jung BH, Chung BC. Measurement of corticoids in the patients with clinical features indicative of mineralocorticoid excess. Clin Chim Acta 2002;320(1–2):95–9.

Christakoudi S, Cowan DA, Taylor NF. Sodium ascorbate improves yield of urinary steroids during hydrolysis with Helix pomatia juice. Steroids 2008;73(3):309–19.

Christakoudi S, Cowan DA, Taylor NF. Steroids excreted in urine by neonates with 21-hydroxylase deficiency: characterization, using GC-MS and GC-MS/MS, of the D-ring and side chain structure of pregnanes and pregnenes. Steroids 2010a;75(1):34–52.

Christakoudi S, Cowan DA, Taylor NF. A new marker for early diagnosis of 21-hydroxylase deficiency: 3beta,16alpha,17alpha-trihydroxy-5alpha-pregnane-7,20-dione. J Steroid Biochem Mol Biol 2010b;121 (3–5):574–81.

Christakoudi S, Cowan DA, Taylor NF. Steroids excreted in urine by neonates with 21-hydroxylase deficiency. 3. Characterization, using GC-MS and GC-MS/MS, of androstanes and androstenes. Steroids 2012a;77(13):1487–501.

Christakoudi S, Cowan DA, Taylor NF. Steroids excreted in urine by neonates with 21-hydroxylase deficiency. 2. Characterization, using GC-MS and GC-MS/MS, of pregnanes and pregnenes with an oxogroup on the A- or B-ring. Steroids 2012b;77(5):382–93.

Christakoudi S, Cowan DA, Christakudis G, Taylor NF. 21-Hydroxylase deficiency in the neonate—trends in steroid anabolism and catabolism during the first weeks of life. J Steroid Biochem Mol Biol 2013a;138:334–47.

Christakoudi S, Cowan DA, Taylor NF. Steroids excreted in urine by neonates with 21-hydroxylase deficiency. 4. Characterization, using GC-MS and GC-MS/MS, of 11oxo-pregnanes and 11oxo-pregnenes. Steroids 2013b;78(5):468–75.

Cimpoiu C, Hosu A, Hodisan S. Analysis of some steroids by thin-layer chromatography using optimum mobile phases. J Pharm Biomed Anal 2006;41(2):633–7.

Dai W, Huang Q, Yin P, Li J, Zhou J, Kong H, et al. Comprehensive and highly sensitive urinary steroid hormone profiling method based on stable isotope-labeling liquid chromatography-mass spectrometry. Anal Chem 2012;84(23):10245–51.

de Kock N, Acharya SR, Ubhayasekera SJKA, Bergquist J. A novel targeted analysis of peripheral steroids by ultra-performance supercritical fluid chromatography hyphenated to tandem mass spectrometry. Sci Rep 2018;8(1):16993.

Denver N, Khan S, Homer NZM, MacLean MR, Andrew R. Current strategies for quantification of estrogens in clinical research. J Steroid Biochem Mol Biol 2019;192:105373.

Desfontaine V, Capetti F, Nicoli R, Kuuranne T, Veuthey JL, Guillarme D. Systematic evaluation of matrix effects in supercritical fluid chromatography versus liquid chromatography coupled to mass spectrometry for biological samples. J Chromatogr B Analyt Technol Biomed Life Sci 2018a;1079:51–61.

Desfontaine V, Losacco GL, Gagnebin Y, et al. Applicability of supercritical fluid chromatography—mass spectrometry to metabolomics. I—Optimization of separation conditions for the simultaneous analysis of hydrophilic and lipophilic substances. J Chromatogr A 2018b;1562:96–107.

Domínguez-Romero JC, García-Reyes JF, Molina-Díaz A. Comparative evaluation of seven different sample treatment approaches for large-scale multiclass sport drug testing in urine by liquid chromatography-mass spectrometry. J Chromatogr A 2014;1361:34–42.

Doué M, Dervilly-Pinel G, Pouponneau K, Monteau F, Le Bizec B. Analysis of glucuronide and sulfate steroids in urine by ultrahigh-performance supercritical-fluid chromatography hyphenated tandem mass spectrometry. Anal Bioanal Chem 2015;407(15):4473–84.

du Toit T, Bloem LM, Quanson JL, Ehlers R, Serafin AM, Swart AC. Profiling adrenal 11β-hydroxyandrostenedione metabolites in prostate cancer cells, tissue and plasma: UPC(2)-MS/MS quantification of 11β-hydroxytestosterone, 11keto-testosterone and 11keto-dihydrotestosterone. J Steroid Biochem Mol Biol 2017;166:54–67.

du Toit T, Stander MA, Swart AC. A high-throughput UPC2-MS/MS method for the separation and quantification of C19 and C21 steroids and their C11-oxy steroid metabolites in the classical, alternative, backdoor and 11OHA4 steroid pathways. J Chromatogr B Analyt Technol Biomed Life Sci 2018;1080:71–81.

du Toit T, van Rooyen D, Stander MA, Atkin SL, Swart AC. Analysis of 52 C19 and C21 steroids by UPC2-MS/MS: characterising the C11-oxy steroid metabolome in serum. J Chromatogr B Analyt Technol Biomed Life Sci 2020;1152, 122243.

Dury AY, Ke Y, Gonthier R, Isabelle M, Simard JN, Labrie F. Validated LC-MS/MS simultaneous assay of five sex steroid/neurosteroid-related sulfates in human serum. J Steroid Biochem Mol Biol 2015;149:1–10.

Eberlein WR, Bongiovanni AM. A paper chromatographic method for the measurement of pregnanediol in urine. J Clin Endocrinol Metab 1958;18(3):300–9.

Ebner MJ, Corol DI, Havlíková H, Honour JW, Fry JP. Identification of neuroactive steroids and their precursors and metabolites in adult male rat brain. Endocrinology 2006;147(1):179–90.

Ellingboe J, Nyström E, Sjövall J. A versatile lipophilic Sephadex derivative for "reversed-phase" chromatography. Biochim Biophys Acta 1968;152(4):803–5.

Ellingboe J, Nyström E, Sjövall J. Liquid-gel chromatography on lipophilic-hydrophobic Sephadex derivatives. J Lipid Res 1970;11(3):266–73.

Enderle Y, Foerster K, Burhenne J. Clinical feasibility of dried blood spots: analytics, validation, and applications. J Pharm Biomed Anal 2016;130:231–43.

Esparza C, Borisov RS, Polovkov NY, Zaikin VG. Post-chromatographic fixed-charge derivatization for the analysis of hydroxyl-containing compounds by a combination of thin-layer chromatography and matrix-assisted laser desorption/ionization mass spectrometry. J Chromatogr A 2018;1560:97–103.

Esquivel A, Matabosch X, Kotronoulas A, Balcells G, Joglar J, Ventura R. Ionization and collision induced dissociation of steroid bisglucuronides. J Mass Spectrom 2017;52(11):759–69.

Exley D, Chamberlain J. Properties of steroidal 3 enol heptafluorobutyrates. Steroids 1967;10(5):509–26.

Fahlbusch FB, Heussner K, Schmid M, Schild R, Ruebner M, Huebner H, et al. Measurement of **amniotic fluid** steroids of midgestation via LC-MS/MS. J Steroid Biochem Mol Biol 2015;152:155–60.

Fanelli F, Belluomo I, Di Lallo VD, Cuomo G, De Iasio R, Baccini M, et al. Serum steroid profiling by isotopic dilution-liquid chromatography-mass spectrometry: comparison with current immunoassays and reference intervals in healthy adults. Steroids 2011;76(3):244–53.

Fenske M. Method development for cortisol and cortisone by micellar liquid chromatography using sodium dodecyl sulphate: application to urine samples of rugby players. J Chromatogr Sci 2006;44(9):579–80.

Fenske M. Determination of cortisol in human plasma by thin-layer chromatography and fluorescence derivatization with isonicotinic acid hydrazide. J Chromatogr Sci 2008;46(1):1–3.

Fiet J, Le Bouc Y, Guéchot J, Hélin N, Maubert MA, Farabos D, et al. A liquid chromatography/tandem mass spectometry profile of 16 serum steroids, including 21-deoxycortisol and 21-deoxycorticosterone, for management of congenital adrenal hyperplasia. J Endocr Soc 2017;1(3):186–201.

Fotsis T. The multicomponent analysis of estrogens in urine by ion exchange chromatography and GC-MS—II. Fractionation and quantitation of the main groups of estrogen conjugates. J Steroid Biochem 1987;28(2):215–26.

Fotsis T, Adlercreutz H. The multicomponent analysis of estrogens in urine by ion exchange chromatography and GC-MS—I. Quantitation of estrogens after initial hydrolysis of conjugates. J Steroid Biochem 1987;28(2):203–13.

Fotsis T, Adlercreutz H, Järvenpää P, Setchell KD, Axelson M, Sjövall J. Group separation of steroid conjugates by DEAE-Sephadex anion exchange chromatography. J Steroid Biochem 1981;14 (5):457–63.

Frey AJ, Wang Q, Busch C, Feldman D, Bottalico L, Mesaros CA, et al. Validation of highly sensitive simultaneous targeted and untargeted analysis of keto-steroids by Girard P derivatization and stable isotope dilution-liquid chromatography-high resolution mass spectrometry. Steroids 2016;116:60–6.

Furuta T, Namekawa T, Shibasaki H, Kasuya Y. Simultaneous determination of tetrahydrocortisol and tetrahydrocortisone in human plasma and urine by stable isotope dilution mass spectrometry. J Chromatogr B Biomed Sci Appl 1998;706(2):181–90.

Furuta T, Eguchi N, Shibasaki H, Kasuya Y. Simultaneous determination of endogenous and 13C-labelled cortisols and cortisones in human plasma by stable isotope dilution mass spectrometry. J Chromatogr B Biomed Sci Appl 2000;738(1):119–27.

Gagné S, Laterreur J, Mahrouche L, et al. Selective isolation of in vitro phase II conjugates using a lipophilic anionic exchange solid phase extraction method. J Chromatogr B Analyt Technol Biomed Life Sci 2008;863(2):242–8.

Galuska CE, Hartmann MF, Sánchez-Guijo A, Bakhaus K, Geyer J, Schuler G, et al. Profiling intact steroid sulfates and unconjugated steroids in biological fluids by liquid chromatography-tandem mass spectrometry (LC-MS-MS). Analyst 2013;138(13):3792–801.

Gao W, Kirschbaum C, Grass J, Stalder T. LC-MS based analysis of endogenous steroid hormones in human hair. J Steroid Biochem Mol Biol 2016;162:92–9.

Gaskell SJ, Brooks CJ. t-Butyldimethylsilyl derivatives in the gas chromatography-mass spectrometry of steroids. Biochem Soc Trans 1976;4(1):111–3.

Gaudl A, Kratzsch J, Bae YJ, Kiess W, Thiery J, Ceglarek U. Liquid chromatography quadrupole linear ion trap mass spectrometry for quantitative steroid hormone analysis in plasma, urine, saliva and hair. J Chromatogr A 2016;1464:64–71.

Giton F, Guechot J, Fiet J. New resuable Celite/ethylene glycol cartridges for selective chromatography of steroids before immunoassay. Clin Biochem 2009;42:1735–8.

Griffiths WJ, Shackleton CHL, Sjövall J. Steroid analysis. In: Caprioli R, editor. The encyclopedia of mass spectrometry, vol. 3. Amsterdam: Elsevier; 2006. p. 447–73.

Gröschl M. Saliva: a reliable sample matrix in bioanalytics. Bioanalysis 2017;9(8):655–68 [review].

Guillarme D, Desfontaine V, Heinisch S, Veuthey JL. What are the current solutions for interfacing supercritical fluid chromatography and mass spectrometry? J Chromatogr B Analyt Technol Biomed Life Sci 2018;1083:160–70.

Guo T, Chan M, Soldin SJ. Steroid profiles using liquid chromatography-tandem mass spectrometry with atmospheric pressure photoionization source. Arch Pathol Lab Med 2004;128(4):469–75.

Häkkinen MR, Heinosalo T, Saarinen N, Linnanen T, Voutilainen R, Lakka T, Jääskeläinen J, Poutanen M, Auriola S. Analysis by LC-MS/MS of endogenous steroids from human serum, plasma, endometrium and endometriotic tissue. J Pharm Biomed Anal 2018;152:165–72.

Halket JM, Zaikin VG. Derivatization in mass spectrometry—1. Silylation. Eur J Mass Spectrom 2003;9(1):1–21.

Halket JM, Zaikin VV. Derivatization in mass spectrometry-3. Alkylation (arylation). Eur J Mass Spectrom 2004;10(1):1–19.

Halket JM, Zaikin VG. Review: derivatization in mass spectrometry—5. Specific derivatization of monofunctional compounds. Eur J Mass Spectrom 2005;11(1):127–60.

Halket JM, Zaikin VG. Derivatization in mass spectrometry—7. On-line derivatisation/degradation. Eur J Mass Spectrom 2006;12(1):1–13.

Hannen RF, Michael AE, Jaulim A, Bhogal R, Burrin JM, Philpott MP. Steroid synthesis by primary human keratinocytes; implications for skin disease. Biochem Biophys Res Commun 2011;404(1):62–7.

Havlíková L, Vlčková H, Solich P, Nováková L. HILIC UHPLC-MS/MS for fast and sensitive bioanalysis: accounting for matrix effects in method development. Bioanalysis 2013;5(19):2345–57.

Heftmann E, Ko ST, Bennett RD. Response of steroids to sulfuric acid in thin-layer chromatography. J Chromatogr 1966;21(3):490–4.

Heikkinen R, Fotsis T, Adlercreutz H. 5. Analytical methods: physiochemical assays. Use of ion exchange chromatography in steroid analysis. J Steroid Biochem 1983;19(1A):175–80.

Hennig K, Antignac JP, Bichon E, Morvan ML, Miran I, Delaloge S, et al. Steroid hormone profiling in human breast adipose tissue using semi-automated purification and highly sensitive determination of estrogens by GC-APCI-MS/MS. Anal Bioanal Chem 2018;410 (1):259–75.

Hennion MC. Solid-phase extraction: method development, sorbents, and coupling with liquid chromatography. J Chromatogr A 1999;856(1–2):3–54.

Higashi T, Ogawa S. Chemical derivatization for enhancing sensitivity during LC/ESI-MS/MS quantification of steroids in biological samples: a review. J Steroid Biochem Mol Biol 2016;162:57–69.

Higashi T, Yamagata K, Kato Y, Ogawa Y, Takano K, Nakaaze Y, et al. Methods for determination of fingernail steroids by LC/MS/MS and differences in their contents between right and left hands. Steroids 2016;109:60–5.

Hill M, Hána Jr V, Velíková M, Pařízek A, Kolátorová L, Vítků J, et al. A method for determination of one hundred endogenous steroids in human serum by gas chromatography-tandem mass spectrometry. Physiol Res 2019;68(2):179–207.

Hobkirk R, Musey P, Nilsen M. Chromatographic separation of estrone and 17 beta-estradiol conjugates on DEAE-Sephadex. Steroids 1969;14 (2):191–206.

Honour JW, Brooks CJ, Shackleton CH. Degradation of steroid derivatives at a gas chromatograph mass spectrometer interface. Biomed Mass Spectrom 1982;9:505–9.

Horning MG, Moss AM, Horning EC. Formation and gas-liquid chromatographic behavior of isometric steroid ketone methoxime derivatives. Anal Biochem 1968;22(2):284–94.

Impens S, De Wasch K, De Brabander H. Determination of anabolic steroids with gas chromatography-ion trap mass spectrometry using hydrogen as carrier gas. Rapid Commun Mass Spectrom 2001;15 (24):2409–14.

Ito Y. C-20-reduced metabolites of cortisol in normal human urine. Endocrinol Jpn 1968;15(1):13–9.

Izquierdo-Hornillos R, Gonzalo-Lumbreras R. Optimization of the separation of a complex mixture of natural and synthetic anabolic steroids by micellar liquid chromatography. J Chromatogr B Analyt Technol Biomed Life Sci 2003;798(1):69–77.

Janzen N, Sander S, Terhardt M, Peter M, Sander J. Fast and direct quantification of adrenal steroids by tandem mass spectrometry in serum and dried blood spots. J Chromatogr B Analyt Technol Biomed Life Sci 2008;861(1):117–22.

Joustra M, Söderqvist B, Fischer L. Gel filtration in organic solvents. J Chromatogr 1967;28(1):21–5.

Junker J, Chong I, Kamp F, Steiner H, Giera M, Müller C, et al. Comparison of strategies for the determination of sterol sulfates via GC-MS leading to a novel deconjugation-derivatization protocol. Molecules 2019;24 (13). pii: E2353.

Karvaly G, Kovács K, Mészáros K, Kocsis I, Patócs A, Vásárhelyi B. The comprehensive characterization of adrenocortical steroidogenesis using two-dimensional ultra-performance liquid chromatography—electrospray ionization tandem mass spectrometry. J Pharm Biomed Anal 2018;153:274–83.

Ke Y, Gonthier R, Isabelle M, Bertin J, Simard JN, Dury AY, Labrie F. A rapid and sensitive UPLC-MS/MS method for the simultaneous quantification of serum androsterone glucuronide, etiocholanolone glucuronide, and androstan-3α, 17β diol 17-glucuronide in postmenopausal women. J Steroid Biochem Mol Biol 2015;149:146–52.

Keller BO, Sui J, Young AB, Whittal RM. Interferents and contaminants encountered in modern mass spectrometry. Anal Chim Acta 2008;627:71–81.

Khan MA, Wang Y, Heidelberger S, Alvelius G, Liu S, Sjövall J, et al. Analysis of derivatised steroids by matrix-assisted laser desorption/ionisation and post-source decay mass spectrometry. Steroids 2006;71(1):42–53.

Kim J, Abdulwahab S, Choi K, Lafrenière NM, Mudrik JM, Gomaa H, et al. A microfluidic technique for quantification of steroids in core needle biopsies. Anal Chem 2015a;87(9):4688–95.

Kim B, Lee MN, Park HD, Kim JW, Chang YS, Park WS, et al. Dried blood spot testing for seven steroids using liquid chromatography-tandem mass spectrometry with reference interval determination in the Korean population. Ann Lab Med 2015b;35(6):578–85.

Kornel L. Studies on steroid conjugates. 3. Separation of conjugated steroids by means of high voltage paper electrophoresis. J Clin Endocrinol Metab 1964;24:956–64.

Kornel L, Miyabo S, Saito Z. New paper chromatographic systems for the separation of conjugated corticosteroids. J Chromatogr 1975;111 (1):200–5.

Kuehnbaum NL, Britz-McKibbin P. Comprehensive profiling of free and conjugated estrogens by capillary electrophoresis-time of flight/mass spectrometry. Anal Chem 2011;83(21):8063–8.

Kulikov AU, Galat MN. Comparison of C18 silica bonded phases selectivity in micellar liquid chromatography. J Sep Sci 2009;32(9):1340–50.

Kulle AE, Welzel M, Holterhus PM, Riepe FG. Principles and clinical applications of liquid chromatography—tandem mass spectrometry for the determination of adrenal and gonadal steroid hormones. J Endocrinol Invest 2011;34(9):702–8.

Kulle AE, Welzel M, Holterhus PM, Riepe FG. Implementation of a liquid chromatography tandem mass spectrometry assay for eight adrenal C-21 steroids and pediatric reference data. Horm Res Paediatr 2013;79(1):22–31.

Kushnir MM, Rockwood AL, Roberts WL, Pattison EG, Owen WE, Bunker AM, et al. Development and performance evaluation of a tandem mass spectrometry assay for 4 adrenal steroids. Clin Chem 2006;52(8):1559–67.

Li T-S, Li J-T, Li H-Z. Modified and convenient preparation of silica impregnated with silver nitrate and its application to the separation of steroids and triterpenes. J Chromatogr A 1995;715:372–5.

Li XS, Li S, Kellermann G. Simultaneous determination of three estrogens in human saliva without derivatization or liquid-liquid extraction for routine testing via miniaturized solid phase extraction with LC-MS/MS detection. Talanta 2018;178:464–72.

Lien SK, Kvitvang HF, Bruheim P. Utilization of a deuterated derivatization agent to synthesize internal standards for gas chromatography-tandem mass spectrometry quantification of silylated metabolites. J Chromatogr A 2012;1247:118–24.

Lindner JM, Vogeser M, Grimm SH. Biphenyl based stationary phases for improved selectivity in complex steroid assays. J Pharm Biomed Anal 2017;142:66–73.

Lisboa BP. Structural analysis of steroids by association of thin-layer chromatography techniques with gas chromatography-mass spectrometry. J Chromatogr 1970;48(2):364–71.

Little JL. Artifacts in trimethylsilyl derivatization reactions and ways to avoid them. J Chromatogr A 1999;844(1–2):1–22.

Liu S, Griffiths WJ, Sjövall J. On-column electrochemical reactions accompanying the electrospray process. Anal Chem 2003;75(4):1022–30.

Luo G, Li Y, Bao JJ. Development and application of a high-throughput sample cleanup process based on 96-well plate for simultaneous determination of 16 steroids in biological matrices using liquid chromatography-triple quadrupole mass spectrometry. Anal Bioanal Chem 2016;408(4):1137–49.

Ma P, Kanizaj N, Chan SA, Ollis DL, McLeod MD. The *Escherichia coli* glucuronylsynthase promoted synthesis of steroid glucuronides: improved practicality and broader scope. Org Biomol Chem 2014;12 (32):6208–14.

Magnisali P, Dracopoulou M, Mataragas M, Dacou-Voutetakis A, Moutsatsou P. Routine method for the simultaneous quantification of 17alpha-hydroxyprogesterone, testosterone, dehydroepiandrosterone, androstenedione, cortisol, and pregnenolone in human serum of neonates using gas chromatography-mass spectrometry. J Chromatogr A 2008;1206(2):166–77.

Makin HL. The gas-liquid chromatography of steroid formates; an application in congenital adrenal hyperplasia. J Endocrinol 1970;47(1):55–64.

Makin HLJ, Honour JW, Griffiths WJ, Shackleton CHL. Extraction purification and measurement of steroids by high-performance liquid chromatography, gas-liquid chromatography and mass spectrometry. In: Makin HLJ, editor. Steroid analysis. London: Blackie; 2010.

Manaf NA, Saad B, Mohamed MH, Wilson LD, Latiff AA. Cyclodextrin based polymer sorbents for micro-solid phase extraction followed by liquid chromatography tandem mass spectrometry in determination of endogenous steroids. J Chromatogr A 2018;1543:23–33.

Mason PA, Fraser R. Estimation of aldosterone, 11-deoxycorticosterone, 18-hydroxy-11-deoxy-corticosterone, corticosterone, cortisol and 11-deoxycortisol in human plasma by gas-liquid chromatography with electron capture detection. J Endocrinol 1975;64(2):277–88.

Matsuzaki Y, Yoshida S, Honda A, Miyazaki T, Tanaka N, Takagiwa A, et al. Simultaneous determination of dehydroepiandrosterone and its 7-oxygenated metabolites in human serum by high-resolution gas chromatography—mass spectrometry. Steroids 2004;69(13–14):817–24.

Mattox VR, Goodrich JE, Litwiller RD. Liquid ion exchangers in paper chromatography of steroidal glucosiduronic acids, glucosiduronic esters and free steroids. Influence of concentration of exchanger and counterion. J Chromatogr 1975a;108(1):23–5.

Mattox VR, Litwiller RD, Goodrich JE. Liquid ion exchangers in paper chromatography of steoidal glucosiduronic acids. Influence of different exchangers on the mobility in chloroform-formamide and correlation of chromatographic data. J Chromatogr 1975b;109(1):129–47.

Matysik S, Liebisch G. Quantification of steroid hormones in human serum by liquid chromatography-high resolution tandem mass spectrometry. J Chromatogr A 2017;1526:112–8.

Meng LJ, Griffiths WJ, Sjövall J. The identification of novel steroid N-acetylglucosaminides in the urine of pregnant women. J Steroid Biochem Mol Biol 1996;58(5–6):585–98.

Methlie P, Hustad SS, Kellmann R, Almås B, Erichsen MM, Husebye E, et al. Multisteroid LC-MS/MS assay for glucocorticoids and androgens, and its application in Addison's disease. Endocr Connect 2013;2(3):125–36.

Miękus N, Konieczna L, Kowiański P, Moryś J, Bączek T. HILIC-MS rat brain analysis, a new approach for the study of ischemic attack. Transl Neurosci 2017;8:70–5.

Mitamura K, Satoh née Okihara R, Kamibayashi M, Sato K, Iida T, Ikegawa S. Simultaneous determination of 18 tetrahydrocorticosteroid sulfates in human urine by liquid chromatography/electrospray ionization-tandem mass spectrometry. Steroids 2014;85:18–29.

Monder C. Stability of corticosteroids in aqueous solutions. Endocrinology 1968;82(2):318–26.

Moon JY, Jung HJ, Moon MH, Chung BC, Choi MH. Heat-map visualization of gas chromatography-mass spectrometry based quantitative signatures on steroid metabolism. J Am Soc Mass Spectrom 2009;20(9):1626–37.

Moon JY, Ha YW, Moon MH, Chung BC, Choi MH. Systematic error in gas chromatography-mass spectrometry-based quantification of hydrolyzed urinary steroids. Cancer Epidemiol Biomarkers Prev 2010;19(2):388–97.

Moros G, Chatziioannou AC, Gika HG, Raikos N, Theodoridis G. Investigation of the derivatization conditions for GC-MS metabolomics of biological samples. Bioanalysis 2017;9(1):53–65.

Murphy BEP, D'Aux RC. The use of Sephadex LH-20 column chromatography to separate unconjugated steroids. J Steroid Biochem 1975;6:233–7.

Napoli KL, Hammett-Stabler C, Taylor PJ, Lowe W, Franklin ME, Morris MR, et al. Multi-center evaluation of a commercial Kit for tacrolimus determination by LC/MS/MS. Clin Biochem 2010;43(10–11):910–20.

Nicolis GL, Wotiz HH, Gabrilove JL. Measurement of urinary tetrahydroaldosterone by gas chromatography with electron capture detection. J Clin Endocrinol Metab 1968;28(4):547–57.

Nováková L, Perrenoud AG, Francois I, West C, Lesellier E, Guillarme D. Modern analytical supercritical fluid chromatography using columns packed with sub-2 μm particles: a tutorial. Anal Chim Acta 2014;824:18–35.

Nováková L, Rentsch M, Grand-Guillaume Perrenoud A, et al. Ultra high performance supercritical fluid chromatography coupled with tandem mass spectrometry for screening of doping agents. II: analysis of biological samples. Anal Chim Acta 2015;853:647–59.

Nozaki O, Ohata T, Ohba Y, Moriyama H, Kato Y. Determination of serum cortisol by reversed-phase liquid chromatography using precolumn sulphuric acid-ethanol fluorescence derivatization and column switching. J Chromatogr 1991;570(1):1–11.

Nozaki O, Ohata T, Ohba Y, Moriyama H, Kato Y. Determination of urinary free cortisol by high performance liquid chromatography with sulphuric acid-ethanol derivatization and column switching. Biomed Chromatogr 1992;6(3):109–14.

O'Hare MJ, Nice EC, Capp M. Reversed- and normal-phase high-performance liquid chromatography of 18-hydroxylated steroids and their derivatives. Comparison of selectivity, efficiency and recovery from biological samples. J Chromatogr 1980;198(1):23–39.

Ostler MW, Porter JH, Buxton OM. Dried blood spot collection of health biomarkers to maximize participation in population studies. J Vis Exp 2014;(83), e50973.

Owen LJ, Adaway JE, Davies S, Neale S, El-Farhan N, Ducroq D, Evans C, Rees DA, MacKenzie F, Keevil BG. Development of a rapid assay for the analysis of serum cortisol and its implementation into a routine service laboratory. Ann Clin Biochem 2013;50(Pt 4):345–52.

Parr MK, Wüst B, Teubel J, Joseph JF. Splitless hyphenation of SFC with MS by APCI, APPI, and ESI exemplified by steroids as model compounds. J Chromatogr B Analyt Technol Biomed Life Sci 2018;1091:67–78.

Peña-García-Brioles D, Gonzalo-Lumbreras R, Izquierdo-Hornillos R, Santos-Montes A. Method development for betamethasone and dexamethasone by micellar liquid chromatography using cetyl trimethyl ammonium bromide and validation in tablets. Application to cocktails. J Pharm Biomed Anal 2004;36(1):65–71.

Plager JE. Extraction and purification of steroid conjugates with ion exchange resins: measurement of androsterone sulfate and dehydroepiandrosterone sulfate in plasma. J Clin Endocrinol Metab 1966;26(12):1275–81.

Pranata A, Fitzgerald CC, Khymenets O, et al. Synthesis of steroid bisglucuronide and sulfate glucuronide reference materials: unearthing neglected treasures of steroid metabolism. Steroids 2019;143:25–40.

Qaiser MZ, Dolman DEM, Begley DJ, et al. Uptake and metabolism of sulphated steroids by the blood-brain barrier in the adult male rat. J Neurochem 2017;142(5):672–85.

Qin F, Zhao YY, Sawyer MB, Li XF. Hydrophilic interaction liquid chromatography-tandem mass spectrometry determination of estrogen conjugates in human urine. Anal Chem 2008;80(9):3404–11.

Quanson JL, Stander MA, Pretorius E, Jenkinson C, Taylor AE, Storbeck KH. High-throughput analysis of 19 endogenous androgenic steroids by ultra-performance convergence chromatography tandem mass spectrometry. J Chromatogr B Analyt Technol Biomed Life Sci 2016;1031:131–8.

Randazzo GM, Tonoli D, Hambye S, Guillarme D, Jeanneret F, Nurisso A, et al. Prediction of retention time in reversed-phase liquid chromatography as a tool for steroid identification. Anal Chim Acta 2016;916:8–16.

Rauh M. Steroid measurement with LC-MS/MS. Application examples in pediatrics. J Steroid Biochem Mol Biol 2010;121(3–5):520–7.

Rauh M, Gröschl M, Rascher W, Dörr HG. Automated, fast and sensitive quantification of 17 alpha-hydroxy-progesterone, androstenedione and testosterone by tandem mass spectrometry with on-line extraction. Steroids 2006;71(6):450–8.

Ray JA, Kushnir MM, Palmer J, Sadjadi S, Rockwood AL, Meikle AW. Enhancement of specificity of aldosterone measurement in human serum and plasma using 2D-LC-MS/MS and comparison with commercial immunoassays. J Chromatogr B Analyt Technol Biomed Life Sci 2014;970:102–7.

Reepmeyer JC, Brower JF, Ye H. Separation and detection of the isomeric equine conjugated estrogens, equilin sulfate and delta 8,9-dehydroestrone sulfate, by liquid chromatography—electrospray-mass spectrometry using carbon-coated zirconia and porous graphitic carbon stationary phases. J Chromatogr A 2005;1083(1–2):42–51.

Rossi SA, Johnson JV, Yost RA. Short-column gas chromatography/tandem mass spectrometry for the detection of underivatized anabolic steroids in urine. Biol Mass Spectrom 1994;23(3):131–9.

Rossi C, Calton L, Hammond G, Brown HA, Wallace AM, Sacchetta P, et al. Serum steroid profiling for congenital adrenal hyperplasia using liquid chromatography-tandem mass spectrometry. Clin Chim Acta 2010;411(3–4):222–8.

Ruiz-Ángel MJ, Carda-Broch S, García-Álvarez-Coque MC. Micellar liquid chromatography in doping control. Bioanalysis 2009;1(7):1225–41.

Salameh WA, Redor-Goldman MM, Clarke NJ, Mathur R, Azziz R, Reitz RE. Specificity and predictive value of circulating testosterone assessed by tandem mass spectrometry for the diagnosis of polycystic ovary syndrome by the National Institutes of Health 1990 criteria. Fertil Steril 2014;101(4): 1135–41.e2.

Salentijn GI, Grajewski M, Verpoorte E. Reinventing (bio)chemical analysis with paper. Anal Chem 2018;90(23):13815–25.

Saugy M, Cardis C, Robinson N, Schweizer C. Test methods: anabolics. Baillieres Best Pract Res Clin Endocrinol Metab 2000;14(1):111–33.

Schöneshöfer M, Kage A, Weber B, Lenz I, Köttgen E. Determination of urinary free cortisol by "on-line" liquid chromatography. Clin Chem 1985;31(4):564–8.

Schöneshöfer M, Kage A, Eisenschmid B, Heilmann P, Dhar TK, Weber B. Automated liquid chromatographic determination of the 20-dihydro isomers of cortisol and cortisone in human urine. J Chromatogr 1986;380(2):267–74.

Sear JW, Holly JM, Trafford DJ, Makin HL. Plasma concentrations of alphaxalone by gas chromatography: comparison with other gas chromatographic methods and gas chromatography-mass spectrometry. J Pharm Pharmacol 1980;32(5):349–52.

Setchell KD, Shackleton CH. The group separation of plasma and urinary steroids by column chromatography on sephadex LH-20. Clin Chim Acta 1973;47(3):381–8.

Setchell KD, Almé B, Axelson M, Sjövall J. The multicomponent analysis of conjugates of neutral steroids in urine by lipophilic ion exchange chromatography and computerised gas chromatography-mass spectrometry. J Steroid Biochem 1976;7:615–29.

Shackleton CH. Derivatization of estrogen conjugates for analysis by capillary gas chromatography. Steroids 1981;38(5):485–94.

Shackleton CH. Profiling steroid hormones and urinary steroids. J Chromatogr 1986;379:91–156.

Shackleton CHL. Genetic disorders of steroid metabolism diagnosed by mass spectrometry. In: Blau N, editor. Laboratory guide to the methods in biochemical genetics. Berlin: Springer-Verlag; 2008 [chapter 5.3].

Shackleton CH. Role of a disordered steroid metabolome in the elucidation of sterol and steroid biosynthesis. Lipids 2012;47:1–10.

Shackleton CH, Honour JW. Simultaneous estimation of urinary steroids by semi-automated gas chromatography. Investigation of neo-natal infants and children with abnormal steroid synthesis. Clin Chim Acta 1976;69(2):267–83.

Shackleton CH, Biglieri EG, Roitman E, Honour JW. Metabolism of radio-labeled corticosterone in an adult with the 17 alpha-hydroxylase deficiency syndrome. J Clin Endocrinol Metab 1979;48(6):976–82.

Shackleton CH, Roitman E, Monder C, Bradlow HL. Gas chromatographic and mass spectrometric analysis of urinary acidic metabolites of cortisol. Steroids 1980;36(3):289–98.

Shackleton CH, Mattox VR, Honour JW. Analysis of intact steroid conjugates by secondary ion mass spectrometry (including FABMS) and by gas chromatography. J Steroid Biochem 1983;19(1A):209–17.

Shimada K, Tanaka T, Nambara T. CL. Separation of catechol estrogens by high-performance liquid chromatography with electrochemical detection. J Chromatogr 1979;178(1):350–4.

Siiteri PK. A universal chromatographic system for the separation of steroid hormones and their metabolites. Methods Enzymol 1975;36:485–9.

Siiteri PK, Murai JT, Hammond GL, Nisker JA, Raymoure WJ, Kuhn RW. The serum transport of steroid hormones. Recent Prog Horm Res 1982;38:457–510.

Sjövall J, Axelson M. General and selective isolation procedures for GC/MS analysis of steroids in tissues and body fluids. J Steroid Biochem 1979;11(1A):129–34.

Sjövall J, Axelson M. Newer approaches to the isolation, identification, and quantitation of steroids in biological materials. Vitam Horm 1982;39:31–144.

Sjövall J, Axelson M. Sample work-up by column techniques. J Pharm Biomed Anal 1984;2(2):265–80.

Sjövall J, Rafter J, Larsen G, Egestad B. Lipophilic ion exchangers for group separation of conjugated metabolites of xenobiotics. J Chromatogr 1983;276(1):150–6.

Smy L, Straseski JA. Measuring estrogens in women, men, and children: recent advances 2012-2017. Clin Biochem 2018;62:11–23.

Søeborg T, Frederiksen H, Johannsen TH, Andersson AM, Juul A. Isotope-dilution TurboFlow-LC-MS/MS method for simultaneous quantification of ten steroid metabolites in serum. Clin Chim Acta 2017;468:180–6.

Stevenson BJ, Waller CC, Ma P, et al. *Pseudomonas aeruginosa* arylsulfatase: a purified enzyme for the mild hydrolysis of steroid sulfates. Drug Test Anal 2015;7(10):903–11.

Stone JA, Fitzgerald RL. Liquid chromatography-mass spectrometry education for clinical laboratory scientists. Clin Lab Med 2018;38(3):527–37.

Storbeck KH, Gilligan L, Jenkinson C, et al. The utility of ultra-high performance supercritical fluid chromatography-tandem mass spectrometry (UHPSFC-MS/MS) for clinically relevant steroid analysis. J Chromatogr B Analyt Technol Biomed Life Sci 2018;1085:36–41.

Strahm E, Baume N, Mangin P, Saugy M, Ayotte C, Saudan C. Profiling of 19-norandrosterone sulfate and glucuronide in human urine: implications in athlete's drug testing. Steroids 2009;74(3):359–64.

Suhr AC, Vogeser M, Grimm SH. Isotope inversion experiment evaluating the suitability of calibration in surrogate matrix for quantification via LC-MS/MS-exemplary application for a steroid multi-method. J Pharm Biomed Anal 2016;124:309–18.

Swart AC, Storbeck KH. 11β-Hydroxyandrostenedione: downstream metabolism by 11βHSD, 17βHSD and SRD5A produces novel substrates in familiar pathways. Mol Cell Endocrinol 2015;408:114–23.

Szarka S, Nguyen V, Prokai L, Prokai-Tatrai K. Separation of dansylated 17β-estradiol, 17α-estradiol, and estrone on a single HPLC column for simultaneous quantitation by LC-MS/MS. Anal Bioanal Chem 2013;405(10):3399–406.

Tang DQ, Zou L, Yin XX, Ong CN. HILIC-MS for metabolomics: an attractive and complementary approach to RPLC-MS. Mass Spectrom Rev 2016;35(5):574–600.

Tetsuo M, Axelson M, Sjövall J. Selective isolation procedures for GC/MS analysis of ethynyl steroids in biological material. J Steroid Biochem 1980;13(8):847–60.

Teubel J, Wüst B, Schipke CG, Peters O, Parr MK. Methods in endogenous steroid profiling—a comparison of gas chromatography mass spectrometry (GC-MS) with supercritical fluid chromatography tandem mass spectrometry (SFC-MS/MS). J Chromatogr A 2018;1554:101–16.

Travers S, Martinerie L, Bouvattier C, Boileau P, Lombès M, Pussard E. Multiplexed steroid profiling of gluco- and mineralocorticoids pathways using a liquid chromatography tandem mass spectrometry method. J Steroid Biochem Mol Biol 2017;165(Pt B):202–11.

Tscherne RJ, Capitano G. High-pressure liquid chromatographic separation of pharmaceutical compounds using a mobile phase containing silver nitrate. J Chromatogr 1977;136(2):337–41.

Turpeinen U, Hämäläinen E, Stenman UH. Determination of aldosterone in serum by liquid chromatography-tandem mass spectrometry. J Chromatogr B Analyt Technol Biomed Life Sci 2008;862(1–2):113–8.

Uusijärvi J, Egestad B, Sjövall J. Manual and automated enrichment procedures for biological samples using lipophilic gels. J Chromatogr 1989;488(1):87–104.

Vallano PT, Shugarts SB, Woolf EJ, Matuszewski BK. Elimination of autosampler carryover in a bioanalytical HPLC-MS/MS method: a case study. J Pharm Biomed Anal 2005;36(5):1073–8.

van de Calseyde JF, Scholtis RJ, Schmidt NA, Kuypers AM. Assay of pregnanediol and pregnanetriol in low-titre urines by gas chromatography. Clin Chim Acta 1970;27(1):139–43.

van der Berg CL, Venter G, van der Westhuizen FH, Erasmus E. Data on the optimisation of a solid phase extraction method for fractionating estrogen metabolites from small urine volumes. Data Brief 2020;29:105222. Published 2020 Feb 3 https://doi.org/10.1016/j.dib.2020.105222.

Vanluchene E, Vandekerckhove D. Group fractionation of free and conjugated steroids by means of disposable silica-based anion-exchange columns. J Chromatogr 1988;456(1):175–82.

Wagner M, Tonoli D, Varesio E, Hopfgartner G. The use of mass spectrometry to analyze dried blood spots. Mass Spectrom Rev 2016;35(3):361–438.

Wang Q, Mesaros C, Blair IA. Ultra-high sensitivity analysis of estrogens for special populations in serum and plasma by liquid chromatography-mass spectrometry: assay considerations and suggested practices. J Steroid Biochem Mol Biol 2016;162:70–9.

Wang R, Hartmann MF, Tiosano D, Wudy SA. Characterizing the steroidal milieu in amniotic fluid of mid-gestation: a GC-MS study. J Steroid Biochem Mol Biol 2019a;193, 105412.

Wang R, Tiosano D, Sánchez-Guijo A, Hartmann MF, Wudy SA. Characterizing the steroidal milieu in amniotic fluid of mid-gestation: a LC-MS/MS study. J Steroid Biochem Mol Biol 2019b;185:47–56.

Watson DJ, Bartosik D. Thin-layer partition chromatography of steroids using volatile stationary phases. J Chromatogr 1971;54(1):91–5.

West C, Melin J, Ansouri H, Mengue Metogo M. Unravelling the effects of mobile phase additives in supercritical fluid chromatography. Part I: polarity and acidity of the mobile phase. J Chromatogr A 2017;1492:136–43.

Wudy SA, Choi MH. Steroid LC-MS has come of age. J Steroid Biochem Mol Biol 2016;162:1–3.

Wudy SA, Dörr HG, Solleder C, Djalali M, Homoki J. Profiling steroid hormones in amniotic fluid of midpregnancy by routine stable isotope dilution/gas chromatography-mass spectrometry: reference values and concentrations in fetuses at risk for 21-hydroxylase deficiency. J Clin Endocrinol Metab 1999;84(8):2724–8.

Wudy SA, Hartmann M, Svoboda M. Determination of 17-hydroxyprogesterone in plasma by stable isotope dilution/benchtop liquid chromatography-tandem mass spectrometry. Horm Res 2000;53(2):68–71.

Wudy SA, Hartmann M, Solleder C, Homoki J. Determination of 17alpha-hydroxypregnenolone in human plasma by routine isotope dilution mass spectrometry using benchtop gas chromatography-mass selective detection. Steroids 2001;66(10):759–62.

Wudy SA, Hartmann M, Homoki J. Determination of 11-deoxycortisol (Reichstein's compound S) in human plasma by clinical isotope dilution mass spectrometry using benchtop gas chromatography-mass selective detection. Steroids 2002;67(10):851–7.

Xu X, Roman JM, Veenstra TD, Van Anda J, Ziegler RG, Issaq HJ. Analysis of fifteen estrogen metabolites using packed column supercritical fluid chromatography-mass spectrometry. Anal Chem 2006;78(5):1553–8.

Yamashita K, Takahashi M, Tsukamoto S, Numazawa M, Okuyama M, Honma S. Use of novel picolinoyl derivatization for simultaneous quantification of six corticosteroids by liquid chromatography-electrospray ionization tandem mass spectrometry. J Chromatogr A 2007;1173(1–2):120–8.

Zaffaroni A, Burton RB, Keutmann EH. Adrenal cortical hormones; analysis by paper partition chromatography and occurrence in the urine of normal persons. Science 1950;111(2871):6–8.

Zaikin VG, Halket JM. Review: derivatization in mass spectrometry-6. Formation of mixed derivatives of polyfunctional compounds. Eur J Mass Spectrom 2005;11(6):611–36.

Zaikin VG, Halket JM. Derivatization in mass spectrometry—8. Soft ionization mass spectrometry of small molecules. Eur J Mass Spectrom 2006;12(2):79–115.

Zarzycki PK, Włodarczyk E, Zarzycka MB, Głód BK. Optimization of a solid-phase extraction protocol for fractionation of selected steroids using retention data from micro thin-layer chromatography. Anal Sci 2009;25(7):935–9.

Zhang YV, Rockwood A. Impact of automation on mass spectrometry. Clin Chim Acta 2015;450:298–303.

Zhao Y, Boyd JM, Sawyer MB, Li XF. Liquid chromatography tandem mass spectrometry determination of free and conjugated estrogens in breast cancer patients before and after exemestane treatment. Anal Chim Acta 2014;806:172–9.

Books

Bush IE. The chromatography of steroids. Pergamon; 1961.

Eik-Nes KB, Horning EC. Gas phase chromatography of steroids. Berlin: Springer-Verlag; 1968.

Ettre LS. Open tubular columns in gas chromatography. Plenum Press; 1965.

Makin HLJ, Gower DB. Steroid analysis. 2nd ed. Springer; 2010, ISBN:978 1 4020 9774 4.

Neher R. Steroid chromatography. Elsevier; 1964.

Snyder LR, Kirkland JJ, Dolan JW. Introduction to modern liquid chromatography. Wiley; 2010, ISBN:978 0 470 16754 0.

Chapter 2.4

Quantitative analysis of steroids

2.4.1 Introduction

Historically, before 1970, steroids had to be determined by crude bioassay and chemical tests. It is rarely necessary these days to refer to these results and they will not be discussed in this book. The technique called **immunoassay** was introduced for steroids in the 1970s which enabled quantification of steroids at low concentrations in biological samples and revolutionized endocrine investigations. Initially steroids were extracted from the sample into an organic solvent and in some cases extracts were further purified by chromatography. With experience of the immunoassay technique, and improved quality of antibodies by introduction of monoclonal antibodies, direct assays became popular. Along with other changes in demand, the automation of immunoassays enables sample processing in parallel with other chemical tests on automated platforms generating hundreds of results every hour. The reduced costs were attractive to hospital administrations. Specificity of the assays however were questioned. The **combination of chromatography with mass spectrometry** has now become the method of choice for many steroids particularly when more than one steroid is determined in an assay.

There have been some attempts to quantify activity of steroids using biological tests with cell lines that have been transfected with genes for a specific response (androgen receptor for example) linked with a reporter for signal generation. This work has not been extensive because of the inherent variation in such bioassays and in some cases lack of specificity. So even though an androgen receptor gene has been used, the assay still had glucocorticoid or progestogenic response because the DNA cis-regulatory elements that bind to AR share sequence similarity with cis-regulatory elements for glucocorticoid, mineralocorticoid and progesterone receptors (GR, MR and PR, respectively).

Normative data of most assays are now available for age of healthy infants and children. Caution is needed however in the interpretation of results in sick premature babies and also children with delayed or advanced puberty at the peri-pubertal ages where age-specific references ranges may not be appropriate. The choice of biochemical investigations should be based on the clinical picture, and good communication between the clinicians and laboratory scientists is important. The interpretation of results must take into account the laboratory methods used and the patient's age and sex. The reduced precision of immunoassays at low concentrations can be a problem in pediatrics, the menopause and clinical situations of steroid suppression.

2.4.2 Examination of column, thin layer and paper chromatograms

2.4.2.1 Ultra-violet absorption

The column eluate can be passed into a spectrophotometer. The α,β-unsaturated ketone in the A-ring of naturally occurring steroid hormones absorbs ultraviolet light with maximum around 240 nm and extinction coefficients 12,000–20,000. Isolated carbonyl groups absorb with a maximum around 280 (275–285) nm and molar extinction coefficients of 17–155. The natural estrogens have peak absorption at 280 nm due to the aromatic A-ring. Underivatized phenolic steroids can be detected with sensitivity limits of 100–10 pg/mL (follicular phase concentrations). Although steroids can absorb UV below 200 nm, in practice it is difficult to achieve a clear signal distinguishable from noise without a reduction in sensitivity particularly when solvent gradients are used to elute the steroids. With some gradient elution systems it is necessary to correct for baseline variation by comparison of the response of the eluate from the analytical column with the flow of solvent alone through a reference cell.

The identification of material in chromatograms is usually assumed from a homogenous peak with elution time which coincides with that of the reference compound under similar conditions. This assumption may however be dangerous since it is not always possible to recognize homogeneity by inspection. Since the detectors currently in use are not selective for distinct classes of substances some further demonstration of specificity is required. The use of a photodiode array detector has a considerable advantage, over observation by UV at one wavelength in this context, in that the spectrum can be compared with a reference spectrum allowing inspection of the peak to ensure homogeneity.

Steroids in the Laboratory and Clinical Practice. https://doi.org/10.1016/B978-0-12-818124-9.00010-3

The sample can also be analyzed separately to achieve purification with a column of differing polarity or selectivity or using a different gradient elution system. Should elution times in each system coincide with those of a standard it is highly probable that each chromatogram reflects the same steroid content. These criteria have not been rigidly applied in the published work relating to steroids.

Retention indices have been widely used for recording and comparing retentions, for identifications and as the basis for prediction methods. Standards such as 1-[4-(2,3-dihydroxypropoxy)phenyl]-1-alkanones have been used in RP-HPLC for steroids with photodiode array detection (Kuronen et al., 1998). In some work, identification is enforced by a second separation of quite different selectivity (orthogonal separation). This may sometimes reveal a further peak that was masked in the first separation.

Chromatograms of HPLC methods often show peaks that are difficult to reproduce and can be hard to trace back. Artifact peaks can arise due to contamination by the injector septum and sampling equipment leading to misinterpretation of impurities and erroneous quantification. A number of investigations may be needed to locate the source of the problems.

2.4.2.2 Derivative formation

For steroids which do not absorb in the UV, a narrow side strip of the aluminum-backed TLC plate can be removed and the steroids located in that strip. The strip can be sprayed and/or heated with a variety of reagents which may or may not be specific for particular types of steroids (17-ketosteroids, 17-hydroxycorticoids, etc.). The position of standards which have been run together with the samples of interest can be located. The 17-ketosteroids are detected by alkaline m-dinitrobenzene, reducing steroids with alkaline blue tetrazolium and certain hydroxylated substances with 10% phosphomolybdic acid.

2.4.2.3 Immunoassay

In many early steroid assays using sample purification with paper chromatography, the solid phase is cut up and eluted for radioimmunoassay as shown in Fig. 2.4.1 for aldosterone. This shows how tedious RIA methods with chromatography could be. The schematic from 1975 would compete with graphical abstracts being used in 2020 (Wudy et al., 2018).

2.4.2.4 Radioactivity

In a TLC analysis, radioactive steroids can be identified by placing a TLC plate in contact with an X-ray film or photostimulated luminescence film from Fuji (https://

www.fujifilm.com/products/medical/technologies/focused_phosphor_technology/) producing an autoradiogram. Several oxidized products of cortisol were identified in urine in this way after intravenous administration of ^{14}C cortisol (Chapter 2.2, Fig. 2.2.2). The aluminum foil plate can be cut with scissors into 1 cm bands for subsequent liquid scintillation counting. The instability of cortisol in aqueous solutions was addressed by TLC (Chapter 2.2, Fig. 2.2.3). The paper in 1968 has important messages for the analyst today on scrupulous attention to cleanliness of reagents and glassware. Oxidation of cortisol to 21-dehydrocortisol is demonstrated in carbonate more than phosphate buffers. The 11-hydroxylated androgens have attracted much interest since around 2015.

Flow-through radioactivity detectors for a column eluates are potentially useful for examining the products of reactions with labeled substrates. The short residence time of the sample components in the counting chamber limits sensitivity. Several publications have demonstrated the variety and complexity of intermediates and products formed when radioactive steroids are incubated with steroid metabolizing tissues. Current detection limits for tritium are 10,000 dpm with flow cells incorporating scintillant (around 1% efficiency) to 1000 dpm (50% efficiency) when the column effluent is mixed with liquid scintillant before passing through a cell.

Many immunoassays used gamma emitting isotopes that could be counted with Geiger counters that could accommodate one or in some cases 16 tubes simultaneously. Disposal of materials for such assays were easier than when beta-emitting isotopes were used and counted in scintillation cocktails.

2.4.2.5 Fluorescence detection

Postcolumn derivative formation can be incorporated for example when 9-anthroyl cyanides of hydroxyls are formed after HPLC separation then fluorescence detection. Estrogens are detected through fluorescence as 2-(4-carboxy)-5,6-dimethylbenzimidazole derivatives (Ishida et al., 1988). A fluorescent derivative of the carboxylic acid of the glucuronides has been used for direct estrogen conjugate analysis (Fujiwara et al., 2020). A highly sensitive and selective fluorescence method of indirect detection is achieved using the enzyme 3α-hydroxysteroid dehydrogenase which oxidizes the relevant steroids after their HPLC separation. The overall reaction involves conversion of the 3α-hydroxy group on the steroid nucleus to a keto group with concomitant reduction of NAD to NADH which is detected fluorimetrically (Takagi et al., 1988). The enzyme is expensive and in its crude form may not be specific. In order to reduce the cost of the enzyme, it can be immobilized on an aminopropyl support in a column in series with the analytical column. Detection of the compounds after the

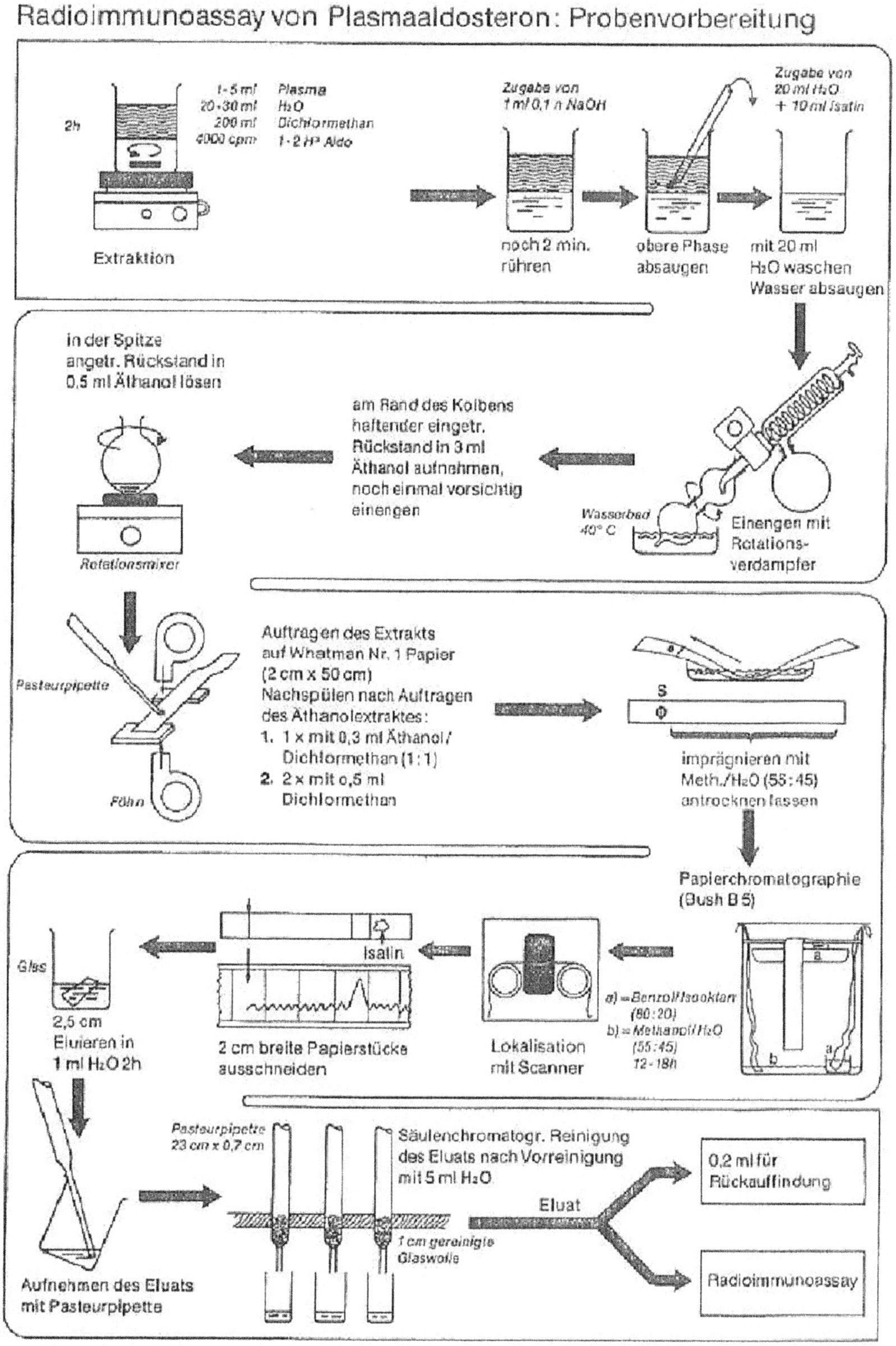

FIG. 2.4.1 Plasma and tritiated aldosterone was diluted with water and extracted into dichloromethane with stirring. The vessel was allowed to stand then the aqueous layer was sucked off. The dichloromethane was washed with water which was removed before the dichloromethane was dried in a rotary evaporator. The residue was dissolved in 0.5 mL of ethanol. The extract was applied as a spot near the end of Whatman No. 1 filter paper (2 × 50 cm). Solvent was evaporated with a hair dryer. The flask was washed with (1) 0.3 mL ethanol:dichloromethane 1:1 then (2) two further washes with 0.5 mL dichloromethane. The filter paper was impregnated with methanol:water (55:45) then transferred to paper chromatography tank. The chromatogram was equilibrated with Bush B5 solvent then developed with (a) benzene:isooctane (50:20) then (b) methanol:water (55:45) over 12–18 h. Radioactivity was detected with scanner. A 2 cm wide strip of paper was cut out and eluted with 1 mL of water for 2 h. Take up eluate with Pasteur pipette and transfer to Pasteur pipette as a column plugged with washed glass wool. Take 0.2 mL for recovery test and use remainder for radioimmunoassay. *(From Vecsei P, Gless K-H. Aldosterone—radioimmunoassay. Ferdinand Eke Verlag Stuttgart; 1975.)*

postcolumn reaction is not uniform. This is influenced by the time the steroids spend in the reactor and the affinity of the enzyme for the limited range of 3α-reduced steroids encountered in biological fluids. Quantification of a number of steroids requires optimization and control of flow rates as well as calibration curves to compensate for the differences in response. Over a period of time the baseline and response characteristics of the enzyme column will vary.

2.4.3 GLC detectors

2.4.3.1 Flame ionization

Steroids separated by GLC are detected by one of three main methods, the commonest of which is **flame ionization detection (FID)** which responds to all steroids with varying response factors (Shackleton and Honour, 1976). In order to use GC-FID for quantitation of steroids, therefore it is necessary to set up a standard curve of peak heights or peak areas against concentration or establish a response factor for the steroid of interest. In the FID, hydrogen and air are mixed with the carrier gas and the mixture is burnt at the tip of a jet (Fig. 2.4.2). A current is applied to the tip and a collector electrode is cited above to complete an electrical circuit.

Since cylinders or supplies of compressed gases are needed the lifetime cost of a GC-FID system may exceed that of a GC-MS, and may not be a good choice. The GC retention time is the only measure of specificity. The use of capillary gas chromatography for the analysis of urinary steroid profiles is a very good example of the valuable use of GLC-FID using trimethylsilyl ether-*O*-methyloxime derivatives and when such systems are also linked to an MS, they can be of immense value.

2.4.3.2 Nitrogen detector

Compounds containing nitrogen atoms can be detected with a **nitrogen-phosphorus detector** (NPD) but as most steroids commonly encountered do not contain nitrogen, to use this detection system requires formation of nitrogen containing derivatives such as methyloximes. This detection system can then be quite useful for the selective measurement and detection of steroids containing oxo groups. Few papers have been described for steroids. Cyanoethyldimethylsilyl derivatives of hydroxysteroids gave derivatives that were suitable for nitrogen detector with good mass spectrometric properties (Bertrand et al., 1990).

2.4.3.3 Electron capture detector

The third method of detection involves the use of **electron capture** which is extremely sensitive and detects steroids containing halogen groups. Halogenated silyl ether derivatives have been used in conjunction with an electron capture detector for the measurement of DHEA in plasma after formation of iodomethyldimethylsilyl ethers. Reports of methods using GLC with electron capture are rare as these detection systems are difficult to use and are susceptible to detector contamination. Pentafluorobenzyl and pentafluoropropionate derivatives have been used as electron capturing derivatives for estrogens by GC-NI-CI (Singh et al., 2000). GLC-EC, although extremely sensitive, is also very difficult to use quantitatively and it has largely been replaced by interfacing the **GLC with a mass spectrometer** which when operated correctly can be as sensitive as electron capture and more selective. Some infrequently used reagents for this purpose produce flophemesyl ethers, *t*-buflophemesyl ethers, pentafluorophenylhydrazone derivatives and halogen substituted aromatic boronic acids (Poole et al., 1980).

GLC with a mass spectrometer is however now the best option detector since it enables retention time, mass and structural information but all applications require the formation of derivatives to survive the high temperatures used in the GC. Only a carrier gas is required.

2.4.4 Detectors for HPLC

2.4.4.1 Ultraviolet detection

HPLC can be coupled to UV detection (Wei et al., 1990, 1992). Results for reference steroids are shown in Fig. 2.4.3. The sensitivity is low but the technique has been used to detect cortisol or other steroids when in excess in urine of patients with defects in cortisol synthesis or when a normal subject has adrenal stimulation.

The Flame Ionisation Detector

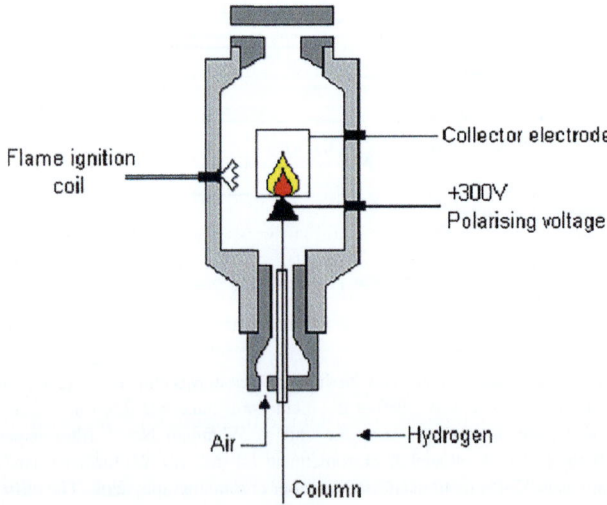

FIG. 2.4.2 Flame ionization detector for gas chromatography.

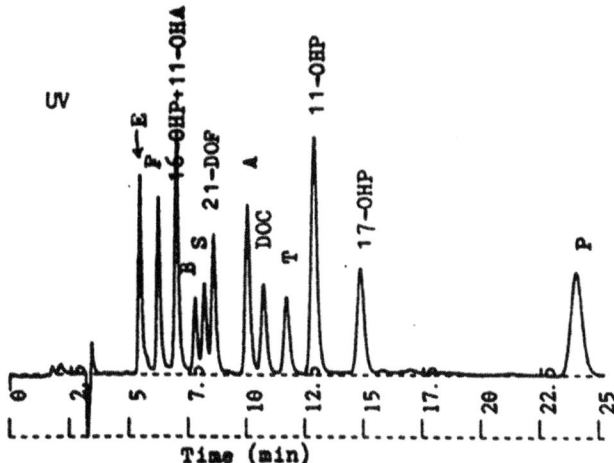

FIG. 2.4.3 UV absorption (254 nm) chromatogram of standard solution cortisone (E), cortisol (F)-16-OH Progesterone (16-OHP), 11-deoxycortisol (S), DOC, testosterone (T), progesterone (P), 11-OHP, and 17OHP from octacylsilane bonded phase column (50 × 2.1 mm) eluted with methanol:tetrahydrofuran:water (26/18/56). *(From Wei JQ, Wei JL, Zhou XT, Cheng JP. Isocratic reversed phase high performance liquid chromatography determination of twelve natural corticosteroids in serum with on-line ultraviolet and fluorescence detection. Biomed Chromatogr 1990;4(4): 161–4. Fig. 1 p. 162.)*

2.4.4.2 Derivatization for HPLC

Most of the urinary steroid metabolites do not have natural absorbance in the UV region. Reactive groups have been utilized in order to make derivatives for spectrophotometric detection. HPLC has thus been used to separate individual oxo-steroids after conversion to phenylhydrazone derivatives. In some cases, it has been necessary to react the steroids in the column eluate with reagents to form UV absorbing derivatives. Postcolumn derivatization methods are however restricted to very fast reactions limiting the scope of application.

Steroid hydrazones formed by reaction of ketosteroids with 2,4-dinitrophenylhydrazine have strong UV absorption (maximum 260 nm and extinction around 10,000) as well as visible absorbance (maximum 350 nm, extinction 10,000) giving detection limits for dehydroepiandrosterone sulfate (DHEAS) of 80 ng/mL. The measurement of 17-oxosteroid conjugates in urine and serum is possible by HPLC of dansyl hydrazine derivatives coupled with a fluorescence detector (Kawasaki et al., 1982). Phenylhydrazine has also been used (Kreutzmann and Silink, 1987). Derivatives have proved even more important to increase sensitivity for steroids at low concentrations because of their greater ability to ionize when the LC is coupled with a mass spectrometer.

Cortisol, cortisone and their tetrahydrometabolites have been determined after derivative formation with 9-anthroyl nitrile before SPE purification and RP HPLC separation (Główka et al., 2010). Fluorescence is measured at 460 nm emission wavelength with excitation at 360 nm wavelength.

2.4.5 Immunoassay of steroids

Measurement of steroids using binding proteins have been the principal methods for steroid immunoassay analysis since the 1970s following the recognition of binding proteins as assay reagents for insulin (Yalow and Berson, 1959) and cortisol by Beverley Murphy in 1963. Roger Ekins at the Middlesex Hospital, in London, England was among the founders of immunoassay (IA) for thyroid hormones (Ekins, 1960) although Rosalyn Yallow was awarded the Nobel Prize for her work using antisera to measure concentrations of insulin. Ekins (1969) coined the term "saturation analysis" to indicate the concentration of binding protein was insufficient to bind all of the analyte. The recognition of steroids by antibodies was the crucial step in the introduction of immunoassay. Steroids themselves are too small to be antigenic and so methods were developed to couple steroids to protein molecules as haptens. If a labeled steroid analyte were introduced to an antibody solution, this competed with the endogenous steroid and the percentage of analyte bound to the antibody was inversely related to the concentration of unlabeled analyte (Fig. 2.4.4). Data can be plotted in a number of ways, e.g., (y) Free/bound (B); amount bound against (x) concentration. Results and concentration can be linear or logarithmic. Initially radioactive steroids were used for the labels and the technique was called radioimmunoassay (RIA) but this will not be covered extensively because new developments are unlikely. The reader should seek reviews from the 1970s for greater detail or the immunoassay handbook of Wild (2013) or Steroid Analysis by Makin & Gower. In general, steroid assays measure the total steroid in serum after displacement from binding proteins. This can be achieved by heating the sample to distort the 3D structure of the proteins, by change in pH of the sample, by adding a competing steroid or by addition of a chemical such 8-anilino-2-naphthalene sulfonic acid (ANSA). Commercial IA kits do not state how total steroids are released from proteins.

At present, most hospital laboratories use automated immunoassays for hormone assays with potential limitations of reduced specificity especially for steroids by cross-reactivity with molecules of similar structures in the biological sample. Most of the analytical platforms use heterogeneous immunoassays where the bound label is physically separated from the unbound label prior to measuring the signal. The separation is often done magnetically using paramagnetic particles, and after separation of bound from free using a washing step, the bound label is reacted with other reagents to generate the signal. This is the mechanism in many chemiluminescent immunoassays (CLIA) where the label may be a small molecule that generates a chemiluminescent signal. Examples of immunoassay systems where the chemiluminescent labels generate signals by chemical reaction are the ADVIA Centaur from Siemens and the

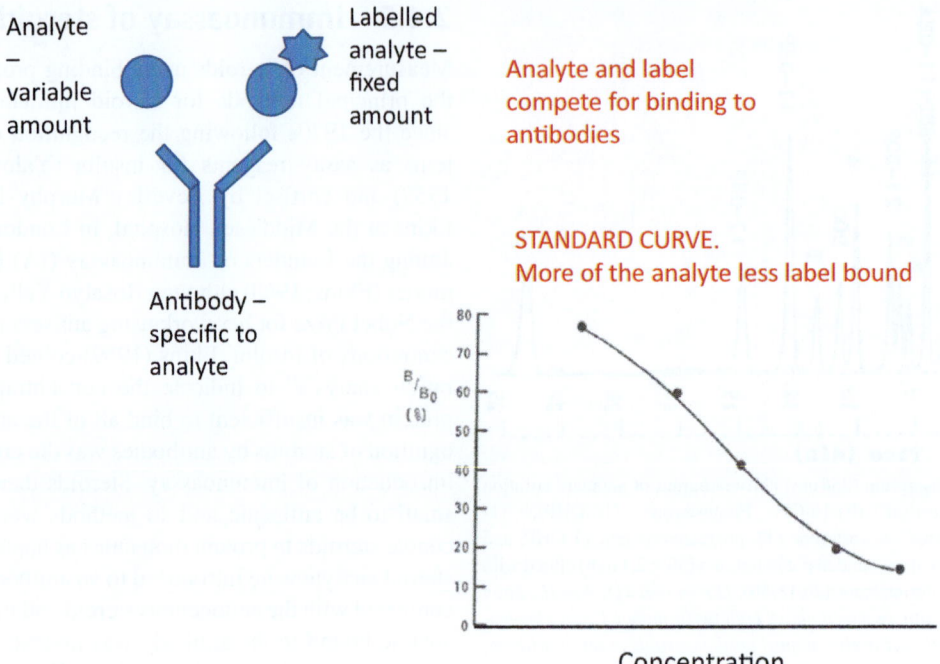

FIG. 2.4.4 Schematic for competitive immunoassay of steroid and simple standard curve to reflect displacement of radiolabel from antibody by steroid in the sample.

Architect from Abbott. An example where the small label is activated electrochemically is the ELECSYS automated immunoassay system from Roche Diagnostics. The label may also be an enzyme (enzyme-linked immunosorbent assay, ELISA) that generates chemiluminescent, fluorometric, or colorimetric signals depending on the enzyme substrates used. Examples of commercial automated assay systems using ELISA technology and chemiluminescent labels are Immulite (Siemens) and ACCESS from Beckman-Coulter. Another type of heterogenous immunoassay uses micro-sized polystyrene particles and the assay is called micro-particle enhanced immunoassay (MEIA).

Most immunoassays have poor precision and sensitivity for steroids at low concentrations (for example, estradiol below 100 pmol/L), which can be problematic with samples from children, adolescents and women postmenarche, as hormone concentrations are often below the limits of detection. The laboratory should be informed of drugs the patient is receiving that can influence the concentrations of the analyte or interfere in the assay.

There is a complex mixture of closely related steroids in biological fluids. The use of a biological technique such as IA for quantitative determination of steroids needs careful consideration both when setting up assays and when maintaining a method routinely. The laboratory has an important responsibility in validating the methods with a full understanding of the principles and pitfalls of any method. The theory of IA will not be discussed in detail here. There are many reviews and books on this subject (Wild, 2013;

Pratt, 1978; James and Jeffcoate, 1974). Although many laboratories also use commercial kits the validation needs to be as thorough as if setting up an assay with in-house reagents. Each stage of an IA is discussed here. Specialist laboratories, particularly in children's hospitals, are more likely to use assays with an extraction step prior to the IA. In practice, fewer laboratories now use IA and chromatography with mass spectrometry is being used for the commonly determined steroid hormones and increasingly for newer steroids (C11-oxygenated androgens, for example).

2.4.5.1 Extraction of steroids prior to IA

Although many current immunoassays are performed directly with biological samples, steroids can be extracted, when required, from biological fluids by simple solvent extraction. Steroid conjugates are poorly extracted into the organic solvents so that some potential immunoassay interfering steroids are eliminated, in particular the adrenal steroid sulfates from the fetal adrenal cortex which can interfere in steroid assay performance in samples from newborn children. For example, 17-hydroxypregnenolone sulfate will interfere in the assay of 17-hydroxyprogesterone unless a solvent extraction of the serum precedes the IA (Wong et al., 1992). All samples from newborn infants should be treated in this way.

For a period in the 1970s to 1980s, the eluates of an HPLC column were collected and dried before reconstitution in a buffer that was taken to individual immunoassays.

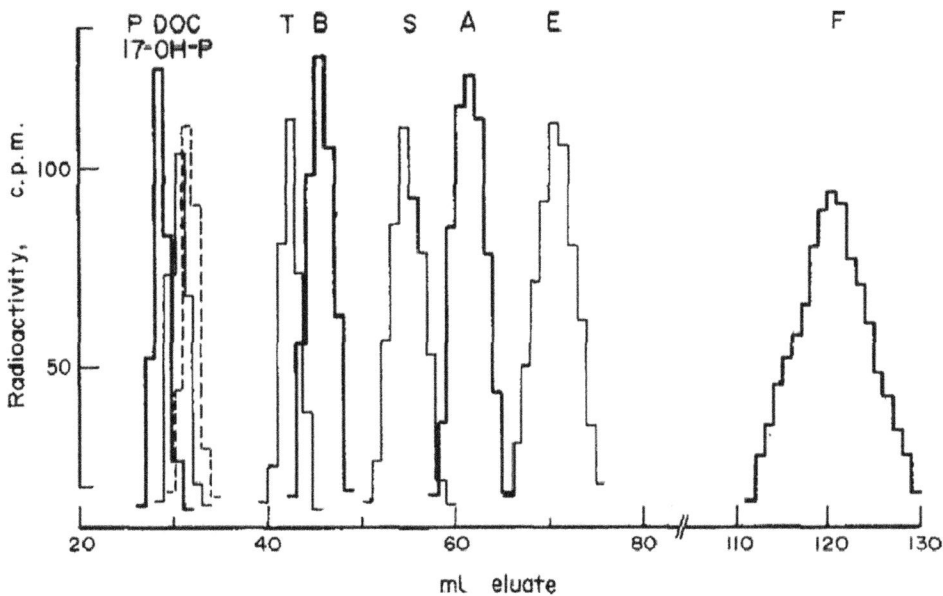

FIG. 2.4.5 Chromatogram of tritiated steroids eluted from 60 cm Sephadex LH-20 column using methylene chloride-methanol (98:2 v/v) at 60 mL/h. Eight steroids in plasma samples were measured by separate immunoassays. *(From Sippell WG, Bidlingmaier F, Becker H, Brünig T, Dörr H, Hahn H, Golder W, Hollmann G, Knorr D. Simultaneous radioimmunoassay of plasma aldosterone, corticosterone, 11-deoxycorticosterone, progesterone, 17-hydroxyprogesterone, 11-deoxycortisol, cortisol and cortisone. J Steroid Biochem 1978;9(1):63–74. Fig. 1 p. 63.)*

This has no place in the laboratory today for steroid measurement but is useful for detecting cross reacting steroids by analysis of eluates from the chromatography. Manfred Schoneshofer developed several methods along those lines. Sippell et al. (1978) developed a multicolumn Sephadex LH-20, mechanized system for steroid separation prior to radioimmunoassay of the eluted fractions for progesterone, deoxycorticosterone, 17-hydroxyprogesterone, corticosterone, 11-deoxycortisol cortisone and cortisol (Fig. 2.4.5). These procedures are far too laborious for the modern laboratory.

2.4.5.2 Standardization of assays

Pure, solid steroids are available from a number of sources and are best stored at 4°C in a desiccator. A stock standard solution is made up in ethanol and stored at −20°C. From this stock, a working standard is made up by dilution in ethanol, in assay buffer or in a steroid-free serum prepared by removal of steroids from serum, e.g., by charcoal adsorption. This stripping procedure is not specific for steroids and the removal by charcoal of other components may have profound effects on the overall sample matrix. Serum from subjects with suppressed hormone levels are more appropriate than stripped serum (e.g., for cortisol assays, serum is taken from subjects after a dose of dexamethasone). There are currently a number of limitations to the source of human sera because of the need to test the serum for HIV virus/antigen before use. Some laboratories therefore use horse serum. The working standards can usually be kept at 4°C for about 4 weeks. For longer storage,

addition of preservatives (such as sodium azide) or freezing may be necessary and the standard curve and quality control results should be watched for evidence of concentration or deterioration of the standards. The assays on automated platforms for multiple analytes have predetermined calibration curves supplied by the manufacturer for the batches/lots of reagents in use. Progress is being made with assay standardization (Stenman, 2013).

2.4.5.3 Antisera

2.4.5.3.1 Polyclonal antibodies

Steroids are not antigenic and need to be coupled as haptens to larger molecules in order to raise antisera (Erlanger, 1980). The injection of animals is with a steroid conjugated to a carrier protein like bovine serum albumin through a site on the steroid molecule distant from the important functional groups of the particular steroid being assayed (Hoffmann et al., 1975). A reactive group such as a hemisuccinate or an oxime group has to be introduced to the steroid molecule which can react with the carrier protein usually by linkage with the amide groups of lysine residues (Fig. 2.4.6). Bovine serum albumin (BSA) is widely used, while linkage to Keyhole limpet (*Megathura crenulata*) hemocyanin increases the number of steroid groups per protein molecule and gives antisera of greater affinity (El-Gamal et al., 1987).

The position on the steroid nucleus, at which the linkage to the antigenic protein is made, is important to the specificity of the resultant antisera (Lacorn et al., 2005). Making

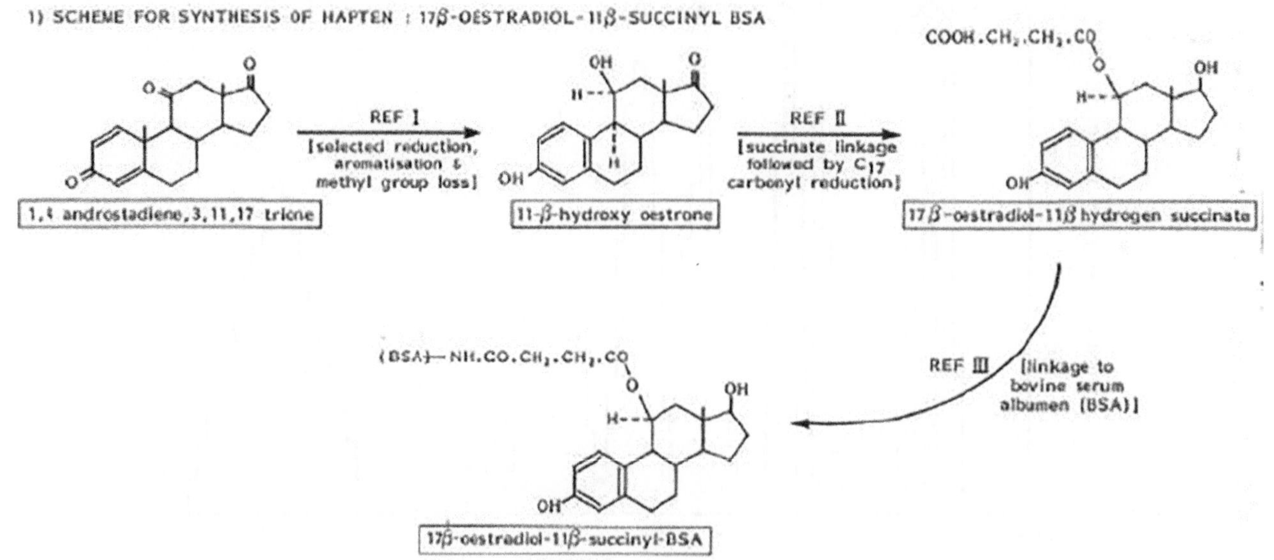

1) SCHEME FOR SYNTHESIS OF HAPTEN : 17β-OESTRADIOL-11β-SUCCINYL BSA

FIG. 2.4.6 Scheme for synthesis of hapten for estradiol RIA (Holownia P and Honour JW, unpublished).

a link from C17 makes the chemistry of the A and B ring open for binding to an antibody, whereas a link from C6 is better if detecting steroids with different functional groups on the D-ring and to a lesser extent the A-ring. This will be clearer to the reader by looking at the steroid chemical illustrations in the Chemistry Chapter. Antisera to estradiol-17β-succinyl-BSA, for example, cross-react with estrone, estradiol-17α and estriol. Antibodies to estradiol-3-succinyl BSA will variably recognize certain androgens, e.g., testosterone. Highly specific antisera to estradiol have been prepared with haptens linked at C-6 and C-11. Of these, a C-6 linked estradiol antigen produces the most specific antisera. Only C-6 substituted estrogens cross-react significantly, but these steroids are only likely to be at low concentration in serum from patients. When determining serum estradiol concentrations in patients being treated with estrone sulfate and estrogen conjugates, the method needs careful evaluation because interference with estrone and with the conjugates may have profound effects under these circumstances. Estrogen conjugates, prescribed for patients requiring hormone replacement, may be detected in certain assays for estradiol. The antisera raised to a C-6 linked hapten seem to be the best for such assays in these particular patients because of lower interference from the administered compound or its metabolites. Extraction of steroids from plasma into an organic solvent will also reduce the cross reaction with conjugates.

The antisera raised to the 11α-hydroxyprogesterone hemisuccinyl conjugate with BSA have good specificity for progesterone, radioactive iodine labeled 11β-hydroxyprogesterone is used as the label. Antisera have been raised to testosterone conjugated to carrier proteins at C1, 3, 6α, 6β, 7, 11, 17 or C-19 on the steroid molecule. The C-19 hapten yields antisera with the least cross-reaction with 5α-dihydrotestosterone although cross-reactions with the 19-norsteroids used in oral contraceptives might need to be considered. The C-19 hapten is the basis of antisera in direct methods, whereas antisera raised against C3 haptens with albumin are generally used for antisera in extraction assays.

Direct assays have provided some misleading results. For example, using one commercial kit several laboratories found testosterone concentrations greater than 5 nmol/L in women (Heald et al., 2006). Such a result is normally suggestive of a tumor unless clinicians are aware of the assay. After solvent extraction, the results were acceptable. On further investigation, these results were partly attributed to interference by the variable SHBG concentrations of samples and differences in binding by the antisera of native steroids. The best antisera for cortisol assays are raised against cortisol-3-carboxymethyloxime-BSA hapten. Prednisolone will however cross-react variably with such antisera and so there is little clinical value in measuring serum cortisol concentrations in patients known to be on prednisolone therapy. A patient with suspected adrenal insufficiency may have an ACTH stimulation test even when on prednisone so as to evaluate adrenal reserve. Linkage of cortisol to BSA through C-21 using a hemisuccinate derivative produces antisera with very poor specificity. For all cortisol assays, the cross-reaction with 11-deoxycortisol needs to be considered particularly when patients are receiving metyrapone (11β-hydroxylase inhibitor).

The process for polyclonal antibody formation is lengthy procedure over many months (Fig. 2.4.7). Rabbits, goats and sheep are commonly used.

Several injections are needed and a crude assay needs to be developed to test the titer of generated antibodies. Animals are bled at frequent intervals. Rabbits may be

PREPARATION OF ANTISERA (vs HAPTEN : 17β-OESTRADIOL-11β-SUCCINYL-BSA)

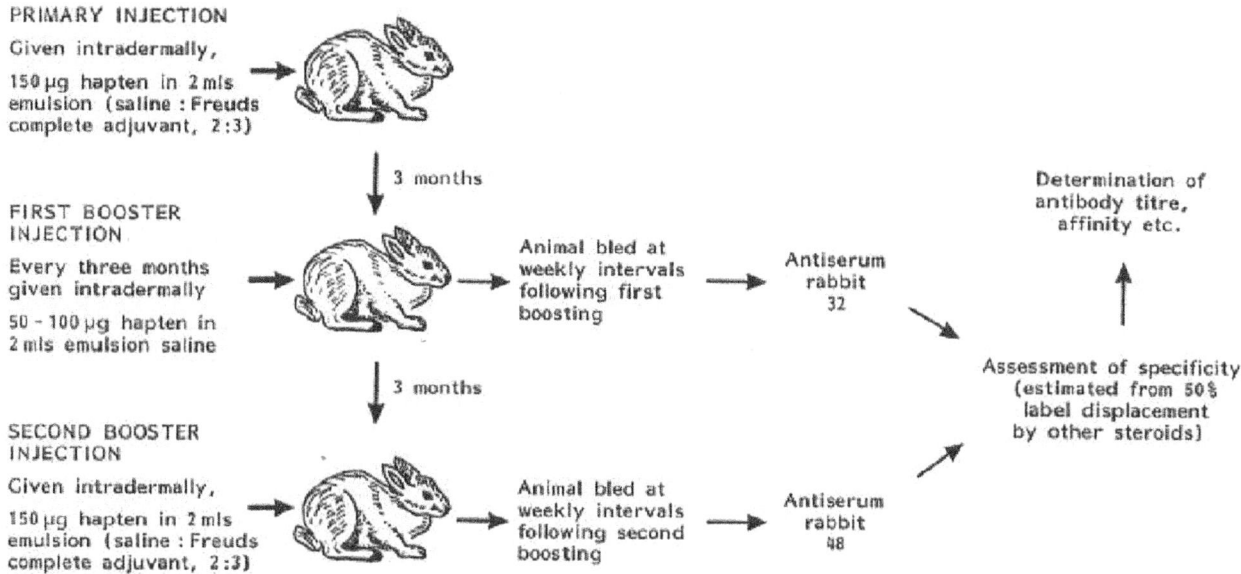

FIG. 2.4.7 Schematic of antibody preparation for estradiol RIA (Holownia P and Honour JW, unpublished).

sacrificed at the end of the program to maximize the blood volume recovered. The resulting antisera can be diluted 1 in 10,000 even 1 in 100,000 with good results.

2.4.5.3.2 Monoclonal antibodies

A polyclonal antiserum will have antibodies that range in specificity and affinity. Some types of assays require better antibody specificity and affinity higher than can be obtained using a polyclonal antiserum. There has been much progress to improve the specificity of the antibodies such that all of the antibodies bind with high affinity to a single epitope. This high specificity can be provided by **monoclonal antibodies (mAbs)** produced in vitro using tissue-culture techniques. Antibodies are produced by immunizing an animal, often a mouse, multiple times with a specific antigen. B cells from the spleen of the immunized animal are then removed. B cells are fused with immortal, cancerous B cells called myeloma cells, to yield **hybridoma** cells. All of the cells are then placed in a selective medium that allows only the hybridomas to grow. The hybridomas, which are capable of growing continuously in culture, are then screened for the desired mAb. Those producing the desired mAb are cloned and grown in tissue culture; the culture medium is harvested periodically and mAbs are purified from the medium (Fig. 2.4.8). This is still a very expensive and time-consuming process. It may take weeks of culturing and many liters of media to provide enough mAbs for an assay. The supply of antibody however is near immortal but retention of activity with time has to be confirmed. Patients who keep mice as pets may have heterophilic antibodies.

Antiidiotypic antibodies have been proposed as the basis for noncompetitive IA (Barnard and Kohen, 1990; Mares et al., 1995) but the cost of preparing the antibodies has inhibited much development of this procedure.

2.4.5.4 Tracers

Many steroid RIA assays were developed using tritium labeled hormones. Except in research laboratories, these were largely replaced in favor of iodine-labeled ligands. This was dictated by the problems of charcoal separation, the disposal of scintillation fluids and the time required to count the isotope in large number of tubes in a typical assay batch. The steroids are usually labeled with tritium at C1, 2, 6 and 7 giving steroids with specific activity of 2–4 TBq/mmol (40–100 Ci/mmol). About 0.37 kBq (0.01 µCi) per tube are used in the assay. Decomposition of the label is usually indicated in the assay by a change in the nonspecific binding or by a change in the shape of the standard curve. Steroids randomly labeled with tritium at up to six positions can have specific activities up to 7 TBq/mmol (180 Ci/mmol). These are required for high sensitivity assays, but the label requires frequent purification. A tritium labeled steroid is immunologically identical to the analyte.

In order to label most steroid hormones with radioactive iodine, some form of derivative is needed. The affinity of the antibody for the label may then be different than to the native steroid so that competition for binding sites is not equal. The site of linkage and the size of the iodinated group may greatly affect the affinity and specificity of the antibody reactions. The preferred tracer for direct serum cortisol assay is

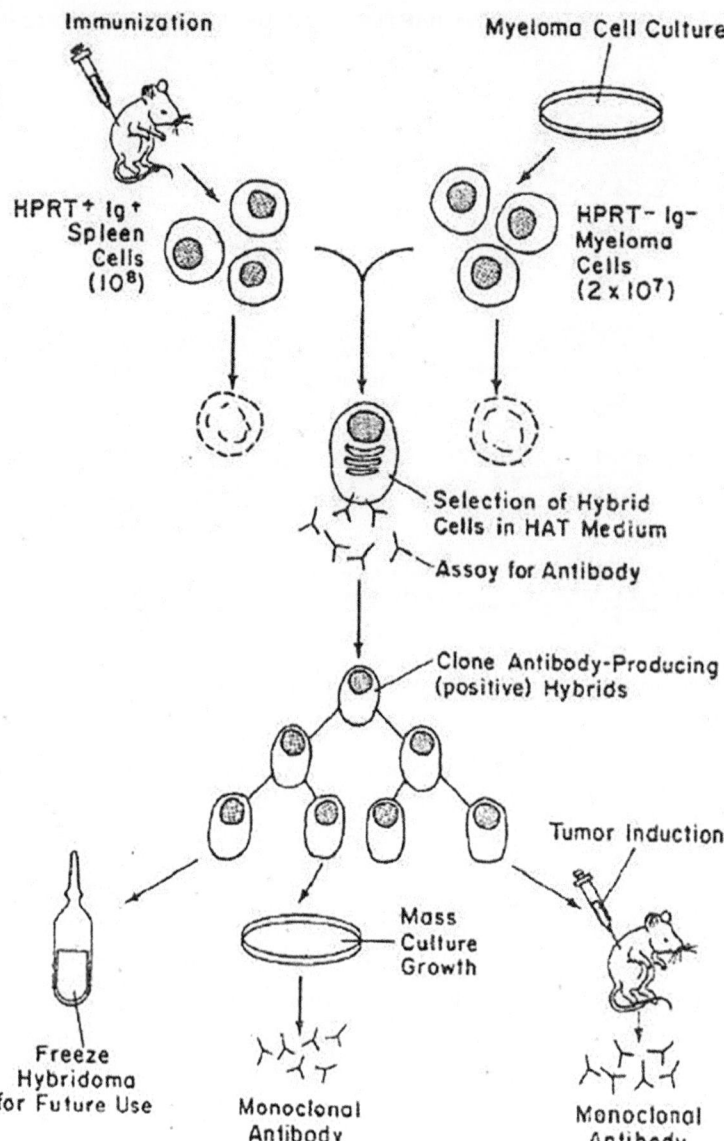

FIG. 2.4.8 General scheme for monoclonal antibody formation (Holownia P and Honour JW, unpublished).

cortisol-3-*O*-carboxymethyloxime (CMO) iodohistamine. The CMO steroid is linked to the iodohistamine by a mixed anhydride reaction. For some progesterone assays, the antisera are raised to 11β-hemisuccinyl conjugate and the label is progesterone glucuronyl iodotyramine.

Estrogens can be labeled with ^{125}I in the phenolic A-ring but if used as the ligand in an RIA there may be some sacrifice of immunoreactivity with this label (Fránek et al., 1983). Estrogen derivatives are therefore used more commonly, for example, 11β-hydroxyestradiol-11β-hemisuccinate coupled with tyrosine methyl ester before iodination of the tyrosine residue (Fig. 2.4.9). Estradiol-6-CMO coupled with ^{125}I-tyramine has also been used.

Most steroid RIA incorporating iodinated tracers coupled to the steroid molecule employ homologous systems where the same steroid derivative is used for both the immunogenic hapten used to raise the antibodies and as a ligand in the assay. These assays are appreciably less sensitive than the corresponding tritiated systems because the antibody can have a higher affinity for the tracer than the native steroid. The sensitivity of assays for steroids can be considerably improved by using different derivatives for the labeled conjugate and the immunogen. The bridge for the ligand and hapten may be similar (homologous) or different (heterologous) (Fig. 2.4.10).

The heterologous assay approach has been successfully applied to assays for progesterone (Karir et al., 2006), androstenedione, testosterone and estradiol (as above for estradiol) (Allen and Redshaw, 1978; Corrie et al., 1981; Rao and Taraporewala, 1992; Rassaie et al., 1992; Fránek et al., 1983).

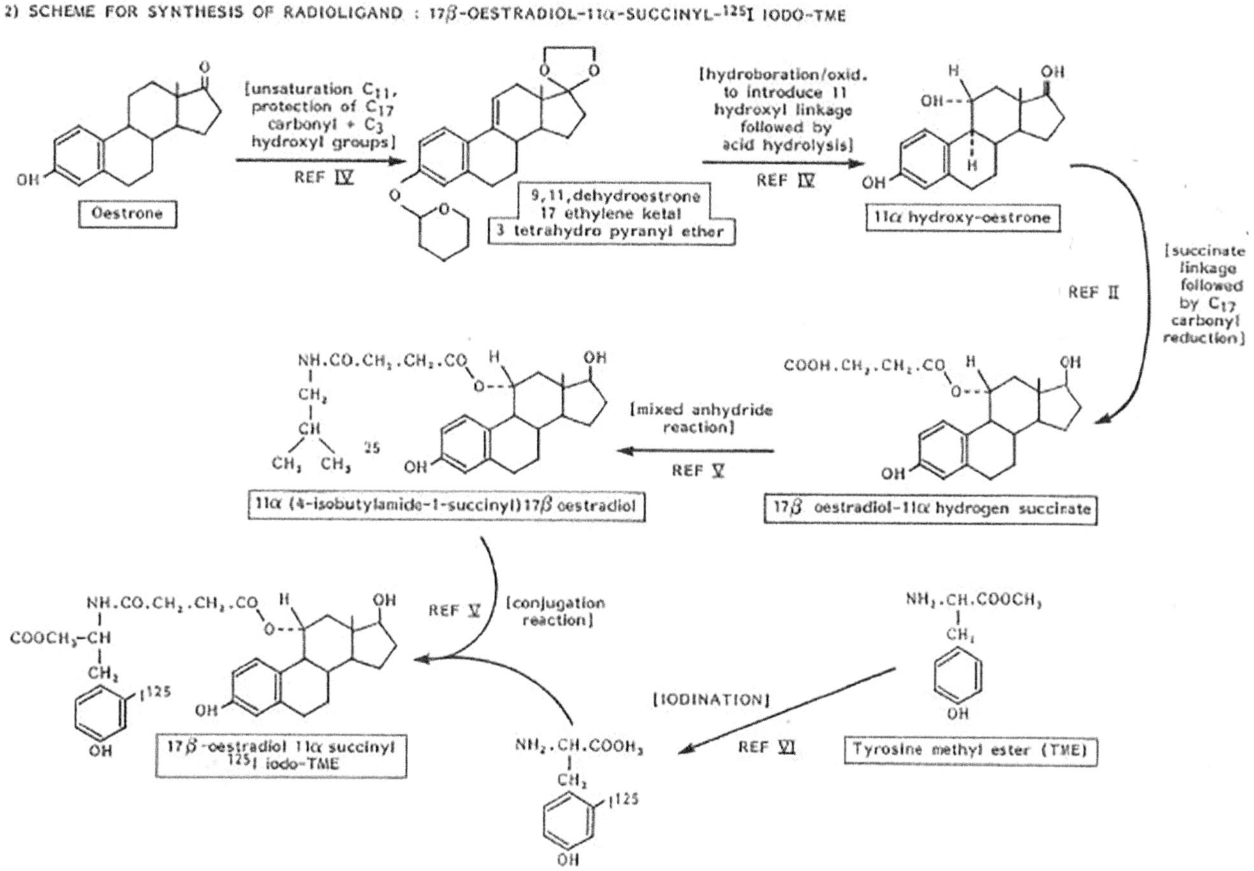

2) SCHEME FOR SYNTHESIS OF RADIOLIGAND : 17β-OESTRADIOL-11α-SUCCINYL-^{125}I IODO-TME

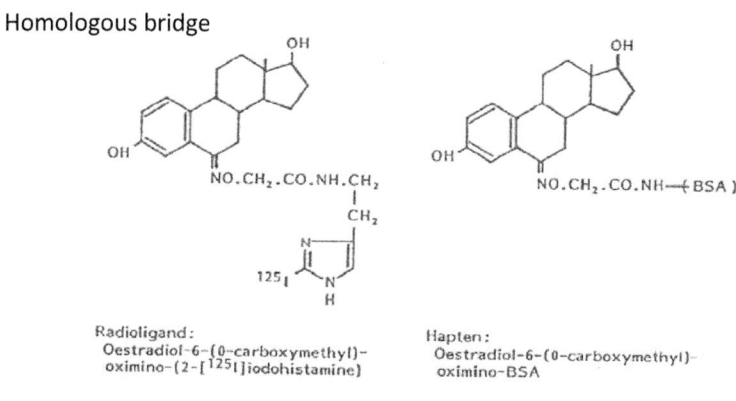

FIG. 2.4.9 Scheme for synthesis of radioligand for estradiol RIA (Holownia P and Honour JW, unpublished).

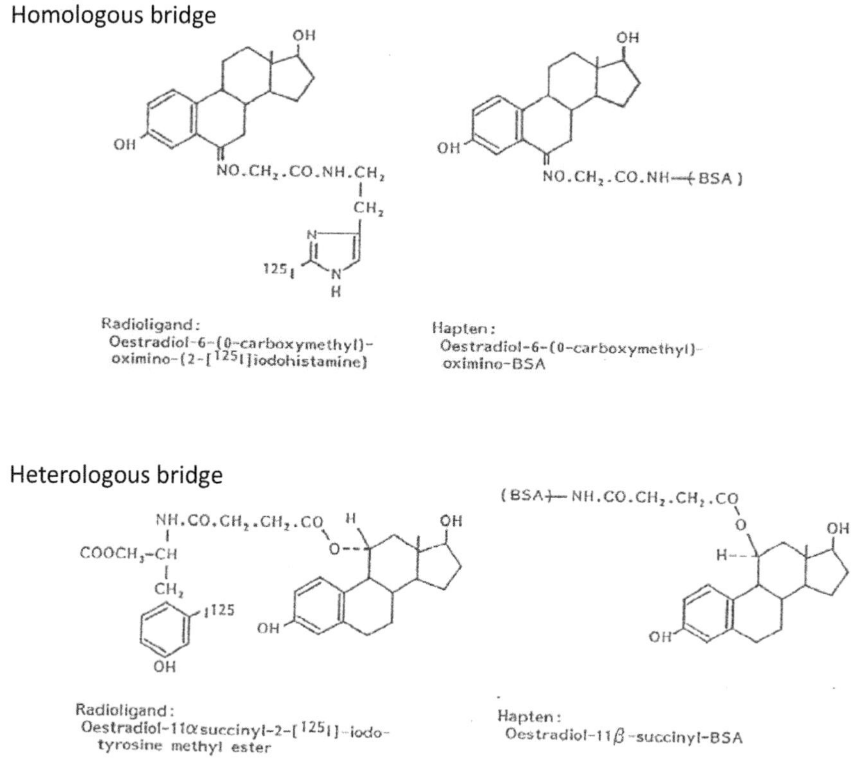

Homologous bridge

Radioligand:
Oestradiol-6-(0-carboxymethyl)-
oximino-(2-[^{125}I]iodohistamine)

Hapten:
Oestradiol-6-(0-carboxymethyl)-
oximino-BSA

Heterologous bridge

Radioligand:
Oestradiol-11α-succinyl-2-[^{125}I]-iodo-
tyrosine methyl ester

Hapten:
Oestradiol-11β-succinyl-BSA

FIG. 2.4.10 Principles of homologous and heterologous antibodies. *(Author original.)*

The mathematical and statistical theory for the optimization and interpretation of radioimmunoassays has been extensively developed. The mathematical theory can be used to compute the optimum combination of tracer and antibody in order to provide the best sensitivity or best precision across the concentration range important for clinical decisions. In general, the labels used for immunoassays have progressed from radioactivity in RIAs to the current systems based on enzymes, luminescence, chemiluminescence and fluorescence.

2.4.5.5 Separation techniques for free and bound label

Several procedures are available for separating the antibody bound steroid from the free steroid. **Dextran-coated charcoal** absorbs the free steroid which is brought to a pellet by centrifugation preferably at low temperature. The bound fraction in the supernatant is decanted for counting. High affinity antisera are needed so that the charcoal does not compete with the antibody for steroid binding. Manual assays are limited to 100 tubes to prevent drift in the results from first to last sample because of the time involved.

Antibody-bound steroid can be precipitated by addition of **polyethylene glycol (PEG 6000)** or with saturated ammonium sulfate. These nonspecific precipitation techniques can lead to high poor precision. These two problems are improved if precipitation of the protein antibody, is achieved by the addition of a solution containing an antiserum (**second antibody**, e.g., donkey anti rabbit gamma globulin) directed against the first antibody (e.g., rabbit antitestosterone) is used. Beads coated with protein A (protein found in wall of *Staphylococcus aureus*) or dead Staphylococci will precipitate IgG. Incubation times for the antibody-antigen reaction are shortened in PEG solution. The efficiency of precipitation can be further increased by the addition of normal serum from the same species in which the first antiserum was raised.

Immobilized antisera are the basis of newer forms of separation. The antisera are coupled to cellulose particles, magnetic particles, nylon beads or to the walls of the assay tubes themselves. These procedures improve the convenience of such assays and are advantageous where rapid results assist clinical decisions. In the calibration curve, the binding at zero concentration of analyte (zero binding) is often low with immobilized antisera particularly when coated tubes are used.

2.4.5.6 Quality of immunoassays

To test specificity of an IA, serial dilutions of potential cross reacting steroids are assayed for displacement of label from the antisera which is plotted against concentration. The resulting curve is then compared with that given by the analyte in the test sample. The calibration curve can be transformed to a linear presentation, that can extend the working range of the assay depending on the imprecision at low and high concentrations. Manual immunoassays were usually performed in duplicate or replicates which allowed the precision of the assay to be determined. The imprecision profile for an assay can be plotted as coefficient of variation against concentration (Fig. 2.4.11). Precision less than 10% is desirable across the working range of the assay. High and unacceptable precision is seen at high and low concentrations. The practice of duplication is becoming obsolete due to the cost implications of extra reagents in automated methods, and duplicates are sadly rarely used now. It is now more common to quote imprecision.

In order to establish interferences in the assay from structurally related steroids, the commonest way of presenting the result is to compare the amount of cross reacting steroid which displaces 50% of the label with the amount of analyte which displaces 50% of the label, the ratio is expressed as a percentage (Dittadi et al., 2019) (Fig. 2.4.12).

To take an extreme example of cross reactivity, in order to exclude an effect of say DHEAS at physiological concentration (8 mmol/L) in an estradiol assay (50% binding at 800 pmol/L) the cross reaction needs to be <0.001. This approach is in many cases circumvented, limited to testing the effect of cross reacting steroids at only one point in the standard curve and in the presence only of tracer quantities of the analyte. Further experiments should assess cross reaction at different levels or response and in the presence of physiological and pathological concentrations of the analyte.

Some commercial kits report cross reaction at a single level as a ratio of the tested analyte concentration measured to the concentration of the cross-reacting steroid added to the assay which is at a concentration above what might be expected in any samples. This may not test the presence of pathological amounts of a cross reactant and gives no measure of cross reaction at low and high levels of the analyte. Even if a laboratory does not prepare its own antisera the quality of antisera should be considered carefully when choosing a commercial antiserum or kit. The assay performance of standards added to such stripped plasma from human or other species may not be identical to the assay of the analyte in patient samples. Immunoassays can detect a picogram of steroid.

The specificity of an immunoassay can also be confirmed by chromatography of the sample and immunoassay of the eluates. An example has already been given with neonatal samples in cases of congenital adrenal hyperplasia where 17-hydroxypregnenolone sulfate affected the results for 17-hydroxyprogesterone. In another example, an immunoassay for cortisol in urine showed a weak reaction for a metabolite, 5α-THF (Fig. 2.4.13A) which was exaggerated if the urine was hydrolyzed (Fig. 2.4.13B) (Horie et al., 2007). Cross reactions were determined in several immunoassays showing that the effects are very variable (Table 2.4.1).

The binding affinities of antibodies and antigens can be tested in Biacore system from Sweden. The instruments monitor molecular interactions in real time, using a

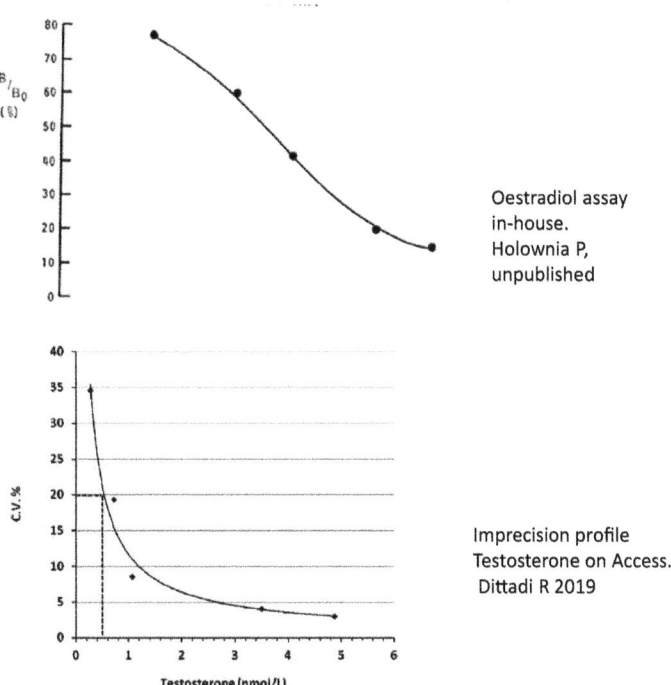

FIG. 2.4.11 Standard curve for estradiol RIA (Holownia P and Honour J, unpublished) and precision profile for testosterone by chemiluminescent assay on Abbott Access. *(From Dittadi R, Polesello V, Zivi A, Carraro P. Evaluation of the possible interference of abiraterone therapy on testosterone immunoassay. Clin Chem Lab Med 2019 ;57(10):e253–4. Fig. 1 p. 3.)*

A.   **B.**

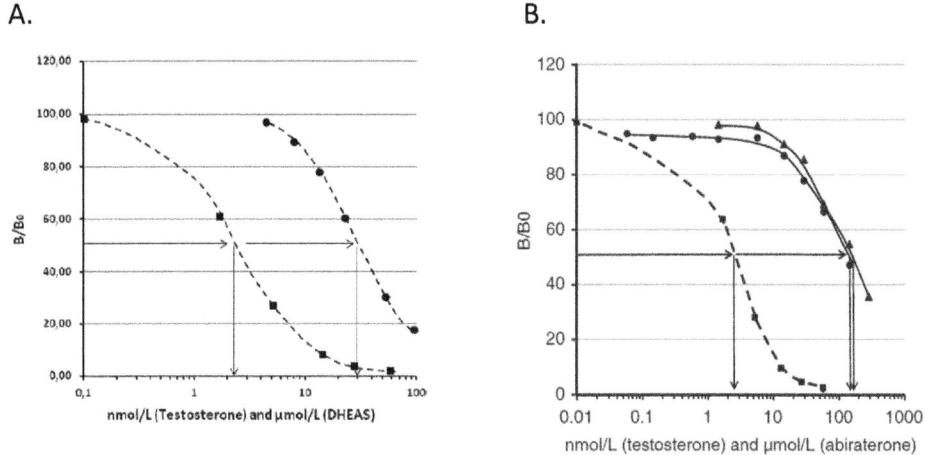

FIG. 2.4.12 (A) Cross-reactivity between testosterone and DHEAS. The standard curves of testosterone (*filled squares*) and DHEAS prepared in testosterone-free matrix (*filled circles*) were assayed with testosterone reagent and the results were reported as percentage of Bound/B0. The cross-reactivity was defined as a percentage of the concentration of DHEAS at the point where the reduction in signal reach the 50% of the binding capacity (50% of the signal achieved in the absence of analyte), respect to the concentration of testosterone giving the same fall in signal. The concentrations are indicated by the *arrows*. (B) Cross-reactivity between testosterone and abiraterone. The standard curves of testosterone (*filled squares*) and abiraterone prepared in testosterone-free matrix (*filled circles* and *triangles*) were assayed with testosterone reagent and the results were reported as percentage of bound(B)/B0. *((A) From Dittadi R, Matteucci M, Meneghetti E, Ndreu R. Reassessment of the access testosterone chemiluminescence assay and comparison with LC-MS method. J Clin Lab Anal 2018;32(3):e22286. Fig. 2 p. 3; (B) From Dittadi R, Polesello V, Zivi A, Carraro P. Evaluation of the possible interference of abiraterone therapy on testosterone immunoassay. Clin Chem Lab Med 2019 ;57(10):e253–4. Fig. 2.)*

noninvasive label-free technology that responds to changes in the concentration of molecules on the chip sensor surface. Surface plasmon resonance (SPR) occurs when light is reflected from the metal-coated interface (Wilson, 2002) and used for detection of testosterone interaction (Zhang

et al., 2014). The angle of reflection of UV light changes when an interaction between the immobilized antibody and a steroid in solution occurs. The detection is sensitive to changes in refractive index within about 150 nm from the sensor surface (Fig. 2.4.14).

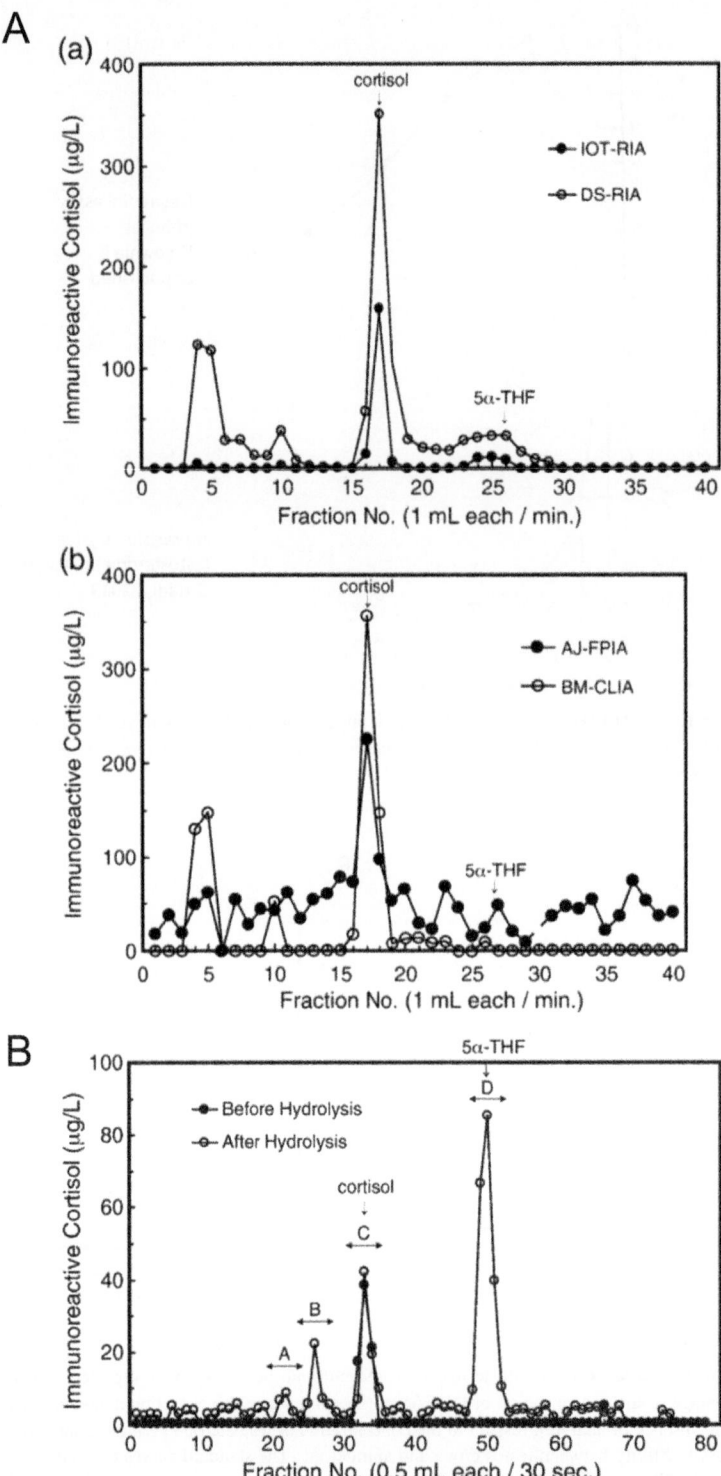

FIG. 2.4.13 Steroids after Sep Pak C18 extraction from urine were separated on a C18 reverse phase silica column eluted with 25% aqueous acetonitrile. (A) Immunoreactive cortisol concentration in each HPLC fraction of urine samples measured with commercial kits (a) IOT—Immunotech; and DS Dia-Sorin (b) AJ Abbott Japan fluorescence polarization IA; Bayer Medical chemiluminescence IA. (B) Immunoreactive cortisol concentrations in each fraction of a urine sample from a healthy adult before (filled circle) and after (open circle) hydrolysis with glucuronidase; cortisol concentrations in fractions were measured with IOT-RIA. *Data (A) From Horie H, Kidowaki T, Koyama Y, Endo T, Homma K, Kambegawa A, Aoki N. Specificity assessment of immunoassay kits for determination of urinary free cortisol concentrations. Clin Chim Acta 2007;378(1–2):66–70. Fig. 2 p. 68; (B) From Horie H, Kidowaki T, Koyama Y, Endo T, Homma K, Kambegawa A, Aoki N. Specificity assessment of immunoassay kits for determination of urinary free cortisol concentrations. Clin Chim Acta 2007;378(1–2):66–70. Fig. 3 p. 68.*

TABLE 2.4.1 Cross reactions of cortisol metabolites and other steroids in the four kits used in the study.

Steroid	Immunotech-RIA	Abbott-FPIA	Dia-Sorin-RIA	Bayer ACS-CLIA
Cortisone	1.0%	1.3%	5.8%	54.5%
11-Deoxycortisol	16.7%	12.9%	8.8%	110.0%
Corticosterone	3.6%	4.5%	0.7%	5.0%
5β-DHF	2.5%	11.3%	5.8%	30.0%
5α-THF	28.5%	30%	12.9%	20.0%
5β-THF	0.2%	1.4%	0.2%	–
5α-THE	<0.1%	<0.9%	<0.1%	0.8%
5β-THE	<0.1%	–	<0.1%	–
5β-THE-3-glucuronide	<0.1%	–	<0.1%	–
6βOHF	<0.1%	<0.9%	3.9%	6.7%
6β-OHE	<0.1%	<0.9%	0.2%	1.5%
20α-Cortol	<0.1%	<0.9%	<0.1%	<0.1%
20β-Cortol	<0.1%	0.0%	–	ND
6α-Methyltestosterone	0.1%	1.3%	8.9%	30.0%
Prednisolone	2.5%	36.0%	38.7%	54.5%
Prednisone	0.1%	<0.9%	2.9%	42.9%

–, no data; *N.D.*, not determined. Cross reaction calculated from the IC50 of each dose response curve compared with the standard cortisol curve. IOT Immunotech; AJ Abbott; DS Dia-Sorin; BM Bayer ACS.

From Horie H, Kidowaki T, Koyama Y, Endo T, Homma K, Kambegawa A, Aoki N. Specificity assessment of immunoassay kits for determination of urinary free cortisol concentrations. Clin Chim Acta 2007;378(1–2):66–70. Table 1 p. 67.

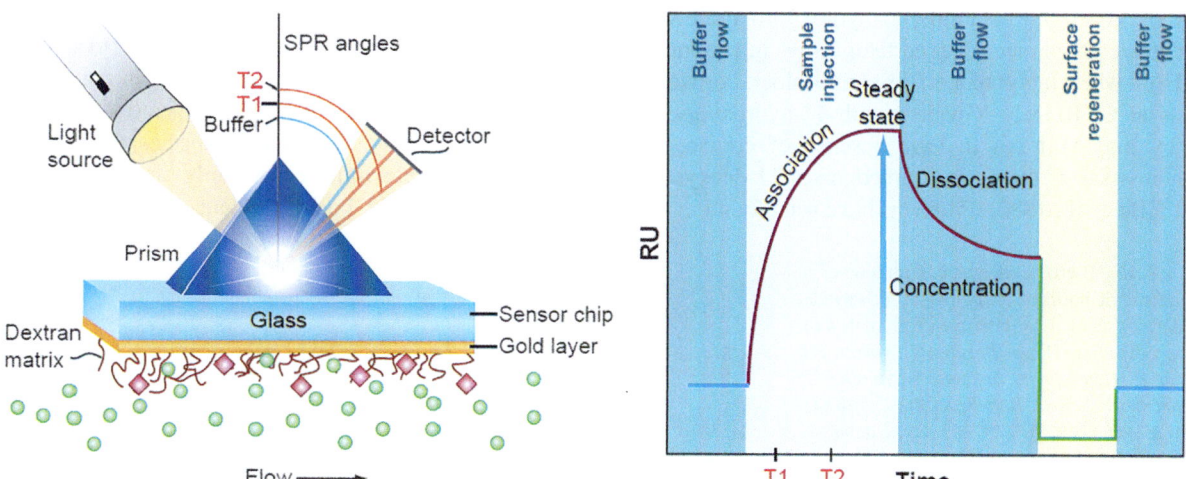

FIG. 2.4.14 Principles of surface plasmon resonance spectroscopy. Left, an SPR optical unit and a sensor chip detect the P molecules (*green spheres*) in the flow solution, which passes by the T (*pink diamonds*) linked to the dextran matrix. The *blue SPR angle* defines the position of the reduced-intensity beam. Time points T1 and T2, shown in the schematic sensorgram (right) correspond to the two *red SPR angles*, which shift as P binds to T over time. As the concentration of bound P increases (*arrow*), the RU response approaches saturation. The complex dissociates upon reintroduction of the buffer. The response to the injection solution will fall below the baseline if its refractive index is lower than that of the buffer. *(From Wilson WD. Tech.Sight. Analyzing biomolecular interactions. Science 2002;295(5562):2103–5. Fig. 1 p. 2103.)*

To study the interaction between two binding partners, one partner is attached to the surface and the other is passed over the surface in a continuous flow of sample solution. The SPR response is directly proportional to the change in mass concentration close to the surface. Biacore systems can be used to study interactions involving (in principle) any kind of molecule, from organic drug candidates to proteins, nucleic acids, glycoproteins and even viruses and whole cells. Since the response is a measure of the change in mass concentration, the response per molar unit of interactant is proportional to the molecular weight (smaller molecules give lower molar responses). The practical lower limit for detection of small molecules with today's instrumentation is about 100 Da. The detection principle does not require any of the interactants to be labeled, and measurements can be performed on complex mixtures such as cell culture supernatants or cell extracts as well as purified interactants. The identity of the interactant monitored in a complex sample matrix is determined by the interaction specificity of the partner attached to the surface. The SPR detection principle is noninvasive and works equally well on clear and colored or opaque samples. This technology can be used to compare binding strengths of antibodies in order to select the best antibody.

2.4.5.7 Interferences in immunoassay

A list of published papers where steroid immunoassays have been subject to interferences is included in Appendix. The commonest problems are due to cross reaction of related steroids particularly synthetic steroids but also endogenous steroids when there is a defect of steroid production due to genetic defect or drug administration (reviewed Jones and Honour, 2006). For example, 17-hydroxyprogesterone and 11-deoxycortisol can affect cortisol assays in patients with CAH or when metyrapone is given to block cortisol synthesis at CYP11B1. A patient with 11-hydroxylase deficiency may still not to have absence of cortisol in an immunoassay if the antibody recognizes 11-deoxycortisol (Ratcliffe et al., 1982). Problems have been found in assays

of sex steroids in postmenopausal women and women treated with aromatase inhibitors and progestogens (Berger et al., 2016).

Some clues to an assay interference can come from clinical testing. Androstenedione can be raised in plasma of patients with polycystic ovaries. The hirsutism can be treated with antiandrogens such as spironolactone (which is primarily a mineralocorticoid antagonist). An increase in immunoreactive androstenedione when a patient was on spironolactone was attributed to cross reaction of spironolactone metabolites (Honour et al., 2010). Similar observations have been seen with abiraterone in patients treated for breast cancer.

Problems with commercial kits are particularly difficult to resolve because the precise formulation of the reagents are not revealed by the manufacturers. For example, the exact nature of the hapten used to raise the antisera (point and nature of bridge to protein) and the chemistry of the label (point and nature of label). Antibodies recognize the bridge as well as the hapten, in addition to the endogenous and exogenous steroids present in the IA itself.

People that keep animals can develop species antibodies that may react in a sandwich immunoassay (a **heterophilic antibody** response) (Bolstad et al., 2013) (Fig. 2.4.15). A clinician may find an elevated test results for ACTH, or LH, or FSH that is not consistent with the clinical picture. This needs to be discussed with the laboratory. Heterophilic antibodies (antianimal antibodies in patients who have mice in the home) have been less of a problem for steroid assays than with the peptide and thyroid hormones (Bolstad et al., 2013). Interference from heterophilic antibodies depends on the species used for antibody preparation (rabbit, sheep are common), the monoclonal antibodies are generated with mice spleen cells.

Dilution of the sample can be used in tests to assess interference giving appropriate lower results in the absence of interference. Antiheterophilic antibodies (blockers) can be added to the sample. These can be species specific or broad spectrum. If **heterophilic antibody** interference is suspected, the sample could be sent for assay at another

FIG. 2.4.15 (Left) A schematic illustration of an immunometric hormone assay; (Right) Falsely elevated result caused by interference from heterophilic antibodies. *(From Bolstad N, Warren DJ, Nustad K. Heterophilic antibody interference in immunometric assays. Best Pract Res Clin Endocrinol Metab 2013;27(5):647–61. Fig. 1 p. 648.)*

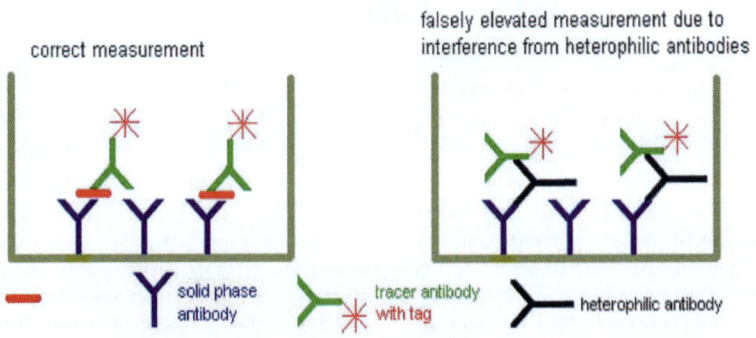

laboratory with a different format. If a patient sample is thought to contain heterophilic antibodies that cross-link mouse IgG1 assay antibodies, the addition of mouse IgG1 to the sample can neutralize the heterophilic antibodies and prevent interference. Aggregated antibodies, either heat-treated or chemically aggregated, are more potent blockers than nonaggregated antibodies. This is most likely explained by the improved ability of antibody aggregates to form complexes with heterophilic antibodies, typically with 2 (IgG) or 10 (IgM) antigen-binding sites, due to the number and proximity of epitopes on the aggregates. Several blocking reagents and blocking tubes, which contain a pellet of blocking reagent, provide an attractive and safe alternative to clinical laboratories in need of an easy-to-perform interference test.

Clinicians have a responsibility to report to the laboratory any unusual results. Interferences can be identified or the laboratory can arrange for samples to be analyzed by a different method at a sister or network laboratory.

2.4.5.8 Immunoassays which replace radioactive labels

There was considerable social pressure in 1980/1990 period to find substitutes for the radioisotopes used in RIA. The health hazards associated with radioactivity plus the logistic and quality control problems arising from the need for regular preparation of radiolabeled reagents which have limited shelf-life. The costs of disposal of radioactive materials were also a factor. When radioactive materials were replaced the alternatives had different risks and required new instrumentation. The appeal for moving to alternative labels arises from the predicted greater sensitivity of such assays and the capability to circumvent separation of the bound and free components (homogeneous assays). The decisions to adopt one system may have been dictated more by considerations of cost, the range of assays available and the necessity for automated equipment than by scientific performance. Nonisotopic tracers have the advantage of

long shelf-life and freedom from the hazards of radioactivity. A number of assays have been reported in the literature and are being offered in commercial kits and automated platforms.

The development of nonradioactive labels has depended on sensitivity which ultimately is achieved by increasing the signal or reducing the background. In that regard, enzymes are widely used because a single molecule of enzyme can react with thousands of molecules of substrate and increase the signal strength. The incubation of enzyme and substrate is sensitive to time, temperature and pH and inhibitory substances, which can include effects on any co-factors. The most commonly used enzyme labels are **horseradish peroxidase, glucose oxidase** (Hosoda et al., 1987) **and alkaline phosphatase**. The enzymes are covalently linked to the antigen or the antibody (Fig. 2.4.16).

The substrates can be incorporated into a colorimetric or fluorometric end-points. The detection limits for most fluorescent tracers is limited by background interference which can be avoided if there is a time gap between excitation and measurement of the emitted light. The DELFIA system is based on time resolved fluorescence with new lanthanides. Greater sensitivity has also been achieved with chemiluminescence and enhancements, such as in the Vitros immunodiagnostic products. Acridinium esters (Richardson et al., 1985) (Fig. 2.4.17) and sulfonamides are chemiluminescent labels (Fig. 2.4.18) that can be linked directly to antigens or antibodies (Kohen et al., 1987).

There are many similarities now in the technologies of immunoassays in the automated platform assays (Table 2.4.2). Separation of components is achieved through centrifugation or magnetism.

At the signal generation stage in ACS 180 and ACS Centaur (Chiron diagnostics), the ester link is cleaved to release unstable N-methylacridone which decomposes with the emission of a flash of light. Electrochemiluminescent substances such as tris-pyridyl ruthenium are used in the Roche system where the detection step takes place in a flow-through electrochemical cell (Elecsys).

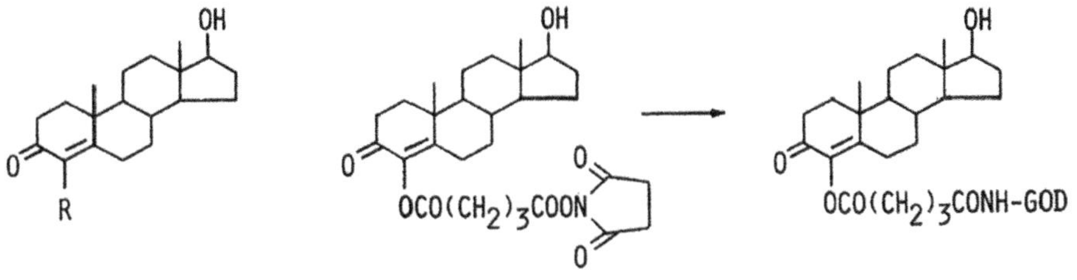

testosterone: R = H
I: R = OCO(CH$_2$)$_3$COOH
II
III

FIG. 2.4.16 Preparation of glucose oxidase labeled antigen for use in enzyme immunoassay. Testosterone was linked to glucose oxidase by the succinimidyl ester method. *(From Hosoda H, Tsukamoto R, Shishido M, Takasaki W, Nambara T. Enzyme labeling of steroids by the N-succinimidyl ester method. Preparation of glucose oxidase-labeled antigen for use in enzyme immunoassay. Chem Pharm Bull (Tokyo) 1987;35(12):4856–61. Chart 1 p. 4857.)*

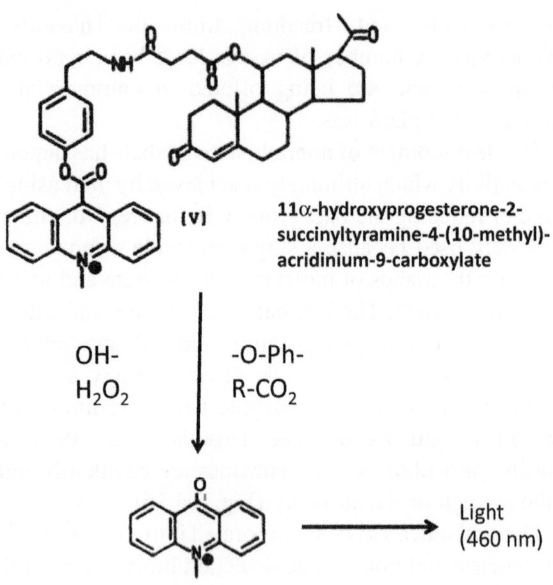

[v]

**11α-hydroxyprogesterone-2-
succinyltyramine-4-(10-methyl)-
acridinium-9-carboxylate**

OH-
H_2O_2

-O-Ph-
R-CO$_2$

Light
(460 nm)

FIG. 2.4.17 Acridinium ester of progesterone. *(From Richardson AP, Kim JB, Barnard GJ, Collins WP, McCapra F. Chemiluminescence immunoassay of plasma progesterone, with progesterone-acridinium ester used as the labeled antigen. Clin Chem 1985;31(10):1664–8.)*

Automated immunoassay systems are standalone modules or part of the string of platform instruments in the larger laboratories (Fig. 2.4.19).

Streptavidin and biotin have been used to develop generic capture systems. Biotin is used in place of the label. The antibody is immobilized on a solid phase such as a microtiter plate. At the end of the assay, streptavidin linked to a signal generating system such as an enzyme is added. **Analytical interference caused by biotin** is increasingly recognized to affect major platforms, including analyzers from Beckman Coulter, Immunodiagnostic Systems, Vitros, Siemens, as well as Roche (Stieglitz and Korpi-Steiner, 2020; Avery, 2019; Luong et al., 2019) (Table 2.4.3).

Problems have unexpectedly arisen because of changes in medical and complementary medicine practices, hence there are problems with patients taking **biotin supplements** that then affect hormone assays that use biotin-streptavidin reagent system. Some patients taking synthetic alkaline phosphatase for treatment of hypophosphatasia have had testosterone assays and abnormal results have been found where the assays use the enzyme in an ELISA (Sofronescu et al., 2018). Sex differences and changes in steroid binding proteins have also been recognized in the published interferences.

2.4.5.9 Current and short-term future position of IA

Many commercial and reference center immunoassays have now been shown to perform as well as MS methods for the

patient samples, reference materials and Proficiency Test materials that were tested. Unfortunately, there are no laboratory tests with immunoassays that detect accuracy in any one clinical sample because specificity and degree of an interference is unknown. Astute clinical interpretation of results has questioned assay performance and prompted investigation of the causes of interference (Warade, 2017; Jones and Honour, 2006). In my laboratory, we have published the effects of drugs (spironolactone, abiraterone) and steroid precursor metabolites (17-hydroxyprogesterone raised due to high concentrations of 11-deoxycortisol) (Ratcliffe et al., 1982) on assay performances. DHEAS has been shown to interfere in testosterone assays and estrone-3-sulfate leads to overestimation of estrone and estradiol. These are probably bridge recognition issues. Interferences are also seen in some endocrine tests with heterophilic antibodies and biotin supplements.

The advantages of Immunoassays are low cost, low sample volume, simplicity, robust, high throughput, ease of automation and adequate sensitivity. The analyzers are compatible with the equipment for many other analytes, whereas currently MS methods are standalone systems. IA will therefore continue to be used as screening tests giving acceptable results in a high proportion of tests. IA can be unreliable for testosterone, DHEAS, androstenedione and 5α-dihydrotestosterone at concentrations below 5 nmol/L and estrogens below 100 pmol/L. Immunoassays are compared with LC-MS/MS methods partly to show industry IA systems perform as well (Fanelli et al., 2011; Taieb et al., 2002, 2003a,b). There are many sources of error when using steroid assays in patients with disorders of steroid synthesis and metabolism. False positive and negative results can also be encountered from patients being treated with cortisol and corticosteroids. Few laboratories offer tests for 11-oxygenated androgens, pregnenolone, 17-hydroxypregnenolone and 18-hydroxy corticosteroids.

2.4.6 Mass spectrometry

A mass spectrometer produces ions from a molecule, separates the ions as a function of their mass to charge ratio (m/z) and records the relative abundance of these ions. There are now many ways of producing ions and these have been subdivided into *hard* and *soft* depending on the amount of energy which is put into the molecule undergoing ionization. The original ionization technique used in GC-MS instruments was **electron impact** (EI) and it still dominates to this day. Electrons produced from a filament impact the molecule (M) displacing an electron, converting it into a positive ion with an odd number of electrons, called the molecular ion ($M^{+\cdot}$) (Fig. 2.4.20). This is a hard ionization technique since the energy of the bombarding electrons is greater than the ionization energy of the molecule and the excess energy causes many of the molecules to fragment

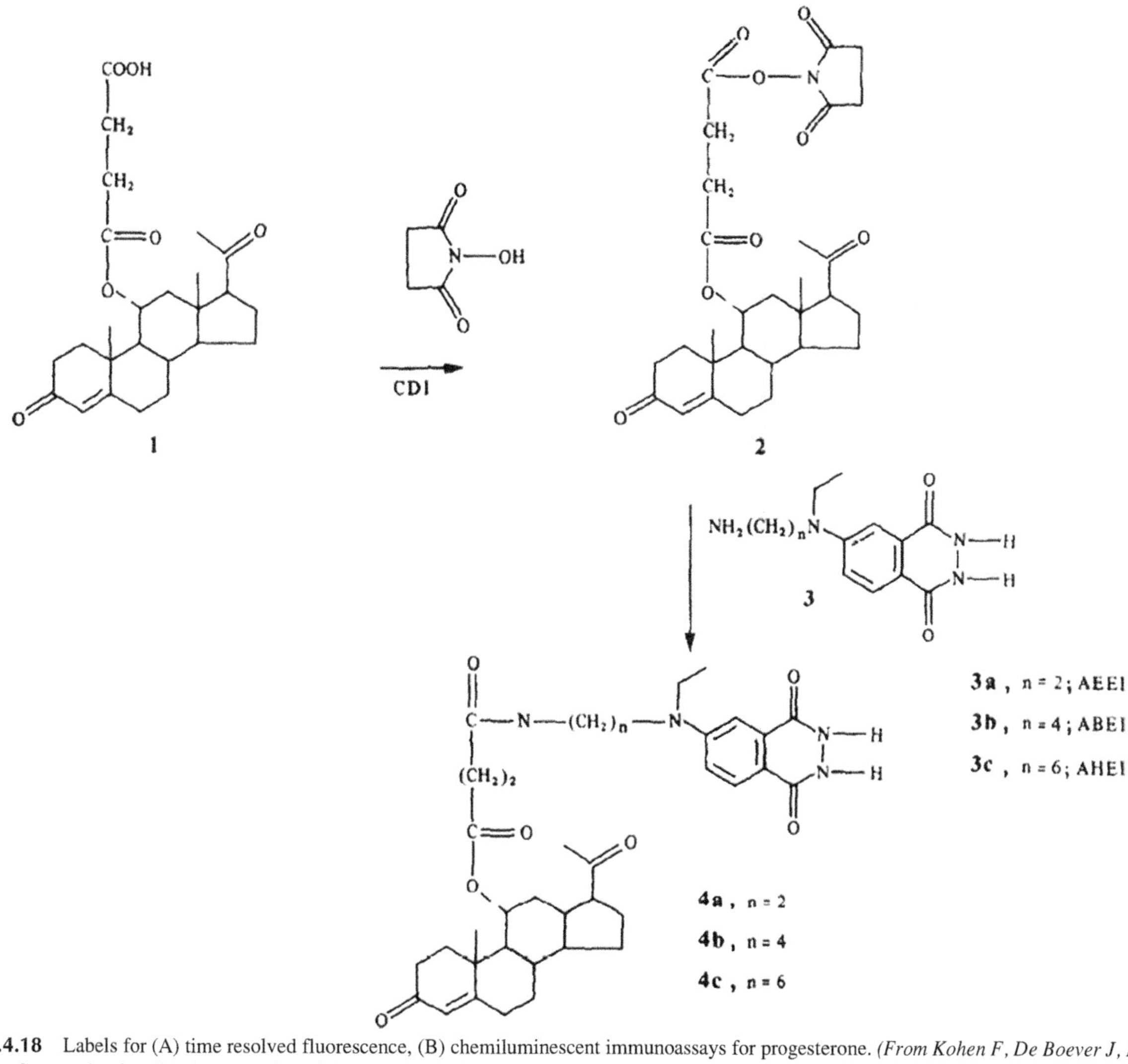

FIG. 2.4.18 Labels for (A) time resolved fluorescence, (B) chemiluminescent immunoassays for progesterone. *(From Kohen F, De Boever J, Kim JB. Recent advances in chemiluminescence-based immunoassays for steroid hormones. J Steroid Biochem 1987;27(1–3):71–9. Fig. 1 p. 72.)*

into structurally informative fragments. Thus, this technique is the most useful for characterizing unknown steroids or confirming the structure of known compounds by comparison with reference materials.

The mass spectrometer can be operated in full scan mode across the mass range 99–800. This avoids the ions at 73 and 75 from TMS groups that carry a high proportion of the ion current from steroids as MO-TMSE derivatives on electron impact. In selected ion, monitoring a targeted analysis of the steroids can be performed by detection of a quantifier and a qualifier ion for each steroid at appropriate time windows.

The need to produce molecular ion species of greater stability motivated the development of less energetic (or softer) ionization techniques, the first of which was **chemical ionization** (CI) (see Fig. 2.4.21).

In this technique, ions are formed of components in the sample by interaction with other ions, the usual source of which are gases introduced into the ion source. Initial

bombardment with electrons produces reagent gas ions (e.g., for methane CH_5^+) which in turn collide with the sample molecules to produce protonated molecular ions (MH^+) through proton transfer. A CI spectrum exhibits strong dependence on the type of reagent gas and the pressure in the ionization region. The usefulness of chemical ionization lies in the fact that the formation of MH^+ can be accomplished with a small transfer of energy. This soft ionization results in the production of large amounts of MH^+ relative to fragment ions thus the technique is less useful than EI for characterizing structurally unknown molecules or confirming the identity of known analytes. CI is however particularly useful in determination of molecular weight and in quantitative methods using selected ion monitoring (SIM) since a greater part of the ion abundance is concentrated in one ion, the protonated molecular ion. Negative ions can also be produced in CI and generally a negative ion SIM assay has greatly

TABLE 2.4.2 Automated immunoassay platforms.

Manufacturer	Model	Detection principle
Abbott	Alinity	CLIA microP
	Archtect	
Beckman	Coulter	CLIA, MagP
Biomerieux	Vidas	ELFA
Chiron	ACS180	
IDS		CLIA
Roche	Modular	ELCLIA
	Elecsys	ELCIA
Siemens	Atelica	CLIA
	Advia	
	Centaur	
Vitros		CLIA
Wallac	Delfia	Time resolved fluorimetry

CIA, chemiluminescence; *EL*, enzyme-linked; *ELFA*, electrofluoroimmunoassay; *FA*, fluoroimmunoassay; *MagP*, Magnetic particles; *microP*, microparticles.

enhanced sensitivity. This has been useful for analysis of estrogens where concentrations in serum can be 2–10 pg/mL.

Once the ions are formed they pass into the *analyzer* which usually means passing the ions through magnetic or electrostatic fields (or both, in *double-focusing* instruments) to filter them on the basis of mass and charge, or through a quadrupole mass filter in which mass separation is accomplished solely by using an electric field. While magnetic and double-focusing instruments are widely used, greater instrument cost restricts their current use to more intractable problems in biochemical analysis so **quadrupole is standard in** the steroid analytical field (Fig. 2.4.22).

In quadrupole instruments, opposing pairs of electrodes in a quadrupole filter system maintain an electrostatic field (DC) upon which is superimposed a radio frequency field (RF). The unresolved ion beam enters these fields parallel to the direction of the electrodes and undergoes an oscillating trajectory of increasing amplitude and are finally collected on the electrodes. However, for selected rod voltages a particular mass is transmitted through the filter to be detected. The success of the instrument stems from the ease with which the masses can be manipulated by complete control of the RF and DC power supplies. The disadvantage of quadrupole instruments lies in their lower mass range (2000–4000 max.) and the lower resolution achievable compared with magnetic instruments.

The widely used **ion-trap** (IT) is a three-dimensional analog of a quadrupole mass filter (Fig. 2.4.23). It consists of a doughnut-shaped central ring electrode and two end-cap electrodes. The field is generated by applying a RF voltage to the central ring electrode and maintaining the end-cap electrode at ground potential. Ionization is achieved through pulsing an electron beam through one of the end electrodes to ionize the gaseous molecules present in the trap.

Automated laboratory

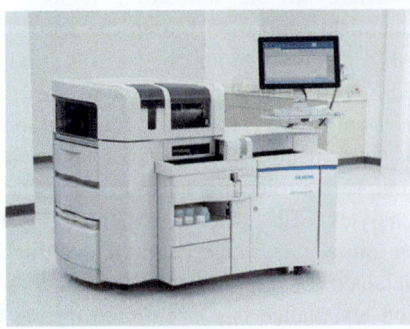

Centaur

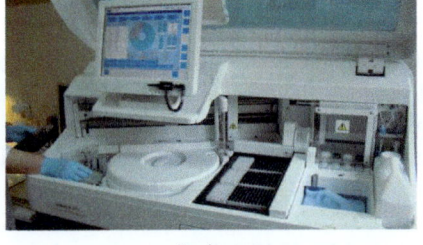

Roche
Elecsys

FIG. 2.4.19 Typical hospital laboratory with automated analyzers.

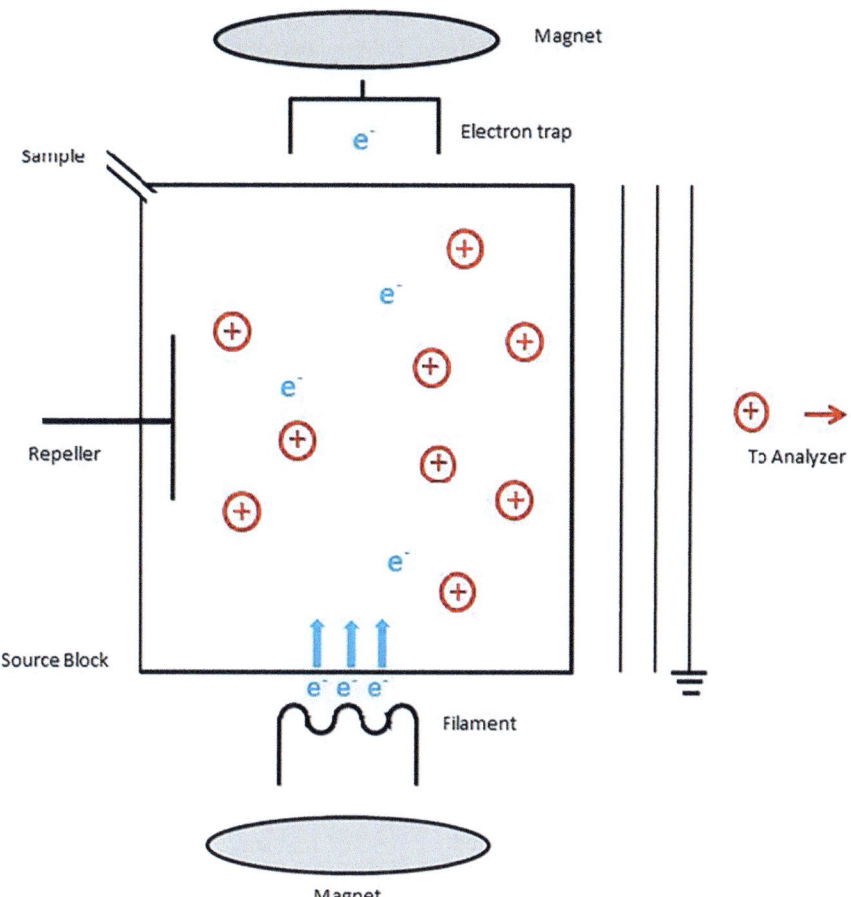

FIG. 2.4.20 Schematic for electron impact ionization and chemical ionization in mass spectrometer interface. CI is a softer ionization so less fragmentation of the molecule. *(Author original.)*

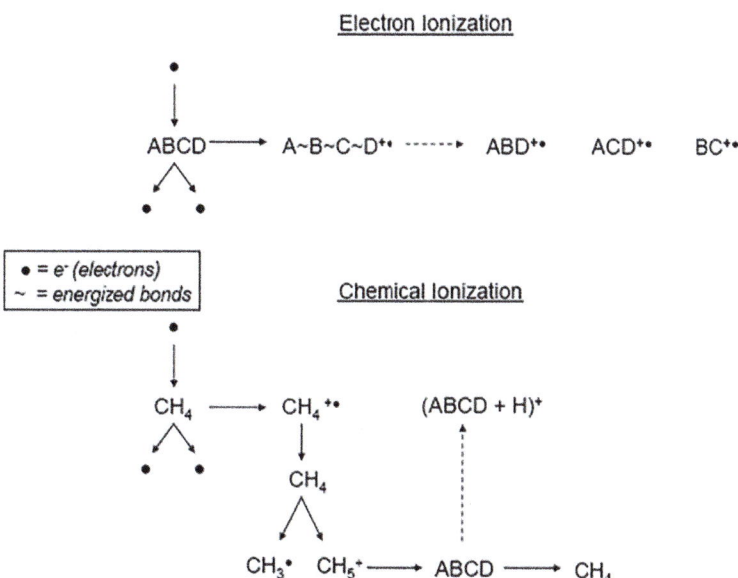

FIG. 2.4.21 Schematic for sample (ABCD) bombardment by electron impact and ionization of the background gas (CH$_4$) to charged particle that instigates softer chemical ionization of the sample in the MS interface. *(Author original.)*

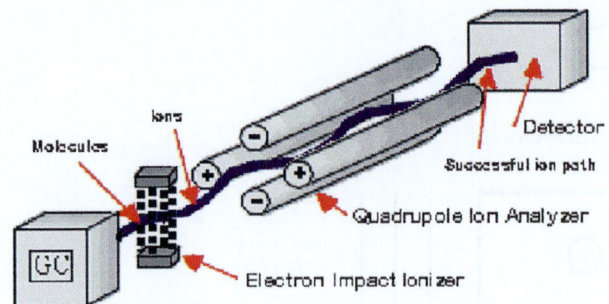

FIG. 2.4.22 Schematic of quadrupole mass spectrometer.

At certain RF amplitudes, all ions above a specific *m/z* are confined within the region bounded by the electrode. Increasing the voltage destabilizes the ion trajectories and ejects the ions sequentially from the ion-trap region through perforations in the other end-cap electrode. The ejected ions are detected to provide a mass spectrum. Ion-trap instruments are becoming increasingly used because like their simpler cousins, the quadrupoles, they have low cost, high scan speed, high ion transmission and are easy to use.

Data produced by mass spectrometers are either in the form of mass scans (mass spectra, usually for qualitative analysis) or selected-ion-monitoring (SIM). The facility with which a computer handles large volumes of data has encouraged the use of *repetitive scanning* as repeatedly scanning the mass range during the time that a sample is introduced into the instrument ensures that data collection is maximized. This is particularly important for LC-MS and GC-MS applications where thousands of spectra may be acquired during a run.

In the SIM mode, selected individual ion masses are monitored continuously or in timed windows to extend the number of analytes in the analysis. When several ions are to be measured each is detected in sequence for a fixed period of time. The sensitivity achievable in this way is greatly increased over a mass scan where each ion in a spectrum is recorded for an exceedingly short time. Quadrupole instruments are particularly suited to SIM as the rod voltage to transmit selected ions can be changed accurately and rapidly.

2.4.6.1 Tandem mass spectrometry (MS-MS or TMS)

As the name implies, MS-MS uses two stages of mass analysis and can in some cases provide a high-resolution measurement of mass. Each MS unit is abbreviated to Q. Ions are separated in the first MS unit (Q1) and delivered to the second unit which is a **collision cell** (abbreviated to Q2 or q2) where further fragmentation of parent or precursor ions takes place in a gas under pressure (Fig. 2.4.24). Those fragment ions selected in the third MS unit (Q3) are sensed by the detector. Each unit can be used to affect all the ions or selected ions. In a neutral loss strategy, Q1 will scan the mass range, q2 will fragment all the ions in the cell and Q3 will scan across a range that covers a fragmentation induced loss of one specific mass. In a precursor ion scan, Q1 will scan across a range, q2 will fragment all ions and Q3 will select one product ion from the second quadrupole.

The first stage of the TMS performs mass selection of a primary ion from the mixture of ions formed in the ion source. This ion undergoes dissociation in an intermediate zone (by collision with inert gas atoms) and the second analyzer records ions formed in this region. The major advantage in MS-MS is its molecular specificity which is the result of the direct link between the precursor ion and all the product ions. This technique carries particular importance when soft ionization methods are used (e.g., CI, LSI-MS, LC-MS spray techniques) since primary spectra

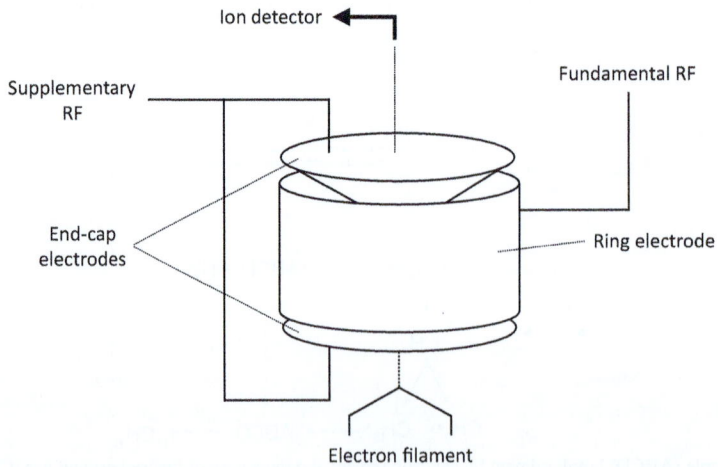

FIG. 2.4.23 Schematic of ion trap mass spectrometer.

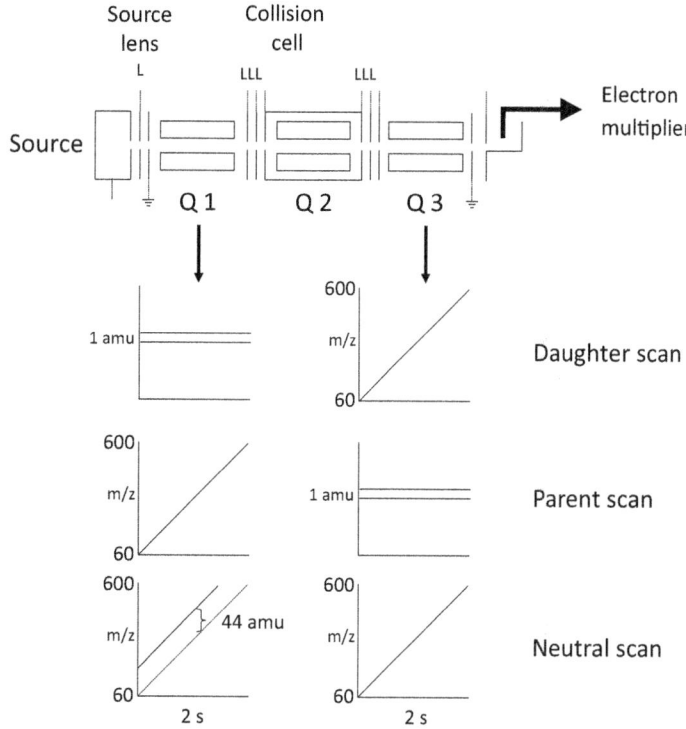

FIG. 2.4.24 Schematic for LC-MS/MS.

are usually simple and give little structural information apart from protonated (MH^+) or deprotonated ($M-H^-$) molecular ions.

Although MS-MS can be carried out using a variety of instrument configurations employing electrostatic, magnetic and quadrupole sectors, the triple quadrupole instrument developed specifically for MS-MS has unique advantages. In this instrument, three quadrupoles are arranged in sequence the first (Q1) and third (Q3) functioning in the normal way, whereas the center quadrupole (Q2) operates in the Rf mode only and serves as an ion-containment region where low energy dissociation takes place usually by collision with a gas (e.g., helium) atoms. In a "triple quadrupole," Q1 is adjusted to allow mass selection of the precursor ion and Q3 is scanned. This procedure is reversed when the precursor ion spectrum is required (i.e., Q3 is used to mass select the desired product ion and Q1 is scanned). The MS is operated in SIM or MRM modes.

The recent commercialization of the tandem ion trap instrument is an exciting development that has cut at least 2/3 off the cost of doing MS-MS. In principle, in triple quadrupoles the different stages of MS-MS analysis are separated in space, whereas in Ion Trap-MS they are tandem in time. Ion formation, excitation and detection are all performed in sequence. The primary ion is formed and stored in the ion-trap chamber and all other ions are ejected. An excitation pulse is applied to the primary ions at a slightly later

time causing them to collide with inert gas and the resulting daughter ions are stored and finally mass-analyzed.

In some cases, full cyclic scan operation is possible, even at high resolution, giving spectra that can afford structural information on steroids. Neutral loss scans and targeted product ion scans have been used particularly when the presence of synthetic anabolic and corticosteroids are suspected. Few staff in clinical laboratories have studied mass spectrometry in depth to interpret these data and a toxicology or drug testing laboratory will have to be consulted.

2.4.6.2 High-resolution mass spectrometry

The **resolving power** of a mass spectrometer is the difference in m/z ions that can be separated divided into the specific m/z value (Fig. 2.4.25). The terms **high resolution** and **accurate mass** are often used interchangeably.

Accurate mass measurement means mass measurement performed to a sufficient number of significant figures to allow for unambiguous determination of an elemental composition. For underivatized steroids, with mass less than 500 Da, an accuracy of 0.0025 Da should enable elemental composition to be determined. In mass spectrometers that separate ions using quadrupole fields, the resolution is constant throughout the scale. Peaks representing two pairs of ions that differ by 1 m/z unit will have the same separation at $mz/100$ and 101 as at mz 2000 and 2001. For accurate mass determination, resolution needs to be greater than

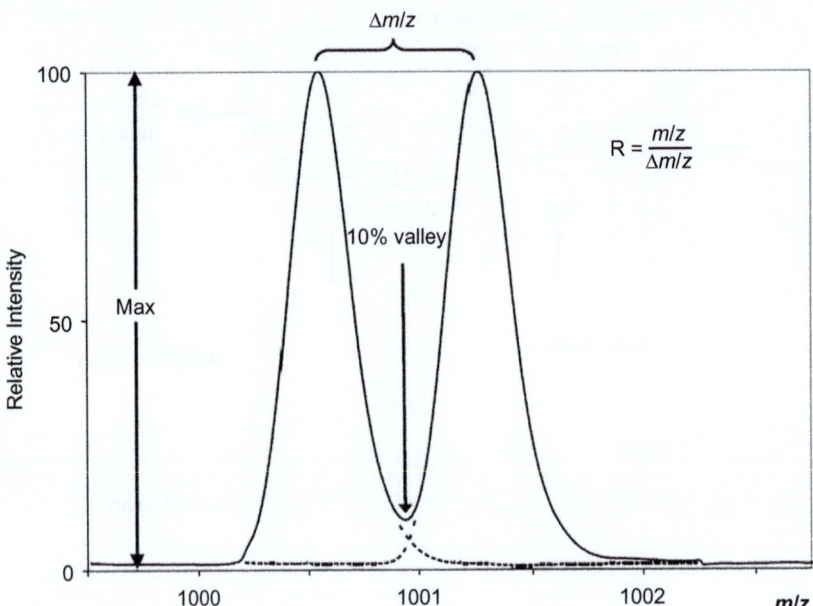

FIG. 2.4.25 Resolution in a mass spectrometer.

10,000 with mass accuracy to 5 ppm. Scan speed will be greater than 5 Hz. Examples of the need for a high resolving power is shown (Fig. 2.4.26.) when the m/z value represents three compounds at m/z 130.

A separation of these ions requires a resolving power of 10,000 which is achieved in a time-of-flight (TOF) analyzer. A representation of a TOF mass spectrometer is shown in Fig. 2.4.27.

The resolution of the mass spectrometer determines the accuracy of mass measurements (see earlier Figs. 2.4.25 and 2.4.26). New generation instruments are now available, allowing a full-scan analysis at ultra-high resolution without remarkable loss of sensitivity, that can enable the analysis of a wide range of compounds. The Orbitrap mass spectrometer confers the typical advantages of HRMS, such as: (a) high selectivity for the target compound, due to its

mass resolution till 100,000 full width at half maximum (FWHM) at 200 m/z; (b) high mass accuracy, with mass error ≤ 2 ppm of amu (as stated by the manufacturer), using as a reference molecular mass (lockmass) the di-isooctyl phthalate, normally present from plastics in a laboratory environment. Estrogens, oxidized metabolites and other endogenous steroids have been determined with a bench top instrument (Franke et al., 2011) but otherwise the instrument has not been greatly used.

HRMS has been used for steroid hormones and metabolites including intact conjugates (Badoud et al., 2011). There is improved specificity with HRMS which will be of value with clinical assays but the instruments are very expensive, the laboratory will have to balance the gains and losses. A quantitative assay in HR-LC-MS/MS can spread in concentrations through three orders of magnitude.

FIG. 2.4.26 Examples of high-resolution mass spectrometer to distinguish three related compounds. R is the resolution.

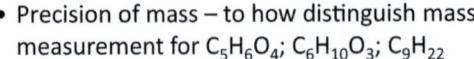

• Precision of mass – to how distinguish mass measurement for $C_5H_6O_4$; $C_6H_{10}O_3$; C_9H_{22}

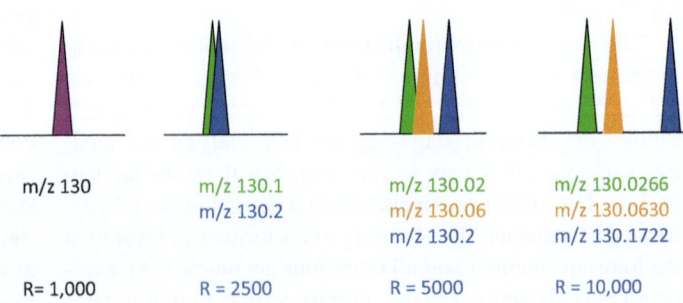

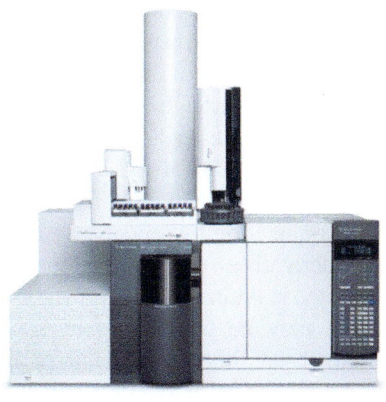

Agilent 7250 GC-QTOF-MS

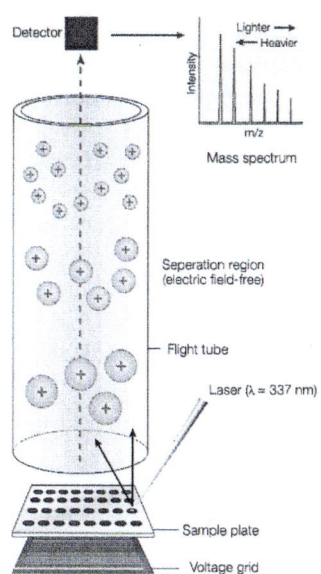

FIG. 2.4.27   Schematic for Agilent time of flight high-resolution mass spectrometer.

2.4.6.3   Quantification

The height or area of the chromatographic peak is measured manually or with the aid of an integrator and ideally the response of the analyte is compared with the response of an appropriate internal standard. The ratios of response for the analyte to the signal from the internal standard are plotted for the concentration range of interest. The concentration of an unknown amount of steroid in the sample is determined from a calibration curve. There are a large number of synthetic steroids available which can be used as internal standards. Since a number of steroid based drugs are widely used in hospital patients the use of two very different internal standards, e.g., 19-nortestosterone and 6α-methyl prednisolone, prevents erroneous results in the case of medication by either one of the steroids selected as internal standard, provided that they behave in a similar way to the analyte during the analytical procedure. When internal standards are not used the extraction and injection must be carefully controlled before peak response can be reliably derived from a calibration curve on injected standards. A deferred standard technique can be adopted in which a known amount of the analyte is injected in pure form sometime after but during the chromatographic run of the unknown sample.

2.4.6.4   Gas chromatography-mass spectrometry (GC-MS)

For specific detection of steroids, the GC is coupled to a mass spectrometer. Steroids were among the first compounds analyzed on the preeminent commercial GC-MS

instrument, the LKB9000. In Stockholm, Sjövall and colleagues separated plasma steroids as trimethylsilyl ethers and methyloxime-trimethylsilyl ethers by polar (QFI) and nonpolar (SE30) packed chromatography columns and characterized previously unidentified components. The wide bore columns had gas flow typically of 60 mL/min so the gas from the column eluates had to be removed in a separator that was designed and implemented also to concentrate the compounds in the gas. Important GC-MS studies at the Karolinska continue to this day. Many of the ancillary chemical techniques necessary for carrying out GC-MS have not changed much within the last 40 years. Methyloxime-trimethylsilyl ethers (MO-TMS) are most often used for profiling urinary steroids after their release from conjugation. The steroids in this "profile" include the metabolites of androgens, cortisol precursors and cortisol. Coupling of the GC with MS is not difficult these days because the narrower capillary GC columns have low gas flow rates which the vacuum system in the MS can cope with so the columns merely passes through a coupling and directs the GC gases into the source of the MS, where ions are created by a beam of electrons and a process of electron bombardment (see earlier Fig. 2.4.20).

GC-MS is commonly used in scan mode to analyze metabolites of steroid hormones and their precursors from urine samples. It is nonselective and will display the entire spectrum of steroids in an individual's metabolome. However, the procedure is relatively time-consuming and expensive because the steroids need to be converted to thermo-stable derivatives. The availability of simple bench-top MS systems has reduced the cost of GC-MS. Electron impact is most commonly used in the MS to ionize and fragment the analytes. Sometimes chemical ionization has been used (Lindberg et al., 1982), particularly for steroids at low concentrations where reducing fragmentation produced more of the molecular ion forms and less fragmentation. This has been used for anabolic steroids in doping control to extend the period of detectability of drugs but not in clinical work. Some methods not requiring deconjugation have been described.

Fewer studies are being published on the GC-MS of steroids because overall the technology has matured and liquid chromatography with MS/MS (LC-MS/MS) for many steroid applications avoids the derivative step making the procedure less tedious and time-consuming. The Shackleton laboratory method for a urinary steroid profile has been applied for clinical diagnosis for more than 45 years (Shackleton and Honour, 1976). The current method measures around 40 steroids excreted as free steroids, steroid sulfates and steroid glucuronides (Krone et al., 2010). The method includes SPE extraction, enzyme hydrolysis and formation of MO-TMS ether derivatives. Retention times of steroids in the GC analysis can be documented in a number of ways. Comparison with a series of *n*-alkanes to give **methylene units** (see earlier Fig. 2.3.19) has proved useful. The mass spectrometer can be operated in full scan or selected ion monitoring (SIM) mode where quantification relies on one ion per component, although specificity is usually confirmed with a second ion. This method profiles metabolites of all the major steroid groups, i.e., the androgens, estrogens, pregnanes, aldosterone and metabolites of corticosterone and cortisol.

The GC-MS conditions for urinary steroid analysis on the automated Hewlett-Packard 5970 MSD or Shimadzu QP2010 are as follows:

- Split/splitless valve closed; Injection volume 2 μL.
- injection carried out with injection block at 260°C and the column oven is held at 50°C for 3 min.
- The split/splitless valve is opened at 2 min.
- After 3 min, the oven temperature is increased at around 27°C/min. For 3 min to reach a starting temperature (for the steroid separation) of 230°C.
   The oven is then programmed at 2°C/min. to a final temperature of 310°C.

GC-MS techniques for synthetic steroids continue to be reported due to the frequent misuse in human and animal competitive sports and in animal husbandry. Some examples are:

- identification of metabolites of corticosteroids in bovine tissues.
- GC-MS-SIM has been used for the measurement of estradiol fatty acid esters in human serum, fat and breast

cyst fluid (Larner et al., 1992). These esters are naturally occurring precursors of biologically potent steroids that can elicit prolonged responses in target tissues because the esters are resistant to catabolism and to their sustained release of estradiol through the action of esterases. A [16, 16 17α-^{2}H$_3$] estradiol stearate was the internal standard although the final quantitative analysis was carried out on free estradiol (as TMS derivatives) released from conjugation by saponification.

- Lipoidal (fatty acid) derivatives of pregnenolone and DHEA have been detected in brain tissue and other types of steroidal conjugate may be present (Liere et al., 2004). These have not been finally characterized but they appear to be sulfalipids with possible structure St–O–SO$_2$–O–L where St is the steroid moiety and L is a lipid. The GC-MS properties of intact estradiol fatty acid esters have been investigated although they have extremely long retention times (40–49 methylene units) on nonpolar capillary columns but have been analyzed as their TMS derivatives with short 5 m capillary columns.
- GC-MS has been used extensively as a reference technique for validating serum assays of steroids. This has been used for testosterone and 17-hydroxyprogesterone by Wudy et al. (2000); cortisol and progesterone by Thienpont laboratory (Thienpont et al., 1991) and for cholesterol by the groups of Siekmann (1991), Eckfeldt et al. (1991) and Ellerbe et al. (1989).
- A combination of flophemesyl and TMS derivatives enabled improved detection of testosterone and pregnenolone in finger nails to 0.1 and 0.2 pg/g, respectively, when normal ranges were 0.24–5.80 pg/g (Choi et al., 2001).

2.4.6.4.1 TMS and MO-TMS derivatives

The most popular derivatization method for steroids is the formation of steroid trimethylsilyl (TMS) ethers. Several reagents with different silyl donor abilities are available. To ensure complete derivatization even of sterically hindered tertiary hydroxyl groups *N*-trimethylsilylimidazole (TSIM) with an auxiliary base like pyridine is used. In the area of sports doping control, an established silylation mixture for complete derivatization of secondary and tertiary hydroxyl groups even at room temperature is *N*-methyl-*N*-trimethylsilyltrifluoroacetamide (MSTFA) with 10% TSIM. A known difficulty in TMS derivatization is the presence of keto groups, because the formation of enol TMS ethers can be observed under these conditions (Junker et al., 2019). There can be artifacts formed during derivative formation (Little, 1999; Meunier-Solère et al., 2005; Moros et al., 2017). The spectra for pregnenolone as mono-TMS and bis-TMS are shown in Fig. 2.4.28A and the structures of the chemical forms in Fig. 2.4.28B.

In sports doping laboratories, TMS derivatives are used because a single derivative reaction makes for a faster method for analysis. In the reaction, carbonyl groups are converted to enol-TMS derivatives. A routine method for serum steroids (17OHP, testosterone, DHEA, androstenedione, cortisol, pregnenolone) in neonates is one of few publications to use enol-TMS derivatives (Magnisali et al., 2008).

Clinical laboratories convert the carbonyl groups to oxime derivatives (usually methyl oximes). The mass spectra of steroids as MO-TMS derivatives are relatively easy to interpret. Most steroids give a protonated molecular ion with losses of 31 from the methyloxime group and 90 from the TMS group (see Table 2.4.3). Taken in combination with GC retention times this provides targeted analysis of steroids.

Spectra of TMS and MO-TMS derivatives ethers show typical peaks from the silyl groups at m/z 73 and 75 (Marcos and Pozo, 2015) (Fig. 2.4.29). This is best avoided

by scanning from m/z 99 since for some steroids there is a characteristic fragment at m/z 100.

In general, the electron impact energy at 20 eV gives structurally more informative spectra of TMS derivatives due to a lesser fragmentation of diagnostically important ions into ambiguous low-mass ions. These ions carry a large proportion of the ion current.

The mass spectra of MO-TMSE derivatives at 70 eV still provide much structural information. They show the molecular ion peaks and characteristic fragmentations. For example, pregnenolone MO-TMS ether (Fig. 2.4.30) the molecular ion $[M]^+$ is observable, and the base peak m/z $[M-15-16]^+$ clearly indicates the fragmentation of the MO moiety (Ebner et al., 2006).

The ion m/z $[M-90]^+$ is typical for the loss of trimethylsilanol and the ions m/z $[M-129]^+$ as well as m/z 129 are characteristic for $\triangle$ 5-sterol TMS ethers referring to the loss of trimethylsilanol from C-3 together with C-1, C-2 and C-3. TMS ethers of steroids with a primary hydroxyl group (at C-18, C-19, or C-21) give ions at m/z 103 $[CH_2OSi(CH_3)_3]^+$ and/or $[M-103]^+$. There are notable ions for certain

TABLE 2.4.3 Features of electron impact mass spectra from GC-MS analysis of steroids as methyloxime-trimethylsilyl ether derivatives.

O-TMS	C=N—O—Me	M Wt Pregnane	Major fragment ions	Examples
1	1	419	388,398	Pregnanolone
2		464 (weak)	117,449	Pregnanediol
2	1	507	476,386,296	Hydroxypregnanolone
3		552	435,345,255	Pregnanetriol
3	1	595	574,474,384	Tetrahydro-11 deoxycortisol
3	1	585	288,275,574	Tetrahydrocorticosterone
3	1; + C=O	609	578,488,398	Tetrahydrocortisone
3	2	638	607	Tetrahydroaldosterone
4		640	550,560,470	Pregnanetriol
4	1	683	652,562,482	Tetrahydrocortisol
4	1; + C=O	654	449,359,269	Cortolone

O-TMS	C=N—O—Me	M Wt Androstane	Major fragment ions	Examples
1	1	391	360,270	Androsterone, etiocholanolone
1	1; 1C=O	405		11-Oxoandrosterone
2		436	421,346,331,256	Androstanediol
2	1	479	448,358,268	11-Hydroxyandrosterone
3		524	434,344	Androstanetriol

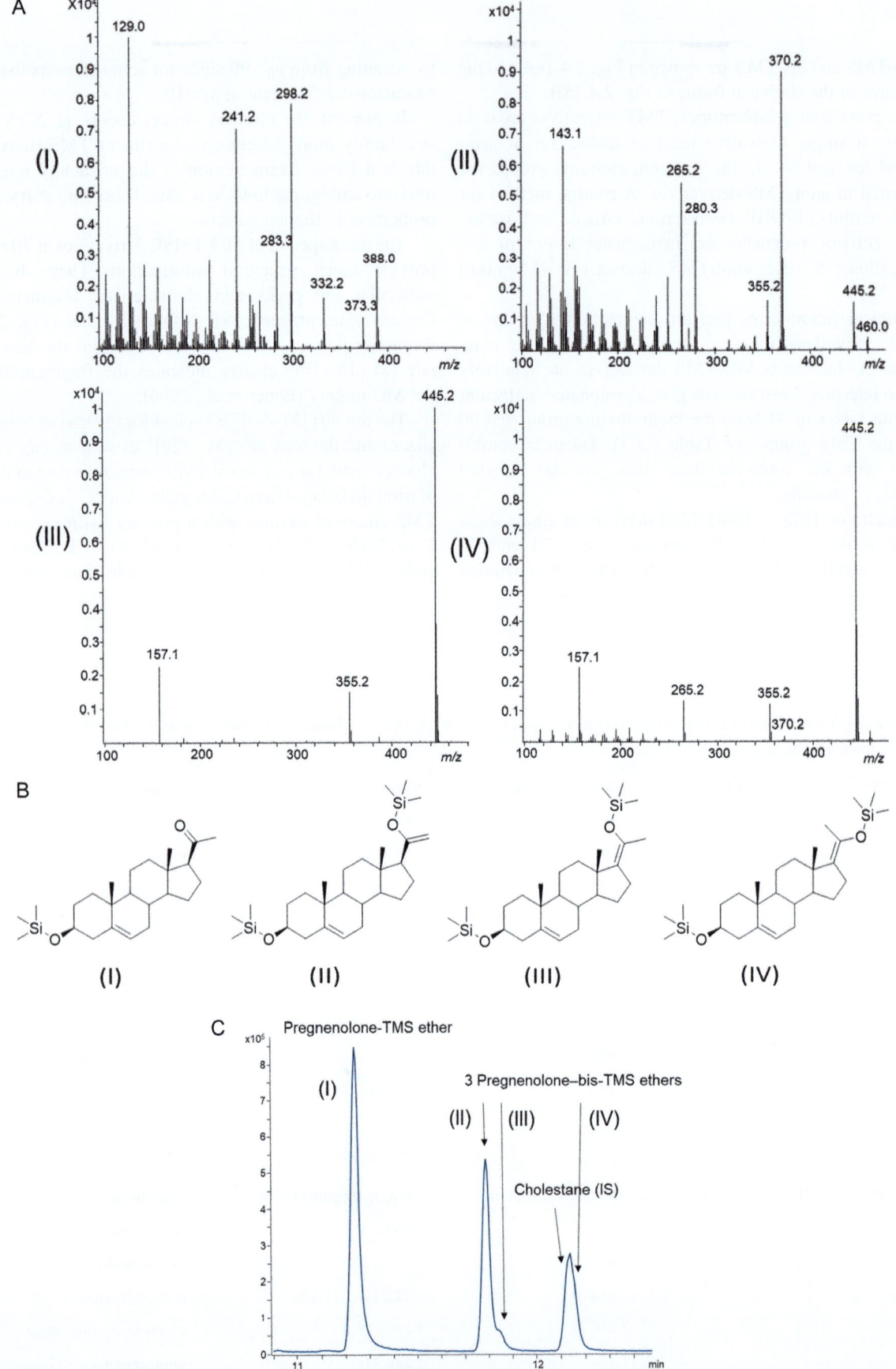

FIG. 2.4.28 (A) Partial total ion current chromatogram of pregnenolone TMS derivatives with cholestane internal standard. (B) Structures of pregnenolone mono and bis-TMS ether derivatives. (C) Low resolution mass spectra of pregnenolone-TMS derivatives after reaction with MSTFA and TSIM (9:1). *(From Junker J, Chong I, Kamp F, Steiner H, Giera M, Müller C, Bracher F. Comparison of strategies for the determination of sterol sulfates via GC-MS leading to a novel deconjugation-derivatization protocol. Molecules 2019;24(13):2353. Fig 3 P5.)*

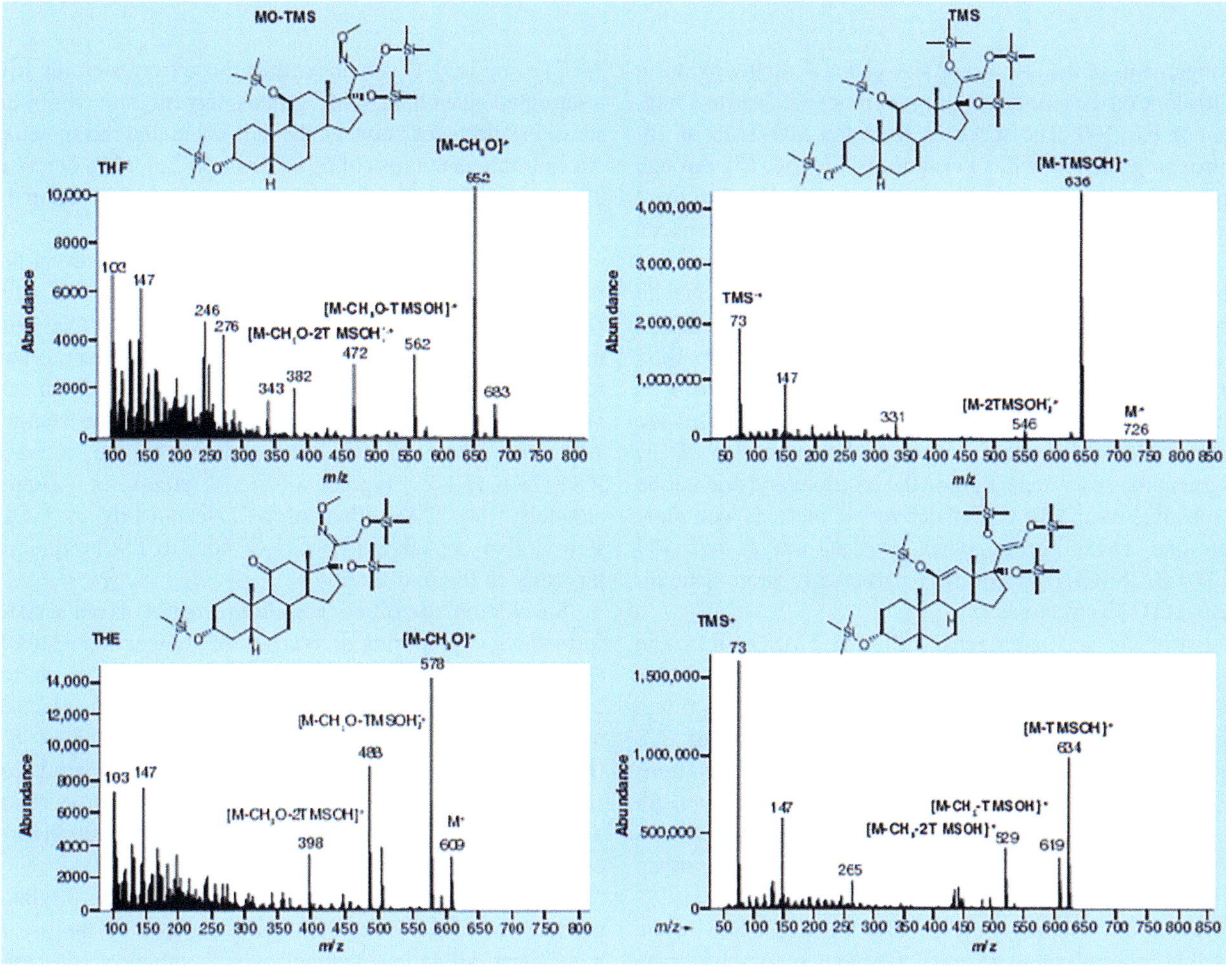

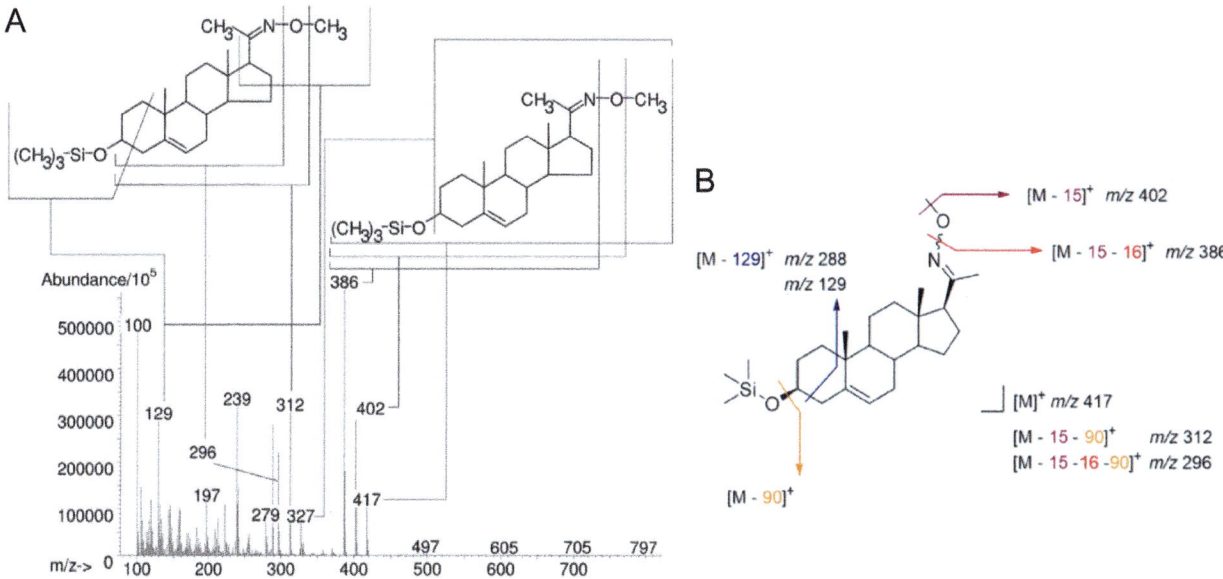

FIG. 2.4.29 Chemical structures and GC-ESI-MS mass spectra of tetrahydrocortisol and tetrahydrocortisone as MO-TMS derivatives. *(From Marcos J, Pozo OJ. Derivatization of steroids in biological samples for GC-MS and LC-MS analyses. Bioanalysis 2015;7(19):2515–36. Fig. 3 p. 2523.)*

FIG. 2.4.30 Mass spectra of pregnenolone as methyloxime trimehylsilyl ether derivatives (molecular weights 417). *Data from (A) Ebner MJ, Corol DI, Havlíková H, Honour JW, Fry JP. Identification of neuroactive steroids and their precursors and metabolites in adult male rat brain. Endocrinology 2006;147(1):179–90. Fig 1 P185. (B) Fragmentation of pregnenolone MO-TMS. From Junker J, Chong I, Kamp F, Steiner H, Giera M, Müller C, Bracher F. Comparison of Strategies for the determination of sterol sulfates via GC-MS leading to a novel deconjugation-derivatization protocol. Molecules 2019;24(13):2353.*

arrangements of the D-ring and side chain. A methyloxime at C-20 alone on the side chain (progesterone) will lead to a fragment at m/z 100. The spectrum from the MO-TMS of 16-hydroxprogesterone will show ions at 117, 156, 157 through d-ring cleavage. 15-hydroxyprogesterone will give a fragment at m/z 258. Loss of the side chain and parts of the D-ring can be seen with ions 103, 175 and 188 in corticosterone metabolites. A prominent ion at M-117 or M-205 is seen with loss of MO-TMS of C-20 and C-21 diol, glyceraldehyde and glycerol side chains, e.g., 5β-pregnane-3α-17α-20α-triol, cortol and cortolone. Loss of the A-ring can also give ions useful for the localization of substituents. These losses are seen in different combinations and the intensities of the fragment ions vary depending on the positions and orientation of substituents and the type of derivative. Steroids with more than one TMSO group often give an ion at m/z 147 $[(CH_3)_3SiOSi(CH_3)_2]^+$ which is particularly prominent for vicinal OTMS structures.

Estrogens and androgens with two TMSO groups in positions 15–18 give an ion at m/z 191 $[(CH_3)_3SiO—CH=O^+Si(CH_3)_3]$ due to migration of a TMSO group. For example, m/z 191 is the base peak in spectra of C18 and C19 steroids with three TMSO groups at positions 15α, 16α, and 17β. A peak at m/z 191 is seen in spectra of many steroids with closely located TMSO groups but the steric requirements for the formation of the ion remain to be determined.

Positional isomers are usually distinguished when suitable reference spectra are available (except in the case of double bonds that may isomerize in the ion source). Stereoisomers often give spectra in which relative intensities of common ions differ reproducibly, see Fig. 2.4.31 for spectra of eight androstanediol isomers (Le Bizec et al., 2005).

The mass spectra are almost identical so retention times are the only way to distinguish them (Fig. 2.4.32).

Information concerning stereochemistry can often be obtained from the retention index in GC-MS analyses as described above to methylene units.

Mass spectra of the TMS ethers of polyhydroxysteroids (Fig. 2.4.33) and steroids with vicinal hydroxyl groups often lose the last trimethylsiloxy group without a hydrogen $[(CH_3)_3SiO, -89]$.

In most cases, an ABCD-ring ion is seen at an m/z value determined by the number of double bonds after loss of all hydroxyl functions and the nature of remaining substituents on the ring system (e.g., at m/z 257, 255, 253, and 251 with one to four double bonds, respectively). Free oxo groups are more clearly seen (giving discrete ABCD-ring ion peaks) and localized than their methyloxime derivatives. Steroids without a side chain (androgens, estrogens) will give ABCD-ring ions 1 Da heavier than those from steroids with a saturated side chain. TMS groups may migrate within the steroid skeleton or between the side chain and the skeleton. An example is the loss of 56 Da from M$^{·+}$ of TMS ethers of 3β-hydroxy-Δ5-steroids also having a carbonyl group in the D-ring or side chain.

TMS ethers of α and β isomers of 20-hydroxy-5α-pregnan-3-one give fragment ions at $[M-44]^+$ and $[M-59]^+$ due to loss of CH_3CHO and $CH_3CHO^+CH_3$ with migration of the TMS group to the charged fragment. A corresponding Δ16 steroid loses the entire side chain and D-ring with migration of the TMS group to the charged ABC ring $[M-156-73]^+$. In addition to the m/z 103 $[CH_2OSi(CH_3)_3]^+$, typical of TMS ethers of primary alcohols; the TMS ether of a 17-oxo-18-hydroxy-C19 steroid gives a peak at $[M-CH_2O]^+$ due to TMS migration probably to the oxo group.

Shackleton identified epi-allopregnenol as an artifact formed when preparing derivatives of pregnenolone sulfate (Fig. 2.4.34) (Shackleton et al., 2004). Steroid degradation by contact of the GC outlet with metal prior to electron impact has also been identified (Honour et al., 1982) (Fig. 2.4.35) and a mechanism for thermal degradation (Fig. 2.4.35C) was proposed for tetrahydrocortisol. Tetrahydro-DOC, tetrahydrocorticosterone and tetrahydrocortisone were also affected.

The formation of mixtures of mono- and bis-silylated products of bis-hydroxy steroids is avoided by the use of a stronger silylating reagent which enhances the enol TMS formation. For this purpose, trimethyliodosilane is used. This reactive reagent is generated in situ from a mixture of MSTFA and ammonium iodide. A reaction with steroids produces enol-TMS derivatives of carbonyls and TMS ethers of hydroxyls with a single reagent (Chambaz et al., 1972; Donike et al., 1983). Upon decomposition of TMSI, iodine is formed that can add to the steroid nucleus. To prevent this side-reaction, the iodine is reduced by the addition of dithioerythritol to form hydrogen iodide. A reducing agent such as mercaptoethanol can also be added in order to avoid undesired formation of iodine. This method requires much effort for optimization depending on the analytes of interest. In addition, with this procedure artifacts can be observed, resulting from incorporation of mercaptoethanol.

The formation of enol TMS ethers of keto sterols is avoided with the two-step derivatization protocol to MO-TMS derivatives. The keto groups are first converted into methyloxime (synonym: oxime methyl ether; MO) derivatives, typically using 2% O-methoxyamine hydrochloride (m/v) in pyridine. The hydroxyl groups are then selectively transformed into TMS ethers This method

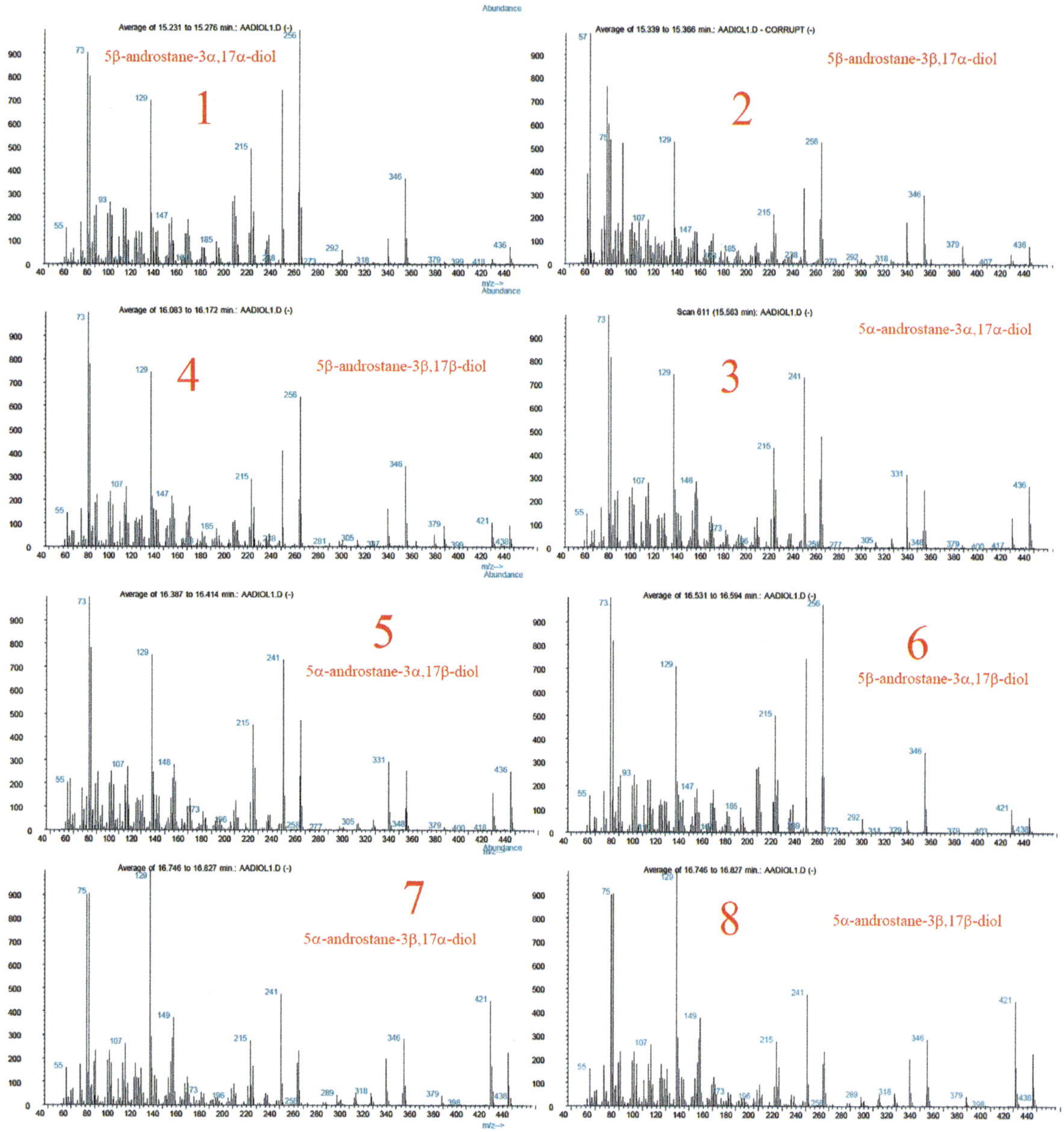

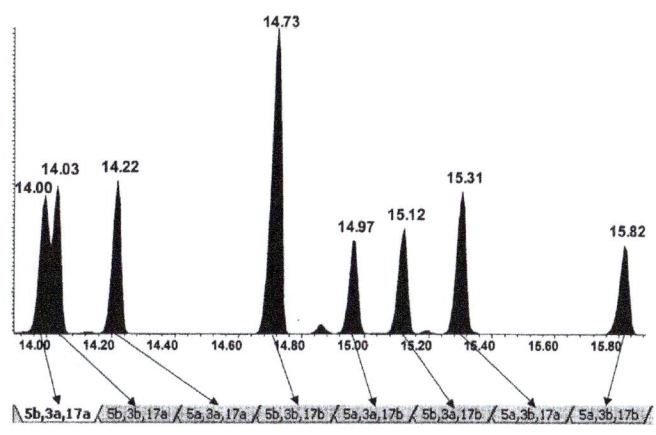

FIG. 2.4.31 Typical GC/EIMS spectra of androstanediol isomers. Molecular weights are 436 and major ions are variably at 421 (M-15); 346 (M-90); 256 (M-90-90); 241 (M-15-90-90), 129—A or D-ring fragmentation. *(Data from Le Bizec B, Antignac JP, Bertrand D, Qannari el M, Andre F. Multidimensional statistical analysis applied to electron ionization mass spectra to determine steroid stereochemistry. Rapid Commun Mass Spectrom 2005;19 (4):509–18. Fig 2 P510.)*

GC-MS androstanediols as TMSE derivatives

FIG. 2.4.32 Partial total ion current chromatogram of eight androstanediol isomers by GC/MS of TMS derivatives. *(From Le Bizec B, Antignac JP, Bertrand D, Qannari El M, Andre F. Multidimensional statistical analysis applied to electron ionization mass spectra to determine steroid stereochemistry. Rapid Commun Mass Spectrom 2005;19(4):509–18. Fig. 1 p. 510.)*

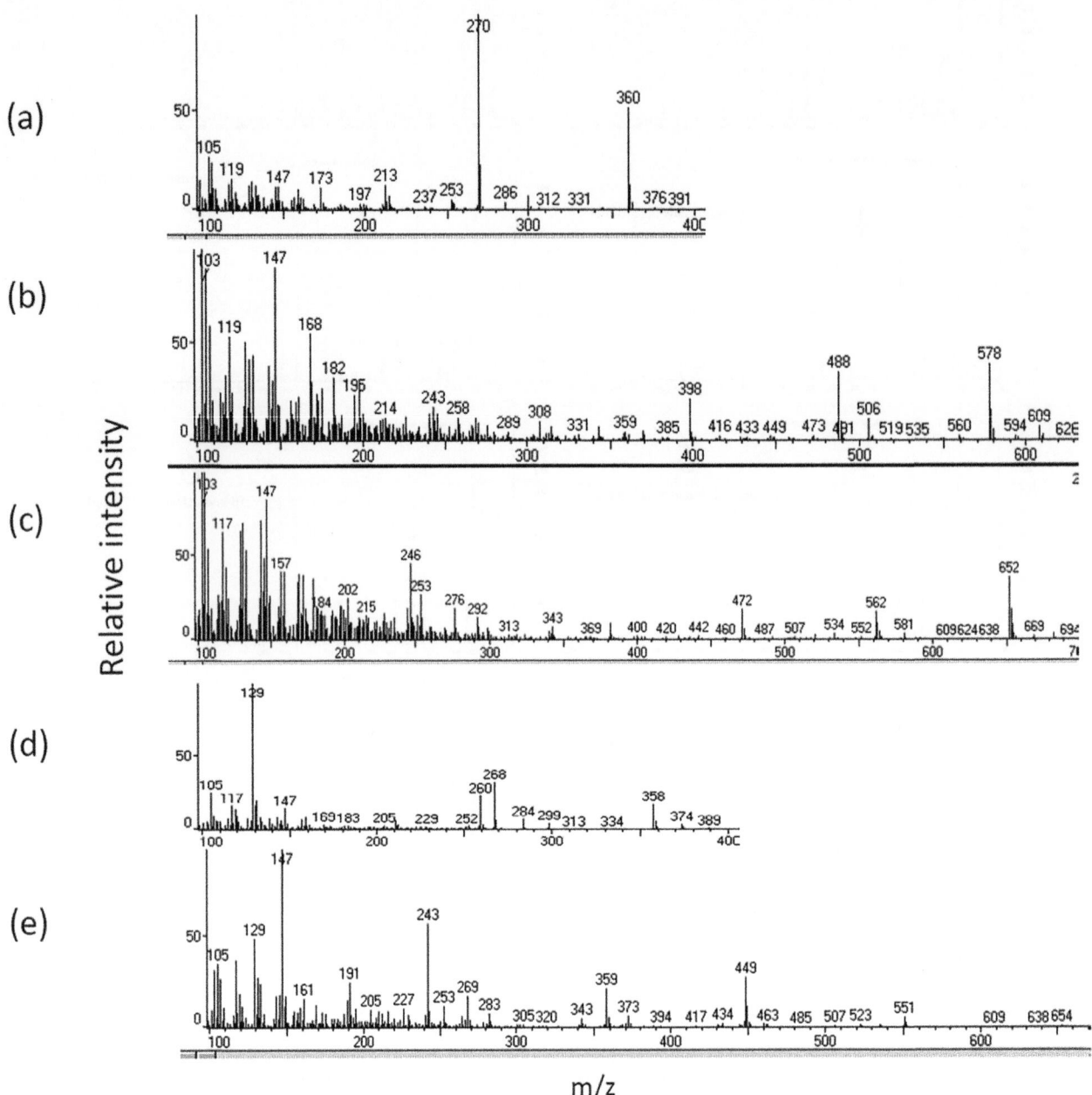

FIG. 2.4.33 Partial mass spectra of urinary steroid metabolites as MO-TMS derivatives (A) is androsterone, (B) is THE, (C) THF, (D) cortol and (E) cortolone. *(From Honour JW, Conway E, Hodkinson R, Lam F. The evolution of methods for urinary steroid metabolomics in clinical investigations particularly in childhood. J Steroid Biochem Mol Biol 2018;181:28–51. Fig. 2 p. 36.)*

produces two isomeric MO derivatives (syn, anti), which are partially or fully separated by GC giving two peaks with the same fragmentation patterns. For most of the steroids in urine, the GC-MS chromatogram shows only one peak for each sterol derivative and no additional peaks or peak shoulders due to syn-/antiisomers of the MO residues are observed except for, but not limited to tetrahydrocorticosterone, 5α-dihydroprogesterone and 16-hydroxy-DHEA. The keto sterol

sulfates are effectively converted into their respective MO-TMS derivatives after solvolysis. Sulfate groups from DHEAS, pregnenolone sulfate and other sulfates are removed during reaction with methoxyamine and the resulting hydroxyl reacts with trimethylsilylimidazole to give the TMS derivative (Fig. 2.4.36).

At least 8h is required to complete the reaction (Shackleton et al., 1983) (Fig. 2.4.37). When making

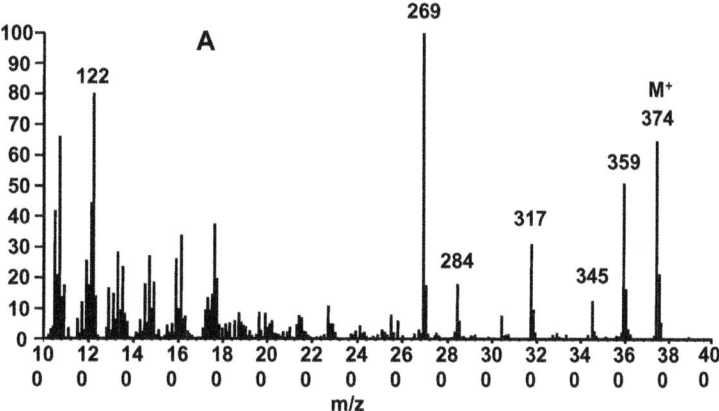

FIG. 2.4.34 (A) The formation of epi-allopregnenol artifact from epi-allo-pregnanediol-20-sulfate during formation of TMSE derivative and the mass spectrum of epiallopregnenol-TMS. (B) In the analysis of urine from patients with POR deficiency, (17E)-5α-pregn-17(20)-en-3β-ol (epiallopregnenol) was identified by mass spectrometric analysis of the TMS derivative. That steroid is formed from epiallopregnanediol-20-sulfate (5α-pregnane-3β, 20α-diol-20-sulfate) by desulfation and dehydration during the derivatization process. Epiallopregnanediol-20-sulfate arises from the enzymatic hydrolysis (*Helix pomatia* beta-glucuronidase and sulfatase) of urinary epiallopregnanediol-3,20-disulfate. Sulfate esters of the 20-hydroxyl of steroids are resistant to this enzymatic hydrolysis so are still present at the time of TMS derivatization. The excreted epiallopregnanediol has two origins, the disulfate conjugate probably being of fetal origin while a glucuronidated or 3-monosulfate form is likely a both a fetal metabolite and a placental progesterone metabolite. *(From Shackleton C, Marcos J, Arlt W, Hauffa BP. Prenatal diagnosis of P450 oxidoreductase deficiency (ORD): a disorder causing low pregnancy estriol, maternal and fetal virilization, and the Antley-Bixler syndrome phenotype. Am J Med Genet A 2004;129A(2):105–12. Fig. 1 p. 106.)*

MO-TMS derivatives of a total urine steroid extract the sulfated steroids elute from GC separation with retention times of free steroids as MO-TMS derivatives, whereas the derivatized glucuronide conjugates elute later.

If a methyl ester of glucuronides is made before the MO-TMS combination the derivatives can be run on a thermostable GC column that can now be run to 350°C. The methyl ester MO-TMS derivatives of steroids have retention times up to methylene units of 44 carbons (Fig. 2.4.38).

Methyl ester TMS of rat plasma steroid glucuronides have been analyzed on a stainless steel capillary column (Choi et al., 2000) (Fig. 2.4.39). The spectra are shown in Fig. 2.4.40 and mechanisms for fragmentation in Fig. 2.4.41. The C-11 oxygenated androgens have become of considerable interest not just as side chain cleavage products of cortisol but as metabolites of androgens.

Reaction times and temperatures for derivatives vary from hours up to a day which is not desirable in clinical laboratories. There are interesting developments in the use of microwaves to accelerate the reactions shortening times from hours to minutes (Lee et al., 2021; Zhao et al., 2016; Casals et al., 2014).

Where isotope labeled internal standards are not available then an isotope labeled derivatizing reagent can be used to introduce an isotope tag to the investigated metabolite group on the internal standard. This is sometimes called **isotope coded analysis** (Higashi and Ogawa, 2016a,b) or **differential isotope labeling** (Higashi et al., 2016). Isotopically labeled analogs with deuterium or C[13] incorporation into the reagents are available.

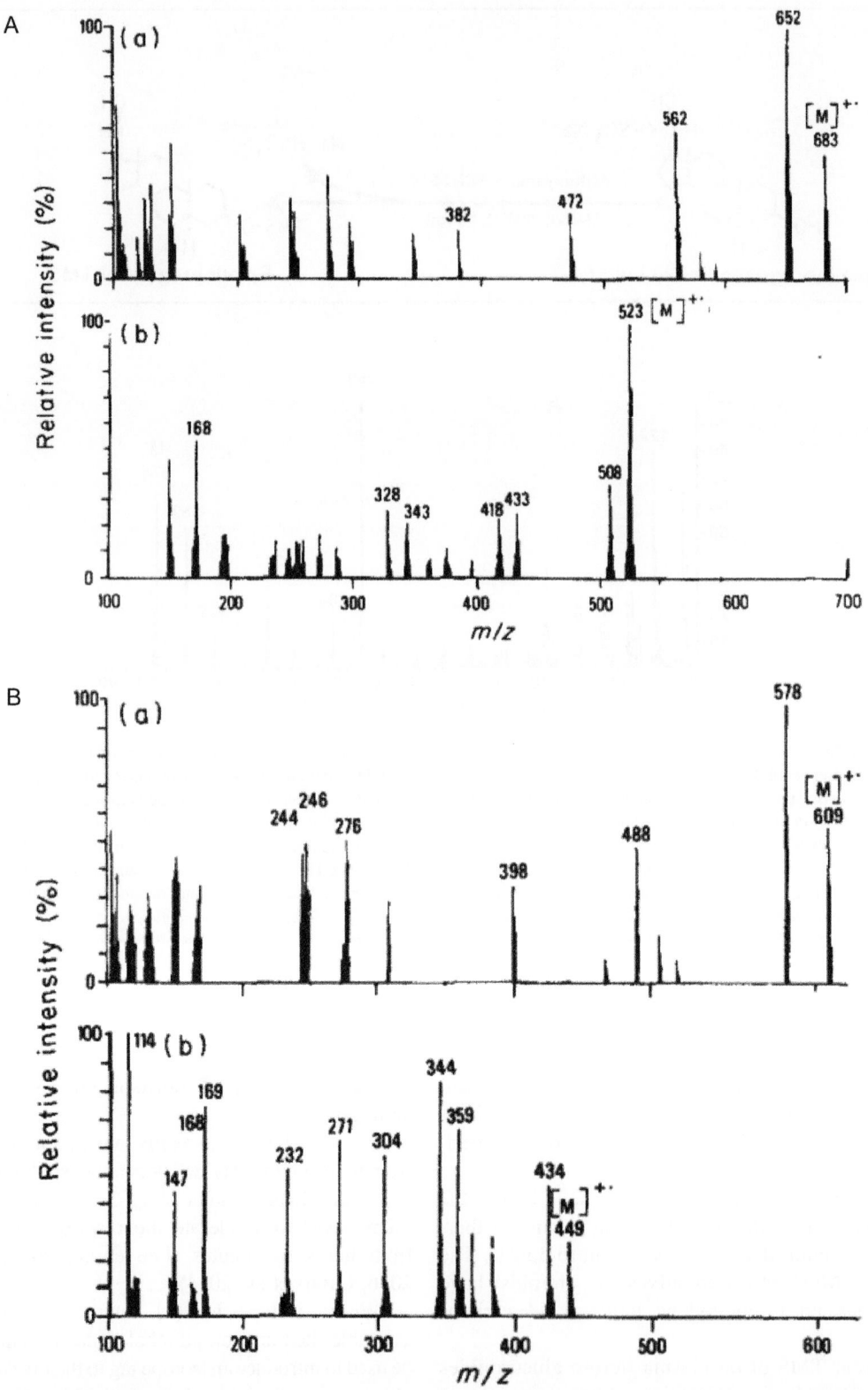

FIG. 2.4.35 Degradation of steroids at the GC-MS interface. (A) Mass spectra of tetrahydrocortisol (MO-TMS) on electron impact of (a) the intact derivative (b) the degradation product formed in the GCMS interface. (B) Proposed mechanism of thermal degradation tetrahydrocortisone-MOTMS electron impact of (a) the intact derivative (b) the degradation product formed in the GCMS interface. *(Data from Honour JW, Brooks CJ, Shackleton CH. Degradation of steroid derivatives at a gas chromatograph mass spectrometer interface. Biomed Mass Spectrom 1982;9:505–9. Fig. 3 p. 507; Fig. 5 p. 508; Fig. 6 p. 508; Scheme 1 p. 507 (Author original).)*

(Continued)

C

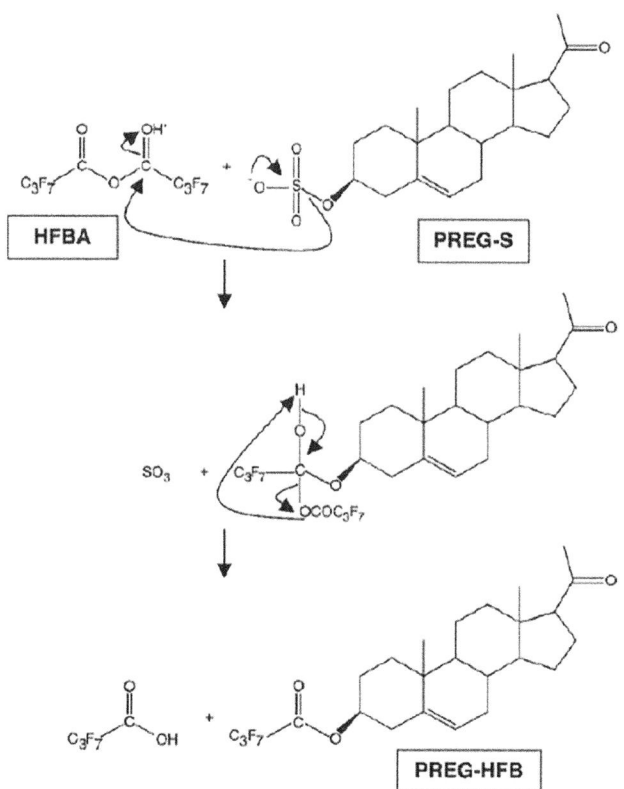

FIG. 2.4.35, CONT'D (C) Proposed mechanism of thermal degradation tetrahydrocortisol-MOTMS. *(From Honour JW, Brooks CJ, Shackleton CH. Degradation of steroid derivatives at a gas chromatograph mass spectrometer interface. Biomed Mass Spectrom 1982;9:505–9. Fig 3 P 507; Fig 5 P 508; Fig 6 P508; Scheme 1 P 507 (Author original) J Steroid Biochem Mol Biol 181:28.)*

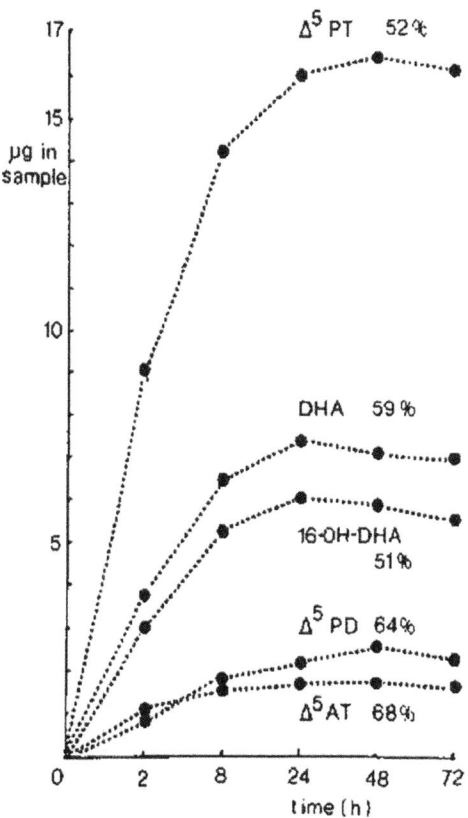

FIG. 2.4.36 Reaction mechanism for the simultaneous hydrolysis/derivative formation of pregnenolone-sulfate during reaction with the acylation reagent heptafluorobutyric acid. *(From Liere P, Pianos A, Eychenne B, Cambourg A, Liu S, Griffiths W, Schumacher M, Sjövall J, Baulieu EE. Novel lipoidal derivatives of pregnenolone and dehydroepiandrosterone and absence of their sulfated counterparts in rodent brain. J Lipid Res 2004;45(12):2287–302. Fig. 2 p. 2290.)*

FIG. 2.4.37 Time course of sulfate-silyl group exchange during direct trimethylsilylation of steroid sulfate. Δ^5PT, pregnenetriol; Δ^5PD, pregnenediol; Δ^5AT, androstenetriol. *(From Shackleton CH, Straub KM. Direct analysis of steroid conjugates: the use of secondary ion mass spectrometry. Steroids 1982;40(1):35–51. Fig. 9 p. 216.)*

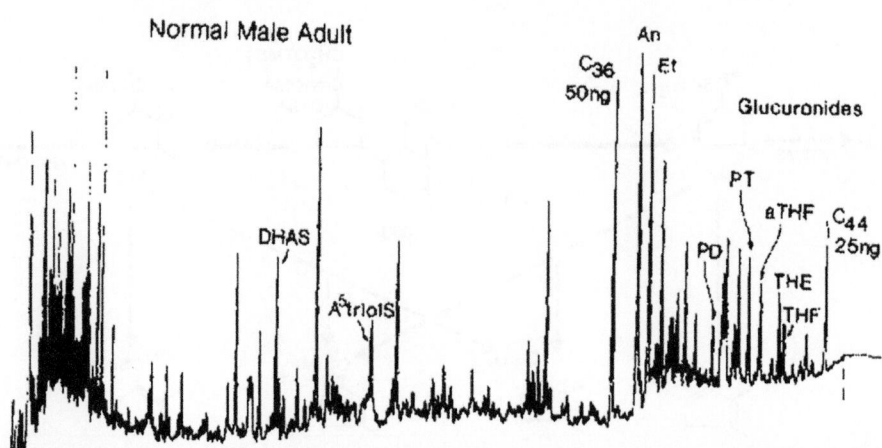

FIG. 2.4.38 Gas chromatographic profile (total ion current) of steroids from adult urine after formation of methyl ester, methyloxime and trimethylsilyl ether derivatives. Steroid sulfates elute as MO-TMS of free steroids, the glucuronides have longer retention times than the derivatives of free steroids. A GC column stable to 380°C is needed for this separation. *A5triolS*, androstenetriol; *An*, androsterone; *DHAS*, dehydroepiandrosterone from DHAS sulfate; *Et*, etiocholanolone; *PD*, pregnanediol; *PT*, pregnanetriol. *(From Shackleton CH, Mattox VR, Honour JW. Analysis of intact steroid conjugates by secondary ion mass spectrometry (including FABMS) and by gas chromatography. J Steroid Biochem 1983;19(1A):209–17. Fig. 6 p. 215.)*

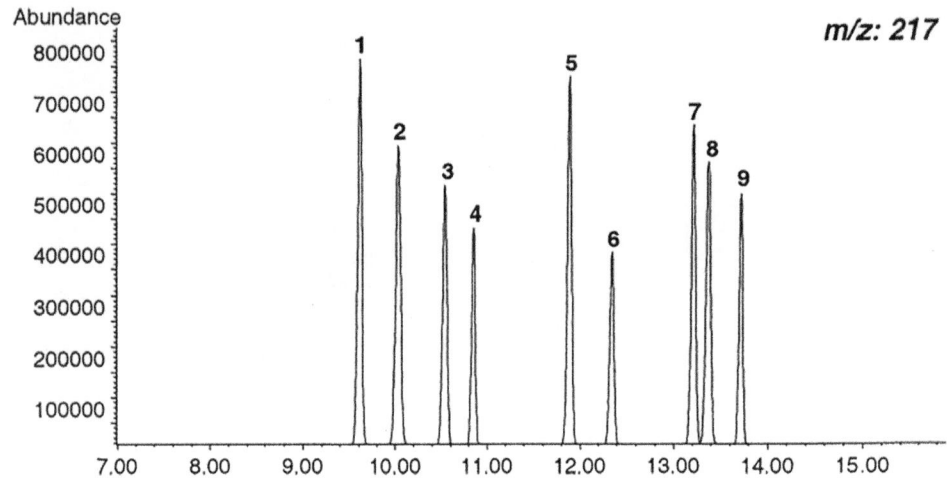

FIG. 2.4.39 Extracted ion chromatogram of 9 steroid glucuronides as Me-TMS derivatives on 30 m MXT-1 GC column temperature from 300 to 332 degrees at 2 degrees per minute. (1) Androsterone-3-glucuronide; (2) etiocholanolone-3-glucuronide; (3) 11-ketoandrosterone-3-glucuronide; (4) 11-ketoetiocholanolone-3-glucuronide; (5) 11β-hydroxyandrosterone-3-glucuronide; (6) 11β-hydroxyetiocholanolone-3-glucuronide; (7) DHT-17-glucuronide; (8) DHEA-3-glucuronide; (9) testosterone-17-glucuronide. *(From Choi MH, Kim KR, Chung BC. Simultaneous determination of urinary androgen glucuronides by high temperature gas chromatography-mass spectrometry with selected ion monitoring. Steroids 2000;65(1):54–9. Fig. 1 p. 56.)*

2.4.6.4.2 *Heptafluorobutyrate derivatives (HFB)*

In order to measure steroids in blood samples after GC separation, flame ionization detection was not sufficiently sensitive and electron capture derivatives were developed. Stefan Wudy developed several methods for determining steroids by GC-MS which culminated in a method to determine androstenedione, testosterone, DHEAS, pregnenolone, 17-hydroxyprogesterone, 11-deoxycortisol and 17-hydroxypregnenolone (Wudy et al., 1992).

Testosterone forms the 3,5-dien-3,17β-di-HFB ester. The mass spectrum of this compound reveals a molecular ion at *m/z* 680.2 of high relative intensity (Fig. 2.4.42A). The spectrum of 17-hydroxyprogesterone forms a 3,5-dien-3-HFB ester. Its spectrum gives a base peak at 465.1 Da that corresponds to the steroid nucleus after loss of the side chain and the 17α-hydroxyl group (Fig. 2.4.42B).

The spectrum of 4-androstenedione gives a thermally stable 3,5-dien-3-mono-HFB ester. Its spectrum shows a very intense molecular ion peak at *m/z* 482.1 (Fig. 2.4.43). The 3β-hydroxy-5-ene steroids DHEA and 17-hydroxypregnenolone (not shown) form the 3-mono-HFB derivatives. DHEA has a base peak at *m/z* 270 which is M-214 from the molecular ion at *m/z* 470.

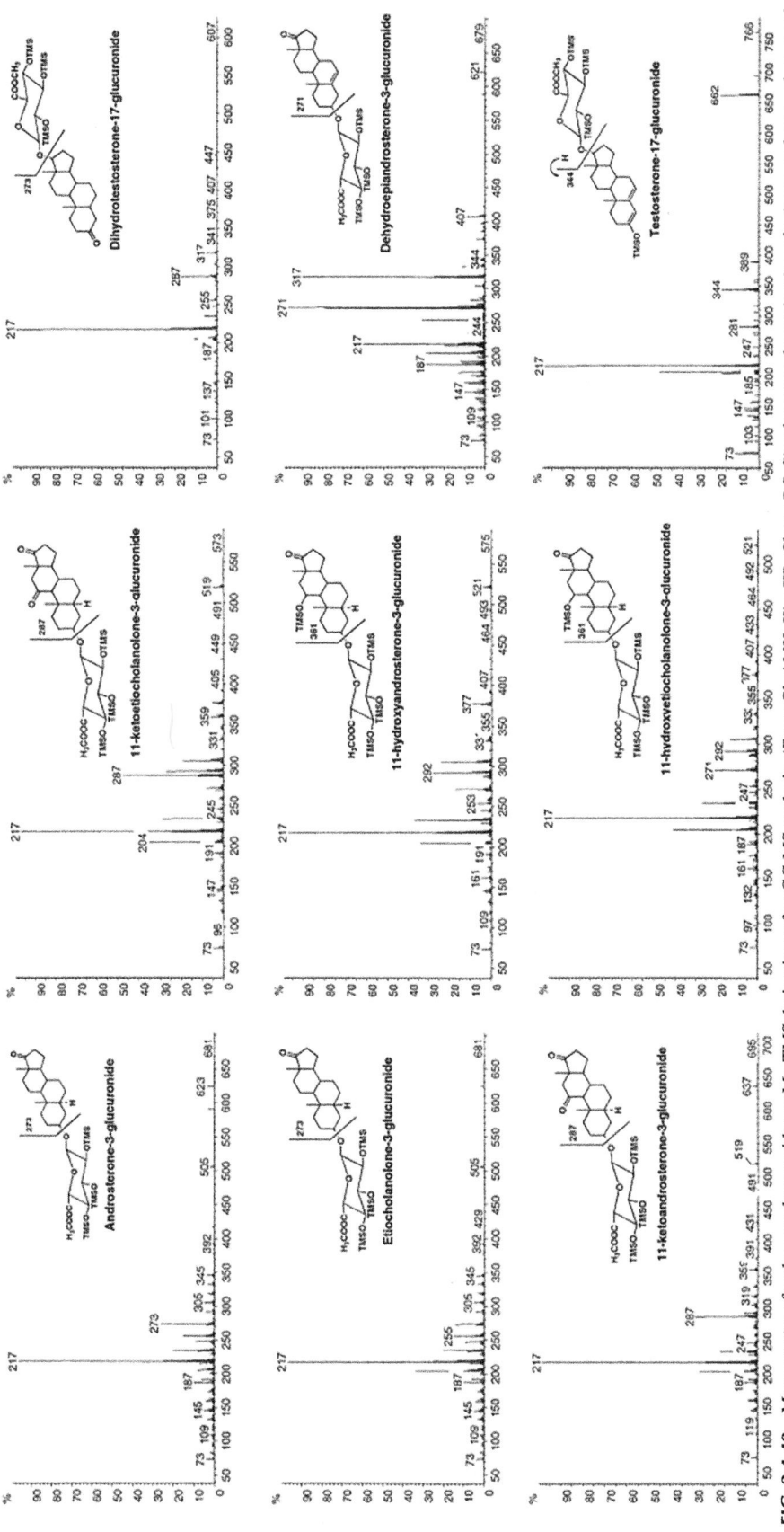

FIG. 2.4.40 Mass spectra of androgen glucuronides as Me-TMS derivatives after GC-MS analysis. (*From Choi MH, Kim KR, Chung BC. Simultaneous determination of urinary androgen glucuronides by high temperature gas chromatography-mass spectrometry with selected ion monitoring. Steroids 2000;65(1).54–9. Fig. 2 p. 56.*)

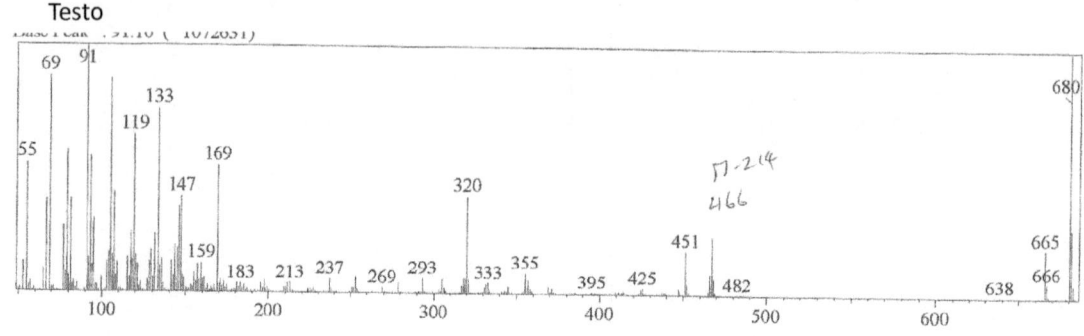

FIG. 2.4.41 Fragmentation in GC-MS analysis of androgen glucuronides as Me-TMS derivatives. *(From Choi MH, Kim KR, Chung BC. Simultaneous determination of urinary androgen glucuronides by high temperature gas chromatography-mass spectrometry with selected ion monitoring. Steroids 2000;65(1):54–9. Fig. 3 p. 57.)*

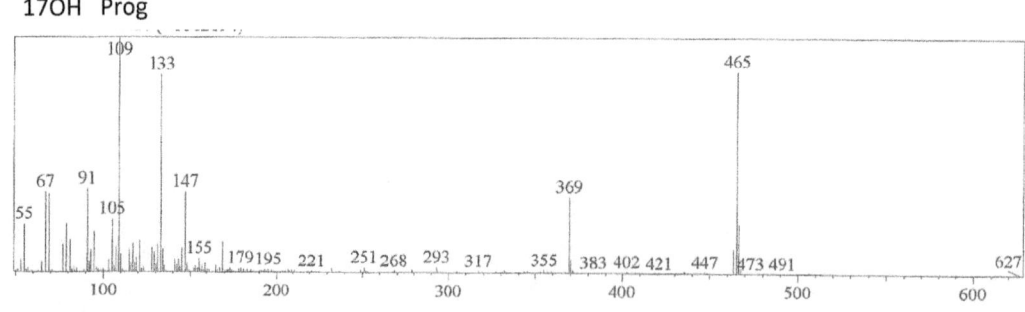

FIG. 2.4.42 Mass spectra of heptafluorobutyrates of testosterone and 17-hydroxyprogesterone (Hodkinson R and Honour JW, unpublished).

5α-Dihydrotestosterone (DHT) gives the 17β-mono-HFB derivative and the spectrum has a fragment suitable for SIM at ion m/z 414 [M−72] (Fig. 2.4.44).

The spectrum for 21-deoxycortisol shows different fragmentation to 11-deoxycortisol with major ions at 463 (M-214) and 465 (M-255) (Fig. 2.4.45).

Quantitative GC-MS of 11-deoxycortisol is based on monitoring the ion at m/z 465 which (like 17-hydroxyprogesterone) is the dominant ion of the spectrum after loss of the side chain. The 17-hydroxyl group is cleaved during derivatization and leads to loss of water. The methyl group at C-13 migrates to C-17 by a Wagner-Meerwein rearrangement. 11-deoxycortisol gives a base peak at m/z-255. The spectra for Cortisol and corticosterone are shown in Fig. 2.4.46.

The molecular ion for cortisol at m/z 718 is not visible. The derivative is a C3-diene enol HFB. There is an additional 9(11) double bond.

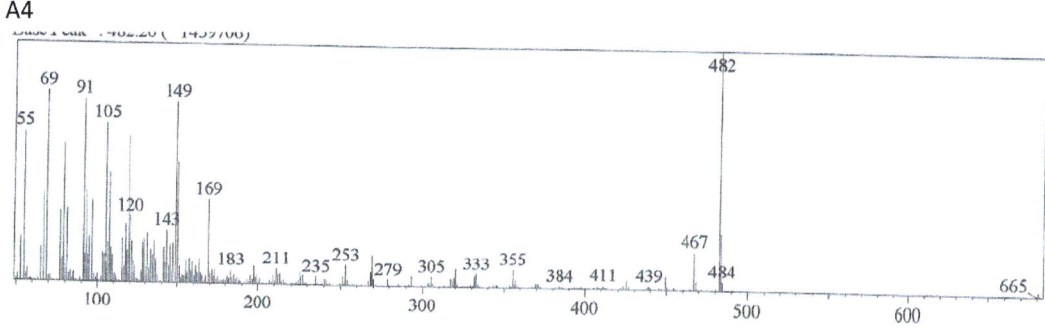

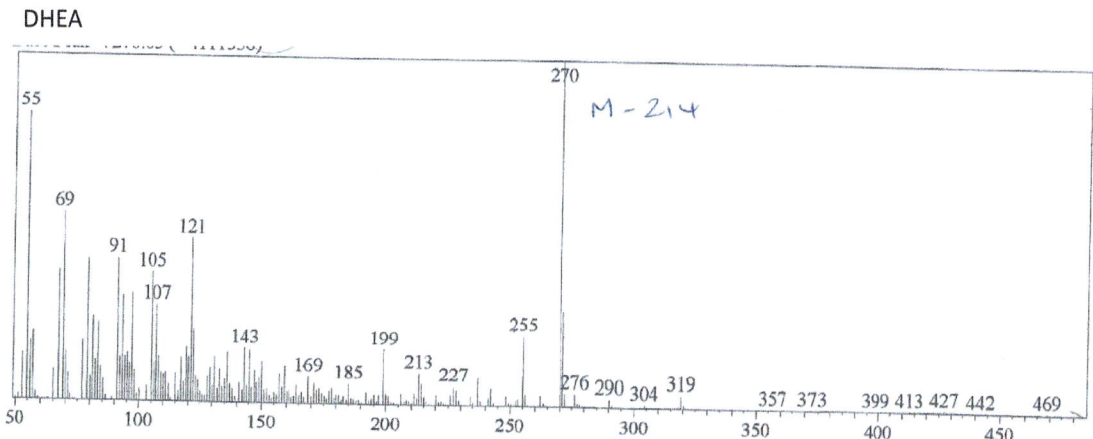

FIG. 2.4.43 Mass spectra of heptafluorobutyrates of androstenedione (A4) and DHEA (Hodkinson R and Honour JW, unpublished).

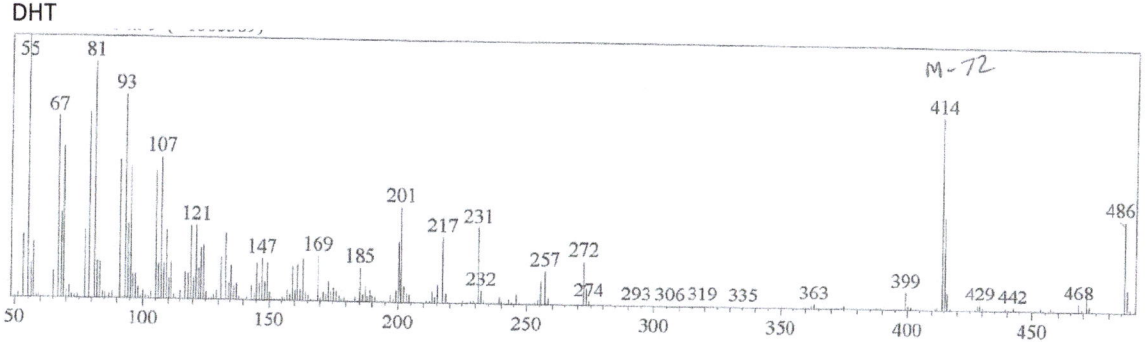

FIG. 2.4.44 Mass spectrum of heptafluorobutyrate of dihydrotestosterone (Hodkinson R and Honour JW, unpublished).

Selected ion monitoring panels too has been used to quantify up to eight steroids in a plasma extract in a single run (Wudy et al., 1992) (Fig. 2.4.47).

The method has been applied by Wudy to amniotic fluid analysis (Wang et al., 2019) and by others to the analysis of rat plasma (Chen et al., 2009) where much smaller sample sizes are available (Fig. 2.4.48).

The mechanisms for fragmentation of some of the HFB derivatives (Fig. 2.4.49) are less predictable than MO-TMS derivatives. DHEA-HFB loses a molecule of HFB to give a strong ion at 270 amu. Some molecular ions are seen in the 700 mass range. Fragments are often due to cleavage of the D-ring especially when a side chain is present. In some cases, the B and C rings are split.

2.4.6.4.3 GC-combustion and GC-isotope ratio MS

When metabolites of radioactive steroids are being analyzed, the GC effluent can be passed through a tube filled

21-DF

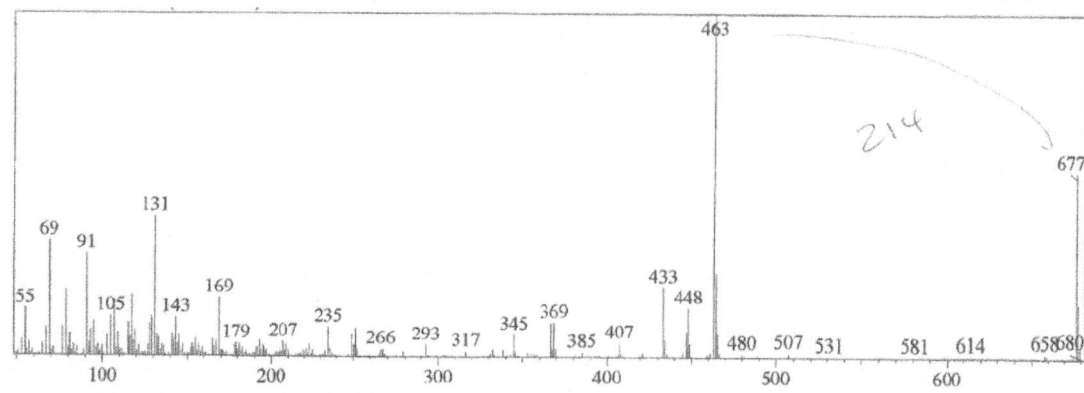

11-DF

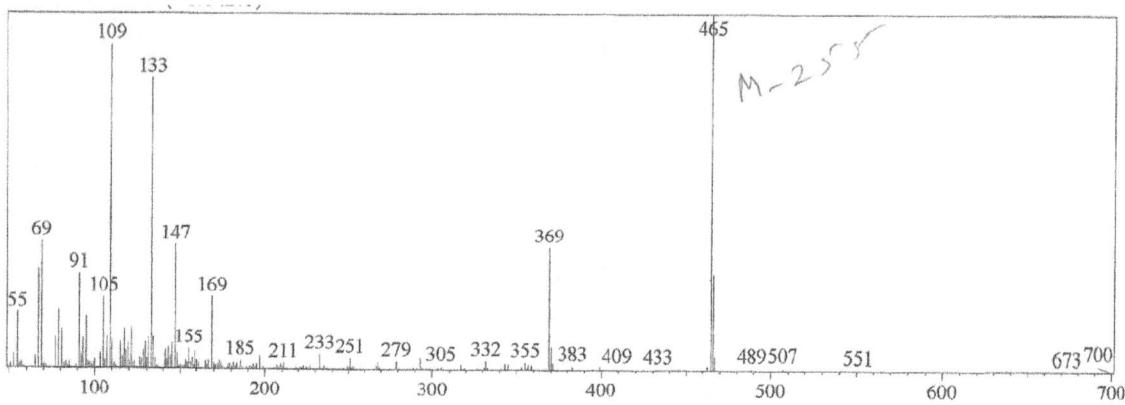

FIG. 2.4.45 Mass spectra of heptafluorobutyrates of 21-deoxycortisol and 11-deoxycortisol (Hodkinson R and Honour JW, unpublished).

with copper oxide in a furnace. Hydrogen is added to the outflow of that furnace and the combined gases passed to a tube of iron filings in another furnace (Fig. 2.4.50). The radioactive gases then pass to a proportional counter. All organic compounds are burnt to CO_2 and water in the first furnace tube, the water is reduced to hydrogen by iron filings. Depending on the capability of the counter, both ^{14}C and ^{3}H could be detected.

Many versions of this system used to be available including ability to collect GC effluent into scintillation fluid in tubes on a fraction collector. Radiogas chromatography was used to confirm metabolic steps in vivo in animals and humans. For instance, 21-deoxymetabolites from corticosterone were confirmed in a patient with congenital adrenal hyperplasia (Shackleton et al., 1979) This conversion had already been demonstrated in rats and negated in germ free rats (Eriksson, 1970, 1971). These were early studies on what today is called the microbiome. Radiogas chromatography of cortisol metabolites was used in a faster method for measurements of cortisol production rate. The use of radioactive steroids in this manner is now rarely used.

In order to uncover the illicit administration of endogenous anabolic steroids, such as testosterone itself, doping control samples undergo an initial testing procedure that provides the so-called steroid profile. Suspicious test results originating, e.g., from elevated urinary steroid concentrations, atypical steroid concentration ratios, or significant variations in the athletes' longitudinal blood profiles (Athlete Biological Passport, steroidal module) can initiate requests for confirmatory analyses.

One such confirmatory analyses is based on carbon isotope ratio (CIR) mass spectrometry. The ^{13}C values of exogenous testosterone are commonly depleted compared with endogenously produced testosterone because steroids are commercially produced from plant sterols that have a different ^{13}C content. The carbon isotope signature of the latter depends on the individual's diet. In order to compensate for this biological variability, $\Delta\delta13C$ values were established, for which thresholds were introduced by WADA as a doping control measure. In addition, the $\Delta\delta13C$ values are compared with reference population-based thresholds. $\Delta\delta13C$ are defined as the difference between the $\delta13C$ values of an endogenous reference compound (ERC) and a target compound (TC). CIRs are expressed as $\delta13C$ values traced on the Vienna Pee Dee Belemnite (VPDB) standard. Besides testosterone, its

Cortisol

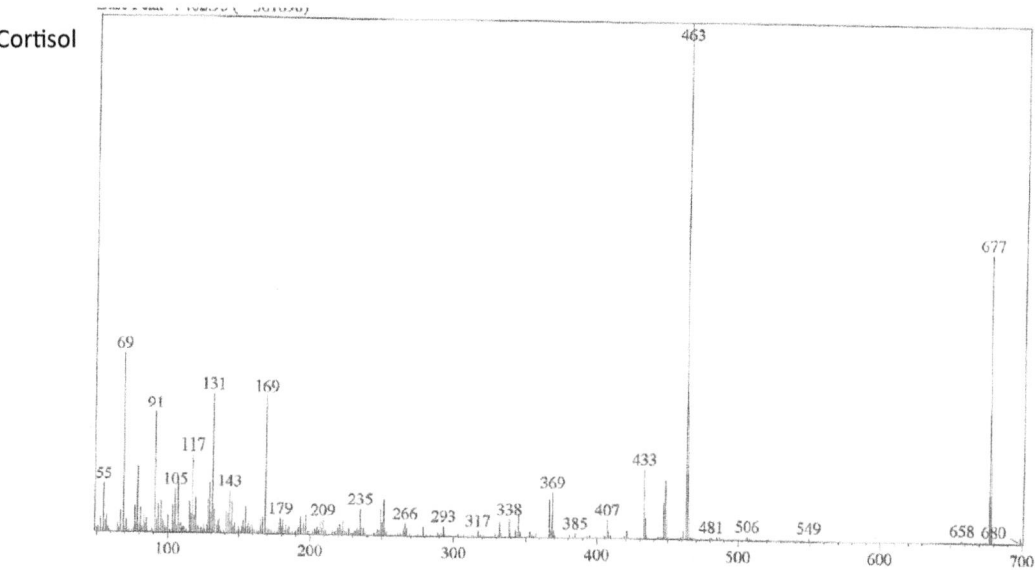

Corticosterone

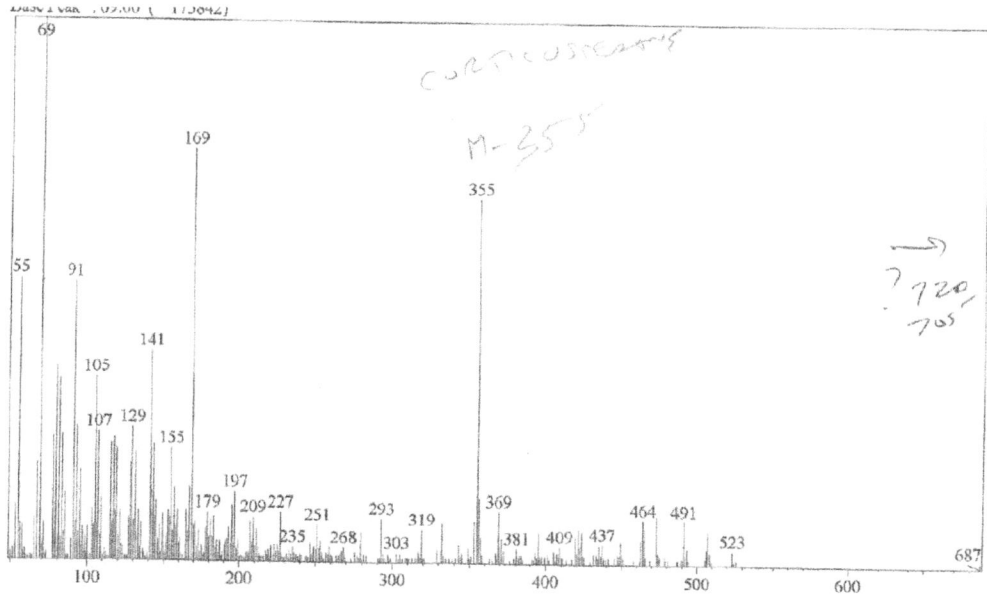

FIG. 2.4.46 Mass spectra of heptafluorobutyrates of cortisol and corticosterone (Hodkinson R and Honour JW, unpublished).

metabolites 5α-androstane-3α,17β-diol (5α), 5β-androstane-3α, 17β-diol (5β), etiocholanolone (AETIO), androsterone (A), and epiandrosterone (EPIA) are referred to as target compounds. Pregnanediol (PD), 5α-androst-16-en-3α-ol (16EN), 11-ketoetiocholanolone (11K) and dehydroepiandrosterone (DHEA) are used as endogenous reference compounds as they are not affected by the testosterone metabolism. While most ERCs and TCs are predominantly excreted as glucuronides, DHEA and EPIA are mainly sulfoconjugated. The appropriate sample preparation for IRMS measurements in complex matrices requires sophisticated approaches. For adequate accuracy and precision, isotopic fractionation has to be avoided,

and high peak purity is required as the analyte combustion results in the loss of any structural information.

Current methods for combustion isotope ratio mass spectrometry (CIRMS) are based on up to two time-consuming HPLC purification steps (Brailsford et al., 2012). Heart-cutting multidimensional gas chromatography (MDGC) consists of two GC columns with different chromatographic selectivity connected by a transfer device, which allows cutting a selected segment on the second dimension in order to improve separation capability (Casilli et al., 2016). With the intention to reduce the manual workload and to accelerate the sample preparation, a method by means of heart-cutting (MDGC-C-IRMS) for the analysis

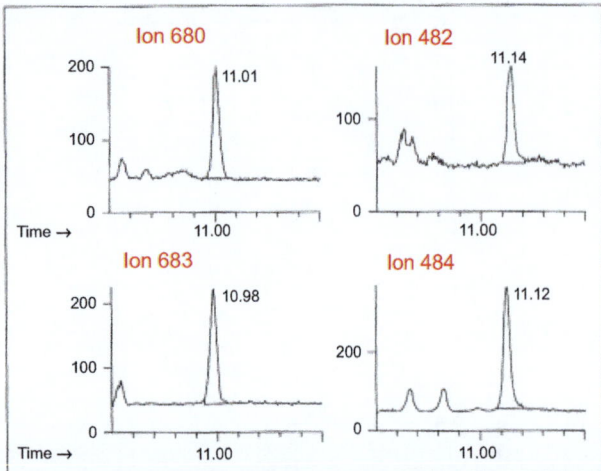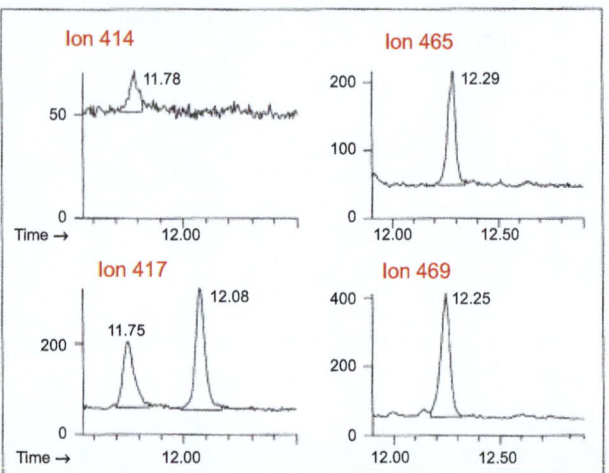

FIG. 2.4.47 Selected ion monitoring of a 1/12 aliquot of a processed extract of 2 mL of female plasma to which 1ng of the corresponding deuterated internal standards had been added. Heptafluorobutyrate derivatives were prepared and analysed by GC-MS with Agilent 5973N mass selective detector. Ion traces of steroid and internal standard are superimposed (testosterone, m/z 680; [^{2}H$_3$]testosterone, m/z 683; 4-androstenedione, m/z 482; [^{2}H$_2$]androst-4-ene-3,17-dione; m/z 484; 5α-dihydrotestosterone; m/z 414; 5α-[^{2}H$_3$]-dihydrotestosterone, m/z 417; 17α-hydroxyprogesterone, m/z 465; 17α-hydroxy-[^{2}H$_4$]-progesterone; m/z 469). The steroids 5α-androstan-3α,17β-diol, dehydroepiandrosterone, 17-hydroxypregnenolone and 11-deoxycortisol were also determined in the panel from responses of ions 470, 270, 467, and 465 and their equivalent ions from their stable labelled steroids. *(Redrawn from Wudy SA, Wachter UA, Homoki J, Teller WM, Shackleton CH. Androgen metabolism assessment by routine gas chromatography/mass spectrometry profiling of plasma steroids: part 1, unconjugated steroids. Steroids 1992;57(7):319–24. Fig 2, p 321.)*

of urinary steroids was developed. Steroids are separated by GC which is heart split into a combustion chamber filled with copper oxide wires at 850°C (Fig. 2.4.51).

The combustion gases are passed through a membrane water removal trap. The remaining CO_2 is analyzed with an isotope mass spectrometer with electron impact source. The ionized CO_2 gas is focused by a magnet onto three Faraday collectors. The ions collected were at masses 44, 45 and 46. The ^{13}Cδ values were calculated for successive peaks from the GC. GC is interfaced with the MS via a combustion chamber that converts steroids to CO_2. Heart-cutting MDGC is approved for complex samples, but publications dealing with heart-cutting MDGC hyphenated to IRMS are scarce. Similar methods for application addressing environmental, authentication of food and biosynthetic studies have been published.

Besides MDGC, comprehensive 2D GC was recently reported as a useful tool in IRMS. In 2D GC, the sample is sequentially trapped and transferred to the second dimension, which allows obtaining a full profile of the sample. The method has been adopted and refined regarding sample preparation and chromatography. The method is intended to be used as an alternative rapid initial testing procedure complementing an HPLC-based confirmation method. It is less time-consuming and necessitates reduced manual workload compared with HPLC-based methods. The method should be performed in specialist laboratories with regular experience of this test. The method could be used in clinical work where prescription steroids have been

taken or when nutritional supplements and herbal medicines are taken that have been spiked with natural steroids. The hospital laboratory should contact the nearest WADA accredited antidoping laboratory.

Uniform steroid isotopic standards (SIS) have been created for use in calibration of GC-C-IRMS and harmonization of results from the carbon isotopic ratio test among the anti-doping laboratories for the detection of synthetic steroid use in sport (Zhang et al., 2012). The SIS were prepared at high volume per ampoule (20mg androsterone, androsterone acetate and 5α-cholestane) in order to allow their use as daily working standards. If one assumes 10 analyses a day of 100ng each standard, and accounting for waste of half sample due to the nature of GC auto-sampler vials, each ampoule of standard would last about 1000 days. The protocol used to prepare the steroid isotopic standards in ampoule containers maintains isotopic integrity to a uniformity within 0.07‰ from ampoule to ampoule. The standards reported on here are available for distribution to other interested laboratories, through the National Measurement Institute of Australia.

2.4.6.4.4 GC tandem mass spectrometry

The term MS/MS is used when two elements of mass spectrometry are applied in tandem MS instrument. GC-MS/MS has not been extensively used although Norman Taylor's group used GC-MS/MS to review the hundreds of steroids in urine of newborn infants with congenital adrenal

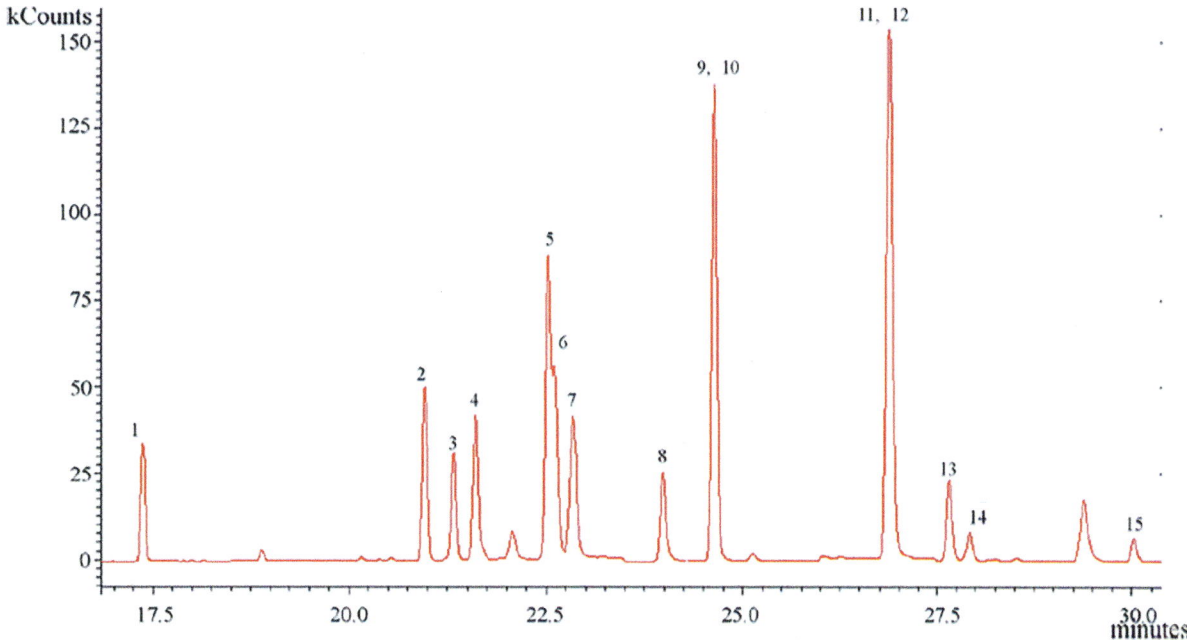

FIG. 2.4.48 Ion current chromatogram constructed from selected ion monitoring for rat plasma steroids ($n = 15$) after HFB derivative formation. Steroids were isolated with Oasis HLB cartridges and derivatives separated with 30 m capillary column coated with 5% phenylmethylsiloxane and 95% dimethylpolysiloxane, temperature programmed. Two or three fragment ions from electron impact ionization were monitored for each steroid (Chen et al., 2009). (1) Aldosterone; (2) testosterone; (3) estradiol; (4) 1,4-androstadien-3,17-dione; (5) androstenedione; (6) dehydroepiandrosterone; (7) estrone; (8) dihydrotestosterone; (9) 17-hydroxypregnenolone; (10) 17-hydroxyprogesterone; (11) progesterone; (12) pregnenolone; (13) 11-deoxycortisol; (14) cortisol and (15) 11-deoxycorticosterone. *(From Chen J, Liang Q, Hua H, Wang Y, Luo G, Hu M, Na Y. Simultaneous determination of 15 steroids in rat blood via gas chromatography-mass spectrometry to evaluate the impact of emasculation on adrenal. Talanta 2009;80(2):826–32. Fig. 1 p. 827.)*

hyperplasia (CAH) due to 21-hydroxylase deficiency (Christakoudi et al., 2010). Tentative identification was based on MS/MS characteristics and GC retention times of the steroids. The structures of the steroids were based on retention times, retention time shifts for specific modifications to a steroid (16-hydroxylation; androstene, androstane, pregnene or pregnane, side chain structure) and fragmentations in the collision cell (Q2) of the tandem MS analysis (Christakoudi et al., 2012a,b). Fragments of steroids with hydroxyl groups at C2, 6 and 11 are shown in Fig. 2.4.52A with 7 oxygenated steroid fragments (Fig. 2.4.52B).

Hydroxylation of steroids at carbons 1, 2, 6, 15 and 18 are uniquely seen in the neonatal period for fetal zone steroids and glucuronide conjugated fractions. The identities of very few of the steroids were however confirmed by comparison of properties with a reference preparation. A series of 6-hydroxy cortisol and 6-hydroxy precursor metabolites were synthesized by Kraan group in Holland and several papers described syntheses of a number of C-15-hydroxylated steroids (Joannou). One steroid was characterized, by its mass spectrum (Fig. 2.4.53) and GC retention shift, as 3β,16α,17α-trihydroxy-5α pregnan-7,20-dione and because it was found in the urine of a newborn infant it was designated as a new early marker for CAH due to 21-hydroxylase deficiency (Christakoudi et al., 2010).

GC-MS/MS along with LC-MS/MS has been used to identify new sulfate metabolites of metandienone in doping control (Gómez et al., 2013). Two laboratories have used the GC-MS/MS approach to produce reference ranges for urine steroids for each decade of life in adults aged 20–79 years (Robles et al., 2017; de Jong et al., 2017). Four internal standards, 11-keto-etiocholanolone-d5, pregnenolone-d4, THE-d5 and DHEA-d6 were used (de Jong et al., 2017) to cover the range of polarities and minimize the costs. The precursor and product ions were used as quantifier and qualifier ion for the analysis (Table 2.4.4). Retention times ranged from 14.3 to 32.8 min. The relative peak heights of quantifier ions are inconsistent because of the different proportion of the ions monitored in the total ion current chromatogram (Fig. 2.4.54). In order to have visualization of the quantities, the concentrations of steroids in a sample are translated into heat maps (this will be addressed later in data interpretation see Section 2.4.12.2).

Calibration is an immense undertaking. The results were compared with published, limited reference ranges. Visual comparison of the results suggest similarity with results from other laboratories using GC-MS. The paper by de Jong et al. reported results in μmol/24 h, some results were higher than a GC-MS comparator and some were significantly lower values.

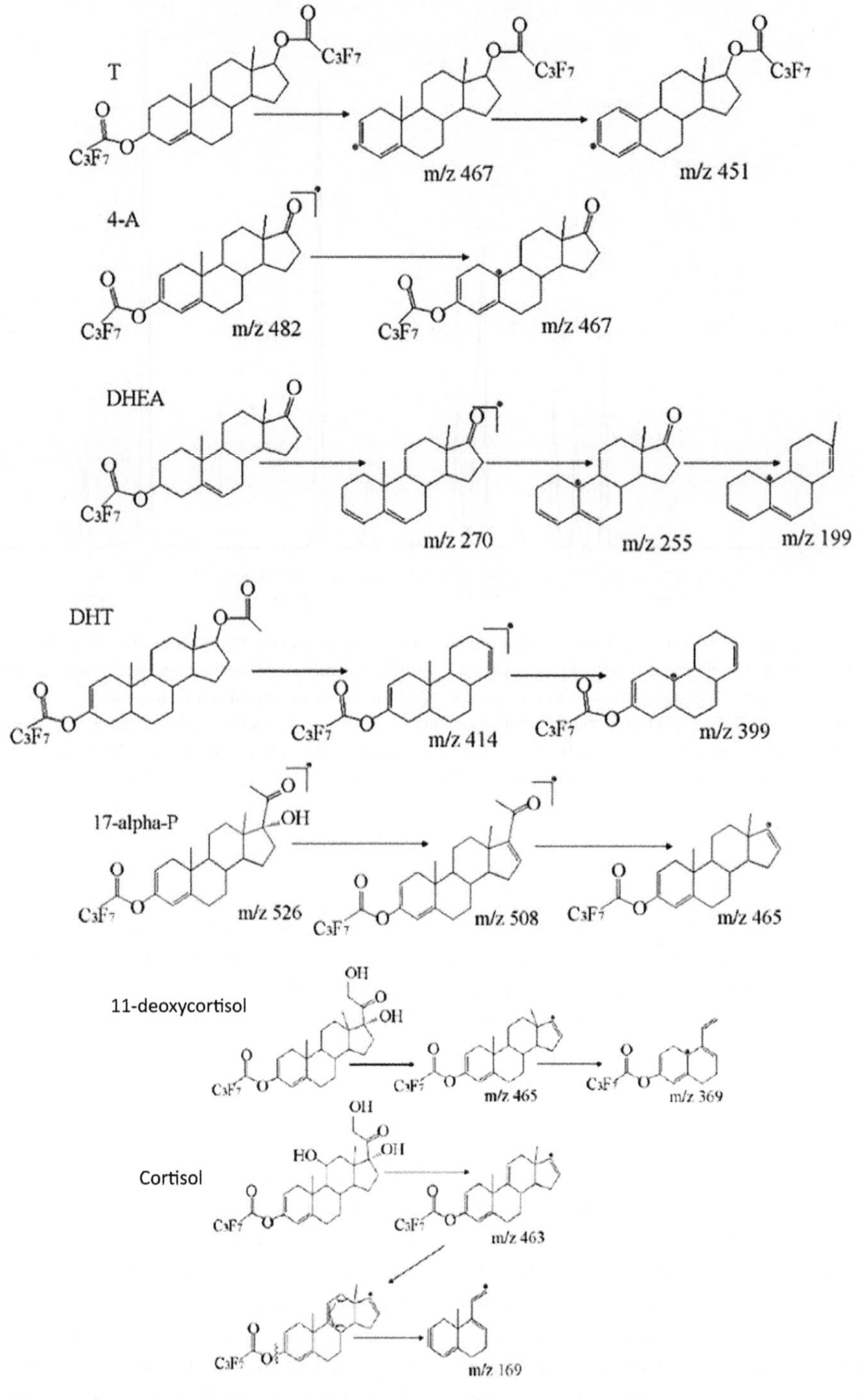

FIG. 2.4.49 Fragmentations in GC/MS of steroid heptafluorobutyrates. *(From Chen J, Liang Q, Hua H, Wang Y, Luo G, Hu M, Na Y. Simultaneous determination of 15 steroids in rat blood via gas chromatography-mass spectrometry to evaluate the impact of emasculation on adrenal. Talanta 2009;80 (2):826–32. Fig. 2 p. 828.)*

Robles et al. in 2017 quantified, by GC-MS/MS, metabolites of androgens, estrogens and pregnenolone in a glucuronide fraction (deconjugated by *E. coli* glucuronidase) and a combined sulfate and glucuronide (deconjugated by *Helix pomatia* sulfatase and glucuronidase) fraction. Enol-TMS derivatives were prepared. Two transitions were monitored as qualifier and quantifier ions for each steroid. Internal standards were added at concentrations appropriate for steroids less than 10 ng/mL, 10–200 ng/mL and greater than 200 ng/mL. This investigation complemented the

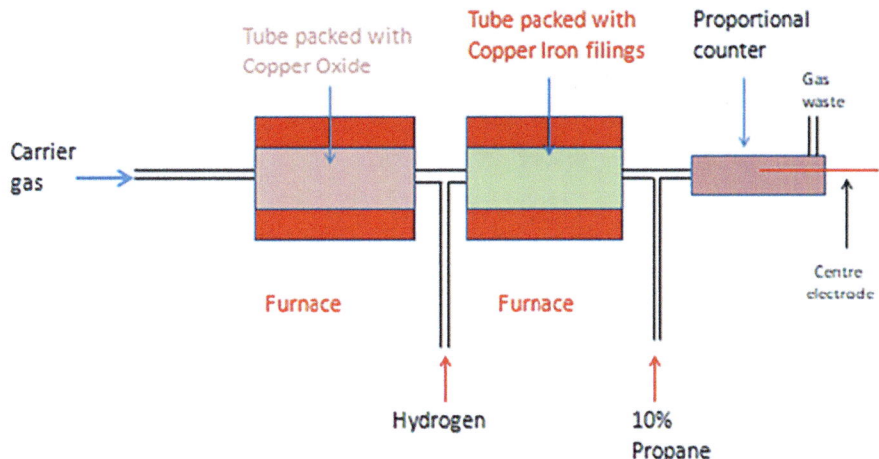

FIG. 2.4.50 Schematic of radioactivity detection following separation of tritium labeled steroids by GC. The column eluate passes into a furnace containing a tube of copper oxide heated to 800 degrees. Hydrogen is added to the oxidized sample and this moves to a tube of copper with iron filings to dry the gas which passes to a proportional counter.

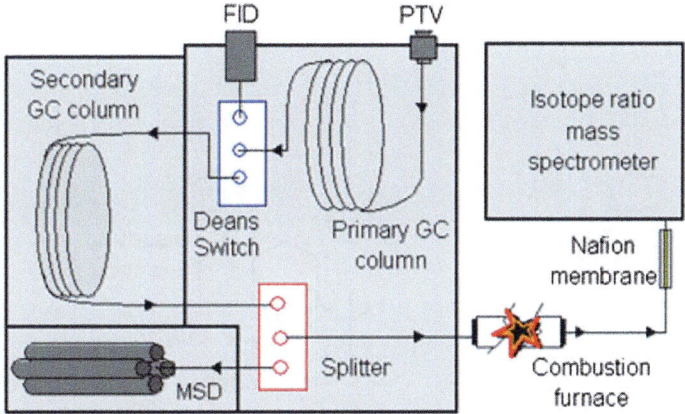

FIG. 2.4.51 2D GC with heart cutting isotope ratio MS. *(From Brailsford AD, Gavrilović I, Ansell RJ, Cowan DA, Kicman AT. Two-dimensional gas chromatography with heart-cutting for isotope ratio mass spectrometry analysis of steroids in doping control. Drug Test Anal 2012;4(12):962–9.)*

antidoping in sport service and so was primarily here concerned with abuse of androgenic-anabolic steroids. Steroids in blood could be detected to 0.2 ng/mL.

One hundred steroids (58 unconjugated and 42 polar conjugates) were quantified in serum in and women and during pregnancy (Hill et al., 2019). For 32 of the steroids, 3 transitions could be monitored (examples are shown in Table 2.4.5) and for 57 steroid 2 transitions were monitored therefore high specificity is demonstrable in this analysis. Considerable development is required to optimize such a method.

2.4.6.5 Liquid chromatography-mass spectrometry (LC-MS)

A significant advantage of LC-MS over the use of GC-MS for steroid analysis is that the compound of interest does not to be derivatized to be in the vapor phase to be analyzed by the technique. In many cases, steroids can be rendered

volatile and remains thermally stable by derivatization but this adds an additional step to the analytical procedures and some steroid types, e.g., glucuronide or sulfate conjugates are not readily amenable to such analysis. HPLC, in contrast to GC, is normally performed at near ambient temperatures and polar compounds and steroid conjugates can be analyzed without degradation.

Since detection selectivity and structural information of analytes separated by HPLC is a requirement, a concerted effort has been made over the last decades to interface the HPLC and mass spectrometric techniques. When using LC-MS/MS it is important that isobaric steroids (those with same molecular weight) are separated in the LC stage. The composition of the mobile phase including any additives will be important since these factors will influence the interactions with mobile phase and stationary phase. This task is complicated because mass spectrometers operate under vacuum, and unlike GC whose components are eluted in inert vapor phase, LC necessitates the removal of the mobile

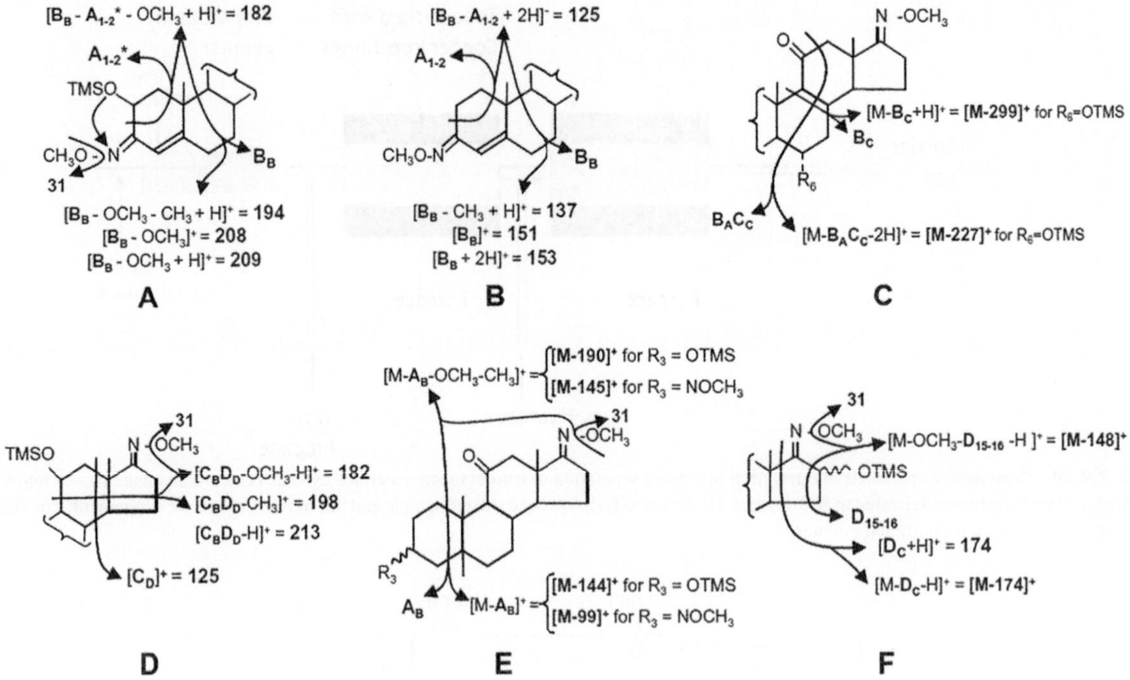

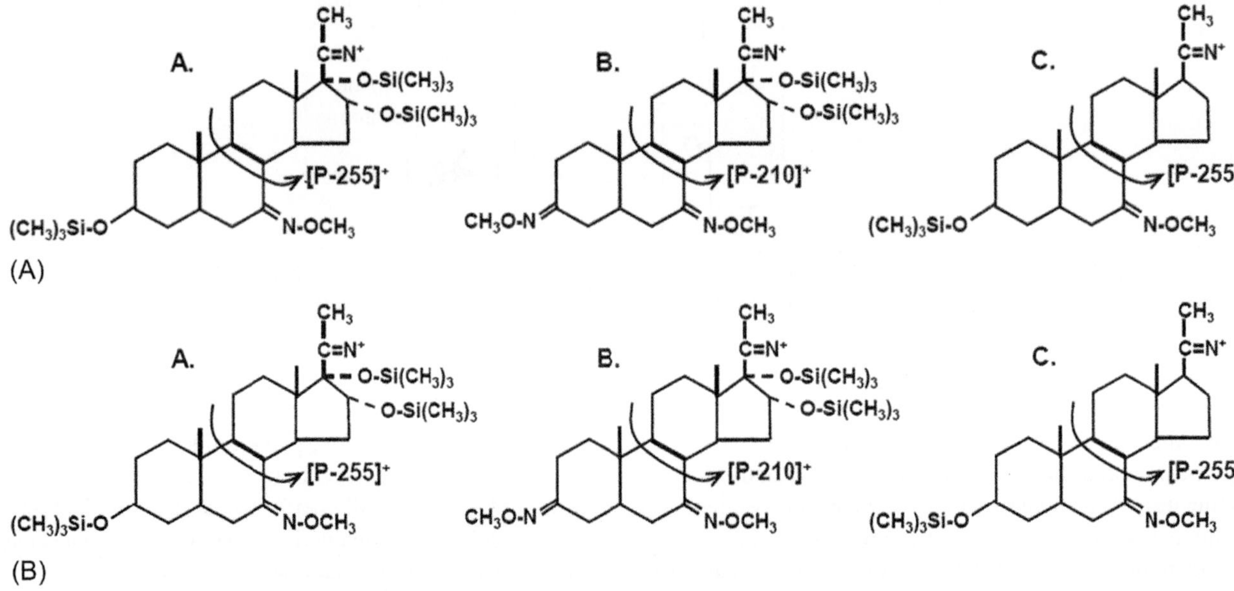

FIG. 2.4.52 (A) Fragmentations in GC-MS/MS of C-2, 6 and 11-oxygenated androgens. (B) Fragmentations in GC-MS/MS of C-7-oxygenated C21 steroids with B-ring cleavage. ((A) From Christakoudi S, Cowan DA, Taylor NF. Steroids excreted in urine by neonates with 21-hydroxylase deficiency. 3. Characterization, using GC-MS and GC-MS/MS, of androstanes and androstenes. Steroids 2012a;77(13):1487–501. Fig. 3 p. 1494; (B) From Christakoudi S, Cowan DA, Taylor NF. A new marker for early diagnosis of 21-hydroxylase deficiency: 3beta,16alpha,17alpha-trihydroxy-5alpha-pregnane-7,20-dione. J Steroid Biochem Mol Biol 2010;121(3–5):574–81. Fig. 4 p. 578.)

phase that can flow at rates up to 2 mL/min. Since the early 1980s the thermospray (TSP) interface developed by Vestal has been used. Interfaces that rely on production of a fine spray of solvent containing analytes have since dominated LC-MS.

A LC-MS method for steroids should not be too expensive when priced per sample. If the instrumentation is leased over a period (3–5 years say) the cost per sample can be very low. Reagents (solvents) are very cheap compared with immunoassay kits although only the highest

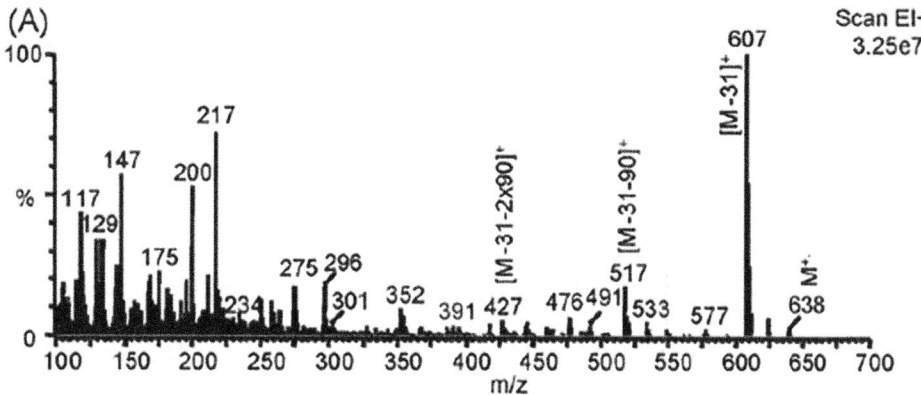

FIG. 2.4.53 GC-MS spectrum of the MO-TMS derivative of 3β,16α,17α-trihydroxy-5α-pregnane-17,20-dione. A marker for 21-hydroxylase deficiency in the neonatal period. *(From Christakoudi S, Cowan DA, Taylor NF. A new marker for early diagnosis of 21-hydroxylase deficiency: 3beta,16alpha,17alpha-trihydroxy-5alpha-pregnane-7,20-dione. J Steroid Biochem Mol Biol 2010;121(3–5):574–81. Fig. 2 p. 576.)*

grade solvents should be used. Low quality solvents can affect long-term costs such as downtime and trouble shooting. Mobile phase additives should also be prepared with high quality chemicals. When changing solvents (particularly if running different tests on the same instrument to optimize use) the system should be flushed with at least five volumes of water to remove any salts that can precipitate in the lines and pumps. Laboratories should consider possibilities of microbial growth in the solvents. Solvent reservoirs should be sealed and mobile phases should be discarded regularly. The containers should be thoroughly cleaned when emptied. Samples should be inspected for particles that may cause blockages. Solvent should be pumped through the system at high volume to clean the flow path in normal and reverse direction. It is important that the system is equilibrated prior to injections. A system suitability test should be used to check retention times, peak shapes, peak area/height of the compounds, contamination, and signal to noise in the analysis whenever solvents, columns and fittings are changed. Precolumn components (frits, filters) and columns should be changed often. A staff training and equipment maintenance program should be in place and the source and probe should be cleaned according to the number of injections as well as any loss in sensitivity (Stone and Fitzgerald, 2018).

HPLC columns have become shorter packed with smaller particles so analytical times are now short (typically less than 10 min). Performance has been improved with turbo flow columns that increase the path length for analytes moving through the particles (Søeborg et al., 2017). Methods with electrospray have been developed for cortisol and cortisone in serum and plasma (Kushnir et al., 2004) and saliva (Higashi et al., 2007; Bae et al., 2016; Jones et al., 2012; Perogamvros et al., 2009) which is a useful test for patients with a genetic defect in HSD11B2.

2.4.6.5.1 Thermospray mass spectrometry

The thermospray interface when introduced around 1986 provided the first reliable method of introducing liquid samples into a mass spectrometer. In LC-MS mode, the sample is passed into the instrument following HPLC separation using conventional columns (4–6 mm id) and solvent flow-rates of 0.5–1.5 mL/min. As the name implies, the technique depends on production of a spray of fine droplets by heating, this being achieved by direct electrical heating of the capillary through which the sample passes (Fig. 2.4.55). These fine droplets continue to vaporize as they pass into the heated ion source. A portion of this vapor and ions produced in the ion source escape into the vacuum system of the mass spectrometer through a sampling cone and the remainder of the excess vapor is pumped away by mechanical vacuum pumps.

The principal form of ionization for LC-MS is termed *direct ion evaporation* which involves evaporation into the gas phase of ions present in solution. Although analytes that are ionized in water or methanol can be analyzed directly, sensitivity is much improved by addition of a volatile buffer such as ammonium acetate at concentration of 0.01–0.1 M. Thermospray MS can be used in positive or negative ion modes. In positive ion mode, the production of MH^+ is most desirable but $M+NH_4^+$ ions are often seen because of ammonium acetate presence in the solvent system and $M+Na^+$ or $M+K^+$ ions are frequently produced due to the ubiquitous presence of these ions in samples, solvents and tubing, often leached from glass containers. Negative ion thermospray is most useful for analyzing steroid conjugates since the dominant presence of $M-H^-$ ions effectively excludes the counter ions. Sensitivity is best when the mobile phase is predominantly aqueous. Although the technique can be used with essentially 100% methanol,

TABLE 2.4.4 Mass spectrometric settings for steroid metabolites and internal standards.

Steroid metabolite	Precursor ion m/z	Product ion m/z
[D5]11-Ketoetiocholanolone	305.1	258.2
[D6] DHEA	364.2	274
[D5] THE	583.3	403.2
[D4] Pregnenolone	390.2	300.1
11-HA	448.2	358.2
11-HE	448.2	268.2
11-KE	300.1	254.2
16K-A'2	446.2	356.2
16-OH-P-ol	474.3	156
17-P3	433.2	253.3
A	360.2	270.2
aP2	269.2	187
aPDL	476.3	386.3
α-Cortol	343.3	199
α-CTLN	449.2	359.3
β-CTLN	449.2	359.3
DHEA	358.2	268
E	360.2	270.2
Estriol	504.3	414.1
P2	269.2	187
P3	435.3	255.2
PDL	476.3	386.3
Polone	388.2	298.2
PTL	449.2	359.3
TH-DOC	476.3	241.2
THE	578.3	398.2
THS	564.3	474.2

the sensitivity is often marginal if the water fraction is less than 20% (Nabarro et al., 1957).

A secondary type of ionization is termed *filament-on operation* that uses a filament to initiate chemical ionization (CI) which is often desirable when samples contain a high proportion of organic solvents, as in normal phase HPLC separation. This is also a useful technique because the presence of salts (e.g., ammonium acetate in the HPLC solvent system) is to be avoided although an alternative way around this is to continue to use "direct ion evaporation" but to introduce the ammonium acetate solution *after* the HPLC separation but before the sample enters the thermospray probe.

2.4.6.5.2 Electrospray mass spectrometry

The invention of electrospray was an important step in generating ions in a spray through ion-evaporation from droplets. Ions do not evaporate from solutions however unless some energy is supplied. Thermospray uses heat but an alternative technique, **electrospray**, uses an electric field. In this technique, the eluant is infused at low flow rates (<10 mL/min) from a 100 mm id fused silica line to the end of the nebulizer where it is exposed to a high electrical potential (Fig. 2.4.56). Compressed air is the source of nebulization resulting in a stable spray of charged droplets. In the electrospray process, ions in solution are desorbed from rapidly shrinking, highly charged droplets by ion-evaporation resulting in abundant molecular ions, MH^+ in positive ion mode and $M-H^-$ in negative ion mode.

A number of other reactions may take place during electrospray including ion suppression that affect signal intensity (Fig. 2.4.57).

Since the interface is operated at ambient temperature, the technique is even more gentle than thermospray. The whole ion-evaporation process is carried out at atmospheric pressure and only after ionization do the ions enter the mass spectrometer after passing through a "curtain" gas which prevents the solvent and buffer molecules from entering the mass analyzer. It is obvious that the low flow-rates (1–5 mL/min) optimal for the interface are in sharp contrast to the thermospray technique, so the use of microbore or capillary HPLC columns is desirable.

A technique somewhat intermediate between thermospray and electrospray has also been extensively used for small molecule studies. It is termed **heated nebulization**. In this technique, eluant from a conventional LC column (at >200 mL/min) is passed into a sprayer where it is converted into fine mist by a high-velocity jet of air (nebulization). A make-up gas sweeps the droplets through a heated quartz tube (225°C) that vaporizes the droplets. The heated gas is typically only at about 100°C, which is high enough for evaporation but low enough to prevent thermal degradation. The gaseous sample mixture flows into an atmospheric pressure reaction region where it is chemically ionized by reagent ions produced by a discharge element. Typically, analyte molecules are ionized by proton transfer from solvated protons in positive-ion mode and by electron transfer from, or proton transfer to oxygen atoms in the negative-ion mode. As in electrospray the ions produced pass through a nitrogen curtain gas and into the evacuated part of the mass spectrometer.

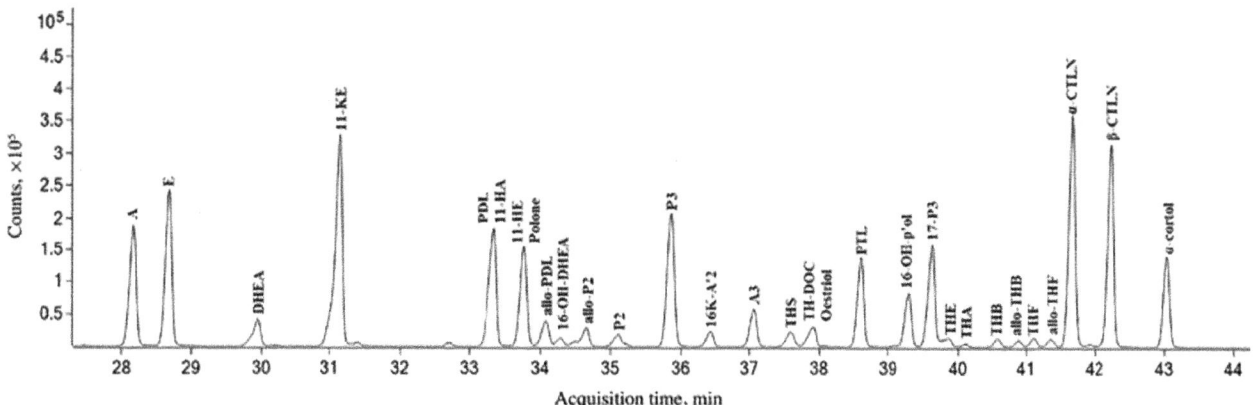

FIG. 2.4.54 Total ion current chromatogram from GC-MS/MS of calibration standard mix used in identification of steroid metabolome. *A*, Androsterone; *E*, etiocholanolone; *DHEA*, dehydroepiandrosterone; *11KE*, 11-ketoE; *11HA*, 11hydroxyA; *11-HE*; *polone*, epi-pregnanolone; *16-OH-DHEA*; *aP2*, allo-pregnanediol; *P2*, pregnanediol; *P3*, pregnanetriol; *16KA2*, 16-ketoandrostenediol; *A3*, androstenetriol; *THS*, tetrahydrodeoxycortisol; *THDOC*, 11-deoxytetrahydrocorticosterone; *PTL*, pregnanetriolone; *16-OH-POL*, 16a hydroxypregnenolone; *17-P3*, 17-hydroxypregnenetriol; *THE*, tetrahydrocortisone; *THA*, 11-dehydrotetrahydrocorticosterone; *THB*, tetrahydrocorticosterone; *alloTHB*; *THF*, tetrahydrocortisol; *alloTHF*; α-*CTLN*, α-cortolone; β-*CTLN*; *PDL*, pregnanediolone; *aPDL*; α-cortol; β-cortol. A precursor and product ion were monitored for each steroid. *(From de Jong WHA, Buitenwerf E, Pranger AT, Riphagen IJ, Wolffenbuttel BHR, Kerstens MN, Kema IP. Determination of reference intervals for urinary steroid profiling using a newly validated GC-MS/MS method. Clin Chem Lab Med 2017 ;56(1):103–12. Fig. 1 p. 105.)*

TABLE 2.4.5 Representative GC-MS/MS analysis of steroids in pregnancy urine. MRM acquisition retention times, three transitions and optimum collision energies for representative steroids.

Steroid	Retention time	MRM transition 1	MRM transition 2	MRM transmission 3
5α-Pregnan-3α,20α-diol	8.41	269 > 187 (12)	269 > 161 (12)	269 > 105 (30)
5β-Pregnane-3α,20α-diol	8.46	269 > 187 (12)	269 > 161 (12)	269 > 105 (30)
Estradiol	8.61	416 > 285 (15)	416 > 326 (6)	285 > 205 (15)
Epiandrosterone	8.63	360 > 270 (9)	360 > 84 (18)	360 > 82 (21)
DHEA	8.64	358 > 84 (18)	268 > 82 (21)	260 > 213 (6)
Allo-pregnanolone	8.96	388 > 298 (15)	388 > 173 (18)	388 > 70 (15)
Testosterone	8.98	389 > 268 (9)	389 > 137 (12)	389 > 125 (9)
Pregnanolone	9.03	388 > 298 (15)	388 > 173 (18)	388 > 70 (18)
17-Hydroxypregnenolone	9.24	472 > 294 (9)	474 > 225 (12)	474 > 157 (21)
Pregnenolone	9.43	402 > 239 (12)	312 > 239 (9)	239 > 157 (18)
Androstenedione	9.77	344 > 313 (9)	344 > 137 (24)	344 > 125 (15)
20α-Dihydroprogesterone	9.99	417 > 117 (12)	301 > 286 (9)	301 > 138 (15)
5α-Dihydroprogesterone	10.27	343 > 244 (24)	343 > 272 (18)	288 > 159 (18)
16α-Hydroxyprogesterone	10.53	429 > 370 (15)	429 > 156 (18)	156 > 73 (15)
17α-Hydroxypregnanolone	8.48	429 > 172 (24)	476 > 386 (12)	476 > 296 (15)
17α,20α-Dihydroxy-4-pregnen-3-one	10.0	388 > 298 (9)	388 > 267 (12)	298 > 145 (15)
Cortisol	10.70	605 > 514 (12)	605 > 143 (21)	515 > 425 (15)
5α-Androstane-3α,17β-diol	6.69	421 > 255 (9)	346 > 256 (6)	346 > 241 (6)
Androsterone	8.05	360 > 270 (9)	270 > 213 (9)	270 > 157 (21)
Etiocholanolone	8.13	360 > 270 (9)	270 > 213 (9)	270 > 157 (21)
5-Androstene-3β,7β,17β-triol	8.17	432 > 327(15)	432 > 233 (21)	432 > 209 (18)

Based on Hill M, Hána V Jr, Velíková M, Pařízek A, Kolátorová L, Vítků J, Škodová T, Šimková M, Šimják P, Kancheva R, Koucký M, Kokrdová Z, Adamcová K, Černý A, Hájek Z, Dušková M, Bulant J, Stárka L. A method for determination of one hundred endogenous steroids in human serum by gas chromatography-tandem mass spectrometry. Physiol Res 2019 ;68(2):179–207. Table 1 p. 2019.

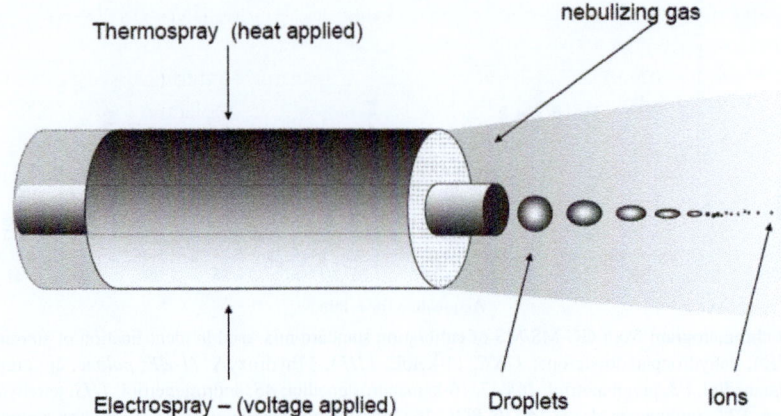

FIG. 2.4.55 Schematic of thermospray LC-MS/MS interface.

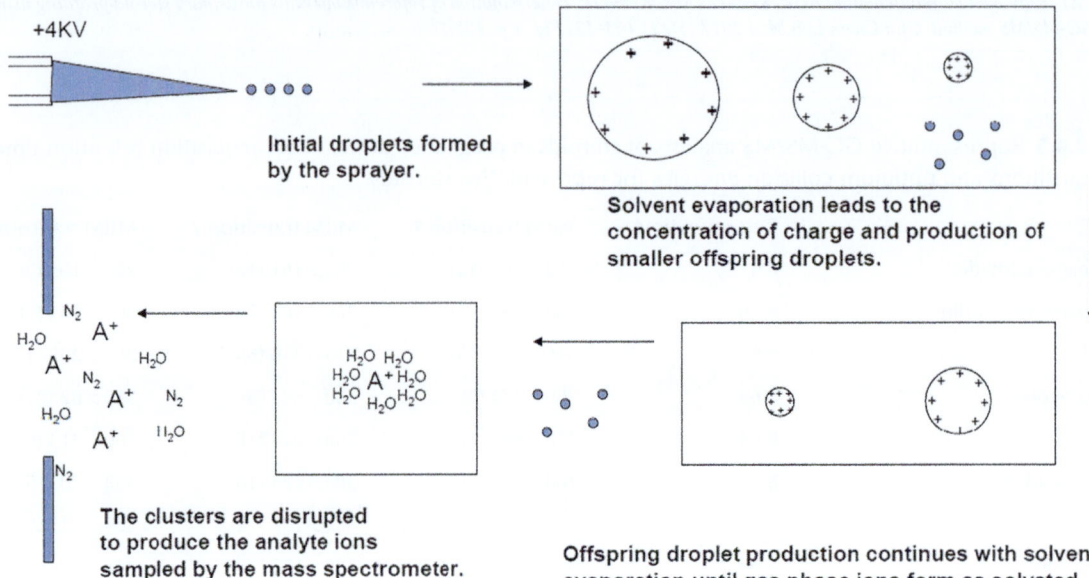

FIG. 2.4.56 Principles of electrospray ionization.

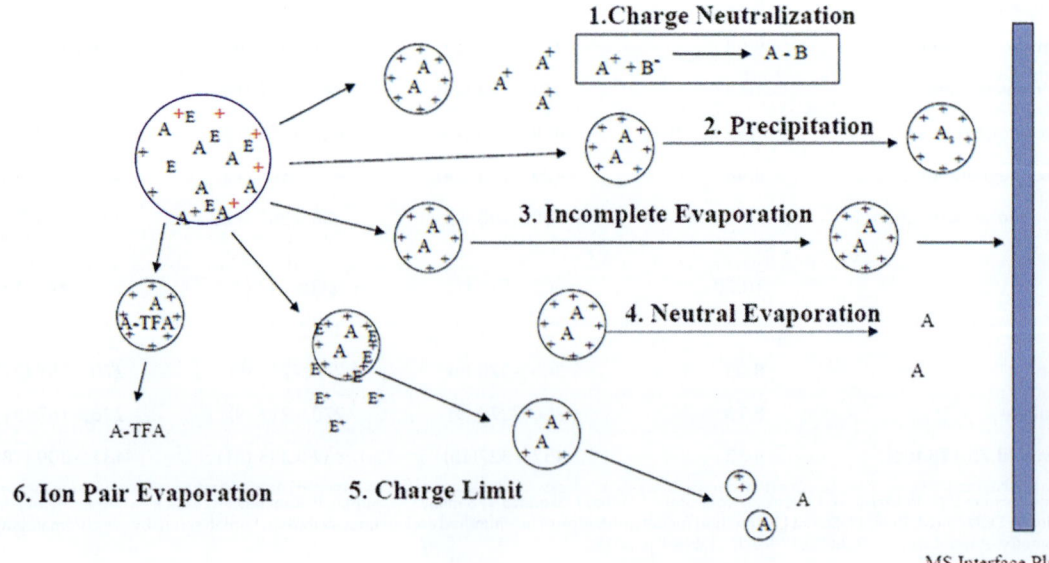

FIG. 2.4.57 Possible causes of electrospray ionization suppression.

In clinical laboratories thermospray, heated nebulization and electrospray interfaces, in order of increasing popularity, are usually used with quadrupole mass spectrometers although they are rarely used with magnetic sector instruments. A perceived disadvantage of soft-ionization has been a relative lack of fragmentation which reduces the specificity of the technique as a detection system and gives little information on structural features of unknown compounds. One of the great advantages of EI GC-MS of derivatized steroids is that such modified steroids fragment in very distinct and well documented ways, facilitating identification. Increased structural information in LC-MS can be obtained if the molecular ions of the analytes are induced to fragment by collisionally induced disassociation. In both thermospray and electrospray, this can be achieved in the collision cell of a triple quadrupole instrument and in electrospray this can be also achieved in the ion source itself by increasing the nozzle/skimmer voltage.

Since both thermospray and electrospray techniques are soft-ionization, steroid spectra from each technique can be virtually identical. Although LC-MS/MS is attractive because the derivative step in GC-MS has been negated there are advantages in LC-MS/MS from using derivatives. The sensitivities of many methods are improved by using derivatives (Higashi et al., 2016; Denver et al., 2019).

Electrochemical reactions have been located to the capillary of the electrospray as the result of the high potential applied to the capillary. On-column solvent electrolysis is proposed to generate free radicals which can subsequently initiate analyte oxidation. Steroid sulfates with a double bond between C-5 and C-6 can be oxidized (Liu et al., 2003). The analyst needs to be vigilant to recognize such possibilities with other analytes including peptides with a site of unsaturation.

2.4.6.5.3 Atmospheric pressure ionization (API)

In the **atmospheric pressure chemical ionization (APCI)** ion source, the mobile phase is evaporated in a heated region of the ion source and analytes get ionized through gas phase reactions in a corona discharge region, the **APCI** interface (Parr et al., 2018; Sun et al., 2018; Storbeck et al., 2008) (Fig. 2.4.58).

In APCI, vaporization is separate from ionization (Fig. 2.4.59).

The corona pin is set to 5–10 μA, the cone voltage to 30 V, and the APCI probe temperature to 450°C. All other settings need to be optimized to obtain the strongest signal possible.

The steroids PREG and DHEA exhibit a high abundance of protonated molecules due to the loss of water ($[M-H_2O+H]^+$), The ionization of pregnenolone is further increased by reducing the cone voltage to 15 V. 17-OHPREG exhibits high abundance of protonated molecules due to the loss of both a single water molecule ($[M-H_2O+H]^+$) and two

water molecules ($[M-2H_2O+H]^+$) (Storbeck et al., 2008) (Fig. 2.4.60).

Δ4 steroids exhibit mass spectra with a high abundance of protonated molecules ($[M+H]^+$) for progesterone, 16-hydroxyprogesterone, 17-hydroxyprogesterone and androstenedione. APCI has been used for steroid quantitative analysis in SIM mode (Higashi et al., 2005). 3β-Hydroxy-5-ene steroids were converted to 3-keto Δ4 steroids with cholesterol oxidase.

Steroids in APCI give stronger signals than ESI as illustrated in three studies (Ceglarek et al., 2009a; Gibson et al., 2019) (Fig. 2.4.61A–C). Matrix effects are lower in APCI than with ESI. The ESI and APCI have been mainly used in steroid analysis.

A method for a panel of 13 steroids (Fig. 2.4.62) has been reported and the results shown to be useful for discriminating patients with adrenal cortical carcinoma (ACC) from other adrenal lesions. 11-Deoxycortisol was markedly increased in all cases of ACC (Taylor et al., 2017).

Atmospheric pressure photo-ionization (APPI) is a newer ionization technique with photo-ionization of a dopant that ionizes nonpolar molecules through proton transfer and charge exchange. The APPI source is similar to the APCI source which vaporizes mobile phase with a heated nebulizer (350–500°C) to generate a dense cloud of gas-phase analytes with minimal thermal decomposition. A dopant, usually toluene, delivered from a HPLC pump is also vaporized in the nebulizer gas within the source chamber. Ionization uses a krypton discharge lamp to emit 10 eV photons to form dopant radical cations. The photo-ions, formed in large quantity from the dopant initiate a cascade of ion-molecule reaction with the analytes and solvent molecules leading to the production of MH^+ of a given analyte M (by proton transfer if the proton affinity (PA) of the analyte is larger than the protonated solvent clusters and the PA of the solvent cluster is larger than that of the dopant radical cation), or M+(by charge exchange if ionization energy (IE) of the analyte is less than the combination energy of the dopant radical cation) in the positive ion mode. For the APPI-MS, source parameters such as temperature and ion transfer voltage have a strong impact on the assay detection limit. Ion transfer voltage, an offset potential between the probe ion deflector and the mass spectrometer curtain plate, is responsible for transferring the analyte ions formed in the ionization region to the mass spectrometer (Gomez et al., 2014; Jeilani et al., 2011; Antignac et al., 2000; Sun et al., 2005).

In some hands (notably Steve Soldin group), APPI is more reliable for steroid measurement than APCI and ESI with results in some cases 50% of immunoassay results because of improved specificity (Stolze et al., 2016; Guo et al., 2004). This claim is due to the softer (less energetic) ionization afforded by APPI, which ionizes the steroids

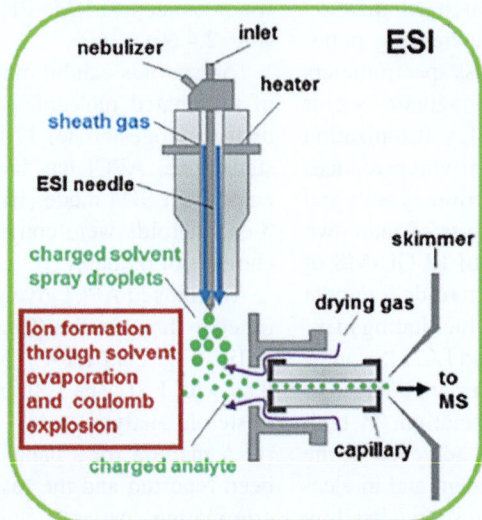

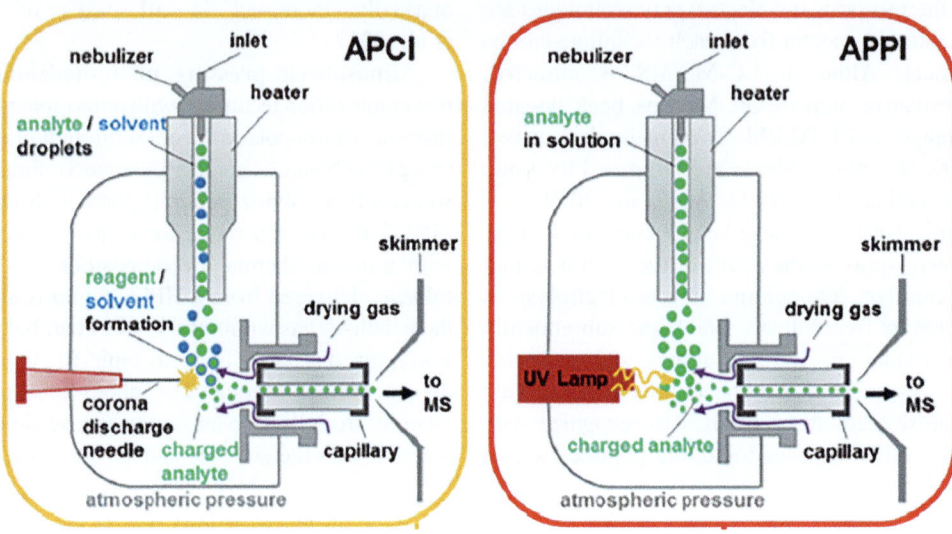

FIG. 2.4.58 Schematic of ESI, APPI and APCI. *(From Parr MK, Wüst B, Teubel J, Joseph JF. Splitless hyphenation of SFC with MS by APCI, APPI, and ESI exemplified by steroids as model compounds. J Chromatogr B Analyt Technol Biomed Life Sci 2018;1091:67–78. Fig. 1 p. 68.)*

more specifically than the far more energetic APCI and ESI sources that simultaneously ionize many different molecular species. APPI has been shown to produce an improved signal-to-noise ratio than APCI by a factor of 3 to 10. For steroid measurement, the ionization suppression provided by ESI is even greater compared with APCI. Given the similar cost-effectiveness and ease of maintenance for APPI, APCI, and ESI, Soldin recommends that clinical laboratories replace APCI/ESI with APPI in steroid measurement.

LC-MS is an important tool for the identification of steroids. Validation of the method is a lengthy process. Disadvantages in the use of mass spectrometry with under-ivatized steroids arise from the absence of significant molecular ions and the ease with which steroids dehydrate at high temperatures. Soft ionization techniques may reveal

more information than from the fragmentation pattern derived by electron impact mass spectrometry and should be particularly beneficial in the case of steroid conjugates. The potential of HPLC-MS has been recognized for some time and is now becoming standard technology. The combination of micro-bore column HPLC with thermospray or electrospray MS seems to be most promising systems with APPI in some cases. Mass spectrometry is currently considered the most specific method for steroid measurement. More recent advances have involved two stages of mass spectrometry with intermediate fragmentation, often called tandem-mass spectrometry (MS/MS).

Liquid chromatography tandem mass spectrometry (LC-MS/MS) measurements are targeted to single or multiples of steroids simultaneously (panel or profile) from a small plasma volume. This is becoming more widely

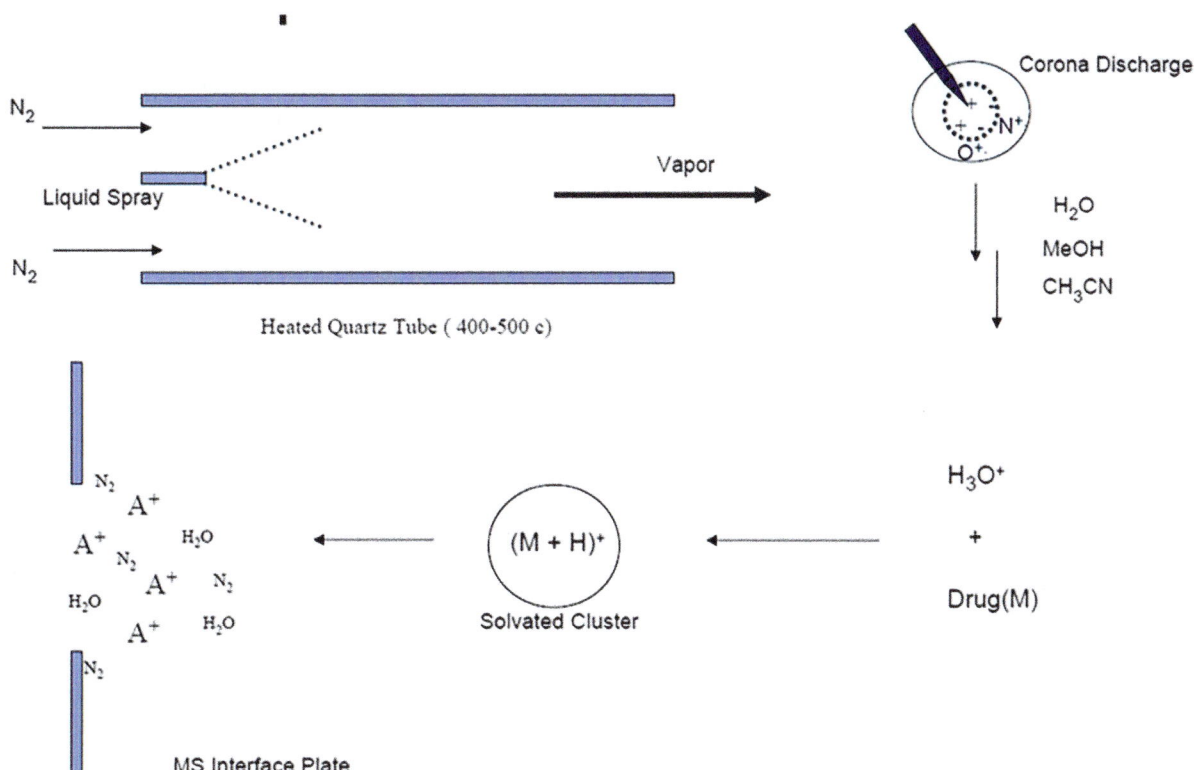

FIG. 2.4.59 Schematic of APCI interface. Vaporization and ionization are separate steps in APCI but not ESI.

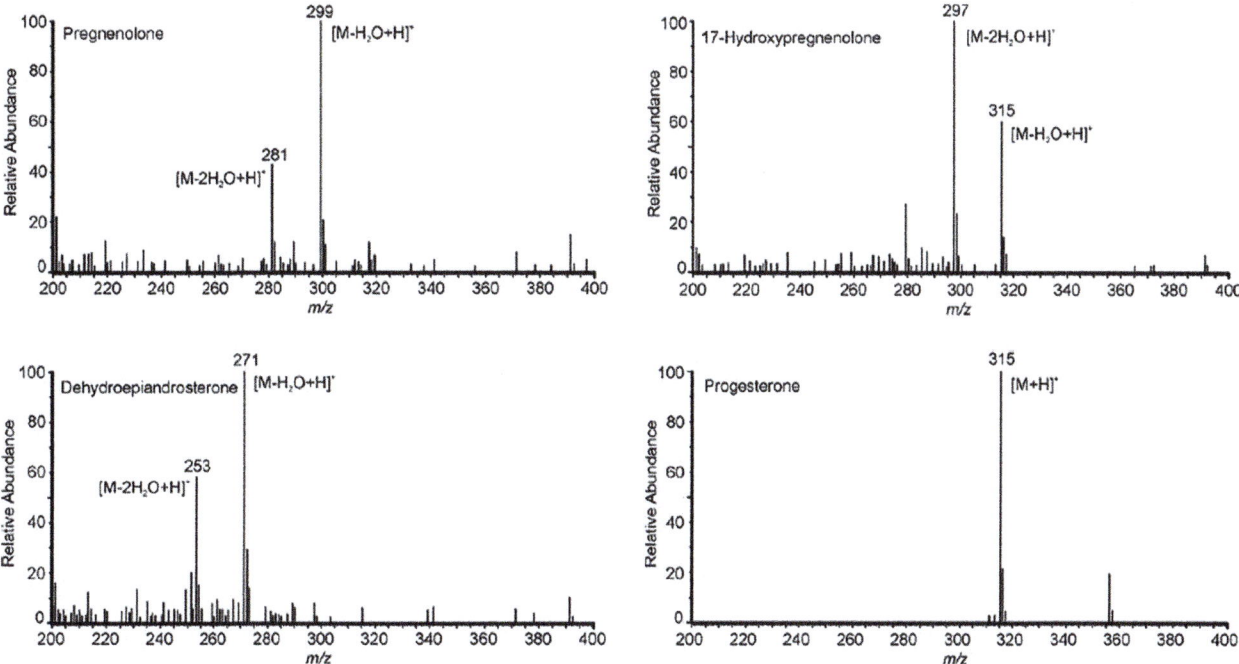

FIG. 2.4.60 LC-APCI-MS mass spectra of steroid standards. *(From Storbeck KH, Kolar NW, Stander M, Swart AC, Prevoo D, Swart P. The development of an ultra performance liquid chromatography-coupled atmospheric pressure chemical ionization mass spectrometry assay for seven adrenal steroids. Anal Biochem 2008;372(1):11–20. Fig. 2 p. 15.)*

available for steroid hormone analysis, with reference values published for androstenedione, dehydroepiandrosterone sulfate (DHEAS), 17-hydroxyprogesterone (17OHP), and testosterone in the pediatric age range as well as adults and many more steroids across age ranges. The main advantage of LC-MS/MS and GC/MS, is that compounds of similar structures can be separated to reduce the problems with cross-reactivity experienced with immunoassays. This has to be balanced however against cost considerations. It is important to interpret hormone results against reference ranges that are appropriate for age and stage of development. There are marked differences in results with different methodologies particularly for androgens and proteins such as sex hormone-binding globulin. The LC-MS interface has two functions. Firstly to remove the mobile phase and secondly to create ions in a vacuum that enter the mass spectrometer.

2.4.7 Analysis of unconjugated neutral steroids

Most steroids are analyzed in positive ion MS mode although estrogens and aldosterone tend to be determined in negative ion mode. For unconjugated steroids in thermospray and electrospray, protonated molecular ions dominate for simple steroids with up to three functional groups and little fragmentation is seen, but in steroids with more hydroxyl groups, losses of water (M-18) are common. In the 17-hydroxycorticoid series, cleavage of the side chain gives rise to M-60 (dihydroxyacetone side-chain) or M-62 (glycerol side-chain) ions as the base peaks. With water loss in the ionization process the subsequent $[M-H_2O+H]^+$ is isobaric with $[M+H]^+$ of another steroid with one less hydroxyl group although this steroid should be separated in the LC stage. The number of water losses that occur is related to the steroid structure. Thermospray mass spectra of more than 30 unconjugated 3-oxo-4-ene steroids has been reported and common fragmentations have been described. Sodium adducts $(M+Na^+)$ are common due to the almost ubiquitous presence of sodium, although rigorous attention to excluding sodium from samples during work-up procedures is one way of reducing the contribution of such ions. Frequent renewal of solvents is recommended.

Electrospray-MS essentially does not give rise to fragmentation of cortisol in contrast to thermospray. This suggests that the heat involved in the TSP technique largely causes the side-chain cleavage. Fragmentation in ES-MS is actively encouraged by both increasing the source nozzle-skimmer voltage from 65v to 350v (*source* fragmentation) and by MS-MS in the collision cell (collision energy 20, argon gas pressure 7.5×10^{-4} mbar). In neither cases, significant side chain fragments were produced. While lack of collision fragmentation is a disadvantage in structural characterization, it gives rise to greater sensitivity in SIM HPLC-MS assays.

Separation of some of the major urinary 3-oxo-4-ene metabolites by microbore HPLC-MS has been attempted. Thermospray MS is an excellent method for the determination of cortisol production rates in humans following infusion of $[9,12,12-^2H_3]$ cortisol (Derks et al., 1977; Esteban et al., 1991). LC-MS/MS has been used for the

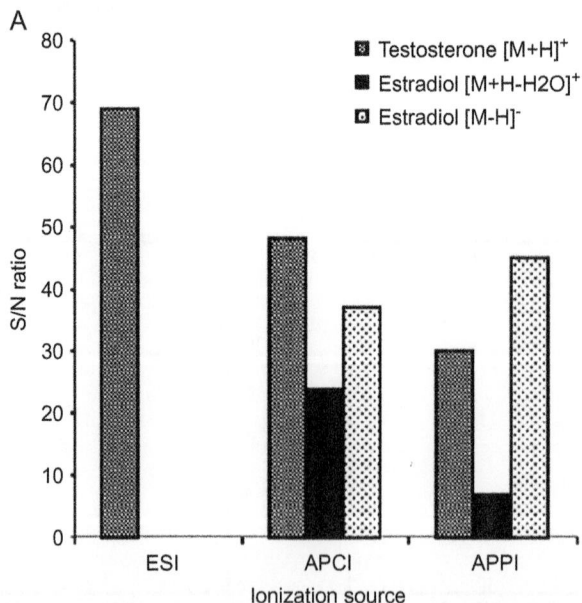

FIG. 2.4.61 (A) Signal/noise ratios of different ion species for testosterone and estradiol dependent on the ionization source and charge; *APCI*, atmospheric pressure chemical ionization; *APPI*, atmospheric pressure photoionization; *ESI*, electrospray ionization.

(Continued)

B

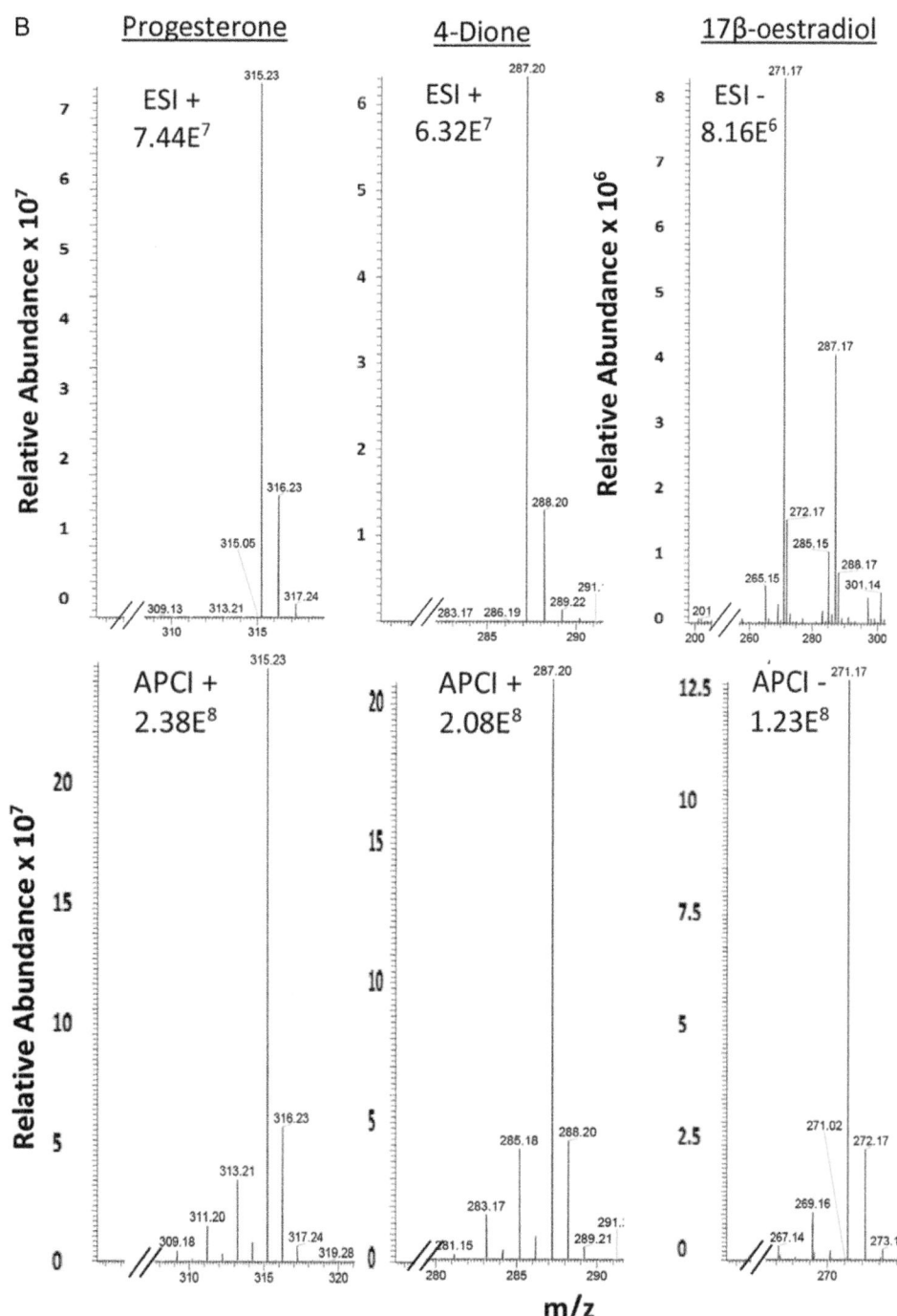

FIG. 2.4.61, CONT'D (B) Comparative relative abundances from ESI (upper) and APCI (lower) of progesterone (m/z 315), androstenedione (m/z 287) and estradiol when 2 mg/mL standards were infused into UPLC flow to MS.

(Continued)

analysis of natural and synthetic corticosteroids related to cortisol in patients being investigated for Cushing's syndrome (Djedovic and Rainbow, 2011) to detect the misuse of such compounds by sport participants, in health supplements (Gosetti et al., 2013) and river water (Tölgyesi et al., 2010). Cortisol, cortisone and typically prednisone,

prednisolone, DOC, corticosterone, 11α-hydroxyprogesterone, betamethasone, triamcinolone, and triamcinolone acetonide are detectable. By using this technique most of the corticosteroids are separated in less than 10 min. Acetylation of the C-21 hydroxyl of cortisol gives a dramatic increase in sensitivity when the hormone is being measured

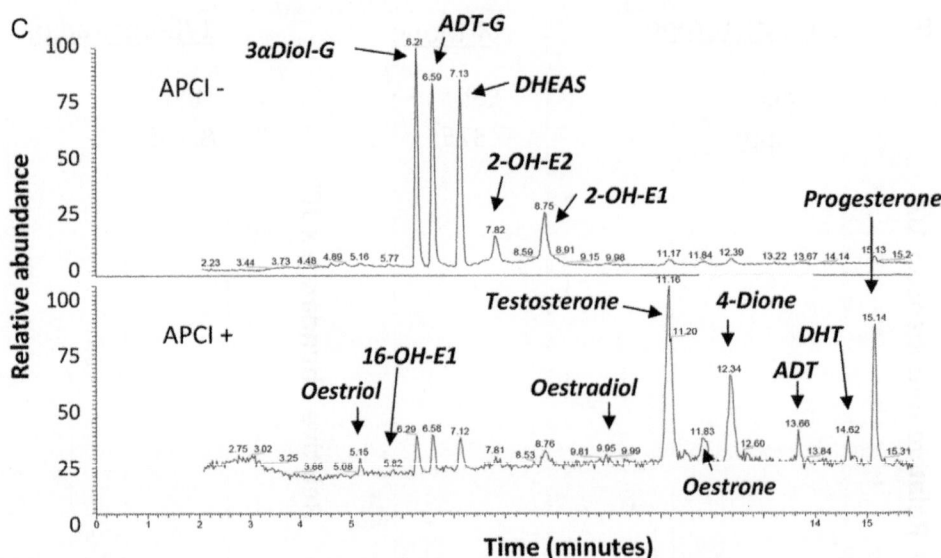

FIG. 2.4.61, CONT'D (C) UPLC-MS total ion chromatogram from alternating negative (upper) and positive (lower) APCI scanning of a mixture of 14 sex steroids to be found in tears. 3αDiol-G for 5α-androstane-3α,17β-diol-glucuronide; ADT-G for androsterone glucuronide; 2-OH-E2 for 2-hydroxyestradiol; 2-OH-E1 for 2-hydroxyestrone. *((A) From Ceglarek U, Leichtle A, Brügel M, Kortz L, Brauer R, Bresler K, Thiery J, Fiedler GM. Challenges and developments in tandem mass spectrometry based clinical metabolomics. Mol Cell Endocrinol 2009a;301(1–2):266–71. Fig. 2 p. 267; (B) From Gibson EJ, Bucknall MP, Golebiowski B, Stapleton F. Comparative limitations and benefits of liquid chromatography—mass spectrometry techniques for analysis of sex steroids in tears. Exp Eye Res 2019;179:168–78. Fig. 2 p. 173; (C) From Gibson EJ, Bucknall MP, Golebiowski B, Stapleton F. Comparative limitations and benefits of liquid chromatography—mass spectrometry techniques for analysis of sex steroids in tears. Exp Eye Res 2019;179:168–78. Fig. 3 p. 174.)*

by thermospray MS with SIM, principally because the fragmentation is decreased.

Thermospray MS has been used to study the metabolites of 4-hydroxyandrost-4-ene-3, 17-dione in prostate cancer patients (Poon et al., 1992). This steroid was under clinical trial for the treatment of patients with estrogen and androgen dependent cancers. At least 6 metabolites were found in glucuronidase-treated urine and structures were proposed.

In general, for LC-MS/MS, clinical laboratories are using quadrupole and ion trap instruments after soft ionization techniques with collision induced dissociation. The first step is to identify steroids in LC-MS/MS is to find the quasimolecular and adduct ions. Mass spectra and fragmentation pathways of testosterone, progesterone, cortisol and 11-deoxycortisol in positive ion mode as representative steroids (Thevis et al., 2011; Gomez et al., 2014; Jeilani et al., 2011; Antignac et al., 2000) are shown in Fig. 2.4.63A–C and estradiol in negative ion mode (Sun et al., 2005) (Fig. 2.4.63D).

There are now large databases on-line such as KEGG pathway, HMDB, Massbank and Metlin. Information on the application of chemometrics to metabolomics has been developed. The links are: https://www.genome.jp/kegg/pathway.html; https://hmdb.ca/; https://massbank.eu/ and https://metlin.scripps.edu/.

A data-dependent program that scans all precursors ions and collects their MS/MS spectra sequentially is called

SWATH (sequential window acquisition of all theoretical fragment ion spectra) can provide information for quantitative analysis and structure (Drotleff et al., 2018).

In LC-MS/MS, the scan mode is only used to resolve an observation suspecting a co-eluting component. The analyst must be cognizant of co-eluting isobaric steroids as seen for 20α- and 20β-cortolone with cortisol in saliva and urine which was resolved by improved chromatography with a biphenyl column (Israelsson et al., 2018) (Fig. 2.4.64). The technology has been used with plasma, urine, saliva, tears, tissues, etc. The use for specific clinical issues are discussed in the clinical chapters.

2.4.7.1 Derivatives in LC-MS/MS

Chemical derivatization in LC-MS has been developed from the 1980s. Since then, there has been a steady growth of derivatization-based LC-MS techniques, which provide a promising strategy that has solved many analytical problems. As with GC-MS the extra time consuming step is not desirable in clinical laboratories. Derivatization aims to modify the structure of the target compounds and, as a consequence, the chemical and physical properties (Santa, 2011). The advantages of integrating derivatization with LC-MS analysis however include:

(1) improvement of selectivity and separation;
(2) enhancement of ionization efficiency;

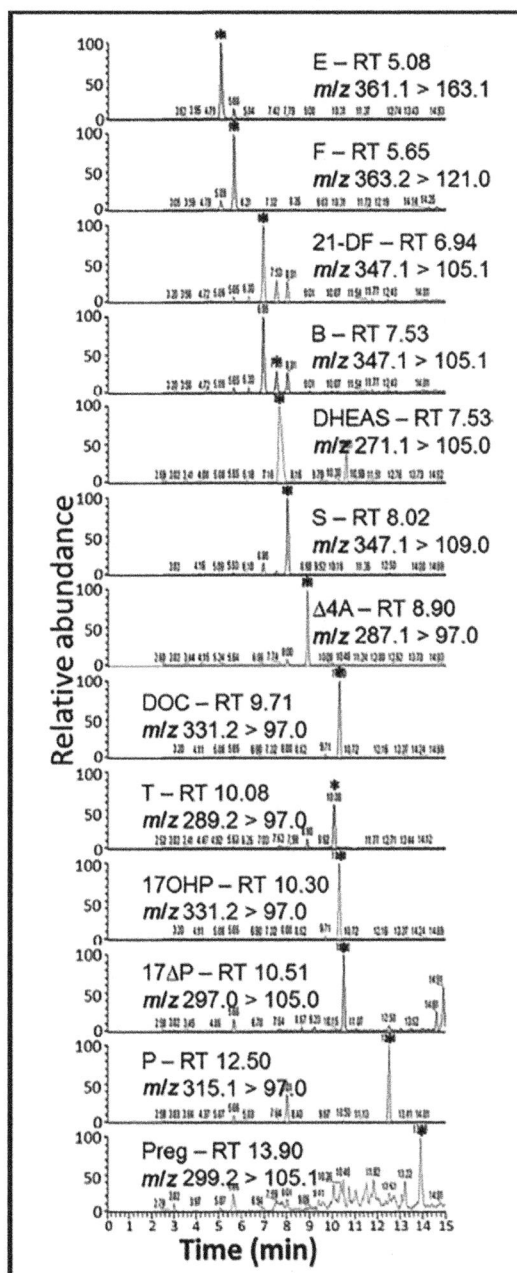

FIG. 2.4.62 Single extracted ion chromatograms showing retention times (RT) and mass to charge (*m/z*) ratio of quantifier ion for 13 steroids used in detection of adrenal steroids in serum of patients with adrenocortical carcinoma—*B*, corticosterone; *DA4*, androstenedione; *DOC*, 11-deoxycorticosterone; *E*, cortisone; *F*, cortisol; *P*, progesterone; *Preg*, pregnenolone; *S*, 11-deoxycortisol; *T*, testosterone; *17DP*, 17-hydroxypregnenolone; *17OHP*, 17-hydroxyprogesterone; *21-DF*, 21-deoxycortisol. An Accucore reverse phase C18 column (100 × 2.1 mm) was eluted with a gradient of water and methanol with 0.1% formic acid. *(From Taylor DR, Ghataore L, Couchman L, Vincent RP, Whitelaw B, Lewis D, Diaz-Cano S, Galata G, Schulte KM, Aylwin S, Taylor NF. A 13-steroid serum panel based on LC-MS/MS: use in detection of adrenocortical carcinoma. Clin Chem 2017;63(12):1836–46. Fig. 1 p. 1839.)*

(3) improvement of structural elucidation;

(4) removal of endogenous interference; and.

(5) facilitation of isomer separation.

The basic idea for enhancing the detection sensitivity in the positive ESI-MS is the introduction of a proton affinitive or permanently charged moiety to the target steroid. The proton-affinitive derivatization mostly involves the introduction of basic nitrogens, such as amino groups. The representative permanently charged moieties are the quaternary ammonium and pyridinium salts.

Dansyl chloride is a popular reagent for esterifying the hydroxyl groups to enhance sensitivity of estrogen analysis. Pentafluorobenzyl, pyridinyl, dansyl, picolinyl, pyridinyl-3-sulfonyl derivatives enhance ESI signal. Shackleton developed a method for testosterone fatty acid esters derivatized with Girard reagent (GP,1-(carboxymethyl) pyridinium chloride hydrazide; Shackleton et al., 1997). The ESI spectra were essentially comprised of molecular ion and $[M-59]^+$ and $[M-87]^+$ product ions. Many reagents suitable for LC-MS/MS derivatives are commercially available but some have been tailor made to meet requirement. A profile of urinary steroid hormones was achieved after derivative formation with dimethylaminobenzoic acid (Dai et al., 2012).

Estrogens are usually analyzed after derivatization due to their low ESI efficiency and occurrence at low concentrations in biological fluids and tissues. Dansylation (Fig. 2.4.65) is now the reference and most practical derivatization procedure for the LC/ESI-MS/MS analyses of estrogens (Huang et al., 2011; Li and Franke, 2015; Higashi and Ogawa, 2016a,b).

Dansyl chloride (DNSCl) rapidly and quantitatively reacts with the phenolic hydroxyl group of estrogens to form the sulfone ester. The dansylation introduces a basic tertiary amino group, which is readily ionized in the commonly used acidic LC mobile phase. The aromatic ring-sulfur atom bond is readily cleaved by the low energy CID to give an intense and characteristic product ion at *m/z*. The generation of pentafluoropropionyl (PFP) derivatives is the most commonly reported approach for estrone and estradiol in serum but the sample preparation is cumbersome. Mobile phase modifiers such ammonium fluoride (Narduzzi et al., 2019) and ammonium hydroxide promote the formation of negative ions.

A number of selective derivative reagents (2-hydrazino-1-methylpyridine and 1,2-dimethyl imidazole-5-sulfonyl chloride) have been used to gain sensitivity for estrogen analysis (Faqehi et al., 2021; Handelsman et al., 2020) although methods without derivatives are being published (Yi et al., 2016). In order to reduce variability of estradiol measurements, immunoassays should not be used for males or postmenopausal women (Handelsman et al., 2014) standardization is recommended (Vesper et al., 2014).

D9-MSTFA has been used for silylation of steroids and GC-MS analysis (Lien et al., 2012).

The measurement of androgens at low concentrations in biological samples including prostatic tissue and saliva is very important for the diagnosis, monitoring the therapeutic efficacy and pathophysiological analysis for diseases of those glands. The methods have been reviewed (Higashi et al., 2016). Hydrazine-based reagents, which react with a keto group of the androgens, have been used for enhancing the sensitivity of the LC/ESI-MS/MS assays for androgens. Girard reagent T (carboxymethyltri-methylammonium chloride hydrazide) (Blair, 2010) (GT, Fig. 2.4.66) is an already-known derivatization reagent for carbonyl compounds to form hydrazones with a quaternary ammonium moiety, and was used for the profiling of androgens in serum.

Griffiths and co-workers developed methods for analysis of the oxysterols (Griffiths and Wang, 2018).

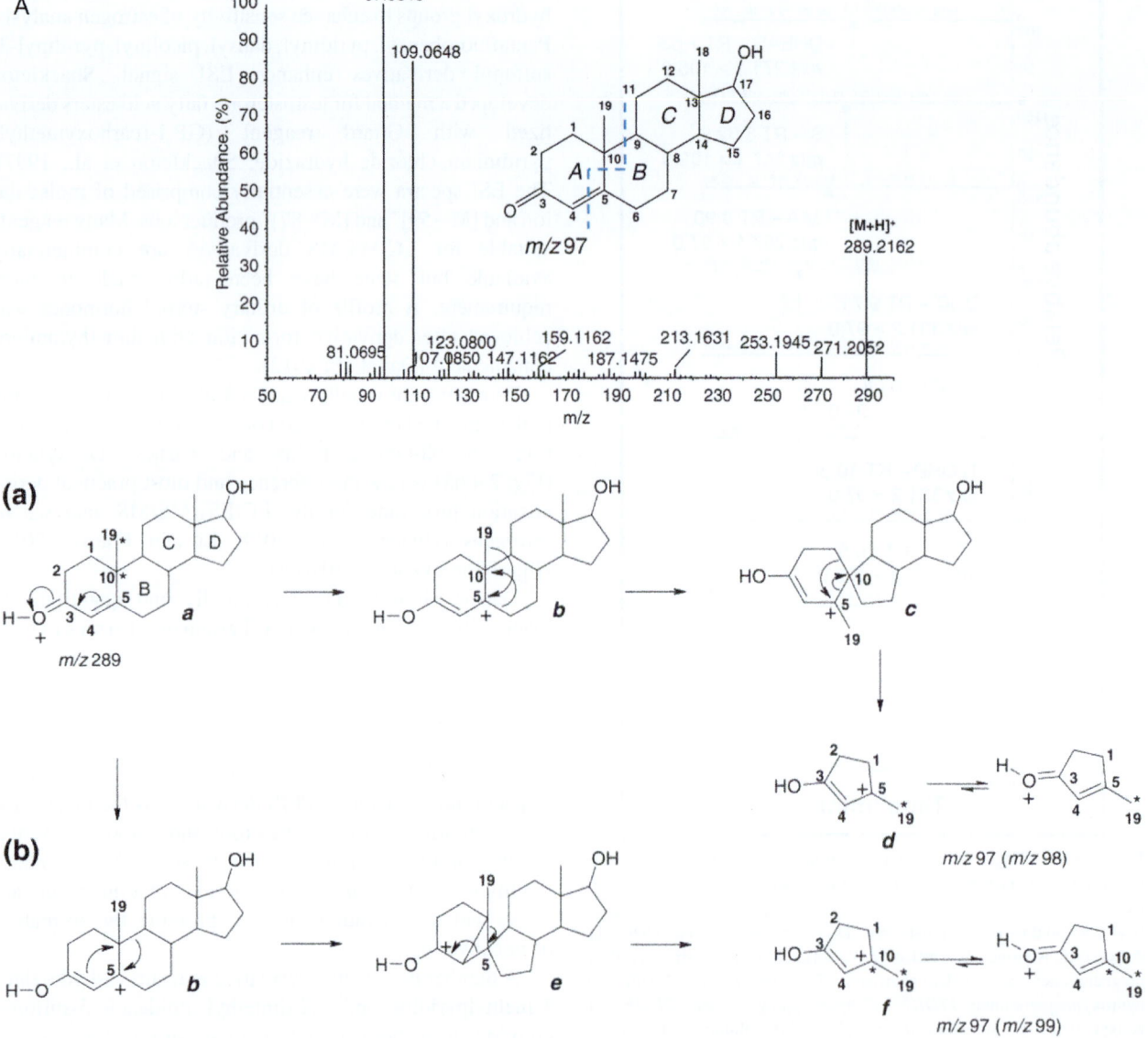

FIG. 2.4.63 (1) ESI product ion spectrum and fragmentation pattern of protonated testosterone at *m/z* 289. The model substance 8a-methyl-3,4,8,8a-tetrahydro-1.6 (2H, 7H) was analysed for confirmation. *Isotopically labeled carbon atoms. (2) Structure elucidation of the diagnostic product ion at *m/z* 97 derived from androst-4-en-3-one-based steroids by ESI-CID and IRMPD spectroscopy. *From Thevis M, Beuck S, Höppner S, Thomas A, Held J, Schäfer M, Oomens J, Schänzer W. Structure elucidation of the diagnostic product ion at m/z 97 derived from androst-4-en-3-one-based steroids by ESI-CID and IRMPD spectroscopy. J Am Soc Mass Spectrom. 2012;23(3):537–46. Fig 1 P 537 and Scheme 2 P544.*

(Continued)

B

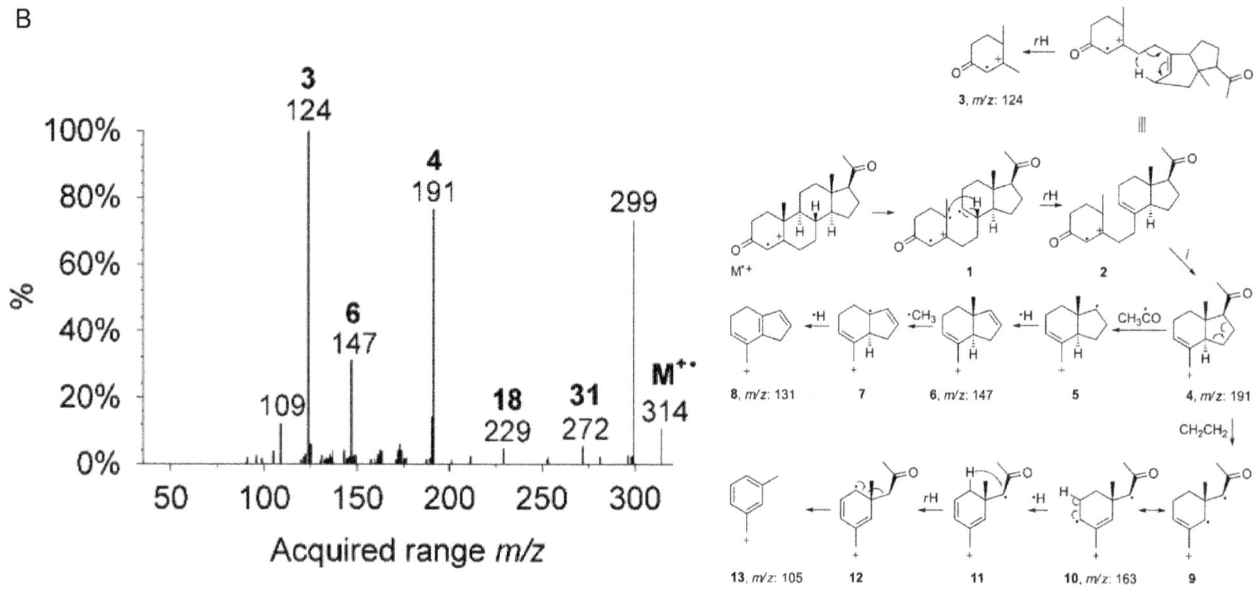

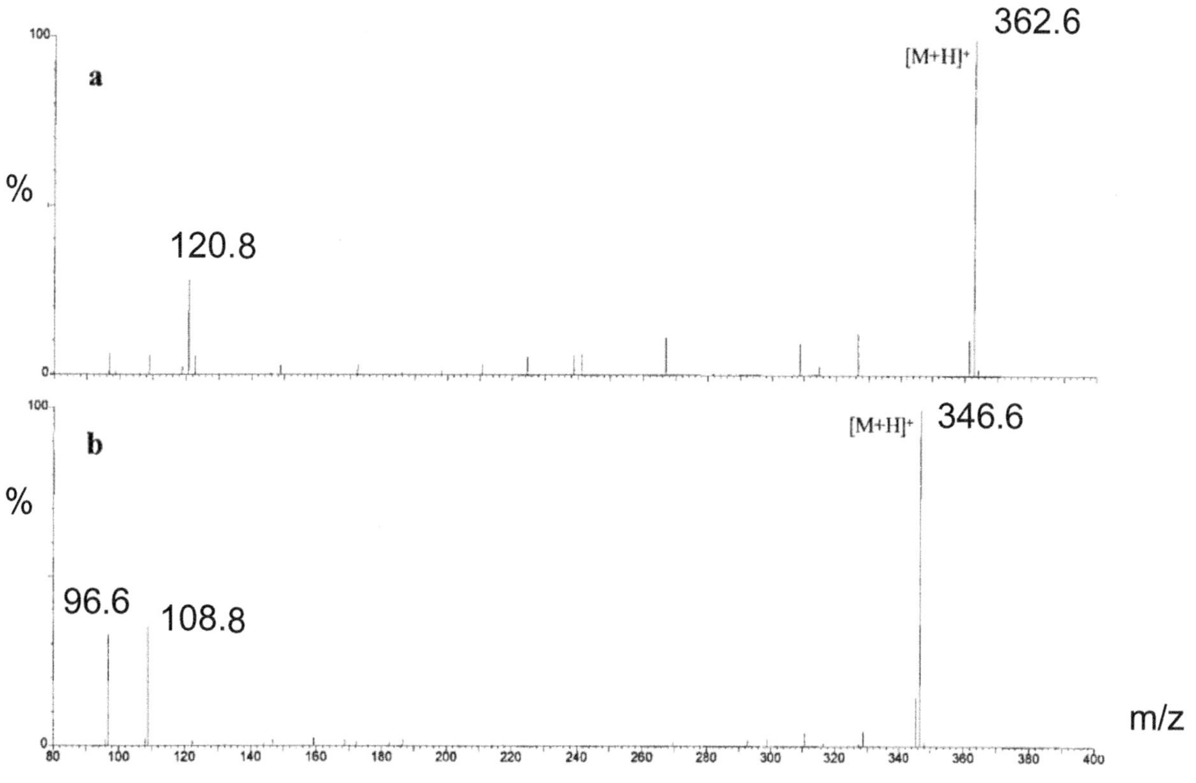

FIG. 2.4.63, CONT'D   Product ion spectrum of progesterone with molecular ion at m/z 314 (upper panel) and ring B cleavage mechanism (lower panel). *(Redrawn from Jeilani YA, Cardelino BH, Ibeanusi VM. Hydrogen rearrangement and ring cleavage reactions study of progesterone by triple quadrupole mass spectrometry and density functional theory. J Mass Spectrom. 2011;46(7):625–34. Redrawn from Fig 2 p 627 and scheme 1 p 629.)* (C) ESI positive MS//MS of [M + H]⁺ spectra of cortisol (upper panel) and deoxycortisone (lower panel) for 20 V collision energy. (b) Differs from (a) only by a 11-hydroxyl group that favors the m/z 97 and 109 fragment ions. *(From Antignac JP, Le Bizec B, Monteau F, Poulain F, André F. Collision-induced dissociation of corticosteroids in electrospray tandem mass spectrometry and development of a screening method by high performance liquid chromatography/tandem mass spectrometry. Rapid Commun Mass Spectrom. 2000;14(1):33–9. Fig 3 p35.)*

(Continued)

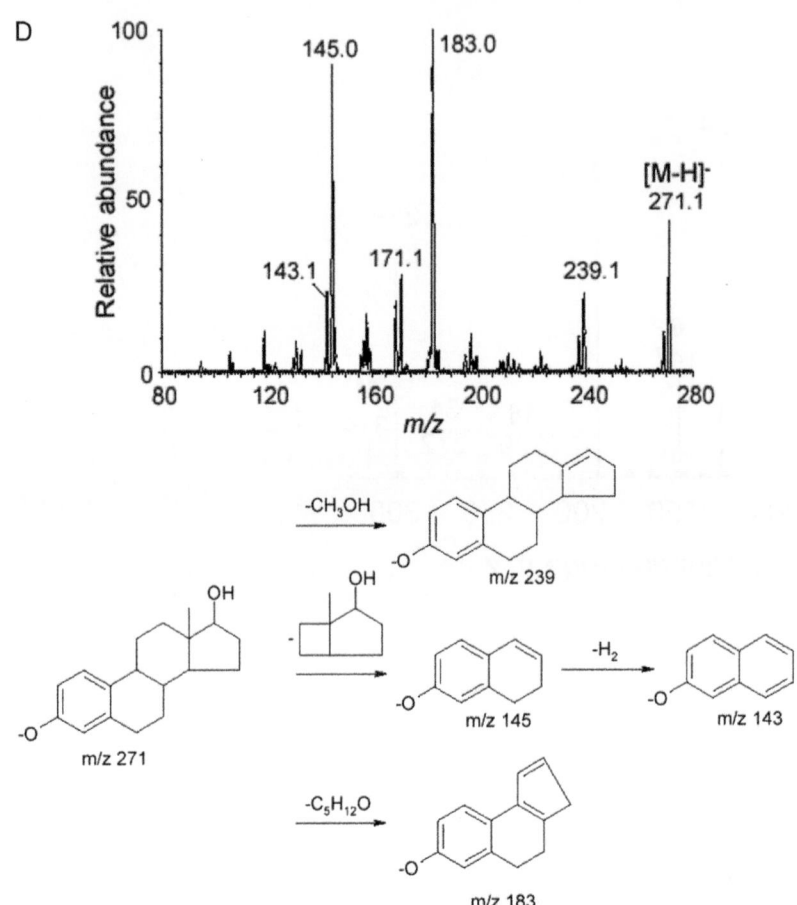

FIG. 2.4.63, CONT'D (D) Negative ion electrospray with CID of estradiol (upper panel) and proposed fragmentation pathways of the deprotonated molecule of E2 (lower panel). *From Sun Y, Gu C, Liu X, Liang W, Yao P, Bolton JL, van Breemen RB. Ultrafiltration tandem mass spectrometry of estrogens for characterization of structure and affinity for human estrogen receptors. J Am Soc Mass Spectrom 2005;16(2):271–9. Redrawn for Fig 4 P 13; Scheme 2 P 15.*

The methods included the derivatization of oxysterols to the hydrazones with Girard reagent P (1-(carboxymethyl) pyridinium chloride hydrazide)). The method was refined so that the 3-ketosterols were derivatized with $^2H_0$-GP, whereas those having the 3β-hydroxy groups were enzymatically oxidized and derivatized with $^2H_5$-GP, and the $^2H_0$- and $^2H_5$-derivatives were mixed and subjected to LC/ESI-MS/MS. $^2H_5$-GP was synthesized by reacting $^2H_5$-pyridine with ethyl bromoacetate, then hydrazine. The oxime-derivatives with hydroxylamine and methoxylamine have also been examined for T analysis.

The salivary concentration of steroids is used as a noninvasive alternative to the blood when monitoring treatment such as 17(OH)P4 in CAH monitoring. A major disadvantage in the use of saliva is the low steroid concentrations, but the sensitivity problem has been resolved for 17-OHP by 2-hydrazino-pyridine-derivatization (LLOQ, 5 pg/mL) (Shibayama et al., 2008). For the analyses of neuroactive androgens, androsterone and 3α,5α-Adiol, in the **rat brain** and serum, isonicotinoyl azide was used as the derivatization reagent (Higashi et al.,

2008a,b). The resulting carbamates showed good fragmentation characteristics during the low-energy CID. In contrast to other reported derivatization reagents, that have been used to enhance the sensitivities for estrogens or androgens, this method is comprehensive with the capability of covering hydroxyl-containing androgens, estrogens, corticoids, and progestogens. Furthermore, the nonderivatized steroid hormones (e.g., 17α-hydroxyprogesterone, progesterone, and androstenedione) are not destroyed during the derivatization process, and their levels could still be obtained in one LC-MS run. Twenty four hormones at sub ng/mL levels (the lower limit of detection could reach 5 pg/mL for estrone and 16α-hydroxy estrone, which is equivalent to 0.1 pg on column) with maximum sensitivity enhancement factors of more than 103- to 104-fold after derivatization. The method was successfully applied to the measurement of free (unconjugated) steroid hormones in urine samples of males, females, and pregnant women (Dai et al., 2012).

As with GC-MS where stable isotope labeled internal standards are not available isotope labeled analogs of the

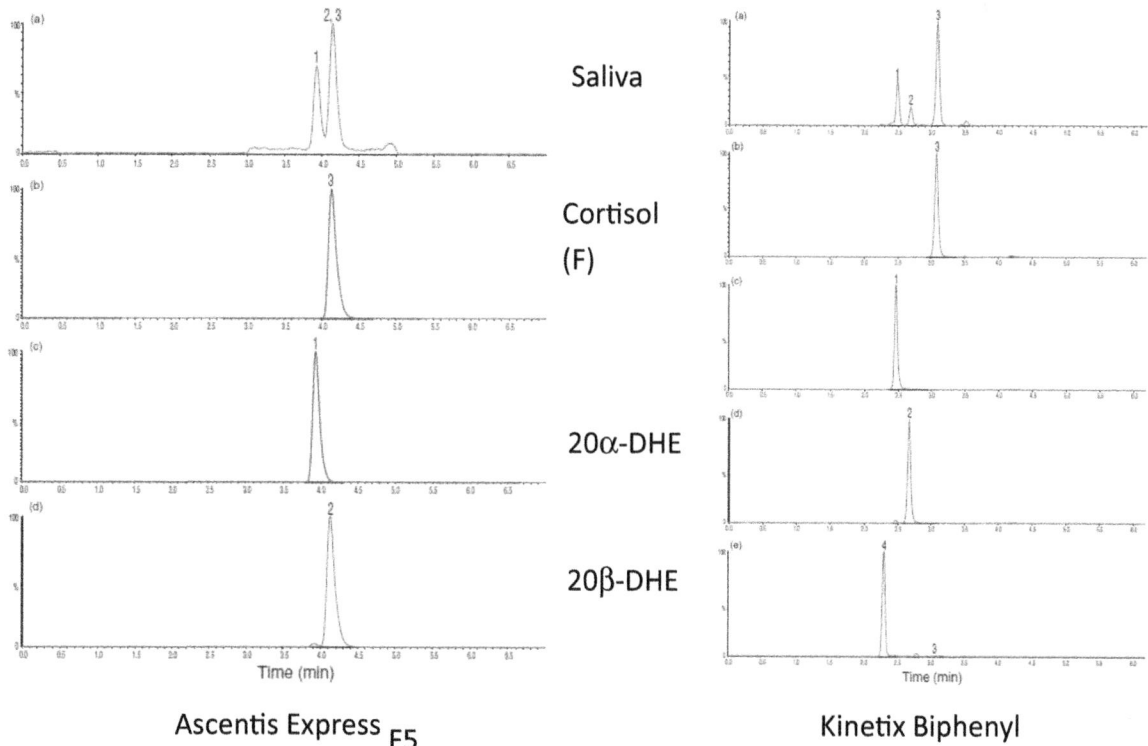

FIG. 2.4.64 Separation of dihydrocortisones from cortisol in saliva on LC with (left panel) Ascentis Express F5 eluted with water (A) and methanol (B) both containing 0.1% formic acid in gradient from 70:30 A/B to 44:56 A/B and (right panel) Kinetex Biphenyl eluted with water and acetonitrile (C) both with 0.1 formic acid in gradient 95:5 A/C to 76:24 A/C. The chromatograms show transition m/z 363.2 > 121.1. 1 represents 20α-dihydrocortisone; 2 is 20β-dihydrocortisone and 3 represents cortisol. *(From Israelsson M, Brattsand R, Brattsand G. 20α- and 20β-dihydrocortisone may interfere in LC-MS/MS determination of cortisol in saliva and urine. Ann Clin Biochem 2018;55(3):341–7. Fig. 3 Fig. 5).*

derivatizing reagent (stable isotope coded) can be used to create labeled internal standards although the label is not actually in the steroid itself. A novel targeted metabolic profiling method is based on the introduction of an easily protonated stable isotope tag to a hydroxyl-containing steroid hormone with a synthesized derivatization reagent, such as deuterium 4-(dimethylamino)-benzoic acid (d4-DMBA) (Liu et al., 2003), and liquid chromatography-mass spectrometry (LC-MS). Griffiths used stable isotope coded Girard reagent (see above, Fig. 2.4.66). d4-DMBA and d5 Girard reagent P to create internal standards for hydroxyl groups and ketone groups, respectively, when measuring free steroids in urine (Lai et al., 2001; Dai et al., 2012).

2.4.8 Plasma steroid profiles

Tandem mass spectrometry (LC-MS/MS) is now the most attractive method for analysis of steroids with little pre-treatment of the sample but the validation of the method requires extensive tests to prove the specificity of the analysis. Although the costs of such apparatus and staff at the present time are high and the level of expertise needed to maintain the equipment seems to mitigate against its

routine use, the reagent costs are low compared with immu-noassay kits and this methodology is becoming more widely used.

In quantitative analysis, tandem mass spectrometers can be used in selected ion monitoring (SIM) and multiple reaction monitoring (MRM) modes. The latter compounds with a common structure can be detected using neutral loss and precursor ion scan modes. Ions for the free steroid are weak. The various molecular ion species for the free steroid hormones have to be considered, e.g., $[M+H]^+$, $[M+NH_4]^+$, $[M+H-nH_2O]^+$, $[M+H+MeOH]^+$ in positive mode and in negative mode $[M-H]^-$, $[M-2H]^-$, $[M-3H]^-$, $[2M-H]^-$, in negative mode with adducts $[M+CH_2O_2-H]^-$, $[M+Na-H]^-$, $[M+Cl]^-$, $[M+K-2H]$; and $[2M+CH_2O_2]^-$ determined by the LC experimental conditions. A typical analysis of standards (Koal et al., 2012) is shown in Fig. 2.4.67 using the Biocrates SteroIDQ kit which provides a calibrator set, QC samples at three levels, internal standard mix, calibration matrix and three solutions for the water in mobile phase A; mobile phase B is acetonitrile/methanol/water.

Recent research on androgens has suggested that 11-oxygenated androgens are significant drivers of androgenic activity in addition to classical androgens

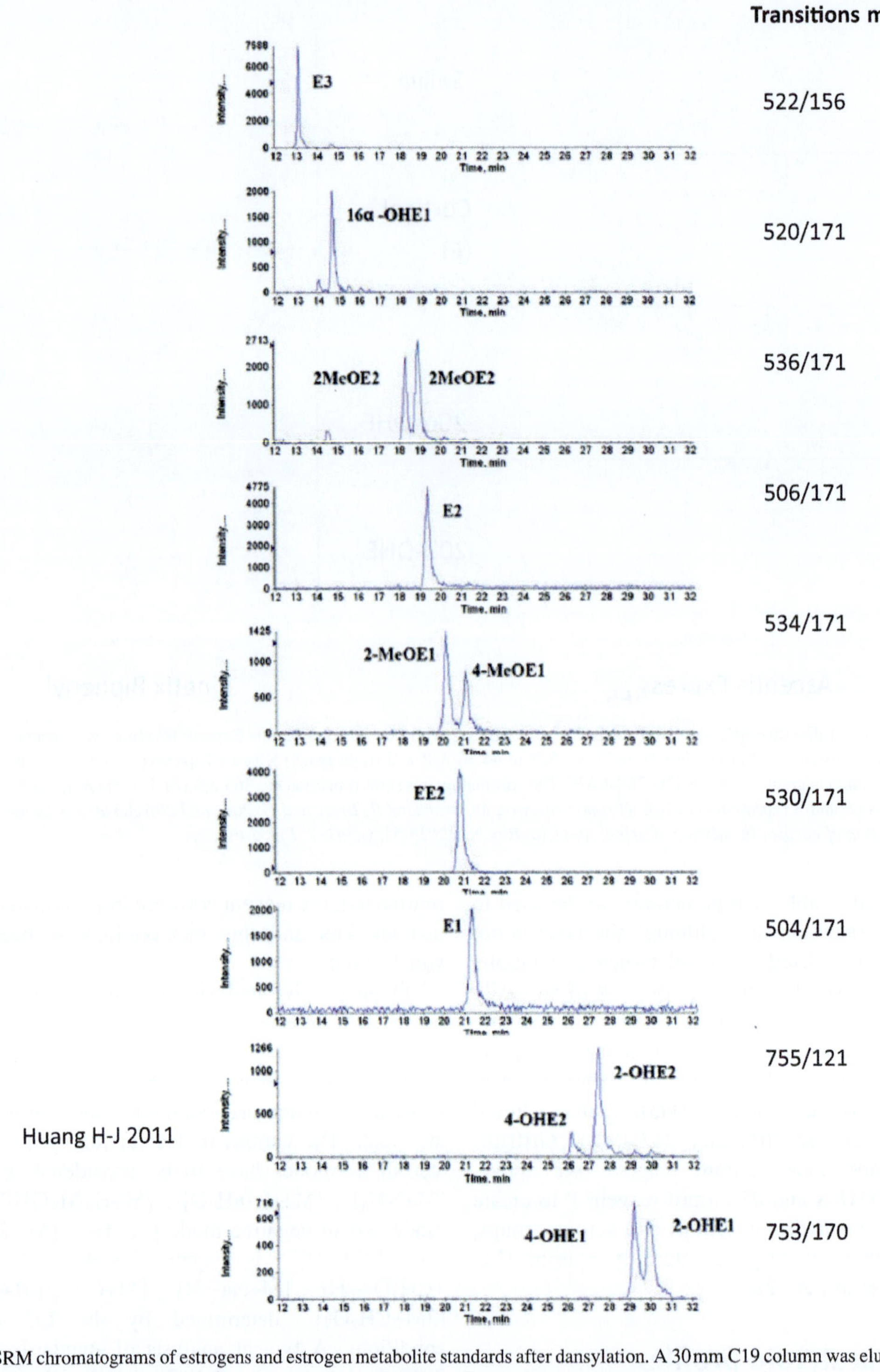

Transitions m/z

FIG. 2.4.65 SRM chromatograms of estrogens and estrogen metabolite standards after dansylation. A 30 mm C19 column was eluted with gradient from mixtures of water and acetonitrile with 0.1% formic acid. *(From Huang HJ, Chiang PH, Chen SH. Quantitative analysis of estrogens and estrogen metabolites in endogenous MCF-7 breast cancer cells by liquid chromatography-tandem mass spectrometry. J Chromatogr B Analyt Technol Biomed Life Sci 2011;879(20):1748–56. Fig. 2 p. 1750.)*

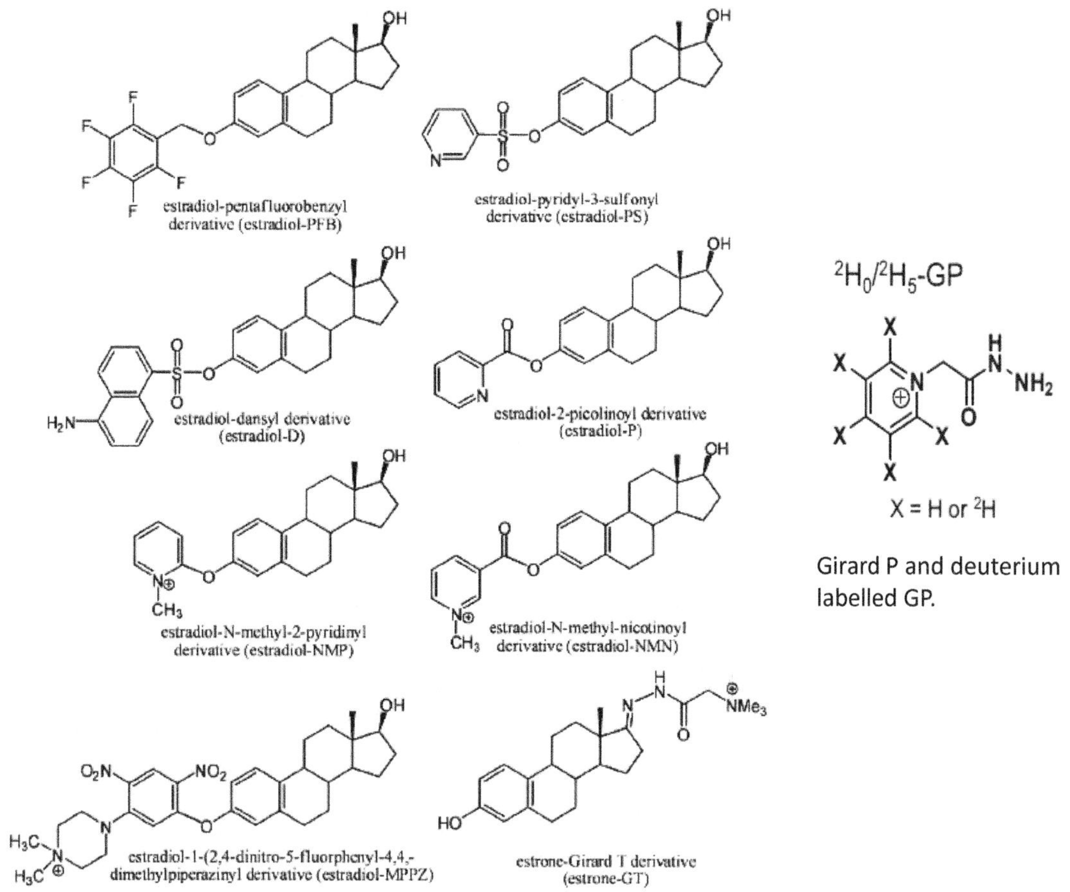

FIG. 2.4.66 Examples of derivatives to enhance sensitivity of estrogen determinations with GC-MS/MS and LC-MS/MS On the right is Girard P reagent which is used to derivatize ketone groups. A coded internal standard has five deuterium atoms at X. *(From Blair IA. Analysis of estrogens in serum and plasma from postmenopausal women: past present, and future. Steroids 2010;75(4–5):297–306. Fig. 4 p. 20.)*

(Auer et al., 2021; Schiffer et al., 2019a,b). Profiles have increased in number so that 16 steroids are measured with 13 isotopomeric internal standards (Koal et al., 2012). For etiocholanolone, the response is related to that from labeled d_3 androsterone; DHEA uses d_4-estrone and deoxycorticosterone is related to d_8-17-hydroxyprogesterone (Table 2.4.6).

That is not ideal because there was no internal standard to compensate for any sample matrix effects beyond those three steroids. The Biocrates commercial kit for steroids has been validated for 16 steroids and shown to achieve LLOQ 0.01–32 ng/mL and ULOQ at 5.0–8000 ng/mL for testosterone and DHEAS, respectively; estradiol was 0.02–20 ng/mL. The analysis time was less than 10 min.

In a second example (Table 2.4.7), the internal standards are a mixture of deuterium labeled, ^{13}-carbon labeled and some with both ^{13}C and deuterium (Boggs et al., 2016).

The methods monitored similar transitions to other examples and are typical of transitions used for steroid paneling giving similar profiles (Fig. 2.4.68).

Although many methods have been published for the analysis of individual steroids by LC-MS/MS, most laboratories will soon be measuring several steroids in one analysis. This can be 5, 10, 17 even up to 100 steroids depending on the need for the results. Most methods use ESI but a few have used APCI or less often APPI. The most reliable data is with the monitoring of two transitions called a quantifier and a qualifier ion. Transitions losing water should be avoided because the reaction is nonspecific. Many transitions include release of fragments from the A-ring which are not unique. Most methods use low resolution, but high-resolution analysis is also being used and has higher specificity. Examples are progesterone metabolites in newborn urine (Bileck et al., 2019), antidoping (Abushareeda et al., 2017, 2018), mixture of 44 steroids including metabolites of progesterone, cortisol, natural and synthetic androgens (Kaabia et al., 2018); 24 steroids in urine suitable for detecting adrenal disorders (Hines et al., 2017); steroids in serum (Matysik and Schmitz, 2015); testosterone, epitestosterone and androstenedione in urine for antidoping purpose (Frey et al., 2016).

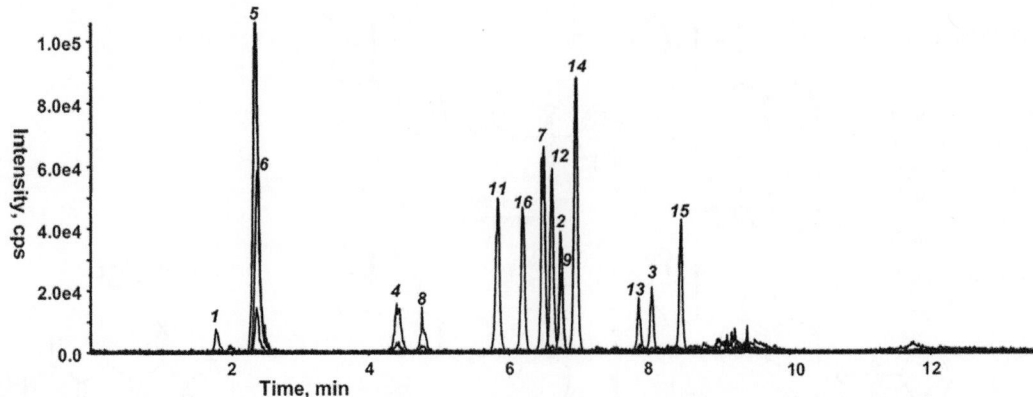

FIG. 2.4.67 Extracted ion chromatogram from HPLC-MS/MS of a calibrator mixture. (1) Aldosterone (0.2 ng/mL); (2) androstenedione (0.13 ng/mL); (3) androsterone (0.24 ng/mL); (4) corticosterone (0.6 ng/mL); (5) cortisol (20 ng/mL); (6) cortisone (2.0 ng/mL); (7) 11-deoxycorticosterone (0.6 ng/mL); (8) 11-deoxycortisol (0.2 ng/mL); (9) DHEA (0.48 ng/mL); (10) DHEAS (128 ng/mL); (11) estradiol (0.4 ng/mL); (12) estrone (0.4 ng/mL); (13) etiocholanolone (0.24 ng/mL); (14) 17OHP (1 ng/mL); (15) progesterone (0.24 ng/mL); (16) testosterone (0.2 ng/mL). *(From Koal T, Schmiederer D, Pham-Tuan H, Röhring C, Rauh M. Standardized LC-MS/MS based steroid hormone profile-analysis. J Steroid Biochem Mol Biol 2012;129 (3–5):129–38. Fig. 2 p. 132.)*

TABLE 2.4.6 Analytical parameters of the SteroIDQ kit: 16 analytes and deuterium-labeled internal standards pairs, MRM of analytes, analyte retention times, calibration ranges.

No.	Analyte	Internal standard	Analyte MRM	Retention time (min)	Calibration range (ng/mL)
1	Aldosterone	D7-aldosterone	361.2/343.3	1.77	0.05–5
2	Androstenedione	D3-androstenedione	287.2/97.1	6.74	0.03–8
3	Androsterone	D4-androsterone	273.2/255.1	8.05	0.06–6
4	Corticosterone	D8-corticosterone	347.2/329.1	4.39	0.03–30
5	Cortisol	D4-cortisol	363.2/345.1	2.34	1–1000
6	Cortisone	D7-cortisone	361.2/163.1	2.37	0.0.–100
7	11-Deoxycorticosterone	D8-17-OHP	331.2/109.1	6.49	0.03–15
8	11-Deoxycortisol	D5-11-deoxycortisol	347.2/109.1	4.75	0.01–10
9	DHEA	D4-E1	271.2/253.2	6.75	0.12–30
10	DHEAS	D5-DHEAS	271.2/253.2	2.78	32–8000
11	E2	D3 E2	255.2/159.1	5.84	0.02–20
12	E1	D4 E1	271.2/253.2	6.62	0.03–15
13	Etiocholanolone	D4 androsterone	273.2/255.1	7.86	0.06–6
14	17-OHP	D8 17-OHP	331.2/109.1	6.96	0.05–50
15	Progesterone	D9 progesterone	315.2/109.1	8.46	0.06–15
16	Testosterone	D5 testosterone	289.2/97.0	6.19	0.01–10

From Koal T, Schmiederer D, Pham-Tuan H, Röhring C, Rauh M. Standardized LC-MS/MS based steroid hormone profile-analysis. J Steroid Biochem Mol Biol 2012;129(3–5):129–38.

TABLE 2.4.7 Parameters for steroid analysis using liquid chromatography mass spectrometry.

Steroid name	Abbreviation	Absolute mass	Ion reaction	Precursor ion (m/z)	Product ion (m/z)	DP (V)	EP (V)	CE (V)	CXP (V)	Retention time (min)
Androstenedione	AE	286.4	$[M+H]^+$	287.2	97.2	100	15	30	12	17.33
Androsterone $^{13}C_3$	AE $^{13}C_3$	289.4	$[M+H]^+$	290.6	100.2	75	10	40	5	17.33
DHEA	DHEA	288.4	$[M+H-2H_2O]^+$	253.3	197.3	75	10	25	10	11.21
Dihydrotestosterone	DHT	290.4	$[M+H]^+$	291.3	255.4	75	10	25	10	13.62
Testosterone	T	288.4	$[M+H]^+$	289.1	97.2	100	15	30	12	14.11
Testosterone $^{13}C_3$	T $^{13}C_3$	291.6	$[M+H]^+$	292.6	112.2	75	10	40	5	14.08
Estradiol	E_2	272.4	$[M+H-H_2O]^+$	255.3	159.2	75	10	25	10	10.91
Estradiol $^{13}C_3$	E_2 $^{13}C_3$	275.4	$[M+H-H_2O]^+$	258.2	162.2	50	10	35	10	10.91
Estriol	E_3	288.4	$[M+H-H_2O]^+$	271.3	133.4	75	10	25	10	10.01
Estrone	E_1	270.4	$[M+H]^+$	271.1	150.3	75	10	30	10	12.84
17-OH-Preg	17-OH P_5	332.5	$[M+H-H_2O]^+$	315.4	297.3	40	5	20	5	10.97
17-OH Prog	17-OH P_4	330.4	$[M+H]^+$	331.2	97.2	75	10	25	10	13.85
17-OH Prog $^{13}C_3$	17-OHP$_4$ $^{13}C_3$	333.4	$[M+H]^+$	334.6	100.1	40	5	40	5	13.88
Preg	P_5	316.5	$[M+H-H_2O]^+$	299.6	161.4	75	15	25	10	13.37
Preg $^{13}C_2$-d$_2$	P_5-$^{13}C_2$-d$_2$	320.5	$[M+H-H_2O]^+$	303.6	161.3	40	5	40	10	13.36
Prog	P_4	314.5	$[M+H]^+$	315.1	97.2	75	15	25	10	19.21
Prog $^{13}C_2$	P_4 $^{13}C_2$	316.5	$[M+H]^+$	317.6	99.1	75	10	45	15	19.18
Prog $^{13}C_3$	P_4 $^{13}C_3$	317.5	$[M+H]^+$	318.4	112.2	50	10	35	5	19.18
11-Deoxycorticosterone	DOC	330.5	$[M+H]^+$	331.7	109.3	75	10	35	5	17.10
11-Deoxycortisol	S	346.5	$[M+H]^+$	347.3	97.1	75	15	30	10	12.97
Corticosterone	B	346.5	$[M+H]^+$	347.2	121.3	40	5	20	15	13.18
Cortisol	F	362.5	$[M+H]^+$	363.1	121.1	75	10	25	10	11.36
Cortisol-d4	F-d4	366.5	$[M+H]^+$	367.4	121.3	75	10	25	10	11.32

From Boggs AS, Bowden JA, Galligan TM, Guillette LJ Jr, Kucklick JR. Development of a multi-class steroid hormone screening method using liquid chromatography/tandem mass spectrometry (LC-MS/MS). Anal Bioanal Chem 2016;408(15):4179–90.

2.4.8.1 Conjugates

The phase II metabolites contain acidic or basic moieties in their structures and therefore, these chemical centers guide the ionization process. The ionization of conjugated steroids is therefore less dependent of the steroid structure and is similar to the behavior observed for the same type of conjugates, independently of the rest of the chemical structure. Adduct formation can be a problem with the steroid conjugate ions having the addition of groups from the solvent mixture or dopant.

Steroid sulfates are almost exclusively determined in negative mode as $[M-H]^-$ (Mitamura et al., 2014) (Fig. 2.4.69).

The efficient ionization of sulfates increases the sensitivity compared with the unconjugated steroids. The spectra for sulfated steroids have a strong ion for the sulfate moiety

at 97 (96.96 in HR-MS) amu. The CID spectra of THF and allo-THF di sulfates are shown in Fig. 2.4.70.

The proposed fragmentations are shown in Fig. 2.4.71.

The ionization of steroid glucuronides as Me-TMS derivatives in negative ionization mode is guided by the carboxylic acid of the glucuronide moiety (Choi et al., 2000). Consequently, the ionization behavior of steroids having a glucuronide group conjugated at different positions showed the same ionization pattern for all metabolites in negative electrospray ionization. A prominent $[M-H]^-$ is found in the spectra of all of glucuronides together with other minor ions such as the $[M-H+Na+HCOO]^-$ ion for androgens (Fig. 2.4.72) (Cho et al., 2009) and corticosteroids (Fig. 2.4.73) (Ikegawa et al., 2009).

When analyzing the glucuronides in negative ion mode the spectra can show $[M-H]^-$, $[M-H-Glu]^-$, $[M-H_2O]$

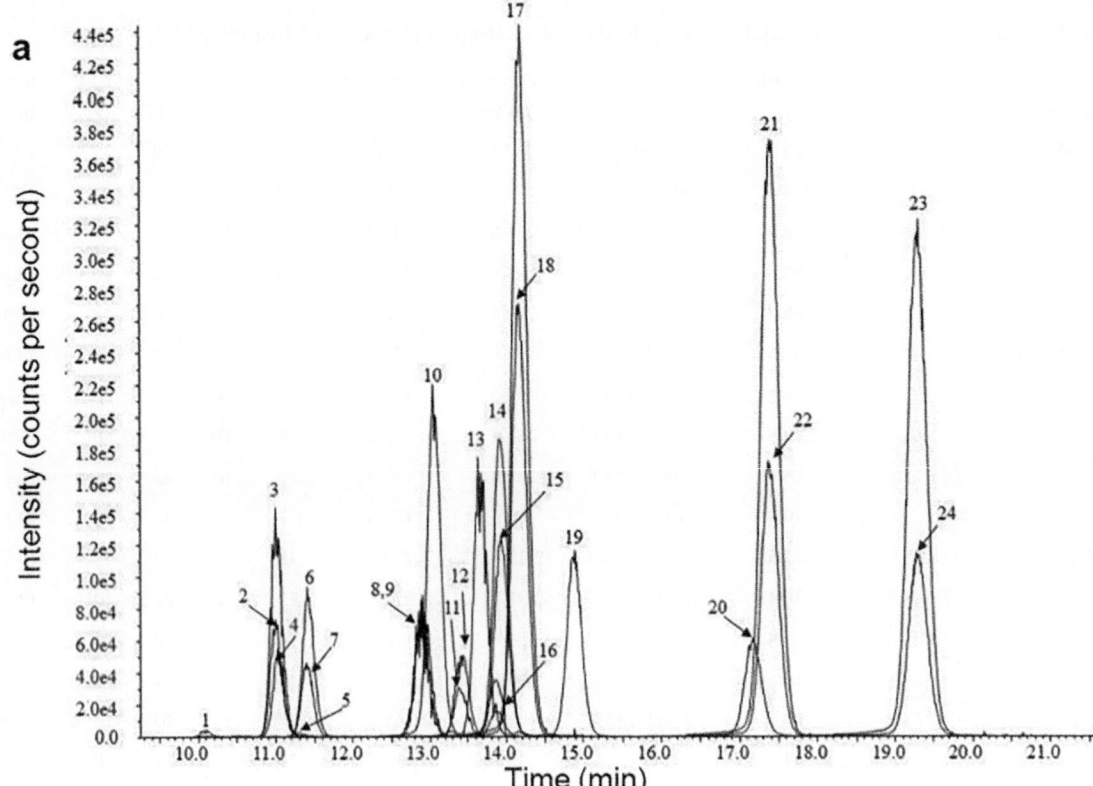

FIG. 2.4.68 Chromatograms for separation of steroids in 100 ng calibration mixture. A Restek Ultra Biphenyl column 250×4.6 mm was used and eluted with gradient using methanol and acetonitrile with 0.1% formic acid at 80% to 55%. The MS was an AB Sciex API4000 QTRAP. (1) Estriol (E$_3$), (2) estradiol-13C$_3$ (E2-^{13}C$_3$), (3) estradiol (E$_2$), (4) 17-OH-pregnenolone (17-OHP5), (5) dehydroepiandrosterone (DHEA), (6) cortisol (F), (7) cortisol-d$_4$ (F-d$_4$), (8) equilin (EQ), (9) estrone (E1), (10) 11-deoxycortisol (S), (11) pregnenolone-^{13}C$_2$d$_2$ (P5-^{13}C$_2$d$_2$), (12) pregnenolone (P5), (13) dihydrotestosterone (DHT), (14) 17-OH-Progesterone (17-OH-P4), (15) 17-OHProgesterone-^{13}C$_3$ (17-OHP4-^{13}C$_3$), (16) corticosterone (B), (17) testosterone (T), (18) testosterone-^{13}C$_3$ (T-^{13}C$_3$), (19) adrenosterone (AT), (20) 11-deoxycorticosterone (DOC), (21) androstenedione (AE), (22) androstenedione-13C$_3$ (AE-^{13}C$_3$), (23) progesterone (P4), (24) progesterone-^{13}C$_2$ (P4-^{13}C$_2$). *(From Boggs AS, Bowden JA, Galligan TM, Guillette LJ Jr, Kucklick JR. Development of a multi-class steroid hormone screening method using liquid chromatography/tandem mass spectrometry (LC-MS/MS). Anal Bioanal Chem 2016;408(15):4179–90. Fig. 2 p. 4186.)*

or [M−H−CH$_2$O]. The mechanisms for fragmentations of THF-3-glucuronide is shown in Fig. 2.4.74 and THF-21-glucuronide in Fig. 2.4.75.

Spectra for androgens can vary with collision energy (Figs. 2.4.76 and 2.4.77), mobile phase (with and without dopant) and ion mode (Table 2.4.8) (Fabregat et al., 2013a,b; Pozo et al., 2008a,b).

Some glucuronides have been analyzed by LC-MS/MS after picolinyl derivatization (Yamashita et al., 2008) (Fig. 2.4.78).

Steroidal bisconjugates have long been known as minor components of the urine steroid profile. In the past, analysis required laborious chromatographic fractionation and hydrolysis, followed by GC-MS detection. Recent developments in both chemical synthesis and LC-MS analytical technology have created avenues for the direct detection of these minor metabolites. In the scan MS with 70 V cone voltage, the mono-anion [M−H]$^−$ was the major ion observed from **bisglucuronides** with some minor [M−H-

gluc]$^−$ in-source fragmentation (where "gluc" was the dehydrated glucuronic acid moiety (C$_6$H$_8$O$_6$) 176 Da) (Esquivel et al., 2017) (Fig. 2.4.79).

The highest relative abundance for the [M−H-gluc]$^−$ ion appeared for estradiol bisglucuronide. In addition, estradiol bisglucuronide also showed another in-source fragment [M−H-2gluc]$^−$ m/z 271 (15%) in the scan MS. In contrast, scan MS with 26 V cone voltage formed the di-anion [M−2H] as the major ion, while still forming mono-anion [M−H]$^−$ with 50%–100% relative abundance. In-source fragmentation was only observed for estradiol bisglucuronide, giving 5% [M−H-gluc]$^−$. A library of crude chromatographically resolved steroid bisglucuronides is now available. The approach has been to form either mono- or di-anion precursors of the bisglucuronides for subsequent MS/MS studies.

In positive mode, **steroid glucuronides** with a conjugated 3-keto function, exhibit [M+H]$^+$ as the major ion. In contrast, steroid glucuronides with a 17-keto function,

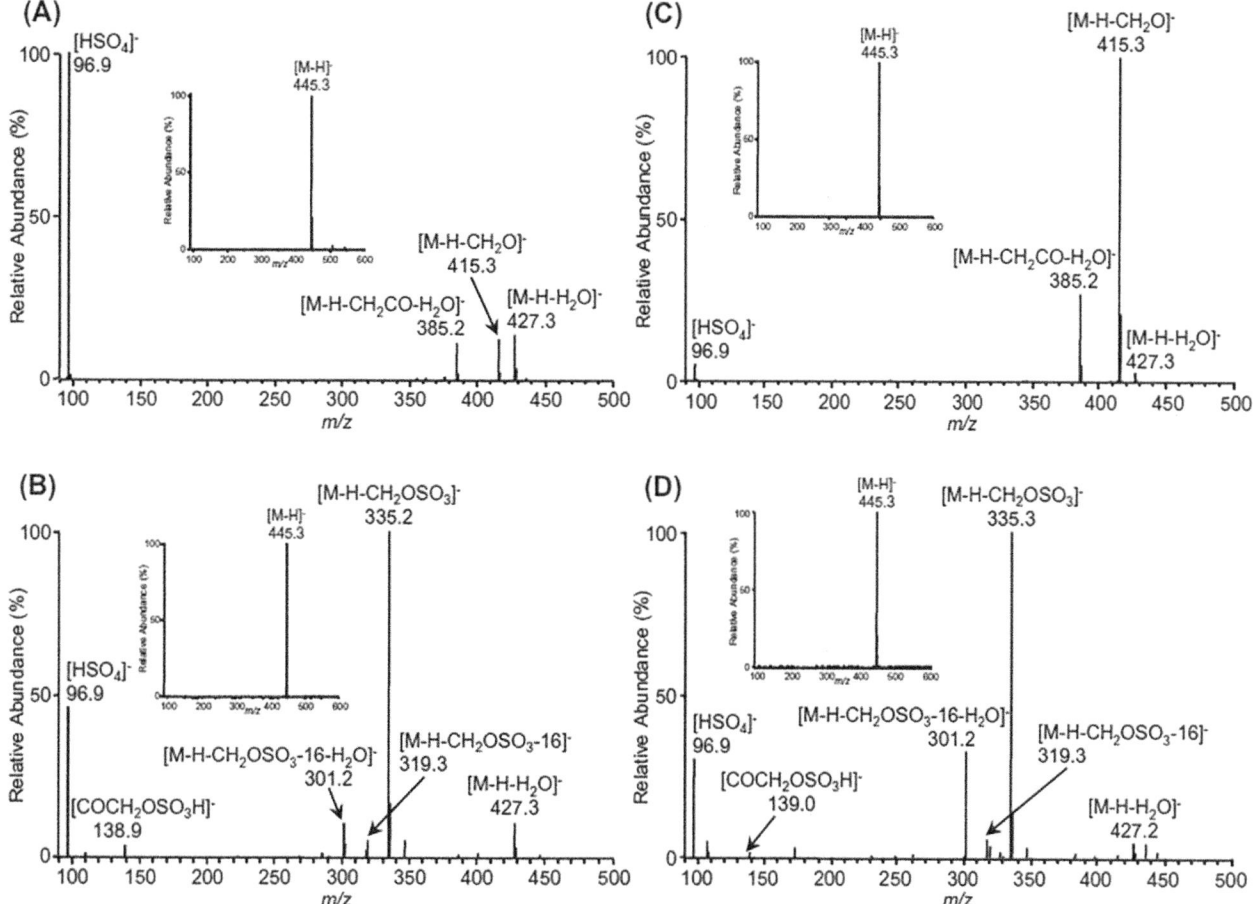

FIG. 2.4.69 Negative ion ESI-MS and MS/MS spectra of (A) THF-3S, (B) THF-21S and (C) alloTHF-3S and allo-THF-21S by CID of [M−H]⁻ *(From Mitamura K, Satoh née Okihara R, Kamibayashi M, Sato K, Iida T, Ikegawa S. Simultaneous determination of 18 tetrahydrocorticosteroid sulfates in human urine by liquid chromatography/electrospray ionization-tandem mass spectrometry. Steroids 2014;85:18–29. Fig. 2 p. 20.)*

an unconjugated 3-keto moiety or without any keto function show the adduct $[M+NH_4]^+$ as the major ion. Other minor ions were obtained for all compounds independently of its structure, like adducts $[M+Na]^+$ and $[M+K]^+$. Collision induced dissociation shows some common patterns in positive and negative modes. In negative mode, ions at m/z 175,157,113, 85 and 73 come from the dehydrated glucuronide (gluc), $[gluc−H]^-$, $[gluc−H−H_2O]^-$, $[gluc−H−H_2O−CO_2]^-$, $[gluc−H−H_2O−CO_2−CO]^-$, and $HOCH_2CO_2^-$. A steroid with a 3-keto group shows a neutral loss of 176D. A CID spectrum does not provide structural information. There is a lack of commercially available standards to establish LC-MS/MS behavior of metabolites. Many papers use the pseudomolecular ion as the precursor ion and ions from fragmentation of the A-ring as product ions. This is not ideal for many steroids because the pattern of fragmentation is common and therefore not specific to one steroid.

Bisglucuronides ionized as $[M+NH_4]^+$ in positive mode, and as $[M−H]^-$ and $[M−2H]^{2-}$ in negative mode. The most specific product ions of steroid bisglucuronides in positive

mode resulted from the neutral losses of 387 and 405 Da (corresponding to $[M+NH_4−NH_3−2gluc−H_2O]^+$ and $[M+NH_4−NH_3−2gluc−2H_2O]^+$, respectively, being "gluc" a dehydrated glucuronide moiety), and in negative mode, the fragmentation of $[M−2H]^{2-}$ showed ion losses of m/z 175 and 75 (gluc− and $HOCH_2CO_2^-$, respectively).

Collision induced dissociation of steroids gives rise to transitions but the mechanisms are complicated (Esquivel et al., 2017) (Fig. 2.4.80). The analyst does not need to understand the mechanisms although it can be useful when selecting the most appropriate ions for a new analysis.

Bis-sulfates ionized preferentially as the dianion ($[M−2H]^{2-}$) with a small contribution of the monoanion ($[M−H]^-$) (McLeod et al., 2017). Product ion spectra generated from the $[M−2H]^{2-}$ precursor ions were dominated by the loss of HSO_4^- to generate two product ions, that is, the ion at m/z 97 (HSO_4^-) and the ion corresponding to the remaining monosulfate fragment (McLeod et al., 2017) (Fig. 2.4.81).

Other product ions were found to be specific for some structures. As an example, the loss of $[CH_3+SO_3]^-$ was

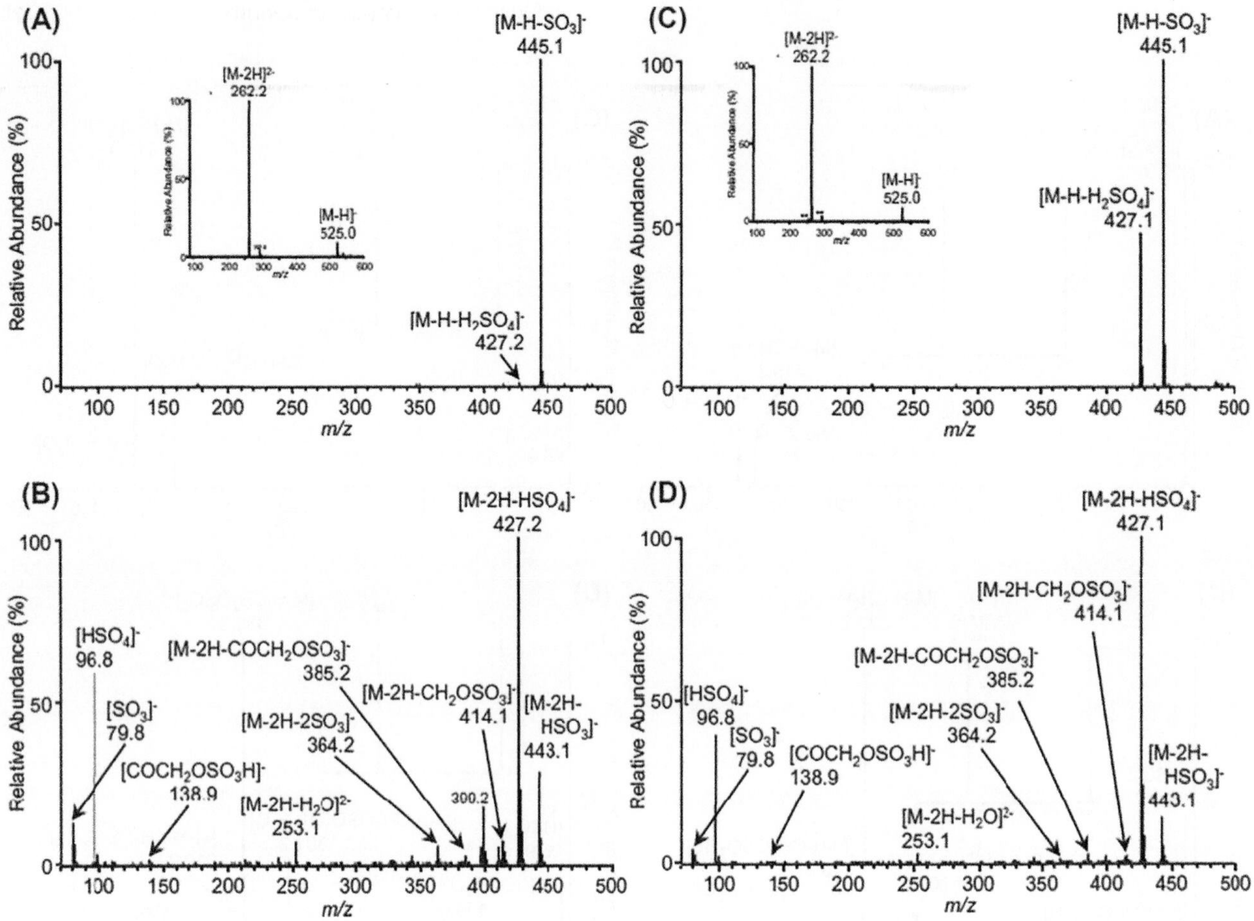

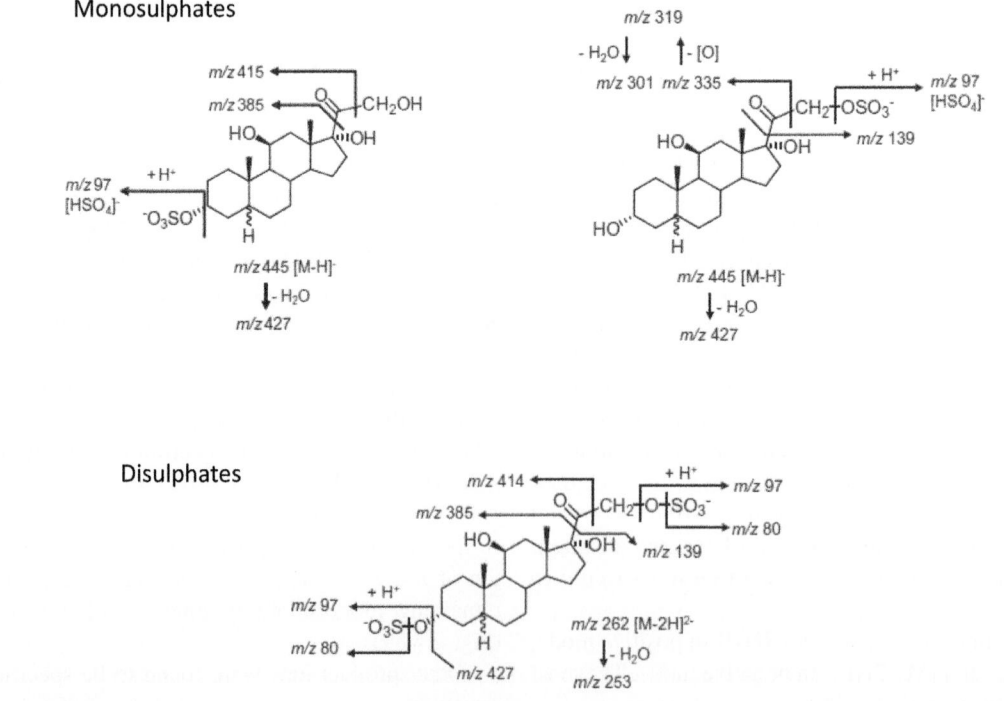

FIG. 2.4.70 Negative ion ESI-MS and MS/MS spectra of (A and B) THF-3,21-di sulfate and (C and D) allo-THF-3,21-disulfate obtained by CID of (A and C) [M−H]⁻ and (B and D) [M−2H]²⁻. *(From Mitamura K, Satoh née Okihara R, Kamibayashi M, Sato K, Iida T, Ikegawa S. Simultaneous determination of 18 tetrahydrocorticosteroid sulfates in human urine by liquid chromatography/electrospray ionization-tandem mass spectrometry. Steroids 2014;85:18–29. Fig. 3 p. 21.)*

FIG. 2.4.71 Proposed major fragments THE and THF monosulfates and THF di-sulfate. *(From Mitamura K, Satoh née Okihara R, Kamibayashi M, Sato K, Iida T, Ikegawa S. Simultaneous determination of 18 tetrahydrocorticosteroid sulfates in human urine by liquid chromatography/electrospray ionization-tandem mass spectrometry. Steroids 2014;85:18–29.)*

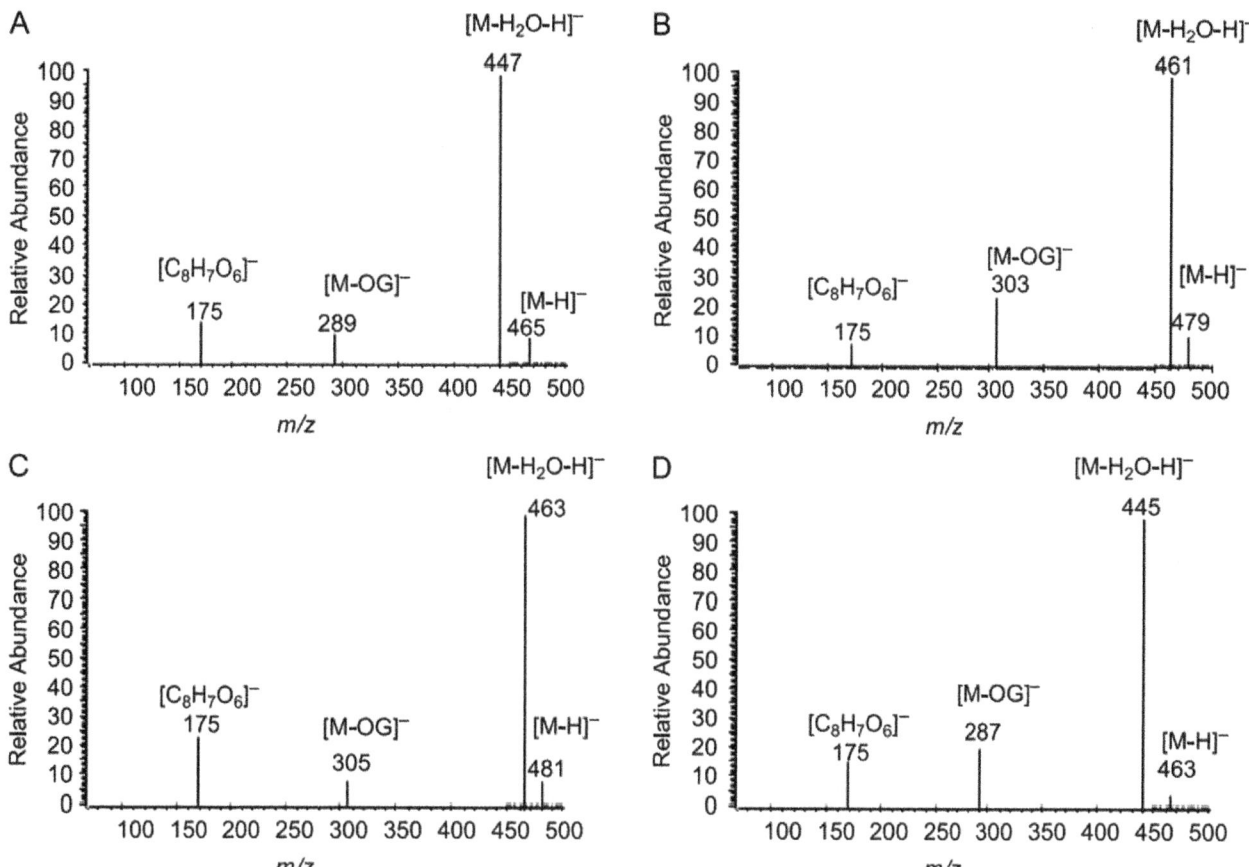

FIG. 2.4.72 MS/MS spectra of (A) An-3 glucuronide; (B) 11-keto-An-3 glucuronide; (C) 11b-OHAn-3 glucuronide; (D) DHEA-3 glucuronide. Spectra for T-17 glucuronide; epi T-17 glucuronide; DHT-17 glucuronide are also shown in original. *(From Cho SH, Lee J, Choi MH, Lee WY, Chung BC. Determination of urinary androgen glucuronides by capillary electrophoresis with electrospray tandem mass spectrometry. Biomed Chromatogr 2009;23(4):426–33.)*

found to be important for several compounds with unsaturation adjacent to the sulfate.

Estrogens conjugated at C-3 present a different fragmentation scheme from the other conjugated steroids, due to the high stability of the aromatic A ring (Indapurkar et al., 2019). Indeed, instead of being on the conjugated group, the charge is kept on the A ring and induces a cleavage of the sulfate or of the glucuronide group. For sulfate compounds, the fragment m/z 97 corresponding to the specific ion $[HSO_4]^-$ can be used as product ion for the quantification signals, except for the sulfated-3-estrogens for which the m/z corresponding to the loss of the sulfate group $[M-SO_3]^-$ is monitored (Indapurkar et al., 2019) (Fig. 2.4.82).

The confirmation ion was the fragment $[SO_3]^-$, m/z 80. For the deuterated internal standard compound, the m/z 98 was observed, revealing the location of one deuterium atom on the sulfate group. For glucuronide forms, the fragment m/z 113 corresponding to the specific ion [glucuronide–CO_2–H_2O]$^-$ was the most abundant fragment obtained in

daughter scan under the conditions described before, except for the estrogens 3-glucuronide for which the fragment [M−glucuronide]$^-$ was the major one. The fragment used for confirmation was the m/z 85 corresponding to the ion [glucuronide–CO–CO_2–H_2O]$^-$, also specific of this conjugated family. Even if these compounds are not chromatographically separated, the MS detection and particularly the MRM mode based on their different specific transitions allowed for an efficient analysis.

Cysteinyl and *N*-acetylcysteinyl conjugates can be identified through their spectra (Fabregat et al., 2013b, 2014). The amine group of the amino acid conjugate promotes the formation of a $[M+H]^+$ irrespective of the steroid structure. (Fig. 2.4.83).

2.4.9 Steroids in urine

In some instances, a qualitative analysis may be sufficient to meet the objectives of the determination. The detection in urine of metabolites of a synthetic anabolic steroid is thus

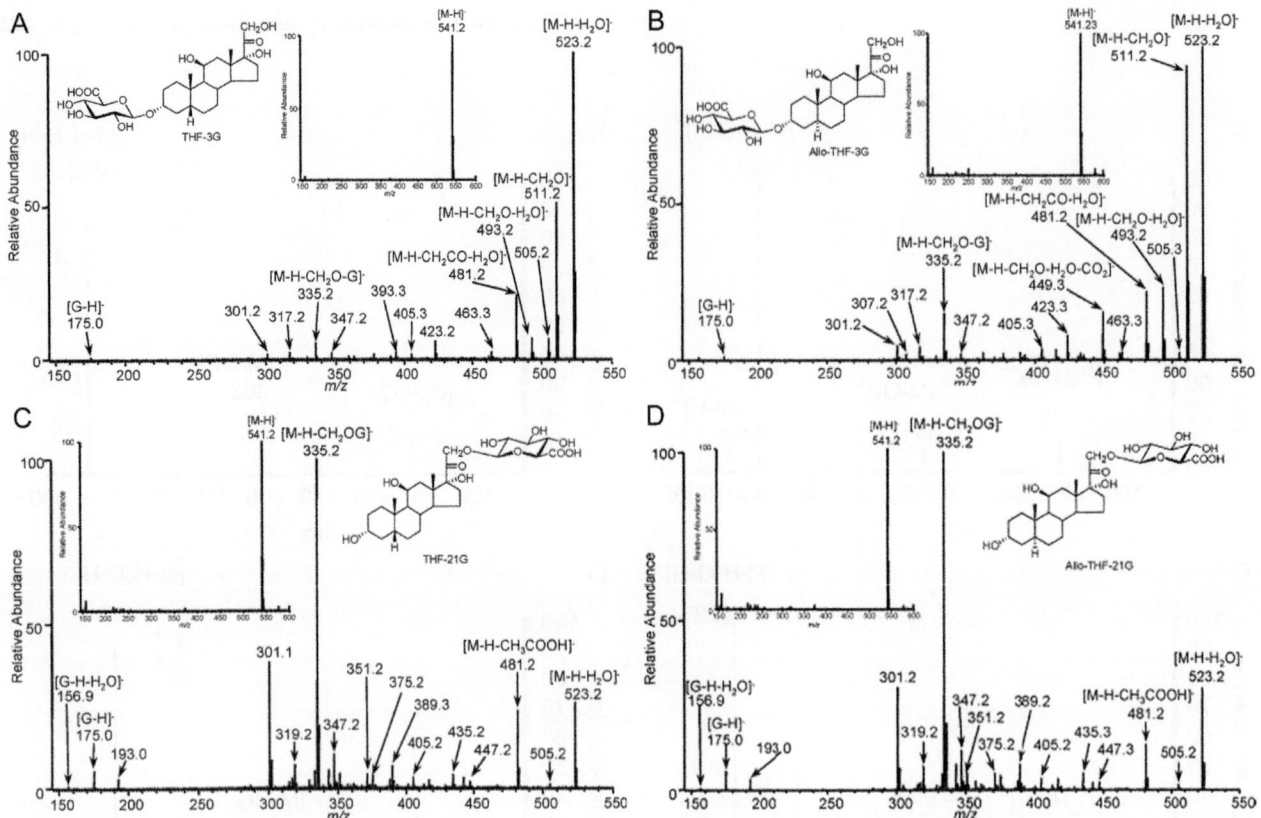

FIG. 2.4.73 Negative ion ESI-MS (main spectra) and CID spectra (small inserts). (A) THF-3G; (B) allo-THF-3G; (C) THF-21G; (D) allo-THF-21G. The [M–H]⁻ ions at m/z 541 were collisionally activated. *(Based on Ikegawa S, Hasegawa M, Okihara R, Shimidzu C, Chiba H, Iida T, Mitamura K. Simultaneous determination of twelve tetrahydrocorticosteroid glucuronides in human urine by liquid chromatography/electrospray ionization-linear ion trap mass spectrometry. Anal Chem 2009;81(24):10124–35. Fig 2 P 10128.)*

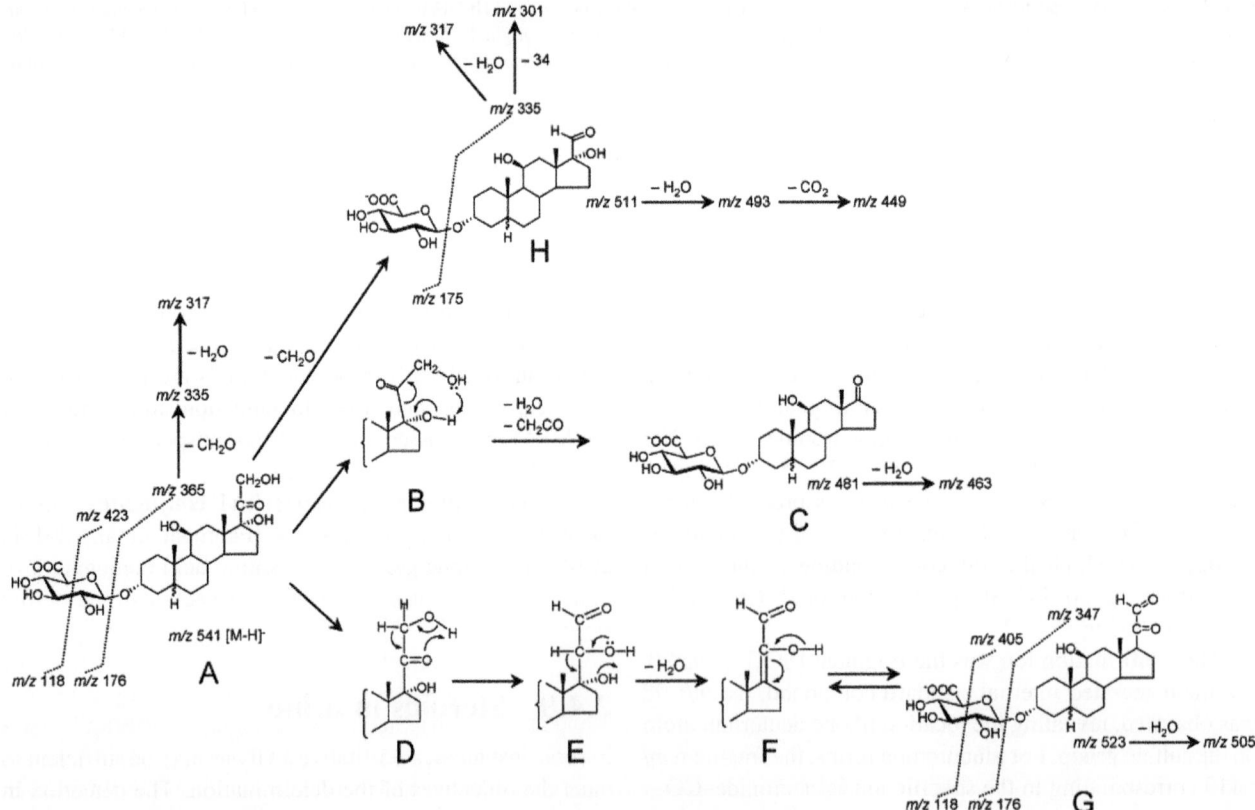

FIG. 2.4.74 Proposed fragmentation of THF-3-glucuronide. *(Redrawn from Ikegawa S, Hasegawa M, Okihara R, Shimidzu C, Chiba H, Iida T, Mitamura K. Simultaneous determination of twelve tetrahydrocorticosteroid glucuronides in human urine by liquid chromatography/ electrospray ionization-linear ion trap mass spectrometry. Anal Chem 2009;81(24):10124–35. Fig. 4 p. 10131.)*

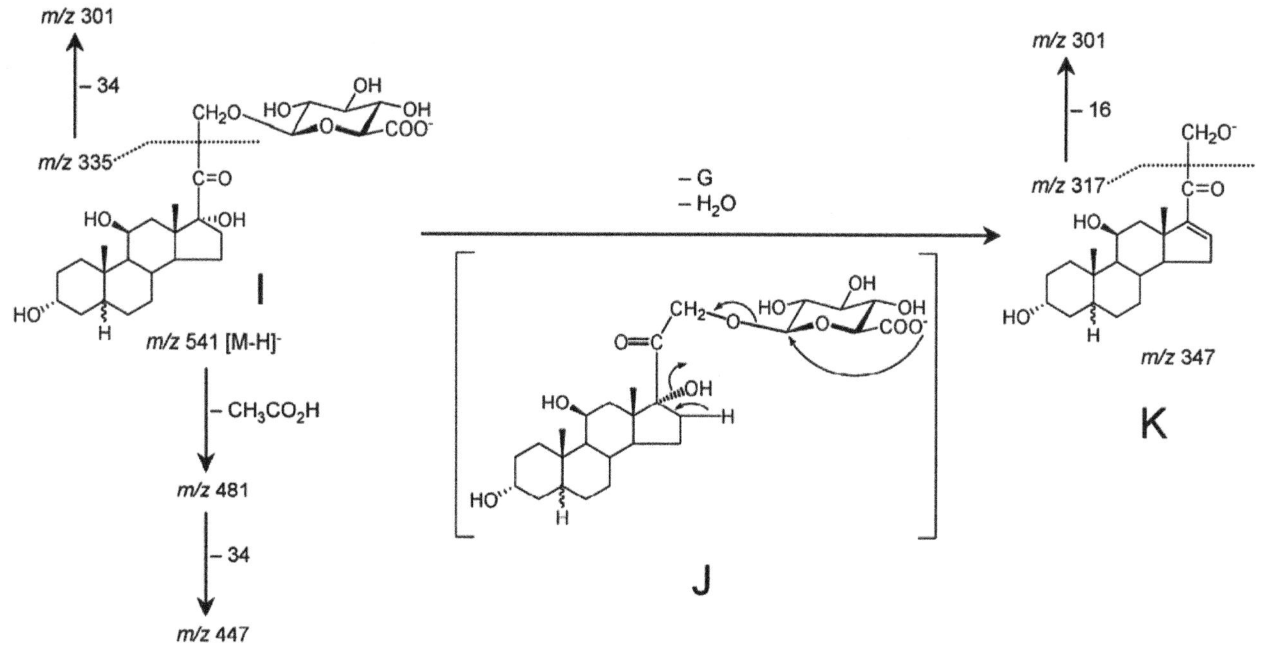

FIG. 2.4.75 Proposed fragmentation THF-21-glucuronide. The formation of K from I can result in the neutral loss of the glucuronic acid moiety via a charge mediated mechanism (J), accompanied by loss of the 17a-hydroxy group as a water molecule. *(From Ikegawa S, Hasegawa M, Okihara R, Shimidzu C, Chiba H, Iida T, Mitamura K. Simultaneous determination of twelve tetrahydrocorticosteroid glucuronides in human urine by liquid chromatography/ electrospray ionization-linear ion trap mass spectrometry. Anal Chem 2009;81(24):10124–35. Fig. 5 p. 10131.)*

taken as sufficient to define abuse of anabolic steroids in sport. The abuse of testosterone and naturally occurring steroids such as 19-nortestosterone (Nandrolone) is a difficult matter requiring a quantitative analysis or determination of ratios of steroids. Carbon isotope ratios can then also be used. It is not uncommon to get adducts of aliphatic and phenolic hydroxyl groups yielding doubly or triply charged species. The tetrahydro-metabolites are ionized as $[M+H-H_2O]^+$ or $[M+NH_4]^+$.

As the basis for studying steroid disorders cortisol-d_4, 6β-hydroxycortisol-d_4, testosterone-d_3, etiocholanolone-d_5 and androsterone glucuronide-d_4 were used as internal standards to quantify 67 steroids in urine (Marcos et al., 2014) (Fig. 2.4.84). Depending on the expected concentration of each steroid they were divided into three categories according to expected concentration. Group A steroids were normally below 10 ng/mL; Group B between 10 and 200 ng/mL and Group C higher than 200 ng/mL. For each category, a calibration was prepared at 9 concentrations 1–200; 2–1000 and 20–5000. Each chromatogram has to be interpreted against the calibration curve in order to quantify the steroid in the urine sample. The excretion rates for all of the steroids in the Marcos paper were in ng/mL although one table has results in μg/24 h which were comparable with published results by Cedric Shackleton.

A constant ion loss (CIL method using the loss of HSO_4^- scan) method for the direct and untargeted detection of urinary steroid bis(sulfate) metabolites has been reported

(McLeod et al., 2017; Pozo et al., 2008a,b, 2018). The CIL scan method has been employed to study the endogenous steroid bis(sulfate) profile, including during pregnancy, to identify markers associated with sports doping, and for the analysis of maternal urine to provide discriminating prenatal diagnosis of inborn errors of steroid biosynthesis associated with Smith-Lemli-Opitz syndrome (SLOS), steroid sulfatase deficiency (STSD), and cytochrome P450 oxidoreductase deficiency (PORD) (see later Chapter 3.3.1).

Reference materials of estradiol 3-sulfate-17 glucuronide were prepared by a five step chemical synthesis from estradiol (Zhao et al., 2014; Doué et al., 2015). Using the glucuronylsynthase approach, sulfation, reduction and glucuronylation of estrone afforded the estradiol 3-sulfate 17 glucuronide in three steps. Further, selective sulfation of estradiol followed by glucuronylation afforded estradiol 17-sulfate 3-glucuronide in just two steps

2.4.9.1 Urine steroids by LC-high-resolution mass spectrometry

Liquid-chromatography, high-resolution, accurate mass, mass spectrometry (HRAM LC-MS) is rapidly gaining in popularity for quantitative clinical analysis of small endogenous molecules, anabolic drug testing, and proteins. HRAM can resolve all steroids and their metabolites except isomers, allowing the use of liquid chromatography,

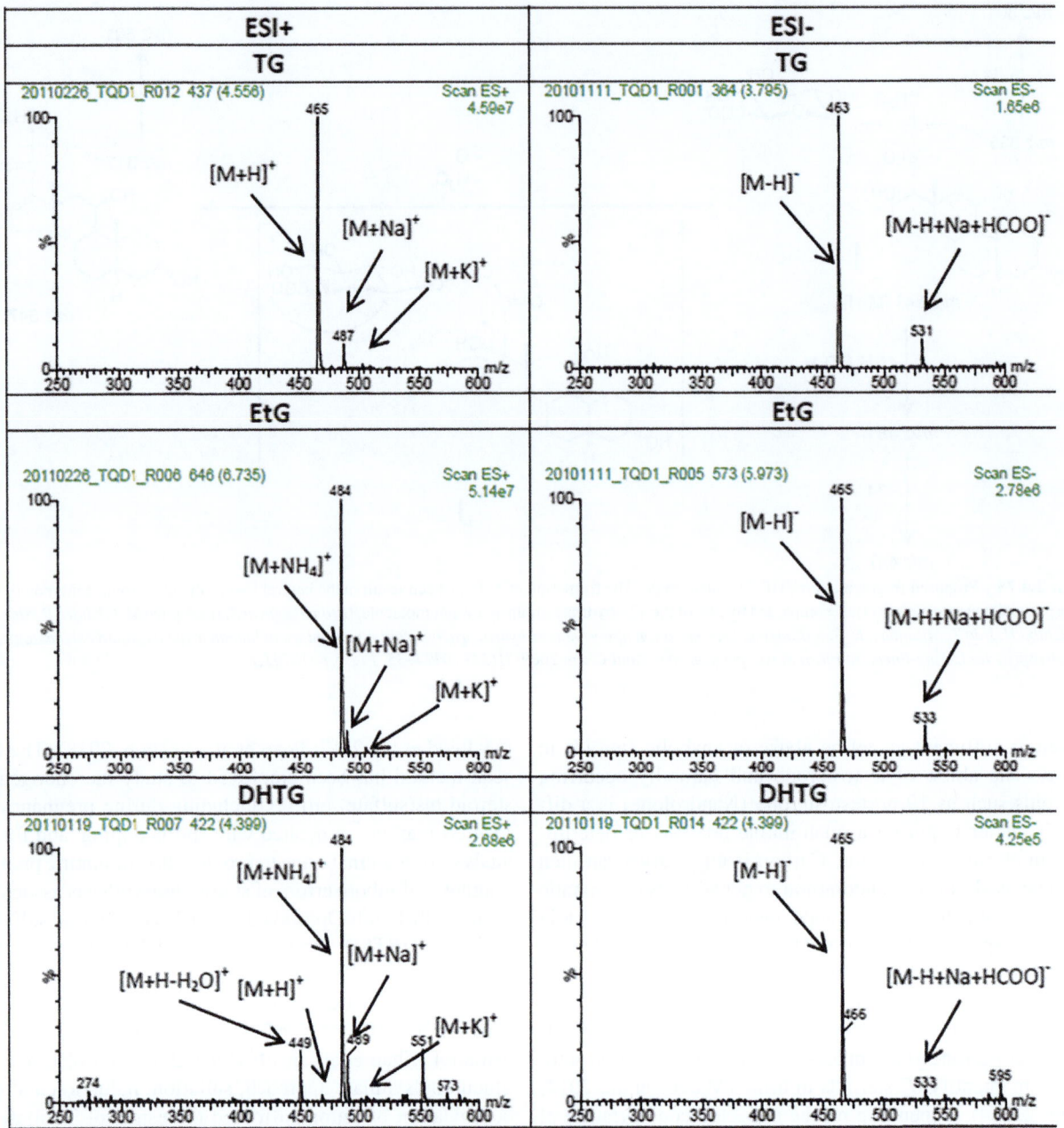

FIG. 2.4.76 Full scan mass spectra in positive and negative ion modes for testosterone, etiocholanolone and dihydrotestosterone glucuronides. *(From Fabregat A, Pozo OJ, Marcos J, Segura J, Ventura R. Use of LC-MS/MS for the open detection of steroid metabolites conjugated with glucuronic acid. Anal Chem. 2013;85(10):5005–14.)*

including multiplexed liquid chromatography setups, as a front-end instead of gas chromatography.

An HRAM LC-MS analysis has been described for a 26-analyte, urine-based steroid panel (Hines et al., 2017) (Table 2.4.9). Sex- and age-based control reference intervals were established and a limited clinical evaluation of the assay in a cohort of patients with different adrenal diseases was tested. Variable signal intensities are recorded for the transitions monitored (Fig. 2.4.85).

2.4.10 Assay validation

Any assay requires validation to agreed standards. A validation must provide objective evidence in order to declare the method as being fit-for-purpose, as well as exhibiting appropriate performance characteristics. There are many published guidelines, probably the most popular is the FDA document (Fig. 2.4.86). Others have been produced by CLSI, Commission for Laboratory

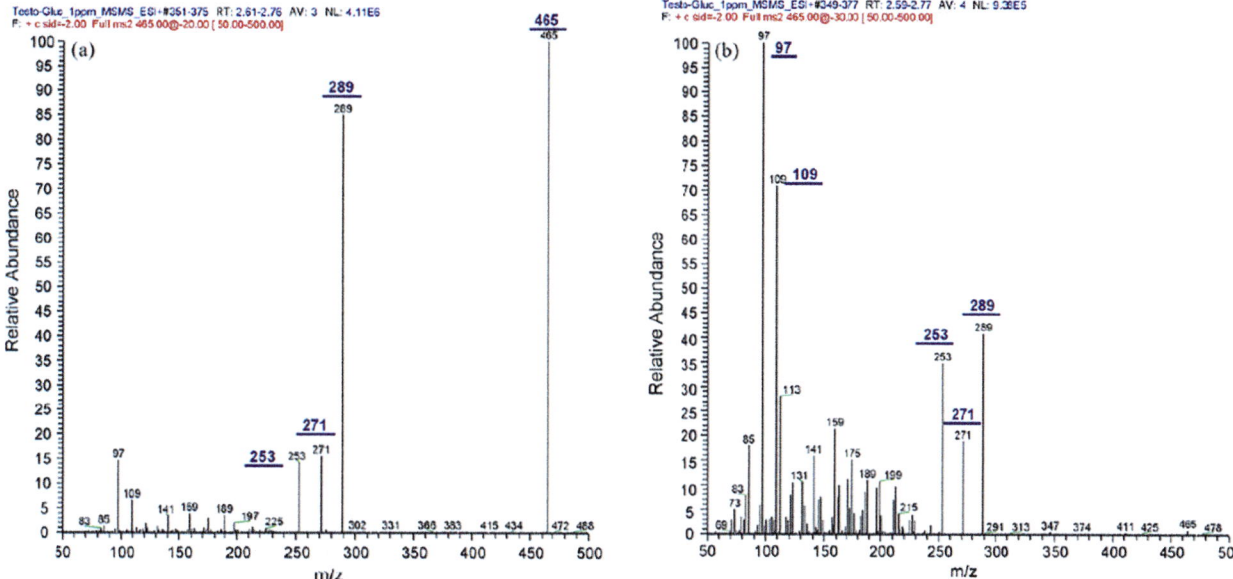

FIG. 2.4.77 Product ion spectra of testosterone glucuronide at 20 eV and 30 eV collision energy. *(From Pozo OJ, Van Eenoo P, Van Thuyne W, Deventer K, Delbeke FT. Direct quantification of steroid glucuronides in human urine by liquid chromatography-electrospray tandem mass spectrometry. J Chromatogr A 2008a;1183(1–2):108–18. Fig. 2 p. 113.)*

TABLE 2.4.8 Major product ions obtained for selected analytes in positive and negative ionization modes.

Analyte	Ionization	Precursor ion	Structure	Product ion	Collision energy	Structure
TG	ESI⁺	465	[M+H]⁺	289	20	[M+H-Gluc]
				271	25	[M+H−H_2O−Gluc]
				253	25	[M+H−2H_2O−Gluc]
				109	30	C_7H_9O (Aring)
				97	30	C_6H_9O (Aring)
	ES⁻	463	[M−H]⁻	287	30	[M−H−Gluc]⁻
				113	35	[Gluc−H−2H_2O−CO_2]⁻
				85	35	[Gluc−H−2H_2O−2CO_2]⁻
				75	35	$HOCH_2CO_2^-$ (gluc)
AG	ESI⁺	484	[M+NH₄]⁺	291	10	[M+H−Gluc]⁺
				273	20	[M+H−H_2O−Gluc]⁺
				255	20	[M+H−H_2O−Gluc]⁺
				177	20	[Gluc+H]⁺
				159	25	[Gluc+H−H_2O]⁺
				141	30	[Gluc+H−2H_2O]⁺
	ESI⁻	465	[M−H]⁻	289	30	[M−H−Gluc]⁻
				113	35	[Gluc−H−2H_2O−CO_2]⁻
				85	35	[Gluc−H−2H_2O−CO_2]⁻
				75	35	$HOCH_2CO_2^-$ (gluc)

Continued

TABLE 2.4.8 Major product ions obtained for selected analytes in positive and negative ionization modes—cont'd

Analyte	Ionization	Precursor ion	Structure	Product ion	Collision energy	Structure
ETG	ESI⁺	484	[M+NH₄]⁺	291	10	[M+H−Gluc]⁺
				273	20	[M+H−H₂O−Gluc]⁺
				255	20	[M+H−2H₂O−Gluc]
				177	15	[Gluc+H]⁺
				59	25	[Gluc+H−H₂O]⁺
				141	30	[Gluc+H−H₂O]⁺
	ESI⁻	465	[M−H]⁻	289	30	[M−H−Gluc]⁻
				113	35	[Gluc−H−2H₂O−CO₂]⁻
				85	35	[Gluc−H−H₂O−2CO₂]⁻
				75	35	OCH₂CO₂⁻ (Gluc)

From Pozo OJ, Van Eenoo P, Van Thuyne W, Deventer K, Delbeke FT. Direct quantification of steroid glucuronides in human urine by liquid chromatography-electrospray tandem mass spectrometry. J Chromatogr A. 2008a;1183(1–2):108–18.

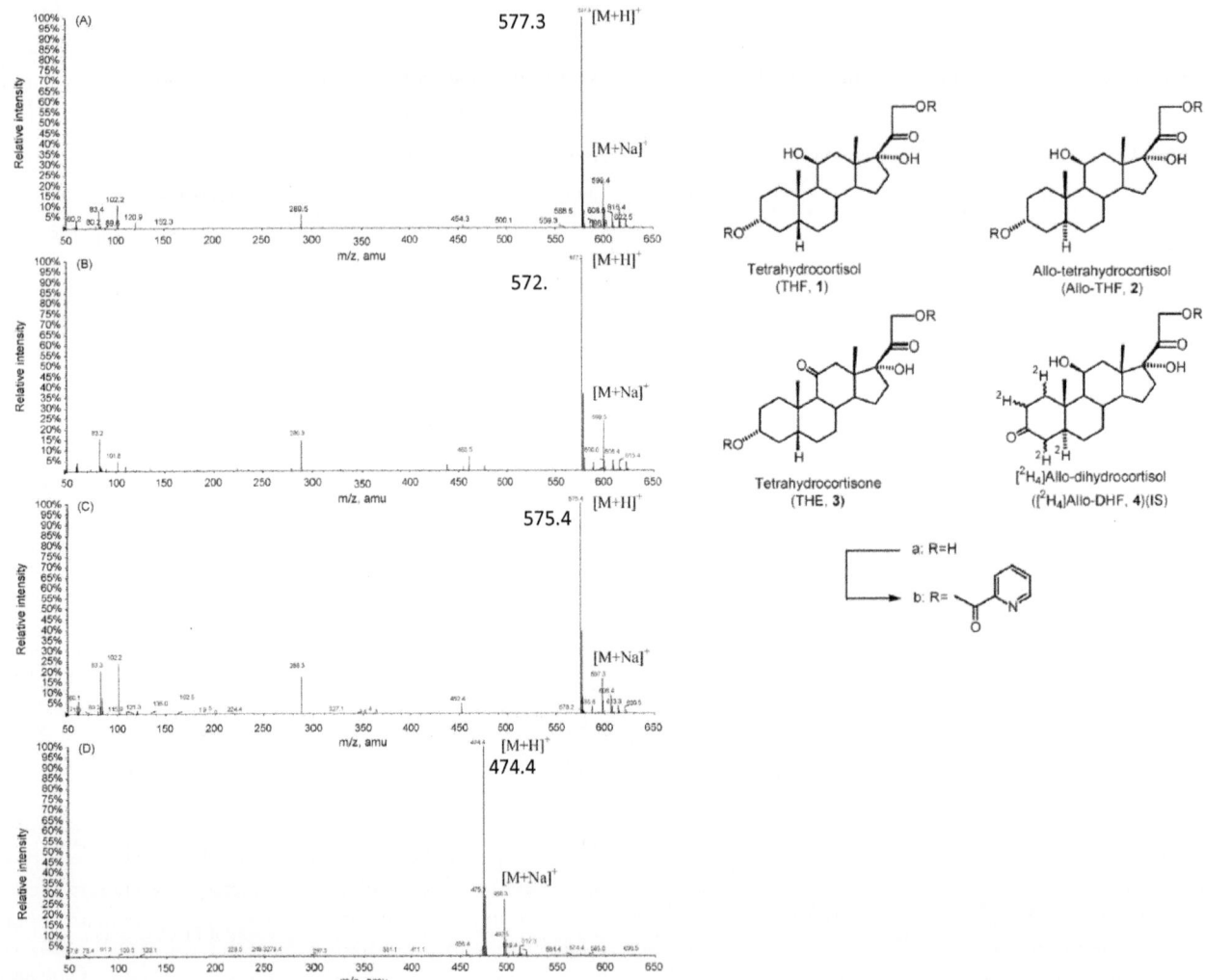

FIG. 2.4.78 Mass spectra of picolinyl derivatives of (A) THF; (B) allo-THF; (C) THE; (D) ²H₄ allo-DHF. *(From Yamashita K, Nakagawa R, Okuyama M, Honma S, Takahashi M, Numazawa M. Simultaneous determination of tetrahydrocortisol, allotetrahydrocortisol and tetrahydrocortisone in human urine by liquid chromatography-electrospray ionization tandem mass spectrometry. Steroids 2008;73(7):727–37. Fig. 3 p. 732.)*

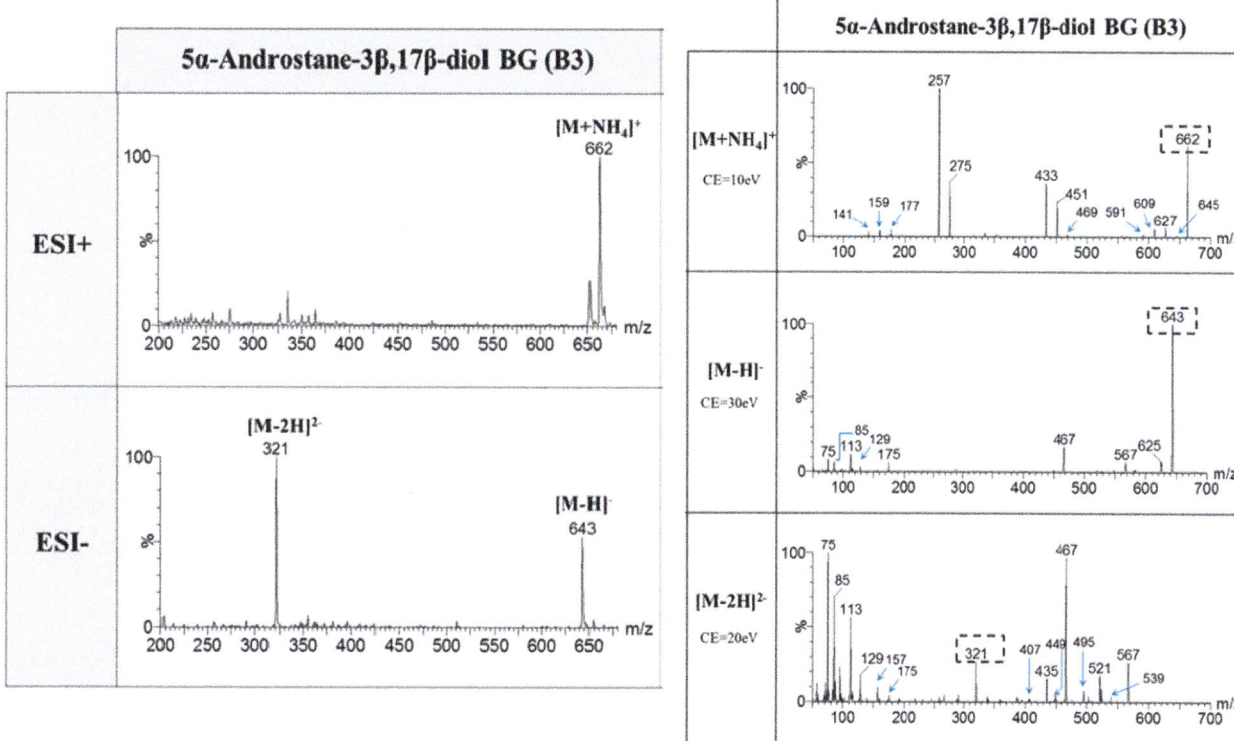

FIG. 2.4.79 Full-scan analyses (left panel) in positive (top) and negative (bottom) modes of bisglucuronide B3 and product ion scan mass spectra (right panel) of [M+NH₄]⁺, [M−H]⁻ and [M−2H]²⁻ of bisglucuronide B3. *(From Esquivel A, Matabosch X, Kotronoulas A, Balcells G, Joglar J, Ventura R. Ionization and collision induced dissociation of steroid bisglucuronides. J Mass Spectrom 2017;52(11):759–69. Fig. 2 p. 765.)*

accreditation, Society for forensic toxicologists, European Commission, European Medicines Agency, Scientific working group for forensic toxicology, Food Drug Administration.

The measurements must be supported with levels of confidence on the results. LC/MS/MS methods allow verification beyond the procedures normally accepted for routine clinical assays (for example, using immunoassay). Each step in the standard operating procedure (SOP) needs to be tested to determine the extent any variable affects the quality of the measurement (Honour, 2011). The validation is a lengthy process requiring full documentation of the SOP through demonstrations of key elements that must include:

- Selectivity,
- Specificity,
- Calibration performance,
- Accuracy,
- Sensitivity,
  1. Lower limit of detection LLOD and
  2. Lower limit of quantification LLOQ.
- Imprecision,
- Stability (reagents, samples, dilution),
- Carry-over and
- Comparison with existing method.

Usually one analyte is determined, but it is becoming more common to measure more than one analyte. The principles of validation apply to each analyte in the panel of interest. During method validation, a blank biological sample will be spiked with the analyte using reference standards to prepare calibration standards. The reference standard and the internal standard must be obtained from authentic and traceable source. A certificate of analysis will include information on purity and storage conditions to be followed. Expiration date and batch number should be recorded.

2.4.10.1 Selectivity

For selectivity, there must be proof of what is being measured. Confirmation of analyte identity must be demonstrated. Three standards of peak shape and position in the chromatogram are used, as follows, and critical examination of the data is important.

Firstly peak shapes should be examined for retention time and symmetry. Fronting, tailing and variable peak widths are evidence of interferences or injection problems. Retention times of the analyte and internal standard should be reproducible. The variation of retention times in samples and calibration curves should not exceed 2.5%. Variation

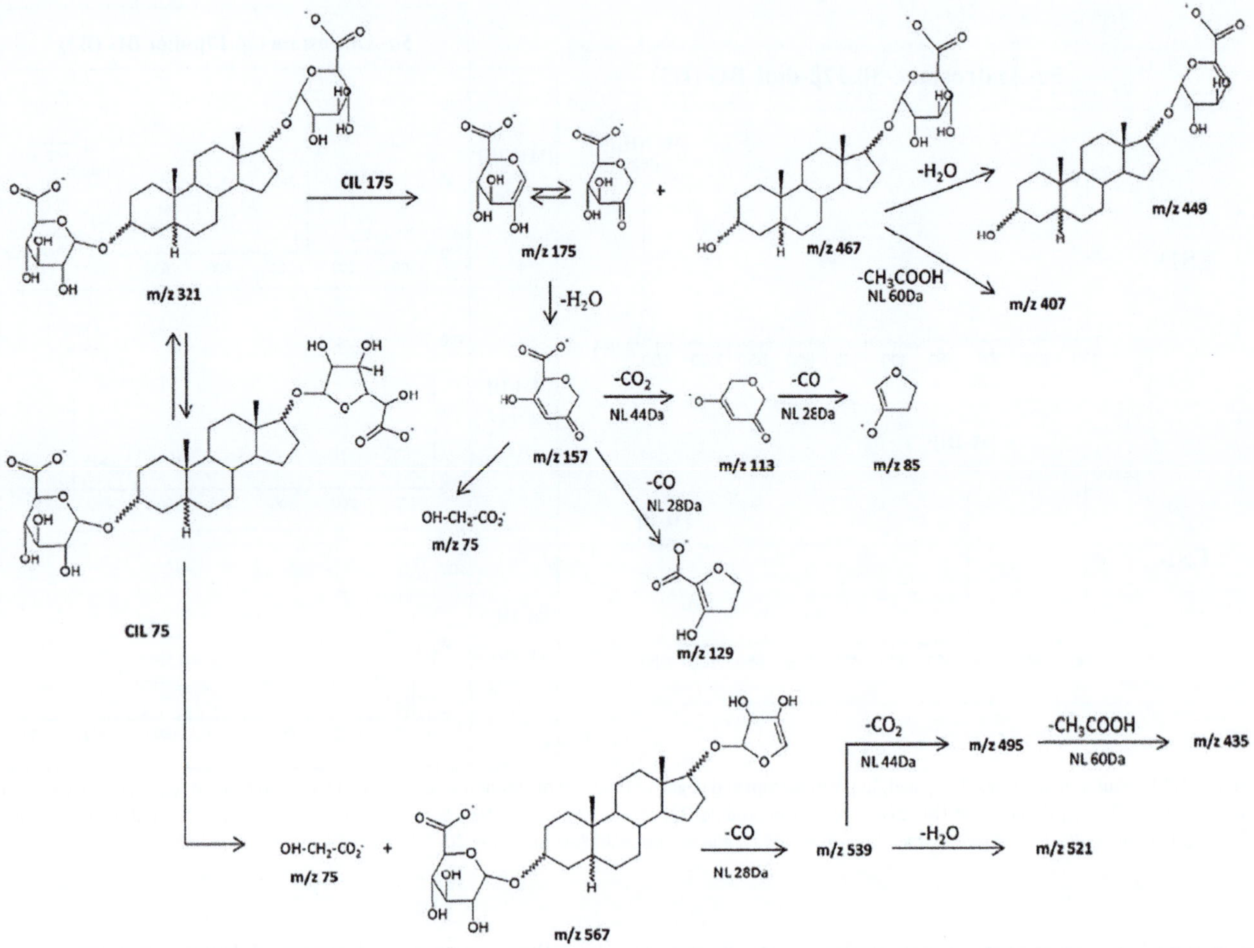

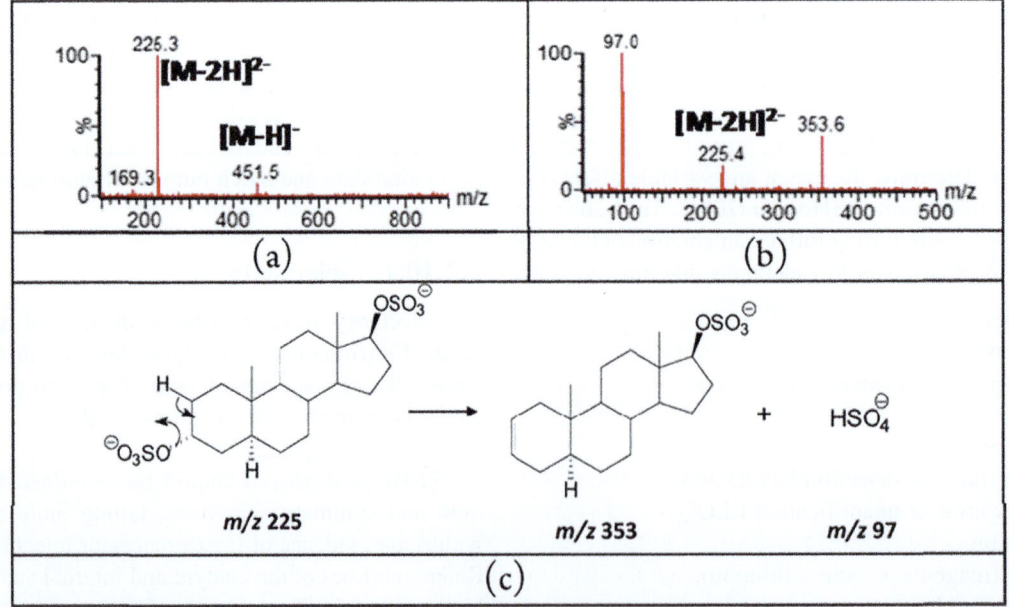

FIG. 2.4.80 Fragmentation scheme for ion $[M-2H]^{2-}$ of androstanediol bisglucuronide. *(From Esquivel A, Matabosch X, Kotronoulas A, Balcells G, Joglar J, Ventura R. Ionization and collision induced dissociation of steroid bisglucuronides. J Mass Spectrom 2017;52(11):759–69. Fig. 3 p. 767.)*

FIG. 2.4.81 MS of androstanediol bissulfate (A) scan in ESI mode (B) product ion scan at 20 eV of precursor ion $[M-2H]^{2-}$ and (C) proposed pathway for ions resulting from ion loss of *m/z* 97. *(From McLeod MD, Waller CC, Esquivel A, Balcells G, Ventura R, Segura J, Pozo OJ. Constant ion loss method for the untargeted detection of Bis-sulfate metabolites. Anal Chem 2017;89(3):1602–9. Fig. 2 p. 1605.)*

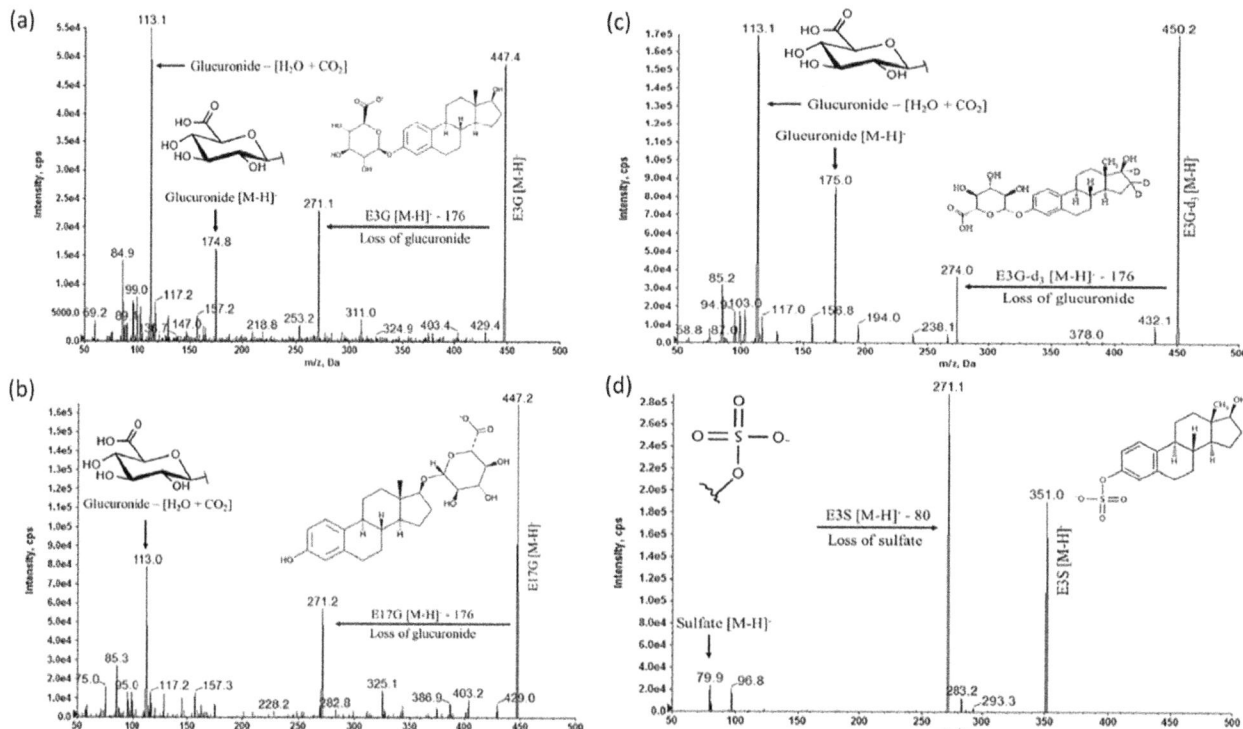

FIG. 2.4.82 Product ion mass spectra with fragmentation patterns of (A) estradiol-3-glucuronide 16,16,17-d3; and (B) and estradiol-3-sulfate. *(From Indapurkar A, Hartman N, Patel V, Matta MK. Simultaneous UHPLC-MS/MS method of estradiol metabolites to support the evaluation of Phase-2 metabolic activity of induced pluripotent stem cell derived hepatocytes. J Chromatogr B Analyt Technol Biomed Life Sci 2019;1126–1127:121765. Fig. 2 p. 5.)*

outside that limit can be due to injector or column problems. The maintenance log should be checked for any changes to tubing or connections and when the column/guard columns were last changed.

The second approach is to monitor the responses of multiple mass transitions. The calculation of their relative intensities (relative ratio) is valuable data for proof of identity. The one ion response ratio for each quantifier and qualifier transitions should be calculated and should not vary more than 15%.

Finally mass transitions under different fragmentation conditions can be studied. Collision gas pressure could be changed for example.

The internal standard needs to be assessed as well. This is often forgotten. In most cases, deuterium labeled compounds elute marginally before their equivalent analyte. For example, d_3-testosterone and testosterone have retention times of around 2.92 and 2.94 min, respectively, in the method reported by Cawood et al. in 2005 (Cawood et al., 2005). Differences in physicochemical properties are attributed to the fact that deuterium atoms have a stronger binding to carbon than hydrogen. The ratios of retention times of the analyte and the internal standard should be determined. The variation in several determinations should be less than 15%.

2.4.10.2 Calibration

The calibration is the reference point for comparison of results from biological samples; see Table 1 in Clinical Laboratory Standards Institute (CLSI EP6) "Evaluation of the Linearity of Quantitative Measurement Procedures: A Statistical Approach, 1st Edition" (https://clsi.org).

Ideally the standard curve should be prepared in the sample matrix with the standard at a minimum of five concentrations in addition to the blank matrix (matrix without added internal standard) including a zero standard (matrix with added internal standard only). The reportable concentration range of the assay should encompass the target analyte concentrations expected in patient investigations. The blank matrix can also be an **analyte-free biological sample**, for example a sample stripped of analyte by absorption to activated charcoal. This is not specific and a range of compounds will be absorbed to the charcoal. There is a matrix difference between fresh whole blood and frozen/lyophilized whole blood. The **sample can be subjected to immunoabsorption** in a precipitation tube or column chromatographic technique. If an equivalent biological sample is not available then **a near physiological buffer** can be used (8% bovine serum albumin for example). The amount of internal standard is usually set at

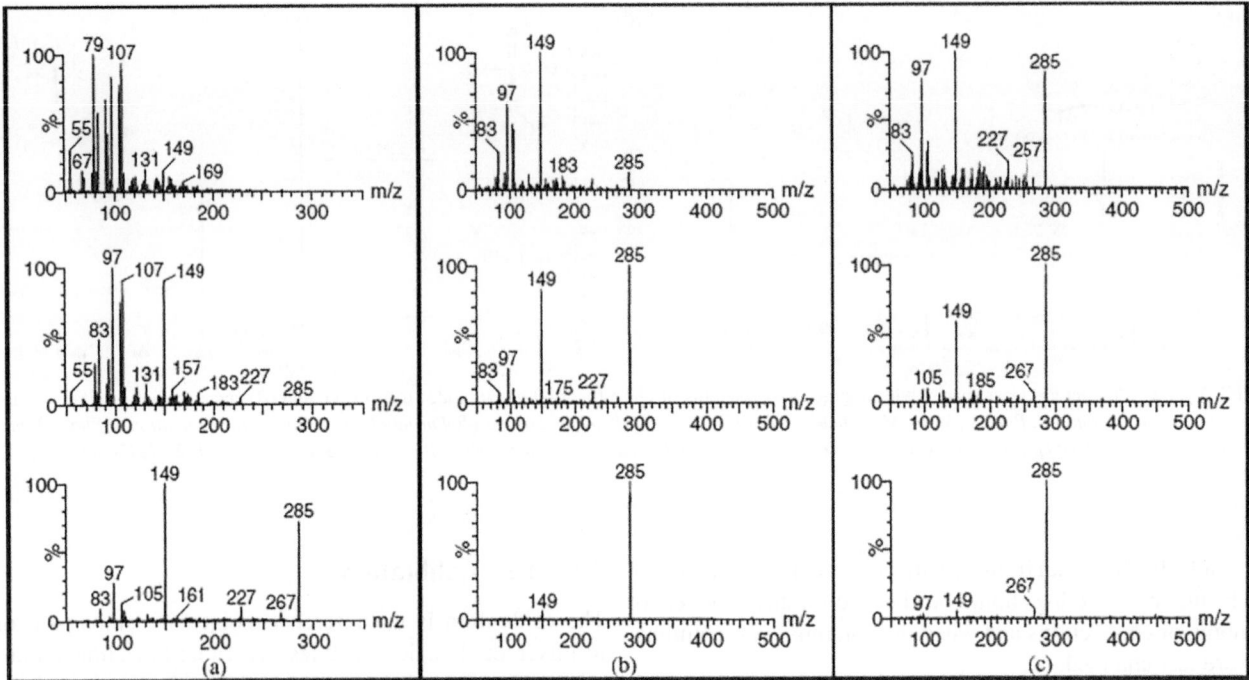

FIG. 2.4.83   Product ion spectra in positive ESI for (A) 4,6-AD-cys (precursor ion *m/z* 285); (B) 4,6-AD-Cys (precursor ion *m/z* 406; (C) 4,6-AD-NAC (precursor ion *m/z* 448 acquired at 20 eV, 30 eV and 40 eV. *(From Fabregat A, Kotronoulas A, Marcos J, Joglar J, Alfonso I, Segura J, Ventura R, Pozo OJ. Detection, synthesis and characterization of metabolites of steroid hormones conjugated with cysteine. Steroids 2013b;78(3):327–36. Fig. 2 p. 331.)*

concentration near the middle of the calibration curve (or at the upper reference or therapeutic limit).

A standard curve of relative response signal of quantifier ions for the analyte at the five concentrations to internal standard response is plotted against relative concentrations. Each calibration standard should be analyzed in triplicate. If there is a lack of reproducibility, the method needs to be re-examined. A minimum of three curves generated during validation should be reported. The back calculated concentrations of the calibration standards should also be examined. Results should be within 15% of the nominal value for at least 75% of the standards.

The response according to concentration of analyte and goodness of fit should be calculated. The blank and zero samples should not be taken into account to calculate the calibration line parameters. The linear response range

for the assay should be determined from least-squares regression analysis. The response can be assumed to be linear if the correlation coefficient (*r*-squared) is greater than 0.99. The plot should not be forced through the origin, some calculations use $1/x$ weighting. A plot of least residuals can be performed. In some cases, a quadratic relation between response and concentration is found. Nonlinear calibration curves can be the result of dimer and cluster formation. This can be prevented by the addition of cationic modifiers to the electrospray solution, and is worth investigating since staff in nonexpert laboratories may not cope with 2nd-order curves.

The calibration is not always linear. If the internal standard has a mass increment of two then this may be detected in the second isotope increment ($^{13}C_2$) of the analyte. Isotopic impurity of the internal standard may work in the

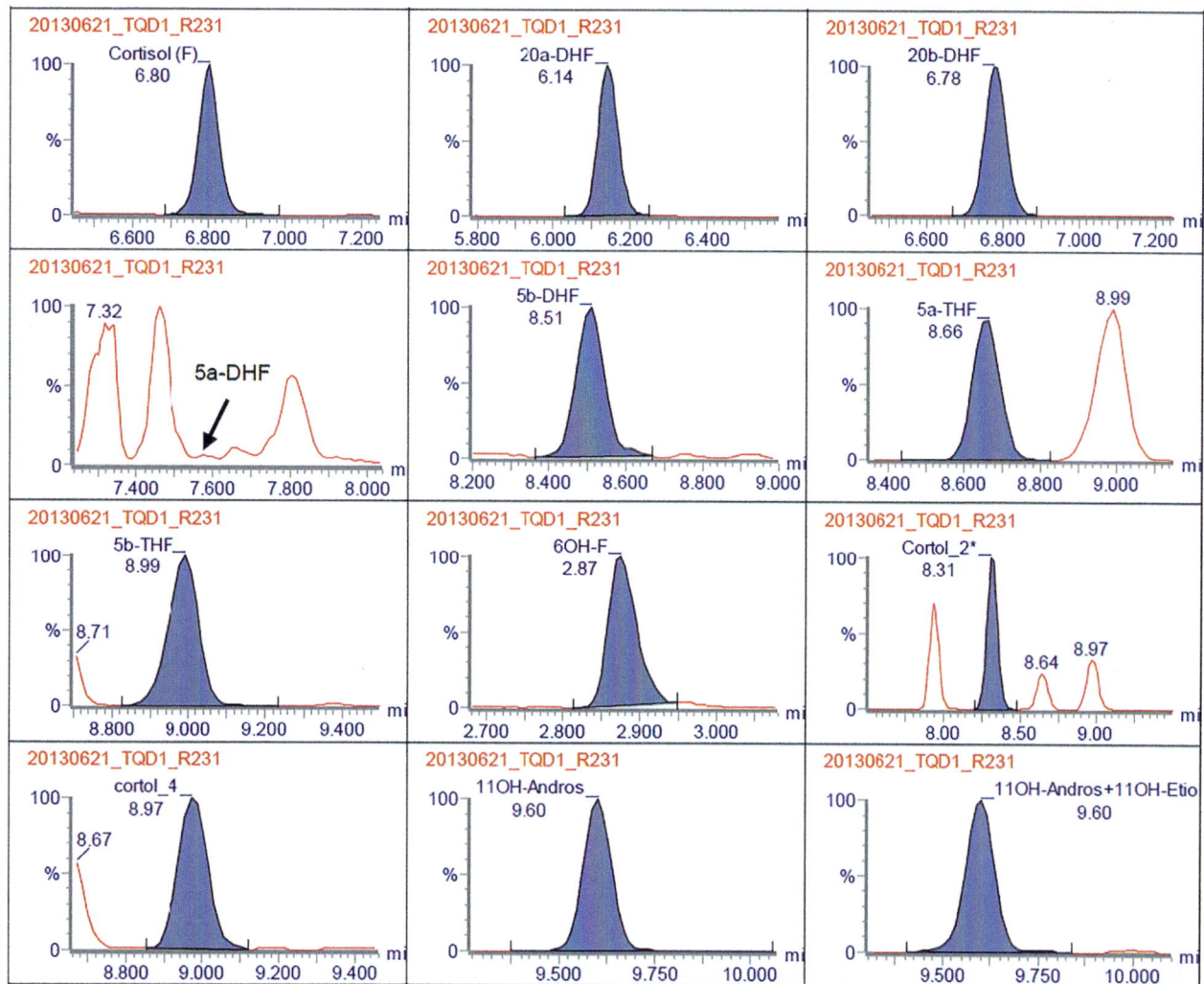

FIG. 2.4.84 LC-MS/MS chromatograms for a real urine sample containing cortisol (88 ng/mL), 20α-DHF (182 ng/mL), 20β-DHF (117 ng/mL), 5β-DHF (23 ng/mL), 5α-THF (3610 ng/mL), 5β-THF (4125 ng/mL), 6βOHF (441 ng/mL), α-cortol (411 ng/mL), β-cortol (155 ng/mL), 11βOH-Andros (1311 ng/mL), 11βOH-Etio (321 ng/mL). Not shown are the signals for cortisone (139 ng/mL), 20α-DHE (97 ng/mL), 20β-DHE (32 ng/mL), 5β-DHE (12 ng/mL), 5α-THE (141 ng/mL), 5β-THE (3650 ng/mL), 6βOHE (6 ng/mL), α-cortolone (2641 ng/mL), β-cortolone (1094 ng/mL), 11CO-Andros +11CO-Etio (252 ng/mL), corticosterone (4 ng/mL), 20α-DHB (5 ng/mL), 20β-DHB (9 ng/mL), 5α-THB (1107 ng/mL), 5β-THB (411 ng/mL), 17-deoxy-β-cortol (5 ng/mL), 11-dehydrocorticosterone (34 ng/mL), 20β-DHA (13 ng/mL), 5α-THA (24 ng/mL), 5β-THA (45 ng/mL), 17-deoxy-β-cortolone (15 ng/mL), DOC (0.7 ng/mL), 20β-DHS (4 ng/mL), 5β-THS (89 ng/mL), α-11-deoxycortolone (25 ng/mL), β-11-deoxycortolone (11 ng/mL), 17OHP (2 ng/mL), 20α-DH17OHP (2 ng/mL), 20β-DH17OHP (0.6 ng/mL), 17HP (421 ng/mL), PT (1463 ng/mL), PTONE (23 ng/mL), 21-deoxy-20α-cortol (86 ng/mL) and 21-deoxy-20β-cortol (65 ng/mL). Guidelines for assay validation. *(From Marcos J, Renau N, Casals G, Segura J, Ventura R, Pozo OJ. Investigation of endogenous corticosteroids profiles in human urine based on liquid chromatography tandem mass spectrometry. Anal Chim Acta 2014;812:92–104. Fig. S1.)*

opposite direction—an unlabeled compound in the internal standard being detected in the ion for the analyte. If a result is above the measurable range the validity of the result should be checked by appropriate dilution to check if the result in the calibration range, but this is not straightforward.

2.4.10.3 Interferences

Experiments need to be performed, particularly in LC-MS/MS methods to challenge if there are there any interferences in the analysis. In any biological sample, there will be a mixture of steroids and if chromatography is used the retention times of other steroids can be established which excludes possibility of detection by the MS. Natural and synthetic steroids need to be tested. Papers from the ARUP laboratory in Salt Lake City tested ranges of compounds in the analysis of adrenal steroids in serum and plasma (Kushnir et al., 2010). Few studies have gone to this extent (Table 2.4.10).

Interferences in tandem mass spectrometric methods cause ion suppression (Annesley, 2003, 2007) leading to changes in the efficiency of analyte ionization and are in general called **matrix effects**. A procedure is needed to reduce matrix effects (Chambers et al., 2007).

TABLE 2.4.9 Selection of steroid hormones with molecular formula and mass. **Note in some cases, duplication of molecular formula and mass of related steroids.**

Peak #	Retention time	Analyte	Formula	Ion formation	Quantifier ion
1	2.70	6β-Hydroxycortisol	$C_{21}H_{30}O_6$	$[M+H]^+$	379.212
2	8.44	Cortisol	$C_{21}H_{30}O_5$	$[M+H]^+$	363.217
3	8.63	Cortisone	$C_{21}H_{28}O_5$	$[M+H]^+$	361.201
4	8.67	β-Cortol	$C_{21}H_{36}O_5$	$[M+H-2H_2O]^+$	333.242
5	9.02	α-Cortolone	$C_{21}H_{34}O_5$	$[M+H-2H_2O]^+$	331.267
6	9.21	16α-Hydroxy-DHEA	$C_{19}H_{28}O_3$	$[M+H-H_2O]^+$	287.201
7	9.30	β-Cortolone	$C_{21}H_{34}O_5$	$[M+H-2H_2O]^+$	331.227
8	9.41	5α-Tetrahydrocortisol	$C_{21}H_{34}O_5$	$[M+H-2H_2O]^+$	331.227
9	9.60	Tetrahydrocortisol	$C_{21}H_{34}O_5$	$[M+H-2H_2O]^+$	331.227
10	10.30	Tetrahydrocortisone	$C_{21}H_{32}O_5$	$[M+H]^+$	365.232
11	10.81	Pregnanetriolone	$C_{21}H_{34}O_4$	$[M+H-2H_2O]^+$	315.232
12	11.12	Tetrahydrocorticosterone	$C_{21}H_{34}O_4$	$[M+H-2H_2O]^+$	315.232
13	11.24	11β-Hydroxyetiocholanolone	$C_{19}H_{30}O_3$	$[M+H-2H_2O]^+$	271.206
14	11.36	Pregnenetriol	$C_{21}H_{34}O_3$	$[M+H-2H_2O]^+$	299.237
15	11.40	11β-Hydroxyandrosterone	$C_{19}H_{30}O_3$	$[M+H-2H_2O]^+$	271.206
16	12.04	11-Oxoetiocholanolone	$C_{19}H_{28}O_3$	$[M+H-H_2O]^+$	287.201
17	13.56	Tetrahydro-11-deoxycortisol	$C_{21}H_{34}O_4$	$[M+H]^+$	351.253
18	14.16	DHEA Dehydroepiandrosterone	$C_{19}H_{28}O_2$	$[M+H-H_2O]^+$	271.206
19	15.14	Pregnanetriol	$C_{21}H_{36}O_3$	$[M+H-2H_2O]^+$	301.252
20	15.54	Tetrahydrodeoxycorticosterone	$C_{21}H_{36}O_3$	$[M+H-2H_2O]^+$	317.248
21	15.76	Pregnenediol	$C_{21}H_{34}O_2$	$[M+H-2H_2O]^+$	283.242
22	15.90	5α-Tetrahydro-11-dehydrocorticosterone	$C_{21}H_{34}O_3$	$[M+H-H_2O]^+$	317.248
23	16.30	Etiocholanolone	$C_{19}H_{30}O_2$	$[M+H-H_2O]^+$	273.221
24	16.79	Androsterone	$C_{19}H_{30}O_2$	$[M+H-H_2O]^+$	273.221
25	17.03	17α-Hydroxypregnanolone	$C_{21}H_{34}O_3$	$[M+H-2H_2O]$	299.237
26	18.23	Pregnanediol	$C_{21}H_{36}O_2$	$[M+H-2H_2O]$	285.258

There are two common methods to assess matrix effects: the **postcolumn infusion method**, defined by Bonfiglio et al. (1999), and **the postextraction spike method**, proposed by Matuszewski et al. (2003) and Matuszewski (2006). The postcolumn infusion method provides a qualitative assessment of matrix effects, identifying chromatographic regions most likely to experience matrix effects (Fig. 2.4.87).

Briefly, an infusion pump delivers a constant amount of analyte into the LC stream entering the ion source of the mass spectrometer. The mass spectrometer is run in **SRM mode** to follow the infused analyte. Blank sample extract is injected on the LC column under conditions chosen for the assay. Since the analyte is infused into the MS at a constant flow, a steady ion response is obtained as a function of time. Any endogenous compound that elutes from the column and causes a variation in ESI response of the infused analyte is seen as a suppression or enhancement in the response of the infused analyte. This approach does not provide a measure of the level of matrix effect observed for specific analytes. In addition, if several compounds are determined in one method, all compounds should be infused separately to investigate possible matrix effects for every analyte. Moreover, analytes are infused at concentrations higher than LLOQ. Matrix effects are therefore not investigated for low concentrated samples.

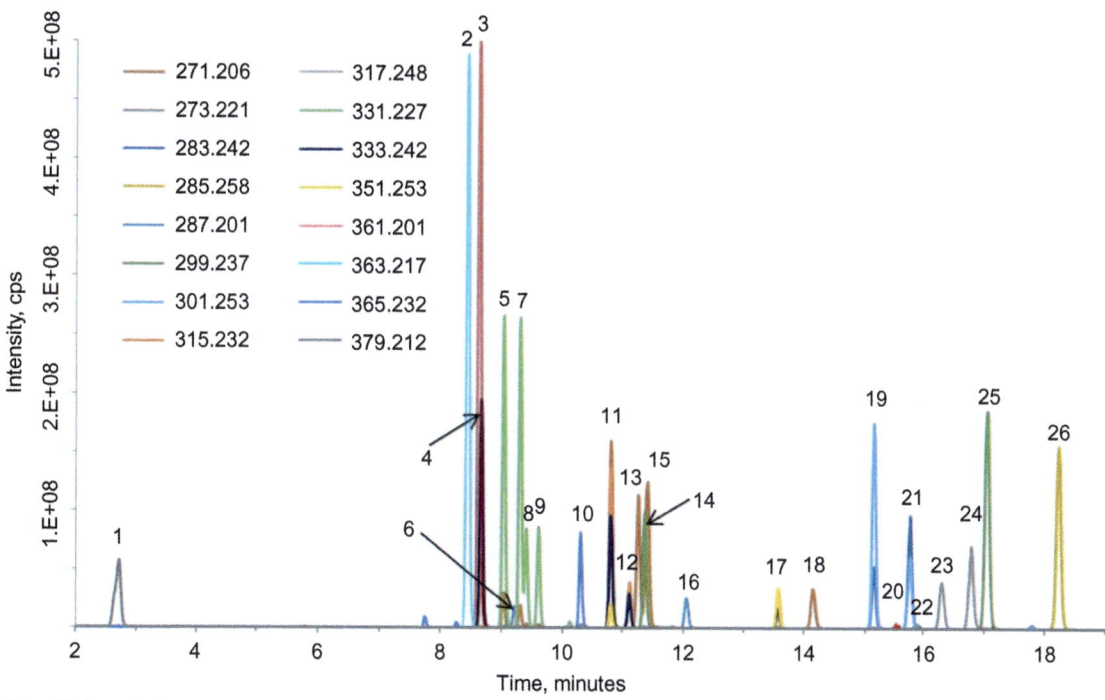

FIG. 2.4.85 High-resolution mass spectrometry of urinary steroids. A Zorbax Extend C-18 column, 1.8 μm particles, 2.1 × 50 mm eluted with a gradient from 10% acetonitrile with 0.1% formic acid to 90% acetonitrile. A Thermo Q Exactive Plus HRAM hybrid quadruple/orbital trap MS was used at 70,000 resolution. Representative chromatogram indicating retention times and color-coded *m/z* used for analysis of steroid standards. The absence of a signal between peak #1 and peak #2 reflects flow diversion from MS by valve at the interface. Peaks are numbered as in Table 2.4.9. *(From Hines JM, Bancos I, Bancos C, Singh RD, Avula AV, Young WF, Grebe SK, Singh RJ. High-resolution, accurate-mass (HRAM) mass spectrometry urine steroid profiling in the diagnosis of adrenal disorders. Clin Chem 2017;63(12):1824–35. Fig. 1 p. 1827.)*

Guideline on bioanalytical method validation

ISO 15193:2009
In vitro diagnostic medical devices —
Measurement of quantities in samples of biological origin — Requirements for content and presentation of reference measurement procedures

FIG. 2.4.86 Guidelines for assay validations.

SOFT / AAFS

FORENSIC TOXICOLOGY LABORATORY GUIDELINES

COMMISSION ON LABORATORY ACCREDITATION

Laboratory Accreditation Program

CHEMISTRY AND TOXICOLOGY CHECKLIST

TABLE 2.4.10 Steroids evaluated for interferences.

1-(2-Chlorophenyl)-1-(chlorophenyl)-2,2,-dichloroethane	Cyprosterone	Hermaphrodiol	Nandrolone
17α-Methylandrostan-17β-ol-30n3	Danazole	Isoandrosterone	Norethandrolone
19-Nortestosterone-17-decanoate	Dihydroepiandrosterone	Isotestosterone	Norethindrone
3β-Hydroxy-5α-androstan-17-one	Epiandrosterone	Levonorgestrel	Norprogesterone
3β-Hydroxyetioallocholan-17-one	Epitestosterone	Medroxyprogesterone	Oxandrolone
5α-Dihydrotestosterone	Ethylestrenole	Mesterolone	Oxymethelone
5-Androstenediol	Ethynodiol diacetate	Methandrostenolone	Stanozolol
5α-Dihydroprogesterone	Etiocholan-17β-ol-3-one	Methenolone	Testosterone
5β-Dihydroprogesterone	Etiocholanolone	Methylandrostanolone	Transandrosterone
Androstenediol	Fluoxymesterone	Methyldihydrotestosterone	Transdehydrotestosterone
Androsterone	Flutamide	Methyltestosterone	

Data from Kushnir MM, Rockwood AL, Roberts WL, Pattison EG, Owen WE, Bunker AM, Meikle AW. Development and performance evaluation of a tandem mass spectrometry assay for 4 adrenal steroids. Clin Chem. 2006;52(8):1559–67. Suppl 1.

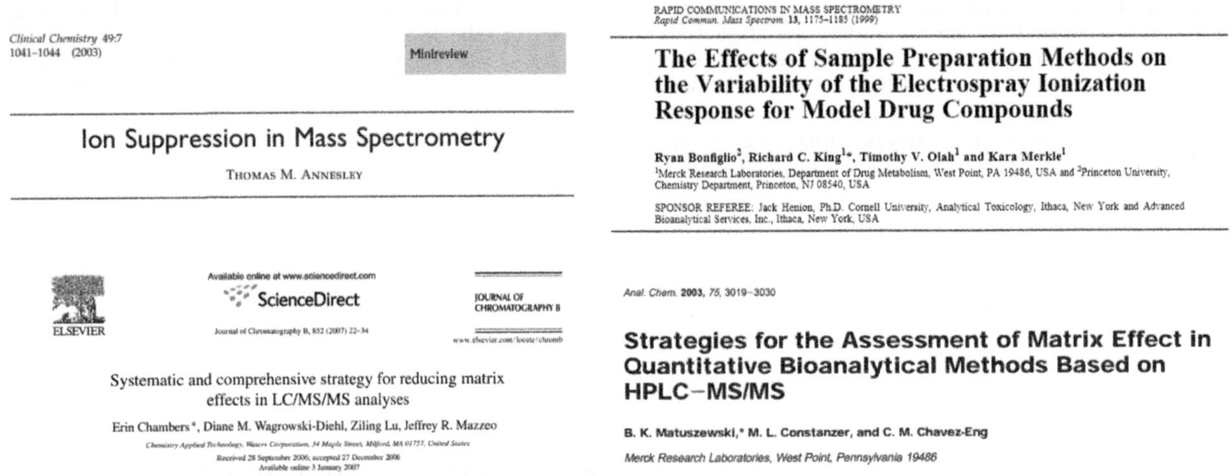

Clinical Chemistry 49:7
1041–1044 (2003)

Minireview

Ion Suppression in Mass Spectrometry

THOMAS M. ANNESLEY

RAPID COMMUNICATIONS IN MASS SPECTROMETRY
Rapid Commun. Mass Spectrom. 13, 1175–1185 (1999)

The Effects of Sample Preparation Methods on the Variability of the Electrospray Ionization Response for Model Drug Compounds

Ryan Bonfiglio[2], Richard C. King[1]*, Timothy V. Olah[1] and Kara Merkle[1]
[1]Merck Research Laboratories, Department of Drug Metabolism, West Point, PA 19486, USA and [2]Princeton University, Chemistry Department, Princeton, NJ 08540, USA

SPONSOR REFEREE: Jack Henion, Ph.D. Cornell University, Analytical Toxicology, Ithaca, New York and Advanced Bioanalytical Services, Inc., Ithaca, New York, USA

Available online at www.sciencedirect.com

ScienceDirect

JOURNAL OF
CHROMATOGRAPHY B

ELSEVIER

Journal of Chromatography B, 852 (2007) 22–34

www.elsevier.com/locate/chromb

Systematic and comprehensive strategy for reducing matrix effects in LC/MS/MS analyses

Erin Chambers *, Diane M. Wagrowski-Diehl, Ziling Lu, Jeffrey R. Mazzeo
Chemistry Applied Technology, Waters Corporation, 34 Maple Street, Milford, MA 01757, United States

Received 28 September 2006; accepted 27 December 2006
Available online 3 January 2007

Anal. Chem. 2003, 75, 3019–3030

Strategies for the Assessment of Matrix Effect in Quantitative Bioanalytical Methods Based on HPLC−MS/MS

B. K. Matuszewski,* M. L. Constanzer, and C. M. Chavez-Eng

Merck Research Laboratories, West Point, Pennsylvania 19486

FIG. 2.4.87 Important papers describing ion suppression and matrix effects. *(Modified from Annesley TM. Ion suppression in mass spectrometry. Clin Chem 2003;49(7):1041–4; Bonfiglio R, King RC, Olah TV, Merkle K. The effects of sample preparation methods on the variability of the electrospray ionization response for model drug compounds. Rapid Commun Mass Spectrom 1999;13(12):1175–85; Chambers E, Wagrowski-Diehl DM, Lu Z, Mazzeo JR. Systematic and comprehensive strategy for reducing matrix effects in LC/MS/MS analyses. J Chromatogr B Analyt Technol Biomed Life Sci 2007;852 (1–2):22–34; Matuszewski BK, Constanzer ML, Chavez-Eng CM. Strategies for the assessment of matrix effect in quantitative bioanalytical methods based on HPLC-MS/MS. Anal Chem 2003;75(13):3019–30.)*

In contrast, the **postextraction spike method** (Lindner et al., 2017) quantitatively assesses matrix effects by comparing the response of an analyte in neat solution to the response of the analyte spiked into a blank matrix sample that has been carried through the sample preparation process (Fig. 2.4.88).

In order to prove the correct quantification of unknown in human body fluid samples based on a calibration in surrogate matrix, a new approach called **Isotope Inversion**

Experiment (Suhr et al., 2016) can be included in the basic validation protocol (Fig. 2.4.89). The nonlabeled steroids are used as the so called inverse internal standard, whereas the labeled steroids are used as the analyte to be quantified. Inverse calibrators are prepared by spiking SIL compound into surrogate mix.

In a calibration, the analyte is used at one concentration with a range of concentrations of the internal standard. Calibrations should be prepared in authentic matrix and in a surrogate matrix. Inverse quality control samples can also be prepared with a fixe concentration of the analyte and a range of concentrations of the label.

Care should be taken to demonstrate absence of "late eluting" interferences. These are recognized usually as very broad peaks.

Phospholipids in sample extracts are well known interferents, and can be monitored with SRM, without fragmentation in Q2, of 184 to 184 and 104 to 104 transitions. Extraction and chromatography conditions can be changed to keep phospholipids away from the quantitative area for any analyte. The use of a stable isotope labeled internal standard (IS), should overcome matrix ionization effects because the IS is chemically like the analyte and therefore will chromatograph and ionize in the same way as the target analyte, but will have a different mass. Deuterium labeled steroids will elute fractionally ahead of its unlabeled form, this is marginal when three deuterium atoms are substituted but is significant if seven atoms or more are substituted as in some IS used.

Having defined the calibration, the validity of the method needs to be demonstrated with patient samples. There are specific tests needed to demonstrate freedom from interferents. In the first experiments, samples need to come from patients in at least five different sources for example out-patients or, different hospital disciplines depending on the clinical relevance of the analyte under determination. Samples from surgical patients may be more suitable than medical patients. The samples need to be tested with and without fortification with the analyte.

Concomitant medications may be a problem not limited to the drugs themselves because the dosing vehicle and packers can also lead to signal changes in the analysis. Samples from patients being prescribed typical drug treatments for any relevant clinical condition should therefore be included. Use of over-the-counter drugs such as aspirin, ascorbic acid, dextromethorphan, pseudoephedrine, salicylic acid, ibuprofen can lead to interferences in some assays. New therapies in the patient population can introduce ion suppression effects that were not present during method validation.

The next set of samples needs to look at patient samples with high concentrations of analyte tested at dilution with matrix or low level sample for **parallelism**. A sample needs to be identified with analyte concentration close to upper limit of assay range. Another sample needs to be identified with analyte concentration close to lower limit of assay range. The two samples are mixed to give equally spaced dilutions. Mixtures can be prepared as shown here from the CLSI EP7A document. Four replicates of each dilution are analyzed. Values are recorded and standard errors determined. A polynomial regression is performed. The 2nd and 3rd order polynomial variables should not be significantly different from zero.

The t-value should be less than critical value at 95% confidence level.

Further samples need to be obtained in order to test for possible interferences: Matrix samples with and without addition of potential interferents can be tested. For steroids, this may involve testing of many steroids and their isomers of the same molecular weights (isobaric) although fragmentations may be different so product ions may not be a feature of all. Metabolites of steroid drugs should also be tested. For example, unlike cortisol prednisolone has a double bond in the A-ring and so a mass 2 less, whereas Prednisone has a carbonyl group instead of an hydroxyl group at C-11. Tetrahydroprednisone and dihydroprednisolone metabolites have the same transitions as cortisol or the internal standard.

The possibility of interference from a number of **common drugs** needs to be eliminated by collection of samples from patients known to be taking the drug or by addition of the drug to samples. Aspirin, ascorbic acid, dextromethorphan, pseudoephidrine, salicylic acid and ibuprofen are among the common drugs that need to be excluded (Keller et al., 2008). The effects of different **anticoagulants, lipemia and hemolysis of samples** should also be tested.

Searches are available on the internet to identify potential interfering substances (http://chemfinder.cambridgesoft.com). Many "hits" are drugs or metabolites. The search needs to cover a range $\pm 70\,m$ Daltons of target compound and allow for potential adducts (Na, MeOH, MeCN, NH_4) and common losses (HCl, CH_3, H_2O). As an example for the protonated 25-hydroxy vitamin D2 ion at m/z 412.6 ± 0.5 (Maunsell et al., 2005) a total of 56 isobaric compounds for 400.6 and 59 compounds 406.8. There were 82 isobaric compounds including 1α-hydroxy vitamin D3 (alfacalcidiol), 7α-hydroxy-4-cholesten-3-one, campesterol. When incorporated into a standard mixture vitamin D3 (alfacalcidiol), 7α-hydroxy-4-cholesten-3-one eluted later in the analysis than 25-hydroxy-vitamin D3. C3-epimers of 25-hydroxy vitamin D2 and 25-hydroxy vitamin D3 can occur variably in samples and these need to be tested to ensure chromatographic separation from the major vitamin Ds. In a candidate reference method for testosterone developed at the National Institution of Standards and Technology in the United States (Tai et al., 2007), a number (19) of structural analogs of molecular weights 286, 288 and 290 and 2 conjugated metabolites were tested for interference in transitions m/z 289 to 97 for testosterone and m/z 292 to 97 for the D_3 internal standard.

2.4.10.4 Sensitivity

Sensitivity (see Fig. 2.4.90) is the lower limit of assay performance achieved by following recommendations in CLSI EP17.

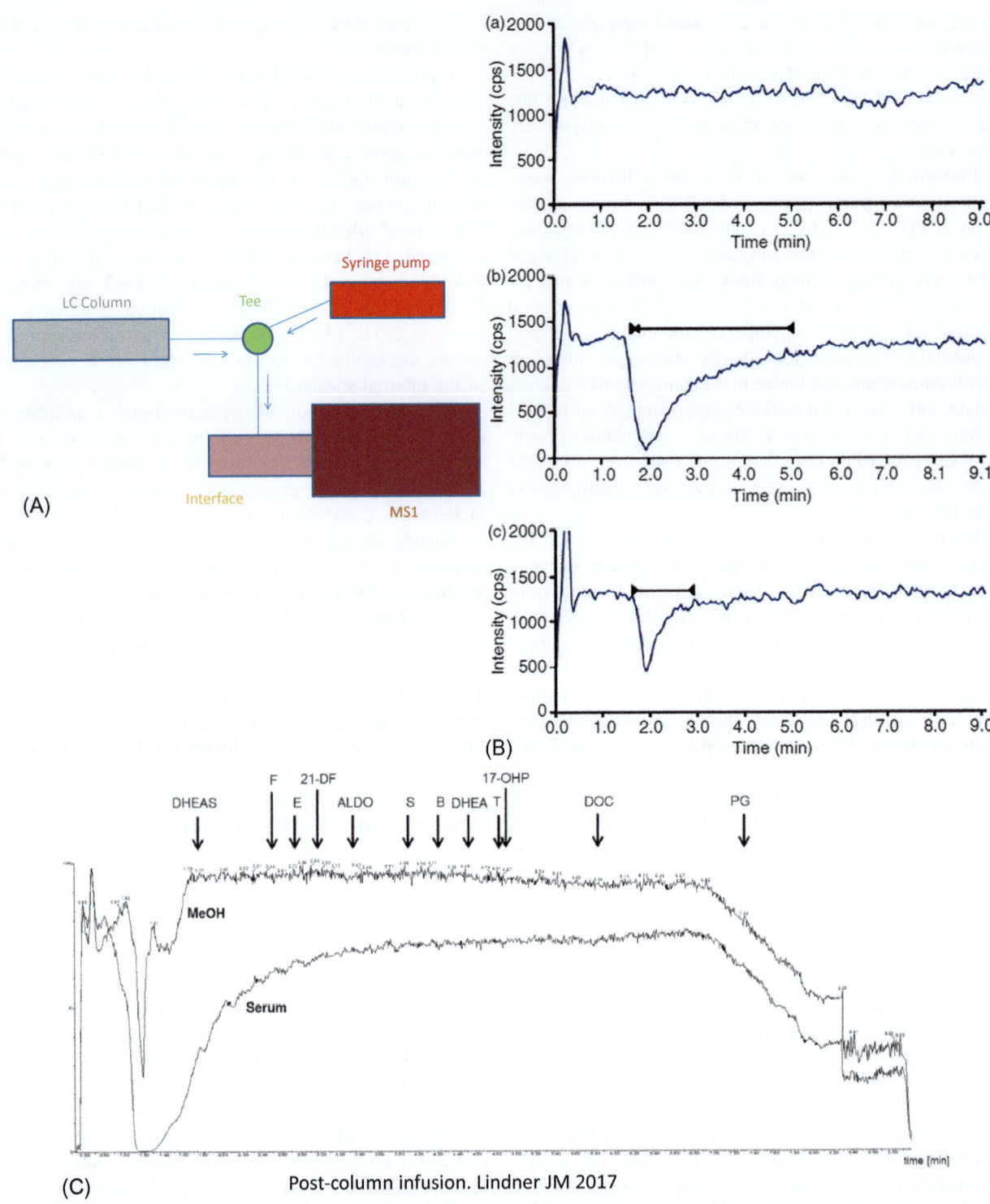

FIG. 2.4.88 (A) Schematic of postcolumn infusion experiment-pure analyte from the syringe is continuously mixed with effluent of a sample eluting from the LC column. (B) Chromatograms when analyte infused continuously postcolumn with injections into LC of: (a) no sample (b) sample extract by protein precipitation (c) a sample after SPE. The *solid line* indicates where ionization is influenced by matrix effects; the analyte should elute after this region. (C) The infusion chromatograms of a MeOH injection and a processed serum sample are shown. Run times were 9 min in total. The result of the postcolumn infusion experiment indicates ion suppression for all analytes of the matrix samples compared to a solvent injection. ((A, B) Redrawn from *Honour JW. Development and validation of a quantitative assay based on tandem mass spectrometry. Ann Clin Biochem 2011;48(Pt 2):97–111. Fig. 2 p. 103, Fig. 3 p. 104; (C) From Lindner JM, Vogeser M, Grimm SH. Biphenyl based stationary phases for improved selectivity in complex steroid assays. J Pharm Biomed Anal 2017;142:66–73. Fig. 3 p. 72.)*

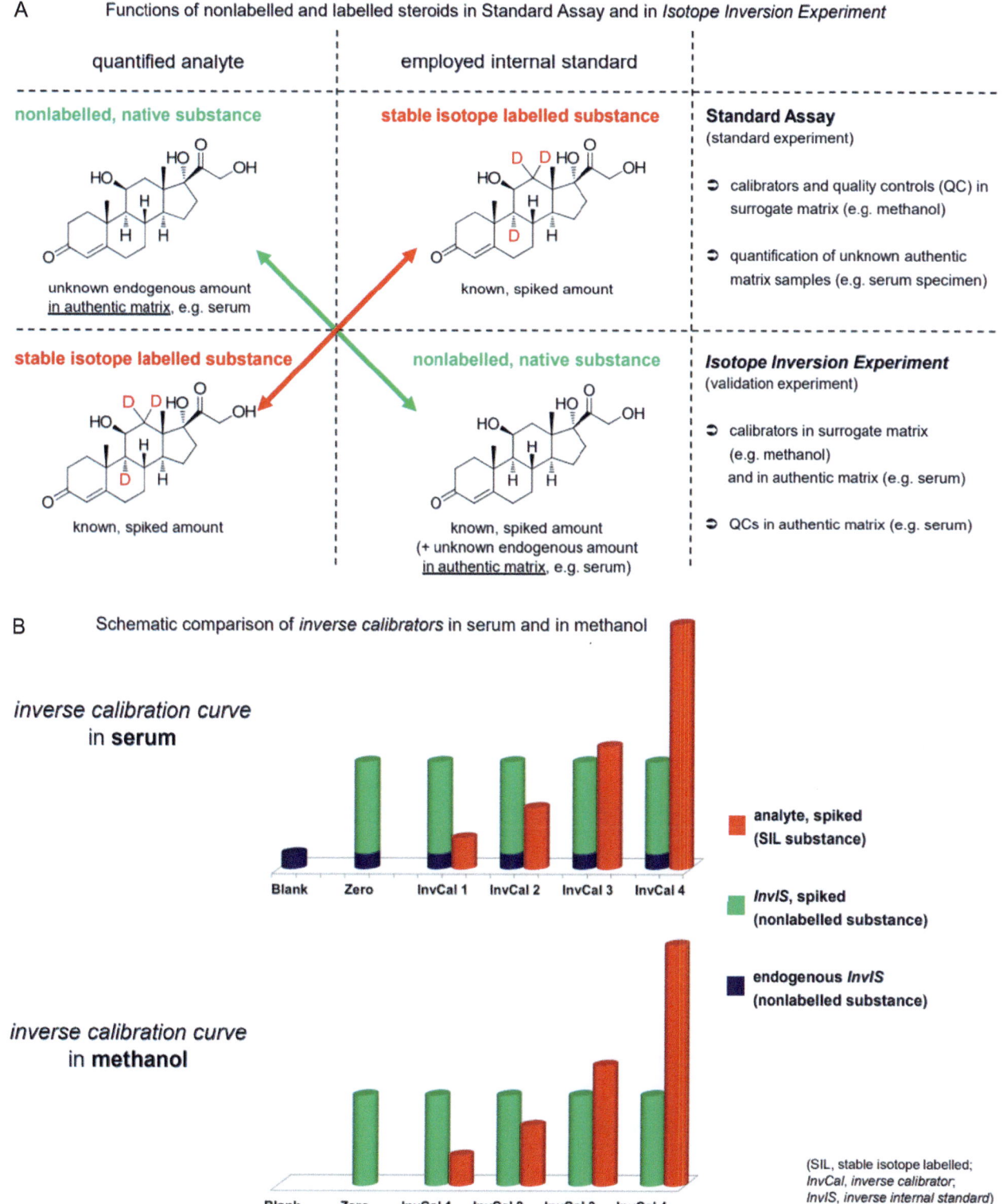

FIG. 2.4.89 Procedure for isotope inversion experiment. *(From Suhr AC, Vogeser M, Grimm SH. Isotope inversion experiment evaluating the suitability of calibration in surrogate matrix for quantification via LC-MS/MS-exemplary application for a steroid multi-method. J Pharm Biomed Anal 2016;124:309–318. Table 2 p. 312.)*

(Continued)

C

Schematic depiction of the comparison of slopes of calibration curves prepared in authentic matrix and in surrogate matrix (Standard Assay)

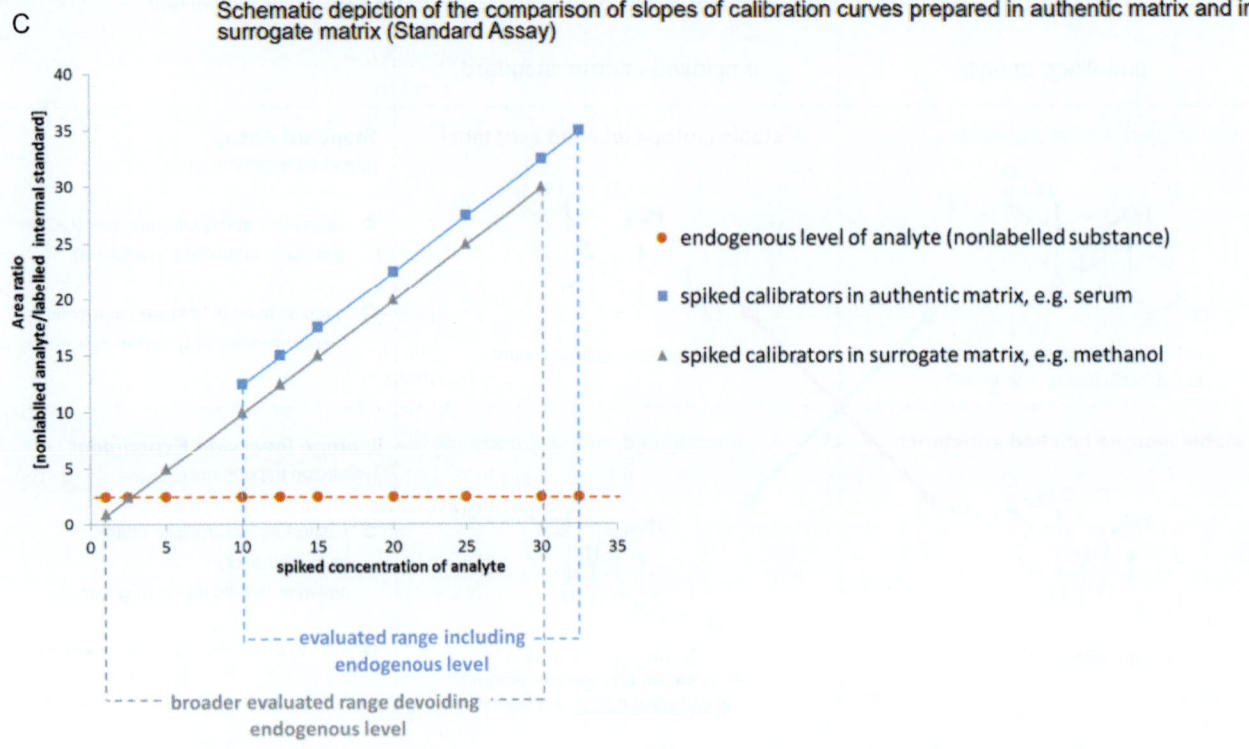

- endogenous level of analyte (nonlabelled substance)
- spiked calibrators in authentic matrix, e.g. serum
- spiked calibrators in surrogate matrix, e.g. methanol

D

Schematic illustration of the influence of the endogenous amount of nonlabelled compounds (contributing to *inverse internal standard*) on the slope of *inverse calibration curves*

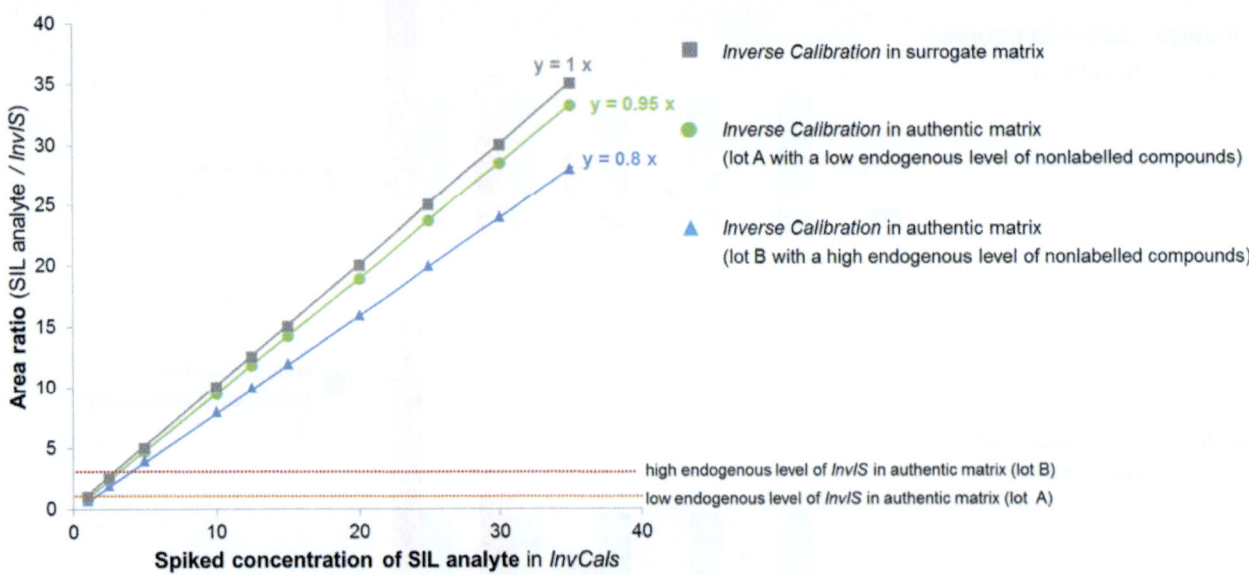

- *Inverse Calibration* in surrogate matrix
- *Inverse Calibration* in authentic matrix (lot A with a low endogenous level of nonlabelled compounds)
- *Inverse Calibration* in authentic matrix (lot B with a high endogenous level of nonlabelled compounds)

(SIL, stable isotope labelled; *InvIS, inverse internal standard; InvCal, inverse calibrator*)

FIG. 2.4.89, CONT'D

The limit of detection is the point where an analyte can be detected but not necessarily quantified. For the lowest standard in the calibration, the analyte response over five injections should be distinct from the results for the blank. The lower limit of detection (**LLOD**) is the lowest concentration at which the peak for the analyte is detected with a defined signal to noise ratio. This is usually 3. The mean result of the lowest standard is above the plus 3 standard deviations of the blank results. The lower limit of quantification (**LLOQ**) is the analyte concentration

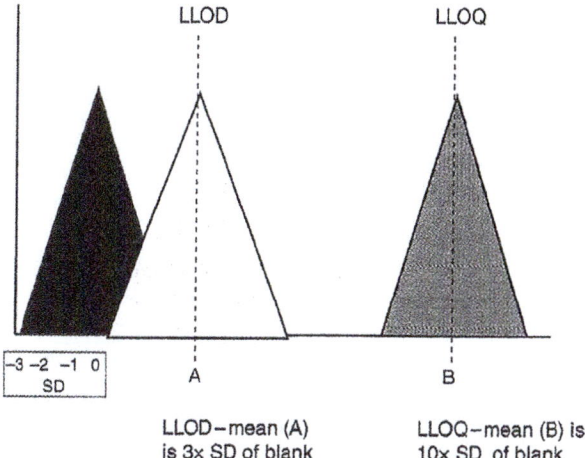

FIG. 2.4.90 Lower limits of detection and quantification (LLOD and LLOQ). SD is the standard deviation of the results. Ranges of results are illustrated for blank (*black triangle*), low level sample at LLOD (*white triangle*) and sample at LLOQ (*gray triangle*). *(From Honour JW. Development and validation of a quantitative assay based on tandem mass spectrometry. Ann Clin Biochem 2011;48(Pt 2):97–111. Fig. 4 p. 104.)*

when the signal to noise ratio is 5 with less than 20% imprecision and 80%–120% accuracy. The lowest standard is diluted and tested for the standard.

Sensitivity of assays can be improved by changes in ionization technique and potentials, detection of adducts, switching modality of ions, and derivative formation. Changes in cone voltage can increase sensitivity. Compare results following solvent extraction, SPE, protein precipitation.

2.4.10.5 Repeatability

The next set of tests concern the demonstration of repeatability of the results. The imprecision of a method is the closeness of individual results when the procedure is applied repeatedly to extracts of a sample. A preliminary precision test in CLSI EP5 suggests analysis of 20 replicates of a mid-range sample, then determine standard deviation and % CV. An acceptance criterion would be results with <10% CV.

For a thorough precision evaluation, again from CLSI EP5, the intra-assay precision of the assay is determined from five replicates QC samples at low (L), medium (M) and high (H) concentrations in one assay or batch. These should be at concentrations within the range of normal and pathological results. The order of analysis of these samples should be mixed so can be M, H, L, M, M, L, L, H, H, M as indicated in CLSI EP10. Intra assay variation is reported from standard deviation and % CV. An acceptance criterion would be results with <10% CV.

The interassay precision is determined over five consecutive assays or batches on different occasions again using the high, medium and low QC samples. Interassay variation should be less than 10% at each concentration.

An internal quality assurance program should be initiated. There are many issues to be considered (see Dudzik et al., 2018 for review). On each day for 5 days, separate assays should be performed with QC and 10 patient samples in duplicate. The order of analysis should be changed. After 2 days, a preliminary QC charts with mean, one standard deviation, standard deviations and 3 standard deviations should be constructed. The subsequent results are recorded daily or by batch on the charts. The data should be examined for outliers—if a result is outside the ±3 SD the sample be should investigated; and if outside ±4 SD the result is rejected. When there are 5 sets of results the imprecision should be calculated. Test acceptance criteria would be met when results within 15% CV. Where possible an external quality assurance scheme should be entered and the performance reviewed regularly.

2.4.10.6 Accuracy

Accuracy is the agreement of results to an accepted reference value. This can be tested in one or more ways. The results can be compared with results from a reference method. The results for the new method should be compared with the existing method. For example, results with an LC-MS/MS can be compared with a GC-MS method for 17-hydroxyprogesterone in Passing Bablok regression (Fig. 2.4.91).

GC-MS is probably a more accurate basis for a reference method because of the greater resolution with GC than LC. A certified reference material (CRM) can be analyzed but availability is limited in number and the material is very expensive for small laboratories to absorb in the cost of

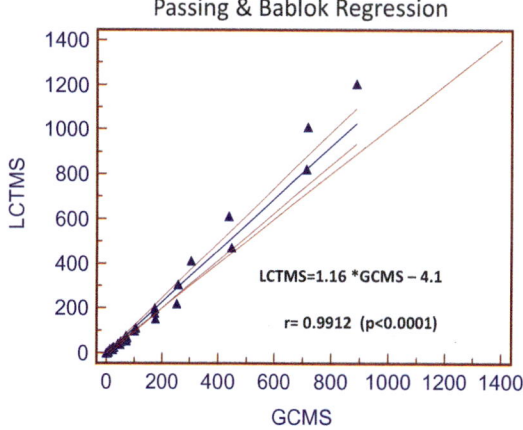

FIG. 2.4.91 Passing Bablock regression for 17-hydroxyprogesterone by GC-MS of heptafluorobutyrate (*y*-axis) (Hodkinson R and Honour JW, unpublished) and LCTMS (*x*-axis). *(Courtesy of Dr C.S. Ho, Hong Kong.)*

validation. Replicates of the CRM should be analyzed 6 times. The mean value should be within 15% of the target unless near LLOQ when 20% is acceptable.

2.4.10.7 Comparison of calibrators

This depends on the source and quality of reagents. For some analytes, calibration can be checked against physicochemical properties such as UV absorption. Blank samples can be spiked and the percentage of recovery can be determined. Ideally the blank should be free of the analyte. Carbon stripped material is not a good choice because the charcoal absorbs other components from the sample. In endocrine studies, samples from patients in a suppressed state can be useful but require approval of ethics committees to obtain. Patient or study samples can be spiked with known amounts of analyte. Six patient samples with low analyte concentration could each be spiked to 1, 2 and 4 times concentration of analyte. All samples should then be assayed in triplicate. Mean recoveries can be calculated. Acceptance criteria are met when results are 90%–110%.

Recovery as a measure of accuracy is assessed by comparing the concentrations of analyte in blank matrix before and after addition of known amounts of the analyte. The percentage of recovery should be calculated. Mean values should be within 5%–10% of the target values. Difference plots, Bland-Altman plots and Deming regression analysis may also be used (as above). When spiked EQA samples are used it is important to check that the samples are a set with the same matrix. Patient samples with low concentrations of analyte can be spiked with known amounts of the analyte—say 1, 2 and 4 times the endogenous amount. Spiked samples are assayed in triplicate. Mean recovery is calculated for each sample. Results should be within 10% of the added material. High and low samples can be mixed. Results should be within 10% of predicted concentrations. The measured against expected can be plotted and regression calculated, when R^2 should be 0.9. In a Altman Bland plot, the measured minus mean of expected is plotted against the mean of expected is divided by measured (Fig. 2.4.92).

2.4.10.8 Reference ranges

Results by mass spectrometry can be significantly lower concentrations than immunoassay methods owing to the improved selectivity of the MS technique. It is therefore not wise to use a reference range from an immunoassay for an MS method. Results with a reference method are more likely to be applicable (CLSI C28; 33, 34). Ideally for a new method the range of concentrations found under normal circumstances need to be established. This is a time-consuming and costly process because volunteers have

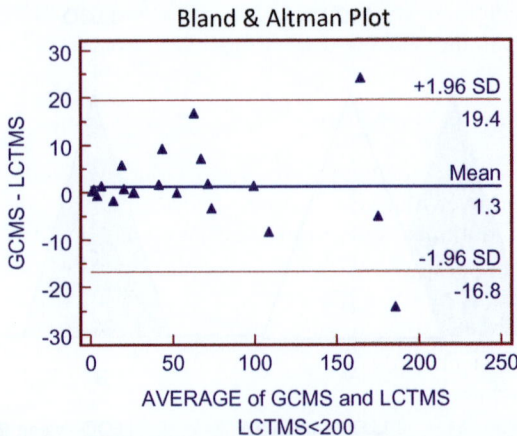

FIG. 2.4.92 Bland Altman plot for 17-hydroxyprogesterone by GC-MS of heptafluorobutyrate (Hodkinson R and Honour JW, unpublished) and LC-MS/MS (Dr C.S. Ho, Hong Kong). The difference between the two methods is plotted against the average of the two methods.

to be selected and recruited to agree to provide the necessary samples. Up to 120 samples may be needed to define a reference range for a normal group, and the data should be examined by parametric and nonparametric statistics. A range may be defined as the limits of the results, mean $\pm 2SD$ or a range from 5% to 95% confidence limits.

2.4.10.9 System stability

The system suitability check sample should be used regularly (batch, daily) to confirm stability of peak area, peak height, peak width and signal-to-noise ratio (basically to ensure that the system is working within the expected parameters) for one or more samples. FDA guidelines are worth consulting.

2.4.10.10 Carry over

The function of the autosampler should be checked when injecting a series of samples (Vallano et al., 2005). Carry over from one sample to the next can be assessed by injecting a high level sample before a blank at several points in the analytical run. Carryover should be understood, and interinjection wash cycles should be adjusted to ensure that any carryover does not significantly affect the determination. A carry over test would need a 10-fold dilution of a sample that has an analyte concentration close to upper limit of analytical range. Run the diluted sample repeatedly for 10 injections. Then run the diluted and reference sample injected alternately 10 times each. Using the independent sample t-test compare the mean of the results for the diluted sample injected sequentially with the samples injected alternately. The calibration curve can be run from low to high concentrations then high to low concentrations to see if the outcomes are the same. Discrepancies in the two approaches should be investigated.

2.4.10.11 Proficiency testing

Performance of internal QA and external QA should be charted over time. For internal quality control (IQC), two or more clinically relevant concentrations should be run throughout the validation. The performance should be charted. Criteria for acceptance are results within two standard deviations of the means. The assay should be tested in an external quality assurance program when available. When there are insufficient users of the method for group data the results should be compared with a reference method or GC-MS target values.

The long-term performance of the methods needs evaluation for ruggedness. The LC columns, reagents, standards will need to be changed with frequencies that depend on work load and cleanliness of injected material. Guard columns and standard LC columns should be changed regularly on the basis of numbers of samples to avoid column failure in the middle of a batch of samples. Using $0.2\,\mu m$ filters before the column and back-flushing the columns with high proportion of organic solvent can maintain performance for longer periods. Procedures need to be in place to document batches and suppliers of all items and check for differences in results when there are changes. Additionally, the use of different production batches of columns, standards and solvents should be checked to ensure consistency of results. A number of situations have been reported in which changing of solvent batches has caused significant problems owing to interferences from the solvents. Instability of standards in solution have been revealed (e.g., ascomycin in acetonitrile). Methanol used in HPLC eluting solutions can contain interferents that cause ion suppression, note when batches are changed.

Performances of some steroids assays can be compared with reference materials. NIST supply two vials of standards for hormones in human serum (SRM971) of males and premenopausal females with two levels of progesterone, estradiol, cortisol and testosterone. The Community Bureau de Reference (BCR) supply lyophilized serum using steroids with metrological traceability and GC-MS calibration. WEQAS offers materials for the four steroids analyzed by candidate reference methods. QA for mass spectrometry in metabolomics has been comprehensively reviewed (Dudzik et al., 2018).

2.4.10.12 Stability of samples

Tests should be conducted to confirm no changes in samples during the preparation stages. For example, when using 96 well plates there can be interaction of solvents with the plate materials and distortion by heat

during derivatization or evaporation of solvents. When frozen samples need to be transported to the laboratory then stability from time of dispatch to receipt in the laboratory needs to be confirmed. The stability of processed samples before LC/MS analysis needs to be tested in case of nonavailability of the LC-MS/MS. When ready for analysis there may be instrument problems that prevent immediate processing. A validation experiment should be performed where a set of samples is divided for immediate analysis and injection after storage to confirm stability (see above). Keep samples on the autosampler for 24 h and compare results with samples injected after processing.

Sometimes samples sit around awaiting analysis or reanalysis. The validation needs to include a mock test for stability of the sample under these situations. Prepared samples should be tested for stability while they are awaiting analysis in the system autosampler. Stability should be ensured for longer, say $2\times$ the anticipated storage time, to ensure that samples do not degrade prior to analysis, e.g., prepare sufficient (pooled) samples for 1000 injections and run them in a single run, looking for decay in response of internal standard or target analyte signal (also monitor the ratio) over time.

To test long-term stability, three patient pools should be prepared and divided into five aliquots before freezing aliquots. Samples are thawed at intervals of 1 week and 1 month. Assay four replicates in each batch. Perform Student t-test on results. Accept result where change $<\pm 10\%$ of initial. Results are not acceptable when $>\pm 15\%$ of initial.

2.4.10.13 Stability of stock solutions

Radiation from light, especially direct natural sunlight, will rapidly degrade steroids Care must be taken when evaporating solutions of many compounds to prevent losses and adsorption to container walls. Temperatures above 35°C for extended periods should be avoided. Stability of the internal standards also needs to be tested. Deuterium-labeled compounds are prone to exchange of deuterium for hydrogen and this may compromise their suitability as internal standards. The deuterium should not be substituted in hydroxyl or amino groups. Deuterium at the carbon of an aldehyde group will not exchange but if attached to the carbon adjacent to the carbon with an aldehyde or ketone group can exchange.

2.4.10.14 Validation of dilution protocol

Samples that give results that fall outside the range of the calibration will need a smaller quantity of the primary sample or a dilution to be put through the analysis again. The original samples should be diluted with analyte-free, similar matrix (stripped plasma, negative tissue) or dissimilar matrix (for example buffer or water or saline with and without albumin). If there is insufficient sample to re-run the injectable laboratory sample can be diluted.

2.4.10.15 Method uncertainty

Medical testing gives results that are subject to a degree of variability due to factors such as poor specimen collection, delays in transport, clerical and reporting errors. It is important to identify and minimize these factors. The laboratory should regularly assay quality control samples in order to ensure the performance of the assays. The variability of IQC results over time can be used to quantify the analytical error of the assay. This assessment is likely to include routine changes in reagent and calibrator batches, different operators, scheduled instrument maintenance. From this the ranges of acceptable results at particular, clinically relevant concentrations can be stated. An external QC assessment compares the results with other laboratories and demonstrates any bias to other methods. The analytical error can be expressed as a percentage and the laboratory can state with 95% certainty the result is a true value.

2.4.10.16 Assay interferences

Clinical laboratory medicine is involved in 60%–70% of critical decisions in patient care and it is the responsibility of the laboratory for quality of the results it produces. Precise and accurate results need to reflect the situation when the sample was collected. Interferences can affect the result at any stage in the preanalytical and postanalytical steps. The laboratory must have processes in place to recognize abnormal results outside the clinical process. The laboratory and clinicians need to work together on this. If interference is suspected then processes need to be in place to investigate the issue. Typically these include:

- Check sample for hemolysis, icterus and lipemia.
- Exclude sample carry-over in an automated assay.
- Check again for any missed medication—prescribed, over the counter or herbal medicine, contact with animals.
- Analysis using a different platform or different reagents.
- Analysis with a reference method.
- Repeat analysis with serial dilutions.
- Add immunoassay blocking agents to the sample and repeat analysis.
- Attempt sample clean up (chromatography, selective solvent extraction).
- Spike sample with analyte to check recovery.
- Spike sample with suspected interferent.

There have been few publications to 2020 reporting interferences in LC-MS/MS methods but undoubtedly with time there will be reports. If users are inexperienced with interpreting chromatographic and mass spectrometric data they may miss clues to problems.

2.4.10.17 Training and documentation

Once the method is established and operational the laboratory needs to establish a training program and a scheme for ensuring competencies. New users of the technology have to master basics in the principles and application of the LC and sample preparation issues and be trained in data interpretation and how to recognize problems within the LC and MS/MS. Recognizing the site of a problem is a key step so that downtime and costs are kept low. Training is also needed for the maintenance of assay performance.

A method can be judged to be viable when the method validation findings meet the standards described in this chapter. Many published methods come up to this standard and more papers follow the published guidelines (Fig. 2.4.86). Several LC-MS/MS papers on tacrolimus, immunosuppressants, drugs, steroid hormones and vitamin D determinations have included certified reference samples and standards.

To standardize the outcome of the process, laboratories must summarize the information in a formal document designed for the purpose of assay verification. The validation document should start with purpose and scope of the test. This will include location of the test, sample matrix, nature and source of materials, training of operators. Reagents need to be defined with respect to source, grade, purity and safety. The equipment needs to be described in its nature and performance with maintenance program. The type and frequency of system checks and quality controls. The report should fully document that the analysis is fit for purpose and can be repeated. The document should include examples of data output such as spectra and chromatograms.

For each method, the laboratory should always participate in an external proficiency testing schemes, preferably a scheme with several technology users. The laboratory should also address competencies and monitor staff performance.

Many new users of the technology have to master basics in the principles and application of the LC and sample preparation issues but problems within the MS/MS may be more difficult for them to recognize. Trace analysis of compounds at pg/mL concentrations needs a different mindset from practices appropriate for the μg/mL concentrations for most clinical analytes. Compliance with accreditation standards may be challenging for some laboratories. Inspectors may raise questions about standards they do not feel have been covered. There can be issues with calibration of bulb pipettes when used by young staff not familiar with such kit. All pipettes need regular checks on performance. Fault finding will be a challenge for staff using tandem mass spectrometry and collaboration with manufacturers and user groups is to be encouraged. The user is advised to check one element at a time.

A laboratory needs to be prepared for instrument failure and have a contingency plan to run samples on a different instrument. Even using an identical model however there may be differences in the results obtained. Portability of the method between systems should therefore be tested in order to maintain a service for clinically critical assays (therapeutic drug monitoring for example) or a change in sample workload.

Manufacturers of immunoassay kits and automated platforms are making a serious effort to demonstrate that their results can approach accuracy of reference and mass spectrometric methods but will always suffer from the inability with every patient sample to demonstrate specificity. The use of commercial kits for LC-MS/MS methods should lead to improvements in interlaboratory performance although there are differences in performances of the same instruments such that optimization of the particular instrumentation is required. LC-MS/MS has the potential to overcome problems with immunoassay. Traceable reference materials and internal standards will be important for high quality assays and harmonization of results. The METHOD VALIDATION must demonstrate that the accuracy and specificity of the analysis is acceptable for use. The costs of LC-MS/MS tests can compete with immunological assays because the reagent costs for LC-MS/MS are considerably less.

Finally if there are any changes to the validated method then these must be tested and justified through the correct process for the laboratory, performed, validated and authorized by agreed parties. New lots/batches of reagents should be tested for satisfactory performance before use in the method on patient samples. This means having new materials to test before existing material is depleted. Method validation is a time consuming process but the laboratory will have contributed to improved health care, by providing more accurate data from patient samples.

2.4.11 Selected applications

2.4.11.1 Basal plasma androgen determinations

In the past 40 years or so, immunoassays (RIA) were widely used. When investigating the neonate it is important to remember that there are high blood concentrations of steroid sulfates (dehydroepiandrosterone, 17-hydroxypregnenolone) from the fetal adrenal cortex that can interfere in the

immunoassay methods which are still frequently in use. Some of these methods, however, use extraction of the steroids into an organic solvent and chromatography (e.g., HPLC) prior to RIA. These combined techniques provide sufficient sensitivity and specificity for many clinical conditions in pediatric endocrinology, especially in neonates, compared with RIA based on neat sera alone. However, this method is cumbersome, time-consuming and therefore expensive and restricted to only few remaining specialized pediatric laboratories. Furthermore, differences in the techniques used in the different laboratories and changing specificities of RIA antibodies hamper comparability and thus implementation of widely usable and comparable normal ranges and cut-offs for diseases associated with testicular endocrine function. Problems have inevitably been encountered with automated methods in the laboratory. In immunoassays, results for testosterone in newborns have been reported to be grossly elevated due to interferences from fetal and placental steroids.

GC-MS has also been found to be a very reliable quantitative method for plasma androgens because of the high specificity through separation of steroids in the sample extract. Comparisons of GC-MS results with clinical samples assayed by direct and extracted RIA methods show close agreement with the extracted RIAs. In the meantime, different techniques based on liquid chromatography tandem mass spectrometry (LC MS/MS) have been developed to detect A, T and DHT. In contrast to immunoassays, these methods are largely independent of matrix effects or cross reactivity. Robust LC-MS/MS methods for sensitive, specific and high-throughput (180 samples per 24 h) determination of A, T and DHT simultaneously from 100 µL of plasma or serum as well as the accompanying reference values covering the whole pediatric age range (Xu et al., 2017) (Fig. 2.4.93). From 2016, there has been a growing interest in the 11-oxygenated androgens.

2.4.11.2 Direct mass spectrometry of samples on a probe

Soft ionization techniques such as LSIMS (liquid secondary ion mass spectrometry) of which FAB (**fast atom bombardment**) is an example have been used for analysis of steroids inserted into the MS on a probe. In principle, the analyte is placed on a probe in a glycerol matrix and a beam of ions (CS^+) or atoms ($Xe°$) is directed upon it. Analyte ions (MH^+, or $M-H^-$) are "sputtered" from the surface and are mass-analyzed (Fig. 2.4.94). The technique is carried out at near ambient temperature and is so soft that steroids such as underivatized steroid conjugates and polar steroids are readily analyzed.

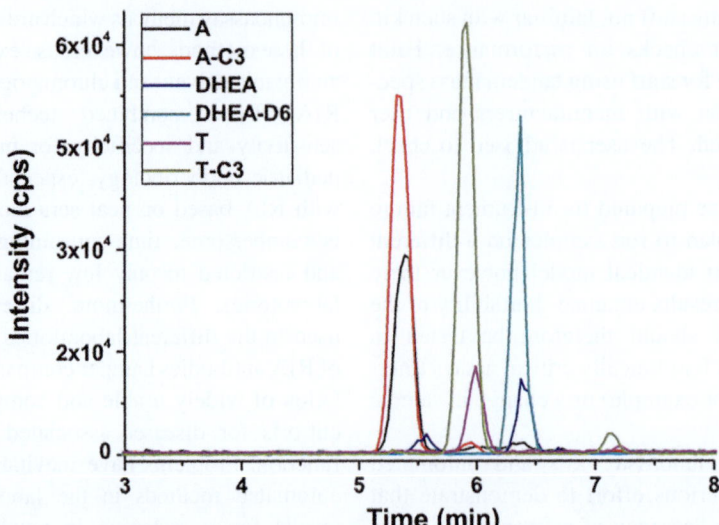

FIG. 2.4.93 Rapid method for serum androgens. The MRM chromatogram of androstenedione, testosterone and DHT with deuterium labeled internal standards is shown. The separation was performed on a Luna C18, 100 Å pore size column 84.0 × 2.0 mm. Mobile phase was 50% aqueous methanol to 80% aqueous ethanol. The MS was a Shimadzu Q-trap 4500 model with APCI in positive mode. *(From Xu W, Li H, Guan Q, Shen Y, Cheng L. A rapid and simple liquid chromatography-tandem mass spectrometry method for the measurement of testosterone, androstenedione, and dehydroepiandrosterone in human serum. J Clin Lab Anal 2017;31(5) e22102. Fig. 1 p. 4.)*

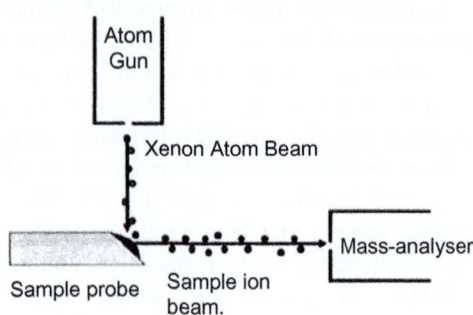

FIG. 2.4.94 Schematic of fast atom bombardment (FAB) MS. *(From Shackleton CH, Mattox VR, Honour JW. Analysis of intact steroid conjugates by secondary ion mass spectrometry (including FABMS) and by gas chromatography. J Steroid Biochem 1983;19(1A):209–17.)*

The SIMS spectra of steroid glucuronides are similar. In the negative ion spectrum of pregnanediol-3-glucuronide, the base peak *m/z* 495 represent [M−H]⁻ with a small peak at *m/z* corresponding to [M+Na]⁻ and several fragments of low mass (Shackleton et al., 1983) (Fig. 2.4.95A). Positive ion spectra are generally less suitable for analysis of steroid conjugate mixtures. In the negative ion spectrum of estriol-16-glucuronide, the pseudomolecular ion is at *m/z* 463.

In urine of a normal pregnancy in second trimester, the major ions represent pregnanediol glucuronide, pregnanetriol, 17-hydroxyprogesterone and androgens with a small peak for estriol. In the spectrum produced from urine of a patient with placental sulfatase deficiency, the major steroid are sulfates (Fig. 2.4.95B).

By utilizing a *continuous flow probe* the technique can be interfaced with HPLC or capillary-zone electrophoresis but there are very few examples of these latter techniques being used in steroid analysis. The relative simplicity of SIMS for steroid conjugate analysis is negated by the rapid contamination of the MS source.

2.4.11.3 Steroid acids

HPLC-MS is a particularly suitable technique for analyzing underivatized steroid acids which give excellent mass spectra in negative ion mode. Steroid acids can also be derivatized prior to analysis. Novel pyrenyl methyl esters were used for the separation and quantification of the four cortolic acids (Iohan et al., 1991). This derivative is particularly useful since it permits UV detection of these steroids which are normally difficult to detect by HPLC (Fig. 2.4.96). The analysis and function of cortolic acids has not been extensive and awaits further investigations.

2.4.11.4 Urinary steroid profile (USP)

The technique of urinary steroid profiling usually measured by capillary GC-FID or GC-MS, provides qualitative and quantitative data on excretion of a wide spectrum of adrenal steroid metabolites simultaneously. The test is available in specialized reference laboratories but could be offered from other centers where there is experience of chromatography and MS analysis.

The steroids in urine are hydrolyzed and derivatized and analyzed in a consistent fashion with strict adherence to enzyme concentration, reaction timings and temperatures. The relative ion currents produced from each steroid in a complex mixture remain constant over a period of weeks providing the glass insert in the splitless injection device is replaced once a week. Calibration is required for each steroid. The responses for 40 steroids is documented for the method used by Groessl and colleagues (Ackermann et al., 2019) in Bern, some examples are shown in Fig. 2.4.97 for progesterones and androgens (A) and for corticosterones and glucocorticoids (B).

This research group has published reference values for males and females with age and compared the results with six other laboratories doing urinary steroid metabolomics. The group subsequently used two-dimensional GC with high-resolution mass spectrometry and confirmed 40 out of 67 steroids (see GC HR MS later).

Some steroids are prone to give variable "ion response" with time of use as column couplings become contaminated. This is particularly a problem for the 18-oxygenated steroids of which 18-hydroxycortisol is a good example. The use of stable isotope labeled internal standards gives rise to a great improvement in accuracy and reproducibility but

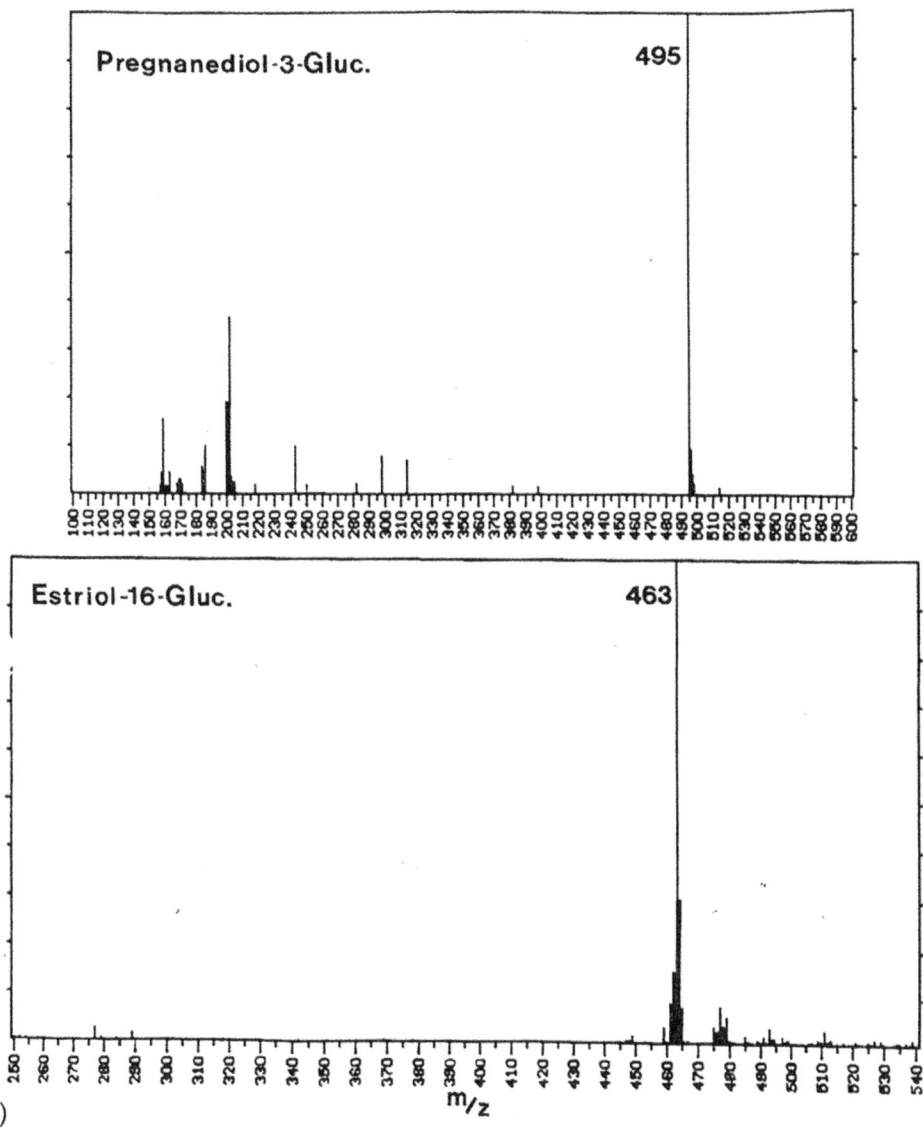

(A)

FIG. 2.4.95 Negative ion spectra in secondary ion FAB/MS for (A) pregnanediol-3-glucuronide (range 100–600 m/z) and estriol-3-glucuronide (range 250–540 m/z). *(Based from Shackleton CH, Mattox VR, Honour JW. Analysis of intact steroid conjugates by secondary ion mass spectrometry (including FABMS) and by gas chromatography. J Steroid Biochem 1983;19(1A):209–17. Fig. 1 p. 212.)*

(Continued)

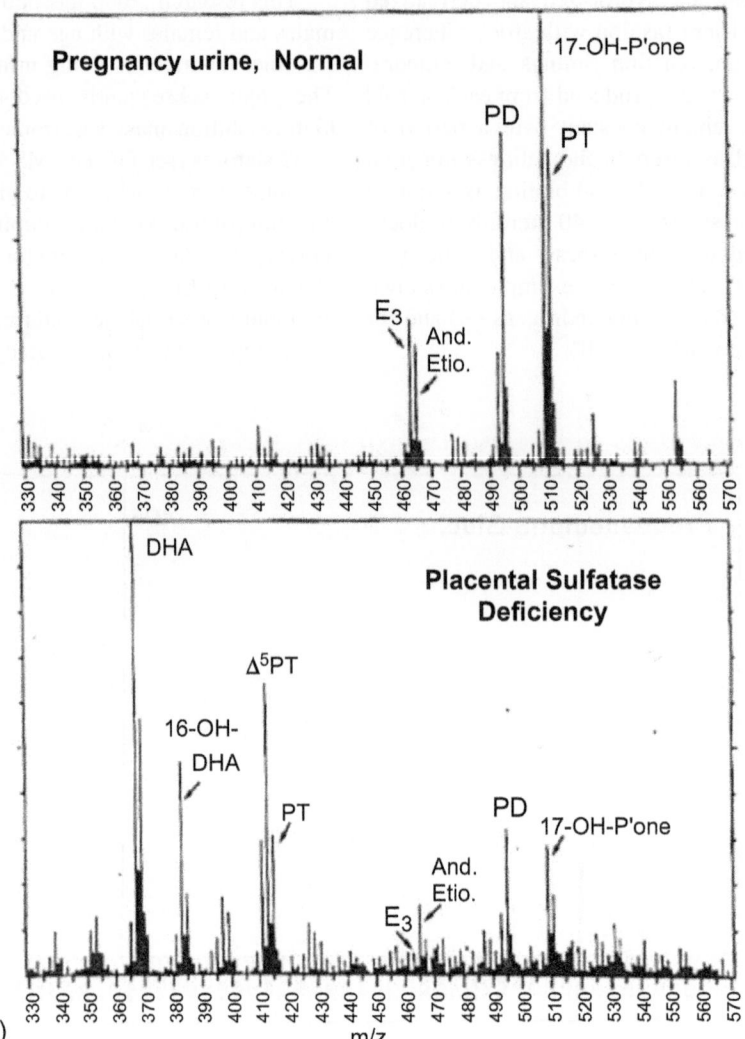

FIG. 2.4.95, CONT'D (B) Negative ion spectra (mass range 330–570) by secondary ion FAB/MS of urine extracts from pregnancy (upper panel) and from a patient with sulfatase deficiency (lower panel) that prevents sulfate cleavage of fetal adrenal androgens DHA, 16-hydroxy-DHA and pregnenetriol. *(Based from Shackleton CH, Mattox VR, Honour JW. Analysis of intact steroid conjugates by secondary ion mass spectrometry (including FABMS) and by gas chromatography. J Steroid Biochem 1983;19(1A):209–17. Fig. 3 p. 213.)*

few standards are available. A method based on selected ion monitoring (SIM) method is used for analyzing tetrahydroaldosterone (THAldo) can be extended to include 18-hydroxycortisol, 18-hydroxytetrahydrocompound A and tetrahydro-18-oxo cortisol. 3β,5α-tetrahydroaldosterone was used as internal standard in the original method (Honour and Shackleton, 1977). This internal standard separates from the natural isomer in the GC analysis. Stable isotope labeled internal standards are now incorporated and this is the best method to date for quantifying these important metabolites. An overnight methyloxime derivative is required to have THALdo with free aldehyde group derivatized (Honour and Shackleton, 1977) (Fig. 2.4.98 lower panel). In a usual steroid profile, most of the THALdo after a 3h MO reaction will be in the hemiacetal form giving

a molecular ion at 609 and a prominent M-103 at 506 amu (Fig. 2.4.98 upper panel).

The isotope dilution 18-oxygenated corticosterone and cortisol method can be adapted to include several other urinary unconjugated cortisol metabolites viz. cortisol, cortisone, 20α-dihydrocortisol, 20β-dihydrocortisol, 18-oxo-cortisol and 6β-hydroxycortisol. Cortisone, cortisol and 18-hydroxycortisol are also measured by isotope dilution (using deuterated internal standards) and the other metabolites are measured by absolute response or response against the usual stigmasterol and cholesteryl butyrate standards by employing the necessary calibrations.

DHEA and DHEAS have been analyzed without derivative formation in GC ion trap MS combination in selected reaction mode (Zemaitis and Kroboth, 1998). An androgen

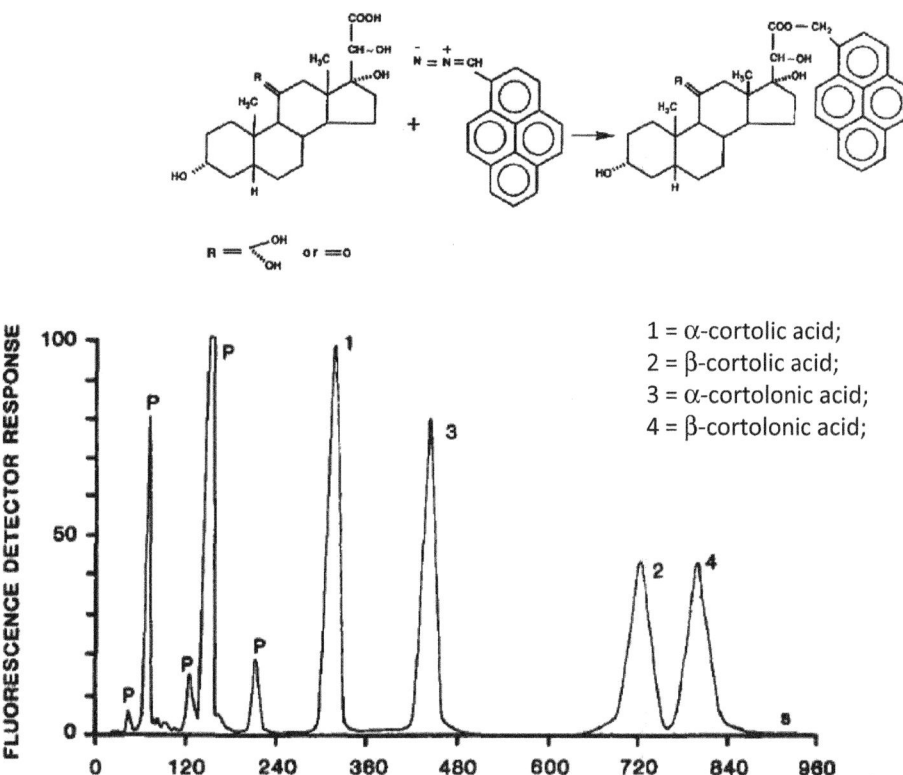

FIG. 2.4.96 HPLC Separation of cortolic and cortolonic acids as pyrenyl esters on Hypersil C18 50 mm × 4.6 mm 3 mm spheres eluted with acetonitrile: methanol (1:1 v/v)—0.03% aqueous phosphoric acid (46:54). *(From Iohan F, Vincze I, Monder C, Cohen S. High-performance liquid chromatographic determination of cortolic and cortolonic acids as pyrenyl ester derivatives. J Chromatogr 1991;564(1):27–41. Fig. 3 p. 32; Fig. 5 p. 36.)*

profile of plasma can now be achieved in a single chromatogram containing both free (for example testosterone, androstenedione, dihydrotestosterone) and conjugated (5α-androstane-3α, 17β-diol-17β-glucuronide and dehydroepiandrosterone sulfate) steroids. A method for plasma steroids has been extended for amniotic fluid utilizing heptafluorobutyrate derivatives and stable isotope labeled internal standards (Wudy et al., 1999).

The odorous components of axillary secretions and axillary hair have been separated with GC-MS of the pentafluorobenzyl derivatives of the 3-ketone and tertbutyldimethylsilyl (tBDMS) derivatives of the 3-hydroxyl containing species. The five steroids quantified were: 5α-androst-16-en-3-one, 4,16-androstedien-3-one, 5α-androst-16-ene-3α-ol and 5,16-androstadien-3β-ol (Nixon et al., 1988). Androgens and 16-androstenes with testosterone and androsta-4,16-dien-3-one have been separated by GC then subject to high-resolution SIM using pentadeutero internal standards produced by deuterium exchange. The internal standards were stable from back-exchange providing samples were always kept slightly alkaline. The steroids gave excellent EI mass spectra with dominant molecular ions. The increased mass spectrometric

resolution used gave chromatograms with higher signal-to-noise ratios and androstadienone was confirmed in human plasma with mean concentration ($n = 11$) of 2.05 pmol/mL, a figure near female plasma testosterone concentrations.

The highest urinary excretion rate of cortisol metabolites is between 10:00 and 18:00 h (Jerjes et al., 2006a). There is an increase in androgen sulfate excretion and other major androgen metabolites in premature adrenarche, whereas very distinctive profiles are seen in cases of genetic adrenal disorders or adrenocortical tumors (Honour et al., 2018) (see Chapters 3.1.1 and 3.3.1). In 21-hydroxylase deficiency, the commonest form of CAH, there is an increase in 17-OHP and 21-deoxycortisol metabolites with low cortisol and corticosterone metabolites. A urinary steroid profile enables differentiation of premature adrenarche or precocious puberty from other serious underlying pathology, i.e., inborn errors of steroid metabolism, and steroid-producing tumors in patients with signs of premature virilization.

Supercritical fluid chromatography (SFC) uses carbon dioxide as the mobile phase has been used in a number of applications for plasma (de Kock et al., 2018) urine steroids (Desfontaine et al., 2016) steroid conjugates

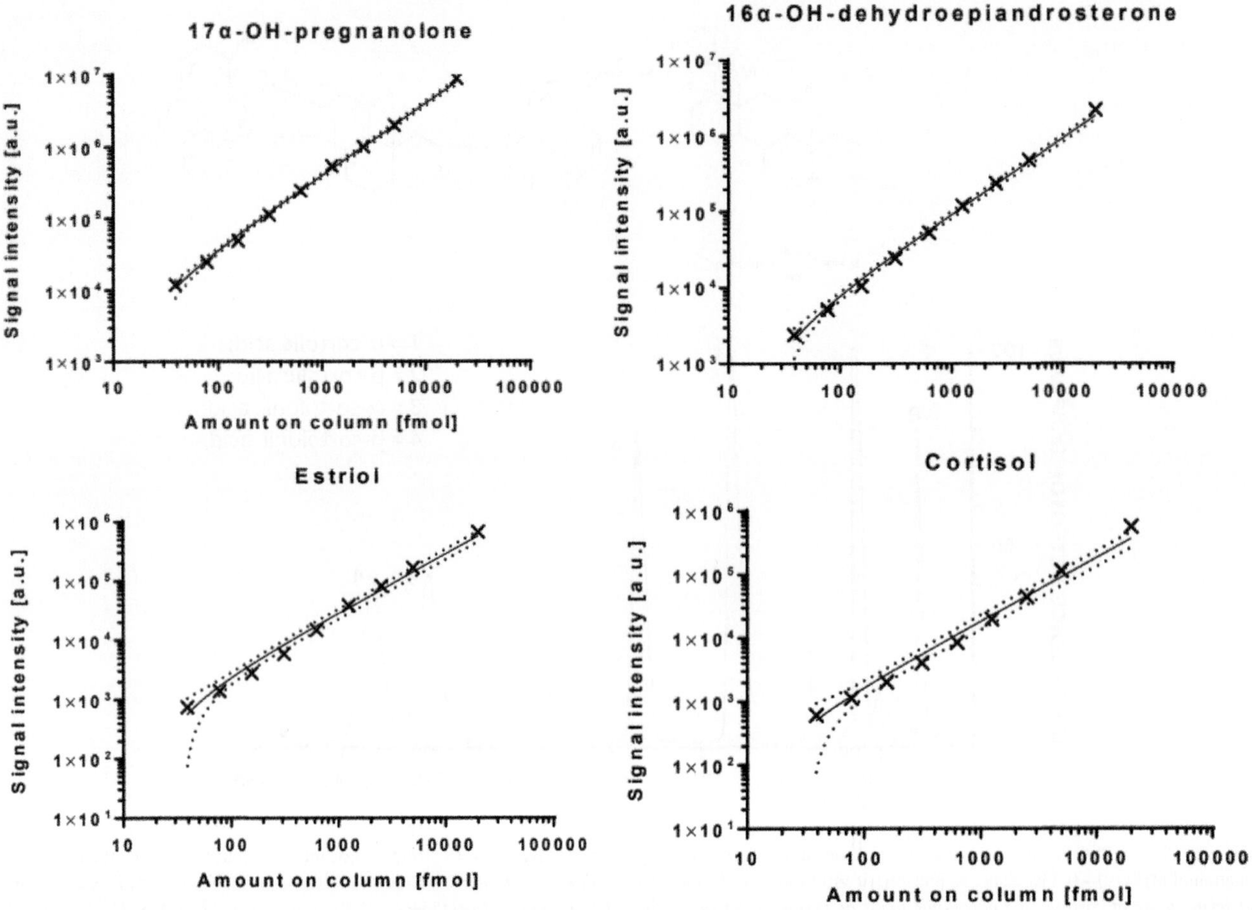

17α-OH-pregnanolone

16α-OH-dehydroepiandrosterone

Estriol

Cortisol

FIG. 2.4.97 Calibration curves (data points and linear regression) for analytes injected on column in the range 39–20,000 fmol. *(From Ackermann D, Groessl M, Pruijm M, Ponte B, Escher G, d'Uscio CH, Guessous I, Ehret G, Pechère-Bertschi A, Martin PY, Burnier M, Dick B, Vogt B, Bochud M, Rousson V, Dhayat NA. Reference intervals for the urinary steroid metabolome: the impact of sex, age, day and night time on human adult steroidogenesis. PLoS One 2019;14(3):e0214549. Supporting Fig. 1.)*

FIG. 2.4.98 Mass spectra of tetrahydroaldosterone as MO-TMS derivatives of hemiacetal (top) and free aldehyde forms (lower). *(From Honour JW, Shackleton CH. Mass spectrometric analysis for tetrahydroaldosterone. J Steroid Biochem 1977;8(4):299–305. Fig. 2 p. 301.)*

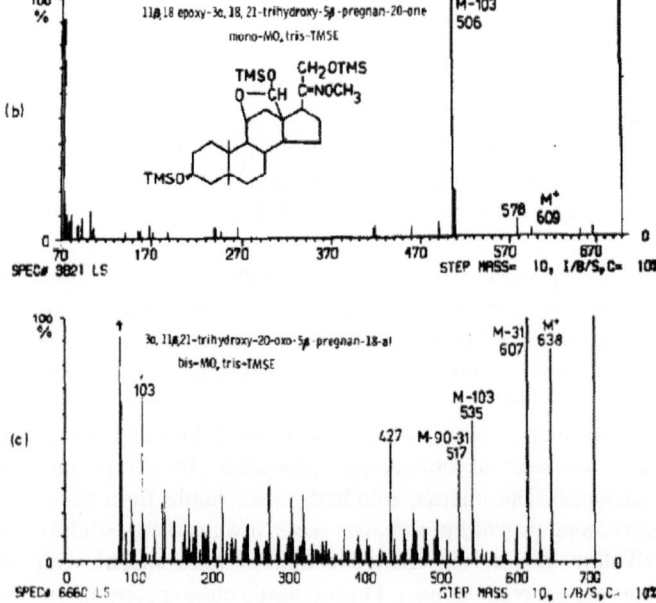

(Doué et al., 2015) doping agents (Nováková et al., 2015a, b). Performance was compared with GC-MS (Storbeck et al., 2018; Teubel et al., 2018). SFC is not yet within the repertoire of clinical laboratories.

2.4.12 Data analysis of MS data

2.4.12.1 Qualitative analysis

GC-MS in scan mode produces hundreds of mass spectra during the analysis, so unusual steroids can be identified through library search routines in the equipment software, manual interpretation of the mass spectra becomes easier with experience. The MS software will normally determine the characteristics of a peak from the slopes, symmetry and discrimination of closely eluting peaks. The analyst may need to action where integration of a peak starts and finishes. The analyst has the opportunity to override the computer determined points subject to local rules described in the standard operating procedure. Any adjustment by the analyst must be recorded. The analyst needs to observe any GC retention time shifts of peaks of interest. As part of the current and evolving metabolomics programs a vast database of metabolites can be found at HMDB (Human metabolome database at www.hmdb.ca).

Interpretation of the GC-MS data is more demanding than for many tests in the clinical laboratory but certainly manageable in laboratories with trained staff running tests for organic acid metabolites and newborn screening. The peak shapes and retention times can give clues to operational difficulties. There will be some differences in performance between instruments and GC column conditions. The analyst must pay particular attention to the positioning of the GC column in the injector and the mass spectrometer. Gas flow control systems will be very different between instruments. Many laboratories use hydrogen instead of helium and hydrogen generators get over the need for cylinders of hydrogen in the laboratory. In a targeted profile analysis by GC-MS, a total of 18 steroids covers the sex steroids and cortisol metabolites to be found in the urine of a normal child, a further 22 steroids are at low levels in normal subjects but when elevated become markers for metabolic blocks in steroid production or action. The analyst may see in the ion chromatogram from scan data other peaks that will need to be identified in a nontargeted analysis. These may be missed in a SIM based method.

The method can take 1–3 days to complete largely depending on the duration of the conjugate hydrolysis and derivative steps. USP therefore does not conform with the rapid sample throughput of routine hospital laboratory tests. The methodology does not yet lend itself to a routine hospital laboratory. Many patients with adrenal disorders have life threatening (salt-loss, adrenal insufficiency) or socially sensitive issues (uncertain gender at birth) so the laboratory should endeavor to produce results as soon as possible. A degree of technical skill is required. As laboratories make more use of mass spectrometry for steroid analysis in plasma samples the USP may be established in more hospitals. The breadth of information on androgens, cortisol metabolites and precursors in urine in one analysis makes the test economical against separate plasma steroid hormone determinations but plasma profiling is now possible with GC-MS/MS and LC-MS/MS in various forms and biological samples.

The total ion current chromatogram from the steroid profile of urine from an 8 year old girl is shown in Fig. 2.4.99. The molecular weights and principal fragments of the main steroids in urine from 389 to 728 Da are shown in Table 2.4.3.

The steroids elute from the low polarity C_{18} bonded GC column over 40 min with a trend to higher mass, hence from $C_{19}O_2$ steroids to $C_{19}O_5$ overlapping with $C_{21}O_2$ to $C_{21}O_6$ steroids, so androgens elute before intermediate metabolites then cortisol metabolites. The stigmasterol peak size is an indicator of a successful TMS derivative, if the peak is reduced in size by more than 80% the derivative step should be repeated.

Steroid retention times are more consistently recorded by relative position to the nearest peaks of a co-injected series of alkanes (C_{22}, C_{24}, C_{26}, C_{28}, C_{30}, C_{32}, C_{34}) as methylene units. In the mass spectra in each GC peak, molecular ions (M^+) for each steroid are often less than 20% relative intensity. Many steroids such as androsterone as MO-TMS derivative show a prominent ion at M-31 representing loss of the methyloxime group (Fig. 2.4.33A). Each hydroxyl group is lost as $OSiMe_3$ with clear fragment losses of 90 mass as for THE and THF (see Fig. 2.4.33B and C). Dehydroepiandrosterone (DHEA) has a base peak at m/z 129 from a fragmentation of the A-ring, the molecular weight is 389 and M-31 is at 358 (Fig. 2.4.33D). Pregnanediol spectrum has a base peak at 117 and the M+ (464 Da) is only 1% or 2% relative intensity, with a weak ion for an M-15 fragment at 449 Da (spectrum not shown).

Pregnanetriol has a weak molecular ion (552) and the ion at 435 amu represents loss of the methyl group at C21 and C20 plus derivative of hydroxyl (CH_3-$CHOSiMe_3$) as M-117. For α-cortolone, there is an M-205 loss to m/z 435 due to loss of carbons 20 and 21 as CH_2OSiMe_3-$CHOSiMe_3$ (Fig. 2.4.33E). The library comparison software may confuse these steroids but retention times are different. For C_{21} steroids with no hydroxyl at C-17 (corticosterone metabolites), the mass spectra show prominent ions at m/z 188 and 175 due to fragmentation across the D-ring and side chain. Identification of the major steroids in urine is straightforward. Steroids such as 16-hydroxy-DHEA and corticosterone metabolites give two peaks because of syn- and antiisomeric forms of the methyloximes. The system software will confirm identities of the steroids by

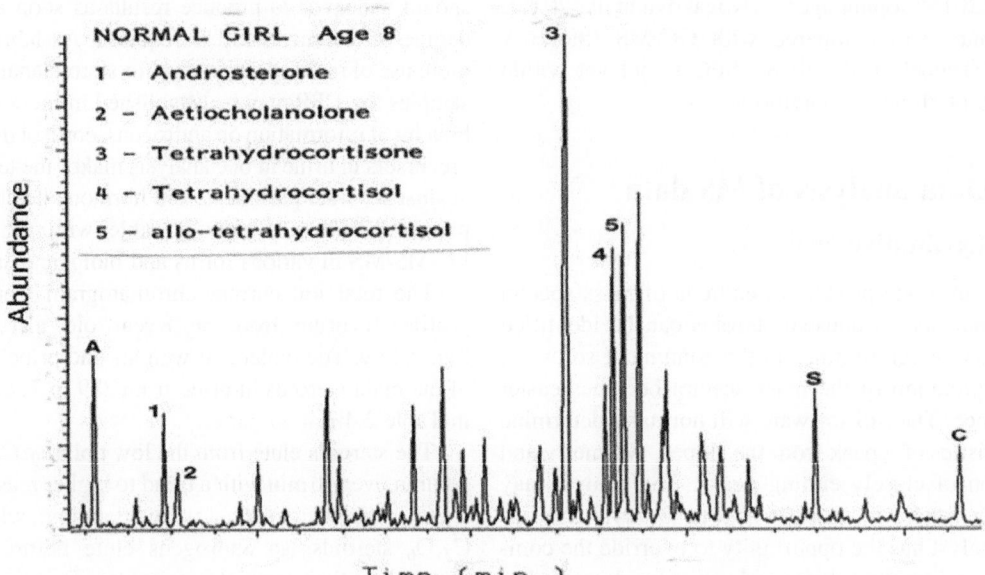

FIG. 2.4.99 Total ion current chromatogram from GC-MS analysis of urinary steroids of normal 8 year old girl. A, S and C are internal standards (androstanediol, stigmasterol and cholesterol butyrate). The androgens are at lower level than later but would increase over the ensuing 4 years with the development of the adrenarche before puberty. *(Author original.)*

comparison with the library of steroids and can be programmed to label peaks in the chromatogram and tabulate results.

2.4.12.2 Quantitative analysis

2.4.12.2.1 GC-MS

The quantitative analysis of steroids in GC-MS is based on peak height or peak area of quantifier ion for each steroid compared with the signal from the designated internal standard. This is better than the process used in early reports based on total ion current chromatograms which did not use the specific data from the MS. Stigmasterol is often used as an internal standard to check efficiency of the TMS derivatives. When the response for stigmasterol is low the derivative should be repeated. Stigmasterol-TMS elutes after all of the urinary steroid metabolites and other compounds that elute closer to the steroids could be used. The excretion rates of androsterone plus etiocholanolone and dehydroepiandrosterone can be used as an androgen index. The outputs of the major cortisol metabolites 11-hydroxyandrosterone, 11-hydroxyetiocholanolone, THE, THF, allo-THF, α- and β-cortolone, and α- and β-cortol are summed to give total cortisol metabolites (TCM).

2.4.12.2.2 LC-MS/MS data

The analyst needs to carefully look at the data for retention times and quantifier/qualifier ratios for each peak. A very much higher degree of technical skill is required particularly to maintain good chromatographic

separation. Positioning of the column at injector and mass spectrometer entry point will be critical. The mixing of solvents, lengths of tubing, couplings at joints are points in the LC system that need attention and regular maintenance particularly to avoid gas in the system. The analyst needs to take note of pressure changes that are monitored in the system, the amount of noise, ghost peaks, missing peaks since they are clues to maintenance requirements.

An LC separation of steroids on small LC columns may be performed in less than 10 min compared with the 30 min or longer for GC separation so it is important to know retention times of a large number of steroid hormones and the metabolites of cortisol, cortisone, corticosterone, 11-deoxycortisol, 11-deoxycorticosterone, 17-hydroxyprogesterone. Commercial kits are available for plasma steroid profiling but not for urine.

Compounds of the same molecular weight may co-elute in the LC and depending upon the fragments detected after fragmentation in the collision cell can interfere with the determinations, these are termed isobaric interferences. In an analysis for cortisol in urine, the steroids 20α and 20β dihydrocortisone were found to affect the performance of the method (Israelsson et al., 2018). This was resolved by changing the LC pentafluorophenylpropyl to C8 silica with biphenyl end-capping column where separation was achieved before the MS/MS step. The 20α-corticosteroids elute earlier than the 20β counterparts in C18 columns. This might have been overcome by selecting different fragments of the C and D ring and side chain rather than the A and B ring which is the same in the three steroids. Biphenyl, phenylbiphenyl, and pentafluorophenyl rather than C8 or C18 columns have been lauded for separation of isomers (Keevil, 2016).

The recent interest in metabolomics has led to development of machine learning methods for data analysis (Moon et al., 2009). Concentrations may need logarithmic transformation because of distribution of the results. Principal component analysis, partial least squares discriminant analysis, hierarchical cluster analysis, peer group normalization, heat map visualizations and Z-scores have been used in several reports. One example of this approach from GC-MS analysis of urinary steroids showed that adrenal tumors secreting aldosterone from primary adrenal tumors also produced excess of cortisol metabolites (Di Dalmazi et al., 2015; Hána et al., 2019).

2.4.12.2.3 Quality of USP

The method needs to be fully validated with respect to reproducibility (typically <10%), recovery (80%–120%) and imprecision (<15%). In good practice, quality control samples will be processed regularly—aliquots of urine from say an adult male are stored frozen and one is processed at prescribed intervals as. Laboratories can register to the international UCLH/SKML EQA scheme (University College London Hospitals, London, United Kingdom and Stichting Kwaliteitsbewaking Medische Laboratoriumdiagnostiek, Nijmegen, The Netherlands). Contact at Clinical Biochemistry-HSL Analytics LLP, Floor 2, 1 Mabledon Place, London WC1H 9AX, United Kingdom. The USP EQA laboratory has a dedicated email address of steroidprofile.eqa@hslpathology.com. Performance of the scheme was reviewed (Phillips et al., 2004). The data from

the external quality assurance scheme run by University College London and SKML (UCL/SKML USP EQA scheme) show wide variation in excretion rates in the results reported on clinical samples and reference ranges used by the laboratories (Fig. 2.4.100).

The results are from USP of a patient with CAH due to 21-hydroxylase deficiency and shows the expected high excretion rates of androgens (peaks #1 and #2) pregnanetriol and 11-oxopregnanetriol (steroid #10 and #20, respectively) and low outputs of cortisol metabolites (steroids # 11–18). Laboratories can compare their results with others in histograms and performance can be judged from the multiple of median score MOM provided with the interpretation of the EQA data from all participants by the organizer.

There is much research centered around steroid profiling in parallel with molecular genetic testing. A European group is addressing the harmonization of investigations of disorders (or differences) of sex development (DSD) not only from a clinical point of view but also for the biochemical and genetic investigations. The EQA of analytical methods will be important in the context of DSD. An EQA for plasma dihydrotestosterone has been proposed and more laboratories performing USP should register with the UCLH/SKML EQA scheme.

2.4.12.2.4 Interferences

The GC-MS method for USP is subject to surprisingly little interference. Quinine metabolites elute with retention time similar to DHEA but are recognizable by the mass spectra. Several metabolite peaks can be seen after morphine use, for catechins from drinking tea (Fig. 2.4.101) plant-derived lignans (Fig. 2.4.102) and reflect drug use, diet, nutrition,

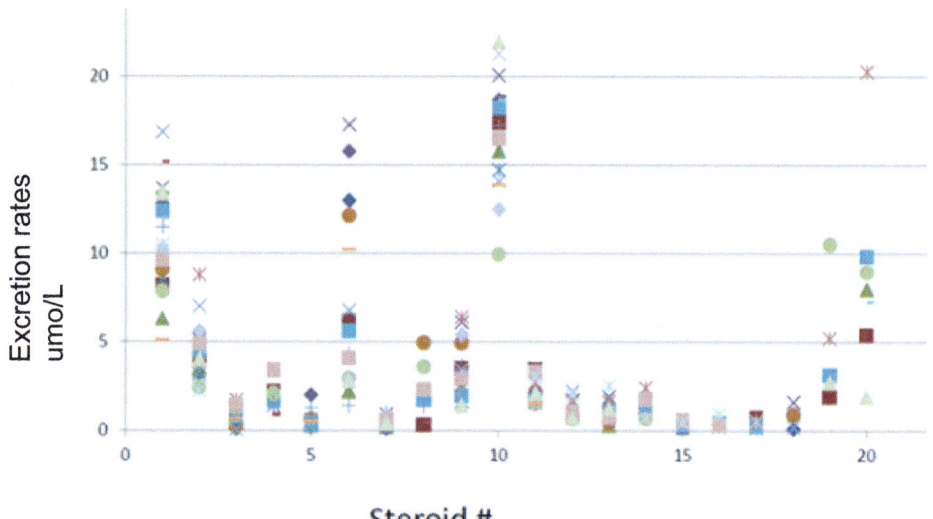

FIG. 2.4.100 Extract of a typical sample circulated to 23 laboratories in the UCLH/SKML EQA scheme for 20 urinary steroids. Laboratories are scored as a multiple of median results. *(I am grateful to Dr. F Lam and Richard Hodkinson for these data.)*

?Catechin

-present in green tea, cocoa, some berries

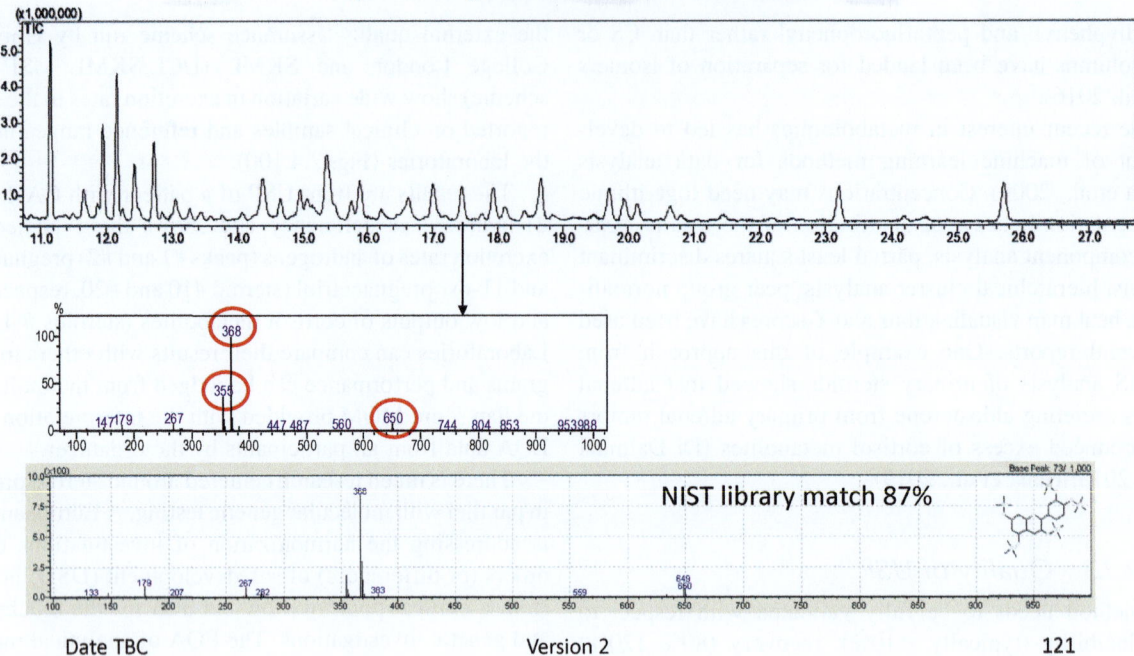

FIG. 2.4.101 Total ion current chromatogram of urinary steroids. One peak was attributed to catechin from green tea in the urinary steroid profile. The spectrum was compared with library data for confirmation and gave 76% match. (*I am grateful to Dr. F Lam for this data unpublished data.*)

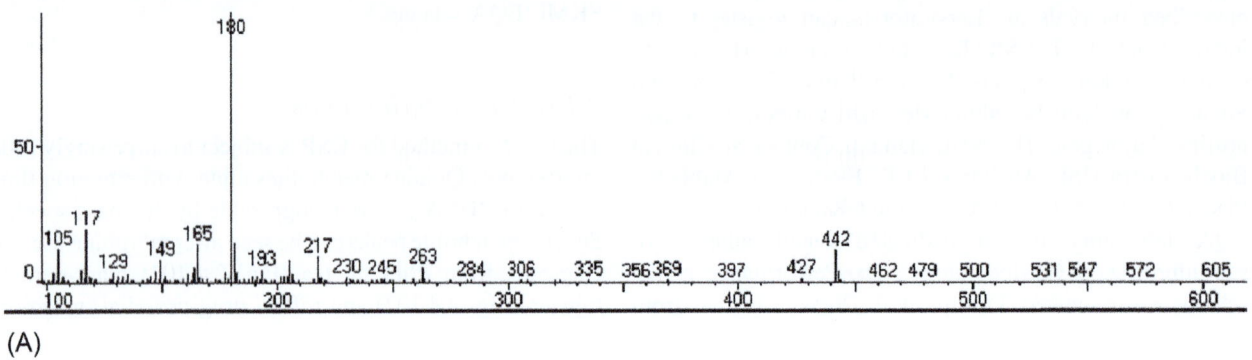

(A)

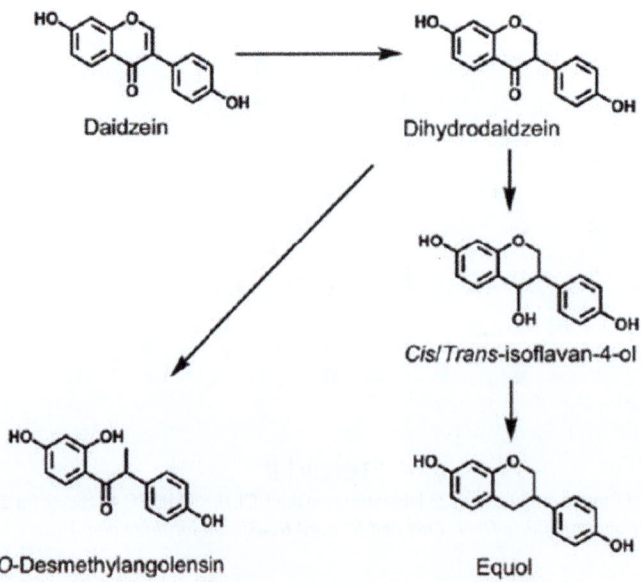

(B) *O*-Desmethylangolensin

Equol

FIG. 2.4.102 (A) Identification of equol and phytosteroids in urinary steroid GC-MS analysis. (B) Equol is one of several soy products formed in the intestine that are seen as contaminant in the analysis. (*Author original.*)

soya foods and tea drinking. Soy products are formed during the enterohepatic circulation through a number of bacterial reactions including deconjugation, demethylation, dehydroxylation reduction and ring cleavage.

Synthetic steroid metabolites can also be problematic. Using the mass spectral library in the software, most of these are identified. A full scan is required so the analyst is blinded to these if SIM is used. Spectral libraries are more complete with GC/MS than LCMS.

Many compounds in biological fluids can threaten the accuracy of any LC-MS analytical method in clinical practice. Interferences can arise from phospholipids, drugs, nutritional compounds, plasma expanders, metabolites arising from pathological conditions, alcohol, drugs of abuse, nutritional supplements, anticoagulants, preservatives, stabilizers, serum separators, hand cream, rubber gloves, collection tube stoppers, hemolysis, icterus, and lipemia. Carbamazepine metabolites, fenofibrate and piperacillin interferences to cortisol assays by LC LC/MS/MS have been reported and with time no doubt there will be many more reports. The analyst must be vigilant at all times. Matrix effects often arise in the interface between the LC and the MS from components in the sample and substances added to the sample during handling and preparation. The deuterium labeled internal standard will be affected to the same extent as the analyte so compensate for many, but not all, of the interferences. Steroids with ^{13}C substitution co-elute with the endogenous steroids, whereas the steroids with deuterium labeling separate in the chromatography. Any test for interference to the analyte should also consider the possibility of ion strength change, negative or positive, to the signal from the internal standard particularly if the analyte and IS do not co-elute as often happens with multilabeled compounds.

Interference testing falls into two categories:-

1. direct testing of the effects of specific substances on analyte concentration—The Clinical and Laboratory Standards Institute guideline EP7-A2 gives recommendations for specific compounds—hemolysis, lipemia, over-the-counter and prescription drugs, dietary supplements. A potential interferent is added to sample pool and the target analyte concentration bias is evaluated against the unspiked pool.
2. evaluation of unidentified interferences
   a. quantitative matrix effect study and
   b. qualitative postcolumn infusion study.

One issue in quantitative analysis of steroids is the choice of matrix for the calibration graph. In order to ensure there are no matrix effects, an interesting new approach involves inversion of the nonlabeled analyte and the stable-isotope labeled analyte.

2.4.13   Steroid conjugates (glucuronides and sulfates)

The steroid conjugates give excellent negative ion mass spectra dominated by the molecular anion (M−H$^-$). Negative-ion spectra are very simple, but depending on the buffer in the solvent, the positive ion spectra give ammoniated adduct molecular ions and show extensive fragmentation. The separate measurement 3α- and 17β-glucuronides of androstanediol was required and thermospray offered a method of achieving this. With the advent of ES-MS, and trideutero androstanediol-17β-glucuronide was synthesized as internal standard so as to permit plasma quantification of this important marker of high androgens in clinical conditions in women.

Microbore HPLC-ES-MS has been used for the analysis of the urinary steroid glucuronides without hydrolysis (Wang et al., 2021). The scope of the analysis was limited by the limited commercial availability of labeled internal standards for each steroid If necessary, this method could form the basis of a quick and easy method for specific diagnosis of steroid disorders. Three androgen glucuronides have been separated and detected by such a technique in serum of postmenopausal women (Ke et al., 2015). Kalogera et al. (2013) reviewed the potential for measuring androgen glucuronides in diagnosis and treatment of malignancies.

Steroid sulfates can be separated by HPLC-TSP-MS in negative ion mode (Shackleton et al., 1990). Frequently, analysis of steroid sulfates (and glucuronides) in the positive ion mode is needlessly confusing because of the sodium and ammonium ions present. The negative ion TSP mass spectra of steroid monosulfates are extremely simple having only a high mass ion at m/z (M−H) and sulfate ions at m/z 97 and 80. Electrospray MS gives almost identical data although less fragmentation results (e.g., sulfate release) is seen presumably due to the lack of heat compared with the TSP-MS technique. TSP-MS could be the basis of a method for measuring intact DHEA-sulfate using [7,7-^{2}H$_2$] DHEA sulfate as internal standard. This technique could even be a candidate reference method for quantification of this important analyte. The methodology reported would be greatly enhanced if ES-MS was utilized instead of thermospray or API MS. Ion-spray was developed by Henion for steroid sulfate analysis to aid in the identification of metabolites of anabolic steroid drugs in urine.

Serum concentrations for steroid sulfates have been determined in patients with sulfatase deficiency (Idkowiak et al., 2016; Sánchez-Guijo et al., 2016; Galuska et al., 2013). The quantification of serum cholesterol sulfate is useful for the diagnosis of the disorder Recessive X-linked Ichthyosis (RXLI). The underlying cause of this disorder is familial sulfatase deficiency and patients (invariably male) with the disorder have levels of cholesterol sulfate in serum in excess of 2000 mg/100 mL while normal individuals have levels between 100 and 200 mg/100 mL. The internal standard used was [$^{13}C_2$] cholesterol sulfate. The ES-MS technique which has higher sensitivity is now used. In pregnancy can be recognized when maternal estriol precursors (particularly 16α-hydroxy-DHEA sulfate) produced by the fetus cannot be converted to estrogen by the placenta which lacks ability to remove the 3-sulfate, a necessary prerequisite for aromatization. Thus, the urine profile would have reduced amounts of estriol glucuronide and elevated amount of DHEA sulfate.

ES-MS was used in a study by Shackleton aimed at determining the nature of a serum component responsible for increasing the measured values of 17-OH-progesterone by RIA in the neonatal period (Wong et al., 1992). The steroid sulfate fraction from newborn serum was separated on a conventional 3.9 × 150 mm C$_{18}$ column housed in an HPLC apparatus attached on-line to a ScieX API ion-spray mass spectrometer. The program was initially 50 mL/L (v/v) methanol in water for 2 min, followed by a linear gradient to 100% methanol at 52 min (then maintained for 20 min.). The flow rate was 0.8 mL/min and the eluant was split and 40 mL/min was introduced into the ion-spray quadrupole instrument. The remaining sample was fraction collected for further analysis of individual samples by 17-OH-progesterone RIA and ES-MS and GC-MS. The steroid sulfates were analyzed by scanning over the whole mass range in negative-ion mode so both a TIC recording and individual spectra were obtained. The spectra were those expected for a series of plasma steroid sulfates, various derivatives of dehydroepiandrosterone (DHEA) and pregnenolone. The fraction which had the highest RIA activity was subjected to individual ES-MS analysis which showed it to be a "hydroxy"—pregnenolone sulfate. Unfortunately, little specific information on the structure of a steroid sulfate can be obtained by ES-MS even when MS-CID is carried out on a tandem MS instrument. Thus, it was not possible, for example, to differentiate 16α- from 17-hydroxypregnenolone. Hydrolysis of the fraction and GC-MS of the derivatized free steroids showed that 17-hydroxypregnenolone sulfate was the compound which gave rise to the apparently high levels of 17-hydroxyprogesterone.

Although applications of negative-ion "spray" techniques for the analysis of glucuronides and sulfates have been considered separately, these conjugates can of course be simultaneously analyzed in the same sample since HPLC solvent systems can readily be developed which allows their separation.

There is the potential to develop selective MS-based methods for the direct and untargeted detection of these metabolite families. Such studies are likely to reveal the extent of steroidal bisconjugates in the steroid profile, which may turn out to be neglected treasures of steroidal metabolism. This may be particularly valuable in detecting metabolites cleared later after drug administration and benefit especially the identification of anabolic steroid abuse.

2.4.14 New techniques

2.4.14.1 GC high-resolution MS

The determination of C7-oxygenated metabolites of DHEA in human serum was shown possible in 2004 with GC coupled to a high-resolution mass spectrometer (Jeol JMS-SX 102) (Matsuzaki et al., 2004) when not detectable by GC-MS. Methyloxime-dimethylisopropylsilyl (MO-DMIPS) ether derivatives were prepared. More recent work has been mainly in the detection of anabolic steroids in urine from sportspersons who may have taken orally or injected anabolic steroids or ingested them because of contamination of a nutritional supplement. One clinical study with high-resolution mass spectrometry of steroids in urine (Hines et al., 2017) is discussed elsewhere in this chapter in Section 2.4.9.1.

A comprehensive analysis of urinary steroids as MO-TMS derivatives by two-dimensional gas chromatography coupled to a time of flight MS GCxGC-MS was undertaken by Groessl and colleagues at the Inselspital in Bern, Switzerland (Bileck et al., 2018). The GC × GC set-up employs two columns with complementary stationary phases, to improve the separation power for accurate mass detection with mass errors in the low ppm range. A double focusing loop modulator was mounted in the GC oven between the first (GC1) and second (GC2) column. The eluate of the first column was trapped in the modulation loop with a cold jet of nitrogen (−80°C). A hot jet of nitrogen was then used for rapid desorption at a modulation period of 6 s. For GC1, a 15 m × 0.25 mm i.d. × 0.25 µm column (crossbond dimethyl polysiloxane, Restek Corporation, United States), for GC2 a 2 m × 0.1 mm i.d. × 0.1 µm column (50% phenyl polysilphenylene-siloxane, SGE, United States) was used. From GC2, 1 m was utilized for the modulation loop. GC oven temperature was held at 50°C for 1 min followed by a first ramp to 220°C at a rate of 30°C/min and a second ramp to 300°C at a rate of 2°C/min and then held at 320°C for 5 min. For the hot jet, an initial temperature of 120°C for 1 min was followed by a first ramp to 240°C at a rate of 30°C/min and a second ramp to 320°C at a rate of 2.5°C/min before returning to 120°C. A constant flow of helium at a rate of 0.8 mL/min was used as carrier gas with an initial head pressure of 1.5 bar. 1 µL of each sample was injected in a split/splitless inlet held at 280°C in pulsed splitless mode. The GC2 column was directly coupled to the MS using a feedthrough block held at 275°C. The ion source temperature was set to 280°C. TOF MS enabled untargeted analysis by recording full mass spectra at high rates. MS analyses were conducted at 100 Hz in a mass range of 45–670 Th. Electron ionization was performed at an electron energy of 70 eV. The mass spectrum was recalibrated at the beginning of each modulation period using pentafluorophenol (PFP) as an internal standard. The ions monitored with retention times in GC1 and GC2 are shown in Table 2.4.11.

TABLE 2.4.11 Steroid list for targeted analysis. Listed ions (nominal mass) were used for quantification.

Steroid name	Abbreviation	Ion (m/z)	GC1 (min)	GC1RSTD (%)	GC2 (s)	RSTD (%)
11β-Hydroxyandrosterone (12)	11β-OH Andro	268	25.75	0.20	3.21	0.64
11β-Hydroxyetiocholanolone (13)	11β-OH-etio	268	26.05	0.20	3.32	0.59
11-Oxo-etiocholanolone (6)	11-oxo-etio	269	23.95	0.22	4.40	0.60
16α-Hydroxy DHEA (15)	16-OH DHEA	266	26.70	0.25	2.92	3.10
17β-Estradiol (7)	17β E2	416	24.57	0.24	3.20	1.00
17-Hydroxypregnanolone (14)	17-HP	476	26.51	0.37	2.90	6.59
18-Hydroxycortisol (42)	18-OH F	344	41.59	0.08	2.98	1.55
20α-Dihydrocortisone (38)	20α-DHE	402	41.96	0.12	4.04	0.80
20α-Dihydrocortisol (46)	20α-DHF	296	42.90	<0.01	2.99	0.61
20β-Dihydrocortisone (39)	20β-DHE	402	41.00	<0.01	3.99	0.66
20β-Dihydrocortisol (45)	20β-DHF	296	41.79	0.06	2.93	0.69
5α-Dihydrotestosterone (8)	5α-DHT	391	24.01	0.37	3.99	2.38
5α-Tetrahydrocorticosterone (28)	5α-THB	564	33.77	0.21	2.98	2.87
5α-Tetrahydrocortisol (30)	5α-THF	652	33.99	0.08	2.44	0.62
5β-Tetrahydrocorticosterone (27)	5β-THB	564	33.25	0.16	3.08	0.70
Tetrahydrocortisol (29)	5β-THF	652	34.20	<0.01	2.44	0.72
Pregnenetriol (24)	5-PT	253	32.60	<0.01	2.22	0.82
6β-Hydroxycortisol (43)	6β-OH F	513	41.68	0.10	2.99	0.88
α-Cortol (35)	α-CTL	343	36.70	<0.01	1.89	0.50
α-Cortolone (31)	α-CTLN	449	34.80	<0.01	2.53	0.54
Androstenediol (5)	AD	239	23.51	0.47	2.18	0.99
Androsterone (1)	Andro	270	21.50	0.30	3.45	0.73
Androstenetriol (18)	AT	432	29.33	0.16	2.48	1.10
β-Cortol (32)	β-CTL	343	35.57	0.14	1.90	0.67
β-Cortolone (33)	β-CTLN	449	35.61	0.17	2.62	0.61
Dihydroandrosterone (3)	DHAndro	331	21.88	0.19	2.34	1.02
Dehydroepiandrosterone (4)	DHEA	268	23.08	0.18	3.81	0.63
Cortisone	E	531	39.12	0.17	5.34	5.50
Estriol (21)	Estriol	504	30.42	0.18	3.35	0.73
Etiocholanolone (2)	Etio	270	21.77	0.21	3.60	0.57
Cortisol (37)	F	605	40.45	0.13	4.01	0.73
Pregnanediol (16)	PD	269	27.57	0.17	2.65	0.73

Continued

TABLE 2.4.11 Steroid list for targeted analysis. Listed ions (nominal mass) were used for quantification—cont'd

Steroid name	Abbreviation	Ion (*m/z*)	GC1 (min)	GC1RSTD (%)	GC2 (s)	RSTD (%)
Pregnanetriol (17)	PT	435	28.35	0.18	2.28	0.79
Pregnanetriolone (22)	Ptone	449	31.25	0.24	2.667	2.22
Tetrahydro-11-dehydrocorticosterone (25)	THA	490	32.93	0.15	4.33	0.72
Tetrahydroaldosterone	THAldo	506	34.80	<0.01	2.86	0.82
Tetrahydrodeoxycorticosterone (20)	THDOC	476	29.97	0.27	3.10	0.91
Tetrahydrocortisone (23)	THE	578	32.50	<0.01	3.35	0.35
Tetrahydro-11-deoxycortisol (19)	THS	564	29.75	0.17	2.71	0.60
Testosterone (9)	TST	389	25.27	0.18	3.88	0.57

Retention times are displayed for the first dimension (GC1) in minutes and for the second dimension (GC2) in seconds including the relative standard deviations of RT variations over 15 measurements.

A TOFWERK EI-TOF was coupled with 2D gas chromatography (GC × GC) to analyze urinary steroids from adults. The group of steroids were quantified by GCqMS as used to screen for adrenal disorders. Many analytes co-elute in the first separation, giving rise to broad and overlapping peaks in the one-dimensional chromatogram (inset). With the addition of the second GC dimension, the final chromatogram is said to show baseline separation of all peaks (Fig. 2.4.103) when looking at mixtures of reference steroids. Further results from the analysis of urinary steroids of children and adults failed in some cases to confirm the identities of many steroids and the data are much less clear (see clinical Chapter 3.1.1).

2.4.14.2 Capillary electrophoresis-MS/MS

Capillary electrophoresis mass spectrometry (CE-MS) in theory represents a promising platform in metabolomics research that is particularly useful for simultaneous resolution of polar metabolites that vary widely in their intrinsic polarity using a simple aqueous buffer system. Published CE methods for analysis of free estrogens and their conjugates have used micellar electrokinetic chromatography with UV detection that is only directly compatible to ESI-MS when using partial filling strategies. The analysis of urinary androgen glucuronides has been demonstrate by CE ESI-MS/MS (Cho et al., 2009); however, the method required a dynamically coated capillary under reverse polarity that suffers from poor stability due to anodic corrosion of the stainless steel needle.

CE-TOF/MS offers a complementary platform to LCMS for resolution of free estrogens together with their intact anionic conjugates with excellent specificity and low matrix-induced suppression effects in human urine (Kuehnbaum and Britz-McKibbin, 2011). The separation mechanism for estrogen metabolites by CE relate to their intrinsic physicochemical parameters. CE TOF/MS enables resolution of anionic estrogen conjugates and their positional isomers that is particularly useful for strongly ionic phytoestrogen disconjugates and glutathionylated estrogen adducts. Given the limited sensitivity of full-scan negative ion mode TOF/MS for detection of endogenous urinary estrogen metabolites, future work is needed to enhance concentration sensitivity when using tandem MS (e.g., Q-TOF), in conjunction with on-line sample preconcentration techniques and/or recent low flow/sheathless interfaces for CEESI-MSI.

The sample size is limited by the dimensions of the capillary column such that only high concentrations of the steroids can be analyzed. A UV detector is limited to detection of the steroid hormones with conjugated A-ring. Mass spectrometry is the better option because of the broader detection capability but detection limits of 100 attomoles on column for neutral steroids and 4 femtomoles for steroid conjugates (Que et al., 2000). On the evidence, to date capillary electrophoresis has little potential for clinical requirements.

2.4.14.3 Ion mobility mass spectrometry

Mass spectrometry has so far detected ions by their mass to charge ratio. Ion mobility mass spectrometry (IMS) provides an option for detection from GC and LC effluents, allowing separation of ions in the gas phase according to their size and shape as well as charge by collisional cross section (Romero and Fernandez-Maestre, 2018; Chouinard et al., 2016; Ahonen et al., 2013; Kanu et al., 2008). The vapor sample can be directly introduced or semivolatile

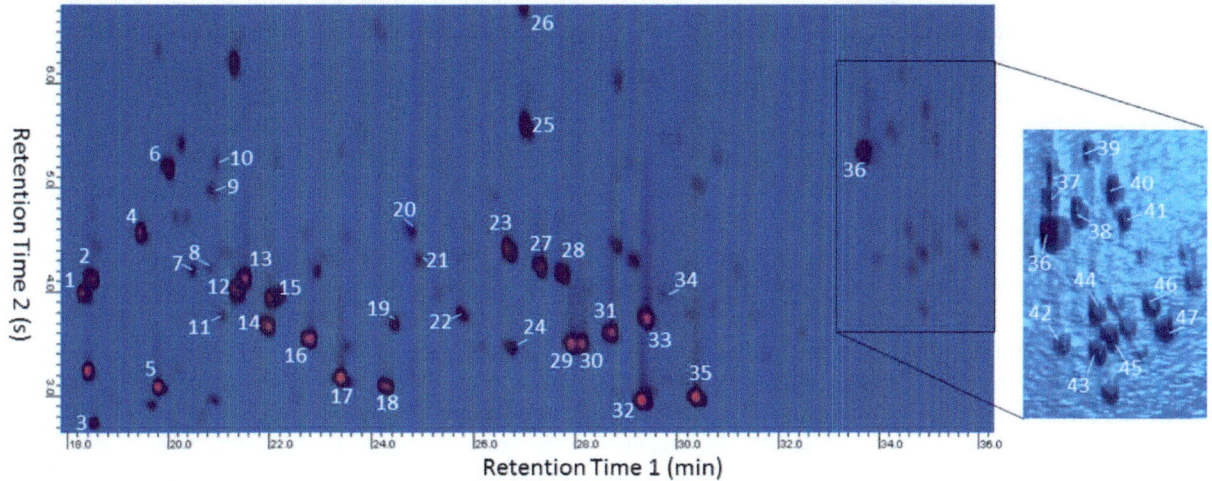

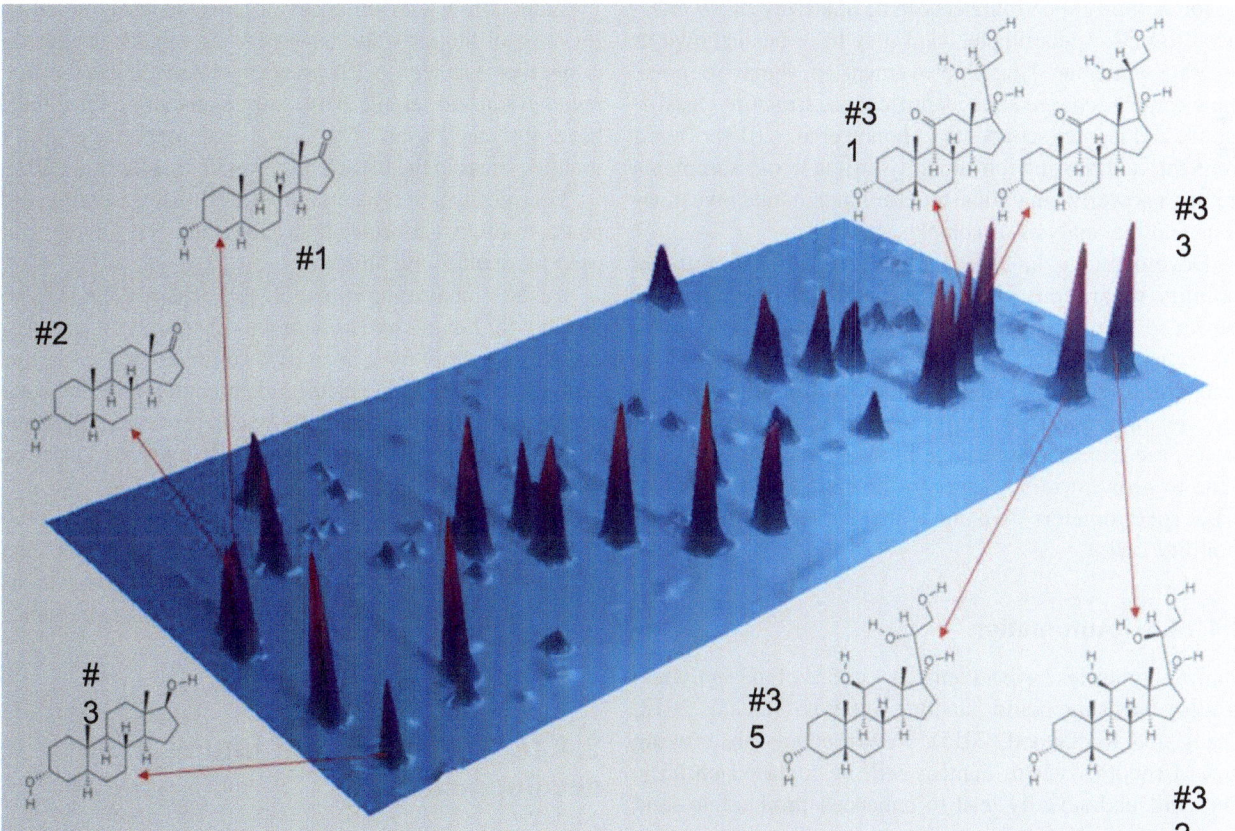

FIG. 2.4.103 (Upper panel) Two dimensional GC×GC-TOF heat plot from total ion current data where the retention time on GC1 is plotted against retention on GC2 is displayed for 47 steroids. There is baseline separation of the urinary steroids. The assignment of numbered blobs is shown in parentheses after name in Table 2.4.11. (Lower panel) 3D plot of the GC×GC-TOF MS chromatogram showing the separation of analytes that co-elute in GC1 alone on GC-MS. *(Adapted from Papadopoulos G, Dick B, Vogt B, Groessl M. Analysis of endogenous urinary steroids by GC×GC-TOFMS. Application Note GC×GCTOF 2015-1, TOFWERK, Switzerland. Kindly provided from personal communication with Michael Groessl at Tofwerk.com/Groessl.)*

compounds can be thermally desorbed from collection filters or traps. Liquid samples can be directly infused into an IMS component (Rister and Dodds, 2019). The most common ionization source is a radioactive ionization source (which is not desirable for health reasons) but corona

discharge, photoionization and secondary electrospray are in use. Discrimination is based on mobility differences in either high vs low electric fields and is dependent on their collisional cross sections. There are three main forms of IMS, drift tube (DT-IMS), traveling wave (TW-IMS) and field asymmetric/differential (FA/DIMS). There is also a low-resolution aspiration AIMS.

DIMS has not been extensively used as of 2020 but has been applied successfully to tissue samples for separation of estrone, estradiol and estriol from tissues of American eels, endogenous steroids in blood serum and testosterone conjugates in urine. Ion mobility allows efficient separation of structural isomers and epimers while reducing background noise over conventional ionization methods. For quantitation, IMS in principle may allow better signal to noise within the detector since species creating contemporary noise maybe separated from analytes through differential mobility. Isomeric estrogens have extremely similar mobility but the potential exists for derivatization to exaggerate structural difference and subsequently increasing separation between structural isomers.

Ion mobility spectrometers can be relatively easily interfaced to mass spectrometers and they have been shown to generate valuable data. Improvement in signal to noise enhancement, charge-state identification, structure classification, and isomer separation (Ahonen et al., 2013) offered from the addition of ion mobility to mass measurements creates a powerful analytical tool for fundamental investigations and the analysis of complex mixtures.

Developments in higher-order (ms^n) differential ion mobility separations and ionization sources may lead to improved sensitivity, quantification, resolving power and separation selectivity. As IMS instruments become more commercially available and move from the research and development laboratory to the applications laboratory, the mobility advantage in mass analyses will continue to grow in importance. In the course of time, most mass spectrometers will have the option to fit an ion mobility cell.

2.4.14.4 Automation

Many laboratories have automated sections of the analytical process by using liquid handling systems (Rochat, 2018; Zhang and Rockwood, 2015). New products are moving toward robotics where humans will be replaced entirely. This will undoubtedly lead to enhanced productivity and increase reproducibility and accuracy. The instruments can orientate bulk-fed tubes and vials and place them in the same position at up to 50 vials per minute. A bar code can be applied to each tube so the tubes can be identified and moved to racks or trays. Selection of equipment is a critical step in the incorporation of these improvements to the laboratory process.

2.4.15 Point of care testing

There is a move toward patient involvement in their health care package. In endocrinology, the best example is in the

care of diabetic patients who for several years have tested their blood sugar concentrations to help optimize the use of insulin. This has progressed to wearing a patch that pierces the skin to access interstitial tissue. The concentration of glucose is determined and coupled with a pump for insulin delivery. In time, the monitoring of cortisol concentrations in patients needing hydrocortisone replacement is envisaged.

Patients are encouraged to self-monitor blood pressure (Fletcher et al., 2016a,b) and there are now many prototype wearable devices to monitor temperature, heart rate, electrocardiography. The classical method for blood pressure measurement is to monitor the pressure in a cuff at which blood flow is stopped, nowadays the blood flow color is sensed by its color (plethysmography PPG) which relates to the pressure.

Lateral flow tests are in development in research settings for cortisol though none have reached regular use beyond prototypes. Devices to 2014 were summarized in a thorough review (Kaushik et al., 2014). Aptamers and microfluidics have also need used. There is a real need to be able to monitor stress in the broadest sense (Vavrinsky et al., 2021).

Eccrine sweat concentrations of glucose, lactate, phosphate, sodium, chloride, potassium pH, NH_4, Orexin are used particularly by athletes.

Tests to monitor hormones in the menstrual cycle can be suitable for home use. Daily mucus characteristic and basal body temperature have been used for many years. The concentrations of estrone glucuronide and pregnanediol glucuronide are monitored but differences between cycles have been seen raising doubts on the reliability of such tests (Blackwell et al., 2018a,b).

The measurement of cortisol in adrenal venous blood at the time of access would shorten and assure the anesthetist of successful sample collection, just as parathyroidectomy can be monitored with PTH measurement at the bedside.

2.4.16 Summary and future developments

Steroid analysis has advanced over the past 50 years from labor intensive, manual assays, one analyte at a time for a few samples to the processing of hundreds of samples for multiple analytes contemporaneously. A number of papers have reviewed the evolution of steroid techniques and these are included in the References section (see particularly Honour et al., 2018; Storbeck et al., 2018; Schiffer et al., 2019a,b). Currently, several technologies are being used and this creates some confusion with the units for concentrations found and the reference ranges being used. Immunoassay for steroids will be phased out and mass spectrometric methods will predominate likely using commercial kits containing all the reagents. Hopefully variation in the results will be reduced but this is not guaranteed.

There will be more use in the future of mass spectrometry at high-resolution, time of flight, ion mobility and charge remote fragmentation to gain more structural information on compounds at low concentrations needed for the breadth of metabolomics studies. Since some of these applications require derivatives to improve ionization and sensitivity they are less attractive. A precursor ion can be searched in databases if accurate mass from high-resolution (HR)MS is available.

The MS technology will continue to advance to improve compatibility with automation in clinical laboratories and ease of use. New reference preparations of steroid conjugates and more stable isotope labeled internal standards are needed. Some new mass spectrometric techniques may be adopted in clinical laboratories. Ion mobility in a number of forms is under development and may prove a key factor in enhancing structural confirmation in cases where shared product ions arise. The ability to connect more than one GC to an MS would increase the uptime of MS use, one column can be providing the steroids for the MS analysis during temperature programming while the other column is returning to starting point.

There are important developments in MS analysis of tissues and organs/tumors that can be sampled during a surgical operation by rapid evaporative ionization (so-called gamma knife) (St John et al., 2017). Appearing at the beginning of the 21st century, ambient mass spectrometry (AMS) allows rapid and direct analysis of various samples in open air with little or no sample preparation. Direct analysis in real time (DART) was first introduced and DART is now considered as one of the most prevalent and established AMS techniques. Alternative ambient mass spectrometry techniques such as desorption atmospheric pressure photoionization (DAPPI) (Vaikkinen et al., 2015), desorption electrospray ionization (DESI), and atmospheric solids analysis probe (ASAP) have already been successfully reported to characterize steroid esters in oils and tablets. AMS will be useful in the doping field for rapid detection of abused substances but probably has little application in clinical work. The analysis of the activity of single cells is said to be within reach (Zhang et al., 2020).

Endocrine biochemical and genetic testing are complementary. The latter techniques are advancing rapidly and it is difficult at this time to predict at what stage in the investigation of a patient that one will dominate over the other. Whole exome sequencing seems to be used more often (e.g., Chan et al., 2015; Kremen and Chan, 2019). In many cases, now genetics are applied to confirm a diagnosis made with biochemical tests that direct a particular fault. Genetic tests can however now seek mutations in a panel of candidate genes preempting some biochemical tests.

The data from steroid profiles will move further to many other analytes in the same sample (so called Metabolomics). Mastering a steroid profile would well prepare an analyst into the metabolomic era. Following from metabolomics and genomics improved, personalized, treatments for medical conditions is the next step (Pharmacogenomics). All of the developments will improve patient care.

An age of "big-data" and computer systems to interpret the data are with us. Big data analysis will help healthcare improve diagnosis, the development of a personalized medicine and enable automated external and internal reporting of patient data. The level of data generated within healthcare systems is not trivial. Wearable technologies like the equipment for continuous glucose monitoring, will continue to increase but along with other developments the volume and accuracy of data will become difficult to interpret. Human inspection at the big data scale is near impossible. There are risks for individual rights, privacy and autonomy, to transparency and trust.

2.4.17 Further reading

2.4.17.1 Immunoassay

Berson and Yalow (2006), Dittadi et al. (2018), Kawasaki et al. (1983), Murphy and Pattee (1963), Rao et al. (1982), van Helden and Weiskirchen (2018), Webb et al. (1985), and Wynne-Roberts (1974).

2.4.17.2 HPLC

Neufeld et al. (1998).

2.4.17.3 GC/MS

Moon et al. (2010).

2.4.17.4 LC/MS assays

Antignac et al. (2005), Gaudl et al. (2016), Kuuranne et al. (2000), Liu et al. (2018), Matysik and Liebisch (2017), Rister et al. (2019), Taylor (2005), Yamashita et al. (2007), and Yan et al. (2014).

2.4.17.5 Steroid profiles

Kulle et al. (2013), Olesti et al. (2021), Quanson et al. (2016), Rodriguuez-Morato et al. (2018), Schoneshofer (1981), Shibata et al. (1998), Storbeck et al. (2019), and Wang et al. (2020).

2.4.17.6 11-oxygenated androgens

du Toit et al. (2017), Bileck et al. (2020), and Ceglarek et al. (2009b).

2.4.17.7 Mass spec imaging

Cobice et al. (2013).

2.4.17.8 Clinical studies

Fink et al. (2002), Gika et al. (2014), Jerjes et al. (2006b), Kulle et al. (2011), Kushnir et al. (2010), Rauh et al. (2006, 2010), Sancez-Guijo et al. (2015), Shackleton (1982, 2010, 2012, 2018), Smy and Straseski (2018), Son et al. (2020), Sun et al. (2005), and Wang et al. (2016).

Appendix. References for interference in steroid immunoassays

Akin L, Kurtoglu S, Kendirci M, et al. Hook effect: a pitfall leading to misdiagnosis of hypoaldosteronism in an infant with pseudohypoaldosteronism. Horm Res Paediatr 2010;74:72–5.

Aldea ML, Barallat J, Martín MA, Rosas I, Pastor MC, Granada ML. Sodium interference in the determination of urinary aldosterone. Clin Biochem 2016;49:295–7.

Bartolone L, Smedile G, Arcoraci V, Trimarchi F, Benvenga S. Extremely high levels of estradiol and testosterone in a case of polycystic ovarian syndrome. Hormone and clinical similarities with the phenotype of the alpha estrogen receptor null mice. J Endocrinol Invest 2000;23:467–72.

Beastall GH, Ratcliffe WA, Thomson M, Semple CG. Trilostane interference with steroid assays. Lancet 1981;1:727–8.

Bell DA, Florkowski CM, Lewis JG, Crooke MJ. Plasma poltergeists: a negative cortisol interference leading to a false diagnosis of adrenal insufficiency. Clin Chim Acta 2012;413:1298–300.

Boddu SK, Madhavan S. High aldosterone and cortisol levels in salt wasting congenital adrenal hyperplasia: a clinical conundrum. J Pediatr Endocrinol Metab 2017;30:1327–31.

Bolland MJ, Chiu WW, Davidson JS, Croxson MS. Heterophile antibodies may cause falsely lowered serum cortisol values. J Endocrinol Invest 2005;28:643–5.

Boscaro M, Barzon L, Sonino N. The diagnosis of Cushing's syndrome: atypical presentations and laboratory shortcomings. Arch Intern Med 2000;160:3045–53.

Brossaud J, Barat P, Gualde D, Corcuff JB. Cross reactions elicited by serum 17-OH progesterone and 11-desoxycortisol in cortisol assays. Clin Chim Acta 2009;407:72–4.

Cao Z, Swift TA, West CA, Rosano TG, Rej R. Immunoassay of estradiol: unanticipated suppression by unconjugated estriol. Clin Chem 2004;50:160–5.

Cheng I, Norian JM, Jacobson JD. Falsely elevated testosterone due to heterophile antibodies. Obstet Gynecol 2012;120:455–8.

Chittawar S, Dutta D, Khandelwal D, Singla R. Neonatal endocrine labomas-pitfalls and challenges in reporting neonatal hormonal reports. Indian Pediatr 2017;54:757–62.

Choy KW, Teng J, Wijeratne N, Tan CY, Doery JC. Immunoassay interference complicating management of Cushing's disease: the onus is on the clinician and the laboratory. Ann Clin Biochem 2017;54:183–4.

Clerico A, Belloni L, Carrozza C, et al. A document endorsed by the Italian Section of the European Ligand Assay Society (ELAS). A Black Swan in clinical laboratory practice: the analytical error due to interferences in immunoassay methods. Clin Chem Lab Med 2018;56:397–402.

Cook DM, Allen JP, Kendall JW, Swanson R. Interference of 21-deoxycortisol with cortisol assay methods. J Clin Endocrinol Metab 1973;36:608–10.

Crowley RK, Broderick D, O'Shea T, et al. Spironolactone interference in the immunoassay of androstenedione in a patient with a cortisol secreting adrenal adenoma. Clin Endocrinol (Oxf) 2014;81:629–30.

Cummings EA, Salisbury SR, Givner ML, Rittmaster RS. Testolactone-associated high androgen levels, a pharmacologic effect or a laboratory artifact? J Clin Endocrinol Metab 1998;83:784–7.

Dupret J, Grenot C, Rolland de Ravel M, Mappus E, Cuilleron CY. Improvement of specificity of anti-testosterone and anti-5 alpha-dihydrotestosterone rabbit antibodies by immunotolerance techniques. J Steroid Biochem 1984;20:1345–52.

Dupret J, Grenot C, Rolland de Ravel M, Mappus E, Cuilleron CY. Improvement of specificity of anti-testosterone (T) and anti-5 alpha-dihydrotestosterone (DHT) rabbit antibodies by immunotolerance techniques. J Steroid Biochem 1984;20:479–85.

Fairfax BP, Morgan RD, Protheroe A, Shine B, James T. Abiraterone acetate: a potential source of interference in testosterone assays. Clin Chem Lab Med 2018;56: e138–40.

Findling JW Pseudohypercortisoluria: spurious elevation of urinary cortisol due to carbamazepine. Endocrinologist 1998; 8:51–54.

Fink RS, Pierre LN, Daley-Yates PT, Richards DH, et al. Hypothalamic-pituitary-adrenal axis function after inhaled corticosteroids: unreliability of urinary free cortisol estimation. J Clin Endocrinol Metab 2002;87:4541–6.

Gordon DL, Holmes E, Kovacs EJ, Brooks MH. A spurious markedly increased serum estradiol level due to an IgA lambda. Endocr Pract 1999;5:80–3.

Gray G, Shakerdi L, Wallace AM. Poor specificity and recovery of urinary free cortisol as determined by the Bayer ADVIA Centaur extraction method. Ann Clin Biochem 2003;40:563–5.

Hamer HM, Finken MJJ, van Herwaarden AE, et al. Falsely elevated plasma testosterone concentrations in neonates: importance of LC-MS/MS measurements. Clin Chem Lab Med 2018;56:e141–3.

Haning RV Jr, Carlson IH, Cortes J, Nolten WE, Meier S. Danazol and its principal metabolites interfere with binding of testosterone, cortisol, and thyroxin by plasma proteins. Clin Chem 1984;28:696–8

Heijboer AC, Savelkout AK, Blankenstein MA. Inaccurate first-generation testosterone binding assays are influenced by sex hormone-binding globulin concentrations. J Appl Lab Med 2016;1:194–201.

Holtkamp U, Klein J, Sander J, et al. EDTA in dried blood spots leads to false results in neonatal endocrinologic screening. Clin Chem 2008;54:602–5.

Jeffery J, MacKenzie F, Beckett G, Perry L, Ayling R. Norethisterone interference in testosterone assays. Ann Clin Biochem 2014;51:284–8.

Jones JC, Carter GD, MacGregor GA. Interference by polar metabolites in a direct radioimmunoassay for plasma aldosterone. Ann Clin Biochem 1981;18:54–9.

Kelly CJ, Ogilvie A, Evans JR, et al. Raised cortisol excretion rate in urine and contamination by topical steroids. BMJ 2001;322:594.

Kishino Y, Tanaka Y, Naitoh S, Kamisako T. Danazol falsely increased values in estradiol immunoassays. Clin Chim Acta 2010;411:1380–1.

Kline GA, Buse J, Krause RD. Clinical implications for biochemical diagnostic thresholds of adrenal sufficiency using a highly specific cortisol immunoassay. Clin Biochem 2017;50:475–80.

Konishi H, Minouchi T, Yamaji A. Interference by danazol with the Porter-Silber method for determination of urinary 17-hydroxycorticosteroids. Ann Clin Biochem 2001;38:277–9.

Koshida H, Miyamori I, Miyazaki R, Tofuku Y, Takeda R. Falsely elevated plasma aldosterone concentration by direct radioimmunoassay in chronic renal failure. J Lab Clin Med 1989;114:294–300.

Krasowski MD, Drees D, Morris CS, et al. Cross-reactivity of steroid hormone immunoassays: clinical significance and two-dimensional molecular similarity prediction. BMC Clin Pathol 2014;14:33.

Kroll MH, Jackson AJ, Elin RJ. Cefoxitin interferes with the "Clini-Skreen" column method for urinary 17-hydroxycorticosteroids. Clin Chem 1987;33):1219–22.

Kuwahara A, Kamada M, Irahara M, et al. Autoantibody against testosterone in a woman with hypergonadotropic hypogonadism. J Clin Endocrinol Metab 1998; 83:14–6.

Langlois F, Moramarco J, He G, Carr BR. Falsely Elevated Steroid Hormones in a Postmenopausal Woman Due to Laboratory Interference. J Endocr Soc 2017;1: 1062–6.

Lee C. Urinary 6 beta-hydroxycortisol in humans: analysis, biological variations, and reference ranges. Clin Biochem 1995;28:49–54.

Lee, C. 6 beta-hydroxycortisol interferes with immunoassay of urinary free cortisol. Clin Chem 1996;42: 1290–1.

Lee C, Goeger DE. Interference of 6 beta-hydroxycortisol in the quantitation of urinary free cortisol by immunoassay and its elimination by solid phase extraction. Clin Biochem 1998;31:229–33.

Legro RS, Schlaff WD, Diamond MP, et al. Total testosterone assays in women with polycystic ovary syndrome: precision and correlation with hirsutism. J Clin Endocrinol Metab 2010;95:5305–13.

Lim YY, Ong L, Loh TP, et al. A diagnostic curiosity of isolated androstenedione elevation due to autoantibodies against horseradish peroxidase label of the immunoassay. Clin Chim Acta 2018;476:103–6.

Mahlab-Guri K, Asher I, Gradstein S, et al. Inhaled fluticasone causes iatrogenic Cushing's syndrome in patients treated with Ritonavir. J Asthma 2011;48:860–3.

Mandic S, Kratzsch J, Mandic D, et al. Falsely elevated serum oestradiol due to exemestane therapy. Ann Clin Biochem 2017;54:402–5.

Masters AM, Hähnel R. Investigation of sex-hormone binding globulin interference in direct radioimmunoassays for testosterone and estradiol. Clin Chem 1989;35:979–84.

Mebes I, Graf M, Kellner M, Keck C, Segerer SE. High estradiol levels during postmenopause—pitfalls in laboratory analysis. Geburtshilfe Frauenheilkd 2015;75: 941–4.

Meikle AW, Findling J, Kushnir MM, Rockwood AL, Nelson GJ, Terry AH. Pseudo-Cushing syndrome caused by fenofibrate interference with urinary cortisol assayed by high-performance liquid chromatography. J Clin Endocrinol Metab 2003;883521–4.

Micallef JV, Hayes MM, Latif A, Ahsan R, Sufi SB. Serum binding of steroid tracers and its possible effects on direct steroid immunoassay. Ann Clin Biochem 1995;32:566–74.

Middle JG. Dehydroepiandrosterone sulphate interferes in many direct immunoassays for testosterone. Ann Clin Biochem 2007;44:173–7.

Mitchell JS, Lowe TE. Matrix effects on an antigen immobilized format for competitive enzyme immunoassay of salivary testosterone. J Immunol Methods 2009 30;349:61–6.

Murakami K, Nakagawa T, Yamashiro G, Araki K, Akasofu K. Adrenal steroids in serum during danazol therapy, taking into account cross-reactions between danazol metabolites and serum androgens. Endocr J 1993;40: 659–64.

Ocvirk R, Bisson JM, Murphy BE. Alcohols which have been in contact with any plastics may interfere in radioimmunoassays of progesterone. J Steroid Biochem Mol Biol 2009;113:150–4.

Ohlsson C, Nilsson ME, Tivesten A, et al. EMAS Study Group Comparisons of immunoassay and mass spectrometry measurements of serum estradiol levels and their influence on clinical association studies in men. J Clin Endocrinol Metab 2013;98:E1097–102.

Owen LJ, Halsall DJ, Keevil BG. Cortisol measurement in patients receiving metyrapone therapy. Ann Clin Biochem 2010;47:573–5.

Pugeat M, Plotton I, de la Perrière AB, et al. Hyperandrogenic states in women: pitfalls in laboratory diagnosis. Eur J Endocrinol 2018;178:R141–54.

Ramaeker D, Brannian J, Egland K, McCaul K, Hansen K. When is elevated testosterone not testosterone? When it is an immunoassay interfering antibody. Fertil Steril 2008;90:886–8.

Rash JM, Jerkunica I, Sgoutas D. Mechanisms of interference of nonesterified fatty acids in radioimmunoassays of steroids. Clin Chim Acta 1979;93:283–94.

Rash JM, Jerkunica I, Sgoutas DS. Lipid interference in steroid radioimmunoassay. Clin Chem 1980;26: 84–8.

Rowe C, Rabet S. Potential interference in the Abbott Architect 2nd generation total testosterone assay. Ann Clin Biochem 2018;55:621–2.

Salameh WA, Redor-Goldman MM, Clarke NJ, Reitz RE, Caulfield MP. Validation of a total testosterone assay using high-turbulence liquid chromatography tandem mass spectrometry: total and free testosterone reference ranges. Steroids 2010;75:169–75.

Saleem M, Lewis JG, Florkowski CM, et al. A patient with pseudo-Addison's disease and falsely elevated thyroxine due to interference in serum cortisol and free thyroxine immunoassays by two different mechanisms. Ann Clin Biochem 2009;46:172–5.

Sinicco A, Raiteri R, Rossati A, Savarino A, Di Perri G. Efavirenz interference in estradiol ELISA assay. Clin Chem 2000;46:734–5.

Slaats EH, Kennedy JC, Kruijswijk H. Interference of sex-hormone binding globulin in the "Coat-A-Count" testosterone no-extraction radioimmunoassay. Clin Chem 1987;33:300–2.

Stieglitz HM, Korpi-Steiner N, Katzman B, Mersereau JE, Styner M. Suspected Testosterone-Producing Tumor in a Patient Taking Biotin Supplements. J Endocr Soc 2018;2:563–9.

Stokes FJ, Bailey LM, Ganguli A, Davison AS. Assessment of endogenous, oral and inhaled steroid cross-reactivity in the Roche cortisol immunoassay. Ann Clin Biochem 2014;51:503–6.

Takahashi K, Karino K, Kurioka H, et al. Estradiol-17beta measurement in women receiving conjugated estrogens. Dissociation between two commercial methods. Clin Chim Acta 1999;284:69–79.

Tejada F, Cremades A, Monserrat F, Peñafiel R. Interference of the antihormone RU486 in the determination of testosterone and estradiol by enzyme-immunoassay. Clin Chim Acta 1998;275:63–9.

Terai I, Yamano K, Ichihara N, Arai J, Kobayashi K. Influence of spironolactone on neonatal screening for congenital adrenal hyperplasia. Arch Dis Child Fetal Neonatal Ed 1999;81:F179–83.

Torjesen PA, Bjøro T. Antibodies against [125I] testosterone in patient's serum: a problem for the laboratory and the patient. Clin Chem 1996;42):2047–8.

Tuhan HU, Catli G, Anik A, et al. Cross-reactivity of adrenal steroids with aldosterone may prevent the accurate diagnosis of congenital adrenal hyperplasia. J Pediatr Endocrinol Metab 2015;28:701–4.

Villanueva RB, Brett E, Gabrilove JL. A cluster of cases of factitious Cushing's syndrome. Endocr Pract 2000;6: 143–7.

Wang C, Shiraishi S, Leung A, et al. Validation of a testosterone and dihydrotestosterone liquid chromatography tandem mass spectrometry assay: Interference and comparison with established methods. Steroids 2008;73:1345–52. Erratum in: Steroids 2018;135:108.

Warner MH, Kane JW, Atkin SL, Kilpatrick ES. Dehydroepiandrosterone sulphate interferes with the Abbott Architect direct immunoassay for testosterone. Ann Clin Biochem 2006;43:196–9.

White A, Gray C, Corrie JE. Monoclonal antibodies to testosterone: the effect of immunogen structure on specificity. J Steroid Biochem 1985;22:169–75.

Willeman T, Casez O, Faure P, Gauchez AS. Evaluation of biotin interference on immunoassays: new data for troponin I, digoxin, NT-Pro-BNP, and progesterone. Clin Chem Lab Med 2017;55:e226–9.

Yarrow JF, Beck DT, Conover CF, et al. Invalidation of a commercially available human 5α-dihydrotestosterone immunoassay. Steroids 2013;78:1220–5.

References

Abushareeda W, Lyris E, Kraiem S, Wahaibi AA, Alyazidi S, Dbes N, et al. Gas chromatographic quadrupole time-of-flight full scan high resolution mass spectrometric screening of human urine in antidoping analysis. J Chromatogr B Analyt Technol Biomed Life Sci 2017;1063:74–83.

Abushareeda W, Vonaparti A, Saad KA, Almansoori M, Meloug M, Saleh A, et al. High resolution full scan liquid chromatography mass spectrometry comprehensive screening in sports antidoping urine analysis. J Pharm Biomed Anal 2018;151:10–24.

Ackermann D, Groessl M, Pruijm M, Ponte B, Escher G, d'Uscio CH, et al. Reference intervals for the urinary steroid metabolome: the impact of sex, age, day and night time on human adult steroidogenesis. PLoS One 2019;14(3), e0214549.

Ahonen L, Fasciotti M, Gennäs GB, Kotiaho T, Daroda RJ, Eberlin M, et al. Separation of steroid isomers by ion mobility mass spectrometry. J Chromatogr A 2013;1310:133–7.

Allen RM, Redshaw MR. The use of homologous and heterologous 125I-radioligands in the radioimmunoassay of progesterone. Steroids 1978;32(4):467–86.

Annesley TM. Ion suppression in mass spectrometry. Clin Chem 2003; 49(7):1041–4.

Annesley TM. Methanol-associated matrix effects in electrospray ionization tandem mass spectrometry. Clin Chem 2007;53(10):1827–34.

Antignac JP, Le Bizec B, Monteau F, Poulain F, André F. Collision-induced dissociation of corticosteroids in electrospray tandem mass spectrometry and development of a screening method by high performance liquid chromatography/tandem mass spectrometry. Rapid Commun Mass Spectrom 2000;14(1):33–9.

Antignac JP, Brosseaud A, Gaudin-Hirret I, André F, Bizec BL. Analytical strategies for the direct mass spectrometric analysis of steroid and corticosteroid phase II metabolites. Steroids 2005;70(3):205–16.

Auer MK, Paizoni L, Neuner M, Lottspeich C, Schmidt H, Bidlingmaier M, et al. 11-oxygenated androgens and their relation to hypothalamus-pituitary-gonadal-axis disturbances in adults with congenital adrenal hyperplasia. J Steroid Biochem Mol Biol 2021;212, 105921.

Avery G. Biotin interference in immunoassay: a review for the laboratory scientist. Ann Clin Biochem 2019;56(4):424–30.

Badoud F, Grata E, Boccard J, Guillarme D, Veuthey JL, Rudaz S, et al. Quantification of glucuronidated and sulfated steroids in human urine by ultra-high pressure liquid chromatography quadrupole time-of-flight mass spectrometry. Anal Bioanal Chem 2011;400 (2):503–16.

Bae YJ, Gaudl A, Jaeger S, Stadelmann S, Hiemisch A, Kiess W, et al. Immunoassay or LC-MS/MS for the measurement of salivary cortisol in children? Clin Chem Lab Med 2016;54(5):811–22.

Barnard G, Kohen F. Idiometric assay: noncompetitive immunoassay for small molecules typified by the measurement of estradiol in serum. Clin Chem 1990;36(11):1945–50.

Berger D, Waheed S, Fattout Y, Kazlauskaite R, Usha L. False increase of estradiol levels in a 36-year-old postmenopausal patient with estrogen receptor-positive breast cancer treated with fulvestrant. Clin Breast Cancer 2016;16(1):e11–3.

Berson SA, Yalow RS. General principles of radioimmunoassay. 1968. Clin Chim Acta 2006;369(2):125–43.

Bertrand MJ, Carazzato D, Sarrasin B. Selective gas chromatographic analysis of monohydroxysteroids as their cyanosyl derivatives. J Chromatogr Sci 1990;28(4):194–9.

Bileck A, Verouti SN, Escher G, Vogt B, Groessl M. A comprehensive urinary steroid analysis strategy using two-dimensional gas chromatography—time of flight mass spectrometry. Analyst 2018;143 (18):4484–94.

Bileck A, Fluck CE, Dhayat N, Groessl M. How high-resolution techniques enable reliable steroid identification and quantification. J Steroid Biochem Mol Biol 2019;186:74–8.

Bileck A, Frei S, Vogt B, Groessl M. Urinary steroid profiles: comparison of spot and 24-hour collections. J Steroid Biochem Mol Biol 2020;200, 105662.

Blackwell LF, Cooke DG, Brown S. Identifying ovulatory cycles and the day of ovulation by the mis-use of pregnanediol glucuronide excretion rate thresholds. Eur J Contracept Reprod Health Care 2018a;23 (5):390–1.

Blackwell LF, Cooke DG, Brown S. The use of estrone-3-glucuronide and pregnanediol-3-glucuronide excretion rates to navigate the continuum of ovarian activity. Front Public Health 2018b;6:153. https://doi.org/ 10.3389/fpubh.2018.00153. 29904626. PMC5990994.

Blair IA. Analysis of estrogens in serum and plasma from postmenopausal women: past present, and future. Steroids 2010;75(4–5): 297–306.

Boggs AS, Bowden JA, Galligan TM, Guillette Jr LJ, Kucklick JR. Development of a multi-class steroid hormone screening method using liquid chromatography/tandem mass spectrometry (LC-MS/MS). Anal Bioanal Chem 2016;408(15):4179–90.

Bolstad N, Warren DJ, Nustad K. Heterophilic antibody interference in immunometric assays. Best Pract Res Clin Endocrinol Metab 2013;27(5):647–61.

Bonfiglio R, King RC, Olah TV, Merkle K. The effects of sample preparation methods on the variability of the electrospray ionization response for model drug compounds. Rapid Commun Mass Spectrom 1999;13(12):1175–85.

Brailsford AD, Gavrilović I, Ansell RJ, Cowan DA, Kicman AT. Two-dimensional gas chromatography with heart-cutting for isotope ratio mass spectrometry analysis of steroids in doping control. Drug Test Anal 2012;4(12):962–9.

Casals G, Marcos J, Pozo OJ, Alcaraz J, Martínez de Osaba MJ, Jiménez W. Microwave-assisted derivatization: application to steroid profiling by gas chromatography/mass spectrometry. J Chromatogr B Analyt Technol Biomed Life Sci 2014;960:8–13.

Casilli A, Piper T, de Oliveira FA, Padilha MC, Pereira HM, Thevis M, et al. Optimization of an online heart-cutting multidimensional gas chromatography clean-up step for isotopic ratio mass spectrometry and simultaneous quadrupole mass spectrometry measurements of endogenous anabolic steroid in urine. Drug Test Anal 2016;8 (11–12):1204–11.

Cawood ML, Field HP, Ford CG, Gillingwater S, Kicman A, Cowan D, et al. Testosterone measurement by isotope-dilution liquid chromatography-tandem mass spectrometry: validation of a method for routine clinical practice. Clin Chem 2005;51(8):1472–9.

Ceglarek U, Leichtle A, Brügel M, Kortz L, Brauer R, Bresler K, et al. Challenges and developments in tandem mass spectrometry based clinical metabolomics. Mol Cell Endocrinol 2009a;301(1–2): 266–71.

Ceglarek U, Kortz L, Leichtle A, Fiedler GM, Kratzsch J, Thiery J. Rapid quantification of steroid patterns in human serum by on-line solid phase extraction combined with liquid chromatography-triple quadrupole linear ion trap mass spectrometry. Clin Chim Acta 2009b;401 (1–2):114–8.

Chambaz EM, Madani C, Ros A. TMS-enol-TMS: a new type of derivative for the gas phase study of dihydroxyacetone side chain saturated corticosteroid metabolites. J Steroid Biochem 1972;3(4):741–7.

Chambers E, Wagrowski-Diehl DM, Lu Z, Mazzeo JR. Systematic and comprehensive strategy for reducing matrix effects in LC/MS/MS analyses. J Chromatogr B Analyt Technol Biomed Life Sci 2007;852(1–2):22–34.

Chan LF, Campbell DC, Novoselova TV, Clark AJ, Metherell LA. Whole-exome sequencing in the differential diagnosis of primary adrenal insufficiency in children. Front Endocrinol 2015;6:113.

Chen J, Liang Q, Hua H, Wang Y, Luo G, Hu M, et al. Simultaneous determination of 15 steroids in rat blood via gas chromatography-mass spectrometry to evaluate the impact of emasculation on adrenal. Talanta 2009;80(2):826–32.

Cho SH, Lee J, Choi MH, Lee WY, Chung BC. Determination of urinary androgen glucuronides by capillary electrophoresis with electrospray tandem mass spectrometry. Biomed Chromatogr 2009;23(4):426–33.

Choi MH, Kim KR, Chung BC. Simultaneous determination of urinary androgen glucuronides by high temperature gas chromatography-mass spectrometry with selected ion monitoring. Steroids 2000;65(1):54–9.

Choi MH, Yoo YS, Chung BC. Measurement of testosterone and pregnenolone in nails using gas chromatography-mass spectrometry. J Chromatogr B Biomed Sci Appl 2001;754(2):495–501.

Chouinard CD, Wei MS, Beekman CR, Kemperman RH, Yost RA. Ion mobility in clinical analysis: current progress and future perspectives. Clin Chem 2016;62(1):124–33.

Christakoudi S, Cowan DA, Taylor NF. A new marker for early diagnosis of 21-hydroxylase deficiency: 3beta,16alpha,17alpha-trihydroxy-5alpha-pregnane-7,20-dione. J Steroid Biochem Mol Biol 2010;121 (3–5):574–81.

Christakoudi S, Cowan DA, Taylor NF. Steroids excreted in urine by neonates with 21-hydroxylase deficiency. 3. Characterization, using GC-MS and GC-MS/MS, of androstanes and androstenes. Steroids 2012a;77(13):1487–501.

Christakoudi S, Cowan DA, Taylor NF. Steroids excreted in urine by neonates with 21-hydroxylase deficiency. 2. Characterization, using GC-MS and GC-MS/MS, of pregnanes and pregnenes with an oxo-group on the A- or B-ring. Steroids 2012b;77(5):382–93.

Cobice DF, Mackay CL, Goodwin RJ, McBride A, Langridge-Smith PR, Webster SP, et al. Mass spectrometry imaging for dissecting steroid intracrinology within target tissues. Anal Chem 2013;85(23): 11576–84.

Corrie JE, Hunter WM, Macpherson JS. A strategy for radioimmunoassay of plasma progesterone with use of a homologous-site 125I-labeled radioligand. Clin Chem 1981;27(4):594–9.

Dai W, Huang Q, Yin P, Li J, Zhou J, Kong H, et al. Comprehensive and highly sensitive urinary steroid hormone profiling method based on stable isotope-labeling liquid chromatography-mass spectrometry. Anal Chem 2012;84(23):10245–51.

de Jong WHA, Buitenwerf E, Pranger AT, Riphagen IJ, Wolffenbuttel BHR, Kerstens MN, et al. Determination of reference intervals for urinary steroid profiling using a newly validated GC-MS/MS method. Clin Chem Lab Med 2017;56(1):103–12.

de Kock N, Acharya SR, Ubhayasekera SJKA, Bergquist J. A novel targeted analysis of peripheral steroids by ultra-performance supercritical fluid chromatography hyphenated to tandem mass spectrometry. Sci Rep 2018;8(1):16993.

Denver N, Khan S, Homer NZM, MacLean MR, Andrew R. Current strategies for quantification of estrogens in clinical research. J Steroid Biochem Mol Biol 2019;192, 105373.

Derks HJ, Muskiet FA, Wolthers BG, Thijssen JH, Drayer NM. Cortisol production rate determined by radio gas-chromatography. Clin Chem 1977;23(3):518–21.

Desfontaine V, Nováková L, Ponzetto F, Nicoli R, Saugy M, Veuthey JL, et al. Liquid chromatography and supercritical fluid chromatography as alternative techniques to gas chromatography for the rapid screening of anabolic agents in urine. J Chromatogr A 2016;1451: 145–55.

Di Dalmazi G, Fanelli F, Mezzullo M, Casadio E, Rinaldi E, Garelli S, et al. Steroid profiling by LC-MS/MS in nonsecreting and subclinical cortisol-secreting adrenocortical adenomas. J Clin Endocrinol Metab 2015;100(9):3529–38.

Dittadi R, Matteucci M, Meneghetti E, Ndreu R. Reassessment of the access testosterone chemiluminescence assay and comparison with LC-MS method. J Clin Lab Anal 2018;32(3), e22286.

Dittadi R, Polesello V, Zivi A, Carraro P. Evaluation of the possible interference of abiraterone therapy on testosterone immunoassay. Clin Chem Lab Med 2019;57(10):e253–4.

Djedovic NK, Rainbow SJ. Detection of synthetic glucocorticoids by liquid chromatography-tandem mass spectrometry in patients being investigated for Cushing's syndrome. Ann Clin Biochem 2011;48(Pt 6):542–9.

Donike M, Bärwald KR, Klostermann K, et al. Nachweis von exogenem testosteron. In: Heck H, Hollmann W, Leisen H, editors. Testosterone in Sport: Leistung und Gesundheit. Köln: Deutsche Artze-Verlag; 1983. p. 293–300.

Doué M, Dervilly-Pinel G, Pouponneau K, Monteau F, Le Bizec B. Analysis of glucuronide and sulfate steroids in urine by ultra-high-performance supercritical-fluid chromatography hyphenated tandem mass spectrometry. Anal Bioanal Chem 2015;407(15): 4473–84.

Drotleff B, Hallschmid M, Lämmerhofer M. Quantification of steroid hormones in plasma using a surrogate calibrant approach and UHPLC-ESI-QTOF-MS/MS with SWATH-acquisition combined with untargeted profiling. Anal Chim Acta 2018;1022:70–80.

du Toit T, Bloem LM, Quanson JL, Ehlers R, Serafin AM, Swart AC. Profiling adrenal 11β-hydroxyandrostenedione metabolites in prostate cancer cells, tissue and plasma: UPC(2)-MS/MS quantification of 11β-hydroxytestosterone, 11keto-testosterone and 11keto-dihydrotestosterone. J Steroid Biochem Mol Biol 2017;166:54–67.

Dudzik D, Barbas-Bernardos C, García A, Barbas C. Quality assurance procedures for mass spectrometry untargeted metabolomics. A review. J Pharm Biomed Anal 2018;147:149–73.

Ebner MJ, Corol DI, Havlíková H, Honour JW, Fry JP. Identification of neuroactive steroids and their precursors and metabolites in adult male rat brain. Endocrinology 2006;147(1):179–90.

Eckfeldt JH, Lewis LA, Belcher JD, Singh J, Frantz Jr ID. Determination of serum cholesterol by isotope dilution mass spectrometry with a benchtop capillary gas chromatograph/mass spectrometer: comparison with the National Reference System's definitive and reference methods. Clin Chem 1991;37(7):1161–5.

Ekins RP. The estimation of thyroxine in human plasma by an electrophoretic technique. Clin Chim Acta 1960;5:453–9.

Ekins R. Saturation analysis. Clin Sci 1969;37(2):570.

El-Gamal BA, Landon J, Abuknesha RA, Gallacher G, Perry LA. The production and characterisation of antisera to 17-hydroxyprogesterone. J Steroid Biochem 1987;26(3):375–82.

Ellerbe P, Meiselman S, Sniegoski LT, Welch MJ, White V E. Determination of serum cholesterol by a modification of the isotope dilution mass spectrometric definitive method. Anal Chem 1989;61 (15):1718–23.

Eriksson H. Steroids in germfree and conventional rats. Unconjugated metabolites of [4-14C]pregnenolone and [4-14C]corticosterone in faeces from female rats. Eur J Biochem 1970;16(2):261–7.

Eriksson H. Steroids in germfree and conventional rats. Metabolites of (4-14C)pregnenolone and (4-14C)corticosterone in urine and faeces from male rats. Eur J Biochem 1971;18(1):86–93.

Erlanger BF. The preparation of antigenic hapten-carrier conjugates: a survey. Methods Enzymol 1980;70(A):85–104.

Esquivel A, Matabosch X, Kotronoulas A, Balcells G, Joglar J, Ventura R. Ionization and collision induced dissociation of steroid bisglucuronides. J Mass Spectrom 2017;52(11):759–69.

Esteban NV, Loughlin T, Yergey AL, Zawadzki JK, Booth JD, Winterer JC, et al. Daily cortisol production rate in man determined by stable isotope dilution/mass spectrometry. J Clin Endocrinol Metab 1991;72(1):39–45.

Fabregat A, Pozo OJ, Marcos J, Segura J, Ventura R. Use of LC-MS/MS for the open detection of steroid metabolites conjugated with glucuronic acid. Anal Chem 2013a;85(10):5005–14.

Fabregat A, Kotronoulas A, Marcos J, Joglar J, Alfonso I, Segura J, et al. Detection, synthesis and characterization of metabolites of steroid hormones conjugated with cysteine. Steroids 2013b;78(3):327–36.

Fabregat A, Marcos J, Garrostas L, Segura J, Pozo OJ, Ventura R. Evaluation of urinary excretion of androgens conjugated to cysteine in human pregnancy by mass spectrometry. J Steroid Biochem Mol Biol 2014;139:192–200.

Fanelli F, Belluomo I, Di Lallo VD, Cuomo G, De Iasio R, Baccini M, et al. Serum steroid profiling by isotopic dilution-liquid chromatography-mass spectrometry: comparison with current immunoassays and reference intervals in healthy adults. Steroids 2011;76(3): 244–53.

Faqehi AM, Denham SG, Naredo G, Cobice DF, Khan S, Simpson JP, et al. Derivatization with 2-hydrazino-1-methylpyridine enhances sensitivity of analysis of 5α-dihydrotestosterone in human plasma by liquid chromatography tandem mass spectrometry. J Chromatogr A 2021;1640, 461933.

Fink RS, Pierre LN, Daley-Yates PT, Richards DH, Gibson A, Honour JW. Hypothalamic-pituitary-adrenal axis function after inhaled corticosteroids: unreliability of urinary free cortisol estimation. J Clin Endocrinol Metab 2002;87(10):4541–6.

Fletcher BR, Hinton L, Hartmann-Boyce J, Roberts NW, Bobrovitz N, McManus RJ. Self-monitoring blood pressure in hypertension, patient and provider perspectives: a systematic review and thematic synthesis. Patient Educ Couns 2016a;99(2):210–9.

Fletcher BR, Hinton L, Bray EP, Hayen A, Hobbs FR, Mant J, et al. Self-monitoring blood pressure in patients with hypertension: an internet-based survey of UK GPs. Br J Gen Pract 2016b;66(652):e831–7.

Fránek M, Bursa J, Hruska K. Characteristics of antibodies against 17 beta-oestradiol in homologous and heterologous systems of radioimmunoassay. J Steroid Biochem 1983;19(3):1371–4.

Franke AA, Custer LJ, Morimoto Y, Nordt FJ, Maskarinec G. Analysis of urinary estrogens, their oxidized metabolites, and other endogenous steroids by benchtop orbitrap LCMS versus traditional quadrupole GCMS. Anal Bioanal Chem 2011;401(4):1319–30.

Frey AJ, Wang Q, Busch C, Feldman D, Bottalico L, Mesaros CA, et al. Validation of highly sensitive simultaneous targeted and untargeted

analysis of keto-steroids by Girard P derivatization and stable isotope dilution-liquid chromatography-high resolution mass spectrometry. Steroids 2016;116:60–6.

Fujiwara T, Inoue R, Ohtawa T, Tsunoda M. Liquid-chromatographic methods for carboxylic acids in biological samples. Molecules 2020; 25(21):4883.

Główka FK, Kosicka K, Karaźniewicz-Łada M. HPLC method for determination of fluorescence derivatives of cortisol, cortisone and their tetrahydro- and allo-tetrahydro-metabolites in biological fluids. J Chromatogr B Analyt Technol Biomed Life Sci 2010;878 (3–4):283–9.

Galuska CE, Hartmann MF, Sánchez-Guijo A, Bakhaus K, Geyer J, Schuler G, et al. Profiling intact steroid sulfates and unconjugated steroids in biological fluids by liquid chromatography-tandem mass spectrometry (LC-MS-MS). Analyst 2013;138(13):3792–801.

Gaudl A, Kratzsch J, Bae YJ, Kiess W, Thiery J, Ceglarek U. Liquid chromatography quadrupole linear ion trap mass spectrometry for quantitative steroid hormone analysis in plasma, urine, saliva and hair. J Chromatogr A 2016;1464:64–71.

Gibson EJ, Bucknall MP, Golebiowski B, Stapleton F. Comparative limitations and benefits of liquid chromatography—mass spectrometry techniques for analysis of sex steroids in tears. Exp Eye Res 2019;179:168–78.

Gika HG, Theodoridis GA, Plumb RS, Wilson ID. Current practice of liquid chromatography-mass spectrometry in metabolomics and metabonomics. J Pharm Biomed Anal 2014;87:12–25.

Gómez C, Pozo OJ, Marcos J, Segura J, Ventura R. Alternative long-term markers for the detection of methyltestosterone misuse. Steroids 2013;78(1):44–52.

Gomez C, Fabregat A, Pozo OJ, Marcos J, Segura L, Ventura R. Analytical strategies based on mass spectrometric techniques for the study of steroid metabolism. Trends Anal Chem 2014;53:106–16.

Gosetti F, Mazzucco E, Gennaro MC, Marengo E. Ultra high performance liquid chromatography tandem mass spectrometry determination and profiling of prohibited steroids in human biological matrices. A review. J Chromatogr B Analyt Technol Biomed Life Sci 2013;927: 22–36.

Griffiths WJ, Wang Y. An update on oxysterol biochemistry: new discoveries in lipidomics. Biochem Biophys Res Commun 2018;504(3):617–22.

Guo T, Chan M, Soldin SJ. Steroid profiles using liquid chromatography-tandem mass spectrometry with atmospheric pressure photoionization source. Arch Pathol Lab Med 2004;128(4):469–75.

Hána V, Ježková J, Kosák M, Kršek M, Hána V, Hill M. Novel GC-MS/MS technique reveals a complex steroid fingerprint of subclinical hypercortisolism in adrenal incidentalomas. J Clin Endocrinol Metab 2019;104(8):3545–56.

Handelsman DJ, Newman JD, Jimenez M, McLachlan R, Sartorius G, Jones GR. Performance of direct estradiol immunoassays with human male serum samples. Clin Chem 2014;60(3):510–7.

Handelsman DJ, Gibson E, Davis S, Golebiowski B, Walters KA, Desai R. Ultrasensitive serum estradiol measurement by liquid chromatography-mass spectrometry in postmenopausal women and mice. J Endocr Soc 2020;4(9), bvaa086.

Heald AH, Butterworth A, Kane JW, Borzomato J, Taylor NF, Layton T, et al. Investigation into possible causes of interference in serum testosterone measurement in women. Ann Clin Biochem 2006;43(Pt 3):189–95.

Higashi T, Ogawa S. Chemical derivatization for enhancing sensitivity during LC/ESI-MS/MS quantification of steroids in biological samples: a review. J Steroid Biochem Mol Biol 2016a;162:57–69.

Higashi T, Ogawa S. Isotope-coded ESI-enhancing derivatization reagents for differential analysis, quantification and profiling of metabolites in biological samples by LC/MS: a review. J Pharm Biomed Anal 2016b;130:181–93.

Higashi T, Takayama N, Shimada K. Enzymic conversion of 3beta-hydroxy-5-ene-steroids and their sulfates to 3-oxo-4-ene-steroids for increasing sensitivity in LC-APCI-MS. J Pharm Biomed Anal 2005;39(3–4):718–23.

Higashi T, Shibayama Y, Shimada K. Determination of salivary dehydroepiandrosterone using liquid chromatography—tandem mass spectrometry combined with charged derivatization. J Chromatogr B Analyt Technol Biomed Life Sci 2007;846(1–2):195–201.

Higashi T, Yokoi H, Nagura Y, Nishio T, Shimada K. Studies on neurosteroids XXIV. Determination of neuroactive androgens, androsterone and 5alpha-androstane-3alpha,17beta-diol, in rat brain and serum using liquid chromatography-tandem mass spectrometry. Biomed Chromatogr 2008a;22(12):1434–41.

Higashi T, Ninomiya Y, Shimada K. Studies on neurosteroids XX. Liquid chromatography-tandem mass spectrometric method for simultaneous determination of testosterone and 5alpha-dihydrotestosterone in rat brain and serum. J Chromatogr Sci 2008b;46(7):653–8.

Higashi T, Aiba N, Tanaka T, Yoshizawa K, Ogawa S. Methods for differential and quantitative analyses of brain neurosteroid levels by LC/MS/MS with ESI-enhancing and isotope-coded derivatization. J Pharm Biomed Anal 2016;117:155–62.

Hill M, Hána Jr V, Velíková M, Pařízek A, Kolátorová L, Vítků J, et al. A method for determination of one hundred endogenous steroids in human serum by gas chromatography-tandem mass spectrometry. Physiol Res 2019;68(2):179–207.

Hines JM, Bancos I, Bancos C, Singh RD, Avula AV, Young WF, et al. High-resolution, accurate-mass (HRAM) mass spectrometry urine steroid profiling in the diagnosis of adrenal disorders. Clin Chem 2017;63(12):1824–35.

Hoffmann K, Samarafeewa P, Smith ER, Kellie AE. Radioimmunoassay of steroids the role of the "bridge" linking the steroid hapten to the protein carrier. J Steroid Biochem 1975;6(2):91–4.

Honour JW. Development and validation of a quantitative assay based on tandem mass spectrometry. Ann Clin Biochem 2011;48(Pt 2):97–111.

Honour JW, Shackleton CH. Mass spectrometric analysis for tetrahydroaldosterone. J Steroid Biochem 1977;8(4):299–305.

Honour JW, Brooks CJ, Shackleton CH. Degradation of steroid derivatives at a gas chromatograph mass spectrometer interface. Biomed Mass Spectrom 1982;9:505–9.

Honour JW, Tsilchorozidou T, Conway GS, Dawnay A. Spironolactone interference in the immunoassay of androstenedione. Ann Clin Biochem 2010;47(Pt 6):564–6.

Honour JW, Conway E, Hodkinson R, Lam F. The evolution of methods for urinary steroid metabolomics in clinical investigations particularly in childhood. J Steroid Biochem Mol Biol 2018;181:28–51.

Horie H, Kidowaki T, Koyama Y, Endo T, Homma K, Kambegawa A, et al. Specificity assessment of immunoassay kits for determination of urinary free cortisol concentrations. Clin Chim Acta 2007;378(1–2):66–70.

Hosoda H, Tsukamoto R, Shishido M, Takasaki W, Nambara T. Enzyme labeling of steroids by the N-succinimidyl ester method. Preparation of glucose oxidase-labeled antigen for use in enzyme immunoassay. Chem Pharm Bull(Tokyo) 1987;35(12):4856–61.

Huang HJ, Chiang PH, Chen SH. Quantitative analysis of estrogens and estrogen metabolites in endogenous MCF-7 breast cancer cells by liquid chromatography-tandem mass spectrometry. J Chromatogr B Analyt Technol Biomed Life Sci 2011;879(20):1748–56.

Idkowiak J, Taylor AE, Subtil S, O'Neil DM, Vijzelaar R, Dias RP, et al. Steroid sulfatase deficiency and androgen activation before and after puberty. J Clin Endocrinol Metab 2016;101(6):2545–53.

Ikegawa S, Hasegawa M, Okihara R, Shimidzu C, Chiba H, Iida T, et al. Simultaneous determination of twelve tetrahydrocorticosteroid glucuronides in human urine by liquid chromatography/electrospray ionization-linear ion trap mass spectrometry. Anal Chem 2009;81(24):10124–35.

Indapurkar A, Hartman N, Patel V, Matta MK. Simultaneous UHPLC-MS/MS method of estradiol metabolites to support the evaluation of Phase-2 metabolic activity of induced pluripotent stem cell derived hepatocytes. J Chromatogr B Analyt Technol Biomed Life Sci 2019;1126–1127, 121765.

Iohan F, Vincze I, Monder C, Cohen S. High-performance liquid chromatographic determination of cortolic and cortolonic acids as pyrenyl ester derivatives. J Chromatogr 1991;564(1):27–41.

Ishida J, Kai M, Ohkura Y. Determination of oestrogens in pregnancy urine by high-performance liquid chromatography with fluorescence detection. J Chromatogr 1988;431(2):249–57.

Israelsson M, Brattsand R, Brattsand G. 20α- and 20β-dihydrocortisone may interfere in LC-MS/MS determination of cortisol in saliva and urine. Ann Clin Biochem 2018;55(3):341–7.

James VH, Jeffcoate SL. Steroids. Br Med Bull 1974;30:50–4. N.B Whole issue on radioimmunoassay and saturation analysis.

Jeilani YA, Cardelino BH, Ibeanusi VM. Hydrogen rearrangement and ring cleavage reactions study of progesterone by triple quadrupole mass spectrometry and density functional theory. J Mass Spectrom 2011;46(7):625–34.

Jerjes WK, Cleare AJ, Peters TJ, Taylor NF. Circadian rhythm of urinary steroid metabolites. Ann Clin Biochem 2006a;43(Pt 4):287–94.

Jerjes WK, Peters TJ, Taylor NF, Wood PJ, Wessely S, Cleare AJ. Diurnal excretion of urinary cortisol, cortisone, and cortisol metabolites in chronic fatigue syndrome. J Psychosom Res 2006b;60(2):145–53.

Jones AM, Honour JW. Unusual results from immunoassays and the role of the clinical endocrinologist. Clin Endocrinol (Oxf) 2006;64(3):234–44. A review of interferences in immunoassays.

Jones RL, Owen LJ, Adaway JE, Keevil BG. Simultaneous analysis of cortisol and cortisone in saliva using XLC-MS/MS for fully automated online solid phase extraction. J Chromatogr B Analyt Technol Biomed Life Sci 2012;881–882:42–8.

Junker J, Chong I, Kamp F, Steiner H, Giera M, Müller C, et al. Comparison of strategies for the determination of sterol sulfates via GC-MS leading to a novel deconjugation-derivatization protocol. Molecules 2019;24(13). pii: E2353.

Kaabia Z, Laparre J, Cesbron N, Le Bizec B, Dervilly-Pinel G. Comprehensive steroid profiling by liquid chromatography coupled to high resolution mass spectrometry. J Steroid Biochem Mol Biol 2018;183:106–15.

Kalogera E, Pistos C, Provatopoulou X, Athanaselis S, Spiliopoulou C, Gounaris A. Androgen glucuronides analysis by liquid chromatography tandem-mass spectrometry: could it raise new perspectives in the diagnostic field of hormone-dependent malignancies? J Chromatogr B Analyt Technol Biomed Life Sci 2013;940:24–34.

Kanu AB, Dwivedi P, Tam M, Matz L, Hill Jr HH. Ion mobility-mass spectrometry. J Mass Spectrom 2008;43(1):1–22.

Karir T, Samuel G, Kothari K, Sivaprasad N, Venkatesh M. Studies on the influence of the structural modifications in the tracer on the immunoassay of progesterone. J Immunoassay Immunochem 2006;27(2):151–71. https://doi.org/10.1080/15321810600573085.

Kaushik A, Vasudev A, Arya SK, Pasha SK, Bhansali S. Recent advances in cortisol sensing technologies for point-of-care application. Biosens Bioelectron 2014;53:499–512.

Kawasaki T, Maeda M, Tsuji A. Determination of 17-oxosteroid glucuronides and sulfates in urine and serum by fluorescence high-performance liquid chromatography using dansyl hydrazine as a prelabeling reagent. J Chromatogr 1982;233:61–8.

Kawasaki T, Maeda M, Tsuji A. Immobilized 3 alpha-hydroxysteroid dehydrogenase and dansyl hydrazine as a pre-labeling reagent for high-performance liquid chromatography with fluorescence detection of bile acids. J Chromatogr 1983;272(2):261–8.

Ke Y, Gonthier R, Isabelle M, Bertin J, Simard JN, Dury AY, et al. A rapid and sensitive UPLC-MS/MS method for the simultaneous quantification of serum androsterone glucuronide, etiocholanolone glucuronide, and androstan-3α, 17β diol 17-glucuronide in postmenopausal women. J Steroid Biochem Mol Biol 2015;149:146–52.

Keevil BG. LC-MS/MS analysis of steroids in the clinical laboratory. Clin Biochem 2016;49(13–14):989–97.

Keller BO, Sui J, Young AB, Whittal RM. Interferences and contaminants encountered in modern mass spectrometry. Anal Chim Acta 2008;627 (1):71–81. Supplementary Excel files for positive ions, negative ions, repeating units, masses, adducts and solvents.

Koal T, Schmiederer D, Pham-Tuan H, Röhring C, Rauh M. Standardized LC-MS/MS based steroid hormone profile-analysis. J Steroid Biochem Mol Biol 2012;129(3–5):129–38.

Kohen F, De Boever J, Kim JB. Recent advances in chemiluminescence-based immunoassays for steroid hormones. J Steroid Biochem 1987;27(1–3):71–9.

Kremen J, Chan YM. Genetic evaluation of disorders of sex development: current practice and novel gene discovery. Curr Opin Endocrinol Diabetes Obes 2019;26(1):54–9.

Kreutzmann DJ, Silink M. Determination of 17-oxosteroid glucuronides and sulphates in urine by liquid chromatography using 2,4-dinitrophenylhydrazine as a prelabelling reagent for spectrophotometric detection. J Chromatogr 1987;415(2):253–60.

Krone N, Hughes BA, Lavery GG, Stewart PM, Arlt W, Shackleton CH. Gas chromatography/mass spectrometry (GC/MS) remains a pre-eminent discovery tool in clinical steroid investigations even in the era of fast liquid chromatography tandem mass spectrometry (LC/MS/MS). J Steroid Biochem Mol Biol 2010;121(3–5):496–504.

Kuehnbaum NL, Britz-McKibbin P. Comprehensive profiling of free and conjugated estrogens by capillary electrophoresis-time of flight/mass spectrometry. Anal Chem 2011;83(21):8063–8.

Kulle AE, Welzel M, Holterhus PM, Riepe FG. Principles and clinical applications of liquid chromatography—tandem mass spectrometry for the determination of adrenal and gonadal steroid hormones. J Endocrinol Invest 2011;34(9):702–8.

Kulle AE, Welzel M, Holterhus PM, Riepe FG. Implementation of a liquid chromatography tandem mass spectrometry assay for eight adrenal C-21 steroids and pediatric reference data. Horm Res Paediatr 2013;79(1):22–31.

Kuronen P, Volin P, Laitalainen T. Reversed-phase high-performance liquid chromatographic screening method for serum steroids using retention index and diode-array detection. J Chromatogr B Biomed Sci Appl 1998;718(2):211–24.

Kushnir MM, Neilson R, Roberts WL, Rockwood AL. Cortisol and cortisone analysis in serum and plasma by atmospheric pressure photoionization tandem mass spectrometry. Clin Biochem 2004;37(5):357–62.

Kushnir MM, Blamires T, Rockwood AL, Roberts WL, Yue B, Erdogan E, et al. Liquid chromatography-tandem mass spectrometry assay for androstenedione, dehydroepiandrosterone, and testosterone with pediatric and adult reference intervals. Clin Chem 2010;56(7):1138–47. Erratum in: Clin Chem. 2011 Jul;57(7):1084.

Kuuranne T, Vahermo M, Leinonen A, Kostiainen R. Electrospray and atmospheric pressure chemical ionization tandem mass spectrometric behavior of eight anabolic steroid glucuronides. J Am Soc Mass Spectrom 2000;11(8):722–30.

Lacorn M, Fleischer K, Willig S, Gremmel S, Steinhart H, Claus R. Use of biotinylated 17beta-estradiol in enzyme-immunoassay development: spacer length and chemical structure of the bridge are the main determinants in simultaneous streptavidin-antibody binding. J Immunol Methods 2005;297(1–2):225–36.

Lai CC, Tsai CH, Tsai FJ, Lee CC, Lin WD. Rapid monitoring assay of congenital adrenal hyperplasia with microbore high-performance liquid chromatography/electrospray ionization tandem mass spectrometry from dried blood spots. Rapid Commun Mass Spectrom 2001;15(22):2145–51.

Larner JM, Shackleton CH, Roitman E, Schwartz PE, Hochberg RB. Measurement of estradiol-17-fatty acid esters in human tissues. J Clin Endocrinol Metab 1992;75(1):195–200.

Le Bizec B, Antignac JP, Bertrand D, Qannari EM, Andre F. Multidimensional statistical analysis applied to electron ionization mass spectra to determine steroid stereochemistry. Rapid Commun Mass Spectrom 2005;19(4):509–18.

Lee W, Lee H, Kim YL, Lee YC, Chung BC, Hong J. Profiling of steroid metabolic pathways in human plasma by GC-MS/MS combined with microwave-assisted derivatization for diagnosis of gastric disorders. Int J Mol Sci 2021;22(4):1872.

Li X, Franke AA. Improved profiling of estrogen metabolites by orbitrap LC/MS. Steroids 2015;99(Pt A):84–90.

Lien SK, Kvitvang HF, Bruheim P. Utilization of a deuterated derivatization agent to synthesize internal standards for gas chromatography-tandem mass spectrometry quantification of silylated metabolites. J Chromatogr A 2012;1247:118–24.

Liere P, Pianos A, Eychenne B, Cambourg A, Liu S, Griffiths W, et al. Novel lipoidal derivatives of pregnenolone and dehydroepiandrosterone and absence of their sulfated counterparts in rodent brain. J Lipid Res 2004;45(12):2287–302.

Lindberg C, Jönsson S, Hedner P, Gustafsson A. Radioimmunoassay and chemical ionization/mass spectrometry compared for plasma cortisol determination. Clin Chem 1982;28(1):174–7.

Lindner JM, Vogeser M, Grimm SH. Biphenyl based stationary phases for improved selectivity in complex steroid assays. J Pharm Biomed Anal 2017;142:66–73.

Little JL. Artifacts in trimethylsilyl derivatization reactions and ways to avoid them. J Chromatogr A 1999;844(1–2):1–22.

Liu S, Griffiths WJ, Sjövall J. On-column electrochemical reactions accompanying the electrospray process. Anal Chem 2003;75(4):1022–30.

Liu C, Sheng X, Wang Y, Yin J, Huang W, Fan Y, et al. A sensitive approach for simultaneous quantification of carbonyl and hydroxy steroids using 96-well SPE plates based on stable isotope coded derivatization UPLC-MRM: method development and application. RSC Adv 2018;8:19713.

Luong JHT, Male KB, Glennon JD. Biotin interference in immunoassays based on biotin-strept(avidin) chemistry: an emerging threat. Biotechnol Adv 2019;37(5):634–41.

Magnisali P, Dracopoulou M, Mataragas M, Dacou-Voutetakis A, Moutsatsou P. Routine method for the simultaneous quantification of 17alpha-hydroxyprogesterone, testosterone, dehydroepiandrosterone, androstenedione, cortisol, and pregnenolone in human serum of neonates using gas chromatography-mass spectrometry. J Chromatogr A 2008;1206(2):166–77.

Marcos J, Pozo OJ. Derivatization of steroids in biological samples for GC-MS and LC-MS analyses. Bioanalysis 2015;7(19):2515–36.

Marcos J, Renau N, Casals G, Segura J, Ventura R, Pozo OJ. Investigation of endogenous corticosteroids profiles in human urine based on liquid chromatography tandem mass spectrometry. Anal Chim Acta 2014;812:92–104.

Mares A, De Boever J, Osher J, Quiroga S, Barnard G, Kohen F. A direct non-competitive idiometric enzyme immunoassay for serum oestradiol. J Immunol Methods 1995;181(1):83–90.

Matsuzaki Y, Yoshida S, Honda A, Miyazaki T, Tanaka N, Takagiwa A, et al. Simultaneous determination of dehydroepiandrosterone and its 7-oxygenated metabolites in human serum by high-resolution gas chromatography—mass spectrometry. Steroids 2004;69(13–14):817–24.

Matuszewski BK. Standard line slopes as a measure of a relative matrix effect in quantitative HPLC-MS bioanalysis. J Chromatogr B Analyt Technol Biomed Life Sci 2006;830(2):293–300.

Matuszewski BK, Constanzer ML, Chavez-Eng CM. Strategies for the assessment of matrix effect in quantitative bioanalytical methods based on HPLC-MS/MS. Anal Chem 2003;75(13):3019–30.

Matysik S, Liebisch G. Quantification of steroid hormones in human serum by liquid chromatography-high resolution tandem mass spectrometry. J Chromatogr A 2017;1526:112–8.

Matysik S, Schmitz G. Determination of steroid hormones in human plasma by GC-triple quadrupole MS. Steroids 2015;99(Pt B):151–4.

Maunsell Z, Wright DJ, Rainbow SJ. Routine isotope-dilution liquid chromatography-tandem mass spectrometry assay for simultaneous measurement of the 25-hydroxy metabolites of vitamins D2 and D3. Clin Chem 2005;51(9):1683–90.

McLeod MD, Waller CC, Esquivel A, Balcells G, Ventura R, Segura J, et al. Constant ion loss method for the untargeted detection of Bis-sulfate metabolites. Anal Chem 2017;89(3):1602–9.

Meunier-Solère V, Maume D, André F, Le Bizec B. Pitfalls in trimethylsi-lylation of anabolic steroids. New derivatisation approach for residue at ultra-trace level. J Chromatogr B Analyt Technol Biomed Life Sci 2005;816(1–2):281–8.

Mitamura K, Satoh née Okihara R, Kamibayashi M, Sato K, Iida T, Ikegawa S. Simultaneous determination of 18 tetrahydrocorticosteroid sulfates in human urine by liquid chromatography/electrospray ionization-tandem mass spectrometry. Steroids 2014;85:18–29.

Moon JY, Jung HJ, Moon MH, Chung BC, Choi MH. Heat-map visualization of gas chromatography-mass spectrometry based quantitative signatures on steroid metabolism. J Am Soc Mass Spectrom 2009;20(9):1626–37.

Moon JY, Ha YW, Moon MH, Chung BC, Choi MH. Systematic error in gas chromatography-mass spectrometry-based quantification of hydrolyzed urinary steroids. Cancer Epidemiol Biomarkers Prev 2010;19(2):388–97.

Moros G, Chatziioannou AC, Gika HG, Raikos N, Theodoridis G. Investigation of the derivatization conditions for GC-MS metabolomics of biological samples. Bioanalysis 2017;9(1):53–65.

Murphy BP, Pattee CJ. A study of the binding capacity of corticosteroid-binding globulin in plasma. J Clin Endocrinol Metab 1963;23:459–64.

Narduzzi L, Royer AL, Bichon E, Guitton Y, Buisson C, Le Bizec B, et al. Ammonium fluoride as suitable additive for HILIC-based LC-HRMS metabolomics. Metabolites 2019;9(12):292.

Neufeld E, Chayen R, Stern N. Fluorescence derivatisation of urinary corticosteroids for high-performance liquid chromatographic analysis. J Chromatogr B Biomed Sci Appl 1998;718(2):273–7.

Nixon A, Mallet AI, Gower DB. Simultaneous quantification of five odorous steroids (16-androstenes) in the axillary hair of men. J Steroid Biochem 1988;29(5):505–10.

Nováková L, Grand-Guillaume Perrenoud A, Nicoli R, Saugy M, Veuthey JL, Guillarme D. Ultra high performance supercritical fluid chromatography coupled with tandem mass spectrometry for screening of doping agents. I: investigation of mobile phase and MS conditions. Anal Chim Acta 2015a;853:637–46.

Nováková L, Rentsch M, Grand-Guillaume Perrenoud A, Nicoli R, Saugy M, Veuthey JL, et al. Ultra high performance supercritical fluid chromatography coupled with tandem mass spectrometry for screening of doping agents. II: analysis of biological samples. Anal Chim Acta 2015b;853:647–59.

Olesti E, Boccard J, Visconti G, González-Ruiz V, Rudaz S. From a single steroid to the steroidome: trends and analytical challenges. J Steroid Biochem Mol Biol 2021;206, 105797.

Parr MK, Wüst B, Teubel J, Joseph JF. Splitless hyphenation of SFC with MS by APCI, APPI, and ESI exemplified by steroids as model compounds. J Chromatogr B Analyt Technol Biomed Life Sci 2018;1091:67–78.

Perogamvros I, Owen LJ, Newell-Price J, Ray DW, Trainer PJ, Keevil BG. Simultaneous measurement of cortisol and cortisone in human saliva using liquid chromatography-tandem mass spectrometry: application in basal and stimulated conditions. J Chromatogr B Analyt Technol Biomed Life Sci 2009;877(29):3771–5.

Phillips IJ, Conway EM, Hodkinson RA, Honour JW. External quality assessment of urinary steroid profile analysis. Ann Clin Biochem 2004;41(Pt 6):474–8.

Poole CF, Zlatkis A, Sye WF, Singhawangcha S, Morgan ED. The determination of steroids with and without natural electrophores by gas chromatography and electron-capture detection. Lipids 1980;15(9):734–44.

Poon GK, Jarman M, McCague R, Davies JH, Heeremans CE, van der Hoeven RA, et al. Identification of 4-hydroxyandrost-4-ene-3,17-dione metabolites in prostatic cancer patients by liquid chromatography-mass spectrometry. J Chromatogr 1992;576(2):235–44.

Pozo OJ, Van Eenoo P, Van Thuyne W, Deventer K, Delbeke FT. Direct quantification of steroid glucuronides in human urine by liquid chromatography-electrospray tandem mass spectrometry. J Chromatogr A 2008a;1183(1–2):108–18.

Pozo OJ, Van Eenoo P, Deventer K, Grimalt S, Sancho JV, Hernández F, et al. Collision-induced dissociation of 3-keto anabolic steroids and related compounds after electrospray ionization. Considerations for structural elucidation. Rapid Commun Mass Spectrom 2008b;22(24):4009–24.

Pozo OJ, Marcos J, Khymenets O, Pranata A, Fitzgerald CC, McLeod MD, et al. Sulfation pathways: alternate steroid sulfation pathways targeted by LC-MS/MS analysis of disulfates: application to prenatal diagnosis of steroid synthesis disorders. J Mol Endocrinol 2018;61(2):M1–M12.

Pratt JJ. Steroid immunoassay in clinical chemistry. Clin Chem 1978;24:1869–90.

Quanson JL, Stander MA, Pretorius E, Jenkinson C, Taylor AE, Storbeck KH. High-throughput analysis of 19 endogenous androgenic steroids by ultra-performance convergence chromatography tandem mass spectrometry. J Chromatogr B Analyt Technol Biomed Life Sci 2016;1031:131–8.

Que AH, Palm A, Baker AG, Novotny MV. Steroid profiles determined by capillary electrochromatography, laser-induced fluorescence detection and electrospray-mass spectrometry. J Chromatogr A 2000;887(1–2):379–91.

Rao PN, Taraporewala IB. A sensitive enzyme-linked immunosorbent assay (ELISA) for testosterone: use of a novel heterologous hapten conjugated to penicillinase. Steroids 1992;57(4):154–61.

Rao PN, Damodaran KM, Moore Jr PH, Desjardins C, Garza R. Synthesis of new steroid haptens for radioimmunoassay. VII. 19-O-Carboxymethyl ether derivative of androstenedione. Specific antiserum for measurement of androstenedione in plasma. J Steroid Biochem 1982;17(5):523–7.

Rassaie MJ, Kumari GL, Rao PN, Shrivastav TG, Pandey HP. Influence of different combinations of antibodies and penicillinase-labeled testosterone derivatives on sensitivity and specificity of immunoassays. Steroids 1992;57(3):112–8.

Ratcliffe WA, McClure JP, Auld WH, Honour JW, Fraser R, Ratcliffe JG. Precocious pseudopuberty due to a rare form of congenital adrenal hyperplasia. Biochemical investigation and pitfalls in interpretation of hormone assays. Ann Clin Biochem 1982;19:145–50.

Rauh M. Steroid measurement with LC-MS/MS. Application examples in pediatrics. J Steroid Biochem Mol Biol 2010;121(3–5):520–7.

Rauh M, Gröschl M, Rascher W, Dörr HG. Automated, fast and sensitive quantification of 17 alpha-hydroxy-progesterone, androstenedione and testosterone by tandem mass spectrometry with on-line extraction. Steroids 2006;71(6):450–8.

Richardson AP, Kim JB, Barnard GJ, Collins WP, McCapra F. Chemiluminescence immunoassay of plasma progesterone, with progesterone-acridinium ester used as the labeled antigen. Clin Chem 1985;31(10):1664–8.

Rister AL, Dodds ED. Steroid analysis by ion mobility spectrometry. Steroids 2019;153:108531. https://doi.org/10.1016/j.steroids.2019.108531. [Epub ahead of print] Review 31672629.

Rister AL, Martin TL, Dodds ED. Formation of multimeric steroid metal adducts and implications for isomer mixture separation by traveling wave ion mobility spectrometry. J Mass Spectrom 2019;54(5):429–36.

Robles J, Marcos J, Renau N, Garrostas L, Segura J, Ventura R, et al. Quantifying endogenous androgens, estrogens, pregnenolone and progesterone metabolites in human urine by gas chromatography tandem mass spectrometry. Talanta 2017;169:20–9.

Rochat B. Fully-automated systems and the need for global approaches should exhort clinical labs to reinvent routine MS analysis? Bioanalysis 2018;10(14):1129–41.

Rodríguez-Morató J, Pozo ÓJ, Marcos J. Targeting human urinary metabolome by LC-MS/MS: a review. Bioanalysis 2018;10(7):489–516.

Romero KI, Fernandez-Maestre R. Ion mobility spectrometry: the diagnostic tool of third millennium medicine. Rev Assoc Med Bras (1992) 2018;64(9):861–8.

Sánchez-Guijo A, Oji V, Hartmann MF, Traupe H, Wudy SA. Simultaneous quantification of cholesterol sulfate, androgen sulfates, and progestogen sulfates in human serum by LC-MS/MS. J Lipid Res 2015;56(9):1843–51.

Sánchez-Guijo A, Neunzig J, Gerber A, Oji V, Hartmann MF, Schuppe HC, et al. Role of steroid sulfatase in steroid homeostasis and characterization of the sulfated steroid pathway: evidence from steroid sulfatase deficiency. Mol Cell Endocrinol 2016;437:142–53.

Santa T. Derivatization reagents in liquid chromatography/electrospray ionization tandem mass spectrometry. Biomed Chromatogr 2011;25(1–2):1–10.

Schiffer L, Adaway JE, Arlt W, Keevil BG. A liquid chromatography-tandem mass spectrometry assay for the profiling of classical and 11-oxygenated androgens in saliva. Ann Clin Biochem 2019a;56(5):564–73.

Schiffer L, Barnard L, Baranowski ES, Gilligan LC, Taylor AE, Arlt W, et al. Human steroid biosynthesis, metabolism and excretion are differentially reflected by serum and urine steroid metabolomes: a comprehensive review. J Steroid Biochem Mol Biol 2019b;194, 105439.

Schöneshöfer M, Fenner A, Dulce HJ. Assessment of eleven adrenal steroids from a single serum sample by combination of automatic high-performance liquid chromatography and radioimmunoassay (HPLC-RIA). J Steroid Biochem 1981;14(4):377–86.

Shackleton C. Clinical steroid mass spectrometry: a 45-year history culminating in HPLC-MS/MS becoming an essential tool for patient diagnosis. J Steroid Biochem Mol Biol 2010;121(3–5):481–90.

Shackleton CH. Role of a disordered steroid metabolome in the elucidation of sterol and steroid biosynthesis. Lipids 2012;47(1):1–12.

Shackleton CH, Honour JW. Simultaneous estimation of urinary steroids by semi-automated gas chromatography. Investigation of neo-natal infants and children with abnormal steroid synthesis. Clin Chim Acta 1976;69(2):267–83.

Shackleton CH, Straub KM. Direct analysis of steroid conjugates: the use of secondary ion mass spectrometry. Steroids 1982;40(1):35–51.

Shackleton CH, Biglieri EG, Roitman E, Honour JW. Metabolism of radiolabeled corticosterone in an adult with the 17 alpha-hydroxylase deficiency syndrome. J Clin Endocrinol Metab 1979;48(6):976–82.

Shackleton CH, Mattox VR, Honour JW. Analysis of intact steroid conjugates by secondary ion mass spectrometry (including FABMS) and by gas chromatography. J Steroid Biochem 1983;19(1A):209–17.

Shackleton CH, Kletke C, Wudy S, Pratt JH. Dehydroepiandrosterone sulfate quantification in serum using high-performance liquid chromatography/mass spectrometry and a deuterated internal standard: a technique suitable for routine use or as a reference method. Steroids 1990;55(10):472–8.

Shackleton CH, Phillips A, Chang T, Li Y. Confirming testosterone administration by isotope ratio mass spectrometric analysis of urinary androstanediols. Steroids 1997;62(4):379–87.

Shackleton C, Marcos J, Arlt W, Hauffa BP. Prenatal diagnosis of P450 oxidoreductase deficiency (ORD): a disorder causing low pregnancy estriol, maternal and fetal virilization, and the Antley-Bixler syndrome phenotype. Am J Med Genet A 2004;129A(2):105–12.

Shackleton C, Pozo OJ, Marcos J. GC/MS in recent years has defined the normal and clinically disordered steroidome: will it soon be surpassed by LC/tandem MS in this role? J Endocr Soc 2018;2(8):974–96.

Shibata N, Hayakawa T, Takada K, Hoshino N, Minouchi T, Yamaji A. Simultaneous determination of glucocorticoids in plasma or urine by high-performance liquid chromatography with precolumn fluorimetric derivatization by 9-anthroyl nitrile. J Chromatogr B Biomed Sci Appl 1998;706(2):191–9.

Shibayama Y, Higashi T, Shimada K, Kashimada K, Onishi T, Ono M, et al. Liquid chromatography-tandem mass spectrometric method for determination of salivary 17alpha-hydroxyprogesterone: a noninvasive tool for evaluating efficacy of hormone replacement therapy in congenital adrenal hyperplasia. J Chromatogr B Analyt Technol Biomed Life Sci 2008;867(1):49–56.

Siekmann L. Reference methods for total cholesterol and total glycerol. Eur J Clin Chem Clin Biochem 1991;29(4):277–9.

Singh G, Gutierrez A, Xu K, Blair IA. Liquid chromatography/electron capture atmospheric pressure chemical ionization/mass spectrometry: analysis of pentafluorobenzyl derivatives of biomolecules and drugs in the attomole range. Anal Chem 2000;72(14):3007–13.

Sippell WG, Bidlingmaier F, Becker H, Brünig T, Dörr H, Hahn H, et al. Simultaneous radioimmunoassay of plasma aldosterone, corticosterone, 11-deoxycorticosterone, progesterone, 17-hydroxyprogesterone, 11-deoxycortisol, cortisol and cortisone. J Steroid Biochem 1978;9(1): 63–74.

Smy L, Straseski JA. Measuring estrogens in women, men, and children: recent advances 2012-2017. Clin Biochem 2018;62:11–23.

Søeborg T, Frederiksen H, Johannsen TH, Andersson AM, Juul A. Isotope-dilution TurboFlow-LC-MS/MS method for simultaneous quantification of ten steroid metabolites in serum. Clin Chim Acta 2017;468:180–6.

Sofronescu AG, Ross M, Rush E, Goldner W. Spurious testosterone laboratory results in a patient taking synthetic alkaline phosphatase (asfotase alfa). Clin Biochem 2018;58:118–21.

Son HH, Yun WS, Cho SH. Development and validation of an LC-MS/MS method for profiling 39 urinary steroids (estrogens, androgens, corticoids, and progestins). Biomed Chromatogr 2020;34(2), e4723.

St John ER, Balog J, McKenzie JS, Rossi M, Covington A, Muirhead L, et al. Rapid evaporative ionisation mass spectrometry of electrosurgical vapours for the identification of breast pathology: towards an intelligent knife for breast cancer surgery. Breast Cancer Res 2017;19(1):59.

Stenman UH. Standardization of hormone determinations. Best Pract Res Clin Endocrinol Metab 2013;27(6):823–30.

Stieglitz HM, Korpi-Steiner N. Characterization of biotin interference in 21 Vitros 5600 immunoassays and risk mitigation for patient safety at a large academic medical center. Clin Biochem 2020;75:53–61.

Stolze BR, Gounden V, Gu J, Elliott EA, Masika LS, Abel BS, et al. An improved micro-method for the measurement of steroid profiles by APPI-LC-MS/MS and its use in assessing diurnal effects on steroid concentrations and optimizing the diagnosis and treatment of adrenal insufficiency and CAH. J Steroid Biochem Mol Biol 2016;162:110–6.

Stone JA, Fitzgerald RL. Liquid chromatography-mass spectrometry education for clinical laboratory scientists. Clin Lab Med 2018;38(3): 527–37.

Storbeck KH, Kolar NW, Stander M, Swart AC, Prevoo D, Swart P. The development of an ultra performance liquid chromatography-coupled atmospheric pressure chemical ionization mass spectrometry assay for seven adrenal steroids. Anal Biochem 2008;372(1):11–20.

Storbeck KH, Gilligan L, Jenkinson C, Baranowski ES, Quanson JL, Arlt W, et al. The utility of ultra-high performance supercritical fluid chromatography-tandem mass spectrometry (UHPSFC-MS/MS) for clinically relevant steroid analysis. J Chromatogr B Analyt Technol Biomed Life Sci 2018;1085:36–41.

Storbeck KH, Schiffer L, Baranowski ES, Chortis V, Prete A, Barnard L, et al. Steroid metabolome analysis in disorders of adrenal steroid biosynthesis and metabolism. Endocr Rev 2019;40(6):1605–25.

Suhr AC, Vogeser M, Grimm SH. Isotope inversion experiment evaluating the suitability of calibration in surrogate matrix for quantification via LC-MS/MS-exemplary application for a steroid multi-method. J Pharm Biomed Anal 2016;124:309–18.

Sun Y, Gu C, Liu X, Liang W, Yao P, Bolton JL, et al. Ultrafiltration tandem mass spectrometry of estrogens for characterization of structure and affinity for human estrogen receptors. J Am Soc Mass Spectrom 2005;16(2):271–9.

Sun Q, Gu J, Stolze BR, Soldin SJ. Atmospheric pressure chemical ionization is a suboptimal ionization source for steroids. Clin Chem 2018;64(6):974–6.

Tai SS-C, Xu B, Welch MJ, Phinney KW. Development and evaluation of a candidate reference measurement procedure for the determination of testosterone in human serum using isotope dilution liquid chromatography/tandem mass spectrometry. Anal Bioanal Chem 2007;388:1087–94.

Taieb J, Benattar C, Birr AS, Lindenbaum A. Limitations of steroid determination by direct immunoassay. Clin Chem 2002;48(3):583–5.

Taieb J, Benattar C, Diop R, Birr AS, Lindenbaum A. Use of the architect-i2000 estradiol immunoassay during in vitro fertilization. Clin Chem 2003a;49(1):183–6.

Taieb J, Mathian B, Millot F, Patricot MC, Mathieu E, Queyrel N, et al. Testosterone measured by 10 immunoassays and by isotope-dilution gas chromatography-mass spectrometry in sera from 116 men, women, and children. Clin Chem 2003b;49(8):1381–95.

Takagi K, Okumura K, Morikawa N, Okuyama S, Wu MC, Narita O. Simultaneous assay for individual sulphated 3 alpha- and beta-hydroxysteroids in serum using high-performance liquid chromatography combined with 3 alpha- and beta-hydroxysteroid dehydrogenases immobilized on one column. J Chromatogr 1988;432:47–56.

Taylor PJ. Matrix effects: the Achilles heel of quantitative high-performance liquid chromatography-electrospray-tandem mass spectrometry. Clin Biochem 2005;38(4):328–34.

Taylor DR, Ghataore L, Couchman L, Vincent RP, Whitelaw B, Lewis D, et al. A 13-steroid serum panel based on LC-MS/MS: use in detection of adrenocortical carcinoma. Clin Chem 2017;63(12):1836–46.

Teubel J, Wüst B, Schipke CG, Peters O, Parr MK. Methods in endogenous steroid profiling—a comparison of gas chromatography mass spectrometry (GC-MS) with supercritical fluid chromatography tandem mass spectrometry (SFC-MS/MS). J Chromatogr A 2018;1554:101–16.

Thienpont L, Siekmann L, Lawson A, Colinet E, De Leenheer A. Development, validation, and certification by isotope dilution gas chromatography-mass spectrometry of lyophilized human serum reference materials for cortisol (CRM 192 and 193) and progesterone (CRM 347 and 348). Clin Chem 1991;37(4):540–6.

Tölgyesi A, Verebey Z, Sharma VK, Kovacsics L, Fekete J. Simultaneous determination of corticosteroids, androgens, and progesterone in river water by liquid chromatography-tandem mass spectrometry. Chemosphere 2010;78(8):972–9. https://doi.org/10.1016/j.chemosphere.2009.12.025. Epub 2010 Jan 13 20071003.

Vaikkinen A, Rejšek J, Vrkoslav V, Kauppila TJ, Cvačka J, Kostiainen R. Feasibility of desorption atmospheric pressure photoionization and desorption electrospray ionization mass spectrometry to monitor urinary steroid metabolites during pregnancy. Anal Chim Acta 2015;880:84–92.

Vallano PT, Shugarts SB, Woolf EJ, Matuszewski BK. Elimination of autosampler carryover in a bioanalytical HPLC-MS/MS method: a case study. J Pharm Biomed Anal 2005;36(5):1073–8.

van Helden J, Weiskirchen R. Cross-method comparison of serum androstenedione measurement with respect to the validation of a new fully automated chemiluminescence immunoassay. Clin Biochem 2018;62:32–8.

Vavrinsky E, Stopjakova V, Kopani M, Kosnacova H. The concept of advanced multi-sensor monitoring of human stress. Sensors (Basel) 2021;21(10):3499.

Vesper HW, Botelho JC, Vidal ML, Rahmani Y, Thienpont LM, Caudill SP. High variability in serum estradiol measurements in men and women. Steroids 2014;82:7–13.

Wang Q, Mesaros C, Blair IA. Ultra-high sensitivity analysis of estrogens for special populations in serum and plasma by liquid chromatography-mass spectrometry: assay considerations and suggested practices. J Steroid Biochem Mol Biol 2016;162:70–9.

Wang R, Hartmann MF, Tiosano D, Wudy SA. Characterizing the steroidal milieu in amniotic fluid of mid-gestation: a GC-MS study. J Steroid Biochem Mol Biol 2019;193, 105412.

Wang Z, Wang H, Peng Y, Chen F, Zhao L, Li X, et al. A liquid chromatography-tandem mass spectrometry (LC-MS/MS)-based assay to profile 20 plasma steroids in endocrine disorders. Clin Chem Lab Med 2020;58(9):1477–87.

Wang R, Hartmann MF, Wudy SA. Targeted LC-MS/MS analysis of steroid glucuronides in human urine. J Steroid Biochem Mol Biol 2021;205, 105774.

Warade J. Retrospective approach to evaluate interferences in immunoassay. EJIFCC 2017;28:224–32.

Webb R, Baxter G, McBride D, Nordblom GD, Shaw MP. The measurement of testosterone and oestradiol-17 beta using iodinated tracers and incorporating an affinity chromatography extraction procedure. J Steroid Biochem 1985;23(6A):1043–51.

Wei JQ, Wei JL, Zhou XT, Cheng JP. Isocratic reversed phase high performance liquid chromatography determination of twelve natural corticosteroids in serum with on-line ultraviolet and fluorescence detection. Biomed Chromatogr 1990;4(4):161–4.

Wei JQ, Wei JL, Lucarelli C, Zhou XT, Wang DQ, Dai WJ, et al. Serum steroid hormonal profiles by reversed-phase liquid chromatography in patients with 17-hydroxylase deficiency and in an affected family. Clin Chem 1992;38(1):76–82.

Wild D. The immunoassay handbook. 4th ed. Elsevier; 2013, ISBN:978-0-08-097037-0.

Wilson WD. Tech.Sight. Analyzing biomolecular interactions. Science 2002;295(5562):2103–5.

Wong T, Shackleton CH, Covey TR, Ellis G. Identification of the steroids in neonatal plasma that interfere with 17 alpha-hydroxyprogesterone radioimmunoassays. Clin Chem 1992;38(9):1830–7.

Wudy SA, Wachter UA, Homoki J, Teller WM, Shackleton CH. Androgen metabolism assessment by routine gas chromatography/mass spectrometry profiling of plasma steroids: part 1, unconjugated steroids. Steroids 1992;57(7):319–24.

Wudy SA, Dörr HG, Solleder C, Djalali M, Homoki J. Profiling steroid hormones in amniotic fluid of midpregnancy by routine stable isotope dilution/gas chromatography-mass spectrometry: reference values and concentrations in fetuses at risk for 21-hydroxylase deficiency. J Clin Endocrinol Metab 1999;84(8):2724–8.

Wudy SA, Hartmann M, Svoboda M. Determination of 17-hydroxyprogesterone in plasma by stable isotope dilution/benchtop liquid chromatography-tandem mass spectrometry. Horm Res 2000; 53(2):68–71.

Wudy SA, Schuler G, Sánchez-Guijo A, Hartmann MF. The art of measuring steroids: principles and practice of current hormonal steroid analysis. J Steroid Biochem Mol Biol 2018;179:88–103.

Wynne-Roberts CR. The evaluation of hydrocortisone antibodies produced in rabbits and sheep. I Radioimmunoassay. Immunology 1974;26(1): 97–113.

Xu W, Li H, Guan Q, Shen Y, Cheng L. A rapid and simple liquid chromatography-tandem mass spectrometry method for the measurement of testosterone, androstenedione, and dehydroepiandrosterone in human serum. J Clin Lab Anal 2017;31(5), e22102.

Yalow RS, Berson SA. Assay of plasma insulin in human subjects by immunological methods. Nature 1959;184(Suppl. 21):1648–9.

Yamashita K, Takahashi M, Tsukamoto S, Numazawa M, Okuyama M, Honma S. Use of novel picolinoyl derivatization for simultaneous quantification of six corticosteroids by liquid chromatography-electrospray ionization tandem mass spectrometry. J Chromatogr A 2007;1173(1–2):120–8.

Yamashita K, Nakagawa R, Okuyama M, Honma S, Takahashi M, Numazawa M. Simultaneous determination of tetrahydrocortisol, allotetrahydrocortisol and tetrahydrocortisone in human urine by liquid chromatography-electrospray ionization tandem mass spectrometry. Steroids 2008;73(7):727–37.

Yan Y, Ubukata M, Cody RB, Holy TE, Gross ML. High-energy collision-induced dissociation by MALDI TOF/TOF causes charge-remote fragmentation of steroid sulfates. J Am Soc Mass Spectrom 2014;25(8): 1404–11.

Yi X, Leung EKY, Bridgman R, Koo S, Yeo KJ. High-sensitivity micro LC-MS/MS assay for serum estradiol without derivatization. J Appl Lab Med 2016;1(1):14–24.

Zemaitis MA, Kroboth PD. Simplified procedure for measurement of serum dehydroepiandrosterone and its sulfate with gas chromatography-ion trap mass spectrometry and selected reaction monitoring. J Chromatogr B Biomed Sci Appl 1998;716(1–2):19–26.

Zhang YV, Rockwood A. Impact of automation on mass spectrometry. Clin Chim Acta 2015;450:298–303.

Zhang Y, Tobias HJ, Sacks GL, Brenna JT. Calibration and data processing in gas chromatography combustion isotope ratio mass spectrometry. Drug Test Anal 2012;4(12):912–22.

Zhang Q, Jing L, Zhang J, Ren Y, Wang Y, Wang Y, et al. Surface plasmon resonance sensor for femtomolar detection of testosterone with water-compatible macroporous molecularly imprinted film. Anal Biochem 2014;463:7–14.

Zhang L, Allworth LL, Vertes A. Identification of metabolites in single cells by ion mobility separation and mass spectrometry. Methods Mol Biol 2020;2064:9–18.

Zhao Y, Boyd JM, Sawyer MB, Li XF. Liquid chromatography tandem mass spectrometry determination of free and conjugated estrogens in breast cancer patients before and after exemestane treatment. Anal Chim Acta 2014;806:172–9.

Zhao XE, Yan P, Wang R, Zhu S, You J, Bai Y, et al. Sensitive determination of cholesterol and its metabolic steroid hormones by UHPLC-MS/MS via derivatization coupled with dual ultrasonic-assisted dispersive liquid-liquid microextraction. Rapid Commun Mass Spectrom 2016;30(Suppl. 1):147–54.

Books

Leaver N. A practical guide to implementing clinical mass spectrometry systems. ILM Publications; 2011, ISBN:978-1-906799-10-6.

Makin HL, Gower DB. Steroid assays. 2nd ed. London: Springer; 2010, ISBN:978-1-4020-9775-1.

Nair H, Clarke W. Mass spectrometry for the clinical laboratory. Academic Press; 2017, ISBN:978-0-128008713.

Watson TJ, Sparkman OD. Introduction to mass spectrometry. 4th ed. Wiley; 2007, ISBN:978-0-470-51634-8.

Chapter 2.5

Normal concentrations of steroids and their regulators in blood

2.5.1 Introduction

In blood plasma or serum, the concentrations of steroids in basal state vary across six orders of magnitude from picomole/L concentrations for estradiol to micromolar amounts per liter for dehydroepiandrosterone-sulfate. This influences a number of factors in the analytical and preanalytical stages. The sensitivity of the quantitative test is one issue to be addressed by the laboratory, although the size of the sample of the biological fluid collected can be adjusted to supply more of the analyte to the detector. It is often not possible from newborn children to get plasma samples of more than 500 µL, so blood spotted onto papers then dried, like the Guthrie newborn screening tests, may then be an alternative to venous samples. The quality of the assays must also have been determined by the laboratory in terms of imprecision and repeatability. The majority of steroids are bound to protein in the blood and in general the steroids are displaced from the proteins before analysis, with or without an extraction and purification steps and the concentration of the free and bound steroid are determined as a total steroid concentration.

For many steroid measurements, the time when the sample is taken is critical because of innate variations by time of day, age and gender of the subject, time in the menstrual cycle and even seasons of the year. Measurements may also be affected by **medications** that affect concentrations of **steroid binding proteins** and steroid metabolism or influence the actual assay method through some **interference**. The ranges of quantitative analytical tests for steroids have been addressed in the previous chapter (Chapter 2.4) and the outcomes for reference ranges mainly by immunoassays reviewed by Fanelli in 2018 is a good reference point and there are other useful papers (Bae et al., 2019; Adeli et al., 2017; Tavita and Greaves, 2017; Karbasy et al., 2016; Raizman et al., 2015; Kelly et al., 2015; Kyriakopoulou et al., 2013; Bailey et al., 2013). The data in the Fanelli paper highlights problems with differences in reference ranges which affects interpretation of results over time and commutability of results. The units in which results are expressed complicate the matter. The preference in Europe is for the *Systeme International*, whereas in the United States mass units are the most used but to different volumes (liter, mL, or dL from µg/dL for cortisol to pg/mL for estradiol). Steroids are now measured more often by mass spectrometric techniques but still in United Kingdom the SI system of reporting results is preferred. Where steroid results are reported in SI and mass units, conversion factors are needed between the two and these are summarized in Table 2.5.1. In one study, multiples of median (MOM) transformed reference ranges were developed for 10 steroids for age and sex and shown to be effective in characterizing patients with congenital adrenal hyperplasia due to 11-hydroxylase or 21-hydroxylase deficiencies (Zalas et al., 2018).

Most of the steroid results in the literature were derived in **immunoassays**, often by direct immunoassay on a platform in the chemistry/pathology laboratory. The performances of many of these assays has been validated against certified reference materials and shown to give comparable results to MS methods for the tested samples (Cao et al., 2019; van Helden and Weiskirchen, 2018). That does not mean that performance is the same across all clinical samples and lack of specificity by cross reaction can still be encountered. Today the use of liquid chromatography coupled with mass spectrometry (LC-MS) or tandem mass spectrometry (LC-MS/MS) is being introduced in many hospital laboratories because of superior specificity (Jeanneret et al., 2016). Initially one steroid at a time was measured but there is the potential to measure typically around 15–30 steroids but easily up to 100 steroids at little extra analytical cost. The service provision implementation progresses from one steroid (testosterone, 17-hydroxyprogesterone or cortisol say) to a panel of androgens or corticosteroids and then to a full profile of adrenal and gonadal steroids (Lee et al., 2020; Fiet et al., 2017). There are LC-MS/MS based normative data of A, T and DHT (van der Veen et al., 2019; Büttler et al., 2015a,b; Mouritsen et al., 2014) to age, Tanner stage based on breast development and testicular volume. Plasma cortisol and its precursors are now measured together in LC-MS/MS methods in positive ion mode making easier

Steroids in the Laboratory and Clinical Practice. https://doi.org/10.1016/B978-0-12-818124-9.00003-6

TABLE 2.5.1 Conversion table for SI and mass units.

Hormone	Unit you know	Multiply	Unit to find
ACTH	pg/mL	0.222	pmol/L
Aldosterone	ng/dL	27.747	pmol/L
Androstanediol	ng/dL	34.199	pmol/L
Androstanediol glucuronide	ng/dL	21.345	pmol/L
Angiotensin I	pg/mL	0.772	pmol/L
Angiotensin II	pg/mL	0.956	pmol/L
Corticosterone	ng/dL	28.086	pmol/L
18-Hydroxy corticosterone	ng/dL	27.594	pmol/L
Cortisol	µg/dL	27.586	nmol/L
Cortisone	µg/dL	27.739	nmol/L
DHEA	ng/dL	34.674	pmol/L
DHEAS	µg/dL	27.211	nmol/L
DOC	ng/dL	30.257	pmol/L
11-Deoxycortisol (S)	ng/dL	28.868	pmol/L
Dihydrotestosterone	ng/dL	34.435	pmol/L
Estradiol	ng/dL	36.711	pmol/L
Insulin	µU/mL	7.175	pmol/L
Prednisolone	ng/dL	27.739	pmol/L
17-Hydroxypregnenolone	ng/dL	30.075	pmol/L
Progesterone	ng/dL	31.796	pmol/L
17-Hydroxy progesterone	ng/dL	30.257	pmol/L
Plasma renin activity	ng/mL/h	0.2778	ng/L/s
SHBG	µg/dL	34.674	nmol/L
Testosterone	ng/dL	34.674	pmol/L

the diagnosis of adrenal disorders from ratios of substrate to product of an enzyme (Storbeck et al., 2019). Aldosterone and estrogens are at the lowest concentrations of the steroids in blood and many MS methods use negative ion mode (Guo et al., 2018; Handelsman et al., 2020). In some methods, steroid derivatives that give stronger ions that improve sensitivity are measured (Gravitte et al., 2021; Denver et al., 2019). Steroid panels of 5–10 or more steroids are now reported from many clinical laboratories (Taylor et al., 2017); in research this can be around 100 or more steroids (Hill et al., 2010, 2014) which is a real challenge to validate the method. Puberty, pregnancy and the menopause are important life events that affect steroid concentrations in blood. Many females will be taking oral contraceptives that affect the steroid concentrations (Søeborg et al., 2014).

Inhaled corticosteroids are widely taken for asthma and hay fever but can suppress the secretion of ACTH.

Reference ranges can be reported in different ways and if not normally distributed by mean ± SD, median and range, upper and lower limit, 2.5 and 97.5 percentiles and presented in tables and graphs. On a personal note, I do not find mean ± SD is useful, many papers do not prove a normal distribution and I find clinicians will look at mean ± 2xSD as the range and find negative results for lower values. Clearly that presentation does not represent the true variation of the data so median and range is much better. At least 120 subjects are needed for a reliable reference range; volunteers are better than a hospitalized patient group. Inclusions and exclusions need to be carefully defined. A female group is difficult to standardize unless all taking one particular contraceptive preparation. The group should only include individuals of normal body weight.

2.5.2 Practices for hormone measurements

The choice of biochemical investigations should be based on the clinical picture, and good communication between the clinicians and laboratory scientists is important. The interpretation of results must take into account the laboratory methods used and the patient's age and gender. Concentrations for steroids quoted in this chapter are for guidance only and depend on the particular assay in use (LC-MS/MS results have been used here wherever possible). The references cited in this chapter are not comprehensive but selected to cover specific issues; they are also based largely on reviewing recent literature and thus a stepping point for further exploration of the literature. Normative data of some assays are now available for age and development of healthy infants and children. However, caution is needed in the interpretation of results in sick premature babies and also children with delayed or advanced puberty at the peri-pubertal ages where age-specific references ranges may not be appropriate. Steroid hormone concentrations are often determined at the same time as the regulatory proteins.

2.5.2.1 Immunoassays

Most hospital laboratories use automated immunoassays, despite the potential limitations of reduced specificity caused especially for steroids by cross-reactivity with molecules of similar structures. There have been efforts by the industry to standardize and harmonize these tests to reference material and quality control samples, such as in the Canadian Laboratory Initiative on Pediatric Reference Intervals (CALIPER) programme. Some immunoassays have poor precision particularly at low concentrations (estradiol for example), which can be problematic in children, adolescents and women postmenarche as hormone concentrations are often below the limits of detection. The user of commercial kits needs to look out for lot-to-lot differences in results due to changes by the manufacturer in the reagents and calibration.

Immunometric (two-site) assays with CLIA and ELISA are most used for the peptide hormones, but MS methods will become available. Plasma renin protein is now measured as concentration (PRC) by immunoassay with antibodies for N and C terminal fragments (Rossi et al., 2016), or activity (PRA) which in some laboratories will involve determining angiotensin I generation by immunoassay and now mass spectrometry (Chappell et al., 2012). Assays for aldosterone and renin are now available on automated chemiluminescence devices (AUTO LUMO A2000

(info@autobio-diagnostics.com); DIA-Sorin LIAISON and IDS-iSys). Reference values (upper limit; lower limit) from the Canadian laboratory initiative (CALIPER) have been summarized for steroids, peptides and other analytes on the five major analytical platforms (Abbott, Beckman, Ortho, Roche and Siemens) (Adeli et al., 2017) although these values need to be verified for local populations. There has been increased demand for verification of laboratory results including diagnostic sensitivity, diagnostic specificity, predictive value, negative predictive value, likelihood ratio and expected values in normal and affected populations.

The user should look out in immunoassay results for the possibility of cross reaction through unusual steroid results, particularly if other exogenous steroids are in the body, which can include oral contraceptives, inhaled steroids and anabolic steroids. Some interference problems will be discussed with each steroid in the course of this chapter. The laboratory is not always informed about current prescription drugs and assay problems can be encountered with the use of herbal medicines and nutritional supplements. The effects of steroids on the regulatory system can be seen, the peptide hormone assays are subject to different problems such as heterophilic antibody interferences if the patient keeps pets (rats, rabbit, and mice) (Moerman and Delanghe, 2020; Stieglitz and Korpi-Steiner, 2020) or farm animals (sheep). The use of biotin as a nutritional supplement can affect performance of Roche Elecsys and Ortho Vitros immunoassay systems (Stieglitz and Korpi-Steiner, 2020) that use streptavidin-biotin detection systems with steroid assays falsely elevated in competitive assays while the peptide sandwich assays can give low results. Assays with Abbott Architect and Aliny and Diasorin Liaison are not affected. Healthcare professionals as well as laboratory staff need to be aware of this problem and have procedures in place to recognize and make corrections for abnormal results (Bowen et al., 2019). Patients need to be guided on use of supplements and re-testing, clinicians should check if patients are taking supplements, the laboratory should have samples retested on different platforms.

2.5.2.2 Mass spectrometry

Mass spectrometry is currently considered the most specific method for steroid measurement. More recent advances have involved two-stages of mass spectrometry with intermediate fragmentation, often called tandem-mass spectrometry (MS/MS). Traditional methods of gas chromatography and high performance liquid chromatography (HPLC, or LC for short) are increasingly combined with mass spectrometry (GC-MS, GC-MS/MS or LC-MS/MS). In a few cases, high-resolution, accurate mass, mass spectrometry (HRAMS) has been used. LC-MS/MS measurements are targeted to single or multiple (panel to profile) of steroids simultaneously from a small plasma volume. This is becoming more widely available for steroid hormone analysis, with reference values in several papers published for cortisol, androstenedione, dehydroepiandrosterone sulfate (DHEAS), 17-hydroxyprogesterone (17OHP), DHT and testosterone (and more) in the pediatric age range and adults.

The main advantage of LC-MS/MS, and more so with HRAMS, is that it firstly allows compounds of similar structures to be separated in the chromatography stage, and then detected more specifically and hence reduces the problems with cross-reactivity experienced with immunoassays (Bileck et al., 2019; Hines et al., 2017). Sample preparation can also be simplified. This has to be balanced, however, against cost considerations. In many laboratories, costs for results by LC-MS/MS methods should be competitive with immunoassay methods because after the initial outlay on equipment and it's maintenance costs the reagent running costs are low (solvents and LC columns). GC-MS has higher chromatographic separation but requires time consuming derivative formation in the sample processing procedure.

Commercial kits for LC-MS/MS methods are available for serum steroid profiles. The user needs to confirm the best MS parameters (ion spray voltage, gas flow, desolvation temperature, etc.) and the transitions to be monitored in tandem mode for the equipment in use since there are differences in MS performances even with the same model of a manufacturer. Performance must be confirmed for the particular laboratory for linearity, accuracy, imprecision, sensitivity, recovery, matrix effects, specificity, carryover, stability and dilutions. It is important to interpret hormone results against reference ranges that are appropriate for method of analysis, age and stage of development. There are marked differences in results with different methodologies particularly for androgens and proteins such as sex hormone-binding globulin. Mass spectrometry is being used to quantify steroids in many laboratories and in some plasma renin activity is so determined. The results will invariably be lower than what has been experienced with immunoassay. The laboratory must consider the level of service in the event of breakdown which could mean a requirement for a back-up instrument or arrangement with another laboratory to perform the tests.

2.5.3 Hypothalamic pituitary adrenal axis (HPA)

The adrenal cortex, hypothalamus and pituitary gland and their interactions constitute the HPA axis (Fig. 2.5.1) (for detail see Chapter 1.4) working together to control response to stress and to regulate the immune system, cardiovascular system, energy expenditure and storage, mood and

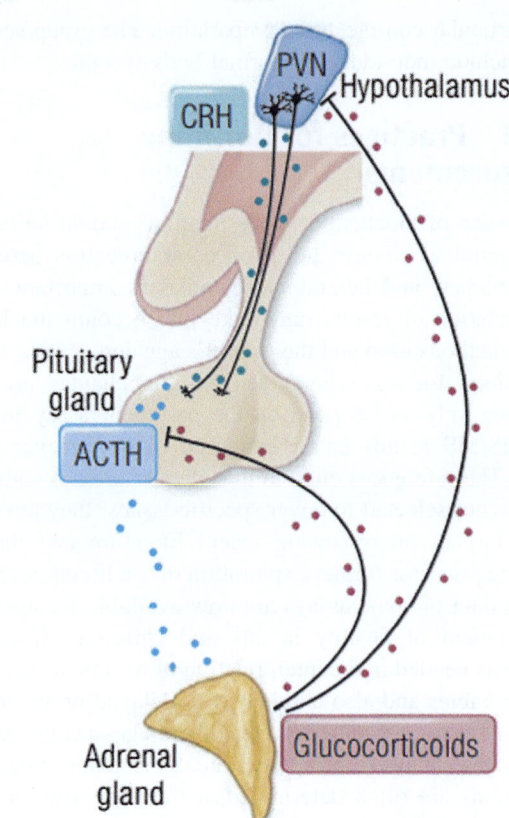

FIG. 2.5.1 A schematic of the HPA axis. Corticotropin-releasing hormone (CRH) and arginine vasopressin (AVP) stimulate the release of adrenocorticotropic hormone (ACTH) from the pituitary. ACTH in turn stimulates the adrenal glands to synthesize CORT, which further regulates its own synthesis through an intra-adrenal feedback loop. Within the HPA axis, CORT acts to inhibit ACTH in the pituitary as well as CRH and AVP in the hypothalamus, creating a dual negative-feedback loop. *(Based on Lightman SL, Birnie MT, Conway-Campbell BL. Dynamics of ACTH and cortisol secretion and implications for disease. Endocr Rev 2020;41(3): bnaa002. Fig 1 P3.)*

emotions. Cortisol, corticotrophin releasing factor (CRF, CRH) and adrenocorticotrophin (ACTH) hormones are involved in the process. The release of CRH from the median eminence of the hypothalamus is influenced by stress, the blood concentrations of cortisol, physical activity and illness. CRF is transported to the pituitary through blood in the hypophyseal stalk to stimulate secretion of ACTH by the pituitary gland into peripheral blood thence stimulating cortisol secretion from the adrenal cortex. Cortisol will negatively feedback to inhibit both the hypothalamus and pituitary.

2.5.3.1 Cortisol and adrenocorticotrophic hormone (ACTH)

The lowest cortisol values (less than 140 nmol/L) are found around midnight, and, following a number of peaks of increasing amplitude in cortisol and ACTH secretion during

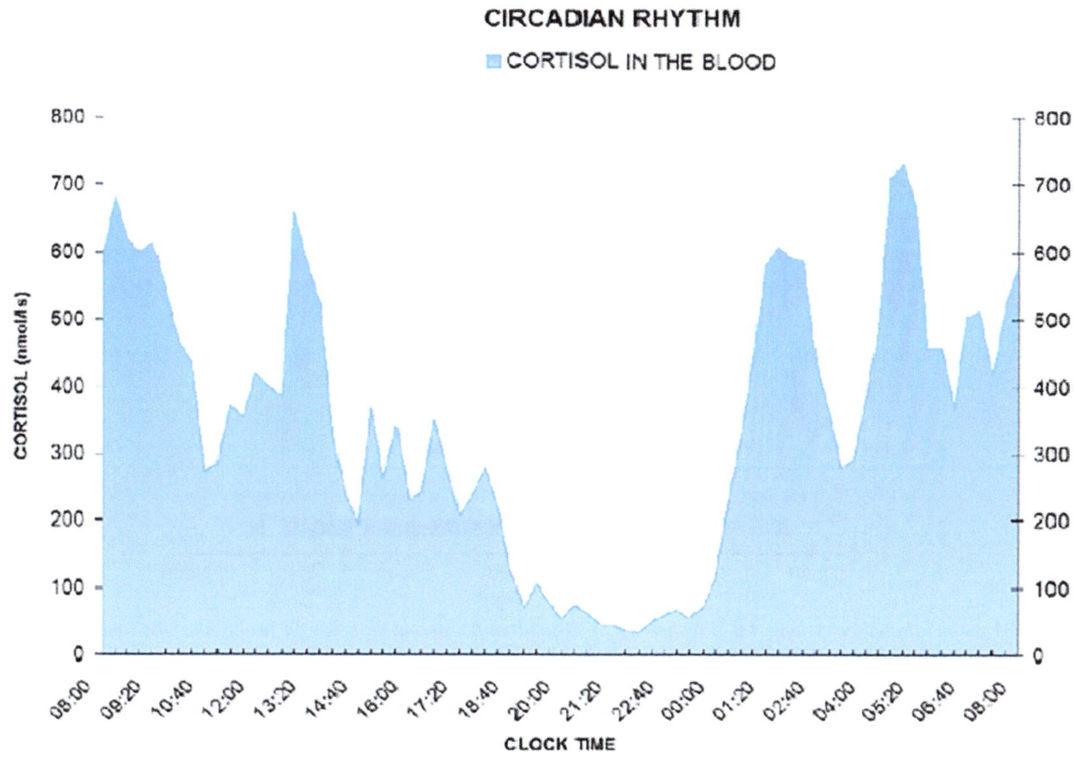

CIRCADIAN RHYTHM

■ CORTISOL IN THE BLOOD

FIG. 2.5.2 Circadian rhythm cortisol. Blood samples were taken at 20 min intervals. *(Reproduced with permisssion of Peter Hindmarsh.)*

the night, to maximal cortisol values near 0800h of around 700 nmol/L (Fig. 2.5.2). The results of limited blood sampling of many subjects shows the HPA as a **diurnal rhythm** in adults and children apart from the newborn infant. The pattern however depends on the frequency of blood sampling over 24 h. If blood is taken at 10 or 20 min intervals ACTH and cortisol concentrations reflect secretion by the pituitary in irregular bursts throughout the day and night and plasma cortisol concentrations tend to rise and fall in consort, an ultradian pattern. One way to assess the HPA is to measure cortisol at 20 min intervals for 2–3 h from 0700 h. The ACTH pulses are more frequent in the early morning and less frequent in the evening. ACTH measured by RIA are normally less than 80 ng/L and the values by immunoradiometric assays (IRMA) are lower with peak concentrations of 10–20 ng/L which reflect both improved sensitivity and specificity of IRMA (Fig. 2.5.3). Several peaks are seen in the 24 h period.

2.5.3.1.1 Basal cortisol determinations

Cortisol is released from the adrenal gland in a pulsatile manner under the stimulus of ACTH which itself is responsive to CRH which is released in an episodic manner. The popular concept of cortisol concentrations in blood is a diurnal rhythm which is the interpretation of the wide range of patterns in individuals that are out of phase (Fig. 2.5.4). In

an individual, therefore, it is difficult to interpret a cortisol concentration in a single sample because the time could be anywhere in the individual between a peak and trough in their ultradian pulsatile pattern.

The performances of cortisol assays were reviewed (Gatti et al., 2009; El-Farhan et al., 2017) although some assays have been reformulated. Many labs use chemiluminescent methods (Roche Elecsys; ACS180; Siemens Centaur; Immulite) for cortisol and this is likely to continue because of the convenience in results along with other analytes. The manufacturers have endeavored to prove satisfactory performances of the assays by comparison with a mass spectrometric standard. In some methods (Roche Elecsys), a new, more specific antibody was incorporated in the second generation assay. This has lowered results such that morning cortisol is 166–507 nmol/L (Vogeser et al., 2017), whereas previously the range seen with this and other assays was 200–800 nmol/L.

Repeated blood sampling at 5–10 min intervals for an hour or 90 min is the only way to get an accurate result near a peak but in reality is a totally impractical approach except in research situations. A microdialysis system has been validated however for continuous blood sampling (Bhake et al., 2020). One study looked at the effects of stress of inserting a catheter at 0800 h or at 1000 h on later steroid concentrations. Values just after inserting the cannula were significantly higher than values when calm. The results

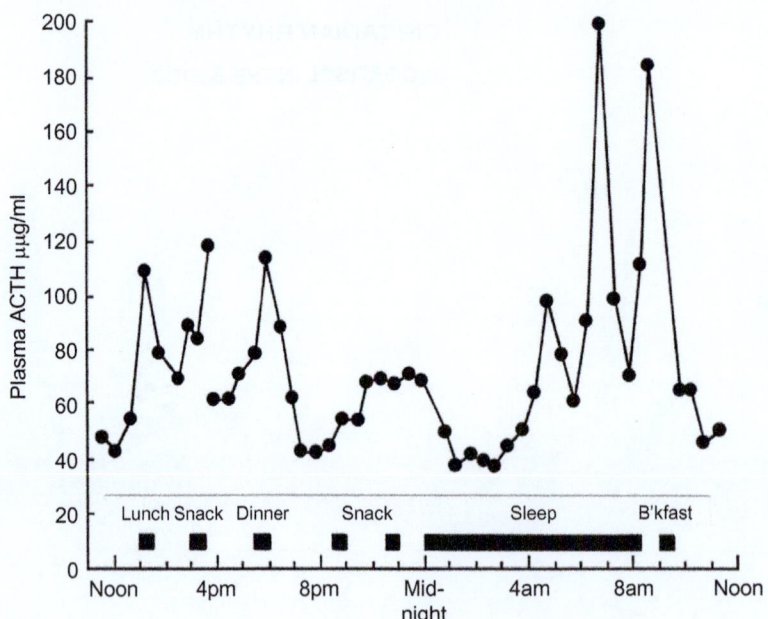

FIG. 2.5.3 ACTH pulses. *(Based on Krieger DT. Rhythms of ACTH and corticosteroid secretion in health and disease, and their experimental modification. J Steroid Biochem 1975;6(5):785–91.)*

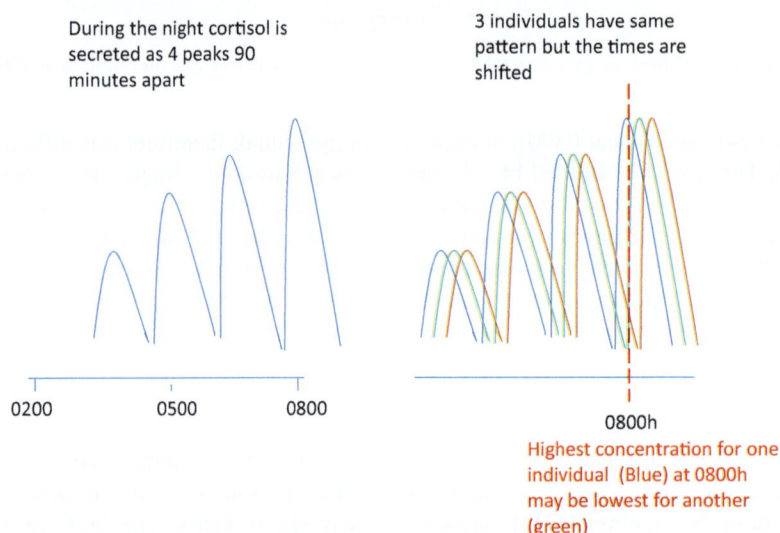

FIG. 2.5.4 Schematic of cortisol 24h profiles from four individuals out of phase. An 0800h cortisol can be between peak and trough values. *(Author original.)*

indicated that blood drawing can invoke variable levels of stress (Fig. 2.5.5) (Dušková et al., 2015). Higher levels of plasma cortisol lasted at least 1 h after the first sampling. The results also indicated that merely the knowledge of the blood sampling could stimulate higher cortisol levels, apart from the needle insertion itself. Cortisone and corticosterone levels were also increased, similar to the cortisol response.

Unlike in the investigations of thyroid function the measurement of free cortisol concentrations in plasma is not yet a routine procedure. Equilibrium dialysis is the method of choice to obtain the free cortisol which is only about 5%–10% of the total. The results agree with calculated free cortisol (Fig. 2.5.6) (Vogeser et al., 2007).

More than 90% of circulating cortisol is bound to proteins, in particular to the corticosteroid-binding protein (CBG) and to a far lesser extent to albumin.

Since cortisol receptors in target cells are located within the cytosol, and because it is generally thought that only unbound cortisol can enter this space by free diffusion,

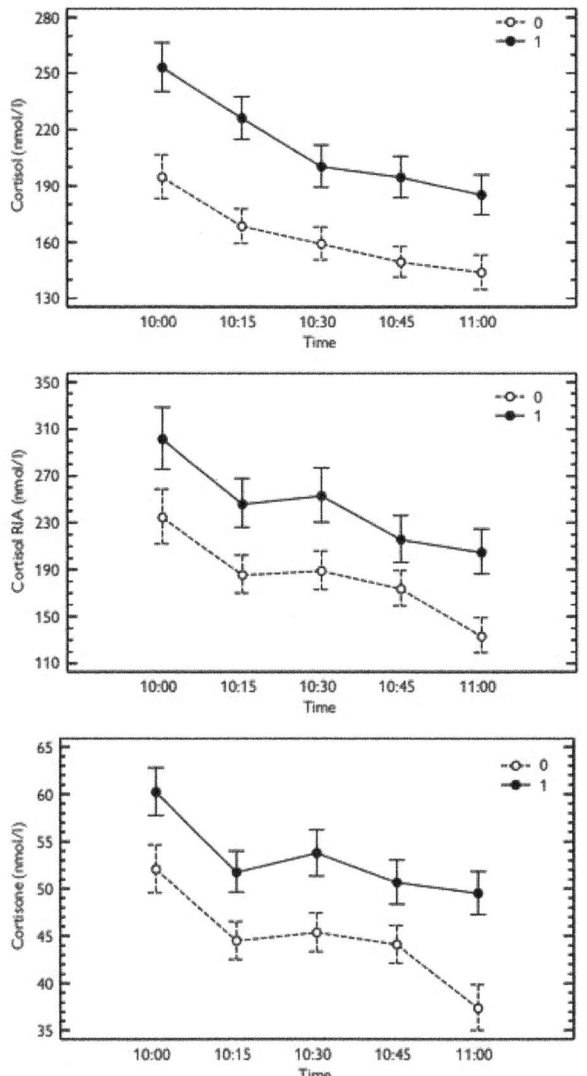

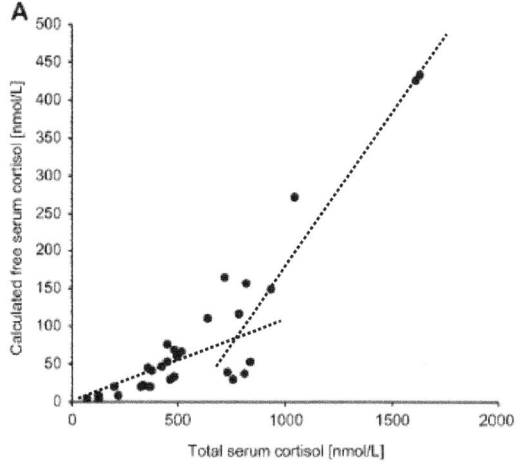

FIG. 2.5.6 Free cortisol measured vs calculated. *(From Vogeser M, Möhnle P, Briegel J. Free serum cortisol: quantification applying equilibrium dialysis or ultrafiltration and an automated immunoassay system. Clin Chem Lab Med 2007;45(4):521–5. Fig. 1 p. 726.)*

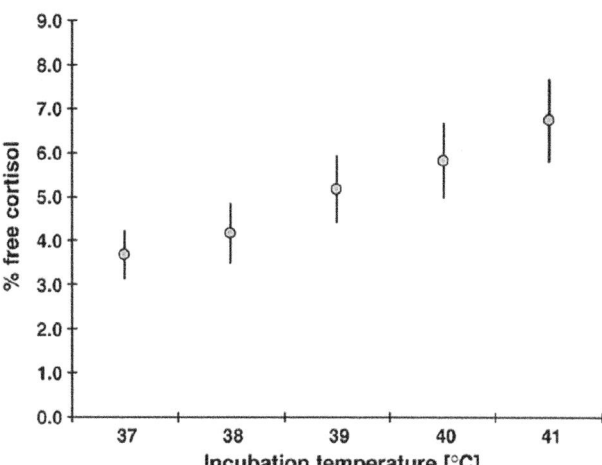

FIG. 2.5.5 Influence of the time of blood drawing after cannulization on cortisol levels (measured LC-MS/MS and RIA) and cortisone, corticosterone and aldosterone (LC-MS/MS) (0—cannula inserted at 8:00; 1—cannula inserted at 10:00). *(Modified from Dušková M, Sosvorová L, Vítků J, Jandíková H, Rácz B, Chlupáčová T, De Cordeiro J, Stárka L. Changes in the concentrations of corticoid metabolites—the effect of stress, diet and analytical method. Prague Med Rep 2015;116(4):268–78. Fig. 1 p. 272.)*

FIG. 2.5.7 Cortisol free-to-total ratio (% free) observed by equilibrium dialysis at different incubation temperatures from sera of 10 healthy persons (mean and standard deviation). *(From Vogeser M, Kratzsch J, Ju Bae Y, Bruegel M, Ceglarek U, Fiers T, Gaudl A, Kurka H, Milczynski C, Prat Knoll C, Suhr AC, Teupser D, Zahn I, Ostlund RE. Multicenter performance evaluation of a second generation cortisol assay. Clin Chem Lab Med 2017;55(6):826–35. Fig. 2 p. 726.)*

the nonprotein bound fraction of circulating cortisol is considered to be biologically active. The free cortisol increases substantially with increasing sample temperatures in vitro (Fig. 2.5.7) (Vogeser et al., 2017). To raise the temperature from 37°C to 41°C results in an increase of nearly 50% in serum-free cortisol.

During acute illness, cortisol action is of essential importance to modulate metabolic, immunological and circulatory responses of the body. In this situation—which is typically associated with fever—decreased protein binding of cortisol seems to increase active cortisol concentrations

in addition to a high adrenal synthesis rate. The oral contraceptive ethinyl estradiol increases CBG production and raises serum cortisol (Panton et al., 2019; Dichtel et al., 2019) (Fig. 2.5.8).

Women on the combined pill had higher total cortisol and CBG but lower free cortisol than nonpill users (Dichtel et al., 2019); patients with cirrhosis had lower albumin and CBG giving lower total cortisol and high free cortisol. The authors are cautious about interpretation of cortisol in sick patients because of complex cortisol dynamics. Progesterone will also bind to CBG, and in the

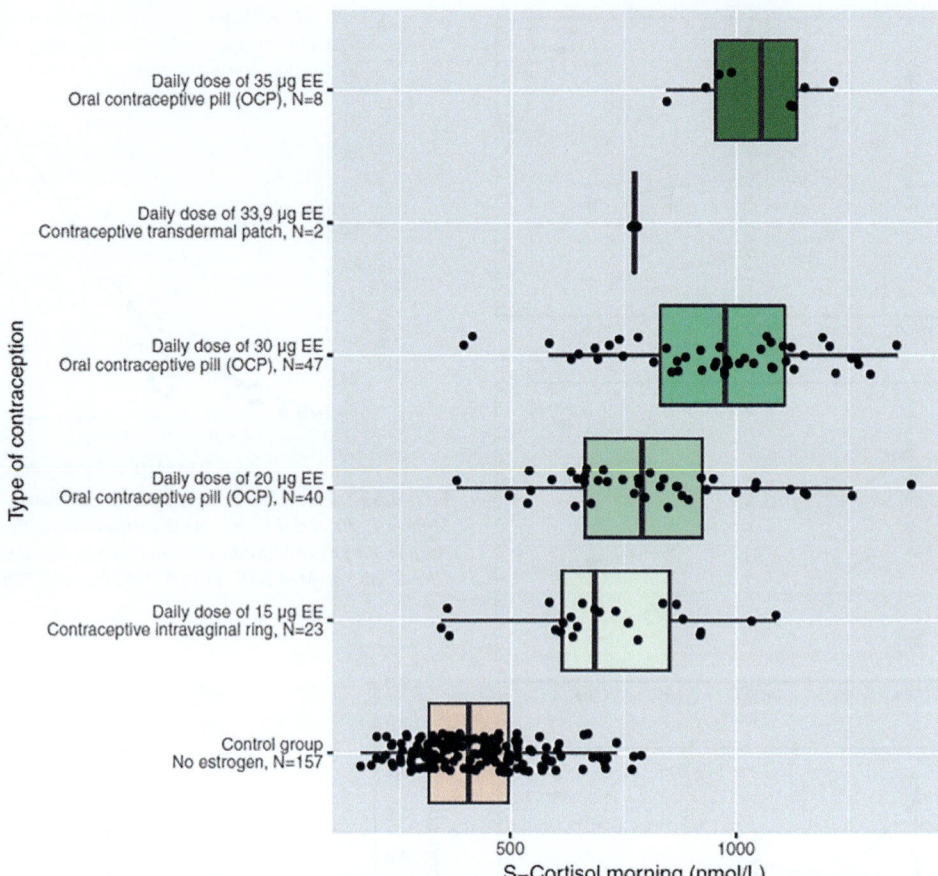

FIG. 2.5.8 Serum cortisol in women taking oral contraceptives. Results shown by subgroup. The group of women using an estrogen containing contraception was divided into subgroups depending on the daily dose of ethinyl estradiol (EE) in the contraception. In the subgroups of women using an oral contraceptive pill or a contraceptive intravaginal ring, the median value of s-cortisol increase with increasing daily dose of EE. *N*, number of women in the subgroup; *S-Cortisol*, serum concentration of cortisol. *(From Panton KK, Mikkelsen G, Irgens WØ, Hovde AK, Killingmo MW, Øien MA, Thorsby PM, Åsberg A. New reference intervals for cortisol, cortisol binding globulin and free cortisol index in women using ethinyl estradiol. Scand J Clin Lab Invest 2019;79(5):314–9. Fig. 1 p. 316.)*

middle and late stages of pregnancy when progesterone is very high and CBG concentration is increased through the high estrogen concentrations, the cortisol binding (in computer modeling) is reduced (Dunn et al., 1981) (Fig. 2.5.9). Changes in CBG concentrations would thus make difficult the interpretation of free cortisol (Hammond, 2016).

Factors affecting CBG concentration are summarized in Table 2.5.2. There are a number of diseases that alter CBG particularly through loss of protein through the kidney in conditions such as nephrotic syndrome. Drugs can also alter CBG concentration particularly the estrogen component of the oral contraceptive pill so total cortisol concentration is higher than might be expected. In patients receiving intravenous gamma-globulin therapy, there is an increased amount of immunoglobulin in the blood which affects the reaction matrix for assays.

Calculations of free cortisol had high bias and imprecision when judged against total cortisol (Molenaar et al.,

2015), whereas measurement on selected samples by equilibrium dialysis had shown the test was effective (Kirchhoff et al., 2011). More research is needed on the value of free cortisol determination.

2.5.3.2 Precursor steroids

Measurement of adrenal hormones, other than cortisol, may be necessary in patients with signs of adrenal insufficiency, hyperandrogenization and possible gonadotrophin-independent precocious puberty (see Clinical chapters in Part 3). Adrenal causes of androgen excess include androgen-secreting tumors and enzymatic defects in the adrenal biosynthesis pathway. Some patients may present with discordant development such as the presence of earlier appearance of pubic hair without testicular enlargement in boys and virilization in girls without breast development. In addition to basal plasma cortisol, other baseline steroid

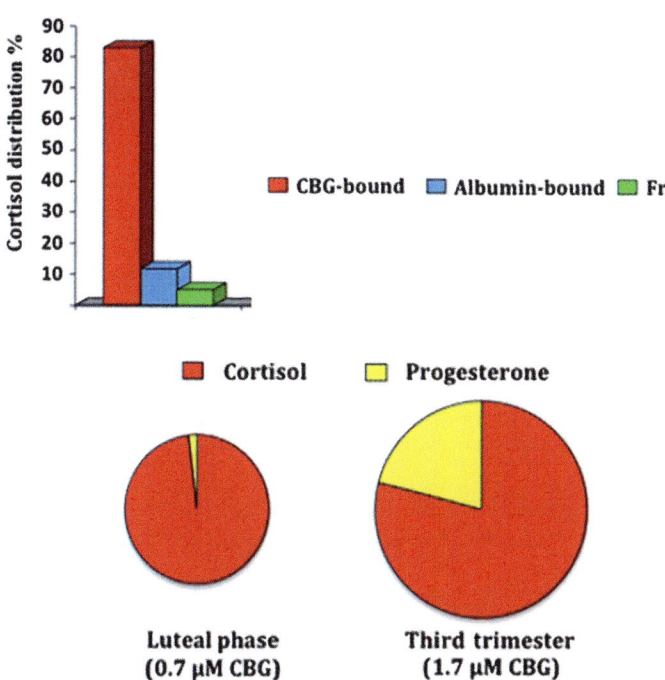

FIG. 2.5.9 Influence of CBG on the plasma distribution of cortisol. (B) Estimated proportional occupancy of plasma CBG by its major ligands, cortisol and progesterone, in blood samples taken from women before and during pregnancy, and the intervillous compartment of the placenta at term. (A) The plasma distribution of cortisol in individuals with normal CBG. (B) Proportional occupancy of CBG in serum from women during the luteal phase of the menstrual cycle vs the third trimester of pregnancy as estimated computationally from data of serum CBG, cortisol and progesterone levels. *(From Hammond GL. Plasma steroid-binding proteins: primary gatekeepers of steroid hormone action. J Endocrinol 2016;230(1):R13–25. Fig. 1 p. R815.)*

TABLE 2.5.2 Factors affecting cortisol binding globulin.

Increase	Decrease
Estrogens	Cirrhosis
Chronic active hepatitis	Obesity
Anticonvulsants	Hypothyroidism
Pregnancy	Nephrosis and other protein losing states
Diabetes mellitus	Multiple myeloma

concentrations may include testosterone, androstenedione, and dihydrotestosterone (DHT) as well as 17-OHP, and 11-deoxycortisol (11-DOC). It is now possible to measure several steroids in the same run of the LC-MS/MS so panels and profiles of steroids can be quantified simultaneously. Patients with equivocal results require provocative tests (see Chapter 2.7 Endocrine tests) and/or a urine steroid profile or genetic tests to confirm the underlying diagnosis.

2.5.3.2.1 17α-Hydroxyprogesterone (17-OHP)

Samples of blood for 17α-hydroxyprogesterone (17-OHP) determination are in the main only useful when a genetic defect of cortisol production is questioned. 21-Hydroxylase deficiency is the commonest disorder leading to congenital adrenal hyperplasia (CAH). 17-OHP is the substrate for the

enzyme. In newborn screening programmes, reference ranges for the assay method are needed for 17-OHP in **dried blood spots (dbs)** (by age after birth, gestation age and birth weight) (Table 2.5.3 has typical results) (Hayashi et al., 2017).

Later, around puberty in girls, once menses start, 17-OHP shows a similar cyclical pattern to progesterone with levels in plasma of 1–3 nmol/L through most of the cycle and peak concentrations less than 15 nmol/L in the luteal phase (Ballerini et al., 2014). A late onset-form of 21-hydroxylase deficiency exists due to mutations that lead to a protein with reduced activity so normal reference ranges are needed for 17-OHP to account for age, gender, and menstrual stage (Turcu et al., 2020a). Some clinicians monitor CAH treatment, by following the degree of 17-OHP suppression through treatment with glucocorticoids, by the use of dbs 17-OHP collected at intervals throughout the day (see Part III chapter on adrenal insufficiency). Monitoring the outcome of the treatment can be dangerous because the highs and lows of the treatment steroid give information on risks of over treatment and under treatment and the clinical consequences with obesity, cardiovascular risk, adrenal insufficiency and mortality (Greaves et al., 2018).

2.5.3.2.2 11-Deoxycortisol

A request for analysis of 11-deoxycortisol will be rare, often only in investigations of causes of hypertension, adrenal tumors or when excluding CAH due to 11-hydroxylase deficiency. Normal concentrations are typically less than

TABLE 2.5.3 Neonatal 17-hydroxyprogesterone (nmol/L) concentrations by Auto DELFIA according to birth weight (BW) and dried blood spot sample collection age.

	<72 h			>72 h			
Weight (g)	Median (IQ)	Range	Cutoffs 99.5th/99.8th	Median (IQ)	Range	Cutoffs 99.5th/99.8th	P
<1500	38.7 (18.9–79.8)	1.5–155.5	330/360	40.2 (23.1–75.3)	0.7–200.2	441/519	<0.001
1501–2000	24.6 (15.9–40.2)	0.8–122	168/213	26.7 (16.8–44.4)	0.4–262.6	207/270	<0.001
2001–2500	20.1 (14.4–29.5)	0.8–125.3	96/117	20.7 (13.5–33.3)	0.4–138.7	144/198	<0.001
>2500	15.6 (11.7–21)	0.5–169	51/60	11.7 (8.1–16.5)	0.3–198.0	60/75	<0.001

From Hayashi GY, Carvalho DF, de Miranda MC, Faure C, Vallejos C, Brito VN, et al. Neonatal 17-hydroxyprogesterone levels adjusted according to age at sample collection and birthweight improve the efficacy of congenital adrenal hyperplasia newborn screening. Clin Endocrinol (Oxf) 2017;86(4):480–7.

12 nmol/L. 11-Deoxycortisol will be measured when increased in a metyrapone test of the HPA axis which inhibits the adrenal 11-hydroxylase enzyme (see Endocrine test, Chapter 2.6), lowers cortisol and increases ACTH which in turn stimulates the adrenal cortex to increase 11-deoxycortisol output. Recently 11-deoxycortisol has been found from studies of the urine steroid metabolome to be elevated in patients with adrenal carcinoma and 11-deoxycortisol is part of many serum steroid profiles for that purpose (Schweitzer et al., 2019; Taylor et al., 2017).

2.5.3.3 Adrenal androgens

2.5.3.3.1 Dehydroepiandrosterone (DHEA) and DHEA-sulfate (DHEAS)

The adrenal glands of the fetus secrete increasing amounts of DHEAS during pregnancy and this capacity continues into the first year of life. The fetal zone involutes after birth over the first 6 months of life. DHEAS is then virtually undetectable until 5–9 years of age when a new zone of cells appears at the boundary of the adrenal cortex with the medulla. This *zona reticularis* has low activity of 3β-hydroxysteroid dehydrogenase and DHEAS is the principal product. This period of life is called the **adrenarche**. In seniors years, there is a decline in DHEAS production which is called the **adrenopause**. The concentrations of DHEAS with age are shown in Fig. 2.5.10 (Bae et al., 2019).

Many studies continue to measure DHEAS by immunoassay. The high concentrations of DHEAS relative to other steroids make interference minimal. LC-MS/MS methods are used usually within a panel or profile of steroids (Campi et al., 2018; Büttler et al., 2015a,b; Kushnir et al., 2010).

In children less than 6 years of age, plasma DHEAS concentrations are in the range 0.2–0.4 μmol/L. The adrenal cortex then grows and develops a *zona reticularis*, outside the adrenal medulla, which secretes androgens. DHEAS is

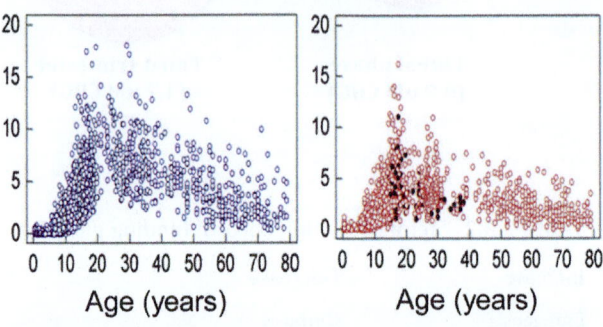

DHEAS (μmol/L)

FIG. 2.5.10 DHEAS concentrations change throughout the life-time in males and females. *(From Bae YJ, Zeidler R, Baber R, Vogel M, Wirkner K, Loeffler M, Ceglarek U, Kiess W, Körner A, Thiery J, Kratzsch J. Reference intervals of nine steroid hormones over the life-span analyzed by LC-MS/MS: effect of age, gender, puberty, and oral contraceptives. J Steroid Biochem Mol Biol 2019;193:105409. Fig. 1 p. 3.)*

thus produced in increasing amounts between age 7 and 15 years and leads to an increase in the circulating levels. At the adrenarche, DHEAS has androgenic activity and is responsible for appearance of pubic hair and sweat gland activity (Goodarzi et al., 2015). Peripheral formation of testosterone, androstenedione and DHT are also involved. Plasma DHEAS concentrations respond to a lesser extent than cortisol to changes in ACTH production with lower levels during the day and a nadir around midnight. Adults have DHEAS concentrations around 2–12 μmol/L during the day. Plasma concentrations increase to a peak at the end of puberty then slowly decline to around 70 years. DHEAS is usually measured in plasma since it is a marker for adrenal tumors and part of the differential diagnosis of hyperandrogenism.

2.5.3.3.2 Androstenedione

Androstenedione is mainly an adrenal androgen though some can be formed from T, DHEA and DHEAS in the

Androstenedione (nmol/L)

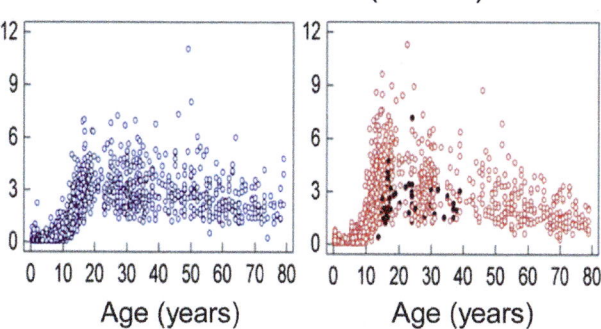

FIG. 2.5.11 Androstenedione concentrations change throughout the life-time in males and females. *(From Bae YJ, Zeidler R, Baber R, Vogel M, Wirkner K, Loeffler M, Ceglarek U, Kiess W, Körner A, Thiery J, Kratzsch J. Reference intervals of nine steroid hormones over the life-span analyzed by LC-MS/MS: effect of age, gender, puberty, and oral contraceptives. J Steroid Biochem Mol Biol 2019;193:105409. Fig. 1 P.)*

Serum androgens (nmol/L)

T	0.3 (0.2–0.5)
A4	5.9 (3.3–9.2)
DHEA	7.1 (4.2–11.8)
DHEAS (μmol/L)	6.0 (3.4–9.6)
FAI	0.6 (0.3–0.9)
11OHA4	6.8 (4.9–12.5)
11KA4	2.7 (2.0–3.9)
11OHT	0.2 (0.1–0.3)
11KT	1.5 (1.2–1.8)

Controls

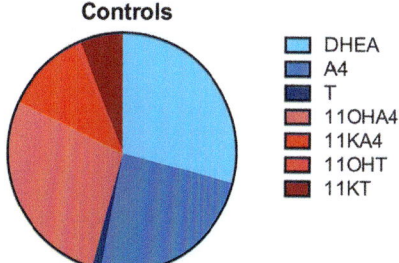

- DHEA
- A4
- T
- 11OHA4
- 11KA4
- 11OHT
- 11KT

FIG. 2.5.12 The 11-oxygenated C19 steroids (11OHA4, 11KA4, 11OHT, 11KT; shades of *red*) contributes to the total circulating androgenic steroid pool healthy female control subjects (*n* = 49). Baseline serum androgens in normal females and relative contribution (median; %) of the classic androgen pathway (DHEA, A4, T; *shades of blue*) and11-oxygenated C19 steroid pathway. *(From O'Reilly MW, Kempegowda P, Jenkinson C, Taylor AE, Quanson JL, Storbeck KH, Arlt W. 11-Oxygenated C19 steroids are the predominant androgens in polycystic ovary syndrome. J Clin Endocrinol Metab 2017;102(3):840–8. Fig. 3 p. 844.)*

periphery. There are several reports of assay interferences with immunoassays and mass spectrometric measurements are usually part of a panel of androgens or a steroid profile. Spironolactone interference has been reported several times in adult females being treated for hyperandrogenism (Honour et al., 2010). Plasma androstenedione concentrations as measured by LC-MS/MS, can be below 1 nmol/L during most childhood except for a period up to 3 months after birth when levels up to 3 nmol/L are seen in line with raised testosterone during this minipuberty (see Section 2.5.5.1 Testosterone). The concentrations of androstenedione begin to rise during late childhood, before the rise in testosterone, indicating adrenal origin due to adrenarche, reaching highest values of up to 8 nmol/L at the age of 13–15 years (Bae et al., 2019) (Fig. 2.5.11). There is a diurnal variation of androstenedione in the adult which is synchronous with cortisol. In girls, androstenedione concentrations rise during the menstrual cycle up to 2–10 nmol/L at the time of ovulation.

2.5.3.3.3 11-Oxygenated androgens

In the past decade, there has been excitement over the relevance of 11-oxygenated androgens. Like the investigations of neurosteroids, LC-MS/MS methods have been developed to quantify the steroids (Storbeck et al., 2019; du Toit et al., 2018; O'Reilly et al., 2017; Rege et al., 2013) and proof for the action of enzymes and the steroid products has been sought. The 11-oxygenated androgen concentrations in serum increase with age in children to adult concentrations (Fig. 2.5.12) (O'Reilly et al., 2017) in parallel with the rise in adrenal androgens. A diurnal rhythm is observed in measurements over 24 h, with lowest concentrations at midnight.

Girl's age 4–7 years with premature adrenarche have adrenal androgen concentrations of girls aged 9–10 year of age. The plasma concentrations of 11-oxygenated

androgens decline at the menopause (Fig. 2.5.13) (Nanba et al., 2019; Turcu et al., 2020b).

The functions of 11-oxyandrogens in human beings in normal physiology and pathologic states remain poorly understood. There is sufficient evidence to identify 11OHA4 as a major product of the adrenal gland and 11KT as the dominant active androgen metabolite of 11OHA4. The contributions of 11-oxyandrogens to androgen excess disorders in women and children awaits further definition

2.5.3.3.4 7-Hydroxy androgens

Some 7-hydroxylated metabolites of dehydroepiandrosterone have been investigated as possible neuroprotective and immunomodulatory steroids. Measurements of this group of steroids are nearly all based on gas liquid chromatography and mass spectrometry by the group of Martin Hill in Prague. Serum concentrations are less than 2 nmol/L so a sensitive method is essential and in some cases derivatives like 2-hydrazinopyridine have been used. Concentrations of 7α hydroxy DHEA and 7β-hydroxy DHEA are less than 5 nmol/L and less than 2.5 nmol/L, respectively, by GC-MS of TMS derivatives (Fig. 2.5.14) (Hill et al., 2005). For LC-MS/MS, the steroids were reacted with 1-amino-4-methyl piperazine (Ke et al., 2016).

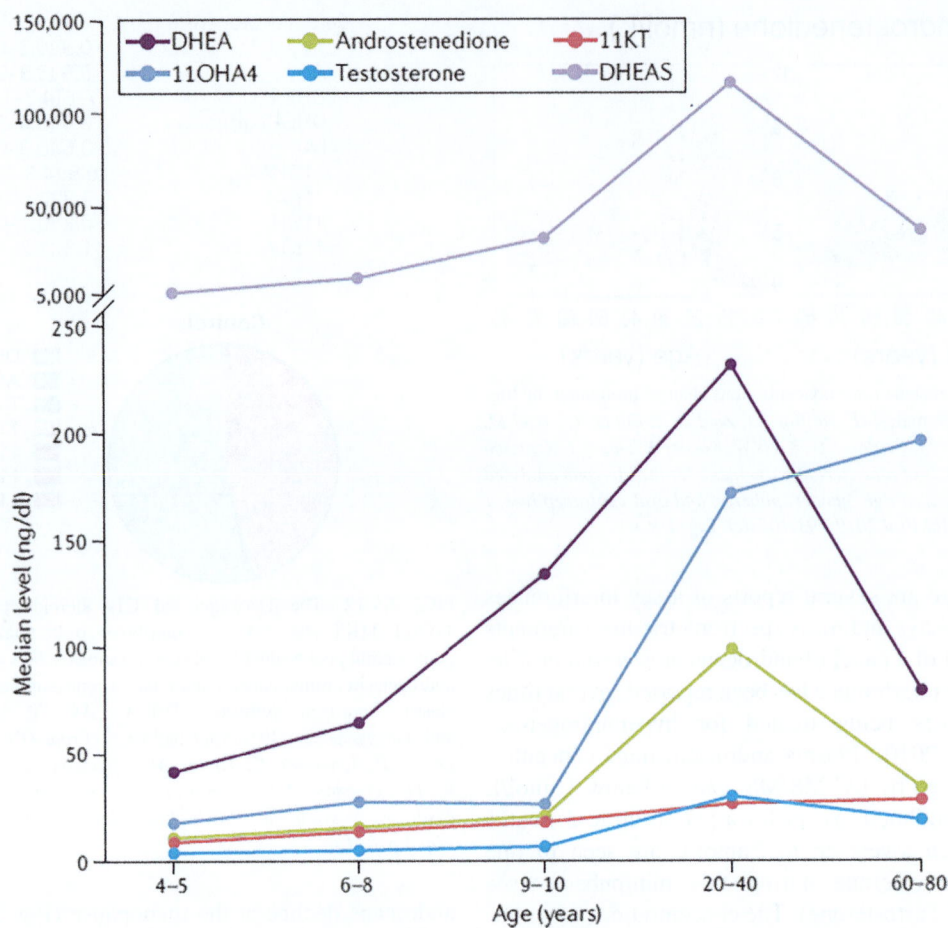

FIG. 2.5.13 Distribution of key C19 steroids across the human female lifespan. Concentrations of 11β-hydroxyandrostenedione (11OHA4) and 11-ketotestosterone (11KT) increase with age and remain elevated in postmenopausal women. Data are expressed in medians. Total $n = 283$ girls and women (aged 4–5 years, $n = 22$; aged 6–8 years, $n = 38$; aged 9–10 years, $n = 23$; aged 20–40 years, $n = 100$; aged 60–80 years, $n = 100$). *(From Turcu AF, Rege J, Auchus RJ, Rainey WE. 11-Oxygenated androgens in health and disease. Nat Rev Endocrinol 2020b;16(5):284-296. Fig. 3 p. 289.)*

The interconversion was from secreted 7α-hydroxy steroid through a 7-oxo derivative. For each of 7α-hydroxy and 7-oxo-DHEA, a single equivalent 3α-deuterium labeled steroid was developed as an internal standard for their endogenous steroid determinations. There is interest in the steroids in psychiatric disorders through effects of the steroids on GABA receptors.

2.5.3.3.5 Androgen conjugates

Several C$_{19}$ steroid conjugates of 5α-reduced steroids are found in serum and have been used as markers of androgen excess. The most used has been 5α-androstane-3α,17β-diol-glucuronide (ADG) although androsterone glucuronide (ADT-G), androsterone sulfate (ADT-S) and AD-S have also been reported. Measurements in peripheral blood, ovarian vein and adrenal vein blood have by deduction concluded that peripheral metabolism is the principal site of formation. Most of the literature derives from immunoassays but new data is appearing with LC-MS/MS methodology

(Kalogera et al., 2013). Mean concentration in normal females are ADG-17G 1.17 ng/mL, ADG-3G 0.76 ng/mL, ADT-G 60.3, and DHT-S 0.40 ng/mL which are comparable with testosterone and DHT mean concentrations 0.26 and 0.13 ng/mL (Trabert et al., 2016). Reference ranges for androstanediol glucuronide by age and gender (Fabregat-Cabello et al., 2019) were similar and agreed with results by Endocrine Sciences; males 112–1046 ng/dL and females 11–249 ng/dL.

2.5.3.3.6 Dehydroepiandrosterone fatty acyl esters

Long chain fatty acids can be conjugated with steroids by reaction catalyzed by lecithin-cholesterol acyltransferase. DHEA-fatty acyl esters (DHEA-FAE) are transported in blood by high density lipoprotein (HDL) and transferred into peripheral cells after binding to lipoprotein receptors. DHEA can be released by hydrolysis to serve as precursor for formation of androst-5-ene-3β,17β-diol,

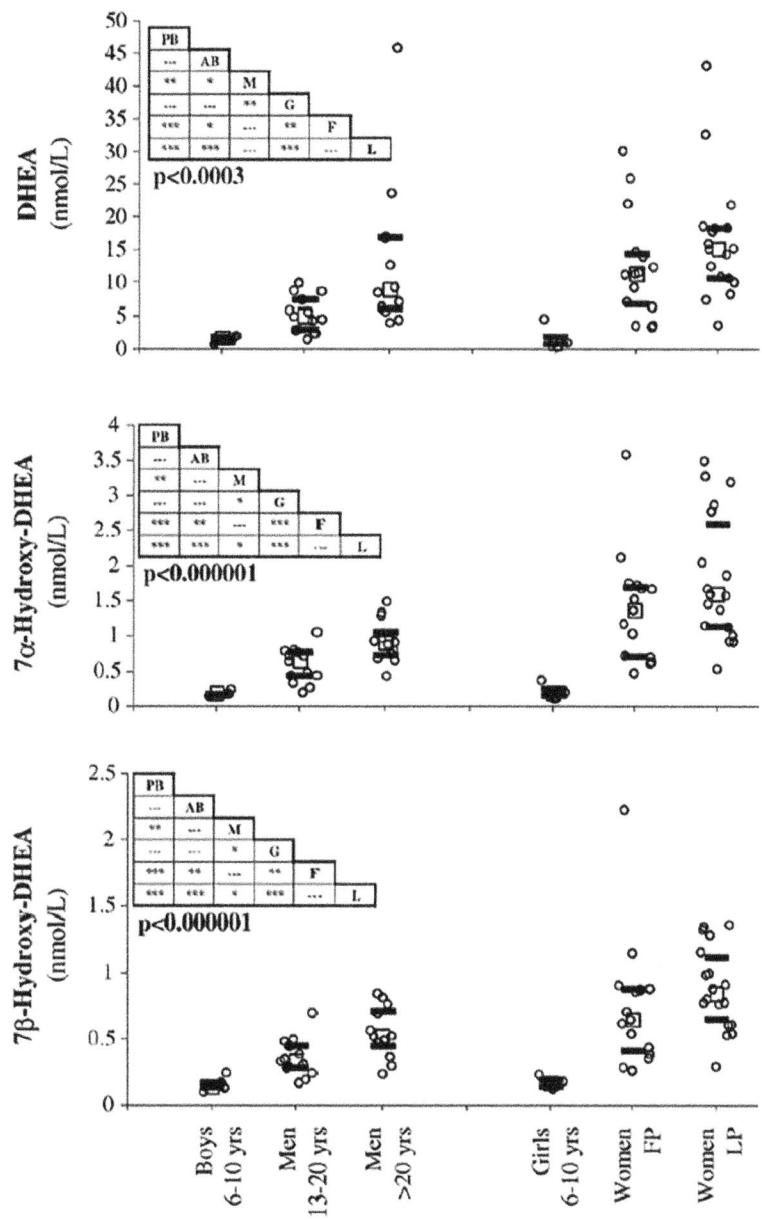

FIG. 2.5.14 Serum levels of DHEAS and its 7-hydroxy-metabolites as evaluated by one-way Kruskall-Wallis ANOVA followed by Kruskall-Wallis multiple comparisons; the *P-value at the insert denotes the significance* of Kruskall-Wallis ANOVA; the table insets represents individual multiple comparisons: *(*) P < .05, (**) P < .01, (***) P < .001, (—) not significant*; the *empty circles* represent the individual subjects, while the *empty squares* with *bold bars* denote the group medians with quartiles; on the *x-axis: FP*, follicular phase of the menstrual cycle; *LP*, luteal phase of the menstrual cycle; *PB*, prepubertal boys 6–10 years; *AB*, adolescent boys 13–20 years; *M*, men older than 20 years; *G*, prepubertal girls 6–10 years; *F*, women in the follicular phase; *F*, women in the luteal phase. *(From Hill M, Havlíková H, Vrbíková J, Kancheva R, Kancheva L, Pouzar V, Cerný I, Stárka L. The identification and simultaneous quantification of 7-hydroxylated metabolites of pregnenolone, dehydroepiandrosterone, 3beta,17beta-androstenediol, and testosterone in human serum using gas chromatography-mass spectrometry. J Steroid Biochem Mol Biol 2005;96(2):187–200. Fig. 7 p. 193.)*

5α-androstanedione and androstenedione particularly in adipose tissues. Mean serum concentrations of DHEA-FAE are 1.4 pmol/L which is around 10% of total DHEA in serum which contrasts with estradiol-FAE that is around 80% of total estradiol in serum. DHEA-FAE is undetectable in adipose tissue, whereas Oe-FAE is 80% of total estradiol (Wang et al., 2011).

2.5.3.4 Adrenocorticotrophin

ACTH is the drive for cortisol production by zona fasciculata cells. ACTH is a 39 amino acid protein produced from processing of pro-opiomelanocortin (POMC, 266 amino acids) (Fig. 2.5.15). In the anterior pituitary gland, proprotein convertase 1 (PC1) cleaves the POMC to an N-terminal fragment (N-POC) and β-lipotrophin (β-LPH).

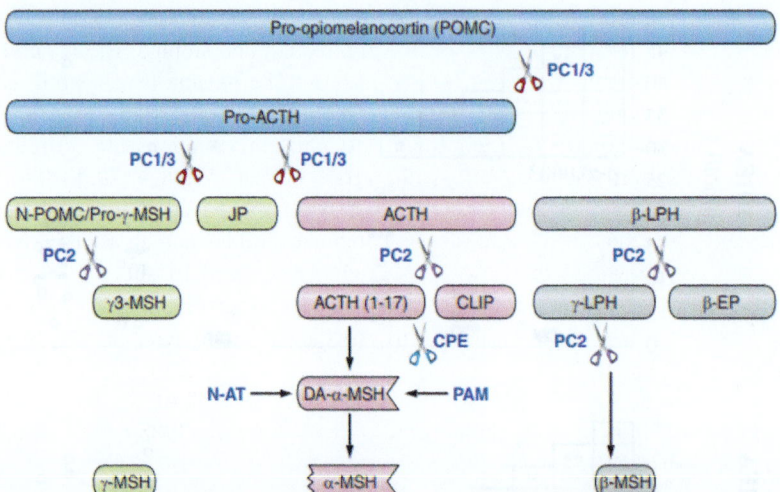

FIG. 2.5.15 Processing of human pro-opiomelanocortin (POMC) in humans. Pro-hormone convertase 1/3 (PC1/3) sequentially cleaves POMC to pro-ACTH then to adrenocorticotropic hormone (ACTH). In hypothalamus, skin, and pars intermedia of the pituitary, ACTH is further cleaved by PC2 to produce ACTH (1–17) and corticotropin-like intermediate peptide (CLIP). Carboxypeptidase E (CPE) then cleaves basic amino acid residues from the COOH terminal, allowing amidation by peptidyl-glycine α-amidating monooxygenase (PAM) to form des-acetyl α-MSH (DA-α-MSH). N-acetyltransferase (N-AT) finally acetylates DA-α-MSH to produce α-MSH. PC2 cleaves β-lipotropic hormone (β-LPH) to β-endorphin (β-EP) and γ-LPH, which is further cleaved to β-MSH. The NH2-terminal peptide N-POMC has dibasic amino acids at the NH2 terminal of γ-MSH which are thought to be cleaved by PC2. *(Redrawn from Harno E, Gali Ramamoorthy T, Coll AP, White A. POMC: the physiological power of hormone processing. Physiol Rev 2018;98(4):2381–430. Fig 1 P 2383.)*

In the neuro-intermediate lobe, PC1 also acts on N-POC to cleave out pro-γ-melanocyte stimulating hormone (γ-MSH), adrenocorticotropin (ACTH) and a joining peptide (JP). γ-Lipotrophin and β-endorphin are products of the action of PC2 on β-LPH. ACTH can also be cleaved to α-MSH and corticotrophin like intermediate lobe peptide (CLIP). Immunometric assays for ACTH with monoclonal antibodies to the N and C terminals can detect ACTH to 0.5 ng/mL. Assay specificity may be a problem in patients with ectopic sources of ACTH because POMC processing can be affected and numerous POMC fragments may cross react in the ACTH assay.

Measurement of plasma ACTH has evolved from bioassays on guinea pig or rat adrenal cells through the milestone introduction of RIA for radioactive and later nonisotopic forms to immunometric assays. The use of ACTH IAs has enabled considerable advances to be made in understanding the physiology of ACTH secretion in certain diseases. The early assays had serious limitations in terms of sensitivity and specificity. Specificity is antibody-dependent, with differing results obtained from N- and C-terminal ACTH-directed antisera used singly or in pairs (sandwich assay) (Dobson et al., 1986). Good assay specificity is also vital due to the presence of the circulating ACTH precursor molecules and fragments devoid of ACTH bioactivity.

An immunometric assay relies on two antibodies binding two different epitopes of the ACTH peptide. In general, antibodies have been raised to N and C-terminal regions of ACTH in order to ensure that ACTH 1–39 is detected rather than fragments derived from ACTH, such as αMSH and CLIP. Assays have been developed which use enzyme or europium labels. IRMAs have overcome the need for iodinating ACTH and, because they employ excess antibody (excess reagent or noncompetitive assay), they overcome the problem of requiring a high-affinity antibody as is needed for RIA.

Comparisons of the performance of new ACTH assays against existing assays have usually revealed good correlation and comparable clinical sensitivity, although widely different results or limited agreement between two assays have in practice also been observed (Talbot et al., 2003) and there are few publications in recent years, several of the assays may still be similar in nature and performance. Reference range (2.5–97.5 percentiles) for the Elecsys ACTH immunoassay are 5.6–76 ng/L. EQA data shows a marked variability in assay performance with both constant and proportional bias among assays tested from major manufacturers. Most assays show adequate repeatability, i.e., intra-assay CV <15%, but assay performance in routine conditions have been consistently less than precision profiles reported in assay technical sheets. Some differences have been seen in IQA depending on the material in use with Roche material preferred to BioRad (Wu and Xu, 2019). The differences in ACTH assay agreement are best summarized by Bland-Altman statistics, which shows a good mean agreement with the expected value although there can be an unacceptably large range of agreement for each measurement. From a clinical viewpoint, this translates into an acceptable agreement in the assignment of results obtained in patients with either normal or high ACTH values, while large excursions among measurements do not necessarily alter their clinical significance, but can cause frequent result misclassification among patients with low ACTH levels. Assay performance is a major concern for ACTH assay kits and, in some cases, has been the reason for assay withdrawal. Absolute ACTH values are commonly used for the distinction of ACTH-independent Cushing's syndrome and ACTH-dependent Cushing's disease, for the differential diagnosis of adrenal insufficiency and to assess the effects of treatment or during follow-up of patients with adrenal or pituitary disorders. In these contexts, interassay and interlaboratory agreement among ACTH measurements becomes mandatory.

There can be several explanations for assay variability, starting with the different ACTH biological standards used in the assays. Not all ACTH (1–39) formulations are equally recognized by assay antibodies. The different potency of ACTH biological standards cannot be rectified without an International Standard for human ACTH. The only International Standard for corticotropin is derived from pig pituitaries, and assays use synthetic or pituitary-derived purified preparations.

ACTH absorbs to unsiliconized glass tubes and, some say, degrades rapidly at room temperature; thus, the use of EDTA-coated plastic tubes, rapid sample processing, and cool storage prior to assay is mandatory (Toprak et al., 2016). These factors have additional, confounding roles in everyday field ACTH measurements. Diagnostic flowcharts commonly exclude adrenal Cushing's syndrome (autonomous or exogenous cortisol) if ACTH values are above 1–2 or 5–10 pg/mL. Many ACTH results however fall in the normal range in patients in whom ACTH secretion should be suppressed. This raises the issue of whether this pattern is due to technical problems or whether the pathophysiology of glucocorticoid negative feedback and secondary adrenal insufficiency should be revisited. ACTH results can be low with sandwich assays using streptavidin-biotin assays if the patient is taking biotin to harden finger nails or treat multiple sclerosis. Elevated ACTH results on Siemens Immulite (Greene et al., 2019) led to unnecessary testing and potentially harmful invasive procedures, including pituitary surgery. Because the ACTH (Immulite) assay is used by most clinical laboratories, this presents a substantial problem for the endocrine community. The Siemens assay is sensitive to heterophilic antibody interference. It is imperative that physicians know which ACTH assay is in use in their practice. An alternate ACTH assay such as the ACTH (Roche Cobas or Tosoh AIA) should be used in the diagnosis and differential diagnosis of patients with suspected disorders of pituitary adrenal function.

There used to be a debate on the best methods for sample collection and storage. Blood should be centrifuged within 4 h and the plasma frozen until assay (Wu and Xu, 2017; Christensen et al., 2016). In normal subjects, mean plasma ACTH concentrations are 30 ng/L at 0730 and 15 ng/L at 1630 at rest (White et al., 1987). The Immulite assay is subject to interference due to heterophilic antibodies in patient samples and an alternative assay (Roche or Tosoh) should be used to check for more reliable ACTH measurement (Greene et al., 2019). Samples in EDTA are stable for 6 h at room temperature (Nandakumar et al., 2020; Donegan et al., 2019; Chakera et al., 2017).

Precursors of ACTH have been determined by IRMA (Crosby et al., 1988) but these assays are not widely available.

2.5.3.5 Corticotrophin releasing hormone (CRH)

The hypothalamus secretes CRH in response to low plasma cortisol concentrations. CRH is a 41 amino acid peptide circulating at low concentrations in plasma (<50 pmol/L) and, in general practice, assays for CRH are not available. The specificity of the assays has been problematic because CRH is processed from a large precursor molecule with many fragments that can cross react in the assay of CRH. The normal range for plasma CRF is 20–110 ng/L (Cunnah et al., 1987). In pregnancy, plasma concentrations of CRH increase to 200–300 pmol/L because CRH is secreted by the placenta (Sanderson et al., 2000). The first assay for CRF was developed in 1983 (Vale et al., 1983) but a binding protein in plasma was found to affect assay performance (Linton et al., 1995). Most methods included chromatographic purification and were too tedious for routine use, so CRH concentrations are rarely measured.

2.5.4 Aldosterone and renin

Plasma aldosterone concentrations are influenced by many factors including age, sodium balance, body posture and activity. Renin release from the kidney is normally stimulated on standing and by volume changes of the vascular compartment and by sodium depletion (Fig. 2.5.16).

In the first weeks of life, normal aldosterone concentrations may be up to 5000 pmol/L because PRA in the normal newborn infant can be up to 50 pmol/mL/h. PRA and aldosterone then decline over the first 18 months of life and from then on are close to adult ranges. Reference ranges for PRA and aldosterone with age using IA are shown in Fig. 2.5.17 (Fiselier et al., 1983).

Most of the literature for aldosterone is around immunoassay, particularly with now some chemiluminescent methods (Deng et al., 2018; Dorrian et al., 2010) but GC and LC methods with and without mass spectrometry are being used (van der Gugten and Holmes, 2016a,b;

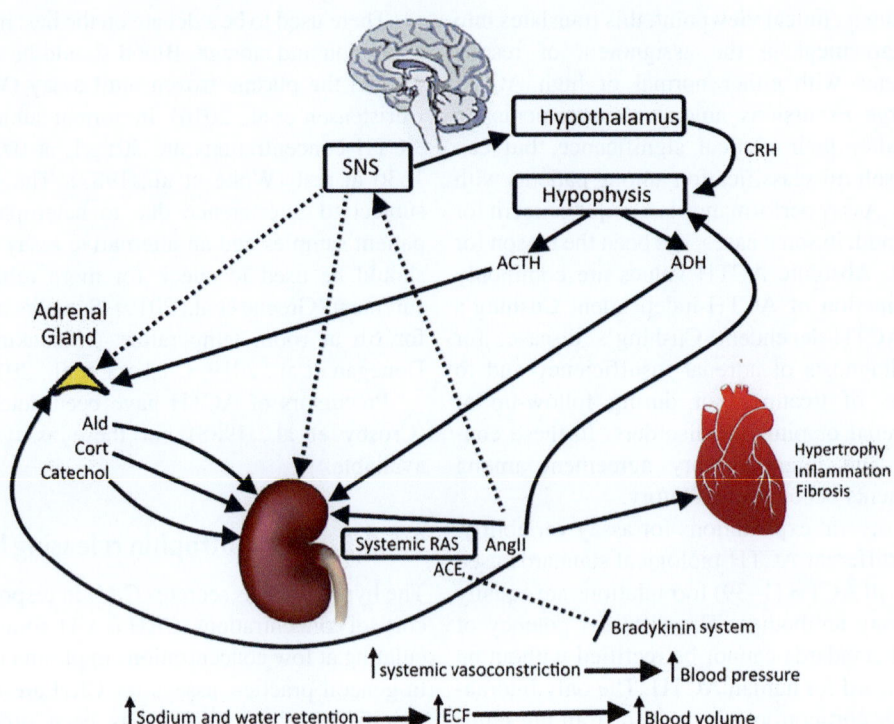

FIG. 2.5.16 Renin-angiotensin system, adrenal gland, sympathetic nervous system (SNS), hypothalamus and hypophysis. The image illustrates the positive interactions among systemic renin angiotensin system and vital organs. There are positive feedbacks and negatives (not displayed) to maintain homeostasis, body water and osmolarity in equilibrium. *Thick arrows* indicate main regulator. *ACE*, angiotensin-converting enzyme; *ACTH*, adrenocorticotropic hormone; *ADH*, antidiuretic hormone; *Ald*, aldosterone; *AngII*, angiotensin II; *Catechol*, catecholamines; *Cort*, cortisol; *CRH*, corticotrophin-releasing hormone; *ECF*, extracellular fluid; *systemic RAS*, systemic renin angiotensin system. *(From Vargas Vargas RA, Varela Millán JM, Fajardo Bonilla E. Renin-angiotensin system: basic and clinical aspects-a general perspective. Endocrinol Diabetes Nutr 2022 ;69(1):52–62. Fig. 2 p. 56.)*

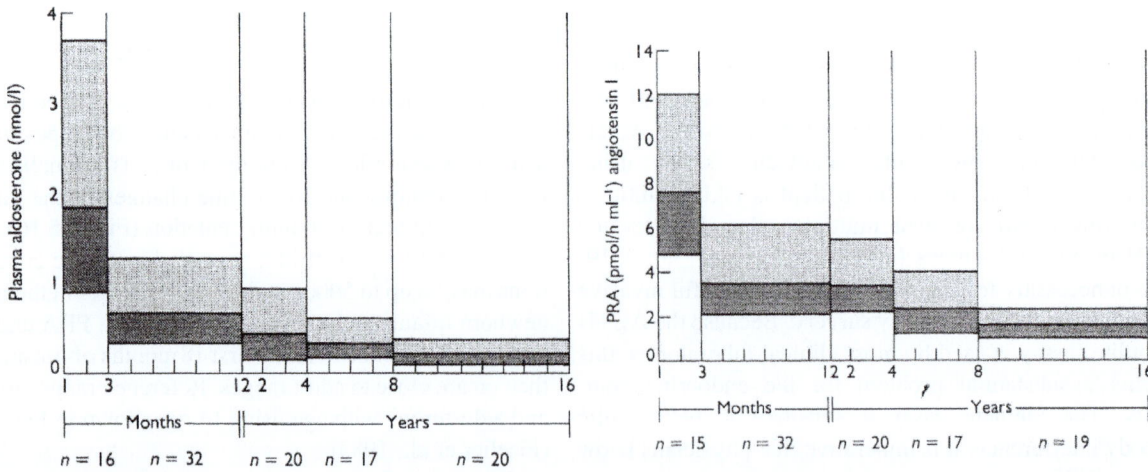

FIG. 2.5.17 Plasma aldosterone and renin activity in different age groups of children. Model median and standard deviation are shown. Conversion traditional to SI units: aldosterone 0.36 ng = 1 pmol; angiotensin I 1297 ng = 1 nmol. *(From Fiselier TJ, Lijnen P, Monnens L, van Munster P, Jansen M, Peer P. Levels of renin, angiotensin I and II, angiotensin-converting enzyme and aldosterone in infancy and childhood. Eur J Pediatr 1983;141(1): 3–7. Fig. 1 and 2 p. 4.)*

TABLE 2.5.4 Effects of drugs on aldosterone/renin ratios.

Drug	Aldosterone	Renin	Effect on ARR
Calcium channel blocker (Nifedipine, Amlodipine)	Minimal decrease	Minimal increase	No effect
Calcium channel blocker (Verapamil)	Small decrease	Minimal increase	No major effect
α-Blocker (Prazosin, doxazosin)	Minimal	Minimal	No effect
Vasodilators (Diazoxide, hydralazine)	No effect	No effect	No effect
ACE inhibitors (Captopril)	Decreased	Increased	False negative
Diuretics (Furosemide—loop, hydrochlorothiazide—distal tubule)	Increased	Increased markedly	False negative
Minoxidil—vasodilator/diuretic (baldness)	Minimal	Increased	False negative
All receptor blocker (Sartans)	Decreased	Increased	False negative
Spironolactone (K sparing diuretic)	Increase	Large increase	False negative
Beta-blockers (Metoprololol, atenolol)	Minimal	Decreased	False positive
Methyl-DOPA	Minimal decrease	Decreased	False positive
Direct renin inhibitor (aliskiren)	Decreased	Decreased	False positive

Hypokalemia—inhibits aldosterone (false negative). Potassium loading (false positive). Estrogens (pregnancy, HRT, O/C) increase renin substrate (false negative). NSAID's retain sodium, reduce PRA (false positive). Renovascular and malignant hypertension, increase renin activity (false negative).
Author original.

Ray et al., 2014; Hinchliffe et al., 2013; Owen and Keevil, 2013). Aldosterone in solution can be in free aldehyde, hemiacetal and hemiketal forms which can form derivatives that can be separated in the chromatography stage. This effect has not been sought in recent methods.

In adults, when lying down for at least 30 min the serum aldosterone is 50–400 pmol/L. After standing for 4 h this usually increases to between 200 and 700 pmol/L. This response to posture is due to increased renin secretion through changed renal blood flow although in very young children, largely bed bound, this is not practical.

A commonly used immunoassay method for renin activity (Dia-Sorin) was withdrawn around 2015. Plasma renin concentration is determined by two-site immunoassay with chemiluminescence (Deng et al., 2018; Dorrian et al., 2010; Rossi et al., 2016; de Bruin et al., 2004). LC-MS/MS methods for renin activity have been in use since 2010 (Bystrom et al., 2010; Owen et al., 2014; van der Gugten and Holmes, 2016a,b). A method with MALDI does not require digestion of the peptide since Angiotensin I has molecular weight of 1295.7 Da (Li et al., 2017). Normal plasma renin activity when recumbent (0800 h, aged 19–40 years) is typically 1.14–2.65 nmol/h/L and this increases on standing to 2.82–4.49 nmol/h/L. Plasma renin concentration is typically 3.3–92.7 mIU/mL in males and 3.7–99.8 mIU/L in females.

Clinicians need to have in place procedures for collection of any samples. Firstly patients with hypertension can be taking a number of drugs to reduce blood pressure

and these can affect the ARR at different points and can affect plasma potassium concentrations which affect the final sites in aldosterone synthesis (Table 2.5.4). Ideally patients need to be off any of the standard treatments (spirolactone, β-adrenergic blockers, ACE inhibitors, angiotensin receptor blockers, renin inhibitors, and calcium channel antagonists) for at least 4 weeks. During that time, the patients should be encouraged to take salt liberally. Blood pressure can be controlled in this time with the calcium channel antagonist verapamil (90–120 mg twice daily), the vasodilator hydralazine (10 increasing to 12.5 mg twice daily), or alpha adrenergic blockers prazosin (0.5–1 mg twice daily), doxazocin (1–2 mg once daily) or terazosin (1–2 mg daily). This can be monitored in general practice. Evaluating the measurements of renin and aldosterone is complicated by a number of factors notably in preparation of the patient for the test. The withdrawal of certain antihypertensive drugs may not always be possible (O'Shea et al., 2016).The patients should not use any products containing licorice root that can inhibit cortisol inactivation to cortisone and lead to suppression of renin and hence aldosterone.

Plasma aldosterone increases during the menstrual cycle being highest in the luteal phase (day 21) and progressively falls from the luteal phase to the menses (Table 2.5.5) (Ahmed et al., 2015; Fommei et al., 2009). Oral contraceptives impact on measurements of aldosterone and renin through actions on binding proteins and angiotensinogen concentrations. Treatment with oral ethinyl estradiol plus

TABLE 2.5.5 Hormones, potassium and blood pressure values during the ovarian cycle.

	7th day	14th day	21st day	28th day	Friedman's test	Significant multiple comparisons (Bonferroni-adjusted Wilcoxon test)
FSH (mUI/mL)	7.2 (6.1–11.8)	7.5 (6.5–13.4)	4.3 (2.9–6.6)	7.0 (5.5–9.6)	0.002	7th vs 21st (=0.012)
						14th vs 21st (<0.001)
LH (mUI/mL)	5.0 (4.0–6.6)	8.4 (6.5–17.9)	3.6 (2.2–6.1)	4.5 (2.9–9.1)	0.002	14th vs 21st (<0.001)
E2 (pg/mL)	93 (43–132)	121 (80–204)	119 (81–187)	57 (38–103)	0.004	21st vs 28th (=0.024)
P (ng/mL)	0.7 (0.6–0.9)	1.2 (0.6–2.3)	7.8 (0.9–12.4)	1.3 (0.9–3.8)	<0.001	7th vs 21st (=0.006)
						7th vs 28th (<0.001)
						14th vs 21st (=0.012)
						21st vs 28th (=0.018)
PRA (ng/mL/h)	0.23 (0.20–0.37)	0.40 (0.20–0.50)	0.35 (0.20–0.63)	0.31 (0.20–0.45)	0.004	7th vs 21st (=0.012)
						7th vs 28th (=0.036)
ALDO (ng/100 mL)	11.2 (7.9–18.5)	15.7 (10.0–22.3)	17.8 (14.9–26.7)	16.2 (11.1–23.8)	0.004	7th vs 21st (=0.001)
ARR	42.5 (28.1–60.8)	45.0 (26.8–66.2)	50.2 (31.4–82.4)	50.0 (36.5–64.6)	NS	–
K (mEq/L)	3.7 (3.6–4.1)	3.8 (3.5–4.0)	4.0 (3.7–4.3)	3.8 (3.7–4.0)	NS	–
SBP (mmHg)	147±17	143±18	142±14	140±17	NS	–
DBP (mmHg)	90±10	85±11	88±6	82±11	NS	–

Abbreviations: *ALDO*, aldosterone; *ARR*, aldosterone/renin ratio; *E2*, 17β estradiol; *P*, progesterone. Values are medians and, in brackets, 25th and 75th percentiles, except for systolic (SBP) and diastolic (DBP) blood pressure, which are expressed as means ± s.d., and in which comparisons were made by ANOVA. From Fommei E, Ghione S, Ripoli A, Maffei S, Di Cecco P, Iervasi A, Turchi S. The ovarian cycle as a factor of variability in the laboratory screening for primary aldosteronism in women. J Hum Hypertens 2009;23(2):130–5. Table 1 p. 132.

drospirenone is associated with increase in aldosterone above baseline and PRA but decreases in DRC leading to increases in ARR calculated with DRC (Fig. 2.5.18) (Ahmed et al., 2011). The combined oral contraceptive is thus capable of significantly increasing ARR with risk of false positive results during screening for primary aldosteronism but only if DRC is used to calculate the ratio.

Factors affecting PRA are summarized earlier in Table 2.5.4.

Samples for aldosterone and renin usually require rapid transport to the laboratory but recent investigations find that requirement has been unnecessarily restrictive and samples are stable for 6 h at room temperature (Chakera et al., 2017). Plasma renin showed significant change with repeated freeze-thaw cycles (Hillebrand et al., 2017).

2.5.4.1 Angiotensin II

The active component of the RAAS on the zona glomerulosa of the adrenal cortex is angiotensin II. An LC-MS/MS method for angiotensin II was introduced and concentrations after a 1 h incubation of serum at 37°C represent concentrations at equilibrium of formation and degradation (Fig. 2.5.19) (Bernstone et al., 2021). The test is under evaluation in wider clinical practice and an aldosterone: eq AngII ratio (AA2R) may replace the current aldo:PRA or aldo:PRC tests.

2.5.4.2 18-Hydroxy steroids

The analysis of steroids in urine was a catalyst for several discoveries in the field of clinical steroid endocrinology.

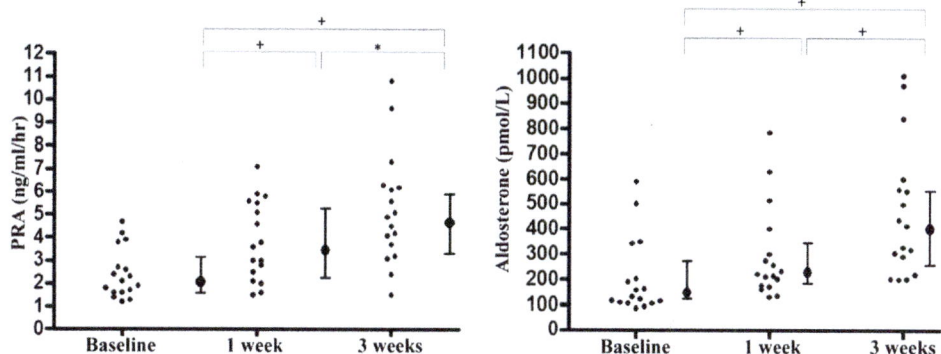

FIG. 2.5.18 The effects of 1 and 3 weeks of oral contraception [ethinyl estradiol (EE) plus drospirenone (D)] on renin and aldosterone. *Error bars* indicate interquartile ranges. *, *P* < .01; _, *P* < .001; #, *P* < .05. The *horizontal line* in A and B indicates the upper limit of the normal range for the ARR. *(From Ahmed AH, Gordon RD, Taylor PJ, Ward G, Pimenta E, Stowasser M. Effect of contraceptives on aldosterone/renin ratio may vary according to the components of contraceptive, renin assay method, and possibly route of administration. J Clin Endocrinol Metab 2011;96(6): 1797–804. Fig. 1 p. 1800.)*

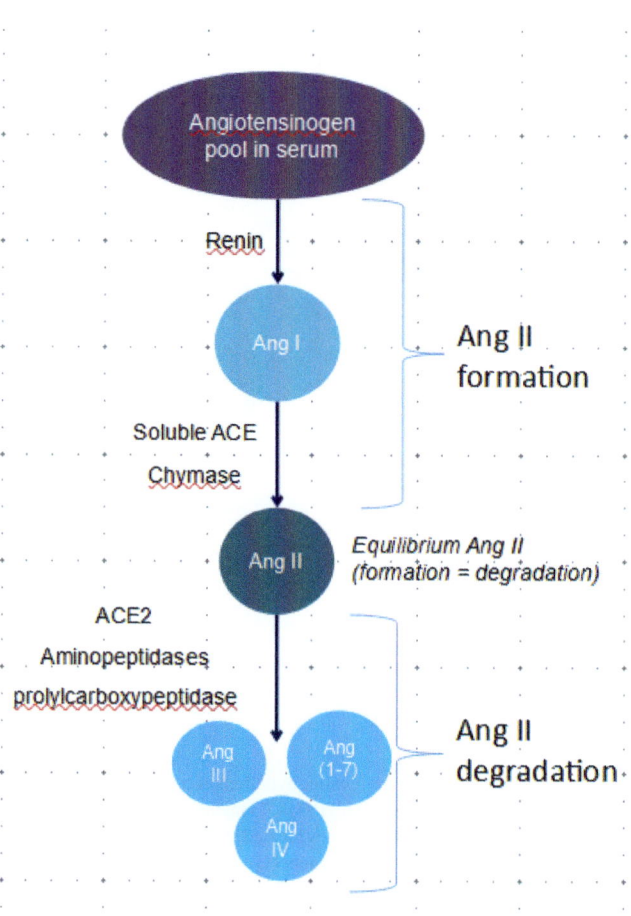

FIG. 2.5.19 Schematic representation of possible enzymatic reactions contributing to the eqAng II concentration in serum samples incubated in vitro. In contrast to analyses renin activity with protease-inhibited plasma, equilibrium analysis uses plasma without protease inhibition that is equilibrated at 37°C ex vivo. Owing to the excess of angiotensinogen, the kinetic equilibrium between formation and degradation of angiotensin metabolites is maintained by activities of circulating angiotensin-processing enzymes resulting in stable equilibrium angiotensin levels. This analysis has the potential to better reflect the in vivo angiotensin levels. *ACE*, angiotensin converting enzyme; *ACE2*, angiotensin-converting enzyme 2; *AGT*, angiotensinogen; *Ang I*, angiotensin I—angiotensin 1-10; *Ang II*, angiotensin II—angiotensin 1-8; *Ang (1-7)*, angiotensin 1-7; *Ang III*, angiotensin 2-8; *Ang IV*, angiotensin 3-8; *APA*, aminopeptidase A; *AT1R*, AngII type 1 receptors; *cAng I and cAngII*, circulating angiotensin I and II concentrations; *eqAngI and eqAngII*, equilibrium angiotensin I and II concentrations; *NEP*, neutral endopeptidase. *(Based on Bernstone L, Adaway JE, Keevil BG. An LC-MS/MS assay for analysis of equilibrium angiotensin II in human serum. Ann Clin Biochem 2021;58(5):422–33. Suppl Fig. S1.)*

18-Hydroxycortisol (18-OHF) in 1982 was first found by Mick Chu and Stanley Ulick in the urine of a patient with primary aldosteronism. The identity was based on GC-MS data. Immunoassays for plasma and urinary 18-OHF (Corrie et al., 1985) and urinary 18-oxo-F have used monoclonal and polyclonal antibodies (Nakamura et al., 2011).

The assays for 18-OHF have given very different results in urine (Lenders et al., 2018) and thus likely in plasma. The two steroids are part of a plasma panel of steroid by LC-MS/MS (Eisenhofer et al., 2017). Normal plasma ranges have been 0.45–4.3 nmol/L by LC-MS/MS. So far the determination of these steroids has only been required when investigating patients with hypertension to discriminate PA and FH-type 1. In such patients, 18-OHF has been measured in adrenal venous blood as well as peripheral blood to lateralize an aldosterone secreting tumor. The measurement of plasma 18-hydroxycorticosterone (18-OHB) (normally less than 30 nmol/L or less than 1.46 nmol/L depending on the method) (Riepe et al., 2003; Freel et al., 2004; Travers et al., 2017) is required in children with hyponatremia due to a genetic condition of aldosterone resistance (aldosterone receptor defect) or a defect in CYP11B2 where 18-hydroxycorticosterone is not oxidized to aldosterone and plasma 18-OHB concentrations were five times higher.

2.5.4.3 Vasopressin

Arginine vasopressin (AVP) is a nonapeptide (M Wt 1084 Da) synthesized in the hypothalamus with Neurophysin I and secreted from the posterior pituitary gland under osmoregulatory control. Plasma concentrations of AVP increase in response to water deficit and decrease in response to a water overload. Measurement of AVP concentrations can help in the investigation of patients with polyuria and altered plasma osmolality (El-Farhan et al., 2013). Polyuria can be due to insufficient AVP or reduced sensitivity to AVP or excessive water intake. Pituitary release of ACTH is, to a lesser extent than CRH, stimulated by vasopressin (AVP) from the neural lobe of the hypothalamus. The pituitary receptor (V1b) is different to the receptor that accounts for the pressor activity (V1a). The first assays for vasopressin were based on bioassays. When immunoassays were introduced, problems were encountered. High molecular weight factors in raw plasma (probably from neurophysin) interfered with the binding of antibodies to AVP, leading to high and erroneous readings. When these interfering factors were removed by extraction of plasma samples, immunoassays gave measurements more consistent with bioassays (Edwards et al., 1972), with measures of turnover and with the sensitivity of target tissues to exogenous hormone. Attempts to measure plasma levels without extracting the samples have generated results impossibly high and wholly erroneous. Vasopressin results differ significantly between assays (Leng and Sabatier, 2016) and it would appear at the time of writing that few or no readily available method exists. Normal plasma concentrations of AVP are 0.25–4 pg/mL (El-Farhan et al., 2013). In pregnancy, plasma concentrations in 10 women by a commercial RIA increased from mean 10 pg/mL at 24–28 weeks to 23 pg/mL near term (Makrigiannakis et al., 2007). HPLC-MS/MS methods do not report normal concentrations (Zhang et al., 2014; Tsukazaki et al., 2016).

Copeptin is derived from the same precursor as vasopressin but is more reliably measured with methods by IRMA and RIA (Agorastos et al., 2020; Timper et al., 2015; Fenske et al., 2009). This may become a reliable surrogate for vasopressin. Concentrations are less than 5 pmol/L that responds in parallel with ACTH to administration of metyrapone and dexamethasone.

2.5.5 Hypothalamic pituitary gonadal axis

The principal steroids from the gonads are testosterone from the male testes and estradiol and progesterone from the ovaries. The activities of the gonads are regulated by luteinizing hormone (LH) and follicle stimulating hormone (FSH) (Fig. 2.5.20) in the hypothalamic-pituitary-gonadal (HPG) axis.

GnRH is released from the hypothalamus in intermittent secretory bursts resulting in a pulsatile secretion of LH and FSH. The pulsatile mode of GnRH release is critical for physiological function of the gonadotrophs and a prerequisite for normal reproductive function. Normally, in adults the peak interval is 90–120 min and the difference between

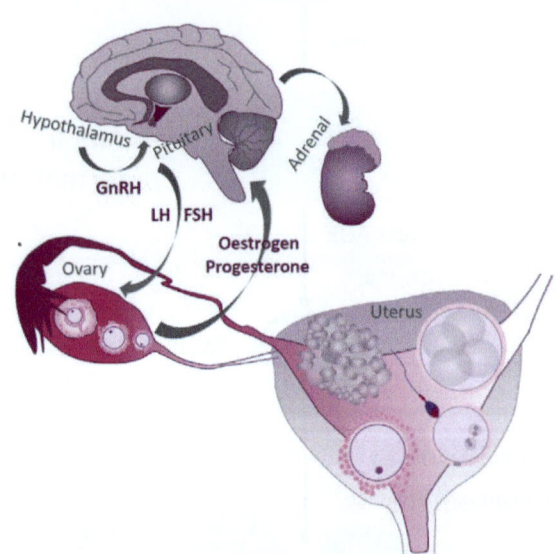

FIG. 2.5.20 The human female hypothalamic gonadal axis. Reproduction is controlled by the HPG axis. Gonadotropin-releasing hormone (GnRH), secreted by GnRH neurons located at the hypothalamus, stimulate the release of gonadotropin hormones [luteinizing hormone (LH) and follicle-stimulating hormone (FSH)] from the pituitary. The release of gonadotropins critically depends on GnRH pulsatile dynamics that are driven by hypothalamic neuronal networks. Gonadotropins act on the gonads, initiating processes involved in gametogenesis and ovulation and triggering the release of sex steroids (estradiol, testosterone, and progesterone) that feedback on the brain and pituitary gland to modulate GnRH and LH/FSH secretion dynamics *(Redrawn from Volozonoka L, Miskova A, Kornejeva L, Kempa I, Bargatina V, Gailite L. A systematic review and standardized clinical validity assessment of genes involved in female reproductive failure. Reproduction 2022;163(6):351–63. Fig 3 P 357.)*

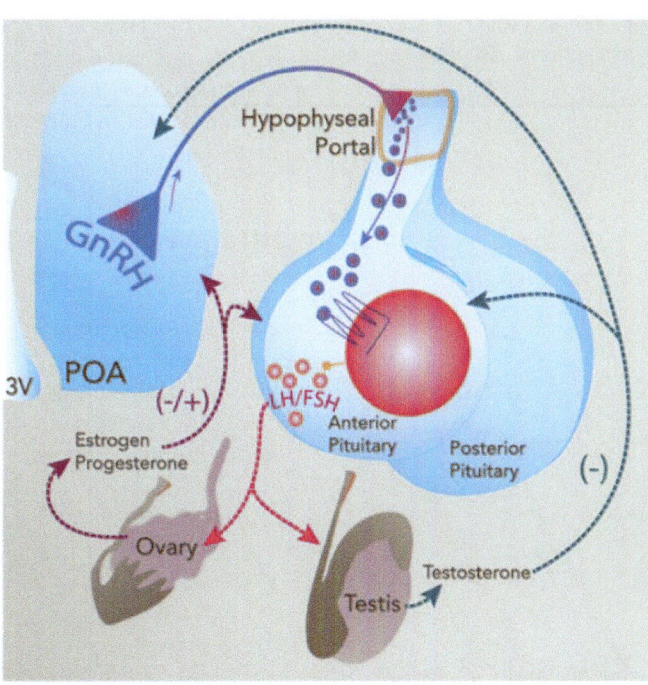

FIG. 2.5.21 Schematic diagram showing the hypothalamic-pituitary-gonadal (HPG) axis and factors involved in its regulation. Hypophysiotropic neurons in the hypothalamus secrete gonadotropin-releasing hormone (GnRH) into the hypothalamo-hypophyseal portal vasculature at the median eminence, where it is carried to the anterior pituitary. Once in the anterior pituitary, GnRH stimulates gonadotroph cells to synthesize and secrete luteinizing hormone (LH) and follicular stimulating hormone (FSH). LH and FSH act on the ovary and testis to regulate steroidogenesis and gametogenesis. Once released, steroid hormones signal through the anterior pituitary and steroid-sensitive hypothalamic neurons to feedback upon the HPG axis to regulate its activity and to other brain areas thereby controlling reproductive behaviors and functions, including HPA axis activity. *(Redrawn from Oyola MG, Handa RJ. Hypothalamic-pituitary-adrenal and hypothalamic-pituitary-gonadal axes: sex differences in regulation of stress responsivity. Stress 2017;20(5):476–94.)*

nadir and peak concentrations is often more than threefold. In fertile women, the concentrations additionally vary during the menstrual cycle.

LH stimulates testosterone production by testicular Leydig cells and ovarian thecal cells, whereas FSH stimulates AMH and inhibin by Sertoli cells and granulosa cells (Fig. 2.5.21).

LH and FSH are glycoproteins of two subunits, each with an alpha-subunit common to LH, FSH and thyroid stimulating hormone (TSH), and similar to human chorionic gonadotrophin (hCG), and a beta-subunit unique to each hormone.

The first day of menstruation is counted as Day 1 of a menstrual cycle which lasts around 28 days. On Day 1, the concentrations of estradiol (<50 pmol/L), progesterone (<2 nmol/L), LH (<15 IU/L) and FSH (<10 IU/L) are their lowest. The ovary contains only small tertiary follicles of less than 5 mm diameter, along with several atretic follicles and thousands of follicles less than 5 mm. By Day 3, LH and FSH have risen and some tertiary follicles have enlarged to about 10 mm in diameter. Some of the tertiary follicles in each ovary increase in size and by Day 10–12 a few reach 15–20 mm in diameter. Some follicles have degenerated (atretic follicles) but by Day 13 one follicle in one ovary undergoes rapid growth to resemble a blister on the surface of the ovary. Plasma estradiol concentrations are below 100 pmol/L in the menstrual phase of the cycle but rise in the follicular phase to peak on Day 12 at 500–3000 pmol/L in response to FSH that increases from

10 to 60 U/L. A surge of then LH peaks on day 15 at 20 U/L. FSH and LH fall in the luteal phase to 10 U/L and 10 U/L. In the lead oocyte, meiosis resumes and ovulation occurs about 12 h after the LH peak. Progesterone is 5 nmol/L at the time of ovulation and rises to 20–30 nmol/L in the mid-luteal phase. During the follicular and luteal phases, the LH concentrations vary between 3 and 10 IU/L and during the LH surge from 21 to 74 IU/L (Fig. 2.5.22) (Roos et al., 2015). These values are not seen with all assays, and the results for other assays are between 60% and 150% of those obtained with the Architect assay (La'ulu et al., 2018).

The gonadotrophins are complex molecules with variable glycosylation and fragmentation and not surprisingly there is variation in immunoassays probably linked to the lack of specificity of antibodies. LH and FSH are commonly measured by sandwich immunoassays which are limited by the lack of sensitivity at low concentrations, and overlapping values in prepubertal and pubertal children. Studies have investigated the use of a single LH measurement in the differentiation between the prepubertal and pubertal HPG-axis, but variations in cut-offs have been reported. Assay interferences have been found with autoimmunity, heterophilic antibodies and biotin. Many of these tests are performed on automated clinical chemistry platform analyzers using immunoassay mostly now with chemiluminescence detection because of the convenience within the large workloads for the laboratories to run in parallel to other analytes.

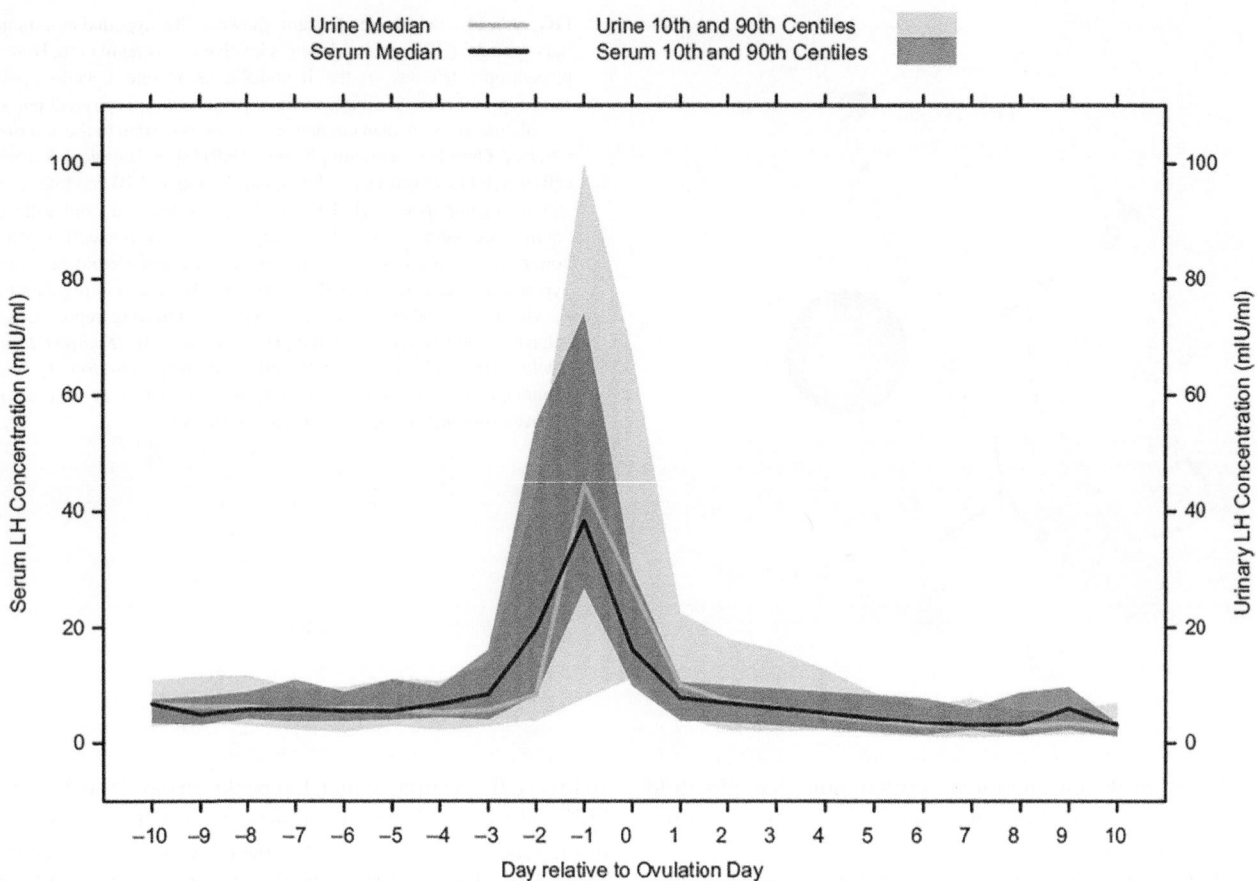

FIG. 2.5.22 Luteinizing hormone (LH) ranges in females referenced to the day of objective ovulation (median and 10th–90th percentile range) Serum (*dark gray*, mIU/mL) and urinary (*light gray*, mIU/mL). *(From Roos J, Johnson S, Weddell S, Godehardt E, Schiffner J, Freundl G, Gnoth C. Monitoring the menstrual cycle: comparison of urinary and serum reproductive hormones referenced to true ovulation. Eur J Contracept Reprod Health Care 2015;20 (6):438–50. Fig. 1 p. 444.)*

In large institutions, mass spectrometry is being introduced for endocrine tests because of the specificity of the tests and the ability to determine several steroids in the same cycle of tests after chromatographic separation of the components. Many steroids have a diurnal rhythm. The age differences and gender differences for the sex steroids need to be taken into account when interpreting test results. The costs per sample by MS methods now compete with IA especially if using the commercial kits containing all the reagents for the assay. In the MS methods, including the costs of leasing equipment over a few years, the running costs are then largely due to the LC solvents, chromatography columns and internal standards which are used around the midpoint of the calibration.

2.5.5.1 Testosterone

During childhood from 6 months to around 7 years of age, testosterone concentrations are below 1 nmol/L which may be near the detection limit of some IA but which is still well within the detection limit of LC-MS/MS methods (0.1 nmol/L). In prepuberty, testosterone is still below 1 nmol/L, initially secretion of testosterone is nocturnal so blood tests during the day will not be useful. As testicular size increases at the start of puberty (see Chapter 1.7—Physiological effects) the LH pulses are in the day and night and testosterone concentrations in blood increase and, from around the age of 10 years, rises continuously up to 10–30 nmol/L in the adult male. In normal menstruating girls, serum testosterone is between 0.5 and 2.5 nmol/L with a small circadian variation and a modest increase in mid-cycle.

2.5.5.1.1 Males

In newborn male children, serum testosterone can be at concentrations of up to 10 nmol/L which is at the lowest level of adult males (10–30 nmol/L) (Fig. 2.5.23). This initially reflects stimulation of the testes by human chorionic gonadotrophin (HCG) remaining in circulation after detachment of the infant from the placenta.

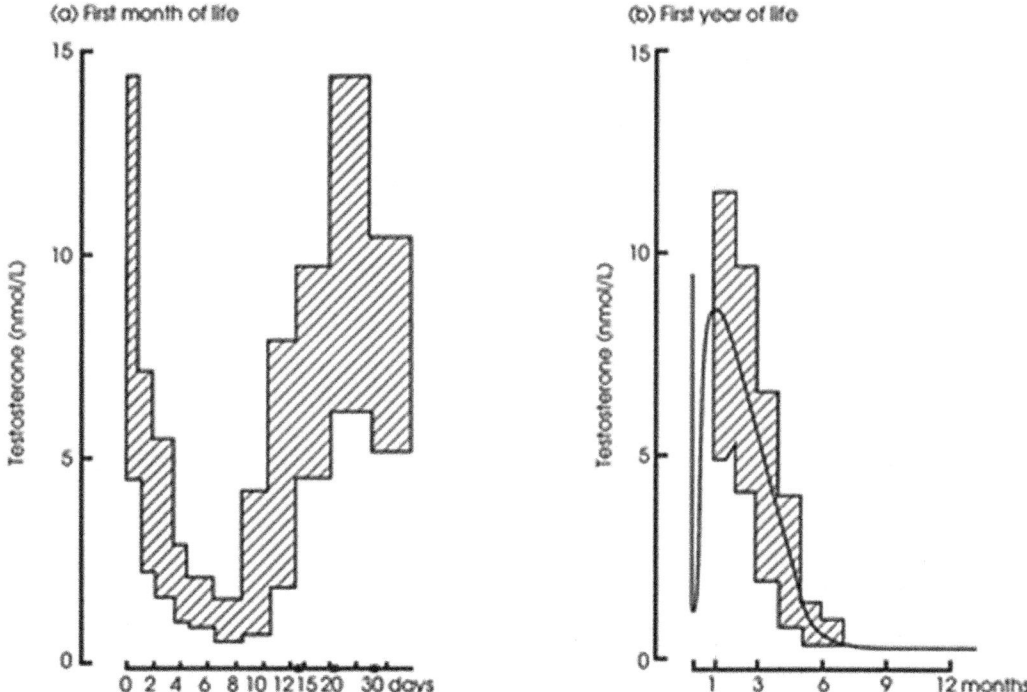

FIG. 2.5.23   Plasma testosterone in male infants as a function of age in the first month of life and first year of life. *Shaded area* is the mean ± 2SD. *(Redrawn from Forest MG, de Peretti E, Bertrand J. Testicular and adrenal androgens and their binding to plasma proteins in the perinatal period: developmental patterns of plasma testosterone, 4-androstenedione, dehydroepiandrosterone and its sulfate in premature and small for date infants as compared with that of full-term infants. J Steroid Biochem 1980;12:25–36. Fig. 2 p. 27.)*

The hCG and serum testosterone declines over the first week of life but in response to an increase in gonadotrophin secretion (by negative feedback) there is a rise again to concentrations of 2–10 nmol/L over the second week which continues for up to 5 months after birth (Forest et al., 1974). This is now called the mini-puberty because it is gonadotrophin dependent. From aged 6 months to around 8 years of age, the concentrations of gonadotrophins are low and serum testosterone remain below 1 nmol/L and is mostly of adrenal origin. The mechanism for suppression of the HPG axis in this period of childhood is unclear.

The onset of puberty LH secretion occurs initially during the night (Boyar et al., 1974) with some dependence on sleep pattern such as the transition from no-rapid eye movement (NREM) to rapid eye movement (REM). In early childhood, there are occasional peaks of gonadotrophins of low amplitude. At the onset of puberty, there is regular pulsatile LH secretion during the night with peak plasma concentrations of 2–5 U/L (Fig. 2.5.24) (Wu et al., 1996). A rise in plasma testosterone occurs within 60–90 min of the initial pulse of LH.

The 24 h pattern of testosterone in the period before puberty shows the highest testosterone concentrations near daybreak increase progressively to be around 10 nmol/L. This drops to half that level by 0900 h and throughout the day may be below 2 nmol/L. The measurement of a random serum testosterone during the day therefore has little value in boys during early puberty. During the day

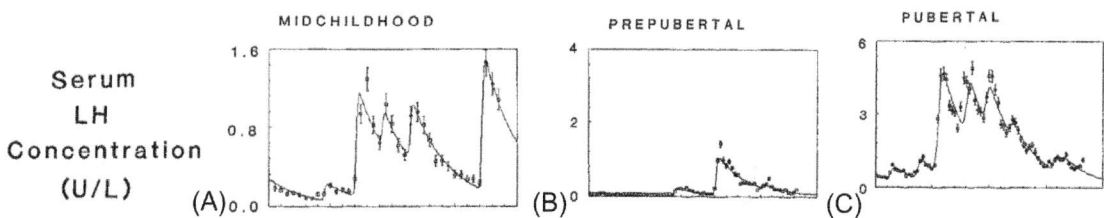

FIG. 2.5.24   Serial plasma LH concentrations measured by DELFIA in blood samples collected at 20 min intervals for 12 h from 2000 h to 0800 h the next morning. The continuous lines through the observed LH data are computer-calculated fits using the multiple parameter convolution integral. (A) Boys aged 4–8 years. (B) Boys aged 11–13 years. (C) Boys aged 13–16 years. *(Redrawn from Wu FC, Butler GE, Kelnar CJ, Huhtaniemi I, Veldhuis JD. Ontogeny of pulsatile gonadotropin releasing hormone secretion from midchildhood, through puberty, to adulthood in the human male: a study using deconvolution analysis and an ultrasensitive immunofluorometric assay. J Clin Endocrinol Metab 1996;81(5):1798–805. Fig. 1 p. 1800; Fig. 2 p. 1801; Fig. 3 p. 1801; Fig. 4 p. 1802.)*

gonadotrophins may be 1–4 U/L and at night, there are peak concentrations around 10 U/L. FSH at night exhibits less obvious pulsatility. As puberty progresses, there is a rise in the amplitude of gonadotrophin pulses. Pulse frequency is approximately 2 hourly and the duration extends from the night only to a pattern throughout the 24 h day. Between 9 and 17 years the testes increase in size, from less than 3 mL before puberty, to the adult range of 12–25 mL. At this time, there is pulsatility in LH throughout the day and night. When 10 mL testes are achieved, plasma testosterone levels rise into the normal adult range. It is not helpful to have reference ranges for testosterone concentrations with age. Each boy progresses through puberty at his own pace (Fig. 2.5.25). Only when testosterone remains low after the age of 16–18 years is an abnormality worth investigating. At the end of puberty, testosterone concentrations are 10–30 nmol/L

and the morning testosterone concentrations may be 20%–40% higher than in the same subject in the evening.

Immunoassays in platform systems now claim to be using better antibodies but concentrations in females may still not be accurate (La'ulu et al., 2018). A comparison of four clinically validated LC-MS/MS methods for testosterone showed good agreement for the results (French et al., 2018). There has been much effort for standardization and harmonization of steroid results, so far testosterone has been the only steroid where consensus results have been established (Travison et al., 2017). The complexity of the onset of puberty has now been known for almost 50 years and overnight blood sampling is an expensive and difficult procedure. The onset of puberty is now charted largely on the basis of morning plasma LH and testosterone or nocturnal urinary LH.

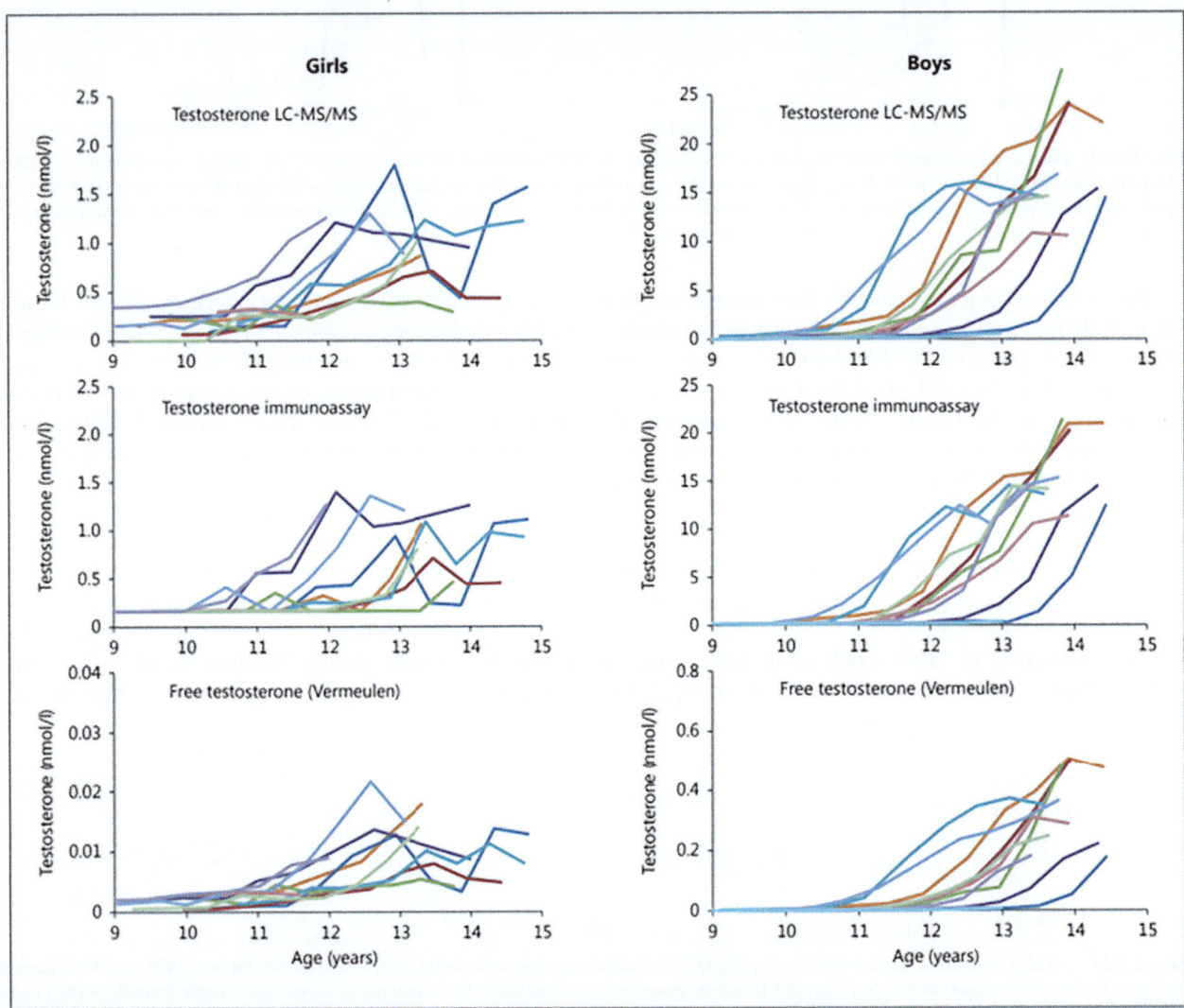

FIG. 2.5.25 Serum concentrations of testosterone according to age in boys (*blue*) and girls (*red*). Each line follows a child. *(Based on Mouritsen A, Søeborg T, Hagen CP, Mieritz MG, Johannsen TH, Frederiksen H, Andersson AM, Juul A. Longitudinal changes in serum concentrations of adrenal androgen metabolites and their ratios by LC-MS/MS in healthy boys and girls. Clin Chim Acta 2015;450:370–5. Fig 1 P 372.)*

Testosterone in the circulation is 97% bound to proteins, principally to sex hormone binding globulin (SHBG) and weakly to albumin. Therefore, testosterone not bound to SHBG can be considered free and bioavailable. Free testosterone may be measured directly by equilibrium dialysis and ultrafiltration which are reference methods, immunoassay, or calculated mathematically based on concentrations of total testosterone, SHBG, and albumin. Total testosterone concentrations show significant diurnal rhythm with highest levels in the morning which is the preferred time for all sample collections. Concentrations can be suppressed by food intake and glucose. Total testosterone concentrations above 320 ng/dL (11 nmol/L) in men are required to maintain lean body mass and muscle size and strength. Obesity, diabetes and the use of androgens lead to reduced concentrations of sex hormone binding globulin. Variability in testosterone concentrations between laboratories is attributed to calibrator, preanalytical and methodological differences. The Endocrine Society working with the Center for Disease Control (CDC) introduced a Harmonization programme and many assays have been certified to these standards, although this will not cure all assay anomalies in patient samples. The current status of testosterone assays is illustrated in Fig. 2.5.26 (Montagna et al., 2018; Cao et al., 2019). First generation automated IA were subject to the effects of SHBG concentration (Heijboer et al.,

2016) but this does not now appear to be an issue with subsequent assays.

Interferences in immunoassays for testosterone have been shown by steroid cross reaction (DHEAS, abiraterone, norethisterone, testolactone, RU486, and danazol). In rare samples, antibodies have been found to testosterone, to iodine labeled testosterone through a histamine bridge. Rabbit antibodies in patient samples (heterophilic) can affect assays based on antibodies raised in rabbits. Biotin and alkaline phosphatase antibodies are new issues arising from supplements and recombinant protein treatment (Stieglitz and Korpi-Steiner, 2020; Maharjan et al., 2019).

GC-MS has also been found to be a very reliable quantitative method for plasma androgens because of the high specificity through separation of steroids in the sample extract. Solvent extraction is adequate for androgens. Heptafluorobutyrate derivatives are the commonest forms used (Wudy et al., 2001, 2002). Electron impact and chemical ionization has been used. Comparisons of GC-MS results with clinical samples assayed by direct and extracted RIA methods show close agreement with the extracted RIA's. In the meantime, different techniques based on liquid chromatography tandem mass spectrometry (LC-MS/MS) have been developed to detect a panel of androstenedione, testosterone and dihydrotestosterone (A, T and DHT) (Wang et al., 2020). In contrast to immunoassays, these methods

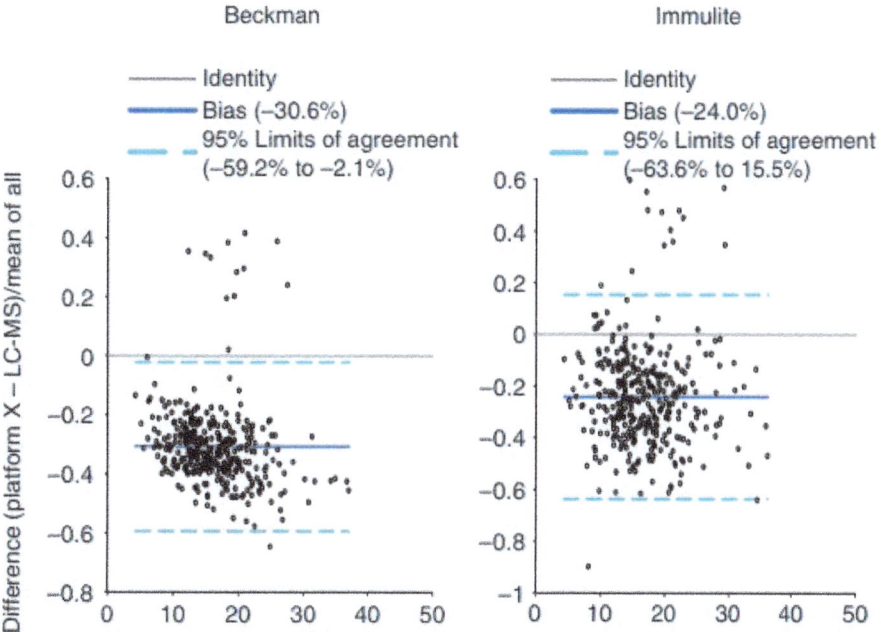

FIG. 2.5.26 Testosterone assays performance on **representa**tive platform CLIA methods compared with LC-MS/MS assay. Differences in serum testosterone levels between commercial IMAs and LC-MS/MS. Bland-Altman plots (A): *Y*-axis depicts the % difference between values of the two measurements (IMA—LC-MS/MS/mean of all). Bias and limits of agreement are depicted as continuous and *dashed line*, respectively. The bias was −30.6% (95% CI −32.2 to −29.1) for Beckman, −24% (95% CI −26.2 to −21.9) for Immulite, −12.4% (95% CI −14.2 to −10.6) for Roche (not shown here) and 0.6% (95% CI −0.7 to 1.9) for Abbott (not shown here). *(Redrawn from Montagna G, Balestra S, D'Aurizio F, Romanelli F, Benagli C, Tozzoli R, Risch L, Giovanella L, Imperiali M. Establishing normal values of total testosterone in adult healthy men by the use of four immunometric methods and liquid chromatography-mass spectrometry. Clin Chem Lab Med 2018;56(11):1936–44. Fig. 1 p. 1939.)*

TABLE 2.5.6 Serum androgens and estrogens in boys determined by GC-MS/MS in relation to testicular volume.

Testicular volume (mL)	1–2	3–6	8–12	15–25
Androstenedione (nmol/L)	0.5 (0.1–0.9)	0.8 (0.3–1.4)	1.9 (0.7–2.9)	2.2 (1.4–2.9)
Testosterone (nmol/L)	0.2 (0.1–0.4)	0.4 (0.2–2.6)	11.0 (2.2–22.9)	19.6 (11.1–23.6)
Dihydrotestosterone (pmol/L)	56 (<27–227)	220 (80–310)	636 (184–1242)	1047 (645–1363)
Estrone (pmol/L)	12 (<9–26)	17 (10–62)	50 (19–80)	53 (27–78)
Estradiol (pmol/L)	3 (<2–14)	7 (<2–22)	33 (3–71)	50 (25–95)
N	9	9	14	9

Data is presented as median (5th–95th percentiles). *$P<.05$, **$P<.01$, ***$P<.001$ compared with closest lower pubertal stage. From Ankarberg-Lindgren C, Dahlgren J, Andersson MX. High-sensitivity quantification of serum androstenedione, testosterone, dihydrotestosterone, estrone and estradiol by gas chromatography-tandem mass spectrometry with sex- and puberty-specific reference intervals. J Steroid Biochem Mol Biol 2018;183:116–24. Table 1 4 p. 122.

are shown by appropriate tests to be independent of matrix effects or cross reactivity. LC-MS/MS methods enable for sensitive, specific and high-throughput (180 samples per 24 h) determination of A, T and DHT with estrone and estradiol simultaneously from 200 μL of plasma or serum as well as the accompanying reference values covering the pediatric age range and sex development (Table 2.5.6) (Ankarberg-Lindgren et al., 2018). Electrospray is the commonest interface with positive ion transitions detected for androgens and negative ions for estrogens. The LC analysis can be achieved in less than 8 min.

Androgen deficiency contributes to many adverse conditions such as diabetes, metabolic syndrome, reduced bone and muscle mass, impaired sexual function, infertility and poor sleep quality. Treatment of hypogonadism has some controversies from what concentration of testosterone defines the condition to how, when and how much steroid is replaced. Low total testosterone is observed frequently in men with abdominal and/or visceral obesity and the metabolic syndrome. A decrease in waist circumference is seen in response to testosterone. Particularly if the men had low levels of testosterone and high BMI (Fig. 2.5.27) (Eisenhofer et al., 2017).

Since many women will be using oral contraceptives the impact on steroid assays need to be factored into investigations involving testosterone measurements (see Zimmerman et al., 2014 for meta-analysis). In women with signs of androgen excess, higher testosterone and free testosterone levels are frequent correlates of increased abdominal and/or visceral fat accumulation that is not the case in nonhyperandrogenic women. Steroid-converting enzymes expressed in adipose tissues are involved in androgen-mediated modulation of body fat distribution (Tchernof et al., 2015, 2018). Plasma concentrations of testosterone decline in men with advancing age leading to many metabolic and pathophysiological changes such as increase in fat mass, decreased bone and muscle mass, sexual function, frailty, depression, insulin resistance and diabetes and cardiovascular risk. The changes are attributed to alterations in Leydig cell function. In the Heritage family study, hundreds of men and women were examined for correlations of steroid hormones with body composition, fat distribution and adiposity before and after an exercise programme (He et al., 2018).

Rodent studies suggest a reduction in the capacity to transfer cholesterol into mitochondria and an increase in free radical formation that affects steroidogenic enzymes. Little is known about the underlying mechanisms and studies with genetically modified mice are likely to bring answers. Mice with increased or decreased expression of antioxidant enzymes will be useful models to elucidate how excessive oxidative stress contributes to the decline in steroidogenesis with age.

2.5.5.2 Estradiol

2.5.5.2.1 Females

In girls under 12 months of age, serum estradiol concentrations are less than 300 pmol/L. Thereafter until the onset of puberty estradiol concentrations are less than 60 pmol/L. These concentrations are difficult to achieve accurately by IA. Basal gonadotrophin levels in prepubertal girls are almost always low (less than 5 U/L). In early puberty, the

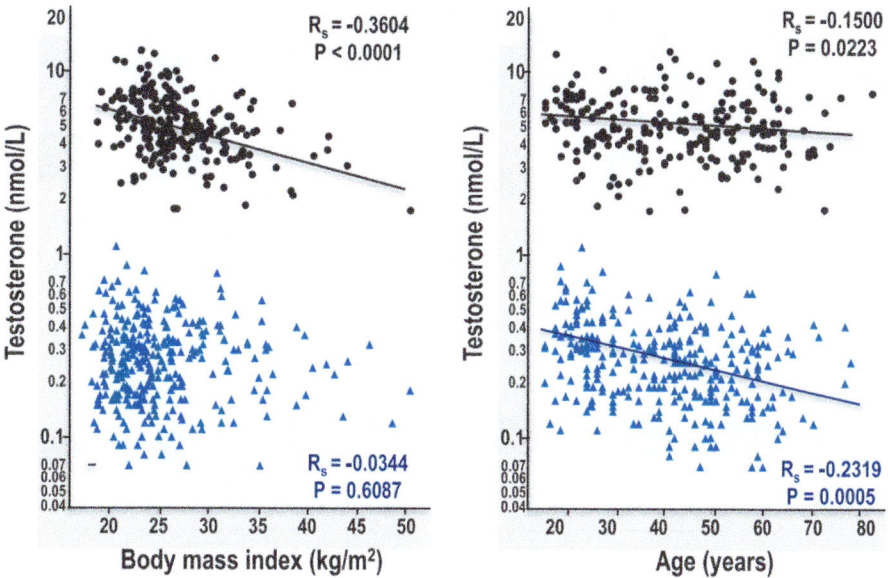

FIG. 2.5.27 Distributions of plasma concentrations (logarithmic scale) of testosterone according to body mass index and age for men (●) and women (▲). Regression lines are shown for relationships that are significant ($P < .05$). *(Based on Eisenhofer G, Peitzsch M, Kaden D, Langton K, Pamporaki C, Masjkur J, Tsatsaronis G, Mangelis A, Williams TA, Reincke M, Lenders JWM, Bornstein SR. Reference intervals for plasma concentrations of adrenal steroids measured by LC-MS/MS: impact of gender, age, oral contraceptives, body mass index and blood pressure status. Clin Chim Acta 2017;470: 115–24. Fig. 1 p. 119.)*

nocturnal release of gonadotrophins (FSH more so than LH) leads to a slow rise in serum estradiol concentrations which peak around mid-day at 500 pmol/L. In the prepubertal girl, the ovary appears on ultrasound to have a number of cysts less than 4 mm in diameter. Even when the gonadotrophins pulse throughout the 24 h period, ovulation cannot occur until an LH surge is sustained for 36 h. This relates to the gradual increase in sex steroid secretion by the ovaries in response to gonadotrophins. LH, FSH and estradiol are monitored during the process of puberty. Breast development is due to the increase in circulating estradiol concentrations (Table 2.5.6) (Ankarberg-Lindgren et al., 2018).

Comprehensive reference ranges for simultaneous quantification of estradiol and estrone is a valuable tool for investigations in clinical practice (Frederiksen et al., 2020; Skiba et al., 2019). Interpretation of results in females need to consider changes with age, body size, menstrual cycle phase, oral contraceptive use, SHBG concentrations, body size and blood pressure.

Once menses begin, there are marked changes in the frequencies of gonadotrophin secretion (Sun et al., 2019) and in the ratio of LH to FSH. In the first half of the normal cycle, LH pulses occur at 1–2 hourly intervals but slow to 4 hourly intervals in the mid and late luteal phases. In more mature girls with menstrual cycles, the secretion of estrogen is episodic but fluctuations in serum estradiol concentrations over a 24 h period are only discernible in the periovulatory period (Skiba et al., 2019) (Fig. 2.5.28).

Serum estradiol may be moderately raised in patients with obesity, hyperthyroidism and liver disease, reflecting increased peripheral production of estrogens from circulating androgens.

Since 1950 estradiol has been measured by a number of techniques (bioassay, immunoassay GC-MS, and UV absorbance) with preference for a solvent extraction of the steroid from the sample. Immunoassays depend on a mechanism to release of the steroid from binding proteins. In commercial assays, this process is rarely defined, Danazol or salicylate is used in some systems. Estradiol acts on many tissue including bone, muscle, skin, blood vessels, liver, kidney, the GI tract, lung and brain so has been implicated in a number of disease processes for its absence and excess. Estradiol impacts sexual development and secondary sex characteristics. Reference ranges are desirable for age, gender, puberty, the menstrual cycle and the menopause. There is a consistent pattern of change across the menstrual cycle (Fig. 2.5.29) (Verdonk et al., 2019). Estradiol shows two areas of increased activity corresponding to the follicular and luteal phases of the cycle around ovulation.

There are a number of demands to be met from an estradiol service assay.

- Concentrations can range across three orders of magnitude from less than 100 pmol/L in children and less than 50 pmol/L in postmenopausal women, then up to 2–3000 pmol/L in pregnancy and women in in vitro fertilization programmes involving ovarian hyperstimulation
- Precision to enable detection of suppression in postmenopausal women taking aromatase inhibitors and children prescribed puberty blockers

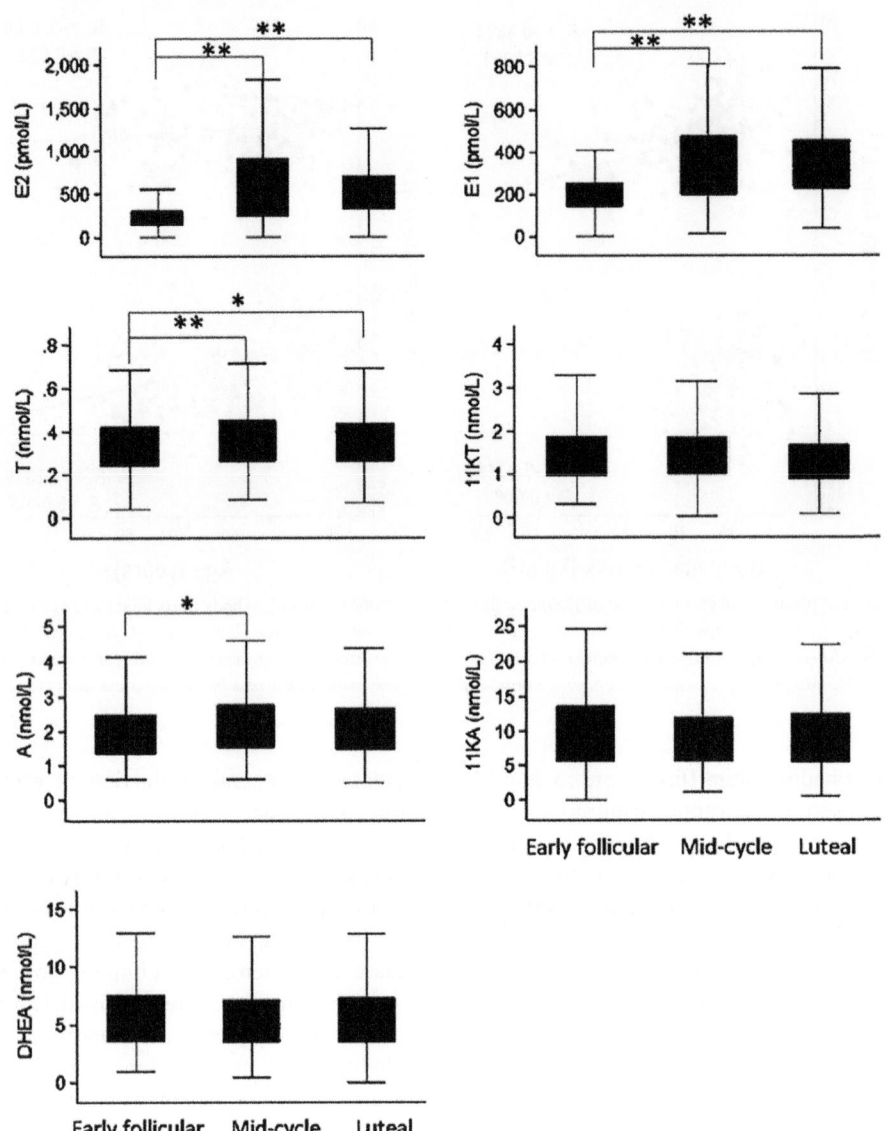

FIG. 2.5.28 Sex steroid levels across the menstrual cycle. In the box and whisker plots, the box represents the interquartile range (IQR); the line in the box is the median. The whiskers extend to the upper and lower adjacent values. The upper adjacent value was defined as the largest data point <75th percentile $+1.5 \times$ IQR. The lower adjacent value was defined as the smallest data point <25th percentile $-1.5 \times$ IQR. After factorial ANOVA, adjusted for BMI and cycle stage, with Bonferroni correction: $*P < .05$; $**P < .01$. *A*, androstenedione; *11KT*, 11-ketotestosterone; *11KA*, 11-ketoandrostenedione. *(From Skiba MA, Bell RJ, Islam RM, Handelsman DJ, Desai R, Davis SR. Androgens during the reproductive years: what is normal for women? J Clin Endocrinol Metab 2019;104(11):5382–92. Fig. 3 p. 5387.)*

- Have high specificity in the presence of a large number of natural estrogen metabolites and conjugates including derivatives of equine estrogens when prescribed. Estrone sulfate is at relatively higher concentrations than estradiol.
- Comparability of results between laboratories will be difficult to achieve when all the preanalytical and methodological variables are considered.

Immunoassays are known to be unreliable in children and postmenopausal females. Estrogens are extracted from samples most often with MTBE or ethyl acetate but diethyl ether and dichloromethane are used. SPE with C18

cartridges or 96 well plates are also used. Pentafluoropropionyl or trimethylsilyl derivatives are used before GC-MS analysis and a number of derivatives have been used before LC-MS/MS analysis (see Denver et al., 2019 for review) where reaction with dansyl chloride is the commonest. With GC-MS/MS both electron impact and chemical ionization have been used and LOQ of less than 20 pg/mL has been achieved. A turboflow column has been found useful for sex steroid separation after liquid-liquid extraction (Frederiksen et al., 2020), reference ranges for estradiol and estrone total and calculated free

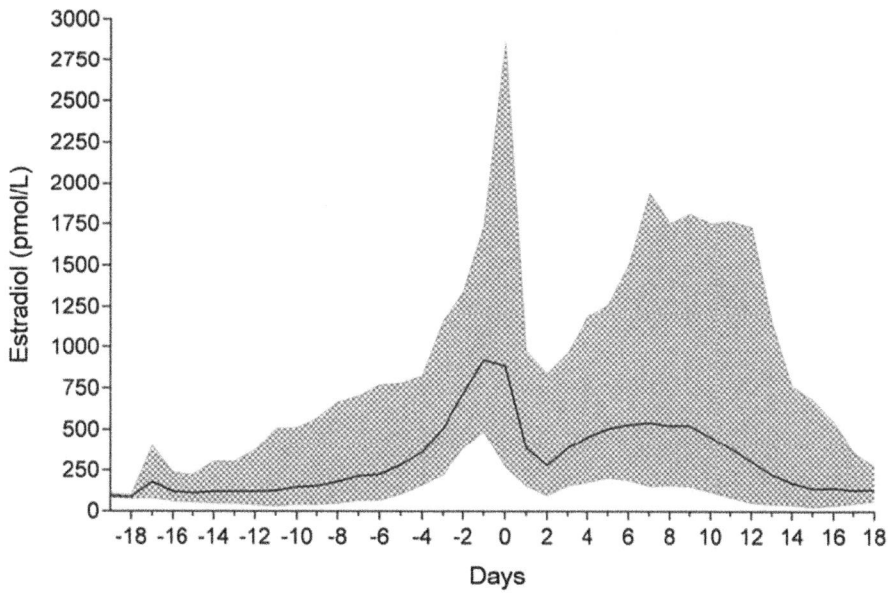

FIG. 2.5.29 Daily mean estradiol concentration in 30 premenopausal women. *Solid line*: presents the mean concentration of estradiol. The *gray areas* around the *solid line*: present the reference interval (the 95% range). To convert to pg/mL, divide the estradiol concentration by 3.67. *(From Verdonk SJE, Vesper HW, Martens F, Sluss PM, Hillebrand JJ, Heijboer AC. Estradiol reference intervals in women during the menstrual cycle, postmenopausal women and men using an LC-MS/MS method. Clin Chim Acta 2019;495:198–204. Fig. 2 p. 201.)*

concentrations were determined. LOD was 4 pmol/L for estradiol. Electrospray and positive mode is favored for derivatives of estrogens although APCI and APPI have been reported. Mobile phase modifiers include formic acid, ammonium formate or acetic acid. Typically 0.1–0.5 mL of serum is used and concentrations as low as 0.5 pg/mL can be detected. Without derivatization and negative ion mode selection, 1 pg/mL can be detected. Ion suppression is less with APCI than APPI and lower concentrations of estrone and estradiol. The performances of immunoassays has been compared with mass spectrometric assays for children (Rahhal et al., 2008) and postmenopausal women (Key et al., 2015) confirming inaccuracy of immunoassay particularly when the assay was performed directly on serum without an extraction step. The reagents for several of these assays have changed so comparative performances may not still be the same. A comparison of methods to determine low serum estradiol in postmenopausal women concluded that at a minimum extraction assays with immunoassay should be used but mass spectrometry is preferred (Lee et al., 2006). Derivatives with 2-fluoro-1-methylpyridinium-*p*-toluene sulfate enabled detection to 0.2 pg (Faqehi et al., 2016).

Much of the activity can be monitored by abdominal ultrasound. Measurements of plasma estradiol concentrations are a useful indicator of reproductive health and to confirm time of ovulation in the cycle so as to optimize timing of intercourse in the hope of a pregnancy (Roos et al., 2015). Reference ranges for LH and estradiol for Elecsys and progesterone for Cobas E801 were established to improve clinical decision making in women with fertility issues (Anckaert et al., 2021).

Prepubertal girls have significantly higher estradiol concentrations compared with boys, when measured using IA, GC-MS or LC-MS/MS and this may reflect their earlier onset of puberty. Prepubertal concentrations of estradiol levels below 50 pg/mL (180 pmol/L) are often undetectable in immunoassays. It should be borne in mind that measurements of sex steroids and gonadotrophins are not interpretable in adolescent girls taking oral contraceptive pills. There is much variation in the timings and duration of the menstrual cycles (Fig. 2.5.30).

Depending on the immunoassay used, interferences have been described due to the presence of antibovine alkaline phosphatase antibodies (Beckman UniCel Dx1 assay), heterophilic antibodies, IgA lamda, inflammatory markers (possibly CRP or related factor), steroid cross reaction with Danazol, exemestane (Mandic et al., 2017), Efavirenz, Fulvestrant, RU486 and conjugated estrogens (see Appendix Chapter 2.4). Aromatase inhibitors are prescribed for patients with breast cancer and the laboratory needs to be informed of this treatment since estradiol results can be compromised.

2.5.5.2.2 Estrone sulfate

E1S is the most abundant estrogen circulating in plasma of pre and postmenopausal women although requests by clinicians for the determination are rare. The literature on concentrations in tissues is confusing because of the wide ranges and differences between methods. An indirect immunoassay involves prior extraction and chromatography preferably including the addition of a radiolabel to quantify

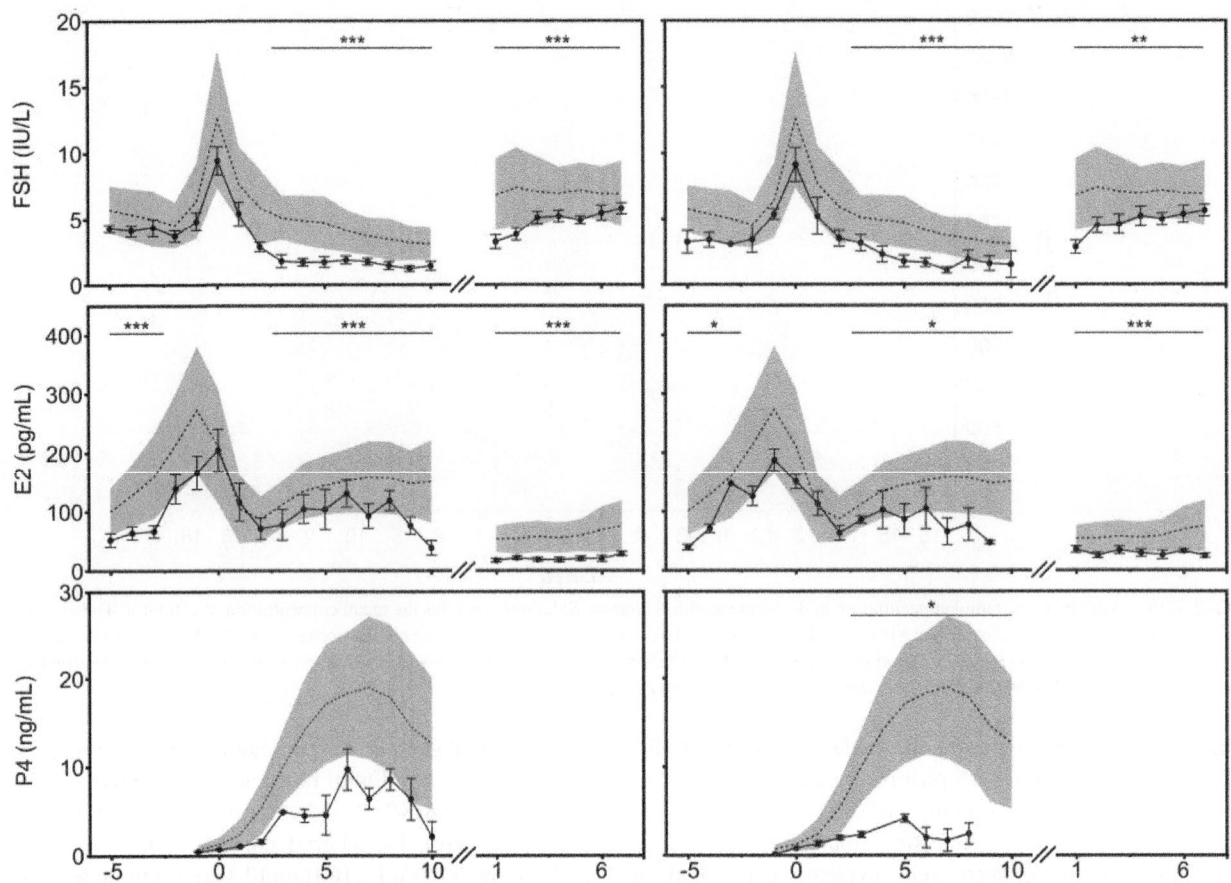

FIG. 2.5.30 FSH, E2, and P4 levels during two consecutive menstrual cycles in adolescent girls with either normal OV (left, $n = 11$) or short OV (right, $n = 5$) cycles. Levels represent serum or serum equivalents as determined from DBS or urine samples. Data for LH not shown here. Cycle days are centered to the MC LH peak of cycle 1 (day 0) and shown during the first week of cycle 2 (day 1 = menses). Adolescents: filled circles are arithmetic mean values and error bars are 61 SE. Both groups are compared with historic adult controls ($n = 65$). Adults: *Dashed line* denotes arithmetic mean level and *shaded area* is 61 SD. *$P \leq .05$; **$P \leq .01$; ***$P \leq .001$ for cycle phase (i.e., LFP, MC, LP, EFP). To convert E2 to SI units (pmol/L), multiply by 3.67; for P4 (nmol/L), multiply by 3.18. *(From Sun BZ, Kangarloo T, Adams JM, Sluss PM, Welt CK, Chandler DW, Zava DT, McGrath JA, Umbach DM, Hall JE, Shaw ND. Healthy post-menarchal adolescent girls demonstrate multi-level reproductive axis immaturity. J Clin Endocrinol Metab 2019;104(2):613–23. Fig. 1 p. 618.)*

recovery (Giton et al., 2002). Direct immunoassays are also used (Giton et al., 2010). Estrone sulfate is produced mainly in peripheral tissues. Sulfatases can transform the biologically inactive sulfate to estrone. Plasma concentrations fluctuate in the menstrual cycle from mean of 0.96 μg/L in follicular phase to 1.74 μg/L in luteal phase similar pattern to estradiol and estrone. In pregnancy, mean plasma concentrations are 19, 66 and 105 μg/L in the three trimesters, respectively (Rezvanpour and Don-Wauchope, 2017; Stanway et al., 2007).

2.5.5.2.3 Estradiol in males

Estradiol in boys is not produced significantly by the testes or adrenal cortex but largely due to the action of aromatase on testosterone in adipose tissue. Obesity will therefore add to this situation. Estradiol in boys is usually less than 100 pmol/L and IA are not precise at that level. If

LC-MS/MS is available for estradiol then more accurate concentrations can be achieved (Table 2.5.6) (Ankarberg-Lindgren et al., 2018).

Estradiol is now known to be physiologically significant in men especially for bone health. Estradiol concentrations in men are 10-fold lower than in premenopausal women and the impact of metabolites on assay performance is largely unknown. Indications for serum estradiol measurements include evaluation of gynaecomastia, adrenal tumors, hepatic tumors, genetic aromatase deficiency or excess, androgen insensitivity, obesity, fertility and even osteoporosis.

2.5.5.3 Progesterone

Plasma **progesterone** concentrations are less than 2 nmol/L in young girls and in pubertal girls during the menstrual cycle prior to the LH surge. Although the ovarian follicles

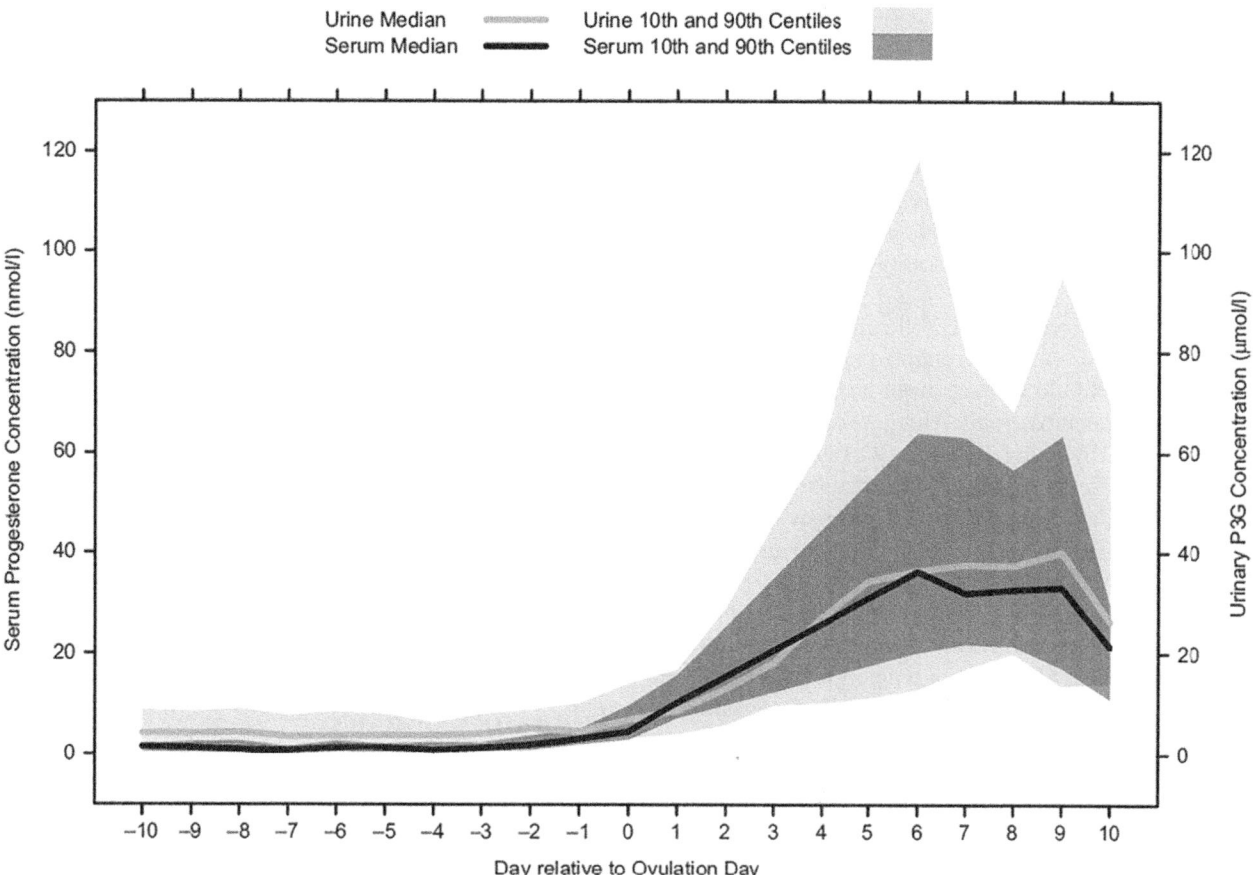

FIG. 2.5.31 Plasma progesterone (*dark gray*, nmol/L) and urinary P3G (*light gray*, μmol/L) ranges referenced to the day of objective ovulation (median and 10th–90th percentile range). *(From Roos J, Johnson S, Weddell S, Godehardt E, Schiffner J, Freundl G, Gnoth C. Monitoring the menstrual cycle: comparison of urinary and serum reproductive hormones referenced to true ovulation. Eur J Contracept Reprod Health Care 2015;20(6):438–50. Fig. 3 p. 446.)*

contain progesterone, most of the circulating progesterone during the follicular phase is adrenal in origin. After the menarche progesterone rises in serum each cycle at 7–9 days after the LH surge in ovulatory cycles with a peak at 30–80 nmol/L (Fig. 2.5.31) (Roos et al., 2015). Then the high progesterone concentration in serum reflects the output of the hormone by the corpus luteum. If a pregnancy starts then the corpus luteum persists and progesterone remains.

Progesterone can be measured directly or after extraction of the steroid into hexane or pentane, methyl tertiary-butyl ether (MTBE) or ethyl acetate. For many years, progesterone has been measured directly on automated chemistry platforms and this may well continue because of the convenience of the test. Progesterone concentrations can be in range 1–3 nmol/L (314–942 pg/mL) by immunoassay but 5.7–91.7 pg/mL in postmenopausal women by LC-MS/MS. Raised concentrations persists if pregnancy is established and at 5–8 weeks increases through secretion by the placenta. In a female under investigation for infertility or failure to progress in puberty, a persistently raised progesterone is a useful indicator for congenital adrenal hyperplasia due to 17-hydroxylase deficiency where

sex steroid production fails. Such a defect is best detected with a plasma steroid panel or urinary steroid profile (Lee et al., 2020) that, respectively, display the hormones or their metabolites in excess (progesterone and corticosterone) while sex steroids and cortisol metabolites are low.

2.5.5.4 Dihydrotestosterone (DHT)

DHT generally follows the excursions of testosterone at a lower concentration level and shows a peak plasma concentration of up to 3 nmol/L at Tanner genital stage P5 (Table 2.5.6). Androgen assays by LC-MS/MS now show good agreement between laboratories although some assays show differences in standardization and others have high variation (Büttler et al., 2015b).

DHT is more potent than T and when the 5-alpha reductase fails through genetic disorders, masculinization of the fetus in utero does not progress and the child will be reared as female. At puberty, there is a surge in testosterone in these patients sufficient to produce some masculinization in body shape.

2.5.6 Neurosteroids

The term neurosteroid was created when steroids were found to be produced in the nervous system. Steroids were found to be neuroactive in a short period of time through action at membranes, thus a nonclassical route. Care is needed when interpreting concentrations in the literature before 1995 because the steroids isolated depended on the methodology used and possibility of auto-oxidation of cholesterol.

Radioimmunoassay (RIA) following HPLC separation, specific RIA, GC-MS or LC coupled to mass spectrometric detection (HPLC-MS) have been used for quantitative analysis of the neurosteroids. Although sufficient sensitivity was obtained by using RIA (LLOQ ranging from 15 to 25 pg), the presence of additional endogenous steroids and structurally related lipids decreased accuracy. Higher sensitivity and better selectivity were achieved by using GC-MS or GC-electron capture negative chemical ionization, although these GC-MS methods required significant sample cleanup and preconcentration steps, such as liquid-liquid extraction (LLE), SPE or even longer procedures like preparative HPLC or TLC. HPLC-MS/MS is versatile and selective, and suitable for high-throughput methodologies for the analysis of neurosteroids. The very low concentrations of neurosteroids in serum necessitate accurate analysis, essential in clinical practice and research. The neurosteroids have relatively low ionization efficiencies ESI and atmospheric pressure chemical ionization modes, and an improved level of sensitivity is required for the levels encountered. Therefore, chemical derivatization reactions,

such as, the conversion of poorly or nonionizable compounds into easily ionizable derivatives by the atmospheric pressure ionization (API) have been used. Unconjugated oxosteroids have been analyzed after conversion to oximes (Ke et al., 2017). In case of derivatization, sample handling procedures for LC-MS are easier to perform than for GC-MS, as well as the run analysis time is usually shorter and the selectivity improved when using LC-MS. In ESI, the steroids detection sensitivity is improved when permanently charged moieties or easily ionizable moieties are introduced. Reagents such as hydroxyammonium chloride and 2-hydrazino-1-methylpyridine and 2-hydrzinopyridine have been used for more efficient detection of neutral neurosteroids or neurosteroids sulfates (Lionetto et al., 2017). The concentrations of allo-pregnanolone, epi-allo pregnanolone, pregnanolone are less than 1 ng/mL (Table 2.5.7) (Lionetto et al., 2017), allopregnanolone has a diurnal rhythm in the follicular phase of the menstrual cycle (Fig. 2.5.32) (Kimball et al., 2020; Moller et al., 2016).

2.5.7 Steroid analysis in current laboratory practice

2.5.7.1 Steroid panels

2.5.7.1.1 *Corticosteroids and precursors*

In addition to cortisol and cortisone concentrations, there are clinical situations where precursor measurements will be useful. Sex steroids, progesterone and 17-hydroxyprogesterone are the major needs in pediatric

TABLE 2.5.7 Demographic data and concentration of analytes calculated for each healthy volunteer analyzed in this study.

Individual	Age (years)	Gender	Education (years)	AP (ng/mL)	EAP (ng/mL)	P-value (ng/mL)	DHEA (ng/mL)	DHEA-S (µg/mL)
1	83	F	12	1.13	0.524	0.254	2.63	0.520
2	73	F	13	0.893	0.368	0.235	3.10	0.242
3	77	M	15	0.688	0.288	0.226	2.81	1.45
4	77	M	17	0.988	0.325	0.417	2.96	1.10
5	70	M	5	0.857	0.164	0.197	3.54	1.31
6	75	M	13	0.818	0.484	0.510	1.99	2.09
7	54	M	11	1.07	0.213	0.316	5.90	4.47
8	77	M	5	0.860	0.219	0.179	5.59	2.2
9	74	M	17	0.822	0.287	0.350	6.23	3.91
10	71	F	8	0.403	0.319	0.461	4.22	0.941

AP, allopregnanolone; *DHEA*, dehydroepiandrosterone; *DHEA-S*, dehydroepiandrosterone 3-sulfate; *EAP*, epiallopregnanolone; *F*, female; *M*, male.
Based on Lionetto L, De Andrés F, Capi M, Curto M, Sabato D, Simmaco M, Bossù P, Sacchinelli E, Orfei MD, Piras F, Banaj N, Spalletta G. LC-MS/MS simultaneous analysis of allopregnanolone, epiallopregnanolone, pregnanolone, dehydroepiandrosterone and dehydroepiandrosterone 3-sulfate in human plasma. Bioanalysis 2017;9(6):527–39. Table 1 p. 530.

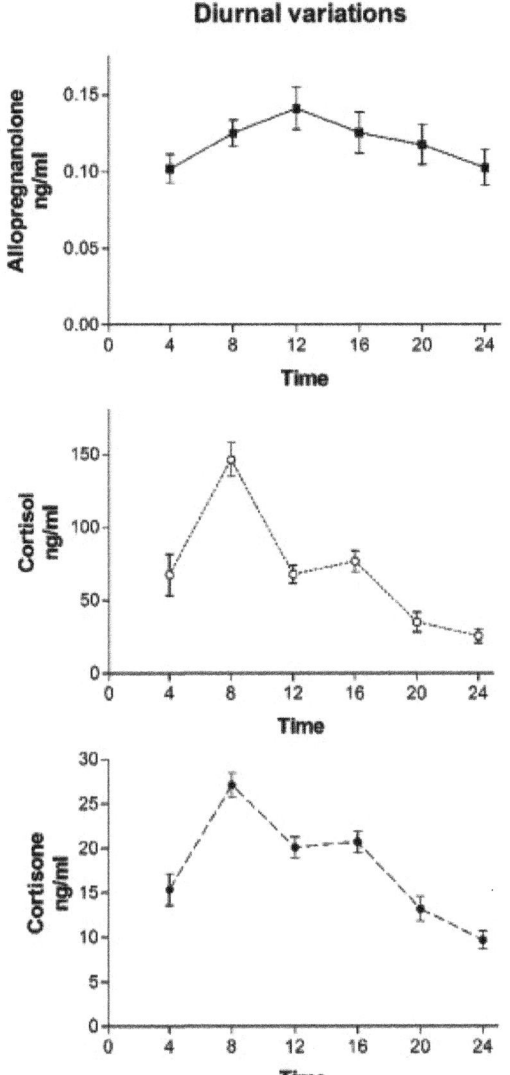

Diurnal variations

FIG. 2.5.32 Serum concentrations of allopregnanolone, cortisol, and cortisone during 24 h presented as mean ± SEM. *(From Moller AT, Backstrom T, Sondergaard HP et al. Diurnal variations of endogenous steroids in the follicular phase of the menstrual cycle. Neurochem Neuropharmacol 2016;2:1–6. Fig. 2 p. 5/6.)*

investigation in this regard (Fig. 2.5.33) (Mouritsen et al., 2015). Many laboratories will then complement those measurements into a full profile. The combination of cortisol and DHEAS is of interest in the investigation of psychoendocrinological issues.

2.5.7.1.2 Androgens

The combination of testosterone, androstenedione, dehydroepiandrosterone and dihydrotestosterone are useful in the investigation of androgen disorders in adults and children. Sample size is now less of a problem, assay sensitivities enable a profile on less than 100 μL of serum/plasma, and with some assays a steroid profile can be

determined from a heel prick blood sample (Boelen et al., 2016). Reference ranges for pubertal stage are useful (Tables 2.5.5 and 2.5.6) (Ankarberg-Lindgren et al., 2018). Gestation appropriate reference ranges during the first 6 weeks of life for infants born preterm have been reported (Greaves et al., 2014). These steroids can be measured individually on immunoassay platforms of chemistry analyzers but specificity is higher with mass spectrometric methods where these steroids can be determined in one analysis with chromatographic separation. Regardless of the method used there are variations in the results between laboratories attributed to differences in standardization.

2.5.7.1.3 Estrogens

Estrogen concentrations vary not only with age but also with sexual development (Table 2.5.8) (Ankarberg-Lindgren et al., 2018) body fat and menstrual status. The clinician will need to consider the relationship of estrogens with indices of estrogen action (uterine volume, endometrial thickness, body fat mass, gynoid fat mass, and breast size). Estrogen panels can include up to 10 metabolites and becoming essential for investigations of patients with breast cancer and postmenopause.

2.5.7.2 Steroid profiles (metabolome)

Several laboratories can now, in one analysis, provide acceptable results for 20 or more steroids in one analysis with stable isotope labeled internal standards ideally for each steroid. Reporting results for 12, 13 or less is more usual. Reference ranges for steroids with sex and age are available (Table 2.5.9 for sources) and compared with a number of variables (age, gender, oral contraceptive use, and pubertal stage) (Tavita and Greaves, 2017; Fanelli et al., 2018).

Comprehensive reference ranges, have been charted and tabulated separately for males and females (Fig. 2.5.34) (Eisenhofer et al., 2017) with comparable ranges for males and females published later (Mezzullo et al., 2020a).

The menstrual cycle can be monitored from assays of estradiol, LH and progesterone (Anckaert et al., 2021; Barrett et al., 2019). The steroids in a profile vary with the clinical interests of the laboratory in question (pediatrics, hypertension, antidoping for examples). The life time issues (mini-puberty, adrenarche, menstrual cycles, oral contraceptives, Tanner staging, menarche, menopause, and body size) are covered. A profile of eight steroids was developed for investigations before the adrenarche (Olisov et al., 2019). Methods are based on isotope dilution for each analyte but calibration curves are not always published. Only 200 μL or less of sample is required Most assays have an extraction step (PPT, solvent, SPE, and SLE) and a few include a derivative step (if GC-MS used

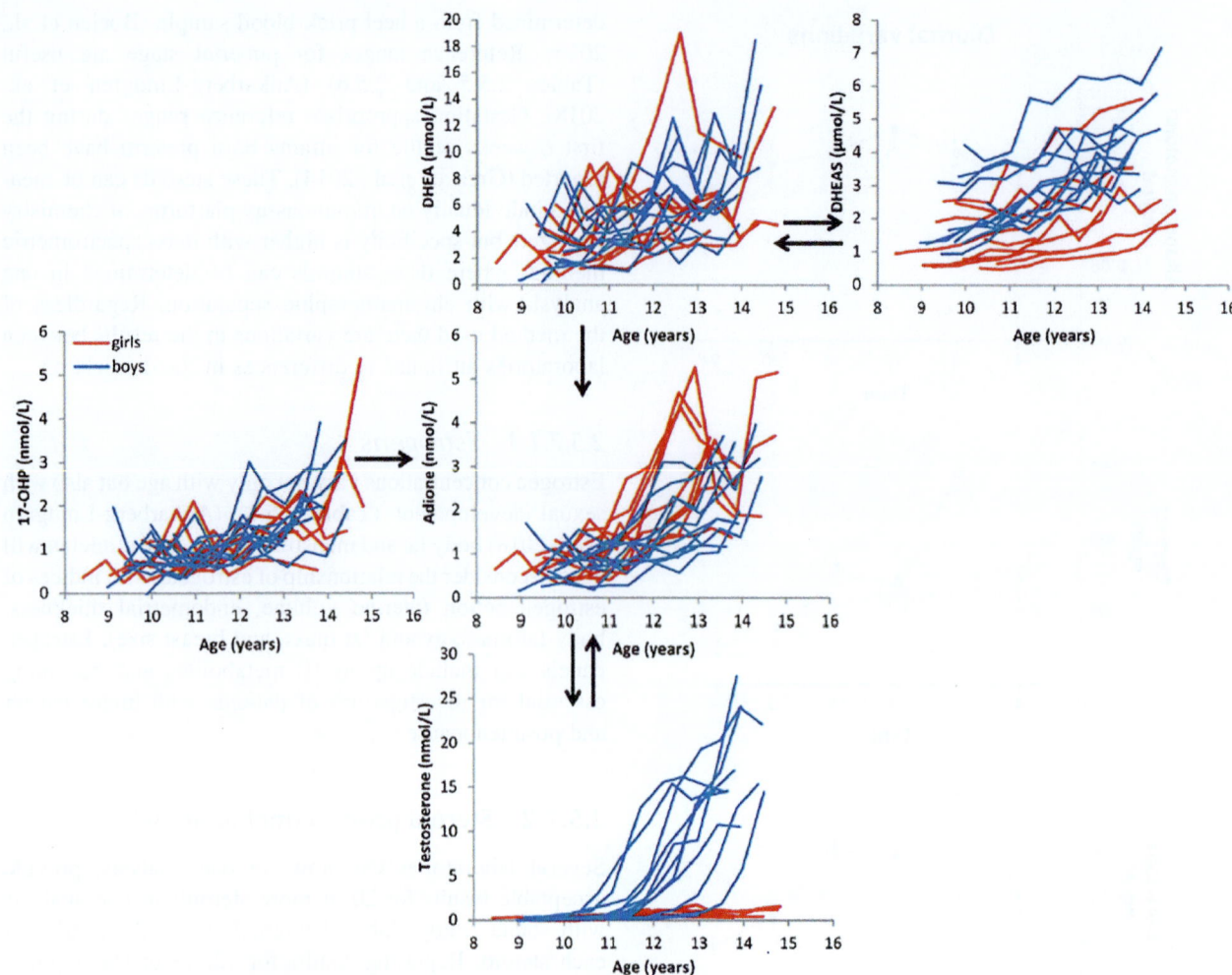

FIG. 2.5.33 Serum concentrations of 17-OHP, DHEA, DHEAS, Adione and T according to age in girls (*red lines*) and boys (*blue lines*). Each line represents a child. (*From Mouritsen A, Søeborg T, Hagen CP, Mieritz MG, Johannsen TH, Frederiksen H, Andersson AM, Juul A. Longitudinal changes in serum concentrations of adrenal androgen metabolites and their ratios by LC-MS/MS in healthy boys and girls. Clin Chim Acta 2015;450:370-5. Fig. 1 p. 372.)*

TABLE 2.5.8 Serum androgens and estrogens in girls determined by GC-MS/MS in relation to breast stage (TB 1–5).

Breast stage	TB1	TB2	TB3	TB 4–5
Androstenedione (nmol/L)	0.6 (0.3–1.3)	0.9 (0.5–2.8)	3.0 (0.9–4.9)	3.0 (0.9–7.6)
Testosterone (nmol/L)	0.2 (0.1–0.5)	0.3 (0.2–0.9)	0.7 (0.3–1.6)	0.8 (0.3–1.5)
Dihydrotestosterone (pmol/L)	52 (<27–187)	80 (28–337)	171 (69–365)	239 (28–528)
Estrone (pmol/L)	16 (<9–32)	31 (18–70)	83 (31–141)	99 (54–556)
Estradiol (pmol/L)	<2 (<2–7)	26 (6–45)	130 (37–589)	367 (89–778)
N	16	15	13	16

Data is presented as median (5th–95th percentiles). *$P<.05$, **$P<.01$, ***$P<.001$ compared with closest lower pubertal stage.
Based on Ankarberg-Lindgren C, Dahlgren J, Andersson MX. High-sensitivity quantification of serum androstenedione, testosterone, dihydrotestosterone, estrone and estradiol by gas chromatography-tandem mass spectrometry with sex- and puberty-specific reference intervals. J Steroid Biochem Mol Biol 2018;183:116–24. Table 1 p. 122.

TABLE 2.5.9 Steroid profile papers that include ref ranges.

Steroids in profile	Subjects	Analysis	Data format	Units	*n*
E1, E2, 17OHP, DHT, 17-OHPreg	**Females:** early follicular, preovulatory, lid-luteal phases. Menopausal. **Men: >17 years**		Median (2.5–97.5 percentiles)	SI units	Mezzullo et al. (2020a)
17-OHPreg, DHEA, 17-OHP, B, S, F, A4, DHT, E1, E2	**Men:** 20–65 years at 5-year intervals		Mean ± SD; Min; Max	SI units	Mezzullo et al. (2020b)
Preg, DHEA, 17OHP, B, A4. Free T		LC-MS/MS CLIA	Median; intervals	Mass	Wang et al. (2020)
F, E, Aldo, DHEAS, Prog, 17OHP, a4, T, E2	Male, female. <1, 1–5, 5–10, 10–15, 15–20, 20–40, 40–60, 60–80 years T1,2,3,4,5		Ref intervals	Mass	Bae et al. (2019)
Preg, DHEA, 17OHP, B, A, T	Boys 6–17 years		Median; intervals	Mass	Cao et al. (2017)
Preg, Prog, 17OHP, DOC, B, Aldo, 18-oxo-F, 21-DF, S, F, E, DHEA, DHEAS, A4, T	Males, females, 20–80 years. Luteal, follicular phases, O/C, BMI, hypertension	LC-MS/MS 10 min	2.5–97.5 percentiles	Mass	Eisenhofer et al. (2017)

Author original.

or for estrogens by LC-MS/MS). APCI is frequently used and rarely APPI. A C18 or phenol column is eluted with methanol which has replaced acetonitrile in gradient elution. Ammonium fluoride is incorporated to enhance the signal because steroid hormones do not easily ionize. Run times are typically less than 10 min. Results are expressed as range (2.5 and 97.5 percentiles) with median or mean ± 1.96 SD.

In order to determine ion suppression or enhancement, a number of experiments are required (see Chapter 2.4 Quantitation). While samples are being processed a continuous flow of test interferent is added after the column to check for change in signal of the analyte. A matrix factor can also be calculated as the percentage ratio between peak areas observed in test samples spiked after extraction and in 50% methanol. A deviation from 100% indicates the presence of ion suppression (<100%) or enhancement (>100%). The overall process efficiency is calculated as the percentage ratio between peak areas observed in test samples spiked before extraction and in 50% methanol. **Matrix effect** is also tested by postcolumn infusion (Mezzullo et al., 2020a). $^{13}C3$-E2 (24.3 nmol/L), $^{13}C3$-E1 (24.5 nmol/L) (both 6.7 ng/mL) and D_3-DHT (45.3 nmol/L, 13.3 ng/mL) were infused at 10 μL/min by syringe pump during LC-MS/MS runs of analyte standards (3.67, 3.70, 3.01 and 3.44 nmol/L for E2, E1, 17OHP5 and DHT, respectively, all 1 ng/mL) (Fig. 2.5.35). For the validation, this experiment was applied in samples from six individual male and female serum extracts (data not shown).

Some commercial kits for MS determination are now available (Chromsystems, Biocrates and Perkin Elmer) with all reagents and columns that are ready to use having been fully validated, although cost may be a deterrent. Details of several components of the kit are proprietary. Surprisingly few publications are based on use of commercial kits. A validation of the SteroIDQ kit was published (Koal et al., 2012) as has a method using Chromsystems certified reference standards (Yuan et al., 2020). The detection limits for some steroids (estradiol, DHT) may not be appropriate for all clinical samples with low concentrations of the steroids. The laboratory is still responsible for confirming the performance of the kit in their hands, and assessing against the cost of validating an in-house assay. Reference ranges need to be established and interferences eliminated. The transitions and best LC and MS conditions will need to be optimized for the instrumentation.

The clinician will need to consider if body size, exercise, use of steroids, misuse of steroids could have affected the results and alert the laboratory to any result that does not fit with the clinical picture in case assay interference should be considered and action taken to resolve the issue. Some steroids are important biomarkers for clinical problems. In addition to investigating clinical problems, there are now issues of hormonal supplementation to be considered in the elderly and patients with gender dysphoria. Since most women in reproductive years are taking oral contraceptives the impact of these steroids on hormone measurements need to be considered.

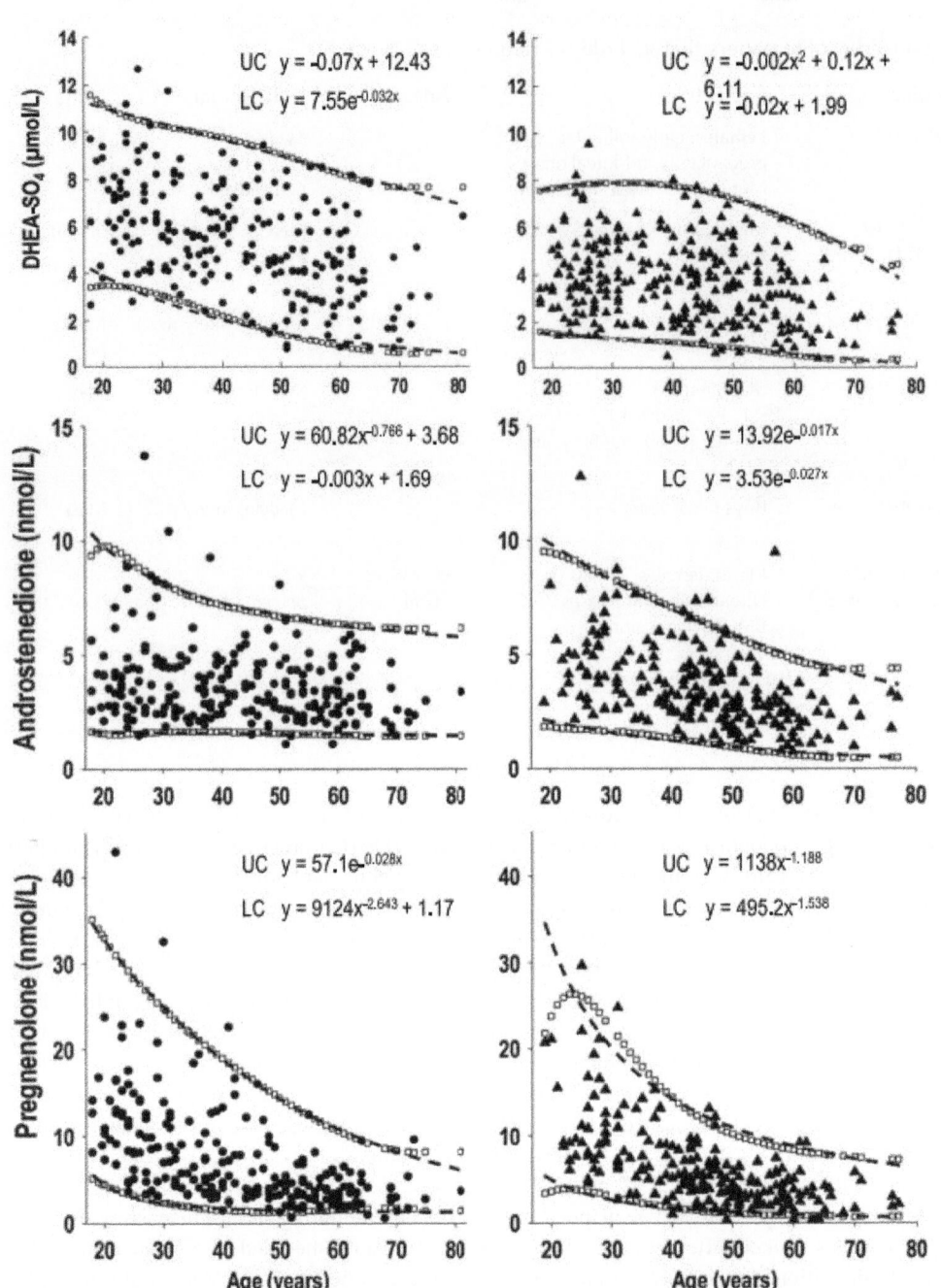

FIG. 2.5.34 Distributions of plasma concentrations of DHEA-SO4, androstenedione and pregnenolone according to age for males *(left panels)* and females *(right panels).* Note that DHEA-SO$_4$ concentrations are shown in µmol/L compared with nmol/L for other steroids. Values for 97.5 and 2.5 percentiles are displayed as *empty square dotted lines* (□□□), whereas best-fit curves for upper and lower cut-offs are displayed by *dashed lines* (– – –) according to the equations shown for age specific upper cut-offs (UC) and lower cut-offs (LC). *(From Eisenhofer G, Peitzsch M, Kaden D, Langton K, Pamporaki C, Masjkur J, Tsatsaronis G, Mangelis A, Williams TA, Reincke M, Lenders JWM, Bornstein SR. Reference intervals for plasma concentrations of adrenal steroids measured by LC-MS/MS: impact of gender, age, oral contraceptives, body mass index and blood pressure status. Clin Chim Acta 2017;470:115–24. Fig. 2 p. 180.)*

Estrogen-containing oral contraceptives increase plasma concentrations of cortisol by increasing corticosteroid-binding globulin, resulting in increased total levels of cortisol. Oral contraceptives decrease plasma concentrations of several steroids, such as progesterone, 17-hydroxyprogesterone, 11-deoxycorticosterone and androstenedione. Oral contraceptive use is thus a major confounder for interpretation of steroid profiles. Many steroids in addition to cortisol show diurnal rhythm due mainly to the changes in the HPA axis (Fig. 2.5.36).

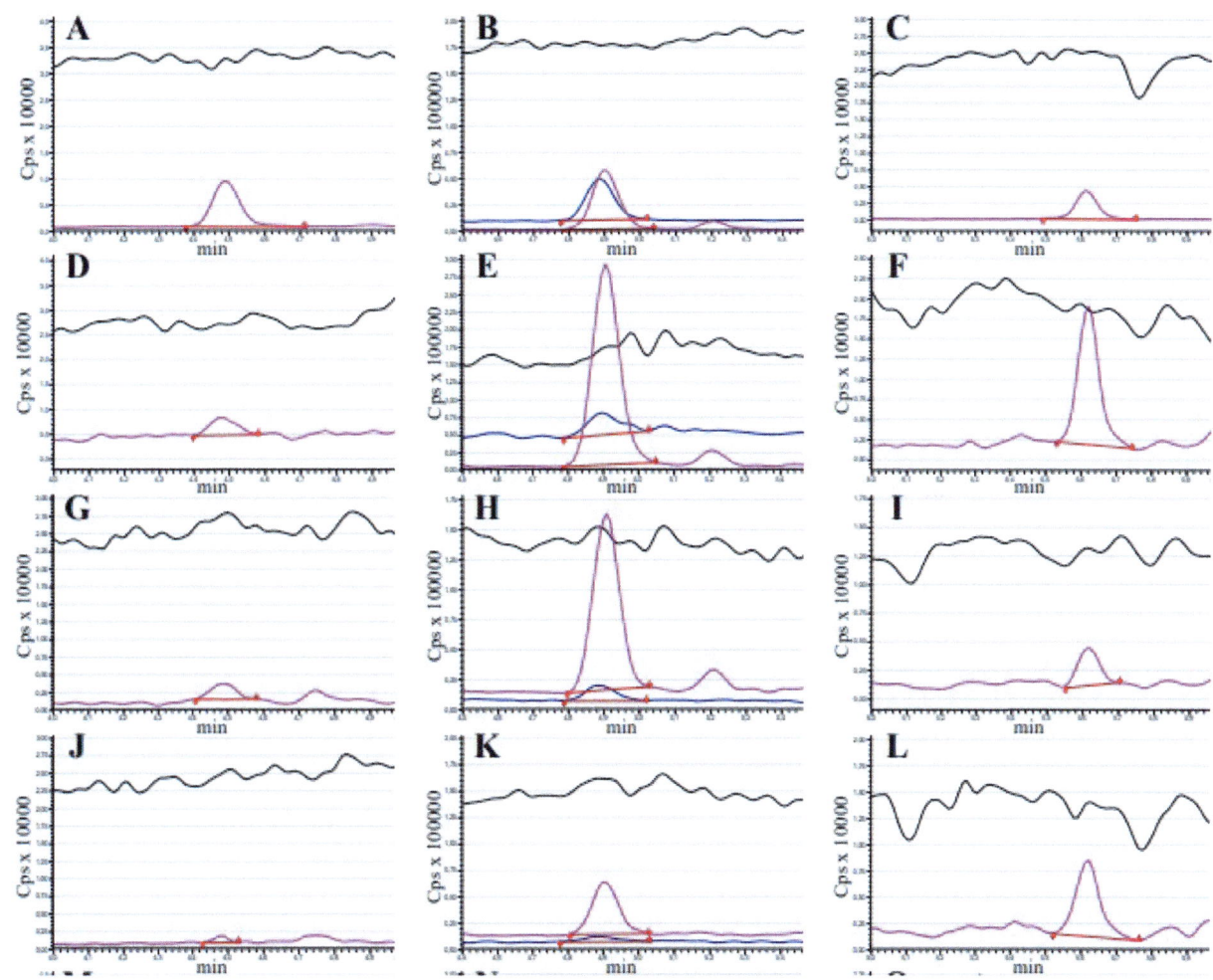

FIG. 2.5.35 Matrix effect evaluation by postcolumn infusion. Signal obtained by postcolumn infusion of $^{13}C_3$-estradiol (*Black line*) (A, D, G, J), $^{13}C_3$-estrone (B, E, H, K) and D_3-dihydrotestosterone (C, F, I, L) during LC-MS/MS run of a mixture of the four analyte standards (A–C), male individual sera (D–L). Data for female individual sera not shown here. *Pink lines*: peaks of estradiol (A, D, G, J), 17OH-pregnenolone (B, E, H, K) and dihydro-testosterone (C, F, I, L). *Light blue line*: peaks of estrone (B, E, H, K). *Cps*, counts per second. *(From Mezzullo M, Pelusi C, Fazzini A, Repaci A, Di Dalmazi G, Gambineri A, Pagotto U, Fanelli F. Female and male serum reference intervals for challenging sex and precursor steroids by liquid chromatography-tandem mass spectrometry. J Steroid Biochem Mol Biol 2020a;197:105538. Suppl 4.)*

2.5.8   Other biochemical tests

2.5.8.1   Sex hormone binding globulin (SHBG)

SHBG is a glycoprotein with high affinity binding for 17β-hydroxysteroid hormones such as testosterone and estradiol. SHBG is measured in girls when gonadal dysfunction may be associated with hyperandrogenism. SHBG concentrations are suppressed by insulin and androgens, and stimulated by estrogen and thyroid hormone. Reduced SHBG concentration is a surrogate marker for insulin resistance. In PCOS patients with abdominal obesity, reduced SHBG concentrations result in a raised free androgen index and hyperandrogenism.

Measurement of plasma sex hormone-binding globulin (SHBG) may indicate a disturbance of testicular function or androgen action. Normally, SHBG levels are around 100 nmol/L in prepubertal boys and fall progressively throughout puberty to around 60 nmol/L. Persistent elevation or a lack of decrease in SHBG following HCG stimulation may indicate a failure of testicular testosterone production, as in anorchia, or a defect of peripheral androgen action, as in AIS. In vivo androgen sensitivity can also be assessed elegantly by measuring the decrease of SHBG in response to oral administration of the synthetic anabolic steroid stanozolol. Stanozolol is a documented transcriptional activator of AR signaling. A decrease of <63% compared with the

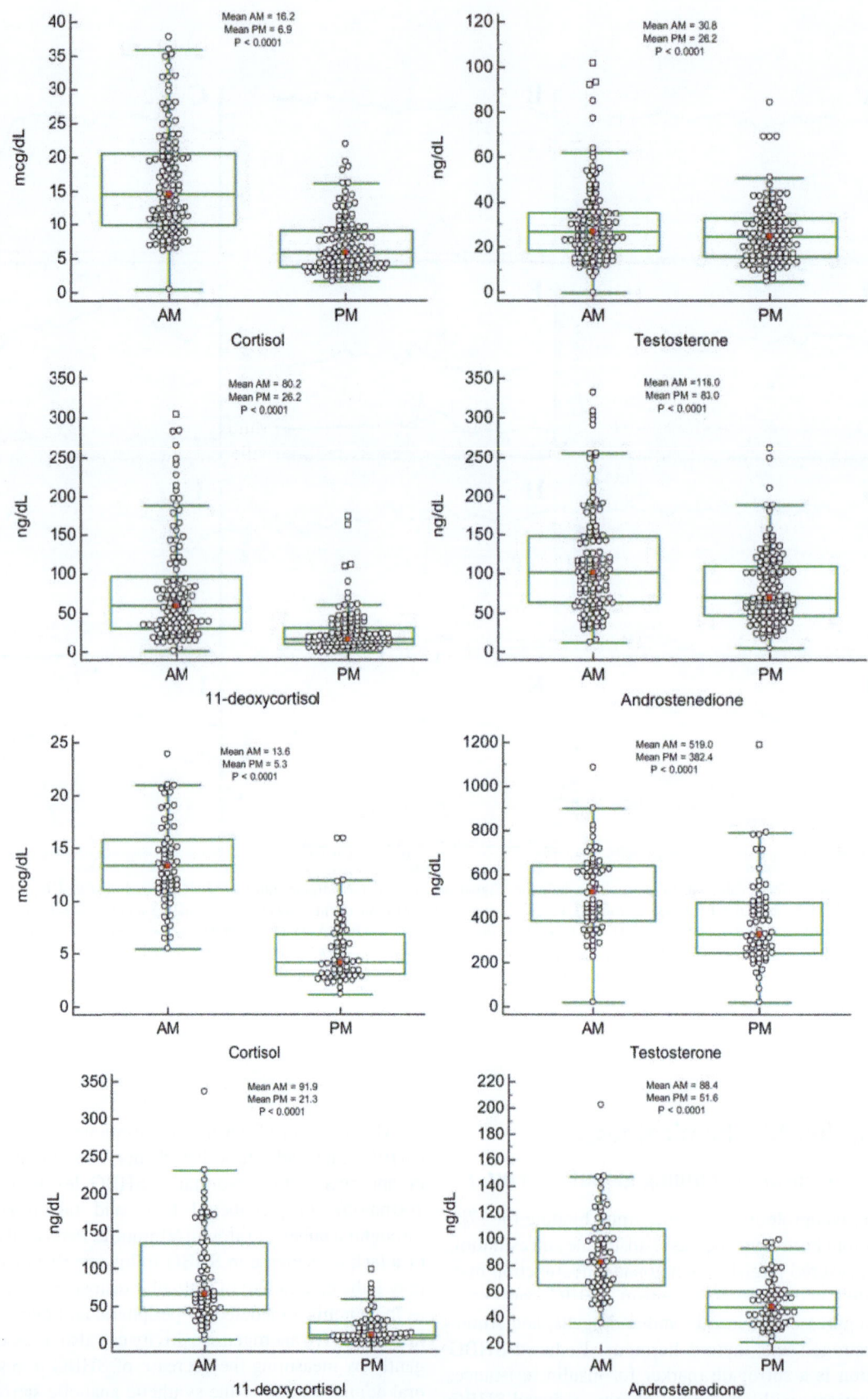

FIG. 2.5.36 Diurnal variation of steroids. Jitter box-and-whisker plots: AM and PM steroid hormone values for female samples ($n = 120$) *(upper panel)* and male samples ($n = 62$) *(lower panel)*. Mean is indicated by *small square markers* and median by *bar* in box, $P < .0001$ ($P < .05$ is considered statistically significant). *(From Parikh TP, Stolze B, Ozarda Y, Jonklaas J, Welsh K, Masika L, Hill M, DeCherney A, Soldin SJ. Diurnal variation of steroid hormones and their reference intervals using mass spectrometric analysis. Endocr Connect 2018;7(12):1354–61. Fig. 1 p. 1356.)*

TABLE 2.5.10 SHBG determination by LC-MS/MS against internal standard with stable isotope label with six additional mass units.

Transition number	Peptide code	AA sequence	Native peptide transitions	Stable isotope-transitions
1.	IAL	IALGGLLFPASNLR	$721.4^{2+}/657.4$	$727.9^{2+}/663.3$
2.	IAL	IALGGLLFPASNLR	$721.4^{2+}/804.4$	$727.9^{2+}/810.4$
3.	IAL	IALGGLLFPASNLR	$721.4^{2+}/917.5$	$727.9^{2+}/923.4$

basal value before medication is considered a normal androgen sensitivity response. In androgen insensitivity syndrome (AIS), there is decreasing downregulation with decreasing receptor function. In complete AIS, SHBG downregulation in response to stanozolol is usually abolished. Of note, patients with somatic mosaicism of an inactivating AR gene mutation may have discordant results.

SHBG has been determined by mass spectrometry after immuno-extraction and tryptic digest of endogenous SHBG (Veldhuis et al., 2014a,b). A carbon isotopically labeled internal standard (IS) is added to each sample digest. Anti-peptide antisera is used to extract a specific peptide called IAL. The stable isotope peptide (IS) was synthesized

in the Mayo Proteomics Research Center (Rochester, MN) on an ACT 396 Multiple Peptide Synthesizer (Advanced ChemTech, Louisville, KY). The peptide was synthesized with two stable-isotope labels at residues Leu (L) position 2, and Pro (P) at position 9, which add six additional mass units. Table 2.5.10 shows the sequences and transitions for native (amino acids 170–183) and isotopically labeled SHBG peptides. The double charged precursor ion at 721.4 2+ and single charge transitions 657.4, 804.4 and 917 were monitored against corresponding internal standard ions (Fig. 2.5.37).

The method gave results higher than immunoassays probably because of calibration issues.

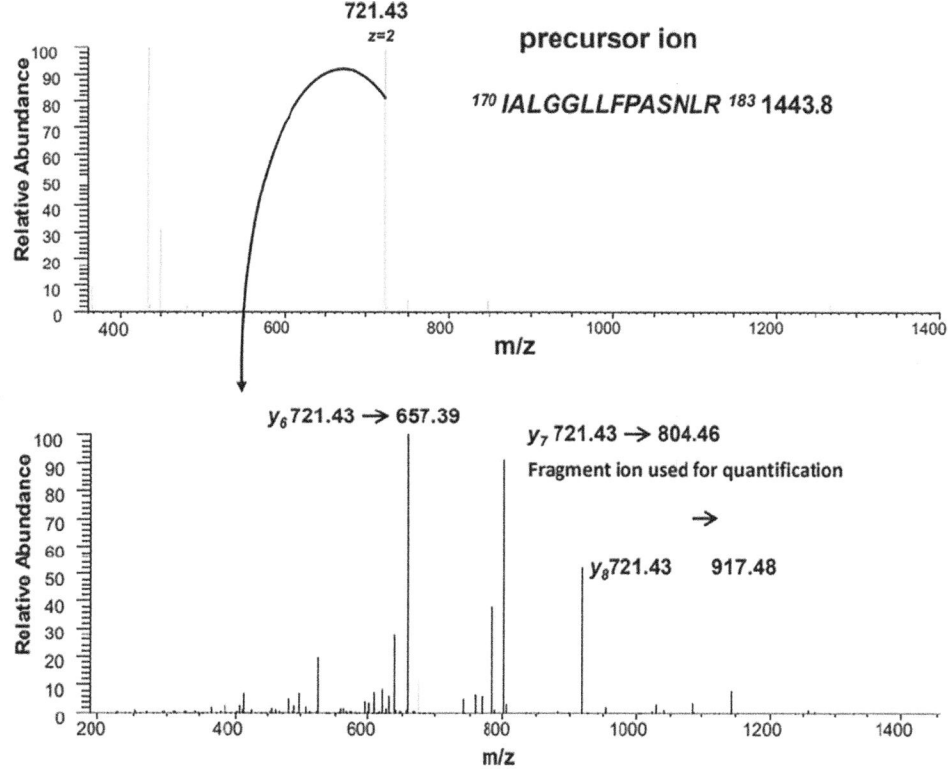

FIG. 2.5.37 Mass spectrum of SHBG-IAL peptide and transition. *(Based on Veldhuis JD, Bondar OP, Dyer RB, Trushin SA, Klee EW, Singh RJ, Klee GG. Immunological and mass spectrometric assays of SHBG: consistent and inconsistent metabolic associations in healthy men. J Clin Endocrinol Metab 2014;99(1):184–93. Suppl Fig 1B.)*

2.5.8.2   Prolactin

Results for prolactin vary considerably depending on methodology and reference preparation of the standard material. Results around 10 ng/mL are typical mean concentrations. Most assays are now performed on the automated IA platforms. Prolactin concentrations may be raised by stress, other systemic disorders and drugs. Prolactin concentrations of prolactin more than twice the reference values are usually pathological, provided the patient is not on antipsychotic medication, hypothyroid, pregnant or recently had a seizure. Hyperprolactinemia (<200 ng/mL) in children may present as delayed puberty and in girls as secondary amenorrhea. Hyperprolactinemia may be caused by either dopamine dis-inhibition from pituitary stalk compression or excess production from a pituitary adenoma. Hyperprolactinemia inhibits GnRH secretion by increasing dopamine release from the hypothalamus and inhibits steroidogenesis.

2.5.8.3   Thyroid hormone

Children with undiagnosed hypothyroidism may present with growth and pubertal delay, which progress normally after initiation of thyroid hormone replacement. Conversely, signs of advanced puberty such as, isolated testicular enlargement in boys and isolated breast development and vaginal bleeding in girls, may be present in children with severe primary hypothyroidism at diagnosis. Baseline tests classically show elevated FSH and estradiol, but suppressed LH. The mechanism is unclear but may be caused by cross-reactivity at the structurally similar FSH and thyroid stimulating hormone (TSH) receptors or direct stimulation of FSH release by elevated TSH.

2.5.8.4   Tumor markers

Tumor marker measurement in precious puberty is indicated when there is suspicion of an oncological process. For example, beta-chorionic gonadotrophin (β-hCG) concentrations in a suspected hCG-secreting tumor that may originate from the gonads, adrenals or liver; alpha-feto protein (AFP) is a marker of germ cell tumor, teratoma or hepatic tumors. The concentrations in males or nonpregnant females are <4.0 IU/L. In pregnancy, hCG is secreted by the placenta. hCG is a glycoprotein related to FSH, LH and TSH. All have similar alpha subunits but characteristic beta subunits. In pregnancy, several forms of the protein are in circulation and in the urine. These can be intact hCG, nicked hCG degradation product, free alpha subunit and free beta subunit. Commercial assays can detect combinations depending on the antibodies used. Pregnancy can usually be detected within the first week of a missed menstrual period depending on the assay and whether serum or urine is tested. In very early pregnancy, serum hCG can be 10–25 IU/L with concentrations >25 IU/L indicating pregnancy. Concentrations peak at 8–12 weeks gestation at 18,000–140,000 IU/L and 940–60,000 in third trimester.

2.5.8.5   AMH

Anti-Müllerian hormone (AMH) is a member of the transforming growth factor-β superfamily. AMH is a 140 kDa homodimer glycoprotein which is composed of two identical 70 kDa monomer subunits linked by disulfide bridges. The biologically active form of AMH in circulation is cleaved at a specific site between a pro-region and mature-region during cytoplasmic transit. Samples for AMH concentration are stable up to 7 days at 4°C, −20°C and −80°C, and 3 days at room temperature. In males, AMH is produced by the Sertoli cells from the eighth week of fetal gestation. In children with normal testicular function, circulating concentrations of AMH peak at 6 months of age and then decline during infancy and subsequently remain stable until puberty, at which point they decline to adult concentrations, which are less than 5% of those in infancy (Johannsen et al., 2020; Yates et al., 2019) (Fig. 2.5.38).

The synthesis of AMH in women comes from follicular granulosa cells in the early stage of follicular development. AMH is the key hormone to regulate follicular maturation. It reflects the number of ovarian antral follicles and preantral follicles (ovarian reserves) so is helpful in evaluating ovarian function. Detection of AMH concentration with fully automated immunoassays has provided more reproducible results with much higher sensitivity than the manual ELISA assays which are time consuming and labor intensive. Automated AMH assay results are reproducible and accurate. At least four automated AMH immunoassays are in use: Elecsys (Roche Diagnostics, Indiana, United States); Access (Beckman Coulter, Inc. California, United States, Ansh Labs, Webster, TX), Lumipulse CLEIA Fujirebio and iFlash (YHLO Biotech, Shenzhen, China). The AFIAs assay can be an alternative to the Roche Elecsys and Beckman Coulter Access 2 AMH assays (Han et al., 2022). Reference intervals for the measurement of serum AMH concentration are critical for the clinical application of AMH in the evaluation of ovarian reserve and IVF treatments (Lotierzo et al., 2021; Anckaert et al., 2016; Jopling et al., 2018) particularly if there are modifications to the assay procedure (Han et al., 2014). The antibodies and antigens used in the immunoassays are not identical, standardization of immunoassays has been a big challenge without with international reference materials. AMH concentrations measured with Elecsys AMH are significantly different from that measured with Access AMH assay (Beckman Coulter).

Female AMH concentrations peak at the age of 20–24 years and gradually decrease to almost undetectable level at menopause. There are differences in results according to

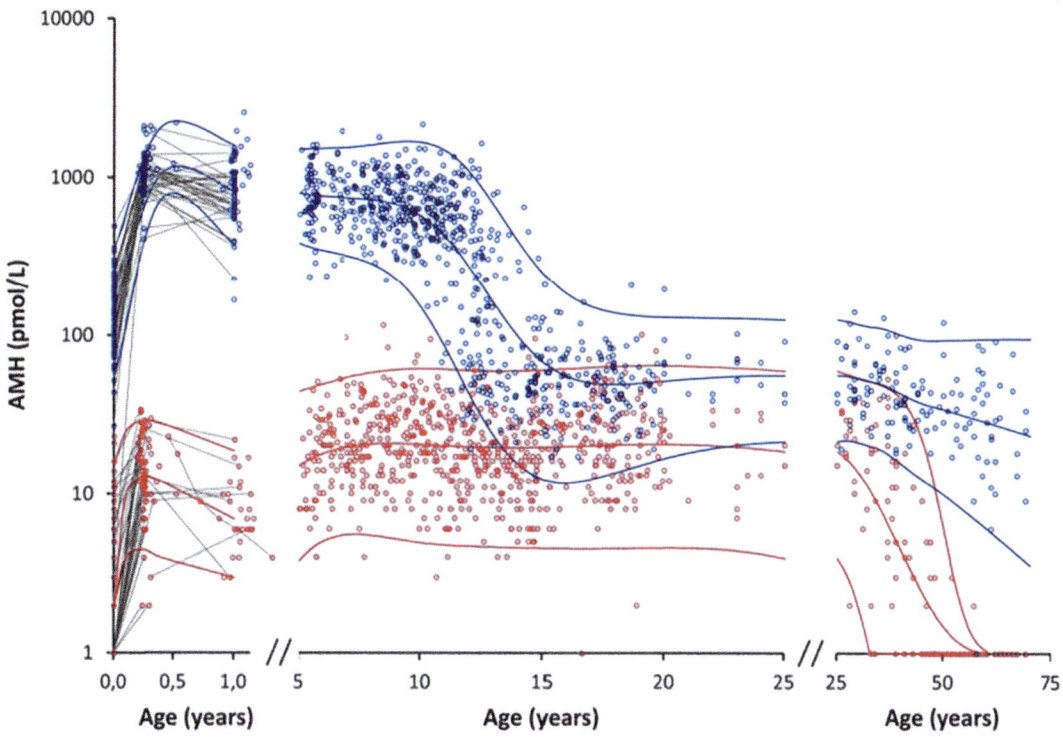

FIG. 2.5.38 Serum concentrations of AMH in 1027 healthy males (*blue*) and 926 healthy females (*red*) throughout life. Longitudinal values during infancy are connected via black lines. *Blue and red lines* mark male and female reference ranges (mean, ±2SD). The concentrations are shown on a $\log_{10}$-transformed y-axis. AMH was measured using a double antibody enzyme-immunometric assay (Immunotech, Beckman Coulter Ltd., Marseilles, France). *(From Johannsen TH, Andersson AM, Ahmed SF, de Rijke YB, Greaves RF, Hartmann MF, Hiort O, Holterhus PM, Krone NP, Kulle A, Ljubicic ML, Mastorakos G, McNeilly J, Pereira AM, Saba A, Wudy SA, Main KM, Juul A. Peptide hormone analysis in diagnosis and treatment of differences of sex development: joint position paper of EU COST Action 'DSDnet' and European Reference Network on Rare Endocrine Conditions. Eur J Endocrinol 2020;182(6):P1–P15. Fig. 2 P3.)*

the detection methods and ethnicity which exact time of AMH peak concentrations (Ortega et al., 2020). Reference intervals for females decline from 1.2 to 10.2 aged 20–24 years to <0.39 at the age of more than 50 years. The range of AMH concentrations at each age group showed significant overlap between groups between 20 and 50 years of age. Clinicians use the serum AMH concentrations as a marker for antral follicle count and ovarian reserve. Serum AMH concentrations can help to estimate how long a pregnancy can be postponed (Deeks, 2015). It can also be used for more accurate prediction of menopause, helping in perimenopausal management.

2.5.8.6 Inhibin B

Inhibins are heterodimeric polypeptide hormones that selectively suppress the secretion of pituitary follicle stimulating hormone (FSH) and also have local paracrine actions on the gonads. The fully processed inhibin molecule has a molecular weight of approximately 32–36 kDa and consists of the two distinct chains (α and β), linked by disulfide bridges (Fig. 2.5.39). Higher molecular weight forms, with precursor forms of the α-subunit, also occur in follicular fluid and serum. In addition, free α-subunit forms,

unassociated with a β-subunit, and lacking inhibin bioactivity, are also present.

Inhibin B consists of an α-subunit and a β-subunit. Inhibin B is produced by the Sertoli cells of the testis in the male and the granulosa cells of the ovary in the female. The primary role of inhibin B appears to be in the regulation of gametogenesis via negative feedback on the production of FSH. Inhibin B is used as an endocrine marker for monitoring the male and female gonadal function (Esposito et al., 2018). The ELISA's uses a pair of antibodies that ideally recognize only the functional dimeric inhibin B molecule and not the free α-subunit forms present in biological fluids. The current assays do not require a sample pretreatment step with hydrogen peroxide to oxidize two methionines in the epitope to the sulfoxide for full immunoreactivity in earlier assays.

In the male, inhibin B supports spermatogenesis. The serum inhibin B level for normal males is usually c. 400 pg/mL. In the female, inhibin B has been studied as a direct, more sensitive and earlier marker of ovarian follicle number as it is secreted directly by the granulosa cells of the small, developing follicles of the ovary. Inhibin B also finds research applications in assisted reproductive technologies (ART). Knowledge of the ovarian reserve is useful prior

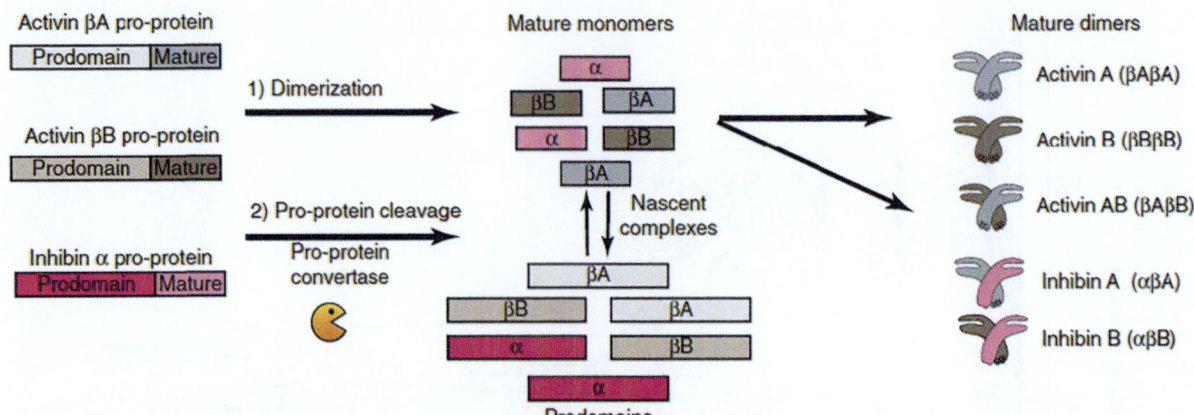

FIG. 2.5.39 Inhibins and activins. Precursor and mature forms of inhibins and activins. The inhibin α and β subunits are produced as larger precursor proteins that are cleaved by proconvertases to form the mature inhibins and activins. Inhibins are heterodimers of inhibin α and β-subunits that assemble via disulfide bridges. Activins are homodimers of two β-subunits. The molecular masses of the subunits are indicated in kilodaltons.

to and during ART and is useful in assessing the level of pharmacological stimulation required in the IVF cycle. Inhibin B levels decline substantially during early perimenopause or transition. As circulating inhibin B levels fall, the suppressive effect on pituitary FSH secretion declines leading to an elevation in circulating FSH. Elevated FSH levels then accelerate follicular recruitment and an overall decline in ovarian reserve as menopause approaches. The calibrators used in the second generation assay (recombinant inhibin B in FBS) are traceable to the WHO inhibin B reference reagent (96/784) using a factor of 2.47. The WHO preparation is composed of a mixture of inhibin forms immunopurified from human follicular fluid. The "Factor" with respect to WHO will vary between methods for the same recombinant calibrator materials due to the unavailability of the pure WHO preparation.

There have been frequent changes in the methods for AMH and inhibin from manual ELISA's to automated platforms with different standardizations (Yeates et al., 2019; Fleming et al., 2018; Jing et al., 2018; de Kat et al., 2017; Li et al., 2016; Kalra et al., 2010). Activins are dimers of the inhibin β-subunit (Wijayarathna and de Kretser, 2016) and not yet routinely measured. Two groups in Europe have produced laboratory guidelines for peptide hormone analysis that may contribute to improved diagnosis and treatment of disorders of sexual development (DSD). Reference ranges for inhibin and AMH concentrations are presented along with FSH and LH (Fig. 2.5.40) (Johannsen et al., 2020).

2.5.9 Catheter studies

Measurements of steroid concentrations in blood from the periphery cannot always be attributed to a particular source and a more direct sample is required. The procedure for obtaining catheter samples requires X-ray monitoring and a skilled radiologist with experience of the procedure. If more than one sample is required then the catheter has to

be cleared of the previous sample and saline flush. Adrenal vein catheterization is the commonest procedure, usually in the context of investigations of primary aldosteronism (see Chapter 3.2.1). The ovary is occasionally examined for androgen excess (Chapter 3.3.1).

2.5.9.1 Adrenal venous blood

Adrenal vein blood is obtained by percutaneous, transfemoral adrenal vein catheterization under imaging guidance. It is deemed to be a difficult procedure and is used only to confirm abnormal production of steroids, most likely a unilateral tumor. Adrenal tumors for cortisol, aldosterone, androgens and estrogens can be localized by adrenal venous sampling (see Chapters 3.1.1 and 3.2.1). Technical difficulty is considerably higher with the right adrenal vein because it directly enters the IVC. It is a smaller and shorter vein than on the left side and lies with a caudal direction in relation to the femoral approach. An antecubital approach has been used in some studies. The success of the procedure has been judged from the steroid determinations. Traditionally cortisol has been measured in samples from patients under investigation for hyperaldosteronism. Some centers have taken intra-operative measurements in samples sent to the laboratory for rapid analysis or using a point-of-care device in the theater (Woods et al., 2000; Page et al., 2018). The use of androstenedione, DHEAS and 11-deoxycortisol measurements to confirm access to the adrenal gland have also been proposed. For a cortisol or aldosterone secreting tumor, it is likely that blood from the contralateral side has lower concentrations than the disease side by virtue of not being stimulated from suppressed concentrations of ACTH or angiotensin II, respectively. The concentrations of most steroids in adrenal vein blood are 30–50 times the concentrations in peripheral blood with the exception of progesterone, cortisone, cortisol and DHEAS. On adrenal stimulation with ACTH, the adrenal steroid concentrations

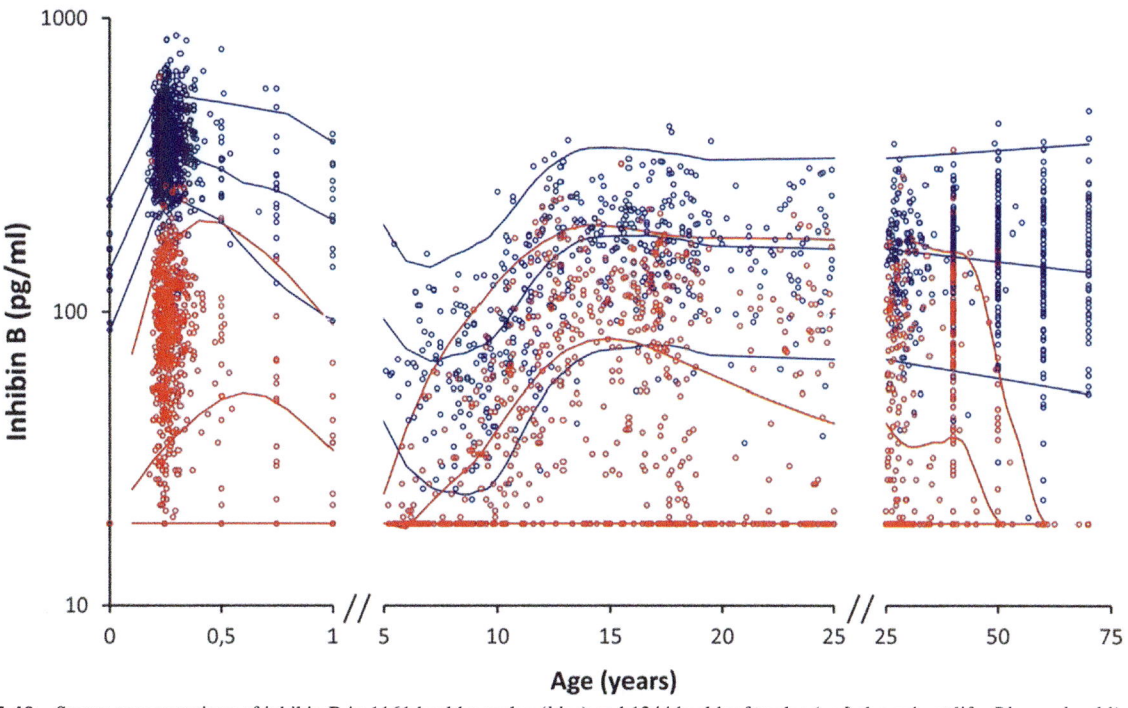

FIG. 2.5.40 Serum concentrations of inhibin B in 1161 healthy males (*blue*) and 1344 healthy females (*red*) throughout life. *Blue and red lines* mark male and female reference ranges (mean, ±2SD). The concentrations are shown on a $\log_{10}$-transformed *y*-axis. Inhibin B was measured using a double antibody enzyme-immunometric assay (Oxford Bio-Innovation, Oxfordshire, United Kingdom; later named Serotec, Oxford, United Kingdom). *(Johannsen TH, Andersson AM, Ahmed SF, de Rijke YB, Greaves RF, Hartmann MF, Hiort O, Holterhus PM, Krone NP, Kulle A, Ljubicic ML, Mastorakos G, McNeilly J, Pereira AM, Saba A, Wudy SA, Main KM, Juul A. Peptide hormone analysis in diagnosis and treatment of differences of sex development: joint position paper of EU COST Action 'DSDnet' and European Reference Network on Rare Endocrine Conditions. Eur J Endocrinol 2020;182(6): P1–P15. Fig. P3.)*

can be increased a further 2–30-fold (see Chapter 2.7 Endocrine tests).

The debate on the best protocol will continue although the cost implications of adrenal venous sampling and poor success rates will probably drive new approaches (see Chapter 3.2.1 Hypertension).

2.5.9.2 Ovarian vein blood

Ovarian vein blood sampling is a rare procedure so literature is mainly case reports where the diagnostic value for tumor location is high. Fast testosterone assays by immunoassay as intra-operative procedure has been an agreed service for "Stat" analysis of samples transferred rapidly to the laboratory in a very few cases to confirm Leydig cell tumor of the ovary (Braun et al., 2013). Testosterone concentration is typically high (105 ng/mL for example) in the vein from the tumor before resection and 0.22 ng/mL (0.78 nmol/L) after the operation while contra-lateral vein testosterone was 5 ng/mL and peripheral concentration was 7 ng/mL (Regnier et al., 2002). Samples from the affected side needed dilution to be within the range of calibration and the laboratory needed to anticipate this to get best of the first sample in the series. Where the source of excess androgen is

totally unknown both ovarian and adrenal veins may be catheterized.

2.5.9.3 Testicular vein blood

Testicular vein samples are collected to localize tumors (Antonio et al., 2019). Access for blood samples is not easy because of the tortuous nature of the veins and thick walls of the arteries. A catheter with holes in the side, near the tip and a wire inside is used to push the blood vessel wall away when aspirating the blood. Concentrations of steroids in testicular vein blood are not much higher (testosterone can be 26 nmol/L or 739 ng/dL) than the peripheral blood but the gradient is not as high as the difference with adrenal blood samples. A comparison of left and right sided venous samples in usually made. Protein hormones markers are also determined in some rare cases. The literature is largely case reports.

2.5.10 Life events

2.5.10.1 Newborn and mini-puberty

In the newborn male infant, directly after birth, normal plasma testosterone concentrations can be up to 15 nmol/L when measured with RIA in an extraction assay. The testosterone concentrations fall over the first week of life as HCG is cleared from the circulation. From 2 weeks of age to 2 months after birth, the concentrations of the androgens testosterone and dihydrotestosterone increase to approach the lower limits of normal adult male concentrations and then decline over the next 3–4 months (see earlier Fig. 2.5.23), LH is the stimulus to T secretion during this period (Johannsen et al., 2018). The period is called "**minipuberty**."

In premature boys, the plasma testosterone concentrations for postconceptional age are similar to those of term babies. The rapid decline in plasma testosterone concentrations in the immediate postnatal period and subsequent rise is related to postnatal age and not to gestational age (Bergadá et al., 2006). The secondary rise lags somewhat behind in premature infants, and peak values are significantly higher and remain elevated for a longer period of time than in term infants.

If serum is assayed for 17-OHP and other steroid hormones in the neonatal period the free steroids must be extracted with a solvent before an IA in order to be separated from fetal adrenal derived steroid sulfates that have been shown to interfere in IA. In preterm infants, the concentrations of 17-hydroxyprogesterone, cortisone and cortisol (Fig. 2.5.41) (Greaves et al., 2015) were lower than reported in term infants (Kulle et al., 2013) reflecting the lesser development of the adrenal cortex.

2.5.10.2 Puberty

Steroids influence a number of factors in development including secondary sex characteristics, height, musculature, pubic hair appearance (adrenarche), beard growth in boys, voice, skin, puberty leading to fertility, immunity, and cognition. Some of these issues are addressed in more detail in Chapter 1.7 (Physiological effects of steroids). Steroid testing needs to take into account sex, age, genital staging (Tanner and Marshall), testicular size (Prader scores), onset of menses (menarche), oral contraceptive use, height, weight, BMI, and thyroid disease. Some of these points are addressed in the review by Fanelli in 2018 and in Further Reading to this chapter though papers include IA and LC-MS/MS data. Polymorphisms in SRD52 and UGT2B17 affect testosterone concentrations (Yeap et al., 2019; Martín-Escudero et al., 2019, respectively). Breast ultrasound and testosterone ultrasound provides a useful assessment of pubertal development (Madsen et al., 2020a,b) in studies of boys and girls with reference ranges for LH, FSH, T and SHBG.

As mentioned more above, in early stages of puberty, testosterone production occurs during the night and only early morning blood samples will be useful. Later in puberty, testosterone is produced throughout the day and night, and timing of blood samples is not critical. Plasma concentrations of testosterone are usually low until pubertal stage 2–3 but can vary considerably between individuals and reach quite high values in single cases. At the clinical level, any defect of androgen production will be accompanied by impaired pubertal virilization. The measurement of plasma androgens is, however, essential in the investigation of hypogonadism in late adolescence and adult life. The finding then of a subnormal adult value indicates the need for investigation and potential androgen replacement.

2.5.10.3 Steroids in pregnancy

Steroidogenesis increases throughout pregnancy to support different physiological demands that determine the success of pregnancy from implantation of the embryo, fetal development and growth, to delivery of a healthy baby. The LH surge is important in driving development and release of an ovum but does not relate to likelihood of spontaneous pregnancy (Johnson et al., 2019). Initially progesterone is made in the corpus luteum but under the influence of hCG the placenta takes over this role (Noyola-Martínez et al., 2019). The most striking steroid changes in pregnancy are the production of estrogens with the principal estrogen, estriol, increasing from 8 weeks to term to 20 ng/mL (Fig. 2.5.42).

Estrogen levels are already raised in the first trimester and may be predictive of development of the axis and estrogen concentrations in the third trimester (Schock et al., 2016) although much can happen in the intervening months to influence those changes. Estriol production is the result of the coordinated activity of the fetal adrenal gland and the placenta. The fetal adrenal secretes DHEAS as the gland grows to be as large as the kidney at term. This adrenal zone lacks HSD3B2 activity. DHEAS is hydroxylated at C-16 in the fetal liver and 16-OH-DHEAS passes to the placenta where a sulfatase cleaves the sulfate group and the action of aromatase (CYP19A1) creates estriol (Pasqualini and Chetrite, 2016). These day's estrogens are rarely measured in pregnancy because fetal viability and development can be monitored with repeat ultrasound scans of the abdomen. Some Downs screening programmes include the measurement of estriol with inhibin A and alpha-fetoprotein. Another hormone that is raised in pregnancy is hCG, secreted by the trophoblast. hCG appears in maternal circulation 9 days after the LH peak of the conception cycle. The hCG concentrations increase from the time of implantation of the embryo to peak around 100 IU/mL around the 8th week of pregnancy, falling to around 10 IU/mL in the 20th week and remaining at a plateau there until term. An LC-MS/MS method for the steroid metabolome has been reported for small volumes of blood (Yuan et al., 2020). Maternal demographics may affect hormone concentrations (Barrett et al., 2019). Unwanted pregnancies can be the result of no, or failed contraception (Yeates et al., 2019).

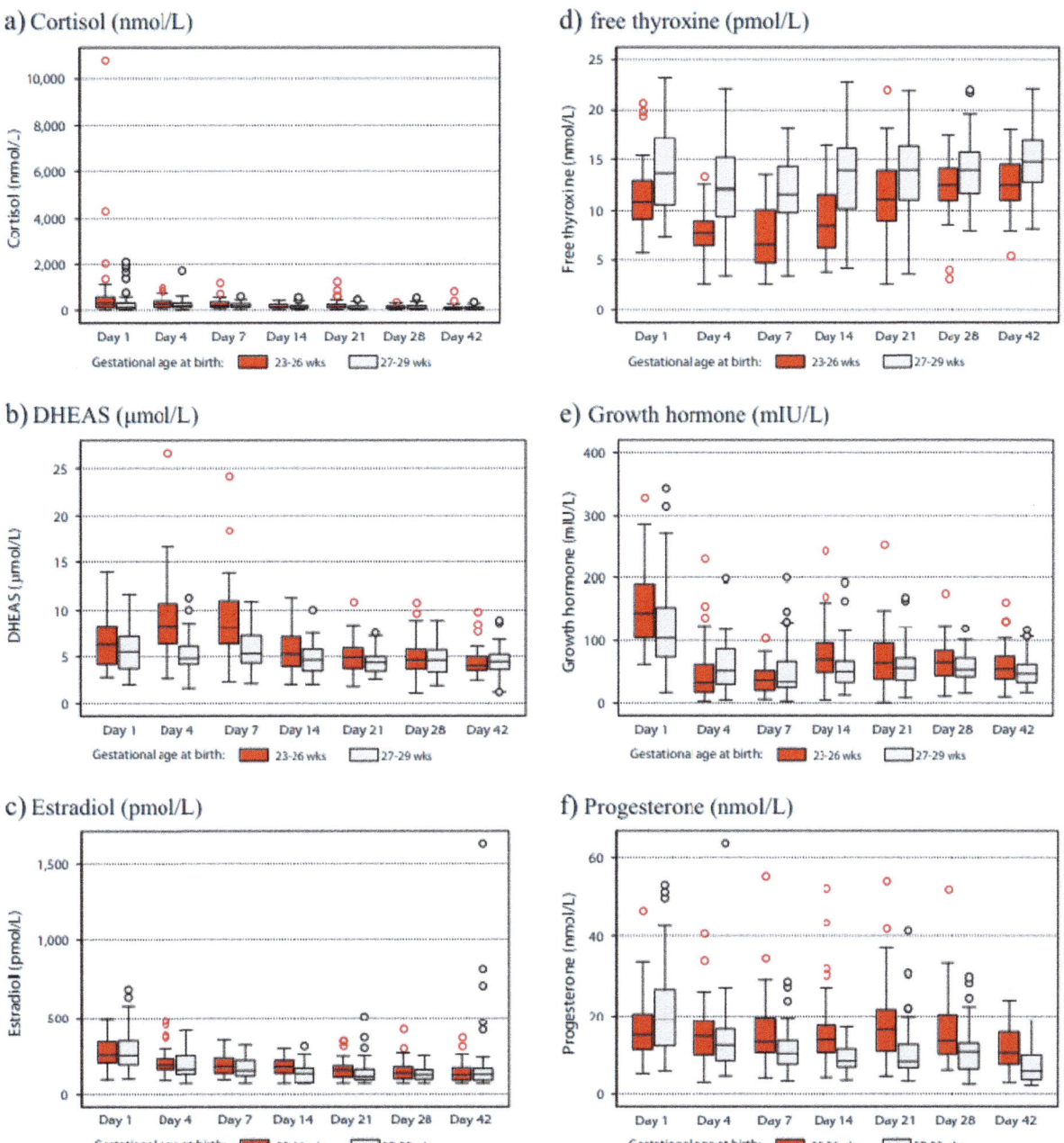

FIG. 2.5.41 Hormone levels of babies born before 32 weeks gestation. (a) Cortisol, (b) DHEAS, (c) estradiol, (d) free thyroxine, (e) growth hormone, (f) progesterone. Box whisker plots representing the "raw" hormone levels for babies born b30 weeks' gestation. The box demonstrates the median and interquartile ranges (IQR; lower i.e., 25th percentile and upper, i.e., 75th percentile). The whiskers represent the minimum and maximum values excluding outliers. The outliers, represented as dots, are values more than 1.5 times the IQR above or below the margins of the box. All results are presented as found (i.e., untransformed). With the exception of free thyroxine, all required log transformation in order to estimate the reference intervals based on the normal distribution. To convert y axis between mass units and SI units the following should be applied: Cortisol nmol/L × 0.03625 = μg/dL; DHEAS μmol/L × 0.3685 = μg/mL; free thyroxine pmol/L × 0.07769 ng/dL; growth hormone mU/L × 0.3846 = ng/mL; progesterone nmol/L × 0.3145 = ng/mL; and E2 pmol/L × 0.2724 = pg/mL. *(Redrawn from Greaves RF, Zacharin MR, Donath SM, Inder TE, Doyle LW, Hunt RW. Establishment of hormone reference intervals for infants born <30 weeks' gestation. Clin Biochem 2014;47(15):101–8. Fig. 2 p. 104.)*

2.5.10.3.1 Umbilical cord blood

There are many obstacles in taking blood samples during pregnancy. Umbilical blood collected at the delivery is noninvasive and provides a near equal mixture of maternal and fetal blood. The steroids in the sample reflect activities of the fetal adrenal and gonads, the placenta and the mother. Cortisol, DHEAS, T, DHT and A4 have been measured in umbilical cord plasma (Allvin et al., 2020; Travers et al., 2018; Hickey et al., 2018; Lundell et al., 2017; Hill et al., 2010). Most studies have used immunoassay of

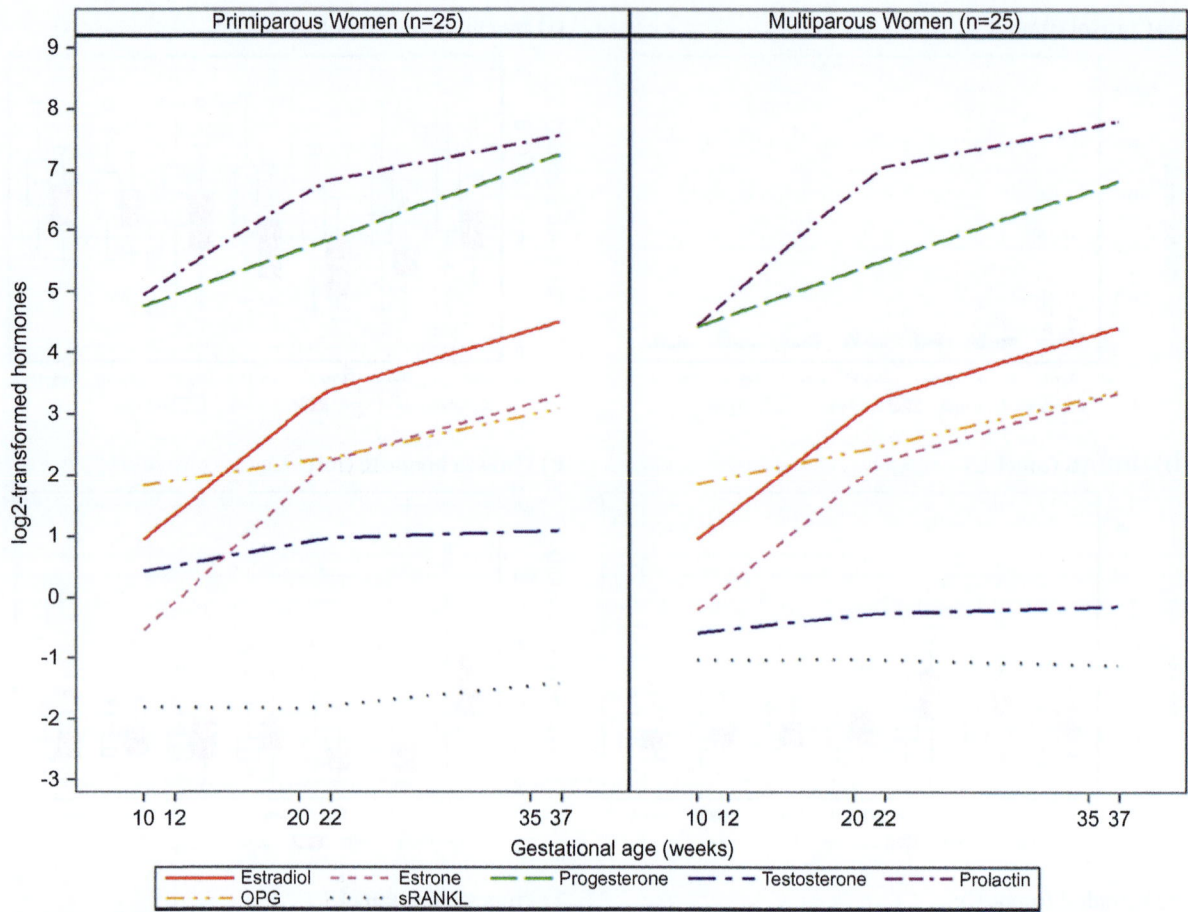

FIG. 2.5.42 Trajectories of hormone concentrations in 25 primiparous and 25 multiparous women. Hormones are log2-transformed to fit in one graph. *(Schock H, Zeleniuch-Jacquotte A, Lundin E, Grankvist K, Lakso HÅ, Idahl A, et al. Hormone concentrations throughout uncomplicated pregnancies: a longitudinal study. BMC Pregnancy Childbirth 2016;16(1):146. Fig. 1 p. 7/11.)*

questionable specificity and will not be discussed further. Few studies have used GC/MS, GC-MS/MS or LC-MS/MS. DHEA and androstenedione are present at the highest concentrations (Fig. 2.5.43). DHT was significantly higher in boys than girls.

Umbilical cord blood levels of cortisol are lower than maternal blood levels. Levels of cortisol in the umbilical cord blood only partly reflect production by the fetal adrenal glands, due to the crossing of maternal cortisol through the placental barrier where, 11β-hydroxysteroid dehydrogenase activity is responsible for the high cortisone/cortisol ratio. DHEA-S levels in umbilical cord blood are higher than maternal blood DHEA-S levels. Umbilical cord steroids have been examined to see whether androgen concentrations correlate with gender, androgen activity (from 2D:4D ratio of fingers) (Hollier et al., 2015), while cortisol has been measured after preterm labor with maternal or fetal inflammation. Umbilical cord samples reflect many tissue activities and correlations in the tests applied have been weak.

2.5.10.4 Steroids at the menopause

From around 45 years of age, the length of menstrual cycles in most women begins to be variable. By this time of life the ovaries are almost depleted of oocytes. Changes in menstrual cycle frequency are a mark of impending menopause (called the peri-menopause). Cycles can vary from 14 to 50 days. Women may complain of headache, joint stiffness, night sweats, hot flushes and breast tenderness. The World Health Organization (WHO) established a workshop for Stages of Reproductive Aging (STRAW) which over some revisions has defined the changes that now stretch from −3 to +2 (Fig. 2.5.44). The first changes are a follicular phase FSH >25 IU/L then one menstrual cycle of more than 60 days, Stage −2 has menstrual cycle variability of 7 days or more. Typical hormone profiles are shown in Fig. 2.5.45.

Detailed hormone analysis is not now recommended practice, a measurement of FSH is the only test to give any value to the clinical assessment of a woman at this time

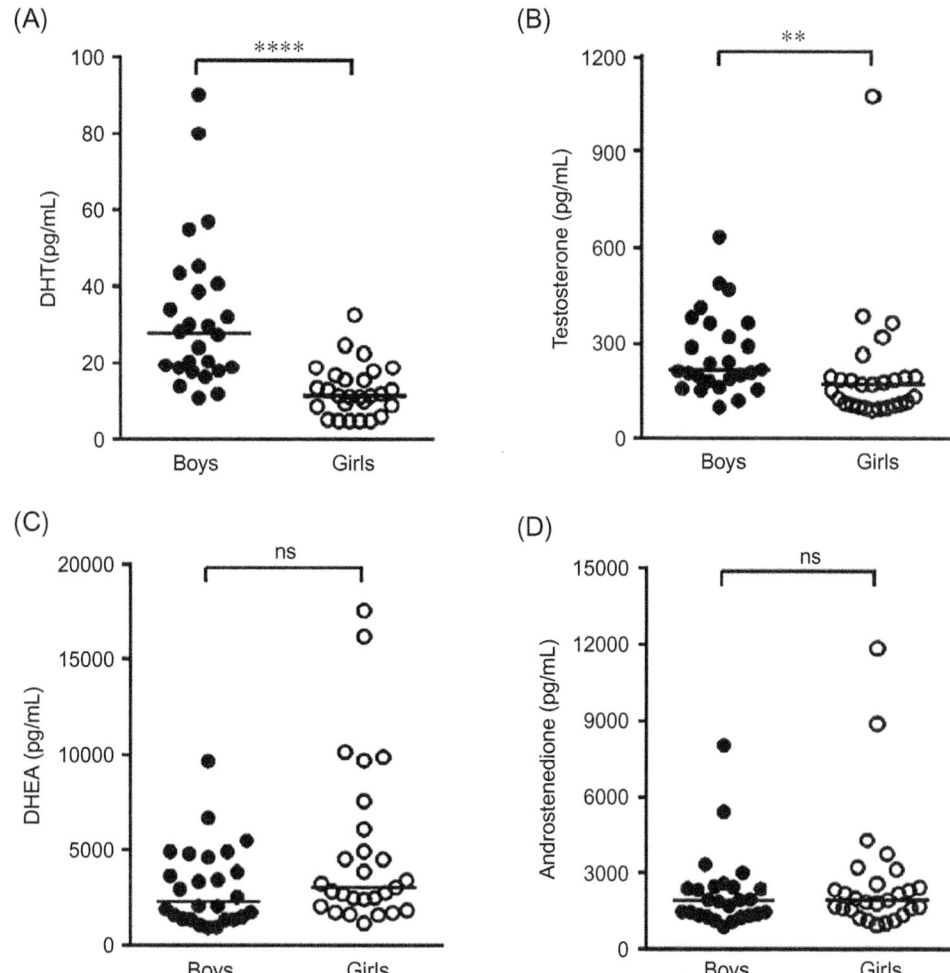

FIG. 2.5.43 Levels of androgens in umbilical cord blood according to sex. Androgens were determined by GC-MS/MS in a cohort of Swedish children (26 boys, 27 girls) born at term. (A) Dihydrotestosterone levels (DHT); (B) testosterone; (C) dehydroepiandrosterone (DHEA); (D) androstenedione. ****P < .01, ****P < .0001** in girls versus boys (Mann-Whitney U test). *ns*, not statistically significant. *(From Lundell AC, Ryberg H, Vandenput L, Rudin A, Ohlsson C, Tivesten Å. Umbilical cord blood androgen levels in girls and boys assessed by gas chromatography-tandem mass spectrometry. J Steroid Biochem Mol Biol 2017;171:195–200. Fig. 1 p. 197.)*

(Honour, 2018). Estrogen deficiency is the basis of many of the signs the woman may experience. Estradiol by LC-MS/MS can be less than 40 pmol/L in menopausal women, whereas by IA the results can be typically three times higher. Body size and composition with and without physical activity can affect interpretation of results (Tin Tin et al., 2020).

Results for steroids by LC-MS/MS are available for 70 year old women and men (Penell et al., 2021; Islam et al., 2021). The data were reported for 2.5th to 97.5th percentiles (Tables; Box/whisker plots) (Fig. 2.5.46) for a Swedish population. Subjects were carefully selected to exclude those with risk of CVD. In a separate study which included 11-oxygenated androgens (Nanba et al., 2019), those steroids did not decline with the menopause unlike other androgens.

2.5.11 Genetic regulation of sex steroid concentrations

Genetic factors strongly contribute to sex hormone concentrations in blood but knowledge of the regulatory mechanisms remains incomplete. A small number of loci have been associated with sex hormone levels in genome-wide-association-studies (GWAS). The studies reported results from the analysis of gene from hundreds of subjects but the results may be specific to, or shared with, different ethnic groups and will need repeating for local populations. A number of genomic regions (SNPS) have been associated with circulating SHBG, LH, FSH concentrations and a novel variant of CYP19A1 associated with estradiol concentrations. Polymorphisms at the LH, FSH and SHBG loci are anticipated (Chen et al., 2013).

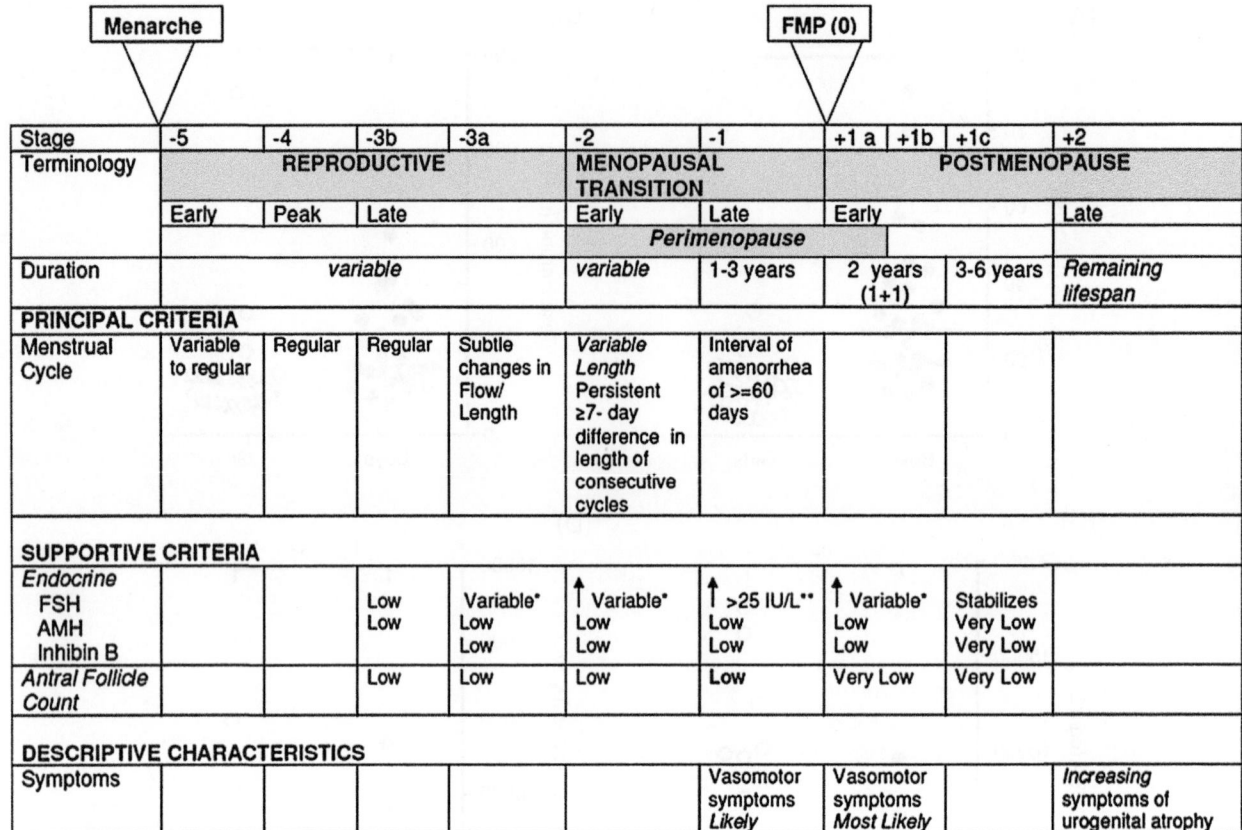

Stage	-5	-4	-3b	-3a	-2	-1	+1 a	+1b	+1c	+2
Terminology	REPRODUCTIVE				MENOPAUSAL TRANSITION		POSTMENOPAUSE			
	Early	Peak	Late		Early	Late	Early			Late
					Perimenopause					
Duration	variable				variable	1-3 years	2 years (1+1)		3-6 years	Remaining lifespan
PRINCIPAL CRITERIA										
Menstrual Cycle	Variable to regular	Regular	Regular	Subtle changes in Flow/ Length	Variable Length Persistent ≥7- day difference in length of consecutive cycles	Interval of amenorrhea of >=60 days				
SUPPORTIVE CRITERIA										
Endocrine FSH AMH Inhibin B			Low Low Low	Variable* Low Low	↑ Variable* Low Low	↑ >25 IU/L** Low Low	↑ Variable* Low Low	Stabilizes Very Low Very Low		
Antral Follicle Count			Low	Low	Low	Low	Very Low	Very Low		
DESCRIPTIVE CHARACTERISTICS										
Symptoms						Vasomotor symptoms *Likely*	Vasomotor symptoms *Most Likely*			*Increasing symptoms of urogenital atrophy*

* Blood draw on cycle days 2-5 ↑ = elevated
**Approximate expected level based on assays using current international pituitary standard

FIG. 2.5.44 The stages of reproductive aging + 10 staging system for reproductive aging in women. *AMH*, anti-Müllerian hormone; *FMP*, final menstrual period; *FSH*, follicle stimulating hormone. FSH was measured on Roche Elecsys and Cobas analyzers and Abbott Architect. *(From Burger HG. The stages of reproductive aging as proposed by workshops held in 2001 and 2010 (STRAW and STRAW + 10): a commentary. Climacteric 2013;16 Suppl. 1:5–7. Fig. 1 p. 6.)*

A polymorphism of SRD5A2 is associated with higher serum T and SHBG. Further discussion is far beyond the extent of knowledge of the author. More data is needed to determine the relevance of sex differences in sex hormone levels.

2.5.12   Summary

Over 50 years the measurements of steroids in blood has changed considerably from immunoassay to chromatographic assays combined with mass spectrometry. The specificities of the assays have improved and reference ranges have become lower. The assays used to measure a single steroid one at a time by immunoassay but now 15 or more (even 100) can be determined in a single LC-MS/MS analysis, saving time and expense per measurement. At one stage, it was suggested that testosterone results were little more than a guess. This prompted many laboratories and kit/instrument manufacturers to look to validation of the methods. More specific antibodies were generated and immunoassays shown to give results equivalent to MS methods. GC-MS methods have the disadvantage of lengthy derivative steps. The LC-MS/MS equipment is now controlled more by the computer than the operator so less skill is required until problems with the assay occur that require sensible isolation and correction of the problem. Maintenance and keeping of logs are essential. In addition to the analysis of steroids, a number of other hormone measurements are needed to understand failings of the HPA, HPG and RAAS. Journals of the American Endocrine Society proposed only accepting papers where reliable steroid results were reported, indicating MS methods or validated alternatives. This was not implemented but could still happen. EQA schemes should be distributing samples that have been measured by reliable methods not judge the results against mean values of the results from participants. There remains a need for standardization of assays (Livingston et al., 2020; Debeljak et al., 2020).

The steroid DHEAS will attract more interest as the relevance of intracrine conversion becomes more accepted. The 11-oxygenated androgens and backdoor route to testosterone are also of more interest. Interpretation of results is

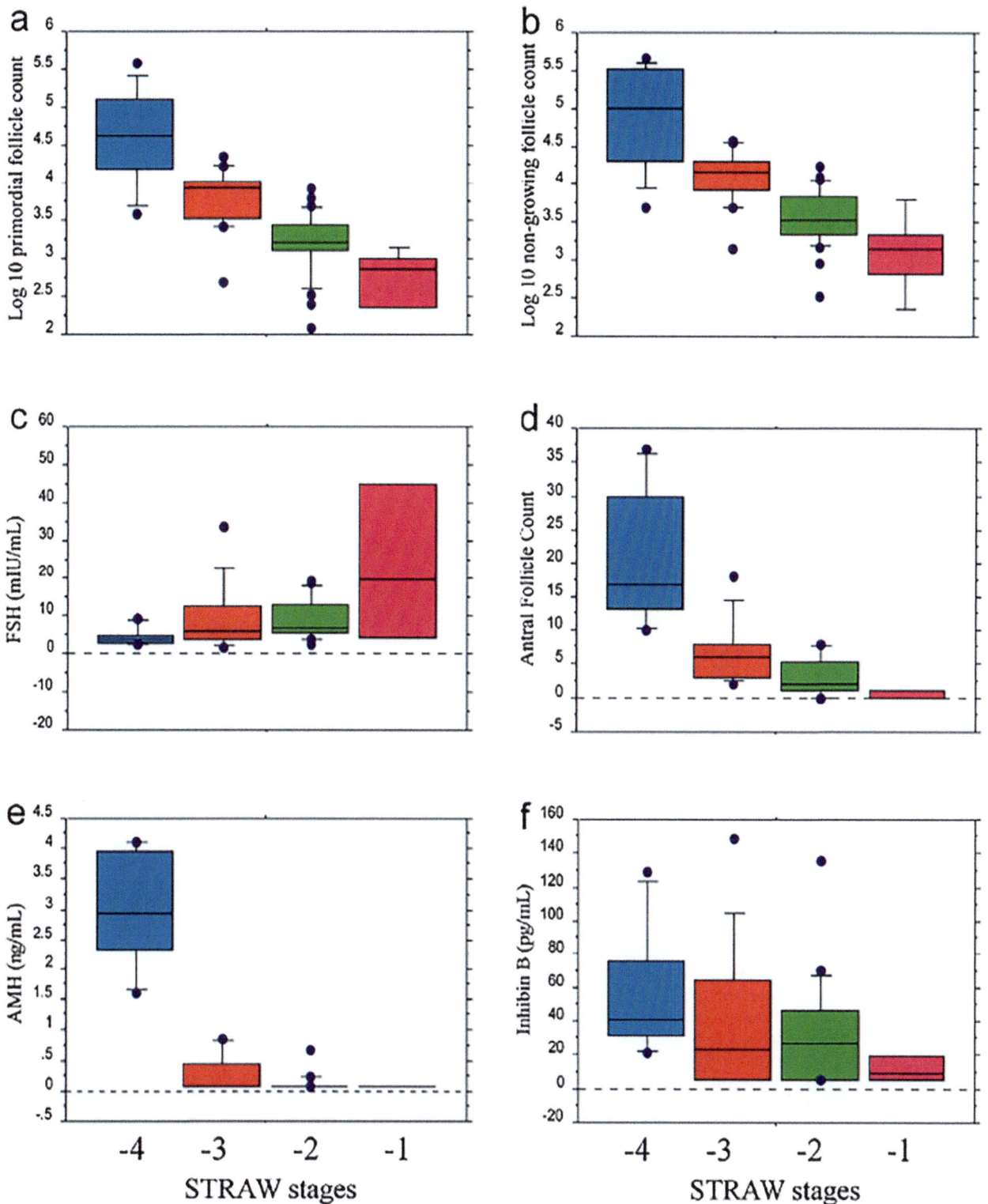

FIG. 2.5.45 Box plots of log₁₀-transformed ovarian primordial and nongrowing follicle counts, total ovarian antral follicle counts, and biomarkers of ovarian reserve for STRAW stages −4 through −1. (A) Log_{10} primordial follicle count: $P < 0.0001$ between all stages except −2/−1, where $P = 0.0074$. (B) Log_{10} nongrowing follicle count: $P < 0.0001$ between all stages except −2/−1, where $P = 0.015$. (C) FSH: $P = 0.0036$ between stages −4 and −1; all others are not significantly different. (D) Antral follicle count: $P < 0.0001$ between all stages except −3/−2, −3/−1, and −2/−1, which are not significantly different. (E) AMH: $P < 0.0001$ between all stages except −3/−2, −3/−1, and −2/−1, which are not significantly different. (F) Inhibin B: ANOVA not significant ($P = 0.23$). *AMH*, anti-Müllerian hormone; *ANOVA*, analysis of variance; *FSH*, follicle-stimulating hormone; *STRAW*, Stages of Reproductive Aging Workshop. *(Based on Hale GE, Robertson DM, Burger HG. The perimenopausal woman: endocrinology and management. J Steroid Biochem Mol Biol 2014;142:121–31. Fig 2 P125.)*

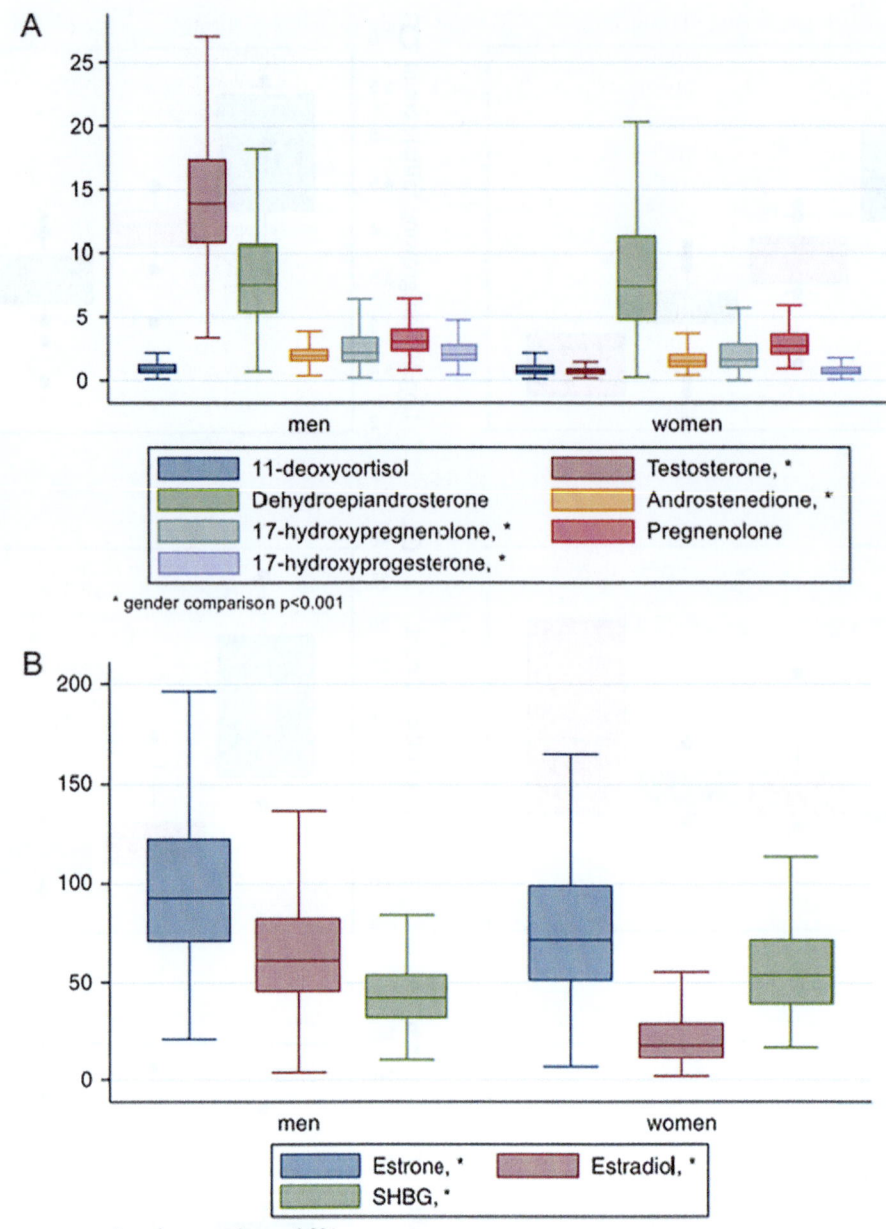

FIG. 2.5.46 Distributions of endogenous steroids in 70-year-old men and women, presented as median and 25th and 75th percentile. (A) Median concentration of testosterone, androstenedione, pregnenolone, 17-hydroxypregnenolone, 17-hydroxyprogesterone, 11-deoxycortisol, and DHEA, in nmol/L with the outline of the boxes representing the interquartile range and the error bars extending to the adjacent values for 452 men and 232 women. (B) Median concentrations of estrone, estradiol (in pmol/L), and SHBG (in nmol/L) with the outline of the boxes representing the interquartile range and the error bars extending to the adjacent values for 452 men and 232 women. *(Based on Penell JC, Kushnir MM, Lind L, Bergquist J, Bergquist J, Lind PM, Naessen T. Concentrations of nine endogenous steroid hormones in 70-year-old men and women. Endocr Connect 2021;10(5):511–20. Fig 2 P 517.)*

not straightforward mainly because of the variations with age, gender, and menstrual cycles. Many of the variables can be controlled through the implementation of operational protocols.

Kisspeptin has emerged as a critical player in the initiation of puberty and reproductive function. In humans, inactivating mutations of the kisspeptin receptor result in hypogonadotrophic hypogonadism and kisspeptin receptor activating mutations cause precocious puberty (Abbara et al., 2021). The serum concentrations of kisspeptin increase across the menstrual cycle but did not correlate with estradiol (Latif and Rafique, 2015). In men, kisspeptin increases LH, FSH and testosterone concentrations (Dhillo et al., 2005). Kisspeptin potently stimulates the release of gonadotrophins predominantly through the release of gonadotrophin-releasing hormone (GnRH). Studies of kisspeptin and infertility are scarce (Trevisan et al., 2018). Multiple sets of genes involve the regulation of puberty initiation (Roberts and Kaiser, 2020). Dysfunctions of these are likely to reveal some useful biomarkers. The Makorin RING finger protein3 (MKRN3) inhibits the secretion of GnRH and is a candidate (Varimo et al., 2016; Hagen et al., 2015). Mutations in MKRN3 gene have been found in central precocious puberty (Abreu et al., 2015). There are many commercial assays for kisspeptin and some for MKRN3 but no publications to confirm validity of measurements. MKRN3 ubiquitin signaling could control both transcriptional and posttranscriptional switches in mammalian puberty (Li et al., 2021). The exact functions of these proteins are unknown.

In the near future, information on clinical phenotyping will be considered in parallel with the biochemical (hormonal) data, the karyotype and the results for candidate genes in an integrative manner (McGrath et al., 2021). In some cases, the genetic analysis will immediately help a diagnosis, but then a multidisciplinary review will be needed, occasionally leading to the discovery of a new disorder.

References

Abbara A, Clarke SA, Dhillo WS. Clinical potential of kisspeptin in reproductive health. Trends Mol Med 2021;27(8):807–23.

Abreu AP, Macedo DB, Brito VN, Kaiser UB, Latronico AC. A new pathway in the control of the initiation of puberty: the MKRN3 gene. J Mol Endocrinol 2015;54(3):R131–9.

Adeli K, Higgins V, Trajcevski K, White-Al Habeeb N. The Canadian laboratory initiative on pediatric reference intervals: a CALIPER white paper. Crit Rev Clin Lab Sci 2017;54(6):358–413.

Agorastos A, Sommer A, Heinig A, Wiedemann K, Demiralay C. Vasopressin surrogate marker copeptin as a potential novel endocrine biomarker for antidepressant treatment response in major depression: a pilot study. Front Psych 2020;11:453.

Ahmed AH, Gordon RD, Taylor PJ, Ward G, Pimenta E, Stowasser M. Effect of contraceptives on aldosterone/renin ratio may vary according to the components of contraceptive, renin assay method, and possibly route of administration. J Clin Endocrinol Metab 2011;96(6):1797–804.

Ahmed AH, Gordon RD, Ward G, Wolley M, Kogovsek C, Stowasser M. Should aldosterone suppression tests be conducted during a particular phase of the menstrual cycle, and, if so, which phase? Results of a preliminary study. Clin Endocrinol (Oxf) 2015;83(3):303–7.

Allvin K, Ankarberg-Lindgren C, Niklasson A, Jacobsson B, Dahlgren J. Altered umbilical sex steroids in preterm infants born small for gestational age. J Matern Fetal Neonatal Med 2020;33(24):4164–70.

Anckaert E, Öktem M, Thies A, Cohen-Bacrie M, Daan NM, Schiettecatte J, et al. Multicenter analytical performance evaluation of a fully automated anti-Müllerian hormone assay and reference interval determination. Clin Biochem 2016;49(3):260–7.

Anckaert E, Jank A, Petzold J, Rohsmann F, Paris R, Renggli M, et al. Extensive monitoring of the natural menstrual cycle using the serum biomarkers estradiol, luteinizing hormone and progesterone. Pract Lab Med 2021;25, e00211.

Ankarberg-Lindgren C, Dahlgren J, Andersson MX. High-sensitivity quantification of serum androstenedione, testosterone, dihydrotestosterone, estrone and estradiol by gas chromatography-tandem mass spectrometry with sex- and puberty-specific reference intervals. J Steroid Biochem Mol Biol 2018;183:116–24.

Antonio L, Albersen M, Billen J, Maleux G, Van Rompuy AS, Coremans P, et al. Testicular vein sampling can reveal gonadotropin-independent unilateral steroidogenesis supporting spermatogenesis. J Endocr Soc 2019;3(10):1881–6.

Bae YJ, Zeidler R, Baber R, Vogel M, Wirkner K, Loeffler M, et al. Reference intervals of nine steroid hormones over the life-span analyzed by LC-MS/MS: effect of age, gender, puberty, and oral contraceptives. J Steroid Biochem Mol Biol 2019;193:105409.

Bailey D, Colantonio D, Kyriakopoulou L, Cohen AH, Chan MK, Armbruster D, et al. Marked biological variance in endocrine and biochemical markers in childhood: establishment of pediatric reference intervals using healthy community children from the CALIPER cohort. Clin Chem 2013;59(9):1393–405.

Ballerini MG, Chiesa A, Morelli C, Frusti M, Ropelato MG. Serum concentration of 17α-hydroxyprogesterone in children from birth to adolescence. Horm Res Paediatr 2014;81(2):118–25.

Barrett ES, Mbowe O, Thurston SW, Butts S, Wang C, Nguyen R, et al. Predictors of steroid hormone concentrations in early pregnancy: results from a multi-center cohort. Matern Child Health J 2019;23(3):397–407.

Bergadá I, Milani C, Bedecarrás P, Andreone L, Ropelato MG, Gottlieb S, et al. Time course of the serum gonadotropin surge, inhibins, and anti-Müllerian hormone in normal newborn males during the first month of life. J Clin Endocrinol Metab 2006;91(10):4092–8.

Bernstone L, Adaway JE, Keevil BG. An LC-MS/MS assay for analysis of equilibrium angiotensin II in human serum. Ann Clin Biochem 2021;58(5):422–33.

Bhake R, Russell GM, Kershaw Y, Stevens K, Zaccardi F, Warburton VEC, et al. Continuous free cortisol profiles in healthy men. J Clin Endocrinol Metab 2020;105(4), dgz002.

Bileck A, Fluck CE, Dhayat N, Groessl M. How high-resolution techniques enable reliable steroid identification and quantification. J Steroid Biochem Mol Biol 2019;186:74–8.

Boelen A, Ruiter AF, Claahsen-van der Grinten HL, Endert E, Ackermans MT. Determination of a steroid profile in heel prick blood using LC-MS/MS. Bioanalysis 2016;8(5):375–84.

Bowen R, Benavides R, Colón-Franco JM, Katzman BM, Muthukumar A, Sadrzadeh H, et al. Best practices in mitigating the risk of biotin interference with laboratory testing. Clin Biochem 2019;74:1–11.

Boyar RM, Rosenfeld RS, Kapen S, Finkelstein JW, Roffwarg HP, Weitzman ED, et al. Human puberty. Simultaneous augmented secretion of luteinizing hormone and testosterone during sleep. J Clin Invest 1974;54(3):609–18.

Braun R, Peter A, Warmann S, Fuchs J, Binder G. Fast intraoperative testosterone assay confirms the location of an ovarian virilizing tumor in a young girl. Horm Res Paediatr 2013;79(2):110–3.

Büttler RM, Martens F, Kushnir MM, Ackermans MT, Blankenstein MA, Heijboer AC. Simultaneous measurement of testosterone, androstenedione and dehydroepiandrosterone (DHEA) in serum and plasma using isotope-dilution 2-dimension ultra high performance liquid-chromatography tandem mass spectrometry (ID-LC-MS/MS). Clin Chim Acta 2015a;438:157–9.

Büttler RM, Martens F, Fanelli F, Pham HT, Kushnir MM, Janssen MJ, et al. Comparison of 7 published LC-MS/MS methods for the simultaneous measurement of testosterone, androstenedione, and dehydroepiandrosterone in serum. Clin Chem 2015b;61(12):1475–83.

Bystrom CE, Salameh W, Reitz R, Clarke NJ. Plasma renin activity by LC-MS/MS: development of a prototypical clinical assay reveals a subpopulation of human plasma samples with substantial peptidase activity. Clin Chem 2010;56(10):1561–9.

Campi B, Frascarelli S, Pietri E, Massa I, Donati C, Bozic R, et al. Quantification of dehydroepiandrosterone in human serum on a routine basis: development and validation of a tandem mass spectrometry method based on a surrogate analyte. Anal Bioanal Chem 2018; 410:407–16.

Cao ZT, Botelho JC, Rej R, Vesper H. Accuracy-based proficiency testing for testosterone measurements with immunoassays and liquid chromatography-mass spectrometry. Clin Chim Acta 2017;469:31–6.

Cao B, Gong C, Wu D, Liang X, Li W, Liu M, et al. A cross-sectional survey of adrenal steroid hormones among overweight/obese boys according to puberty stage. BMC Pediatr 2019;19(1):414.

Chakera AJ, McDonald TJ, Knight BA, Vaidya B, Jones AG. Current laboratory requirements for adrenocorticotropic hormone and renin/aldosterone sample handling are unnecessarily restrictive. Clin Med (Lond) 2017;17(1):18–21.

Chappell DL, McAvoy T, Weiss B, Weiner R, Laterza OF. Development and validation of an ultra-sensitive method for the measurement of plasma renin activity in human plasma via LC-MS/MS. Bioanalysis 2012;4(23):2843–50.

Chen Z, Tao S, Gao Y, Zhang J, Hu Y, Mo L, et al. Genome-wide association study of sex hormones, gonadotropins and sex hormone-binding protein in Chinese men. J Med Genet 2013;50(12):794–801.

Christensen M, Madsen RF, Møller LR, Knudsen CS, Samson MH. Whole blood samples for adrenocorticotrophic hormone measurement can be stored at room temperature for 4 hours. Scand J Clin Lab Invest 2016;76(8):653–6.

Corrie JE, Edwards CR, Budd PS. A radioimmunoassay for 18-hydroxycortisol in plasma and urine. Clin Chem 1985;31(6):849–52.

Crosby SR, Stewart MF, Ratcliffe JG, White A. Direct measurement of the precursors of adrenocorticotropin in human plasma by two-site immunoradiometric assay. J Clin Endocrinol Metab 1988;67(6): 1272–7.

Cunnah D, Jessop DS, Besser GM, Rees LH. Measurement of circulating corticotrophin-releasing factor in man. J Endocrinol 1987;113 (1):123–31.

de Bruin RA, Bouhuizen A, Diederich S, Perschel FH, Boomsma F, Deinum J. Validation of a new automated renin assay. Clin Chem 2004;50(11):2111–6.

de Kat AC, Broekmans FJM, van Westing AC, Lentjes E, Verschuren WMM, van der Schouw YT. A quantitative comparison of anti-Müllerian hormone measurement and its shifting boundaries between two assays. Maturitas 2017;101:12–6.

Debeljak Ž, Marković I, Pavela J, Lukić I, Mandić D, Mandić S, et al. Analytical bias of automated immunoassays for six serum steroid hormones assessed by LC-MS/MS. Biochem Med (Zagreb) 2020;30(3), 030701.

Deeks ED. Elecsys(®) AMH assay: a review in anti-Müllerian hormone quantification and assessment of ovarian reserve. Mol Diagn Ther 2015;19(4):245–9.

Deng L, Xiong Z, Li H, Lei X, Cheng L. Analytical validation and investigation on reference intervals of aldosterone and renin in Chinese Han population by using fully automated chemiluminescence immunoassays. Clin Biochem 2018;56:89–94.

Denver N, Khan S, Homer NZM, MacLean MR, Andrew R. Current strategies for quantification of estrogens in clinical research. J Steroid Biochem Mol Biol 2019;192:105373.

Dhillo WS, Chaudhri OB, Patterson M, Thompson EL, Murphy KG, Badman MK, et al. Kisspeptin-54 stimulates the hypothalamic-pituitary gonadal axis in human males. J Clin Endocrinol Metab 2005;90(12):6609–15.

Dichtel LE, Schorr M, Loures de Assis C, Rao EM, Sims JK, Corey KE, et al. Plasma free cortisol in states of normal and altered binding globulins: implications for adrenal insufficiency diagnosis. J Clin Endocrinol Metab 2019;104(10):4827–36.

Dobson SH, Gray C, Smith H, Baker T, Ratcliffe JG, White A. Selection and optimisation of monoclonal antibodies for a two-site immunoradiometric assay for ACTH. J Immunol Methods 1986;88(1):83–90.

Donegan DM, Algeciras-Schimnich A, Hamidi O, Young WF, Nippoldt T, Bancos I, et al. Corticotropin hormone assay interference: a case series. Clin Biochem 2019;63:143–7.

Dorrian CA, Toole BJ, Alvarez-Madrazo S, Kelly A, Connell JM, Wallace AM. A screening procedure for primary aldosteronism based on the Diasorin Liaison automated chemiluminescent immunoassay for direct renin. Ann Clin Biochem 2010;47(Pt 3):195–9.

du Toit T, Stander MA, Swart AC. A high-throughput UPC²-MS/MS method for the separation and quantification of C_{19} and C_{21} steroids and their C11-oxy steroid metabolites in the classical, alternative, backdoor and 11OHA4 steroid pathways. J Chromatogr B Analyt Technol Biomed Life Sci 2018;1080:71–81.

Dunn JF, Nisula BC, Rodbard D. Transport of steroid hormones: binding of 21 endogenous steroids to both testosterone-binding globulin and corticosteroid-binding globulin in human plasma. J Clin Endocrinol Metab 1981;53(1):58–68.

Dušková M, Sosvorová L, Vítků J, Jandíková H, Rácz B, Chlupáčová T, et al. Changes in the concentrations of corticoid metabolites—the effect of stress, diet and analytical method. Prague Med Rep 2015;116(4):268–78.

Edwards CR, Chard T, Kitau MJ, Forsling ML, Landon J. The development of a radioimmunoassay for arginine-vasopressin: production of antisera and labelled hormone; separation techniques; specificity and sensitivity of the assay in aqueous solution. J Endocrinol 1972; 52(2):279–88.

Eisenhofer G, Peitzsch M, Kaden D, Langton K, Pamporaki C, Masjkur J, et al. Reference intervals for plasma concentrations of adrenal steroids measured by LC-MS/MS: impact of gender, age, oral contraceptives, body mass index and blood pressure status. Clin Chim Acta 2017; 470:115–24.

El-Farhan N, Hampton D, Penney M. Measurement of arginine vasopressin. Methods Mol Biol 2013;1065:129–39.

El-Farhan N, Rees DA, Evans C. Measuring cortisol in serum, urine and saliva—are our assays good enough? Ann Clin Biochem 2017; 54(3):308–22.

Esposito S, Cofini M, Rigante D, Leonardi A, Lucchetti L, Cipolla C, et al. Inhibin B in healthy and cryptorchid boys. Ital J Pediatr 2018;44(1):81.

Fabregat-Cabello N, Peeters SD, Yilmaz T, Cavalier É, Le Goff CM. Establishment of reference intervals for serum concentrations of androstanediol glucuronide by a newly developed LC-MS/MS method. Steroids 2019;143:62–6.

Fanelli F, Baronio F, Ortolano R, Mezzullo M, Cassio A, Pagotto U, et al. Normative basal values of hormones and proteins of gonadal and adrenal functions from birth to adulthood. Sex Dev 2018;12(1–3): 50–94.

Faqehi AMM, Cobice DF, Naredo G, Mak TCS, Upreti R, Gibb FW, et al. Derivatization of estrogens enhances specificity and sensitivity of analysis of human plasma and serum by liquid chromatography tandem mass spectrometry. Talanta 2016;151:148–56.

Fenske W, Störk S, Blechschmidt A, Maier SG, Morgenthaler NG, Allolio B. Copeptin in the differential diagnosis of hyponatremia. J Clin Endocrinol Metab 2009;94(1):123–9.

Fiet J, Le Bouc Y, Guéchot J, Hélin N, Maubert MA, Farabos D, et al. A liquid chromatography/tandem mass spectometry profile of 16 serum steroids, including 21-deoxycortisol and 21-deoxycorticosterone, for management of congenital adrenal hyperplasia. J Endocr Soc 2017;1 (3):186–201.

Fiselier TJ, Lijnen P, Monnens L, van Munster P, Jansen M, Peer P. Levels of renin, angiotensin I and II, angiotensin-converting enzyme and aldosterone in infancy and childhood. Eur J Pediatr 1983;141(1):3–7.

Fleming R, Fairbairn C, Gaudoin M. Objective multicentre performance of the automated assays for AMH and estimation of established critical concentrations. Hum Fertil (Camb) 2018;21(4):269–74.

Fommei E, Ghione S, Ripoli A, Maffei S, Di Cecco P, Iervasi A, et al. The ovarian cycle as a factor of variability in the laboratory screening for primary aldosteronism in women. J Hum Hypertens 2009;23(2):130–5.

Forest MG, Sizonenko PC, Cathiard AM, Bertrand J. Hypophyso-gonadal function in humans during the first year of life. 1. Evidence for testicular activity in early infancy. J Clin Invest 1974;53(3):819–28.

Frederiksen H, Johannsen TH, Andersen SE, Albrethsen J, Landersoe SK, Petersen JH, et al. Sex-specific estrogen levels and reference intervals from infancy to late adulthood determined by LC-MS/MS. J Clin Endocrinol Metab 2020;105(3):754–68.

Freel EM, Shakerdi LA, Friel EC, Wallace AM, Davies E, Fraser R, et al. Studies on the origin of circulating 18-hydroxycortisol and 18-oxocortisol in normal human subjects. J Clin Endocrinol Metab 2004;89(9):4628–33.

French D, Drees J, Stone JA, Holmes DT, van der Gugten JG. Comparison of four clinically validated testosterone LC-MS/MS assays: harmonization is an attainable goal. Clin Mass Spectrom 2018;11:12–20.

Gatti R, Antonelli G, Prearo M, Spinella P, Cappellin E, De Palo EF. Cortisol assays and diagnostic laboratory procedures in human biological fluids. Clin Biochem 2009;42(12):1205–17.

Giton F, Valleix A, Boudou P, Villette JM, Bélanger A, Galons H, et al. Specific radioimmunoassay of estrone sulfate. Application to measurement in male plasma. J Steroid Biochem Mol Biol 2002;81(1): 85–94.

Giton F, Caron P, Bérubé R, Bélanger A, Barbier O, Fiet J. Plasma estrone sulfate assay in men: comparison of radioimmunoassay, mass spectrometry coupled to gas chromatography (GC-MS), and liquid chromatography-tandem mass spectrometry (LC-MS/MS). Clin Chim Acta 2010;411(17–18):1208–13.

Goodarzi MO, Carmina E, Azziz R. DHEA, DHEAS and PCOS. J Steroid Biochem Mol Biol 2015;145:213–25.

Gravitte A, Archibald T, Cobble A, Kennard B, Brown S. Liquid chromatography-mass spectrometry applications for quantification of endogenous sex hormones. Biomed Chromatogr 2021;35(1), e5036.

Greaves RF, Zacharin MR, Donath SM, Inder TE, Doyle LW, Hunt RW. Establishment of hormone reference intervals for infants born <30 weeks' gestation. Clin Biochem 2014;47(15):101–8.

Greaves RF, Pitkin J, Ho CS, Baglin J, Hunt RW, Zacharin MR. Hormone modeling in preterm neonates: establishment of pituitary and steroid hormone reference intervals. J Clin Endocrinol Metab 2015;100(3): 1097–103.

Greaves RF, Ho CS, Loh TP, Chai JH, Jolly L, Graham P, et al. Current state and recommendations for harmonization of serum/plasma 17-hydroxyprogesterone mass spectrometry methods. Clin Chem Lab Med 2018;56(10):1685–97.

Greene LW, Geer EB, Page-Wilson G, Findling JW, Raff H. Assay-specific spurious ACTH results lead to misdiagnosis, unnecessary testing, and surgical misadventure—a case series. J Endocr Soc 2019;3(4):763–72.

Guo Z, Poglitsch M, McWhinney BC, Ungerer JPJ, Ahmed AH, Gordon RD, et al. Aldosterone LC-MS/MS assay-specific threshold values in screening and confirmatory testing for primary aldosteronism. J Clin Endocrinol Metab 2018;103(11):3965–73.

Hagen CP, Sørensen K, Mieritz MG, Johannsen TH, Almstrup K, Juul A. Circulating MKRN3 levels decline prior to pubertal onset and through puberty: a longitudinal study of healthy girls. J Clin Endocrinol Metab 2015;100(5):1920–6.

Hammond GL. Plasma steroid-binding proteins: primary gatekeepers of steroid hormone action. J Endocrinol 2016;230(1):R13–25.

Han X, McShane M, Sahertian R, White C, Ledger W. Pre-mixing serum samples with assay buffer is a prerequisite for reproducible anti-Mullerian hormone measurement using the Beckman Coulter Gen II assay. Hum Reprod 2014;29(5):1042–8.

Han A, Suh B, Yi G, Lee YJ, Kim SE. Comparison of the automated fluorescent immunoassay system with Roche Elecsys and Beckman Coulter Access 2 assays for anti-Müllerian hormone measurement. Ann Lab Med 2022;42(1):47–53.

Handelsman DJ, Desai R, Seibel MJ, Le Couteur DG, Cumming RG. Circulating sex steroid measurements of men by mass spectrometry are highly reproducible after prolonged frozen storage. J Steroid Biochem Mol Biol 2020;197:105528.

Hayashi GY, Carvalho DF, de Miranda MC, Faure C, Vallejos C, Brito VN, et al. Neonatal 17-hydroxyprogesterone levels adjusted according to age at sample collection and birthweight improve the efficacy of congenital adrenal hyperplasia newborn screening. Clin Endocrinol (Oxf) 2017;86(4):480–7.

He Z, Rankinen T, Leon AS, Skinner JS, Tchernof A, Bouchard C. Plasma steroids, body composition, and fat distribution: effects of age, sex, and exercise training. Int J Obes (Lond) 2018;42(7):1366–77.

Heijboer AC, Savelkoul E, Kruit A, Endert E, Blankenstein MA. Inaccurate first-generation testosterone assays are influenced by sex hormone-binding globulin concentrations. J Appl Lab Med 2016;1(2):194–201.

Hickey M, Lawson LP, Marino JL, Keelan JA, Hart R. Relationship between umbilical cord sex hormone binding globulin, sex steroids, and age at menarche: a prospective cohort study. Fertil Steril 2018;110(5):965–73.

Hill M, Havlíková H, Vrbíková J, Kancheva R, Kancheva L, Pouzar V, et al. The identification and simultaneous quantification of 7-hydroxylated metabolites of pregnenolone, dehydroepiandrosterone, 3beta,17beta-androstenediol, and testosterone in human serum using gas chromatography-mass spectrometry. J Steroid Biochem Mol Biol 2005;96(2):187–200.

Hill M, Parízek A, Kancheva R, Dusková M, Velíková M, Kríz L, et al. Steroid metabolome in plasma from the umbilical artery, umbilical vein, maternal cubital vein and in amniotic fluid in normal and preterm labor. J Steroid Biochem Mol Biol 2010;121(3–5):594–610.

Hill M, Pašková A, Kančeva R, Velíková M, Kubátová J, Kancheva L, et al. Steroid profiling in pregnancy: a focus on the human fetus. J Steroid Biochem Mol Biol 2014;139:201–22.

Hillebrand JJ, Heijboer AC, Endert E. Effects of repeated freeze-thaw cycles on endocrine parameters in plasma and serum. Ann Clin Biochem 2017;54(2):289–92.

Hinchliffe E, Carter S, Owen LJ, Keevil BG. Quantitation of aldosterone in human plasma by ultra high performance liquid chromatography tandem mass spectrometry. J Chromatogr B Analyt Technol Biomed Life Sci 2013;913-914:19–23.

Hines JM, Bancos I, Bancos C, Singh RD, Avula AV, Young WF, et al. High-resolution, accurate-mass (HRAM) mass spectrometry urine steroid profiling in the diagnosis of adrenal disorders. Clin Chem 2017;63(12):1824–35.

Hollier LP, Keelan JA, Jamnadass ES, Maybery MT, Hickey M, Whitehouse AJ. Adult digit ratio (2D,4D) is not related to umbilical cord androgen or estrogen concentrations, their ratios or net bioactivity. Early Hum Dev 2015;91(2):111–7.

Honour JW. Biochemistry of the menopause. Ann Clin Biochem 2018;55(1):18–33.

Honour JW, Tsilchorozidou T, Conway GS, Dawnay A. Spironolactone interference in the immunoassay of androstenedione. Ann Clin Biochem 2010;47(Pt 6):564–6.

Islam RM, Bell RJ, Handelsman DJ, Robinson PJ, Wolfe R, Davis SR, et al. Longitudinal changes over three years in sex steroid hormone levels in women aged 70 years and over. Clin Endocrinol (Oxf) 2021;94(3):443–8.

Jeanneret F, Tonoli D, Rossier MF, Saugy M, Boccard J, Rudaz S. Evaluation of steroidomics by liquid chromatography hyphenated to mass spectrometry as a powerful analytical strategy for measuring human steroid perturbations. J Chromatogr A 2016;1430:97–112.

Jing J, Xia F, Ding Z, Chen L, Shao Y, Ge YF, et al. A single-center performance evaluation of the fully automated iFlash anti-Müllerian hormone immunoassay. Clin Chem Lab Med 2018;57(2):e19–22.

Johannsen TH, Main KM, Ljubicic ML, Jensen TK, Andersen HR, Andersen MS, et al. Sex differences in reproductive hormones during mini-puberty in infants with normal and disordered sex development. J Clin Endocrinol Metab 2018;103(8):3028–37.

Johannsen TH, Andersson AM, Ahmed SF, de Rijke YB, Greaves RF, Hartmann MF, et al. Peptide hormone analysis in diagnosis and treatment of differences of sex development: joint position paper of EU COST Action 'DSDnet' and European Reference Network on Rare Endocrine Conditions. Eur J Endocrinol 2020;182(6):P1–P15.

Johnson S, Schiffner J, Freundl G, Bachmann N, Gnoth C. Luteinising hormone profiles in conception and non-conception natural cycles. Eur J Contracept Reprod Health Care 2019;24(2):140–7.

Jopling H, Yates A, Burgoyne N, Hayden K, Chaloner C, Tetlow L. Paediatric anti-Müllerian hormone measurement: male and female reference intervals established using the automated Beckman Coulter Access AMH assay. Endocrinol Diabetes Metab 2018;1(4), e00021.

Kalogera E, Pistos C, Provatopoulou X, Athanaselis S, Spiliopoulou C, Gounaris A. Androgen glucuronides analysis by liquid chromatography tandem-mass spectrometry: could it raise new perspectives in the diagnostic field of hormone-dependent malignancies? J Chromatogr B Analyt Technol Biomed Life Sci 2013;940:24–34.

Kalra B, Kumar A, Patel K, Patel A, Khosravi MJ. Development of a second generation Inhibin B ELISA. J Immunol Methods 2010;362(1-2):22–31.

Karbasy K, Lin DC, Stoianov A, Chan MK, Bevilacqua V, Chen Y, et al. Pediatric reference value distributions and covariate-stratified reference intervals for 29 endocrine and special chemistry biomarkers on the Beckman Coulter Immunoassay Systems: a CALIPER study of healthy community children. Clin Chem Lab Med 2016;54(4):643–57. https://doi.org/10.1515/cclm-2015-0558. 26457782.

Ke Y, Gonthier R, Simard JN, Labrie F. A validated LC-MS/MS method for the sensitive quantitation of serum 7alpha hydroxy-, 7beta hydroxy- and 7-keto-dehydroepiandrosterone using a novel derivatization reagent. Steroids 2016;108:112–7.

Ke Y, Gonthier R, Labrie F. A sensitive and accurate LC-MS/MS assay with the derivatization of 1-amino-4-methylpiperazine applied to serum allopregnanolone, pregnenolone and androsterone in pre- and postmenopausal women. Steroids 2017;118:25–31.

Kelly J, Raizman JE, Bevilacqua V, Chan MK, Chen Y, Quinn F, et al. Complex reference value distributions and partitioned reference intervals across the pediatric age range for 14 specialized biochemical markers in the CALIPER cohort of healthy community children and adolescents. Clin Chim Acta 2015;450:196–202.

Key TJ, Appleby PN, Reeves GK, Travis RC, Brinton LA, Helzlsouer KJ, et al. Steroid hormone measurements from different types of assays in relation to body mass index and breast cancer risk in postmenopausal women: reanalysis of eighteen prospective studies. Steroids 2015;99(Pt A):49–55.

Kimball A, Dichtel LE, Nyer MB, Mischoulon D, Fisher LB, Cusin C, et al. The allopregnanolone to progesterone ratio across the menstrual cycle and in menopause. Psychoneuroendocrinology 2020;112, 104512.

Kirchhoff F, Briegel J, Vogeser M. Quantification of free serum cortisol based on equilibrium dialysis and isotope dilution-liquid chromatography-tandem mass spectrometry. Clin Biochem 2011;44(10–11):894–9.

Koal T, Schmiederer D, Pham-Tuan H, Röhring C, Rauh M. Standardized LC-MS/MS based steroid hormone profile-analysis. J Steroid Biochem Mol Biol 2012;129(3–5):129–38.

Kulle AE, Welzel M, Holterhus PM, Riepe FG. Implementation of a liquid chromatography tandem mass spectrometry assay for eight adrenal C-21 steroids and pediatric reference data. Horm Res Paediatr 2013;79(1):22–31.

Kushnir MM, Blamires T, Rockwood AL, Roberts WL, Yue B, Erdogan E, et al. Liquid chromatography-tandem mass spectrometry assay for androstenedione, dehydroepiandrosterone, and testosterone with pediatric and adult reference intervals. Clin Chem 2010;56(7):1138–47.

Kyriakopoulou L, Yazdanpanah M, Colantonio DA, Chan MK, Daly CH, Adeli K. A sensitive and rapid mass spectrometric method for the simultaneous measurement of eight steroid hormones and CALIPER pediatric reference intervals. Clin Biochem 2013;46(7-8):642–51.

Latif R, Rafique N. Serum kisspeptin levels across different phases of the menstrual cycle and their correlation with serum oestradiol. Neth J Med 2015;73(4):175–8.

La'ulu SL, Kalp KJ, Straseski JA. How low can you go? Analytical performance of five automated testosterone immunoassays. Clin Biochem 2018;58:64–71.

Lee JS, Ettinger B, Stanczyk FZ, Vittinghoff E, Hanes V, Cauley JA, et al. Comparison of methods to measure low serum estradiol levels in postmenopausal women. J Clin Endocrinol Metab 2006;91(10):3791–7.

Lee C, Kim JH, Moon SJ, Shim J, Kim HI, Choi MH. Selective LC-MRM/SIM-MS based profiling of adrenal steroids reveals metabolic signatures of 17α-hydroxylase deficiency. J Steroid Biochem Mol Biol 2020;198, 105615.

Lenders JWM, Williams TA, Reincke M, Gomez-Sanchez CE. Diagnosis of endocrine disease: 18-oxocortisol and 18-hydroxycortisol: is there clinical utility of these steroids? Eur J Endocrinol 2018;178(1):R1–9.

Leng G, Sabatier N. Measuring oxytocin and vasopressin: bioassays, immunoassays and random numbers. J Neuroendocrinol 2016;28(10). https://doi.org/10.1111/jne.12413.

Li HW, Wong BP, Ip WK, Yeung WS, Ho PC, Ng EH. Comparative evaluation of three new commercial immunoassays for anti-Müllerian hormone measurement. Hum Reprod 2016;31(12):2796–802.

Li H, Popp R, Frohlich B, Chen MX, Borchers CH. Peptide and protein quantification using automated immuno-MALDI (iMALDI). J Vis Exp 2017;126:55933.

Li C, Han T, Li Q, Zhang M, Guo R, Yang Y, et al. MKRN3-mediated ubiquitination of poly(A)-binding proteins modulates the stability and translation of GNRH1 mRNA in mammalian puberty. Nucleic Acids Res 2021;49(7):3796–813.

Linton EA, Perkins AV, Hagan P, Poole S, Bristow AF, Tilders F, et al. Corticotrophin-releasing hormone (CRH)-binding protein interference with CRH antibody binding: implications for direct CRH immunoassay. J Endocrinol 1995;146(1):45–53.

Lionetto L, De Andrés F, Capi M, Curto M, Sabato D, Simmaco M, et al. LC-MS/MS simultaneous analysis of allopregnanolone, epiallopregnanolone, pregnanolone, dehydroepiandrosterone and dehydroepiandrosterone 3-sulfate in human plasma. Bioanalysis 2017;9(6):527–39.

Livingston M, Downie P, Hackett G, Marrington R, Heald A, Ramachandran S. An audit of the measurement and reporting of male testosterone levels in UK clinical biochemistry laboratories. Int J Clin Pract 2020;74(11), e13607.

Lotierzo M, Urbain V, Dupuy AM, Cristol JP. Evaluation of a new automated immunoassay for the quantification of anti-Müllerian hormone. Pract Lab Med 2021;25, e00220.

Lundell AC, Ryberg H, Vandenput L, Rudin A, Ohlsson C, Tivesten Å. Umbilical cord blood androgen levels in girls and boys assessed by gas chromatography-tandem mass spectrometry. J Steroid Biochem Mol Biol 2017;171:195–200.

Madsen A, Bruserud IS, Bertelsen BE, Roelants M, Oehme NHB, Viste K, et al. Hormone references for ultrasound breast staging and endocrine profiling to detect female onset of puberty. J Clin Endocrinol Metab 2020a;105(12):e4886–95.

Madsen A, Oehme NB, Roelants M, Bruserud IS, Eide GE, Viste K, et al. Testicular ultrasound to stratify hormone references in a cross-sectional Norwegian study of male puberty. J Clin Endocrinol Metab 2020b;105(6), dgz094. https://doi.org/10.1210/clinem/dgz094. 31697832.

Maharjan AS, Wyness SP, Ray JA, Willcox TL, Seiter JD, Genzen JR. Detection and characterization of estradiol (E2) and unconjugated estriol (uE3) immunoassay interference due to anti-bovine alkaline phosphatase (ALP) antibodies. Pract Lab Med 2019;17, e00131.

Makrigiannakis A, Semmler M, Briese V, Eckerle H, Minas V, Mylonas I, et al. Maternal serum corticotropin-releasing hormone and ACTH levels as predictive markers of premature labor. Int J Gynaecol Obstet 2007;97(2):115–9.

Mandic S, Kratzsch J, Mandic D, Debeljak Z, Lukic I, Horvat V, et al. Falsely elevated serum oestradiol due to exemestane therapy. Ann Clin Biochem 2017;54(3):402–5.

Martín-Escudero P, Muñoz-Guerra JA, García-Tenorio SV, Garde ES, Soldevilla-Navarro AB, Galindo-Canales M, et al. Impact of the UGT2B17 polymorphism on the steroid profile. Results of a crossover clinical trial in athletes submitted to testosterone administration.

Steroids 2019;141:104–13. https://doi.org/10.1016/j.steroids.2018.11. 009. Epub 2018 Nov 29 30503386.

McGrath IM, Mortlock S, Montgomery GW. Genetic regulation of physiological reproductive lifespan and female fertility. Int J Mol Sci 2021;22(5):2556. https://doi.org/10.3390/ijms22052556. 33806348. PMC7961500.

Mezzullo M, Pelusi C, Fazzini A, Repaci A, Di Dalmazi G, Gambineri A, et al. Female and male serum reference intervals for challenging sex and precursor steroids by liquid chromatography-tandem mass spectrometry. J Steroid Biochem Mol Biol 2020a;197, 105538.

Mezzullo M, Di Dalmazi G, Fazzini A, Baccini M, Repaci A, Gambineri A, Vicennati V, Pelusi C, Pagotto U, Fanelli F. Impact of age, body weight and metabolic risk factors on steroid reference intervals in men. Eur J Endocrinol. 2020b;182(5):459-471.

Moerman A, Delanghe JR. Sense and nonsense concerning biotin interference in laboratory tests. Acta Clin Belg 2020;1–7.

Molenaar N, Groeneveld AB, de Jong MF. Three calculations of free cortisol versus measured values in the critically ill. Clin Biochem 2015;48 (16–17):1053–8.

Moller AT, Backstrom T, Sondergaard HP, et al. Diurnal variations of endogenous steroids in the follicular phase of the menstrual cycle. Neurochem Neuropharmacol 2016;2:1–6.

Montagna G, Balestra S, D'Aurizio F, Romanelli F, Benagli C, Tozzoli R, et al. Establishing normal values of total testosterone in adult healthy men by the use of four immunometric methods and liquid chromatography-mass spectrometry. Clin Chem Lab Med 2018; 56(11):1936–44.

Mouritsen A, Søeborg T, Johannsen TH, Aksglaede L, Sørensen K, Hagen CP, et al. Longitudinal changes in circulating testosterone levels determined by LC-MS/MS and by a commercially available radioimmunoassay in healthy girls and boys during the pubertal transition. Horm Res Paediatr 2014;82(1):12–7.

Mouritsen A, Søeborg T, Hagen CP, Mieritz MG, Johannsen TH, Frederiksen H, et al. Longitudinal changes in serum concentrations of adrenal androgen metabolites and their ratios by LC-MS/MS in healthy boys and girls. Clin Chim Acta 2015;450:370–5.

Nakamura Y, Satoh F, Morimoto R, Kudo M, Takase K, Gomez-Sanchez CE, et al. 18-Oxocortisol measurement in adrenal vein sampling as a biomarker for subclassifying primary aldosteronism. J Clin Endocrinol Metab 2011;96(8):E1272–8.

Nanba AT, Rege J, Ren J, Auchus RJ, Rainey WE, Turcu AF. 11-Oxygenated C19 steroids do not decline with age in women. J Clin Endocrinol Metab 2019;104(7):2615–22.

Nandakumar V, Paul Theobald J, Algeciras-Schimnich A. Evaluation of plasma ACTH stability using the Roche Elecsys immunoassay. Clin Biochem 2020;81:59–62.

Noyola-Martínez N, Halhali A, Barrera D. Steroid hormones and pregnancy. Gynecol Endocrinol 2019;35(5):376–84.

O'Reilly MW, Kempegowda P, Jenkinson C, Taylor AE, Quanson JL, Storbeck KH, et al. 11-Oxygenated C19 steroids are the predominant androgens in polycystic ovary syndrome. J Clin Endocrinol Metab 2017;102(3):840–8.

Olisov D, Lee K, Jun SH, Song SH, Kim JH, Lee YA, et al. Measurement of serum steroid profiles by HPLC-tandem mass spectrometry. J Chromatogr B Analyt Technol Biomed Life Sci 2019;1117:1–9.

Ortega MT, Carlson L, McGrath JA, Kangarloo T, Adams JM, Sluss PM, et al. AMH is higher across the menstrual cycle in early postmenarchal girls than in ovulatory women. J Clin Endocrinol Metab 2020;105(4):e1762–71. https://doi.org/10.1210/clinem/dgaa059. 32016427. PMC7082083.

O'Shea P, Brady JJ, Gallagher N, Dennedy MC, Fitzgibbon M. Establishment of reference intervals for aldosterone and renin in a Caucasian population using the newly developed immunodiagnostic systems specialty immunoassay automated system. Ann Clin Biochem 2016;53(Pt 3):390–8.

Owen LJ, Keevil BG. Supported liquid extraction as an alternative to solid phase extraction for LC-MS/MS aldosterone analysis? Ann Clin Biochem 2013;50(Pt 5):489–91.

Owen LJ, Adaway J, Morris K, Lockhart S, Keevil BG. A widely applicable plasma renin activity assay by LC-MS/MS with offline solid phase extraction. Ann Clin Biochem 2014;51(Pt 3):409–11.

Page MM, Taranto M, Ramsay D, van Schie G, Glendenning P, Gillett MJ, et al. Improved technical success and radiation safety of adrenal vein sampling using rapid, semi-quantitative point-of-care cortisol measurement. Ann Clin Biochem 2018;55(5):588–92.

Panton KK, Mikkelsen G, Irgens WØ, Hovde AK, Killingmo MW, Øien MA, et al. New reference intervals for cortisol, cortisol binding globulin and free cortisol index in women using ethinyl estradiol. Scand J Clin Lab Invest 2019;79(5):314–9.

Pasqualini JR, Chetrite GS. The formation and transformation of hormones in maternal, placental and fetal compartments: biological implications. Horm Mol Biol Clin Invest 2016;27(1):11–28.

Penell JC, Kushnir MM, Lind L, Bergquist J, Bergquist J, Lind PM, et al. Concentrations of nine endogenous steroid hormones in 70-year-old men and women. Endocr Connect 2021;10(5):511–20.

Rahhal SN, Fuqua JS, Lee PA. The impact of assay sensitivity in the assessment of diseases and disorders in children. Steroids 2008; 73(13):1322–7.

Raizman JE, Quinn F, Armbruster DA, Adeli K. Pediatric reference intervals for calculated free testosterone, bioavailable testosterone and free androgen index in the CALIPER cohort. Clin Chem Lab Med 2015; 53(10):e239–43. https://doi.org/10.1515/cclm-2015-0027. 25883205.

Ray JA, Kushnir MM, Palmer J, Sadjadi S, Rockwood AL, Meikle AW. Enhancement of specificity of aldosterone measurement in human serum and plasma using 2D-LC-MS/MS and comparison with commercial immunoassays. J Chromatogr B Analyt Technol Biomed Life Sci 2014;970:102–7.

Rege J, Nakamura Y, Satoh F, Morimoto R, Kennedy MR, Layman LC, et al. Liquid chromatography-tandem mass spectrometry analysis of human adrenal vein 19-carbon steroids before and after ACTH stimulation. J Clin Endocrinol Metab 2013;98(3):1182–8.

Regnier C, Bennet A, Malet D, Guez T, Plantavid M, Rochaix P, et al. Intraoperative testosterone assay for virilizing ovarian tumor topographic assessment: report of a Leydig cell tumor of the ovary in a premenopausal woman with an adrenal incidentaloma. J Clin Endocrinol Metab 2002;87(7):3074–7.

Rezvanpour A, Don-Wauchope AC. Clinical implications of estrone sulfate measurement in laboratory medicine. Crit Rev Clin Lab Sci 2017; 54(2):73–86.

Riepe FG, Krone N, Peter M, Sippell WG, Partsch CJ. Chromatographic system for the simultaneous measurement of plasma 18-hydroxy-11-deoxycorticosterone and 18-hydroxycorticosterone by radioimmunoassay: reference data for neonates and infants and its application in aldosterone-synthase deficiency. J Chromatogr B Analyt Technol Biomed Life Sci 2003;785(2):293–301.

Roberts SA, Kaiser UB. Genetics in endocrinology: genetic etiologies of central precocious puberty and the role of imprinted genes. Eur J Endocrinol 2020;183(4):R107–17.

Roos J, Johnson S, Weddell S, Godehardt E, Schiffner J, Freundl G, et al. Monitoring the menstrual cycle: comparison of urinary and serum reproductive hormones referenced to true ovulation. Eur J Contracept Reprod Health Care 2015;20(6):438–50.

Rossi GP, Ceolotto G, Rossitto G, Seccia TM, Maiolino G, Berton C, et al. Prospective validation of an automated chemiluminescence-based assay of renin and aldosterone for the work-up of arterial hypertension. Clin Chem Lab Med 2016;54(9):1441–50.

Sanderson TC, Woods RJ, Kemp CF, Lowry PJ. Detection of N-terminal pro-corticotrophin releasing hormone (CRH) and a 'novel' CRH in human maternal plasma and placenta. Placenta 2000;21(2–3):218–25. https://doi.org/10.1053/plac.1999.0462. 10736245.

Schock H, Zeleniuch-Jacquotte A, Lundin E, Grankvist K, Lakso HÅ, Idahl A, et al. Hormone concentrations throughout uncomplicated pregnancies: a longitudinal study. BMC Pregnancy Childbirth 2016; 16(1):146.

Schweitzer S, Kunz M, Kurlbaum M, Vey J, Kendl S, Deutschbein T, et al. Plasma steroid metabolome profiling for the diagnosis of adrenocortical carcinoma. Eur J Endocrinol 2019;180(2):117–25.

Skiba MA, Bell RJ, Islam RM, Handelsman DJ, Desai R, Davis SR. Androgens during the reproductive years: what is normal for women? J Clin Endocrinol Metab 2019;104(11):5382–92.

Søeborg T, Frederiksen H, Mouritsen A, Johannsen TH, Main KM, Jørgensen N, et al. Sex, age, pubertal development and use of oral contraceptives in relation to serum concentrations of DHEA, DHEAS, 17α-hydroxyprogesterone, Δ4-androstenedione, testosterone and their ratios in children, adolescents and young adults. Clin Chim Acta 2014;437:6–13.

Stanway SJ, Purohit A, Reed MJ. Measurement of estrone sulfate in post-menopausal women: comparison of direct RIA and GC-MS/MS methods for monitoring response to endocrine therapy in women with breast cancer. Anticancer Res 2007;27(4C):2765–7.

Stieglitz HM, Korpi-Steiner N. Characterization of biotin interference in 21 vitros 5600 immunoassays and risk mitigation for patient safety at a large academic medical center. Clin Biochem 2020;75:53–61.

Storbeck KH, Schiffer L, Baranowski ES, Chortis V, Prete A, Barnard L, et al. Steroid metabolome analysis in disorders of adrenal steroid biosynthesis and metabolism. Endocr Rev 2019;40(6):1605–25.

Sun BZ, Kangarloo T, Adams JM, Sluss PM, Welt CK, Chandler DW, et al. Healthy post-menarchal adolescent girls demonstrate multi-level reproductive axis immaturity. J Clin Endocrinol Metab 2019;104 (2):613–23.

Talbot JA, Kane JW, White A. Analytical and clinical aspects of adrenocorticotrophin determination. Ann Clin Biochem 2003;40(Pt 5): 453–71.

Tavita N, Greaves RF. Systematic review of serum steroid reference intervals developed using mass spectrometry. Clin Biochem 2017;50 (18):1260–74.

Taylor DR, Ghataore L, Couchman L, Vincent RP, Whitelaw B, Lewis D, et al. A 13-steroid serum panel based on LC-MS/MS: use in detection of adrenocortical carcinoma. Clin Chem 2017;63 (12):1836–46.

Tchernof A, Mansour MF, Pelletier M, Boulet MM, Nadeau M, Luu-The V. Updated survey of the steroid-converting enzymes in human adipose tissues. J Steroid Biochem Mol Biol 2015;147:56–69.

Tchernof A, Brochu D, Maltais-Payette I, Mansour MF, Marchand GB, Carreau AM, et al. Androgens and the regulation of adiposity and body fat distribution in humans. Compr Physiol 2018;8(4):1253–90.

Timper K, Fenske W, Kühn F, Frech N, Arici B, Rutishauser J, et al. Diagnostic accuracy of copeptin in the differential diagnosis of the polyuria-polydipsia syndrome: a prospective multicenter study. J Clin Endocrinol Metab 2015;100(6):2268–74.

Tin Tin S, Reeves GK, Key TJ. Body size and composition, physical activity and sedentary time in relation to endogenous hormones in pre-menopausal and postmenopausal women: findings from the UK Biobank. Int J Cancer 2020;147(8):2101–15.

Toprak B, Yalcin H, Arı E, Colak A. EDTA interference in electrochemiluminescence ACTH assay. Ann Clin Biochem 2016;53(6):699–701. https://doi.org/10.1177/0004563216636898. Epub 2016 Sep 28 27166315.

Trabert B, Xu X, Falk RT, Guillemette C, Stanczyk FZ, McGlynn KA. Assay reproducibility of serum androgen measurements using liquid chromatography-tandem mass spectrometry. J Steroid Biochem Mol Biol 2016;155(Pt A):56–62.

Travers S, Martinerie L, Bouvattier C, Boileau P, Lombès M, Pussard E. Multiplexed steroid profiling of gluco- and mineralocorticoids pathways using a liquid chromatography tandem mass spectrometry method. J Steroid Biochem Mol Biol 2017;165(Pt B):202–11.

Travers S, Martinerie L, Boileau P, Xue QY, Lombès M, Pussard E. Comparative profiling of adrenal steroids in maternal and umbilical cord blood. J Steroid Biochem Mol Biol 2018;178:127–34.

Travison TG, Vesper HW, Orwoll E, Wu F, Kaufman JM, Wang Y, et al. Harmonized reference ranges for circulating testosterone levels in men of four cohort studies in the United States and Europe. J Clin Endocrinol Metab 2017;102(4):1161–73.

Trevisan CM, Montagna E, de Oliveira R, Christofolini DM, Barbosa CP, Crandall KA, et al. Kisspeptin/GPR54 system: what do we know about its role in human reproduction? Cell Physiol Biochem 2018;49(4):1259–76.

Tsukazaki Y, Senda N, Kubo K, Yamada S, Kugoh H, Kazuki Y, et al. Development of a high-sensitivity quantitation method for arginine vasopressin by high-performance liquid chromatography tandem mass spectrometry, and comparison with quantitative values by radioimmunoassay. Anal Sci 2016;32(2):153–9.

Turcu AF, El-Maouche D, Zhao L, Nanba AT, Gaynor A, Veeraraghavan P, et al. Androgen excess and diagnostic steroid biomarkers for nonclassic 21-hydroxylase deficiency without cosyntropin stimulation. Eur J Endocrinol 2020a;183(1):63–71.

Turcu AF, Rege J, Auchus RJ, Rainey WE. 11-Oxygenated androgens in health and disease. Nat Rev Endocrinol 2020b;16(5):284–96.

Vale W, Vaughan J, Yamamoto G, Bruhn T, Douglas C, Dalton D, et al. Assay of corticotropin releasing factor. Methods Enzymol 1983;103:565–77.

van der Gugten JG, Holmes DT. Quantitation of aldosterone in serum or plasma using liquid chromatography-tandem mass spectrometry (LC-MS/MS). Methods Mol Biol 2016a;1378:37–46.

van der Gugten JG, Holmes DT. Quantitation of plasma renin activity in plasma using liquid chromatography-tandem mass spectrometry (LC-MS/MS). Methods Mol Biol 2016b;1378:243–53.

van der Veen A, van Faassen M, de Jong WHA, van Beek AP, Dijck-Brouwer DAJ, Kema IP. Development and validation of a LC-MS/MS method for the establishment of reference intervals and biological variation for five plasma steroid hormones. Clin Biochem 2019;68:15–23.

van Helden J, Weiskirchen R. Cross-method comparison of serum androstenedione measurement with respect to the validation of a new fully automated chemiluminescence immunoassay. Clin Biochem 2018;62:32–8.

Varimo T., Hero M., Känsäkoski J., Vaaralahti K., Matikainen N., Raivio T. Circulating makorin ring-finger protein-3 (MKRN3) levels in healthy men and in men with hypogonadotropic hypogonadism. Clin Endocrinol (Oxf) 2016;84(4):638 Erratum for: Clin Endocrinol (Oxf). 2016;84(1):151–2.

Veldhuis JD, Dyer RB, Trushin SA, Bondar OP, Singh RJ, Klee GG. Immunologic and mass-spectrometric estimates of SHBG concentrations in healthy women. Metabolism 2014a;63(6):783–92.

Veldhuis JD, Bondar OP, Dyer RB, Trushin SA, Klee EW, Singh RJ, et al. Immunological and mass spectrometric assays of SHBG: consistent and inconsistent metabolic associations in healthy men. J Clin Endocrinol Metab 2014b;99(1):184–93.

Verdonk SJE, Vesper HW, Martens F, Sluss PM, Hillebrand JJ, Heijboer AC. Estradiol reference intervals in women during the menstrual cycle, postmenopausal women and men using an LC-MS/MS method. Clin Chim Acta 2019;495:198–204.

Vogeser M, Möhnle P, Briegel J. Free serum cortisol: quantification applying equilibrium dialysis or ultrafiltration and an automated immunoassay system. Clin Chem Lab Med 2007;45(4):521–5.

Vogeser M, Kratzsch J, Ju Bae Y, Bruegel M, Ceglarek U, Fiers T, et al. Multicenter performance evaluation of a second generation cortisol assay. Clin Chem Lab Med 2017;55(6):826–35.

Wang F, Koskela A, Hämäläinen E, Turpeinen U, Savolainen-Peltonen H, Mikkola TS, et al. Quantitative determination of dehydroepiandrosterone fatty acyl esters in human female adipose tissue and serum using mass spectrometric methods. J Steroid Biochem Mol Biol 2011;124(3–5):93–8.

Wang JL, Cao BY, Gong CX, Wu D, Chen JJ, Wei LY. Reference intervals for steroid hormones in healthy 6- to 15-year-old girls based on liquid chromatography-tandem mass spectrometry in China. Chin Med J (Engl) 2020;133(10):1239–41.

White A, Smith H, Hoadley M, Dobson SH, Ratcliffe JG. Clinical evaluation of a two-site immunoradiometric assay for adrenocorticotrophin in unextracted human plasma using monoclonal antibodies. Clin Endocrinol (Oxf) 1987;26(1):41–51.

Wijayarathna R, de Kretser DM. Activins in reproductive biology and beyond. Hum Reprod Update 2016;22(3):342–57. https://doi.org/10.1093/humupd/dmv058. Epub 2016 Feb 15 26884470.

Woods JJ, Sampson ML, Ruddel ME, Remaley AT. Rapid intraoperative cortisol assay: design and utility for localizing adrenal tumors by venous sampling. Clin Biochem 2000;33(6):501–3.

Wu ZQ, Xu HG. Preanalytical stability of adrenocorticotropic hormone depends on both time to centrifugation and temperature. J Clin Lab Anal 2017;31(5), e22081.

Wu ZQ, Xu HG. Comparison of two commercial quality control sera for adrenocorticotropin (ACTH) used in Elecsys® immunoassay system. J Clin Lab Anal 2019;33(1), e22618.

Wu FC, Butler GE, Kelnar CJ, Huhtaniemi I, Veldhuis JD. Ontogeny of pulsatile gonadotropin releasing hormone secretion from mid-childhood, through puberty, to adulthood in the human male: a study using deconvolution analysis and an ultrasensitive immunofluorometric assay. J Clin Endocrinol Metab 1996;81(5):1798–805.

Wudy SA, Hartmann M, Solleder C, Homoki J. Determination of 17alpha-hydroxypregnenolone in human plasma by routine isotope dilution mass spectrometry using benchtop gas chromatography-mass selective detection. Steroids 2001;66(10):759–62.

Wudy SA, Hartmann M, Homoki J. Determination of 11-deoxycortisol (Reichstein's compound S) in human plasma by clinical isotope dilution mass spectrometry using benchtop gas chromatography-mass selective detection. Steroids 2002;67(10):851–7.

Yates AP, Jopling HM, Burgoyne NJ, Hayden K, Chaloner CM, Tetlow L. Paediatric reference intervals for plasma anti-Müllerian hormone: comparison of data from the Roche Elecsys assay and the Beckman Coulter Access assay using the same cohort of samples. Ann Clin Biochem 2019;56(5):536–47.

Yeap BB, Knuiman MW, Handelsman DJ, Ho KKY, Hui J, Divitini ML, et al. A 5α-reductase (SRD5A2) polymorphism is associated with serum testosterone and sex hormone-binding globulin in men, while aromatase (CYP19A1) polymorphisms are associated with oestradiol and luteinizing hormone reciprocally. Clin Endocrinol (Oxf) 2019; 90(2):301–11.

Yeates SL, Elder CV, Grover SR. Unintended and unwanted pregnancy in Australia: a cross-sectional, national random telephone survey of prevalence and outcomes. Med J Aust 2019;210(8):382–382.e1.

Yuan TF, Le J, Wang ST, Li Y. An LC/MS/MS method for analyzing the steroid metabolome with high accuracy and from small serum samples. J Lipid Res 2020;61(4):580–6.

Zalas D, Reinehr T, Niedziela M, Borzikowsky C, Flader M, Simic-Schleicher G, et al. Multiples of median-transformed, normalized reference ranges of steroid profiling data independent of age, sex, and units. Horm Res Paediatr 2018;89(4):255–64.

Zhang D, Rios DR, Tam VH, Chow DS. Development and validation of a highly sensitive LC-MS/MS assay for the quantification of arginine vasopressin in human plasma and urine: application in preterm neonates and child. J Pharm Biomed Anal 2014;99:67–73.

Zimmerman Y, Eijkemans MJ, Coelingh Bennink HJ, Blankenstein MA, Fauser BC. The effect of combined oral contraception on testosterone levels in healthy women: a systematic review and meta-analysis. Hum Reprod Update 2014;20(1):76–105.

Chapter 2.6

Steroids in urine, other fluids and tissues

2.6.1 Steroid concentrations in samples other than blood

Although blood samples are the most popular in clinical investigations and research other biological samples are used. The safety and feelings of the subject under investigation have to be considered. Some samples can only be obtained under anesthesia. The time and duration of the procedure have to be considered, so a urine sample is noninvasive but if a 24 h urine collection is requested that has to be planned and all voiding have to be remembered to be included in the sample. This is surprisingly difficult for many individuals. A single urine sample analysis reflects endocrine activity in a few hours since the last voiding, whereas a 24 h collection compensates for diurnal changes. Saliva is easy to obtain but the steroid concentration is a snapshot at the time of collection. Hair is a long-term indicator of endocrine activity.

2.6.2 Steroids in urine

Urine collection is noninvasive so it is a useful biological fluid. The concentrations of steroids are often corrected in relation to creatinine excretion (sometimes osmolality) to compensate for the problems in collecting urine, particularly if a 24 h collection is required. This is controversial in general use because of the wide ranges for creatinine concentrations, although in studying an individual on several occasions the creatinine excretion is consistent. Urinary-free cortisol is a frequently requested test for investigations of hypercortisolism, immunoassays and automated immunoassays are commonly used, in general after solvent extraction. The range of steroid metabolites in urine is quantified by GC-MS, GC-MS/MS and LC-MS/MS.

GC-MS in a nonselective full scan mode can reveal the entire spectrum of an individual's major steroid metabolome, even the low level of some steroids can be determined from ion monitoring data or by reconstruction of ion chromatograms from MS analysis. The procedure is relatively time-consuming and expensive as a one-off test but not expensive compared with the total costs of several hormones determined separately. GC-MS of steroids in full scan mode usually produces a chromatogram with a peak for each steroid, whereas when selected ion monitoring is used the results for steroid excretion have to be calculated for each steroid and it is not immediately obvious that the relative amounts are different. The application of the steroid profile (metabolome) technique in diagnostics has now been tested in most adrenal disorders by GC-MS/MS and LC-MS/MS and is a reliable test throughout with few exceptions.

2.6.2.1 Single steroids

2.6.2.1.1 Free cortisol

The concentration of **cortisol in urine** reflects filtration in the kidney of the free cortisol in blood serum. Around 1% of the total cortisol in blood is excreted in the urine until binding proteins in blood are saturated when there is an increase in the free cortisol in urine. The preferred sample is a 24 h collection to avoid any circadian/ultradian rhythm of cortisol or the metabolites that can interfere in an immunoassay for cortisol. Measurement by current immunoassay platforms after extraction of cortisol from urine gives very much higher results than by mass spectrometry mainly due to cross reactions of closely related structures (Fig. 2.6.1, Oßwald et al., 2019).

SPE extraction is preferred to organic solvent (dichloromethane or ethyl acetate) extraction of the cortisol (Gatti et al., 2009). Cross reactions from related steroids in an immunoassay should be less than 5% which is rarely achieved with the synthetic glucocorticoid prednisolone which is among the glucocorticosteroids commonly prescribed in hospital patients. Measurement of the cortisol is a very useful test if investigating **overactivity** of the adrenal gland but should **not be used to determine low or suppressed cortisol production**, the results can be imprecise and misleading particularly if a patient is taking inhaled corticosteroids that can suppress the HPA and may interfere in the cortisol analysis by immunoassay (Fink et al., 2002) (Fig. 2.6.2).

Mass spectrometric methods for cortisol in urine are usually based on stable isotope dilution against deuterium

Steroids in the Laboratory and Clinical Practice. https://doi.org/10.1016/B978-0-12-818124-9.00007-3

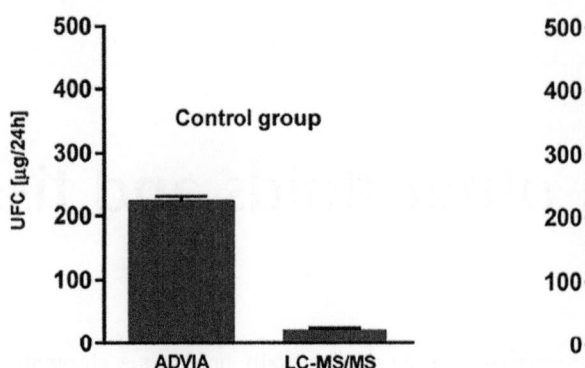

FIG. 2.6.1 Mean ± SEM for urine-free cortisol (UFC) with ADVIA and Liaison immunoassays and LC-MS/MS in patients with Cushings syndrome and patients where Cushing syndrome excluded (control group). *(Form Oßwald A, Wang R, Beuschlein F, Hartmann MF, Wudy SA, Bidlingmaier M, Zopp S, Reincke M, Ritzel K. Performance of LC-MS/MS and immunoassay based 24-h urine free cortisol in the diagnosis of Cushing's syndrome. J Steroid Biochem Mol Biol 2019;190:193–7. Fig. 1 p 195.)*

labeled internal standards. GC-MS methods have used tri-methylsilyl ether (TMS), heptafluorobutyrate (HFB) and rarely other derivatives that make the method for determination of one steroid too time consuming for hospital laboratories. Most LC-MS/MS methods for cortisol use solvent mixtures of water with acetonitrile or methanol, sometimes with a buffer to enhance ionization (ammonium acetate and formic acid) and elution of steroid from a C18 column of 2 μm packings (reviewed by Hawley et al., 2016). Phenyl columns may enable separation of cortisol and prednisolone and isomeric steroids. Electrospray (ESI) in positive ion mode is the most common approach to give $[M+H]^+$ ions although a few methods are based on negative ion mode of adducts for example $[M+HCOO]^-$ that fragment to $[M—H—CH_2O]^-$ ion.

Interferences in urine cortisol measurement have been reported even with MS methods due to lack of LC separation of isomeric steroids (dihydrocortisones) (Israelsson et al., 2018) which emphasizes the need for full validation of the method for specificity and matrix effects. The clearance rate of cortisol by LC-M/MS was apparently enhanced by ion suppression of the internal standard from drugs (Piperacillin) (Danese et al., 2017) or interferes to give high cortisol results by HPLC with UV detection (Carbamazepine; Findling et al., 1998). These are cautionary notes for clinicians to report the use of drugs and for the laboratory to look at the data which can suggest the presence of other steroids.

The reference range for urine cortisol in the literature is 11–70 μg/day (30–193 nmol/day) although results by immunoassay can be 1.5 or more times the MS method results (Bianchi et al., 2019). Typical reference ranges for four separate laboratory LC-MS/MS methods are 17–126, 15–134, 12–118, 25–157 nmol/day (Brossaud et al., 2018). It must be born in mind that the total cortisol metabolites in urine can be 2–15 μmol/day. The mean cortisol production rate determined by stable isotope dilution technology is 7.5 μmol/day (9.9 mg/day) or 15.7 μmol/day/m² body

surface area (5.7 mg/d/M²). Higher UFC in males than females has been reported in many studies (Lamb et al., 1994, 1997; Taylor and Raven, 1996). A recent study (Moffat et al., 2020) from 1814 subjects aged 20–90 years providing 5527 urine collections showed a U-shaped relationship for UFC corrected for creatinine excretion (Fig. 2.6.3). The authors appreciated many contributory variables but made a suggestion that the HPA axis changed with aging leading to the increased cortisol output.

Some studies have seen a bimodal distribution in UFC and dividing the data from a large study into high and low modes found a relationship with cardiovascular disease risk. In children, a relationship of UFC with urinary electrolyte excretion has been reported that was not apparent in their mothers. The data for UFC has much variation but is used as a disease marker in defining clear hypercortisolism (see Chapter 3.3.1, Hypertension).

2.6.2.1.2 Cortisone

There is no need to measure cortisone in urine alone although in saliva there is merit because cortisol is oxidized in the glands. In combination with cortisol, the F/E ratio in urine is an index of HSD11B enzyme activities and thus useful in diagnosis of an apparent mineralocorticoid excess syndrome (AME; see Chapter 3.3.1) and some other clinical situations that will be addressed later in the book. Chromatographic methods are the only practical way to measure cortisone and cortisol. Increased cortisone excretion has been reported in depressed patients by immunoassay and was also found in chronic fatigue syndrome where cortisone and cortisol were measured by a GC-MS method (Weber et al., 2000; Jerjes et al., 2006a).

2.6.2.1.3 Aldosterone-18-glucuronide

This acid labile conjugate of aldosterone in urine was first described around 1957, soon after the discovery of

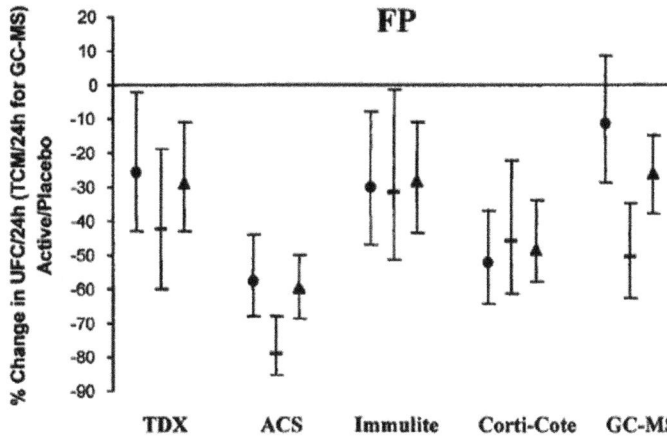

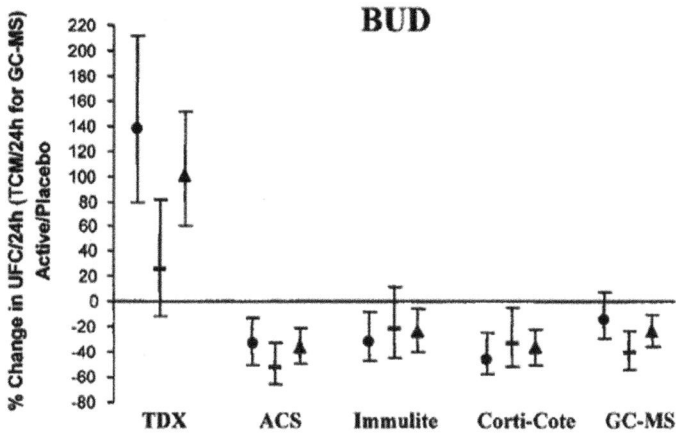

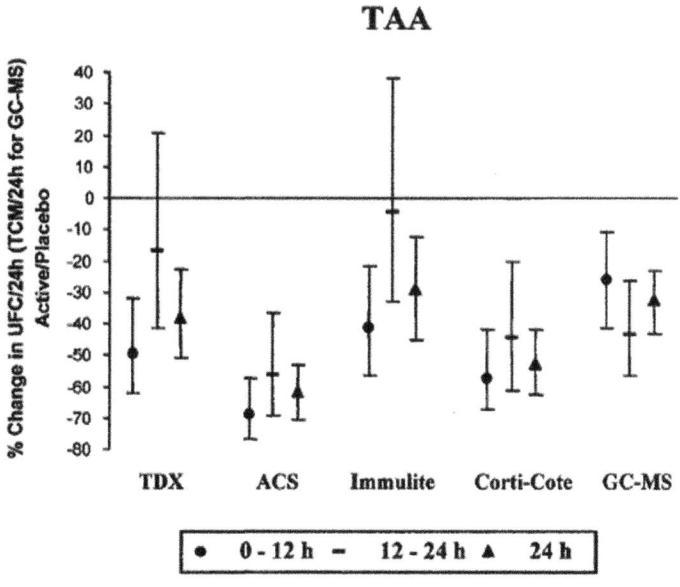

FIG. 2.6.2 Effect of inhaled steroids fluticasone proprionate (FP), budesonide (BUD) and triamcinolone acetate (TAA) on 24h free cortisol by immunoassays and total cortisol metabolite (TCM) excretion rates excretion rates. Values are percent change (95% confidence intervals). *(From Fink RS, Pierre LN, Daley-Yates PT, Richards DH, Gibson A, Honour JW. Hypothalamic-pituitary-adrenal axis function after inhaled corticosteroids: unreliability of urinary free cortisol estimation. J Clin Endocrinol Metab 2002;87(10):4541–6. Fig. 2 p. 4545.)*

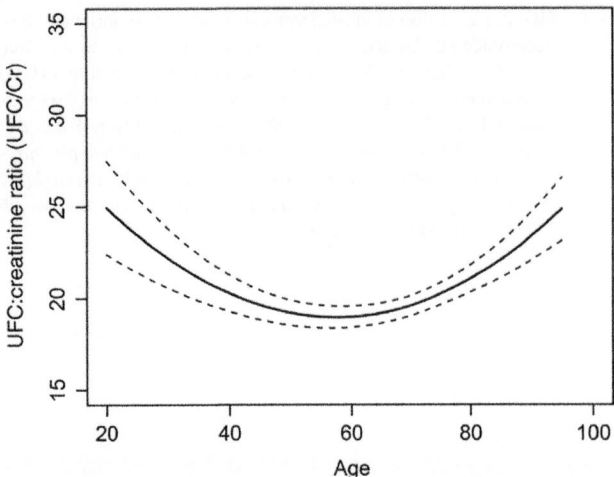

FIG. 2.6.3 Relationship (solid line) (95% confidence intervals dotted lines) between 24 h urinary-free cortisol to creatinine ratio (UFC/Cr) and age. A significant nonlinear effect is found in which UFC/Cr decreased to early in adulthood, was stable in middle adulthood and increased thereafter. *(From Moffat SD, An Y, Resnick SM, Diamond MP, Ferrucci L. Longitudinal change in cortisol levels across the adult life span. J Gerontol A Biol Sci Med Sci 2020;75(2):394–400. Fig. 2 p. 397.)*

aldosterone in London. Early methods for determination were chemical but the literature from 1970 is dominated by immunoassay with a few papers to date using gas chromatography and less with mass spectrometry. Problems in the analysis of aldosterone have been identified and include:

- interferences from other aldosterone metabolites,
- the conditions for hydrolysis around pH 1,
- the need for purification of urine extracts before immunoassay, and
- isomerization of radioactive aldosterone to isoaldosterone.

There is good inverse agreement of steroid excretion with sodium (Fig. 2.6.4). Normal excretion rates are less than 18–5 µg/day from reliable immunoassays. The steroid is measured as an indicator of aldosterone excess.

2.6.2.1.4 Tetrahydroaldosterone (THAldo)

Aldosterone metabolism has been studied since 1963 and THAldo recognized as a metabolite accounting for 7%–15% of the metabolites. Only specialist laboratories have had methods to quantify excretion rates which are less than 100 µg per day in normal subjects which is dependent on sodium intake with low intake stimulating aldosterone excretion. The measurements have proved useful in young children with salt-losing problems where blood sampling is difficult. Other useful results have been in hypertensive patients with primary aldosteronism (Tumor secreting aldosterone; see Chapter 3.3.1). The majority of the literature is based on immunoassay of THAldo after chromatography

although GC-MS has been used for the steroid (Honour and Shackleton, 1977) which can be part of the steroids reported from a urinary steroid profile (McQuarrie et al., 2013). Ackermann et al. (2019) reported THAldo as median; 25–75 percentile in men as 20;12.6–31.5 and for women 17.2; 10.6–30.3 µg/24 h with actual results between 2 and 200 µg/24 h (Fig. 2.6.5).

2.6.2.1.5 DHEA and DHEA-S

Dehydroepiandrosterone sulfate (DHEAS) is the major steroid secreted by the adrenal cortex in adults, the fetus and newborn infant for the first 6 months after birth. DHEAS is a marker of the appearance of the *zona reticularis* in the adrenal cortex before puberty. Although uptake and desulfation of DHEAS in adipose tissue is important only low concentrations of free DHEA are found in urine at less than 37 nmol/L (mean 3 nmol/L). Mean DHEAS can increase from 28 µg/dL to 73 µg/dL as children pass through five Tanner stages of puberty (Table 2.6.1) (Saczawa et al., 2013).

Measurement of DHEAS in urine has been useful in assessing adrenal function of preterm infants where blood is difficult to collect on even one venepuncture and more of a problem when several blood samples are required particularly when samples from catheters should be avoided wherever possible. In pregnancy, DHEAS increases as the fetal adrenal cortex grows but most of the DHEAS is both desulfated and aromatized to estrone and estradiol in the placenta. In clinical investigations, the urine DHEAS test is useful when looking for certain genetic defects of adrenal function where excretion of several precursors will be raised and cortisol production low. DHEAS can also be high from certain adrenocortical tumors. Free DHEA in urine is not usually measured.

2.6.2.1.6 Testosterone/testosterone glucuronide/androsterone

Androgen excretion has been quantified in neonates as testosterone concentration in blood declines over the first week, as hCG is cleared, then T rises in response to LH to low levels of adult males which is sustained for up to 8 months (minipuberty) (up to 100 µg/mmol creatinine; Dhayat et al., 2015) then falls (for unknown reasons) until puberty. Androsterone is the main metabolite of testosterone and output follows that pattern during the first months (Fig. 2.6.6) but that must be interpreted in the context of high DHEAS production until regression of the fetal zone by 6 months of age.

Testosterone in urine is rarely measured in clinical practice but there is immense interest in the context of anti-doping programs in sport. Testosterone glucuronide is determined along with epitestosterone glucuronide, because both steroids are produced in the body but synthetic testosterone does not contain epitestosterone so the T-G to e-G ratio

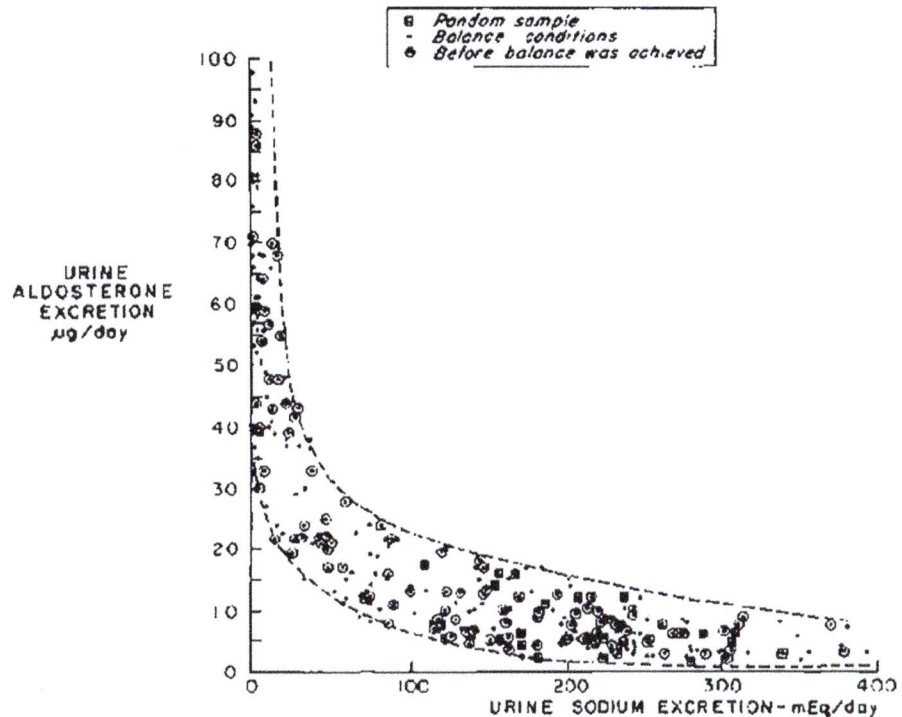

FIG. 2.6.4 Relationship of Aldosterone excretion to daily urine sodium excretion. Aldosterone was measured by radioimmunoassay (*n* = 198) and double isotope dilution (*n* = 75) with overlap of results by two methods across the entire range. No difference was found in measurements from random samples or those collected under balance conditions. *(From Sealey JE, Bühler FR, Laragh JH, Manning EL, Brunner HR. Aldosterone excretion. Physiological variations in man measured by radioimmunoassay or double-isotope dilution. Circ Res 1972;31(3):367–78. Fig. 3 p. 372.)*

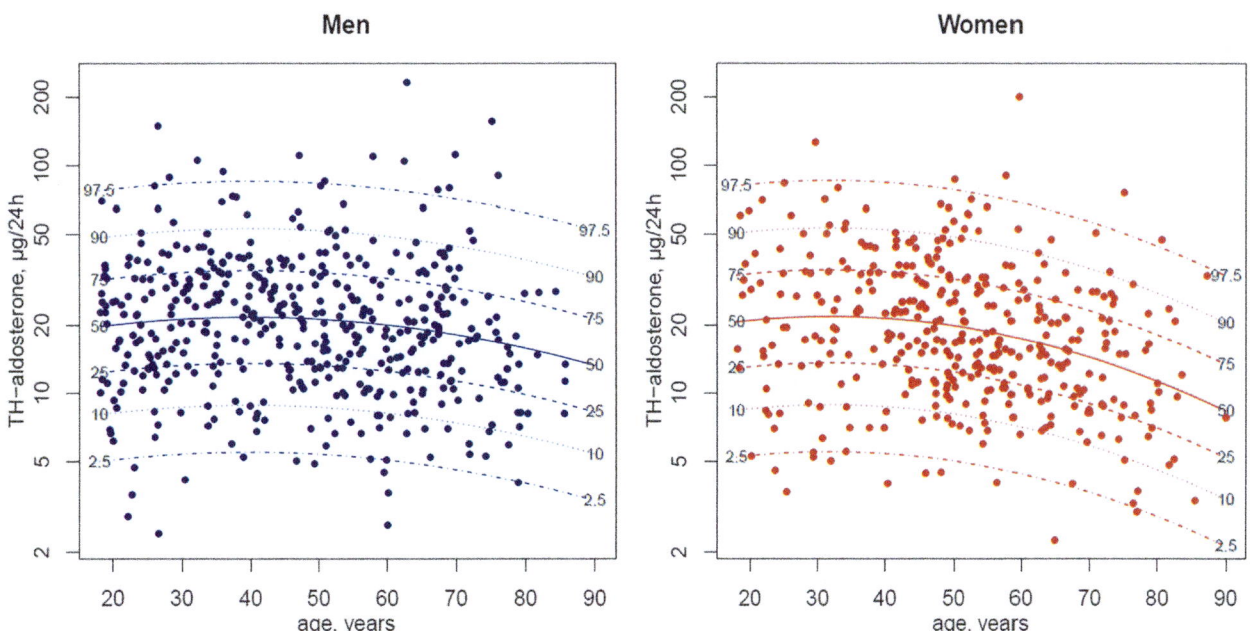

FIG. 2.6.5 Tetrahydroaldosterone excretion in urine by age in men (*blue*) and women (*red*). The percentiles 2.5, 10, 25, 50, 75, 90, and 97.5 of the steroid compounds in function of age and sex are shown on a log-scale. *(From Ackermann D, Groessl M, Pruijm M, Ponte B, Escher G, d'Uscio CH, Guessous I, Ehret G, Pechère-Bertschi A, Martin PY, Burnier M, Dick B, Vogt B, Bochud M, Rousson V, Dhayat NA. Reference intervals for the urinary steroid metabolome: the impact of sex, age, day and night time on human adult steroidogenesis. PLoS ONE 2019;14(3):e0214549. Suppl. Fig. 4).*

TABLE 2.6.1 Urinary excretion (mean) of DHEAS for gender and Tanner pubic hair (TannerPH) stage (units: µg/dl).

DHEAS	Boys. Mean concentration; number	Girls. Mean concentration; number
Tanner PHI	36.938; n=17	25.187; n=136
Tanner PH2	47.816; n=81	43.879; n=74
Tanner PH 3	62.294; n=39	61.093; n=40
Tanner PH4/5	72.288; n=18	88.930; n=49

From Saczawa ME, Graber JA, Brooks-Gunn J, Warren MP. Methodological considerations in use of the cortisol/DHEA(S) ratio in adolescent populations. Psychoneuroendocrinology 2013;38(11):2815–2819. From Table 1 p. 2817.

increases when testosterone in abused. The best measure of testosterone abuse is the determination of $^{12}C/^{13}C$ isotope ratios in metabolites because pharmaceutical testosterone is prepared from plant products that have a different ^{13}C content to human testosterone derived from cholesterol in the body.

2.6.2.1.7 Androstanediol glucuronide

The skin, especially the pilosebaceous unit composed of sebaceous glands and hair follicles, can synthesize androgens de novo from cholesterol or by locally converting circulating weaker androgens to more potent ones. As in other classical steroidogenic organs, the same six major enzyme systems are involved in cutaneous androgen metabolism, namely steroid sulfatase, HSD3B2, HSD17B, SRD5A, HSD3A andCYP19. Steroid sulfatase, together with CYP17 (hydroxylase and lyase) are found to reside in the cytoplasm of sebocytes and keratinocytes. HSD3B has been mainly immunolocalized to sebaceous glands, with the type 1 being the key cutaneous isoenzyme. The

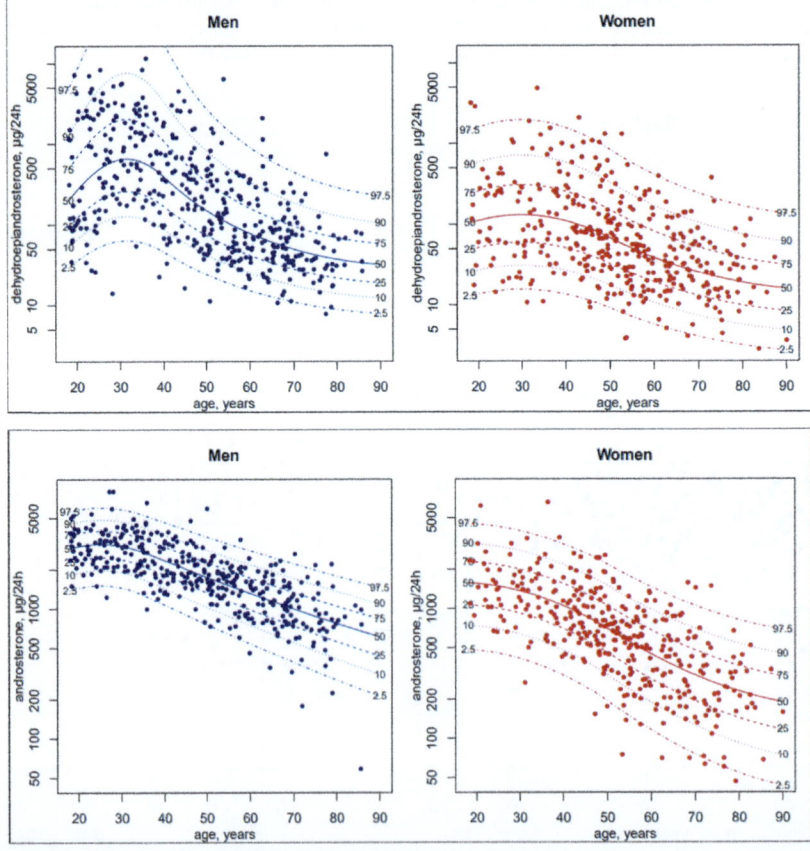

FIG. 2.6.6 Androgen excretion in urine by age in men (*blue*) and women (*blue*). The percentiles 2.5, 10, 25, 50, 75, 90, and 97.5 of the steroid compounds in function of age and sex are shown on a log-scale. (A) Dehydroepiandrosterone; (B) Androsterone. (*Based on Ackermann D, Groessl M, Pruijm M, Ponte B, Escher G, d'Uscio CH, et al. Reference intervals for the urinary steroid metabolome: the impact of sex, age, day and night time on human adult steroidogenesis. PLoS One 2019;14(3):e0214549. Suppl 4.*)

HSD17B2 isoenzyme predominates in sebaceous glands and exhibits greater reductive activity in glands from facial areas compared with areas not prone to acne. In hair follicles, HSD17B is found mainly in outer root sheath cells. The type SRD5A1 mainly occurs in the sebaceous glands, whereby SRD5A2 seems to be localized in the hair follicles. HSD3A converts dihydrotestosterone to 3α-androstanediol, which is conjugated to **3α-androstanediol glucuronide (ADG)**. Concentrations in children have been determined by a heterologous ELISA (Fig. 2.6.7) and recently by LC-MS/MS (Fabregat-Cabello et al., 2019). The results are difficult to compare because different units are used.

High concentrations of ADG reflect a hyperandrogenic state in hirsute women, especially for idiopathic hirsutism. In acne patients, 3α-androstanediol glucuronide and androsterone glucuronide serve as suitable urine and serum markers for measuring androgenicity. Aromatase, localized to sebaceous glands and to both outer and inner root sheath cells of anagen terminal hair follicles, may play a role in removing excess androgens. Pharmacologic development of potent specific isoenzyme antagonists has led to some

clinical benefit and prevention of hair loss and androgen-dependent skin disorders. The measurements have not been widely used in the past 25 years because the results have added little value to clinical judgment. However, there is better understanding of the significance of adrenal steroids, particularly DHEAS, in pathology and the use of androstanediol glucuronide as a biomarker of particular routes of hyperandrogenism may return.

2.6.2.1.8 Pregnanetriol

The 17-hydroxyprogesterone in blood is reduced in the A/B rings with the addition of 4 hydrogen atoms to give 17-hydroxypregnanolone which is reduced at C-20 to form pregnanetriol (5β-pregnane-3α,17α,20α-triol) the main urinary metabolite of 17-hydroxyprogesterone (Fig. 2.6.8).

A number of techniques from immunoassay through GC and GC-MS (Koyama et al., 2012; Homma et al., 2004) have been used to quantify pregnanetriol as marker for diagnosis and monitoring treatment of congenital adrenal hyperplasia due to 21-hydroxlyase deficiency. The hydrolyzed conjugate and the intact conjugate have been analyzed. Concentrations of pregnanetriol around 1.2–2.1 mg/M^2/day and 2.2–3.3 mg/g creatinine have been used as indices of optimal control (Izawa et al., 2007; Kamrath et al., 2016).

2.6.2.1.9 Pregnanetriol-one

The 21-deoxycortisol in blood is reduced with the addition of 6 hydrogen atoms then the 11-hydroxyl group is oxidized to a carbonyl with final formation of pregnanetriolone (full name 5β-pregnane-3α,17α,20α-triol, 11-one) (Fig. 2.6.8).

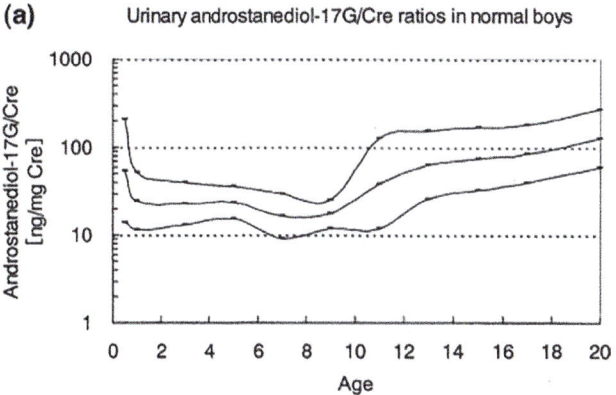
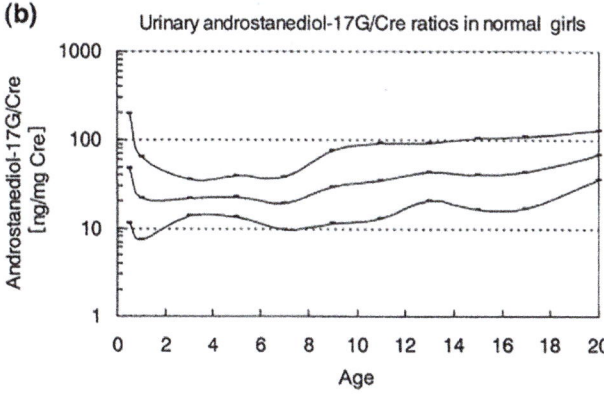

FIG. 2.6.7 Urinary androstanediol 17-glucuronide/creatinine ratios in normal boys (upper panel) and girls (lower panel) (upper limit 97.5 percentiles, mean and lower limit 2.5 percentiles). *(From Onishi T, Takei H, Kambegawa A, et al. A highly specific heterologous enzyme immunoassay for 5 alpha-androstane-3 alpha, 17 beta-diol 17-glucuronide (androstanediol-17G) and developmental patterns of urinary androstanediol-17G excretions. Steroids 2002;67(3–4):175–83. Fig. 7.)*

FIG. 2.6.8 Chemical structures of 17-hydroxyprogesterone and 21-deoxycortisol metabolites. *(Author original.)*

A spot urinary concentration of 0.1 mg/g creatinine is a cut-off value in distinguishing CAH and normal term and preterm infants. This metabolite has also been determined in urine collected into absorbent paper (Alonso-Fernández, 2016).

2.6.2.1.10 Other metabolites

A number of other single urinary steroids have been measured and tests are still available from a few specialist laboratories. In the 1970s, there was much research for markers of hypertension that has not been pursued in recent years. 19-nor steroids have been shown to be highly hypertensinogenic and is among these steroids (Lepenies et al., 2009). The results for metabolites of cortisol and corticosterone are summarized in Table 2.6.2. and the data, using outdated technology, are a guide to measurements in the future as part of metabolomic data analysis that will inevitably consider the steroids.

2.6.3 Panels of steroids

Several pairs of steroids are measured in a sample and the ratios used to assess an enzyme activity or a physiological process with integrated activities.

2.6.3.1 Estrogens and progesterone

2.6.3.1.1 Menstrual cycle

Ovarian hormonal activity can be monitored by steroid determinations with early morning urine samples. Urinary concentrations of estradiol glucuronide and pregnanediol glucuronide correlate with serum concentrations of estradiol and progesterone (Fig. 2.6.9). The excretion of urinary estrone glucuronide (E1G) gives a direct measure of follicular growth, and the postovulatory rise in urinary PdG following an E1G peak provides good evidence of ovulation.

Specific values of the PdG excretion rate can be used to determine whether a cycle is anovulatory with or without a LUF or is ovulatory and infertile or ovulatory and fertile. Laboratory E1G and PdG assays have been in-house ELISA assays while urine progesterone can be determined on automated immunoassay platforms (Fig. 2.6.10). Urine progesterone concentrations on the Abbott Architect show a rise at ovulation but the Roche Cobas assay is rather poor for this discrimination.

The spot urine samples are easy to collect and several can be collected in days during a menstrual cycle to illustrate serial changes of the steroids. The time of ovulation

TABLE 2.6.2 Determinations of individual mineralocorticoid metabolites in urine with methodology and ranges of excretion rates.

Steroid	Range (µg/day)	Mean	No. of subjects	Method	Ref
Tetrahydroaldosterone	9–29	28.2±15.2	5 18	Gas liquid chromatography Immunoassay	Nicolis et al. (1968) Gomez-Sanchez and Holland (1981)
Tetrahydrocorticosterone (THB)	100–360	200	18	Blue tetrazolium	Visser and Cost (1964)
Allo-THB	80–360	200	18	Blue tetrazolium	Visser and Cost (1964)
Tetrahydo-11-dehydro-corticosterone (THA)	60–240	160	18	Blue tetrazolium	Visser and Cost (1964)
Tetrahydro-11-deoxycorticosterone (THDOC)	10–40	24	15	Double isotope derivative	Schambelan and Biglieri (1972)
18-Hydroxy-THA	97–182		9	Double isotope derivative	Ulick, 1976
18-Hydroxy-THDOC		15	27	Porter Silber reaction	Melby et al. (1972)
18-Oxo-cortisol		1.2	37	Immunoassay	Gomez-Sanchez et al. (1989)
18-Hydroxycortisol		111±112	9	Immunoassay	Gomez-Sanchez et al. (1987)
19-Nor-DOC		0.6±0.1	14	Immunoassay	Ehlers et al. (1987)

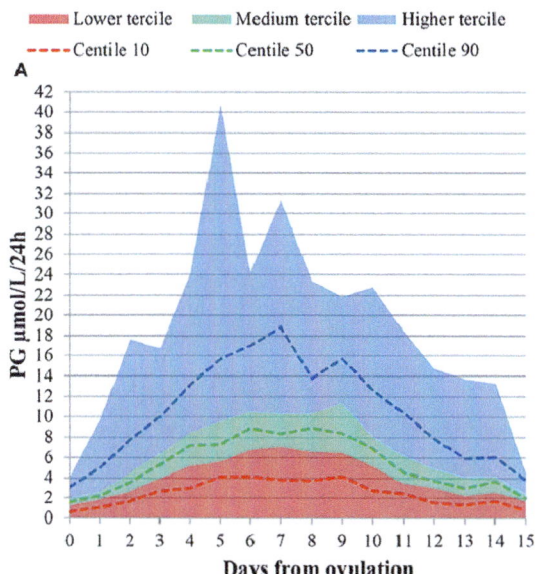

FIG. 2.6.9 Left: Pregnanediol glucuronide (PG) and right estradiol glucuronide (EG) after ovulation. Day 0 was presumed as ovulation (1 day after LH peak). Daily lower, median and upper tercile hormone ranges are shown as well as 10th, 50th and 90th percentiles. *(From Alliende ME, Arraztoa JA, Guajardo U, Mellado F. Towards the clinical evaluation of the luteal phase in fertile women: a preliminary study of normative urinary hormone profiles. Front Public Health 2018;6:147. Fig. 2 p. 6.)*

and the adequacy of the luteal phase can be assessed without repeated blood tests through, temperature charts, clinical examination, abdominal ultrasound and quality of cervical mucus. Home kits are available for instant testing of urinary luteinizing hormone (LH) or pregnanediol glucuronide to reveal results above thresholds for the patient to interpret, or the urine samples can be frozen and sent to the laboratory for analysis. Since the 1980s home urinary ovulation predictor tests (OT) have been used by those women charting the hormone pattern to optimize intercourse to become pregnant. These urinary tests attempt to predict when ovulation is about to occur by measuring the LH surge. Some women report a range of negative experiences when using OTs, they also reported similar negative experiences when trying to conceive without using the tests. There are many

positive themes associated with OT use, including an increased understanding of the menstrual cycle, confirmation of ovulation timing and providing a source of help and support when trying to conceive. Overall, when women are trying to conceive, ensuring they have access to high-quality information, including use of OT, may be of benefit to help address some of the questions and uncertainties that were raised by the participants in this study.

The rise in LH in the urine occurs near the time when ovulation takes place during the menstrual cycle (4–6) and may not be strictly 1 or 2 days before ovulation (Fig. 2.6.11). The LH peak is best described as a wave rather than as a peak with its surge occurring prior to ovulation but LH levels may remain high after ovulation into the luteal phase of the cycle and beyond if pregnancy is established.

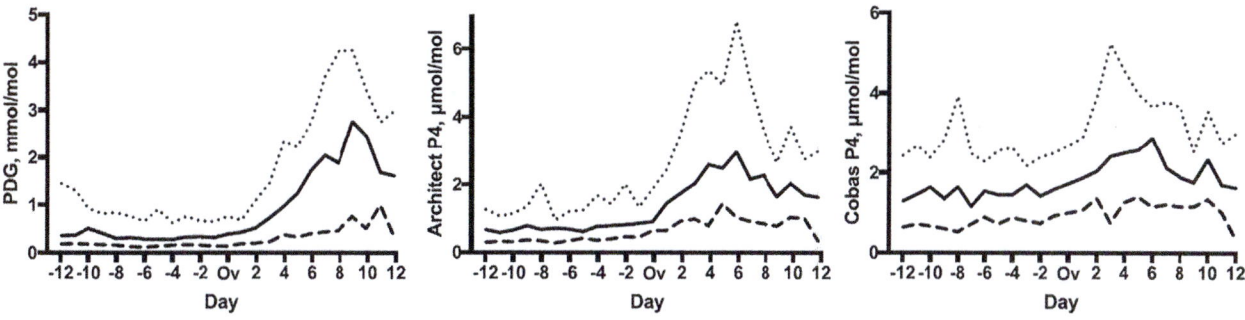

FIG. 2.6.10 Analysis of pregnanediol glucuronide (PDG) and progesterone (P4) in daily urine samples. Solid line—median; dashed line—0th centile; dotted line—90th centile. Day of ovulation—OV. PDG was measured by ELISA. P4 was measured on Abbott Architect chemiluminescent microparticle immunoassay and Cobas; Roche Cobas e 411 electrochemiluminescence immunoassay. *(From Gifford RM, Howie F, Wilson K, Johnston N, Todisco T, Crane M, Greeves JP, Skorupskaite K, Woods DR, Reynolds RM, Anderson RA. Confirmation of ovulation from urinary progesterone analysis: assessment of two automated assay platforms. Sci Rep 2018;8(1):17621. Fig. 2 p. 5.)*

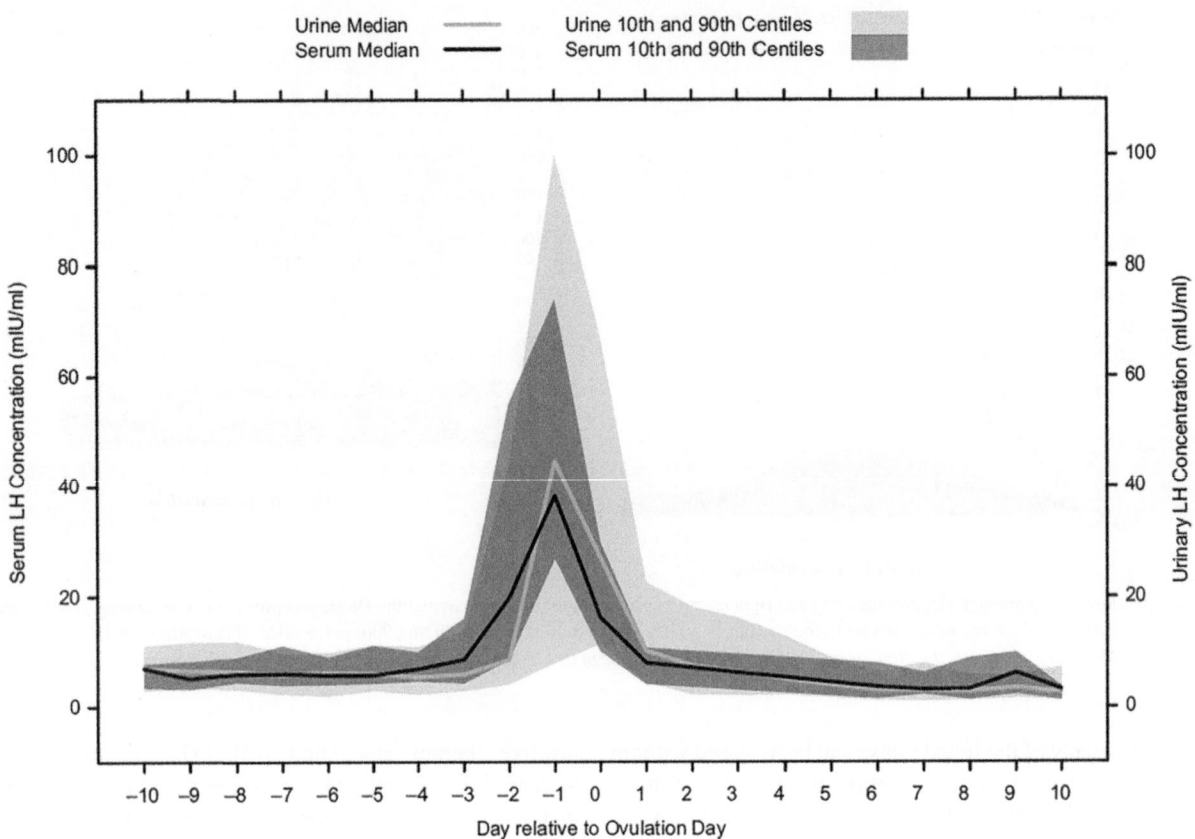

Urine Median ——— Urine 10th and 90th Centiles
Serum Median ——— Serum 10th and 90th Centiles

FIG. 2.6.11 Urinary LH (light gray, mIU/L) and serum LH (*dark gray*, mIU/L) referenced to the day of ovulation (mean, 10th and 90th percentile ranges. *(From Roos J, Johnson S, Weddell S, Godehardt E, Schiffner J, Freundl G, Gnoth C. Monitoring the menstrual cycle: comparison of urinary and serum reproductive hormones referenced to true ovulation. Eur J Contracept Reprod Health Care 2015;20(6):438–50. Fig. 1 p. 444.)*

All of these factors may affect how these tests are interpreted by the user in relation to the day of ovulation. The best scenario to predict ovulation at random was to detect the LH surge within 24 h after the first single positive test. A threshold of 25–30 mIU/ml may present the best predictive value for ovulation within 24 h (Fig. 2.6.11). Using the LH test alone to delineate the end of the fertile window, may provide a false end of the fertile window and may even occur before ovulation in 30% of cycles. The rise in progesterone and pregnanediol glucuronide (PdG) above baseline is a consistent marker of luteinization confirming ovulation. Both LH and progesterone surges delivered clear, sharp signals allowing reliable detection and confirmation of ovulation. Values for LH in urine corrected for urinary creatinine have been determined with the Perkin-Elmer Wallac Delfia assay. There is a large variation between individuals and the concentrations tend to increase between early and late follicular phase being in the range 1–25 IU/g creatinine. During the LH peak, the values of 30–100 IU/g creatinine are reached and after the peak the concentrations slowly decreased to less than 10 IU/g creatinine by the end of the luteal phase. The assay used to detect the LHbcf but this is not detected by the LH assay presently available from this manufacturer.

Studies by Adlercreutz et al. (1986) many years ago looked at the impact of dietary fiber on estrogen metabolism including the role of the enterohepatic circulation. At that time, it was difficult to analyze the bacterial content of the intestine. A repeat of the study (Oh et al., 2015) found fiber and fat intakes associated with patterns of estrogen metabolism. Such studies need to be repeated to determine the role of particular bacteria now that they can be better characterized. Estrogen metabolism is complex but studies to quantify multiple metabolites by GC-MS (Xu et al., 1999) have shown changes in the menstrual cycle. Compared with the early and mid-follicular phases, the ratios of 2-hydroxyestrogens/16-hydroxyestrogens and 2-hydrox/4-hydroxyestrogens were significantly increased during the periovulatory and luteal phases. In a more recent study, taller height and higher current BMI were associated with lower levels of urinary EM in the mid-luteal phase among premenopausal women (Xie et al., 2012) (Fig. 2.6.12).

High levels of physical activity are also associated with lower concentrations of parent estrogens and metabolites of the 16 hydroxylation pathway and preferential metabolism along 2-hydroxylation pathway to catechols (Matthews et al., 2012) (Fig. 2.6.13). In that study, steroids were measured by LC-MS/MS technology.

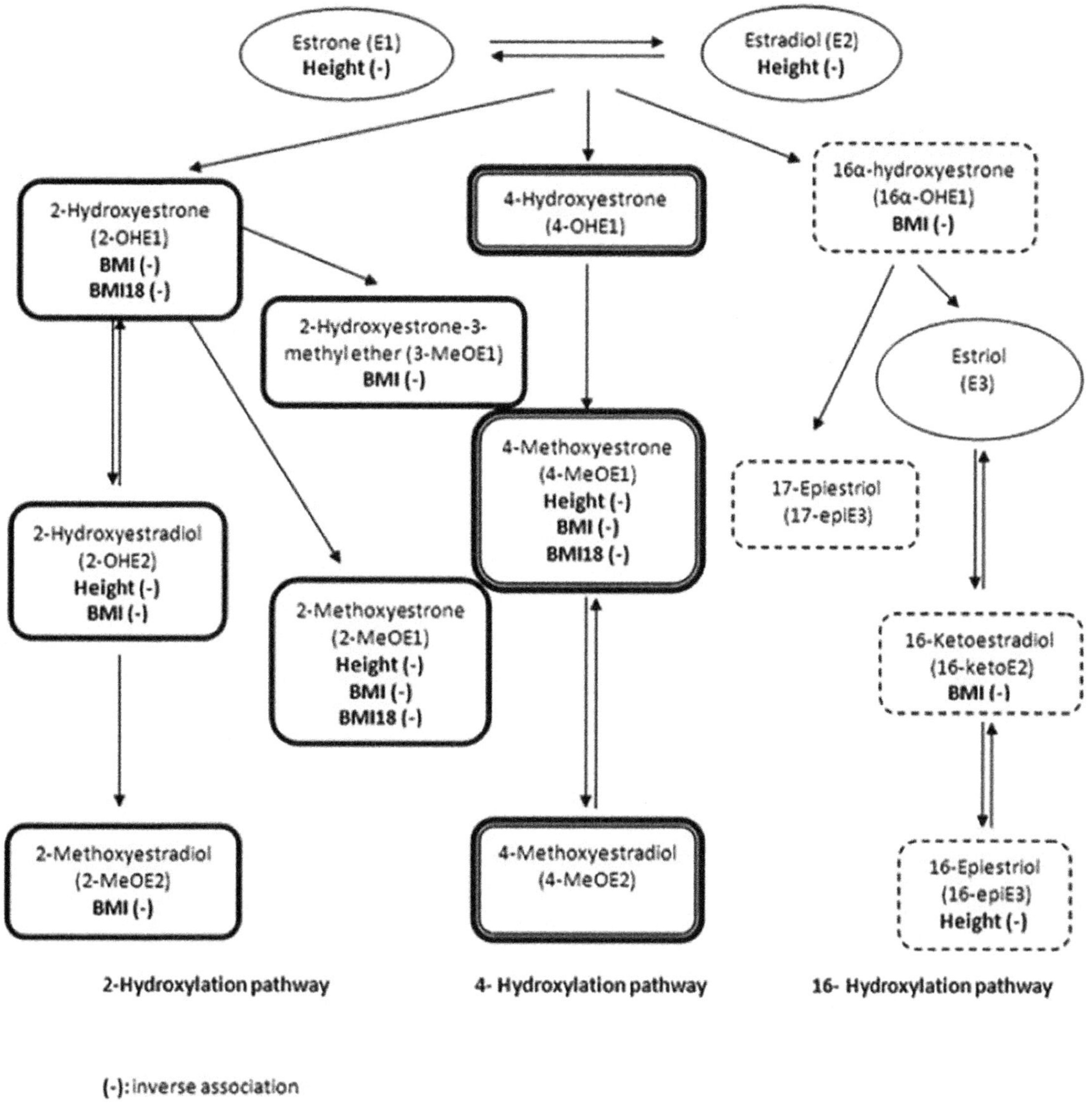

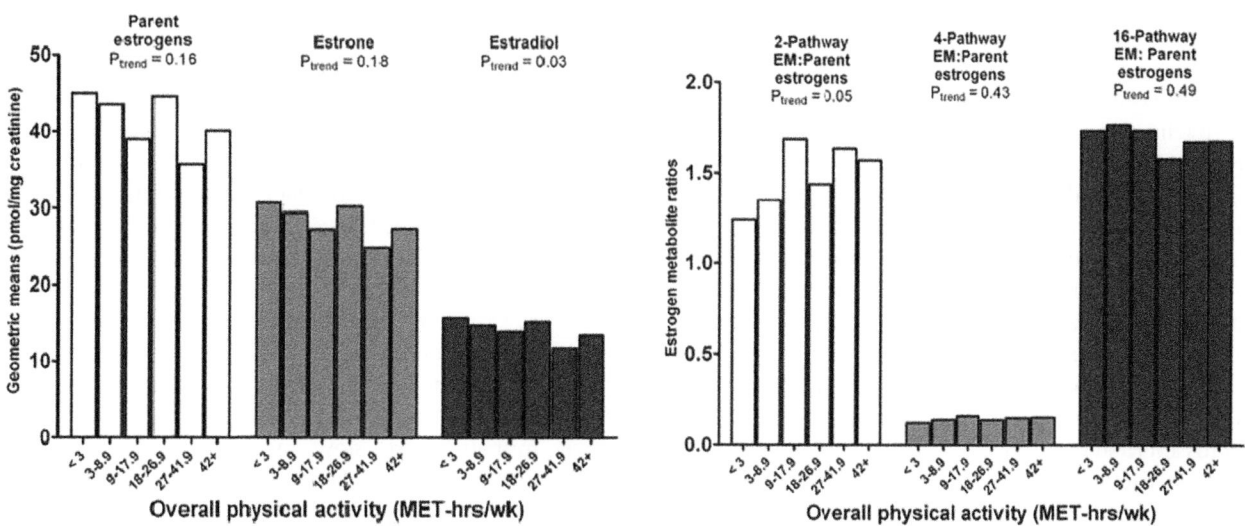

FIG. 2.6.12 Associations of BMI with endogenous estrogen metabolism. Adjusted geometric means of urinary estrogen and estrogen metabolite (EM) measures by BMI at age 18 in women. *(Adapted from Xie J, Eliassen AH, Xu X, Matthews CE, Hankinson SE, Ziegler RG, Tworoger SS. Body size in relation to urinary estrogens and estrogen metabolites (EM) among premenopausal women during the luteal phase. Horm Cancer 2012;3(5–6):249–60. Figure p. 17 and 18.)*

FIG. 2.6.13 Geometric means for urinary concentrations of parent estrogens (pmol per milligram creatinine) by level of overall physical activity in the NHSII (*n* = 603) (Left panel) and for the 2-hydroxylation, 4-hydroxylation and 16-hydroxylation pathways to parent estrogens ratio (right panel). *(From Matthews CE, Fortner RT, Xu X, Hankinson SE, Eliassen AH, Ziegler RG. Association between physical activity and urinary estrogens and estrogen metabolites in premenopausal women. J Clin Endocrinol Metab 2012;97(10):3724–33. Fig. 2 p. 3729.)*

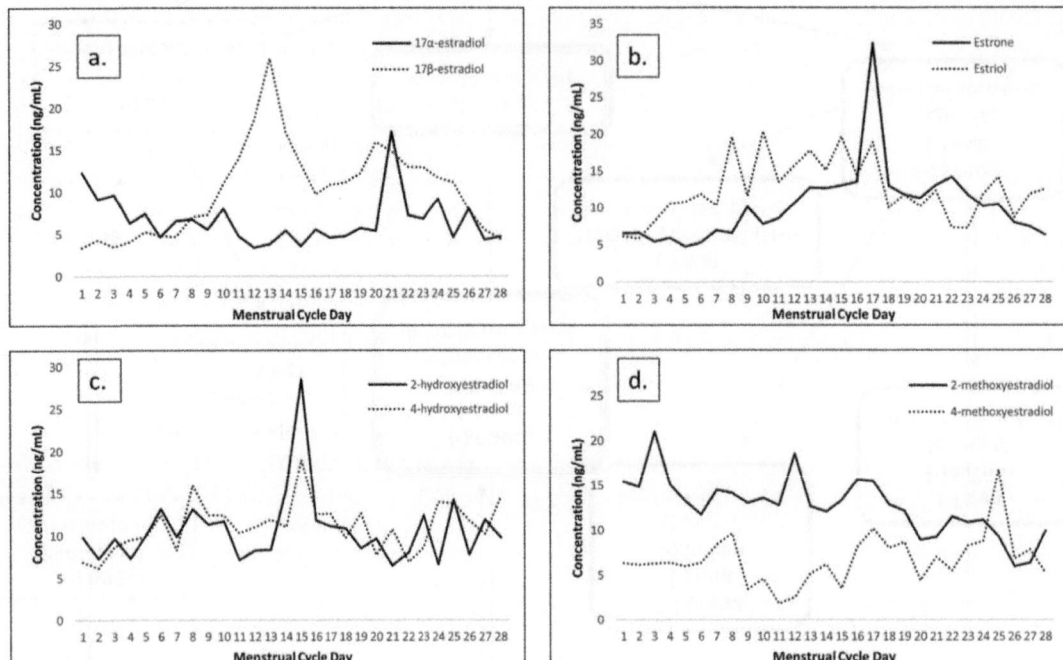

FIG. 2.6.14 Concentration profiles (normalized for creatinine) for estrogen metabolites across the 28 day menstrual cycle. Concentration profile (normalized for the creatinine value) of the target analytes along the 28-day menstrual cycle achieved by applying PARAFAC approach: (A) 17α-estradiol and 17β-estradiol, (B) estrone and estriol, (C) 2-hydroxyestradiol and 4-hydroxyestradiol, (D) 2-methoxyestradiol and 4-methoxyestradiol. Original file includes 2-hydroxyestrone, 4-hydroxyestrone, 2-methoxyestrone, 4-methoxyestrone, 16α-hydroxyestrone, 16-epiestriol, and 17-epiestriol. *(From Bozzolino C, Vaglio S, Amante E, Alladio E, Gerace E, Salomone A, Vincenti M. Individual and cyclic estrogenic profile in women: structure and variability of the data. Steroids 2019;150:108432. Fig. 1 p. 5.)*

The results of physical activity and estrogen metabolism may have to be considered in some future studies investigating the etiology of diseases linked to both physical activity and endogenous estrogen.

The early literature was able to only report variations of estrone and estradiol concentrations across the complete menstrual cycle, whereas now a generalized picture for a broad urinary estrogen panel along the whole menstrual period can be described. It is now possible to determine many more estrogens in a urine sample (hydroxy estradiols and estrones, estriols and catecholestrogens) and to explore the cyclic and interindividual variation in each estrogen (Fig. 2.6.14).

Multivariate data analysis and the application of principal component analysis (PCA) to data from estrogen profiles in the menstrual cycle, provide an easier visualization and efficient partition of the data into three groups, corresponding to the three phases of the menstrual cycle, namely the follicular phase, ovulation and the luteal phase, together with the transitions between the phases. This may provide clearer prospective information in fertility potential studies. Multivariate statistics beyond principal component analysis (Parallel Factor Analysis) was applied to the chronological data and from the three-dimensional data (Bozzolino et al., 2019) it was found that 4-methxyestradiol, 16-epiestriol and

17-epiestriol profiles increased around ovulation and remained stable for 10 days, whereas 2-methoxyestradiol showed a constant decrease through the cycle. This study is an example where machine learning and analysis of large data sets is increasingly becoming an essential aspect of metabolic data analysis.

2.6.4 Urinary steroid profiles

In a urine sample, the range of steroid metabolites can be analyzed by capillary gas chromatography (GC) with flame ionization detector or more specifically by GC-MS even GC-MS/MS and LC-MS/MS and has been a research tool for decades called **Urinary steroid profiling** or recently **metabolome**. This technique provides qualitative and quantitative data on excretion of the wide spectrum of adrenal steroid metabolites simultaneously. For GC analysis, steroids are extracted from the urine, conjugates hydrolyzed enzymatically before derivative formation for the steroids to be stable in the high temperature GC analysis. Methods for intact steroid conjugates without hydrolysis and derivative formation are now available by GC-MS and LC-MS/MS.

There are several fundamental issues to understand about a USP.

1. A USP seems to be quite an expensive test but provides data on metabolites of 40 or more of the steroid hormone metabolites and therefore competes with the cost of several separate hormone tests. Since the common tests for steroids will still initially be performed on automated clinical chemistry analyzers there will be multiple test charges.

2. There is a **circadian rhythm** in the excretion of steroid metabolites with highest excretion rates for cortisol metabolites in samples between 10.00 h and 16.00 h when urine was collected every 3 h (Fig. 2.6.15) (Ackermann et al., 2019; Jerjes et al., 2006b) which contrasts to the plasma profile with peaks around 0800 h time of day.

Differences are seen in concentrations of the steroids in an overnight versus daytime sample collection with highest results for adrenal androgen and cortisol metabolites in the daytime collection (Fig. 2.6.16).

Similar results have been found in the Bern laboratory, reported as mg/hour (Ackermann et al., 2019). Quantitative results from the analysis of spot urine samples must therefore be treated with caution, but ratios of steroids often useful in interpretation of enzyme activities.

3. It is important that the clinical user of a steroid profile understands that androgens are mainly metabolites of DHEAS that is produced in 10–30 mg quantities per day. Even when testosterone production is raised from normal, the metabolites are swamped by the products from DHEAS. Testosterone is excreted as many metabolites, some of which are also products of androstenedione and DHEAS. Urine androgen excretion is low until 5–8 years of age when DHEAS production rises with the development of the zona reticularis in the adrenal cortex outside the adrenal medulla. Appropriate reference ranges for age need to be used in interpretation of the results.

4. In clinical use as detailed later in Part 3 of the book, a urinary steroid profile enables the diagnosis of inborn errors of steroid metabolism, and steroid-producing tumors in patients with signs of premature virilization and permits the differentiation of premature adrenarche or precocious puberty from other serious underlying pathology (see Chapters 3.3.1 and 3.3.2). Very distinctive profiles are seen in genetic disorders and adrenocortical tumors, whereas an earlier increase in androgen sulfate excretion and other major androgen

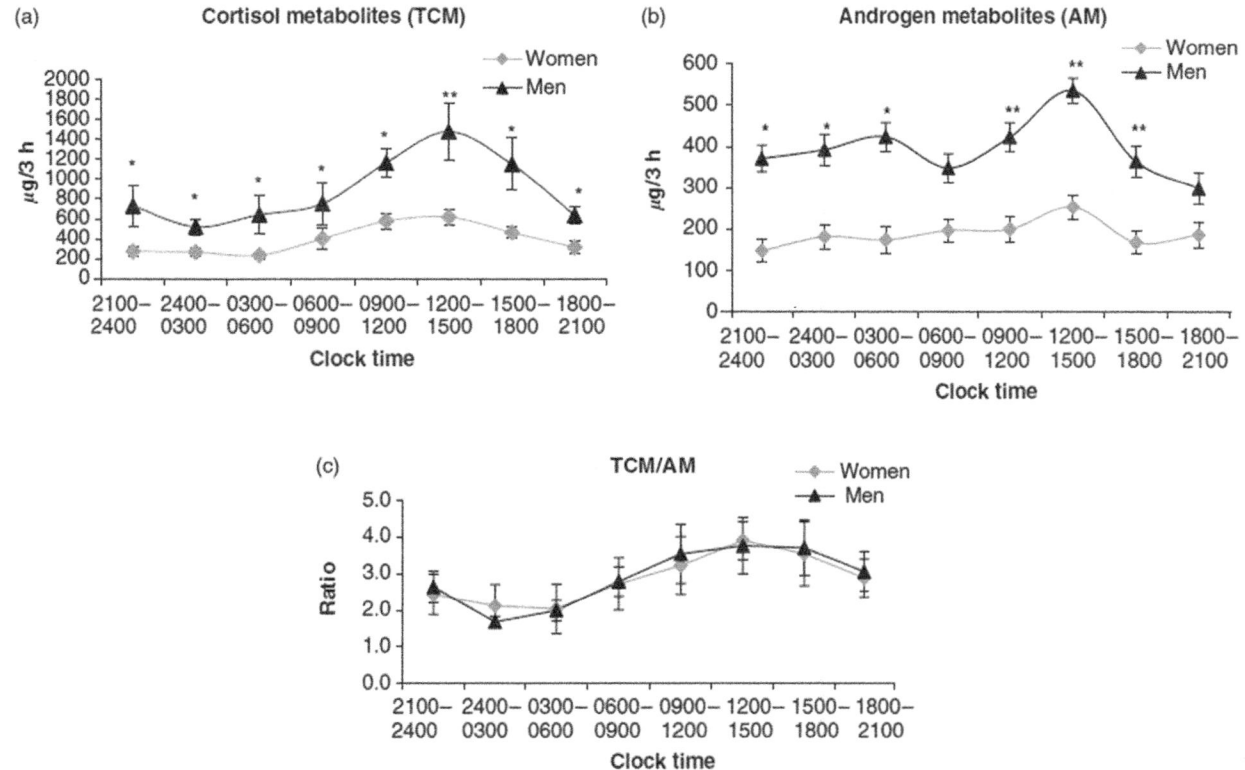

FIG. 2.6.15 Diurnal rhythm of urinary steroids. Mean values and standard errors of (A) total urinary cortisol metabolite (TCM) excretion (B) urinary androgen metabolites (AM) (C) ratio cortisol metabolites/androgen metabolites at 3 h intervals over 24 h. *$P < .05$; **$P < .005$. *(Jerjes WK, Cleare AJ, Peters TJ, Taylor NF. Circadian rhythm of urinary steroid metabolites. Ann Clin Biochem 2006a;43(Pt 4):287–294. Fig. 1. p. 289.)*

FIG. 2.6.16 Excretion rates of steroid in urine collected in the day (0730 to 2300h) and night (2301 to 0729h). *(Author original.)*

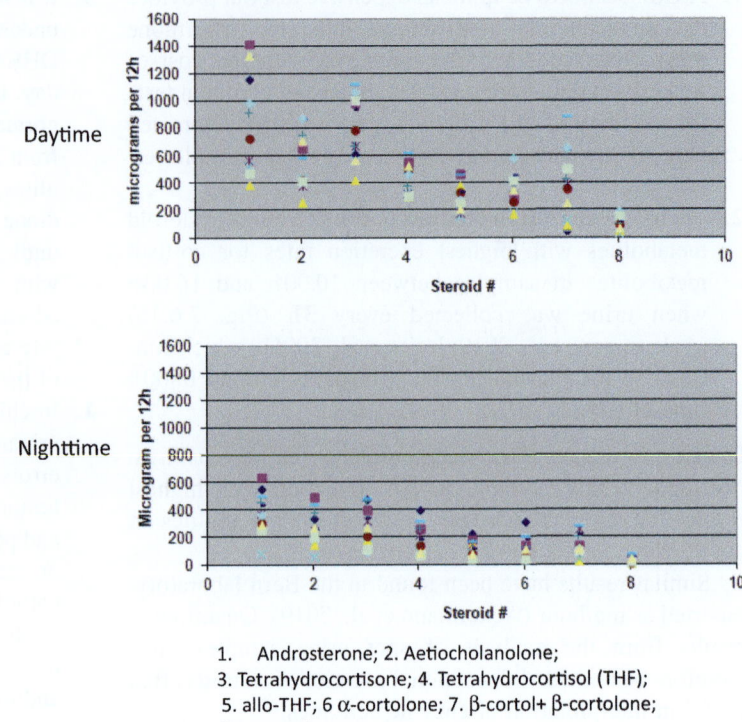

Daytime

Nighttime

1. Androsterone; 2. Aetiocholanolone;
3. Tetrahydrocortisone; 4. Tetrahydrocortisol (THF);
5. allo-THF; 6 α-cortolone; 7. β-cortol+ β-cortolone;
8. α cortol.

2.6.4.1.1   Effects of age and gender

Laboratories report the concentrations or **excretion rates** of around 20 to 40 or even 100 steroids. In addition to excretion rates of the steroids, reference ranges for **steroid ratios** have been developed in some laboratories for the purpose of recognizing where low enzyme activity leads to changes in the ratios of substrate to product. These are summarized in Table.2.6.3.

The application of these ratios will be discussed in more detail in the relevant clinical sections where genetic disorders of the enzymes are detected.

Changes in the outputs of urine steroids with age through childhood to adults have been documented in a number of centers by GC-MS, GC-MS/MS and LC-MS/MS at low and high MS resolution and there are visually significant differences in the reference ranges between laboratories so local ranges should be used. It is difficult to compare the numerical data for urinary steroids because many different units are used, some with correction for creatinine

metabolites are recognized in premature adrenarche. The patient with nonclassical 21-hydroxylase deficiency, the commonest form of CAH, shows an increase in 17-OHP and 21-deoxycortisol metabolites with low cortisol and corticosterone metabolites. A USP has now been tested for nearly all steroid clinical disorders.

output (see later Table 2.6.8, de Jong et al., 2017). Comprehensive reference ranges including neonates and children are few in number (Homma et al., 2003; Wudy and Hartmann, 2011; Shackleton, 1986; Honour et al., 2018; Ackermann et al., 2019; de Jong et al., 2017). The published data on reference ranges for children and adults from nine laboratories were summarized in a Shackleton review in 1986 and for a further 8 laboratories by Ackermann in 2017 so will not be summarized again here (Shackleton, 1986, 1993, 2008, 2010; Shackleton et al., 2018); Bevan et al., 1986 (Taylor lab.); Weykamp et al., 1989; Finken et al., 1999; Shamim et al., 2000 (UCLH); Taylor, 2006, 2013; Chan et al., 2008; de Jong et al., 2017). Results from the UCLH (Tables 2.6.4A and 2.6.4B) are shown as representative data in μg/day units. The results in general for the urinary steroids are higher in males than females partly due to body mass differences.

Differences in urinary cortisol output by males and females are well known (Lamb et al., 1994, 1997; Taylor and Raven, 1996) and are also seen in the outputs of steroid metabolites (Tables 2.6.5A, 2.6.5B, and 2.6.6).

Excretion rates for 11-hydroxyandrosterone, THF, allo-THF, cortolone are significantly lower in females, so median total cortisol metabolites are 6965mg/day in men and 4595mg/day in women ($P = .0005$). Androgen metabolites are also higher in men than women (Table 2.6.7) (Shamim et al., 2000).

TABLE 2.6.3 Summary of ratios of urinary steroids used by various labs.

		de Jong	de Jong	Rhousson/Dhayat	Krone
		M	F	F	
HSD3B2	DHEA/(THE+THF+aTHF)	0–0.8	0–0.7		
	(A+AE)/(THE+THF+aTHF)	0.5–1.9	0.1–2.2		
	5PT/THE			0.031–0.097	0.03–0.08
SRD5A2	E/A	0.3–1.8		0.899–1.36	
	THF/aTHF	0.3–2.8			
	THB/aTHB	0.1–1.2			
HSD11B2	(THF+aTHF)/THE	0.5=2.4			
	(F+E)/(THE+THF+aTHF)			0.757–0.858	
HSD17B3	(A+AE)/(THE+THF+aTHF)	0.5–1.9		0.624–1.24	
POR	(PDL+P3)/(A+Ae)	0.1–0.2			
	(PDL+P3)/((THE+THF+aTHF))	0.1–0.3			
	P2 (THE+THF+aTHF)	0–0.1			
	17OHP+PT/(THE+THF+aTHF)				0.1–0.32
	(THA+THB+aTHB(/(THE+THF+aTHF)				0.1–0.22
CYP11B1	THS/(THE+THF+aTHF)	0–0.02			0.5–2.0
	THS/THE			0.018–0.031	
CYP17	(THA+THB+aTHB)/(THE+THF+aTHF)	0–0.1			
	(THA+THB+aTHB)/THE			0.176–0.279	
	(THA+THB+aTHB)/(A+Ae)	0–0.2			
	PD/(A+Ae)			0.073–0.384	
CYP21	PDL/(THE+ THF+aTHF)	0–0.1			
	P3/(THE+THF+aTHF)	0.1–0.3			0.05–0.2
	PTL/(THE+THF+aTHF)	0–0.004		0.004–0.008	0.06–0.9

Data from de Jong WHA, Buitenwerf E, Pranger AT, Riphagen IJ, Wolffenbuttel BHR, Kerstens MN, Kema IP. Determination of reference intervals for urinary steroid profiling using a newly validated GC-MS/MS method. Clin Chem Lab Med 2017;56(1):103–12; Rousson V, Ackermann D, Ponte B, Pruijm M, Guessous I, d'Uscio CH, et al. Sex- and age-specific reference intervals for diagnostic ratios reflecting relative activity of steroidogenic enzymes and pathways in adults. PLoS One 2021;16(7):e0253975; Krone N, Hughes BA, Lavery GG, Stewart PM, Arlt W, Shackleton CH. Gas chromatography/mass spectrometry (GC/MS) remains a pre-eminent discovery tool in clinical steroid investigations even in the era of fast liquid chromatography tandem mass spectrometry (LC/MS/MS). J Steroid Biochem Mol Biol 2010;121(3–5):496–504.

It is difficult to compare reference ranges when so many different units are reported in absolute terms or corrected for body size and creatinine output.

Using **GC-MS/MS, 33** steroids were measured by Wilhelmina de Jong and colleagues in urine of adults from 20 to 79 years of age. Only four internal standards were used (D_6-DHEA, D_4-pregnenolone, D_5-11KE, and D_5-THE). The gender differences were similar to the UCLH and Bern data by GC/MS. Women using OCP had lower excretions of androgen metabolites in the 20–29 year old age group (16.3 vs. 21.8 µmol/24 h, $P = .04$) (Fig. 2.6.17). Progesterone metabolite excretion was decreased in both age groups using OCP compared with corresponding age groups not using OCP (20–29 years: 1.9 vs. 3.6 µmol/24 h, $P = .006$ and 30–39 years: 2.6 [1.5–4.0] vs. 6.3 [3.4–13.2] µmol/24 h, $P = .001$).

Several cortisol metabolites were excreted in significantly lower amounts in the women using OCP and total cortisol metabolite excretion showed a trend toward lower excretion in women using OCP in both age groups (20–29 years: 13.1 vs. 20.0 µmol/24 h, $P = .06$; Table 2.6.8).

TABLE 2.6.4A Excretion rates for urine steroids (normal ranges) (μg/24 h) by GC-MS (UCLH laboratory data) for children and adults.

Females	<3 yrs	4 yrs	6 yrs	8 yrs	9 yrs	11 yrs	12 yrs	Adults
N=	4	12	6	6	103	24	20	53
Androsterone	<10	<50	30–100	10–150	10–680	160–740	200–700	200–2580
Etiocholanolone	<10	<50	30–100	10–120	10–760	200–540	200–500	210–2740
Dehydropeiandrosterone	<10	<20	<50	<100		20–80	10–400	30–1670
11-Hydroxyandrosterone	<50	10–120	50–150	50–200	20–600	150–550	100–600	120–1550
11-Hydroxyetiocholanolone	<20	10–80	50–200	50–200		<200	50–300	40–570
16-Hydroxy-DMA	<80	<20	10–200	10–200		<200	20–500	40–510
Pregnanediol	<10	<20	<30	<30		70–340	20–600	40–2110
Pregnanetriol	<50	<50	<50	<50		10–250	20–100	40–1810
Androstenetriol	<80	<70	<50	<50		<200	20–200	30–1620
THE	50–250	200–800	300–1000	400–1200	120–2390	350–1660	500–1500	900–3690
THF	10–150	50–250	50–300	100–400	30–1130	200–600	200–800	430–1840
Allo-THF	20–300	100–700	100–700	100–800	40–1110	200–700	200–1100	160–1680
α-Cortolone	<50	50–250	50–250	50–300	20–1510	440–710	200–900	260–1330
β-Cortolone + β-cortol	<50	50–250	50–250	50–250	30–1080	300–580	100–500	200–1950
α-Cortol	<20	<50	50–820	50–1000	10–300	20–200	20–200	120–590
Total cortisol metabs	300–900	630–1730	820–2160	1000–2730	380–6420	1500–3530	2650–4640	2630–8820
Total androgens	<80	<80	40–170	90–140	20–1220	250–790	500–1400	470–4510

TABLE 2.6.4B Excretion rates for urine steroids (normal ranges) (μg/24 h) by GC-MS (UCLH laboratory data) for children and adults.

Males	<3 yrs	4 yrs	6 yrs	8 yrs	9 yrs	10 yrs	12 yrs	Adults
N=	4	12	7	33	141	85	66	23
Androsterone	<10	<50	20–150	10–100	10–800	100–600	100–1250	820–1730
Etiocholanolone	<10	<50	10–200	10–120	10–580	150–550	100–1000	620–2030
Dehydropeiandrosterone	<110	<20	<100	<100		<200	10–500	40–1720
11-Hydroxyandrosterone	<50	10–120	50–200	50–200	10–910	50–400	200–700	210–1240
11-Hydroxyetiocholanolone	<20	10–80	50–200	50–200		50–250	50–400	40–540
16-Hydroxy-DHA	<80	<20	10–200	<30		10–200	20–500	130–1020
Pregnanetriol	<50	<50	<50	<50		<50	20–100	110–520
Androstenetriol	<80	<70	<50	<50		<50	20–250	290–600
THE	50–250	200–800	300–1000	400–1200	120–2270	500–1300	500–2000	1530–5730
THF	10–150	50–250	50–300	100–400	30–920	200–500	200–900	600–2150
Allo-THF	20–300	100–700	100–800	100–900	30–1150	200–900	200–1100	380–1830
α-Cortolone	<50	50–250	50–250	50–300	20–980	50–700	200–800	220–1600
β-Cortolone + β-cortol	<50	50–250	20–250	50–250	20–980	50–500	100–600	410–2600
α-Cortol	<20	< 50	10–50	10–50	10–220	50–200	20–200	100–450
Total cortisol metabs	300–900	630–1730	820–2160	1000–2730	180–6220	1600–3600	2659–4640	4070–12,470
Total androgens	<80	<80	40–300	30–470	20–1380	220–1200	280–1290	1850–4500

TABLE 2.6.5A Reference intervals of urinary steroid metabolites in men (μmol/24h) per decade.

Metabolite, μg/24h	N	Age group, years							CF to nmol/24h
		18–29.9	30–39.9	40–49.9	50–59.9	60–69.9	70–79.9	80–85.7	
Androsterone	381	1450–5821	1456–5839	1073–4644	759–3594	541–2810	390–2224	285–1786	×3.444
11β-OH-androsterone	447	357–1724	373–1771	376–1780	367–1752	345–1687	313–1589	273–1462	×3.264
Etiocholanolone	390	883–5047	794–4743	673–4308	531–3770	385–3165	249–2531	138–1910	×3.444
17β-Estradiol	457	0.7–4.5	0.9–5.5	1–6.1	1.1–63	1–5.9	0.9–5.6	0.9–5.6	×3.671
Estriol	456	1.8–13.8	1.9–14.5	2–15.1	2.2–15.8	2.3–16.5	2.4–17.3	2.5–18.1	×3.467
TH-11-deoxycorticosterone	455	3.2–23.3	3.1–22.7	2.9–21.4	2.7–19.5	2.4–17.3	2.2–16.2	2.2–16.2	×2.990
TH-11-dehydrocorticosterone	452	49.1–277	49.5–278	46.6–267	40.7–242	32.8–209	28.8–191	28.8–191	×2.869
18-OH-TH-11-dehydrocorticosterone	433	14–259	15.1–297	15.7–318	15.7–318	15.1–297	14–259	12.5–212	×2.743
TH-corticosterone	457	52.6–326	58.3–350	59.8–356	57.1–345	50.5–317	48.6–309	60.1–357	×2.853
5α-TH-corticosterone	457	155–891	170–976	147–844	128–738	119–681	115–663	115–663	×2.853
TH-aldosterone	456	5.1–792	5.4–84	5.5–85.6	5.4–83.9	5.1–79.1	4.6–71.7	4–62.5	×2.743
TH-11-deoxycortisol	457	27.5–145	28.4–152	29.4–160	30.4–168	31.5–176	32.6–185	33.8–194	×2.853
Cortisol	457	41.6–243	47.1–282	50–302	49.7–300	47.3–282	48.4–291	54.5–335	×2.759
6β-OH-cortisol	457	35.3–303	35.3–303	35.3–303	353–303	35.3–303	35.3–303	35.3–303	×2.642
18-OH-cortisol	424	40.7–638	40.7–638	40.7–638	40.7–638	40.7–638	40.7–638	40.7–638	×2.642
20α-DH-cortisol	457	19.2–136	19.2–136	19.2–136	19.2–136	19.2–136	19.2–136	19.2–136	×2.743
TH-cortisol	371	714–2724	866–3301	966–3682	992–3781	938–3575	918–3501	1047–3992	×2.729
α-Cortol	452	127–519	157–638	161–658	164–668	169–688	176–717	186–757	×2.714
β-Cortol	453	213–1171	213–1171	213–1171	213–1171	213–1171	213–1171	213–1171	×2.714
11β-OH-etiocholanolone	455	29.7–955	15.1–861	27.7–944	43.7–1029	55–1081	59.1–1099	59.1–1099	×3.264
5α-TH-cortisol	381	524–3424	631–3992	578–3715	527–3436	497–3278	488–3226	488–3226	×2.729
Cortisone	456	69.2–378	74–404	77–420	77.9–425	76.6–418	73.4–401	68.3–373	×2.774
20α-DH-cortisone	457	11.1–63.3	11.1–63.3	11.1–63.3	11.1–63.3	11.1–63.3	11.1–63.3	11.1–63.3	×2.759
20β-DH-cortisone	457	21.8–127	22.7–131	23.6–135	24.5–140	25.5–145	26.5–149	27.5–154	×2.759
TH-cortisone	407	1543–6080	1643–6378	1667–6449	1611–6285	1483–5902	1297–5332	1072–4623	×2.743
α-Cortolone	426	552–2214	587–2338	611–2420	622–2456	618–2443	600–2382	570–2277	×2.729
β-Cortolone	427	366–1387	349–1328	332–1271	315–1216	300–1164	285–1114	271–1065	×2.729
11-Keto-etiocholanolone	455	97.7–1055	67.3–884	85–986	103–1082	109–1113	102–1076	83–975	×3.285

From Ackermann D, Groessl M, Pruijm M, Ponte B, Escher G, d'Uscio CH, et al. Reference intervals for the urinary steroid metabolome: the impact of sex, age, day and night time on human adult steroidogenesis. PLoS ONE 2019;14(3):e0214549. Table 2.

TABLE 2.6.5B Reference intervals of urinary steroid metabolites in men (μmol/24h) per decade.

Metabolite, μg/24h	N	Age group, years							CF to nmol/24h
		18–29.9	30–39.9	40–49.9	50–59.9	60–69.9	70–79.9	80–90.0	
17α-OH-pregnanolone	375	18.2–383	28.4–704	27.5–672	16.5–337	9.7–166	8.2–132	8.2–132	×2.990
Pregnanetriol	360	173–1156	216–1346	200–1275	135–975	80.9–696	56.8–554	48.3–501	×2.972
Pregnanetriol	378	18.7–810	12–661	7.4–535	4.3–429	2.4–341	1.2–268	0.5–207	×2.990
Pregnanetriolone	379	4.1–95.7	3–54.5	3.1–57.9	3.4–67	3.6–77.4	3.9–89.4	4.2–103	×2.853
Pregnanediol	376	629–2180	95.2–5269	94.6–5192	61.7–2101	38.2–830	29.5–520	27.5–460	×3.160
Dehydroepiandrosterone	377	13.8–1623	15.6–1989	13.6–1578	9.1–834	5.7–409	4.1–242	3.2–168	×3.467
16α-OH-dehydroepiandrosterone	379	37.7–1742	33.9–1565	25–1157	15.2–704	8.9–411	5.8–270	4.3–199	×3.285
Androstenediol	378	22.5–873	16.8–571	12.6–380	9.5–257	7.2–177	5.6–123	4.3–86.5	×3.444
Androstenetriol	378	99.7–1351	72–1051	51.4–813	36.2–625	25.2–476	17.3–361	11.7–271	×3.264
Testosterone	367	2.4–55.1	27–66.1	2.6–61.8	2.1–45.3	1.6–30.9	1.2–22.8	1.1–18.3	×3.467
5α-DH-testosterone	377	2.9–75.6	25–66.9	2.2–59.1	2–523	1.8–46.3	1.5–41	1.4–36.2	×3.444
Androstarediol	372	8.9–106	10.5–124	9.7–114	7–83.1	4.9–57.7	4–46.7	3.7–44	×3.420
Androsterone	349	480–4482	414–3978	307–3138	194–2181	116–1462	75.6–1051	543–816	×3.444
11β-OH-androsterone	376	192–1183	192–1183	192–1183	192–1183	192–1183	192–1183	192–1183	×3.264
Etiocholanolone	351	543–3870	448–3483	345–3031	243–2540	153–2037	82.1–1551	33.91106	×3.444
17β-Estradiol	377	0.3–7.6	0.7–26.6	0.9–35.8	0.5–17	0.3–7.1	0.2–5.4	0.2–5.4	×3.671
Estriol	374	0.8–36.3	1.6–91.8	1.7–99.5	1–453	0.6–19.5	0.5–15	0.5–15	×3.467
TH-11-deoxycorticosterone	378	1.8–34.2	25–66.1	2.6–69.5	1.9–39	1.4–215	1.2–17.9	1.2–17.9	×2.990
TH-11-dehydrocorticosterone	379	31.5–240	29.5–228	27.6–216	25.9–205	24.3–194	22.7–184	21.2–174	×2.869
18-OH-TH-11-dehydrocorticosterone	342	10–262	10–262	10–262	10–262	10–262	10–262	10–262	×2.743
TH-corticosterone	379	28–226	34.8–261	39.7–285	41.7–294	40.4–288	36.2–268	29.7–235	×2.853
5α-TH-corticosterone	379	60.6–556	57–535	53.5–514	50.2–493	47–474	44.1–454	41.2–436	×2.853
TH-aldosterone	378	5.3–83	5.6–86.3	5.5–84.9	5.1–78.8	4.5–69.1	3.7–57.3	2.9–44.9	×2.743
TH-11-deoxycortisol	379	15–97.3	17.6–114	19.8–128	21.4–138	22.2–144	22.1–143	21.1–137	×2.853
Cortisol	379	33.7–250	33.7–250	33.7–2.50	33.7–250	33.7–250	33.7–250	33.7–250	×2.759
6β-OH-cortisol	378	25.6–282	25.6–282	25.6–282	25.6–282	25.6–282	25.6–282	25.6–282	×2.642
18-OH-cortisol	344	32.4–516	32.4–516	32.4–516	32.4–516	32.4–516	32.4–516	32.4–516	×2.642
20α-DH-cortisol	379	16.7–176	16.2–169	15.6–163	15.2–156	14.7–150	14.2–144	13.8–138	×2.743
TH-cortisol	340	336–1677	416–2015	490–2319	550–2560	589–2714	603–2766	588–2711	×2.729
α-Cortol	379	92.5–464	95.1–479	97.8–495	101–511	103–528	106–546	109–564	×2.714
β-Cortol	378	125–751	125–751	125–751	125–751	125–751	125–751	125–751	×2.714
11β-OH-etiocholanakine	378	0.5–612	12.2–756	28.4–860	39.1–916	39.3–917	28.9–863	12.7–760	×3.264
5α-TH-cortisol	362	177–1982	177–1982	177–1982	177–1982	177–1982	177–1982	177–1982	×2.729
Cortisone	379	48.3–308	56.3–360	59.4–379	56.7–362	50.2–321	47.9–306	47.9–306	×2.774
20α-DH-cortisone	379	806–52.3	8.7–53.1	8.3–49.9	7.4–43.7	6.4–365	6–34.2	6–34.2	×2.759
20β-DH-cortisone	379	23.9–145	225–136	21.2–129	20–121	18.8–114	17.8–107	16.7–101	×2.759

TABLE 2.6.5B Reference intervals of urinary steroid metabolites in men (μmol/24h) per decade—cont'd

Metabolite, μg/24h	N	18–29.9	30–39.9	40–49.9	50–59.9	60–69.9	70–79.9	80–90.0	CF to nmol/24h
TH-cortisone	360	900–4654	900–4654	900–46.54	900–4654	900–4654	900–4654	900–4654	×2.743
α-Cortolone	362	405–1932	405–1932	405–1932	405–1932	405–1932	405–1932	405–1932	×2.729
β-Cortolone	369	164–854	164–854	164–854	164–854	164–854	164–854	164–854	×2.729
11-Keto-etiocholanolone	379	23.6–652	43.4–778	60.8–870	70.9–918	71–919	61–871	43.6–779	×3.285

Reference intervals have been estimated by the described statistical models and are given as 25th–97.5th percentiles in the unit μg/24h for different age groups. N represents the sample number per analyte included in the statistical model. C.F., conversion factor.

From Ackermann D, Groessl M, Pruijm M, Ponte B, Escher G, d'Uscio CH, et al. Reference intervals for the urinary steroid metabolome: the impact of sex, age, day and night time on human adult steroidogenesis. PLoS ONE 2019;14(3):e0214549.

TABLE 2.6.6 Reference intervals (μg/24 h) for steroid metabolites by GC-MS by men and women.

Steroid	Male Kings range	Male Bern 27–75th centile	Female Kings range	Female Bern 25–75th centile
Androsterone (5α) (391)	490–2570	**1194–2719**	**<1610**	**337–1092**
Etiocholanolone (5β) (391)	180–2424	**1029–2362**	**< 2180**	**493–1399**
Dehydroepiandrosterone **(389)**	, 1750	**54–572**	**<800**	**25–141**
11β-Hydroxyandrosterone (479)	**370–1340**	**646–1086**	**80–980**	**339–630**
11β-Hydroxyetiocholanolone	<1540	**199–546**	**20–650**	**162–452**
16α-Hydroxy-DHA (477)	<1480	**89–495**	**<515**	**44–214**
Pregnanediol (Progesterone) (464)	<1450	**145–294**	**<2430**	**118–388**
Pregnanetriol (17-OH-Prog) (552)	≪1563	**518–920**	**<890**	**226–526**
Androstenetriol (532)	<1630	**236–626**	**<760**	**90–277**
Tetrahydrocortisone (609)	570–5700	**2492–4016**	**370–3510**	**1528–2661**
Allo-tetrahydrocorticosterone (595)	**80–570**	**236–456**	**<422**	**119–257**
Tetrahydrocortisol (683)	310–2220	**1407–2231**	**250–1510**	**881–1455**
Allo-THF (683)	190–2220	**1055–1986**	**90–920**	**409–851**
α-Cortolone (654)	<1520	**984–1525**	**140–1210**	**673–1153**
β-Cortolone + β-cortol	150–1590			
α-Cortol (728)	60–700	**255–421**	**70–340**	**166–290**

Assembled from Ackermann D, Groessl M, Pruijm M, Ponte B, Escher G, d'Uscio CH, et al. Reference intervals for the urinary steroid metabolome: the impact of sex, age, day and night time on human adult steroidogenesis. PLoS One 2019;14(3):e0214549. Table 2 Page 9 Table 3 Page 10 and Taylor NF. Urinary steroid profiling. Methods Mol Biol 2013;1065:259–76. Chapter 9 Table 1.

TABLE 2.6.7 Excretion rates of steroids by normal male and normal females (μg/24 h) after hydrolysis of conjugates, extraction, MO-TMS derivative formation and GC-MS analysis of MO-TMS derivatives.

	Male (n = 16)		Female (n = 14)		
	Median	Range[a]	Median	Range[a]	P-value
Androsterone (5α)	1200	1020–1482	720	522–820	0.0001
Etiocholanolone (5β)	1220	825–1502	955	600–1260	0.01
Dehydroepiandrosterone	460	290–750	100	75–180	0.01
16α-Hydroxyandrosterone	0.4	220–380	200	150–377	0.2
Androstenetriol	460	430–560	320	195–877	0.4
11β-Hydroxyandrosterone	730	520–740	325	220–410	0.003
11β-Hydroxyetiocholanolone	140	97–332	160	112–207	0.39
Tetrahydrocortisone	2290	2115–2715	1485	1140–1990	0.001
Tetrahydrocortisol (THF) (5β)	1080	1025–1350	830	480–950	0.0007
Allo-THF (5α)	830	607–925	400	270–530	0.0002
α-Cortolone	770	717–917	730	350–910	0.1
β-Cortolone + β-cortol	1040	890–1102	765	490–820	0.002
α-Cortol	265	215–365	200	147–262	0.05
Total cortisol metabolites	6965	6480–7420	4595	3371–5525	0.0005
Total androgen metabolites	2660	2030–3215	1850	1310–2110	0.0003

[a]First to third quartile range.

Shamim W, Yousufuddin M, Bakhai A, Coats AJ, Honour JW. Gender differences in the urinary excretion rates of cortisol and androgen metabolites. Ann Clin Biochem. 2000;37 (Pt 6):770–4. Table 2 p. 772.

Steroid output declined in elderly Dutch men and women. The study included reference ranges for the steroids normally only seen in the first 6 months of life presumably as a test of the method for neonatal samples. The highest result were comparable with the outputs of 11-ketoetiocholanolone which is very low in adults. Normal metabolite ratios for substrates and products of the enzymes in steroidogenesis were calculated.

Steroid output declined in elderly Dutch men and women. The study included reference ranges for the steroids

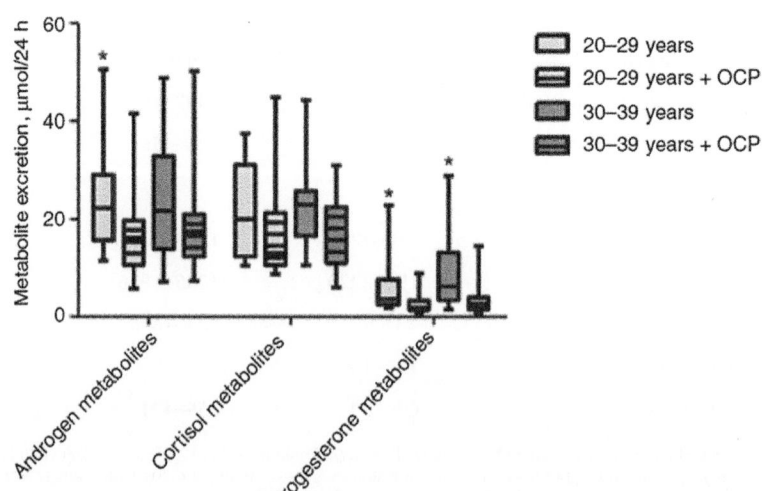

FIG. 2.6.17 Urinary excretion of women and women using oral contraceptives. Boxes represent median with interquartile range. Whiskers are minimum and maximum values. *P < .05 compared with the same group. *(From de Jong WHA, Buitenwerf E, Pranger AT, Riphagen IJ, Wolffenbuttel BHR, Kerstens MN, Kema IP. Determination of reference intervals for urinary steroid profiling using a newly validated GC-MS/MS method. Clin Chem Lab Med 2017;56(1):103–12. Fig. 4 p. 110.)*

TABLE 2.6.8 Reference intervals of urinary steroid metabolites in women and the effects of oral contraceptive use (µmol/24h) per decade.

Steroid metabolite	20–29 OCP− Reference interval	Mean[a]	20–29 OCP+ Reference interval	Mean[a]	30–39 OCP− Reference interval	Mean[a]	30–39 OCP+ Reference interval	Mean[a]	40–49 Reference interval	Mean[a]
Androgen										
A	2.8–20.8	6.7(NP)	0.9–12.2	6.5	0.1–17.3	8.0	1.9–10.9[b]	6.4	2.1–15.4	5.2(NP)
E	1.8–14.4	8.1	1.1–10.4	5.8	1.4–13.2	7.3	1.9–17.8	6.9(NP)	2.6–17.7	6.2(NP)
DHEA	0.2–11.8	1.2(NP)	0.1–2.5[b]	0.7(NP)	0.1–12.2	0.8(NP)	0.2–3.4	0.9(NP)	0.1–5.7	0.7(NP)
11-KE	0.4–3.1	1.0(NP)	0.1–2.3[c]	0.5(NP)	0.5–2.8	1.1(NP)	0.2–3.1[c]	0.8(NP)	0.1–3.1	1.6
11-HA	0.6–6.3	1.7(NP)	0.5–3.1	1.0(NP)	0.6–6.0	2.2(NP)	0.5–3.7[c]	1.5(NP)	0.6–5.2	2.9
11-HE	0.1–2.3	1.2	0.0–2.2[b]	0.6(NP)	0.3–3.8	1.3(NP)	0.1–3.3	0.8(NP)	0.2–3.1	1.6
Estriol	<0.1	<0.1	<0.1	<0.1	<0.1	<0.1	0.0–0.3	<0.1	<0.1	<0.1
Cortisol										
THE	3.1–23.4	7.8(NP)	2.3–11.2[c]	4.7(NP)	2.6–14.4	8.5	1.3–15.1[c]	5.3(NP)	4.7–18.2	9.2(NP)
THF	1.0–7.1	4.1	1.6–5.6	3.0(NP)	2.2–8.7	4.2(NP)	0.4–7.4	3.9	2.8–8.0	4.5(NP)
aTHF	0.4–10.4	3.2(NP)	0.4–5.0[c]	1.3(NP)	0.1–7.6	3.6	0.7–4.2[c]	1.8(NP)	0.9–5.4	3.1
α-CTLN	0.9–7.4	4.1	1.1–5.6	3.4	1.5–7.4	3.3(NP)	1.3–5.4	3.3	1.3–6.6	3.9
β-CTLN	0.1–3.3	1.7	0.4–2.4[b]	0.9(NP)	0.4–2.8	1.6	0.4–2.4[c]	1.2	0.7–4.0	1.6(NP)
α-cortol	0.1–1.1	0.6	0.3–1.0	0.6	0.3–1.3	0.5	0.3–1.5	0.7(NP)	0.3–1.0	0.6
Progesterone										
aP2	0.0–3.5	0.3(NP)	0.0–0.4	0.2(NP)	0.0–3.6	0.9(NP)	0.0–0.4[c]	0.1(NP)	0.1–5.6	0.3(NP)
P2	0.3–14.2	1.5(NP)	0.2–2.1[b]	0.7(NP)	0.4–21.9	2.2(NP)	0.2–4.4[c]	1.0(NP)	0.2–25.6	1.6(NP)
P3	0.6–5.7	2.1(NP)	0.3–2.9[b]	0.8(NP)	0.7–6.5	2.7(NP)	0.3–5.6[c]	1.4(NP)	0.1–5.3	2.6
Polone	0.0–1.3	0.2(NP)	0.0–0.3[c]	0.1(NP)	0.0–2.2	0.2(NP)	0.0–0.4[b]	0.1(NP)	0.0–2.5	0.2(NP)
Aldosterone										
THA	0.1–1.3	0.3(NP)	0.1–0.6	0.3(NP)	0.0–0.5	0.3	0.0–0.8	0.2(NP)	0.2–0.6	0.3(NP)
THB	0.1–1.3	0.4(NP)	0.1–0.7	0.3(NP)	0.1–0.8	0.4(NP)	0.1–1.0	0.3(NP)	0.2–0.7	0.4
aTHB	0.2–3.5	0.8(NP)	0.2–1.5[b]	0.4(NP)	0.1–1.5	0.8	0.2–1.6[b]	0.5(NP)	0.2–1.3	0.8

Continued

TABLE 2.6.8 Reference intervals of urinary steroid metabolites in women and the effects of oral contraceptive use (µmol/24h) per decade—cont'd

Steroid metabolite	20–29 OCP− Reference interval	Mean[a]	20–29 OCP+ Reference interval	Mean[a]	30–39 OCP− Reference interval	Mean[a]	30–39 OCP+ Reference interval	Mean[a]	40–49 Reference interval	Mean[a]
Intermediate										
THS	0.0–0.3	0.1[NP]	0.0–0.3	0.1[NP]	0.0–0.3	0.2[NP]	0.0–0.3	0.1[NP]	0.1–0.4	0.2[NP]
PDL	0.1–1.9	0.4[NP]	0.0–0.6[b]	0.2[NP]	0.1–2.1	0.9	0.0–1.0[c]	0.2[NP]	0.1–2.6	0.6[NP]
PTL	0.0–0.2	0.04[NP]	<0.1[c]	<0.1	0.0–0.2	0.03[NP]	<0.1	<0.1	<0.1	<0.1
aPDL	0.0–0.2	0.1[NP]	<0.1	<0.1	0.0–0.2	<0.1	<0.1	<0.1	0.0–0.2	<0.1
TH-DOC	<0.1	<0.1	<0.1	<0.1	<0.1	<0.1	<0.1	<0.1	<0.1	<0.1

From de Jong WHA, Buitenwerf E, Pranger AT, Riphagen IJ, Wolffenbuttel BHR, Kerstens MN, Kema IP. Determination of reference intervals for urinary steroid profiling using a newly validated GC–MS/MS method. Clin Chem Lab Med 2017;56(1):103–112. Table 4 p. 109.

normally only seen in the first 6 months of life presumably as a test of the method for neonatal samples. The highest result was comparable with the outputs of 11-ketoetiocholanolone which is very low in adults. Normal metabolite ratios for substrates and products of the enzymes in steroidogenesis were calculated. The androgens decrease from 50 to 60 years and are significantly lower in Chinese (Chan et al., 2008), Swiss (Ackermann et al., 2019), and Dutch by 70–79 years (de Jong et al., 2017) (Table 2.6.9).

The total of all cortisol metabolites in a 24 h collection is close to the cortisol production rate The cortisol output for body size is almost constant around 6–8 mg per meter squared reflecting the role of the HPA axis in maintaining cortisol concentrations. The total of cortisol metabolites per day is useful in detecting hypocortisolism as well as hypercortisolism which contrasts to the utility of urine-free cortisol only for excess cortisol production. Cortisol production can be low in patients with genetic defects of cortisol synthesis and when synthetic corticosteroids are used. Prednisolone metabolites can be distinguished in a mass analysis, the adrenal suppression by inhaled corticosteroids can be a particular concern.

Although the secretion of cortisol is different over 24 h with peak around 0800 h and trough at midnight the outputs of cortisol metabolites in urine is highest between 1000 and 1600 h (Ackermann et al., 2019; Jerjes et al., 2006b) when samples are collected for 3 h periods. This is not practical but if urine is collect for longer periods in day and night the outputs of metabolites in the urine are highest in the daytime collection (Table 2.6.10).

TABLE 2.6.9 Reference intervals for daily outputs of urine steroids in elderly men and women by GC-MS.

		n	2.5	50	97.5
Women					
Androsterone	<40	29	431	1083	2037
	40–59 years	47	143	684	1765
	>60	13	100	389	919
Men					
Androsterone	<40	43	792	2312	3514
	40–59 years	30	589	1792	3105
	>60	10	343	957	2372
Etiocholanolone	<40		577	1373	2802
	40–59 years		585	1169	237
	>60		235	828	1362
DHEA	<40		164	1152	3657
	40–59 years		49	889	3105
	>60		15	167	1214
11-OH Andro	<40		233	776	1618
	40–59 years		318	904	1898
	>60		311	527	1059
11-OH aetio	<40		17	142	572
	40–59 years		26	213	482
	>60		24	52	321
16-OH DHEA	<40		253	535	1375
	40–59 years		91	486	1511
	>60		17	226	511
THE	<40		1253	3051	6409
	40–59 years		1301	3117	4917
	>60		1202	1740	3293
aTHB	<40		58	251	653
	40–59 years		48	204	402
	>60		79	179	311
Allo-THF	<40		422	1834	4959
	40–59 years		200	1848	3283
	>60		465	1223	1865
b-cortolone	<40		220	520	1391
	40–59 years		237	493	771
	>60		204	312	571

Redrawn from Chan AO, Taylor NF, Tiu SC, Shek CC. Reference intervals of urinary steroid metabolites using gas chromatography-mass spectrometry in Chinese adults. Steroids 2008;73(8):828–37. Table 3.

TABLE 2.6.10 Day time excretion rates of steroids (0800–2000 h) (2000–80 h) and night time.

	Mean	Min	Max	Mean	Min	Max
Androsterone	904	390	1420	383	90	640
Etiocholanolone	590	260	880	279	70	500
THE	488	420	1130	303	100	1620
THF	441	300	630	191	70	490
aTHF	304	160	470	129	40	230
a-Cortolone	353	170	580	126	40	310
b-Cortolone+ b-cortol	404	60	890	145	30	290
a-Cortol	101	30	190	37	10	80
	Mean	Min	Max	Mean	Min	Max
Androsterone	447	160	860	295	70	650
Etiocholanolone	440	90	1400	274	70	700
THE	667	200	1050	363	40	940
THF	365	140	590	191	10	440
aTHF	241	40	390	141	30	500
a-Cortolone	392	160	1230	180	40	320
b-Cortolone+ b-cortol	277	80	470	147	10	300
a-Cortol	193	10	1230	33	10	60

Author original.

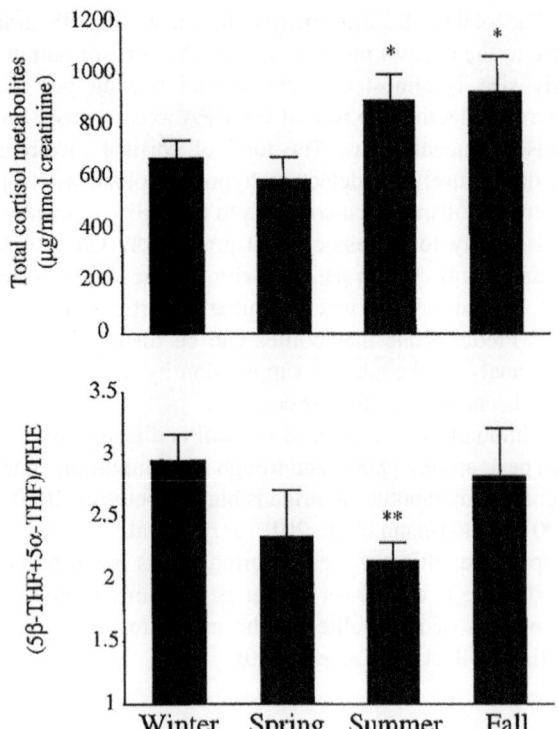

FIG. 2.6.18 Seasonal variation in plasma and urinary corticosteroids. Bars—Mean ± SEM. *THF*, tetrahydrocortisol; *THE*, tetrahydrocortisone. *(From Walker BR, Best R, Noon JP, Watt GC, Webb DJ. Seasonal variation in glucocorticoid activity in healthy men. J Clin Endocrinol Metab 1997;82 (12):4015–9. Fig. 1 p. 4017.)*

TABLE 2.6.11 Reference intervals for detectable steroids of a total of 40 steroid metabolites measured in 24-h urine in men. Steroids were analyzed by LC-MS/MS after glucuronidase hydrolysis and SPE.

1. Steroid	Range (ng/mL)
16α-Hydroxy DHEA	6.6–45.9
11β-Hydroxy androsterone	45.4–1026.4
11β-Hydroxyetiocholanolone	44.3–10,110
Androstenedione	3.7–40.6
T	9.6–129.0
DHEA	176.0–644.9
DHT	1.3–10.2
Androsterone	728.4–7325.76
Etiocholanolone	439.4–2236.1
20α-Hydro E	34.1–128.0
20α-Hidydro F	23.0–220.2
E	122.3–442.2
F	49.7–352.0
11-Dehydro B	28.2–230.7
21-Deoxy F	2.9–280.9
11-Deoxy F	0.3–4.1
17α-Hydroxy prog	14.1 19.0
Prog	0.5–2.5
20α-OH Prog	13.7–69.4
Preg	459.7–1043.7
P triol	106.9–5822.6

Abbreviations: *2-MeO E1*, 2-methoxyestrone; *2-MeO E2*, 2-methoxyestradiol; *4-MeO E2*, 4-methoxyestradiol; *5α-dehydro-11-dehydro-B*, 5α-dehydro-11-dehydrocorticosterone; *5α-dihydro-Prog*, 5α-dihydroprogesterone; *5β-dihydro-Prog*, 5β-dihydroprogesterone; *11-dehydro-B*, 11-dehydrocorticosterone; *11-desoxy-F*, 11-desoxycortisol; *11β-OH An*, 11β-hydroxyandrosterone; *11β-OH et*, 11β-hydroxyetiocholanolone; *16-epi E3*, 16-epiestriol; *16α-OH DHEA*, 16α-hydroxydehydroepiandrosterone; *17-epi E3*, 17-epiestriol; *17α-OH Prog*, 17α-hydroxyprogesterone; *20α-dihydro-E*, 20α-dihydrocortisone; *20α-dihydro-F*, 20α-dihydrocortisol; *20α-OH Prog*, 20α-hydroxyprogesterone; *21-deoxy-F*, 21-deoxycortisol; *Al*, aldosterone; *Allo-P-one*, allopregnanolone; *An*, androsterone; *An-dion*, androstenedione; *B*, corticosterone; *DHEA*, dehydroepiandrosterone; *DHT*, dihydrotestosterone; *E*, cortisone; *E1*, estrone; *E2*, 17β-estradiol; *E3*, estriol; *epi An*, epiandrosterone; *epi T*, epitestosterone; *epi-Preg*, epipregnanolone; *et*, etiocholanolone; *F*, cortisol; *P-one*, pregnanolone; *Preg*, pregnenolone; *Prog*, progesterone; *P-triol*, pregnanetriol; *T*, testosterone.
From Son HH, Yun WS, Cho SH. Development and validation of an LC-MS/MS method for profiling 39 urinary steroids (estrogens, androgens, corticoids, and progestins). Biomed Chromatogr 2020;34(2):e4723.

The production rate of cortisol is high in the summer when the ratio of THF plus allo-THF to THE is lowest (Fig. 2.6.18) (Walker et al., 1997), even though the plasma cortisol concentrations at 0900 h and tissue sensitivity were higher in winter and spring.

This study has not been repeated despite the observation of higher prevalence of infections and disease in winter in Scotland.

A method was developed and thoroughly validated using LC-MS/MS for steroids without any derivatization steps, using both 0.01% formic acid and 1 mM ammonium formate as the mobile phase additives to increase ionization properties and chromatographic resolution. The novel LC-MS/MS method enabled quantitative profile of 39 steroids, including estrogens, androgens, corticoids and progestins, in a single analytical run (Table 2.6.11) (Son et al., 2020).

Results are presented only for ranges because the results as mean and SD were disappointing in that they could not possibly reflect the data. Standard deviations were higher than the mean such that lower limits were negative values.

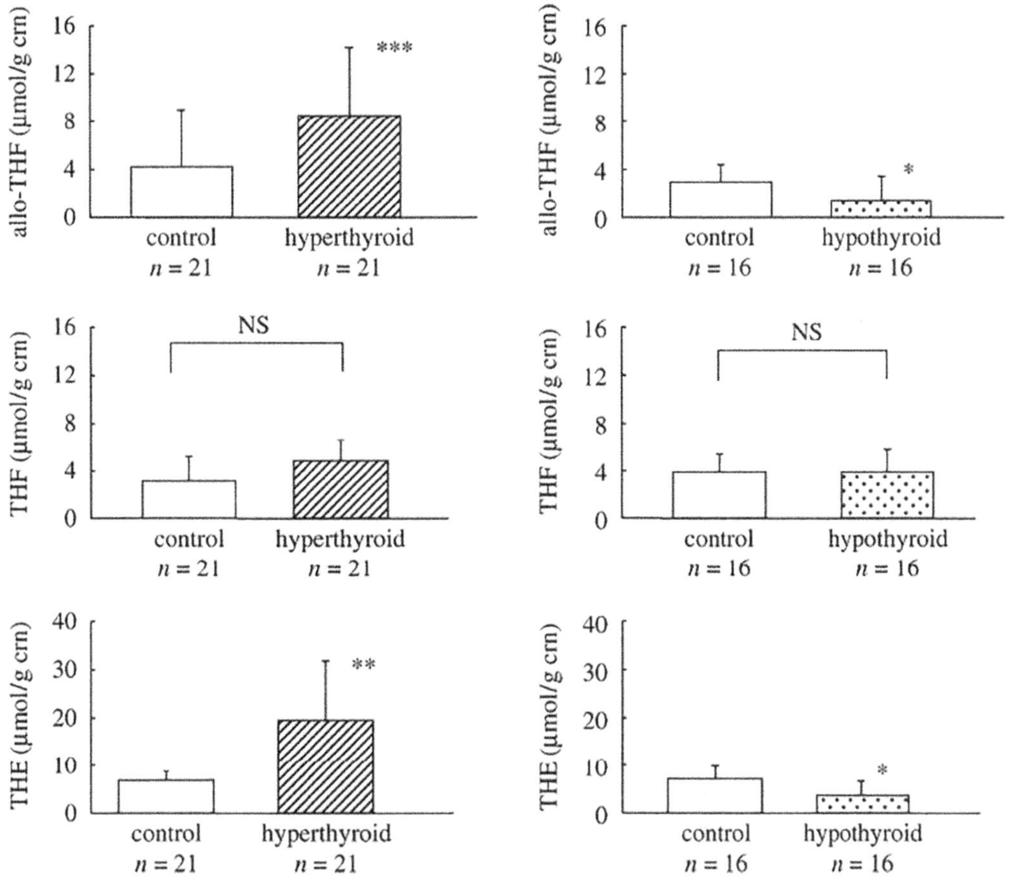

FIG. 2.6.19 Effects of thyroid function on cortisol metabolism Urinary allo-THF, THF and THE in hyperthyroid and hypothyroid patients. The data are expressed as means $\pm$ SD. *P < .05; **P < .001; ***P < .0001; NS, not significant compared with the controls. *(From Hoshiro M, Ohno Y, Masaki H, Iwase H, Aoki N. Comprehensive study of urinary cortisol metabolites in hyperthyroid and hypothyroid patients. Clin Endocrinol (Oxf) 2006;64(1):37–45. Fig. 2 p. 40.)*

The outputs of total cortisol metabolites are raised in **hyperthyroidism** (Fig. 2.6.19) (Hoshiro et al., 2006).

A low urinary allo-THF + THF/THE ratio, reflecting the overall activity of 11β-HSD, found in the hyperthyroid patients was attributed to the marked increase in THE in hyperthyroid patients that reflects suppressed cortisone (E)-to-cortisol (F) conversion (11β-HSD1) and the increased F-to-E conversion (11β-HSD2). An earlier study compared urinary cortisol metabolites with three groups (Vantyghem et al., 1998). (A) overt hypothyroidism with low serum FT3 and FT4 and raised TSH, (B) control group with normal thyroid hormones and (C) subclinical hypothyroid group with moderately increased serum TSH and low/normal FT3 and FT4. The mean THE/(THF + aTHF) and aTHF/THF ratios were lower in overt hypothyroid group (Fig. 2.6.20).

High resolution, accurate mass (HRAM) mass spectrometry has also been applied to steroid profiling (Bileck et al., 2019; Hines et al., 2017). Reference intervals for pre- and postmenopausal women and men are comparable with data from GC/MS and LC-MS/MS. Twenty-seven steroids were determined against 9 internal standards.

This approach is expensive and not really necessary for the particular application but a good test moving into the future for metabolomic studies. The long term problem will then be the lack of reference compounds (the steroid conjugates) and isotope labeled standards.

2.6.4.1.2 Effects of growth hormone

When patients have to be treated with glucocorticoid, the tissue exposure can be supra-physiological for example in hypopituitary patients taking hydrocortisone replacement therapy. Some of the patients also require growth hormone (GH) replacement therapy. Circulating cortisol concentrations are vulnerable to the inhibitory effect of GH on 11βHSD1 such that cortisol and cortisone assessments particularly when taking cortisone acetate after GH therapy to ensure that glucocorticoid replacement remain adequate (Table 2.6.12).

The metabolites of cortisol and cortisone are determined and ratio is used (Fig. 2.6.21) in this and many clinical studies to assess the relative activities of the two enzymes, and these will discussed in Part 3 of the book.

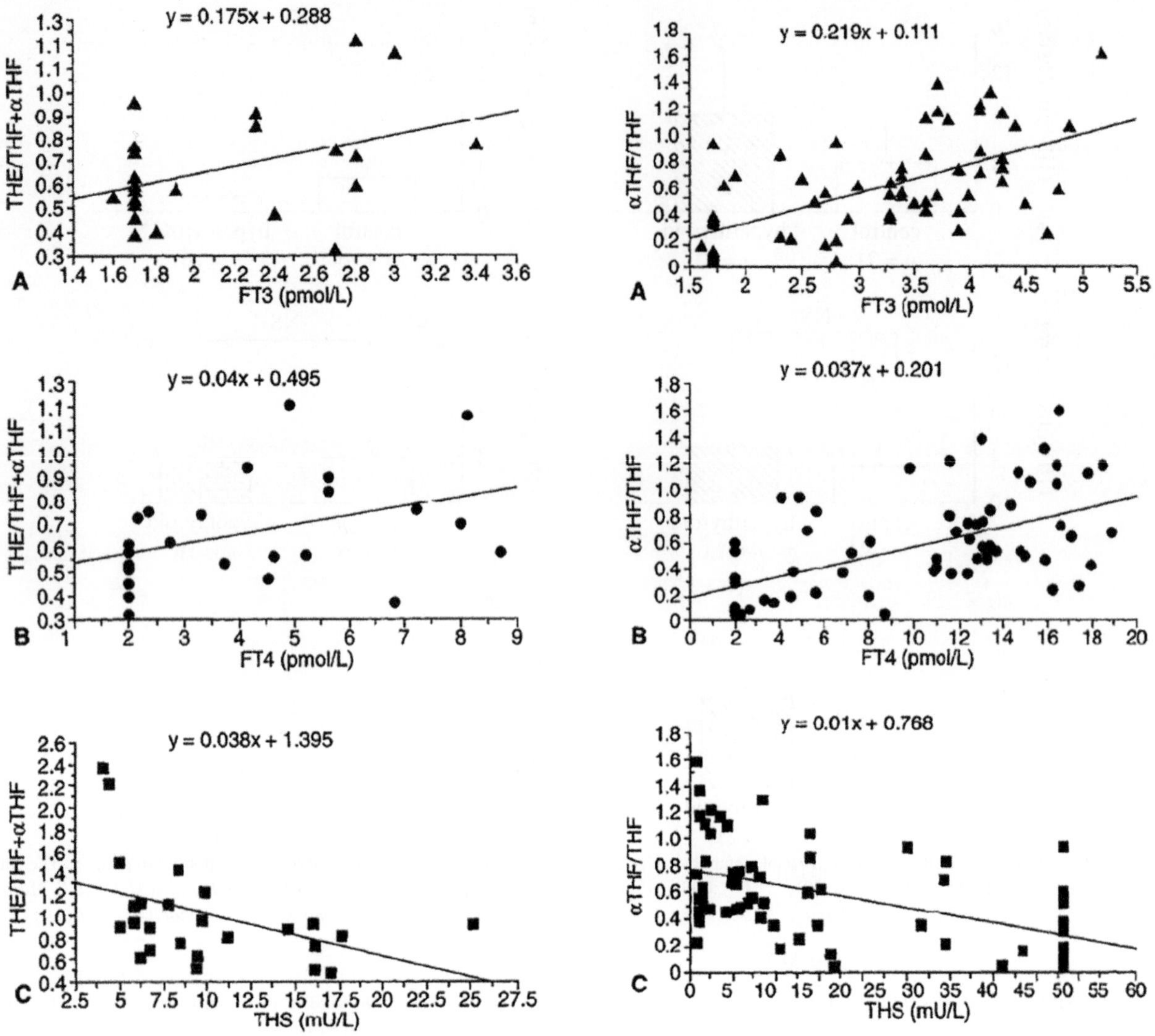

FIG. 2.6.20 Cortisol metabolites in hypothyroidism. Markers for HSD11b (left) and SRD5A2 (right) against (Vantyghem et al., 1998) (A) free T3 (B) free T4 and (C) TSH (not THS as in graphs).

TABLE 2.6.12 Comparison of urinary steroid profiles before and after growth hormone therapy (mg, mean and SD) for hypopituitary patients while taking hydrocortisone or cortisone acetate.

Urinary steroid assessed	Cortisone acetate		Hydrocortisone		Normal subjects	
	Pre-GH	Post-GH	Pre-GH	Post-GH	Male	Female
Total cortisol metabolites (µg/24 h)	12,842 (4478)	13,131 (4823)	10,733 (6321)	10,114 (6001)	7618 (2761)	4752 (1693)
Fm (µg/24 h)	4427 (1399)	4170 (987)	53,790 (2179)	4728 (2458)	3294 (1075)	1863 (631)
Em (µg/24 h)	8486 (1870)	8961 (2931)	5355 (2889)	5386 (2834)	4323 (1794)	2889 (1115)
THF (µg/24 h)	3070 (1898)	1746 (794)	2222 (1002)	1994 (1088)	1275 (481)	878 (314)

TABLE 2.6.12 Comparison of urinary steroid profiles before and after growth hormone therapy (mg, mean and SD) for hypopituitary patients while taking hydrocortisone or cortisone acetate—cont'd

Urinary steroid assessed	Cortisone acetate		Hydrocortisone		Normal subjects	
	Pre-GH	Post-GH	Pre-GH	Post-GH	Male	Female
5αTHF (μg/24h)	2385 (1683)	1614 (558)	2422 (1348)	2070 (2166)	1205 (509)	502 (208)
THE (μg/24h)	7025 (1759)	7512 (2365)	4175 (2433)	4253 (2291)	3137 (1283)	1941 (783)
UFF (nmol/24h)	84 (34)	107 (64)	123 (49)	135 (97)	–	–
UFE (nmol/24h)	101 (57)	209 (105)	177 (67)	171 (90)	–	–
UFF/UFE	0.54 (0.18)	0.52 (0.24)	0.73 (0.25)	0.75 (0.36)	–	–
5α/5βTHF	1.04 (0.8)	1.12 (0.62)	1.18 (0.72)	1.02 (0.61)	1.07 (0.69)	0.58 (0.20)
(THF+5αTllF)/THE	0.53 (0.20)	0.47 (0.20)	1.33 (0.50)	1.03* (0.47)	0.84 (0.27)	0.74 (0.17)
Fm/Em	0.52 (0.09)	0.44* (0.09)	1.17 (0.28)	0.90* (0.25)	0.81 (0.24)	0.66 (0.14)
20OH/20oxo Steroid	0.21 (0.06)	0.20 (0.03)	0.24 (0.07)	0.23 (0.06)	0.36 (0.06)	0.44 (0.09)

Total cortisol metabolites is sum of THE, THF, allo-THF, cortols and cortolones. *Fm*, 11-hydroxylated metabolites; *Em*, sum of 11-oxo steroids; *UFF*, urinary-free cortisol; *UFE*, urinary-free cortisone.

From Swords FM, Carroll PV, Kisalu J, Wood PJ, Taylor NF, Monson JP. The effects of growth hormone deficiency and replacement on glucocorticoid exposure in hypopituitary patients on cortisone acetate and hydrocortisone replacement. Clin Endocrinol (Oxf) 2003;59(5):613–20. Table 2 p. 617.

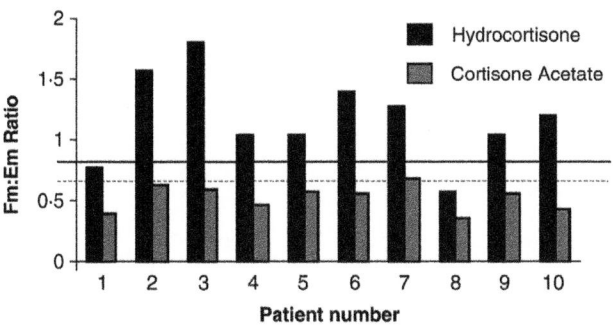

FIG. 2.6.21 GH treatment. Fm/Em ratios are shown for all patients during hydrocortisone treatment (*black bars*) or cortisone acetate treatment (*gray bars*). All measurements were taken prior to initiation of GH therapy. *P* < .001. Values for normal subjects are indicated with horizontal lines (unbroken line males, broken line females). *(From Swords FM, Carroll PV, Kisalu J, Wood PJ, Taylor NF, Monson JP. The effects of growth hormone deficiency and replacement on glucocorticoid exposure in hypopituitary patients on cortisone acetate and hydrocortisone replacement. Clin Endocrinol (Oxf) 2003;59(5):613–20.)*

2.6.4.2 Pediatric data

Paula Midgley working at UCLH and the Middlesex Hospitals in London, UK studied 22 preterm infants (12M, 10F) (gestation age 24–31 weeks). Urine was collected at around weekly intervals from birth until discharge from hospital. Steroids were extracted by SPE, divided by using Sephadex LH-20 chromatography into a fraction of free and glucuronidated steroids and a fraction of sulfated steroids. Conjugates were hydrolyzed with *Helix pomatia* glucuronidase/sulfatase preparation (Sigma H1) and steroids stabilized as MO-TMS derivatives which were subjected to analysis by GC-MS in full scan mode. Chromatograms of the two fractions are shown in Fig. 2.6.22 showing the complexity compared with later in life (see Fig. 2.6.32, later).

The sulfate fraction contained more steroid than the free and glucuronide fraction. The steroids were identified by the GC retention times and mass spectra of the steroids compared with the properties of reference compounds (Table 2.6.13).

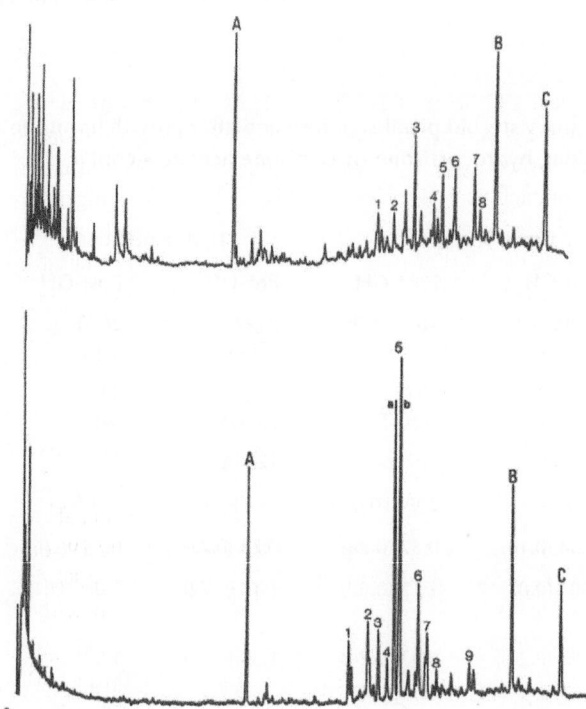

FIG. 2.6.22 The total ion current chromatogram of steroids from newborn urine after separation with Sephadex LH-20 into free plus glucuronide (top chromatogram) and sulfate fractions (bottom chromatogram). Conjugates were hydrolyzed then converted to MO-TMS derivatives for GC-EI-MS analysis. The steroids are: (1) 16α-hydroxy DHEA; (2) 16β-hydroxy DHEA; (3) 16-oxo-androstenediol and 15β,16α-dihydroxy DHEA; (4) androstenetriol; (5) 16,18-dihydrox-DHEA (doublet); (6) 16α-hydroxypregnenolone; (7) androstenetetrol; (8) 6α-hydroxytetrahydrocortisone; (9) 5-pregnane-3β,16α,20α,21-tetrol.

TABLE 2.6.13 Identification of steroids in urine of a neonate by GC retention times of MO-TMS derivatives in methylene units and principal ions on mass spectrometry after electron impact.

Steroid (molecular weight as MO-TMS)	Main fragment ions	Methylene units (MU)
0–30 fraction		
Tetrahydrocortisone (609)	578	29.65
α-Cortolone (654)	449	30.51
6-Hydroxy- THE (697)	666/576	30.80
β-Cortolone (654)	449	30.73
6-Hydroxy-20α-cortolone (742)	537, 243, 639	31.82
6-Hydroxy-20β-cortolone (742)	537, 243, 639	
1-Hydroxy-20β-cortolone	196, 272, 537,652	32.00
30 + fraction		
16α-Hydroxy-DHA (477)	266/446	27.38/27.42
16βα-Hydroxy-DHA (477)	266/446	27.74
(16-Oxo-androstenediol [477]) and 15,16-Dihydroxy-DHA (565)	(446/356) (565, 232)	28.19
Androstenetriol (532)	432	28.46
16,18 Dihydroxy-DHA (565)	384,444,534,283	28.65/28.81
16-Hydroxypregnenolone (505)	474	29.39

A includes free and glucuronide conjugated steroids; Table B includes sulfated steroids. The two groups were achieved by column chromatography with Sephadex LH-20.

Author original.

In the sulfate fraction, for most of the mass spectra (Fig. 2.6.23) the molecular ion is weak and M-15 and M-31 is the first prominent fragmentations.

Loss of 90 mass units is seen for each silanol group. All, except one, of the major 3β-hydroxy-5-ene steroids were 16-hydroxylated apart from 3β,15β,17α,20β-tetrol. Many more related steroids were tentatively identified using GC-MS/MS (Christakoudi et al., 2010, 2012a,b).

The 16-hydroxylated steroids represented about 90% of the total steroids. The steroids identified were metabolites of 16-hydroxy-DHEA and 16-hydroxypregnenolone. Three 18-hydroxylated steroids were identified but represented only a small amount of the total urinary steroids.

In the glucuronide fraction (Fig. 2.6.24), only cortisone and very little cortisol metabolites were seen in the urine of newborn infants. Mass spectra (Fig. 2.10.24) were observed to confirm assignments of steroid identities.

Tetrahydrocortisone (THE) was the main steroid identified along with hydroxy-THE, cortolones and hydroxycortolones with a predominance of 6-hydroxylation over 1-hydroxylation. Cortisone is higher than cortisol excretion in the urine of newborn infants (mean 23.86 vs. 6.12 ng/mL) giving an F to E ratio of 0.28, whereas in adults the 0.8 ratio is found.

From the quantitative analysis, the percentage of 16-hydroxylated steroids in preterm babies declined after term in parallel with the overall decline of 3β-hydroxy-5-ene steroids (Fig. 2.6.25). 16,18-dihydroxy-DHEA and 5-androstene-3β,15β,16α,18-tetrol were the most abundant steroids in the study.

The 3β-hydroxy-5-ene steroids accounted for around 90% of the steroids excreted in the urine until after 112 days (16 weeks) postnatal age, when they started to decline reaching 50% by 196 days (28 weeks). 18-Hydroxylated steroids were about 40% of the total steroids in the first week of life and declined with time. The activities of the adrenal cortex in preterm infants continues to term then declines over 12 weeks in much the same way as a term infant (Fig. 2.6.26).

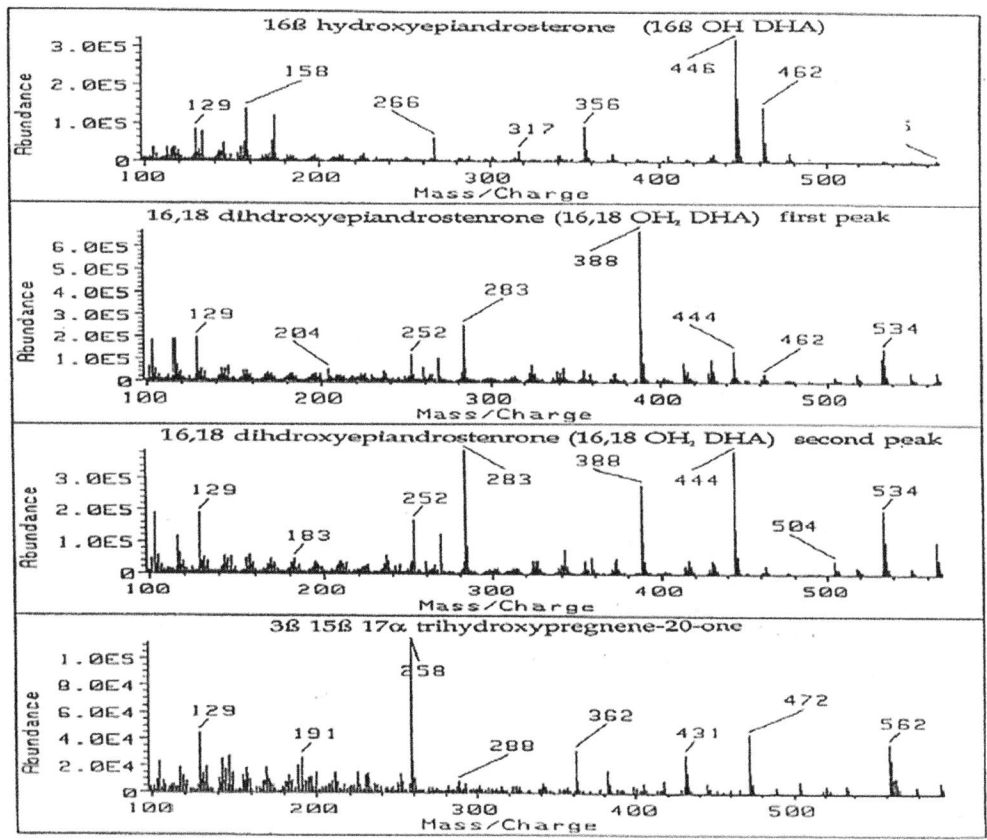

FIG. 2.6.23 Mass spectra of steroids from sulfate fraction of newborn urine after hydrolysis and formation of MO-TMS derivatives which were analyzed by GC-EI-MS.

(Continued)

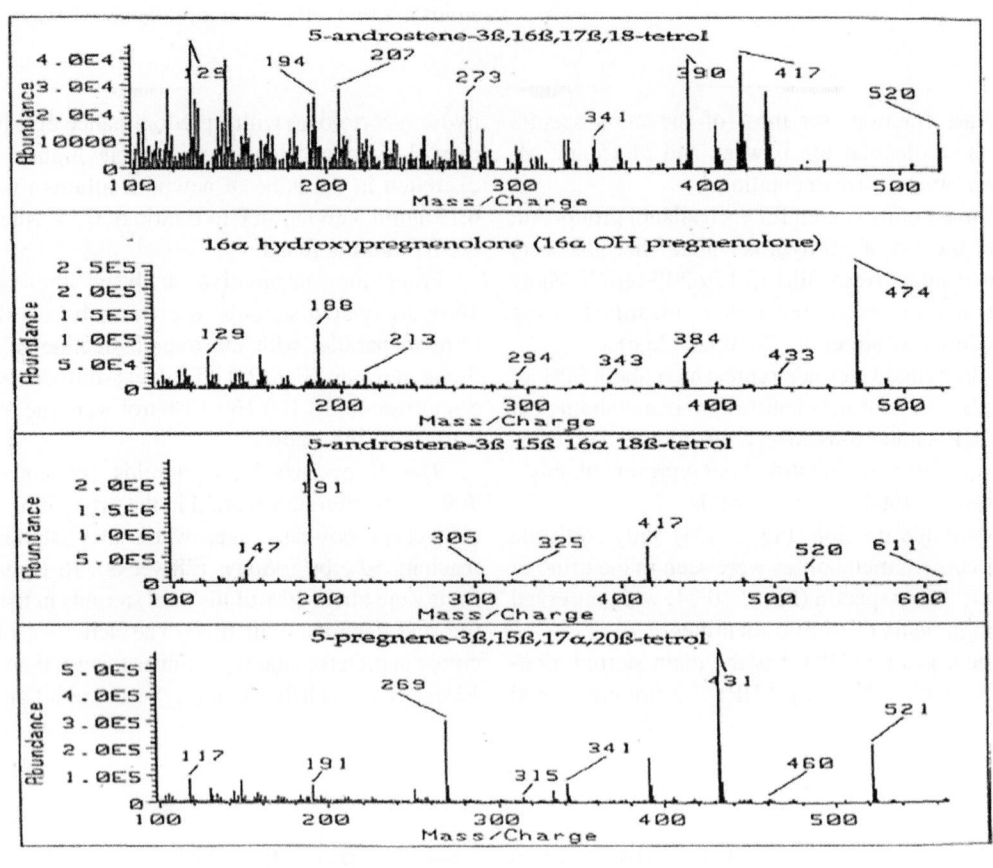

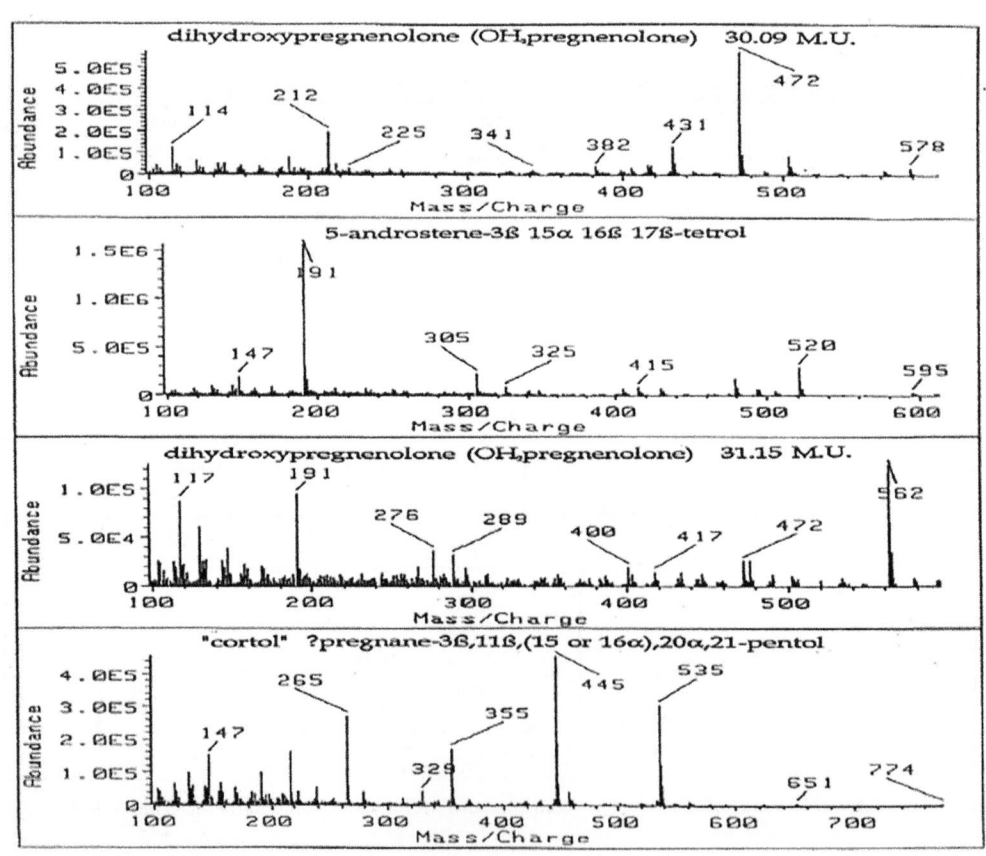

FIG. 2.6.23, CONT'D

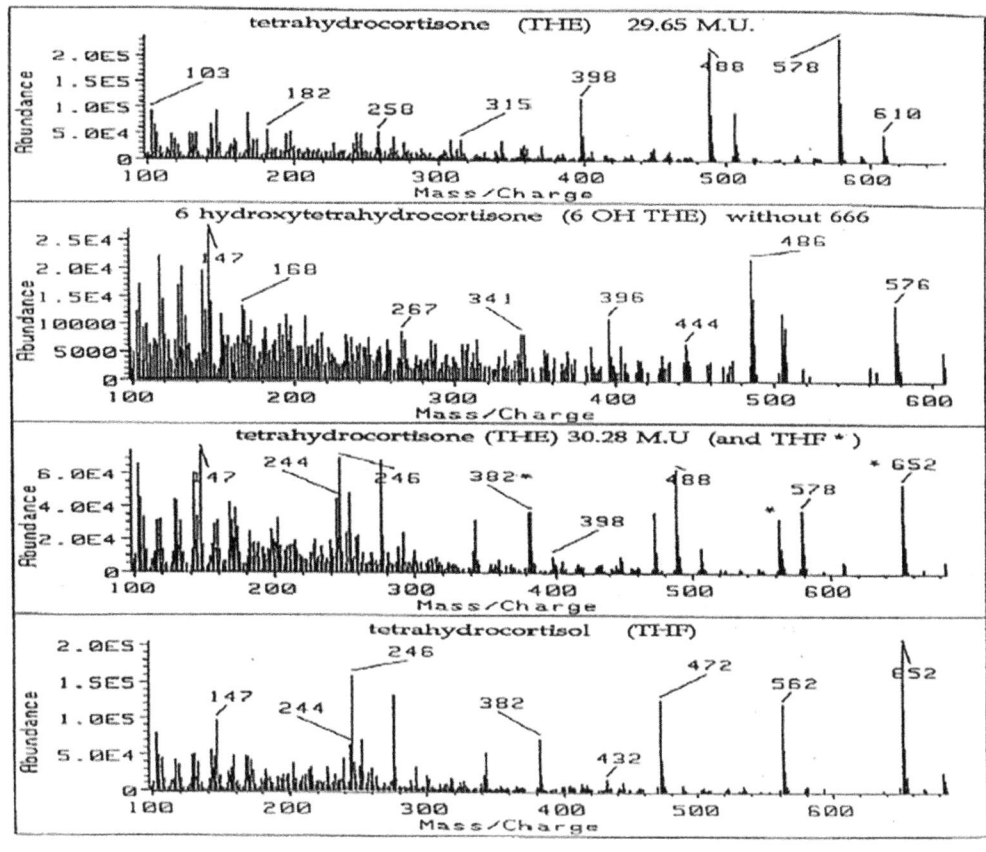

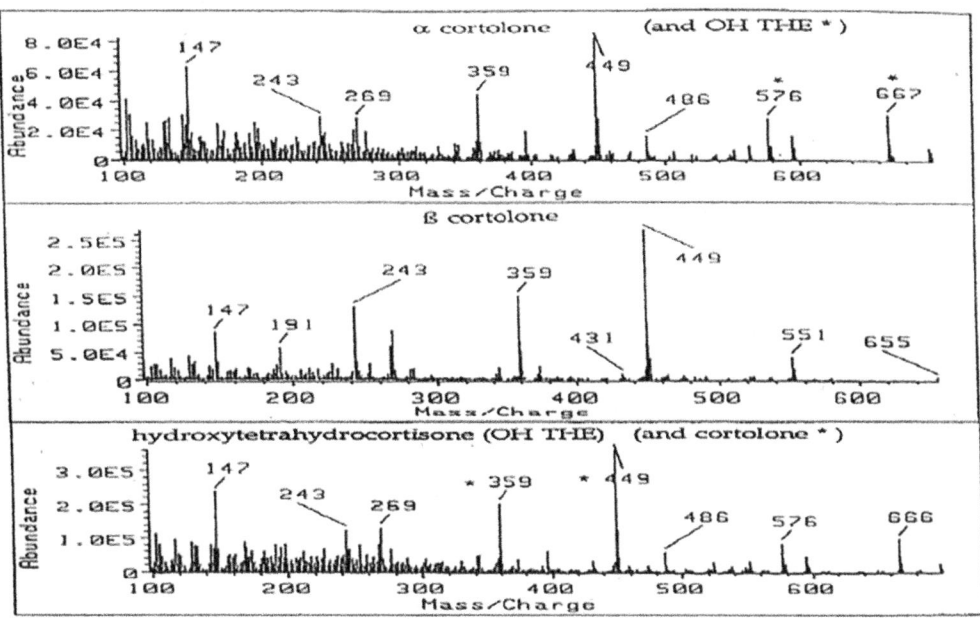

FIG. 2.6.24 Mass spectra of steroids from free and glucuronide fraction of newborn urine after hydrolysis and formation MO-TMS derivatives which were analyzed by GC-EI-MS.

(Continued)

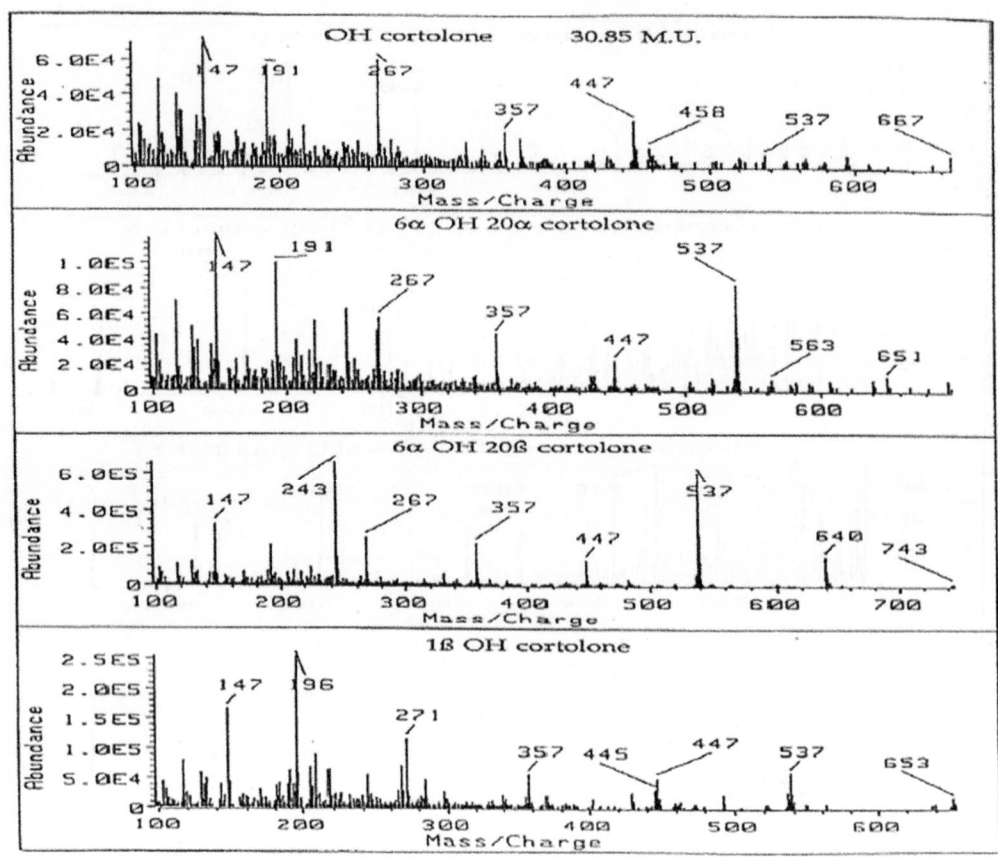

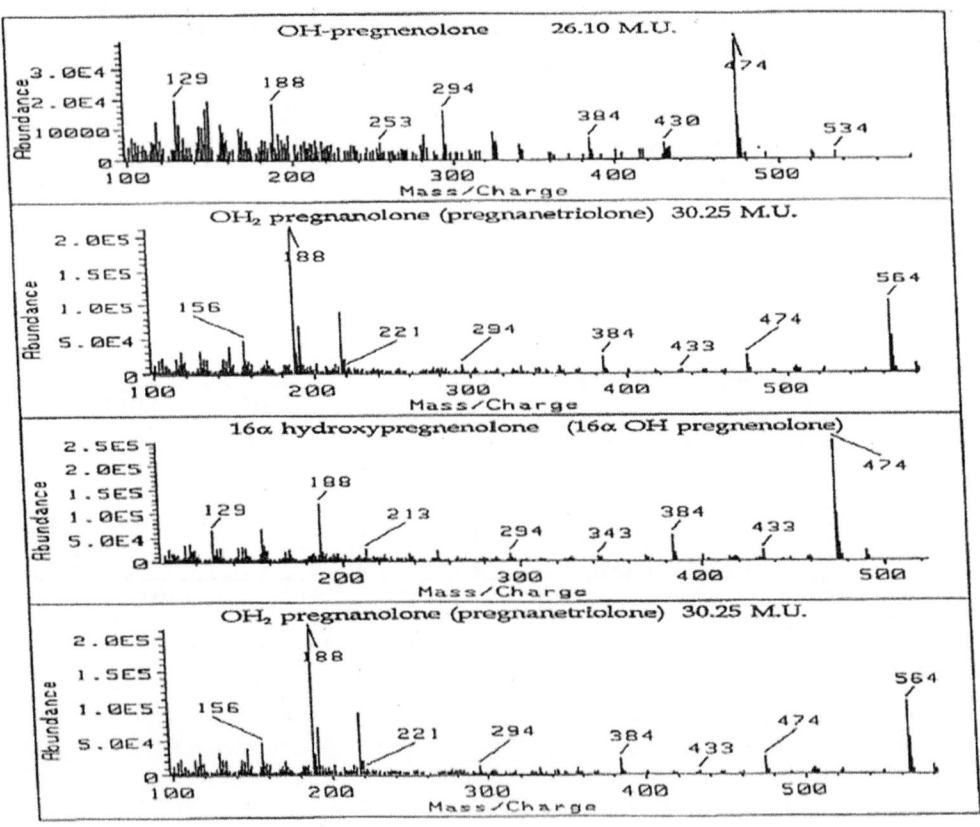

FIG. 2.6.24, CONT'D

(Continued)

FIG. 2.6.24, CONT'D

Overall the steroids in urine reflect high activity of fetal 16-hydroxylase and sulfokinase and low activity of HSD3B2 for many weeks after birth. Restraint of HSD3B2 is therefore not due to the effects of estrogens as inhibitors of the enzyme. If these data are judged by postconceptional age there was persistent excretion of 3β-hydroxy-5-ene steroids until term (40 weeks) after which there was a decline (Figs. 2.6.27 and 2.6.28).

Similar results have been published by the Wudy group (Fig. 2.6.29). The findings also support inactive HSD11B2 that converts cortisone back to cortisol.

The profiles of newborn infants have also been studied in detail (Dhayat et al., 2017) over the first year of life using GC-MS in SIM mode which is more sensitive than full scan and reconstructed ion chromatograms. The total steroid metabolites were analyzed after hydrolysis and derivative formation. They were thus able to find many reduced metabolites of progesterone and corticosterone with excretion rates from 1 up to 200 µmol/mg creatinine that were not reported by Paula Midgley. The Dhayat publication showed 16-hydroxy-DHEA and androstenetriol to be the major steroids but they did not include ions scanning for 16,18-dihydroxy-DHA, androstene-3β,15β,16α,17β-tetrol and 16α-hydroxypregnenolone which are important steroids in the data for term babies in the UCLH laboratory (Table 2.6.14).

The excretion rates of the major androgens reported by Dhayat (16-hydroxy-DHEA and androstenetriol) declined over the first 20 weeks as the fetal zone regresses (Fig. 2.6.30). This decline in fetal androgen output is the same in preterm infants so this pattern of fetal zone regression is related to postconception age.

The concentrations of the androgens were much greater than for the cortisol (Table 2.6.15) where Tetrahydrocortisone (THE) was the dominant steroid (THE up to 2000 µmol/mg creatinine, 200 µg/day) with several hydroxylated cortisone metabolites.

The cortisol metabolites are just detectable in the Dhayat study. Tetrahydrocortisol output for example is less than 200 µmol/mg creatinine for 30 weeks and allo-THF is higher. There are significant levels of 1- and 6-hydroxylated steroids in the first 30 weeks.

Urine-free steroids have been determined by GC-MS in newborn infants following extraction and heptafluorobutyrate derivative formation (Koyama et al., 2013) (Table 2.6.16). The study did not include mineralocorticoid steroids.

Results had conversion factors between the ranges reported in nmol/L and µmol/mol creatinine. The method was established as a step improvement from analysis of cortisol and cortisone before a fuller method for urine steroids was described by the same group.

The steroid profile in newborn period into the first year of life is very different from subsequent profiles as the activity of the fetal adrenal gland disappears through apoptosis. Ackermann and colleagues looked closely at ratios of steroids during the mini-puberty. The excretion of

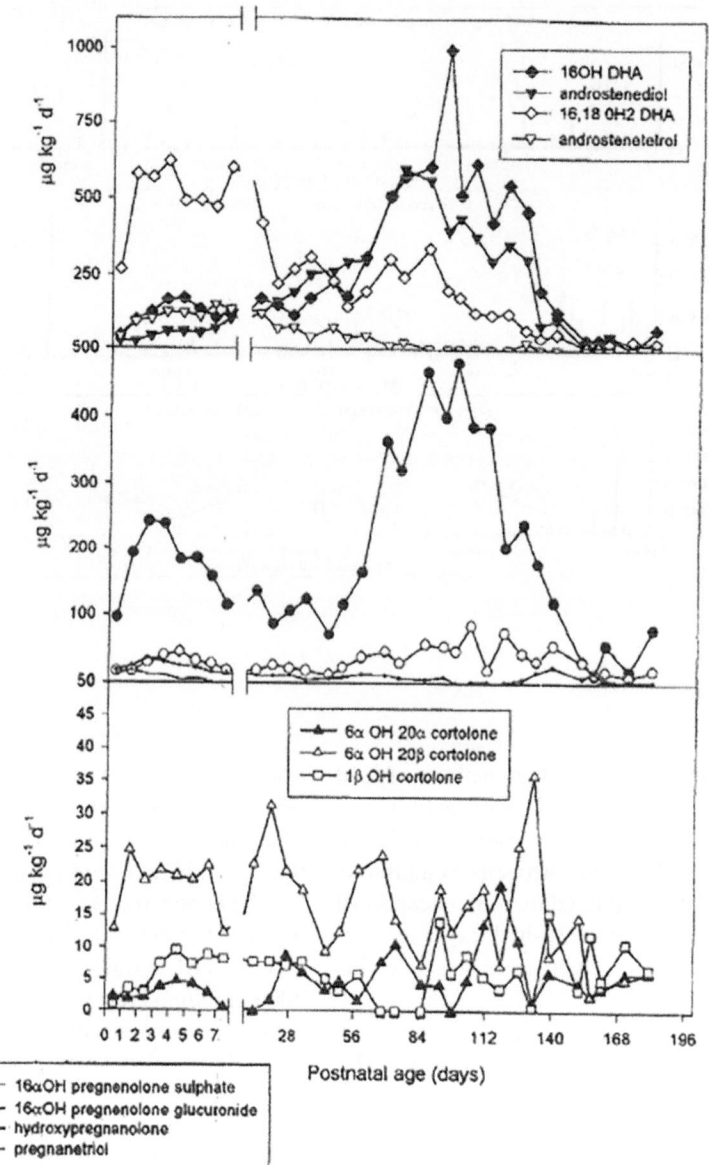

FIG. 2.6.25 Urine excretion steroids preterm infants. *(Based on Midgley P. Adrenal funcion of preterm infants, MD thesis. University of Aberdeen, 1992.)*

androsterone/etiocholanolone ratio was high which supported the backdoor path for dihydrotestosterone. To get a measure of the testicular contribution, they subtracted the excretion rates of steroids in urine of girls and the residue was considered to be the contribution from the testes (Fig. 2.6.31).

Flux calculations for 17,20 lyase activity were also performed, which is essential for any androgen

production. While lyase activity in the Δ5-path leads to the classic androgen biosynthesis, the Δ4-path produces 17-hydroxyprogesterone, which feeds into the backdoor path. In the cohort, lyase activity in the D5-path was extremely high at birth and dropped massively after birth to 10 weeks of age. This likely reflects the involution of the fetal adrenal gland, which produces exclusively DHEA over the D5-path, while postnatally the adult adrenal cortex

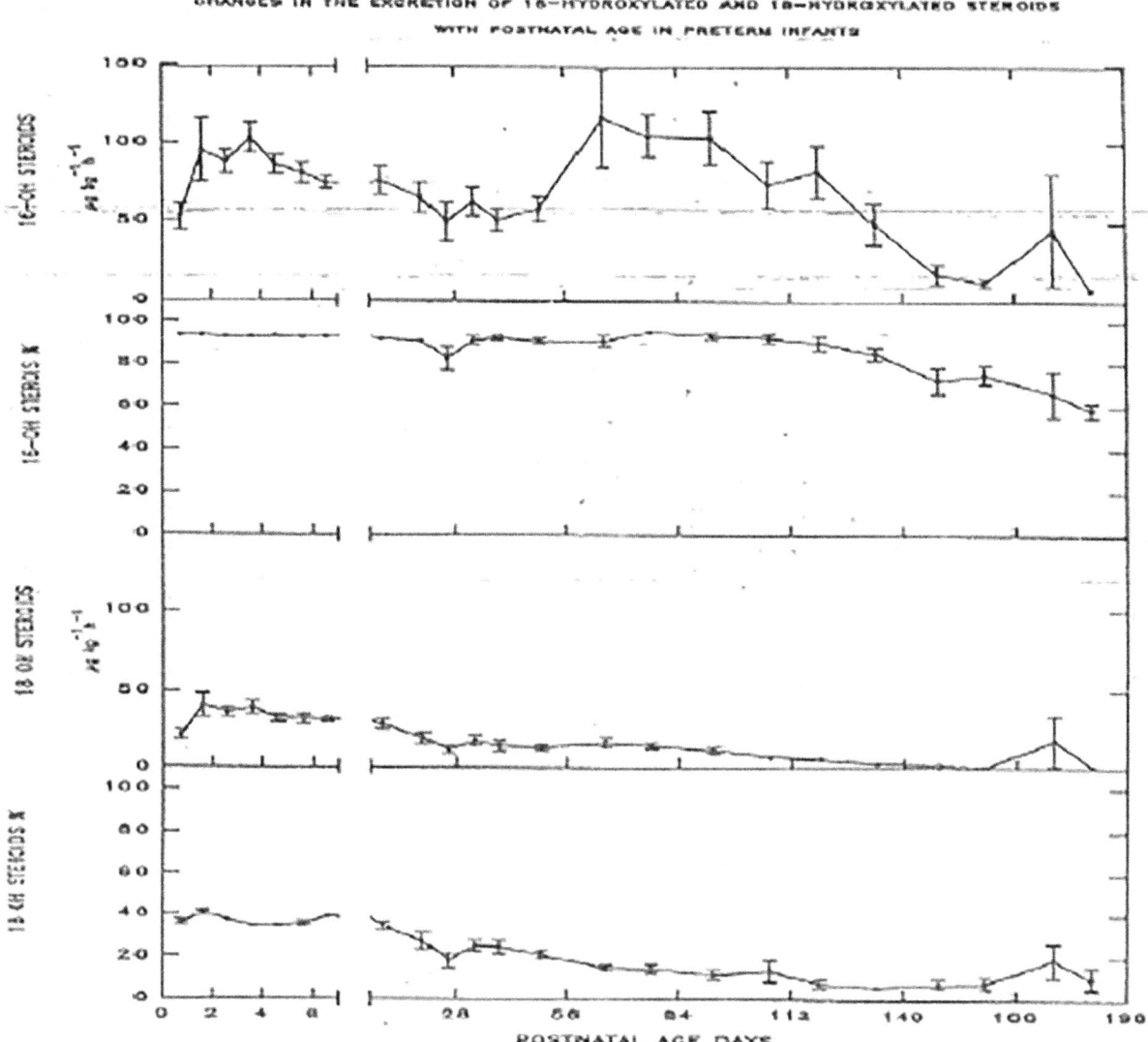

FIG. 2.6.26   Percentage of 16-hydroxy and 18-hydroxy steroids in urine of preterm infants during the early weeks of life in hospital. *(Data from Paula Midgley MD thesis.)*

in the first year of life does not produce androgens. By contrast, a significant rise in lyase activity in the D4-path within the first 10 weeks postnatally in the cohort may reflect higher androgen biosynthesis through the backdoor pathway in the testis during minipuberty. Overall, the data suggested to Dhayat and colleagues that during the first 3 months of life the human testis favors the backdoor over the classic path for androgens.

By 2 years of age the profile (Fig. 2.6.32) is near that seen for the rest of life although the ratios of androgens

changes notably when the zona reticularis develops in the inner layer of the adrenal cortex.

At 5–8 years of age, the metabolites of DHEAS increase in the urine following the development of a zon reticularis outside the adrenal medulla. The steroid profile during childhood to adults and on to senior years is described in detail in Chapter 2.4. The profile essentially reflects the activity of the adrenal cortex, with the contributions from the gonads too low to have any effects on the outputs of sex steroid metabolites in the urine of normal subjects.

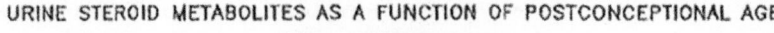

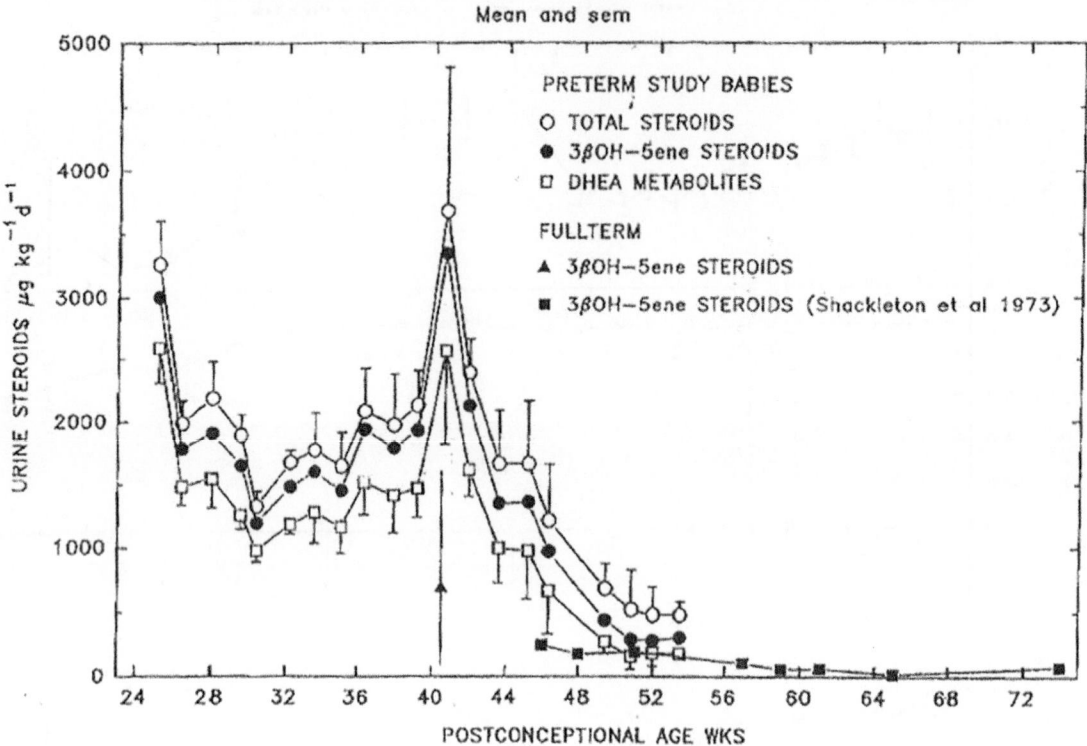

FIG. 2.6.27 Steroids in urine of preterm infants according to postconceptional age. *(Data from Midgley P. Adrenal funcion of preterm infants, MD thesis. University of Aberdeen, 1992.)*

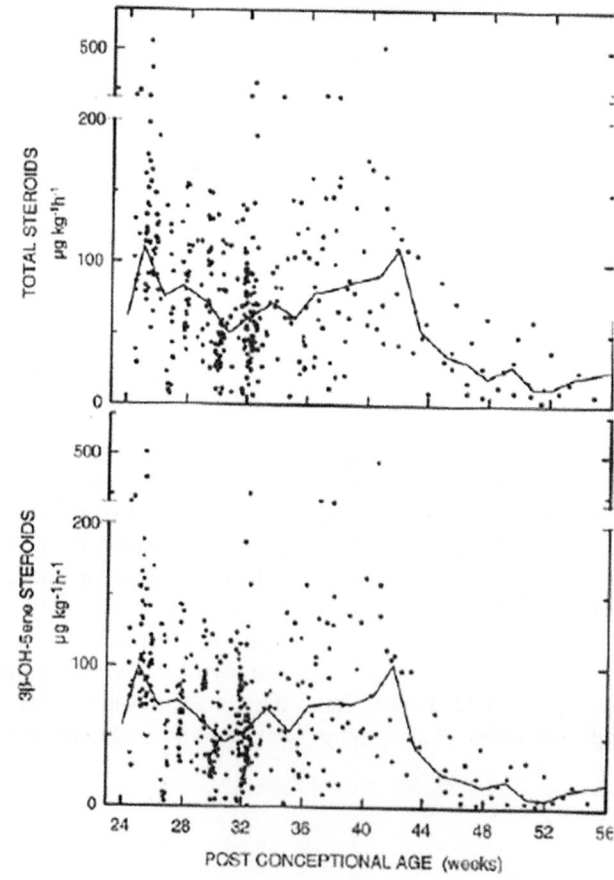

FIG. 2.6.28 Overall A/C activity as judged from urine steroid excretion of preterm infants until past expected date of term delivery. *(Data from Paula Midgley MD thesis.)*

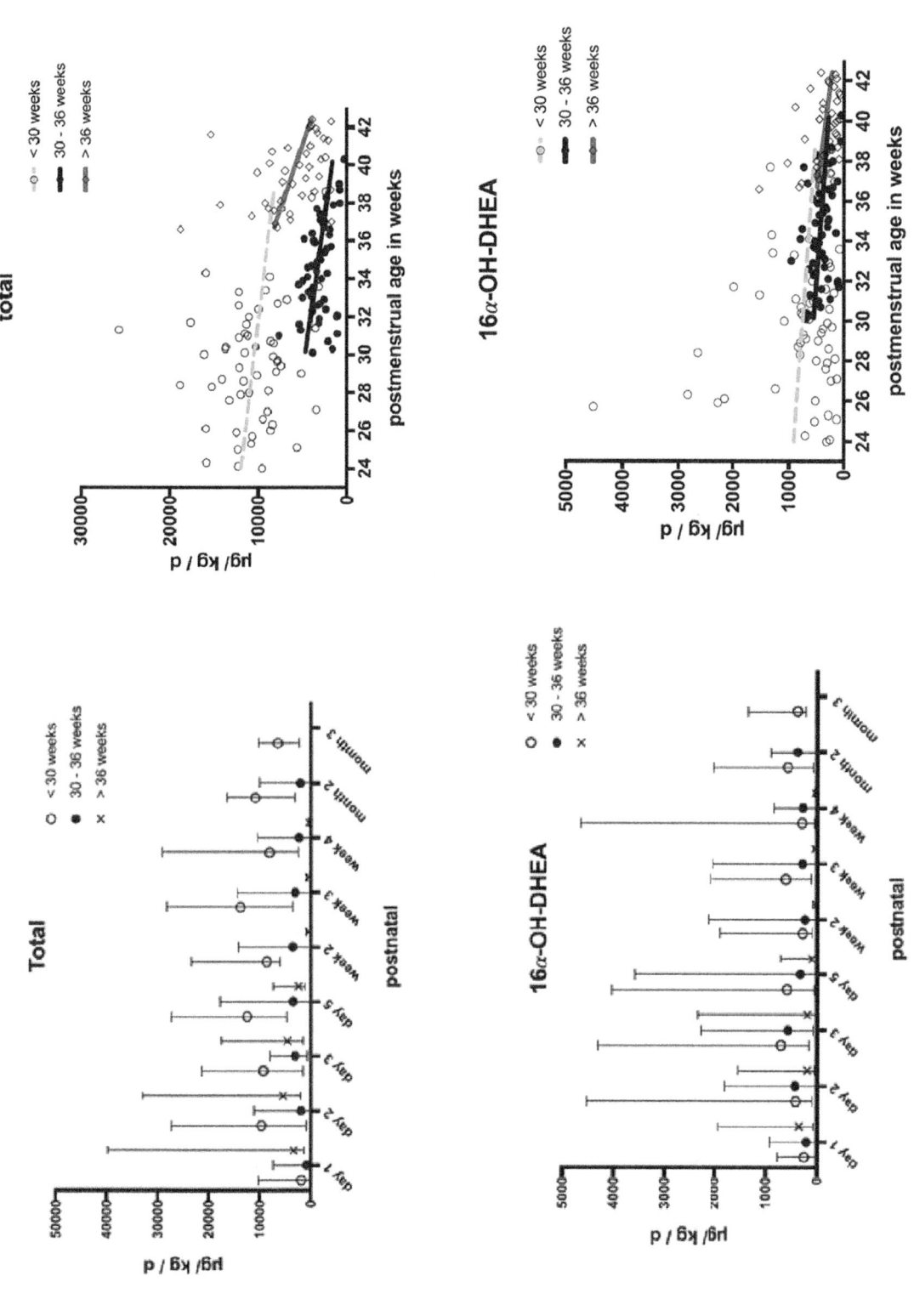

FIG. 2.6.29 Urinary excretion rates (mg/kg/day) of total fetal zone steroids (FZS) (A), 5-androsten-3β,17β-diol (Adiol) (B), DHEA (C) and 16α-OH-DHEA (D) as a function of postnatal age for each gestational age group (*whites dots*: <30 weeks, *black dots*: 30–36 weeks, *black crosses*: term infants, median and range). Excretion rates of total 3β-hydroxy-5-ene-steroids and of steroid metabolites grouped into pregnenolone and 17-OH-pregnenolone metabolites (Δ5 unsaturated C21 steroids) and DHEA and metabolites (Δ5 unsaturated C19 steroids) for each gestational age group as function of postmenstrual age. Gray dashed line represents the correlation line for the gestational age with preterm FZS for infants <30 weeks of gestational age, black line for 30–36 weeks of gestational age and gray line illustrate the association with term infants (>36 weeks of gestational age). (*From Ruhnau J, Hübner S, Sunny D, Ittermann T, Hartmann MF, De Lafollie J, et al. Impact of gestational and postmenstrual age on excretion of fetal zone steroids in preterm infants determined by gas chromatography-mass spectrometry. J Clin Endocrinol Metab 2021;106(9):e3725–38. Fig. 2 p. 730; Fig. 3 p. e3733.*)

TABLE 2.6.14 Excretion rates of steroids in urine of 16 newborn infants.

Steroid	Normal infant Median	Normal infant Range
Fetal adrenal		
16α-Hydroxy DHA[a]	60	10–280
16β-Hydroxy-DHA	40	10–130
16-Oxo-androstenediol +		
15β,16α-Dihydroxy-DHA	70	20–350
Androstenetriol	50	10–150
16α,18-Dihydroxy-DHA	100	20–450
16α-Hydroxypregnenolone	10	30–460
21-Hydroxypregnenolone	20	<5–40
5-Androstene-3β,16α,17β,18-tetrol	10	<5–30
5-Androstene-3β,15β,16α,17β-tetrol	30	<5–70
5-Androstene-3β,15α,16α,17β-tetrol	10	<5–70
5-Androstene-3β,15α,16β,17β-tetrol	5	<5–30
5-Pregnene-3β,20β,21-triol	5	<5–30
5-Pregnene-3β,16α,20α,21-triol	5	<5–20
Total	600	140–1590
Cortisol metabolites		
Tetrahydrocortisone (THE)	30	10–100
α-Cortolone	<5	<5–80
6α-Hydroxy tetrahydrocortisone	20	2–130
β-Cortolone	5	<5–20
6α-Hydroxy-20α-cortolone	5	<5–30
6α-Hydroxy-20β-cortolone	4	<5–20
1α-HYDROXY-20β-cortolone	10	<5–30
Total	90	10–330

Result is in μg/kg/day (UCLH laboratory data) by GC-MS.
[a]DHA, *dehydroepiandrosterone.*
Author original.

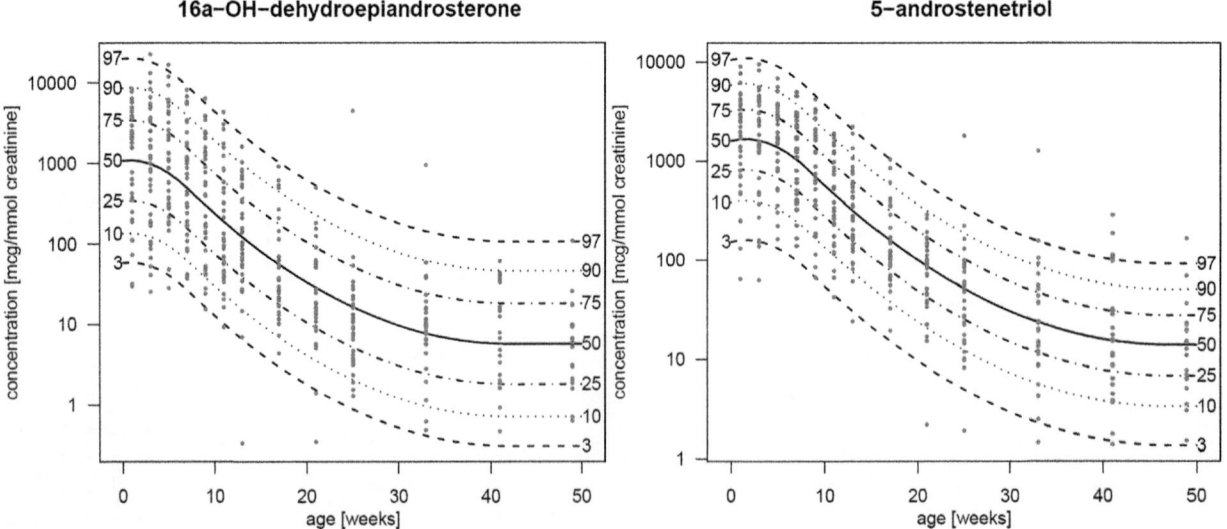

FIG. 2.6.30 Excretion rates of 16-hydroxy-DHEA and androstenetriol during first 6 months of life. *(From Dhayat NA, Dick B, Frey BM, d'Uscio CH, Vogt B, Flück CE. Androgen biosynthesis during minipuberty favors the backdoor pathway over the classic pathway: Insights into enzyme activities and steroid fluxes in healthy infants during the first year of life from the urinary steroid metabolome. J Steroid Biochem Mol Biol. 2017;165(Pt B):312–22. Fig. 35, 36 Suppl.)*

TABLE 2.6.15 Urinary steroids first year Dhayat (10th–50th–90th centiles μg/day) (NB > full data set is for 13 time points).

	Week 1	Week 5	Week 11	Week 17	Week 21	Week 25	Week 49
Pregnanetriol	5–14–39	7–22–60	7–22–60	7–22–60	7–22–60	7–22–60	7–22–60
Pregnenetriol	4–15–58	10–40–156	9–34–132	6–23–89	5–18–71	4–15–59	2–9–36
DHEA	2–9–38	2–6–27	1–4–18				
16-OH DHEA	136–1088–8720	94–757–6070	24–189–1515	7–56–446	4–29–228	2–16–131	1–6–46
Androstenetriol	395–1656–6005	329–1376–4990	110–462–1676	38–159–575	2–7–33	12–52–187	3–14–151
11-OH Andro	2–6–18	5–12–40	10–24–78	11–26–84	11–26–84	11–26–84	1–26–84
Allo-THB	5–28–109	2–10–39	4–23–9	8–46–178	12–66–257	16–90–348	28–153–595
THF	2–5–15	3–7–21	6–16–48	12–32–97	18–47–143	25–64–197	45–118–363
α-Cortol	16–49–145	27–85–249	21–66–192	17–52–153	15–47–137	14–43 – 126	13–40–117
β-Cortol	4–11–24	8–21–47	15–38–85	17–44–100	17–44–100	17–44–100	17–44–100
11-OH Aetio	3–5–111	4–22–159	3–19–37	3–6–116	2–14–103	2–13–94	2–10–75
THE	186–536–1328	226–652–1616	241–697–1726	204–588–1456	181–522–1294	164–472–1169	127–366–906
1β-OH THE	49–152–412	36–112–305	25–77–210	19–59–161	17–52–142	16–48–131	15–47–127
6α-OH THE	942–2679–6011	984–2798–6279	731–2078–4662	456–1295–2906	349–991–2224	277–788–1768	156–442–992
1β-OH β cortolone	106–413–1405	46–179–609	17–65–222	8–32–108	6–23–78	5–19–64	5–18–60
6α-OH-β-cortolone	417–1089–3139	273–712–2052	127–331–953	59–153–442	38–100–289	27–70–202	12–32–93

TABLE 2.6.16 Relative urinary-free steroid concentrations in newborn infants.

Steroid	Urinary concentration (nmol/L)		Relative to creatinine (µmol/mol creatinine)	
P5	6.2	(1.4–34.6)	4.2	(0.7–31.6)
P4	<0.3	(<0.3–1.4)	0.5	(n.d.–0.6)
16OHP4	1.8	(<1.5–14.6)	1.4	(n.d.–10.3)
17OHP4	<1.5	(<1.5–2.8)	1.1	(n.d.–1.9)
21DOE	<1.5	(<1.5–<1.5)	n.d.	(n.d.–n.d.)
21DOF	<14.4	(<14.4–<14.4)	n.d.	(n.d.–n.d.)
DHEA	3.1	(0.8–37.2)	2.2	(0.6–27.3)
AD4	0.9	(<0.3–7.7)	0.7	(n.d.–5.2)
11OHAD4	3.5	(<1.0–40.1)	2.9	(n.d.–26.7)
			(Median [minimum–maximum])	

Steroids are defined as "not detected" (n.d.) when urinary concentration is under the detection limit. P5: µmol/mol creatinine × 2.80 = µg/g creatinine, P4: µmol/mol creatinine × 2.78 = µg/g creatinine, 16OHP4: µmol/mol creatinine × 2.92 = µg/g creatinine, 17OHP4: µmol/mol creatinine × 2.92 = µg/g creatinine, 21DOE: µmol/mol creatinine × 3.04 = µg/g creatinine, 21DOF: µmol/mol creatinine × 3.06 = µg/g creatinine, DHEA: µmol/mol creatinine × 2.55 = µg/g creatinine, AD4: µmol/mol creatinine × 2.53 = µg/g creatinine, 11OHAD4: µmol/mol creatinine × 2.71 = µg/g creatinine.

Koyama Y, Homma K, Miwa M, Ikeda K, Murata M, Hasegawa T. Measurement of reference intervals for urinary free adrenal steroid levels in Japanese newborn infants by using stable isotope dilution gas chromatography/mass spectrometry. Clin Chim Acta 2013;415:302–5. Redrawn from Table 3 p. 305.

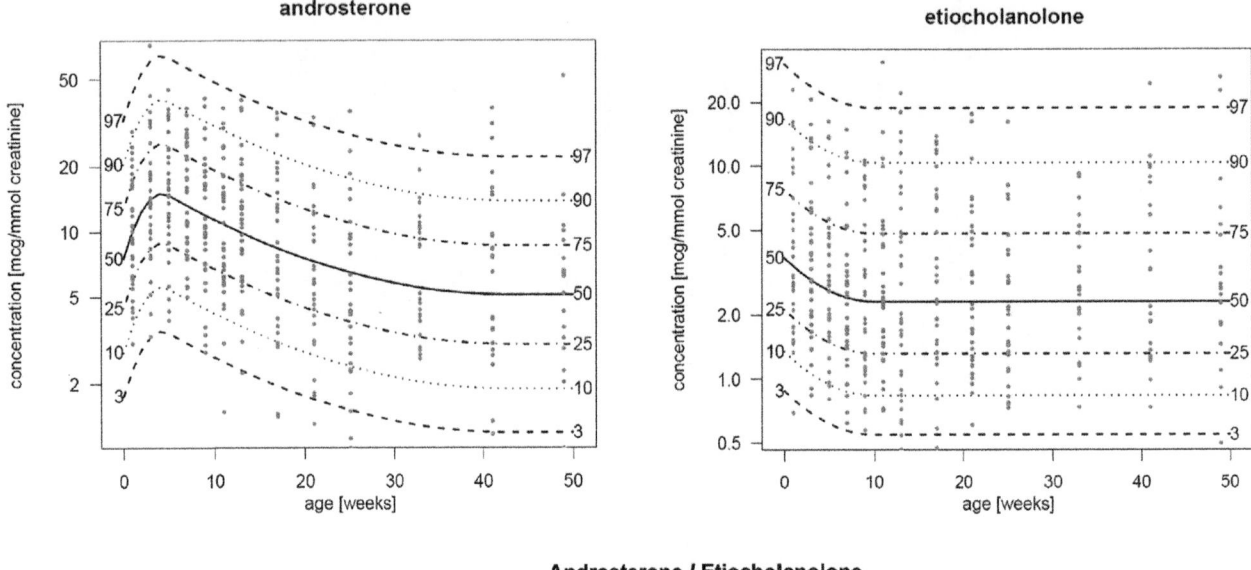

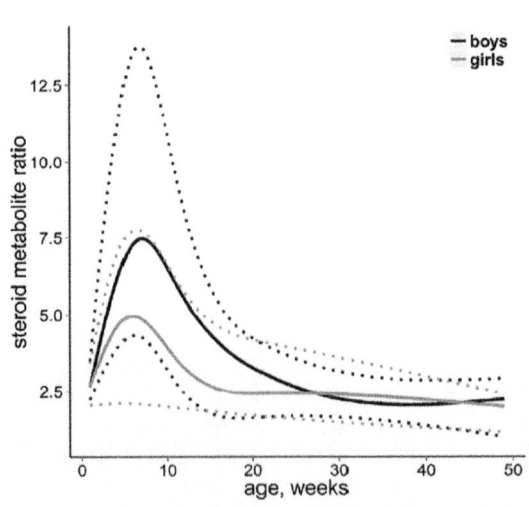

FIG. 2.6.31 Mini puberty Dhayat. Excretion rates of androsterone and etiocholanolone during early weeks of life. *(From Dhayat NA, Dick B, Frey BM, d'Uscio CH, Vogt B, Flück CE. Androgen biosynthesis during minipuberty favors the backdoor pathway over the classic pathway: Insights into enzyme activities and steroid fluxes in healthy infants during the first year of life from the urinary steroid metabolome. J Steroid Biochem Mol Biol. 2017;165 (Pt B):312–22. Fig. 1 p. 315.)*

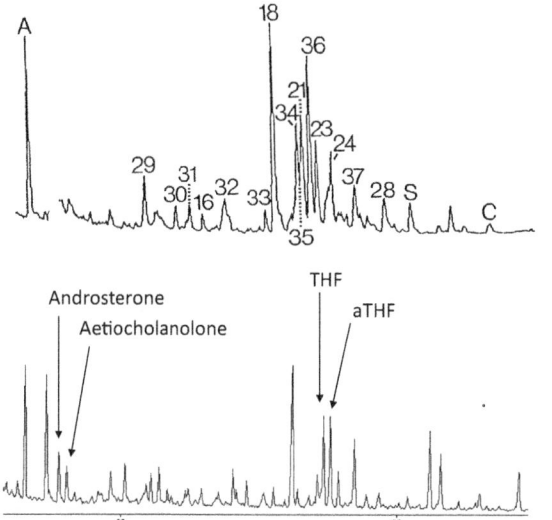

FIG. 2.6.32 Total ion current chromatograms from 2-year-old child and adult female. *(Based on Midgley P. MD thesis, University of Aberdeen. 1992. Adrenal funcion of preterm infants.)*

2.6.4.2.1 Pregnancy

The profile of steroids in urine during pregnancy gets more complicated with each period of gestation due to the increased production of progesterone and estradiol and the multitude of metabolites that arise from those steroids. Hydroxylations at C2, C4, C6 and C16 in particular are seen for all the steroid hormones. The analysis of the steroid hormones in urine provides a view of circulating steroids but has to consider the complex picture from analysis of the metabolites.

Estriol and pregnanediol are major metabolites of estrogen and progesterone and are prominent steroids in urine during pregnancy. Estriol is the produced through the combined actions of the fetal adrenal cortex and liver and of the placenta. The adrenal synthesizes DHEA which is 16-hydroxylated then sulfated in the liver. In the placenta, sulfatase releases 16-hydroxy-DHEA which is the substrate for aromatase and transfer of estriol to the circulation then urine of mother (see Feto-placental unit in biosynthesis chapter). During the first 9 weeks of pregnancy, the corpus luteum is the main contributor to the progesterone metabolite, pregnanediol, measured in urine (Fig. 2.6.33). The placenta then becomes the predominant progesterone source. Progesterone is important for normal feto-placental function.

Estriol and pregnanediol are measured as indices of fetal viability (Schneider et al., 1993), but abdominal ultrasound to visualize the ovaries and endometrium provides the clinician with most of the information needed on pregnancy progression (and more) and has replaced the need for urine testing. Expulsion of the placenta causes a decline in estrogen levels at term. Several clinical conditions are characterized by low estrogen excretion and estriol measurement is included in some Down's syndrome screening programs, other conditions are considered elsewhere in the book (Chapter 3.2.2).

2.6.4.2.2 Conjugates other than glucuronides and sulfates

A series of pregnanediols and pregnanetriols doubly conjugated with n-acetylglucosamine (GlcNAc) and glucuronic

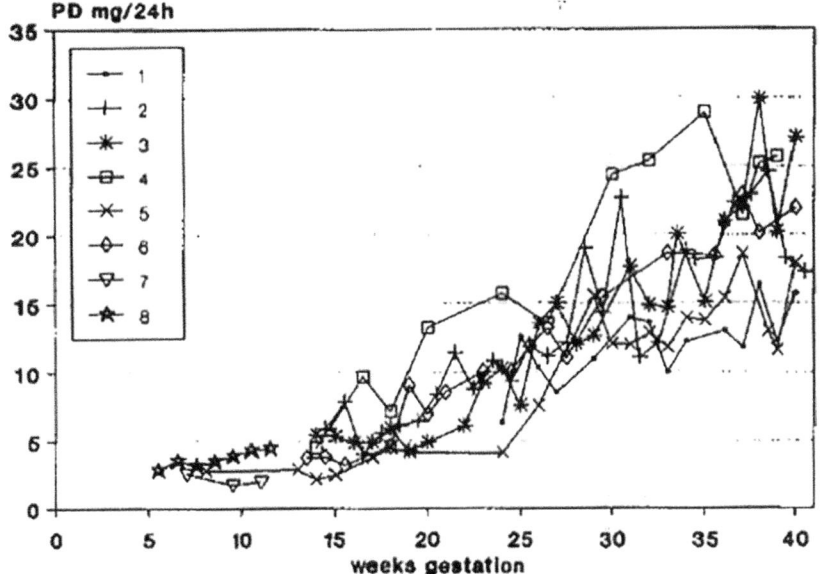

FIG. 2.6.33 Urinary excretion (A) PdG and (B) U1G in pregnancy. Steroid conjugates were hydrolyzed and steroids converted to methyloxime-trimethylsilyl ether derivatives for GC-MS analysis. *From Schneider MA, Davies MC, Honour JW. The timing of placental competence in pregnancy after oocyte donation. Fertil Steril. 1993;59(5):1059–64.*

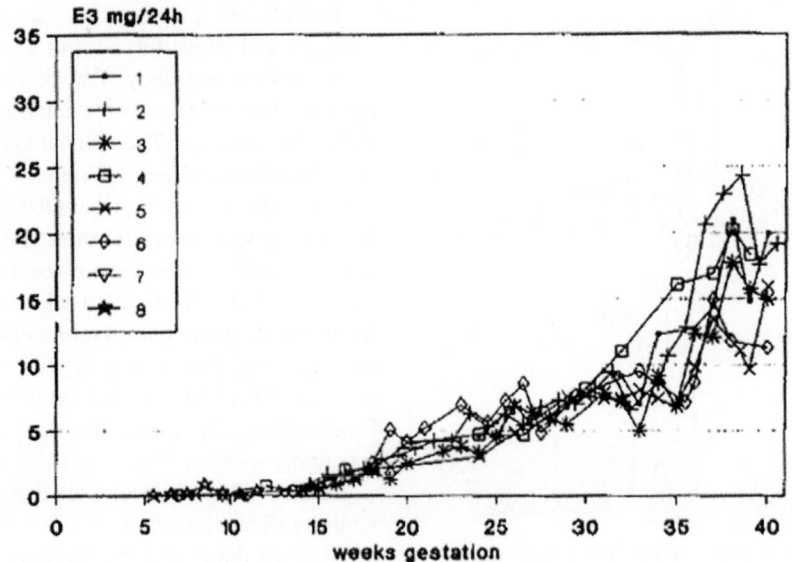

FIG. 2.6.33, CONT'D

16-Cys-Prog   **16-Cys-Preg**

FIG. 2.6.34 Structures of cysteinyl conjugates of progesterone. *(From Marcos J, Pol M, Fabregat A, Ventura R, Renau N, Hanzu FA, Casals G, Marfà S, Barceló B, Barceló A, Robles J, Segura J, Pozo OJ. Urinary cysteinyl progestogens: occurrence and origin. J Steroid Biochem Mol Biol 2015;152:53–61. Modified from Fig. 1 p. 54.)*

acid or sulfuric acid have been identified in urine from pregnant women The concentrations by GC-MS have been reported in mmol/g creatinine so difficult to compare with the cysteinyl conjugates described latterly (Fig. 2.6.34). The GlcNAc metabolites were about 45% of the concentration of 5β-pregnane-3α,20α-diol (Marcos et al., 2015).

Changes in the menstrual cycle and pregnancy are found and some response is also seen with ACTH stimulation (Figs. 2.6.35–2.6.38).

The concentrations of steroids in the first trimester of pregnancy spread across three orders of magnitude (mean concentrations from 0.9 to 1162 ng/mL (Zhou and Cai, 2020) (Table 2.6.17). Only three internal standards were used.

The analysis of up to 100 steroids has been described with techniques from GC-MS to LC-ESI-MS/MS and desorption APPI or APESI with high resolution MS that evades chromatography. Isomeric and isobaric steroids can be resolved in ambient MS methods only if they have

different fragmentation patterns. GC-MS has needed the conjugates to be hydrolyzed and the free steroids converted to thermostable derivatives. LC-MS/MS analysis now permits the intact conjugates to be analyzed after chromatography and ionization or by ambient MS. Gestation specific reference ranges for steroid concentrations (as μg/mmol creatinine) in spot urine collections have been determined with GC-MS (Mistry et al., 2015) (Figs. 2.6.39 and 2.6.40). (The detailed tables of results are not included because of the sizes of the Tables are large and the units used for concentrations are not used by others.)

For three trimesters (Mistry et al., 2015) for the much more extensive list of steroids with reference intervals as mean ±SD and 2.5–97.5 CI, estrogen and progesterone metabolite concentrations increased during pregnancy. Cortisol and corticosterone metabolite excretion decreases during pregnancy but cortisone metabolites remain constant, reflecting inhibition of HSD11B2. In a separate study based on 24h urine collections, the total cortisol metabolites

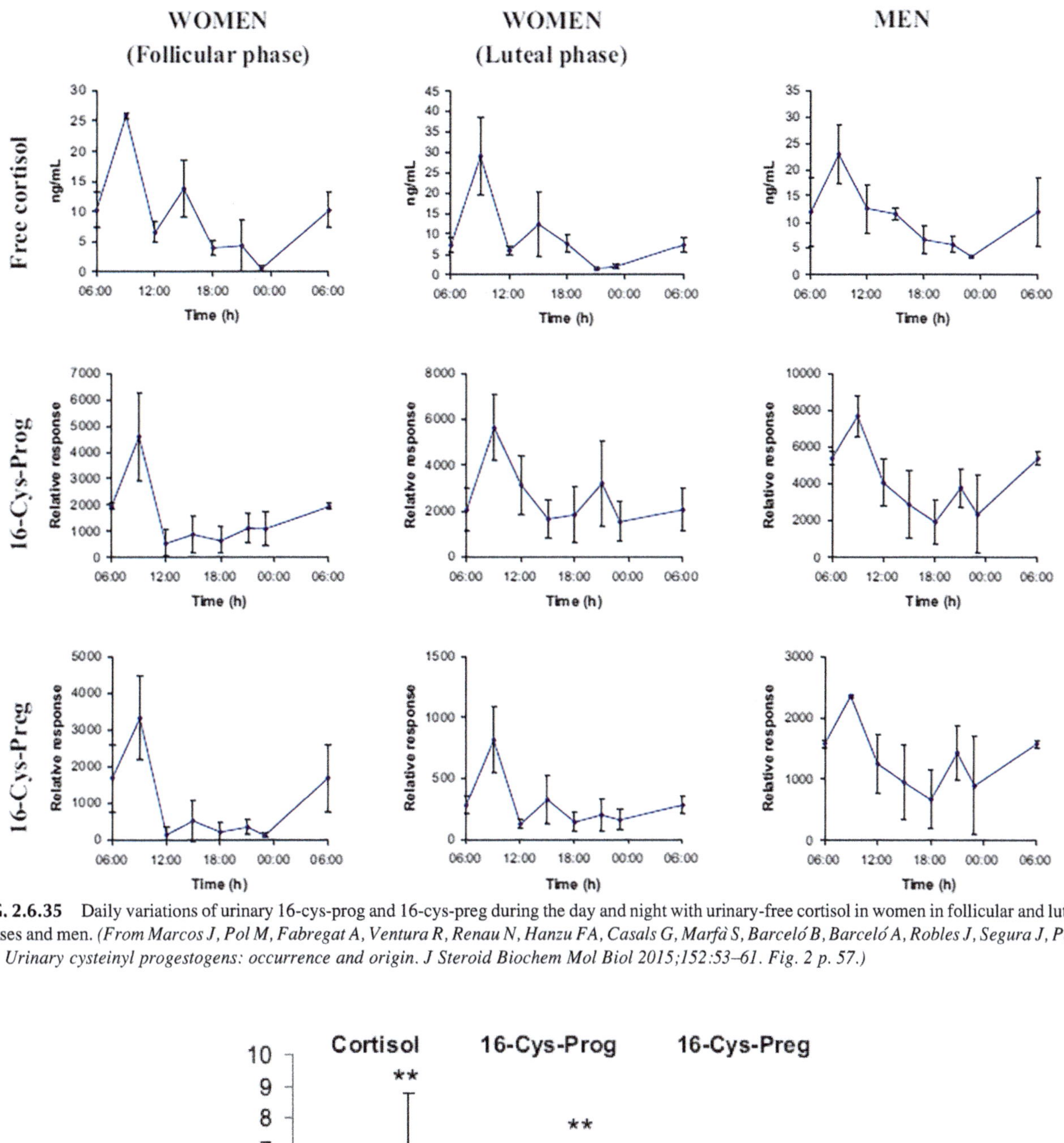

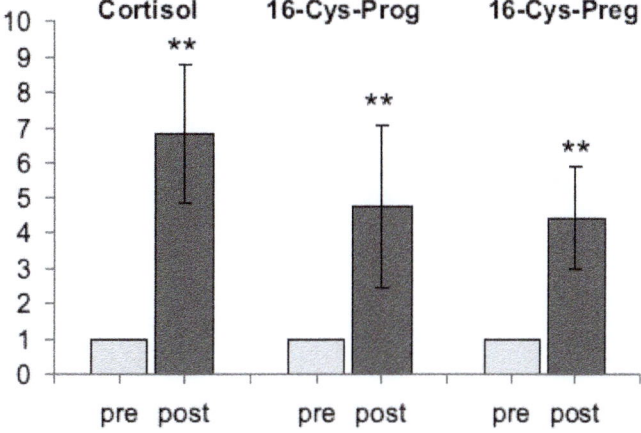

WOMEN (Follicular phase) — WOMEN (Luteal phase) — MEN

FIG. 2.6.35 Daily variations of urinary 16-cys-prog and 16-cys-preg during the day and night with urinary-free cortisol in women in follicular and luteal phases and men. *(From Marcos J, Pol M, Fabregat A, Ventura R, Renau N, Hanzu FA, Casals G, Marfà S, Barceló B, Barceló A, Robles J, Segura J, Pozo OJ. Urinary cysteinyl progestogens: occurrence and origin. J Steroid Biochem Mol Biol 2015;152:53–61. Fig. 2 p. 57.)*

FIG. 2.6.36 Relative effect of ACTH administration on urinary 16-Cys-Prog and 16-Cys-Preg compared with urinary-free cortisol ($n = 8$, **$P < .01$). LC-MS/MS chromatograms in the same signal scale corresponding to the analysis of a sample collected before and after ACTH administration. *(From Marcos J, Pol M, Fabregat A, Ventura R, Renau N, Hanzu FA, Casals G, Marfà S, Barceló B, Barceló A, Robles J, Segura J, Pozo OJ. Urinary cysteinyl progestogens: occurrence and origin. J Steroid Biochem Mol Biol 2015;152:53–61. Fig. 3 p. 58.)*

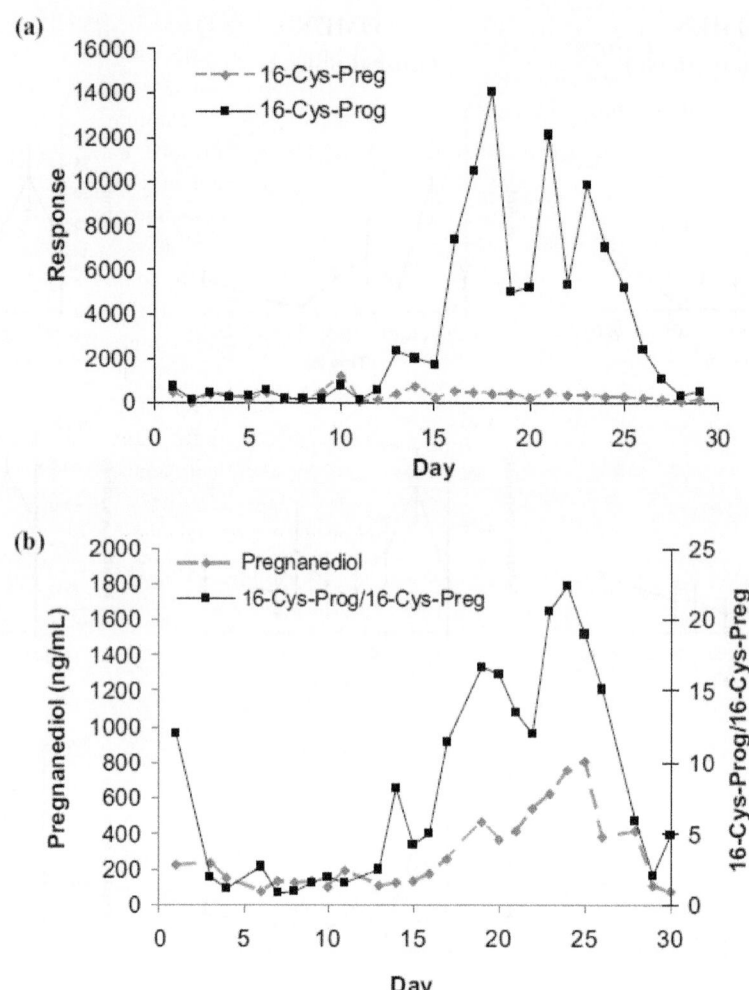

FIG. 2.6.37 Menstrual cycle variation in (A) urinary 16-cys-prog and 16-cys-preg (B) ratio of 16-cys-prog to 16-cys-preg compared with pregnanediol output. *(From Marcos J, Pol M, Fabregat A, Ventura R, Renau N, Hanzu FA, Casals G, Marfà S, Barceló B, Barceló A, Robles J, Segura J, Pozo OJ. Urinary cysteinyl progestogens: occurrence and origin. J Steroid Biochem Mol Biol 2015;152:53–61. Fig. 4 p. 59.)*

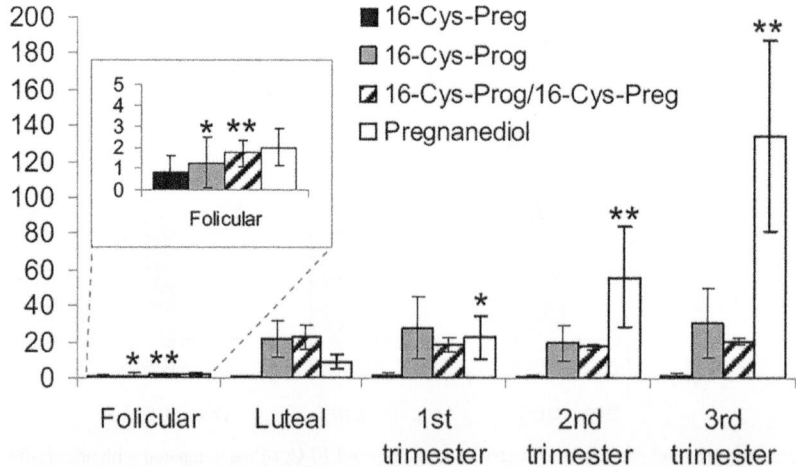

FIG. 2.6.38 Urinary 16-cys-preg and urinary 16-cys-prog excretion and ratio 16-cycs-prog:cys-16-preg during pregnancy compared with urinary pregnanediol. *(From Marcos J, Pol M, Fabregat A, Ventura R, Renau N, Hanzu FA, Casals G, Marfà S, Barceló B, Barceló A, Robles J, Segura J, Pozo OJ. Urinary cysteinyl progestogens: occurrence and origin. J Steroid Biochem Mol Biol 2015;152:53–61. Fig. 5 p. 59.)*

TABLE 2.6.17 The SG-adjusted mean concentrations in ng/mL (±2SD) of hormones in urine samples from 30 women in early pregnancy (<16 weeks of gestation).

Steroid	Mean (2SD)
Cortisone	1161 (2232)
Cortisol	158 (249)
11-Deoxycortisol	0.99 (1.77)
Aldosterone	20.98 (29.31)
Corticosterone	41 (9)
11-Deoxycorticosterone	5.9 (14.80)
Progesterone	6.20 (6.00)
17-OH-Progesterone	4.9 (9.0)
Pregnenolone	18 (42)
Estrone	76 (1142)
Estradiol	28.7 (44.26)
Estriol	229.8 (282)
Testosterone	5.0 (12.8)
Dehydroepiandrosterone	57 (103)

From Zhou Y, Cai Z. Determination of hormones in human urine by ultra-high-performance liquid chromatography/triple-quadrupole mass spectrometry. Rapid Commun Mass Spectrom. 2020;34 Suppl 1:e8583. Table 4 p. 7.

increased in the second and third trimesters of pregnancy (Stoye et al., 2020) (Fig. 2.6.41) (Table 2.6.18).

Analysis of steroids in pregnancy is of interest with regard to psychological and immunological changes, genetic defects of the fetus and the relationship between events in pregnancy and birth that influence presentation of cardiovascular and other disorders in later life. These will be discussed in more detail in the relevant clinical topics.

In conclusion, analysis of urine for steroid content provides valuable information that can be correlated with somatic changes (age, gender, life events, minipuberty, adrenarche, puberty and pregnancy). A 24h collection of urine is the preferred sample in order to integrate the diurnal variation and for more reproducible data. When samples from patients are analyzed in order to demonstrate changes in enzyme activities the reader (see clinical chapters) will observe large changes in results from patients with similar genetic problems compared with normality and spot samples will provide as much diagnostic evidence as from a 24h collection. Using clinical samples from patients with enzyme defects (confirmed by genetic testing), the defects in steroidogenic enzymes such as SRD5A2, CYP17A1, CYP21A2 and CYP11B1 can be diagnosed with high confidence using spot urine only (Bileck et al., 2018). However, the authors of that paper concluded that spot urine collections should not be used for the diagnosis of disorders that are based on absolute excretion values such as adrenal tumors or in some cases of adrenal insufficiency.

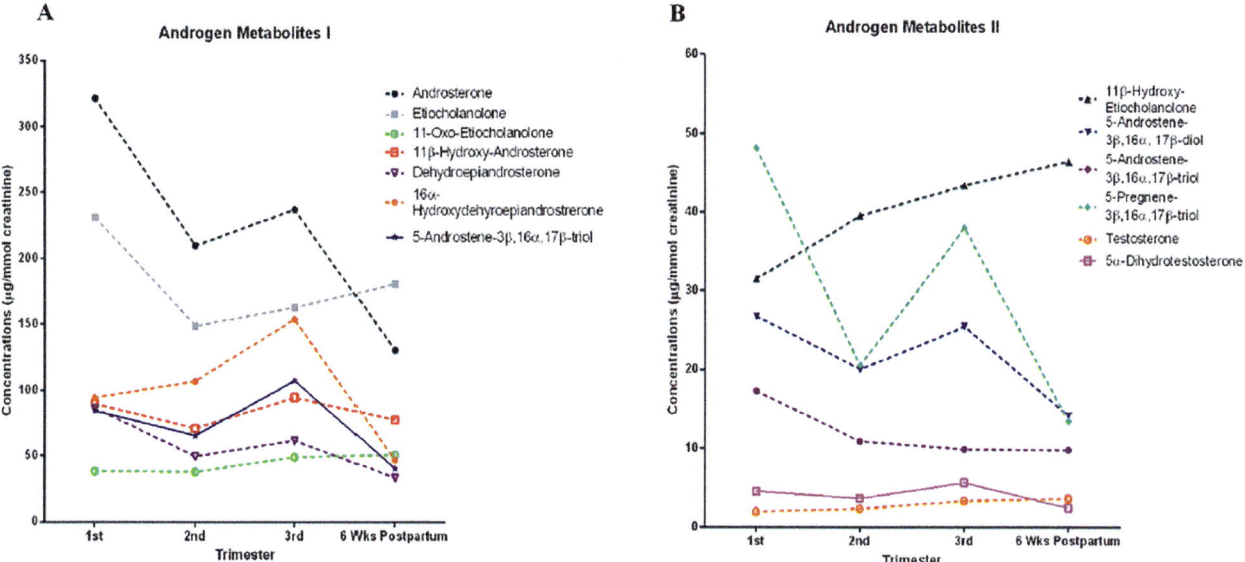

FIG. 2.6.39 Urinary excretions of androgen metabolites in the first, second and third trimesters of pregnancy and 6 weeks postpartum. Each point is the mean excretion (μg) corrected for creatinine (μg/mmol). *(From Mistry HD, Eisele N, Escher G, Dick B, Surbek D, Delles C, Currie G, Schlembach D, Mohaupt MG, Gennari-Moser C. Gestation-specific reference intervals for comprehensive spot urinary steroid hormone metabolite analysis in normal singleton pregnancy and 6 weeks postpartum. Reprod Biol Endocrinol 2015;13:101. Fig. 1 p. 9.)*

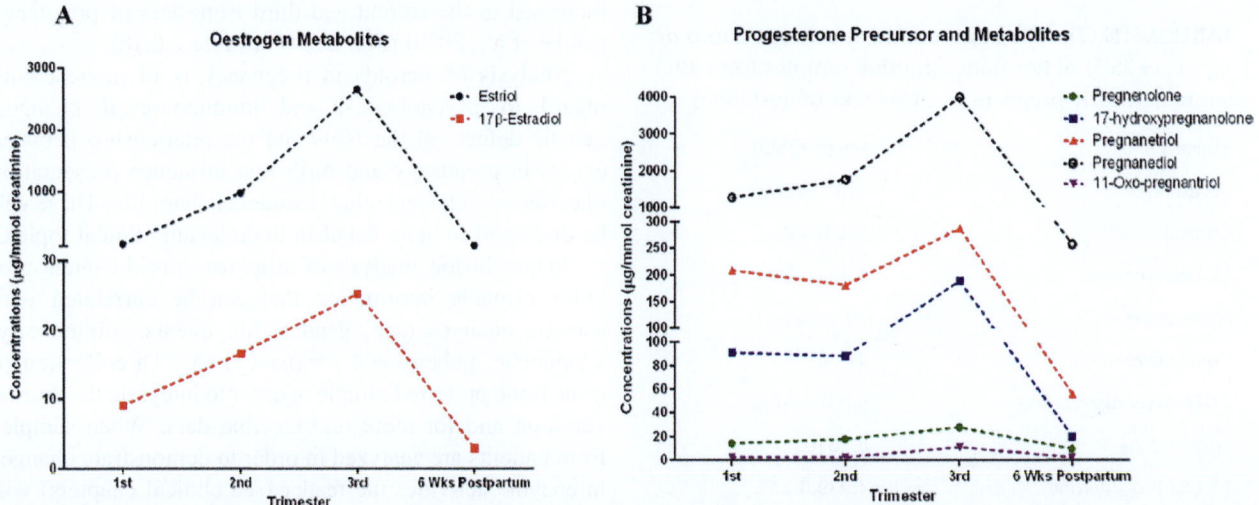

FIG. 2.6.40 Urinary excretions of estrogen and progesterone metabolites in the first, second and third trimesters of pregnancy and 6 weeks post partum. Each point is the mean excretion (µg) corrected for creatinine (µg/mmol). *(From Mistry HD, Eisele N, Escher G, Dick B, Surbek D, Delles C, Currie G, Schlembach D, Mohaupt MG, Gennari-Moser C. Gestation-specific reference intervals for comprehensive spot urinary steroid hormone metabolite analysis in normal singleton pregnancy and 6 weeks postpartum. Reprod Biol Endocrinol 2015;13:101. Fig. 2 p. 9.)*

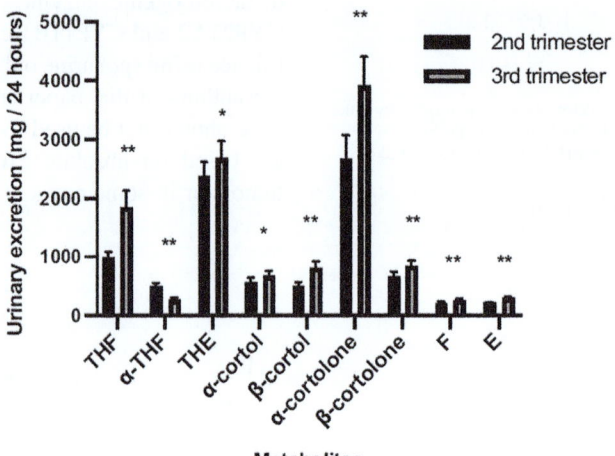

FIG. 2.6.41 Excretion of cortisol metabolites during the second and third trimesters of pregnancy (Boxes are geometric means and bars are 95% confidence limits). *(From Stoye DQ, Andrew R, Grobman WA, et al. Maternal glucocorticoid metabolism across pregnancy: a potential mechanism underlying fetal glucocorticoid exposure. J Clin Endocrinol Metab 2020;105(3):e782-90. Fig. 2 p. e786.)*

TABLE 2.6.18 Changes in urinary metabolites excretion and ratios across pregnancy.

	Second trimester: median (lower quartile-upper quartile)	Third trimester: median (lower quartile-upper quartile)	Change across gestations: RGM (95% CI)
Urinary metabolites, (µg/24 h)			
THF	1043 (691–1397)	1768 (1066–3269)	1.88 (1.65–2.15)[b]
α-THF	494 (331–781)	291 (177–436)	0.55 (0.50–0.61)[b]
THE	2500 (1588–3579)	2799 (1805–4222)	1.13 (1.04–1.23)[a]
α-Cortol	586 (368–917)	641 (455–1140)	1.19 (1.05–1.34)[a]
β-Cortol	545 (259–947)	849 (540–1410)	1.65 (1.45–1.88)[b]
α-Cortolone	2420 (1589–4473)	3685 (2371–6241)	1.46 (1.25–1.71)[b]
β-Cortolone	632 (424–979)	796 (574–1189)	1.29 (1.13–1.47)[b]

TABLE 2.6.18 Changes in urinary metabolites excretion and ratios across pregnancy—cont'd

	Second trimester: median (lower quartile-upper quartile)	Third trimester: median (lower quartile-upper quartile)	Change across gestations: RGM (95% CI)
F	231 (160–315)	272 (215–361)	1.23 (1.13–1.35)[b]
E	228 (171–292)	316 (227–410)	1.36 (1.26–1.48)[b]
Total urinary glucocorticoids	9691 (6157–12,805)	13,523 (8955–18,269)	1.37 (1.22–1.52)[b]
Ratios of metabolites			
11β-HSD2 activity=F/E	0.99 (0.78–1.28)	0.88 (0.73–1.16)	0.90 (0.86–0.95)[b]
11β-HSD total activity=(THF +α-THF)/THE	0.61 (0.52–0.85)	0.76 (0.48–1.23)	1.27 (1.14–1.42)[b]
Relative 5β -reductase and 5α - reductase	1.78 (1.33–2.83)	7.19 (3.64–11.74)	3.41 (3.04–3.83)[b]
Activity = THF/α-THF			
5α-Reductase activity=F/α-THF	0.45 (0.27–0.60)	0.98 (0.61–1.51)	2.24 (2.00–2.50)[b]
5β-Reductase metabolism of F = F/ (THF+α-cortol + β-cortol)	0.10 (0.07–0.14)	0.07 (0.05–0.11)	0.72 (0.65–0.81)[b]
5β-Reductase metabolism of E = E/ (THE + +α-cortolone+ β-cortolone)	0.04 (0.02–0.06)	0.04 (0.03–0.06)	1.05 (0.96–1.15)

Paired t-test (2-tailed) of log-transformed urine values.
Abbreviations: *E*, cortisone; *F*, cortisol; *HSD*, hydroxysteroid dehydrogenase; *HSD2*, hydroxysteroid dehydrogenase type 2; *RGM*, ratio of geometric means; *THE*, tetrahydrocortisone; *THF*, 5β-tetrahydrocortisol.
[a]$P<.01$.
[b]$P<.001$.
From Stoye DQ, Andrew R, Grobman WA, et al. Maternal glucocorticoid metabolism across pregnancy: a potential mechanism underlying fetal glucocorticoid exposure. J Clin Endocrinol Metab 2020;105(3):e782–e790. Table 2 p. e786.

2.6.4.2.3 Performance of steroid assays for urine samples

The steroids in urine are mainly conjugated forms of reduced (tetrahydro-) hormones. There are few steroids that can be purchased as reference compounds and purity will never be 100%. Until recently urinary steroids have been determined after enzyme hydrolysis of the conjugates. The validation of a method will still need as far as possible to comply with analytical guidelines (described in Chapter 2.4). Around 30 laboratories that perform urinary steroid profile analysis participate in an EQA scheme which judges results by multiples of medians of results from all participants (Phillips et al., 2004). The quality of any interpretative comments are judged by an independent panel are assessed on the safety of the advice.

Assay interferences are not easy to determine. Standard addition is a good method (Fig. 2.6.42) (Gomez-Gomez et al., 2020).

In the absence of a true matrix sample, standards are prepared in methanol or buffer. The samples are then spiked with reference steroid preparation at several levels and the results for spiked samples are compared with the standard curve (Gomez-Gomez et al., 2020). A calculated concentration is then compared with the "real" concentration from the graph. The technique is called "standard addition" and would be used in the validation stage of assay development only before the introduction of the assay into routine laboratory testing. This test is limited for urinary steroids by lack of available reference standards. The assay of samples at dilution is another good style. Dilution to 1 in 10 is usually the maximum that can be achieved within the LOD for the analyte as shown elegantly by Häkkinen et al. (2018) for serum, plasma and endometrial tissue.

2.6.5 Other fluids

The collection of urine can sometimes be difficult or impossible so other fluids can be taken in clinical practice which can be in day to day living or in a surgical procedure. Preanalytical factors as well as analytical issues need to be taken into account with appropriate guidelines when procedure is not by or in the presence of healthcare professional. The availability of drug-free matrix in adequate quantities for the calibrators in quantification is sometimes challenging, e.g. in the case of cerebrospinal fluid. In the absence of a true matrix sample, the standards are prepared in methanol or buffer and further tests on the samples are needed to exclude matrix effects. Recently, LC-ESI-MS methods for simultaneous quantitation of 28 steroids in animal fluids have been reported.

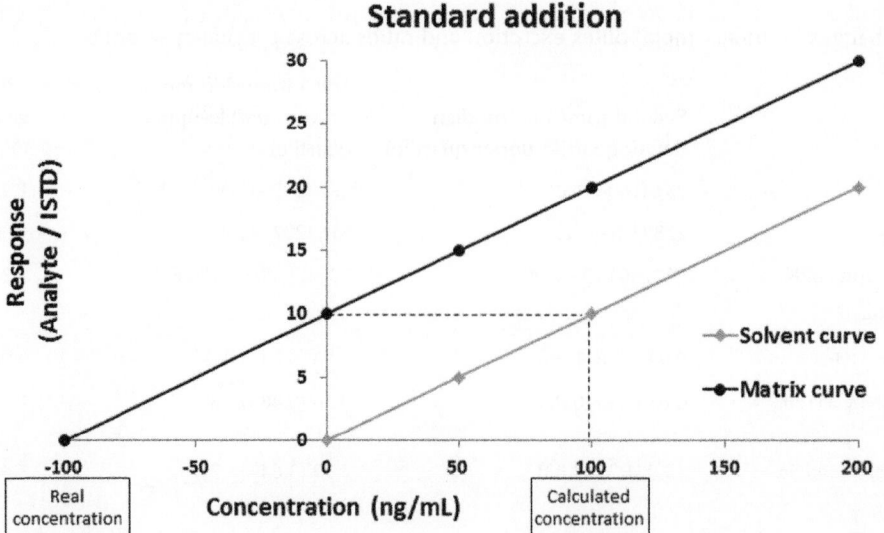

FIG. 2.6.42 Graphical representation of the "real" and "calculated concentrations using the standard addition approach. *(Based on Gomez-Gomez A, Miranda J, Feixas G, Arranz Betegon A, Crispi F, Gratacós E, Pozo OJ. Determination of the steroid profile in alternative matrices by liquid chromatography tandem mass spectrometry. J Steroid Biochem Mol Biol 2020;197:105520. Suppl. Fig. 1.)*

2.6.5.1 Saliva

Saliva collection is a noninvasive procedure but steroid concentrations (cortisol <10 ng/mL; estradiol <500 pg/mL) are very low because, like urine and the kidney, saliva reflects the steroid filtered from the free fraction in blood. The laboratory should produce a strict protocol for the collection procedure. The subject should refrain from eating, drinking, or oral hygiene procedures for at least 1 h prior to the collection. The subject is asked to rinse their mouth out well with distilled water for 1 min either expectorating or swallowing the water. Saliva should be collected in the mouth for a few minutes and discarded then the sample for analysis can be collected. Over a short period of time saliva will collect in the floor of the mouth and can then be transferred to a collection tube (passive drool) or the subject can chew on one of several absorbent devices (Salivette is the most popular for steroids) which is more acceptable with children. Some test devices are chewed on and have an indicator that changes color when sufficient sample has been collected (Quest Oral-Eze is one such device). Saliva flow can be stimulated by chewing paraffin wax but not chewing gum (Büttler et al., 2018). Citric acid or lemon drops also effectively stimulate salivation. Free steroids in blood passively diffuse into saliva via acinar cells. More than 60% of saliva is produced by the submandibular glands. Contamination of the sample with blood should be avoided (Gröschl and Rauh, 2006; Gröschl et al., 2003). Saliva has fewer measurable steroidal compounds than in serum or urine, less protein and fewer cells so is less of an analytical challenge. Some proteins pass from blood into saliva (e.g. insulin) others are produced in the salivary glands (amylase and lipase). Samples for salivary steroid analysis are stable for up to 7 days at room temperature, 1 month or more at 4°C and 3 months or more at −20°C. In order to minimize the viscosity of saliva samples before SPE, c. 0.5 g of sodium sulfate anhydrous (solid) can be added and the mixture vortexed (1 min) and then centrifuged (3000 g, 5 min).

Salivary cortisol concentration is a reliable indicator of serum unbound-cortisol concentration regardless of sex, oral contraceptive usage or pregnancy (Fig. 2.6.43) (Büttler et al., 2013; Vining et al., 1983).

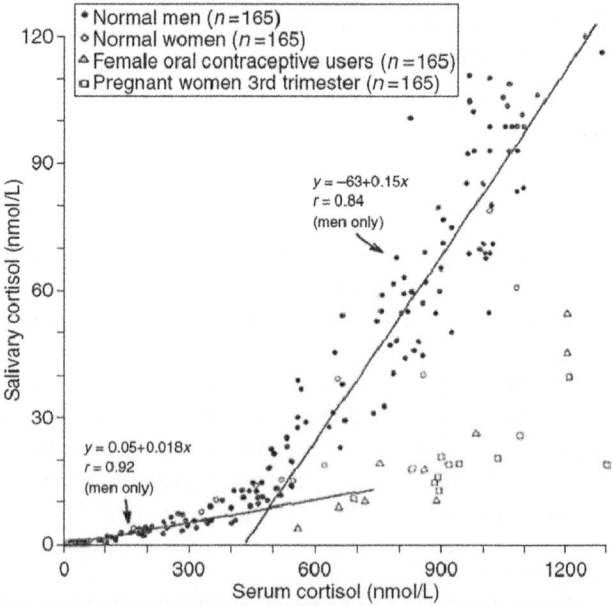

FIG. 2.6.43 Cortisol concentration by radioimmunoassay in time-matched samples of serum and saliva from normal men, normal women, women on oral contraceptives and pregnant women. *(From Vining RF, McGinley RA, Maksvytis JJ, Ho KY. Salivary cortisol: a better measure of adrenal cortical function than serum cortisol. Ann Clin Biochem 1983;20 (Pt 6):329–35. Fig. 2 p. 331.)*

There are considerable variations in steroid results due to timing of sample collection and the analytical methods used (Hollanders et al., 2020; Bakusic et al., 2019). The diurnal rhythm of plasma steroids has been also observed with saliva samples (Fig. 2.6.44) (Mezzullo et al., 2017).

Steroids are usually extracted from the saliva by protein precipitation techniques, solvent extraction or SPE.

Salivary testosterone can be used to assess the course of puberty with concentrations increasing from around 8 years of age (Fig. 2.6.45) (Büttler et al., 2013). Note samples were

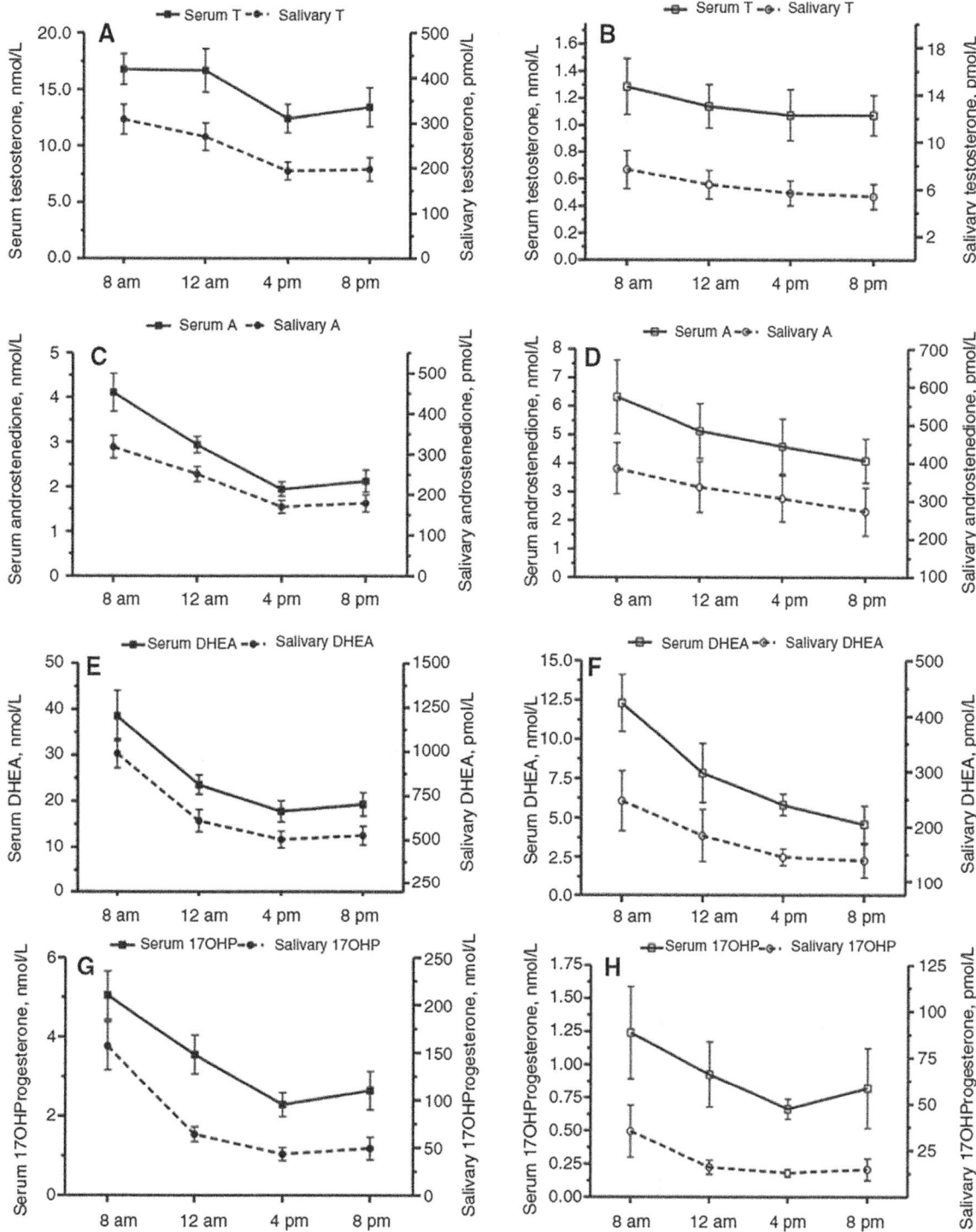

FIG. 2.6.44 Salivary concentrations parallel the serum fluctuation during daytime. Serum (bold line) and saliva (dotted line) diurnal changes of T ($n = 12$), A ($n = 12$), DHEA ($n = 10$) and 17OHP ($n = 9$) in males (A, C, E and G, respectively). *(From Mezzullo M, Fazzini A, Gambineri A, Di Dalmazi G, Mazza R, Pelusi C, Vicennati V, Pasquali R, Pagotto U, Fanelli F. Parallel diurnal fluctuation of testosterone, androstenedione, dehydroepiandrosterone and 17OHprogesterone as assessed in serum and saliva: validation of a novel liquid chromatography-tandem mass spectrometry method for salivary steroid profiling. Clin Chem Lab Med 2017;55(9):1315–23. Fig. 2 p. 1321.)*

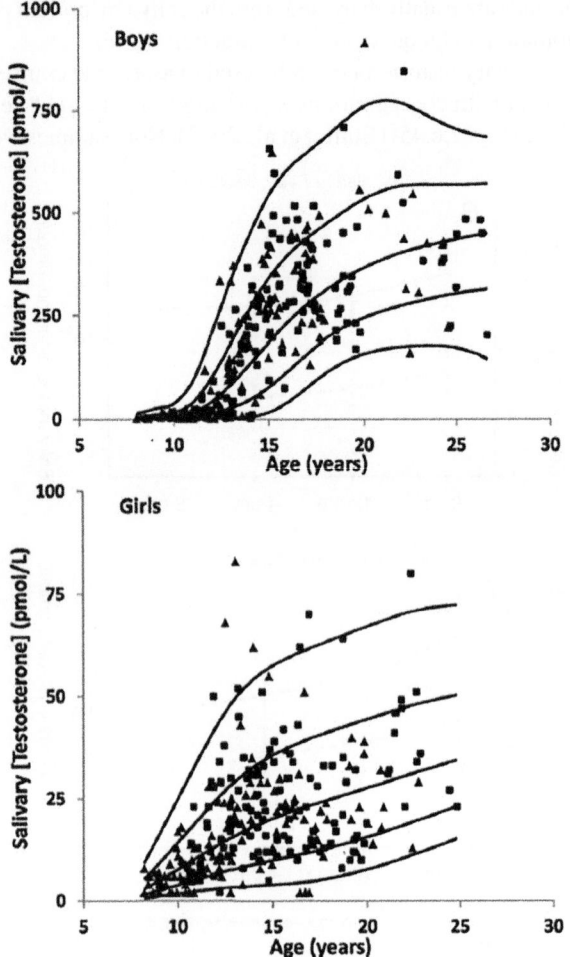

FIG. 2.6.45 Individual points for salivary testosterone and fitted percentile ranges (2.5, 15.9, 50, 84.1, and 97.5 percentiles) for age of boys (n = 123) and girls (n = 131). *(From Büttler RM, Peper JS, Crone EA, Lentjes EGW, Blankenstein MA, Heijboer AC. Reference values for salivary testosterone in adolescent boys and girls determined using isotope-dilution liquid-chromatography tandem mass spectrometry (ID-LC-MS/MS). Clin Chim Acta 2016;456:15–18. Fig. 1 p. 17.)*

collected soon after waking in the morning because activity early in puberty is nocturnal.

Salivary 17-hydroxyprogesterone and androstenedione assays have value in patients with congenital adrenal hyperplasia due to 21-hydroxylase deficiency (Mezzullo et al., 2017) (Fig. 2.6.38) as noninvasive tests for the home monitoring of hydrocortisone replacement therapy (Bacila et al., 2019; Gröschl et al., 2003). Salivary glands have HSD17B activity converting testosterone to androstenedione.

A saliva method for cortisol was validated against ISO15189 and is a useful example of the detailed process required for LC-MS/MS measurements (Antonelli et al., 2018). When assessed against strict criteria, salivary cortisol in evening samples (collected near mid-night) or following dexamethasone suppression provide a reliable and effective screen for Cushing's syndrome (see Chapter 3.1.1, Hypercortisolism) especially for the investigation of suspected cyclical Cushing's syndrome. There is potential for the identification of adrenal insufficiency although the response to Synacthen stimulation should be tested. The test is also used in psychoendocrinology to demonstrate low levels of cortisol particularly in patients with depression and stress but the method must be sensitive to 2 nmol/L. In the parotid glands, cortisol is a substrate for HSD11B2 which is very active producing high concentrations of cortisone which can cross react in many immunoassays, cortisol and cortisone are now measured together.

Testosterone in saliva is a good marker for following pubertal development according to testicular volume or pubic hair signs in boys (Krebs et al., 2019) (Table 2.6.19).

Progesterone (Schiffer et al., 2019a,b) is sometimes used as a marker of ovarian function and possible predictor of spontaneous preterm delivery (Abuelghar et al., 2019) (Fig. 2.6.46).

On a general note, because of the low salivary steroid concentrations, measurements in hypo-states are not recommended. Over the past 10 years saliva assays have moved

TABLE 2.6.19 Saliva and serum steroid concentrations related to testis volume and pubic hair stages.

Analyte	PH1	TV < 3 mL	PH 2–4	TV 4 – 15 mL	PH 5/6	TV > 15 mL
17-OHP ng/L saliva	11.3 (10–11)	11 (10–11)	13 13–14)	13 (13–14)	15 (14–16)	15 (14–16)
17-OHP µg/L serum	0.51 (0.1–1.30)	0.51 (0.1–2.0)	1.10 0.26–3.80)	1.10 (0.26–3.80)	1.25 (0.48–2.70)	1.30 (0.48–2.70)
DHEA ng/L saliva	113.6 (10–633)	115.7 (10–680)	278 (52.6–809)	266 (52.6–756)	414 (171–1275)	455 (171–1274)
DHEAS mg/L serum	0.94 (0.3–3.9)	0.97 (0.3–3.39)	2.2 (0.8–8.1)	2.2 (0.8–8.1)	2.45 (1–5.9)	2.5 (1.0–5.9)
A4 ng/L saliva	32 (20–192)	35.3 (20–192)	63.9 (20–199)	62.5 (20–253)	91.7 (20–263)	88.9 (20–263)
A4 µg/L serum	0.30 (0.30–3.6)	0.30 (0.30–3.6)	0.9 (0.30–2.5)	0.8 (0.30–2.50)	1.25 (0.8–2.30)	1.30 (0.80–2.30)

TABLE 2.6.19 Saliva and serum steroid concentrations related to testis volume and pubic hair stages—cont'd

Analyte	PH1	TV < 3 mL	PH 2–4	TV 4 – 15 mL	PH 5/6	TV > 15 mL
T ng/saliva	11.8 (6.2–63.2)	11.8 (6.2–4.5)	67.9 (10–230.2)	66.3 (10–230.2)	108.7 (36.9–320.1)	110.5 (36.9–320.1)
T µg/L serum	0.12 (0.12–0.35)	0.12 (0.12–0.62)	1.93 (0.12–7.91)	2.28 (0.12–7.91)	4.11 (1.29–5.56)	3.62 (1.29–5.68)

Data are presented as median (range). *PH*, pubic hair; *TV*, testicular volume; *17-OHP*, 17-hydroxyprogesterone; *DHEA*, dehydroepiandrosterone; *DHEAS*, dehydroepiandrosterone-sulfate; *AE*, androstenedione; *TE*, testosterone; *BMI*, body mass index. *U* test: aPH1 to PH2–4: $P < .001$, bPH1 to PH2–4: $P = .005$, cPH1 to PH5–6: $P < .001$, dPH2–4 to PH5–6: $P = .001$, ePH2–4 to PH5–6: $P < .01$, fPH2–4 to PH5–6: $P < .05$, gPH1 to PH2–4: $P < .04$, hPH2–4 to PH5–6: $P < .001$. To convert from ng/L to nmol/L, multiply 17-OHP by 0.00303, AE by 0.00349, DHEA by 0.00347 and TE by 0.00347; to convert from µg/L to nmol/L, multiply 17-OHP by 3.03; AE by 3.49 and TE by 3.47; to convert from mg/L to µmol/L, multiply DHEAS by 2.725.

From Krebs A, Dickhuth K, Mumm R, Stier B, Doerfer J, Grueninger D, Wurm M, Brichta C, Schwab KO. Evaluating the four most important salivary sex steroids during male puberty: testosterone best characterizes pubertal development. J Pediatr Endocrinol Metab. 2019;32(3):287–94. Table 2 p. 289.

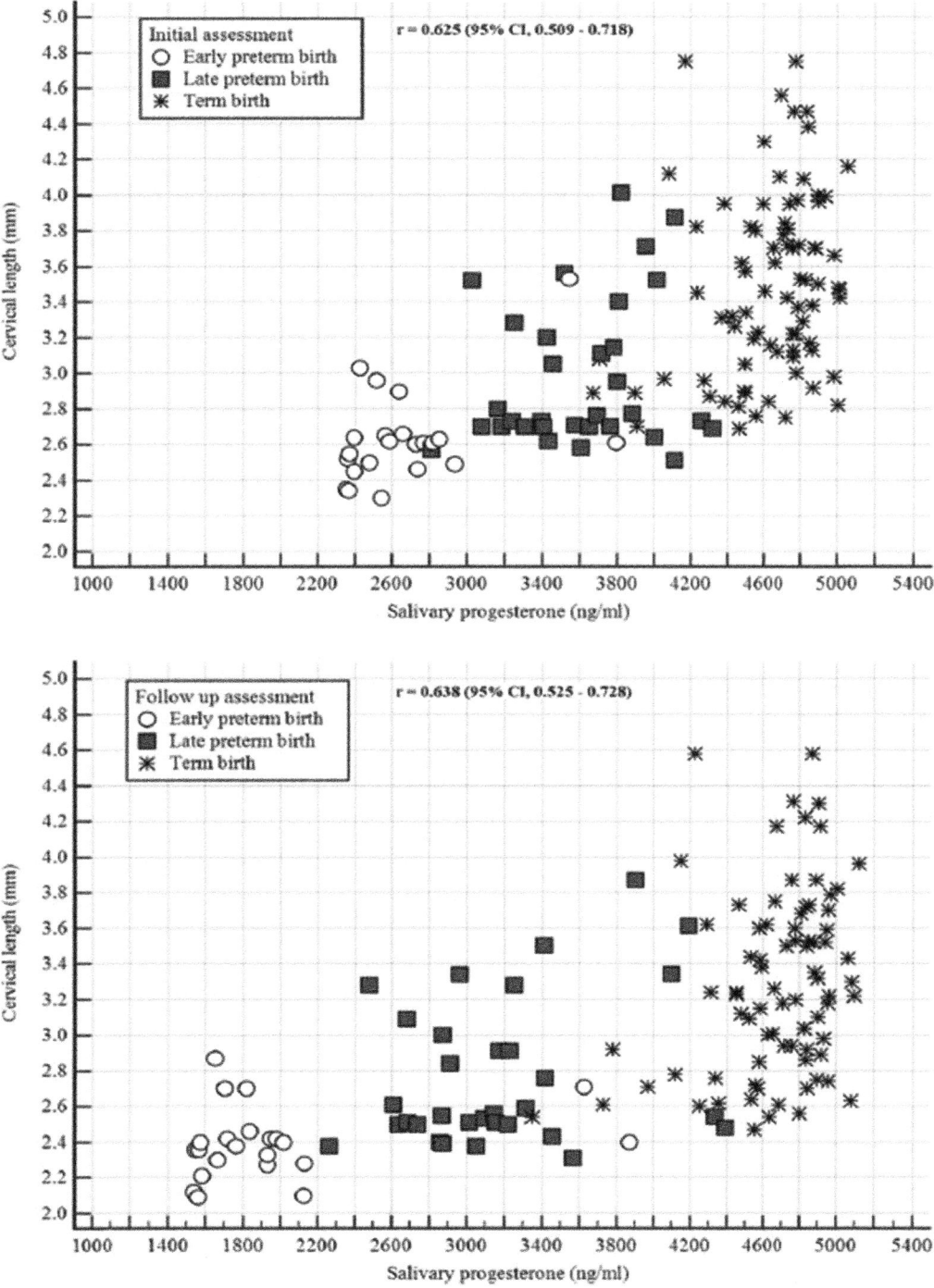

FIG. 2.6.46 Scatter plots illustrating the correlation between the initial (upper) and follow up (lower) measurements of cervical length and salivary progesterone. *(From Abuelghar WM, Ellaithy MI, Swidan KH, Allam IS, Haggag HM. Prediction of spontaneous preterm birth: salivary progesterone assay and transvaginal cervical length assessment after 24 weeks of gestation, another critical window of opportunity. J Matern Fetal Neonatal Med 2019;32(22):3847–58. Fig. 4 p. 3855.)*

from single steroid measurements to profiles of 7–10 steroids including estrogens and aldosterone, extending the clinical applicability of saliva analysis. Results for several steroids were reported for saliva and other tissues (Table.2.6.20).

The diagnostic value of salivary progesterone and 17-OHP is compromised by fluctuations of the steroids in the menstrual cycle (Liu et al., 2018) (Fig. 2.6.47).

2.6.5.2 Breast milk

Typical basal concentrations for steroids in milk are summarized in Table 2.6.21(Gomez-Gomez et al., 2020; Jurgens et al., 2019; Svensson et al., 2019; Liu et al., 2018; Gao et al., 2015).

Considering the amount of milk consumed by humans there is little literature on the steroid content regardless of source. Lactating mothers may get concerned about transfer of drugs in breast milk (Greenberger et al., 1993). Prednisolone however even when taken at 40mg per day has been judged to be a minimal problem for the baby with 0.025% of the steroid passed into the milk (Fig. 2.6.48).

Synthetic steroids such as oral contraceptives can be measured in breast milk for days after a single injection of steroid such as norethindrone. Sahlberg and Axelson (1986) used ion exchange chromatography with modified Sephadex (TEAP, no longer used since availability of SPE) to separate the free and conjugated steroids. For that year, an impressive number of steroids (c 50 steroids as early steroid profile) in different conjugate forms were quantified by GC-MS, and the majority of the steroids were sulfated. In order to distinguish isobaric cortisol/cortisone metabolites, tetrahydrocortisol isomers were detected as ammonia adducts ($[M+NH_4^+]$), whereas cortisol, cortisone and 20-dihydrocortisol metabolites were detected as protonated species ($[M+H+]$) (Gomez-Gomez et al., 2020). A comparison of the ranges for steroid concentrations in milk with saliva reveals higher E to F ratios (Table 2.6.21). Ratios are also used, for example F to E can be up to 5 in breast milk, 8 in amniotic fluid but 0.5 in saliva because of HSD11B1 activity in saliva; F to prog is near 1 in amniotic fluid, 0.25 in saliva and 4 in breast milk. These and other comparisons suggest different transport systems or local production. Some concentrations in breast milk are near concentrations in plasma suggesting diffusion of the steroid into the milk. Cortisone concentrations in

TABLE 2.6.20 Summary of steroid concentrations in saliva.

Saliva	Present study range		Reported range (Polet et al., 2018)	
Analyte	Female	Male	Female	Male
F	0.12–1.8	0.11–1542	0.84–2.5 (n=387)	0.018–2.3 (n=453)
20aDHF	0.041–0.19	0.053–0.068	–	–
20bDHF	0.1–0.69	0.10–0.48	–	–
5aTHF	0.042–0.35	0.047–0.60	–	–
5bTHF	0.046–0.15	0.047–0.12	–	–
E	0.69–1.9	0.54–5.0	4.4–9.3 (n=387)	0.013–9.4 (n=453)
20aDHE	0.29–1.2	0.34–1.7	–	–
20bDHE	0.35–1.5	0.314–1.6	–	–
5bTHE	0.38–2.0	0.39–1.7	–	–
aHHE	0.90–19	1.1–4.5	–	–
bHHE	0.98–27	0.80–5.2	–	–
A	0.48–1.3	0.5–1.8	–	–
AED	0.014–0.10	0.012–0.060	0.025–0.062 (n=387)	0.037–0.069 (n=439)
T	0.020–0.14	0.022–1.4	0.0014–0.028 (n=91)	0.026–76.3 (n=112)
Prog	0.045–0.62	0.053–0.44	0.014–0.049 (n=8)	–

Concentration ranges (n=30) obtained in this study for selected analytes compared with ranges previously reported in the literature.
From Gomez-Gomez A, Miranda J, Feixas G, et al. Determination of the steroid profile in alternative matrices by liquid chromatography tandem mass spectrometry. J Steroid Biochem Mol Biol. 2020;197:105520. Table 3 p. 8.

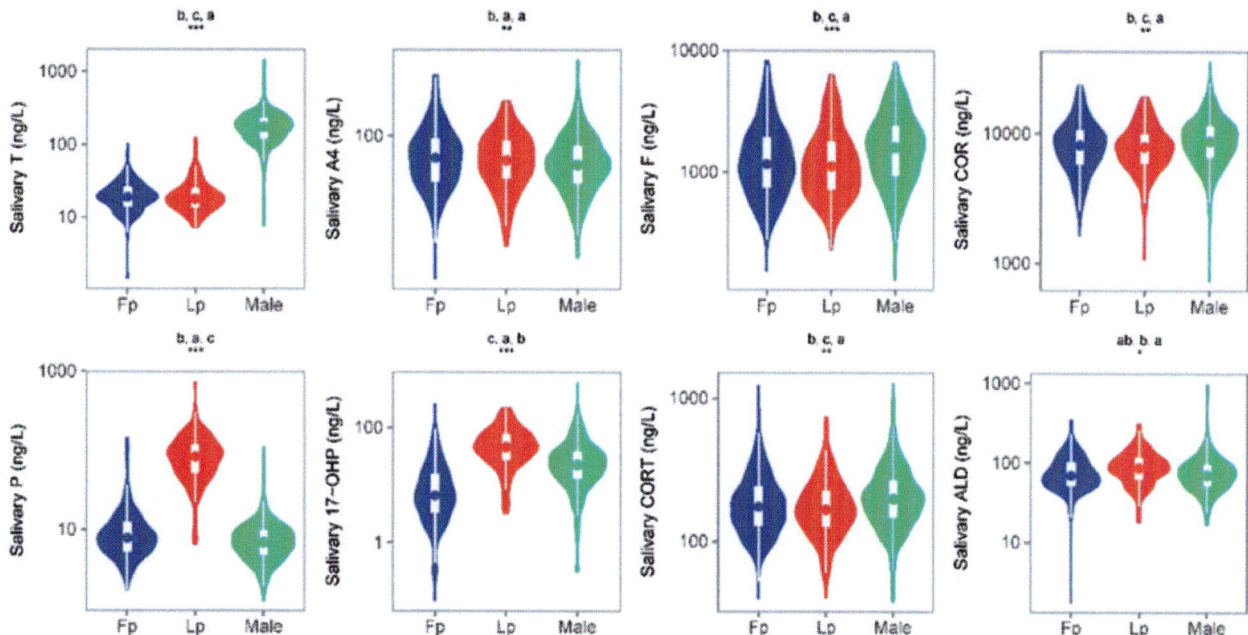

FIG. 2.6.47 Saliva concentrations of 10 salivary steroid hormones between women in the follicular phase (Fp, $n = 310$), women in the luteal phase (Lp, $n = 153$) and men (Male, $n = 421$) (one-way analysis of variance; *$P < .05$, **$P < .01$, ***$P < .001$). The letters above each plot represent the comparison result. For all groups with the same letter, the difference between the means was not statistically significant (Student's t-test, $P < .05$). The groups that do not share the same letters are significantly different. *T*, testosterone; *A4*, androstenedione; *F*, cortisol; *COR*, cortisone; *P*, progesterone; *17-OHP*, 17a-hydroxyprogesterone; *CORT*, corticosterone; *ALD*, aldosterone. *(From Liu J, Qiu X, Wang D, et al. Quantification of 10 steroid hormones in human saliva from Chinese adult volunteers. J Int Med Res 2018;46(4):1414–27. Fig. 1 p. 1421.)*

TABLE 2.6.21 Concentration ranges for selected analytes in breast milk (n30) concentration ranges ($n = 30$) obtained in this study compared with ranges previously the literature reported in the literature.

BM	Present study range[a]		Reported range[a]
Analyte	Colostrum	Mature milk	
F	164–5500	131–6070	144–1400 ($n = 83$)
5aTHF	12–635	34–3385	–
5bTHF	94–1554	102–2578	–
20aDHF	10.1–411	6.7–301	–
20bDHF	76–2568	83–1168	–
60HF	38–450	44–381	–
E	1549–21,200	850–17,400	1100–2100 ($n = 53$)
20aDHE	177–7163	154–2195	–
20bDHE	116–2288	101–1092	–
60HE	30–233	16–123	–
A	5–49	1.3–114	–
Prog	2730–9340	480–6790	1700–4000 ($n = 35$)

[a]*Data from Lu M, Xiao H, Li K, Jiang J, Wu K, Li D. Concentrations of estrogen and progesterone in breast milk and their relationship with the mother's diet. Food Funct 2017;8(9):3306–10; van der Voorn B, de Waard M, van Goudoever JB, Rotteveel J, Heijboer AC, Finken MJ. Breast-milk cortisol and cortisone concentrations follow the diurnal rhythm of maternal hypothalamus-pituitary-adrenal axis activity. J Nutr 2016;146(11):2174–79. https://doi.org/10.3945/jn.116.236349.*
From Gomez-Gomez A, Miranda J, Feixas G, et al. Determination of the steroid profile in alternative matrices by liquid chromatography tandem mass spectrometry. J Steroid Biochem Mol Biol 2020;197:105520. From Table 3 p. 8.

FIG. 2.6.48 Analysis of prednisolone concentrations by immunoassay in plasma and breast milk. Solid lines represent the least-squares fit of the plasma concentrations shown by the solid circles. Data shown by the open circles were excluded from this pharmacokinetic analysis. Prednisolone concentrations in milk and milk volume (in milliliters) are shown by the triangles and associated numbers. Estimates of unbound serum prednisolone concentrations are shown by the broken lines. The shaded area represents the range of unbound prednisolone serum concentrations expected if serum transcortin binding capacity is allowed to vary by ±1 SD from the mean value that was used to construct the broken line. *(From Greenberger PA, Odeh YK, Frederiksen MC, Atkinson AJ Jr. Pharmacokinetics of prednisolone transfer to breast milk. Clin Pharmacol Ther 1993;53(3):324–8. Fig. 1 p. 326.)*

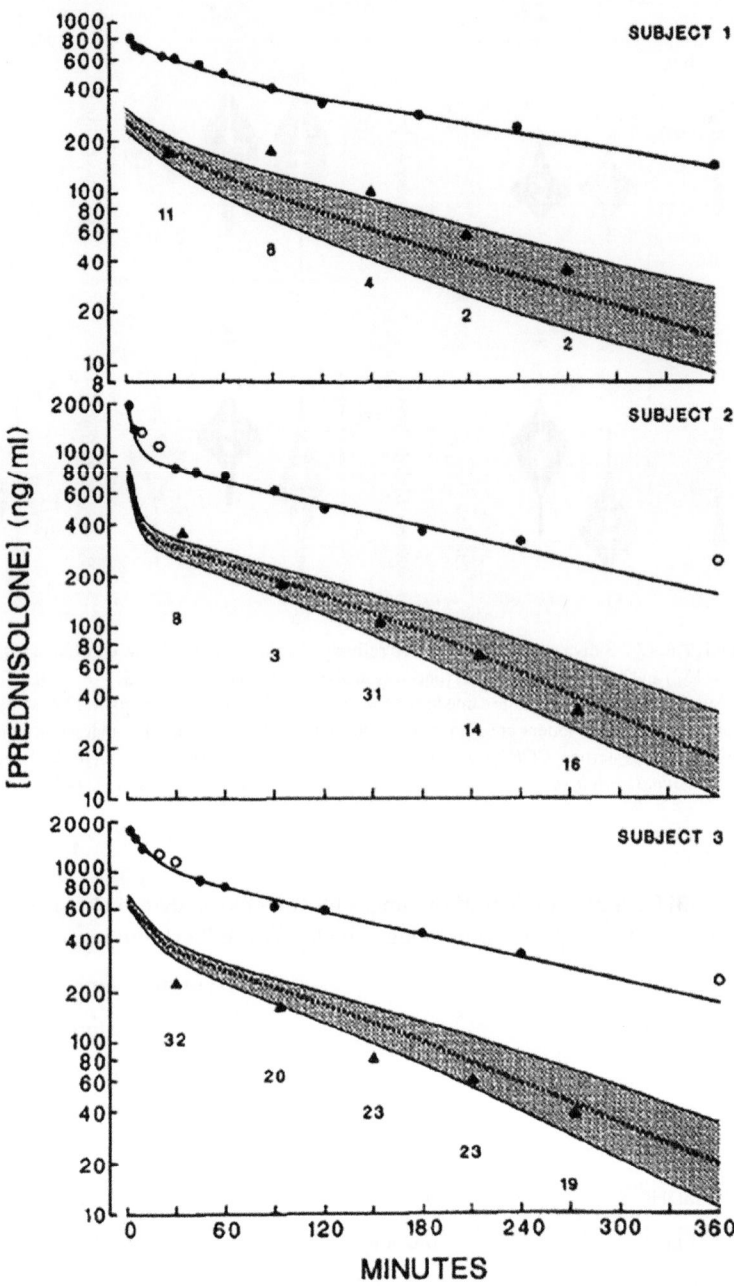

breast milk are higher than cortisol and follow a diurnal rhythm (Fig. 2.6.49) (Pundir et al., 2017). The concentrations of cortisol were lower in milk of mothers that delivered preterm (van der Voorn et al., 2016) (Fig. 2.6.50).

2.6.5.3 Seminal plasma

There seems to be an increase in male infertility across the globe so analysis of steroids in seminal plasma has become of interest. The spermatozoa account for up to 5% of seminal fluid, the rest is seminal plasma containing many organic and some inorganic compounds that nourish sperm development, maturation and survival in the female reproductive tract. Testosterone is produced in the Leydig cells and the seminal plasma reflects activities of the accessory sexual organs namely epididymis, seminal vesicles, prostate and bulbourethral organs. Progesterone is also found in seminal plasma and this may act in nonclassical way on membranes with cyclic AMP production and release of proteolytic enzymes that assist capacitating of spermatozoa and

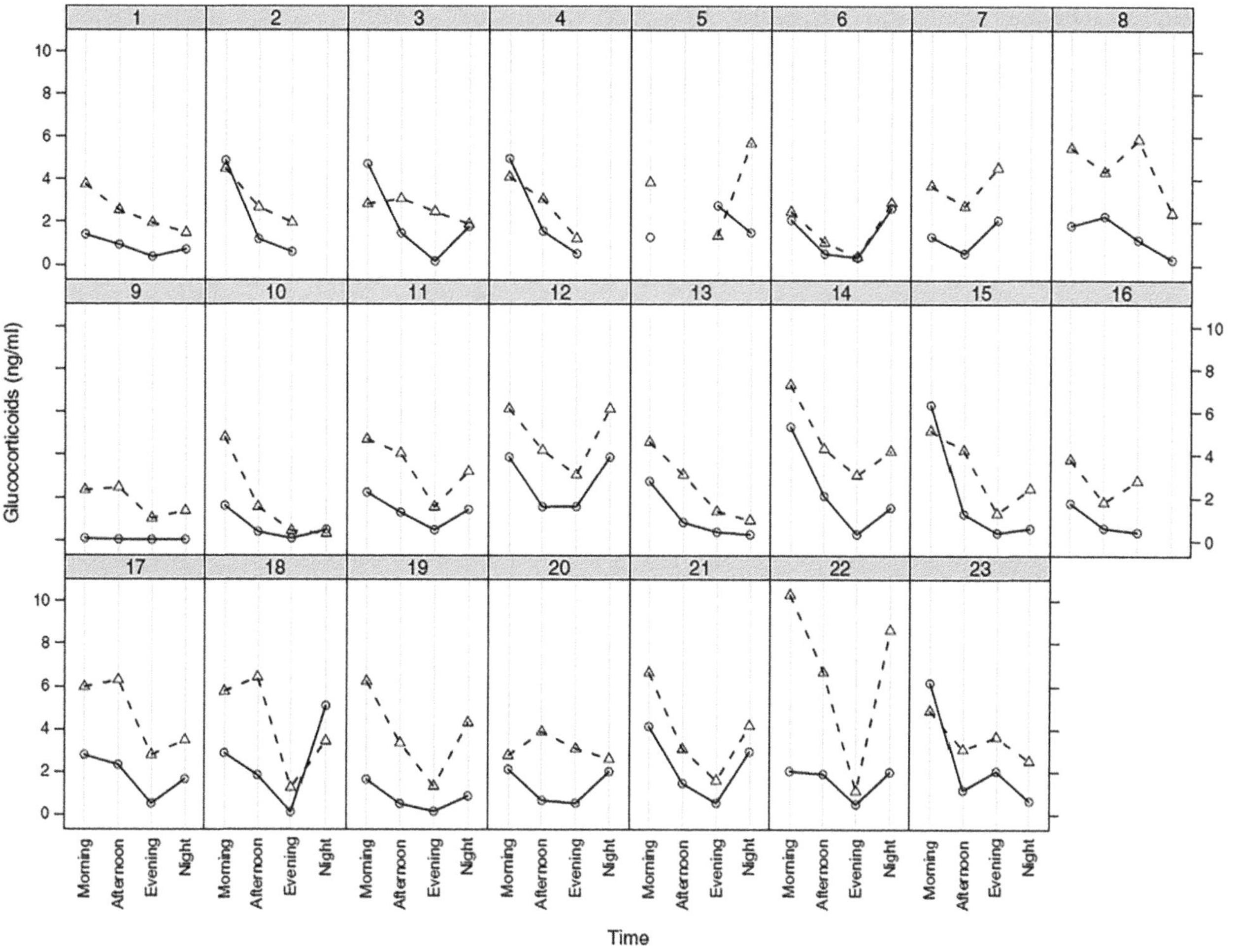

FIG. 2.6.49 Glucocorticoids (cortisol and cortisone) in milk over 24 h in individual mothers ($n = 23$). Each box of the lattice plot indicates a single mother milk glucocorticoid profile with symbol (o) representing cortisol and (Δ) representing cortisone. Each solid line indicates cortisol and dashed line cortisone. Time points are categorized as morning (0401–1000h), afternoon (1001–1600h), evening (1601–2200h) and night (2201–0400h). *(From Pundir S, Wall CR, Mitchell CJ, Thorstensen EB, Lai CT, Geddes DT, Cameron-Smith D. Variation of human milk glucocorticoids over 24 hour period. J Mammary Gland Biol Neoplasia 2017;22(1):85–92. Fig. 1 p. 88.)*

permeation of the ovum. DHEA and 7-oxygenated metabolites in seminal plasma are immunoprotective although 16-hydroxy-DHEA can counteract the effects. Many enzymes in the reproductive tract can act on steroids (Fig. 2.6.51).

The seminal plasma contains basic amines (putrescine and spermidine for example) to neutralize acid in the vaginal environment, also immunomodulatory molecules (prostaglandins, leukotrienes, cytokines and glucocorticoids), nutritional compounds (fructose, citric acid, lactic acid and amino acids), enzymes (proteases and phosphatases), phospholipids, vitamins, antioxidants, bile acids and inorganic ions. Peptide hormones (LH, FSH, hCG, Inhibin B, melatonin, noradrenalin, oxytocin, PTH, prolactin and vasopressin) have been quantified in seminal plasma. Samples are heated at 37 degrees C for liquefaction prior to centrifugation to remove sperm. LLE, SLE and SPE have been used for extraction and part purification of

steroids in the sample. The concentrations of many steroids have been determined but because many different analytical techniques have been used (RIA, ELISA, HPLC-UV, GC-MS, and LC-MS/MS) the wide variations in results and different units make it difficult to put accurate concentrations to the steroids (see Johannsen et al., 2020 for review of steroids and reproductive protein hormones). A brief list from several sources (Table 2.6.22) illustrates some variations (Zufferey et al., 2018; Vitku et al., 2017; Zhao et al., 2004).

A profile of around 50 steroids in seminal plasma by the preferred LC-high resolution MS has measured around 40 of the steroids (Olesti et al., 2020). The extraction and identification of steroids has been optimized, with SPE superior to SLE, but further work is needed since androstenedione-d_7 was the only internal standard and the LOD for steroids ranged from 5 to 400 pg/mL. Samples after vasectomy were

FIG. 2.6.50 Boxplots showing medians and ranges of concentrations of milk cortisol (A) and milk cortisone (B) over 24 h during week 4 postpartum. The circles represent outliers and are labeled with their anonymized study number. $n = 10$ mothers and 92 milk samples. *(From van der Voorn B, de Waard M, van Goudoever JB, Rotteveel J, Heijboer AC, Finken MJ. Breast-milk Cortisol and cortisone concentrations follow the diurnal rhythm of maternal hypothalamus-pituitary-adrenal Axis activity. J Nutr 2016;146(11):2174–79. Fig. 2 p. 2177.)*

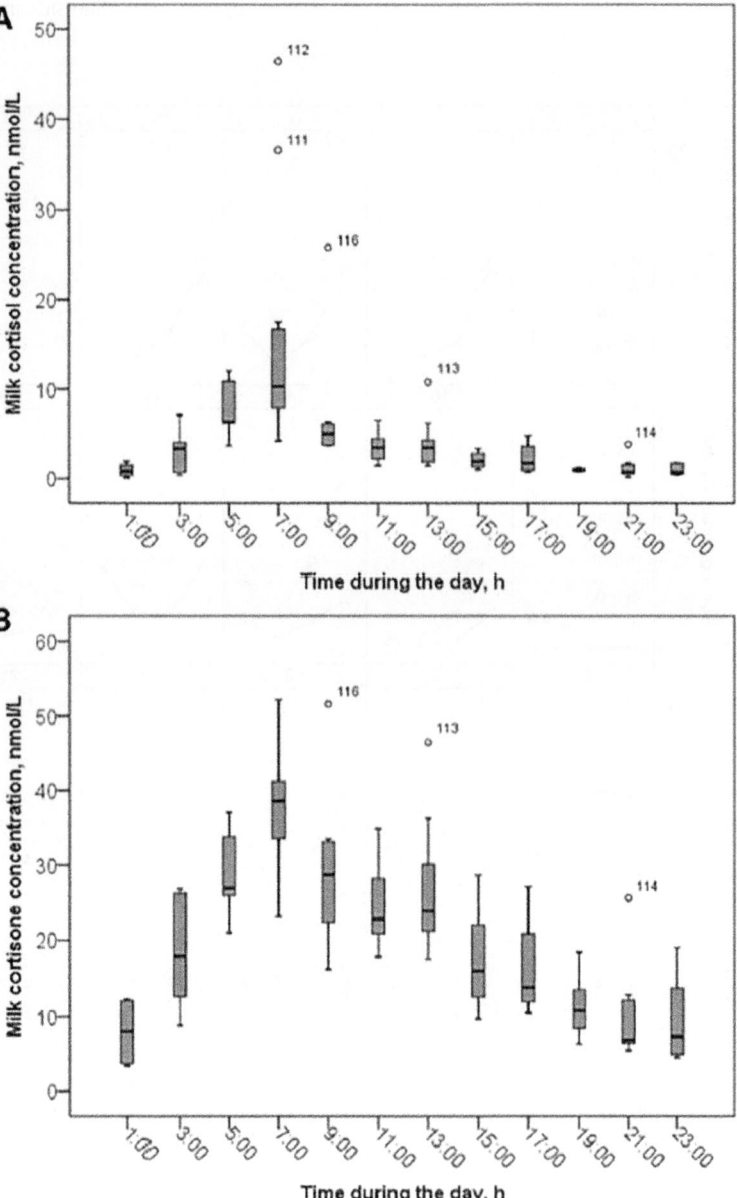

FIG. 2.6.51 Schematic of male urogenital system showing the expression of major enzymes in steroid synthesis in each reproductive organ—testes, prostate, seminal vesicle and Epididymis. 3β-HSD (3β-hydroxysteroid dehydrogenase) transforms dehydroepiandrosterone (from periphery) to androstendione, 17β-HSD (17β-hydroxysteroid dehydrogenase) converts androstendione to testosterone), 5α-reductase transforms testosterone to dihydrotestosterone), aromatase converts testosterone to estradiol, sulfotransferase catalyzes the steroid sulfatation and UDP-glucuronosyltransferase (UGT) catalyzes the steroid glucuronidation. *UGT in epididymis was reported till this time in monkeys only. Enzymes reproductive tract. *(From Vitku J, Kolatorova L, Hampl R. Occurrence and reproductive roles of hormones in seminal plasma. Basic Clin Androl 2017;27:19. Fig. 1 p. 4.)*

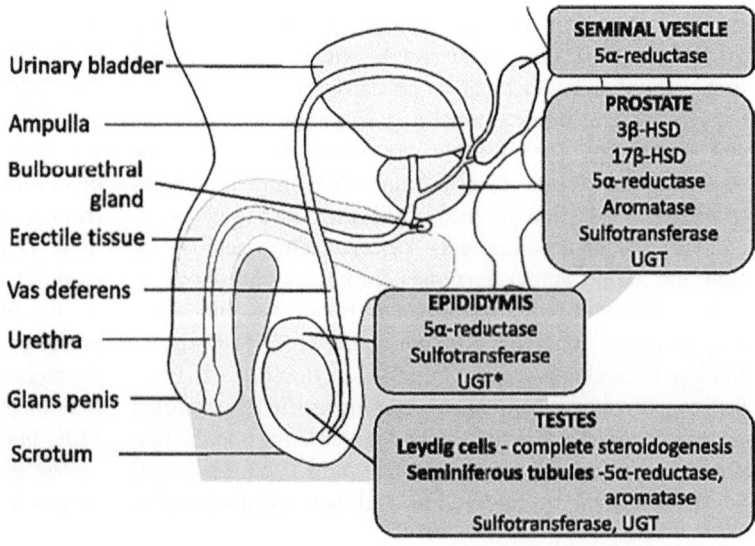

TABLE 2.6.22 Comparison of published concentrations of steroids in seminal plasma using mass spectrometric techniques.

	Olesti et al., 2020 Measured ng/mL LC-HRMS	Olesti calculated nmol/L	Zhao et al., 2004 Measured Ng/mL LC-MS/MS	Zhao et al., 2004 Calculated nmol/L	Vitku et al., 2017 Calculated nmol/L	Vitku et al., 2017 Measured ng/mL Various
11-OH andro	100					
11-Keto aetio	100					
16a,17b-Estriol	200				0.11–0.19	
Androstenedione					0.3–2	0.09, −0.58
17b-Estradiol	150		8–29		0.2–0.6	0.06–0.16
20b-Cortolone	450					
3a,5b-THE	350					
DHT			8–26		1.1–1.9	0.29–0.56
5a-Androstane-3b,7a,16b-triol	300					
5b-androstan-3a,17b-diol			3–26			
6b-Hydroxy T	500					
Cortisol					59–176	21–64
Androst-5-ene-3b,16a,17b-triol	150					
Pregnenolone	400				0.54–0.73	
Testosterone	50		10–100		0.3–4.6	0.09–1.33

From Olesti E, Garcia A, Rahban R, Rossier MF, Boccard J, Nef S, González-Ruiz V, Rudaz S. Steroid profile analysis by LC-HRMS in human seminal fluid. J Chromatogr B Analyt Technol Biomed Life Sci 2020;1136:121929.; Zhao M, Baker SD, Yan X, Zhao Y, Wright WW, Zirkin BR, Jarow JP. Simultaneous determination of steroid composition of human testicular fluid using liquid chromatography tandem mass spectrometry. Steroids 2004;69(11–12):721–6.; Vitku J, Kolatorova L, Hampl R. Occurrence and reproductive roles of hormones in seminal plasma. Basic Clin Androl 2017;27:19.

used for comparison and assessment of matrix effects. Steroid concentrations in seminal plasma do not correlate with blood concentrations so in addition to the effects of enzymes in the tract there may be transport issues that have not yet been identified. Few estrogens or conjugates steroids have been measured and the processing of androgen sulfates has not been considered. The analysis of steroids in seminal plasma will likely be extended to include measurements of endocrine disrupting chemicals that are considered to be involved in infertility.

2.6.5.4 Ovarian follicular fluid

In women of fertile age, ovarian follicles are the main source of biosynthesis of estrogens and progestins. Local steroid production in ovarian follicles is controlled by enzymes expressed in the ovaries, which regulate conversion between the steroids and availability of precursors, with the adrenal cortex serving as a source of androgen biosynthesis. Human ovarian follicular fluid (hFF) is produced during follicular growth and provides the microenvironment for oocyte development. Main origins of hFF are circulating blood, which diffuses through the follicular wall into an antrum of a follicle, and follicular secretions, which are predominantly products of metabolic processes within an oocyte. The steroids in FF provide information on the status of health of the oocyte and potentially serve as diagnostic biomarkers. Follicular steroids are secreted by granulosa and theca cells under the control of gonadotropins; the hormonal microenvironment in the follicles affects the follicular development and the oocyte viability. A higher concentration of estradiol (E2) in FF is associated with healthy mature follicles containing oocytes capable of meiosis, while a higher concentration of androgens is indicative of atretic changes in the follicles.

Exogenous gonadotropins are used in programmes for in-vitro fertilization (IVF) to promote growth of follicles and stimulate growth of multiple follicles simultaneously with the aim of making several oocytes available for fertilization from one invasive procedure. A number of studies have focused on analyzing steroids in FF samples of women undergoing IVF treatment, and assessing the relationship between the steroid concentrations and the follicular development. The published information on the association of FF steroid concentrations with IVF outcomes is conflicting. Lower concentrations of P4, higher E2 and higher androstendione (A4) in FF have been reported in association with pregnancy following IVF treatment, while other studies have demonstrated lower E2, higher P4 or no association. In studies pre 2000 measurements of steroids in FF were performed using IAs.

Dehennin et al. was the first to use mass spectrometry-based methods to analyze steroids in stimulated follicular (sFF) (Dehennin et al., 1987). Dehennin identified deoxycorticosterone, 21-hydroxypregnenolone and 11DOC and suggested existence of a positive gradient between FF and peripheral plasma. The early studies were based on analysis of a few steroids. Kushnir et al. (2012, 2016) % compared concentrations of 17 steroids determined by LC-MS/MS methods and the concentrations determined by LC-MS/MS were typically lower than those determined by IAs. This was especially true for testosterone, suggesting cross-reactivity in the IAs when used for analysis of **stimulated follicular fluid (sFF)**. The potential for cross reactivity and overestimation of concentrations was high when IAs intended for analysis of serum and plasma samples are used for analyzing sFF samples.

Kushnir et al. (2016) determined steroid profiles in sFF in relation to prognosis of IVF treatment (live birth or no pregnancy) with the aim of finding biomarkers that could allow selecting oocytes more likely to result in a successful IVF outcome. In sFF samples of the live birth (LB) group, lower concentrations of Progesterone and pregnenolone, lower ratios of concentrations of Te/DHEA and A4/DHEA and higher ratios of concentrations of E2/Te and E1/A4 were found (Table 2.6.23).

The results suggested increased biosynthesis of estrogens in association with LB. Oocytes that did not result in pregnancy had follicle concentrations of estradiol below

TABLE 2.6.23 Medians and central 80% distribution of concentrations (ng/mL) of steroids in follicluar fluid for groups based on the IVF outcome.

AF	Present study range[a]		Reported range[a,b]
Analyte	Vaginal delivery	Cesarean delivery	
F	12–119	17–87	10–40 (n = 20)
5aTHF	4.6–45	4.4–31	–
5bTHF	4.7–32	4.2–27	–
20aDHF	1.2–16	1.4–17	–
20bDHF	2.1–17	2.2–13	–
5bDHF	0.023–0.132	0.021–0.15	–
6bOHF	9.8–54	4.8–43	–
E	6.4–80	11–61	5–20 (n = 20)
5aTHE	0.033–0.15	0.034–0.090	–
5bTHE	34–112	29–119	–
20aDHE	1.8–17	2.7–30	–
20bDHE	2.1–17	2.1–13	–
6bOHE	1.9–13	1.5–10	–
bHHE	4.4–20	4.9–68	–
B	0.87–9.9	0.84–4.1	1.3 (n = 10)
5bTHB	3.1–140	3.3–114	–
A	0.66–3.9	0.30–3.1	1.3 (n = 10)
S	0.41–3.1	0.29–1.4	–

TABLE 2.6.23 Medians and central 80% distribution of concentrations (ng/mL) of steroids in follicluar fluid for groups based on the IVF outcome—cont'd

AF	Present study range[a]		Reported range[a,b]
Analyte	Vaginal delivery	Cesarean delivery	
T	0.18–3.2	0.19–4.4	0.03–0.29 (n=75)
AED	0.12–1.8	0.35–2.4	–
Andros	5.7–27	6.4–19	–
Etio	1.7–87	1.5–101	–
PT	38–77	51.2–98	–
17OHP	0.4–0.84	0.5–2.7	–
17HP	7.0–29	4.9–23	–
Prog	3.3–136	12–95	–
DOC	0.12–1.9	0.054–2.7	–
Preg	2.8–23	3.4–8.2	–

Comparison between the groups was performed using the nonparametric Wilcoxon rank sum test.
Concentration ranges (n = 30) obtained in this study for selected analytes compared with ranges previously reported in the literature.

From Kushnir MM, Naessén T, Wanggren K, Hreinsson J, Rockwood AL, Meikle AW, Bergquist J. Exploratory study of the association of steroid profiles in stimulated ovarian follicular fluid with outcomes of IVF treatment. J Steroid Biochem Mol Biol 2016;162:126–33. Part of Table 2 p. 129.

the distribution, and androstenedione above the distribution observed in the LB group. There were no significant associations between the follicular volumes and IVF outcomes, while trends of lower concentrations of Pr, and higher concentrations of P4 and E3 with increasing follicular volume were observed. The results were processed as associations and did not determine cause and effect of the observed differences; however, the results agreed with earlier observations and had biological plausibility. The study has some strengths and limitations. In all participants, sFF was sampled from individual follicles, when the outcomes corresponding to each follicle were not known. Steroids were analyzed using high specificity LC-MS/MS methods. One limitation of this study was that the enrolled patients had different underlying causes of the infertility and two ovarian stimulation protocols were utilized. Kirsty Walters in 2019 found none of nine steroids predicted IVF outcomes.

Follicular fluid steroids have been determined in patients with polycystic ovary syndrome to see if insulin resistance contributed to ovarian dysfunction in steroidogenesis. Lewicka et al. (2003) reported that lower concentrations of cortisone in FF and increased cortisol/cortisone ratio have been associated with pregnancy, while Michael et al. observed association with lower concentrations of cortisol and cortisone. Local generation of cortisol by granulosa cells was consistent with elevation of HSD11B1 mRNA (Zhu et al., 2016). Due to the discrepancy in the results from these and other studies, association of concentrations of steroids in FF with IVF outcome remains unclear.

The presence of estradiol-FAE in ovarian follicular fluid remains controversial and may be method dependent (Vihma et al., 2011). Estrone fatty acyl esters are found at low levels in the circulation (c 1 pmol/L) and in the periphery where they can be substrates for release of free steroid and act as a reservoir. The methodology so far has depended on measuring the free steroid after saponification in potassium hydroxide (1 mol/L). Most of the data in the literature is for RIA detection, some papers use GC-MS and some LC-MS/MS. The intact esters have not been measured by RIA, a mass spectrum from LC-MS/MS of estrone-3-oleate have been published.

Inhibin-B has been measured in follicular fluid as a factor in dominant oocyte selection. This needs to be studied along with isoforms of FSH since paracrine and endocrine effects may be involved. Other proteins have been identified in follicular fluid (Kushnir et al., 2012). Apelin may increase IGF-1 induced steroidogenesis (Roche et al., 2016) via a receptor in granulosa cells. Adropin is important in glucose homeostasis and may be involved in the insulin resistance see in polycystic ovary syndrome (Bousmpoula et al., 2018) where it is suppressed by increased TNFα. Better characterization of the hFF metabolome will not only expand the knowledge of the folliculogenesis and advance understanding of roles of proteins role during oocyte development but potentially contribute to the development of new methods of assessment of oocytes for viability of an IVF treatment. This will be discussed in more detail in the clinical section on IVF (Chapter 3.3.2). There is more research needed, on steroids and proteins, to clarify outstanding issues within the ovary. Follicular fluid will usually be from several oocytes after hyperstimulation and will have a range of properties so animal studies will be needed to look at natural development of ovarian function.

2.6.5.5 Amniotic fluid

Amniotic fluid is partly fetal urine and is in fact swallowed by the fetus. The data on steroid concentrations derive from samples collected during early labor or at abortion, otherwise collection by transabdominal amniocentesis when genetic studies were being performed. Sample availability is limited to periods around mid-gestation and near term. Activity of the hypothalamus-pituitary-adrenal axis in the fetus can be detected as early as 8th–12th week of gestation. Corticotrophin releasing hormone (CRH) is produced by fetal hypothalamus and by the placenta. CRH stimulates the pituitary to secrete adrenocorticotrophic hormone (ACTH) which in turn stimulates fetal adrenals to synthesize dehydroepiandrosterone sulfate (DHEAS) from cholesterol. DHEAS from fetal adrenals is thereafter hydroxylated at position 16α in fetal liver by cytochrome P450 CYP3A7. DHEAS and 16OH-DHEAS are main precursors for estrogen production in the placenta. The collections of fetal blood or the amniotic fluid are highly risky in utero. Most published studies on steroids in AF have been carried out by immunoassays. Due to drawbacks of immunoassays such as cross-reactivity, less efficiency and matrix interferences, GC-MS and LC-MS nowadays are the main tools for the qualitative and quantitative analysis of steroids. Until recently only the active steroid forms which pass through cell membrane to interact with nuclear receptors have been considered. Sulfated steroids are water-soluble excretion products that are now known to enter cells by specific uptake carriers and modulate the intracellular milieu following local metabolism.

The first attempt to use mass spectrometry in determination of AF steroids was made by Homoki et al. (1983). In that study, AF steroids were measured, after LH-20 fraction and enzymatic hydrolysis in pooled AF, by gas chromatography followed by mass spectrometric identification (full scan mode). Wudy et al. (1999) used stable isotope dilution/GC-MS to establish reference data on six unconjugated steroids in AF at mid-gestation. Unconjugated steroids were eluted from SepPak C18 column with chloroform and sulfated steroids were eluted by methanol. Apart from DHEAS, no other steroid sulfates had previously been quantified in AF by MS technology. Samples were taken from normal and congenital adrenal hyperplasia (CAH) affected fetuses. Steroid analysis of AF permitted the prenatal diagnosis of 21-hydroxylase deficiency but better methods are available today. For progesterone, the mean

concentrations were similar to data produced by various immunoassays. Hill et al. (2010) reported data for 69 steroids by GC-MS in order to have a near complete metabolome of amniotic fluid steroids. Samples from 12 women were at normal labor and 38 women in early labor usually due to infection.

A larger group of subjects and patients were reported in a subsequent study from the Wudy laboratory (Wang et al., 2019a, b) with more steroids measured. The study was based on targeted steroid measurements with steroids typical for urine so many steroid hormones were not included. AF levels of testosterone and 4-androstenedione were higher in male than female fetuses at mid-gestation. Derivatization was needed to improve the sensitivity of estrogens. Dansyl chloride has been used as a derivatization agent in the determination of estrogens. Sodium bicarbonate usually provided an alkaline environment for the reaction. A 0.1 M $NaHCO_3$ was actually used because a less acidic reconstitution solution was required later for neutralization to facilitate ion detection in positive mode. The E1 and the E2 derivatives only differed by 2 Da (m/z 504.2 and 506.1), and they formed the same product ion in the collision cell of the LC-MS/MS. The product ion m/z 171.0 is formed from all estrogen-dansyl derivatives by cleavage of the S—C bond of the sulfonyl group. Baseline separation was achieved using an Accucore Polar Premium column, instead of a C18 column. The concentrations of steroids in AF of mid-gestation were compared with the available literature as shown in Table 2.6.24).

A study by Fahlbusch et al. (2015) using LC-MS/MS produced values that scattered more widely and in several cases the mean values were 2–3 times higher than reported from the Wudy laboratory using LC-MS/MS (Wang et al., 2019b). Fahlbusch collected amniotic fluid at the time of amniocentesis for genetic studies. The levels of 17OHProg were slightly lower than results from previous LC-MS/MS and GC-MS studies. The mean concentrations of PregS were by 50% lower than those obtained by immunoassay (Forest et al., 1980), attributed to the lack of specificity of the immunoassay even though extraction and chromatographic purification had been used before the immunoassay. 17OHPregS was higher than the value reported in 1980 by RIA although no corresponding internal standard for 17OHPregS was available for the LC-MS/MS study. Fahlbusch reported results for many cortisol metabolites, there were few other reported ranges for comparison of the data.

The concentrations of T showed a significant sex difference of comparable magnitude to levels found in other studies. Epitestosterone (eT) and T are converted from 4 A by 17α-HSD and 17β-HSD, respectively. The levels of eTS to be approximately 6 times higher than the levels of TS. The reaction from 4 A to T is reversible while the conversion of 4 A to eT is irreversible. DHEAS, the precursor for E1 and E2, was comparable with the result from

TABLE 2.6.24 Concentration ranges for selected analytes in amniotic fluid (n 30).

	Live birth (LB)		No pregnancy (NP)		Wilcoxon test P-value	Ratio of the medians NP/LB
	Median	10th–90th percentile	Median	10th–90th percentile		
Pregnenolone (Pr)	194	82–448	293	98–635	0.10[a]	1.51
17OH Pregnenolone (17OH Pr)	2.8	1.4–36	2.9	2.9–9.0	0.41	1.04
Progesterone (P4)	2380	1230–5940	3500	1940–6720	0.67	1.47
17OH Progesterone (17OH P4)	357	160–965	377	118–802	0.71	1.06
11 Deoxycortisol (11 DC)	3.5	0.88–8.74	2.75	0.83–6.61	0.82	0.79
Cortisol (F)	85	37–120	76	29–114	0.39	0.89
Cortisone (E)	20	8.6–31.2	17.5	9.3–25.4	0.84	0.88
DHEA	0.58	0.41–11.2	0.51	0.29–0.91	0.49	0.88
Androstenedione (A4)	1.6	0.9–3.5	2.2	0.74–8.6	0.16	1.29
Testosterone (Te)	0.04	0.02–0.23	0.07	0.04–0.23	0.14	1.75
Estrone (E1)	14	7–50.4	14	4.8–37.3	0.72	1.00
Estradiol (E2)	191	88–324	208	62–306	0.67	1.09
Estriol (E3)	4.8	2.2–12.4	5.95	1.9–9.9	0.62	1.24
Allopregnanolone	5.5	1.8–7.8	6.25	3.1–13	0.27	1.14

Concentration ranges (n = 30) obtained in this study for selected analytes compared with ranges previously reported in the literature.

From Gomez-Gomez A, Miranda J, Feixas G, et al. Determination of the steroid profile in alternative matrices by liquid chromatography tandem mass spectrometry. J Steroid Biochem Mol Biol 2020;197:105520. Table 3 p. 8.

GC analysis. 16OH-DHEAS, the most important precursor for E3, showed the highest level of all sulfated steroids measured. DHTS and AnDiolS were both at the lower concentrations. The concentrations of epiAnS were considerably lower than the concentrations of AnS. In most samples, epiAnS was below LOQ.

E3 was the dominant estrogen at mid-gestation while E3S showed the highest concentrations among all estrogens measured. The concentrations of E1, E2, E2S and E2-17S were mostly below their LOQs. No comparable literature was available for the levels of E1S, E2S and E2-17S in AF of mid-gestation. In the Wang study with GC-MS, many metabolites of DHEA, pregnenolone, 17-hydroxypregnenolone and 17-hydroxyprogesterone were reported since the method was based on urinary steroid profiling of urine. DHEAS correlated highly with 16OH-DHEAS, consistent with the concept of the classical steroid pathway in the feto-placental unit. Strong correlation was further found between 16OH-DHEAS and E3S. 16OHDHEAS is the principal precursor of estriol.

A new LC-MS/MS study (Gomez-Gomez et al., 2020) for amniotic fluid, saliva and breast milk is based on a urinary steroid profile method which is not surprising because the laboratory focuses on antidoping rather than clinical investigations. Amniotic fluid was collected from 30 cases at the time of cesarean section or early vaginal delivery. The concentrations of many cortisol metabolites were reported with no significant differences in the results (Table 2.6.25). Few comparable results have been published.

The analysis of amniotic fluid in a limited number of studies has produced data for concentrations of many steroids but between studies there is little duplication. The nature of the samples is difficult to control and many samples are taken during difficult clinical times so results may be distorted by unnatural circumstances.

TABLE 2.6.25 Steroids in amniotic fluid mid gestation. LC-MS/MS with APCI in positive mode.

	Male (µg/L)	Female (µg/L)	P-value
Androstenedione Mean±SD Median (min–max)	0.89±0.49 0.78 (0.09–3.11)	0.39±0.26 0.33 (0.09–1.65)	<.001
Corticosterone Mean±SD Median (min–max)	0.62±0.4 0.56 (0.05–1.72)	0.53±0.42 0.38 (0.01–2.37)	−.055
Cortisol Mean±SD Median (min–max)	7.34±3.01 6.75 (0.50–16.95)	6.38±2.73 6.32 (0.67–13.60)	−.057
Cortisone Mean±SD Median (min–max)	18.43±6.35 18.3 (1.7–45.4)	17.14±6.52 16.6 (3.3–34.6)	−.206
Deoxycorticosterone Mean±SD Median (min–max)	0.16±0.07 0.15 (0.03–0.34)	0.14±0.06 0.14 (0.01–0.3)	−.112
11-Deoxycortisol Mean±SD Median (min–max)	0.52±0.38 0.42 (0.04–1.71)	0.61±0.46 0.53 (0.03–2.34)	−.354
DHEA Mean±SD Median (min–max)	0.64±0.48 0.54 (0.01–2.39)	0.56±0.36 0.53 (0.04–2.11)	−.47
DHEA-S Mean±SD Median (min–max)	9.73±8.61 7.68 (0.38–65.1)	8.67±6.45 7.17 (0.38–43.3)	−.344
17-OHP Mean±SD Median (min–max)	1.81±0.73 1.73 (0.67–3.64)	1.73±0.87 1.59 (0.06–4.32)	−.248
Progesterone Mean±SD Median (min–max)	98.24±63.76 85.75 (4.27–376.0)	81.19±52.86 66.85 (8.85–341.0)	−.061
Testosterone Mean±SD Median (min–max)	0.30±0.15 0.28 (0.01–0.77)	0.02±0.02 0.01 (0–0.07)	<.001

Fahlbusch FB, Heussner K, Schmid M, Schild R, Ruebner M, Huebner H, Rascher W, Doerr HG, Rauh M. Measurement of amniotic fluid steroids of midgestation via LC-MS/MS. J Steroid Biochem Mol Biol 2015;152:155–60. Table 2 p. 158.

2.6.5.6 Cerebrospinal fluid

CSF is in contact with the brain extracellular fluid and thus can reflect processes in the brain. Some of the steroids may be synthesized de novo in the CNS although a substantial part of the steroid metabolites may be synthesized in the CNS from the steroid precursors or may be transported from the periphery. Collection of CSF from patients is rarely needed and the procedure itself may affect the concentrations of steroids in the fluid in addition to the usual variables of gender, age, circadian and menstrual cycle variations. CSF has lipid content around 10 µg/mL and cholesterol up to 5 µg/mL. The analytical methods that have been used for neuroactive steroids were reviewed by Kaleta et al. (2021). The quantitative analysis of steroids in CSF was thoroughly reviewed by Teubel and Parr (2020), covering a detailed critique of papers published to 2020. In view of the difficulties in sample collection and processing of CSF, the lack of reliability of methods for CSF and high variation in the published results it is difficult to define concentrations to the analytes and will not attempt here. The Teubel review covers sugar, salt and sex steroids as well as the neurosteroids in normal and pathological situations such as neurodegenerative diseases.

There is an interest in measuring oxygenated steroids at C-7 position because of antiglucocorticoid, immunomodulatory, neuroprotective, antioxidant and antiapoptotic effects of 7-hydroxylated DHEA metabolites. DHEA, testosterone and related metabolites as well as 7-oxo and 7-hydroxy DHEA have been quantified in CSF (Table 2.6.26) with lower results by LCMS/MS than earlier immunoassay data (Kancheva et al., 2011; Sosvorova et al., 2015).

7-oxo-DHEA was less than 12 pg/mL which was the detection limit of the assay. The oxidation of 7-hydroxy-DHEA is akin to the effect of HSD11B1 which favors cortisol over cortisone in CSF (Table 2.6.27) (Sosvorova et al., 2015).

TABLE 2.6.26 Concentrations of steroids in the cerebrospinal fluid (CSF) and serum (nmol/L).

Steroid	CSF. free steroids			Serum, free steroids			Serum, polar conjugates		
		Quartile			Quartile			Quartile	
	Median	Lower	Upper	Median	Lower	Upper	Median	Lower	Upper
Pregnenolone	0.0604	0.0448	0.0659	0.417	0.245	0.786	35.3	17.0	54.5
Dehydroepiandrosterone	0.078	0.058	0.154	2.28	1.11	4.47	1544	726	2095
Progesterone	0.235	0.194	0.306	1.68	1.50	2.21			
17-Hydroxy-progesterone	0.0167	0.0077	0.0217	0.77	0.56	1.97			
Androstenedione	0.208	0.123	0.273	1.26	0.67	1.89			
Testosterone	0.231	0.186	0.428	6.8	4.2	12.8			
Allopregnanolone	0.0078	0.0058	0.0110	0.125	0.075	0.149	3.65	1.88	6.57
Isopregnanolone	0.0399	0.0244	0.0584	0.189	0.155	0.210	7.0	4.8	13.0
Androsterone	0.0047	0.0033	0.0135	0.111	0.080	0.215	331	213	1043
Epiandrosterone	0.0044	0.0021	0.0080	0.128	0.067	0.215	109	76	295
7α-Hydroxy-dehydroepiandrosterone	0.300	0.178	0.375	0.79	0.52	1.28			
7β-Hydroxy-dehydroepiandrosterone	0.0369	0.0107	0.0538	0.314	0.171	0.442			
5-Androstene-3β,7α,17β-triol	0.0068	0.0038	0.0177	0.095	0.040	0.167			
5-Androstene-3β,7β, 17β-triol	0.0119	0.0035	0.0205	0.0504	0.0339	0.0829			
16α-Hydroxy-pregnenolone	0.0014	0.0012	0.0041	0.071	0.052	0.223			
16α-Hydroxy-dehydroepiandrosterone	0.0059	0.0044	0.0115	0.180	0.103	0.216	71.8	34.9	94.1
16α-Hydroxy-progesterone	0.072	0.047	0.158	2.51	1.69	3.11			
Cortisol	5.59	2.24	9.15	412	155	607			

From Kancheva R, Hill M, Novák Z, Chrastina J, Kancheva L, Stárka L. Neuroactive steroids in periphery and cerebrospinal fluid. Neuroscience 2011;191:22–7. Table 6 p. 38.

TABLE 2.6.27 Medians and concentration ranges of free steroids in plasma and cerebrospinal fluid samples.

	Plasma		Cerebrospinal fluid	
Analyte	Concentration range	Medians	Concentration range	Medians
DHEA	0.066–2.190	0.83	0.004–0.610	0.03
Cortisol	23.8–184.0	96.75	0.808–20.20	6.54
Cortisone	7.161–26.601	18.85	0.137–5.028	2.0
Cortisol/cortisone	2.72–9.02	5.06	1.15–10.31	2.91
7α-OH-DHEA	0.058–0.380	0.138	0.016–0.315	0.06
7β-OH-DHEA	0.024–0.072	0.049	0.003–0.041	0.01
7-oxo-DHEA	0.009–0.052	0.017	Under LOD	Under LOD
16α-OH-DHEA	0.027–0.068	0.051	0.005–0.027	0.006

LOD means limit of detection. The concentration of steroids are provided in ng/mL.

Sosvorová L, Bičíková M, Mohapl M, Hampl R. Steroids and their metabolites in CSF from shunt as potential predictors of further disease progression in patients with hydrocephalus and the importance of 11β-hydroxysteroid dehydrogenase. Horm Mol Biol Clin Investig 2012;10(3):287–92. Table 4 p. 6.

Estrogens require dansylation to be detectable from CSF, a Polaris Amide-C18 column was used to separate estriol and estrone while a second dimension of two C18 columns was used to separate 17β-estradiol and 17α estradiol (Nguyen et al., 2011). Detection limits were near 30 pg/mL.

The CNS synthesis of neurosteroids and transport from periphery might be complementary in some cases. So brain synthesis might provide minimum level of steroids, which are indispensable for the CNS functions. The CSF steroids need further investigation to further understand neurodegenerative diseases. The source of steroids and transport in and out of the brain need to be clarified.

2.6.6 Tissues

Biological fluids provide a picture of the tissue exposure to steroids but direct analysis of the tissue has benefits although in most cases the procedure to collect samples involves anesthesia. Normal tissue can thus be obtained (with the permission of the patient) at the same time as disease material is being taken. Hair and fingernails are processed because of the ease of sample collection and in some cases are surrogates for blood samples. Adipose tissue is of particular interest because of the fat content and need to more clearly understand the global problem of obesity. Androgens, corticosteroids, estrogens and progestogens control reproduction, secondary sexual characteristics, maturation, gene expression, cardiovascular health and neurological functions, and especially the estrogens are implicated in the development and/or progression of many diseases, such as breast cancer, ovarian cancer, prostate cancer, endometrial cancer, osteoporosis, neurodegenerative diseases, cardiovascular disease and obesity. The immediate levels of exposure of tissues to the steroids is therefore of interest. The cumulative exposure of exogenous and endogenous estrogens increases the risk of breast and other cancers. Breast cancer risk is higher in women with early menarche and late menopause, who have had longer exposure to estrogens.

Bioanalysis of tissue samples using LC-MS/MS comes with unique challenges in terms of sample handling and inconsistent analyte response owing to nonvolatile matrix components causing ion suppression (matrix) or enhancement. Electrospray ionization (ESI) mode of ionization is more affected than atmospheric pressure chemical ionization (APCI) or atmospheric pressure photoionization (APPI). Nonvolatile matrix components include drug formulation excipients and phospholipids eluting at the same time as the analyte of interest. Most of the phospholipids show poor ionization in APCI mode when compared with ESI mode. Nevertheless, ESI-LC-MS/MS mode is more popular among bioanalytical researchers than APCI-LC-MS/MS, since the steroids ionize well in ESI mode and generate good signals. A "matrix-matching" approach was preferred by

analysts, but this approach adds lots of samples to improve the integrity and accuracy of generated data in ESI mode.

Physiological steroid levels in many tissues are generally low and vary among individuals with sex, gender, age, etc. Immunoassay based measurements of steroids are simple and sensitive but raise concerns about reproducibility, cross reactivity, dynamic range, and matrix effects and cannot analyze multiple steroids in a single assay. GC/MS and LC-MS methods have large dynamic range, and several determinations of the steroids can be done simultaneously. The availability of steroid-free matrix in adequate quantities for quantification is sometimes challenging, e.g. in the case of brain, bone marrow and other rare matrices.

The accurate determination of tissue steroids was addressed by Maeda et al. (2013). The steroids were fractionated by absorption from an acetonitrile extract with 2% formic acid onto WAX cartridge and elution of the waste to a Hybrid SPE phospholipid cartridge. The water soluble conjugates on the WAX cartridge were eluted with 10 mM triethylamine/methanol (1:9) before deconjugation of steroids. The corticoids (corticosterone) and phospholipids on the phospholipid SPE were eluted with 2% citric acid/acetonitrile (1/9). Brain (cerebrum, cerebellum and hippocampus) as well as testis, adrenal glands, liver and muscle were analyzed by TOF-MS and LC-MS/MS.

2.6.6.1 Hair

Hair samples are collected for cortisol analysis in order to act as a biomarker for stress, in forensic and toxicology studies, in antidoping programs hair has been tested as an alternative biological matrix to urine or blood for sex steroid concentrations. Human hair grows at the rate of on average of about 1 cm per month but this can vary from 0.6 to 3.4 cm/month. Hair grows in cycles through new growth (anagen), transition (catagen) and quiescence (Telogen) phases with incorporation of steroid into the hair mainly in the anagen phase. In animal studies, it has been common practice to shave the hair then conduct the experiment in the re-growth phase that is mainly in anagen phase. Human hair at the back of the head in the *vertex posterior* site has the most uniform growth rate. A typical sample is a lock of hair from the scalp about the thickness of a pencil and 3 cm in length. Any analysis is therefore a retrospective study over time. When subjects are bald, hair can be taken from the chest, arms, legs and pubic region but smaller sample size may be a problem for sensitivity. For samples from men with Afro-textured hair, electric clippers are used at scalp level at front, back and sides of the head to get 50–300 mg hair (Doyle and Brindle, 2019).

Steroids can get into hair by passive diffusion from the bloodstream, incorporation via sweat, from sebum and possibly by local production in the hair follicle because skin has all the requisite enzymes. The uptake of cortisol into hair has been validated with radioactive cortisol into the hair of rhesus monkeys (Kapoor et al., 2018). Radioactivity was detected in 14 day hair but not 28 day hair samples supporting the hypothesis that cortisol gets in to the hair matrix then advances with growth of the hair. Cortisone and cortisol metabolites were detected.

The assays for steroids in hair vary in technical aspects (Voegel et al., 2020; Grova et al., 2020; Liu and Doan, 2019; Greff et al., 2019; Binz et al., 2018). Most protocols include one or more washing steps prior to the actual extraction to prevent interference from sweat, hair care products or other environmental surface materials. Washes with isopropanol or methylene chloride are the washes of choice. Other variables then include the amount of hair milled or minced and in what device. If an internal standard is to be used this will be added to the hair at this milling stage when the extractant is added. Buffers (Sorensen) and solvents are used for the extraction step, methanol being the most used. The length and condition of incubation, sonication/shaking, extraction agent (solvent/ buffer and time, temperature) varies between methods. The properties of antibodies used in IA and full details of the

chromatographic-mass spectrometric method need to be disclosed. The experimental approach with MS methods needs to circumvent matrix effects. In MS methods, calibration standards and sample extracts have been prepared in solutions containing melanin, in hair washed in methanol for a 5 days and hair from the tips of long hair that are presumed to have been leached of steroids. Another approach is the method of standard addition to the samples to confirm recovery.

Most of the literature on concentration of steroids in hair is based on immunoassay data. Now the preferred method of choice will be MS. Concentrations of cortisol in the range 5–90 pg/mg have been determined in human hair. Scalp hair is preferred. The effects of hair pigment, BMI, sex, hair treatments (dying and perming), oral contraceptive use, frequency of hair washing have been considered (Greff et al., 2019; van den Heuvel et al., 2020). The use of hair from chest, arms and legs is not recommended (Voegel et al., 2020) (Fig. 2.6.52). The amount of hair processed has varied from 1.25 to 20 mg with LC-MS/MS detection but larger samples are needed with IA detection.

Lower results for cortisol have been reported in post-partum depression. Reference ranges for children have

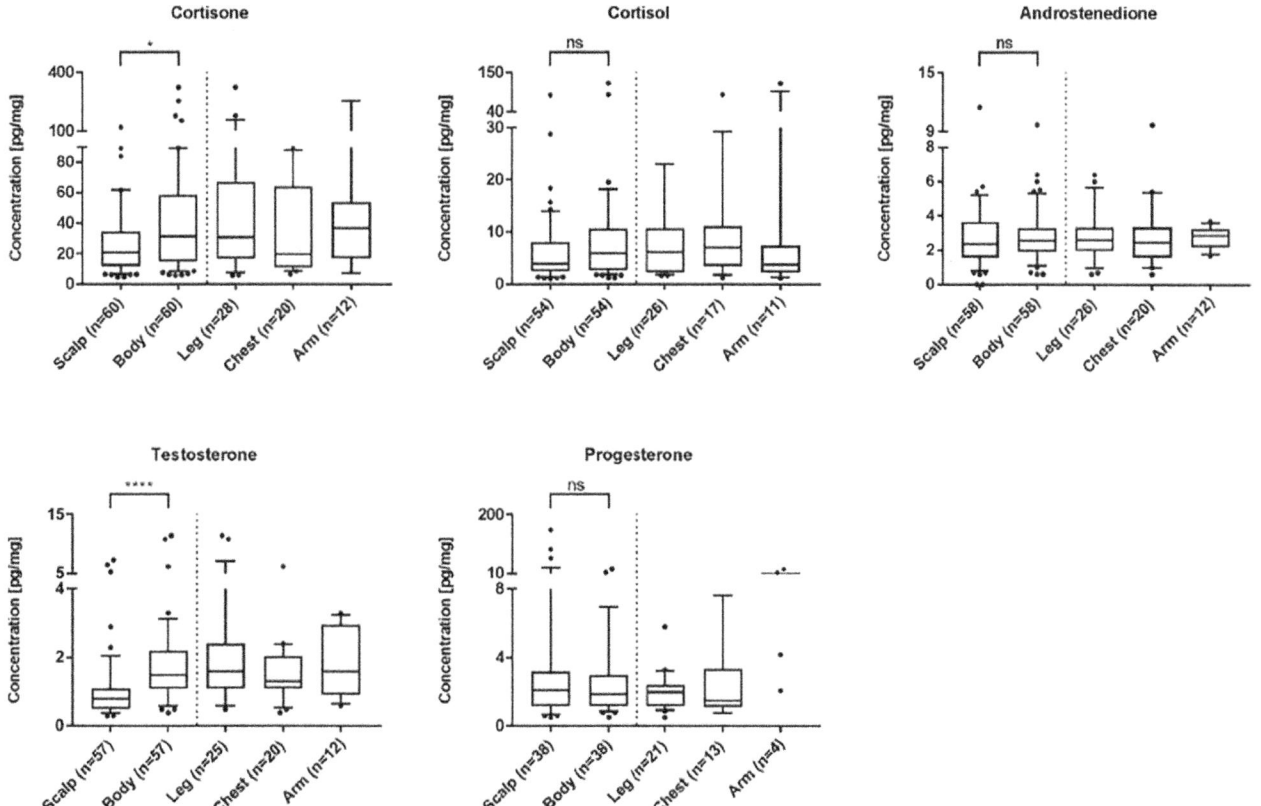

FIG. 2.6.52 Comparative analysis of steroid concentrations in scalp and body hair and representative box plots of leg, chest and arm hair. The *black line* in middle represents the median. The boxes represent the 25% and 75% percentile. The whiskers represent the 10% and 90% percentile. Statistical analysis was done by Wilcoxon-signed rank test. Significance level is indicated with asterisks. *P*-values > .05 were considered as not significant (ns), *P* < .05 (*) as significant, *P* < .0001 (****) as extremely significant. *n* = number of pairs. Only pairs with values over LOQ were considered for statistical analysis. *(From Voegel CD, Hofmann M, Kraemer T, Baumgartner MR, Binz TM. Endogenous steroid hormones in hair: investigations on different hair types, pigmentation effects and correlation to nails. Steroids 2020;154:108547. Fig. 1 p. 5.)*

not been established for all steroid hormones. Androstenedione, testosterone, progesterone and cortisone are typically 1 to 5; 0.3 to 3; 0.9 to 50; 8 to 140 pg/mg so very sensitive methods are needed, 0.2 pg/mg being the required level of sensitivity for testosterone. Cortisol, androstenedione and 17-hydoxyprogesterone have been used to monitor treatment in patients with congenital adrenal hyperplasia (Noppe et al., 2016). Cortisol concentrations in hair increase during pregnancy in parallel with the findings in other samples (Marceau et al., 2020). Estradiol, when measured, has been derivatized with dansyl chloride for example. Cortisol and DHEA in hair have been measured where the steroids have been replaced in patients with adrenal insufficiency (Binder et al., 2009) and in studies in the field of endocrine pathology and psychoneuroendocrinology (Greff et al., 2019).

2.6.6.2 Finger nails

Some religious groups object to taking samples of scalp hair but fingernails are acceptable material for investigations of stress. Nails are keratinized tissue-like hair but melanin is the main protein in hair. Growth rate of nails (3 mm per month) is less than hair and there is variation with site (hands or feet) and between digits. Processing of the sample and interpretation of the results is challenging. The nail samples (10 mg) are fragmented to fine powder in a ball milling

machine (e.g., Retsch type MM400). Concentrations of cortisol, cortisone progesterone and androstenedione are typically 1.2–2.9; 6.2–34.3; 1.3–65.9 and 1.1–15 pg/mg (Voegel et al., 2018). Studies of cortisol concentrations in nails of newborn infants have been associated with some degree of stress in utero but data are limited in the low numbers studied. A poor correlation between hair and nail concentrations has been found (Fig. 2.6.53), so the use of only one matrix is recommended (Voegel et al., 2020).

The concentrations of cortisol in nails from the thumb can be lower than the ring finger and more so the index finger, there are some differences in samples from left vs. right hands (Higashi et al., 2016). A profile of steroids has been achieved using ^{13}C-labeled steroids as internal standards from less than 10 mg nail clipping (Voegel et al., 2020).

2.6.6.3 Adipose tissue

There is an increasing interest in adipose tissue in relation to obesity where fat is increased in and between cells with differences in abdominal and visceral compartments. Fat deposition varies with sex, diet, ethnicity, pre- and postmenopause and the site (abdomen vs. gluteofemoral) and type (brown or white). Samples of abdominal fat are taken during intestinal surgery and subcutaneous fat during breast surgery, but visceral fat samples are not easily taken,

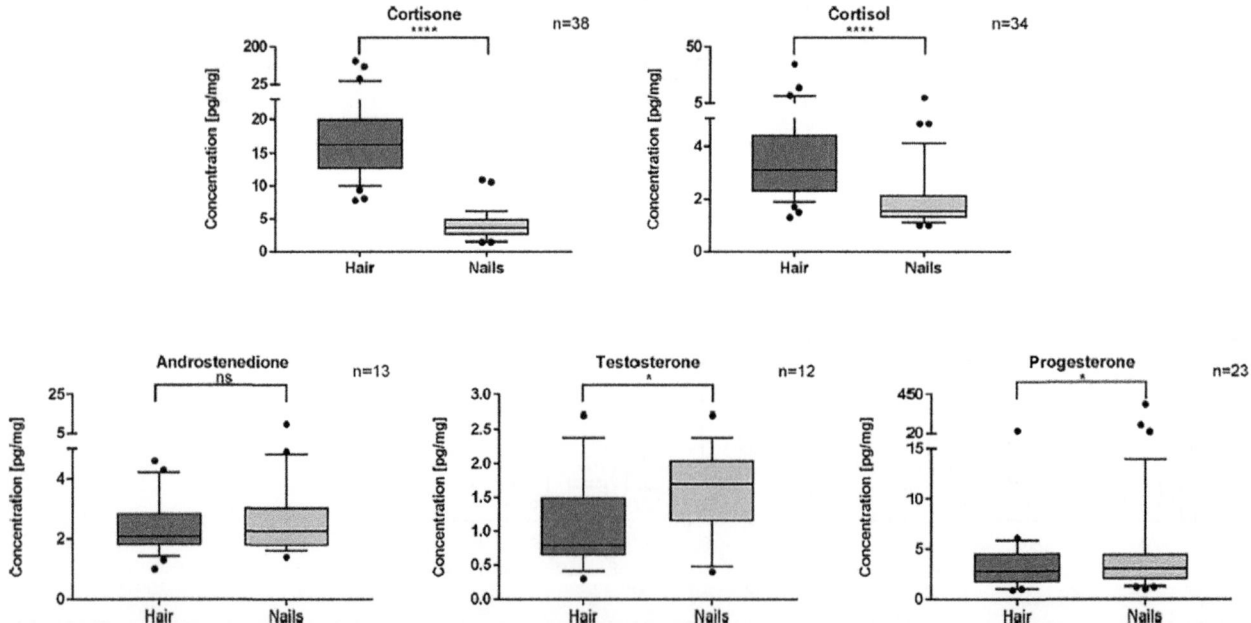

FIG. 2.6.53 Comparative statistical analysis of steroids in hair and nails. The *black line* in middle represents the median. The boxes represent the 25% and 75% percentile. The whiskers represent the 10% and 90% percentile. Statistical analysis was done by Wilcoxon-signed rank test. Significance level is indicated with asterisks. *P* values >.05 were considered as not significant (ns), *P* <.05 (*) as significant, *P* <.0001 (****) as extremely significant. *n* = number of pairs. *(From Voegel CD, Hofmann M, Kraemer T, Baumgartner MR, Binz TM. Endogenous steroid hormones in hair: investigations on different hair types, pigmentation effects and correlation to nails. Steroids 2020;154:108547. Fig. 2 p. 6.)*

biopsies have been taken and cells cultured in order to study the molecular biology of the steroidogenic enzymes. In adipose tissue, many enzymes of steroid metabolism are expressed and contribute to local and then circulating levels of steroids. The intracellular creation of steroids is now called an intracrine process. Fat can take up chylomicrons, VLDL and steroid hormones from the circulation and create steroids through classical and nonclassical pathways (backdoor to DHT for example). Lipoprotein lipase, sulfatase and aromatase are key activities. Fat cells and adipocytes secrete a variety of paracrine and endocrine substances including lipids, cytokines, miRNA and peptide hormones. There is an over-expression of aromatase in adipose tissue. The resulting increase in estradiol contributes to cancer initiation and progression. Adipose tissue has immune, metabolic and endocrine functions. Obesity and associated insulin resistance is a global problem. Males have less body fat overall and accumulate fat in the abdominal (visceral) compartment, whereas women store in the glutofemoral depot. In postmenopausal women, adipose tissue is an important source of sex steroids. Adipose tissue is a pool of glucocorticoid derived from the circulation and local production leading to increased lipolysis.

Leptin, adiponectin, apelin, omentin, vaspin, adipolin subfatin and visfatin are newly characterized proteins that improve insulin sensitivity in rodent models (Booth et al., 2016), epidemiological studies however suggests obesity is associated with insulin resistance. These insulin sensitizing factors may only be effective during periods of rapid adipose tissue growth in order to maintain glucose homeostasis in the face of an increase in the intake of calories. In chronic obesity, the beneficial effects may be lost are lost with chronic obesity because of in-balance as adiponectin and omentin concentrations (helpful adipokines) decrease, or increase, e.g., leptin and apelin, or become less efficient through resistance due to lower binding affinity or increased degradation.

In tissues, steroids are present at low picomolar concentrations distributed around the blood, cells and fat. Few studies have analyzed the steroid content of adipose tissue by reliable methods. Blouin et al. (2009) used thin layer chromatography in several studies for crude assessment of androgen conversion. Bélanger et al. (2006) used gas chromatography with chemical ionization mass spectrometry to compare estrogen concentrations in adipose tissue of obese men (Table 2.6.28).

Regional differences in adipose tissue steroid levels were observed for dihydrotestosterone, androstenedione and dehydroepiandrosterone levels which were all significantly more concentrated in omental versus subcutaneous fat. The study illustrated the need for sensitive and precise methods to determine the steroid content of adipose tissue starting with effective extraction of the steroids from the samples.

Equilibration of internal standards with the tissue elements is difficult, usually circumvented by extending the duration of this step after tissue disruption. The sample, usually frozen, is pulverized/homogenized in solvent rather than water or buffer because of the lipid content of adipose tissue. Purification of sample extracts is required and SPE is the most often used today, although Sephadex LH-20 has been used (Vihma et al., 2016). In order to determine estrogens and cortisol simultaneously, the relative signal from the estrogens needed to be amplified. The phenolic group of estrogens was reacted with 1-(2,4-dinitro-5-fluorophenyl)-4-methylpiperazine (PPZ) and methyl iodide to generate charged 1-(2,4-dinitro-5-fluorophenyl)-4,4-dimethylpiperazine (MPPZ) derivatives (Fig. 2.6.54) (Laforest et al., 2019) thus enabling simultaneous determination of estrogens and glucocorticoids.

TABLE 2.6.28 Regional differences in omental and subcutaneous adipose tissue **steroid concentrations of men.**

Steroid	Omental adipose tissue Mean (SD)	Subcutaneous adipose tissue Mean (SD)	Significance
Estrone (ng/g)	0.45 (0.21)	0.58 (0.60)	NS
Testosterone (ng/g)	2.55 (1.10)	2.47 (0.94)	NS
Dihydrotestosterone (ng/g)	0.18 (0.11)	0.12 (0.06)	.005
Androstenedione (ng/g)	6.60 (2.39)	3.62 (1.20)	.0001
Dehydroepiandrosterone (ng/g)	19.2 (14.2)	15.0 (6.9)	.05

Based on Bélanger C, Hould FS, Lebel S, Biron S, Brochu G, Tchernof A. Omental and subcutaneous adipose tissue steroid levels in obese men. Steroids 2006;71(8):674–682. Table 2 P 677.

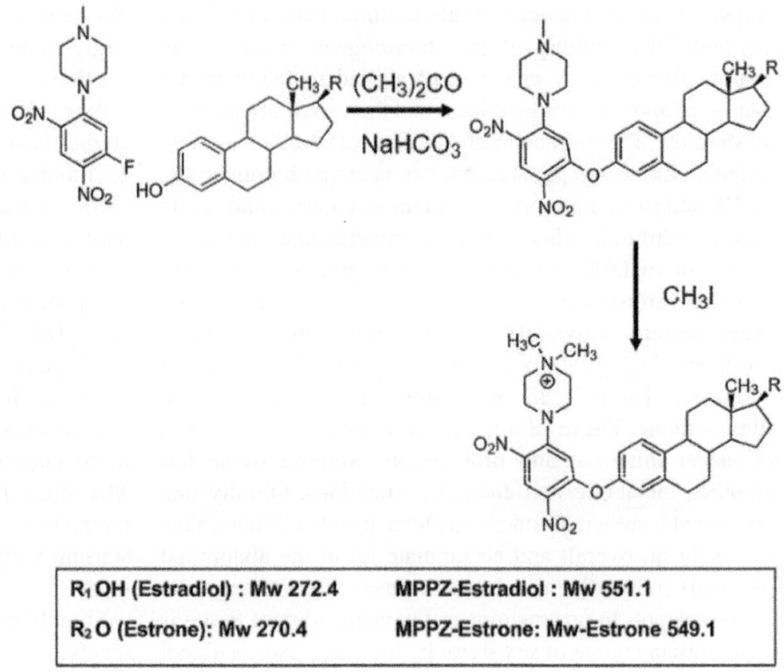

FIG. 2.6.54 Formation of derivatives of estrogens from adipose tissue. *(From Laforest S, Pelletier M, Denver N, Poirier B, Nguyen S, Walker BR, Durocher F, Homer NZM, Diorio C, Tchernof A, Andrew R. Simultaneous quantification of estrogens and gluco-corticoids in human adipose tissue by liquid-chromatography-tandem mass spectrometry. J Steroid Biochem Mol Biol 2019;195:105476. Fig. 1 p. 4.)*

R₁ OH (Estradiol) : Mw 272.4   MPPZ-Estradiol : Mw 551.1

R₂ O (Estrone): Mw 270.4   MPPZ-Estrone: Mw-Estrone 549.1

A C18 PFP column was used for the separation but, in validation of the method, problems were encountered with lipid in the extract or with residues of the derivative reaction and concentrations did not compare well with published results by GC-MS (Hennig et al., 2018) (Table 2.6.29). In that study, adipose tissues were sampled around breast cancer tumors, so they may not be pure adipose tissue samples.

Steroid FAE bound to lipoproteins can be taken up into adipose tissues or in some cases produced by the action of esterifying enzymes which so far is unknown apart from plasma and ovarian follicles where lecitihin: cholesterol acyltransferase (LCAT) has been shown to be involved. The fatty acylated form of estradiol constitutes 80% of total E2 in adipose tissue in postmenopausal women, but the DHEA-FAE concentration is less than 5% of total DHEA (Table 2.6.30). Up to ten times more free DHEA is stored in adipose tissue compared with serum. DHEA-S needs to be desulfated to form free DHEA or,

TABLE 2.6.29 Concentrations of the steroid hormones determined in the analyzed adipose tissue samples ($n = 102$; 51 in proximity and 51 distant to the tumor).

Compound	Detection rate (%)	Min (ng/g lipids)	Max (ng/g lipids)	Median (ng/g lipids)	1st quartile (ng/g lipids)	3rd quartile (ng/g lipids)	P values[a]	Percent of total steroids[b]
DHEA	100	1.0	156.0	14.2	6.6	30.4	0.463	35
AEDIONE	100	0.7	26.5	5.4	3.7	8.5	0.569	13.3
β-TESTO	88	0.1	7.4	0.3	0.2	0.4	0.252	0.7
PROG	100	0.2	197.6	0.8	0.5	1.2	0.602	2.1
OH-PROG	64	0.1	6.1	0.8	0.5	1.3	0.777	2.1
PREG	100	1.6	48.3	11.1	6.3	20.0	0.052	27.4
OH-PREG	97	0.3	18.8	2.8	1.3	6.1	0.442	7.0
ALLOP[c]	90	0.2	31.2	1.0	0.6	2.3	0.144	2.5
ANDR[d]	47	0.6	39.1	2.6	1.5	4.0	0.919	3.8
ETIO	83	0.1	41.4	0.5	0.	1.0	0.253	1.2

TABLE 2.6.29 Concentrations of the steroid hormones determined in the analyzed adipose tissue samples (*n* 2pt= 102; 51 in proximity and 51 distant to the tumor).—cont'd

Compound	Detection rate (%)	Min (ng/g lipids)	Max (ng/g lipids)	Median (ng/g lipids)	1st quartile (ng/g lipids)	3rd quartile (ng/g lipids)	*P* values	Percent of total steroids
EPIANDR[d]	33	0.2	5.8	0.6	0.4	0.7	0.685	1.2
βαβ-ADIOL	17	0.3	975.3	1.1	0.7	5.8	1.000	2.6
EI	100	0.02	4.3	0.4	0.21	0.72	0.908	1.0
β-E2	75	0.005	0.4	All = 0.04 Adip 1 = 0.05 Adpi2 = 0.04	0.02	0.09	0.027	0.1

For comparative reasons, concentrations were also calculated for values between LOD and LOQ.

[a]*Statistical difference of steroid concentrations in proximity and distant to the tumor was tested with Wilcoxon rank test.*

[b]*Based on the sum of the median of all the steroids.*

[c]*The method validation demonstrated an underestimation of 50% of ALLOP. values are not quantitative.*

[d]*Only data with confirmed identification criteria were included in the data analysis.*

From Hennig K, Antignac JP, Bichon E, et al. Steroid hormone profiling in human breast adipose tissue using semi-automated purification and highly sensitive determination of estrogens by GC-APCI-MS/MS. Anal Bioanal Chem 2018;410(1):259-75. Table 7 p. 269.

TABLE 2.6.30 Concentrations of E2-FAE and E2 in subcutaneous abdominal and visceral adipose tissue and in serum in obese men and women and in control women by TR-FIA [median (range)].

	E_2-FAE[a]	E_2
Obese men (*n* = 14)		
sc fat (pmol/kg)	251 (45–804)[b]	262 (183–421)
Visceral fat (pmol/kg)	262 (44–562)[c]	267 (143–361)
Serum (pmol/L)	79 (21–130)[d,k]	102 (55–138)[b]
Obese women (*n* = 21)		
sc fat (pmol/kg)	255 (57–1001)[e,k]	422 (117–1514)[l]
Visceral fat (pmol/kg)	216 (52–1803) (*n* = 19)[f,j]	398 (102–1649) (*n* = 19)
Serum (pmol/L)	51 (16–156) (*n* = 20)[g,k]	245 (48–954) (*n* = 20)[m]
Control women (*n* = 8)		
sc fat (pmol/kg)	148 (121–277)[h,j]	609 (175–1673)[l]
Visceral fat (pmol/kg)	212 (122–328)[i,j]	586 (203–1156)[m]
Serum (pmol/L)	52 (35–85)[i]	256 (57–625)[m]

[a]*Concentration expressed as picomoles per kilogram or picomoles per liter estradiol.*

[b]*Reference value for serum E_2 in men is 0–130pmol/L by immunofluorometric method at HUSLAB, Helsinki University Central Hospital.*

[c]*The concentration of E_2-FAE was below the limit of quantification (151pmol/kg in adipose tissue or 30pmol/L in serum) in the following numbers of subjects: [b]n = 5;[c]n = 4;[d]n = 2;[e]n = 5;[f]n = 10,[g]n = 8;[h]n = 5;[i]n = 2.*

[d]*E_2-FAE vs. E_2 (Wilcoxon signed ranks test): [j]P < .05; [k]P < .01.*

[e]*E_2 in women vs. E_2 in men (Mann-Whitney U test):[l]P < .05; [m]P < .01.*

Wang F, Vihma V, Soronen J, Turpeinen U, Hämäläinen E, Savolainen-Peltonen H, Mikkola TS, Naukkarinen J, Pietiläinen KH, Jauhiainen M, Yki-Järvinen H, Tikkanen MJ. 17β-Estradiol and estradiol fatty acyl esters and estrogen-converting enzyme expression in adipose tissue in obese men and women. J Clin Endocrinol Metab 2013;98(12):4923–31. Table 3 p. 4926.

to a lesser extent, DHEA-FAE after hydrolysis, is available for conversions to hormonally active androgenic and estrogenic steroids in adipose tissue. This represents a mechanism to produce estrogens after the cessation of ovarian steroid production.

2.6.6.4 Prostate tissue

Prostate cancer is close to a leading cause of cancer in men which is androgen dependent so it is important to understand if this is due to circulating or locally produced steroids (endocrine or paracrine effects).

An analytical method based on liquid extraction and LC-MS/MS has been developed and validated for the quantification of 21 steroids (T, A4, DHA4, DHEA, DHT, 11KDHT, 11KT, 11bOHA4, 11OHT, E1, 17OHP5, 17OHP4, P4, P5, F, E, B, S, 21OHP4, E2 and 11aOHA4) from serum, plasma and prostatic tissue homogenate samples. The analysis of all of the 23 steroids in 11 min was achieved within a single run from 150 μL sample using toluene for LLE, oxime derivatization with hydroxylamine in the sample preparation and isotope-labeled internal standards. Simultaneous analysis of the most important adrenal and gonadal steroids and their metabolites was possible was from a small sample volume in the pmol/L or pg/g range, as required for several steroids, including DHT, in tissue, plasma and serum samples. The method was used to examine steroid hormone profiles in serum and prostate tissue homogenates (Hakkinen et al.) (Table 2.6.31).

Several steroids (all of the 11-oxygenated androgens in prostate tissue) were below the LLOQ for each analyte. Most steroids had low intraday precision in plasma and prostate but interday precision was unacceptable for 11bOHA4 steroids in prostate tissue. Cortisol in plasma across 3 days had high standard deviations (6194, 15,645, 19,702 pM). The method was validated against reference serum samples for cortisol (ERM-DA-192), testosterone (ERM-DA346 from LGC), estradiol (BCR-577) and progesterone (ERM-DA347). Concentrations for several androgens were at lower concentrations in prostate tissue than in plasma.

2.6.6.5 Endometrial tissue

Tissues of the endometrium can be taken in the investigation of pathological processes, with a control sample of biopsy material taken from another site nearby or from another procedure. In the investigation of endometriosis, for example, endometrial tissue from an endometriosis patient can be compared with an endometrial biopsy taken during tubule ligation. Endometriosis is a clinical condition characterized by ectopic tissue that causes a range of symptoms including chronic pain and menstrual irregularity. Increased estrogen and progesterone production with or without progesterone

resistance are associated with the disease. The tissue can be homogenized in an aqueous solution and because the tissue has high lipid content, steroid extraction with toluene can be the best solvent, although MTBE is a better solvent for the more hydrophilic steroids such as cortisol. The concentrations of steroids in endometrial tissue (Table 2.6.32) are in picomolar concentrations and about 1000 times less than in plasma.

The menstrual cycle is one of several factors that will influence the measurement of steroid concentrations. Recent guidelines highlight the necessity going forward for standardization of clinical and surgical data and biological sampling (Becker et al., 2014; Vitonis et al., 2014; Rahmioglu et al., 2014; Fassbender et al., 2014; Anupa et al., 2019).

2.6.6.6 Brain

Material is rarely available before death and the dying process may influence concentrations of steroids in the tissue. The literature on steroids in brain must be examined critically because the methods used to isolate the steroids have influenced what is measured being different from what is actually present. The earliest papers reported "pregnenolone-sulfate" and "DHEA-sulfate" in rat brain and Baulieu and colleagues called them "neurosteroids" but the findings have not been upheld in later studies. A review of methodology showed that auto-oxidation of cholesterol was behind the early findings. The fault lie in isolating "conjugates" then applying solvolysis to release free steroid (thus presumed to derive from steroid sulfates). Papers since 2010 are more likely to have reliable data. The methodology should now include removal of cholesterol before isolation of steroid sulfates, fatty acid esters and glucuronides by ion exchange chromatography then analysis of the intact conjugates or the freed steroids after hydrolysis or saponification. Ebner and colleagues at UCL in London (Ebner et al., 2006) in 2006 were the first to adopt such a complete approach for screening and identification of steroids in brain, finding DHEAS at 1.04 ng/g but not pregnenolone sulfate (detection limit 0.05 ng/g). Analysis of steroids by LC-MS/MS without derivatization has measured steroids, steroid sulfates and steroid glucuronides. Steroids in the brain are involved in the modulation of several receptors including GABAa, N-methyl-D-aspartate (NMDA) and sigma receptors. Steroids influence brain development, behavior, cognition, neuroplasticity and neuroinflammation. In view of the concerns on the nature of the steroids in brain, it is safer to avoid quoting concentrations at this time as stated for CSF analysis (Teubel and Parr, 2020).

2.6.7 Summary

Urine cortisol is an important indicator for cortisol excess states, as is demonstrated later in the clinical chapters of the

TABLE 2.6.31 Endogenous steroid concentrations (pM) for plasma and prostate tissue homogenates.

	Plasma					Prostate				
	Day 1 (n = 6)	Day 2 (n = 6)	Day 3 (n = 5)	Intra-day precision	Interday precision	Day 1 (n = 6)	Day 2 (n = 4)	Day 3 (n = 6)	Intra-day precision	Interday precision
A[a]	208±22	136±18	192±24	11/14/13	21	<LLOQ	<LLOQ	<LLOQ		
F[b]	242,118±6194	254,465±15,645	243,744±19,702	3/6/8	6	1889±154	1892±165	2026±138	8/9/7	8
E	64,550±4265	62,327±3078	59,106±4874	7/5/8	7	357±26	346±37	321±21	7/11/7	9
11aOHA4	<LLOQ	<LLOQ	<LLOQ			<LLOQ	<LLOQ	<LLOQ		
11KA4	519±89	558±84	351±49	17/15/14	24	<LLOQ	<LLOQ	<LLOQ		
11KT	809±59	822±81	889±41	7/10/5	8	<LLOQ	<LLOQ	<LLOQ		
B[b]	8618±632	9333±901	8820±327	7/10/4	8	27±2	32±1	31±2	6/4/7	10
11KDHT	15±3	17±2	13±0	18/10/2	15	30±5	30±3	23±1	15/10/5	17
11OHT	370±15	389±24	323±19	4/6/6	9	<LLOQ	<LLOQ	<LLOQ		
11 bOHA4[b]	2791±105	2778±207	2631±118	4/7/4	6	15±2	9±2	15±2	11/19/14	23
E1	82±4	83±6	92±3	5/7/4	7	13±2	12±1	11±2	15/4/15	15
S	346±13	344±13	327±8	4/4/2	4	<LLOQ	<LLOQ	<LLOQ		
17OHP5	2268±119	2190±122	2170±36	5/6/2	5	136±16	124±9	123±9	12/7/7	10
E2	<LLOQ	<LLOQ	<LLOQ			<LLOQ	<LLOQ	<LLOQ		
DHEA	5752±178	5497±358	5409±95	3/6/2	5	4277±483	3925±205	3840±313	11/5/8	10
A4	1715±77	1726±114	1720±79	4/7/5	5	17±3	16±0	15±1	16/2/5	12
21OHP4	55±3	58±3	53±2	5/5/3	6	<LLOQ	<LLOQ	<LLOQ		
T[b]	13,966±450	13,823±404	13,158±223	3/3/2	4	34±2	31±1	31±2	6/4/7	7
DHT	1273±35	1301±54	1263±21	3/4/2	3	1004±63	882±17	839±22	6/2/3	9
DHA4	147±12	155±15	158±3	8/10/2	8	100±7	94±3	90±5	7/3/6	7
17OHP4	2144±64	2167±120	1979±60	3/6/3	6	<LLOQ	<LLOQ	<LLOQ		
P5	1908±76	1609±91	1780±16	4/6/1	8	548±64	472±24	489±36	12/5/7	11
P4	150±2	152±11	146±3	2/7/2	4	10±1	11±0	11±1	6/4/14	10

[a]A: Matrix disturbs the analysis, no better than a semiquantitative result.
[b]F: ULOQ=133,900. B: LLOQ=33.5 pM. 11 bOHA4: LLOQ=33.5. T: ULOQ=13,293. Plasma F was » ULOQ. Prostate B and 11bOHA4 were below LLOQ.

From Häkkinen MR, Murtola T, Voutilainen R, Poutanen M, Linnanen T, Lakka T, Jääskeläinen J, Auriola S. Simultaneous analysis by LC-MS/MS of 22 ketosteroids with hydroxylamine derivatization and underivatized estradiol from human plasma, serum and prostate tissue. J Pharm Biomed Anal. 2019;164:642–52.

TABLE 2.6.32 Endogenous steroid concentrations (pmol), intraday and interday precisions for plasma and tissue homogenates.

Plasma

	Day 1 (n=5)	Day 2 (n=3–4)	Day 3 (n=6)	Intraday	Interday
F[a]	264,571±19,256	284,197±16,092	272,304±28,955	7%	8%
E	39,251±1478	41,246±625	39,681±4059	4%	7%
E2	61±4	54±2	62±5	6%	9%
A	69±9	69±3	64±7	12%	10%
17OHP5	10,397±234	10,981±228	11,054±877	2%	6%
S	1533±62	1593±34	1496±117	4%	6%
B[b,bb]	10,054±415	11,376±254	10,058±662	4%	7%
E1	86±5	86±1	84±7	5%	6%
DHEA	19,732±1657	32,873±2055	30,750±5340	8%	25%
T	11,541±515	11,047±235	11,092±829	4%	6%
17OHP4[c]	1893±54	1687±107	1799±83	3%	6%
A4	2833±83	2968±155	2939±211	3%	6%
21OHP4	168±10	170±5	158±10	6%	6%
P4	170±8	167±18	159±13	5%	8%

Endometrium

	Day 1 (n=3)	Day 2 (n=6)	Intraday	Interday
F[a]	254±19	235±7	8%	6%
E	940±103	878±36	11%	7%
E2	11±1	11±1	13%	12%
A	<LLOQ	<LLOQ		
17OHP5	NA	NA		
S	<LLOQ	<LLOQ		
B	19±0	18±1	2%	4%
E1	10±0	9±0	2%	6%
DHEA	4267±690	3437±549	16%	19%
T	10±1	9±0	11%	8%
17OHP4	14±1	13±1	10%	7%
A4	33±1	32±1	2%	4%
21OHP4	<LLOQ	<LLOQ		
P4	201±26	181±10	13%	10%

Deep endometriosis

	Day 1 (n=3)	Day 2 (n=5)	Intraday	Interday
F[a]	11,343±442	11,702±160	4%	3%
E	3434±128	3599±70	4%	3%
E2	59±4	60±3	7%	6%
A	<LLOQ	<LLOQ		
17OHP5	980±36	1043±29	4%	4%
S	102±3	100±3	3%	3%
B[b]	1485±37	1383±19	2%	4%
E1	24±1	23±1	3%	5%
DHEA	9176±1087	8225±1122	12%	13%
T	153±5	164±3	3%	4%
17OHP4[c]	109±4	108±3	4%	3%
A4	354±12	355±8	3%	3%
21OHP4	37±2	38±2	5%	6%
P4	160±9	159±12	6%	6%

Ovarian endometriosis

	Day 1 (n=3)	Day 2 (n=5)	Intraday	Interday
F[a]	2607±222	2506±24	9%	5%
E	464±9	461±9	2%	2%
E2	29±2	28±2	6%	6%
A	<LLOQ	<LLOQ		
17OHP5	101±2	101±2	2%	2%
S	26±1	29±1	5%	6%
B[b]	229±3	213±7	1%	4%
E1	18±1	16±1	4%	6%
DHEA	2869±154	2319±315	5%	15%
T	22±0	23±1	1%	3%
17OHP4[c]	27±3	28±2	11%	9%
A4	96±1	79±10	1%	14%
21OHP4	<LLOQ	<LLOQ		
P4	47±2	58±3	4%	11%

Mean values ± standard deviation (n=3–6). <LLOQ=concentration below the lower limit of quantification. NA data not available, matrix disturbs the analysis.

[a] Plasma F > ULOQ.

[b] Endometrium B < LLOQ.

[c] Endometrium and Ovarian endometriosis 17OHP4 < LLOQ.

From Häkkinen MR, Heinosalo T, Saarinen N, Linnanen T, Voutilainen R, Lakka T, Jääskeläinen J, Poutanen M, Auriola S. Analysis by LC-MS/MS of endogenous steroids from human serum, plasma, endometrium and endometriotic tissue. J Pharm Biomed Anal. 2018;152:165–72.

book. Immunoassay will continue to be used for this analysis as front-line test ideally after a solvent extraction. A few other individual steroid measurements have some merit but since the preferred technique for measuring cortisol in urine involves chromatography and mass spectrometry, it is now becoming standard practice to measure more than one steroid at a time in a panel or profile. In using the profile techniques for diagnostic use, the concentrations of steroids and the relative amounts need to be examined when judging the contribution of the enzymes. A number of steroid ratios also prove useful; these effects will be discussed in the clinical sections of the book in the disorders relating to sugar, salt and sex.

Collection of samples of hair, finger nails, saliva, semen and breast milk are minimally invasive. Cerebrospinal fluid, amniotic fluid, endometrial tissue is confined to specific clinical investigations but provide information on the exposure of tissues to steroids particularly when brain cannot be sampled when the patient is living. Standardization of the source material will be important. Tissue samples and fluids require a lot of preparation because of the protein and lipid content that can affect the analysis of the steroids. Small amounts of some tissues (endometrium and breast cancer) are collected by biopsy and need sensitive tests for the steroid content. Concentrations of steroids in hair represent an integrated measure of synthesis over months, whereas saliva is near the levels at a short period in time. Hair steroid measurements therefore do not replace concentrations in saliva or plasma but have limited clinical application. Many methods of analysis have been used and in reviewing a particular application it is essential that the validity of the method for the matrix is demonstrated. Many results will be based on kits developed for human plasma or saliva and used with extracts of other fluids or tissues claiming to be surrogate material. Proof is not always demonstrated and it beholds referees of such papers to be satisfied that the validation is complete (Hillebrand et al., 2021).

GC/MS is now been extensively used in steroid analysis; however, the steroid measurements from biological specimens require elaborate and tedious sample preparation procedures. Prior to GC/MS analysis, steroid derivatization is required as a part of sample preparation to increase the volatility of the steroid and thermal stability of the molecules and to improve chromatographic separation and detection. In recent years, LC-MS has been shown to be very useful in determining steroid levels in the biological samples. Electrospray ionization (ESI) and atmospheric pressure chemical ionization (APCI) are routinely used in qualitative and quantitative analysis of steroids. Many new techniques, like MS-based imaging tools, including matrix-assisted laser desorption and liquid extraction surface sampling, have become available to study tissue distribution of steroids. These high-end techniques are expensive and semi-quantitative but will have merit, complementing other data.

Analysis of gene expression will also provide valuable data. All methods for quantitative analysis must be subjected to quality management processes (Vogeser and Seger, 2016). The method of standard addition was lauded for urine steroid analysis (Gomez-Gomez et al., 2020) and analysis of serum, plasma and endometrial tissue samples at 1 in 3 and 1 in 10 dilution (Häkkinen et al., 2019) for differences <20% to support precision is limited by the LOD of analytes and limited availability of reference compounds.

2.6.8   Further reading

The following references were not cited in the chapter but may be of value in further research of a topic.

2.6.8.1   Methodology

Abdel-Khalik et al. (2013), Junker et al. (2019), Kim et al. (2015), Koopman et al. (1986), Pedersen et al. (2017), and Wudy et al. (2018).

2.6.8.2   LC/MS

Bae et al. (2019a), Cao et al. (2019), and Vaikkinen et al. (2015).

2.6.8.3   Urine

Binnie et al. (2017), Christakoudi et al. (2012a,b, 2013), Cuzzola et al. (2014), Doué et al. (2015), Haas et al. (2019), Izawa et al. (2008), Onishi et al. (2002), Ulick et al. (1991), Wood et al. (2008), Yoshitake et al. (1992), Zhang et al. (2011), and Zheng et al. (2016, 2019).

2.6.8.4   Fluids (saliva, etc.)

Bae et al. (2019b), Büttler et al. (2016), Galuska et al. (2013), Gröschl (2017), Li et al. (2018), Raff and Phillips (2019), Sahlberg (1987), Saito-Abe et al. (2020), and Schiffer et al. (2019a).

2.6.8.5   Tissues, hair

Ben Khelil et al. (2011), Gao et al. (2016), Gray et al. (2018), Schiffer et al. (2018), Stalder et al. (2017), Wang et al. (2011), Wester and van Rossum (2015), and Yding (2017).

2.6.8.6   Reproduction

Alliende et al. (2018), Blackwell et al. (2016), Blaschka et al. (2018), Dehennin (1990), Gibson et al. (2016, 2018), Gifford et al. (2018), Huhtinen et al. (2014), Jäntti et al. (2013), Konings et al. (2018), Mäkelä et al. (2020),

Makin (1996), Maybin et al. (2017), Michael et al. (2013), Roos et al. (2015), and Walters et al. (2019).

2.6.8.7 Metabolomics

Ceglarek et al. (2010), Gaikwad (2013), Pussard et al. (2020), and Storbeck et al. (2019).

2.6.8.8 Paediatrics

Idkowiak et al. (2016), Kamrath et al. (2014), Lee et al. (2016), Midgley (1994), Patil et al. (2020), Pozo et al. (2018), Remer et al. (2005), Reynolds (2013), Ruhnau et al. (2021), Shackleton (2012), and Singh et al. (2015).

2.6.8.9 Neurosteroids

Gaignard et al. (2017), Melcangi et al. (2014), Jäntti et al. (2010), Kancheva et al. (2010), Liere et al. (2009), Martin et al. (2019), Meng et al. (1996), Qaiser et al. (2017), Taves et al. (2011), and Trent et al. (2014).

2.6.8.10 Clinical

Sealey et al. (1972), Swords et al. (2003), and Torres et al. (2019).

References

Abdel-Khalik J, Björklund E, Hansen M. Development of a solid phase extraction method for the simultaneous determination of steroid hormones in **H295R cell line** using liquid chromatography-tandem mass spectrometry. J Chromatogr B Analyt Technol Biomed Life Sci 2013;935:61–9.

Abuelghar WM, Ellaithy MI, Swidan KH, Allam IS, Haggag HM. Prediction of spontaneous preterm birth: salivary progesterone assay and transvaginal cervical length assessment after 24 weeks of gestation, another critical window of opportunity. J Matern Fetal Neonatal Med 2019;32(22):3847–58.

Ackermann D, Groessl M, Pruijm M, Ponte B, Escher G, d'Uscio CH, et al. Reference intervals for the urinary steroid metabolome: the impact of sex, age, day and night time on human adult steroidogenesis. PLoS ONE 2019;14(3), e0214549.

Adlercreutz H, Fotsis T, Bannwart C, Hämäläinen E, Bloigu S, Ollus A. Urinary estrogen profile determination in young Finnish vegetarian and omnivorous women. J Steroid Biochem 1986;24(1):289–96.

Alliende ME, Arraztoa JA, Guajardo U, Mellado F. Towards the clinical evaluation of the luteal phase in fertile women: a preliminary study of normative urinary hormone profiles. Front Public Health 2018;6:147.

Alonso-Fernández JR. Pregnanetriolone in paper-borne urine for neonatal screening for 21-hydroxylase deficiency: the place of urine in neonatal screening. Mol Genet Metab Rep 2016;8:99–102.

Antonelli G, Sciacovelli L, Aita A, Padoan A, Plebani M. Validation model of a laboratory-developed method for the ISO15189 accreditation: the example of salivary cortisol determination. Clin Chim Acta 2018;485:224–8.

Anupa G, Sharma JB, Roy KK, Sengupta J, Ghosh D. An assessment of the multifactorial profile of steroid-metabolizing enzymes and steroid receptors in the eutopic endometrium during moderate to severe ovarian endometriosis. Reprod Biol Endocrinol 2019;17(1):111.

Bacila I, Adaway J, Hawley J, et al. Measurement of salivary adrenal-specific androgens as biomarkers of therapy control in 21-hydroxylase deficiency. J Clin Endocrinol Metab 2019;104(12):6417–29.

Bae YJ, Zeidler R, Baber R, Vogel M, Wirkner K, Loeffler M, et al. Reference intervals of nine steroid hormones over the life-span analyzed by LC-MS/MS: effect of age, gender, puberty, and oral contraceptives. J Steroid Biochem Mol Biol 2019a;193:105409.

Bae YJ, Reinelt J, Netto J, et al. Salivary cortisone, as a biomarker for psychosocial stress, is associated with state anxiety and heart rate. Psychoneuroendocrinology 2019b;101:35–41.

Bakusic J, De Nys S, Creta M, Godderis L, Duca RC. Study of temporal variability of salivary cortisol and cortisone by LC-MS/MS using a new atmospheric pressure ionization source. Sci Rep 2019;9(1):19313.

Becker CM, Laufer MR, Stratton P, et al. World endometriosis research foundation endometriosis phenome and biobanking harmonisation project: I. Surgical phenotype data collection in endometriosis research [published correction appears in Fertil Steril. 2015;104(4):1047] [published correction appears in Fertil Steril 2015;104(4):1047]. Fertil Steril 2014;102(5):1213–22.

Bélanger C, Hould FS, Lebel S, Biron S, Brochu G, Tchernof A. Omental and subcutaneous adipose tissue steroid levels in obese men. Steroids 2006;71(8):674–82.

Ben Khelil M, Tegethoff M, Meinlschmidt G, Jamey C, Ludes B, Raul JS. Simultaneous measurement of endogenous cortisol, cortisone, dehydroepiandrosterone, and dehydroepiandrosterone sulfate in nails by use of UPLC-MS-MS. Anal Bioanal Chem 2011;401(4):1153–62.

Bevan BR, Savvas M, Jenkins JM, Baker K, Pennington GW, Taylor NF. Abnormal steroid excretion in gestational trophoblastic disease complicated by ovarian theca-lutein cysts. J Clin Pathol 1986;39(6):627–34.

Bianchi L, Campi B, Sessa MR, De Marco G, Ferrarini E, Zucchi R, et al. Measurement of urinary free cortisol by LC-MS-MS: adoption of a literature reference range and comparison with our current immunometric method. J Endocrinol Invest 2019;42(11):1299–305.

Bileck A, Verouti SN, Escher G, Vogt B, Groessl M. A comprehensive urinary steroid analysis strategy using two-dimensional gas chromatography-time of flight mass spectrometry. Analyst 2018;143(18):4484–94.

Bileck A, Fluck CE, Dhayat N, Groessl M. How high-resolution techniques enable reliable steroid identification and quantification. J Steroid Biochem Mol Biol 2019;186:74–8.

Binder G, Weber S, Ehrismann M, Zaiser N, Meisner C, Ranke MB, et al. Effects of dehydroepiandrosterone therapy on pubic hair growth and psychological well-being in adolescent girls and young women with central adrenal insufficiency: a double-blind, randomized, placebo-controlled phase III trial. J Clin Endocrinol Metab 2009;94(4):1182–90.

Binnie JE, Cooke DG, Blackwell LF. Establishment of a reference ELISA for measurement of universal thresholds of pregnanediol glucuronide excretion rates using urine samples diluted to a constant volume per unit time. J Immunoassay Immunochem 2017;38(2):202–20.

Binz TM, Gaehler F, Voegel CD, Hofmann M, Baumgartner MR, Kraemer T. Systematic investigations of endogenous cortisol and cortisone in nails by LC-MS/MS and correlation to hair. Anal Bioanal Chem 2018;410(20):4895–903.

Blackwell LF, Vigil P, Alliende ME, Brown S, Festin M, Cooke DG. Monitoring of ovarian activity by measurement of urinary excretion rates using the Ovarian Monitor, Part IV: the relationship of the pregnanediol glucuronide threshold to basal body temperature and cervical mucus as markers for the beginning of the post-ovulatory infertile period. Hum Reprod 2016;31(2):445–53.

Blaschka C, Schuler G, Sánchez-Guijo A, Zimmer B, Feller S, Kotarski F, et al. Occurrence of sulfonated steroids and ovarian expression of steroid sulfatase and SULT1E1 in cyclic cows. J Steroid Biochem Mol Biol 2018;179:79–87.

Blouin K, Veilleux A, Luu-The V, Tchernof A. Androgen metabolism in adipose tissue: recent advances. Mol Cell Endocrinol 2009;301 (1–2):97–103.

Booth A, Magnuson A, Fouts J, Foster MT. Adipose tissue: an endocrine organ playing a role in metabolic regulation. Horm Mol Biol Clin Invest 2016;26(1):25–42.

Bousmpoula A, Kouskouni E, Benidis E, et al. Adropin levels in women with polycystic ovaries undergoing ovarian stimulation: correlation with lipoprotein lipid profiles. Gynecol Endocrinol 2018;34(2):153–6.

Bozzolino C, Vaglio S, Amante E, Alladio E, Gerace E, Salomone A, et al. Individual and cyclic estrogenic profile in women: structure and variability of the data. Steroids 2019;150:108432.

Brossaud J, Leban M, Corcuff JB, Boux de Casson F, Leloupp AG, Masson D, et al. LC-MSMS assays of urinary cortisol, a comparison between four in-house assays. Clin Chem Lab Med 2018;56(7):1109–16.

Büttler RM, Kruit A, Blankenstein MA, Heijboer AC. Measurement of dehydroepiandrosterone sulphate (DHEAS): a comparison of isotope-dilution liquid chromatography tandem mass spectrometry (ID-LC-MS/MS) and seven currently available immunoassays. Clin Chim Acta 2013;424:22–6.

Büttler RM, Peper JS, Crone EA, Lentjes EGW, Blankenstein MA, Heijboer AC. Reference values for salivary testosterone in adolescent boys and girls determined using isotope-dilution liquid-chromatography tandem mass spectrometry (ID-LC-MS/MS). Clin Chim Acta 2016;456:15–8.

Büttler RM, Bagci E, Brand HS, Heijer MD, Blankenstein MA, Heijboer AC. % Testosterone, androstenedione, cortisol and cortisone levels in human unstimulated, stimulated and parotid saliva. Steroids 2018; 138:26–34.

Cao ZT, Wemm SE, Han L, Spink DC, Wulfert E. Noninvasive determination of human cortisol and dehydroepiandrosterone sulfate using liquid chromatography-tandem mass spectrometry. Anal Bioanal Chem 2019;411(6):1203–10.

Ceglarek U, Shackleton C, Stanczyk FZ, Adamski J. Steroid profiling and analytics: going towards sterome. J Steroid Biochem Mol Biol 2010; 121(3–5):479–80.

Chan AO, Taylor NF, Tiu SC, Shek CC. Reference intervals of urinary steroid metabolites using gas chromatography-mass spectrometry in Chinese adults. Steroids 2008;73(8):828–37.

Christakoudi S, Cowan DA, Taylor NF. Steroids excreted in urine by neonates with 21-hydroxylase deficiency: characterization, using GC-MS and GC-MS/MS, of the D-ring and side chain structure of pregnanes and pregnenes. Steroids 2010 Jan;75(1):34–52.

Christakoudi S, Cowan DA, Taylor NF. Steroids excreted in urine by neonates with 21-hydroxylase deficiency. 2. Characterization, using GC-MS and GC-MS/MS, of pregnanes and pregnenes with an oxo-group on the A- or B-ring. Steroids 2012a;77(5):382–93.

Christakoudi S, Cowan DA, Taylor NF. Steroids excreted in urine by neonates with 21-hydroxylase deficiency. 3. Characterization, using GC-MS and GC-MS/MS, of androstanes and androstenes. Steroids 2012b;77(13):1487–501.

Christakoudi S, Cowan DA, Christakudis G, Taylor NF. 21-Hydroxylase deficiency in the neonate—trends in steroid anabolism and catabolism during the first weeks of life. J Steroid Biochem Mol Biol 2013;138:334–47.

Cuzzola A, Mazzini F, Petri A. A comprehensive study for the validation of a LC-MS/MS method for the determination of free and total forms of urinary cortisol and its metabolites. J Pharm Biomed Anal 2014 Jun;94:203–9.

Danese E, Salvagno GL, Guzzo A, Scurati S, Fava C, Lippi G. Urinary free cortisol assessment by liquid chromatography tandem mass spectrometry: a case study of ion suppression due to unacquainted administration of piperacillin. Biochem Med (Zagreb) 2017;27(3), 031001.

de Jong WHA, Buitenwerf E, Pranger AT, Riphagen IJ, Wolffenbuttel BHR, Kerstens MN, et al. Determination of reference intervals for urinary steroid profiling using a newly validated GC-MS/MS method. Clin Chem Lab Med 2017;56(1):103–12.

Dehennin L. Estrogens, androgens, and progestins in follicular fluid from preovulatory follicles: identification and quantification by gas chromatography/mass spectrometry associated with stable isotope dilution. Steroids 1990;55(4):181–4.

Dehennin L, Nahoul K, Scholler R. Steroid 21-hydroxylation by human preovulatory follicles from stimulated cycles: a mass spectrometrical study of deoxycorticosterone, 21-hydroxypregnenolone and 11-deoxycortisol in follicular fluid. J Steroid Biochem 1987;26(3):337–43.

Dhayat NA, Frey AC, Frey BM, d'Uscio CH, Vogt B, Rousson V, et al. Estimation of reference curves for the urinary steroid metabolome in the first year of life in healthy children: tracing the complexity of human postnatal steroidogenesis. J Steroid Biochem Mol Biol 2015;154:226–36.

Dhayat NA, Dick B, Frey BM, d'Uscio CH, Vogt B, Flück CE. Androgen biosynthesis during minipuberty favors the backdoor pathway over the classic pathway: Insights into enzyme activities and steroid fluxes in healthy infants during the first year of life from the urinary steroid metabolome. J Steroid Biochem Mol Biol 2017;165(Pt B):312–22.

Doué M, Dervilly-Pinel G, Pouponneau K, Monteau F, Le Bizec B. Analysis of glucuronide and sulfate steroids in urine by ultra-high-performance supercritical-fluid chromatography hyphenated tandem mass spectrometry. Anal Bioanal Chem 2015;407(15):4473–84.

Doyle JA, Brindle E. Development and validation of hair specimen collection methods among extremely short-length Afro-textured hair. Am J Hum Biol 2019;31(3), e23222.

Ebner MJ, Corol DI, Havlíková H, Honour JW, Fry JP. Identification of neuroactive steroids and their precursors and metabolites in adult male rat brain. Endocrinology 2006;147(1):179–90.

Ehlers ME, Griffing GT, Wilson TE, Melby JC. Elevated urinary 19-Nordeoxycorticosterone glucuronide in Cushing's syndrome. J Clin Endocrinol Metab 1987;64(5):926–30.

Fabregat-Cabello N, Peeters SD, Yilmaz T, Cavalier É, Le Goff CM. Establishment of reference intervals for serum concentrations of androstanediol glucuronide by a newly developed LC-MS/MS method. Steroids 2019;143:62–6.

Fahlbusch FB, Heussner K, Schmid M, et al. Measurement of amniotic fluid steroids of midgestation via LC-MS/MS. J Steroid Biochem Mol Biol 2015;152:155–60.

Fassbender A, Rahmioglu N, Vitonis AF, et al. World endometriosis research foundation endometriosis phenome and biobanking harmonisation project: IV. Tissue collection, processing, and storage in endometriosis research. Fertil Steril 2014;102(5):1244–53.

Findling JW, Pinkstaff SM, Shaker JL, Raff H, Nelson JC. Pseudohypercortisoluria: spurious elevation of urinary cortisol due to carbamazepine. The Endocrinologist 1998;8:51–4.

Fink RS, Pierre LN, Daley-Yates PT, Richards DH, Gibson A, Honour JW. Hypothalamic-pituitary-adrenal axis function after inhaled corticosteroids: unreliability of urinary free cortisol estimation. J Clin Endocrinol Metab 2002;87(10):4541–6.

Finken MJ, Andrews RC, Andrew R, Walker BR. Cortisol metabolism in healthy young adults: sexual dimorphism in activities of A-ring reductases, but not 11beta-hydroxysteroid dehydrogenases. J Clin Endocrinol Metab 1999;84(9):3316–21.

Forest MG, de Peretti E, Lecoq A, Cadillon E, Zabot MT, Thoulon JM. Concentration of 14 steroid hormones in human amniotic fluid of midpregnancy. J Clin Endocrinol Metab 1980;51(4):816–22.

Gaignard P, Liere P, Thérond P, Schumacher M, Slama A, Guennoun R. Role of sex hormones on brain mitochondrial function, with special reference to aging and neurodegenerative diseases. Front aging Neurosci 2017;9:406.

Gaikwad NW. Ultra performance liquid chromatography-tandem mass spectrometry method for profiling of steroid metabolome in human tissue. Anal Chem 2013;85(10):4951–60.

Galuska CE, Hartmann MF, Sánchez-Guijo A, Bakhaus K, Geyer J, Schuler G, et al. Profiling intact steroid sulfates and unconjugated steroids in **biological fluids** by liquid chromatography-tandem mass spectrometry (LC-MS-MS). Analyst 2013;138(13):3792–801.

Gao W, Stalder T, Kirschbaum C. Quantitative analysis of estradiol and six other steroid hormones in human saliva using a high throughput liquid chromatography-tandem mass spectrometry assay. Talanta 2015;143:353–8.

Gao W, Kirschbaum C, Grass J, Stalder T. LC-MS based analysis of endogenous steroid hormones in **human hair**. J Steroid Biochem Mol Biol 2016;162:92–9.

Gatti R, Antonelli G, Prearo M, Spinella P, Cappellin E, De Palo EF. Cortisol assays and diagnostic laboratory procedures in human biological fluids. Clin Biochem 2009;42(12):1205–17.

Gibson DA, Simitsidellis I, Saunders PT. Regulation of androgen action during establishment of pregnancy. J Mol Endocrinol 2016;57(1):R35–47.

Gibson DA, Simitsidellis I, Collins F, Saunders PTK. Endometrial intracrinology: estrogens, androgens and endometrial disorders. Int J Mol Sci 2018;19(10):3276.

Gifford RM, Howie F, Wilson K, Johnston N, Todisco T, Crane M, et al. Confirmation of ovulation from urinary progesterone analysis: assessment of two automated assay platforms. Sci Rep 2018;8(1):17621.

Gomez-Gomez A, Miranda J, Feixas G, et al. Determination of the steroid profile in alternative matrices by liquid chromatography tandem mass spectrometry. J Steroid Biochem Mol Biol 2020;197:105520.

Gomez-Sanchez CE, Holland OB. Urinary tetrahydroaldosterone and aldosterone-18-glucuronide excretion in white and black normal subjects and hypertensive patients. J Clin Endocrinol Metab 1981;52(2):214–9.

Gomez-Sanchez CE, Upcavage RJ, Zager PG, Foecking MF, Holland OB, Ganguly A. Urinary 18-hydroxycortisol and its relationship to the excretion of other adrenal steroids. J Clin Endocrinol Metab 1987;65(2):310–4.

Gomez-Sanchez CE, Zager PG, Foecking MF, Holland OB, Ganguly A. 18-oxocortisol: effect of dexamethasone, ACTH and sodium restriction. J Steroid Biochem 1989;32(3):409–12.

Gray NA, Dhana A, Van Der Vyver L, Van Wyk J, Khumalo NP, Stein DJ. Determinants of hair cortisol concentration in children: a systematic review. Psychoneuroendocrinology 2018;87:204–14.

Greenberger PA, Odeh YK, Frederiksen MC, Atkinson Jr AJ. Pharmacokinetics of prednisolone transfer to breast milk. Clin Pharmacol Ther 1993;53(3):324–8.

Greff MJE, Levine JM, Abuzgaia AM, Elzagallaai AA, Rieder MJ, van Uum SHM. Hair cortisol analysis: an update on methodological considerations and clinical applications. Clin Biochem 2019;63:1–9.

Gröschl M. **Saliva**: a reliable sample matrix in bioanalytics. Bioanalysis 2017;9(8):655–68 [Review].

Gröschl M, Rauh M. Influence of commercial collection devices for saliva on the reliability of salivary steroids analysis. Steroids 2006;71(13–14):1097–100.

Gröschl M, Rauh M, Dörr HG. Circadian rhythm of salivary cortisol, 17alpha-hydroxyprogesterone, and progesterone in healthy children. Clin Chem 2003;49(10):1688–91.

Grova N, Wang X, Hardy EM, Palazzi P, Chata C, Appenzeller BMR. Ultra performance liquid chromatography-tandem mass spectrometer method applied to the analysis of both thyroid and steroid hormones in human hair. J Chromatogr A 2020;1612:460648.

Haas AV, Hopkins PN, Brown NJ, et al. Higher urinary cortisol levels associate with increased cardiovascular risk. Endocr Connect 2019;8(6):634–40. https://doi.org/10.1530/EC-19-0182.

Häkkinen MR, Heinosalo T, Saarinen N, et al. Analysis by LC-MS/MS of endogenous steroids from human serum, plasma, endometrium and endometriotic tissue. J Pharm Biomed Anal 2018;152:165–72.

Häkkinen MR, Murtola T, Voutilainen R, Poutanen M, Linnanen T, Koskivuori J, et al. Simultaneous analysis by LC-MS/MS of 22 ketosteroids with hydroxylamine derivatization and underivatized estradiol from human plasma, serum and prostate tissue. J Pharm Biomed Anal 2019;164:642–52.

Hawley JM, Owen LJ, Lockhart SJ, Monaghan PJ, Armston A, Chadwick CA, et al. Serum cortisol: an up-to-date assessment of routine assay performance. Clin Chem 2016;62(9):1220–9.

Hennig K, Antignac JP, Bichon E, et al. Steroid hormone profiling in human breast adipose tissue using semi-automated purification and highly sensitive determination of estrogens by GC-APCI-MS/MS. Anal Bioanal Chem 2018;410(1):259–75.

Higashi T, Yamagata K, Kato Y, Ogawa Y, Takano K, Nakaaze Y, et al. Methods for determination of **fingernail** steroids by LC/MS/MS and differences in their contents between right and left hands. Steroids 2016;109:60–5.

Hill M, Parízek A, Kancheva R, Dusková M, Velíková M, Kríz L, et al. Steroid metabolome in plasma from the umbilical artery, umbilical vein, maternal cubital vein and in amniotic fluid in normal and preterm labor. J Steroid Biochem Mol Biol 2010;121(3–5):594–610.

Hillebrand JJ, Wickenhagen WV, Heijboer AC. Improving science by overcoming laboratory pitfalls with hormone measurements. J Clin Endocrinol Metab 2021;106(4):e1504–12.

Hines JM, Bancos I, Bancos C, Singh RD, Avula AV, Young WF, et al. High-resolution, accurate-mass (HRAM) mass spectrometry urine steroid profiling in the diagnosis of adrenal disorders. Clin Chem 2017;63(12):1824–35.

Hollanders JJ, van der Voorn B, de Goede P, et al. Biphasic glucocorticoid rhythm in one-month-old infants: reflection of a developing HPA-Axis? J Clin Endocrinol Metab 2020;105(3), dgz089.

Homma K, Hasegawa T, Masumoto M, Takeshita E, Watanabe K, Chiba H, et al. Reference values for urinary steroids in Japanese newborn infants: gas chromatography/mass spectrometry in selected ion monitoring. Endocr J 2003;50(6):783–92.

Homma K, Hasegawa T, Takeshita E, Watanabe K, Anzo M, Toyoura T, et al. Elevated urine pregnanetriolone definitively establishes the diagnosis of classical 21-hydroxylase deficiency in term and preterm neonates. J Clin Endocrinol Metab 2004;89(12):6087–91.

Homoki J, Roitman E, Shackleton CH. Characterization of the major steroids present in amniotic fluid obtained between the 15th and 17th weeks of gestation. J Steroid Biochem 1983;19(2):1061–8.

Honour JW, Shackleton CH. Mass spectrometric analysis for tetrahydroaldosterone. J Steroid Biochem 1977;8(4):299–305.

Honour JW, Conway E, Hodkinson R, Lam F. The evolution of methods for urinary steroid metabolomics in clinical investigations particularly in childhood. J Steroid Biochem Mol Biol 2018;181:28–51.

Hoshiro M, Ohno Y, Masaki H, Iwase H, Aoki N. Comprehensive study of urinary cortisol metabolites in hyperthyroid and hypothyroid patients. Clin Endocrinol (Oxf) 2006 Jan;64(1):37–45.

Huhtinen K, Saloniemi-Heinonen T, Keski-Rahkonen P, et al. Intra-tissue steroid profiling indicates differential progesterone and testosterone metabolism in the endometrium and endometriosis lesions. J Clin Endocrinol Metab 2014;99(11):E2188–97.

Idkowiak J, Taylor AE, Subtil S, et al. Steroid sulfatase deficiency and androgen activation before and after puberty. J Clin Endocrinol Metab 2016;101(6):2545–53.

Israelsson M, Brattsand R, Brattsand G. 20α- and 20β-dihydrocortisone may interfere in LC-MS/MS determination of cortisol in saliva and urine. Ann Clin Biochem 2018;55(3):341–7.

Izawa M, Aso K, Higuchi A, Ariyasu D, Hasegawa Y. Pregnanetriol in the range of 1.2-2.1 mg/m(2)/day as an index of optimal control in CYP21A2 deficiency. Clin Pediatr Endocrinol 2007;16(2):45–52.

Izawa M, Aso K, Higuchi A, Ariyasu D, Hasegawa Y. The range of 2.2-3.3 mg/gCr of pregnanetriol in the first morning urine sample as an index of optimal control in CYP21 deficiency. Clin Pediatr Endocrinol 2008;17(3):75–80.

Jäntti SE, Tammimäki A, Raattamaa H, Piepponen P, Kostiainen R, Ketola RA. Determination of steroids and their intact glucuronide conjugates in mouse brain by capillary liquid chromatography-tandem mass spectrometry. Anal Chem 2010;82(8):3168–75.

Jäntti SE, Hartonen M, Hilvo M, et al. Steroid and steroid glucuronide profiles in urine during pregnancy determined by liquid chromatography-electrospray ionization-tandem mass spectrometry. Anal Chim Acta 2013;802:56–66.

Jerjes WK, Cleare AJ, Peters TJ, Taylor NF. Circadian rhythm of urinary steroid metabolites. Ann Clin Biochem 2006a;43(Pt 4):287–94.

Jerjes WK, Taylor NF, Peters TJ, Wessely S, Cleare AJ. Urinary cortisol and cortisol metabolite excretion in chronic fatigue syndrome. Psychosom Med 2006b;68(4):578–82.

Johannsen TH, Andersson AM, Ahmed SF, de Rijke YB, Greaves RF, Hartmann MF, et al. Peptide hormone analysis in diagnosis and treatment of Differences of Sex Development: joint position paper of EU COST Action 'DSDnet' and European Reference Network on Rare Endocrine Conditions. Eur J Endocrinol 2020;182(6):1–15.

Junker J, Chong I, Kamp F, et al. Comparison of strategies for the determination of sterol sulfates via GC-MS leading to a novel Deconjugation-derivatization protocol. Molecules 2019;24(13):2353.

Jurgens E, Knaven EJ, Hegeman EC, van Gemert MV, Emmen JM, Willemen I, et al. Quantitative profiling of seven steroids in saliva using LC-MS/MS. J Appl Biophys 2019;5:34–45.

Kaleta M, Oklestkova J, Novák O, Strnad M. Analytical methods for the determination of neuroactive steroids. Biomolecules 2021;11(4):553.

Kamrath C, Hartmann MF, Boettcher C, Wudy SA. Reduced activity of 11β-hydroxylase accounts for elevated 17α-hydroxyprogesterone in preterms. J Pediatr 2014;165(2):280–4.

Kamrath C, Hartmann MF, Boettcher C, Zimmer KP, Wudy SA. Diagnosis of 21-hydroxylase deficiency by urinary metabolite ratios using gas chromatography-mass spectrometry analysis: reference values for neonates and infants. J Steroid Biochem Mol Biol 2016;156:10–6.

Kancheva R, Hill M, Novák Z, et al. Peripheral neuroactive steroids may be as good as the steroids in the cerebrospinal fluid for the diagnostics of CNS disturbances. J Steroid Biochem Mol Biol 2010;119(1–2):35–44.

Kancheva R, Hill M, Novák Z, Christina J, Kancheva L, Stárka L. Neuroactive steroids in periphery and cerebrospinal fluid. Neuroscience 2011;191:22–7.

Kapoor A, Schultz-Darken N, Ziegler TE. Radiolabel validation of cortisol in the hair of rhesus monkeys. Psychoneuroendocrinology 2018;97:190–5.

Kim J, Abdulwahab S, Choi K, Lafrenière NM, Mudrik JM, Gomaa H, et al. A microfluidic technique for quantification of steroids in **core needle biopsies**. Anal Chem 2015;87(9):4688–95.

Konings G, Brentjens L, Delvoux B, et al. Intracrine regulation of estrogen and other sex steroid levels in endometrium and non-gynecological tissues; pathology, physiology, and drug discovery. Front Pharmacol 2018;9:940.

Koopman BJ, Lokerse IJ, Verweij H, et al. Improved gas chromatographic-mass fragmentographic assay for tetrahydroaldosterone and aldosterone in urine. J Chromatogr 1986;378(2):283–92.

Koyama Y, Homma K, Fukami M, Miwa M, Ikeda K, Ogata T, et al. Two-step biochemical differential diagnosis of classic 21-hydroxylase deficiency and cytochrome P450 oxidoreductase deficiency in Japanese infants by GC-MS measurement of urinary pregnanetriolone/tetrahydroxycortisone ratio and 11β-hydroxyandrosterone. Clin Chem 2012;58(4):741–7.

Koyama Y, Homma K, Miwa M, Ikeda K, Murata M, Hasegawa T. Measurement of reference intervals for urinary free adrenal steroid levels in Japanese newborn infants by using stable isotope dilution gas chromatography/mass spectrometry. Clin Chim Acta 2013;415:302–5.

Krebs A, Dickhuth K, Mumm R, Stier B, Doerfer J, Grueninger D, et al. Evaluating the four most important salivary sex steroids during male puberty: testosterone best characterizes pubertal development. J Pediatr Endocrinol Metab 2019;32(3):287–94.

Krone N, Hughes BA, Lavery GG, Stewart PM, Arlt W, Shackleton CH. Gas chromatography/mass spectrometry (GC/MS) remains a pre-eminent discovery tool in clinical steroid investigations even in the era of fast liquid chromatography tandem mass spectrometry (LC/MS/MS). J Steroid Biochem Mol Biol 2010;121(3–5):496–504.

Kushnir MM, Naessén T, Wanggren K, Rockwood AL, Crockett DK, Bergquist J. Protein and steroid profiles in follicular fluid after ovarian hyperstimulation as potential biomarkers of IVF outcome. J Proteome Res 2012;11(10):5090–100.

Kushnir MM, Naessén T, Wanggren K, Hreinsson J, Rockwood AL, Meikle AW, et al. Exploratory study of the association of steroid profiles in stimulated ovarian follicular fluid with outcomes of IVF treatment. J Steroid Biochem Mol Biol 2016;162:126–33.

Laforest S, Pelletier M, Denver N, Poirier B, Nguyen S, Walker BR, et al. Simultaneous quantification of estrogens and glucocorticoids in human adipose tissue by liquid-chromatography-tandem mass spectrometry. J Steroid Biochem Mol Biol 2019;195:105476.

Lamb EJ, Noonan KA, Burrin JM. Urine-free cortisol excretion: evidence of sex-dependence. Ann Clin Biochem 1994 Sep;31(Pt 5):455–8.

Lamb EJ, Noonan KA, Burrin JM. Origins of the sex difference in human urinary free cortisol excretion. Ann Clin Biochem 1997;34 (Pt 3):326.

Lee SH, Kim SH, Lee WY, Chung BC, Park MJ, Choi MH. Metabolite profiling of sex developmental steroid conjugates reveals an association between decreased levels of steroid sulfates and adiposity in obese girls. J Steroid Biochem Mol Biol 2016;162:100–9.

Lepenies J, Stewart PM, Quinkler M. The hypertensiogenetic steroid 19-nor-progesterone does not influence cortisol inactivation by 11beta-hydroxysteroid dehydrogenase type 2. Clin Exp Hypertens 2009;31 (4):376–9.

Lewicka S, von Hagens C, Hettinger U, Grunwald K, Vecsei P, Runnebaum B, et al. Cortisol and cortisone in human follicular fluid and serum and the outcome of IVF treatment. Hum Reprod 2003;18(8):1613–7.

Li XS, Li S, Kellermann G. Simultaneous determination of three estrogens in human **saliva** without derivatization or liquid-liquid extraction for routine testing via miniaturized solid phase extraction with LC-MS/MS detection. Talanta 2018;178:464–72.

Liere P, Pianos A, Eychenne B, et al. Analysis of pregnenolone and dehydroepiandrosterone in rodent brain: cholesterol autoxidation is the key. J Lipid Res 2009;50(12):2430–44.

Liu CH, Doan SN. Innovations in biological assessments of chronic stress through hair and nail cortisol: conceptual, developmental, and methodological issues. Dev Psychobiol 2019;61(3):465–76.

Liu J, Qiu X, Wang D, et al. Quantification of 10 steroid hormones in human saliva from Chinese adult volunteers. J Int Med Res 2018;46 (4):1414–27.

Maeda N, Tanaka E, Suzuki T, et al. Accurate determination of tissue steroid hormones, precursors and conjugates in adult male rat. J Biochem 2013;153(1):63–71.

Mäkelä R, Leinonen A, Suominen T. Analysis of luteinizing hormone (LH): validation of a commercial ELISA kit for LH analysis and quantification in doping control samples. Drug Test Anal 2020;12(2):239–46.

Makin HL. Origins of the sex difference in human urinary free cortisol excretion. Ann Clin Biochem 1996;33(Pt 5):471–2.

Marceau K, Wang W, Robertson O, Shirtcliff EA. A systematic review of hair cortisol during pregnancy: reference ranges and methodological considerations. Psychoneuroendocrinology 2020;122:104904.

Marcos J, Pol M, Fabregat A, Ventura R, Renau N, Hanzu FA, et al. Urinary cysteinyl progestogens: occurrence and origin. J Steroid Biochem Mol Biol 2015;152:53–61.

Martin J, Plank E, Jungwirth B, Hapfelmeier A, Podtschaske A, Kagerbauer SM. Weak correlations between serum and cerebrospinal fluid levels of estradiol, progesterone and testosterone in males. BMC Neurosci 2019;20(1):53.

Matthews CE, Fortner RT, Xu X, Hankinson SE, Eliassen AH, Ziegler RG. Association between physical activity and urinary estrogens and estrogen metabolites in premenopausal women. J Clin Endocrinol Metab 2012;97(10):3724–33.

Maybin JA, Thiruchelvam U, Madhra M, Saunders PTK, Critchley HOD. Steroids regulate CXCL4 in the human endometrium during menstruation to enable efficient endometrial repair. J Clin Endocrinol Metab 2017;102(6):1851–60.

McQuarrie EP, Freel EM, Mark PB, Fraser R, Connell JM, Jardine AG. Urinary sodium excretion is the main determinant of mineralocorticoid excretion rates in patients with chronic kidney disease. Nephrol Dial Transplant 2013;28(6):1526–32.

Melby JC, Dale SL, Grekin RJ, Gaunt R, Wilson TE. 18-Hydroxy-11-deoxycorticosterone (18-OH-DOC) secretion in experimental and human hypertension. Recent Prog Horm Res 1972;28:287–351.

Melcangi RC, Giatti S, Calabrese D, et al. Levels and actions of progesterone and its metabolites in the nervous system during physiological and pathological conditions. Prog Neurobiol 2014;113:56–69.

Meng LJ, Griffiths WJ, Sjövall J. The identification of novel steroid N-acetylglucosaminides in the urine of pregnant women. J Steroid Biochem Mol Biol 1996;58(5–6):585–98.

Mezzullo M, Fazzini A, Gambineri A, Di Dalmazi G, Mazza R, Pelusi C, et al. Parallel diurnal fluctuation of testosterone, androstenedione, dehydroepiandrosterone and 17OHprogesterone as assessed in serum and saliva: validation of a novel liquid chromatography-tandem mass spectrometry method for salivary steroid profiling. Clin Chem Lab Med 2017;55(9):1315–23.

Michael AE, Glenn C, Wood PJ, Webb RJ, Pellatt L, Mason HD. Ovarian 11β-hydroxysteroid dehydrogenase (11βHSD) activity is suppressed in women with anovulatory polycystic ovary syndrome (PCOS): apparent role for ovarian androgens. J Clin Endocrinol Metab 2013;98 (8):3375–83.

Midgley P. Adrenal function in preterm infants [MD thesis]. Aberdeen University; 1994.

Mistry HD, Eisele N, Escher G, Dick B, Surbek D, Delles C, et al. Gestation-specific reference intervals for comprehensive spot urinary steroid hormone metabolite analysis in normal singleton pregnancy and 6 weeks postpartum. Reprod Biol Endocrinol 2015;13:101.

Moffat SD, An Y, Resnick SM, Diamond MP, Ferrucci L. Longitudinal change in Cortisol levels across the adult life span. J Gerontol A Biol Sci Med Sci 2020;75(2):394–400.

Nguyen HP, Li L, Gatson JW, Maass D, Wigginton JG, Simpkins JW, et al. Simultaneous quantification of four native estrogen hormones at trace levels in human cerebrospinal fluid using liquid chromatography-tandem mass spectrometry. J Pharm Biomed Anal 2011;54(4):830–7.

Nicolis GL, Wotiz HH, Gabrilove JL. Measurement of urinary tetrahydroaldosterone by gas chromatography with electron capture detection. J Clin Endocrinol Metab 1968;28(4):547–57.

Noppe G, de Rijke YB, Koper JW, van Rossum EF, van den Akker EL. Scalp hair 17-hydroxyprogesterone and androstenedione as a long-term therapy monitoring tool in congenital adrenal hyperplasia. Clin Endocrinol (Oxf) 2016;85(4):522–7.

Oh H, Smith-Warner SA, Tamimi RM, Wang M, Xu X, Hankinson SE, et al. Dietary fat and Fiber intakes are not associated with patterns of urinary estrogen metabolites in premenopausal women. J Nutr 2015;145(9):2109–16.

Olesti E, Garcia A, Rahban R, Rossier MF, Boccard J, Nef S, et al. Steroid profile analysis by LC-HRMS in human seminal fluid. J Chromatogr B Analyt Technol Biomed Life Sci 2020;1136:121929.

Onishi T, Takei H, Kambegawa A, et al. A highly specific heterologous enzyme immunoassay for 5 alpha-androstane-3 alpha, 17 beta-diol 17-glucuronide (androstanediol-17G) and developmental patterns of urinary androstanediol-17G excretions. Steroids 2002;67(3–4):175–83.

Oßwald A, Wang R, Beuschlein F, Hartmann MF, Wudy SA, Bidlingmaier M, et al. Performance of LC-MS/MS and immunoassay based 24-h urine free cortisol in the diagnosis of Cushing's syndrome. J Steroid Biochem Mol Biol 2019;190:193–7.

Patil AS, Gaikwad NW, Grotegut CA, Dowden SD, Haas DM. Alterations in endogenous progesterone metabolism associated with spontaneous very preterm delivery. Hum Reprod Open 2020;2020(2), hoaa007.

Pedersen M, Frandsen HL, Andersen JH. Optimised deconjugation of androgenic steroid conjugates in bovine urine. Food Addit Contam Part A Chem Anal Control Expo Risk Assess 2017;34(4):482–8.

Phillips IJ, Conway EM, Hodkinson RA, Honour JW. External quality assessment of urinary steroid profile analysis. Ann Clin Biochem 2004;41(Pt 6):474–8.

Polet M, De Wilde L, Van Renterghem P, Van Gansbeke W, Van Eenoo P. Potential of saliva steroid profiling for the detection of endogenous steroid abuse: reference thresholds for oral fluid steroid concentrations and ratios. Anal Chim Acta 2018;999:1–12. Table 4 p. 6.

Pozo OJ, Marcos J, Khymenets O, Pranata A, Fitzgerald CC, McLeod MD, et al. Sulfation pathways: alternate steroid sulfation pathways targeted by LC-MS/MS analysis of disulfates: application to prenatal diagnosis of steroid synthesis disorders. J Mol Endocrinol 2018;61(2):M1–M12.

Pundir S, Wall CR, Mitchell CJ, Thorstensen EB, Lai CT, Geddes DT, et al. Variation of human milk glucocorticoids over 24 hour period. J Mammary Gland Biol Neoplasia 2017;22(1):85–92.

Pussard E, Travers S, Bouvattier C, et al. Urinary steroidomic profiles by LC-MS/MS to monitor classic 21-hydroxylase deficiency. J Steroid Biochem Mol Biol 2020;198:105553.

Qaiser MZ, Dolman DEM, Begley DJ, et al. Uptake and metabolism of sulphated steroids by the blood-brain barrier in the adult male rat. J Neurochem 2017;142(5):672–85.

Raff H, Phillips JM. Bedtime salivary cortisol and cortisone by LC-MS/MS in healthy adult subjects: evaluation of sampling time. J Endocr Soc 2019;3(8):1631–40.

Rahmioglu N, Fassbender A, Vitonis AF, et al. World endometriosis research foundation endometriosis phenome and biobanking harmonization project: III. Fluid biospecimen collection, processing, and storage in endometriosis research. Fertil Steril 2014;102(5):1233–43.

Remer T, Boye KR, Hartmann MF, Wudy SA. Urinary markers of adrenarche: reference values in healthy subjects, aged 3-18 years. J Clin Endocrinol Metab 2005;90(4):2015–21.

Reynolds RM. Glucocorticoid excess and the developmental origins of disease: two decades of testing the hypothesis—2012 Curt Richter award winner. Psychoneuroendocrinology 2013;38(1):1–11.

Roche J, Ramé C, Reverchon M, et al. Apelin (APLN) and Apelin receptor (APLNR) in human ovary: expression, signaling, and regulation of steroidogenesis in primary human luteinized granulosa cells. Biol Reprod 2016;95(5):104.

Roos J, Johnson S, Weddell S, Godehardt E, Schiffner J, Freundl G, et al. Monitoring the menstrual cycle: comparison of urinary and serum reproductive hormones referenced to true ovulation. Eur J Contracept Reprod Health Care 2015;20(6):438–50.

Ruhnau J, Hübner S, Sunny D, Ittermann T, Hartmann MF, De Lafollie J, et al. Impact of gestational and postmenstrual age on excretion of fetal zone steroids in preterm infants determined by gas chromatography-mass spectrometry. J Clin Endocrinol Metab 2021;106(9):e3725–38.

Saczawa ME, Graber JA, Brooks-Gunn J, Warren MP. Methodological considerations in use of the cortisol/DHEA(S) ratio in adolescent populations. Psychoneuroendocrinology 2013;38(11):2815–9.

Sahlberg BL. The characterization of sulphated metabolites of norethindrone in human milk after oral administration of contraceptive steroids. J Steroid Biochem 1987;26(4):481–5.

Sahlberg BL, Axelson M. Identification and quantitation of free and conjugated steroids in milk from lactating women. J Steroid Biochem 1986;25(3):379–91.

Saito-Abe M, Yamamoto-Hanada K, Nakayama SF, Hashimoto Y, Natsume O, Fukami M, et al. Reference values for salivary cortisol in healthy young infants by liquid chromatography-tandem mass spectrometry. Pediatr Int 2020;62(7):785–8.

Schambelan M, Biglieri EG. Deoxycorticosterone production and regulation in man. J Clin Endocrinol Metab 1972;34(4):695–703.

Schiffer L, Arlt W, Storbeck KH. Intracrine androgen biosynthesis, metabolism and action revisited. Mol Cell Endocrinol 2018;465:4–26.

Schiffer L, Adaway JE, Baranowski ES, Arlt W, Keevil BG. A novel high-throughput assay for the measurement of salivary progesterone by liquid chromatography tandem mass spectrometry. Ann Clin Biochem 2019a;56(1):64–71.

Schiffer L, Barnard L, Baranowski ES, Gilligan LC, Taylor AE, Arlt W, et al. Human steroid biosynthesis, metabolism and excretion are differentially reflected by serum and urine steroid metabolomes: a comprehensive review. J Steroid Biochem Mol Biol 2019b;194:105439.

Sealey JE, Bühler FR, Laragh JH, Manning EL, Brunner HR. Aldosterone excretion. Physiological variations in man measured by radioimmunoassay or double-isotope dilution. Circ Res 1972;31(3):367–78.

Shackleton CH. Profiling steroid hormones and urinary steroids. J Chromatogr 1986;379:91–156.

Shackleton CH. Mass spectrometry in the diagnosis of steroid-related disorders and in hypertension research. J Steroid Biochem Mol Biol 1993;45(1–3):127–40.

Shackleton C. Genetic disorders of steroid metabolism diagnosed by mass spectrometry. In: Blau N, Duran M, Gibson KM, editors. Laboratory guide to the methods in biochemical genetics. Berlin Heidelberg: Springer; 2008. p. 549–605.

Shackleton C. Clinical steroid mass spectrometry: a 45-year history culminating in HPLC-MS/MS becoming an essential tool for patient diagnosis. J Steroid Biochem Mol Biol 2010;121(3–5):481–90.

Shackleton CH. Role of a disordered steroid metabolome in the elucidation of sterol and steroid biosynthesis. Lipids 2012;47(1):1–12.

Shackleton C, Pozo OJ, Marcos J. GC/MS in recent years has defined the normal and clinically disordered steroidome: will it soon be surpassed by LC/tandem MS in this role? J Endocr Soc 2018;2(8):974–96.

Shamim W, Yousufuddin M, Bakhai A, Coats AJ, Honour JW. Gender differences in the urinary excretion rates of cortisol and androgen metabolites. Ann Clin Biochem 2000;37(Pt 6):770–4. 11085621.

Singh GK, Balzer BW, Kelly PJ, Paxton K, Hawke CI, Handelsman DJ, et al. Urinary sex steroids and anthropometric markers of puberty—a novel approach to characterising within-person changes of puberty hormones. PLoS ONE 2015;10(11), e0143555.

Son HH, Yun WS, Cho SH. Development and validation of an LC-MS/MS method for profiling 39 urinary steroids (estrogens, androgens, corticoids, and progestins). Biomed Chromatogr 2020;34(2), e4723.

Sosvorova L, Vitku J, Chlupacova T, Mohapl M, Hampl R. Determination of seven selected neuro- and immunomodulatory steroids in human cerebrospinal fluid and plasma using LC-MS/MS. Steroids 2015;98:1–8.

Stalder T, Steudte-Schmiedgen S, Alexander N, et al. Stress-related and basic determinants of hair cortisol in humans: a meta-analysis. Psychoneuroendocrinology 2017;77:261–74.

Storbeck KH, Schiffer L, Baranowski ES, Chortis V, Prete A, Barnard L, et al. Steroid metabolome analysis in disorders of adrenal steroid biosynthesis and metabolism. Endocr Rev 2019;40(6):1605–25.

Stoye DQ, Andrew R, Grobman WA, et al. Maternal glucocorticoid metabolism across pregnancy: a potential mechanism underlying fetal glucocorticoid exposure. J Clin Endocrinol Metab 2020;105(3):e782–90.

Svensson K, Just AC, Fleisch AF, et al. Prenatal salivary sex hormone levels and birth-weight-for-gestational age. J Perinatol 2019;39(7):941–8.

Swords FM, Carroll PV, Kisalu J, Wood PJ, Taylor NF, Monson JP. The effects of growth hormone deficiency and replacement on glucocorticoid exposure in hypopituitary patients on cortisone acetate and hydrocortisone replacement. Clin Endocrinol (Oxf) 2003;59(5):613–20.

Taves MD, Ma C, Heimovics SA, Saldanha CJ, Soma KK. Measurement of steroid concentrations in brain tissue: methodological considerations. Front Endocrinol (Lausanne) 2011;2:39.

Taylor NF. Urinary steroid profiling. In: Wheeler MJ, Hutchinson JSM, editors. Methods in molecular biology: hormone assays in biological fluids. Totowa, NJ: Humana Press Inc; 2006. p. 159–75.

Taylor NF. Urinary steroid profiling. Methods Mol Biol 2013;1065:259–76.

Taylor N, Raven P. Origins of the sex difference in human urinary free cortisol excretion. Ann Clin Biochem 1996;33(Pt 2):174–5.

Teubel J, Parr MK. Determination of neurosteroids in human cerebrospinal fluid in the 21st century: a review. J Steroid Biochem Mol Biol 2020;204:105753.

Torres SJ, Grimes C, Nowson CA, Jayasinghe SU, Bruce CR, Mason SA, et al. Urinary sodium is positively associated with urinary free cortisol and total cortisol metabolites in a cross-sectional sample of Australian schoolchildren aged 5–12 years and their mothers. Br J Nutr 2019;121(2):164–71.

Trent S, Fry JP, Ojarikre OA, Davies W. Altered brain gene expression but not steroid biochemistry in a genetic mouse model of neurodevelopmental disorder. Mol Autism 2014;5(1):21.

Ulick S. Diagnosis and nomenclature of the disorders of the terminal portion of the aldosterone biosynthetic pathway. J Clin Endocrinol Metab 1976;43(1):92–6.

Ulick S, Chan CK, Wang JZ. Measurement of 4 urinary C-18 oxygenated corticosteroids by stable isotope dilution mass fragmentography. J Steroid Biochem Mol Biol 1991;38(1):59–66.

Vaikkinen A, Rejšek J, Vrkoslav V, Kauppila TJ, Cvačka J, Kostiainen R. Feasibility of desorption atmospheric pressure photoionization and desorption electrospray ionization mass spectrometry to monitor urinary steroid metabolites during pregnancy. Anal Chim Acta 2015;880:84–92.

van den Heuvel LL, Acker D, du Plessis S, Stalder T, Suliman S, Thorne MY, et al. Hair cortisol as a biomarker of stress and resilience in South African mixed ancestry females. Psychoneuroendocrinology 2020;113:104543.

van der Voorn B, de Waard M, van Goudoever JB, Rotteveel J, Heijboer AC, Finken MJ. Breast-milk cortisol and cortisone concentrations follow the diurnal rhythm of maternal hypothalamus-pituitary-adrenal Axis activity. J Nutr 2016;146(11):2174–9.

Vantyghem MC, Ghulam A, Hober C, et al. Urinary cortisol metabolites in the assessment of peripheral thyroid hormone action: overt and subclinical hypothyroidism. J Endocrinol Invest 1998;21(4):219–25.

Vihma V, Koskela A, Turpeinen U, et al. Are there endogenous estrone fatty acyl esters in human plasma or ovarian follicular fluid? J Steroid Biochem Mol Biol 2011;127(3–5):390–5.

Vihma V, Wang F, Savolainen-Peltonen H, et al. Quantitative determination of estrone by liquid chromatography-tandem mass spectrometry in subcutaneous adipose tissue from the breast in postmenopausal women. J Steroid Biochem Mol Biol 2016;155(Pt A):120–5.

Vining RF, McGinley RA, Maksvytis JJ, Ho KY. Salivary cortisol: a better measure of adrenal cortical function than serum cortisol. Ann Clin Biochem 1983;20(Pt 6):329–35.

Visser HK, Cost WS. A new hereditary defect in the biosynthesis of aldosterone: urinary c21-corticosteroid pattern in three related patients with a salt-losing syndrome, suggesting an 18-oxidation defect. Acta Endocrinol (Copenh) 1964;47:589–612.

Vitku J, Kolatorova L, Hampl R. Occurrence and reproductive roles of hormones in seminal plasma. Basic Clin Androl 2017;27:19.

Vitonis AF, Vincent K, Rahmioglu N, et al. World endometriosis research foundation endometriosis phenome and biobanking harmonization project: II. Clinical and covariate phenotype data collection in endometriosis research. Fertil Steril 2014;102(5):1223–32.

Voegel CD, La Marca-Ghaemmaghami P, Ehlert U, Baumgartner MR, Kraemer T, Binz TM. Steroid profiling in nails using liquid chromatography-tandem mass spectrometry. Steroids 2018;140:144–50.

Voegel CD, Hofmann M, Kraemer T, Baumgartner MR, Binz TM. Endogenous steroid hormones in hair: investigations on different hair types, pigmentation effects and correlation to nails. Steroids 2020;154:108547.

Vogeser M, Seger C. Quality management in clinical application of mass spectrometry measurement systems. Clin Biochem 2016;49(13–14):947–54.

Walker BR, Best R, Noon JP, Watt GC, Webb DJ. Seasonal variation in glucocorticoid activity in healthy men. J Clin Endocrinol Metab 1997;82(12):4015–9.

Walters KA, Eid S, Edwards MC, Thuis-Watson R, Desai R, Bowman M, et al. Steroid profiles by liquid chromatography-mass spectrometry of matched serum and single dominant ovarian follicular fluid from women undergoing IVF. Reprod Biomed Online 2019;38(1):30–7.

Wang F, Koskela A, Hämäläinen E, Turpeinen U, Savolainen-Peltonen H, Mikkola TS, et al. Quantitative determination of dehydroepiandrosterone fatty acyl esters in human female adipose tissue and serum using mass spectrometric methods. J Steroid Biochem Mol Biol 2011;124(3–5):93–8.

Wang R, Hartmann MF, Tiosano D, Wudy SA. Characterizing the steroidal milieu in amniotic fluid of mid-gestation: a GC-MS study. J Steroid Biochem Mol Biol 2019a;193:105412.

Wang R, Tiosano D, Sánchez-Guijo A, Hartmann MF, Wudy SA. Characterizing the steroidal milieu in amniotic fluid of mid-gestation: a LC-MS/MS study. J Steroid Biochem Mol Biol 2019b;185:47–56.

Weber B, Lewicka S, Deuschle M, Colla M, Vecsei P, Heuser I. Increased diurnal plasma concentrations of cortisone in depressed patients. J Clin Endocrinol Metab 2000;85(3):1133–6.

Wester VL, van Rossum EF. Clinical applications of cortisol measurements in hair. Eur J Endocrinol 2015;173(4):M1–M10.

Weykamp CW, Penders TJ, Schmidt NA, Borburgh AJ, van de Calseyde JF, Wolthers BJ. Steroid profile for urine: reference values. Clin Chem 1989;35(12):2281–4.

Wood L, Ducroq DH, Fraser HL, Gillingwater S, Evans C, Pickett AJ, et al. Measurement of urinary free cortisol by tandem mass spectrometry and comparison with results obtained by gas chromatography-mass spectrometry and two commercial immunoassays. Ann Clin Biochem 2008;45(Pt 4):380–8.

Wudy SA, Hartmann MF. Mass spectrometry in the diagnosis of steoid related disorders: clinical applications. In: Ranke MB, Mullis PE, editors. Diagnostics of endocrine function in children and adolescents. Karger; 2011. p. 379–401.

Wudy SA, Dörr HG, Solleder C, Djalali M, Homoki J. Profiling steroid hormones in amniotic fluid of midpregnancy by routine stable isotope dilution/gas chromatography-mass spectrometry: reference values and concentrations in fetuses at risk for 21-hydroxylase deficiency. J Clin Endocrinol Metab 1999;84(8):2724–8.

Wudy SA, Schuler G, Sánchez-Guijo A, Hartmann MF. The art of measuring steroids: principles and practice of current hormonal steroid analysis. J Steroid Biochem Mol Biol 2018;179:88–103.

Xie J, Eliassen AH, Xu X, Matthews CE, Hankinson SE, Ziegler RG, et al. Body size in relation to urinary estrogens and estrogen metabolites (EM) among premenopausal women during the luteal phase. Horm Cancer 2012;3(5–6):249–60.

Xu X, Duncan AM, Merz-Demlow BE, Phipps WR, Kurzer MS. Menstrual cycle effects on urinary estrogen metabolites. J Clin Endocrinol Metab 1999;84(11):3914–8.

Yding AC. Inhibin-B secretion and FSH isoform distribution may play an integral part of follicular selection in the natural menstrual cycle. Mol Hum Reprod 2017;23(1):16–24.

Yoshitake T, Ishida J, Sonezaki S, Yamaguchi M. High performance liquid chromatographic determination of 3 alpha, 5 beta-tetrahydroaldosterone and cortisol in human urine with fluorescence detection. Biomed Chromatogr 1992;6(5):217–21.

Zhang Y, Tobias HJ, Auchus RJ, Brenna JT. Comprehensive 2-dimensional gas chromatography fast quadrupole mass spectrometry (GC × GC-qMS) for urinary steroid profiling: mass spectral characteristics with chemical ionization. Drug Test Anal 2011;3(11–12):857–67.

Zhao M, Baker SD, Yan X, Zhao Y, Wright WW, Zirkin BR, et al. Simultaneous determination of steroid composition of human testicular fluid using liquid chromatography tandem mass spectrometry. Steroids 2004;69(11–12):721–6.

Zheng N, Christopher LJ, Ma X, et al. UHPLC-MS/MS bioanalysis of urinary DHEA, cortisone and their hydroxylated metabolites as potential biomarkers for CYP3A-mediated drug-drug interactions. Bioanalysis 2016;8(23):2429–43.

Zheng Y, Zhao H, Zhu L, Cai Z. Comprehensive identification of steroid hormones in human urine based on liquid chromatography-high resolution mass spectrometry. Anal Chim Acta 2019;1089:100–7.

Zhou Y, Cai Z. Determination of hormones in human urine by ultra-high-performance liquid chromatography/triple-quadrupole mass spectrometry. Rapid Commun Mass Spectrom 2020;34(Suppl. 1), e8583.

Zhu Q, Zuo R, He Y, Wang Y, Chen ZJ, Sun Y, et al. Local regeneration of cortisol by 11β-HSD1 contributes to insulin resistance of the granulosa cells in PCOS. J Clin Endocrinol Metab 2016;101(5):2168–77.

Zufferey F, Rahban R, Garcia A, Gagnebin Y, Boccard J, Tonoli D, et al. Steroid profiles in both blood serum and seminal plasma are not correlated and do not reflect sperm quality: study on the male reproductive health of fifty young Swiss men. Clin Biochem 2018; 62:39–46.

Chapter 2.7

Confirmatory tests in steroid endocrinology

2.7.1 Introduction

The dynamics of steroid production is such that no single measurement is likely to have a particular diagnostic meaning. The concentration of a steroid hormone at a particular time is subject to much variation that include:

- time of day,
- day of month in women,
- age,
- dependence upon trophic hormones,
- sex,
- pregnancy,
- levels of binding proteins,
- rate of clearance from the body and
- use of steroids medically.

The measurement of more than one steroid at a time may be needed in some clinical investigations. For example, when investigating a patient with a possible genetic defect of an enzyme in a steroidogenic pathway, the concentrations of substrate and product of one or more enzymes will be measured. It is also often useful to measure the concentrations of the regulatory hormones. Normal ranges for blood concentrations and urine excretion of steroids are needed in order to interpret test results. In some cases, ranges for age and gender are needed. The clinician is trained to recognize a number of features of the patient and will be alerted to medical problems by differences from normal in physical attributes and behavior which can be challenging with newborn infants (Shah et al., 2020).

A number of **provocative tests** are also used to **stimulate** or **suppress** components of an endocrine axis under likely circumstances. A plan is needed by the clinical team which should include preparation of the patient. The clinician may need advice from the laboratory for the best type of samples to be taken. Blood is an invasive measure at a point in time whereas urine concentrations will be an integer of the time period of collection since last voiding and saliva is an index of free hormone in the circulation. Most of the further confirmatory tests are described as dynamic tests. In some cases, "out of normal" tests are requested as static

procedures in exceptional cases. Arrangements may be needed if samples are to be forwarded to academic or specialist centers.

Dose response relationships are commonly used and the dose can be administered in many different routes (oral, vaginal, rectal for tablets) (intravenous, intramuscular, intraperitoneal, infusion, intranasal for liquids). In animal experiments, implantable mini-pumps meet a number of research needs where a drug is administered over time, the dose and rate can be adjusted. Alzet pumps are small, infusion pumps inserted under the skin of small animals such as mice and rates for accurate and continuous dosing without restraining the laboratory animals. They are a convenient and reliable alternative to frequent injections that save time by eliminating the need for frequent animal handling.

It is good practice with any procedures involving multiple samples to have sample tubes labeled before the testing, with spares in case of any mishap along the way. Tubes must have requisite identifiers for the patient, confirmatory procedure name and timings of samples in the procedure.

2.7.2 Dynamic adrenal cortex tests

Profound changes in steroid hormone concentrations can be due to disorders of steroidogenesis (primary) or the result of changes in the hormones that regulate steroidogenesis (secondary) (Horrocks et al., 1990) or to a defect in a receptor for the steroid. There are many reasons for results that are inconclusive and further dynamic and static tests are requested that afford some control to the test. Further dynamic tests involve manipulation of the activity of a steroidogenic organ or of the regulatory path for the organ. Both stimulation and suppression tests are used (Karaca et al., 2021; Giordano et al., 2008). In these tests, blood concentrations of the peptide hormones controlling the adrenal cortex and gonads can be increased by exogenous or endogenous means. Reduction of a peptide hormone is used to reveal autonomous output of a steroid. The timings of tests

Steroids in the Laboratory and Clinical Practice. https://doi.org/10.1016/B978-0-12-818124-9.00006-1

are important and the frequency of sampling can be dictated by patterns of hormone secretion which can be in short pulses over a few minutes.

2.7.2.1 Adrenal stimulation tests

2.7.2.1.1 Synacthen, short test

When basal cortisol concentrations are repeatedly low, the first confirmatory test is for the response of the adrenal cortex to exogenous ACTH. This is particularly required:

1. To diagnose Addison's disease where patients may present with a hypotensive crisis, an inappropriately low plasma cortisol (<200 nmol/L) is near diagnostic (Barthel et al., 2019; Husebye et al., 2014).
2. Diagnosis of late-onset congenital adrenal hyperplasia (LOCAH) (Kulle et al., 2015). Cortisol and steroid precursors may be measured during the test when it is used for this purpose. This use is subject to development and evaluation.

The newer assays may have cut-off points lower than conventionally quoted.

Adrenal output is stimulated by an injection of synthetic ACTH (Synacthen), the 1–24 amino acids of the 39 amino acid full protein hence the name tetracosactrin. Cortisol is measured in plasma or serum but the adrenal output of other steroids may be measured as required. All corticosteroid therapy other than dexamethasone or betamethasone interferes with the immunoassay of cortisol. Hydrocortisone treatment should be stopped for at least 12 h but prednisone, prednisolone or other likely interfering therapy should be stopped for at least 3 days. Immediately after the test, glucocorticoid cover may be provided by an appropriate dose of dexamethasone (e.g., total of 2 mg/day in four doses). Cases where Addison's disease is newly suspected may have the stimulation test performed immediately, including in A&E/Casualty. The test may then be followed by therapeutic use of steroids before the results are known. Other steroid measurements may be requested on the samples, as may be agreed with the laboratory or by special arrangement (androgens, for example, if adrenal tumor

suspected). The procedure is summarized in Protocol 2.7.1, although the timings of blood samples will vary.

There are many pitfalls in interpretation of the results (Burgos et al., 2019). An increase from baseline in serum cortisol concentration of 150 nmol/L, to high-point above 450 nmol/L (even 350 nmol/L with an LC-MS/MS method) is a normal result, depending on the assay used. An impaired response does not distinguish between adrenal failure and pituitary failure although a long Synacthen test may do so. Patients with pituitary failure may have additional abnormalities of gonadal and thyroid function so further pituitary function tests are recommended. In patients who have been on long-term steroid therapy, the adrenals may not be stimulated during the short Synacthen test or they may be stimulated only partially (Barthel et al., 2019; Burgos et al., 2019; Lindner et al., 2018). A response indicates that the adrenals are active, but the test may be difficult to interpret. Oral contraceptives can stimulate CBG synthesis so total cortisol concentrations in plasma can be higher basally and on ACTH stimulation (Fig. 2.7.1) and this must be taken into account when interpreting the results. Cortisol will still largely be measured on analytical chemistry platforms although more specifically LC-MS/MS is being used (Ueland et al., 2018).

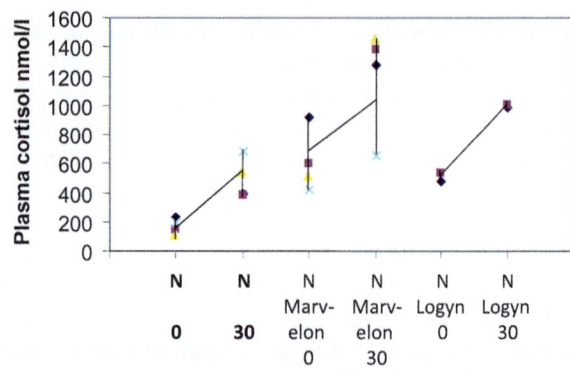

Plasma cortisol at zero and 30 mins after ACTH in normal subjects not taking and taking oral contraceptives

FIG. 2.7.1 Plasma cortisol concentrations and effects of oral contraceptives. *(Author original.)*

Protocol 2.7.1 Short Synacthen test

1. Obtain Synacthen (dose 250 μg for injection) from Pharmacy.
2. The patient should be resting quietly.
3. Take blood for cortisol (Basal—0 min). Record time of day of test and medications.[a]
4. Inject Synacthen intramuscularly.
5. Take bloods for cortisol at 30 and 60 min after Synacthen.
6. Send samples to the laboratory. Note these are not assayed urgently.

[a]If ACTH measurements are required for further diagnosis, e.g., of Addison's disease, then take blood for this also (state time of day). Blood for ACTH must be delivered to the laboratory immediately for urgent preparation. Approved, local, near-patient preparation may be available to separate off plasma and freeze paired aliquots of the sample being taken.

The standard Synacthen test entails stimulation of the adrenal glands by a high dose (250 μg in adults, reduced on the basis of body size for children) of synthetic ACTH (1–24) administered either intravenously or intramuscularly, which results in a large output of cortisol and other adrenal steroids. A normal cortisol response is demonstrated by an adequate peak concentration of between 350 and 500 nmol/L, depending on local laboratory assays, and 17OHP to less than 10 nmol/L at 30 or 60 min post-Synacthen administration. The response of intermediates is not proportional. A 10-fold increase in corticosterone was reported (Lindner et al., 2018). The Synacthen dose can be adjusted for body size.

The test is indicated in suspected adrenal insufficiency and disorders of the steroidogenesis pathway such as CAH. The Synacthen test, if performed as part of workup for abnormal puberty due to suspected steroidogenesis defects, should include interpretation of responses of cortisol and its precursors (17-OHP, 11-deoxycortisol (S), corticosterone, DHEA, 17-hydroxypregnenolone) (Bacila et al., 2019; Travers et al., 2017). Patients with nonclassical CAH show a 17-OHP increment of >20 nmol/L while heterozygotes have intermediate response somewhere between the normal range and classical disease group (Fig. 2.7.2).

11β-Hydroxylase deficiency is suggested by a rise of 11-deoxycortisol to greater than 60 nmol/L, an increase in 11-DOC to cortisol ratio and decreased plasma renin activity (Tran et al., 2018), defects of 17-hydroxylase (Lee et al., 2020) and 3β-hydroxysteroid dehydrogenase (Guran et al., 2020) have also been characterized with steroid profiling. Reference ranges for plasma steroid concentrations for the rarer genetic disorders are available and the overall patterns give clues to the diagnosis. For example, elevation in 17-hydroxypregnenolone and DHEA in 3β-hydroxysteroid deficiency (Guran et al., 2020),

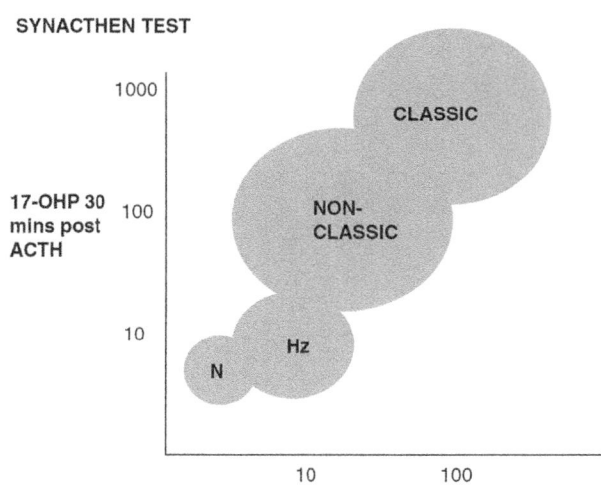

FIG. 2.7.2 The response of plasma 17-hydroxyprogesterone to ACTH. *(Author original.)*

Collection of **saliva samples** at zero and 30 min after the ACTH can also be used to assess the adrenal response to ACTH. Because cortisol in saliva is converted to cortisone by HSD11B2 in the salivary glands measurements of F or E or both have been examined (Perogamvros et al., 2010; Chao et al., 2020; Vaiani et al., 2019). Saliva cortisol in the morning increased from 8.3 nmol/L (3.2–14.8) to 33.7 (12.8–68.6) whereas cortisone increased from 34.6 (18.8–47) to 65.1 (26.8–102.4) (Perogamvros et al., 2010). The difficulty is that the response observed depends on time of day, the effectiveness of the cortisol/cortisone shuttle, the prevailing CBG concentration and salivary flow.

A long acting ACTH preparation of the 34 amino acid porcine formulation (Acton Prolongatum, Ferring Pharmaceuticals) has been available in India and has acceptable stimulation properties (Nair et al., 2019; George et al., 2020). If available globally, this drug may become an economical alternative to synthetic ACTH tetracosapeptide.

A nasal preparation of Synacthen (500 μg **Nasacthin003, Archimedes Pharma, Nottingham**, United Kingdom) became available in 2019 and has undergone preliminary tests with promising results (Elder et al., 2020). A drug enhancer (Chitosan, FMC Biopolymer AS, Sandvika, Norway) was used to optimize nasal absorption Nasacthin atomized from 0.1 mL solution. Peak cortisol response at 60 min for serum and saliva was comparable with 250 μg intravenous Synacthen. If further testing is acceptable, there is the opportunity for noninvasive testing (Auchus, 2020).

2.7.2.1.2 Long Synacthen test

When a longer stimulus to adrenal function than can be provided by the short Synacthen test is required adrenal steroid output is stimulated by 250 μg Synacthen given daily over 3 days (Patti et al., 2018; Chanson et al., 2017). This test will assist the investigations if there is diagnostic uncertainty following the short Synacthen test. All corticosteroid therapy other than dexamethasone or betamethasone interferes with the assay of cortisol. Hydrocortisone should have been stopped for at least 12 h. Prednisone, prednisolone or other interfering therapy should have been stopped for at least 3 days. Cover may be provided by an appropriate dose of dexamethasone. This protocol is optimized for outpatient use. Four clear working days are required to complete the protocol (Protocol 2.7.2).

Admission to hospital is desirable but not essential, the blood samples should be taken around the same time of day each time.

An **increase** in plasma cortisol concentration of 150 nmol/L to a peak at least 400 nmol/L is an adequate response. Dormant adrenals due to pituitary failure or long term steroid therapy should be stimulated by the long Synacthen test even if they were not stimulated by the short Synacthen test.

Protocol 2.7.2 Long Synacthen test

Obtain Synacthen (1 mg dose for injection) from Pharmacy. Three doses are required.

   The test is liable to cause fluid retention which can be dangerous in some patients. The test commences on Day 1 at 0900 h.

1. Take blood for cortisol (Basal Day 1 ideally 0900 h).
2. Inject depot Synacthen.
3. On day 2, repeat the blood collection for cortisol at 0900 h, inject depot Synacthen.
4. On day 3, repeat the blood collection for cortisol at 0900 h, inject depot Synacthen.
5. On day 4, repeat the blood collection for cortisol at 0900 h.

Send each of the blood samples to the laboratory as it is collected. State the timing clearly on each request.

2.7.2.1.3 Low-dose Synacthen test

The 250 μg Synacthen dose is greatly in excess of the dose needed for the test, and low-dose tests have been used with variable success rates due to differences in timings and refinements to procedures (Fig. 2.7.3). The Synacthen protein may stick to surfaces it contacts (Burgos et al., 2019). The best procedure for preparing the dose of Synacthen is to inject 1 mL of Synacthen into a 500-mL bag of 5% dextrose (Cross et al., 2018). After gently mixing the contents of the bag, 2 mL of the diluted dextrose is injected. Blood for cortisol measurement can be taken at 0, 15, 20, 25, 30 and 35 min (Crowley et al., 1991, 1993; Bridges et al., 1998). The response to the low dose is similar to the early response to the standard ACTH with peak being between 15 and 35 min (Fig. 2.7.4). If other steroids are to be measured, then ideally blood should be taken at 5 min intervals throughout the 40 min postinjection, since intermediates have earlier peak times normally (Fig. 2.7.5) but continue to rise when a sign of a defective enzyme in the cortisol pathway. Aldosterone also responds to low doses of ACTH (Honour et al., 2008) (Fig. 2.7.6). Most papers

in the literature focus on blood samples at zero, 30 and 60 min, ignoring the above observations on correct timing of samples (Gill et al., 2019; Vaiani et al., 2019). The term "peak" should not be used since the exact time of the highest level will be unknown (Hindmarsh and Honour, 2020). Cortisol is measured on automated IA or by LC-MS/MS (Ueland et al., 2018).

2.7.2.2 Adrenal suppression

2.7.2.2.1 Dexamethasone suppression test (overnight)

The aim of dexamethasone administration is to reduce ACTH secretion by feedback of the synthetic glucocorticoid to the pituitary and thus lower adrenal synthesis of cortisol. In normal subjects, the morning peak of blood cortisol (around 0800 h) is abolished by oral dexamethasone given the previous night (Sasaki et al., 2017). This test is part of the preliminary investigations following high cortisol results suggestive of possible Cushing's disease (Berlińska et al., 2020; Mojtahedzadeh et al., 2018). The

FIG. 2.7.3 Response of cortisol to low-dose Synacthen compared with high dose.

Standard ACTH stim 250μg bloods at 5 min intervals for 90 minutes vs low dose ACTH stim 500ng ACTH
Usually measure only at zero, 30 and 60 minutes

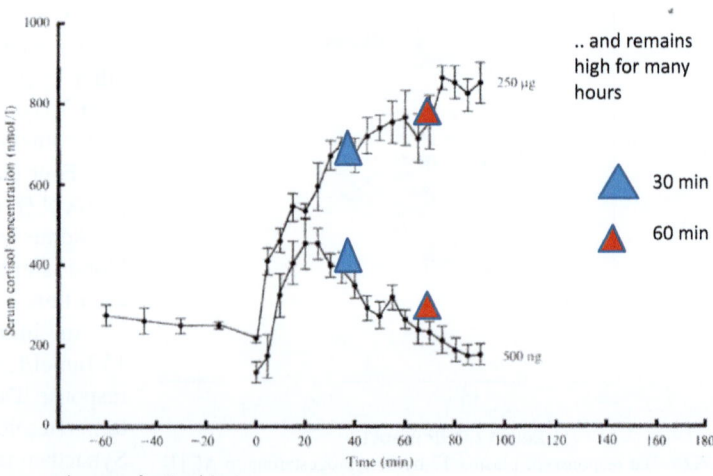

Means and ranges for 5 subjects

Cortisol response to low dose Synacthen

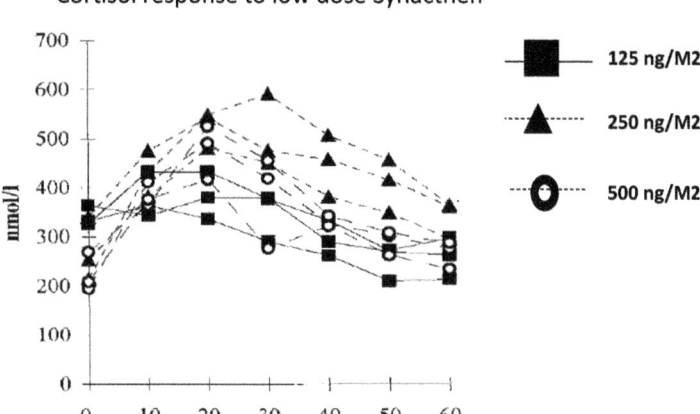

FIG. 2.7.4 Responses of cortisol to low doses of Synacthen.

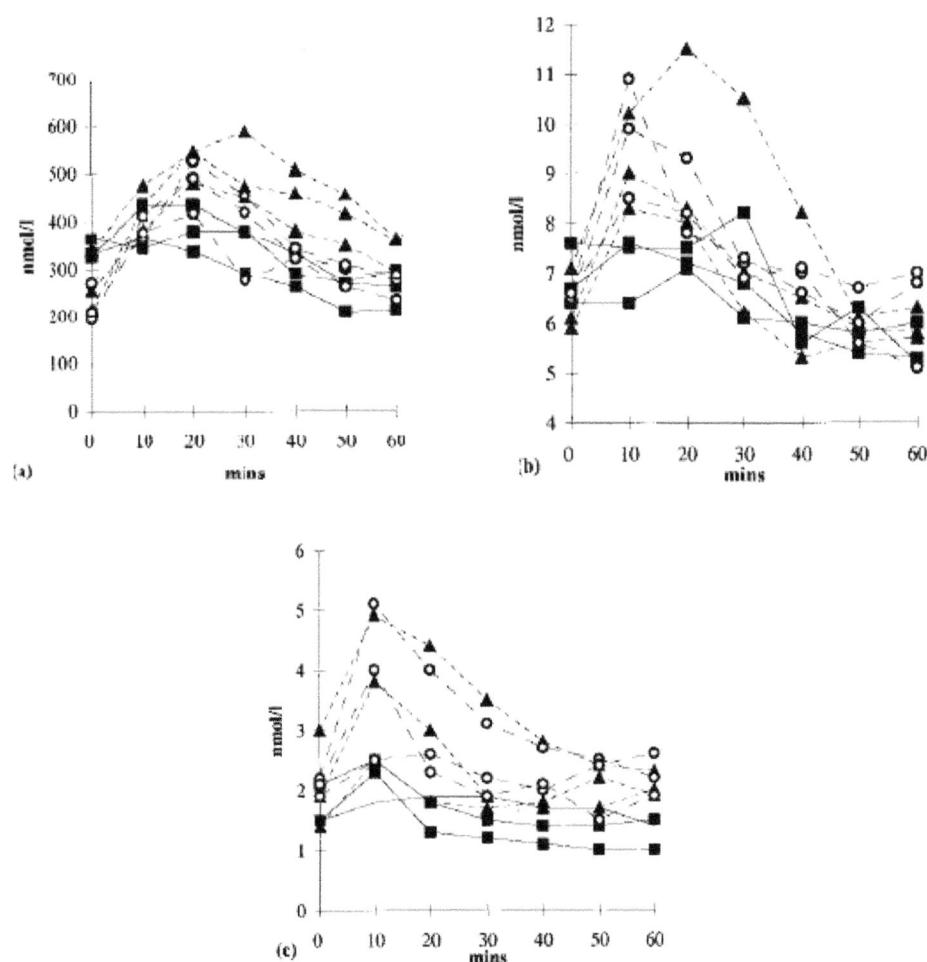

FIG. 2.7.5 Responses of (A) cortisol, (B) androstenedione and (C) 17-hydroxyprogesterone. *(From Bridges NA, Hindmarsh PC, Pringle PJ, Honour JW, Brook CG. Cortisol, androstenedione (A4), dehydroepiandrosterone sulphate (DHEAS) and 17 hydroxyprogesterone (17OHP) responses to low doses of (1–24)ACTH. J Clin Endocrinol Metab. 1998;83(10):3750–3. Erratum in: J Clin Endocrinol Metab 1999;84(8):2972. J Clin Endocrinol Metab 1999;84 (11):4177. 1998 p. 3753; 1999 p. 2972; 1999 p. 4177.)*

FIG. 2.7.6 Responses of aldosterone to low doses of Synacthen. *(From Honour JW, Bridges NA, Conway-Phillips E, Hindmarsh PC. Plasma aldosterone response to the low-dose adrenocorticotrophin (ACTH 1-24) stimulation test, Clin Endocrinol (Oxf), 2008;68 (2):299–303. Fig. 1 p. 300.)*

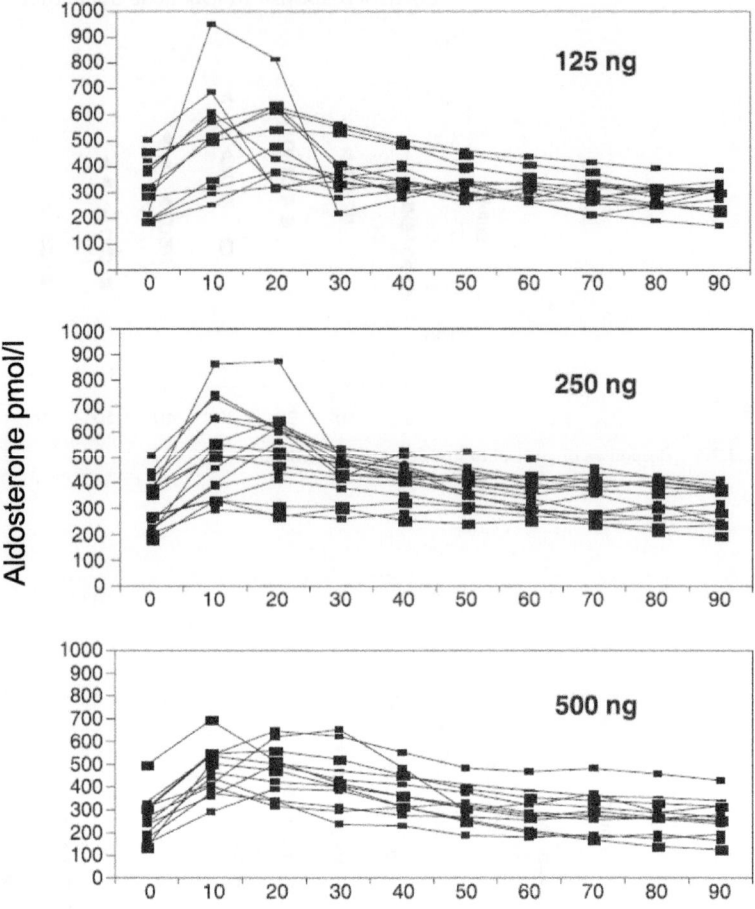

patient should not be on steroid therapy and have no recent major infection and no severe psychological stress. The test is administered as an out-patient procedure. When used as a first investigation, there will be no prior knowledge of the patient's cortisol rhythm.

The patient is prescribed 1 mg of dexamethasone as a single dose late at night. The dose for pediatric patients is 0.3 mg per sq. meter body surface area. The patient must attend hospital or G.P. early the following morning for blood sampling for measurement of plasma cortisol in the laboratory. After the single night-time dose of dexamethasone, suppression of cortisol to below 140 nmol/L is normal. Patients with Cushing's syndrome/disease or ectopic ACTH syndrome fail to suppress adequately (serum cortisol generally above 400 nmol/L). Patients with depression may also fail to suppress and this procedure is not ideal as a diagnostic test for them (Berlińska et al., 2020). Oral contraceptives should be stopped 6 weeks before the test (Vastbinder et al., 2016). A supplementary file (Berlińska et al., 2020) considers the impact of organ dysfunction on dexamethasone suppression tests and provides a useful list of drugs that induce or inhibit CYP3A4/activity.

2.7.2.2.2 Dexamethasone suppression test (long)

This test is used for a detailed diagnosis and evaluation of possible Cushing's disease (Sasaki et al., 2017). There should be a reasonable suspicion of Cushing's disease, e.g., from clinical state, and preliminary investigations. The suppression or otherwise of serum cortisol is evaluated during prolonged dexamethasone administration (Mojtahedzadeh et al., 2018). False positive results (i.e., failure of suppression) can occur in patients taking enzyme inducing drugs, e.g., rifampicin or phenytoin which induce cytochrome activity affecting dexamethasone metabolism and clearance. This test should be performed as an in-patient admission. Sample preparation is required for the night-time ACTH. The tests should be conducted where an approved ward-based preparation area is available. The amount of steroid used is contraindicated in patients with diabetes mellitus or in patients with psychiatric symptoms of Cushing's disease (Berlińska et al., 2020).

The test proceeds in three phases each lasting 2 days (Protocol 2.7.3).

Protocol 2.7.3 Dexamethasone suppression with urine sampling

Basal phase:

    Take blood samples 8–9 a.m. and 11–12 p.m., each time for cortisol and ACTH.

    Collect 24 h urine sample from Day 2 (for example, 12 a.m.–12 a.m.) for urine cortisol.

    During the 2 day **low-dose dex**. Phase:

    Give oral dexamethasone 0.5 mg qds starting midnight.

    Take blood samples at 8–9 a.m. and 11–12 p.m.

    Collect 24 h urine in second day for urine cortisol.

    During the 2 day **high-dose dexamethasone**. Phase:

    Give oral dexamethasone 2 mg qds starting midnight.

    Take blood samples 8–9 a.m. and 11–12 p.m.

    Collect 24 h urine for urine cortisol.

Given the limitations on admission to hospital of patients for investigations, a more restricted procedure may still give adequate information. Without admission, late night blood samples will not be possible. Sampling of bloods must be limited to mornings, and to the collection of 24 h urine samples. The high-dose dexamethasone phase should not be used on out-patients. It may be thought useful to continue low-dose dexamethasone for a third day (Box 2.7.1).

ACTH is perhaps more stable than used to be thought, but plasma should be separated as soon as possible and samples should be kept cold. The morning samples must be sent to the laboratory without delay. The midnight samples must be centrifuged and paired aliquots (ACTH and cortisol) of the plasma refrigerated or frozen as soon as possible. Cortisol is stable in blood, the night time samples need to have serum or plasma separated and kept cold and sent to the laboratory the following morning. All request forms for samples in this procedure must clearly show the correct date and time of each sample. N.B. the sample time for midnight is 2400 h on the day just ended and not 0000 h on the day just starting. Forms must also show how samples relate to the overall procedure (Basal day 1, high-dose dexamethasone day 2). Note: A day in this procedure starts after one night time sample and continues to include the next night-time sample.

The laboratory will analyze the plasma cortisol samples while holding the ACTH samples. The cortisol data will be reviewed to determine which, if any, of the ACTH samples will give the best further information. Immediate suppression of the serum cortisol, i.e., at 0800 h on the first low-dose day demonstrates the absence of any abnormality.

BOX 2.7.1 Dexamethasone suppression over 3 days with blood sampling

Admit patient.

The test proceeds in three phases—Basal, low-dose and high-dose dexamethasone: each phase lasting 2 days.

During each day of the 2-day basal phase:

    Take blood samples 8 a.m.–9 a.m. and 11 p.m.–12 mn.

        Each time for cortisol (serum) and ACTH (EDTA plasma).

    Optional: Collect 24 h urine samples *11 p.m.–11 p.m.* for urine cortisol (i.e., the first urine collection starts before the first blood sample).

During each day of the 2-day low-dose phase:

    Give oral dexamethasone 0.5 mg qds, starting *midnight*.

    Take blood samples 8 a.m.–9 a.m. and 11 p.m.–12 mn.

        Each time for cortisol (serum) and ACTH (EDTA plasma).

    Optional: Collect 24 h urine samples *11 p.m.–11 p.m.* for urine cortisol.

During each day of the 2-day high-dose phase:

    Give oral dexamethasone 2 mg qds, starting *midnight*.

    Take blood samples 8 a.m.–9 a.m. and 11 p.m.–12 mn.

        Each time for cortisol (serum) and ACTH (EDTA plasma).

    Optional: Collect 24 h urine samples *11 p.m.–11 p.m.* for urine cortisol.

All samples for ACTH a must be sent to the laboratory individually without delay.

All orders for samples in this procedure must clearly show the correct date and time of each sample. The sample time for midnight is 24.00 h on the day just ending (and NOT 00.00 h on the day just starting).

Patients with all forms of Cushing's syndrome show loss of the circadian rhythm in serum cortisol in the basal period. Overall the results may not be very high but the loss of the lower night-time level is notable. Patients with the pituitary dependent Cushing's disease usually show some or complete suppression of serum cortisol (<50% of basal) during the test. The suppression is not immediate but appears progressively as the time/dose of dexamethasone continues. This suppression occurs because the pituitary retains some ability to lower ACTH in response to the steroid load. A paradoxical later rise (escape) of cortisol during dexamethasone is also characteristic.

Patients with adrenal Cushing's syndrome or ectopic ACTH production show no ACTH suppression although a reduction of cortisol during the procedure can occur also because of lowering of ACTH. Suppression from high levels is therefore presumptive evidence of raised basal ACTH.

In some cases, it may be necessary to have samples analyzed for dexamethasone concentrations (Genere et al., 2022; Vogg et al., 2021; de Graaf et al., 2019; Hawley et al., 2018; Ueland et al., 2017; Hempen et al., 2012; Meikle, 1982). This can be part of a plasma steroid profile if available (Methlie et al., 2013). The dexamethasone should be greater than 5 nmol/L if compliance or rapid clearance is to be challenged, calculations on the theoretically expected dexamethasone concentrations, for example, 6.5 nmol/L for a person of 60 kg and 4.8 nmol/L for a person of 80 kg.

2.7.2.3 Stimulation of H-P-A axis

2.7.2.3.1 Insulin tolerance (stress) test

An insulin test is the gold standard for adrenal stimulation but is not a pleasant procedure (Tuchelt et al., 2000; Nye et al., 1999). The change in ACTH secretion is assessed indirectly via measurement of plasma cortisol concentrations. A day-case admission is feasible (Protocol 2.7.4).

Arrangements are needed for the patient to have a meal after the procedure. A rapid blood glucose test on finger-prick blood samples should be available, and the operator must be trained in its use. Resuscitation equipment should be available. This test is hazardous both due to the production of hypoglycemia, and to its possible treatment. The test should be carried out in the presence of medical staff trained in the procedure; and only on the instruction of a consultant. The insulin test should not be performed on patients over 65 years old or on those with epilepsy, cardiovascular or cerebrovascular disease; or on those with known Addison's disease, hypothyroidism or panhypopituitarism. Patients should have been screened adequately for these conditions, and affected patients may be stressed with glucagon instead of insulin. Pituitary function of ACTH can be evaluated in relation to growth hormone (GH) and TSH if a broader effect of pituitary function is suspected. The secretions of GH and ACTH are stimulated by hypoglycemia, induced by intravenous insulin.

Protocol 2.7.4 Insulin tolerance test (ITT)

Fast the patient overnight: water is permitted. Weigh the patient. Insert a venous cannula before the test. Keep patent with small amounts of saline or dilute heparin solution.

Prepare and label the necessary number of sample tubes according to Table 2.7.1. (Times are relative to the injection of Insulin.)

Where the table shows (✓) the sample collection is optional. To maintain a complete set, at least label a tube and send it empty to the laboratory.

Prepare the dose of Insulin. Calculations and doses must be fully cross-checked by a second qualified person.

Insulin: Standard dose: 0.15 U/kg intravenously.

N.B. Where a prior glucose (meter) test is ≤3.6 mmol/L, use a reduced dose of 0.10 U/kg.

GREAT CARE MUST BE TAKEN IN CALCULATING THE DOSE. IF THE DOSE EXCEEDS 10 units, STOP AND RECHECK. Draw up the dose only in a syringe with adequately detailed markings for the volume required.

Take the required Pre 20 min and Basal samples (Table 2.7.1). Inject the Insulin, with flushing of the syringe.

Hypoglycemia (Glucose <2.2 mmol/L on laboratory analysis or "Point of Care Test" meter) should occur at between 20 and 40 min after Insulin.

The patient should eat a light snack once symptoms of hypoglycemia are established.

If there are no symptoms of hypoglycemia, then the samples may still be collected. Repeating the procedure is not recommended. If subsequent laboratory testing shows no hypoglycemia, then the procedure may have to be repeated on another day.

If symptoms of hypoglycemia are severe, then treatment must be given. Standard treatment of hypoglycemia is with intravenous Glucose. Infuse 10% dextrose to a total of 2 mL/kg over 3 min. Measure the blood Glucose using a meter. Only if the symptoms require, restart the infusion of 10% dextrose at not more than 0.1 mL/kg/min. Continue to check the blood Glucose which should not be allowed to exceed 8.0 mmol/L. The object of the treatment is to raise the blood glucose, but not excessively.

TABLE 2.7.1 Timings of blood samples for insulin tolerance tests.

Time	Glucose	Cortisol	GH
	[fluor.]	[serum]	[serum]
Pre 20 min	(✓)	✓	✓
Basal	✓	✓	✓
20 min	✓	(✓)	(✓)
30 min	✓	✓	✓
60 min	✓	✓	✓
90 min	✓	✓	✓
120 min	✓	✓	✓

Where the table shows (✓) the sample collection is optional. To maintain a complete set, at least label a tube and send it empty to the laboratory.

2.7.2.3.2 Hypoglycemia during the ITT

Hypoglycemia (defined as blood glucose <2.2 mmol/L on laboratory analysis) should occur between 20 and 40 min after insulin. Note the GH and cortisol response is maximal at glucose concentration of 2.6 mmol/L. The blood glucose may be checked with a stick test or meter. The patient is encouraged to report symptoms. Milder symptoms, due to catecholamine release, pass off in 10–15 min as the blood glucose rises. This sequence indicates a satisfactory response and the sample collections should continue according to Table 2.7.1.

If there are no symptoms of hypoglycemia, then the samples may still be collected. The blood glucose may be checked using a meter. Experienced operators may review their metered glucose results, have medical approval to giving more insulin and prolong the collection of samples. If subsequent laboratory testing shows no hypoglycemia, then the procedure may have to be repeated on another day. If symptoms of hypoglycemia are severe, then glucose treatment must be given. The object of that treatment is to raise the blood glucose but not excessively.

Standard treatment of hypoglycemia uses intravenous glucose. Infuse 10% dextrose to a total of 2 mL/kg over 3 min. Stop and assess the symptoms. Measure the blood glucose using a meter. Only if symptoms require, restart the infusion of 10% dextrose at not more than 0.1 mL/kg/min. Continue to check the blood glucose which should not exceed 8.0 mmol/L. If venous access is lost, an alternative treatment for hypoglycemia is to inject glucagon subcutaneously. Effectiveness depends upon adequate liver glycogen. The blood glucose concentrations should be checked, and further action taken as required.

Hypoglycemia is not expected if glucagon is used as initial stress although there can be a later rebound hypoglycemia. Collect the blood samples for glucose, growth hormone and cortisol according to Table 2.7.1.

All of the symptoms and treatment must be recorded in the patients' notes.

At the end of the procedures, the patient should have an adequate meal. After taking the last blood sample some protocols specify that hydrocortisone 50 mg should be given intravenously either if hypopituitarism is suspected or if hypoglycemia has been treated.

Hypoglycemia has been achieved after insulin if the glucose falls to 2.2 mmol/L or below at any time. If the patient did not become hypoglycemic, then the growth hormone and cortisol results are not relevant. This requirement does not apply to the use of glucagon as a stress.

2.7.2.3.3 Glucagon tolerance test

As an alternative to insulin a 1-mg dose of glucagon can be given subcutaneously (Weintrob et al., 2018). For patients >90 kg, the dose is increased to 1.5 mg glucagon. Take the required basal samples (see Table 2.7.1 for insulin tolerance). Inject the glucagon (dose and route as above). Glucagon leads to an increase in blood glucose but can be accompanied later in the test at 150–180 min by hypoglycemia due to enhanced insulin release by the glucagon stimulated glucose elevation. Overall, glucagon is not as predictable a stimulus for cortisol than the ITT is. In response to glucagon, cortisol should peak between 60 and 90 min into the test if hypoglycemia has been achieved (peak 30–45 min after hypoglycemia). An increment of cortisol of 200 nmol/L or an absolute peak above 450 nmol/L can be interpreted as an acceptable response. A suboptimal response indicates pituitary or adrenal insufficiency. The normal growth hormone response is arbitrarily considered to be an increase to 20 mU/L. An increase to 9 mU/L or less is diagnostic of GH deficiency but there is a spectrum of severity and intermediate levels which suggest partial GH deficiency (Giordano et al., 2008).

2.7.2.3.4 Metyrapone test

Metyrapone blocks 11-beta hydroxylase in the adrenal cortex and results in the inhibition of conversion of 11-deoxycortisol to cortisol. Serum levels of cortisol decrease and concentration of 11-deoxycortisol increases, however 11-deoxycortisol does not downregulate ACTH (Creemers et al., 2015). Therefore in a normally functioning HPA axis there is an increase in ACTH and hence 11-deoxycortisol. This metabolite can be directly measured in the serum or saliva or rarely these days measured in the urine as 17-OH corticosteroids. This test will be abnormal in either primary adrenal deficiency or ACTH deficiency. LC-MS/MS is available as a biochemical test in these studies (Monaghan et al., 2011).

The test can be performed as an overnight test. Metyrapone is given orally (30 mg/kg body weight or 2 g for <70 kg, 2.5 g for 70–90 kg and 3 g for >90 kg body weight) at midnight with a glass of milk or a small snack. Serum 11-deoxycortisol and cortisol are measured at 8 a.m. the next morning; plasma ACTH can also be measured if available. After metyrapone at night the 8 a.m. serum 11-deoxycortisol concentrations should be >200 nmol/L (7 µg/dL) with serum cortisol less than 140 nmol/L (5 µg/dL), confirming adequate metyrapone blockade. The plasma ACTH concentration at 8 a.m. should exceed 17 pmol/L (75 pg/mL), confirming that any increases in serum 11-deoxycortisol concentrations are ACTH-dependent, thereby separating primary from secondary adrenal insufficiency.

The concurrent use of corticosteroids will interfere with the test. Any medications that the patient is taking that increases the P450 enzymes (such as rifampin, phenobarbital or phenytoin) will increase the metabolism and clearance of metyrapone. Grapefruit juice should be avoided for the same reason. Similarly, hypothyroidism or hyperthyroidism will affect clearance of metyrapone and the adrenal responsiveness. Thyroid function tests should therefore be measured prior to performing this test. Measurement of 11-deoxycortisol, like cortisol itself is dependent on CBG binding, drugs such as estrogens and oral contraceptives will falsely increase the concentrations of 11-deoxycortisol. The test may lead to headache, dizziness, sedation; Allergic rash; Nausea, vomiting, abdominal discomfort or pain; decreased white blood cell count or bone marrow suppression is rare (Nieman, 2018).

In pregnancy, this test should only be used if essential. Subnormal response may occur in pregnant women and the fetal pituitary may be affected. The extent of excretion of the metyrapone in breast milk is unknown therefore the advice is to use with caution in lactation.

2.7.2.3.5 CRF test with peripheral vein sampling

This procedure is used to evaluate the source of known excessive ACTH secretion (Iwanaga et al., 2017). Tumors providing ectopic sources of ACTH have no CRF receptors and do not respond. Following intravenous CRF, the ACTH response is monitored via the changes in plasma cortisol (Newell-Price et al., 2002).

Obtain CRF (100 µg dose) for I.V. injection from pharmacy (CRF, Bissendorf, Germany). This is an unlicensed product obtainable on a named patient basis against the signature of the clinical consultant organizing the procedure. The patient may be admitted on a day-patient basis.

The patient may notice mild facial flushing following the CRF injection. An in-dwelling catheter may be used.

1. Take blood for serum cortisol (Basal, pre-CRF T-15 min).
2. Take blood for serum cortisol (Basal, 0 min) and then inject CRF intravenously.
3. Take further blood samples for cortisol at 15, 30, 45, 60, 90 and 120 min after CRF.

Interpretation:

Normal: Cortisol rises to a maximum above 550 nmol/L. In patients with pituitary disease, cortisol rises and exceeds 700 nmol/L and for ectopic tumors secreting ACTH, cortisol is high and shows no rise after CRF.

2.7.2.3.6 Desmopressin

Desmopressin is a vasopressin analog acting at V receptors to mediate water retention. Desmopressin, by consensus, has negligible or no ACTH-releasing activity in the context of a normal H-P-A. The literature shows variability between studies that can be attributed to dose, bolus or continuous mode of administration, the time of investigation and criteria for positive response (Vassiliadi and Tsagarakis, 2018). When used in patients with Cushing's syndrome there are high concentrations of V3 receptors in neoplastic tissue and a significant response of cortisol is recorded. The desmopressin test is thus best used in differential diagnosis of ACTH dependent Cushing's syndrome (Vassiliadi et al., 2015). A dose of 5 or 10 µg of desmopressin is injected and blood samples for cortisol and separately for ACTH are drawn 15, 30, 45, 60, 90 and 120 min after the injection (Malerbi et al., 1993). A 21%–380% increase in cortisol was seen in patients with Cushing's disease.

2.7.3 Aldosterone and renin

To investigate for mineralocorticoid activity related to a suspected adrenal disorder, plasma renin activity and aldosterone concentration are measured. **Plasma renin activity (PRA)** is determined as Angiotensin I (AI) generation over time with the detection of AI by immunoassay or mass spectrometry (Bystrom et al., 2010), or the **plasma renin concentration (PRC)** can now be measured by two-site ELISA. The mass of AI is 1296 Da which is within the range for detection by mass spectrometry when detecting multiply charged ions according to the m/z convention. Some laboratories now use measurements of renin concentration with immunoradiometric assays. Appropriate reference ranges for age must be used as children have higher renin activity and concentrations than adults (Fiselier et al., 1983a, b, 1984a, b) (see in earlier chapter Fig. 2.5.17).

In congenital adrenal hyperplasia aldosterone concentrations are low when there is a defect of 21-hydroxylase or 3β-hydroxysteroid dehydrogenase so plasma renin activity or concentration is raised. On the contrary, in CAH due to 11-hydroxylase and 17-hydroxylase defects, high concentrations of the mineralocorticoid deoxycorticosterone leads to sodium retention and renin suppression and aldosterone concentrations are low (Tran et al., 2018).

2.7.3.1 Aldosterone/renin ratio (diagnosis of Conn's syndrome)

Renin and aldosterone are measured in the basal state and after provocative testing (Morera and Reznik, 2019; Wu et al., 2019; Funder et al., 2016; Rehan et al., 2015; Vaughan et al., 1981). Conn's syndrome is a treatable cause of hypertension due to hyperaldosteronism and is suspected when there is hypokalemia (unless caused by diuretics). Normokalemia however should not now prevent further investigation. Random renin results are not diagnostic. The basal assessment may be repeated after **sodium loading**. Without the salt loading procedure, admission is not required overnight for the determination of the response to assuming upright posture from a prone state. With salt loading longer admission is required. The patient must be off all antihypertensive treatment. ACE inhibitors and spironolactone must be stopped for at least 4 weeks, any others for at least 2 weeks. Bethanidine, Nifedipine or Prazosin may be given if necessary. The patient should receive potassium supplements if hypokalemic. The diet should offer 100–300 mmol/24 h sodium and 50–100 mmol/24 h potassium, inclusive of dietary sources for 3 days before the test. All supplements should be stopped before taking the samples.

The patient is admitted overnight. No food or drink except water is allowed after midnight. The patient must remain recumbent all night (and use bedpan rather than getting up to use the toilet). This must continue until the first samples are taken. At 8 a.m. with the patient basal and still recumbent, take blood samples for plasma (1) U and E (2) recumbent renin activity/concentration and aldosterone. The patient should then get up. A 24-h urine collection is started for urea, electrolytes and creatinine. The patient must walk around for at least 30 min. At 8.30 a.m. (30 min after getting up), take blood sample for ambulant renin activity/concentration. Send the samples for the two available times to the laboratory. At 12 noon, take blood sample for ambulant aldosterone concentration. Send to the laboratory.

Salt-loading (optional) involves additional salt to a normal diet (Thuzar et al., 2020; Stowasser et al., 2018; Schirpenbach et al., 2006). Give 170 mmol "Slow Sodium" per day (10 mmol sodium per tablet) divided in six hourly doses (5 tablets, then 4, 4, 4) together with fludrocortisone

TABLE 2.7.2 Aldosterone/renin ratios from plasma aldosterone concentration (PAC), plasma renin activity (PRA) and direct plasma renin concentration (DRC) according to units of measurement.

	PRA (ng/mL/h)	PRA (pmol/L/min)	DRC (mU/L)	DRC (ng/L)
PAC (ng/dL)	20	1.6	2.4	3.8
PAC (ng/dL)	30	2.5	3.7	5.7
PAC (pmol/L)	40	3.1	4.9	7.7
PAC (pmol/L)	750	60	91	144
	1000	80	122	192

0.5 mg once daily and continue for 3 days. Continue 24-h urine collections for urea, electrolytes and creatinine. On Day 3, repeat the recumbent and ambulant renin and aldosterone levels.

The literature can be very confusing because of the different units in use for the aldosterone concentration and PRA or PRC. This means there is no number for a clinician that is easy to remember other than the local range. Cut-offs were summarized by the Endocrine Society (Table 2.7.2). It is often not possible to test the patient off all medications that affect the RAA that can affect the results through primary affects and hence ratios. If it is essential to perform the test, then the impact of drugs on ARR are summarized in Table 2.7.3. Oral contraceptives can affect the ARR (Ahmed et al., 2011) depending on timing and assay used (see earlier chapter Fig. 2.5.18).

2.7.3.2 Tests of the renin-angiotensin-aldosterone system

If an initial ARR raises suspicion of a clinical issue, then further tests are recommended as confirmatory tests.

1. Aldosterone suppression

Normal subjects will show suppression of the serum aldosterone concentrations after **volume expansion** with isotonic saline. The test is done after an overnight fast. An **infusion of 2 L of 0.9% sodium chloride** is given intravenously by an infusion pump over 4 h into the recumbent patient. At the end of the infusion, blood is taken for measurement of serum aldosterone concentration which in normal subjects decreases to <300 pmol/L (<10 ng/dL). During this test, a patient requires careful monitoring and because of the risks of

TABLE 2.7.3 Effects of drugs on aldosterone renin ratios.

Drug	Aldosterone	Renin	Effect on ARR
Calcium channel blocker (Nifedipine, Amlodipine)	Minimal decrease	Minimal increase	No effect
Calcium channel blocker (Verapamil)	Small decrease	Minimal increase	No major effect
α-Blocker (Prazosin, doxazosin)	Minimal	Minimal	No effect
Vasodilators (Diazoxide, hydralazine)	No effect	No effect	No effect
ACE inhibitors (Captopril)	Decreased	Increased	False negative
Diuretics (Furosemide—loop, hydrochlorothiazide—distal tubule)	Increased	Increased markedly	False negative
Minoxidil—vasodilator/diuretic (baldness)	Minimal	Increased	False negative
AII receptor blocker (Sartans)	Decreased	Increased	False negative
Spironolactone (K sparing diuretic)	Increase	Large increase	False negative
Beta-blockers (Metoprolol, atenolol)	Minimal	Decreased	False positive
Methyl-DOPA	Minimal decrease	Decreased	False positive
Direct renin inhibitor (aliskiren)	Decreased	Decreased	False positive

Hypokalemia—inhibits aldosterone (False negative); potassium loading (False positive). Estrogens (Pregnancy, HRT, O/C) increase renin substrate (False negative). NSAIDs retain sodium and reduce PRA (False positive). Renovascular and malignant hypertension increase renin activity (False negative).

hypertension or heart failure the test is rarely used (Thuzar et al., 2020; Stowasser et al., 2018).

2. **Stimulation of renin**

   PRA levels can normally be stimulated by **a change in posture**. A 4-h period in the upright posture when on a sodium restricted diet will raise PRA to 3 pmol/mL/h (3 ng/mL/h) whereas patients with primary aldosteronism will have PRA which remains <2 pmol/mL/h (3 ng/mL/h) (Funder et al., 2016).

3. **Converting enzyme inhibition—Captopril**

   After 25 mg Captopril, plasma aldosterone falls by more than 50% over 2 h. Similar results are found however in essential hypertension. In primary aldosteronism, there is no suppression of plasma aldosterone in this test (Funder et al., 2016).

2.7.4 Gonadal function tests

The clinician is alerted to gonadal disorders by changes in endocrine sensitive tissues (see Chapter 1.7) which need to be recognized against normal changes in development before and after birth (Shah et al., 2020) and during puberty. The cause of gonadal disease can be partly characterized from the measurements of a steroid hormone and the pituitary regulatory hormones. The hypothalamic pituitary gonadal axis involves principally LH, FSH, estradiol or testosterone (Corradi et al., 2016) but several other proteins are now known to be involved although the whole axis is still not completely understood. An overlap with adrenal disease

can be excluded by a dexamethasone test. The measurements of inhibin B and anti-Müllerian hormone are increasingly useful (Wei et al., 2017).

2.7.4.1 HPG stimulation tests

Patients with clinical signs of precocious puberty but low concentrations of basal LH require further investigations of the HPG axis for clarification. The luteinizing hormone releasing hormone (LHRH) or the gonadotrophin hormone releasing hormone analog (GnRHa) tests are useful to identify whether the HPG-axis is activated to a pubertal level. Different LHRH and GnRHa test protocols have been described.

2.7.4.1.1 LHRH test

Determinations of the concentrations of plasma gonadotrophins are unlikely to give diagnostic information in newborn infants with ambiguous genitalia. In the pubertal child with primary testicular failure, basal LH and follicle-stimulating hormone (FSH) levels may be elevated (Pasternak et al., 2012; Houk et al., 2009). In testicular dysgenesis, FSH is a more sensitive index of seminiferous tubular damage. When plasma basal gonadotrophins are elevated, the response to luteinizing hormone-releasing hormone (LHRH) is unlikely to provide any further diagnostic information. In androgen insensitivity syndrome (AIS), testosterone and gonadotrophins are usually within or above the normal range. In pubertal patients with hypogonadotrophic hypogonadism, testicular function may be low because of lack of stimulation from gonadotrophin-releasing hormone (GnRH) (Rey et al., 2013; Harrington and Palmert, 2012).

A common LHRH protocol involves measurements of FSH and LH concentrations at baseline, then 20 and 60 min after an intravenous bolus of 2.5 µg/kg or 100 µg/m^2 LHRH (maximum 100 µg). FSH and LH concentrations increase at 20–30 min and decrease again at 60 min. A prepubertal response includes a low LH with FSH predominance (i.e., FSH peak greater than LH). International guidelines suggested peak LH of >5 IU/L, and/or stimulated LH/FSH ratio of >0.66 as a cut-off for pubertal response during LHRH testing. However, there are variations in peak LH responses depending on local assays (Özalkak et al., 2020), and lower peak LH responses have been reported in those with obesity. Basal and stimulated FSH concentrations alone are not useful due to the large variability and overlapping results between prepubertal and postpubertal children. Patients with nonprogressive thelarche may show some rise to LHRH in LH with a lower LH/FSH ratio than gonadotrophin-dependent precocious puberty.

An LHRH test can localize the primary defect to the hypothalamus. In some patients, the distinction between hypogonadotrophic hypogonadism and constitutional delay of puberty may be difficult. In these cases, a GnRH agonist test (e.g., Nafarelin) is useful. A maximum LH peak after 4 h is <6.2 U/L and/or a testosterone increase after 24 h of <1 nmol/L support hypogonadotrophic hypogonadism. If the result is negative, repeated bolus injections of GnRH will prime the pituitary and may normalize gonadotrophin release in patients with hypogonadotrophic hypogonadism patients with hypothalamic but not pituitary defects. The response to GnRH is variable and some consider this test is useless.

2.7.4.1.2 GnRHa (analog) tests

The gonadotrophin releasing hormone analog (GnRHa) test with Leuprolide or Triptorelin is used as an alternative to the LHRH test in countries where LHRH is not available (Yazdani et al., 2012). However, large variations in the timing and concentration of peak LH which define puberty have been reported. After Leuprolide administration, LH and FSH increase for 2 and 4 h, respectively, and remain high for 24 h. Peaks for testosterone and estradiol occur later (around 16 h). Reference ranges have been published (Bang et al., 2017; Bahaeldein and Brassill, 2018).

In polycystic ovary syndrome, dysregulation of androgen synthesis has been defined by an increase in 17-hydroxyprogesterone on GnRHa (Nafarelin) stimulation (Rosenfield et al., 1994) (Fig. 2.7.7). The test is performed with overnight dexamethasone suppression of the adrenal cortex. The basis of the results is still under investigation (Homer et al., 2019).

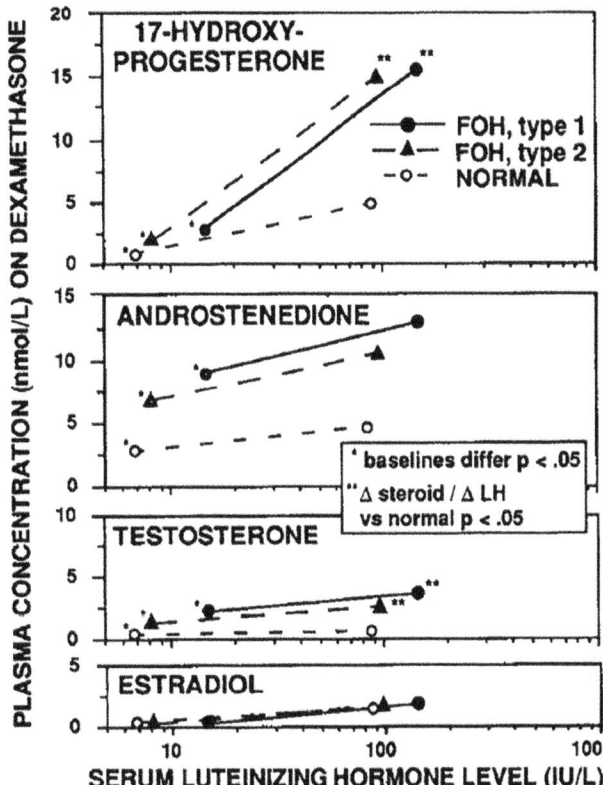

FIG. 2.7.7 Response of 17-OHP to GnRH. Apparent dose-response relationships between baseline and peak blood levels of LH and steroids in women with ovarian hyperandrogenism (FOH) compared with those in normal subjects. Mean baseline and peak blood levels of LH are plotted against these respective parameters for steroids. Baseline 17-PROG, AD and T levels differ among groups independently of LH according to ANCOVA (*). In addition, the apparent slopes of the LH—17PROG and LH—T dose-response curves (Δ steroid/Δ LH) are significantly greater in both types of FOH than normal (**). The data-in nafarelin-negative subjects (not shown) are similar to those in normal subjects for 17-PROG and E and are greater than but parallel to normal for AD and T. *(From Rosenfield RL, Barnes RB, Ehrmann DA. Studies of the nature of 17-hydroxyprogesterone hyperresonsiveness to gonadotropin-releasing hormone agonist challenge in functional ovarian hyperandrogenism. J Clin Endocrinol Metab 1994;79(6):1686–92. Fig. 1 p. 1691.)*

In patients with precocious puberty, a pubertal response from the LHRH or GnRHa test will be observed in gonadotrophin-dependent precocious puberty, but absent in gonadotrophin-independent precocious puberty. False positive results may be present in premature thelarche and ovarian failure.

In patients with delayed puberty, the LHRH stimulation test may not differentiate between CDGP and IHH as there is significant overlap of LH and FSH responses between these patients (Chan et al., 2018). Alternative protocols of the LHRH test involving administration of small repetitive pulses of LHRH over 36 h has demonstrated lower LH responses in IHH than CDGP, but the protocol is complicated, invasive with unsatisfactory diagnostic accuracy. The GnRHa test has shown better discriminatory values in differentiating IHH from CDGP which may be due to the greater stimulation of GnRHa in patients CDGP by awakening the primed normal gonadotrophic cells. However, studies are limited by small patient numbers, male gender only and lack of consistent diagnostic thresholds (Turra et al., 2015).

2.7.4.1.3 Beta human chorionic gonadotrophin test (hCG)

The hCG test assesses function of testicular tissue and defects of testosterone biosynthesis (Anderson et al., 2018; Bertelloni et al., 2018). In the pediatric patient aged 6 months to 8 years, basal gonadal steroids are frequently undetectable in plasma, and gonadal function can only be assessed by Leydig cell stimulation using hCG. Many different protocols are used for **hCG administration**. Excellent Leydig cell stimulation has been demonstrated using a single injection of $5000 IU/m^2$ body surface area with plasma testosterone measured 24 or 72 h later. Leydig cell desensitization occurs following multiple hCG administrations (Bang et al., 2017).

The hCG test must be performed after the LHRH test if they are both being done on the same day to avoid contamination of the LHRH test with the hCG response. The hCG stimulates testicular testosterone production from Leydig cells over a prolonged period of time via stimulation of LH receptors. A typical protocol for the short hCG test to investigate for biosynthetic disorders involves daily intramuscular injection of hCG 500–2000 IU over 3 days. A prolonged test over 3 weeks with two injections per week should be performed particularly if gonadotrophins deficiency is suspected to allow a response by receptors which may have been downregulated due to the absence of their ligand. A normal testosterone response will show a rise from baseline level but reference ranges are age and assay

dependent. The testosterone response is blunted if there is poor testicular function or absent functioning testicular tissues (Roli et al., 2017).

The normal testosterone response to hCG depends on the age of the patient. Of note, the values given here have not been *re*-validated with recent LC-MS/MS technology but they are based on RIA methods. In infancy, a normal testosterone increment after hCG may vary from 2-fold to 10- or even 20-fold. During childhood, the increment is between 5- and 10-fold. Increments in plasma testosterone are typically between 2.0 and 8.5 nmol/L with peak values between 2.3 and 8.7 nmol/L. During puberty, as the basal concentration is higher, the increment is less, i.e., 2- to 3-fold. If there is no response to a short hCG test, a prolonged test over 2 weeks (Table 2.7.4) should be performed, and a 5- to 10-fold increment from the basal testosterone level will be a normal response. If an ovotestis is suspected, estradiol should be measured in a human menopausal gonadotrophin test.

Measurement of hCG concentration is sometimes used to distinguish a hypothalamic/pituitary source of hCG from a gonadal cause in boys with delayed puberty who have coexisting abnormalities of testicular development such as bilateral cryptorchidism, gonadal dysgenesis, anorchia ("vanishing testis syndrome") or other forms of disorders of sexual development, puberty and sexual development (Sturgeon and McAllister, 1998). Steroids may also be measured so androstenedione concentrations and androstenedione to testosterone ratio in 17β-hydroxysteroid dehydrogenase type III deficiency (Yang et al., 2017), and testosterone to dihydrotestosterone ratio in 5α-reductase type II deficiency (Fan et al., 2020). Elevation of androstenedione in relation to testosterone suggests a diagnosis of 17β-hydroxysteroid dehydrogenase deficiency in most but not all cases. Exaggerated testosterone to dihydrotestosterone ratio is seen in 5α-reductase deficiency.

TABLE 2.7.4 Responses to transdermal dihydrotestosterone of 4 patients with 5-alpha reductase deficiency.

Patient	Age (yr)	DHT dose (mg/day)	Duration (weeks)	LH (IU/L) before	LH (IU/L) after	FSH (IU/L) before	FSH (IU/L) after	SPL (cm) before	SPL (cm) after
1	11	2.5	8	1.4	<0.2	3.8	1.7	3.3	4.5
2	10 to 11	2.5	16	<0.2	<0.2	0.6	<0.2	22.2	5
3	8	12.5	8	<0.2	<0.2	1.2	<0.2	2.8	4.2
4	4	12.5	16	<0.2	<0.2	0.5	<0.2	2	3.5

DHT, dihydrotestosterone; SPL, stretched penile length.

(From Sasaki G, Ishii T, Hori N, Amano N, Homma K, Sato S, Hasegawa T. Effects of pre- and post-pubertal dihydrotestosterone treatment on penile length in 5α-reductase type 2 deficiency. Endocr J 2019;66(9):837–42. Table 2 p. 839; Table 3 p. 841.)

2.7.5 Clinical tests

Not all tests require steroid/hormone measurements. The clinical examination and history is always an important start to investigations of patients. The response to some provocative tests can be judged by a clinical response such as ovulation and pregnancy or by abdominal ultrasound to observe changes in the ovary and uterus.

In addition to endocrine tests based on the measurements of circulating hormones, the response of the female reproductive tract is amenable to direct clinical assessment. Hence, the morphology of the cells from the vaginal epithelium changes in response to estrogen. Cervical mucous secretion ceases when progesterone production increases.

2.7.5.1 Ultrasound

These days abdominal ultrasound is widely used to measure in the ovaries the follicle number and development, the dimensions of the uterus and the endometrial thickness. Transvaginal ultrasound is a procedure free from radiation. Pelvic ultrasound examination of the uterus is used to measure longitudinal and transverse lengths and the anterior-posterior diameter (Kelsey et al., 2016). Uterine volume is then calculated by multiplying those measurements and further multiplying by 0.52 based on an ellipse (Gilligan et al., 2019) (Fig. 2.7.8). There is a direct correlation between uterine dimensions and serum estradiol concentrations.

2.7.5.2 Ovulation induction

Clomiphene, a nonsteroidal derivative of triphenylethylene, is one of the most effective agents for inducing ovulation. By opposing estrogen action at the pituitary, clomiphene can be used to stimulate gonadotrophin secretion. At a typical dose of 100 mg/day for 5 days (days 5 through 9 in the cycle), this leads to an increase in FSH on day 10 of the cycle. The resulting rise in circulating estrogen triggers an LH surge that induces ovulation 5–10 days after the last dose of Clomid (Hendriks et al., 2006). FSH concentrations are measured at the start of Clomid treatment then from 10 days after the treatment. Intercourse should take place at this stage. Blood tests can also be performed for progesterone from 14 to 16 days after the last Clomid. Pregnancy may follow. The effects are variable and this may be due to the fact that the Clomiphene preparation is an isomeric mixture of two forms one isomer being an antiestrogen the other is a weak estrogen that are sensed differently at the pituitary. It is difficult to put accurate results to the FSH concentrations because of assay variability and changes in carbohydrate content of the protein. This test has no value in children. Clomiphene increases LH and FSH in males with testosterone stimulation and an increase in sperm mobility.

2.7.5.3 Response to testosterone/hCG

In a newborn 46XY individual with micropenis, the administration of testosterone leads to growth of the stretched penis (Fig. 2.7.9). This test is rarely used now to exclude hormone resistance. In boys with delayed puberty, oral TU is administered in increasing doses; starting with 40 mg once daily rapidly increasing to 40 mg twice daily and subsequently 80 mg twice daily. Over 1 year, pubertal progression was assessed through changes in growth (height and BMI), testicular volume, gonadotrophins, sex steroids, SHBG and inhibin B (Lawaetz et al., 2015).

The speed of dose increments depended on the individual boy (age and maturity). If basal LH and/or testicular

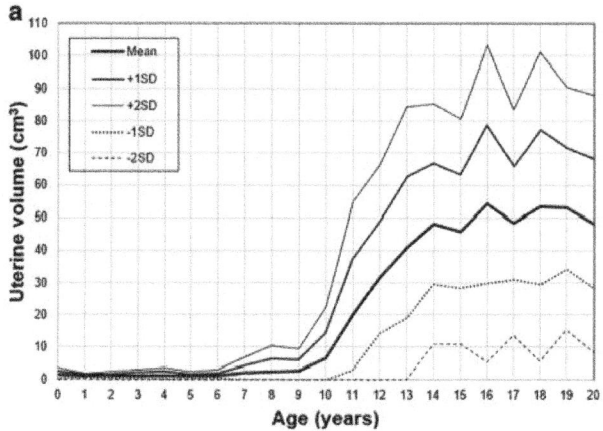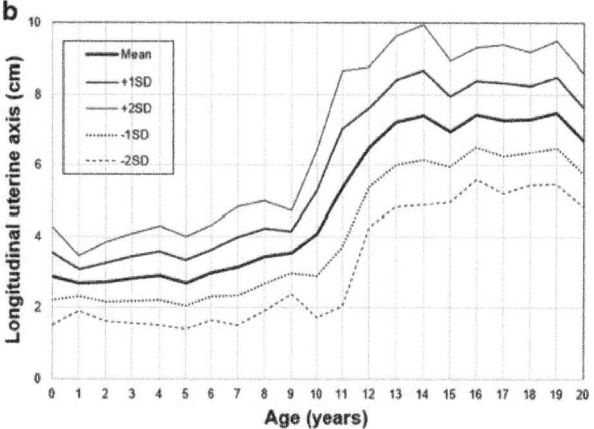

FIG. 2.7.8 Mean uterine volume (A) and uterine longitudinal axis (B). In the original figure anteroposterior axis, transverse axis, left ovarian volume, right ovarian volume, and endometrial stripe thickness were reported. *(From Gilligan LA, Trout AT, Schuster JG, Schwartz BI, Breech LL, Zhang B, Towbin AJ. Normative values for ultrasound measurements of the female pelvic organs throughout childhood and adolescence. Pediatr Radiol 2019;49(8):1042–50. Fig. 2 p. 1046.)*

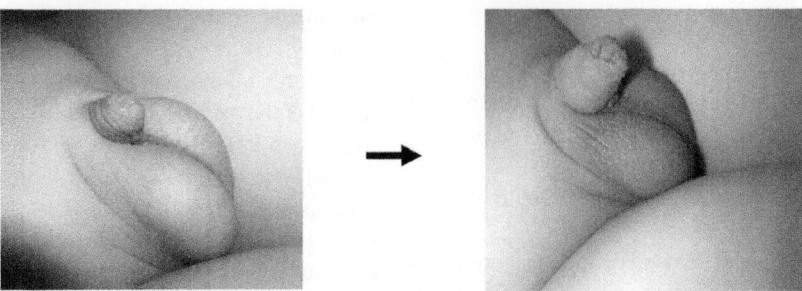

FIG. 2.7.9 Effect of dihydrotestosterone on stretched penile length in a prepubertal 5αRD2 patient. Patient was treated with 25 mg per day of dihydrotestosterone (DHT) for 8 weeks at the age of 10–11 years. The stretched penile length (SPL) was enlarged from 2.2 cm (−3.1 SD, left) to 3.2 cm (−1.9 SD, right). Then additional 8-week application of DHT extended to 5.0 cm (+0.3 SD) in SPL. Thickened band of scrotum is also presented. *(From Sasaki G, Ishii T, Hori N, Amano N, Homma K, Sato S, Hasegawa T. Effects of pre- and post-pubertal dihydrotestosterone treatment on penile length in 5α-reductase type 2 deficiency. Endocr J 2019;66(9):837–42. Fig. 1 p. 840.)*

volume increased significantly during oral TU, therapy was paused for 2 weeks followed by measurement of LH and testosterone. Treatment was stopped if a significant increase in LH and testosterone was observed. In patients treated with testosterone supplementation, the type of product, dose and duration of treatment were registered. Oral TU treatment was in the form of Andriol (MSD). Clinical and biochemical values were recorded at baseline and at each year thereafter (0.2 years). The response was satisfactory in a number of cases. (N.B. Restandol is no longer available in the United Kingdom.)

2.7.6 Catheter studies

In order to localize a suspected tumor, it may sometimes be necessary to insert a catheter into likely sites. With modern imaging techniques the use of contrasts is rarely required when following the location of the catheter (Page et al., 2018; Monroe et al., 2017; see Chapter 3.1.1 for cortisol secreting tumors and ACTH secreting tumors; Chapter 3.3.1 for aldosterone secreting tumors).

2.7.6.1 Adrenal venous sampling is used to localize an aldosterone secreting tumor

This test helps to distinguish unilateral primary aldosteronism from hyperplasia (Turcu et al., 2020; Blondin et al., 2015). Access to the adrenal vein is usually gained from the femoral vein and inferior vena cava, occasionally from the antecubital vein (Jiang et al., 2017), radiography is required to follow the course of the catheter. The test is conducted sometimes after ACTH stimulation (Deinum et al., 2019; Nakamura et al., 2012) (Fig. 2.7.10). Note the concentrations of the steroids in comparison with circulating concentrations, for most ACTH stimulation samples a dilution will be required for steroid results to be within calibration ranges.

If seeking an aldosterone secreting tumor, the usual practice is to measure the concentration of cortisol as well. The success of the catheterization is based on the concentrations of steroids in the samples. At UCLH in 89 adrenal catheter studies, that plasma of blood from left adrenal vein was 1750–58,320 nmol/L in 85 of the cases. On the right side, only 64 of the cases had grossly elevated cortisol concentrations. Samples from the inferior vena cava were 460–2077 nmol/L in the whole group. There have been attempts to have the cortisol measurement during the procedure by using a simple point-of-care device near the patient who needs a quick centrifugation of the sample or using a whole blood method or by sending the sample to the laboratory for stat analysis in the laboratory (Cesari et al., 2017; Yoneda et al., 2016). A very quick response is required to produce cortisol results quick enough to influence the radiologists approach to judge if sampling was in the correct site. If looking at tests retrospectively in some cases, catecholamines were also measured. Now that several steroids can be measured in the sample using LC-MS/MS technology other steroids have also been used including DHEAS and 11-deoxycortisol (Eisenhofer et al., 2016; Nilubol et al., 2017; Ceolotto et al., 2017; Peitzsch et al., 2015). The AVS has also been performed with metoclopramide stimulation (Rossitto et al., 2018)—APA patients show an enhanced response. A unilateral tumor should result in low concentrations of steroid in the contralateral samples (Kline et al., 2015).

In order to **localize an ectopic ACTH secreting tumor**, many sites for sampling blood are required (Fig. 2.7.11). In the search for sites of excess androgens, the adrenal and ovary are catheterized.

2.7.6.2 CRF test with petrosal vein sampling

The site of excessive ACTH secretion from the pituitary can be localized with petrosal sinus sampling with or without

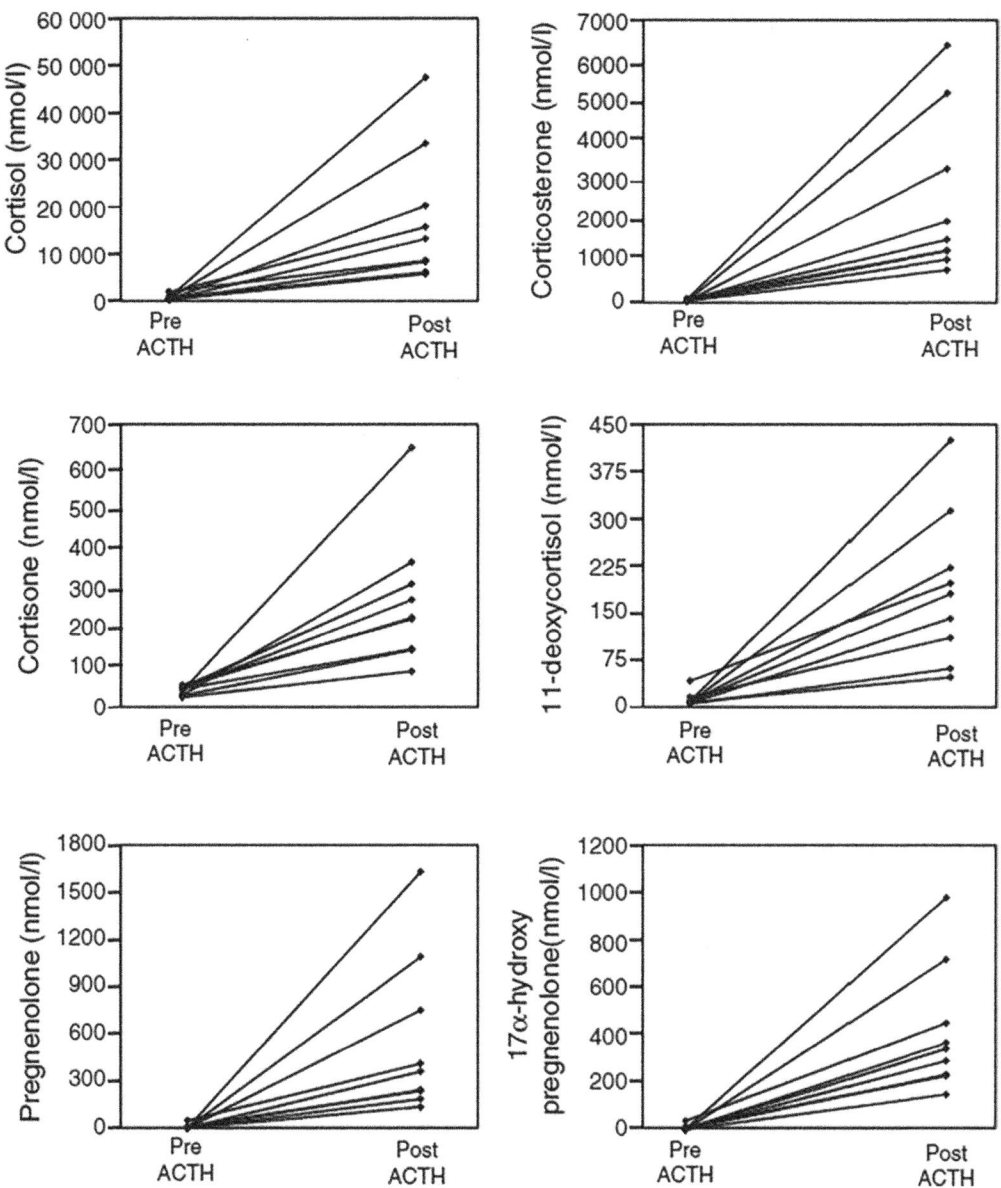

FIG. 2.7.10 Adrenal venous sampling. Concentrations (nmol/L) were determined using LC-MS/MS for cortisol, corticosterone, cortisone, 11-deoxycortisol, pregnenolone and 17α-hydroxypregnenolone in the adrenal vein serum of nine subjects pre- and post-ACTH stimulation (15 min). Each line denotes the steroid increase in an individual ($n=9$, 3 men and 6 women). The line graphs denote the concentrations (nmol/L) of cortisol, corticosterone, cortisone, 11-deoxycortisol, pregnenolone and 17α-hydroxypregnenolone in the adrenal vein serum of nine subjects pre- and post-ACTH stimulation (15 min), with each line denoting one individual. Individual subjects are denoted by the same line color in each panel; allowing the marked variation in production the relative steroid production varied between subjects. *(From Nakamura Y, Rege J, Satoh F, Morimoto R, Kennedy MR, Ahlem CN, Honma S, Sasano H, Rainey WE. Liquid chromatography-tandem mass spectrometry analysis of human adrenal vein corticosteroids before and after adrenocorticotropic hormone stimulation. Clin Endocrinol (Oxf) 2012;76(6):778–84. Fig. 2 p. 12.)*

injection of corticotrophin releasing factor (CRF) (Wang et al., 2020; Pecori Giraldi et al., 2015; Utz and Biller, 2007; Colao et al., 2001; Fig. 2.7.12).

The test should be performed only when abnormal increased ACTH secretion has already been established, but the results from dexamethasone and CRF testing are equivocal (Landolt et al., 1986). The test should not be performed on a patient on steroid therapy. This procedure requires careful and complete planning with the consultant in charge of the case and with a radiologist who will be positioning the catheters. This is an interventional procedure requiring a skilled radiologist working to a standard operating procedure. Complete clinical details and the results of all previous investigations must be

1	Right aorta
	R high internal jugular ;
	R low internal jugular;
	L high internal jugular;
	Thymic;
	R subclavian;
	Superior vena cava
2	Hepatic
3	High IVC
4	Right adrenal
5	Right renal
6	Left adrenal
7	Distal L renal
8	Proximal L renal
	L gonadal vein exits here
9	Left gonadal vein
10	Low IVC
11	L internal iliac
12	R external iliac

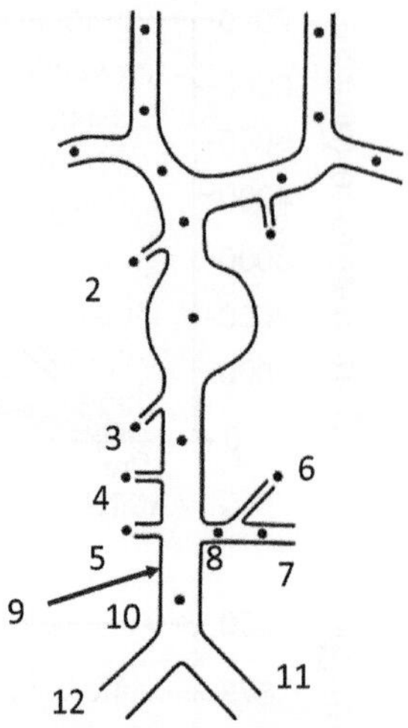

FIG. 2.7.11 Sites for catheterization in the vascular tree to find sites of hormone secretion.

made available for a multidisciplinary discussion of the plan. The plan should confirm with the laboratory that all the samples can be received and processed urgently as required. If the samples are to be delivered unprocessed, then ice buckets and a "runner" are required. Alternatively, near-patient sample preparation may be possible.

Obtain CRF (100 μg dose) for I.V. injection from pharmacy (CRF, Bissendorf, Germany). This is an unlicensed product obtainable on a named patient basis against the signature of the clinical consultant organizing the procedure.

The ACTH response to intravenous CRF is monitored via the changes in plasma ACTH and cortisol. The samples are taken from the venous drainage of the pituitary.

Prepare suitably labeled heparin tubes for all the specimens required (Table 2.7.5). These must show the patient's name (and date of birth, Hospital #) with sample site and timing.

TABLE 2.7.5 Samples during inferior petrosal sinus sampling (IPSS) with corticotrophin releasing factor (CRF).

Table of samples				
Time		Peripheral	Left	Right
Basal		✓	Jugular	Jugular
Basal		✓	Petrosal	Petrosal
CRF	3 min	✓	Petrosal	Petrosal
	8 min	✓	Petrosal	Petrosal
	15 min	✓	Petrosal	Petrosal

Collect samples from all three catheters (Peripheral, Left, Right) at 3–5 min, at 8–10 min and at 13–15 min following the CRF injection.
Total 15 samples; each 5 mL EDTA, for ACTH and Cortisol.
For simplicity, Cortisol will be measured on the EDTA sample, and serum is not required.

FIG. 2.7.12 Schematic Coronal view of the pituitary venous drainage. *(From Utz A, Biller BM. The role of bilateral inferior petrosal sinus sampling in the diagnosis of Cushing's syndrome. Arq Bras Endocrinol Metabol 2007;51(8):1329–38.)*

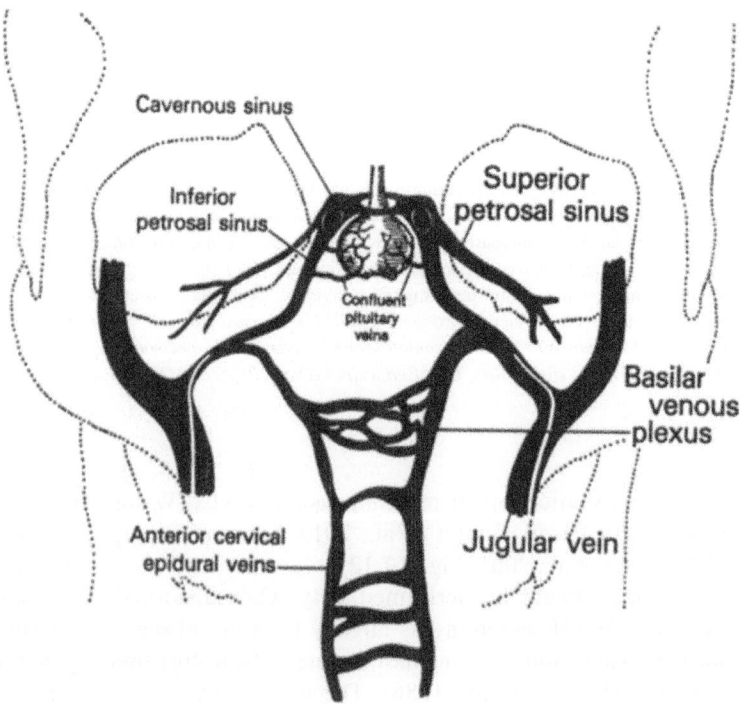

If samples are to be held on ice, make sure the labels and any writing will not soak off.

Sampling catheters are first positioned

1. in a peripheral vein,
2. in the left internal jugular vein and
3. in the right internal jugular vein.

Samples are collected from all three catheters. Each blood sample should be a minimum of 5 mL. The catheters may require flushing out before obtaining a sample.

The jugular catheters are advanced to the petrosal sinuses. Samples are again collected from all three catheters.

The CRF is injected intravenously into a peripheral vein. Keep the catheters in the same places. Samples are collected from all three catheters at 3–5 min, at 8–10 min and at 13–15 min following the CRF injection. Total 15 samples, each 5 mL in heparin for cortisol and ACTH.

The pattern of high ACTH results in the petrosal vein samples should localize the tumor. The peripheral results provide indices of control. Cortisol should confirm the exaggerated response. However the effective direct removal of ACTH by the sampling process may limit any cortisol rise if the procedure is prolonged. Hormone concentrations in small vessels may differ because blood flow becomes an important factor so a "localized" results may reflect blood flow rather than actual localization. A way around this is to administer another stimulant say GHRH or TRH and measure GH or TSH. If the results are same as ACTH 10 suggests change more likely due to flow.

A more potent stimulation in this test is achieved by the addition of desmopressin (Castinetti et al., 2007; Tsagarakis et al., 2007).

2.7.6.3 Ovarian vein catheterization

PCOS is the commonest cause of hyperandrogenism. An ovarian contribution to androgen secretion in women with hyperandrogenism is not affected by the administration of dexamethasone. The ovarian component to the hyperandrogenism can be assessed by evaluating the degree of androgen suppression to estrogen or GnRH-analog administration; however, gonadotrophin-dependent ovarian-steroid secreting tumors have been described. Although ultrasonography and computed tomography (CT), have been extensively used to demonstrate adrenal or ovarian pathology in hyperandrogenic women they may occasionally fail to localize small ovarian tumors embedded within the ovary. Magnetic resonance imaging (MRI) sometimes prove to be useful. Some tumors have only been localized successfully by biochemical analysis of blood obtained from the adrenal and ovarian veins during venous catheterization (Bailey et al., 2012; Elhadd et al., 1996).

High concentrations of 5α-reduced androgens have been found in ovarian vein catheter studies although suppression with dexamethasone suggested an adrenal component (Stanczyk et al., 2014). Studies such as this need repeating with LC-MS/MS based steroid profiling (Kalogera et al., 2013). Venous catheterization and sampling should be reserved for patients in whom uncertainty remains after extensive investigations (Duan et al., 2019), as the presence of a small ovarian tumor can sometimes not be excluded on biochemical and imaging studies. Use should be restricted to units with expertise in this area (Kaltsas et al., 2003).

2.7.6.4 Testicular veins

When testosterone concentrations are high and gonadotrophins are suppressed but no mass detected by palpation or imaging the spermatic veins can be selectively catheterized (Richter-Unruh et al., 2002). In one patient, testosterone concentration has been shown in the left vein (c. 130 nmol/L) compared with the opposite side (1400 pmol/L) and the patient was referred for left orchidopexy (Antonio et al., 2019). The test should only be performed in units with an experienced radiologist.

2.7.7 Further potential static tests

Pituitary hormones (follicle stimulating hormone and luteinizing hormone) are regulators of gonadal function. These protein concentrations are measured in blood during initial patient investigations. A number of other hormones are now measured as indices of gonadal function (Namwanje and Brown, 2016). Anti-Müllerian hormone (AMH) also known as Müllerian inhibitory substance (MIS) has been known for some time. Inhibin and activin are members of the transforming growth factor-B (TGF-β) family with combinations of α- and β-subunits (see Fig. 2.7.13) also known to be produced by the gonads.

Inhibins are heterodimers (32–34 kDa) consisting of a common alpha subunit (c. 20 kDa) linked by disulfide bridges to a beta-A (inhibin A) or beta-B subunit (inhibin B).

Inhibin was defined by its activity to suppress FSH. The monomeric free alpha subunits are in the circulation at much higher concentrations than the biologically active dimers. Inhibins bind reversibly with α2-macroglobulin. There have been frequent changes in the methods from manual ELISAs (Groome et al., 1996) to automated platforms with different standardizations for AMH (Yates et al., 2019; Fleming et al., 2018; Jing et al., 2018; de Kat et al., 2017; Li et al., 2016) and inhibin (Kalra et al., 2010). Activins are dimers of the inhibin β-subunit (Wijayarathna and de Kretser, 2016) and not yet routinely measured.

AMH has two cleavage sites at amino acids 229 and 451 giving an N-terminal domain and a biologically active

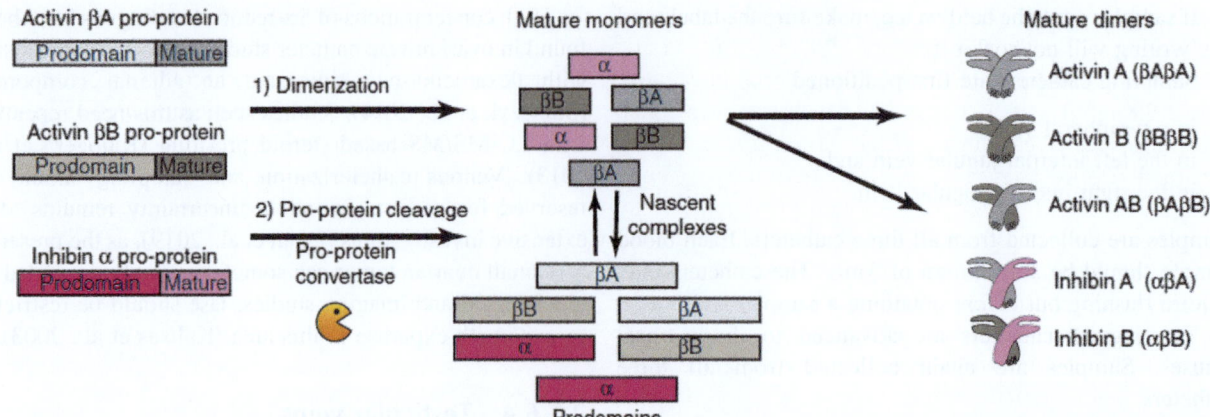

FIG. 2.7.13 Activins and inhibins. Activin bA, activin bB and inhibin A are synthesized as pro-proteins that comprise a prodomain and mature domain. The pro-proteins associate to form homo- or heterodimers, which are ultimately processed into activins A, B, AB and inhibins A and B. The junctions of the pro- and mature domains are cleaved by pro-protein convertases, resulting in dimer complexes that retain the noncovalently linked prodomains. *(From Namwanje M, Brown CW. Activins and Inhibins: roles in development, physiology, and disease. Cold Spring Harb Perspect Biol 2016;8(7):a021881. Fig. 1. p. 3.)*

12.5 kDa C-terminal domain. The nature of the material used to raise antisera is proprietary and the extent to which epitopes are formed at all stages before analysis and recognized in the assay are unknown. Variations in AMH test performance with early plate immunoassay tests were found to be due to the specificity of the antibodies, antibody pairs, assay process (enzyme-linked immunoabsorbent or automated platform) and the nature of the calibrators (reviewed by Punchoo and Bhoora, 2021). Complement was found to interfere in the early assays but is not found with the two commonest used assay now (Roche Elecys and Beckman Access; Turner et al., 2020).

Reference AMH material is available from World Health Organization (WHO) and National Institute of Standards and Technology (NIST). A recombinant preparation (SS-581) is under investigation (Ferguson et al., 2018). A collaborative study at 16 laboratories showed a wide variation in the estimates of the AMH content of a new product (16/190) with estimates ranging from 282 to 1157 ng/ampoule (ng per amp), resulting in a geometric mean content estimate from valid assays of 511 ng/amp (95% CI: 426–612, $n = 16$, GCV 42%) and a robust geometric mean of 489 ng/amp (Ferguson et al., 2020). A second sample (B) with half the content of sample a gave proportional results (B/A geometric mean 0.53 and 95% confidence limits 0.50–0.56 at 12% CV). The product16/190 is classified as a WHO international reference material because it did not meet the criteria to be an international standard at 489 ng/ampoule

Laboratory guidelines for peptide hormone analysis seek to improve diagnosis and treatment of disorders of sexual development (DSD). Reference ranges for inhibin and AMH concentrations are presented along with FSH and LH (Johannsen et al., 2020). Lifetime results are available for males (Aksglaede et al., 2010) and females (Hagen et al.,

2010). AMH concentrations increase in early infancy in males from 1 month after birth peaking around 6 months of age but remain low in females. AMH in males are up to 40-fold higher in males than females. As males pass through puberty the AMH concentrations decline but increase in females so that in adults there is overlap in the data (see Section 2.5.8).

2.7.7.1 Kisspeptin

Kisspeptin and neurokinin B are members of a family of tachykinins that have emerged as master regulators of GnRH and LH secretion. Tachykinins and kisspeptins (Leon and Navarro, 2019) are involved in fertility through spermatozoa in males and ovarian follicles in females, respectively. Tachykinins act through specific receptors (Tacr3/NK3R; Zhang et al., 2020). Kisspeptin is an essential gatekeeper of puberty and is the upstream regulator of GnRH neurons. The mRNA expression of the ovarian Kiss1 system correlates well with the serum luteinizing hormone (LH) levels. Kiss1 neurons are sensitive to the energy state of the organism, and a negative energy balance induces the suppression of the hypothalamic Kiss1 system. Kisspeptin (KP), an RF-amide coded by the *KISS*1 gene, is a major neuroendocrine regulator of reproduction, that It plays a focal part in the control of the HPG axis neuronal activity (highly sensitive to the body energy stores and nutritional status) to trigger puberty as well as affects fertility (Abbara and Dhillo, 2020). KP not only triggers onset of puberty (Navarro, 2020a) but also relays information about body's energy stores to the central nervous system by modulating negative and positive feedback of gonadal steroids. The expression of *KISS*1 is thus influenced by nutritional status where KP neurons function as a gatekeeper of reproductive function at pivotal times in the human lifespan, downregulating fertility at times of physical strain such as overexercise or weight loss (Navarro, 2020b; Dhillo et al., 2005). Median values by ELISA in prepubertal boys have been 13.8 pg/mL with higher concentrations (median 45 pg/mL) in delayed puberty (Chan et al., 2020). In the menstrual cycle, there is an increase in kisspeptin (Varikasuvu et al., 2019) from early follicular phase to the preovulatory phase 264 (28) (mean ± SD) vs. 472 (18) nmol/L and from preovulatory phase to the luteal phase 724 (37) (Latif and Rafique, 2015). Kisspeptin concentrations are significantly higher in PCOS then in matched women with unexplained infertility and male factor infertility (Kaya et al., 2019) with mean concentrations (±SD) at 525 (116); 281 (74) and 354 (111) ng/mL, respectively.

Kisspeptin infusion significantly increases LH in some cases of delayed puberty (Narayanaswamy et al., 2016; Dhillo et al., 2005; Chan et al., 2020).

2.7.7.2 Neurokinin

Neurokinin B is coexpressed with kisspeptin in the arcuate nucleus and synchronizes the pulsatile secretion of kisspeptin (Navarro, 2020a,b; Zhang et al., 2020). NKB neurons convey a stimulatory effect via the release of kisspeptin to GnRH neurons, which express both kisspeptin and NKB receptors. Plasma concentrations are above 50 pmol/L or 50 pg/mL with neuroendocrine tumors. The role of neurokinin B (NKB) is not widely known in precocious puberty (PP), but it may influence LH dysregulation in PCOS (Skorupskaite et al., 2020).

2.7.7.3 Melatonin

Melatonin (*N*-acetylmethoxytryptamine) is a hormone produced by the pineal gland that regulates the sleep wake cycle, and sometimes it is taken as a supplement for jet lag. The pharmacological and physiological properties of melatonin suggest this could be useful against ovarian cancer (OC) progression and metastasis. In normal cells, melatonin has potent antioxidant and antiapoptotic actions. OC has the highest mortality rate of all gynecological cancers, and most patients develop resistance after first-line treatments. Despite recent advances, the 5-year relative survival is around 45% for all OC subtypes, while invasive epithelial OC has only a 17% survival rate when diagnosed at a late stage. Identification of new biomarkers represents important opportunities in the treatment of OC. Conversely, in many cancer types including hormone-dependent cancers, melatonin has pro-oxidant as well as antiproliferative, antiangiogenic and immunomodulatory properties. Melatonin receptors have been identified in OC cells, but the exact mechanism of anticancer activities is still incompletely understood. Clinical studies have failed to find a positive correlation between aggression of OC and serum concentrations of melatonin. Many studies suggest melatonin controls cellular signals associated with malignancy and may be used in combination with other OC treatments (Zharinov et al., 2020). Melatonin is measured in laboratories with an interest in the peptide hormone and LC-MS/MS will now become the principal mode of analysis these days in plasma, saliva and urine (Greendale et al., 2020; Shin et al., 2021; van Faassen et al., 2017). There will be developments in the measurement of melatonin with potential clinical benefits.

2.7.7.4 Cytokines

Body weight, fat content and overall metabolic state are modifiers for puberty and fertility. Overnutrition and undernutrition is frequently linked to reproductive disorders. Adipocytokines may be the link between energy balance and reproduction, but our understanding of the neurobiological basis for this is still incomplete. Cytokines are involved in the maturation of oocytes and in the progression of puberty and pregnancy. Adipokines (i.e., cytokine/hormones or enzymes secreted from adipose tissue) are now under investigation as a missing link. Most adipokines do not act on the HPG axis, despite their role in reproduction and metabolic regulations has great potential. The measurements of cytokines are not routine practice yet because of the uncertainties about the analysis and interpretation of the results. The tests are usually by ELISA methods based on competitive enzyme immunoassay and there will be issues about the standards used for the method and the quality of the antibodies not to mention the interactions of the proteins and their precursors and fragments. This is an exciting and evolving area of steroid endocrinology with new molecules yet to be accounted for. The research will lead to new treatment regimens. The assays are used for infertility investigations and for recognizing activities of neuroendocrine tumors.

2.7.7.4.1 Leptin

Leptin is an adipokine secreted by both the adipose tissue and endometrial cells. Its role in obesity and reproductive regulation are well known. Leptin is a 167 amino acid that acts on receptors in the lateral hypothalamus to inhibit hunger and the medial hypothalamus to stimulate satiety. Leptin counteracts the effects of neuropeptide Y, a potent hunger promoter secreted by cells in the gut and in the hypothalamus. Leptin counteracts the effects of anandamide which is another potent hunger promoter. In the medial hypothalamus, leptin stimulates satiety by promoting the synthesis of α-MSH, a hunger suppressant. Obesity has been shown to impact the timing of puberty. Leptin is known to play a major role in the metabolic gating of pubertal maturation, that is mainly conducted at hypothalamic levels.

Leptin alone however cannot trigger early puberty. Serum leptin concentrations can be determined by ELISA kit methods. Normal results have been 5.6±0.9 ng/mL or 4.62±10.2 pmol/L (Kelesidis et al., 2010).

2.7.7.5 GnSAF gonadotrophin surge attenuating factor

At this time, GnSAF has not been characterized so dynamic tests around the effects of the hormone are not available, a situation that is bound to change in the years to come to become part of future investigations (Messinis et al., 2018; Tsutsui et al., 2018). A bioassay is needed to determine activity but is not widely available. There is evidence from in vivo and in vitro experiments for a gonadotrophin surge attenuating factor (GnSAF) that is a nonsteroidal substance in the ovary that plays an important role during the normal menstrual cycle (Dimitraki et al., 2014). The bioactivity of this factor is high in the early- to midfollicular phase, decreasing gradually thereafter until the midcycle. The factor antagonizes the sensitizing effect of estradiol on the pituitary response to GnRH during the greater part of the follicular phase. Activity decreases in the late follicular phase, when GnSAF facilitates the full expression of the positive feedback mechanism and the occurrence of a normal midcycle LH surge. GnSAF is proposed to be the "missing link" between the ovaries and the hypothalamo-pituitary system, to maintain the pituitary in a state of low responsiveness to GnRH pulses in the early- to midfollicular phase, limiting the release of LH to low but adequate amounts for normal folliculogenesis. This factor appears to affect the pulsatile aspect of LH release that is controlled centrally by GnRH and modulated by the ovarian steroids. GnSAF therefore provides a negative action in the context of the positive feedback loop.

2.7.7.6 Other hormones

In the investigation of patients with suspected Cushing's syndrome, **big ACTH, POMC fragments, lipotrophin and calcitonin** may need to be measured (Drury et al., 1982; Howlett et al., 1985; Solcia et al., 1987; Himsworth et al., 1977) (see Chapter 3.1.1).

2.7.7.7 Autoantibody determination

A number of auto-antibodies to steroidogenic enzymes have been described (Hellesen et al., 2018; Flück, 2017). The commonest is the 21-hydroxylase antibody (Falorni et al., 2015) for which there are binding tests, immunofluorescence or immunoprecipitation (Betterle et al., 1999) and an ELISA (del Pilar Larosa et al., 2018). Less common are auto-antibodies to 17-hydroxylase, side chain cleavage enzyme (CYP11A1) and 3β-hydroxysteroid dehydrogenase

(HSD3B). The function of the adrenal cortex, and in some case the ovary (de Bellis et al., 2017), can be compromised causing Addison's disease and premature ovarian failure (Naletto et al., 2019). Samples may have to be referred to specialist laboratories.

2.7.7.8 Steroid receptor assays

2.7.7.8.1 Androgen receptor

In many cases of 46XY individuals with high testosterone with no mutations of the androgen receptor, the cause of DSD is unknown. These patients have been assigned to a group called Partial androgen insensitivity syndrome (PAIS). Some of those cases have later been found to have defects of 5-alpha reductase (Berra et al., 2011) or 17-ketosteroid reductase (Phelan et al., 2015). Other possibilities include defects of coregulators (Appari et al., 2009; Adachi et al., 2000). Genital skin fibroblasts can be serially subcultured until enough cells are obtained for androgen receptor-binding studies. Total receptor concentration, receptor-binding affinity and the effects of increased incubation temperature and prolonged androgen preincubation of cells on specific androgen receptor with DHT binding are measured (Evans et al., 1984; Evans and Hughes, 1985).

A recombinant adenovirus has been used to deliver an androgen-responsive reporter gene (mouse mammary tumor virus luciferase) to fibroblasts (McPhaul et al., 1997). Over the past 25 years, several bioassays have been developed using different methods. One of the first assays developed relied on a chloramphenicol acetyltransferase (CAT) reporter model (Deslypere et al., 1992). This system was limited by experimental variation due to the transient nature of transgene expression. Other reporter cell lines were developed but most of them were transiently transfected. Transient transfection assays can provide similar information to stable assays but may not reflect endogenous levels of receptor. A stable expression of AR in monkey kidney CV1 cells eliminates the need for repetitious transient transfections, reduces the variability associated with these transient assays and moreover be utilized for high-throughput studies (Campana et al., 2016). The cells did not respond to aldosterone, cortisol or progesterone.

The development of receptor assays has not been extensive because of the inherent variation in such bioassays and in some cases lack of specificity. So even though an androgen receptor gene was active, the assay could still have glucocorticoid or progestogenic response because the DNA cis-regulatory elements that bind to AR share sequence similarity with cis-regulatory elements for glucocorticoid, mineralocorticoid and progesterone receptors (GR, MR and PR, respectively). Normative data of assays are available for age of healthy infants and children. Caution is needed however in the interpretation of results in sick premature babies and also children with delayed or advanced puberty at the peripubertal ages where age-specific references ranges may not be appropriate and there is potential for interferences from the fetal adrenal zone steroids. These methods have been restricted to patients having molecular defects in the hormone-binding domain.

Skin fibroblasts from the scrotum have also been shown to show significant differences of gene expression compared with fibroblasts from foreskin (Holterhus et al., 2007). When DHT induced gene expression of scrotal skin fibroblasts was examined in the presence the demethylating agent 5-aza-deoxy-cytidine significant upregulation of apolipoprotein D (APOD) in cells from normal individuals but severely reduced or absent in PIAS and CAIS (Appari et al., 2009; Hornig et al., 2016). Other such tests may follow in the course of time such as for detection of anabolic steroid use (Zierau et al., 2008).

2.7.7.8.2 GR, PR, ER and MR

Similar tests for receptor number, binding affinity, stability of the complex, activity and transfection can be used for other steroid receptors. Much is now known about the relevant genes in relation to protein structure and function and many mutations have been defined so genetic tests are now used at an earlier stage in the diagnosis of receptor defects.

2.7.7.8.3 Tests of smell

Some central endocrine disorders affecting olfactory nerve development or function of smell which may now be more reliably tested (Dżaman et al., 2017). The test is under evaluation in the United Kingdom. A Sniffin Sticks tests kit (now called Odofin) looks to be a convenient way to test for responses to rose, eucalyptus, lemon, lime, clove, apple, banana, butter, coffee, grass, fish, rose, lemon, orange, lemon, onion, peach and strawberry. The screening test and taste strips are supplied with multiple choice cards. The test is based on a number of felt-tip pen-like odor dispensing devices or scratchcards which are presented to the patient for detection and identification of an odorant substance (Hummel et al., 1997). It comprises three tests for olfactory function, namely odor threshold, odor discrimination and odor identification. The odor identification uses 16 common odorants and asks the subject to verbally identify the smell from a selection of four options. Initially developed in Germany, the Sniffin stick test was validated on a large number of local subjects and it is used in many European countries. Odor identification is strongly dependent on familiarity with the odors presented in the Sniffin sticks. Cultural differences and age might prevent odor identification and as a result limit the applicability of this olfactory test in a region of the world. A study is in progress to establish that the odors and normal values for the test are valid in a UK population.

Each stick is labeled with a "best before date" and can be used for many persons before this date provided that the tip of a pen is never touched with fingers or nose. That will contaminate the stick and should not be used anymore because the quality of the odor of the stick cannot be guaranteed. The shelf life for the screening 12 test is approximately 1–1.5 years.

2.7.8 Summary

A number of provocative tests are available to gain confidence in a particular diagnosis of a steroid related disorder. A low basal result will need a stimulation test in the confirmation process and a high result will need a suppression test. The references for HPA axis stimulation are summarized in one review (Karaca et al., 2021). An ACTH or hCG test are the most common procedures beyond the basal hormone determinations. The dynamic test procedures do not have agreed protocols and these do need to be standardized allowing for factors such as age of the patients. A learned body should take responsibility for this in the way that steroid assays are being harmonized. Optimal timing of hormone measurements is also important to produce reliable results. Not all centers have access to all of the steroid assays required (DHT, for example). Some tests can be performed by immunoassay on analytical platforms and increasingly LC-MS/MS tests and profile are being introduced. The harmonization of tests started with a focus on testosterone assays but is continuing with development of certified reference materials, reference methods, reference laboratories and reference intervals to improve assay quality. An EQA scheme for DHT has been introduced (Greaves, 2017). Databases are under development (www.orpha.net and www.dsd.net) in order to improve access to tests for rare disorders.

Some clinical situations are still difficult to resolve. The mechanisms of cholesterol delivery to the steroidogenic machinery, the biochemistry of androgen synthesis, the mini-puberty, the regulation and biological role of adrenarche, fetal adrenal development and involution, the roles of steroids made in "extraglandular" cells, Cushing's syndrome are still important questions. Over the past 25 years or so, there have been an increasing number of biochemical and genetic tests that can identify loss or gain of function effects and copy number variants. New disorders are being discovered all the time. Biochemical tests will still have an important part in the steps to diagnosis of a problem in a patient then to monitoring the treatment. Standardization of assays such as AMH, inhibin and activin is needed. This chapter has not addressed the genetic tests being used for diagnosis of steroid disorders. This has changed over the past 30 years from location of mutations by single strand conformational polymorphisms (SSCP) testing to gene sequencing.

Note added in proof.

A useful test procedure to measure one analyte in multiple samples involves derivative formation with different Girard reagents such that they can all be measured by GC/MS in one sample because of differences in masses (Ohta et al., 2022).

References

Abbara A, Dhillo WS. Makorin rings the kisspeptin bell to signal pubertal initiation. J Clin Invest 2020;130(8):3957–60.

Adachi M, Takayanagi R, Tomura A, Imasaki K, Kato S, Goto K, et al. Androgen-insensitivity syndrome as a possible coactivator disease. N Engl J Med 2000;343(12):856–62.

Ahmed AH, Gordon RD, Taylor PJ, Ward G, Pimenta E, Stowasser M. Effect of contraceptives on aldosterone/renin ratio may vary according to the components of contraceptive, renin assay method, and possibly route of administration. J Clin Endocrinol Metab 2011;96(6):1797–804.

Aksglaede L, Sørensen K, Boas M, Mouritsen A, Hagen CP, Jensen RB, et al. Changes in anti-Müllerian hormone (AMH) throughout the life span: a population-based study of 1027 healthy males from birth (cord blood) to the age of 69 years. J Clin Endocrinol Metab 2010;95 (12):5357–64.

Anderson RC, Newton CL, Anderson RA, Millar RP. Gonadotropins and their analogs: current and potential clinical applications. Endocr Rev 2018;39(6):911–37.

Antonio L, Albersen M, Billen J, Maleux G, Van Rompuy AS, Coremans P, et al. Testicular vein sampling can reveal gonadotropin-independent unilateral steroidogenesis supporting spermatogenesis. J Endocr Soc 2019;3(10):1881–6.

Appari M, Werner R, Wünsch L, Cario G, Demeter J, Hiort O, et al. Apolipoprotein D (APOD) is a putative biomarker of androgen receptor function in androgen insensitivity syndrome. J Mol Med (Berl) 2009;87(6):623–32.

Auchus RJ. An innovative approach to noninvasive dynamic adrenal testing. J Clin Endocrinol Metab 2020;105(10), dgaa455.

Bacila IA, Elder C, Krone N. Update on adrenal steroid hormone biosynthesis and clinical implications. Arch Dis Child 2019;104(12):1223–8.

Bahaeldein E, Brassill MJ. Utilisation of gonadotrophin-releasing hormone (GnRH) analogue to differentiate ovarian from adrenal hyperandrogenism in postmenopausal women. Endocrinol Diabetes Metab Case Rep 2018;2018.

Bailey AP, Schutt AK, Carey RM, Angle JF, Modesitt SC. Hyperandrogenism of ovarian etiology: utilizing differential venous sampling for diagnosis. Obstet Gynecol 2012;120(2 Pt 2):476–9.

Bang AK, Nordkap L, Almstrup K, Priskorn L, Petersen JH, Rajpert-De Meyts E, et al. Dynamic GnRH and hCG testing: establishment of new diagnostic reference levels. Eur J Endocrinol 2017;176(4):379–91.

Barthel A, Benker G, Berens K, Diederich S, Manfras B, Gruber M, et al. An update on Addison's disease. Exp Clin Endocrinol Diabetes 2019;127(2-03):165–75.

Berlińska A, Świątkowska-Stodulska R, Sworczak K. Factors affecting dexamethasone suppression test results. Exp Clin Endocrinol Diabetes 2020;128(10):667–71.

Berra M, Williams EL, Muroni B, Creighton SM, Honour JW, Rumsby G, et al. Recognition of 5α-reductase-2 deficiency in an adult female 46XY DSD clinic. Eur J Endocrinol 2011;164(6):1019–25.

Bertelloni S, Russo G, Baroncelli GI. Human chorionic gonadotropin test: old uncertainties, new perspectives, and value in 46,XY disorders of sex development. Sex Dev 2018;12(1–3):41–9.

Betterle C, Volpato M, Pedini B, Chen S, Smith BR, Furmaniak J. Adrenal-cortex autoantibodies and steroid-producing cells autoantibodies in patients with Addison's disease: comparison of immunofluorescence and immunoprecipitation assays. J Clin Endocrinol Metab 1999;84 (2):618–22.

Blondin D, Quack I, Haase M, Kücükköylü S, Willenberg HS. Indication and technical aspects of adrenal blood sampling. Rofo 2015;187 (1):19–28.

Bridges NA, Hindmarsh PC, Pringle PJ, Honour JW, Brook CG. Cortisol, androstenedione (A4), dehydroepiandrosterone sulphate (DHEAS) and 17 hydroxyprogesterone (17OHP) responses to low doses of (1–24) ACTH. J Clin Endocrinol Metab 1998;83(10):3750–3.

Burgos N, Ghayee HK, Singh-Ospina N. Pitfalls in the interpretation of the cosyntropin stimulation test for the diagnosis of adrenal insufficiency. Curr Opin Endocrinol Diabetes Obes 2019;26(3):139–45.

Bystrom CE, Salameh W, Reitz R, Clarke NJ. Plasma renin activity by LC-MS/MS: development of a prototypical clinical assay reveals a subpopulation of human plasma samples with substantial peptidase activity. Clin Chem 2010;56(10):1561–9.

Campana C, Rege J, Turcu AF, Pezzi V, Gomez-Sanchez CE, Robins DM, et al. Development of a novel cell based androgen screening model. J Steroid Biochem Mol Biol 2016;156:17–22.

Castinetti F, Morange I, Dufour H, Jaquet P, Conte-Devolx B, Girard N, et al. Desmopressin test during petrosal sinus sampling: a valuable tool to discriminate pituitary or ectopic ACTH-dependent Cushing's syndrome. Eur J Endocrinol 2007;157(3):271–7.

Ceolotto G, Antonelli G, Maiolino G, Cesari M, Rossitto G, Bisogni V, et al. Androstenedione and 17-α-hydroxyprogesterone are better indicators of adrenal vein sampling selectivity than cortisol. Hypertension 2017;70(2):342–6.

Cesari M, Ceolotto G, Rossitto G, Maiolino G, Seccia TM, Rossi GP. The intra-procedural cortisol assay during adrenal vein sampling: rationale and design of a randomized study (I-Padua). High Blood Press Cardiovasc Prev 2017;24(2):167–70.

Chan YM, Lippincott MF, Kusa TO, Seminara SB. Divergent responses to kisspeptin in children with delayed puberty. JCI Insight 2018;3(8), e99109.

Chan YM, Lippincott MF, Sales Barroso P, Alleyn C, Brodsky J, Granados H, et al. Using kisspeptin to predict pubertal outcomes for youth with pubertal delay. J Clin Endocrinol Metab 2020;105(8):e2717–25.

Chanson P, Guignat L, Goichot B, Chabre O, Boustani DS, Reynaud R, et al. Group 2: adrenal insufficiency: screening methods and confirmation of diagnosis. Ann Endocrinol (Paris) 2017;78(6):495–511.

Chao CS, Shi RZ, Kumar RB, Aye T. Salivary cortisol levels by tandem mass spectrometry during high dose ACTH stimulation test for adrenal insufficiency in children. Endocrine 2020;67(1):190–7.

Colao A, Faggiano A, Pivonello R, Pecori Giraldi F, Cavagnini F, Lombardi G, et al. Inferior petrosal sinus sampling in the differential diagnosis of Cushing's syndrome: results of an Italian multicenter study. Eur J Endocrinol 2001;144(5):499–507.

Corradi PF, Corradi RB, Greene LW. Physiology of the hypothalamic pituitary gonadal axis in the male. Urol Clin North Am 2016;43(2): 151–62.

Creemers SG, Hofland LJ, Lamberts SW, Feelders RA. Cushing's syndrome: an update on current pharmacotherapy and future directions. Expert Opin Pharmacother 2015;16(12):1829–44.

Cross AS, Helen Kemp E, White A, Walker L, Meredith S, Sachdev P, et al. International survey on high- and low-dose Synacthen test and assessment of accuracy in preparing low-dose synacthen. Clin Endocrinol (Oxf) 2018;88(5):744–51.

Crowley S, Hindmarsh PC, Holownia P, Honour JW, Brook CG. The use of low doses of ACTH in the investigation of adrenal function in man. J Endocrinol 1991;130(3):475–9.

Crowley S, Hindmarsh PC, Honour JW, Brook CG. Reproducibility of the cortisol response to stimulation with a low dose of ACTH(1-24): the effect of basal cortisol levels and comparison of low-dose with high-dose secretory dynamics. J Endocrinol 1993;136(1):167–72.

Dżaman K, Zborowska-Piskadło K, Pietniczka-Załęska M, Kantor I. Kallmann syndrome in pediatric otorhinolaryngology practice—case report and literature review. Int J Pediatr Otorhinolaryngol 2017;100:149–53.

de Bellis A, Bellastella G, Falorni A, Aitella E, Barrasso M, Maiorino MI, et al. Natural history of autoimmune primary ovarian insufficiency in patients with Addison's disease: from normal ovarian function to overt ovarian dysfunction. Eur J Endocrinol 2017;177(4):329–37.

de Graaf AJ, Mulder AL, Krabbe JG. Retrospective analysis of repeated dexamethasone suppression tests—the added value of measuring dexamethasone. Ann Clin Biochem 2019;, 4563219870834. https://doi.org/10.1177/0004563219870834 [Epub ahead of print] 31370673.

de Kat AC, Broekmans FJM, van Westing AC, Lentjes E, Verschuren WMM, van der Schouw YT. A quantitative comparison of anti-Müllerian hormone measurement and its shifting boundaries between two assays. Maturitas 2017;101:12–6.

Deinum J, Groenewoud H, van der Wilt GJ, Lenzini L, Rossi GP. Adrenal venous sampling: cosyntropin stimulation or not? Eur J Endocrinol 2019;181(3):D15–26.

del Pilar Larosa M, Chen S, Steinmaus N, Macrae H, Guo L, Masiero S, et al. A new ELISA for autoantibodies to steroid 21-hydroxylase. Clin Chem Lab Med 2018;56(6):933–8.

Deslypere JP, Young M, Wilson JD, McPhaul MJ. Testosterone and 5 alpha-dihydrotestosterone interact differently with the androgen receptor to enhance transcription of the MMTV-CAT reporter gene. Mol Cell Endocrinol 1992;88(1–3):15–22.

Dhillo WS, Chaudhri OB, Patterson M, Thompson EL, Murphy KG, Badman MK, et al. Kisspeptin-54 stimulates the hypothalamic-pituitary gonadal axis in human males. J Clin Endocrinol Metab 2005;90(12):6609–15.

Dimitraki M, Messini CI, Dafopoulos K, Gioka T, Koutlaki N, Garas A, et al. Attenuating activity of the ovary on LH response to GnRH during the follicular phase of the cycle. Clin Endocrinol (Oxf) 2014;80(3):439–43.

Drury PL, Ratter S, Tomlin S, Williams J, Dacie JE, Rees LH, et al. Experience with selective venous sampling in diagnosis of ACTH-dependent Cushing's syndrome. Br Med J (Clin Res Ed) 1982;284(6308):9–12.

Duan L, Yang Y, Gu Y, Zhang X, Mao Q, Pan B, et al. The utility of adrenal and ovarian venous sampling in a progesterone-producing adrenal tumor and review of the literature. Endocrine 2019;66(2):319–25.

Eisenhofer G, Dekkers T, Peitzsch M, Dietz AS, Bidlingmaier M, Treitl M, et al. Mass spectrometry-based adrenal and peripheral venous steroid profiling for subtyping primary aldosteronism. Clin Chem 2016;62(3):514–24.

Elder CJ, Vilela R, Johnson TN, Taylor RN, Kemp EH, Keevil BG, et al. Pharmacodynamic studies of nasal tetracosactide with salivary glucocorticoids for a noninvasive Short Synacthen test. J Clin Endocrinol Metab 2020;105(8), dgaa323.

Elhadd TA, Connolly V, Cruickshank D, Kelly WF. An ovarian lipid cell tumour causing virilization and Cushing's syndrome. Clin Endocrinol (Oxf) 1996;44(6):723–5.

Evans BA, Hughes IA. Augmentation of androgen-receptor binding in vitro: studies in normals and patients with androgen insensitivity. Clin Endocrinol (Oxf) 1985;23(5):567–77.

Evans BA, Jones TR, Hughes IA. Studies of the androgen receptor in dispersed fibroblasts: investigation of patients with androgen insensitivity. Clin Endocrinol (Oxf) 1984;20(1):93–105.

Falorni A, Bini V, Betterle C, Brozzetti A, Castaño L, Fichna M, et al. Determination of 21-hydroxylase autoantibodies: inter-laboratory concordance in the Euradrenal International Serum Exchange Program. Clin Chem Lab Med 2015;53(11):1761–70.

Fan L, Song Y, Polak M, Li L, Ren X, Zhang B, et al. Clinical characteristics and genotype-phenotype correlations of 130 Chinese children in a high-homogeneity single-center cohort with 5α-reductase 2 deficiency. Mol Genet Genomic Med 2020;8(10), e1431.

Ferguson JM, Pépin D, Duru C, Matejtschuk P, Donahoe PK, Burns CJ. Towards international standardization of immunoassays for Müllerian inhibiting substance/anti-Müllerian hormone. Reprod Biomed Online 2018;37(5):631–40.

Ferguson J, Hockley J, Rigsby P, Burns C. Establishment of a WHO reference reagent for anti-Mullerian hormone. Reprod Biol Endocrinol 2020;18(1):86.

Fiselier T, Monnens L, Moerman E, Van Munster P, Jansen M, Peer P. Influence of the stress of venepuncture on basal levels of plasma renin activity in infants and children. Int J Pediatr Nephrol 1983a;4(3):181–5.

Fiselier TJ, Lijnen P, Monnens L, van Munster P, Jansen M, Peer P. Levels of renin, angiotensin I and II, angiotensin-converting enzyme and aldosterone in infancy and childhood. Eur J Pediatr 1983b;141(1):3–7.

Fiselier T, Derkx F, Monnens L, Van Munster P, Peer P, Schalekamp M. The basal levels of active and inactive plasma renin concentration in infancy and childhood. Clin Sci (Lond) 1984a;67(4):383–7.

Fiselier T, Monnens L, van Munster P, Jansen M, Peer P, Lijnen P. The renin-angiotensin-aldosterone system in infancy and childhood in basal conditions and after stimulation. Eur J Pediatr 1984b;143(1):18–24.

Fleming R, Fairbairn C, Gaudoin M. Objective multicentre performance of the automated assays for AMH and estimation of established critical concentrations. Hum Fertil (Camb) 2018;21(4):269–74.

Flück CE. Mechanisms in endocrinology: update on pathogenesis of primary adrenal insufficiency: beyond steroid enzyme deficiency and autoimmune adrenal destruction. Eur J Endocrinol 2017;177(3):R99–R111.

Funder JW, Carey RM, Mantero F, Murad MH, Reincke M, Shibata H, et al. The management of primary aldosteronism: case detection, diagnosis, and treatment: an endocrine society clinical practice guideline. J Clin Endocrinol Metab 2016;101(5):1889–916.

Genere N, Kaur RJ, Athimulam S, Thomas MA, Nippoldt T, Van Norman M, et al. Interpretation of abnormal dexamethasone suppression test is enhanced with use of synchronous free cortisol assessment. J Clin Endocrinol Metab 2022;107(3):e1221–30.

George GS, Jabbar PK, Jayakumari C, John M, Mini M, Thekkumkara Surendran Nair A, et al. Long-acting porcine ACTH stimulated salivary cortisol in the diagnosis of adrenal insufficiency. Clin Endocrinol (Oxf) 2020;93(6):652–60.

Gill H, Barrowman N, Webster R, Ahmet A. Evaluating the low-dose ACTH stimulation test in children: ideal times for cortisol measurement. J Clin Endocrinol Metab 2019;104(10):4587–93.

Gilligan LA, Trout AT, Schuster JG, Schwartz BI, Breech LL, Zhang B, et al. Normative values for ultrasound measurements of the female pelvic organs throughout childhood and adolescence. Pediatr Radiol 2019;49(8):1042–50.

Giordano R, Picu A, Bonelli L, Balbo M, Berardelli R, Marinazzo E, et al. Hypothalamus-pituitary-adrenal axis evaluation in patients with hypothalamo-pituitary disorders: comparison of different provocative tests. Clin Endocrinol (Oxf) 2008;68(6):935–41.

Greaves RF. The central role of external quality assurance in harmonisation and standardisation for laboratory medicine. Clin Chem Lab Med 2017;55(4):471–3.

Greendale GA, Witt-Enderby P, Karlamangla AS, Munmun F, Crawford S, Huang M, et al. Melatonin patterns and levels during the human menstrual cycle and after menopause. J Endocr Soc 2020;4(11), bvaa115.

Groome NP, Illingworth PJ, O'Brien M, Pai R, Rodger FE, Mather JP, et al. Measurement of dimeric inhibin B throughout the human menstrual cycle. J Clin Endocrinol Metab 1996;81(4):1401–5.

Guran T, Kara C, Yildiz M, Bitkin EC, Haklar G, Lin JC, et al. Revisiting classical 3β-hydroxysteroid dehydrogenase 2 deficiency: lessons from 31 pediatric cases. J Clin Endocrinol Metab 2020;105(3), dgaa022.

Hagen CP, Aksglaede L, Sørensen K, Main KM, Boas M, Cleemann L, et al. Serum levels of anti-Müllerian hormone as a marker of ovarian function in 926 healthy females from birth to adulthood and in 172 Turner syndrome patients. J Clin Endocrinol Metab 2010;95 (11):5003–10.

Harrington J, Palmert MR. Distinguishing constitutional delay of growth and puberty from isolated hypogonadotropic hypogonadism: critical appraisal of available diagnostic tests. J Clin Endocrinol Metabol 2012;97:3056–67.

Hawley JM, Owen LJ, Debono M, Newell-Price J, Keevil BG. Development of a rapid liquid chromatography tandem mass spectrometry method for the quantitation of serum dexamethasone and its clinical verification. Ann Clin Biochem 2018;55(6):665–72.

Hellesen A, Bratland E, Husebye ES. Autoimmune Addison's disease—an update on pathogenesis. Ann Endocrinol (Paris) 2018;79(3):157–63.

Hempen C, Elfering S, Mulder AH, van den Bergh FA, Maatman RG. Dexamethasone suppression test: development of a method for simultaneous determination of cortisol and dexamethasone in human plasma by liquid chromatography/tandem mass spectrometry. Ann Clin Biochem 2012;49(Pt 2):170–6.

Hendriks DJ, Mol BW, Bancsi LF, te Velde ER, Broekmans FJ. The clomiphene citrate challenge test for the prediction of poor ovarian response and nonpregnancy in patients undergoing in vitro fertilization: a systematic review. Fertil Steril 2006;86(4):807–18.

Himsworth RL, Bloomfield GA, Coombes RC, Ellison M, Gilkes JJ, Lowry PJ, et al. 'Big ACTH' and calcitonin in an ectopic hormone secreting tumour of the liver. Clin Endocrinol (Oxf) 1977;7(1):45–62.

Hindmarsh PC, Honour JW. Would cortisol measurements be a better gauge of hydrocortisone replacement therapy? Congenital adrenal hyperplasia as an exemplar. Int J Endocrinol 2020;2020:2470956.

Holterhus PM, Deppe U, Werner R, Richter-Unruh A, Bebermeier JH, Wünsch L, et al. Intrinsic androgen-dependent gene expression patterns revealed by comparison of genital fibroblasts from normal males and individuals with complete and partial androgen insensitivity syndrome. BMC Genomics 2007;8:376.

Homer MV, Toloubeydokhti T, Lawson MA, Garzo G, Duleba AJ, Chang RJ. Individual 17-hydroxyprogesterone responses to hCG are not correlated with follicle size in polycystic ovary syndrome. J Endocr Soc 2019;3(4):687–98.

Honour JW, Bridges NA, Conway-Phillips E, Hindmarsh PC. Plasma aldosterone response to the low-dose adrenocorticotrophin (ACTH 1-24) stimulation test. Clin Endocrinol (Oxf) 2008;68(2):299–303.

Hornig NC, Ukat M, Schweikert HU, Hiort O, Werner R, Drop SL, et al. Identification of an AR mutation-negative class of androgen insensitivity by determining endogenous AR activity. J Clin Endocrinol Metab 2016;101(11):4468–77.

Horrocks PM, Jones AF, Ratcliffe WA, Holder G, White A, Holder R, et al. Patterns of ACTH and cortisol pulsatility over twenty-four hours in normal males and females. Clin Endocrinol (Oxf) 1990;32(1): 127–34.

Houk CP, Kunselman AR, Lee PA. Adequacy of a single unstimulated luteinizing hormone level to diagnose central precocious puberty in girls. Pediatrics 2009;123:e1059–63.

Howlett TA, Price J, Hale AC, Doniach I, Rees LH, Wass JA, et al. Pituitary ACTH dependent Cushing's syndrome due to ectopic production of a bombesin-like peptide by a medullary carcinoma of the thyroid. Clin Endocrinol (Oxf) 1985;22(1):91–101.

Hummel T, Sekinger B, Wolf SR, Pauli E, Kobal G. Sniffin' sticks': olfactory performance assessed by the combined testing of odor identification, odor discrimination and olfactory threshold. Chem Senses 1997;22(1):39–52.

Husebye ES, Allolio B, Arlt W, Badenhoop K, Bensing S, Betterle C, et al. Consensus statement on the diagnosis, treatment and follow-up of patients with primary adrenal insufficiency. J Intern Med 2014;275 (2):104–15.

Iwanaga K, Yamamoto A, Matsukura T, Niwa F, Kawai M. Corticotrophin-releasing hormone stimulation tests for the infants with relative adrenal insufficiency. Clin Endocrinol (Oxf) 2017;87(6):660–4.

Jiang X, Dong H, Peng M, Che W, Zou Y, Song L, et al. A novel method of adrenal venous sampling via an antecubital approach. Cardiovasc Intervent Radiol 2017;40(3):388–93.

Jing J, Xia F, Ding Z, Chen L, Shao Y, Ge YF, et al. A single-center performance evaluation of the fully automated iFlash anti-Müllerian hormone immunoassay. Clin Chem Lab Med 2018;57(2):e19–22.

Johannsen TH, Andersson AM, Ahmed SF, de Rijke YB, Greaves RF, Hartmann MF, et al. Peptide hormone analysis in diagnosis and treatment of differences of sex development: joint position paper of EU COST Action 'DSDnet' and European Reference Network on Rare Endocrine Conditions. Eur J Endocrinol 2020;182(6):P1–P15.

Kalogera E, Pistos C, Provatopoulou X, Athanaselis S, Spiliopoulou C, Gounaris A. Androgen glucuronides analysis by liquid chromatography tandem-mass spectrometry: could it raise new perspectives in the diagnostic field of hormone-dependent malignancies? J Chromatogr B Analyt Technol Biomed Life Sci 2013;940:24–34. https://doi.org/10.1016/j.jchromb.2013.09.022. Epub 2013 Sep 27 24140653.

Kalra B, Kumar A, Patel K, Patel A, Khosravi MJ. Development of a second generation inhibin B ELISA. J Immunol Methods 2010;362(1–2): 22–31.

Kaltsas GA, Mukherjee JJ, Kola B, Isidori AM, Hanson JA, Dacie JE, et al. Is ovarian and adrenal venous catheterization and sampling helpful in the investigation of hyperandrogenic women? Clin Endocrinol (Oxf) 2003;59(1):34–43.

Karaca Z, Grossman A, Kelestimur F. Investigation of the hypothalamo-pituitary-adrenal (HPA) axis: a contemporary synthesis. Rev Endocr Metab Disord 2021;22(2):179–204.

Kaya C, Alay İ, Babayeva G, Gedikbaşı A, Ertaş Kaya S, Ekin M, et al. Serum Kisspeptin levels in unexplained infertility, polycystic ovary syndrome, and male factor infertility. Gynecol Endocrinol 2019;35 (3):228–32.

Kelesidis T, Kelesidis I, Chou S, Mantzoros CS. Narrative review: the role of leptin in human physiology: emerging clinical applications. Ann Intern Med 2010;152(2):93–100.

Kelsey TW, Ginbey E, Chowdhury MM, Bath LE, Anderson RA, Wallace WH. A validated normative model for human uterine volume from birth to age 40 years. PLoS One 2016;11(6), e0157375.

Kline GA, Chin A, So B, Harvey A, Pasieka JL. Defining contralateral adrenal suppression in primary aldosteronism: implications for diagnosis and outcome. Clin Endocrinol (Oxf) 2015;83(1):20–7.

Kulle AE, Riepe FG, Hedderich J, Sippell WG, Schmitz J, Niermeyer L, et al. LC-MS/MS based determination of basal- and ACTH-stimulated plasma concentrations of 11 steroid hormones: implications for detecting heterozygote CYP21A2 mutation carriers. Eur J Endocrinol 2015;173(4):517–24.

Landolt AM, Valavanis A, Girard J, Eberle AN. Corticotrophin-releasing factor-test used with bilateral, simultaneous inferior petrosal sinus blood-sampling for the diagnosis of pituitary-dependent Cushing's disease. Clin Endocrinol (Oxf) 1986;25(6):687–96.

Latif R, Rafique N. Serum kisspeptin levels across different phases of the menstrual cycle and their correlation with serum oestradiol. Neth J Med 2015;73(4):175–8.

Lawaetz JG, Hagen CP, Mieritz MG, Blomberg Jensen M, Petersen JH, Juul A. Evaluation of 451 Danish boys with delayed puberty: diagnostic use of a new puberty nomogram and effects of oral testosterone therapy. J Clin Endocrinol Metab 2015;100(4):1376–85.

Lee C, Kim JH, Moon SJ, Shim J, Kim HI, Choi MH. Selective LC-MRM/SIM-MS based profiling of adrenal steroids reveals metabolic signatures of 17α-hydroxylase deficiency. J Steroid Biochem Mol Biol 2020;198:105615.

Leon S, Navarro VM. Novel biology of tachykinins in gonadotropin-releasing hormone secretion. Semin Reprod Med 2019;37(3):109–18.

Li HW, Wong BP, Ip WK, Yeung WS, Ho PC, Ng EH. Comparative evaluation of three new commercial immunoassays for anti-Müllerian hormone measurement. Hum Reprod 2016;31(12):2796–802.

Lindner JM, Suhr AC, Grimm SH, Möhnle P, Vogeser M, Briegel J. The dynamics of a serum steroid profile after stimulation with intravenous ACTH. J Pharm Biomed Anal 2018;151:159–63.

Malerbi DA, Mendonça BB, Liberman B, Toledo SP, Corradini MC, Cunha-Neto MB, et al. The desmopressin stimulation test in the differential diagnosis of Cushing's syndrome. Clin Endocrinol (Oxf) 1993;38(5):463–72.

McPhaul MJ, Schweikert HU, Allman DR. Assessment of androgen receptor function in genital skin fibroblasts using a recombinant adenovirus to deliver an androgen-responsive reporter gene. J Clin Endocrinol Metab 1997;82(6):1944–8.

Meikle AW. Dexamethasone suppression tests: usefulness of simultaneous measurement of plasma cortisol and dexamethasone. Clin Endocrinol (Oxf) 1982;16(4):401–8.

Messinis IE, Messini CI, Anifandis G, Garas A, Daponte A. Gonadotropin surge-attenuating factor: a nonsteroidal ovarian hormone controlling GnRH-induced LH secretion in the normal menstrual cycle. Vitam Horm 2018;107:263–86.

Methlie P, Hustad SS, Kellmann R, Almås B, Erichsen MM, Husebye E, et al. Multisteroid LC-MS/MS assay for glucocorticoids and androgens, and its application in Addison's disease. Endocr Connect 2013;2(3):125–36.

Mojtahedzadeh M, Shaesteh N, Haykani M, Tran JLA, Mangubat M, Shahinian HK, et al. Low-dose and standard overnight and low dose-two day dexamethasone suppression tests in patients with mild and/or episodic hypercortisolism. Horm Metab Res 2018;50(6):453–61.

Monaghan PJ, Owen LJ, Trainer PJ, Brabant G, Keevil BG, Darby D. Comparison of serum cortisol measurement by immunoassay and liquid chromatography-tandem mass spectrometry in patients receiving the 11β-hydroxylase inhibitor metyrapone. Ann Clin Biochem 2011;48(Pt 5):441–6.

Monroe EJ, Carney BW, Ingraham CR, Johnson GE, Valji K. An Interventionist's guide to endocrine consultations. Radiographics 2017;37 (4):1246–67.

Morera J, Reznik Y. Management of endocrine disease: the role of confirmatory tests in the diagnosis of primary aldosteronism. Eur J Endocrinol 2019;180(2):R45–58.

Nair A, Jayakumari C, George GS, Jabbar PK, Das DV, Jessy SJ, et al. Long acting porcine sequence ACTH in the diagnosis of adrenal insufficiency. Eur J Endocrinol 2019;181(6):639–45.

Nakamura Y, Rege J, Satoh F, Morimoto R, Kennedy MR, Ahlem CN, et al. Liquid chromatography-tandem mass spectrometry analysis of human adrenal vein corticosteroids before and after adrenocorticotropic hormone stimulation. Clin Endocrinol (Oxf) 2012;76(6): 778–84.

Naletto L, Frigo A, Ceccato F, Sabbadin C, Scarpa R, Presotto F, et al. The natural history of autoimmune Addison's disease from the detection of autoantibodies to development of the disease: a long follow-up study on 143 patients. Eur J Endocrinol 2019. https://doi.org/10.1530/EJE-18-0313. pii: EJE-18-0313.R3. [Epub ahead of print] 30608902.

Namwanje M, Brown CW. Activins and Inhibins: roles in development, physiology, and disease. Cold Spring Harb Perspect Biol 2016;8(7): a021881.

Narayanaswamy S, Jayasena CN, Ng N, Ratnasabapathy R, Prague JK, Papadopoulou D, et al. Subcutaneous infusion of kisspeptin-54 stimulates gonadotrophin release in women and the response correlates with basal oestradiol levels. Clin Endocrinol (Oxf) 2016;84(6):939–45.

Navarro VM. Tachykinin signaling in the control of puberty onset. Curr Opin Endocr Metab Res 2020a;14:92–6.

Navarro VM. Metabolic regulation of kisspeptin—the link between energy balance and reproduction. Nat Rev Endocrinol 2020b;16(8): 407–20.

Newell-Price J, Morris DG, Drake WM, Korbonits M, Monson JP, Besser GM, et al. Optimal response criteria for the human CRH test in the differential diagnosis of ACTH-dependent Cushing's syndrome. J Clin Endocrinol Metab 2002;87(4):1640–5.

Nieman LK. Recent updates on the diagnosis and management of Cushing's syndrome. Endocrinol Metab (Seoul) 2018;33(2):139–46.

Nilubol N, Soldin SJ, Patel D, Rwenji M, Gu J, Masika LS, et al. 11-Deoxycortisol may be superior to cortisol in confirming a successful adrenal vein catheterization without cosyntropin: a pilot study. Int J Endocr Oncol 2017;4(2):75–83.

Nye EJ, Grice JE, Hockings GI, Strakosch CR, Crosbie GV, Walters MM, et al. Comparison of adrenocorticotropin (ACTH) stimulation tests and insulin hypoglycemia in normal humans: low dose, standard high dose, and 8-hour ACTH-(1-24) infusion tests. J Clin Endocrinol Metab 1999;84(10):3648–55.

Ohta A, Hobo W, Ogawa S, Sugiura Y, Nishikawa T, Nishimoto K, et al. A method for determination of aldosterone concentrations of six adrenal venous serum samples during a single LC/ESI-MS/MS run using a sextet of Girard reagents. J Pharm Biomed Anal 2022;207:114423.

Özalkak Ş, Çetinkaya S, Budak FC, Erdeve ŞS, Aycan Z. Evaluation of gonadotropin responses and response times according to two different cut-off values in luteinizing hormone releasing hormone stimulation test in girls. Indian J Endocrinol Metab 2020;24(5):410–5.

Page MM, Taranto M, Ramsay D, van Schie G, Glendenning P, Gillett MJ, et al. Improved technical success and radiation safety of adrenal vein sampling using rapid, semi-quantitative point-of-care cortisol measurement. Ann Clin Biochem 2018;55(5):588–92.

Pasternak Y, Friger M, Loewenthal N, Haim A, Hershkovitz E. The utility of basal serum LH in prediction of central precocious puberty in girls. Eur J Endocrinol 2012;166:295–9.

Patti G, Guzzeti C, Di Iorgi N, Maria Allegri AE, Napoli F, Loche S, et al. Central adrenal insufficiency in children and adolescents. Best Pract Res Clin Endocrinol Metab 2018;32(4):425–44.

Pecori Giraldi F, Cavallo LM, Tortora F, Pivonello R, Colao A, Cappabianca P, et al. The role of inferior petrosal sinus sampling in ACTH-dependent Cushing's syndrome: review and joint opinion statement by members of the Italian Society for Endocrinology, Italian Society for Neurosurgery, and Italian Society for Neuroradiology. Neurosurg Focus 2015;38(2):E5.

Peitzsch M, Dekkers T, Haase M, Sweep FC, Quack I, Antoch G, et al. An LC-MS/MS method for steroid profiling during adrenal venous sampling for investigation of primary aldosteronism. J Steroid Biochem Mol Biol 2015;145:75–84.

Perogamvros I, Owen LJ, Keevil BG, Brabant G, Trainer PJ. Measurement of salivary cortisol with liquid chromatography-tandem mass spectrometry in patients undergoing dynamic endocrine testing. Clin Endocrinol (Oxf) 2010;72(1):17–21.

Phelan N, Williams EL, Cardamone S, Lee M, Creighton SM, Rumsby G, et al. Screening for mutations in 17β-hydroxysteroid dehydrogenase and androgen receptor in women presenting with partially virilised 46,XY disorders of sex development. Eur J Endocrinol 2015;172 (6):745–51.

Punchoo R, Bhoora S. Variation in the measurement of anti-Müllerian hormone—what are the laboratory issues? Front Endocrinol (Lausanne) 2021;12, 719029.

Rehan M, Raizman JE, Cavalier E, Don-Wauchope AC, Holmes DT. Laboratory challenges in primary aldosteronism screening and diagnosis. Clin Biochem 2015;48(6):377–87.

Rey RA, Grinspon RP, Gottlieb S, Pasqualini T, Knoblovits P, Aszpis S, et al. Male hypogonadism: an extended classification based on a developmental, endocrine physiology-based approach. Andrology 2013;1 (1):3–16.

Richter-Unruh A, Jorch N, Wessels HT, Weber EA, Hauffa BP. Venous sampling can be crucial in identifying the testicular origin of idiopathic male luteinising hormone-independent sexual precocity. Eur J Pediatr 2002;161(12):668–71.

Roli L, Santi D, Belli S, Tagliavini S, Cavalieri S, De Santis MC, et al. The steroid response to human chorionic gonadotropin (hCG) stimulation in men with Klinefelter syndrome does not change using immunoassay or mass spectrometry. J Endocrinol Invest 2017;40(8):841–50.

Rosenfield RL, Barnes RB, Ehrmann DA. Studies of the nature of 17-hydroxyprogesterone hyperresonsiveness to gonadotropin-releasing hormone agonist challenge in functional ovarian hyperandrogenism. J Clin Endocrinol Metab 1994;79(6):1686–92.

Rossitto G, Maiolino G, Lenzini L, Bisogni V, Seccia TM, Cesari M, et al. Subtyping of primary aldosteronism with adrenal vein sampling: hormone-and side-specific effects of cosyntropin and metoclopramide. Surgery 2018;163(4):789–95.

Sasaki Y, Katabami T, Asai S, Fukuda H, Tanaka Y. In the overnight dexamethasone suppression test, 1.0 mg loading is superior to 0.5 mg loading for diagnosing subclinical adrenal Cushing's syndrome based on plasma dexamethasone levels determined using liquid chromatography-tandem mass spectrometry. Endocr J 2017;64 (9):833–42.

Schirpenbach C, Seiler L, Maser-Gluth C, Rüdiger F, Nickel C, Beuschlein F, et al. Confirmatory testing in normokalaemic primary aldosteronism: the value of the saline infusion test and urinary aldosterone metabolites. Eur J Endocrinol 2006;154(6):865–73.

Shah R, Alshaikh B, Schall JI, Kelly A, Ford E, Zemel BS, et al. Endocrine-sensitive physical endpoints in newborns: ranges and predictors. Pediatr Res 2020. https://doi.org/10.1038/s41390-020-0950-2.

Shin S, Oh H, Park HR, Joo EY, Lee SY. A sensitive and specific liquid chromatography-tandem mass spectrometry assay for simultaneous quantification of salivary melatonin and cortisol: development and comparison with immunoassays. Ann Lab Med 2021;41(1):108–13.

Skorupskaite K, George JT, Veldhuis JD, Millar RP, Anderson RA. Kisspeptin and neurokinin B interactions in modulating gonadotropin secretion in women with polycystic ovary syndrome. Hum Reprod 2020;35(6):1421–31.

Solcia E, Usellini L, Buffa R, Rindi G, Villani L, Zampatti C, et al. Endocrine cells producing regulatory peptides. Experientia 1987;43 (7):839–50.

Stanczyk FZ, Saxena T, Lobo RA. Dexamethasone suppressibility and adrenal and ovarian venous effluents of 5α-reduced C19 conjugates in hyperandrogenic women. J Steroid Biochem Mol Biol 2014; 139:73–7.

Stowasser M, Ahmed AH, Cowley D, Wolley M, Guo Z, McWhinney BC, et al. Comparison of seated with recumbent saline suppression testing for the diagnosis of primary aldosteronism. J Clin Endocrinol Metab 2018;103(11):4113–24.

Sturgeon CM, McAllister EJ. Analysis of hCG: clinical applications and assay requirements. Ann Clin Biochem 1998;35(Pt 4):460–91.

Thuzar M, Young K, Ahmed AH, Ward G, Wolley M, Guo Z, et al. Diagnosis of primary aldosteronism by seated saline suppression test-variability between immunoassay and HPLC-MS/MS. J Clin Endocrinol Metab 2020;105(3), dgz150.

Tran MT, Tran NA, Nguyen PM, Vu CD, Tran MD, Ngo DN, et al. 11β-Hydroxylase deficiency detected by urine steroid metabolome profiling using gas chromatography-mass spectrometry. Clin Mass Spectrom 2018;7:1–8.

Travers S, Martinerie L, Bouvattier C, Boileau P, Lombès M, Pussard E. Multiplexed steroid profiling of gluco- and mineralocorticoids pathways using a liquid chromatography tandem mass spectrometry method. J Steroid Biochem Mol Biol 2017;165(Pt B):202–11.

Tsagarakis S, Vassiliadi D, Kaskarelis IS, Komninos J, Souvatzoglou E, Thalassinos N. The application of the combined corticotropin-releasing hormone plus desmopressin stimulation during petrosal sinus sampling is both sensitive and specific in differentiating patients with Cushing's disease from patients with the occult ectopic adrenocorticotropin syndrome. J Clin Endocrinol Metab 2007;92(6):2080–6.

Tsutsui K, Osugi T, Son YL, Ubuka T. Review: structure, function and evolution of GnIH. Gen Comp Endocrinol 2018;264:48–57.

Tuchelt H, Dekker K, Bähr V, Oelkers W. Dose-response relationship between plasma ACTH and serum cortisol in the insulin-hypoglycaemia test in 25 healthy subjects and 109 patients with pituitary disease. Clin Endocrinol (Oxf) 2000;53(3):301–7. Erratum in: Clin Endocrinol (Oxf) 2001;54(1):135.

Turcu AF, Wannachalee T, Tsodikov A, Nanba AT, Ren J, Shields JJ, et al. Comprehensive analysis of steroid biomarkers for guiding primary aldosteronism subtyping. Hypertension 2020;75(1):183–92.

Turner KA, Larson BL, Kreofskys NC, et al. Assessment of complement interference in anti-Mullerian hormone immunoassays. Clin Chem Lab Med 2020;58:e8–e10.

Turra J, Granzotto M, Gallea M, Faggian D, Conte L, Litta P, et al. Sertoli-Leydig cell tumors: hormonal profile after dynamic test with GnRH analogue: triptorelin represents a useful tool to evaluate tumoral hyperandrogenism. Gynecol Endocrinol 2015;31(1):18–21.

Ueland GÅ, Methlie P, Kellmann R, Bjørgaas M, Åsvold BO, Thorstensen K, et al. Simultaneous assay of cortisol and dexamethasone improved diagnostic accuracy of the dexamethasone suppression test. Eur J Endocrinol 2017;176(6):705–13.

Ueland GÅ, Methlie P, Øksnes M, Thordarson HB, Sagen J, Kellmann R, et al. The short cosyntropin test revisited: new normal reference range using LC-MS/MS. J Clin Endocrinol Metab 2018;103(4):1696–703.

Utz A, Biller BM. The role of bilateral inferior petrosal sinus sampling in the diagnosis of Cushing's syndrome. Arq Bras Endocrinol Metabol 2007;51(8):1329–38.

Vaiani E, Lazzati JM, Ramirez P, Costanzo M, Gil S, Dratler G, et al. The low-dose ACTH test: usefulness of combined analysis of serum and salivary maximum cortisol response in pediatrics. J Clin Endocrinol Metab 2019;104(10):4323–30.

van Faassen M, Bischoff R, Kema IP. Relationship between plasma and salivary melatonin and cortisol investigated by LC-MS/MS. Clin Chem Lab Med 2017;55(9):1340–8.

Varikasuvu SR, Prasad VS, Vamshika VC, Satyanarayana MV, Panga JR. Circulatory metastin/kisspeptin-1 in polycystic ovary syndrome: a systematic review and meta-analysis with diagnostic test accuracy. Reprod Biomed Online 2019;39(4):685–97.

Vassiliadi DA, Tsagarakis S. Diagnosis of endocrine disease: the role of the desmopressin test in the diagnosis and follow-up of Cushing's syndrome. Eur J Endocrinol 2018;178(5):R201–14.

Vassiliadi DA, Tzanela M, Tsatlidis V, Margelou E, Tampourlou M, Mazarakis N, et al. Abnormal responsiveness to dexamethasone-suppressed CRH test in patients with bilateral adrenal incidentalomas. J Clin Endocrinol Metab 2015;100(9):3478–85.

Vastbinder M, Kuindersma M, Mulder AH, Schuijt MP, Mudde AH. The influence of oral contraceptives on overnight 1 mg dexamethasone suppression test. Neth J Med 2016;74(4):158–61.

Vaughan NJ, Jowett TP, Slater JD, Wiggins RC, Lightman SL, Ma JT, et al. The diagnosis of primary hyperaldosteronism. Lancet 1981;1 (8212):120–5.

Vogg N, Kurlbaum M, Deutschbein T, Gräsl B, Fassnacht M, Kroiss M. - Method-specific cortisol and dexamethasone thresholds increase clinical specificity of the dexamethasone suppression test for Cushing syndrome. Clin Chem 2021;67(7):998–1007.

Wang H, Ba Y, Xing Q, Cai RC. Differential diagnostic value of bilateral inferior petrosal sinus sampling (BIPSS) in ACTH-dependent Cushing syndrome: a systematic review and meta-analysis. BMC Endocr Disord 2020;20(1):143.

Wei C, Davis N, Honour J, Crowne E. The investigation of children and adolescents with abnormalities of pubertal timing. Ann Clin Biochem 2017;54(1):20–32.

Weintrob N, Davidov AS, Becker AS, Israeli G, Oren A, Eyal O. Serum free cortisol during glucagon stimulation test in healthy short-statured children and adolescents. Endocr Pract 2018;24(3):288–93.

Wijayarathna R, de Kretser DM. Activins in reproductive biology and beyond. Hum Reprod Update 2016;22(3):342–57.

Wu S, Yang J, Hu J, Song Y, He W, Yang S, et al. Confirmatory tests for the diagnosis of primary aldosteronism: a systematic review and meta-analysis. Clin Endocrinol (Oxf) 2019;90(5):641–8.

Yang Z, Ye L, Wang W, Zhao Y, Wang W, Jia H, et al. 17β-Hydroxysteroid dehydrogenase 3 deficiency: three case reports and a systematic review. J Steroid Biochem Mol Biol 2017;174:141–5.

Yates AP, Jopling HM, Burgoyne NJ, Hayden K, Chaloner CM, Tetlow L. Paediatric reference intervals for plasma anti-Müllerian hormone: comparison of data from the Roche Elecsys assay and the Beckman Coulter Access assay using the same cohort of samples. Ann Clin Biochem 2019;56(5):536–47.

Yazdani P, Lin Y, Raman V, Haymond M. A single sample GnRHa stimulation test in the diagnosis of precocious puberty. Int J Pediatr Endocrinol 2012;2012:23.

Yoneda T, Karashima S, Kometani M, Usukura M, Demura M, Sanada J, et al. Impact of new quick gold nanoparticle-based cortisol assay during adrenal vein sampling for primary aldosteronism. J Clin Endocrinol Metab 2016;101(6):2554–61.

Zhang WW, Wang Y, Chu YX. Tacr3/NK3R: beyond their roles in reproduction. ACS Chem Nerosci 2020;11(19):2935–43.

Zharinov GM, Bogomolov OA, Chepurnaya IV, Neklasova NY, Anisimov VN. Melatonin increases overall survival of prostate cancer patients with poor prognosis after combined hormone radiation treatment. Oncotarget 2020;11(41):3723–9.

Zierau O, Lehmann S, Vollmer G, Schänzer W, Diel P. Detection of anabolic steroid abuse using a yeast transactivation system. Steroids 2008;73(11):1143–7.

Guidelines

American College of Obstetricians and Gynecologists. Primary ovarian insufficiency in adolescents and young women. Committee Opinion No. 605. Obs Gynecol 2014;123:193–7.

Part III

Steroids in clinical practice

A patient with a steroid disorder may present with symptoms as broad as physical signs, (hirsutism, buffalo hump, hypertension, osteoporosis, anorexia, fatigue, weight loss, fatigue, weight gain, infertility, hyperpigmentation, vomiting, diarrhea, and rapid growth) or biochemical changes (hypoglycemia, hyponatremia, and hyperkalemia) and opportunistic findings (adrenal mass on CT scan). There has been a definite increase in the numbers of patients with adrenal incidentaloma with a prevalence of up to 2% of the population. Congenital adrenal hyperplasia can have an incidence of 1 in 10,000 to 1 in 15,000, whereas Addison's disease and Cushing disease can have the prevalence of less than 150 or 70, respectively, per million of the population. The prevalence of primary aldosteronism may be up to 6% of the population but if so is likely grossly underdiagnosed. Polycystic ovary syndrome affects 4%–25% of the population in countries worldwide. Puberty is said to be earlier, and age of first pregnancy is increasing to 30 years of age and more as women establish a career first.

Whatever the clinical picture and the age of the patient, the clinician has to decide a route of investigations in order to make a diagnosis and treatment plan. Any dysregulation of steroid precursor uptake, steroid hormone synthesis, transport, action, metabolism, or excretion can lead to the development of a human disease so the diagnosis is a challenge with any affected patient. The measurements of steroids have been, and still are, the basis of investigations of patients with suspected steroid disorders. Many factors outside the performance of the tests need to be considered such as

- nature and timing of samples
- consulting guidelines and practice notes
- patient preparation
- notification to the laboratory in some cases (if rapid results are needed)
- ordering supplies and items from the pharmacy (ACTH, for example)

Continuing the arrangement of the book, the sugar, salt, and sex themes will be considered in turn in relation to underactive (hypo) and overactive (hyper) gland situations. The corticosteroids normally control sugar and salt concentrations, hence the glucocorticoids and mineralocorticoids distinction and thus mainly about cortisol and aldosterone, respectively, and their actions mainly on glucose and sodium chloride homeostasis. Abnormalities in male sex steroid production can be the result of defects in adrenal synthesis of cortisol and aldosterone or problems in gonadal function and will be addressed in both categories. In addition to the clinical consequences of abnormal steroid production, defects are now known for the actions and metabolism of steroids. Diagnosis will depend on the demonstration of unusual steroid concentrations although increasingly genetic tests are undertaken because defects in the relevant genes have been found which in many cases account for changes in a protein activity. The reader will not be bombarded with the genetic mutations but is directed to the most recent literature and reviews for the detail of all mutations found in particular genes.

The steroid assays must have been fully validated, and the clinicians need to be kept aware of changes in reference ranges as the laboratory migrates between methods. The assays need to be reliable and, as near as possible, specific for the analyte. The laboratory is likely to have or may need to adopt new technology. The method used needs to accommodate the clinical needs above and below the reference ranges for each analyte. Ideally, reference ranges for each test will be established locally, but this is a big undertaking and a published reference range for the method may have to be used. The clinician and the laboratory need to work together in establishing and judging the suitability of results.

Over the past 70 years, there have been huge changes in how the laboratory serves the clinicians in terms of the number of analytes and the specificity of the measurement of each. Assay interferences are likely to be less of a problem with LC-MS methods than with automated immunoassay platforms but not impossible, the use of a new drug may interfere in a method directly or indirectly in a matrix effect for instance. Some results can be obtained in less than an hour, but others may take days or weeks. Nevertheless, satisfying the patient must be key to the outcome of the service.

Lab investigations

Care should be taken with the age of the patient at the time when blood samples are taken for hormone measurements.

At birth plasma testosterone concentrations can be up to 15 nmol/L in an extraction assay and sometimes up to 100 nmol/L in direct assays. The T concentrations fall over the first week of life as human chorionic gonadotrophin (hCG) is cleared from the circulation, so that at 7 days testosterone concentrations are less than 3 nmol/L. In response to an increase in LH concentrations, plasma testosterone rise over the second week and remain up to 12 nmol/L for up to 6 months (the minipuberty) when LH concentrations decline and plasma testosterone concentrations fall to <3 nmol/L by 6 months. These changes in absolute concentrations and methodological interference must be understood, otherwise results will not be interpreted correctly particularly if gonadal stimulation is attempted.

When assessing whether there is a normal cortisol production, it must be remembered that the circadian rhythm of cortisol does not emerge until a child is 2–6 months of age. A single low concentration result does not exclude normal cortisol production, and it may be necessary to perform an ACTH stimulation test for certainty. In newborn infants, the dose of synacthen can be considerably reduced to 1 μg with blood taken for cortisol measurements at 15 and 30 min. The laboratory should be clear of the extent to which cortisone is measured in the assay since this is more important in the newborn infant than cortisol itself and will affect the cortisol measurement. The further investigation of males with poor androgenization and normal cortisol will depend upon the initial findings for testosterone and gonadotrophins. The immunoassays in use for androgens have been inaccurate and manufacturers have endeavored to make changes and prove results are comparable to MS though not in young children. Improved, simultaneous androgen assays (testosterone, dihydrotestosterone, androstenedione, DHEA, and DHEAS) by tandem mass spectrometry will aid in the diagnosis of DSD. The hCG test will, in many cases, be necessary to unmask or accentuate an enzyme deficiency affecting testosterone production. The test may be necessary when there is doubt about the presence of testes. There are many protocols for hCG stimulation tests and mostly involve repeated injections. In addition to preparations from human pregnancy urine, recombinant hCG preparations are now available. A common protocol in a newborn is 1,000 international units of hCG daily for 3 days. In older children, hCG (1000 IU) is given twice a week for 3 weeks.

A range of targeted and untargeted genetic tests are now available but fall outside the remit of the author although the reader is directed to recent publications to guide further reading.

Requests for laboratory tests

The typical request form or computer request needs information on age, sex, id numbers, tests required, and clinical information. All of these should be filled in as far as possible. The sex can be a problem when unknown at the time of birth pending investigations and later in life there can be issues around gender dysphoria and transgender states particularly if the patient or family see the request form. There are problems over the lab reporting results without an appropriate reference range.

When patients have to undergo any investigation, they must be fully informed of what will happen and what will be the risks. Where the test are performed by non-healthcare professionals (e.g., phlemotomists), they must be trained and have standard working practices, and they are crucial to the lab getting the right samples in the correct state for analysis.

Steroid analysis and reporting

The lab can offer most help when informed of the clinical background to the testing requirements, particularly with complex endocrine investigations around steroids. Reference ranges must be provided by the lab if clinicians are to make the best of interpretation of the results. The author is averse to using mean and standard deviation. First, this assumes normal distribution of results that is rarely the case. Second, if 2 standard deviations are calculated from the mean the range of results will usually be from a negative value and does not reflect the range of the data. Median and range are thus the preferred expression. In some tests such as a urine steroid profile, the lab must talk the clinician through the data with charts. Currently, labs do not report in a harmonized manner particularly with units for excretion rates so the literature is very difficult to interpret unless local standards are used.

The author has reported to a template based on sex steroids, cortisol metabolites, and intermediates with reference ranges for 24 h urine collection in micrograms that is how the analysis is based (Fig. 3.1). Additional lines can be added where unusual steroids are quantified. This can be converted to a histogram (Fig. 3.2). The Shackleton lab has standardized the breakdown with colors (Fig. 3.3) that are consistent from synthetic pathway to results and is to be applauded for helping clinicians. This approach was probably first used by Homma et al. for infants with enzyme defects (Fig. 3.4). Where a 24 h collection is not provided labs use a number of ways to express results. At UCLH, the mean 24 h volume for age is used in calculations, and other laboratories determine creatinine excretion rates and report steroid output against the creatinine (again units can be SI or mass).

The expression of results as ratios is a help to interpreting the data for enzyme defects using the products of substrate and product and does not rely upon the need for a 24 h collection. Many ratios are in use but seem to be in

PATIENT NAME:

AGE: DOB:

REGISTRATION NUMBER:

CONSULTANT:

HOSPITAL

SAMPLE DATE   TOTAL VOLUME   mL

BODY SURFACE AREA:

CLINICAL INFORMATION:

Patients results (μg/day) against medians and ranges for normal adult males,
n = 10, aged 19–45 years.

STEROID	PATIENT	NORMAL MALE MEDIAN	NORMAL MALE RANGE
ADRENAL ANDROGENS			
Androsterone (5α)		1200	1040–1620
Aetiocholanolone (5β)		940	620–2030
Dehydroepiandrosterone (DHA)		210	<50–1110
11β-Hydroxyandrosterone		630	450–780
11β-Hydroxyaetiocholanolone		160	<50–310
16α-Hydroxy-DHA		230	130–400
INTERMEDIATES			
Pregnanediol		110	80–220
Pregnanetriol		270	110–520
Androstenetriol		390	290–600
CORTISOL METABOLITES			
Tetrahydrocortisone (THE)		1890	1600–3330
Tetrahydrocortisol (THF) (5β)		990	600–1420
Allo-THF (5α)		640	380–910
α-Cortolone		650	220–1100
β-Cortolone + β-cortol		870	410–1370
α-Cortol			

COMMENT

FIG. 3.1   Style of report for steroid excretion in urine by GC profile.

near agreement between labs (Fig. 3.5). Some examples are given and more will be in the appropriate text to follow in clinical chapters. Interpretation is variable from brief (normal/abnormal) to an essay depending on the lab. At UCLH, we devised a series of codes that could be assembled to the complete interpretation.

First, the clinical question must be addressed (hirsutism, salt loss) reply androgens/mineralocorticoids raised/not raised.

Second, the highs and lows of the results should be highlighted (e.g., increased output of x and Y...., low Q and Z...).

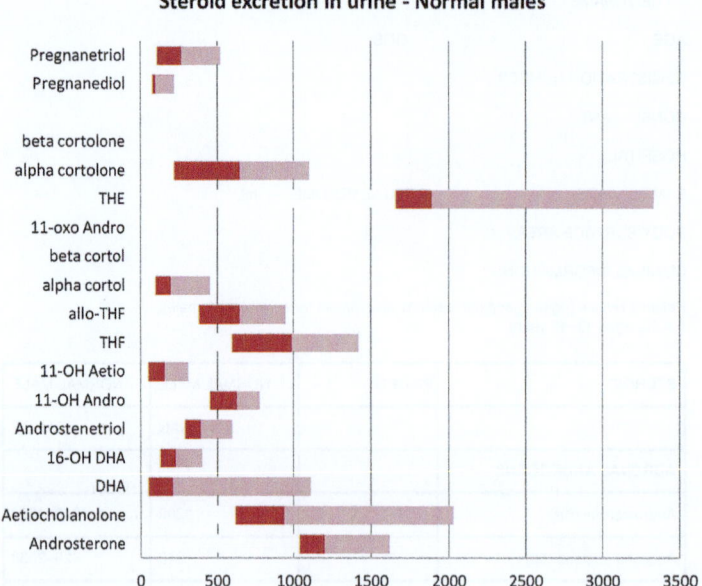

FIG. 3.2 Report in histogram format for steroid excretion in urine. Median and range are shown for 30 adult males.

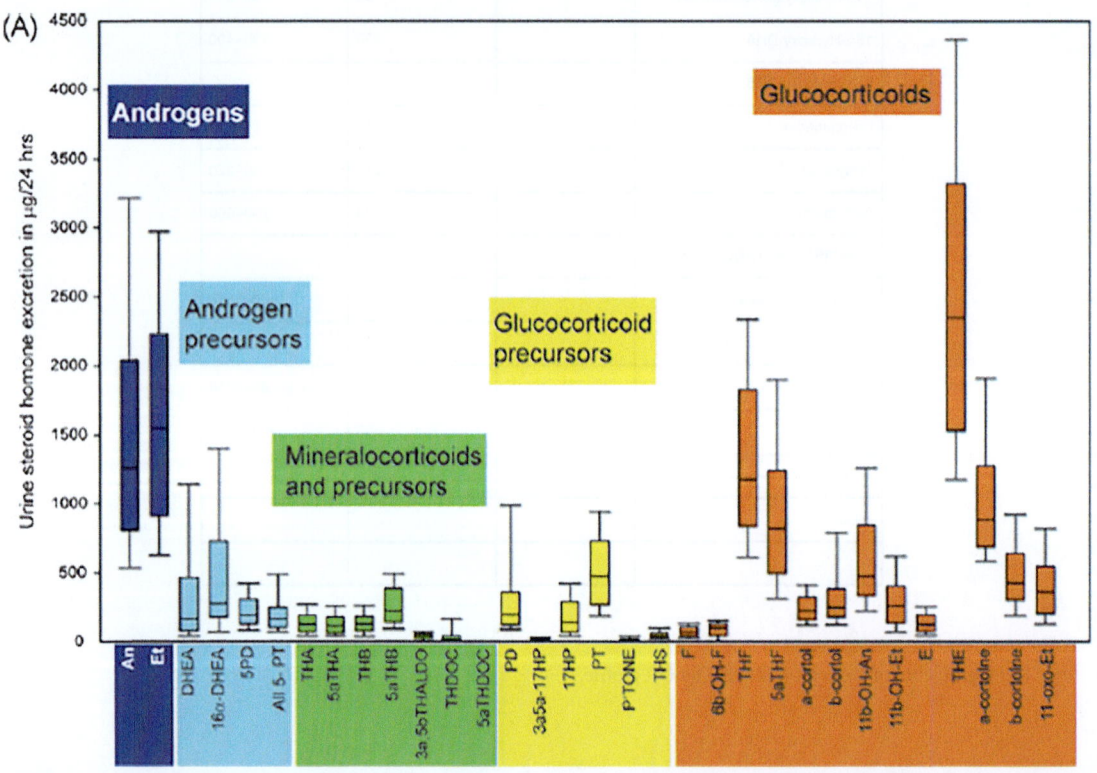

FIG. 3.3 Report of urinary steroid excretion as μg/24 h. This panel shows the normal excretion of metabolites in adults. Mean and 95th percentiles are shown. *From Krone N, Hughes BA, Lavery GG, Stewart PM, Arlt W, Shackleton CH. Gas chromatography/mass spectrometry (GC/MS) remains a preeminent discovery tool in clinical steroid investigations even in the era of fast liquid chromatography tandem mass spectrometry (LC/MS/MS). J Steroid Biochem Mol Biol 2010;121(3–5):496–504.*

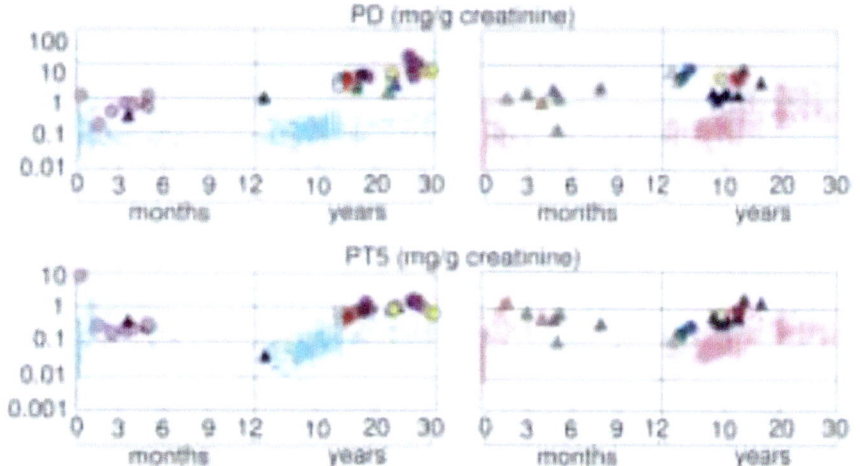

FIG. 3.4 Representative results of urine steroid profile analysis for patients with cytochrome P450 oxidoreductase (POR) deficiency. The results are for the assessment of the conventional front door pathway. The triangles indicate POR patients with two missense mutations, and the circles represent patients with one missense and one non-missense mutation. The light blue circles and the pink circles show the data obtained from 854 control males and 909 control females, respectively. Urine samples from the control subjects were obtained before 12 months of age in 425 males and 351 females, and after 1 year of age in the remaining 429 males and 558 females. Note that all the data are expressed using a logarithm scale. *From Homma K, Hasegawa T, Nagai T, Adachi M, Horikawa R, Fujiwara I, Tajima T, Takeda R, Fukami M, Ogata T. Urine steroid hormone profile analysis in cytochrome P450 oxidoreductase deficiency: implication for the backdoor pathway to dihydrotestosterone. J Clin Endocrinol Metab 2006;91(7):2643–2649.*

11β-hydroxylase deficiency

100*(THS)/(THE+THF+5aTHF)

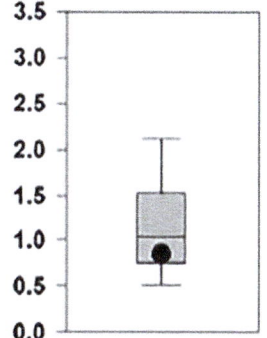

FIG. 3.5 Diagnostic ratio for 11-hydroxylase deficiency. The diagnostic ratio is of tetrahydro-11-deoxycortisol over THE + THF + alloTHF. The normal mean with 5th and 95th percentiles are shown. The block filled circle is for one patent with 11-hydroxylase deficiency.

Third, What is the inference of the results if add up to a biosynthetic defect or a conclusion. Report low cortisol, metabolites of X raised consistent with congenital adrenal hyperplasia due to defect of y enzyme.

Finally recommend follow-up investigations. Suggest measure ….. in blood. Check plasma renin activity. Check ACTH. Suggest genetic testing.

Genetic tests

The author is not an expert in genetic testing, and this information has been tackled in the book with that in mind. Some recent references are included to lead the interested reader to search the literature. It is difficult to judge at this stage the extent to which the adoption off genetic tests will replace biochemical tests in clinical investigations.

Pharmacology of steroids

Steroids are prescribed where production is deficient although replacement to match the natural rhythms is difficult and creates many problems to clinicians. Steroid action can be blocked at the synthetic level or the receptor level. Steroids are used as immunosuppressants for asthma and rheumatoid arthritis, fertility regulators in the normal menstrual cycle or hormonal cancers.

Chapter 3.1.1

Cortisol excess states

3.1.1.1 Cortisol hypersecretion

Plasma cortisol concentrations in a normal subject are within the range of 200–800 nmol/L (7.3–29 μg/dL) in the morning but less than 200 nmol/L around midnight. The actual timing of the nadir varies with age. In children and adolescents, the nadir occurs at 2300h, in young adults at 0200h and in older people midnight. The pulsatile nature of this secretion, particularly in the night from around 0300h, has already been emphasized in the book, so the time when a blood or saliva sample is collected for analysis is important for interpretation of the result. The assays for steroids have become more specific as methods have moved from polyclonal to monoclonal antibodies in immunoassays and as chromatography with mass spectrometry is being adopted. There are issues of interferences in immunoassays due to heterophilic antibodies in the sample (Choy et al., 2017). Clinicians need to ask the patients about domestic pets and use of nutritional supplements where biotin can interfere with certain tests depending on the assay configuration. Progress in assay methodology over the past 20 years means that the clinician is faced with wide a variable ranges of published steroid concentrations that may be difficult to harmonize. Cortisol concentrations above the normal range for the test should be judged by the clinician in the context of patient presentation and examination. A number of screening tests eliminate some confounding issues but further investigations are needed to ascertain the cause and formulate treatment of the actual problem. Careful interpretation of the patient data is mandatory. In addition to the methodological considerations of quantifying free cortisol in urine, there are the problems of obtaining complete 24h urine collections especially in children. Patients need clear instructions from the clinical team.

High cortisol production can be ACTH dependent or independent. A pituitary adenoma secreting **adrenocorticotrophic hormone (ACTH)** was first described by Harvey Cushing (**Cushing's disease**) and is the commonest noniatrogenic cause of the consequential cortisol excess. An adenoma is a growth in epithelial tissue. Less common causes are autonomous cortisol secretion from an adrenal neoplasm and ACTH secretion from an extra-pituitary tumor (ectopic ACTH syndrome). All causes of cortisol excess are classified under the generic name Cushing's syndrome. Symptoms may also be caused by the over use of exogenous glucocorticoids, depression and alcoholism. The literature is becoming confusing because of differences in the biochemical tests that are used, with immunoassay dominant before 2017 and increasingly since then by mass spectrometric methods. There are therefore many reference ranges. The incidence of CS and CD is variable between populations but less than 3 cases per million of the population, so care is needed when requesting biochemical tests and interpreting the results.

The detection rate of adrenal tumors has increased in recent years through the use of MRI that has revealed adrenal masses during abdominal scanning. Some of these masses are nonfunctional in terms of adrenal steroid secretion and are called adrenal incidentalomas. In some cases, there is mild autonomous cortisol secretion (MACS).

3.1.1.2 Cushing's disease/syndrome

Cushing's syndrome is the symptoms and signs which result from an increase in the amount of cortisol produced, with a reduction in or loss of the normal circadian rhythm of cortisol. Chronic cortisol excess induces alterations in protein, lipid and carbohydrate metabolism leading to an increased cardiometabolic burden (Vega-Beyhart et al., 2021).

The clinical examination should take into account a number of issues. High plasma concentrations of cortisol leads to combinations of a number of typical signs—full moon-shaped face, buffalo hump, and striae (Fig. 3.1.1.1). The rounded face of a patient with Cushing's syndrome (Fig. 3.1.1.2) can sometimes also show hirsutism.

Cushing's disease caused by a pituitary adenoma secreting adrenocorticotrophic hormone (ACTH) was first described in 1932 by Harvey Cushing. This disease form is more frequent in males than females. The high production of other steroids (more typical of adrenal tumors) accounts for acne, hirsutism, and in some cases for hypertension. Other causes of cortisol excess are classified under the generic name of Cushing's syndrome (Table 3.1.1.1).

Steroids in the Laboratory and Clinical Practice. https://doi.org/10.1016/B978-0-12-818124-9.00019-X

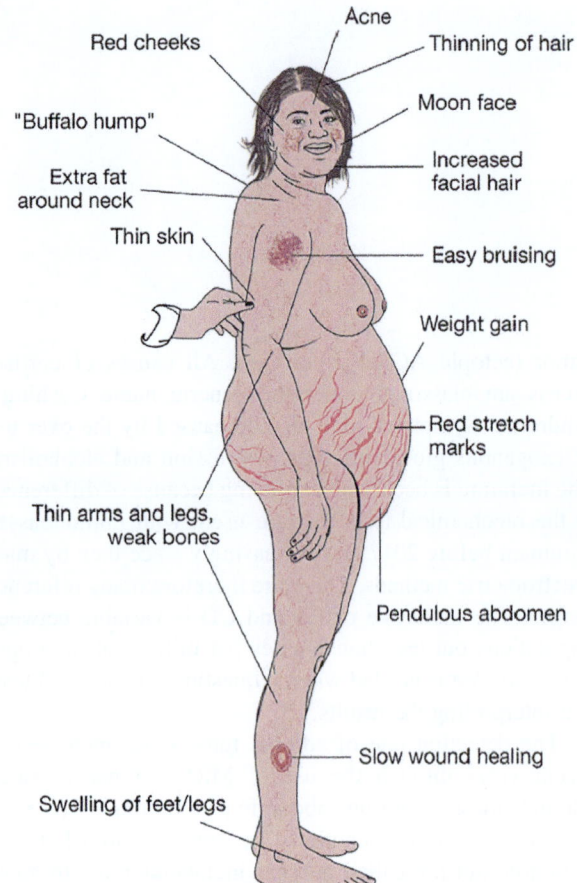

FIG. 3.1.1.1   Signs and symptoms of Cushing's syndrome.

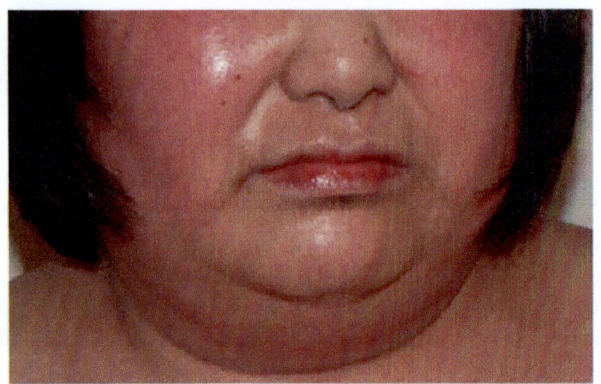

FIG. 3.1.1.2   Full-moon shaped face. Hirsutism (hair on upper lip) is often a problem.

TABLE 3.1.1.1 Signs and symptoms Cushing's syndrome.

Symptoms	Signs
Weight gain/growth failure	Reduction in height SDS and increased BMI SDS
Acne	Facial appearance
Striae	Purple striae lower abdomen
Hirsutism	Facial appearance
Virilization	Hirsutism, clitormegaly
Lethargy/depression	Muscle weakness
Emotional lability	Osteoporosis
Headache	Buffalo hump
Bruising	Carney complex
Family history	
Lentigines/freckles	

The use of exogenous corticosteroids is the commonest iatrogenic cause of such problems and this should be excluded before investigations are taken further. The high production of other steroids (more typical of adrenal tumors) accounts for acne, hirsutism, and in some cases for hypertension although it is difficult to exclude direct effects of cortisol on blood pressure. Clinicians should be alert to the possibility that herbal medicines may be supplemented with glucocorticoids (Fig. 3.1.1.5), so patients will need to be questioned on any such practice plus all other medications.

Cushing's syndrome can thus be due to primary cortisol excess (Adrenal tumor), primary ACTH secreting tumor (Cushing's disease), CRF excess, ectopic ACTH or exogenous corticosteroids (Fig. 3.1.1.6).

In order to complete baseline investigations, the measurements of Free T4, TSH, Prolactin, LH, FSH, testosterone or estradiol, androstenedione, DHEAS and SHBG should be included The picture is complicated by HPA activation seen also in:

- obesity,
- polycystic ovary syndrome (PCOS),
- poorly controlled diabetes mellitus,
- chronic alcoholism, and
- psychiatric disorders.

Circulating concentrations of the adipose tissue hormone leptin are often increased in obesity, in some cases due to leptin resistance. PCOS is primarily a state of hyperandrogenism and is dealt with separately (Chapter 3.2.2). The diagnosis of **alcohol-induced pseudo-Cushing's syndrome** is assisted with raised serum levels of gamma glutamyl

Growth retardation and weight gain are the more likely presentation in a child with Cushing's syndrome (Fig. 3.1.1.3). Care is needed as an acceleration of weight gain accompanied by a decline of growth velocity down to stagnation of growth (Fig. 3.1.1.4) can be observed in severe acquired hypothyroidism. The diagnosis of the disease in the early phases of these growth chart changes is a real challenge.

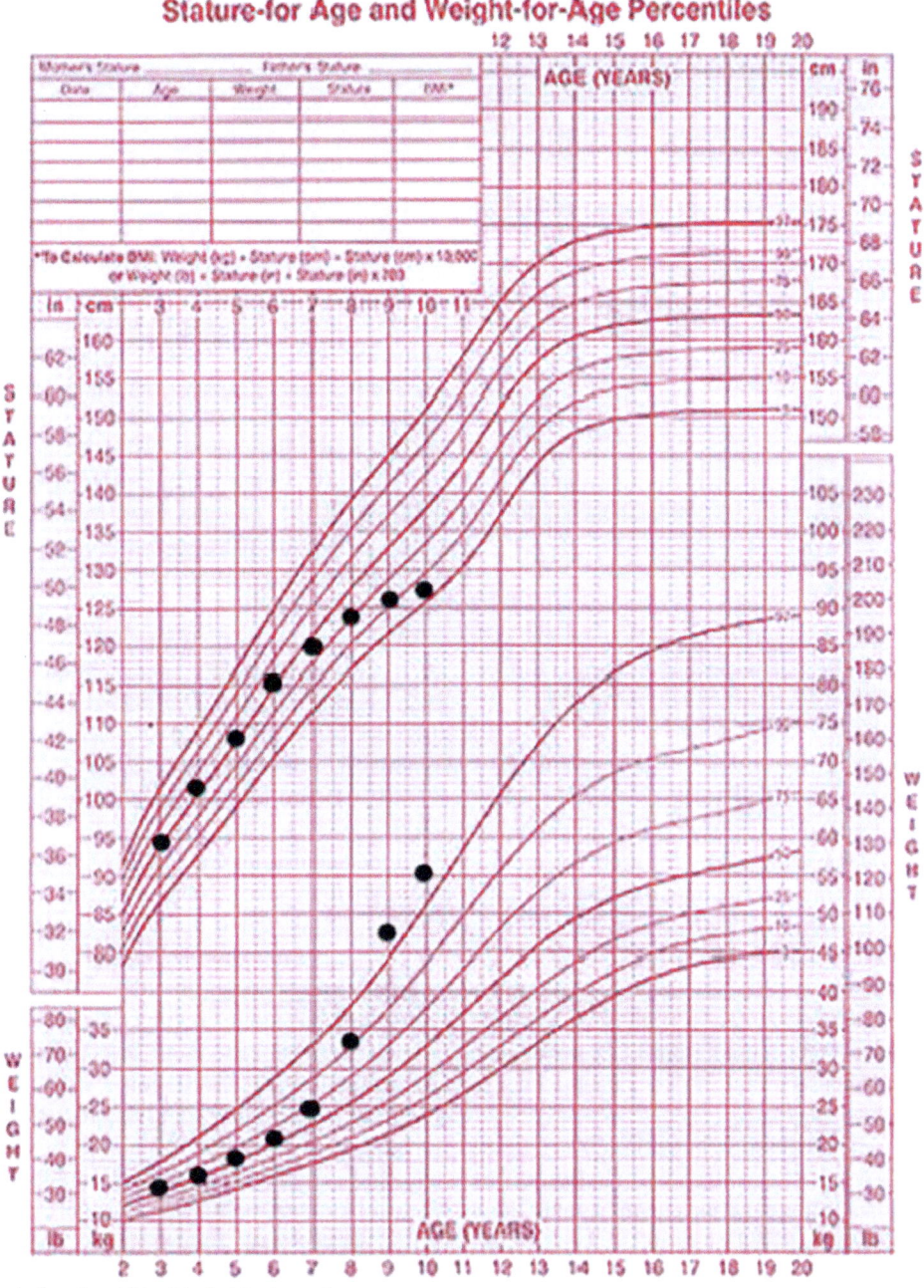

FIG. 3.1.1.3 Growth chart for child with Cushing's syndrome.

transpeptidase, aspartate aminotransferase, carbohydrate deficient transferrin, elevated mean corpuscular volume or occasionally by blood alcohol determination. Depression may be difficult to distinguish from Cushing's syndrome, though the cortisol response to insulin induced hypoglycemia remains intact in the former but is usually impaired in the latter condition. The biochemical features rapidly become normal when alcohol is stopped or depression is treated. Urine collections for 24h cortisol determination should be repeated three times. Blood or saliva cortisol concentrations at 0800, 1800 and 2400h are needed.

3.1.1.2.1 Demonstration of high cortisol concentrations in body fluids

Most of the recent literature will report total **plasma cortisol concentrations** from automated chemistry platform analyzers which are still used because of low cost and convenience in the laboratory test flow organization. The means of displacement of cortisol from binding proteins (CBG and albumin) is often not declared in commercial kit and platform methods. **Total plasma cortisol concentration** can be raised if CBG is elevated for example by exogenous

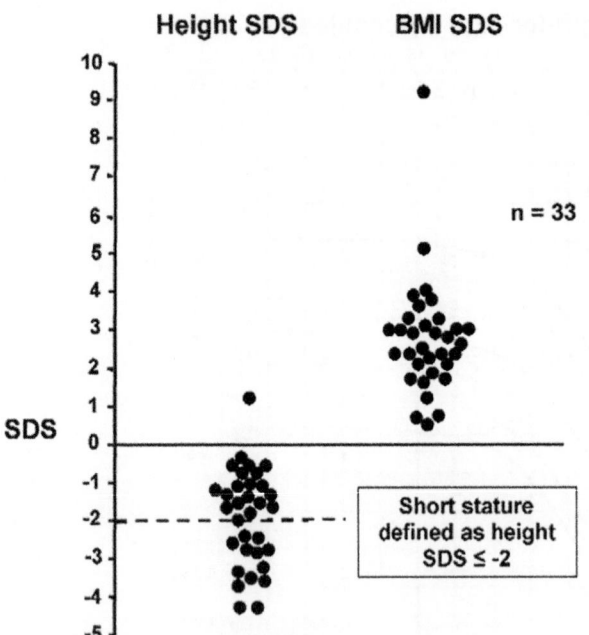

FIG. 3.1.1.4 Height SDS and BMI SDS in 33 patients with CD. Short stature defined as Height SDS ≤ -2 is shown as a dotted line. *(From Chan LF, Storr HL, Grossman AB, Savage MO. Pediatric Cushing's syndrome: clinical features, diagnosis, and treatment. Arq Bras Endocrinol Metabol. 2007a;51(8):1261–71.)*

estrogens (oral contraceptives) or stressful events (critical illness, pregnancy, depression, hospitalization and poorly controlled diabetes mellitus). The most commonly used antibodies are raised against cortisol-3-(O-carbxymethyl)-oxime conjugates. The performances of cortisol assays are variable (El-Farhan et al., 2017) which affects reference ranges and potential for interferences. High cross reactions are seen in immunoassays for cortisol from 11-deoxycortisol, prednisolone and prednisone but not dexamethasone or beclomethasone.

The pattern of cortisol secretion in Cushing's syndrome is such that there is a **loss of the circadian rhythm of the plasma total cortisol concentrations** with notably **high levels (>300 nmol/L (11 µg/dL)** in samples taken at **midnight** when concentrations are <100 nmol/L (3.6 µg/dL) in normal subjects. This investigation must be performed with careful attention to detail for correct interpretation. The patient needs ideally to be stabilized in hospital for at least 48 h and asleep prior to a stress-free venepuncture.

In adults, **saliva samples** collected at home are therefore being used instead of blood sampling (Mezzullo et al., 2016) and this is an attractive option for investigation with children who dislike needles. Salivary cortisol concentrations fluctuate in the 24 h period just like plasma concentrations change, with morning peak and trough around midnight (Casals et al., 2011; Dom et al., 2007). In some cases, saliva has been collected between 2200 and 2300 h before bed rather than set an alarm clock for midnight.

Samples taken at bedtime after 2230 h will give cortisol results that are as effective as midnight samples in demonstrating low cortisol in normal subjects and raised concentrations in Cushing's patients (Raff and Phillips, 2019; Elamin et al., 2008). Results by immunoassay are less accurate than by LC-MS/MS (Bae et al., 2016). Setting an alarm to wake up at midnight counts as a stress event so samples must be collected quickly. Analysis for cortisol concentrations of samples collected over the evening show a nadir before midnight (Chan et al., 2007a,b; Carroll et al., 2009). Normal **late night salivary cortisol concentrations** are typically less than 16 nmol/L (0.58 ng/dL) by a chemiluminescent assay (Ceccato et al., 2013) and 20 to 120 nmol/L (0.72–4.35 ng/mL) in CS. Lower results are found with Salimetrics EIA with results closer to LC-MS/MS (Deming regression slope 0.65) (Raff and Phillips, 2019). Patients need instructions for late night saliva collection with respect to time and the brushing of teeth. When determined by LC-MS/MS cortisol reference ranges and cut-offs (3.6 nmol/L; 1 ng/mL) (Bäcklund et al., 2020) are lower than with chemiluminescence (Ceccato et al., 2019). Cortisone concentrations in saliva are higher than cortisol (14.4–57.5 nmol/L at 0800 h and 1.5–13.5 nmol/L at 2300 h) because salivary glands express high HSD11B2 activity (Bäcklund et al., 2020). Results by LC-MS/MS are lower than with EIA because of improved specificity (Fig. 3.1.1.7).

Blood contamination of saliva from heavy tooth brushing (Kivlighan et al., 2004), smoking or chewing tobacco (Badrick et al., 2007) should be avoided. Eating liquorice can increase saliva cortisol and decrease cortisone concentrations (Räikkönen et al., 2010). The measurement of salivary cortisone alone or with cortisol is now offered in some laboratories (Ponzetto et al., 2020; Ceccato et al., 2018a,b; Kannankeril et al., 2020; Jones et al., 2012). Mass spectrometry is becoming the preferred technique for cortisol measurement regardless of biological fluid because of the avoidance of interfering steroids and the ability to simultaneously measure cortisone and other steroids (Olesti et al., 2021; Casals and Hanzu, 2020; %Rossi et al., 2020; Harrison et al., 2019; Antonelli et al., 2015). Saliva concentrations of cortisol can now be measured in less than 100 µL of sample (Ueland et al., 2021).

Urinary cortisol is a measure of free cortisol in blood since this is the filtered fraction by the kidney. A urine-free cortisol (UFC) determination is commonly requested in the investigation of an obese patient. In addition to the methodological considerations of quantifying free cortisol in urine, there are the problems of obtaining and hence relying on complete 24 h urine collections especially in children. The patient should be given a protocol for the 24 h sample collection. UFC is preferably determined on a solvent extract of the sample. A raised 24 h UFC is a good indicator for Cushing's syndrome (Ceccato et al., 2019; Baid et al.,

FIG. 3.1.1.5 Nutritional supplements. *(From Thrower SL, Taylor NF, Buckley DA. Cushing syndrome and rebound generalized pustular psoriasis from Chinese herbs. Br J Dermatol. 2013;169(6):1367–9. Fig. 1 p. 1368.)*

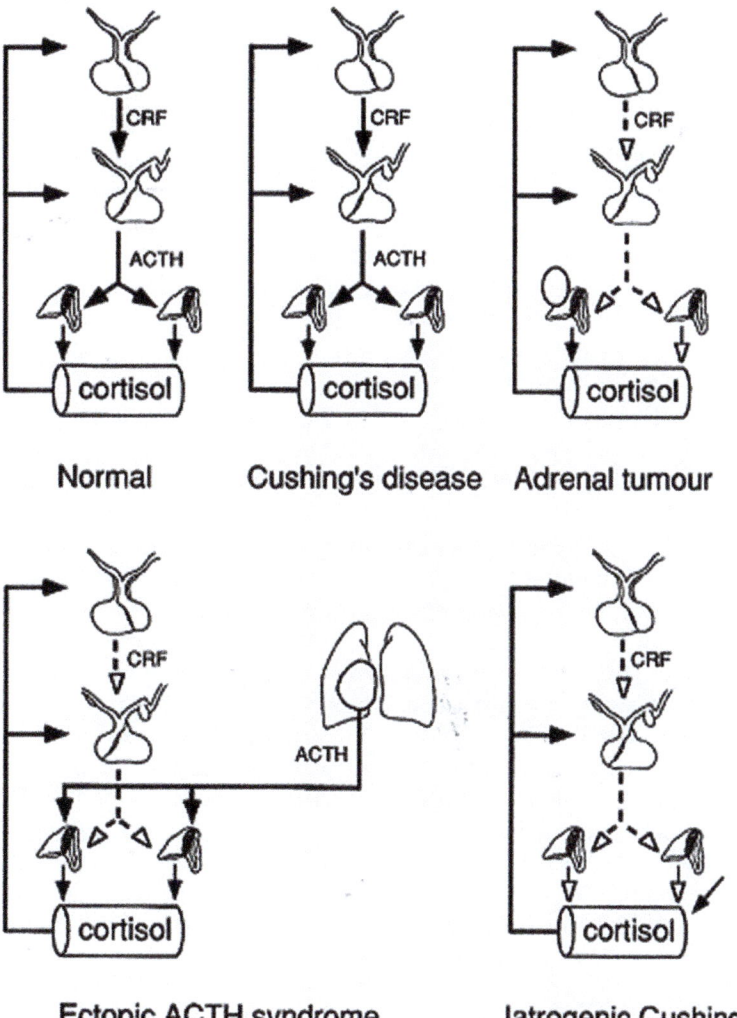

FIG. 3.1.1.6 Adrenal steroid excess due to adrenal tumors secreting excess cortisol; pituitary or ectopic ACTH secreting tumors or exogenous use of glucocorticoid steroids. *(Based on Chan LF, Storr HL, Grossman AB, Savage MO. Pediatric Cushing's syndrome: clinical features, diagnosis, and treatment. Arq Bras Endocrinol Metabol. 2007a;51(8):1261–71.))*

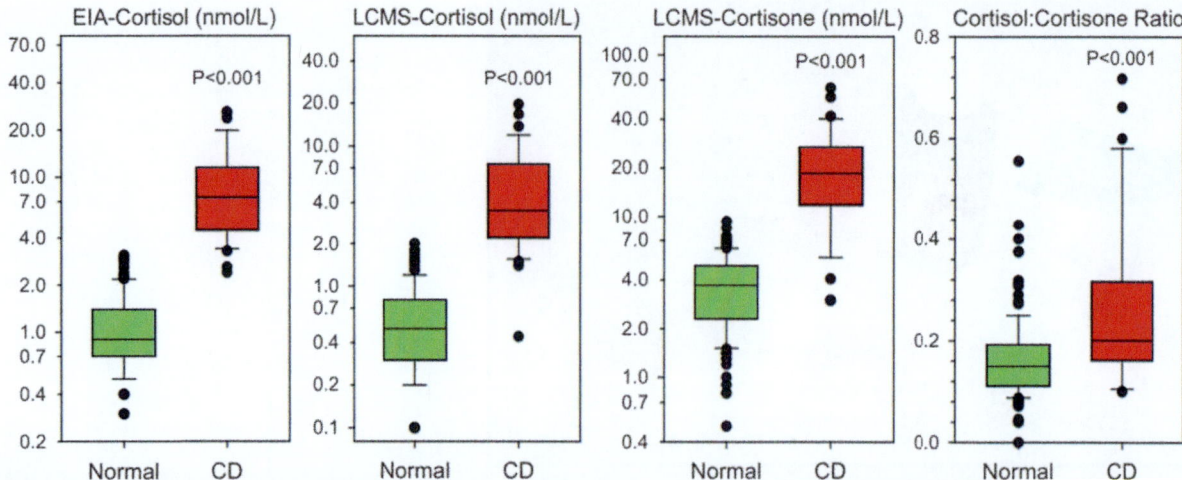

FIG. 3.1.1.7 Late-night salivary cortisol by enzyme immunoassay (EIA-Cortisol); liquid chromatography-tandem mass spectrometry (LCMS) cortisol and cortisone, and the ratio of LCMS cortisol to cortisone in 35 patients with Cushing's disease compared with the 121 patients randomly chosen with all normal salivary steroid results and without the diagnosis of Cushing's syndrome of any type. Horizontal line is the median; box indicates 25th to 75th percentile, whisker indicates 10th and 90th percentiles, and outliers are indicated by circles. *P* values are from Mann-Whitney rank sum tests. *(From Kannankeril J, Carroll T, Findling JW, Javorsky B, Gunsolus IL, Phillips J, Raff H. Prospective evaluation of late-night salivary cortisol and cortisone by EIA and LC-MS/MS in suspected Cushing syndrome. J Endocr Soc 2020;4(10):bvaa107. Fig. 2 p. 7.)*

2007) but immunoassay platforms even after dichloromethane extraction of the steroids have cut-offs different to newer results with LC-MS/MS (330 μg/24h Advia; 234 Liaison; 51 LC-MS/MS) (Oßwald et al., 2019). A solvent extract will need to be dried and material dissolved in a buffer or assay zero standard. A high cortisol excretion rate may reflect stress on the day of collection, although in children the effects of alcoholism and psychiatric disorders seen in adults are unlikely. The practice now is to repeat the test 3 times to assess consistency with respect to the cortisol result and the creatinine excretion. Results are compromised when the 24h urine collection exceeds 5L. When synthetic corticosteroids are prescribed for the patients (Budesonide, Fluticasone, and Triamcinolone) there is variable suppression of **urine-free cortisol (UFC)** when compared with other tests of adrenal function such as urine steroid profile (Fink et al., 2002) (Fig. 3.1.1.8). See Table 3.1.1.2 for other drug effects on urinary-free cortisol.

A number of factors contribute to the variability in urine cortisol output including compliance with the procedure, stress, social practices (use of alcohol) during the collection period. Two studies have shown that a high intake of sodium can increase urine cortisol output (47% higher with liberal salt intake (Chen et al., 2020) and total cortisol metabolite excretion by GC-MS (55% and 37% higher with high salt) (Baudrand et al., 2014). The latter study unfortunately did not examine the cortisol/cortisone metabolite ratios. Urine-free cortisol was measured by Coat-A-Count radioimmunoassay and Immulite, respectively. A number of studies have shown that higher urinary sodium excretion levels were associated with an increased likelihood of

overweight and central obesity in adults, and adolescents (Lee et al., 2018).

In a patient with high clinical suspicion of Cushings, but normal urine-free cortisol excretion rate on the first occasion, repeated 24h urine collections over many weeks can reveal **cyclical** form of the disease which has been characterized, at least in adults (Humayun et al., 2017; Atkinson and Mullan, 2011). Studies in this way showed peak frequency over weeks to months (Wędrychowicz et al., 2019; Alexandraki et al., 2009).

Steroid profiling by GC-MS of urine steroids after deconjugation, and more recently the plasma steroids by LC-MS/MS, has not been extensively used in the investigation of patients with Cushing's syndrome. The first study by George Phillipou in 1982 showed a dominance of 11-hydroxycorticosteroids and 5β reduced steroids in urine in pituitary (*n* = 4) and ectopic ACTH Cushing's syndrome (*n* = 2). The GC-MS pattern was similar to the steroids in urine collected after a female given 50mg/day of hydrocortisone (Fig. 3.1.1.9). In contrast, dehydroepiandrosterone and pregnanetriol were markedly elevated in one case of adrenal carcinoma (Phillipou, 1982). Cushing's syndrome is caused by excess ACTH from a pituitary tumor or an ectopic source (often the lungs).

The pattern of steroids in urine is of cortisol excess, just as seen with autonomous adrenal Cushing's syndrome.

3.1.1.2.1.1 Steroid analysis

In view of the uncertainty in the actual steroids in excess, a general assessment such as produced by urinary or plasma

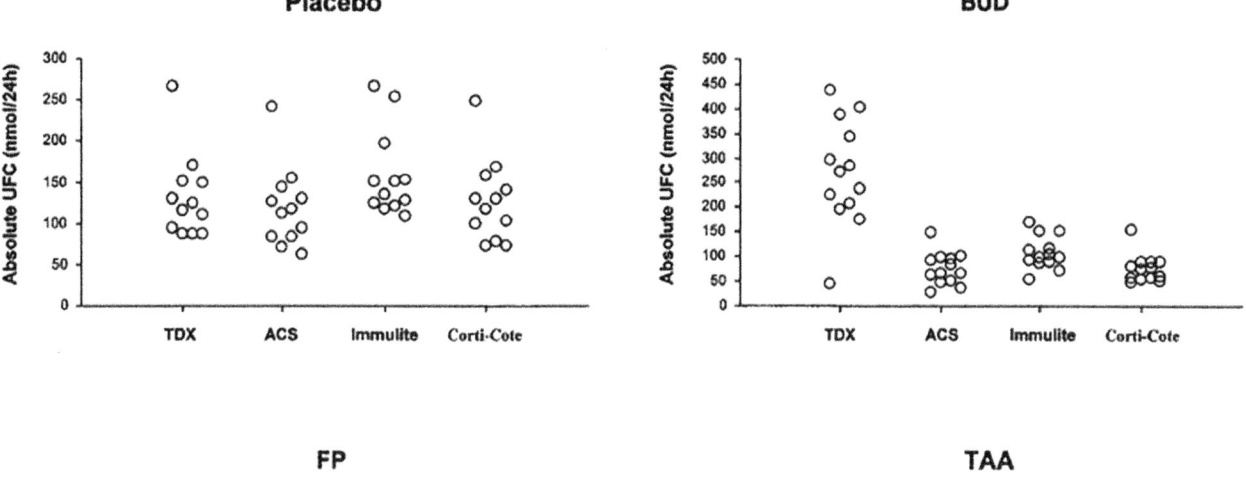

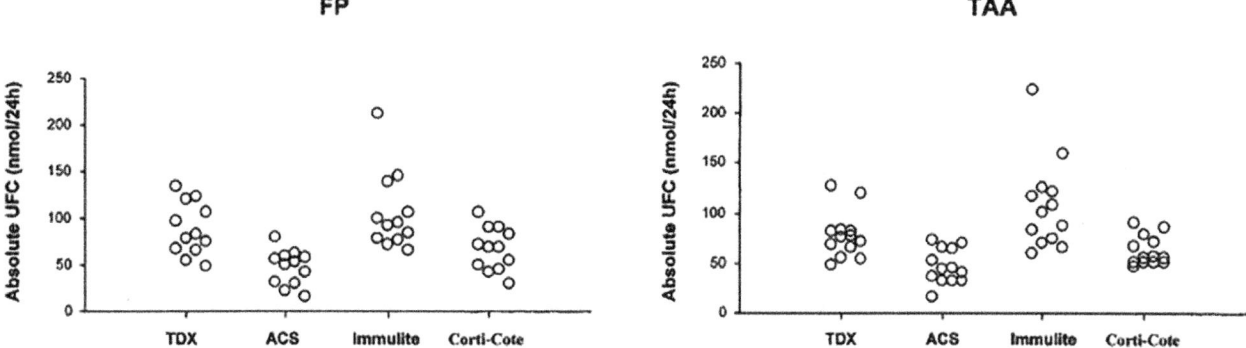

FIG. 3.1.1.8 Individual excretion rates over 24 h of urinary-free cortisol (UFC) after placebo and inhaled corticosteroids (BUD, FP and TAA) as measured with TDX, ACS, Immulite and CortiCote assays. Lower reference limit is 50 nmol/24 h. *(From Fink RS, Pierre LN, Daley-Yates PT, Richards DH, Gibson A, Honour JW. Hypothalamic-pituitary-adrenal axis function after inhaled corticosteroids: unreliability of urinary free cortisol estimation. J Clin Endocrinol Metab 2002;87(10):4541–6. Fig 1 p. 4543.)*

TABLE 3.1.1.2 Drugs that interfere with evaluation of urinary-free cortisol in diagnosis of Cushing's syndrome.

Induce CYP3A4

- Phenobarbital
- Phenytoin
- Carbamazepine
- Primidone
- Rifampin
- Rifapentine
- Etthosuximide
- Pioglitazone

Inhibit CYP3A4

- Aprepitant/fosaprepitant
- Itraconazole
- Ritinovir

TABLE 3.1.1.2 Drugs that interfere with evaluation of urinary-free cortisol in diagnosis of Cushing's syndrome.—cont'd

- Fluoxetine
- Diltiazem
- Cimetidine

Increase CBG

- Estrogens
- Mitotane

Increase UFC

- Carbamazepine
- Fenofibrate
- Synthetic glucocorticoids
- Inhibitors HSD11B2

Continued

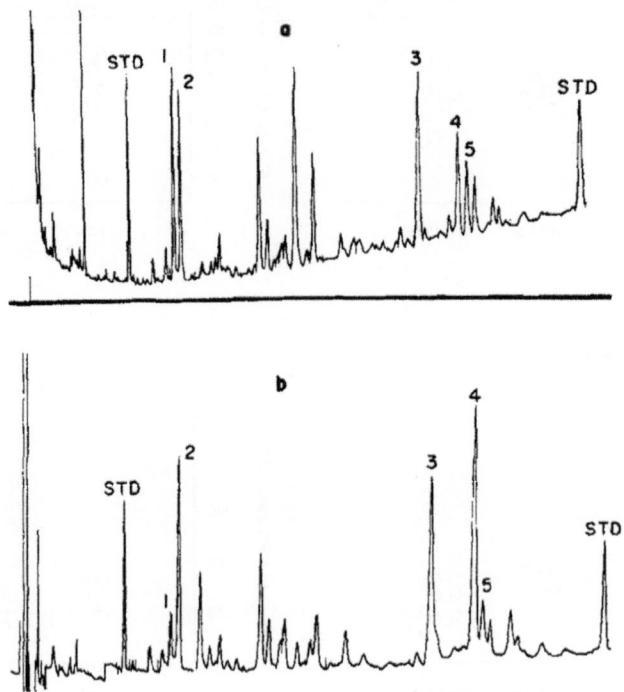

FIG. 3.1.1.9   Neutral steroid urinary GC profile of a. normal subject; b. patient diagnosed as having Cushing's disease. The pattern was similar in the analysis of urine from anoth patient receiving exogenous cortisol therapy (data not shown here). Steroids: 1 = andro; 2 = aetio; 3. THE; 4. THF; 5. alloTHF. *(From Phillipou G. Investigation of urinary steroid profiles as a diagnostic method in Cushing's syndrome. Clin Endocrinol (Oxf) 1982;16(5):433–9. Fig. 1 p. 435.)*

steroid profile analysis of a panel of steroids, may be the best approach with capillary column gas chromatography (GC-MS) or high pressure liquid chromatography coupled with mass spectrometry (LC-MS/MS). This test can also confirm normal adrenal function should a mass be **an incidentaloma**. A 24 h urine collection is essential for these investigations requiring absolute excretion rates to be determined without any regard for circadian variations (Bileck et al., 2020).

High excretion of 11-deoxycortisol or its metabolite THS has been found in several GC-MS reports of **adrenal carcinoma** (Fig. 3.1.1.10) (Bancos et al., 2020; Chortis et al., 2020; Velikanova et al., 2016; Arlt et al., 2011).

This test can also confirm normal or abnormal adrenal function should a mass be seen in an abdominal CT or MRI scan for nonendocrine reasons (so called incidentaloma) may reveal an adrenal tumor or display hyperplasia. In GC-MS analysis, raised androgens, glucocorticoid and precursor metabolites and THS were increased in patients with ACC compared with adenoma cases (Table 3.1.1.3) and a group with hormonally nonactive adenomas (data not shown here) (Velikanova et al., 2016).

Examination of the excretion rates of steroids plotted from a pool of patient material showed crude associations for representative steroids with the diagnosis (Fig. 3.1.1.11). Further analysis of the data by machine learning and principal component analysis was then applied to the data to

distinguish results for Cushing's syndrome (Wilkes et al., 2018) (Fig. 3.1.1.12).

Urine steroid profiling of 26 steroids utilizing GC and **high-resolution, accurate- mass, mass spectrometry** LC-MS assay was tested in 71 patients with adrenal disorders (8 with Cushing's syndrome; Hines et al., 2017). Patients with pituitary CS had increased androgen and progestogen metabolites [androsterone, etiocholanolone (ET), dehydroepiandrosterone, 16-hydroxydehydroepiandrosterone, pregnanetriol, and pregnanediol] and glucocorticoids (cortisol, 6βOH-cortisol, THF, 5α-THF, β-cortol, 11β-OH-AN, 11β-OH-ET, cortisone, THE, α-cortolone, β-cortolone, and 11-OXO-ET), whereas patients with adrenal CS had suppressed androgens and increased glucocorticoids. Steroid profiling demonstrated significant differences in patients with ACC when compared with patients with ACAs for 11 steroids, most notably in THS, PT and Etio (Fig. 3.1.1.13). PD, 5PD, DHEA, and 17HP also showed significant distinction from ACA with Z scores ranging from 1.1 to 3.3.

Patients with adrenocorticotropic hormone-dependent pituitary hypercortisolism (Cushing's disease) had increase of similar range of steroids, whereas patients with cortisol producing adrenal adenomas had suppressed androgens and increased glucocorticoids, although to a lesser degree than Cushing's disease patients. The difference in 3 analytes achieved statistical significance: Etio, 11B-OHAN, and a-cortolone. In LC-MS analysis

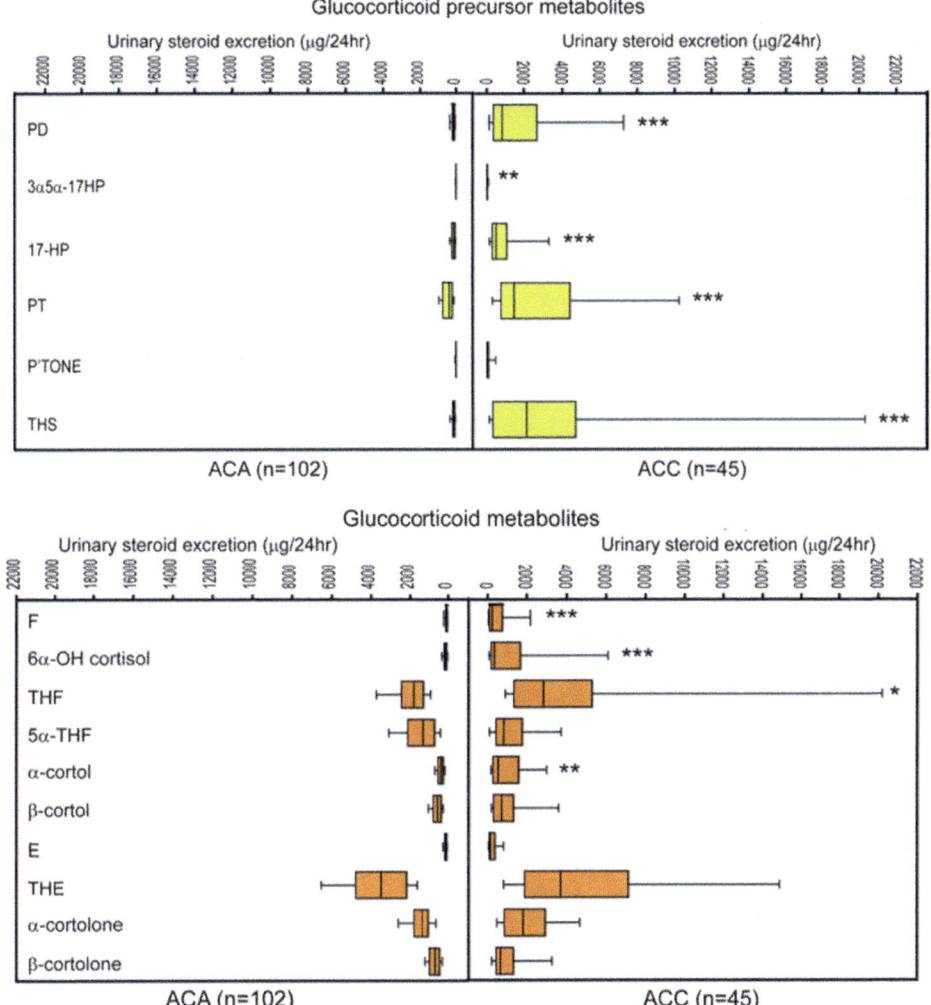

FIG. 3.1.1.10 Steroid metabolite excretion in adrenocortical adenoma ACA (n _ 102) and carcinoma ACC (n _ 45) according to steroid classes. Metabolites of adrenal androgen precursors and active androgens and the metabolites of mineralocorticoids and their precursors are shown in the original figure. (Top panel) metabolites of glucocorticoid precursors; (lower panel) cortisol and cortisone metabolites. Box plots represent median and interquartile ranges; the whiskers represent 5th and 95th percentile, respectively. *, P _ .05; **, P _ .01; ***, P _ .001 comparing ACA with ACC. *(From Arlt W, Biehl M, Taylor AE, Hahner S, Libé R, Hughes BA, Schneider P, Smith DJ, Stiekema H, Krone N, Porfiri E, Opocher G, Bertherat J, Mantero F, Allolio B, Terzolo M, Nightingale P, Shackleton CH, Bertagna X, Fassnacht M, Stewart PM. Urine steroid metabolomics as a biomarker tool for detecting malignancy in adrenal tumors. J Clin Endocrinol Metab 2011;96(12):3775–84. Fig. 2 p. 3779.)*

(Velikanova et al., 2019) of urine steroids, UFF, UFE and 6-hydroxy F were higher in ACC compared with hormonally nonactive adenomas.

3.1.1.2.2 Tumors

A comprehensive analysis of urinary steroids as MO-TMS derivatives by **two dimensional gas chromatography coupled to a time of flight MS (GCxGC-MS)** was undertaken by Groessl and colleagues (Bileck et al., 2019). GC × GC employs two columns consisting of complementary stationary phases, to improve the separation power for accurate mass detection with mass errors in the low ppm range. A double focusing loop modulator was mounted in the GC oven between the first (GC1) and second (GC2) column. The eluate of the first column was trapped in the modulation loop with a cold jet of nitrogen (−80°C). A hot jet of nitrogen was then used for rapid desorption of the trapped sample at a modulation period of 6 s. For GC1, a 15 m × 0.25 mm i.d. × 0.25 μm column (crossbond dimethyl polysiloxane, Restek Corporation, USA), for GC2 a 2 m × 0.1 mm i.d. × 0.1 μm column (50% phenyl polysilphenylene-siloxane, SGE, USA) was used. Of GC2, 1 m was utilized for the modulation loop. GC oven temperature was held at 50°C for 1 min followed by a first ramp to 220°C at a rate of 30°C min^{-1} and a

TABLE 3.1.3 Common features of urine steroid patterns for adrenocortical carcinoma in patients with adrenocortical carcinoma without cortisol and its metabolite hypersecretion (ACC) and in patients with adrenocortical carcinoma and Cushing's syndrome (ACC-CS) by gas chromatography-mass spectrometry.

	Median/lower and upper quartiles (µg/24h)			
	Adrenocortical adenomas		Patients with adrenocortical carcinoma	
Name of steroids	Hormonally non-active adenomas (n = 52)	Cushing's syndrome (n = 44)	ACC (n = 18)	ACC-CS (n = 13)
Androgens				
Etiocholanolone	240/148–440	279/129–412	1464/554–2476***	723/365–8215* P=.005
Androstendio-17β	58/37–91	75/25–115	705/348–1673***	1010/54–3142 P=.03
Dehydroepiandrosterone (DMEA)	40/32–55	11/9–14*	3407/776–11171***	3283/20–10235 P=.005
16-Hydroxy-DHEA	150/36–212	153/65–269	1851/953–9837*	2113/245–7379 P=.02
11-Hydioxy-etiocholanolone	227/73–377	313/136–701	672/214–968*	1720/839–2229*** P=.02
Androstenetriol	133/41–177	118/51–234	1630/492–4462*	1322/299–1248* P=.003
16-Oxo-androstendiol	27/14–40	32/23–56	533/387–659*	1232/504–2189* P=.005
Androgen and glucocorticoid precursor and their metabolites				
17-Hydroxy-pregnanolone	172/70–185	86/48–109	355/275–1237*	1253/696–3139** P=.002
Pregnanediol (P2)	228/186–495	483/181–628*	2356/1097–3528***	3278/2803–6864*** P=.001
Pregnanetriol (P3)	458/283–705	523/256–745	1195/739–2200**	3167/1612–5479*** P=.002
11-Oxo-pregnanetriol	37/33–46	66/36–101	150/99–227*	305/165–721* P=.04
Pregnenediol	430/181–558	386/236–688	2530/1540–3214***	3669/2176–5981*** P=.006
5-Pregnen,3α, 16α,20α-triol	75/51–149	121/83–208	957/306–1299***	2130/1184–9722*** P=.02
5-Pregnen,3α, l7α, 20α-triol	146/91–314	168/118–266	1554/1112–2576***	2489/524–11235*** P=.002
6-Hydroxy-pregnanolone	33/15–55	19/15–43	113/27–210*	198/102–312* P=.03
Tetrahydro-11-deoxycortisol (THS)	93/49–171	411/100–539	858/131–1355*	1081/691–3732*** P=.04
Hexahydro-11-deoxycortisol (HMS)	75/27–143	52/25–183	272/132–1370*	622/160–8165 P=.02
21-Deoxy-tetrahydrocortisol	48/22–54	131/110–217	192/80–203*	1036/881–1258** P=.02
Mineralocorticoid metabolite				
Tetrahydro-11-deoxycorticosterone	54/14–74	66/35–89	110/88–168*	176/148–205* P=.04

*P < .05, **P < .001, ***P < .0001—comparison of each group of patients with hormonally non-active adenomas; P—comparison of ACC-CS with ACA-CS.

From Velikanova LI, Shafigullina ZR, Lisitsin AA, Vorokhobina NV, Grigoryan K, Kukhianidze EA, Strelnikova EG, Krivokhizhina NS, Krasnov LM, Fedorov EA, Sablin IV, Moskvin AL, Bessonova EA. Different types of urinary steroid profiling obtained by high-performance liquid chromatography and gas chromatography-mass spectrometry in patients with adrenocortical carcinoma. Horm Cancer 2016;7(5–6):327–35. Table 3 p. 330.

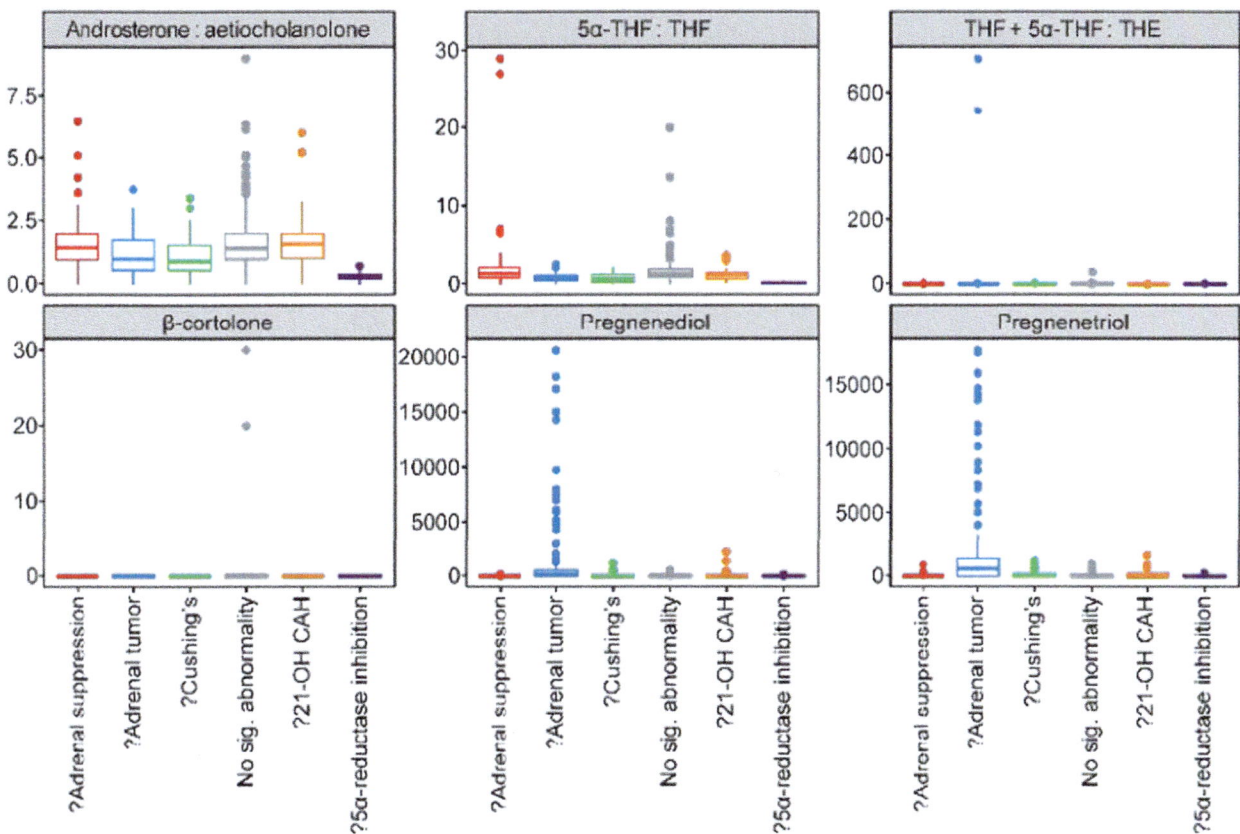

FIG. 3.1.1.11 Feature distributions of steroids for each diagnostic class. Boxes represent the median +/− IQR, whiskers represent 1.5 times the IQR and individual points represent data points > 1.5 time the IQR. Colors represent the six diagnostic classes. *(From Wilkes EH, Rumsby G, Woodward GM. Using machine learning to aid the interpretation of urine steroid profiles. Clin Chem 2018;64(11):1586–1595. Supplement S2.)*

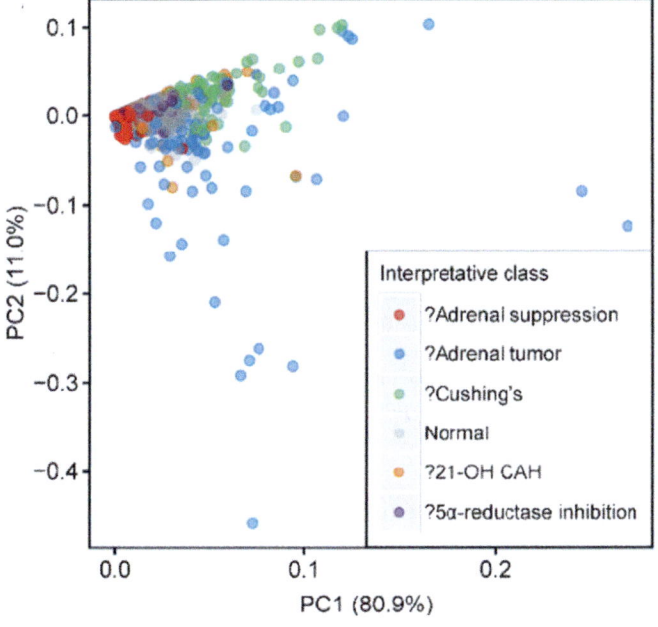

FIG. 3.1.1.12 Principal component analysis PC1 vs PC2. Points are colored according to six interpretative classes. *(Based on Wilkes EH, Rumsby G, Woodward GM. Using machine learning to aid the interpretation of urine steroid profiles. Clin Chem 2018;64(11): 1586–1595. Supplement S1.)*

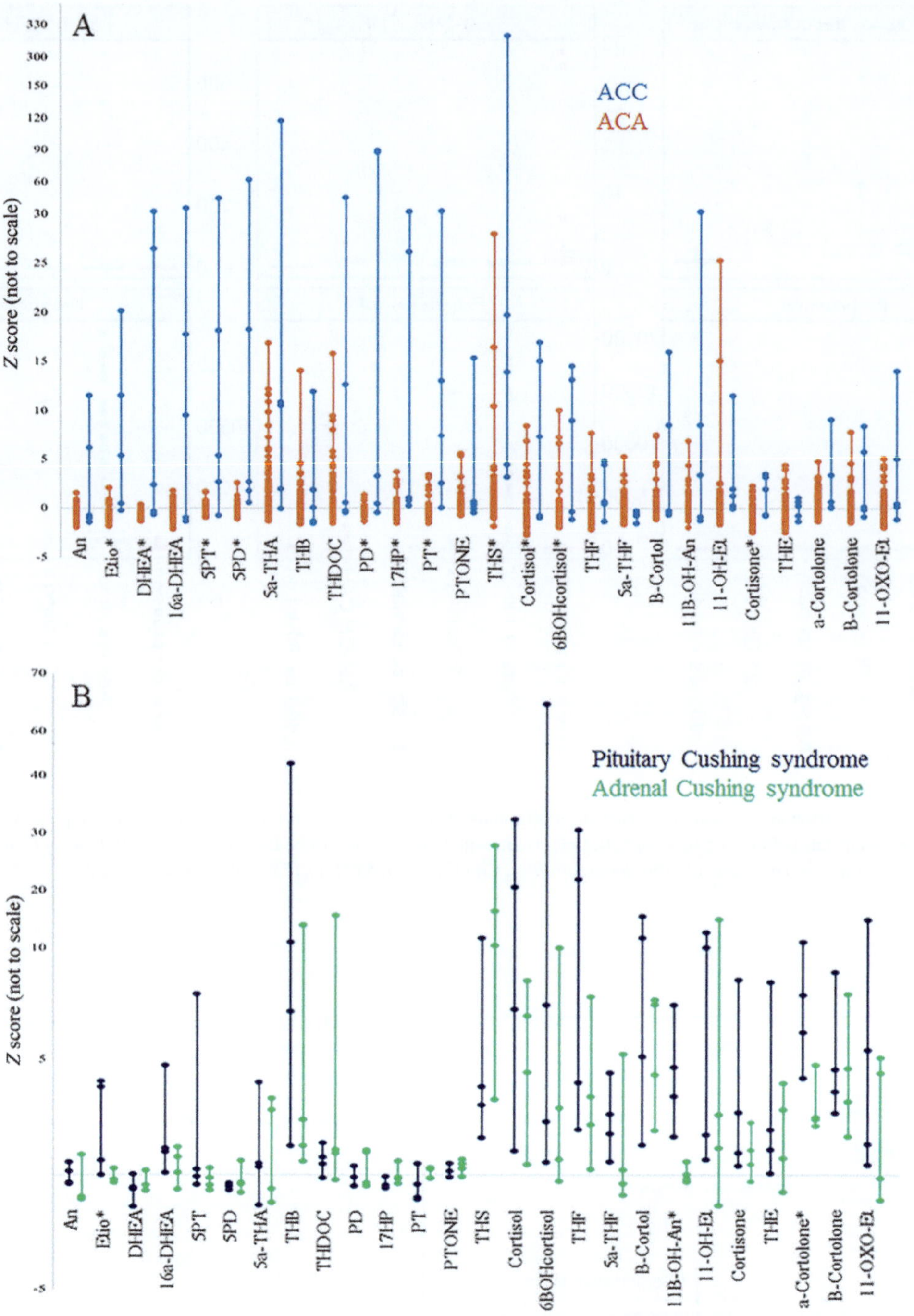

FIG. 3.1.1.13 Clinical significance of steroid metabolite profiling in adrenal disease states. Comparison of Z score (y axis) between ACC and ACA (A) using HRAM LC-MS steroid profiling established 11 of 26 metabolites as statistically significant (*) in distinguishing ACC from ACA. Adrenal CS versus pituitary CS (B) showed 3 analytes with statistical significance (*) steroids. *(From Hines JM, Bancos I, Bancos C, Singh RD, Avula AV, Young WF, Grebe SK, Singh RJ. High-resolution, accurate-mass (HRAM) mass spectrometry urine steroid profiling in the diagnosis of adrenal disorders. Clin Chem 2017;63(12):1824–1835. Fig. 2 p. 1832.)*

second ramp to 300°C at a rate of $2°C\,min^{-1}$ and then held at 320°C for 5 min. For the hot jet, an initial temperature of 120°C for 1 min was followed by a first ramp to 240°C at a rate of $30°C\,min^{-1}$ and a second ramp to 320°C at a rate of $2.5°C\,min^{-1}$ before returning to 120°C. A constant flow of

helium at a rate of $0.8\,mL\,min^{-1}$ was used as carrier gas with an initial head pressure of 1.5 bar. 1 µL of each sample was injected in a split/splitless inlet held at 280°C in pulsed splitless mode. The GC2 column was directly coupled to the MS using a feedthrough block held at 275°C. The

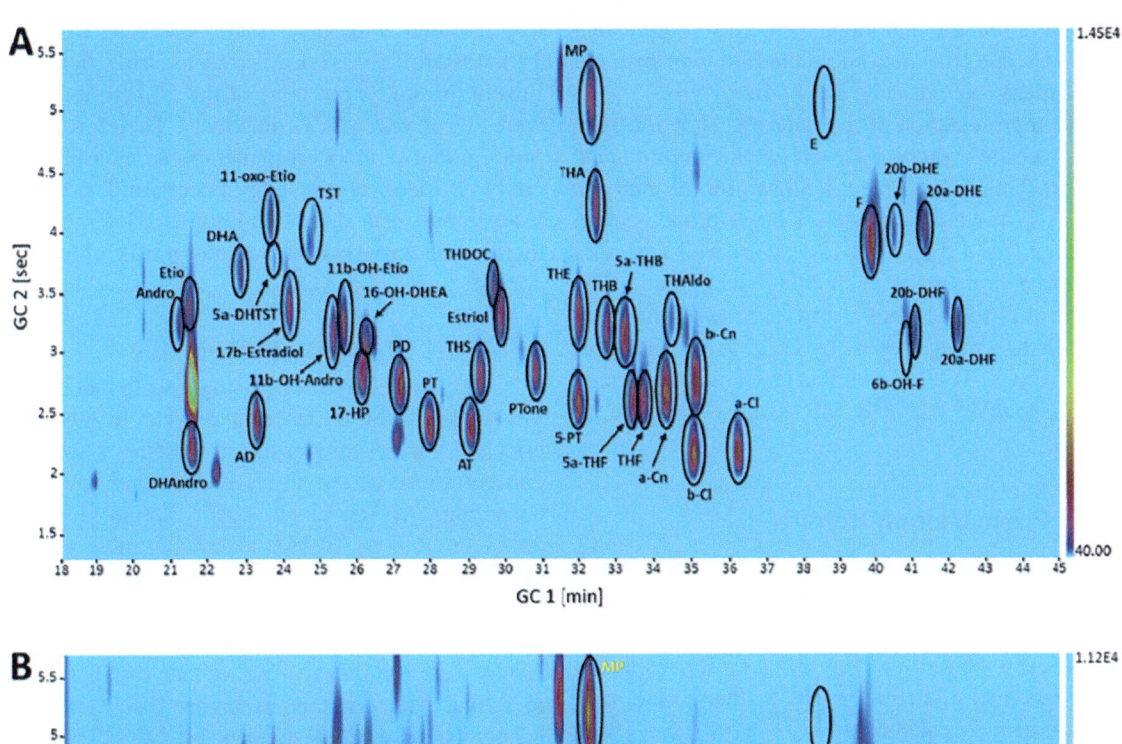

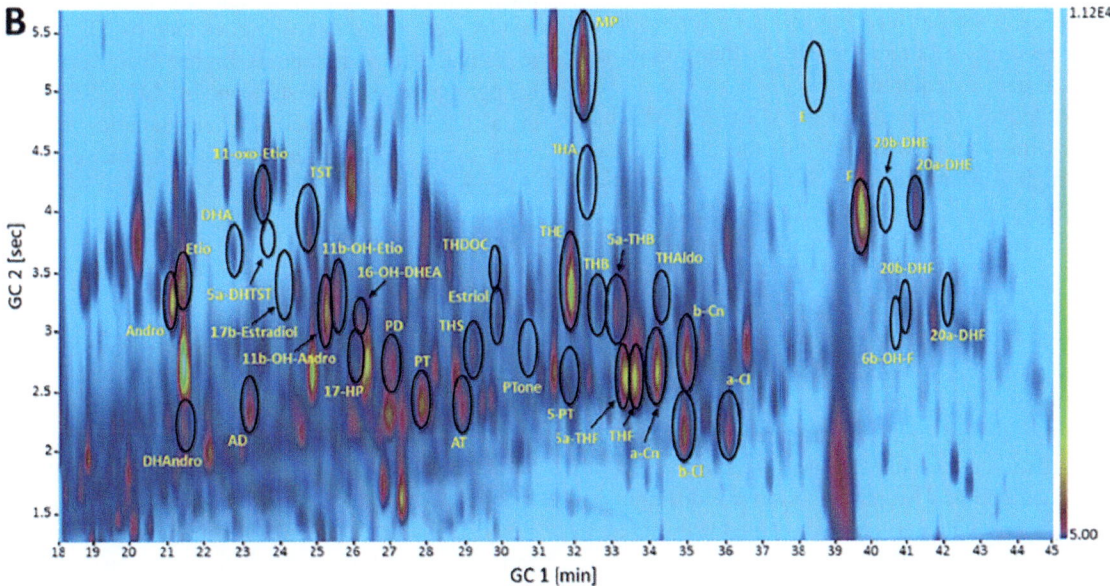

FIG. 3.1.1.14 GC × GC-TOF MS for targeted analysis. Total ion chromatograms of (A) an equimolar standard solution consisting of 40 steroids and two internal standards and (B) a human 24 h urine sample. *(Reproduced from Bileck A, Verouti SN, Escher G, Vogt B, Groessl M. A comprehensive urinary steroid analysis strategy using two-dimensional gas chromatography—time of flight mass spectrometry. Analyst. 2018;143(18):4484–94.)*

ion source temperature was set to 280°C. TOF MS enabled untargeted analysis by recording full mass spectra at high rates. MS analyses were conducted at 100 Hz in a mass range of 45–670 Th. Electron ionization was performed at an electron energy of 70 eV. The mass spectrum was recalibrated at the beginning of each modulation period using pentafluorophenol (PFP) as an internal standard (Fig. 3.1.1.14).

At the Inselspital in Bern, Switzerland, a TOFWERK EI-TOF was coupled with 2D gas chromatography (GCxGC) to analyze urinary steroids from adults. The group of steroids had been measured by GCqMS to screen for adrenal disorders. Many analytes co-elute in the first separation, giving rise to broad and overlapping peaks in the one-dimensional chromatogram (inset). With the addition of the second GC dimension, the final chromatogram is said to show baseline separation of all peaks in standard mixture (Fig. 3.1.1.14A). The analysis of urine from a 43-year-old female showed high outputs of cortisol metabolites suggestive of an adrenal tumor which was located and surgically removed. Steroids in GCqMS were confirmed in the GCXGC TOF method but 40 other steroids were seen and high resolution mass spectra obtained. The chromatogram looks incredibly complicated (Fig. 3.1.1.16) and the analysis failed to confirm the identities of many steroids (Fig. 3.1.1.14B).

A clearer representation of the data was shown on the Company website (https://www.tofwerk.com/products/ei-tof-for-gc/). This improvement is shown for the analytes designated as 1–4, which appear as a single broad peak in the 1D chromatogram (Fig. 3.1.1.15), but are well separated in the 2D projection (lower chart blue background). The steroids were unfortunately not identified. One steroid identified as 17-hydroxyprogesterone by GCqMS was not confirmed on GCxGC-TOFMS. Further work is anticipated although LC coupled with HR MS is more likely to be exploited.

3.1.1.2.3 Plasma/serum steroids

Analysis of steroids in serum or plasma by LC-MS/MS produces a chromatogram where excess of cortisol is less obvious because the relative intensities of the specific fragment ion are different for each steroid (see Fig. 3.1.1.16 for responses from equal amounts of steroid injected). Interpretation of the numerical data is required.

The profiles of serum steroids by LC-MS/MS have recapitulated the abnormal findings of the metabolites in urine with high concentrations of S, DHEA, pregnenolone and 17-hydroxpregnenolone (Schweitzer et al., 2019; Rege et al., 2018; Eisenhofer et al., 2018; Eisenhofer et al., 2017; Taylor et al., 2017) (Fig. 3.1.1.17).

The urine of patients with autonomous cortisol secretion (ACS) showed higher cortisol and lower androgen concentrations than those with adrenal incidentaloma (Di Dalmazi et al., 2015a,b). In dexamethasone tests, patients with ACS hyperplasia had reduced suppression of cortisol, 11-deoxycortisol and corticosterone (Di Dalmazi et al., 2019). (ACS) Subclinical hypercortisolism was distinguished by low androgens when 83 steroids were determined by GC-MS (Hána Jr et al., 2019).

Not every laboratory can afford mass spectrometric methodology so one study used radioimmunoassays for the measurements of 12 steroids in blood and urine of patients with adrenal carcinoma (ACC) or adenoma (Suzuki et al., 2020) (Fig. 3.1.1.18). The authors recommend the measurement of 17-hydroxypregnenolone (glucocorticoid and androgen precursor), 11-deoxycorticosterone (mineralocorticoid precursor), or the combination of DHEAS and androstenedione (androgen precursor) for the diagnosis of cortisol-producing ACC.

Comparisons of steroids measured by immunoassay and LC-MS/MS have confirmed the positive bias in immunoassays (Debeljak et al., 2020). The combination of DHEAS and androstenedione could be a valuable tool to differentiate ACC from CPA is of extreme relevance, since these are two widely available markers, and can be implemented without a major increase in costs. Only 17-ketosteroids were measured as urine steroid metabolites in the study, but none of the 17-ketosteroid fractions were sensitive to ACC diagnosis. Therefore, the authors propose that serum steroid metabolite measurement, especially 11-deoxycortisol (glucocorticoid precursor) and testosterone, as a simple and noninvasive method for predicting the progression and prognosis of patients with cortisol-producing ACC.

3.1.1.2.3.1 Cortisol in hair

The analysis of cortisol in a single scalp hair sample about the thickness of a pencil and 3 cm length provides is effective for the diagnosis of Cushing's syndrome with concentrations of ten times greater than normal (107 vs 8.4 pg/mg) (Wester et al., 2017; Slominski et al., 2015). At 31.1 ng/mg by ELISA technique, there was detection of CS at 93% sensitivity and 90% specificity. Results for four laboratories using immunoassay showed good agreement (Hodes et al., 2018; Russell et al., 2015), although results were significantly higher than by LC-S/MS. Similar to saliva, hair cortisone is higher than cortisol (13.9 vs 8.7 pg/mg on normal subjects by LC-MS/MS; Brossaud et al., 2021). Cortisol in hair samples from patients with cyclic CS corresponded with the clinical course (Manenschijn et al., 2012).

3.1.1.2.3.2 Investigation of the cause of cortisol hypersecretion

When the clinician is satisfied of raised cortisol concentration in a fluid, further biochemical tests and imaging procedures are justified to determine CS or CD. The protocol will vary between hospitals depending upon the laboratory services, clinical practice and experience and according to the availability of specialized tests, e.g., ACTH assay, CT scanning, MRI scanning and venous catheterization. Twenty-one published papers on these tests have been reviewed (Scaroni et al., 2020) although all depended on cortisol measurements with chemiluminescent methods. Further detailed biochemical tests with mass spectrometric detection are needed.

Testing for Cushing's syndrome first requires establishing the degree of autonomy of this production (Ferrigno et al., 2021; Casals and Hanzu, 2020; Bansal et al., 2015; Nieman et al., 2008). The plasma concentrations of potassium and ACTH should be checked since hypokalemia and very high ACTH values by RIA are characteristic of the ectopic ACTH tumor. Other dynamic tests should be considered if ACTH assays are not available to gain further indication of the pathology in a patient with high cortisol production (Fig. 3.1.1.19).

While severe hypokalemic alkalosis and hypertension strongly supports an ectopic source, electrolyte changes

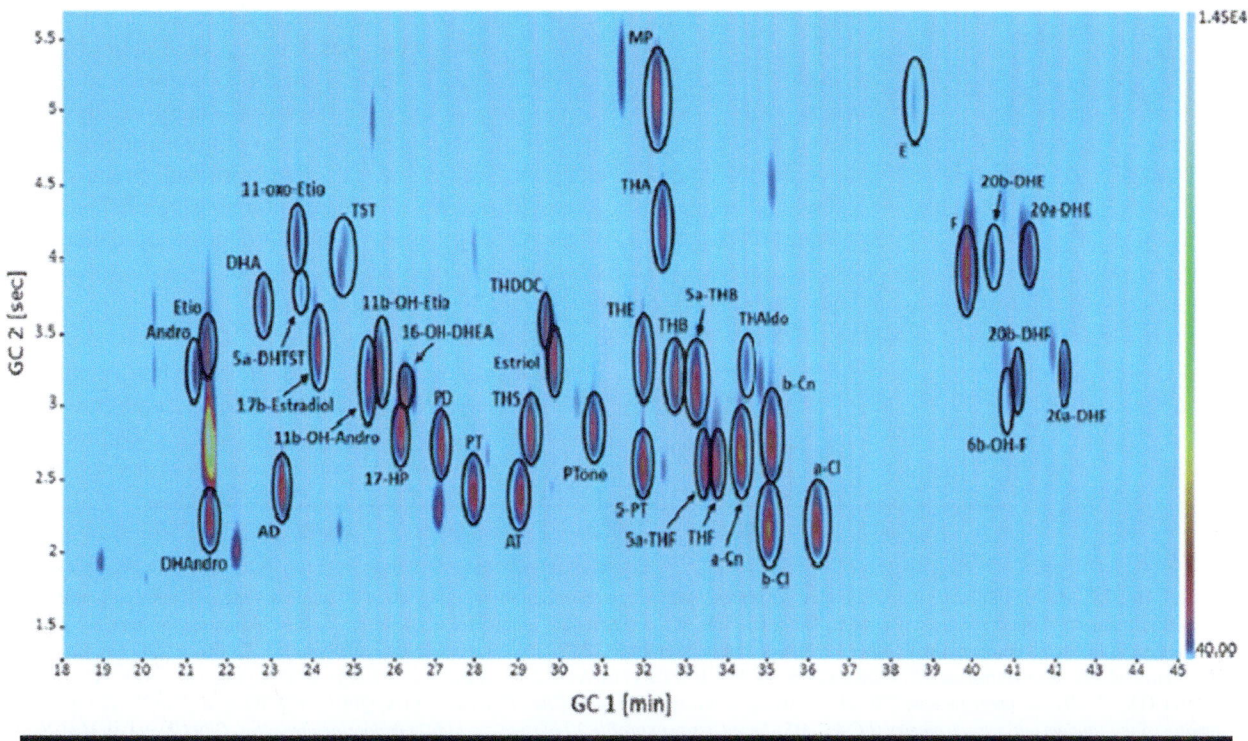

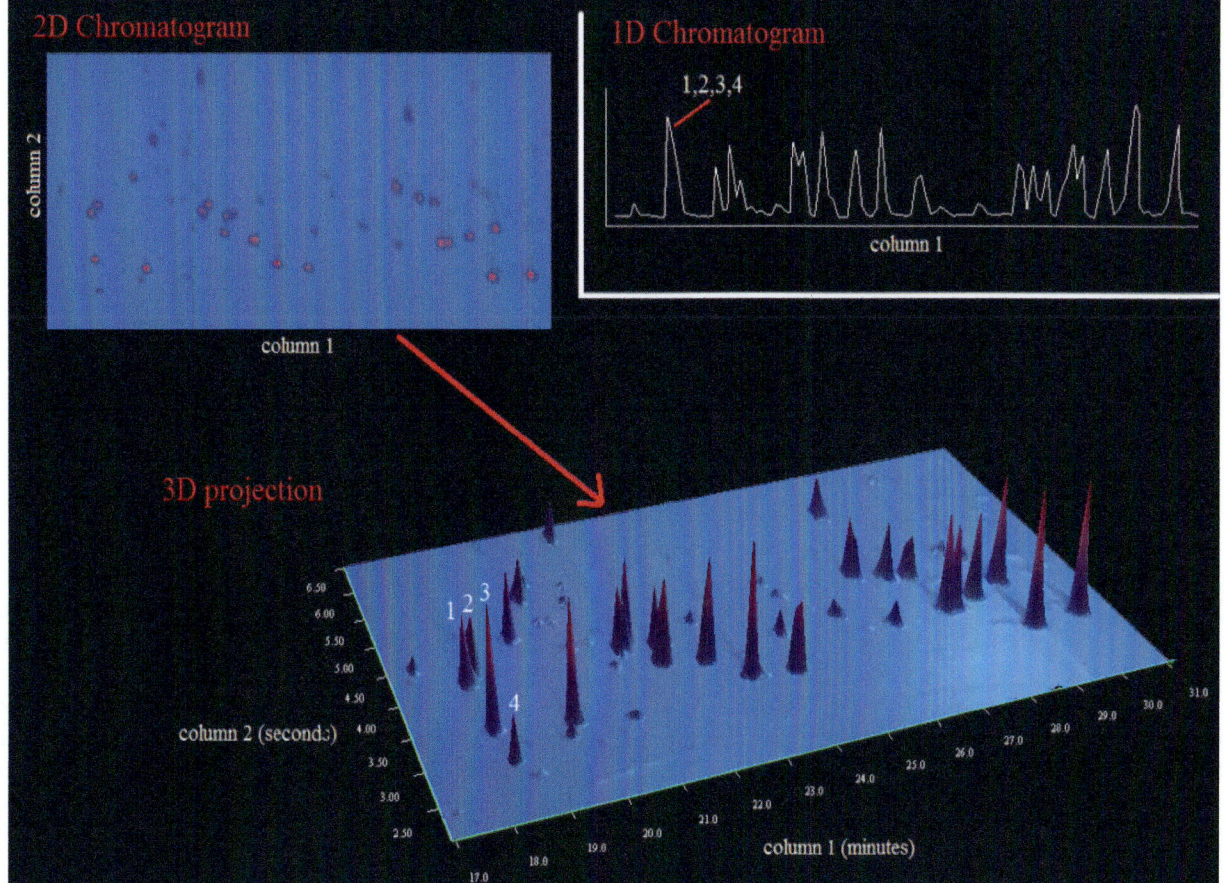

FIG. 3.1.1.15   2D plot of data.

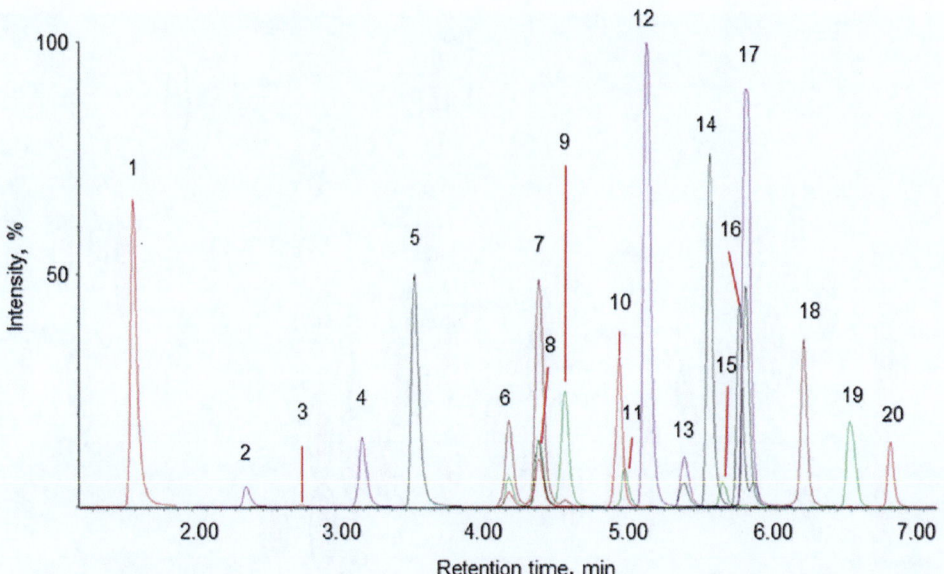

FIG. 3.1.1.16 LC-MS/MS of plasma steroids. Figure is a representative chromatogram showing signals of 20 steroids. Fragment ions in the collision are monitored in the second mass spectrometer stage. The intensities of the signals are unique for each steroid so the chromatogram does not obviously display a steroid in excess compared with other steroids as can be seen in GC-MS. The 20 steroids are melatonin (1), E3 (2), ALDO (3), E (4), F (5), 21-deoxycortisol (6), dexamethasone (7), CORT (8), S (9), E1 (10), E2 (11), A4 (12), 21-OH-P (13), T (14), DHEA (15), 17-OH-PR (16), 17-OH-P (17), DHT (18), P (19) and pregnenolone (20). *(From Wang Z, Wang H, Peng Y, Chen F, Zhao L, Li X, Qin J, Li Q, Wang B, Pan B, Guo W. A liquid chromatography-tandem mass spectrometry (LC-MS/MS)-based assay to profile 20 plasma steroids in endocrine disorders. Clin Chem Lab Med 2020;58 (9):1477–1487. Fig. 1 p. 1480.)*

FIG. 3.1.1.17 Steroid heterogeneity in adrenocortical carcinoma revealed by serum steroid paneling. Data for each adrenocortical carcinoma case are expressed as multiples of the median value for the non-adrenocortical carcinoma adrenal lesion group. *(From Taylor DR, Ghataore L, Couchman L, Vincent RP, Whitelaw B, Lewis D, et al. A 13-steroid serum panel based on LC-MS/MS: use in detection of adrenocortical carcinoma. Clin Chem 2017;63(12):1836–46.)*

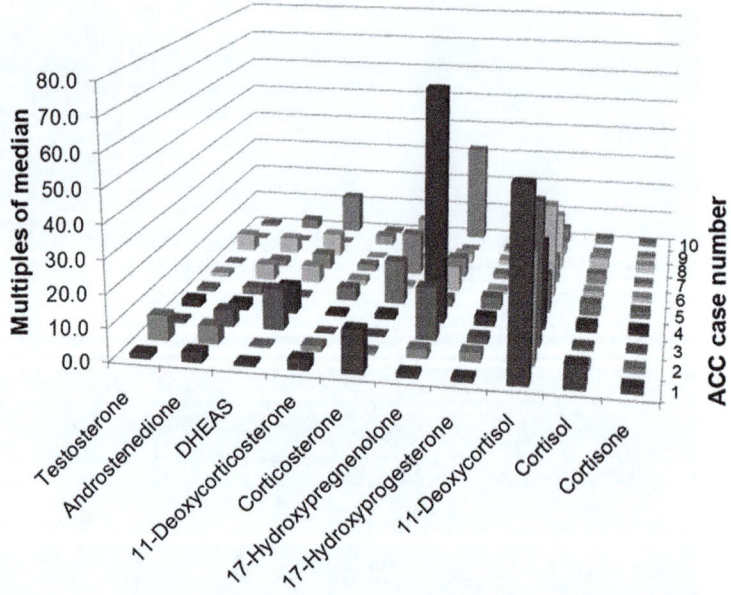

are not uniform and mild hypokalemia may also occur occasionally in Cushing's disease. Other ectopic tumors such as carcinoid (neuroendocrine tumor) may not always be associated with hypokalemia and ACTH may be only moderately raised. These cases are rare especially in children. The adrenals are enlarged on CT imaging and cortisol production is usually grossly elevated. A patient with an ectopic ACTH tumor may present with rapid onset weight loss,

edema and muscle weakness which is not the classical clinical picture of Cushing's syndrome.

3.1.1.2.4 Low-dose dexamethasone test

Even if the UFC excretion is clearly elevated above 400 nmol/24 h (>140 μg/day), in the majority of patients tested, the urine-free cortisol excretion will suppress to

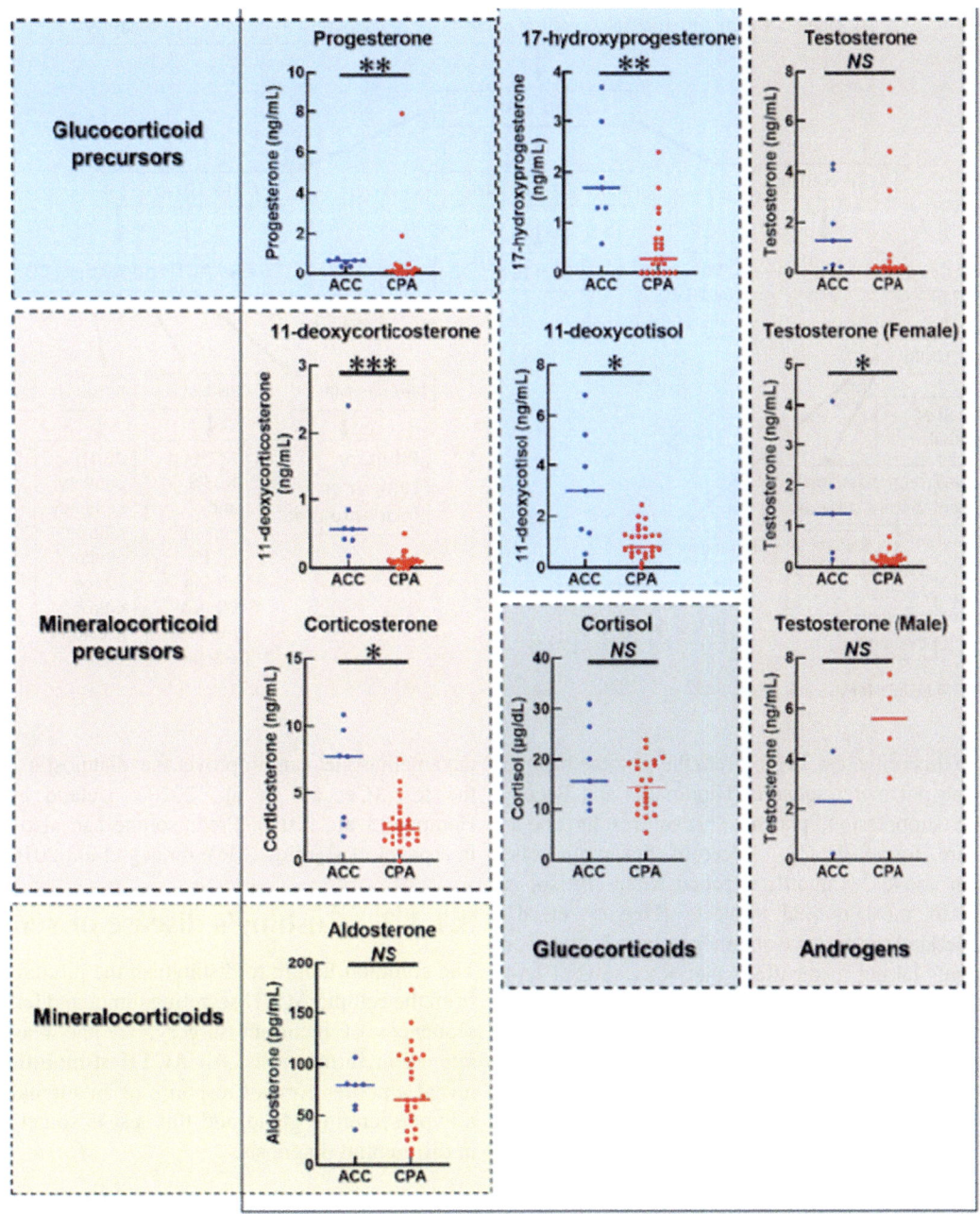

FIG. 3.1.1.18 Comparison of serum concentrations of steroids by radioimmunoassays and urinary 17-ketosteroid excretion between ACC and CPA. DHEAS: dehydroepiandrosterone sulfate. ACC: adrenocortical carcinomas, CPA: cortisol-producing adenomas. $P < .05$ was defined as significant. *: <.05, **: <.01, ***: <.001, NS: nonsignificant. *(From Suzuki S, Minamidate T, Shiga A, Ruike Y, Ishiwata K, Naito K, Ishida A, Deguchi H, Fujimoto M, Koide H, Tatsuno I, Ikeda JI, Yamazaki Y, Sasano H, Yokote K. Steroid metabolites for diagnosing and predicting clinicopathological features in cortisol-producing adrenocortical carcinoma. BMC Endocr Disord 2020;20(1):173. Fig. 1 p. 177.)*

the lower region of the normal range after taking 0.5 mg of dexamethasone every 6 h for 2 days (<70 nmol/day; 20 μg/day). The urine collection is made on the second day of treatment after the fifth dose. Children should get an adjusted dexamethasone dose of 25 micrograms per kg per day (Ferrigno et al., 2021) or 30 mg per kg (Magiakou and Chrousos, 2002). Children less than 40 kg should

get a dose of 30 micrograms per kg per day. The test can be performed as an out-patient procedure (Castro et al., 1999). Saliva samples can be collected (Perogamvros et al., 2010) as an alternative.

After taking 1.0 mg of dexamethasone at 2300 h, serum or plasma cortisol concentrations can normally be <50 nmol/L (<1.8 μg/dL) measured at 0800–0900 h as an

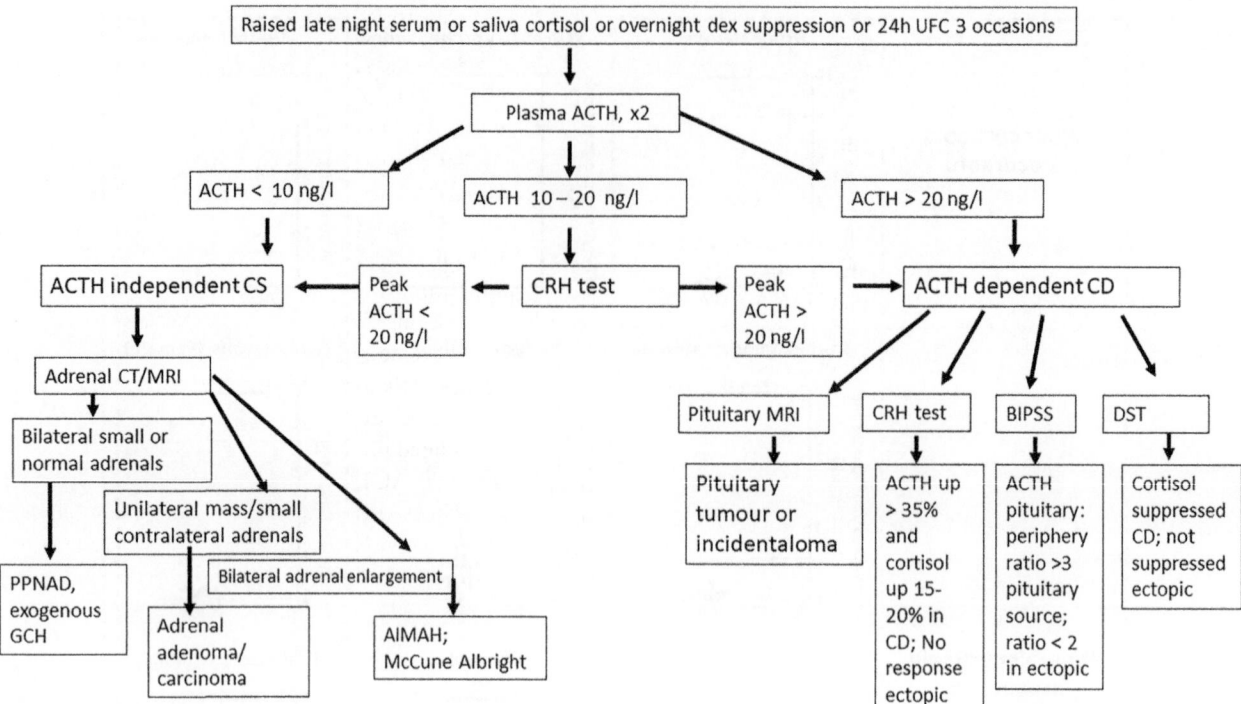

FIG. 3.1.1.19 Schedule for dynamic tests.

alternative to this urine test. In children, the dexamethasone dose is 0.3 mg per meter squared (Hindmarsh and Brook, 1985). After suppression, plasma cortisol results above 50 nmol/L are found in CS. A serum dexamethasone concentration above 3.3 nmol/L is found when cortisol is <50 nmol/L in most normal subjects (Hawley et al., 2018). Accelerated dexamethasone metabolism is however seen in patients taking medications that activate CYP3A4 enzyme (barbiturates, ethosuximide, phenytoin, pioglitazone, primidone, and rifampicin) thus giving a false positive test for CS. Other drugs inhibit CYP3A4 that affects dexamethasone clearance (aprepitant, cimetidine, diltiazem, itraconazole, fluoxetine, and ritonavir).

3.1.1.2.5 High-dose dexamethasone

The diagnosis of pituitary dependent Cushing's disease is confirmed when previously high cortisol production is substantially suppressed with **high doses of dexamethasone** (2 mg, 6 hourly for 2 days). Plasma cortisol, ACTH and UFC (or the total cortisol metabolites) all fall to less than 50% of their basal values in the majority of affected patients when on high-dose dexamethasone (Athimulam et al., 2021; Genere et al., 2021; Pinelli et al., 2021). Up to 10% of patients with ectopic ACTH secreting tumors can however suppress with these high doses of dexamethasone.

The use of synthetic corticosteroids can lead to false negative CS test results (Berlińska et al., 2020). In this regard, the simultaneous measurement of cortisol and

dexamethasone can improve the diagnostic precision of the test (Ceccato et al., 2020c; Ueland et al., 2017; Hempen et al., 2012). Prednisolone can also be included in a panel of steroids (McWhinney et al., 2010).

3.1.1.3 Cushing's disease or syndrome?

The common failure to distinguish the pituitary-dependent from the ectopic ACTH secreting tumor and fear of the consequences of incorrect surgery, are the reasons for the interest in further tests. An **ACTH stimulation test** may reveal a poorer cortisol response of an adrenal tumor than a hypersecreting gland and this test is sometimes helpful in differential diagnosis.

3.1.1.3.1 ACTH measurement

If a validated and specific **ACTH assay** is available, plasma ACTH measurements at midnight and 0800 h will readily distinguish ACTH dependent from independent causes. Circulating ACTH is low (<10 pg/mL; 2.2 pmoL/L) in patients with iatrogenic Cushing's or an adrenal tumor secreting cortisol and further investigations of the adrenal gland in these patients should be directed at imaging the abdomen to determine the site and possible spread of a tumor. Typically plasma ACTH levels are high normal or elevated in patients with pituitary dependent Cushing's disease, and even higher and above 250 pg/mL (55 pmoL/L) in the ectopic ACTH syndrome. However, there is considerable overlap and ACTH levels do not always resolve the main clinical problem of distinguishing Cushing's disease and an occult ectopic source of ACTH.

In considering assays for ACTH, the specificity of the RIA or IRMA should be known with respect to the measurement of ACTH itself, POMC and pro-ACTH. The precursors can be measured by relatively specific IRMA's and are found at high concentrations in patients with ectopic tumors (Page-Wilson et al., 2014; Oliver et al., 2003).

ACTH results on the Immulite platform can be falsely elevated due to interfering antibodies (Greene et al., 2019). A patient was suspected of alcohol induced hypercortisolism on the basis of ACTH results with the Immulite system but normal when samples were assayed with Cobas platform and the apparent hypercortisolism on the Immulite results was resolved with abstinence. Chromogranin A may prove to be useful in the differentiation of ectopic tumors.

An intact ACTH analysis by LC-MS/MS has been reported and was the arbiter when immunoassay results were clinically discordant (Shi et al., 2019). An Aeris Peptide column XB-C18 was eluted with gradient 3% (B 0.1% formic acid in acetonitrile and A 0.1% formic acid in water) to 97% B in A. Three transitions were monitored. The internal standard was Mouse ACTH (pG26V and pD29N). The assay was linear from 9 to 1938 pg/mL (2–427 pmoL/L).

3.1.1.3.2 Venous sampling

Whole body venous sampling with ACTH measurements may be useful in locating a tumor if an ectopic source is suspected but not identified radiologically (Fig. 3.1.1.20). The venous sampling techniques require special experience and expertise.

Imaging of ectopic tumors may be assisted by isotope scans after injection of labeled somatostatin analogues. CT scanning may localize an adrenal tumor or display hyperplasia. Benign adrenal lesions associated with Cushing's syndrome are linked to dysregulation of cyclic AMP signaling (Hannah-Shmouni et al., 2016), whereas tumors are linked to insulin-like growth factor II, p53 protein and related proteins.

If an ectopic ACTH secreting tumor is suspected the pancreas and chest are the most likely sites and should be sought by CT scanning. Whole body venous catheter studies for the localization of tumors producing ACTH may be needed (Drury et al., 1982) (Fig. 3.1.1.20B).

A number of sites of ectopic ACTH tumors have been located (Table 3.1.1.4). The measurement of tumor markers (e.g., calcitonin, gastrin, somatostatin, and carcinoembryonic antigen) should be considered (Howlett et al., 1985; Himsworth et al., 1977) although this needs further investigations since, using immunohistochemistry of tumors, variable

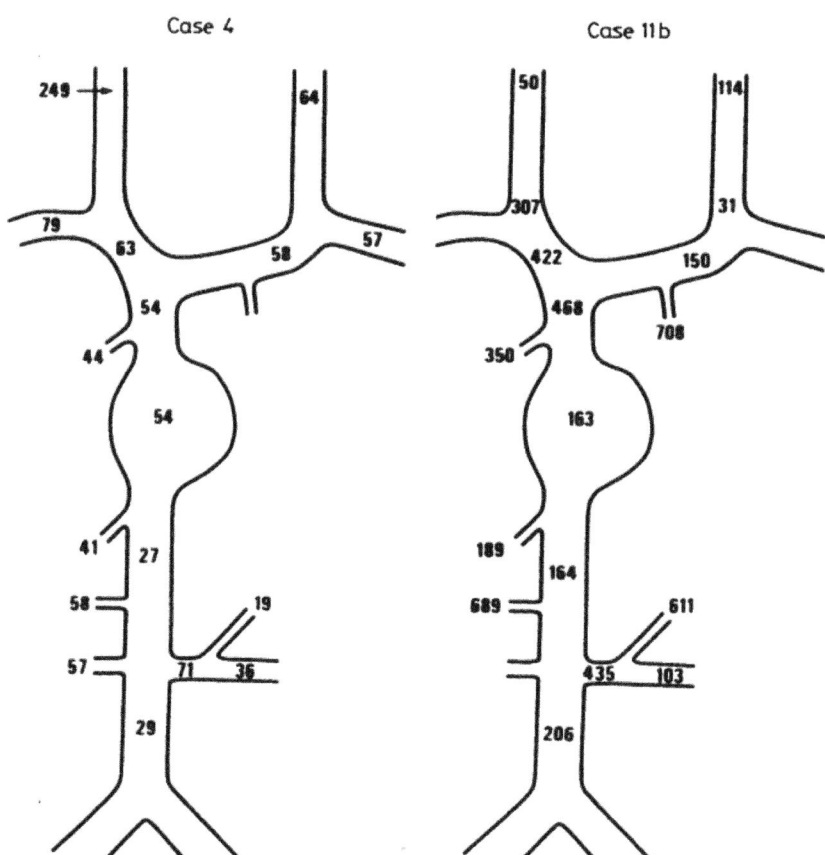

FIG. 3.1.1.20 Plasma ACTH concentrations (ng/L) at sites sampled during catheterizations in case with pituitary-dependent disease *(left)* and case with proven thymic tumor *(right)*. See Fig. 2.7.11 for key to sampling sites. *(Data of Drury PL, Ratter S, Tomlin S, Williams J, Dacie JE, Rees LH, Besser GM. Experience with selective venous sampling in diagnosis of ACTH-dependent Cushing's syndrome. Br Med J (Clin Res Ed) 1982;284(6308):9–12.)*

TABLE 3.1.1.4 Sites ectopic tumors.

Small cell carcinoma lung
Carcinoma sigmoid colon
Pancreatic endocrine tumor
Carcinoid of gall bladder
Bronchial carcinoid
Thymic carcinoid
Phaeochromocytoma

responses with antibodies to C-POMC, α-MSH, CRF, bombesin, somatostatin, calcitonin, α-hcg and β-hcg were documented in 18 cases (Coates et al., 1986).

3.1.1.3.3 CRF test

If a pituitary adenoma is suspected, a **CRF test** may be informative in many cases (Ceccato et al., 2020b; Nishioka and Yamada, 2019; Chan et al., 2007a). After an intravenous injection of 100 μg CRF (1 μg per kg in children), blood samples are taken at 15 min before the CRF then at 0, 15, 30, 45, 60, 90 and 120 min. In Cushing's disease, there is often an increase in cortisol above the normal response to the CRF (Fig. 3.1.1.21). The measurement of ACTH response was less helpful than the cortisol response (Newell-Price et al., 2002).

The combination of high-dose dexamethasone and the CRF test with measurement of serum cortisol is superior to either test alone in the differential diagnosis of Cushing's syndrome (Pecori Giraldi et al., 2015). The interpretation of this test in children and adults is confounded by severe

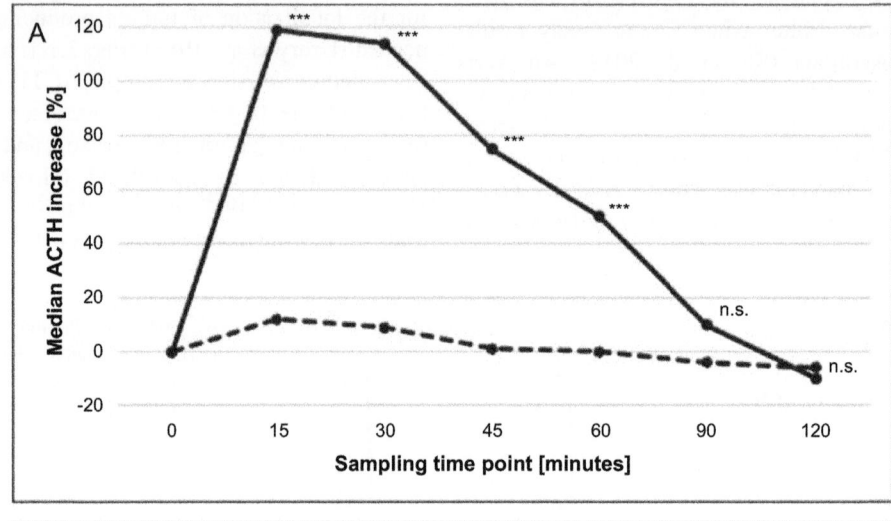

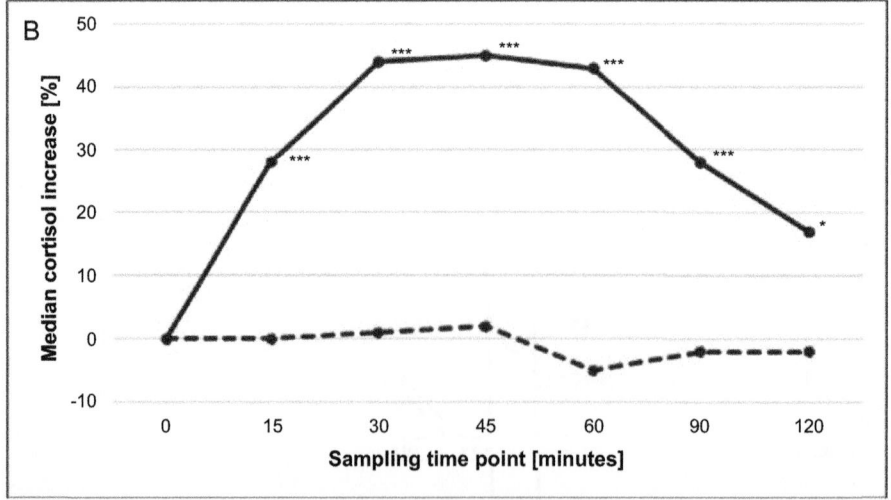

FIG. 3.1.1.21 Median %-increases of (A) ACTH and (B) cortisol during the CRH stimulation test in patients with CD *(solid line)* and ECS *(dashed line)*. Stars indicate statistical significant differences between both sub-entities (* <0.05; *** <0.001). ACTH, adrenocorticotropin; CD, Cushing's disease; CRH, corticotropin-releasing hormone; ECS, ectopic Cushing's syndrome. n.s., not significant. *(From Detomas M, Ritzel K, Nasi-Kordhishti I, Wolfsberger S, Quinkler M, Losa M, Tröger V, Kroiss M, Fassnacht M, Vila G, Honegger JB, Reincke M, Deutschbein T. Outcome of CRH stimulation test and overnight 8 mg dexamethasone suppression test in 469 patients with ACTH-dependent Cushing's syndrome. Front Endocrinol (Lausanne) 2022;13:955945. Fig 1 p 5.)*

obesity (Saiegh et al., 2016; Batista et al., 2008). Simultaneous measurements of ACTH and cortisol may improve the sensitivity and specificity of the test (Arnaldi et al., 2009; Oliver et al., 2003;).

3.1.1.3.4 DDAVP test

The ACTH and or cortisol responses to an intravenous bolus of DDAVP (a vasopressin analogue) have been found to be a reliable diagnostic marker for ACTH secreting tumors (Vassiliadi and Tsagarakis, 2018; Rollin et al., 2015). The response is absent in obesity, depressive illness and healthy subjects (Tsagarakis et al., 1999) (Fig. 3.1.1.22).

3.1.1.3.4.1 Petrosal sinus sampling ACTH

If there is still doubt on the diagnosis after HDDS and CRF tests, petrosal sinus sampling for ACTH in a specialized center may be helpful (Fig. 3.1.1.23). The discriminating power of the test is increased by combining petrosal sinus and simultaneous peripheral venous sampling with CRH stimulation.

The sampling of blood bilaterally from inferior petrosal sinuses (BIPSS) for simultaneous assessment of ACTH concentrations can be used to differentiate Cushing's disease from the syndrome and for lateralization of a pituitary microadenoma (Tortora et al., 2020; Chen et al., 2019;

Lienhardt et al., 2001). An intriguing observation from these studies has been the parallel increased secretion of prolactin, growth hormone, TSH and glycoprotein α-subunit. These findings may reflect changes in the vasculature or a paracrine effect of β-endorphin from the tumor on adjacent tissue.

3.1.1.3.4.2 ACTH precursors assay

Processing of the ACTH precursor, pro-opiomelanocortin (POMC), differs in pituitary and extra-pituitary tumors. The application of a two site immunoradiometric assay for ACTH precursors (POMC and pro-ACTH) suggests that precursor levels are higher in the ectopic ACTH syndrome than in Cushing's disease and can distinguish the two conditions better than ACTH assays (Page-Wilson et al., 2014). The source of ACTH may still be elusive despite extensive investigation of Cushing's syndrome. It may then be necessary to render the patient eucorticoid medically and repeat investigations after several weeks. POMC can be measured using an ELISA based on capture antibodies to ACTH 10-118 and detection antibody to γ-MSH. The normal range is 7–32 fmol/L. Raised Agouti related protein (AgRP), chromogranin and calcitonin may prove to be useful in the differentiation of ectopic tumors (Granberg, 2015; Page-Wilson et al., 2014; Glinicki et al., 2013; Zemskova et al., 2009).

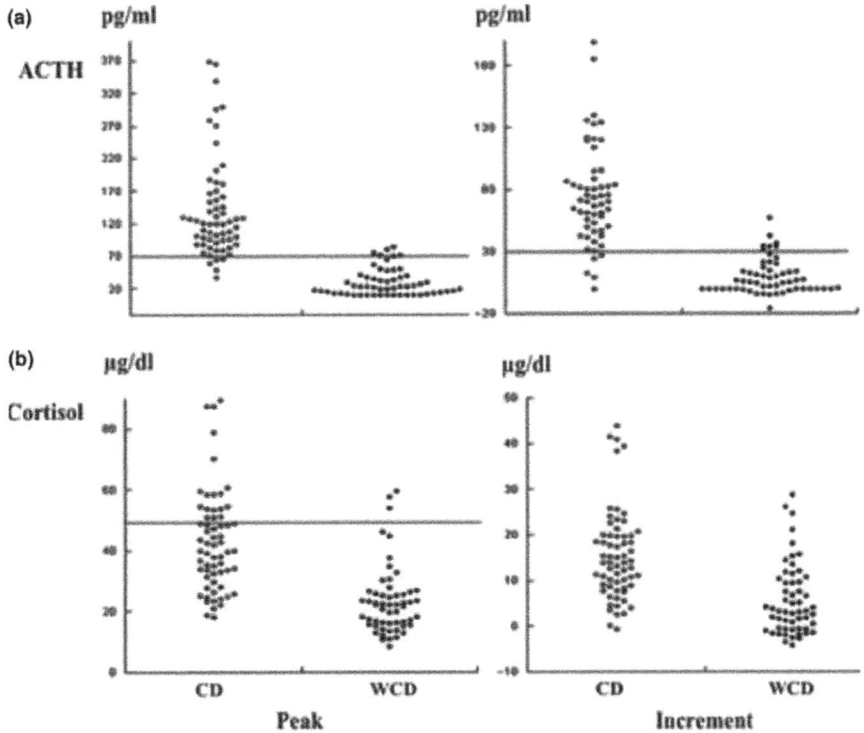

FIG. 3.1.1.22 Evaluation of peak and increment after **DDAVP administration**. ACTH (A) and cortisol (B) CD: Cushing's Disease, WCD: Without Cushing's Disease. To convert serum cortisol 1g/dL to SI units, multiply by 27_6. To convert plasma ACTH pg/mL to SI units, multiply by 0.22. *(From Rollin GA, Costenaro F, Gerchman F, Rodrigues TC, Czepielewski MA. Evaluation of the DDAVP test in the diagnosis of Cushing's disease. Clin Endocrinol (Oxf) 2015;82(6):793–800. Fig. 3. p. 798.)*

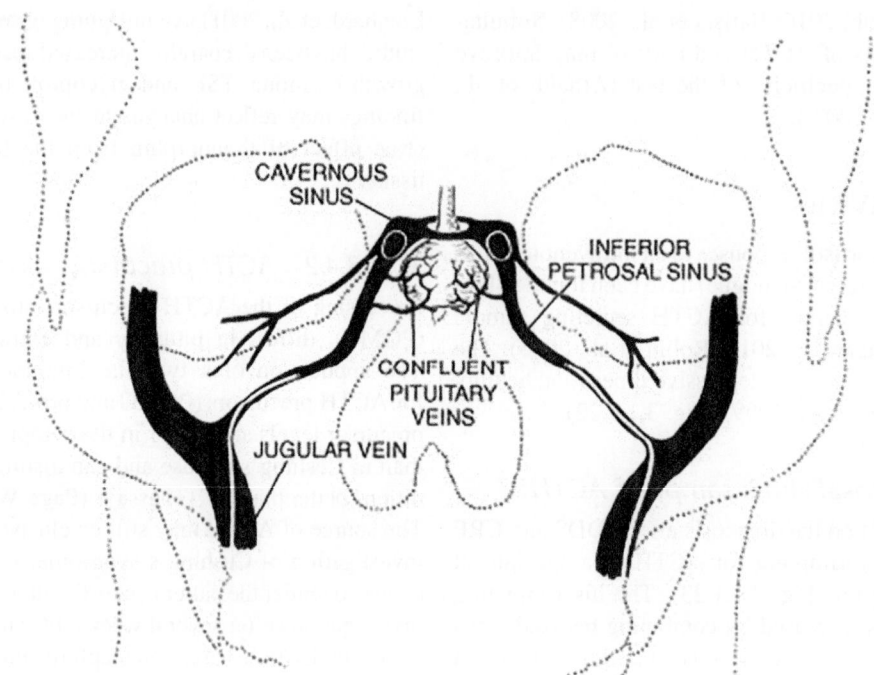

FIG. 3.1.1.23 Schematic of petrosal sinus sampling. *(Based on Utz A, Biller BM. The role of bilateral inferior petrosal sinus sampling in the diagnosis of Cushing's syndrome. Arq Bras Endocrinol Metabol. 2007;51(8):1329–38. Fig 2 p. 1332.)*

3.1.1.3.5 Metyrapone test

In order to examine the entire HPA axis, the cortisol response to hypoglycemia induced with insulin can be used but the test procedure is demanding for patients and medical staff. Overnight single-dose metyrapone provides a simple alternative (Fiad et al., 1994). Metyrapone is given orally (30 mg/kg body weight, or 2 g for <70 kg, 2.5 g for 70–90 kg, and 3 g for >90 kg body weight) at midnight with a glass of milk or a small snack. The 8 AM serum 11-deoxycortisol concentrations should then be >200 nmol/L (7 µg/dL) with serum cortisol less than 140 nmol/L (5 µg/dL), confirming adequate metyrapone blockade. The plasma ACTH concentration at 8 AM should exceed 17 pmol/L (75 pg/mL), confirming that any increases in serum 11-deoxycortisol concentrations are ACTH-dependent, thereby separating primary from secondary adrenal insufficiency. The test fell out of favor because of the difficulties in steroid and ACTH measurements. Ideally cortisol and 11-deoxycortisol should be measured together by LC-MS/MS since immunoassays are unreliable (Monaghan et al., 2011; Owen et al., 2010). This test may be resurrected now that 11-deoxycortisol is more easily measured in a steroid panel by LC-MS/MS.

3.1.1.3.6 Cortisone reductase deficiency

Cortisone reductase deficiency is a rare cause of increased cortisol production because cortisone is not reduced to cortisol in the periphery. Patients present with a polycystic ovary picture without weight gain. Adrenal androgen output is raised because of the adrenal hyperplasia from high ACTH stimulation. **Mutations** can be found in the **HSD11B1 gene** and in the **hexose-6-phosphate dehydrogenase (H6PDH) gene** which is involved in NADPH regeneration for the enzyme (Draper et al., 2003; Lavery et al., 2013; Phillipov et al., 1996; Phillipou and Higgins, 1985). A urine steroid profile shows higher excretions of 11-oxo than 11-hydroxy cortisol metabolites (high ratio of THE plus cortolones to THF, allo-THF and cortols) (Fig. 3.1.1.24).

A cortisone acetate challenge test is recommended to provide evidence to support the defects (Jamieson et al., 1999). Cortisol generation from the cortisone is lower than normal in patients with reductase defect.

3.1.1.3.7 Familial glucocorticoid resistance

High plasma cortisol concentrations and production rate can be seen in familial glucocorticoid resistance. The variable clinical phenotype of this sometimes so-called Chrousos syndrome, including chronic fatigue, mild hypertension and hyperandrogenism, in association with the difficulties encountered in establishing the correct diagnosis on biochemical grounds, may account for the low reported prevalence of the condition.

Patients may display signs of androgen and mineralocorticoid excess but without the glucocorticoid effects (van Rossum and Lamberts, 2006; Charmandari et al., 2008). Teenagers with resistant hypertension can show biochemical features of cortisol resistance (Shamim et al., 2001) and the genetic basis of this should be undertaken. The syndrome has

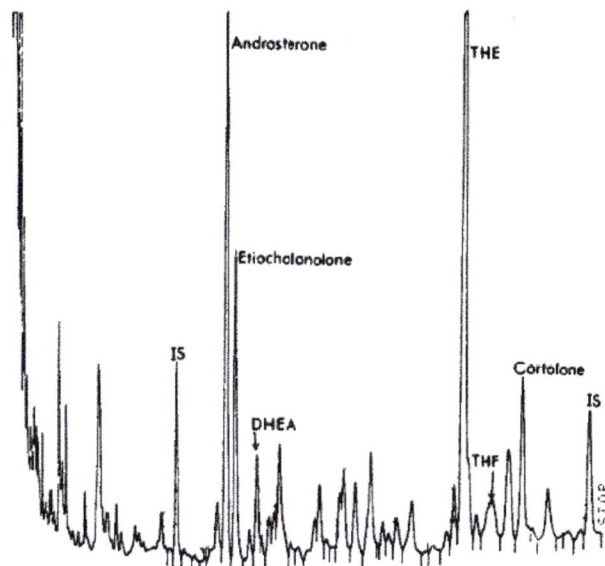

FIG. 3.1.1.24 Urinary steroid profile for suspected cortisone to cortisol defect. Steroid chromatographed as methyloxime-TMS derivatives on 15 m × 0.3 mm OV-101 vitreous silica capillary column. IS = internal standard, *n*-tetracosane and *n*-dotriacontane, respectively. *(From Phillipou G, Higgins BA. A new defect in the peripheral conversion of cortisone to cortisol. J Steroid Biochem 1985;22(3):435–6. Fig. 1 p. 435.)*

been associated with genetic defects in the glucocorticoid receptor (NR3C1) gene although some patients with clinical manifestations of this condition did not have any mutations, deletions or insertions in the gene encoding the NR3C1 (Nicolaides and Charmandari, 2021 for review of variants). The molecular basis of Chrousos syndrome has been attributed to inherited or de novo genetic defects in the NR3C1 gene, which are depicted in Fig. 3.1.1.25. The search for further molecular mechanisms in patients will enhance our understanding of glucocorticoid signaling.

3.1.1.3.8 Adrenal masses

A finding of an adrenal mass on CT or MRI scans of the abdomen is now quite common and investigations of function are obligatory. A mass can be seen when 3 mm or more, up to several centimeters which may then be

palpable. Bilateral adrenal masses do not always equate with malignancy. In some cases, no abnormality of steroids or catecholamines can be identified and the mass is called an **incidentaloma**. The literature on this topic was reviewed in 2020 (Sherlock et al., 2020) with prevalence rates in general less than 3% but approaching 10% in the elderly. The adrenal mass can secrete steroids independently of ACTH and can be benign (adrenal adenoma ACA) or malignant (adrenal carcinoma ACA). The pattern of steroid output can be for cortisol only, cortisol and androgens, androgens only but the literature is difficult to review without an effective classification. The risk of malignancy was a driving factor for surgery and thorough biochemical investigations were not always conducted. The distinction of adrenal tumors was retrospective based on histology that included the presence of mitoses and clear cell components on histology and evidence in surgery of venous and/or sinusoidal invasion. In the published literature, infants will have presented with symptoms (virilization, obesity) rather than imaging, whereas older patients can be asymptomatic. This could also bias the biochemistry which has progressed from group steroid measurements (17 ketosteroids, 17-hydroxycorticosteroids), through individual steroid measurements by "specific" immunoassays to steroid profiles in urine by GC-MS (Arlt et al., 2011) and plasma profiles by LC-MS/MS (Taylor et al., 2004; Wang et al., 2020; Hannah-Shmouni and Stratakis, 2020). In addition to the measurement of steroids in biological fluids, the extent of steroid derangement can now be assessed by immunochemical techniques for the enzymes in the tissue after surgical removal (Uchida et al., 2017; Pereira et al., 2020).

3.1.1.4 Genetic studies

At surgical removal, adrenal tumors are distinguished on size, number and pigmentation. The tissue around nodules should be examined for hyperplasia or atrophy. Lipofuscin may be present as brown auto-fluorescent granularity and a marker of lysosomal residual bodies containing end products of oxidative damage to lipids. The patient should be examined for spread of the tumor to lymph nodes and for the presence of metastases.

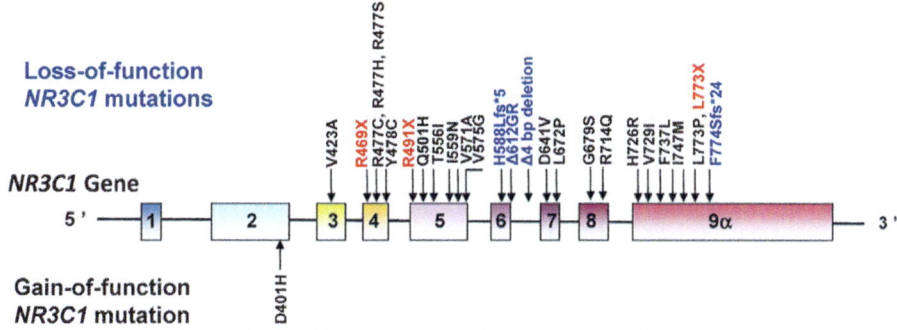

FIG. 3.1.1.25 GR loss of function mutations and gain of function mutations in the human NR3C1 gene. *(Redrawn from Vitellius G, Trabado S, Bouligand J, Delemer B, Lombès M. Pathophysiology of glucocorticoid signaling. Ann Endocrinol (Paris). 2018;79(3):98–106. Fig. 2 p. 102.)*

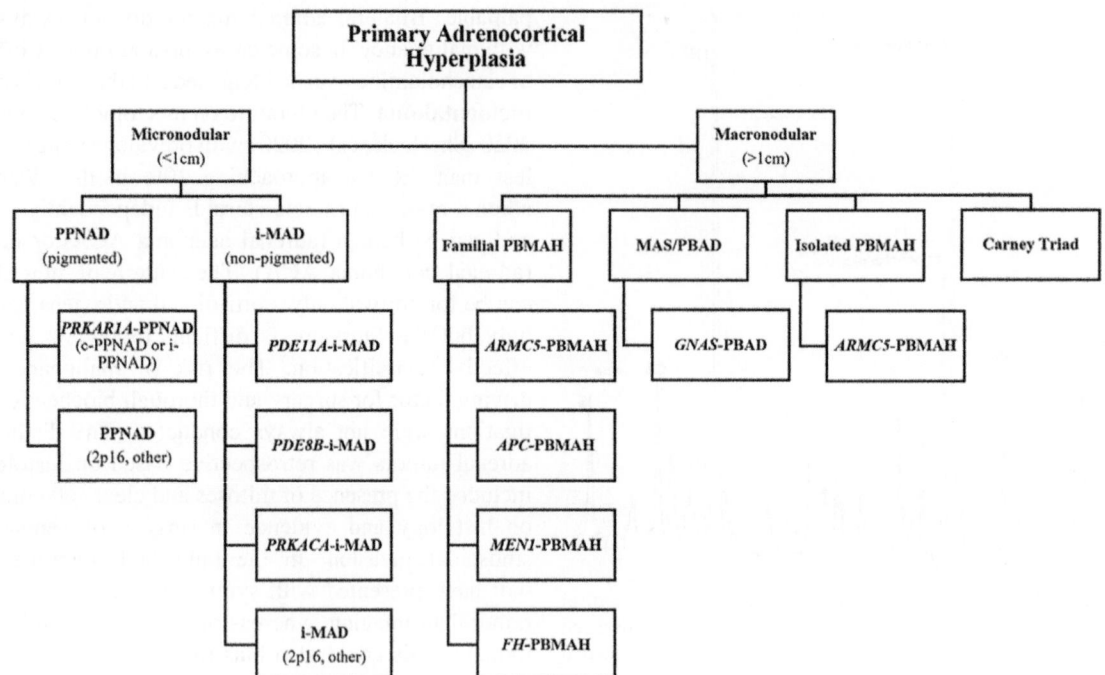

FIG. 3.1.1.26 Path for investigation of adrenal tumors with autonomous cortisol secretion. *(From Kamilaris CDC, Stratakis CA, Hannah-Shmouni F. Molecular genetic and genomic alterations in Cushing's syndrome and primary aldosteronism. Front Endocrinol 2021;12:632543. Fig. 3 p. 6.)*

A scheme for investigations of adrenal tumors is based on size of tumors and frequency of findings (Fig. 3.1.1.26).

In normal adrenal zona fasciculata, pituitary-derived ACTH binds to the membrane G protein coupled receptor MC2R which in turn results in activation of adenylate cyclase that leads to an increased cyclic AMP production (Fig. 3.1.1.27) (Vaduva et al., 2020).

Protein kinase A (PKA) has two R and two C subunits encoded by type 1 alpha regulatory protein (PRKAR1A) and PRKACA. The tetrameric PKA takes on four molecules of cAMP and releases the catalytic subunits to phosphorylate nuclear and cytoplasmic targets including the transcription factor CREB (cyclic AMP response element binding protein) to stimulate transcription of several cAMP dependent genes. The cAMP is degraded by phosphodiesterases (PDE). Mutations in any gene of all of these components have been found in adrenal tumors.

Primary pigmented nodular adrenal disease (PPNAD) is characterized by multiple micronodules of less than 1cm while **primary bilateral macronodular adrenal hyperplasia (PBMAH)** has nodules more than 1cm. PPNAD is found in children with Carney complex (CNC) (Table 3.1.1.5) (Berthon and Bertherat, 2020).

Benign adrenal lesions (usually less than 1 cm) associated with Cushing's syndrome are linked with genetic alterations in GNAS, PRKAR1A, PRKACA, PRKACB, PDE11A, and PDE8B, that lead to **dysregulation of cyclic AMP signaling** (Hannah-Shmouni and Stratakis, 2020) (Fig. 3.1.1.28).

A GNAS activating mutation was found in a patient that presented with a mild cortisol and androgen secreting tumor.

Administration of dexamethasone had paradoxically increased cortisol concentrations (Sakaguchi et al., 2019). Micronodular adrenocortical hyperplasia can be divided into two subclasses—**pigmented adrenocortical nodules (PPNAD)** and **isolated micronodular adrenocortical disease (i-MAD) that** can be familial, autosomal dominant inherited or sporadic. PPNAD is more common and has pigmented nodules and atrophic surrounding tissues. CNC is attributed to a defect in a gene coding for a component of the PKA enzyme. **Inactivating mutations of PRKAR1A** are found leading to constitutive activation of the cAMP/PKA pathway. **Copy number variants of the PRKCAB locus** on chromosome 1 encode the catalytic subunit of PKA.

PPNAD can present with other forms of Carney complex due to triplication of 1p31.1 gene location including PRKACB; duplication of gene 19p region including PRKACA; deleterious variants PDE11A; base substitution PDE8A. For some cases of PPNAD, beta-catenin accumulation is found. Intraadrenal ACTH production has also been demonstrated in PBMAH (Louiset et al., 2013).

Defects in genes encoding **phosphodiesterase** and the catalytic subunits of PKA have been found in cases with iMAD which has little nodular pigment with hyperplasia of the zona fasciculata. Inactivating defects of **PDE11A** which encodes phosphodiesterase type 11A is the most frequent followed by **PDE8**. Inactivating defects in PDE11A have been implicated in the development of prostate cancer and testicular germ cell tumors (Hannah-Shmouni et al., 2016). Germline copy number gains resulting in amplification of PRKACA that codes the catalytic subunit of

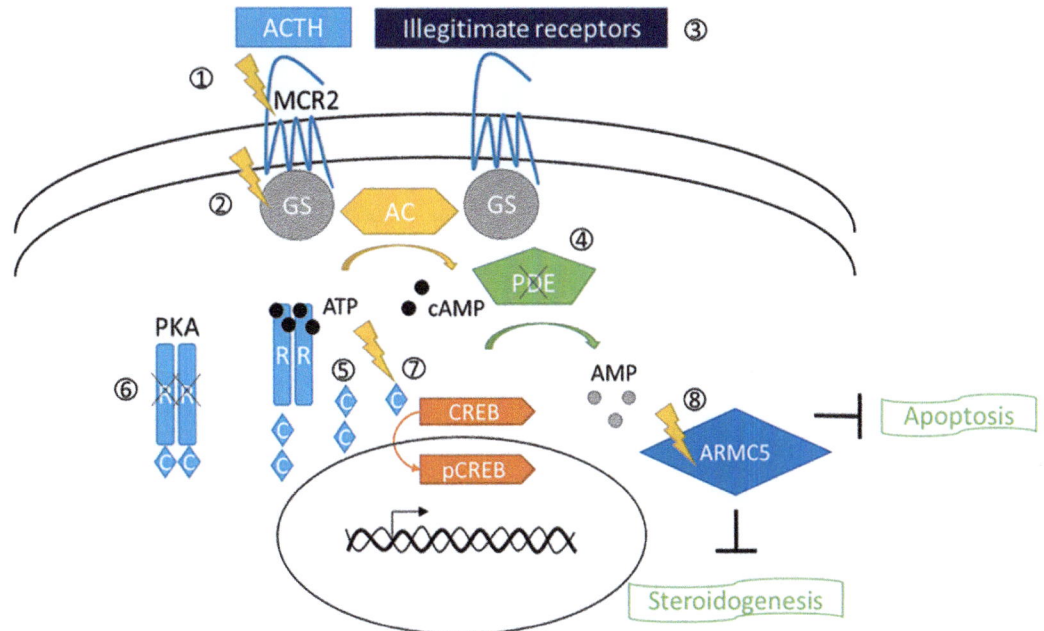

FIG. 3.1.1.27 Signaling pathways and molecular alterations in adrenal Cushing's syndrome. Physiologically, adrenocorticotropic hormone (ACTH) binds to a G protein-coupled receptor, the melanocortin receptor (MC2R), resulting in Gs protein activation. This in turn activates adenylate cyclase (AC) leading to cyclic adenosine monophosphate (cAMP) production. Four cAMP molecules bind to the protein kinase A (PKA) regulatory subunits dimer, which allows the release and activation of the 2 catalytic subunits of PKA. The free catalytic subunits will phosphorylate the transcription factor CREB (cAMP response element-binding protein), stimulating the transcription of several cAMP-dependent genes. Phosphodiesterases (PDE) involved in cAMP degradation are negative regulators of this pathway. *(From Vaduva P, Bonnet F, Bertherat J. Molecular basis of primary aldosteronism and adrenal Cushing syndrome. J Endocr Soc 2020;4(9):bvaa075. Fig. 1 p. 2.)*

TABLE 3.1.1.5 Carney complex.

- Cardiac tumor (myxoma)

- Bumps on skin (skin myxoma)

- Spotted areas on skin (lentiginosis)

- Multiple moles (blue nevi)

- Primary pigmented nodular adrenocortical disease (PPNAD)

- Testicular tumors

- Acromegaly

- Thyroid tumors

- Or an melanotic schwannoma (nerve sheath tumor)

- Osteochondromyxoma (bone tumors)

PKA (Cα) have been reported. Abnormalities of the Wnt-β-catenin pathway may finally act as a genetic modifier causing i-MAD (Almeida and Stratakis, 2011) (Fig. 3.1.1.29).

Germline hereditary factors are implicated with the development of larger tumors (>>1 cm) in primary bilateral macronodular adrenocortical hyperplasia (Hannah-Shmouni

and Stratakis, 2020; Liu et al., 2018; Assié et al., 2013). PBMAH is a rare cause of CS, not diagnosed until senior years. Type 1 PBMAH is characterized by internodular atrophy and the more common type 2 is diffusely hyperplastic without internodular tissue. Dysregulation of PKA signaling has been found in the tumors (Espiard et al., 2014) and aberrant ectopic expression of G-protein coupled receptors has been identified (Lacroix, 2010). Another subtype of macronodular bilateral adrenocortical hyperplasia disease, PBAD, is due to aberrant expression of the G-protein coupled receptor.

Defects in ARMC5 have been characterized in isolated PBMAH (Berthon et al., 2017). These defects ultimately result in abnormal activation of the cAMP-PKA signaling pathway and clinical hypersensitivity to ACTH. Germline **ARMC5** defects are familial (Assié et al., 2013). The precise role of the armadillo repeat containing 5 protein (AMCR5) is not clear but certainly acts as a tumor suppressor gene (Chevalier et al., 2021) (Fig. 3.1.1.30) leading to reduced expression of steroidogenic enzymes. The hypersecretion of cortisol is likely due to the increased adrenocortical mass.

AMCR5 is highly polymorphic with several variants in the population that may produce genetic modifiers for PKRAR1A (Maria et al., 2020). ARMC5 mutations are associated with more severe hypertension due to hyperaldosteronism (Albiger et al., 2017). Germline inactivating

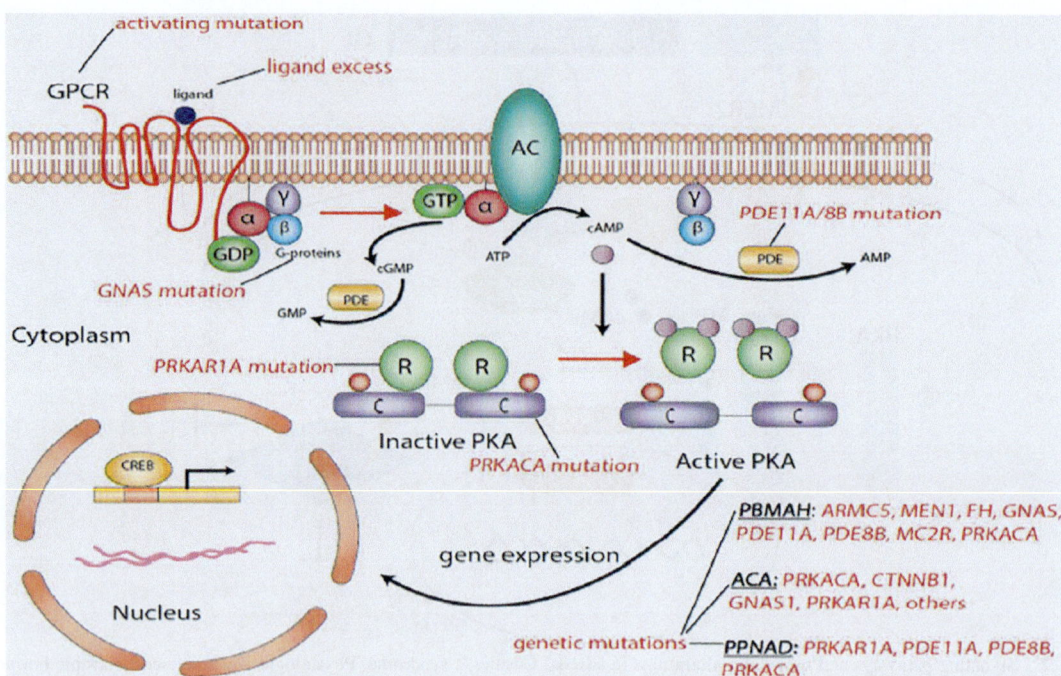

FIG. 3.1.1.28 The cyclic AMP-signaling pathways in primary adrenocortical cortisol-producing neoplasms. The G-protein coupled receptor for ACTH (ACTHR), also known as melanocortin 2 receptor (MC2R), is a seven-transmembrane receptor that undergoes extensive conformational changes in response to its ligand, ACTH. This leads to activation of adenylyl cyclase (AC) through the G proteins and the generation of cyclic AMP, activating protein kinase A (PKA), a holoenzyme that consists of a tetramer of two homo- or heterodimers regulatory subunits (R1α, R1β, R2α and R2β), and catalytic subunits (Cα, Cβ, Cγ and PRKX) that are encoded by the respective genes. This in turn enables phosphorylation of PKA targets, including gene expression to mediate cell growth, differentiation and hormone production (e.g., cortisol and its precursors). Genetic alterations in key genes of the cAMP-signaling pathway leads to the formation of various primary adrenocortical lesions. *AC*, adenylyl cyclase; *C*, catalytic subunit of protein kinase A; *cAMP*, cyclic AMP; *CREB*, cyclic AMP response element binding protein, a transcription factor; *GPCR*, G-protein-coupled receptor; *Gsα*, stimulatory subunit α of the G-protein; *PDE11A*, phosphodiesterase 11A; *PKA*, cAMP-dependent protein kinase; *R*, regulatory subunit. *(From Hannah-Shmouni F, Stratakis CA. A gene-based classification of primary adrenocortical hyperplasias. Horm Metab Res 2020;52(3):133–141. Fig. 1 p. 134.)*

defects in the adenomatous polyposis coli (**APC**) cause FAH. The APC protein is a component of the β-catenin pathway. Inactivating defects of the tumor suppressor MEN1, that is implicated in the regulation of transcription, genome stability, cell division and cell proliferation, have been characterized but the mode of action in adrenal hyperplasia is unclear. Finally germline defects in **fumarate hydratase (FH)** lead to hereditary leiomyomatosis and enal cell carcinoma (HLRCC). Leiomyomatosis are benign tumors in smooth muscle of the skin. FH is a component of the Krebs tricarboxylic acid cycle The gene is located at 1q42.3-q43 but the molecular pathogenesis has not been elucidated. About 8% of patients with HLRCC develop adrenal tumors (Matyakhina et al., 2005).

Adrenocortical carcinoma has been linked primarily to aberrant p53 signaling and/or Wnt-β-catenin signaling, with frequent genetic alterations in TP53, ZNRF3, CTNNB1, and chromosome region 11p15. Tumors are linked to overexpression of **insulin-like growth factor II, p53 protein and related proteins** (Kamilaris et al., 2020; Stratakis and Boikos, 2007). Table 3.1.1.6 summarizes the genetic basis for micronodular and macronodular adrenocortical hyperplasias.

3.1.1.4.1   Pituitary adenoma

The removal of pituitary tumors produces only small amounts of material for analysis which hampered investigations on the basis of these lesions. In 2015, recurrent somatic mutations in a hotspot region of the **Ubiquitin Specific Peptidase 8 (USP8) gene** in CD were found for the first time-using next-generation exome sequencing (Reincke et al., 2015). Other studies using exome sequencing (Sbiera et al., 2019) or targeted sequencing (Perez-Rivas et al., 2015; Ballmann et al., 2018; Sesta et al., 2020) confirmed the fact that USP8 hotspot mutations are responsible for around half of the CD tumors and identified further recurrent mutations. Mutations in the somatic ubiquitin-specific protease 8 (USP8) gene are now a common cause of CD in children (Treppiedi et al., 2021; Faucz et al., 2017; Perez-Rivas and Reincke, 2016; Reincke et al., 2015). Increased de-ubiquitinating activity alters the degradation process of endocytosed EGFR, with elevated EGFR recycling on the plasma membrane. POMC gene transcription is strengthened by high EGFR level leading to increased ACTH secretion and corticotroph tumor development (Albani et al., 2019) (Fig. 3.1.1.31).

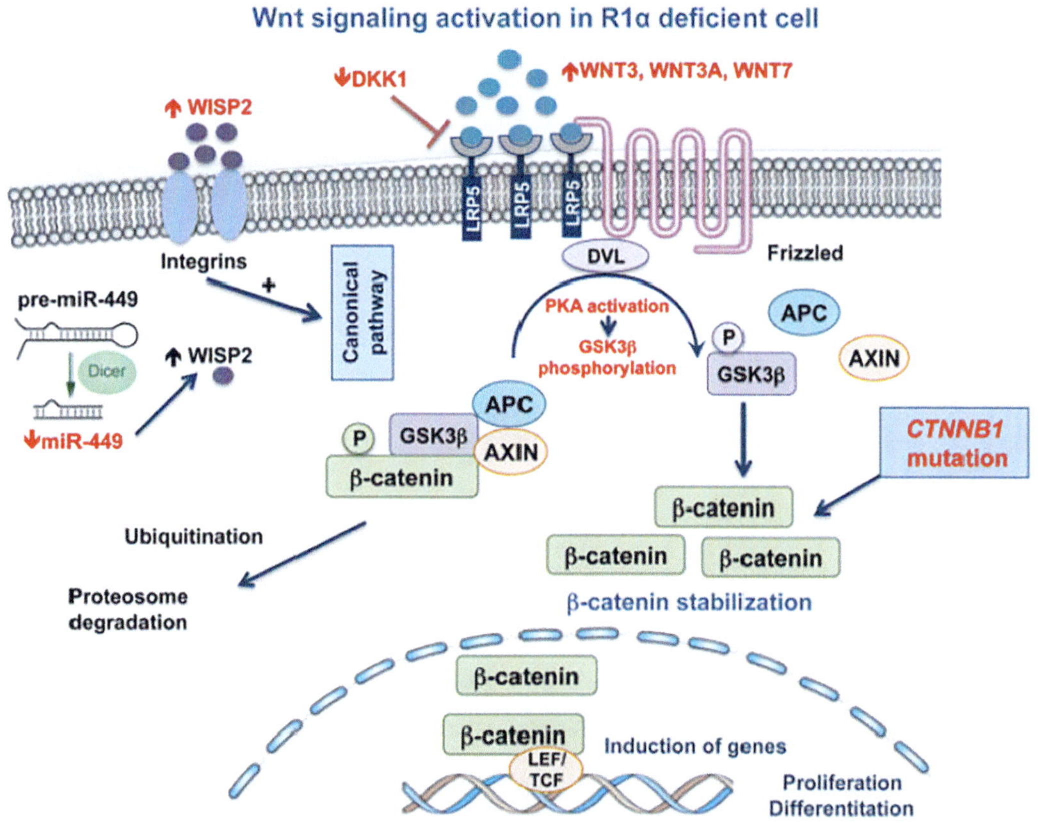

FIG. 3.1.1.29 Schematic representation of Wnt signaling pathway activation in R1α deficient tumor cells. *Wnt signaling enrichment was demonstrated in human and mouse tumors associated with* cAMP/PKA activation. Up-regulation of WNT3, WNT3A, WNT7 and WISP2 induces the activation of the β-catenin-dependent signaling (canonical pathway). Binding of Wnt ligands to the frizzled, LRP5, and LRP6 receptors inhibits the degradation of β-catenin cytoplasmic complex and leads to nuclear accumulation of β-catenin. *CTNNB1 mutations in adrenal* tumors also promote β-catenin stabilization. miR-449 is one of the highest down-regulated microRNAs in PPNAD and regulates *WISP2 expression. WNT, Wingless-type MMTV* integration site family; *DKK,* Dickkopf; *WISP2,* WNT1-inducible signaling pathway protein 2; *LRP5,* low-density lipoprotein receptor-related protein 5; *DVL,* disheveled; *APC,* activated protein C; *GSK3β,* glycogen synthase kinase 3 beta; *LEF/TCF,* lymphoid enhancer factor/T-cell factor; *PKA,* protein kinase A; *PPNAD,* primary pigmented nodular adrenocortical disease. *(From Almeida MQ, Stratakis CA. How does cAMP/protein kinase A signaling lead to tumors in the adrenal cortex and other tissues? Mol Cell Endocrinol 2011;336(1–2):162–8. Fig. 3 p. 166.)*

Ubiquitination is a posttranslational protein modification that directs proteolytic degradation at the 26S proteasome complex. De-ubiquitinating enzymes can remove ubiquitin from target proteins to prevent degradation. USP8 activity is modulated through the binding of **14-3-3 proteins**, first identified in the mid-1960s as a family of abundant acidic proteins in brain that eluted from DEAE chromatography in the 14th fraction that was further purified by gel electrophoresis (fraction 3.3). Function was attributed 20 years later, as **activator of tyrosine and tryptophan hydroxylases involved** in the biosynthesis of neurotransmitters. Later on 14-3-3 proteins were shown to inhibit the protein kinase C and to activate Raf (from originally virus rapidly accelerated fibrosarcoma). In 1996, the 14-3-3 proteins were found to bind to specific phosphorylated motifs in protein targets, and subsequently to numerous kinases such as the MEK kinases (mitogen enhanced protein kinase) as well as phosphatases. The 14-3-3 proteins are now regarded as a major class of molecular chaperones, with more than 200 proteins in networks for cell signaling.

Researchers in China examined tumor tissue samples from 22 patients with pituitary adenomas but a normal *USP8* gene (Chen et al., 2018). Four genes that were recurrently mutated, including—*BRAF* and *USP48*—had not previously been reported in CD. Then in 91 samples from patients, *BRAF* mutations were found in 17% of cases and *USP48* mutations in 23% of patients. These mutations were also found in patients with *USP8*-mutant pituitary tumors, but at a much lower rate—5.1% for *BRAF* and 1.2% for *USP48* mutations. Mutations in these two genes were not seen in patients with pituitary tumors producing other hormones, suggesting they are genetic signatures of ACTH-producing adenomas. The researchers found that *BRAF* and *USP48* mutations activate signaling pathways that lead to the production of proopiomelanocortin (POMC), the precursor of ACTH (Fig. 3.1.1.32).

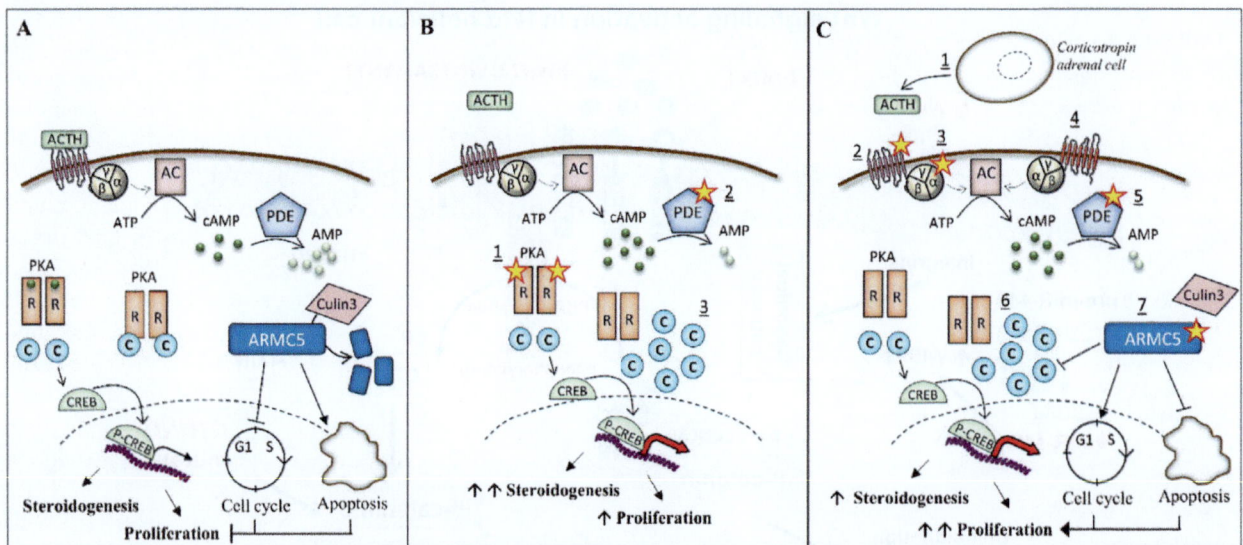

FIG. 3.1.1.30 Alteration of protein kinase A (PKA) pathway and ARMC5 in bilateral adrenal hyperplasia. (A) In normal adrenocortical cells, ACTH activates the MC2R receptor, leading to the activation of the Gα subunits of the G protein. The latter activates the adenylate cyclase (AC), which converts the ATP in cAMP. The phosphodiesterases (PDE) inactivates cAMP in AMP. The regulatory (R) subunits of the PKA bind the cAMP, leading to the release of the catalytic (C) subunits. The catalytic subunits phosphorylate their targets, including the cAMP Response Element-Binding protein (CREB), which activates genes involved in steroidogenesis. ARMC5 blocks the cell cycle in G1 phase and induces apoptosis. ARMC5 is degraded by Culin3. (B) In PPNAD and iMAD, the PKA pathway is activated by (1) mutations in the regulatory subunit R1α of PKA, (2) mutations in phosphodiesterases genes, and (3) duplication of the catalytic subunit Cα have also been described. (C) In PBMAH, the PKA pathway is activated by (1) ACTH locally produced by clusters of corticotropin adrenal cells, (2) mutations in the gene coding for MC2R, (3) mutations in gene GNAS coding for Gα, (4) aberrant expression of G-coupled protein receptors, (5) mutations in phosphodiesterase genes, (6) duplication of the catalytic subunit Cα, and (7) ARMC5 mutations, which lead to the activation of the cell cycle and the loss of apoptosis. Moreover, some mutations prevent its binding to Culin3 and its subsequent degradation. In addition, ARMC5 decreases the PKA activity. *(From Chevalier B, Vantyghem MC, Espiard S. Bilateral adrenal hyperplasia: pathogenesis and treatment. Biomedicines. 2021;9(10):1397.)*

Three cases of CD with somatic inactivating missense mutations in the tumor suppressor gene TP53 have been described; two were carcinomas and one an invasive corticotropinoma. The authors reported TP53 mutations in 33% of USP8-negative tumors and raised the hypothesis that p53 protein is important for regulation of apoptosis (one of the main pathways affected in corticotropinomas) or the BRCA1 mediated DNA-repair in corticotroph cells (Tanizaki et al., 2007). Further studies are needed to confirm this finding.

Pituitary corticotroph adenomas in several studies have USP8, USP48, and BRAF genes mutated causing CD by enhancing the promoter activity and transcription of the gene encoding POMC, which is the precursor of ACTH (Sbiera et al., 2019). Further studies have shown that pituitary adenomas with mutated USP8 display an increased incidence of EGFR expression (Mizuno et al., 2005), EGFR protein abundance and the mRNA expression levels of POMC, indicating that EGFR plays an important role in adenoma development.

Among the potential USP8 substrates, GLi1 (glioma-associated oncogene) is the downstream mediator of sonic hedgehog signaling (SHH) that is down regulated in corticotroph tumors. CRH signaling is potentiated by the SHH/Gli1 pathway to stimulate ACTH synthesis (Sbiera et al., 2019) (Fig. 3.1.1.33).

3.1.1.4.2 Rare forms of CD

CD can occur in children with DICER 1 syndrome and a pituitary blastoma which is an aggressive tumor more usually found in lungs, thyroid, gonads and kidney. DICER gene encodes an endoribonuclease responsible for functional miRNA production and loss of function mutations initiate tumor development (Sahakitrungruang et al., 2014).

Pathogenic cases with CABLES1 variants have been identified in 2% of cases in one cohort. The CABLES1 (Cdk5 and ABL enzyme substrate 1) gene (18q11.2) is a negative regulator of cell cycle progression that is activated in corticotroph cells in response to glucocorticoid. The physiological negative feedback exerted by glucocorticoids on the corticotroph cells is often impaired in corticotropinomas, and, concordantly, CABLES1 protein expression is lost in around half of such tumors (Hernández-Ramírez et al., 2017).

CD has been described with RET (REarranged during Transfection gene) mutations as part of MEN2A (medullary thyroid carcinoma and bilateral phaeochromocytoma) in a 68-year-old male (Naziat et al., 2013) or MEN2B (metastatic medullary thyroid cancer) in a 21-year-old male (Kasturi et al., 2017). The cases are worth reading because they illustrate some of the problems with the diagnosis of Cushing's disease. The RET protein is a transmembrane

TABLE 3.1.1.6 Gene-based classification for primary cortisol producing adrenocortical hyperplasia according to tumor size.

Gene-based diagnostic algorithm for primary cortisol-producing adrenocortical hyperplasias

Micronodular

PPNAD, primary pigmented nodular adrenocortical disease

PRKAR1A protein kinase, cAMP-dependent, catalytic, alpha regulatory subunit

i-MAD, isolated micronodular adrenocortical disease

PDE8B, phosphodiesterase 8B gene

PDE11A, phosphodiesterase 11A gene

Protein kinase A catalytic subunit (PRKACA)

Macronodular

PBMAH, primary bilateral macronodular adrenocortical hyperplasia

MAS, McCune-Albright syndrome

Carney triad Carney complex-associated primary pigmented nodular adrenocortical disease

ARMC5, armadillo repeat-containing protein 5

APC, adenomatous polyposis coli gene

MEN1, multiple endocrine neoplasia type 1

FH, fumarate hydratase

GNAS, gene coding for the stimulatory subunit a of the G-protein (Gsa)

PBAD, primary bimorphic adrenocortical disease

protein, with one end of the protein inside the cell and the other end projecting from the outer surface of the cell. When molecules that stimulate growth and development (growth factors) attach to the RET protein, a complex cascade of chemical reactions inside the cell is triggered. These reactions instruct the cell to undergo certain changes, such as dividing or maturing to take on specialized functions. The mutations responsible for multiple endocrine neoplasia type 2 result in an overactive RET protein that can alter growth factors without first attaching to growth factors outside the cell. The overactive protein likely triggers cells to grow and divide abnormally, which leads to the formation of tumors in the endocrine system and other tissues.

Familial isolated pituitary adenoma (FIPA) due to aryl hydrocarbon receptor interacting protein (AIP) gene mutations is an autosomal dominant disease with incomplete penetrance. FIPA recognizes the presence of pituitary adenomas in two or more members of the same family in the absence of other clinical features, with autosomal dominant inheritance and incomplete penetrance. This accounts for about 2.5% of all pituitary adenomas (Vierimaa et al., 2006). One-fifth of the FIPA cases are due to germline loss-of function mutations in the AIP gene (11q13.2). AIP mutations are also detected in a subset of sporadic pituitary adenomas affecting young patients, and in one-third of cases of gigantism (Hernández-Ramírez et al., 2015). The genetic screening of patients with other sporadic, young-onset pituitary adenoma with no evidence of other endocrine tumors should be focused on AIP mutations in first instance in cases of GH excess (with or without PRL cosecretion) and on MEN1 mutations, especially in cases of prolactinoma (Cuny et al., 2013) because this can be the first manifestation of MEN1 (Thakker et al., 2012).

Aryl hydrocarbon receptor interacting protein (AIP) is a tumor suppressor gene that has a complex effect as a negative regulator of the cAMP/PKA pathway and of the downstream effects of a Gi protein-coupled receptor, probably a somatostatin receptor (SSTR) (Hernández-Ramírez et al., 2017). In the somatotroph cells, AIP mutations lead to tumors secreting growth hormone. More than 100 variants in AIP have been identified, of which the most frequent mutation occurs in the p.R304 locus (Tuominen et al., 2015). Approximately 15%–30% of familial isolated PAs harbor germline mutations in AIP. PAs with mutations in AIP are predominantly somatotropinomas and prolactinomas; however, studies have revealed that AIP mutations may also rarely occur in CD (Beckers et al., 2013).

3.1.1.4.3 Multiple endocrine neoplasia

CD is rare in multiple endocrine neoplasia (MEN1) which also affects parathyroid, enteropancreas and is inherited in an autosomal dominant manner (Brandi et al., 2021). The MENIN protein has roles in multiple pathways (TGF-β/BNP; Wnt/β-catenin and hedgehog) associated with cell proliferation.

Primary nodular adrenocortical hyperplasia is an important cause of Cushing's syndrome in children and is usually associated with tumors of endocrine glands (**multiple endocrine neoplasia** type 1, MEN1) or **Carney complex** (Horvath and Stratakis, 2008; Stratakis, 2009). A paradoxical increase in UFC in the second phase of the high-dose dexamethasone suppression test can be diagnostic of PPNAD (primary, pigmented nodular adrenocortical disease) (Fig. 3.1.1.34) (Stratakis et al., 1999; Gunther et al., 2004; Louiset et al., 2009).

Patients will need careful examination of the skin for lentiginosis (freckles) (Fig. 3.1.1.35) and echocardiogram for cardiac myxoma.

Blood tests for insulin-like growth factor 1 (IGF-1) and prolactin will reflect hormones from affected pancreas and

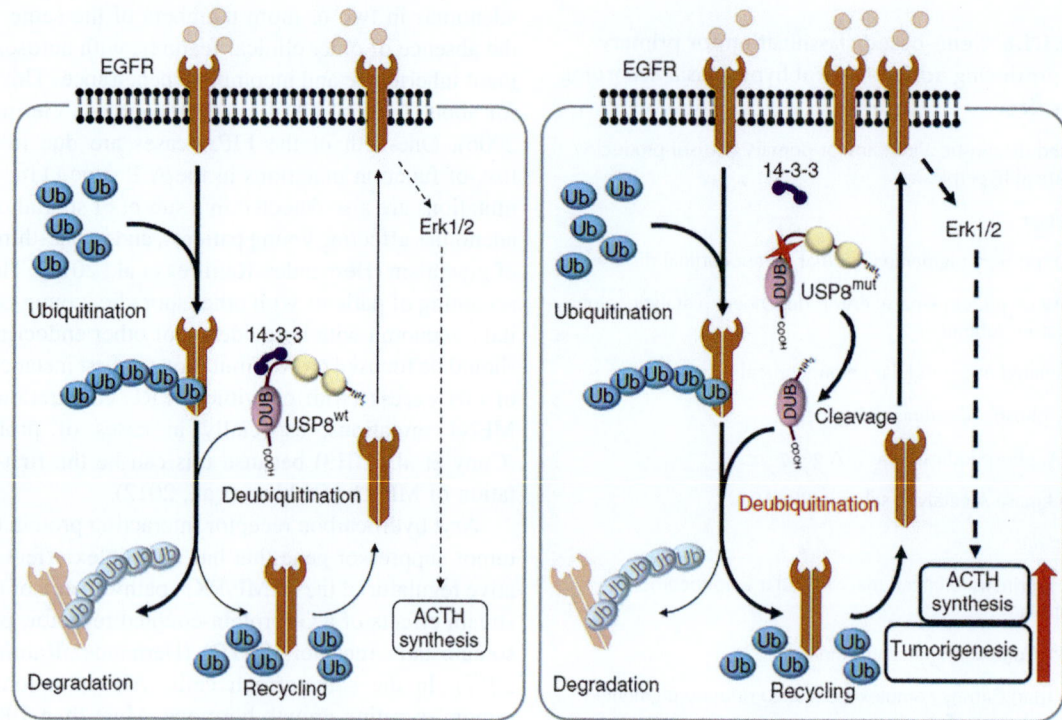

FIG. 3.1.1.31 USP8 wild-type vs USP8 mutant cell. USP8 potentiates EGFR signaling in corticotroph tumor cells. Upon ligand binding, EGFR gets ubiquitinated and targeted to the lysosome. USP8 deubiquitinates EGFR, rescuing it from lysosomal degradation and promoting its recycling to the cell membrane where it is able to bind EGF and transduce its stimulatory action on ACTH synthesis. In corticotroph tumor cells with wild-type USP8 (USP8wt), (A) 14-3-3 proteins bind to USP8, blocking its accessibility to proteases and preventing its full activation. In contrast, (B) the presence of a mutation in the hotspot prevents 14-3-3 from binding to USP8, rendering it accessible to proteases that cleave it to a peptide fragment with high deubiquitinase activity. These highly potent USP8 mutants effectively deubiquitinate EGFR, potentiating its signaling and subsequently ACTH synthesis. *(From Albani A, Berr CM, Beuschlein F, Treitl M, Hallfeldt K, Honegger J, et al. A pitfall of bilateral inferior petrosal sinus sampling in cyclic Cushing's syndrome. BMC Endocr Disord. 2019;19(1):105.)*

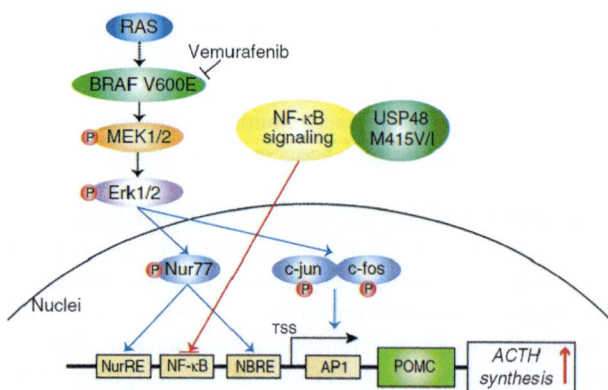

FIG. 3.1.1.32 USP48 mutants potentiate POMC promoter activity and ACTH production. Schematic representation showing the mechanisms by which BRAF V600E and USP48 mutants promote POMC transcription in corticotroph adenomas. *(Chen J, Jian X, Deng S, Ma Z, Shou X, Shen Y, Zhang Q, Song Z, Li Z, Peng H, Peng C, Chen M, Luo C, Zhao D, Ye Z, Shen M, Zhang Y, Zhou J, Fahira A, Wang Y, Li S, Zhang Z, Ye H, Li Y, Shen J, Chen H, Tang F, Yao Z, Shi Z, Chen C, Xie L, Wang Y, Fu C, Mao Y, Zhou L, Gao D, Yan H, Zhao Y, Huang C, Shi Y. Identification of recurrent USP48 and BRAF mutations in Cushing's disease. Nat Commun 2018;9(1):3171. Fig. 1f p. 5.)*

pituitary gland. Identification of two or more of the symptoms in typical fashion is indicative of Carney complex. Carney complex is associated with a particular mutation of PRKRAR1A gene (Protein kinase RI alpha regulatory protein) causing dysregulation of cyclic AMP/protein kinase A pathway.

A few cases with or without neuroendocrine features of MEN 4 had mutations in CDKN1B that is the cyclin-dependent kinase inhibitor 1B gene (*CDKN1B*, 12p13.1) (Chasseloup et al., 2020). The protein normally acts as a key negative regulator of cell cycle progression from G1 to S phase; its function is often impaired in human neoplasms.

3.1.1.4.4 McCune-Albright syndrome (MAS)

McCune-Albright syndrome presents with Cushing's syndrome in children along with polyostotic fibrous dysplasia, café-au-lait skin pigmentation and precocious puberty (Tufano et al., 2020). Cushing's syndrome in children along with polyostotic fibrous dysplasia (scar-like tissues in bones) and café-au-lait skin pigmentation (usually on

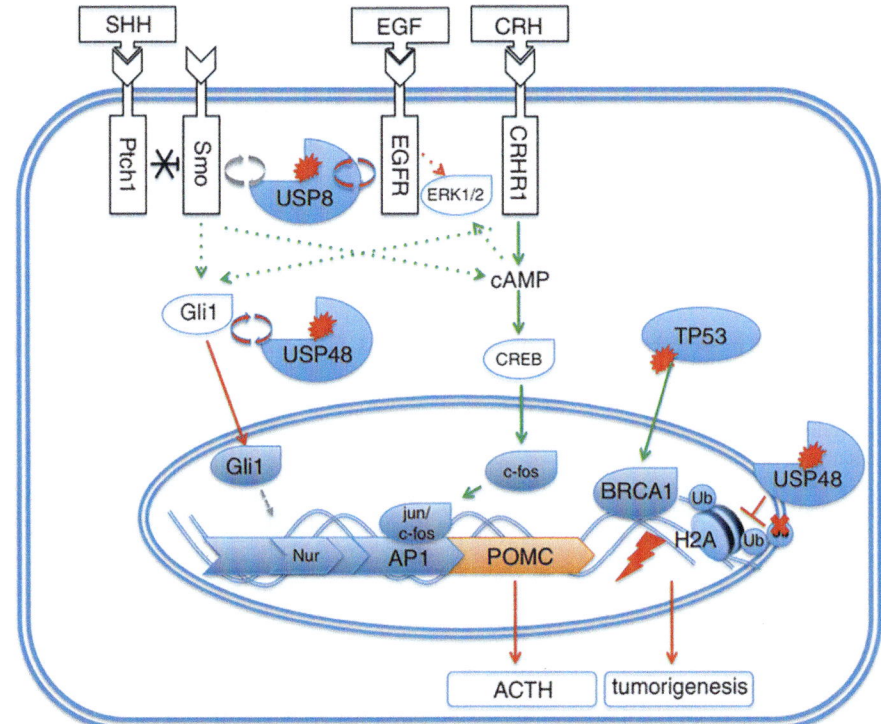

Sbiera S 2019 driver mutations USP8

FIG. 3.1.1.33 Hypothetical mechanisms leading to corticotroph tumor formation involving the different recurrently mutated genes discovered in CD. The 3 genes most frequently mutated in CD (USP8, USP48, and TP53) are shown in the context of patho) physiological corticotroph cell regulation. CRH acts on Nur and AP1 binding elements on the POMC promoter to activate transcription downstream to the cAMP/extracellular signal-regulated kinase (ERK) signaling pathways. Activated epidermal growth factor receptor acts via ERK1/2 to stimulate POMC transcription; activating USP8 mutations potentiate this effect by deubiquitinating and rescuing the receptor from lysosomal degradation.13 SHH binding to Patched (Ptch1) relieves suppression of Smoothened (Smo) allowing for Gli1 activation; SHH and CRH crosstalk at Gli1 level to stimulate POMC transcription and ACTH secretion.19 We hypothesize that activating mutations in the USP48 deubiquitinase lead to increased levels of Gli1, which enhances basal and CRH-induced POMC transcription via an unknown at present mechanism. In Drosophila S2 cells, USP8 was shown to prevent Smo ubiquitination.32 Activating USP48 mutations may also lead to increased deubiquitination of histone H2A and thus to a decreased recruitment of DNA repair factors in case of DNA damage and increased tumorigenesis potential. This effect can also be triggered by inactivating mutations in TP53. 40 *Green* lines represent physiological, red lines pathological activity, and gray lines mechanism of action shown in other cell systems but not yet proven in corticotroph tumor cells. Dotted lines present indirect effects. *(From Sbiera S, Kunz M, Weigand I, Deutschbein T, Dandekar T, Fassnacht M. The new genetic landscape of Cushing's disease: deubiquitinases in the spotlight. Cancers 2019;11(11):1761. Fig. 4 p. 1281.)*

one side of the body) and precocious puberty (Salpea and Stratakis, 2014; Dumitrescu and Collins, 2008) are signs of McCune-Albright syndrome. The syndrome is caused by activating mutations of the G protein alpha subunit (**GNAS1**) (Lumbroso et al., 2004).

3.1.1.4.4.1 Neurofibromatosis

In this condition, tumors occur along nerve tissue and can be anywhere in brain, spine and periphery most noticeable in the skin. Cafe-au-lait markings are also seen, the cause for which is unknown. Mutations in the *NF1* gene lead to the production of abnormal protein called neurofibromin. This protein is produced in many cells and acts as a tumor suppressor, the nonfunctional version of neurofibromin cannot regulate cell growth and division. A complex profile

of steroids in urine by GC-MS was seen in a 49-year-old female who had an 8 cm adrenocortical carcinoma. Dehydroepiandrosterone (DHEA), 17-hydroxypregnenolone, THS and other steroids were elevated in addition to cortisol metabolites (Fig. 3.1.1.36) (Menon et al., 2014). DNA analysis showed loss of heterozygosity (LOH) at the NF1 locus.

Adrenal incidentaloma

A problem with abdominal MRI scanning is the high prevalence of adrenal masses that prompts investigations of function. In the majority of cases, no increase in steroids or catecholamines can be detected. Some cases have a low degree of hypercortisolism which raises questions for the endocrinologists. Is this a subtle (early) Cushing's

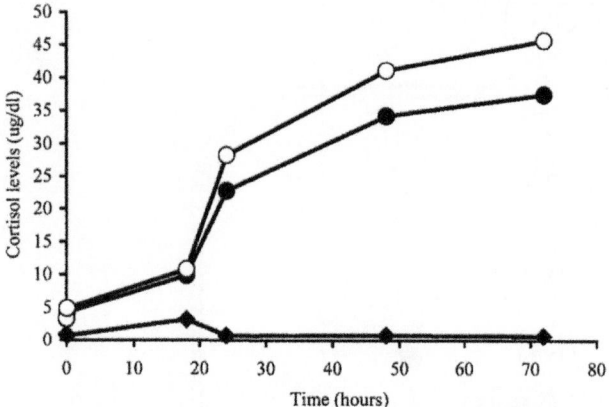

FIG. 3.1.1.34 Cortisol secretion from cultured slices of PPNAD response to dexamethasone. The addition of dexamethasone leads to an increase in cortisol secretion in right (*closed circle*) and left (*open circle*) adrenal glands from patient 3. In the absence of dexamethasone (*closed diamond*) no significant variation in cortisol levels was observed over 72 h. (*From Bourdeau I, Lacroix A, Schürch W, Caron P, Antakly T, Stratakis CA. Primary pigmented nodular adrenocortical disease: paradoxical responses of cortisol secretion to dexamethasone occur in vitro and are associated with increased expression of the glucocorticoid receptor. J Clin Endocrinol Metab. 2003;88(8):3931–7. Fig. 1 p. 3933.*)

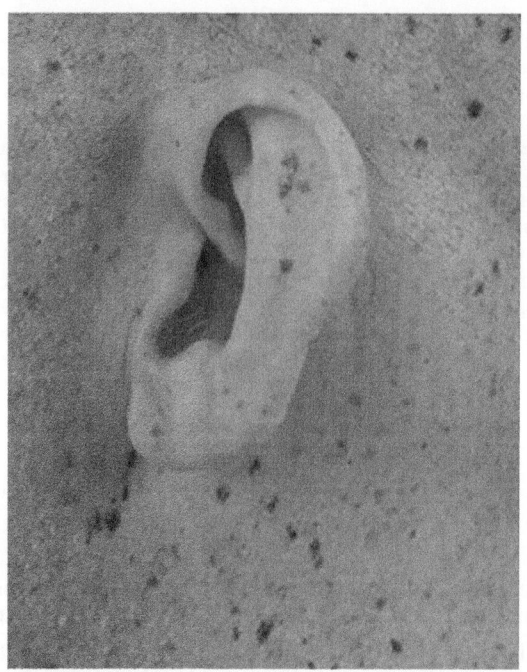

FIG. 3.1.1.35 Lentigines on the face, earlobes and neck and elsewhere are seen in patients with Carney complex. (*From Kamilaris CDC, Faucz FR, Voutetakis A, Stratakis CA. Carney complex. Exp Clin Endocrinol Diabetes. 2019;127(2–3):156–64.*)

syndrome? Is the mass malignant? Are there risks from the hypercortisolemia? Is treatment required? and further questions. This state is labeled Subclinical Cushing's Syndrome (SCS) or dysregulated hypercortisolism and remains controversial.

The adrenal condition is best defined with an overnight dexamethasone suppression test although the dose can be 1, 2 or 8 mg and the 0800 h cortisol cut-off is 1.8, 2.2 or 5 µg/dL (Araujo-Castro et al., 2021). As seen earlier the plasma concentration of cortisol varies with the method of determination (El-Farhan et al., 2017). It is now apparent that dexamethasone should be determined at the same time to be sure that ACTH is suppressed and the concentration of cortisol truly reflects autonomous output of the adenoma (Vogg et al., 2021).

3.1.1.5 Subclinical Cushing's syndrome

SCS does not present with full clinical symptoms of CS (striae, infections, and proximal muscle weakness) although there is a high prevalence of hypertension, obesity, diabetes mellitus, dyslipidemia and osteoporosis (Di Dalmazi et al., 2015a,b). Patients with SCS are at risk of cardiovascular events. The extent to which metabolic parameters affect outcome are unknown. The initial approach for management is "wait and see" with regular tests for exacerbation of the Cushingoid state. Some cases have undergone adrenalectomy which has been found to improve cardiovascular impairment (Iacobone et al., 2015) although replacement of adrenal steroids can never match normal function. A learned panel of experts from the European Society of Endocrinology produced a clinical practice guideline (Fassnacht et al., 2016) which fully covers the debate for the study of patients with adrenal tumors, When surgery was not appropriate or declined there are several medical treatments available (Favero et al., 2021).

Late night salivary cortisol by LCMSMS above 2.8 nmol/L had 83.3% specificity for SCS and 31.3% sensitivity for predicting subclinical hypercortisolism. The tests had 85.2% specificity and 55.6% SN for predicting hypertension, T2DM and osteoporosis while the saliva test (with cortisol >1.4 nmoL/L) when combined with 1 mg dexamethasone suppression test (cortisol >50 nmoL/L) had an 88.9% SN and 85.2% SP (Palmieri et al., 2013).

A plasma steroid profile of the adrenal venous effluent, bilaterally, is ideally required (Athimulam et al., 2021; Nakamura et al., 2012). Urinary steroid profile analysis has been applied to patients with adrenal tumors (not all incidentally discovered), the majority of which were tumors >4 cm (Bancos et al., 2020). The data analysis was extremely complicated based on machine learning of the data and so specific to the laboratory rather than general use. No doubt some clarity will come with time. A previous study had revealed a metabolite of 11-deoxycortisol to be a marker of malignancy (Arlt et al., 2011). When the steroid content of adrenal tissue was examined cortisol, cortisone, 11-deoxycortisol, corticosterone, progesterone and 17-hydroxyprogesterone in the adenoma were present at higher concentrations than adjacent normal tissue. Furthermore corticosterone, 18-hydroxycorticosterone, 21-deoxycortisol, progesterone, and 17-hydroxyprogesterone concentrations

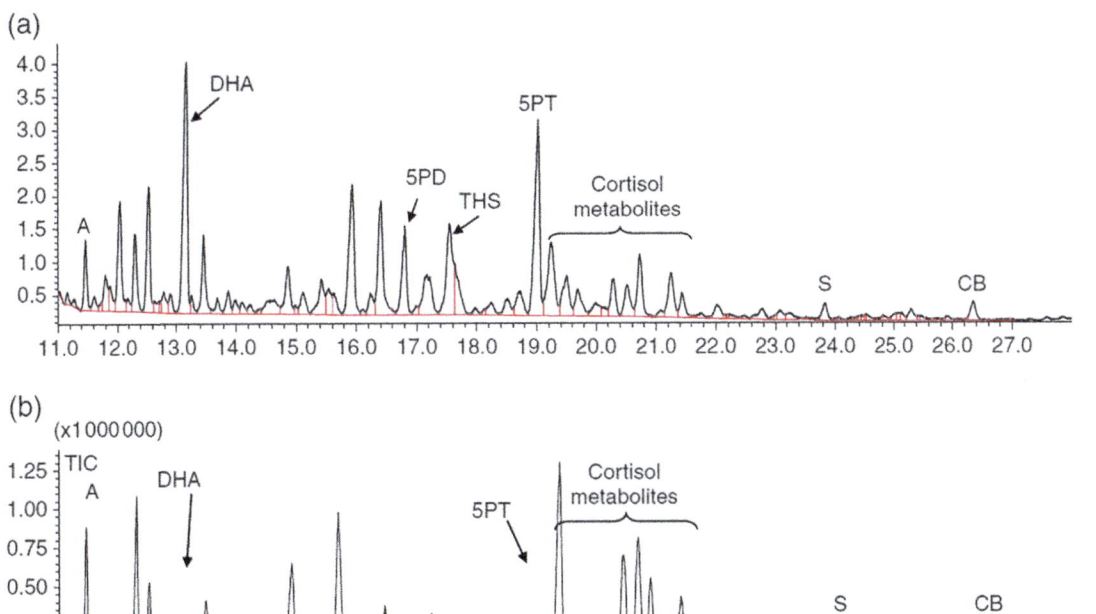

FIG. 3.1.1.36 A 24-h urinary steroid profile: (A) preadrenalectomy compared with (B) the normal control. The profile is dominated by the D5 steroids, in particular pregnanetriol (5PT), DHEA (DHA) and 5PD, and the glucocorticoid precursor, THS. Peaks A, S, and CB are internal standards.

were higher in adenoma than overt Cushing's syndrome (Teuber et al., 2021). Similar findings were reported in comparing plasma steroid profiles of subclinical and overt Cushing's syndrome (Masjkur et al., 2019). Clearly more work is needed relating steroid profiles to the genetics behind abnormal tissues. The profile techniques generate large amounts of data such that multivariate data analysis and chemometrics are used to interpret clinical studies (Kotłowska, 2014). This technology is evolving to metabolomics which is the global assessment of metabolites beyond steroids to select biomarkers for diagnosis and progression of the disease process and pathophysiology.

To date, adrenal imaging with labeled cholesterol [^{75}Se]-6α-methyl19-nor-cholesterol or ^{18}FGD PET has been used to reveal the extent of steroid activity.

3.1.1.5.1 Pituitary incidentaloma

Tumors of the pituitary gland less than 1 cm can also be seen as an incidental finding when scanning the head. This is rare, but can be associated with SCS by virtue of ACTH secretion and increased cortisol production. The tumors may be more common than the 10% frequency at autopsy. Some corticotroph tumors do not clinically express Cushing's disease which is thought to be due to impaired processing of POMC or the tumor transitioning between microfunction and silent state (García-Martínez et al., 2019). The masses can be nonfunctional in terms of no demonstrable excess peptide hormone production, although

some may show ACTH immunoactivity on histology and called silent corticotroph adenomata (Alwani et al., 2014; Baldeweg et al., 2005). Some patients only present with headaches or loss of visual fields. Microadenomas may cause hypopituitarism due to blocking the natural production of peptide hormones or mild hyperprolactinemia if the pituitary stalk is affected.

Many strategies have been proposed to investigate the differential diagnosis (Pinelli et al., 2021).

The Endocrine Society guidelines (Freda et al., 2011) suggest that for large macro-incidentalomas, the laboratory should ideally measure prolactin on diluted serum samples to ensure that levels are not falsely low due to the high-dose hook effect. Screening for GH-secreting tumors should be done using IGF-1. If this is elevated, further evaluation for GH excess is suggested. Screening for ACTH hypersecretion should be undertaken when there is clinical suspicion. Routine measurement of plasma ACTH levels is not recommended. For possible hypopituitarism free T4, TSH, morning cortisol, testosterone, LH+FSH and IGF-1 testing may be considered as a broad approach to detect deficits in the gonadal, cortisol and GH axes. If these baseline measurements suggest hypopituitarism, further stimulation tests of the adrenal or GH-IGF-1 axes should be performed.

Removal of the masses by transphenoidal surgery is recommended when visual fields are affected or there are signs of neurological damage due to compression of structures (Freda et al., 2011). Surgery does not guarantee complete remission.

3.1.1.6 Food dependent Cushings/GIP dependent

The action of glucose dependent insulinotropic polypeptide (GIP; formerly gastric inhibitory peptide) in Cushing's syndrome was first considered when a patient with typical symptoms of CS was found to have an abnormal circadian rhythm of cortisol with high levels in the afternoon instead of declining values (Hamet et al., 1987; Lacroix et al., 1992; Lecoq et al., 2017; Messidoro et al., 2009; Reznik et al., 1992). The observation was further refined to show cortisol secretion was stimulated by food. Glucose, protein and fat ingestion all increased cortisol, but the meal and glucose induced response was blocked with somatostatin. The adrenal glands show bilateral macronodular hyperplasia. Like ACTH, GIP receptor signaling in rats activated cyclic AMP (Mazzocchi et al., 1999). Ectopic overexpression in the adrenal cortex or mutated GIP receptors was proposed and the overexpression was confirmed by RNA sequencing (N'Diaye et al., 1998). Further studies made little progress in knowledge until 2017 when somatic duplications in chromosome region 19q13.32 containing the GIPR locus in the adrenocortical lesions derived from 3 patients were found (Lecoq et al., 2017). In 2 adenoma samples, the duplicated 19q13.32 region was rearranged with other chromosome regions, whereas a single tissue sample with hyperplasia had a 19q duplication only. Then the authors demonstrated that juxtaposition with cis-acting regulatory sequences such as glucocorticoid response elements in the newly identified genomic environment drives abnormal expression of the translocated GIPR allele in adenoma cells. The breakthrough came with the finding of pathogenic variants in KDN1A that encode for a histone lysine demethylase. Methylation of lysine residues can be associated with activation or repression of transcription. KDM1A promotes a chromatin state refractory to gene expression by demethylating histone H3 on lysine 4, usually linked to active gene transcription. Additionally, KDM1A has been shown to affect methylation of nonhistone proteins involved in tumorigenesis, such as the cellular tumor antigen p53, RB1, and STAT3.8. Both mechanisms could be important in the pathogenesis of GIP-dependent primary bilateral macronodular adrenal hyperplasia with Cushing's syndrome. Persistent histone methylation, secondary to loss of KDM1A function, can result in aberrant transcriptional activation and absence of KDM1A interaction with oncogenic proteins, which can lead to cell cycle dysregulation and consequently adrenal tumorigenesis. The loss of KDM1A profoundly modifies the epigenetic landscape of adrenal cells and generates a distinct cellular phenotype. It is now thought a two hit activation of KDN1A ablates the tumor suppression gene. For those who carry KDM1A pathogenic variants, patients should have clinical examination and biochemical screening with measurements including fasting and postprandial plasma cortisol, urinary-free cortisol excretion, and serum protein electrophoresis to detect monoclonal gammopathy (Chasseloup et al., 2021).

3.1.1.6.1 Treatment of Cushing's syndrome

If left untreated, patients with CS can die from infections and cardiovascular consequences of cortisol excess. Surgery is the optimal treatment, but recurrence is common (Bileck et al., 2020) so more information on the etiology of the disease will assist the search for targeted medical treatment. Transphenoidal surgery is needed for pituitary tumors with access through the nose and roof of the mouth. If the surgeon is uncertain about total resection of the tumor then radiation may be given. Postoperative cortisol measurements are needed. A DDAVP tests may predict late recurrence (Barbot et al., 2013). In some cases, glucocorticoid replacement will be needed which complicates biochemical follow-up testing. Close surveillance is required postoperatively (Bansal et al., 2017).

The results of genetic tests will increasingly help to direct treatment. The genes involved in Cushings syndrome are summarized cartoon form in Fig. 3.1.1.38 with more detail seen in Figs. 3.1.1.29 and 3.1.1.30. PRKAR1A gene defects in adrenal CS and the involvement of abnormal cAMP-PKA activity in several adrenal disorders leading to excess cortisol production, led to identification of additional genes of the cAMP-PKA pathway which are causative or contributory to adrenal-related hypercortisolemia. PDE11A (and possibly PDE8B) mutations contribute to a variety of pathologic adrenal lesions. Phosphodiesterases (PDEs) are enzymes involved in the hydrolysis of cAMP and defects in these genes lead to increased levels of cAMP and aberrant PKA signaling (Fig. 3.1.1.38 left panel). Amplifications and activating mutations of the PRKACA gene, coding for the C alpha catalytic subunit of PKA, have also been involved in the pathogenesis of adrenal CS, Somatic mutations of β-catenin gene (CTNNB1) explain up to half of the cases where Wnt signaling is found increased (see Figs. 3.1.1.30 and 3.1.1.38 right panel).

In Cushings disease pituitary surgery has been the treatment of choice but medical treatments are increasingly under investigation depending on findings of genetic tests. TP53 mutations are found in 33% of USP8-negative tumors. Thep53 protein is likely important for regulation of apoptosis (one of the main pathways affected in corticotropinomas) or the BRCA1 mediated DNA repair in corticotroph cells. Two independent groups in 2015 independently reported somatic variants in the USP8 gene in corticotropinomas. BRAF and USP48 mutations activate signaling pathways that lead to the production of proopiomelanocortin (POMC), the precursor of ACTH.

In case of FIPA, screening for AIP mutations should be considered, as mutations are identified in about 10% of unselected families and 20% of those with familial acromegaly. CABLES1 was recently identified as a rare cause of corticotropinomas.

The combination of pituitary adenomas, pheochromocytomas (PHEO) and/or paragangliomas (PGL) in members of the same family (three P association or 3PA). Most cases are caused by succinate dehydrogenase (SDH)-related genes (SDHA, SDHAF2, SDHB, SDHC, SDHD), while additional genes have been more recently reported (VHL, MEN1, RET, and MAX). Succinate dehydrogenase (SDH) is part of the complex II of the mitochondria involved in energy production and the respiratory chain (see Fig. 3.1.1.33).

Food-dependent CS is a rare variant of ACTH-independent CS, where cortisol secretion is stimulated by the postprandial state, most probably by the effect of glucose-dependent insulinotropic polypeptide (GIP). Somatic duplications in chromosome 19q13.3 containing GIPR gene in three patients with food-dependent CS.

3.1.1.6.1.1 Nelson's syndrome

After surgical removal of both adrenal glands, an ACTH producing tumor can develop (Reincke et al., 2021; Papakokkinou et al., 2021; Fountas and Karvitaki, 2020). This can grow upwards and compress the optic nerves, and outwards to reduce the secretion of other peptide hormones (GH, TSH, FSH, and LH). The high concentrations of ACTH act on cells in the skin to increase pigmentation, this defines Nelson's syndrome. The residual tumor can be treated medically or surgically (Chan et al., 2007a,b; Bansal et al., 2017).

3.1.1.6.1.2 Medical treatments

The blockade of cortisol synthesis is usually the first line of treatment because they are effective (Favero et al., 2021; Daniel and Newell-Price, 2015). Ketoconazole (400–1600 mg/day orally), osilodrostat (4–14 mg per day orally) or metyrapone (500 mg to 6 g per day, orally) are preferred for mild hypercortisolism (Daniel and Newell-Price, 2015; Weber et al., 1993; Figs. 3.1.1.37–3.1.1.40).

Cabergoline is in use but is less effective (Barbot et al., 2014). Osilodrostat and metyrapone are 11-hydroxylase inhibitors that act quickly but there is the risk of adrenal insufficiency. If these drugs are not effective than bilateral adrenalectomy should be considered. Some centers use a combination of ketoconazole and metyrapone. Metyrapone blocks the action of 11-hydroxylase and reduces cortisol production (Daniel et al., 2015). The blood concentrations of 11-deoxycortisol can be used to monitor the drug effects. There can be excess androgen production. Metyrapone is not approved in the United States. Ketoconazole is an antifungal agent that is another inhibitor of steroidogenesis (Daniel and Newell-Price, 2015). There are side effects including liver toxicity. Etomidate blocks 11-hydroxylase and lowers serum cortisol concentrations (Carroll et al., 2018).

Mifepristone (300–1200 mg per day orally) is a potent glucocorticoid receptor antagonist, approved for the treatment of hyperglycemia (Brown et al., 2020; Debono et al., 2013). Patients suffer nausea, headaches, peripheral edema and hypokalemia so tolerance is a distinct problem. There is no marker for the treatment. Several agents can be used to block steroidogenesis. Mitotane can be used alone or in combination with other drugs to block steroid synthesis (Kamenický et al., 2011). Mitotane (500 mg to 4 g per day orally) increases the clearance of cortisol probably by induction of CYP3A4 which

Genes involved in adrenal related Cushing syndrome

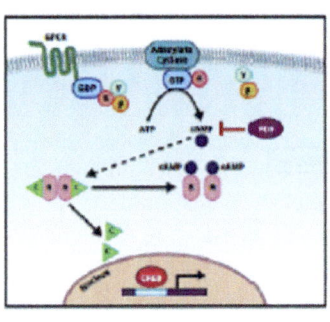

PRKAR1A
PDE11A
PDE8A
PRKACA

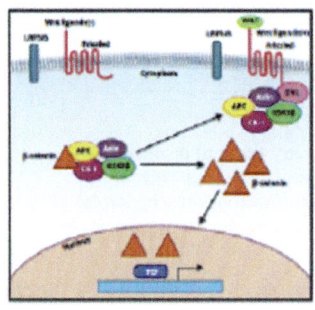

CTNBB1
WNT

MEN1   **and others**
FH
GNAS

FIG. 3.1.1.37 Summary of genes involved in Cushings syndrome. *(Reproduced from Tatsi C., Flippo C., Stratakis C.A. Cushing syndrome: old and new genes. Best Pract Res Clin Endocrinol Metab 2020;34(2):101418.)*

Genes involved in Cushing disease

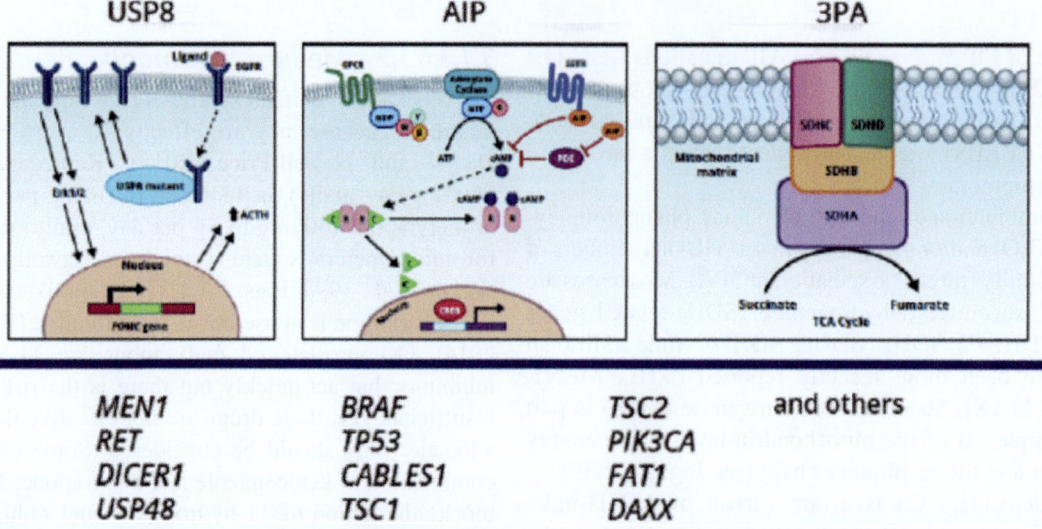

USP8   AIP   3PA

MEN1   BRAF   TSC2   and others
RET   TP53   PIK3CA
DICER1   CABLES1   FAT1
USP48   TSC1   DAXX

FIG. 3.1.1.38 Summary of genes in Cushings disease. *(Reproduced from Tatsi C., Flippo C., Stratakis C.A. Cushing syndrome: old and new genes. Best Pract Res Clin Endocrinol Metab 2020;34(2):101418.)*

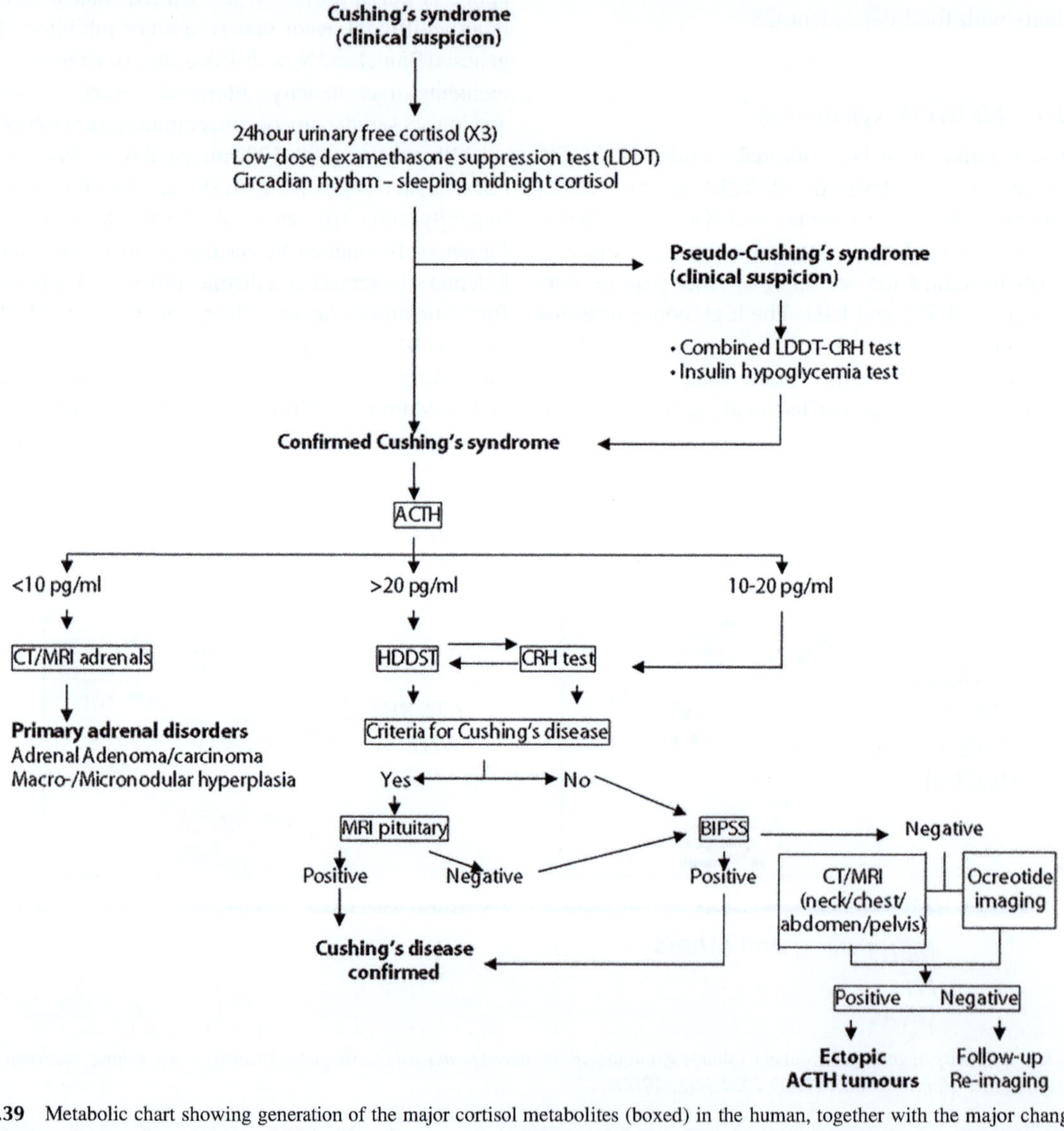

Cushing's syndrome (clinical suspicion)

24 hour urinary free cortisol (X3)
Low-dose dexamethasone suppression test (LDDT)
Circadian rhythm – sleeping midnight cortisol

Pseudo-Cushing's syndrome (clinical suspicion)

• Combined LDDT-CRH test
• Insulin hypoglycemia test

Confirmed Cushing's syndrome

ACTH

<10 pg/ml   >20 pg/ml   10-20 pg/ml

CT/MRI adrenals   HDDST ← CRH test

Primary adrenal disorders
Adrenal Adenoma/carcinoma
Macro-/Micronodular hyperplasia

Criteria for Cushing's disease

Yes   No

MRI pituitary   BIPSS ──→ Negative

Positive   Negative   Positive

                              CT/MRI (neck/chest/abdomen/pelvis)   Ocreotide imaging

Cushing's disease confirmed

Positive   Negative

Ectopic ACTH tumours   Follow-up Re-imaging

FIG. 3.1.1.39 Metabolic chart showing generation of the major cortisol metabolites (boxed) in the human, together with the major changes brought about by mitotane. Crosses show pathways inhibited by mitotane and upward arrows show pathways stimulated by mitotane. Names shaded white are for steroids that show decreases after mitotane. *(From Ghataore L, Chakraborti I, Aylwin SJ, Schulte KM, Dworakowska D, Coskeran P, Taylor NF. Effects of mitotane treatment on human steroid metabolism: implications for patient management. Endocr Connect 2012;1(1):37–47. Fig. 1 p. 38.)*

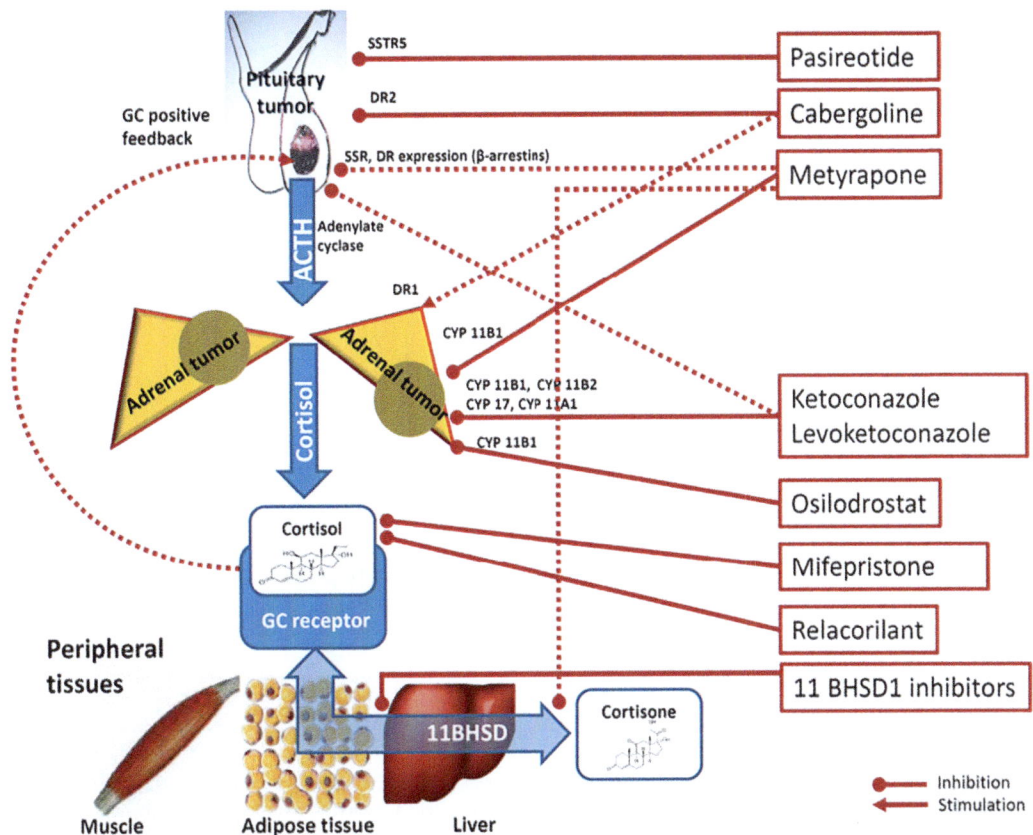

FIG. 3.1.1.40 Mechanisms underlying the effects of medical therapy of hypercortisolism. Footnotes: Cabergoline inhibit pituitary corticotrophs via dopamine receptor type 2 (DR2). Cabergoline is also thought to exert a stimulatory role on cortisol secretion on adrenal cortex cells via DR1 dopamine receptor type 2. Pasireotide inhibits pituitary corticotrophs via somatostatine receptors type 5 (SSTR5). Metyrapone inhibits adrenal 11β-hydroxylase and, to a lesser degree, 18-hydroxylase (CYP11B1). Metyrapone is also hypothesized to reduce at the peripheral target tissues (i.e., muscle, adipocytes and liver) the conversion of cortisone into the more active cortisol via modulation of 11beta-hydroxysteroid-dehydrogenase (11BHSD) and to inhibit pituitary corticotrophs via reduction of GC-driven positive-feedback and via SSTR5 and DR2 receptors expression via the modulation of β-arrestin 1 and β-arrestin 2 expression. Ketoconazole and levoketoconazole act on the adrenal steroidogenesis via 11-βhydroxylase (CYP11B1) inhibition, 18-hydroxylase (CYP11B2) inhibition, 20,22-desmolase (CYP11A1) inhibition and on 17a-hydroxylase and 17,20-desmolase (CYP17) inhibition. Ketoconazole and levo-ketoconazole are also hypothesized to inhibit pituitary corticotroph inhibition by impairing adenylate cyclase activation. Osilodrostat is a steroidogenesis inhibitor acting on CYP11B1) via 11-βhydroxylase inhibition. Mifepristone inhibits the peripheral effects of glucocorticoids (GC) by nonselectively antagonizing the GC receptor. Relacorilant is a selective inhibitor of GC receptor. Finally, some 11 beta-hydroxysteroid dehydrogenase (11BHSD) type 1 (11BHSD1) inhibitors (for example INCB13739, S-707106 and chenodeoxycholic acid) have been suggested to decrease cortisone-to-cortisol conversion. Dotted lines are used for not clearly demonstrated pathways. *(From Favero V, Cremaschi A, Falchetti A, Gaudio A, Gennari L, Scillitani A, Vescini F, Morelli V, Aresta C, Chiodini I. Management and medical therapy of mild hypercortisolism. Int J Mol Sci. 2021;22(21):11521. Fig. 1 p. 5.)*

explains the increased urinary output of 6-hydroxylated cortisol metabolites (Ghataore et al., 2012). The drug is cytotoxic. The effects can be monitored with urinary steroid analysis (Ghataore et al., 2012) (Fig. 3.1.1.41). In patients with recurrence after adrenal surgery, there was an increase in 6-hydroxylated steroids and reduction of 5-alpha reduced steroids with mitotane.

Inhibitors of HSD11B1 are under investigation as part of the treatment of Cushing's disease. Cortisol needs to be converted to cortisone by the action of HSD11B2 in order to prevent action on the MR. The aim of inhibiting HSD11B1 is then to block recycling of cortisone to cortisol. Studies in mouse HSD11B1 knockouts have demonstrated proof of concept (Morgan et al., 2014). So far investigations of HSD11B1 inhibition in man have been shown to modestly improve glucose control, insulin sensitivity and lipid profiles (Fig. 3.1.1.42).

Cabergoline (0.5–7 mg per week orally) inhibits pituitary corticotrophs through dopamine receptor type 2 (DR2), whereas pasireotide (0.6–1.8 mg/mL subcutaneous twice a day) acts via somatostatin receptors type 5 (SSTR5).

Knowledge of the outcome of MEN signaling has influenced new treatments (Brandi et al., 2021) including receptor tyrosine kinase (RTKs) inhibitors, novel mechanistic target of rapamycin (mTOR) inhibitors, β-catenin antagonists, epigenetic modulators, and thrombospondin analogues.

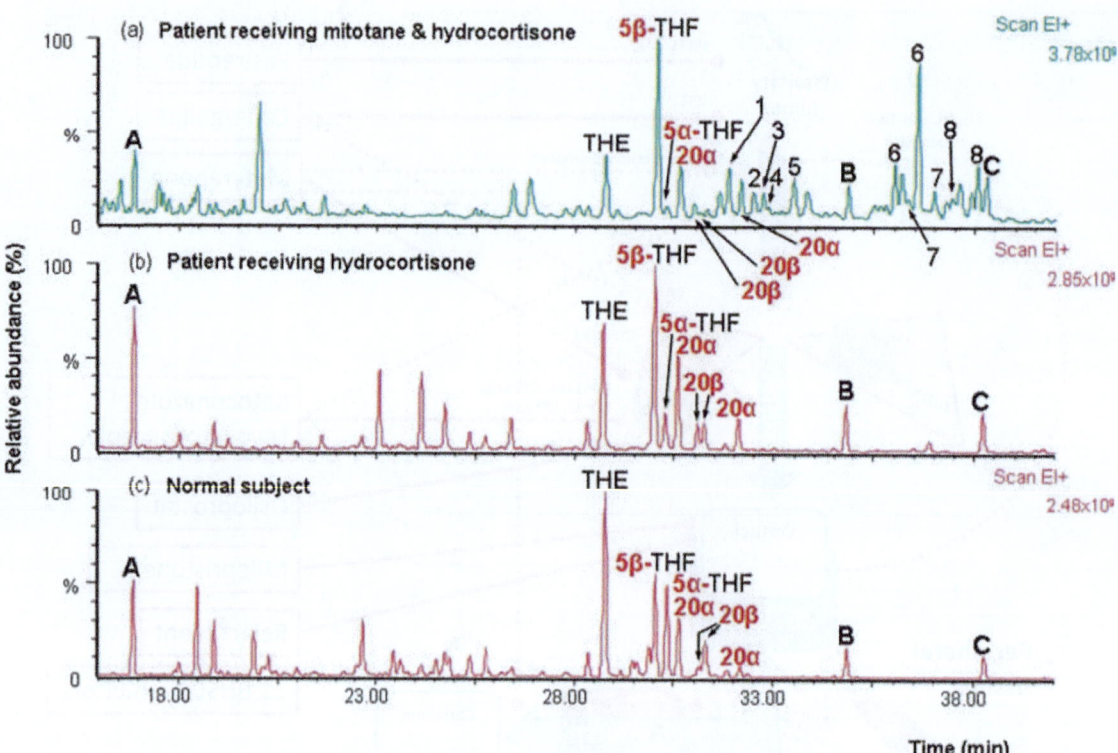

FIG. 3.1.1.41 Total ion current chromatograms from GC–MS analysis of urinary steroid metabolites excreted (A) by a patient after surgery when receiving mitotane and hydrocortisone, (B) the same patient on a previous occasion when receiving hydrocortisone only and (C) a normal subject. A, B, C, internal standards: androstanediol, stigmasterol and cholesterol butyrate; THE, tetrahydrocortisone; THF, tetrahydrocortisol (5α- and 5β-reduced epimers shown); 20α- and 20β-, 20-reduced steroids in the order α-cortolone, β-cortol, β-cortolone, α-cortol. Numbered peaks (repeat numbers show syn/antipairs of the methyl oxime derived from a single steroid) are 1, 6α-hydroxytetrahydrocortisol; 2, 6α-hydroxy-α-cortolone; 3, 1β-hydroxy-α-cortolone; 4, 6α-hydroxy-β-cortolone; 5, 1β-hydroxy-bcortolone; 6, 6β-hydroxycortisol; 7, 6β-hydroxy-20β-dihydrocortisol and 8, 6β-hydroxy-20α-dihydrocortisol. The steroids marked were identified by comparison of mass spectra and retention times with those of authentic standards. *(From Ghataore L, Chakraborti I, Aylwin SJ, Schulte KM, Dworakowska D, Coskeran P, Taylor NF. Effects of mitotane treatment on human steroid metabolism: implications for patient management. Endocr Connect 2012;1(1):37–47. Fig. 2 p. 41.)*

MEN1 gene replacement is in preclinical studies for efficacy in MEN1 patients, and somatostatin (SST) analogues may have chemopreventative efficacy. A USP8 inhibitor is a novel strategy to target USP8 mutated tumors (Treppiedi et al., 2021).

3.1.1.6.2 Polycystic ovary syndrome

This condition is common among women of reproductive age. PCOS is generally diagnosed from the presence of two of three key features from hyperandrogenism, anovulation and presence of polycystic ovaries. Androgen excess in women with polycystic ovary syndrome (PCOS) may be ovarian and/or adrenal in origin, and one proposed contributing mechanism is altered cortisol metabolism (see Chapter on Hyperandrogenism). Increased peripheral metabolism of cortisol may occur by enhanced inactivation of cortisol by 5α-reductase (5α-R) or impaired reactivation of cortisol from cortisone by 11β-hydroxysteroid dehydrogenase type 1 (11β-HSD1) resulting in decreased negative

feedback suppression of ACTH secretion maintaining normal plasma cortisol concentrations at the expense of androgen excess. Urinary excretion of steroid metabolites by gas chromatography/mass spectrometry and fasting metabolic and hormone profiles. Urinary excretion of androgens, and C19 steroid sulfates, cortisone metabolites and total cortisol metabolites (TCM) are raised in lean PCOS subjects when compared with controls (Tsilchorozidou et al., 2003). Changes in 5α-R, 11β-HSD1, and 20α/β-HSD enzyme activities observed in PCOS may contribute to the increased production rates of cortisol and androgens, supporting the concept of a widespread dysregulation of steroid metabolism in a pseudo-Cushings state.

3.1.1.6.3 Pregnancy

The diagnosis of hypercortisolism in pregnancy would be rarely suspected because symptoms such as hypertension and striae are normal features. Striae in pregnancy are white but pink in CS. The condition before the pregnancy is

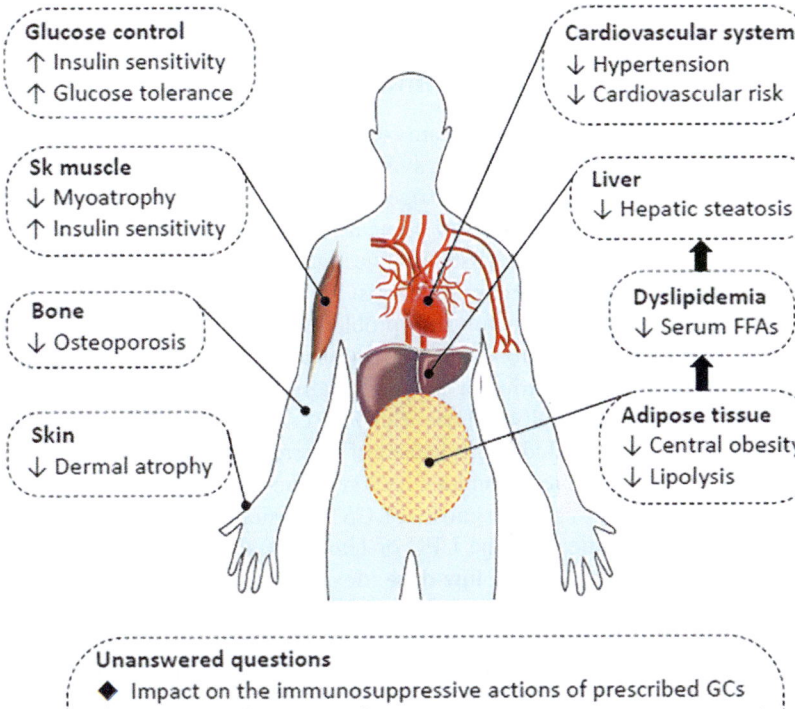

Potential Benefits of Selectively Targeting 11β-HSD1 for the Treatment of Cushing's Syndrome/disease

Glucose control
↑ Insulin sensitivity
↑ Glucose tolerance

Sk muscle
↓ Myoatrophy
↑ Insulin sensitivity

Bone
↓ Osteoporosis

Skin
↓ Dermal atrophy

Cardiovascular system
↓ Hypertension
↓ Cardiovascular risk

Liver
↓ Hepatic steatosis

Dyslipidemia
↓ Serum FFAs

Adipose tissue
↓ Central obesity
↓ Lipolysis

Unanswered questions
◆ Impact on the immunosuppressive actions of prescribed GCs
◆ Impact on the acute inflammatory response
◆ Long-term consequences of suppressed local GC metabolism

FIG. 3.1.1.42 The therapeutic potential of using a selective 11β-HSD1 inhibitor to abrogate the systemic metabolic complications associated with circulatory GC excess in both exogenous and endogenous Cushing's syndrome. *FFA*, free fatty acid. *(From Wamil M, Seckl JR. Inhibition of 11beta-hydroxysteroid dehydrogenase type 1 as a promising therapeutic target. Drug Discov Today 2007;12(13–14):504–520. Fig 3 p 516.)*

unlikely because ovarian function is compromised with CS particularly when androgen production is raised. If suspected, a diagnosis is desirable because the high cortisol concentrations can lead to diabetes mellitus, preeclampsia, infections and preterm birth (meta-analysis Caimari et al., 2017). Increased estrogen increases CBG glycosylation preventing CBG cleavage and increasing concentrations 2- or 3-fold (Nenke et al., 2017). Total cortisol is normally raised in pregnancy, a 24h urinary-free cortisol or late-night salivary cortisol (Lopes et al., 2016) are the best measurements for demonstrating CS in pregnancy so the elevation with CS would have to be significant (2–3 x normal) for reliable detection (Nieman et al., 2008). The placenta secretes increasing amounts of CRF which is another confounder (Linton et al., 1993). Abdominal ultrasound and MRI can be used without contrast in pregnancy; small adenomas in the pituitary may not be detected because of poor discrimination from normal tissue. The use of contrast with MRI is not acceptable and IPSS would involve excessive radiation exposure. Pregnancy induced Cushing's syndrome can be due to hCG-stimulated transformation of LHCGR (Plöckinger et al., 2017).

A multidisciplinary team will need to decide on the significance of a number of issues including the effects from, hypercortisolism, the risks to mother and unborn child, the likelihood of malignancy, medical or surgical intervention and other questions. Metyrapone, ketoconazole and cabergoline can be used in pregnancy but mitotane is teratogenic (Pivonello et al., 2015). Surgery would lead to the need for replacement therapy which itself is not without problems (see Chapter 3.1.2 Adrenal Insufficiency).

3.1.1.6.4 Exogenous corticosteroids and herbal medicines

The commonest cause of Cushing's syndrome is the use of glucocorticoids particularly high potency ones for the management of inflammatory conditions such as arthritis and for the prevention of organ transplant rejection. These glucocorticoids can be taken by the oral, topical, rectal, intraarticular and nasal routes. These steroids are prescribed for the treatment of eczema, asthma, psoriasis and inflammation and have variable potency. A few cases have been reported of iatrogenic Cushing's syndrome from corticosteroid use on infants for diaper rash (Alkhuder and Mawlawi, 2019). A self-reported, high incidence of glucocorticoid use was found in patients attending an obesity clinic (Savas et al., 2017). Increasingly glucocorticoids are used as skin blanching agents. Dexamethasone and Clobetasol propionate are the most frequently used glucocorticoids. Herbal medicines have on chemical analysis been often found to contain unlisted corticosteroids (Ching et al., 2018; Chong et al., 2015; DiNicola et al., 2013). There is a risk of infections when taking such products and adrenal insufficiency on withdrawal after long term adrenal suppression (Kempegowda et al., 2019). The steroids can be detected after solvent extraction of the products (Franke et al., 2017; Giaccone et al., 2017; Golubović et al., 2015) or analysis of samples from the patients (Taylor et al., 2004). A low excretion rate of urinary-free cortisol may be an indicator of adrenal suppression during use of these glucocorticoids (van der Valk et al., 2019).

3.1.1.6.4.1   Other effects of high doses of GCH

When excessive GCH levels result from high doses of exogenous hydrocortisone there are effects on many systems.

- hyperglycemia and insulin resistance.
- reduction in the synthesis and production of collagen, mucopolysaccharides and muscle protein synthesis in skin and connective tissue.
- GCH stimulate the deposition of visceral or central adipose tissue.
- inhibition of osteoblast function leading to low bone density, osteopenia and more severe osteoporosis (Sharma et al., 2019).
- inhibition of intestinal calcium absorption and increased renal calcium excretion leading to an increase in parathyroid hormone secretion. There is inhibition of bone formation, suppression of calcium absorption and delayed wound healing (Hachemi et al., 2018; Hartmann et al., 2016).
- inhibition of insulin like growth factor (IGF-1) in children can lead to suppression of linear growth.

An understanding of the role of GCH in glucose homeostasis, protein and lipid metabolism will help treatment of metabolic disease and the development of pharmacotherapy. A reduction of GCH will elevate the body's inflammatory response. The antagonism of GCH to insulin is critical during stress to maintain glucose concentrations in the brain. GCH suppress uptake and oxidation of glucose in hypothalamus and hippocampus. The ability of GCH to inhibit glucose oxidation has been linked to apoptosis in leukemia cells. GCH excess can cause depression, euphoria, psychosis, apathy and lethargy. Some aspects of cognitive function are known to both stimulate GCH

secretion and be influenced by GCH (Wheelan et al., 2018). Fear-inducing stimuli lead to secretion of GCH from the adrenal gland, and treatment of phobic individuals with GCH prior to a fear-inducing stimulus can blunt the fear response.

3.1.1.7   Summary

Signs and symptoms of CS overlap with common diseases such as metabolic syndrome, obesity and depression or can be the result of exogenous corticosteroids. Patients present with a variety of clinical features and get referred to different specialists for investigations—an orthopedic surgeon for fracture, gynecologist for menstrual problems, and dermatologist for skin problems and psychiatrist for mental health problems. The measurement of cortisol excretion rate in urine or late night salivary cortisol has satisfactory sensitivity and specificity as screening tests. Clinicians need to exclude other causes of hypercortisolism prior to provocative tests and localization studies (Pappachan et al., 2017). An early diagnosis of CS is desirable to prevent complications. If the UFC or late night saliva cortisol is clearly elevated, a **low-dose dexamethasone suppression test (LDDS)** is used to identify patients requiring further investigation. In normal subjects, dexamethasone (1 mg orally at 2300h) will suppress serum cortisol levels to below 100 nmol/L at 0800 to 0900h the following morning. Normal suppression effectively excludes Cushing's syndrome (i.e., few false negatives). This version of the LDDS test is more convenient than the classical test which involves taking 0.5 mg of dexamethasone every 6 h for 2 days with a 24 h urine collection being made through the second day of treatment. Drugs which increase the clearance rate of dexamethasone (e.g., anticonvulsants) can result in false positives but the simultaneous measurement of cortisol and dexamethasone is not required routinely.

In a patient with high clinical suspicion of Cushing's syndrome, but normal urine UFC excretion rate on the first occasion, repeated 24 h urine collections over many weeks may be necessary since cyclical forms of the syndrome (due usually to Cushing's disease) occur occasionally.

The diagnosis of pituitary dependent Cushing's disease is confirmed when previously high cortisol production is substantially suppressed with **high doses of dexamethasone** (2 mg 6 hourly for 2 days). Plasma cortisol, ACTH and urine-free cortisol (or better of total cortisol metabolites) should all fall to less than 50% of their basal values in the majority of affected patients. In a meta-analysis of three tests (UFC, dexamethasone suppression and midnight cortisol alone and in combination) the suppression test was most effective in the diagnosis of Cushing's syndrome (Casals and Hanzu, 2020; Galm et al., 2020). Up to 10% of patients with **ectopic ACTH secreting tumors** can however suppress with the high dose of dexamethasone. The plasma concentrations of potassium and ACTH should be checked since hypokalemia and very high ACTH values by RIA are characteristic of the ectopic tumor. Other ectopic tumors such as neuroendocrine **carcinoid** may not always be associated with hypokalemia and ACTH may be only moderately raised in early stages of slow tumor growth rate. These cases are rare in children. The adrenals are enormous on CT and cortisol production is usually grossly elevated. A patient with an ectopic ACTH tumor may present with rapid onset weight loss, edema and muscle weakness (Himsworth et al., 1977) which is not the classical clinical picture of Cushing's syndrome. The diagnosis of Cushing's syndrome is difficult even for experienced endocrinologists often taking many months (Rubinstein et al., 2020). There is much variation between countries in the tests used (Valassi et al., 2017) especially when obesity is the presenting problem (van der Valk et al., 2019). There has been an effort to agree to a consensus for diagnosis and management of Cushing's disease (Fleseriu et al., 2021).

The genetic bases of pituitary and adrenal tumors has received much attention in the past two decades and many associations have been found (Kamilaris et al., 2021; Tatsi et al., 2020). There is much overlap in a gene mutation and clinical pictures. The classification of these disorders will need revision.

Over several decades, great efforts have been made in discovering novel and reliable diagnostic and prognostic biomarkers including microRNAs, steroid profiling, circulating tumor cells, circulating tumor DNAs and radiomics. Tumor registries enable diagnostic and treatment options to be investigated for low incidence states such as adrenocortical carcinoma (Crona et al., 2021).

3.1.1.8 Further reading

The reader may find other useful references not cited in the chapter.

3.1.1.7.1 Further tests

Barbot et al. (2016), Ceccato et al. (2014), Ceccato et al. (2015a,b), Ceccato et al. (2017a,b), Ceccato et al. (2018), Ceccato et al. (2020a,b,c), Fanelli and Di Dalmazi (2019), Invitti et al. (1999), Louiset et al. (2010), Martinez de la Piscina et al. (2021), Raff et al. (2014), and Stratakis (2012).

3.1.1.7.2 Disease state

Räikkönen et al. (2020) and Ceccato et al. (2020a).

3.1.1.7.3 Post surgery

Braun et al. (2020).

3.1.1.7.4 Hair

Brossaud et al. (2021).

References

Albani A, Berr CM, Beuschlein F, Treitl M, Hallfeldt K, Honegger J, et al. A pitfall of bilateral inferior petrosal sinus sampling in cyclic Cushing's syndrome. BMC Endocr Disord 2019;19(1):105.

Albiger NM, Regazzo D, Rubin B, Ferrara AM, Rizzati S, Taschin E, et al. A multicenter experience on the prevalence of ARMC5 mutations in patients with primary bilateral macronodular adrenal hyperplasia: from genetic characterization to clinical phenotype. Endocrine 2017; 55(3):959–68.

Alexandraki KI, Kaltsas GA, Isidori AM, Akker SA, Drake WM, Chew SL, et al. The prevalence and characteristic features of cyclicity and variability in Cushing's disease. Eur J Endocrinol 2009;160(6):1011–8.

Alkhuder L, Mawlawi H. Infantile iatrogenic Cushing syndrome due to topical steroids. Case Rep Pediatr 2019;2019, 2652961.

Almeida MQ, Stratakis CA. How does cAMP/protein kinase A signaling lead to tumors in the adrenal cortex and other tissues? Mol Cell Endocrinol 2011;336(1–2):162–8.

Alwani RA, Schmit Jongbloed LW, de Jong FH, van der Lely AJ, de Herder WW, Feelders RA. Differentiating between Cushing's disease and pseudo-Cushing's syndrome: comparison of four tests. Eur J Endocrinol 2014;170(4):477–86.

Antonelli G, Ceccato F, Artusi C, Marinova M, Plebani M. Salivary cortisol and cortisone by LC-MS/MS: validation, reference intervals and diagnostic accuracy in Cushing's syndrome. Clin Chim Acta 2015;451 (Pt B):247–51.

Araujo-Castro M, Parra Ramírez P, Robles Lázaro C, García Centeno R, Gracia Gimeno P, Fernández-Ladreda MT, et al. Accuracy of the dexamethasone suppression test for the prediction of autonomous cortisol secretion-related comorbidities in adrenal incidentalomas. Hormones (Athens) 2021;20(4):735–44.

Arlt W, Biehl M, Taylor AE, Hahner S, Libé R, Hughes BA, et al. Urine steroid metabolomics as a biomarker tool for detecting malignancy in adrenal tumors. J Clin Endocrinol Metab 2011;96 (12):3775–84.

Arnaldi G, Tirabassi G, Papa R, Furlani G, Trementino L, Cardinaletti M, et al. Human corticotropin releasing hormone test performance in the differential diagnosis between Cushing's disease and pseudo-Cushing state is enhanced by combined ACTH and cortisol analysis. Eur J Endocrinol 2009;160(6):891–8.

Assié G, Libé R, Espiard S, Rizk-Rabin M, Guimier A, Luscap W, et al. ARMC5 mutations in macronodular adrenal hyperplasia with Cushing's syndrome. N Engl J Med 2013;369(22):2105–14.

Athimulam S, Grebe S, Bancos I. Steroid profiling in the diagnosis of mild and overt Cushing's syndrome. Best Pract Res Clin Endocrinol Metab 2021;35(1), 101488.

Atkinson B, Mullan KR. What is the best approach to suspected cyclical Cushing syndrome? Strategies for managing Cushing's syndrome with variable laboratory data. Clin Endocrinol (Oxf) 2011;75 (1):27–30.

Bäcklund N, Brattsand G, Israelsson M, Ragnarsson O, Burman P, Edén Engström B, et al. Reference intervals of salivary cortisol and cortisone and their diagnostic accuracy in Cushing's syndrome. Eur J Endocrinol 2020;182(6):569–82.

Badrick E, Kirschbaum C, Kumari M. The relationship between smoking status and cortisol secretion. J Clin Endocrinol Metab 2007;92(3): 819–24.

Bae YJ, Gaudl A, Jaeger S, Stadelmann S, Hiemisch A, Kiess W, et al. Immunoassay or LC-MS/MS for the measurement of salivary cortisol in children? Clin Chem Lab Med 2016;54(5):811–22.

Baid SK, Sinaii N, Wade M, Rubino D, Nieman LK. Radioimmunoassay and tandem mass spectrometry measurement of bedtime salivary cortisol levels: a comparison of assays to establish hypercortisolism. J Clin Endocrinol Metab 2007;92(8):3102–7.

Baldeweg SE, Pollock JR, Powell M, Ahlquist J. A spectrum of behaviour in silent corticotroph pituitary adenomas. Br J Neurosurg 2005;19 (1):38–42.

Ballmann C, Thiel A, Korah HE, Reis AC, Saeger W, Stepanow S, et al. *USP8* mutations in pituitary Cushing adenomas-targeted analysis by next-generation sequencing. J Endocr Soc 2018;2(3):266–78.

Bancos I, Taylor AE, Chortis V, Sitch AJ, Jenkinson C, Davidge-Pitts CJ, et al. Urine steroid metabolomics for the differential diagnosis of adrenal incidentalomas in the EURINE-ACT study: a prospective test validation study. Lancet Diabetes Endocrinol 2020;8(9):773–81.

Bansal V, El Asmar N, Selman WR, Arafah BM. Pitfalls in the diagnosis and management of Cushing's syndrome. Neurosurg Focus 2015;38(2):E4.

Bansal P, Lila A, Goroshi M, Jadhav S, Lomte N, Thakkar K, et al. Duration of post-operative hypocortisolism predicts sustained remission after pituitary surgery for Cushing's disease. Endocr Connect 2017;6 (8):625–36.

Barbot M, Albiger N, Koutroumpi S, Ceccato F, Frigo AC, Manara R, et al. Predicting late recurrence in surgically treated patients with Cushing's disease. Clin Endocrinol (Oxf) 2013;79(3):394–401.

Barbot M, Albiger N, Ceccato F, Zilio M, Frigo AC, Denaro L, et al. Combination therapy for Cushing's disease: effectiveness of two schedules of treatment: should we start with cabergoline or ketoconazole? Pituitary 2014;17(2):109–17.

Barbot M, Trementino L, Zilio M, Ceccato F, Albiger N, Daniele A, et al. Second-line tests in the differential diagnosis of ACTH-dependent Cushing's syndrome. Pituitary 2016;19(5):488–95.

Batista DL, Courcoutsakis N, Riar J, Keil MF, Stratakis CA. Severe obesity confounds the interpretation of low-dose dexamethasone test combined with the administration of ovine corticotrophin-releasing hormone in childhood Cushing syndrome. J Clin Endocrinol Metab 2008;93 (11):4323–30.

Baudrand R, Campino C, Carvajal CA, Olivieri O, Guidi G, Faccini G, et al. High sodium intake is associated with increased glucocorticoid production, insulin resistance and metabolic syndrome. Clin Endocrinol (Oxf) 2014;80(5):677–84.

Beckers A, Aaltonen LA, Daly AF, Karhu A. Familial isolated pituitary adenomas (FIPA) and the pituitary adenoma predisposition due to mutations in the aryl hydrocarbon receptor interacting protein (AIP) gene. Endocr Rev 2013;34(2):239–77.

Berlińska A, Świątkowska-Stodulska R, Sworczak K. Factors affecting dexamethasone suppression test results. Exp Clin Endocrinol Diabetes 2020;128(10):667–71.

Berthon A, Bertherat J. Update of genetic and molecular causes of adrenocortical hyperplasias causing Cushing syndrome. Horm Metab Res 2020;52(8):598–606.

Berthon A, Faucz F, Bertherat J, Stratakis CA. Analysis of ARMC5 expression in human tissues. Mol Cell Endocrinol 2017;441:140–5.

Bileck A, Fluck CE, Dhayat N, Groessl M. How high-resolution techniques enable reliable steroid identification and quantification. J Steroid Biochem Mol Biol 2019;186:74–8.

Bileck A, Frei S, Vogt B, Groessl M. Urinary steroid profiles: comparison of spot and 24-hour collections. J Steroid Biochem Mol Biol 2020;200, 105662.

Brandi ML, Agarwal SK, Perrier ND, Lines KE, Valk GD, Thakker RV. Multiple endocrine neoplasia type 1: latest insights. Endocr Rev 2021;42(2):133–70.

Braun LT, Rubinstein G, Zopp S, Vogel F, Schmid-Tannwald C, Escudero MP, et al. Recurrence after pituitary surgery in adult Cushing's disease: a systematic review on diagnosis and treatment. Endocrine 2020;70 (2):218–31.

Brossaud J, Charret L, De Angeli D, Haissaguerre M, Ferriere A, Puerto M, et al. Hair cortisol and cortisone measurements for the diagnosis of overt and mild Cushing's syndrome. Eur J Endocrinol 2021;184 (3):445–54.

Brown DR, East HE, Eilerman BS, Gordon MB, King EE, Knecht LA, et al. Clinical management of patients with Cushing syndrome treated with mifepristone: consensus recommendations. Clin Diabetes Endocrinol 2020;6(1):18.

Caimari F, Valassi E, Garbayo P, Steffensen C, Santos A, Corcoy R, et al. Cushing's syndrome and pregnancy outcomes: a systematic review of published cases. Endocrine 2017;55(2):555–63.

Carroll T, Raff H, Findling JW. Late-night salivary cortisol for the diagnosis of Cushing syndrome: a meta-analysis. Endocr Pract 2009; 15(4):335–42.

Carroll TB, Peppard WJ, Herrmann DJ, Javorsky BR, Wang TS, Patel H, et al. Continuous etomidate infusion for the management of severe Cushing Syndrome: validation of a standard protocol. J Endocr Soc 2018;3(1):1–12.

Casals G, Hanzu FA. Cortisol measurements in Cushing's syndrome: immunoassay or mass spectrometry? Ann Lab Med 2020;40(4):285–96.

Casals G, Foj L, de Osaba MJ. Day-to-day variation of late-night salivary cortisol in healthy voluntaries. Clin Biochem 2011;44(8–9):665–8.

Castro M, Elias PC, Quidute AR, Halah FP, Moreira AC. Out-patient screening for Cushing's syndrome: the sensitivity of the combination of circadian rhythm and overnight dexamethasone suppression salivary cortisol tests. J Clin Endocrinol Metab 1999;84(3):878–82.

Ceccato F, Barbot M, Zilio M, Ferasin S, Occhi G, Daniele A, et al. Performance of salivary cortisol in the diagnosis of Cushing's syndrome, adrenal incidentaloma, and adrenal insufficiency. Eur J Endocrinol 2013;169(1):31–6.

Ceccato F, Antonelli G, Barbot M, Zilio M, Mazzai L, Gatti R, et al. The diagnostic performance of urinary free cortisol is better than the cortisol:cortisone ratio in detecting de novo Cushing's syndrome: the use of a LC-MS/MS method in routine clinical practice. Eur J Endocrinol 2014;171(1):1–7.

Ceccato F, Barbot M, Zilio M, Ferasin S, De Lazzari P, Lizzul L, et al. Age and the metabolic syndrome affect salivary cortisol rhythm: data from a community sample. Hormones (Athens) 2015a;14(3):392–8.

Ceccato F, Barbot M, Zilio M, Frigo AC, Albiger N, Camozzi V, et al. Screening tests for Cushing's syndrome: urinary free cortisol role measured by LC-MS/MS. J Clin Endocrinol Metab 2015b;100 (10):3856–61.

Ceccato F, Trementino L, Barbot M, Antonelli G, Plebani M, Denaro L, et al. Diagnostic accuracy of increased urinary cortisol/cortisone ratio to differentiate ACTH-dependent Cushing's syndrome. Clin Endocrinol (Oxf) 2017a;87(5):500–7.

Ceccato F, Antonelli G, Frigo AC, Regazzo D, Plebani M, Boscaro M, et al. First-line screening tests for Cushing's syndrome in patients with adrenal incidentaloma: the role of urinary free cortisol measured by LC-MS/MS. J Endocrinol Invest 2017b;40(7):753–60.

Ceccato F, Barbot M, Albiger N, Antonelli G, Zilio M, Todeschini M, Regazzo D, Plebani M, Lacognata C, Iacobone M, Mantero F, Boscaro M, Scaroni C. Daily salivary cortisol and cortisone rhythm in patients with adrenal incidentaloma. Endocrine 2018;59(3):510–19.

Ceccato F, Zilio M, Barbot M, Albiger N, Antonelli G, Plebani M, et al. Metyrapone treatment in Cushing's syndrome: a real-life study. Endocrine 2018b;62(3):701–11.

Ceccato F, Marcelli G, Martino M, Concettoni C, Brugia M, Trementino L, et al. The diagnostic accuracy of increased late night salivary cortisol for Cushing's syndrome: a real-life prospective study. J Endocrinol Invest 2019;42(3):327–35.

Ceccato F, Lizzul L, Barbot M, Scaroni C. Pituitary-adrenal axis and peripheral cortisol metabolism in obese patients. Endocrine 2020a;69(2):386–92.

Ceccato F, Artusi C, Barbot M, Lizzul L, Pinelli S, Costantini G, Niero S, Antonelli G, Plebani M, Scaroni C. Dexamethasone measurement during low-dose suppression test for suspected hypercortisolism: threshold development with and validation. J Endocrinol Invest 2020b;43(8):1105–13.

Ceccato F, Tizianel I, Vedolin CK, Boscaro M, Barbot M, Scaroni C. Human corticotropin-releasing hormone tests: 10 years of real-life experience in pituitary and adrenal disease. J Clin Endocrinol Metab 2020c;105(11):dgaa564.

Chan LF, Storr HL, Grossman AB, Savage MO. Pediatric Cushing's syndrome: clinical features, diagnosis, and treatment. Arq Bras Endocrinol Metabol 2007a;51(8):1261–71.

Charmandari E, Kino T, Ichijo T, Chrousos GP. Generalized glucocorticoid resistance: clinical aspects, molecular mechanisms, and implications of a rare genetic disorder. J Clin Endocrinol Metab 2008;93(5):1563–72.

Chasseloup F, Pankratz N, Lane J, Faucz FR, Keil MF, Chittiboina P, et al. Germline CDKN1B loss-of-function variants cause pediatric Cushing's disease with or without an MEN4 phenotype. J Clin Endocrinol Metab 2020;105(6):1983–2005.

Chasseloup F, Bourdeau I, Tabarin A, Regazzo D, Dumontet C, Ladurelle N, et al. Loss of KDM1A in GIP-dependent primary bilateral macronodular adrenal hyperplasia with Cushing's syndrome: a multicentre, retrospective, cohort study. Lancet Diabetes Endocrinol 2021;9(12):813–24.

Chen J, Jian X, Deng S, Ma Z, Shou X, Shen Y, et al. Identification of recurrent USP48 and BRAF mutations in Cushing's disease. Nat Commun 2018;9(1):3171.

Chen S, Chen K, Lu L, Zhang X, Tong A, Pan H, et al. The effects of sampling lateralization on bilateral inferior petrosal sinus sampling and desmopressin stimulation test for pediatric Cushing's disease. Endocrine 2019;63(3):582–91.

Chen AX, Haas AV, Williams GH, Vaidya A. Dietary sodium intake and cortisol measurements. Clin Endocrinol (Oxf) 2020;93(5):539–45.

Chevalier B, Vantyghem MC, Espiard S. Bilateral adrenal hyperplasia: pathogenesis and treatment. Biomedicines 2021;9(10):1397.

Ching CK, Chen SPL, Lee HHC, Lam YH, Ng SW, Chen ML, et al. Adulteration of proprietary Chinese medicines and health products with undeclared drugs: experience of a tertiary toxicology laboratory in Hong Kong. Br J Clin Pharmacol 2018;84(1):172–8.

Chong YK, Ching CK, Ng SW, Mak TW. Corticosteroid adulteration in proprietary Chinese medicines: a recurring problem. Hong Kong Med J 2015;21(5):411–6.

Chortis V, Bancos I, Nijman T, Gilligan LC, Taylor AE, Ronchi CL, et al. Urine steroid metabolomics as a novel tool for detection of recurrent adrenocortical carcinoma. J Clin Endocrinol Metab 2020;105(3):e307–18.

Choy KW, Teng J, Wijeratne N, Tan CY, Doery JC. Immunoassay interference complicating management of Cushing's disease: the onus is on the clinician and the laboratory. Ann Clin Biochem 2017;54(1):183–4.

Coates PJ, Doniach I, Howlett TA, Rees LH, Besser GM. Immunocytochemical study of 18 tumours causing ectopic Cushing's syndrome. J Clin Pathol 1986;39(9):955–60.

Crona J, Baudin E, Terzolo M, Chrisoulidou A, Angelousi A, Ronchi CL, et al. ENSAT registry-based randomized clinical trials for adrenocortical carcinoma. Eur J Endocrinol 2021;184(2):R51–9.

Cuny T, Pertuit M, Sahnoun-Fathallah M, Daly A, Occhi G, Odou MF, et al. Genetic analysis in young patients with sporadic pituitary macroadenomas: besides AIP don't forget MEN1 genetic analysis. Eur J Endocrinol 2013;168(4):533–41.

Daniel E, Newell-Price JD. Diagnosis of Cushing's disease. Pituitary 2015;18(2):206–10.

Daniel E, Aylwin S, Mustafa O, Ball S, Munir A, Boelaert K, et al. Effectiveness of metyrapone in treating Cushing's syndrome: a retrospective multicenter study in 195 patients. J Clin Endocrinol Metab 2015;100(11):4146–54.

Debeljak Ž, Marković I, Pavela J, Lukić I, Mandić D, Mandić S, et al. Analytical bias of automated immunoassays for six serum steroid hormones assessed by LC-MS/MS. Biochem Med 2020;30(3), 030701.

Debono M, Chadarevian R, Eastell R, Ross RJ, Newell-Price J. Mifepristone reduces insulin resistance in patient volunteers with adrenal incidentalomas that secrete low levels of cortisol: a pilot study. PLoS ONE 2013;8(4), e60984.

Di Dalmazi G, Fanelli F, Zavatta G, Ricci Bitti S, Mezzullo M, Repaci A, et al. The steroid profile of adrenal incidentalomas: subtyping subjects with high cardiovascular risk. J Clin Endocrinol Metab 2019;104(11):5519–28.

DiNicola C, Kekevian A, Chang C. Integrative medicine as adjunct therapy in the treatment of atopic dermatitis—the role of traditional Chinese medicine, dietary supplements, and other modalities. Clin Rev Allergy Immunol 2013;44(3):242–53.

Dom LD, Lucke JF, Loucke TL, Berga SH. Salivary cortisol reflects serum cortisol: analysis of circadian profiles. Ann Clin Biochem 2007;44:281–4.

Draper N, Walker EA, Bujalska IJ, Tomlinson JW, Chalder SM, Arlt W, et al. Mutations in the genes encoding 11beta-hydroxysteroid dehydrogenase type 1 and hexose-6-phosphate dehydrogenase interact to cause cortisone reductase deficiency. Nat Genet 2003;34(4):434–9.

Drury PL, Ratter S, Tomlin S, Williams J, Dacie JE, Rees LH, et al. Experience with selective venous sampling in diagnosis of ACTH-dependent Cushing's syndrome. Br Med J (Clin Res Ed) 1982;284(6308):9–12.

Dumitrescu CE, Collins MT. McCune-Albright syndrome. Orphanet J Rare Dis 2008;3:12.

Eisenhofer G, Peitzsch M, Kaden D, Langton K, Pamporaki C, Masjkur J, et al. Reference intervals for plasma concentrations of adrenal steroids measured by LC-MS/MS: impact of gender, age, oral contraceptives, body mass index and blood pressure status. Clin Chim Acta 2017;470:115–24.

Eisenhofer G, Masjkur J, Peitzsch M, Di Dalmazi G, Bidlingmaier M, Grüber M, et al. Plasma steroid metabolome profiling for diagnosis and subtyping patients with Cushing syndrome. Clin Chem 2018;64 (3):586–96.

El-Farhan N, Rees DA, Evans C. Measuring cortisol in serum, urine and saliva—are our assays good enough? Ann Clin Biochem 2017;54 (3):308–22.

Elamin MB, Murad MH, Mullan R, Erickson D, Harris K, Nadeem S, et al. Accuracy of diagnostic tests for Cushing's syndrome: a systematic review and metaanalyses. J Clin Endocrinol Metab 2008;93(5). 1553–62.

Espiard S, Ragazzon B, Bertherat J. Protein kinase A alterations in adreno-cortical tumors. Horm Metab Res 2014;46(12):869–75.

Fanelli F, Di Dalmazi G. Serum steroid profiling by mass spectrometry in adrenocortical tumors: diagnostic implications. Curr Opin Endocrinol Diabetes Obes 2019;26(3):160–5.

Fassnacht M, Arlt W, Bancos I, Dralle H, Newell-Price J, Sahdev A, et al. Management of adrenal incidentalomas: European society of endocri-nology clinical practice guideline in collaboration with the European network for the study of adrenal tumors. Eur J Endocrinol 2016;175 (2):G1–G34. https://doi.org/10.1530/EJE-16-0467. 27390021.

Faucz FR, Tirosh A, Tatsi C, Berthon A, Hernández-Ramírez LC, Settas N, et al. Somatic USP8 gene mutations are a common cause of pediatric Cushing disease. J Clin Endocrinol Metab 2017;102(8):2836–43.

Favero V, Cremaschi A, Falchetti A, Gaudio A, Gennari L, Scillitani A, et al. Management and medical therapy of mild hypercortisolism. Int J Mol Sci 2021;22(21):11521.

Ferrigno R, Hasenmajer V, Caiulo S, Minnetti M, Mazzotta P, Storr HL, et al. Paediatric Cushing's disease: epidemiology, pathogenesis, clinical management and outcome. Rev Endocr Metab Disord 2021;22(4):817–35.

Fiad TM, Kirby JM, Cunningham SK, McKenna TJ. The overnight single-dose metyrapone test is a simple and reliable index of the hypothalamic-pituitary-adrenal axis. Clin Endocrinol (Oxf) 1994;40 (5):603–9.

Fink RS, Pierre LN, Daley-Yates PT, Richards DH, Gibson A, Honour JW. Hypothalamic-pituitary-adrenal axis function after inhaled corticoste-roids: unreliability of urinary free cortisol estimation. J Clin Endo-crinol Metab 2002;87(10):4541–6.

Fleseriu M, Auchus R, Bancos I, Ben-Shlomo A, Bertherat J, Biermasz NR, et al. Consensus on diagnosis and management of Cushing's disease: a guideline update. Lancet Diabetes Endocrinol 2021;9(12):847–75.

Fountas A, Karavitaki N. Nelson's syndrome: an update. Endocrinol Metab Clin North Am 2020;49(3):413–32.

Franke V, Scholtens WF, von Rosenstiel IA, Walenkamp MJ. Exogenous Cushing's syndrome due to a Chinese herbalist's prescription of ointment containing dexamethasone. BMJ Case Rep 2017;2017, bcr2016218721.

Freda PU, Beckers AM, Katznelson L, Molitch ME, Montori VM, Post KD, et al. Pituitary incidentaloma: an endocrine society clinical practice guideline. J Clin Endocrinol Metab 2011;96(4):894–904.

Galm BP, Qiao N, Klibanski A, Biller BMK, Tritos NA. Accuracy of lab-oratory tests for the diagnosis of Cushing syndrome. J Clin Endocrinol Metab 2020;105(6), dgaa105.

García-Martínez A, Cano DA, Flores-Martínez A, Gil J, Puig-Domingo M, Webb SM, et al. Why don't corticotroph tumors always produce Cushing's disease? Eur J Endocrinol 2019;181(3):351–61.

Genere N, Kaur RJ, Athimulam S, Thomas MA, Nippoldt T, Van Norman M, et al. Interpretation of abnormal dexamethasone suppression test is enhanced with use of synchronous free cortisol assessment. J Clin Endocrinol Metab 2021;dgab724. https://doi.org/10.1210/clinem/dgab724. Epub ahead of print. PMID 34648626.

Ghataore L, Chakraborti I, Aylwin SJ, Schulte KM, Dworakowska D, Coskeran P, et al. Effects of mitotane treatment on human steroid metabolism: implications for patient management. Endocr Connect 2012;1(1):37–47.

Giaccone V, Polizzotto G, Macaluso A, Cammilleri G, Ferrantelli V. Determination of ten corticosteroids in illegal cosmetic products by a simple, rapid, and high-performance LC-MS/MS method. Int J Anal Chem 2017;2017, 3531649.

Glinicki P, Jeske W, Bednarek-Papierska L, Kasperlik-Załuska A, Rosło-nowska E, Gietka-Czernel M, et al. Chromogranin A (CgA) in adrenal tumours. Endokrynol Pol 2013;64(5):358–62.

Golubović JB, Otašević BM, Protić AD, Stanković AM, Zečević ML. Liquid chromatography/tandem mass spectrometry for simultaneous determination of undeclared corticosteroids in cosmetic creams. Rapid Commun Mass Spectrom 2015;29(24):2319–27.

Granberg D. Biochemical testing in patients with neuroendocrine tumors. Front Horm Res 2015;44:24–39.

Greene LW, Geer EB, Page-Wilson G, Findling JW, Raff H. Assay-specific spurious ACTH results lead to misdiagnosis, unnecessary testing, and surgical misadventure—a case series. J Endocr Soc 2019;3(4):763–72.

Gunther DF, Bourdeau I, Matyakhina L, Cassarino D, Kleiner DE, Griffin K, et al. Cyclical Cushing syndrome presenting in infancy: an early form of primary pigmented nodular adrenocortical disease, or a new entity? J Clin Endocrinol Metab 2004;89(7):3173–82.

Hachemi Y, Rapp AE, Picke AK, Weidinger G, Ignatius A, Tuckermann J. Molecular mechanisms of glucocorticoids on skeleton and bone regeneration after fracture. J Mol Endocrinol 2018;61(1):R75–90.

Hamet P, Larochelle P, Franks DJ, Cartier P, Bolte E. Cushing syndrome with food-dependent periodic hormonogenesis. Clin Invest Med 1987;10(6):530–3.

Hána Jr V, Ježková J, Kosák M, Kršek M, Hána V, Hill M. Serum steroid profiling in Cushing's syndrome patients. J Steroid Biochem Mol Biol 2019;192, 105410.

Hannah-Shmouni F, Stratakis CA. A gene-based classification of primary adrenocortical hyperplasias. Horm Metab Res 2020;52(3):133–41.

Hannah-Shmouni F, Faucz FR, Stratakis CA. Alterations of phosphodies-terases in adrenocortical tumors. Front Endocrinol 2016;7:111.

Harrison RF, Debono M, Whitaker MJ, Keevil BG, Newell-Price J, Ross RJ. Salivary cortisone to estimate cortisol exposure and sampling frequency required based on serum cortisol measurements. J Clin Endocrinol Metab 2019;104(3):765–77.

Hartmann K, Koenen M, Schauer S, Wittig-Blaich S, Ahmad M, Baschant U, et al. Molecular actions of glucocorticoids in cartilage and bone during health, disease, and steroid therapy. Physiol Rev 2016;96(2):409–47.

Hawley JM, Owen LJ, Debono M, Newell-Price J, Keevil BG. Devel-opment of a rapid liquid chromatography tandem mass spectrometry method for the quantitation of serum dexamethasone and its clinical verification. Ann Clin Biochem 2018;55(6):665–72.

Hempen C, Elfering S, Mulder AH, van den Bergh FA, Maatman RG. Dexamethasone suppression test: development of a method for simultaneous determination of cortisol and dexamethasone in human plasma by liquid chromatography/tandem mass spectrometry. Ann Clin Biochem 2012;49(Pt. 2):170–6.

Hernández-Ramírez LC, Gabrovska P, Dénes J, Stals K, Trivellin G, Tilley D, et al. Landscape of familial isolated and young-onset pituitary adenomas: prospective diagnosis in AIP mutation carriers. J Clin Endocrinol Metab 2015;100(9):E1242–54.

Hernández-Ramírez LC, Gam R, Valdés N, Lodish MB, Pankratz N, Balsalobre A, et al. Loss-of-function mutations in the *CABLES1* gene are a novel cause of Cushing's disease. Endocr Relat Cancer 2017;24 (8):379–92.

Himsworth RL, Bloomfield GA, Coombes RC, Ellison M, Gilkes JJ, Lowry PJ, et al. 'Big ACTH' and calcitonin in an ectopic hormone secreting tumour of the liver. Clin Endocrinol (Oxf) 1977;7(1):45–62.

Hindmarsh PC, Brook CG. Single dose dexamethasone suppression test in children: dose relationship to body size. Clin Endocrinol (Oxf) 1985;23 (1):67–70.

Hines JM, Bancos I, Bancos C, Singh RD, Avula AV, Young WF, et al. High-resolution, accurate-mass (HRAM) mass spectrometry urine steroid profiling in the diagnosis of adrenal disorders. Clin Chem 2017;63(12):1824–35.

Hodes A, Meyer J, Lodish MB, Stratakis CA, Zilbermint M. Mini-review of hair cortisol concentration for evaluation of Cushing syndrome. Expert Rev Endocrinol Metab 2018;13(5):225–31.

Horvath A, Stratakis CA. Unraveling the molecular basis of micronodular adrenal hyperplasia. Curr Opin Endocrinol Diabetes Obes 2008;15 (3):227–33.

Howlett TA, Price J, Hale AC, Doniach I, Rees LH, Wass JA, et al. Pituitary ACTH dependent Cushing's syndrome due to ectopic production of a bombesin-like peptide by a medullary carcinoma of the thyroid. Clin Endocrinol (Oxf) 1985;22(1):91–101.

Humayun MA, Hart T, Richardson T. Cyclical Cushing's: how best to catch the ups and downs. BMJ Case Rep 2017;2017, bcr2016218451.

Iacobone M, Citton M, Scarpa M, Viel G, Boscaro M, Nitti D. Systematic review of surgical treatment of subclinical Cushing's syndrome. Br J Surg 2015;102(4):318–30.

Invitti C, Pecori Giraldi F, de Martin M, Cavagnini F. Diagnosis and management of Cushing's syndrome: results of an Italian multicentre study. Study group of the Italian society of endocrinology on the pathophysiology of the hypothalamic-pituitary-adrenal axis. J Clin Endocrinol Metab 1999;84(2):440–8.

Jamieson PM, Chapman KE, Seckl JR. Tissue- and temporal-specific regulation of 11beta-hydroxysteroid dehydrogenase type 1 by glucocorticoids in vivo. J Steroid Biochem Mol Biol 1999;68(5–6): 245–50.

Jones RL, Owen LJ, Adaway JE, Keevil BG. Simultaneous analysis of cortisol and cortisone in saliva using XLC-MS/MS for fully automated online solid phase extraction. J Chromatogr B Analyt Technol Biomed Life Sci 2012;881–882:42–8.

Kamenický P, Droumaguet C, Salenave S, Blanchard A, Jublanc C, Gautier JF, et al. Mitotane, metyrapone, and ketoconazole combination therapy as an alternative to rescue adrenalectomy for severe ACTH-dependent Cushing's syndrome. J Clin Endocrinol Metab 2011;96 (9):2796–804.

Kamilaris CDC, Hannah-Shmouni F, Stratakis CA. Adrenocortical tumorigenesis: lessons from genetics. Best Pract Res Clin Endocrinol Metab 2020;34(3), 101428.

Kamilaris CDC, Stratakis CA, Hannah-Shmouni F. Molecular genetic and genomic alterations in Cushing's syndrome and primary aldosteronism. Front Endocrinol 2021;12, 632543.

Kannankeril J, Carroll T, Findling JW, Javorsky B, Gunsolus IL, Phillips J, et al. Prospective evaluation of late-night salivary cortisol and cortisone by EIA and LC-MS/MS in suspected Cushing syndrome. J Endocr Soc 2020;4(10), bvaa107.

Kasturi K, Fernandes L, Quezado M, Eid M, Marcus L, Chittiboina P, et al. Cushing disease in a patient with multiple endocrine neoplasia type 2B. J Clin Transl Endocrinol Case Rep 2017;4:1–4.

Kempegowda P, Quinn L, Shepherd L, Kauser S, Johnson B, Lawson A, et al. Adrenal insufficiency from steroid-containing complementary therapy: importance of detailed history. Endocrinol Diabetes Metab Case Rep 2019;2019(1):1–4.

Kivlighan KT, Granger DA, Schwartz EB, Nelson V, Curran M, Shirtcliff EA. Quantifying blood leakage into the oral mucosa and its effects on the measurement of cortisol, dehydroepiandrosterone, and testosterone in saliva. Horm Behav 2004;46(1):39–46.

Kotłowska A. Application of chemometric techniques in search of clinically applicable biomarkers of disease. Drug Dev Res 2014;75 (5):283–90.

Lacroix A. Approach to the patient with adrenocortical carcinoma. J Clin Endocrinol Metab 2010;95(11):4812–22.

Lacroix A, Bolté E, Tremblay J, Dupré J, Poitras P, Fournier H, et al. Gastric inhibitory polypeptide-dependent cortisol hypersecretion—a new cause of Cushing's syndrome. N Engl J Med 1992;327 (14):974–80.

Lavery GG, Idkowiak J, Sherlock M, Bujalska I, Ride JP, Saqib K, et al. Novel H6PDH mutations in two girls with premature adrenarche: 'apparent' and 'true' CRD can be differentiated by urinary steroid profiling. Eur J Endocrinol 2013;168(2):K19–26.

Lecoq AL, Stratakis CA, Viengchareun S, Chaligné R, Tosca L, Deméocq V, et al. Adrenal GIPR expression and chromosome 19q13 microduplications in GIP-dependent Cushing's syndrome. JCI Insight 2017;2(18), e92184.

Lee J, Hwang Y, Kim KN, Ahn C, Sung HK, Ko KP, et al. Associations of urinary sodium levels with overweight and central obesity in a population with a sodium intake. BMC Nutr 2018;4:47.

Lienhardt A, Grossman AB, Dacie JE, Evanson J, Huebner A, Afshar F, et al. Relative contributions of inferior petrosal sinus sampling and pituitary imaging in the investigation of children and adolescents with ACTH-dependent Cushing's syndrome. J Clin Endocrinol Metab 2001;86(12):5711–4.

Linton EA, Perkins AV, Woods RJ, Eben F, Wolfe CD, Behan DP, et al. Corticotropin releasing hormone-binding protein (CRH-BP): plasma levels decrease during the third trimester of normal human pregnancy. J Clin Endocrinol Metab 1993;76(1):260–2.

Liu Q, Tong D, Xu J, Yang X, Yi Y, Zhang D, et al. A novel germline ARMC5 mutation in a patient with bilateral macronodular adrenal hyperplasia: a case report. BMC Med Genet 2018;19(1):49.

Lopes LM, Francisco RP, Galletta MA, Bronstein MD. Determination of nighttime salivary cortisol during pregnancy: comparison with values in non-pregnancy and Cushing's disease. Pituitary 2016;19(1):30–8.

Louiset E, Stratakis CA, Perraudin V, Griffin KJ, Libé R, Cabrol S, et al. The paradoxical increase in cortisol secretion induced by dexamethasone in primary pigmented nodular adrenocortical disease involves a glucocorticoid receptor-mediated effect of dexamethasone on protein kinase A catalytic subunits. J Clin Endocrinol Metab 2009;94(7):2406–13.

Louiset E, Gobet F, Libé R, Horvath A, Renouf S, Cariou J, et al. ACTH-independent Cushing's syndrome with bilateral micronodular adrenal hyperplasia and ectopic adrenocortical adenoma. J Clin Endocrinol Metab 2010;95(1):18–24.

Louiset E, Duparc C, Young J, Renouf S, Tetsi Nomigni M, Boutelet I, et al. Intraadrenal corticotropin in bilateral macronodular adrenal hyperplasia. N Engl J Med 2013;369(22):2115–25.

Lumbroso S, Paris F, Sultan C, European Collaborative Study. Activating Gsalpha mutations: analysis of 113 patients with signs of McCune-Albright syndrome—a European Collaborative Study. J Clin Endocrinol Metab 2004;89(5):2107–13.

Magiakou MA, Chrousos GP. Cushing's syndrome in children and adolescents: current diagnostic and therapeutic strategies. J Endocrinol Invest 2002;25(2):181–94.

Manenschijn L, Koper JW, van den Akker EL, de Heide LJ, Geerdink EA, de Jong FH, et al. A novel tool in the diagnosis and follow-up of (cyclic) Cushing's syndrome: measurement of long-term cortisol in scalp hair. J Clin Endocrinol Metab 2012;97(10):E1836–43.

Maria AG, Tatsi C, Berthon A, Drougat L, Settas N, Hannah-Shmouni F, et al. ARMC5 variants in PRKAR1A-mutated patients modify cortisol levels and Cushing's syndrome. Endocr Relat Cancer 2020;27(9):509–17.

Martínez de LaPiscina I, Portillo Najera N, Rica I, Gaztambide S, Webb SM, Santos A, et al. Clinical and genetic characteristics in patients under 30 years with sporadic pituitary adenomas. Eur J Endocrinol 2021;185(4):485–96.

Masjkur J, Gruber M, Peitzsch M, Kaden D, Di Dalmazi G, Bidlingmaier M, et al. Plasma steroid profiles in subclinical compared with overt adrenal Cushing syndrome. J Clin Endocrinol Metab 2019;104(10):4331–40.

Matyakhina L, Freedman RJ, Bourdeau I, Wei MH, Stergiopoulos SG, Chidakel A, et al. Hereditary leiomyomatosis associated with bilateral, massive, macronodular adrenocortical disease and atypical Cushing syndrome: a clinical and molecular genetic investigation. J Clin Endocrinol Metab 2005;90(6):3773–9.

Mazzocchi G, Rebuffat P, Meneghelli V, Malendowicz LK, Tortorella C, Gottardo G, et al. Gastric inhibitory polypeptide stimulates glucocorticoid secretion in rats, acting through specific receptors coupled with the adenylate cyclase-dependent signaling pathway. Peptides 1999;20 (5):589–94.

McWhinney BC, Briscoe SE, Ungerer JP, Pretorius CJ. Measurement of cortisol, cortisone, prednisolone, dexamethasone and 11-deoxycortisol with ultra high performance liquid chromatography-tandem mass spectrometry: application for plasma, plasma ultrafiltrate, urine and saliva in a routine laboratory. J Chromatogr B Analyt Technol Biomed Life Sci 2010;878(28):2863–9.

Menon RK, Ferrau F, Kurzawinski TR, Rumsby G, Freeman A, Amin Z, et al. Adrenal cancer in neurofibromatosis type 1: case report and DNA analysis. Endocrinol Diabetes Metab Case Rep 2014;2014, 140074.

Messidoro C, Elte JW, Castro Cabezas M, van Agteren M, Lacroix A, de Herder WW. Food-dependent Cushing's syndrome. Neth J Med 2009;67(5):187–90.

Mezzullo M, Fanelli F, Fazzini A, Gambineri A, Vicennati V, Di Dalmazi G, et al. Validation of an LC-MS/MS salivary assay for glucocorticoid status assessment: evaluation of the diurnal fluctuation of cortisol and cortisone and of their association within and between serum and saliva. J Steroid Biochem Mol Biol 2016;163:103–12.

Mizuno E, Iura T, Mukai A, Yoshimori T, Kitamura N, Komada M. Regulation of epidermal growth factor receptor down-regulation by UBPY-mediated deubiquitination at endosomes. Mol Biol Cell 2005;16(11):5163–74.

Monaghan PJ, Owen LJ, Trainer PJ, Brabant G, Keevil BG, Darby D. Comparison of serum cortisol measurement by immunoassay and liquid chromatography-tandem mass spectrometry in patients receiving the 11β-hydroxylase inhibitor metyrapone. Ann Clin Biochem 2011;48(Pt. 5):441–6.

Morgan SA, McCabe EL, Gathercole LL, Hassan-Smith ZK, Larner DP, Bujalska IJ, et al. 11β-HSD1 is the major regulator of the tissue-specific effects of circulating glucocorticoid excess. Proc Natl Acad Sci USA 2014;111(24):E2482–91.

N'Diaye N, Tremblay J, Hamet P, De Herder WW, Lacroix A. Adrenocortical overexpression of gastric inhibitory polypeptide receptor underlies food-dependent Cushing's syndrome. J Clin Endocrinol Metab 1998;83(8):2781–5.

Nakamura Y, Rege J, Satoh F, Morimoto R, Kennedy MR, Ahlem CN, et al. Liquid chromatography-tandem mass spectrometry analysis of human adrenal vein corticosteroids before and after adrenocorticotropic hormone stimulation. Clin Endocrinol (Oxf) 2012;76(6): 778–84.

Naziat A, Karavitaki N, Thakker R, Ansorge O, Sadler G, Gleeson F, et al. Confusing genes: a patient with MEN2A and Cushing's disease. Clin Endocrinol (Oxf) 2013;78(6):966–8.

Nenke MA, Zeng A, Meyer EJ, Lewis JG, Rankin W, Johnston J, et al. Differential effects of estrogen on corticosteroid-binding globulin forms suggests reduced cleavage in pregnancy. J Endocr Soc 2017;1 (3):202–10.

Newell-Price J, Morris DG, Drake WM, Korbonits M, Monson JP, Besser GM, et al. Optimal response criteria for the human CRH test in the differential diagnosis of ACTH-dependent Cushing's syndrome. J Clin Endocrinol Metab 2002;87(4):1640–5.

Nicolaides NC, Charmandari E. Primary generalized glucocorticoid resistance and hypersensitivity syndromes: a 2021 update. Int J Mol Sci 2021;22(19):10839.

Nieman LK, Biller BM, Findling JW, Newell-Price J, Savage MO, Stewart PM, et al. The diagnosis of Cushing's syndrome: an endocrine society clinical practice guideline. J Clin Endocrinol Metab 2008;93 (5):1526–40.

Nishioka H, Yamada S. Cushing's disease. J Clin Med 2019;8(11):1951.

Olesti E, Boccard J, Visconti G, González-Ruiz V, Rudaz S. From a single steroid to the steroidome: trends and analytical challenges. J Steroid Biochem Mol Biol 2021;206, 105797.

Oliver RL, Davis JR, White A. Characterisation of ACTH related peptides in ectopic Cushing's syndrome. Pituitary 2003;6(3):119–26.

Oßwald A, Wang R, Beuschlein F, Hartmann MF, Wudy SA, Bidlingmaier M, et al. Performance of LC-MS/MS and immunoassay based 24-h urine free cortisol in the diagnosis of Cushing's syndrome. J Steroid Biochem Mol Biol 2019;190:193–7.

Owen LJ, Halsall DJ, Keevil BG. Cortisol measurement in patients receiving metyrapone therapy. Ann Clin Biochem 2010;47 (Pt. 6):573–5.

Page-Wilson G, Freda PU, Jacobs TP, Khandji AG, Bruce JN, Foo ST, et al. Clinical utility of plasma POMC and AgRP measurements in the differential diagnosis of ACTH-dependent Cushing's syndrome. J Clin Endocrinol Metab 2014;99(10):E1838–45.

Palmieri S, Morelli V, Polledri E, Fustinoni S, Mercadante R, Olgiati L, et al. The role of salivary cortisol measured by liquid chromatography-tandem mass spectrometry in the diagnosis of subclinical hypercortisolism. Eur J Endocrinol 2013;168(3):289–96.

Papakokkinou E, Piasecka M, Carlsen HK, Chantzichristos D, Olsson DS, Dahlqvist P, et al. Prevalence of Nelson's syndrome after bilateral adrenalectomy in patients with cushing's disease: a systematic review and meta-analysis. Pituitary 2021;24(5):797–809.

Pappachan JM, Hariman C, Edavalath M, Waldron J, Hanna FW. Cushing's syndrome: a practical approach to diagnosis and differential diagnoses. J Clin Pathol 2017;70(4):350–9.

Pecori Giraldi F, Cavallo LM, Tortora F, Pivonello R, Colao A, Cappabianca P, et al. The role of inferior petrosal sinus sampling in ACTH-dependent Cushing's syndrome: review and joint opinion statement by members of the Italian Society for Endocrinology, Italian Society for Neurosurgery, and Italian Society for Neuroradiology. Neurosurg Focus 2015;38(2), E5.

Pereira SS, Costa MM, Gomez-Sanchez CE, Monteiro MP, Pignatelli D. Incomplete pattern of steroidogenic protein expression in functioning adrenocortical carcinomas. Biomedicine 2020;8(8):256.

Perez-Rivas LG, Reincke M. Genetics of Cushing's disease: an update. J Endocrinol Invest 2016;39(1):29–35.

Perez-Rivas LG, Theodoropoulou M, Ferraù F, Nusser C, Kawaguchi K, Stratakis CA, et al. The gene of the ubiquitin-specific protease 8 is frequently mutated in adenomas causing Cushing's disease. J Clin Endocrinol Metab 2015;100(7):E997–E1004.

Perogamvros I, Owen LJ, Keevil BG, Brabant G, Trainer PJ. Measurement of salivary cortisol with liquid chromatography-tandem mass spectrometry in patients undergoing dynamic endocrine testing. Clin Endocrinol (Oxf) 2010;72(1):17–21.

Phillipou G. Investigation of urinary steroid profiles as a diagnostic method in Cushing's syndrome. Clin Endocrinol (Oxf) 1982;16(5):433–9.

Phillipou G, Higgins BA. A new defect in the peripheral conversion of cortisone to cortisol. J Steroid Biochem 1985;22(3):435–6.

Phillipov G, Palermo M, Shackleton CH. Apparent cortisone reductase deficiency: a unique form of hypercortisolism. J Clin Endocrinol Metab 1996;81(11):3855–60.

Pinelli S, Barbot M, Scaroni C, Ceccato F. Second-line tests in the diagnosis of adrenocorticotropic hormone-dependent hypercortisolism. Ann Lab Med 2021;41(6):521–31.

Pivonello R, De Leo M, Cozzolino A, Colao A. The treatment of Cushing's disease. Endocr Rev 2015;36(4):385–486.

Plöckinger U, Chrusciel M, Doroszko M, Saeger W, Blankenstein O, Weizsäcker K, et al. Functional implications of LH/hCG receptors in pregnancy-induced Cushing syndrome. J Endocr Soc 2017;1(1):57–71.

Ponzetto F, Settanni F, Parasiliti-Caprino M, Rumbolo F, Nonnato A, Ricciardo M, et al. Reference ranges of late-night salivary cortisol and cortisone measured by LC-MS/MS and accuracy for the diagnosis of Cushing's syndrome. J Endocrinol Invest 2020;43(12):1797–806.

Raff H, Phillips JM. Bedtime salivary cortisol and cortisone by LC-MS/MS in healthy adult subjects: evaluation of sampling time. J Endocr Soc 2019;3(8):1631–40.

Raff H, Sharma ST, Nieman LK. Physiological basis for the etiology, diagnosis, and treatment of adrenal disorders: Cushing's syndrome, adrenal insufficiency, and congenital adrenal hyperplasia. Compr Physiol 2014;4(2):739–69.

Räikkönen K, Seckl JR, Heinonen K, Pyhälä R, Feldt K, Jones A, et al. Maternal prenatal licorice consumption alters hypothalamic-pituitary-adrenocortical axis function in children. Psychoneuroendocrinology 2010;35(10):1587–93.

Räikkönen K, Gissler M, Kajantie E. Associations between maternal antenatal corticosteroid treatment and mental and behavioral disorders in children. JAMA 2020;323(19):1924–33.

Rege J, Nanba AT, Auchus RJ, Ren J, Peng HM, Rainey WE, et al. Adrenocorticotropin acutely regulates pregnenolone sulfate production by the human adrenal in vivo and in vitro. J Clin Endocrinol Metab 2018;103(1):320–7.

Reincke M, Sbiera S, Hayakawa A, Theodoropoulou M, Osswald A, Beuschlein F, et al. Mutations in the deubiquitinase gene USP8 cause Cushing's disease. Nat Genet 2015;47(1):31–8.

Reincke M, Albani A, Assie G, Bancos I, Brue T, Buchfelder M, et al. Corticotroph tumor progression after bilateral adrenalectomy (Nelson's syndrome): systematic review and expert consensus recommendations. Eur J Endocrinol 2021;184(3):P1–P16.

Reznik Y, Allali-Zerah V, Chayvialle JA, Leroyer R, Leymarie P, Travert G, et al. Food-dependent Cushing's syndrome mediated by aberrant adrenal sensitivity to gastric inhibitory polypeptide. N Engl J Med 1992;327(14):981–6.

Rollin GA, Costenaro F, Gerchman F, Rodrigues TC, Czepielewski MA. Evaluation of the DDAVP test in the diagnosis of Cushing's disease. Clin Endocrinol (Oxf) 2015;82(6):793–800.

Rossi C, Cicalini I, Verrocchio S, Di Dalmazi G, Federici L, Bucci I. The potential of steroid profiling by mass spectrometry in the management of adrenocortical carcinoma. Biomedicine 2020;8(9):314.

Rubinstein G, Osswald A, Hoster E, Losa M, Elenkova A, Zacharieva S, et al. Time to diagnosis in Cushing's syndrome: a meta-analysis based on 5367 patients. J Clin Endocrinol Metab 2020;105(3), dgz136.

Russell E, Kirschbaum C, Laudenslager ML, Stalder T, de Rijke Y, van Rossum EF, et al. Toward standardization of hair cortisol measurement: results of the first international interlaboratory round robin. Ther Drug Monit 2015;37(1):71–5.

Sahakitrungruang T, Srichomthong C, Pornkunwilai S, Amornfa J, Shuangshoti S, Kulawonganunchai S, et al. Germline and somatic DICER1 mutations in a pituitary blastoma causing infantile-onset Cushing's disease. J Clin Endocrinol Metab 2014;99(8):E1487–92.

Saiegh L, Keren D, Rainis T, Sheikh-Ahmad M, Reut M, Nakhleh A, et al. Dexamethasone-suppressed corticotropin-releasing hormone stimulation test in morbid obese adults. Obes Res Clin Pract 2016;10(3):275–82.

Sakaguchi C, Ashida K, Kohashi K, Ohe K, Fujii Y, Yano S, et al. A case of autonomous cortisol secretion in a patient with subclinical Cushing's syndrome, GNAS mutation, and paradoxical cortisol response to dexamethasone. BMC Endocr Disord 2019;19(1):13.

Salpea P, Stratakis CA. Carney complex and McCune Albright syndrome: an overview of clinical manifestations and human molecular genetics. Mol Cell Endocrinol 2014;386(1–2):85–91.

Savas M, Wester VL, Staufenbiel SM, Koper JW, van den Akker ELT, Visser JA, et al. Systematic evaluation of corticosteroid use in obese and non-obese individuals: a multi-cohort study. Int J Med Sci 2017;14(7):615–21.

Sbiera S, Kunz M, Weigand I, Deutschbein T, Dandekar T, Fassnacht M. The new genetic landscape of Cushing's disease: deubiquitinases in the spotlight. Cancers 2019;11(11):1761.

Scaroni C, Albiger NM, Palmieri S, Iacuaniello D, Graziadio C, Damiani L, et al. Approach to patients with pseudo-Cushing's states. Endocr Connect 2020;9(1):R1–R13.

Schweitzer S, Kunz M, Kurlbaum M, Vey J, Kendl S, Deutschbein T, et al. Plasma steroid metabolome profiling for the diagnosis of adrenocortical carcinoma. Eur J Endocrinol 2019;180(2):117–25.

Sesta A, Cassarino MF, Terreni M, Ambrogio AG, Libera L, Bardelli D, et al. Ubiquitin-specific protease 8 mutant corticotrope adenomas present unique secretory and molecular features and shed light on the role of ubiquitylation on ACTH processing. Neuroendocrinology 2020;110(1–2):119–29.

Shamim W, Yousufuddin M, Francis DP, Gualdiero P, Honour JW, Anker SD, et al. Raised urinary glucocorticoid and adrenal androgen precursors in the urine of young hypertensive patients: possible evidence for partial glucocorticoid resistance. Heart 2001;86(2):139–44.

Sharma AK, Shi X, Isales CM, McGee-Lawrence ME. Endogenous glucocorticoid signaling in the regulation of bone and marrow adiposity: lessons from metabolism and cross talk in other tissues. Curr Osteoporos Rep 2019;17(6):438–45.

Sherlock M, Scarsbrook A, Abbas A, Fraser S, Limumpornpetch P, Dineen R, et al. Adrenal Incidentaloma. Endocr Rev 2020;41(6):775–820.

Shi J, Dhaliwal P, Zi Zheng Y, Wong T, Straseski JA, Cervinski MA, et al. An intact ACTH LC-MS/MS assay as an arbiter of clinically discordant immunoassay results. Clin Chem 2019;65(11):1397–404.

Slominski R, Rovnaghi CR, Anand KJ. Methodological considerations for hair cortisol measurements in children. Ther Drug Monit 2015;37 (6):812–20.

Stratakis CA. New genes and/or molecular pathways associated with adrenal hyperplasias and related adrenocortical tumors. Mol Cell Endocrinol 2009;300(1–2):152–7.

Stratakis CA. Research and development in the molecular genetics of pituitary adenomas and related tumors. Expert Rev Endocrinol Metab 2012;7(6):593–8.

Stratakis CA, Boikos SA. Genetics of adrenal tumors associated with Cushing's syndrome: a new classification for bilateral adrenocortical hyperplasias. Nat Clin Pract Endocrinol Metab 2007;3(11):748–57.

Stratakis CA, Sarlis N, Kirschner LS, Carney JA, Doppman JL, Nieman LK, et al. Paradoxical response to dexamethasone in the diagnosis of primary pigmented nodular adrenocortical disease. Ann Intern Med 1999;131(8):585–91.

Suzuki S, Minamidate T, Shiga A, Ruike Y, Ishiwata K, Naito K, et al. Steroid metabolites for diagnosing and predicting clinicopathological features in cortisol-producing adrenocortical carcinoma. BMC Endocr Disord 2020;20(1):173.

Tanizaki Y, Jin L, Scheithauer BW, Kovacs K, Roncaroli F, Lloyd RV. P53 gene mutations in pituitary carcinomas. Endocr Pathol 2007;18 (4):217–22.

Tatsi C, Flippo C, Stratakis CA. Cushing syndrome: old and new genes. Best Pract Res Clin Endocrinol Metab 2020;34(2), 101418.

Taylor RL, Grebe SK, Singh RJ. Quantitative, highly sensitive liquid chromatography-tandem mass spectrometry method for detection of synthetic corticosteroids. Clin Chem 2004;50(12):2345–52.

Taylor DR, Ghataore L, Couchman L, Vincent RP, Whitelaw B, Lewis D, et al. A 13-steroid serum panel based on LC-MS/MS: use in detection of adrenocortical carcinoma. Clin Chem 2017;63(12):1836–46.

Teuber JP, Nanba K, Turcu AF, Chen X, Zhao L, Else T, et al. Intratumoral steroid profiling of adrenal cortisol-producing adenomas by liquid chromatography- mass spectrometry. J Steroid Biochem Mol Biol 2021;212, 105924.

Thakker RV, Newey PJ, Walls GV, Bilezikian J, Dralle H, Ebeling PR, et al. Clinical practice guidelines for multiple endocrine neoplasia type 1 (MEN1). J Clin Endocrinol Metab 2012;97(9):2990–3011.

Tortora F, Negro A, Briganti F, Del Basso De Caro ML, Cavallo LM, Solari D, et al. Pituitary magnetic resonance imaging vs. bilateral inferior petrosal sinus sampling: comparison between non-invasive and invasive diagnostic techniques for Cushing's disease—a narrative review. Gland Surg 2020;9(6):2260–8.

Treppiedi D, Barbieri AM, Di Muro G, Marra G, Mangili F, Catalano R, et al. Genetic profiling of a cohort of Italian patients with ACTH-secreting pituitary tumors and characterization of a novel USP8 gene variant. Cancers 2021;13(16):4022.

Tsagarakis S, Vasiliou V, Kokkoris P, Stavropoulos G, Thalassinos N. Assessment of cortisol and ACTH responses to the desmopressin test in patients with Cushing's syndrome and simple obesity. Clin Endocrinol (Oxf) 1999;51(4):473–7.

Tsilchorozidou T, Honour JW, Conway GS. Altered cortisol metabolism in polycystic ovary syndrome: insulin enhances 5alpha-reduction but not the elevated adrenal steroid production rates. J Clin Endocrinol Metab 2003;88(12):5907–13.

Tufano M, Ciofi D, Amendolea A, Stagi S. Auxological and endocrinological features in children with McCune Albright syndrome: a review. Front Endocrinol 2020;11:522.

Tuominen I, Heliövaara E, Raitila A, Rautiainen MR, Mehine M, Katainen R, et al. AIP inactivation leads to pituitary tumorigenesis through defective Gαi-cAMP signaling. Oncogene 2015;34(9):1174–84.

Uchida T, Nishimoto K, Fukumura Y, Asahina M, Goto H, Kawano Y, et al. Disorganized steroidogenesis in adrenocortical carcinoma, a case study. Endocr Pathol 2017;28(1):27–35.

Ueland GÅ, Methlie P, Kellmann R, Bjørgaas M, Åsvold BO, Thorstensen K, et al. Simultaneous assay of cortisol and dexamethasone improved diagnostic accuracy of the dexamethasone suppression test. Eur J Endocrinol 2017;176(6):705–13.

Ueland GÅ, Kellmann R, Jørstad Davidsen M, Viste K, Husebye ES, Almås B, et al. Bedtime salivary cortisol as a screening test for Cushing syndrome in children. J Endocr Soc 2021;5(5), bvab033.

Vaduva P, Bonnet F, Bertherat J. Molecular basis of primary aldosteronism and adrenal Cushing syndrome. J Endocr Soc 2020;4(9), bvaa075.

Valassi E, Franz H, Brue T, Feelders RA, Netea-Maier R, Tsagarakis S, et al. Diagnostic tests for Cushing's syndrome differ from published guidelines: data from ERCUSYN. Eur J Endocrinol 2017;176 (5):613–24.

van der Valk ES, van den Akker ELT, Savas M, Kleinendorst L, Visser JA, Van Haelst MM, et al. A comprehensive diagnostic approach to detect underlying causes of obesity in adults. Obes Rev 2019;20 (6):795–804.

van Rossum EF, Lamberts SW. Glucocorticoid resistance syndrome: a diagnostic and therapeutic approach. Best Pract Res Clin Endocrinol Metab 2006;20(4):611–26.

Vassiliadi DA, Tsagarakis S. DIAGNOSIS OF ENDOCRINE DISEASE: the role of the desmopressin test in the diagnosis and follow-up of Cushing's syndrome. Eur J Endocrinol 2018;178(5):R201–14.

Vega-Beyhart A, Iruarrizaga M, Pané A, García-Eguren G, Giró O, Boswell L, et al. Endogenous cortisol excess confers a unique lipid signature and metabolic network. J Mol Med (Berl) 2021;99(8):1085–99.

Velikanova LI, Shafigullina ZR, Lisitsin AA, Vorokhobina NV, Grigoryan K, Kukhianidze EA, et al. Different types of urinary steroid profiling obtained by high-performance liquid chromatography and gas chromatography-mass spectrometry in patients with adrenocortical carcinoma. Horm Cancer 2016;7(5–6):327–35.

Velikanova LI, Shafigullina ZR, Vorokhobina NV, Malevanaya EV. Gas chromatography-mass spectrometry analysis of urinary steroid metabolomics for detection of early signs of adrenal neoplasm malignancy in patients with Cushing's syndrome. Bull Exp Biol Med 2019;167(5):676–80.

Vierimaa O, Georgitsi M, Lehtonen R, Vahteristo P, Kokko A, Raitila A, et al. Pituitary adenoma predisposition caused by germline mutations in the AIP gene. Science 2006;312(5777):1228–30.

Vogg N, Kurlbaum M, Deutschbein T, Gräsl B, Fassnacht M, Kroiss M. Method-specific cortisol and dexamethasone thresholds increase clinical specificity of the dexamethasone suppression test for Cushing syndrome. Clin Chem 2021;67(7):998–1007.

Wędrychowicz A, Hull B, Tyrawa K, Kalicka-Kasperczyk A, Zieliński G, Starzyk J. Cushing disease in children and adolescents—assessment of the clinical course, diagnostic process, and effects of the treatment—experience from a single paediatric centre. Pediatr Endocrinol Diabetes Metab 2019;25(3):127–43.

Wang Z, Wang H, Peng Y, Chen F, Zhao L, Li X, et al. A liquid chromatography-tandem mass spectrometry (LC-MS/MS)-based assay to profile 20 plasma steroids in endocrine disorders. Clin Chem Lab Med 2020;58(9):1477–87 [s].

Weber MM, Lang J, Abedinpour F, Zeilberger K, Adelmann B, Engelhardt D. Different inhibitory effect of etomidate and ketoconazole on the human adrenal steroid biosynthesis. Clin Investig 1993;71(11):933–8.

Wester VL, Reincke M, Koper JW, van den Akker ELT, Manenschijn L, Berr CM, et al. Scalp hair cortisol for diagnosis of Cushing's syndrome. Eur J Endocrinol 2017;176(6):695–703.

Wheelan N, Kenyon CJ, Harris AP, Cairns C, Al Dujaili E, Seckl JR, et al. Midlife stress alters memory and mood-related behaviors in old age: role of locally activated glucocorticoids. Psychoneuroendocrinology 2018;89:13–22.

Wilkes EH, Rumsby G, Woodward GM. Using machine learning to aid the interpretation of urine steroid profiles. Clin Chem 2018;64(11):1586–95.

Zemskova MS, Nylen ES, Patronas NJ, Oldfield EH, Becker KL, Nieman LK. Diagnostic accuracy of chromogranin A and calcitonin precursors measurements for the discrimination of ectopic ACTH secretion from Cushing's disease. J Clin Endocrinol Metab 2009;94(8):2962–5.

Chapter 3.1.2

Cortisol deficiency

3.1.2.1 Introduction

Cortisol deficiency can be due to failure of the adrenal cortex to produce the steroid hormone either because the adrenal functions fail (**primary adrenal insufficiency**) or there are problems with the production or action of pituitary ACTH (**secondary adrenal insufficiency**) and is life-threatening (Younes et al., 2021; Uçar et al., 2016; Shulman et al., 2007). The commonest cause of AI in children and adults is the result of adrenal suppression by **exogenous glucocorticoids** (e.g., treatment of asthma). Adrenal insufficiency is seen in genetic disorders of the enzymes in cortisol production. Activation of the H-P-A axis through cortisol deficiency then leads to (congenital) adrenal hyperplasia from the action of raised ACTH and in children often presents with symptoms due to excess or deficiency of sex steroids (DHEA and androstenedione) or mineralocorticoids (aldosterone) (Fig. 3.1.2.1). Once the low cortisol state has been determined, a measurement of **ACTH** should be requested to have evidence to distinguish primary or secondary AI.

Hyponatremia is a frequent finding in primary adrenal insufficiency due to loss of mineralocorticoid steroids. The low plasma sodium is associated with hyperkalemia and a raised plasma urea. In the urine, there will be low potassium loss.

Although Addison in 1855 described nodules in the adrenal glands due to **tuberculosis infection**, today **auto-immune adrenalitis** which spares the adrenal medulla is more common unless TB and HIV infections are endemic. Other rare, acquired causes of adrenal insufficiency include adrenal infarction or hemorrhage and adrenal hypoplasia (particularly in newborn infants). Later, **adrenal destruction by metastases, sarcoidosis, histoplasmosis and amyloidosis** may be found. The differential diagnosis of hypoglycemia in childhood should include adrenal cortical insufficiency. This can be due to congenital adrenal hypoplasia (Flück, 2017; Buonocore et al., 2021) due to a primary defect in adrenal development or defects in ACTH synthesis, processing, release and response (familial glucocorticoid deficiency) (Heshmatzad et al., 2020; Dias et al., 2010).

Secondary adrenal insufficiency may result from impairment of the pituitary-adrenal axis (Bornstein, 2009). In some cases, the adrenal insufficiency may be part of more general hypopituitarism (Prodam et al., 2021). If hypocortisolism is due to ACTH deficiency resulting from pituitary or hypothalamic disease such as tumors, infarction, trauma, there are usually other signs of deficiency of other hormones (e.g., loss of body hair). There will be no pigmentation. In the absence of ACTH, the production of aldosterone under the stimulus of the renin-angiotensin system will usually be unaffected and the blood pressure will probably be normal. Hyponatremia can be seen with secondary AI because in the absence of cortisol, there is loss of the ability to excrete water. Hypopituitarism can lead to partial or complete pituitary hormone deficiency that requires complicated treatment regimens and life-long follow-up by endocrinologists (Prabhakar and Shalet, 2006).

Tertiary adrenal insufficiency reflects blockade of the hypothalamus from exogenous glucocorticoids. Patients with asthma who have long-term treatment with glucocorticoids can have adrenal function compromised. Long-term treatment with glucocorticoids should never be stopped abruptly and reductions in dosage should be undertaken with medical supervision. A recent problem has arisen from the use of immune checkpoint inhibitors in cancer treatments that cause hypophysitis. Replacement therapy is the cornerstone of treating adrenal insufficiency which in practice is challenging.

Adrenal insufficiency can affect nearly all systems in the body (Fig. 3.1.2.2).

Classical signs of AI are nonspecific—**a**ppetite loss, **d**iscoloration of the skin, **d**ehydration, **i**ncreased thirst and need to urinate frequently, **s**alt craving, **o**ligomenorrhea, **n**o energy or motivation, **s**ore/painful muscles and joints with first letters an acronym for Addison. There can be weight loss, vomiting, hypotension, abdominal pain, anemia and pigmentation in the axillae and/or the oral mucosa, areas of skin bleaching and loss of pubic hair being seen. Electrolytes may show hyponatremia, hyperkalemia and glucose concentration may be low. AI is not just the picture of feeling tired, exhausted, irritable and unhappy, sold as "adrenal fatigue."

Steroids in the Laboratory and Clinical Practice. https://doi.org/10.1016/B978-0-12-818124-9.00012-7

The hypothalamic–pituitary–adrenal axis

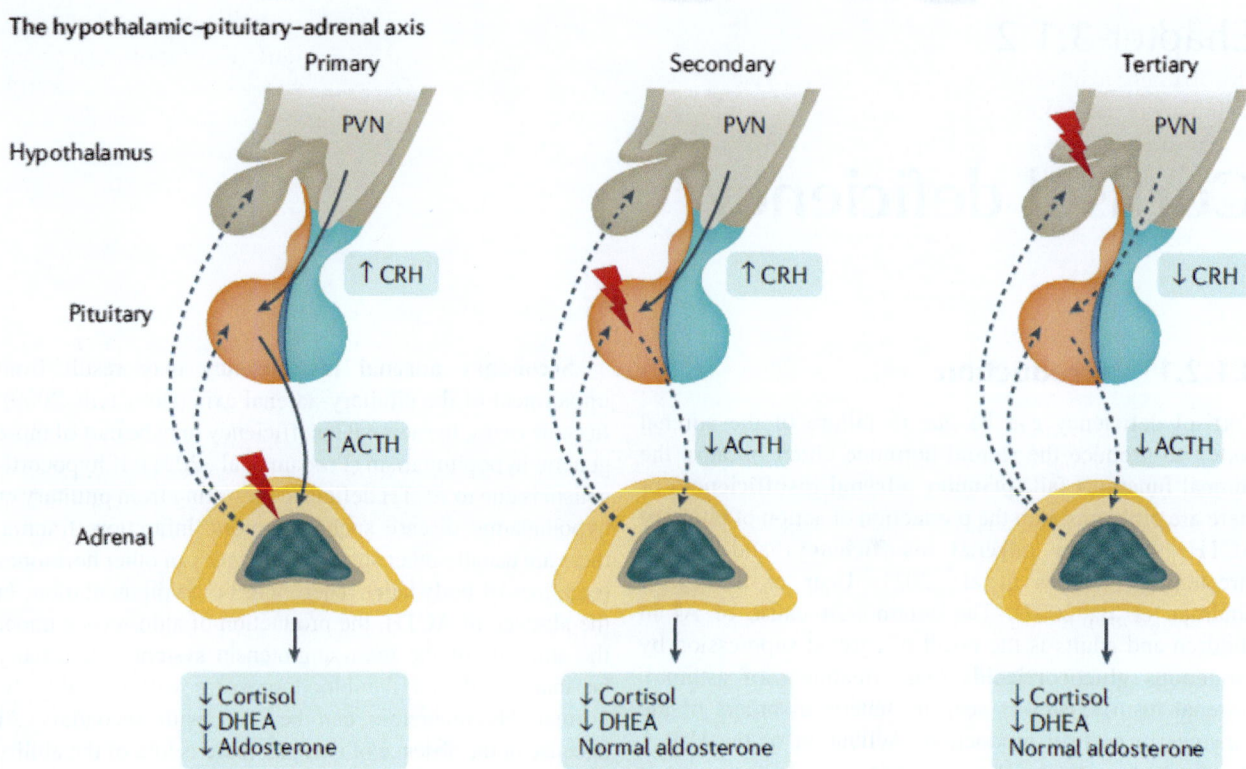

FIG. 3.1.2.1 HPA axis in primary, secondary and tertiary adrenal insufficiency.

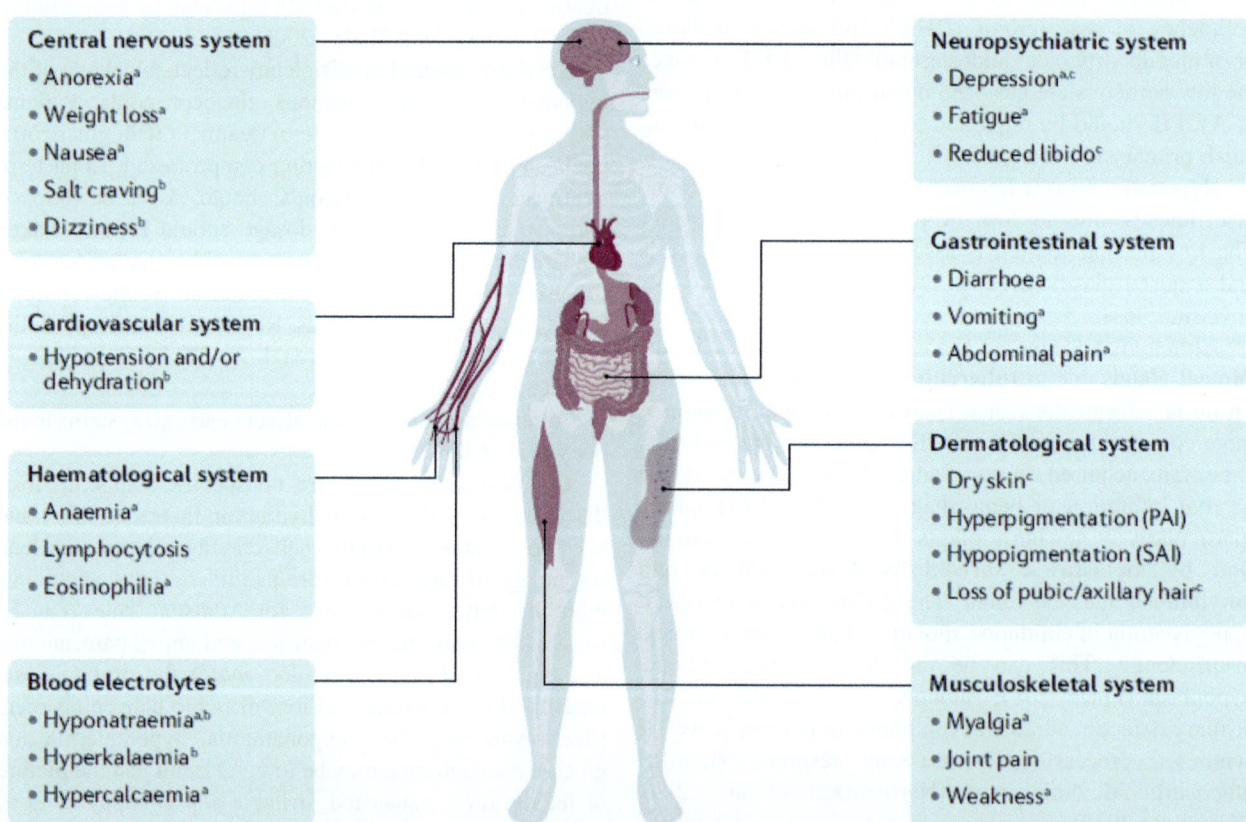

FIG. 3.1.2.2   Clinical manifestations of adrenal insufficiency. Deficiency of adrenocortical hormones affects all systems in the human body leading to a wide range of symptoms. Frequent symptoms are shown and their main associations with the respective deficiency of glucocorticoids, mineralocorticoids and adrenal androgens are demonstrated. *PAI*, primary adrenal insufficiency; *SAI*, secondary adrenal insufficiency. (a) Symptoms specific to glucocorticoid deficiency; (b) Symptoms specific to mineralocorticoid deficiency; (c) Symptoms specific to adrenal androgen deficiency. *(From Hahner S, Ross RJ, Arlt W, Bancos I, Burger-Stritt S, Torpy DJ, Husebye ES, Quinkler M. Adrenal insufficiency. Nat Rev Dis Primers 2021;7(1):19. Fig. 2 p. 2.)*

3.1.2.2 Identification of AI

The first steps to investigating a suspected case are a physical examination and family history. The most common causes of PAI are **autoimmune destruction** of the adrenal cortex in adults and **congenital adrenal hyperplasia (CAH)** in children. Tests for 21-hydroxylase antibodies (Wolff et al., 2021; Napier et al., 2020; Chen et al., 1996) and a baseline serum 17-hydroxyprogesterone level will aid diagnosis of these (Fig. 3.1.2.3). **17-Hydroxyprogesterone** is the substrate for the commonest form of primary AI due to congenital adrenal hyperplasia from 21-hydroxylase deficiency.

Newborn screening programmes include 17-OHP and some include the volatile long chain fatty acid (C26:0) for adrenoleukodystrophy. The measurement of **DHEAS** and DHEA provide measures of adrenal destruction with infection and autoimmune disease.

Since the presentation can require urgent investigations, the blood will be taken at any time of day and not at the preferred times around 0800 h and midnight. There may be difficulties in interpreting cortisol concentrations. The time of the blood sample should be recorded since electronic timing of receipt in the laboratory may be some hours later. If cortisol is below 100 nmol/L, there is clear indication of AI

requiring confirmation, the figure is a guide and the laboratory may have established different cut-offs according to the assay used. The range can spread up to 250 nmol/L (Abbott Architect and Alinity), with lower cut-offs when LC-MS/MS is used and higher levels with automated chemiluminescent or the similar platform assays (Ramadoss et al., 2021). There is considerable variation in the methods used to diagnose and investigate AI (Wallace et al., 2009).

A **Synacthen test** should be ordered if there is still doubt about the cause, again with problems in interpretation of cortisol response depending on time of day of the test (30 and 60 min concentrations 366 and 418 nmol/L) (Ramadoss et al., 2021; Bornstein, 2009). The cortisol response can be monitored by LC-MS/MS measurements in saliva (Chao et al., 2020). Aldosterone can also be measured in the ACTH stimulation test. A stimulated peak plasma aldosterone of <5 ng/dL (0–14 nmol/L) 30 or 60 min after 250 μg Cortrosyn accurately distinguished between PAI and SAI in 88% of the patients (Fig. 3.1.2.4) (Abraham et al., 2015).

Hydrocortisone treatment can be started as soon as all blood tests have been completed. The patient may also need corrective measures initially for hypoglycemia and hyponatremia.

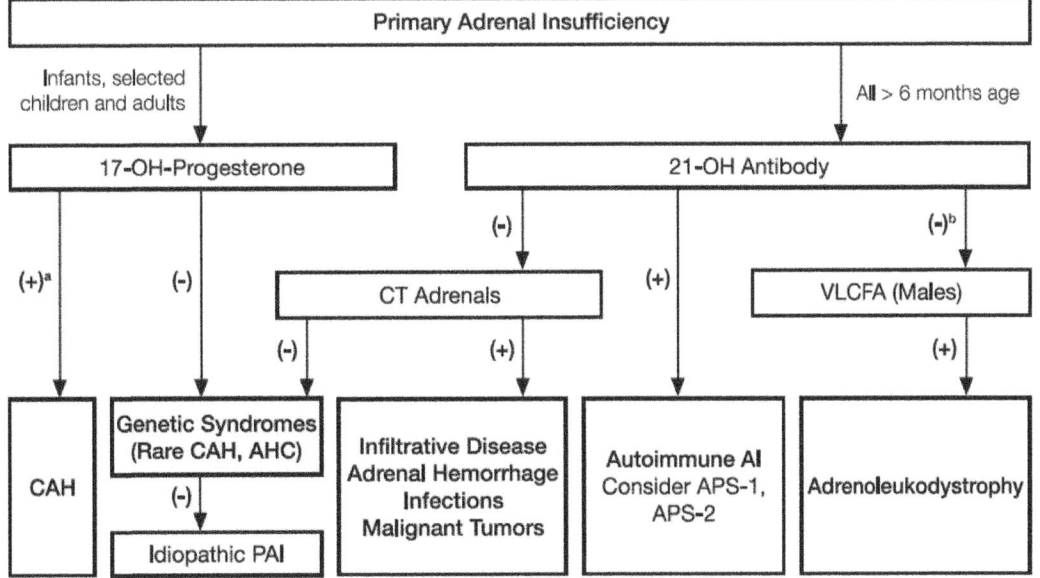

FIG. 3.1.2.3 Algorithm for the initial diagnostic approach to the patient with PAI. The most common causes of PAI are autoimmune destruction of the adrenal cortex in adults and CAH in children. Tests for 21-hydroxylase antibodies and a baseline serum 17-hydroxyprogesterone level will aid diagnosis of these. Males with negative 21-hydroxylase antibodies should be tested for adrenoleukodystrophy with plasma VLCFAs. If these diagnoses are excluded, a CT scan of the adrenals may reveal evidence of adrenal infiltrative processes or metastases. The clinical picture and family history may render some steps in the algorithm redundant or suggest specific genetic syndromes. The latter includes subtypes of autoimmune polyglandular syndromes or specific rare genetic disorders where adrenal failure is part of a broader phenotype. *AHC*, adrenal hypoplasia congenita; *AI*, adrenal insufficiency; *VLCFA*, very long chain fatty acid. [a]17-OH progesterone >1000 ng/dL is diagnostic for 21-OH deficiency, [b]VLCFA should be measured in the initial evaluation of preadolescent boys. *(From Bornstein SR, Allolio B, Arlt W, Barthel A, Don-Wauchope A, Hammer GD, Husebye ES, Merke DP, Murad MH, Stratakis CA, Torpy DJ. Diagnosis and treatment of primary adrenal insufficiency: an Endocrine Society clinical practice guideline. J Clin Endocrinol Metab 2016;101(2):364–89. Fig. 1 p. 372.)*

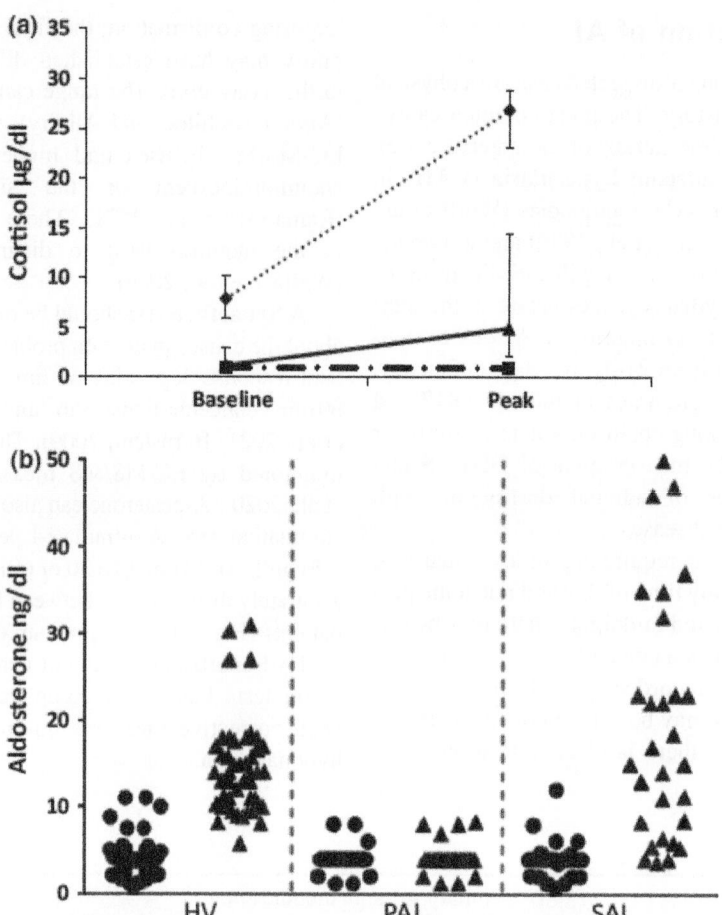

FIG. 3.1.2.4 (A) Baseline and peak cortisol values in healthy volunteers (HV—*dotted line*), and subjects with primary (PAI—*long/short dashed line*) and secondary (SAI—*solid black line*) adrenal insufficiency (B) baseline (*circles*) and Cortrosyn-stimulated peak (*triangles*) aldosterone levels in healthy volunteers (HV) and subjects with primary (PAI) and secondary (SAI) adrenal insufficiency. *(From Abraham SB, Abel BS, Sinaii N, Saverino E, Wade M, Nieman LK. Primary vs secondary adrenal insufficiency: ACTH-stimulated aldosterone diagnostic cut-off values by tandem mass spectrometry. Clin Endocrinol (Oxf) 2015;83(3):308–14. Fig. 1 p. 311.)*

3.1.2.3 Primary adrenal insufficiency

The presentation of adrenal insufficiency can be in the newborn period or later depending on whether there is a genetic basis or other cause. Regardless of the basis, AI is an important clinical problem requiring urgent medical care and endocrine tests although there is no agreement on which should be performed and which is the best biological fluid for the investigations. Morning plasma cortisol (Sbardella et al., 2017) and the cortisol response to an injection of ACTH are often used. The low-dose ACTH test can give false positive results due to timings of the first blood sample, the assay used and the cut-off value (Mongioì et al., 2019). Samples should be taken at 10 min intervals from 20 to 40 min after the dose. Urine-free cortisol determination is a poor test of adrenal suppression although measurements of total cortisol metabolites are of value (Fink et al., 2002; Priftis et al., 1990). Saliva is also being used to assess steroid status (Langelaan et al., 2018), although cortisol concentrations are <20 nmol/L and the assay needs appropriate low precision. Cortisone concentrations in saliva

are higher and can be used to assess AI (Perogamvros et al., 2009).

Genetic disorders of the adrenal cortex can affect development and function from synthesis and trafficking of cholesterol to mitochondria and the involvement of endoplasmic reticulum, peroxisomes and endosomes in steroidogenesis. The adrenal gland can also be damaged in very rare cases by hemorrhage, infarction (Herndon et al., 2018) and anticoagulation as well as infection by cytomegalovirus, HIV, coccidiomycosis, histiocytosis, blastomycosis and cryptococcosis (Flück, 2017).

3.1.2.4 Disorders of cortisol synthesis

3.1.2.4.1 Congenital adrenal hyperplasia

One of the several disorders of genes in the adrenal cortex with autosomal recessive hereditary pattern may be defective. The genes code for steroidogenic proteins in the synthesis of cortisol from cholesterol and when defective there is low cortisol production leading to excess secretion of ACTH which causes adrenal hyperplasia. Most of the enzymes are cytochrome P450 proteins. Several of the genes are duplicated on a chromosome to give active and inactive pairs (CYP21A2 and pseudogene CYP21B2), enzymes of different function (11-hydroxylase for cortisol synthesis—CYP11B1 and for aldosterone synthesis—CYP11B2) and enzymes with different tissues expression (HSD3B1 in periphery and HSD3B2 in adrenals and gonads). Care must be taken in the genetic tests to demonstrate the gene location of mutations. The enzymes are located in the endoplasmic reticulum or mitochondria and have distinct co-factor requirements. Congenital adrenal hyperplasia due to defects of cholesterol side chain cleavage enzyme (CYP11A1) (Kallali et al., 2020; Kim et al., 2008; Hiort et al., 2005) can be lethal at an early stage but there are exceptions (Rubtsov et al., 2009) such that some cases present with hypospadias. The infants usually present as phenotypic females and have an adrenal crisis in the newborn period. Detailed analysis of the genes involved in CAH has been critically reviewed (Balsamo et al., 2020; Baranowski et al., 2018).

Although **congenital adrenal hyperplasia** is a common cause of primary adrenal insufficiency especially in children, the diagnosis of AI can be missed because there are several possibilities (Shulman et al., 2007; Perry et al., 2005; Miller, 2008; Krone and Arlt, 2009; Claahsen-van der Grinten et al., 2022) and risks of misinterpretation of steroid results. **Newborn screening** in many countries (not United Kingdom) includes determinations of blood spot 17-hydroxyprogesterone concentrations that should pick up the 21-hydroxylase deficiency especially when mass spectrometric tests are used in the second tier. Some countries have screens for adrenal disorders beyond 21-hydroxylase deficiency with the addition of DHEAS and 11-deoxycortisol measurements (Janzen et al., 2012). A rapid tests programme is needed to return an abnormal result ideally before the onset of an adrenal crisis (presenting usually from 5 to 10 days after birth). Hyperkalemia before hyponatremia can be one of the early signs of some forms of CAH. Prolonged jaundice in a newborn infant may also be a sign of adrenal insufficiency. Patients may have few or no symptoms of adrenal cortical insufficiency until they suffer a physical stress such as trauma, surgery or infection when they present with tiredness, weakness, lethargy, anorexia, nausea, weight loss, dizziness and hypoglycemia. Pigmentation will sometimes be noted as a smokey brown coloration which affects the buccal mucosa (inside cheeks, gums and lips), on skin creases, scars, genitalia and papillae. This is a reflection of melanocyte stimulating hormone action which is a product of the proopiomelanocortin secreted by the pituitary along with ACTH. The patient will probably have postural hypotension with high plasma renin activity (PRA) but low aldosterone.

Adrenal insufficiency is seen from disorders of cortisol production. Activation of the H-P-A axis by low circulating cortisol concentrations leads to (congenital) adrenal hyperplasia in children who often present with symptoms due to excess or deficiency of sex steroids from blockade in synthesis of cortisol. Defects of 21-hydroxylase, 11β-hydroxylase and 3β-hydroxysteroid dehydrogenase are found with decreasing frequency and will be considered with further detail in other chapters because of the associated changes in sex steroids and mineralocorticoids.

3.1.2.4.1.1 21-Hydroxylase deficiency

A defect of 21-hydroxylase prevents aldosterone and cortisol production with accumulation of the substrates, pregnenolone, progesterone, 17-hydroxypregnenolone and 17-hydroxyprogesterone (Fig. 3.1.2.5). The side chain of pregnenolone is cleaved to dehydroepiandrosterone which in peripheral tissues can be converted to androstenedione and testosterone.

The effects of excess androgens are evident in 46XX girls presenting with masculinization of the genitalia at birth such that a male sex may be assigned. From the first week of life and beyond the signs of adrenal insufficiency become apparent. In severe cases (salt-losers SW), vomiting, diarrhea, weakness, dehydration, low blood pressure reflect the absence of cortisol and aldosterone. Boys may present later with an adrenal crisis or acne and a rapid growth rate (simple virilizers SV). Nonclassical forms can be detected at any age with raised 17-OHP, exaggerated on ACTH stimulation. Hirsutism and acne are signs of excess androgens.

A raised blood test result for 17-hydroxyprogesterone concentration suggests the diagnosis of 21-hydroxylase deficiency. Cortisol concentrations will be low; androstenedione, renin and ACTH raised. Normal ranges for aldosterone and renin are very much higher in children than adults which must be considered when interpreting patient results. A plasma steroid panel or urine steroid profile should confirm the defect due to appropriate elevation of cortisol precursors. The measurement of 21-deoxycortisol is a useful addition to the diagnosis. Certainly, the urine excretion of pregnanetriol-11-one is prominent in the steroid profile (Fig. 3.1.2.6), together with 17-hydroxypregnanolone and pregnanetriol.

The steroid profile may show evidence of increased 17-hydroxyprogesterone that is metabolized to DHT in the backdoor pathway through androstanedione or androsterone (Sumińska et al., 2020).

The analysis of the 21-hydroxylase gene is complicated by the duplication of the chromosome region. Mutations can be classified according to pseudogene origin (Kolahdouz et al., 2015). This is discussed in more detail in the chapter on Hypogonadism (Section 3.2.2). The 21-hydroxylase gene (*CYP21A2*) is located in the highly polymorphic region

FIG. 3.1.2.5 Defect of 21-hydroxylase reduces cortisol synthesis with elevation of 17-hydroxyprogesterone and androgens. If aldosterone synthesis is also affected, salt loss will add to adrenal insufficiency. *(Author original.)*

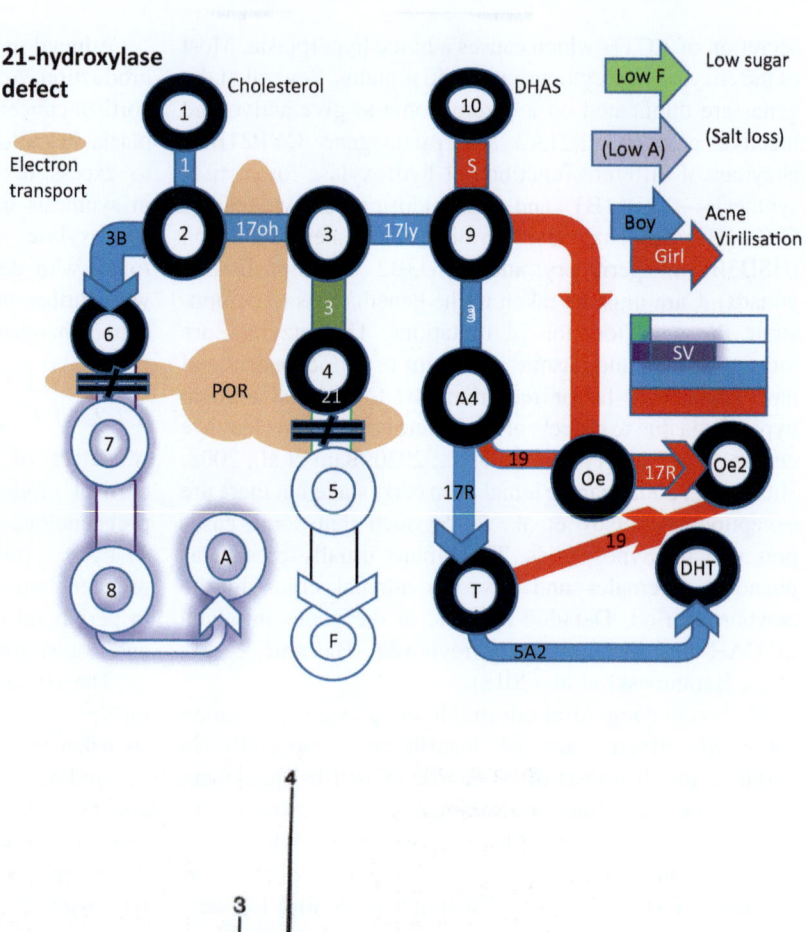

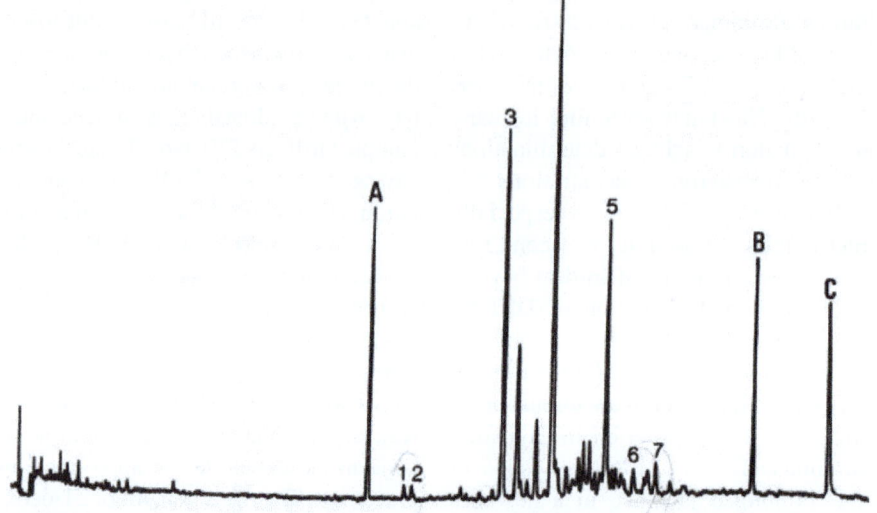

FIG. 3.1.2.6 Urinary steroid profile by GC-MS of a patient with 21-hydroylase deficiency. (1) Androsterone; (2) etiocholanolone; (3) 17-hydroxypregnanolone; (4) pregnanetriol; (5) 11-oxo-pregnaneriol; (6) tetrahydrocortisone; (7) tetrahydrocortisol. *(Author original.)*

of the major histocompatibility complex (HLA) on the short arm of chromosome 6, close to a pseudogene (*CYP21P*) which has 98% homology (Carvalho et al., 2021; Merke and Auchus, 2020). Most mutations are generated through misalignment of the CYP21A2 and the CYP21A1P and gene conversion during meiosis of the homologous genes. One or more of 10 deleterious sequences in *CYP21A1P* can relocate into *CYP21A2*. Twenty to 30% of *CYP21A2* classic mutations are 30-kb deletions, usually associated with a null mutation. Approximately 3% of deletions retain

partial 21-hydroxylase activity because of the location of the junction sites and are associated with a milder phenotype. Most affected persons have different mutations on each allele (compound heterozygotes) and a phenotype corresponding to the milder gene defect. New genetic patterns in patients are still possible (Iezzi et al., 2021).

Mutations in *CYP21A2* in Turkish children were summarized and found to be comparable with other populations (Karaoglan, 2019; Baş et al., 2009). Restriction enzyme digests, Southern blotting, polymerase chain reactions,

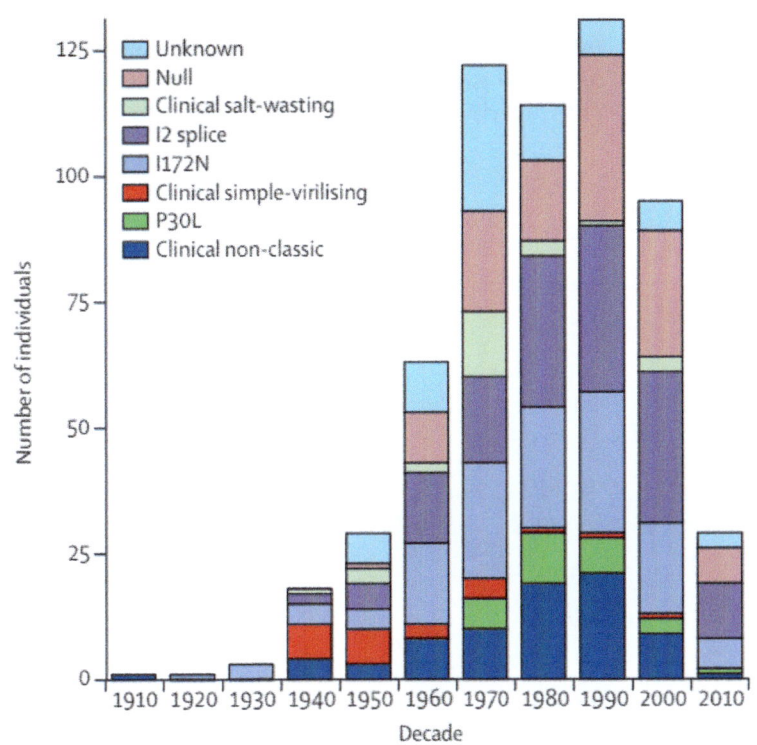

FIG. 3.1.2.7 **Distribution of congenital adrenal hyperplasia patients by genotype or phenotype when genotype was not available.** *(Redrawn from Gidlöf S, Falhammar H, Thilén A, von Döbeln U, Ritzén M, Wedell A, Nordenström A. One hundred years of congenital adrenal hyperplasia in Sweden: a retrospective, population-based cohort study. Lancet Diabetes Endocrinol 2013;1(1):35–42.)*

specific oligonucleotide hybridization techniques, exome sequencing, next-generation sequencing and Sanger sequencing have been used. In view of the obvious virilizing effects of androgens in excess, more detailed description of the genetics can be found in Chapter 3.2.1.

The ability to detect 21-hydroxylase deficiency has changed over past five decades due to technological advances (steroid assays, newborn screening and molecular genetics) and clinical experience (recognition of adrenal crisis and salt loss) (Fig. 3.1.2.7).

Genotyping has shown that the majority of defects are due to gene deletions, gene conversion or pseudogene derived mutations in most of the 490 cases (of a total population of 606 cases) in Sweden (Gidlöf et al., 2013) (Table 3.1.2.1) and small study in China (Liu et al., 2020).

3.1.2.4.1.2 HSD3B2

Steroids are more active when the 3β-hydroxy-5ene structure is converted to the 3-keto-4-ene configuration by the action of 3β-hydroxysteroid dehydrogenase (HSD3B2). A deficiency of HSD3B2 thus has profound effects on all steroid hormones (Fig. 3.1.2.8). The typical presentation of HDS3B2 deficiency is ambiguous genitalia and salt loss.

A 46XY infant will be poorly masculinized because of low testosterone/DHT and a 46XX patient may have some virilization due to weak androgen effects of DHEA that can be converted to androstenedione by peripheral HSD3B1. This defect will be covered also in Chapters 3.2.1 (androgen excess) and 3.3.2 (mineralocorticoid deficiency).

There may be a false elevation of 17-hydroxyprogesterone because a related enzyme, HSD3B1, can convert 17-hydroxypregnenolone to 17-OHP. A raised 17-OHP in a blood test or newborn screen needs confirmation by steroid profile/panel or genetic tests for 21-hydroxylase, P450 oxidoreductase and HSD3B2. The steroid profile plasma/serum LC/MS using Euraka kit (Ancona, Italy) (Guran et al., 2020) found highly significant elevation of the 17-hydroxypregnenolone/cortisol ratio (Table 3.1.2.2).

Other important characteristics of the profile are increased concentrations of pregnenolone, DHEA, DHEAS and lower aldosterone, corticosterone, progesterone, 21-deoxycortisol, cortisol, cortisone, androstenedione and testosterone. The urine steroid profile by GC-MS shows high DHEA and pregnenetriol (Fanis et al., 2020) as well as further metabolites of DHEA such 16-hydroxy-DHEA and androstenetriol.

3.1.2.4.1.3 CYP 17

A defect of 17-hydroxylase prevents cortisol and androgen production while raised corticosterone provides some glucocorticoid cover (Fig. 3.1.2.9) and the presentation does not usually include adrenal insufficiency (Breder et al., 2018).

The increase in production of 11-deoxycorticosterone (Lee et al., 2020a,b,c) leads to sodium retention through mineralocorticoid activity so renin is suppressed and aldosterone production is low. The usual presentation is primary

TABLE 3.1.2.1 Frequency of underlying mutations in the congenital adrenal hyperplasia population due to 21-hydroxylase deficiency in Sweden.

	Number of patients ($N = 490$)
CYP21A2 deletion or large gene conversion	220 (18%)
Pseudogene-derived mutations	
I2 splice	218 (27%)
I172N	135 (17%)
V281L	62 (8%)
Q318X	31 (4%)
R356W	25 (3%)
P30L	21 (3%)
Cluster E6	7 (1%)
P453S	7 (1%)
L307insT	6 (1%)
Combinations of pseudogene-derived mutations	
Q318X + R356W	5 (1%)
I2 splice and Q318X (two genes on one chromosome)	4 (1%)
I172N + P453S	4 (1%)
L307insT + Q318X	3 (<0.5%)
Cluster E6 + V281L	2 (<0.5%)
V281L + L307insT	2 (<0.5%)
I172N + ClusterE6 + V281L + L307insT + Q318X + R356W	2 (<0.5%)
I172N + ClusterE6 + L307insT + Q318X	2 (<0.5%)
I2 splice + I172N	1 (<0.5%)
I172N + ClusterE6 + V281L + L307insT	1 (<0.5%)
P30L + Q318X	1 (<0.5%)
H62L + P453S	1 (<0.5%)
I2 splice + P453S	1 (<0.5%)
V281L + R356W + A487P	1 (<0.5%)
Nonpseudogene-derived mutations	
R483GGtoC	11 (1%)
R483P	2 (<0.5%)
R341W	2 (<0.5%)
R233G	2 (<0.5%)
R354H	1 (<0.5%)

TABLE 3.1.2.1 Frequency of underlying mutations in the congenital adrenal hyperplasia population due to 21-hydroxylase deficiency in Sweden—cont'd

	Number of patients ($N = 490$)
W22X	1 (<0.5%)
I1 splice	1 (<0.5%)
G291S	1 (<0.5%)
I7 splice	1 (<0.5%)
R356P	1 (<0.5%)
W405X	1 (<0.5%)
R356Q	1 (<0.5%)
T295N	1 (<0.5%)
G424S	1 (<0.5%)
C147R	1 (<0.5%)
V139E	1 (<0.5%)
R426H	1 (<0.5%)
P482S	1 (<0.5%)
R444X	1 (<0.5%)
F404C	1 (<0.5%)
T52P	1 (<0.5%)
M283V	1 (<0.5%)
R408C	1 (<0.5%)
Unknown despite gene sequencing:	2 (<0.5%)

Data from Gidlöf S, Falhammar H, Thilén A, von Döbeln U, Ritzén M, Wedell A, Nordenström A. One hundred years of congenital adrenal hyperplasia in Sweden: a retrospective, population-based cohort study. Lancet Diabetes Endocrinol 2013;1(1):35–42. Table 2 P 38.

amenorrhea and hypertension (Table 3.1.2.3) (Wijaya et al., 2019; Unal et al., 2020) and further discussion will be found in Chapters 3.3.1 (mineralocorticoid excess) and 3.2.2 (infertility).

A steroid profile in urine (Honour et al., 1978) or blood (Lee et al., 2020a,b,c; Petri et al., 2014) shows increased outputs of progesterone and corticosterone and of note 21-deoxycorticosterone metabolites were identified (Honour et al., 1978, 1985) (Fig. 3.1.2.10) and like the rat attributed to bacterial metabolism in the enterohepatic circulation (Eriksson et al., 1968, 1969; Eriksson and Gustafsson, 1970; Eriksson, 1971).

Corticosterone is reduced in the liver with saturation of the A and B ring then conjugated before the metabolites are excreted in bile. Bacteria in the intestine can remove conjugate groups. Among many bacterial reactions, the

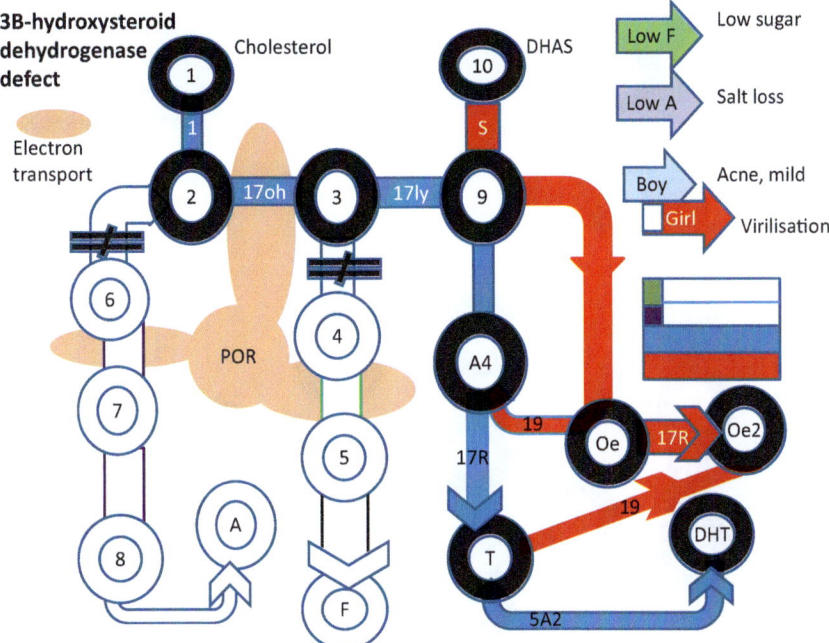

FIG. 3.1.2.8 Schematic for steroid biosynthesis in HSD3B2 deficiency. *(Author original.)*

TABLE 3.1.2.2 Adrenocortical steroids measurements in plasma of patients with HSD3B2 defects compared with controls.

Metabolite nmol/l	Patients IQR	Patients (Median)	Controls (Median)	Controls (IQR)
Aldosterone	0.11 - 0.41	0.16	0.36	0.19 - 1.6
Corticosterone	0.34 - 3.49	0.6	11.01	5.78 - 31.79
Pregnenolone	27.9 - 98.58	49.6	1.86	1.14 - 9.3
Progesterone	0.06 - 0.66	0.28	0.09	0.06 - 0.31
17-OH-Prog	0.27 - 7.73	2.38		0.15 - 1.26
17-OH-Preg	45.4 - 125	83.3	0.63	0.36 - 1.89
21-deoxycortisol	0.05 - 0.46	0.14	0.02	0.0 - 0.02
11-deoxycortisol	0.37 - 2.85	1.09	1.44	0.57 - 4.20
Cortisol	1.78 - 113	21.1	253	198 - 555
Cortisone	0.13 - 54.2	39.1	109	58.1 - 140
DHEA	8.01 - 105	39.9	5.37	2.3 - 17
DHEAS	200 - 4444	2116	197	59 - 2025
Androstenedione	0.27 - 2.82	0.97	0.27	0.13 - 0.76
Testosterone	0.17 - 1.94	0.55	0.1	0.06 - 0.2

Redrawn from
Guran et al 2020

From Guran T, Kara C, Yildiz M, Bitkin EC, Haklar G, Lin JC, Keskin M, Barnard L, Anik A, Catli G, Guven A, Kirel B, Tutunculer F, Onal H, Turan S, Akcay T, Atay Z, Yilmaz GC, Mamadova J, Akbarzade A, Sirikci O, Storbeck KH, Baris T, Chung BC, Bereket A. Revisiting Classical 3β-hydroxysteroid Dehydrogenase 2 Deficiency: Lessons from 31 Pediatric Cases. J Clin Endocrinol Metab 2020;105(3):dgaa022.

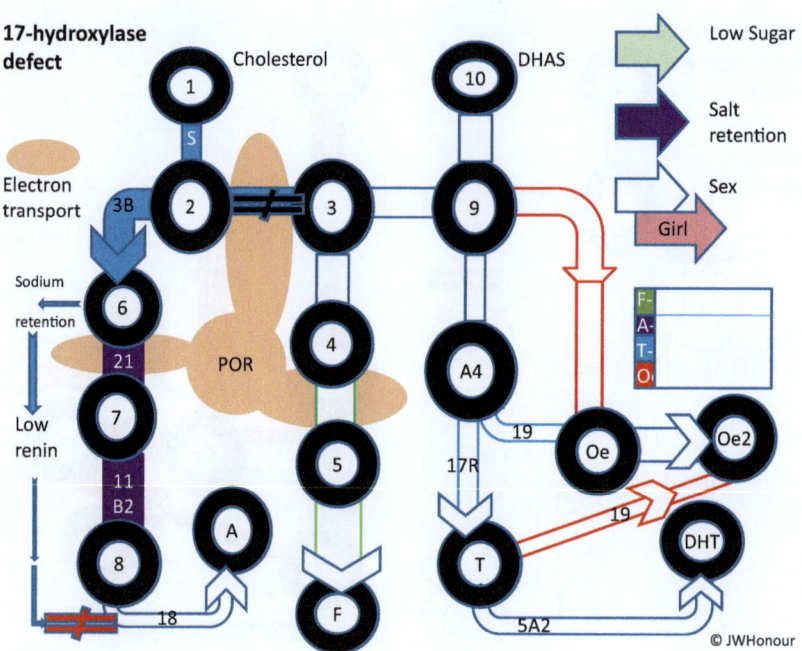

FIG. 3.1.2.9 Schematic for steroid biosynthesis in 17-hydroxylase deficiency. *(Author original.)*

TABLE 3.1.2.3 Single center experience of patients with CAH due to 17-hydroxylase deficiency.

Patient no	Sex phenotype	Sex genotype	Age, year	Exon	Variant	Signs/symptoms
P4	Female	46, XX	15.67	8	C.1459_1467delGACTCTTTC	Amenorrhea, lack of secondary sexual development, normotension, normokalemia
P5	Female	46, XX	11	8	C.1459_1467delGACTCTTTC	Severe fatigue, hypokalemic hypertension
P6	Female	46, XY	14	8	C.1459_1467delGACTCTTTC	Amenorrhea, lack of secondary sexual development, hypokalemic hypertension
P7	Female	46, XY	14.75	6	Y329K/418X	Lack of secondary sexual development, hypokalemic hypertension
P8	Male	46, XY	12.83	6	C.1073G>A	Hypospadias, gynecomastia, normokalemic normotension
				7	C.1169C>G	

From Wijaya M, Huamei M, Jun Z, Du M, Li Y, Chen Q, Chen H, Song G. Etiology of primary adrenal insufficiency in children: a 29-year single-center experience. J Pediatr Endocrinol Metab 2019;32(6):615–22. Table 3 p. 618.

21-hydroxyl group is removed and 21-deoxy steroids reabsorbed and excreted in urine (Fig. 3.1.2.11).

During an ACTH stimulation test, changes in excretion of 21-deoxysteroids were delayed due to the time of steroid excretion in bile, bacterial metabolism and reabsorption before urinary excretion (Shackleton et al., 1979) (Fig. 3.1.2.12).

When rats were given antibiotics to sterilize the intestine the blood pressure increased with corticosterone (Honour, 1982, 2015; Honour et al., 1985). Studies of anaerobic

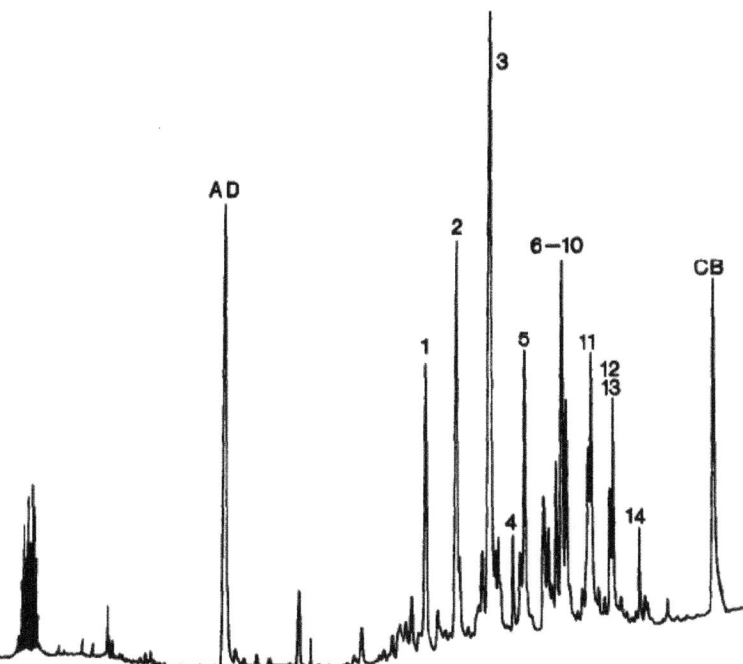

FIG. 3.1.2.10 Urinary steroid profile by GC-MS of patient with CYP17A1 deficiency. Androgen metabolites would normally elute as peaks between internal standard AD and steroid #1. Major steroids are: (1) 5β-pregnane-3α,20α-diol; (2) 5-pregnene-3α,20α-diol; (3) 3α,20α-dihydroxy-5α-pregnan-11-one; (4) 5β-pregnane-3α,11β,20α-triol; (5) 5α-pregnane-3α,11β,20α-triol; (6) 3α,21-dihydroxy-5β-pregnane-11,20-dione; (7) 3α,21-dihydroxy-5α-11,20-dione; (8) tetrahydrocorticosterone; (9) allo-tetrahydrocorticosterone; (10) partially characterised as 1,3,20-trihydroxy-5-prenan-11-one; (11) 3α,20α,21-trihydroxy-5α-pregnane-3β,16α,20α,21-tetrol; (12,13) 5α(and β)-pregnane-3α,11β,20α,21-tetrol. Peaks 1 and 2 are metabolites of progesterone. Steroids 3, 4 and 5 are formed by bacterial 21-dehydroxylation of 6, 7, 8, 9, 11, 12 and 13.

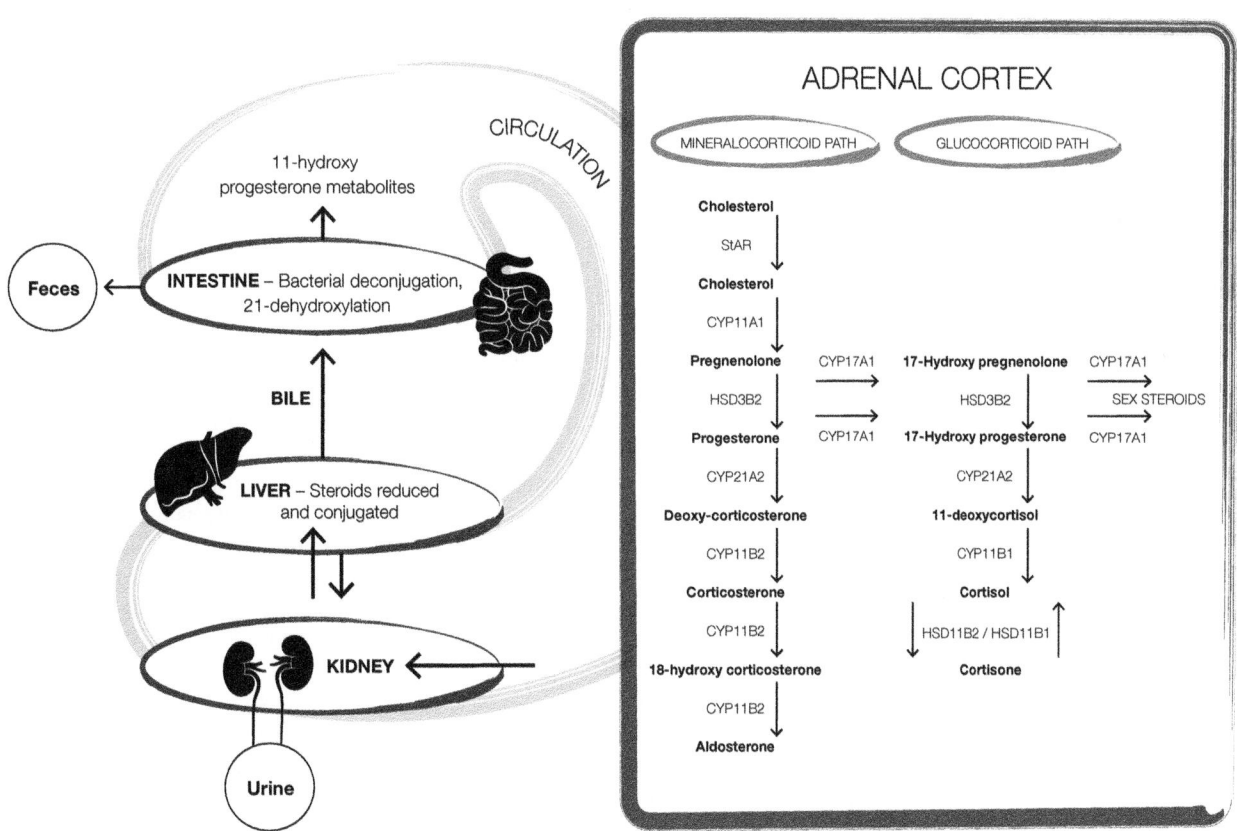

FIG. 3.1.2.11 Metabolism of corticosterone in liver, intestine and enterohepatic circulation. Most steroids are excreted in urine after reduction in the liver by A-ring and conjugation with glucuronic acid. Corticosterone and sex steroids are excreted in bile. Steroids in the intestine are subject to bacterial metabolism. Tetrahydrocorticosterone can lose the 21-hydroxyl group and 21-deoxysteroid metabolites, in effect metabolites of 11-hydroxyprogesterone, are reabsorbed from the intestine. Deoxycorticosterone excess causes hypertension. Progesterone can also be converted to 11-hydroxyprogesterone by the action of CYP11B2. *CYP11A1*, side chain cleavage; *CYP11B1*, 11B-hydroxylase; *CYP11B2*, 11 B-hydroxylase and aldosterone synthase; *CYP17A1*, 17-hydroxylase; *CYP21A2*, 21-hydroxylase; *HSD11B*, 11B-hydroxysteroid dehydrogenase; *HSD17B*, 17-hydroxysteroid dehydrogenase; *HSD3B2*, 3-beta hydroxysteroid dehydrogenase; *StAR*, steroidogenic acute regulatory protein. *(From Honour JW. Historical perspective: gut dysbiosis and hypertension. Physiol Genomics 2015;47(10):443–6. Fig. 1 p. 444.)*

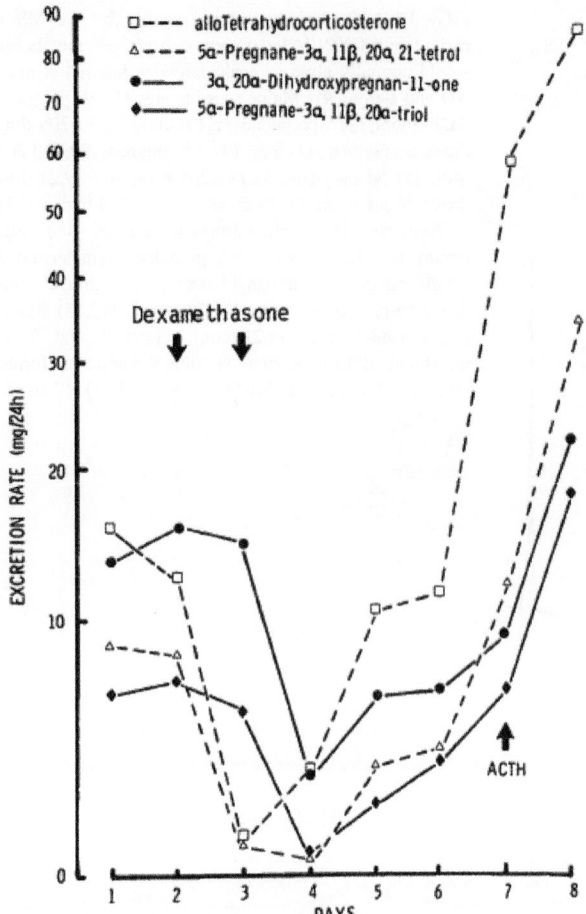

FIG. 3.1.2.12 Effects of dexamethasone then ACTH on urinary steroid excretion of corticosterone metabolites in patient with defect of 17-hydroxylase. Changes in excretion of 21-deoxy steroids delayed due to time in the enterohepatic circulation. *(Based on Honour J, Millar G, Roitman E, Shackleton C. Steroid excretion in urine during suppression and stimulation of adrenals in the 17 alpha-hydroxylase deficiency syndrome. J Clin Endocrinol Metab 1981;52(5):1039–42.)*

bacteria were difficult before 2010 when bacterial 16S RNA analysis was introduced. More research around the microbiome and steroid metabolism is anticipated. Some steroids may act as glycyrrhetinic-like acids factors (Morris et al., 2014).

3.1.2.4.1.4   CYP11B1

Deficiency of 11-hydroxylase also prevents synthesis of cortisol but sex steroids are produced so ambiguous genitalia with clitoromegaly and fused labia are seen in 46XX patients but boys have normal genitalia (Breil et al., 2019). The steroid 11-deoxycortisol is the substrate for the enzyme and this is elevated in plasma with a high ratio of 11-deoxycortisol to cortisol (Khattab et al., 2017) and raised tetrahydro-11-deoxycortisol in urine (Tran et al., 2019; Ratcliffe et al., 1982). Deoxycorticosterone is stimulated by ACTH and causes sodium retention, renin suppression and low aldosterone (Fig. 3.1.2.13).

The usual presentation in boys is precocious puberty with acne due to excess androgens, rapid growth and hypertension is a feature from puberty (see Hypergonadism and Mineralocorticoid excess chapters). At puberty, girls can also develop hirsutism and menstrual irregularities.

3.1.2.4.1.5   Cytochrome P450 oxidoreductase (POR) deficiency

An apparent combined deficiency of 21-hydroxylase and 17-hydroxylase is attributed to a defect in a common electron transporter called **cytochrome P450 oxidoreductase** (POR; ORD) (Fan et al., 2019; Hao et al., 2018; Reardon et al., 2000; Kelley et al., 2002; Miller et al., 2009). This defect can be associated with appearance of the Antley-Bixler syndrome (Fig. 3.1.2.14).

A raised 17-hydroxyprogesterone is suggestive of 21-hydroxylase, other steroids (progesterone and corticosterone) need to be checked if a POR defect is likely from facial features. A urine steroid profile is a valuable test to cover these possibilities (Fig. 3.1.2.15) (Lee et al., 2020a, b,c; Krone et al., 2012; Fukami et al., 2009; Shackleton et al., 2004; Wudy et al., 2004). POR is the electron transporter of NADH and serves two enzymes in steroid biosynthesis.

3.1.2.4.1.6   Congenital lipoid adrenal hyperplasia (CLAH)

Lipoid CAH was first described pathologically with cholesterol ester accumulation in the adrenal glands and gonads and later with impairment of steroidogenesis. Cholesterol uptake into mitochondria requires **Steroidogenic acute regulatory protein** (StAR) so when the protein is deficient steroids are absent in blood and urine. Investigations of patients with familial glucocorticoid deficiency failed to find mutations in the ACTH receptor but a genetic locus was found on Chromosome 8. Defects in steroidogenic acute regulatory protein StAR gene at 8p11.2 were found. Mutations have been found in the **Steroidogenic acute regulatory protein (StAR)** (Fig. 3.1.2.16) (Zhang et al., 2021; Fujieda et al., 2003; Miller, 2007).

There may be a family history of boys and neonatal death. Some patients with partial inactivation of StAR present after 2 years of age with mild hypoglycemia and hyperreninemia or later with compromised fertility in females and azoospermia in males (Ishii et al., 2020; Metherell et al., 2009). A c.772C>T mutation is common in Chinese patients from founder effect in Korea (Zhang et al., 2021; Huang et al., 2016). Skin pigmentation is a sign of nonclassic CLAH where gene mutations lead to protein with partial activity of StAR (Bae et al., 2020).

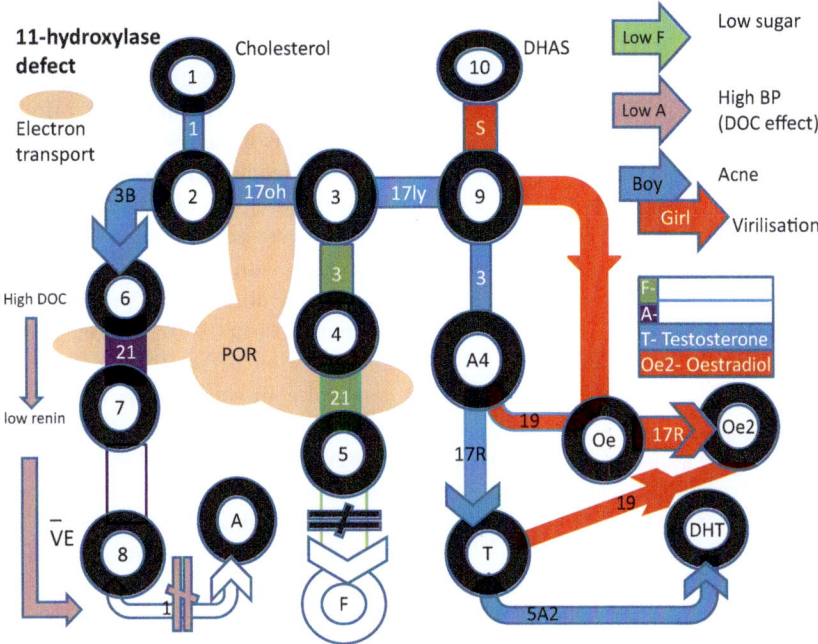

FIG. 3.1.2.13 In 11-hydroxylase deficiency—suppression renin by DOC. *(Author original.)*

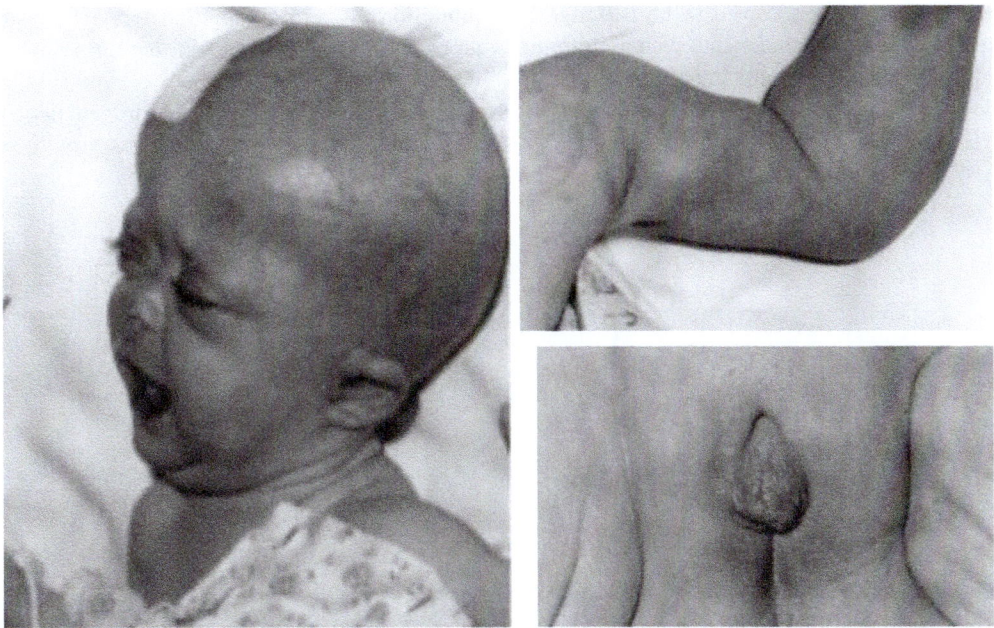

FIG. 3.1.2.14 Patient with Antley-Bixler syndrome showing craniofacial features, characteristic fixed elbow, and genital abnormalities (clitoromegaly and hooded prepuce). *(Modified from Reardon W, Smith A, Honour JW, Hindmarsh P, Das D, Rumsby G, Nelson I, Malcolm S, Adès L, Sillence D, Kumar D, DeLozier-Blanchet C, McKee S, Kelly T, McKeehan WL, Baraitser M, Winter RM. Evidence for digenic inheritance in some cases of Antley-Bixler syndrome? J Med Genet 2000;37(1):26–32. Fig 1. P27.)*

3.1.2.4.2 Treatment of congenital adrenal hyperplasia

The treatment of CAH is based on the replacement of cortisol and aldosterone and on suppression of excess androgens. This is not straightforward and varies considerably.

Cortisol was thought to be produced at 18 mg/day and early treatments were based on two thirds of the requirement in the morning and one third at night. In some centers, the dose was divided into three with additional dose mid-day. More recently, four times per day dosing has been introduced.

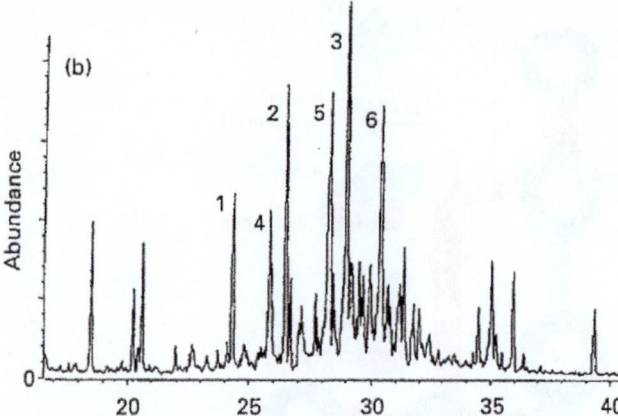

FIG. 3.1.2.15 Urinary steroid profile by GC-MS of patient with cytochrome P450 reductase deficiency. 1. 17-hydroxypregnanolone; 2. Pregnanetriol; 3. 11-oxopregnanetriol; 4. Pregnanediol; 5. 11-oxo-pregnanediol 6. Allo-tetrahydrocorticosterone. *(Author original.)*

Replacement of the natural ultradian secretion is not possible; taking the morning dose at 0400h was one approach to having cortisol in circulation before 0800h. The dose of hydrocortisone has been reduced since production rates by stable isotope technology are accepted to be 6–9 mg/M²/day. There have been other developments such as administration of hydrocortisone by insulin pump (Bryan et al., 2009) and delayed release tablets that are in Phase 3 clinical trial.

In some countries, lower doses of hydrocortisone are available in the prescription products. Two items based on hydrocortisone granules (1mg, Alkindi or Infacort) or coded tablets of 1, 5 and 10mg offer more flexible treatment plans. Cortisol was thought to be produced at 18mg/day and early treatments were based on two thirds of the requirement in the morning and one third at night. In adults, this protocol can be reversed. In some centers, the dose was divided into three with additional dose mid-day. Replacement of the natural ultradian secretion is not possible; taking the morning dose at 0400h was one approach to having cortisol in circulation before 0800h (the late JDH [Willie] Slater personal communication). The dose of hydrocortisone has been reduced since production rates by stable isotope technology became accepted to be 6–9 mg/M²/day. There have been other developments such as administration of hydrocortisone by insulin pump (Bryan et al., 2009) which can provide the best 24h cortisol profile and delayed release tablets that are in various stages of clinical trial development (Plenadren, Efmody, Chronocort) (Prete et al., 2021). Abiraterone acetate is an inhibitor of CYP17 that reduces steroid production at an early stage, reducing androgen production but the risk of hypertension. Blocking steroid synthesis by inhibition of sterol-*O*-acyltransferase (Nevanibe) failed at an early stage of clinical trial but does not exclude further investigation of drugs to block steroid synthesis. Mitotane which is normally used to androgen dependent cancers was found to be useful to suppress growth of adrenal rest

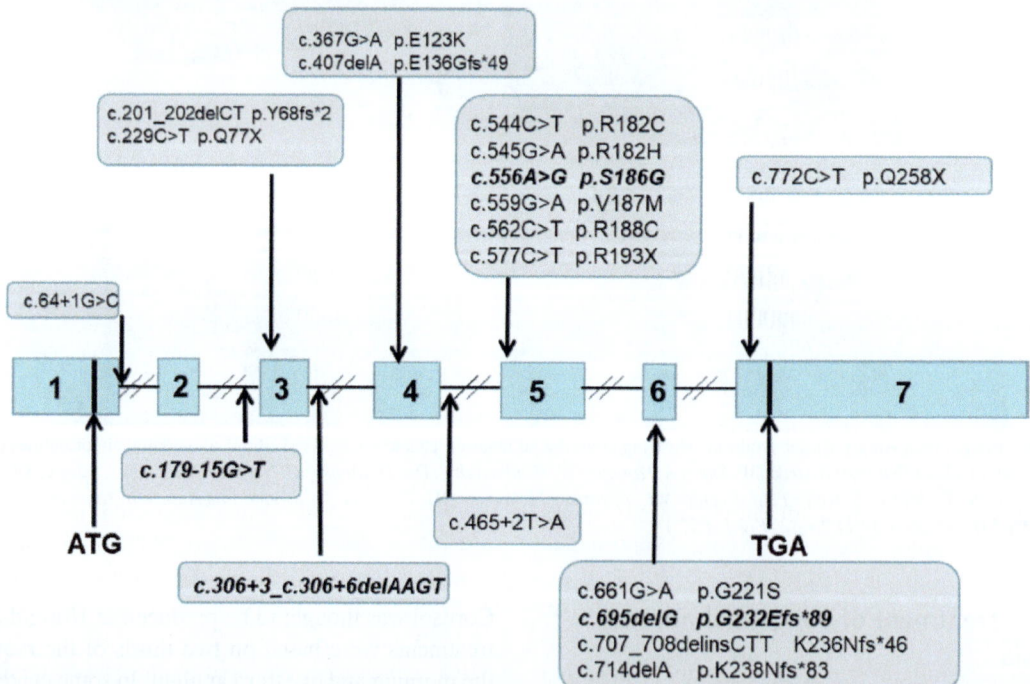

FIG. 3.1.2.16 Diagram of the STAR gene showing the location of mutations identified in patients with LCAH. Novel variants are in **bold**. *(From Zhang T, Ma X, Wang J, Jia C, Wang W, Dong Z, Ye L, Sun S, Hu R, Ning G, Li C, Lu W. Clinical and molecular characterization of thirty Chinese patients with congenital lipoid adrenal hyperplasia. J Steroid Biochem Mol Biol 2021;206:105788. Fig. 1 p. 6.)*

tumors that are a side effect of long-term exposure to excess ACTH in poorly treated CAH. Adrenal rests are a problem if bilateral adrenalectomy is used as a possible cure for CAH.

With any treatment so far there are risks to the patients with too much or too little hydrocortisone causing features of Cushing's syndrome or osteoporosis and for androgens causing virilization or short stature and infertility. Although 17-OHP, cortisol and androstenedione are dependent on ACTH, changes in their concentrations do not always go together. This makes monitoring concentrations difficult to interpret.

Prices of drugs have changed for a number of commercial reasons and one center changed hydrocortisone with a single dose of prednisolone at night to mimic the diurnal rhythm of cortisol. This left the patient without glucocorticoid cover for many hours in the daytime.

To combat androgen excess spironolactone (or eplerenone) that blocks the androgen receptor as well as the MR was introduced but the side effect was gynecomastia that is treatable with an aromatase inhibitor. This has been replaced with Flutamide but this needs to be in conjunction with glucocorticoid. A number of drugs targeted to CRF1 receptor antagonism are in development (Crinecerfont; Tildacerfont).

Growth hormone or Leuprolide has been used in children to improve growth but did not improve outcome compared with controlled GCH and MCH treatment. Other treatments in the development are anti-ACTH monoclonal antibodies and ACTH receptor antagonists.

Production of epinephrine in the adrenal medulla is dependent on cortisol in the blood from the cortex. Patients with CAH have adrenomedullary insufficiency. The physiologic responses of glucose, insulin and leptin pathways are dysregulated during exercise in patients lacking both cortisol and epinephrine. Epinephrine deficiency may contribute to hypoglycemia during illnesses with fever, especially young children, and impair the stress response. Decreased epinephrine production has been observed in newborns with classic CAH compared with controls. Epinephrine replacement or supplementation has not been studied. It is not known whether a compensatory norepinephrine response is sufficient.

An adenoviral gene therapy transiently restored enzyme activity in a mouse model of 21-hydroxylase deficiency (Tajima et al., 1999) further animal research is ongoing. Intravenous injection of an adenoviral-Cyp21a1 vector in similar mice showed enzyme expression in adrenal tissue, resulting in weight gain, reduction in progesterone concentrations and improved stress response that lasted for more than 15 weeks (Perdomini et al., 2017) a response that was not repeated in another laboratory where the effect only lasted for 8 weeks (Markmann et al., 2018). Auto-transplantation of Cyp21A1-expressing fibroblasts into 21-hydroxylase-deficient into mouse subcutaneous

tissue or direct injection of adenovirus-Cyp21A1 constructs into mouse muscle demonstrated reduction in the ratio of progesterone to deoxycorticosterone for nearly 8 weeks when the animals were sacrificed (Naiki et al., 2016).

The ability to generate donor-specific and functional adrenocortical-like cells would facilitate: (1) the next generation of cell-based treatments for AI; (2) the modeling of adrenal-specific diseases and (3) the testing of personalized interventions on cells derived from patients (Bornstein et al., 2020). Permanent correction of mutations causing CAH is desirable and has been tested in by culturing cells from urine and selection of cells for ability to express steroidogenesis (Zhou et al., 2012; Ruiz-Babot et al., 2018). Cells were isolated from urine of patients with CAH then programmed to express native CYP21A1 using a lentiviral vector. Cells were created that corrected the loss of CYP21A2 in a patient's own adrenal stem cells. When cells in a fibrinogen/thrombin mixture were transplanted into mouse adrenal gland or kidney capsule enzyme activity was preserved. This approach would theoretically cure CAH and supplant imperfect steroid replacement.

3.1.2.4.2.1 Auto-immune adrenal insufficiency

AI can be due to **specific auto-antibodies** against steroidogenic enzymes, 21-hydroxylase being the commonest (Wolff et al., 2021) but 17-hydroxylase, side chain cleavage (Chen et al., 1996) and HSD3B2 (Arif et al., 2001) have also been found. Immunoprecipitation (Colls et al., 1995) and ELISA assays are used for autoantibody tests (Del Pilar Larosa et al., 2018). Commercial kits seem to perform better than "in house" tests (Falorni et al., 2015). The diagnosis of APS types is difficult because of rarity and range of presentation. Antibodies are needed against cytokines, enzymes and specific organs (Savvateeva et al., 2021).

Adrenal insufficiency can be part of **autoimmune polyendocrine syndromes (APSs)** involving other endocrine glands causing type 1 diabetes, thyroid disease and ovarian failure as well as other clinical issues such as candidiasis and hypoparathyroidism. The autoimmune conditions tend to cluster within families. There are five types of APS (or PAS), where type 1 is adrenal insufficiency, hypoparathyroidism (CH), mucocutaneous candidiasis (CMC); type 2 is Addison's disease (AD) with thyroid disease (AITD) and/or type 1 diabetes. Type 4 includes autoimmune conditions other than AITD and T1D such as chronic atrophic gastritis (CAG) and vitiligo. APS3 is associated with premature ovarian insufficiency, Hashimoto-type thyroid antibodies and 21-hydroxylase antibodies.

In a large study of 158 patients, mutations in the **AIRE gene** are associated with APS-1 (Garelli et al., 2021), the presentation fitted the classical triad of CMC (Table 3.1.2.4) (76% at 9.1 years), CH (86% at 11 years) and AD (77% at 16.3 years).

TABLE 3.1.2.4 Autoimmune polyendocrine syndrome type 1.

Presentation: classic triad (*n*=158)

- Chronic mucocutaneous candidiasis 76%
  Median age presentation 9.1 years
- Chronic hypoparathyroidism 86%
  Median age presentation 11 years
- Addison's disease 77%
  Median age at presentation 16.3 years

Data from Garelli S, Dalla Costa M, Sabbadin C, Barollo S, Rubin B, Scarpa R, Masiero S, Fierabracci A, Bizzarri C, Crinò A, Cappa M, Valenzise M, Meloni A, De Bellis AM, Giordano C, Presotto F, Perniola R, Capalbo D, Salerno MC, Stigliano A, Radetti G, Camozzi V, Greggio NA, Bogazzi F, Chiodini I, Pagotto U, Black SK, Chen S, Rees Smith B, Furmaniak J, Weber G, Pigliaru F, De Sanctis L, Scaroni C, Betterle C. Autoimmune polyendocrine syndrome type 1: an Italian survey on 158 patients. J Endocrinol Invest 2021;44 (11):2493–2510.

The range of mutations probably reflects founder genes from different migrant populations. In addition to AIRE mutations, the study concluded that other genetic, environmental or stochastic factors may impact the phenotypes of APS-1.

3.1.2.4.3 Glucocorticoid resistance

Patients with hypercortisolism without signs of Cushing's syndrome may have glucocorticoid resistance. There can be signs of mineralocorticoid and androgen excess so there can be hypertension and hypokalemic alkalosis, hirsutism, acne and male pattern hair loss. Morning cortisol concentrations can be seven times normal. Urine-free cortisol should be determined on at least 2 consecutive days and can be grossly elevated (50 times normal). ACTH will be raised. There will be suppression of cortisol with dexamethasone, both steroids should be measured to exclude increased clearance or reduced absorption of the dexamethasone. The NR3C1 gene should be sequenced. The molecular mutations and clinical manifestations of patients with resistance were reviewed by Nicolaides and Charmandari (2021). Inherited and de novo defects have been found.

3.1.2.5 Adrenal dysgenesis/hypoplasia

A number of disorders affecting adrenal and gonadal development have been characterized in recent years. They present clinically and biochemically with inadequate or absence of adrenal and/or gonadal steroids. These include in boys defects of the nuclear receptors DAX-1 and SF-1. (More detail on these conditions can be found in the chapter on Hypogonadism.)

3.1.2.5.1 X-linked adrenal hypoplasia congenita

Mutations in the NROB1 gene are found in the dosage sensitive sex reversal **adrenal hypoplasia congenita** on the X-chromosome gene 1 (DAX-1) that lead to neonatal adrenal insufficiency and failure to undergo puberty with hypogonadotrophic hypogonadism (Butz et al., 2021; Zhang et al., 2015; Yu et al., 1998; Calliari et al., 2007; Ozer et al., 2009; Yang et al., 2009; Landau et al., 2010). This was originally described as the cytomegalic form of CAHo. Until puberty, gonadotrophin concentrations are normally suppressed so evidence for hypogonadotrophic hypogonadism is not apparent without a stimulation test.

In a 29-year experience of 434 cases of primary adrenal insufficiency, 83% were congenital adrenal hyperplasias and 20 cases with mutations in NROB1 (or sometimes called DAX-1) were identified (Table 3.1.2.5) (Wijaya et al., 2019).

In the group studied, 5 presented with an adrenal crisis before 2 years of age, 10 presented with salt craving, presented with pigmentation and 2 patients presented with premature puberty.

DAX-1 protein is related to nuclear receptors although it lacks the zinc finger DNA-binding domain that is highly conserved in other member of the family. DAX-1 has been shown to repress the transcription of genes that are regulated by SF-1, another nuclear receptor (Yu et al., 1998).

A miniature adult form of CAHo is due to contiguous gene deletion of Xp21.2-p21.3 with **glycerol kinase** and **Duchenne muscular dystrophy**. Infants present with hypoglycemic acidosis, older patients have hyperpigmentation, metabolic acidosis and hyponatremia (Rathnasiri et al., 2021). Elevated glycerol concentrations are seen in urine analysis of organic acids and elevated triglycerides in blood give serum which is cloudy.

3.1.2.5.2 SF-1/NR5A1

Steroidogenic factor 1 (SF-1), a nuclear receptor, is expressed in adrenals and gonads and mutations in the NR5A1 gene are found in some patients who present with adrenal failure and 46XY DSD (de-Souza et al., 2006; Köhler et al., 2009; Hoivik et al., 2010). The NR5A1 gene is located at 9q33.3 and SF-1 regulates the promoters of CYP19A, DAX, CYP11A1 and AMH. Heterozygous defects in SF1 range from adrenal insufficiency, sex reversal, to ovarian failure or azoospermia (Fabbri-Scallet et al., 2020; Wijaya et al., 2019; An et al., 2021; Köhler et al., 2008; Köhler and Achermann, 2010).

TABLE 3.1.2.5 Patients with adrenal hypoplasia congenita.

Patient no	Sex	Age at onset	Gene mutation	Variant	Signs/symptoms	HH	Diagnosis
P11	Male	1 month	NROB1	c.1168+1_116 +2dupGT	Salt craving	+	X-AHC
P12	Male	10 days	NROB1	c.791_793delAGA	Salt craving	+	X-AHC
P13	Male	1 month	NROB1	c.848A>C	AC	+	X-AHC
P14	Male	2 years	NROB1	c.814_816delTTG	AC	+	X-AHC
P15	Male	2.5 years	NROB1	c.604dupT	AC		X-AHC
P16	Male	15 days	NROR1	c.460A>T	Salt craving		X-AHC
P17	Male	10 days	NROB1	c.332_333delCT	Salt craving		X-AHC
P18	Male	6 years	NROB1	c.754C>T	Skin pigmentation	+	X-AHC
P19	Male	5 years	NROB1	c.460A>T	Skin pigmentation		X-AHC
P20	Male	1 month	NROB1	c.1231_1234delCTCA	Salt craving		X-AHC
P21	Male	1.67 years	NROB1	c.604dupT	AC		X-AHC
P22	Male	5 days	NROB1	c.332_333delCT	Salt craving		X-AHC
P23	Male	1 month	NROB1	E1-2 del	Salt craving		X-AHC
P24	Male	2 month	NROB1	c.913C>T	Premature puberty		X-AHC
P25	Male	5 days	NROB1	c.585_595del11	Salt craving		X-AHC
P26	Male	2 years	NROB1	c.848A>C	Premature puberty		X-AHC
P27	Male	2 years	NROB1	c.848A>C	AC		X-AHC
P28	Male	1 month	NROB1	c.838delC	Salt craving		X-AHC
P30	Male	5 years	NROB1	c.793delA	Skin pigmentation		X-AHC
P30	Male	1 month	NROB1	del, hom	PAI, intellectual disability, hypertriglyceridemia	+	Xp21 contiguous gene deletion
P31	Female (46, XY)	1 month	NR5A1	c.616_758 del	Poor weight gain, skin pigmentation		SF-1 gene mutation

AC, adrenal crisis; *HH*, hypogonadotropin hypogonadism; *PAI*, primary adrenal insufficiency; *X-AHC*, X-linked adrenal hypoplasia congenital.
From Wijaya M, Huamei M, Jun Z, Du M, Li Y, Chen Q, Chen H, Song G. Etiology of primary adrenal insufficiency in children: a 29-year single-center experience. J Pediatr Endocrinol Metab 2019;32(6):615–22. Table 4 p. 619.

3.1.2.6 Scheme for biochemical investigation of PAI

The steroid enzyme disorders have detectable changes in the steroid hormones (or their metabolites in urine) (summarized in Fig. 3.1.2.17) that enable early detection if samples are processed and interpreted promptly. Genetic testing is required for other disorders.

Plasma cortisol concentrations between 7 and 9 am that are repeatedly less than 170 nmol/L are suggestive of adrenal cortical insufficiency. In a sick child, cortisol concentrations of 300 nmol/L that do not increase on Synacthen should be regarded as inappropriately low and worth investigating. This may not be seen however, until the course of the disease has advanced. Low concentrations of cortisol in saliva also reflect Addison's disease (<15 nmol/L) (Kim et al., 2020). A short Synacthen test should be performed to assess adrenal reserve. Blood for serum cortisol is taken before and at 30 and 60 min after an intravenous injection of 250 μg of soluble Synacthen. Lower doses of Synacthen

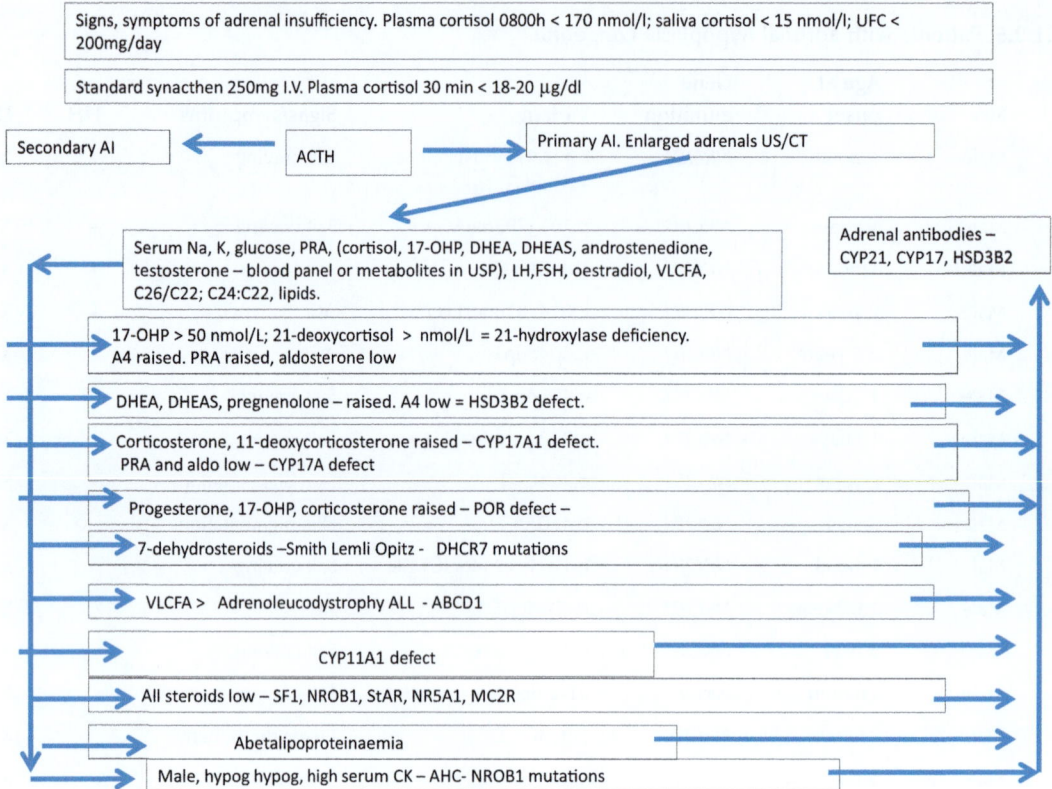

FIG. 3.1.2.17 Biochemical investigations for adrenal insufficiency. *(Author original.)*

(62.5 or 125 μg) should be used in young children but bloods should be taken at 15, 20, 25 and 30 min because the cortisol peak is earlier than with high-dose stimulation. A normal response is characterized by an increment in cortisol of at least 200 nmol/L or a rise to levels above 500 nmol/L. If an assay for cortisol is specific enough to exclude cross reaction with prednisolone, this synthetic steroid can be given immediately after the basal blood has been taken so as to afford glucocorticoid cover without affecting the adrenal response to exogenous ACTH. If there is pigmentation, the patient has primary adrenal failure (Addison's disease). Plasma ACTH measurements will be raised in Addison's disease and normal or low in secondary adrenal failure. A metyrapone test or insulin tolerance test can be helpful in distinguishing primary from secondary adrenal insufficiency. Circulating antibodies to the adrenal cortex suggest an autoimmune process and other endocrine tests may be required to look for an extension of the autoimmune process to other hormonal tissues. Cortisol should be replaced at 9–12 mg/M^2/24 h and it is useful to check cortisol concentrations throughout the day in plasma samples taken at 30 min intervals over 2 h after a morning dose of hydrocortisone then at 2–3 h intervals, or better with hourly sampling, throughout the day. Depending on the frequency of sampling and the assay used, cortisol concentrations can be between 500 and 550 nmol/L within an appropriate pattern reflects adequate replacement therapy.

3.1.2.7   Disorders associated with syndromic features

3.1.2.7.1   Adrenoleukodystrophy

An adrenal defect with an X-linked transmission is adrenoleukodystrophy. The presence of neurological signs is not constant. Adrenal insufficiency appears after the age of 4 or 5 years. The measurement of long chain fatty acid in high concentrations allows the diagnosis of adrenoleukodystrophy. Indications of Addison disease in young males should prompt consideration of adrenoleukodystrophy (Moser et al., 2007; Chang et al., 2008) which is a peroxisomal disorder. Elevated levels of very long chain fatty acids (VLCFA) are found in plasma. Neurologic problems may or may not be present and can progress to an early demise. Hematopoietic stem cell transplantation has so far been the only treatment. Newborn screening programmes in many states in the USA include measurement of the C26:0 very long chain fatty acid (Tortorelli et al., 2016). The concentration of C26:0 (1.38–10.28 μmol/L; normal 0.45–1.32) can be determined or an elevated ratio of C26:0 to C22:0 is indicative of ALD. The disease is caused by defects in the gene for the **adenosine triphosphate (ATP)-binding cassette protein, subfamily D (ABCD1)** that encodes the peroxisomal transporter of very-long-chain fatty acids (VLCFAs).

MIRAGE syndrome is characterized by **m**yelodys-plasia, **i**nfections, **r**estriction of **g**rowth, **a**drenal **hypoplasia, g**enital phenotypes and **e**nteropathy. Other anomalies include, but not limited to, microcephaly, crypt-orchidism, hypospadias and mild facial dysmorphism (Table 3.1.2.6).

Many cases of Mirage syndrome are born preterm. Most of the published cases had died within 2 years of birth (Viaene and Harding, 2020). Heterozygous, activating mutations of **SAMD9 (sterile alpha motif domain containing 9)** result in increased antiproliferative effects that restrict growth in multiple systems. In the years prior to recognition of the syndrome (Narumi, 2018; Narumi et al., 2016), an 11-year-old boy had undergone investigations for immune thrombocytopenia purpura, anemia, proteinuria, hypertension, repeat noninfectious diarrhea and recurrent airway disease (Perisa et al., 2019). At 10 years of age, a plasma cortisol (3.4 mg/dL; normal range 2–18 mg/dL) and poor response to Synacthen (peak 9.7 mg/dL; normal >15.5) suggested adrenal insufficiency. A missense mutation in SAMD9 was reported. This may be a variant of MIRAGE.

TABLE 3.1.2.6 Patient characteristics MIRAGE syndrome.

Clinical feature	Patient characteristics
Karyotype	46 XY
SAMD9 mutation	C 2318 T>C /p 1773T
Gestational age at delivery	34 weeks
Intrauterine growth restriction (IUGR)	Yes
Birth weights (mean and standard deviation score)	1843 g—2.5
Urogenital abnormalities	Cryptorchidism; unilateral vesicoureteral reflux; hypospadias
Anemia	Yes
Thrombocytopenia	Yes
Adrenal insufficiency	Yes
Recurrent infections	Yes, with hypogammaglobulinemia
Other phenotypic changes	Microcephaly, brachydactyly 5th digits, attention deficit hyperactive disorder (ADHD), mild developmental delay

From Perisa MP, Rose MJ, Varga E, Kamboj MK, Spencer JD, Bajwa RPS. A novel SAMD9 variant identified in patient with MIRAGE syndrome: further defining syndromic phenotype and review of previous cases. Pediatr Blood Cancer 2019;66(7):e27726. Table 1 p. 204.

3.1.2.7.2 IMAGe

Intrauterine growth restriction, metaphyseal dysplasia, adrenal hypoplasia congenita and genital abnormalities are the features of the acronym **IMAGe** syndrome. On physical examination, there are several dysmorphic features—small, low set ears, craniosynostosis, cleft palate, scoliosis and cryptorchidism. Some features overlap with Russell Syndrome and isolated growth hormone deficiency. Bilateral sensorineural hearing loss was described later. The clinical features of published cases have been reviewed (Balasubramanian et al., 2010) and epiphyseal dysplasia was a more striking feature than metaphyseal dysplasia. Gain of function mutations in **cyclin-dependent kinase inhibitor 1C gene (CDKN1C)** were found in these patients (Buonocore et al., 2021).

A number of other syndromes are extremely rare and are not described in detail here (see Table 3.1.2.7). In addition to disorders of steroidogenesis, there can be defects of cholesterol trafficking and mitochondrial cholesterol uptake.

3.1.2.8 Cholesterol disorders

In stress, there is a need to mobilize cholesterol rapidly to meet the demands of cortisol synthesis. Cholesterol for steroidogenesis has multiple sources including free cholesterol, cholesterol esters, HDL and LDL. Some defects in cholesterol transfer affect adrenal and gonadal functions.

3.1.2.8.1 Steroid synthesis

Smith-Lemli-Opitz syndrome may present as ambiguous genitalia and distinctive facial features, small head size (microcephaly). Malformations of the heart, lungs, kidneys and GI tract are also common. Infants with Smith-Lemli-Opitz syndrome have weak muscle tone hypotonia, experience feeding difficulties and tend to grow more slowly than other infants. Most affected individuals have fused second and third toes (syndactyly), and some have extra fingers or toes (polydactyly). A defect in cholesterol synthesis through 7-dehydrocholesterol reductase is reflected in elevation of delta-7 steroids/sterols in blood and urine (Ryan et al., 1998; Shackleton et al., 1999) (Fig. 3.1.2.18).

3.1.2.8.2 Cholesterol transport

Dietary cholesterol binds to Niemann-Pick protein (NPC2) in endosomes to meet lysosomal acid lipase (LAL) that releases free cholesterol. NPC1 then trafficks free cholesterol to the mitochondria. **Niemann-Pick disorders** are now classified as an acid sphingomyelinase disorder. There are three types, with type C involved in cholesterol movement, A and B are sphingomyelinases. Defects of NPC-C lead to accumulation of cholesterol and

TABLE 3.1.2.7 Extremely rare syndromes with adrenal insufficiency.

Syndrome	Gene affected	Gene location	Inheritance	Clinical features
Galloway-Mowat	WDR73	15q25.2	AR	Early onset extreme encephalopathy, epilepsy, nephrotic syndrome, microcephaly, optic atrophy, hiatal hernia
Hydrolethalus	HYLS1	11q24.2	AR	Prenatal-onset severe hydrocephalus, polydactyly, micrognathia, congenital heart and pulmonary defects
Meckel	MKS1	17q22	AR	CNS malformation, polycystic kidney, hepatic fibrosis, polydactyly
Pallister-Hall syndrome	GLI3	7p14.1	AD	Hypothalamic hamartoblastoma, imperforate anus, polydactyly, cleft palate
Pena-Shokier	OK7	4p16.3	AD	Arthrogryposis, fetal akinesia, IUGR, cystic hygroma, pulmonary hypoplasia, cleft palate, cryptorchidism, cardiac defects, intestinal malrotation
Pseudotrisomy 13	RAPSN			Holoprosencephaly, facial abnormalities, postaxial polydactyly
Townes-Brocks syndrome	SALL1	16q12.1	AD	Imperforate anus, triphalangeal and supernumerary thumbs, malformed ears, sensorineural deafness

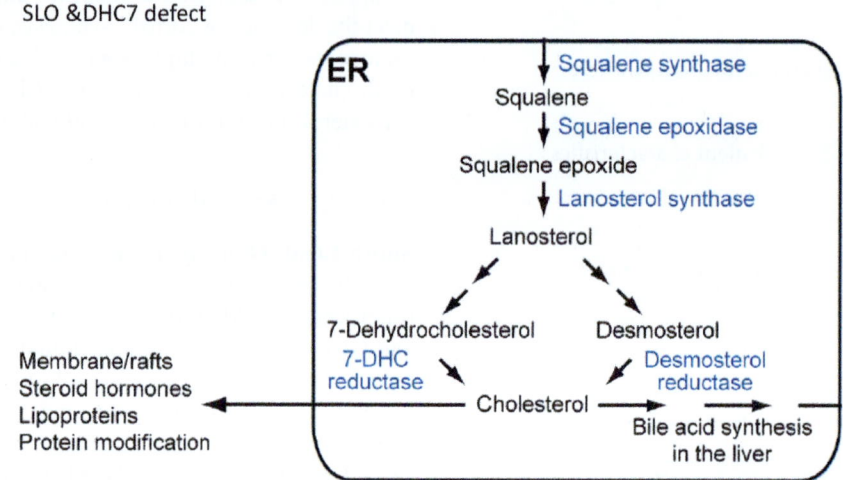

FIG. 3.1.2.18 Paths for cholesterol synthesis from squalene. Defects of 7-Dehydrocholesterol reductase (DHCR7) lead to Smith-Lemli-Opitz syndrome (SLO).

glycososphingolipids in endosomes causing neurodegeneration and death. Infants and older children present with yellow discoloration of the skin and eyes, difficulty in walking, feeding difficulties, learning disabilities, loss of previously learned speech and an enlarged liver or spleen. Cholestane-3β,5α,6β triol is a marker for NPC deficiency (Geberhiwot et al., 2018). Genetic analysis of NPC should be undertaken to confirm the defect. NPC-C deficiency is incredibly rare and patients require close observation as the disease progresses.

Low-density lipoproteins are the richest source of cholesterol and enter the adrenal cortex and gonadal cells through **LDL receptor** mediated endocytosis in clathrin coated pits. Familial hypercholesterolemia (FH) is caused by mutations in the gene that encodes the LDL receptor (LDLR) at 19p13.3. FH is associated with high levels of LDL protein cholesterol (LDL-C), inherited in an autosomal dominant manner. Mutations of LDL-R lead to decreased LDL-C uptake but defects in the gene for apolipoprotein B (ApoB) have also been found. Further mutations have been found in **proprotein convertase subtilisin/kexin type 9 (PCSK9)** that normally degrades LDL-R in lysosomes (Sharifi et al., 2017).

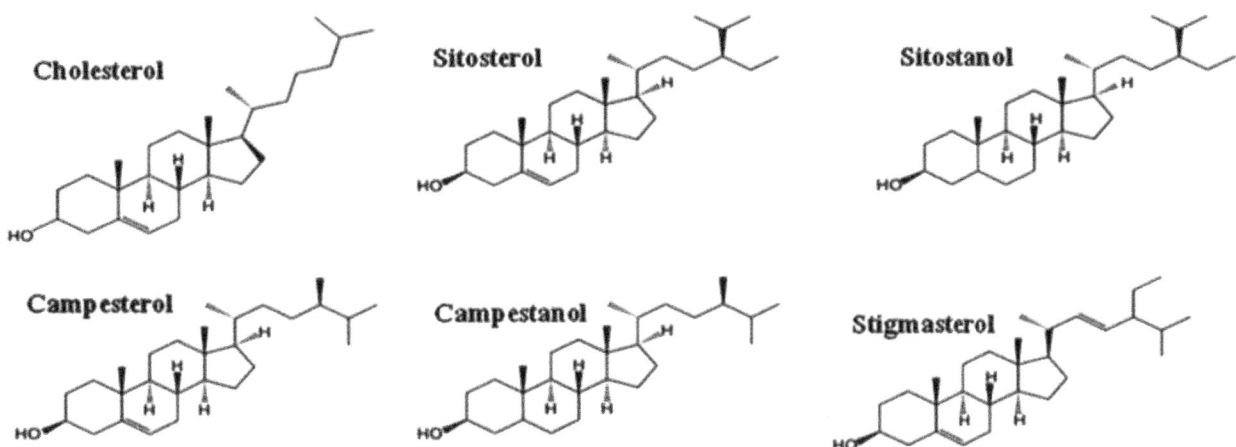

FIG. 3.1.2.19 Structures of plant sterols and stanols compared with cholesterol. Plant sterols have the same squalene ring nucleus as cholesterol, but differ mainly in the side-chain configuration. All plant sterols have a side-chain of variable length at C24. In addition, sitostanol and campestanol lack a double bond between C5 and C6 compared with their unsaturated counterparts; stigmasterol has a double bond between C22 and C23; avenosterol has a double bond between C24 and C28; brassicasterol has a double bond between C22 and C23; and ergosterol differs by the presence of a double bond between C7 and C8. *(From Izar MC, Tegani DM, Kasmas SH, Fonseca FA. Phytosterols and phytosterolemia: gene-diet interactions. Genes Nutr 2011;6(1):17–26. Fig. 1. P 18.)*

Sitosterolemia is due to the accumulation of dietary plant sterols (Fig. 3.1.2.19) in blood and tissues due to increased intestinal absorption and decreased biliary excretion of plant sterols.

Cardiovascular disease, anemia, arthritis and liver failure may be seen. Skin xanthomas (yellow, fat-filled nodules) may be the only abnormality and diagnosis can be confused with hypercholesterolemia since routine enzymatic cholesterol assays also detect phytosterols. The apparent hypercholesterolemia is reduced with low cholesterol diet but not with statins (Yoo, 2016).

Bile acid micelles facilitate the solubilization of dietary and endogenous sterols in the proximal small intestine (Fig. 3.1.2.20).

Niemann-Pick C-type L1 NPC1L1 facilitates uptake of cholesterol and phytosterols into intestinal enterocytes. Cholesterol is incorporated into chylomicrons, delivered to the plasma compartment through the lymphatic system and cleared by the liver. ABCG5 ABCG8 also promotes cholesterol secretion into the intestinal lumen. Phytosterols are poorly absorbed and largely returned to the intestinal lumen by ABCG5 ABCG8. Bile acids are reabsorbed in the distal small intestine, stimulate FXR dependent expression of FGF15/19 and are returned to the liver through the portal system. In the liver, bile acids stimulate ABCG5 ABCG8 catalytic activity and promote the formation of bile acid micelles that serve as acceptors for ABCG5 ABCG8 mediated biliary cholesterol secretion. Cholesterol metabolites (oxysterols), through LXR and bile acids, through FXR and in cooperation with FGF15/19, activate ABCG5 ABCG8. The half transporters heterodimerize, traffic to the canalicular surface and promote biliary phytosterol and cholesterol secretion. Excess cholesterol,

phytosterols and bile acids that are not absorbed/reabsorbed are eliminated from the body.

Sitosterol should be measured by GC-MS (Lee et al., 2020a,b,c; Zein et al., 2019; Kidambi and Patel, 2008). Variants in genes for ATP cassette-binding proteins ABCG5 (Sterolin-1) or ABCG8 (Sterolin-2) were found by whole-exome sequencing (Kiss et al., 2020; Berge et al., 2000). The highly homologous genes are mapped to chromosome 2p21. About 60 variants each of ABCG5 and ABCG8 have now been reported (Williams et al., 2021) which has increased in number in the ensuing 15 years (Fig. 3.1.2.20) (Kidambi and Patel, 2008). These are predicted loss of function variants, and there are many more missense variants. ABCG5 and ABCG8 are thought to act as an obligate heterodimer, based on the fact that complete mutations on either gene can cause sitosterolemia. ABCG5 and ABCG8 Sterolin-1 (ABCG5) and sterolin-2 (ABCG8) belong to the ATP-binding cassette (ABC) transporter superfamily and the "G" subfamily. They localize to the apical membrane of the enterocyte and the canalicular membrane of the biliary tract. Each ABCG5 and ABCG8 gene is composed of 13 exons; they are arranged in a head-to-head configuration on chromosome 2p21, with about 140 bases separating their first exons (Fig. 3.1.2.21).

Their genomic organization suggests a gene duplication event during evolution. A classification has been proposed based on nonsense, frameshift and deletion mutations (Class I), maturation (II), activity (III), stability (IV) and no detectable defect (V). Further study of this disorder will reveal new knowledge on the function of transporters. Treatment is currently based on diet to restrict plant sterol intake (Escolà-Gil et al., 2014).

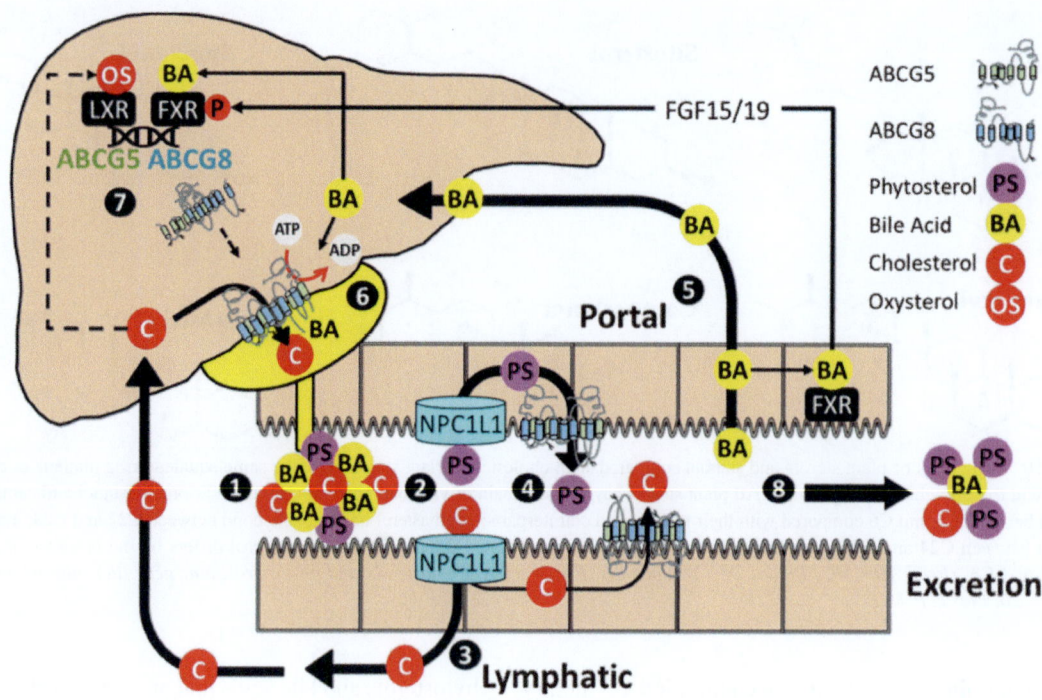

Sitosterolaemia 20 years

FIG. 3.1.2.20 Sterol transport in Sitosterolemia showing enterohepatic sterol flux and regulation of ABCG5 ABCG8. (1) Bile acid micelles facilitate the solubilization of dietary and endogenous sterols in the proximal small intestine. Phospholipids not depicted. (2) Niemann-Pick C-type L1 NPC1L1 facilitates uptake of cholesterol and phytosterols into intestinal enterocytes. (3) Cholesterol is incorporated into chylomicrons, delivered to the plasma compartment through the lymphatic system and cleared by the liver. ABCG5 ABCG8 also promotes cholesterol secretion into the intestinal lumen. (4) Phytosterols are poorly absorbed and largely returned to the intestinal lumen by ABCG5 ABCG8. (5) Bile acids are reabsorbed in the distal small intestine, stimulate FXR dependent expression of FGF15/19 and are returned to the liver through the portal system. (6) In the liver, bile acids stimulate ABCG5 ABCG8 catalytic activity and promote the formation of bile acid micelles that serve as acceptors for ABCG5 ABCG8 mediated biliary cholesterol secretion. (7) Cholesterol metabolites (oxysterols), through LXR, and bile acids, through FXR and in cooperation with FGF15/19, activate ABCG5 ABCG8. The half transporters heterodimerize, traffic to the canalicular surface and promote biliary phytosterol and cholesterol secretion. (8) Excess cholesterol, phytosterols and bile acids that are not absorbed/reabsorbed are eliminated from the body. *(From Williams K, Segard A, Graf GA. Sitosterolemia: twenty years of discovery of the function of ABCG5ABCG8. Int J Mol Sci 2021;22(5):2641. Fig. 1 p. 2.)*

3.1.2.8.3 Abetalipoproteinemia

Abetalipoproteinemia is a rare disorder due to defect in synthesis of Apoprotein B such that there is absence of chylomicrons, VLDL and LDL. High-density lipoprotein is also reduced by about 50% of normal the disease is suspected in infants that have steatorrhea, vomiting and failure to thrive. Adults present with neuromuscular, ophthalmological and hepatic anomalies (Table 3.1.2.8). HDL production requires the actions of ATP-binding cassette (ABCA1) on ApoA-1 and, lecithin:cholesterol acyltransferase (LCAT) (Fig. 3.1.2.22).

Apolipoprotein A-1 formed in the liver and intestine acquires lipids in the form of cholesterol or phospholipids by the action of ATP-binding cassette transporter A1 (ABCA1) to generate pre-b-HDL. Lecithin: cholesterol acyltransferase (LCAT) using APoA-1 as a cofactor converts lecithin to lysolecithin and cholesterol to cholesterol ester which is then sequestered into HDL. Scavenger protein class B Type 1 (SR-B1) promotes the uptake of mature HDL by the adrenal cells for the synthesis of steroids and in the liver for secretion in the bile.

Males with low levels of HDL-C can have mutations in the LDL-receptor gene, the ApoB gene, the **ATP-binding cassette transporter 1 (ABCA1)** gene or in **lecithin-cholesterol acyltransferase (LCAT)** gene (Kosmas et al., 2018). Patients with abetalipoproteinemia were found to have lower urinary excretion of 17-ketosteroids than controls but normal 17-hydroxycorticosteroids (Bochem et al., 2013). The response to ACTH was normal. These findings were confirmed with urinary steroid profiles (Fig. 3.1.2.23).

In females with molecularly defined low HDL-C, the response of cortisol to ACTH was not different to normal

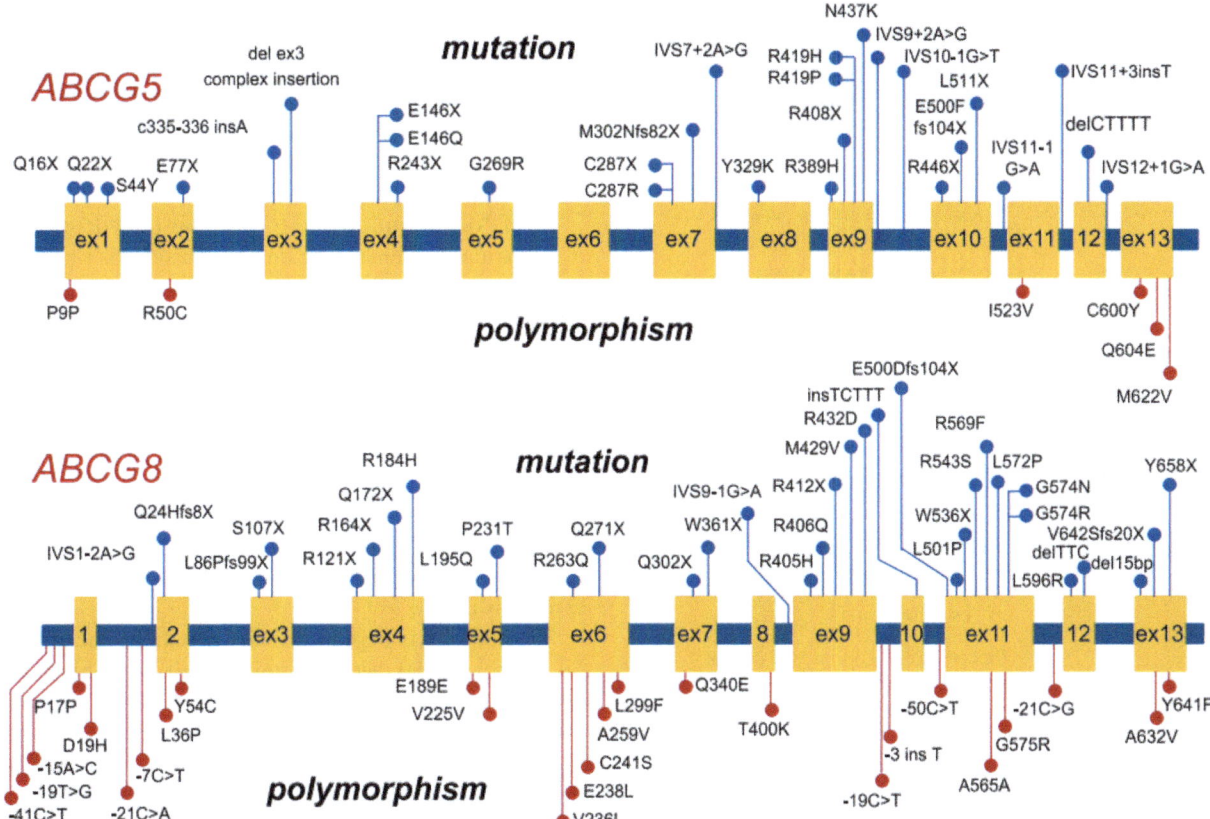

FIG. 3.1.2.21 Mutations and polymorphisms in the ABCG5 and ABCG8 genes. Mutations are illustrated as blue. Polymorphisms are illustrated as red. Upper panel: Mutations and polymorphisms in the ABCG5 gene. Lower panel: Mutations and polymorphisms in the ABCG8 gene. *(Sun W, Zhang T, Zhang X, Wang J, Chen Y, Long Y, Zhang G, Wang Y, Chen Y, Fang T, Chen M. Compound heterozygous mutations in ABCG5 or ABCG8 causing Chinese familial Sitosterolemia. J Gene Med. 2020;22(8):e3185.)*

females (Bochem et al., 2014). Males with abetalipoproteinemia have normal basal cortisol but poor response to Synacthen with peak at 12h of 24.7 µg/dL (normal c. 33 µg/dL at 12h continuing to rise to 42 µg/dL at 24h) (Illingworth et al., 1984). Normal cortisol responses are seen in patients with heterozygous hypobetalipoproteinemia but when patients present with steatorrhea the early response to ACTH was lower, but overall adrenal function was normal (Arem et al., 1997).

Mutations in apoA-1, ATP-binding cassette (ABCA1) underlie familial hypoalphaproteinemia that lowers LDL-c and reduces risk of atherosclerosis. FHA is associated with increased risk of cardiovascular disease. Ultrasound determination of the carotid artery intima media thickness (IMT) is a surrogate for atheroclerotic vascular disease. Four families with FHA types (Hovingh et al., 2005). The IMT was measured along with lipid fractions. The IMT increased with decreasing cellular efflux capacity of the patients (Fig. 3.1.2.24).

Microsomal triglyceride transfer protein (MTTP) transfers triglyceride and cholesterol ester to apoB and is essential for formation of VLDL and chylomicrons

(Fig. 3.1.2.25). This results in malabsorption of dietary fat and fat-soluble vitamins.

Measurements of lipid levels are needed for a diagnosis (Table 3.1.2.9) (Rodríguez Gutiérrez et al., 2021) and analysis of MTTP gene (Takahashi et al., 2021). MTTP has two subunits, a 97-kDa subunit of 894 amino acids and a 55-kDa unit. MMT is coded by 18 exons to give a protein of 3 domains. At least 74 mutations have been reported in the MTTP gene (Fig. 3.1.2.26) (Takahashi et al., 2021).

3.1.2.8.4 Cholesterol storage disease

Cholesterol ester storage disease, also known as **familial hypercholesterolemia and lysosomal acid lipase deficiency**, presents in childhood. Morbidity depends on severity of vascular, hepatic and gastrointestinal problems. Mutations in LIPA gene Lysosomal acid lipase deficiency is reduced but not absent as in Wolman's syndrome. Hypercholesterolemia is frequent leading to atherosclerosis. The most common variant is a splice site mutation in exon (c894G>A) which encodes LAL protein with 3%–8% of normal activity.

TABLE 3.1.2.8 Diagnostic criteria abetalipoproteinemia.

A. Laboratory entry criterion	Plasma LDL-C < 15 mg/dL AND/OR plasma apoB < 15 m/dL
B. Clinical manifestations	1. Gastrointestinal—steatorrhea, chronic diarrhea, vomiting, failure to thrive
	2. Neuromuscular—ataxia, spastic paralysis, hypoesthesia (numbness to touch), deep tendon reflexes
	3. Ophthalmological—retinitis pigmentosa (vision loss); poor night vision, constriction of visual field
C. Bloods	Acanthocytosis (red blood cells with spurs)
D. Differential diagnosis	Familial hypobetalipoproteinemia
E. Genetic test	Mutation in MTTP gene
Diagnosis	Definite ABL—A plus one of B or C and exclusion of D and E
	Probable ABL—A plus two of B or C and exclusion of D

From Takahashi M, Okazaki H, Ohashi K, Ogura M, Ishibashi S, Okazaki S, Hirayama S, Hori M, Matsuki K, Yokoyama S, Harada-Shiba M. Current diagnosis and management of abetalipoproteinemia. J Atheroscler Thromb 2021;28(10):1009–19. Table 2 p. 1015.

3.1.2.8.5 Lysosomal defect

3.1.2.8.5.1 Wolman's disease

The complete absence of **lysosomal acid lipase (LAL)** usually presents in the first year of life and is known as Wolman disease. It causes infiltration of liver, spleen, adrenals, lymph nodes and intestinal organs with macrophages filled with cholesteryl esters by 1–2 months of age, leading to symptoms of malabsorption, failure to thrive from nutritional deficiencies, hepatomegaly, progressive liver failure and early death in infancy if missed or untreated. **Lysosomal acid lipase (LAL)** deficiency presents with hepatomegaly and abnormal liver enzyme concentrations. Cholesterol, triglycerides and LDL-C are low but HDL-C is elevated. Screening a laboratory database for low LDL-c and aspartases only found 2 likely cases of LDL deficiency in 1825 patients (Reynolds et al., 2018). Adrenal gland calcification results in adrenal cortical insufficiency. Reduced enzyme activity can be demonstrated in fibroblasts and white blood cells (Cappuccio et al., 2019). The substrate is 4-methyl umbelliferyl palmitate with incorporation of Lalistat 2 as inhibitor of related enzymes (Lukacs et al., 2017). Unless successfully treated with hematopoietic stem cell transplantation (HSCT), recombinant or sebelipase alfa, infants with classic Wolman disease do not survive beyond age 1 year. Mutations have been found in all 10 exons of the LIPA gene (Pericleous et al., 2017).

3.1.2.9 Mitochondrial abnormalities

Hyperpigmentation and hyponatremia can be presenting symptoms of adrenal insufficiency with very rare mitochondrial disorders. Impaired ATP production with oxidative stress may be the explanation. The age at presentation can be from 1 to 30+ years.

NNT Nicotinamide nucleotide transdehydrogenase is an antioxidant defense gene on chromosome 5. Under most circumstances, NNT uses energy from the mitochondrial proton gradient to generate NADPH which reduces thioredoxin and glutathione (Fig. 3.1.2.27).

Targeted exome sequencing of patients with glucocorticoid deficiency led to the identification of NNT variants (Meimaridou et al., 2012; Jazayeri et al., 2015). Mutations are scattered throughout the gene (Fig. 3.1.2.28) (Roucher-Boulez et al., 2016) with changes to the protein on both sides of the mitochondrial membrane. Patients present with hyperpigmentation and glucocorticoid deficiency after 12 months of age.

Thioredoxin is kept in a reduced state by the action of the mitochondrial selenoprotein TXNRD2 **Thioredoxin reductase** that transfers electrons from NADPH. The action of the protein together with the glutathione system is to eliminate superoxides which are generated by steroidogenesis. A mutation in TXNRD2 was found in a family with adrenal insufficiency (Prasad et al., 2014).

Pearson syndrome classically presents in the first year of life with bone marrow failure and exocrine pancreas insufficiency. Patients get abdominal pain and anorexia with recurrent pancreatitis. Most have lactic acidosis and hypoglycemia (Williams et al., 2012). Large mitochondrial DNA deletions are found (Liu et al., 2021; Wild et al., 2020). The etiology of adrenal insufficiency is unclear.

MELAS is the acronym for **mitochondrial encephalomyopathy, lactic acidosis and stroke.** Patients also have failure to thrive, progressive deafness, migraine-like headaches, vomiting and seizures (Fig. 3.1.2.29) (Fan et al., 2021).

There are no clear indications of adrenal insufficiency but in mitochondria glucose and protein metabolism produces acetyl-CoA which in peroxisomes is taken to farnesyl-pyrophosphate that is transferred to the endoplasmic reticulum for the conversion of squalene to cholesterol. L-Arginine and glucocorticosteroids are used in the treatment of the condition. **A mutation in mitochondrial DNA** at A3243G is common (Fan et al., 2021) and large deletions have been found (Liu et al., 2021).

Kearns-Sayre (KSS) disease presents before 20 years of age with pigmentary retinopathy and progressive ophthalmoplegia (weakness/paralysis of eye muscles) and hearing loss. The syndrome is due to deletions of mitochondrial DNA. Children with KSS show failure to thrive and toward the end of puberty develop drooping eyelid (a ptosis),

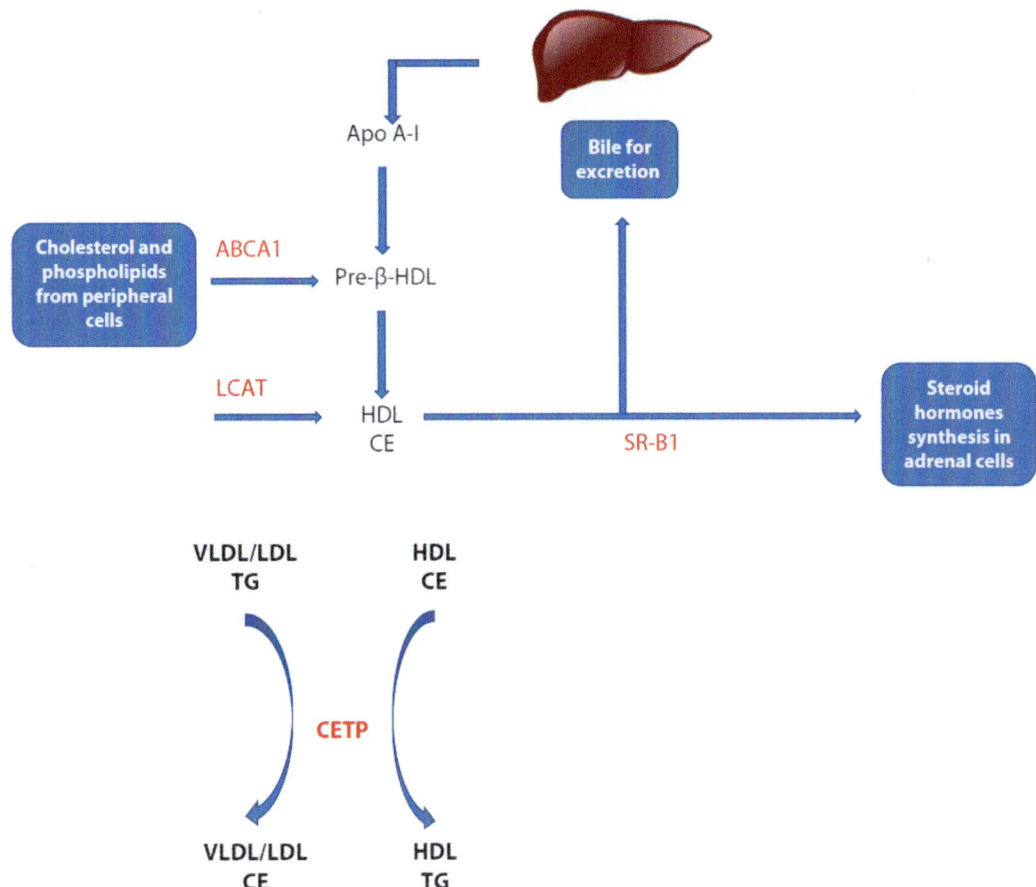

FIG. 3.1.2.22 HDL metabolism.

hearing impairment and neurological symptoms. Mitochondrial DNA rearrangements with especially large deletions (6600 bp) are found (Guo et al., 2020).

3.1.2.10 Peroxisome disorders

Peroxisomes are organelles within cells important in cellular metabolism of fatty acids, phospholipids, and reactive oxygen species through about 50 enzymes. Some disorders are associated with biochemical evidence of adrenal insufficiency although the cause is not clear (Table 3.1.2.10) (Klouwer et al., 2016).

Clinical symptoms of the disorders involve the nervous system, the eye, the auditory nerve, the liver and skeletal system and include neonatal hypotonia, bilateral cataracts, development regression and retinitis pigmentosa with sensorineural hearing loss. Presentation varies with age and from biochemical characteristics as revealed by enzyme activity measurements in plasma, erythrocytes, blood spots and fibroblasts.

Refsum syndrome is another leukodystrophy with accumulation of lipids. There are few symptoms before 10 years of age when loss of night vision and weakness in arms and legs are apparent. Presentation may include

retinitis pigmentosum, ataxia, chronic polyneuropathy, deafness, anosmia, ichthyosis, cardiac abnormalities and skeletal abnormalities (metatarsal shortening). Onset is during late childhood or later. Adults with Refsum's disease accumulate large stores of phytanic acid in their blood and tissues. This frequently leads to peripheral polyneuropathy, cerebellar ataxia, retinitis pigmentosa, anosmia and hearing loss. Refsum's disease is rare with an incidence around 1 in 1000,000. Phytanic acid (or 3, 7, 11, 15-tetramethyl hexadecanoic acid) is a C_{20} branched chain fatty acid from theconsumption of dairy products, ruminant animal fats and certain fish (van den Brink and Wanders, 2006). Phytol is cleaved from chlorophyll and oxidized to phytanic acid (Fig. 3.1.2.30).

For humans, the intake of phytanic acid in food is more important than from chlorophyll and phytanic acid then undergoes α-oxidation in the peroxisome (Verhoeven et al., 1998), where it is converted into pristanic acid by the removal of one carbon. Pristanic acid can undergo further β-oxidation in the peroxisome to form medium chain fatty acids that can be converted to carbon dioxide and water in mitochondria (Fig. 3.1.2.31).

Mutations in phytanoyl-CoA hydroxylase (PHYH) that lead to accumulation of phytanic acid have been

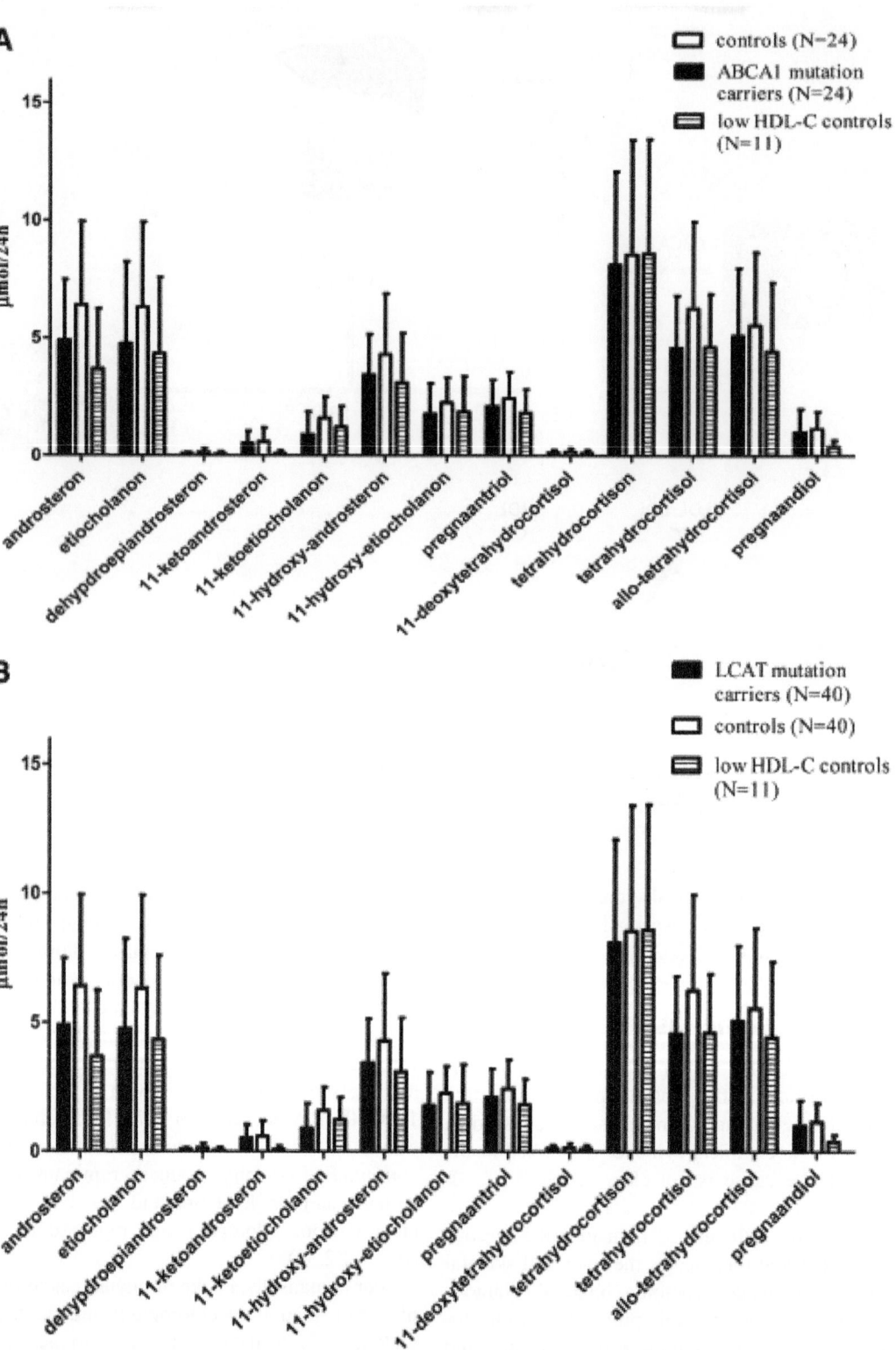

FIG. 3.1.2.23 Urinary steroid metabolites in male *ABCA1 (A) and LCAT (B) mutation carriers compared with* age-matched male controls. Data are presented as mean ± SD. *(From Bochem AE, Holleboom AG, Romijn JA, Hoekstra M, Dallinga-Thie GM, Motazacker MM, Hovingh GK, Kuivenhoven JA, Stroes ESG. High density lipoprotein as a source of cholesterol for adrenal steroidogenesis: a study in individuals with low plasma HDL-C. J Lipid Res 2013;54(6):1698–1704.; Fig. 2 p. 1702.)*

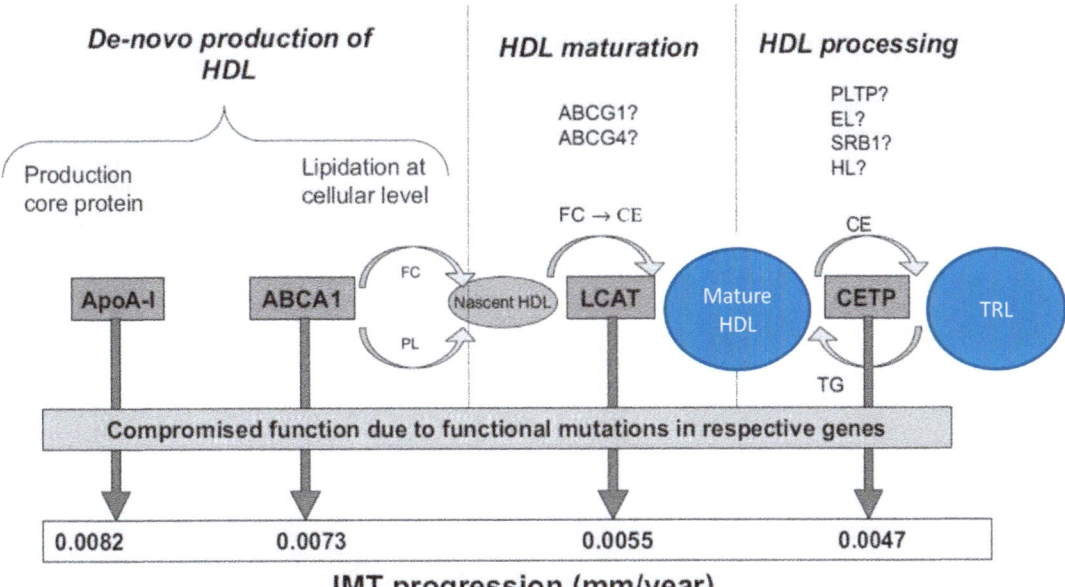

FIG. 3.1.2.24 **HDL metabolism and inherited defects causing atherosclerosis. Intima media thickness was measured as an index of disease pro-gression.** *ABCA1*, ATP-binding cassette transporter A1; *ApoA-I*, apolipoprotein A-I; *CE*, cholesteryl ester; *CETP*, cholesteryl ester transfer protein; *EL*, endothelial lipase; *FC*, free cholesterol; *HL*, hepatic lipase; *IMT*, intima media thickness; *LCAT*, lecithin:cholesterol acyltransferase; *PL*, phospholipid; *PLTP*, phospholipid transfer protein; *SRB1*, scavenger-receptor BI; *TG*, triglyceride; *TRL*, triglyceride rich lipoprotein. *(From Hovingh GK, de Groot E, van der Steeg W, Boekholdt SM, Hutten BA, Kuivenhoven JA, Kastelein JJ. Inherited disorders of HDL metabolism and atherosclerosis. Curr Opin Lipidol 2005;16(2):139–45. Fig. 2 p. 144.)*

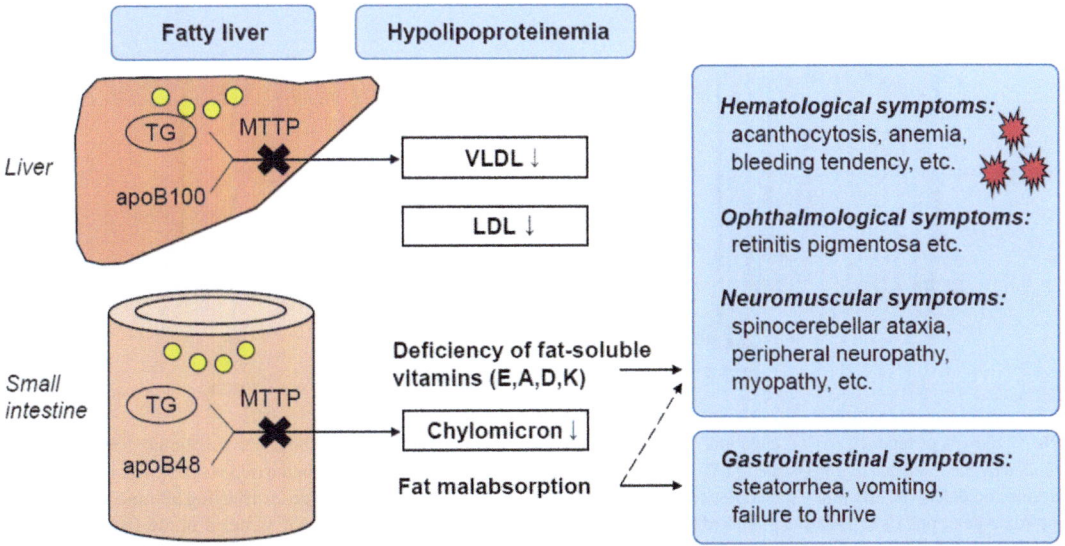

FIG. 3.1.2.25 MTTP is a requisite for the assembly of VLL and CM by the liver and small intestine, respectively. Homozygous MTTP deficiency causes fat malabsorption, steatorrhea, vomiting, failure to thrive, hypolipoproteinemia, fatty liver as well as symptoms related to deficiency of fat soluble vitamins. *(From Takahashi M, Okazaki H, Ohashi K, Ogura M, Ishibashi S, Okazaki S, Hirayama S, Hori M, Matsuki K, Yokoyama S, Harada-Shiba M. Current diagnosis and management of abetalipoproteinemia. J Atheroscler Thromb 2021;28(10):1009–19. Fig. 1 p. 1011.)*

documented in cases of **Refsum syndrome** (Elghawy et al., 2021).

The **Zellweger syndrome** spectrum (ZSS) is of three clinical phenotypes:

- the cerebrohepatorenal syndrome or Zellweger syndrome (ZS),

- neonatal adrenoleukodystrophy (NALD) and
- infantile Refsum's disease (IRD) (Cheillan, 2020).

These were originally described as separate clinical entities but they are now considered different presentations within the same clinical and biochemical continuum with ZS being the most and IRD the least severe. Patients at the severe side

TABLE 3.1.2.9 Biochemical findings of Mexican patient with abetalipoproteinemia and her relatives.

	Proband	Mother	Father	Sister I	Sister II	Reference ranges
Age (years)	6	30	28	8	10	
Glucose (mg/dL)	81	87	65	78	80	72–108
Urea (mg/dL)	7	16	26	21	11	12–54
Creatinine (mg/dL)	0.3	0.6	0.9	0.4	0.4	0.6–1.2
TC (mg/dL)	<50	169	141	116	124	104–220
HDL-C (mg/dL)	25	38	55	30	36	35–85
LDL-C (mg/dL)	20	119.2	79.5	74.1	75.9	70–120
TG (mg/dL)	<10	144	111	73	53	60–150
apoB (mg/dL)	<35	91.1	65.7	61.5	64	55–125

From Rodríguez Gutiérrez PG, González García JR, Castillo De León YA, Zárate Guerrero JR, Magaña Torres MT. A novel p.Gly417Valfs*12 mutation in the MTTP gene causing abetalipoproteinemia: Presentation of the first patient in Mexico and analysis of the previously reported cases. J Clin Lab Anal. 2021;35(3):e23672. Table 2 p. 5 of 10.

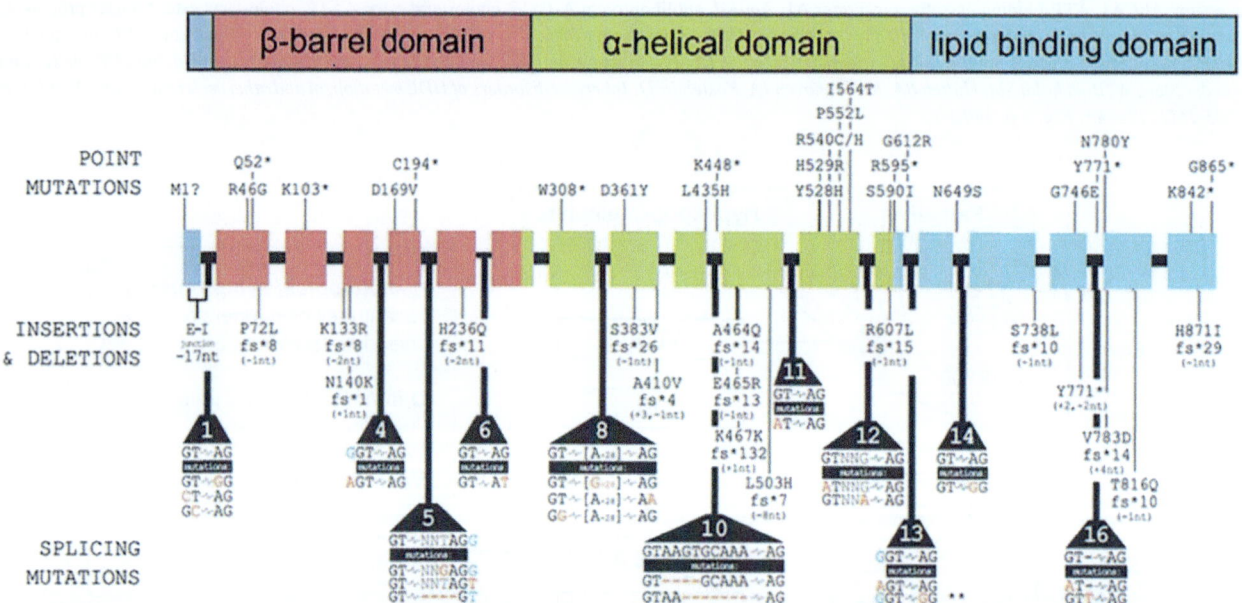

FIG. 3.1.2.26 Synopsis of reported MTTP mutations causing abetalipoproteinemia. The top panel shows the functional domains of the MTP protein, which are encoded by the indicated regions of the MTTP gene depicted with its 18 exons and 17 introns in the middle of the figure. Point mutations, frame-shifting insertions and deletions, and splicing mutations are indicated and occur throughout the MTTP gene, affecting all functional domains of the MTP protein. Large-scale deletions (≥25 base pairs) and MTTP mutations linked to other conditions are not included in the figure. (*From Vlasschaert C, McIntyre AD, Thomson LA, Kennedy BA, Ratko S, Prasad C, Hegele RA. Abetalipoproteinemia due to a novel splicing variant in MTTP in 3 siblings. J Investig Med High Impact Case Rep 2021;9:23247096211022484. Fig 3 P3.*)

of the continuum usually come to clinical attention in the newborn period, while patients with an IRD presentation are often recognized later in childhood. Infants with the ZS presentation usually die during the first year of life, often without having made any developmental progress. ZS classically presents with severe hypotonia and with characteristic craniofacial features that include a large anterior fontanel, a prominent forehead, shallow orbital ridges, epicanthal folds, a high arched palate, a broad nasal bridge and

a small nose with anteverted nares (Fig. 3.1.2.32) (Klouwer et al., 2016). Cataracts, glaucoma and corneal clouding are common ocular abnormalities.

Patients with ZS have seizures, renal cysts and neonatal jaundice with elevation in liver function tests. Chondrodysplasia punctata of the patellae and other long bones may occur. Patients with the NALD and IRD can present with developmental delay, hypotonia, liver dysfunction, sensorineural hearing loss, retinal dystrophy and vision

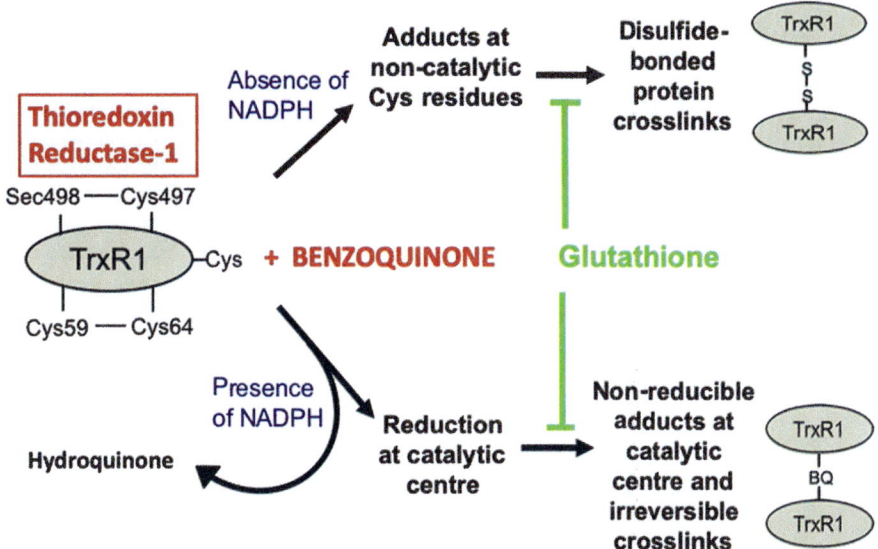

FIG. 3.1.2.27 Role of NNT in free radical metabolism in the mitochondria. *ETC*, electron transport chain; *GPX*, glutathione peroxidase; *GR*, glutathione reductase; *GSH*, glutathione; *GSSG*, glutathione disulfide; *PRDX3*, peroxiredoxin 3; *TXNRD2*, thioredoxin reductase. *(From Shu N, Cheng Q, Arnér ESJ, Davies MJ. Inhibition and crosslinking of the selenoprotein thioredoxin reductase-1 by p-benzoquinone. Redox Biol 2020;28:101335. Graphical abstract.)*

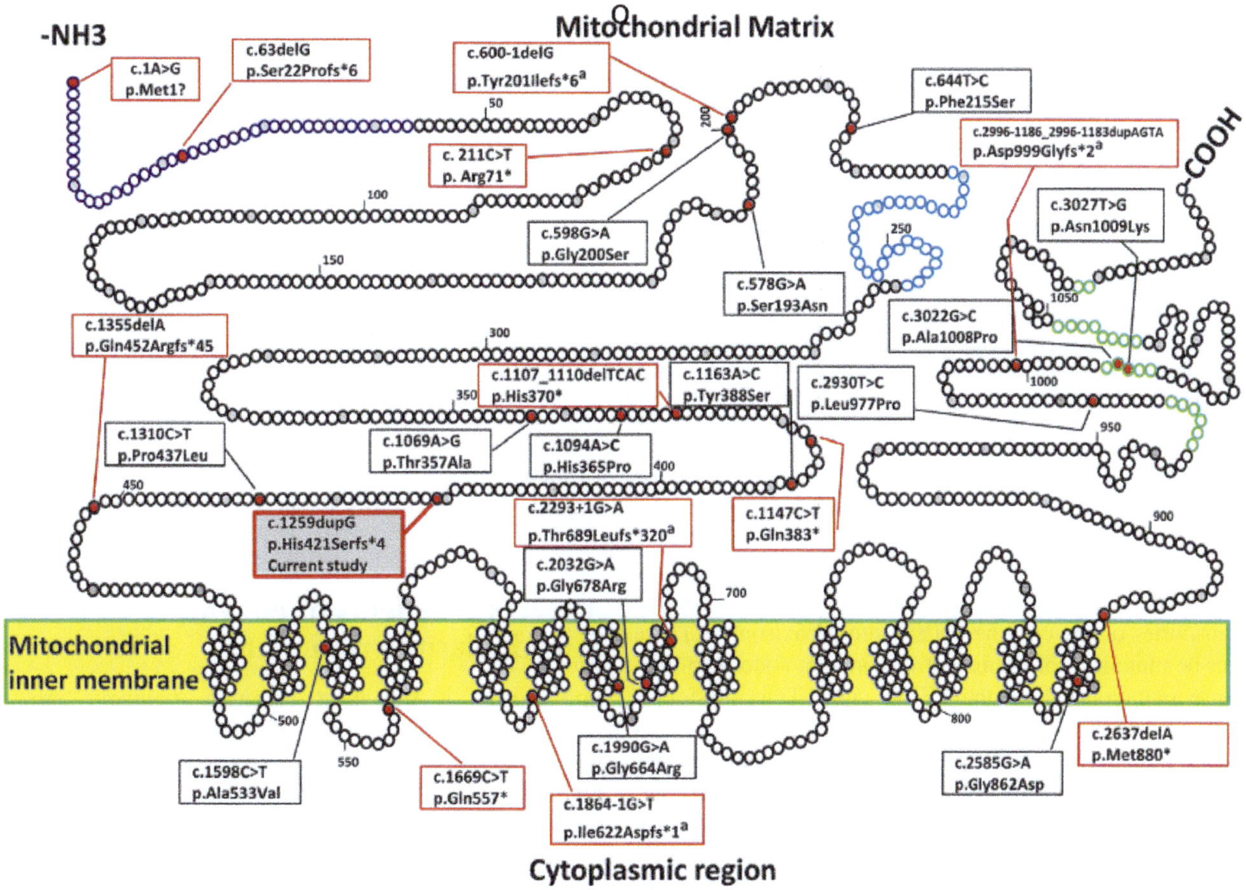

FIG. 3.1.2.28 Schematic representation of the NNT protein. Amino acid positions and predicted protein domains have been adapted from UniProtKB database (Q13423). *Purple circles* represent the transit peptide. *Blue circles* represent NADH⁻ and *green circles* NADPH-binding sites. *Red circles* indicate the amino acids where mutations have been reported. Truncating mutations are shown in *red boxes* and nontruncating mutations in *black boxes*. [a]Predicted result if exon skipped. *(From Jazayeri O, Liu X, van Diemen CC, Bakker-van Waarde WM, Sikkema-Raddatz B, Sinke RJ, Zhang J, van Ravenswaaij-Arts CM. A novel homozygous insertion and review of published mutations in the NNT gene causing familial glucocorticoid deficiency (FGD). Eur J Med Genet 2015;58(12):642–9. Fig. 1 p. 643.)*

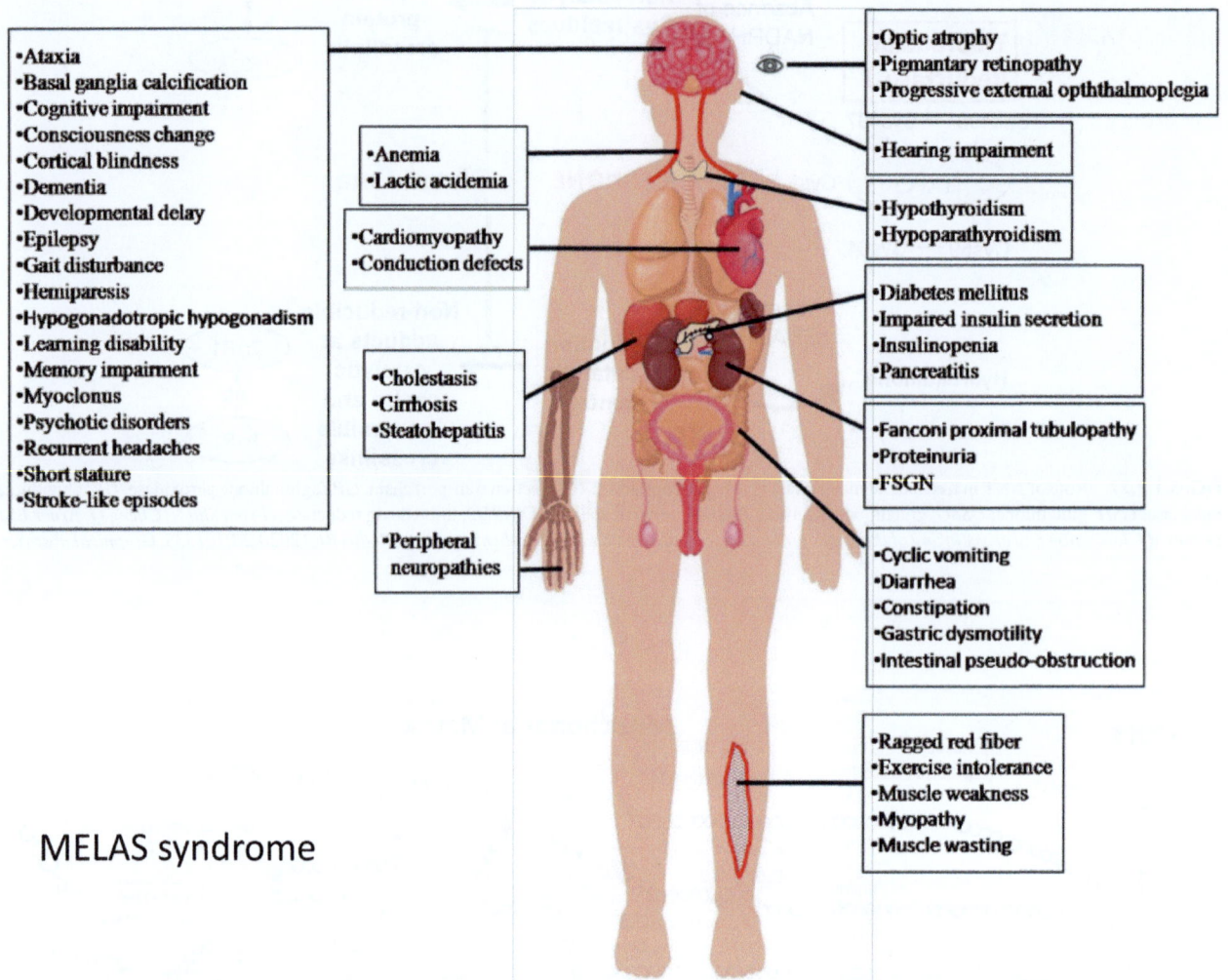

Ataxia
- Ataxia
- Basal ganglia calcification
- Cognitive impairment
- Consciousness change
- Cortical blindness
- Dementia
- Developmental delay
- Epilepsy
- Gait disturbance
- Hemiparesis
- Hypogonadotropic hypogonadism
- Learning disability
- Memory impairment
- Myoclonus
- Psychotic disorders
- Recurrent headaches
- Short stature
- Stroke-like episodes

- Anemia
- Lactic acidemia

- Cardiomyopathy
- Conduction defects

- Cholestasis
- Cirrhosis
- Steatohepatitis

- Peripheral neuropathies

- Optic atrophy
- Pigmantary retinopathy
- Progressive external opththalmoplegia

- Hearing impairment

- Hypothyroidism
- Hypoparathyroidism

- Diabetes mellitus
- Impaired insulin secretion
- Insulinopenia
- Pancreatitis

- Fanconi proximal tubulopathy
- Proteinuria
- FSGN

- Cyclic vomiting
- Diarrhea
- Constipation
- Gastric dysmotility
- Intestinal pseudo-obstruction

- Ragged red fiber
- Exercise intolerance
- Muscle weakness
- Myopathy
- Muscle wasting

MELAS syndrome

FIG. 3.1.2.29 Manifestation of MELAS syndrome. The clinical features are not specific and are variable between patients, including neurological and nonneurological presentations. MELAS is a progressive syndrome where patients can recover from one phenotype and develop others later. Subjects with mtDNA mutations can be asymptomatic or have multiorgan involvement. *(From Fan HC, Lee HF, Yue CT, Chi CS. Clinical characteristics of mitochondrial encephalomyopathy, lactic acidosis, and stroke-like episodes. Life (Basel) 2021;11(11):1111. Fig. 2 p. 4 of 24.)*

impairment. Children with NALD may reach their teens, but patients with IRD reach adulthood. Loss of vision and hearing progresses slowly at variable rates.

The related **Zellweger** syndrome is due to mutations in the **peroxin gene 1 (PEX1)** that normally codes for the transporter of the enzyme PHYH into peroxisomes. There can be mutations of premature termination codons to missense mutations along the PEX-1 gene. Transmission is through autosomal recessive patter. Most investigations of PEX1 mutations to date have focused on coding regions of the gene. The identification of polymorphisms that regulate PEX1 gene expression suggest that factors other than coding mutations may impact PEX1 activity, either directly or as a consequence of changes to the steady-state interactions of PEX1 with other peroxins. The identification and characterization of other factors that regulate PEX1 activity may hold the key to better elucidating the phenotypic

variation found in Zellweger spectrum patients. Sanger sequencing is the most reliable way to detect mutations.

3.1.2.11 Defects of enzyme in endoplasmic reticulum

Sphingosine is the most common sphingoid base in mammals. Ceramides contain sphingosine in amide linkage with a fatty acid of variable chain length and saturation. Sphingosine can be phosphorylated, giving rise to the bioactive signaling molecule, sphingosine-1-phosphate (S1P). Sphingomyelin is formed when a phosphocholine head (Choi and Saba, 2019) (Fig. 3.1.2.33).

In a patient cohort of more than 350 cases up to 2017, the genetic cause of AI was unknown in nearly 40% of cases (Choi and Saba, 2019; Settas et al., 2019; Prasad et al.,

TABLE 3.1.2.10 Classification and nomenclature of peroxisome disorders.

Disorder	Abbreviation	Gene name	OMIM no.
Zellweger spectrum disorders	ZSD	PEX genes	
- Zellweger syndrome	ZS		214100
- Neonatal adrenoleucodystrophy	NALD		202370
- Infantile Refsum disease	IRD		266510
Rhizomelic chondrodysplasia punctata Type 1	RCDP1	PEX7	215100
Rhizomelic chondrodysplasia punctata Type 5	RCDP5	PEX5	
Rhizomelic chondrodysplasia punctata Type 2	RCDP2	GNPAT	602744
Rhizomelic chondrodysplasia punctata Type 3	RCDP3	AGPS	600121
Rhizomelic chondrodysplasia punctata Type 4	RCDP4	FAR1	616107
X-linked adrenoleucodystrophy	X-ALD	ABCD1	300100
Peroxisomal acyl-CoA oxidase 1	ACOX1	ACOX	264470
D-bifunctional protein deficiency	DBP	HSD1B4	261515
Refsum disease (classic)	RD (CRD/ARD)	PHYH/PEX7	266500
α-methylacyl-CoA-racemase deficiency	AMACR	AMACR	604489

Peroxisome Biogenesis disorders

Single Peroxisome Biogenesis disorders

Klouwer

From Klouwer FC, Huffnagel IC, Ferdinandusse S, Waterham HR, Wanders RJ, Engelen M, Poll-The BT. Clinical and biochemical pitfalls in the diagnosis of peroxisomal disorders. Neuropediatrics 2016;47(4):205–20. Fig 1 P206.

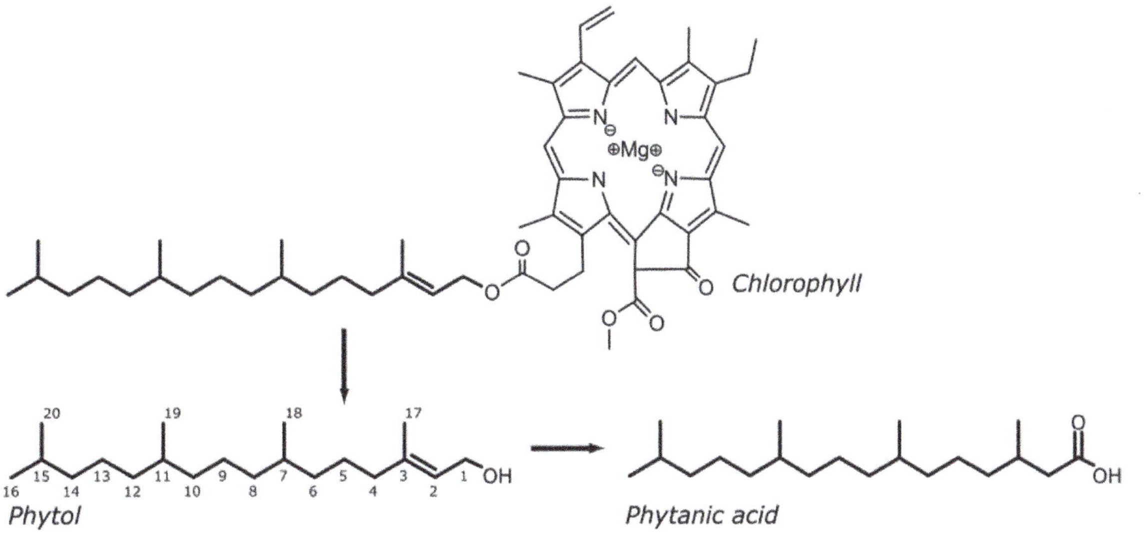

FIG. 3.1.2.30 The structures of chlorophyll, phytol, and phytanic acid. *(Based on van den Brink DM, van Miert JN, Dacremont G, Rontani JF, Wanders RJ. Characterization of the final step in the conversion of phytol into phytanic acid. J Biol Chem 2005;280(29):26838–44. Fig 1 P 26839.)*

2017; Lovric et al., 2017). Whole-exome sequencing revealed families with mutations in **sphingosine-1-phosphate lyase (SGPL1)**. Primary adrenal insufficiency was found in conjunction with steroid resistant nephrotic syndrome. Some of the patients presented with other features including ichthyosis, primary hypothyroidism, neurological symptoms and cryptorchidism (Table 3.1.2.11) (Maharaj et al., 2020a). Knockout mice had disrupted adrenocortical zonation and defective expression of steroidogenic enzymes.

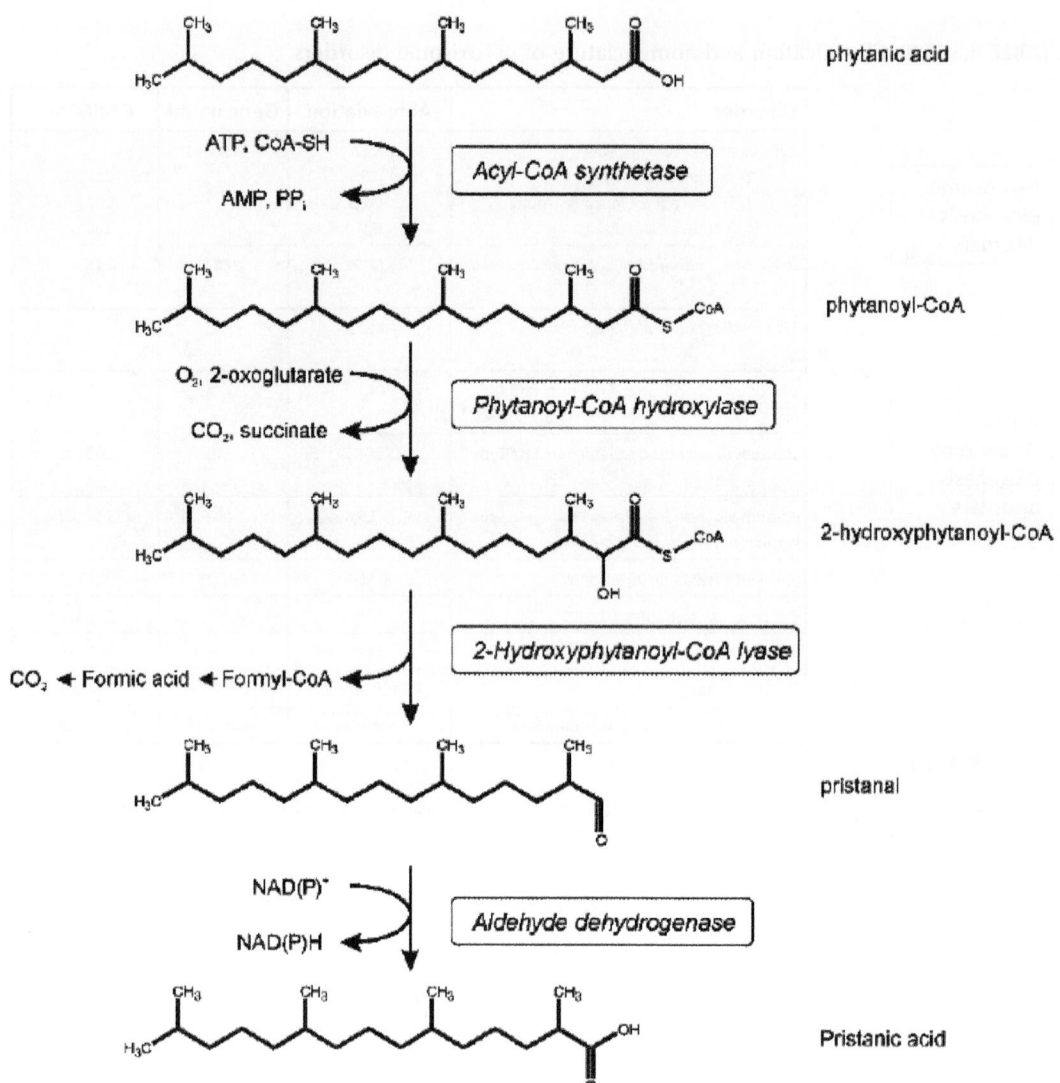

FIG. 3.1.2.31 Pathway of phytanic acid α-oxidation. Following activation of phytanic acid to phytanoyl-CoA, 2-hydroxyphytanoyl-CoA is formed by phytanoyl-CoA hydroxylase. Thereafter, formyl-CoA is released and pristanal is formed. Pristanal is converted to pristanic acid, which, after activation, is degraded by peroxisomal β-oxidation. *(Based on Jansen GA, Waterham HR, Wanders RJ. Molecular basis of Refsum disease: sequence variations in phytanoyl-CoA hydroxylase (PHYH) and the PTS2 receptor (PEX7). Hum Mutat 2004;23(3):209–18. GFig. 1. P 211.)*

SGPL1 is an intracellular enzyme that carries out an irreversible step in degradation of sphingosine (Fig. 3.1.2.34). Frameshift and splice site mutations predicted markedly truncated mRNA and loss of enzyme activity. Imaging showed calcification of enlarged adrenals.

Degeneration of nerves and skin reflect harmful accumulation of lysosomal sphingolipid species. Ceramide and sphingosine can reduce steroidogenesis. Sphingosine interacts with nuclear receptor SF-1 to keep SF-1 in an inactive conformation (Urs et al., 2007).

S1P binds to S1PR1 and/or SIPR3 and activates Gai. The Gai couples to PLC, thereby increasing intracellular IP3 and subsequently cytosolic Ca^{2+}. Intracellular Ca^{2+} accumulation activates CamKII, which in turn mediates the phosphorylation of ERK1/2. Acute activation of the S1P signaling pathway culminates in an induction of StAR gene expression and an increase in cortisol secretion. S1P also stimulates PKA-mediated phosphorylation of HSL at Ser563, which increases cholesterol ester hydrolysis and substrate availability for cortisol production (Lucki et al., 2012) (Fig. 3.1.2.35).

Some patients with SPLIS may respond to vitamin B6 supplementation, which may overcome the low binding affinity of mutant SPL proteins and/or may exert a chaperone function that enables mutant SPL proteins to achieve proper folding, enhancing their enzyme activity (Fig. 3.1.2.36). However, only patients with a susceptible missense SGPL1 allele are likely to benefit from this treatment strategy.

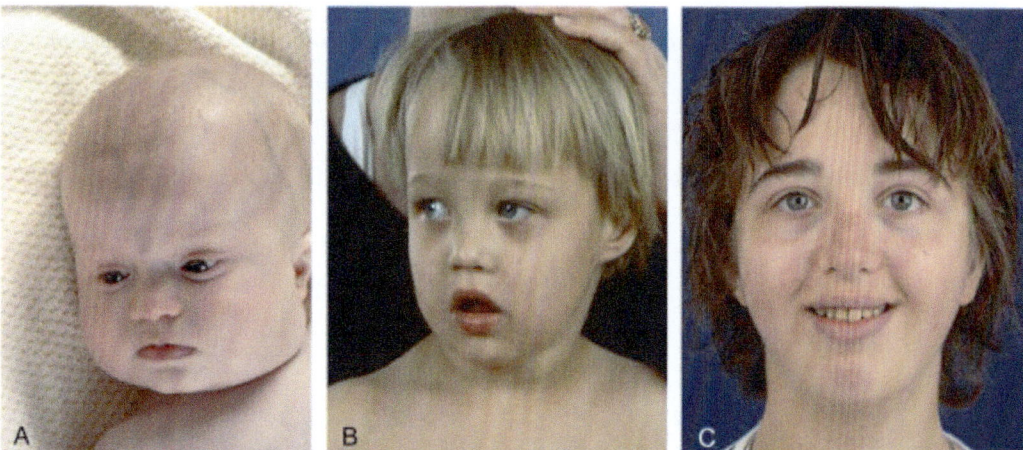

FIG. 3.1.2.32 Different facial appearances in Zellweger spectrum disorders. (A) Photograph of a 6-month-old girl with typical craniofacial dysmorphia. Note the high forehead, epicanthal folds, broad nasal bridge, hypoplastic supraorbital ridges and low-set ear on the left side (right not shown). (B) A 5-year-old girl with less pronounced craniofacial dysmorphism. A high forehead is seen, along with anteverted nares and a broad nasal bridge. Note the downwards turned mouth. (C) Photograph of an 18-year-old woman without evident craniofacial dysmorphism. However, yellow discoloration of the teeth is noticeable. (A written informed consent was obtained from patient [C] and from the parents of patients [A] and [B]). *(From Verhoeven NM, Wanders RJ, Poll-The BT, Saudubray JM, Jakobs C. The metabolism of phytanic acid and pristanic acid in man: a review. J Inherit Metab Dis 1998;21 (7):697–728.)*

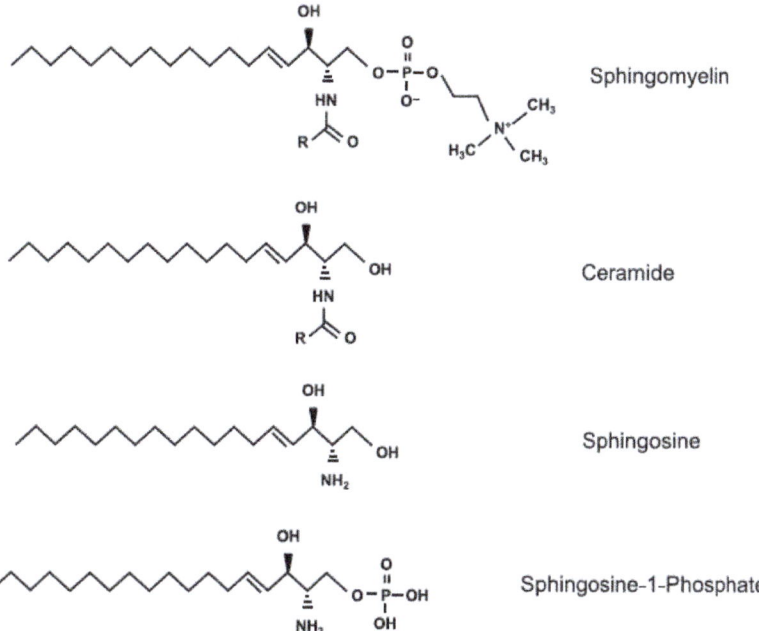

FIG. 3.1.2.33 Basic structures of sphingolipids. **Sphingolipids contain a long chain amino base, also known as a sphingoid base. Sphingosine is the most** common sphingoid base in mammals. Ceramides contain sphingosine in amide linkage with a fatty acid of variable chain length and saturation. Sphingosine can be phosphorylated, giving rise to the bioactive signaling molecule, sphingosine-1-phosphate (S1P). Sphingomyelin is formed when a phosphocholine head. *(From Choi YJ, Saba JD. Sphingosine phosphate lyase insufficiency syndrome (SPLIS): a novel inborn error of sphingolipid metabolism. Adv Biol Regul 2019;71:128–140. Fig. 1 p. 129.)*

In contrast, AAV-mediated SGPL1 gene therapy represents a universal treatment for SPLIS that would address the root cause of the condition by replacing the defective SGPL1 gene, ideally resulting in the production of a functional SPL enzyme in its proper subcellular context (Saba et al., 2021).

Inactivation of SPL has resulted from recessive mutations that translate to a range of molecular effects including protein truncation, frameshifts, nonsense mediated mRNA decay and splicing defects as well as missense mutations that would be predicted to influence protein localization, stability and/or catalytic activity (Fig. 3.1.2.37). The

TABLE 3.1.2.11 Main clinical features of group of patients with genetically confirmed sphingosine-1-phosphate lyase (SGPL1) deficiency.

Feature	Patient numbers	Patient details
Sex M:F	21:14	
Deaths	15	Postnatally with fetal hydrops to 9 years after septic episodes
Steroid resistant nephrotic syndrome (SRNS)	29	Congenital to 19 years with SRNS
Adrenal insufficiency	23	Glucocorticoid deficiency ± mineralocorticoid deficiency postnatally 11 years
Adrenal calcification on imaging	5	
Hypothyroidism	12 (Not all investigated)	Primary hypothyroidism with raised TSH
Gonadal dysfunction	7	Males with cryptorchidism ± microphallus. Raised gonadotrophins.
Ichthyosis	12	
Neurological/developmental delay	18	Varied from microcephaly, seizures, sensorineural deafness and later abnormal gait, peripheral neuropathy
Immunodeficiency		Lymphopenia some with additional hypogammaglobulinemia

From Maharaj A, Theodorou D, Banerjee II, Metherell LA, Prasad R, Wallace D. A sphingosine-1-phosphate lyase mutation associated with congenital nephrotic syndrome and multiple endocrinopathy. Front Pediatr 2020a;8:151. Table 2 p. 6.

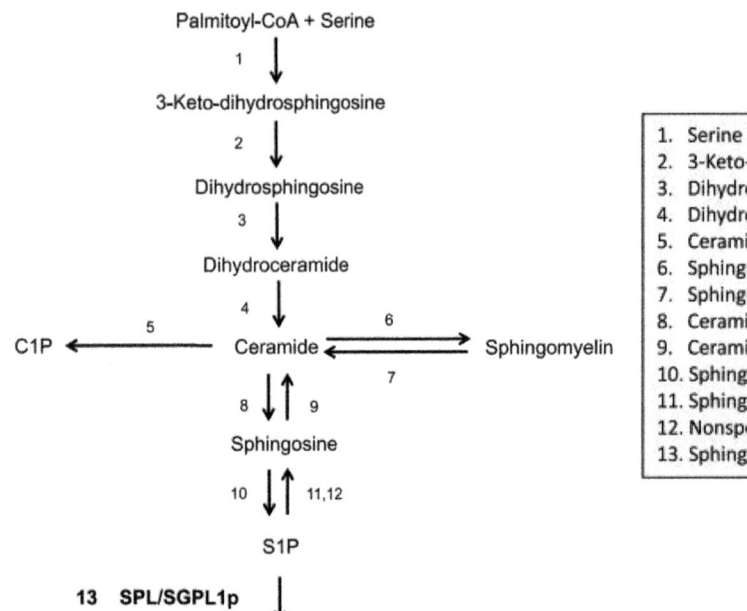

FIG. 3.1.2.34 **Sphingolipid metabolic pathway**. Enzymes catalyzing the conversion steps are listed numerically. Of note, serine palmitoyltransferase (SPT) catalyzes the condensation reaction between serine and palmitoyl-CoA, the rate limiting induction step in sphingolipid biosynthesis. Sphingosine phosphate lyase (SPL) catalyzes the irreversible cleavage of sphingosine-1-phosphate (S1P), the final and essential step in sphingolipid catabolism. *(From Choi YJ, Saba JD. Sphingosine phosphate lyase insufficiency syndrome (SPLIS): a novel inborn error of sphingolipid metabolism. Adv Biol Regul 2019;71:128–140. Fig. 2 p. 130.)*

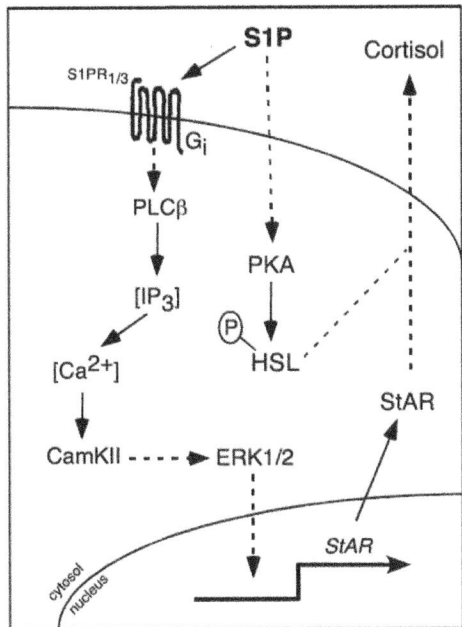

FIG. 3.1.2.35 Proposed model for S1P-mediated cortisol biosynthesis in H295R cells. S1P binds to S1PR1 and/or S1PR3 and activates Gαi. Gαi couples to PLC, thereby increasing intracellular IP3 and subsequently cytosolic Ca^{2+}. Intracellular Ca^{2+} accumulation activates CamKII, which in turn mediates the phosphorylation of ERK1/2. Acute activation of the S1P signaling pathway culminates in an induction of StAR gene expression and an increase in cortisol secretion. S1P also stimulates PKA-mediated phosphorylation of HSL at Ser563, which increases cholesterol ester hydrolysis and substrate availability for cortisol production. *(Based on Lucki NC, Li D, Sewer. Sphingosine-1-phosphate rapidly increases cortisol biosynthesis and the expression of genes involved in cholesterol uptake and transport in H295R adrenocortical cells. Mol Cell Endocrinol 2012;348(1):165–75.)*

severity in SPLIS presentations could ultimately factor into decisions about treatment. Patient derived dermal fibroblasts and CRISP engineered SGLP-1 knockout HeLa cells have been used to demonstrate aspects of mitochondrial dysfunction from gene mutations (Maharaj et al., 2020b).

3.1.2.11.1 ACTH resistance

Hereditary resistance to ACTH is characterized as Familial Glucocorticoid deficiency (FGD) and is divided into two types on the basis of receptor binding or transport defects. The ACTH receptor gene (**MC2R-melanocortin 2 receptor**) has been found to have mutations in infants that present with glucocorticoid but not usually mineralocorticoid failure (Fig. 3.1.2.38) (Abuduxikuer et al., 2019; Lin et al., 2007; Cooray et al., 2008; Clark et al., 1993).

Less than 50 cases have been identified, some present with dysmorphic features. Mutations, that inactivate pro-opiomelanocortin (POMC), have been described in children with adrenal insufficiency, early onset obesity and sometimes along with fair skin with hyperpigmentation and red hair (Gregoric et al., 2021; Clément et al., 2008). Deficiency of cortisol causes hypoglycemia, convulsions and prolonged jaundice in neonates whereas adults present

with repeat infections, lethargy and pigmentation. Plasma ACTH concentrations are high (250–66,888 ng/L). Hyponatremia can be due to vomiting.

MC2R is transported across the adrenal cell membrane by **MRAP** (melanocortin receptor accessory protein). Splicing defects and nonsense mutation in MRAP production prevent MC2R movement and patients present in a similar manner to those with mutations in MC2R gene (Fig. 3.1.2.39) (Novoselova et al., 2019; Meimaridou et al., 2013; Chung et al., 2010). Plasma ACTH were 108–4500 ng/L, cortisol,17-OHP, androstenedione and DHEA were low. The MC2R is degraded by proteasomal enzymes.

A good algorithm for the investigation of primary adrenal insufficiency in children was published from the Department of Pediatrics in Montreal, Canada (Perry et al., 2005) and updated in 2017 (Kirkgoz and Guran, 2018; Guran et al., 2016; Fig. 3.1.2.40). Another case series for neonates was recently published (Gao and Chen, 2020).

A recent diagnostic workup for inherited primary adrenal insufficiency in children (Kirkgoz and Guran, 2018) has been broken into three to be legible (Fig. 3.1.2.41A–C).

After the initial tests basal cortisol, ACTH stimulated cortisol, ACTH measurement further tests include electrolytes, spectrum of steroids and adrenal imaging. In males, only VLCFA, CK and serum lipids are determined. Karyotype. Then adrenal autoantibodies. For further testing, a decision has to be made which steroids are high or low (Fig. 3.1.2.41A).

An etiologic diagnosis is established on the basis of a combination of clinical, laboratory, radiologic and molecular findings. Karyotype can be excluded in female phenotype patients whenever pelvic US is informative for the presence of normal ovaries and Mullerian structures. Assessment of karyotype-matched normal external and internal genitalia and gonads is crucial to decide about gonadal sex steroid production. Measurement of gonadotropins in minipuberty or pubertal cases with non-CAH PAI shows the coexistence or type of hypogonadism. Non-CAH PAI associated with hypogonadotropic hypogonadism would suggest NR0B1 or NR5A1 gene defects while the ones with hypergonadotropic hypogonadism strongly suggest STAR or CYP11A1 mutations.

If adrenal steroid precursors are high, then defects of cortisol production at CYP21A2, CYP11B1, CYP17A1, HSD3B2 and POR can be recognized by precursor elevated (Fig. 3.1.2.41B). A group with dysmorphic features can also be dissected. If adrenal steroid precursors are low, further distinction can be achieved on the basis isolated GC deficiency, GC and MC deficiency or GC plus MC and sex steroid deficiency (Fig. 3.1.2.41C). Genetic studies can also be conducted.

A long-term review of managing patients with adrenal insufficiency now presents clues to a diagnosis according to age (Fig. 3.1.2.42) (Buonocore et al., 2021) and anticipating changes with time. So boys with X-linked adrenal hypoplasia are at risk of developing hypogonadotrophic

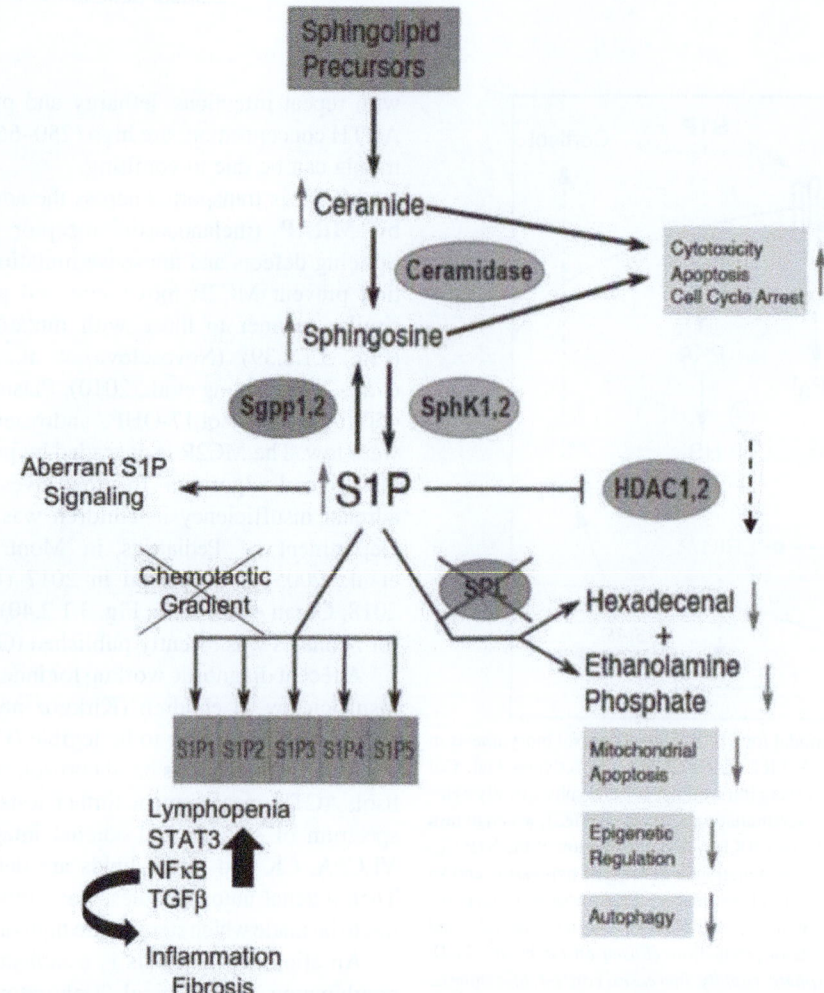

FIG. 3.1.2.36 Features of SPLIS. Some patients with SPLIS may respond to vitamin B6 supplementation, which may overcome the low binding affinity of mutant SPL proteins and/or may exert a chaperone function that enables mutant SPL proteins to achieve proper folding, enhancing their enzyme activity. However, only patients with a susceptible missense SGPL1 allele are likely to benefit from this treatment strategy. In contrast, AAV-mediated SGPL1 gene therapy represents a universal treatment for SPLIS that would address the root cause of the condition by replacing the defective SGPL1 gene, ideally resulting in the production of a functional SPL enzyme in its proper subcellular context. *(From Saba JD, Keller N, Wang JY, Tang F, Slavin A, Shen Y. Genotype/phenotype interactions and first steps toward targeted therapy for sphingosine phosphate lyase insufficiency syndrome. Cell Biochem Biophys 2021;79(3):547–59. Fig. 1 p. 548.)*

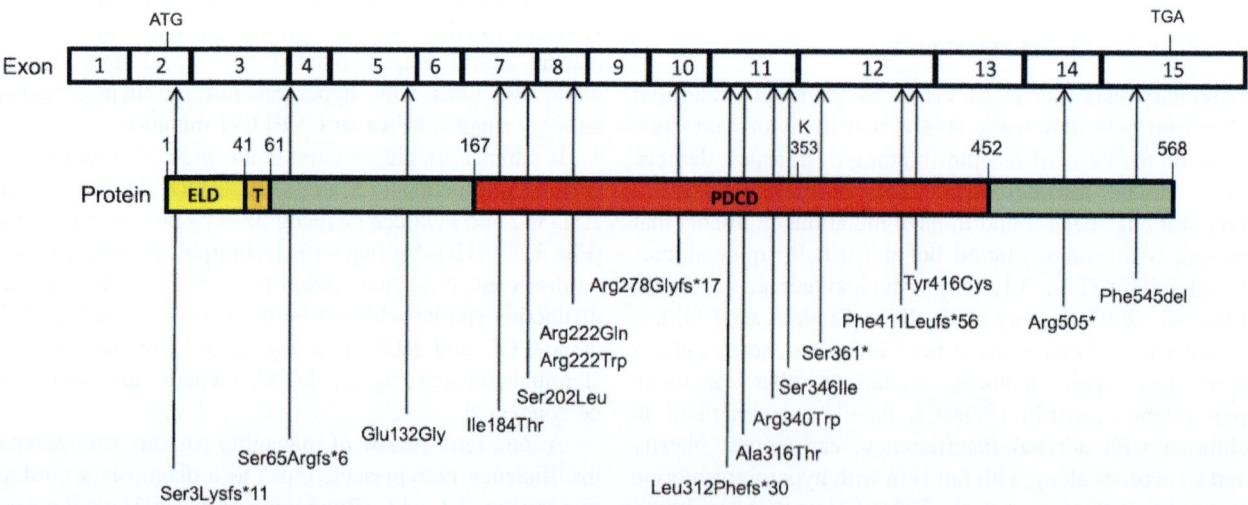

FIG. 3.1.2.37   Sites of all identified *SGPL1* mutations in exon and protein structures. *SGPL1 cDNA contains 15 exons, and the coding sequence is 568 amino acids long. Pyridoxal-dependent decarboxylase conserved domain (PDCD) contains active site residue K353 which binds to pyridoxal 5′-phosphate (PLP). ELD, endoplasmic lumenal domain; T, transmembrane domain; cytoplasmic domain in green. (From Choi YJ, Saba JD. Sphingosine phosphate lyase insufficiency syndrome (SPLIS): a novel inborn error of sphingolipid metabolism. Adv Biol Regul 2019;71:128–40. Fig. 3 p. 131.)*

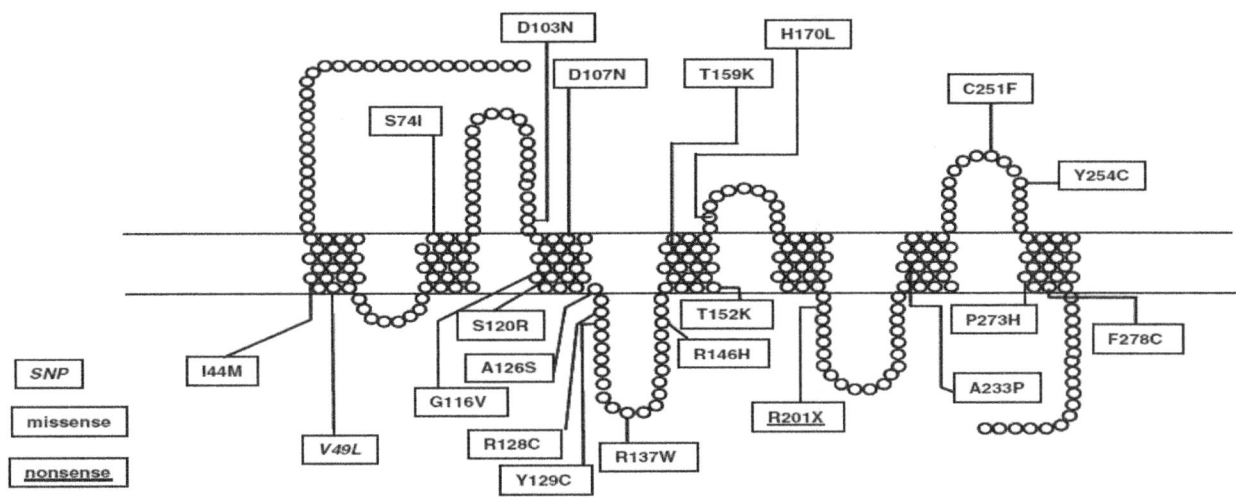

FIG. 3.1.2.38 Schematic of MC2R gene and mutations. *(Chung TT, Chan LF, Metherell LA, Clark AJ. Phenotypic characteristics of familial glucocorticoid deficiency (FGD) type 1 and 2. Clin Endocrinol (Oxf). 2010;72(5):589–94.) Fig. 1 P591.*

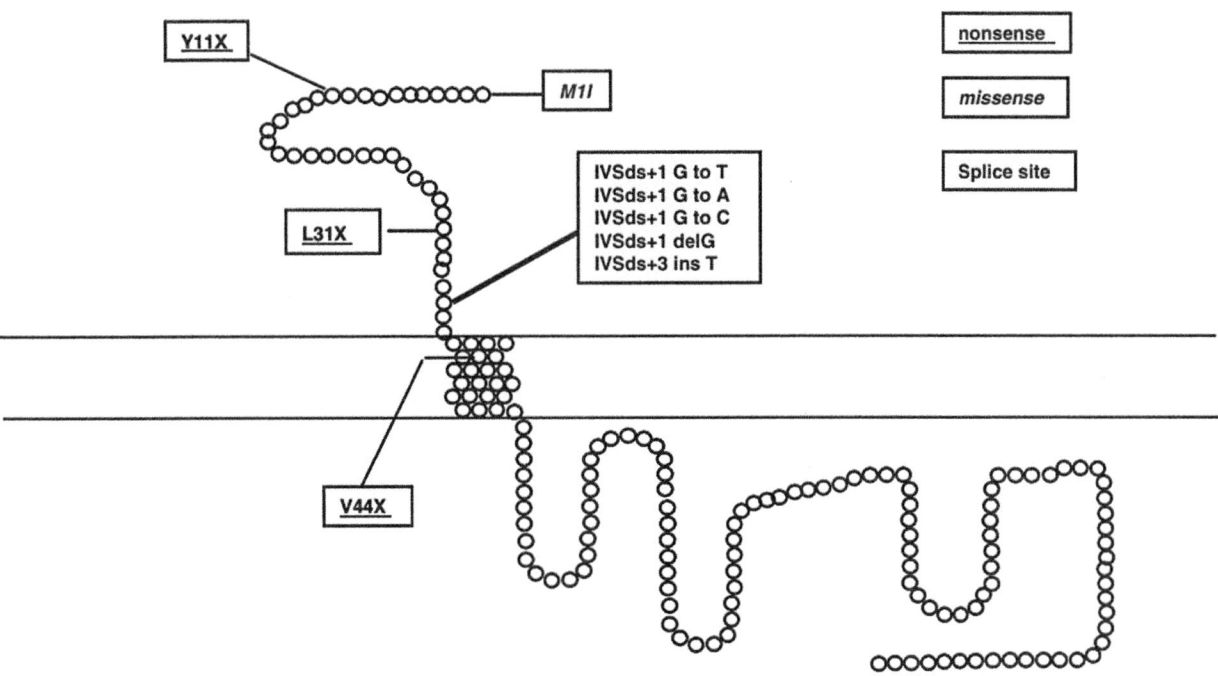

FIG. 3.1.2.39 Schematic of MRAP gene and mutations. *(Chung TT, Chan LF, Metherell LA, Clark AJ. Phenotypic characteristics of familial glucocorticoid deficiency (FGD) type 1 and 2. Clin Endocrinol (Oxf). 2010;72(5):589–94.)*

hypogonadism in adolescence and infertility with defects of StAR or CYP11A1.

There are a number of acquired causes of secondary adrenal insufficiency (Table 3.1.2.12).

3.1.2.11.2 Secondary adrenal insufficiency

Secondary adrenal insufficiency occurs when the adrenal cortex is deprived of ACTH stimulation. An expanding pituitary tumor which leads to impairment of pituitary function usually affects the pituitary hormones in a sequential manner over a period of time. Growth hormone secretion

fails first followed by gonadotrophins, TSH and ACTH (Mullarkey et al., 2021; Higham et al., 2016).

3.1.2.11.2.1 Triple A syndrome

The **achalasia-Addisonian-alacrima syndrome (AAA) or Triple A syndrome** (sometimes called Allgrove syndrome), which appears to be autosomal recessive, is another example of combined adrenal and neurologic (autonomic) involvement (Patt et al., 2017; Toromanovic et al., 2009). With achalasia the esophageal sphincter does not relax making it difficult to swallow food, with alacrima there is inability to secrete tears. A gene for **aladin** on chromosome

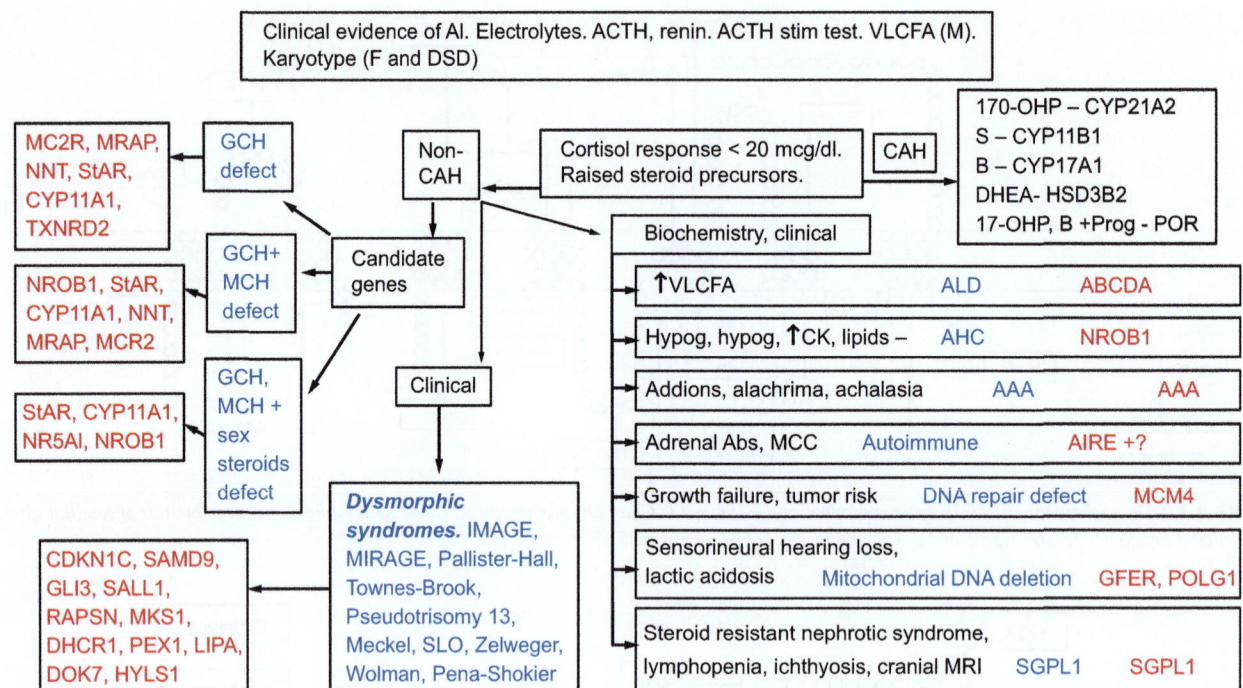

FIG. 3.1.2.40   Algorithm for primary adrenal failure. This chart proposes a diagnostic workup to an appropriate gene. If steroid analysis shows 17OHP is elevated (indicating defect of CYP21A2), then further steroid analysis may reveal markers for the less frequent forms of CAH. Other biochemical findings, clinical observation and targetted genetic testing are needed to determine diagnosis of AI. *Abs*, antibodies; *AHC*, adrenal hypoplasia congenita; *AIRE*, autoimmune regulator; *ALD*, adrenoleucodystrophy; *APS-2*, autoimmune polyglandular syndrome type 2; *CNKN1C*, cyclin-dependent kinase inhibitor 1C; *CYP11A1*, cholesterol side chain cleavage; *DHCR1*, 7-dehydrocholesterol reductase 1; DM, diabetes mellitus; *DOK7*, downstream of tyrosine kinase 7; *GCH*, glucocorticoid; *GFER*, growth factor erv-1; *GLI3*, glioma associated oncogene 3; *HYLS1*, hydrolethalus syndrome protein; *LIPA*, lipase A, lysosomal acid type; *MCC*, mucocutaneous candidiasis; *MCH*, mineralocorticoid; *MC2R*, melanocortin 2 receptor; *MCM4*, minichromosome maintenance deficient 4 gene; *MKS1*, Meckel syndrome 1; *MRAP*, melanocortin 2 receptor accessory protein; *NNT*, nicotinamide nucleotide transdehydrogenase; *NR5A1*, steroidogenic factor 1; *NROB1*, dax-1 gene; *PEX1*, peroxisome biogenesis factor 1; *POLG1*, DNA polymerase subunit G; *RAPSN*, receptor associated protein of the synapse; *SALL1*, spalt like gene 1; *SAMD9*, sterile alpha motif domain containing protein 9; *SGPL1*, sphingosine phosphate lyase; *StAR*, steroidogenic acute regulatory protein; *TXNRD2*, thioredoxin reductase 2; *VLCFA*, very-long-chain fatty acids. *(Author original.)*

12q3 is involved (Salehi et al., 2005). A multidisciplinary team is needed to manage the systems involved beyond adrenal function (neurology, gastroenterology, ophthalmology, dentistry and genetics) (Flokas et al., 2019). Genetic analysis of the AAS gene may facilitate an earlier diagnosis (Li et al., 2015).

3.1.2.11.2.2   DNA repair defect: MCM4 deficiency

Among the Irish traveler community a number of children more than 4 years of age presented to hospital with an acute illness, failure to thrive, hyperpigmentation and hypoglycemia. They had some dysmorphic features (thin upper lip, cupids bow lip and prominent forehead) and short stature (O'Riordan et al., 2008). They had raised ACTH and low cortisol, normal renin and aldosterone suggestive of familial glucocorticoid deficiency although screening

tests at 1 year had been normal. Low counts of NK cells were found. No mutations were found in StAR, MC2R or MRAP so a degenerative disorder of the adrenals was predicted. A high prevalence in the community was related to consanguinity. Children had evidence of chromosome fragility on testing with diepoxybutane. Targeted exome sequencing identified a variant of **minichromosome maintenance 4 (MCM4)** leading to a severely truncated protein (Hughes et al., 2012). MCM4 is one part of a MCM2–7 complex which is a replicative helicase essential for DNA replication and stability. Patients may be at increased risk of neoplastic change and need monitoring in anticipation.

This chapter has covered some of the many gene variants that can cause adrenal insufficiency. The encoded proteins can be in the cell membrane or the organelles—mitochondria, endoplasmic reticulum, nuclear membrane (Fig. 3.1.2.43) and other sites within the cell that are not depicted here, such as peroxisomes.

3.1.2.12 Tertiary adrenal insufficiency

The commonest cause is after steroid therapy. Glucocorticosteroids of varying potency are used topically and systemically (Prete and Bancos, 2021). Steroids are often taken orally but also by inhalation, topical, nasal, intra-articular and intravenous routes. High-dose glucocorticoid treatment is used in the management of rheumatic diseases and glomerular diseases and prolonged exposure lead to suppression of the HPA (Karangizi et al., 2019). If treatment with steroids is to be stopped, then stepwise reduction of the steroids is essential to allow recovery of the HPA. This may take many months and has to be balanced against reactivation of the condition for which the glucocorticoids were used. As a rough guide the dose is halved every 2–4 weeks (Pelewicz and Miśkiewicz, 2021). If prednisolone is in use, clinicians may have to convert to hydrocortisone equivalents based on potency estimates that differ between patients. It may be

easier to switch to hydrocortisone which can be better monitored by hormone measurements (Sagar et al., 2021). A basal serum cortisol (above 350 nmol/L for Siemens and Abbott analysis but 500 nmol/L for Roche II assay) excludes adrenal suppression (Woods et al., 2015; Sbardella et al., 2017). A common variation in the platelet derived growth factor (PDGFD) gene locus is associated with an increased risk of adrenal suppression (Hawcutt et al., 2018).

Many patients presenting at the emergency room have hyponatremia that can get mismanaged with respect to sodium and fluids and possible syndrome of inappropriate antidiuretic hormone (SIADH). Patients with serum sodium <135 mmol/L and urine sodium >30 mmol/L with no clinical signs of dehydration (sunken eyes, loss of skin turgor, dry oral mucosa and hypotension) or evidence of fluid excess (presacral or pedal edema or ascites) can be loosely classified as having euvolemic hyponatremia. A low basal cortisol and poor Synacthen stimulated cortisol

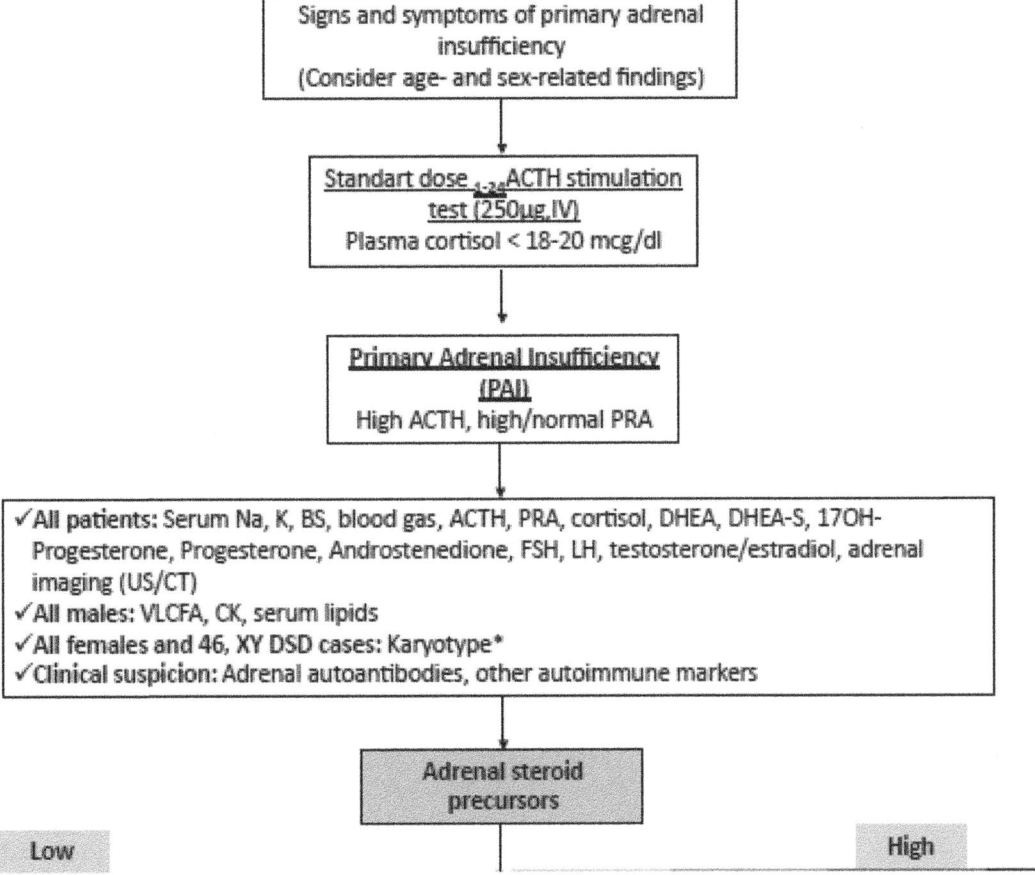

FIG. 3.1.2.41   (A) Diagnostic workup for inherited primary adrenal insufficiency in children is established on the basis of a combination of clinical, laboratory, radiologic and molecular findings. The original chart has been broken into 3 to be legible. After the initial tests basal cortisol, ACTH stimulated cortisol, ACTH measurement further tests include electrolytes, spectrum of steroids and adrenal imaging. In males, only VLCFA, CK and serum lipids are determined. Karyotype. Then adrenal autoantibodies. For further testing, a decision has to be made which steroids are high or low. Abbreviations: *ACTH*, adrenocorticotropin; *BS*, blood sugar; *CAH*, congenital adrenal hyperplasia; *CK*, creatine kinase; *CT*, computerized tomography; *DHEA*, dehydroepiandrosterone; *FSH*, follicle-stimulating hormone; *K*, potassium; *LH*, luteinizing hormone; *Na*, sodium; *PRA*, plasma renin activity; *US*, ultrasound; *VLCFA*, very-long-chain fatty acids.

(Continued)

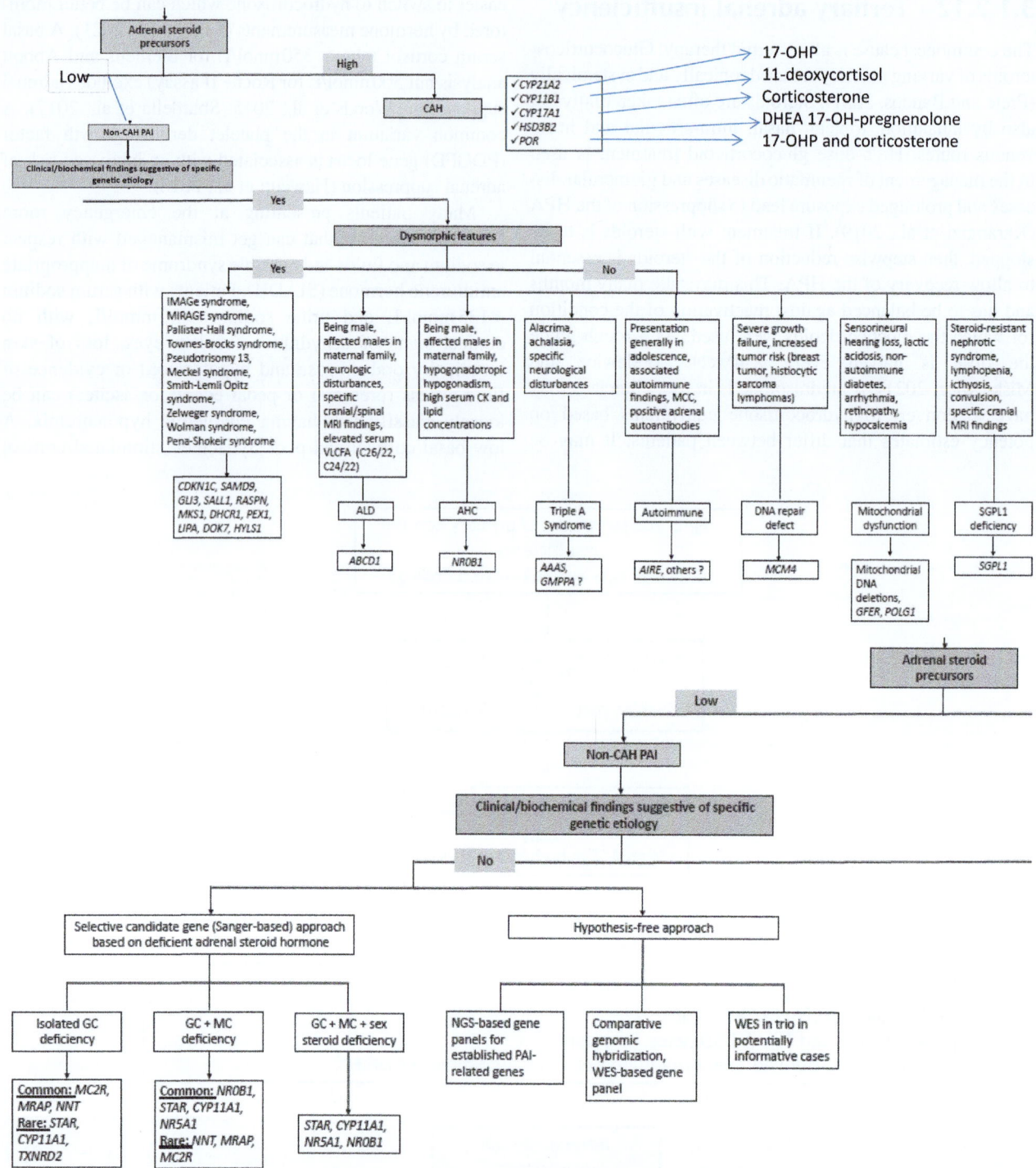

FIG. 3.1.2.41, CONT'D   (B) Although this arm of the algorithm is called "non-CAH PAI" and "no clinical/biochemical findings suggestive of specific etiology" in fact defects of GC, MC and sex steroids can be clues to a diagnosis. Many of the disorders will be confirmed by genetic analysis. Abbreviations: *ALD*, adrenoleukodystrophy; *AHC*, adrenal hypoplasia congenita; *MCC*, mucocutaneous candidiasis; *MRI*, magnetic resonance imaging; *SGPL1*, sphingosine-1-phosphate lyase. (C) *Karyotype can be excluded in female phenotype patients whenever pelvic US is informative for the presence of normal ovaries and Mullerian structures. Assessment of karyotype-matched normal external and internal genitalia and gonads is crucial to decide about gonadal sex steroid production. Measurement of gonadotropins in minipuberty or pubertal cases with non-CAH PAI shows the coexistence or type of hypogonadism. Non-CAH PAI associated with hypogonadotropic hypogonadism would suggest NR0B1 or NR5A1 gene defects while the ones with hypergonadotropic hypogonadism strongly suggest STAR or CYP11A1 mutations. Abbreviations: *GC*, glucocorticoid; *MC*, mineralocorticoid; *NGS*, next-generation sequencing; *WES*, whole-exome sequencing. *(From Kirkgoz T, Guran T. Primary adrenal insufficiency in children: diagnosis and management. Best Pract Res Clin Endocrinol Metab 2018;32(4):397–424. Fig. 2 p. 416.)*

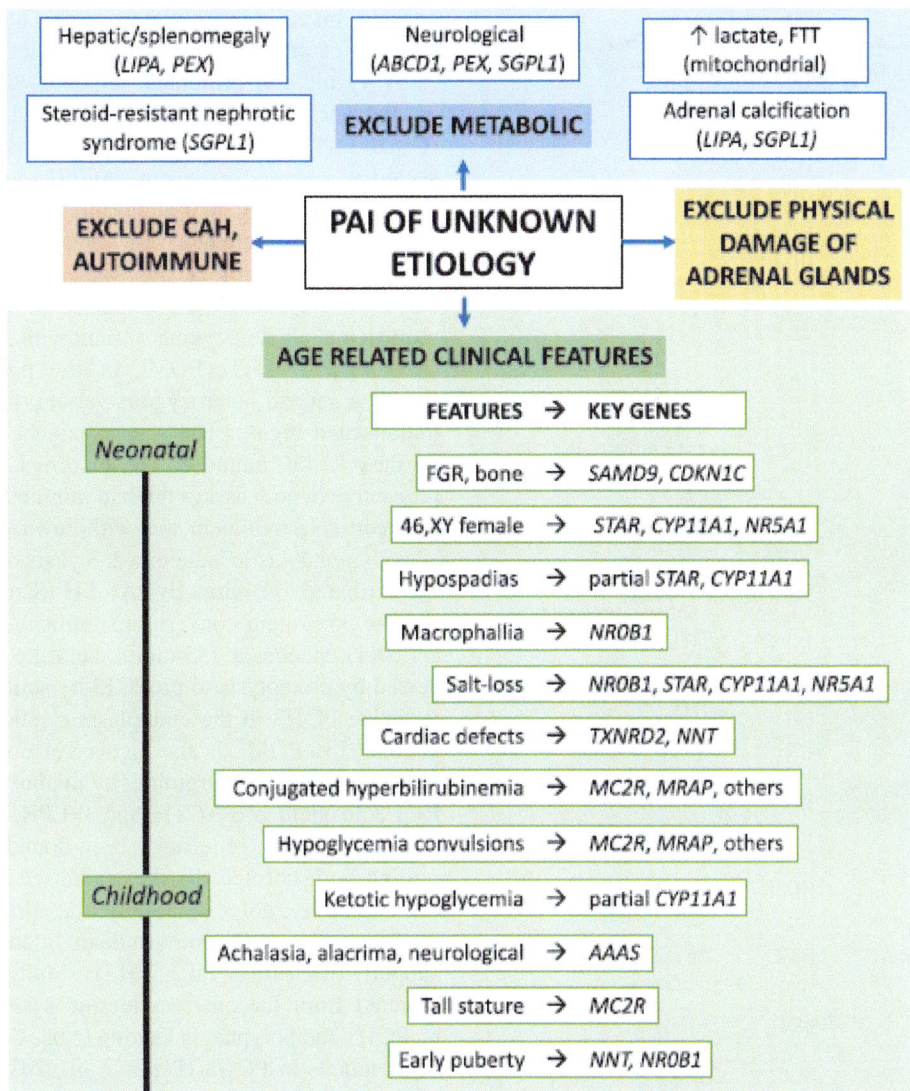

FIG. 3.1.2.42 Clinical features associated with different genetic etiologies of PAI. *(From Buonocore F, Maharaj A, Qamar Y, Koehler K, Suntharalingham JP, Chan LF, Ferraz-de-Souza B, Hughes CR, Lin L, Prasad R, Allgrove J, Andrews ET, Buchanan CR, Cheetham TD, Crowne EC, Davies JH, Gregory JW, Hindmarsh PC, Hulse T, Krone NP, Shah P, Shaikh MG, Roberts C, Clayton PE, Dattani MT, Thomas NS, Huebner A, Clark AJ, Metherell LA, Achermann JC. Genetic analysis of pediatric primary adrenal insufficiency of unknown etiology: 25 years' experience in the UK. J Endocr Soc 2021;5 (8):bvab086.)*

will confirm adrenal insufficiency although prevalence is likely to be low (Kumar et al., 2021; Pelewicz and Miśkiewicz, 2021; McLaughlan and Barth, 2017).

In some populations, corticosteroid creams are used as skin lightening agents which can lead to central adrenal insufficiency (Benn et al., 2016, 2019; Lee et al., 2015; Petit et al., 2006).

Hypothalamic function may not be normal in inflammatory disorders, trauma, radiation therapy (postsurgery for pituitary Cushings), surgery, tumors pressing on gland, infiltrative disease such as sarcoidosis and histocytosis X.

People living with human immunodeficiency virus (HIV) infection (PLHIV) can have endocrinological abnormalities including adrenal insufficiency. Fatigue is a common presenting problem as well as weight loss, anorexia, nausea and vomiting. The picture is complicated by the effects of the infection itself, by other infections and drug use for pain, muscle wasting and infection (Mifsud et al., 2021). AI is characterized by a persistently low plasma cortisol and poor response to Synacthen. Raised renin and low aldosterone is also supportive of AI. Adrenal androgen output can be low (Honour et al., 1995), as seen with tuberculosis infection (Rook et al., 1996) and histoplasmosis (Madhavan et al., 2020). If there is a normal cortisol and elevated ACTH, this defines subclinical AI. If cortisol is raised but the response to Synacthen is poor, then relative AI can be diagnosed. When cortisol is raised and there is good response to Synacthen, the diagnosis is glucocorticoid resistance. ACTH will be raised but is not suppressed with dexamethasone. All situations are seen with PLHIV.

TABLE 3.1.2.12 Causes of secondary adrenal insufficiency due to pituitary disorders.

	Gene	
Acquired causes		
Steroid withdrawal		
Opioids		
Tumor		
Trauma		
Pituitary apoplexy (Sheehan's syndrome)		
Congenital causes		
Apoplasia or hypoplasia		
PROP deficiency	PROP1	Defects also of growth hormone, prolactin, TSH, LH, FSH
LHX4 deficiency	LHX4	Defects also of GH and TSH
SOX3 deficiency	SOX3	Defects of pituitary hormones
Isolated ACTH deficiency		
TBX19 deficiency	TBX19	Severe neonatal adrenal insufficiency (AI)
Proopiomelanocortin	POMC	AI, early onset obesity, red hair
Proprotein convertase 1	PCSK1	Hypoglycaemia, malabsorption, hypog hypog

Based on Husebye ES, Pearce SH, Krone NP, Kämpe O. Adrenal insufficiency. Lancet 2021;397(10274):613–29.

3.1.2.12.1   Pituitary insufficiency

POMC is a precursor for ACTH and other peptides (alpha MSH) and deficiency in the synthesis of POMC leads to early adrenal insufficiency, pale skin and red hair (Krude et al., 1998). The disease due to **POMC deficiency** leads to early onset of obesity due to uncontrolled polyphagia. Hydrocortisone treatment and diet are needed. Twenty two cases in the literature were reviewed (Graves et al., 2021; Gregoric et al., 2021). Neonatal presentation was usually hypoglycemia and hyperbilirubinemia. Red hair is attributed to the absence of MC1 that is stimulated by alpha-MSH. MC1R shifts expression of red pigment (pheomelanin) to dark pigment (eumelanin). The absence of ACTH leads to

adrenal insufficiency. Some cases have further deficiencies of anterior pituitary hormones.

A 4y old girl presented in coma with hypoglycemia, normal electrolytes and low cortisol, high ACTH by IRMA (Cis-Bio). The adrenals could not be visualized. A boy presented as a generalize seizure. ACTH was raised (Immulite 2000). Both had red hair and primary adrenal insufficiency was suspected. Patients were treated with glucocorticoid and mineralocorticoid. Genes known to cause ACTH resistance (MC2R and MRAP) were normal. After whole-exome sequencing, a heterozygous variant with missense mutation encoding p.R1415C in POMC and thus p.R8C in ACTH was found. A second heterozygous variant was found in the 5′ untranslated region. The second patient was homozygous for the p.R145C mutation. The abnormal ACTH was immunoreactive in both assays but had minimal bioactivity. Mineralocorticoid treatment was withdrawn. Case 2 developed acute lymphoblastic leukemia at 3 years of age. These cases are attributed to **Bioinactive ACTH** (Samuels et al., 2013).

The proprotein convertase subtilisin/kexin-type 1 gene (PCSK1) encodes a 753-amino acid precursor that is processed by proteolysis to proPC13 by action of PCSK1 then to active PC1/3 in the endoplasmic reticulum (Turpeinen et al., 2013). POMC is also cleaved at a dibasic amino acid sequence (lysine and arginine) by **prohormone convertase PC1/3** to yield pro-ACTH and β-LPH. Loss of function mutations in the gene have been found in children presenting with chronic diarrhea, adults may have hypoglycemia, obesity, polydipsia, polyuria, adrenal insufficiency, hypogonadism and hypothyroidism. Infants need nutritional support (Martín et al., 2013). Multiple prohormones secreted from the enteroendocrine system are processed by PC1/3 and polyphagia leading to obesity is characteristic of mutations in PC1/3 (Pépin et al., 2019; Ramos-Molina et al., 2016).

Hyponatremia is an important presenting feature of pituitary disease due to nonfunctioning pituitary adenoma, intracranial aneurysm, lymphocytic hypophysitis and traumatic brain injury (Miljic et al., 2017; Diederich et al., 2003). Other pituitary transcription factor genes can be affected (PROP1, HESX1, OTX2, LHX4, SOX3 and TBX19) (see Bornstein, 2009 for summary at that time), and mutations of others may come to light in the course of time. Patients with obesity due to POMC or MC4R mutations were found to have lower cardiac muscle mass than other obese subjects (Puder et al., 2021) and stored Na^+ within subcutaneous fat.

The **Prader-Willi Syndrome (PWS)** is a rare genetic disorder resulting from the loss of imprinted gene expression in the paternal chromosome 15q11-q13. Children have a long, narrow head (dolichocephaly), small mouth with downturned corners and almond shaped eyes. PWS is characterized by several endocrine abnormalities, such as growth hormone (GH) deficiency, hyperphagia

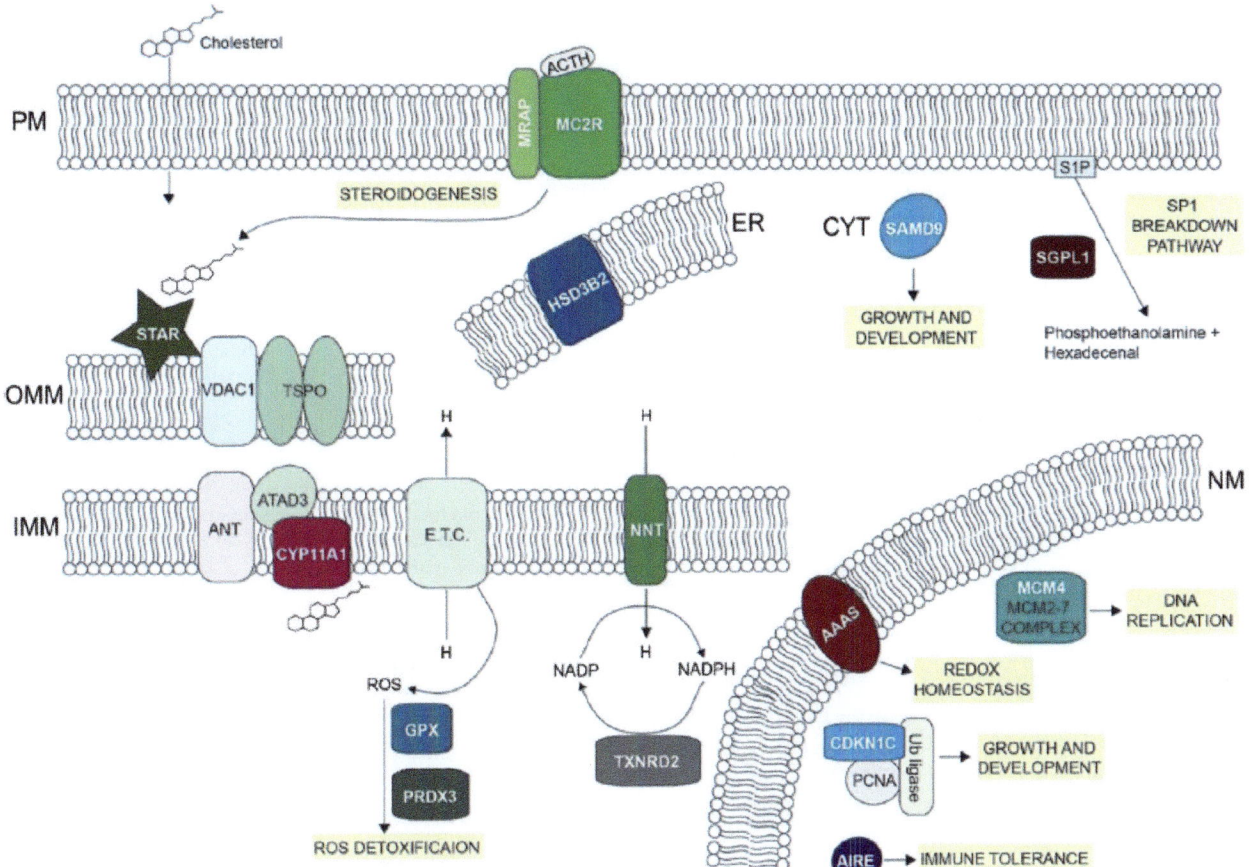

FIG. 3.1.2.43 Summary of genetic causes of isolated and syndromic glucocorticoid deficiency and associated molecular mechanisms. Abbreviations (with mutated genes in *white*): *AAAS*, achalasia-Addisonianism-alacrimia (triple A) syndrome gene; *AIRE*, autoimmune regulator; *ANT*, adenine nucleotide translocator; *ATAD3*, ATPase family AAA domain-containing protein 3; *CDKN1C*, cyclin-dependent kinase inhibitor 1C; *CYP11A1*, cholesterol side-chain cleavage enzyme encoding-gene; *ER*, endoplasmic reticulum; *etc.*, electron transport chain; *GPX1*, glutathione peroxidase 1; *HSD3B2*, 3-beta-hydroxysteroid dehydrogenase; *IMM*, inner mitochondrial membrane; *MC2R*, melanocortin receptor type-2; *MCM*, minichromosome maintenance complex component; *MRAP*, melanocortin receptor accessory protein; *NM*, nuclear membrane; *NNT*, nicotinamide nucleotide transhydrogenase; *OMM*, outer mitochondrial membrane; *PCNA*, proliferating cell nuclear antigen; *PM*, plasma membrane; *PRDX3*, thioredoxin-dependent peroxide reductase; *SGPL1*, sphingosine-1-phosphate lyase 1; *SMAD9*, sterile alpha motif domain-containing protein 9; *SP-1*, sphingosine 1 phosphate; *STAR*, steroidogenic acute regulatory protein; *TSPO*, translocator protein; *TXNRD2*, thioredoxin reductase 2; *Ub*, ubiquitin; *VDAC*, voltage-dependent anion-selective channel 1. *(From Maharaj A, Maudhoo A, Chan LF, Novoselova T, Prasad R, Metherell LA, Guasti L. Isolated glucocorticoid deficiency: genetic causes and animal models. J Steroid Biochem Mol Biol 2019;189:73–80.)*

causing obesity, central adrenal insufficiency (rare), hypothyroidism, hypogonadism and complex behavioral and intellectual difficulties (Ramos-Molina et al., 2018). PWS individuals may present with sleep related breathing problems, scoliosis, constipation, dental issues and coagulation disorders. Management is mainly through caloric restriction to 900 kcal/day, daily aerobic exercises and postural therapy (Heksch et al., 2017). Treatment with recombinant growth hormone is recommended starting as soon as the diagnosis is made. A few need hydrocortisone replacement. A metyrapone test is the best insult of the HPA for adrenal insufficiency. Surgery is a challenge especially for anesthetists.

Congenital causes of tertiary AI include septo-optic dysplasia and CRF deficiency (Table 3.1.2.13).

3.1.2.12.1.1 Drugs

Besides use for replacement therapy where cortisol production is impaired, glucocorticoids are prescribed for a number of medical conditions such as asthma, rheumatoid arthritis, autoimmune disorders as immunosuppressants. Steroids can be administered by oral, intravenous, intramuscular injection, inhalation, intra-articular, percutaneous and rectal routes. The dose and potency of the steroids used are variable. Features of Cushing's syndrome can develop with high exposure and care must be taken on withdrawal of the treatment to avoid an Addisonian crisis (Laugesen et al., 2019; Prete and Bancos, 2021 for a thorough review).

Several drugs inhibit cortisol synthesis (Table 3.1.2.14). Inhibition of 11-hydroxylase through use of **etomidate** is

TABLE 3.1.2.13 Congenital causes of tertiary AI.

Cause	Signs and symptoms
Acquired	
Corticosteroid withdrawal	Endogenous due to Cushing's syndrome or exogenous cortisol for more than 2 weeks
Opioids	Hypogonadotrophic hypogonadism
Inflammatory disorders	Abscess, meningitis, encephalitis
Radiation therapy	Craniospinal irradiation in leukemia
Tumor	Craniopharyngioma, glioma, meningioma, ependymoma, germinoma and intrasellar or suprasellar metastases
Infiltrative diseases	Sarcoidosis, histiocytosis X and hemochromatosis
Trauma	Pituitary stalk lesions, battered child, vehicular trauma
HIV/AIDS (Mifsud et al., 2021)	
Histoplasmosis (Madhavan et al., 2020)	Mass left maxilla. Swelling, pain, hyponatremia
Genetic	
Septo-optic dysplasia, mutation HESX1	Optic nerve hypoplasia, combined pituitary hormone defects, midline defects
Corticotrophic releasing factor	ACTH deficiency

Modified from Husebye ES, Pearce SH, Krone NP, Kämpe O. Adrenal insufficiency. Lancet 2021;397(10274):613–29. Table 3 p. 617.

TABLE 3.1.2.14 Drug induced adrenal insufficiency.

Drug	Mechanism
Aminoglutethimide	Aromatase inhibitor
Anticoagulants	Hemorrhage
Antidepressant (imipramine)	Inhibit glucocorticoid gene transcription
Antimycotic drugs (ketoconazole, fluconazole)	Mitochondrial cytochrome P450 enzyme inhibition
Antipsychotic drugs (chlorpromazine)	Inhibit glucocorticoid gene transcription
Etomidate	Mitochondrial cytochrome P450 enzyme inhibition—CYP11B1
Glucocorticoid treatment (ciclesonide, fluticasone, hydrocortisone, megestrol acetate, prednisolone, prednisone)	Suppression of CRH and ACTH synthesis
Immunosuppressant drugs (Pembrolizumab)	Hypophysitis
Mifepristone	Binds glucocorticoid receptor
Phenobarbital	Induction CYP450-dependent metabolizers of cortisol (CYP2B1 and CYP2B2)
Phenytoin	Induction CYP450-dependent metabolizers of cortisol (CYP3A4)
Rifampin	Induction CYP450-dependent metabolizers of cortisol (CYP3A4)
Trilostane	Inhibit HSD3B2
Troglitazone	Induction CYP450-dependent metabolizers of cortisol (CYP3A4)

not uncommon as this short acting intravenous anesthetic agent is used to cover procedures such as intubation.

The drug should not be used when the patient may have an adrenal problem. **Ketoconazole** is an antifungal agent that can also blocks 11-hydroxylase activity in the adrenal gland (Weber et al., 1993). The related **posaconazole** was prescribed for a boy with cystic fibrosis and an allergic broncho-pulmonary aspergillosis who became hypertensive. Analysis of steroids in urine (Agarwal et al., 2020) revealed an increased excretion of the 11-deoxycortisol metabolite, tetrahydro-11-deoxycortisol (261 μg/mmol creatinine) and the 11-deoxycorticosterone metabolite, tetrahydro-11- deoxycorticosterone (32 μg/mmol creatinine) (Table 3.1.2.15) which raised the possibility of CAH secondary to 11β-hydroxylase deficiency.

Cortisol metabolites were present, and there was no relative increase in the androgen metabolites (DHEA < 0.4 μmol/L, androstenedione < 0.4 nmol/L, and testosterone < 0.7 nmol/L), which would be consistent with a mild form of such a defect.

Three children with SLO symptoms but without mutations in 7DHCR had been exposed to **aripiprazole** (an antipsychotic) or **trazodone** (antidepressant) in the pregnancy. Both drugs can be metabolized to piperazine derivatives that inhibit DHCR7 (Hall et al., 2013).

Secondary adrenal insufficiency is also seen in people taking opioids for noncancer pain (Li et al., 2020; Lamprecht et al., 2018). Adrenal insufficiency is an uncommon side-effect of new cancer treatments (Pembrolizumab, etc.) based on antiprogrammed cell death therapy

TABLE 3.1.2.15 Clinical and biochemical parameters on and off posaconazole treatment.

Parameters	On posaconazole	Off posaconazole	Normal reference range
Blood pressure (mmHg)	150/84	92/56	95/56
Urine 11-deoxycorticosterone metabolite (μg/mmol creatinine)	32	1.0	<1
Urine 11-deoxycortisol metabolite (μg/mmol creatinine)	261	11.7	<15
Plasma renin (mU/L)	2.0	N/A	15.8–100.8
Plasma aldosterone (pmol/L)	45.0	N/A	80–970
Serum cortisol on SST[a] (nmol/L)	420[b]	735[b]	>500
Serum cortisone on SST[a] (nmol/L)	41[b]	55[b]	19.7–77.5
Cortisol:cortisone ratio	12[b]	3.5[b]	1.0–10.5
17-Hydroxyprogesterone on SST[a] (nmol/L)	5[b]	7.5[b]	<30
Serum 11-deoxycortisol on SST[a] (nmol/L)	102[b]	5[b]	<2.7
Serum 11-deoxycorticosterone on SST (nmol/L)	41[b]	1.6[b]	<1.4
Serum corticosterone on SST[a] (nmol/L)	71[b]	106[b]	3.5–59.2
Posaconazole assay (mg/L)	6	N/A	1–3.75

N/A, not available.
[a]*Standard Synacthen test.* [b]*Peak concentration quoted.*
From Agarwal N, Apperley L, Taylor NF, Taylor DR, Ghataore L, Rumsby E, Treslove C, Holt R, Thursfield R, Senniappan S. Posaconazole-Induced Hypertension Masquerading as Congenital Adrenal Hyperplasia in a Child with Cystic Fibrosis. Case Rep Med. 2020;2020:8153012. Table 1 p. 3.

(Hattersley et al., 2021). This was first noticed as eosinophilia on blood tests and radiological evidence of pituitary inflammation during follow-up. Some cases had hypotension and hyponatremia. Fatigue, anorexia, weight loss and hypoglycemia may also be seen (Iglesias et al., 2021; Ariyasu et al., 2017).

Clinicians need a high degree of clinical suspicion but this is not always easy when faced with patients with infections who are likely to have symptoms of the infection similar to AI and will be prescribed with a range of drugs. Delay in treatment with steroids could result in life threatening consequence.

3.1.2.12.2 Adrenal insufficiency in pregnancy

The commonest of cases of AI in pregnancy would be caused by an existing autoimmune disorder causing adrenal destruction or a patient taking exogenous glucocorticoids for Addison's disease, congenital adrenal hyperplasia (Reisch, 2019), asthma or rheumatism that suppresses pituitary ACTH secretion. Pregnancy is not easy to achieve with these conditions so mothers need careful monitoring during the process. Successful deliveries have been reported from frozen thawed embryo transfer for 17-hydroxylase deficiency (Bianchi et al., 2016), 17,20 lyase deficiency (Blumenfeld and Koren, 2021) and P450 oxidoreductase

deficiency (Pan et al., 2021). Clinical information such as well-being, weight gain, and blood pressure and plasma potassium concentrations should be monitored. Cortisol production during normal pregnancy increases two- to threefold partly as CBG is increased through the action of estrogens and partly through upregulation of the HPA axis. There is an exaggerated response to ACTH (Suri et al., 2006).

The results of a short Synacthen test that would normally be used to assess AI may be difficult to interpret. Peak cortisol concentrations near 30 min after intravenous ACTH increase to 700, 800 then 900 nmol/L in the trimesters (Bornstein et al., 2016; Suri et al., 2006). The activity of HSD11B2 in the placenta that inactivates cortisol is also increased during pregnancy (Pasqualini and Chetrite, 2016). A rise in cortisol toward term is essential for the mother for the metabolic demands of labor. Clinicians need to ensure patients understand the changing needs during pregnancy and how to adjust to stressful situations such as infection (Bothou et al., 2020).

Corticosteroid replacement for patients with AI should be increased in pregnancy. The suggested management of GCH treatment in pregnancy have been reviewed recently (Pofi and Tomlinson, 2020; Woodcock et al., 2020; Bornstein et al., 2016) taking into account the nuances of the HPA axis in pregnancy, the normal rise in cortisol across the trimesters, the changes in CBG and CRF production from mid-pregnancy to term. High activity of HSD11B2 in the placenta has a significant effect on cortisol metabolism. Glucocorticoid replacement should be increased 20%–40% in the third trimester. At delivery, 100 mg hydrocortisone i/m is recommended when labor starts and 200 mg every 24 h (Pofi and Tomlinson, 2020). The dose is reduced for 2–4 days after delivery to double the dose used before pregnancy.

After the delivery, some mothers can get hypopituitarism (Sheehan's syndrome postpartum pituitary necrosis). There is little concern over transfer of cortisol to the baby in breast milk though there is little data to support that. Prednisolone, measured by LC-MS/MS, is only marginally excreted in milk (Ryu et al., 2018). Breathing difficulties for preterm babies are common. Ultrasound images of the fetal lung can be used to assess maturity (Xia et al., 2021). Steroids are used for lung maturation of babies likely to be delivered early (Deshmukh and Patole, 2021; Roberts et al., 2017; Gyamfi-Bannerman et al., 2016). Dexamethasone (4×6 mg 12 h apart) and betamethasone (2×12 mg 12 h apart) are used. Uptake of this practice across the globe is variable. Some maternal screening programmes in pregnancy for Down's syndrome measure concentration of unconjugated estriol during pregnancy. The results can reflect fetal disorders such as sulfatase deficiency and adrenal insufficiency (Langlois et al., 2009; Weintrob et al., 2006; Malpuech et al., 1988).

3.1.2.13 Summary

Some of the disorders of adrenal insufficiency present a real clinical challenge because of similarities in presentation with inborn errors of metabolism. Dialogues between clinicians and families and clinicians and the laboratory are important in difficult life-threatening situations. Testing for adrenal insufficiency is becoming easier as more accurate and more precise steroid tests are becoming more widely available, particularly when steroid panel and profile tests are accessible. Most of the genetic disorders present in the younger patient. Consanguinity is a contributing factor to many cases. Most of the disorders of steroid synthesis arise from mutation sin genes for enzymes in the mitochondria and endoplasmic reticulum but rare forms are now due to enzymes in other cell compartment such as lysosomes and peroxisomes and sites in the hypothalamic-pituitary-adrenal axis. Genetic testing will have an increasing importance as panels of genes and sequencing become more widely available.

Most disorders can be treated with steroid replacement and there have been many developments in modes of treatment aside from taking tablets. All patients with adrenal insufficiency should wear a medical alert bracelet.

References

Abraham SB, Abel BS, Sinaii N, Saverino E, Wade M, Nieman LK. Primary vs secondary adrenal insufficiency: ACTH-stimulated aldosterone diagnostic cut-off values by tandem mass spectrometry. Clin Endocrinol (Oxf) 2015;83(3):308–14.

Abuduxikuer K, Li ZD, Xie XB, Li YC, Zhao J, Wang JS. Novel melanocortin 2 receptor variant in a Chinese infant with familial glucocorticoid deficiency type 1, case report and review of literature. Front Endocrinol (Lausanne) 2019;10:359.

Agarwal N, Apperley L, Taylor NF, Taylor DR, Ghataore L, Rumsby E, et al. Posaconazole-induced hypertension masquerading as congenital adrenal hyperplasia in a child with cystic fibrosis. Case Rep Med 2020;2020:8153012.

An M, Liu Y, Zhang M, Hu K, Jin Y, Xu S, et al. Targeted next-generation sequencing panel screening of 668 Chinese patients with non-obstructive azoospermia. J Assist Reprod Genet 2021;38(8):1997–2005.

Arem R, Ghusn H, Ellerhorst J, Comstock JP. Effect of decreased plasma low-density lipoprotein levels on adrenal and testicular function in man. Clin Biochem 1997;30(5):419–24.

Arif S, Varela-Calvino R, Conway GS, Peakman M. 3 Beta hydroxysteroid dehydrogenase autoantibodies in patients with idiopathic premature ovarian failure target N- and C-terminal epitopes. J Clin Endocrinol Metab 2001;86(12):5892–7.

Ariyasu R, Horiike A, Yoshizawa T, Dotsu Y, Koyama J, Saiki M, et al. Adrenal insufficiency related to anti-programmed death-1 therapy. Anticancer Res 2017;37(8):4229–32.

Baş F, Kayserili H, Darendeliler F, Uyguner O, Günöz H, Yüksel Apak M, et al. CYP21A2 gene mutations in congenital adrenal hyperplasia: genotype-phenotype correlation in Turkish children. J Clin Res Pediatr Endocrinol 2009;1(3):116–28.

Bae H, Kim MS, Park H, Jang JH, Choi JM, Lee SM, et al. Nonclassic congenital lipoid adrenal hyperplasia diagnosed at 17 months in a Korean boy with normal male genitalia: emphasis on pigmentation as a diagnostic clue. Ann Pediatr Endocrinol Metab 2020;25(1):46–51.

Balasubramanian M, Sprigg A, Johnson DS. IMAGe syndrome: case report with a previously unreported feature and review of published literature. Am J Med Genet A 2010;152A(12):3138–42.

Balsamo A, Baronio F, Ortolano R, Menabo S, Baldazzi L, Di Natale V, et al. Congenital adrenal hyperplasias presenting in the newborn and Young infant. Front Pediatr 2020;8, 593315.

Baranowski ES, Arlt W, Idkowiak J. Monogenic disorders of adrenal steroidogenesis. Horm Res Paediatr 2018;89(5):292–310.

Benn EK, Alexis A, Mohamed N, Wang YH, Khan IA, Liu B. Skin bleaching and dermatologic health of African and Afro-Caribbean populations in the US: New directions for methodologically rigorous, multidisciplinary, and culturally sensitive research. Dermatol Ther (Heidelb) 2016;6(4):453–9.

Benn EKT, Deshpande R, Dotson-Newman O, Gordon S, Scott M, Amarasiriwardena C, et al. Skin bleaching among African and Afro-Caribbean women in New York City: primary findings from a P30 pilot study. Dermatol Ther (Heidelb) 2019;9(2):355–67.

Berge KE, Tian H, Graf GA, Yu L, Grishin NV, Schultz J, et al. Accumulation of dietary cholesterol in sitosterolemia caused by mutations in adjacent ABC transporters. Science 2000;290(5497):1771–5.

Bianchi PH, Gouveia GR, Costa EM, Domenice S, Martin RM, de Carvalho LC, et al. Successful live birth in a woman with 17α-hydroxylase deficiency through IVF frozen-thawed embryo transfer. J Clin Endocrinol Metab 2016;101(2):345–8.

Blumenfeld Z, Koren I. Successful delivery in 17,20-lyase deficiency. J Clin Endocrinol Metab 2021;106(7):1882–6. https://doi.org/10.1210/clinem/dgab222. 33824988.

Bochem AE, Holleboom AG, Romijn JA, Hoekstra M, Dallinga-Thie GM, Motazacker MM, et al. High density lipoprotein as a source of cholesterol for adrenal steroidogenesis: a study in individuals with low plasma HDL-C. J Lipid Res 2013;54(6):1698–704.

Bochem AE, Holleboom AG, Romijn JA, Hoekstra M, Dallinga GM, Motazacker MM, et al. Adrenal function in females with low plasma HDL-C due to mutations in ABCA1 and LCAT. PLoS One 2014;9(5), e90967.

Bornstein SR. Predisposing factors for adrenal insufficiency. N Engl J Med 2009;360(22):2328–39.

Bornstein SR, Allolio B, Arlt W, Barthel A, Don-Wauchope A, Hammer GD, et al. Diagnosis and treatment of primary adrenal insufficiency: an Endocrine Society clinical practice guideline. J Clin Endocrinol Metab 2016;101(2):364–89.

Bornstein SR, Malyukov M, Heller C, Ziegler CG, Ruiz-Babot G, Schedl A, et al. New horizons: novel adrenal regenerative therapies. J Clin Endocrinol Metab 2020;105(9):3103–7.

Bothou C, Anand G, Li D, Kienitz T, Seejore K, Simeoli C, et al. Current management and outcome of pregnancies in women with adrenal insufficiency: experience from a multicenter survey. J Clin Endocrinol Metab 2020;105(8):e2853–63.

Breder ISS, Garmes HM, Mazzola TN, Maciel-Guerra AT, de Mello MP, Guerra-Júnior G. Three new Brazilian cases of 17α-hydroxylase deficiency: clinical, molecular, hormonal, and treatment features. J Pediatr Endocrinol Metab 2018;31(8):937–42.

Breil T, Yakovenko V, Inta I, Choukair D, Klose D, Mittnacht J, et al. Typical characteristics of children with congenital adrenal hyperplasia due to 11β-hydroxylase deficiency: a single-centre experience and review of the literature. J Pediatr Endocrinol Metab 2019;32(3):259–67.

Bryan SM, Honour JW, Hindmarsh PC. Management of altered hydrocortisone pharmacokinetics in a boy with congenital adrenal hyperplasia using a continuous subcutaneous hydrocortisone infusion. J Clin Endocrinol Metab 2009;94(9):3477–80.

Buonocore F, Maharaj A, Qamar Y, Koehler K, Suntharalingham JP, Chan LF, et al. Genetic analysis of pediatric primary adrenal insufficiency of unknown etiology: 25 Years' experience in the UK. J Endocr Soc 2021;5(8), bvab086.

Butz H, Nyírő G, Kurucz PA, Likó I, Patócs A. Molecular genetic diagnostics of hypogonadotropic hypogonadism: from panel design towards result interpretation in clinical practice. Hum Genet 2021;140(1):113–34.

Calliari LE, Longui CA, Rocha MN, Faria CD, Kochi C, Melo MR, et al. A novel mutation in DAX1 gene causing different phenotypes in three siblings with adrenal hypoplasia congenita. Genet Mol Res 2007;6(2):277–83.

Cappuccio G, Donti TR, Hubert L, Sun Q, Elsea SH. Opening a window on lysosomal acid lipase deficiency: biochemical, molecular, and epidemiological insights. J Inherit Metab Dis 2019;42(3):509–18.

Carvalho B, Marques CJ, Santos-Silva R, Fontoura M, Carvalho D, Carvalho F. Congenital adrenal hyperplasia due to 21-hydroxylase deficiency: an update on genetic analysis of CYP21A2 gene. Exp Clin Endocrinol Diabetes 2021;129(7):477–81.

Chang YC, Huang CC, Huang SC, Hung FC. Neonatal adrenoleukodystrophy presenting with seizure at birth: a case report and review of the literature. Pediatr Neurol 2008;38(2):137–9.

Chao CS, Shi RZ, Kumar RB, Aye T. Salivary cortisol levels by tandem mass spectrometry during high dose ACTH stimulation test for adrenal insufficiency in children. Endocrine 2020;67(1):190–7.

Cheillan D. Zellweger syndrome disorders: from severe neonatal disease to atypical adult presentation. Adv Exp Med Biol 2020;1299:71–80.

Chen S, Sawicka J, Betterle C, Powell M, Prentice L, Volpato M, et al. Autoantibodies to steroidogenic enzymes in autoimmune polyglandular syndrome, Addison's disease, and premature ovarian failure. J Clin Endocrinol Metab 1996;81(5):1871–6.

Choi YJ, Saba JD. Sphingosine phosphate lyase insufficiency syndrome (SPLIS): a novel inborn error of sphingolipid metabolism. Adv Biol Regul 2019;71:128–40.

Chung TT, Chan LF, Metherell LA, Clark AJ. Phenotypic characteristics of familial glucocorticoid deficiency (FGD) type 1 and 2. Clin Endocrinol (Oxf) 2010;72(5):589–94.

Claahsen-van der Grinten HL, Speiser PW, Ahmed SF, Arlt W, Auchus RJ, Falhammar H, et al. Congenital adrenal hyperplasia—current insights in pathophysiology, diagnostics and management. Endocr Rev 2022;43(1):91–159. bnab016.

Clark AJ, McLoughlin L, Grossman A. Familial glucocorticoid deficiency associated with point mutation in the adrenocorticotropin receptor. Lancet 1993;341(8843):461–2.

Clément K, Dubern B, Mencarelli M, Czernichow P, Ito S, Wakamatsu K, et al. Unexpected endocrine features and normal pigmentation in a young adult patient carrying a novel homozygous mutation in the POMC gene. J Clin Endocrinol Metab 2008;93(12):4955–62.

Colls J, Betterle C, Volpato M, Prentice L, Smith BR, Furmaniak J. Immunoprecipitation assay for autoantibodies to steroid 21-hydroxylase in autoimmune adrenal diseases. Clin Chem 1995;41(3):375–80.

Cooray SN, Chan L, Metherell L, Storr H, Clark AJL. Adrenocorticotropin resistance syndromes. Endocr Dev 2008;13:99–116.

de-Souza BF, Lin L, Achermann JC. Steroidogenic factor-1 (SF-1) and its relevance to pediatric endocrinology. Pediatr Endocrinol Rev 2006;3(4):359–64. 16816804.

Del Pilar Larosa M, Chen S, Steinmaus N, Macrae H, Guo L, Masiero S, et al. A new ELISA for autoantibodies to steroid 21-hydroxylase. Clin Chem Lab Med 2018;56(6):933–8.

Deshmukh M, Patole S. Antenatal corticosteroids for impending late preterm (34-36+6 weeks) deliveries-A systematic review and meta-analysis of RCTs. PLoS One 2021;16(3), e0248774.

Dias RP, Chan LF, Metherell LA, Pearce SH, Clark AJ. Isolated Addison's disease is unlikely to be caused by mutations in MC2R, MRAP or STAR, three genes responsible for familial glucocorticoid deficiency. Eur J Endocrinol 2010;162(2):357–9.

Diederich S, Franzen NF, Bähr V, Oelkers W. Severe hyponatremia due to hypopituitarism with adrenal insufficiency: report on 28 cases. Eur J Endocrinol 2003;148(6):609–17.

Elghawy O, Zhang AY, Duong R, Wilson WG, Shildkrot EY. Ophthalmic diagnosis and novel management of infantile refsum disease with combination docosahexaenoic acid and cholic acid. Case Rep Ophthalmol Med 2021;2021:1345937.

Eriksson H. Absorption and enterohepatic circulation of neutral steroids in the rat. Eur J Biochem 1971;19(3):416–23. https://doi.org/10.1111/j.1432-1033.1971.tb01331.x. 5554229.

Eriksson H, Gustafsson JA. Steroids in germfree and conventional rats. Sulpho- and glucuronohydrolase activities of caecal contents from conventional rats. Eur J Biochem 1970;13(1):198–202.

Eriksson H, Gustafsson JA, Sjövall J. Steroids in germfree and conventional rats. 4. Identification and bacterial formation of 17 alpha-pregnane derivatives. Eur J Biochem 1968;6(2):219–26.

Eriksson H, Gustafsson JA, Sjövall J. Steroids in germfree and conventional rats. 21-dehydroxylation by intestinal microorganisms. Eur J Biochem 1969;9(4):550–4.

Escolà-Gil JC, Quesada H, Julve J, Martín-Campos JM, Cedó L, Blanco-Vaca F. Sitosterolemia: diagnosis, investigation, and management. Curr Atheroscler Rep 2014;16(7):424.

Fabbri-Scallet H, de Sousa LM, Maciel-Guerra AT, Guerra-Júnior G, de Mello MP. Mutation update for the NR5A1 gene involved in DSD and infertility. Hum Mutat 2020;41(1):58–68.

Falorni A, Bini V, Betterle C, Brozzetti A, Castaño L, Fichna M, et al. Determination of 21-hydroxylase autoantibodies: inter-laboratory concordance in the Euradrenal International Serum Exchange Program. Clin Chem Lab Med 2015;53(11):1761–70.

Fan L, Ren X, Song Y, Su C, Fu J, Gong C. Novel phenotypes and genotypes in Antley-Bixler syndrome caused by cytochrome P450 oxidoreductase deficiency: based on the first cohort of Chinese children. Orphanet J Rare Dis 2019;14(1):299.

Fan HC, Lee HF, Yue CT, Chi CS. Clinical characteristics of mitochondrial encephalomyopathy, lactic acidosis, and stroke-like episodes. Life (Basel) 2021;11(11):1111.

Fanis P, Neocleous V, Kosta K, Karipiadou A, Hartmann MF, Wudy SA, et al. Late diagnosis of 3β-hydroxysteroid dehydrogenase deficiency: the pivotal role of gas chromatography-mass spectrometry urinary steroid metabolome analysis and a novel homozygous nonsense mutation in the HSD3B2 gene. J Pediatr Endocrinol Metab 2020;34(1):131–6.

Fink RS, Pierre LN, Daley-Yates PT, Richards DH, Gibson A, Honour JW. Hypothalamic-pituitary-adrenal axis function after inhaled corticosteroids: unreliability of urinary free cortisol estimation. J Clin Endocrinol Metab 2002;87(10):4541–6.

Flokas ME, Tomani M, Agdere L, Brown B. Triple A syndrome (Allgrove syndrome): improving outcomes with a multidisciplinary approach. Pediatric Health Med Ther 2019;10:99–106.

Flück CE. Mechanisms in endocrinology: update on pathogenesis of primary adrenal insufficiency: beyond steroid enzyme deficiency and autoimmune adrenal destruction. Eur J Endocrinol 2017;177(3):R99–R111.

Fujieda K, Okuhara K, Abe S, Tajima T, Mukai T, Nakae J. Molecular pathogenesis of lipoid adrenal hyperplasia and adrenal hypoplasia congenita. J Steroid Biochem Mol Biol 2003;85(2–5):483–9.

Fukami M, Nishimura G, Homma K, Nagai T, Hanaki K, Uematsu A, et al. Cytochrome P450 oxidoreductase deficiency: identification and characterization of biallelic mutations and genotype-phenotype correlations in 35 Japanese patients. J Clin Endocrinol Metab 2009;94(5):1723–31.

Gao J, Chen L. Primary adrenocortical insufficiency case series in the neonatal period: genetic etiologies are more common than expected. Front Pediatr 2020;8:464.

Garelli S, Dalla Costa M, Sabbadin C, Barollo S, Rubin B, Scarpa R, et al. Autoimmune polyendocrine syndrome type 1: an Italian survey on 158 patients. J Endocrinol Invest 2021;44(11):2493–510.

Geberhiwot T, Moro A, Dardis A, Ramaswami U, Sirrs S, Marfa MP, et al. International Niemann-Pick Disease Registry (INPDR). Consensus clinical management guidelines for Niemann-Pick disease type C. Orphanet J Rare Dis 2018;13(1):50.

Gidlöf S, Falhammar H, Thilén A, von Döbeln U, Ritzén M, Wedell A, et al. One hundred years of congenital adrenal hyperplasia in Sweden: a retrospective, population-based cohort study. Lancet Diabetes Endocrinol 2013;1(1):35–42.

Graves LE, Khouri JM, Kristidis P, Verge CF. Proopiomelanocortin deficiency diagnosed in infancy in two boys and a review of the known cases. J Paediatr Child Health 2021;57(4):484–90.

Gregoric N, Groselj U, Bratina N, Debeljak M, Zerjav Tansek M, Suput Omladic J, et al. Two cases with an early presented proopiomelanocortin deficiency-a long-term follow-up and systematic literature review. Front Endocrinol (Lausanne) 2021;12, 689387.

Guo L, Wang X, Ji H. Clinical phenotype and genetic features of a pair of Chinese twins with Kearns-Sayre syndrome. DNA Cell Biol 2020;39(8):1449–57.

Guran T, Buonocore F, Saka N, Ozbek MN, Aycan Z, Bereket A, et al. Rare causes of primary adrenal insufficiency: genetic and clinical characterization of a large nationwide cohort. J Clin Endocrinol Metab 2016;101(1):284–92.

Guran T, Kara C, Yildiz M, Bitkin EC, Haklar G, Lin JC, et al. Revisiting classical 3β-hydroxysteroid dehydrogenase 2 deficiency: lessons from 31 pediatric cases. J Clin Endocrinol Metab 2020;105(3), dgaa022.

Gyamfi-Bannerman C, Thom EA, Blackwell SC, Tita AT, Reddy UM, Saade GR, et al. Antenatal betamethasone for women at risk for late preterm delivery. N Engl J Med 2016;374(14):1311–20.

Hall P, Michels V, Gavrilov D, Matern D, Oglesbee D, Raymond K, et al. Aripiprazole and trazodone cause elevations of 7-dehydrocholesterol in the absence of Smith-Lemli-Opitz syndrome. Mol Genet Metab 2013;110(1–2):176–8.

Hao C, Guo J, Guo R, Qi Z, Li W, Ni X. Compound heterozygous variants in POR gene identified by whole-exome sequencing in a Chinese pedigree with cytochrome P450 oxidoreductase deficiency. Pediatr Investig 2018;2(2):90–5.

Hattersley R, Nana M, Lansdown AJ. Endocrine complications of immunotherapies: a review. Clin Med (Lond) 2021;21(2):e212–22.

Hawcutt DB, Francis B, Carr DF, Jorgensen AL, Yin P, Wallin N, et al. Susceptibility to corticosteroid-induced adrenal suppression: a genome-wide association study. Lancet Respir Med 2018;6(6):442–50.

Heksch R, Kamboj M, Anglin K, Obrynba K. Review of Prader-Willi syndrome: the endocrine approach. Transl Pediatr 2017;6(4):274–85.

Herndon J, Nadeau AM, Davidge-Pitts CJ, Young WF, Bancos I. Primary adrenal insufficiency due to bilateral infiltrative disease. Endocrine 2018;62(3):721–8.

Heshmatzad K, Mahdieh N, Rabbani A, Didban A, Rabbani B. The genetic perspective of familial glucocorticoid deficiency: in silico analysis of two novel variants. Int J Endocrinol 2020;2020, 2190508.

Higham CE, Johannsson G, Shalet SM. Hypopituitarism. Lancet 2016;388(10058):2403–15.

Hiort O, Holterhus PM, Werner R, Marschke C, Hoppe U, Partsch CJ, et al. Homozygous disruption of P450 side-chain cleavage (CYP11A1) is associated with prematurity, complete 46,XY sex reversal, and severe adrenal failure. J Clin Endocrinol Metab 2005;90(1):538–41.

Hoivik EA, Lewis AE, Aumo L, Bakke M. Molecular aspects of steroidogenic factor 1 (SF-1). Mol Cell Endocrinol 2010;315(1–2):27–39.

Honour J. The possible involvement of intestinal bacteria in steroidal hypertension. Endocrinology 1982;110(1):285–7.

Honour JW. Historical perspective: gut dysbiosis and hypertension. Physiol Genomics 2015;47(10):443–6.

Honour JW, Tourniaire J, Biglieri EG, Shackleton CH. Urinary steroid excretion in 17 alpha-hydroxylase deficiency. J Steroid Biochem 1978;9(6):495–505.

Honour JW, Borriello SP, Ganten U, Honour P. Antibiotics attenuate experimental hypertension in rats. J Endocrinol 1985;105(3):347–50. https://doi.org/10.1677/joe.0.1050347. 2987388.

Honour JW, Schneider MA, Miller RF. Low adrenal androgens in men with HIV infection and the acquired immunodeficiency syndrome. Horm Res 1995;44(1):35–9.

Hovingh GK, de Groot E, van der Steeg W, Boekholdt SM, Hutten BA, Kuivenhoven JA, et al. Inherited disorders of HDL metabolism and atherosclerosis. Curr Opin Lipidol 2005;16(2):139–45.

Huang Z, Ye J, Han L, Qiu W, Zhang H, Yu Y, et al. Identification of five novel STAR variants in ten Chinese patients with congenital lipoid adrenal hyperplasia. Steroids 2016;108:85–91.

Hughes CR, Guasti L, Meimaridou E, Chuang CH, Schimenti JC, King PJ, et al. MCM4 mutation causes adrenal failure, short stature, and natural killer cell deficiency in humans. J Clin Invest 2012;122(3):814–20.

Iezzi ML, Varriale G, Zagaroli L, Lasorella S, Greco M, Iapadre G, et al. A case of salt-wasting congenital adrenal hyperplasia with triple homozygous mutation: review of literature. J Pediatr Genet 2021;10(1):57–62.

Iglesias P, Sánchez JC, Díez JJ. Isolated ACTH deficiency induced by cancer immunotherapy: a systematic review. Pituitary 2021;24(4):630–43.

Illingworth DR, Alam NA, Lindsey S. Adrenocortical response to adrenocorticotropin in heterozygous familial hypercholesterolemia. J Clin Endocrinol Metab 1984;58(1):206–11.

Ishii T, Tajima T, Kashimada K, Mukai T, Tanahashi Y, Katsumata N, et al. Clinical features of 57 patients with lipoid congenital adrenal hyperplasia: criteria for nonclassic form revisited. J Clin Endocrinol Metab 2020;105(11), dgaa557.

Janzen N, Riepe FG, Peter M, Sander S, Steuerwald U, Korsch E, et al. Neonatal screening: identification of children with 11β-hydroxylase deficiency by second-tier testing. Horm Res Paediatr 2012;77(3):195–9.

Jazayeri O, Liu X, van Diemen CC, Bakker-van Waarde WM, Sikkema-Raddatz B, Sinke RJ, et al. A novel homozygous insertion and review of published mutations in the NNT gene causing familial glucocorticoid deficiency (FGD). Eur J Med Genet 2015;58(12):642–9.

Kallali W, Gray E, Mehdi MZ, Lindsay R, Metherell LA, Buonocore F, et al. Long-term outcome of partial P450 side-chain cleavage enzyme deficiency in three brothers: the importance of early diagnosis. Eur J Endocrinol 2020;182(3):K15–24.

Karangizi AHK, Al-Shaghana M, Logan S, Criseno S, Webster R, Boelaert K, et al. Glucocorticoid induced adrenal insufficiency is common in steroid treated glomerular diseases - proposed strategy for screening and management. BMC Nephrol 2019;20(1):154.

Karaoglan M. The distribution of intrafamilial CYP21A2 mutant alleles and investigation of clinical features in Turkish children and their siblings in Southeastern Anatolia. J Pediatr Endocrinol Metab 2019;32(12):1311–20.

Kelley RI, Kratz LE, Glaser RL, Netzloff ML, Wolf LM, Jabs EW. Abnormal sterol metabolism in a patient with Antley-Bixler syndrome and ambiguous genitalia. Am J Med Genet 2002;110(2):95–102.

Khattab A, Haider S, Kumar A, Dhawan S, Alam D, Romero R, et al. Clinical, genetic, and structural basis of congenital adrenal hyperplasia due to 11β-hydroxylase deficiency. Proc Natl Acad Sci U S A 2017;114(10). E1933-E19s R40.

Kidambi S, Patel SB. Sitosterolaemia: pathophysiology, clinical presentation and laboratory diagnosis. J Clin Pathol 2008;61(5):588–94.

Kim CJ, Lin L, Huang N, Quigley CA, AvRuskin TW, Achermann JC, et al. Severe combined adrenal and gonadal deficiency caused by novel mutations in the cholesterol side chain cleavage enzyme, P450scc. J Clin Endocrinol Metab 2008;93(3):696–702.

Kim YJ, Kim JH, Hong AR, Park KS, Kim SW, Shin CS, et al. Stimulated salivary cortisol as a noninvasive diagnostic tool for adrenal insufficiency. Endocrinol Metab (Seoul) 2020;35(3):628–35.

Kirkgoz T, Guran T. Primary adrenal insufficiency in children: diagnosis and management. Best Pract Res Clin Endocrinol Metab 2018;32(4):397–424.

Kiss S, Lee JY, Pitt J, MacGregor D, Wallace J, Marty M, et al. Dig deeper when it does not make sense: juvenile xanthomas due to sitosterolemia. JIMD Rep 2020;56(1):34–9.

Klouwer FC, Huffnagel IC, Ferdinandusse S, Waterham HR, Wander J, Engelen M, et al. Clinical and biochemical pitfalls in the diagnosis of peroxisomal disorders. Neuropediatrics 2016;47(4):205–20.

Köhler B, Achermann JC. Update—steroidogenic factor 1 (SF-1, NR5A1). Minerva Endocrinol 2010;35(2):73–86.

Köhler B, Lin L, Ferraz-de-Souza B, Wieacker P, Heidemann P, Schröder V, et al. Five novel mutations in steroidogenic factor 1 (SF1, NR5A1) in 46,XY patients with severe underandrogenization but without adrenal insufficiency. Hum Mutat 2008;29(1):59–64.

Köhler B, Lin L, Mazen I, Cetindag C, Biebermann H, Akkurt I, et al. The spectrum of phenotypes associated with mutations in steroidogenic factor 1 (SF-1, NR5A1, Ad4BP) includes severe penoscrotal hypospadias in 46,XY males without adrenal insufficiency. Eur J Endocrinol 2009;161(2):237–42.

Kolahdouz M, Mohammadi Z, Kolahdouz P, Tajamolian M, Khanahmad H. Pitfalls in molecular diagnosis of 21-hydroxylase deficiency in congenital adrenal hyperplasia. Adv Biomed Res 2015;31(4):189.

Kosmas CE, Silverio D, Sourlas A, Garcia F, Montan PD, Guzman E. Primary genetic disorders affecting high density lipoprotein (HDL). Drugs Context 2018;7, 212546.

Krone N, Arlt W. Genetics of congenital adrenal hyperplasia. Best Pract Res Clin Endocrinol Metab 2009;23(2):181–92.

Krone N, Reisch N, Idkowiak J, Dhir V, Ivison HE, Hughes BA, et al. Genotype-phenotype analysis in congenital adrenal hyperplasia due to P450 oxidoreductase deficiency. J Clin Endocrinol Metab 2012;97(2):E257–67.

Krude H, Biebermann H, Luck W, Horn R, Brabant G, Grüters A. Severe early-onset obesity, adrenal insufficiency and red hair pigmentation caused by POMC mutations in humans. Nat Genet 1998;19(2):155–7.

Kumar A, Ghosh M, Jacob JJ. Prevalence of adrenal insufficiency among patients with euvolemic hyponatremia. Endocr Connect 2021;, EC-21-0500.R1. https://doi.org/10.1530/EC-21-0500. Epub ahead of print 34788227.

Lamprecht A, Sorbello J, Jang C, Torpy DJ, Inder WJ. Secondary adrenal insufficiency and pituitary dysfunction in oral/transdermal opioid users with non-cancer pain. Eur J Endocrinol 2018;179(6):353–62.

Landau Z, Hanukoglu A, Sack J, Goldstein N, Weintrob N, Eliakim A, et al. Clinical and genetic heterogeneity of congenital adrenal hypoplasia due to NR0B1 gene mutations. Clin Endocrinol (Oxf) 2010;72(4):448–54.

Langelaan MLP, Kisters JMH, Oosterwerff MM, Boer AK. Salivary cortisol in the diagnosis of adrenal insufficiency: cost efficient and patient friendly. Endocr Connect 2018;7(4):560–6.

Langlois S, Armstrong L, Gall K, Hulait G, Livingston J, Nelson T, et al. Steroid sulfatase deficiency and contiguous gene deletion syndrome amongst pregnant patients with low serum unconjugated estriols. Prenat Diagn 2009;29(10):966–74.

Laugesen K, Petersen I, Sørensen HT, Jørgensen JOL. Clinical indicators of adrenal insufficiency following discontinuation of oral glucocorticoid therapy: A Danish population-based self-controlled case series analysis. PLoS One 2019;14(2), e0212259.

Lee AS, Perera NJ, Chua EL. Hypoadrenalism secondary to topical corticosteroid-containing skin-lightening cream: danger of over-the-counter cosmetic agents. Med J Aust 2015;203(7):287.

Lee C, Kim JH, Moon SJ, Shim J, Kim HI, Choi MH. Selective LC-MRM/SIM-MS based profiling of adrenal steroids reveals metabolic signatures of 17α-hydroxylase deficiency. J Steroid Biochem Mol Biol 2020a;198, 105615.

Lee JH, Song DY, Jun SH, Song SH, Shin CH, Ki CS, et al. High prevalence of increased sitosterol levels in hypercholesterolemic children suggest underestimation of sitosterolemia incidence. PLoS One 2020b;15(8), e0238079.

Lee Y, Choi JH, Oh A, Kim GH, Park SH, Moon JE, et al. Clinical, endocrinological, and molecular features of four Korean cases of cytochrome P450 oxidoreductase deficiency. Ann Pediatr Endocrinol Metab 2020c;25(2):97–103.

Li W, Gong C, Qi Z, Wu DI, Cao B. Identification of AAAS gene mutation in Allgrove syndrome: a report of three cases. Exp Ther Med 2015;(4):1277–82.

Li T, Cunningham JL, Gilliam WP, Loukianova L, Donegan DM, Bancos I. Prevalence of opioid-induced adrenal insufficiency in patients taking chronic opioids. J Clin Endocrinol Metab 2020;105(10), dgaa499.

Lin L, Hindmarsh PC, Metherell LA, Alzyoud M, Al-Ali M, Brain CE, et al. Severe loss-of-function mutations in the adrenocorticotropin receptor (ACTHR, MC2R) can be found in patients diagnosed with salt-losing adrenal hypoplasia. Clin Endocrinol (Oxf) 2007;66(2):205–10.

Liu Y, Zheng J, Liu N, Xu X, Zhang X, Zhang Y, et al. The spectrum of CYP21A2 gene mutations in patients with classic salt wasting form of 2l-hydroxylase deficiency in a Chinese cohort. Mol Genet Genomic Med 2020;8(11), e1501.

Liu R, Mo GL, Song YZ. Identification of a novel large deletion of the mitochondrial DNA in an infant with Pearson syndrome: a case report. Transl Pediatr 2021;10(1):204–8.

Lovric S, Goncalves S, Gee HY, Oskouian B, Srinivas H, Choi WI, et al. Mutations in sphingosine-1-phosphate lyase cause nephrosis with ichthyosis and adrenal insufficiency. J Clin Invest 2017;127(3):912–92. MB 8.

Lucki NC, Li D, Sewer. Sphingosine-1-phosphate rapidly increases cortisol biosynthesis and the expression of genes involved in cholesterol uptake and transport in H295R adrenocortical cells. Mol Cell Endocrinol 2012;348(1):165–75.

Lukacs Z, Barr M, Hamilton J. Best practice in the measurement and interpretation of lysosomal acid lipase in dried blood spots using the inhibitor Lalistat 2. Clin Chim Acta 2017;471:201–5.

Madhavan P, Nallu R, Luthra P. Histoplasmosis: An unusual cause of adrenal insufficiency. AACE Clin Case Rep 2020;7(1):29–31.

Maharaj A, Theodorou D, Banerjee II, Metherell LA, Prasad R, Wallace D. A sphingosine-1-phosphate lyase mutation associated with congenital nephrotic syndrome and multiple endocrinopathy. Front Pediatr 2020a;8:151.

Maharaj A, Williams J, Bradshaw T, Güran T, Braslavsky D, Casas J, et al. Sphingosine-1-phosphate lyase (SGPL1) deficiency is associated with mitochondrial dysfunction. J Steroid Biochem Mol Biol 2020b;202, 105730.

Malpuech G, Vanlieferinghen P, Dechelotte P, Gaulme J, Labbé A, Guiot F. Isolated familial adrenocorticotropin deficiency: prenatal diagnosis by maternal plasma estriol assay. Am J Med Genet 1988;29(1):125–30.

Markmann S, De BP, Reid J, Jose CL, Rosenberg JB, Leopold PL, et al. Biology of the adrenal gland cortex obviates effective use of adeno-associated virus vectors to treat hereditary adrenal disorders. Hum Gene Ther 2018;29(4):403–12.

Martín MG, Lindberg I, Solorzano-Vargas RS, Wang J, Avitzur Y, Bandsma R, et al. Congenital proprotein convertase 1/3 deficiency causes malabsorptive diarrhea and other endocrinopathies in a pediatric cohort. Gastroenterology 2013;145(1):138–48.

McLaughlan E, Barth JH. An analysis of the relationship between serum cortisol and serum sodium in routine clinical patients. Pract Lab Med 2017;8:30–3.

Meimaridou E, Kowalczyk J, Guasti L, Hughes CR, Wagner F, Frommolt P, et al. Mutations in NNT encoding nicotinamide nucleotide transhydrogenase cause familial glucocorticoid deficiency. Nat Genet 2012;44(7):740–2.

Meimaridou E, Hughes CR, Kowalczyk J, Guasti L, Chapple JP, King PJ, et al. Familial glucocorticoid deficiency: new genes and mechanisms. Mol Cell Endocrinol 2013;371(1–2):195–200.

Merke DP, Auchus RJ. Congenital adrenal hyperplasia due to 21-hydroxylase deficiency. N Engl J Med 2020;383(13):1248–61.

Metherell LA, Naville D, Halaby G, Begeot M, Huebner A, Nürnberg G, et al. Nonclassic lipoid congenital adrenal hyperplasia masquerading as familial glucocorticoid deficiency. J Clin Endocrinol Metab 2009;94(10):3865–71.

Mifsud S, Gauci Z, Gruppetta M, Mallia Azzopardi C, Fava S. Adrenal insufficiency in HIV/AIDS: a review. Expert Rev Endocrinol Metab 2021;16(6):351–62.

Miljic D, Doknic M, Stojanovic M, Nikolic-Djurovic M, Petakov M, Popovic V, et al. Impact of etiology, age and gender on onset and severity of hyponatremia in patients with hypopituitarism: retrospective analysis in a specialised endocrine unit. Endocrine 2017;58(2):312–9.

Miller WL. Steroidogenic acute regulatory protein (StAR), a novel mitochondrial cholesterol transporter. Biochim Biophys Acta 2007;1771(6):663–76.

Miller WL. Steroidogenic enzymes. Endocr Dev 2008;13:1–18.

Miller WL, Huang N, Agrawal V, Giacomini KM. Genetic variation in human P450 oxidoreductase. Mol Cell Endocrinol 2009;300(1–2):180–4.

Mongioì LM, Condorelli RA, Barbagallo F, Cannarella R, La Vignera S, Calogero AE. Accuracy of the low-dose ACTH stimulation test for adrenal insufficiency diagnosis: a re-assessment of the cut-off value. J Clin Med 2019;8(6):806.

Morris DJ, Latif SA, Brem AS. An alternative explanation of hypertension associated with 17α-hydroxylase deficiency syndrome. Steroids 2014;79:44–8.

Moser HW, Mahmood A, Raymond GV. X-linked adrenoleukodystrophy. Nat Clin Pract Neurol 2007;3(3):140–51.

Mullarkey EM, Iyer A, Ihuoma A. Lessons of the month: a challenging presentation of hypopituitarism secondary to an intracerebral aneurysm. Clin Med (Lond) 2021;21(2):e228–30.

Naiki Y, Miyado M, Horikawa R, Katsumata N, Onodera M, Pang S, et al. Extra-adrenal induction of Cyp21a1 ameliorates systemic steroid metabolism in a mouse model of congenital adrenal hyperplasia. Endocr J 2016;63(10):897–904.

Napier C, Allinson K, Gan EH, Mitchell AL, Gilligan LC, Taylor AE, et al. Natural history of adrenal steroidogenesis in autoimmune Addison's disease following diagnosis and treatment. J Clin Endocrinol Metab 2020;105(7):2322–30.

Narumi S. Rare monogenic causes of primary adrenal insufficiency. Curr Opin Endocrinol Diabetes Obes 2018;25(3):172–7.

Narumi S, Amano N, Ishii T, Katsumata N, Muroya K, Adachi M, et al. SAMD9 mutations cause a novel multisystem disorder, MIRAGE syndrome, and are associated with loss of chromosome 7. Nat Genet 2016;48(7):792–7.

Nicolaides NC, Charmandari E. Primary generalized glucocorticoid resistance and hypersensitivity syndromes: a 2021 update. Int J Mol Sci 2021;22(19):10839.

Novoselova TV, King PJ, Guasti L, Metherell LA, Clark AJL, Chan LF. ACTH signalling and adrenal development: lessons from mouse models. Endocr Connect 2019;8(7):R122–30.

O'Riordan SM, Lynch SA, Hindmarsh PC, Chan LF, Clark AJ, Costigan C. A novel variant of familial glucocorticoid deficiency prevalent among the Irish traveler population. J Clin Endocrinol Metab 2008;93(7):2896–9.

Ozer EA, Kaya A, Yildirimer M, Guler O, Can S, Aydinlioglu H. A novel DAX1 gene mutation in a Turkish infant with X-linked adrenal hypoplasia congenita. Eur J Pediatr 2009;168(3):367–9.

Pan P, Zheng L, Chen X, Huang J, Yang D, Li Y. Successful live birth in a Chinese woman with P450 oxidoreductase deficiency through frozen-thawed embryo transfer: a case report with review of the literature. J Ovarian Res 2021;14(1):22.

Pasqualini JR, Chetrite GS. The formation and transformation of hormones in maternal, placental and fetal compartments: biological implications. Horm Mol Biol Clin Investig 2016;27(1):11–28.

Patt H, Koehler K, Lodha S, Jadhav S, Yerawar C, Huebner A, et al. Phenotype-genotype spectrum of AAA syndrome from Western India and systematic review of literature. Endocr Connect 2017;6(8):901–13.

Pelewicz K, Miśkiewicz P. Glucocorticoid withdrawal-an overview on when and how to diagnose adrenal insufficiency in clinical practice. Diagnostics (Basel) 2021;11(4):728.

Pépin L, Colin E, Tessarech M, Rouleau S, Bouhours-Nouet N, Bonneau D, et al. A new case of PCSK1 pathogenic variant with congenital pro-protein convertase 1/3 deficiency and literature review. J Clin Endocrinol Metab 2019;104(4):985–93.

Perdomini M, Dos Santos C, Goumeaux C, Blouin V, Bougnères P. An AAVrh10-CAG-CYP21-HA vector allows persistent correction of 21-hydroxylase deficiency in a Cyp21$^{-/-}$ mouse model. Gene Ther 2017;24(5):275–81.

Pericleous M, Kelly C, Wang T, Livingstone C, Ala A. Wolman's disease and cholesteryl ester storage disorder: the phenotypic spectrum of lysosomal acid lipase deficiency. Lancet Gastroenterol Hepatol 2017;2(9):670–9.

Perisa MP, Rose MJ, Varga E, Kamboj MK, Spencer JD, Bajwa RPS. A novel SAMD9 variant identified in patient with MIRAGE syndrome: further defining syndromic phenotype and review of previous cases. Pediatr Blood Cancer 2019;66(7):e27726.

Perogamvros I, Owen LJ, Newell-Price J, Ray DW, Trainer PJ, Keevil BG. Simultaneous measurement of cortisol and cortisone in human saliva using liquid chromatography-tandem mass spectrometry: application in basal and stimulated conditions. J Chromatogr B Analyt Technol Biomed Life Sci 2009;877(29):3771–5.

Perry R, Kecha O, Paquette J, Huot C, Van Vliet G, Deal C. Primary adrenal insufficiency in children: twenty years experience at the Sainte-Justine Hospital, Montreal. J Clin Endocrinol Metab 2005;90(6):3243–50.

Petit A, Cohen-Ludmann C, Clevenbergh P, Bergmann JF, Dubertret L. Skin lightening and its complications among African people living in Paris. J Am Acad Dermatol 2006;55(5):873–8.

Petri C, Wudy SA, Riepe FG, Holterhus PM, Siegel J, Hartmann MF, et al. 17α-Hydroxylase deficiency diagnosed in early infancy caused by a novel mutation of the CYP17A1 gene. Horm Res Paediatr 2014;81(5): 350–5.

Pofi R, Tomlinson JW. Glucocorticoids in pregnancy. Obstet Med 2020;13 (2):62–9.

Prabhakar VK, Shalet SM. Aetiology, diagnosis, and management of hypopituitarism in adult life. Postgrad Med J 2006;82(966):259–66.

Prasad R, Chan LF, Hughes CR, Kaski JP, Kowalczyk JC, Savage MO, et al. Thioredoxin reductase 2 (TXNRD2) mutation associated with familial glucocorticoid deficiency (FGD). J Clin Endocrinol Metab 2014;99(8):E1556–63.

Prasad R, Hadjidemetriou I, Maharaj A, Meimaridou E, Buonocore F, Saleem M, et al. Sphingosine-1-phosphate lyase mutations cause primary adrenal insufficiency and steroid-resistant nephrotic syndrome. J Clin Invest 2017;127(3):942–53.

Prete A, Bancos I. Glucocorticoid induced adrenal insufficiency. BMJ 2021;374:n1380. Erratum in: BMJ. 2021;374:n1936.

Prete A, Auchus RJ, Ross RJ. Clinical advances in the pharmacotherapy of congenital adrenal hyperplasia. Eur J Endocrinol 2021;186(1): R1–R14.

Priftis K, Milner AD, Conway E, Honour JW. Adrenal function in asthma. Arch Dis Child 1990;65(8):838–40.

Prodam F, Caputo M, Mele C, Marzullo P, Aimaretti G. Insights into non-classic and emerging causes of hypopituitarism. Nat Rev Endocrinol 2021;17(2):114–29.

Puder L, Roth S, Krabusch P, Wiegand S, Opitz R, Bald M, et al. Cardiac phenotype and tissue sodium content in adolescents with defects in the melanocortin system. J Clin Endocrinol Metab 2021;106(9): 2606–16.

Ramadoss V, Lazarus K, Prevost AT, Tan T, Meeran K, Choudhury S. Improving the interpretation of afternoon cortisol levels and SSTs to prevent misdiagnosis of adrenal insufficiency. J Endocr Soc 2021;5(11), bvab147.

Ramos-Molina B, Martin MG, Lindberg I. PCSK1 variants and human obesity. Prog Mol Biol Transl Sci 2016;140:47–74.

Ramos-Molina B, Molina-Vega M, Fernández-García JC, Creemers JW. Hyperphagia and obesity in Prader⁻Willi syndrome: *PCSK1* deficiency and beyond? Genes (Basel) 2018;9(6):288.

Ratcliffe WA, McClure JP, Auld WH, Honour JW, Fraser R, Ratcliffe JG. Precocious pseudopuberty due to a rare form of congenital adrenal hyperplasia. Biochemical investigation and pitfalls in interpretation of hormone assays. Ann Clin Biochem 1982;19(3):145–50.

Rathnasiri A, Senarathne U, Arunath V, Hoole T, Kumarasiri I, Muthukumarana O, et al. A rare co-occurrence of duchenne muscular dystrophy, congenital adrenal hypoplasia and glycerol kinase deficiency due to Xp21 contiguous gene deletion syndrome: case report. BMC Endocr Disord 2021;21(1):214.

Reardon W, Smith A, Honour JW, Hindmarsh P, Das D, Rumsby G, et al. Evidence for digenic inheritance in some cases of Antley-Bixler syndrome? J Med Genet 2000;37(1):26–32.

Reisch N. Pregnancy in congenital adrenal hyperplasia. Endocrinol Metab Clin North Am 2019;48(3):619–41.

Reynolds TM, Mewies C, Hamilton J, Wierzbicki AS. PATHFINDER Project Collaboration group. Identification of rare diseases by screening a population selected on the basis of routine pathology results-the PATHFINDER project: lysosomal acid lipase/cholesteryl ester storage disease substudy. J Clin Pathol 2018;71(7):608–13.

Roberts D, Brown J, Medley N, Dalziel SR. Antenatal corticosteroids for accelerating fetal lung maturation for women at risk of preterm birth. Cochrane Database Syst Rev 2017;3(3):CD004454. Update in: Cochrane Database Syst Rev. 2020 Dec 25;12:CD004454.

Rodríguez Gutiérrez PG, González García JR, Castillo De León YA, Zárate Guerrero JR, Magaña Torres MT. A novel p.Gly417Valfs*12 mutation in the MTTP gene causing abetalipoproteinemia: presentation of the first patient in Mexico and analysis of the previously reported cases. J Clin Lab Anal 2021;35(3):e23672.

Rook GAW, Honour J, Kon OM, Wilkinson RJ, Davidson R, Shaw RJ. Urinary adrenal steroid metabolites in tuberculosis—a new clue to pathogenesis? QJM 1996;89:333–42.

Roucher-Boulez F, Mallet-Motak D, Samara-Boustani D, Jilani H, Ladjouze A, Souchon PF, et al. NNT mutations: a cause of primary adrenal insufficiency, oxidative stress and extra-adrenal defects. Eur J Endocrinol 2016;175(1):73–84.

Rubtsov P, Karmanov M, Sverdlova P, Spirin P, Tiulpakov A. A novel homozygous mutation in CYP11A1 gene is associated with late-onset adrenal insufficiency and hypospadias in a 46,XY patient. J Clin Endocrinol Metab 2009;94(3):936–9.

Ruiz-Babot G, Balyura M, Hadjidemetriou I, Ajodha SJ, Taylor DR, Ghataore L, et al. Modeling congenital adrenal hyperplasia and testing interventions for adrenal insufficiency using donor-specific reprogrammed cells. Cell Rep 2018;22(5):1236–49.

Ryan AK, Bartlett K, Clayton P, Eaton S, Mills L, Donnai D, et al. Smith-Lemli-Opitz syndrome: a variable clinical and biochemical phenotype. J Med Genet 1998;35(7):558–65.

Ryu RJ, Easterling TR, Caritis SN, Venkataramanan R, Umans JG, Ahmed MS, et al. Prednisone pharmacokinetics during pregnancy and lactation. J Clin Pharmacol 2018;58(9):1223–32.

Saba JD, Keller N, Wang JY, Tang F, Slavin A, Shen Y. Genotype/phenotype interactions and first steps toward targeted therapy for sphingosine phosphate lyase insufficiency syndrome. Cell Biochem Biophys 2021;79(3):547–59.

Sagar R, Mackie S, Morgan AW, Stewart P, Abbas A. Evaluating tertiary adrenal insufficiency in rheumatology patients on long-term systemic glucocorticoid treatment. Clin Endocrinol (Oxf) 2021; 94(3):361–70.

Salehi M, Houlden H, Sheikh A, Poretsky L. The diagnosis of adrenal insufficiency in a patient with Allgrove syndrome and a novel mutation in the ALADIN gene. Metabolism 2005;54(2):200–5.

Samuels ME, Gallo-Payet N, Pinard S, Hasselmann C, Magne F, Patry L, et al. Bioinactive ACTH causing glucocorticoid deficiency. J Clin Endocrinol Metab 2013;98(2):736–42.

Savvateeva EN, Yukina MY, Nuralieva NF, Filippova MA, Gryadunov DA, Troshina EA. Multiplex autoantibody detection in patients with autoimmune polyglandular syndromes. Int J Mol Sci 2021;22(11): 5502.

Sbardella E, Isidori AM, Woods CP, Argese N, Tomlinson JW, Shine B, et al. Baseline morning cortisol level as a predictor of pituitary-adrenal reserve: a comparison across three assays. Clin Endocrinol (Oxf) 2017;86(2):177–84.

Settas N, Persky R, Faucz FR, Sheanon N, Voutetakis A, Lodish M, et al. SGPL1 deficiency: a rare cause of primary adrenal insufficiency. J Clin Endocrinol Metab 2019;104(5):1484–90.

Shackleton CH, Biglieri EG, Roitman E, Honour JW. Metabolism of radiolabeled corticosterone in an adult with the 17 alpha-hydroxylase deficiency syndrome. J Clin Endocrinol Metab 1979;48(6):976–82.

Shackleton CH, Roitman E, Kelley R. Neonatal urinary steroids in Smith-Lemli-Opitz syndrome associated with 7-dehydrocholesterol reductase deficiency. Steroids 1999;64(7):481–90.

Shackleton C, Marcos J, Malunowicz EM, Szarras-Czapnik M, Jira P, Taylor NF, et al. Biochemical diagnosis of Antley-Bixler syndrome by steroid analysis. Am J Med Genet A 2004;128A(3):223–31. https://doi.org/10.1002/ajmg.a.30104. PMID: 15216541.

Sharifi M, Futema M, Nair D, Humphries SE. Genetic architecture of familial hypercholesterolaemia. Curr Cardiol Rep 2017;19(5):44.

Shulman DI, Palmert MR, Kemp SF, Lawson Wilkins Drug and Therapeutics Committee. Adrenal insufficiency: still a cause of morbidity and death in childhood. Pediatrics 2007;119(2):e484–94.

Sumińska M, Bogusz-Górna K, Wegner D, Fichna M. Non-classic disorder of adrenal steroidogenesis and clinical dilemmas in 21-hydroxylase deficiency combined with backdoor androgen pathway. Mini-review and case report. Int J Mol Sci 2020;21(13):4622.

Suri D, Moran J, Hibbard JU, Kasza K, Weiss RE. Assessment of adrenal reserve in pregnancy: defining the normal response to the adrenocorticotropin stimulation test. J Clin Endocrinol Metab 2006;91(10): 3866–72.

Tajima T, Okada T, Ma XM, Ramsey W, Bornstein S, Aguilera G. Restoration of adrenal steroidogenesis by adenovirus-mediated transfer of human cytochrome P450 21-hydroxylase into the adrenal gland of 21-hydroxylase-deficient mice. Gene Ther 1999;6(11):1898–903.

Takahashi M, Okazaki H, Ohashi K, Ogura M, Ishibashi S, Okazaki S, et al. Current diagnosis and management of abetalipoproteinemia. J Atheroscler Thromb 2021;28(10):1009–19.

Toromanovic A, Tahirovic H, Milenkovic T, Koehler K, Kind B, Zdravkovic D, et al. Clinical and molecular genetic findings in a 6-year-old Bosnian boy with triple A syndrome. Eur J Pediatr 2009;168(3): 317–20.

Tortorelli S, Turgeon CT, Gavrilov DK, Oglesbee D, Raymond KM, Rinaldo P, et al. Simultaneous testing for 6 lysosomal storage disorders and X-adrenoleukodystrophy in dried blood spots by tandem mass spectrometry. Clin Chem 2016;62(9):1248–54.

Tran MT, Tran NA, Nguyen PM, Vu CD. 11B-hydroxylase deficiency detected by urine steroid metabolome profiling using gas chromatography-mass spectrometry. Clin Mass Spec 2019;7:1–5.

Turpeinen H, Ortutay Z, Pesu M. Genetics of the first seven proprotein convertase enzymes in health and disease. Curr Genomics 2013;14(7): 453–67.

Uçar A, Baş F, Saka N. Diagnosis and management of pediatric adrenal insufficiency. World J Pediatr 2016;12(3):261–74.

Unal E, Yıldırım R, Taş FF, Tekin S, Ceylaner S, Haspolat YK. A rare cause of delayed puberty in two cases with 46,XX and 46,XY karyotype: 17 α-hydroxylase deficiency due to a novel variant in CYP17A1 gene. Gynecol Endocrinol 2020;36(8):739–42.

Urs AN, Dammer E, Kelly S, Wang E, Merrill Jr AH, Sewer MB. Steroidogenic factor-1 is a sphingolipid binding protein. Mol Cell Endocrinol 2007;265–266:174–8.

van den Brink DM, Wanders RJ. Phytanic acid: production from phytol, its breakdown and role in human disease. Cell Mol Life Sci 2006;63 (15):1752–65.

Verhoeven NM, Wanders RJ, Poll-The BT, Saudubray JM, Jakobs C. The metabolism of phytanic acid and pristanic acid in man: a review. J Inherit Metab Dis 1998;21(7):697–728.

Viaene AN, Harding BN. The neuropathology of MIRAGE syndrome. J Neuropathol Exp Neurol 2020;79(4):458–62.

Wallace I, Cunningham S, Lindsay J. The diagnosis and investigation of adrenal insufficiency in adults. Ann Clin Biochem 2009;46(Pt 5): 351–67.

Weber MM, Lang J, Abedinpour F, Zeilberger K, Adelmann B, Engelhardt D. Different inhibitory effect of etomidate and ketoconazole on the human adrenal steroid biosynthesis. Clin Investig 1993;71(11):933–8.

Weintrob N, Drouin J, Vallette-Kasic S, Taub E, Marom D, Lebenthal Y, et al. Low estriol levels in the maternal triple-marker screen as a predictor of isolated adrenocorticotropic hormone deficiency caused by a new mutation in the TPIT gene. Pediatrics 2006;117(2):e322–7.

Wijaya M, Huamei M, Jun Z, Du M, Li Y, Chen Q, et al. Etiology of primary adrenal insufficiency in children: a 29-year single-center experience. J Pediatr Endocrinol Metab 2019;32(6):615–22.

Wild KT, Goldstein AC, Muraresku C, Ganetzky RD. Broadening the phenotypic spectrum of Pearson syndrome: five new cases and a review of the literature. Am J Med Genet A 2020;182(2):365–73.

Williams TB, Daniels M, Puthenveetil G, Chang R, Wang RY, Abdenur JE. Pearson syndrome: unique endocrine manifestations including neonatal diabetes and adrenal insufficiency. Mol Genet Metab 2012;106 (1):104–7.

Williams K, Segard A, Graf GA. Sitosterolemia: twenty years of discovery of the function of ABCG5ABCG8. Int J Mol Sci 2021;22(5):2641.

Wolff AB, Breivik L, Hufthammer KO, Grytaas MA, Bratland E, Husebye ES, et al. The natural history of 21-hydroxylase autoantibodies in autoimmune Addison's disease. Eur J Endocrinol 2021;184(4):607–15.

Woodcock T, Barker P, Daniel S, Fletcher S, Wass JAH, Tomlinson JW, et al. Guidelines for the management of glucocorticoids during the peri-operative period for patients with adrenal insufficiency: Guidelines from the Association of Anaesthetists, the Royal College of Physicians and the Society for Endocrinology UK. Anaesthesia 2020;75 (5):654–63. Erratum in: Anaesthesia. 2020;75(9):1252.

Woods CP, Argese N, Chapman M, Boot C, Webster R, Dabhi V, et al. Adrenal suppression in patients taking inhaled glucocorticoids is highly prevalent and management can be guided by morning cortisol. Eur J Endocrinol 2015;173(5):633–42.

Wudy SA, Hartmann MF, Draper N, Stewart PM, Arlt W. A male twin infant with skull deformity and elevated neonatal 17-hydroxyprogesterone: a prismatic case of P450 oxidoreductase deficiency. Endocr Res 2004;30(4):957–64.

Xia TH, Tan M, Li JH, Wang JJ, Wu QQ, Kong DX. Establish a normal fetal lung gestational age grading model and explore the potential value of deep learning algorithms in fetal lung maturity evaluation. Chin Med J (Engl) 2021;134(15):1828–37.

Yang F, Hanaki K, Kinoshita T, Kawashima Y, Nagaishi J, Kanzaki S. - Late-onset adrenal hypoplasia congenita caused by a novel mutation of the DAX-1 gene. Eur J Pediatr 2009;168(3):329–31.

Yoo EG. Sitosterolemia: a review and update of pathophysiology, clinical spectrum, diagnosis, and management. Ann Pediatr Endocrinol Metab 2016;21(1):7–14.

Younes N, Bourdeau I, Lacroix A. Latent adrenal insufficiency: from concept to diagnosis. Front Endocrinol (Lausanne) 2021;12, 720769.

Yu RN, Achermann JC, Ito M, Jameson JL. The role of DAX-1 in reproduction. Trends Endocrinol Metab 1998;9(5):169–75.

Zein AA, Kaur R, Hussein TOK, Graf GA, Lee JY. ABCG5/G8: a structural view to pathophysiology of the hepatobiliary cholesterol secretion. Biochem Soc Trans 2019;47(5):1259–68.

Zhang Z, Feng Y, Ye D, Li CJ, Dong FQ, Tong Y. Clinical and molecular genetic analysis of a Chinese family with congenital X-linked adrenal hypoplasia caused by novel mutation 1268delA in the DAX-1 gene. J Zhejiang Univ Sci B 2015;16(11):963–8.

Zhang T, Ma X, Wang J, Jia C, Wang W, Dong Z, et al. Clinical and molecular characterization of thirty Chinese patients with congenital lipoid adrenal hyperplasia. J Steroid Biochem Mol Biol 2021;206, 105788.

Zhou T, Benda C, Dunzinger S, Huang Y, Ho JC, Yang J, et al. Generation of human induced pluripotent stem cells from urine samples. Nat Protoc 2012;7(12):2080–9.

Hypergonadism

3.2.1.1 Introduction

Investigations of androgen excess are quite common in the endocrine laboratory with infertility, hirsutism, acne, hair loss being common clinical problems. In pediatrics, the presentation can be severe such as ambiguous genitalia in newborn infants, to mild as in advanced growth rate and acne in childhood. For convenience, the issues are discussed by age since the investigations of androgen excess have different objectives depending on the age and karyotype of the patient. During childhood, boys with precocious puberty may have space occupying lesions of the brain or congenital adrenal hyperplasia (CAH) or very rarely tumors of the adrenals and gonads. Imaging of the brain and abdomen may be needed. Polycystic ovary syndrome is the commonest problem. It is important to stress that a multidisciplinary team is required in the management of these conditions (Cools et al., 2018; Brain et al., 2010). In terms of the laboratories, there has been progress to improve specificities of the steroid assays and there are continual changes to the genetic tests involved. The initial determinations of testosterone, estradiol, gonadotrophins (LH, FSH) and dehydroepiandrosterone sulfate DHEAS may continue to be performed on automated immunoassay platforms but investigation beyond this will largely fall within the capability for LC-MS/MS methods. The sample sizes needed for many assays has fallen as the sensitivity of the tests has improved, in some cases the utility of dried blood spots goes beyond newborn screening (Malsagova et al., 2020). Clinicians need to advise the laboratory of abnormal results so that the possibility of assay interferences can investigated (Ghazal et al., 2022; Jones and Honour, 2006).

3.2.1.2 Androgen excess of newborn infant

The birth of a child with genitalia different from normal of either sex makes for the clinical dilemma that any uncertainty is extremely distressing for the family and in some cases life threatening for the child. There have been many changes in attitude to this complex situation. The parents need counseling and assurance on the likely delays in resolving the diagnosis. Urgent investigations are required

depending upon the genotype. Care should be taken in the laboratory to select assay methods as well as interpretation of the results against appropriate pediatric reference ranges, which can change even during the neonatal period. The family history should be taken and an obstetric history taken of any abnormalities in the pregnancy and use of drugs. Precise information in the family history can be cursory but any recollection of neonatal deaths, precocious puberty, amenorrhea, infertility can be useful information, Ultrasound images during pregnancy should be reviewed for anomalies and the report included in the child's notes.

The physical examination of the baby must focus on visual appearance and palpation for testes which can be in the inguinal canal, inguinal pouch and upper scrotum. The size of the clitoris should be measured against normal values of penis/clitoris and position of perineal opening recorded. Examine the labia majora and labia minora. An ultrasound should be performed to locate/exclude a possible uterus. A neonate with bilateral unpalpable testes and severe hypospadias needs to be investigated as a candidate for a 46XX DSD and in situ hybridization of the SRY gene along with the karyotype should be requested immediately. An external masculinizing score can be applied (Ahmed et al., 2000). The skeleton should be examined for facial abnormalities (shape of head, cleft lip, position of eyes and ears, deformities of the pinna), fused fingers or toes (see Délot et al., 2017 supplement for a suggested physical examination form, intake form).

The clinical management of a **DSD (Disorder of Sex Determination or Difference in Sex Development)** requires understanding of normal sexual development. Studies of DSD patients without a complete diagnosis will provide useful material for the discovery of new defects as has been very effective in the past.

A DSD (ambiguous genitalia) at birth is rare with a frequency of 1 in 5000 to 10,000 births and is noted in various ways:

Impaired testicular descent (cryptorchidism). The location and number of testes and appearance and content of the scrotum should be noted. Unilateral undescended testes are one of the most frequent congenital abnormalities in humans, involving 1%–2% of live births. Bilateral undescended testes are less common. **The size and structure of**

the phallus can vary from a normal clitoris through to a normal penis with the urethral opening anywhere along the length (**hypospadias**). A normal clitoris is around 5 mm at birth, whereas a normal penis can be 3–4 cm. Internally, it may be necessary to verify the presence of uterus, fallopian tubes, epididymis and *vas deferens*.

The clinicians may be unable to assign gender at birth because of the sometimes subtle differences from normal in the appearance of the genitalia. These cases are now classified in a **DSD** (differences in sex development) group. Families will be asked by family, relatives, neighbors and others "Is it a boy or a girl" so for them to have no answer is very distressing and needs psychosocial support. The laboratory also needs to bear this in mind and expedite testing.

A full karyotype should be ordered but the quickest test is to establish whether or not the chromosomes includes the Y type can be done by Fluorescence in situ hybridization (FISH) analysis. The **genotype** is required since virilization of a female (46XX) will have clinical similarities to the male (46XY) with incomplete masculinization. If the Y chromosome is not present, the diagnosis is for a virilized girl 46XX. Very rare cases of ambiguous genitalia are the consequence of abnormal karyotype, with or without mosaics, and mutations in genes for proteins in the hypothalamic-pituitary-gonadal axis affecting sex determination.

The investigation of a 46XY DSD individual (undervirilized male) is considered in Chapter 3.2.2.

The commonest cause of ambiguous genitalia in the newborn with **46XX karyotype** is due to inherited enzyme **defects of cortisol synthesis** with diversion of intermediates to androgen production (Table 3.2.1.1) (Miller, 2018; Krone and Arlt, 2009) and raised adrenocorticotrophic hormone (ACTH).

The synthesis of aldosterone can also be reduced because of the steroid enzyme defect or as the result of renin and angiotensin suppression (CYP11B defect where deoxycorticosterone production is raised). There can therefore be

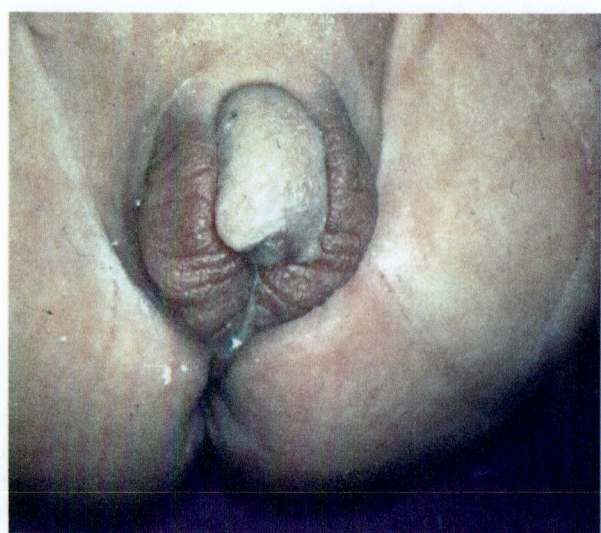

FIG. 3.2.1.1 External genitalia of 46XX infant with virilization of the clitoris and labia such that the appearance is barely distinguishable from a normal male except the urethra is not at the tip of the penis and there are no palpable testes. (Author original.)

further medical complications with salt loss being a common symptom and adrenal insufficiency where life is then at risk. Some infants will show signs of **pigmentation**, commonly of the labia. The clinical picture is of an **enlarged clitoris** (clitoromegaly) with fused labia majora can make the overall appearance near that of a normal male (Fig. 3.2.1.1). The urethra is not at the tip of the clitoris as would be the normal case in a male and the urethra may merge with the vagina (**hypospadias**).

An abdominal ultrasound should determine the internal reproductive tract (uterus, no testes). The extent of masculinization will indicate the degree of androgen exposure supporting onset at 8–10 weeks of gestation or later differentiating endogenous from exogenous androgen exposure. The ano-genital distance is a further measure of androgen excess.

A historical post mortem examination in 1865 of a Giuseppe Marzo by Professor Luigi de Crechio shows a male-looking individual externally (Fig. 3.2.1.2) but clearly with a uterus and ovaries suggesting the earliest account of a patient with 46XX DSD. This predates knowledge of the adrenal glands above the kidneys which might have been seen to be enlarged.

Several congenital defects of steroid synthesis in the infant can cause the clinical picture, whereas the mother, or environment, may rarely these days be the source of the excess androgens.

Reports in the literature of combined enzyme defects can be pretty much now negated because defects in the genes would now have to be assigned to separate chromosomes. However, an apparent combined deficiency of 21-hydroxylase and 17-hydroxylase on biochemical

TABLE 3.2.1.1 Ambiguous genitalia with 46XX chromosomes.

Congenital adrenal hyperplasia
21-Hydroxylase deficiency
11β-Hydroxylase deficiency
3β-Hydroxysteroid dehydrogenase deficiency
Androgen exposure
Maternal ingestion of androgens or progestogens
Adrenal tumor of mother
Ovarian tumor in mother
Aromatase deficiency

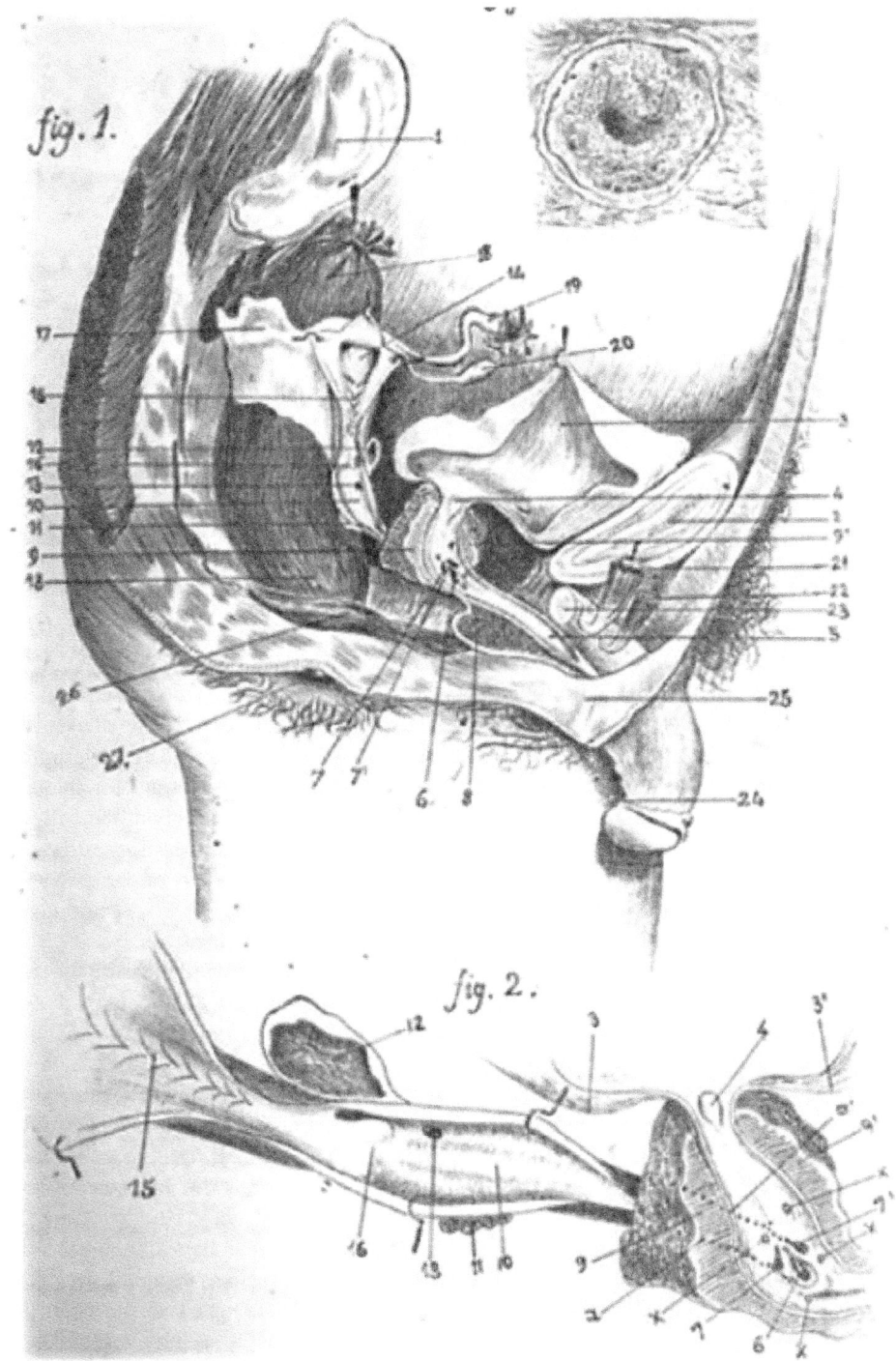

FIG. 3.2.1.2 Drawing from De Crecchio's paper, showing a longitudinal view of the pelvic contents: 5. Urethra (opened); 10. Vagina; 14. Uterus (opened); 15. Cervix; 17. Right broad ligament, containing the right ovary and fallopian tube; 19. Left fallopian tube and fimbriae; 20. Left ovary; 23. Right corpus cavernosum (cut); 24. Urethral meatus. *(Modified from Miller WL, White PC. A brief history of congenital adrenal hyperplasia. Horm Res Paediatr 2022;95(6):529–45. Fig 2 p531.)*

grounds is due to a defect in the electron transport path to these two enzymes which have normal genes (see later for P450 oxidoreductase deficiency) (Miller, 2021). The above enzymes are normally part of the steroid metabolic pathways that link intermediates between cholesterol

and cortisol, aldosterone and androgens (Fig. 3.2.1.3). The division of steroid synthesis to these products reflects the functional activities of the zones of the adrenal cortex.

The outer *zona glomerulosa* principally secretes aldosterone because CYP17 is not expressed but CYP11B1

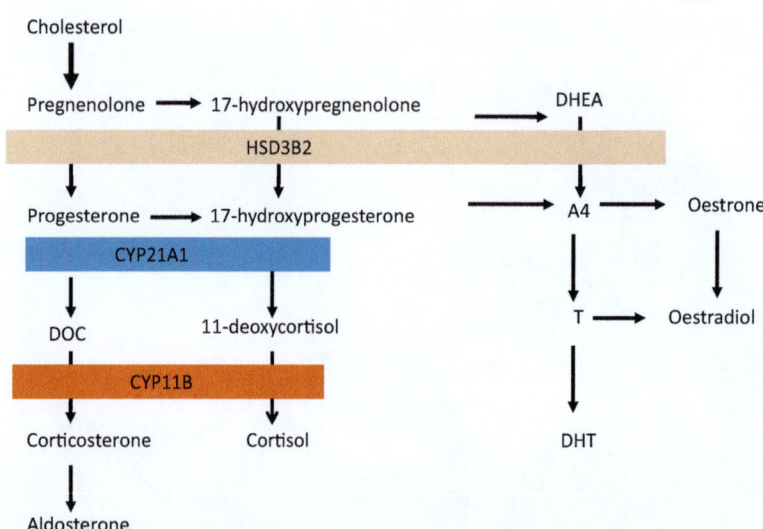

FIG. 3.2.1.3   Disorders of enzymes in steroid biosynthesis. Defects in CYP21, CYP11B and HSD3B2 cause androgen excess. *(Author original.)*

and CYP11B2 are active. The *zona fasciculata* is the site of cortisol production and the *zona reticularis* secretes androgens. The zona reticularis develops in mid-childhood so between 4 and 8 years of age. In the absence, or lowered potential, of cortisol production through defects in any of the steroidogenic enzymes, the ACTH concentrations are high leading to adrenal hyperplasia and in some cases excess androgen production hence **congenital adrenal hyperplasia (CAH)**.

Early biochemical investigations are complicated by the changes in the hypothalamic-pituitary-gonadal (HPG) axis and reduction in the size and function of the fetal adrenal cortex during the first weeks of life. Plasma testosterone in a male at birth can be as high as 10–15 nmol/L (depending on the assay) due to incomplete clearance of chorionic gonadotrophin. By 7 days, testosterone concentration will decline to approaching zero, then the HPG negative feedback is activated and testosterone will increase to around 10 nmol/L by days 12–14 in response to gonadotrophins and this level will continue for 4 or 5 months. This period of active testicular steroidogenesis is called the **mini-puberty**. The HPG is then not active from 6 months of age until entering puberty.

The usual clinical approach to this patient is to seek evidence for the commonest defect first, so investigations are directed to a **defect of the cytochrome P450 21-hydroxylase (*CYP21A1*)** (Speiser et al., 2018). Cortisol, testosterone, DHEAS and androstenedione would have been measured on a platform immunoassay, whereas 17-hydroxyprogesterone assay would have been a radioimmunoassay (Wallace et al., 1987) or time resolved fluorescence assay (Fingerhut, 2009). This approach has been changing as plasma steroid panels (4 steroids) (Kushnir et al., 2006) or profiles (up to 16 steroids) (Guo et al., 2006;

Magnisali et al., 2011; Fiet et al., 2017; Cicalini et al., 2019) have become available and if included 21-deoxycortisol will also be raised. In many countries, a marker for this disorder is included in the panels of tests in newborn screening (NBS) programmes. **17-Hydroxyprogesterone** is the substrate for this enzyme and blood tests can be started from about the fifth day of life to look for elevated concentrations in plasma. In many cases, the female patient is likely to be under investigation for ambiguous genitalia, or adrenal insufficiency before screening results are back.

Many laboratories will measure the required steroid hormones in **blood spots** collected on a Guthrie card for newborn screening from 2 days of age (but in some countries later). For many logistical reasons, there is a delay in when the first sample is obtained. Reference ranges have been published according to age, gestation and body weight. It is important to check if results are for blood or plasma since blood concentrations are half the plasma concentrations (Honour, 2014). Since 21-hydroxylase deficiency is confined to the adrenal cortex, andostenedione is the most appropriate androgen to be measuring although many clinicians might think to have testosterone measured, although T can be raised as a by-product of the androstenedione synthetic route.

The infant needs to be monitored for evidence of urinary salt loss, this should be by determinations of plasma sodium and potassium concentrations. Hyperkalemia can precede hyponatremia and can be evident from day 5 after birth (Mullis et al., 1990). Aldosterone and plasma renin should also be measured although few laboratories will have their own reference ranges in the newborn period and literature values will be needed (see Chapter 2.5). In the newborn infant and child, aldosterone and renin are normally very much higher than in the adolescent and adult.

Specificity in steroid analysis is crucial in the newborn period because plasma has high concentrations of androgen and pregnenolone sulfates from the adrenal cortex that can interfere in immunoassays. In some cases, the raised 17-OHP is transient with normal results on repeat testing (Cavarzere et al., 2009; Fingerhut, 2009). Some compensation is achieved if free steroids are extracted from the plasma into an organic solvent leaving the sulfates behind. The immunoassay for 17-OHP and androstenedione is then performed on the organic extract after drying under vacuum or under a gas flow and reconstitution into a suitable matrix for the immunoassay (buffer with albumin or zero standard for the assay). A profile of steroids has clear advantages in a diagnosis of CAH when all the requisite intermediates, androgens and cortisol can be determined in one test to cover the several enzyme defects (Sarathi et al., 2019; Fiet et al., 2017; Stolze et al., 2016). Since 17-hydroxypregnenolone was a likely cause of interference in immunoassay an LC-MS/MS was set up for measurements in dried blood spots of 17-OHP and 17-OH-preg (Higashi et al., 2008). Genetic tests (Buonocore et al., 2020; Pignatelli et al., 2019; Krone and Arlt, 2009) have become easier and available in regional and specialist centers.

3.2.1.2.1.1 Congenital adrenal hyperplasia (CAH)

21-hydroxylase deficiency

The learned societies in endocrinology have published practice guidelines for this condition (Speiser et al., 2018; Kulle et al., 2017) that cover a range of issues, although there will be differences in endocrine centers in how they are interpreted depending on the availability of services. A defect of the steroid 21-hydroxylase accounts for 90% of cases of virilization of a female and therefore should be excluded before proceeding to assign other causes for ambiguous genitalia unless there are other clinical features that are clues to this and other diagnoses. CAH due to 21-hydroxylase deficiency causes variable virilization. In Europe, 60% of all cases of steroid 21-hydroxylase deficiency will then present in the first 10 days with a salt-losing crisis ("salt wasting form" SW) due to low production of aldosterone. Patients without salt-loss are usually called "simple virilizer" (SV). Some cases may be 46XX girls with such severe masculinization (Prader stage IV or V) that the child in a quick examination has been assigned to the male sex.

As 17-hydroxyprogesterone (17-OHP) is a biosynthetic precursor of cortisol, in patients with deficiency of 21-hydroxylase the production of 17OHP increases and plasma concentrations are elevated through raised ACTH from cortisol deficiency. The measurements of 17OHP in serum or plasma or blood spots (dbs) are used to assist the diagnosis of this disorder. The AutoDelphia assay may be subject to interference in dbs (Han et al., 2019). In many countries, this is part of a newborn screening programme that can be extended to LCMS/MS measurements of seven steroids (Kim et al., 2015). Reference ranges are needed, adjusted according to age at sample collection and birthweight (Hayashi et al., 2017). The timing of blood sampling is very important in order that the laboratory can interpret the findings. In healthy newborn infants, 17-OHP concentrations in serum may be above 100 nmol/L on the first day of life. After 36 h, there is usually good discrimination of 17-OHP in affected cases (100–800 nmol/L) from normal infants (<15 nmol/L). In term babies, but not necessarily in preterm babies, the 17-OHP concentrations fall in the first 3 days to 1 week to less than 5 nmol/L. Results of a directly measured 17-OHP, are higher than those measured in an organic extract of serum prior to immunoassay (Wallace et al., 1987) particularly in preterm infants. Steroids in blood (probably steroid sulfates from the fetal adrenal) (Wong et al., 1992) such as 17-hydroxypregnenolone sulfate interfere in direct immunoassay and these steroids contribute to the abnormally high 17-OHP results. A high 17-OHP concentration in a direct immunoassay should be verified in an assay that at least involves solvent extraction before the immunoassay and preferably chromatography with mass spectrometric confirmation. Screening data leads to a high rate of false positive results so confirmatory testing is needed (Janzen et al., 2007; Higashi et al., 2008; Schwarz et al., 2009; Seo et al., 2014; Bialk et al., 2019), with the best options being GC-MS or LC-MS/MS but this is not available in all countries. One objection to GC-MS analysis of steroids has been the time for derivative formation but using microwave irradiation this has been reduced to 120 s (Deng et al., 2005). Incidence figures vary between countries (1 in 10,000 to 1 in 25,000), in urban and rural settings, not all are able to follow-up cases after an initial screen, there are difficulties in applying birth weight and gestational age cut-offs (Dabas et al., 2020). There are cultural reasons in some populations to conceal genital ambiguity and access to clinics and medicines is not uniform across the globe. Screening programmes aim to detect classical forms of 21-hydroxylase deficiency (Held et al., 2020), whether nonclassic CAH is detectable is for debate and proof of principle. For MS determination of 17-OHP by isotope dilution, the use of 17-OHP-[2,3,4-^{13}C] and 17-OHP-[2,2,4,6,6,21,21,21-^{2}H] internal standards gave equivalent results (Loh et al., 2020).

In the United Kingdom, the practice is for a physical examination plus blood tests. The examination includes screening for problems with eyes, heart, hips and in boys the testes. Three common disorders are screened namely sickle cell disease, cystic fibrosis and congenital hypothyroidism. Tandem mass spectrometry is used to detect phenylketonuria (PKU), medium-chain-acyl-CoA dehydrogenase deficiency (MCADD), isovaleric acidemia (IVA), glutaric aciduria type 1 (GA1) and homocystinuria pyridoxine unresponsive (HCU). The UK Government quote frequencies of 1 in 10,000 for each of PKU and MCADD and 1 in 100,000 for GA1. MSUD is generally thought to have an incidence of 1 in 185,000. The decision to include a disease in screening is based on incidence and ease of treatment and lifelong cost of survival with quality of life.

The determination of 21-deoxycortisol has been suggested by some authorities to replace 17-OHP as the marker for 21-hydroxylase deficiency (Miller, 2019) although principal component analysis of profile data does not support that claim (Lasarev et al., 2020). In an ACTH stimulation test 21-deoxycortisol by LC-MS/MS was superior to 17-OHP in discriminating nonclassic patients for 21-OHD with results of median 1260 (302–3619) compared with 81 ng/dL (<24–296) for classic and nonclassic carriers and controls median <24 (<24–39) (Costa-Barbosa et al., 2010). The steroid profile of plasma shows a clear pattern of 17-OHP, androgen and 21-deoxycortisol excess. The steroid profile of plasma will need to target cortisol, 17-OHP, androstenedione for this diagnosis. The plasma analysis may suggest the predominance of other steroids as substrates for other enzyme defects if 11-deoxycortisol, DHEAS, corticosterone and testosterone, androstenedione and dihydrotestosterone are included in the steroid panel in order to suggest defects of CYP11B1, HSD3B2 and POR (Stolze et al., 2016). The 11-oxygenated androgens are promising tools for a more precise diagnosis (Balsamo et al., 2020).

A provisional diagnosis of CAH by immunoassay screening needs to be confirmed or refuted as soon as possible so that the medical team can facilitate counseling of the parents who will be anxious about the ambiguity over sex of their newborn child. A 17-OHP analysis of the mother and partner should be undertaken if further children are considered and mutation analysis of the gene in both partners will be useful to predict chances of CAH in a future child.

Urinary steroid analysis is a reliable diagnostic test, although pregnanetriol is not necessarily the most important neonatal metabolite of 17-OHP. A steroid profile of urine by GC-MS will show the predominance of 21-deoxy steroids (pregnanetriol, metabolite from 17-hydroxyprogesterone and pregnanetriolone from 21-deoxycortisol) if the defect is 21-hydroxylase. The structures of a large number of unique metabolites have been tentatively elucidated with the help of tandem mass spectrometry (GC-MS/MS) and a characteristic, very complicated urine steroid profile in the first week of life

can be attributed to a vast number of steroids not having a 21-hydroxyl group. In the newborn period, one particular steroid (3β,16α,17α-trihydroxy-5α-pregnane-7,20-dione) is a biomarker for the defect (Christakoudi et al., 2010; Christakoudi et al., 2013). Pregnanetriolone in CAH was 10 (0.079–360 mg/g creatinine (median and range) in 21OHD and 1.5 (0.42–2.6 in PORD (Koyama et al., 2016). The pregnanetriolone/THE ratio in 21-OHD was 2.5 (0.14–15) and 0.18 (0.051–0.23) in PORD and 0.00038 (0.000068–0.0083) I controls. The Wudy laboratory found the ratio of pregnanetriolone to 6-hydroxy-tetrahydrocortisone was the best discriminator (Kamrath et al., 2016). A diagnostic steroid profile is recognized from day 3 after birth. In the GC-MS analysis of urinary steroids at 2–6 months after birth, three steroids become the dominant products (17-hydroxypregnanolone, pregnanetriol and 11-oxo-pregnanetriol) (Fig. 3.2.1.4). Reference ranges have been reported by the Wudy laboratory (Kamrath et al., 2016). A result (including mass spectrometric analysis) can, and should be, obtained and reported within 36 h of getting the sample to the specialist laboratory. A rapid hydrolysis of urinary steroid conjugates and shortened derivative step may be needed to achieve this target. A blood steroid profile will take less time to perform.

All children diagnosed with 21-hydroxylase deficiency based on direct blood tests should preferably have a urine or plasma steroid profile by GC-MS or LC-MS/MS as a confirmatory test since these tests characterize without doubt the steroids in excess (Travers et al., 2020). In some countries, genetic mutation analysis is used for confirmation but there be will some cases where the mutation cannot be found with the tests available locally or nationally. In some of the national newborn screening programmes, an immunoassay revealing raised 17-OHP will go on to second tier tests for steroid profile by LC-MS/MS (de Hora et al., 2020; Bialk et al., 2019; Schwarz et al., 2009). Other steroids are being measured in the context of diagnosis and management of 21-hydroxylase deficiency, so, 21-deoxycorticosterone is a precursor for aldosterone and thus useful in patients with salt wasting CAH (SW-CAH) (Travers et al., 2020; Fiet et al., 2017). In a research setting, 11-oxygenated androgens have some additional benefit for monitoring androgen excess (Jha et al., 2020; Kamrath et al., 2018) and response to continuous subcutaneous glucocorticoids compared with oral hydrocortisone (Turcu et al., 2020b). 17-Hydroxprogesterone can be converted to dihydrotestosterone through a "backdoor path" including androsterone (Sumińska et al., 2020; Kamrath et al., 2012) because of low 5-alpha reductase and high 17,20 lyase activities. Measurements of plasma renin activity (PRA) and aldosterone help to define the extent of 21-hydroxylase enzyme blockage in the mineralocorticoid pathway and these tests can be used to monitor efficacy of treatment in CAH.

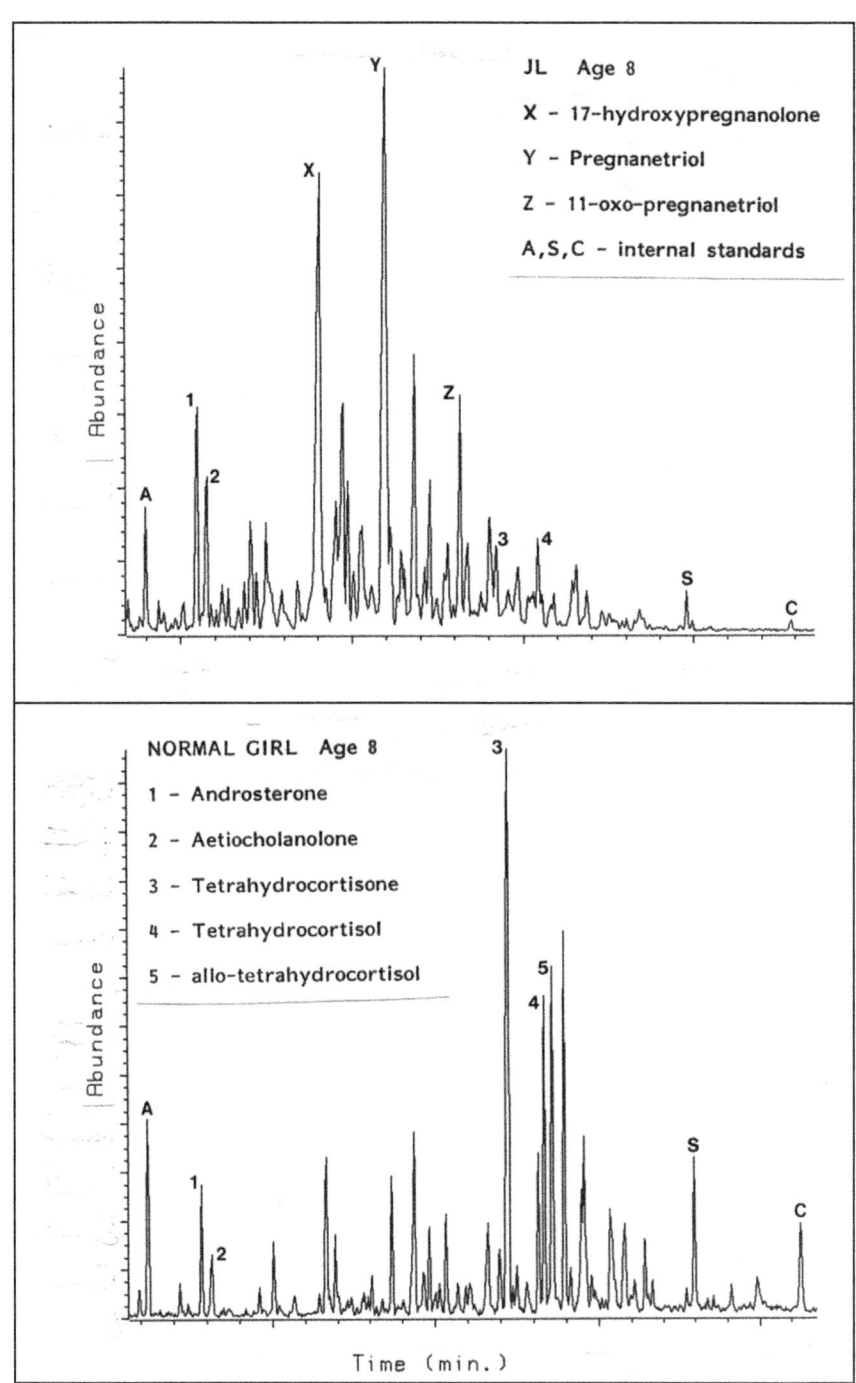

FIG. 3.2.1.4 Urinary steroid profile (USP) by GC_MS in 21-hydroxylase deficiency. After quantitative analysis, the cortisol metabolites are low, androgens are higher than cortisol but the metabolites 17-hydroxypegnanolone, pregnanetriol and 11-oxo-pregnanetriol are grossly elevated. *(Author original.)*

Genetics of 21-hydroxylase deficiency The gene encoding 21-hydroxylase (*CYP21B*) has been characterized along with a 98% identical pseudogene (*CYP21A*). Located on the short arm of chromosome 6 in the midst of the class III HLA region, CYP21B is closely linked to the highly polymorphic genes encoding HLA-B and HLADR. Almost all of the mutations characterized so far in the genes of patients with 21-hydroxylase deficiency appear to result from recombinations between the *CYP21B* and *CYP21A*. These are either deletions caused by unequal crossing-over during meiosis or apparent transfers of deleterious mutations due to a phenomenon called gene conversion

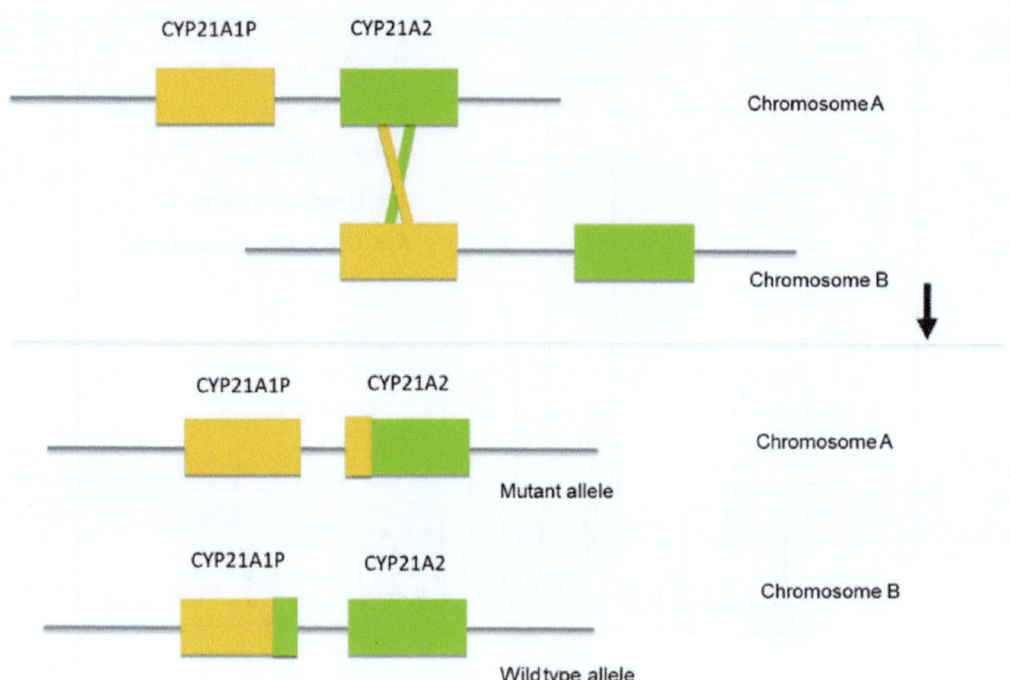

FIG. 3.2.1.5 Schematic representation of the mechanism of gene conversion, where a misalignment between the two DNA sequences results in a recombination between the CYP21A2 gene and the CYP21A1P pseudogene. *(From Pignatelli D, Carvalho BL, Palmeiro A, Barros A, Guerreiro SG, Macut D. The complexities in genotyping of congenital adrenal hyperplasia: 21-hydroxylase deficiency. Front Endocrinol 2019;10:432. Fig. 1, p. 5.)*

(Carvalho et al., 2021; Simonetti et al., 2018). There is some agreement between genotype and phenotype but since compound heterozygotes are frequent there are always exceptions. The genetic testing is beyond the scope of this book, recent reviews and databases should be consulted for up to date information.

The experience of Gillian Rumsby working with the author is that studies of the genetics of 21-hydroxylase is extremely complicated and best in the domain of experts. The frequent existence of copy number variations together with the large number of possible genetic variants makes the characterization of CYP21A2 alleles rather difficult. Pathogenic variants have been identified both in the coding and noncoding regions of the gene inclusively in the 5′UTR and the 3′UTR regions. Consequently, it is important to screen all coding exons, as well as intron-exon boundaries of the gene.

CYP21A2 pathogenic alterations Due to close proximity of the CYP21A2 gene and pseudogene and the highly polymorphic complexity of the gene region, recombination events are the major cause of CYP21A2 pathogenic variants. Two types of recombination are found:

- one is the result of an unequal crossing over during meiosis, with the production of large rearrangements and
- the other, consists of smaller gene conversions where a segment of the functional CYP21A2 gene is replaced by a segment copied from the CYP21A1P pseudogene

(Fig. 3.2.1.5). The segment of the converted CYP21A2 gene will carry either a few nucleotides from CYP21A1P (microconversions) or a short sequence affecting one or more exons. The converted sequences will translate to inactive or at least significantly modified activity protein (Pignatelli et al., 2019).

More than 95% of the pathogenic variants causing 21OHD are due to intergenic recombinations. **Large gene conversions** and large deletions, sometimes involving C4B and CYP21A2 with the formation of CYP21A1P/CYP21A2 chimeric genes are found in around 20% of the pathogenic variants. A 26 or 32Kb deletion (depending on whether C4B is the long or short gene), involving the 3′ end of CYP21A1P, all of the C4B gene, and the 5′ end of the CYP21A2 gene, produces a single nonfunctional chimeric gene with its 5′ and 3′ ends corresponding to CYP21A1P and CYP21A2, respectively (Fig. 3.2.1.6). Several different chimeric CYP21A1P/CYP21A2 genes have been found and characterized.

Around 75% of the deleterious variants are transferred during meiosis by **small conversions** from the pseudogene. These conversions can involve a single point variant called "microconversions" or more pseudogene variants (Fig. 3.2.1.7).

There are four promoter mutations, two frameshift mutations (exons 3 and 7), one intronic mutation (intron 2) and eight single base mutations (exons 1, 4, 6, 7, and 8).

A pathogenic variant in intron 2 is characterized by the substitution of A or C nucleotide at 13 bp before the end of

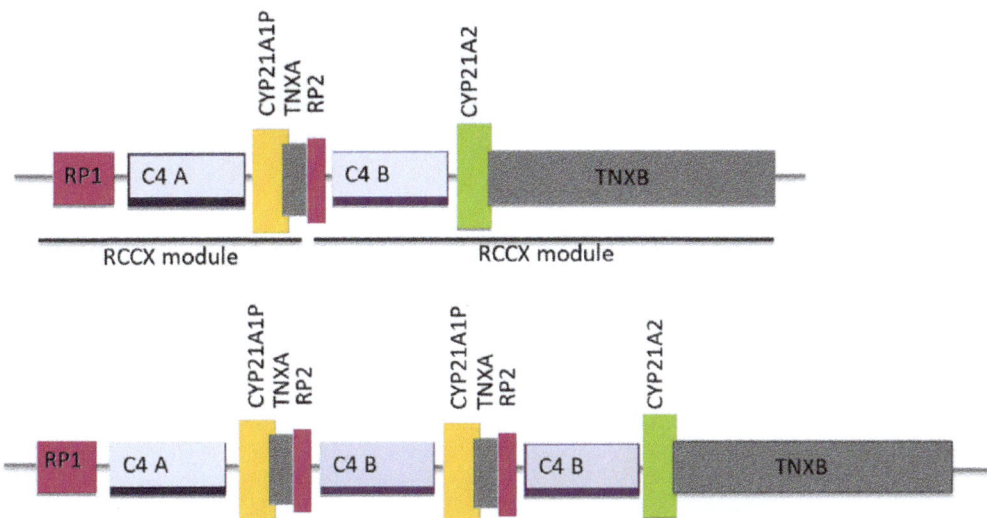

FIG. 3.2.1.6 Schematic diagram of the organization of the RCCX modules, one with the CYP21A1P pseudogene and the other with the CYP21A2 gene for the most common bimodular haplotype and for the three modular haplotype with two modules harboring the CYP21A1P pseudogene and one the CYP21A2 gene. *(From Pignatelli D, Carvalho BL, Palmeiro A, Barros A, Guerreiro SG, Macut D. The complexities in genotyping of congenital adrenal hyperplasia: 21-hydroxylase deficiency. Front Endocrinol 2019;10:432. Fig. 2 p. 5.)*

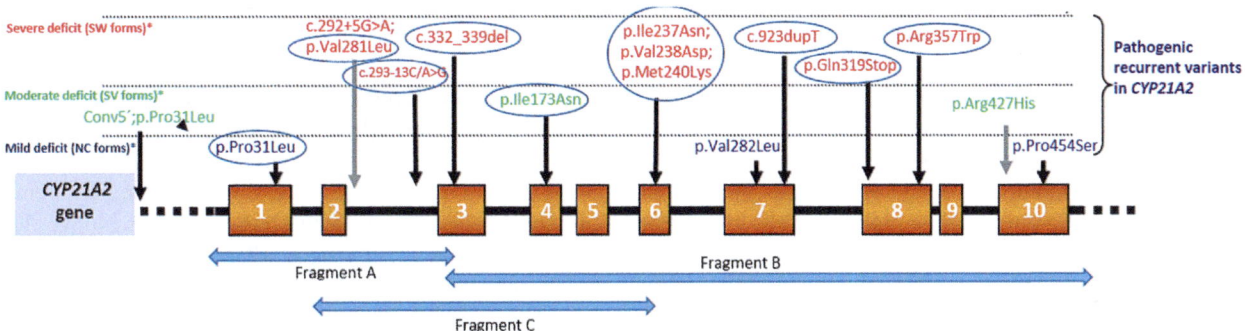

FIG. 3.2.1.7 Recurrent variants grouped according to how they affect the enzymatic functionality: severely (red), moderately severe (green), or mildly (blue). Recurrent variants in all populations are circled. The complete nomenclature of each variant including the cDNA position (NM_000500.9) would be: c.92C>T [p.Pro31Leu], c.292+5C>A, c.293-13C>G, c.332-339del, c.518T>A [p.Ile173Asn], c.(710T>A; 713T>A; 719T>A) p.[Ile237Asn; Val238Glu; Met240Lys], c.844G>T [p.Val282Leu], c.923dupT, c.955C>T [p.Gln319*], c.1069C>T [p.Arg357Trp], c.1280G>A [p.Arg427His] and c.1360C>T [p.Pro454Ser]). The arrows "Fragment A," "Fragment B," and "Fragment C" represents specific amplicons for CYP21A2 amplification. *(From Arriba M, Ezquieta B. Molecular diagnosis of steroid 21-hydroxylase deficiency: a practical approach. Front Endocrinol (Lausanne) 2022;13:834549. Fig. 1 p. 3.)*

intron 2 (nt 656) to G (c.293-13A/C > G) that causes aberrant splicing of intron 2 with retention of 19 nucleotides normally spliced out of mRNA, resulting in a shift in the translational reading frame (Pignatelli et al., 2019). Other rearrangements, such as a deletion of 10 nucleotides in exon 8 and a duplication of 16 nucleotides in exon 9 have also been reported.

CAH in the remaining 20%–25% of the cases is due to misalignment owing to unequal crossing over during meiosis and gene deletions, gene duplications and deletions involving CYP21A2 and the other contiguous genes. In rare cases, CAH can also be caused by uniparental isodisomy.

More than 1300 genetic variants have been reported but only 230 affecting human health of which 67% of variants result in the classic CAH form. Some genetic variants have been described in the nontranslated regions of the gene resulting in promoter conversions. Around 66% of the genetic variants are missense mutations in all forms of the disease while nonsense and frameshift mutations always result in the classic forms.

Polymorphisms A missense pathogenic variant results from a transition of a CCC to a TCC and was initially described as not present in the pseudogene. CYP21A1P

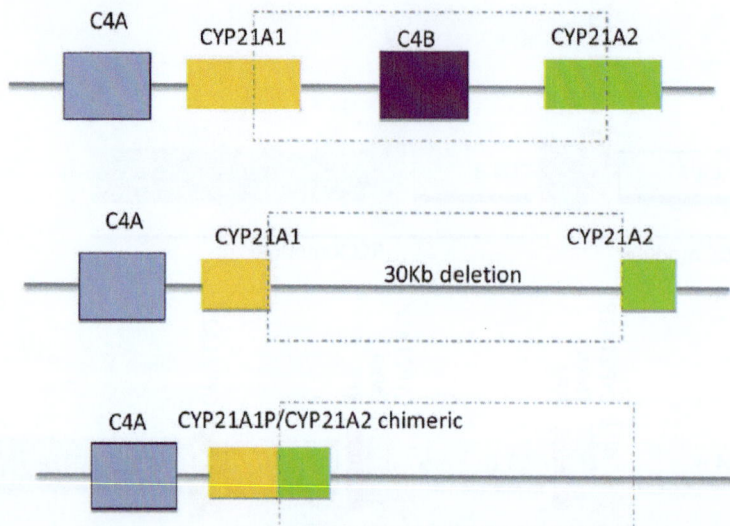

FIG. 3.2.1.8   Schematic diagram of the formation of chimeric genes by large gene deletions. *(From Pignatelli D, Carvalho BL, Palmeiro A, Barros A, Guerreiro SG, Macut D. The complexities in genotyping of congenital adrenal hyperplasia: 21-hydroxylase deficiency. Front Endocrinol 2019;10:432. Fig. 4 p. 6.)*

may carry P453S as an occasional polymorphism and that this pathogenic variant *P453S: Pro-453Ser (p.(Pro454Ser))* is transferred to CYP21A2 in the same way as the other pathogenic variants frequently causing 21-hydroxylase deficiency. The functional mechanism is not clearly explained although it corresponds to a decrease of 50%–70% of 17-OHP and 20%–45% of progesterone utilization and occurs in a number of different populations. A chimeric gene due to misalignment during meiosis results in deletion of CYP21A2 gene (Fig. 3.2.1.8).

Other pathogenic variants   More than 200 other different pathogenic variants have been described and this number is increasing as the techniques for molecular diagnosis improve (see http://www.cypalleles.ki.se/cyp21.htm and http://www.hgmd.cf.ac.uk). Some of these pathogenic variants have been reported in several cases, but most of them were described only in one family. Except for the nonsense, frameshift and rearrangement alterations that are deduced as severe, most of these pathogenic variants are missense, and require functional studies to be classified. Less than 5% of the pathogenic variants in the CYP21A2 gene are not caused by gene conversions and possibly are not present in the pseudogene.

Rare CYP21A2 pathogenic variants are associated with the particular CAH types:

- severe SW form (L167P, G291S, G292D, and R354H),
- the SV form (I77T, E320K, R341P, and G424S), and
- and with the NC form (I230T and R233K).

Some of these pathogenic variants are associated with different phenotypes depending on if there is another pathogenic variant. This synergistic effect that results in a different phenotype has also been described for other pathogenic variants, such as H62L, R339H, or P105L with P453S.

Molecular biology is useful in genetic counseling for the family, with an affected child, considering more children, since the testing affords the opportunity of antenatal diagnosis. If the mutation is known, gene amplification techniques can be used at the start of a pregnancy with small amounts of tissue taken with chorionic villus biopsy. This tissue can be obtained around 12 weeks of the pregnancy and the tests can be completed within days. A full karyotype should be performed with the biopsy material to exclude chromosomal abnormalities such as Down syndrome. Dexamethasone administration to the mother can prevent severe virilization of an affected female by suppression of the fetal pituitary-adrenal axis but may have long-term effects on the child but is not now recommended (Miller, 2015). This treatment must be started as early as possible in pregnancy but only if the affected fetus is female, should steroid treatment be continued to term. There may be side-effects of concern from the dexamethasone for the mother and the child. Only one child in four pregnancies will need to be treated throughout pregnancy and 3 cases will get treatment for 8–12 weeks without reason. In vitro fertilization with preimplantation genetic testing of the embryo is the way forward.

Rare patients with classic CAH deficiency have a "contiguous gene syndrome" named **Ehlers-Danlos syndrome**. This a connective tissue disorders affecting joints, tendons, skin, gastrointestinal tract and the heart (Rymen et al., 2019; Miller and Merke, 2018). Most common is a bipolar RP-C4-CYP21-TNX region involving the tenascin gene (TNX) (Fig. 3.2.1.9).

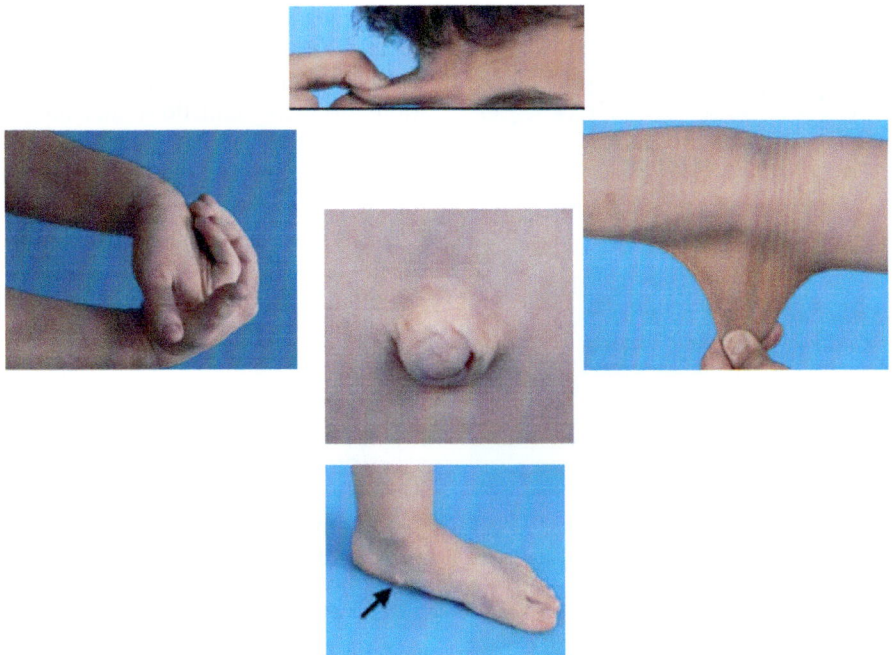

FIG. 3.2.1.9   Clinical manifestations of Ehlers-Danlos syndrome (CAH-X). Patients with CAH-X due to heterozygosity of a TNXA/TNXB chimera commonly have hypermobility of small joints (left middle row) and large joints (not shown). Pes planus and piezogenic papules (arrow) (bottom image) are frequently observed. Approximately 25% of patients have a congenital cardiac defect such as a quadricuspic aortic valve (not shown) heterozygous for CAH-X CH-1. Hernias are most often observed in patients heterozygous for CAH-X CH-2 (central image) or biallelic for CAH-X. Hyperextensible skin is observed with CAH-X CH-2, and most severe with biallelic CAH-X (middle right, top). *(From Miller WL, Merke DP. Tenascin-X, congenital adrenal hyperplasia, and the CAH-X syndrome. Horm Res Paediatr 2018;89(5):352–61. Fig. 3, p. 356.)*

This condition is likely under investigated and should be considered in patients particularly those *with a* 30kB deletion of CYP21A12 (Narasimhan and Khattab, 2019) or symptoms of a connective tissue disorder (Kolli et al., 2019). A TNXB c12463+2T > C variant is associated with moderate EDS manifestations (Lao et al., 2021).

11β-hydroxylase deficiency

In cases of CAH aside from 21-hydroxylase deficiency, the defect is due to the absence of the steroid 11β-hydroxylase and the serum concentration of 11-deoxycortisol (compound S) (Khattab et al., 2017; Nimkarn and New, 2008) should be determined. The extent of elevation of 11-deoxycorticosterone is not always apparent in publications since few laboratories were able to measure the steroid, but this is changing. Hypertension is not always a sign of the condition in the affected newborn and blood pressure is often normal in children until around the time of puberty when high blood pressure is seen. Many clinicians are conditioned to consider this form of CAH as the hypertensive form but this is not the case in practice and the term should not be used for neonates.

Girls present with virilization affecting the genitals, but boys are often not diagnosed until much later with acne or rapid growth rate. Bone age advancement is seen in both sexes but after puberty the older children can have short stature (see Fig. 3.2.1.10).

The incidence of the disorder may vary with ethnicity and geography (Baronio et al., 2019) but seems to be more common in the Middle East and North Africa. As with the interpretation of 17-OHP results in the newborn, care should be taken in assigning significance to a raised 11-deoxycortisol concentration in serum. In known cases of 11β-hydroxylase deficiency, plasma S exceeds 500 nmol/L. A modestly elevated value for 17-hydroxyprogesterone by a direct radioimmunoassay may be the result of interference in the assay and should be confirmed after extraction of steroids from plasma to an organic solvent. The urine steroid profile in the newborn infant will show a relatively simple pattern with a high excretion of 6-hydroxy-tetrahydro-11-deoxycortisol as well as tetrahydro-S (Hughes et al., 1986). This is different from a profile later in life that will show THS and hexahydro-S metabolites (Fig. 3.2.1.11).

Androstenedione measurements when elevated are helpful in the diagnosis and management of CAH due to 11β-hydroxylase defects as with 21-hydroxylase. Concentrations of androstenedione using an extraction method can approach 10 nmol/L at birth of a normal infant but decline during the first week of life in normal males and females. In CAH, androstenedione concentrations will rise well above 10 nmol/L after the first week.

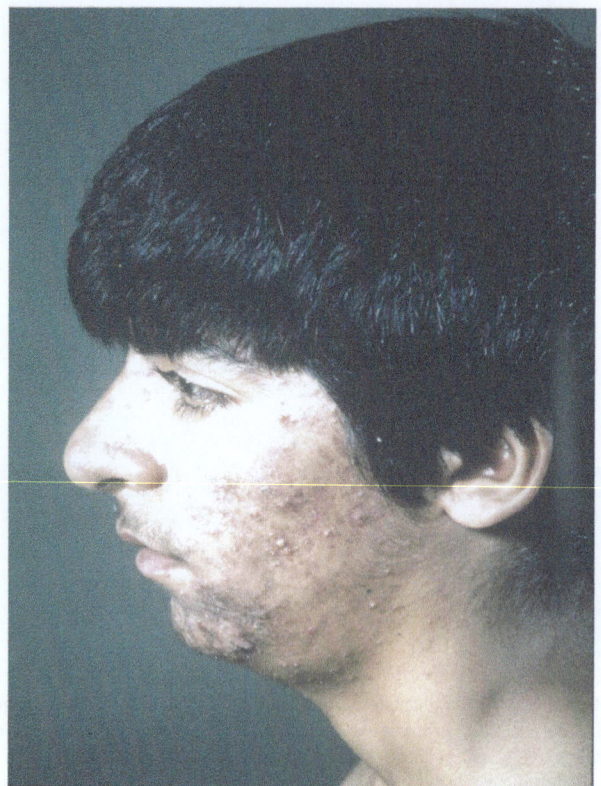

FIG. 3.2.1.10 Youth with 11-hydroxylase deficiency presenting with severe acne. *(Author original.)*

In 11β-hydroxylase deficiency, with a high production of deoxycorticosterone (DOC), a potent mineralocorticoid, there is suppression of PRA that will rise on adequate suppression of adrenals with treatment. Paradoxically with over treatment there can be loss of sodium in the urine and hyponatremia. Assays for DOC are not readily available, urine metabolites of DOC however can be detected in a urine steroid profile (Nguyen et al., 2016). The implementation of a blood steroid profile by LC-MS/MS enables the diagnosis of a patient with 11-hydroxylase deficiency (Bulsari and Falhammar, 2017; Sarathi et al., 2019). It is important when validating a steroid profile that isobaric steroids are resolved in the chromatographic stage otherwise a false diagnosis can be made. For example, 11-deoxycortisol, corticosterone and 21-deoxycortisol have the same molecular weight and in a short LC separation may not be resolved. A diagnosis of 11-hydroxylase deficiency may be confused with 17-hydroxylase deficiency if 11-deoxycortisol is not resolved from corticosterone (Burns et al., 2020).

Genetics of 11-hydroxylase deficiency

Steroid 11β-OHD CAH is the result of mutations in CYP11B1 gene (OMIM #202010) as an autosomal recessive trait. The nine exonic CYP11B1 gene is located on the long arm of chromosome 8, and the highly homologous CYP11B2, that is 95% and 97% identical in coding and noncoding regions, respectively, is 40 Kb upstream. More than 150 mutations including missense/nonsense, splicing, small/gross deletions, insertions and complex rearrangements have been described over the entire encoding region, in the CYP11B1 gene (Baronio et al., 2019). These CYP11B1 mutations tend to cluster in exons 2, 5, 6, 7, and 8. The R448H mutation and 8 other mutations are the most frequently reported and account for approximately 40% of all cases (Fig. 3.2.1.12).

There is significant ethnic specificity in CYP11B1 mutations (Khattab et al., 2017; Wang et al., 2015). A variant p.R448H among Sephardic Jews of Moroccan ancestry and p.Q356X is common among sub-Saharan Africans and African-Americans and p.G379V among Tunisian patients. In Turkey, 13 different CYP11B1 mutations were found among 28 patients from 25 families (Baş et al., 2018b). The T318 M mutation is most common among Yemenis (Motaghedi et al., 2005), and mutations such as c.53_54insT, G206 V, W260X, R448P, and H465L are often found in Saudi Arabs (Alzahrani et al., 2017). The R454C mutation has only been reported among the Chinese (Wang et al., 2015). In one case, a compound heterozygous mutations Y195H and R453Q located in exons 3 and 8, respectively, that are mutation hotspots (Wang et al., 2018). A R453Q mutation in 3 Chinese individuals and is likely to be more common among the Chinese. The residue R453 is located in the L-helix and is adjacent to the Cys-pocket motif. This domain is highly conserved in the P450 family of enzymes and causes 11-hydroxylase activity to decrease to approximately 1% of the wild-type activity.

Patients carrying the R448H mutation are the most numerous. The normal structure of the protein from this gene region is essential for maintaining normal enzymatic activity. Point mutations in exon 8 result in severe reduction of enzymatic activity, thereby resulting in classical 11βOHD. There are many mutations in exon 3, most cause a partial reduction in enzymatic activity and thus result in nonclassical 11βOHD.

Genetic recombination between CYP11B1 and CYP11B2, crossing over during meiotic reduction, do occur (Duan et al., 2018; Menabò et al., 2016; Hampf et al., 2001). An unequal recombination between the CYP11B1 and CYP11B2 genes has been found in 11β-OHD CAH patients. A recombination event where the CYP11B1 gene is under the control of the CYP11B2 promoter such that it responds to angiotensin II and not to ACTH has been described resulting clinically in classical 11-OHD (Hampf et al., 2001; Menabò et al., 2016).

Some dinucleotides are prone to methylation of the cytosine followed by deamination. Most of the known mutations completely abolish the activity of the enzyme,

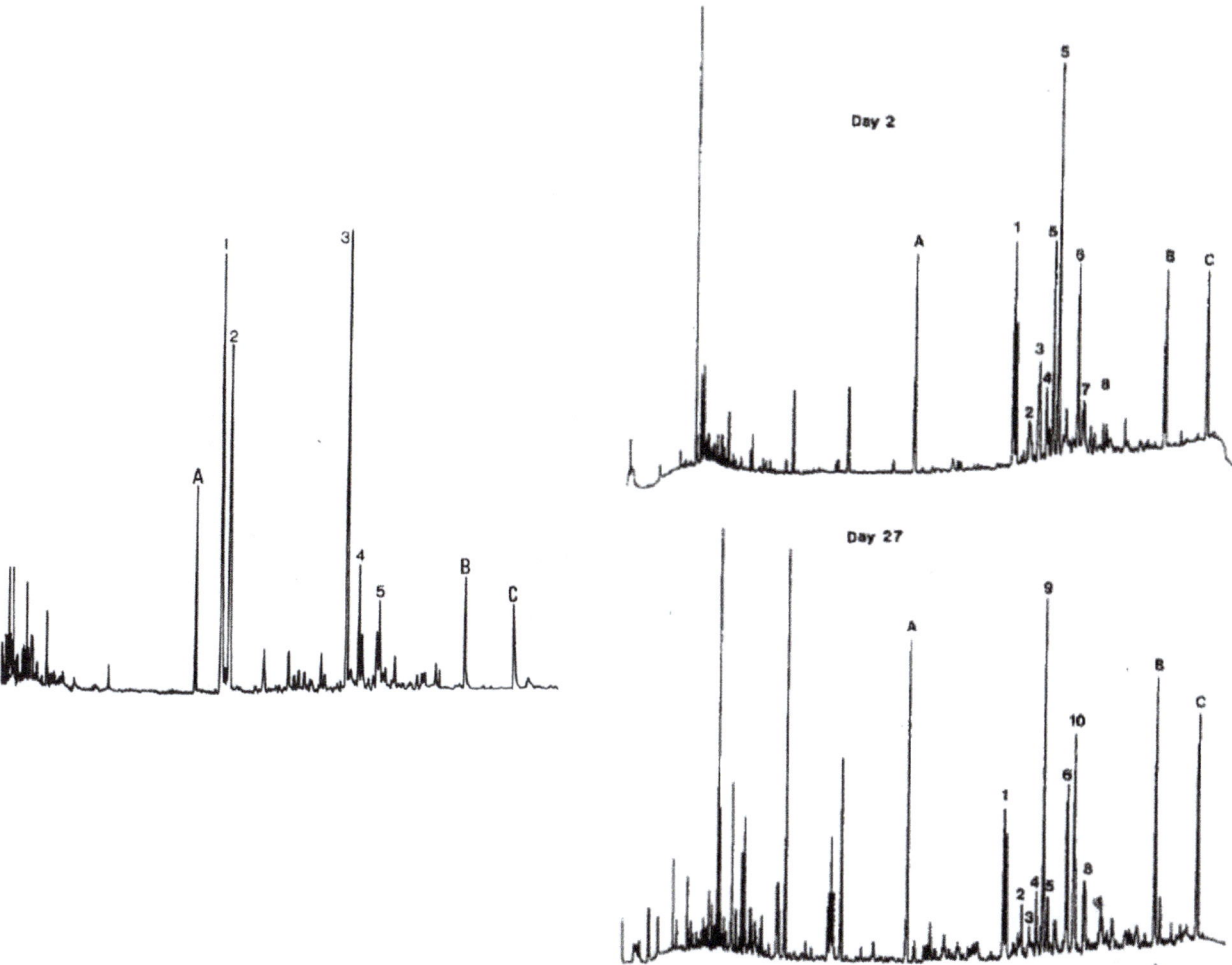

FIG. 3.2.1.11 USP in CYP11B1 deficiency. Newborn (*top right*) and day 27 (*bottom right*). The steroids identified were (1) 16α-hydroxy-DHEA; (2) 16β-hydroxy-DHEA; (3) 16-oxo-androstene-3β,17α-diol; (4) 5-androstene-3β,16α,17β-triol; (5) 16,18-dihydroxy-DHEA; (6) 16α-hydroxypregne-nolone; (7) 21-hydroxypregnenolone; (8) 5-androstene-3β,15,16,17 tetrol; (9) tetrahdro-11-deoxycortisol (THS); and (10) hydroxy-THS. On left is results from 6-year-old patient with (3) THS; (4) HHS; and (5) HHS. *(Left: From Honour JW, Anderson JM, Shackleton CH. Difficulties in the diagnosis of congenital adrenal hyperplasia in early infancy: the 11 beta-hydroxylase defect. Acta Endocrinol (Copenh) 1983;103(1):101–9. Right: Author original.)*

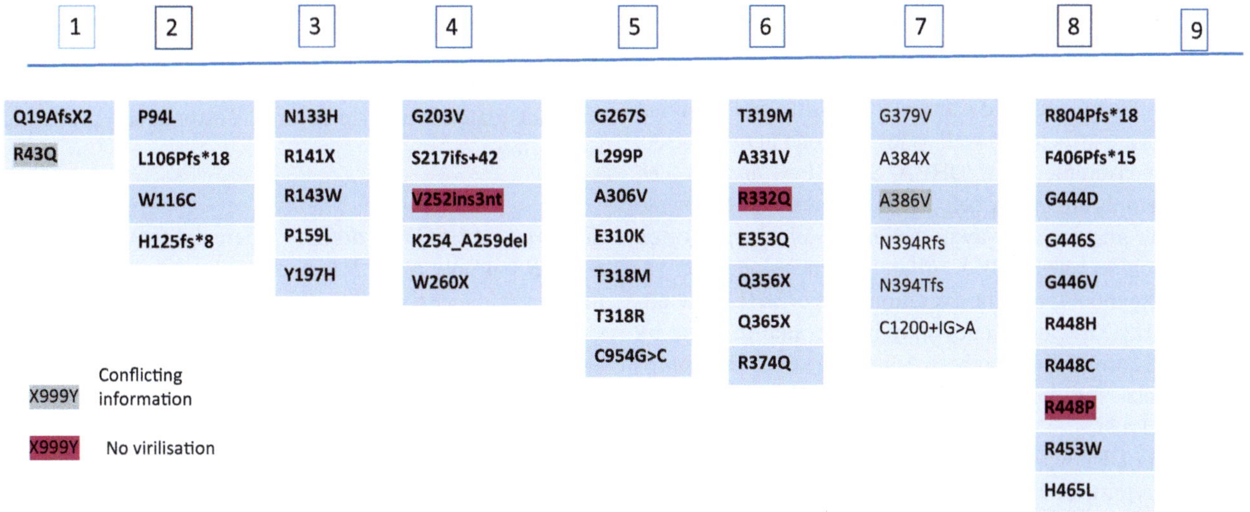

1	2	3	4	5	6	7	8	9
Q19AfsX2	P94L	N133H	G203V	G267S	T319M	G379V	R804Pfs*18	
R43Q	L106Pfs*18	R141X	S217ifs+42	L299P	A331V	A384X	F406Pfs*15	
	W116C	R143W	V252ins3nt	A306V	R332Q	A386V	G444D	
	H125fs*8	P159L	K254_A259del	E310K	E353Q	N394Rfs	G446S	
		Y197H	W260X	T318M	Q356X	N394Tfs	G446V	
				T318R	Q365X	C1200+IG>A	R448H	
				C954G>C	R374Q		R448C	
							R448P	
							R453W	
							H465L	

X999Y Conflicting information

X999Y No virilisation

FIG. 3.2.1.12 Structure of the human CYP11B1 gene with the pathogenic variants that have been described in 46,XX patients. The nomenclature of the pathogenic variations is given according to historical references cited in a supplement to the paper. The nomenclature and the color of the boxes of each pathogenic variant are in accordance with the phenotypic characteristics of each patient, the variant present on the second allele and the references cited are in the supplement. *(From Baronio F, Ortolano R, Menabò S, Cassio A, Baldazzi L, Di Natale V, Tonti G, Vestrucci B, Balsamo A. 46,XX DSD due to androgen excess in monogenic disorders of steroidogenesis: genetic, biochemical, and clinical features. Int J Mol Sci 2019;20(18):4605. Fig. 5, p. 6 of 35.)*

but the clinical manifestations of the disorder vary substantially. The genotype phenotype correlation for hypertension is less than found for the clinical variables with CYP21A21 even though there is similar gene duplication.

Seven novel CYP11B1 mutations were transiently transfected in COS7 cells and the conversion of 11-deoxycortisol to cortisol was monitored. Three mutations detected in patients with classic 11OHD (p.W116G, p.A165D, p.K254_A259del) had absent enzymatic activity. Only the p.A165D mutation showed some conversion for 11-deoxycortisol to cortisol when these three mutations were incubated with a lower 11-deoxycortisol concentration (250 nmol/L) for 24 h. The other four mutations (p.M88I, p.Pro159Leu, p.R366C, and p.T401A) detected in patients with mild 11OHD and heterozygous carriers resulted in partial reduction of 11β-hydroxylase activity. The p.M88I mutation reduced activity to $39.8 \pm 6.4\%$ of wild-type. The transfected p.P159L and p.R366C showed a similar impairment with $25.8 \pm 3.3\%$ and $23.0 \pm 3.1\%$ of wild-type activity, respectively, whereas p.T401A had $37.5 \pm 3.8\%$ of the normal 11β-hydroxylase activity. Determination of kinetic constants showed similar Km values for p.M88I, p.R366C, and p.T401A with significantly impaired Vmax compared with wild-type. The p.P159L mutation did not reach substrate saturation under the established reaction conditions (Parajes et al., 2010).

3β-Hydroxysteroid dehydrogenase (HSD3B2) deficiency

This rare defect will present with variable degrees of virilization of a 46XX child and is difficult to confirm biochemically in a newborn child because the markers for the defect (dehydroepiandrosterone (DHEA), DHEAS, pregnenolone and pregnenolone sulfate) are normal products of the adrenals at this time. Salt loss, hypoglycemia and hypogonadism are also seen. Mineralocorticoid deficiency is not apparent in one third of cases (Guran et al., 2020). Steroidogenic pathways to aldosterone, cortisol and adrenal androgens are affected as well as testosterone in the gonads. Plasma measurements of DHEA_S, DHEA, and ACTH, before steroids are given, help in the diagnosis. The enzyme is however inactive anyway in the large fetal adrenal cortex zone which secretes DHEA sulfate, pregnenolone and 17-hydroxypregnenolone. The excretion rates of DHEA metabolites in HSD3B2 deficiency may be elevated in urine by GC-MS if corrected for body size and compared with appropriate data for healthy infants (Fanis et al., 2020) (Fig. 3.2.1.13).

Plasma DHEAS concentrations are particularly high in normal preterm infants compared with infants delivered at term. Infants with 3β-hydroxysteroid dehydrogenase

deficiency can become very sick and this is the more likely presentation of a girl with the defect. Mild clitoromegaly may be apparent on closer examination than in the delivery suite although only one third of females reviewed by Guran et al. had ambiguous genitalia (Guran et al., 2020). In many cases, steroid treatment is likely to have been started before biochemical tests have been performed. Plasma 17-OHP results may well appear to be elevated in this condition because 17-hydroxypregnenolone (and sulfate) production that cross-reacts in many 17-OHP assays will be high. Newborn screening may produce a moderately elevated 17-OHP result and the diagnosis made by second tier screening (Levy-Shraga and Pinhas-Hamiel, 2016; Araújo et al., 2014; Janzen et al., 2012). A profile of 19 steroids by LC-MS/MS reveals high baseline 17OH-preg to cortisol ratio with low 11-oxyandrogen concentrations (Guran et al., 2020). Testosterone was only slightly elevated. It was suggested that DHT synthesis was achieved through the backdoor path although androsterone excretion was low and DHT was not measured.

If a child suspected of having a HSD3B2 defect is maintained with synthetic steroids (plus fludrocortisone if necessary for electrolyte balance) then given daily injections of depot ACTH (Synacthen) the markers for the defect (pregnenetriol, DHEA) are clearly elevated in urine steroid profile. In the short term, dexamethasone treatment allows measurements of endogenous steroids during an ACTH stimulation test without the risk of assay interference (Lutfallah et al., 2002). The synthetic steroids have little effect on the steroid profile due to their lower excretion rate and different metabolism to the endogenous steroids. Hydrocortisone (15–20 mg/M^2/day) has been the preferred long-term treatment of a child with CAH although lower doses are now used since cortisol production rates (PR) based on stable isotope tests have shown the PR to be 6–9 mg per meter squared BSA. HSD3B2 deficiency also leads to low aldosterone production, consequently mineralocorticoid replacement (fludrocortisone 150 μg/M^2/day) is needed. PRA is elevated prior to treatment but normalizes with adequate electrolyte control. Testosterone production is also low.

Genetics of HSD3B2 defects There are two related genes *HSD3B* but only *HSD3B2* is active in adrenals and gonads for steroidogenesis, *HSD3B1* is expressed in peripheral tissues (Naelitz and Sharifi, 2020). Activity of HSD3B1 is another reason why steroid hormone assays can be affected because steroids such 17-hydroxypregnenolone can be converted to 17-hydroxyprogesterone. More than 50 mutations throughout the HSD3B2 gene were reviewed recently in more than 60 families (Baquedano et al., 2018) and 82 patients in a more recent study (Guran et al., 2020).

A

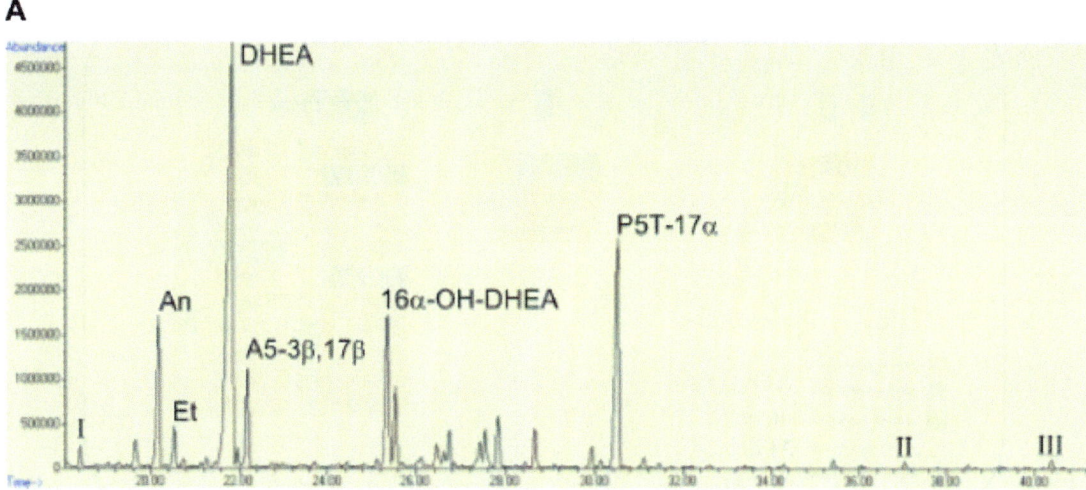

B

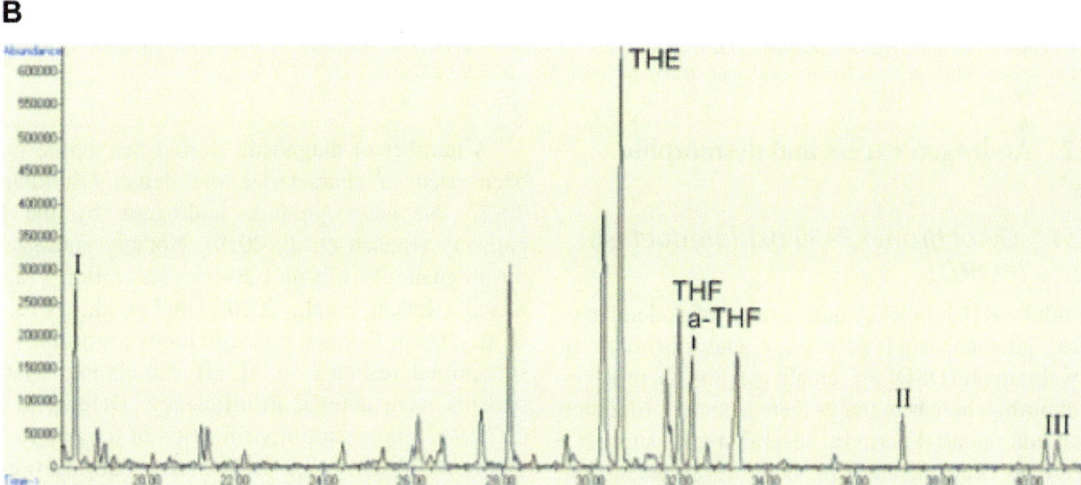

FIG. 3.2.1.13 Scans of the gas chromatography-mass spectrometry (GC-MS) steroid metabolome analysis of the patient (A) as well as a healthy control (B). I, II, II: internal standards (5α-androstane-3α,17α-diol, stigmasterol, cholesteryl butyrate); An (androsterone, 5α-androstane-3α-ol17-one), Et (etiocholanolone, 5β-androstane-3α-ol-17-one), DHEA (dehydroepiandrosterone, 5-androstene-3β-ol-17-one), A5-3β,17β (androstenediol, 5-androstene-3β,17β-diol), 16α-OH-DHEA (5-androstene-3β,16α-diol-17-one), P5T-17α (pregnenetriol, 5-pregnene3β,17α,20α-triol). Cortisol metabolites: THE (tetrahydrocortisone, 5β-pregnane-3α,17α,21-triol-11,20-dione), THF (tetrahydrocortisol, 5β-pregnane-3α,11β,17α,21-tetrol-20-one) and a-THF (5α-tetrahydrocortisol, 5α-pregnane-3α,11β,17α,21-tetrol-20-one). *(From Fanis P, Neocleous V, Kosta K, Karipiadou A, Hartmann MF, Wudy SA, et al. Late diagnosis of 3β-hydroxysteroid dehydrogenase deficiency: the pivotal role of gas chromatography-mass spectrometry urinary steroid metabolome analysis and a novel homozygous nonsense mutation in the* HSD3B2 *gene. J Pediatr Endocrinol Metab 2020;34(1):131–6. Fig. 1, p. 134.)*

Frame shift, nonsense, in-frame deletions, splicing and missense mutations have been reported (Fig. 3.2.1.14).

The proteins have been unstable when transfected into cells so it has been difficult to determine levels of activity. In vitro data such as in silico pathogenicity prediction suggests there can be a molecular explanation for the disease state. The gene is expressed in the adrenals and gonads so the synthesis of cortisol, aldosterone, progesterone, androgens and estrogens are affected to varying degrees. Patients can present with salt-wasting and or a mild DSD. 46XX children can have mild to severe virilization at birth and salt wasting or present later with premature pubarche.

Labial fusion and or clitoral enlargement are seen depending on the activity of mutant genes.

Mutations in DAX-1

X-linked congenital adrenal hyperplasia (AHC) is usually associated with hypogonadotrophic hypogonadism but precocious puberty has been reported in some patients (Nagel et al., 2019). The authors propose an intrinsic gonadotrophin and ACTH independent activation of steroidogenesis without conversion of adrenal androgens to testosterone. The source of uncontrolled testosterone production is unclear.

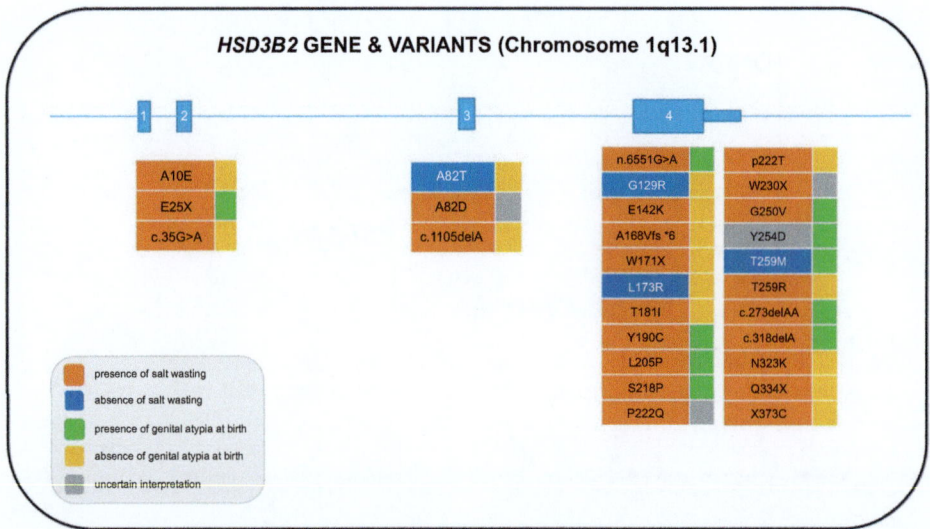

HSD3B2 GENE & VARIANTS (Chromosome 1q13.1)

FIG. 3.2.1.14 Structure of the human HSD3B2 gene with the pathogenic variants described in 46,XX. The nomenclature and the color of the boxes of each pathogenic variant are in accordance with the phenotypic characteristics described in each patient and the variant present on the second allele. *(From Baronio F, Ortolano R, Menabó S, Cassio A, Baldazzi L, Di Natale V, Tonti G, Vestrucci B, Balsamo A. 46,XX DSD due to androgen excess in monogenic disorders of steroidogenesis: genetic, biochemical, and clinical features. Int J Mol Sci 2019;20(18):4605. Fig 2 p 3.)*

3.2.1.2.2 Androgen excess and dysmorphic features

3.2.1.2.2.1 Cytochrome P450 oxidoreductase deficiency (PORD)

Patients with PORD deficiency have a range of skeletal malformations, glucocorticoid deficiency and disorders of sexual development (DSD). In female patients, there is virilization at birth, whereas in males there is undervirilization in the fetal and pubertal periods. Several studies suggest a role of POR in drug metabolism, bone development and retinoic acid metabolism. Skeletal malformations in PORD affect the face (midface hypoplasia), cranium (craniosynostosis), hands and feet (arachnodactyly, talipes), large joints (radiohumeral synostosis), femurs (bowing, fractures), and other areas (e.g., scoliosis, pectus excavatum) (Fig. 3.2.1.15). Antley-Bixler syndrome is a genetically heterogeneous condition caused by mutations not only in POR but also in FGFR2 and CYP26B1 (Azoury et al., 2017; Ko, 2016; Ko et al., 2009).

Cytochrome P450 enzymes in the endoplasmic reticulum get electrons from NADPH through the action of POR (Miller, 2021) for 21-hydroxylase and 17-hydroxylase (Fig. 3.2.1.16). Each pathway of steroidogenesis is affected to a different extent (depending on the locations of the POR mutations), resulting in high variability of the clinical presentation of PORD.

High concentrations of pregnenolone, progesterone, 17-hydroxyprogesterone and corticosterone are found in plasma and urine metabolites thereof in urine (Fig. 3.2.1.17) (Reisch et al., 2019; Ono et al., 2018; Homma et al., 2006; Flück et al., 2004).

A number of diagnostic steroid metabolite ratios have been used to characterize the defect (Burkhard et al., 2017). Neonates produce androgens by the backdoor pathway (Reisch et al., 2019). Notably androsterone and 5α-pregnane-3α,17α-diol-20-one excretion rates were raised (Reisch et al., 2019; Ono et al., 2018; Homma et al., 2006). Cortisol concentrations are usually low with suboptimal response to ACTH stimulation. Many of the patients have adrenal insufficiency. Defects in placental *CYP19A1* may result in virilization of the pregnant mother in around 20% of cases (Dean et al., 2020). Low estriol and increased metabolites of pregnenolone can be found in urine and amniotic fluid (Cragun et al., 2004).

Genetics of PORD

The genetic cause of this form of CAH was discovered in 2004 and explained the previous description of "apparent pregnene hydroxylation deficiency" (Shackleton and Malunowicz, 2003), "apparent associated 17α- and 21-hydroxylase deficiency" (Flück et al., 2004; Huang et al., 2017). The POR gene is located on the long arm of chromosome 7 (7q11.2). More than 80 female patients affected by PORD and with more than 30 different pathogenic variants have been reported to date (Fig. 3.2.1.18) (Dean et al., 2020; Lee et al., 2020a; Parween et al., 2019; Krone et al., 2012; Idkowiak et al., 2011; Huang et al., 2017). POR variants must be studied for each potential P450 target enzyme. POR mutations can impair a large number of enzymes that rely on POR for electron transfer.

Impairments in enzymes involved in cholesterol synthesis (CYP51A1 and squalene epoxidase) and retinoic acid metabolism (CYP26 isozymes) are believed to cause

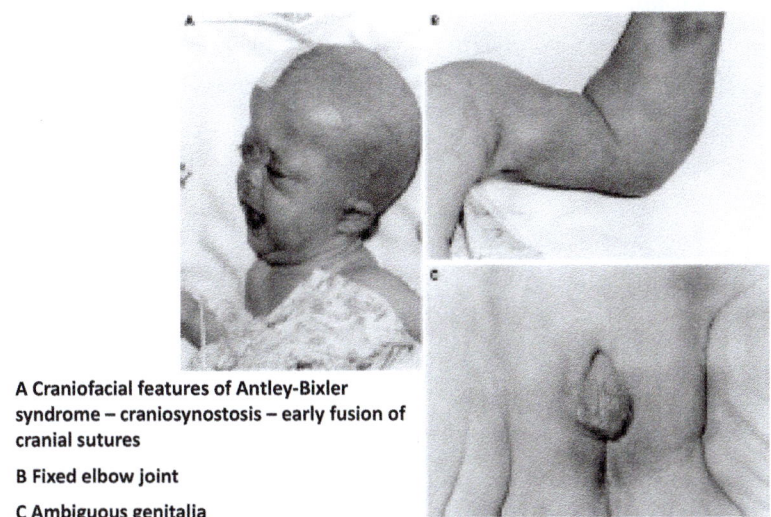

A Craniofacial features of Antley-Bixler
syndrome – craniosynostosis – early fusion of
cranial sutures

B Fixed elbow joint

C Ambiguous genitalia

POR defects associated with Antley Bixler syndrome in some cases

FIG. 3.2.1.15 Characteristic craniofacial features of patient with Antley-Bixler syndrome (A) with craniosynostosis due to early fusion of cranial sutures. The characteristic fixed elbow joint of Antley-Bixler syndrome is shown (B). The genital malformations are illustrated to document the clitoromegaly (C) and hooded prepuce. *(Photographs reproduced with permission. From Reardon W, Smith A, Honour JW, Hindmarsh P, Das D, Rumsby G, Nelson I, Malcolm S, Adès L, Sillence D, Kumar D, DeLozier-Blanchet C, McKee S, Kelly T, McKeehan WL, Baraitser M, Winter RM. Evidence for digenic inheritance in some cases of Antley-Bixler syndrome? J Med Genet 2000;37(1):26–32. Fig. 1 p. 27.)*

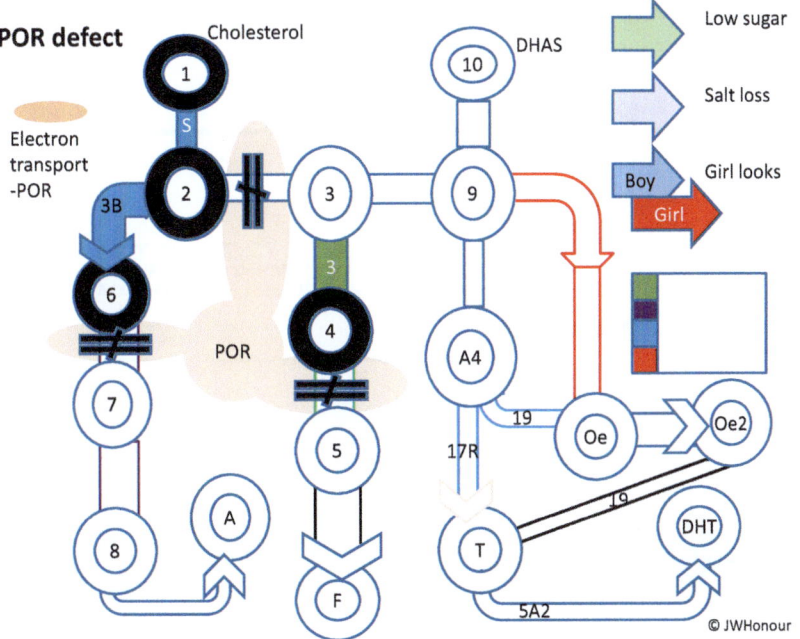

FIG. 3.2.1.16 Site of POR in steroid metabolism pathway. POR transfers electrons to CYP21 and CYP17. *(Author original.)*

skeletal malformations. Loss of function of CYP17A1 is associated with DSD, specifically male undervirilization (from decreased androgen production) and female virilization (from use of the backdoor pathway for androgen synthesis in fetal life). Computational analysis has been used to study the expression of mutant genes (Burkhard et al.,

2017). The phenotypic result in genetic females characteristically depends on the POR pathogenic variant (Miller, 2021) that may reduce the CYP17A1, CYP21A2 and CYP19A1 activities differently and thus influence, adrenal insufficiency and differences in sex development and skeletal development. Loss of function in the fetus may lead

P450 oxidoreductase deficiency (ORD)

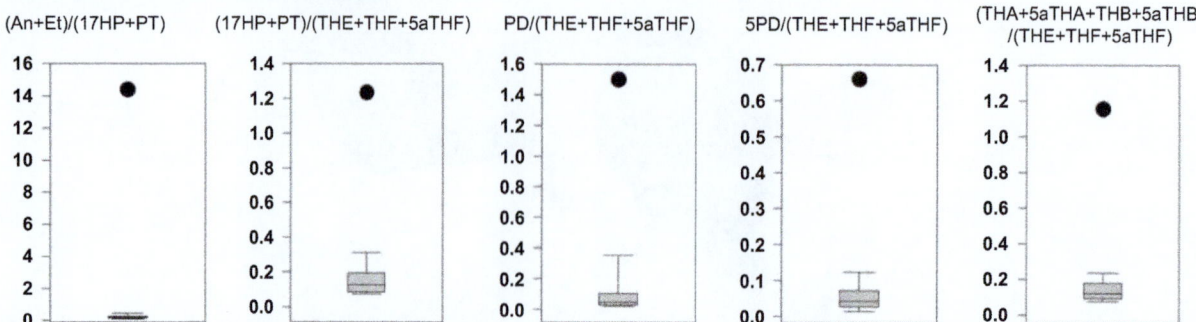

FIG. 3.2.1.17 Diagnostic ratio panel for CAH. Quantitative data *(black dot)* for ORD patient compared to normative data appropriate for several forms of CAH. Using the designated ratios the diagnosis can be accurately given. The ORD ratios are all abnormal. *(From Krone N, Hughes BA, Lavery GG, Stewart PM, Arlt W, Shackleton CH. Gas chromatography/mass spectrometry (GC/MS) remains a pre-eminent discovery tool in clinical steroid investigations even in the era of fast liquid chromatography tandem mass spectrometry (LC/MS/MS). J Steroid Biochem Mol Biol 2010;121(3–5):496–504. Fig 5 p 503.)*

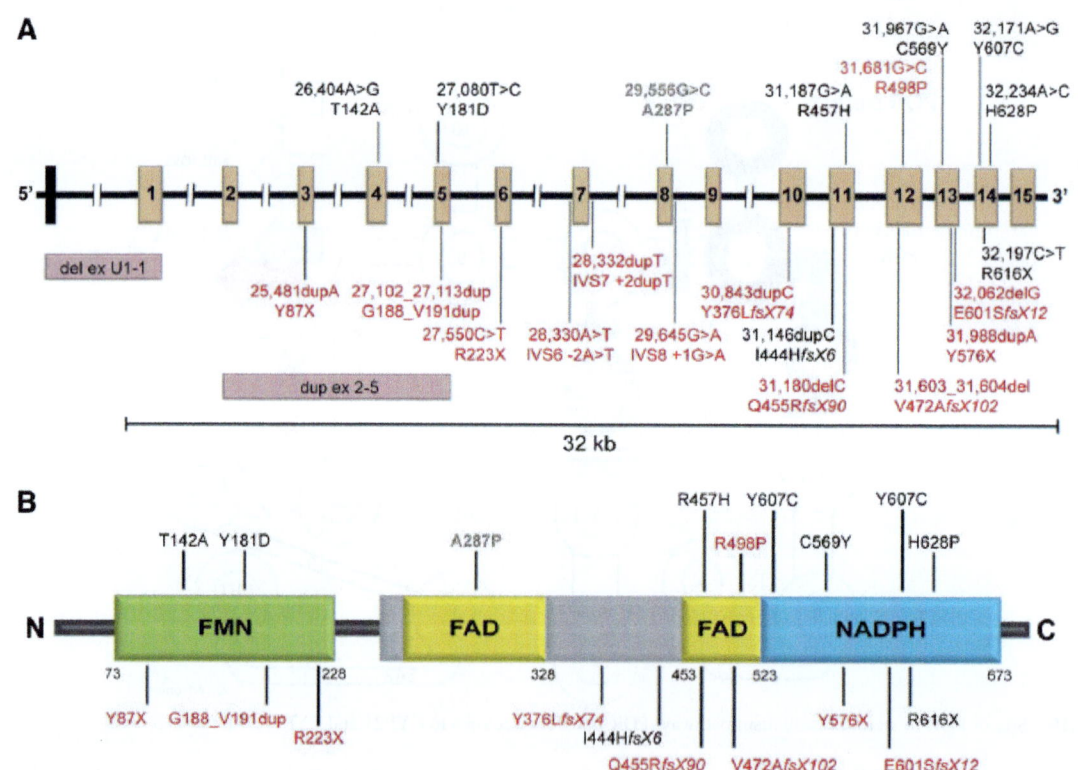

FIG. 3.2.1.18 Localization of POR mutations at the DNA and protein level. (A) Schematic representation of the POR gene. The untranslated exon U1 is given as a *black box* and coding exons are numbered. Novel mutations are given in *red font*, and the most common mutation p.A287P in exon 8 is marked in *blue*. Mutations cluster in the 3 region without any additional common mutation apart from the p.A287P mutation. The partial deletion del ex U1–1 and the partial duplication dup ex2–5 that were identified by MLPA are represented by labeled *red boxes*. (B) Schematic representation of the POR protein. Mutant residues are located in various locations all over the three functional domains of the POR protein. *(From Krone N, Reisch N, Idkowiak J, Dhir V, Ivison HE, Hughes BA, Rose IT et al, Genotype-phenotype analysis in congenital adrenal hyperplasia due to P450 oxidoreductase deficiency. J Clin Endocrinol Metab 2012;97(2):E257–67. Fig. 1 p. e261.)*

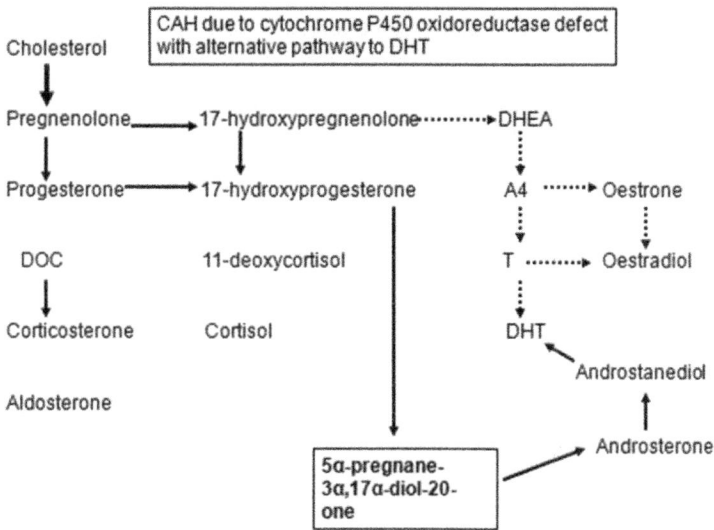

FIG. 3.2.1.19 Backdoor path to dihydrotestosterone (DHT). *(Author original.)*

to sufficient excess androgen to cause virilization of the mother.

The severity of skeletal malformations varies according to the distribution of POR mutation pairs. The greater the severity to protein function of the involved mutations, greater the severity of the "skeletal" phenotype (Krone et al., 2012). Adrenal insufficiency is attributed to null and missense mutations, although R457H/R457H mutations are associated with milder adrenal insufficiency (Dean et al., 2020). DSD are common in patients with PORD, across all types of mutations, but less common in males (46XY) with R457H/R457H mutations. The POR mutation p.R457H allows for an increase in androgens in the maternal circulation and virilizes the mother. Deviation of steroidogenesis towards the "backdoor pathway" (Fig. 3.2.1.19) contributes to the virilization of mothers and affected female fetuses (Fukami et al., 2005).

Severe phenotypes can lead to stillbirth or early neonatal deaths (Reisch et al., 2013).

The POR mutation p.A287P, in European patients alters CYP17A1 activity but not CYP21A2 activity or CYP19A1 activity, whereas the POR mutation p.R457H, which is frequent in Japanese patients (Fukami et al., 2005; Huang et al., 2017) predominantly reduces CYP19A1 activity (Pandey et al., 2007). The ethnic variability and the high polymorphism of the POR gene are reflected in the high recurrence of p.a503v sequence variation in mildly affected patients of Chinese origin, but not in those of African origin (Miller, 2021; Pandey and Sproll, 2014).

3.2.1.2.2.2 Smith-Lemli-Opitz syndrome

Smith-Lemli-Opitz syndrome is a rare autosomal recessive disorder of cholesterol biosynthesis also characterized by multiple congenital malformations. An infant can present with ambiguous genitalia (bifid scrotum, micropenis) or less frequently with an acute adrenal crisis (Jayamanne et al., 2018), circulatory collapse and low blood pressure. Early infancy can be complicated by poor feeding, episodes of loose stools, failure to thrive, and several episodes of unexplained drowsiness. Growth measures (weight, height, and occipitofrontal circumference) can be very low. The infant can have soft dysmorphic features that include microcephaly, bitemporal narrowing, upward slanting eyes, epicanthal folds (eyelid covering corner of the eye), partial ptosis (drooping of the eyelid), broad nasal bridge, low set posteriorly rotated ears, high arched palate and short neck (Fan et al., 2019). Some of these features are seen in ultrasound images of the fetus in utero (Schoner et al., 2020) and a diagnosis may often now be made at mid-gestation. Some have fusion of toes (syndactyly) (Fig. 3.2.1.20).

After birth, pigmentation of the skin may be seen around the mouth (Jayamanne et al., 2018), buccal and palmar areas. In later years, a dentist may note dentofacial anomalies (Rojare et al., 2019). Some cases have holoprosencephaly (failure in forebrain to divide into hemispheres) although chromosomal anomalies (aneuploides—abnormal number) are more common presentation of this feature. Children with SLO syndrome have developmental delay. Mild variants have also been reported (Temple et al., 2020; Tucci et al., 2016). Blood glucose and serum bicarbonate levels are low in SLO syndrome and serum electrolytes reveal hyponatremia with hyperkalemia. Hydrocortisone brings about a marked clinical response. Serum spot cortisol level is low normal and 17-hydroxyprogesterone levels are low. Cholesterol may be low.

Diagnosis of Smith-Lemli-Opitz syndrome and associated adrenal crisis is made based on clinical and

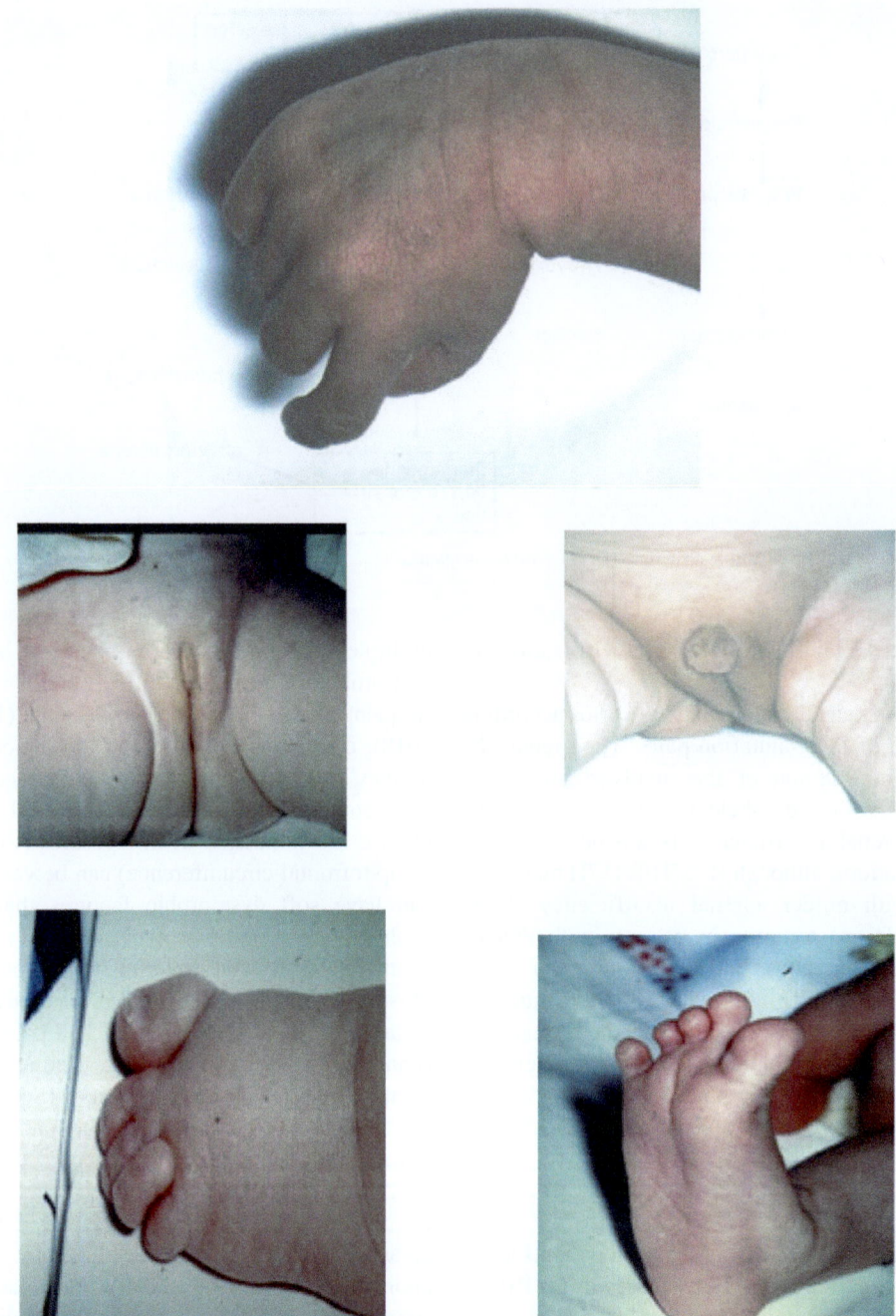

FIG. 3.2.1.20 Newborn infant with Smith Lemli Opitz syndrome has specific craniofacial dysmorphism, microcephaly, growth restriction of prenatal onset and internal organ malformations. Hypoplastic external genitalia in a male and 2–3 syndactyly of the toes are shown. *(Author original.)*

biochemical features. 7-Dehydrocholesterol can be raised in blood and urine. In urine of the infant, a monosulfate and disulfate fraction was analyzed as MO-TMS derivatives by GC-MS (Fig. 3.2.1.21). Metabolites of 7-dehydrosteroids were found such as 16-hydroxy-7-dehydro-DHEA (peak 2), 16-hydroxy-7-dehydropregnenolone (peaks 8 and 10) and 21-hydroxy-dehydropregnenolone (peaks 17 and 18)

where the double bond in the B-ring is at position 7 or 8 (9). The identities were confirmed from characteristic mass spectra (Fig. 3.2.1.22). The pairs of steroids with similar spectra were attributed to double bonds at C5,6 or C7,8 (9).

Further 7-dehydro steroid metabolites have been described by others in urine notably of 16-hydroxy-DHEA,

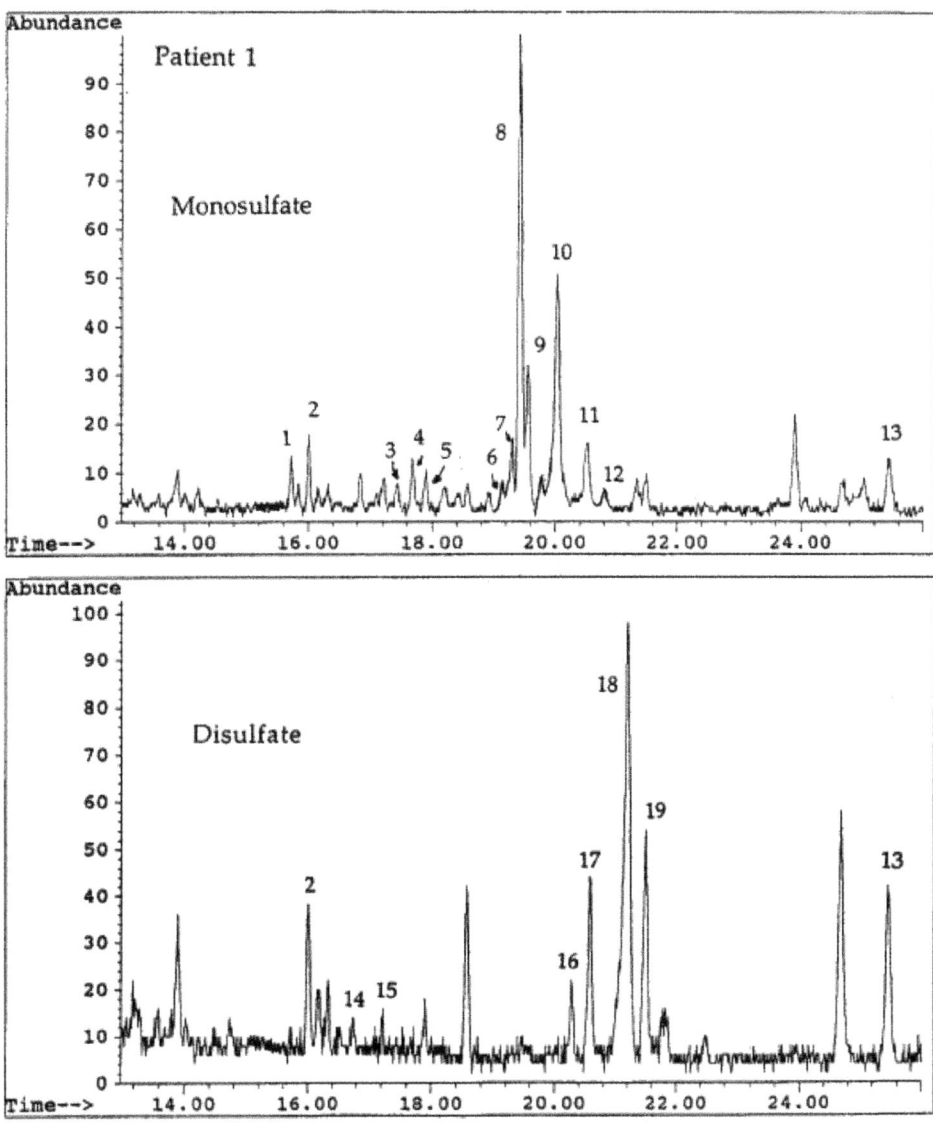

FIG. 3.2.1.21 USP for patient with Smith Lemli Opitz syndrome. Gas chromatography/mass spectrometry total ion current recording of the methyloxime-trimethysilyl ethers derivatives of the monosulfate and disulfate fractions of SLO Patient. Representative steroids have been labeled: (1) 16α-hydroxyDHEA; (2) 3β,16α-dihydroxy-5,8(9)-androstadien-17-one; (3) 5-androstene-3β,16α,17α-triol; (4) and (5) androst-5,8(9)-diene-3β,16α,17β-triols; (6) 16α-hydroxypregnenolone; (7) 3β,16α-dihydroxypregnatrien-20-one; (8) 3β,16α-dihydroxy-5,8(9)-pregnadien-20-one; (9) 5,8 (9)-pregnadiene-3β,16α,20α-triol; (10) 3β,16α-dihydroxy-5,7-pregnadien-20-one; (11) 5,7-pregnadiene-3β,17α,20α-triol; (12) 5,7-pregnadiene-3β,16α,20α-triol; (13) Stigmasterol (int. std.); (14) 3β,16β-dihydroxyandrostadien-17-one; (15) 3β,17β-dihydroxyandrostadien-16-one; (16) 21-hydroxypregnenolone plus 3β,21-dehydroxypregnatrien-20-one; (17) 3β,21-dihydroxy-5,8(9)-pregnadien-20-one; (18) 3β,21-dihydroxy-5,7-pregnadien-20-one, and (19) separated *syn-* or antiform of the methoxime derivatives of compounds 17 and 18. *(From Shackleton CH, Roitman E, Kelley R. Neonatal urinary steroids in Smith-Lemli-Opitz syndrome associated with 7-dehydrocholesterol reductase deficiency. Steroids 1999;64(7):481–90. Fig. 1, p. 483.)*

pregnenetriol and pregnanetriol (Jezela-Stanek et al., 2020; Jezela-Stanek et al., 2015; Shackleton et al., 1999b).

A defect in reduction of 7-dehydrocholesterol is the cause of SLOS such that 7-dehydropregnenolone is the result of the action of side chain cleavage acting on 7-dehydrocholesterol and 16-hydroxy-7-dehydropregnenolone is the product of 17 and 16 hydroxylases acting on 7-dehydropregnenolone (Fig. 3.2.1.23) with metabolites also of 21-hydroxy-7-dehydropregnenolone.

In pregnancy, 7-dehydropregnenetriol and 7-dehydroestriol are the main steroids that can be biomakers in a prenatal diagnosis of SLO.

Genetics of SLO

Mutations are found in the **DHCR7 gene**, which encodes 7-dehydrocholesterol reductase, the enzyme that catalyzes the final step in cholesterol biosynthesis. The DHCR7

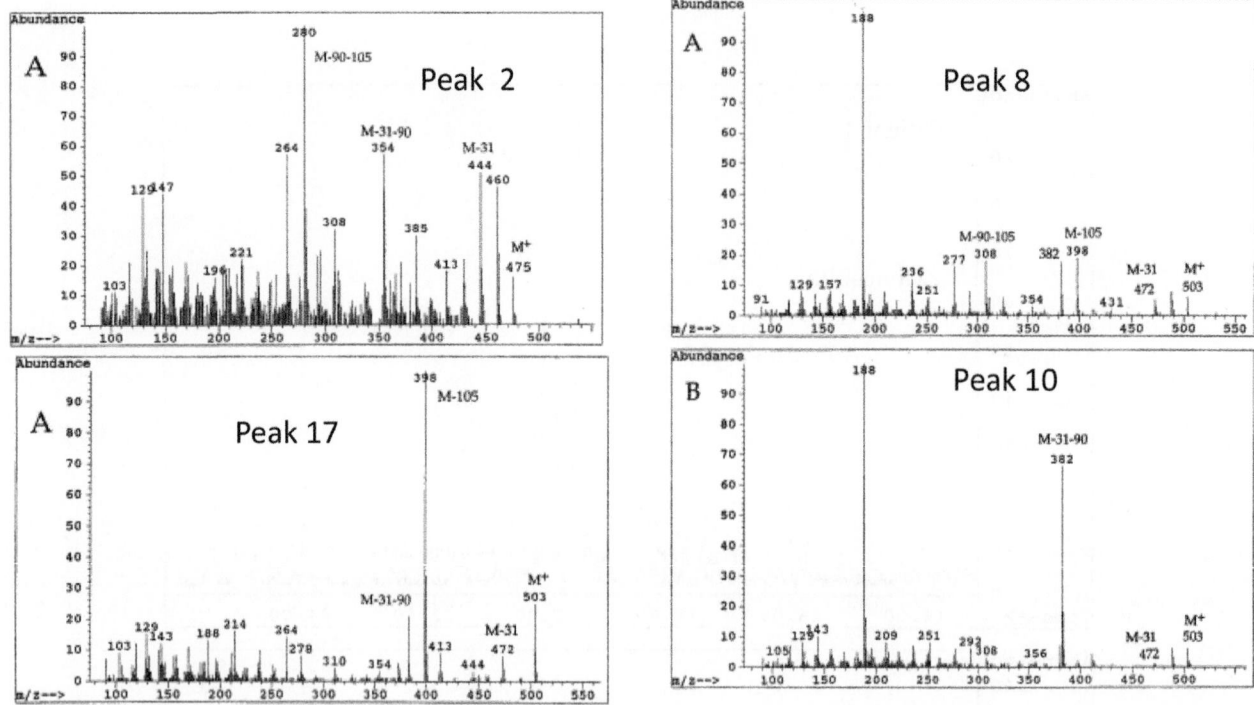

FIG. 3.2.1.22 Mass spectra of methyloxime-trimethysilyl ethers derivative of compounds identified as 3β,16α-dihydroxy-5,8(9)-pregnadien-20-one (A) and 3β,16α-dihydroxy-5,7-pregnadien-20-one (B). *(From Shackleton CH, Roitman E, Kelley R. Neonatal urinary steroids in Smith-Lemli-Opitz syndrome associated with 7-dehydrocholesterol reductase deficiency. Steroids 1999;64(7):481–90. Fig. 3, p. 485.)*

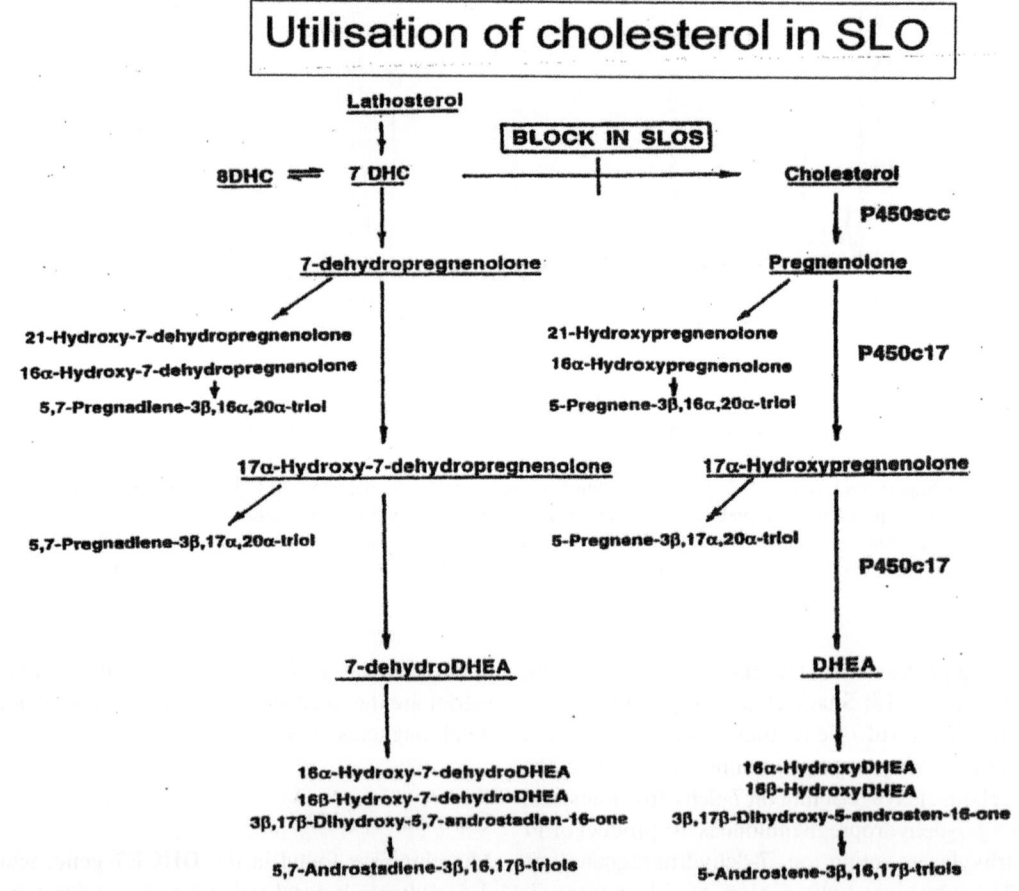

FIG. 3.2.1.23 C_{19} steroid synthesis from 7-dehydrocholesterol (7 DHC) in normal and infants with Smith-Lemli-Opitz syndrome. The major biosynthetic intermediates are underlined, and urinary metabolites are not. An equivalent biosynthetic and catabolic pathway also operates for 8-dehydrocholesterol (8DHC). *(From Shackleton CH, Roitman E, Kelley R. Neonatal urinary steroids in Smith-Lemli-Opitz syndrome associated with 7-dehydrocholesterol reductase deficiency. Steroids 1999;64(7):481–90. Fig. 9, p. 489.)*

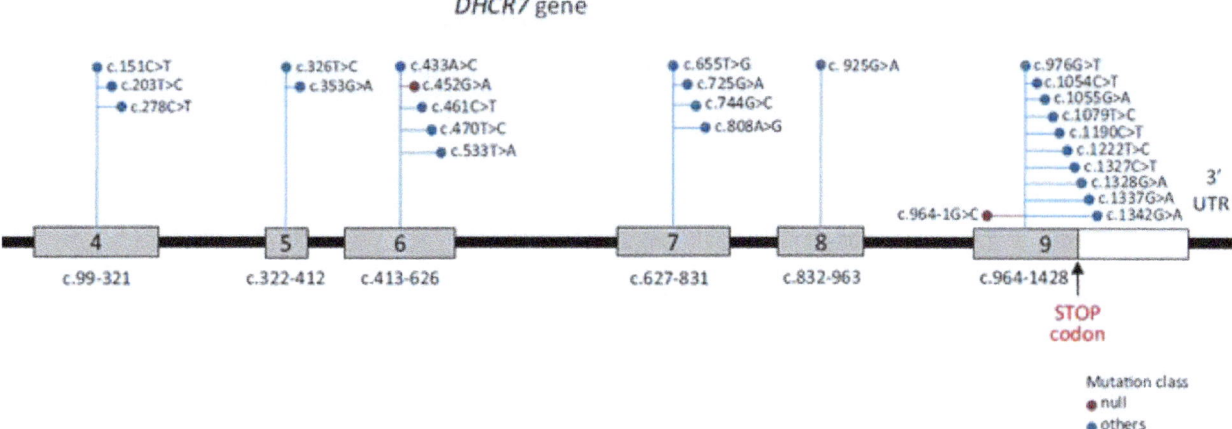

FIG. 3.2.1.24 Mutations in DHCR7 gene. 1 The DHCR7 gene scheme with mutation distribution and the topology of the cytosol loops (CL), the C-terminal domain (CT), ER lumen (LL), N-terminus (NT), and transmembrane domains (TM) in DHCR7 protein structure. Amino acid (aa) of each TM domain is also given. SRM, sterol reductase motif; NADPH—binding site of the coactivator NADPH. *(From Różdżyńska-Świątkowska A, Ciara E, Halat-Wolska P, Krajewska-Walasek M, Jezela-Stanek A. Anthropometric characteristics of 65 Polish Smith-Lemli-Opitz patients. J Appl Genet. 2021;62 (3):469–475. Erratum in: J Appl Genet. 2021 May 27. Fig. 1. p. 471.)*

gene, is located at chromosome 11q13.2-q13.5, spans 14 kb and contains nine exons with the translation initiation codon located in exon 3 (Fig. 3.2.1.24). The gene produces two main DHCR7 mRNAs (c 2.9 and 2.3 kb), which vary in length of the 30 noncoding regions encoded by exon 9.

The protein is localized to the endoplasmic reticulum with nine transmembrane helices and a sterol sensing domain. There are more than 160 different mutations, mainly missense ($n = 130/154$) and rarely nonsense, deletions, insertions, indel and splice site mutations. Many mutations are found in single or only a few patients, whereas other mutations are found frequently only in selected populations due to founder effects (Waterham and Hennekam, 2012). The 154 different mutations described therein were 130 missense, 8 nonsense, 8 deletions, 2 insertions, 1 indel and 5 splice site mutations. Several polymorphisms have been found in the DHCR7 gene.

Treatment of CAH

CAH due to 21-hydroxylase deficiency leads to a number of clinical problems (Fig. 3.2.1.25) that need to be addressed in the treatment plan. Hypoglycemia and salt wasting will need replacement cortisol and aldosterone, respectively. Cortisol may aid adrenal medullary function by supporting adrenalin synthesis. Suppression of ACTH reduces the risks of tumor formation especially **testicular adrenal rests (TART)**. Psychological effects will improve unless corticosteroid replacement is excessive, final height and fertility are important issues (Ng et al., 2020). Androgen production will be lowered although surgery may be required to the enlarged clitoris of an affected female.

Once CAH has been confirmed, lifelong cortisol treatment must be commenced. The aim is to mimic the activity of the adrenal cortex, the assessment of which is covered in several places in this book because of gross misunderstandings and variations in clinical practice. Oral hydrocortisone is usually prescribed at 9–12 mg/M²/day with 2/3 of the dose in the morning and 1/3 in the evening. Some centers give hydrocortisone four times per day with 35% at 6 a.m., 25% at lunchtime, 10% late afternoon and 30% at midnight. Exogenous corticosteroid will suppress ACTH and reduce adrenal production of androgens and prevent formation of adrenal rest tumors. Fludrocortisone should be given to control salt loss. It must be remembered that this steroid is a potent glucocorticoid itself and the dose should not exceed 0.15 mg/M² to a maximum of 0.1 mg. Sodium supplements may be needed at times in infancy (for more detail see Chapter 3.3.2).

Compliance is best assessed in children in the long term by following growth (Sellick et al., 2018; Appan et al., 1989). Height and weight should be followed at 3 monthly intervals in the first 2 years of life then at 6 monthly visits. Bone age is checked yearly from X-rays of the wrist and hands. The replacement therapy should be adjusted according to body size. Bone mineral density in adult patients with CAH can be lower than normal probably due to over treatment (Rangaswamaiah et al., 2020).

Electrolytes can be measured periodically but for long-term assessment of mineralocorticoid replacement the measurement of plasma renin activity is advisable. The measurement of 17-OHP and androgens in blood (or saliva) taken at regular intervals will define the adrenal steroid output in relation to treatment (Pussard et al., 2020) but in practice has little effect on a patients drug taking habits.

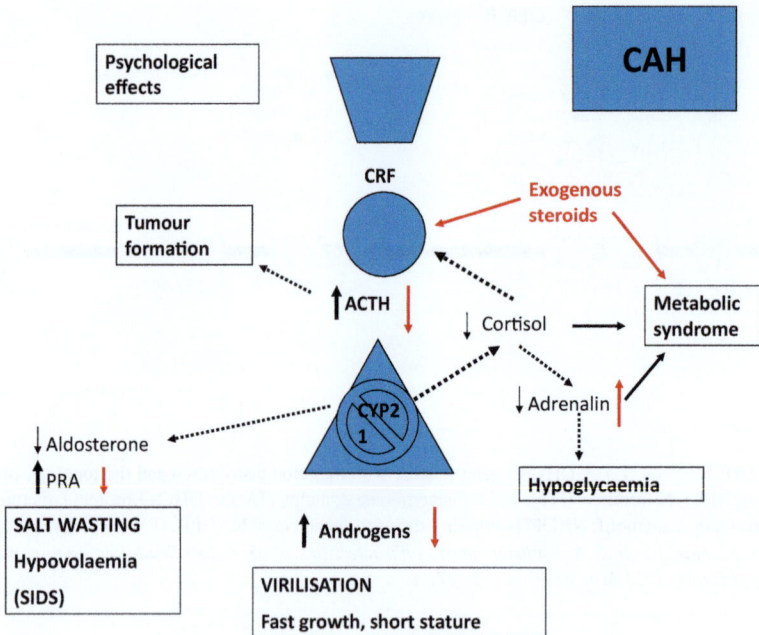

FIG. 3.2.1.25 Effects of CAH due to 21-hydroxylase deficiency requiring treatment. *(Author original.)*

Androstenedione measurements may be helpful in the management of CAH due to 21-hydroxylase and 11β-hydroxylase defects. Patients with 21-hydroxylase and some with 3β-hydroxysteroid dehydrogenase defects manifest elevated PRA while in the defects with mineralocorticoid excess (11β-hydroxylase) PRA is suppressed. PRA is normalized with effective treatment. In the case of 21-hydroxylase deficiency and 3β-hydroxysteroid dehydrogenase defects, treatment is improved with addition of fludrocortisone. A 24-h plasma cortisol and 17-hydroxyprogesterone profiles from hourly blood samples are helpful in getting the hydrocortisone matched to a normal cortisol circadian rhythm. In 4 groups of patients treated with prednisolone at 1–2.5, 2.5–5, 5–7.5, and 7.5–15 mg/day, a number of metabolic parameters have been studied in plasma using HILIC interaction chromatography with high-resolution mass spectrometry (Alwashih et al., 2017). From 382 metabolites studied, 24 metabolites of free fatty acids, bile acids and amino acids were significantly different between treated groups. A combination of seven metabolites readily discriminated replacement dose effects.

There are several tests that assess effectiveness of therapy for CAH. Auxological data have been thought to be the gold standard over the long term. Among the biochemical tests urinary pregnanetriol is a marker over the short term. Ranges of 2.3–3.3 mg/g creatinine (Izawa et al., 2008) and 1.2–2.1 mg per m²/day have been proposed. Morning and night-time plasma or saliva 17-OHP and/or androstenedione have been used. It is important not only measure 17-OHP but also cortisol since some treatment

practices leave the patient with hours of adrenal insufficiency (Hindmarsh and Honour, 2020).

The Wudy laboratory established target values for urinary steroid excretion rates based on growth rate (Kamrath et al., 2019 then devised four unique metabotypes based on urinary steroid excretion patterns of treated CAH patients. Metabotype 1 ($N = 21$ (19%)) revealed adequate metabolic control with low cortisol metabolites and suppressed androgen and 17α-hydroxyprogesterone (17OHP) metabolites. Metabotype 2 ($N = 23$ (21%)) showed overtreatment consisting of a constellation of elevated urinary cortisol metabolites and low metabolites of androgens and 17OHP. Metabotype 3 ($N = 32$ (29%)) demonstrated under-treated patients with low cortisol metabolites and elevated metabolites of androgens and 17OHP. Metabotype 4 ($N = 33$ (30%)) presented patients with treatment failure reflected by unsuppressed androgen- and 17OHP metabolites despite elevated urinary cortisol metabolites (Kamrath et al., 2020).

New treatment regimens are being tested (Schröder and Claahsen-van der Grinten, 2022). Subcutaneous delivery of HC via a programmed pump achieved significant reduction in adrenal androgens in eight adults with CAH and improved quality of life with less fatigue (Bryan et al., 2009; Nella et al., 2016). Pump management is complex but the approach works with motivated patients. An early trial with a once a day, modified release oral HC preparation (Chronocort, Diurnal, Cardiiff, UK) decreased adrenal precursors (Mallappa et al., 2015). Chronocort can be taken at night, then after a delay in release of the HC an early morning peak of HC can be achieved. Subsequent Phase 3 trial apparently failed to demonstrate superior treatment to standard HC treatment and this potential new treatment is on hold. In Europe, a different modified release oral HC (Plenadren, Shire, London, UK) is approved for use in adrenal insufficiency but not yet for CAH. Plenadren has an immediate release outer layer coating and an extended release core.

A newly recognized issue with CAH is that the lack of cortisol synthesis will influence the activity of enzymes in the adrenal medulla (Merke et al., 2000; Weise et al., 2004b; Riepe et al., 2007; Kim et al., 2014) such that there will be **epinephrine deficiency** which may contribute to hypoglycemia during febrile illness and impair the response to stress. Replacement of epinephrine has not yet been reported.

Testicular adrenal rest tumors (TARTs) are rare and benign tumors of the testis often seen secondary to CAH. Ultrasound provides diagnostic information. The tumors are found in young (>4 years) patients with CAH (Kim et al., 2019) and common in a survey of CAH patients aged 3–23 years (Werneck et al., 2019). The tumors are likely the outcome of poor treatment of the CAH and should be considered at any age (Mendes-Dos-Santos et al., 2018). A PET/CT scan after 11C-metomidate is a sensitive method for detecting TARTS (Burman et al., 2021).

The use of DEX during pregnancy to prevent or diminish the risk of virilization of the young girl is now not recommended because of the potential long-term effects of such molecules on brain function or metabolism (Bachelot et al., 2017).

TABLE 3.2.1.2 Steroid concentrations in patients with nonclassic 21-hydroxylase deficiency (NC21OHD) and controls (median and interquartile range) (Ueland et al., 2022).

Steroid (nmol/L)	NC21OHD	Control	Ratio of medians Patient/ controls
17OHP*	1.1. (0.5–2.7)	1.6 (0.8–4.3)	6.5
16OHP*	2.4 (1.4–4.9)	0.6 (0.4–0.9)	4.1
21DF*	1.8 (1.3–5.3)	0 (0–0.3)	
Prog	0.8 (0.5–1.4)	0.4 (0.3–0.8)	1.8
Cortisol	202.5 (125.9–442.8)	249.1 (208.9–373.7)	0.8
Corticosterone*	1.6 (0.8–3.5)	5.9 (3.5–13.6)	0.3
A	4.3 (2.0–8.1)	3.9 (2.2–8.5)	1.1
T	1.1 (0.5–2.7)	1.6 (0.8–4.3)	0.7
11-OHA*	9.6 (5.1–25.9)	4.5 (2.9–7.1)	2.1
11KA*	1.1 (0.7–1.7)	0.6 (0.4–0.8)	1.7
11OHT*	1.0 (0.3–1.7)	0.4 0.3–0.6)	2.2
11KT*	1.8 (1.2–3.3)	0.9 (0.6–1.3)	2.0

*$P < .0001$.

3.2.1.2.2.3 Nonclassical CAH (NCCAH)

Genital virilization is not a feature of NCCAH patients and the earliest time for concerns of hyperandrogenism is usually from the age of 5 years (Turcu et al., 2020a; Dörr et al., 2020; Kurtoğlu and Hatipoğlu, 2017; Lin-Su et al., 2008). Premature pubarche (appearance of pubic hair) is the first issue although other signs include, but not limited to, increased penile length, clitoromegaly, increased bone age, axillary apocrine odor, acne. Adolescents and adults females can present with hirsutism and menstrual cycle changes.

A mildly elevated concentration of 17-hydroxyprogesterone is an index of nonclassical CAH due to defect of CYP21A1 in the adrenal cortex (Turcu et al., 2020b; Papadakis et al., 2019) (Table 3.2.1.2).

The serum concentrations by LC-MS/MS of 16-hydroxyprogesterone, 21-deoycortisol and 11-oxygenated androgens were also significantly raised. The results for 17-hydroxyprogesterone need to be judged against reference ranges for the method in the laboratory. Ideally the 17-OHP assay should include mass spectrometry (Honour, 2014).

An ACTH stimulation test is an important test for NCCAH from other causes—CCAH and tumors of adrenals or gonads. The responses to the ACTH of 17-OHP, DHEA and 11-deoxycortisol are required in order to detect NCCAH of CYP21, HSD3B2 and CYP11B1, respectively (Table 3.2.1.3) (Costa-Barbosa et al., 2010). The addition of 21-deoxycortisol has been shown by others to improve detection of carriers (Kulle et al., 2015). There are distinct differences in clinical presentation and the response of steroids to ACTH according geographical location. It can be difficult to distinguish PCOS from NCCAH. Specific genetic mutations have been assigned to NCCAH for CYP21A and CYP11B but not HSD3B2 and that remains unclear.

A synacthen test may be ordered to validate the result. In a group of PCOS with immunoassay basal 17-OHP above 2 ng/mL (6 nmol/L), nonclassic CAH was detected with

TABLE. 3.2.1.3 PCOS.

NIH definition 1990	Rotterdam definition 2004
Less inclusive 1 and 2 needs to be met	More inclusive 2 of 3 need to be met

1. Hyperandrogenism
   - Clinical (hirsutism, acne, frontal balding)
   - biochemical (high serum androgen concentrations
	1. Hyperandrogenism
	   - Clinical or biochemical

2. Menstrual irregularity
   - Chronic anovulation
   - Oligomeorrhea >35 days
	2. Menstrual irregularity

	3. Polycystic ovaries*

For both: Exclude other causes (hyperprolactinemia, nonclassic CAH, thyroid disorder).
Key difference from NIH.

high degree of specificity when 21-deoxycortisol by LC-MS/MS was >0.087 ng/mL (>0.26 nmol/L) (Oriolo et al., 2020). The existence of a nonclassic form of 3β-hydroxysteroid dehydrogenase (HSD3B2) is controversial (Al Alawi et al., 2019; Mermejo et al., 2005; Pang et al., 1983).

The ovarian steroid response to GnRH after naferelin shows elevated levels of delta 4 steroids androstenedione and 17-OH suggestive of dysregulation of CYP17 (Barnes et al., 1993). Late onset 11-hydroxylase deficiency is incredibly rare, recognized by mildly elevated levels of cortisol precursors (deoxycorticosterone and 11-deoxycortisol) (Reisch et al., 2013; Mooij et al., 2015; Joehrer et al., 1997).

3.2.1.2.2.4   Genetics of NC_CAH

CYP21 A mutation at V281L: Val-281Leu (p.(Val282Leu))

This pathogenic variant results in an enzyme with 50% of normal activity when 17-OHP is the substrate but only 20% of normal activity for progesterone. This variant occurs in the majority of patients with nonclassic 21-hydroxylase deficiency who carry the HLA haplotype B14; DR1, which is likely a founder effect. Overall, about 70% of all nonclassic alleles carry the V281L pathogenic variant.

CYP21A mutation at P30L: Pro-30Leu (p.(Pro31Leu))

The enzyme from this pathogenic variant has 20%–60% of normal activity when expressed in cultured cells. Patients carrying this pathogenic variant tend to have more severe signs of androgen excess than patients carrying the more common nonclassic pathogenic variant V281L (p. (Val282Leu)). This pathogenic variant is found in

approximately one-sixth of alleles in patients with nonclassic disease although it may comprise a higher percentage in Japan. The pathogenic variant **I230T is** responsible for a NC CAH form unless associated with the V281L pathogenic variant that results in a more severe phenotype.

CYP11B1 mutation c449C > T

This mutation puts a change of pSer150Leu with reduced activity of $19.2 \pm 1.4\%$ (mean $\pm$ S.D.) and $14.7 \pm 0.5\%$ (mean $\pm$ S.D.) for the conversion to B and F compared with WT, respectively (Polat et al., 2014).

Hydrocortisone treatment of NCCAH should reduce androgen effects (acne, hirsutism, regular menstrual cycles) depending on the age of the patient. In the case of severe hirsutism or acne, the prescription of an oral contraceptive and or antiandrogens may be needed (Livadas and Bothou, 2019).

3.2.1.2.3   Other causes

Virilization of the genitalia has been attributed to **maternal ingestion of progestogens and androgens**. This is now less likely than in the past because most of the prescription drugs of these types (e.g., Danazol) (Rosa, 1984) are now much less potent and are discontinued in pregnancy. Other causes of a virilized female at birth are the result of exposure to androgens from an **adrenal or ovarian tumor in the mother** (Morris et al., 2011; Bustamante et al., 2017; Blake et al., 2014). In these cases, the child will have normal endocrinology after birth but need corrective surgery on the external genitalia. The mother will need assessment of the tumor, initially by imaging of the abdomen and androgen measurements most likely leading to surgery. Virilization of the fetus was reported in one case attributed to increased production of 11-oxygenated C_{19} steroids by an adrenal tumor in the mother (Nagasaki et al., 2020).

Clitoral hypertrophy in premature or low birth weight infants may suggest the diagnosis of congenital adrenal hyperplasia (Midgley et al., 1990; Greaves et al., 2004; Heckmann et al., 2006; Greaves et al., 2008). The clinical appearance (Fig. 3.2.1.26) may reflect the relative underdevelopment of the labia. Some cases have required surgical clitoral reduction.

Hyponatremia can be due to low sodium intake or renal loss due to tubular immaturity. In very immature preterm babies (less than 30 weeks gestation), virilization has been attributed to exposure to androgens (DHA) due to persistence of the fetal adrenal zone for months after birth and the problem often resolves as the infant gains weight.

A girl with **aromatase deficiency** may have abnormal genitalia with Prader 3 to 4 stage virilization (Jones et al.,

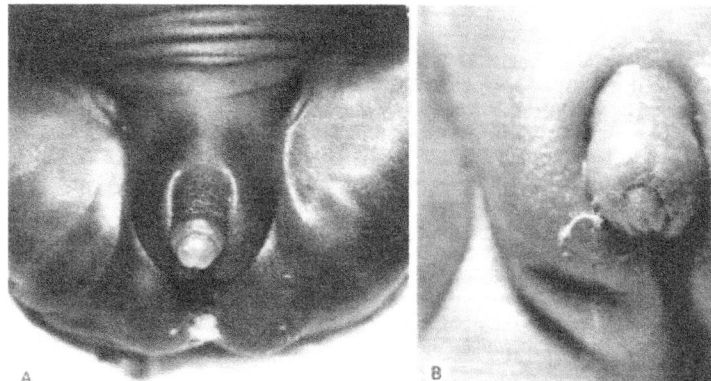

FIG. 3.2.1.26 Clitoromegaly in infant 1 month of age, born preterm at 24 weeks gestation. *(From Midgley PC, Azzopardi D, Oates N, Shaw JC, Honour JW. Virilisation of female preterm infants. Arch Dis Child. 1990;65(7 Spec No):701–3. Fig. 1 p. 702.)*

2007a) but many cases are raised as female (Özen et al., 2020). At birth, serum estrogens are very low, androstenedione and FSH concentrations are significantly raised. Virilization of the mother during pregnancy may be an indicator for **aromatase** deficiency in the fetus (Alsaleem et al., 2019). Men with the disorder become very tall with eunochoid proportions, delayed epiphyaeal closure and osteoporosis. Few males with mutations in the **estrogen receptor gene** seem to have been reported since the original case (Smith et al., 1994).

Another external causes of a virilized female are attributed to **ovotesticular DSD** (Melardi et al., 2020; Paula et al., 2015; Maciel-Guerra et al., 2008). This is a true state of hermaphroditism, a term no-longer used. There are a number of combinations of testis and ovary in terms of lateralization, number and size of the gonads (Deng et al., 2019). Variants in the sequence of the A-box of **SF-1** gene have been found. A basic arginine residue (Arg, R) in wild-type NR5A1 (left panel) is replaced by a neutral tryptophan residue (Trp, W) that has a bulky indole side chain (Bashamboo and McElreavey, 2016).

N.B. Sulfatase deficiency and aromatase deficiency are examples where low estrogen in pregnancy provide a clue to disorders of the fetus. Estrogens are these days rarely measured because ultrasound is used to detect many relevant developmental disorders of the fetus. Smith-Lemli-Opitz is another example although the method for measuring estradiol or estriol would need to characterize the 7-dehydro steroids.

3.2.1.2.4 Approach to 46XX neonate, ambiguous genitalia

The diagnosis of a DSD in newborn infants requires a systematic approach. Many algorithms have been constructed. An approach based largely on biochemistry is shown in Fig. 3.2.1.27. After a karyotype, 17-OHP is measured in blood to suggest or exclude the commonest CAH due to 21-hydroxylase deficiency. A confirmatory test is required.

Over the first week or so, the child may have begun to show signs of adrenal insufficiency, measurements of electrolytes, cortisol, renin and aldosterone may confirm a salt-losing state with low cortisol and aldosterone. The results of a urine steroid profile or plasma steroid panel will confirm 21-hydroxylase or may reflect defects of HSD3B2, CP11B1 or POR. If CAH is then unlikely investigations will be needed for maternal androgen excess, aromatase deficiency and Y remnants.

3.2.1.2.5 Androgen excess in childhood

Acne and advanced growth rate in childhood need to be investigated as signs of early puberty or adrenal disease. The main sign to suspect the onset of puberty is breast tissue growth in girls and testicular enlargement above 4 mL in boys (Cheuiche et al., 2021). Early appearance of pubic hair (before 6 years in girls and 8 years in boys) is called **premature adrenarche** and usually thought to be benign (Novello and Speiser, 2018) but needs to be distinguished from more sinister problems (such as a tumor).

During childhood boys with precocious puberty may have:

- space occupying lesions of the brain,
- CAH or
- very rarely tumors of the adrenals or gonads.

Late presentation of CAH may include an adrenal crisis under stress and signs of androgen excess particularly with boys. A few girls with hirsutism and/or acne, with or without menstrual disturbance, may have nonclassic congenital adrenal hyperplasia or very rarely have excess secretion of androgens of ovarian or adrenal origin due to tumors. Polycystic ovaries are commonly found in mild hyperandrogenization in adolescence that may progress to infertility in adult life.

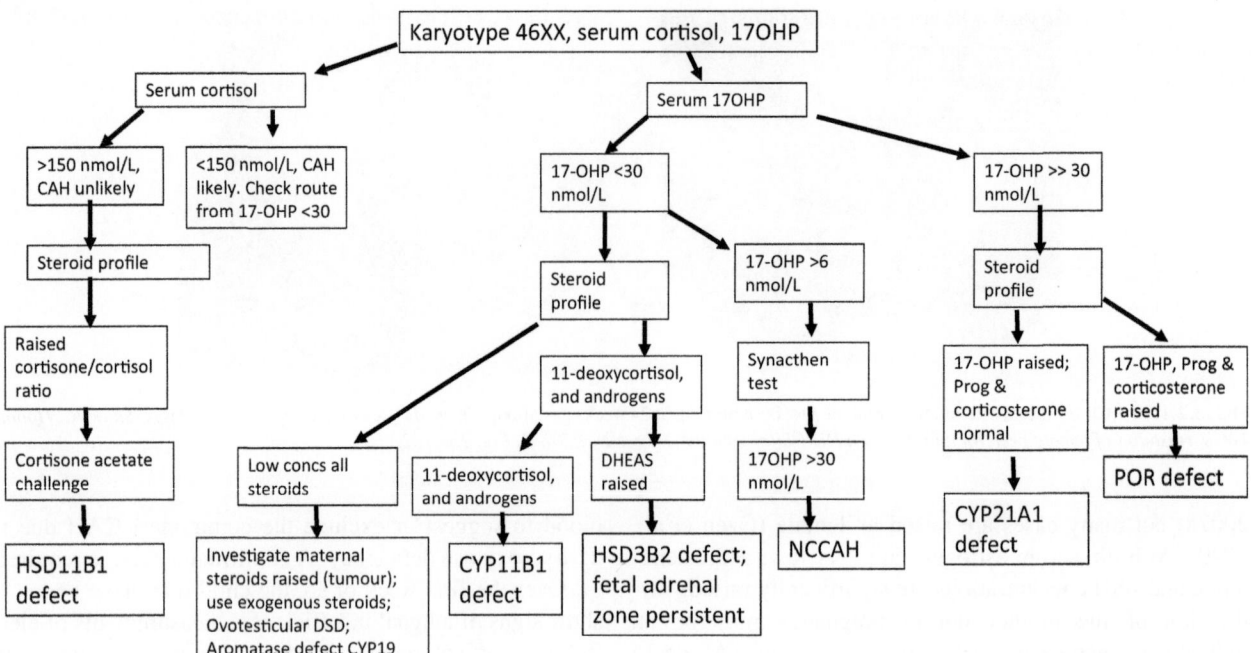

FIG. 3.2.1.27 Biochemical approach to diagnosis new born female with ambiguous genitalia. Serum cortisol and 17-hydroxyprogesterone are first steps since CAH due to 21-hydroxylase is the commonest problem. A urine or plasma steroid profile will look for steroids in excess as markers of other forms of CAH. *(Author original.)*

FIG. 3.2.1.28 Adrenarche: Plots for adrenarche patients of andro sterone plus aetiocholanolone (micrograms per day) for age against reference range. *(Author original.)*

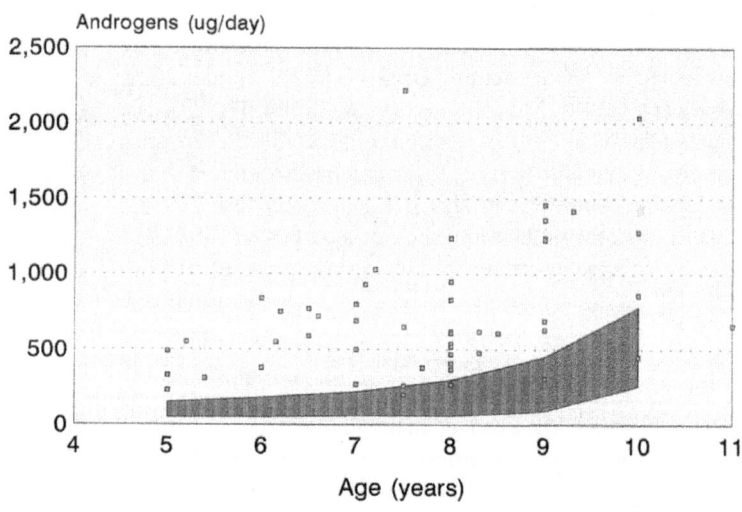

Investigations include a steroid profile in blood or urine to look for excess precursors and androgen metabolites. Increased outputs of cortisol and androgens are indicative of glucocorticoid resistance, and if the output is not suppressible with dexamethasone confirming that hypothesis. Pubic hair growth before breast development or testicular enlargement may be the outcome of increased secretion of DHEA and DHEAS from the adrenal cortex due to early differentiation of the zona reticularis. This is a benign disorder called *premature adrenarche* not requiring treatment. DHEA and DHEAS concentrations in serum should be interpreted against normal ranges for age. Testosterone may be slightly elevated for age due to peripheral conversion of the adrenal androgens. The 24-h urine excretion of androsterone, etiocholanolone as well as cortisol metabolites are above the normal range for age and body size (Fig. 3.2.1.28) reflecting advance adrenal growth.

Premature adrenarche is more common in Asian and Afro-Caribbean children than Caucasians. A state of glucocorticoid resistance explains some features of children with premature adrenarche (Panayiotopoulos et al., 2020), higher doses of dexamethasone (4 or more milligrams of dexamethasone) (Honour, unpublished) are required to suppress 24 h outputs of cortisol metabolites.

If a precursor is marginally elevated a Synacthen test should be performed to identify a nonclassic CAH with mild enzyme normalities. Abdominal CT or ultrasound should seek any tumors. Androgen analysis should include DHEAS, testosterone, androstenedione, DHT, 11-hydroxyandrostenedione, 11-deoxycortisol and pregnenolone in blood samples or their metabolites in urine.

3.2.1.2.6 Androgen excess around puberty

3.2.1.2.6.1 *Early puberty*

Early puberty can cause significant anxiety to families although in girls it may represent a variation of normal development. However, particularly in boys it can be the presentation of serious pathology. Children may be reluctant to disclose symptoms so parents need to be vigilant. Early puberty can present with significant behavior problems, psychological issues, reduced educational attainment and also result in **early epiphyseal closure** and short adult stature. It therefore warrants investigation and may need treatment to ameliorate social and psychological problems and optimize growth. Parents still need reassurance when the presentation actually represents a variation of normal development. Boys under 9 years or girls under 8 years presenting with signs of androgen activity (enlarged testes), and growth acceleration or girls under 8 years presenting with signs of estrogen activity (breast development) (Carel and Léger, 2008), virilization and growth acceleration should be investigated (Santi et al., 2021; Cheuiche et al., 2021; Farello et al., 2019; Sultan et al., 2018). Whether a brain MRI scan is needed is under debate (Canton et al., 2021) with opposition based on costs, sedation, contrast agents and incidental findings.

Early puberty can be divided into two groups (Fig. 3.2.1.29).

1. **gonadotrophin dependent, "central precocious puberty"** presenting with concordant clinical signs of puberty
   - boys: testicular enlargement followed by penile development and pubic hair;
   - girls: breast buds (thelarche) followed by later menstruation with pubic hair just before or after) or
2. **gonadotrophin independent "peripheral precocious puberty"** presenting with discordant signs of puberty
   - boys: penile development and pubic hair without testicular development and

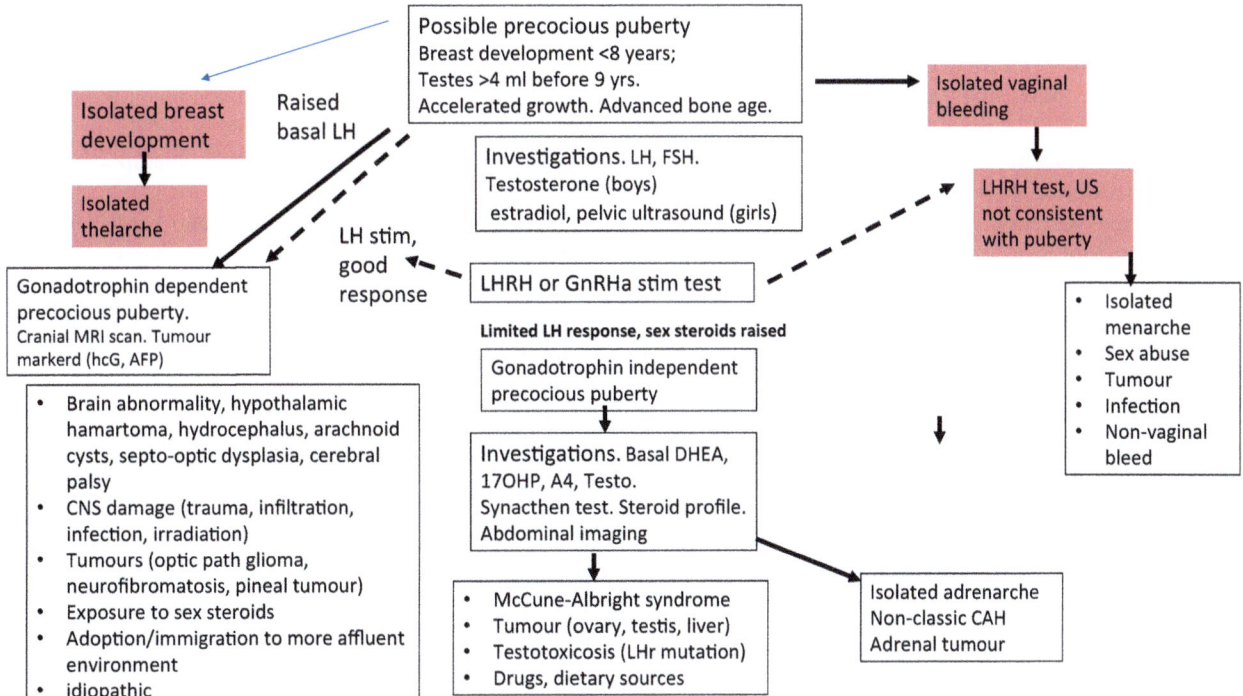

FIG. 3.2.1.29 Precocious puberty. A full clinical assessment with auxology and bone age determination is followed with endocrine measurements to establish if sex steroid production is gonadotrophin dependent or adrenal in origin. *(From Wei C, Davis N, Honour J, Crowne E. The investigation of children and adolescents with abnormalities of pubertal timing. Ann Clin Biochem 2017;54(1):20–32. Fig. 1 p. 21.)*

• girls: signs of virilization such as pubic hair without breast development, or marked virilization such as clitoromegaly, or menstruation without mature breast development).

Signs of puberty before 8 years in a girl or 9 years in a boy require assessment and are of particular significance in boys (Farello et al., 2019; Haddad and Eugster, 2019). All children with clinical evidence of early puberty need a full medical history to include factors such as family history (Durand et al., 2016), ethnicity and weight issues plus previous medical history (e.g., neurological condition or injury, previous medical conditions or exposure to sex steroids). The age at pubertal onset seems to decline with the passage of time and this should be borne in mind when assessing the younger patients (Perry et al., 2015). Hair loss in adolescents is associated with a strong family history (Griggs et al., 2019). Parents frequently comment on change in behavior with moodiness or tearfulness. Children need careful auxology (height, weight, pubertal staging) and full examination including neurological examination. Examination and pubertal staging with gonadotrophin measurements will indicate whether the clinical picture suggests gonadotrophin dependent (see Fig. 3.2.1.29) or gonadotrophin independent precocious puberty and thus points to the relevant investigations.

3.2.1.2.6.2 Gonadotrophin-dependent (central) precocious puberty

Gonadotrophin-dependent precocious puberty indicates central activation of the hypothalamic-pituitary-gonadal axis and can be either idiopathic (in the majority of girls but very rare in boys) or secondary to intracranial pathology or damage. A child with clinical precocious puberty who has pubertal gonadotrophin levels and augmented nocturnal gonadotrophin secretion has central precocious puberty. LH results should be treated with caution because of interference from heterophilic antibodies and high results may have to be confirmed using a second assay designed to avoid such interference (Segal et al., 2003). Assay formats can be modified to circumvent this problem (Andersson et al., 2003). Girls should have a bone age and LHRH test (Özalkak et al., 2020) and a pelvic ultrasound to confirm gonadotrophin dependent precocious puberty. Failure to find a pelvic mass by palpation or ultrasound scan reduces the likelihood of a rare ovarian tumor. If confirmed, a cranial MRI with hypothalamus/pituitary views to look for any intracranial pathology is indicated to exclude tumors in the hypothalamic-pituitary area. This is mandatory in boys and for the younger girls (<7 years) or those girls with any neurological symptoms or signs such as seizures. CNS tumors (hamartomas, third ventricular cysts, astrocytomas or gliomas) can be recognized by MRI. The most common brain lesion is a **hypothalamic hamartoma** which can

present as early as 12 months of age or later with short stature, breast development, pubic hair presence, maturation of sexual reproductive organs, deepening voice and acne (Alomari et al., 2020). Some of the tumors (**dysgerminomas, hepatoblastomas**) secrete **hCG so** this hormone should be measured in plasma (Huang et al., 2017). A cerebral tumor is relatively more common in boys with central precocious puberty than in girls. Exact mechanisms are not known.

If the clinical picture is associated with poor growth this should alert the clinician to think of either **primary hypothyroidism** or additional hypothalamic/pituitary dysfunction with growth hormone deficiency Along with an increase in TSH there can be a concomitant increase in FSH and prolactin that presumably lead to sexual precocity (Riaz et al., 2020; Salerno et al., 2016).

At least four monogenic causes of CPP have been reported (Roberts and Kaiser, 2020; Gohil and Eugster, 2020). Activating mutations of Kisspeptin (Kiss1) and its receptor Kiss1R) have been found but rarely. Loss of function mutations of **Makorin ring finger protein 3 (MKRN3)** are found in CPP with early puberty (Ge et al., 2020; Maione et al., 2020; Li et al., 2021 Naulé and Kaiser, 2019; Abreu et al., 2015). The MKRN3 is inhibitor of puberty initiation, maternally inherited and paternally expressed gene so the phenotype is inherited for the father. Studies in MKRN3 knockout mice indicate mechanisms for epigenetic silencing of GnRH expression (Li et al., 2021). Loss of function of the **delta-like homolog 1 gene (DLK1)** with similar inheritance to MKRN3 and only females who inherited the mutation from their father had CPP. The DLK1 mutations are probably a rare cause of CPP. The precise role of DLK1 in pubertal onset is unclear other than acting as a preadipocyte differentiation inhibitor.

Missense mutations in NOTCH2 and HerC2 found by exome sequencing are new candidate genes with roles in pubertal development (Lee et al., 2020b). More studies will be necessary to confirm or refute the role of the Notch signaling pathway in pubertal development. Notch signaling may promote gonadotropin differentiation and function.

3.2.1.2.6.3 Management of gonadotrophin dependent (central) precocious puberty

Management of early puberty involves first identification and management of any underlying pathology then careful explanation about the normal timing and sequence of events, reassurance and practical advice for parents (Kang et al., 2016). This is key in all cases as early puberty can cause significant distress at home and issues at school (see information leaflets http://www.bsped.org.uk). In those who are approaching the normal age range for puberty without underlying pathology, for instance girls over 7 years, explanation and advice is the key part of management. Parents may still wish to explore options to stop puberty in older girls, particularly if there are other issues such as neurological handicap or learning difficulties or marked psychological factors.

For younger girls and boys and those with underlying pathology causing early puberty and requiring management, LHRH analogue treatment can be used to suppress pituitary gonadotrophin pulsatility and is given as a depot injection every 28 days or 12 weeks and should result in suppression of signs of puberty (although they may not resolve altogether), slower growth and avoidance of mood swings. Treatment can be monitored clinically (growth rate and signs of puberty) and biochemically (once treatment is established a repeat LHRH test prior to an injection will confirm gonadotrophins are suppressed). Dose interval may need to be reduced for older, larger children and girls whose puberty may be harder to suppress than boys. Compliance with regular treatment schedule is essential as delayed injections can result in paradoxical stimulation. Treatment can continue until an appropriate age for puberty and should optimize remaining growth potential. In older girls, there is little published evidence that treatment improves final height and so indications for treatment to suppress puberty relates to psychosocial concerns outweighing the impact of treatment. Medroxyprogesterone has an important place in managing PP when GnRHa therapy is too expensive (Kumar et al., 2015).

3.2.1.2.6.4 Gonadotrophin independent (peripheral) precocious puberty

Gonadotrophin independent precocious puberty indicates sex hormone production from the adrenals, gonads or even a tumor that is not directly under pituitary control and is pathological for instance **late onset CAH**, **adrenal tumors**, **McCune-Albright syndrome** or **testotoxicosis**. One striking feature of the McCune-Albright syndrome at the clinical examination of the patient is the cafe-au-lait skin markings on head or body (Anderson, 2020; Tufano et al., 2020) (Fig. 3.2.1.30).

The main differential diagnoses are with variations of normal pubertal development, i.e., adrenarche (the onset of adrenal androgen production) (Idkowiak et al., 2019). Boys may have acne, increased body odor and behavioral problems and may become taller than their peers. As a result of rapid skeletal maturation there is early epiphyseal fusion and the outcome is for short stature in adulthood.

The more common scenario is presentations with signs of androgen excess and the main differential is between adrenarche and late onset CAH and with rare instances of

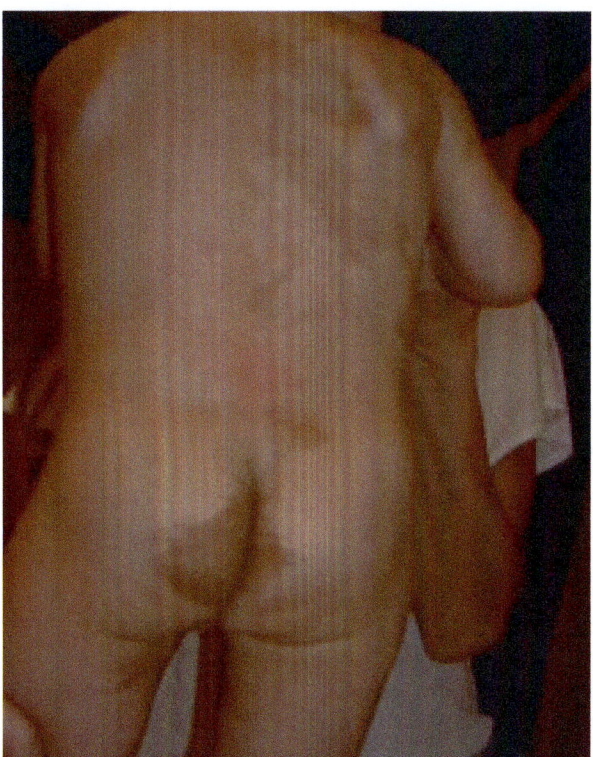

FIG. 3.2.1.30 Coast of Maine cafe-au-lait macules. Irregularly bordered cafe-au-lait macules that tend not to cross the midline are characteristic of McCune-Albright syndrome. *(From Anderson S. Café au Lait macules and associated genetic syndromes. J Pediatr Health Care 2020;34(1):71–81. Fig. 2 p 73.)*

adrenal tumors. Children with CAH are likely to have a longer history of virilization, whereas the rare **adrenal tumors** are likely to have a shorter and more rapidly progressing history of virilization. Clinical and growth assessment, bone age and then measurement of blood androgen levels or a urine steroid profile are required (Wolthers et al., 1999; Honour et al., 1984). If CAH or a tumor is identified further assessment of adrenal function will be necessary (plasma renin and synacthen test) and imaging is required if a tumor is suspected.

3.2.1.2.6.5 Rare causes of gonadotrophin independent (peripheral) precocious puberty

Tumors can be secreting sex steroids from gonadal, adrenal or liver tumors or **HCG** from pineal gland, liver or testes. There will be clinical evidence of sex steroid action but suppressed LH or FSH response in an LHRH test. The most common pediatric adrenal tumor reported in the literature secretes **DHEAS**. In general, the tumors have been quite large. A number of cases of tumors secreting other androgens have been reported and using urine steroid profile analysis the secretion of **11β-hydroxyandrostenedione** has been defined on the basis of high excretion of 11β-hydroxyandrosterone (Małunowicz et al.,

1995; Wolthers et al., 1999; Honour et al., 1984). Those reports predate by several decades a recent interest in 11-oxygenated androgens. In those reported cases, androgen excretion was not grossly elevated and without scanning of the adrenals a tumor may have been dismissed. FSH and LH are suppressed to within prepubertal ranges. The secretion of the androgens by adrenal tumors is not suppressed by giving dexamethasone. This is an important test for autonomous secretion. A testicular mass with grossly elevated 17-OH-P usually indicates **CAH with adrenal rests** in the testes (testicular adrenal rest tumors TARTS). **Leydig cell tumors** however may also produce elevated 17-OHP which is not suppressed with dexamethasone. 21-Deoxycortisol is not raised in these cases unlike the situation with CAH.

A few recent reports of premature pubarche describe children with degrees of hyperandrogenism and short stature who had low plasma DHEAS due to defect of the sulfotransferase SULT2A1 through a defect in sulfate delivery from PAPS through *inactivating mutations in PAPSS2* (Eltan et al., 2019; Mueller et al., 2018; Oostdijk et al., 2015; Noordam et al., 2009). Plasma DHEA/DHEAS ratio can be less than 20 (normal range 31–345).

Acne, hirsutism, menstrual disturbance can be due to *nonclassical forms of CAH* and a mild defect of the steroid 21-hydroxylase. In many such patients, the basal 17-hydroxyprogesterone concentrations are above the normal range (>5 nmol/L). In the plasma steroid profile, the excess of 17-OHP is clear in an unstimulated state along with elevation of 16-hydroxyprogesterone, 21-deoxycortisol (earlier Table 3.2.1.2).

An injection of ACTH will lead to an increase in serum 17-OHP at 30 and 60 min after the trophic hormone injection and affected subjects will have an increment in 17-OHP greater than seen in normal subjects. Fifty percent of patients with nonclassic or late onset form of CAH have HLA B14, DR1 in association since this is coded nearby on the short arm of chromosome 6 and specific mutations in exon 1 or exon 7 CYP21A2 (V281L; P453S).

The nonclassic form of the 21-hydroxylase disorder is due to mutations in the gene producing a protein with less than 20% activity. A late onset form of 3β-hydroxysteroid dehydrogenase deficiency was suggested on the basis of exaggerated DHEA or 17-hydroxypregnenolone response to ACTH but mutation analysis has not supported this etiology (Pang, 1998; Sakkal-Alkaddour et al., 1996). Boys can present with severe acne and growth acceleration due to late onset 11-hydroxylase (Wang et al., 2018; Reisch et al., 2013).

Tumor markers (beta-HCG, alphafetoprotein) may be raised. Imaging of the likely source site is required. Treatment is of the underlying tumor and so will involve referral to the appropriate specialist multidisciplinary team.

Endocrine follow-up will be required, however, because there is also always the risk of a central precocious puberty subsequently following exposure to sex steroids or thyroid hormones. Germinomas spread suddenly and rapidly along the optic tracts leading to irreversible, profound visual loss, and in rare circumstances, instead of leading to a panhypopituitary state, cause a paradoxical sexual precocity due to secretion of human chorionic gonadotrophin, alongside deficiency of other pituitary hormones, save for prolactin, which may be moderately raised due to dopamine deficiency (Goswami et al., 2013).

Plasma gonadotrophin concentrations are often pubertal but may in some cases be prepubertal. Symptoms may wax and wane and on ultrasound this can be attributed to the appearance and regression of unilateral ovarian cysts. This condition is caused by a postzygotic activating missense mutations in the **α-subunit gene of the stimulatory G protein (GNAS1)** that regulates cell function by coupling hormone and other receptors to adenylyl cyclase.

Hyperpigmented birthmarks (cafe-au-lait macules) can be associated with a number of problems including neurofibromatosis type 1 and rare ring chromosome syndromes (Anderson, 2020). Cafe-au-lait markings are seen in McCune-Albright syndrome with precocious puberty, polyostotic dysplasia of bone and other endocrine abnormalities (Tufano et al., 2020). Missense mutations in codon 201 of GNAS1 were first identified in **McCune-Albright syndrome** (Weinstein et al., 1991). Replacement of arginine by histidine at codon 201 of GNAS1 exon 8 is the most frequent mutation, but substitution by cysteine in this same codon may occur in a few cases. More rare mutations in codon 224 of exon 9 have also been described. These mutations result in longer activation of Gsα and stimulation of endocrine system. Intracellular levels of cyclic adenosine monophosphate (cAMP) in osteoprogenitor cells increase as a consequence of Gsα activation.

The mutations may be found in affected tissues, including the ovaries and bone lesions or peripheral lymphocytes (Cho et al., 2016). Estrogen production occurs from large ovarian cysts, independent of gonadotrophin action.

Testotoxicosis (familial male-limited precocious puberty) is due to an **activating mutation of the luteinizing hormone (LH/CGR) receptor**, leading to increased levels of sex steroids, with enlarging testes from 3 to 4 years of age in the context of low LH levels (Gurnurkar et al., 2021; Daussac et al., 2020; Qiao and Han, 2019). There may be a family history involving early puberty in males. Around 20 activating mutations of the LHCGR gene located at chromosome 2p21 region have been reported from different ethnic backgrounds. In some cases, the mutation is found in DNA from a testicular biopsy but not from blood (Daussac et al., 2020).

Medroxyprogesterone, ketoconzole and cyproterone have been used to reduce synthesis and action of testosterone.

3.2.1.2.6.6 *Management of Gonadotrophin Independent (peripheral) precocious puberty*

Management involves Identification and management of any underlying pathology—in CAH or tumor-related conditions this will address the over-production of sex steroids. Specific treatment of excess sex steroid production will be required, for example, in Testotoxicosis and McCune-Albright Syndrome and this requires drugs to block sex steroid biosynthesis. The family need careful explanation of what is a very complex area. Finally management of secondary central precocious puberty may also be required.

3.2.1.3 **Pseudo-precocious puberty**

Exposure to *exogenous sex steroids* can cause a pseudo-precocious puberty. Both contraceptive pill and androgenic/anabolic steroids (AAS) are reported offenders. AAS are taken not only to enhance performance in sport but also to improve male physique for social reasons.

3.2.1.3.1 **Sex reversal**

A change in direction of the phenotype is called sex reversal. This has been seen when children raised as female become masculinized (Fig. 3.2.1.31). This is seen where androgens that have failed to act in fetal life because of a

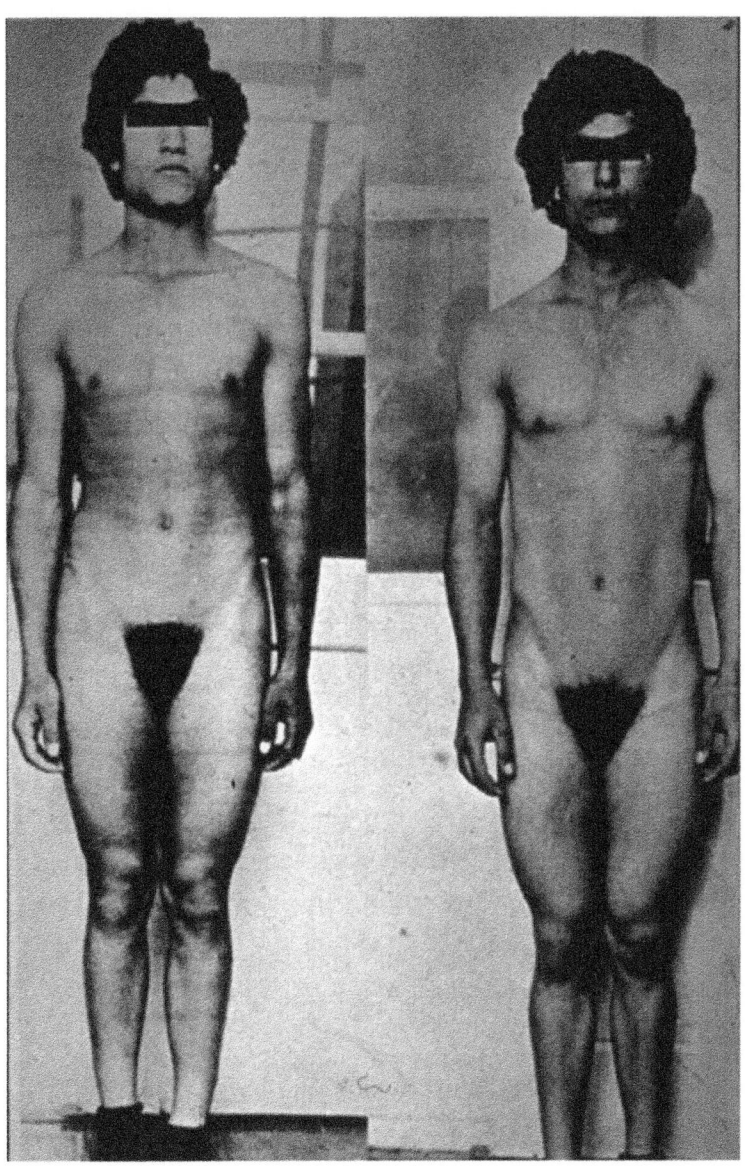

FIG. 3.2.1.31 Young adults raised as girls with SRD5A2 defect and masculinization at puberty.

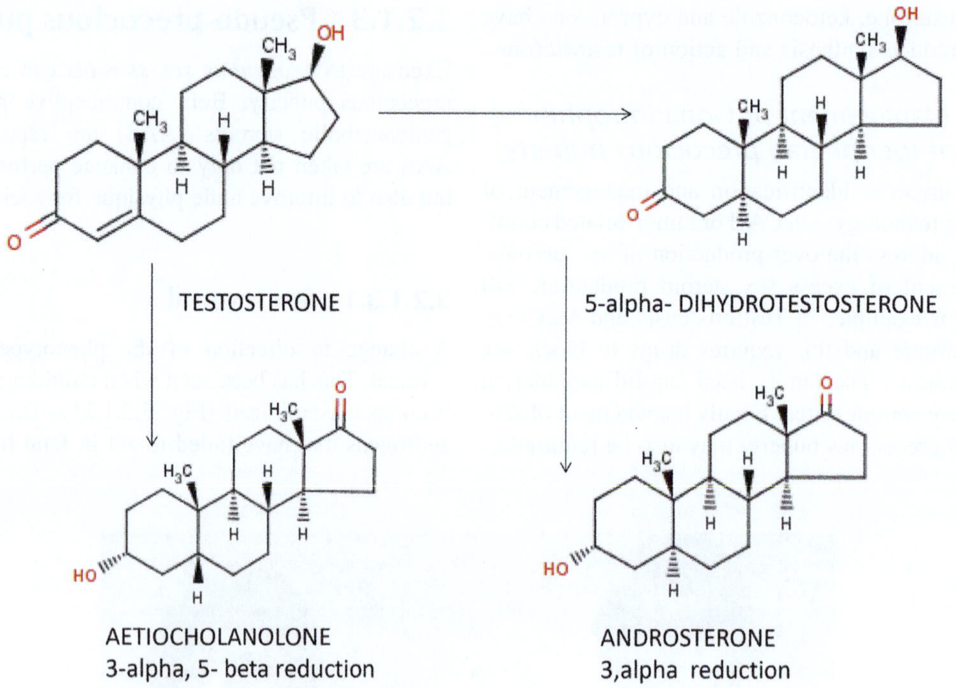

TESTOSTERONE

5-alpha- DIHYDROTESTOSTERONE

AETIOCHOLANOLONE
3-alpha, 5- beta reduction

ANDROSTERONE
3,alpha reduction

FIG. 3.2.1.32 T to DHT and metabolites. *(Author original.)*

failing in androgen synthesis but activation of the gonads approaching or at puberty has produced a significant degree of masculinization.

Two disorders of androgen synthesis have been described based on ratios of substrate to product for defective enzymes in androgen synthesis. Dihydrotestosterone production from testosterone requires **5-alpha reductase** (SRD5A2) and the ratio of T to DHT is above 3 with plasma determinations. An abnormally high ratio of 5β- to 5α-reduced androgens (etiocholanolone to androsterone) and cortisol (THF to allo-THF) are seen in urine. Children aged beyond 3 months of age only show the abnormality of cortisol metabolites because in the newborn period only cortisone metabolites are found (Berra et al., 2011). Many of these patients in an adult intersex clinic had been classified as having partial androgen insensitivity in years when there was little biochemical support (pre 1980).

3.2.1.3.1.1 5α-Reductase (SRD5A2) deficiency

Raised as girls, they virilize at a pubertal age with development of male muscularity and habitus, testicular descent and spermatogenesis, but have markedly reduced facial hair, acne and frontal hair recession. Male psychosexual characteristics develop after puberty through exposure to testosterone. In the Dominican Republic, a highly consanguineous community where this gene mutation is common, most affected people switch at puberty to a male gender role (Imperato-McGinley et al., 1974). The process was accepted as normal and because of the concomitant testicular descent was termed "guevedoces": eggs at twelve.

Although rare 5-alpha reductase deficiency seems to occur more frequently in consanguineous communities, there is failure of conversion of testosterone to dihydrotestosterone (Fig. 3.2.1.32) and external genital virilization is impaired (Fig. 3.2.1.33).

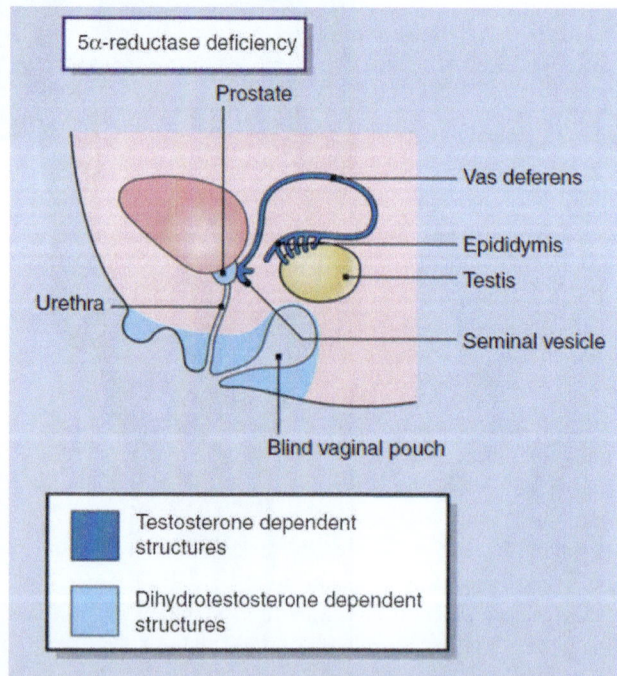

FIG. 3.2.1.33 Differentiation of male external genitalia through 5α-reductase activity converting testosterone to dihydrotestosterone. *(From Belchetz PE, Barth JH, Kaufman JM. Biochemical endocrinology of the hypogonadal male. Ann Clin Biochem 2010;47(Pt 6):503–15. Fig. 3 p. 506.)*

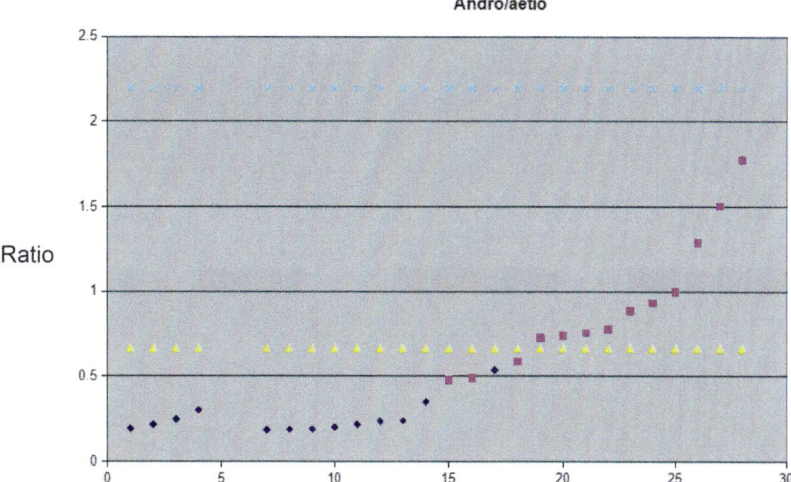

FIG. 3.2.1.34 Ratios of urinary androsterone to aetiocholanolone in a cohort of 29 adults with DSD and uncertain diagnosis (13 blue diamonds with SRD5A2 mutations; no SRD5A2 mutations). Normal range (low yellow diamonds; high blue crosses). The ratios of A/Ae in the patients are ordered according to the ascending values for each ratio SRD5A2 gene mutations. Four subjects raised as males with SRD5A2 deficiency are shown at the left followed by females with 46XY DSD. *(Author original figure.)*

In the first six months of life, blood samples for androgens can be taken because LH secretion is not suppressed. An HCG test can contribute to the diagnosis by demonstrating elevated testosterone to dihydrotestosterone ratio in plasma or an elevated ratio of 5β to 5α metabolites (etiocholanolone to androsterone) (Fig. 3.2.1.34) in urine.

The steroid profile may also show abnormalities in the distribution of cortisol metabolites (high ratio of tetrahydrocortisol to allo-tetrahydrocortisol) (data not shown). In newborns, however, the majority of cortisol metabolites reflect the strong oxidation to cortisone for which there is no isomeric pair of metabolites to measure.

A raised etiocholanolone to androsterone ratio is nevertheless useful for assessing the defect in suspected cases aged 1 month or older when the excretion of androgens is sufficient to make an assessment from the ratio (Fig. 3.2.1.34). The increased ratio of 5β- to 5α-reduced cortisol metabolites can be detected around 1 month of age if GC-MS or LC-MS/MS is used to look at the ratio for low 5α-THF to THF which are present at too low a concentration to be measured accurately by GC analysis alone. The ambiguity of genitalia may be so mild that the child may be reared as a female until puberty when there is some development of male musculature (not as marked as in 17-KSR deficiency) and psychosexual orientation reflecting the pubertal surge of testosterone.

Missense and nonsense mutations in the coding region of the gene encoding 5α-reductase type 2 have been identified as the cause of the disease and techniques are now available to screen for mutations (Fig. 3.2.1.35).

Increased ratios of 5α to 5β reduced androgens have been found in association with **Serkal syndrome** (sex reversal with dysgenesis of kidneys, adrenals and lungs) possibly reflecting increased 5α-reductase and loss of function of WNT4 signaling.

Defects in the 5α-reductase type 2 enzyme arise from mutations in the SRD5A2 gene. This gene is made up of five exons and four introns and allelic variants have been reported in the whole gene. Impairment in the 5α-reductase type 2 enzymatic activity results from either homozygous or compound heterozygous allelic variants. Initially, this disorder was reported in clusters around the world in individuals from specific ethnic groups. There is growing evidence reporting affected individuals with a variety of ethnic backgrounds and coming from several geographical areas, suggesting that 5α-reductase type 2 deficiency has a worldwide distribution (Batista and Mendonca, 2020). The 5α-reductase type 2 defects have been reported in individuals with several degrees of under-virilization, ranging from typical female external genitalia to hypospadias or isolated micropenis. The causes of divergent phenotypes are still unclear.

Mendonca found there were 70% of homozygous allelic variants and 30% compound heterozygous. Most were missense variants (76%). However, small indels (11%), splicing (5%) and large deletions (4%) were all reported. They were distributed along all exons with exon 1 (33%) and exon 4 (25%) predominance. Allelic variants in the exon 4 (NADPH-binding domain) resulted in lower virilization ($P < .0001$). The codons 55, 65, 196, 235, and 246 are hotspots making up 25% of all allelic variants. Most of them (76%) were located at conserved aa. However, allelic variants at nonconserved aa were more frequently indels (28% vs. 6%; $P < .01$).

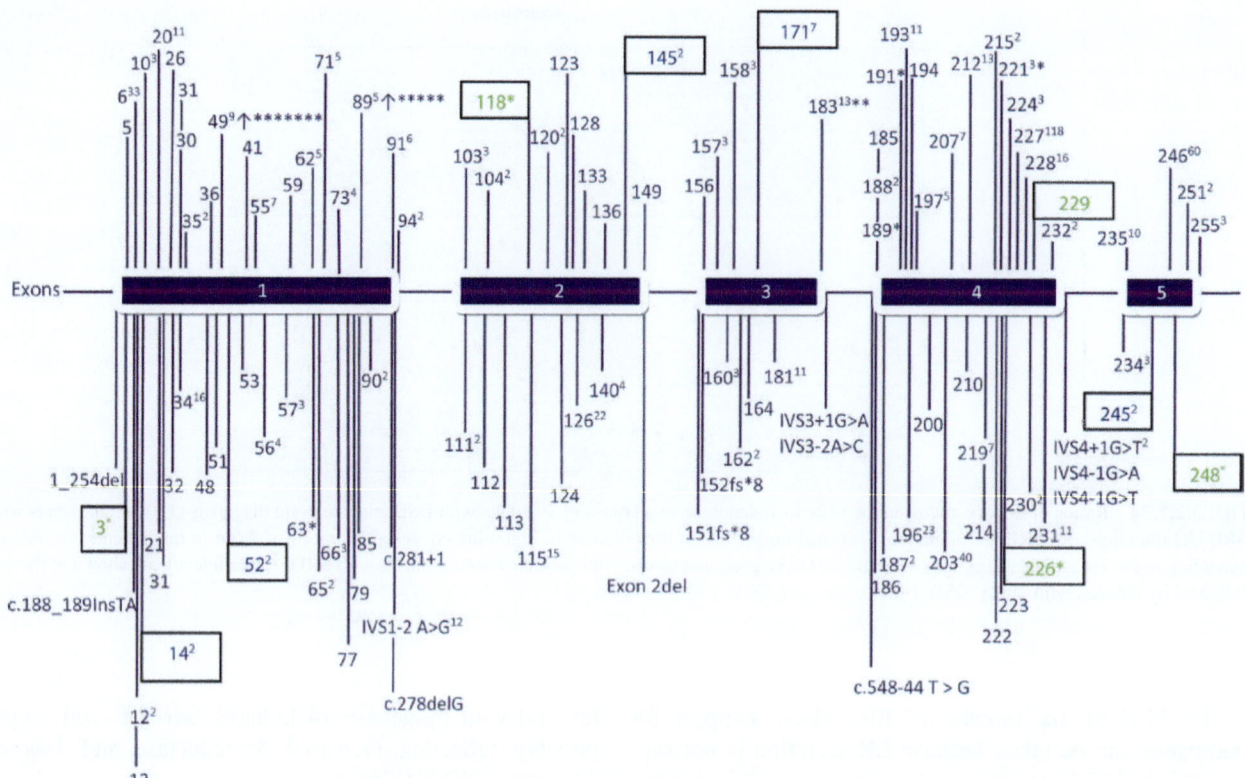

FIG. 3.2.1.35 Molecular identification of SRD5A2 gene mutations. Numbers indicate the amino acid substitutions reported in the literature and blue numbers (3, 14, 52, 118, 145, 171, 226, 245, and 248) show the mutations studied by kinetic analysis in this study. The green numbers (3, 118, 226, 229, and 248) indicates a novel substitutions associated with steroid 5α-reductase 2 deficiency. The superscript numbers denote the times that the amino acid has been reported with mutations. Upwards arrows designate the polymorphic variants (49 and 89) and asterisks indicate the codons that have possible associations with prostate cancer (63, 118, 183, 189, 191, 221, 226, and 248). *(From Ramos L, Vilchis F, Chávez B, Mares L. Mutational analysis of SRD5A2: from gene to functional kinetics in individuals with steroid 5α-reductase 2 deficiency. J Steroid Biochem Mol Biol 2020;200:105691. Fig. 6 p. 10.)*

3.2.1.3.1.2 17-Ketosteroid reductase (17β-hydroxysteroid dehydrogenase) (17HSD3) deficiency

Testosterone synthesis from androstenedione is catalyzed by **17-ketosteroid reductase** (HSD17B3) which when defective leads to low ratio of testosterone to androstenedione (Fig. 3.2.1.36).

In urine, a high ratio of androsterone to etiocholanolone is often seen which may reflect activation of the backdoor path to DHT. A deficiency of 17-βHSD3 is considered the commonest DSD due to a biosynthetic testosterone defect in some cultures (Mendonca et al., 2017), but it is very rare at least in European countries and China (Yang et al., 2017). A frequency of 1:147,000 was estimated in a Dutch study (Boehmer et al., 1999). In Italy, only 11 patients were collected in a multicentric study of the Italian DSD Study Group, which is an association connecting the main Italian centers involved in the care of DSD. The large dsd-LIFE study collecting data from 14 centers in six European countries (France, Germany, the Netherlands, Poland, Sweden, United Kingdom) also identified 11 patients with 17-βHSD3 deficiency (~0.4% of all 46,XY DSD group).

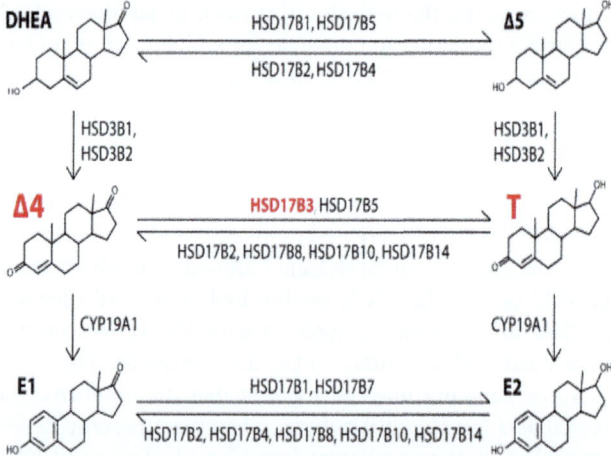

FIG. 3.2.1.36 Eleven human HSD17B isozymes control key steps of sex steroid synthesis. A total of 14 HSD17B isozymes have been identified in vertebrates, of which 11 are found in humans. *DHEA*, dehydroepiandrosterone; *Δ4*, androstenedione; *Δ5*, androstenediol; *T*, testosterone; *E1*, estrone; *E2*, estradiol; *HSD3B1 and HSD3B2*, 3β-hydroxysteroid dehydrogenase; *CYP19A1*, aromatase. *(From Khattab A, Yuen T, Yau M, Domenice S, Frade Costa EM, Diya K, Muhuri D, Pina CE, Nishi MY, Yang AC, de Mendonça BB, New MI. Pitfalls in hormonal diagnosis of 17-beta hydroxysteroid dehydrogenase III deficiency. J Pediatr Endocrinol Metab 2015;28(5–6):623–8. Fig. 1 p. 624.)*

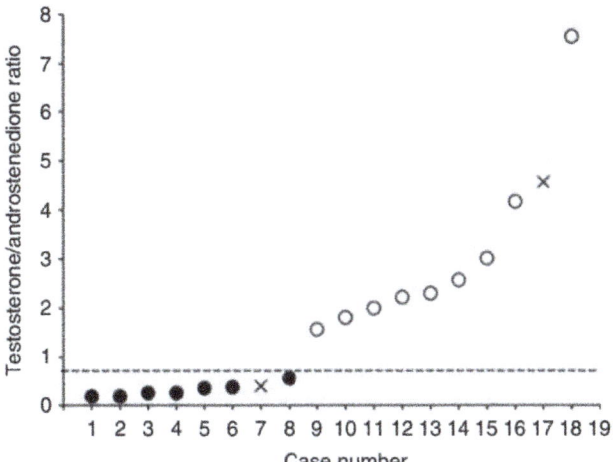

FIG. 3.2.1.37 Basal serum testosterone/ androstenedione ratios in partially virilized adult patients in DSD clinic. A defect in HSD17B (closed circles) or SRD5A2 (open circles) was confirmed by gene analysis. Only 46XY patients with gonads in situ were studied. *(Auther original.)*

Infants with this defect are born usually with female looking external genitalia and a phallus closely resembling a normal clitoris. On closer examination, there may be mild posterior fusion of the labioscrotal folds and a urogenital sinus. There is a high prevalence of this disorder within the Arab population of the Gaza strip (Rösler et al., 1996). This defect is confirmed when the ratio of androstenedione to testosterone is elevated above 10 after HCG stimulation. In urine, the high ratio of androsterone to

etiocholanolone is seen which may reflect activation of the backdoor to DHT (Fig. 3.2.1.37) where androsterone is an intermediate as shown in congenital adrenal hyperplasia (Kamrath et al., 2012).

However, the androsterone/etiocholanolone ratio in urine and the androstenedione/testosterone ratio in blood are not reliable indicators of the 17KSR defect. In humans, 11 forms of 17KSR have been identified. Serum levels of Δ4 and T are regulated not only by HSD17B3 but also by HSD17B5, HSD17B2, HSD17B8, HSD17B10 and HSD17B14 (Fig. 3.2.1.38). HSD17B5 catalyzes the reduction of Δ4 to T and is expressed in the testis, prostate, adrenals and liver. HSD17B2, HSD17B8, HSD17B10, and HSD17B14 convert T to Δ4 and estradiol (E2) to estrone (E1) and have widespread expression in various tissues, including the testis, prostate, liver, kidney, and the central nervous system. Individual and temporal variability in the activities of these HSD17B isozymes may affect serum Δ4/T ratio (Khattab et al., 2015).

Outside the neonatal period infants with the disorder may present with swelling in the groin (hernia) or labia and there is marked virilization at puberty. Genetic testing of 17BHSD3 may be needed. Mutations in 17β-hydroxysteroid dehydrogenase type-3 explain the disease (Phelan et al., 2015) although phenotypic variation can occur within families with the same mutations. A male with this defect can be reared as a boy but will need testosterone treatment.

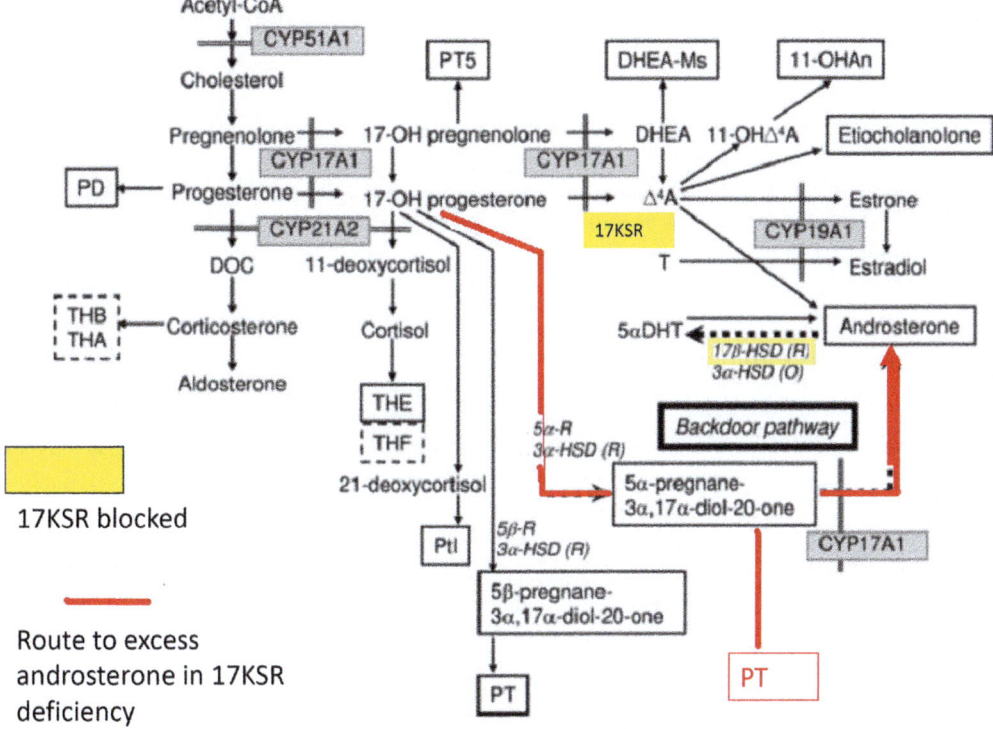

FIG. 3.2.1.38 Proposed route for increased urinary androsterone in defects of HSD17B3.

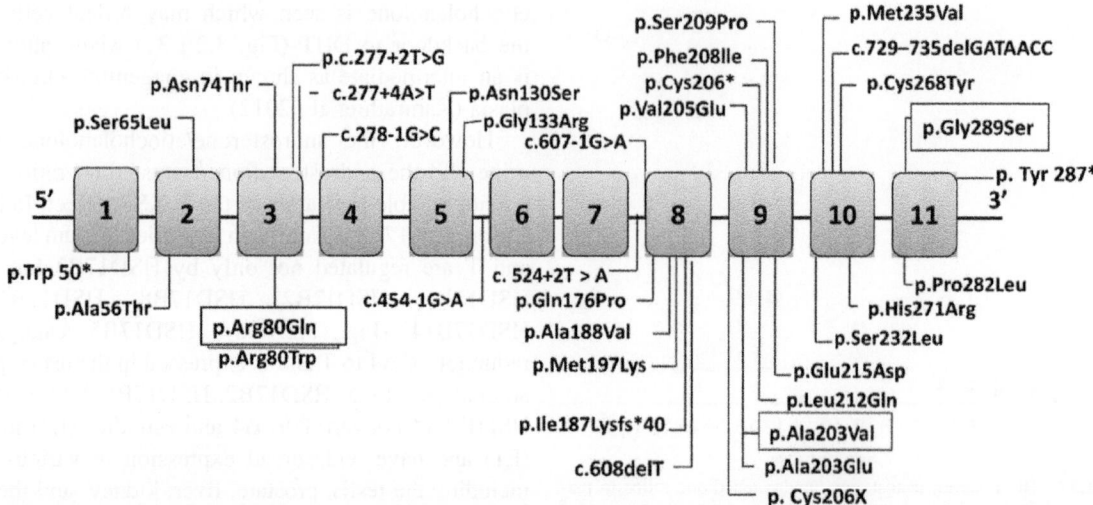

FIG. 3.2.1.39 Mutations found in HSD17B3 gene in patients with 46,XY DSD due to 17β-HSD3 deficiency. The recurrent mutations are represented in framed boxes. Duplication of the exons 3–10 was not included. *(From Mendonca BB, Gomes NL, Costa EM, Inacio M, Martin RM, Nishi MY, Carvalho FM, Tibor FD, Domenice S. 46,XY disorder of sex development (DSD) due to 17β-hydroxysteroid dehydrogenase type 3 deficiency. J Steroid Biochem Mol Biol 2017;165(Pt. A):79–85. Fig. 1 p. 84.)*

About 70 mutations have been identified (https://www.hgmd.cf.ac.uk/ac/gene.php?gene=HSD17   B3). These mutations include intronic splice sites, exonic deletions, missense and nonsense mutations, four of which are small deletions (Fig. 3.2.1.39) (Phelan et al., 2015). The c.277 +4A > T in intron 3 has been previously reported in several patients of various countries and it may be spread worldwide by North European common founders. The c.277+4A > T variant disrupts normal splicing. Although mutations throughout the gene have been described, a mutation cluster region in exon nine with complete elimination of 17β-HSD3 activity was identified in many populations.

46XX sex reversal (male sex assignment) with **gonadal dysgenesis** is a distinct or dysmorphic syndrome usually present later in life with lack of pubertal development, growth delay and infertility and will not be considered in this chapter which focuses on disorders of sex differentiation with androgen excess.

3.2.1.3.2   Androgen excess and unwanted hair growth

Androgen excess causes unwanted body hair in females on chin, back and legs. Since 1961 when Ferriman and Gallway published a classification of excess hair patterns it has been usual practice to score hairs in nine body areas (upper lip, chin, chest, arm, upper abdomen, lower abdomen, upper back, lower back and thighs) with scores from 1 (minimal terminal hairs present) to 4 (Fig. 3.2.1.40) (equivalent to male hair patter). If no terminal hairs are observed in the body area being examined the score is zero (left blank).

Clinically terminal hairs can be distinguished from vellus hairs primarily by their length (i.e., greater than 0.5 cm) and the fact that they are usually pigmented. The extent can be calibrated against standards set by Ferrimen and Gallway although the original cartoons are now available as color photographs (Fig. 3.2.1.41) (Yildiz et al., 2010). A total score above 6–8 signifies hirsutism. The interpretation of the photographs is easier to interpret than the F&G sketches.

3.2.1.3.3   Androgen excess in the adult

Several conditions due to hyperandrogenism present in adults, females particularly and simultaneous determination of several androgens is a starting point for investigations (Martin et al., 2018). Androsterone is raised in polycystic ovary syndrome (PCOS) and adrenocortical carcinoma, testosterone is raised in CAH, adrenal adenoma (Zhou et al., 2019), sertoli cells in males (Giglio et al., 2003) and ovarian hyperthecosis (Elhassan et al., 2018). Estrogens and gonadotrophins should also be measured. New markers have been defined for adrenocortical carcinoma (Mizdrak et al., 2021). Studies with androgen receptor knockout mice have confirmed that AR mediated androgen action plays a key role in regulating the female HPG axis. Androgens in excess play a key role in the origin of polycystic ovaries (Walters et al., 2019).

3.2.1.3.3.1   Polycystic ovaries (PCO)

In a female, when LH is raised in the presence of low or normal FSH the most likely diagnosis is *polycystic ovaries* (rarely indicative of pregnancy due to reaction of hCG in the LH assay). Patients with PCO may have oligomenorrhea and have a characteristic ovarian picture on ultrasound

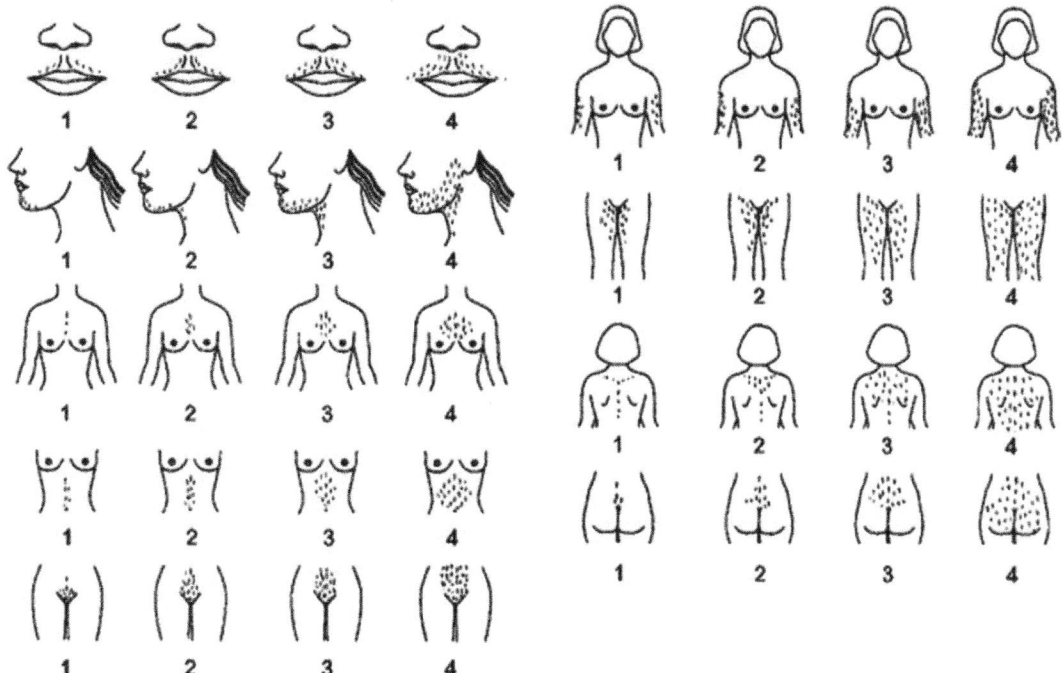

FIG. 3.2.1.40 Unwanted hair and modified Ferriman and Gallway scoring system. Nine body areas (upper lip, chin, chest, arm, upper abdomen, lower abdomen, upper back, lower back and thighs) are scored from 1 (minimal terminal hairs present) to 4 (equivalent to a hairy man). If no terminal hairs are observed in the body area being examined the score is zero (left blank). Clinically terminal hair hairs can be distinguished from vellus hairs primarily by their length (i.e., greater than 0.5 cm) and the fact that they are usually pigmented. *(From Yildiz BO, Bolour S, Woods K, Moore A, Azziz R. Visually scoring hirsutism. Hum Reprod Update 2010;16(1):51–64. Fig. 2, p. 57.)*

and may have elevated LH concentrations such that the LH to FSH ratio is 3 or more.

An abdominal ultrasound will reveal PCO in 80% of the females presenting with hirsutism, acne or alopecia. A ring of 20–20 cystic follicles of 2–9 mm in diameter around the rim of the ovary is characteristic of PCOS. PCO syndrome (PCOS, Stein-Leventhal) is the picture of polycystic ovaries (Fig. 3.2.1.42) accompanied with anovulation, menstrual irregularity and infertility.

There is an increased risk of dyslipidemia, obesity, hypertension and cardiovascular disease. In Europe, abdominal ultrasound is used to image follicle numbers and location. Up to 80% of women with hirsutism will have an endocrine disorder. One of **CAH, adrenal or ovarian tumors, hyperprolactinemia, thyroid disorders and Cushing's syndrome** may be a precise diagnosis that needs to be excluded and with the passage time further causes will be identified.

Standards for recognition of PCOS have been published by three learned groups (Table 3.2.1.4). They agree that 2 out of 3 criteria are met from chronic anovulation, androgen excess and polycystic ovaries but have a different emphasis.

An international consensus guideline has subsequently been published with a focus on refining specific diagnostic features (Teede et al., 2018). Diagnosis of PCOS during adolescence is challenging and the situation should be re-

evaluated once menstrual cycle pattern has established (Peña et al., 2020; Witchel and Plant, 2020). The choice of investigations in adults will depend on funds and resources in terms of both equipment and staff. The patient needs to be assessed in terms of menstrual cycle frequency and dysfunction, extent of body hair growth towards the male pattern. A blood sample will be needed to investigate the nature of androgen excess and ability to exclude CAH, hyperprolactinemia and thyroid dysfunction. The patient should ideally be off medical contraception for 6 weeks and provide a blood sample near 0800 h and declare the date of the last menstrual period. A single raised prolactin in a 0800 h blood sample should not be regarded as proof of hyperprolactinemia until further measurements throughout the day have been performed since prolactin does have a circadian rhythm and can be raised following a painful venepuncture.

When primary amenorrhea is associated with hirsutism measurements of testosterone, DHAS, 17α-hydroxyprogesterone, cortisol and prolactin are required to identify or exclude major causes of abnormal androgen production. A serum testosterone >5 nmol/L is indicative of a rare ovarian tumor. DHAS > 20 μmol/L suggest the possibility of an adrenal androgen secreting tumor. 17α-Hydroxyprogesterone >5 nmol/L is an indication of some degree of adrenal 21-hydroxylase deficiency. An ACTH stimulation test with exaggerated response of 17OH-P is required to confirm this disease. The patient may be a heterozygote for the classical disease or may have a milder form sometimes referred to as late onset CAH (Papadakis et al., 2019). In very specialized laboratories, 21-deoxycortisol is a better marker for a defective 21-hydroxylase. These patients can be treated with glucocorticoids.

Measurement of plasma free testosterone is regarded as the best index of testosterone excess. Ideally an equilibrium dialysis with then a mass spectrometric analysis of free testosterone in the dialysate is the preferred method. Analogue

displacement assays are not reliable. If free testosterone by dialysis and MS is not available the next best test is plasma total testosterone by LC-MS/MS. The algorithms derived for calculating free T, from plasma testosterone and SHBG concentrations, are inaccurate because they were founded on models of testosterone binding to SHBG that are not relevant in all populations however they can still give clinically useful results (Keevil and Adaway, 2019; Keevil et al., 2018). Testosterone determinations on chemistry platform systems have improved and give results similar to MS determinations but for patients can still be subject to interference from components in the sample. Extraction

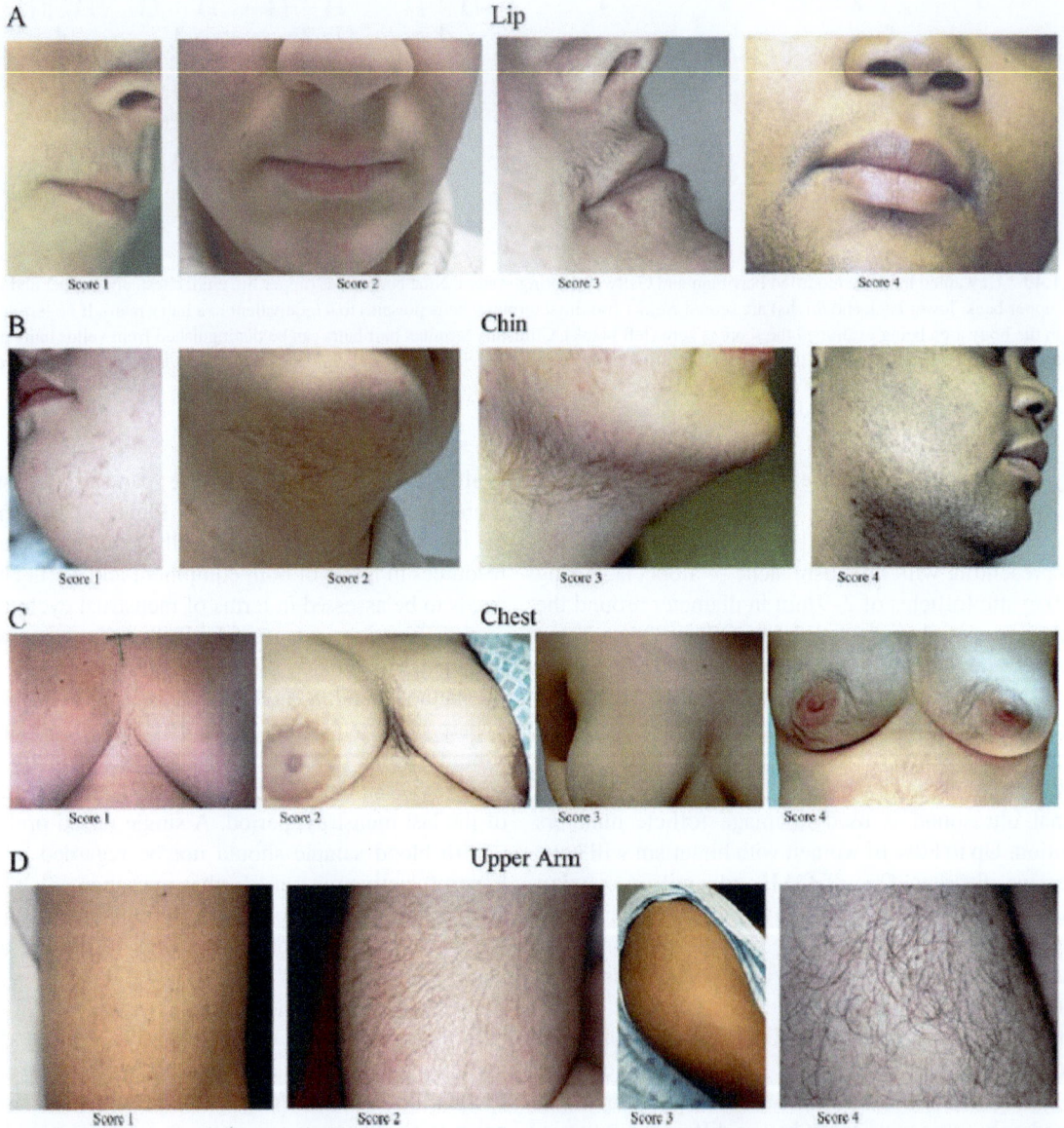

FIG. 3.2.1.41 Photographs depicting facial and body terminal hair growth scored according to the modified FG method. All were taken on women who had not used laser or electrolysis for at least 3 months, not depilated or waxed for at least 4 weeks, not shaved or plucked for at least 5 days before the photograph. The photographs depict scores of 1 through 4 for the upper lip (A), chin (B), chest (C), arm (D), upper abdomen

(Continued)

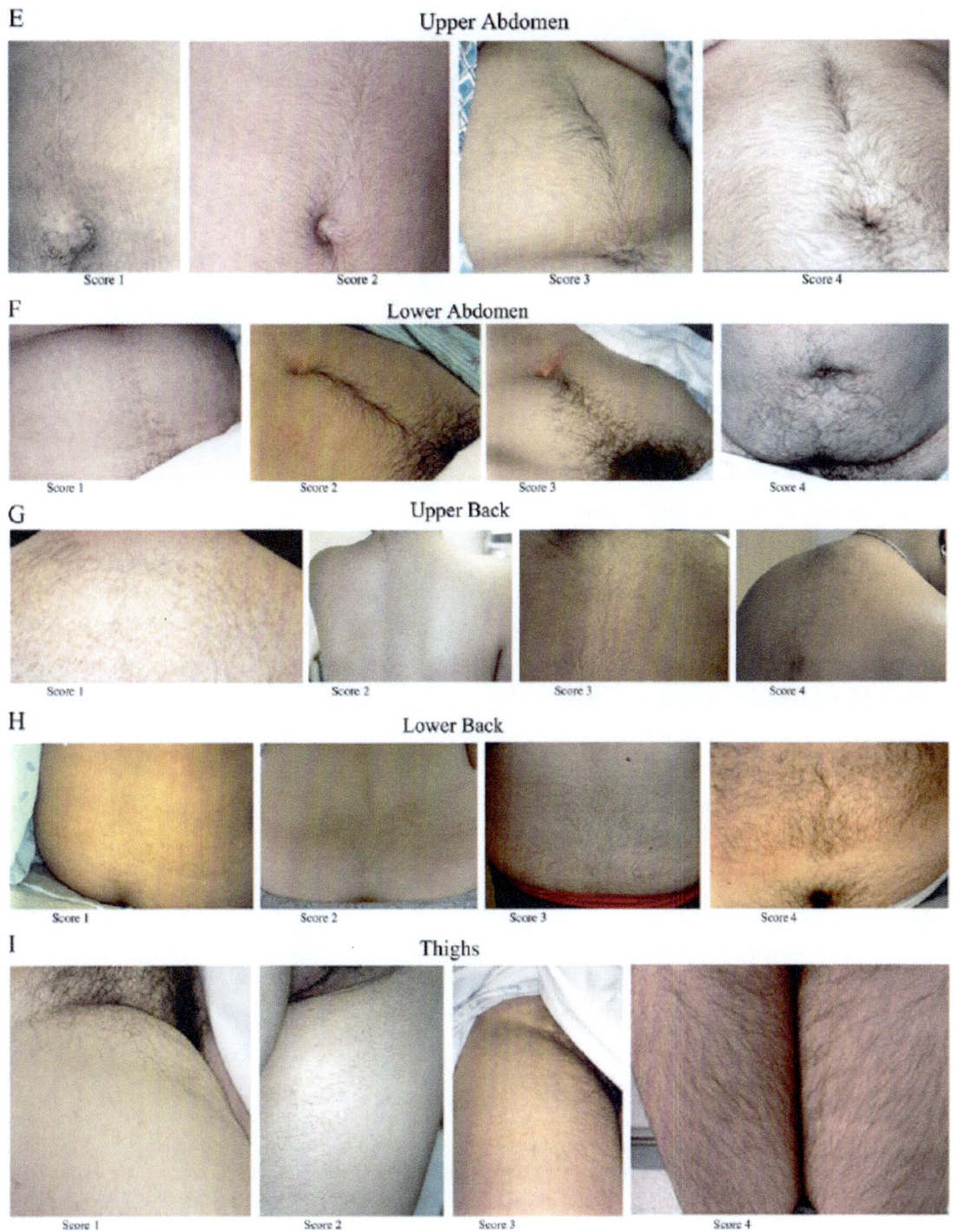

FIG. 3.2.1.41, CONT'D (E), lower abdomen (F), upper back (G), lower back (H) and thighs (I). The areas were photographed with a standard single-lens reflex camera (Nikon N50, Nikon Corp, Melville, NY, USA) equipped with a macro lens (Vivitar 50 or 100 mm Auto Focus Macro, Vivitar Corp, Newbury Park, CA) and ring flash (Vivitar Macroflash 5000, Vivitar Corp). For film, Kodacolor VR 200 ISO film (Eastman Kodak Co, Rochester, NY, USA) was used. Representative areas were selected. All photographs of hair were anonymized and all identifying information removed, meeting current Institutional Review Board for Human Use and Health Insurance Portability and Accountability Act of 1996 standards. *(From Yildiz BO, Bolour S, Woods K, Moore A, Azziz R. Visually scoring hirsutism. Hum Reprod Update 2010;16(1):51–64. Fig. 3, p. 58 and 59.)*

FIG. 3.2.1.42   Ultrasound PCOS.

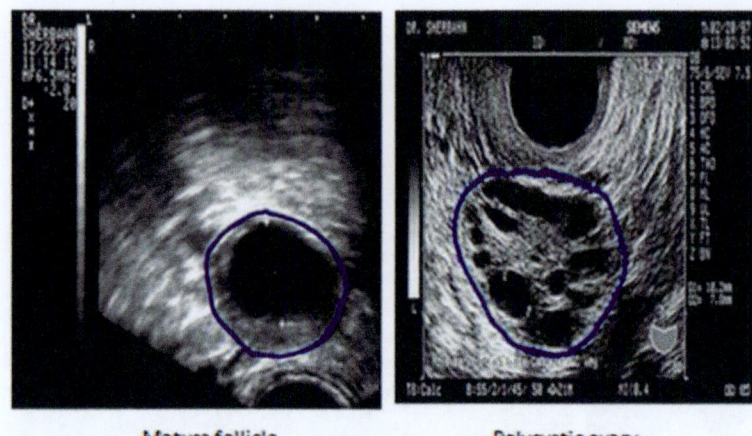

Mature follicle   Polycystic ovary

TABLE 3.2.1.4 Androgens in idiopathic hirsutism.

3α,17β-Androstanediol-glucuronide	Horton et al. (1982)
3α-Androstenediol-gluc, T, DHT (Celite RIA)	Toscano et al. (1982)
5α-Androstane-3α,17β-diol glucuronide, T, DHT (Celite chrom, RIA)	Gompel et al. (1986)
3α,17β-Androstanediol-glucuronide	Greep et al. (1986)
C-19 Steroids and 5-alpha reduced glucuronides	Brochu et al. (1987)
3α-Diol-G, fT, T, A (Celite chrom, hydrolysis RIA)	Kirschner et al. (1987)
A,T, DHT,AD, ADG, DHEA, DHEAS (TLC and RIA), conversion rates for ADG and DS	Giagulli et al. (1991)
Androsterone sulfate 3.5 µmol/L (RIA androsterone after extraction and solvolysis)	Zwicker and Rittmaster (1993)
Serum A-diol and A_diol-G (GC-MS)	Wudy et al. (1996)
Serum T, Oe2, 17-OHP, DHEA, DHEAS, A4	Escobar-Morreale et al. (1997)
Serum 3alpha diol -G, T fT, DHEAS (RIAs), SHBG	Meczekalski et al. (2007)
Serum DHEAS, T, free T, A4,17-OHP (commercial kits immunoassays), SHBG (IRMA) and genes expression	Caglayan et al. (2011)
Serum DHEAS, T, A4, estradiol (immunoassays), SHBG (IRMA) and genes expression	Taheri et al. (2015)

of testosterone into an organic solvent before immunoassay increases the specificity of the analysis but is an extra time consuming test.

If ovarian ultrasound is not used, measurements of progesterone will be raised after ovulation but this will be needed on several days when cycle frequency is irregular.

An LC-MS/MS assay of 13 steroids overcame the uncertainties from multiple individual determinations (Saito et al., 2016) and showed excessive testosterone in PCOS could be attributed to conventional pathways and DHT was likely produced in the backdoor path. Over production of DHEA was due to oxidative stress phosphorylation of p38α and enhanced 17,20 lyase activity (Zhu et al., 2019b).

A panel of steroid measurements by MS will be financially viable for most laboratories when compared with four or more separate immunoassays.

In urine, a reproducible finding has been high ratios of 5α to 5β reduced steroids suggesting that an increased peripheral metabolism of cortisol may occur by enhanced chemical reduction of cortisol by an **increase of 5α-reductase** (SRD5A2) (Fig. 3.2.1.43) (Torchen et al., 2016; Vassiliadi et al., 2009; Fassnacht et al., 2003; Tsilchorozidou et al., 2003) or impaired reactivation of cortisol from cortisone by **11β-hydroxysteroid dehydrogenase type 1 (HSD11B1)** (Rodin et al., 1994) resulting in decreased negative feedback suppression of ACTH

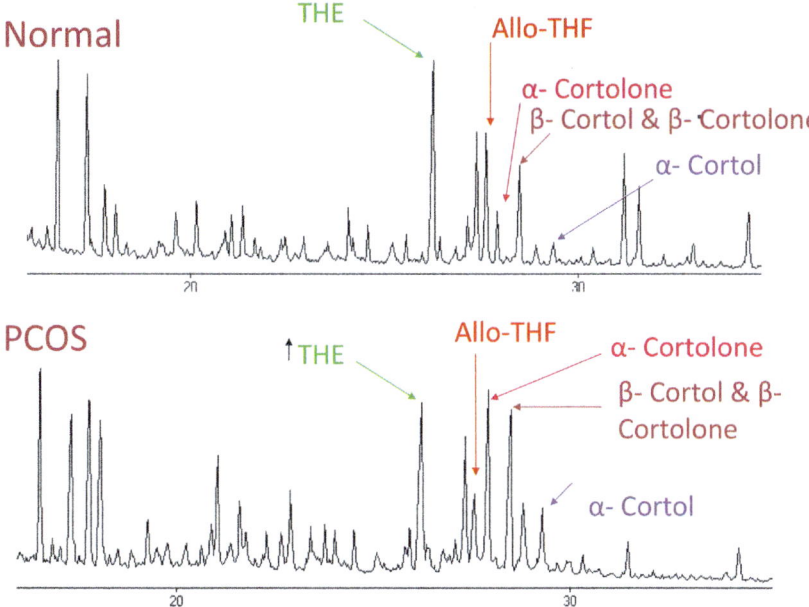

Normal

THE
Allo-THF
α- Cortolone
β- Cortol & β- Cortolone
α- Cortol

PCOS

↑THE
Allo-THF
α- Cortolone
β- Cortol & β- Cortolone
α- Cortol

FIG. 3.2.1.43 Steroid profile analysis of urine from normal adult female (A) and patient with PCOS (B). A, S and C are internal standards. The numbered steroids are: (1) androsterone, (2) etio-cholanolone, (3) 11-hydroxyandrosterone, (4) THE, (5) THF, (6) alloTHF, (7) -cortolone, (8) -cortolone-cortol (mainly cortolone in B and C) and (9) -cortol. The progressive increase in steroid peak heights from sample A to C relative to the internal standards reflects increased excretion (cortisol production) rates due to raised THE and cortolones. (*From Tsilchorozidou T, Honour JW, Conway GS. Altered cortisol metabolism in polycystic ovary syndrome: insulin enhances 5alpha-reduction but not the elevated adrenal steroid production rates. J Clin Endocrinol Metab 2003;88 (12):5907–13. Fig. 2 p. 5912.*)

secretion maintaining normal plasma cortisol concentrations at the expense of androgen excess.

Androstenedione is also mainly produced by the adrenal cortex so when 17-OHP concentration is raised androstenedione should be measured as an androgen index of adrenal disease. The analysis of androstenedione, 17-hydroxyprogesterone, 11-deoxycortisol and dehydroepiandrosterone sulfate should cover possibility of any of the late onset forms of CAH (Wang et al., 2018; Carmina et al., 2017; Reisch et al., 2013). The determinations should use mass spectrometric methods particularly as all can be measured together in one assay (Keevil, 2019; Keefe et al., 2014). Plasma concentrations of 11-hydroxyandrostenedione by immunoassay were higher in PCOS compared with women with normal ovaries (Owen et al., 1992). The measurement of 11-oxyandrogens by LC-MS/MS may become useful biomarkers (Turcu et al., 2017; Kempegowda et al., 2020) when further studies confirm initial findings. High activity of StAR was shown in ovarian follicles which may promote androgen synthesis from early in the pathway (Kahsar-Miller et al., 2001).

Hyperandrogenism is characterized by enhanced response of 17OHP to gonadotropin stimulation in many cases (Rosenfield and Ehrmann, 2016) through an intrinsic abnormality in the normal mechanism for down-regulation of the steroidogenic response to LH (Fig. 3.2.1.44). The ovary can be the source of mildly excess 17-OHP. In such a case, a urine steroid profile may provide evidence to confirm the ovarian origin. In the profile, 17-hydroxpregnanolone and pregnanetriol are raised but 11-oxo-pregnanetriol is normal because the ovary cannot form 21-deoxycortisol. DHEAS should be measured and if raised an abdominal CT for adrenal tumor organized.

The goal of research into PCOS pathophysiology is ultimately to understand the biochemical and genetic subtypes of the condition. A number of individual androgens have been measured since 1982 (Table.3.2.1.5).

Several of those steroids could now be included in a steroid panel to exclude the obvious defects (Pasquali et al., 2016). Hyperandrogenism seems to develop as a complex trait from interactions between predisposing congenital factor(s) and provocative environmental factor(s). The most common provocative factors seem to be obesity and insulin resistance, that occur in about half of cases and can have heritable components. Obesity up-regulates ovarian androgen production primarily via insulin-resistant hyperinsulinemia and to some extent via inflammatory cytokines. Testosterone measurements by LC-MS/MS differentiated more cases of PCOS with raised concentrations then when a CLIA was used (Yang et al., 2019).

Excessive production of anti-Müllerian hormone (AMH) is seen in PCOS due to the increased number of growing follicles and other factors (Dewailly et al., 2020). There have been several changes in the assays available in format, antibodies and standards such that cut-off concentrations have varied between 5.6 ng/mL (40 pmol/L) for manual ELISA and 4.2 ng/mL (30 pmol/L) for automated platform assays (Dumont et al., 2015). A homozygous gene variant of AMH may be less immunogenic and not well detected by Ash AMH ELISA (Hoyos et al., 2020).

PCOS is a syndrome with heterogeneous characteristics; therefore, there are different pathways, hormonal, reproductive, metabolic that may be involved in its etiology. Hyperandrogenism causes its adverse effects on follicular development and negative feedback mechanisms.

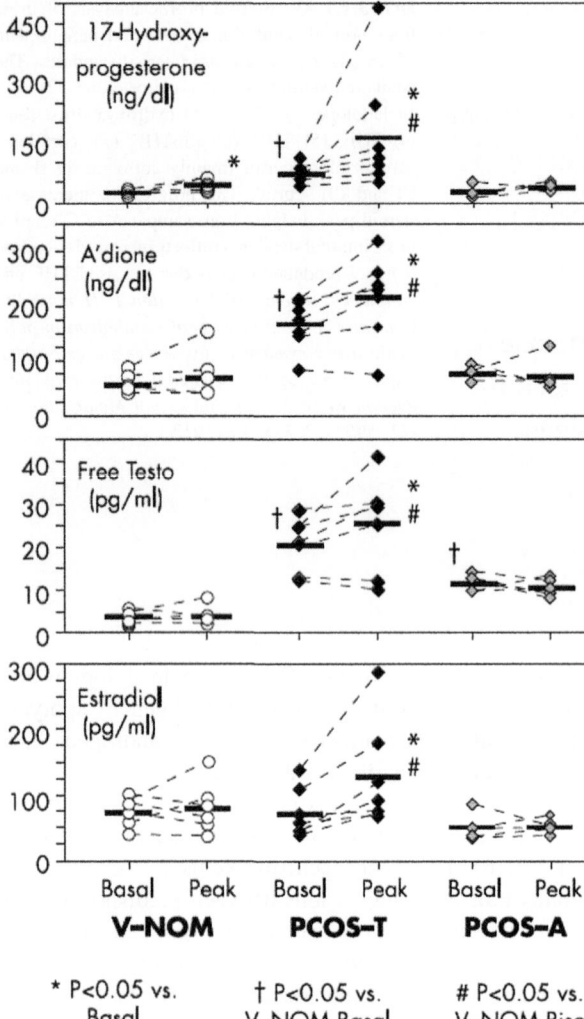

FIG. 3.2.1.44 Response of 17-OHP to half-maximal hCG stimulation during overnight dexamethasone suppression of normal and PCOS subjects. After bedtime, dexamethasone 0.25 mg/m² , 500-IU hCG was administered im at 8 AM; a basal blood sample was drawn before hCG, and the peak response to hCG was sampled after a repeat dexamethasone dose 24 h later. Subjects were healthy volunteers with normal ovarian morphology (V-NOMs), PCOS patients with functionally typical FOH (i.e., 17OHP hyperresponsiveness to GnRHag; PCOS-T), and PCOS patients with functionally atypical FOH (i.e., normal 17OHP responsiveness to GnRHag; PCOS-A). V-NOM had a small but significant rise in serum 17OHP but not in other steroids. PCOS-T had hyperresponsiveness of all steroids. PCOS-A, a heterogenous group, had elevated basal serum free testosterone, but normal hCG-responses of all steroids. To convert to SI units, multiply 17OHP by 0.0303 (nM), androstenedione (A'dione) by 0.0340 (nM), free testosterone by 3.47 (pM) and estradiol by 3.67 (pM). *(From Rosenfield RL, Ehrmann DA. The pathogenesis of polycystic ovary syndrome (PCOS): the hypothesis of PCOS as functional ovarian hyperandrogenism revisited. Endocr Rev 2016;37(5):467–520. Fig. 3 p. 473.)*

For instance,

(a) hormonal imbalances such as hyperandrogenism increased LH/ follicle stimulating hormone (FSH) ratio, increased estrogen levels and decreased serum progesterone. The elevation of frequency and amplitude of

gonadotropin releasing hormone (GnRH) release and subsequent increase in LH secretion is an important pathophysiological feature of PCOS (Coyle and Campbell, 2019; Lu et al., 2018). GnRH secretion is impaired due to a dysfunction of gonadotropin-inhibitory hormone (GnIH). LH then affects ovarian theca cells with increased synthesis of androgens,

(b) anovulation and menstrual irregularities,

(c) impaired glucose tolerance and insulin resistance, obesity, cardiovascular disease and type 2 diabetes. Hyperinsulinemia increases the secretion of androgens with different effects on ovary, adrenal, pituitary, LH receptor, sex hormone-binding globulin (SHBG) protein, and

(d) changes in serum lipid parameters,

are all encountered in this complex disorder. A holistic approach is needed to manage the condition (Barber et al., 2019) particularly when the incidence of obesity is increasing. There are several reasons for the hyperandrogenism that can disrupt normal activity of ovary and interfere with menstrual cycle.

The genetic evaluation of PCOS has been mainly conducted through the candidate gene approach focusing gene selection based on an assumed role in the syndrome. A summary of genes studied to 2019 were published by Shaaban et al. (2019). This approach has failed in complex syndromes such as PCOS, due to incomplete understanding of the pathophysiology of the condition. The selected genes were primarily responsible for the synthesis of proteins that belong to six categories,

(a) gonadotrophin secretion and actions,

(b) steroid hormones biosynthesis and functions,

(c) insulin secretion and signaling,

(d) insulin resistance and type 2 diabetes mellitus,

(e) obesity and dyslipidemia, and

(f) chronic inflammatory reactions.

Four hypotheses cover the impact of peripheral hormones on brain function in pathogenesis of PCOS (Fig. 3.2.1.45) (Shaaban et al., 2019).

The first hypothesis is hyperinsulinemia that elevates the activity of GnRH neurons or pituitary responsiveness to GnRH.

The second hypothesis is that low levels of serum progesterone from the anovulation removes the influence of negative feedback by progesterone on GnRH release. The third hypothesis is hyperandrogenism that changes the neuronal circuits for negative feedback of steroid hormones.

A recent hypothesis from studies in rats and mice exposed to continuous light found that the activity of **GnRH inhibitors** (GnIH) was reduced by low mRNA expression of hypothalamic RFRP-3 neuropeptide that is a **GnIH**

TABLE 3.2.1.5 Selected hormonal changes in mother with aromatase deficiency during pregnancy.

Serum steroid	36 weeks pregnancy—Patient	36 weeks pregnancy—Control	38 weeks pregnancy—patient	38 weeks pregnancy—control
Estrone[a]	1028	36,620±13,316	183	40,489±11,837
Estradiol[a]	1690	102,790±32,670	1650	103,160±46,620
Estriol[a]	213	55,173±24,473	134	55,173±19,085
Progesterone[b]			700	571±154
Testosterone[b]	41.6	4±0.5	33	4±0.5
DHT[b]			7.8	1.1±1.1
Androstenedione[b]	83.8	20.2±2.3	160.8	20.2±2.3
DHEA[b]			20.5	59.5±7.0
DHEAS mmol/L			1.1	0.9±0.2

[a]*Picomoles per liter.* [b]*Nanomoles per liter.*

From Shozu M, Akasofu K, Harada T, Kubota Y. A new cause of female pseudohermaphroditism: placental aromatase deficiency. J Clin Endocrinol Metab 1991;72(3):560–6. Table 1 p. 562.

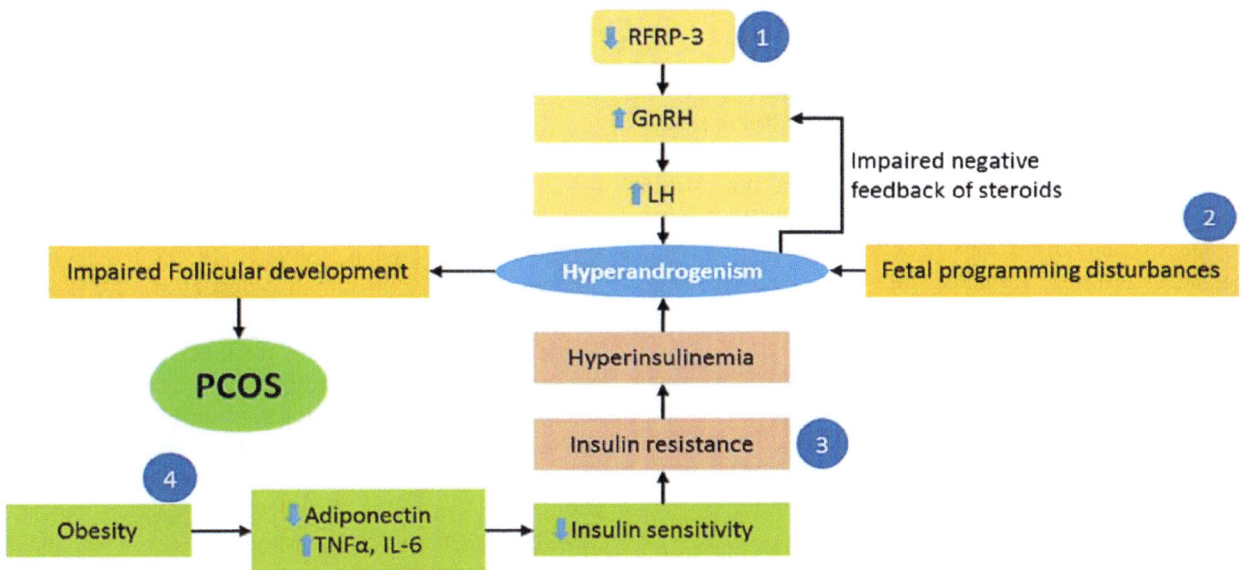

FIG. 3.2.1.45 Mechanism PCOS. The pathophysiological pathways that are presumed to mediate polycystic ovary syndrome (PCOS) formation are expressed. Four different pathophysiological pathways lead to PCOS. Abbreviation: *RFRP-3*, arginine-phenylalanine-amide (RFamide)-related peptide 3; *GnRH*, gonadotropin releasing hormone; *LH*, luteinizing hormone; *TNFα*, tumor necrosis factor-α; *IL-6*, Interleukin-6. *(From Shaaban Z, Khoradmehr A, Jafarzadeh Shirazi MR, Tamadon A. Pathophysiological mechanisms of gonadotropins- and steroid hormones-related genes in etiology of polycystic ovary syndrome. Iran J Basic Med Sci 2019;22(1):3–16. Fig. 1, p. 4.)*

(Lima et al., 2014). This did not however affect kisspeptin expression (Nooranizadeh et al., 2018). There is no research on alteration of GnIH neuronal activity in the human PCOS condition. Genetic alterations in *GnRH* and its receptor (*GnRHR*) genes can interfere with PCOS. Although, susceptible variants as a risk factor for PCOS in this gene have not yet been determined. Studies in animals with knockout genes will help with these investigations (Lima et al., 2014).

Using a nontargeted metabolomics approach genes beyond those coding for steroidogenic enzymes, that affected metabolic alterations were significantly involved (Chang et al., 2017). PCOS was most associated with elevated branched chain amino acids, essential amino acids and the lysine metabolite α-aminoadipic acid. Patients with PCOS exhibit several mitochondrial gene mutations, deletions and nucleotide variations (Zhang et al., 2019; Zeng et al., 2020) that have a role in energy metabolism. The gut flora

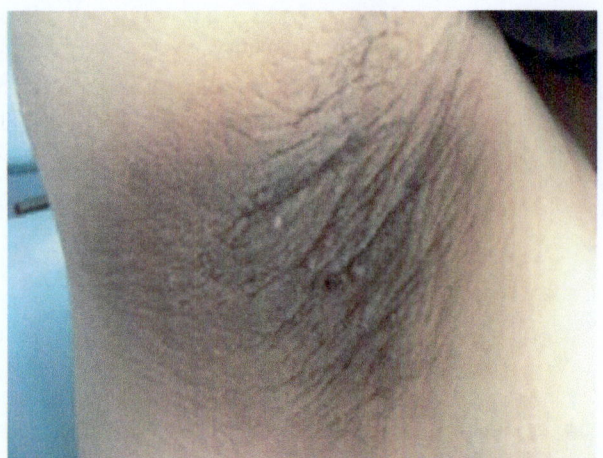

FIG. 3.2.1.46 Acanthosis nigricans. Left—Male youth facing to right with neck showing coarse thickened and hyperpigmented skin with accentuation of skin marking. Right—young girl facing left with hyperpigmented velvety plaques on the sides of the neck. Acanthosis nigricans. *(From Ng HY. Acanthosis nigricans in obese adolescents: prevalence, impact, and management challenges. Adolesc Health Med Ther 2016;8:1–10. Fig. 3, p. 3; Das A, Datta D, Kassir M, Wollina U, Galadari H, Lotti T, Jafferany M, Grabbe S, Goldust M. Acanthosis nigricans: a review. J Cosmet Dermatol 2020;19(8):1857–1865. Fig. 2 p. 1858.)*

may contribute to metabolic changes and this aspect will receive renewed interest (Sun et al., 2012). Further investigations of the causative role of hyperandrogenism in occurrence of PCOS and evaluation of its upstream and downstream pathways drives is necessary (Abbott et al., 2019).

Obesity and insulin resistance (IR) often associate with a cutaneous disorder, acanthosis nigricans, characterized by velvety hyperpigmented plaques on axilla, neck and groin (Fig. 3.2.1.46).

Fasting insulin and glucose should be measured. There are several calculations in the literature but the Homeostasis model assessment-insulin resistance (HOMA-IR) is a good tool for assessment of the insulin resistance (Das et al., 2020).

HOMA-IR = [fasting glucose (mmol/L) × fasting insulin (mU/mL)] divided by 22.5.
IR is diagnosed if HOMA-IR is >2.7

Dysregulation of adrenocortical steroidogenesis seems to account for the associated hyperandrogenism found in about one-quarter of cases. Insulin-resistant hyperinsulinism is often an important aggravating factor for androgens in PCOS pathogenesis. Triglycerides and HDL cholesterol as metabolic markers, individually and as a ratio were raised significantly in PCOS (Blum et al., 2021) The hyperinsulinemia sensitizes ovarian theca cells to LH stimulation. Stimulation of adipogenesis and lipogenesis and inhibition of lipolysis by insulin excess also appear to contribute to the obesity of PCOS. About half of women with PCOS have an abnormal degree of insulin resistance for BMI. The

hyperandrogenism is associated with a significantly higher prevalence of glucose intolerance in the presence of increased insulin resistance, which suggests a relationship with pancreatic-cell failure. The insulin resistance of PCOS is independent of obesity. Women with PCOS have elevated levels of advanced glycation end products that could affect granulosa and theca cell functions (Mouanness and Merhi, 2022).

There is a lack of knowledge about the normal and abnormal regulation of ovarian steroidogenesis, homologous desensitization to LH, and folliculogenesis as well as the links between mechanisms underpinning the associated abnormalities seen in PCOS. Untransformed human cell lines for the tissues that can be dysfunctional in PCOS: theca, granulosa, adrenocortical zona reticularis and preadipocyte cells need to be developed as a complementary approach. One important unanswered question is what links ovarian hyperandrogenism, obesity and insulin resistance? Considerable basic research will be necessary to clarify this.

Kisspeptin levels are higher in the PCOS population, which supports the hypothesis that an over-active KISS1 system causes HPG-axis overactivity, leading to irregular menstrual cycles and excessive androgen release (Witchel and Tena-Sempere, 2013). In animal studies, however, kisspeptin levels are not increased in all subtypes of PCOS models which might reflect the way the model is generated rather than enhancing our understanding of pathophysiologic changes in PCOS in humans.

Metabolomics will be applied to PCOS in order to seek biomarkers for the condition (Alesi et al., 2021; Rajska et al., 2020; Zhao et al., 2012). There is limited evidence for derangement of pathways for lipids (mainly glycerophospholipids and sphingolipids) (Jové et al., 2017), fatty acids (Zou et al., 2018), carbohydrates (Sun et al., 2012), and amino acids (Alesi et al., 2021). Protein expression was examined by two dimensional difference gel electrophoresis and matrix assisted laser desorption ionization MS/MS in PCOS with and without IR (Li et al., 2020b). Twenty proteins were expressed differently of which 4 were identified namely afamin, serotransferrin, complement C3 and apolipoprotein C3. The expression level of APOC3 was positively associated with IR. Gut dysbiosis may aggravate metabolic and endocrine malfunction (Han et al., 2021). MicroRNA may affect the symptoms of PCOS and become biomarkers (Abdalla et al., 2020; Sørensen et al., 2019). Much further work is anticipated in this area.

Genetic variation of ***FSHR*** gene (Shi et al., 2012; Chen et al., 2014) and ***LHCGR*** gene (Almawi et al., 2015) may be risk factors for the PCOS. The possible role of these and other genes in PCOS development needs to be evaluated with respect to any association in different populations. The genetic evaluation of **follistatin** (Jones et al., 2007b) should be performed with regard to its association with raised androgens in in PCOS and studied in different ethnic backgrounds of participants.

The origins of hyperandrogenemia can be the steroidogenic cells of ovary and adrenal, which have similar effective enzymes for steroidogenesis. In addition, the biological impacts of androgens are mediated by the **androgen receptor (AR)** (Schüring et al., 2012), and **SHBG** regulates the serum level of free androgen (Zhu et al., 2019a; Moran et al., 2013); so, all of these may be involved in the PCOS pathophysiology. The roles of StAR (Kahsar-Miller et al., 2001) for cholesterol availability, as well as CYP19A (Mehdizadeh et al., 2017), CYP17A1 (Comim et al., 2013), HSD17B (Jones et al., 2009) mediating androgen biosynthesis may be susceptibility genes in PCOS development but further research is needed. There is not enough evidence that HSD3B (Carbunaru et al., 2004), CYP17 (Comim et al., 2013) or CYP11A genes (Shen et al., 2014) are risk factors for PCOS.

The promise of molecular genetic approaches to understanding the cause of PCOS is illustrated by the recent identification by genome-wide association screening of a previously unrecognized protein variant in androgen-producing cells, DENND1A.V2, as a facilitator of steroidogenesis (McAllister et al., 2014). Increased expression of a splice variant of DENND1A RNA, DENND1A.V2, is characteristic of PCOS. The **DENN** (**differentially expressed in neoplastic vs. normal cells**) domain is a protein module conserved during evolution. In human genome, there are 18 genes encoding DENN domain proteins, which can be categorized into eight families. The DENND1A-1C family, also known as connecdenn, has been identified as guanine nucleotide exchange factor (GEF) for a small GTPase Rab35. DENND1A/connecdenn 1 functions in the endocytic pathway, which binds Rab35 in its GDP-loaded conformation, mediating the switch from GDP-bound to GTP-bound forms, to promote endocytic recycling of various cargoes on clathrin-coated vesicles (CCVs) after internalization. Forced expression of DENND1A.V2 in normal theca cells increases CYP17A1 and CYP11A1 gene expression and converts the cells to a PCOS phenotype of augmented androgen and progestin biosynthesis. In contrast, knock-down of DENND1A.V2 with silencing shRNA plasmids or lentivirus in PCOS theca cells reverts the cells to a normal phenotype of reduced CYP17A1 and CYP11A1 gene expression and androgen and progestin biosynthesis. These observations suggest that DENND1A is involved in a signaling cascade that augments transcription of steroidogenic genes that subsequently results in increased androgen production (McAllister et al., 2014).

Genetic wide association studies will continue to provide clues towards candidate genes for PCOS (Dadachanji et al., 2018). The possible involvement of epigenetics and microRNA in PCOS has only been sparsely

investigated (Chen et al., 2019; Dumesic et al., 2015, 2020; Rosenfield, 2020; Duică et al., 2021) and more research is needed before conclusions can be drawn. The expression of miRNA needs to be studied further in granulosa cells, theca cells, adipose tissue, follicular fluid and serum. Dysglycemia may contribute to cognitive dysfunction in women of reproductive age (Jarrett et al., 2019) and will attract interest for further research.

Treatment of PCOS

The patient is first directed to changes in lifestyle treatment (LST) including a calorie restricted diet and exercise (Elias et al., 2004a,b). If the patient is insulin resistant then metformin can be prescribed (Unluhizarci et al., 2021). Bariatric surgery can be recommended in some cases (Ahmed et al., 2021) for weight loss. Oral contraceptives, cyproterone and drospironone can be prescribed. Metformin, an insulin sensitizing drug used alone or in combination with pioglitozone or spironolactone to improve hyperandrogenism (Zeng et al., 2020), Photoepilation and topical eflornithine can be used for hirsutism. Menstrual cycles are controlled with letrozole, clomiphene or gonadotrophins. Ovarian diathermy has been used for many years (Armar et al., 1990) and drilling is occasionally used (Hafizi et al., 2020; Seow et al., 2020) as replacements for the earlier surgical wedge resection. For IVF, a GnRH analogue may be needed to control the cycle with or without metformin (Raperport et al., 2021). Insulin tolerance is diagnosed with an oral glucose tolerance test. The patient must be monitored for cardiovascular disease risk, psychiatric disturbance (Standeven et al., 2021) and sleep apnea (Zhou et al., 2021). There is a strong need for continued research to improve quality of life for patients with PCOS, a number of animal models have been studied in this regard (Rodriguez Paris and Bertoldo, 2019).

3.2.1.3.3.2 CAH—Late onset

Hirsutism and amenorrhea can be a sign of NCCAH and measurements of 17-OHP, DHEA and 11-deoxycortisol should be instigated basally and after ACTH stimulation as discussed earlier.

3.2.1.3.4 Other chromosomal problems

The biochemistry for DSDs is well covered for proteins and gene products that in 2020/1 are currently known. New abnormal genes will be discovered and this may open new biochemical avenues. Exposure to androgens in utero during fetal development can result in PCOS phenotype in adulthood. During development of the fetus, the embryo may be exposed to additional androgens for four reasons, with epigenetic changes leading to polycystic ovary syndrome (PCOS) in the adult. First, the mother has PCOS and the placenta is also unable to aromatize androgens that lead to an increase in the concentration of SHBG, hence the mother receiving androgen via the placenta. Second, the fetal undifferentiated ovary is the source of excess androgen production. The third reason is abnormalities of tissues producing androgens including the adrenals of the fetus, congenital adrenal hyperplasia for example. The fourth reason is H-P-G disorders during embryonic development simultaneously with evolution of this system that may increase androgen production. Androgen excess in both embryonic stage and adulthood may play important roles in initiating PCOS, although the hypothalamus–pituitary axis and insulin pathways are valuable.

3.2.1.3.5 Adrenal tumors

In the past, biochemistry has not figured high in the investigations of patients with adrenal masses. Adrenocortical carcinoma (ACC) is a rare malignant disease with a poor prognosis so the surgical priority was to remove the tumor and rely on a histologist to characterize the tumor on the basis of cell types, necrosis, calcification, hemorrhage and mitoses to indicate malignancy. A Weiss score is used to stage adrenal tumors. Tumors are also classified on histology with the Ki67 score. Expression of the Ki67 protein (pKi67) is associated with the proliferative activity of intrinsic cell populations in malignant tumors, allowing it to be used as a marker of tumor aggressiveness.

There are difficulties in predicting the malignancy potential. Adrenal cancers are aggressive with incidence of around 1% and variable size tumors with evidence of metastases and local invasion at presentation. Tumors show high uptake of 2-deoxy-2-[^{18}F]-fluoro-D-glucose on PET scan. Despite radical resection around 80% show recurrence of the local disease often with concurrent metastases. Adrenolytic therapy with mitotane, administered alone or in combination with cytotoxic agents, is currently the primary (palliative) treatment for patients with advanced ACC.

Where steroids were analyzed in blood or urine, DHEAS excess was the commonest adrenal tumor secreting androgens and the determination of other steroids was not undertaken or involved analysis of cortisol or broadly 17-ketosteroids, 17-hydroxycorticosteroids and 17-ketogenic steroids. When steroids in urine were determined by GC-MS, other dominant steroids were found such as 11-hydroxyandrosterone, 16-hydroxy-DHEA and pregnenolone in children (Fig. 3.2.1.47) (Honour et al., 1984; Małunowicz et al., 1995) and more precursors including androsterone, etiocholanolone and tetrahydro-11-deoxycortisol in adults (Gröndal et al., 1990; Velikanova et al., 2016; Elhassan et al., 2018; Chortis et al., 2020).

In 1990, Shackleton demonstrated high production of androgens including 11-hydroxyandrostenedione by NCI-H295 adrenocortical carcinoma cells in culture (Gazdar et al., 1990). Using histochemistry significant

evidence demonstrated lower expression of CYP11B1, StAR and CYP17A1 in ACC when compared to adenomas presenting with Cushing's syndrome (ACAc). An incomplete pattern of steroidogenic enzyme expression could justify the increased secretion of steroid metabolite precursors witnessed. CYP11B1 was shown to be a highly accurate molecular marker for differential diagnosis between ACC and ACAc (Pereira et al., 2020). Many patients have evidence of cortisol excess and/or DHEAS secretion, although biochemistry has not been thorough in this group. IGF2 is the most frequently over-expressed gene in ACC along with structural abnormalities in the 11p15 gene region. The IGF1R is also over expressed and all of these genes have altered DNA methylation patterns.

Tumors secreting cortisol or aldosterone are considered in detail in the relevant steroid excess chapters (Chapters 3.1 and 3.3).

The most common mutations in adrenocortical carcinoma are mutations or deletions of TP53 and ZNRF3 or CTNNB1 that alter the p53 or Wnt/β-catenin pathways. Mutations affecting cortisol excess have been described but not for gonadal hormone excess.

3.2.1.3.5.1 Incidentaloma/adenoma

With the widespread use of MRI and CT scanning, there has been an increase in the finding of tumors with no symptoms of excess hormone production. These are called incidentalomas. Several types based on biochemistry of dominant steroids have been found including androgen, cortisol and precursor steroids (see Chapter 3.1.1). Malignant adrenal tumors that do not secrete excess cortisol cannot be reliably identified by CT scan. Prior to removal of the tumor a steroid profile has proved very effective in identifying abnormalities in function (Bancos et al., 2020). A patient in the neurofibromatosis clinic at UCLH had an adrenal mass on ultrasound (Fig. 3.2.1.48). The urine steroid profile demonstrated elevated levels of androgen precursors (pregnenediol, pregnenetriol) DHEAS as well as glucocorticoid precursor metabolite (THS) (Menon et al., 2014).

This incidentaloma was shown to be an adrenal cancer on the basis of ^{18}fluoro-deoxy glucose uptake. DNA analysis demonstrated loss of heterozygosity (LOH) at the NF1 locus in the adrenal cancer, supporting the hypothesis of an involvement of the NF1 gene in the pathogenesis of ACC. Loss of heterozygosity (LOH) analysis of the tumor suggests that the loss of neurofibromin in the adrenal cells may lead to tumor formation. Several other studies have shown raised 11-deoxycortisol in adrenal carcinoma (Bancos et al., 2020; Velikanova et al., 2016). An increased excretion of THS is the steroid most relevant in urine for the differentiation of ACC from ACA, followed by the pregnenolone and 17-hydroxypregnenolone metabolites 5-PD and 5-PT (Velikanova et al., 2016) similar to the findings of

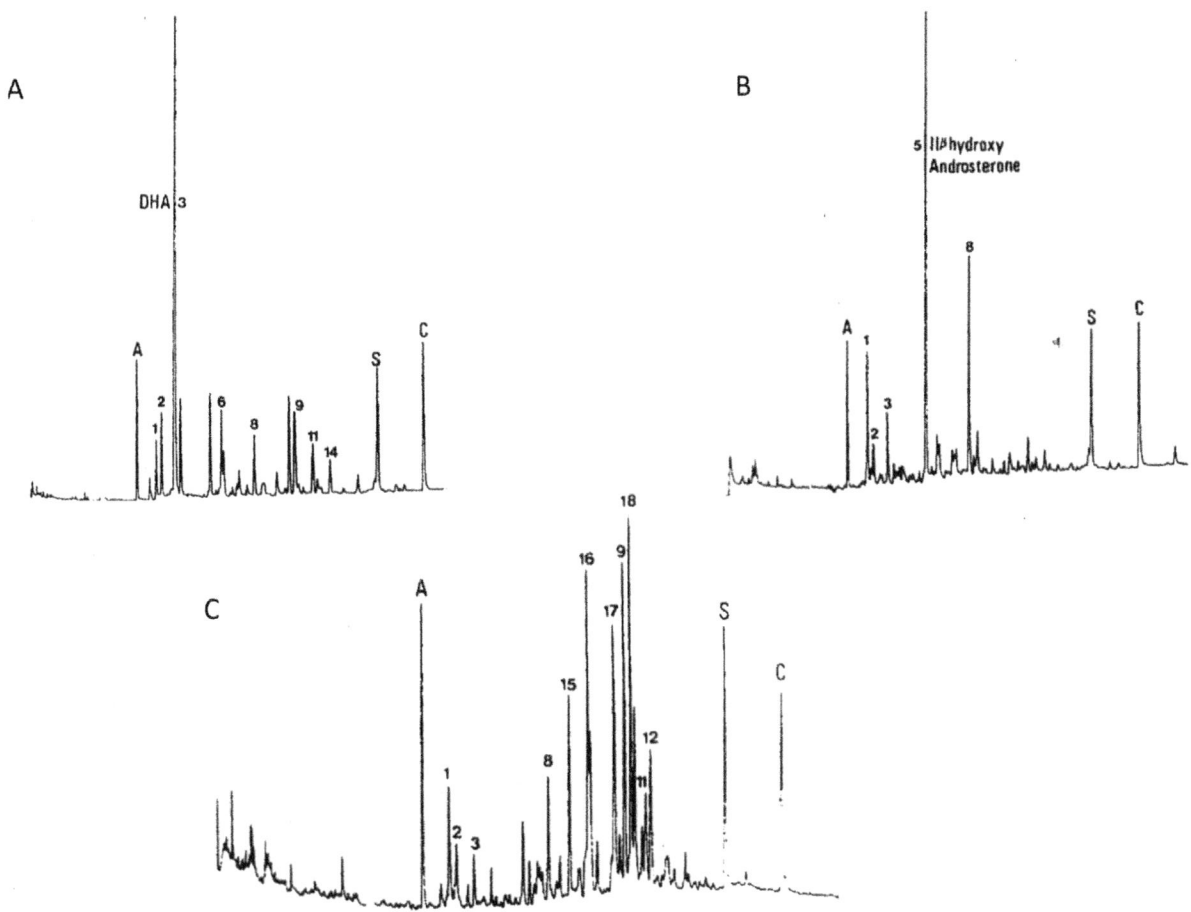

FIG. 3.2.1.47 GC profiles of steroids in urine of children with androgen secreting tumors: (A) 3 years old girl with 4 cm adenoma DHEA secreting tumor. (B) 1 yearr old girl with 3 cm adenoma and high excretion of 11β-hydroxyandrosterone. (C) 2 years old boy with 6 cm adenoma secreting pregnenolone. A 20 m open tubular capillary column with solid injection and flame ionization detector of the steroids was used. Numbered steroids are (1) androsterone, (2) etiocholanolone (3) DHA, (6) 16α-hydroxy-DHA, (8) 5-androstene-3β, 16α,17β-triol, (9) tetrahydrocortisone (11) tetrahydrocortisol, (12) 5α-tetrahydrocortisol, (14) β−cortolone with β-cortol, (15) 5-pregnene-3β,20α-diol, (16) 5-pregnene-3β,16α,20α-triol (17) 16α-hydroxypregnenolone, (18) 5-pregnene-3β, 16α,20β-triol. *(From Honour JW, Price DA, Taylor NF, Marsden HB, Grant DB. Steroid biochemistry of virilising adrenal tumours in childhood. Eur J Pediatr 1984;142(3):165–9. Fig. 1 p. 167.)*

Gröndal et al. (1990) and others. Steroids in plasma are in agreement with the urine observations (Chortis, 2019). Similar patterns have been seen in urine patients with incidentalomas. A machine learning analysis determined the 11-deoxycortisol metabolite THS as the single most important steroid metabolite in differentiating postrecurrence urine samples from samples provided by nonrecurred patients, followed by the mineralocorticoid precursor metabolite tetrahydrocorticosterone, the pregnenolone metabolite pregnenediol and the androgen metabolite etiocholanolone (Chortis et al., 2020).

3.2.1.3.6 Ovarian tumors

Ovarian tumors can arise from sex cord cells which surround the oocytes as stromal cell or Leydig cell tumors. They are often less than 1 cm in diameter and difficult to locate with ultrasound. MRI or laparoscopy may be better

(Klimek et al., 2020; Chen et al., 2018). Ovarian tumors can secrete androgens and other patterns. Tumors are often found after the menopause possibly because of an increase in free androgens as the decline in estrogen leads to reduction in sex hormone binding globulin. Rapid progression of hirsutism indicates a tumor etiology. Other signs can be forehead baldness, acne, coarser voice and clitoromegaly. Ovarian and adrenal veins sampling may be needed in some cases. Access to the right ovarian vein can be difficult due to anatomical factors (Tong and Tai, 2020). Markedly elevated testosterone concentrations are found. Oophorectomy is the usual treatment (Escobar-Morreale et al., 2012).

3.2.1.3.7 Androgenic alopecia

Hair loss (alopecia) is a sign of androgen excess in men but less so in women although presentation is more common at

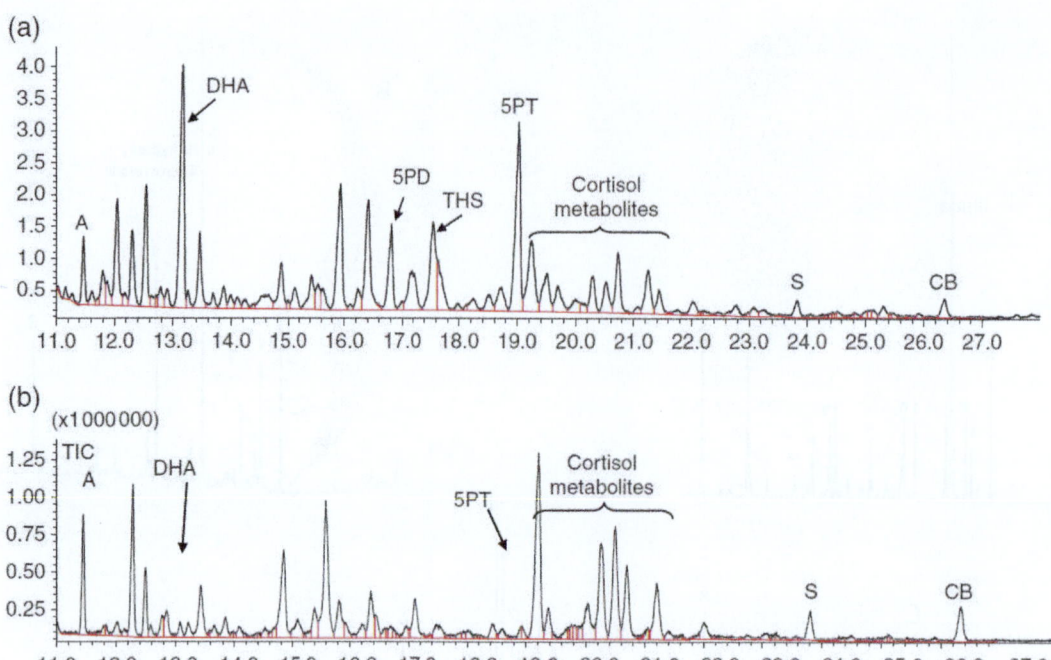

FIG. 3.2.1.48 Total ion current chromatograms (GC-MS) (A) 49 years female patient under follow-up at specialist neurofibromatosis clinic with novel heterozygous mutation in NF-1 gene (B) matched control. A 24-h urinary steroid profile is shown of the patient preadrenalectomy compared with a normal control. The profile is dominated by the Δ5 steroids, in particular pregnenetriol (5PT), DHEA (DHA) and 5PD and the glucocorticoid precursor, THS. Peaks A, S, and CB are internal standards. *(From Menon RK, Ferrau F, Kurzawinski TR, Rumsby G, Freeman A, Amin Z, Korbonits M, Chung TT. Adrenal cancer in neurofibromatosis type 1: case report and DNA analysis. Endocrinol Diabetes Metab Case Rep 2014;2014:140074. Fig. 3 p. 3.)*

FIG. 3.2.1.49 Hair loss pattern men. *(From Olsen EA. Current and novel methods for assessing efficacy of hair growth promoters in pattern hair loss. J Am Acad Dermatol 2003;48(2):253–62. Fig. 1 p. 255.)*

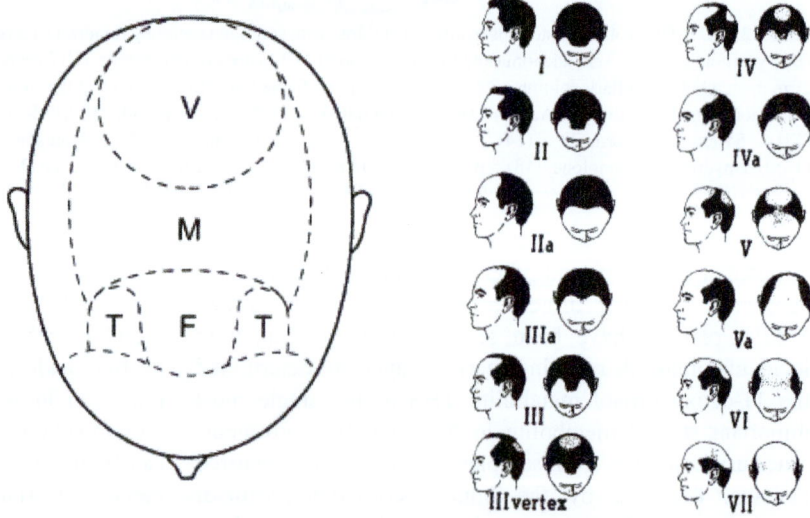

puberty and around the menopause. Women with shorter (<15) CAG and GGC repeats in the androgen receptor gene are at an increased risk of developing female pattern hair loss (FPHL).

Male androgenic alopecia and female pattern hair loss show different patterns of balding and are graded on different scales (Carmina et al., 2019; Fabbrocini et al., 2018; Dinh and Sinclair, 2007). Male hair loss is graded on the Hamilton-Norwood scale. A receding hairline in men may progress with accentuated temporal regression, and a bald spot on the vertex of the scalp. Eventually the fronto-parietal scalp may become completely bald (Fig. 3.2.1.49).

Women with FPHL first notice increased shedding of hair before the onset of a thinning scalp (Shapiro, 2007). FPHL is graded on multiple scales, such as the Ludwig scale, and the newest is the Sinclair scale (Fig. 3.2.1.50).

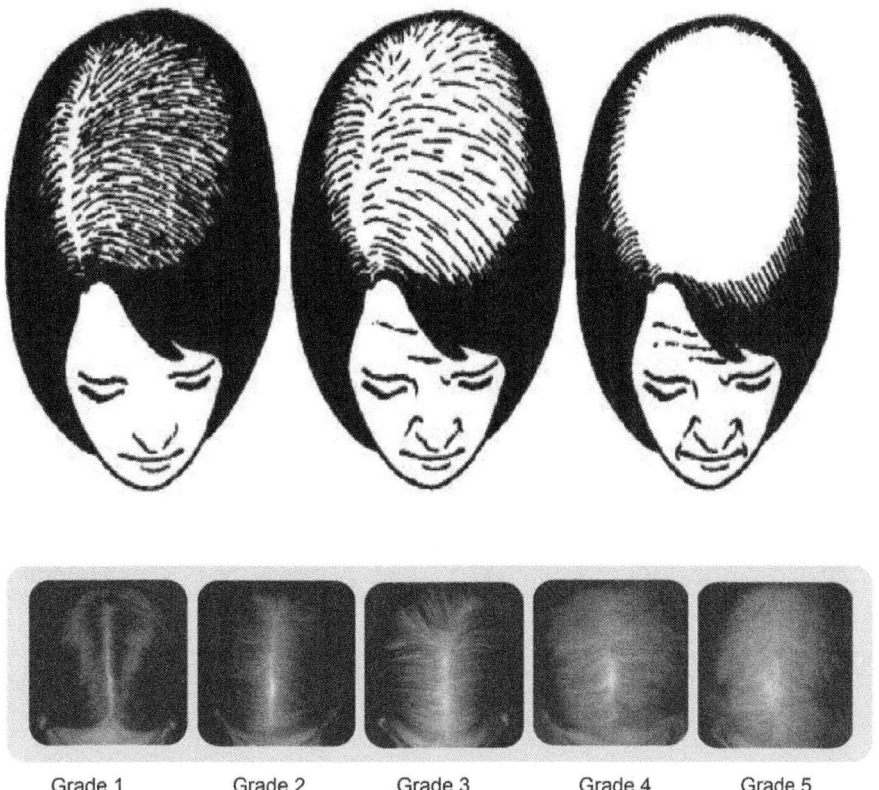

Grade 1   Grade 2   Grade 3   Grade 4   Grade 5

FIG. 3.2.1.50 Ludwig scale (upper images). Grade I: Perceptible thinning of the hair on the crown, limited in the front by a line situated 1–3 cm behind the frontal hair line. Grade II: Pronounced rarefaction of the hair on the crown within the area seen in Grade I. Grade III: Full baldness (total denudation) within the area seen in Grades I and II. Sinclair scale (lower images). Grade 1: is normal. This pattern is found in all girls prior to puberty but in only forty-five percent of women aged eighty or over. Grade 2: shows a widening of the central part. Grade 3: shows a widening of the central part and thinning of the hair on either side of the central part. Grade 4: reveals the emergence of a diffuse hair loss over the top of the scalp. Grade 5: indicates advanced hair loss. *(Modified from Dinh QQ, Sinclair R. Female pattern hair loss: current treatment concepts. Clin Interv Aging 2007;2(2):189–99. Fig 1 p190; Fig 2 p191.)*

FPHL is associated with an increased risk for cardiovascular and metabolic diseases, specifically PCOS, diabetes mellitus and carotid atherosclerosis. Male pattern alopecia has an association with dihydrotestosterone and inhibition of SRD5A2 is an effective treatment for many cases. In females, low concentrations of estrogens may be the basis for hair loss.

The investigations effects of androgen excess are shown in Fig. 3.2.1.51. A raised testosterone suggests a tumor, congenital adrenal hyperplasia or PCOS and is followed with MRI and CT scan and venous catheterization or measurement of 17-hydroxyprogesterone.

3.2.1.3.8 Androgen excess in pregnancy

Signs of androgen excess appearing in pregnancy range from acne, hirsutism and sometimes hair loss to clitormegaly and voice deepening and nearly always are due to circumstances within the pregnancy. High levels of testosterone are found because of the estrogen stimulated increase in SHBG concentrations. A tumor in the ovary (Luteoma) or adrenals may secrete androgens, raised androstenedione/testosterone can be due to CAH and estrogen synthesis

can be blocked due to sulfatase or aromatase deficiency (Hakim et al., 2017).

Hyperreactio luteinalis is a rare condition that stems from theca cell hyperplasia in the ovaries due to the high levels of human chorionic gonadotropin during gestation. It occurs commonly in pregnant patients with trophoblastic disease, occasionally in multiple pregnancies and rarely in normal singleton pregnancy. Subjects with **ovarian cysts** show pronounced excesses in urine of androgen metabolites, 17α-hydroxypregnanolone, pregnanediol and pregnanetriol a pattern similar to CAH with 21-hydroxylase deficiency (Fig. 3.2.1.52) (Bevan et al., 1986).

A **luteoma** is a large solid, nontumor mass near the ovary that secretes androgens in about 30% of cases (Fig. 3.2.1.53). Sometimes the child is virilized, lactation is delayed and the luteoma regresses in the first months after the pregnancy. The excess androgen may not always cause severe symptoms because the functions of the fetoplacental unit convert the androgens to estrogens. The luteoma may be recognized on ultrasound imaging during the pregnancy. An extremely rare adrenal tumor in the mother may also be discovered in this way.

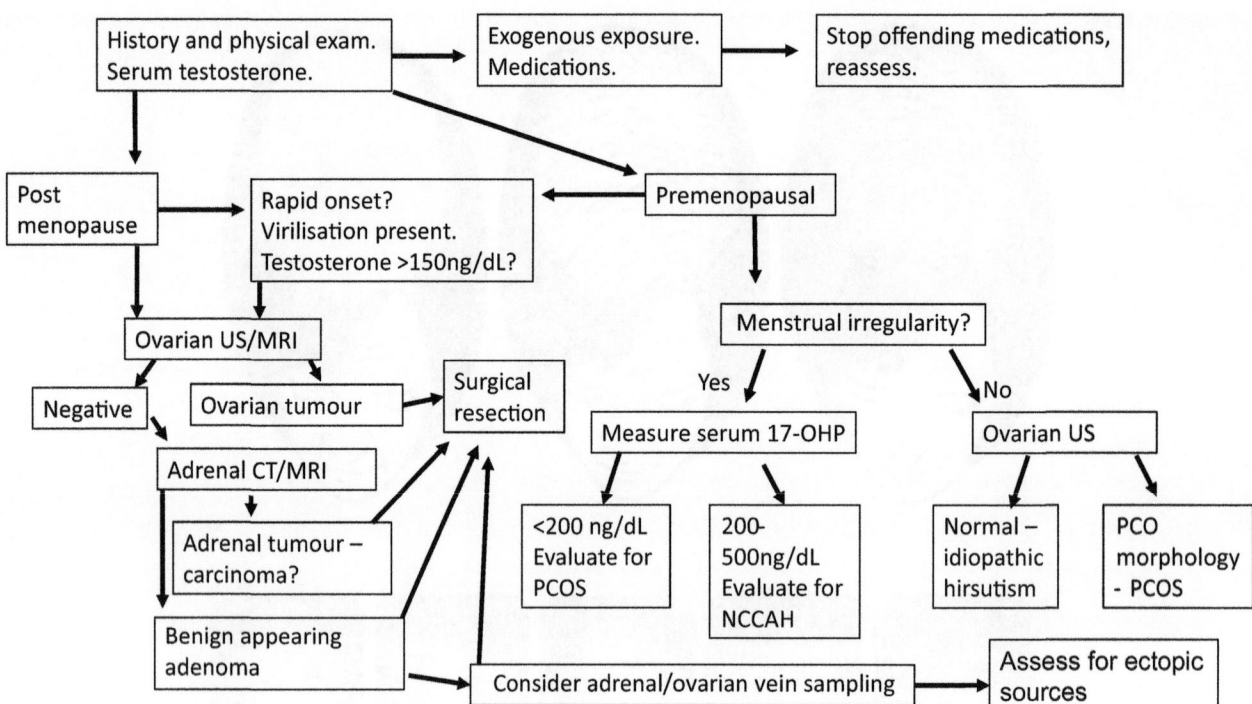

FIG. 3.2.1.51   Investigation of hyperandrogenism.

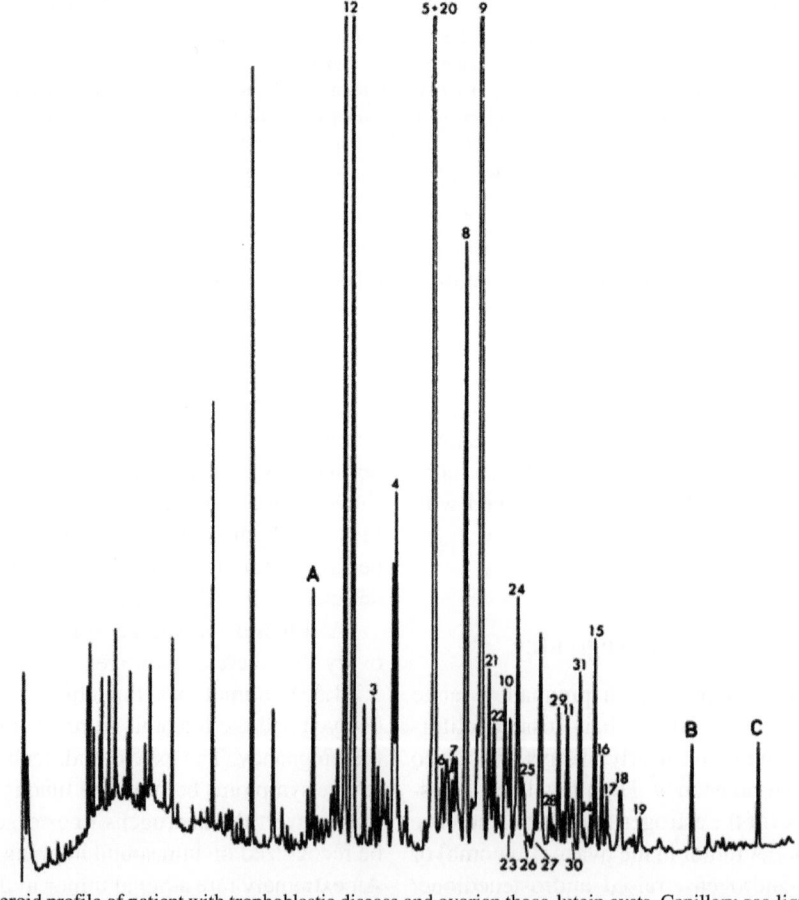

FIG. 3.2.1.52   Urinary steroid profile of patient with trophoblastic disease and ovarian theca-lutein cysts. Capillary gas-liquid chromatogram of urinary steroids in case 2. The major numbered steroids are 1. Androsterone 2 etiocholanolone 4. 11-oxo-etiocholanolone 5. 11-hydroxyandrosterone 20. 17-hydroxypregnanolone 8. Pregnanediol 9. Pregnanetriol. The peaks ABC are internal standards = 5a-androstane-3a-diol, stigmasterol and cholesteryl butyrate, respectively. *(Courtesy of Dr. Norman Taylor, Kings College Hospital, London, England.)*

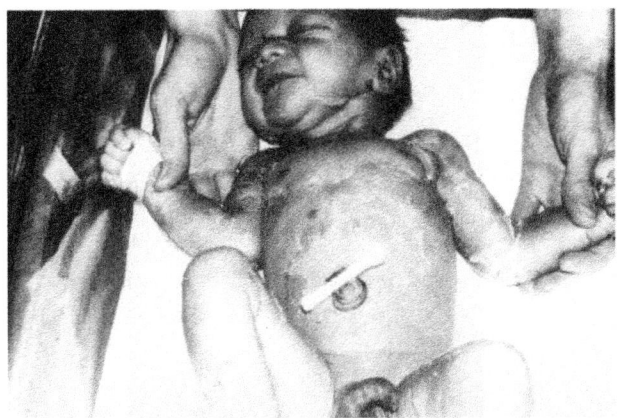

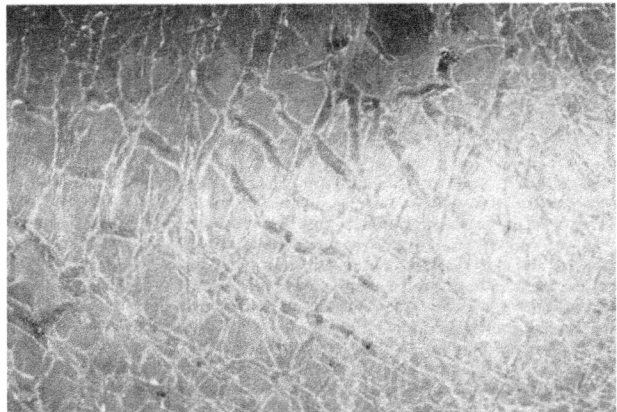

FIG. 3.2.1.53 (A) Parchment-like skin on 3 day old infant with sulfatase deficiency. (B) Ichthyosis with large dark scales and sparing of the flexures at 6 months. *(Author original.)*

Laboratory examination of androgen excess is focused on determining the level of testosterone, androstenedione, dehydroepiandrosterone and its sulfate, dihydrotestosterone, SHBG, cortisol and estrogens (estradiol, estriol). Ultrasound examination of the adrenal glands and ovaries enables an assessment of the size of the formation, distinguish their solid or cystic nature, unilateral and bilateral affliction, and sometimes reveal changes in the genitalia of the fetus.

3.2.1.3.8.1 Aromatase deficiency

In rare cases of aromatase deficiency from mid gestation, a mother can begin to suffer from progressive virilization, including severe acne on her face, pigmentation, lowering and breaking of her voice, clitoral hypertrophy, and masculinizing changes in her face (Kaňová and Bičíková, 2011; Jones et al., 2007a; Morishima et al., 1995; Shozu et al., 1991). These signs of virilization worsen progressively until delivery. Aromatase deficiency has however been reported in the absence of a history of maternal virilization (Bouchoucha et al., 2014; Özen et al., 2020; Dursun and

Ceylaner, 2019) but this is a rare finding. In the last week of pregnancy, estrogens in serum are low and DHEAS and DHEA are raised (Table 3.2.1.6), these concentrations fall dramatically in the first week after the delivery. The maternal manifestations of virilization disappear over months after the delivery although this is longer than might be predicted due to persistently elevated hCG (Riedl et al., 2013).

At birth, the child has masculine-appearing external genitalia: a greatly enlarged phallus (Prader V), complete fusion of posterior scrotolabial folds, rugation (wrinkles as in a scrotum) of the scrotolabial folds and a single meatus on the top of a phallus. Following a clinical examination 46XX DSD is suspected, and a comprehensive workup initiated. The karyotype of the patient is 46,XX. To rule out the possibility of 46,XX true hermaphroditism, the infant should be given 1500 U human CG, im daily for 3 days. Serum free T does not increase. Further, the infant is SRY negative, and the plasma concentration of the anti-Müllerian hormone is less than the sensitivity of the assay. These observations are consistent with the absence of testicular tissue. A blood screening test for 21-hydroxylase deficiency indicates normal concentration of 17-OH progesterone; a urine steroid profile is normal, and no evidence for a virilizing form of congenital adrenal hyperplasia is found. The diagnosis of a nonadrenal form of DSD is made. At 2 months of age, FSH and testosterone are elevated, consistent with aromatase deficiency (Unal et al., 2018). Ultrasonographic examination of the patient at the age of 6 months reveals slightly enlarged ovaries and a vagina that entered a persistent urogenital sinus.

Girls may present later with primary amenorrhea and osteopenia. Some cases may show signs of virilization due to lack of aromatization of androgens. Affected males do not present with obvious defects at birth but later in life have tall stature, delayed skeletal maturation, delayed epiphyseal closure, bone pains, eunuchoid body proportions and adiposity (Rochira and Carani, 2009) (see Chapter 3.2.2).

Approximately 40 variants of CYP19A1 have been reported including missense and nonsense mutations, deletions, insertions and splicing variants (http://www.hmgd.cf.ac.uk/ac/index.php).

3.2.1.3.8.2 Sulfatase deficiency

This condition is also suspected prenatally when low estriol is found in a triple screening test in pregnancy or the **mother shows clinical signs of virilization**. The low estrogens in pregnancy urine (Table 3.2.1.6) are the result of sulfatase deficiency in the fetus and placenta so C_{19} steroid sulfates are raised in the urine.

TABLE 3.2.1.6 Excretion of steroid sulphates in pregnancy urine (μg/24 h).

	Normal (n = 10) (mean ± SE)	Placental sulfatase deficiency (n = 12) (mean)
Gestation age at collection (weeks)	36	37.5
16α-hydroxy DHA	476 ± 99	8990
16-oxo-androstenetriol + 15β,16α-dihydoxy DHA	153 ± 61	1809
Androstenetriol	326 ± 93	5676
16,18-dihydroxy DHA	<100	2195
16α-hydroxypregnanolone	84 ± 84	3284

From Harkness RA, Taylor NF, Bowman PR, Gordon H, Cummins M, Valman HB. The causes of low oestrogen excretion in pregnancy: the development of diagnostic methods for the antenatal detection of familial congenital adrenocortical hypoplasia. Clin Endocrinol (Oxf) 1980;12(5):453–60.

This is an X-linked condition due to deletions or X-Y translocation of the Xp22.3 locus. There may be delay in delivery and induction of labor may be required (Dreyer et al., 2018). Data from California show the overall prevalence is 1 in 1500 males (Craig et al., 2010). The newborn infant may show ichthyosis (flaky skin scales) due to accumulation of cholesterol sulfate in the epidermis (Elias et al., 2004a) and cryptorchidism depending on the extent of chromosome deletion affecting the Kallmann's syndrome locus.

Gonadotrophins are low but respond to LHRH stimulation. Urine steroids show high excretion rates of sulfated C$_{19}$ steroids (Sánchez-Guijo et al., 2016; Marcos et al., 2009) and low estrogens (Fig. 3.2.1.54) A urine steroid profile of maternal urine resembles the pattern of a newborn infant because the placenta without sulfatase activity produces no substrate for aromatase activity.

Most X-linked ichthyosis patients have extensive deletions (complete or partial) of the *STS* gene (Ben Khelifa et al., 2013); however, point mutations may result in complete STS deficiency. Female carriers of *STS* gene do not exhibit any manifestations because the gene is localized to a region of the X-chromosome that does not undergo X-inactivation. De novo *STS* mutation also occurs (Wells and Kerr, 1965; Elias et al., 2004a). A subset of patients will have a deletion associated with microdeletions/contiguous gene syndromes that can cause a more severe phenotype. For example, deletion of the VCX3A gene on Xp22.3 has been shown to cause XLI with an abnormal neurocognitive (intellectual disability) in some persons (Zhu et al., 2016; Winge et al., 2011; Ben Khelifa et al., 2013). The prepubertal rise in DHEAS (adrenarche) was absent in STSD patients (Idkowiak et al., 2016) but an increase in 5α-reductase activity might enhance peripheral androgen activation.

3.2.1.3.9 Androgen excess after the menopause

Some postmenopausal ladies can present with acne and sometimes hair loss as signs of androgen excess. Increased hair growth may also be on chin, lips and abdomen which can be scored using Ferriman-Gallway scoring. Some women have lowering of the voice and clitoromegaly. Many cases will be obese or have acne (Khunger and Mehrotra, 2019). Headaches and visual field defects should be investigated. The onset of symptoms may be short or longstanding before or after the menopause. The patient should be investigated for cortisol and androgen excess by blood and urine steroid analysis. A dexamethasone suppression test should be performed. Reference ranges pre and post menopause are needed to interpret the endocrine results (Elhassan et al., 2018). Ultrasound scan should look for adrenal or ovarian tumors. Ovarian tumors that present with hyperandrogenism include Leydig cell tumors, Sertoli cell tumors and ovarian hyperthecosis (Elhassan et al., 2018; Sharma et al., 2018; Czyzyk et al., 2017). Basal testosterone overlaps between various etiologies. Histopathological findings are the gold standard for ovarian hyperandrogenism (Yance et al., 2017). GnRH analogue stimulation tests and ovarian/adrenal catheter studies are used but rarely (Tng and Tam, 2020). Ultrasound will reveal most tumors but small ovarian tumors may need MRI scan (Zou et al., 2021).

3.2.1.4 Iatrogenic causes of hyperandrogenism

Antiepileptic drug treatment can affect serum concentrations of sex hormones especially if they induce liver enzymes. Polycystic ovaries, hyperandrogenism, menstrual

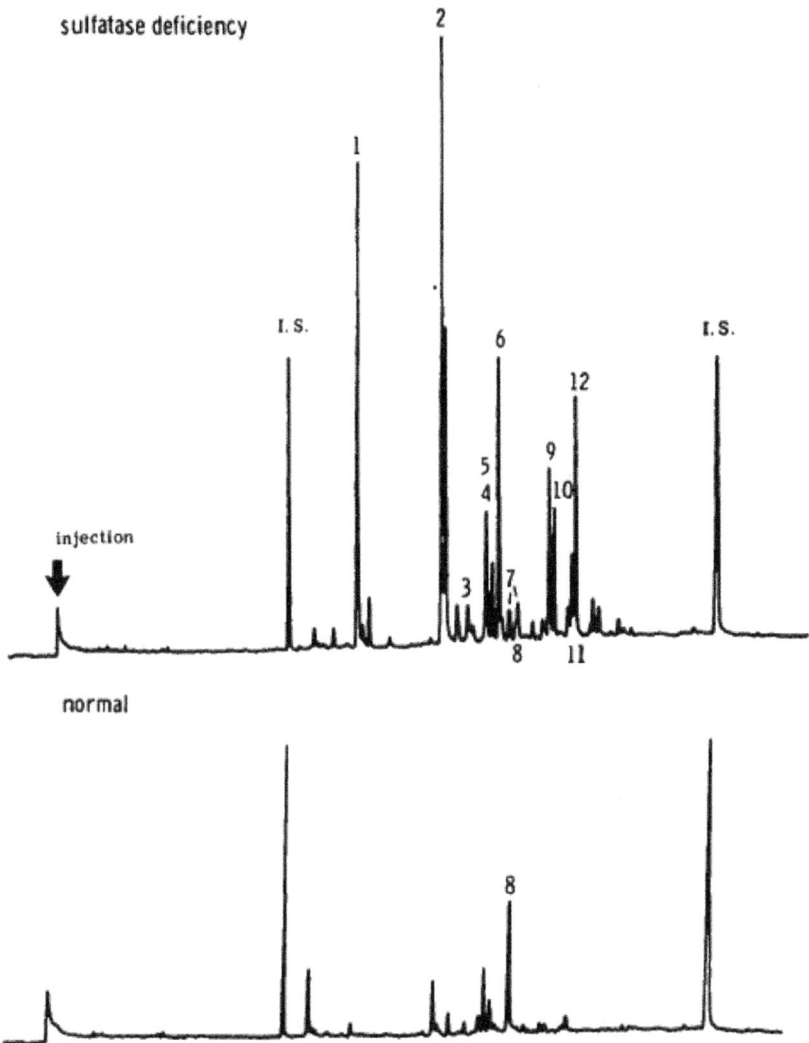

FIG. 3.2.1.54 Urine steroids normal pregnant woman and pregnant woman with PSD (A) Glass capillary column GC of methyloxime trimethylsilyl ether derivatives of the steroid monosulfate fractions of urine from normal woman and pregnant woman with PSD. I.S., Internal standards androstanediol and cholesteryl butyrate; (1) DHEA; (2) 16α-hydroxyDHEA (two peaks); (3) 16 β-hydroxy-DHEA; (4) 16-oxo-androstenediol (not resolved in this system); (5) 15β,16α-dihydroxy-DHEA (not resolved in this system); (6) androstenetriol; (7) 16ε,18-dihydroxy-DHEA (two peaks); (8) estriol; (9) 16α-hydroxypregnenolone; (10) 16α-hydroxypregnanolone; (11) 5-pregnene-3β,16α,20α-triol; (12) 5α-pregnane-3β,16α,20α-triol. *(From Shackleton CH, Roitman E, Kratz LE, Kelley RI. Midgestational maternal urine steroid markers of fetal Smith-Lemli-Opitz (SLO) syndrome (7-dehydrocholesterol 7-reductase deficiency). Steroids 1999;64(7):446–52.)*

disturbances and hirsutism are seen in patients with epilepsy on sodium valproate (Isojärvi et al., 1993). Amenorrhea is a side effect of valproate treatment. Serum testosterone concentrations are raised but serum LH is normal. The effect of valproate on ovarian function and androgen production is not caused by altered pituitary LH secretion. Carbamazepine does not have effects on menstrual cycles. Valproate is not used as much now and also risk of fetal malformation to kept away from females.

A female or child may come in contact with testosterone prescribed to a male for hypogonadism. DHEAS is available without prescription as a nutritional supplement. The use of androgenic anabolic steroids has to be excluded. These can

be taken to enhance performance in sport or for cosmetic effects. Side effects include acne (Fig. 3.2.1.55). Testosterone and DHEA are taken by women to improve libido and overtreatment is possible.

In men, there can be suppression of the HPG axis, reduced sperm count, lower testicular volume, aggressive behavior and personality disorders. These effects are reversible. In women, there can be menstrual abnormalities, acne, hirsutism, clitoromegaly, shrinkage of the breasts and deepening of the voice. If anabolic steroids are taken by young persons growth may be halted. Damage to the liver (jaundice, tumors) is associated with a 17α-alkylated steroid taken orally. Left ventricular hypertrophy

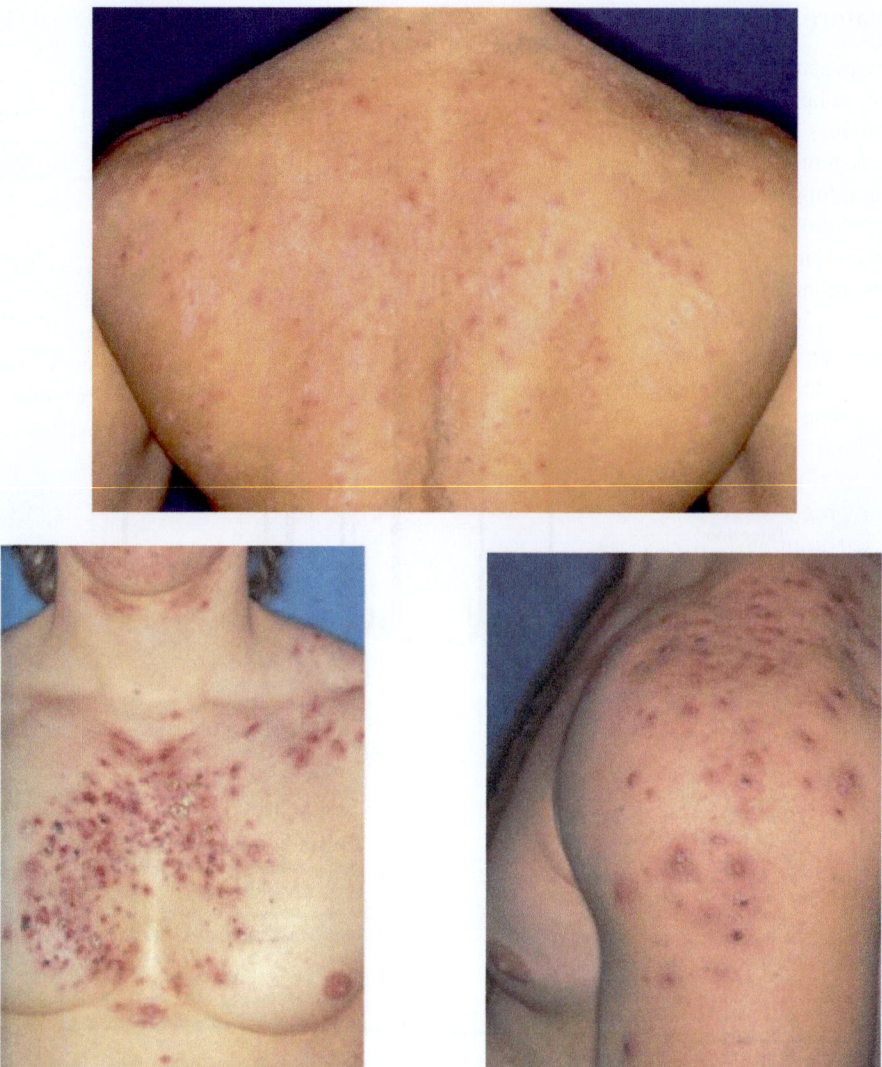

FIG. 3.2.1.55 Severe acne conglobata induced by anabolic-androgenic steroids. Upper image shows lesions include papules, pustules, abscesses and deep ulcerations. In the lower image is of the patient after 6 weeks of antiseptic-antibiotic therapy (C). *(From Gerber PA, Kukova G, Meller S, Neumann NJ, Homey B. The dire consequences of doping. Lancet 2008;372(9639):656. Fig. 1 p. 656.)*

is a distinct risk with anabolic steroid use. Gonadotrophins will be suppressed and the steroids may be detected indirectly by cross reaction in an immunoassay or directly by GC-MS or LC-MS/MS of steroids in urine. A WADA antidoping laboratory may be needed to assist here because of their broad experience. No patient preparation or sample preservative is required, which is not seen as good practice in hospital practice but accepted in forensics and toxicology.

Exposure to endocrine-disrupting chemicals such as 1,1,1-trichloro-2,2-bis(4-chlorophenyl)ethane (DDT), monobutyl phthalate (MBP), n-nonyl phenol (n-NP), t-octylphenol (t-OP) and isofavones like equol, genistein and daidzein can influence hormonal dysregulation leading to PP. Premature thelarche and prepubertal gynecomastia

have been attributed to continuous exposure to lavender fragranced products (Ramsey et al., 2019) or flame retardants of polybrominated diphenyl ethers (Deodati et al., 2016).

3.2.1.5 Estrogen excess

Males can present with rapidly developing gynecomastia (breasts) if excess estrogen is encountered. A rare adrenal carcinoma may be the cause (Vogt et al., 2021) or testicular cancer (Hassan et al., 2008). Exposure to chemicals as "endocrine disruptors" may also be an issue with personal care products (Hawkins et al., 2020), environmental and industrial exposure as well as natural sources such as lavender oil (Henley et al., 2007).

3.2.1.5.1 Premature thelarche (PreT)

Girls can present with isolated breast development before the normal age of puberty which can be unilateral or bilateral and may regress or persist. Breasts can be Tanner stage 2 or 3 and can be tender to palpation with no discharge. Dietary intake (including substances in breast milk) or exposure to excess phyto-estrogens, lavender oil, tea tree oil, fennel tea, organo-halogen pesticides (Deodati et al., 2016) and flame retardants should be excluded. PreT was found in several children adopted from developing countries and likely exposed to toxic chemicals. Recombinant receptor assays have been used to assess estrogenic activity with variable results. The breast enlargement can wax and wane in cycles of about six weekly intervals. Growth and epiphyseal maturation is normal. In girls less than 2 years, the disorder does not progress to central precocious puberty and is essentially self-limiting. Further investigations are unnecessary unless on follow-up, breast enlargement had persisted for more than 6 months.

In older girls, there can be cyclical uterine withdrawal bleeding which is of considerable concern to families already alarmed by the breast changes. A temporary activation of the HPG axis with raised gonadotrophins, predominantly FSH can be demonstrated in variant cases. According to LHRH stimulation tests with central precocious puberty basal LH exceeds 0.5 IU/L, peak LH is greater than 5.0 IU/L and peak LH/FSH is more than 0.6. Congenital adrenal hyperplasia, estrogen secreting tumors and aromatase excess syndrome should be excluded. Ovarian morphology on ultrasound is usually characterized by isolated cyst formation. A meta-analysis of published papers with ultrasound findings teased out a uterine length greater than 3.2 cm exhibited diagnostic accuracy in differentiating risk of CPP (Nguyen et al., 2021). Ovarian parameters were inferior markers. An experienced radiologist is required since interpretation of scans is challenging. There is at this time insufficient data to opine on value of measurements of Makorin ring finger protein (MKRN3 gene) although some suppression of circulating protein in PT has been found and greater suppression of results in CPP (Ge et al., 2020).

3.2.1.5.2 Gynecomastia

In spite of being rare, gynecomastia is encountered in clinical practice as the expression of estrogen excess with benign proliferation of glandular tissue in a male. Pubertal signs are common, usually as a transient idiopathic event due to hormone imbalance. Trauma can be another cause, for example, a 10-year-old boy in a farming family who was carrying water melons from land to the transport vehicle using his dominant left arm causing irritation to the pectoral region (Laimon et al., 2021) and mainly hypertrophy of fibro-fatty tissue on ultrasound examination.

Excessive consumption of dietary soy was reported in one case that resolved on exclusion of dietary products containing phyto-estrogens (Sea et al., 2020). It is imperative that the clinician is vigilant for underlying malignancy (Ali et al., 2018).

On physical examination, patients can have Tanner 2–4 stage breasts, children may lose height and be short. Most patients have normal puberty although testes can be small and facial hair may be sparse. Adult men have features of hypogonadism such as reduced libido and erectile dysfunction. The patient should be tested for serum testosterone and estradiol (by methods with mass spectrometry), SHBG, LH, FSH, prolactin, hCG, free thyroxine, and TSH. Increased estrogen concentrations result in increased SHBG that in turn leads to lowering of free testosterone concentrations to increasing the estrogen:androgen ratio. If serum estradiol is above 150 pmol/L by LC-MS/MS (173–287 pmol/L depending on immunoassay) (Handelsman et al., 2014) or hCG is elevated a testicular ultrasound should be performed and further imaging of adrenals. Gonadotrophin levels may be normal or low. FSH levels if low respond poorly to GnRH stimulation. Tumors secreting hCG affect men in third or fourth decade of life and about 10% of cases have gynecomastia. Klinefelter syndrome needs to be excluded.

Gynecomastia may be secondary to chronic liver or kidney disease, hypogonadism, hyperthyroidism and some medications. Gynecomastia can be secondary to Sertoli cell tumor (Gourgari et al., 2012; Giglio et al., 2003) or adrenal tumor (Moreno et al., 2006; Balakumar et al., 1997) that may be accompanied by excess glucocorticoid. Leydig cell tumors can be seen in prepubertal boys or adult men. Some patients with androgen insensitivity syndrome and 5α-reductase deficiency can have gynecomastia after pubertal age because of raised testosterone that is aromatized to estradiol. Patients with 17-ketosteroid reductase deficiency do not get gynecomastia because estrone from androstenedione is not reduced to estradiol. AMH is elevated in hyperestrogenic states (Valeri et al., 2020). Treatment with an aromatase inhibitor such as anastrazole should ameliorate progression of the gynecomastia and the patient may have to undergo mastectomy.

3.2.1.6 Progesterone resistance

Progesterone in the menstrual cycle prepares the endometrium for blastocyst implantation and then is important for the maintenance of pregnancy. The corpus luteum is the major sources of progesterone during the second half of the menstrual cycle and the placenta is the source at the beginning of pregnancy. Luteinizing hormone (LH) is responsible for stimulating progesterone synthesis by the corpus luteum of the menstrual cycle and chorionic gonadotropin for the corpus luteum of pregnancy. Reproductive competence would be impossible with **complete end-organ resistance to progesterone**. In males, progesterone has no known function and would not be affected by hormone resistance. Failure of the uterus to respond to progesterone would lead to a proliferative endometrium incompatible with blastocyst implantation. Partial resistance to progesterone would be expected to be associated with various degrees of luteal phase defect with incomplete maturation of the endometrium leading to infertility or early abortions. Plasma progesterone concentrations would be normal or elevated.

In normal endometrium, progesterone acts via the PR to increase transcription of proteins such as HSD17B2. In **endometriosis,** affecting 10% of women of reproductive age, there is estrogen dependent growth of endometrial tissue outside the uterus causing pelvic pain (Bulun et al., 2019). Endometrial cells fail to downregulate genes needed for cell cycle regulation and decidualization. There are many possible explanations for the condition around stem cells, retrograde cells from the uterus, stress, estrogen action, inflammation and lack of progesterone action (Susheelamma et al., 2018). The pathogenesis likely originates from eutopic endometrium. Genetic causes of progesterone resistance under investigation include polymorphisms, altered mRNA expression and epigenetic modifications of transcribed products.

Using LC-MS/MS normal endometrial concentrations of progesterone and T mainly correspond to serum concentrations, whereas impaired local synthesis and metabolism determine the tissue sex steroid concentrations in endometriosis (Huhtinen et al., 2014). The elevated tissue P4 and T concentrations in endometriosis did not show cycle-dependent changes, so it was concluded that these concentration differences, together with higher tissue E2 concentration, are critical for the pathogenesis of endometriosis. In addition, there was weak but consistent changes in tissue steroid concentrations and related gene expression in eutopic endometrium, which remains to be further characterized. These novel data suggest that local steroid metabolism is a key step in the regulation of steroid hormone action in endometriosis. Progesterone and estrogen signaling are disrupted, commonly resulting in progesterone resistance and estrogen dominance. This hormone imbalance leads to heightened inflammation and may also increase the pelvic pain of the disease and decrease endometrial receptivity to embryo implantation (Marquardt et al., 2019). Two thorough reviews of the literature has revealed that there is increasing evidence for intracrine metabolism of DHEA sulfate to the estrogens accumulated in endometriosis (Konings et al., 2018; Piccinato et al., 2018). There is a strong expression of sulfotransferase (STS) in human endometrial tissue and in ectopic tissue from patients with endometriosis. The higher STS activity in human ectopic

endometrial tissue correlates with worse severity of disease. Furthermore, inhibition of STS has been shown to decrease the size of human endometrial explants tissue that has been implanted in mice as a model of endometriosis (Foster, 2021). Oral contraceptives, GnRH antagonists, aromatase inhibitors and antiprogestins are used in treatment of the condition to block menstruation along with pain control (Donnez and Dolmans, 2021).

3.2.1.7 Summary

In the past 10 years, there has been a huge improvement in the abilities of laboratories to support investigations of hyperandrogenism. Immunoassay is still important for measurements of gonadotrophins, cortisol, testosterone, DHEAS and rarely 17-OHP concentrations. Assay qualities, particularly for testosterone, have improved and many interferences have been addressed or more clearly understood by the laboratory and clinicians. Many laboratories now offer simultaneous analysis of several steroids in a panel or profile in blood or urine using GC-MS (Holst et al., 2007; Wudy et al., 2018) or LC-MS/MS (Holst et al., 2007; Guo et al., 2006; Kushnir et al., 2006; Rauh, 2009). These tests enable several forms of CAH and other adrenal disorders to be detected. Newborn screening is variable in countries across the globe in terms of number of tests, timing of blood spot collection and quality of assays. Tandem mass spectrometry (Lai et al., 2020; Janzen et al., 2008; Janzen et al., 2007; Higashi et al., 2008; Schwarz et al., 2009) complements immunoassay in NBS.

Despite an immense amount of research there are still difficulties in finding a diagnosis for hyperandrogenism. For example, up to 80% of girls and 40% of boys with precocious puberty may have as yet no identifiable cause and are termed idiopathic. Genetic studies suggest that PP has an autosomal dominant mode of inheritance. Mutations in genes that are involved in sexual development such as MKRN3, KISS1, GPR54, CYP19A1, and LHCGR contribute to PP. Apart from genetic factors, intrauterine growth retardation and low birth weight are linked to early menarche. It is unclear if childhood obesity is cause or effect of PP. Through the application of genomics, proteomics and metabolomics the situation will change with time. There are already gene panels such as HaloPlex (Agilent) which enable screening for many genes (Yang et al., 2019). For the patients, there are many support groups around the world.

The diagnosis of the cause of hyperandrogenism, like many areas in medicine, is becoming more and more complicated and there will be limits to knowledge clinicians can retain. A dialogue between laboratory scientists and clinicians is essential and in the future the interaction will become more important.

3.2.1.7.1 Androgen excess

The investigations of androgen excess have different objectives depending on the age and karyotype of the patient. Thus 46,XX girls present with ambiguous genitalia usually due to inherited metabolic disease (congenital adrenal hyperplasia, CAH) of the adrenal cortex. During childhood boys with precocious puberty may have space occupying lesions of the brain, CAH or very rarely tumors of the adrenals or gonads. A few girls with hirsutism and/or acne with or without menstrual disturbance may have nonclassic congenital adrenal hyperplasia or very rarely have excess secretion of androgens of ovarian or adrenal origin due to tumors. Polycystic ovaries are commonly found in mild hyperandrogenization in adolescence that may progress to infertility in adult life.

In newborn girls with ambiguous genitalia (46XX DSD), the commonest cause is due to enzyme defects of cortisol synthesis with diversion of intermediates to androgen production (Miller, 2008; Krone and Arlt, 2009). The genotype should be confirmed by chromosome analysis. A pelvic ultrasound of the pelvis may also be helpful but results need to be treated with some caution.

Other external causes of a virilized female are attributed to aromatase deficiency (Jones et al., 2007a) and an ovotesticular DSD (Maciel-Guerra et al., 2008), maternal ingestion of progestogens (now very rare) or androgens or to maternal production of androgens by an adrenal or ovarian tumor. In these cases, the child will have normal endocrinology after birth but may need surgery on the external genitalia if androgenization is severe. A reduction in steroid 21-hydroxylase activity or absence of 11β-hydroxylase or 3β-hydroxysteroid dehydrogenase can be the cause of congenital adrenal hyperplasia of which 21-hydroxylase deficiency is by far the commonest cause of CAH (90% of cases). In Europe, 60% of all cases of steroid 21-hydroxylase deficiency will present in the newborn period with a salt-losing crisis. This reflects the nature of the enzyme block. If there is a salt-losing crisis this will need immediate treatment. An increase in serum potassium may be seen prior to a fall in body weight and hyponatremia.

17α-Hydroxyprogesterone is a biosynthetic precursor of cortisol and in patients with deficiency of 21-hydroxylase the production of 17-OHP increases and serum levels are elevated. The reduced capacity to produce cortisol however leads to high ACTH levels which causes adrenal hyperplasia. There is also a high adrenal secretion of androgens which cause virilization. The measurement of 17-OHP in serum, plasma and blood spots is used for the diagnosis of this disorder. The timing of blood sampling is very important in order that the laboratory can interpret the findings. In all newborn infants, 17-OHP concentrations in serum are high on the first day of life (>100 nmol/L) and the levels fall over the first week to below 50 nmol/L. After day 3, there is usually good discrimination of the 17-OHP in affected cases (100–800 nmol/L) from normal infants (<15 nmol/L). Differences in results with a direct and an extraction method would suggest that steroids in blood (probably steroid sulfates from the fetal adrenal) cross react in the RIA giving higher results in the direct assay (Wallace et al., 1987; Wong et al., 1992). The suggestion that steroid sulfates affect the quality of the assay is supported on the observation of greater discrepancy between the results obtained by the 2 methods in premature and low birth weight babies both of which have sustained fetal adrenal activity after birth. Plasma 17-OHP can be mildly elevated in defects of HSD3B2, POR, and CYP11B1.

In rare cases of CAH, the defect is due to low activity of the 11β-hydroxylase enzyme (Nimkarn and New, 2008). This defect is best identified by a raised serum concentration of 11-deoxycortisol or by a urine steroid profile. A high excretion of 6-hydroxy-tetrahydro-11-deoxycortisol (6-hydroxy-THS) is a better marker of the defect in the newborn than THS which is elevated in older patients (>120 nmol/L) but not so clearly raised in the newborn period (Hughes et al., 1986). This again emphasizes the requirement to involve specialized laboratories with relevant experience.

The cause of ambiguous genitalia in a newborn child is required as soon as possible in order to counsel anxious parents and start treatment. A 17-OHP assay which involves solvent extraction before the RIA is essential. Now that specific urine metabolites of 17-OHP have been recognized, the diagnosis by GC analysis of steroids in urine is reliable. Pregnanetriol is one marker for the disorder in the newborn. A characteristic urine steroid pattern can be recognized of which the most informative steroid is 17α-hydroxypregnanolone. It is essential that a laboratory offering this analysis can provide a rapid service (<2 days). In order to achieve this, the identity of the steroids in the GC analysis must be confirmed by a further analysis with GC coupled to a mass spectrometer. Some laboratories now offer simultaneous analysis of several steroids by using GC-MS (Wudy and Choi, 2016) or LC coupled with tandem mass spectrometry (Holst et al., 2007; Guo et al., 2006; Kushnir et al., 2006; Rauh, 2009). These tests enable one of several forms of CAH to be detected and will become important on a regional or national basis. Newborn screening has been based on immunoassay of steroids in blood spots. Repeat testing has eliminated the disease when the concentrations of 17-OHP were lower than an initial high level that suggested 21-hydroxylase deficiency (Cavarzere et al., 2009). This approach is stressful to the family. False positive results are reduced by the use of cut-off levels adjusted for gestation age or birth weight and by solvent extraction of steroids (Fingerhut, 2009). Screening will also improve with more specific technology based on tandem mass spectrometry (Janzen et al., 2007; Higashi et al., 2008; Schwarz et al., 2009).

Once CAH has been confirmed and lifelong cortisol treatment has commenced, compliance is best assessed in children by following growth (Appan et al., 1989). Height and weight should be followed at 3 monthly intervals in the first 2 years of life then at 6 monthly visits. Bone age is checked yearly from X-rays of the wrist and hands. The replacement therapy should be adjusted according to body size. Hydrocortisone is usually prescribed at 9–12 mg/M^2/day with 2/3 of the dose in the morning and 1/3 in the evening. Some centers give hydrocortisone four times per day with 35% at 6 a.m., 25% at lunchtime, 10% late afternoon and 30% at midnight to achieve cortisol concentrations nearer the circadian rhythm. If fludrocortisone is given to control salt loss it must be remembered that this steroid is a potent glucocorticoid itself and the dose should not exceed 0.15 mg/m squared to a maximum of 0.1 mg per day. Any higher dose needed should only take place after careful consideration of sodium balance and renal function. Sodium supplements may be needed in infancy. Electrolytes can be measured periodically but for long-term assessment of mineralocorticoid replacement the measurement of plasma renin activity is advisable. The measurement of 17-OHP and androgens in blood (or saliva) taken at regular intervals will define the adrenal steroid output in relation to treatment but in practice has little effect on a patients drug taking habits. Androstenedione measurements may be helpful in the management of treatment for CAH due to 21-hydroxylase and 11β-hydroxylase defects. Measurement of PRA can be used to monitor efficacy of treatment in salt-losing CAH. Patients with 21-hydroxylase and some with 3β-hydroxysteroid dehydrogenase defects manifest elevated PRA while in the defects with mineralocorticoid excess (17α-hydroxylase and 11β-hydroxylase) PRA is suppressed. PRA is normalized with effective treatment. In the case of 21-hydroxylase deficiency and 3β-hydroxysteroid dehydrogenase defects, treatment is improved with addition of fludrocortisone.

When sexual maturation appears before 8 years in girls and 9 years in boy's puberty is considered precocious. The diagnosis of precocious puberty has been reviewed (Berberoğlu, 2009; Traggiai and Stanhope, 2003). Normal puberty is initiated by an increase in pulsatile secretion of GnRH at night although the mechanism and timing of this initiation is still not understood. Other factors need to be considered such as the contribution of adipose tissue to steroid synthesis, gastrointestinal function, energy sensing, physical and psychological stress, fetal and early life stress and exposure to endocrine disruptors (Livadas and Chrousos, 2019).

Failure to find a pelvic mass by palpation or ultrasound scan reduces the likelihood of a rare ovarian tumor. Central precocious puberty (gonadotrophin dependent precocious puberty) reflects early activation of the gonadotrophic drive to increased gonadal function. This is more common in girls than in boys. A child with clinical precocious puberty who has pubertal gonadotrophin levels and augmented nocturnal gonadotrophin secretion has central precocious puberty (Traggiai and Stanhope, 2003; Carel and Léger, 2008). LH results should be treated with caution because of interference from heterophilic antibodies and high results should be confirmed using a second assay (Segal et al., 2003). Assays have been devised to avoid such interference (Andersson et al., 2003). In girls, pelvic ultrasound is useful for the assessment of central precocious puberty (Traggiai and Stanhope, 2003). Tumors in the CNS (hamartomas, third ventricular cysts, astrocytomas or gliomas) can be recognized by MRI. Some of the tumors (dysgerminomas, hepatoblastomas) secrete hCG. A cerebral tumor is relatively more common in boys with central precocious puberty than in girls.

Gonadotrophin independent precocious puberty is due to inappropriate production of gonadal or adrenal hormones which affect secondary sex characteristics. Children may have acne and behavioral problems and may become taller than their peers. As a result of rapid skeletal maturation there is early epiphyseal fusion and the outcome is for short stature in adulthood. Hypothyroidism should be excluded. With the increase in TSH there is a concomitant increase in FSH and prolactin that presumably lead to sexual precocity. In patients with the McCune-Albright syndrome, precocious puberty is associated with cafe-au-lait skin marks and polyostotic fibrous dysplasia (Dumitrescu and Collins, 2008). Plasma gonadotrophin concentrations are often pubertal but may in some cases be prepubertal. Symptoms may wax and wane and on ultrasound this can be attributed to the appearance and regression of unilateral ovarian cysts. Up to 80% of girls and 40% of boys with precocious puberty may have no responsible cause and are termed idiopathic.

Pseudo-precocious puberty can be the result of exposure to exogenous sex steroids such as the contraceptive pill and anabolic steroids. Abnormal sex steroid secretion from tumors is also a cause. The most common adrenal tumor reported in the literature secretes DHAS. In general, the tumors have been quite large. A number of cases of tumors secreting other androgens have been reported and using urine steroid profile analysis the secretion of 11β-hydroxyandrostenedione has been defined on the basis of high excretion of 11β-hydroxyandrosterone (Honour et al., 1984; Wolthers and Schou, 2005). In these reported cases, androgen excretion was not grossly elevated and without scanning of the adrenals a tumor may have been dismissed. FSH and LH are suppressed to within prepubertal ranges. The secretion of androgens by adrenal tumors is not suppressed by giving dexamethasone. A testicular mass with grossly elevated 17-OH-P usually indicates CAH with adrenal rests in the testes. Leydig cell tumors however may produce elevated 17-OH-P which is not suppressed with dexamethasone. Plasma 21-deoxycortisol is not raised in these cases unlike the situation with CAH. One recent report of premature pubarche described a child with hyperandrogenic, anovulation who had low plasma DHAS due to defect of the sulfotransferase SULT2A1 through a defect in sulfate delivery from PAPS through inactivating mutations in PAPSS2 (Noordam et al., 2009).

Pubic hair growth before breast development or testicular enlargement may be the outcome of increased secretion of DHA and DHAS from the adrenal cortex due to early differentiation of the zona reticularis. This is a benign disorder called premature adrenarche not requiring treatment. DHA and DHAS concentrations in serum should be interpreted against normal ranges for age. Testosterone may be slightly elevated for age due to peripheral conversion of the adrenal androgens. The 24-h urine excretion of androsterone, etiocholanolone as well as cortisol metabolites are above the normal range for age and body size reflecting advance adrenal growth. Premature adrenarche is more common in Asian and Afro-Caribbean children than Caucasians.

Acne, hirsutism, menstrual disturbance can be due to nonclassical forms of CAH and a mild defect of the steroid 21-hydroxylase. In many such patients, the basal 17-hydroxyprogesterone concentrations are above the normal range (>5 nmol/L). An injection of ACTH will lead to an increase in serum 17-OHP at 30 and 60 min after the trophic hormone injection and affected subjects will have an increment in 17-OHP greater than seen in normal subjects (Bidet et al., 2009). Fifty percent of patients with nonclassic or late onset form of CAH have HLA B14, DR1 in association since this is coded nearby on the short arm of chromosome 6 and specific mutations in exon 1 or exon 7 CYP21A2 (V281L; P453S) (Lin-Su et al., 2008). A late onset form of 3β-hydroxysteroid dehydrogenase deficiency was suggested on the basis of exaggerated DHA or 17-hydroxypregnenolone response to ACTH (Pang et al., 1985) but mutation analysis does not support this etiology (Carbunaru et al., 2004; Mermejo et al., 2005).

Some girls present with isolated breast development (premature thelarche) and no growth acceleration (Borges et al., 2008). Increased estrogen production in girls with premature thelarche is a benign condition not to be confused with central precocious puberty. There is no increase in growth rate. Central precocious puberty is more serious and causes progressive breast development with pubic hair growth, accelerated growth rate and bone maturation and early epiphyseal fusion. A pelvic ultrasound may show a follicle in the ovary. The two conditions have been resolved by repeated blood sampling at 15 min intervals throughout the night (Traggiai and Stanhope, 2003). In patients with thelarche, the FSH is higher (2–7 iu/L) than the LH (1–3 iu/L) which contrasts with precocious puberty where LH secretion predominates. A GnRH stimulation test may sometimes be needed to distinguish the 2 disorders again by revealing differences in the prodominant gonadotrophin but interpretation may be assay dependent (Zevenhuijzen et al., 2004; Resende et al., 2007). Various doses of GnRH are used with 10 and 100 μm the most common and blood samples drawn at zero, 20 and 60 min. Peak LH usually occurs at 20 min and peak FSH at 60 min after injection of GnRH.

References

Abbott DH, Dumesic DA, Levine JE. Hyperandrogenic origins of polycystic ovary syndrome—implications for pathophysiology and therapy. Expert Rev Endocrinol Metab 2019;14(2):131–43.

Abdalla M, Deshmukh H, Atkin SL, Sathyapalan T. miRNAs as a novel clinical biomarker and therapeutic targets in polycystic ovary syndrome (PCOS): a review. Life Sci 2020;259, 118174.

Abreu AP, Macedo DB, Brito VN, Kaiser UB, Latronico AC. A new pathway in the control of the initiation of puberty: the MKRN3 gene. J Mol Endocrinol 2015;54(3):R131–9.

Ahmed SF, Khwaja O, Hughes IA. The role of a clinical score in the assessment of ambiguous genitalia. BJU Int 2000;85(1):120–4.

Ahmed B, Ammori BJ, Akhtar K, Senapati S, New JP, Syed AA. Weight loss and metabolic outcomes in women with or without polycystic ovarian syndrome after Roux-en-Y gastric bypass: a case-matched study. Surgeon 2021. S1479-666X(21)00062-7. Epub ahead of print 33863670.

Al Alawi AM, Nordenström A, Falhammar H. Clinical perspectives in congenital adrenal hyperplasia due to 3β-hydroxysteroid dehydrogenase type 2 deficiency. Endocrine 2019;63(3):407–21.

Alesi S, Ghelani D, Mousa A. Metabolomic biomarkers in polycystic ovary syndrome: a review of the evidence. Semin Reprod Med 2021. https://doi.org/10.1055/s-0041-1729841. Epub ahead of print 33946122.

Ali SN, Jayasena CN, Sam AH. Which patients with gynaecomastia require more detailed investigation? Clin Endocrinol (Oxf) 2018;88(3):360–3.

Almawi WY, Hubail B, Arekat DZ, Al-Farsi SM, Al-Kindi SK, Arekat MR, et al. Leutinizing hormone/choriogonadotropin receptor and follicle stimulating hormone receptor gene variants in polycystic ovary syndrome. J Assist Reprod Genet 2015;32(4):607–14.

Alomari SO, Houshiemy MNE, Bsat S, Moussalem CK, Allouh M, Omeis IA. Hypothalamic hamartomas: a comprehensive review of the literature—part 1: neurobiological features, clinical presentations and advancements in diagnostic tools. Clin Neurol Neurosurg 2020;197, 106076.

Alsaleem M, Miller DE, Saadeh L, Majumdar I. Aromatase deficiency: a rare cause of maternal virilisation and ambiguous genitalia in neonates. BMJ Case Rep 2019;12(6), e231267.

Alwashih MA, Watson DG, Andrew R, Stimson RH, Alossaimi M, Blackburn G, et al. Plasma metabolomic profile varies with glucocorticoid dose in patients with congenital adrenal hyperplasia. Sci Rep 2017;7(1):17092.

Alzahrani AS, Alswailem MM, Murugan AK, Alhomaidah DS, Capper CP, Auchus RJ, et al. A high rate of novel CYP11B1 mutations in Saudi Arabia. J Steroid Biochem Mol Biol 2017;174:217–24.

Anderson S. Café au Lait macules and associated genetic syndromes. J Pediatr Health Care 2020;34(1):71–81.

Andersson M, Rönnmark J, Areström I, Nygren PA, Ahlborg N. Inclusion of a non-immunoglobulin binding protein in two-site ELISA for quantification of human serum proteins without interference by heterophilic serum antibodies. J Immunol Methods 2003;283(1–2): 225–34.

Appan S, Hindmarsh PC, Brook CG. Monitoring treatment in congenital adrenal hyperplasia. Arch Dis Child 1989;64(9):1235–9.

Araújo VG, Oliveira RS, Gameleira KP, Cruz CB, Lofrano-Porto A. 3β-Hydroxysteroid dehydrogenase type II deficiency on newborn screening test. Arq Bras Endocrinol Metabol 2014;58(6):650–5.

Armar NA, McGarrigle HH, Honour J, Holownia P, Jacobs HS, Lachelin GC. Laparoscopic ovarian diathermy in the management of anovulatory infertility in women with polycystic ovaries: endocrine changes and clinical outcome. Fertil Steril 1990;53(1):45–9.

Azoury SC, Reddy S, Shukla V, Deng CX. Fibroblast growth factor receptor 2 (FGFR2) mutation related syndromic craniosynostosis. Int J Biol Sci 2017;13(12):1479–88.

Baş F., Toksoy G., Ergun-Longmire B., Uyguner Z.O., Abalı Z.Y., Poyrazoğlu Ş., et al. Prevalence, clinical characteristics and long-term outcomes of classical 11 β-hydroxylase deficiency (11BOHD) in Turkish population and novel mutations in CYP11B1 gene. J Steroid Biochem Mol Biol 2018b;181:88–97.

Bachelot A, Grouthier V, Courtillot C, Dulon J, Touraine P. Management of endocrine disease: congenital adrenal hyperplasia due to 21-hydroxylase deficiency: update on the management of adult patients and prenatal treatment. Eur J Endocrinol 2017;176(4):R167–81.

Balakumar T, Perry LA, Savage MO. Adrenocortical adenoma—an unusual presentation with hypersecretion of oestradiol, androgens and cortisol. J Pediatr Endocrinol Metab 1997;10(2):227–9.

Balsamo A, Baronio F, Ortolano R, Menabo S, Baldazzi L, Di Natale V, et al. Congenital adrenal hyperplasias presenting in the newborn and young infant. Front Pediatr 2020;8:593315.

Bancos I, Taylor AE, Chortis V, Sitch AJ, Jenkinson C, Davidge-Pitts CJ, et al. Urine steroid metabolomics for the differential diagnosis of adrenal incidentalomas in the EURINE-ACT study: a prospective test validation study. Lancet Diabetes Endocrinol 2020;8 (9):773–81.

Baquedano MS, Guercio G, Costanzo M, Marino R, Rivarola MA, Belgorosky A. Mutation of HSD3B2 gene and fate of dehydroepiandrosterone. Vitam Horm 2018;108:75–123.

Barber TM, Hanson P, Weickert MO, Franks S. Obesity and polycystic ovary syndrome: implications for pathogenesis and novel management strategies. Clin Med Insights Reprod Health 2019;13. 117955811 9874042.

Barnes RB, Ehrmann DA, Brigell DF, Rosenfield RL. Ovarian steroidogenic responses to gonadotropin-releasing hormone agonist testing with nafarelin in hirsute women with adrenal responses to adrenocorticotropin suggestive of 3 beta-hydroxy-delta 5-steroid dehydrogenase deficiency. J Clin Endocrinol Metab 1993;76(2):450–5.

Baronio F, Ortolano R, Menabò S, Cassio A, Baldazzi L, Di Natale V, et al. 46,XX DSD due to androgen excess in monogenic disorders of steroidogenesis: genetic, biochemical, and clinical features. Int J Mol Sci 2019;20(18):4605.

Bashamboo A, McElreavey K. Mechanism of sex determination in humans: insights from disorders of sex development. Sex Dev 2016;10(5–6): 313–25.

Batista RL, Mendonca BB. Integrative and analytical review of the 5-alpha-reductase type 2 deficiency worldwide. Appl Clin Genet 2020;14 (13):83–96.

Ben Khelifa H, Soyah N, Ben-Abdallah-Bouhjar I, Gritly R, Sanlaville D, Elghezal H, et al. Xp22.3 interstitial deletion: a recognizable chromosomal abnormality encompassing VCX3A and STS genes in a patient with X-linked ichthyosis and mental r etardation. Gene 2013;527(2): 578–83.

Berberoğlu M. Precocious puberty and normal variant puberty: definition, etiology, diagnosis and current management. J Clin Res Pediatr Endocrinol 2009;1(4):164–74.

Berra M, Williams EL, Muroni B, Creighton SM, Honour JW, Rumsby G, et al. Recognition of 5α-reductase-2 deficiency in an adult female 46XY DSD clinic. Eur J Endocrinol 2011;164(6):1019–25.

Bevan BR, Savvas M, Jenkins JM, Baker K, Pennington GW, Taylor NF. Abnormal steroid excretion in gestational trophoblastic disease complicated by ovarian theca-lutein cysts. J Clin Pathol 1986;39(6): 627–34.

Bialk ER, Lasarev MR, Held PK. Wisconsin's screening algorithm for the identification of newborns with congenital adrenal hyperplasia. Int J Neonatal Screen 2019;5(3):33.

Bidet M, Bellanné-Chantelot C, Galand-Portier MB, Tardy V, Billaud L, Laborde K, et al. Clinical and molecular characterization of a cohort of 161 unrelated women with nonclassical congenital adrenal hyperplasia due to 21-hydroxylase deficiency and 330 family members. J Clin Endocrinol Metab 2009;94(5):1570–8.

Blake EA, Carter CM, Kashani BN, Kodama M, Mabuchi S, Yoshino K, et al. Feto-maternal outcomes of pregnancy complicated by ovarian sex-cord stromal tumor: a systematic review of literature. Eur J Obstet Gynecol Reprod Biol 2014;175:1–7.

Blum MR, Popat RA, Nagy A, Cataldo NA, McLaughlin TL. Using metabolic markers to identify insulin resistance in premenopausal women with and without polycystic ovary syndrome. J Endocrinol Invest 2021. https://doi.org/10.1007/s40618-020-01430-2. Epub ahead of print 33687700.

Boehmer AL, Brinkmann AO, Sandkuijl LA, Halley DJ, Niermeijer MF, Andersson S, et al. 17Beta-hydroxysteroid dehydrogenase-3 deficiency: diagnosis, phenotypic variability, population genetics, and worldwide distribution of ancient and de novo mutations. J Clin Endocrinol Metab 1999;84(12):4713–21.

Borges MF, Pacheco KD, Oliveira AA, Rita CV, Pacheco KD, Resende EA, et al. Premature thelarche: clinical and laboratorial assessment by immunochemiluminescent assay. Arq Bras Endocrinol Metabol 2008;52(1):93–100.

Bouchoucha N, Samara-Boustani D, Pandey AV, Bony-Trifunovic H, Hofer G, Aigrain Y, et al. Characterization of a novel CYP19A1 (aromatase) R192H mutation causing virilization of a 46,XX newborn, undervirilization of the 46,XY brother, but no virilization of the mother during pregnancies. Mol Cell Endocrinol 2014;390(1–2):8–17.

Brain CE, Creighton SM, Mushtaq I, Carmichael PA, Barnicoat A, Honour JW, et al. Holistic management of DSD. Best Pract Res Clin Endocrinol Metab 2010;24(2):335–54.

Brochu M, Bélanger A, Tremblay RR. Plasma levels of C-19 steroids and 5 alpha-reduced steroid glucuronides in hyperandrogenic and idiopathic hirsute women. Fertil Steril 1987;48(6):948–53.

Bryan SM, Honour JW, Hindmarsh PC. Management of altered hydrocortisone pharmacokinetics in a boy with congenital adrenal hyperplasia using a continuous subcutaneous hydrocortisone infusion. J Clin Endocrinol Metab 2009;94(9):3477–80.

Bulsari K, Falhammar H. Clinical perspectives in congenital adrenal hyperplasia due to 11β-hydroxylase deficiency. Endocrine 2017;55(1): 19–36.

Bulun SE, Yilmaz BD, Sison C, Miyazaki K, Bernardi L, Liu S, et al. Endometriosis. Endocr Rev 2019;40(4):1048–79.

Buonocore F, McGlacken-Byrne SM, Del Valle I, Achermann JC. Current insights into adrenal insufficiency in the newborn and young infant. Front Pediatr 2020;8:619041.

Burkhard FZ, Parween S, Udhane SS, Flück CE, Pandey AV. P450 oxidoreductase deficiency: analysis of mutations and polymorphisms. J Steroid Biochem Mol Biol 2017;165(Pt. A):38–50.

Burman P, Falhammar H, Waldenström E, Sundin A, Bitzén U. 11C-metomidate PET/CT detected multiple ectopic adrenal rest tumors in a woman with congenital adrenal hyperplasia. J Clin Endocrinol Metab 2021;106(2):e675–9.

Burns AD, Taylor NF, Taylor DR, Bhake RC, Rahman F. A curious case of primary amenorrhea. Clin Chem 2020;66(9):1150–4.

Bustamante C, Hoyos-Martínez A, Pirela D, Díaz A. In utero virilization secondary to a maternal Krukenberg tumor: case report and review of literature. J Pediatr Endocrinol Metab 2017;30(7):785–90.

Caglayan AO, Dundar M, Tanriverdi F, Baysal NA, Unluhizarci K, Ozkul Y, et al. Idiopathic hirsutism: local and peripheral expression of aromatase (CYP19A) and 5α-reductase genes (SRD5A1 and SRD5A2). Fertil Steril 2011;96(2):479–82.

Canton APM, Krepischi ACV, Montenegro LR, Costa S, Rosenberg C, Steunou V, et al. Insights from the genetic characterization of central precocious puberty associated with multiple anomalies. Hum Reprod 2021;36(2):506–18.

Carbunaru G, Prasad P, Scoccia B, Shea P, Hopwood N, Ziai F, et al. The hormonal phenotype of nonclassic 3 beta-hydroxysteroid dehydrogenase (HSD3B) deficiency in hyperandrogenic females is associated with insulin-resistant polycystic ovary syndrome and is not a variant of inherited HSD3B2 deficiency. J Clin Endocrinol Metab 2004;89(2): 783–94.

Carel JC, Léger J. Clinical practice. Precocious puberty. N Engl J Med 2008;358(22):2366–77.

Carmina E, Dewailly D, Escobar-Morreale HF, Kelestimur F, Moran C, Oberfield S, et al. Non-classic congenital adrenal hyperplasia due to 21-hydroxylase deficiency revisited: an update with a special focus on adolescent and adult women. Hum Reprod Update 2017;23(5): 580–99.

Carmina E, Azziz R, Bergfeld W, Escobar-Morreale HF, Futterweit W, Huddleston H, et al. Female pattern hair loss and androgen excess: a report from the multidisciplinary androgen excess and PCOS committee. J Clin Endocrinol Metab 2019;104(7):2875–91.

Carvalho B, Marques CJ, Santos-Silva R, Fontoura M, Carvalho D, Carvalho F. Congenital adrenal hyperplasia due to 21-hydroxylase deficiency: an update on genetic analysis of CYP21A2 gene. Exp Clin Endocrinol Diabetes 2021;129(7):477–81.

Cavarzere P, Samara-Boustani D, Flechtner I, Dechaux M, Elie C, Tardy V, et al. Transient hyper-17-hydroxyprogesteronemia: a clinical subgroup of patients diagnosed at neonatal screening for congenital adrenal hyperplasia. Eur J Endocrinol 2009;161(2):285–92.

Chang AY, Lalia AZ, Jenkins GD, Dutta T, Carter RE, Singh RJ, et al. Combining a nontargeted and targeted metabolomics approach to identify metabolic pathways significantly altered in polycystic ovary syndrome. Metabolism 2017;71:52–63.

Chen DJ, Ding R, Cao JY, Zhai JX, Zhang JX, Ye DQ. Two follicle-stimulating hormone receptor polymorphisms and polycystic ovary syndrome risk: a meta-analysis. Eur J Obstet Gynecol Reprod Biol 2014;182:27–32.

Chen M, Zhou W, Zhang Z, Zou Y, Li C. An ovarian Leydig cell tumor of ultrasound negative in a postmenopausal woman with hirsutism and hyperandrogenism: a case report. Medicine 2018;97(10), e0093.

Chen B, Xu P, Wang J, Zhang C. The role of MiRNA in polycystic ovary syndrome (PCOS). Gene 2019;706:91–6.

Cheuiche AV, da Silveira LG, de Paula LCP, Lucena IRS, Silveiro SP. Diagnosis and management of precocious sexual maturation: an updated review. Eur J Pediatr 2021. https://doi.org/10.1007/s00431-021-04022-1. Epub ahead of print 33745030.

Cho EK, Kim J, Yang A, Ki CS, Lee JE, Cho SY, et al. Clinical and endocrine characteristics and genetic analysis of Korean children with McCune-Albright syndrome: a retrospective cohort study. Orphanet J Rare Dis 2016;11(1):113.

Chortis V. The role of steroid metabolome analysis for the diagnosis and follow-up of adrenocortical tumors. Minerva Endocrinol 2019;44(1): 19–24.

Chortis V, Bancos I, Nijman T, Gilligan LC, Taylor AE, Ronchi CL, et al. Urine steroid metabolomics as a novel tool for detection of recurrent adrenocortical carcinoma. J Clin Endocrinol Metab 2020;105(3): e307–18.

Christakoudi S, Cowan DA, Taylor NF. A new marker for early diagnosis of 21-hydroxylase deficiency: 3beta,16alpha,17alpha-trihydroxy-5alpha-pregnane-7,20-dione. J Steroid Biochem Mol Biol 2010; 121(3–5):574–81.

Christakoudi S, Cowan DA, Christakudis G, Taylor NF. 21-hydroxylase deficiency in the neonate—trends in steroid anabolism and catabolism during the first weeks of life. J Steroid Biochem Mol Biol 2013;138:334–47.

Cicalini I, Tumini S, Guidone PI, Pieragostino D, Zucchelli M, Franchi S, et al. Serum steroid profiling by liquid chromatography-tandem mass spectrometry for the rapid confirmation and early treatment of congenital adrenal hyperplasia: a neonatal case report. Metabolites 2019;9(12):284.

Comim FV, Teerds K, Hardy K, Franks S. Increased protein expression of LHCG receptor and 17α-hydroxylase/17-20-lyase in human polycystic ovaries. Hum Reprod 2013;28(11):3086–92.

Cools M, Nordenström A, Robeva R, Hall J, Westerveld P, Flück C, et al. COST action BM1303 working group 1. Caring for individuals with a difference of sex development (DSD): a consensus statement. Nat Rev Endocrinol 2018;14(7):415–29.

Costa-Barbosa FA, Tonetto-Fernandes VF, Carvalho VM, Nakamura OH, Moura V, Bachega TA, et al. Superior discriminating value of ACTH-stimulated serum 21-deoxycortisol in identifying heterozygote carriers for 21-hydroxylase deficiency. Clin Endocrinol (Oxf) 2010;73(6):700–6.

Coyle C, Campbell RE. Pathological pulses in PCOS. Mol Cell Endocrinol 2019;498:110561.

Cragun DL, Trumpy SK, Shackleton CH, Kelley RI, Leslie ND, Mulrooney NP, et al. Undetectable maternal serum uE3 and postnatal abnormal sterol and steroid metabolism in Antley-Bixler syndrome. Am J Med Genet A 2004;129A(1):1–7.

Craig WY, Roberson M, Palomaki GE, Shackleton CH, Marcos J, Haddow JE. Prevalence of steroid sulfatase deficiency in California according to race and ethnicity. Prenat Diagn 2010;30(9):893–8.

Czyzyk A, Latacz J, Filipowicz D, Podfigurna A, Moszynski R, Jasinski P, et al. Severe hyperandrogenemia in postmenopausal woman as a presentation of ovarian hyperthecosis. Case report and mini review of the literature. Gynecol Endocrinol 2017;33(11):836–9.

Dabas A, Bothra M, Kapoor S. CAH newborn screening in India: challenges and opportunities. Int J Neonatal Screen 2020;6(3):70.

Dadachanji R, Shaikh N, Mukherjee S. Genetic variants associated with hyperandrogenemia in PCOS pathophysiology. Genet Res Int 2018;2018:7624932.

Das A, Datta D, Kassir M, Wollina U, Galadari H, Lotti T, et al. Acanthosis nigricans: a review. J Cosmet Dermatol 2020;19(8):1857–65.

Daussac A, Barat P, Servant N, Yacoub M, Missonier S, Lavran F, et al. Testotoxicosis without testicular mass: revealed by peripheral precocious puberty and confirmed by somatic *LHCGR* gene mutation. Endocr Res 2020;45(1):32–40.

de Hora MR, Heather NL, Patel T, Bresnahan LG, Webster D, Hofman PL. Measurement of 17-hydroxyprogesterone by LCMSMS improves newborn screening for CAH due to 21-hydroxylase deficiency in New Zealand. Int J Neonatal Screen 2020;6(1):6.

Dean B, Chrisp GL, Quartararo M, Maguire AM, Hameed S, King BR, et al. P450 oxidoreductase deficiency: a systematic review and meta-analysis of genotypes, phenotypes, and their relationships. J Clin Endocrinol Metab 2020;105(3):dgz255.

Délot EC, Papp JC, Genetics Workgroup DSD-TRN, Sandberg DE, Vilain E. Genetics of disorders of sex development: the DSD-TRN experience. Endocrinol Metab Clin North Am 2017;46(2):519–37.

Deng C, Ji J, Zhang L, Zhang X. Diagnosis of congenital adrenal hyperplasia by rapid determination of 17alpha-hydroxyprogesterone in dried blood spots by gas chromatography/mass spectrometry following microwave-assisted silylation. Rapid Commun Mass Spectrom 2005;19(20):2974–8.

Deng S, Sun A, Chen R, Yu Q, Tian Q. Gonadal dominance and internal genitalia phenotypes of patients with ovotesticular disorders of sex development: report of 22 cases and literature review. Sex Dev 2019;13(4):187–94.

Deodati A, Sallemi A, Maranghi F, Germani D, Puglianiello A, Baldari F, et al. Serum levels of polybrominated diphenyl ethers in girls with premature Thelarche. Horm Res Paediatr 2016;86(4):233–9.

Dewailly D, Barbotin AL, Dumont A, Catteau-Jonard S, Robin G. Role of anti-Müllerian hormone in the pathogenesis of polycystic ovary syndrome. Front Endocrinol (Lausanne) 2020;11:641.

Dinh QQ, Sinclair R. Female pattern hair loss: current treatment concepts. Clin Interv Aging 2007;2(2):189–99.

Donnez J, Dolmans MM. Endometriosis and medical therapy: from progestogens to progesterone resistance to GnRH antagonists: a review. J Clin Med 2021;10(5):1085.

Dörr HG, Schulze N, Bettendorf M, Binder G, Bonfig W, Denzer C, et al. Genotype-phenotype correlations in children and adolescents with nonclassical congenital adrenal hyperplasia due to 21-hydroxylase deficiency. Mol Cell Pediatr 2020;7(1):8.

Dreyer FE, Abdulrahman GO, Waring G, Hinshaw K. Placental steroid sulphatase deficiency: an approach to antenatal care and delivery. Ann Saudi Med 2018;38(6):445–9.

Duan L, Shen R, Song L, Liao Y, Zheng H. A novel chimeric CYP11B2/CYP11B1 combined with a new p.L340P CYP11B1 mutation in a patient with 11OHD: case report. BMC Endocr Disord 2018;18(1):23.

Duică F, Dănilă CA, Boboc AE, Antoniadis P, Condrat CE, Onciul S, et al. Impact of increased oxidative stress on cardiovascular diseases in women with polycystic ovary syndrome. Front Endocrinol 2021;12: 614679.

Dumesic DA, Oberfield SE, Stener-Victorin E, Marshall JC, Laven JS, Legro RS. Scientific statement on the diagnostic criteria, epidemiology, pathophysiology, and molecular genetics of polycystic ovary syndrome. Endocr Rev 2015;36(5):487–525.

Dumesic DA, Abbott DH, Sanchita S, Chazenbalk GD. Endocrine-metabolic dysfunction in polycystic ovary syndrome: an evolutionary perspective. Curr Opin Endocr Metab Res 2020;12:41–8.

Dumitrescu CE, Collins MT. McCune-Albright syndrome. Orphanet J Rare Dis 2008;3:12.

Dumont A, Robin G, Catteau-Jonard S, Dewailly D. Role of anti-Müllerian hormone in pathophysiology, diagnosis and treatment of polycystic ovary syndrome: a review. Reprod Biol Endocrinol 2015;13:137.

Durand A, Bashamboo A, McElreavey K, Brauner R. Familial early puberty: presentation and inheritance pattern in 139 families. BMC Endocr Disord 2016;16(1):50.

Dursun F, Ceylaner S. A novel homozygous *CYP19A1* gene mutation: aromatase deficiency mimicking congenital adrenal hyperplasia in an infant without obvious maternal virilisation. J Clin Res Pediatr Endocrinol 2019;11(2):196–201.

Elhassan YS, Idkowiak J, Smith K, Asia M, Gleeson H, Webster R, et al. Causes, patterns, and severity of androgen excess in 1205 consecutively recruited women. J Clin Endocrinol Metab 2018;103(3):1214–23.

Elias PM, Crumrine D, Rassner U, Hachem JP, Menon GK, Man W, et al. Basis for abnormal desquamation and permeability barrier dysfunction in RXLI. J Invest Dermatol 2004a;122(2):314–9.

Elias SG, Onland-Moret NC, Peeters PH, Rinaldi S, Kaaks R, Grobbee DE, et al. Urinary endogenous sex hormone levels in postmenopausal women after caloric restriction in young adulthood. Br J Cancer 2004b;90(1):115–7.

Eltan M, Yavas Abali Z, Arslan Ates E, Kirkgoz T, Kaygusuz SB, Türkyılmaz A, et al. Low DHEAS concentration in a girl presenting with short stature and premature pubarche: a novel PAPSS2 gene mutation. Horm Res Paediatr 2019;92(4):262–8.

Escobar-Morreale HF, Serrano-Gotarredona J, García-Robles R, Sancho J, Varela C. Mild adrenal and ovarian steroidogenic abnormalities in hirsute women without hyperandrogenemia: does idiopathic hirsutism exist? Metabolism 1997;46(8):902–7.

Escobar-Morreale HF, Samino S, Insenser M, Vinaixa M, Luque-Ramírez M, Lasunción MA, et al. Metabolic heterogeneity in polycystic ovary syndrome is determined by obesity: plasma metabolomic approach using GC-MS. Clin Chem 2012;58(6):999–1009.

Fabbrocini G, Cantelli M, Masarà A, Annunziata MC, Marasca C, Cacciapuoti S. Female pattern hair loss: a clinical, pathophysiologic, and therapeutic review. Int J Womens Dermatol 2018;4(4):203–11.

Fan L, Ren X, Song Y, Su C, Fu J, Gong C. Novel phenotypes and genotypes in Antley-Bixler syndrome caused by cytochrome P450 oxidoreductase deficiency: based on the first cohort of Chinese children. Orphanet J Rare Dis 2019;14(1):299.

Fanis P, Neocleous V, Kosta K, Karipiadou A, Hartmann MF, Wudy SA, et al. Late diagnosis of 3β-hydroxysteroid dehydrogenase deficiency: the pivotal role of gas chromatography-mass spectrometry urinary steroid metabolome analysis and a novel homozygous nonsense mutation in the *HSD3B2* gene. J Pediatr Endocrinol Metab 2020; 34(1):131–6.

Farello G, Altieri C, Cutini M, Pozzobon G, Verrotti A. Review of the literature on current changes in the timing of pubertal development and the incomplete forms of early puberty. Front Pediatr 2019;7:147.

Fassnacht M, Schlenz N, Schneider SB, Wudy SA, Allolio B, Arlt W. Beyond adrenal and ovarian androgen generation: increased peripheral 5 alpha-reductase activity in women with polycystic ovary syndrome. J Clin Endocrinol Metab 2003;88(6):2760–6. https://doi.org/10.1210/jc.2002-021875. PMID: 12788885.

Fiet J, Le Bouc Y, Guéchot J, Hélin N, Maubert MA, Farabos D, et al. A liquid chromatography/tandem mass spectometry profile of 16 serum steroids, including 21-deoxycortisol and 21-deoxycorticosterone, for management of congenital adrenal hyperplasia. J Endocr Soc 2017;1(3):186–201.

Fingerhut R. False positive rate in newborn screening for congenital adrenal hyperplasia (CAH)-ether extraction reveals two distinct reasons for elevated 17alpha-hydroxyprogesterone (17-OHP) values. Steroids 2009;74(8):662–5.

Flück CE, Tajima T, Pandey AV, Arlt W, Okuhara K, Verge CF, et al. Mutant P450 oxidoreductase causes disordered steroidogenesis with and without Antley-Bixler syndrome. Nat Genet 2004;36(3):228–30.

Foster PA. Steroid sulphatase and its inhibitors: past, present, and future. Molecules 2021;26(10):2852.

Fukami M, Horikawa R, Nagai T, Tanaka T, Naiki Y, Sato N, et al. Cytochrome P450 oxidoreductase gene mutations and Antley-Bixler syndrome with abnormal genitalia and/or impaired steroidogenesis: molecular and clinical studies in 10 patients. J Clin Endocrinol Metab 2005;90(1):414–26.

Gazdar AF, Oie HK, Shackleton CH, Chen TR, Triche TJ, Myers CE, et al. Establishment and characterization of a human adrenocortical carcinoma cell line that expresses multiple pathways of steroid biosynthesis. Cancer Res 1990;50(17):5488–96.

Ge W, Wang HL, Shao HJ, Liu HW, Xu RY. Evaluation of serum Makorin ring finger protein 3 (MKRN3) levels in girls with idiopathic central precocious puberty and premature thelarche. Physiol Res 2020;69(1):127–33.

Ghazal K, Brabant S, Prie D, Piketty ML. Hormone immunoassay interference: a 2021 update. Ann Lab Med 2022;42(1):3–23.

Giagulli VA, Giorgino R, Vermeulen A. Is plasma 5 alpha-androstane 3 alpha, 17 beta-diol glucuronide a biochemical marker of hirsutism in women? J Steroid Biochem Mol Biol 1991;39(1):55–61.

Giglio M, Medica M, De Rose AF, Germinale F, Ravetti JL, Carmignani G. Testicular sertoli cell tumours and relative sub-types. Analysis of clinical and prognostic features. Urol Int 2003;70(3):205–10.

Gohil A, Eugster EA. Delayed and precocious puberty: genetic underpinnings and treatments. Endocrinol Metab Clin North Am 2020;49(4):741–57.

Gompel A, Wright F, Kuttenn F, Mauvais-Jarvis P. Contribution of plasma androstenedione to 5 alpha-androstanediol glucuronide in women with idiopathic hirsutism. J Clin Endocrinol Metab 1986;62(2):441–4.

Goswami S, Chakraborty PP, Bhattacharjee R, Roy A, Thukral A, Selvan C, et al. Precocious puberty: a blessing in disguise! Indian J Endocrinol Metab 2013;17(Suppl. 1):S111–3.

Gourgari E, Saloustros E, Stratakis CA. Large-cell calcifying Sertoli cell tumors of the testes in pediatrics. Curr Opin Pediatr 2012;24(4):518–22.

Greaves R, Kanumakala S, Read A, Zacharin M. Genital abnormalities mimicking congenital adrenal hyperplasia in premature infants. J Paediatr Child Health 2004;40(4):233–6.

Greaves R, Hunt RW, Zacharin M. Transient anomalies in genital appearance in some extremely preterm female infants may be the result of foetal programming causing a surge in LH and the over activation of the pituitary-gonadal axis. Clin Endocrinol (Oxf) 2008;69(5):763–8.

Greep N, Hoopes M, Horton R. Androstanediol glucuronide plasma clearance and production rates in normal and hirsute women. J Clin Endocrinol Metab 1986;62(1):22–7.

Griggs J, Burroway B, Tosti A. Pediatric androgenetic alopecia: a review. J Am Acad Dermatol 2019. S0190-9622(19)32565-4.

Gröndal S, Eriksson B, Hagenäs L, Werner S, Curstedt T. Steroid profile in urine: a useful tool in the diagnosis and follow up of adrenocortical carcinoma. Acta Endocrinol 1990;122(5):656–63.

Guo T, Taylor RL, Singh RJ, Soldin SJ. Simultaneous determination of 12 steroids by isotope dilution liquid chromatography-photospray ionization tandem mass spectrometry. Clin Chim Acta 2006;372(1–2):76–82.

Guran T, Kara C, Yildiz M, Bitkin EC, Haklar G, Lin JC, et al. Revisiting classical 3β-hydroxysteroid dehydrogenase 2 deficiency: lessons from 31 pediatric cases. J Clin Endocrinol Metab 2020;105(3):dgaa022.

Gurnurkar S, DiLillo E, Carakushansky M. A case of familial male-limited precocious puberty with a novel mutation. J Clin Res Pediatr Endocrinol 2021;13(2):239–44. https://doi.org/10.4274/jcrpe.galenos.2020.2020.0067.

Haddad NG, Eugster EA. Peripheral precocious puberty including congenital adrenal hyperplasia: causes, consequences, management and outcomes. Best Pract Res Clin Endocrinol Metab 2019;33(3), 101273.

Hafizi L, Amirian M, Davoudi Y, Jaafari M, Ghasemi GH. Comparison of laparoscopic ovarian drilling success between two standard and dose-adjusted methods in polycystic ovary syndrome: a randomized clinical trial. Int J Fertil Steril 2020;13(4):282–8.

Hakim C, Padmanabhan V, Vyas AK. Gestational hyperandrogenism in developmental programming. Endocrinology 2017;158(2):199–212.

Hampf M, Dao NT, Hoan NT, Bernhardt R. Unequal crossing-over between aldosterone synthase and 11beta-hydroxylase genes causes congenital adrenal hyperplasia. J Clin Endocrinol Metab 2001;86(9):4445–52.

Han L, Tavakoli NP, Morrissey M, Spink DC, Cao ZT. Liquid chromatography-tandem mass spectrometry analysis of 17-hydroxyprogesterone in dried blood spots revealed matrix effect on immunoassay. Anal Bioanal Chem 2019;411(2):395–402.

Han Q, Wang J, Li W, Chen ZJ, Du Y. Androgen-induced gut dysbiosis disrupts glucolipid metabolism and endocrinal functions in polycystic ovary syndrome. Microbiome 2021;9(1):101.

Handelsman DJ, Newman JD, Jimenez M, McLachlan R, Sartorius G, Jones GR. Performance of direct estradiol immunoassays with human male serum samples. Clin Chem 2014;60(3):510–7.

Hassan HC, Cullen IM, Casey RG, Rogers E. Gynaecomastia: an endocrine manifestation of testicular cancer. Andrologia 2008;40(3):152–7.

Hawkins J, Hires C, Dunne E, Baker C. The relationship between lavender and tea tree essential oils and pediatric endocrine disorders: a systematic review of the literature. Complement Ther Med 2020;49:102288.

Hayashi GY, Carvalho DF, de Miranda MC, Faure C, Vallejos C, Brito VN, et al. Neonatal 17-hydroxyprogesterone levels adjusted according to age at sample collection and birthweight improve the efficacy of congenital adrenal hyperplasia newborn screening. Clin Endocrinol (Oxf) 2017;86(4):480–7.

Heckmann M, Hartmann MF, Kampschulte B, Gack H, Bödeker RH, Gortner L, et al. Persistent high activity of the fetal adrenal cortex in preterm infants: is there a clinical significance? J Pediatr Endocrinol Metab 2006;19(11):1303–12.

Held PK, Bird IM, Heather NL. Newborn screening for congenital adrenal hyperplasia: review of factors affecting screening accuracy. Int J Neonatal Screen 2020;6(3):67.

Henley DV, Lipson N, Korach KS, Bloch CA. Prepubertal gynecomastia linked to lavender and tea tree oils. N Engl J Med 2007;356(5):479–85.

Higashi T, Nishio T, Uchida S, Shimada K, Fukushi M, Maeda M. Simultaneous determination of 17alpha-hydroxypregnenolone and 17alpha-hydroxyprogesterone in dried blood spots from low birth weight infants using LC-MS/MS. J Pharm Biomed Anal 2008;48(1): 177–82.

Hindmarsh PC, Honour JW. Would cortisol measurements be a better gauge of hydrocortisone replacement therapy? Congenital adrenal hyperplasia as an exemplar. Int J Endocrinol 2020;, 2470956.

Holst JP, Soldin SJ, Tractenberg RE, Guo T, Kundra P, Verbalis JG, et al. Use of steroid profiles in determining the cause of adrenal insufficiency. Steroids 2007;72(1):71–84.

Homma K, Hasegawa T, Nagai T, Adachi M, Horikawa R, Fujiwara I, et al. Urine steroid hormone profile analysis in cytochrome P450 oxidoreductase deficiency: implication for the backdoor pathway to dihydrotestosterone. J Clin Endocrinol Metab 2006;91(7):2643–9.

Honour JW. 17-Hydroxyprogesterone in children, adolescents and adults. Ann Clin Biochem 2014;51(Pt. 4):424–40.

Honour JW, Price DA, Taylor NF, Marsden HB, Grant DB. Steroid biochemistry of virilising adrenal tumours in childhood. Eur J Pediatr 1984;142(3):165–9.

Horton R, Hawks D, Lobo R. 3 alpha, 17 beta-androstanediol glucuronide in plasma. A marker of androgen action in idiopathic hirsutism. J Clin Invest 1982;69(5):1203–6.

Hoyos LR, Visser JA, McLuskey A, Chazenbalk GD, Grogan TR, Dumesic DA. Loss of anti-Müllerian hormone (AMH) immunoactivity due to a homozygous AMH gene variant rs10417628 in a woman with classical polycystic ovary syndrome (PCOS). Hum Reprod 2020;35(10):2294–302.

Huang H, Wang C, Tian Q. Gonadal tumour risk in 292 phenotypic female patients with disorders of sex development containing Y chromosome or Y-derived sequence. Clin Endocrinol (Oxf) 2017;86(4):621–7.

Hughes IA, Arisaka O, Perry LA, Honour JW. Early diagnosis of 11 beta-hydroxylase deficiency in two siblings confirmed by analysis of a novel steroid metabolite in newborn urine. Acta Endocrinol 1986;111(3):349–54.

Huhtinen K, Saloniemi-Heinonen T, Keski-Rahkonen P, Desai R, Laajala D, Ståhle M, et al. Intra-tissue steroid profiling indicates differential progesterone and testosterone metabolism in the endometrium and endometriosis lesions. J Clin Endocrinol Metab 2014;99(11):E2188–97.

Idkowiak J, O'Riordan S, Reisch N, Malunowicz EM, Collins F, Kerstens MN, et al. Pubertal presentation in seven patients with congenital adrenal hyperplasia due to P450 oxidoreductase deficiency. J Clin Endocrinol Metab 2011;96(3):E453–62.

Idkowiak J, Taylor AE, Subtil S, O'Neil DM, Vijzelaar R, Dias RP, et al. Steroid sulfatase deficiency and androgen activation before and after puberty. J Clin Endocrinol Metab 2016;101(6):2545–53.

Idkowiak J, Elhassan YS, Mannion P, Smith K, Webster R, Saraff V, et al. Causes, patterns and severity of androgen excess in 487 consecutively recruited pre- and post-pubertal children. Eur J Endocrinol 2019;180 (3):213–21.

Imperato-McGinley J, Guerrero L, Gautier T, Peterson RE. Steroid 5alpha-reductase deficiency in man: an inherited form of male pseudohermaphroditism. Science 1974;186(4170):1213–5.

Isojärvi JI, Laatikainen TJ, Pakarinen AJ, Juntunen KT, Myllylä VV. Polycystic ovaries and hyperandrogenism in women taking valproate for epilepsy. N Engl J Med 1993;329(19):1383–8.

Izawa M, Aso K, Higuchi A, Ariyasu D, Hasegawa Y. The range of 2.2-3.3 mg/gCr of pregnanetriol in the first morning urine sample as an index of optimal control in CYP21 deficiency. Clin Pediatr Endocrinol 2008;17(3):75–80.

Janzen N, Peter M, Sander S, Steuerwald U, Terhardt M, Holtkamp U, et al. Newborn screening for congenital adrenal hyperplasia: additional steroid profile using liquid chromatography-tandem mass spectrometry. J Clin Endocrinol Metab 2007;92(7):2581–9.

Janzen N, Sander S, Terhardt M, Peter M, Sander J. Fast and direct quantification of adrenal steroids by tandem mass spectrometry in serum and dried blood spots. J Chromatogr B Analyt Technol Biomed Life Sci 2008;861(1):117–22.

Janzen N, Riepe FG, Peter M, Sander S, Steuerwald U, Korsch E, et al. Neonatal screening: identification of children with 11β-hydroxylase deficiency by second-tier testing. Horm Res Paediatr 2012;77(3):195–9.

Jarrett BY, Vantman N, Mergler RJ, Brooks ED, Pierson RA, Chizen DR, et al. Dysglycemia, not altered sex steroid hormones, affects cognitive function in polycystic ovary syndrome. J Endocr Soc 2019;3(10): 1858–68.

Jayamanne C, Sandamal S, Jayasundara K, Saranavabavananthan M, Mettananda S. Smith-Lemli-Opitz syndrome presenting as acute adrenal crisis in a child: a case report. J Med Case Reports 2018;12(1):217.

Jezela-Stanek A, Małunowicz E, Anna S, Kucharczyk M, Goryluk-Kozakiewicz B, Sodowska H, et al. Trends in prenatal diagnosis of non-specific multiple malformations disorders with reference to the own experience and research study on Smith-Lemli-Opitz syndrome. Ginekol Pol 2015;86(8):598–602.

Jezela-Stanek A, Siejka A, Kowalska EM, Hosiawa V, Krajewska-Walasek M. GC-MS as a tool for reliable non-invasive prenatal diagnosis of Smith-Lemli-Opitz syndrome but essential also for other cholesterolopathies verification. Ginekol Pol 2020;91(5):287–93.

Jha S, Turcu AF, Sinaii N, Brookner B, Auchus RJ, Merke DP. 11-Oxygenated androgens useful in the setting of discrepant conventional biomarkers in 21-hydroxylase deficiency. J Endocr Soc 2020;5(2): bvaa192.

Joehrer K, Geley S, Strasser-Wozak EM, Azziz R, Wollmann HA, Schmitt K, et al. CYP11B1 mutations causing non-classic adrenal hyperplasia due to 11 beta-hydroxylase deficiency. Hum Mol Genet 1997;6 (11):1829–34.

Jones AM, Honour JW. Unusual results from immunoassays and the role of the clinical endocrinologist. Clin Endocrinol (Oxf) 2006;64(3):234–44.

Jones ME, Boon WC, McInnes K, Maffei L, Carani C, Simpson ER. Recognizing rare disorders: aromatase deficiency. Nat Clin Pract Endocrinol Metab 2007a;3(5):414–21.

Jones MR, Wilson SG, Mullin BH, Mead R, Watts GF, Stuckey BG. Polymorphism of the follistatin gene in polycystic ovary syndrome. Mol Hum Reprod 2007b;13(4):237–41. https://doi.org/10.1093/molehr/gal120. Epub 2007 Feb 6 17284512.

Jones MR, Mathur R, Cui J, Guo X, Azziz R, Goodarzi MO. Independent confirmation of association between metabolic phenotypes of polycystic ovary syndrome and variation in the type 6 17beta-hydroxysteroid dehydrogenase gene. J Clin Endocrinol Metab 2009;94(12):5034–8.

Jové M, Pradas I, Naudí A, Rovira-Llopis S, Bañuls C, Rocha M, et al. Lipidomics reveals altered biosynthetic pathways of glycerophospholipids and cell signaling as biomarkers of the polycystic ovary syndrome. Oncotarget 2017;9(4):4522–36.

Kaňová N, Bičíková M. Hyperandrogenic states in pregnancy. Physiol Res 2011;60(2):243–52.

Kahsar-Miller MD, Conway-Myers BA, Boots LR, Azziz R. Steroidogenic acute regulatory protein (StAR) in the ovaries of healthy women and those with polycystic ovary syndrome. Am J Obstet Gynecol 2001;185(6):1381–7.

Kamrath C, Hochberg Z, Hartmann MF, Remer T, Wudy SA. Increased activation of the alternative "backdoor" pathway in patients with 21-hydroxylase deficiency: evidence from urinary steroid hormone analysis. J Clin Endocrinol Metab 2012;97(3):E367–75.

Kamrath C, Hartmann MF, Boettcher C, Zimmer KP, Wudy SA. Diagnosis of 21-hydroxylase deficiency by urinary metabolite ratios using gas chromatography-mass spectrometry analysis: reference values for neonates and infants. J Steroid Biochem Mol Biol 2016;156:10–6.

Kamrath C, Wettstaedt L, Boettcher C, Hartmann MF, Wudy SA. Androgen excess is due to elevated 11-oxygenated androgens in treated children with congenital adrenal hyperplasia. J Steroid Biochem Mol Biol 2018;178:221–8.

Kamrath C, Wettstaedt L, Hartmann MF, Wudy SA. Height velocity defined metabolic control in children with congenital adrenal hyperplasia using urinary GC-MS analysis. J Clin Endocrinol Metab 2019;2019(4):4214–24.

Kamrath C, Hartmann MF, Pons-Kühnemann J, Wudy SA. Urinary GC-MS steroid metabotyping in treated children with congenital adrenal hyperplasia. Metabolism 2020;112:154354.

Kang E, Cho JH, Choi JH, Yoo HW. Etiology and therapeutic outcomes of children with gonadotropin-independent precocious puberty. Ann Pediatr Endocrinol Metab 2016;21(3):136–42.

Keefe CC, Goldman MM, Zhang K, Clarke N, Reitz RE, Welt CK. Simultaneous measurement of thirteen steroid hormones in women with polycystic ovary syndrome and control women using liquid chromatography-tandem mass spectrometry. PLoS ONE 2014;9(4), e93805.

Keevil B. Steroid mass spectrometry for the diagnosis of PCOS. Med Sci 2019;7(7):78.

Keevil BG, Adaway J. Assessment of free testosterone concentration. J Steroid Biochem Mol Biol 2019;190:207–11.

Keevil BG, Adaway J, Fiers T, Moghetti P, Kaufman JM. The free androgen index is inaccurate in women when the SHBG concentration is low. Clin Endocrinol (Oxf) 2018;88(5):706–10.

Kempegowda P, Melson E, Manolopoulos KN, Arlt W, O'Reilly MW. Implicating androgen excess in propagating metabolic disease in polycystic ovary syndrome. Ther Adv Endocrinol Metab 2020;11. 2042018820934319.

Khattab A, Yuen T, Yau M, Domenice S, Frade Costa EM, Diya K, et al. Pitfalls in hormonal diagnosis of 17-beta hydroxysteroid dehydrogenase III deficiency. J Pediatr Endocrinol Metab 2015;28(5–6):623–8.

Khattab A, Haider S, Kumar A, Dhawan S, Alam D, Romero R, et al. Clinical, genetic, and structural basis of congenital adrenal hyperplasia due to 11β-hydroxylase deficiency. Proc Natl Acad Sci U S A 2017;114(10):E1933–40.

Khunger N, Mehrotra K. Menopausal acne—challenges and solutions. Int J Womens Health 2019;11:555–67.

Kim MS, Ryabets-Lienhard A, Bali B, Lane CJ, Park AH, Hall S, et al. Decreased adrenomedullary function in infants with classical congenital adrenal hyperplasia. J Clin Endocrinol Metab 2014;99(8):E1597–601.

Kim B, Lee MN, Park HD, Kim JW, Chang YS, Park WS, et al. Dried blood spot testing for seven steroids using liquid chromatography-tandem mass spectrometry with reference interval determination in the Korean population. Ann Lab Med 2015;35(6):578–85.

Kim MS, Koppin CM, Mohan P, Goodarzian F, Ross HM, Geffner ME, et al. Absence of testicular adrenal rest tumors in newborns, infants, and toddlers with classical congenital adrenal hyperplasia. Horm Res Paediatr 2019;92(3):157–61. Erratum in: Horm Res Paediatr. 2019 Dec 12:1. Erratum in: Horm Res Paediatr. 2021 Apr 7:1.

Kirschner MA, Samojlik E, Szmal E. Clinical usefulness of plasma androstanediol glucuronide measurements in women with idiopathic hirsutism. J Clin Endocrinol Metab 1987;65(4):597–601.

Klimek M, Radosz P, Lemm M, Szanecki W, Dudek A, Pokładek S, et al. Leydig cell ovarian tumor—clinical case description and literature review. Prz Menopauzalny 2020;19(3):140–3.

Ko JM. Genetic syndromes associated with craniosynostosis. J Korean Neurosurg Soc 2016;59(3):187–91.

Ko JM, Cheon CK, Kim GH, Yoo HW. A case of Antley-Bixler syndrome caused by compound heterozygous mutations of the cytochrome P450 oxidoreductase gene. Eur J Pediatr 2009;168(7):877–80.

Kolli V, Kim H, Torky A, Lao Q, Tatsi C, Mallappa A, et al. Characterization of the CYP11A1 nonsynonymous variant p.E314K in children presenting with adrenal insufficiency. J Clin Endocrinol Metab 2019;104(2):269–76.

Konings G, Brentjens L, Delvoux B, Linnanen T, Cornel K, Koskimies P, et al. Intracrine regulation of estrogen and other sex steroid levels in endometrium and non-gynecological tissues; pathology, physiology, and drug discovery. Front Pharmacol 2018;9:940.

Koyama Y, Homma K, Fukami M, Miwa M, Ikeda K, Ogata T, et al. Classic and non-classic 21-hydroxylase deficiency can be discriminated from P450 oxidoreductase deficiency in Japanese infants by urinary steroid metabolites. Clin Pediatr Endocrinol 2016;25(2):37–44.

Krone N, Arlt W. Genetics of congenital adrenal hyperplasia. Best Pract Res Clin Endocrinol Metab 2009;23(2):181–92.

Krone N, Reisch N, Idkowiak J, Dhir V, Ivison HE, Hughes BA, et al. Genotype-phenotype analysis in congenital adrenal hyperplasia due to P450 oxidoreductase deficiency. J Clin Endocrinol Metab 2012;97(2):E257–67.

Kulle AE, Riepe FG, Hedderich J, Sippell WG, Schmitz J, Niermeyer L, et al. LC-MS/MS based determination of basal- and ACTH-stimulated plasma concentrations of 11 steroid hormones: implications for detecting heterozygote CYP21A2 mutation carriers. Eur J Endocrinol 2015;173(4):517–24.

Kulle A, Krone N, Holterhus PM, Schuler G, Greaves RF, Juul A, et al. EU COST action. Steroid hormone analysis in diagnosis and treatment of DSD: position paper of EU COST action BM 1303 'DSDnet'. Eur J Endocrinol 2017;176(5):P1–9.

Kumar M, Mukhopadhyay S, Dutta D. Challenges and controversies in diagnosis and management of gonadotropin dependent precocious puberty: an Indian perspective. Indian. J Endocrinol Metab 2015;19(2):228–35.

Kurtoğlu S, Hatipoğlu N. Non-classical congenital adrenal hyperplasia in childhood. J Clin Res Pediatr Endocrinol 2017;9(1):1–7.

Kushnir MM, Rockwood AL, Roberts WL, Pattison EG, Owen WE, Bunker AM, et al. Development and performance evaluation of a tandem mass spectrometry assay for 4 adrenal steroids. Clin Chem 2006;52(8):1559–67.

Lai F, Srinivasan S, Wiley V. Evaluation of a two-tier screening pathway for congenital adrenal hyperplasia in the New South Wales newborn screening programme. Int J Neonatal Screen 2020;6(3):63.

Laimon W, El-Hawary A, Aboelenin H, Elzohiri M, Abdelmaksoud S, Megahed N, et al. Prepubertal gynecomastia is not always idiopathic: case series and review of the literature. Eur J Pediatr 2021;180(3):977–82.

Lao Q, Mallappa A, Rueda Faucz F, Joyal E, Veeraraghavan P, Chen W, et al. A TNXB splice donor site variant as a cause of hypermobility type Ehlers-Danlos syndrome in patients with congenital adrenal hyperplasia. Mol Genet Genomic Med 2021;9(2):e1556.

Lasarev MR, Bialk ER, Allen DB, Held PK. Application of principal component analysis to newborn screening for congenital adrenal hyperplasia. J Clin Endocrinol Metab 2020;105(8):dgaa371.

Lee Y, Choi JH, Oh A, Kim GH, Park SH, Moon JE, et al. Clinical, endocrinological, and molecular features of four Korean cases of cytochrome P450 oxidoreductase deficiency. Ann Pediatr Endocrinol Metab 2020a;25(2):97–103.

Lee HS, Jeong HR, Rho JG, Kum CD, Kim KH, Kim DW, et al. Identification of rare missense mutations in NOTCH2 and HERC2 associated with familial central precocious puberty via whole-exome sequencing. Gynecol Endocrinol 2020b;36(8):682–6.

Levy-Shraga Y, Pinhas-Hamiel O. High 17-hydroxyprogesterone level in newborn screening test for congenital adrenal hyperplasia. BMJ Case Rep 2016;2016. bcr2015213939.

Li L, Zhang J, Zeng J, Liao B, Peng X, Li T, et al. Proteomics analysis of potential serum biomarkers for insulin resistance in patients with polycystic ovary syndrome. Int J Mol Med 2020b;45(5):1409–16.

Li M, Chen Y, Liao B, Tang J, Zhong J, Lan D. The role of kisspeptin and MKRN3 in the diagnosis of central precocious puberty in girls. Endocr Connect 2021;10(9):1147–54.

Lima CJ, Cardoso SC, Lemos EF, Zingler E, Capanema C, Menezes LD, et al. Mutational analysis of the genes encoding RFamide-related peptide-3, the human orthologue of gonadotrophin-inhibitory hormone, and its receptor (GPR147) in patients with gonadotrophin-releasing hormone-dependent pubertal disorders. J Neuroendocrinol 2014;26(11):817–24.

Lin-Su K, Nimkarn S, New MI. Congenital adrenal hyperplasia in adolescents: diagnosis and management. Ann N Y Acad Sci 2008;1135:95–8.

Livadas S, Bothou C. Management of the female with non-classical congenital adrenal hyperplasia (NCCAH): a patient-oriented approach. Front Endocrinol 2019;10:366.

Livadas S, Chrousos GP. Molecular and environmental mechanisms regulating puberty initiation: an integrated approach. Front Endocrinol (Lausanne) 2019;10:828.

Loh TP, Ho CS, Hartmann MF, Zakaria R, Lo CWS, van den Berg S, et al. Influence of isotopically labeled internal standards on quantification of serum/plasma 17α-hydroxyprogesterone (17OHP) by liquid chromatography mass spectrometry. Clin Chem Lab Med 2020;58(10):1731–9.

Lu C, Hutchens EG, Farhy LS, Bonner HG, Suratt PM, McCartney CR. Influence of sleep stage on LH pulse initiation in the normal late follicular phase and in polycystic ovary syndrome. Neuroendocrinology 2018;107(1):60–72.

Lutfallah C, Wang W, Mason JI, Chang YT, Haider A, Rich B, et al. Newly proposed hormonal criteria via genotypic proof for type II 3beta-hydroxysteroid dehydrogenase deficiency. J Clin Endocrinol Metab 2002;87(6):2611–22.

Małunowicz EM, Ginalska-Malinowska M, Romer TE, Ruszczyńska-Wolska A, Dura M. Heterogeneity of urinary steroid profiles in children with adrenocortical tumors. Horm Res 1995;44(4):182–8.

Maciel-Guerra AT, de Mello MP, Coeli FB, Ribeiro ML, Miranda ML, Marques-de-Faria AP, et al. XX maleness and XX true hermaphroditism in SRY-negative monozygotic twins: additional evidence for a common origin. J Clin Endocrinol Metab 2008;93(2):339–43.

Magnisali P, Chalioti MB, Livadara T, Mataragas M, Paliatsiou S, Malamitsi-Puchner A, et al. Simultaneous quantification of 17α-OH progesterone, 11-deoxycortisol, Δ4-androstenedione, cortisol and cortisone in newborn blood spots using liquid chromatography-tandem mass spectrometry. J Chromatogr B Analyt Technol Biomed Life Sci 2011;879(19):1565–72.

Maione L, Naulé L, Kaiser UB. Makorin RING finger protein 3 and central precocious puberty. Curr Opin Endocr Metab Res 2020;14:152–9.

Mallappa A, Sinaii N, Kumar P, Whitaker MJ, Daley LA, Digweed D, et al. A phase 2 study of Chronocort, a modified-release formulation of hydrocortisone, in the treatment of adults with classic congenital adrenal hyperplasia. J Clin Endocrinol Metab 2015;100 (3):1137–45.

Malsagova K, Kopylov A, Stepanov A, Butkova T, Izotov A, Kaysheva A. Dried blood spot in laboratory: directions and prospects. Diagnostics 2020;10(4):248.

Marcos J, Craig WY, Palomaki GE, Kloza EM, Haddow JE, Roberson M, et al. Maternal urine and serum steroid measurements to identify steroid sulfatase deficiency (STSD) in second trimester pregnancies. Prenat Diagn 2009;29(8):771–80.

Marquardt RM, Kim TH, Shin JH, Jeong JW. Progesterone and estrogen signaling in the endometrium: what goes wrong in endometriosis? Int J Mol Sci 2019;20(15):3822.

Martin KA, Anderson RR, Chang RJ, Ehrmann DA, Lobo RA, Murad MH, et al. Evaluation and treatment of hirsutism in premenopausal women: an endocrine society clinical practice guideline. J Clin Endocrinol Metab 2018;103(4):1233–57.

McAllister JM, Modi B, Miller BA, Biegler J, Bruggeman R, Legro RS, et al. Overexpression of a DENND1A isoform produces a polycystic ovary syndrome theca phenotype. Proc Natl Acad Sci U S A 2014;111(15):E1519–27.

Meczekalski B, Slopien R, Warenik-Szymankiewicz A. Serum levels of 3alpha-androstanediol glucuronide in young women with polycystic ovary syndrome, idiopathic hirsutism and in normal subjects. Eur J Obstet Gynecol Reprod Biol 2007;132(1):88–92.

Mehdizadeh A, Kalantar SM, Sheikhha MH, Aali BS, Ghanei A. Association of SNP rs.2414096 CYP19 gene with polycystic ovarian syndrome in Iranian women. Int J Reprod Biomed 2017;15(8):491–6.

Melardi JW, Cunha DFS, Steinmetz L, Damiani D. Puberty in patients with ovotesticular DSD: evaluation of 20 patients and review of the literature. Pediatr Endocrinol Rev 2020;17(3):243–9.

Menabò S, Boccassini S, Gambineri A, Balsamo A, Pasquali R, Prontera O, et al. Improving the diagnosis of 11β-hydroxylase deficiency using home-made MLPA probes: identification of a novel chimeric CYP11B2/CYP11B1 gene in a Sicilian patient. J Endocrinol Invest 2016;39(3):291–5.

Mendes-Dos-Santos CT, Martins DL, Guerra-Júnior G, Baptista MTM, de-Mello MP, de Oliveira LC, et al. Prevalence of testicular adrenal rest tumor and factors associated with its development in congenital adrenal hyperplasia. Horm Res Paediatr 2018;90(3):161–8.

Mendonca BB, Gomes NL, Costa EM, Inacio M, Martin RM, Nishi MY, et al. 46,XY disorder of sex development (DSD) due to 17β-hydroxysteroid dehydrogenase type 3 deficiency. J Steroid Biochem Mol Biol 2017;165(Pt. A):79–85.

Menon RK, Ferrau F, Kurzawinski TR, Rumsby G, Freeman A, Amin Z, et al. Adrenal cancer in neurofibromatosis type 1: case report and DNA analysis. Endocrinol Diabetes Metab Case Rep 2014;2014, 140074.

Merke DP, Chrousos GP, Eisenhofer G, Weise M, Keil MF, Rogol AD, et al. Adrenomedullary dysplasia and hypofunction in patients with classic 21-hydroxylase deficiency. N Engl J Med 2000;343 (19):1362–8.

Mermejo LM, Elias LL, Marui S, Moreira AC, Mendonca BB, de Castro M. Refining hormonal diagnosis of type II 3beta-hydroxysteroid dehydrogenase deficiency in patients with premature pubarche and hirsutism based on HSD3B2 genotyping. J Clin Endocrinol Metab 2005;90(3):1287–93.

Midgley PC, Azzopardi D, Oates N, Shaw JC, Honour JW. Virilisation of female preterm infants. Arch Dis Child 1990;65:701–3 [7 Spec No].

Miller WL. Steroidogenic enzymes. Endocr Dev 2008;13:1–18.

Miller WL. Fetal endocrine therapy for congenital adrenal hyperplasia should not be done. Best Pract Res Clin Endocrinol Metab 2015; 29(3):469–83.

Miller WL. Mechanisms in endocrinology: rare defects in adrenal steroidogenesis. Eur J Endocrinol 2018;179(3):R125–41.

Miller WL. Congenital adrenal hyperplasia: time to replace 17OHP with 21-deoxycortisol. Horm Res Paediatr 2019;91(6):416–20.

Miller WL. Steroidogenic electron-transfer factors and their diseases. Ann Pediatr Endocrinol Metab 2021;26(3):138–48.

Miller WL, Merke DP. Tenascin-X, congenital adrenal hyperplasia, and the CAH-X syndrome. Horm Res Paediatr 2018;89(5):352–61.

Mizdrak M, Tičinović Kurir T, Božić J. The role of biomarkers in adrenocortical carcinoma: a review of current evidence and future perspectives. Biomedicine 2021;9(2):174.

Mooij CF, Parajes S, Rose IT, Taylor AE, Bayraktaroglu T, Wass JA, et al. Characterization of the molecular genetic pathology in patients with 11β-hydroxylase deficiency. Clin Endocrinol (Oxf) 2015;83(5): 629–35.

Moran LJ, Teede HJ, Noakes M, Clifton PM, Norman RJ, Wittert GA. Sex hormone binding globulin, but not testosterone, is associated with the metabolic syndrome in overweight and obese women with polycystic ovary syndrome. J Endocrinol Invest 2013;36(11):1004–10.

Moreno S, Guillermo M, Decoulx M, Dewailly D, Bresson R, Proye C. Feminizing adreno-cortical carcinomas in male adults. A dire prognosis. Three cases in a series of 801 adrenalectomies and review of the literature. Ann Endocrinol (Paris) 2006;67(1):32–8.

Morishima A, Grumbach MM, Simpson ER, Fisher C, Qin K. Aromatase deficiency in male and female siblings caused by a novel mutation and the physiological role of estrogens. J Clin Endocrinol Metab 1995; 80(12):3689–98.

Morris LF, Park S, Daskivich T, Churchill BM, Rao CV, Lei Z, et al. Virilization of a female infant by a maternal adrenocortical carcinoma. Endocr Pract 2011;17(2):e26–31.

Motaghedi R, Betensky BP, Slowinska B, Cerame B, Cabrer M, New MI, et al. Update on the prenatal diagnosis and treatment of congenital adrenal hyperplasia due to 11beta-hydroxylase deficiency. J Pediatr Endocrinol Metab 2005;18(2):133–42.

Mouanness M, Merhi Z. Impact of dietary advanced glycation end products on female reproduction: review of potential mechanistic pathways. Nutrients 2022;14(5):966.

Mueller JW, Idkowiak J, Gesteira TF, Vallet C, Hardman R, van den Boom J, et al. Human DHEA sulfation requires direct interaction between PAPS synthase 2 and DHEA sulfotransferase SULT2A1. J Biol Chem 2018;293(25):9724–35.

Mullis PE, Hindmarsh PC, Brook CG. Sodium chloride supplement at diagnosis and during infancy in children with salt-losing 21-hydroxylase deficiency. Eur J Pediatr 1990;150(1):22–5.

Naelitz BD, Sharifi N. Through the looking-glass: reevaluating DHEA metabolism through HSD3B1 genetics. Trends Endocrinol Metab 2020;31(9):680–90.

Nagasaki K, Takase K, Numakura C, Homma K, Hasegawa T, Fukami M. Foetal virilisation caused by overproduction of non-aromatisable 11-oxygenated C19 steroids in maternal adrenal tumour. Hum Reprod 2020;35(11):2609–12.

Nagel SA, Hartmann MF, Riepe FG, Wudy SA, Wabitsch M. Gonadotropin- and adrenocorticotropic hormone-independent precocious puberty of gonadal origin in a patient with adrenal hypoplasia congenita due to DAX1 gene mutation—a case report and review of the literature: implications for the pathomechanism. Horm Res Paediatr 2019;91(5):336–45.

Narasimhan ML, Khattab A. Genetics of congenital adrenal hyperplasia and genotype-phenotype correlation. Fertil Steril 2019;111(1):24–9.

Naulé L, Kaiser UB. Evolutionary conservation of MKRN3 and other Makorins and their roles in puberty initiation and endocrine functions. Semin Reprod Med 2019;37(4):166–73.

Nella AA, Mallappa A, Perritt AF, Gounden V, Kumar P, Sinaii N, et al. A phase 2 study of continuous subcutaneous hydrocortisone infusion in adults with congenital adrenal hyperplasia. J Clin Endocrinol Metab 2016;101(12):4690–8.

Ng SM, Stepien KM, Krishan A. Glucocorticoid replacement regimens for treating congenital adrenal hyperplasia. Cochrane Database Syst Rev 2020;3(3), CD012517.

Nguyen HH, Eiden-Plach A, Hannemann F, Malunowicz EM, Hartmann MF, Wudy SA, et al. Phenotypic, metabolic, and molecular genetic characterization of six patients with congenital adrenal hyperplasia caused by novel mutations in the CYP11B1 gene. J Steroid Biochem Mol Biol 2016;155(Pt. A):126–34.

Nguyen NN, Huynh LBP, Do MD, Yang TY, Tsai MC, Chen YC. Diagnostic accuracy of female pelvic ultrasonography in differentiating precocious puberty from premature thelarche: a systematic review and meta-analysis. Front Endocrinol (Lausanne) 2021;12:735875.

Nimkarn S, New MI. Steroid 11beta- hydroxylase deficiency congenital adrenal hyperplasia. Trends Endocrinol Metab 2008;19(3):96–9.

Nooranizadeh MH, Mogheiseh A, Kafi M, Sepehrimanesh M, Vaseghi H. Induction of superovulation in mature mice and rats using serum of spayed female dogs. Lab Anim Res 2018;34(4):211–5.

Noordam C, Dhir V, McNelis JC, Schlereth F, Hanley NA, Krone N, et al. Inactivating PAPSS2 mutations in a patient with premature pubarche. N Engl J Med 2009;360(22):2310–8. Erratum in: N Engl J Med. 2009;361(2):217.

Novello L, Speiser PW. Premature adrenarche. Pediatr Ann 2018;47(1): e7–e11.

Ono H, Numakura C, Homma K, Hasegawa T, Tsutsumi S, Kato F, et al. Longitudinal serum and urine steroid metabolite profiling in a 46,XY infant with prenatally identified POR deficiency. J Steroid Biochem Mol Biol 2018;178:177–84.

Oostdijk W, Idkowiak J, Mueller JW, House PJ, Taylor AE, O'Reilly MW, et al. PAPSS2 deficiency causes androgen excess via impaired DHEA sulfation—in vitro and in vivo studies in a family harboring two novel PAPSS2 mutations. J Clin Endocrinol Metab 2015;100(4): E672–80.

Oriolo C, Fanelli F, Castelli S, Mezzullo M, Altieri P, Corzani F, et al. Steroid biomarkers for identifying non-classic adrenal hyperplasia due to 21-hydroxylase deficiency in a population of PCOS with suspicious levels of 17OH-progesterone. J Endocrinol Invest 2020;43 (10):1499–509.

Owen EJ, Holownia P, Conway GS, Jacobs HS, Honour JW. 11 beta-hydroxyandrostenedione in plasma, follicular fluid, and granulosa cells of women with normal and polycystic ovaries. Fertil Steril 1992; 58(4):713–8.

Özalkak Ş, Çetinkaya S, Budak FC, Erdeve ŞS, Aycan Z. Evaluation of gonadotropin responses and response times according to two different cut-off values in luteinizing hormone releasing hormone stimulation test in girls. Indian J Endocrinol Metab 2020;24(5):410–5.

Özen S, Atik T, Korkmaz Ö, Onay H, Gökşen D, Özkınay F, et al. Aromatase deficiency in two siblings with 46,XX karyotype raised as different genders: a novel mutation (p.R115X) in the CYP19A1 gene. J Clin Res Pediatr Endocrinol 2020;12(1):109–12.

Panayiotopoulos A, Bhangoo A, Khurana D, Ten S, Michl J, Ghanny S. Glucocorticoid resistance in premature adrenarche and PCOS: from childhood to adulthood. J Endocr Soc 2020;4(9):bvaa111.

Pandey AV, Sproll P. Pharmacogenomics of human P450 oxidoreductase. Front Pharmacol 2014;5:103.

Pandey AV, Kempná P, Hofer G, Mullis PE, Flück CE. Modulation of human CYP19A1 activity by mutant NADPH P450 oxidoreductase. Mol Endocrinol 2007;21(10):2579–95.

Pang S. The molecular and clinical spectrum of 3beta-hydroxysteroid dehydrogenase deficiency disorder. Trends Endocrinol Metab 1998;9(2): 82–6.

Pang S, Levine LS, Stoner E, Opitz JM, Pollack MS, Dupont B, et al. Nonsalt-losing congenital adrenal hyperplasia due to 3 beta-hydroxysteroid dehydrogenase deficiency with normal glomerulosa function. J Clin Endocrinol Metab 1983;56(4):808–18.

Pang SY, Lerner AJ, Stoner E, Levine LS, Oberfield SE, Engel I, et al. Late-onset adrenal steroid 3 beta-hydroxysteroid dehydrogenase deficiency. I. A cause of hirsutism in pubertal and postpubertal women. J Clin Endocrinol Metab 1985;60(3):428–39.

Papadakis G, Kandaraki EA, Tseniklidi E, Papalou O, Diamanti-Kandarakis E. Polycystic ovary syndrome and NC-CAH: distinct characteristics and common findings. A systematic review. Front Endocrinol 2019;10:388.

Parajes S, Loidi L, Reisch N, Dhir V, Rose IT, Hampel R, et al. Functional consequences of seven novel mutations in the CYP11B1 gene: four mutations associated with nonclassic and three mutations causing classic 11{beta}-hydroxylase deficiency. J Clin Endocrinol Metab 2010;95(2):779–88.

Parween S, Rojas Velazquez MN, Udhane SS, Kagawa N, Pandey AV. Variability in loss of multiple enzyme activities due to the human genetic variation P284T located in the flexible hinge region of NADPH cytochrome P450 oxidoreductase. Front Pharmacol 2019;10:1187.

Pasquali R, Zanotti L, Fanelli F, Mezzullo M, Fazzini A, Morselli Labate AM, et al. Defining hyperandrogenism in women with polycystic ovary syndrome: a challenging perspective. J Clin Endocrinol Metab 2016;101(5):2013–22.

Paula GB, Ribeiro Andrade JG, Guaragna-Filho G, Sewaybricker LE, Miranda ML, Maciel-Guerra AT, et al. Ovotesticular disorder of sex development with unusual karyotype: patient report. J Pediatr Endocrinol Metab 2015;28(5–6):677–80.

Peña AS, Witchel SF, Hoeger KM, Oberfield SE, Vogiatzi MG, Misso M, et al. Adolescent polycystic ovary syndrome according to the international evidence-based guideline. BMC Med 2020;18(1):72.

Pereira SS, Costa MM, Gomez-Sanchez CE, Monteiro MP, Pignatelli D. Incomplete pattern of steroidogenic protein expression in functioning adrenocortical carcinomas. Biomedicine 2020;8(8):256.

Perry JR, Murray A, Day FR, Ong KK. Molecular insights into the aetiology of female reproductive ageing. Nat Rev Endocrinol 2015;11(12):725–34.

Phelan N, Williams EL, Cardamone S, Lee M, Creighton SM, Rumsby G, et al. Screening for mutations in 17β-hydroxysteroid dehydrogenase and androgen receptor in women presenting with partially virilised 46,XY disorders of sex development. Eur J Endocrinol 2015;172 (6):745–51.

Piccinato CA, Malvezzi H, Gibson DA, Saunders PTK. Sulfation pathways: contribution of intracrine oestrogens to the aetiology of endometriosis. J Mol Endocrinol 2018;61(2):T253–70.

Pignatelli D, Carvalho BL, Palmeiro A, Barros A, Guerreiro SG, Macut D. The complexities in genotyping of congenital adrenal hyperplasia: 21-hydroxylase deficiency. Front Endocrinol 2019;10:432.

Polat S, Kulle A, Karaca Z, Akkurt I, Kurtoglu S, Kelestimur F, et al. Characterisation of three novel CYP11B1 mutations in classic and non-classic 11β-hydroxylase deficiency. Eur J Endocrinol 2014;170 (5):697–706.

Pussard E, Travers S, Bouvattier C, Xue QY, Cosson C, Viengchareun S, et al. Urinary steroidomic profiles by LC-MS/MS to monitor classic 21-hydroxylase deficiency. J Steroid Biochem Mol Biol 2020;198, 105553.

Qiao J, Han B. Diseases caused by mutations in luteinizing hormone/chorionic gonadotropin receptor. Prog Mol Biol Transl Sci 2019; 161:69–89.

Rajska A, Buszewska-Forajta M, Rachoń D, Markuszewski MJ. Metabolomic insight into polycystic ovary syndrome—an overview. Int J Mol Sci 2020;21(14):4853.

Ramsey JT, Li Y, Arao Y, Naidu A, Coons LA, Diaz A, et al. Lavender products associated with premature thelarche and prepubertal gynecomastia: case reports and endocrine-disrupting chemical activities. J Clin Endocrinol Metab 2019;104(11):5393–405.

Rangaswamaiah S, Gangathimmaiah V, Nordenstrom A, Falhammar H. Bone mineral density in adults with congenital adrenal hyperplasia: a systematic review and meta-analysis. Front Endocrinol 2020;11:493.

Raperport C, Chronopoulou E, Homburg R. Effects of metformin treatment on pregnancy outcomes in patients with polycystic ovary syndrome. Expert Rev Endocrinol Metab 2021;16(2):37–47.

Rauh M. Steroid measurement with LC-MS/MS in pediatric endocrinology. Mol Cell Endocrinol 2009;301(1–2):272–81.

Reisch N, Högler W, Parajes S, Rose IT, Dhir V, Götzinger J, et al. A diagnosis not to be missed: nonclassic steroid 11β-hydroxylase deficiency presenting with premature adrenarche and hirsutism. J Clin Endocrinol Metab 2013;98(10):E1620–5.

Reisch N, Taylor AE, Nogueira EF, Asby DJ, Dhir V, Berry A, et al. Alternative pathway androgen biosynthesis and human fetal female virilization. Proc Natl Acad Sci U S A 2019;116(44):22294–9.

Resende EA, Lara BH, Reis JD, Ferreira BP, Pereira GA, Borges MF. Assessment of basal and gonadotropin-releasing hormone-stimulated gonadotropins by immunochemiluminometric and immunofluorometric assays in normal children. J Clin Endocrinol Metab 2007;92(4):1424–9.

Riaz M, Ibrahim MN, Laghari TM, Hanif MI, Raza J. Van Wyk Grumbach syndrome. J Coll Physicians Surg Pak 2020;30(12):1332–4.

Riedl S, Springer A, Häusler G, Price G, Richter-Unruh A, Stener-Victorin E, et al. Hypothesis: persistently elevated hCG causes gestational ovarian overstimulation associated with prolonged postpartum hyperandrogenism in mothers of aromatase-deficient babies. J Clin Endocrinol Metab 2013;98(8):3115–20.

Riepe FG, Sippell WG. Recent advances in diagnosis, treatment, and outcome of congenital adrenal hyperplasia due to 21-hydroxylase deficiency. Rev Endocr Metab Disord 2007;8(4):349–63.

Roberts SA, Kaiser UB. Genetics in endocrinology: genetic etiologies of central precocious puberty and the role of imprinted genes. Eur J Endocrinol 2020;183(4):R107–17.

Rochira V, Carani C. Aromatase deficiency in men: a clinical perspective. Nat Rev Endocrinol 2009;5(10):559–68.

Rodin A, Thakkar H, Taylor N, Clayton R. Hyperandrogenism in polycystic ovary syndrome. Evidence of dysregulation of 11 beta-hydroxysteroid dehydrogenase. N Engl J Med 1994;330(7):460–5.

Rodriguez Paris V, Bertoldo MJ. The mechanism of androgen actions in PCOS etiology. Med Sci (Basel) 2019;7(9):89.

Rojare C, Opdenakker Y, Laborde A, Nicot R, Mention K, Ferri J. The Smith-Lemli-Opitz syndrome and dentofacial anomalies diagnostic: case reports and literature review. Int Orthod 2019;17(2):375–83.

Rosa FW. Virilization of the female fetus with maternal danazol exposure. Am J Obstet Gynecol 1984;149(1):99–100.

Rosenfield RL. Current concepts of polycystic ovary syndrome pathogenesis. Curr Opin Pediatr 2020;32(5):698–706.

Rosenfield RL, Ehrmann DA. The pathogenesis of polycystic ovary syndrome (PCOS): the hypothesis of PCOS as functional ovarian hyperandrogenism revisited. Endocr Rev 2016;37(5):467–520.

Rösler A, Silverstein S, Abeliovich D. A (R80Q) mutation in 17 beta-hydroxysteroid dehydrogenase type 3 gene among Arabs of Israel is associated with pseudohermaphroditism in males and normal asymptomatic females. J Clin Endocrinol Metab 1996;81(5):1827–31.

Rymen D, Ritelli M, Zoppi N, Cinquina V, Giunta C, Rohrbach M, et al. Clinical and molecular characterization of classical-like Ehlers-Danlos syndrome due to a novel *TNXB* variant. Genes (Basel) 2019; 10(11):843.

Saito K, Matsuzaki T, Iwasa T, Miyado M, Saito H, Hasegawa T, et al. Steroidogenic pathways involved in androgen biosynthesis in eumenorrheic women and patients with polycystic ovary syndrome. J Steroid Biochem Mol Biol 2016;158:31–7.

Sakkal-Alkaddour H, Zhang L, Yang X, Chang YT, Kappy M, Slover RS, et al. Studies of 3 beta-hydroxysteroid dehydrogenase genes in infants and children manifesting premature pubarche and increased adrenocorticotropin-stimulated delta 5-steroid levels. J Clin Endocrinol Metab 1996;81(11):3961–5.

Salerno M, Capalbo D, Cerbone M, De Luca F. Subclinical hypothyroidism in childhood—current knowledge and open issues. Nat Rev Endocrinol 2016;12(12):734–46.

Sánchez-Guijo A, Neunzig J, Gerber A, Oji V, Hartmann MF, Schuppe HC, et al. Role of steroid sulfatase in steroid homeostasis and characterization of the sulfated steroid pathway: evidence from steroid sulfatase deficiency. Mol Cell Endocrinol 2016;437:142–53.

Santi M, Graf S, Zeino M, Cools M, Van De Vijver K, Trippel M, et al. Approach to the virilizing girl at puberty. J Clin Endocrinol Metab 2021;106(5):1530–9.

Sarathi V, Atluri S, Pradeep TVS, Rallapalli SS, Rakesh CV, Sunanda T, et al. Utility of a commercially available blood steroid profile in endocrine practice. Indian J Endocrinol Metab 2019;23(1):97–101.

Schoner K, Witsch-Baumgartner M, Behunova J, Petrovic R, Bald R, Kircher SG, et al. Smith-Lemli-Opitz syndrome—fetal phenotypes with special reference to the syndrome-specific internal malformation pattern. Birth Defects Res 2020;112(2):175–85.

Schröder MAM, Claahsen-van der Grinten HL. Novel treatments for congenital adrenal hyperplasia. Rev Endocr Metab Disord 2022;23 (3):631–45.

Schüring AN, Welp A, Gromoll J, Zitzmann M, Sonntag B, Nieschlag E, et al. Role of the CAG repeat polymorphism of the androgen receptor gene in polycystic ovary syndrome (PCOS). Exp Clin Endocrinol Diabetes 2012;120(2):73–9.

Schwarz E, Liu A, Randall H, Haslip C, Keune F, Murray M, et al. Use of steroid profiling by UPLC-MS/MS as a second tier test in newborn screening for congenital adrenal hyperplasia: the Utah experience. Pediatr Res 2009;66(2):230–5.

Sea JL, Abramyan M, Chiu HK. Prepubescent unilateral gynecomastia secondary to excessive soy consumption. J Pediatr Endocrinol Metab 2020;34(4):521–5.

Segal DG, DiMeglio LA, Ryder KW, Vollmer PA, Pescovitz OH. Assay interference leading to misdiagnosis of central precocious puberty. Endocrine 2003;20(3):195–9.

Sellick J, Aldridge S, Thomas M, Cheetham T. Growth of patients with congenital adrenal hyperplasia due to 21-hydroxylase in infancy, glucocorticoid requirement and the role of mineralocorticoid therapy. J Pediatr Endocrinol Metab 2018;31(9):1019–22.

Seo JY, Park HD, Kim JW, Oh HJ, Yang JS, Chang YS, et al. Steroid profiling for congenital adrenal hyperplasia by tandem mass spectrometry as a second-tier test reduces follow-up burdens in a tertiary care hospital: a retrospective and prospective evaluation. J Perinat Med 2014;42(1):121–7.

Seow KM, Chang YW, Chen KH, Juan CC, Huang CY, Lin LT, et al. Molecular mechanisms of laparoscopic ovarian drilling and its therapeutic effects in polycystic ovary syndrome. Int J Mol Sci 2020;21 (21), 8147.

Shaaban Z, Khoradmehr A, Jafarzadeh Shirazi MR, Tamadon A. Pathophysiological mechanisms of gonadotropins- and steroid hormones-related genes in etiology of polycystic ovary syndrome. Iran J Basic Med Sci 2019;22(1):3–16.

Shackleton C, Malunowicz E. Apparent pregnene hydroxylation deficiency (APHD): seeking the parentage of an orphan metabolome. Steroids 2003;68(9):707–17.

Shackleton CH, Roitman E, Kelley R. Neonatal urinary steroids in Smith-Lemli-Opitz syndrome associated with 7-dehydrocholesterol reductase deficiency. Steroids 1999b;64(7):481–90.

Shapiro J. Clinical practice. Hair loss in women. N Engl J Med 2007; 357(16):1620–30.

Sharma A, Kapoor E, Singh RJ, Chang AY, Erickson D. Diagnostic thresholds for androgen-producing tumors or pathologic hyperandrogenism in women by use of total testosterone concentrations measured by liquid chromatography-tandem mass spectrometry. Clin Chem 2018;64(11):1636–45.

Shen W, Li T, Hu Y, Liu H, Song M. Common polymorphisms in the CYP1A1 and CYP11A1 genes and polycystic ovary syndrome risk: a meta-analysis and meta-regression. Arch Gynecol Obstet 2014;289 (1):107–18.

Shi Y, Zhao H, Shi Y, Cao Y, Yang D, Li Z, et al. Genome-wide association study identifies eight new risk loci for polycystic ovary syndrome. Nat Genet 2012;44(9):1020–5.

Shozu M, Akasofu K, Harada T, Kubota Y. A new cause of female pseudohermaphroditism: placental aromatase deficiency. J Clin Endocrinol Metab 1991;72(3):560–6.

Simonetti L, Bruque CD, Fernández CS, Benavides-Mori B, Delea M, Kolomenski JE, et al. CYP21A2 mutation update: comprehensive analysis of databases and published genetic variants. Hum Mutat 2018;39(1):5–22.

Smith EP, Boyd J, Frank GR, Takahashi H, Cohen RM, Specker B, et al. Estrogen resistance caused by a mutation in the estrogen-receptor gene in a man. N Engl J Med 1994;331(16):1056–61. Erratum in: N Engl J Med 1995;332(2):131.

Sørensen AE, Udesen PB, Maciag G, Geiger J, Saliani N, Januszewski AS, et al. Hyperandrogenism and metabolic syndrome are associated with changes in serum-derived microRNAs in women with polycystic ovary syndrome. Front Med 2019;6:242.

Speiser PW, Arlt W, Auchus RJ, Baskin LS, Conway GS, Merke DP, et al. Congenital adrenal hyperplasia due to steroid 21-hydroxylase deficiency: an endocrine society clinical practice guideline. J Clin Endocrinol Metab 2018;103(11):4043–88. Erratum in: J Clin Endocrinol Metab 2019;104(1):39–40.

Standeven LR, Olson E, Leistikow N, Payne JL, Osborne LM, Hantsoo L. Polycystic ovary syndrome, affective symptoms, and neuroactive steroids: a focus on allopregnanolone. Curr Psychiatry Rep 2021; 23(6):36.

Stolze BR, Gounden V, Gu J, Elliott EA, Masika LS, Abel BS, et al. An improved micro-method for the measurement of steroid profiles by APPI-LC-MS/MS and its use in assessing diurnal effects on steroid concentrations and optimizing the diagnosis and treatment of adrenal insufficiency and CAH. J Steroid Biochem Mol Biol 2016;162:110–6.

Sultan C, Gaspari L, Maimoun L, Kalfa N, Paris F. Disorders of puberty. Best Pract Res Clin Obstet Gynaecol 2018;48:62–89.

Sumińska M, Bogusz-Górna K, Wegner D, Fichna M. Non-classic disorder of adrenal steroidogenesis and clinical dilemmas in 21-hydroxylase deficiency combined with backdoor androgen pathway. Mini-review and case report. Int J Mol Sci 2020;21(13), 4622.

Sun L, Hu W, Liu Q, Hao Q, Sun B, Zhang Q, et al. Metabonomics reveals plasma metabolic changes and inflammatory marker in polycystic ovary syndrome patients. J Proteome Res 2012;11(5):2937–46.

Susheelamma CJ, Pillai SM, Asha NS. Oestrogen, progesterone and stem cells: the discordant trio in endometriosis? Expert Rev Mol Med 2018;20:e2.

Taheri S, Zararsiz G, Karaburgu S, Borlu M, Ozgun MT, Karaca Z, et al. Is idiopathic hirsutism (IH) really idiopathic? mRNA expressions of skin steroidogenic enzymes in women with IH. Eur J Endocrinol 2015;173 (4):447–54.

Teede HJ, Misso ML, Costello MF, Dokras A, Laven J, Moran L, et al. Recommendations from the international evidence-based guideline for the assessment and management of polycystic ovary syndrome. Hum Reprod 2018;33(9):1602–18.

Temple SEL, Sachdev R, Ellaway C. Familial *DHCR7* genotype presenting as a very mild form of Smith-Lemli-Opitz syndrome and lethal holoprosencephaly. JIMD Rep 2020;56(1):3–8.

Tng EL, Tan JMM. Gonadotropin-releasing hormone analogue stimulation test versus venous sampling in postmenopausal hyperandrogenism. J Endocr Soc 2020;5(1):bvaa172.

Tong CV, Tai YT. Hyperandrogenism caused by ovarian Leydig cell tumour: finding the needle in a haystack. BMJ Case Rep 2020;13 (12), e238012.

Torchen LC, Idkowiak J, Fogel NR, O'Neil DM, Shackleton CH, Arlt W, et al. Evidence for Increased 5α-reductase activity during early childhood in daughters of women with polycystic ovary syndrome. J Clin Endocrinol Metab 2016;101(5):2069–75.

Toscano V, Sciarra F, Adamo MV, Petrangeli E, Foli S, Caiola S, et al. Is 3 alpha-androstanediol a marker of peripheral hirsutism? Acta Endocrinol 1982;99(2):314–20.

Traggiai C, Stanhope R. Disorders of pubertal development. Best Pract Res Clin Obstet Gynaecol 2003;17(1):41–56.

Travers S, Bouvattier C, Fagart J, Martinerie L, Viengchareun S, Pussard E, et al. Interaction between accumulated 21-deoxysteroids and mineralocorticoid signaling in 21-hydroxylase deficiency. Am J Physiol Endocrinol Metab 2020;318(2):E102–10.

Tsilchorozidou T, Honour JW, Conway GS. Altered cortisol metabolism in polycystic ovary syndrome: insulin enhances 5alpha-reduction but not the elevated adrenal steroid production rates. J Clin Endocrinol Metab 2003;88(12):5907–13.

Tucci A, Ronzoni L, Arduino C, Salmin P, Esposito S, Milani D. The p. Phe174Ser mutation is associated with mild forms of Smith Lemli Opitz syndrome. BMC Med Genet 2016;17, 22.

Tufano M, Ciofi D, Amendolea A, Stagi S. Auxological and Endocrinological features in children with McCune Albright syndrome: a review. Front Endocrinol 2020;11:522.

Turcu AF, Mallappa A, Elman MS, Avila NA, Marko J, Rao H, et al. 11-Oxygenated androgens are biomarkers of adrenal volume and testicular adrenal rest tumors in 21-hydroxylase deficiency. J Clin Endocrinol Metab 2017;102(8):2701–10.

Turcu AF, El-Maouche D, Zhao L, Nanba AT, Gaynor A, Veeraraghavan P, et al. Androgen excess and diagnostic steroid biomarkers for nonclassic 21-hydroxylase deficiency without cosyntropin stimulation. Eur J Endocrinol 2020;183(1):63–71.

Turcu AF, Rege J, Auchus RJ, Rainey WE. 11-Oxygenated androgens in health and disease. Nat Rev Endocrinol 2020b;16(5):284–96.

Ueland GÅ, Dahl SR, Methlie P, Hessen S, Husebye ES, Thorsby PM. Adrenal steroid profiling as a diagnostic tool to differentiate polycystic ovary syndrome from nonclassic congenital adrenal hyperplasia: pinpointing easy screening possibilities and normal cutoff levels using liquid chromatography tandem mass spectrometry. Fertil Steril 2022;118(2):384–91.

Unal E, Yıldırım R, Taş FF, Demir V, Onay H, Haspolat YK. Aromatase deficiency due to a novel mutation in *CYP19A1* gene. J Clin Res Pediatr Endocrinol 2018;10(4):377–81.

Unluhizarci K, Karaca Z, Kelestimur F. Role of insulin and insulin resistance in androgen excess disorders. World J Diabetes 2021;12(5):616–29.

Valeri C, Lovaisa MM, Racine C, Edelsztein NY, Riggio M, Giulianelli S, et al. Molecular mechanisms underlying AMH elevation in hyperoestrogenic states in males. Sci Rep 2020;10(1):15062.

Vassiliadi DA, Barber TM, Hughes BA, McCarthy MI, Wass JA, Franks S, et al. Increased 5 alpha-reductase activity and adrenocortical drive in women with polycystic ovary syndrome. J Clin Endocrinol Metab 2009;94(9):3558–66.

Velikanova LI, Shafigullina ZR, Lisitsin AA, Vorokhobina NV, Grigoryan K, Kukhianidze EA, et al. Different types of urinary steroid profiling obtained by high-performance liquid chromatography and gas chromatography-mass spectrometry in patients with adrenocortical carcinoma. Horm Cancer 2016;7(5–6):327–35.

Vogt EC, Hammerling K, Sorbye H, Heie A, Sulen A, Ueland G, et al. Feminizing adrenal tumor identified by plasma steroid profiling. Endocrinol Diabetes Metab Case Rep 2021;2021, 21-0104.

Wallace AM, Beesley J, Thomson M, Giles CA, Ross AM, Taylor NF. Adrenal status during the first month of life in mature and premature human infants. J Endocrinol 1987;112(3):473–80.

Walters KA, Rodriguez Paris V, Aflatounian A, Handelsman DJ. Androgens and ovarian function: translation from basic discovery research to clinical impact. J Endocrinol 2019;242(2):R23–50.

Wang X, Nie M, Lu L, Tong A, Chen S, Lu Z. Identification of seven novel CYP11B1 gene mutations in Chinese patients with 11β-hydroxylase deficiency. Steroids 2015;100:11–6.

Wang D, Wang J, Tong T, Yang Q. Non-classical 11β-hydroxylase deficiency caused by compound heterozygous mutations: a case study and literature review. J Ovarian Res 2018;11(1):82.

Waterham HR, Hennekam RC. Mutational spectrum of Smith-Lemli-Opitz syndrome. Am J Med Genet C Semin Med Genet 2012;160C(4):263–84.

Weinstein LS, Shenker A, Gejman PV, Merino MJ, Friedman E, Spiegel AM. Activating mutations of the stimulatory G protein in the McCune-Albright syndrome. N Engl J Med 1991;325(24):1688–95.

Weise M, Drinkard B, Mehlinger SL, Holzer SM, Eisenhofer G, Charmandari E, et al. Stress dose of hydrocortisone is not beneficial in patients with classic congenital adrenal hyperplasia undergoing short-term, high-intensity exercise. J Clin Endocrinol Metab 2004b;89(8):3679–84.

Wells RS, Kerr CB. Genetic classification of ichthyosis. Arch Dermatol 1965;92(1):1–6.

Werneck G, Rodrigues EMR, Mantovani RM, Lane JSS, Silva IN. Testicular adrenal rest tumors in patients with congenital adrenal hyperplasia: 6 years of follow-up. J Pediatr Endocrinol Metab 2019;32(5):519–26.

Winge MC, Hoppe T, Liedén A, Nordenskjöld M, Vahlquist A, Wahlgren CF, et al. Novel point mutation in the STS gene in a patient with X-linked recessive ichthyosis. J Dermatol Sci 2011;63(1):62–4.

Witchel SF, Plant TM. Intertwined reproductive endocrinology: puberty and polycystic ovary syndrome. Curr Opin Endocr Metab Res 2020;14:127–36.

Witchel SF, Tena-Sempere M. The Kiss1 system and polycystic ovary syndrome: lessons from physiology and putative pathophysiologic implications. Fertil Steril 2013;100(1):12–22.

Wolthers OD, Schou AJ. Short-term growth after withdrawal of exogenous glucocorticoids. Horm Res 2005;64(3):116–8.

Wolthers OD, Cameron FJ, Scheimberg I, Honour JW, Hindmarsh PC, Savage MO, et al. Androgen secreting adrenocortical tumours. Arch Dis Child 1999;80(1):46–50.

Wong T, Shackleton CH, Covey TR, Ellis G. Identification of the steroids in neonatal plasma that interfere with 17 alpha-hydroxyprogesterone radioimmunoassays. Clin Chem 1992;38(9):1830–7.

Wudy SA, Choi MH. Steroid LC-MS has come of age. J Steroid Biochem Mol Biol 2016;162:1–3.

Wudy SA, Wachter UA, Homoki J, Teller WM. 5 alpha-androstane-3 alpha, 17 beta-diol and 5 alpha-androstane-3 alpha, 17 beta-diol-glucuronide in plasma of normal children, adults and patients with idiopathic hirsutism: a mass spectrometric study. Eur J Endocrinol 1996;134(1):87–92.

Wudy SA, Schuler G, Sánchez-Guijo A, Hartmann MF. The art of measuring steroids: principles and practice of current hormonal steroid analysis. J Steroid Biochem Mol Biol 2018;179:88–103.

Yance VRV, Marcondes JAM, Rocha MP, Barcellos CRG, Dantas WS, Avila AFA, et al. Discriminating between virilizing ovary tumors and ovary hyperthecosis in postmenopausal women: clinical data, hormonal profiles and image studies. Eur J Endocrinol 2017;177(1):93–102.

Yang Z, Ye L, Wang W, Zhao Y, Wang W, Jia H, et al. 17β-hydroxysteroid dehydrogenase 3 deficiency: three case reports and a systematic review. J Steroid Biochem Mol Biol 2017;174:141–5.

Yang Y, Ding M, Di N, Azziz R, Yang D, Zhao X. Close correlation between hyperandrogenism and insulin resistance in women with polycystic ovary syndrome-based on liquid chromatography with tandem mass spectrometry measurements. J Clin Lab Anal 2019;33(3): e22699.

Yildiz BO, Bolour S, Woods K, Moore A, Azziz R. Visually scoring hirsutism. Hum Reprod Update 2010;16(1):51–64.

Zeng X, Xie YJ, Liu YT, Long SL, Mo ZC. Polycystic ovarian syndrome: correlation between hyperandrogenism, insulin resistance and obesity. Clin Chim Acta 2020;502:214–21.

Zevenhuijzen H, Kelnar CJ, Crofton PM. Diagnostic utility of a low-dose gonadotropin-releasing hormone test in the context of puberty disorders. Horm Res 2004;62(4):168–76.

Zhang J, Bao Y, Zhou X, Zheng L. Polycystic ovary syndrome and mitochondrial dysfunction. Reprod Biol Endocrinol 2019;17(1):67.

Zhao Y, Fu L, Li R, Wang LN, Yang Y, Liu NN, et al. Metabolic profiles characterizing different phenotypes of polycystic ovary syndrome: plasma metabolomics analysis. BMC Med 2012;10:153. https://doi.org/10.1186/1741-7015-10-153. 23198915. PMCID: PMC3599233.

Zhou WB, Chen N, Li CJ. A rare case of pure testosterone-secreting adrenal adenoma in a postmenopausal elderly woman. BMC Endocr Disord 2019;19(1):14.

Zhou X, Jaswa E, Pasch L, Shinkai K, Cedars MI, Huddleston HG. Association of obstructive sleep apnea risk with depression and anxiety symptoms in women with polycystic ovary syndrome. J Clin Sleep Med 2021. https://doi.org/10.5664/jcsm.9372. Epub ahead of print 33983110.

Zhu WJ, Cheng T, Zhu H, Han B, Fan MX, Gu T, et al. Aromatase deficiency: a novel compound heterozygous mutation identified in a Chinese girl with severe phenotype and obvious maternal virilization. Mol Cell Endocrinol 2016;433:66–74.

Zhu JL, Chen Z, Feng WJ, Long SL, Mo ZC. Sex hormone-binding globulin and polycystic ovary syndrome. Clin Chim Acta 2019a;499: 142–8.

Zhu W, Han B, Fan M, Wang N, Wang H, Zhu H, et al. Oxidative stress increases the 17,20-lyase-catalyzing activity of adrenal P450c17 through p38α in the development of hyperandrogenism. Mol Cell Endocrinol 2019b;484:25–33.

Zou Y, Zhu FF, Fang CY, Xiong XY, Li HY. Identification of potential biomarkers for urine metabolomics of polycystic ovary syndrome based on gas chromatography-mass spectrometry. Chin Med J (Engl) 2018;131(8):945–9.

Zou M, Chen R, Wang Y, He Y, Wang Y, Dong Y, et al. Clinical and ultrasound characteristics of virilizing ovarian tumors in pre- and postmenopausal patients: a single tertiary center experience. Orphanet J Rare Dis 2021;16(1):426.

Zwicker H, Rittmaster RS. Androsterone sulfate: physiology and clinical significance in hirsute women. J Clin Endocrinol Metab 1993;76(1): 112–6.

Further reading

Askari M, Rastari M, Seresht-Ahmadi M, McElreavey K, Bashamboo A, Razzaghy-Azar M, et al. A missense mutation in NR5A1 causing female to male sex reversal: a case report. Andrologia 2020;52(6): e13585.

Azziz R, Carmina E, Dewailly D, Diamanti-Kandarakis E, Escobar-Morreale HF, Futterweit W, et al. The androgen excess and PCOS society criteria for the polycystic ovary syndrome: the complete task force report. Fertil Steril 2009;91(2):456–88.

Baş F, Abalı ZY, Toksoy G, Poyrazoğlu Ş, Bundak R, Güleç Ç, et al. Precocious or early puberty in patients with combined pituitary hormone deficiency due to POU1F1 gene mutation: case report and review of possible mechanisms. Hormones (Athens) 2018a;17(4):581–8.

Baş F, Toksoy G, Ergun-Longmire B, Uyguner ZO, Abalı ZY, Poyrazoğlu Ş, et al. Prevalence, clinical characteristics and long-term outcomes of classical 11 β-hydroxylase deficiency (11BOHD) in Turkish population and novel mutations in CYP11B1 gene. J Steroid Biochem Mol Biol 2018b;181:88–97.

Belchetz PE, Barth JH, Kaufman JM. Biochemical endocrinology of the hypogonadal male. Ann Clin Biochem 2010;47(Pt. 6):503–15.

Bertrand-Delepine J, Manouvrier-Hanu S, Cartigny M, Paris F, Mallet D, Philibert P, et al. In cases of familial primary ovarian insufficiency and disorders of gonadal development, consider NR5A1/SF-1 sequence variants. Reprod Biomed Online 2020;40(1):151–9.

Bhasin S, Storer TW, Berman N, Callegari C, Clevenger B, Phillips J, et al. The effects of supraphysiologic doses of testosterone on muscle size and strength in normal men. N Engl J Med 1996;335(1):1–7.

Binder G, Ziegler J, Schweizer R, Habhab W, Haack TB, Heinrich T, et al. Novel mutation points to a hot spot in CDKN1C causing Silver-Russell syndrome. Clin Epigenetics 2020;12(1):152.

Breil T, Yakovenko V, Inta I, Choukair D, Klose D, Mittnacht J, et al. Typical characteristics of children with congenital adrenal hyperplasia due to 11β-hydroxylase deficiency: a single-centre experience and review of the literature. J Pediatr Endocrinol Metab 2019;32(3):259–67.

Butler AE, Ramachandran V, Sathyapalan T, David R, Gooderham NJ, Benurwar M, et al. microRNA expression in women with and without polycystic ovarian syndrome matched for body mass index. Front Endocrinol 2020;11:206.

Castro-Magana M, Angulo M, Canas JA, Mazur B, Sarrantonio M, Vitollo P, et al. Characterization of zona glomerulosa function in patients with classic and non-classic forms of congenital adrenal hyperplasia due to 11 beta-hydroxylase deficiency. J Pediatr Endocrinol Metab 1995; 8(1):19–25.

de Peretti E, Forest MG. Pitfalls in the etiological diagnosis of congenital adrenal hyperplasia in the early neonatal period. Horm Res 1982; 16(1):10–22.

Delle Piane L, Rinaudo PF, Miller WL. 150 years of congenital adrenal hyperplasia: translation and commentary of De Crecchio's classic paper from 1865. Endocrinology 2015;156(4):1210–7.

Di Dalmazi G. Hyperandrogenism and adrenocortical tumors. Front Horm Res 2019;53:92–9.

Domenice S, Machado AZ, Ferreira FM, Ferraz-de-Souza B, Lerario AM, Lin L, et al. Wide spectrum of NR5A1-related phenotypes in 46,XY and 46,XX individuals. Birth Defects Res C Embryo Today 2016;108(4):309–20.

Escobar-Morreale HF, Sanchón R, San Millán JL. A prospective study of the prevalence of nonclassical congenital adrenal hyperplasia among women presenting with hyperandrogenic symptoms and signs. J Clin Endocrinol Metab 2008;93(2):527–33.

Fanelli F, Gambineri A, Mezzullo M, Vicennati V, Pelusi C, Pasquali R, et al. Revisiting hyper- and hypo-androgenism by tandem mass spectrometry. Rev Endocr Metab Disord 2013;14(2):185–205.

Flück CE, Meyer-Böni M, Pandey AV, Kempná P, Miller WL, Schoenle EJ, et al. Why boys will be boys: two pathways of fetal testicular androgen biosynthesis are needed for male sexual differentiation. Am J Hum Genet 2011;89(2):201–18. Erratum in: Am J Hum Genet. 2011;89(2):347.

Garg D, Merhi Z. Relationship between advanced glycation end products and steroidogenesis in PCOS. Reprod Biol Endocrinol 2016;14(1):71.

Gerber PA, Kukova G, Meller S, Neumann NJ, Homey B. The dire consequences of doping. Lancet 2008;372(9639):656.

Goursaud C, Mallet D, Janin A, Menassa R, Tardy-Guidollet V, Russo G, et al. Aberrant splicing is the pathogenicity mechanism of the p. Glu314Lys variant in CYP11A1 gene. Front Endocrinol 2018;9:491.

Gunness A, Pazderska A, Ahmed M, McGowan A, Phelan N, Boran G, et al. Measurement of selected androgens using liquid chromatography-tandem mass spectrometry in reproductive-age women with type 1 diabetes. Hum Reprod 2018;33(9):1727–34.

Hamed SA. The effect of epilepsy and antiepileptic drugs on sexual, reproductive and gonadal health of adults with epilepsy. Expert Rev Clin Pharmacol 2016;9(6):807–19.

Han Z, Elliott MS. Neglected issues concerning teaching human adrenal steroidogenesis in popular biochemistry textbooks. Biochem Mol Biol Educ 2017;45(6):469–74.

Hao C, Guo J, Guo R, Qi Z, Li W, Ni X. Compound heterozygous variants in POR gene identified by whole-exome sequencing in a Chinese pedigree with cytochrome P450 oxidoreductase deficiency. Pediatr Investig 2018;2(2):90–5.

Herlin MK, Petersen MB, Brännström M. Mayer-Rokitansky-Küster-Hauser (MRKH) syndrome: a comprehensive update. Orphanet J Rare Dis 2020;15(1):214.

Honour JW, Anderson JM, Shackleton CH. Difficulties in the diagnosis of congenital adrenal hyperplasia in early infancy: the 11 beta-hydroxylase defect. Acta Endocrinol 1983;103(1):101–9.

Hurwitz A, Brautbar C, Milwidsky A, Vecsei P, Milewicz A, Navot D, et al. Combined 21- and 11 beta-hydroxylase deficiency in familial congenital adrenal hyperplasia. J Clin Endocrinol Metab 1985;60(4): 631–8.

Ibrahim F, Giton F, Boudou P, Villette JM, Julien R, Galons H, et al. Plasma 11beta-hydroxy-4-androstene-3,17-dione: comparison of a time-resolved fluoroimmunoassay using a biotinylated tracer with a radioimmunoassay using a tritiated tracer. J Steroid Biochem Mol Biol 2003;84(5):563–8.

Idkowiak J, Randell T, Dhir V, Patel P, Shackleton CH, Taylor NF, et al. A missense mutation in the human cytochrome b5 gene causes 46,XY disorder of sex development due to true isolated 17,20 lyase deficiency. J Clin Endocrinol Metab 2012;97(3):E465–75.

Inoue T, Nakamura A, Iwahashi-Odano M, Tanase-Nakao K, Matsubara K, Nishioka J, et al. Contribution of gene mutations to Silver-Russell syndrome phenotype: multigene sequencing analysis in 92 etiology-unknown patients. Clin Epigenetics 2020;12(1):86.

Kallali W, Gray E, Mehdi MZ, Lindsay R, Metherell LA, Buonocore F, et al. Long-term outcome of partial P450 side-chain cleavage enzyme deficiency in three brothers: the importance of early diagnosis. Eur J Endocrinol 2020;182(3):K15–24.

Kamrath C. Beyond the adrenals: organ manifestations in inherited primary adrenal insufficiency in children. Eur J Endocrinol 2020;182(3): C9–C12.

Karlekar MP, Sarathi V, Lila A, Rai K, Arya S, Bhandare VV, et al. Expanding genetic spectrum and discriminatory role of steroid profiling by LC-MS/MS in 11β-hydroxylase deficiency. Clin Endocrinol (Oxf) 2021;94(4):533–43.

Karnak I, Senocak ME, Göğüş S, Büyükpamukçu N, Hiçsönmez A. Testicular enlargement in patients with 11-hydroxylase deficiency. J Pediatr Surg 1997;32(5):756–8.

Kok RC, Timmerman MA, Wolffenbuttel KP, Drop SL, de Jong FH. Isolated 17,20-lyase deficiency due to the cytochrome b5 mutation W27X. J Clin Endocrinol Metab 2010;95(3):994–9.

Koyama Y, Homma K, Fukami M, Miwa M, Ikeda K, Ogata T, et al. Two-step biochemical differential diagnosis of classic 21-hydroxylase deficiency and cytochrome P450 oxidoreductase deficiency in Japanese infants by GC-MS measurement of urinary pregnanetriolone/ tetrahydroxycortisone ratio and 11β-hydroxyandrosterone. Clin Chem 2012;58(4):741–7.

Kreutzmann DJ, Cowell CT, Howard NJ, De Souza M, Silink M. Congenital adrenal hyperplasia family studies using the short ACTH test. Aust Paediatr J 1989;25(6):340–5.

Li H, Zang Y, Wang C, Li H, Fan A, Han C, et al. The interaction between microorganisms, metabolites, and immune system in the female genital tract microenvironment. Front Cell Infect Microbiol 2020a;10: 609488.

Li L, Zhang J, Zeng J, Liao B, Peng X, Li T, et al. Proteomics analysis of potential serum biomarkers for insulin resistance in patients with polycystic ovary syndrome. Int J Mol Med 2020b;45(5):1409–16.

Lizneva D, Gavrilova-Jordan L, Walker W, Azziz R. Androgen excess: investigations and management. Best Pract Res Clin Obstet Gynaecol 2016;37:98–118.

Long Y, Han S, Zhang X, Zhang X, Chen T, Gao Y, et al. The combination of a novel 2 bp deletion mutation and p.D63H in CYP11B1 cause congenital adrenal hyperplasia due to steroid 11β-hydroxylase deficiency. Endocr J 2016;63(3):301–10.

Maharaj A, Buonocore F, Meimaridou E, Ruiz-Babot G, Guasti L, Peng HM, et al. Predicted benign and synonymous variants in *CYP11A1* cause primary adrenal insufficiency through missplicing. J Endocr Soc 2018;3(1):201–21.

McCartney CR. Maturation of sleep-wake gonadotrophin-releasing hormone secretion across puberty in girls: potential mechanisms and relevance to the pathogenesis of polycystic ovary syndrome. J Neuroendocrinol 2010;22(7):701–9.

Meczekalski B, Szeliga A, Maciejewska-Jeske M, Podfigurna A, Cornetti P, Bala G, et al. Hyperthecosis: an underestimated nontumorous cause of hyperandrogenism. Gynecol Endocrinol 2021;37(8):677–82.

Menabò S, Polat S, Baldazzi L, Kulle AE, Holterhus PM, Grötzinger J, et al. Congenital adrenal hyperplasia due to 11-beta-hydroxylase deficiency: functional consequences of four CYP11B1 mutations. Eur J Hum Genet 2014;22(5):610–6. Erratum in: Eur J Hum Genet. 2020; 28(5):692.

Merke DP, Mallappa A, Arlt W, Brac de la Perriere A, Lindén Hirschberg A, Juul A, et al. Modified-release hydrocortisone in congenital adrenal hyperplasia. J Clin Endocrinol Metab 2021;106(5):e2063–77.

Miller WL, Tee MK. The post-translational regulation of 17,20 lyase activity. Mol Cell Endocrinol 2015;408:99–106.

Na X, Mao Y, Tang Y, Jiang W, Yu J, Cao L, et al. Identification and functional analysis of *fourteen* NR5A1 variants in patients with the 46 XY disorders of sex development. Gene 2020;760, 145004.

Newfield RS. ACTH receptor blockade: a novel approach to treat congenital adrenal hyperplasia, or Cushing's disease. Med Hypotheses 2010;74(4):705–6.

Ng HY. Acanthosis nigricans in obese adolescents: prevalence, impact, and management challenges. Adolesc Health Med Ther 2016;8:1–10.

Nimkarn S, New MI. Prenatal diagnosis and treatment of congenital adrenal hyperplasia due to 21-hydroxylase deficiency. Mol Cell Endocrinol 2009;300(1–2):192–6.

Nishikido A, Okamura T, Nakajima Y, Ishida E, Miyamoto T, Toki AK, et al. Regulation of the KCNJ5 gene by SF-1 in the adrenal cortex: complete genomic organization and promoter function. Mol Cell Endocrinol 2020;501, 110657.

Nixon M, Upreti R, Andrew R. 5α-reduced glucocorticoids: a story of natural selection. J Endocrinol 2012;212(2):111–27.

Olsen EA. Current and novel methods for assessing efficacy of hair growth promoters in pattern hair loss. J Am Acad Dermatol 2003; 48(2):253–62.

Papadimitriou DT, Bothou C, Zarganis D, Karantza M, Papadimitriou A. Heterozygous mutations in the cholesterol side-chain cleavage enzyme gene (CYP11A1) can cause transient adrenal insufficiency and life-threatening failure to thrive. Hormones (Athens) 2018;17(3): 419–21.

Pugeat M, Plotton I, de la Perrière AB, Raverot G, Déchaud H, Raverot V. Management of endocrine disease hyperandrogenic states in women: pitfalls in laboratory diagnosis. Eur J Endocrinol 2018;178(4): R141–54.

Ramos L, Vilchis F, Chávez B, Mares L. Mutational analysis of SRD5A2: from gene to functional kinetics in individuals with steroid 5α-reductase 2 deficiency. J Steroid Biochem Mol Biol 2020;200, 105691.

Reardon W, Smith A, Honour JW, Hindmarsh P, Das D, Rumsby G, et al. Evidence for digenic inheritance in some cases of Antley-Bixler syndrome? J Med Genet 2000;37(1):26–32.

Rinonapoli G, Ruggiero C, Meccariello L, Bisaccia M, Ceccarini P, Caraffa A. Osteoporosis in men: a review of an underestimated bone condition. Int J Mol Sci 2021;22(4):2105.

Różdżyńska-Świątkowska A, Ciara E, Halat-Wolska P, Krajewska-Walasek M, Jezela-Stanek A. Anthropometric characteristics of 65 Polish Smith-Lemli-Opitz patients. J Appl Genet 2021;62(3): 469–75. Erratum in: J Appl Genet. 2021 May 27.

Rothman MS, Wierman ME. How should postmenopausal androgen excess be evaluated? Clin Endocrinol (Oxf) 2011;75(2):160–4.

Roucher-Boulez F, Mallet D, Chatron N, Dijoud F, Gorduza DB, Bretones P, et al. Reversion SAMD9 mutations modifying phenotypic expression of MIRAGE syndrome and allowing inheritance in a usually *de novo* disorder. Front Endocrinol 2019;10:625.

Royer-Pokora B, Busch MA, Tenbusch S, Schmidt M, Beier M, Woods AD, et al. Comprehensive biology and genetics compendium of Wilms tumor cell lines with different *WT1* mutations. Cancers 2020;13(1):60.

Saari A, Pokka J, Mäkitie O, Saha MT, Dunkel L, Sankilampi U. Early detection of abnormal growth associated with juvenile acquired hypothyroidism. J Clin Endocrinol Metab 2021;106(2):e739–48.

Schüring AN, Busch AS, Bogdanova N, Gromoll J, Tüttelmann F. Effects of the FSH-β-subunit promoter polymorphism -211G->T on the hypothalamic-pituitary-ovarian axis in normally cycling women indicate a gender-specific regulation of gonadotropin secretion. J Clin Endocrinol Metab 2013;98(1):E82–6.

Shackleton CH, Roitman E, Kratz LE, Kelley RI. Midgestational maternal urine steroid markers of fetal Smith-Lemli-Opitz (SLO) syndrome (7-dehydrocholesterol 7-reductase deficiency). Steroids 1999a;64(7): 446–52.

Shackleton CH, Roitman E, Kratz L, Kelley R. Dehydro-oestriol and dehydropregnanetriol are candidate analytes for prenatal diagnosis of Smith-Lemli-Opitz syndrome. Prenat Diagn 2001;21(3):207–12.

Shammas C, Byrou S, Phelan MM, Toumba M, Stylianou C, Skordis N, et al. Genetic screening of non-classic CAH females with hyperandrogenemia identifies a novel CYP11B1 gene mutation. Hormones (Athens) 2016;15(2):235–42.

Shirazi FKH, Khodamoradi Z, Jeddi M. Insulin resistance and high molecular weight adiponectin in obese and non-obese patients with polycystic ovarian syndrome (PCOS). BMC Endocr Disord 2021;21(1):45.

Siviero-Miachon AA, Kizys MM, Ribeiro MM, Garcia FE, Spinola-Castro AM, Dias da Silva MR. Cosegregation of a novel mutation in the sixth transmembrane segment of the luteinizing/choriogonadotropin hormone receptor with two Brazilian siblings with severe testotoxicosis. Endocr Res 2017;42(2):117–24.

Spoudeas HA, Slater JD, Rumsby G, Honour JW, Brook CG. Deoxycorticosterone, 11 beta-hydroxylase and the adrenal cortex. Clin Endocrinol (Oxf) 1993;39(2):245–51.

Taylor NF, Shackleton CH. Gas chromatographic steroid analysis for diagnosis of placental sulfatase deficiency: a study of nine patients. J Clin Endocrinol Metab 1979;49(1):78–86.

Travers S, Martinerie L, Boileau P, Lombès M, Pussard E. Alterations of adrenal steroidomic profiles in preterm infants at birth. Arch Dis Child Fetal Neonatal Ed 2018;103(2):F143–51.

Vakili R. Precocious puberty: an unusual presentation of juvenile hypothyroidism. Ann Saudi Med 2004;24(2):145–7.

Valadares LP, Pfeilsticker ACV, de Brito Sousa SM, Cardoso SC, de Moraes OL, Gonçalves de Castro LC, et al. Insights on the phenotypic heterogenity of 11β-hydroxylase deficiency: clinical and genetic studies in two novel families. Endocrine 2018;62(2):326–32.

van der Straaten S, Springer A, Zecic A, Hebenstreit D, Tonnhofer U, Gawlik A, et al. The external genitalia score (EGS): a European multicenter validation study. J Clin Endocrinol Metab 2020;105(3):dgz142.

Wang F, Cai J, Wang J, He M, Mao J, Zhu K, et al. A novel WT1 gene mutation in a chinese girl with denys-drash syndrome. J Clin Lab Anal 2021a;35(5):e23769.

Wang W, Han R, Yang Z, Zheng S, Li H, Wan Z, et al. Targeted gene panel sequencing for molecular diagnosis of congenital adrenal hyperplasia. J Steroid Biochem Mol Biol 2021b;211, 105899.

Wei C, Davis N, Honour J, Crowne E. The investigation of children and adolescents with abnormalities of pubertal timing. Ann Clin Biochem 2017;54(1):20–32.

Weise M, Mehlinger SL, Drinkard B, Rawson E, Charmandari E, Hiroi M, et al. Patients with classic congenital adrenal hyperplasia have decreased epinephrine reserve and defective glucose elevation in response to high-intensity exercise. J Clin Endocrinol Metab 2004a;89(2):591–7.

Xie W, Zhou H, Zhou L, Gong Y, Lin J, Chen Y. Clinical features and genetic analysis of two Chinese families with X-linked ichthyosis. J Int Med Res 2020;48(10). 300060520962292.

Xu L, Xia W, Wu X, Wang X, Zhao L, Nie M. Chimeric CYP11B2/CYP11B1 causing 11β-hydroxylase deficiency in Chinese patients with congenital adrenal hyperplasia. Steroids 2015;101:51–5.

Zadik Z, Kahana L, Kaufman H, Benderli A, Hochberg Z. Salt loss in hypertensive form of congenital adrenal hyperplasia (11-beta-hydroxylase deficiency). J Clin Endocrinol Metab 1984;58(2):384–7.

Zhang M, Huang H, Lin N, He S, An G, Wang Y, et al. X-linked ichthyosis: molecular findings in four pedigrees with inconspicuous clinical manifestations. J Clin Lab Anal 2020;34(5):e23201.

Zhou Q, Wang D, Wang C, Zheng B, Liu Q, Zhu Z, et al. Clinical and molecular analysis of four patients with 11β-hydroxylase deficiency. Front Pediatr 2020;8:410.

Chapter 3.2.2

Hypogonadism

3.2.2.1 Introduction

Hypogonadism results from inadequate testicular or ovarian function which can affect sexual differentiation in fetal life, lack of puberty and during adult life lead to infertility, reduced muscle mass, bone abnormalities, psychological effects and changes beyond normal old age (Acién and Acién, 2020; Grinspon et al., 2020). Testosterone (T) and dihydrotestosterone (DHT) are the key androgens, estradiol and progesterone are ovarian steroids in females and the placenta during pregnancy. In the menstrual cycle, estradiol is secreted by the Graafian follicle until ovulation then with progesterone from the corpus luteum. Peak estradiol concentrations in the menstrual cycle are 16–420 pg/mL (250–1400 pmol/L) and at the end of pregnancy are 5–30 ng/mL (18–100 nmol/L). In hypogonadism, the sex steroids are low or ineffective. Symptoms of hypogonadism relate to changes in functions of the reproductive system but can also affect bones, muscles, hair and vasomotor effects. Some patients have unusual facial appearance with additional dysmorphic features. Certain congenital anomalies may be seen on fetal ultrasound although this is not yet standard practice (Corsten-Janssen et al., 2020).

Androgen deficiency may be investigated at birth or early in the postnatal period, when a **micropenis, hypospadias and cryptorchidism** are noticed. Some abnormal features might have been reported from the ultrasound in pregnancy. The testes may be in the abdomen or inguinal canal and can be located by palpation or ultrasound imaging. Facial (cleft lip, overall shape) and skeletal abnormalities may be noticed at birth, earlier or later. Digits should be counted and examined for fusion and low ratios of finger lengths (2D to 4D). In some cases, hyponatremia and hypoglycemia due to defects of the adrenal cortex may necessitate early hospital admission. After **chromosome analysis** and some genetic tests, the presence of an SRY gene will focus investigations on a failure of sex steroid production or to receptor resistance to these hormones.

Investigations of a hernia in childhood may reveal abdominal testes that are linked to a defect of androgen synthesis or action. Low estrogens and androgens are seen with **delayed puberty and infertility**. Testosterone production shows a steady decline with age but the change is not as

severe as the loss of estrogen and progesterone at the menopause. The laboratory should provide relevant reference ranges for the tests.

3.2.2.2 Hypogonadism in infant with ambiguous genitalia

In hypogonadism, the appearance of genitalia at birth can vary from micropenis to normal clitoris, with or without one or both gonads palpable on examination. The picture can vary from normal female to normal male. The genital scores should be taken into account such as phallic length although the overlap is considerable (Fig. 3.2.2.1) (Finlayson et al., 2018). A careful clinical examination and family history is necessary. For the younger patients with ambiguous genitalia the parents should be probed about a consanguineous relationship, virilization of mother in pregnancy, drugs taken and family history of ambiguous genitalia.

Many patients with hypogonadism have other malformations so the clinician must take careful note of face shape, fingers, toes, body habitus and height, hair pattern in addition to the external, and as far as possible, internal genitalia and anus (Hutson et al., 2014). In some cases, a complex type of anorectal malformation is characterized by the urological, genital, and intestinal tracts opening through a single common channel in the perineum (cloaca).

A karyotype or array comparative genomic hybridization test (CGH testing) is essential. Biochemical testing should start with measurements of basal concentrations of LH, FSH, testosterone, estradiol, and cortisol bearing in mind the sensitivities and imprecision of steroid assays at low concentrations. Later measurements of AMH, and Inhibin B and in some cases dihydrotestosterone, cortisol, aldosterone, renin and potassium are needed.

3.2.2.3 Clinical recognition of causes of hypogonadism

The early presentation of hypogonadism includes a male with poor phallic development, unfused labia with perineal urethral opening and testicular maldescent. A premature

Steroids in the Laboratory and Clinical Practice. https://doi.org/10.1016/B978-0-12-818124-9.00018-8

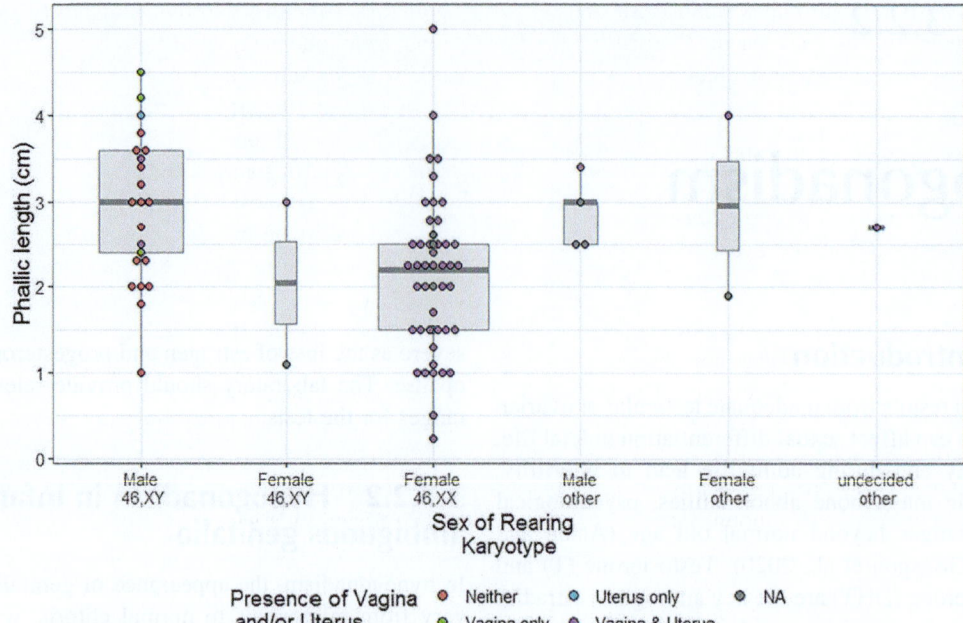

FIG. 3.2.2.1 Predictive phenotypic factors for sex designation: phallic length and presence of vagina and uterus. *(Finlayson C, Rosoklija I, Aston CE, Austin P, Bakula D, Baskin L, et al. Baseline characteristics of infants with atypical genital development: phenotypes, diagnoses, and sex of rearing. J Endocr Soc 2018;3(1):264–72. Fig. 1. p. 269. J Endocr Soc 3:264.)*

female baby may have a prominent clitoris, but this is possibly due to the raised concentration of androgens from the fetal adrenal cortex rather than a case of hypogonadism (Midgley et al., 1990; Greaves et al., 2008). **Cryptorchism** is when one or both the testes are still in the abdomen or along the normal route of descent (Fig. 3.2.2.2) and is one of the most common congenital malformations in boys.

In some cases, ultrasound is needed to locate the testes in the abdomen. Descent can occur during the minipuberty so that by 1 year more than 98.5% of males have scrotal testes. Rodprasert et al. (2020) reviewed the conditions where cryptorchidism is found, although the causes are often not elucidated. This situation will change as new defective genes are identified. Testicular hormones and insulin-like peptide 3 (INSL-3) are involved in descent of the testes. Cryptorchidism can, long-term, affect Leydig cell and Sertoli cell function. The condition can be part of Down and Noonan syndrome (Moniez et al., 2018) which have characteristic facial features. Down syndrome has an extra chromosome, usually trisomy 21.

In boys with **anorchia**, undetectable plasma concentrations of AMH and Inhibin B and an elevated FSH with 46XY complement are sufficient to make a diagnosis (Brauner et al., 2011).

Hypospadias is a variant of fetal development where urine does not exit from the urethra at the head of the penis. The phenotype of hypospadias is seen in 1:200 to 1:300 live births but the etiology is unknown in most cases. In some males with defective testicular descent, a uterus and

Testicular locations

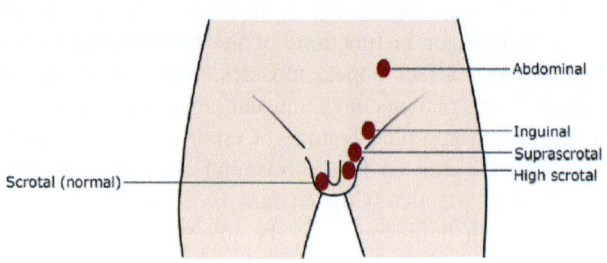

Testicular locations. Normally, both testes locate at the bottom of the scrotum. In cryptorchidism, one or both testes do not stay at the normal position, but anywhere along the normal path of testicular descent as illustrated in the figure Rodprasert 2020

FIG. 3.2.2.2 Location of undescended testes in abdomen. *(Rodprasert W, Virtanen HE, Mäkelä JA, Toppari J. Hypogonadism and cryptorchidism. Front Endocrinol. 2020;10:906. Fig. 1 p. 2.)*

Fallopian tubes are found during surgery to correct cryptorchidism; the phenotypic male is a true hermaphrodite. An ovotestis may be found during the operation. This can be with an ovary and testis on opposite or the same side (Kim et al., 2021). **Persistent Müllerian ducts** may also be detected which is usually due either to mutations in genes coding for anti-Müllerian hormone production or the hormone receptor (Josso et al., 2005, 2012). The majority of affected subjects are infertile when adults.

Baby boys with an abnormally small penis and one or both undescended testes may have **Prader-Willi syndrome.** This has several distinctive features, including:

- almond-shaped eyes,
- a narrow forehead at the temples,
- narrow bridge of the nose,
- a thin upper lip and a downturned mouth,
- unusually fair hair, skin and eyes, and
- small hands and feet.

Floppiness (hypotonia) caused by weak muscles is noticed soon after birth and the baby does not have full range of movement, a weak cry and poor reflexes. The baby cannot suck milk properly and by 1 year of age is underweight.

Bardet-Biedl syndrome is another disorder where boys have small genitalia at birth; other signs of the condition become apparent in childhood although extra toes or fingers will be seen in a newborn case. Obesity due to fat deposition in the abdomen is a feature. Progressive vision loss, kidney problems and learning difficulties are encountered.

3.2.2.4 Chromosomal defects in gonadal development (gonadal dysgenesis)

Where all or parts of the X and or Y chromosome are absent or mutated a disorder in development of the reproductive system (gonadal dysgenesis) is found (Anon, 2020). Three distinct phenotypes are encountered based on 46XX, 46XY and mixed gonadal dysgenesis.

Babies with **Down syndrome** tend to share certain physical features such as a flat facial profile, an upward slant to the eyes, small ears, and a protruding tongue. Chromosome tests reveal trisomy of chromosome 21. Most cases are detected during pregnancy by demonstrating nuchal translucency (NT) by ultrasound or screening tests depending on age of the mother and stage of pregnancy. For a fetus with Downs syndrome (DS) in first trimester PAPP-A (pregnancy associated plasma protein A) is lower and hCG (human chorionic gonadotrophin) higher and NT is increased, in the second trimester alpha fetoprotein (AFP), uE3 (urinary estriol) lower than normal and hCG higher. In a quadruple test, inhibin-A is added and this is increased in DS. Cell free DNA is also being used (ACOG Bulletin # 226) and further genetic tests are anticipated. Second trimester ultrasound can detect many congenital anomalies of skull, skeleton, heart, and kidneys that can be followed up with rapid exome sequencing of DNA from chorionic villi or amniotic fluid. Cases of **MIRAGE, CHARGE, Noonan** syndromes and **congenital adrenal hyperplasia** were also detected by rapid exome sequencing (rES) of fetal DNA after an ultrasound with anomalies (Corsten-Janssen et al., 2020). The inclusion criteria for rES were: (a) two or more independent major fetal

anomalies, (b) either hydrops fetalis or bilateral renal cysts alone, or (c) one major fetal anomaly and a first-degree relative with the same anomaly.

Turner syndrome is a rare disorder due to partial or complete loss of one of the second sex chromosomes (45X or 45O). In some cases, this abnormality may not be seen in all cells (mosaicism). A ring chromosome can also be found where the ends of a chromosome break off and the long and short arms join (Tzancheva et al., 1999). Abnormalities of eyes, ears, skeleton, heart and kidneys develop through childhood (Fig. 3.2.2.3). A short neck with a webbed appearance, low hairline at neck and narrow digits are common. Lymphedema due to fluid accumulation in a number of organs can even be seen at the neck on a fetal ultrasound.

Noonan syndrome is a genetic disorder that at birth is characterized by a wide spectrum of physical features that vary greatly in range and severity. In many affected individuals, there is a distinctive facial appearance; a broad or webbed neck, widely set eyes (ocular hypertelorism); skin folds that may cover the eyes' inner corners (epicanthal folds); drooping of the upper eyelids (ptosis); a small jaw (micrognathia); a depressed nasal root; a short nose with broad base; and low-set, posteriorly rotated ears (pinnae) (Fig. 3.2.2.4). The testes have not descended into the scrotum.

Most of the genes mutated in Noonan syndrome cause dysregulated RAS-MaPK signaling. *PTPN11* mutations have been found (50% of affected individuals) with *KRAS, SOS1, RIT1* (5%), and *RAF m*utations each in around 5% of cases (Bertola et al., 2020). Other gene defects have been seen in *NRAS, BRAF, MEK2, RRAS, RASA2, A2ML1,* and *SOS2.*

46XY gonadal dysgenesis is due to mutations in the SRY region of the Y chromosome. These include SRY, SOX9, WT1, SF1 and DHH genes. At birth the phenotype is female or ambiguous. Recent application of next-generation sequencing has found variants in 46XY complete gonadal dysgenesis of SRY (11%), DMRT1 (3.7%), NR5A1 (3.7%), and DHH (3.7%) accounting for 22% of cases (Buonocore et al., 2019).

Denys-Drash syndrome presents with ambiguous genitalia and nephrotic syndrome (protein in urine usually in the first year) and often a Wilm's kidney tumor (not always apparent at birth but palpable later) in 46XY DSD. If nephropathy is absent the diagnosis of WAGR should be considered (see below). If genitalia are female and both Müllerian and Wolffian ducts are present then Denys-Drash is more likely (Mueller, 1994). Mutations have been found in the **WT1 gene** mainly in exons 8 and 9 of the gene in the 11p13 chromosome region (Baird et al., 1992; Royer-Pokora et al., 2020; Wang et al., 2021). WT1 protein is a transcription factor with four zinc fingers that acts as a tumor suppressor. WT1 has a primary role in differentiation

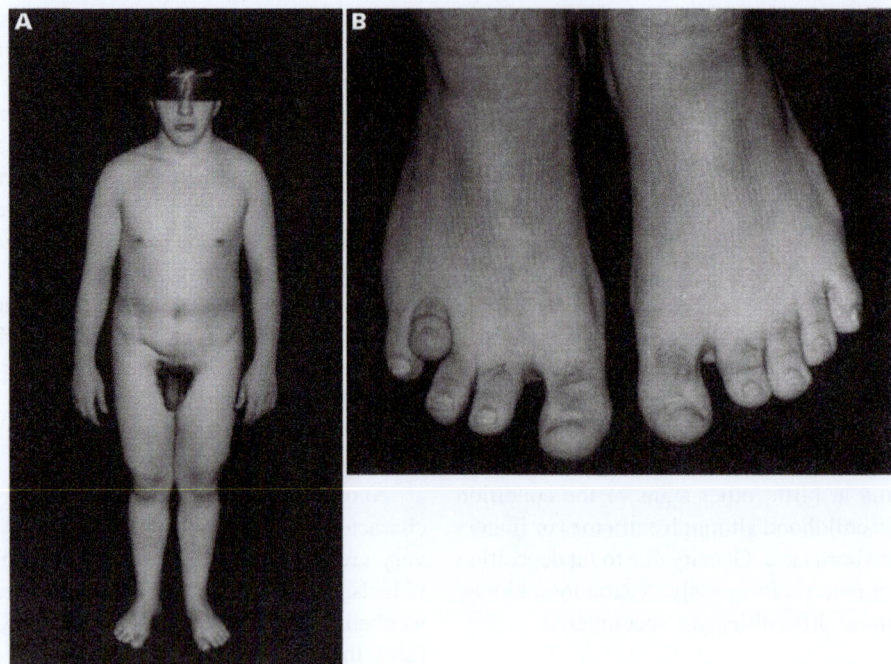

FIG. 3.2.2.3 Phenotype of Turner syndrome showing full stigmata. Note short right fourth toe and metatarsal bone and hypoplastic nails. *(Tzancheva M, Kaneva R, Kumanov P, Williams G, Tyler-Smith C. Two male patients with ring Y: definition of an interval in Yq contributing to Turner syndrome. J Med Genet 1999;36(7):549-53.)*

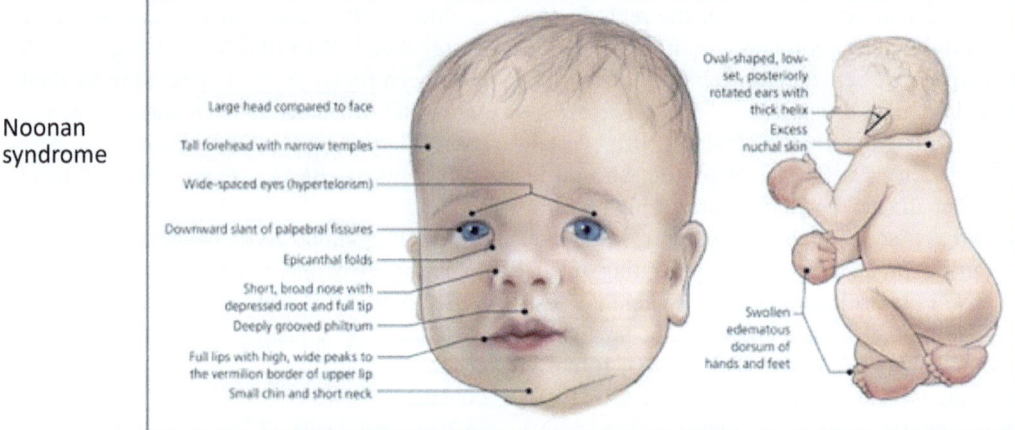

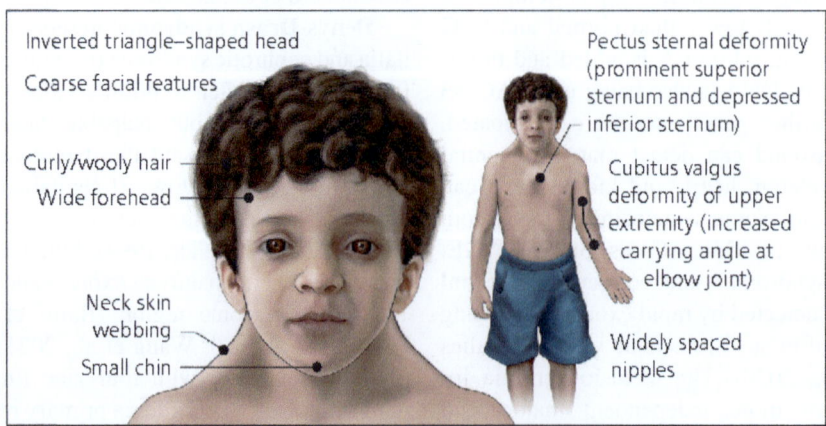

FIG. 3.2.2.4 Noonan syndrome (Public domain images National Human Genome Research Institute). *(Zenker M, Edouard T, Blair JC, Cappa M. Noonan syndrome: improving recognition and diagnosis. Arch Dis Child 2022:archdischild-2021-322858.)*

in the kidney. Loss of WT function leads to unrestrained growth of metanephric blastemal cells leading to Wilms tumor (Rauscher, 1993; Liu and Suson, 2020).

A number of syndromes have an increased risk of Wilm's tumor (Bhutani et al., 2021). Anirhdia and hemihypertrophy are further complications with Wilm's tumor. **WAGR** stands for **Wilms tumor, aniridia, genitourinary abnormalities** and leads to mental retardation. The syndrome is associated with a deletion on the short arm of chromosome 11. Breakpoints differ in individual cases, but the minimum deletion involves both PAX6 and WT1, which are B700 kb apart in the distal half of band 11p13.

The most common signs and symptoms of a related **Frasier syndrome** include normal female to ambiguous genitalia with proteinuria. Mutations responsible for Frasier syndrome predominantly occur in intron 9 of the *WT1* gene, specifically nucleotide substitutions that influence a splice site (Lavi et al., 2020).

3.2.2.4.1 Smith-Lemli-Opitz syndrome

A child with Smith-Lemli Opitz syndrome has prenatal growth retardation, a characteristic facial appearance with small head (microcephaly), short nose, pyloric stenosis and cleft palate, genital abnormalities, 2–3 toe fusion (syndactyly) (see Hypergonadism chapter).

The presence of 7-dehydro steroids can be found in the urine of an infant with this condition (Donoghue et al., 2018) consistent with a defect in cholesterol synthesis at the 7-dehydrocholesterol reduction step (Fig. 3.2.2.5).

The disorder may be detected by anomalies in fetal ultrasound. Postnatal growth is slower than normal.

3.2.2.5 Biochemical investigations at birth

3.2.2.5.1 Initial tests

Some investigations of hypogonadism can be undertaken locally but specialized centers may be called upon to resolve the cause of hypogonadism in many cases. Following confirmation of chromosomes, the serum concentrations of **cortisol** and **testosterone or estradiol with progesterone** should be measured, usually on a platform based immunoassay. From these results some clear decisions can be made. Defects of sex steroid production can be separated into:

- those affecting cortisol and androgens and
- those relating to androgens alone.

Time of day for blood sample collection is less important in hypogonadism because the circadian rhythm of T will not be

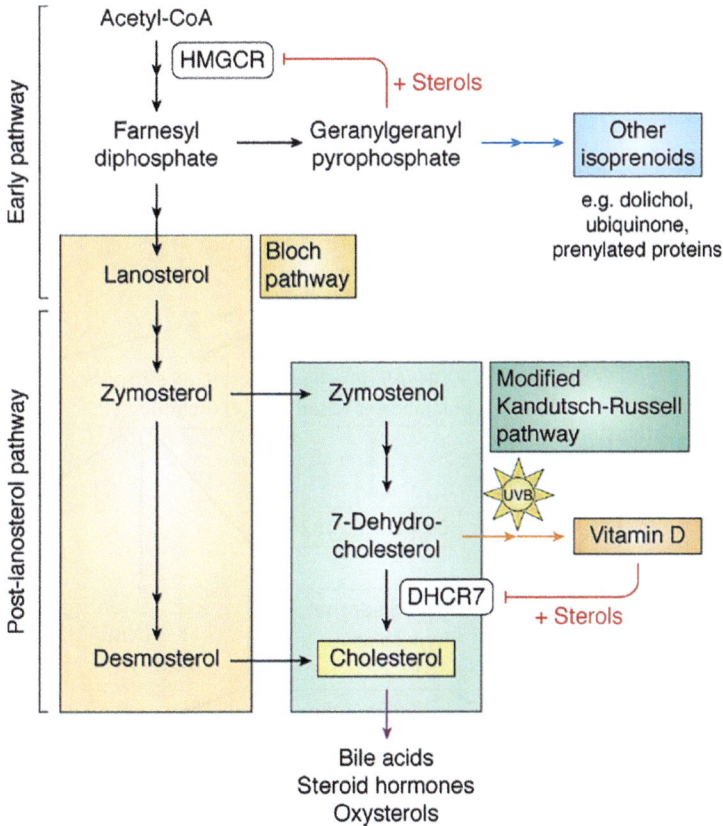

FIG. 3.2.2.5 Site of SLO defect cholesterol synthesis and DHCR7.

apparent above the assay imprecision. Gonadotrophin measurements will help distinguish primary and secondary causes. An hCG test and T measurement will determine if Leydig cells are present. Since cortisol production can be affected in some cases then testing should respect the circadian rhythm of adrenocortical activity although this is not fully established until 3 months post term.

A 46XX infant with ambiguous genitalia may have some masculine signs such that they are investigated for hyperandrogenism (see Chapter 3.2.1). A raised 17OHP is a further clue to this direction of investigation. Some patients with a normal female phenotype can be later shown to have hypogonadism. Failure of male (46XY) sexual differentiation can occur due to defects in T production from the fetal testis (9th to 14th week of gestation), a defect in the HPG axis or failure of androgen receptor response.

3.2.2.5.1.1   Defects of sex steroid production

Low androgen production in a poorly virilized male can be attributed to rare forms of congenital adrenal hyperplasia. When cortisol is low defects of five enzymes should be considered (Fig. 3.2.2.6):

- cholesterol uptake (defect of steroidogenic acute regulatory protein StAR),
- cholesterol 20,22 desmolase (side-cleavage, CYP11A1 chain),

- 17α-hydroxylase and 17–20 lyase (cytochrome *b*5 co factor delivery) (Very rare in neonates),
- isolated 17–20 lyase (very rare in neonates), and
- 3β-hydroxysteroid dehydrogenase (salt loss common presentation in neonate).

In the 2 weeks of life after birth, there are difficulties in interpreting serum steroid results because of the fall in testosterone over the first week as hCG is cleared from the system then the rise in testoterone over the second week as LH production increases as a result of negative feedback effects. From the third week of life for the next 3–6 months is a good period (the minipuberty) in which to evaluate the patient with hypogonadism.

Consanguiniuity is a risk for DSDs (Alswailem et al., 2021). These enzymes affect the production of all important adrenal and gonadal steroids and disorders of the enzymes can be fatal for the child because of adrenal insufficiency so that few cases are documented in the literature. A patient with **17-hydroxylase deficiency** rarely presents in the newborn period and will be discussed further under investigations for primary amenorrhea (see Section 3.2.2.7) and hypocortisolism Chapter 3.1.2). The activities of the enzymes can be affected in some cases by **defects in cytochrome *b*5** cofactor delivery.

In this group, AMH measurements are useful and intermediates in the cortisol pathway can be observed in a blood panel or urine profile. Corticosterone, DHEAS, 17-

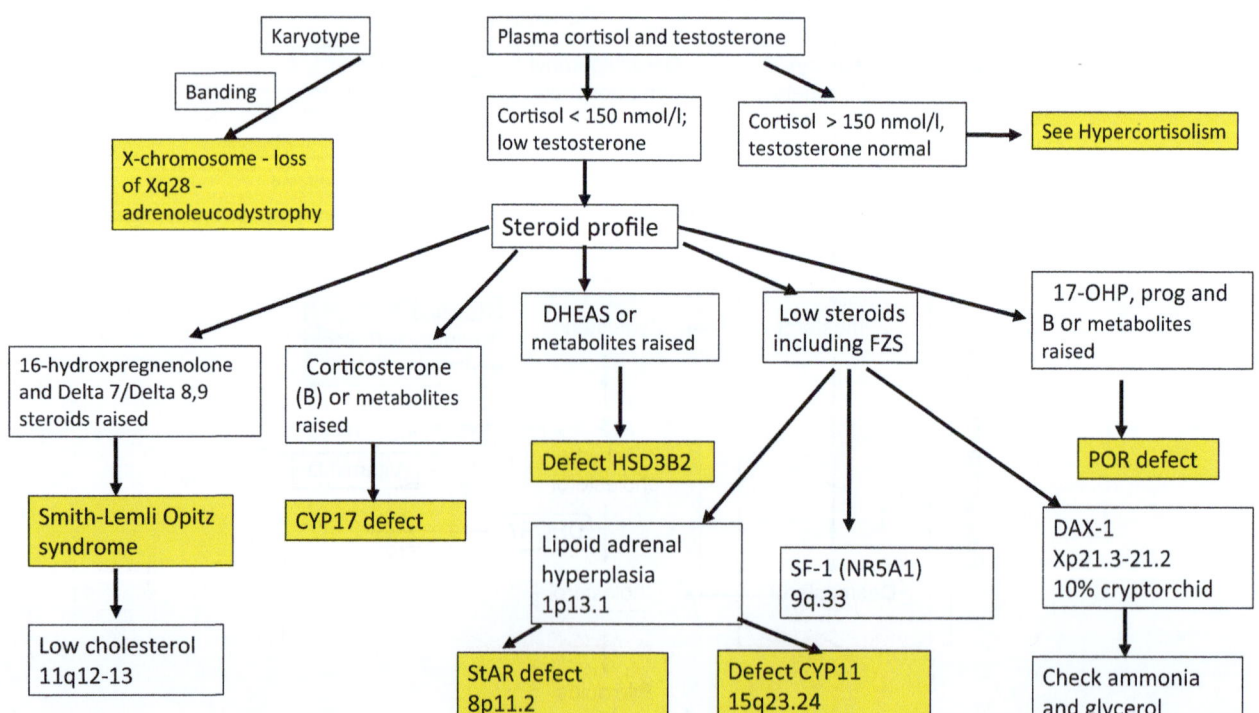

FIG. 3.2.2.6   Algorithm for poorly virilized patient with defects in the synthesis of cortisol and androgens. *(Author original.)*

hydroxyprogesterone and delta-7 oxidized equivalents in blood or their metabolites in urine are important.

3.2.2.5.2 Androgen defects

The causes of hypogonadism include inherited mutations of critical enzymes involved in steroid biosynthesis affecting T and sometimes adrenocortical function. They include adrenal 3β-hydroxysteroid dehydrogenase II (HSD3B2) deficiency and 17α-hydroxylase/17,20-lyase deficiency (CYP17A1) and gonadal 5α-reductase (SRD5A2) and 17β-hydroxysteroid dehydrogenase type 3 (HSD17B3) deficiency in 46,XY individuals (Auchus and Miller, 2012; Miller and Chung, 2016; Miller 2017). These patients have female phenotype and they are not usually seen in the young child except when there are signs of salt loss and adrenal insufficiency, notably in patients with defect of HSD3B2.

When cortisol production is normal, defects in sex steroids alone affect gonadal development and function (Fig. 3.2.2.7). Defects of sex steroid production with lack of testosterone synthesis during the critical period of gonadal development can be separated into two groups according to whether testosterone is low (less than 2 nmol/L) or normal to slightly raised (more than 2 nmol/L).

AMH and urine steroid profile are not helpful in the group with very low testosterone and of limited use in the newborn infant where testosterone is greater than 2 nmol/L. A defect of **5α-reductase** is unlikely to be detected in the newborn infant because gonadotrophins and testosterone are normal. The urine steroid profiles will however, show a characteristic pattern of raised THF to allo-THF after 6 months (Akgun et al., 1986) (Fig. 3.2 Hyperandrogenism). A raised androstenedione is a clue to **17-ketosteroid reductase** defect. **Androgen insensitivity** does not usually present at birth because the external genitalia are so near female as not to raise concern. Some cases of hypospadias and or cryptorchidism need to be investigated starting with gonadotrophin and testosterone measurements. These three disorders are discussed further in Chapter 3.2.1 with virilization at puberty.

3.2.2.6 Investigations of hypogonadism in newborn infant

Following confirmation of chromosomes, the serum concentrations of cortisol and testosterone should be checked. Genital skin tissue for fibroblast generation may be needed for androgen receptor analysis. DNA can be prepared from leucocytes. From the gonadotrophin, cortisol, testosterone/estradiol and 17-hydroxyprogesterone results some clear decisions can be made that lead to a diagnosis.

3.2.2.6.1 Enzyme defects

Defects in **cytochrome P450 oxidoreductase (POR)** (Fig. 3.2.2.6) and **cytochrome *b*5A** (Fig. 3.2.2.7) genes are first recognized as children in some cases seen especially with consanguinity.

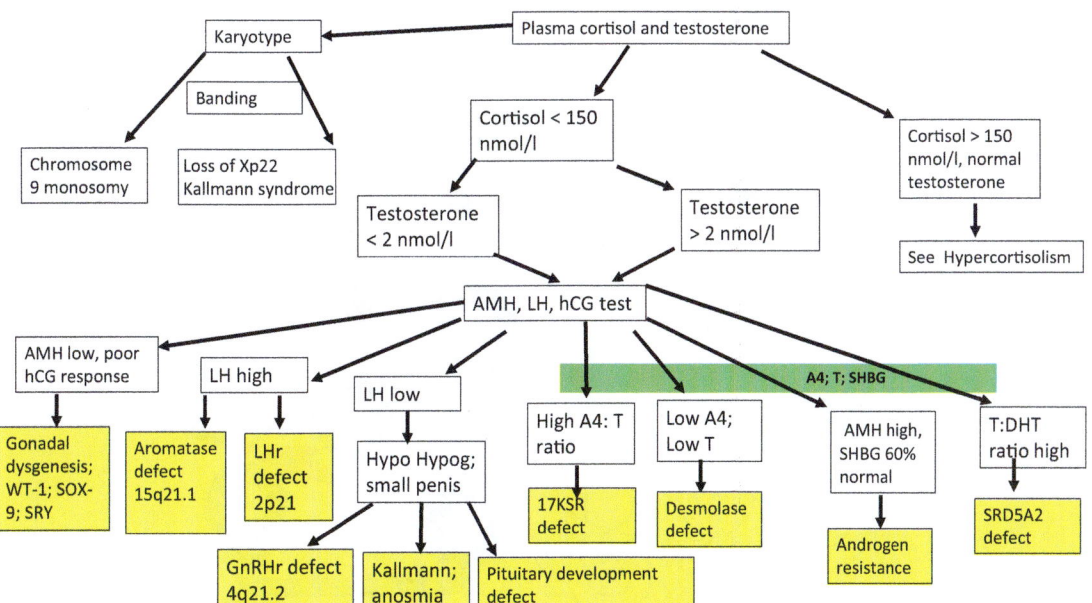

FIG. 3.2.2.7 Algorithm for patient with low testosterone but normal cortisol. *(Author original.)*

3.2.2.6.1.1 3β-Hydroxysteroid dehydrogenase (HSD3B2) deficiency

The synthesis of all active steroid hormones requires the action of HSD3B2 because it is the lead enzyme step in all pathways to cortisol, aldosterone and sex steroids (Fig. 3.2.2.8). Two steps are involved, namely dehydrogenation at C3 and isomerization of the double bond at C5 to C4.

Two isoforms are found, HSD3B1 is in the placenta and HSD3B2 is in the adrenals and gonads. The backdoor path to 11-oxygenated androgens and dihydrotestosterone is excluded. Genitalia of patients with HSD3B2 deficiency at birth are ambiguous and signs of adrenal insufficiency can develop in the first weeks. In the absence of testosterone production, the high concentrations of DHEA and DHEAS from the adrenal cortex can produce some effects on the genitalia so the clitoris can be enlarged.

In practice, HSD3B2 deficiency is difficult to confirm in a newborn child because the markers for the defect (DHEA and pregnenolone) are normal products of the adrenal in the newborn due to inactivity of this enzyme in the fetal adrenal cortex. Since these infants are usually very sick, even in the first week of life, the diagnosis is not straightforward. A 24-h urine collection with determination of excretion rates of all steroids is ideal but in practice difficult to achieve unless diapers are collected. Elevated 17-hydroxyprogesterone has been found in newborn screening that may be real or apparent depending on the assay used (Probst-Scheidegger et al., 2016). A raised plasma 17-hydroxypregenenolone concentration and ratio of 17-hydroxypregnenolone to cortisol in blood were criteria for diagnosis in newborn (Lutfallah et al., 2002) and older patients (Fanis et al., 2020; Mermejo et al., 2005). Few laboratories offer analysis of 17-hydroxypregnenolone. Cases up to 3 months of age were studied because of ambiguous genitalia with and without salt loss (Lutfallah et al., 2002). Older patients present with premature pubarche and hirsutism so will be discussed in Chapter 3.2.1 on Hypergonadism. In urine, the ratios of pregnenetriol (metabolite of 17-hydroxypregnenolone) to cortisol metabolites and DHEA to cortisol are raised (see Fig. 3.2.1.13 in Hypergonadism chapter) (Fanis et al., 2020; Krone et al., 2010). Excretion rates of DHA and 16-hydroxy-DHA are also elevated. Reference ranges for about 70 steroids in the urine of Japanese infants were published using milligrams per gram of creatinine units (Homma et al., 2003) and for Swiss babies using microgram per micromol creatinine (Dhayat et al., 2015).

Plasma measurements of DHEAS, pregnenolone and ACTH may be better tests before treatment but in the face of an adrenal crisis replacement therapy may be needed before the necessary tests are completed. Raised 17-hydroxypregnenolone concentrations after ACTH stimulation have been diagnostic in some cases (Al Alawi et al., 2019). That paper includes a thorough literature review. To overcome difficulties in interpretation of basal results dexamethasone can be given as temporary treatment during the ACTH stim test because it does not interfere in many steroid assays. If a child suspected of having 3β-hydroxysteroid dehydrogenase deficiency is switched to dexamethasone and fludrocortisone then given daily injections of depot

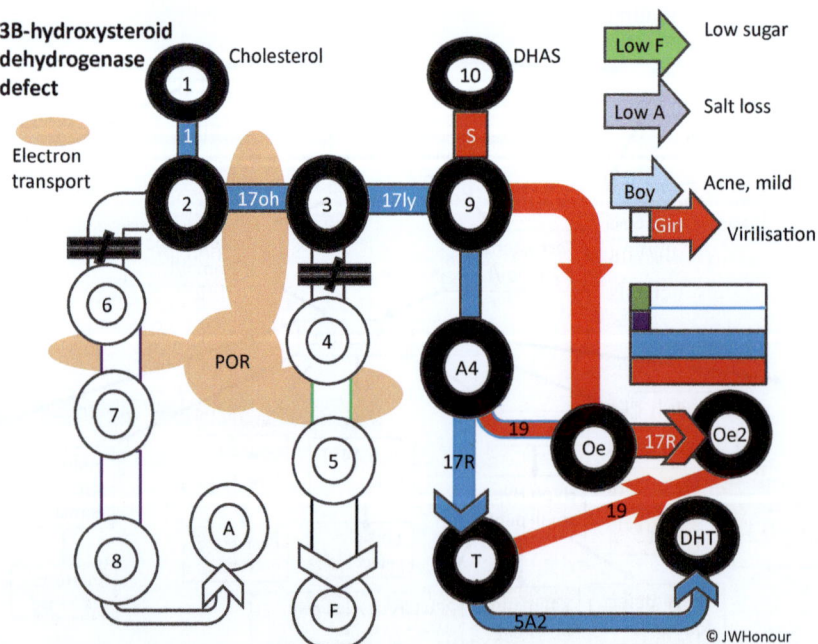

FIG. 3.2.2.8 Site HSD3B2 in steroidogenesis. *(Author original.)*

ACTH (Synacthen) the markers for the defect can be displayed in the urine steroid profile without interference from dexamethasone. This approach has been used successfully in confirming 3β-OHSDH deficiency although ACTH may be needed for several days before the suppressed adrenal secreted sufficient steroid to be detected in the urine steroid assay. In HSD3B2 deficiency, PRA is elevated. Treatment can be monitored to some extent by normalization of PRA.

HSD3B2 gene, which encodes a protein of 371 amino acids (Fig. 3.2.2.9), shares 93.5% identity with the HSD3B type 1 and is almost exclusively expressed in the adrenals and gonads. The enzyme has distinct functional regions.

Mutations are located in HSD3B2 in regions coding for domains crucial for enzyme function (Fig. 3.2.2.10) (Guran et al., 2020). More than 50 mutations have been found including frameshift, nonsense, in-frame deletion, splicing and missense mutations (Baquedano et al., 2018). Males with defects of cortisol and androgen production can be reared as boys needing adrenal steroid replacement with the addition of androgens at puberty.

Salt loss is a real complication with this condition and treatment of the disorder has not proved straightforward (Benkert et al., 2015). Testicular adrenal rests (TARTS) are common (Lolis et al., 2018).

3.2.2.6.1.2 Side chain cleavage enzyme deficiency

Cholesterol side chain cleavage enzyme (CYP11A1) catalyzes the first step in steroidogenesis so failure of the adrenals and gonads is predicted (Fig. 3.2.2.11). This is a rarer defect than STAR deficiency.

Less than 50 patients have been described usually presenting with signs of adrenal insufficiency, failure to thrive, hyponatremia and raised ACTH often causing hyperpigmentation (Maharaj et al., 2019) (Fig. 3.2.2.12).

Mutations were transfected into V29 cells and pregnenolone formation from 22R-hydroxycholesterol was determined using tandem mass spectrometry and shown to be reduced (Table 3.2.2.1). Two of the mutations did not show time-dependent increase in pregnenolone.

If the diagnosis is made in young children and they survive to pubertal age they will have hypergonadotrophism. This disorder is considered in more detail in chapter on cortisol deficiency. The diagnosis is important to ensure counseling and management particularly with regard to the risk of testicular adrenal rests (TARTS) (Kallali et al., 2020).

Cytochrome P450 oxidoreductase (POR) deficiency

At birth, and in some cases in fetal life, skeletal abnormalities are noted with features similar to the Antley-Bixler syndrome. The head is an unusual shape (craniosynostosis due to early fusion of bones in skull) with mid-face hypoplasia and there are, hand and feet malformations (see Chapter 3.2.1 Hyperandrogenism). A defect in POR prevents electron donation from NADPH to enzymes in steroidogenesis notably CYP17A1, CYP21A2, and CYP19A1 (Krone et al., 2012; Idkowiak et al., 2010, 2011) when 17-OHP, progesterone and corticosterone are raised in blood or their metabolites in urine.

Cortisol and aldosterone are low with raised ACTH and renin. The extent of failure to synthesize cortisol and sex steroids is variable (Dean et al., 2020, review of 211 patients; Idkowiak et al., 2011; Idkowiak et al., 2010). The first results of an affected male patient at 6 months of age were described as a defect of steroid biosynthetic microsomal mixed function oxidases (Peterson et al., 1985). A urine steroid profile analysis with gas chromatography alone revealed obvious increase in the metabolites of progesterone, 17-hydroxyprogesterone and corticosterone with hormones elevated in blood. In newborn infants, pregnanediol, pregnanetriolone and pregnanetriol are elevated and pregnenetriol tends to be increased. Cortisol metabolites are low. The 17-hydroxyprogesterone metabolite, pregnanetriol, is the dominant steroid in the profile

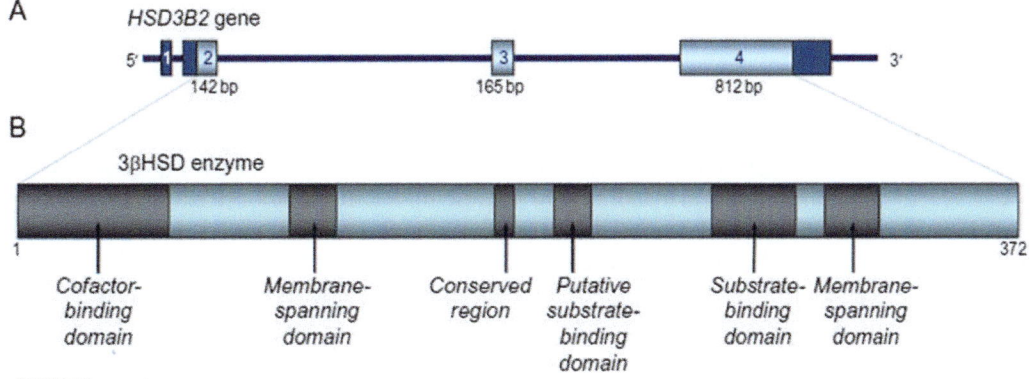

FIG. 3.2.2.9 HSD3B2 protein.

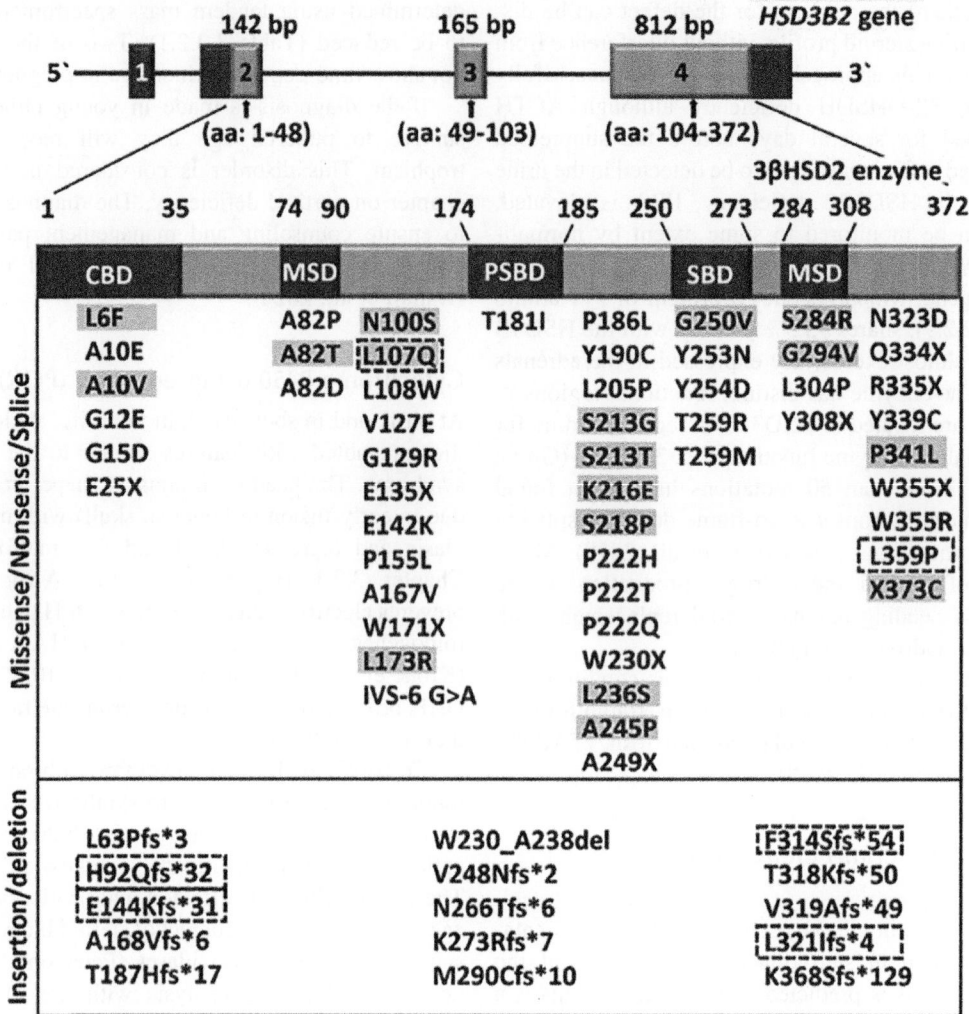

FIG. 3.2.2.10 Schematic presentation of *HSD3B2 with all known mutations and novel mutations*. Upper panel represents genomic organization for the *HSD3B2 gene. Light and dark gray boxes and lines represent exons*, noncoding exonic segments, and introns, respectively. Lower panel illustrates functional domains of the 3βHSD2 enzyme, including cofactor binding domain (CBD), membrane-spanning domain (MSD), putative substrate-binding domain (PSBD), and substrate-binding domains (SBD). Most of the previously reported mutations are located in exon 4. Pathogenic missense/nonsense/ splice mutations are shown in relation to functional domains (above the line in the box). Insertions and deletions are indicated below the line in the box. Novel mutations reported by Guran et al. (2020) are shown in boxes with dashed borders. Mutations with >5% residual 3βHSD2 activity in vitro are highlighted in gray boxes. *(Guran T, Kara C, Yildiz M, Bitkin EC, Haklar G, Lin JC, Keskin M, Barnard L, Anik A, Catli G, Guven A, Kirel B, Tutunculer F, Onal H, Turan S, Akcay T, Atay Z, Yilmaz GC, Mamadova J, Akbarzade A, Sirikci O, Storbeck KH, Baris T, Chung BC, Bereket A. Revisiting classical 3β-hydroxysteroid dehydrogenase 2 deficiency: lessons from 31 pediatric cases. J Clin Endocrinol Metab. 2020;105(3):dgaa022. Fig. b1 p. e1723.)*

and From 3 months of age corticosterone metabolites become more important (Fig. 3.2.2.13).

There is high excretion of a pregnenolone sulfate which undergoes enzyme desulfation at C3 and dehydration in the derivative stage to form the artifact metabolite epi-allopregnanediol which elutes before androsterone in the GC analysis and is recognizable by its mass spectrum (Fig. 3.2.2.14).

Many diagnostic ratios are considered for defects of androgen synthesis and are summarized in Table 3.2.2.2, and in the case of POR deficiency confirmed defects

of 21-hydroxylase and 17-hydroxylase. Results can be plotted on histograms with reference ranges for ratios (Fig. 3.2.2.15). Androsterone excretion was raised suggesting increased T and DHT production which was the cause of virilization of the mother (Ono et al., 2018; Shackleton et al., 2004).

POR is a substrate for estrogen (CYP19A1) and drug metabolism by actions of CYP1A2, CYP2D6, CYP2C19, and CYP3A4 so bone and xenobiotic metabolism can be affected differently depending on the gene mutation (Burkhard et al., 2017).

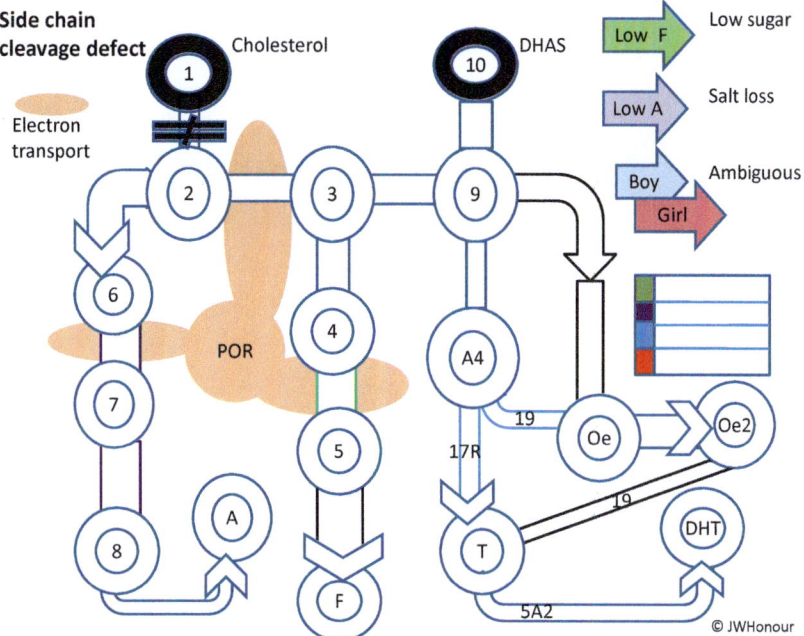

FIG. 3.2.2.11 CYP11A1 scc.

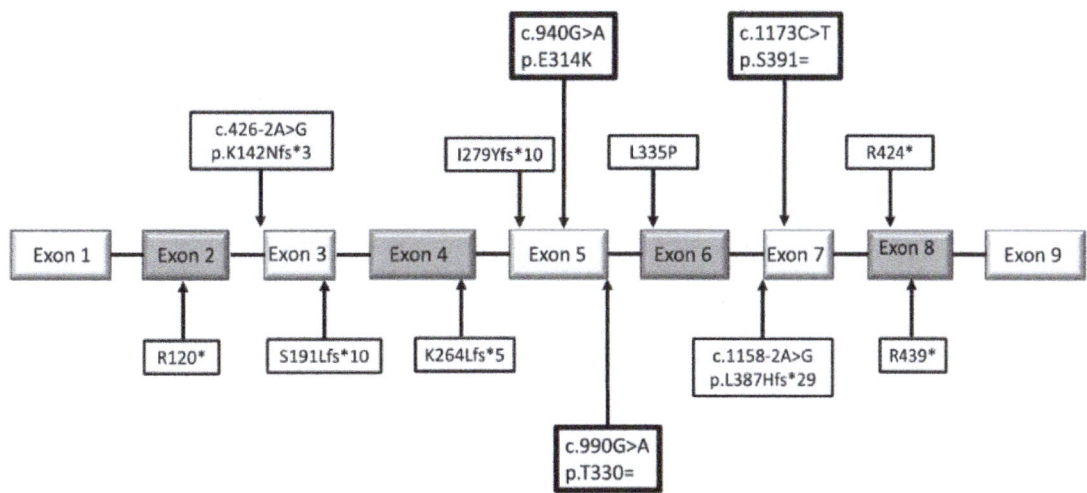

FIG. 3.2.2.12 CYP11A1 mutations. Position of variants in CYP11A1 genomic/pre-mRNA sequence found in patients with PAI. Boxed in bold, the three predicted benign or synonymous variants assessed for their effect on splicing; SNP rs6161 (c.940G.A, p.Glu314Lys) is 110 bp from the start and 51 bp from the end of exon 5, c.990G.A (p.Thr330 =) occurs at the last base of exon 5, and c.1173C.T (p.Ser391 =) is 16 bp from the start of exon 7. *(Maharaj A, Buonocore F, Meimaridou E, Ruiz-Babot G, Guasti L, Peng HM, Capper CP, Burgos-Tirado N, Prasad R, Hughes CR, Maudhoo A, Crowne E, Cheetham TD, Brain CE, Suntharalingham JP, Striglioni N, Yuksel B, Gurbuz F, Gupta S, Lindsay R, Couch R, Spoudeas HA, Guran T, Johnson S, Fowler DJ, Conwell LS, McInerney-Leo AM, Drui D, Cariou B, Lopez-Siguero JP, Harris M, Duncan EL, Hindmarsh PC, Auchus RJ, Donaldson MD, Achermann JC, Metherell LA. Predicted benign and synonymous variants in CYP11A1 cause primary adrenal insufficiency through missplicing. J Endocr Soc. 2018; 3(1):201–221. Fig. 1. p. 213.)*

TABLE 3.2.2.1 Activity of CYP11A1 (scc) when transfected to V29 cells.

	Pregnenolone formed (pg/well 6 h)	Pregnenolone formed (pg/well 12 h)
Wild type CYP11A1	23,238 ± 3919	48,113 ± 10,369
Mutant R460W	105 ± 28	388 ± 226
Mutant L335P	84 ± 15	73 ± 19
Mutant E314K	83 ± 9	86 ± 30

Maharaj A, Maudhoo A, Chan LF, Novoselova T, Prasad R, Metherell LA, Guasti L. Isolated glucocorticoid deficiency: genetic causes and animal models. J Steroid Biochem Mol Biol 2019;189:73–80.. Table 3 p. 219.

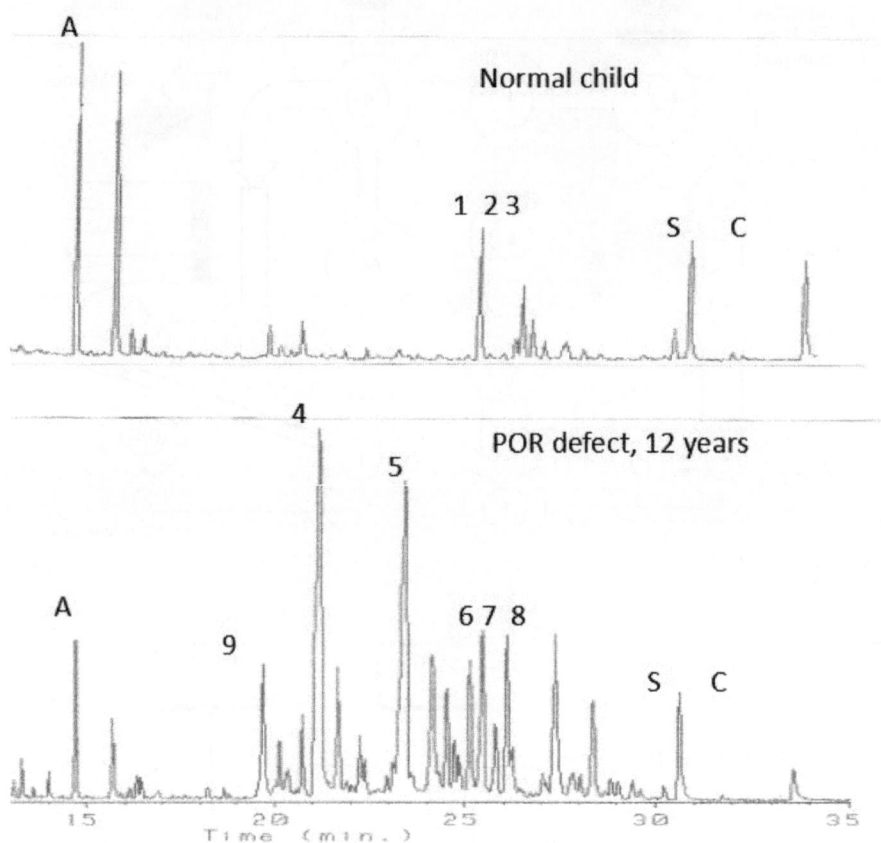

FIG. 3.2.2.13 Urine steroid profile of normal child and patient with cytochrome P450 oxidoreductase (POR) defect. Note increased outputs of metabolites of progesterone, 17-hydroxyprogesterone and corticosterone. Key to steroids: A, S, C Internal standards. (1) Tetrahydrocortisone; (2) tetrahydrocortisol; (3) allo-tetrahydrocortisol; (4) pregnanediol; (5) 11-oxo-pregnanediol; (6) tetrahydro-11-dehydrocorticosterone; (7) allo-tetrahydrocorticosterone; (8) hexahydro-corticosterone; and (9) 17-hydroxypregnanolone. *(Author original.)*

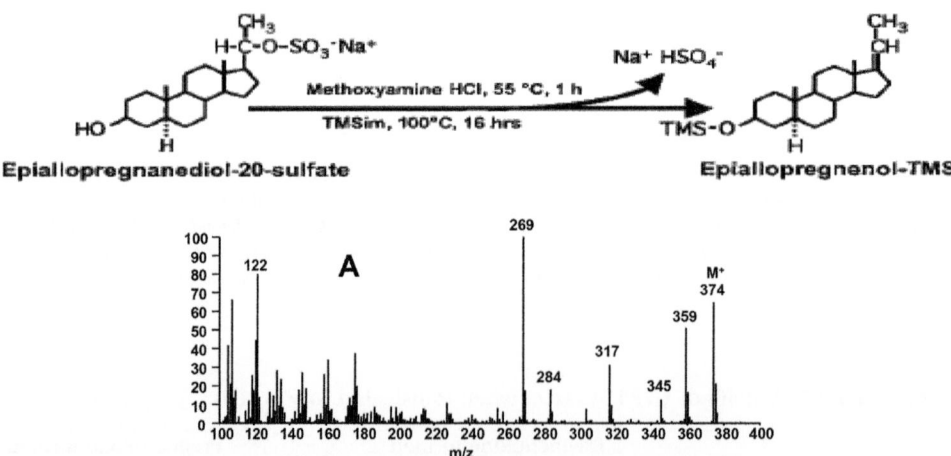

FIG. 3.2.2.14 Analysis by GC-MS of urine from patients with POR deficiency (17E)-5α-pregn-17(20)-en-3β-ol (epiallopregnenol) was identified by mass spec analysis of the TMS derivative (lower panel). That steroid is formed from epiallopregnanediol-20-sulfate (5α-pregnane-3β, 20α-diol-20-sulfate) by desulfation and dehydration during the derivatization process (upper panel). Epiallopregnanediol-20-sulfate arises from the enzymatic hydrolysis (*Helix pomatia* β-glucuronidase and sulfatase) of urinary epiallopregnanediol-3,20-disulfate. Sulfate esters of the 20-hydroxyl of steroids are resistant to enzymatic hydrolysis so are still present at the time of TMS derivatization. The excreted epiallopregnanediol has two origins, the disulfate conjugate probably being of fetal origin while a glucuronidated or 3-monosulfate form is likely both a fetal metabolite and a placental progesterone metabolite. *(Shackleton C, Marcos J, Arlt W, Hauffa BP. Prenatal diagnosis of P450 oxidoreductase deficiency (ORD): a disorder causing low pregnancy estriol, maternal and fetal virilization, and the Antley-Bixler syndrome phenotype. Am J Med Genet A 2004;129A(2):105–12. Fig. 1 p. 106.)*

TABLE 3.2.2.2 Diagnostic ratios of steroid metabolites.

Enzyme defect	Diagnostic ratios—urine	Diagnostic serum
21-Hydroxylase (CYP21A2)	PT/(THE + THF + αTHF); PTONE/(THE + THF + αTHF)	17-OHP/F; 21DF/F
11β-Hydroxylase (CYP11B1)	THS/(THE + THF + αTHF)	S/F
3β-Hydroxysteroid dehydrogenase (HSD3B2)	DHEA/(THE + THF + αTHF); Δ5-PT/(THE + THF + αTHF)	DHEA/F; Pregnenolone (Preg)/F; 17-OH-preg/F
17α-Hydroxylase (CYP17A1)	(THA + THB + αTHB)/(THE + THF + αTHF)	
Cytochrome P450 oxidoreductase (POR)	(17HP + PT)/(THE + THF + αTHF); PD/(THE + THF + αTHF); (THA + THB + αTHB)/(THE + THF + αTHF)	Progesterone; 17-OHP; B
5α-Reductase (SRD5A2)	Aetio/andro; THF/5αTHF	T/DHT
17β-Hydroxysteroid dehydrogenase (HSD17B3)	Andro/aetio	A4/T

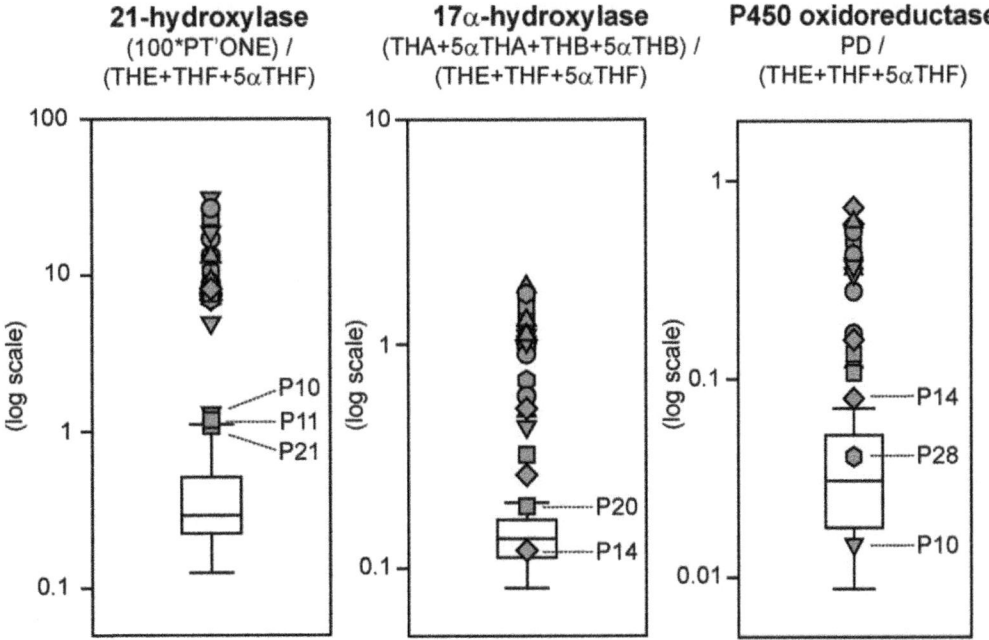

FIG. 3.2.2.15 In vivo steroidogenic enzyme activities in PORD patients (n23) as determined by urinary steroid profiling. Diagnostic **steroid substrate over product ratios** reflective of distinct steroidogenic enzyme activities and measured by GC/MS is shown in comparison with an age-matched reference cohort. Box plots represent the interquartile ranges (25th to 75th percentile), whiskers the fifth and 95th percentiles, respectively, of the reference cohort; each PORD case is represented by specific symbols. *(Krone N, Reisch N, Idkowiak J, Dhir V, Ivison HE, Hughes BA, Rose IT, O'Neil DM, Vijzelaar R, Smith MJ, MacDonald F, Cole TR, Adolphs N, Barton JS, Blair EM, Braddock SR, Collins F, Cragun DL, Dattani MT, Day R, Dougan S, Feist M, Gottschalk ME, Gregory JW, Haim M, Harrison R, Olney AH, Hauffa BP, Hindmarsh PC, Hopkin RJ, Jira PE, Kempers M, Kerstens MN, Khalifa MM, Köhler B, Maiter D, Nielsen S, O'Riordan SM, Roth CL, Shane KP, Silink M, Stikkelbroeck NM, Sweeney E, Szarras-Czapnik M, Waterson JR, Williamson L, Hartmann MF, Taylor NF, Wudy SA, Malunowicz EM, Shackleton CH, Arlt W. Genotype-phenotype analysis in congenital adrenal hyperplasia due to P450 oxidoreductase deficiency. J Clin Endocrinol Metab. 2012;97(2):E257–67. Fig. 2 p. e264.)*

3.2.2.6.2 StAR defect

StAR ensures delivery of cholesterol to the mitochondria and absence of activity impairs testicular steroidogenesis with fetal androgen deficiency leading to under-virilization of the fetus (Miller and Strauss, 1999; Flück et al., 2011a,b; Ishii et al., 2020; Galano et al., 2021; Zhang et al., 2021) as well as cortisol synthesis. Survival rates with this condition are poor because of primary adrenal insufficiency. Salt loss will be evident in the first week and severe hypoglycemia may also occur. This condition is likely under-diagnosed. Areas of skin pigmentation may be noted. StAR deficiency is more frequent in Arabia and in East Asia. StAR deficiency is usually only diagnosed at post mortem with histological appearance of large, lipid droplet filled adrenal glands as a consequence of defective StAR protein that cannot transport cholesterol esters into the mitochondria, hence another name is **lipoid adrenal hyperplasia (LCAH)**. This condition can be confused with a defect of 20,22 desmolase (side-chain cleavage of cholesterol to pregnenolone). All steroids in blood will be at low concentrations, gonadotrophins, ACTH and PRA are raised (Zhang et al., 2021), several mutations are described. Splicing mutations of the **StAR (NR5a1)** gene cause severe impairment of protein function and phenotype. There are pQ258.X, pR182L and p.L260P founder mutations. The c772C > T is the commonest mutation in Chinese population (Fig. 3.2.2.16) (Zhang et al., 2021).

3.2.2.6.3 SF-1 defects

Steroidogenic factor 1 (SF-1) has a crucial role in adrenal and gonadal development by modulating expression of SRY and SOX9 in the bipotential gonad, AMH during testicular differentiation, StAR in cholesterol mobilization and SF1 in regression of the mesonephric duct (Bashamboo et al., 2016). Deficiency can vary through hypospadias, ambiguous genitalia or complete female appearance (Tuhan et al., 2017; Lin and Achermann, 2008). A 46XY individual with atypical or female external genitalia with clitoromegaly and palpable gonads may have a defect in SF-1. Müllerian ducts are absent. The diagnosis is more often suspected in early childhood (Mönig et al., 2022; Domenice et al., 2016) usually because of investigations of adrenal insufficiency and spontaneous male puberty of a phenotypic female or concerns of a DSD. Testosterone is not always low and often in the normal range but LH is consistently raised (Faienza et al., 2019; Mönig et al., 2022) (Fig. 3.2.2.17).

The human SF-1 protein spans 461 amino acids and the positions of the substituted amino acid or nonsense mutation in the 2 studies are shown in Fig. 3.2.2.18.

Typical of nuclear receptors, the SF-1 protein contains an amino terminal core DBD with a primary proximal box (P-box), an accessory DBD (Ftz-F1 box/A-box) and a carboxy terminal LBD. The DBD and LBD flank a hinge region. The LBD and its carboxy terminal activation

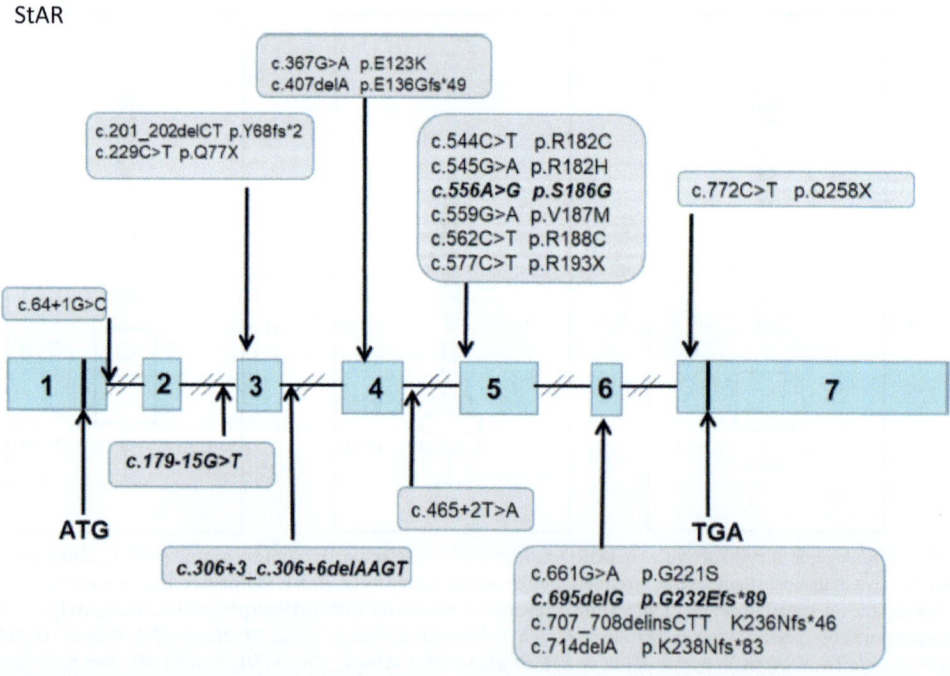

FIG. 3.2.2.16 STAR gene mutations in LCAH patients. Diagram of the STAR gene showing the location of mutations identified in patients with LCAH and novel variants are in bold. *(Zhang T, Ma X, Wang J, Jia C, Wang W, Dong Z, Ye L, Sun S, Hu R, Ning G, Li C, Lu W. Clinical and molecular characterization of thirty Chinese patients with congenital lipoid adrenal hyperplasia. J Steroid Biochem Mol Biol 2021;206:105788. Fig. 1 p. 6.)*

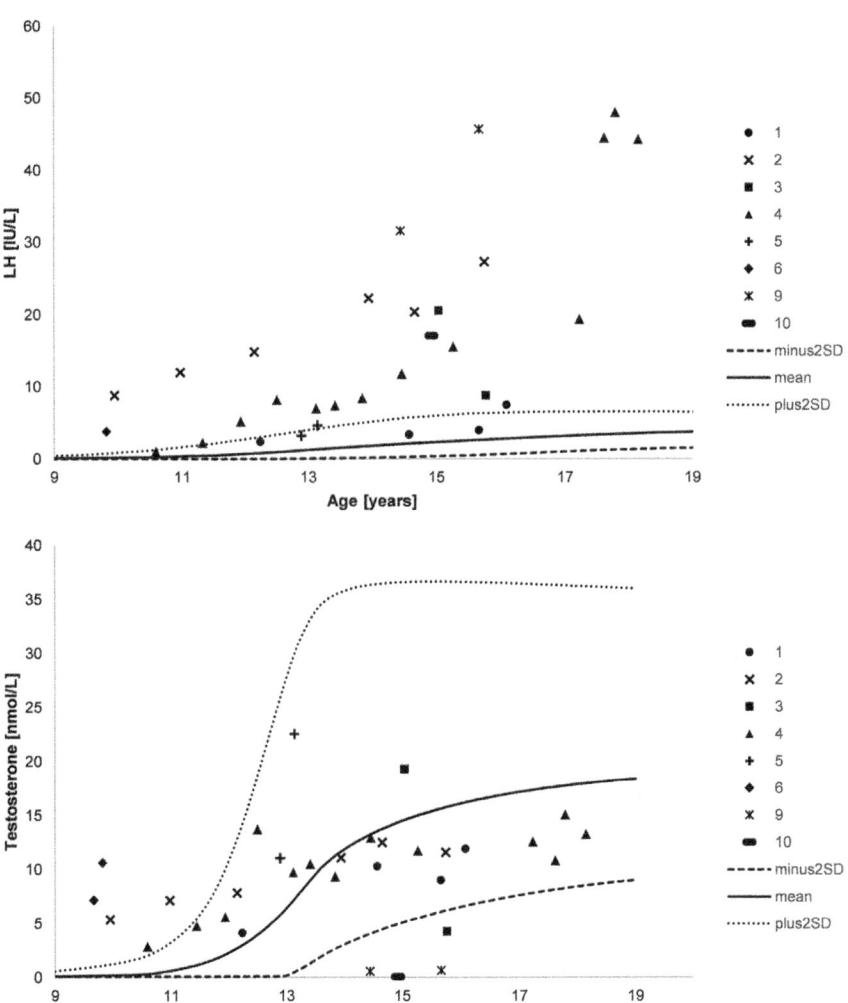

FIG. 3.2.2.17 T and LH characteristics in patients with NR5A1 during course of puberty and loss of SF1 function. The solid lines are means and broken lines plus and minus 2 standard deviations. *(Mönig I, Schneidewind J, Johannsen TH, Juul A, Werner R, Lünstedt R, Birnbaum W, Marshall L, Wünsch L, Hiort O. Pubertal development in 46,XY patients with NR5A1 mutations. Endocrine 2022;75(2):601–13. Fig. 1 p. 6. Endocrine 75:601.)*

function 2 (AF2) domain are sites of interaction for cofactors (Ito et al., 1998; Park et al., 2007) and putative phospholipid ligands (Krylova et al., 2005; Li et al., 2005; Wang et al., 2005).

Mutations were found in patients presenting with hypospadias (Allali et al., 2011). Targeted next-generation sequencing, comprising 163 candidate genes involved in sexual differentiation and development is used for the investigation of likely patients (Pan et al., 2020) followed by the functional evaluation of the NR5A1 mutations.

Many mutations map to functional domains of the protein or the extracellular domains. More than 100 mutations have been reported. Many SF-1 mutants are predicted in silico to alter DNA, ligand or cofactor binding (Sreenivasan et al., 2018).

There is considerable phenotypic variability among individuals with *NR5A1* mutations. Neither the mutation in *NR5A1* nor its reduction in SF1 activity is a good indicator of a patient's phenotype or clinical outcome. Inherited mutations in DSD tend to be rare as fertility is often affected; incomplete penetrance is often observed with *NR5A1* variants in DSD patients. Additional genes, polymorphisms, and contributing environmental factors have been studied and extensive research will be needed to establish how mutations of them play roles in the phenotype. Variants of MPK3 (Cheng et al., 2021) AMH, StAR and ZFPM2/FOG2 have been revealed (Martínez de LaPiscina et al., 2020). Using whole exome sequencing 19 potentially deleterious mutations(one to seven per patient) were found in 18 genes considered to interact with SF1 (for example AKR1C3, CACNG4, NAV1, FBLN2, SRA1, SMAD6, ZDHHC11, GDNF, SOX30) (Camats et al., 2018) suggesting an oligogenic basis for the defect. Very large-scale studies that compare a range of *NR5A1* mutations and patient phenotypes will likely reveal the complexities of these interactions. Further discussion on the defect can be found later and in the chapter on adrenal insufficiency (Chapter 3.1.2).

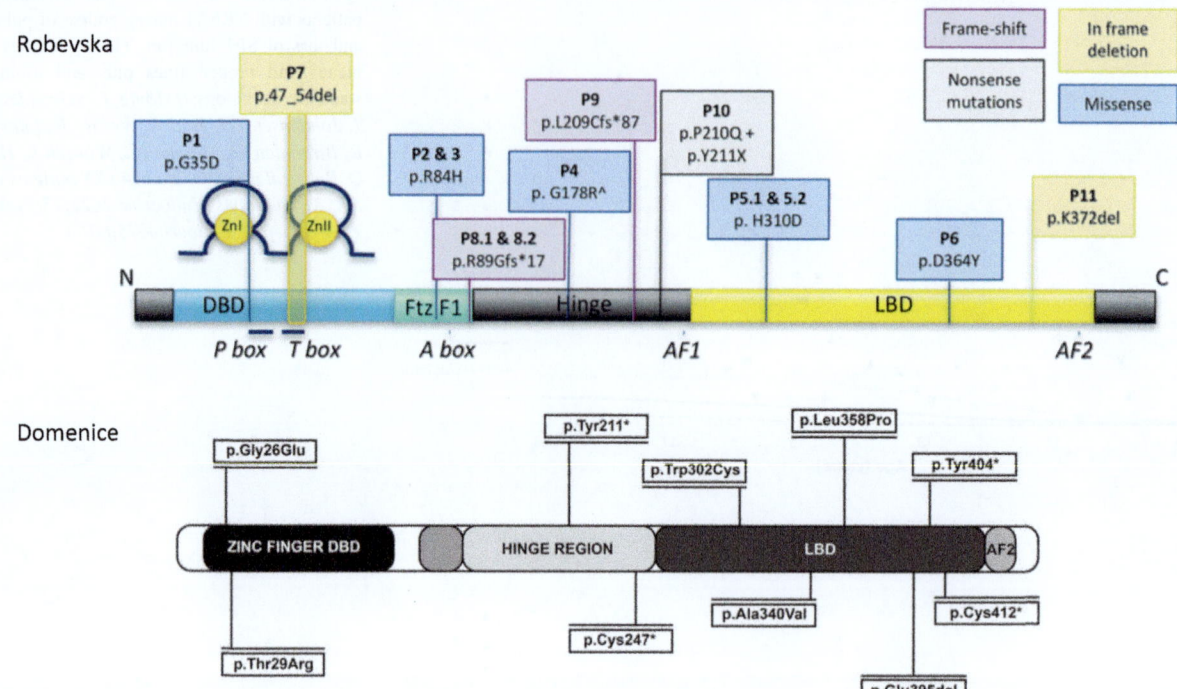

FIG. 3.2.2.18 Structure of wild-type SF-1 protein showing location of 20 mutants found in 46,XYDSD patients. Each amino acid substitution is coded according to severity of developmental phenotype (black for developmentally mild phenotypes with under-masculinized male genitalia; underlined for severe phenotypes with ambiguous or female genitalia). Numbers represent the first and last amino acid of the SF-1 protein. Functional domains of SF-1 are labeled as follows: *DBD*, DNA binding domain; *P-box*, proximal box; *NLS*, nuclear localization signal; *A-box*, accessory box; *Hinge*, hinge domain; *LBD*, ligand binding domain; *AF2*, activation function 2. *(Robevska G, van den Bergen JA, Ohnesorg T, Eggers S, Hanna C, Hersmus R, Thompson EM, Baxendale A, Verge CF, Lafferty AR, Marzuki NS, Santosa A, Listyasari NA, Riedl S, Warne G, Looijenga L, Faradz S, Ayers KL, Sinclair AH. Functional characterization of novel NR5A1 variants reveals multiple complex roles in disorders of sex development. Hum Mutat 2018;39(1):124–39; Domenice S, Machado AZ, Ferreira FM, Ferraz-de-Souza B, Lerario AM, Lin L, Nishi MY, Gomes NL, da Silva TE, Silva RB, Correa RV, Montenegro LR, Narciso A, Costa EM, Achermann JC, Mendonca BB. Wide spectrum of NR5A1-related phenotypes in 46,XY and 46,XX individuals. Birth Defects Res C Embryo Today 2016;108(4):309–20.)*

A low basal testosterone with elevated gonadotrophins and a poor androgen response to HCG suggest either **Leydig cell hypoplasia** (LCH) or an **androgen biosynthetic defect**. A patient with SRD5A2 or HSD3B2 defects, leading to impaired DHT production in utero, will at birth only sometimes have ambiguous external genitalia, clitoral-like phallus, bifid scrotum, blind vaginal pouch and inguinal or labial testes but usually the phenotype can be near female. There are androgenic issues at puberty (see Chapter 3.1 Hyperandrogenism).

With the conditions most often encountered there will be no Müllerian structures because AMH production is not affected. Patients with defects in the androgen receptor (androgen insensitivity syndrome AIS proven by mutations in the androgen receptor) do not show the postnatal rise in LH and testosterone (Bouvattier et al., 2002).

Inactivating mutations in the LH receptor gene in phenotypic girls usually present at a pubertal age with lack of full breast development and amenorrhea. There is a blind ending vagina, no uterus/tubes. LH is raised,

estradiol and progesterone concentrations do not increase without ovulation (Qiao and Han, 2019; Huhtaniemi, 2006; Latronico et al., 1996). This disorder is very rare, sometimes revealed during assisted reproduction due to failure of estradiol to increase on gonadotropin stimulation. Mutations in the LH receptor gene are responsible for LCH (Xu et al., 2018; Kossack et al., 2013; Qiao et al., 2009). Mild forms can present with hypospadias, micropenis or cryptorchidism although these patients usually present at a pubertal age with lack of breast development and amenorrhea. Missense, nonsense mutations, insertions and deletions are scattered throughout the whole gene. Some mutations can affect translocation of protein to the cell surface, but mutations have been reported in the promoter region, the extracellular domain, transmembrane areas and cytoplasmic loops (Qiao et al., 2009) (Fig. 3.2.2.19). Genetically modified mouse models are used in studies of LH action (Huhtaniemi et al., 2006).

During childhood, **undescended testes** may lead to painful **inguinal hernias** and be the first sign of

Mutations LHr gene 2006

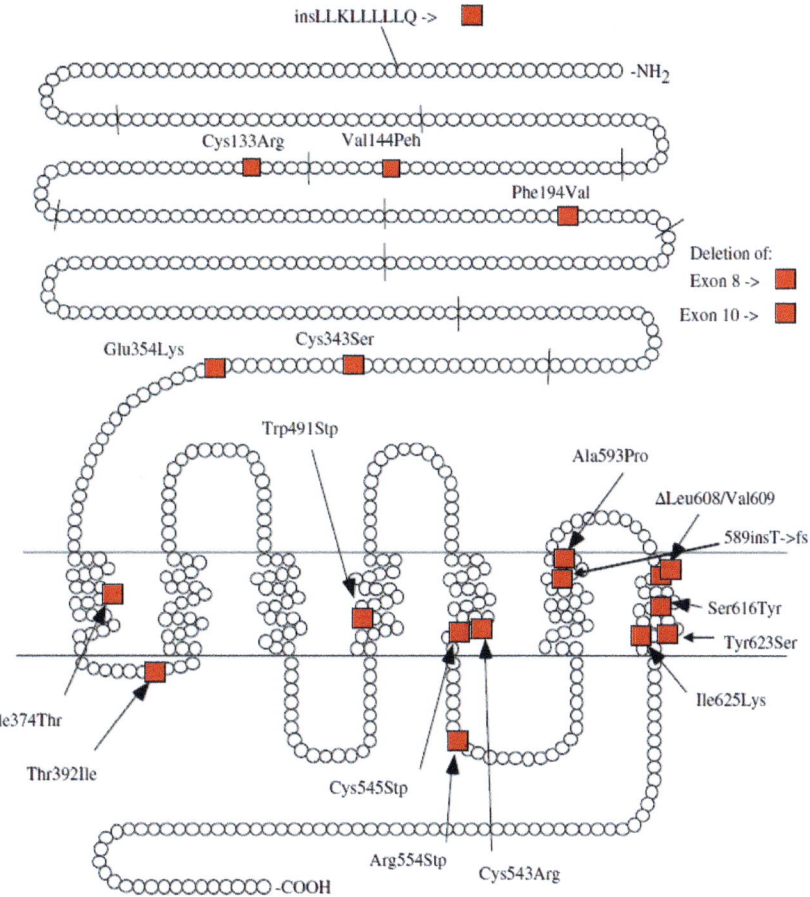

FIG. 3.2.2.19 Currently identified mutations in LHCGR. Inactivating (*red squares* of boxes) and activating (*green circles*) are marked according to their amino acid position. *(Rivero-Müller A, Huhtaniemi I. Genetic variants of gonadotrophins and their receptors: Impact on the diagnosis and management of the infertile patient. Best Pract Res Clin Endocrinol Metab. 2022;36(1):101596. Fig. 3 p. 10.)*

cryptorchidism requiring surgery, where other anomalies can be revealed. Undescended testes are associated in adults with infertility, due to the raised temperature within the abdominal cavity (Gurney et al., 2017). Early surgical correction improves the chances of fertility. Female reproductive structures (streak gonads) characteristic of **persistent Müllerian duct syndrome** (PMDS) or **ovotestis** may be found in the operation.

3.2.2.6.4 Presentation more likely in childhood/puberty

Later presentations of hypogonadism may be of **delayed puberty**, defined by the absence of sexual maturation by an age 2–2.5 standard deviations later than the mean of the population; typically no breast development by 13 years of age in girls, and testicular development by 14 years in boys. Final height may be normal and later some will be infertile.

Delayed puberty can be caused by conditions which induce.

(i) primary/hypergonadotropic hypogonadism such as **Turner** syndrome, **Klinefelter** syndrome or chemotherapy-induced gonadal toxicity,

(ii) secondary/hypogonadotropic hypogonadism such as **Kallmann's** syndrome or high-dose central nervous system irradiation, or

(iii) transient hypogonadotropic hypogonadism as induced by chronic disease including inflammatory bowel disease or anorexia.

For children, teenagers and adults with hypogonadism the parents or patient should be probed in detail about symptoms of chronic disease, the use of medications, substance abuse, symptoms indicative of deficits in or excess of other hormones, previous treatment or surgery, age of onset of puberty in the parents, plus their final heights and ethnic background. Any history of abnormal eating such as anorexia should be taken into account, especially in girls in whom delayed puberty is more unusual than in boys, as well as any intense exercise programmes. Childhood is a period of sexual quiescence but hypogonadism manifests when the normal pattern and timing of puberty fails to occur unless there are syndromes with characteristic somatic or developmental abnormalities. In girls, delay in the onset of breast development and absence of growth acceleration are conspicuous and menarche is late in the process. The growth rate of hypogonadal boys often falls behind that of his peers and if the underlying problem persists, the long bones continue growing, leading to the span exceeding the standing height by 5 cm or more that constitutes eunuchoidal proportions. The failure of virilization becomes obvious, often with psychosocial effects with loss of confidence and self-esteem. Boys are often reluctant to get medical help. Complete absence of clinical signs in girls by the fifteenth year warrant investigation. Pubertal delay can be familial.

Some disorders of sex may only be questioned clinically as the onset of puberty is delayed. When the history is taken it is important to observe the timings of any changes in secondary sex characteristics (breasts, testes, and pubic hair) in comparison with normal standards that have taken including any events since birth and childhood. Is there a growth chart on the kitchen refrigerator or elsewhere? Skeletal abnormalities, facial appearance and ability to smell should be tested. A Sniffin Sticks tests kit looks to be a convenient way to test for responses to apple, banana, butter, coffee, grass, fish, rose, lemon, orange, lemon, onion, peach and strawberry. Where possible the family history should record puberty timing of family members as well as issues of infertility. The diet and activity of the child should be questioned. The clinician should recognize signs of particular syndromes associated with delayed puberty—small or tall stature (Turners syndrome or Kallmann's syndrome) texture of testes, gynecomastia (Klinefelters), shape of skull, face and teeth (Smith Lemli Opitz syndrome, Antley Bixler syndrome).

The causes of delayed puberty are many (Honour, 2015) (Fig. 3.2.2.20). Laboratory tests should start with LH and FSH. Gonadotrophin measurements will distinguish primary and secondary forms (Wei et al., 2017). Measurements of cortisol, DHEAS, estradiol or testosterone, AMH, prolactin, free T4 and TSH may be needed. In exceptional cases a chromosomal microarray or G-banded karyotype, bone density and brain MRI. Adrenal insufficiency should be considered in boys and girls with low or high gonadotrophins.

Delayed puberty in boys requires similar investigations of the hypothalamic-pituitary-gonadal axis. Height and testicular volume should be measured. Height is often short due to the delay in the onset of the pubertal growth spurt.

Lesser degrees of T failure of production or insensitivity are linked with a range of intermediate or ambiguous sexual phenotypes with poor virilization up to minor degrees of

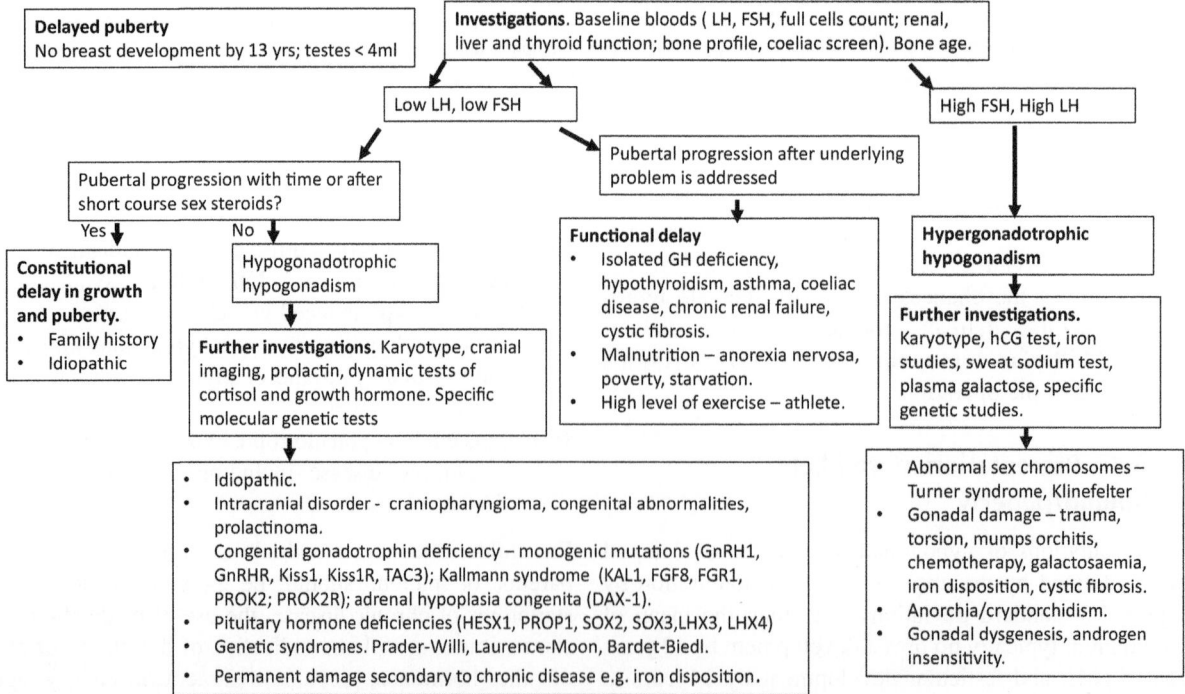

FIG. 3.2.2.20 Flow diagram for the investigation of delayed puberty. *(Wei C, Davis N, Honour J, Crowne E. The investigation of children and adolescents with abnormalities of pubertal timing. Ann Clin Biochem 2017;54(1):20–32. Fig. 2 p. 23.)*

hypospadias or some forms of idiopathic male infertility (Wilson et al., 1974). Serum testosterone should be measured early in the morning. Testicular failure can be due to torsion, surgical damage or inflammation.

Raised gonadotrophins confirm primary gonadal failure. A boy with low gonadotrophins should have an hCG test and brain MRI to exclude a pituitary defect.

AMH is high when there are defects in testosterone synthesis and complete androgen insensitivity unless Sertoli cell function is not normal (Josso and Rey, 2020) (Fig. 3.2.2.21).

Inhibin B is at its highest concentrations in infancy (100–500 ng/mL), low during childhood (<50 ng/mL) then rises in the early stages of puberty along with the increase in serum testosterone and testicular volume. Inhibin B in hypogonadism is raised in androgen insensitivity syndrome (Fig. 3.2.2.22) (Kubini et al., 2000).

3.2.2.6.4.1 Turner syndrome 45XO

High FSH is a feature of Turner syndrome, autoimmune destruction of ovaries and gonadal dysgenesis (Fig. 3.2.2.23).

Almost all females with Turner syndrome are short with no growth spurt in puberty. Some girls show signs of breast

development and menarche then periods cease. Haplo insufficiency of the SHOX gene is a major contributor to short stature, and mutations in TIMP1 and TIMP3 are linked to problems in cardiac valves (Corbitt et al., 2018).

In **46XX gonadal dysgenesis**, there are streak ovaries or nonfunctional tissues. Growth delay in a girl should prompt a karyotype determination. An abdominal ultrasound will reveal details of the reproductive tract. Estrogen production is low so there is failure of pubertal development, lack of breast development and no menarche. Gonadotrophins are raised.

Abnormalities in the FSHr can be uncovered through gene sequencing (Fig. 3.2.2.24). A GnRH stimulation test with lower inhibin response may help distinguish congenital hypogonadotropic hypogonadism from constitutional delay of growth and puberty although not reliably (Mosbah et al., 2020).

Boys with gonadal dysgenesis show low AMH and low inhibin B levels consistent with the absence or reduced numbers of Sertoli cells and germ cells that is characteristic of the dysgenetic gonad (Fig. 3.2.2.25). Individuals without NR5A1 mutations and either ambiguous external genitalia or hypospadias showed a much more variable aspect with plasma AMH and inhibin B levels below, within or above the normal range for age. Interestingly all individuals carrying heterozygote NR5A1 mutations had plasma AMH

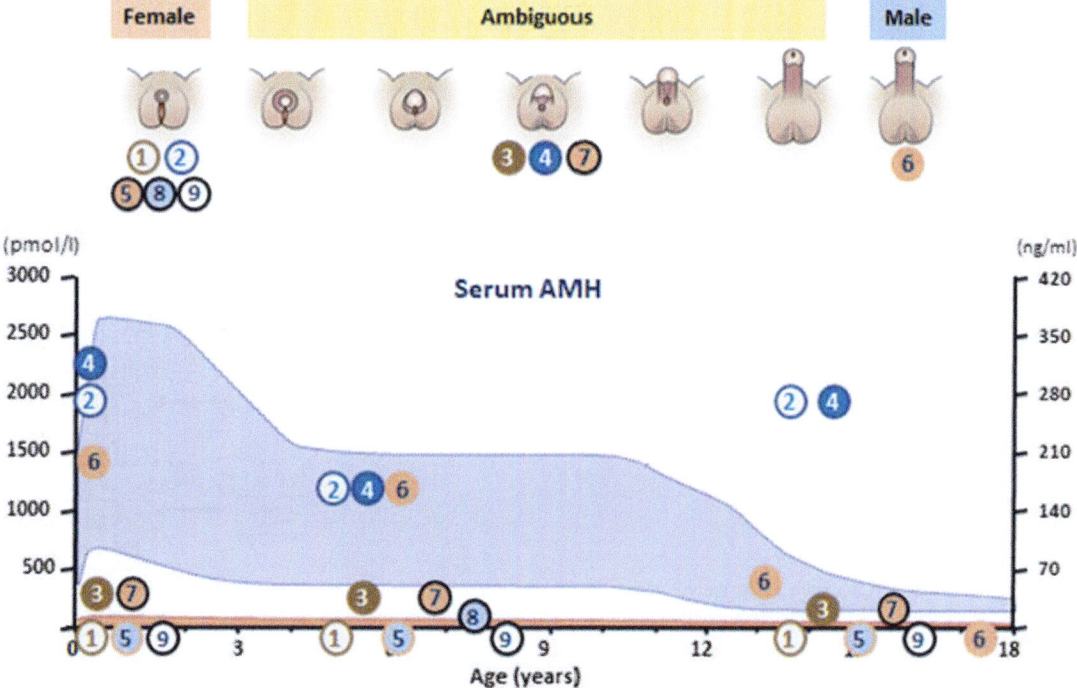

FIG. 3.2.2.21 Schematic of AMH levels in various types of disorders of sex development (DSD) in relationship to the aspect of the external genitalia and age. The shaded area represents reference levels for AMH from Grinspon et al. (2011). (1) 46XY complete gonadal dysgenesis; (2) 46XY Leydig cell aplasia/complete steroidgenesis defects; (3) 46XY chromosomal DSD/partial testicular dysgenesis; (4) 46XY Leydig cell hypoplasia/partial steroidogenic defect; (5) 46XX CAH/aromatase deficiency/androgenic tumor or drugs; (6) 46XX male/47 XXY Klinefelter syndrome; (7) 46XX ovotesticular DSD; (8) 45× or variants—Turner syndrome with ovarian tissue; (9) 45× or variants—Turner syndrome without ovarian tissue. *(Josso N, Rey RA. What does AMH tell us in pediatric disorders of sex development? Front Endocrinol. 2020;11:619. Fig. 6 p. 9.)*

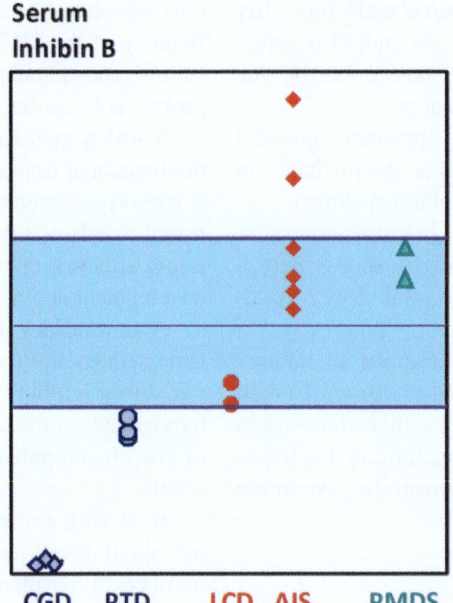

FIG. 3.2.2.22   Inhibin B levels in patients with DSD. The shaded area represents the normal male level; thin line = 50th centile, thick lines = 5th and 95th centiles. *AIS*, androgen insensitivity syndrome; *CGD*, complete gonadal dysgenesis; *LCD*, Leydig cell disorders (aplasia, hypoplasia); *PMDS*, persistent Müllerian duct syndrome; *PTD*, partial testicular dysgenesis. *(From Freire AV, Grinspon RP, Rey RA. Importance of Serum Testicular Protein Hormone Measurement in the Assessment of Disorders of Sex Development. Sex Dev 2018;12(1–3):30–40. Fig 3 P35.)*

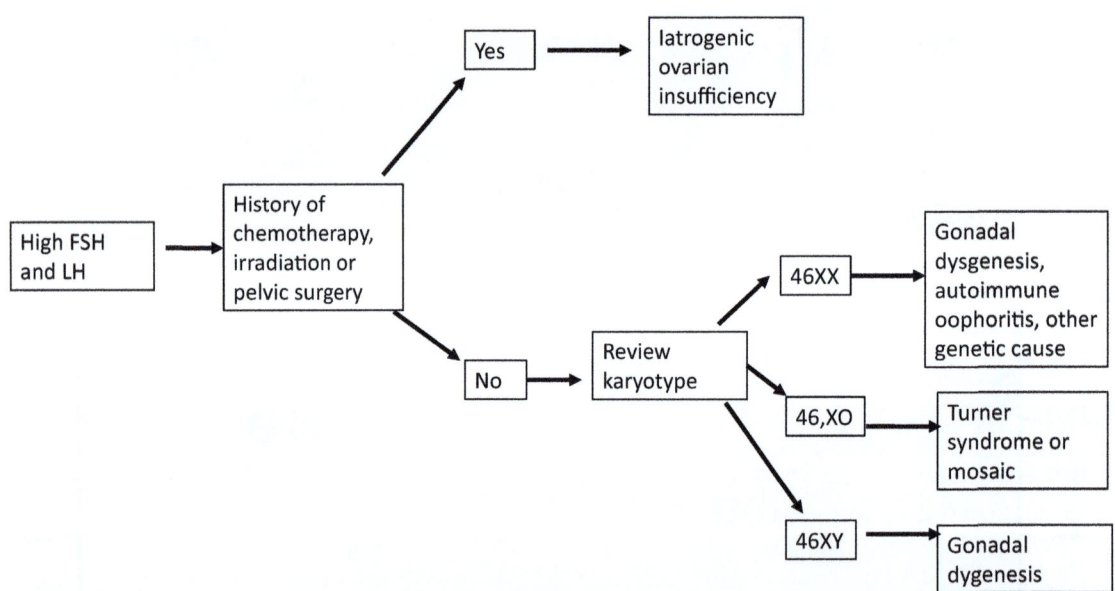

FIG. 3.2.2.23   Algorithm for the approach to the patient with primary amenorrhea and hypergonadotropic hypogonadism with diagnosis of Turner syndrome and gonadal dysgenesis. *FSH*, follicle-stimulating hormone; *LH*, luteinizing hormone; *POI*, premature ovarian insufficiency.

and inhibin B levels that were under or at the lower limit of normal range irrespective of the phenotype.

Pathogenic variants of **DEAH-box 37 Helicase (DHX37)** are a newly recognized cause of an autosomal dominant form of 46XY DSD with gonadal dysgenesis and testicular regression syndrome distinguished by female genitalia, Müllerian derivatives and streak gonads versus micropenis, partially developed Müllerian structures and no gonadal tissue (da Silva et al., 2019; McElreavey et al., 2020; Zidoune et al., 2021).

Mutations FSHr gene 2006

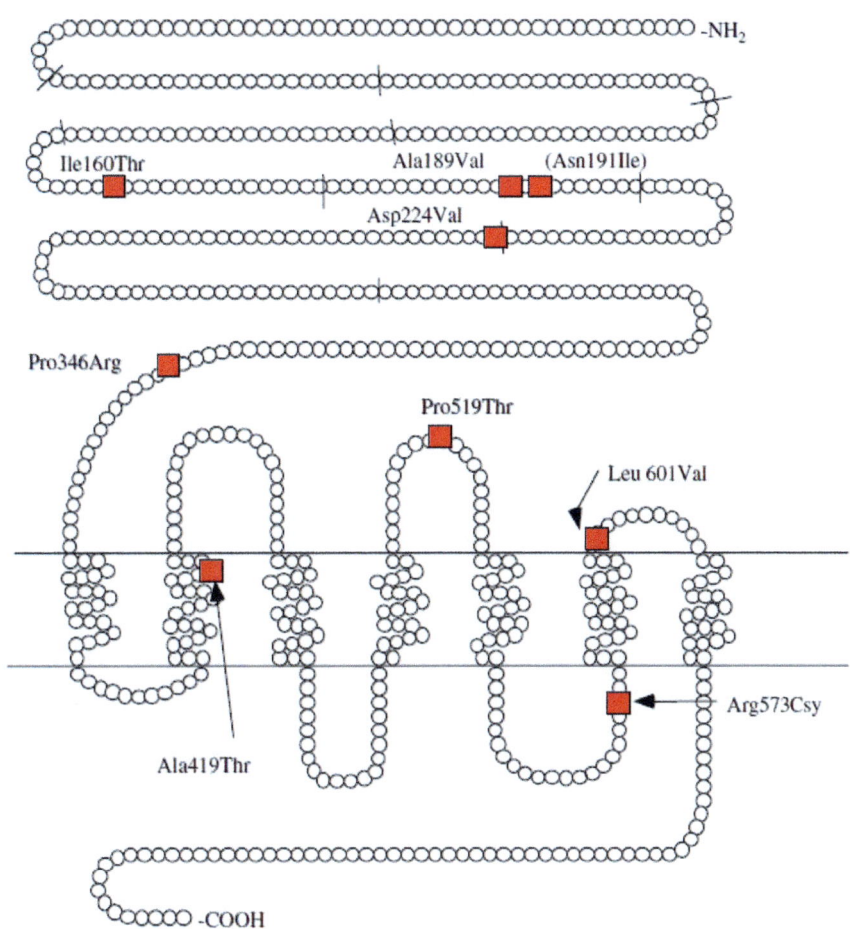

FIG. 3.2.2.24 Currently identified mutations of FSHR. Inactivating (*red squares* or boxes), activating (*green circles*) and promiscuous (*blue circle*) mutations are shown in their corresponding positions. *(Rivero-Müller A, Huhtaniemi I. Genetic variants of gonadotrophins and their receptors: Impact on the diagnosis and management of the infertile patient. Best Pract Res Clin Endocrinol Metab. 2022;36(1):101596. Fig. 4 p. 11.)*

Girls with **Noonan syndrome** are short with distinctive skeletal malformations, such as abnormalities of the breastbone (sternum), curvature of the spine (kyphosis and/or scoliosis), and outward deviation of the elbows (cubitus valgus). Many also have heart (cardiac) defects, such as pulmonary valvular stenosis and thickening of the ventricular heart muscle (hypertrophic cardiomyopathy). Additionally, the girls may have malformations of certain blood and lymph vessels, blood clotting and platelet deficiencies, learning difficulties or mild intellectual disability.

Primary amenorrhea and tall stature are features of **Swyer's syndrome** where chromosomes are 46XY but there are mutations in the SRY gene probably in combination with other gene defects (Racca et al., 2016). A small uterus without ovaries is found on investigation of primary amenorrhea, low sex steroid concentrations. Inheritance is variable. Patients with this disorder of mixed gonadal dysgenesis are at high risk of gonadal tumors (Bumbuliené et al., 2020).

Cryptorchid testes are prone to malignancy. Persistent Mullerian duct syndrome (PMDS) can be due to mutations in the AMH gene or the AMH receptor type II (AMHR-II) (Brunello and Rey, 2022; Josso et al., 2005). Males have a uterus and Fallopian tubes. The level of circulating AMH is normal for age in AMHR-II and low or undetectable in AMH gene defects (Fig. 3.2.2.26).

Mutations in the AMH gene of 64 patients were reviewed in 2017 (Picard et al., 2017). The inheritance was autosomal recessive. The AMH gene is on the short arm of chromosome 19. Missense mutations are the most frequent and are particularly common in exons 1 and 5. In the same paper, mutations in AMHRII were summarized. The AMHRII gene is on the long arm of chromosome 12.

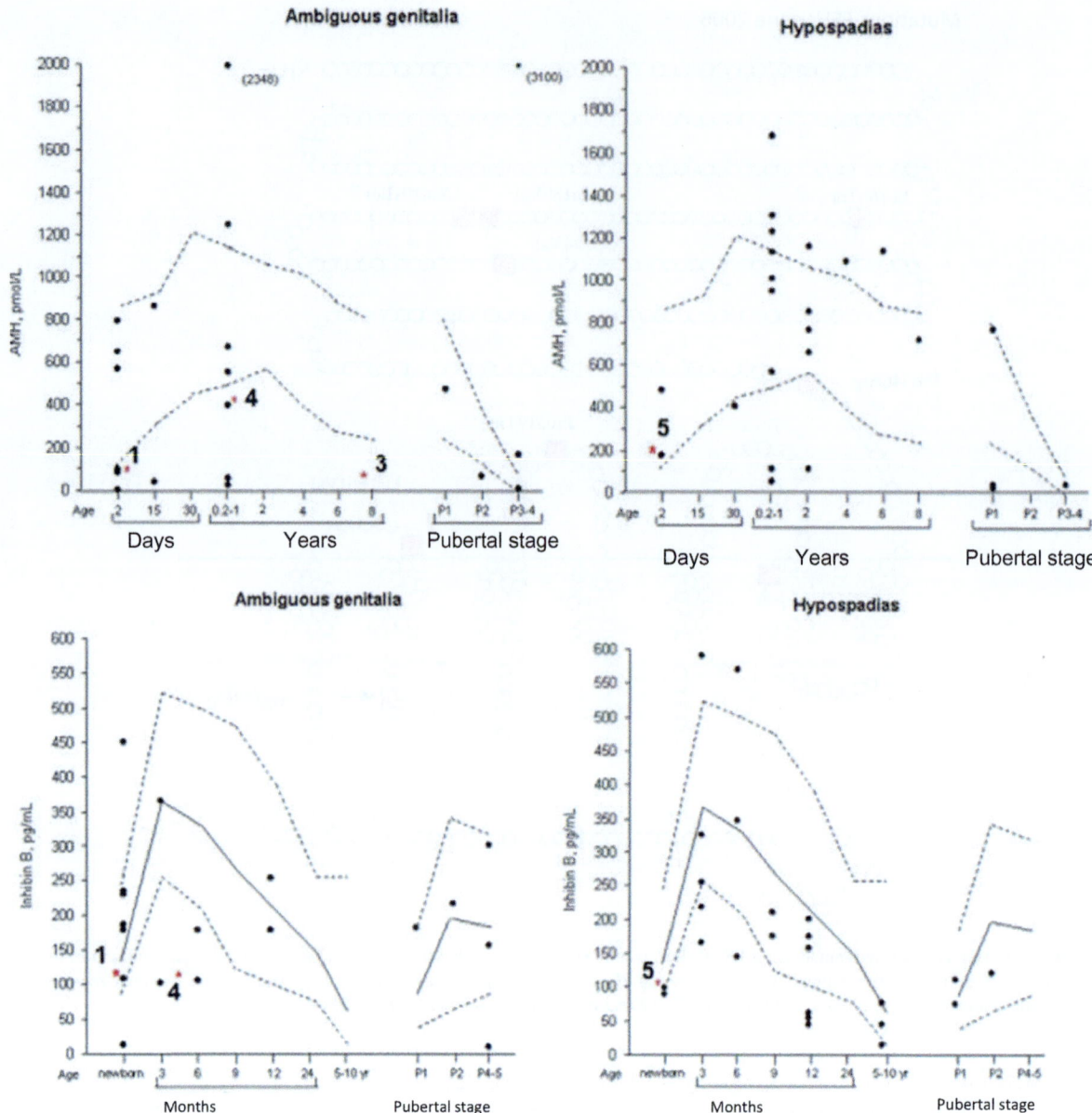

FIG. 3.2.2.25 Plasma AMH and inhibin B concentrations in patients with gonadal dysgenesis or ambiguous genitalia. In each AMH graph, the broken lines correspond to the upper and lower limits of the normal range. For inhibin B, the solid line corresponds to the median and the broken lines to the 5th and 95th percentiles. The red asterisk indicates individuals carrying an NR5A1 heterozygote mutation and the numbers indicate particular patients. *(Allali S, Muller JB, Brauner R, Lourenço D, Boudjenah R, Karageorgou V, Trivin C, Lottmann H, Lortat-Jacob S, Nihoul-Fékété C, De Dreuzy O, McElreavey K, Bashamboo A. Mutation analysis of NR5A1 encoding steroidogenic factor 1 in 77 patients with 46, XY disorders of sex development (DSD) including hypospadias. PLoS ONE 2011;6(10):e24117. Fig. 2 p. 5.)*

All types of variants have been reported in homozygous and heterozygous state (Fig. 3.2.2.27).

3.2.2.7 Pubertal failure or arrest

The investigations for delayed puberty are dependent on the clinical picture. Many children with mild symptoms will have constitutional delay and do not need extensive tests. Karyotyping will exclude girls with Turner syndrome.

Patients with dysmorphic features and learning disabilities can have specific disorders such as Klinefelter, Prader-WIlli or Bardet-Biedl syndrome. Pubertal delay with hypogonadotrophic hypogonadism will need neuroimaging and tests for other pituitary hormone deficiencies. Pubertal arrest is defined as lack of pubertal progression for more than 2 years after spontaneous onset such as failure to achieve menarche from the onset of breast development or attainment of testes greater than 15 mL from 4 mL over 5 years.

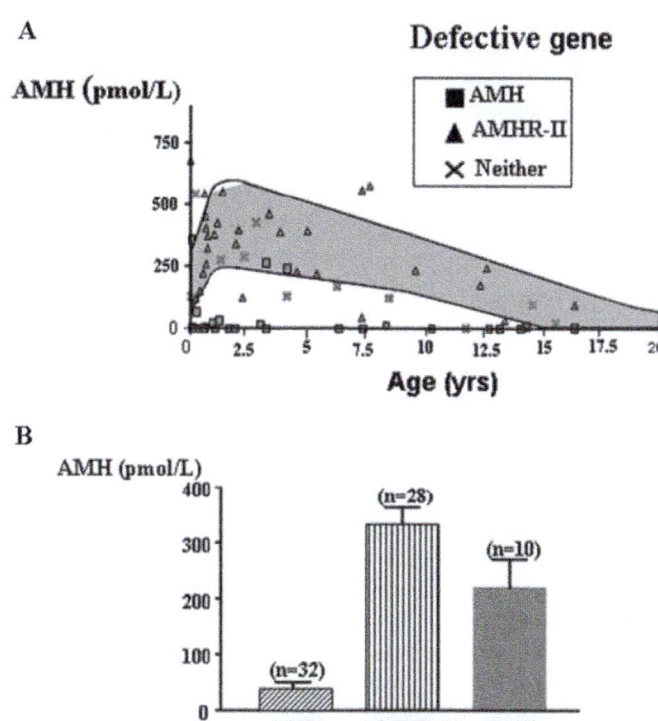

FIG. 3.2.2.26 AMH concentrations in patients with defect of AMH gene or AMHr gene. *(Brunello FG, Rey RA. AMH and AMHR2 involvement in congenital disorders of sex development. Sex Dev 2022;16(2–3):138–46. https://doi.org/10.1159/000518273. Epub ahead of print. PMID: 34515230. Fig. 2 p. 3.)*

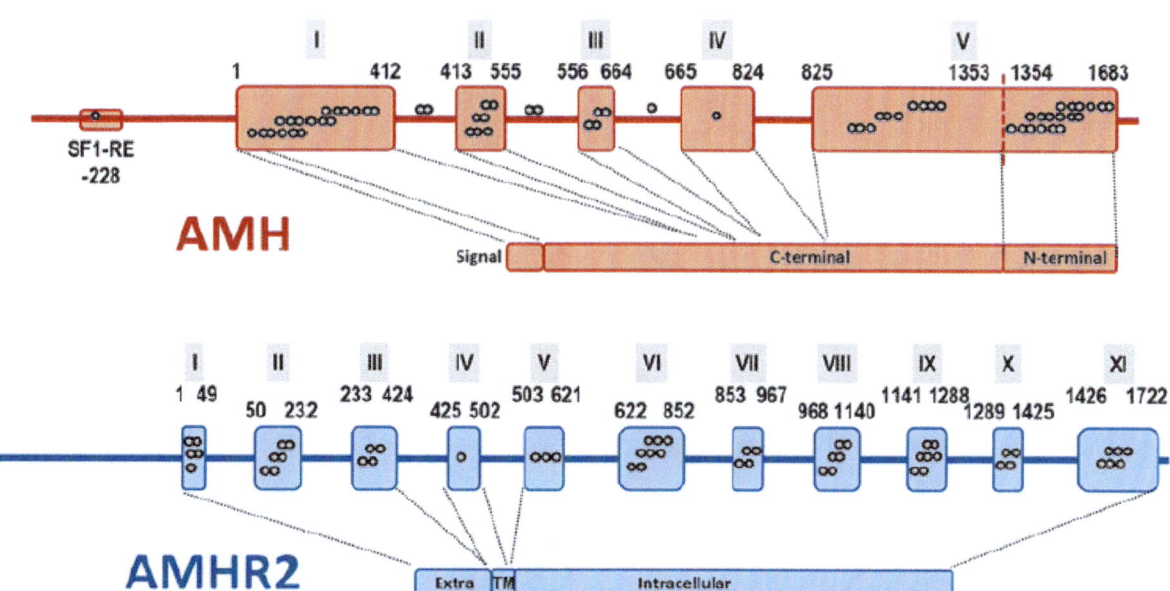

FIG. 3.2.2.27 Mutations AMH and AMHr. Schematic of the *AMH* and *AMHR2* genes and proteins. Roman numerals indicate exons, arabic numerals indicate initial and last base number of each exon. SF1-RE, SF1 response element at position −228 of the *AMH* promoter; Extra, extracellular domain of AMHR2; TM, transmembrane domain of AMHR2. Dots indicate independent mutations described. Deletions are not included in this figure. *(Brunello FG, Rey RA. AMH and AMHR2 involvement in congenital disorders of sex development. Sex Dev 2022;16(2–3):138–46. https://doi.org/10.1159/000518273. Epub ahead of print. PMID: 34515230. Fig. 3 p. 4.)*

3.2.2.7.1   Somatic issues

Nutrition is important and eating disorders are common. This can be anorexia or binge eating, bulimia to obesity which represent ends of a spectrum of problems. Anorexia is loss of appetite or inability to eat which can be due to infection, depression and sometimes cancer. Anorexia nervosa (AN) is a more complex psychological disorder with extreme limit of food consumption or vomiting to maintain abnormally low body weight. Girls with AN can have a fear of gaining weight and a distorted self-image. Obesity is the result of high food intake through practices such as binge eating in some cases with vomiting to ensure no weight gain. Amenorrhea is a common problem. Counseling is needed to restore energy balance; medical treatment may also be needed such as antidepressants, antipsychotics and leptin administration. Obesity treatment with lifestyle changes and dieting is not always successful and severely obese patients may require bariatric surgery.

For the obese a reduction in bone mass due to estrogen deficiency can lead to fractures. Bone mass can be measured by densitometry (dual X-ray absorptiometry DEXA scan). Bone turnover markers can be measured in blood. Elevated bone resorption markers (C-telopeptide, and urinary N-telopeptide) can be normal or elevated and bone formation markers (ostoecalcin, procollagen type-1 N-propeptide and bone specific alkaline phosphatase) normal or decreased. These are not always reliable at the outset but are effective biomarkers when treatment is successful.

A term "female athlete triad" was proposed more than 20 years ago to encompass disordered eating, menstrual disturbance and low bone mass. The syndrome is attributed to low energy intake relative to the energy expenditure in sport and is now called "relative energy deficiency in sport" or RED-s. The hypothalamic release of pulstile GnRH is reduced leading to hypogonadotrophism, amenorrhea, estrogen deficiency and reduced bone mass (Fig. 3.2.2.28) (Dipla et al., 2021). Leptin, IGF-1 and thyroid hormones are raised. A similar state is seen in male athletes (Dwyer et al., 2019).

3.2.2.7.1.1   Gender dysphoria

Patients can present with their "feeling to be in the wrong body". Girls with this state need firstly to be investigated for a DSD as the underlying cause. A patient with 5α-reductase or 17-hydroxysteroid dehydrogenase deficiencies or androgen insensitivity syndrome can genuinely fall in this category and biochemistry with genetic tests will exclude these diagnoses. There are several times in a lifetime where there is an imbalance between sex steroids (minipuberty, entry to puberty, and stress situations) that can influence a perception of being the wrong gender. Clinical and

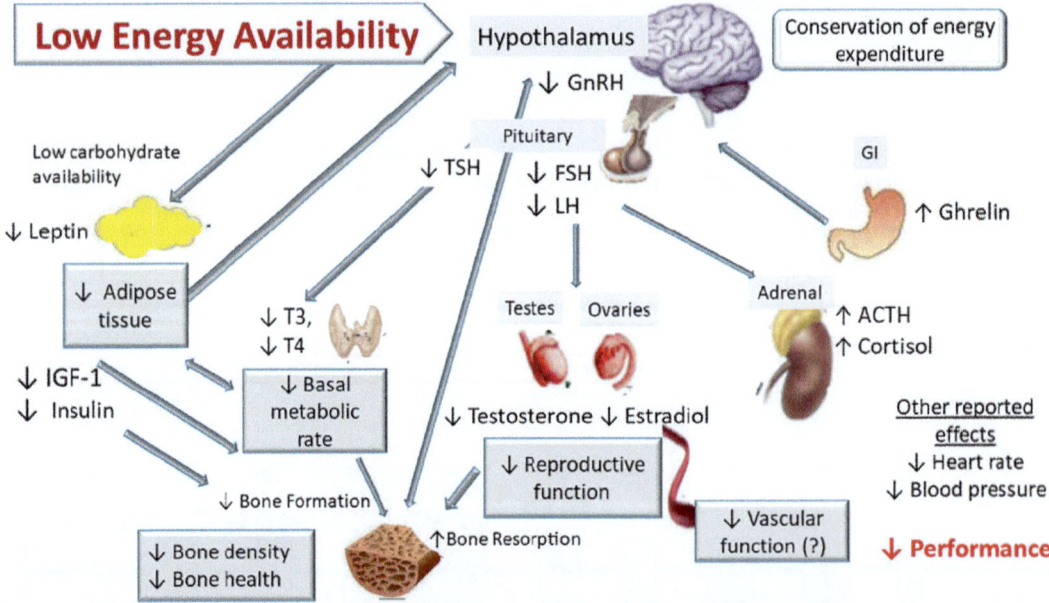

FIG. 3.2.2.28   Athlete triad. Stressors such as a low energy availability (LEA) activate the sympathetic nervous system and the hypothalamic pituitary adrenal (HPA) axis, resulting in a plethora of neuroendocrine alterations to conserve energetic expenditure. The interplay between the brain, the peripheral endocrine organs, and the bone has been shown. The brain and the gonads act as regulators of skeletal homeostasis. The bone also influences brain functions via the release of molecules that are able to cross the blood-brain barrier and regulate neurotransmitter release. Leptin, a hormone primarily secreted by adipocytes, not only affects energy metabolism but also, via the hypothalamus and the central nervous system, indirectly inhibits bone accrual. GnRH, gonadotropin-releasing hormone; IGF-1, insulin-like growth factor-1 (currently, insulin growth factor, IGF); TSH, thyroid-stimulating hormone; LH, luteinizing hormone; FSH, follicle-stimulating hormone; ACTH, and adrenocorticotropic hormone. *(Dipla K, Kraemer RR, Constantini NW, Hackney AC. Relative energy deficiency in sports (RED-S): elucidation of endocrine changes affecting the health of males and females. Hormones (Athens) 2021;20(1):35–47. Fig. 1 p. 39.)*

psychological evaluation of the patients is essential in an understanding environment. Depending on age and sexual maturity, a number of options are available these days including hormones for enhancing (gender affirming) or repressing secondary sex characteristics, (puberty blockers) and in certain cases surgery for complete transition. This is a long, slow process and at the end it is not unusual for the outcome to feel wrong again.

Children with **Prader-Willi** syndrome experience puberty later than usual and may not go through full development into an adult. Boys may still have a high pitched voice and may not have much facial and body hair. Girls will often not start periods until in the third decade and the breasts will not fully develop. When periods do start they are usually irregular with light menstruation.

Obesity is a characteristic of autosomal recessive **Bardet-Biedl syndrome**. Vision loss is one of the major features. Impaired speech and delayed development of motor skills as well as behavioral problems are additional signs. The syndrome can result from mutations in at least 15 genes that play critical roles in cell structures called cilia.

17α-Hydroxylase deficiency (CYP17)α

There are four elements to this enzyme complex involving (1) 17-hydroxylation, (2) 17,20 lyase action, (3) the role of cytochrome P450 oxidoreductase (POR) as electron donor, (4) the action of cytochrome $b5$. The actions of (1) and (2) can be combined or the components can be in isolation. The deficiencies constitute a rare form of autosomal recessive disorder.

17α-Hydroxylase deficiency is identified biochemically by demonstrating high serum concentrations (Sun et al., 2021; Lee et al., 2020; Burns et al., 2020; Wei et al., 2006) and urine metabolites of progesterone and corticosterone (Lee et al., 2020; Tiosano et al., 2008; Rosa et al., 2007; Honour et al., 1978) (Fig. 3.2.2.29).

The production rate of corticosterone can be much higher than cortisol due to the lower capacity for negative feedback to the HPA and patients do not experience cortisol deficiency. Corticosterone is significantly raised in LC-MS/MS analysis provided that the validation had assured separation of the isobaric steroids 11-deoxycortisol and 21-deoxycortisol. Even if the steroids are not resolved

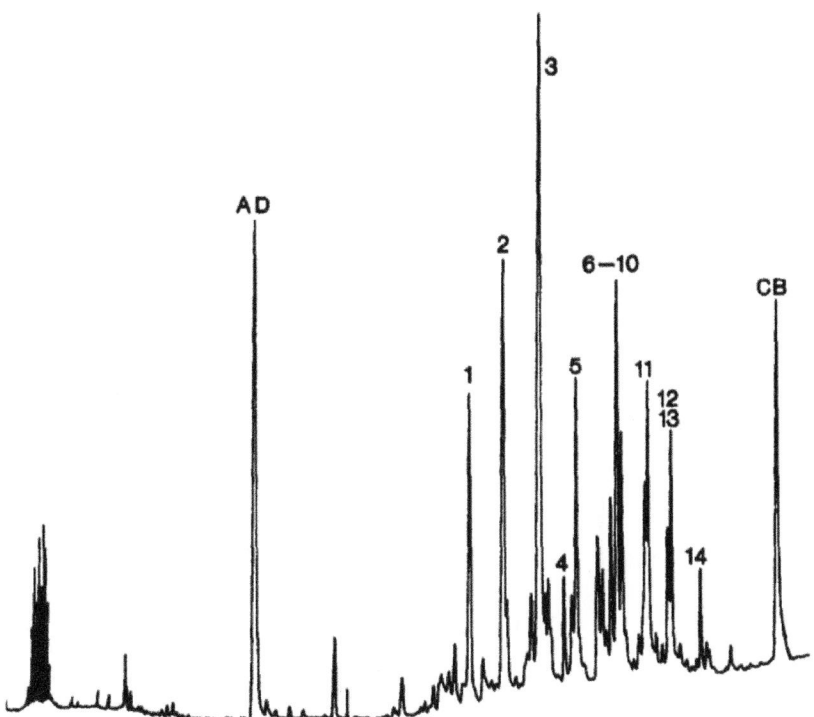

FIG. 3.2.2.29 Gas chromatogram of the major steroids excreted by the patient with 17α-hydroxylase deficiency. Steroids in urine were deconjugated before formation of methoxime trimethylsilyl derivatives. A 20-m OV-101 WCOT column was used for this analysis. Some of the major steroids are indicated. (1) 5β-Pregnane-3α,20α-dial: (2) 5-pregnene-3α,20α-diol; (3) 3α,20α-dihydroxy-5α (and β)-pregnan-11-one; (4) 5β-pregnane-3α,11β, 20α–triol; (5) 5α-pregnane-3α,11β,20α-triol and 5-pregnene-3β,16α,20α-triol; (6) 3α,21-dihydroxy-5β-pregnane-11,20-dione; (7) 3α,21-dihydroxy-5α-pregnane-11,20-dione; (8) tetrahydrocorticosterone; (9) allotetrahydrocorticosterone; (10) partially characterized 1,3,20-trihydroxy-5-pregnan-11-one; (11) 3α,20α,21-trihydroxy-5α(and β)-pregnan-11-one and second peaks of 6–9; (12, 13) 5α(and β)-pregnane-3α,11β,20α,21-tetrol; (14) 5-pregnene-3β,16α,20α,21-tetrol and hydroxytetrahydrocorticosterone. AD (5α-androstane-3α,17α-diol) and CB (cholesteryl butyrate) were internal standards. *(Honour JW, Tourniaire J, Biglieri EG, Shackleton CH. Urinary steroid excretion in 17 alpha-hydroxylase deficiency. J Steroid Biochem 1978; 9(6):495–505. Fig. 1 p. 497.)*

corticosterone and 11-deoxycortisol have similar 347 to 109/97 transitions but 21-deoxycortisol shows different response for 347 to 105/121 transitions compared with corticosterone and 11-deoxycortisol (Burns et al., 2020). A urine steroid profile will confirm absence of cortisol and androgens with raised corticosterone metabolites (Honour et al., 1978). Several corticosterone metabolites were 21-dehydroxylated. The impact of the enterohepatic circulation and the microbiome on corticosterone clearance was then recognized, decades before the microbiome attracted real interest in health and disease (Fig. 3.2.2.30).

A 46XX individual with this defect is phenotypically normal in childhood and most cases are not detected until hypertension or failure of puberty or infertility is investigated (Sun et al., 2021). Lack of breast development and absence of pubic hair at 14 years of age should be investigated but because the patients are usually taller than peers a primary health care provider might conclude that the patient is healthy because she was growing well. A 46XY patients can present with inguinal or abdominal hernia.

The experience with urine steroid profile analysis in such cases in the newborn period is limited because at this age the disorder is usually without symptoms. The pattern of steroids in urine of one neonatal case, later confirmed to have CYP17 deficiency, showed unusually high neonatal excretion rates of 16α-hydroxypregnenolone. At 15 months of age the child excreted excess corticosterone metabolites just as is found in urine of adults with this disease (Dean et al., 1984).

The usual presentation of this disorder is delayed puberty and primary amenorrhea. High concentrations of deoxycorticosterone lead to hypernatremia, hypokalemia and later in life to hypertension. CYP17 deficiency is associated with low PRA. Ultrasound of internal genital structures provides additional information, sometimes before karyotype and endocrine results are available (Hryhorczuk et al., 2022).

An ovary is identified by the presence of anechoic follicles interspersed with hypoechoic stroma. A testicle is homogeneously hypoechoic with a hyperechoic mediastinal band. An ovotestis has a mixture of ovarian and testicular features. A streak gonad is difficult to distinguish from surrounding tissues. The presence of a uterus and ovaries would suggest a 46XX karyotype while absence of a uterus

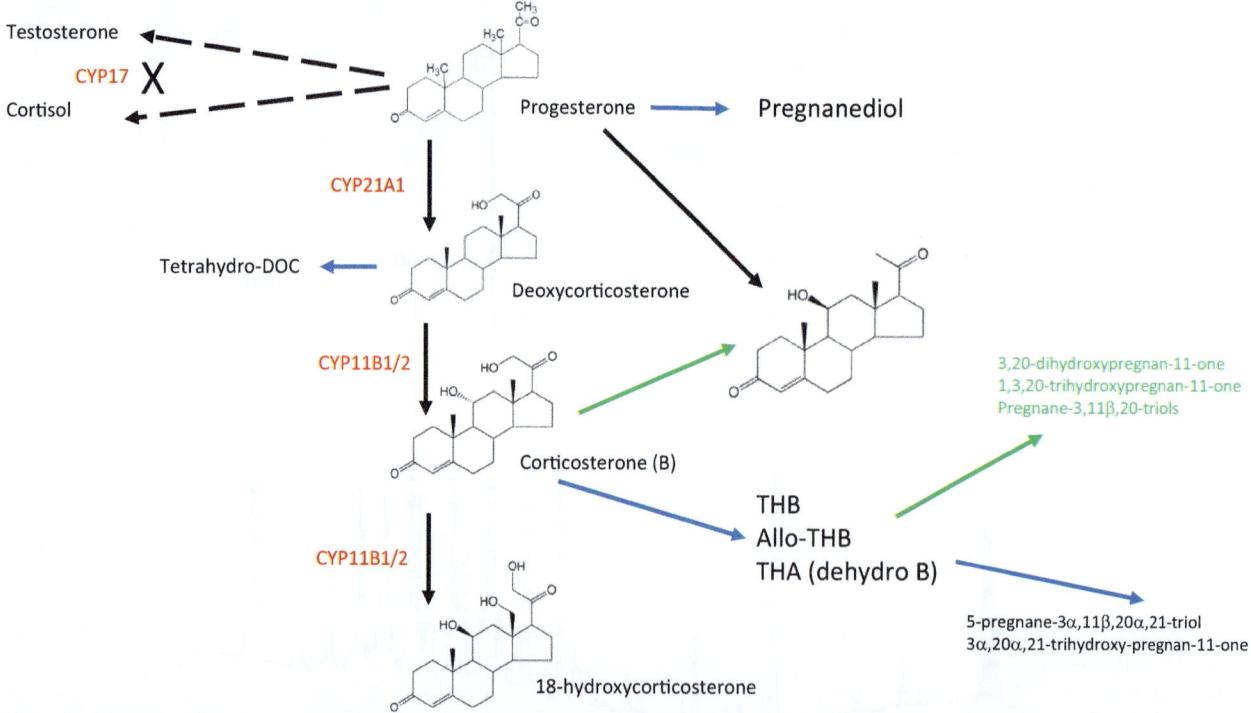

FIG. 3.2.2.30 Biosynthesis and metabolism of adrenal steroids. Progesterone can be converted to 11-hydroxyprogesterone by the action of CYP11B2. Hydroxylation at C21 lead to formation of corticosterone. 18-Hydroxycorticosterone is a precursor of aldosterone. Cortisol and testosterone synthesis is blocked in the absence of CYP17. Most steroids are excreted in urine after reduction of the A-ring (blue arrows) and conjugation with glucuronic acid. Corticosterone and sex steroids are excreted in bile. Steroids in the intestine are subject to bacterial metabolism (green arrows). Tetrahydrocorticosterone can lose the 21-hydroxyl group and 21-deoxysteroid metabolites, in effect metabolites of 11-hydroxyprogesterone, are reabsorbed from the intestine. Rats do not produce cortisol. Deoxycorticosterone excess causes hypertension. *StAR*, steroidogenic acute regulatory protein; *CYP11A1*, side chain cleavage; *HSD3B2*, 3-beta hydroxysteroid dehydrogenase; *CYP17A1*, 17-hydroxylase; *CYP21A2*, 21-hydroxylase; *CYP11B1*, 11B-hydroxylase; *CYP11B2*, 11B-hydroxylase and aldosterone synthase; *HSD11B*, 11B-hydroxysteroid dehydrogenase; *HSD17B*, 17-hydroxysteroid dehydrogenase. *Modified from Honour JW. Historical perspective; gut dysbiosis and hypertension. Physiol Genomics 2015;47:443–6. Fig 1 P444.*

would indicate a 46XY karyotype. Ovarian cysts are a feature of 17-hydroxylase deficiency attributed to the increased gonadotrophins. Testes may be seen on imaging of the abdomen. Androgens, estrogens and cortisol are low. **A persistent raised progesterone**, is in the experience of the author, an **important clue to the defect** with the common tests. The tall stature relates to the estrogen deficiency, a symptom of aromatase deficiency.

Gene sequencing analysis has shown base deletions and duplications leading to stop and termination codons that have explained several of the deficiencies in the enzyme in patients with CYP17 deficiency (Rumsby et al., 1993). A P409R mutation was predicted to abolish heme incorporation with the enzyme (Lam et al., 2001). Missense and nonsense mutation, small and gross deletions, small and complex insertions or deletions, splice site variants and mutations in regulatory elements are recorded at http://hgmd.org (requires registration). Over 100 mutations in the CYP17A1 gene are known (Auchus, 2017) (www.hgmd.cf.ac.uk). Many mutations affect the protein near the C-terminus. There are distinct ethno-graphic distributions in some cultures (Wu et al., 2017), families and their descendants around the globe. Some individuals with low-renin hypertension that is responsive to mineralocorticoid-antagonists may have mild, unrecognized P450c17 deficiencies linked to polymorphisms in the gene.

17,20 desmolase deficiency

A defect of 17,20-desmolase is to some extent related to the CYP17 defect. Ser/Thr phosphorylation of CYP17A1 appears to promote POR-CYP17AI interaction in vitro (Tee and Miller, 2013). The association between the two defects is due to the fact that these enzyme activities are coded by the same gene. The external genitalia may vary in appearance from female through male with perineal hypospadias to hypoplastic male (Miller and Tee, 2015). At puberty there may be raised 17-hydroxyprogesterone, 17-hydroxypregnenolone and gonadotrophins. HCG stimulation produces a marked rise in plasma concentrations of these steroids without change in the already low androgens. There is a similar response to ACTH showing the absence of the defect in the adrenal as well as the gonads. Testicular tissue fails to metabolize C21-steroids to androgens.

Urinary steroid profiles have been reported for a patient and families (Tiosano et al., 2008; Fernández-Cancio et al., 2018). Activation of the backdoor path to dihydrotestosterone may explain some of the findings.

A few mutations in CYP17 explain the defect of desmolase only (Miller, 2012). Isolated 17,20-lyase deficiency appears to be secondary to variants in protein coded by CYP17A1. Four mutations in CYP17A1 have been described (pArg347His, pArg347Cys, pArg538Gln and pGlu305Gly) (Sherbet et al., 2003; Van Den Akker et al., 2002; Tiosano et al., 2008). Some mutations such R347H

in CYP17A1 can affect the interactions of the protein and cytochrome b5 (Geller et al., 1999).

3.2.2.7.1.2   AKR1C2 and AKR1C4 mutations

An early case of 17,20 desmolase deficiency (Zachmann et al., 1972) based on early steroid analysis by capillary gas chromatography then available was re-examined many years later, using a candidate gene approach to reveal mutations in **AKR1C2 and AKR1C2/4** which are elements of the backdoor path to dihydrotestosterone (Fig. 3.2.2.31).

AKR1C1-4 on chromosomal region 10p14-p15 encodes four enzymes in C15 aldo-keto reductases that act as 3-, 17-, and 20-ketosteroid reductases but only AKR1C2 and AKR1C4 oxidize 3α-diol to DHT. AKR1C2 is primarily a reductive enzyme using the readily available NAD and low levels of NADPH. AKR1C4 is catalytically the most efficient, is abundant in liver and expressed in testes and adrenals. In elegant work, the Christa Fluck team showed mutations in the coding region of AKR1C2 and a splicing mutation in AKR1C4 (Flück et al., 2011b). Both mutations were found in family 1 but only AKR1C2 in family 2. Mutations in AKR1C4 were not found in other 46XY DSD patients. Mutations in AKR1C2 are sufficient for disease manifestation. It was concluded that 5α DHP needs to be reduced to allopregnanolone for synthesis of 3α diol in the backdoor path to DHT, a path complementary to the classical path in humans (Fig. 3.2.2.32).

Cytochrome b5 deficiency

In the defect of cytochrome b5, cases were presumed to be female at birth although the clitoris can be judged to be mildly enlarged and labia majora scrotalized. The skinfold between the labia, urethral orifice and vaginal opening cannot be measured. An affected male sibling was the stimulus to investigate some newborn infants and older relatives (Idkowiak et al., 2012). There is no deficit of cortisol or aldosterone which excludes sole 17-hydroxylase deficiency. Four genetic mutations have been reported affecting protein; a H44L, W27X, Tyr35Ter (2 cases), and Tyr11Ter (Shaunak et al., 2020) sequence change. **Methemaglobinemia** is associated with isolated 17,20 lyase deficiency due to the abnormal cytochrome b5 protein. Urine steroid analysis in the first week of life showed only a relative increase in 16-hydroxypregnenolone (Author, observation in one case in Shaunak cases). A urine steroid profile can provide evidence to support the defect (Leung et al., 2019). The excretion of DHEAS, corticosterone and cortisol metabolites was normal. In cases after puberty, 17-hydroxyprogesterone metabolites were increased and androgens were low. In plasma, 17-OHP is stimulated by hCG but not ACTH.

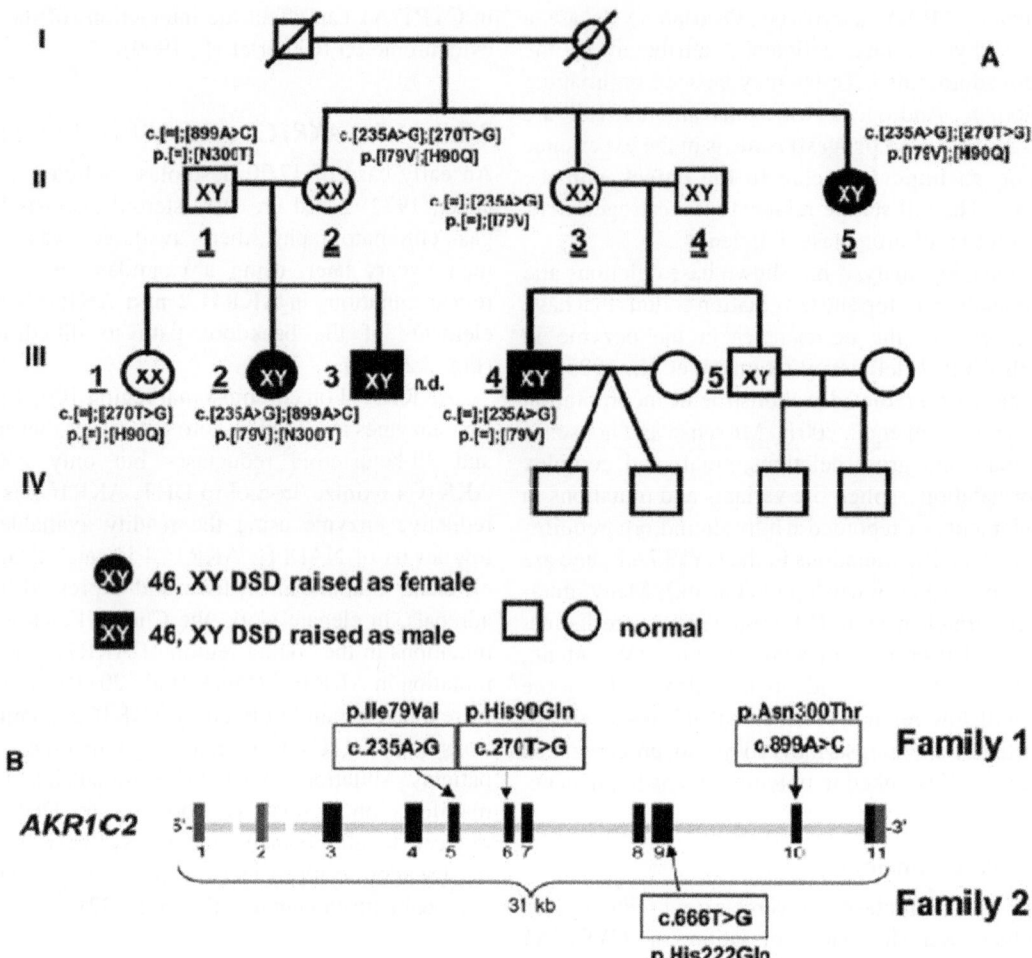

FIG. 3.2.2.31   Identification of AKR1C2 mutations in 2 families. (Upper panel) The pedigree of the index family (family 1). The AKR1C2 genotype and the amino acid exchange are given adjacent to the symbols. Individuals II.4 and III.5 had a normal genotype, which is not indicated. Family generations are indicated in Roman numerals, and relevant individuals within each generation are numbered. Individuals whose numbers are underlined were analyzed. Individual III.3 refused genetic testing (n.d. is abbreviation for not done). The sexes of rearing are indicated as square for a male and a circle for a female. The filled symbols represent individuals with a DSD, all had 46XY chromosomes. The proband is individual III.2. Primary amenorrhea and testes in inguinal canal were in II.5 and III.2, whereas III.3, III.4 and all had absent uterus. (Lower panel) Diagram of AKR1C2 on chromosomal region 10p14-p15 encoding 3α-HSD3. Coding exons 3–11 are shown as *black boxes*, noncoding exons 1 and 2 are shown in *gray*. The locations of the four identified missense mutations are shown. *(Flück CE, Pandey AV, Dick B, Camats N, Fernández-Cancio M, Clemente M, et al. Characterization of novel StAR (steroidogenic acute regulatory protein) mutations causing non-classic lipoid adrenal hyperplasia. PLoS ONE 2011b;6(5):e20178. Fig. 2 p. 203.)*

3.2.2.7.1.3   Aromatase deficiency

In males with tall stature, delayed skeletal maturation without fusion of epiphyses, osteoporosis, elevated gonadotrophins, raised androgens, low estradiol a defect in aromatase may be causative. A female with this condition will have primary amenorrhea, no breast development, elevated gonadotrophins and low estradiol (Bulun, 2014). Girls show signs of virilization at puberty indicating the importance of adrenal androgen production as well as the increase in T before the block (Praveen et al., 2020). Hyperandrogenism is discussed in more detail in chapter on Hypergonadism (Chapter 3.2.2.1). Detail of the mutations reported has been summarized (Praveen et al., 2020).

DAX-1 mutation

The DAX-protein (coded by NROB1) is an orphan nuclear receptor and transcription factor in adrenal and gonadal development. **Mutations in the NROB1 gene for DAX-1**, a member of the steroid receptor superfamily, are found in boys that fail to undergo puberty with hypogonadotrophic hypogonadism and history of neonatal adrenal insufficiency often with a normal female appearance without noticing ambiguous genitalia (Wu et al., 2020; Nagel et al., 2019; Rojek et al., 2016; Suntharalingham et al., 2015; Cho et al., 2021). The condition is called adrenal hypoplasia congenita (AHC) and is also known as dosage sensitive sex reversal gene syndrome (DSS). Delayed puberty is

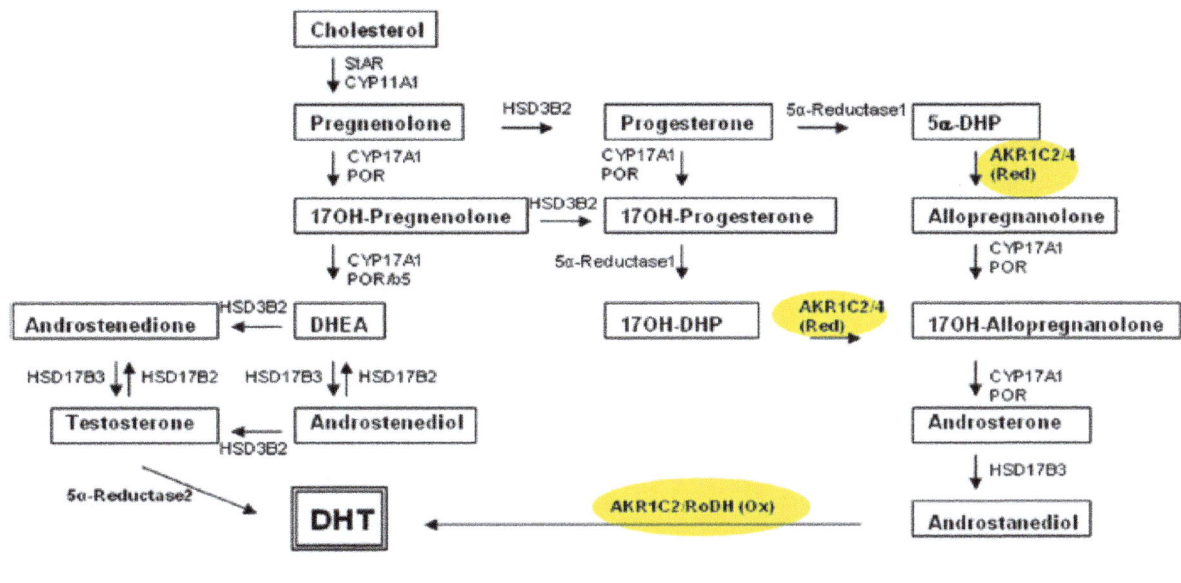

"Classic" Pathway   Alternative/Backdoor Pathway

FIG. 3.2.2.32 Synthesis of dihydrotestosterone via the classic and alternative (backdoor) pathways. The classic pathway of steroidogenesis leading to dihydrotestosterone is shown on the left, and the alternative pathway is shown on the right. The factors in the classic pathway are CYP11A1 (cholesterol side-chain cleavage enzyme, P450scc), StAR (steroidogenic acute regulatory protein), CYP17A1 (17α-hydroxylase/17,20-lyase, P450c17), HSD3B2 (3β-hydroxysteroid dehydrogenase, type 2), HSD17B3(17β-HSD3 [17β-hydroxysteroid dehydrogenase, type 3] and 5α-reductase, type 2 [5α-reductase 2, encoded by SRD5A2]). The alternative pathway is characterized by the presence of additional enzymes: 5α-reductase, type 1 (5α-reductase 1, encoded by SRD5A1), AKR1C2 3 (3α-reductase, type 3) and possibly AKR1C4 (3α-reductase, type 1) and RoDH (3-hydroxyepimerase, encoded by HSD17B6). Most steroids are identified by their trivial names; 17-hydroxy-dihydroprogesterone (17OH-DHP) is 5α-pregnane-17α-ol-3,20-dione; 17-hydroxy-allopregnanolone (17OH-allo) is 5α-pregnan-3α,17α-diol-20-one; 5α-dihydroprogesterone (5α-DHP) is 5α-pregnane-3,20-dione, and allopregnanolone is 3α-hydroxy-dihydroprogesterone (3α-OH-DHP) or 5α-pregnane-3α-ol-20-one. *(Flück CE, Pandey AV, Dick B, Camats N, Fernández-Cancio M, Clemente M, et al. Characterization of novel StAR (steroidogenic acute regulatory protein) mutations causing non-classic lipoid adrenal hyperplasia. PLoS ONE 2011b;6(5):e20178. Fig. 1 p. 202.)*

relatively common, often familial and as adults these males are infertile with azoospermia. Müllerian ducts can be present. Precocious puberty has also been reported (Nagel et al., 2019). Many patients respond to hCG treatment. Older patients can present with tumors sometimes with distinct abdominal mass secondary to duplication of DAX-1 (García-Acero et al., 2019). Sperm banking is recommended (Vargas et al., 2020).

3.2.2.7.1.4 SF-1 (NR5A1)

Mutations in SF-1 have been found in boys with hypospadias or **delayed puberty** with history of adrenal failure (Köhler and Achermann, 2010; Achermann et al., 1999) but thereafter the majority of cases present with a reproductive phenotype only (Camats et al., 2018). In fact, the majority of cases reported have been 46XY. There is a broad spectrum in the phenotype at presentation. **Mutations in SF-1** should be screened in cases of suspected partial androgen insensitivity when no mutations are found in the androgen receptor gene (Coutant et al., 2007). A total of 163 gonad-related target genes (and more) can be screened by targeted NGS (Pan et al., 2020). Heterozygous missense, nonsense,

splicing mutations, small deletions and insertions have been reported leading to severe gonadal dysplasia, differentiation of testes and ovaries, abnormal transcription, subcellular location of SF-1. The range of presentation has prompted research into the possibility of additional affected genes (Martínez de LaPiscina et al., 2020; Neocleous et al., 2020).

CDKN1C (IMAGE syndrome)

The combination of Intrauterine growth restriction, Metaphyseal dysplasia (malformation of ends of long bones), Adrenal hypoplasia and GEnitourinary anomalies (IMAGE) is attributed to missense variants with gain of function in cyclin-dependent kinase inhibitor 1C (CDKN1C) which is a negative cell cycle regulator (Vilain et al., 1999; Kerns et al., 2014; Cabrera-Salcedo et al., 2017; Suntharalingham et al., 2019; Berland et al., 2020). Some infants have distinct facial features and curvature of the spine (scoliosis). Glucocorticoid and mineralocorticoid insufficiency are variable, infections are common. A young patient can present in the first week with hypotension, high urine output, hyponatremia, hyperkalemia and raised ACTH and PRA. Genital abnormalities occur in males with micropenis, cryptorchidism and hypospadias.

A gain of function mutation in CDKN1C is a cause of the related but less frequent **Silver-Russell syndrome** without adrenal insufficiency (Binder et al., 2020) or genital ambiguity. A mutation was found in SAMD9 (Kim et al., 2018) and a loss of function mutation in **Polymerase epsilon (POL1)** that leads to delayed S-phase progression and cell division with unknown mechanism (Logan et al., 2018; Schmit and Bielinsky, 2021).

SAMD9 (MIRAGE syndrome)

Preterm infants with growth restriction and genitalia between hypospadias and female genitalia can develop the signs linked to the acronym MIRAGE for Myelodysplasia (type of blood cancer), Infections, Respiratory distress, Anemia, Gonadal anomalies and Enteropathy. There can be thrombocytopenia (low blood platelet count) and hydrocephalus (fluid on brain affecting head size, fontanelle and eyes). Gain of function variants in Sterile Alpha Motif Binding Domain containing 9 (SAMD9) have been found (Narumi et al., 2016; Shima et al., 2018; Csillag et al., 2019; Mengen et al., 2020; Onuma et al., 2020).

3.2.2.7.2 Defects of steroid action

3.2.2.7.2.1 Androgen insensitivity syndrome (AIS)

In boys with poor masculinization of the external genitalia that have androgen resistance, basal testosterone and gonadotrophin concentrations are often within age-related reference ranges. In cases of complete androgen insensitivity, there is no increase in LH and T usually seen between 2 weeks and 3 months called the minipuberty (Bouvattier et al., 2002). In children and teenagers with CAIS, the response to hCG results in stimulated testosterone concentrations that can be 10 times basal and DHT typically 2.2-fold the basal concentration. There is an exaggerated LH response to LHRH. An additional in vivo test for AIS is to examine the decline in SHBG in response to oral anabolic steroids. In normal subjects, there is a 40%–60% decline in SHBG but in AIS the SHBG is 65%–90% of basal (Krause et al., 2004). Girls with inguinal hernias should have karyotype checked and an hCG stimulation test performed. Older patients will seek help when menstruation does not occur despite normal breast development. High plasma AMH measurement is a useful marker for androgen resistance (Liu et al., 2020b).

Androgen receptor defects can influence the number and stability of androgen receptors. Genital skin can be taken and sent to a specialist laboratory. Before further steroid binding experiments can be undertaken the cells have to be increased in number by culturing through several passages in tissue culture. Androgen receptor binding studies have been developed for the fibroblasts from genital skin but do not really provide results early enough to assist important clinical decisions (Ahmed et al., 2000). Three types of abnormalities have been identified according to the binding affinity for testosterone with and without NADPH (Hughes and Evans, 1987). Androgen receptor defects can influence the number and stability of androgen receptors.

Receptor gene analysis has been reported from a few centers (Boehmer et al., 2001; Barbaro et al., 2007; Jeske et al., 2007; Martínez-Garza et al., 2008; Deeb et al., 2008; Philibert et al., 2010). The risk of malignancy in CAIS is low and gonadectomy can be delayed until after pubertal age (Barros et al., 2021). The androgen receptor gene is located on the X chromosome. Mutation screening and linkage analysis can now be used but nearly 400 mutations have been described in the androgen receptor gene (Arya et al., 2020; Costagliola et al., 2021; Naumova et al., 2020; Philibert et al., 2010; Deeb et al., 2008; Jeske et al., 2007; Barbaro et al., 2007). Complete absence of the gene, large deletions and point mutations affect the protein sequence. Deletions in the LBD abolish hormone binding completely but deletions in the NTD and DNA-binding domain do not affect hormone binding. Androgen receptor variants have been shown to regulate similar patterns of gene expression to the full-length hormone-bound receptor. Deletion of the LBD leads to a constitutively active AR protein with trans-activation capacity like the full-length AR. The hormone binding domain seems to act as a repressor of the trans-activation function in the absence of hormone. Receptor positive patients may have mutations which affect transcription activation, nuclear localization and phosphorylation domains.

The detailed biochemical and genetic tests are time consuming and expensive and are not provided by service laboratories and only available in specialized laboratories with variable (often too long) turnaround times. Failure to respond to a clinical trial of testosterone treatment is a useful predictive test of androgen response and assists decisions about how best to bring up the child. Stanozolol suppresses SHBG in patients classified as PAIS but not with CAIS (Sinnecker et al., 1997).

In the past, many patients have been classified with partial forms of the disease (PAIS) with poor correlation of phenotype and genotype (not of AR) suggesting that other factors are involved in the phenotype. Two studies found mutations in SRD5A2 and HSD17B3 from this patient group (Berra et al., 2011; Phelan et al., 2015). There has been much confusion in the diagnosis of the undervirilized 46XY DSD male. A partially virilized (pvDSD) designation may be appropriate (Buonocore et al., 2019).

In one case of CAIS and no mutation in the coding region, the abnormality was attributed to a co-activator through decreased transmission of a transactivation signal from the AF1 region of the AR (Adachi et al., 2000). An extended family through 3 generations showed X-linked inheritance of under virilization at birth (Batista and Mendonca, 2020). Five individuals were reared as female and four assigned as male at birth. Whole exon sequencing of the AR gene was normal. A 1100 base pair insertion was found in the 5′ untranslated region due to retrotransposon insertion of an LINE1 inherited condition and constitute CAIS type II. A recent assay using the androgen regulated gene, apolipoprotein D (APOD), as a marker for androgen insensitivity at the cellular level, demonstrated that one third of genital fibroblasts from PAIS patients had APOD induction below the cut-off (Hornig and Holterhus, 2021). Hypermethylation of the AR promoter is proposed, caused by cofactors in androgen signaling. The molecular diagnosis of PAIS thus remains a challenge.

3.2.2.7.2.2 Estrogen resistance

A mutation in the estrogen receptor ERα gene led to a tall male who had grown continuously without a pubertal growth spurt. For many years only 2 publications had described individuals with this condition, a 28-year-old male with genu valgum (knock knees) and an 18-year-old female with primary amenorrhea (Smith et al., 1994; Quaynor et al., 2013). Gonadotrophins, testosterone and estradiol were raised and genitalia were normal. A point mutation in exon 2 was found in the first patient and a loss of function mutation in the second patient. In 2017, a consanguineous family with a high incidence of pubertal failure was described (Bernard et al., 2017). Two females and one male had markedly raised estradiol and gonadotrophins concentrations with homozygous R394H missense mutation in helix 5 of the LBD of ERα. The SHBG concentrations were very low consistent with absence of a hepatic effect of estrogen. The females (25 years and 21 years) had primary amenorrhea; the male (18 years) had Tanner stage 1 development with cryptorchid right testis. Like androgen insensitivity the basis for estrogen resistance is a challenge and the detection of more cases is anticipated.

3.2.2.7.2.3 Gonadotrophin deficiency or resistance

Leydig cell hypoplasia is a familial disorder where the external genitalia are ambiguous or female and gonads are not palpable. There are no Müllerian structures. Serum gonadotrophins are elevated and testosterone concentrations are low. Mutations have been found that affect the extracellular domain of the LHr gene (Qiao et al., 2009; Qiao and Han, 2019; Xu et al., 2018; Hassan et al., 2020).

Deficiency of GnRH results from failure of embryonic migration of cells from the olfactory lobe to the forebrain leading to congenital hypogonadotrophic hypogonadism. This is associated with anosmia in most cases (**Kallmann's syndrome**) but there is tremendous phenotypic and genotypic heterogeneity (Millar et al., 2021). Several genes have been implicated in CHH through mutation analysis of ANOS1 (formerly KAL-1), SOX-10, IL17RD, TAC3 and TCC3R, FGFR1, FGF8, PROKR2 and CHD7; ANOS1 (Zhou et al., 2018; Amato et al.,

2019; Wen et al., 2019; Ying et al., 2020; Gach et al., 2020a, 2020b; Neocleous et al., 2020; Danda et al., 2021) (Table 3.2.2.3). The prevalence varies with population and size of group studied (Cangiano et al., 2021).

Other defective genes are coming to light and genetic testing is essential if effective treatment is to be realized (Rivero-Müller and Huhtaniemi, 2022). A mutation in the GnRH receptor gene has been described (Bouligand et al., 2009).

The **CHARGE** syndrome associated with congenital deficiency of GnRH manifests as failure of puberty and infertility. In a minority of patients, other pituitary defects are found. The presentation is variable from micropenis and cryptorchidism to the full features of **Coloboma, Heart defects, choanal Atresia, Retardation of growth and development, Gonadal defects, and Ear/hearing abnormalities.**

Mutations in CHD7 gene affect ATPase and nucleosome remodeling (Li et al., 2020; Balasubramanian and Crowley, 2017). The precise role of CHD7 is so far unknown.

In summary, the investigation of delayed puberty depends on gonadotrophin, cortisol and sex steroid measurements. Several conditions are associated with abnormalities of the skull and skeleton. After taking a medical history through an algorithm for boys and girls, the differential diagnosis of the most common causes of the problem can be resolved in a proportion of cases. Increasingly genetic tests are required.

3.2.2.7.3 Arrested puberty

This is usually a very significant presentation which requires prompt investigation. The history should indicate that puberty has started but failed to progress (allowing for variations in pubertal tempo-puberty may take 2–5 years to complete) or pubertal signs seem to have regressed and this may be associated with growth problems. A child with breast development and primary amenorrhea would fall into this presentation (Morawiecka-Pietrzak et al., 2021). There may be obvious significant ill health—acute or chronic or history of weight loss which may explain delay in pubertal progression but there are a number of potential significant pathologies that can cause pubertal arrest. The investigations of arrested puberty (Seppä et al., 2021) are indicated in Fig. 3.2.2.33 depending on the status of gonadotrophins. If concentrations of LH and FSH are low, an MRI scan of the brain and further biochemistry are required as the next steps to focus on pituitary abnormalities. Androgen secreting tumors may explain low gonadotrophins and evidence of androgen excess. If gonadotrophins are normal the uterus in a girl without menarche needs to be examined for structural defects and a karyotype should be performed.

Measurements of LH, FSH, (DHEAS, testosterone, androstenedione if signs of hyperandrogenism) or estradiol, SHBG and thyroid function are key to the resolution of this problem.

TABLE 3.2.2.3 Associated phenotypic features of selected gene mutations in hypogonadotrophic hypogonadism.

Affected genes	Associated features
ANOS (KAL1)	Cryptorchidism,small testes, renal agenesis (Costa-Barbosa et al., 2013)
SOX10	Waardenburg syndrome, sensorineural deafness; skin, hair and iris pigmentation (Pingault et al., 2013)
IL17RD	Hearing loss (Miraoui et al., 2013)
TAC3 and TAC3R	Microphallus and cryptorchidism (Gianetti et al., 2010)
FGFR1	Cleft lip or palate, dental agenesis, bimanual synkinesis, iris coloboma, unilateral hearing loss, digital malformations (Dodé et al., 2003)
FGF8	Hearing loss, high arched palate, cleft lip/palate, severe osteoporosis, microphallus, cryptorchidism, flat nasal bridge (Falardeau et al., 2008)
CHD7	Charge syndrome (Pitteloud et al., 2007)

Chromosomes should be reviewed for causes of Müllerian agenesis. Outflow tract problems should be investigated when menarche is absent. Prolactin should always be measured to exclude prolactinomas although the hormone concentration can be raised with pregnancy or prescription of antipsychotic, antidepressant, opioid and hypertensive drugs.

3.2.2.7.4 Hypogonadism in adult life

The presentation of hypogonadism in adults is of a decline in sexual function and general somatic symptoms, especially fatigue or weakness, and osteoporosis. The spectrum varies in severity and speed of onset, which relates to the underlying cause. In the complete forms, there is in the adult patient absence of secondary sexual hair, breast development and rudimentary nipples. As with hypogonadism with onset in puberty body size and exercise are issues in adults. Anorexia and binge eating can affect fertility partly corrected by getting to a normal body size although in some cases the damage is not reversible. The gut microbiota is going to be an area of new research. In the laboratory of the author, the effect of a vegetarian diet on bone density was studied. Osteopenia was found. The results are summarized in Chapter 1.8 (see Table 1.8.2 for fuller details).

Acquired hypogonadism in adult life may have widely differing effects on secondary sexual characteristics and performance. The main outcome of hypogonadism is loss of libido in either sex, often associated with erectile dysfunction in males although this is commonly not overtly associated with endocrine dysfunction. Depression and anxiety, often exacerbate the clinical situation and require sympathetic handling. Men with acquired hypogonadism may experience hot flushes and vasomotor disturbances like postmenopausal women that may be severe if the cause is of rapid onset. Hypogonadism may be discovered during investigation of infertility and may be asymptomatic. Fatigue and weakness can have many causes. Low T can lead to loss of skeletal muscle mass and strength, combined with tendency to central obesity. Anemia may also contribute to symptoms. Sleep patterns can be disrupted and sleep apnea may be a problem associated with obesity, particularly affecting increase in shirt collar size. Depression can reflect an awareness of body and facial hair loss, muscle bulk and height as well as loss of libido. The skin typically becomes sallow and thin. Osteoporosis can develop. Loss of bone mass correlates with the bio-available estradiol concentration (most of which is derived from peripheral aromatization of T) (Khosla et al., 2008). The management of hypogonadism in older men and women needs a discussion of hormone treatments. In extreme cases, adult men with undescended testes can present with abdominal pain and on surgery found to have female reproductive structures (streak gonads) characteristic of persistent Müllerian duct syndrome.

A serum prolactin greater than 120 ng/mL (normal range <20 ng/mL) is found when patients have a prolactin secreting tumor (microprolactinoma) and macroprolactinoma (>10 mm) when prolactin is greater than 250 ng/mL. Gonadotrophins are suppressed leading to amenorrhea and infertility in women and loss of libido, oligospermia in men. Pituitary imaging is needed to confirm a pituitary mass. Prolactinomas can be treated with a dopamine agonist that shrinks the tumor and reduces prolactin with restoration of fertility. Cabergoline has replaced bromocryptine treatment. In some cases, pituitary surgery is needed for macroprolactinoma.

Loss of menses in women before 40 years of age is called **premature ovarian insufficiency (POI)**, premature ovarian failure or premature menopause. Typical symptoms are amenorrhea or oligomenorrhea, hot flushes, night sweats, pain on intercourse. An earlier age for the menopause due to premature ovarian failure (POF) can be due to stress, eating disorders or obesity, excessive exercise, dieting and from medication for fibroids or endometriosis. POI can be due to chromosomal disorders such as Fragile X or Turner syndrome, and genetic disorders are coming to light. Environmental pollution is a new threat to health including POF.POI is diagnosed when FSH is above 25 IU/L (ESHRE et al., 2016). Females of reproductive age that present with lower abdominal pain should be offered a urine hCG test to exclude the possibility of pregnancy. Aberrations of both **FSHR** and **LHR** genes have

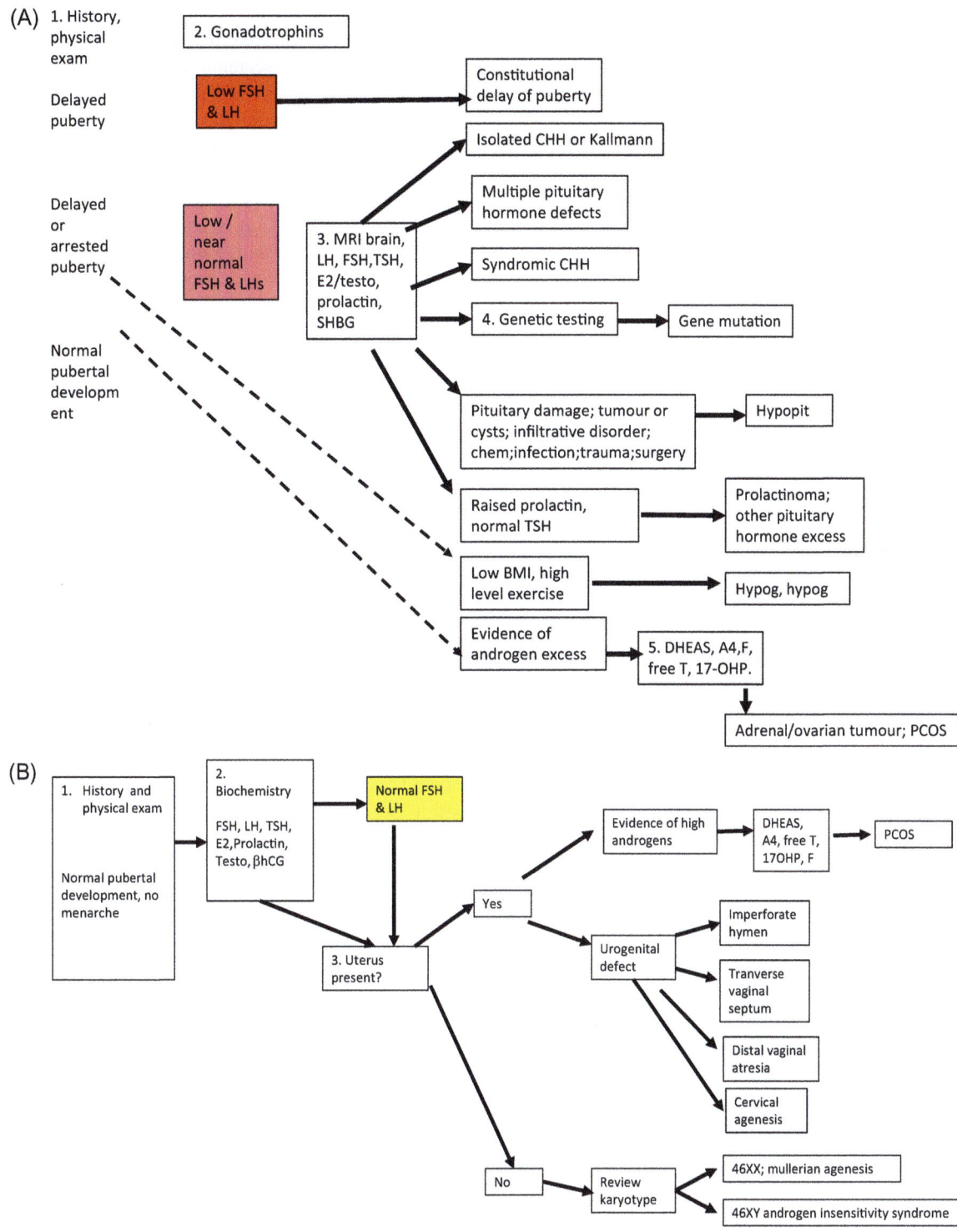

FIG. 3.2.2.33 Approach to the patient with primary amenorrhea and hypogonadotropic (A) or normogonadotropic (B) hypogonadism. *CBC*, complete blood count; *CHH*, congenital hypogonadotropic hypogonadism; *E2*, estradiol; *FSH*, follicle-stimulating hormone; *FT4*, free thyroxine; *β-hCG*, beta-human chorionic gonadotropin; *LH*, luteinizing hormone; *NCAH*, nonclassic congenital adrenal hyperplasia; *PCOS*, polycystic ovary syndrome; *TSH*, thyroid-stimulating hormone; *17-OHP*, 17-hydroxyprogesterone.

been identified in women with premature ovarian insufficiency (POI) (Aittomäki et al., 1995; Touraine et al., 1999; Rannikko et al., 2002; Gheorghiu, 2019). NR5A1 is critical for steroidogenesis and is expressed in both theca and granulosa cells; mutations in this gene have been identified in women with POI (Domenice et al., 2016; Camats et al., 2012). Both BMP15 and GDF9 are genes that code for members of the transforming growth factor-b superfamily and are both required for ovarian folliculogenesis. Mutations in a number of genes have been identified in women with POI (Wesevich et al., 2020; Ferrarini et al., 2021; Santos et al., 2019; Sanfins et al., 2018). INHA codes for inhibin alpha, which is a protein produced by granulosa cells required for folliculogenesis, mutations of which have been identified in POI in New Zealand (Shelling et al., 2000) but not confirmed in other populations for example Brazil (Christofolini et al., 2017). FOXL2 is a transcription factor that is a member of the forkhead box superfamily that plays a role in folliculogenesis. Mutations in FOXL2 have been found in women with abnormalities of the eyelids (Grzechocińska et al., 2019).

Disordered ovarian function presents with **primary amenorrhea**. In some patients, this can co-exist with signs of androgenization. Body size and degree of exercise are important clinical considerations. The initial endocrine assessment of amenorrhea includes measurements of prolactin, FSH, LH, 17-hydroxprogesterone, estradiol, TSH and thyroxine (Fig. 3.2.2.34). A pregnancy test should be conducted as a precaution.

The most common problem is *isolated gonadotrophin deficiency.* Occasionally, *Hyperthyroidism* is associated with amenorrhea. When the serum prolactin exceeds 1000 IU/L a prolactin secreting adenoma needs to be confirmed by appropriate scanning techniques. *Hyperprolactinemia* is a cause of arrested puberty. A **prolactinoma** or **pituitary hormone deficiency** may be uncovered and successful medical or surgical treatment will restore regular cycles. Plasma growth hormone concentration should be measured and if low along with low concentrations of FSH, LH, TSH would suggest a pituitary disorder. A cranial MRI scan may be needed. Failure to show growth of the uterus on ultrasound also provides valuable diagnostic information. Failure to have a menstrual bleed following a progestogen challenge test is usually a further indication of low estrogen production (poor follicular growth).

If FSH is high a karyotype will exclude 45XO (*Turner syndrome (45XO) and Turner mosaic*) (see earlier Fig. 3.2.2.23). Chromosomal analysis may suggest gonadal dysgenesis. The presence of ovarian antibodies should be excluded.

Inactivating mutations of the FSH receptor have been found in a few patients with hypergonadotrophic hypogonadism (Sassi et al., 2020; Liu et al., 2017; Aittomäki et al., 1995; Tapanainen et al., 1997). Low serum LH and FSH suggests *hypogonadotrophic hypogonadism* and these patients can respond well to pulsatile GnRH treatment.

Infertility is defined by WHO as inability to achieve pregnancy after at least 12 months of regular unprotected sexual intercourse. Male infertility constitutes up to 30% of cases. Clinicians will look at the patient medical and reproductive histories. Many health conditions can affect fertility, so a thorough evaluation of both partners is needed

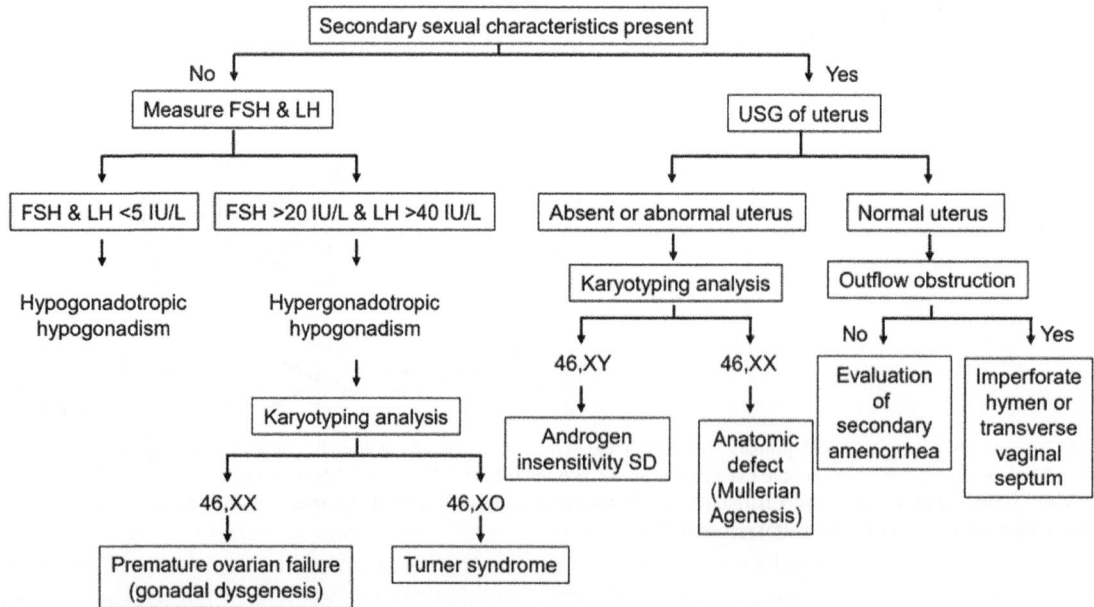

FIG. 3.2.2.34 Algorithm for the investigation of primary amenorrhea with hypergonadotropic hypogonadism. FSH, follicle-stimulating hormone; LH, luteinizing hormone; POI, premature ovarian insufficiency.

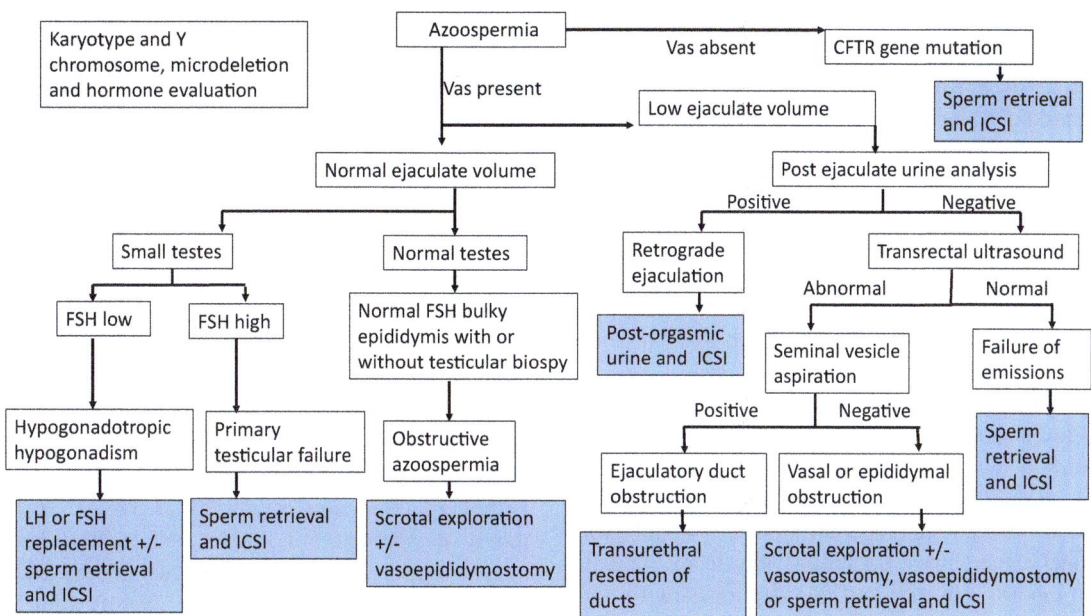

FIG. 3.2.2.35 Classification of azoospermia. *FSH*, follicle stimulating hormone; *ICSI*, intracytoplasmic sperm injection. *(Redrawn from Agarwal A, Baskaran S, Parekh N, Cho CL, Henkel R, Vij S, Arafa M, Panner Selvam MK, Shah R. Male infertility. Lancet 2021;397(10271):319–33. Fig.2. p 327.)*

to identify treatable or reversible lifestyle factors or medical conditions (Sharma et al., 2021).

Standard **semen analysis** comprises the analysis of macroscopic and microscopic parameters—volume, pH, and sperm count motility. Where the semen volume is less than 1.5 mL the patient is assessed for retrograde ejaculation, ejaculatory duct obstruction and epididymal obstruction (Fig. 3.2.2.35) and if the volume is normal then endocrine investigations proceed with small testes. Normal testes with normal FSH are investigated for obstructive azoospermia.

Sperm concentration should be more than 15 million per mL with at least 32% motile. The nomenclature for semen quality includes 14 groups depending on number and quality of the semen from aspermia (no semen) through azoospermia (no spermatozoa), Normozoospermia (number of motility equal or greater than lower reference limits) to oligozoospermia (total number less lower reference) and teratozoospermia (percentage morphologically normal spermatozoa less than lower reference). There are now technologies for some tests of semen quality to be conducted at home for sperm count and mobility (Gonzalez et al., 2021; Yu et al., 2018) where patients are reluctant to hospital tests. This is a useful step but may not provide enough detail to completely avoid hospital tests. An advanced sperm function test comprises the determination of reactive oxygen species (ROS), sperm DNA fragmentation, acrosome reaction and mitochondrial membrane potential using different techniques. The obesity, hypertension, diabetes, cystic fibrosis and metabolic syndrome should be considered. Lifestyle factors such as smoking, alcohol

intake and recreational drug use are relevant to male infertility. In the physical examination, size and shape of the phallus and testes should be assessed. The epididymes should be palpated and spermatic cords checked for varicoeles, LH, FSH, testosterone and prolactin. The testosterone concentration cut-off is 300 ng/dL in USA but 230 ng/dL (8 nmol/L) in UK and Europe. Increased FSH is seen in cases of defective spermatogenesis. The most common karyotype defect is Klinefelter's syndrome (47XXY). Y chromosome microdeletion is indicated in patients with azoospermia and sperm count is below one million per ml semen. Much research is needed in the genetics of azoospermia (Peña et al., 2020).

3.2.2.7.4.1 Causes of adult hypogonadism

The underlying causes of hypogonadism are divided into a primary group, where the defect lies at the gonadal level, and secondary causes due to deficient secretion or action of pituitary gonadotrophins. A number of conditions act at more than one site to produce hypogonadism. These are often chronic, multisystem diseases and rare chromosomal disorders. Adults with signs of adrenal insufficiency and hypogonadotrophic hypogonadism should be tested for DAX-1 mutations (Kyriakakis et al., 2017). The suggested mechanisms for infertility in men with undescended testes have included the lower abdomino-scrotal temperature gradient. Abnormalities of thermoregulation and circulation are cited as the cause of impaired spermatogenesis in occupations where long hours are seated (taxi and lorry drivers) and spinal injuries leading to paraplegia or quadriplegia (Naderi and Safarinejad, 2003).

Elevation of SHBG can reduce free T concentration without markedly changing total T, is often seen in aging men, due to obesity but also with liver disease, especially due to high alcohol intake. In men with hyperthyroidism, the total T may be elevated due to SHBG rise, yet free T is low (La Vignera et al., 2017). SHBG can be raised in men with prostate cancer given oral estrogens; germ cell tumors may secrete estrogens, estrogens can arise from disturbed steroid metabolism in men with cirrhosis. Abnormalities of the veins in the scrotum (varicocele) are often blamed for subfertility or mild feminization such as gynecomastia. Prescribed drugs should be reviewed since some drugs interfere with androgen synthesis include aminoglutethimide, etomidate and ketoconazole. Drugs that block the action at the androgen receptor include spironolactone, cyproterone acetate and flutamide. Physicians underestimate osteoporosis in men and this requires new research endeavor (Rinonapoli et al., 2021).

Microdeletions of the Y chromosome can be the cause of male infertility through loss of genes important in sexual development (spermatogenesis, primary spermatocyte development). Mutations of the cystic fibrosis transmembrane conductance regulator (CFTCR) can be sought in men with no palpable vas deferens and the ejaculate is primarily prostatic (Peña et al., 2020). GWAS is a poor choice for gene discovery in male infertility. The future of clinical genetic testing is likely to be with WGS (Witherspoon and Flannigan, 2021; Thirumavalavan et al., 2019).

Primary hypogonadism

Testicular damage can also arise from trauma or vascular causes such as testicular torsion, or in the attempted surgical correction of cryptorchidism and as an inadvertent consequence of inguinal hernia repair. Bacterial epididymitis and orchitis may result from infections of ascending urethra, bladder or prostate. Mumps is the commonest cause of orchitis (Davis et al., 2010; Choi et al., 2020). Patients with acquired immunodeficiency syndrome (AIDS) frequently develop hypogonadism and infertility. In tropical countries, leprosy is a common cause of testicular damage (Leal and Foss, 2009). Bacterial infections of urethra or bladder are usually sexually transmitted. Autoimmune involvement is seen in patients with forms of polyglandular syndrome and is especially associated with primary hypoadrenalism due to sex steroid cell antibodies directed at common epitopes in both glands (Perniola et al., 2021).

Pathogenic variants in **desert hedgehog homologue (DHH)** are a rare cause of autosomal recessive 46XY DSD causing partial or complete gonadal dysgenesis (Canto et al., 2004; Neocleous et al., 2019). Homozygous and compound heterozygotes have been described and classified as a ribosome-opathy (Ayers et al., 2019; Abramyan, 2019). Four novel mutations were found in patients with oligoasthenozoospermic, bilateral and unilateral undescended testes, and gonadal dysgenesis of a total 202 patients so mutations in the DHH gene impact reproduction with mild mutations affecting fertility (Mehta et al., 2021).

3.2.2.7.5 Testicular cancer

Men with testicular cancer can present with low testosterone at the time of diagnosis, or can experience a decrease in testosterone as a side effect of surgery or chemotherapy. Tumors in the testes are incredibly rare and diverse. Ultrasonography is the first line investigation for scrotal pathology (Hermann et al., 2022). Germ cell (predominantly postpubertal), nongerm cell (mainly prepubertal), sex cord and adrenal rests are encountered. New genetic information has broadened the classification based on molecular alterations and the interested reader should refer to recent reviews (Al-Obaidy and Idrees, 2021). Testosterone, beta-chorionic gonadotrophin (b-HCG), alpha fetoprotein (AFP), lactate dehydrogenase (LDH), and placental alkaline phosphatase (PALP) are biomarkers. Yolk sac tumors secreting AFP are the most type in boys up to 4 years of age. Choriocarcinoma secreting HCG is an aggressive tumor in adults. High serum concentrations of LDH are seen in patients with seminoma. PALP has little specificity for any given neoplasm but provides valuable information on response to chemotherapy or tumor recurrence.

3.2.2.7.5.1 Chromosomal disorders

Down's syndrome (trisomy 21) is usually associated with infertility not hypogonadism (Salzano et al., 2016). Tall men may have **Klinefelter's** syndrome which is typically caused by a 47 XXY karyotype. There are small firm testes with azoospermia (Fig. 3.2.2.36). Testicular biopsies show atrophic seminiferous tubules on histology and may recover occasional spermatozoa at least in younger adults (Rohayem et al., 2016; Masterson 3rd et al., 2020).

Patients are often tall and display female fat distribution and gynecomastia as well as eunuchoid proportion, but others are thin and narrow chested. Some may look quite normal apart from their small testes (Lanfranco et al., 2004). Klinefelter's syndrome is a recognized cause of leg ulcers in men. Men with the 47 XXY/46 XY mosaicism present with less severe features and much more severe with increasing supernumerary X chromosomes. Diagnosis of Klinefelters before puberty is rare apart from cases suspected on the basis of problems in psycho-social development or of long limbs. Puberty may fail to develop or undergo early arrest, normal T concentrations are sometimes seen in early adult life that then decline with

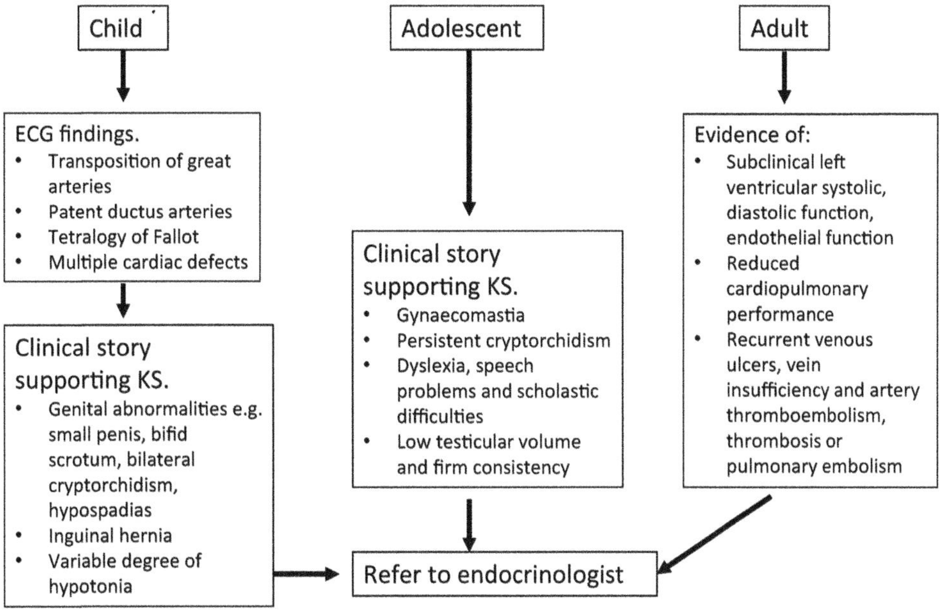

FIG. 3.2.2.36 Suggested flow-chart for patients with cardiovascular and other abnormalities associated with KS needing referral to cardiologists.

development of hypogonadism a decade later. The phenotype is also affected by the number of CAG triplet repeats in the androgen receptor gene.

3.2.2.7.5.2 Cancer treatment

The alkylating agents used in the treatment of cancers such as cyclophosphamide, busulfan and chlorambucil notoriously damage seminiferous tubules, causing infertility, rather than causing impaired T production. High doses even to children can lead to permanent infertility and azoospermia (van Dorp et al., 2018; Parekh et al., 2020; Shimazaki et al., 2020). Other cytotoxic drug types are much less harmful, usually leading, at most, only to transitory suppression of spermatogenesis (Panner Selvam et al., 2021).

Ionizing radiation is very toxic to the germinal epithelium, resembling the effects of alkylating drugs. Radioiodine, in the doses used for treatment of thyrotoxicosis, can lead to short lived depression of the sperm count. High doses used in patients with thyroid cancer are more subject to adverse effects, especially with repeated administration (Rosário et al., 2006).

3.2.2.7.5.3 Secondary hypogonadism

Hyperprolactinemia with **pituitary prolactinomas** may only be revealed during investigations for infertility or headaches. Large pituitary tumors can compress the optic chiasm and even cause cerebrospinal fluid leak (rhinorrhea). Prolactin concentrations can be raised by distortion of the pituitary stalk by a large pituitary tumor, that removes the usual inhibition of prolactin secretion exerted by dopamine secreted from the hypothalamus into the portal veins running down to the anterior pituitary gland. Gonadotrophin secretion may be inhibited because the frequency of the hypothalamic pulse generator for gonadotrophins is slowed. Hyperprolactinemia can suppress libido. In rare cases, a reduction in gonadotrophin secretion may result from earlier neurosurgery or pituitary radiotherapy.

In the male with infertility, gonadotrophin deficiency can be due to syndromes such as ***Kallmann's syndrome***. There can be loss of smell (anosmia) and cleft palate and mirror movements. A commercial "Sniffin sticks" test kit may be helpful although children may not be familiar with the sources of cloves, peppermint, turpentine and anise among the full test of 16 components (Liu et al., 2020a). A number of genetic mutations have been identified in a minority of cases. Failure of migration of GnRH cell bodies from the olfactory placode in the embryonic forebrain to the medial-basal hypothalamus, or failure to connect with the hypothalamo-pituitary portal blood vessels is the basis of the problem. Kallmann's syndrome is associated with anosmia. During the past 20 years, mutations in many genes in GnRH neuronal embryogenesis (ANOS1 {KAL1}, GnRHR, FGFR1, GPR54, PROK2, PROK2R, FGF8, CHD7, TAC3, TAC3R, and NELF) (Zhang et al., 2020; Ying et al., 2020; Jin et al., 2018; de Castro et al., 2017; Hardelin and Dodé, 2008; Jongmans et al., 2009) and pituitary development (Prop1, HESX1, POUF1, LHX3, LHX4, SOX2, and SOX3) have been identified (Cohen, 2012). Treatment of Kallmann aims to improve secondary sex characteristics and bone health (Berges-Raso et al., 2017).

A craniopharyngioma is a frequent cause of hypopituitarism in children but can be present at any time to old age. A variety of hypothalamic tumors occur, often malignant, and including metastases from extra-cranial primaries (Alter et al., 2021; Greuter et al., 2021; Karavitaki et al., 2005). Head injury is capable of producing hypopituitarism, by damage to the pituitary, hypothalamus and/or pituitary stalk leading to a combined deficiency of anterior and posterior lobe hormones and raised concentrations of prolactin (Darzy and Shalet, 2009; Behan et al., 2008).

Hemochromatosis and **β-thalassemia** leads to iron deposition in the testes and also the liver and anterior pituitary, where there is a preference for the gonadotroph cells, leading to a variably picture of primary and secondary hypogonadism (Gabrielsen et al., 2018). Patients with thalassemia require multiple blood transfusions, and for those with hemochromatosis, the treatment is venesection to correct hemoglobin and hematocrit values.

Impaired spermatogenesis and testicular damage often leads to infertility in **chronic renal failure**, with abnormalities at all levels of the hypothalamo-pituitary-gonadal axis and prolactin (Grossman et al., 2016; Gómez et al., 1980). Sexual steroid concentrations are low in patients on dialysis and increase after renal transplantation (Reinhardt et al., 2018; Grossman et al., 2016).

In **type 2 diabetes** and males with the metabolic syndrome, erectile dysfunction is common and testosterone deficiency is associated with increased cardiovascular risk, which is reduced with testosterone treatment (Alwani et al., 2021; Gopal et al., 2010). Moderate obesity (BMI 30–35) is associated with lower serum total T, resulting from decreased serum SHBG, morbidly obese men (BMI >40) also tend to have decreased free T as a consequence of diminished gonadotrophin secretion (Lima et al., 2000).

3.2.2.8 Hypogonadism in pregnancy

3.2.2.8.1 Sulfatase deficiency pregnancy

The condition is suspected prenatally when low estriol is found in triple screening (see Chapter Hypergonadism Fig. 3.2.1.52) (Marcos et al., 2009). Low estrogens in pregnancy are also the result of sulfatase deficiency in the fetus and steroids sulfates in urine of the mother are raised (see earlier Table 3.2.1.5). The profile is very similar to the pattern of steroids in urine of a newborn infant.

The STS gene is mapped to the short arm of the X-chromosome (band Xp22.3) and deletion of the gene is the most common basis of the condition. A few patients can have a microdeletion/contiguous gene syndrome. The rate of spontaneous labor and delivery are reduced as the enzyme deficiency is linked with failure of cervical dilatation. Should further pregnancy be contemplated then a new pregnancy can be monitored with urine estriol analysis if the fetus is male and counseled of increased risk of miscarriage or need for caesarean section delivery.

3.2.2.8.2 Smith-Lemli-Opitz syndrome

Estrogen production is affected when the fetus has Smith-Lemli-Opitz syndrome (see earlier for defect of 7-dehydrocholesterole reductase deficiency). The steroids 5β-pregn-7(or 8)-ene-3α,17α,20α-triol, 5β-pregn-7(or 8)-ene-3α,16α,20α-triol, and 5α-pregn-7(or 8)-ene-3,16α,20α-triol have been identified in maternal urine from mid-gestation (Shackleton et al., 2001; Shackleton et al., 1999a,b). The steroid 7-dehydro-equilinin was identified by its mass spectrum in GC-MS analysis of trimethylsily ether derivatives with mass 2 units less than estriol (Fig. 3.2.2.37). Diagnostic steroids ratios were established.

3.2.2.9 The menopause

The female menopause is a period of life when the ovaries are depleted of oocytes and the cyclical activity of gonadotrophins, peptides and steroids is lost. Age at the menopause reflects the complex interactions of health and socio-economic factors including for example ethnicity, diet, education, oral contraceptive use, weight, employment, exposure to endocrine disrupting chemicals, alcohol use, smoking and physical activity. Some of the changes at the menopause are attributed to estrogen and or progesterone depletion.

At birth the ovaries can have one million oocytes that fall off rapidly to around 200,000 by the end of puberty. At the onset of menstruation (menarche) the pattern of menstrual cycles can be irregular with frequency up to 70 days. There may be short menstrual cycles, sometimes interspersed with normal cycles. Only about 350 follicles actually ovulate during the reproductive years. Further follicle loss is largely by apoptosis. A regular cycle frequency is around 30 days between the ages of 20 and 45 years. By the last cycle, there may only be about 1000 follicles in the ovaries.

In perimenopausal women, an increase in the cyclical early follicular concentrations of FSH is a consistent finding. Concentrations of estrogens and inhibins are lower than seen in earlier years which correlate with the depletion of follicles in the ovaries and the resistance of the ovaries to gonadotrophins with a reduction in estrogen and inhibin production. The menopause transition goes through four phases (Fig. 3.2.2.38) from normal ovarian activity (called D), through a stage with prolonged follicular phase and no luteal activity (B) (Hale et al., 2014). There is a stage with normal follicular activity and insufficient luteal phase (C), and finally the menopause with low estrogens and progesterone (A).

Hormone levels were monitored across the menopause transition. Data were gathered from multiple, annual blood

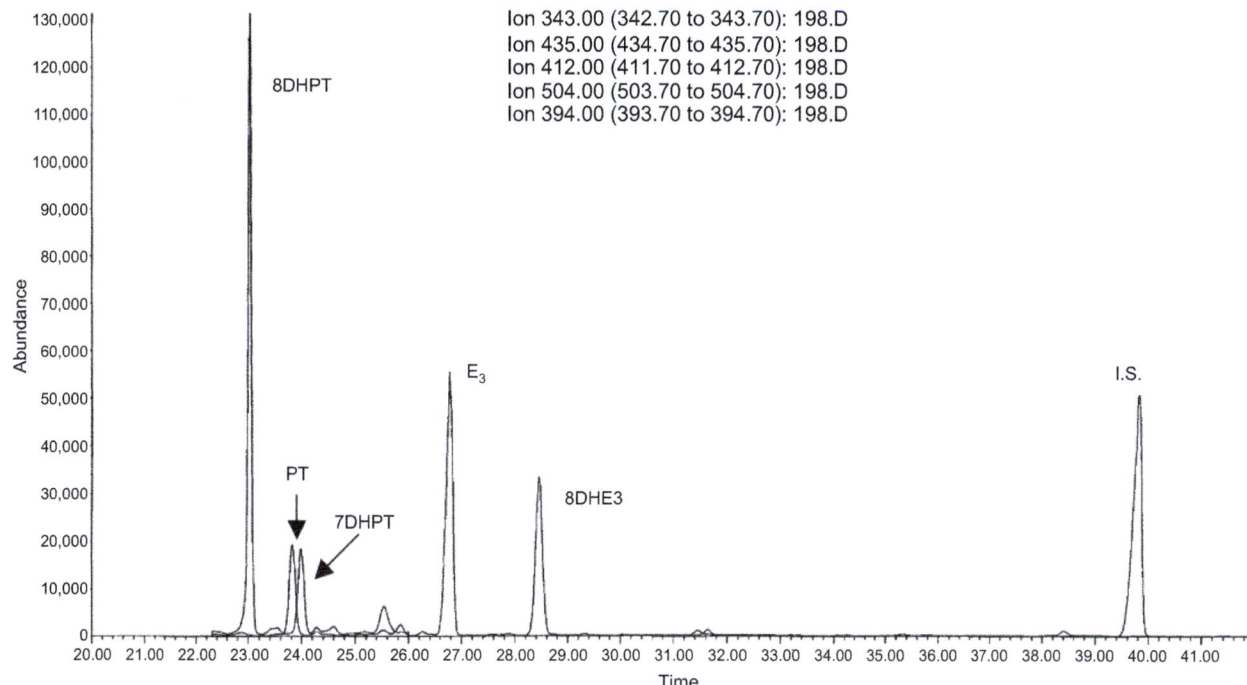

FIG. 3.2.2.37 GC/MS-SIM diagnosis of SLOS-positive fetus at 18 weeks of gestation; identification of the selected ions from: 7 and 8DHPT m/z 343; 8DHE3 m/z 412; PT m/z 435; E3 m/z 504; stigmasterol (I.S.) m/z 394. *DHPT*, dehydropregnanetriol; *GC*, gas chromatography; *MS*, mass spectrometry; *PT*, pregnanetriol; *SIM*, selecting ion monitoring. *(From Jezela-Stanek A, Małunowicz EM, Ciara E, Popowska E, Goryluk-Kozakiewicz B, Spodar K, Czerwiecka M, Jezuita J, Nowaczyk MJ, Krajewska-Walasek M. Maternal urinary steroid profiles in prenatal diagnosis of Smith-Lemli-Opitz syndrome: first patient series comparing biochemical and molecular studies. Clin Genet. 2006;69(1):77–85. Fig 4 p 83.)*

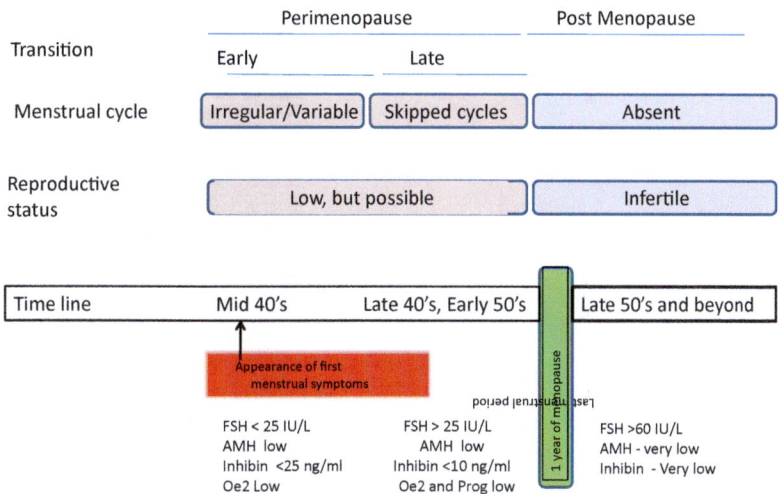

FIG. 3.2.2.38 Staging menopause. *(Hale GE, Robertson DM, Burger HG. The perimenopausal woman: endocrinology and management. J Steroid Biochem Mol Biol 2014;142:121–31.)*

tests in 13 women for between 4 and 9 years up to the menopause (Santoro and Randolph Jr., 2011). Blood samples were taken three times a week for a 4-week period in each year. B cycles that became longer were delayed ovulatory cycles or anovulatory cycles. As the process moved on, FSH concentrations were higher and inhibins lower than

in the D follicular phases. Intervals of 24 up to 38 days separate menses in early phase and the interval can be greater than 60 days in later phase. FSH concentrations above 25 IU/L are seen in the late transitional period. Once menstruation has ceased FSH is above 60 IU/L with LH raised (10–45 IU/L), but estradiol (<200 pmol/L), progesterone and inhibins (B < 25; A < 10 ng/L) concentrations are low. Hormone measurements other than FSH during the perimenopause are generally considered to be of little diagnostic value. The transition may take 4 or more years.

In a separate study, blood samples were collected pre- and postmenopause (Rothman et al., 2011). Estradiol testosterone and DHT were significantly lower and less significantly lower for estrone and free testosterone (Fig. 3.2.2.39). Estrone was much higher than estradiol after the menopause. FSH was significantly higher after the menopause.

Hot flushes, night sweats and vaginal dryness are common symptoms of the menopause from estrogen withdrawal. A hot flush is a sensation of heat of unknown cause in the upper body, face and neck lasting for 3–5 min accompanied by higher heart rate and peripheral blood flow with a rise in skin temperature. With estrogen loss there is a reduction in bone density and an increase in cardiovascular risk, although the latter may be related more to concentrations of high-density lipoprotein cholesterol. The current guidelines from the National Institute

FIG. 3.2.2.39 (A) Estradiol (E2), estrone E1), SHBG, FSH, testosterone (T) and dihydrotestosterone (DHT) during early follicular phase (EFP) in premenopausal women compared with postmenopausal women. E2 and E1 were measured by LC-MS/MS (**P < .0001). SHBG and FSH levels were measured by fluoroimmunometric assay (*P < .05 and **P < .0001). (B) T and DHT levels during EFP in premenopausal women compared with postmenopausal women. T and DHT were measured by LC-MS/MS (**P < .001). (C) Free T during EFP in premenopausal women compared with postmenopausal women. Free T was calculated as T/SHBG x 100 (*P < .05). *(Rothman MS, Carlson NE, Xu M, Wang C, Swerdloff R, Lee P, Goh VH, Ridgway EC, Wierman ME. Reexamination of testosterone, dihydrotestosterone, estradiol and estrone levels across the menstrual cycle and in postmenopausal women measured by liquid chromatography-tandem mass spectrometry. Steroids 2011;76 (1–2):177–82. Fig. 2 p. 180.)*

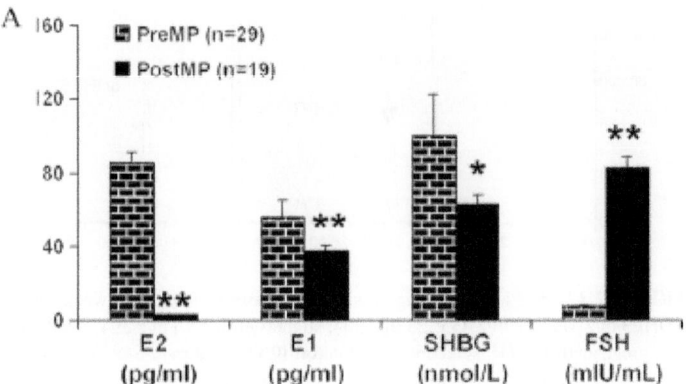

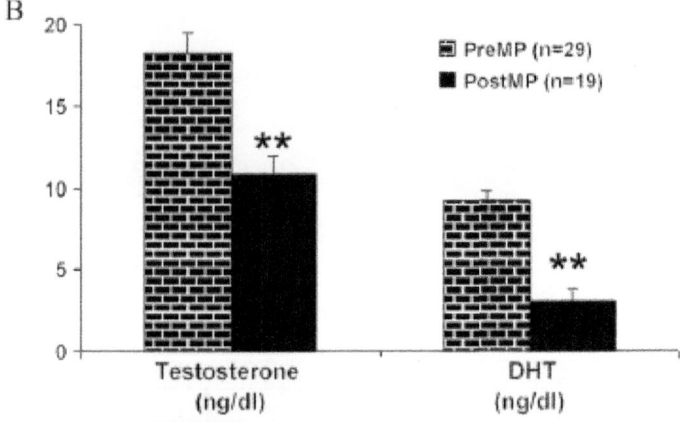

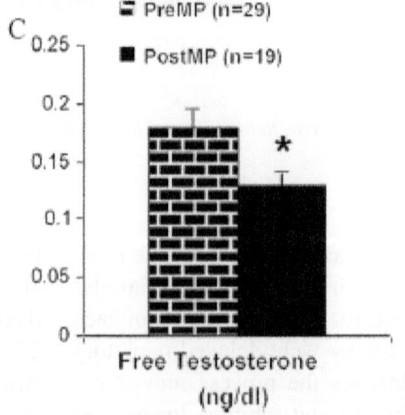

Box 3.2.2.1 Extract of NICE guideline for menopause.

1.2. Diagnosis of perimenopause and menopause

    1.2.1. Diagnose the following without laboratory tests in otherwise healthy women aged over 45 years with menstrual symptoms
- perimenopause based on vasomotor symptoms and irregular periods,
- menopause in women who have not had a period for at least 12 months and are not using hormonal contraception, and
- menopause based on symptoms in women without a uterus.

    1.2.2. Take into account that it can be difficult to diagnose menopause in women who are taking hormonal treatment for example for the treatment of heavy periods

    1.2.3. Do not use the following laboratory and imaging tests to diagnose perimenopause or menopause in women age over 45 years
- anti-Müllerian hormone,
- inhibin A,
- inhibin B, and
- estradiol.

    1.2.4. Do not use a serum follicle-stimulating hormone (FSH) test to diagnose menopause in women using estrogen and progestogen contraception or high dose progestogen

    1.2.5. Consider using a FSH test to diagnose menopause only
- in women aged 40 to 45 years with menopausal symptoms including a change in their menstrual cycle
- in women aged under 40 years in whom menopause is suspected

---

for Clinical Excellence (NICE) for tests in the menopause are shown in Box 3.2.2.1.

3.2.2.9.1 Treatment of hypogonadism

Hormone replacement therapy is desirable in many cases of hypogonadism mainly to prevent cardiovascular and skeletal disease. For women with POI, HRT alleviates vasomotor and genitourinary symptoms (Ishizuka, 2021).

3.2.2.10 Summary

Hypogonadism is heterogeneous group of congenital conditions with chromosomal, gonadal and anatomical issues. Gonadal development used to be considered the result of testosterone and anti-Müllerian hormone actions from the Y chromosome. A number of other genetic networks acting on the bipotential are now being investigated in normal development of ovary or testis (Gomes et al., 2020). Hypogonadism can be a feature in any of the 3 groups—sex chromosomal DSD, 46XY DSD and 46XX DSD. These include defects of adrenal steroid biosynthesis where diagnosis is usually made in the newborn period because of the side-effects of dominant steroids and loss of cortisol and/or aldosterone, and defects in gonadal steroids synthesis which are not usually clinically suspected until puberty. Each condition has a spectrum of genital phenotype depending on the underlying variant and other modifying factors including the chromosomes and steroid receptors. An early biochemical test (gonadotrophins, sex steroids and biomarkers [17-hydroxyprogesterone, hCG, prolactin and steroid panel in

plasma or urine]) or genetic finding by gene panel testing can reduce subjecting patients to uninformative clinical tests. The clinician will have to make important decisions on when and how to instigate genetic tests.

Genetic testing is becoming increasingly important. A panel-based approach is effective for known genes but whole exome or genome sequencing is better for gene discovery (Mazen et al., 2021). Specialist centers have reviewed their long-term experience in primary adrenal insufficiency (Buonocore et al., 2021; Guran et al., 2016) DSDs (Ata et al., 2021) and pubertal delay (Saengkaew et al., 2021). Gene testing already needs around 200 genes to be included in a panel test and this will increase further as mechanisms for sex determination and kidney development, smell, alternative biosynthetic pathways including the 11-oxygenated androgens are elucidated. Some further classification may be needed as the genetic basis becomes clearer. Several conditions can be suspected depending on the age of the patient, clinical and biochemical grounds. Further genetic testing of patients with renal (Wilms tumor) and cardiac problems is likely to reveal new mutated genes. Further studies will improve the diagnosis of hypogonadism by investigating the contribution of epigenetics and other routes. The many divergences in results may be associated with the different ethnicities of the population studied. The clinician may be faced with interpretation of the findings which may be new proteins or mechanisms. The multidisciplinary team discussions will become even more important.

Artificial intelligence is being used for diagnosis of genetic conditions based on facial recognition (Porras et al., 2021). A diagnosis of a condition is important in determining treatment which can vary from hormone replacement to gonadectomy. A risk for malignancy and cardiovascular disease must be made and acted upon. A good relationship between the patient and clinicians in several disciplines must be established for the medical and psychological needs of the patient. Gone are the days when "testes were removed because they were bad ovaries" without telling the patient they were 46XY males, as was the case with some of the patient histories encountered by the author. In many conditions, there are now good expectations for fertility. Uterine transplants and cryopreservation are options available today or soon. Many conditions of hypogonadism present with facial and skeletal abnormalities and anosmia. Some patients will need surgery to correct, genital, renal, pituitary/hypothalamus, and facial abnormalities. Research is in progress for regenerative and more curative strategies than corticoid replacement therapy (Bornstein et al., 2020; Ruiz-Babot et al., 2018).

References

Abramyan J. Hedgehog signaling and embryonic craniofacial disorders. J Dev Biol 2019;7(2):9.

Achermann JC, Ito M, Ito M, Hindmarsh PC, Jameson JL. A mutation in the gene encoding steroidogenic factor-1 causes XY sex reversal and adrenal failure in humans. Nat Genet 1999;22(2):125–6.

Acién P, Acién M. Disorders of sex development: classification, review, and impact on fertility. J Clin Med 2020;9(11):3555.

Adachi M, Takayanagi R, Tomura A, Imasaki K, Kato S, Goto K, et al. Androgen-insensitivity syndrome as a possible coactivator disease. N Engl J Med 2000;343(12):856–62. Erratum in: N Engl J Med 2001;344(9):696.

Ahmed SF, Cheng A, Dovey L, Hawkins JR, Martin H, Rowland J, et al. Phenotypic features, androgen receptor binding, and mutational analysis in 278 clinical cases reported as androgen insensitivity syndrome. J Clin Endocrinol Metab 2000;85(2):658–65.

Aittomäki K, Lucena JL, Pakarinen P, Sistonen P, Tapanainen J, Gromoll J, et al. Mutation in the follicle-stimulating hormone receptor gene causes hereditary hypergonadotropic ovarian failure. Cell 1995;82(6):959–68.

Akgun S, Ertel NH, Imperato-McGinley J, Sayli BS, Shackleton C. Familial male pseudohermaphroditism due to 5-alpha-reductase deficiency in a Turkish village. Am J Med 1986;81(2):267–74.

Al Alawi AM, Nordenström A, Falhammar H. Clinical perspectives in congenital adrenal hyperplasia due to 3β-hydroxysteroid dehydrogenase type 2 deficiency. Endocrine 2019;63(3):407–21.

Al-Obaidy KI, Idrees MT. Testicular tumors: a contemporary update on morphologic, immunohistochemical and molecular features. Adv Anat Pathol 2021;28(4):258–75.

Allali S, Muller JB, Brauner R, Lourenço D, Boudjenah R, Karageorgou V, et al. Mutation analysis of NR5A1 encoding steroidogenic factor 1 in 77 patients with 46, XY disorders of sex development (DSD) including hypospadias. PLoS ONE 2011;6(10), e24117.

Alswailem M, Alsagheir A, Abbas BB, Alzahrani O, Alzahrani AS. Molecular genetics of disorders of sex development in a highly consanguineous population. J Steroid Biochem Mol Biol 2021;208:105736.

Alter CA, Shekdar KV, Cohen LE. Pituitary tumors in children. Adv Pediatr 2021;68:211–25.

Alwani M, Yassin A, Talib R, Al-Qudimat A, Aboumarzouk O, Al-Zoubi RM, et al. Cardiovascular disease, hypogonadism and erectile dysfunction: early detection, prevention and the positive effects of long-term testosterone treatment: prospective observational, real-life data. Vasc Health Risk Manag 2021;17:497–508.

Amato LGL, Montenegro LR, Lerario AM, Jorge AAL, Guerra Junior G, Schnoll C, et al. New genetic findings in a large cohort of congenital hypogonadotropic hypogonadism. Eur J Endocrinol 2019;181(2):103–19.

Anon. Screening for fetal chromosomal abnormalities: ACOG Practice Bulletin Summary, Number 226. Obstet Gynecol 2020;136(4):859–67.

Arya S, Tiwari A, Lila AR, Sarathi V, Bhandare VV, Kumbhar BV, et al. Homozygous p.Val89Leu plays an important pathogenic role in 5α-reductase type 2 deficiency patients with homozygous p.Arg246Gln in SRD5A2. Eur J Endocrinol 2020;183(3):275–84.

Ata A, Özen S, Onay H, Uzun S, Gökşen D, Özkınay F, et al. A large cohort of disorders of sex development and their genetic characteristics: 6 novel mutations in known genes. Eur J Med Genet 2021;64(3):104154.

Auchus RJ. Steroid 17-hydroxylase and 17,20-lyase deficiencies, genetic and pharmacologic. J Steroid Biochem Mol Biol 2017;165(Pt. A):71–8.

Auchus RJ, Miller WL. Defects in androgen biosynthesis causing 46,XY disorders of sexual development. Semin Reprod Med 2012;30(5):417–26.

Ayers K, van den Bergen J, Robevska G, Listyasari N, Raza J, Atta I, et al. Functional analysis of novel desert hedgehog gene variants improves the clinical interpretation of genomic data and provides a more accurate diagnosis for patients with 46,XY differences of sex development. J Med Genet 2019;56(7):434–43.

Baird PN, Santos A, Groves N, Jadresic L, Cowell JK. Constitutional mutations in the WT1 gene in patients with Denys-Drash syndrome. Hum Mol Genet 1992;1(5):301–5.

Balasubramanian R, Crowley Jr WF. Reproductive endocrine phenotypes relating to CHD7 mutations in humans. Am J Med Genet C Semin Med Genet 2017;175(4):507–15.

Baquedano MS, Guercio G, Costanzo M, Marino R, Rivarola MA, Belgorosky A. Mutation of HSD3B2 gene and fate of dehydroepiandrosterone. Vitam Horm 2018;108:75–123.

Barbaro M, Oscarson M, Almskog I, Hamberg H, Wedell A. Complete androgen insensitivity without Wolffian duct development: the AR-A form of the androgen receptor is not sufficient for male genital development. Clin Endocrinol (Oxf) 2007;66(6):822–6.

Barros BA, Oliveira LR, Surur CRC, Barros-Filho AA, Maciel-Guerra AT, Guerra-Junior G. Complete androgen insensitivity syndrome and risk of gonadal malignancy: systematic review. Ann Pediatr Endocrinol Metab 2021;26(1):19–23.

Bashamboo A, Donohoue PA, Vilain E, Rojo S, Calvel P, Seneviratne SN, et al. A recurrent p.Arg92Trp variant in steroidogenic factor-1 (NR5A1) can act as a molecular switch in human sex development. Hum Mol Genet 2016;25(23):5286. Erratum for: Hum Mol Genet. 2016;25(16):3446–3453.

Batista RL, Mendonca BB. Integrative and analytical review of the 5-alpha-reductase type 2 deficiency worldwide. Appl Clin Genet 2020;13:83–96.

Behan LA, Phillips J, Thompson CJ, Agha A. Neuroendocrine disorders after traumatic brain injury. J Neurol Neurosurg Psychiatry 2008;79(7):753–9.

Benkert AR, Young M, Robinson D, Hendrickson C, Lee PA, Strauss KA. Severe salt-losing 3β-hydroxysteroid dehydrogenase deficiency: treatment and outcomes of HSD3B2 c.35G>A homozygotes. J Clin Endocrinol Metab 2015;100(8):E1105–15.

Berges-Raso I, Giménez-Palop O, Gabau E, Capel I, Caixàs A, Rigla M. Kallmann syndrome and ichthyosis: a case of contiguous gene deletion syndrome. Endocrinol Diabetes Metab Case Rep 2017; 2017, EDM170083.

Berland S, Haukanes BI, Juliusson PB, Houge G. Deep exploration of a *CDKN1C* mutation causing a mixture of Beckwith-Wiedemann and IMAGe syndromes revealed a novel transcript associated with developmental delay. J Med Genet 2020;59:155–64. jmedgenet-2020-107401.

Bernard V, Kherra S, Francou B, Fagart J, Viengchareun S, Guéchot J, et al. Familial multiplicity of estrogen insensitivity associated with a loss-of-function ESR1 mutation. J Clin Endocrinol Metab 2017;102(1):93–9.

Berra M, Williams EL, Muroni B, Creighton SM, Honour JW, Rumsby G, et al. Recognition of 5α-reductase-2 deficiency in an adult female 46XY DSD clinic. Eur J Endocrinol 2011;164(6):1019–25.

Bertola DR, Castro MAA, Yamamoto GL, Honjo RS, Ceroni JR, Buscarilli MM, et al. Phenotype-genotype analysis of 242 individuals with RASopathies: 18-year experience of a tertiary center in Brazil. Am J Med Genet C Semin Med Genet 2020;184(4):896–911.

Bhutani N, Kajal P, Sharma U. Many faces of wilms tumor: recent advances and future directions. Ann Med Surg (Lond) 2021;64, 102202.

Binder G, Ziegler J, Schweizer R, Habhab W, Haack TB, Heinrich T, et al. Novel mutation points to a hot spot in CDKN1C causing Silver-Russell syndrome. Clin Epigenetics 2020;12(1):152.

Boehmer AL, Brinkmann O, Brüggenwirth H, van Assendelft C, Otten BJ, Verleun-Mooijman MC, et al. Genotype versus phenotype in families with androgen insensitivity syndrome. J Clin Endocrinol Metab 2001;86(9):4151–60.

Bornstein SR, Malyukov M, Heller C, Ziegler CG, Ruiz-Babot G, Schedl A, et al. New horizons: novel adrenal regenerative therapies. J Clin Endocrinol Metab 2020;105(9):3103–7.

Bouligand J, Ghervan C, Tello JA, Brailly-Tabard S, Salenave S, Chanson P, et al. Isolated familial hypogonadotropic hypogonadism and a GNRH1 mutation. N Engl J Med 2009;360(26):2742–8.

Bouvattier C, Carel JC, Lecointre C, David A, Sultan C, Bertrand AM, et al. Postnatal changes of T, LH, and FSH in 46,XY infants with mutations in the AR gene. J Clin Endocrinol Metab 2002;87(1):29–32.

Brauner R, Neve M, Allali S, Trivin C, Lottmann H, Bashamboo A, et al. Clinical, biological and genetic analysis of anorchia in 26 boys. PLoS One 2011;6(8), e23292.

Brunello FG, Rey RA. AMH and AMHR2 involvement in congenital disorders of sex development. Sex Dev 2022;16(2–3):138–46. https://doi.org/10.1159/000518273. Epub ahead of print 34515230.

Bulun SE. Aromatase and estrogen receptor α deficiency. Fertil Steril 2014;101(2):323–9.

Bumbulienė Ž, Varytė G, Geimanaitė L. Dysgerminoma in a prepubertal girl with complete 46XY gonadal dysgenesis: case report and review of the literature. J Pediatr Adolesc Gynecol 2020;33(5):599–601.

Buonocore F, Clifford-Mobley O, King TFJ, Striglioni N, Man E, Suntharalingham JP, et al. Next-generation sequencing reveals novel genetic variants (SRY, DMRT1, NR5A1, DHH, DHX37) in adults with 46, XY DSD. J Endocr Soc 2019;3(12):2341–60.

Buonocore F, Maharaj A, Qamar Y, Koehler K, Suntharalingham JP, Chan LF, et al. Genetic analysis of pediatric primary adrenal insufficiency of unknown etiology: 25 years' experience in the UK. J Endocr Soc 2021;5(8), bvab086.

Burkhard FZ, Parween S, Udhane SS, Flück CE, Pandey AV. P450 Oxidoreductase deficiency: analysis of mutations and polymorphisms. J Steroid Biochem Mol Biol 2017;165(Pt. A):38–50.

Burns AD, Taylor NF, Taylor DR, Bhake RC, Rahman F. A curious case of primary amenorrhea. Clin Chem 2020;66(9):1150–4.

Cabrera-Salcedo C, Kumar P, Hwa V, Dauber A. IMAGe and related undergrowth syndromes: the complex Spectrum of gain-of-function CDKN1C mutations. Pediatr Endocrinol Rev 2017;14(3):289–97.

Camats N, Pandey AV, Fernández-Cancio M, Andaluz P, Janner M, Torán N, et al. Ten novel mutations in the NR5A1 gene cause disordered sex development in 46,XY and ovarian insufficiency in 46, XX individuals. J Clin Endocrinol Metab 2012;97(7):E1294–306.

Camats N, Fernández-Cancio M, Audí L, Schaller A, Flück CE. Broad phenotypes in heterozygous NR5A1 46,XY patients with a disorder of sex development: an oligogenic origin? Eur J Hum Genet 2018;26(9): 1329–38.

Cangiano B, Swee DS, Quinton R, Bonomi M. Genetics of congenital hypogonadotropic hypogonadism: peculiarities and phenotype of an oligogenic disease. Hum Genet 2021;140(1):77–111.

Canto P, Söderlund D, Reyes E, Méndez JP. Mutations in the desert hedgehog (DHH) gene in patients with 46,XY complete pure gonadal dysgenesis. J Clin Endocrinol Metab 2004;89(9):4480–3.

Cheng Y, Chen J, Zhou X, Yang J, Ji Y, Xu C. Characteristics and possible mechanisms of 46, XY differences in sex development caused by novel compound variants in NR5A1 and MAP3K1. Orphanet J Rare Dis 2021;16(1):268.

Cho CY, Tsai WY, Lee CT, Liu SY, Huang SY, Chien YH, et al. Clinical and molecular features of idiopathic hypogonadotropic hypogonadism in Taiwan: a single center experience. J Formos Med Assoc 2021;121:218–26.

Choi HI, Yang DM, Kim HC, Kim SW, Jeong HS, Moon SK, et al. Testicular atrophy after mumps orchitis: ultrasonographic findings. Ultrasonography 2020;39(3):266–71.

Christofolini DM, Cordts EB, Santos-Pinheiro F, Kayaki EA, Dornas MCF, Santos MC, et al. How polymorphic markers contribute to genetic diseases in different populations? The study of inhibin A for premature ovarian insufficiency. Einstein (Sao Paulo) 2017;15(3):269–72.

Cohen LE. Genetic disorders of the pituitary. Curr Opin Endocrinol Diabetes Obes 2012;19(1):33–9.

Corbitt H, Morris SA, Gravholt CH, Mortensen KH, Tippner-Hedges R, Silberbach M, et al. TIMP3 and TIMP1 are risk genes for bicuspid aortic valve and aortopathy in Turner syndrome. PLoS Genet 2018;14(10), e1007692.

Corsten-Janssen N, Bouman K, Diphoorn JCD, Scheper AJ, Kinds R, El Mecky J, et al. A prospective study on rapid exome sequencing as a diagnostic test for multiple congenital anomalies on fetal ultrasound. Prenat Diagn 2020;40(10):1300–9.

Costa-Barbosa FA, Balasubramanian R, Keefe KW, Shaw ND, Al-Tassan N, Plummer L, et al. Prioritizing genetic testing in patients with Kallmann syndrome using clinical phenotypes. J Clin Endocrinol Metab 2013;98(5):E943–53.

Costagliola G, Cosci O di Coscio M, Masini B, Baldinotti F, Caligo MA, Tyutyusheva N, et al. Disorders of sexual development with XY karyotype and female phenotype: clinical findings and genetic background in a cohort from a single centre. J Endocrinol Invest 2021;44 (1):145–51.

Coutant R, Mallet D, Lahlou N, Bouhours-Nouet N, Guichet A, Coupris L, et al. Heterozygous mutation of steroidogenic factor-1 in 46,XY subjects may mimic partial androgen insensitivity syndrome. J Clin Endocrinol Metab 2007;92(8):2868–73.

Csillag B, Ilencikova D, Meissl M, Webersinke G, Laccone F, Narumi S, et al. Somatic mosaic monosomy 7 and UPD7q in a child with MIRAGE syndrome caused by a novel SAMD9 mutation. Pediatr Blood Cancer 2019;66(4), e27589.

da Silva TE, Gomes NL, Lerário AM, Keegan CE, Nishi MY, Carvalho FM, et al. Genetic evidence of the association of DEAH-box helicase 37 defects with 46,XY gonadal dysgenesis spectrum. J Clin Endocrinol Metab 2019;104(12):5923–34.

Danda VSR, Paidipelly SR, Verepula M, Lodha P, Thaduri KR, Konda C, et al. Exploring the genetic diversity of isolated hypogonadotropic hypogonadism and its phenotypic spectrum: a case series. J Reprod Infertil 2021;22(1):38–46.

Darzy KH, Shalet SM. Hypopituitarism following radiotherapy. Pituitary 2009;12(1):40–50.

Davis NF, McGuire BB, Mahon JA, Smyth AE, O'Malley KJ, Fitzpatrick JM. The increasing incidence of mumps orchitis: a comprehensive review. BJU Int 2010;105(8):1060–5.

de Castro F, Seal R, Maggi R. Group of HGNC consultants for KAL1 nomenclature. ANOS1: a unified nomenclature for Kallmann syndrome 1 gene (KAL1) and anosmin-1. Brief Funct Genomics 2017;16(4):205–10.

Dean HJ, Shackleton CH, Winter JS. Diagnosis and natural history of 17-hydroxylase deficiency in a newborn male. J Clin Endocrinol Metab 1984;59(3):513–20.

Dean B, Chrisp GL, Quartararo M, Maguire AM, Hameed S, King BR, et al. P450 oxidoreductase deficiency: a systematic review and meta-analysis of genotypes, phenotypes, and their relationships. J Clin Endocrinol Metab 2020;105(3):dgz255.

Deeb A, Jääskeläinen J, Dattani M, Whitaker HC, Costigan C, Hughes IA. A novel mutation in the human androgen receptor suggests a regulatory role for the hinge region in amino-terminal and carboxy-terminal interactions. J Clin Endocrinol Metab 2008;93(10):3691–6.

Dhayat NA, Frey AC, Frey BM, d'Uscio CH, Vogt B, Rousson V, et al. Estimation of reference curves for the urinary steroid metabolome in the first year of life in healthy children: tracing the complexity of human postnatal steroidogenesis. J Steroid Biochem Mol Biol 2015;154:226–36. Erratum in: J Steroid Biochem Mol Biol. 2018; 183:238.

Dipla K, Kraemer RR, Constantini NW, Hackney AC. Relative energy deficiency in sports (RED-S): elucidation of endocrine changes affecting the health of males and females. Hormones (Athens) 2021;20(1): 35–47.

Dodé C, Levilliers J, Dupont JM, De Paepe A, Le Dû N, Soussi-Yanicostas N, et al. Loss-of-function mutations in FGFR1 cause autosomal dominant Kallmann syndrome. Nat Genet 2003;33(4):463–5.

Domenice S, Machado AZ, Ferreira FM, Ferraz-de-Souza B, Lerario AM, Lin L, et al. Wide spectrum of NR5A1-related phenotypes in 46,XY and 46,XX individuals. Birth Defects Res C Embryo Today 2016;108(4):309–20.

Donoghue SE, Pitt JJ, Boneh A, White SM. Smith-Lemli-Opitz syndrome: clinical and biochemical correlates. J Pediatr Endocrinol Metab 2018;31(4):451–9.

Dwyer AA, Chavan NR, Lewkowitz-Shpuntoff H, Plummer L, Hayes FJ, Seminara SB, et al. Functional hypogonadotropic hypogonadism in men: underlying neuroendocrine mechanisms and natural history. J Clin Endocrinol Metab 2019;104(8):3403–14.

European Society for Human Reproduction and Embryology (ESHRE) Guideline Group on POI, Webber L, Davies M, Anderson R, Bartlett J, Braat D, et al. ESHRE guideline: management of women with premature ovarian insufficiency. Hum Reprod 2016;31(5):926–37.

Faienza MF, Chiarito M, Baldinotti F, Canale D, Savino C, Paradies G, et al. NR5A1 gene variants: variable phenotypes, new variants, different outcomes. Sex Dev 2019;13(5–6):258–63.

Falardeau J, Chung WC, Beenken A, Raivio T, Plummer L, Sidis Y, et al. Decreased FGF8 signaling causes deficiency of gonadotropin-releasing hormone in humans and mice. J Clin Invest 2008;118(8): 2822–31.

Fanis P, Neocleous V, Kosta K, Karipiadou A, Hartmann MF, Wudy SA, et al. Late diagnosis of 3β-hydroxysteroid dehydrogenase deficiency: the pivotal role of gas chromatography-mass spectrometry urinary steroid metabolome analysis and a novel homozygous nonsense mutation in the *HSD3B2* gene. J Pediatr Endocrinol Metab 2020; 34(1):131–6.

Fernández-Cancio M, Camats N, Flück CE, Zalewski A, Dick B, Frey BM, et al. Mechanism of the dual activities of human CYP17A1 and binding to anti-prostate cancer drug abiraterone revealed by a novel V366M mutation causing 17,20 lyase deficiency. Pharmaceuticals (Basel) 2018;11(2):37.

Ferrarini E, De Marco G, Orsolini F, Gianetti E, Benelli E, Fruzzetti F, et al. Characterization of a novel mutation V136L in bone morphogenetic protein 15 identified in a woman affected by POI. J Ovarian Res 2021;14(1):85.

Finlayson C, Rosoklija I, Aston CE, Austin P, Bakula D, Baskin L, et al. Baseline characteristics of infants with atypical genital development: phenotypes, diagnoses, and sex of rearing. J Endocr Soc 2018;3(1): 264–72.

Flück CE, Meyer-Böni M, Pandey AV, Kempná P, Miller WL, Schoenle EJ, et al. Why boys will be boys: two pathways of fetal testicular androgen biosynthesis are needed for male sexual differentiation. Am J Hum Genet 2011a;89(2):201–18. Erratum in: Am J Hum Genet 2011 Aug 12;89(2):347.

Flück C.E., Pandey A.V., Dick B., Camats N., Fernández-Cancio M., Clemente M., et al. Characterization of novel StAR (steroidogenic acute regulatory protein) mutations causing non-classic lipoid adrenal hyperplasia. PLoS ONE 2011b;6(5):e20178.

Gabrielsen JS, Lamb DJ, Lipshultz LI. Iron and a man's reproductive health: the good, the bad, and the ugly. Curr Urol Rep 2018;19(8):60.

Gach A, Pinkier I, Sałacińska K, Szarras-Czapnik M, Salachna D, Kucińska A, et al. Identification of gene variants in a cohort of hypogonadotropic hypogonadism: diagnostic utility of custom NGS panel and WES in unravelling genetic complexity of the disease. Mol Cell Endocrinol 2020a;517:110968.

Gach A, Pinkier I, Szarras-Czapnik M, Sakowicz A, Jakubowski L. Expanding the mutational spectrum of monogenic hypogonadotropic hypogonadism: novel mutations in ANOS1 and FGFR1 genes. Reprod Biol Endocrinol 2020b;18(1):8.

Galano M, Li Y, Li L, Sottas C, Papadopoulos V. Role of constitutive STAR in leydig cells. Int J Mol Sci 2021;22(4):2021.

García-Acero M, Molina M, Moreno O, Ramirez A, Forero C, Céspedes C, et al. Gene dosage of DAX-1, determining in sexual differentiation: duplication of DAX-1 in two sisters with gonadal dysgenesis. Mol Biol Rep 2019;46(3):2971–8.

Geller DH, Auchus RJ, Miller WL. P450c17 mutations R347H and R358Q selectively disrupt 17,20-lyase activity by disrupting interactions with P450 oxidoreductase and cytochrome b5. Mol Endocrinol 1999; 13(1):167–75.

Gheorghiu ML. Actualities in mutations of luteinizing hormone (LH) and follicle-stimulating hormone (FSH) receptors. Acta Endocrinol (Buchar) 2019;5(1):139–42.

Gianetti E, Tusset C, Noel SD, Au MG, Dwyer AA, Hughes VA, et al. TAC3/TACR3 mutations reveal preferential activation of gonadotropin-releasing hormone release by neurokinin B in neonatal life followed by reversal in adulthood. J Clin Endocrinol Metab 2010;95(6):2857–67.

Gomes NL, Chetty T, Jorgensen A, Mitchell RT. Disorders of sex development-novel regulators, impacts on fertility, and options for fertility preservation. Int J Mol Sci 2020;21(7):2282.

Gómez F, de la Cueva R, Wauters JP, Lemarchand-Béraud T. Endocrine abnormalities in patients undergoing long-term hemodialysis. The role of prolactin. Am J Med 1980;68(4):522–30.

Gonzalez D, Narasimman M, Best JC, Ory J, Ramasamy R. Clinical update on home testing for male fertility. World J Mens Health 2021; 39(4):615–25.

Gopal RA, Bothra N, Acharya SV, Ganesh HK, Bandgar TR, Menon PS, et al. Treatment of hypogonadism with testosterone in patients with type 2 diabetes mellitus. Endocr Pract 2010;16(4):570–6.

Greaves R, Hunt RW, Zacharin M. Transient anomalies in genital appearance in some extremely preterm female infants may be the result of foetal programming causing a surge in LH and the over activation of the pituitary-gonadal axis. Clin Endocrinol (Oxf) 2008;69(5):763–8.

Greuter L, Guzman R, Soleman J. Typical pediatric brain tumors occurring in adults-differences in management and outcome. Biomedicine 2021;9(4):356.

Grinspon RP, Bedecarrás P, Ballerini MG, Iñiguez G, Rocha A, Mantovani Rodrigues Resende EA, et al. Early onset of primary hypogonadism revealed by serum anti-Müllerian hormone determination during infancy and childhood in trisomy 21. Int J Androl 2011;34(5 Pt 2):e487–98.

Grinspon RP, Bergadá I, Rey RA. Male hypogonadism and disorders of sex development. Front Endocrinol (Lausanne) 2020;11:211.

Grossman A, Koren R, Tirosh A, Michowiz R, Shohat Z, Rahamimov R, et al. Prevalence and clinical characteristics of adrenal incidentalomas in potential kidney donors. Endocr Res 2016;41(2):98–102.

Grzechocińska B, Warzecha D, Wypchło M, Ploski R, Wielgoś M. Premature ovarian insufficiency as a variable feature of blepharophimosis, ptosis, and epicanthus inversus syndrome associated with c.223C > T p.(Leu75Phe) FOXL2 mutation: a case report. BMC Med Genet 2019;20(1):132.

Guran T, Buonocore F, Saka N, Ozbek MN, Aycan Z, Bereket A, et al. Rare causes of primary adrenal insufficiency: genetic and clinical characterization of a large nationwide cohort. J Clin Endocrinol Metab 2016;101 (1):284–92.

Guran T, Kara C, Yildiz M, Bitkin EC, Haklar G, Lin JC, et al. Revisiting classical 3β-hydroxysteroid dehydrogenase 2 deficiency: lessons from 31 pediatric cases. J Clin Endocrinol Metab 2020;105(3):dgaa022.

Gurney JK, McGlynn KA, Stanley J, Merriman T, Signal V, Shaw C, et al. Risk factors for cryptorchidism. Nat Rev Urol 2017;14(9):534–48.

Hale GE, Robertson DM, Burger HG. The perimenopausal woman: endocrinology and management. J Steroid Biochem Mol Biol 2014; 142:121–31.

Hardelin JP, Dodé C. The complex genetics of Kallmann syndrome: KAL1, FGFR1, FGF8, PROKR2, PROK2, et al. Sex Dev 2008;2(4–5):181–93.

Hassan HA, Essawi ML, Mekkawy MK, Mazen I. Novel mutations of the LHCGR gene in two families with 46,XY DSD causing Leydig cell hypoplasia I. Hormones (Athens) 2020;19(4):573–9.

Hermann AL, L'Herminé-Coulomb A, Irtan S, Audry G, Cardoen L, Brisse HJ, et al. Imaging of pediatric testicular and para-testicular tumors: a pictural review. Cancers (Basel) 2022;14(13):3180.

Homma K, Hasegawa T, Masumoto M, Takeshita E, Watanabe K, Chiba H, et al. Reference values for urinary steroids in Japanese newborn infants: gas chromatography/mass spectrometry in selected ion monitoring. Endocr J 2003;50(6):783–92.

Honour JW. Historical perspective: gut dysbiosis and hypertension. Physiol Genomics 2015;47(10):443–6.

Honour JW, Tourniaire J, Biglieri EG, Shackleton CH. Urinary steroid excretion in 17 alpha-hydroxylase deficiency. J Steroid Biochem 1978;9(6):495–505.

Hornig NC, Holterhus PM. Molecular basis of androgen insensitivity syndromes. Mol Cell Endocrinol 2021;523:111146.

Hryhorczuk AL, Phelps AS, Yu RN, Chow JS. The radiologist's role in assessing differences of sex development. Pediatr Radiol 2022; 52(4):752–64.

Hughes IA, Evans BA. Androgen insensitivity in forty-nine patients: classification based on clinical and androgen receptor phenotypes. Horm Res 1987;28(1):25–9.

Huhtaniemi I. Mutations along the pituitary-gonadal axis affecting sexual maturation: novel information from transgenic and knockout mice. Mol Cell Endocrinol 2006;254–255:84–90.

Huhtaniemi I, Ahtiainen P, Pakarainen T, Rulli SB, Zhang FP, Poutanen M. Genetically modified mouse models in studies of luteinising hormone action. Mol Cell Endocrinol 2006;252(1–2):126–35.

Hutson JM, Baskin LS, Risbridger G, Cunha GR. The power and perils of animal models with urogenital anomalies: handle with care. J Pediatr Urol 2014;10(4):699–705.

Idkowiak J, Malunowicz EM, Dhir V, Reisch N, Szarras-Czapnik M, Holmes DM, et al. Concomitant mutations in the P450 oxidoreductase and androgen receptor genes presenting with 46,XY disordered sex development and androgenization at adrenarche. J Clin Endocrinol Metab 2010;95(7):3418–27.

Idkowiak J, O'Riordan S, Reisch N, Malunowicz EM, Collins F, Kerstens MN, et al. Pubertal presentation in seven patients with congenital adrenal hyperplasia due to P450 oxidoreductase deficiency. J Clin Endocrinol Metab 2011;96(3):E453–62.

Idkowiak J, Randell T, Dhir V, Patel P, Shackleton CH, Taylor NF, et al. A missense mutation in the human cytochrome b5 gene causes 46,XY disorder of sex development due to true isolated 17,20 lyase deficiency. J Clin Endocrinol Metab 2012;97(3):E465–75.

Ishii T, Tajima T, Kashimada K, Mukai T, Tanahashi Y, Katsumata N, et al. Clinical features of 57 patients with lipoid congenital adrenal hyperplasia: criteria for nonclassic form revisited. J Clin Endocrinol Metab 2020;105(11):dgaa557.

Ishizuka B. Current understanding of the etiology, symptomatology, and treatment options in premature ovarian insufficiency (POI). Front Endocrinol (Lausanne) 2021;12, 626924.

Ito M, Yu RN, Jameson JL. Steroidogenic factor-1 contains a carboxy-terminal transcriptional activation domain that interacts with steroid receptor coactivator-1. Mol Endocrinol 1998;12(2):290–301.

Jeske YW, McGown IN, Cowley DM, Oley C, Thomsett MJ, Choong CS, et al. Androgen receptor genotyping in a large Australasian cohort with androgen insensitivity syndrome; identification of four novel mutations. J Pediatr Endocrinol Metab 2007;20(8):893–908.

Jin BF, Ji ZY, Su ZY, Mei LB, Huang XJ, Lin SB, et al. Identification of a novel mutation in FGFR1 gene in patients with Kallmann syndrome by high throughput sequencing. Syst Biol Reprod Med 2018;64 (3):202–6.

Jongmans MC, van Ravenswaaij-Arts CM, Pitteloud N, Ogata T, Sato N, Claahsen-van der Grinten HL, et al. CHD7 mutations in patients initially diagnosed with Kallmann syndrome—the clinical overlap with CHARGE syndrome. Clin Genet 2009;75(1):65–71.

Josso N, Rey RA. What does AMH tell us in pediatric disorders of sex development? Front Endocrinol (Lausanne) 2020;11:619.

Josso N, Belville C, di Clemente N, Picard JY. AMH and AMH receptor defects in persistent Müllerian duct syndrome. Hum Reprod Update 2005;11(4):351–6.

Josso N, Rey R, Picard JY. Testicular anti-Müllerian hormone: clinical applications in DSD. Semin Reprod Med 2012;30(5):364–73. https:// doi.org/10.1055/s-0032-1324719. Epub 2012 Oct 8 23044872.

Kallali W, Gray E, Mehdi MZ, Lindsay R, Metherell LA, Buonocore F, et al. Long-term outcome of partial P450 side-chain cleavage enzyme deficiency in three brothers: the importance of early diagnosis. Eur J Endocrinol 2020;182(3):K15–24.

Karavitaki N, Brufani C, Warner JT, Adams CB, Richards P, Ansorge O, et al. Craniopharyngiomas in children and adults: systematic analysis of 121 cases with long-term follow-up. Clin Endocrinol (Oxf) 2005; 62(4):397–409.

Kerns SL, Guevara-Aguirre J, Andrew S, Geng J, Guevara C, Guevara-Aguirre M, et al. A novel variant in CDKN1C is associated with intrauterine growth restriction, short stature, and early-adulthood-onset diabetes. J Clin Endocrinol Metab 2014;99(10):E2117–22.

Khosla S, Amin S, Singh RJ, Atkinson EJ, Melton 3rd LJ, Riggs BL. Comparison of sex steroid measurements in men by immunoassay versus mass spectroscopy and relationships with cortical and trabecular volumetric bone mineral density. Osteoporos Int 2008;19(10):1465–71.

Kim YM, Seo GH, Kim GH, Ko JM, Choi JH, Yoo HW. A case of an infant suspected as IMAGE syndrome who were finally diagnosed with MIRAGE syndrome by targeted Mendelian exome sequencing. BMC Med Genet 2018;19(1):35.

Kim HI, Lee I, Kim SH, Lee YS, Han SW, Yun BH. Ovotesticular disorder of sex development in Korean children: a single-center analysis over a 30-year period. J Pediatr Adolesc Gynecol 2021;34(5):626–30. https:// doi.org/10.1016/j.jpag.2021.02.105. Epub 2021 Mar 2 33667640.

Köhler B, Achermann JC. Update–steroidogenic factor 1 (SF-1, NR5A1). Minerva Endocrinol 2010;35(2):73–86.

Kossack N, Troppmann B, Richter-Unruh A, Kleinau G, Gromoll J. Aberrant transcription of the LHCGR gene caused by a mutation in exon 6A leads to Leydig cell hypoplasia type II. Mol Cell Endocrinol 2013;366(1):59–67.

Krause A, Sinnecker GH, Hiort O, Thamm B, Hoepffner W. Applicability of the SHBG androgen sensitivity test in the differential diagnosis of 46,XY gonadal dysgenesis, true hermaphroditism, and androgen insensitivity syndrome. Exp Clin Endocrinol Diabetes 2004;112(5):236–40.

Krone N, Hughes BA, Lavery GG, Stewart PM, Arlt W, Shackleton CH. Gas chromatography/mass spectrometry (GC/MS) remains a preeminent discovery tool in clinical steroid investigations even in the era of fast liquid chromatography tandem mass spectrometry (LC/ MS/MS). J Steroid Biochem Mol Biol 2010;121(3–5):496–504.

Krone N, Reisch N, Idkowiak J, Dhir V, Ivison HE, Hughes BA, et al. Genotype-phenotype analysis in congenital adrenal hyperplasia due to P450 oxidoreductase deficiency. J Clin Endocrinol Metab 2012;97(2):E257–67.

Krylova IN, Sablin EP, Moore J, Xu RX, Waitt GM, MacKay JA, et al. Structural analyses reveal phosphatidyl inositols as ligands for the NR5 orphan receptors SF-1 and LRH-1. Cell 2005;120(3):343–55.

Kubini K, Zachmann M, Albers N, Hiort O, Bettendorf M, Wölfle J, et al. Basal inhibin B and the testosterone response to human chorionic gonadotropin correlate in prepubertal boys. J Clin Endocrinol Metab 2000;85(1):134–8.

Kyriakakis N, Shonibare T, Kyaw-Tun J, Lynch J, Lagos CF, Achermann JC, et al. Late-onset X-linked adrenal hypoplasia (DAX-1, NR0B1): two new adult-onset cases from a single center. Pituitary 2017;20 (5):585–93.

La Vignera S, Vita R, Condorelli RA, Mongioì LM, Presti S, Benvenga S, et al. Impact of thyroid disease on testicular function. Endocrine 2017;58(3):397–407.

Lam CW, Arlt W, Chan CK, Honour JW, Lin CJ, Tong SF, et al. Mutation of proline 409 to arginine in the meander region of cytochrome p450c17 causes severe 17 alpha-hydroxylase deficiency. Mol Genet Metab 2001;72(3):254–9.

Lanfranco F, Kamischke A, Zitzmann M, Nieschlag E. Klinefelter's syndrome. Lancet 2004;364(9430):273–83.

Latronico AC, Anasti J, Arnhold IJ, Rapaport R, Mendonca BB, Bloise W, et al. Brief report: testicular and ovarian resistance to luteinizing hormone caused by inactivating mutations of the luteinizing hormone-receptor gene. N Engl J Med 1996;334(8):507–12.

Lavi E, Zighan M, Abu Libdeh A, Klopstock T, Weinberg-Shukron A, Renbaum P, et al. A unique presentation of XY gonadal dysgenesis in Frasier syndrome due to WT1 mutation and a literature review. Pediatr Endocrinol Rev 2020;17(4):302–7.

Leal AM, Foss NT. Endocrine dysfunction in leprosy. Eur J Clin Microbiol Infect Dis 2009;28(1):1–7.

Lee C, Kim JH, Moon SJ, Shim J, Kim HI, Choi MH. Selective LC-MRM/ SIM-MS based profiling of adrenal steroids reveals metabolic signatures of 17α-hydroxylase deficiency. J Steroid Biochem Mol Biol 2020;198:105615.

Leung MT, Cheung HN, Iu YP, Choi CH, Tiu SC, Shek CC. Isolated 17,20-lyase deficiency in a *CYB5A* mutated female with normal sexual development and fertility. J Endocr Soc 2019;4(2):bvz016.

Li Y, Choi M, Cavey G, Daugherty J, Suino K, Kovach A, et al. Crystallographic identification and functional characterization of phospholipids as ligands for the orphan nuclear receptor steroidogenic factor-1. Mol Cell 2005;17(4):491–502.

Li JD, Wu J, Zhao Y, Wang X, Jiang F, Hou Q, et al. Phenotypic spectrum of idiopathic hypogonadotropic hypogonadism patients with CHD7 variants from a large Chinese cohort. J Clin Endocrinol Metab 2020;105(5):dgz182.

Lima N, Cavaliere H, Knobel M, Halpern A, Medeiros-Neto G. Decreased androgen levels in massively obese men may be associated with impaired function of the gonadostat. Int J Obes Relat Metab Disord 2000;24(11):1433–7.

Lin L, Achermann JC. Steroidogenic factor-1 (SF-1, Ad4BP, NR5A1) and disorders of testis development. Sex Dev 2008;2(4–5):200–9.

Liu EK, Suson KD. Syndromic Wilms tumor: a review of predisposing conditions, surveillance and treatment. Transl Androl Urol 2020;9(5): 2370–81.

Liu H, Xu X, Han T, Yan L, Cheng L, Qin Y, et al. A novel homozygous mutation in the FSHR gene is causative for primary ovarian insufficiency. Fertil Steril 2017;108(6):1050–1055.e2.

Liu D.T., Welge-Lissen A., Besser G., Mueller C.A., Renner B. Assessment of odor hedonic perception: the Sniffin' sticks parosmia test (SSParoT). Sci Rep 2020a;10:18019.

Liu Q, Yin X, Li P. Clinical, hormonal and genetic characteristics of androgen insensitivity syndrome in 39 Chinese patients. Reprod Biol Endocrinol 2020b;18(1):34.

Logan CV, Murray JE, Parry DA, Robertson A, Bellelli R, Tarnauskaitė Ž, et al. DNA polymerase epsilon deficiency causes IMAGe syndrome with variable immunodeficiency. Am J Hum Genet 2018;103(6): 1038–44.

Lolis E, Juhlin CC, Nordenström A, Falhammar H. Extensive bilateral adrenal rest testicular tumors in a patient with 3β-hydroxysteroid dehydrogenase type 2 deficiency. J Endocr Soc 2018;2(6): 513–7.

Lutfallah C, Wang W, Mason JI, Chang YT, Haider A, Rich B, et al. Newly proposed hormonal criteria via genotypic proof for type II 3beta-hydroxysteroid dehydrogenase deficiency. J Clin Endocrinol Metab 2002;87(6):2611–22.

Maharaj A, Maudhoo A, Chan LF, Novoselova T, Prasad R, Metherell LA, et al. Isolated glucocorticoid deficiency: genetic causes and animal models. J Steroid Biochem Mol Biol 2019;189:73–80.

Marcos J, Craig WY, Palomaki GE, Kloza EM, Haddow JE, Roberson M, et al. Maternal urine and serum steroid measurements to identify steroid sulfatase deficiency (STSD) in second trimester pregnancies. Prenat Diagn 2009;29(8):771–80.

Martínez de LaPiscina I, Mahmoud RA, Sauter KS, Esteva I, Alonso M, Costa I, et al. Variants of STAR, AMH and ZFPM2/FOG2 May Contribute towards the Broad Phenotype Observed in 46,XY DSD Patients with Heterozygous Variants of NR5A1. Int J Mol Sci 2020;21(22): 8554.

Martínez-Garza SG, Gallegos-Rivas MC, Vargas-Maciel M, Rubio-Rubio JM, de Los Monteros-Rodríguez ME, González-Ortega C, et al. Genetic screening in infertile Mexican men: chromosomal abnormalities, Y chromosome deletions, and androgen receptor CAG repeat length. J Androl 2008;29(6):654–60.

Masterson 3rd TA, Nassau DE, Ramasamy R. A clinical algorithm for management of fertility in adolescents with the Klinefelter syndrome. Curr Opin Urol 2020;30(3):324–7.

Mazen I, Mekkawy M, Kamel A, Essawi M, Hassan H, Abdel-Hamid M, et al. Advances in genomic diagnosis of a large cohort of Egyptian patients with disorders of sex development. Am J Med Genet A 2021;185(6):1666–77.

McElreavey K, Jorgensen A, Eozenou C, Merel T, Bignon-Topalovic J, Tan DS, et al. Pathogenic variants in the DEAH-box RNA helicase DHX37 are a frequent cause of 46,XY gonadal dysgenesis and 46,XY testicular regression syndrome. Genet Med 2020;22 (1):150–9.

Mehta P, Singh P, Gupta NJ, Sankhwar SN, Chakravarty B, Thangaraj K, et al. Mutations in the desert hedgehog (DHH) gene in the disorders of sexual differentiation and male infertility. J Assist Reprod Genet 2021;38(7):1871–8.

Mengen E, Küçükçongar Yavaş A, Uçaktürk SA. A rare etiology of 46,XY disorder of sex development and adrenal insufficiency: a case of MIRAGE syndrome caused by mutations in the SAMD9 gene. J Clin Res Pediatr Endocrinol 2020;12(2):206–11.

Mermejo LM, Elias LL, Marui S, Moreira AC, Mendonca BB, de Castro M. Refining hormonal diagnosis of type II 3beta-hydroxysteroid dehydrogenase deficiency in patients with premature pubarche and hirsutism based on HSD3B2 genotyping. J Clin Endocrinol Metab 2005;90(3):1287–93.

Midgley PC, Azzopardi D, Oates N, Shaw JC, Honour JW. Virilisation of female preterm infants. Arch Dis Child 1990;65(7):701–3. Spec No.

Millar AC, Faghfoury H, Bieniek JM. Genetics of hypogonadotropic hypogonadism. Transl Androl Urol 2021;10(3):1401–9.

Miller WL. The syndrome of 17,20 lyase deficiency. J Clin Endocrinol Metab 2012;97(1):59–67.

Miller WL. Disorders in the initial steps of steroid hormone synthesis. J Steroid Biochem Mol Biol 2017;165(Pt. A):18–37.

Miller WL, Chung BC. The first defect in electron transfer to mitochondrial P450 enzymes. Endocrinology 2016;157(3):1003–6.

Miller WL, Strauss 3rd JF. Molecular pathology and mechanism of action of the steroidogenic acute regulatory protein, StAR. J Steroid Biochem Mol Biol 1999;69(1–6):131–41.

Miller WL, Tee MK. The post-translational regulation of 17,20 lyase activity. Mol Cell Endocrinol 2015;408:99–106.

Miraoui H, Dwyer AA, Sykiotis GP, Plummer L, Chung W, Feng B, et al. Mutations in FGF17, IL17RD, DUSP6, SPRY4, and FLRT3 are identified in individuals with congenital hypogonadotropic hypogonadism. Am J Hum Genet 2013;92(5):725–43.

Moniez S, Pienkowski C, Lepage B, Hamdi S, Daudin M, Oliver I, et al. Noonan syndrome males display Sertoli cell-specific primary testicular insufficiency. Eur J Endocrinol 2018;179(6):409–18.

Mönig I, Schneidewind J, Johannsen TH, Juul A, Werner R, Lünstedt R, et al. Pubertal development in 46,XY patients with NR5A1 mutations. Endocrine 2022;75(2):601–13.

Morawiecka-Pietrzak M, Dąbrowska E, Gliwińska A, Góra A, Geisler G, Gawlik A, et al. A rare case of primary amenorrhoea and breast development in a 46,XY 15-year-old girl. Pediatr Endocrinol Diabetes Metab 2021;27(1):62–7.

Mosbah H, Bouvattier C, Maione L, Trabado S, De Filippo G, Cartes A, et al. GnRH stimulation testing and serum inhibin B in males: insufficient specificity for discriminating between congenital hypogonadotropic hypogonadism from constitutional delay of growth and puberty. Hum Reprod 2020;35(10):2312–22.

Mueller RF. The Denys-Drash syndrome. J Med Genet 1994;31(6):471–7.

Naderi AR, Safarinejad MR. Endocrine profiles and semen quality in spinal cord injured men. Clin Endocrinol (Oxf) 2003;58(2):177–84.

Nagel SA, Hartmann MF, Riepe FG, Wudy SA, Wabitsch M. Gonadotropin- and adrenocorticotropic hormone-independent precocious puberty of gonadal origin in a patient with adrenal hypoplasia congenita due to DAX1 gene mutation—a case report and review of the literature: implications for the pathomechanism. Horm Res Paediatr 2019;91(5):336–45.

Narumi S, Amano N, Ishii T, Katsumata N, Muroya K, Adachi M, et al. SAMD9 mutations cause a novel multisystem disorder, MIRAGE syndrome, and are associated with loss of chromosome 7. Nat Genet 2016;48(7):792–7.

Naumova OY, Rychkov SY, Burenkova OV, Solodunova MY, Polyanskaya IV, Arintcina IA, et al. Male pseudohermaphroditism: a case study of 46,XY disorder of sexual development using whole-exome sequencing. Clin Case Rep 2020;8(12):2889–94.

Neocleous V, Fanis P, Cinarli F, Kokotsis V, Oulas A, Toumba M, et al. 46,XY complete gonadal dysgenesis in a familial case with a rare mutation in the desert hedgehog (DHH) gene. Hormones (Athens) 2019;18(3):315–20.

Neocleous V, Fanis P, Toumba M, Tanteles GA, Schiza M, Cinarli F, et al. GnRH deficient patients with congenital hypogonadotropic hypogonadism: novel genetic findings in ANOS1, RNF216, WDR11, FGFR1, CHD7, and POLR3A genes in a case series and review of the literature. Front Endocrinol 2020;11:626.

Ono H, Numakura C, Homma K, Hasegawa T, Tsutsumi S, Kato F, et al. Longitudinal serum and urine steroid metabolite profiling in a 46,XY infant with prenatally identified POR deficiency. J Steroid Biochem Mol Biol 2018;178:177–84.

Onuma S, Wada T, Araki R, Wada K, Tanase-Nakao K, Narumi S, et al. MIRAGE syndrome caused by a novel missense variant (p.Ala1479Ser) in the SAMD9 gene. Hum Genome Var 2020;7:4.

Pan S, Guo S, Liu L, Yang X, Liang H. Functional study of a novel c.630delG (p.Y211Tfs*85) mutation in NR5A1 gene in a Chinese boy with 46,XY disorders of sex development. J Assist Reprod Genet 2020;37(2):477–86.

Panner Selvam MK, Finelli R, Agarwal A, Henkel R. Evaluation of seminal oxidation-reduction potential in male infertility. Andrologia 2021; 53(2), e13610. https://doi.org/10.1111/and.13610.

Parekh NV, Lundy SD, Vij SC. Fertility considerations in men with testicular cancer. Transl Androl Urol 2020;9(Suppl. 1):S14–23.

Park YY, Park KC, Shong M, Lee SJ, Lee YH, Choi HS. EID-1 interacts with orphan nuclear receptor SF-1 and represses its transactivation. Mol Cells 2007;24(3):372–7.

Peña VN, Kohn TP, Herati AS. Genetic mutations contributing to nonobstructive azoospermia. Best Pract Res Clin Endocrinol Metab 2020;34(6), 101479d.

Perniola R, Fierabracci A, Falorni A. Autoimmune Addison's disease as part of the autoimmune Polyglandular syndrome type 1: historical overview and current evidence. Front Immunol 2021;12, 606860.

Peterson RE, Imperato-McGinley J, Gautier T, Shackleton C. Male pseudohermaphroditism due to multiple defects in steroid-biosynthetic microsomal mixed-function oxidases. A new variant of congenital adrenal hyperplasia. N Engl J Med 1985;313(19):1182–91.

Phelan N, Williams EL, Cardamone S, Lee M, Creighton SM, Rumsby G, et al. Screening for mutations in 17β-hydroxysteroid dehydrogenase and androgen receptor in women presenting with partially virilised 46,XY disorders of sex development. Eur J Endocrinol 2015;172 (6):745–51.

Philibert P, Audran F, Pienkowski C, Morange I, Kohler B, Flori E, et al. Complete androgen insensitivity syndrome is frequently due to premature stop codons in exon 1 of the androgen receptor gene: an international collaborative report of 13 new mutations. Fertil Steril 2010;94 (2):472–6.

Picard JY, Cate RL, Racine C, Josso N. The persistent Müllerian Duct Syndrome: an update based upon a personal experience of 157 cases. Sex Dev 2017;11(3):109–25.

Pingault V, Bodereau V, Baral V, Marcos S, Watanabe Y, Chaoui A, et al. Loss-of-function mutations in SOX10 cause Kallmann syndrome with deafness. Am J Hum Genet 2013;92(5):707–24.

Pitteloud N, Quinton R, Pearce S, Raivio T, Acierno J, Dwyer A, et al. Digenic mutations account for variable phenotypes in idiopathic hypogonadotropic hypogonadism. J Clin Invest 2007;117 (2):457–63.

Porras AR, Rosenbaum K, Tor-Diez C, Summar M, Linguraru MG. Development and evaluation of a machine learning-based point-of-care screening tool for genetic syndromes in children: a multinational retrospective study. Lancet Digit Health 2021;3(10):e635–43.

Praveen VP, Ladjouze A, Sauter KS, Pulickal A, Katharopoulos E, Trippel M, et al. Novel *CYP19A1* mutations extend the genotype-phenotype correlation and reveal the impact on ovarian function. J Endocr Soc 2020;4(4):bvaa030.

Probst-Scheidegger U, Udhane SS, l'Allemand D, Flück CE, Camats N. Non-virilizing congenital adrenal hyperplasia in a female patient with a novel HSD3B2 mutation. Sex Dev 2016;10(4):200–4.

Qiao J, Han B. Diseases caused by mutations in luteinizing hormone/chorionic gonadotropin receptor. Prog Mol Biol Transl Sci 2019;161: 69–89.

Qiao J, Han B, Liu BL, Chen X, Ru Y, Cheng KX, et al. A splice site mutation combined with a novel missense mutation of LHCGR cause male pseudohermaphroditism. Hum Mutat 2009;30(9):E855–65.

Quaynor SD, Stradtman Jr EW, Kim HG, Shen Y, Chorich LP, Schreihofer DA, et al. Delayed puberty and estrogen resistance in a woman with estrogen receptor α variant. N Engl J Med 2013;369(2):164–71.

Racca JD, Chen YS, Yang Y, Phillips NB, Weiss MA. Human sex determination at the edge of ambiguity: Inherited XY sex reversal due to enhanced ubiquitination and proteasomal degradation of a master transcription factor. J Biol Chem 2016;291(42):22173–95.

Rannikko A, Pakarinen P, Manna PR, Beau I, Misrahi M, Aittomäki K, et al. Functional characterization of the human FSH receptor with an inactivating Ala189Val mutation. Mol Hum Reprod 2002;8(4): 311–7.

Rauscher 3rd FJ. Tumor suppressor genes which encode transcriptional repressors: studies on the EGR and Wilms' tumor (WT1) gene products. Adv Exp Med Biol 1993;348:23–9. https://doi.org/10.1007/978-1-4615-2942-2_2. 8172019.

Reinhardt W, Kübber H, Dolff S, Benson S, Führer D, Tan S. Rapid recovery of hypogonadism in male patients with end stage renal disease after renal transplantation. Endocrine 2018;60(1):159–66.

Rinonapoli G, Ruggiero C, Meccariello L, Bisaccia M, Ceccarini P, Caraffa A. Osteoporosis in men: a review of an underestimated bone condition. Int J Mol Sci 2021;22(4):2105.

Rivero-Müller A, Huhtaniemi I. Genetic variants of gonadotrophins and their receptors: impact on the diagnosis and management of the infertile patient. Best Pract Res Clin Endocrinol Metab 2022;36(1): 101596.

Rodprasert W, Virtanen HE, Mäkelä JA, Toppari J. Hypogonadism and cryptorchidism. Front Endocrinol 2020;10:906.

Rohayem J, Nieschlag E, Zitzmann M, Kliesch S. Testicular function during puberty and young adulthood in patients with Klinefelter's syndrome with and without spermatozoa in seminal fluid. Andrology 2016;4(6):1178–86.

Rojek A, Krawczynski MR, Jamsheer A, Sowinska-Seidler A, Iwaniszewska B, Malunowicz E, et al. X-linked adrenal hypoplasia congenita in a boy due to a novel deletion of the entire NR0B1 (DAX1) and MAGEB1-4 genes. Int J Endocrinol 2016;2016, 5178953.

Rosa S, Duff C, Meyer M, Lang-Muritano M, Balercia G, Boscaro M, et al. P450c17 deficiency: clinical and molecular characterization of six patients. J Clin Endocrinol Metab 2007;92(3):1000–7.

Rosário PW, Barroso AL, Rezende LL, Padrão EL, Borges MA, Guimarães VC, et al. Testicular function after radioiodine therapy in patients with thyroid cancer. Thyroid 2006;16(7):667–70.

Rothman MS, Carlson NE, Xu M, Wang C, Swerdloff R, Lee P, et al. Reexamination of testosterone, dihydrotestosterone, estradiol and estrone levels across the menstrual cycle and in postmenopausal women measured by liquid chromatography-tandem mass spectrometry. Steroids 2011;76(1–2):177–82.

Royer-Pokora B, Busch MA, Tenbusch S, Schmidt M, Beier M, Woods AD, et al. Comprehensive biology and genetics compendium of Wilms tumor cell lines with different *WT1* mutations. Cancers (Basel) 2020; 13(1):60.

Ruiz-Babot G, Balyura M, Hadjidemetriou I, Ajodha SJ, Taylor DR, Ghataore L, et al. Modeling congenital adrenal hyperplasia and testing interventions for adrenal insufficiency using donor-specific reprogrammed cells. Cell Rep 2018;22(5):1236–49.

Rumsby G, Skinner C, Lee HA, Honour JW. Combined 17 alpha-hydroxylase/17,20-lyase deficiency caused by heterozygous stop codons in the cytochrome P450 17 alpha-hydroxylase gene. Clin Endocrinol (Oxf) 1993;39(4):483–5.

Saengkaew T, Patel HR, Banerjee K, Butler G, Dattani MT, McGuigan M, et al. Genetic evaluation supports differential diagnosis in adolescent patients with delayed puberty. Eur J Endocrinol 2021;185(5): 617–27.

Salzano A, Arcopinto M, Marra AM, Bobbio E, Esposito D, Accardo G, Giallauria F, Bossone E, Vigorito C, Lenzi A, Pasquali D, Isidori AM, Cittadini A. Klinefelter syndrome, cardiovascular system, and thromboembolic disease: review of literature and clinical perspectives. Eur J Endocrinol 2016;175(1):R27–40.

Sanfins A, Rodrigues P, Albertini DF. GDF-9 and BMP-15 direct the follicle symphony. J Assist Reprod Genet 2018;35(10):1741–50. https://doi.org/10.1007/s10815-018-1268-4. Epub 2018 Jul 23 30039232. PMCID: PMC6150895.

Santoro N, Randolph Jr JF. Reproductive hormones and the menopause transition. Obstet Gynecol Clin North Am 2011;38(3):455–66.

Santos M, Cordts EB, Peluso C, Dornas M, Neto FHV, Bianco B, et al. Association of BMP15 and GDF9 variants to premature ovarian insufficiency. J Assist Reprod Genet 2019;36(10):2163–9.

Sassi A, Désir J, Janssens V, Marangoni M, Daneels D, Gheldof A, et al. Novel inactivating follicle-stimulating hormone receptor mutations in a patient with premature ovarian insufficiency identified by next-generation sequencing gene panel analysis. F S Rep 2020;1(3): 193–201.

Schmit M, Bielinsky AK. Congenital diseases of DNA replication: clinical phenotypes and molecular mechanisms. Int J Mol Sci 2021; 22(2):911. https://doi.org/10.3390/ijms22020911. 33477564. PMCID: PMC7831139.

Seppä S, Kuiri-Hänninen T, Holopainen E, Voutilainen R. MANAGEMENT OF ENDOCRINE DISEASE: diagnosis and management of primary amenorrhea and female delayed puberty. Eur J Endocrinol 2021;184(6):R225–42.

Shackleton CH, Roitman E, Kratz LE, Kelley RI. Midgestational maternal urine steroid markers of fetal Smith-Lemli-Opitz (SLO) syndrome (7-dehydrocholesterol 7-reductase deficiency). Steroids 1999a;64(7): 446–52.

Shackleton CH, Roitman E, Kratz LE, Kelley RI. Equine type estrogens produced by a pregnant woman carrying a Smith-Lemli-Opitz syndrome fetus. J Clin Endocrinol Metab 1999b;84(3):1157–9.

Shackleton CH, Roitman E, Kratz L, Kelley R. Dehydro-oestriol and dehydropregnanetriol are candidate analytes for prenatal diagnosis of Smith-Lemli-Opitz syndrome. Prenat Diagn 2001;21(3):207–12.

Shackleton C, Marcos J, Arlt W, Hauffa BP. Prenatal diagnosis of P450 oxidoreductase deficiency (ORD): a disorder causing low pregnancy estriol, maternal and fetal virilization, and the Antley-Bixler syndrome phenotype. Am J Med Genet A 2004;129A(2):105–12.

Sharma A, Minhas S, Dhillo WS, Jayasena CN. Male infertility due to testicular disorders. J Clin Endocrinol Metab 2021;106(2):e442–59.

Shaunak M, Taylor NF, Hunt D, Davies JH. Isolated 17, 20 lyase deficiency secondary to a novel CYB5A variant: comparison of steroid metabolomic findings with published cases provides diagnostic guidelines and greater insight into its biological role. Horm Res Paediatr 2020;93(7–8):483–96.

Shelling AN, Burton KA, Chand AL, van Ee CC, France JT, Farquhar CM, et al. Inhibin: a candidate gene for premature ovarian failure. Hum Reprod 2000;15(12):2644–9.

Sherbet DP, Tiosano D, Kwist KM, Hochberg Z, Auchus RJ. CYP17 mutation E305G causes isolated 17,20-lyase deficiency by selectively altering substrate binding. J Biol Chem 2003;278(49):48563–9.

Shima H, Hayashi M, Tachibana T, Oshiro M, Amano N, Ishii T, et al. MIRAGE syndrome is a rare cause of 46,XY DSD born SGA without adrenal insufficiency. PLoS ONE 2018;13(11), e0206184.

Shimazaki S, Kazukawa I, Mori K, Kihara M, Minagawa M. Factors predicting endocrine late effects in childhood cancer survivors from a Japanese hospital. Endocr J 2020;67(2):131–40.

Sinnecker GH, Hiort O, Nitsche EM, Holterhus PM, Kruse K. Functional assessment and clinical classification of androgen sensitivity in patients with mutations of the androgen receptor gene. German collaborative intersex study group. Eur J Pediatr 1997;156(1):7–14.

Smith EP, Boyd J, Frank GR, Takahashi H, Cohen RM, Specker B, et al. Estrogen resistance caused by a mutation in the estrogen-receptor gene in a man. N Engl J Med 1994;331(16):1056–61.

Sreenivasan R, Ludbrook L, Fisher B, Declosmenil F, Knower KC, Croft B, et al. Mutant NR5A1/SF-1 in patients with disorders of sex development shows defective activation of the SOX9 TESCO enhancer. Hum Mutat 2018;39(12):1861–74.

Sun M, Mueller JW, Gilligan LC, Taylor AE, Shaheen F, Noczyńska A, et al. The broad phenotypic spectrum of 17α-hydroxylase/17,20-lyase (CYP17A1) deficiency: a case series. Eur J Endocrinol 2021;185 (5):729–41.

Suntharalingham JP, Buonocore F, Duncan AJ, Achermann JC. DAX-1 (NR0B1) and steroidogenic factor-1 (SF-1, NR5A1) in human disease. Best Pract Res Clin Endocrinol Metab 2015;29(4):607–19.

Suntharalingham JP, Ishida M, Buonocore F, Del Valle I, Solanky N, Demetriou C, et al. Analysis of *CDKN1C* in fetal growth restriction and pregnancy loss. F1000Res 2019;8:90.

Tapanainen JS, Aittomäki K, Min J, Vaskivuo T, Huhtaniemi IT. Men homozygous for an inactivating mutation of the follicle-stimulating hormone (FSH) receptor gene present variable suppression of spermatogenesis and fertility. Nat Genet 1997;15(2):205–6.

Tee MK, Miller WL. Phosphorylation of human cytochrome P450c17 by p38α selectively increases 17,20 lyase activity and androgen biosynthesis. J Biol Chem 2013;288(33):23903–13. Erratum in: J Biol Chem. 2014;289(41):28137.

Thirumavalavan N, Gabrielsen JS, Lamb DJ. Where are we going with gene screening for male infertility? Fertil Steril 2019;111(5):842–50.

Tiosano D, Knopf C, Koren I, Levanon N, Hartmann MF, Hochberg Z, et al. Metabolic evidence for impaired 17alpha-hydroxylase activity in a kindred bearing the E305G mutation for isolate 17,20-lyase activity. Eur J Endocrinol 2008;158(3):385–92.

Touraine P, Beau I, Gougeon A, Meduri G, Desroches A, Pichard C, et al. New natural inactivating mutations of the follicle-stimulating hormone receptor: correlations between receptor function and phenotype. Mol Endocrinol 1999;13(11):1844–54.

Tuhan H, Anik A, Catli G, Onay H, Aykut A, Abaci A, et al. A novel mutation in steroidogenic factor (SF1/NR5A1) gene in a patient with 46 XY DSD without adrenal insufficiency. Andrologia 2017;49(1).

Tzancheva M, Kaneva R, Kumanov P, Williams G, Tyler-Smith C. Two male patients with ring Y: definition of an interval in Yq contributing to Turner syndrome. J Med Genet 1999;36(7):549–53.

Van Den Akker EL, Koper JW, Boehmer AL, Themmen AP, Verhoef-Post M, Timmerman MA, et al. Differential inhibition of 17alpha-hydroxylase and 17,20-lyase activities by three novel missense CYP17 mutations identified in patients with P450c17 deficiency. J Clin Endocrinol Metab 2002;87(12):5714–21.

van Dorp W, Haupt R, Anderson RA, Mulder RL, van den Heuvel-Eibrink MM, van Dulmen-den Broeder E, et al. Reproductive function and outcomes in female survivors of childhood, adolescent, and young adult cancer: a review. J Clin Oncol 2018;36(21):2169–80. Erratum in: J Clin Oncol. 2020;38(8):847.

Vargas MCC, Moura FS, Elias CP, Carvalho SR, Rassi N, Kunii IS, et al. Spontaneous fertility and variable spectrum of reproductive phenotype in a family with adult-onset X-linked adrenal insufficiency harboring a novel DAX-1/NR0B1 mutation. BMC Endocr Disord 2020;20(1):21.

Vilain E, Le Merrer M, Lecointre C, Desangles F, Kay MA, Maroteaux P, et al. IMAGe, a new clinical association of intrauterine growth retardation, metaphyseal dysplasia, adrenal hypoplasia congenita, and genital anomalies. J Clin Endocrinol Metab 1999;84(12):4335–40.

Wang W, Zhang C, Marimuthu A, Krupka HI, Tabrizizad M, Shelloe R, et al. The crystal structures of human steroidogenic factor-1 and liver receptor homologue-1. Proc Natl Acad Sci U S A 2005;102(21):7505–10.

Wang F, Cai J, Wang J, He M, Mao J, Zhu K, et al. A novel WT1 gene mutation in a chinese girl with denys-drash syndrome. J Clin Lab Anal 2021;35(5), e23769.

Wei JQ, Wei JL, Li WC, Bi YS, Wei FC. Genotyping of five chinese patients with 17alpha-hydroxylase deficiency diagnosed through high-performance liquid chromatography serum adrenal profile: identification of two novel CYP17 mutations. J Clin Endocrinol Metab 2006;91(9):3647–53.

Wei C, Davis N, Honour J, Crowne E. The investigation of children and adolescents with abnormalities of pubertal timing. Ann Clin Biochem 2017;54(1):20–32.

Wen Y, Zhang Z, Li Z, Liu G, Tao G, Song X, et al. The PROK2/PROKR2 signaling pathway is required for the migration of most olfactory bulb interneurons. J Comp Neurol 2019;527(18):2931–47.

Wesevich V, Kellen AN, Pal L. Recent advances in understanding primary ovarian insufficiency. F1000Res 2020;9. F1000 Faculty Rev-1101.

Wilson JD, Harrod MJ, Goldstein JL, Hemsell DL, MacDonald PC. Familial incomplete male pseudohermaphroditism, type 1. Evidence for androgen resistance and variable clinical manifestations in a family with the Reifenstein syndrome. N Engl J Med 1974;290(20):1097–103.

Witherspoon L, Flannigan R. Male factor infertility: Initial workup and diagnosis in primary care. Can Fam Physician 2021;67(4):248–54. https://doi.org/10.46747/cfp.6704248. 33853910. PMCID: PMC8324160.

Wu C, Fan S, Qian Y, Zhou Y, Jin J, Dai Z, et al. 17α-Hydroxylase/17, 20-lyase deficiency: clinical and molecular characterization of eight Chinese patients. Endocr Pract 2017;23(5):576–82.

Wu SM, Gao JZ, He B, Long WJ, Luo XP, Chen L. A novel NR0B1 gene mutation causes different phenotypes in two male patients with congenital adrenal hypoplasia. Curr Med Sci 2020;40(1):172–7.

Xu Y, Chen Y, Li N, Hu X, Li G, Ding Y, et al. Novel compound heterozygous variants in the LHCGR gene identified in a subject with Leydig cell hypoplasia type 1. J Pediatr Endocrinol Metab 2018;31(2):239–45.

Ying H, Sun Y, Wu H, Jia W, Guan Q, He Z, et al. Posttranslational modification defects in fibroblast growth factor receptor 1 as a reason for normosmic isolated hypogonadotropic hypogonadism. Oxid Med Cell Longev 2020;2020, 2358719.

Yu S, Rubin M, Geevarughese S, Pino JS, Rodriguez HF, Asghar W. Emerging technologies for home-based semen analysis. Andrology 2018;6(1):10–9.

Zachmann M, Völlmin JA, Hamilton W, Prader A. Steroid 17,20-desmolase deficiency: a new cause of male pseudohermaphroditism. Clin Endocrinol (Oxf) 1972;1(4):369–85.

Zhang Q, He HH, Janjua MU, Wang F, Yang YB, Mo ZH, et al. Identification of two novel mutations in three Chinese families with Kallmann syndrome using whole exome sequencing. Andrologia 2020;52(7), e13594.

Zhang T, Ma X, Wang J, Jia C, Wang W, Dong Z, et al. Clinical and molecular characterization of thirty Chinese patients with congenital lipoid adrenal hyperplasia. J Steroid Biochem Mol Biol 2021;206:105788.

Zhou C, Niu Y, Xu H, Li Z, Wang T, Yang W, et al. Mutation profiles and clinical characteristics of Chinese males with isolated hypogonadotropic hypogonadism. Fertil Steril 2018;110(3):486–495.e5.

Zidoune H, Martinerie L, Tan DS, Askari M, Rezgoune D, Ladjouze A, et al. Expanding DSD phenotypes associated with variants in the DEAH-box RNA helicase DHX37. Sex Dev 2021;22:1–9.

Mineralocorticoid excess

3.3.1.1 Introduction

3.3.1.1.1 Background physiology

Three paths are delineated for the synthesis of aldosterone, cortisol, and sex steroids. Hydroxylation at C-17 is not needed for aldosterone production but is necessary for the synthesis of cortisol or the sex steroids. Aldosterone is synthesized from cholesterol (Fig. 3.3.1.1) in a number of enzyme catalyzed steps in the *zona glomerulosa* of the adrenal cortex. Hydroxylations of the steroid nucleus at positions C-11 and C-21 are important steps in the generation of aldosterone.

21-Hydroxylase (CYP17) is a microsomal enzyme, whereas two forms of 11-hydroxylase are in the mitochondria. The 11-hydroxylases convert 11-deoxycortisol to cortisol and deoxycorticosterone (DOC) to corticosterone in both *zona fasciculata* and *glomerulosa*. CYP11B1 normally catalyzes the terminal step of cortisol synthesis in the *zona fasciculata* under the primary control of adrenocorticotrophin (ACTH). The enzyme is expressed at high levels throughout the human adrenal cortex. Two 11-hydroxylase enzymes catalyze 18-hydroxylation of 11-deoxycortisol and 11-deoxycorticosterone but only aldosterone synthase (CYP11B2) can carry out the subsequent reaction to produce aldosterone, thereby restricting aldosterone synthesis to the *zona glomerulosa*. The hydroxylases are members of the cytochrome P450 superfamily (shortened to CYP when referring to the enzyme, and *CYP* for the gene) of hemoproteins. The enzymes have active domains for heme binding, substrate recognition, active site, and membrane anchoring to endoplasmic reticulum or mitochondria.

The production of aldosterone is under the control of the renin-angiotensin system with angiotensin II the major stimulus to aldosterone production. The major trigger for renin release is a decrease in renal perfusion pressure as a result of a reduction in extracellular fluid volume after sodium depletion, hemorrhage, or hypotension. Renin, a protease, is produced in active and inactive forms and methods are available for measuring active renin, which may in time replace the measurements of renin activity. Following release, the renin converts angiotensinogen to a decapeptide, angiotensin I, which is then converted to an octapeptide, angiotensin II by the enzyme angiotensin converting enzyme in the lungs (or ACE) (Fig. 3.3.1.2).

Angiotensin II stimulates aldosterone biosynthesis through enhanced transcription of the cholesterol side-chain cleavage enzyme (CYP11A) but also inhibits expression of CYP17 and therefore cortisol biosynthesis in a dose dependent manner. Hyperkalemia can directly stimulate the *zona glomerulosa*; conversely hypokalemia inhibits aldosterone synthesis. Hyperkalemia is believed to act through enhancement of CYP11A transcription but by a process independent of angiotensin II. ACTH in the short term acutely stimulates aldosterone through the enhanced transcription of the genes involved in steroid biosynthesis and mediated by cyclic AMP (cAMP).

The activities and expression of StAR and CYP11B2 are stimulated at higher calcium concentrations in the mitochondria. The cytosolic calcium concentrations of *glomerulosa* cells is about 100–200 nanomolar, whereas the extracellular calcium concentration is 1–2 millimolar. The sodium/calcium exchanger (NCX) and plasma calcium ATPases (PMCA) export calcium to maintain the resting low calcium concentration (Fig. 3.3.1.3).

Calcium influx is controlled by store operated channel (SOC) and two voltage activated calcium channels (CaV) which are activated by depolarization of the cell membrane. T type CaV is partially open at resting potential but the L-type CaV is closed at resting potential and opened at higher levels of membrane depolarization. The membrane potential and cytosolic calcium concentrations depend on the actions of potassium channels (TASK or KCN), calcium channels (CaV), sodium/potassium exchangers (ATP1A1), sodium/calcium exchanger (NCX), sodium/hydrogen exchanger (NHE) and plasma membrane calcium ATPases (PMCA). The specific contributions of these pathways to calcium signaling in *glomerulosa* cells is largely unknown. More detail can be found in a 2014 review by Bollag.

3.3.1.2 Mineralocorticoid hypertension

Hypertension is a major medical problem affecting approximately 15% of the worldwide population. A number of possible mechanisms have been identified (Table 3.3.1.1).

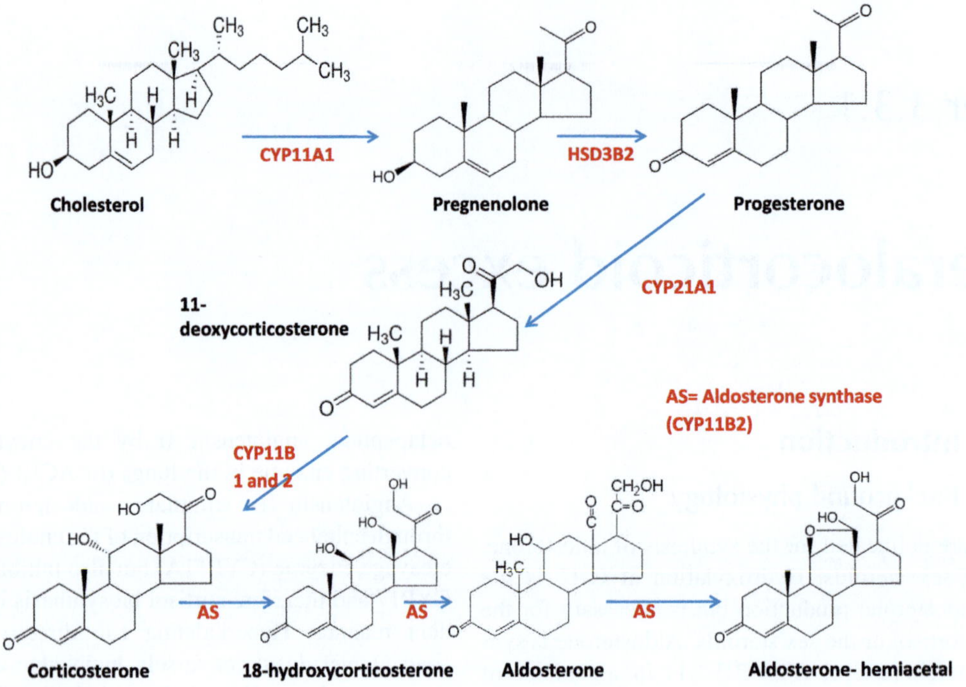

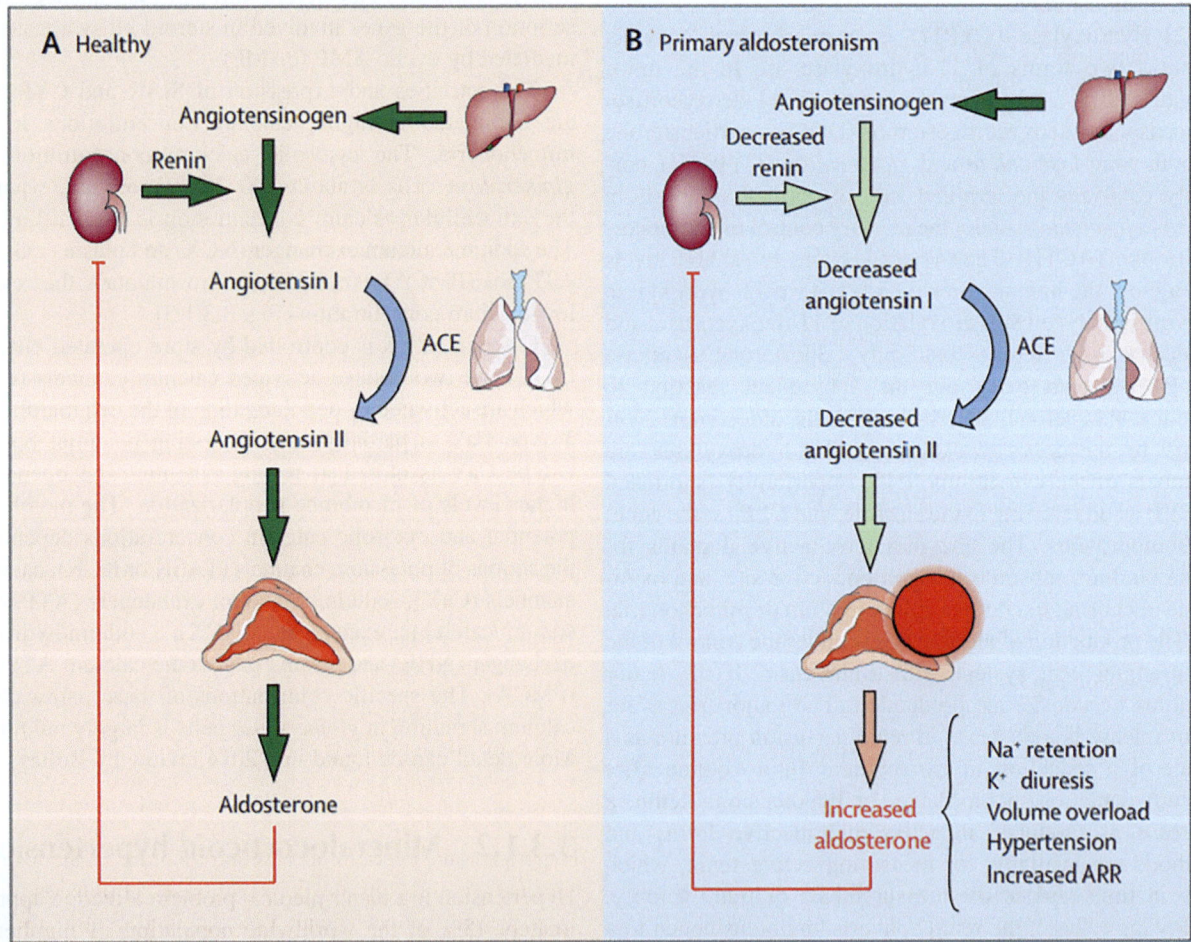

FIG. 3.3.1.1 Aldo synthesis.

FIG. 3.3.1.2 Physiology and pathophysiology of the renin angiotensin aldosterone system Physiological regulation (A). Dysregulation in primary aldosteronism (B). ACE = angiotensin converting enzyme. ARR = aldosterone-renin ratio. *(From Reincke M, Bancos I, Mulatero P, Scholl UI, Stowasser M, Williams TA. Diagnosis and treatment of primary aldosteronism. Lancet Diabetes Endocrinol 2021; 9(12): 876–892. Fig. 1 p. 877. Lancet Diabetes Endocrinol 9:867.)*

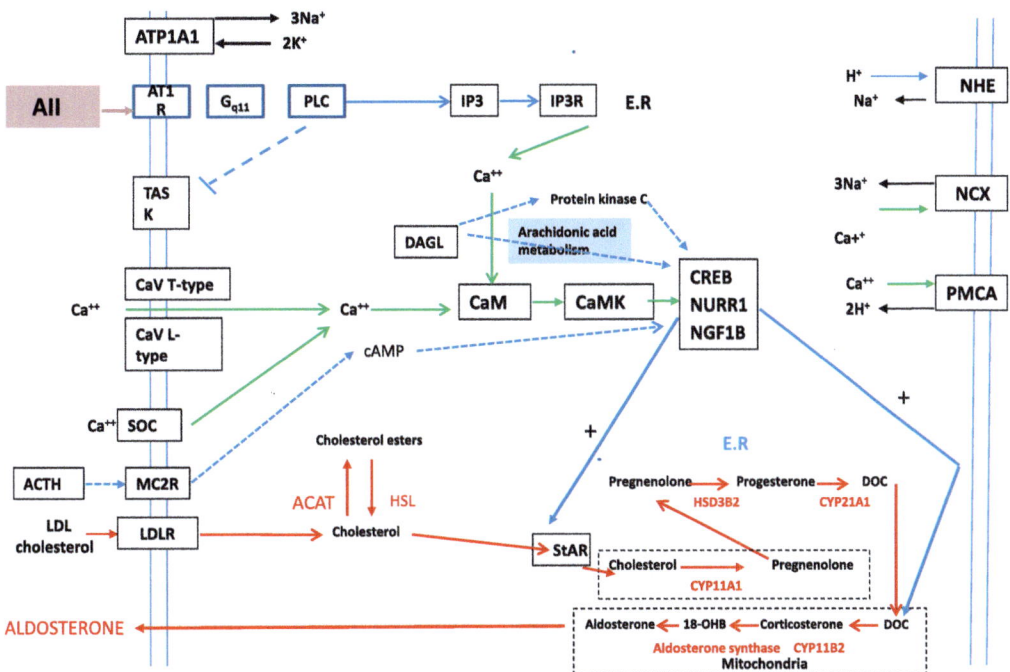

FIG. 3.3.1.3 Control of aldosterone synthesis by angiotensin I through calmodulin and calcium signaling pathways. *AII,* angiotensin II; *AT1R,* angiotensin II receptor type 1; *ATPA1,* sodium/potassium ATPase; $G_{q/11}$, guanine protein subunit alpha 11; *PLC,* phospholipase C; *IP1,* inositol triphosphate; *IP3R,* inositol triphosphate receptor; *DAGL,* diacylglycerol; *CaV-T,* voltage dependent Ca channel type T; *CaV-L,* voltage dependent Calcium channel type L; *CAM,* calmodulin; *CAMK,* calmodulin kinase; *CREB,* cAMP responsive element binding protein 3; *NURR1,* nuclear receptor subfamily 4 member 2; *NGF1B,* nerve growth factor 1B like receptor; *SOC,* calcium release-activated calcium channel protein; *TASK,* potassium two pore domain channel subfamily K member 3 (KCN3); *LDLR,* low density lipoprotein receptor; *NHE,* sodium/hydrogen exchanger; *NCX,* sodium/calcium exchanger; *PMCA,* plasma membrane calcium ATPase; *E.R.,* endoplasmic reticulum; *ACT,* acyl-CoA cholesterol transferase; *HSL,* hormone sensitive lipase; *HSD3B2,* 3β-hydroxysteroid dehydrogenase and Δ5-Δ4-isomerase; *CYP21A1,* 21-hydroxylase; *CYP11A1,* 11β-hydroxylase type 1; *CYP11B2,* 11β-hydroxylase type 2; *AS,* aldosterone synthase; *StAR,* steroidogenic acute regulatory protein; *DOC,* 11-deoxycorticosterone; *ACTH,* adrenocorticotrophic hormone; *MC2R,* melanocortin 2 receptor; *cAMP,* cyclic AMP. *(Author original.)*

Most of the patients are without symptoms for many years and, in the majority, no cause for the elevated blood pressure can be found.

The lowering of high blood pressure however has been shown to reduce the incidence of stroke, coronary artery disease and also renal failure. General practitioners manage patients with drugs that act at sites within the renin-angiotensin-aldosterone system (RAAS). Guidelines for the management of these patients do not suggest steroid measurements until there is poor response to combinations of 3 or 4 drugs which in the opinion of the author is unfortunate. Treatment for high blood pressure is lucrative for the pharmaceutical industry but precise targeting of treatment could be more effective.

Hypertension can result from overactivity of any one or more components of the RAAS (see Chapter 2.3.3.1). In addition to aldosterone, the steroids deoxycorticosterone (DOC) and corticosterone are mineralocorticoids and these become significant in certain forms of hypertension. The classic clinical markers are

- hypertension which is often long-standing and resistant to conventional drugs and
- hypokalemia that can manifest with muscle weakness, cramp, headache, palpitations, nocturia, etc.

Hypertension is a common vascular disorder influenced by environmental factors and racial origin. Hypertension is salt sensitive and reflects renal, vascular, and central mechanisms (Fig. 3.3.1.4). Genetic disorders that cause excesses of mineralocorticoid steroids cause hypertension by

TABLE 3.3.1.1 Mechanisms of hypertension.

Hypertension can be the result of:

- Increased peripheral resistance
- Increased mineralocorticoid production
- Increased overall mineralocorticoid activity
- Increased renal sodium reabsorption by factors which do not directly influence mineralocorticoids but lead to compensatory aldosteronism.

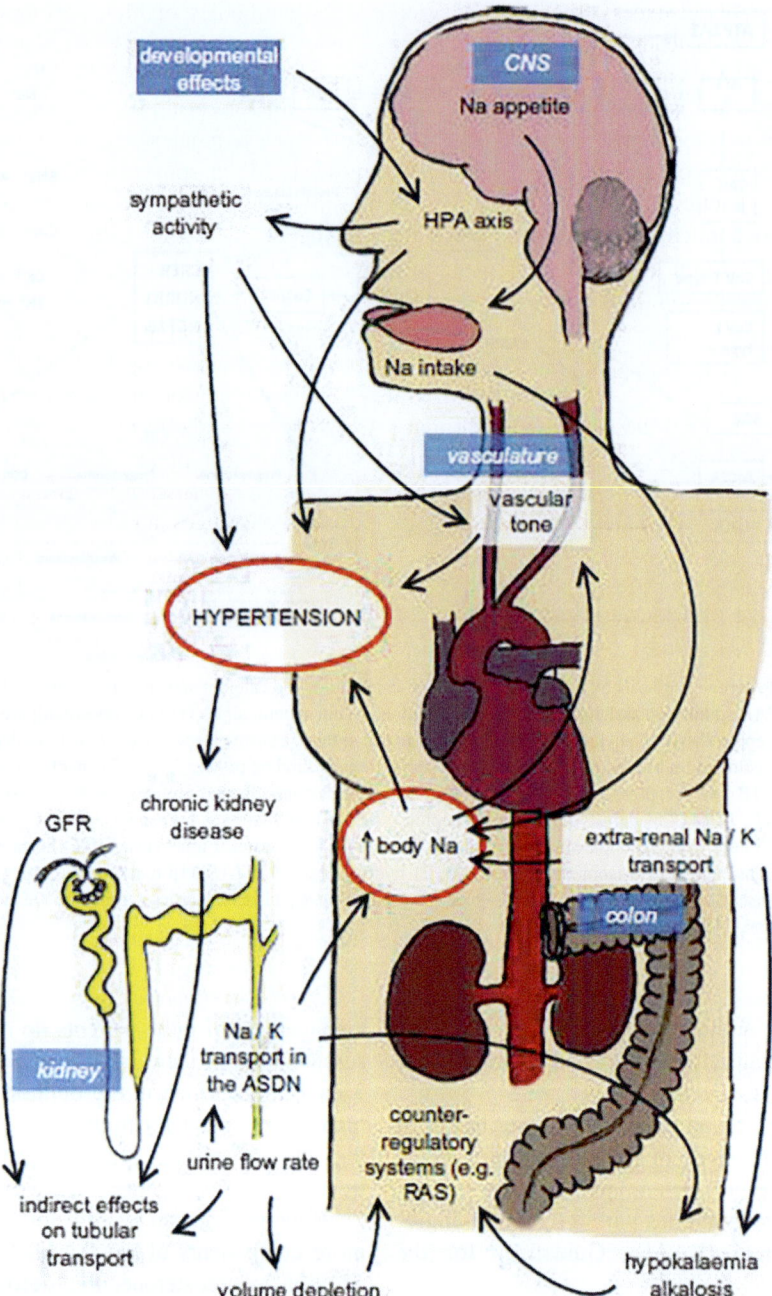

FIG. 3.3.1.4 Mechanisms contributing to systemic arterial hypertension in the metabolic syndrome. Hypertension is salt-sensitive and reflects renal, vascular, and central mechanisms. Hypertension driven by sodium retention arising from renal mineralocorticoid actions is not the whole story, and glucocorticoid receptor blockade is likely to be beneficial (*ASDN*, aldosterone-sensitive distal nephron). *(From Bailey MA. 11β-Hydroxysteroid dehydrogenases and hypertension in the metabolic syndrome. Curr Hypertens Rep 2017; 19(12): 100. Fig. 2 p. 5. Curr Hypertens Rep 19(12):100.)*

facilitating sodium retention at renal tubules. Such a mechanism operates in primary aldosteronism and some forms of congenital adrenal hyperplasia (CAH).

High production rates of cortisol in Cushing's syndrome or the condition of glucocorticoid resistance may provoke hypertension through the action of cortisol on the mineralocorticoid receptor. Some forms of hypertension with endocrine features are now recognized as compensation for mutations in genes for ion transporter proteins. The genetic bases of certain hypertensive conditions have been studied to identify mutations that cause susceptibility. Mutations in at least 10 genes have been shown to alter blood pressure, severally of these acting through a common pathway of changing salt and water reabsorption in the kidney. Such findings in rare disorders may provide insight into mechanisms underlying more common forms of hypertension.

Although mineralocorticoid excess is one explanation for hypertension, a diagnosis leads to the opportunity to render in some cases a surgical cure or the ability to achieve a dramatic pharmacological response in the treatment of hypertension. Exciting developments have included the description of new subtypes of mineralocorticoid excess and elucidation of the genetic basis of certain forms. In addition to classical biochemical measurements, there are now options to examine certain genes in order to reliably screen for causes of hypertension.

3.3.1.3 Primary aldosteronism

In 1955, Jerome Conn described the syndrome of primary aldosteronism (PA), subsequently called Conn's syndrome, in which the increased secretion of aldosterone typically presents with hypertension, hypokalemia, and hyporeninemia. In the case described by Conn, the removal of an **adrenal adenoma** was followed by a normalization of blood pressure and of biochemical parameters. Following the first description it was realized that a similar clinical disease is caused by **bilateral adrenal hyperplasia**. Further cases have been published in which not only the secretion of aldosterone was increased but also of other mineralocorticoids and sometimes of cortisol. Screening for PA is recommended when systolic blood pressure is more than 160–180 mmHg and diastolic greater than 100–110 mmHg, and if blood pressure is greater than 140/90 mm HG when on 3 antihypertensive drugs (Funder et al., 2016; Mulatero et al., 2020). Some adrenal tumors are carcinomas that need urgent surgical removal before turning metastatic. Excess aldosterone may negatively affect insulin action leading to an increased risk for dysglycemia (Grewal et al., 2021). The causal role of insulin action in glucose intolerance associated with PA remains unclear.

In the earliest investigations, blood samples must be taken on several occasions for measurement of electrolytes. Patients with primary hyperaldosteronism have in the past been considered for investigation only when repeated plasma potassium values were below 3.7 mmol/L. This is now not the case. Causes for potassium depletion (diarrhea, purgatives, liquorice ingestion, diuretics) must be excluded when taking case notes. Potassium supplements may be needed to restore blood concentrations to normal. The tests should be conducted under medical supervision. Not all tests have been validated in children and may be difficult.

3.3.1.3.1 Plasma aldosterone

Aldosterone concentration in plasma is now often measured without extraction and without chromatography (Fortunato et al., 2015), although these determinations come with problems for laboratories in which immunoassay (IA) techniques are used. Such assays should only be performed

when antibodies of high specificity are available. There are reference laboratories or research centers which perform such tests using LC-MS/MS (Taylor et al., 2009; Tan et al., 2020; Gibbons et al., 2021). Plasma aldosterone is now measured by homogeneous immunoassays and chemiluminescent IA, in some cases with automation. Aldosterone results can vary by a factor of two between methods (Deng et al., 2018; Schirpenbach et al., 2006; Pizzolo et al., 2006). Assays are subject to interferences with false positive results in nonmass spectrometric methods (Constantinescu et al., 2020; Lopez et al., 2019). The use of antihypertensive agents is often the problem.

Aldosterone concentrations in plasma are extremely sensitive to postural effects but rarely can patients in a clinic be kept for 1h or preferably 2h in the horizontal position. Aldosterone and renin increase on standing after lying for 2h due to changes in renal blood flow and release of renin. Patients with primary aldosteronism have lost this response (Fig. 3.3.1.5) (Fuss et al., 2022).

Plasma aldosterone concentrations are influenced by day to day or period to period fluctuations in aldosterone secretion that can be observed in a 24h profile (Fig. 3.3.1.6) (Hurwitz et al., 2004), although in the case of the plasma aldosterone tests minute to minute variations may cause further difficulties in borderline cases of aldosterone excess.

The diagnosis of primary hyperaldosteronism rests on finding plasma aldosterone concentrations that are high for the corresponding PRA. Aldosterone assays are available in the full range of technologies and direct measurements by radioimmunoassay can be as good as results after solvent extraction (Fig. 3.3.1.7) (Miller et al., 1997), although LC-MS/MS measurements are now preferred.

Since removal of an adenoma corrects the blood pressure, localization of the source of aldosterone assists a decision on the surgical approach for the operation. Primary aldosteronism is rarely discovered in children (Dinleyici et al., 2007). An algorithm for the diagnosis of primary aldosteronism in adults can be followed to reach a diagnosis of a rare pediatric problem (Funder et al., 2016) although taking blood samples is more problematic in children although with fine needles used these days pain is minimal and short-lived.

3.3.1.3.1.1 *Preparation of patients for diagnosis of primary aldosteronism*

The patients have to be prepared for some time prior to the detailed endocrine studies. Primary aldosteronism is more often investigated at a late stage due to resistance to common therapeutic measures, rapid development of hypokalemia after therapy with thiazide derivatives or the discovery of an enlarged adrenal mass by computerized tomography. Since most forms of therapy used in

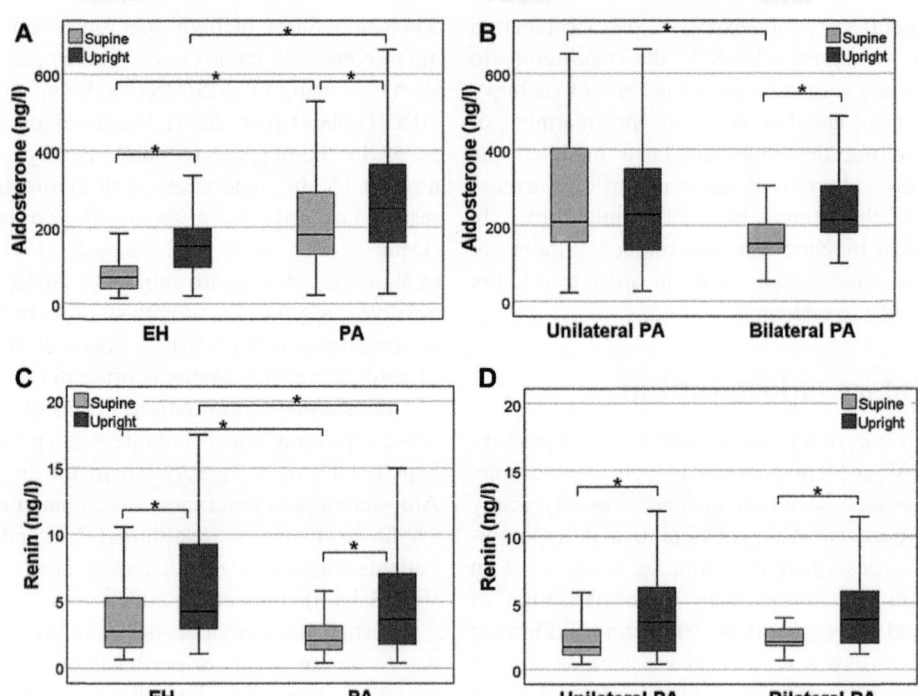

FIG. 3.3.1.5   Differences in aldosterone (A and B); and renin (E and F), levels during postural stimulation testing in patients with essential hypertension (EH, $n = 54$) and primary aldosteronism (PA, $n = 106$), as well as unilateral ($n = 55$) and bilateral ($n = 29$) primary aldosteronism. *P less than .05. Not shown here C and D, cortisol (C and D). *(From Fuss CT, Brohm K, Fassnacht M, Kroiss M, Hahner S. Reassessment of postural stimulation testing as a simple tool to identify a subgroup of patients with unilateral primary aldosteronism. J Clin Endocrinol Meab 2022;107(2):e865–73. Fig. 1 p. e868. JCEM 107:865.)*

hypertensive patients influence aldosterone secretion (Table 3.3.1.2) it is necessary to stop therapy for at least 2 weeks, in some cases up to 6 weeks prior to the collection of samples for diagnostic tests.

Depending on the clinical features, it is often impossible to keep this schedule. The question as to the therapy allowed during this period is not clear. Reserpine is probably the only hypertensive drug which does not influence aldosterone secretion. In the case of other antihypertensive therapy, the 2 week withdrawal period can be shortened. If the plasma or urine aldosterone concentrations are not elevated, then hyperaldosteronism can be excluded.

When patients are being treated with drugs which suppress aldosterone secretion the situation is more difficult, for instance beta-blocking drugs or converting enzyme inhibitors. In these cases, the interruption of treatment is inevitable during which time the high blood pressure may return and also the patient may suffer the beta-blocker withdrawal syndrome. All these doubts stimulate the view that tests to exclude primary aldosteronism have to be performed immediately after detection of hypertension, and preferably prior to the initiation of treatment.

Physical and psychological stresses of different kinds have to be eliminated. Ideally, the patient should be in electrolyte equilibrium for 2 or 3 days prior to the investigations. In the United States and in some other countries with

clinical research wards, sodium balance can be achieved. In other countries, this is less easy, but can be compromised. In Germany, for instance, with the average food of the population, a relative sodium equilibrium is achieved with appropriate advice so that patients excrete sodium and potassium within the normal range over the period of the test. In any case, sodium excretion should be controlled in the days immediately prior to and during the tests for aldosterone production. An additional problem is the body position. A raised plasma aldosterone concentration after 1 h or preferably 2 h in the **supine position** is strongly indicative of primary aldosteronism. However, this can only be assured when patients are controlled by hospital personnel.

3.3.1.3.2   Urinary aldosterone

For the **urinary aldosterone determinations**, a high level of rest during the 24-h collection period is desirable. It is helpful to check urinary creatinine excretion as an indication of a complete collection (Ames et al., 2016) or to standardize results in spot samples. The normal range for creatinine excretion is wide (7–18 mmol/24 h) and it will be ideal to have three collections from the patient to check the personal values, which should agree. For the first screening to exclude primary aldosteronism, methods have

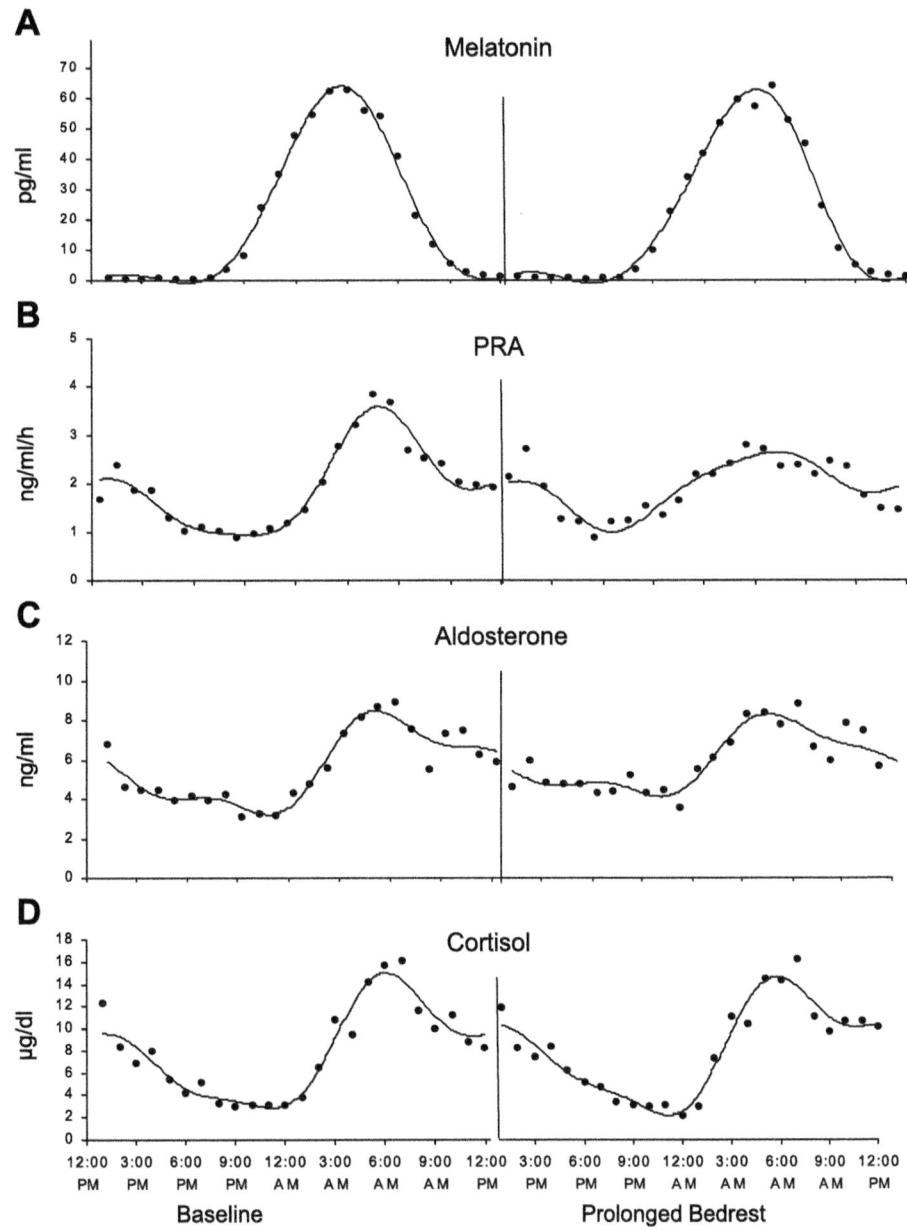

FIG. 3.3.1.6 Patterns of hormone secretion at baseline and after prolonged bed rest. *A:* melatonin. *B: plasma renin activity (PRA).C: aldosterone. D: cortisol. Solid lines, average* of the harmonic regressions that were fit to individual subject data; F, group averages of the observed data. *(From Hurwitz S, Cohen RJ, Williams GH. Diurnal variation of aldosterone and plasma renin activity: timing relation to melatonin and cortisol and consistency after prolonged bed rest. J Appl Physiol (1985) 2004;96(4):1406–14. Fig. 1 p. 1408. J Appl Physiol (1985) 96:1406.)*

for many years been able to measure aldosterone in unprocessed urine (Vecsei et al., 1982a). Highly specific antibodies with high affinity for aldosterone are essential. Pal Vecsei was an expert in this particular area with paper chromatography in most cases to improve the specificities of the assays. Urinary-free aldosterone, the aldosterone-18-glucuronide, and tetrahydroaldosterone can also be measured. For young children, especially girls, a 24-h urine collection is a special problem. For the above reasons, blood samples are preferred.

Urinary aldosterone excretion rates in samples collected for 24h are perhaps more reliably interpreted than plasma steroid concentrations. Of course, in cases with highly elevated plasma or urinary values, these variations cannot seriously influence the interpretation when making a diagnosis. The urinary excretion of the aldosterone metabolite, aldosterone-18-glucuronide is used to evaluate increased aldosterone secretion although this test is not widely used. The conjugate is uniquely hydrolyzed by acid at pH 1 before the quantitative analysis. The estimation of

FIG. 3.3.1.7 Comparison of methods measuring plasma aldosterone concentration: Linear regressions between various immunoassays (RIA and CLIA) and two different LC-MS/MS methods: (A) all range, (B) PAC below 500 pmol/L. *(From Baron S, Amar L, Faucon AL, Blanchard A, Baffalie L, Faucard C, Travers S, Pagny JY, Azizi M, Houillier P. Criteria for diagnosing primary aldosteronism on the basis of liquid chromatography-tandem mass spectrometry determinations of plasma aldosterone concentration. J Hypertens 2018;36(7):1592–1601. Fig. 1 p. 1595.)*

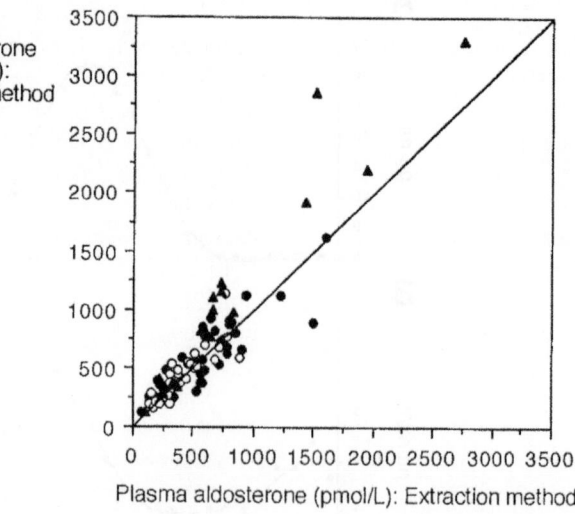

TABLE 3.3.1.2 Drugs and the RAAS.

Drug	Aldosterone	Renin	Effect on ARR
Calcium channel blocker (Nifedipine, Amlodipine)	Minimal decrease	Minimal increase	No effect
Calcium channel blocker (Verapamil)	Small decrease	Minimal increase	No major effect
α-Blocker (Prazosin, doxazosin)	Minimal	Minimal	No effect
Vasodilators (Diazoxide, hydralazine)	No effect	No effect	No effect
ACE inhibitors (Captopril)	Decreased	Increased	False negative
Diuretics (Furosemide-loop, hydrochlorthiazide—distal tubule)	Increased	Increased markedly	False negative
Minoxidil—vasodilator/diuretic (baldness)	Minimal	Increased	False negative
AII receptor blocker (Sartans)	Decreased	Increased	False negative
Spironolactone (K sparing diuretic)	Increase	Large increase	False negative
Beta-blockers (Metoprolol, atenolol)	Minimal	Decreased	False positive
Methyl-DOPA	Minimal decrease	Decreased	False positive
Direct renin inhibitor (aliskiren)	Decreased	Decreased	False positive

Hypokalemia—inhibits aldosterone (FALSE negative); potassium loading (false positive).
Estrogens (pregnancy, HRT, O/C) increase renin substrate (false negative).
NSAIDs retain sodium, reduce PRA (false positive).
Renovascular and malignant hypertension, increase renin activity (false negative).

tetrahydroaldosterone excretion has been shown to be more useful (Abdelhamid et al., 2003). Tetrahydroaldosterone may be subject to 21-dehydroxylation by intestinal bacteria in the course of the entero-hepatic circuit (Winter and Bokkenheuser, 1978; Winter et al., 1984). Urine samples particularly from females should be refrigerated until the steroids analysis to avoid bacterial degradation. A major steroid in primary aldosteronism was identified as 18-hydroxycortisol by accurate mass measurement for $C_{21}H_{38}O_4$ and degradative chemistry to assign hydroxyls at carbons 11β, 18, and 21 (Chu and Ulick, 1982; Miyamori et al., 1992). As with other steroid assays there have been developments through RIA (Corrie et al., 1985), time resolved immunofluorescence (DELFIA) (Morra di Cella et al., 2002, monoclonal antibodies in immunoassay (Kohno et al., 1994), GC-MS (Shackleton, 1983) and LC-MS/MS (Jin et al., 2013). The performances have been with assays showing variable discriminatory value (Lenders et al., 2018; Reynolds et al., 2005).

3.3.1.3.3 Plasma renin activity

Plasma renin activity or concentration should be determined to establish if the raised aldosterone is autonomous. PRA is determined by measuring the rate of angiotensin I production from endogenous angiotensinogen at 37°C and pH 5, usually by immunoassay of angiotensin I. Automated platforms for renin, or aldosterone or both renin and aldosterone have been available for some time (Teruyama et al., 2022; Tamura et al., 2021; Li et al., 2019; Burrello et al., 2016; Manolopoulou et al., 2015; Dorrian et al., 2010). Mass spectrometry has been used for angiotensin measurement since 1999 (Fredline et al., 1999; Bystrom et al., 2010; Owen et al., 2014; Van Der Gugten and Holmes, 2016; Gibbons et al., 2021). Angiotensin I with arginine $^{13}C_6$, $^{15}N_4$ in the angiotensin sequence (DRVYIHPHFL) is used as an internal standard. Renin activity can be reported for minutes or hours. Direct renin concentrations are expressed in units per liter or ng/liter.

RAS equilibrium analysis has recently been reported to quantify different angiotensin metabolites (e.g., AngI, AngII, AngIII, AngIV, angiotensin [1–7], angiotensin [1–5]) simultaneously from 1 blood sample using LC-MS/MS methodology (Binder et al., 2019; Guo et al., 2020; Bernstone et al., 2021). Compared with the measurement of circulating AngII (cAngII), which requires immediate inhibition of angiotensin-processing enzymes by adding a protease inhibitor cocktail to newly collected blood, sample preparation for measuring equilibrium AngII (eqAngII) is more convenient and straightforward. The principle of this assay is to establish an ex vivo equilibrated status of the RAS by incubating prestored frozen serum at 37°C and pH 7.4 for 1 h without blocking any angiotensin-producing or -degrading enzymes. Because AGT exists in a very large "molar excess" in human peripheral blood (in the micromolar range) compared with renin (in the picomolar range), this would allow "adequate" enzyme-substrate interactions for angiotensin formation and degradation during the incubation period. The final eqAngII level is determined by the activity of all soluble angiotensin-processing enzymes present in tested samples and is maintained by the equal formation and degradation rates for each peptide in the status of RAS equilibrium (Guo et al., 2020). This required further validation.

3.3.1.3.4 Aldosterone: Renin ratio (ARR)

The ratio of aldosterone to renin (ARR) is used as the biomarker for PA (Weinberger and Fineberg, 1993) but many different units are used to report each analyte so the cut-off can be from 1.6 to 1000 (Table 3.3.1.3). The most commonly used ARR are 30 when both are expressed in mass units and 750 when both are expressed in SI units.

TABLE 3.3.1.3 Variations in ARR according to units of measurement.

	PRA (ng/mL/h)	PRA (pmol/L/min)	DRC (mU/L)	DRC (ng/L)
PAC (ng/dL)	20	1.6	2.4	3.83
	30	2.5	3.7	5.7
	40	3.1	4.9	7.7
PAC (pmol/L)	750	60	91	144
	1000	80	122	192

Conversion factor for PRA (ng/mL/h) to DRC (mU/L) is 8.2 but may be as high as 12.
There is poor correlation between DRC and PRA when PRA is <1 ng/mL/h which is often the case in primary aldosteronism.
The most commonly adopted cut off values are 30 when using mass units and 750 when PAC is expressed in SI Units.

Females have higher ARR ratios than males, and false positive ratios can occur during the luteal menstrual phase and while taking an oral ethynylestradiol/drospirenone (but not implanted subdermal etonogestrel) contraceptive, but only if calculated using direct renin concentration and not plasma renin activity.

Where feasible, diuretics should be ceased for at least 6 weeks and other interfering medications at least 2 weeks before ARR measurement, substituting noninterfering agents (e.g., with verapamil, slow-release hydralazine and prazosin or doxazosin) when required. Hypokalemia should be corrected and a liberal salt diet encouraged. Collecting blood mid-morning from seated patients following 2–4 h upright posture improves sensitivity.

Although the ARR is the most reliable screening test for primary aldosteronism, false positives and negatives occur. Dietary salt restriction, concomitant malignant or renovascular hypertension, pregnancy, and treatment with diuretics (including spironolactone), dihydropyridine calcium blockers, angiotensin converting enzyme inhibitors, and angiotensin receptor antagonists can produce false negatives by stimulating renin. Selective serotonin reuptake inhibitors lower the ratio. Because potassium regulates aldosterone, uncorrected hypokalemia can lead to false negatives. Beta-blockers, alpha-methyldopa, clonidine, and nonsteroidal antiinflammatory drugs suppress renin, raising the ARR with potential for false positives. False positives may occur in patients with renal dysfunction or advancing age.

The ARR is a screening test only and should be repeated one or more times before deciding whether to proceed to confirmatory suppression testing. Liquid chromatography-tandem mass spectrometry aldosterone assays represent a major advance toward addressing inaccuracies inherent in other available methods.

3.3.1.3.5 Diagnosis of primary aldosteronism

The chemical diagnosis of primary aldosteronism has two different tasks:

(a) answer the question; Does the particular patient have primary aldosteronism or not?
(b) Determine which form of primary aldosteronism or related disease does the patient have?

Precise diagnosis of primary aldosteronism is approached after exclusion of other possible causes of hypertension with estimations of electrolytes. Especially relevant is the circulating potassium concentration. Low serum potassium concentrations were obligatory in primary aldosteronism investigations and this opinion was widely followed in clinical practice. The impression from reviews is that about 20%–30% of patients with primary aldosteronism are occasionally, and around 10% are regularly, normokalemic.

3.3.1.3.5.1 Organization for the diagnosis or exclusion of primary aldosteronism

When the initial screening has revealed an elevated aldosterone value, then confirmation is necessary in a method using separation with column chromatography or HPLC. If it is possible, further steroids should be determined, for example, not only urinary aldosterone 18-glucuronide but also tetrahydroaldosterone, free urinary aldosterone aldosterone, deoxycorticosterone, corticosterone, and 18-hydroxylated precursors (Fig. 3.3.1.8) (Arlt et al., 2017). Few laboratories are equipped for those investigations.

Of course, other biochemical studies such as determination of plasma renin activity and electrolytes are also necessary. When these further tests confirm that aldosterone production is elevated, then steps are necessary to differentiate the various forms of primary aldosteronism.

Technical aspects

Most of the activity around steroid determinations in patients with hypertension took place up to 1985, in many cases using immunoassay with antibodies from the laboratory of Pal Vecsei in Heidelberg who was using paper chromatography for purification of plasma extracts (Abdelhamid et al., 2003). Celso Gomez-Sanchez was another source of good antibodies. Most antibodies were raised to the 3-carboxymethyloxime steroid derivative coupled to albumin (Chandler et al., 1976; Martin et al., 1975). Column chromatography with Sephadex LH-20 before immunoassay (Schöneshöfer et al., 1981) was introduced then high performance liquid chromatography for rat plasma (Imaizumi et al., 1987); but these were assays only

suitable for use in research settings. Aldosterone metabolites could be measured by HPLC with fluorescent (Yoshitake et al., 1992) and chemiluminescent detection (Ishida et al., 1992).

In some cases, high blood pressure can be characterized by finding that steroids other than aldosterone are produced in excess of normal. Thus in addition to aldosterone, both deoxycorticosterone and corticosterone are relatively commonly elevated, even cortisol can be elevated (Arlt et al., 2017) (Fig. 3.3.1.8). Increased concentrations of certain uncommon steroids, such as 19-hydroxy and 19-nor-corticosterone, 19-nor deoxycorticosterone, 19-nor-aldosterone as well as 18-hydroxycortisol and 18-oxocortisol have historically also been found in quantitative analysis by immunoassay, bioassays in adrenalectomized rats of different substrains, toad bladder and GC-MS (Table 3.3.1.4).

19-Nor-deoxycorticosterone is elevated in rats with hypertension induced in certain strains of rat after unilateral adrenalectomy and nephrectomy in combination with demedullation of the contralateral adrenal (Gomez-Sanchez et al., 1983). This latter step not only removes the adrenal medulla, but most of the *zona fasciculata*. The *zona glomerulosa* at first remains untouched. Later however, the remaining adrenal tissue decomposes and new tissue develops. In parallel with the development of the new adrenal tissue, blood pressure becomes elevated particularly when the rats are fed with a salt rich diet. 19-Nor-DOC was found to have comparable hypertensinogenic potency to other 19-nor-steroids such as 19-nor-corticosterone, 19-nor-progesterone and more recently 19-nor-aldosterone (Takeda et al., 1990, 1991, 1992, 1995, 1996). The group of Melby found that in spontaneous hypertensive rats and also in humans with primary aldosteronism the excretion of 19-nor-DOC is increased (Griffing et al., 1983). The role of other 19-nor-steroids such as 19-nor-progesterone, -corticosterone, and -aldosterone as a cause or additional cause in the diseases related to primary aldosteronism have not yet been studied but are now amenable to new technology. Recent publications of steroid profiles include 18-hydroxycorticosterone, 18-oxocortisol and 18-hydroxycortisol (Fig. 3.3.1.9) (Chen et al., 2022; Berke et al., 2021; Eisenhofer et al., 2021; Travers et al., 2017; Mulatero et al., 2012a) by LC-MS/MS.

Some uncommon mineralocorticoids or other steroids, e.g., the possible aldosterone precursors 18-hydroxycorticosterone or 18-hydroxydeoxycorticosterone or 21-deoxyaldosterone and/or bicyclic products (Kelly M1 material) (Lewicka et al., 1988) are produced in large amounts without concomitant increase in the usual aldosterone results. Assay interference was difficult to demonstrate other than by excluding cross reactivity. In those days, steroids were readily available and extensive lists of steroids could be tested. With the demise of the Medical Research

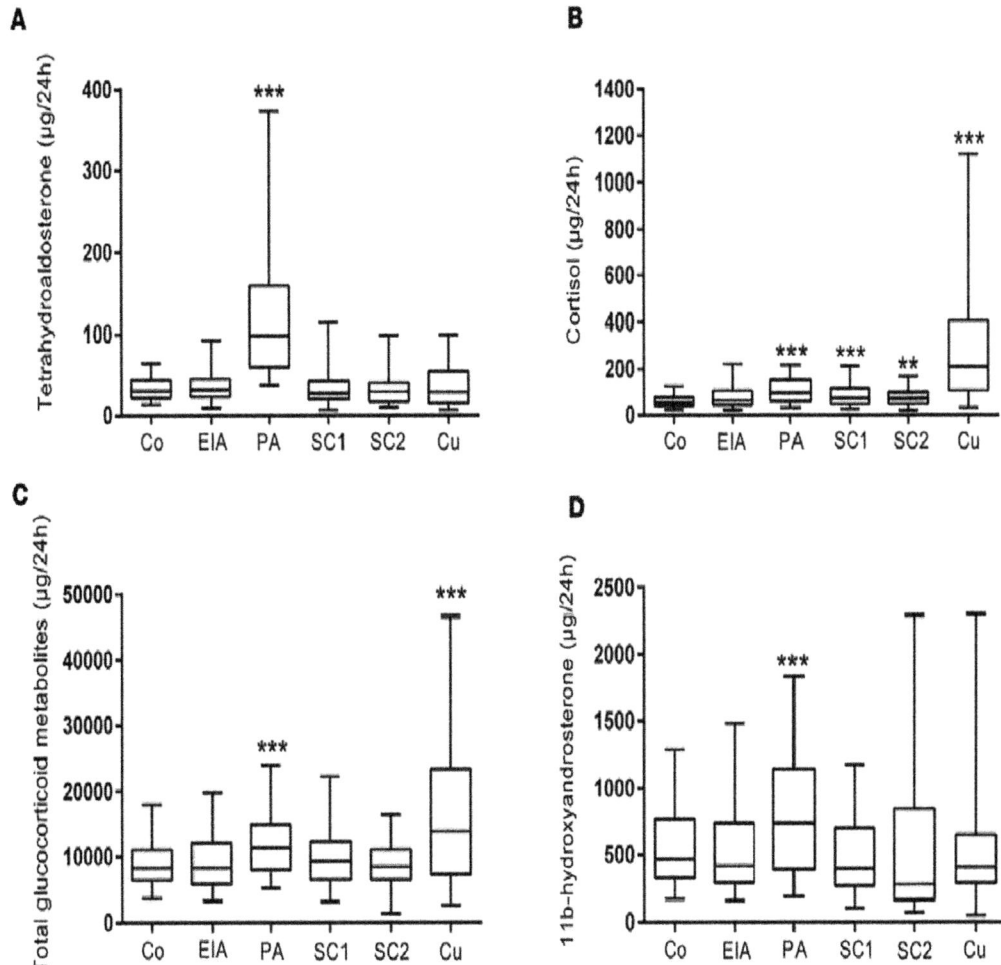

FIG. 3.3.1.8 Steroid metabolite excretion in primary aldosteronism in comparison to healthy controls and patients with endocrine-inactive and cortisol producing adrenal adenomas. The panels show the 24-h urinary excretion of tetrahydroaldosterone (A), cortisol (B), total glucocorticoid metabolites (C), and the major adrenal androgen metabolite 11β-hydroxyandrosterone (D) in primary aldosteronism patients (PA; *n* = 174) in comparison to healthy controls (Co; *n* = 162), patients with endocrine-inactive adrenal adenoma (EIA; *n* = 56), patients with subclinical Cushing's (differentiated into 2 groups: SC1 (*n* = 55), morning cortisol after 1 mg dexamethasone overnight >50 and < 138 nmoL/L; SC2 (*n* = 49), morning cortisol >138 nmoL/L), and overt adrenal Cushing's syndrome patients (Cu; *n* = 47). Boxes represent median and interquartile range, whiskers represent 5th and 95th centiles. **P < .01 versus controls, ***P < .001 versus controls. Comparisons between groups were made with linear regression models to adjust for age and sex in comparisons between all 6 groups. *(From Arlt W, Lang K, Sitch AJ, Dietz AS, Rhayem Y, Bancos I, Feuchtinger A, Chortis V, Gilligan LC, Ludwig P, Riester A, Asbach E, Hughes BA, O'Neil DM, Bidlingmaier M, Tomlinson JW, Hassan-Smith ZK, Rees DA, Adolf C, Hahner S, Quinkler M, Dekkers T, Deinum J, Biehl M, Keevil BG, Shackleton CH, Deeks JJ, Walch AK, Beuschlein F, Reincke M. Steroid metabolome analysis reveals prevalent glucocorticoid excess in primary aldosteronism. JCI Insight 2017;2(8):e93136. Fig. 3 p. 4. JCI insight 2:e93136.)*

Council Steroid Reference Collection an important source of steroids was lost.

The diagnosis of Conn's syndrome related diseases should be considered more often. Due to the rising costs of assays however and the overloading of laboratories, new organizational considerations have to be made for the exclusion of primary aldosteronism.

For many years, few laboratories offered little more than aldosterone assays, maybe with corticosterone and 11-deoxycorticosterone, rarely with 18-hydroxy steroids (Morra di Cella et al., 2002; Riepe et al., 2003). In one publication, a Delfia method for 18-hydroxycortisol had better

discrimination in diagnosis of primary aldosteronism than GC-MS or RIA (Reynolds et al., 2005). Stable isotope labeled internal standards are added to ensure specificity of the LC-MS/MS analysis, since many are not available, surrogates will be needed.

3.3.1.3.5.2 Steroid profiles (metabolome)

The steroid profile of adrenal vein blood (AV) blood has been examined by LC-MS/MS (Rossi et al., 2020; Nakamura et al., 2012; Mulatero et al., 2012a,b; Eisenhofer et al., 2016; Meyer et al., 2018) and GC-MS

TABLE 3.3.1.4 Earlier methods for hybrid steroid determination.

Steroid			Reference
18-Hydroxycorticosterone (18-OHB)	Human	Plasma	Uchida et al. (1989)
19-Noraldosterone; 18,19-dihydroxy B; 18-hydroxy-19 nor B	Human	Urine	Takeda et al. (1992)
19-nor-Aldo	Chemistry	Synthesis	Harnik et al. (1986)
5α-Dihydro-DOC	Rat	Blood pressure	Carroll et al. (1981)
19-nor DOC; 19-nor progesterone	Rat	Kidney, liver bioassays	Wynne et al. (1981)
19-OH-Aldo; 19-norAldo	Toad	Bladder	Morris et al. (1986)
19-nor-DOC	Adrenal_X rats	Blood pressure	Gorsline and Morris, (1985) and Gomez-Sanchez et al. (1979)
19-Hydroxy androstenedione	Rats	Sodium retention	Sekihara (1983b)
18-Hydroxycortisol	Human	Adrenal tissue	Gomez-Sanchez et al. (1987) and Ulick et al. (1983)
18-Hydroxycortisol	Human	Serum, urine	Mosso et al. (2001)
18-Oxo-cortisol	Human	Adrenal tissue	Ulick et al. (1983)
Kelly M1[a]	Human	Urine	Lewicka et al. (1988)
21-Deoxy-Aldo; 21-deoxy-THAldo	Human	Urine	Abdelhamid et al. (1979, 1981)
19-nor-DOC	Human	Urine	Casey et al. (1985)
2-Hydroxy-THF; 6-hydroxy-F	Human	Urine	Bournot et al. (1982)

[a]M1 is 11β,18 (S), 18:20α-diepoxy-5β-pregnan-3α-ol.

(Hána et al., 2019; Hill et al., 2019) technology. Immunoassay methods are being replaced with LC-MS/MS methods in many laboratories because of improved specificity and higher accuracy. The chromatography stages of GC-MS and GC-MS/MS have greater resolution but time to results is longer than LC-MS/MS methods mainly because of the initial derivative step to make steroids stable to high temperature analysis. From these investigations, steroids other than cortisol have been found to be better markers for successful adrenal venous catheterization. Androstenedione, DHEAS, 17-hydroxyprogesterone and 11-deoxycortisol have been superior to cortisol in that regard (Ceolotto et al., 2017; Nilubol et al., 2017). Some kits are available to measure up to 15 steroids, all of the reagents to run the kit are included—SPE plates or sample clean up columns, wash buffer, extraction buffer, elution buffer, reconstitution buffer, analytical column, calibrators, internal standards, 3 level internal quality samples, rinsing solution, equilibration reagents, mobile phase A and B. Before use in a clinical laboratory a full validation is required (Le Goff et al., 2020; Honour, 2011).

Plasma 18-oxo-cortisol was higher in patients with adenoma compared with adrenal hyperplasia (Eisenhofer et al., 2016). When steroid profile results were judged against the tumor size, the concentrations of 18-oxocortisol, 18-hydroxycortisol and aldosterone were higher in patients with macroadenoma than with microadenoma (Yang et al., 2019). The concentrations of 11-deoxycorticosterone, DHEAS and pregnenolone were higher in patients with microadenoma and bilateral adrenal hyperplasia (BAH). In a separate study, higher 18-oxocortisol and 18-hydroxycortisol were higher with adenomas due to KCNJ5 mutations.

Differential diagnosis between the two main forms of primary aldosteronism; adrenal adenoma and hyperplasia

Other biochemical studies contribute to the differential diagnosis. **The response of plasma aldosterone to assumption of the upright body posture is one such test.** This method can indeed be helpful, but according to the literature there are exceptions. In cases with adrenal adenoma, an elevation of plasma aldosterone may occur after orthostasis (Fuss et al., 2022). A second, more reliable differential diagnostic test is the concentration of plasma

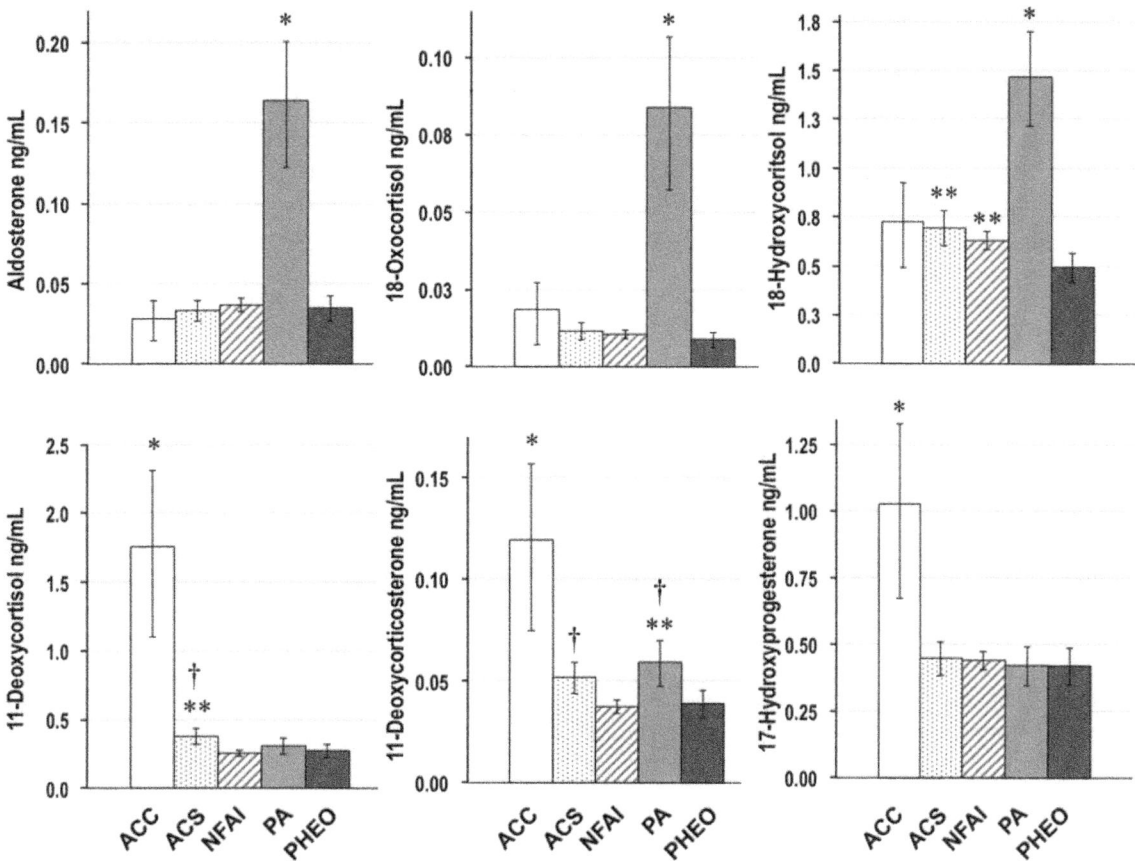

FIG. 3.3.1.9 Plasma concentrations of 6 (out of 12 in original figure) selected steroids in patients with adrenocortical carcinoma (ACC), autonomous cortisol secretion (ACS), nonfunctional adrenal incidentaloma (NFAI), primary aldosteronism (PA), and pheochromocytoma (PHEO). Results are shown as least square geometric means corrected for age and sex with 95% CIs. *P less than .05 higher than all 4 other groups; **P less than .05, higher than PHEO; $^{\dagger}P$ less than .05, higher than NFAI; $^{\S}P$ less than .05 lower than all 4 other groups. For conversion of ng/mL to nmol/mL or mg/mL to mmol/mL, divide by molecular weight. *(Redrawn from Berke K, Constantinescu G, Masjkur J, Kimpel O, Dischinger U, Peitzsch M, Kwapiszewska A, Dobrowolski P, Nolting S, Reincke M, Beuschlein F, Bornstein SR, Prejbisz A, Lenders JWM, Fassnacht M, Eisenhofer G. Plasma steroid profiling in patients with adrenal incidentaloma. J Clin Endocrinol Metab 2022;107(3):e1181–92. Fig. 3 p. 6.)*

18-hydroxycorticosterone at rest (Mulatero et al., 2012a,b). This method was shown to be very useful, but not absolute. The 18-hydroxycorticosterone:cortisol ratio as well as the aldosterone/corticosterone ratios by immunoassays have been used for differential diagnosis of both forms of primary aldosteronism. Basal values but also the change in ratio after an infusion of 1 liter normal **saline solution,** proved to be of a specially high diagnostic power (Arteaga et al., 1985; Kem et al., 1971). For idiopathic hyperaldosteronism, 18OHB/F was <3, whereas the ratio is >3 for primary aldosteronism (Note 18OHB is in ng/dL and F in μg/dL). The protocol of the test has changed over nearly 40 years with variations in posture, steroids measured, assay methods, ARR response, potassium state, and volume of saline so there is confusion in the interpretation of results (Eisenhofer et al., 2021; Rossi et al., 2006).

An unilateral adrenal adenoma occurs more often in Conn's syndrome than hyperplasia. Adrenal hyperplasia is also not a homogeneous disease, at least two types are known; microscopic noduli and hyperplasia with widening of the zona glomerulosa. It seems that the form with noduli reacts better to surgical treatment with bilateral or subtotal adrenalectomy than the other forms. However, in general, the operative results of cases with adrenal adenoma are definitely better than for hyperplasia. Therefore the differential diagnosis between the two forms is essential. This should include **lateralization**. For these purposes, radiological studies of different kinds are available; computerized tomography, venography, etc. and measurements of hormone concentrations in **adrenal venous blood.**

There are cases with the typical clinical syndrome of primary aldosteronism where an adrenal adenoma can be visualized and an oversecretion of aldosterone repeatedly excluded. Before a specialist laboratory is asked to search for these atypical steroids, certain further studies can be informative. The development of steroid assays after high-resolution chromatography will enable these uncommon steroids to be measured as part of a profile of the steroids in plasma or urine. The development of HPLC with MS/MS techniques enables the separation of a number of related steroids from small biological samples prior to specific measurement using mass spectrometry. There is now with LC-MS/MS techniques the potential to cover all (and more) (Taylor et al., 2017; Travers et al., 2017; Turcu et al., 2020). Metabolomic studies must be imminent. There is however limited availability of suitable stable isotope labeled internal standards.

3.3.1.3.5.3 Adrenal venous blood

The collection of adrenal venous blood is not without problems even for an experienced radiologist (Betz and Zech, 2022; So et al., 2022; Quencer, 2021; Wolley et al., 2020; Jakobsson et al., 2018). In general, the single adrenal vein from each gland are catheterized under X-ray control (see earlier chapter Fig. 3.1.1.). However, there are additional veins draining each adrenal gland and the venous system has variations between individuals.

On the right side, the collection of venous blood is particularly difficult. To eliminate mistakes caused by dilution of the adrenal venous blood, the aldosterone and 18-hydroxycorticosterone concentrations are related to cortisol or exceptionally to adrenalin concentrations. The ratio of

cortisol concentrations on the two sides is called the selectivity index. The aldosterone concentrations in each vein normalized to cortisol on each side gives a lateralization index. So the ratio of A left/C left to A right/C right above four indicates a left adrenal tumor.

As a guide to whether the catheters are located correctly, some centers have attempted measurement of cortisol with rapid results using laboratory kit in adjacent room or sent by track to the central laboratory for rapid analysis on automated platforms, e.g., Roche e411 (Chen and Singh, 2018; Betz et al., 2011; Betz and Zech, 2022).

The measurements of cortisol by immunoassay however can be disproportionately high compared with LC-MS/MS results, suggesting much cross reactivity from other steroids which are removed in the LC separation stage of an LC-MS/MS method (Fig. 3.3.1.10) (Ma et al., 2022).

A lateral flow test has been used in a few reports and is ideal for the procedure in the radiology suite (Page et al., 2018). One report used a 6 min immunochromatographic method that agreed well with Roche chemiluminescent method and measured cortisol to 1655 nmol/L (Yoneda et al., 2016; Cesari et al., 2017). A sample of 100 μL plasma was applied. The protocol did not indicate dilution but the method was validated for samples up to 100 fold dilution

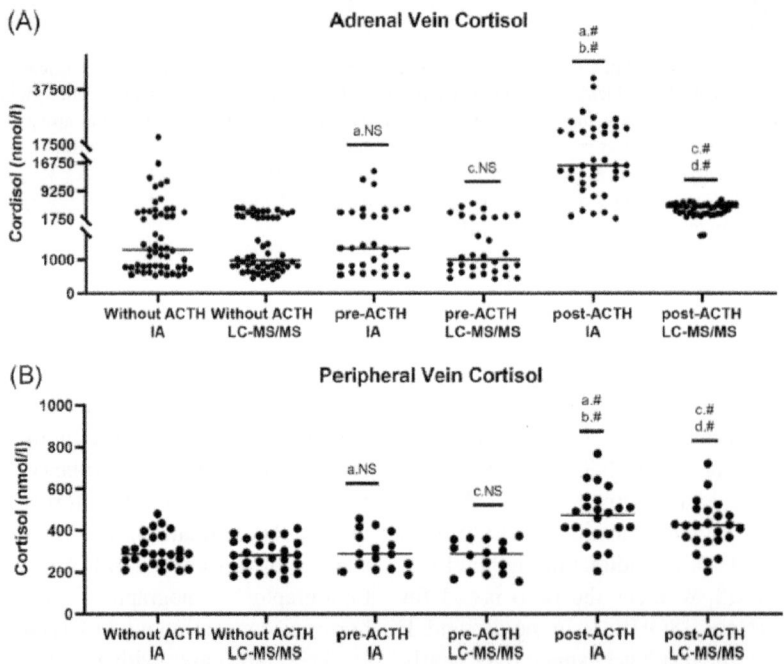

FIG. 3.3.1.10 Cortisol by immunoassay compared with LC-MS/MS determination (A) Cortisol results by immunoassay (IA) and liquid chromatography-mass spectrometry (LC-MS/MS) in the adrenal vein. (A) Comparison with the without adrenocorticotropic hormone (ACTH)-IA group; (B) comparison with the pre-ACTH-IA group; (C) comparison with the without ACTH-LC-MS/MS group; (D) comparison with the pre-ACTH-LC-MS/MS group; NS, not significant; [#]$P < .05$. (B) Cortisol results by both immunoassay (IA) and liquid chromatography-mass spectrometry (LC-MS/MS) in the peripheral vein. (A) Comparison with the without ACTH-IA group; (B) comparison with the pre-ACTH-IA group; (C) comparison with the without ACTH-LC-MS/MS group; (D) Comparison with the pre-ACTH-LC-MS/MS group; NS, not significant; [#]$P < .05$. *(From Ma Y, Chen H, Chen F, Jiang J, Guo W, Li X, Gao X, Lu Z, Zhou B, Zhao L, Li X. Mass spectrometry-based cortisol profiling during adrenal venous sampling reveals misdiagnosis for subtyping primary aldosteronism. Clin Endocrinol (Oxf) 2022;96(5):680–9. Fig. 2 p. 6. Clin Endocrinol 96:680.)*

of the sample. A kit from Trust Medical Support Co., Ltd. Fukuoka, Japan is under evaluation. At UCLH, London, England, cortisol concentrations up to 50,000 nmoL/L were found in AV blood plasma after dilution (Honour & Kurzawinski, personal data, unpublished).

The adrenal venous studies have been extended to analysis of 17 steroids in the samples from patients classified by their genotype (Turcu et al., 2020). Cortisol was found not to be the best marker for successful catheterization, while 11-deoxycortisol, 11-hydroxyandrostenedione and corticosterone showed the highest unstimulated gradient between adrenal vein and the periphery (Selectivity Index) (Fig. 3.3.1.11).

The differential was greater for 16-hydroxyprogesterone and 17-hydroxyprogesterone on ACTH stimulation. The hybrid steroids 18-oxo-cortisol and 18-hydroxy-cortisol were not consistent between Japanese and European participants. The two hybrid steroids were highest in patients with KCNJ5 mutations. By measuring 17 steroids there was however inconsistent lateralization between pre and post ACTH stimulation (data not shown). Distinct profiles were found in cases of unilateral disease.

ACTH can be given as a bolus prior to the catheterization or as a continuous infusion with better lateralization results (Hu et al., 2021). The use of a rapid cortisol assay has enhanced the performance of the whole procedure such that a reduction in radiation exposure during AVS has been achieved (Augustin et al., 2021).

The 18-hydroxycorticosterone values are higher in cases with adrenal adenoma than with bilateral adrenal hyperplasia (Auchus et al., 2007). The methodological error is smaller for 18-hydroxycorticosterone than aldosterone. In any case, two values are always better than one, although one study did not confirm that point (Dekkers et al., 2016).

As with many endocrine patient investigations if a steroid is found at high concentration it is valuable to test suppressibility of secretion and if a steroid is at low concentration a stimulation test is used. The concentrations of plasma steroids in dynamic tests were compared in patients with aldosterone-producing adenoma (APA) and bilateral PA (BPA) (Tezuka et al., 2021b) (Fig. 3.3.1.12. upper panel).

Following dexamethasone suppression, the APA group had larger decrements of 18OHF, 18oxoF, and aldosterone than the BPA group. Despite this suppression, these steroids remained higher in the APA group than in the BPA group following the DST. In contrast, cortisol levels were suppressed similarly in both groups, as were the remaining steroids. In several patients, the aldosterone secretion is stimulated rather than suppressed by dexamethasone as seen by others (Stratakis et al., 1999).

Although not strictly a diagnostic test for primary aldosteronism, all of the steroids measured increased in response to the Synacthen test. Seven steroids had a disproportionally higher increment in the APA group compared with the BPA group (Fig. 3.3.1.12 lower panel).

3.3.1.3.6 Imaging of tumors

A tumor may accumulate radiolabeled cholesterol and be detected by X-rays and recently by CT and PET scans. The latter needs a cyclotron which is expensive equipment and not widely available. Adrenal venous sampling may have to be used to lateralize the lesion. The catheterization of the adrenal veins is, in some centers, a successful procedure in adults but the venous drainage of the adrenal glands is often complex and even an experienced radiographer is not guaranteed of success. In a patient with an

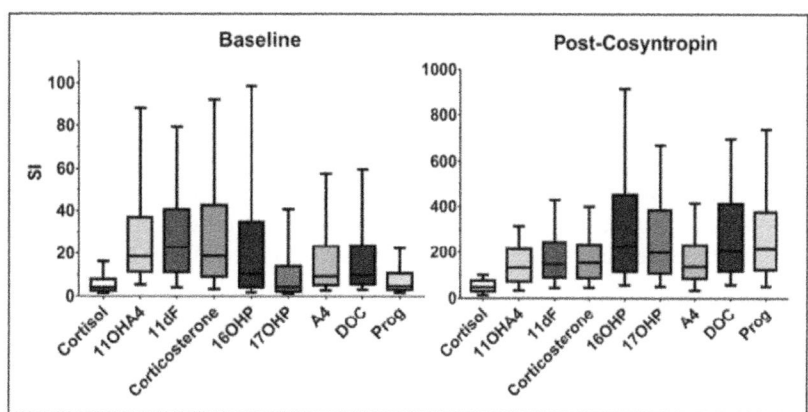

FIG. 3.3.1.11 Comparison of selectivity index (SI) using cortisol and alternative steroids. Figure illustrates the standard-of-care, cortisol, and all steroids with SI higher than cortisol ($P < .01$ for all). The boxes represent the interquartile range, the horizontal line marks the median, and the whiskers mark the 10–90 percentiles. *11dF*, 11-deoxycortisol; *11OHA4*, 11β-hydroxyandrostenedione; *16OHP*, 16-hydroxyprogesterone; *17OHP*, 17α-hydroxyprogesterone; *A4*, androstenedione; *DOC*, 11-deoxycorticosterone; *Prog*, progesterone. *(From Turcu AF, Wannachalee T, Tsodikov A, Nanba AT, Ren J, Shields JJ, O'Day PJ, Giacherio D, Rainey WE, Auchus RJ. Comprehensive analysis of steroid biomarkers for guiding primary aldosteronism subtyping. Hypertension 2020;75(1):183–192. Fig. 2 p. 185. Hypertension 75:183.)*

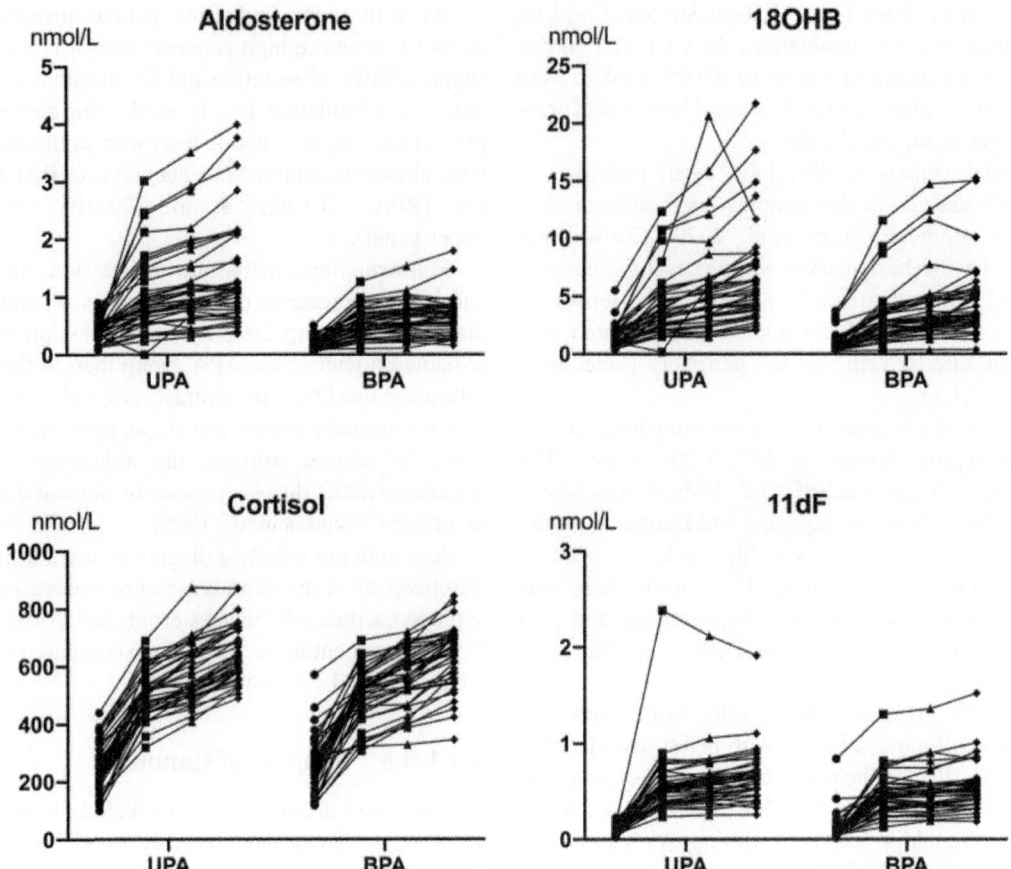

FIG. 3.3.1.12 (Top row) Steroid responses to cosyntropin stimulation in PA subtypes. Changes in steroid concentrations following cosyntropin stimulation in patients with aldosterone-producing adenoma (APA) and bilateral primary aldosteronism (BPA), 40 per group. Filled circles, squares, triangles, and diamonds represent concentrations at baseline, 15, 30, and 60 min after 250 mg cosyntropin injection, respectively. Abbreviations: *11dF*, 11-deoxycortisol; *18OHB*, 18-hydroxycorticosterone. Not shown here data for *18OHF*, 18-hydroxycortisol; *18oxoF*, 18-oxocortisol; *DOC*, 11-deoxycorticosterone. (Bottom row) Steroid responses to dexamethasone suppression in PA subtypes. Steroid concentrations before (circles) and after (squares) overnight 1 mg dexamethasone suppression test in patients with aldosterone-producing adenoma (APA) and bilateral primary aldosteronism (BPA) subtype, 40 per group. Unsuppressed morning steroids served as reference. Abbreviations: 18OHF, 18-hydroxycortisol; 18oxoF, 18-oxocortisol. *((Top) Tezuka Y, Ishii K, Zhao L, Yamazaki Y, Morimoto R, Sasano H, Udager AM, Satoh F, Turcu AF. ACTH stimulation maximizes the accuracy of peripheral steroid profiling in primary aldosteronism subtyping. J Clin Endocrinol Metab 2021;106(10):e3969–78. Fig. 1 p. e3973; (Bottom) Tezuka Y, Ishii K, Zhao L, Yamazaki Y, Morimoto R, Sasano H, Udager AM, Satoh F, Turcu AF. ACTH stimulation maximizes the accuracy of peripheral steroid profiling in primary aldosteronism subtyping. J Clin Endocrinol Metab 2021;106(10):e3969–78. Fig. 2 p. e3974.)*

aldosterone secreting adenoma, the adrenal venous serum from the affected side usually shows a higher concentration of aldosterone compared with either the contralateral adrenal vein or the periphery sampled simultaneously. Aldosterone and cortisol should be measured in all samples to check the authenticity of the site from which the blood is taken. The cortisol concentrations in the adrenal vein samples may be in the range of 400–58,000 nmoL/L or more, so dilution of 5–100 fold for the sample is required to read within the range of most assays.

More precise imaging is based on the label seeking out the activity of the tumor. The first attempts were based on labels that inhibit aldosterone synthase and several variants of etomidate were used with ^{131}I and ^{18}F. These varied in specificity for CYP11B1 or CYP11B2. To circumvent

detection of CYP11B1, the imaging study was performed when cortisol production was suppressed with dexamethasone. A fluorine-18 labeled agent (CDP2230) has higher specificity for CYP11B2 than CYP11B1 (Abe et al., 2016). The CXC chemokine receptor type 4 (CXCR4), which is over-expressed in many types of malignancy, is a new target under investigation (Heinze et al., 2018; Bluemel et al., 2017) with a Gadalinium[68] label. A 18F label based on a direct binding agent for CYP11B2 is also under investigation (Sander et al., 2021). Most radiolabels have short half-lives.

Localization of a unilateral aldosterone secreting tumor can lead to removal of the offending adrenal gland to reveal the canary yellow colored Conns tumor and a cure for the hypertension.

3.3.1.3.6.1 The role of computerized tomography in the diagnosis of primary aldosteronism

CT scanning of the abdomen can often reveal an adrenal mass. Figures from various large CT studies give an occurrence varying from 0.5% and 0.6%. The incidentally found large adrenal masses are more frequent in patients with elevated blood pressure than in patients with normal blood pressure. Further investigations of this problem have revealed that the increase in adrenal mass is due to several abnormalities. Adrenal adenomas with elevation of androgens, aldosterone or cortisol secretion in other words, an androgen producing tumor, Conn's syndrome or Cushing's syndrome—often in the early, clinically asymptomatic stages have been found. Phaeochromocytomas, lipomas, lipomyomas, and adrenal cysts have also been identified. The supporting biochemical studies have not been sufficiently extensive for a clear understanding of any underlying endocrinopathy. The "incidentalomas" are also called "silent" adrenal tumors. The CT figures may also be lower than the true prevalence since the resolutions of scanners varies. Only adrenal tumors bigger than 0.5–1 cm in diameter are visualized in most cases so small tumors can be missed (Fujiwara et al., 2010) because some adrenal adenomas are smaller than this limit.

Post mortem studies have found higher frequency of adrenal adenomas than with computerized tomography (CT). The figures for post mortem identification vary from 2.5% and 9%. The dissociation of the CT data from the histological statistics is understandable because the post mortem studies are able to detect the smallest adrenal adenomas. In some cases of clinical primary aldosteronism in which computerized tomography was unable to detect an adenoma, nuclear magnetic resonance equipment was successful. There will undoubtedly be further developments in the application of this technology with contrast.

Low renin activity with raised aldosterone levels are clues to a primary source of aldosterone or mineralocorticoid production. Primary hyperaldosteronism is also suspected when a hypertensive patient is found to be hypokalemic but this is now not universal. Most patients with primary aldosteronism are 30–60 years of age. Plasma sodium concentrations tend to be in the upper normal range with bicarbonate concentrations often raised. For endocrine assessment, the patient should be on an adequate sodium diet and blood samples should be collected without stasis or hemolysis. Around 30% of patients with essential hypertension have subnormal PRA, but a low PRA is not specific for primary aldosteronism. A high ratio of PAC to PRA above 800 is more suggestive of primary aldosteronism (when the units are in Systeme Internationale units not mass units). Many units are in use which makes it difficult to interpret the results (see earlier Table 3.3.1.3 for the

variation in ARR). The increased aldosterone concentration in blood or urine must be confirmed in the context of normal cortisol production.

If a patient is already being treated with spironolactone, this should be stopped for at least 6 weeks. PRA and PAC measurements should be repeated. If the patient is hypokalemic with low PRA and raised PAC/PRA ratio, an aldosterone suppression testing can be performed on a high salt diet for 3 days with potassium supplements. On the third day, a 24-h urine is collected for measurement of aldosterone, electrolytes, and cortisol. The 24 h sodium intake should be more than 200 mmol to document sodium repletion. High urine aldosterone excretion is consistent with hyperaldosteronism.

Computerized tomography (CT) scanning will pick up and localize many of the primary adrenal abnormalities. If PAC to PRA ratio is less than 400, the patient should be investigated for secondary hyperaldosteronism; and if PAC and PRA are low other causes of mineralocorticoid excess need to be studied.

The morphology of the tumor when removed will define the clinical state (Table 3.3.1.5).

Dissection of the tumor will reveal idiopathic hyperaldosteronism, noduli with diffuse hyperplasia and without noduli (Kumagai, 1980).

The term primary aldosteronism or Conn's syndrome is usually restricted to cases with adrenal adenoma. However, overproduction of other mineralocorticoids and/or hypertensinogenic steroids with and without simultaneous elevation of aldosterone presents a clinical picture indistinguishable from Conn's syndrome. Idiopathic hyperaldosteronism and the dexamethasone suppressible hyperaldosteronism are secondary to defects of the regulatory mechanism. Nevertheless, in accordance with the lature for all these syndromes the common name "primary aldosteronism" is retained. Another nosological variant can be

---

TABLE 3.3.1.5 Morphology of APA tumors.

Morphology of hyperaldosteronism

1. Hyperaldosteronism due to solitary adrenal adenoma

2. Hyperaldosteronism with bilateral adrenal hyperplasia

3. Hyperaldosteronism with unilateral adrenal hyperplasia

4. Hypermineralocorticoidism including hyperaldosteronism due to solitary adrenal adenoma

5. Hypermineralocorticoidism including hyperaldosteronism due to bilateral adrenal aldosteronism due to adrenal carcinoma

6. Dexamethasone remediable hyperaldosteronism

7. Aldosteronism after ovarian carcinoma

8. Adrenal hyperplasia

hypertension with hypoaldosteronism probably due to corticosteroid receptor hypersensitivity (Charmandari et al., 2008).

Classification of tumors is based on the Weiss score that is composed of nine items assigned by a pathologist (three concerning the architecture, three the nucleus, and three the presence of any type of invasion). Each item scores 1 point and the sum of the positive items defines the final score. Staining of thin sections of tumors with hematoxilin-eosin will demonstrate nodules. Immunostaining for Ki67 protein using monoclonal antibody MIB-1 is a measure of cell proliferation. Staining for CYP11B2 can show focal or diffuse distribution for nonneoplastic and hyperplasia, respectively. The process has been reviewed leading to an international consensus (Williams et al., 2021).

The evaluation of the histological statistics is handicapped by differences in interpretation of large adrenal noduli and small adrenal tumors. The differentiation is made by demarcation from the neighboring tissue but this rule is not always applied so that "noduli with signs of demarcation" can be construed as smaller tumors or noduli. These issues need to be resolved, but account for different figures of the histological statistics.

In addition, numerous physiological maneuvres have been described to discern forms of primary aldosteronism (adenoma versus hyperplasia) (Table 3.3.1.6 summarizes these tests).

Bilateral adrenal hyperplasia is not homogeneous pathologically. There are differences in the morphological picture with widening of the *zona glomerulosa* compared with the micronodular hyperplasia. Both forms react differently to surgery. No diagnostic measure has yet been described to satisfactorily differentiate both forms of adrenal hyperplasia.

Prevalence of primary aldosteronism

Primary hyperaldosteronism probably occurs in 1%–2% of all hypertensive subjects. Approximately 70% have a unilateral adenoma, and approximately 30% have bilateral hyperplasia. These are usually treated by adrenalectomy or medically, respectively. The data on the prevalance of primary aldosteronism varies between 0.3% and 2.8% and is difficult to define accurately because of differences in health care practices between countries as well as the facilities available and genetic bias between populations. Screening studies based on investigation for hypokalemia have resulted in lower frequencies. When steroid estimations were also performed on patients with hypertension and normokalemia then primary aldosteronism was more frequently diagnosed. However, the studies from different groups are not comparable because the parameters estimated were different. Hence to date it is impossible to calculate acceptable figures on the prevalence of primary

TABLE 3.3.1.6 Provocative tests for confirmation of primary aldosteronism.

Confirmatory tests

- Oral sodium loading test. Patient is advised to increase sodium intake to about 6 g (>200 mmol) for 3 days after which a 24-h urine is collected for sodium and urine aldosterone measurement. A urinary aldosterone <10 μg/24 h (28 nmol/d) makes PA unlikely. An elevated urinary aldosterone >12 μg/24 h (>33 nmol/d) makes PA highly likely. This test is not recommended in patients with severe uncontrolled hypertension, renal insufficiency, cardiac arrhythmia, or severe hypokalemia.

- Saline infusion test. Patient stays in a recumbent position an hour before and during the infusion of 2 L of 0.9% saline over 4 h in the morning. Bloods are taken for renin, aldosterone, cortisol, and potassium at 0 and after 4 h. BP and heart rate are monitored throughout the test. A post infusion aldosterone <5 ng/dL (140 pmol/L) makes PA less likely and levels >10 ng/dL (280 nmol/L) makes PA more likely. Values between 5 and 10 ng/dL are intermediate. This test is also contraindicated in patients with severe uncontrolled hypertension, renal insufficiency, cardiac arrhythmia, or severe hypokalemia.

- Fludrocortisone suppression test. Patient is given 0.1 mg of oral fludrocortisone every 6 h for 4 days, potassium chloride supplements may be needed to keep plasma potassium close to 4.0 mmol/L, and high sodium diet to maintain urinary sodium excretion rate of at least 3 mmol/Kg body weight. On the fourth day, plasma aldosterone and PRA are measured at 10 a.m. and plasma cortisol at 7 a.m. and 10 a.m. Plasma aldosterone >6 ng/dL (170 nmol/L) on day 4 at 10 a.m. confirms PA provided PRA is <1 ng/mL/h and 10 a.m. plasma cortisol is lower than the 7 a.m. measurement.

- Captopril challenge test. Patient is given 25–50 mg of captopril orally after sitting or standing for at least 1 h. PRA, plasma aldosterone, and cortisol are measured at time 0, and at 1 or 2 h after the challenge, with the patient remaining seated during this period. Plasma aldosterone is normally suppressed by captopril but remains elevated while PRA remains suppressed in patients with PA. False negatives have been reported in patients with APA and in those with IAH.

aldosteronism in unselected hypertensives. Nevertheless, published figures of less than 0.5% seem unrealistically low. Interesting investigations based on measurements of several steroids favor higher figures. These have confirmed the results of previous statistics and recent data obtained by serial computed tomography of the adrenals. The number of cases of hyperaldosteronism is greater than those with adrenal hyperplasia. In surveys of cases, those with adenoma accounted for 75%, around 20% were adrenal hyperplasia and 2% adrenal carcinoma. As the clinical symptoms and biochemical changes are less pronounced in patients with adrenal hyperplasia, unrecognized cases are more likely to be of this type. Primary aldosteronism is more frequently diagnosed in the 4th and 5th decades of life although the diagnosis has been made in other age groups. An adrenal adenoma is more frequently found in females. Hyperplasia with noduli is predominant in aged patients. In diffuse adrenal hyperplasia, no gender preference has been revealed. For the same disorder, the incidence is higher in younger subjects.

Pathophysiology

In the earlier stages of the disorder, the unilateral overproduction of aldosterone can be compensated by suppression of the contralateral and homolateral normal adrenal cortex. Some of the patients in whom an apparently hormone inactive adrenal adenoma was discovered by computed tomography may go on to show the clinical onset of Conn's syndrome. In some cases, it has been shown that in the earlier stages of the disease, steroids other than aldosterone are produced in excess. Abdelhamid et al. published clinical and hormonal data from patients with adrenal adenoma and hypertension in whom 18-hydroxycorticosterone was elevated (Vecsei et al., 1982b). Later high aldosterone levels were found in these subjects. The aldosterone secretion is independent of the renin angiotensin system and of ACTH and other regulatory mechanisms.

The mechanisms by which aldosterone hyperproduction causes and maintains elevated blood pressure are multifactorial. There are many studies performed in patients with primary aldosteronism after withdrawal of spironolactone therapy. In the early stages of the disease, the sodium retention and the expansion of the extracellular volume are the most decisive changes. Later, there is an increase in peripheral resistance. Experimental data in patients with Conn's syndrome previously treated with spironolactone (Wenting et al., 1982) showed in the first weeks an increased heart rate/minute volume in addition to the elevated blood volume accompanied with high blood pressure. With continued treatment both values decreased especially in elderly patients. The peripheral resistance was increased in the later stages but not during the early period of spironolactone withdrawal. An increase in peripheral resistance may be the main mechanism of maintaining elevated blood pressure. The relevance of peripheral resistance was also shown by Miller et al. (1979) in experiments in young pigs with DOC-induced hypertension. The etiology of the increased peripheral resistance is not yet understood. Under certain circumstances the so far hypothetical steroid receptors in the vessel walls may have a role. Six mechanisms have been considered (Table 3.3.1.7).

Much research is still needed to resolve outstanding clinical questions. In view of the rarity of the disease, collaboration between centers will be needed to agreed protocols and responsibilities.

TABLE 3.3.1.7 Six mechanisms of increase peripheral resistance.

Increased peripheral resistance

1. Facilitation of vasoconstriction by hypovolemia

2. Changes in the membrane of the blood vessel muscle cells (increased permeability for catecholamines)

3. Increased vascular resistance to angiotensin and noradrenaline

4. Increased vasopressin secretion and increased sensitivity of the target organs, i.e., vessels, for vasopressin

5. Changes in the function of the central nervous system centers (region anteroventral of the third ventricle of the brain)

6. Increased activity of the sympathetic nervous system: increased adrenalin production

3.3.1.3.6.2 Adrenal carcinoma

An isolated tumor mass is often seen during an MRI scan of the abdomen. Malignancy is recognized by enlargement of lymph nodes locally or remotely. An adrenal carcinoma per se may not be identified without hormonal investigations and genetic examination in a liquid biopsy. The tumor may have no recognizable function and is called an incidentaloma. Conn's syndrome due to adrenal carcinoma is a relatively rare disease (1%-3% of the described cases of Conn's syndrome) encountered when adrenal nodules proliferate and metastases may appear in the abdomen, lungs, bones, etc. When the adrenal carcinoma is not removed early, the prognosis is very bad. Occasionally, an aldosterone producing carcinoma can originate in the ovary. Urine steroid analysis has sought differences between tumors secreting aldosterone (Conn's syndrome), cortisol (Cushing's syndrome), androgens, estrogens, adrenocortical adenoma (ACA) and adrenocortical carcinoma (ACC). The tumor may also be a phaeochromocytoma secreting catecholamines. The clinical picture will often dictate the investigations.

An increase in the excretion of 11-deoxycortisol and DHEA metabolites was seen on GC-MS analysis of steroids (Fig. 3.3.1.13) (Arlt et al., 2011).

This was a significant study including a total of 45 ACC patients and 102 ACA patients. A machine learning algorithm was applied to the data. The findings have been replicated in several studies using GC-MS (Kerkhofs et al., 2015; Shafigullina et al., 2018), LC-MS/MS (Velikanova et al., 2016; Bancos et al., 2020) and LC-HRAM (Hines et al., 2017). LC-MS/MS has been applied to serum samples from patients with nonfunctioning adenoma (NFA, $n = 73$), Cushing's syndrome (CS, $n = 30$) and primary aldosteronism (PA, $n = 40$). The PA group showed decreased levels

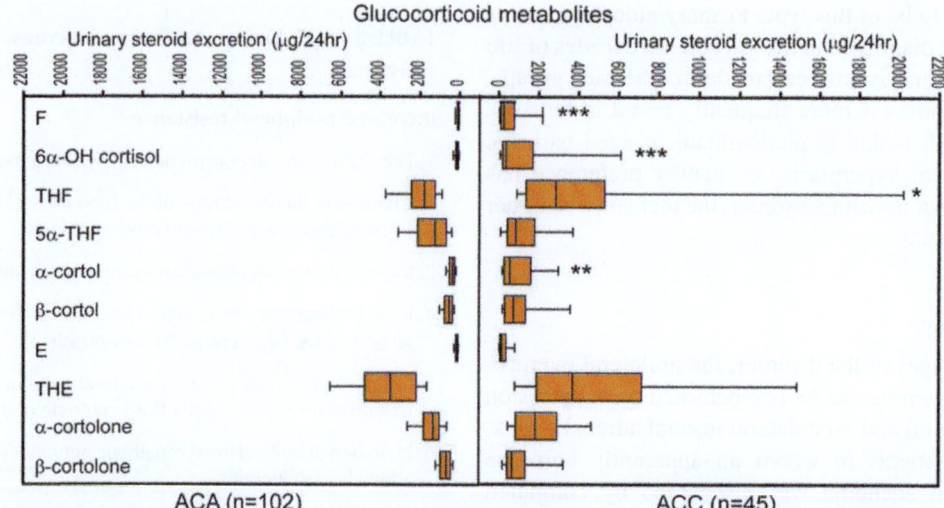

FIG. 3.3.1.13   Steroid metabolite excretion in ACA (n = 102) and ACC (n = 45). Results for cortisol and cortisone metabolites only shown here. Original chart shows results for adrenal androgen precursors and active androgens, metabolites of mineralocorticoids and their precursors and metabolites of glucocorticoid precursors. Box plots represent median and interquartile ranges: the whiskers represent 5th and 95th percentiles, respectively. *P 0.05, **P 0.01, ***P 0.001 comparing ACA with ACC. *(From Arlt W, Biehl M, Taylor AE, Hahner S, Libé R, Hughes BA, Schneider P, Smith DJ, Stiekema H, Krone N, Porfiri E, Opocher G, Bertherat J, Mantero F, Allolio B, Terzolo M, Nightingale P, Shackleton CH, Bertagna X, Fassnacht M, Stewart PM. Urine steroid metabolomics as a biomarker tool for detecting malignancy in adrenal tumors. J Clin Endocrinol Metab. 2011;96(12):3775-84. Fig. 2. p. 3779.)*

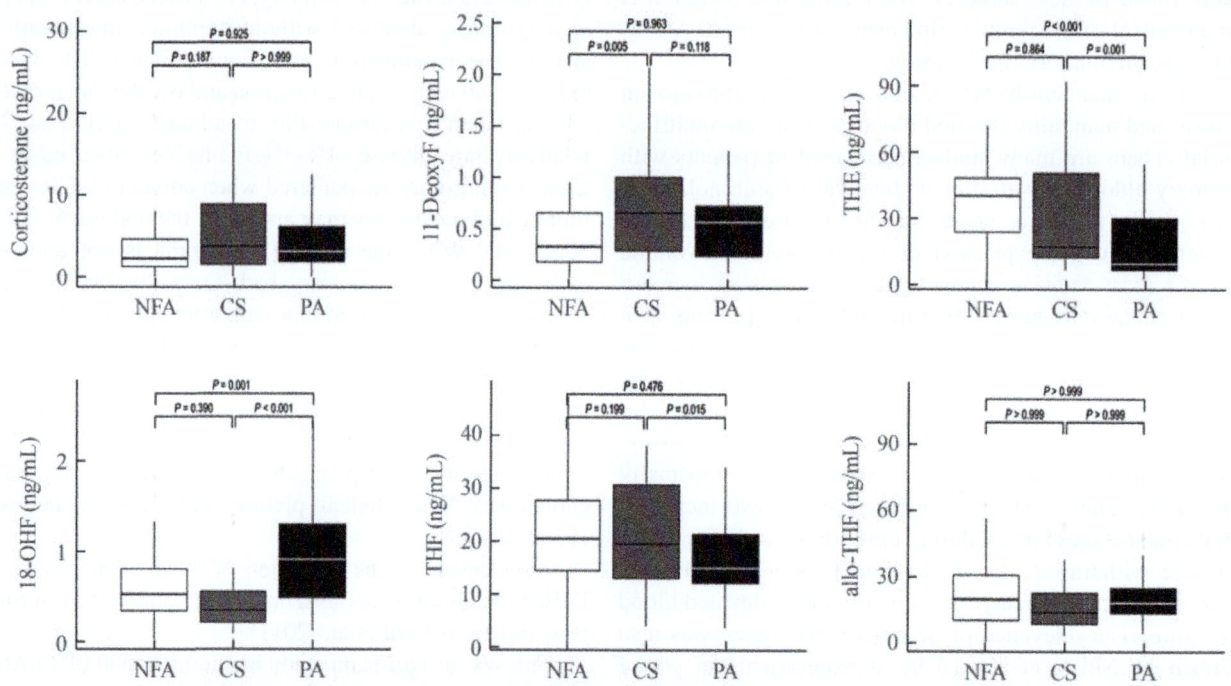

FIG. 3.3.1.14   Comparative serum levels of 6 (out of 15 reported) selected adrenal steroids between patients with nonfunctioning adenoma (NFA), Cushing's syndrome (CS), and primary aldosteronism (PA) after adjustment for age and gender. *11-deoxyF*, 11-deoxycortisol; *THE*, tetrahydrocortisone; *18-OHF*, 18-hydroxycortisol; *THF*, tetrahydrocortisol. *(From Ku EJ, Lee C, Shim J, Lee S, Kim KA, Kim SW, Rhee Y, Kim HJ, Lim JS, Chung CH, Chun SW, Yoo SJ, Ryu OH, Cho HC, Hong AR, Ahn CH, Kim JH, Choi MH. Metabolic subtyping of adrenal tumors: prospective multi-center cohort study in Korea. Endocrinol Metab (Seoul) 2021;36(5):1131–41. Fig. 1 p. 1135.)*

of tetrahydrocortisone and increased 18-hydroxycortisol (Fig. 3.3.1.14). Through two routes in a decision tree analysis, the PA group were distinguished from the NFA group (Fig. 3.3.1.15).

3.3.1.3.7   Genetics of primary aldosteronism

Exome sequencing has revealed somatic mutations in genes that encode ion channels or pumps and the mutations lead

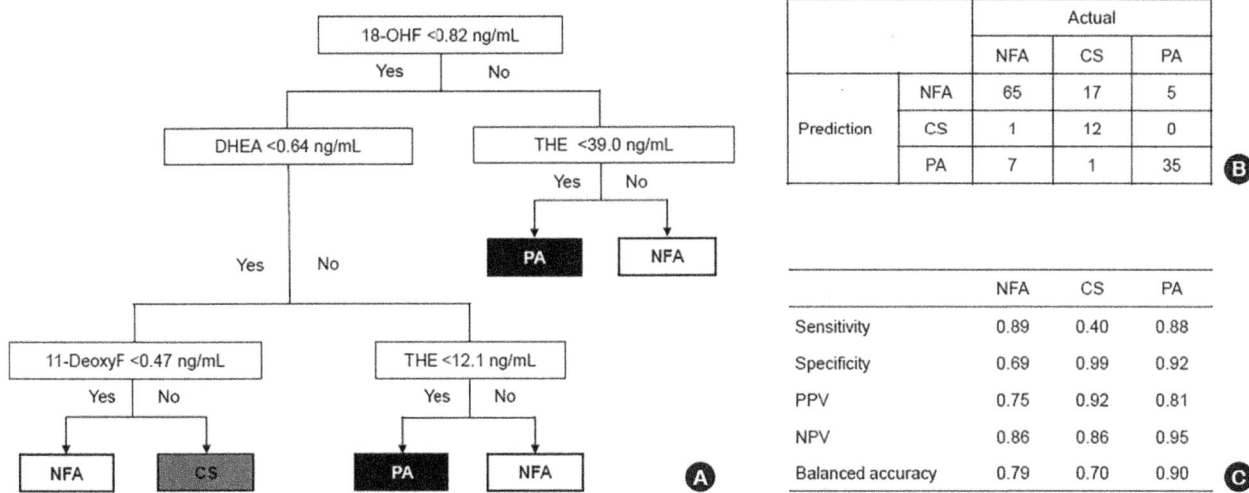

FIG. 3.3.1.15 The decision tree analysis for classification of nonfunctioning adenoma (NFA), Cushing's syndrome (CS), and primary aldosteronism (PA) groups in subjects with adrenal tumors using the multiple steroid panels. (A) The significant features of steroid panels, (B) the confusion matrix for decision tree analysis, (C) the diagnostic performance of decision tree analysis in each group. Accuracy, 0.78 (95% confidence interval, 0.71–0.85); $P < 1.6 \times 1011$. *18-OHF*, 18-hydroxycortisol; *DHEA*, dehydroepiandrosterone; *THE*, tetrahydrocortisone; *11-deoxyF*, 11-deoxycortisol; *PPV*, positive predictive value; *NPV*, negative predictive value. *(From Ku EJ, Lee C, Shim J, Lee S, Kim KA, Kim SW, Rhee Y, Kim HJ, Lim JS, Chung CH, Chun SW, Yoo SJ, Ryu OH, Cho HC, Hong AR, Ahn CH, Kim JH, Choi MH. Metabolic subtyping of adrenal tumors: prospective multi-center cohort study in Korea. Endocrinol Metab (Seoul) 2021;36(5):1131–41. Fig. 2 p. 1136.)*

to depolarization of the cell membrane. Depolarization activates voltage gated Ca^{2+} ion channels and intracellular ion signaling and promotes the response of the adrenal cortex to angiotensin II and transcription of aldosterone synthase resulting in overproduction of aldosterone.

3.3.1.3.8 Genetics of familial hyperaldosteronism

There are four forms of PA with autosomal dominant inheritance of which 3 are associated with genetic mutations.

FH-1 presents with severe hypertension before 20 years of age in patients with a family history of hypertension and cerebrovascular incidents at young age. The disease carries a risk of cerebral hemorrhage. This disease was first described (Sutherland et al., 1966) in a father and son. Hypertension and hypokalemia is corrected by administration of dexamethasone (Dluhy et al., 2001). Parents had presented with hypertension at young age.

Aldosterone production in the *zona fasciculata* is regulated by ACTH in this condition. Bilateral adrenal hyperplasia is seen and hybrid steroids such as 18-hydroxycortisol and 18-oxo-cortisol are found at high concentrations in blood or the metabolites in urine. Estimation of 18-hydroxycortisol is likely to prove useful in screening for this disorder (Nicod et al., 2004; Reynolds et al., 2005) but this assay has limited availability. An increased urine excretion of the steroid can be seen in a urine steroid profile. Genetic testing is available (Lifton et al., 1992). The chimeric gene can be detected in a PCR test.

Low-dose glucocorticoids (dexamethasone or prednisone) is effective treatment in most cases although spironolactone or eplerenone can be added.

The group of Maria New found that the total mineralocorticoid activity of a plasma extract was higher than calculated from the summation of the mineralocorticoid activities of the measured aldosterone, corticosterone and deoxycorticosterone values. 18-hydroxycortisol was found to be elevated (Ulick, 1991). Later, however, it was shown that not only in **glucocorticoid remediable hypertension** but also in other forms of Conn's syndrome the plasma concentrations of 18-hydroxycortisol and 18-oxocortisol may be high. The relevance of these findings is not yet known because the mineralocorticoid activities of these C-18-hydroxy derivatives are relatively low.

Less than 150 cases had been reported worldwide and most of those were picked up owing to clinical features of severe early familial hypertension with signs of aldosteronism such as spontaneous hypokalemia. The hypertension is often severe with a high incidence of premature death from hemorrhagic stroke and ruptured intracranial aneurysms. Clinical evaluation reveals variable degrees of hyperaldosteronism with suppressed renin activity. The disorder is probably underdiagnosed particularly in children but is characterized by the dramatic lowering of blood pressure and hyperaldosteronism with dexamethasone. A post dexamethasone PAC below 4 ng/dL (120 pmol/L) is more effective in detecting GRA than suppression compared with baseline. Screening for intracranial aneurysms by magnetic resonance angiography is now advised for patients with proven GRA (Litchfield et al., 1998; Al Romhain et al., 2015).

In addition to the unusual regulation of aldosterone secretion, patients with GRA secrete high levels of 18-hydroxycortisol and 18-oxocortisol which are also positively regulated by ACTH (Reynolds et al., 2005; Lenders et al., 2018). The occurrence of abnormal steroids in GRA has led to suggestions that GRA might derive from a failure of proper differentiation of the adrenal cortex. An alternative explanation was that the two enzymatic activities are expressed in the wrong part of the gland (Gomez-Sanchez et al., 2017).

3.3.1.3.8.1 Genetics of GRA

The etiology of this disease has been clarified by molecular genetic studies. This disorder was found to be caused by **a chimeric gene formed by unequal cross over between CYP11B1 and CYP11B2** (Lifton et al., 1992) such that aldosterone synthase is controlled by ACTH rather than angiotensin (Fig. 3.3.1.16).

The genomic structure of the two *CYP11B* genes enabled a gene map to be built up for patients with GRA. A chimeric gene created by nonhomologous recombination, has the 5′ end of the CYP11B1 abutted to the 3′ end of CYP11B2. An 11-hydroxylase enzyme acquires aldosterone synthesizing properties under the control of a promoter responsive to ACTH (Lifton et al., 1992). From restriction enzyme, maps with cutting sites in *CYP11B1, CYP11B2* and the chimeric gene, a diagnostic 6.3 kB BamH1 fragment was recognized from the chimeric gene (Fig. 3.3.1.17).

In family studies, hybridization of exons 3–4 of 11-hydroxylase to BamH1-digested DNA reveals a 6.3 kb

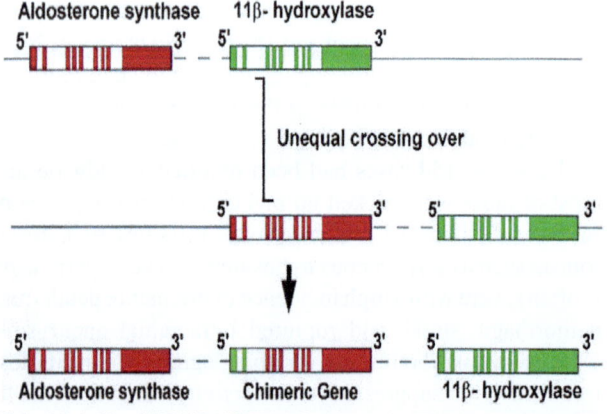

FIG. 3.3.1.16 CYP11B1/CYP11B2 chimera in FH1. Unequal crossing-over of CYP11B1 and CYP11B2 results in a chimeric gene with a 5′ ACT responsive regulatory sequence encoding an enzyme with aldosterone synthase activity and another gene with 11-hydroxylase activity whose expression is regulated by angiotensin II. *(From Lifton RP, Gharavi AG, Geller DS. Molecular mechanisms of human hypertension. Cell 2001;104(4):545–56. Fig. 3 p. 547. Cell 104:545.)*

fragment only in affected pedigree members. The intensity of the fragment is 50% of that observed for either the normal aldosterone synthase at 4.5 kb or 11-hydroxylase fragment at 8.5 kb, compatible with the presence of one additional copy of a chimeric aldosterone synthase/11-hydroxylase gene in affected subjects. The crossover break points have varied between the second intron of 11-hydroxylase and exon 4 of 11-hydroxylase. The activity for 18-hydroxylation is localized to exon 5 so that the exon 4 break point is the limit for GRA. The restriction enzyme test for GRA is reliable and can be performed from a small sample of peripheral blood.

In principle, the treatment of GRA by interference with ectopic aldosterone production or action should provide optimal therapy. Suppression of ACTH with glucocorticoids such as prednisone is supported by mineralocorticoid blockade with spironolactone; and inhibition of mineralocorticoid-sensitive distal tubule sodium channel with amiloride.

FH-3. Subjects present in a similar manner to FH-1 with severe hypertension in childhood and with **severe hypokalemia**. FH-III is very rare. Hyperaldosteronism is caused by bilateral adrenal hyperplasia (Mulatero, 2008; Geller et al., 2008). Patients with FH3 have shown a paradoxical increase in aldosterone when treated with dexamethasone (Stratakis et al., 1999 and others) and have poor responses to spironolactone and amiloride.

After exclusion of the chimeric CYP11B gene, mutations in **KCNJ5**, particularly in exon 2, were found by sequencing (Choi et al., 2011). GIRK4 (Kir 3.4) comprises a pore formed from a tetramer of units each having two membrane spanning domains with intracellular N and C termini. The first mutations found were substitutions at G151R and L168R leading to loss of potassium ion selectivity and increased Na+influx with resulting activated voltage-gated Ca^{2+} channel and autonomous aldosterone production. Increased CYP11B2 expression is mediated by a Ca^{2+}/calmodulin cascade (Mulatero et al., 2013; Oki et al., 2012) (Fig. 3.3.1.18). Gain-in function changes for the encoded **G protein activated inward rectifier potassium channel (GIRK4)** lead to loss of potassium selectivity and increased sodium conductance.

The cause of cell proliferation is uncertain. Cardiovascular complications have been found in some patients with KCNJ5 mutations (Rossi et al., 2014). Increased influx of sodium activates voltage dependent calcium channels and accumulation of calcium which is the main regulator of aldosterone production. Selective blockers for the mutated GIRK4 channel are predicted to benefit patients with FH3. A metaanalysis of 13 studies and 1636 patients found a prevalence of 43% (Lenzini et al., 2015), more common in females and larger tumors albeit at low significance.

FH-2. Patients present with hypertension and hypokalemia.

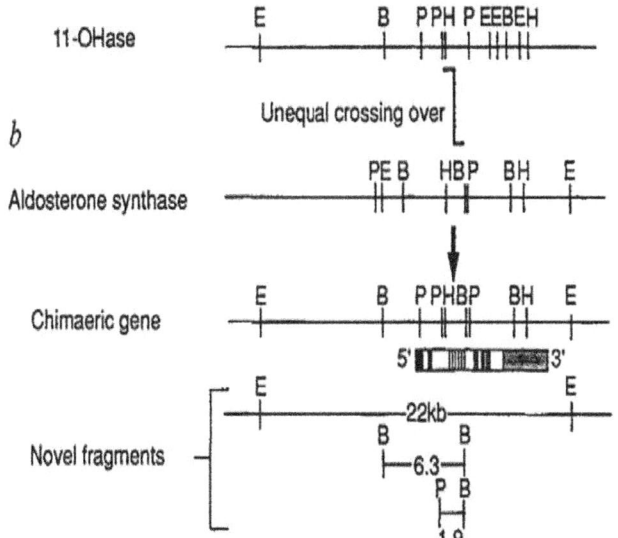

FIG. 3.3.1.17 Location of cleavage sites for restriction endonucleases in 11-0Hase, aldosterone synthase and the hypothesized chimeric gene resulting from unequal crossing over. Sites cut by BamHI (B), *Eco*RI (E). Hindlll (H), and Pvull (P) are shown and the location of the exons of the chimeric gene on the restriction enzyme map are indicated. The map predicts that digests of DNA of affected subjects with the enzymes indicated give 3 novel fragments hybridizing to exon 3–4. All other products of these digests should be identical in size to fragments derived from either 11-0Hase or AldoS. The origin of each of these novel fragments is demonstrated. The map of sites in 11-0Hase and AldoS includes a Pvull site located in intron II of 11-0Hase, a region not sequenced previously. A cluster of 4 Pvull sites and a BamHI site common to the 3′ ends of all 3 genes are not shown. *(From Lifton RP, Dluhy RG, Powers M, Rich GM, Gutkin M, Fallo F, Gill JR Jr, Feld L, Ganguly A, Laidlaw JC, et al. Hereditary hypertension caused by chimaeric gene duplications and ectopic expression of aldosterone synthase. Nat Genet 1992;2(1):66–74. Fig. 3 p. 284. Cell 104:545.)*

Whole exome sequencing using DNA from 3 family members revealed a heterozygous mutation in the voltage gated **CLCN2 gene** (Scholl et al., 2018) (Fig. 3.3.1.19).

The mutations lead to protein that increases chloride conductance at resting potential and opening of the voltage gated calcium channels. Aldosterone production is stimulated at high intracellular calcium concentrations (Fig. 3.3.1.20).

FH-4

This form of PA presents with hypertension before the age of 10 years. Mutations in **CACNA1H** (voltage gated calcium channel T type—Cav 3.2) at chromosome 16p13 were found by whole exome sequencing (Scholl et al., 2015) (Fig. 3.3.1.21).

Beta catenin is encoded by CTNNB1 and mutations in this gene in APA have been found that lead to constitutive activation of Wnt/β-catenin in signaling (Kim et al., 2008). CTNNB1 mutations have also been reported in adrenal adenomas and carcinoma (Boulkroun et al., 2011; Berthon et al., 2014). The concentrations of β-catenin was increased

in adrenal tumors (Tissier et al., 2005). Catenin binds to the promoters for the AT1R, NURR1, and Nur77 to increase expression of the genes which in turn stimulates CYP21 and CYP11B2 leading to raised production of aldosterone (Berthon et al., 2014) (Fig. 3.3.1.22).

In summary, pathogenetic mechanisms involved in familial forms of Primary Aldosteronism and treatment are:

(A) FH-I caused by an unequal crossover between CYP11B1 and CYP11B2 genes.
(B) FH-II caused by mutations in CLCN2 gene that lead to opening of calcium channels.
(C) FH-III caused by mutations in KCNJ5 gene.
(D) FH-IV caused by mutations in CACNA1H gene (Fig. 3.3.1.23).

Mutations of KCNJ5, ATP1A1, and CLCN2 lead to depolarization of the cell membrane due to impairment of ion transport. Depolarization activates voltage-gated Ca^{2+} channels and increases intracellular Ca^{2+} levels. Conversely, mutations of CACNA1D and CACNA1H directly cause an increase in Ca^{2+} conductance. ATP2B3 mutation reduces Ca^{2+} export from the cell. Activated calcium signaling promotes transcription of aldosterone synthase (CYP11B2), resulting in overproduction of aldosterone.

A new study has shown patients with posture unresponsive APA had higher concentrations of hybrid steroids (18-oxocortisol, 18-hydroxy-F), recumbent aldosterone and cortisol than patients with posture responsive aldosterone APA (Guo et al., 2020).

Other forms of primary aldosteronism and related diseases

CACNA1D encodes a calcium channel voltage dependent subunit of six transmembrane segments. Four such units line a pore and mutations shift the voltage dependent gating to increase intracellular calcium levels. Somatic mutations in CACNA1D are found throughout the gene (Scholl et al., 2018). Gain of function mutations in **CACNA1D** L type voltage gated calcium channel—Cav 1.3) on chromosome 3p14.3. have also been found (Scholl et al., 2013) (Fig. 3.3.1.24). Germline mutations have also been reported (Fig. 3.3.1.25) (Ortner et al., 2020).

The syndrome is more fully defined as **primary aldosteronism with seizures and neurological abnormalities (PASNA)**. The mutation leads to increased calcium concentrations and activation of aldosterone production, independent of angiotensin. Family members have variable blood pressure so other factors may constrain the expression of the defect. More than 30 different mutations have been reported in about 9% of sporadic PA. Calcium channel blockers such as amlodipine may be effective (Pinggera et al., 2017).

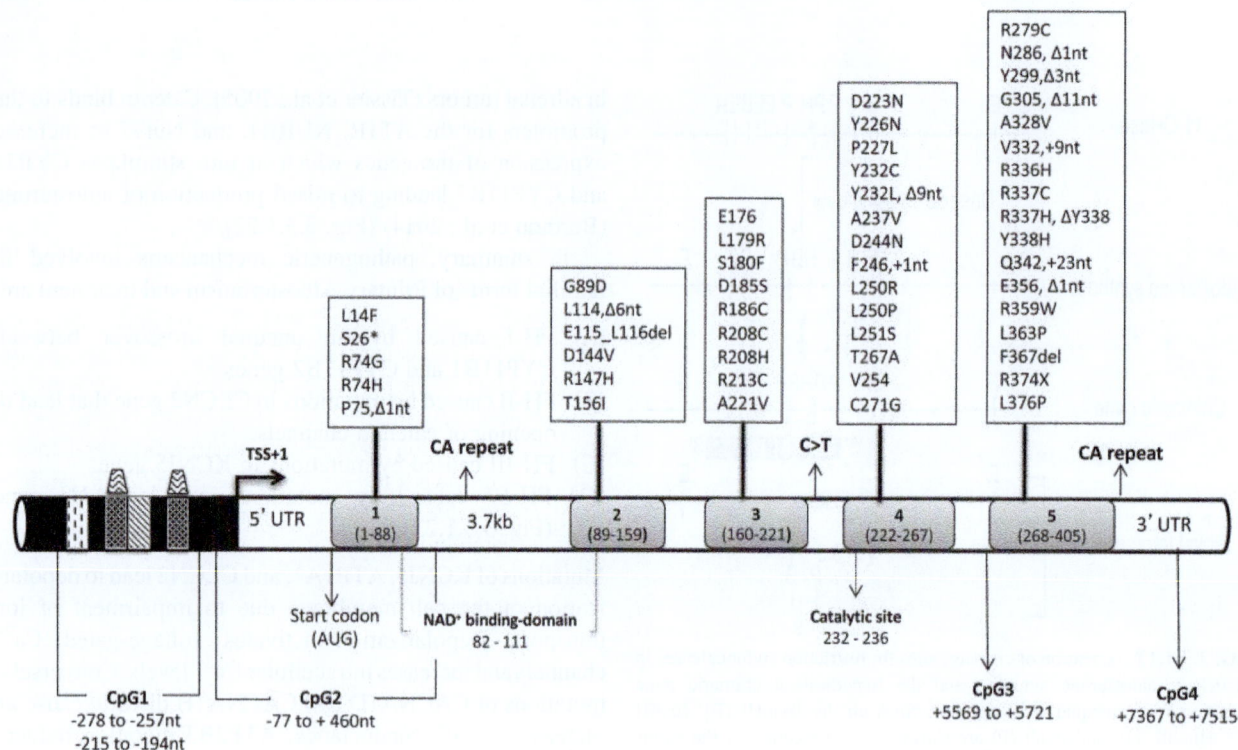

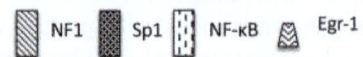

FIG. 3.3.1.18 Genetic variants in the human HSD11B2 gene. The diagram shows the human HSD11B2 gene and the location of identified pathogenic variants causing AME. The HSD11B2 gene is located on chromosome 16 and contains 5 exons, 4 CpG regions, 2 main (CA) repeat regions, and a promoter with SP1/SP3, NF1, NF-κB, and EGR-1 sites. Exons are represented by *gray boxes*. Mutations are listed relative to their position in the gene. *(From Carvajal CA, Tapia-Castillo A, Vecchiola A, Baudrand R, Fardella CE. Classic and nonclassic apparent mineralocorticoid excess syndrome. J Clin Endocrinol Metab. 2020;105(4):dgz315. https://doi.org/10.1210/clinem/dgz315. Erratum in: J Clin Endocrinol Metab. 2021;106(4):e1934. Fig. 2 p. e928. JCEM 106:dgz315.)*

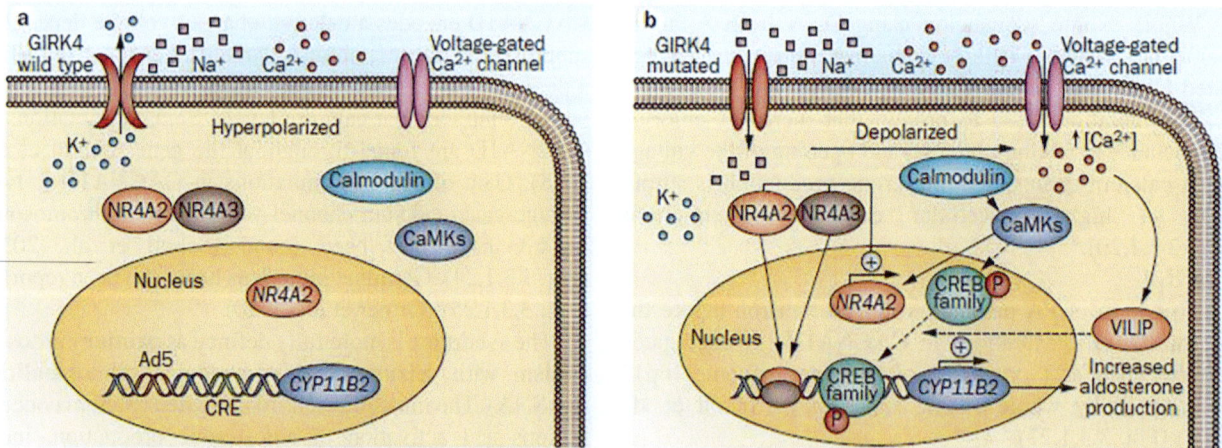

FIG. 3.3.1.19 Effect of *KCNJ5* mutations on adrenal glomerulosa cells. (A) Adrenal glomerulosa cells display a high resting background K+ conductance and the membrane is hyperpolarized. NR4A2 and NR4A3 are transcriptional regulators of the aldosterone synthase gene *CYP11B2*. NR4A2 binds to *cis*-regulatory elements in the *CYP11B2* promoter. (B) Mutations in the selectivity filter region of **GIRK4** (Gly151Arg, Thr158Ala, Leu168Arg, Gly151Glu and Ile157del) result in Na+ influx and membrane depolarization. This depolarization activates voltage-gated Ca2+ channels resulting in Ca2+ influx followed by activation of calmodulin and calmodulin-dependent kinases (CAMKs). Activation of calcium signaling pathways results in increased *CYP11B2* expression, via the transcriptional regulators NR4A2 and NR4A336 and probably the phosphorylation of cyclic AMP regulatory element binding protein (CREB) family members. Increased *CYP11B2* expression is instrumental in causing a deregulation of aldosterone production. *(From Mulatero P, Monticone S, Rainey WE, Veglio F, Williams TA. Role of KCNJ5 in familial and sporadic primary aldosteronism. Nat Rev Endocrinol 2013; 9(2): 104–12. Fig. 1 p. 107. Nat Rev. Endocrinol 9:104.)*

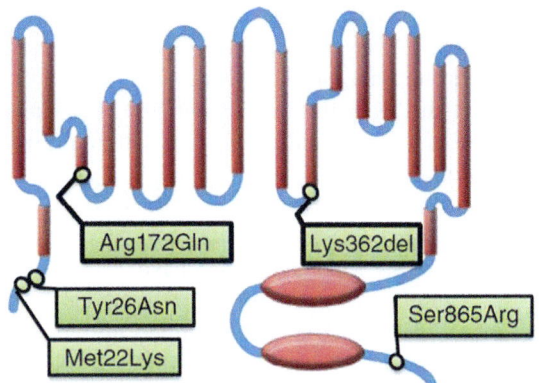

FIG. 3.3.1.20 Position of the variants in the N terminus, D helix, K helix and C terminus of the ClC-2 chloride channel encoded by *CLCN2. Red ellipses represent C-terminal CBS domains. (From Scholl UI, Stölting G, Schewe J, Thiel A, Tan H, Nelson-Williams C, Vichot AA, Jin SC, Loring E, Untiet V, Yoo T, Choi J, Xu S, Wu A, Kirchner M, Mertins P, Rump LC, Onder AM, Gamble C, McKenney D, Lash RW, Jones DP, Chune G, Gagliardi P, Choi M, Gordon R, Stowasser M, Fahlke C, Lifton RP. CLCN2 chloride channel mutations in familial hyperaldosteronism type II. Nat Genet 2018;50(3):349–54. Fig. 1b p. 350. Nat Genet 50:349.)*

3.3.1.3.8.2 ATPAse mutations

A somatic mutation was found in ATP1A1 in about 5% of APA (Azizan et al., 2013) then some in cells that were rich in zona glomerulosa cells (Beuschlein et al., 2013). ATP1A1 encodes the alpha 1 subunit of the Na+/K+ ATPase that transports three Na+ in exchange for two K+ ions. The alpha subunit is a transmembrane protein of 10 units and intracellular N and C termini. The protein from the mutations has lost K+ binding and depolarization of the cell membrane leading to autonomous aldosterone synthesis. The recognition of mutations seems to be dependent on the population (Taguchi et al., 2012) and gene analysis method with NGS more effective than Sanger sequencing (Nanba et al., 2018; Nanba et al., 2019). Somatic mutations have been found in ATP1A1 at chromosome 1p21 that

codes for ATPase Na/K transporting subunit alpha 1, and ATP2B3 on the X chromosome that codes for ATPase plasma membrane calcium transporting 3 (Beuschlein et al., 2013). These proteins play critical roles in intracellular calcium homeostasis (Fig. 3.3.1.26).

The plasma membrane Ca^{2+} ATPase type 3 (PMCA3) derives from the PMCA3 gene and has similar structure to ATP1A1 and exports Ca^+ ions from the cytoplasm. Deletion mutations in PMCA3 (Fig. 3.3.1.27) lead to disruption of calcium balance and depolarization of the cell membrane (Beuschlein et al., 2013). Mutation of *ATP1A1* results in alteration of potassium ion binding and loss of function followed by cell depolarization. Mutation of *ATP2B3 is a*ssociated with impaired clearance of cytoplasmic calcium ions. *CYP11B2*, aldosterone synthase gene expression is raised.

3.3.1.3.8.3 Primary mineralocorticoid excess

DOC secreting tumors are usually large and malignant. Some of the tumors also secrete androgens and estrogens which may cause virilization in women and feminization in men. A high level of plasma DOC or urinary excretion of tetrahydro-DOC will support the diagnosis. A large adrenal tumor on CT scan will confirm the diagnosis. Optimal treatment is surgical removal of the tumor. Medical treatment with o,p-DDD can be effective. Hypertension is seen with Cushing's syndrome particularly of the ectopic ACTH tumor type.

Deoxycorticosterone excess and low PRA may be encountered when mineralocorticoids other than aldosterone are in excess. This can be due to CAH due to 11β-hydroxylase deficiency or 17α-hydroxylase deficiency (11-deoxycorticosterone DOC excess) (Miller, 2008; Krone and Arlt, 2009; Rosa et al., 2007) or with mineralocorticoid secreting tumors (Irony et al., 1987). All are rare.

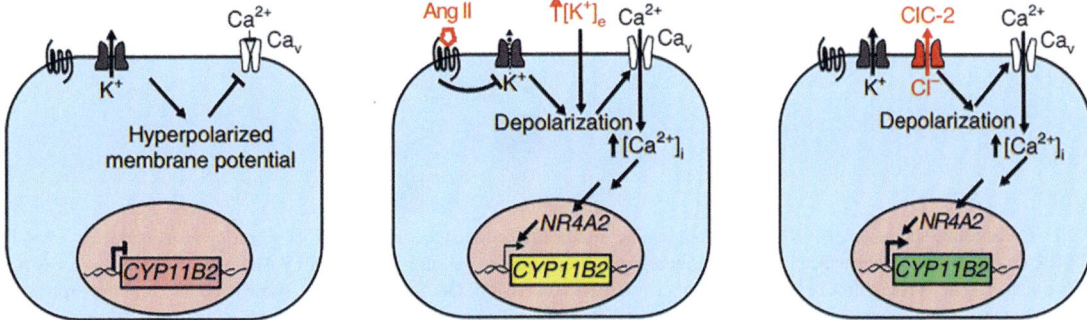

FIG. 3.3.1.21 Model of ClC-2 function in human adrenal glomerulosa. Resting cells are hyperpolarized. AngII and hyperkalemia cause depolarization, activation of voltage-dependent calcium channels, calcium influx and increased *CYP11B2 expression via the transcription factor NR4A2 (NURR1). ClC-2MUT causes increased CYP11B2 expression by membrane depolarization via increased chloride efflux.* $**P < .01$; $***P < .001$; $****P < .0001$. *(From Scholl UI, Stölting G, Schewe J, Thiel A, Tan H, Nelson-Williams C, Vichot AA, Jin SC, Loring E, Untiet V, Yoo T, Choi J, Xu S, Wu A, Kirchner M, Mertins P, Rump LC, Onder AM, Gamble C, McKenney D, Lash RW, Jones DP, Chune G, Gagliardi P, Choi M, Gordon R, Stowasser M, Fahlke C, Lifton RP. CLCN2 chloride channel mutations in familial hyperaldosteronism type II. Nat Genet 2018; 50(3): 349–354. Fig. 4d p. 353. Nat Genet 50:349.)*

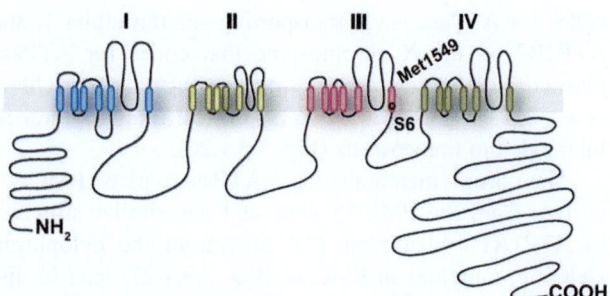

FIG. 3.3.1.22 Transmembrane structure of CaV3.2 (encoded by CACNA1H), the pore-forming subunit of a voltage-gated Ca2+ channel, is shown. These channels have four internal homologous repeats (I–IV), each with six transmembrane segments (S1–S6) and a membrane-associated loop between the pore-forming S5 and S6 segments. The p. Met1549Val mutation is located in S6 of repeat III. *(From Scholl UI, Stölting G, Nelson-Williams C, Vichot AA, Choi M, Loring E, Prasad ML, Goh G, Carling T, Juhlin CC, Quack I, Rump LC, Thiel A, Lande M, Frazier BG, Rasoulpour M, Bowlin DL, Sethna CB, Trachtman H, Fahlke C, Lifton RP. Recurrent gain of function mutation in calcium channel CACNA1H causes early-onset hypertension with primary aldosteronism. Elife 2015;4:e06315. Fig. 1b p. 4. Elife 4:e06315.)*

Unilateral adrenal hyperplasia. There are a few publications in the literature describing this phenomenon. Ganguly et al. described patients in whom removal of one hyperplastic adrenal normalized the blood pressure which then remained low for several years (Ganguly et al., 1980a). The contra-lateral adrenal gland was assumed to be normal, a fact supported by normal radiological findings. The group of Biglieri described three cases in which 18-hydroxycorticosterone values were elevated like that in adrenal adenoma (Banks et al., 1984). The surgical finding

was monolateral adrenal hyperplasia behaving as autonomous tissue. After removing the hyperplastic adrenal, blood pressure normalized (Katayama et al., 2005; Williams et al., 2017). It is worth mentioning that Cushing's syndrome has a similar unilateral form with multiple autonomic noduli.

Adrenal adenoma with elevated 18-hydroxycorticosterone, but normal aldosterone *production.* In the literature, four cases of this kind altogether have been described, three by Vecsei group, two of which had been operated. One further operated case was described by the group of Kanazawa in Japan (Miyamori et al., 1987). These patients were characterized by normal aldosterone, but elevated 18-hydroxycorticosterone values accompanied by somewhat elevated 18-hydroxydeoxycorticosterone. The serum potassium concentrations were in the low normal range, plasma renin activities were low and did not react to exertion. After removing the adrenal adenoma, blood pressure became normal. The etiology of the blood pressure elevation is unknown. 18-hydroxycorticosterone has some mineralocorticoid activity. Nothing is however known about its chronic hypertensinogenic activity. Acutely blood pressure remains unaffected by 18-hydroxycorticosterone.

3.3.1.3.8.4 Nodular hyperplasia

In children, **idiopathic hyperaldosteronism,** in which the adrenals have a micro or macronodular hyperplastic appearance has been described (Oberfield et al., 1984; Goh et al., 2007; Dewez and Bachy, 2009). Surgery was not curative. The best treatment is with the aldosterone antagonist, spironolactone, or with amiloride.

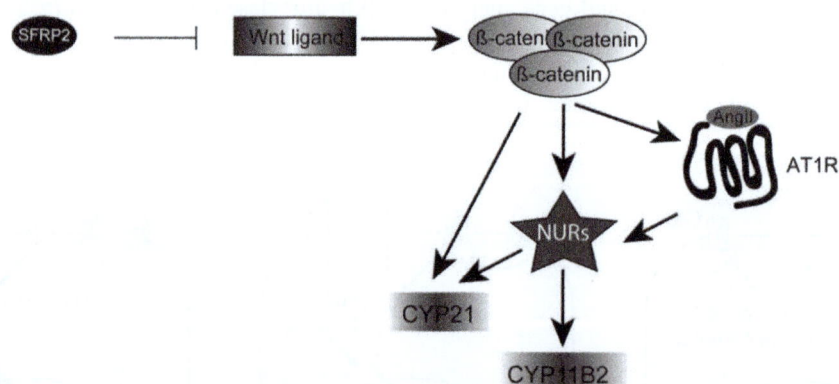

FIG. 3.3.1.23 Constitutive activation of WNT/b-catenin signaling. In a healthy adrenal (not shown), WNT signaling is maintained at a basal level by the action of SFRP2 (and presumably other factors of WNT signaling). Aldosterone secretion is stimulated by Angiotensin II through a cascade involving binding of Angiotensin II to its receptor AT1R, which stimulates NURR1 and NUR77 (NURs) expression. These in turn control the expression of CYP21 and CYP11B2, two enzymes essential for aldosterone production. Whether b-catenin plays a role in the control of aldosterone production in a healthy adrenal is unknown. In aldosterone-producing tumors (panel shown in figure) down-regulation of SFRP2 (or other regulators) results in deregulated WNT/b-catenin activation (70% of APA). b-Catenin is constitutively bound to the promoters of AT1R, NURR1 and NUR77. This results in increased expression of these genes, which in turn stimulates expression of CYP21 and CYP11B2, leading to increased production of aldosterone. *(From Berthon A, Drelon C, Ragazzon B, Boulkroun S, Tissier F, Amar L, Samson-Couterie B, Zennaro MC, Plouin PF, Skah S, Plateroti M, Lefèbvre H, Sahut-Barnola I, Batisse-Lignier M, Assié G, Lefrançois-Martinez AM, Bertherat J, Martinez A, Val P. WNT/β-catenin signalling is activated in aldosterone-producing adenomas and controls aldosterone production. Hum Mol Genet 2014; 23(4): 889–905. Fig. 7 p. 899. Hum Mol Genet 23:889.)*

FH-II

CLCN2

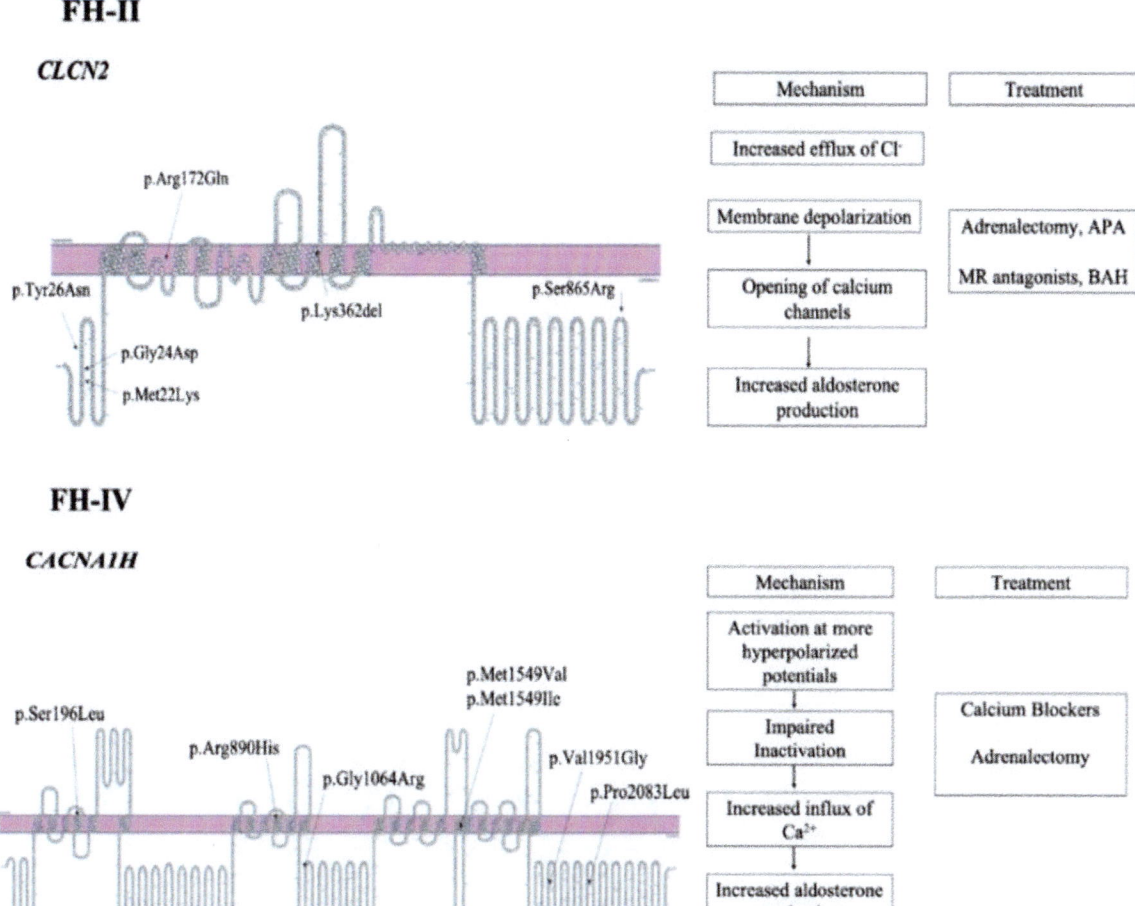

FIG. 3.3.1.24 Pathogenic mechanisms involved in familial forms of primary aldosteronism. *(Mourtzi N, Sertedaki A, Markou A, Piaditis GP, Charmandari E. Unravelling the genetic basis of primary aldosteronism. Nutrients 2021;13(3):875. Fig. 1 p. 7 of 12. Nutrients 13:875.)*

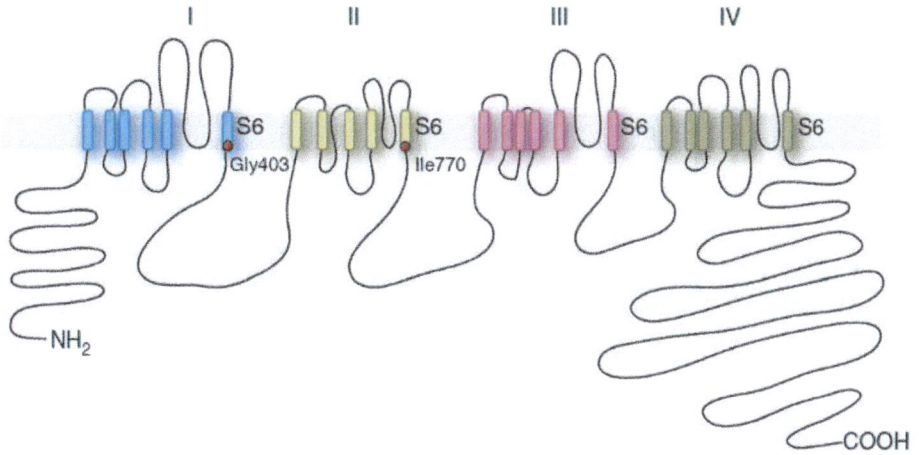

FIG. 3.3.1.25 Transmembrane structure of Cav1.3. *CACNA1D* encodes the pore-forming α1 subunit of a voltage-gated calcium channel. These channels feature four homologous repeats (I–IV) with six transmembrane segments (S1–S6) and a membrane-associated loop between segments S5 and S6. The five APA and two germline *CACNA1D* mutations identified in this study alter residues located at the ends of S6 segments implicated in channel gating. *(From Scholl UI, Goh G, Stölting G, de Oliveira RC, Choi M, Overton JD, Fonseca AL, Korah R, Starker LF, Kunstman JW, Prasad ML, Hartung EA, Mauras N, Benson MR, Brady T, Shapiro JR, Loring E, Nelson-Williams C, Libutti SK, Mane S, Hellman P, Westin G, Åkerström G, Björklund P, Carling T, Fahlke C, Hidalgo P, Lifton RP. Somatic and germline CACNA1D calcium channel mutations in aldosterone-producing adenomas and primary aldosteronism. Nat Genet 2013;45(9):1050–4. Fig. 2 p. 1052. Nat Genet 45:1050.)*

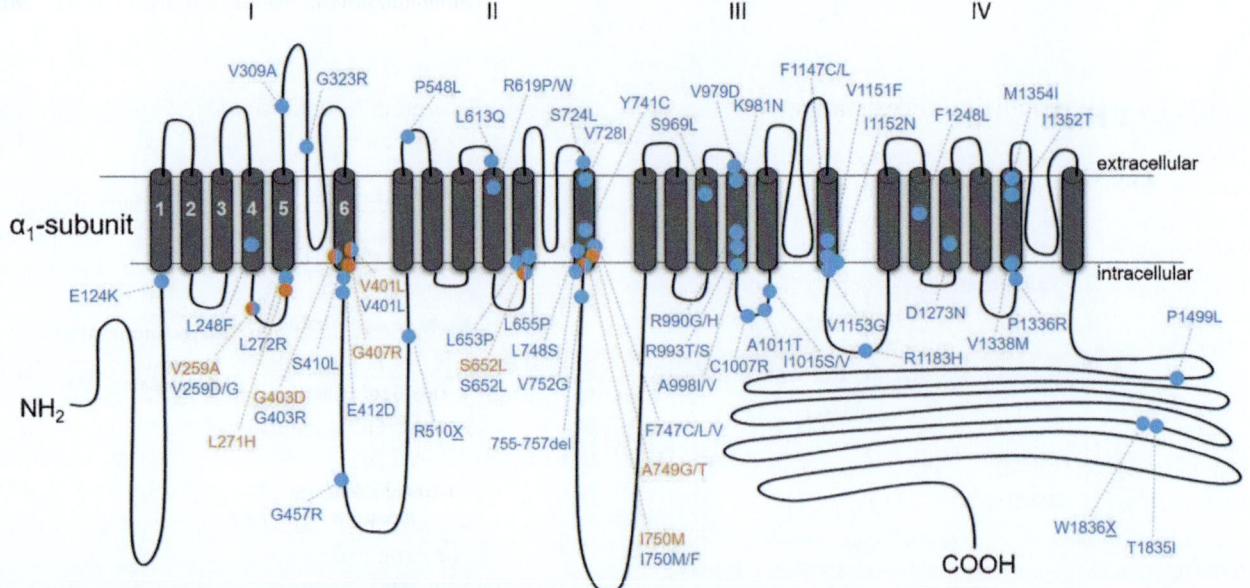

FIG. 3.3.1.26 Scheme of the position of de novo germline and de novo somatic APA/APCC variants within the Cav1.3 α1-subunit of CACNA1D. The α1-subunit consists of four homologous repeats (I–IV), each comprising six transmembrane segments (S1–S6). From each repeat, S1–S4 form the voltage sensors (S4 contains the positive gating charges) and S5–S6 with their connecting loop build the ion conducting pore. The S4–S5 loop links the voltage-sensor movements to the pore opening. Color code: germline (*orange*), somatic (*blue*). *(From Ortner NJ, Kaserer T, Copeland JN, Striessnig J. De novo CACNA1D Ca2+ channelopathies: clinical phenotypes and molecular mechanism. Pflugers Arch 2020;472(7):755–73. Fig. 2 p. 762. Pflugers Arch 472:755.)*

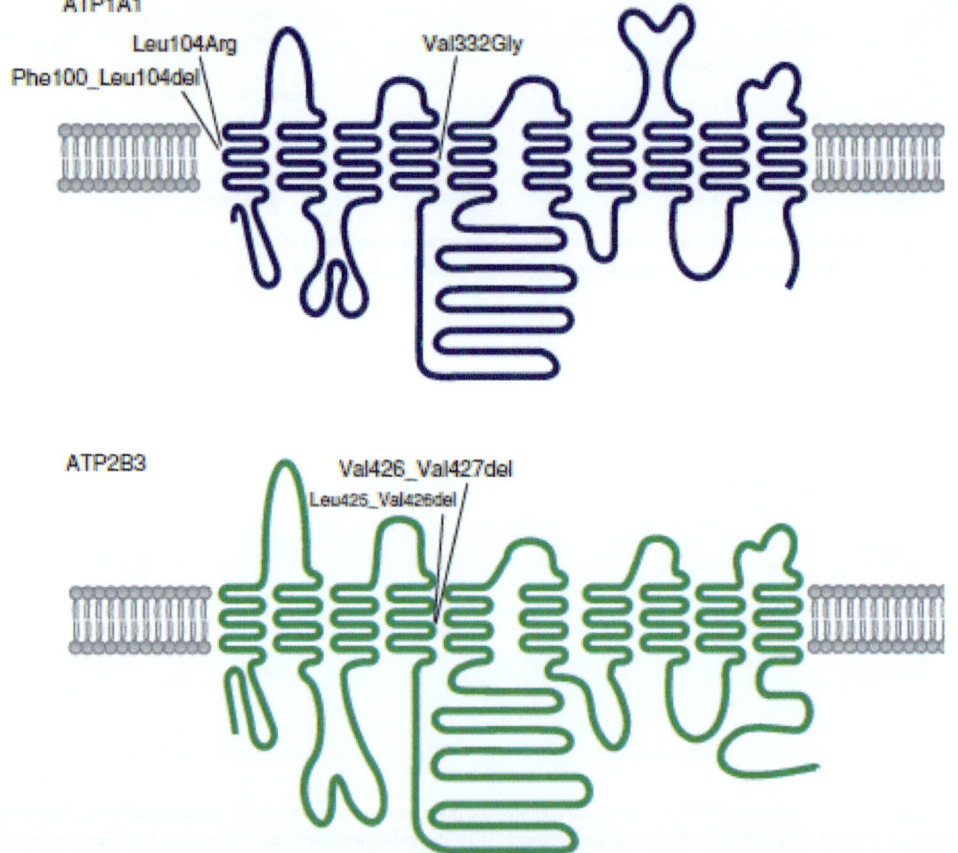

FIG. 3.3.1.27 Summary of somatic alterations in ATP1A1 and ATP2B3 identified in APAs. The ten transmembrane segments (M1-M10) are shown with the most N terminal (M1) on the left. The transported ions are bound by specific residues in M4, M5, M6 and M8, with M4 being of particular importance9,11. ATP and phosphate are bound in the large cytoplasmic loop between M4 and M5. The smaller cytoplasmic loop between M2 and M3 has a major role in conformational changes associated with energy transduction. *(From Beuschlein F, Boulkroun S, Osswald A, Wieland T, Nielsen HN, Lichtenauer UD, Penton D, Schack VR, Amar L, Fischer E, Walther A, Tauber P, Schwarzmayr T, Diener S, Graf E, Allolio B, Samson-Couterie B, Benecke A, Quinkler M, Fallo F, Plouin PF, Mantero F, Meitinger T, Mulatero P, Jeunemaitre X, Warth R, Vilsen B, Zennaro MC, Strom TM, Reincke M. Somatic mutations in ATP1A1 and ATP2B3 lead to aldosterone-producing adenomas and secondary hypertension. Nat Genet 2013;45(4):440–4, [444e1–2]. Fig. 1 p. 441. Nat Genet 45:440.)*

3.3.1.3.9 Glucocorticoid resistance

Primary cortisol resistance is detected when increased cortisol secretion is found in a patient with no evidence of Cushing's syndrome. A range of symptoms are found in children to adults, from hypokalemic alkalosis, hypertension and increased adrenal androgen secretion that characterize the spectrum of primary cortisol resistance. Adults can present with chronic fatigue and depression, infants can have seizures, hypoglycemia and hypertension. Total plasma cortisol should be measured at 0800 h and midnight. ACTH, serum aldosterone, plasma renin activity (lying and standing), androstenedione, testosterone, DHEAS, triglycerides, total cholesterol, HDL and LDL should also be measured. Cortisol shows a normal circadian rhythm but does not suppress with conventional doses of dexamethasone (up to 3 mg of dexamethasone may be required). The compensatory increase in CRF, ACTH and AVP leads to adrenal hyperplasia. Urinary-free cortisol can be up to 50 times normal. Patients with high urinary-free cortisol excretion rates have the least sensitivity of blood pressure to dietary sodium loading. This association may also have a genetic basis possibly through interaction with other genes. High plasma cortisol levels are a feature of glucocorticoid resistance but have been found in patients with salt sensitive hypertension.

The gene is typical of the steroid receptor superfamily. Defects in the glucocorticoid receptor gene NR3C1 affecting the steroid-receptor complex are responsible (Fig. 3.3.1.28) (reviewed by Nicolaides and Charmandari, 2021).

Mutations in the NR3C1 gene are concentrated to some extent around the binding sites in the protein (Fig. 3.3.1.29) and affect translocation of the steroid receptor complex.

Two polymorphisms of the glucocorticoid receptor have been found. The first polymorphism, located in the exon 2 of the *NR3C1* gene, ER22/23EK (rs6189, rs6190) polymorphism corresponds to two linked changes in codons 22 and 23, GGA.GAG to GAA.AAG, leading to one amino acid from ER to EK. The second polymorphism GR-9β (rs6198 or A3669G) is a A→G change, located in the 3′ UTR end of the exon 9β.

Mutations in NR3C1 create a protein with decreased binding to cortisol, as well as alterations in a number of loop conformations due to changes in the motion of the mutant protein (Monteiro et al., 2019). In a second study, in vitro and in silico methods were used to describe the structural biology of the mutant hGR L773P receptor (Kaziales et al., 2021). Significant conformational alterations were found for the binding of the mutant receptor to dexamethasone, to an NCOA-2 co-activator peptide, to DNA, and to HSP90. More information is needed to understand the action of steroids for protein synthesis in normal and pathological states.

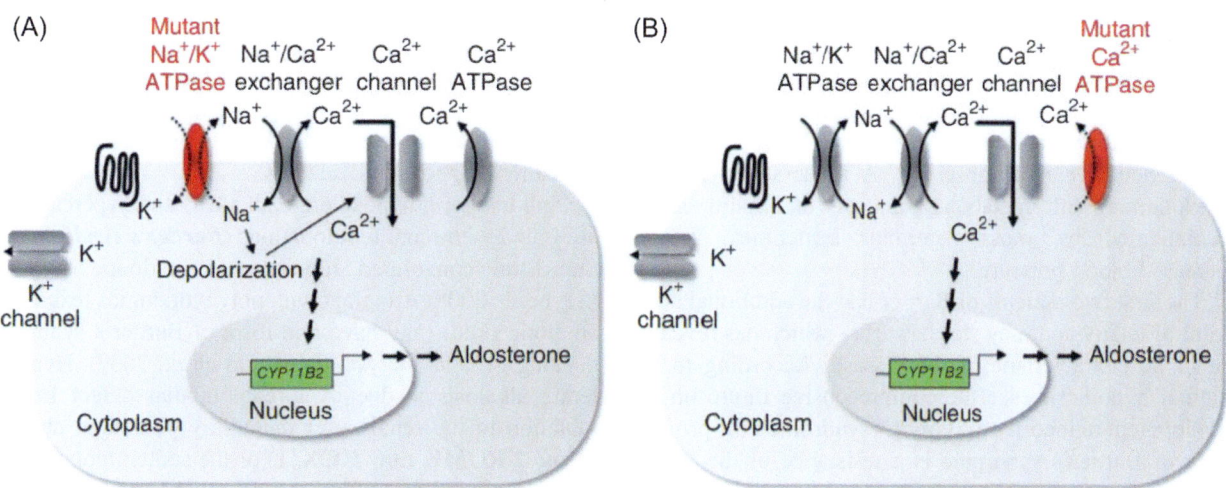

FIG. 3.3.1.28 Proposed mechanism for autonomous aldosterone secretion in APAs with somatic ATPase alterations. (A) Glomerulosa cell with hyperpolarized membrane potential under baseline conditions. (B) Angiotensin II (AngII)-induced inhibition of K+ channels and Na+/K+ ATPases leading to cell depolarization, activation of voltage-gated Ca2+ channels and higher cytoplasmic calcium ion levels. The original paper also presented the impact (i) mutation of ATP1A1 resulting in alteration of potassium ion binding and loss of function followed by cell depolarization. (ii) Mutation of ATP2B3 associated with impaired clearance of cytoplasmic calcium ions. CYP11B2, aldosterone synthase gene. *(Beuschlein F, Boulkroun S, Osswald A, Wieland T, Nielsen HN, Lichtenauer UD, Penton D, Schack VR, Amar L, Fischer E, Walther A, Tauber P, Schwarzmayr T, Diener S, Graf E, Allolio B, Samson-Couterie B, Benecke A, Quinkler M, Fallo F, Plouin PF, Mantero F, Meitinger T, Mulatero P, Jeunemaitre X, Warth R, Vilsen B, Zennaro MC, Strom TM, Reincke M. Somatic mutations in ATP1A1 and ATP2B3 lead to aldosterone-producing adenomas and secondary hypertension. Nat Genet 2013;45 (4):440–4, 444e1–2. Fig 4 Page 443.)*

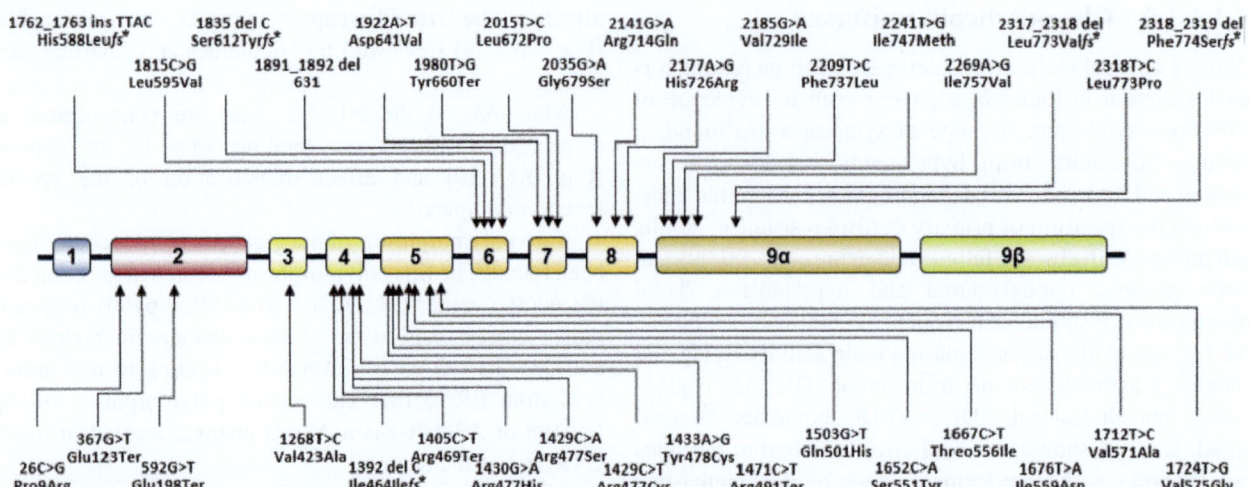

FIG. 3.3.1.29 Schematic representation of the published genetic defects causing Chrousos syndrome in the NR3C1 gene (**A**). *(From Nicolaides NC, Charmandari E. Primary generalized glucocorticoid resistance and hypersensitivity syndromes: a 2021 update. Int J Mol Sci. 2021;22(19):10839. Fig. 3 p. 5. Int J Mol Sci 22;10839.)*

3.3.1.3.10   Renal tubulopathies

Hypertension can be caused by disorders of renal salt reabsorption and potassium and hydrogen excretion through mechanisms independent of aldosterone. Defects in segments of the distal renal tubules are the basis of the hypernatremia.

3.3.1.3.10.1   Bartter's syndromes

In 1962, Bartter described a syndrome which was later named after him. This clinical entity reflects hyperplasia of the juxtaglomerulosa cells in the kidney, elevation of plasma renin activity and hyperaldosteronism. A decreased response of the vasculature to angiotensin II and other vasopressor agents can be observed. A moderate to marked hypokalemia with metabolic alkalosis and kaliuresis is accompanied by mostly normal, sometimes slightly decreased blood pressure.

The first two patients of Bartter had the additional complaint of tetany and they were dwarfs—which has revealed not to be characteristic of the disease. According to the original hypothesis (see the comprehensive figure on the development of theories of Bartter's syndrome), the primary event of Bartter's syndrome is a resistance of the vasculature to angiotensin II which causes a compensatory elevation in renin secretion and angiotensin generation. The elevated angiotensin II stimulates the *zona glomerulosa* of the adrenal, causing a hyperplasia and hyperfunction of this zone. Bartter's syndrome is relatively rare. In time, it has been revealed that the original hypothesis on the mechanism cannot be upheld. For instance, not all cases show the decreased reactivity of the vasculature to pressor agents. On the other hand, indomethacin, a known inhibitor of

prostaglandin biosynthesis, has improved the disease lowers plasma renin and aldosterone. Bradykinin, which was found also to be increased, decreased during treatment when hypokalemia and the previous low urinary kinin levels rose to normal. The marked effect of indomethacin led to a second hypothesis for the pathogenesis of Bartter's syndrome which assumed prostaglandins were the primary cause of the disease. Later however, after extensive studies, this prostaglandin hypothesis had to be discounted.

Nowadays, a defect in chloride reabsorption in the thick ascending limb of Henle's loop is considered as the cause of Bartter's syndrome. All other events were consequences of this basic failure as shown in Fig. 3.3.1.30. Aldosterone production is markedly elevated in Bartter's syndrome and does not cause diagnostic difficulties.

Salt loss with hypokalemic alkalosis and hypercalciuria are found with renal tubulopathies (***Bartter's syndrome***) in the distal convoluted tubules and or loop of Henle (Seyberth, 2008) An infant with polyhydramnios (excessive amniotic fluid) may have one form of Bartter's syndrome needing genetic analysis (Brochard et al., 2009). Hypokalemic alkalosis is due to a renal tubular defect from a mutation in the renal outer medullary potassium channel (**Type 2 ROMK now KCNJ1**) or the sodium/potassium/2 chloride transporter (**Type 1 SLCl2A1 now NKCC2**) (Marcoux et al., 2021; Starremans et al., 2003). The **infantile form** is associated with growth failure and sensorineural deafness. The primary cause for this disorder (**type 3**) is due to mutations in the **chloride channel gene (CLCNKB)** (Brochard et al., 2009). Type 4 has simultaneous mutations in BSND and CLCNKB or CLCNKA. There is resistance of the vasculature to the pressor action of angiotensin II associated with a compensatory increase

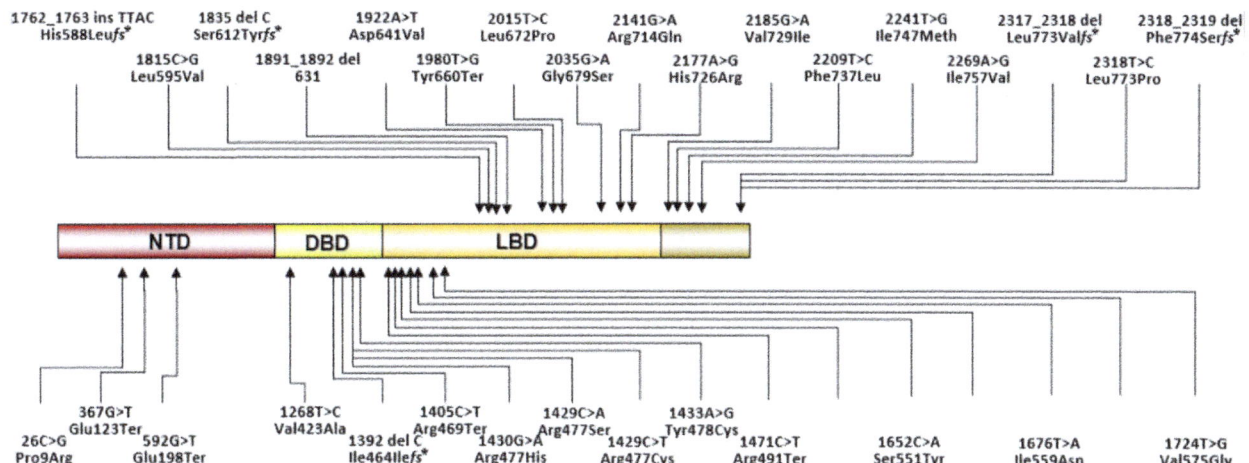

FIG. 3.3.1.30 Schematic representation of the published genetic defects causing Chrousos syndrome in the linearized hGR protein (**B**). *(From Nicolaides NC, Charmandari E. Primary generalized glucocorticoid resistance and hypersensitivity syndromes: a 2021 update. Int J Mol Sci 2021;22 (19):10839. Fig. 3 p. 5. Int J Mol Sci 22;10839.)*

in PRA with juxtaglomerular hyperplasia. The plasma concentrations of both aldosterone and PRA are raised.

This syndrome is a severe tubulopathy causing renal salt loss due to alterations in ion channels located in the TAL of Henle's loop and distal convoluted tubules. The result is hypokalemic, hypochloremic metabolic alkalosis. Mutations have been found in at least five genes coding for transporters that are expressed in the loop of Henle and distal convoluted tubules (Besouw et al., 2020). On a genetic basis, Bartters 1 to 5 are due to gene mutations (Table 3.3.1.8).

In the thick ascending loop of Henle (TAL), the apical sodium potassium dichloride co-transporter (NKCC2) protein reabsorbs 25% of the filtered sodium chloride (Gonzalez-Vicente et al., 2019). NKCC2 is expressed on basolateral membranes of epithelial and nonepithelial cells. Sodium is reabsorbed electroneutrally via NKCC2 together with one potassium and two chloride ions (Caceres and

Ortiz, 2019). The potassium is recycled through the potassium channel ROMK encoded by KCNJ1. SLC12A1 on chromosome 15 codes for NKCC2 and mutations in SLC12A1 (NKCC2) cause Bartters type 1. Mutations in KCNJ1 (ROMK) cause Bartters type 2. Type 2 has been called furosemide/amiloride phenotype. After birth, most patients present with low blood pressure, metabolic alkalosis, secondary aldosteronism, hypercalciuria and nephrocalcinosis.

Defects in NKCC2 and ROMK (Bartter I and II) produce the most severe manifestations initially with polyhydramnios that requires premature delivery. Macula densa cells sense variations in tubular fluid composition including salt content. Signal transduction includes activation of MAP-kinases p38 and pERK1/2, PGE2 synthesis through COX-2 and mPGES. PGE2 through paracrine signaling cause increased renin synthesis and release from juxtaglomerular cells and activation of renin-angiotensin-aldosterone axis (Cunha and Heilberg, 2018; Peti-Peterdi and Harris, 2010). Mutations are distributed throughout the transporter gene (Fig. 3.3.1.31). Impaired NKCC2 reduces the lumen positive potential difference that is necessary for paracellular concentrations of magnesium and calcium. NKCC2 was originally called furosemide-type loop disorder.

Sodium is actively pumped out of the renal TAL by basolateral sodium-potassium ATPase, whereas in response to an intracellular negative voltage chloride leaves the cell through specific chloride channels CLA-Ka in the TAL and DCT and CLC-Kb in the thin ascending loop. For proper function, both CLC require the protein Barttin as a subunit. A lumen positive transepithelial voltage promotes paracellular transport of magnesium and calcium by Claudin 16 and 19. Bartter's syndrome (BSIII) was the classical variant caused by mutations in the CLCKNB gene

TABLE 3.3.1.8 Bartter's syndrome I to V.

Bartter's syndromes

1. SLC12A1gene affecting NKCC2 protein- Na-K-2Cl co-transporter

2. KCNJ1 gene that codes for ROMK

3. CLCNKB basolateral chloride channel

4a. BSND Bartter with sensoneural deafness codes is due to mutations in CLCKNB subunit called Barttin

4b. Heterozygotes for CLCNKA and CLCNKb

5. Gain of function CASR

6. MAGED

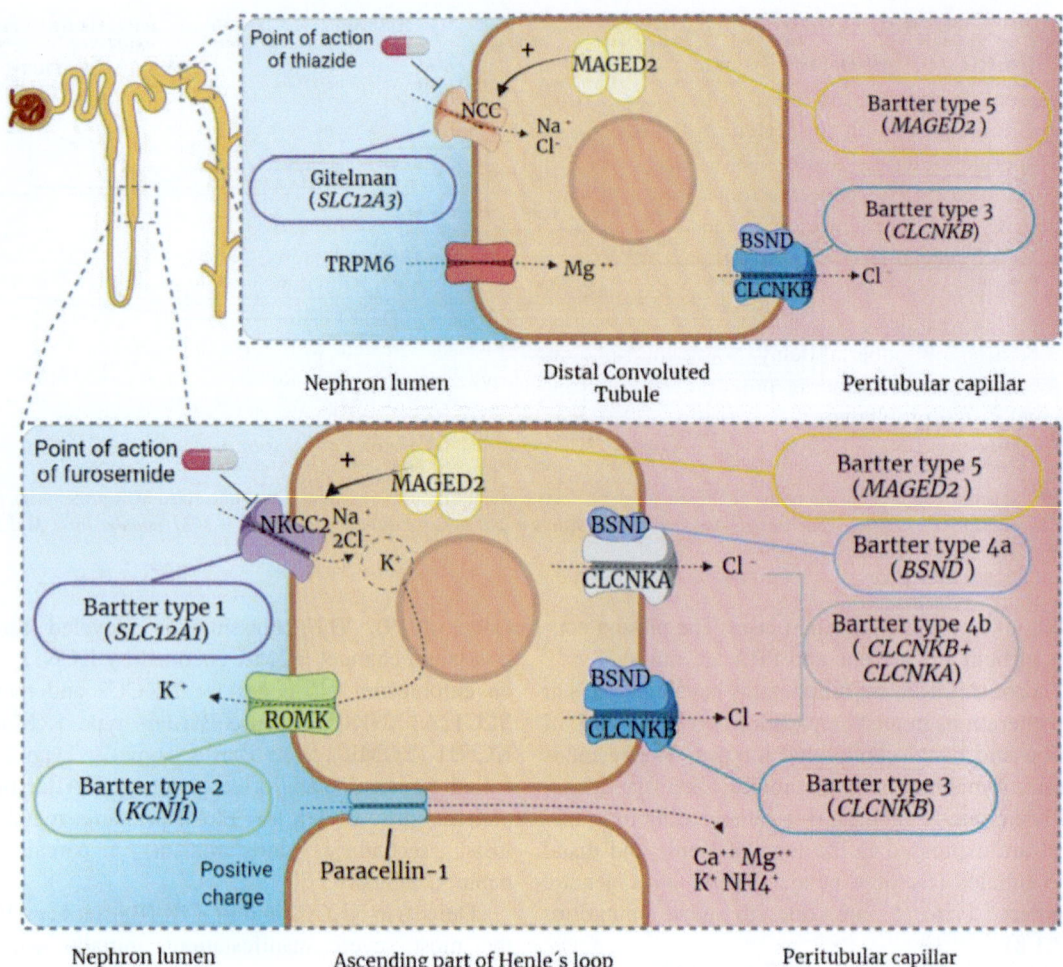

FIG. 3.3.1.31 Proteins and channels implicated in the pathogenesis of Gitelman and Bartter's syndromes. The electrolyte transports of the most important channels for the diseases are represented as well as the channels related to the inhibition by thiazide (NCC) and furosemide diuretics (NKCC2). Each disease is accompanied by the causative gene (in capital letters, brackets and italics), whereas the corresponding protein is indicated above the channel (only in capital letters). NCC: Solute carrier family 12 member 3; MAGED2: Melanoma-associated antigen D2; TRPM6: Transient receptor potential cation channel subfamily M member 6; CLCNKB: Chloride channel protein ClC-Kb; NKCC2: Solute carrier family 12 member 1; BSND: Barttin; CLCNKA: Chloride channel protein ClC-Ka; and ROMK: ATP-sensitive inward rectifier potassium channel 1. The positive charge of the DCT makes an electrochemical gradient from the luminal tubule from the interstitium possible. *(From Nuñez-Gonzalez L, Carrera N, Garcia-Gonzalez MA. Molecular basis, diagnostic challenges and therapeutic approaches of Bartter and Gitelman syndromes: a primer for clinicians. Int J Mol Sci 2021;22 (21):11414. Fig. 1 p. 3. Int J Mol Sci 22:11414.)*

which encodes basolateral chloride channel b variant that responds to thiazide and furosemide. Mutations were expressed in *Xenopus laevis* oocytes. Large deletions, frameshift, nonsense and essential splicing mutations were associated with children who presented at younger age (Seys et al., 2017). CLCKA, CLKCAB and Barttin are also expressed in epithelial cells of the inner ear.

Barrter syndrome type 4 is either a double heterozygote mutation of CLCNKA and CLCNKB or a monogenic mutation of Barttin (Miyamura et al., 2003; García-Nieto et al., 2006; Krämer et al., 2008; Janssen et al., 2009). Hypokalemia and hearing impairment are clues to this defect. Diagnosis is confirmed by genetic testing. Adults can

present with hypokalemia, deafness, secondary hyperparathyroidism and erythrocytosis (Sahbani et al., 2020; Heilberg et al., 2015).

The calcium sensing receptor (CaSR) promotes calcium reabsorption in response to increases of its plasma concentration. Activating mutations of CASR produce Bartter type V (Papadopoulou et al., 2021; Vahe et al., 2017).

Mutations in MAGED - melanoma associated antigen D2 causing X-linked polyhydramnios with an early onset in utero (15 weeks gestation) leading to prematurity and a severe but transient form of antenatal Bartters (Legrand et al., 2018; Laghmani et al., 2016). Mutations were detected by whole genome sequencing.

3.3.1.3.10.2 Gitelman syndrome (Familial hypokalemia-hypomagnesemia)

Patients present from late childhood with hypokalemic, hypochloremic, alkalosis with hypocalciuria and hypermagnesuria. Hypermagnesemia distinguishes GS from BS. Mutations in the SLC12A3 gene are found with abnormal thiazide sensitive sodium chloride cotransporter (NCC) protein (Blanchard et al., 2017). Adolescents or younger children can present variably with salt craving, muscle weakness, fatigue, limited sport ability, fainting, cramps, growth retardation and pubertal delay. Some adult patients develop hypertension, dizziness, polyuria, palpitations, joint pain and visual problems.

Biochemical investigations reveal hypokalemia, hypomagnesemia with renal potassium and magnesium wasting but hypocalciuria. Any electrolyte supplement should be stopped at least 48h prior to these tests. The supplements to the Blanchard paper provide useful information on urine calcium creatinine/ratios in children; foods rich in potassium, magnesium and calories; support groups and sick day rules.

3.3.1.3.10.3 EAST/SeSAME

This rare autosomal recessive syndrome was first described in 2009 (Bockenhauer et al., 2009; Scholl et al., 2012). The acronyms stand for the clinical features of the condition:

(i) epilepsy, ataxia, sensorineural deafness, tubulopathy and
(ii) seizures, sensorineural deafness, ataxia, mental retardation and electrolyte imbalance.

Presentation of seizures early in life allows for recognition and treatment. Mutations in KCNJ10 were found in a highly conserved transmembrane region of the channel that influence activity of Kir4.1 an inwardly rectifying potassium channel expressed in brain, inner ear, kidney and eyes (Bandulik et al., 2011). Children fail to respond to antiepileptic drugs (Valproate, Carbamazepine, Lamotrigine). Laboratory findings include hypokalemia, metabolic alkalosis, hypomagnesemia, hyponatremia, hypochloremia, elevated renin and aldosterone. The task of KCNJ10 is to recycle potassium that is necessary for the function of the active sodium/potassium ATPase.

KCNJ10 is expressed in glial cells in the cerebral cortex and cerebellar cortex and elsewhere, and it is believed to establish the resting membrane potential of neuronal cells. With repeated excitation and repolarization, a neuron takes up substantial amounts of sodium and loses potassium. Potassium accumulates extracellularly, decreasing the membrane potential, facilitating further excitations, and creating a tendency toward seizures. Glial cells presumably take up the extruded potassium and distribute it through gap junctions; KCNJ10 has been implicated in this process.

Absence of fully functional KCNJ10 removes this protective functional and accounts for the seizures in the patients.

3.3.1.3.10.4 HELIX

This is a syndrome acronym encompassing hypohidrosis (sweat glands make no sweat), electrolyte imbalance, lacrimal gland dysfunction (no tears), ichthyosis and xerostomia (dry mouth) (Hadj-Rabia et al., 2018; Bongers et al., 2017). Patients present with renal loss of NaCl with secondary hyperaldosteronism and hypokalemia (Fig. 3.3.1.32). Further tests confirm decreased NaCl absorption in the thick ascending loop of Henle. Hypercalcemia and hypermagnesemia are found but not consistently. The syndrome is distinct from Bartter and Gitelman syndromes. Patients have a preserved aquaporin response to desmopressin and intact response to furosemide. Compound heterozygous mutations in **claudin gene CLDN10b** affect paracellular ion transport in the thick ascending loop with novel tight junction disease characterized by hypokalemic-alkalosis and salt losing phenotype.

For patients with unexplained hypertension and hypokalemia, high-throughput sequencing of 5 genes (CLCKNB, SLC12A1, KCNJ1, BSND, and SLC12A3) can be used to diagnose Bartter or Gitelman disorders (Bao et al., 2019). MAGED2 needs to be added to that panel (Legrand et al., 2018). Newer panels include 37 genes (Ashton et al., 2018). More detailed genetic testing as part of chronic kidney disease is available (Knoers et al., 2022).

Hypertension with low aldosterone and other steroids. This scenario was described and extensively studied by the group of Maria New. Patients had low steroid values, not only of aldosterone and other mineralocorticoids, but even the glucocorticoids were suppressed. The etiological considerations have changed with the passage of time. Originally, **5α-dihydrocortisol**, a cortisol metabolite was taken as a causative factor (Mikami et al., 1980). Indeed this metabolite potentiates the mineralocorticoid effect of aldosterone. 5α-Dihydrocortisol however, does not evoke an increase in blood pressure. Recently, **cortisol receptor hypersensitivity** has been proposed with this glucocorticoid acting as mediator of the disease (Nicolaides et al., 2015).

3.3.1.3.11 Deoxycorticosterone excess—CAH

Low PRA may be encountered when mineralocorticoids other than aldosterone are in excess. This can be due to CAH due to 11β-hydroxylase deficiency or 17α-hydroxylase deficiency (**11-deoxycorticosterone DOC excess**) or with mineralocorticoid secreting tumors (Fig. 3.3.1.33).

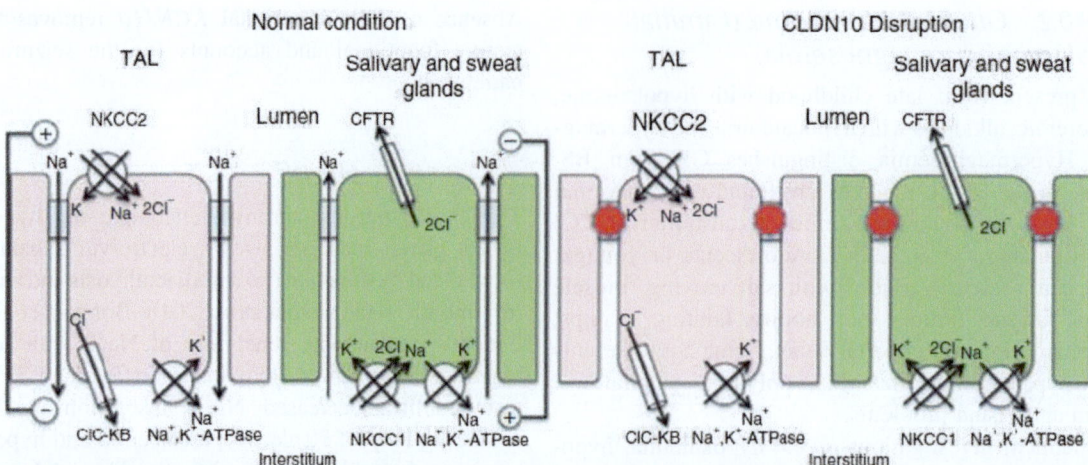

FIG. 3.3.1.32   Suggested model of ion transport across the epithelium of the thick ascending limb of the loop of Henle (TAL), and salivary and sweat glands under normal conditions and in the absence of CLDN10. Under normal conditions (left panel), the TAL (*pink*) absorbs NaCl via the furosemide-sensitive Na-K-2Cl cotransporter NKCC2 expressed at the apical membrane. Na and Cl exit the cell across the basolateral membrane via the Na, K-ATPase and the chloride channel ClC-Kb, respectively. Active NaCl absorption is an electrogenic process, resulting in a lumen-positive transepithelial potential difference, owing for a passive reabsorption of Na along the paracellular pathway. Salivary and sweat glands epithelia (*green*) secrete NaCl: NaCl enters the cell across the basolateral membrane via the Na-K-2Cl cotransporter Na-K-Cl cotransporter1 NKCC1; Cl is secreted into the lumen across the apical chloride channel cystic fibrosis transmembrane conductance regulator, thereby creating a lumen-negative transepithelial potential difference; Na exits the cell across the basolateral membrane via the Na,K-ATPase, and is passively secreted into the lumen along the paracellular pathway. In the absence of functional CLDN10 (right panel), the TAL epithelium still actively reabsorbs NaCl but the paracellular diffusion of Na is impaired, resulting in an overall decrease in NaCl absorption; in the salivary and sweat glands, the passive Na secretion is completely abolished. (*From Hadj-Rabia S, Brideau G, Al-Sarraj Y, Maroun RC, Figueres ML, Leclerc-Mercier S, Olinger E, Baron S, Chaussain C, Nochy D, Taha RZ, Knebelmann B, Joshi V, Curmi PA, Kambouris M, Vargas-Poussou R, Bodemer C, Devuyst O, Houillier P, El-Shanti H. Multiplex epithelium dysfunction due to CLDN10 mutation: the HELIX syndrome. Genet Med 2018;20(2):190–201. Fig. 3e p. 198. Genet Med 20:190.*)

All are rare. A defect in CYP11B1 presents with ambiguous genitalia in girls and precocious puberty and malignant hypertension in boys (Baronio et al., 2019; Hague and Honour, 1983). 17-hydroxylase deficiency is not usually detected until a female is investigated for primary amenorrhea and lack of sexual development. Persistent high progesterone is a useful diagnostic clue to this defect (Martin et al., 2003). The increased plasma corticosterone concentrations have been demonstrated by HPLC with UV detection.

Enzymatic defects in adrenal steroidogenesis result in deficient cortisol secretion and congenital adrenal hyperplasia (CAH). The lack of cortisol feedback inhibition on the hypothalamus and pituitary producers an ACTH-driven accumulation of intermediates in cortisol synthesis proximal to the enzymatic defect. CYP11B1 and CYP17 defects lead to hypertension and hyperkalemia because of hypersecretion of DOC. The sodium retention lowers PRA and hence aldosterone production. The disorders are very rare, but the study of these disorders has however provided a valuable insight into the mechanisms of genetic change and structure-function relationships as described below.

3.3.1.3.11.1   17α-Hydroxylase deficiency

Less than 150 cases of this condition have been described in the literature since first described (Biglieri et al., 1966).

Diagnosis of this enzyme deficiency is rarely made in childhood, the usual presentation is of a phenotypic female investigated for amenorrhea and failure of sexual development who is found to be hypertensive. Other characteristics are tall stature, eunuchoidism, bone age retardation and osteoporosis but other clinical evidence of glucocorticoid deficiency is unusual. The karyotype can be either 46XX or 46XY but since there is no sex steroid production pseudohermaphroditism is rare. Raised progesterone, absence of estrogen and cortisol and often low PRA and PAC are revealed in the initial endocrine tests. When a female is investigated for primary amenorrhea and lack of sexual development, persistent high progesterone is a useful diagnostic test for 17-hydroxylase deficiency. The increased plasma corticosterone concentrations have been demonstrated even by HPLC with UV detection (Wei et al., 2006).

In most laboratories, progesterone will be measured on automated immunoassay platforms but the markers for the defect are raised levels of DOC and corticosterone have been difficult to get measured in blood. Laboratories running LC-MS/MS for a plasma steroid profile now may include these steroids. It is important that isobaric steroids are separated in the chromatographic stage, 11-deoxycortisol and corticosterone are a pair of steroids with the same molecular weight which in a short LC separation may not be resolved. In an analysis of steroids in urine, metabolites of progesterone, 11-hydroxyprogesterone and corticosterone are clearly raised while cortisol and androgen metabolites are absent (Martin et al., 2003; Honour et al., 1978).

The gene for CYP17, CYP17, has been mapped to chromosome 10q24-25. No major deletions have been observed at this locus but a number of smaller deletions and duplications in addition to point mutations have been described in patients with CAH with regional differences (Costa-Santos et al., 2004; Kim et al., 2014; Wang et al., 2019). A cysteine residue involved in hem binding is located at codon 441. Several mutations have been located downstream of this site in the carboxy terminal region indicating that this region, although not believed to play a role in the active site, is important for the function of the enzyme. Genetic tests have some diagnostic value. This disorder is considered in more detail in the chapter on hypoandrogenism (Section 3.2.2) which is the usual context for this condition.

3.3.1.3.11.2 11β-Hydroxylase deficiency

A defect in CYP11B1 presents with ambiguous genitalia in girls and precocious puberty in boys. Defects of 11-hydroxylase and 17-hydroxylase cause hypertension through the action of raised 11-deoxycorticosterone. Examination of the steroid pathways can cause some confusion for students with respect to understanding DOC excess in 11-hydroxylase deficiency (Han and Elliott, 2017). It is important to distinguish the 11-hydroxylase for cortisol synthesis (CYP11B1) in the *zona fasciculata* and the 11-hydroxylase for aldosterone synthesis in the *zona glomerulosa* (CYP11B2). Furthermore, CYP11B1 is under the control of ACTH, whereas CYP11B2 is controlled by angiotensin II.

In 11-hydroxylase deficiency, most of the plasma DOC is synthesized in the *zona fasciculata*, stimulated (Fig. 3.3.1.33B) by ACTH due to low production rate of cortisol. The elevated DOC leads to sodium retention at the kidney and suppression of renin output with low circulating AII and normal/low aldosterone synthesis in the *zona glomerulosa*. During glucocorticoid treatment, the synthesis of steroids in the *zona fasciculata* can be suppressed with paradoxical salt loss in the kidney which then elevates renin and stimulates aldosterone secretion by the preserved activity of the CYP11B2 in the *zona glomerulosa* (Spoudeas et al., 1993; Zadik et al., 1984).

The 11-hydroxylase disorder is sometimes called the nonclassic form of CAH (Shammas et al., 2016) because 21-hydroxylase is the commonest and thus classic form. This description should be discouraged because nonclassic 21-hydroxylase deficiency is a distinct form due to mutations in CYP21A2 gene that express a 21-hydroxylase with mild activity. Females present in early childhood because of ambiguous genitalia and this will be discussed in the chapter on hyperandrogenism (Chapter 3.2.1). In 11-hydroxylase

deficiency, virilization is the typical presentation in the male with severe acne, advanced bone age, and hypokalemia in combination with hypertension (Breil et al., 2019; Bulsari and Falhammar, 2017) although raised blood pressure is not a presenting feature in the very young patient (Fig. 3.3.1.34).

The mineralocorticoid effect of increased circulating levels of DOC lowers PRA and the late steps of aldosterone synthesis are not activated. Blood measurements of 11-deoxycortisol, and DOC are more readily available (Karlekar et al., 2021; Li et al., 2021; Yildiz et al., 2021 although a urine steroid profile by GC-MS has picked up this disorder in cases before the introduction of LC-MS/MS methods, and is still reported (Tran et al., 2018). In a patient with 11-hydroxylase deficiency, cortisol metabolites are absent, there are high androgens and excess intermediate metabolites particularly of 11-deoxycortisol. The condition is treated with glucocorticoid replacement. In the affected neonate, tetrahydro-6-hydroxy-11-deoxycortisol may be dominant (Fig. 3.3.1.35) (Honour et al., 1983a). 6-OH-THS and two peaks for hexahydro-S (HHS) isomers (Hampf et al., 2001; Chabre et al., 2000; Skinner et al., 1996).

In a newborn screening programme in Germany, out of 986,098 tests, 78 cases were positive for 21-hydroxylase. Second tier screening by in-house LC-MS/MS of steroids including 17-OHP, 21-deoxycortisol, androstenedione and 11-deoxycortisol (472 to 1294 nmol/L NR <1 to 43) confirmed 5 cases of 11-hydroxylase deficiency (Janzen et al., 2012). The results for 17-OHP, and 21-deoxycortisol were within normal ranges and androstenedione was elevated. In another center in Germany, the Chromsystems steroid kit was used for LC-MS/MS testing (Breil et al., 2019) 11 deoxycortisol (8.6 to 58 μg/dL NR 0.2 to 0.25) and 17-OHP was 189 to 838 ng/dL (NR 3 to 90) so justifying a modestly raised first screen.

The gene encoding steroid 11β-hydroxylation (*CYP11B1*) is a close relative of aldosterone synthase (*CYP11B2*). The two genes are approximately 40 kb apart on chromosome 8q22, and share 93% base sequence homology in coding regions although only 48% in the 5' promoter region. The genes have identical intron-exon structure. CYP11B1 has an ACTH response element, CYP11B2 has an angiotensin II response element. In patients with 11-hydroxylase deficiency, mutations were initially thought to be concentrated in exons 6, 7 and 8 but as more patients have been studied the mutations are now scattered along the gene (Zhao et al., 2008). C to G transitions by deamination have been common mechanisms for mutation (Fig. 3.3.1.36). Several mutations are related to splicing site mutation.

Genetic tests for this condition are useful in family planning. Gene conversion does not appear to play a major role in the generation of mutations at this locus although regions of apparent gene conversion in intronic sequences

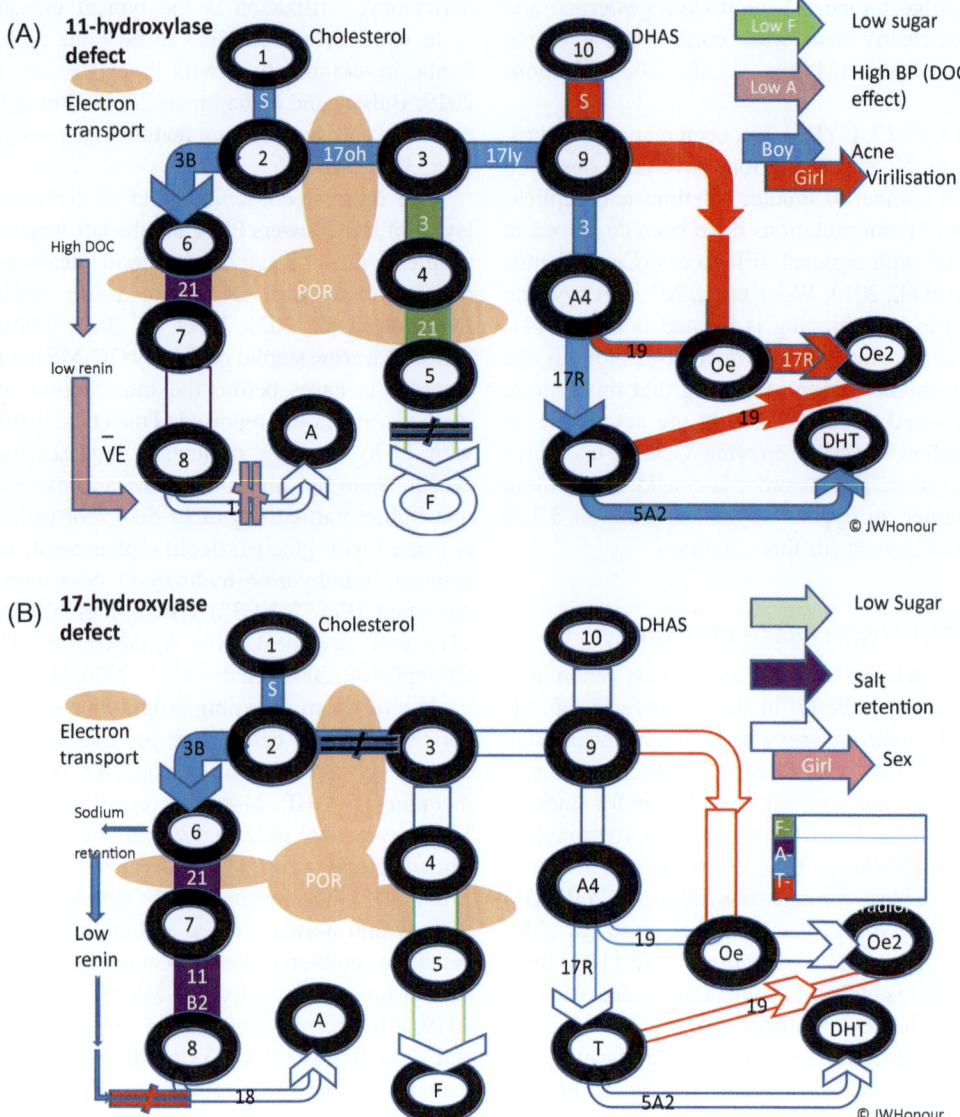

FIG. 3.3.1.33   CAH and raised DOC. (A) 17-Hydroxylase deficiency prevents synthesis of sex steroids and cortisol. The path to aldosterone is available but the production of excess 11-deoxycorticosterone (#6 in chart) causes sodium retention in the kidney and suppression of renin output. Without renin there is no angiotensin stimulation of the final steps in aldosterone production. (B) 11-hydroxylase deficiency prevents the last step in cortisol synthesis. Sex steroid production and deoxycorticosterone are raised, leading to virilization and hypertension. High DOC production again leads to low aldosterone production through lack of angiotensin stimulation. *(Author original.)*

of CYP11B1 and CYP11B2 have been identified. So far correlations between genotype, degree of virilization and severity of hypertension have not been verified as the numbers studied are small. Other factors may contribute to the phenotype.

In a cohort of 108 patients from 11 countries, Maria New and colleagues (Khattab et al., 2017) reported 41 compound heterozygotes or homozygotes for select missense or nonsense mutations causing severe classic CAH (hyperandrogenism). Computational modeling of 25 missense mutations revealed modification in the heme-binding or substrate binding site or alterations in enzyme stability may predict severe disease.

Polymorphisms have been identified in both the CYP11B1 and CYP11B2 genes and there is evidence of linkage disequilibrium between markers. Studies in Finland have shown that the 344CC allele in the CYP11B2 promoter was associated with higher blood pressure and increased left ventricular size (Hautanena et al., 1998). However, higher supine PAC with lower upright PRA was an unexpected result in such studies and such associations do not support a major role for the aldosterone synthase gene in hypertension at least in white populations.

Due to the close location of CYP11B1 and CYP11B2 on chromosome 8q21-22 chimeric genes are encountered (see earlier section on GRA/DSH). In a few cases, the chimera

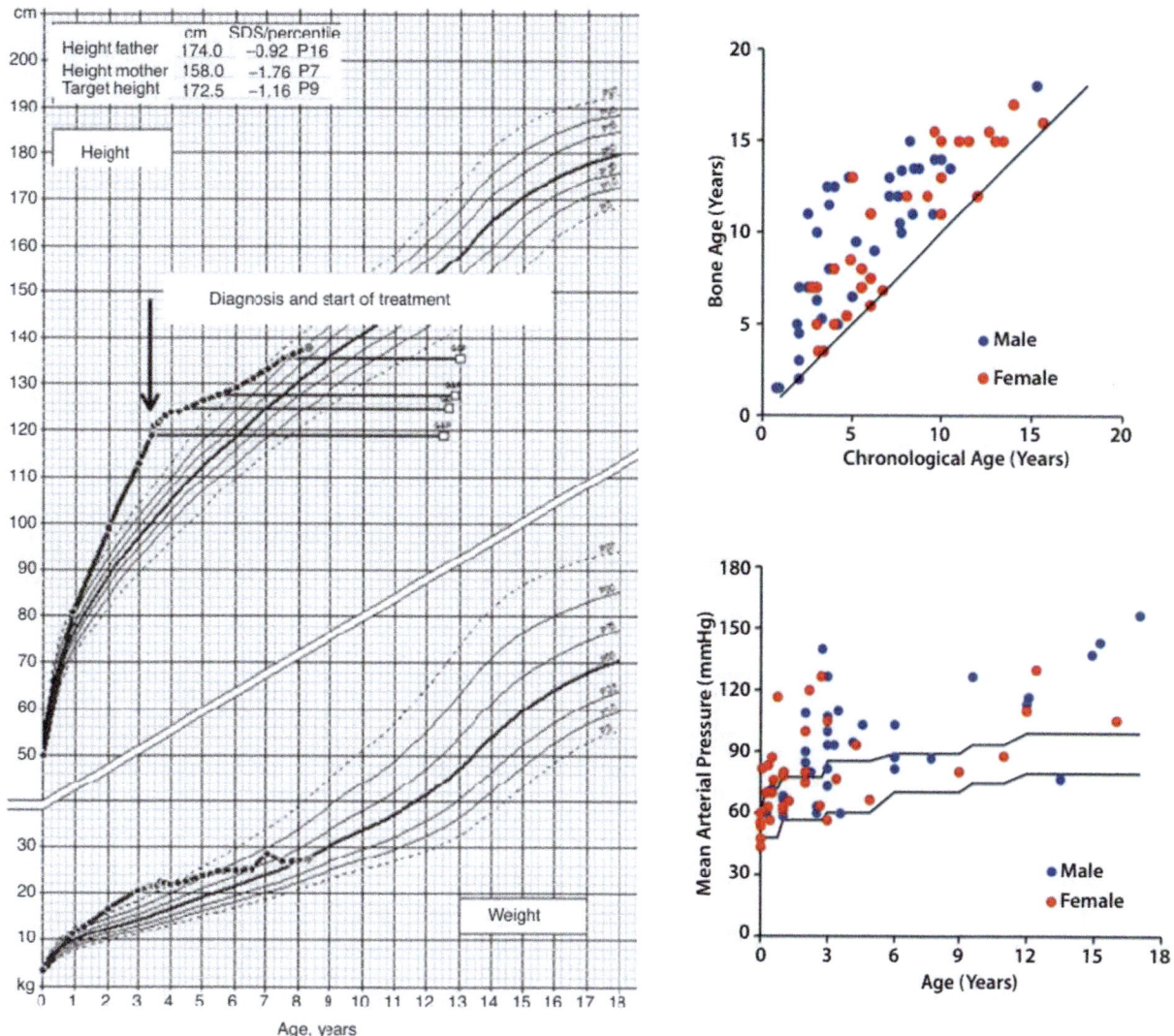

FIG. 3.3.1.34 Presentation of 11-hydroxylase deficiency. (A) Height and weight chart of one patient before and after treatment started. German references according to Reinken and van Oost 1992 and Kromeyer-Hauschild et al. 2001. Repetitive bone ages (Greulich and Pyle 1959 are depicted as arrows from distinct chronological ages to illustrate bone age acceleration by indicating height for bone ages. SDS, standard deviation score. (B) Bone age versus chronological age in male (*blue*) and female (*red*) patients. The black line represents a slope of 1 (bone age = chronological age). Of note is that almost all patients show evidence of advanced bone age. Rapid growth is a feature of 11-hydroxylase deficiency from very early in childhood, hypertension is not seen until 3–5 years of age and beyond. *(From (A) Breil T, Yakovenko V, Inta I, Choukair D, Klose D, Mittnacht J, Schulze E, Alrajab A, Grulich-Henn J, Bettendorf M. Typical characteristics of children with congenital adrenal hyperplasia due to 11β-hydroxylase deficiency: a single-centre experience and review of the literature. J Pediatr Endocrinol Metab 2019; 32(3):259–267. Fig. 1 p. 263. J Pediatr Endocrinol Metab 32:259. (B) Khattab A, Haider S, Kumar A, Dhawan S, Alam D, Romero R, Burns J, Li D, Estatico J, Rahi S, Fatima S, Alzahrani A, Hafez M, Musa N, Razzghy Azar M, Khaloul N, Gribaa M, Saad A, Charfeddine IB, Bilharinho de Mendonça B, Belgorosky A, Dumic K, Dumic M, Aisenberg J, Kandemir N, Alikasifoglu A, Ozon A, Gonc N, Cheng T, Kuhnle-Krahl U, Cappa M, Holterhus PM, Nour MA, Pacaud D, Holtzman A, Li S, Zaidi M, Yuen T, New MI. Clinical, genetic, and structural basis of congenital adrenal hyperplasia due to 11β-hydroxylase deficiency. Proc Natl Acad Sci U S A. 2017; 114(10): E1933–E1940. Fig. 1 p. e1934. PNAS 114:e1933.)*

consists of the promoter, exons 1–6 of CYP11B2 and exons 7–9 of CYP11B1. Lack of CYP11B1 consists under the control of CYP11B2 promoter explains the cortisol defect (Xu et al., 2015; Khattab et al., 2017; Nimkarn and New, 2008). 11-hydroxylase deficiency occurs in the Caucasian population at a prevalence of 1 in 100,000 - twenty times less common than CYP21A1 deficiency.

Liddle's syndrome

The more common systemic form of pseudohypoaldosteronism (PHA) is due to activating mutations of the **epithelial sodium transport channel** (ENaC-α, β-, or γ-) genes (SCNN1B,G and A (Brower et al., 2020; Fan et al., 2018, 2020a; Tetti et al., 2018; Belot et al., 2008; Hanukoglu

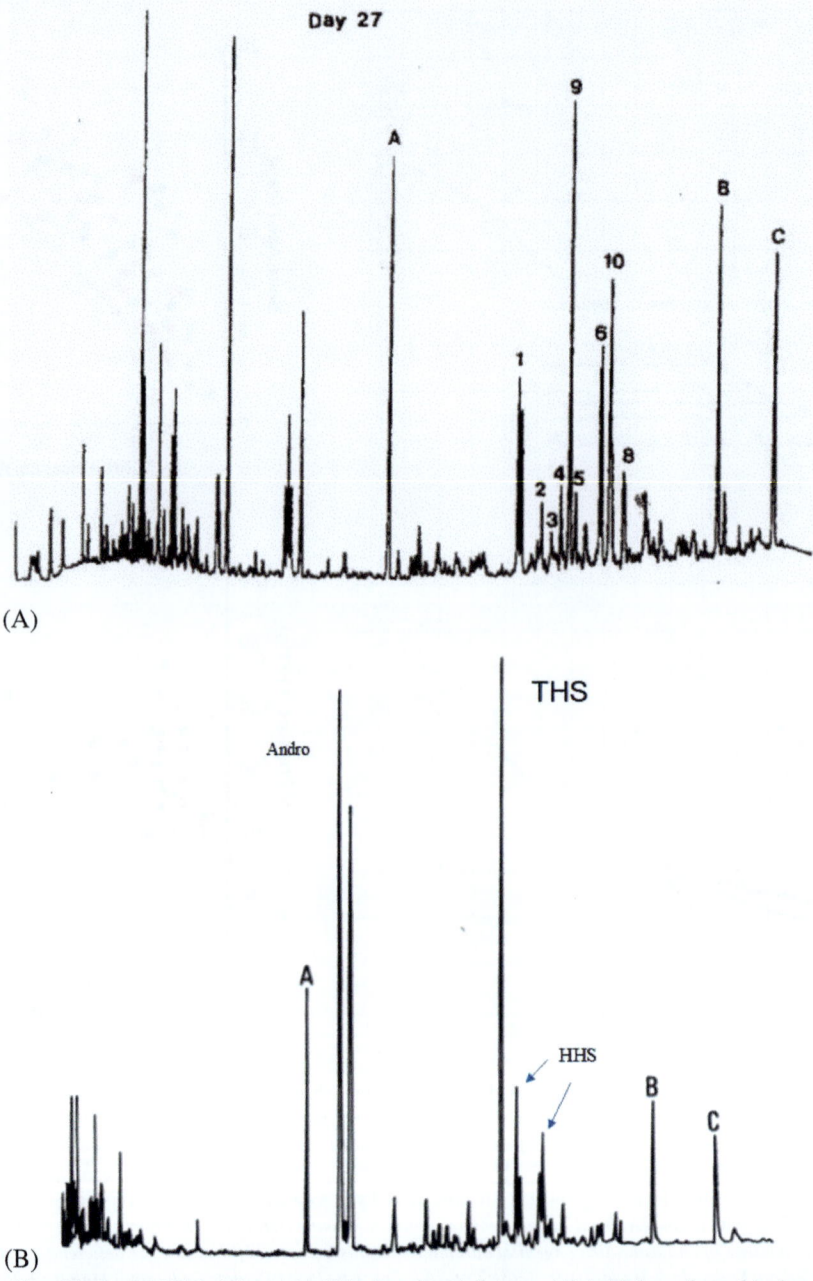

FIG. 3.3.1.35 Urine steroid profiles by GC-MS for infants with 11-hydroxylase deficiency. (A) At 27 days of age, 6-hydroxy-THS (peak 9) is the major steroid in the urine. Peak 1 (doublet is for 16a-hydroxy-DHEA; 2 is 16b-hydroxyDHEA; 6 is 16-hydroxypregnenolone and 10 is hexahydro-11-deoxycortisol (HHS) (B) At 1 year the dominant steroid is THS (tetrahydro-11-deoxycortisol), the androgens androsterone (androsterone) and the next peak for etiocholanolone are also grossly raised. Two isomers of HHS are also seen. *(Honour JW, Anderson JM, Shackleton CH. Difficulties in the diagnosis of congenital adrenal hyperplasia in early infancy: the 11 beta-hydroxylase defect. Acta Endocrinol (Copenh) 1983;103(1):101–9. Fig. 1 p. 102.)*

et al., 2008). Salt loss occurs at the sweat glands, epithelium of lung and intestines in addition to the distal nephron. Potassium restriction is effective treatment (Adachi et al., 2020). Patients with Liddle's syndrome present with early and typically moderate to severe hypertension. Transmission is in autosomal dominant manner. The pathogenesis of hypertension in some cases entails increased sodium reabsorption of salt and water. **Low PRA and PAC** suggests

the effect of other mineralocorticoids but is a consequence of sodium retention and is discussed further in Chapter 3.3.2.

3.3.1.3.12 Treatment of CAH

Treatment of the disease due to defects in CYP11B1 or CYP17A1 is usually achieved with glucocorticoid

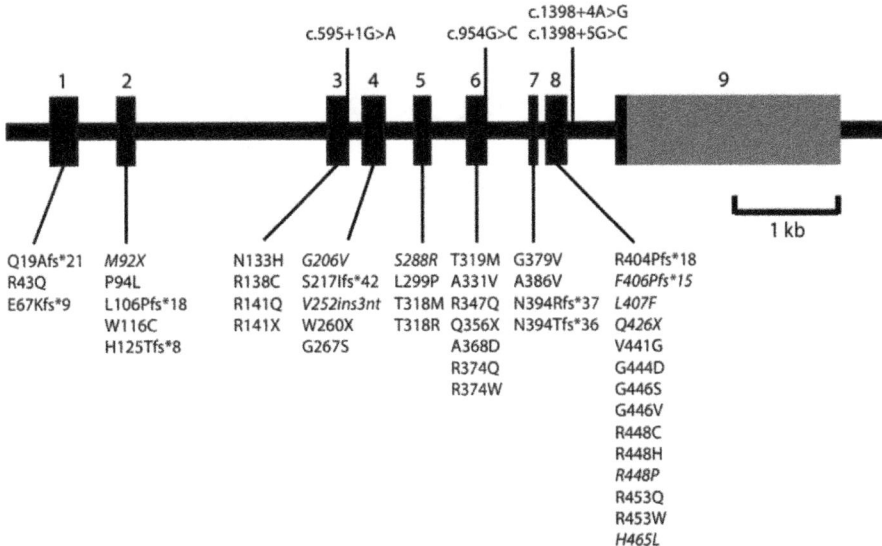

FIG. 3.3.1.36 Genetic mutations in patients with CAH resulting from 11β-hydroxylase deficiency from 13 nations comprising an International Consortium for Rare Steroid Disorders. The human CYP11B1 gene contains nine exons with mutations shown for 108 patients in this international cohort. Previously unreported mutations are shown in italics. *(From Khattab A, Haider S, Kumar A, Dhawan S, Alam D, Romero R, Burns J, Li D, Estatico J, Rahi S, Fatima S, Alzahrani A, Hafez M, Musa N, Razzghy Azar M, Khaloul N, Gribaa M, Saad A, Charfeddine IB, Bilharinho de Mendonça B, Belgorosky A, Dumic K, Dumic M, Aisenberg J, Kandemir N, Alikasifoglu A, Ozon A, Gonc N, Cheng T, Kuhnle-Krahl U, Cappa M, Holterhus PM, Nour MA, Pacaud D, Holtzman A, Li S, Zaidi M, Yuen T, New MI. Clinical, genetic, and structural basis of congenital adrenal hyperplasia due to 11β-hydroxylase deficiency. Proc Natl Acad Sci U S A. 2017;114(10):E1933–40.)*

replacement (Bulsari and Falhammar, 2017; Khattab et al., 2017) as described in detail for 21-hydroxylase deficiency (Chapter 3.1.2). The suppression of ACTH does reduce production of sex steroids in the case of CYP11B1 deficiency (Khattab et al., 2017). Testicular adrenal rests are common (Mazzilli et al., 2019; Engels et al., 2018) when ACTH suppression is not complete. TART are benign but a growing mass can compress the ducts and spermatogonia leading to irreversible damage and infertility. TART can be detected by ultrasound scanning (Ma et al., 2019). Overtreatment with hydrocortisone alone in patients with CYP17A1 or CYP11B1 defects can block production of deoxycorticosterone by the *zona fasciculata*, should DOC production by the *zona glomerulosa* also be low (when PRA is low) a salt losing state can be precipitated. It may then be necessary to add mineralocorticoid treatment (fludrocortisone is used) (Spoudeas et al., 1993; Hochberg et al., 1986). Spironolactone or eplerenone can be used as mineralocorticoid antagonists to control hypertension and will also block sex steroid action and are therefore not ideal treatments.

In most hospitals, these condition will be monitored by measurement of the principal substrate for the defective enzyme (corticosterone for CYP17A1 defects and 11-deoxycortisol for CP11B1 defects). Few laboratories have been able to measure 11-deoxycorticosterone but that is changing with the introduction of steroid profiling. PRA can be measured as an index of DOC suppression or mineralocorticoid activity.

3.3.1.3.13 Apparent mineralocorticoid excess (AME)

The interconversion of cortisol to cortisone particularly in the kidney inactivates cortisol which would otherwise act at the MR (Fig. 3.3.1.37).

Children with a **defect in HSD11B2** present with symptoms of hypokalemia, i.e., thirst, polyuria, failure to thrive and weakness. The hypertension is severe with a 10% mortality rate. Despite an extensive search for a mineralocorticoid responsible for the features of AME, all assays failed to detect activity in the plasma or urine of such patients. It is now clear that cortisol is the mineralocorticoid responsible for the condition. The diagnosis can be made from a characteristic urine steroid profile as measured by gas chromatography-mass spectrometry in full scan mode (Fig. 3.3.1.38) (Shackleton et al., 1980).

There is an increase in the ratio of cortisol metabolites to cortisone metabolites. The data analysis can target the steroids using an ion for a specific fragment of each steroid (Fig. 3.3.1.39).

The plasma cortisol half-life is prolonged but patients do not show features of cortisol excess because there is a fall in daily production. Circulating levels of cortisol are normal. The concentration of urinary-free cortisol is usually elevated. It is generally believed that the interconversion of cortisol to cortisone particularly in the kidney inactivates the cortisol which otherwise competes with aldosterone for binding at the mineralocorticoid receptor.

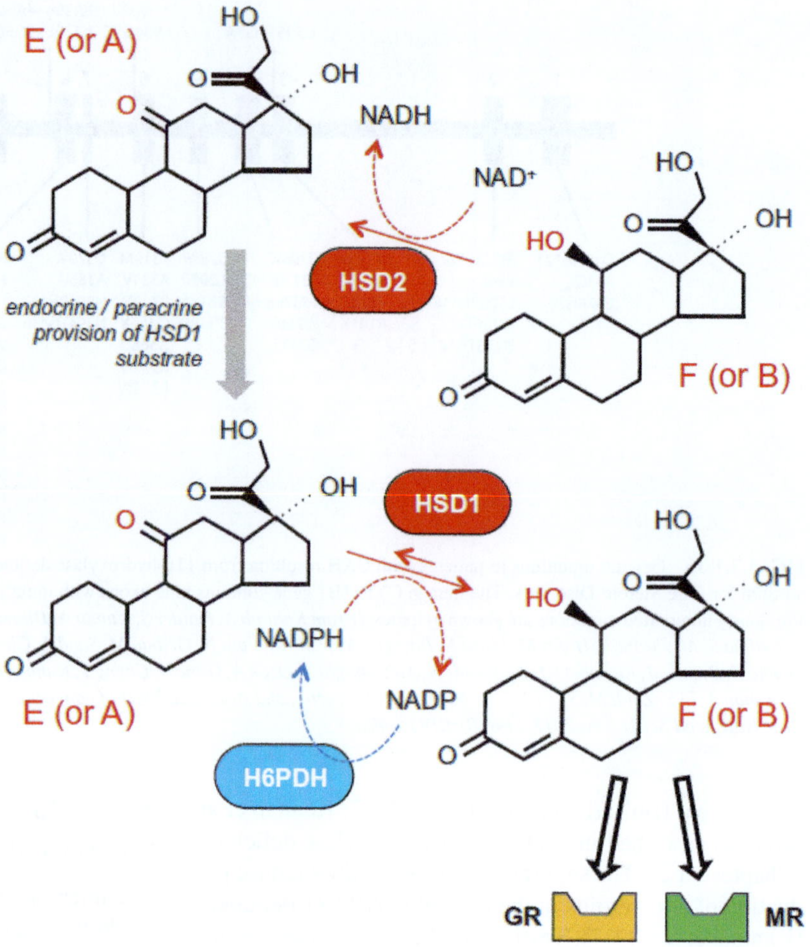

FIG. 3.3.1.37 Actions of the 11βHSD enzymes. The bioactivity of glucocorticoid is regulated by enzymatic modification of the C11 side chain. In humans, the reduced 11-hydroxy form cortisol (F) is physiologically active at the mineralocorticoid receptor; the oxidized 11-keto form cortisone (E) is inert. The same is true in rodents for active corticosterone (B) and inactive 11-dehydrocorticosterone (A). Interconversion between the oxidized and reduced forms is catalyzed by two 11β-hydroxysteroid dehydrogenase (11βHSD) enzymes. 11BHSD1 operates as an NAPDH-dependent reductase, regenerating active glucocorticoids in target tissues. It is co-expressed in the endoplasmic reticulum with hexose-6-phosphate dehydrogenase (H6PDH), which generates NADPH requisite for reductase activity. 11βHSD2 is a high affinity NAD+-dependent dehydrogenase, inactivating glucocorticoids in vivo. The changes in redox potential that accompany NAD+ metabolism may lock MR-cortisol complexes in an inactive state. *(From Bailey MA. 11β-Hydroxysteroid dehydrogenases and hypertension in the metabolic syndrome. Curr Hypertens Rep 2017;19(12):100. Fig. 1 p. 3 of 9. Curr Hypertens Rep 19:100.)*

FIG. 3.3.1.38 Capillary gas chromatographic urinary steroid profile of the patient with apparent mineralocorticoid excess. A, B, and C, Internal standards (5a-androstane-3a,17a-diol, stigmasterol, and cholesteryl butyrate, respectively). Peak I, 11-hydroxyandrosterone; 2, 11-hydroxyaetiocholanolone; 3, pregnanetriol; 4,tetrahydro-Substance S; 5, tetrahydrocortisone; 6, 5a-tetrahydrocorticosterone; 7, tetrahydrocortisol; 8, 5a-tetrahydrocortisol; 9, b-cortol; 10, a-cortol; 11, cortisol (doublet); 12, 20a-dihydrocortisol (doublet). The peak between 11 and 12 is 6b-hydroxycortisol. *(From Shackleton CH, Honour JW, Dillon MJ, Chantler C, Jones RW. Hypertension in a four-year-old child: gas chromatographic and mass spectrometric evidence for deficient hepatic metabolism of steroids. J Clin Endocrinol Metab 1980; 50(4): 786-02. Fig. 2 p. 788. JCEM 50:786.)*

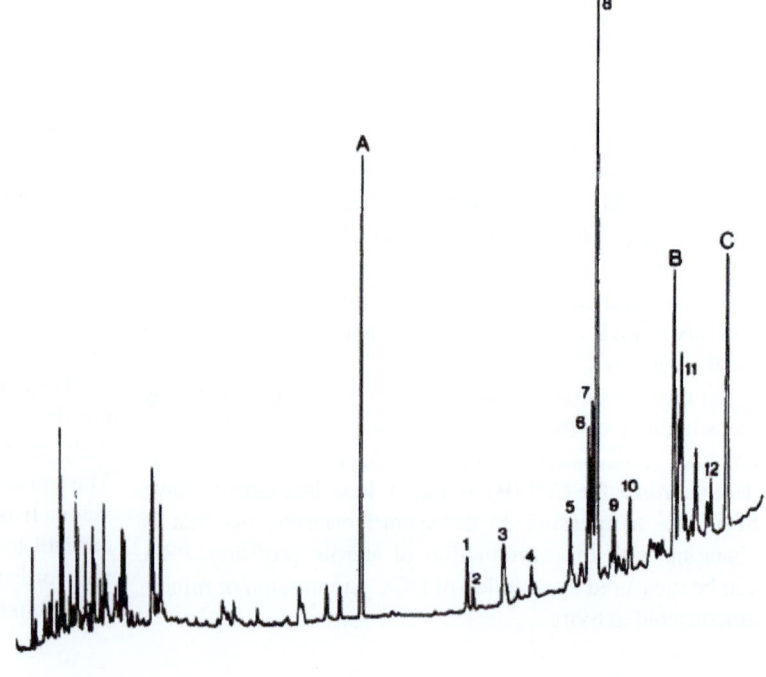

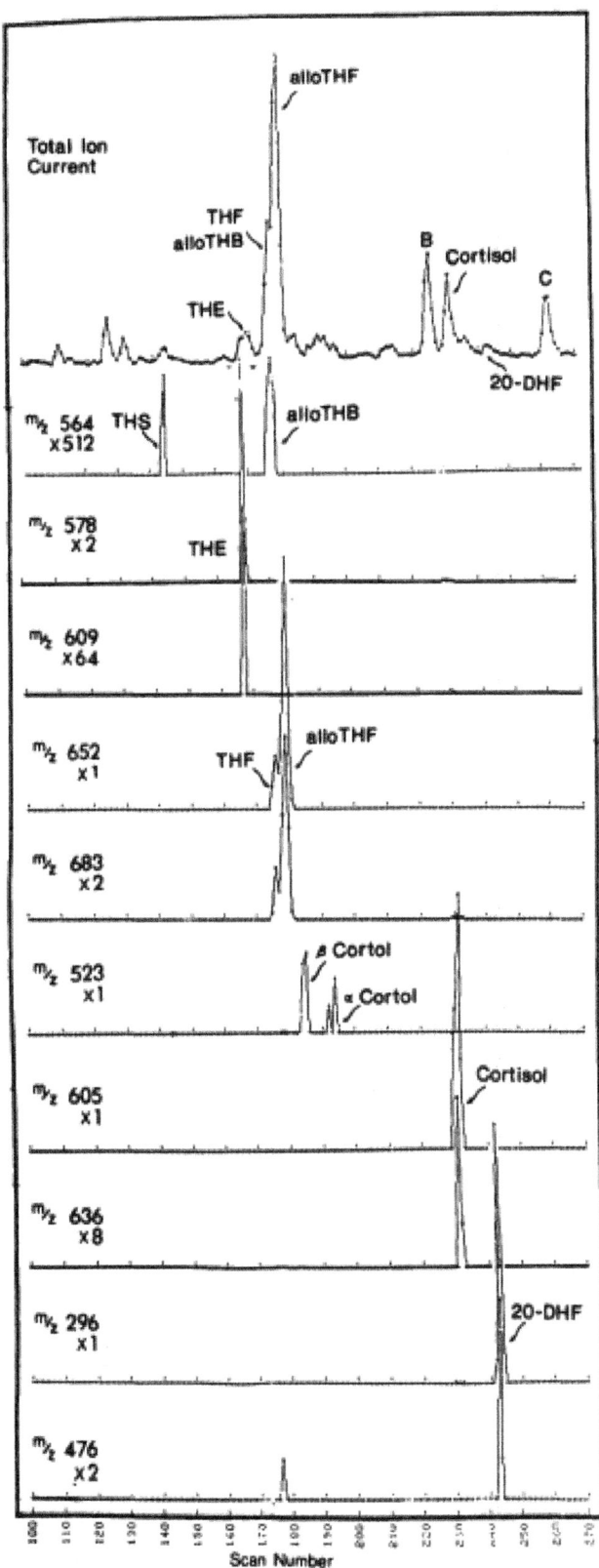

FIG. 3.3.1.39 Mass spectral identification of urinary steroids. Reconstructed selected ion chromatograms from repetitive scanning analysis carried out on an SE-S2 capillary column. The *ordinate is an arbitrary intensity* scale, and the *abscissa represents the scan numbers of the complete* spectrum obtained. The multiplication factor of each of the *m/z* numbers represents the increase in the intensity relative to a normal intensity of unity. The *m/z* values represent the molecular ions or fragments formed by the loss of specific groups from the parent molecule named, viz. *m/z* 564, M - 31 for tetrahydro-Substance S (THS) and 5a-tetrahydrocorticosterone (alloTHB); *m/z* 578, M - 31 for tetrahydrocortisone (THE); *m/z* 609, M for THE; *m/z* 609, M+ for THE; *m/z* 652, M - 31 for tetrahydrocortisol (THF) and 5a-tetrahydrocortisol (alloTHF); *m/z* 683. M+ for THF and alloTHF; *m/z* 523, M - 205 for cortols; *m/z* 605, M – 31 for cortisol; *m/z* 636, M+ of cortisol; *m/z* 296, M - 205 - 90 - 90 of 20a-dihydrocortisol (20-DHF); and *m/z* 476, M - 205 of 20 DHF. *Compounds* Band C are the internal standards stigmasterol and cholesteryl butyrate, respectively. B, Corticosterone. *(From Shackleton CH, Honour JW, Dillon MJ, Chantler C, Jones RW. Hypertension in a four-year-old child: gas chromatographic and mass spectrometric evidence for deficient hepatic metabolism of steroids. J Clin Endocrinol Metab 1980; 50(4): 786-02. Fig. 2 p. 788. JCEM 50:786.)*

Dexamethasone administration induced natriuresis and potassium retention and when cortisol or ACTH is infused a mineralocorticoid excess state can be reproduced.

In a syndrome of apparent mineralocorticoid excess (AME), there is hypertension, hypokalemia and reduced secretion rate of cortisol. The improvement brought about by treatment with spironolactone (aldosterone antagonist) or triamterine (potassium sparing diuretic) suggested the presence of an unidentified mineralocorticoid. The disease is attributed to cortisol acting as both glucocorticoid and mineralocorticoid with a prolonged half-life due to low activity of 11β-hydroxysteroid dehydrogenase type 2 (which normally oxidizes cortisol to inactive cortisone (Hammer and Stewart, 2006). This defect is most easily detected by a saliva cortisol and cortisone ratio (Antonelli et al., 2015; De Palo et al., 2009). A urine steroid profile clearly displays a high excretion of cortisol metabolites relative to cortisone (Morineau et al., 2006). Tandem mass spectrometry methods have enabled simultaneous analysis of cortisol and cortisone in blood (Taylor et al., 2002; Kushnir et al., 2004; Raul et al., 2004) and saliva (De Palo et al., 2009) and of metabolites in urine (Turpeinen et al., 2006). The disease has been found mainly in children with severe hypertension and may be lethal which may explain why very few adult cases have been described.

The mineralocorticoid hormones acting at the MR lead to sodium retention or potassium excretion. Increased aldosterone secretion may be suspected in a patient presenting with muscle weakness due to hypokalemia and headaches due to hypertension. Before the laboratory enters investigations of the renin-angiotensin-aldosterone system, abnormalities of electrolytes and water balance need to be confirmed. Many antihypertensive drugs affect plasma renin and aldosterone (thiazide and loop diuretics, beta-receptor blockers, calcium channel blockers) and these should ideally be stopped 2 weeks before meaningful tests can be carried out.

Hypokalemia with metabolic alkalosis can be seen with rare disorders of ion channels in the renal tubules. Measurements of calcium and magnesium give characteristic features for further renal tubular disorders.

3.3.1.3.13.1 Genetics of HSD11B2 deficiency

Patients with AME may be either homozygous or compound heterozygous carriers of HSD11B2 mutations (Yau et al., 2017; New et al., 2005) Published mutations are mostly missense mutations clustering in exons 3, 4, and 5 (Morineau et al., 2006). A systematic review recorded 54 homozygous or compound heterozygous mutations, including 31 (57%) missense mutations and 42 (78%) mutations located in exons 3–5 (see earlier Fig. 3.3.1.18) (Fan et al., 2020c).

Other genetic aberrancie including nonsense, splicing, insertion, and deletion mutations were also recorded, but at a lower rate. The identified compound heterozygous mutations were situated in exons 2 and 5, and resulted in deletions of amino acid residues causing truncated 11βHSD2.

3.3.1.3.13.2 Hypertension in pregnancy

A distinct form of **hypertension** has been seen in **pregnancy** and attributed to the action of progesterone on the mineralocorticoid receptor (MR). An **activating mutant of the MR** in the hormone binding domain was found in a 15 year old boy resulted in substitution of leucine for serine at codon 810. When 23 relatives were tested, 11 had severe hypertension before 20 years of age (Geller et al., 2000). Plasma and urine aldosterone concentrations were low. Progesterone was an agonist to the mutant MR, furthermore spironolactone was also an agonist precluding this treatment option.

3.3.1.3.13.3 Liquorice abuse

The mineralocorticoid activity of liquorice, originally thought to block the action of aldosterone, is now known to be due to inhibition of the enzyme system 11β-hydroxysteroid dehydrogenase. Patients should be asked about confectionary habits (Dellow et al., 1999; Buhl et al., 2018). Other substances have such glycyrrhetinic acid like factor (GALF) activity (see Morris et al., 2021 for review). The urine steroid profile of a patient who eats liquorice is similar to that seen for patients with apparent mineralocorticoid excess (Table 3.3.1.9), thus confirming the action of liquorice on HSD11B2 rather than the MR as originally thought to be the case.

3.3.1.4 Secondary aldosteronism

Diuretic therapy is the commonest cause of secondary aldosteronism. PRA is raised as a physiological response to hypovolemia (hemorrhage or intestinal fluid loss). Hypokalemia without hypertension is the result of increased aldosterone but certainly in adults can be due to diuretic or laxative abuse or psychogenic vomiting. Renin secretion is increased with renal ischemia, **renal artery stenosis**. Peripheral PRA levels may be only moderately raised but renal vein samples may show a result on the ischemic side 1.5 times or more that from the contralateral normal kidney.

Renal hypertension constitutes about 15% of all cases of hypertension and patients have secondary hyperaldosteronism. Renal hypertension due to renal vascular disease, renal parenchymatous disease have high renin activity and secondary aldosteronism. Treatment of secondary hyperaldosteronism is possible with one of several effective inhibitors of the renin-angiotensin system. Renin output may be reduced by beta blockade with propanolol. The angiotensin converting enzyme can be inhibited by captopril and enalopril. The drug

TABLE 3.3.1.9 Urine steroid profile of patient using liquorice. Results for sodium, potassium, plasma renin activity, and aldosterone are shown.

	On liquorice	On liquorice	Off liquorice	Off liquorice	Normal ranges
Date	02/04/20	15/04/20	20/01/21	14/05/21	
Urine cortisol metabolites					
THE	1140	940	860	1350	980–3240
THF	1290	1120	1120	2090	440–1610
Allo-THF	300	310	410	430	160–640
THF + allo-THF	1590	1450	1420	2520	610–2140
THF + allo-THF/THE	1.39	1.54	1.65	1.85	0.47–0.87
THE/THF	0.72	0.65	0.61		0.79–1.92
Allo-THF/THF	0.23	0.27	0.27		0.29–0.72
Total C21 cortisol metabolites	3330	3440	3260		2630–7300
Blood pressure	180/100	226/120	138/82	136/88	
Electrolytes					
Plasma potassium		2.4	4.7	4.1	3.3–4.8
Plasma sodium		145	132	142	137–145
Aldosterone pmol/L	170	89	295		600–1200
Plasma renin activity nmol/L/h	1.86	1.01	4.21		2.8–4.5

From Dellow EL, Unwin RJ, Honour JW. Pontefract cakes can be bad for you: refractory hypertension and liquorice excess. Nephrol Dial Transplant 1999;14(1):218–20.

saralasin is a competitive inhibitor of angiotensin II at the receptor site.

3.3.1.4.1 Renal artery stenosis

Renin secretion is increased with renal ischemia, renal artery stenosis. Peripheral PRA levels may be only moderately raised but renal vein samples may show a result on the ischemic side 1.5 times or more that from the contralateral normal kidney.

3.3.1.4.2 Pseudo Bartter's syndrome

A clinical syndrome which is very similar to Bartter's syndrome can be caused by **abuse of purgatives and diuretics.** In contrast to Bartter's syndrome, this disease is relatively frequent. Patients, more usually women, are taking chronically purgatives or diuretics. The disease is often a psychosomatic one; the patients are discrete with

respect to their abuse. The main event of the disease is the chronic volume deficit leading to elevated renin and aldosterone. The clinical syndrome is very similar to the true Bartter's syndrome with hypokalemia and normal or slightly decreased blood pressure, sometimes extreme hyperaldosteronism is present. The main threat of this disease is the chronic hypokalemia which may cause renal damage and later failure, therefore, an early diagnosis is important. Detection of the drugs is more of a diagnostic problem than the endocrine studies because patients are often not co-operative.

3.3.1.4.3 High aldosterone concentrations in other situations

3.3.1.4.3.1 Pseudohypoaldosteronism

Salt loss in children can be associated with high aldosterone production suggestive of aldosterone resistance, thus having clinical features of hypoaldosteronism and is considered in more detail in Chapter 3.3.2. Hypertension and severe hyperkalemia are features of Gordon syndrome. Renin and aldosterone are suppressed and thus considered in more detail in Pseudohypoaldosteronism. Patients respond to thiazide diuretics since sodium chloride transporter (NCC) is activated through mutations in WNK1 and WNK4 (Ceccato and Mantero, 2019).

3.3.1.4.3.2   Diabetes

Some diabetic patients have small increase in plasma aldosterone and exhibit 10% increase in CVD mortality relative to diabetics with normal aldosterone concentrations (McCurley and Jaffe, 2012). MR antagonism was found to reduce the risk of sudden death from cardiac causes (Pitt et al., 2003) confirmed in the Eplerenone Postacute myocardial infarction heart failure efficacy and survival study (EPHESUS). There are no specific recommendations of blood pressure targets for diabetics (Arguedas et al., 2013). In line with the American Diabetes Association, a BP of 130/80 mmHg would be a target (Diabetes Care Addendum 10. Cardiovascular disease and risk management: Standards of medical care in diabetes—2022. Diabetes Care 2022; 45 (Suppl 1): S144–74).

3.3.1.4.3.3   Renal disease

Aldosterone has some roles in the progression of renal failure independent of effects on blood pressure. Blockade of the MR with MR antagonists such as spironolactone or eplerenone results in significant reduction in proteinuria in patients with several types of kidney diseases (Minakuchi et al., 2020).

3.3.1.4.3.4   Sickle cell disease

Sickle cell disease is a disorder of hemoglobin that results in abnormal red blood cells and multiple pathological consequences. Kidney damage is a complication of the disease. Renal blood flow is reduced. Patients have lower blood pressures than normal individuals (Brito et al., 2022). Concentrations of angiotensin converting enzyme (ACE) were significantly reduced. Patients with sickle cell disease have chronic kidney disease leading to albuminuria. In these patients, however, treatment with blocking agents of the renin-angiotensin-aldosterone pathway increased the proteinuria and risk of death (Thrower et al., 2019).

3.3.1.4.3.5   Obesity

Obese subjects often have hypertension and cardiovascular and renal diseases. Increased cortisol secretion, even with subclinical Cushing's syndrome and hyperaldosteronism

are related to obesity. A higher percentage of body fat is associated with increased risk of hypertension due to primary hyperaldosteronism (Manosroi and Atthakomol, 2020; Stowasser et al., 2001) and increased serum aldosterone have been related to the development of hypertension (Kawarazaki and Fujita, 2016; Cooper et al., 2013). Twenty-four hour excretion of urinary aldosterone was positively correlated with BMI across all quartiles, whereas plasma aldosterone correlated to BMI only across the first three quartiles (Dudenbostel et al., 2016) (Fig. 3.3.1.40).

Human adipose tissues secrete a variety of endocrinologically active substances such as leptin. IL-6 and TNFα. Human adipocytes secrete potent mineralocorticoid releasing factors when a fat cell conditioned medium was tested with human adrenocortical cells (NCI-H295R) with response comparable to forskolin (2×10^{-5} M) (Ehrhart-Bornstein et al., 2003). The activity was heat sensitive, precipitated with 70% aqueous ammonium sulfate. This work does not appear to have been reproduced and the characters not chemically defined.

MR are expressed in adipose tissue (Yang and Fuller, 2012). Excess calorie intake leads to pathogenic adipose tissue metabolic and immune response (adiposopathy) that promote metabolic disease. If this can be substantiated, a patient can be provided by the clinician with evidence to lose weight (Bays et al., 2008) though this is difficult to achieve and more difficult to maintain. Bariatric surgery is a near last resort.

3.3.1.4.3.6   Epigenetics

A conventional Mendelian genetic approach does not suffice to explain the regulation of BP. Gene-specific DNA methylation is one of the most studied hallmarks among all epigenetic modifications in essential hypertension. Posttranslational histone modifications, such as methylation, acetylation, and phosphorylation, are important epigenetic markers. MicroRNAs also affect blood pressure by regulating master genes such as those implicated in the renin-angiotensin-aldosterone system. Epigenetic modifications, that appear to contribute to the pathophysiology of cardiovascular diseases, may be a promising research area for the development of novel future strategies for hypertension prevention and therapeutics. In the context of this book, a published review will satisfy the reader as a comprehensive resource of near current literature on major epigenetic modifications during essential hypertension, highlighting the potential contribution of DNA methylation, posttranslational histone modifications, and microRNAs (miRNAs) (Arif et al., 2019).

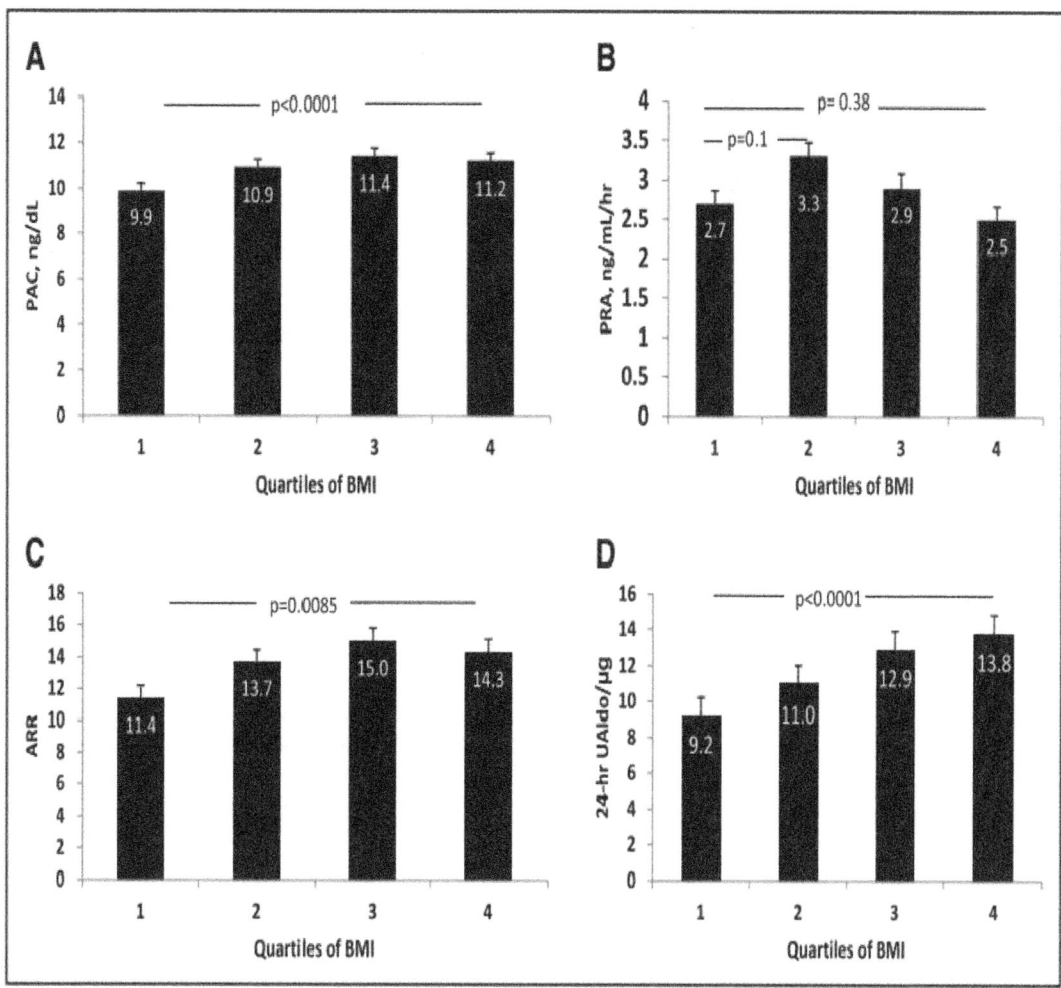

FIG. 3.3.1.40 Increase in mean plasma aldosterone concentration (PAC; A), plasma renin activity (PRA; B), aldosterone:renin ratio (ARR; C), and 24-h urinary aldosterone (24-h UAldo) levels (D) with increasing quartiles of body mass index (BMI) in patients with resistant hypertension. All panels show Bonferroni corrected P values. *(Dudenbostel T, Ghazi L, Liu M, Li P, Oparil S, Calhoun DA. Body mass index predicts 24-hour urinary aldosterone levels in patients with resistant hypertension. Hypertension 2016;68(4):995–1003. Fig. 1 p. 998. Hypertension 68:995.)*

3.3.1.5 Summary

Aldosterone and mineralocorticoid steroids are important in the development of hypertension through salt retention although other mechanism are under investigation. The biochemistry of mineralocorticoids has received little attention in the last 35 years but starting from the evidence generated from 1955 to 1985 this area is open for re-investigation. Many unusual steroids were identified before 1987 but quantitative analyses were only available in research settings. The assays were based on purification of steroid extracts by chromatography before immunoassay and the quality (accuracy, specificity) was likely to be high. From 1985, only assays for aldosterone were generally available, mostly on automated platforms with chemiluminescence. Aldosterone is now determined by LC-MS/MS methods but the addition of 11-deoxycorticosterone, 18-hydroxycorticosterone, 18-oxo-cortisol in one method is anticipated more widely. It is not possible to predict the extent to which other steroids will be added.

11-Hydroxysteroid dehydrogenases have attracted much interest for the protective effect against the action of cortisol at the MR and potential targets for treatment of hyperaldosteronism. GALF's will be further identified when investigating disease mechanisms for forms of low renin hypertension. With the current interest in the microbiome steroids such as 21-deoxycortisol, 21-deoxycorticosterone and 21-deoxyaldosterone may become of interest although limited by the lack of stable isotope labeled internal standards. Studies are needed with extensive steroid measurements in body fluids in the unselected hypertension population to gain a true endocrine picture for establishing the prevalence of primary aldosteronism. Mass spectrometric methods are essential for accuracy and specificity of the determinations.

There is a need to improve the analysis of renin because accurate determination when suppressed is important to the classification of primary and secondary aldosteronism. The concentration of renin can now be determined by LC-MS/MS methods, negating the angiotensin I generation incubation step with activity assays. Several fragments of angiotensin have been defined and measurement of these in clinical situations may reveal further abnormalities of the RAAS and new targeted treatments.

Genetic studies have identified new familial forms of hypertension and revealed genetic abnormalities associated with primary aldosteronism. These abnormalities implicate genes coding for ion channels and ATPases that regulate ion potential and intracellular ion homeostasis. The recognition of other genes is inevitable.

Hypertension and obesity are very important clinical problems involving significant proportions of the population. The frequency of PA is now accepted to affect at least 5% of the population and screening tests should be more widely used particularly when 3 or more drugs fail to control the hypertension when 20% may have PA. An evidence base for treatment is desirable as opposed to the clinical management based on the effectiveness of treatments. Analysis of "big data" has enabled development of a nomogram for diagnosis of PA when ARR, sex, age, antihypertensive medication, plasma potassium and other factors are taken into account (Tezuka et al., 2021a; Wang et al., 2021). A trial of spironolactone or eplerenone would determine if primary aldosteronism can be treated without adrenalectomy in 50% of the patients. Newer treatments are under development including further elements of the RAAS.

Hypokalemia with metabolic alkalosis can be seen with rare disorders of ion channels in the renal tubules. Measurements of calcium and magnesium give characteristic features for further renal tubular disorders.

Low renin hypertension is common and often unknown entity in clinical practice. Guidelines for the management of hypertension do not yet include aldosterone and renin measurements or genetic studies. High blood pressure is treated with a range of drugs acting on the R-A-A-S. As more is known about genetics of hypertension there will arise new treatments targeted to a defect in the patient.

The ability to explore genomic data through the use of new advanced techniques, such as whole genome and exome sequencing, will aid in efficiently examining the genomic regulatory regions and associated epigenetic markers affecting variants involved in hypertension. These approaches will also provide the significant potential to explore beyond the single variant risk model to simultaneously study multiple risk variants of essential hypertension.

Exosomes are small particles released into the extracellular space from many cell types found in body fluids. Exosomes in urine are useful sources for protein and microRNA analysis (Maggio et al., 2021; Hoorn et al., 2005) to characterize potential biomarkers using mass spectrometry (Thongboonkerd, 2020; Barros and Carvajal, 2017).

References

Abdelhamid S, Vecsei P, Haack D, Gless KH, Walb D, Lichtwald K, et al. Dissociation in the excretion of different aldosterone metabolites and unmetabolized ('free') aldosterone in hypertension. Clin Sci (Lond) 1979;57(5):409–14.

Abdelhamid S, Vecsei P, Haack D, Gless KH, Walb D, Fiegel P, et al. Elevated 'free' 18-hydroxy-corticosterone excretion as a possible indicator for early diagnosis of primary aldosteronism. J Steroid Biochem 1981;14(9):913–20.

Abdelhamid S, Blomer R, Hommel G, Haack D, Lewicka S, Fiegel P, et al. Urinary tetrahydroaldosterone as a screening method for primary aldosteronism: a comparative study. Am J Hypertens 2003;16(7):522–30.

Abe T, Naruse M, Young Jr WF, Kobashi N, Doi Y, Izawa A, et al. A Novel CYP11B2-Specific Imaging Agent for Detection of Unilateral Subtypes of Primary Aldosteronism. J Clin Endocrinol Metab 2016;101(3):1008–15.

Adachi M, Tajima T, Muroya K. Dietary potassium restriction attenuates urinary sodium wasting in the generalized form of pseudohypoaldosteronism type 1. CEN Case Rep 2020;9(2):133–7. Erratum in: CEN Case Rep. 2020;9(3):294.

Al Romhain B, Young AM, Battacharya JJ, Suttner N. Intracranial aneurysm in a patient with glucocorticoid-remediable aldosteronism. Br J Neurosurg 2015;29(5):715–7.

Ames MK, Atkins CE, Lantis AC, zum Brunnen J. Evaluation of subacute change in RAAS activity (as indicated by urinary aldosterone:creatinine, after pharmacologic provocation) and the response to ACE inhibition. J Renin Angiotensin Aldosterone Syst 2016;17(1), 1470320316633897.

Antonelli G, Ceccato F, Artusi C, Marinova M, Plebani M. Salivary cortisol and cortisone by LC-MS/MS: validation, reference intervals and diagnostic accuracy in Cushing's syndrome. Clin Chim Acta 2015;451(Pt B):247–51.

Arguedas JA, Leiva V, Wright JM. Blood pressure targets for hypertension in people with diabetes mellitus. Cochrane Database Syst Rev 2013;(10), CD008277.

Arif M, Sadayappan S, Becker RC, Martin LJ, Urbina EM. Epigenetic modification: a regulatory mechanism in essential hypertension. Hypertens Res 2019;42(8):1099–113.

Arlt W, Biehl M, Taylor AE, Hahner S, Libé R, Hughes BA, et al. Urine steroid metabolomics as a biomarker tool for detecting malignancy in adrenal tumors. J Clin Endocrinol Metab 2011;96(12):3775–84.

Arlt W, Lang K, Sitch AJ, Dietz AS, Rhayem Y, Bancos I, et al. Steroid metabolome analysis reveals prevalent glucocorticoid excess in primary aldosteronism. JCI Insight 2017;2(8), e93136.

Arteaga E, Klein R, Biglieri EG. Use of the saline infusion test to diagnose the cause of primary aldosteronism. Am J Med 1985;79(6):722–8.

Ashton EJ, Legrand A, Benoit V, Roncelin I, Venisse A, Zennaro MC, et al. Simultaneous sequencing of 37 genes identified causative mutations in the majority of children with renal tubulopathies. Kidney Int 2018;93(4):961–7.

Auchus RJ, Chandler DW, Singeetham S, Chokshi N, Nwariaku FE, Dolmatch BL, et al. Measurement of 18-hydroxycorticosterone during adrenal vein sampling for primary aldosteronism. J Clin Endocrinol Metab 2007;92(7):2648–51.

Augustin AM, Dalla Torre G, Fuss CT, Fassnacht M, Bley TA, Kickuth R. Reduction of radiation exposure in adrenal vein sampling: impact of the rapid cortisol assay. Rofo 2021;193(12):1392–402.

Azizan EA, Poulsen H, Tuluc P, Zhou J, Clausen MV, Lieb A, et al. Somatic mutations in ATP1A1 and CACNA1D underlie a common subtype of adrenal hypertension. Nat Genet 2013;45(9):1055–60.

Bancos I, Taylor AE, Chortis V, Sitch AJ, Jenkinson C, Davidge-Pitts CJ, et al. Urine steroid metabolomics for the differential diagnosis of adrenal incidentalomas in the EURINE-ACT study: a prospective test validation study. Lancet Diabetes Endocrinol 2020;8(9):773–81.

Bandulik S, Schmidt K, Bockenhauer D, Zdebik AA, Humberg E, Kleta R, et al. The salt-wasting phenotype of EAST syndrome, a disease with multifaceted symptoms linked to the KCNJ10 K+ channel. Pflugers Arch 2011;461(4):423–35.

Banks WA, Kastin AJ, Biglieri EG, Ruiz AE. Primary adrenal hyperplasia: a new subset of primary hyperaldosteronism. J Clin Endocrinol Metab 1984;58(5):783–5.

Bao M, Cai J, Yang X, Ma W. Genetic screening for Bartter syndrome and Gitelman syndrome pathogenic genes among individuals with hypertension and hypokalemia. Clin Exp Hypertens 2019;41(4):381–8.

Baronio F, Ortolano R, Menabò S, Cassio A, Baldazzi L, Di Natale V, et al. 46,XX DSD due to androgen excess in monogenic disorders of steroidogenesis: genetic, biochemical, and clinical features. Int J Mol Sci 2019;20(18):4605.

Barros ER, Carvajal CA. Urinary exosomes and their cargo: potential biomarkers for mineralocorticoid arterial hypertension? Front Endocrinol (Lausanne) 2017;8:230.

Bays HE, González-Campoy JM, Henry RR, Bergman DA, Kitabchi AE, Schorr AB, et al. Adiposopathy working group. Is adiposopathy (sick fat) an endocrine disease? Int J Clin Pract 2008;62(10):1474–83.

Belot A, Ranchin B, Fichtner C, Pujo L, Rossier BC, Liutkus A, et al. Pseudohypoaldosteronisms, report on a 10-patient series. Nephrol Dial Transplant 2008;23(5):1636–41.

Berke K, Constantinescu G, Masjkur J, Kimpel O, Dischinger U, Peitzsch M, et al. Plasma steroid profiling in patients with adrenal incidentaloma. J Clin Endocrinol Metab 2021;, dgab751.

Bernstone L, Adaway JE, Keevil BG. An LC-MS/MS assay for analysis of equilibrium angiotensin II in human serum. Ann Clin Biochem 2021;58(5):422–33.

Berthon A, Drelon C, Ragazzon B, Boulkroun S, Tissier F, Amar L, et al. WNT/β-catenin signalling is activated in aldosterone-producing adenomas and controls aldosterone production. Hum Mol Genet 2014;23(4):889–905.

Besouw MTP, Kleta R, Bockenhauer D. Bartter and Gitelman syndromes: questions of class. Pediatr Nephrol 2020;35(10):1815–24.

Betz MJ, Zech CJ. Adrenal venous sampling in the diagnostic workup of primary aldosteronism. Br J Radiol 2022;95(1129):20210311.

Betz MJ, Degenhart C, Fischer E, Pallauf A, Brand V, Linsenmaier U, et al. Adrenal vein sampling using rapid cortisol assays in primary aldosteronism is useful in centers with low success rates. Eur J Endocrinol 2011;165(2):301–6.

Beuschlein F, Boulkroun S, Osswald A, Wieland T, Nielsen HN, Lichtenauer UD, et al. Somatic mutations in ATP1A1 and ATP2B3 lead to aldosterone-producing adenomas and secondary hypertension. Nat Genet 2013;45(4):440–4 [444e1–2].

Biglieri EG, Herron MA, Brust N. 17-hydroxylation deficiency in man. J Clin Invest 1966;45(12):1946–54.

Binder C, Poglitsch M, Agibetov A, Duca F, Zotter-Tufaro C, Nitsche C, et al. Angs (Angiotensins) of the alternative renin-angiotensin system predict outcome in patients with heart failure and preserved ejection fraction. Hypertension 2019;74(2):285–94.

Blanchard A, Bockenhauer D, Bolignano D, Calò LA, Cosyns E, Devuyst O, et al. Gitelman syndrome: consensus and guidance from a kidney disease: improving global outcomes (KDIGO) controversies conference. Kidney Int 2017;91(1):24–33.

Bluemel C, Hahner S, Heinze B, Fassnacht M, Kroiss M, Bley TA, et al. Investigating the chemokine receptor 4 as potential theranostic target in adrenocortical cancer patients. Clin Nucl Med 2017;42(1):e29–34.

Bockenhauer D, Feather S, Stanescu HC, Bandulik S, Zdebik AA, Reichold M, et al. Epilepsy, ataxia, sensorineural deafness, tubulopathy, and KCNJ10 mutations. N Engl J Med 2009;360(19):1960–70.

Bongers EMHF, Shelton LM, Milatz S, Verkaart S, Bech AP, Schoots J, et al. A novel hypokalemic-alkalotic salt-losing tubulopathy in patients with CLDN10 mutations. J Am Soc Nephrol 2017;28(10):3118–28.

Boulkroun S, Samson-Couterie B, Golib-Dzib JF, Amar L, Plouin PF, Sibony M, et al. Aldosterone-producing adenoma formation in the adrenal cortex involves expression of stem/progenitor cell markers. Endocrinology 2011;152(12):4753–63.

Bournot P, Pitoizet N, Zachmann M, Maume BF. Partial characterization of unusual polar steroids in the urine of a child with low renin hypertension. J Steroid Biochem 1982;16(3):467–77.

Breil T, Yakovenko V, Inta I, Choukair D, Klose D, Mittnacht J, et al. Typical characteristics of children with congenital adrenal hyperplasia due to 11β-hydroxylase deficiency: a single-centre experience and review of the literature. J Pediatr Endocrinol Metab 2019;32(3):259–67.

Brito PL, Dos Santos AF, Chweih H, Favero ME, Gotardo EMF, Silva JAF, et al. Reduced blood pressure in sickle cell disease is associated with decreased angiotensin converting enzyme (ACE) activity and is not modulated by ACE inhibition. PLoS One 2022;17(2), e0263424.

Brochard K, Boyer O, Blanchard A, Loirat C, Niaudet P, Macher MA, et al. Phenotype-genotype correlation in antenatal and neonatal variants of Bartter syndrome. Nephrol Dial Transplant 2009;24(5):1455–64.

Brower RK, Ghlichloo IA, Shabgahi V, Elsholz D, Menon RK, Vyas AK. Liddle syndrome due to a novel c.1713 deletion in the epithelial sodium channel β-subunit in a normotensive adolescent. AACE Clin Case Rep 2020;7(1):65–8.

Buhl LF, Pedersen FN, Andersen MS, Glintborg D. Licorice-induced apparent mineralocorticoid excess compounded by excessive use of terbutaline and high water intake. BMJ Case Rep 2018;2018, bcr2017223918.

Bulsari K, Falhammar H. Clinical perspectives in congenital adrenal hyperplasia due to 11β-hydroxylase deficiency. Endocrine 2017;55(1):19–36.

Burrello J, Monticone S, Buffolo F, Lucchiari M, Tetti M, Rabbia F, et al. Diagnostic accuracy of aldosterone and renin measurement by chemiluminescent immunoassay and radioimmunoassay in primary aldosteronism. J Hypertens 2016;34(5):920–7.

Bystrom CE, Salameh W, Reitz R, Clarke NJ. Plasma renin activity by LC-MS/MS: development of a prototypical clinical assay reveals a subpopulation of human plasma samples with substantial peptidase activity. Clin Chem 2010;56(10):1561–9.

Caceres PS, Ortiz PA. Molecular regulation of NKCC2 in blood pressure control and hypertension. Curr Opin Nephrol Hypertens 2019;28(5): 474–80.

Carroll J, Komanicky P, Melby JC. The relationship between plasma 18-hydroxy-11-deoxycorticosterone levels and production of hypertension in the rat. J Steroid Biochem 1981;14(10):989–95.

Casey ML, Guerami A, Milewich L, Gomez-Sanchez CE, MacDonald PC. Origin of urinary nonconjugated 19-nor-deoxycorticosterone and metabolism of infused radiolabeled 19-nor-deoxycorticosterone in men and women. J Clin Invest 1985;75(4):1335–8.

Ceccato F, Mantero F. Monogenic forms of hypertension. Endocrinol Metab Clin North Am 2019;48(4):795–810.

Ceolotto G, Antonelli G, Maiolino G, Cesari M, Rossitto G, Bisogni V, et al. Androstenedione and 17-α-hydroxyprogesterone are better indicators of adrenal vein sampling selectivity than cortisol. Hypertension 2017;70(2):342–6.

Cesari M, Ceolotto G, Rossitto G, Maiolino G, Seccia TM, Rossi GP. The intra-procedural cortisol assay during adrenal vein sampling: rationale and design of a randomized study (I-Padua). High Blood Press Cardiovasc Prev 2017;24(2):167–70.

Chabre O, Portrat-Doyen S, Vivier J, Morel Y, Defaye G. Two novel mutations in splice donor sites of CYP11B1 in congenital adrenal hyperplasia due to 11beta-hydroxylase deficiency. Endocr Res 2000;26 (4):797–801.

Chandler DW, Tuck M, Mayes DM. The measurement of 18-hydroxy-11-deoxycorticosterone in human plasma by radioimmunoassay. Steroids 1976;27(2):235–46.

Charmandari E, Kino T, Ichijo T, Chrousos GP. Generalized glucocorticoid resistance: clinical aspects, molecular mechanisms, and implications of a rare genetic disorder. J Clin Endocrinol Metab 2008;93(5):1563–72.

Chen LS, Singh RJ. Niche point-of-care endocrine testing—reviews of intraoperative parathyroid hormone and cortisol monitoring. Crit Rev Clin Lab Sci 2018;55(2):115–28.

Chen F, Cheng Z, Wang Z, Peng Y, Wang B, Guo W, et al. Liquid chromatography-tandem mass spectrometry (LC-MS/MS) based assay for the simultaneous quantification of 18-hydroxycorticosterone, 18-hydroxycortisol and 18-oxocortisol in human plasma. J Chromatogr B Analyt Technol Biomed Life Sci 2022;1188:123030.

Choi M, Scholl UI, Yue P, Björklund P, Zhao B, Nelson-Williams C, et al. K+ channel mutations in adrenal aldosterone-producing adenomas and hereditary hypertension. Science 2011;331(6018):768–72.

Chu MD, Ulick S. Isolation and identification of 18-hydroxycortisol from the urine of patients with primary aldosteronism. J Biol Chem 1982;257(5):2218–24.

Constantinescu G, Bidlingmaier M, Gruber M, Peitzsch M, Poitz DM, van Herwaarden AE, et al. Mass spectrometry reveals misdiagnosis of primary aldosteronism with scheduling for adrenalectomy due to immunoassay interference. Clin Chim Acta 2020;507:98–103.

Cooper LA, Marsteller JA, Noronha GJ, Flynn SJ, Carson KA, Boonyasai RT, et al. A multi-level system quality improvement intervention to reduce racial disparities in hypertension care and control: study protocol. Implement Sci 2013;8:60.

Corrie JE, Edwards CR, Budd PS. A radioimmunoassay for 18-hydroxycortisol in plasma and urine. Clin Chem 1985;31(6):849–52.

Costa-Santos M, Kater CE, Auchus RJ, Brazilian Congenital Adrenal Hyperplasia Multicenter Study Group. Two prevalent CYP17 mutations and genotype-phenotype correlations in 24 Brazilian patients with 17-hydroxylase deficiency. J Clin Endocrinol Metab 2004;89 (1):49–60.

Cunha TDS, Heilberg IP. Bartter syndrome: causes, diagnosis, and treatment. Int J Nephrol Renovasc Dis 2018;11:291–301.

De Palo EF, Antonelli G, Benetazzo A, Prearo M, Gatti R. Human saliva cortisone and cortisol simultaneous analysis using reverse phase HPLC technique. Clin Chim Acta 2009;405(1–2):60–5.

Dekkers T, Prejbisz A, Kool LJS, Groenewoud HJMM, Velema M, Spiering W, et al. Adrenal vein sampling versus CT scan to determine treatment in primary aldosteronism: an outcome-based randomised diagnostic trial. Lancet Diabetes Endocrinol 2016;4(9):739–46.

Dellow EL, Unwin RJ, Honour JW. Pontefract cakes can be bad for you: refractory hypertension and liquorice excess. Nephrol Dial Transplant 1999;14(1):218–20.

Deng L, Xiong Z, Li H, Lei X, Cheng L. Analytical validation and investigation on reference intervals of aldosterone and renin in Chinese Han population by using fully automated chemiluminescence immunoassays. Clin Biochem 2018;56:89–94.

Dewez JE, Bachy A. Hyperaldostéronisme primaire chez l'enfant, à propos d'un cas [A case of primary aldosteronism in childhood]. Arch Pediatr 2009;16(1):37–40 [French].

Dinleyici EC, Dogruel N, Acikalin MF, Tokar B, Oztelcan B, Ilhan H. An additional child case of an aldosterone-producing adenoma with an atypical presentation of peripheral paralysis due to hypokalemia. J Endocrinol Invest 2007;30(10):870–2.

Dluhy RG, Anderson B, Harlin B, Ingelfinger J, Lifton R. Glucocorticoid-remediable aldosteronism is associated with severe hypertension in early childhood. J Pediatr 2001;138(5):715–20.

Dorrian CA, Toole BJ, Alvarez-Madrazo S, Kelly A, Connell JM, Wallace AM. A screening procedure for primary aldosteronism based on the Diasorin liaison automated chemiluminescent immunoassay for direct renin. Ann Clin Biochem 2010;47(Pt 3):195–9.

Dudenbostel T, Ghazi L, Liu M, Li P, Oparil S, Calhoun DA. Body mass index predicts 24-hour urinary aldosterone levels in patients with resistant hypertension. Hypertension 2016;68(4):995–1003.

Ehrhart-Bornstein M, Lamounier-Zepter V, Schraven A, Langenbach J, Willenberg HS, Barthel A, et al. Human adipocytes secrete mineralocorticoid-releasing factors. Proc Natl Acad Sci U S A 2003;100(24):14211–6.

Eisenhofer G, Dekkers T, Peitzsch M, Dietz AS, Bidlingmaier M, Treitl M, et al. Mass spectrometry-based adrenal and peripheral venous steroid profiling for subtyping primary aldosteronism. Clin Chem 2016;62(3):514–24.

Eisenhofer G, Kurlbaum M, Peitzsch M, Constantinescu G, Remde H, Schulze M, et al. The saline infusion test for primary aldosteronism: implications of immunoassay inaccuracy. J Clin Endocrinol Metab 2021;, dgab924.

Engels M, Gehrmann K, Falhammar H, Webb EA, Nordenström A, Sweep FC, et al. Gonadal function in adult male patients with congenital adrenal hyperplasia. Eur J Endocrinol 2018;178(3):285–94.

Fan P, Lu CX, Zhang D, Yang KQ, Lu PP, Zhang Y, et al. Liddle syndrome misdiagnosed as primary aldosteronism resulting from a novel frameshift mutation of SCNN1B. Endocr Connect 2018;7(12): 1528–34.

Fan P, Pan XC, Zhang D, Yang KQ, Zhang Y, Tian T, et al. Pediatric liddle syndrome caused by a novel SCNN1G variant in a Chinese family and characterized by early-onset hypertension. Am J Hypertens 2020a;33 (7):670–5.

Fan P, Lu YT, Yang KQ, Zhang D, Liu XY, Tian T, et al. Apparent mineralocorticoid excess caused by novel compound heterozygous mutations in HSD11B2 and characterized by early-onset hypertension and hypokalemia. Endocrine 2020c;70(3):607–15.

Fortunato A, Prontera C, Masotti S, Franzini M, Marchetti C, Giovannini S, et al. State of the art of aldosterone immunoassays. A multicenter collaborative study on the behalf of the Cardiovascular Biomarkers Study Group of the Italian Section of European Society of Ligand Assay (ELAS) and Società Italiana di Biochimica Clinica (SIBIOC). Clin Chim Acta 2015;444:106–12.

Fredline VF, Kovacs EM, Taylor PJ, Johnson AG. Measurement of plasma renin activity with use of HPLC-electrospray-tandem mass spectrometry. Clin Chem 1999;45(5):659–64.

Fujiwara M, Murao K, Imachi H, Yoshida K, Muraoka T, Ohyama T, et al. Misdiagnosis of two cases of primary aldosteronism owing to failure of computed tomography to detect adrenal microadenoma. Am J Med Sci 2010;340(4):335–7.

Funder JW, Carey RM, Mantero F, Murad MH, Reincke M, Shibata H, et al. The management of primary aldosteronism: case detection, diagnosis, and treatment: an endocrine society clinical practice guideline. J Clin Endocrinol Metab 2016;101(5):1889–916.

Fuss CT, Brohm K, Fassnacht M, Kroiss M, Hahner S. Reassessment of postural stimulation testing as a simple tool to identify a subgroup of patients with unilateral primary aldosteronism. J Clin Endocrinol Meab 2022;107(2):e865–73.

Ganguly A, Zager PG, Luetscher JA. Primary aldosteronism due to unilateral adrenal hyperplasia. J Clin Endocrinol Metab 1980a;51 (5):1190–4.

García-Nieto V, Flores C, Luis-Yanes MI, Gallego E, Villar J, Claverie-Martín F. Mutation G47R in the BSND gene causes Bartter syndrome with deafness in two Spanish families. Pediatr Nephrol 2006;21 (5):643–8.

Geller DS, Farhi A, Pinkerton N, Fradley M, Moritz M, Spitzer A, et al. Activating mineralocorticoid receptor mutation in hypertension exacerbated by pregnancy. Science 2000;289(5476):119–23.

Geller DS, Zhang J, Wisgerhof MV, Shackleton C, Kashgarian M, Lifton RP. A novel form of human mendelian hypertension featuring nonglucocorticoid-remediable aldosteronism. J Clin Endocrinol Metab 2008;93(8):3117–23.

Gibbons SM, Field HP, Fairhurst A, Fleming A, Ford C, Williams EL, et al. Clinical evaluation of assays for plasma renin activity and aldosterone measurement by liquid chromatography-tandem mass spectrometry. J Appl Lab Med 2021;6(3):668–78.

Goh BK, Tan YH, Chang KT, Eng PH, Yip SK, Cheng CW. Primary hyperaldosteronism secondary to unilateral adrenal hyperplasia: an unusual cause of surgically correctable hypertension. A review of 30 cases. World J Surg 2007;31(1):72–9.

Gomez-Sanchez CE, Holland OB, Murry BA, Lloyd HA, Milewich L. 19-nor-deoxycorticosterone: a potent mineralcocorticoid isolated from the urine of rats with regenerating adrenals. Endocrinology 1979;105 (3):708–15.

Gomez-Sanchez CE, Gomez-Sanchez EP, Upcavage RJ, Hall EB. Urinary free and serum 19-nor-deoxycorticosterone in adrenal regeneration hypertension. Hypertension 1983;5(2 Pt 2):I32–4.

Gomez-Sanchez CE, Upcavage RJ, Zager PG, Foecking MF, Holland OB, Ganguly A. Urinary 18-hydroxycortisol and its relationship to the excretion of other adrenal steroids. J Clin Endocrinol Metab 1987;65 (2):310–4.

Gomez-Sanchez CE, Qi X, Gomez-Sanchez EP, Sasano H, Bohlen MO, Wisgerhof M. Disordered zonal and cellular CYP11B2 enzyme expression in familial hyperaldosteronism type 3. Mol Cell Endocrinol 2017;439:74–80.

Gonzalez-Vicente A, Saez F, Monzon CM, Asirwatham J, Garvin JL. Thick ascending limb sodium transport in the pathogenesis of hypertension. Physiol Rev 2019;99(1):235–309.

Gorsline J, Morris DJ. The hypertensinogenic activity of 19-nor-deoxycorticosterone in the adrenalectomized spontaneously hypertensive rat. J Steroid Biochem 1985;23(4):535–6.

Grewal S, Fosam A, Chalk L, Deven A, Suzuki M, Correa RR, et al. Insulin sensitivity and pancreatic β-cell function in patients with primary aldosteronism. Endocrine 2021;72(1):96–103.

Griffing GT, Dale SL, Holbrook MM, Melby JC. 19-nor-deoxycorticosterone excretion in primary aldosteronism and low renin hypertension. J Clin Endocrinol Metab 1983;56(2):218–21.

Guo Z, Nanba K, Udager A, McWhinney BC, Ungerer JPJ, Wolley M, et al. Biochemical, histopathological, and genetic characterization of posture-responsive and unresponsive APAs. J Clin Endocrinol Metab 2020;105(9):e3224–35.

Hadj-Rabia S, Brideau G, Al-Sarraj Y, Maroun RC, Figueres ML, Leclerc-Mercier S, et al. Multiplex epithelium dysfunction due to CLDN10 mutation: the HELIX syndrome. Genet Med 2018;20(2):190–201.

Hague WM, Honour JW. Malignant hypertension in congenital adrenal hyperplasia due to 11 beta-hydroxylase deficiency. Clin Endocrinol (Oxf) 1983;18(5):505–10.

Hammer F, Stewart PM. Cortisol metabolism in hypertension. Best Pract Res Clin Endocrinol Metab 2006;20(3):337–53.

Hampf M, Dao NT, Hoan NT, Bernhardt R. Unequal crossing-over between aldosterone synthase and 11beta-hydroxylase genes causes congenital adrenal hyperplasia. J Clin Endocrinol Metab 2001;86(9):4445–52.

Han Z, Elliott MS. Neglected issues concerning teaching human adrenal steroidogenesis in popular biochemistry textbooks. Biochem Mol Biol Educ 2017;45(6):469–74.

Hána V, Ježková J, Kosák M, Kršek M, Hána V, Hill M. Novel GC-MS/MS technique reveals a complex steroid fingerprint of subclinical hypercortisolism in adrenal incidentalomas. J Clin Endocrinol Metab 2019;104(8):3545–56.

Hanukoglu A, Edelheit O, Shriki Y, Gizewska M, Dascal N, Hanukoglu I. Renin-aldosterone response, urinary Na/K ratio and growth in pseudohypoaldosteronism patients with mutations in epithelial sodium channel (ENaC) subunit genes. J Steroid Biochem Mol Biol 2008;111(3–5):268–74.

Harnik M, Kashman Y, Cojocaru M, Rosenthal T, Morris DJ. Synthesis of 19-noraldosterone, a potent mineralocorticoid. J Steroid Biochem 1986 Jun;24(6):1163–9.

Hautanena A, Lankinen L, Kupari M, Jänne OA, Adlercreutz H, Nikkilä H, et al. Associations between aldosterone synthase gene polymorphism and the adrenocortical function in males. J Intern Med 1998;244 (1):11–8.

Heilberg IP, Tótoli C, Calado JT. Adult presentation of Bartter syndrome type IV with erythrocytosis. Einstein (Sao Paulo) 2015;13(4):604–6.

Heinze B, Fuss CT, Mulatero P, Beuschlein F, Reincke M, Mustafa M, et al. Targeting CXCR4 (CXC chemokine receptor type 4) for molecular imaging of aldosterone-producing adenoma. Hypertension 2018;71 (2):317–25.

Hill M, Hána Jr V, Velíková M, Pařízek A, Kolátorová L, Vítků J, et al. A method for determination of one hundred endogenous steroids in human serum by gas chromatography-tandem mass spectrometry. Physiol Res 2019;68(2):179–207.

Hines JM, Bancos I, Bancos C, Singh RD, Avula AV, Young WF, et al. High-resolution, accurate-mass (HRAM) mass spectrometry urine steroid profiling in the diagnosis of adrenal disorders. Clin Chem 2017;63(12):1824–35.

Hochberg Z, Benderly A, Kahana L, Zadik Z. Requirement of mineralocorticoid in congenital adrenal hyperplasia due to 11 beta-hydroxylase deficiency. J Clin Endocrinol Metab 1986;63(1):36–40.

Honour JW. Development and validation of a quantitative assay based on tandem mass spectrometry. Ann Clin Biochem 2011;48(Pt 2):97–111.

Honour JW, Tourniaire J, Biglieri EG, Shackleton CH. Urinary steroid excretion in 17 alpha-hydroxylase deficiency. J Steroid Biochem 1978;9(6):495–505.

Honour JW, Dillon MJ, Levin M, Shah V. Fatal, low renin hypertension associated with a disturbance of cortisol metabolism. Arch Dis Child 1983a;58(12):1018–20.

Hoorn EJ, Pisitkun T, Zietse R, Gross P, Frokiaer J, Wang NS, et al. Prospects for urinary proteomics: exosomes as a source of urinary biomarkers. Nephrology (Carlton) 2005;10(3):283–90.

Hu J, Chen J, Cheng Q, Jing Y, Yang J, Du Z, et al. Comparison of bolus and continuous infusion of adrenocorticotropic hormone during adrenal vein sampling. Front Endocrinol (Lausanne) 2021;12:784706.

Hurwitz S, Cohen RJ, Williams GH. Diurnal variation of aldosterone and plasma renin activity: timing relation to melatonin and cortisol and consistency after prolonged bed rest. J Appl Physiol (1985) 2004;96 (4):1406–14.

Imaizumi N, Yamamoto I, Kamei M, Yoshida I, Miyauchi E, Kigoshi T, et al. High-performance liquid chromatographic separation of corticosteroids in plasma of rats. Its application to the determination of the circadian rhythm of 18-hydroxycorticosterone related to aldosterone and corticosterone. Horm Res 1987;27(1):53–60.

Irony I, Biglieri EG, Perloff D, Rubinoff H. Pathophysiology of deoxycorticosterone-secreting adrenal tumors. J Clin Endocrinol Metab 1987;65(5):836–40.

Ishida J, Sonezaki S, Yamaguchi M, Yoshitake T. High-performance liquid chromatographic determination of 3 alpha,5 beta-tetrahydroaldosterone in human urine with chemiluminescence detection. Analyst 1992;117(11):1719–24.

Jakobsson H, Farmaki K, Sakinis A, Ehn O, Johannsson G, Ragnarsson O. Adrenal venous sampling: the learning curve of a single interventionalist with 282 consecutive procedures. Diagn Interv Radiol 2018;24(2):89–93.

Janssen AG, Scholl U, Domeyer C, Nothmann D, Leinenweber A, Fahlke C. Disease-causing dysfunctions of barttin in Bartter syndrome type IV. J Am Soc Nephrol 2009;20(1):145–53.

Janzen N, Riepe FG, Peter M, Sander S, Steuerwald U, Korsch E, et al. Neonatal screening: identification of children with 11β-hydroxylase deficiency by second-tier testing. Horm Res Paediatr 2012;77 (3):195–9.

Jin S, Wada N, Takahashi Y, Hui SP, Sakurai T, Fuda H, et al. Quantification of urinary 18-hydroxycortisol using LC-MS/MS. Ann Clin Biochem 2013;50(Pt 5):450–6.

Karlekar MP, Sarathi V, Lila A, Rai K, Arya S, Bhandare VV, et al. Expanding genetic spectrum and discriminatory role of steroid profiling by LC-MS/MS in 11β-hydroxylase deficiency. Clin Endocrinol (Oxf) 2021;94(4):533–43.

Katayama Y, Takata N, Tamura T, Yamamoto A, Hirata F, Yasuda H, et al. A case of primary aldosteronism due to unilateral adrenal hyperplasia. Hypertens Res 2005;28(4):379–84.

Kawarazaki W, Fujita T. The role of aldosterone in obesity-related hypertension. Am J Hypertens 2016;29(4):415–23.

Kaziales A, Rührnößl F, Richter K. Glucocorticoid resistance conferring mutation in the C-terminus of GR alters the receptor conformational dynamics. Sci Rep 2021;11(1):12515.

Kem DC, Weinberger MH, Mayes DM, Nugent CA. Saline suppression of plasma aldosterone in hypertension. Arch Intern Med 1971;128 (3):380–6.

Kerkhofs TM, Kerstens MN, Kema IP, Willems TP, Haak HR. Diagnostic value of urinary steroid profiling in the evaluation of adrenal tumors. Horm Cancer 2015;6(4):168–75.

Khattab A, Haider S, Kumar A, Dhawan S, Alam D, Romero R, et al. Clinical, genetic, and structural basis of congenital adrenal hyperplasia due to 11β-hydroxylase deficiency. Proc Natl Acad Sci U S A 2017;114(10):E1933–40.

Kim AC, Reuter AL. Targeted disruption of beta-catenin in Sf1-expressing cells impairs development and maintenance of the adrenal cortex. In: Development, 135. Zubair M, Else T, Serecky; 2008. p. 2593–602.

Kim YM, Kang M, Choi JH, Lee BH, Kim GH, Ohn JH, et al. A review of the literature on common CYP17A1 mutations in adults with 17-hydroxylase/17,20-lyase deficiency, a case series of such mutations among Koreans and functional characteristics of a novel mutation. Metabolism 2014;63(1):42–9.

Knoers N, Antignac C, Bergmann C, Dahan K, Giglio S, Heidet L, et al. Genetic testing in the diagnosis of chronic kidney disease: recommendations for clinical practice. Nephrol Dial Transplant 2022;37(2): 239–54. https://doi.org/10.1093/ndt/gfab218. 34264297. PMCID: PMC8788237.

Kohno H, Sato S, Chiba H, Kobayashi K, Ikegawa S, Kurosawa T, et al. Monoclonal antibodies specific for 18-hydroxycortisol and their use in an enzyme immunoassay for human urinary 18-hydroxycortisol for diagnosis of primary aldosteronism. Clin Biochem 1994;27 (4):277–82.

Krämer BK, Bergler T, Stoelcker B, Waldegger S. Mechanisms of disease: the kidney-specific chloride channels ClCKA and ClCKB, the Barttin subunit, and their clinical relevance. Nat Clin Pract Nephrol 2008;4 (1):38–46.

Krone N, Arlt W. Genetics of congenital adrenal hyperplasia. Best Pract Res Clin Endocrinol Metab 2009;23(2):181–92.

Kumagai A. Results of treatment in 68 patients with idiopathic aldosteronism in Japan. Endocrinol Jpn 1980;27(2):121–7.

Kushnir MM, Neilson R, Roberts WL, Rockwood AL. Cortisol and cortisone analysis in serum and plasma by atmospheric pressure photoionization tandem mass spectrometry. Clin Biochem 2004;37 (5):357–62.

Laghmani K, Beck BB, Yang SS, Seaayfan E, Wenzel A, Reusch B, et al. Polyhydramnios, transient antenatal Bartter's syndrome, and MAGED2 mutations. N Engl J Med 2016;374(19):1853–63.

Le Goff C, Farre-Segura J, Stojkovic V, Dufour P, Peeters S, Courtois J, et al. The pathway through LC-MS method development: in-house or ready-to-use kit-based methods? Clin Chem Lab Med 2020;58 (6):1002–9.

Legrand A, Treard C, Roncelin I, Dreux S, Bertholet-Thomas A, Broux F, et al. Prevalence of novel *MAGED2* mutations in antenatal Bartter syndrome. Clin J Am Soc Nephrol 2018;13(2):242–50.

Lenders JWM, Williams TA, Reincke M, Gomez-Sanchez CE. Diagnosis of endocrine disease: 18-oxocortisol and 18-hydroxycortisol: is there clinical utility of these steroids? Eur J Endocrinol 2018;178(1):R1–9.

Lenzini L, Rossitto G, Maiolino G, Letizia C, Funder JW, Rossi GP. A Meta-analysis of somatic KCNJ5 K(+) channel mutations in 1636 patients with an aldosterone-producing adenoma. J Clin Endocrinol Metab 2015;100(8):E1089–95.

Lewicka S, Vecsei P, Bige K, Fisher T, Winter J, Abdelhamid S, et al. Urinary excretion of aldosterone metabolite Kelly-M1 in patients with adrenal dysfunction. J Steroid Biochem 1988;29(3):333–9.

Li T, Ma Y, Zhang Y, Liu Y, Fu T, Zhang R, et al. Feasibility of screening primary aldosteronism by aldosterone-to-direct renin concentration ratio derived from chemiluminescent immunoassay measurement: diagnostic accuracy and cutoff value. Int J Hypertens 2019;2019:2195796.

Li Z, Liang Y, Du C, Yu X, Hou L, Wu W, et al. Clinical applications of genetic analysis and liquid chromatography tandem-mass spectrometry in rare types of congenital adrenal hyperplasia. BMC Endocr Disord 2021;21(1):237.

Lifton RP, Dluhy RG, Powers M, Rich GM, Gutkin M, Fallo F, et al. Hereditary hypertension caused by chimaeric gene duplications and ectopic expression of aldosterone synthase. Nat Genet 1992;2(1):66–74.

Litchfield WR, Hunt SC, Jeunemaitre X, Fisher ND, Hopkins PN, Williams RR, et al. Increased urinary free cortisol: a potential intermediate phenotype of essential hypertension. Hypertension 1998;31(2):569–74.

Lopez AG, Fraissinet F, Lefebvre H, Brunel V, Ziegler F. Pharmacological and analytical interference in hormone assays for diagnosis of adrenal incidentaloma. Ann Endocrinol (Paris) 2019;80(4):250–8.

Ma L, Xia Y, Wang L, Liu R, Huang X, Ye T, et al. Sonographic features of the testicular adrenal rests tumors in patients with congenital adrenal hyperplasia: a single-center experience and literature review. Orphanet J Rare Dis 2019;14(1):242.

Ma Y, Chen H, Chen F, Jiang J, Guo W, Li X, et al. Mass spectrometry-based cortisol profiling during adrenal venous sampling reveals misdiagnosis for subtyping primary aldosteronism. Clin Endocrinol (Oxf) 2022;96(5):680–9.

Maggio S, Polidori E, Ceccaroli P, Cioccoloni A, Stocchi V, Guescini M. Current methods for the isolation of urinary extracellular vesicles. Methods Mol Biol 2021;2292:153–72.

Manolopoulou J, Fischer E, Dietz A, Diederich S, Holmes D, Junnila R, et al. Clinical validation for the aldosterone-to-renin ratio and aldosterone suppression testing using simultaneous fully automated chemiluminescence immunoassays. J Hypertens 2015;33(12):2500–11.

Manosroi W, Atthakomol P. High body fat percentage is associated with primary aldosteronism: a cross-sectional study. BMC Endocr Disord 2020;20(1):175.

Marcoux AA, Tremblay LE, Slimani S, Fiola MJ, Mac-Way F, Garneau AP, et al. Molecular characteristics and physiological roles of Na$^+$ -K$^+$ -Cl$^-$ cotransporter 2. J Cell Physiol 2021;236(3):1712–29.

Martin VI, Edwards CR, Biglieri EG, Vinson GP, Bartter FC. The development and application of a radioimmunoassay for 18-hydroxy-corticosterone. Steroids 1975;26(5):591–604.

Martin RM, Lin CJ, Costa EM, de Oliveira ML, Carrilho A, Villar H, et al. P450c17 deficiency in Brazilian patients: biochemical diagnosis through progesterone levels confirmed by CYP17 genotyping. J Clin Endocrinol Metab 2003;88(12):5739–46.

Mazzilli R, Stigliano A, Delfino M, Olana S, Zamponi V, Iorio C, et al. The high prevalence of testicular adrenal rest tumors in adult men with congenital adrenal hyperplasia is correlated with ACTH levels. Front Endocrinol (Lausanne) 2019;10:335.

McCurley A, Jaffe IZ. Mineralocorticoid receptors in vascular function and disease. Mol Cell Endocrinol 2012;350(2):256–65.

Meyer LS, Wang X, Sušnik E, Burrello J, Burrello A, Castellano I, et al. Immunohistopathology and steroid profiles associated with biochemical outcomes after adrenalectomy for unilateral primary aldosteronism. Hypertension 2018;72(3):650–7.

Mikami H, Nugent CA, Ogihara T, Naka T, Iwanaga K, Kumahara Y. The effect of 5 alpha-dihydrocortisol on the blood pressure of rats treated with deoxycorticosterone acetate and salt. Endocrinol Jpn 1980;27 (6):769–73.

Miller WL. Steroidogenic enzymes. Endocr Dev 2008;13:1–18.

Miller 2nd AW, Bohr DF, Schork AM, Terris JM. Hemodynamic responses to DOCA in young pigs. Hypertension 1979;1(6):591–7.

Miller MA, Sagnella GA, MacGregor GA. Extraction method and nonextracted kit method compared for measuring plasma aldosterone. Clin Chem 1997;43(10):1995–7. 9342029.

Minakuchi H, Wakino S, Urai H, Kurokochi A, Hasegawa K, Kanda T, et al. The effect of aldosterone and aldosterone blockade on the progression of chronic kidney disease: a randomized placebo-controlled clinical trial. Sci Rep 2020;10(1):16626.

Miyamori I, Yasuhara S, Matsubara T, Okamoto S, Ikeda M, Koshida H, et al. Effects of corticotropin-releasing factor (CRF) on aldosterone and 18-hydroxycorticosterone in essential hypertension and primary aldosteronism. Endocrinol Jpn 1987;34(6):809–19.

Miyamori I, Takeda Y, Takasaki H, Itoh Y, Iki K, Takeda R. Determination of urinary 18-hydroxycortisol in the diagnosis of primary aldosteronism. J Endocrinol Invest 1992;15(1):19–24.

Miyamura N, Matsumoto K, Taguchi T, Tokunaga H, Nishikawa T, Nishida K, et al. Atypical Bartter syndrome with sensorineural deafness with G47R mutation of the beta-subunit for ClC-Ka and ClC-Kb chloride channels, barttin. J Clin Endocrinol Metab 2003;88 (2):781–6.

Monteiro LLS, Franco OL, Alencar SA, Porto WF. Deciphering the structural basis for glucocorticoid resistance caused by missense mutations in the ligand binding domain of glucocorticoid receptor. J Mol Graph Model 2019;92:216–26.

Morineau G, Sulmont V, Salomon R, Fiquet-Kempf B, Jeunemaître X, Nicod J, et al. Apparent mineralocorticoid excess: report of six new cases and extensive personal experience. J Am Soc Nephrol 2006;17 (11):3176–84.

Morra di Cella S, Veglio F, Mulatero P, Christensen V, Aycock K, Zhu Z, et al. A time-resolved fluoroimmunoassay for 18-oxocortisol and 18-hydroxycortisol. Development of a monoclonal antibody to 18-oxocortisol. J Steroid Biochem Mol Biol 2002;82(1):83–8.

Morris DJ, Brem AS, Saccoccio NA, Pacholski M, Harnik M. Mineralocorticoid activity of 19-hydroxyaldosterone, 19-nor-aldosterone, and 3 beta-hydroxy-delta 5-aldosterone: relative potencies measured in two bioassay systems. Endocrinology 1986;118(6):2505–9.

Morris DJ, Brem AS, Odermatt A. Modulation of 11β-hydroxysteroid dehydrogenase functions by the cloud of endogenous metabolites in a local microenvironment: the glycyrrhetinic acid-like factor (GALF) hypothesis. J Steroid Biochem Mol Biol 2021;214:105988.

Mosso L, Gómez-Sánchez CE, Foecking MF, Fardella C. Serum 18-hydroxycortisol in primary aldosteronism, hypertension, and normotensives. Hypertension 2001;38(3 Pt 2):688–91.

Mulatero P. A new form of hereditary primary aldosteronism: familial hyperaldosteronism type III. J Clin Endocrinol Metab 2008;93 (8):2972–4.

Mulatero P, Monticone S, Rainey WE, Veglio F, Williams TA. Role of KCNJ5 in familial and sporadic primary aldosteronism. Nat Rev Endocrinol 2013;9(2):104–12.

Mulatero P, Sechi LA, Williams TA, Lenders JWM, Reincke M, Satoh F, et al. Subtype diagnosis, treatment, complications and outcomes of primary aldosteronism and future direction of research: a position statement and consensus of the Working Group on Endocrine Hypertension of the European Society of Hypertension. J Hypertens 2020;38 (10):1929–36.

Nakamura Y, Rege J, Satoh F, Morimoto R, Kennedy MR, Ahlem CN, et al. Liquid chromatography-tandem mass spectrometry analysis of human adrenal vein corticosteroids before and after adrenocorticotropic hormone stimulation. Clin Endocrinol (Oxf) 2012;76(6): 778–84.

Nanba K, Omata K, Else T, Beck PCC, Nanba AT, Turcu AF, et al. Targeted molecular characterization of aldosterone-producing adenomas in White Americans. J Clin Endocrinol Metab 2018;103 (10):3869–76.

Nanba K, Omata K, Gomez-Sanchez CE, Stratakis CA, Demidowich AP, Suzuki M, et al. Genetic characteristics of aldosterone-producing adenomas in blacks. Hypertension 2019;73(4):885–92.

New MI, Geller DS, Fallo F, Wilson RC. Monogenic low renin hypertension. Trends Endocrinol Metab 2005;16(3):92–7.

Nicod J, Dick B, Frey FJ, Ferrari P. Mutation analysis of CYP11B1 and CYP11B2 in patients with increased 18-hydroxycortisol production. Mol Cell Endocrinol 2004;214(1–2):167–74.

Nicolaides NC, Charmandari E. Primary generalized glucocorticoid resistance and hypersensitivity syndromes: a 2021 update. Int J Mol Sci 2021;22(19):10839.

Nicolaides NC, Lamprokostopoulou A, Polyzos A, Kino T, Katsantoni E, Triantafyllou P, et al. Transient generalized glucocorticoid hypersensitivity. Eur J Clin Invest 2015;45(12):1306–15.

Nilubol N, Soldin SJ, Patel D, Rwenji M, Gu J, Masika LS, et al. 11-Deoxycortisol may be superior to cortisol in confirming a successful adrenal vein catheterization without cosyntropin: a pilot study. Int J Endocr Oncol 2017;4(2):75–83.

Nimkarn S, New MI. Steroid 11beta- hydroxylase deficiency congenital adrenal hyperplasia. Trends Endocrinol Metab 2008;19(3):96–9.

Oberfield SE, Levine LS, Firpo A, Lawrence Sr D, Stoner E, Levy DJ, et al. Primary hyperaldosteronism in childhood due to unilateral macronodular hyperplasia. Case report. Hypertension 1984;6 (1):75–84.

Oki K, Plonczynski MW, Lam ML, Gomez-Sanchez EP, Gomez-Sanchez CE. The potassium channel, Kir3.4 participates in angiotensin II-stimulated aldosterone production by a human adrenocortical cell line. Endocrinology 2012;153(9):4328–35.

Ortner NJ, Kaserer T, Copeland JN, Striessnig J. De novo CACNA1D Ca2+ channelopathies: clinical phenotypes and molecular mechanism. Pflugers Arch 2020;472(7):755–73.

Owen LJ, Adaway J, Morris K, Lockhart S, Keevil BG. A widely applicable plasma renin activity assay by LC-MS/MS with offline solid phase extraction. Ann Clin Biochem 2014;51(Pt 3):409–11.

Page MM, Taranto M, Ramsay D, van Schie G, Glendenning P, Gillett MJ, et al. Improved technical success and radiation safety of adrenal vein sampling using rapid, semi-quantitative point-of-care cortisol measurement. Ann Clin Biochem 2018;55(5):588–92.

Papadopoulou A, Bountouvi E, Karachaliou FE. The molecular basis of calcium and phosphorus inherited metabolic disorders. Genes (Basel) 2021;12(5):734.

Peti-Peterdi J, Harris RC. Macula densa sensing and signaling mechanisms of renin release. J Am Soc Nephrol 2010;21(7):1093–6.

Pinggera A, Mackenroth L, Rump A, Schallner J, Beleggia F, Wollnik B, et al. New gain-of-function mutation shows CACNA1D as recurrently mutated gene in autism spectrum disorders and epilepsy. Hum Mol Genet 2017;26(15):2923–32.

Pitt B, Stier Jr CT, Rajagopalan S. Mineralocorticoid receptor blockade: new insights into the mechanism of action in patients with cardiovascular disease. J Renin Angiotensin Aldosterone Syst 2003;4(3): 164–8.

Pizzolo F, Corgnati A, Guarini P, Pavan C, Bassi A, Corrocher R, et al. Plasma aldosterone assays: comparison between chemiluminescence-based and RIA methods. Clin Chem 2006;52(7):1431–2.

Quencer KB. Adrenal vein sampling: technique and protocol, a systematic review. CVIR Endovasc 2021;4(1):38.

Raul JS, Cirimele V, Ludes B, Kintz P. Detection of physiological concentrations of cortisol and cortisone in human hair. Clin Biochem 2004;37 (12):1105–11.

Reynolds RM, Shakerdi LA, Sandhu K, Wallace AM, Wood PJ, Walker BR. The utility of three different methods for measuring urinary 18-hydroxycortisol in the differential diagnosis of suspected primary hyperaldosteronism. Eur J Endocrinol 2005;152(6):903–7.

Riepe FG, Krone N, Peter M, Sippell WG, Partsch CJ. Chromatographic system for the simultaneous measurement of plasma 18-hydroxy-11-deoxycorticosterone and 18-hydroxycorticosterone by radioimmunoassay: reference data for neonates and infants and its application in aldosterone-synthase deficiency. J Chromatogr B Analyt Technol Biomed Life Sci 2003;785(2):293–301.

Rosa S, Duff C, Meyer M, Lang-Muritano M, Balercia G, Boscaro M, et al. P450c17 deficiency: clinical and molecular characterization of six patients. J Clin Endocrinol Metab 2007;92(3):1000–7.

Rossi GP, Bernini G, Caliumi C, Desideri G, Fabris B, Ferri C, et al. A prospective study of the prevalence of primary aldosteronism in 1,125 hypertensive patients. J Am Coll Cardiol 2006;48(11): 2293–300.

Rossi GP, Cesari M, Letizia C, Seccia TM, Cicala MV, Zinnamosca L, et al. KCNJ5 gene somatic mutations affect cardiac remodelling but do not preclude cure of high blood pressure and regression of left ventricular hypertrophy in primary aldosteronism. J Hypertens 2014;32(7): 1514–21.

Rossi C, Cicalini I, Verrocchio S, Di Dalmazi G, Federici L, Bucci I. The potential of steroid profiling by mass spectrometry in the management of adrenocortical carcinoma. Biomedicine 2020;8(9):314.

Sahbani D, Strumbo B, Tedeschi S, Conte E, Camerino GM, Benetti E, et al. Functional study of novel Bartter's syndrome mutations in ClC-kb and rescue by the accessory subunit Barttin toward personalized medicine. Front Pharmacol 2020;11:327.

Sander K, Gendron T, Cybulska KA, Sirindil F, Zhou J, Kalber TL, et al. Development of [18F]AldoView as the first highly selective aldosterone synthase PET tracer for imaging of primary hyperaldosteronism. J Med Chem 2021;64(13):9321–9.

Schirpenbach C, Seiler L, Maser-Gluth C, Beuschlein F, Reincke M, Bidlingmaier M. Automated chemiluminescence-immunoassay for aldosterone during dynamic testing: comparison to radioimmunoassays with and without extraction steps. Clin Chem 2006;52 (9):1749–55.

Scholl UI, Dave HB, Lu M, Farhi A, Nelson-Williams C, Listman JA, et al. SeSAME/EAST syndrome–phenotypic variability and delayed activity of the distal convoluted tubule. Pediatr Nephrol 2012;27 (11):2081–90.

Scholl UI, Goh G, Stölting G, de Oliveira RC, Choi M, Overton JD, et al. Somatic and germline CACNA1D calcium channel mutations in aldosterone-producing adenomas and primary aldosteronism. Nat Genet 2013;45(9):1050–4.

Scholl UI, Stölting G, Nelson-Williams C, Vichot AA, Choi M, Loring E, et al. Recurrent gain of function mutation in calcium channel CACNA1H causes early-onset hypertension with primary aldosteronism. Elife 2015;4, e06315.

Scholl UI, Stölting G, Schewe J, Thiel A, Tan H, Nelson-Williams C, et al. CLCN2 chloride channel mutations in familial hyperaldosteronism type II. Nat Genet 2018;50(3):349–54.

Schöneshöfer M, Fenner A, Dulce HJ. Assessment of eleven adrenal steroids from a single serum sample by combination of automatic high-performance liquid chromatography and radioimmunoassay (HPLC-RIA). J Steroid Biochem 1981;14(4):377–86.

Sekihara H. 19-Hydroxyandrostenedione: evidence for a new class of sodium-retaining and hypertensinogenic steroids. Endocrinology 1983b;113(3):1141–8.

Seyberth HW. An improved terminology and classification of Bartter-like syndromes. Nat Clin Pract Nephrol 2008;4(10):560–7.

Seys E, Andrini O, Keck M, Mansour-Hendili L, Courand PY, Simian C, et al. Clinical and genetic spectrum of bartter syndrome type 3. J Am Soc Nephrol 2017;28(8):2540–52.

Shackleton CH. Mass spectrometry in the diagnosis of steroid-related disorders and in hypertension research. J Steroid Biochem Mol Biol 1983;45:127–40.

Shackleton CH, Honour JW, Dillon MJ, Chantler C, Jones RW. Hypertension in a four-year-old child: gas chromatographic and mass spectrometric evidence for deficient hepatic metabolism of steroids. J Clin Endocrinol Metab 1980;50(4), 786-02.

Shafigullina ZR, Velikanova LI, Vorokhobina NV, Shustov SB, Lisitsin AA, Malevanaia EV, et al. Urinary steroid profiling by gas chromatography mass spectrometry: early features of malignancy in patients with adrenal incidentalomas. Steroids 2018;135:31–5.

Shammas C, Byrou S, Phelan MM, Toumba M, Stylianou C, Skordis N, et al. Genetic screening of non-classic CAH females with hyperandrogenemia identifies a novel CYP11B1 gene mutation. Hormones (Athens) 2016;15(2):235–42.

Skinner CA, Rumsby G, Honour JW. Single strand conformation polymorphism (SSCP) analysis for the detection of mutations in the CYP11B1 gene. J Clin Endocrinol Metab 1996;81(6):2389–93.

So CB, Leung AA, Chin A, Kline GA. Adrenal venous sampling in primary aldosteronism: lessons from over 600 single-operator procedures. Clin Radiol 2022;77(2):e170–9.

Spoudeas HA, Slater JD, Rumsby G, Honour JW, Brook CG. Deoxycorticosterone, 11 beta-hydroxylase and the adrenal cortex. Clin Endocrinol (Oxf) 1993;39(2):245–51.

Starremans PG, Kersten FF, Knoers NV, van den Heuvel LP, Bindels RJ. Mutations in the human Na-K-2Cl cotransporter (NKCC2) identified in Bartter syndrome type I consistently result in nonfunctional transporters. J Am Soc Nephrol 2003;14(6):1419–26.

Stowasser M, Gordon RD, Rutherford JC, Nikwan NZ, Daunt N, Slater GJ. Diagnosis and management of primary aldosteronism. J Renin Angiotensin Aldosterone Syst 2001;2(3):156–69.

Stratakis CA, Sarlis N, Kirschner LS, Carney JA, Doppman JL, Nieman LK, et al. Paradoxical response to dexamethasone in the diagnosis of primary pigmented nodular adrenocortical disease. Ann Intern Med 1999;131(8):585–91.

Sutherland DJ, Ruse JL, Laidlaw JC. Hypertension, increased aldosterone secretion and low plasma renin activity relieved by dexamethasone. Can Med Assoc J 1966;95(22):1109–19.

Taguchi R, Yamada M, Nakajima Y, Satoh T, Hashimoto K, Shibusawa N, et al. Expression and mutations of KCNJ5 mRNA in Japanese patients with aldosterone-producing adenomas. J Clin Endocrinol Metab 2012;97(4):1311–9.

Takeda Y, Lewicka S, Koch S, Bige K, Vecsei P, Abdelhamid S, et al. Synthesis of 19-nor-aldosterone, 18-hydroxy-19-nor-corticosterone and 18,19-dihydroxycorticosterone in the human aldosterone-producing adenoma. J Steroid Biochem Mol Biol 1990;37(4):599–604.

Takeda Y, Bige K, Iwuanyanwu T, Lewicka S, Vecsei P, Abdelhamid S, et al. Urinary 18,19-dihydroxycorticosterone and 18-hydroxy-19-norcorticosterone excretion in patients with primary and secondary aldosteronism. Steroids 1991;56(11):566–70.

Takeda Y, Miyamori I, Iki K, Takeda R, Vecsei P. Urinary excretion of 19-noraldosterone, 18, 19-dihydroxycorticosterone and 18-hydroxy-19-norcorticosterone in patients with aldosterone-producing adenoma or idiopathic hyperaldosteronism. Acta Endocrinol 1992;126(6): 484–8.

Takeda Y, Miyamori I, Takeda R. Significance of 19-noraldosterone, a new mineralocorticoid, in clinical and experimental hypertension. Steroids 1995;60(1):137–42.

Takeda Y, Miyamori I, Yoneda T, Furukawa K, Hatakeyama H, Inaba S, et al. Effect of adrenocorticotropin stimulation on the synthesis of 19-noraldosterone in man. J Clin Endocrinol Metab 1996;81 (5):1852–5.

Tamura N, Watanabe E, Shirakawa R, Nakatani E, Yamada K, Hatakeyama H, et al. Comparisons of plasma aldosterone and renin data between an automated chemiluminescent immunoanalyzer and conventional radioimmunoassays in the screening and diagnosis of primary aldosteronism. PLoS One 2021;16(7), e0253807.

Tan X, Li F, Wang X, Wang Y. Quantitation and clinical evaluation of plasma aldosterone by ultra-performance liquid chromatography-mass spectrometry. J Chromatogr A 2020;1609:460456.

Taylor RL, Machacek D, Singh RJ. Validation of a high-throughput liquid chromatography-tandem mass spectrometry method for urinary cortisol and cortisone. Clin Chem 2002;48(9):1511–9.

Taylor PJ, Cooper DP, Gordon RD, Stowasser M. Measurement of aldosterone in human plasma by semiautomated HPLC-tandem mass spectrometry. Clin Chem 2009;55(6):1155–62.

Taylor DR, Ghataore L, Couchman L, Vincent RP, Whitelaw B, Lewis D, et al. A 13-steroid serum panel based on LC-MS/MS: use in detection of adrenocortical carcinoma. Clin Chem 2017;63(12):1836–46.

Teruyama K, Naruse M, Tsuiki M, Kobayashi H. Novel chemiluminescent immunoassay to measure plasma aldosterone and plasma active renin concentrations for the diagnosis of primary aldosteronism. J Hum Hypertens 2022;36(1):77–85.

Tetti M, Monticone S, Burrello J, Matarazzo P, Veglio F, Pasini B, et al. Liddle syndrome: review of the literature and description of a new case. Int J Mol Sci 2018;19(3):812.

Tezuka Y, Yamazaki Y, Nakamura Y, Sasano H, Satoh F. Recent development toward the next clinical practice of primary aldosteronism: a literature review. Biomedicine 2021a;9(3):310.

Tezuka Y, Ishii K, Zhao L, Yamazaki Y, Morimoto R, Sasano H, et al. ACTH stimulation maximizes the accuracy of peripheral steroid profiling in primary aldosteronism subtyping. J Clin Endocrinol Metab 2021b;106(10):e3969–78.

Thongboonkerd V. Roles for exosome in various kidney diseases and disorders. Front Pharmacol 2020;10:1655.

Thrower A, Ciccone EJ, Maitra P, Derebail VK, Cai J, Ataga KI. Effect of renin-angiotensin-aldosterone system blocking agents on progression of glomerulopathy in sickle cell disease. Br J Haematol 2019;184 (2):246–52.

Tissier F, Cavard C, Groussin L, Perlemoine K, Fumey G, Hagneré AM, et al. Mutations of beta-catenin in adrenocortical tumors: activation of the Wnt signaling pathway is a frequent event in both benign and malignant adrenocortical tumors. Cancer Res 2005;65(17): 7622–7.

Tran MT, Tran NA, Nguyen PM, et al. 11β-hydroxylase deficiency detected by urine steroid metabolome profiling using gas chromatography-mass spectrometry. Clin Mass Spectrom 2018;7:1–5.

Travers S, Martinerie L, Bouvattier C, Boileau P, Lombès M, Pussard E. Multiplexed steroid profiling of gluco- and mineralocorticoids pathways using a liquid chromatography tandem mass spectrometry method. J Steroid Biochem Mol Biol 2017;165(Pt B):202–11.

Turcu AF, Wannachalee T, Tsodikov A, Nanba AT, Ren J, Shields JJ, et al. Comprehensive analysis of steroid biomarkers for guiding primary aldosteronism subtyping. Hypertension 2020;75(1):183–92.

Turpeinen U, Markkanen H, Sane T, Hämäläinen E. Determination of free tetrahydrocortisol and tetrahydrocortisone ratio in urine by liquid chromatography-tandem mass spectrometry. Scand J Clin Lab Invest 2006;66(2):147–59.

Uchida K, Kigoshi T, Iwasaki R, Morimoto S. Altered responses of plasma 18-hydroxycorticosterone and aldosterone to angiotensin II and adrenocorticotropin in patients with a 18-hydroxycorticosterone-producing tumor. Jpn J Med 1989;28(4):446–51.

Ulick S. Two uncommon causes of mineralocorticoid excess. Syndrome of apparent mineralocorticoid excess and glucocorticoid-remediable aldosteronism. Endocrinol Metab Clin North Am 1991;20(2):269–76.

Ulick S, Land M, Chu MD. 18-oxocortisol, a naturally occurring mineralocorticoid agonist. Endocrinology 1983;113(6):2320–2.

Vahe C, Benomar K, Espiard S, Coppin L, Jannin A, Odou MF, et al. Diseases associated with calcium-sensing receptor. Orphanet J Rare Dis 2017;12(1):19.

Van Der Gugten JG, Holmes DT. Quantitation of plasma renin activity in plasma using liquid chromatography-tandem mass spectrometry (LC-MS/MS). Methods Mol Biol 2016;1378:243–53.

Vecsei P, Benraad TJ, Hofman J, Abdelhamid S, Haack D, Lichtwald K. Direct radioimmunoassays for "aldosterone" and "18-hydroxycorticosterone" in unprocessed urine, and their use in screening to distinguish primary aldosteronism from hypertension. Clin Chem 1982a;28(3):453–6.

Vecsei P, Abdelhamid S, Haack D, Lichtwald K, Lewicka S, von Mittelstädt G. Increased excretion of 18-hydroxycorticosterone in patients with adrenal adenomas and hypertension. Clin Exp Hypertens A 1982b;4(9–10):1759–70.

Velikanova LI, Shafigullina ZR, Lisitsin AA, Vorokhobina NV, Grigoryan K, Kukhianidze EA, et al. Different types of urinary steroid profiling obtained by high-performance liquid chromatography and gas chromatography-mass spectrometry in patients with adrenocortical carcinoma. Horm Cancer 2016;7(5–6):327–35.

Wang M, Wang H, Zhao H, Li L, Liu M, Liu F, et al. Prevalence of CYP17A1 gene mutations in 17α-hydroxylase deficiency in the Chinese Han population. Clin Hypertens 2019;25:23.

Wang MH, Li NF, Luo Q, Wang GL, Heizhati M, Wang L, et al. Development and validation of a novel diagnostic nomogram model to predict primary aldosteronism in patients with hypertension. Endocrine 2021;73(3):682–92.

Wei JQ, Wei JL, Li WC, Bi YS, Wei FC. Genotyping of five chinese patients with 17alpha-hydroxylase deficiency diagnosed through high-performance liquid chromatography serum adrenal profile: identification of two novel CYP17 mutations. J Clin Endocrinol Metab 2006;91(9):3647–53.

Weinberger MH, Fineberg NS. The diagnosis of primary aldosteronism and separation of two major subtypes. Arch Intern Med 1993;153 (18):2125–9.

Wenting GJ, Man in 't Veld AJ, Derkx FH, Schalekamp MA. Recurrence of hypertension in primary aldosteronism after discontinuation of spironolactone. Time course of changes in cardiac output and body fluid volumes. Clin Exp Hypertens A 1982;4(9–10):1727–48.

Williams TA, Lenders JWM, Mulatero P, Burrello J, Rottenkolber M, Adolf C, et al. Outcomes after adrenalectomy for unilateral primary aldosteronism: an international consensus on outcome measures and analysis of remission rates in an international cohort. Lancet Diabetes Endocrinol 2017;5(9):689–99.

Williams TA, Gomez-Sanchez CE, Rainey WE, Giordano TJ, Lam AK, Marker A, et al. International histopathology consensus for unilateral primary aldosteronism. J Clin Endocrinol Metab 2021;106(1):42–54.

Winter J, Bokkenheuser VD. 21-dehydroxylation of corticoids by anaerobic bacteria isolated from human fecal flora. J Steroid Biochem 1978;9(5):379–84.

Winter J, Shackleton CH, O'Rourke S, Bokkenheuser VD. Bacterial formation of aldosterone metabolites. J Steroid Biochem 1984;21(5):563–9.

Wolley M, Thuzar M, Stowasser M. Controversies and advances in adrenal venous sampling in the diagnostic workup of primary aldosteronism. Best Pract Res Clin Endocrinol Metab 2020;34(3):101400.

Wynne KN, Rae ID, O'Keefe DF, Adam WR, Pearce P, Stockigt JR, et al. Mineralocorticoid activity of 21-deoxyaldosterone derivatives: structure-function studies. J Steroid Biochem 1981;14(10):1041–4.

Xu L, Xia W, Wu X, Wang X, Zhao L, Nie M. Chimeric CYP11B2/CYP11B1 causing 11β-hydroxylase deficiency in Chinese patients with congenital adrenal hyperplasia. Steroids 2015;101:51–5.

Yang J, Fuller PJ. Interactions of the mineralocorticoid receptor—within and without. Mol Cell Endocrinol 2012;350(2):196–205.

Yang Y, Burrello J, Burrello A, Eisenhofer G, Peitzsch M, Tetti M, et al. Classification of microadenomas in patients with primary aldosteronism by steroid profiling. J Steroid Biochem Mol Biol 2019;189:274–82.

Yau M, Haider S, Khattab A, Ling C, Mathew M, Zaidi S, et al. Clinical, genetic, and structural basis of apparent mineralocorticoid excess due to 11β-hydroxysteroid dehydrogenase type 2 deficiency. Proc Natl Acad Sci U S A 2017;114(52):E11248–56.

Yildiz M, Isik E, Abali ZY, Keskin M, Ozbek MN, Bas F, et al. Clinical and hormonal profiles correlate with molecular characteristics in patients with 11β-hydroxylase deficiency. J Clin Endocrinol Metab 2021;106 (9):e3714–24.

Yoneda T, Karashima S, Kometani M, Usukura M, Demura M, Sanada J, et al. Impact of New quick gold nanoparticle-based cortisol assay during adrenal vein sampling for primary aldosteronism. J Clin Endocrinol Metab 2016;101(6):2554–61.

Yoshitake T, Ishida J, Sonezaki S, Yamaguchi M. High performance liquid chromatographic determination of 3 alpha, 5 beta-tetrahydroaldosterone and cortisol in human urine with fluorescence detection. Biomed Chromatogr 1992;6(5):217–21.

Zadik Z, Kahana L, Kaufman H, Benderli A, Hochberg Z. Salt loss in hypertensive form of congenital adrenal hyperplasia (11-beta-hydroxylase deficiency). J Clin Endocrinol Metab 1984;58(2):384–7.

Zhao LQ, Han S, Tian HM. Progress in molecular-genetic studies on congenital adrenal hyperplasia due to 11beta-hydroxylase deficiency. World J Pediatr 2008;4(2):85–90.

Further reading

Abdelhamid S, Lewicka S, Vecsei P, Winter J. 21-Deoxytetrahydroaldosterone excretion in primary hyperaldosteronism. Cardiology 1985;72(Suppl 1): 102–6.

Abdelhamid S, Lewicka S, Vecsei P, Haack D, Pahl S, Remberger K, et al. A new subset of mineralocorticoid hypertension with excess of 21-deoxyaldosterone and Kelly's-M1 steroid: clinical and morphological findings. J Clin Endocrinol Metab 1995;80(3):737–44.

Amar L, Plouin PF, Steichen O. Aldosterone-producing adenoma and other surgically correctable forms of primary aldosteronism. Orphanet J Rare Dis 2010;5:9.

Bailey MA. 11β-Hydroxysteroid dehydrogenases and hypertension in the metabolic syndrome. Curr Hypertens Rep 2017;19(12):100.

Bartter FC, Pronove P, Gill Jr JR, Maccardle RC. Hyperplasia of the juxtaglomerular complex with hyperaldosteronism and hypokalemic alkalosis. A new syndrome. Am J Med 1962;33:811–28.

Bokkenheuser VD, Winter J, Dehazya P, Kelly WG. Isolation and characterization of human fecal bacteria capable of 21-dehydroxylating corticoids. Appl Environ Microbiol 1977;34(5):571–5.

Bokkenheuser VD, Winter J, Honour JW, Shackleton CH. Reduction of aldosterone by anaerobic bacteria: origin of urinary 21-deoxy metabolites in man. J Steroid Biochem 1979;11(2):1145–9.

Bollag WB. Regulation of aldosterone synthesis and secretion. Compr Physiol 2014;4(3):1017–55.

Carroll J, Dluhy R, Fallo F, Pistorello M, Bradwin G, Gomez-Sanchez CE, et al. Aldosterone-producing adenomas do not contain glucocorticoid-remediable aldosteronism chimeric gene duplications. J Clin Endocrinol Metab 1996;81(12):4310–2.

Carvajal CA, Tapia-Castillo A, Vecchiola A, Baudrand R, Fardella CE. Classic and nonclassic apparent mineralocorticoid excess syndrome. J Clin Endocrinol Metab 2020;105(4), dgz315. https://doi.org/10.1210/clinem/dgz315. Erratum in: J Clin Endocrinol Metab. 2021;106(4):e1934.

Dluhy RG, Lifton RP. Glucocorticoid-remediable aldosteronism. J Clin Endocrinol Metab 1999;84(12):4341–4.

Eisenhofer G, Durán C, Cannistraci CV, Peitzsch M, Williams TA, Riester A, et al. Use of steroid profiling combined with machine learning for identification and subtype classification in primary aldosteronism. JAMA Netw Open 2020;3(9), e2016209.

Fan P, Zhang D, Pan XC, Yang KQ, Zhang QY, Lu YT, et al. Premature stroke secondary to severe hypertension results from liddle syndrome caused by a novel SCNN1B mutation. Kidney Blood Press Res 2020b;45(4):603–11.

Ganapathipillai S, Laval G, Hoffmann IS, Castejon AM, Nicod J, Dick B, et al. CYP11B2-CYP11B1 haplotypes associated with decreased 11 beta-hydroxylase activity. J Clin Endocrinol Metab 2005;90(2):1220–5.

Ganguly A, Bergstein J, Grim CE, Yum MN, Weinberger MH. Childhood primary aldosteronism due to an adrenal adenoma: preoperative localization by adrenal vein catheterization. Pediatrics 1980b;65(3):605–9.

Honour JW, Anderson JM, Shackleton CH. Difficulties in the diagnosis of congenital adrenal hyperplasia in early infancy: the 11 beta-hydroxylase defect. Acta Endocrinol 1983b;103(1):101–9.

Inoue K, Yamazaki Y, Kitamoto T, Hirose R, Saito J, Omura M, et al. Aldosterone suppression by dexamethasone in patients with KCNJ5-mutated aldosterone-producing adenoma. J Clin Endocrinol Metab 2018;103(9):3477–85.

Ku EJ, Lee C, Shim J, Lee S, Kim KA, Kim SW, et al. Metabolic subtyping of adrenal tumors: prospective multi-center cohort study in Korea. Endocrinol Metab (Seoul) 2021;36(5):1131–41.

Lifton RP, Gharavi AG, Geller DS. Molecular mechanisms of human hypertension. Cell 2001;104(4):545–56.

Lu YT, Fan P, Zhang D, Zhang Y, Meng X, Zhang QY, et al. Overview of monogenic forms of hypertension combined with hypokalemia. Front Pediatr 2021;8:543309.

Melby JC, Dale SL, Holbrook M, Wilson TE, Griffing GT, Arison BH. 19-Nor-corticosteroids in health, in hypertensive states in humans including 17 alpha-hydroxylase deficiency and in the spontaneously hypertensive rat (SHR). Endocr Res. 1984-1985;10(3–4):591–607.

Mourtzi N, Sertedaki A, Markou A, Piaditis GP, Charmandari E. Unravelling the genetic basis of primary Aldosteronism. Nutrients 2021;13(3):875.

Mulatero P, di Cella SM, Monticone S, Schiavone D, Manzo M, Mengozzi G, et al. 18-hydroxycorticosterone, 18-hydroxycortisol, and 18-oxocortisol in the diagnosis of primary aldosteronism and its subtypes. J Clin Endocrinol Metab 2012a;97(3):881–9.

Mulatero P, Tauber P, Zennaro MC, Monticone S, Lang K, Beuschlein F, et al. KCNJ5 mutations in European families with nonglucocorticoid remediable familial hyperaldosteronism. Hypertension 2012b;59 (2):235–40.

Nuñez-Gonzalez L, Carrera N, Garcia-Gonzalez MA. Molecular basis, diagnostic challenges and therapeutic approaches of Bartter and Gitelman syndromes: a primer for clinicians. Int J Mol Sci 2021;22 (21):11414.

Rassi-Cruz M, Maria AG, Faucz FR, London E, Vilela LAP, Santana LS, et al. Phosphodiesterase 2A and 3B variants are associated with primary aldosteronism. Endocr Relat Cancer 2021;28(1):1–13.

Reincke M, Bancos I, Mulatero P, Scholl UI, Stowasser M, Williams TA. Diagnosis and treatment of primary aldosteronism. Lancet Diabetes Endocrinol 2021;9(12):876–92.

Rich GM, Ulick S, Cook S, Wang JZ, Lifton RP, Dluhy RG. Glucocorticoid-remediable aldosteronism in a large kindred: clinical spectrum and diagnosis using a characteristic biochemical phenotype. Ann Intern Med 1992;116(10):813–20.

Schiff MJ, Schwartz JH, Bengele HH, Dale SL, Melby JC, Alexander EA. Mineralocorticoid activity of 19-nor-corticosterone and 19-nor-progesterone in the toad bladder. Endocrinology 1984;115(4):1235–8.

Scholl UI, Lifton RP. New insights into aldosterone-producing adenomas and hereditary aldosteronism: mutations in the K+ channel KCNJ5. Curr Opin Nephrol Hypertens 2013;22(2):141–7.

Sekihara H. 6 beta-Hydroxyandrostenedione: evidence for a new hypertensinogenic agent. Clin Exp Hypertens A 1983a;5(1):1–9.

Spiga F, Lightman SL. Dynamics of adrenal glucocorticoid steroidogenesis in health and disease. Mol Cell Endocrinol 2015;408:227–34.

Vakkalanka S, Zhao A, Samannodi M. Primary hyperaldosteronism: a case of unilateral adrenal hyperplasia with contralateral incidentaloma. BMJ Case Rep 2016;2016, bcr2016216209.

Vorselaars WMCM, van Beek DJ, Postma EL, Spiering W, Borel Rinkes IHM, Valk GD. Vriens MR; international CONNsortium study group. Clinical outcomes after surgery for primary aldosteronism: evaluation of the PASO-investigators' consensus criteria within a worldwide cohort of patients. Surgery 2019;166(1):61–8.

Watanabe F, Ryota T, Kobayashi Y. Enzyme immunoassay for serum 18-hydroxycorticosterone and its clinical application. Steroids 1984;43(5):509–16.

Chapter 3.3.2

Hypoaldosteronism, pseudohypoaldosteronism and salt loss

3.3.2.1 Introduction

Aldosterone has a critical role in regulating concentrations of sodium, potassium and hydrogen ions. It acts at the renal distal convoluted tubules through a chain of tubular cell channels and transporters. The action results in sodium retention and excretion of potassium and hydrogen ions. Hypoaldosteronism is therefore associated with sodium loss, hyperkalemic metabolic acidosis, volume depletion and hypotension. The synthesis of aldosterone is controlled by angiotensin II (AngII) which is produced by the action of renin on angiotensinogen to produce angiotensin I (AngI) a peptide of 10 amino acids that is further cleaved to AngII with 8 amino acids by the action of angiotensin converting enzyme (ACE) (Fig. 3.3.2.1).

Renin, angiotensin and aldosterone (RAAS) act to elevate arterial pressure in response to decreased renal blood pressure, decreased salt delivery to the distal convoluted tubule, and/or beta-agonism.

The synthesis of aldosterone from progesterone requires the actions of three enzymes that are subject to genetic mutations these are 21-hydroxylase (CYP21A1), 11-hydroxylase (CYP11B1) and aldosterone synthase (CYP11B2). The commonest genetic defect is the salt-wasting form of 21-hydroxylase (SWCAH) deficiency that is discussed later in this chapter and is also discussed in Chapter 3.1.2 on cortisol deficiency and Chapter 3.2.1 on sex steroid excess. CYP21 converts progesterone to 11-deoxycorticosterone (DOC), then CYP11B1 or CYP11B2 forms corticosterone from DOC. Aldosterone synthase (AS) (CYP11B2) has three activities.

(1) 11-hydroxylation of DOC to corticosterone (B),
(2) C-18 hydroxylation of B to 18-hydroxy B and
(3) oxidation of C-18 hydroxyl group to aldehyde of aldosterone.

Aldosterone and 18-hydroxycorticosterone in solution are in the form of 11,18-hemiacetal and 18,20-hemiacetal, respectively (Fig. 3.3.2.2) and the chemistry of the interconversions will need to be considered in the storage of the steroids and processing of samples before analysis.

Accurate assessment of aldosterone in biological fluids and renin activity or concentration in plasma is key to the investigation of patients with hyponatremia. Reference ranges for age are essential because of the much higher levels in infants than in adults. Few laboratories have pediatric reference ranges for local assays because of the difficulties in meeting requirements for derivation so literature values are used (my laboratory at UCL used data from Fiselier et al., 1984 for example). The assays have changed dramatically in the past two decades as radioimmunoassay has been almost completely replaced with chemiluminescent assays (Ozeki et al., 2021), often on analytical platforms for clinical chemistry laboratories. There are dedicated instruments for renin and aldosterone concentration and mass spectrometric methods for both analytes. The imprecision of aldosterone results at low concentrations may need to be considered although the really low concentrations of aldosterone in adults with disorders of aldosterone are not encountered in children. Sample size will be more of an issue with pediatric samples. Ideally less than 100 microliters of plasma from children should be needed for the required tests (Suessenbach et al., 2019).

LC-MS/MS methods are available for aldosterone alone (Lin et al., 2021; Tan et al., 2020; Ray et al., 2014), aldosterone in a panel of steroids (Gaudl et al., 2016), renin activity alone (Chappell et al., 2012; Bystrom et al., 2010) and aldosterone with 18-hydroxy-precursors (Chen et al., 2022; Makowski and Burckhardt, 2019; Travers et al., 2017), simultaneous aldosterone with renin activity and angiotensin concentration (Chen et al., 2021). High-resolution mass spectrometry has been used for aldosterone and precursors (Makowski and Burckhardt, 2019). Some ready-to use kits for LC-MS/MS methods include the requirements for determination of aldosterone in panels of up to 19 steroids, complete validation in the laboratory is still required because of differences in performances of the mass spectrometers used (Le Goff et al., 2020). The LOD of the aldosterone assay should be 20–50 pmol/L if accurate determination of hypoaldosteronism is to be demonstrated.

Steroids in the Laboratory and Clinical Practice. https://doi.org/10.1016/B978-0-12-818124-9.00021-8

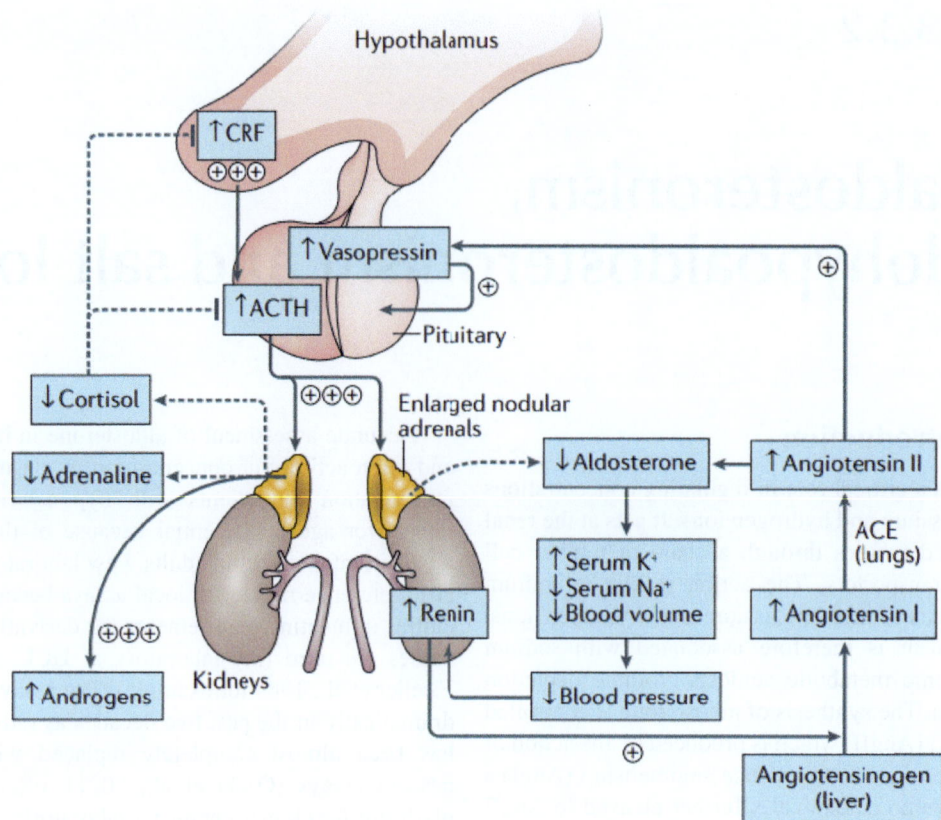

FIG. 3.3.2.1 Regulation of aldosterone. The renin-angiotensin-aldosterone system regulates blood pressure as well as fluid and electrolyte balance and is not directly under the influence of ACTH. Volume depletion and salt loss from aldosterone insufficiency leads to an increase in circulating levels of angiotensin II, which in turn stimulates vasopressin secretion. In congenital adrenal hyperplasia (CAH) due to 21-hydroxylase deficiency, reduced circulating levels of cortisol increase the hypothalamic secretion of corticotropin-releasing factor (CRF) and pituitary production of adrenocorticotrophic hormone (ACTH), and decrease adrenomedullary adrenaline secretion. Elevated ACTH drives adrenocortical hyperplasia and uninhibited synthesis of adrenal androgens. Vasopressin acts synergistically with CRF to augment ACTH release. The *dashed lines* indicate processes that are blunted in CAH. *Plus symbols* indicate processes that are enhanced in CAH; processes with *three plus* symbols are greatly enhanced and those with *one plus symbol* are mildly enhanced. *ACE,* angiotensin-converting enzyme. *(From Mallappa A, Merke DP. Management challenges and therapeutic advances in congenital adrenal hyperplasia. Nat Rev Endocrinol 2022;18(6):337–52. Fig. 1 p. 339.)*

Aldosterone stimulates sodium transport by regulating the expression and activity of ENaC. Aldosterone also stimulates the expression of serum glucocorticoid induced kinase (SGK1) (Valinsky et al., 2019) which directly and indirectly increase the expression and activity of ENaC. SGK1 phosphorylates Nedd4-2 (neuronal precursor cell expressed developmentally downregulated 4 protein), an ubiquitin ligase that ubiquitinates a motif of ENaC and targets it for degradation (Rotin and Staub, 2012, 2021). Upon phosphorylation by SGK1, Nedd4-2 loses its affinity to ENaC, thereby increasing the number of channels in the plasma membrane. SGK1 also phosphorylates WNK4, a negative regulator of ENaC activity. Upon phosphorylation by SGK1 WNK weakens its interaction with ENaC. All of these substances are subject to genetic mutations and with the study of patient material, mutation studies have enabled the recognition of emerging mechanisms of pathogenesis due to dysregulation of protein degradation and accumulation of proteins.

There is abundant evidence that WNK kinases (with no lysine) regulate the renal outer medullary potassium channel (ROMK), the sodium potassium chloride cotransporter (NKCC2), the sodium chloride cotransporter (NCC) and ENaC (Fig. 3.3.2.3) (Hoorn et al., 2011). WNK1 and WNK4 are key regulators of the thiazide sensitive Na-CL cotransporter (NCC) in the distal convoluted tubule (Boyd-Shiwarski et al., 2018). The kinase active forms of WNK1 and WNK4 phosphorylate and activate two downstream homologous kinases SPAK and OSR (oxidative stress response kinase).

SPAK (STE20 (sterile 20)/SPS1-related proline/alanine-rich kinase) is a member of germinal center kinase IV subfamily of STE20-related protein kinases that shares 67% amino acid sequence homology with the other mammalian STE20 kinase—oxidative stress-responsive kinase-1 (OSR1) (Ko et al., 2013). Both SPAK and OSR1 are enriched in transporting epithelia, especially renal tubules. In vitro studies showed that SPAK and OSR1 enhance the

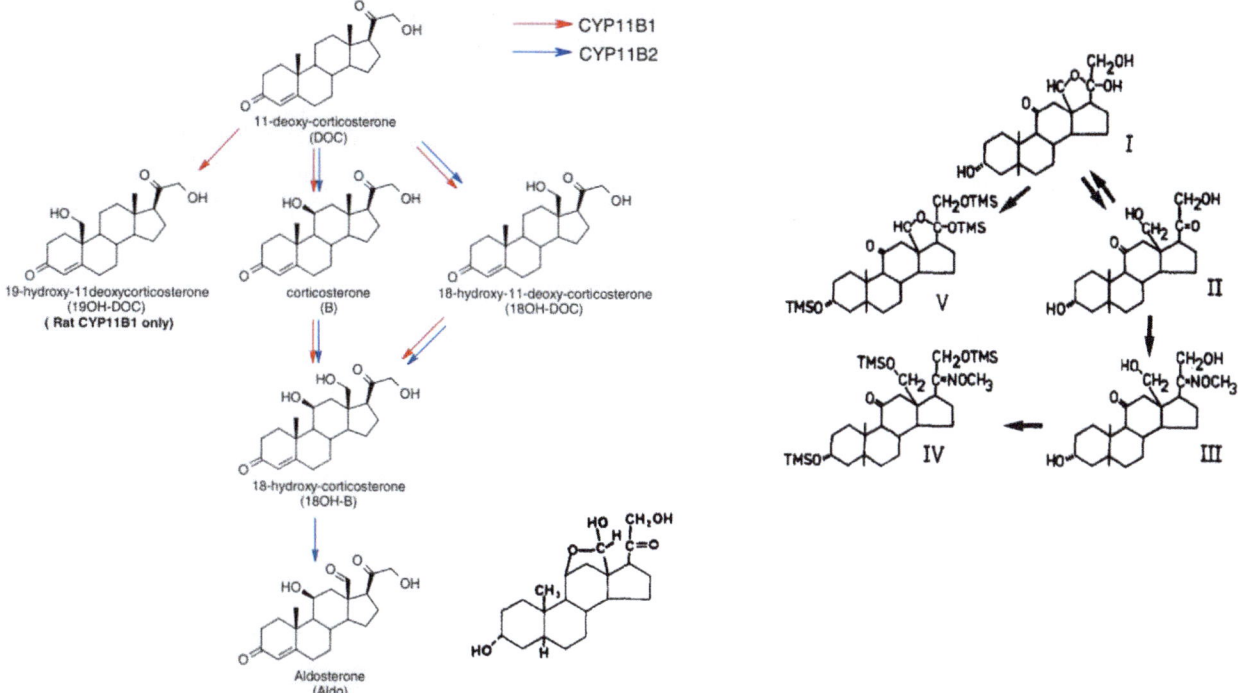

FIG. 3.3.2.2 Aldosterone synthesis and metabolism. The final steps of aldosterone synthesis are performed by CYP11B2. Hydroxylation of deoxycorticosterone at C-11 is followed by hydroxylation at C18 and oxidation to a C-18 aldehyde. The combined effects of A ring reduction at C-3 and reduction of the double bond at C-5 produces the 3α,5β-tetrahydroaldosterone (VI) which in solution will be in a hemiacetal form (VII). 18-Hydroxytetrahydro Compound A in open form (II) is the principal metabolite of 18-hydroxycorticosterone after A-ring reduction and oxidation of the C-11 hydroxyl to a carbonyl group. 18-HydroxyTHA can form a methyloxime derivative (III) and in the 18,20-hemiacetal form (I) forms trimethylsilyl ether derivative (IV) and methyloxime-trimethylsilyl derivative (IV). *(Author original.)*

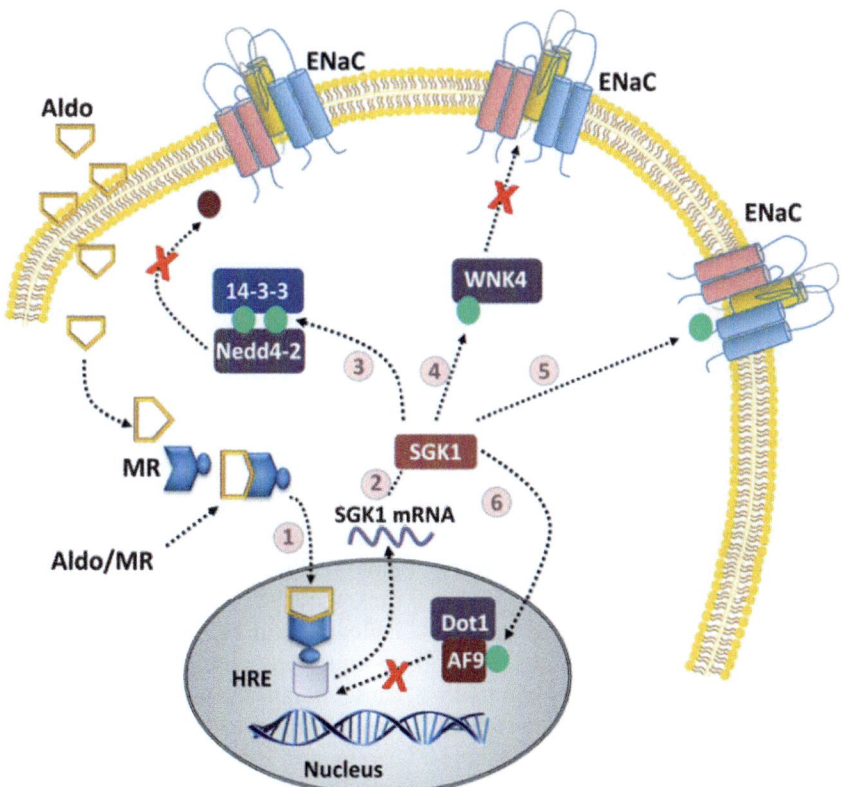

FIG. 3.3.2.3 Schematic of aldosterone, SGK1, and ENaC interactions. Aldosterone freely crosses phospholipid membranes and binds to the cytosolic mineralocorticoid receptor (MR) (1). The aldo/MR complex translocates to the nucleus, binds to specialized hormone response elements (HREs), and promotes the transcription of aldosterone-regulated genes, including SGK1, which is translated into protein (2). Newly synthesized SGK1 upregulates ENaC activity through several distinct pathways that reduce ENaC ubiquitination through bi-phosphorylation of Nedd4-2 (3), prevent ENaC endocytosis by phosphorylation of WNK4 (4), recruit silent ENaC channels to active ones by direct phosphorylation (5), and inhibit the transcriptional repressor complex Dot1-AF9 via phosphorylation of AF9 (6). *(From Valinsky WC, Touyz RM, Shrier A. Aldosterone, SGK1, and ion channels in the kidney. Clin Sci (Lond) 2018;132(2):173–83. Fig. 1 p. 175.)*

activities of Na, Cl cotransporter (NCC) and Na, K, 2Cl cotransporter isoform 1 and 2 (NKCC1/2) via phosphorylating a cluster of threonine and serine residues in the N terminus of these transporters (Cheng et al., 2015). By analysis of proteins that interact with-no-lysine (WNK) kinases, SPAK and OSR1 were found to be downstream substrates of WNK kinases. WNK kinases phosphorylate and activate SPAK and OSR1, which in turn phosphorylate and activate NCC and NKCC1/2. Heterozygous mutations in WNK1 and WNK4 genes cause an autosomal-dominant hypertension known as pseudohypoaldosteronism type II (PHAII). Polymorphisms within introns of the SPAK gene are also associated with hypertension. These findings implicate the importance of WNK-SPAK/OSR1 in renal salt reabsorption (Cheng et al., 2015). Signaling pathways are thus involved in sodium homeostasis and blood pressure (Tsilosani et al., 2022) and more paths may be revealed from further research in this area. The glucocorticoid induced leucine zipper (GILZ) protein is another factor that stimulates expression of ENaC (Soundararajan et al., 2005) and may affect aldosterone levels.

3.3.2.2   Hypoaldosteronism

Hypoaldosteronism is considered in any patient in whom there is no obvious cause such as kidney failure or the patient declares the use of potassium supplements or a potassium-sparing diuretic. In a newborn infant, plasma aldosterone less than the normal range of 1000–3500 pmol/L (40–130 ng/dL) indicates hypoaldosteronism, in older children and adults ambulant aldosterone is in the normal range 600–1200 pmol/L (20–40 ng/dL). Normal renin activity is 4–12 nmol/L/h (5–15 ng/mL/h) in newborn infants and 2–6 nmol/L/h (2.6–7.8 ng/mL/h) later. The differences can be attributed to reduced mineralocorticoid receptor expression in preterm infants. The causes of hypoaldosteronism can be acquired or genetic due to aldosterone deficiency or resistance. In children with SWCAH, renin activity is high. Aldosterone acts at the kidney and elsewhere through an epithelial sodium transport channel (ENaC), which is comprised of three subunits each of which is subject to transformation from gene mutations. In the absence of renin, due to certain forms of congenital adrenal hyperplasia, diabetic kidney disease and certain drugs there is aldosterone deficiency. Angiotensin converting enzyme (ACE) inhibitors, angiotensin II receptor blockers (ARBs) and direct renin inhibitors are widely prescribed and this must be considered when assessing a patient for hypoaldosteronism. **Pseudohypoaldosteronism** is a condition in which serum aldosterone is normal or elevated but the action of aldosterone is deficient.

3.3.2.3   Hypoaldosteronism

Deficiency in aldosterone production through certain steroidogenic enzyme defects or destruction of the adrenal zona glomerulosa constitutes primary hypoaldosteronism. Patients present with salt loss and failure to thrive with hyperkalemia and hyperreninemia. A similar clinical picture is seen when aldosterone is not low if the mineralocorticoid receptor is defective. This clinical state is described as pseudohypoaldosteronism (PHA). Hypoaldosteronism may be secondary to lack of renin-angiotensin stimulation and can be the result of a congenital adrenal hyperplasia where deoxycorticosterone is produced in excess to retain sodium in the kidney. Under those circumstances cortisol production is low.

3.3.2.4   Primary hypoaldosteronism

Classically, patients present with weakness, lethargy, anorexia, nausea, weight loss, and hypotension but they may have few or no symptoms of adrenal cortical insufficiency until they are stressed by trauma, surgery or infection. The patient should be examined for other genetic anomalies and a family history should be taken. A number of biochemical tests are needed but increasingly a genetic diagnosis is becoming more practical. Hyponatremia, hyperkalemia and raised serum urea levels are typical biochemical findings. Hypoglycemia is particularly common in children. In the urine, there will be low potassium loss. Balance studies of electrolytes may be needed. Autoimmune adrenalitis which spares the adrenal medulla is a cause of hypoaldosteronism due to adrenal antibodies.

Plasma should be tested for a growing list of **adrenal antibodies** such as against the steroid synthetic enzymes (21-OH, 17α-hydroxylase (17α-OH), side-chain cleavage enzyme (SCC)), SRY (sex determining region Y) and SOX-10 (SRY-related HMG-box) (Arif et al., 1996; Chen et al., 1996; Uibo et al., 1994).

Hyponatremia in the neonatal period is an urgent diagnostic problem which should consider whether the sodium intake is adequate (>4 mmol/kg/day in term babies, up to 12 mmol/kg/day in preterm infants), the fluid intake is high or there is sodium loss from the gastrointestinal tract or kidneys. Renal salt loss can be due to anatomical abnormalities, obstructive or renal tubular disorders which can include failure to respond to aldosterone. Low production of mineralocorticoid due to adrenal disease is a common cause of salt loss with hyperkalemia in newborns. In particular, urinary tract infections with pyelonephritis are the commonest cause of hyponatremia and pseudohypoaldosteronism in the first 6 months of life. Hyponatremia is often seen in the first weeks of life in preterm infants (<30 weeks gestation) (Honour et al., 1977). This reflects immaturity of renal function as well as in the adrenal production of aldosterone and the antidiuretic effect of increased vasopressin production (Martinerie et al., 2009a,b). Age, gender, menstrual phase, pregnancy, time of day, posture and medication use are factors that affect plasma renin concentration (PRC) and activity PRA and need to be considered when taking samples and interpreting results.

3.3.2.4.1 Primary hypoaldosteronism with defects in aldosterone synthesis

3.3.2.4.1.1 CYP11B2 defect (aldosterone synthase)

If the biosynthetic path to cortisol is normal, the production of aldosterone needs to be evaluated. Defects of aldosterone production and action are recognized. In both cases, PRA will be elevated, the defects are distinguished by the serum concentrations of aldosterone and 18-hydroxycorticosterone or urinary excretion rates of the metabolites. Most of the literature is based on immunoassay of steroids after chromatography. The 18-oxygenated steroids are not determined in all studies and diagnosis is based on demonstrating hyperreninemic hypoaldosteronism. Aldosterone synthase used to be called corticosterone methyl oxidase (CMO) and two types of CMO deficiency are recognized biochemically. CMOI (ASD1) is a defect of first steps with high DOC and B but no C-18 oxygenated steroids (Üstyol et al., 2016; Nguyen et al., 2008). CMOII (ASD2) defect is due to a defect in oxidation of the C-18 hydroxyl to aldehyde so high ratios of B and 18-OHB to aldosterone are found (Turan et al., 2021; Pascoe et al., 1992).

The simultaneous determination in urine of corticosterone, 18-hydroxycorticosterone and aldosterone metabolites will distinguish defects of the two enzymes in the final stages of aldosterone biosynthesis. Corticosterone will be raised if no C-18 hydroxylation and a **defect in the late steps of aldosterone biosynthesis** will be confirmed when B or 18-hydroxycorticosterone metabolite excretion are elevated. In the study of one neonate of 2 weeks only, a corticosterone metabolite (6-hydroxy tetrahydro-11-dehydrocorticosterone) was significantly raised (Fig. 3.3.2.4) (Honour et al., 1982) and assumed (before the advent of genetic studies) to reflect loss of 18-hydroxylation (ASD Type I).

Identification 18-OH THA in urine of a salt losing infant with normal cortisol production was first confirmed by Shackleton et al. (1976) after periodate oxidation and comparison with a reference preparation of the resulting etiolactone. Later a preparation of 18-hydroxycorticosterone was also converted to 3α,5β-tetrahydroaldosterone by incubation with *Clostridium paraputrificum* (Shackleton et al., 1979). 18-OH-corticosterone metabolites in solution occur

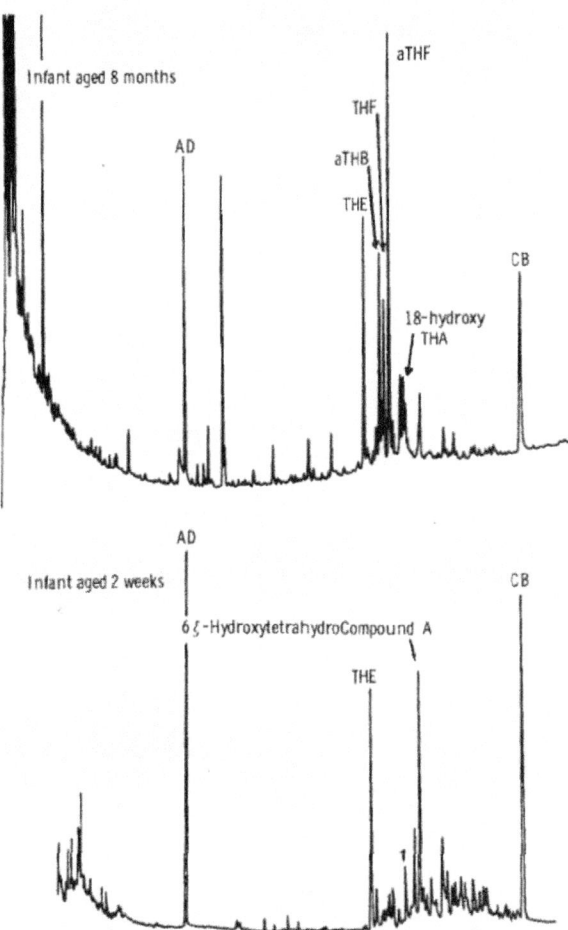

FIG. 3.3.2.4 GC profile of steroids in urine of infants with defect in aldosterone biosynthesis at 8 months (top panel) and 2 weeks (lower panel). Peak 1 was identified as 6-hydroxy tetrahydrocortisone. *(From Honour JW, Dillon MJ, Shackleton CH. Analysis of steroids in urine for differentiation of pseudohypoaldosteronism and aldosterone biosynthetic defect. J Clin Endocrinol Metab 1982;54(2):325–31. Fig. 1 p. 328.)*

in hemiacetal form which in ethanolic solution is converted to the ethyl derivative of the 18,20-hemiacetal (see earlier Fig. 3.3.2.1). The MO-TMS of 18-hydroxyTHB has a molecular weight of 683 and prominent M-152 peak at 531. As MO-TMS derivative, 18-hydroxy-THA has a molecular weight of 609 and M-31 fragment at 578 (Fig. 3.3.2.5). A characteristic fragment is seen at m/z 457 (Shackleton and Honour, 1977). 18-Hydroxy THB and THDOC have molecular weights of 683 and 595, respectively.

The derivatized 18,20-hemiacetal has a molecular weight of 580 with base peak at M-103 (477 amu) (Fig. 3.3.2.6) and the ethyl ketal of the hemiacetal has molecular weight 536 and base peak at 433 (M-103).

In a standard GC-MS analysis of urinary steroids, the main peak for tetrahydroaldosterone has a mass spectrum

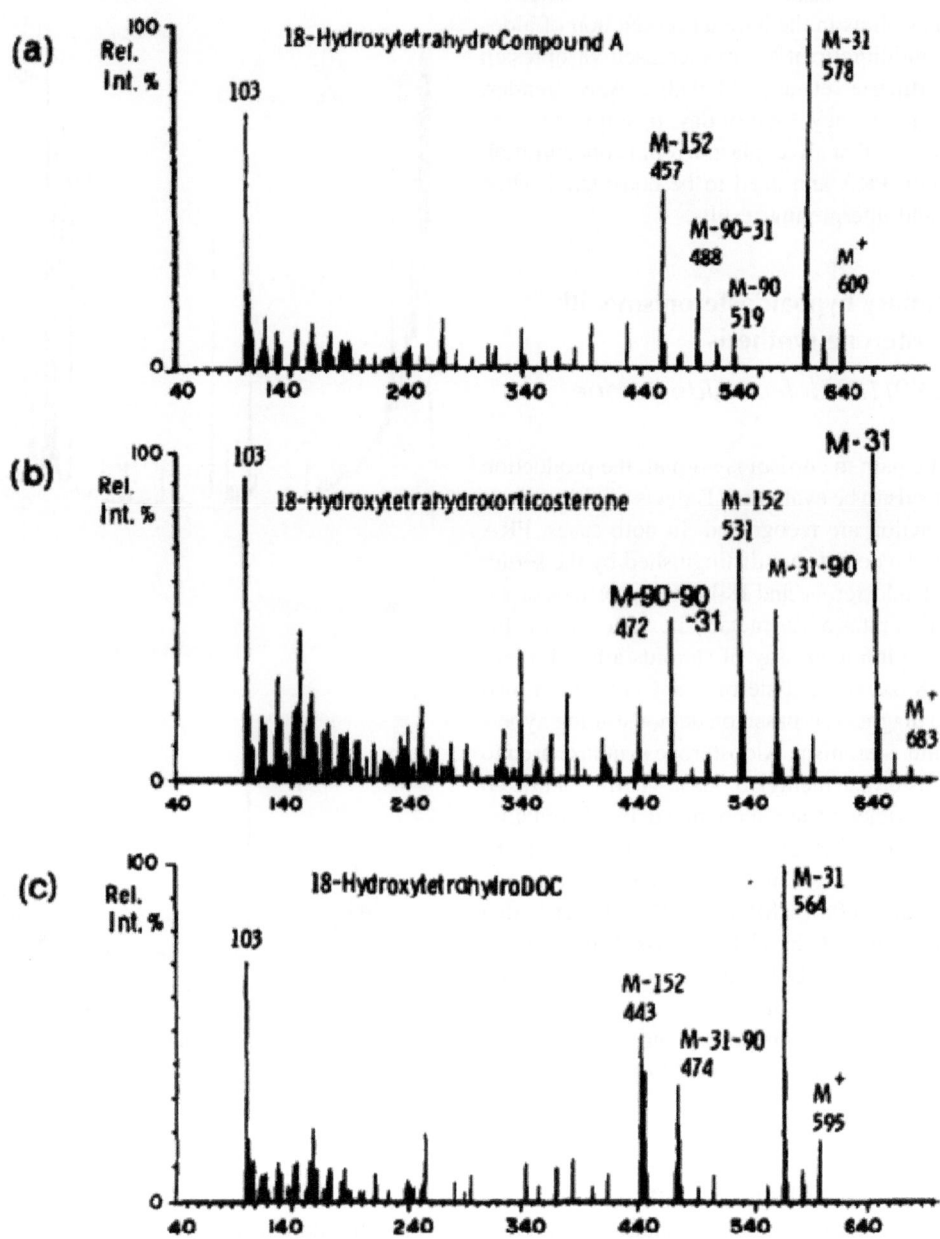

FIG. 3.3.2.5 GC-MS mass spectra of 18-hydroxy-THB, 18-hydroxy-THA and 18-hydroxy THDOC as MO-TMS derivatives. *(Author original.)*

with molecular weight 609 and base peak at 506 (M-103) for the steroid in the hemiacetal form (Honour and Shackleton, 1977) (Fig. 3.3.2.7).

A long reaction with methyloxime hydrochloride is required to have THAldo in the free aldehyde form with molecular weight 638 and base peak at 607 (M-31).

Quantitative results for five cases with 18-oxidation defect show the variation seen with 18-hydroxyTHA, relative to corticosterone and aldosterone (Table 3.3.2.1). Similar results (with mutation analysis) were seen in a separate paper using GC-MS with selected ion monitoring (Nguyen et al., 2010). High plasma renin and low aldosterone are taken as

evidence for the defect in some studies (Merakou et al., 2021). That study demonstrated why it is important to use appropriate reference ranges for renin and aldosterone although this is not always the case (Miao et al., 2019) since a high ratio of serum 18-hydroxycorticosterone to aldosterone (1.25 vs. <0.02) by isotope dilution LC-MS/MS (although the isotopes were not specified) is diagnostic. When the activities of expressed proteins were studied by HPLC with UV detection, two peaks were recorded for the 18-OHB standard but only one for the expression study. The second peak was attributed to formation of a tautomer on storage. This problem was addressed earlier in the book

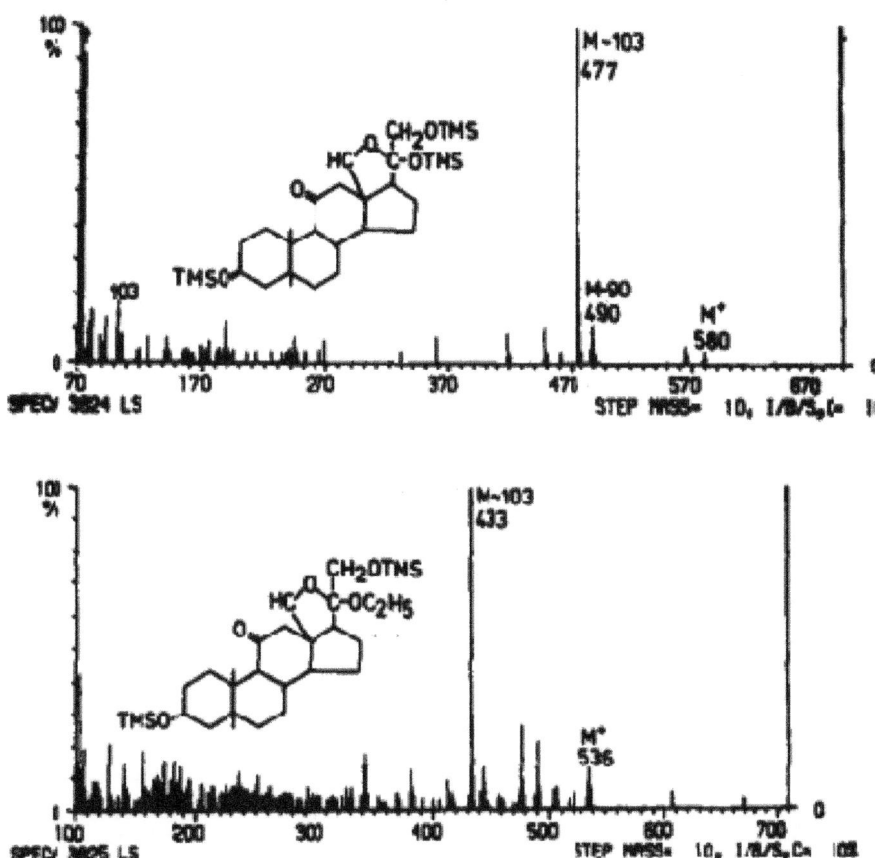

FIG. 3.3.2.6 Mass spectra of 18-hydroxy-THA as hemiacetal and ethyl-ketol derivative. *(Author original.)*

in the Chemistry Chapter 1.1. A high-resolution mass spectrometry method for aldosterone, tetrahydroaldosterone, and 18-hydroxycorticosterone has been validated for 50 μL serum (Makowski and Burckhardt, 2019). Defects in aldosterone synthesis need further biochemical investigations with more detail on concentrations of DOC, corticosterone and 18-hydroxycorticosterone so as to characterize CMO I and CMO II in relation to the genotypes. So far, many studies only had information on corticosterone and aldosterone.

A five base pair deletion has been found in CMO1 (Mitsuuchi et al., 1992) as well as cases with frame shift and premature stop codon. Point mutations have been found with CMOII defect (Turan et al., 2018). A gene conversion was reported where exons 3 and 4 of CYP11B2 were replaced with exon 3 and 4 of CYP11B1 (Fardella et al., 1996). Miao and colleagues reviewed 28 papers between 1992 and 2014 with more than 50 mutations found in *CYP11B2* by sequencing (Miao et al., 2019), with 22 mutations in ASDI in all exons except exon 2 and 15 in ASDII in exons 3–8 with most in exon 3 (Fig. 3.3.2.8). The majority were missense mutations. A study in Greece of 62 patients showed high prevalence of T185I mutations (Merakou et al., 2021).

A growth chart of child with defect of aldosterone biosynthesis shows the growth rate to be low in the first year of

life and improvement after treatment (Miao et al., 2019) (Fig. 3.3.2.9).

A child with an aldosterone biosynthetic defect will have increased excretion of corticosterone metabolites when the defect is in the 18-oxygenation step (Fig. 3.3.2.10).

3.3.2.4.2 Addison's disease

In adults, adrenal insufficiency is often autoimmune or due to tuberculosis infection, whereas in children the commonest cause is congenital adrenal hyperplasia with 21-hydroxylase deficiency by far the commonest (Capalbo et al., 2021).

3.3.2.4.3 Congenital adrenal hyperplasia

The group of inborn errors of adrenal steroid biosynthesis which reduce cortisol production and lead to compensatory hypersecretion of ACTH and hyperplasia of the adrenal cortex are thus named congenital adrenal hyperplasia (CAH) (Fig. 3.3.2.11). Four of the enzymes, 21-hydroxylase, 11β-hydroxylase, side chain cleavage, and 3β-hydroxysteroid dehydrogenase (HSD3B2) are found in the adrenal *zona glomerulosa* as well as adrenal *zona*

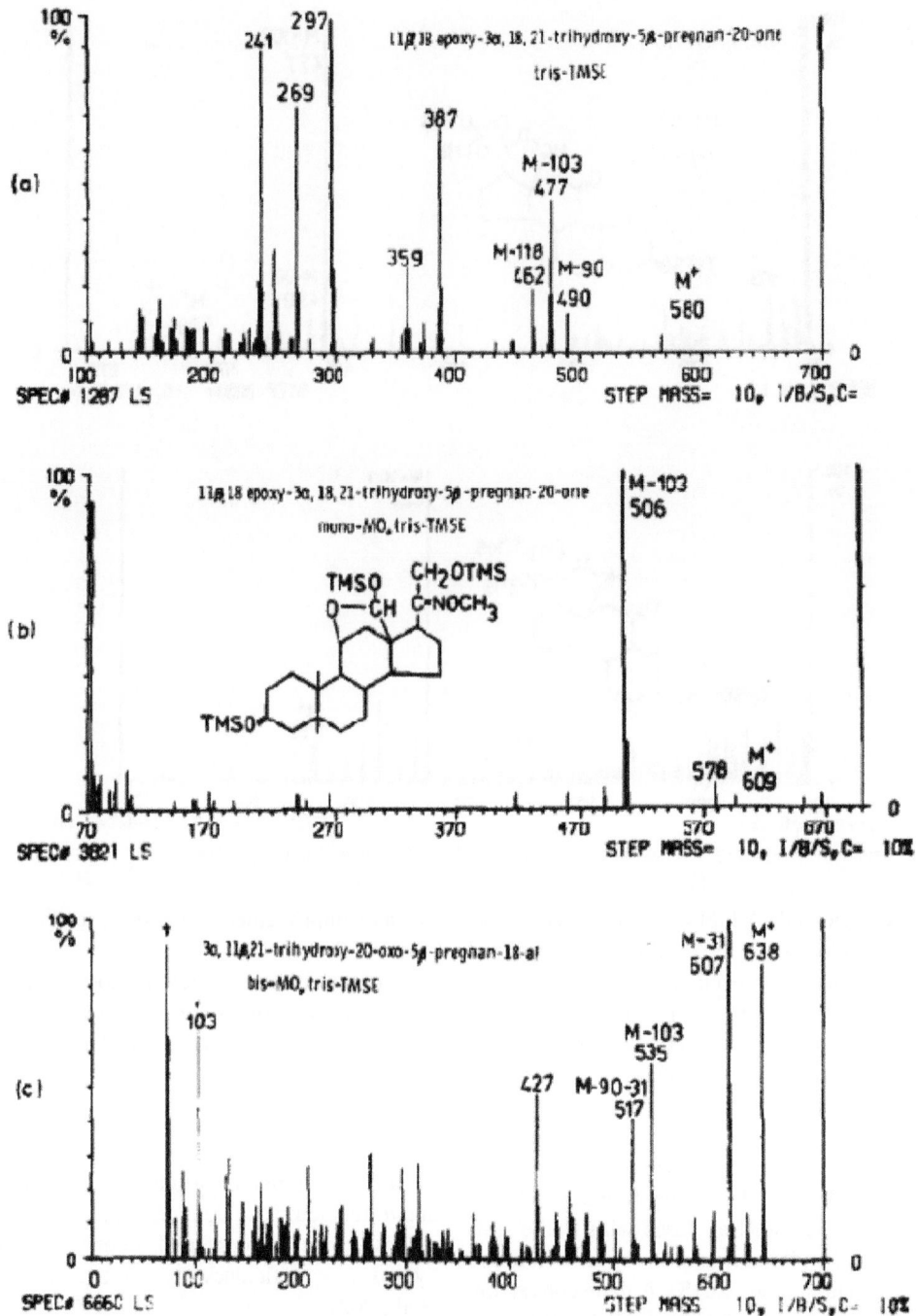

FIG. 3.3.2.7 GC-MS mass spectrum of THAldo in aldehyde, hemiacetal and hemiketal forms as MO-TMS derivatives. *(Author original.)*

fasciculata and deficiency can therefore affect the production of the steroids that influence sodium and glucose concentrations, respectively, and in some cases the sex steroids. Defects in the enzymes which only affect aldosterone production in the glomerulosa zone do not present with hyperplastic adrenal glands but with hyperreninemic hypoaldosteronism.

A defect in the early stages of steroid synthesis can lead to hypoaldosteronism. Defects in **side-chain cleavage** (SCC), **StAR or adrenal hypoplasia** may explain a salt-losing state so other tests should be considered, such as a urine or plasma steroid profile for evidence of cortisol and aldosterone deficiency. A defect of **StAR** prevents steroid production from cholesterol and all steroids in blood and urine will be low. DHEAS will be elevated in blood or urine if the defect is in HSD3B2 and 17-hydroxypregnenolone in blood or pregnenetriol in urine will be detected.

TABLE 3.3.2.1 Urinary steroid excretion by GC-MS of four patients with defects in aldosterone synthase.

Patient (age, years)	Mutations on two alleles	THDOC (μg/24h)	6α-THA	THA	aTHA	THB	allo-THB	18-OH-THA	18-OH-THB	THAldo
1 (2)	p.W260X/p.W260X	69	ND	130	280	60	148	ND	ND	3
2 (0.1)	p.G206WfsX51/p.L496SfsX169	ND	212	101	ND	9	24	ND	ND	ND
3 (3)	p.S315R/p.R374W	45	ND	203	49	49	301	ND	ND	3
4 (0.1)	p.S315R/p.R374W	ND	123	164	57	ND	ND	ND	ND	3
Reference range										
0–3 months		0–5	ND	2–28	3–42	1–10	6–24	3–12	ND	4–12
1–3 years		0–12	ND	12–87	15–98	5–16	12–98	5–25	ND	6–34

From Nguyen HH, Hannemann F, Hartmann MF, Malunowicz EM, Wudy SA, Bernhardt R. Five novel mutations in CYP11B2 gene detected in patients with aldosterone synthase deficiency type I: functional characterization and structural analyses. Mol Genet Metab 2010;100(4):357–64. Table 1 p. 358.

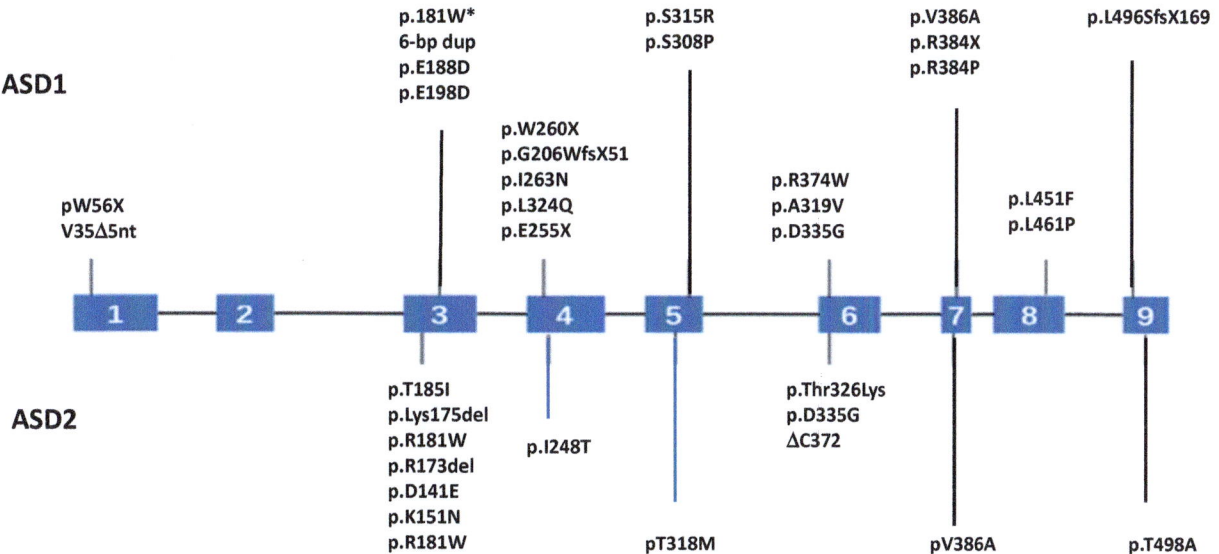

FIG. 3.3.2.8 Mutations of CYP11B2 from literature review. *ASD*, aldosterone synthase deficiency. *(From Miao H, Yu Z, Lu L, Zhu H, Auchus RJ, Liu J, Jiang J, Pan H, Gong F, Chen S, Lu Z. Analysis of novel heterozygous mutations in the CYP11B2 gene causing congenital aldosterone synthase deficiency and literature review. Steroids 2019;150:108448. Fig. 4 p. 7.)*

3.3.2.4.3.1 21-Hydroxylase deficiency

A male child who collapses during the second week of life should be tested for congenital adrenal hyperplasia due to **21-hydroxylase deficiency** because this is the commonest disorder in this situation. The steroids 17-hydroxyprogesterone and 21-deoxycortisol are substrates for the enzyme and when in excess are biomarkers. Sixty percent of all children with this genetic defect have the salt-losing genetic variant of the disease. The enzyme deficiency in cortisol production extends to the synthesis of aldosterone. Plasma renin activity will be elevated and aldosterone will be inappropriately low for age (normal ranges 4–12 nmol/L/h and 1000–3500 pmol/L, respectively, compared with 2.8–4.5 and 600–1200 in adults). Since plasma renin activity is much higher in all newborn infants than in adults it is important to check the activity against a normal range for age of the infant (Fiselier et al., 1984).

An elevated 17-OHP and exaggerated response to synthetic ACTH above the normal response is indicative of **congenital adrenal hyperplasia** due to **21-hydroxylase deficiency** which is the commonest cause of AI (see later). The biochemistry of this enzyme deficiency is considered in more detail in Chapter 3.1.2. A urinary steroid profile will show elevated excretion of 17-hydroxypregnanolone then

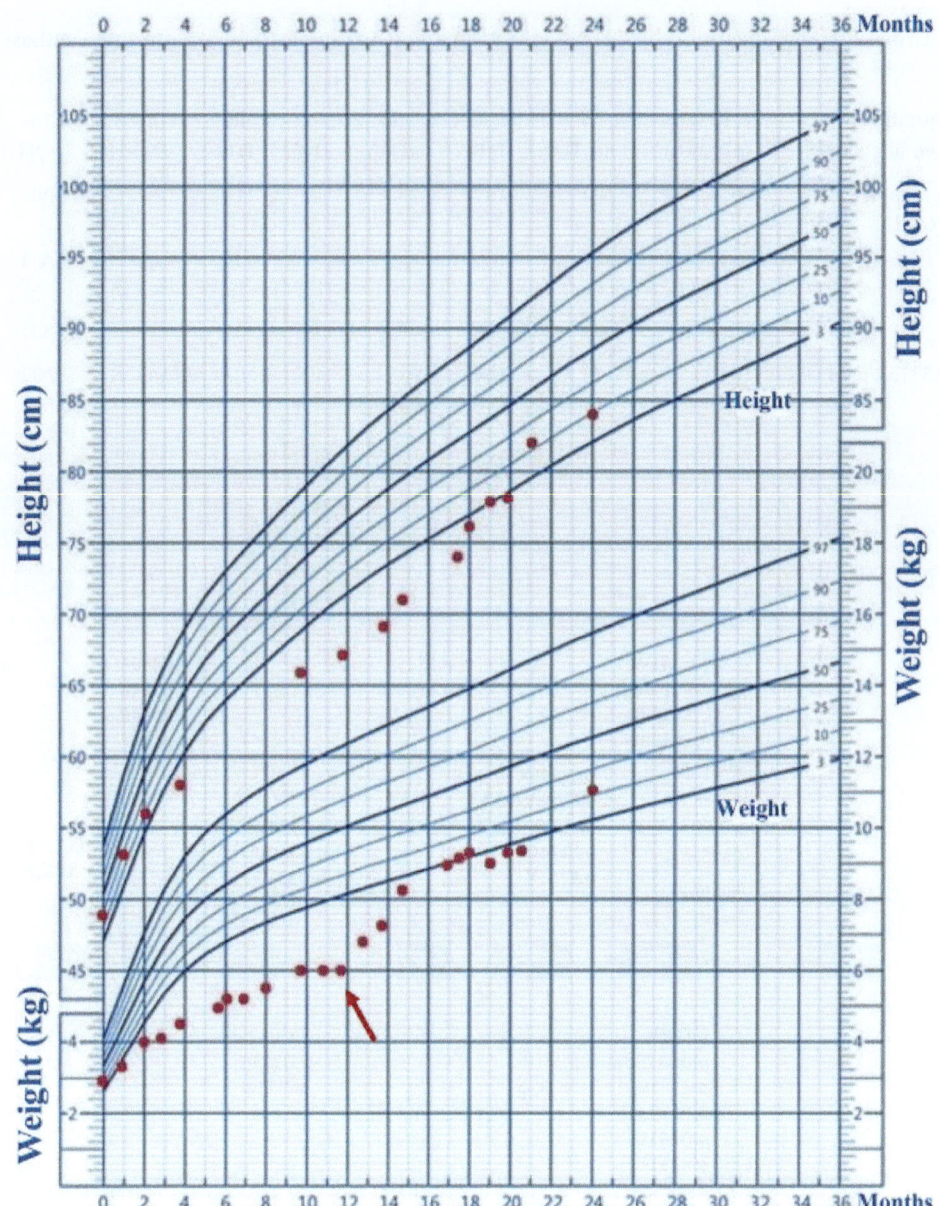

FIG. 3.3.2.9 Growth chart of child with aldosterone biosynthetic defect. Note response to mineralocorticoid treatment (red arrow) that would not be seen in a case of aldosterone resistance. *(From Miao H, Yu Z, Lu L, Zhu H, Auchus RJ, Liu J, Jiang J, Pan H, Gong F, Chen S, Lu Z. Analysis of novel heterozygous mutations in the CYP11B2 gene causing congenital aldosterone synthase deficiency and literature review. Steroids 2019;150:108448. Fig. 1 p. 4.)*

pregnanetriol and 11-oxo-pregnanetriol (Fig. 3.3.2.12) the main metabolites of 17-hydroxyprogesterone and 21-deoxycortisol, respectively, which can be seen in LC-MS/MS of plasma steroids.

Steroid 21-hydroxylase deficiency has an incidence in the order of 1:15,000 in the United Kingdom and therefore a carrier frequency of 1 in 60. The highest prevalence is found in Iceland. The gene *CYP21* is located on the short arm of chromosome 6 within the HLA class 3 region and in humans a single gene is present approximately 30 kb

downstream of a pseudogene, *CYP21P*, with 98% sequence homology in exons and 96% in introns. The pseudogene CYP21A1P is inactive because of the presence of multiple small insertions or deletions and point pathogenic variants that prevent the synthesis of a functional protein. The location and high rate of homology between the two genes facilitates misalignment that results in recombination events between the gene and the pseudogene.

About 70% of disease-causing *CYP21A2* mutations arise from gene conversion with the transfer of deleterious

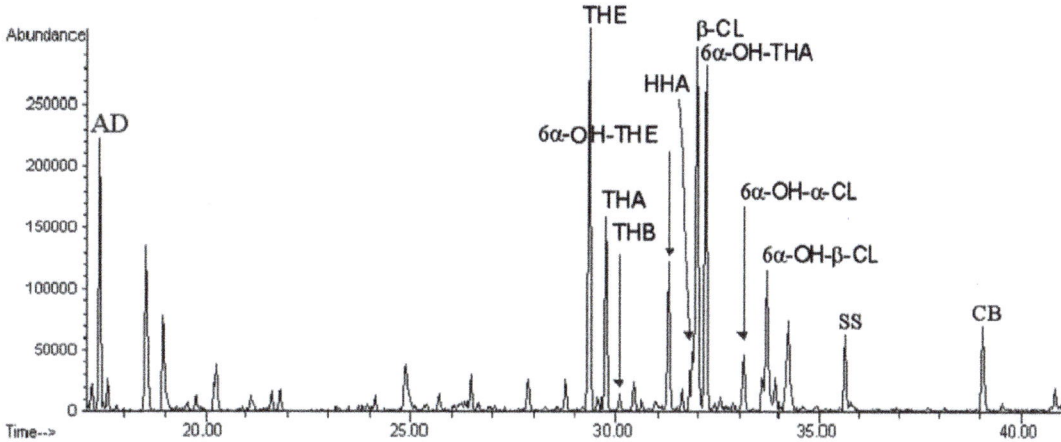

FIG. 3.3.2.10 GC-MS urinary steroid profile in a newborn with aldosterone synthase deficiency type I. The profile was dominated by the huge peaks reflecting highly elevated corticosterone metabolites: THA (tetrahydro-11-dehydrocorticosterone), 6α-OH-THA (6α-hydroxy-tetrahydro-11-dehydrocorticosterone), HHA (hexahydro-11-dehydrocorticosterone), THB (tetrahydrocorticosterone). 18-Oxygenated corticosterone metabolites were not detectable. Excretion of cortisol metabolites (THE, tetrahydrocortisone; 6α-OH-THE, 6α-hydroxy-tetrahydrocortisone; β-CL, β-cortolone; 6α-OH-α-CL, 6α-hydroxy-α-cortolone; 6α-OH-β-CL, 6α-hydroxy-β-cortolone) was normal. AD (5α-androstane-3α, 17α-diol), SS (stigmasterol) and CB (5-cholestene-3β-ol-butyrate) indicate internal standards. *(From Nguyen HH, Hannemann F, Hartmann MF, Wudy SA, Bernhardt R. Aldosterone synthase deficiency caused by a homozygous L451F mutation in the CYP11B2 gene. Mol Genet Metab 2008;93(4):458–67. Fig. 1 p. 461.)*

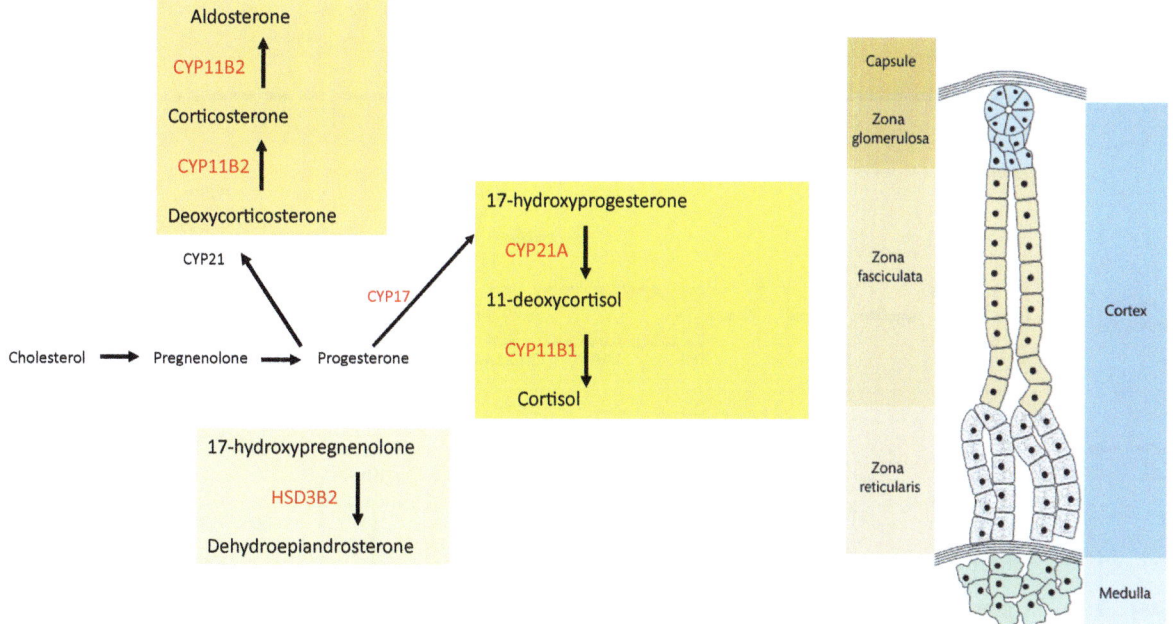

FIG. 3.3.2.11 Biosynthesis of aldosterone in the *zona glomerulosa*, cortisol in the *zona fasciculata* and DHEA in the zona reticularis of the adrenal cortex. *(Author original.)*

mutations from *CYP21A1P*. Over 200 pseudogene-independent mutations are listed in the Human Gene Mutation Database (HGMD, http://www.hgmd.cf.ac.uk) and the Pharmacogene Variation Consortium (https://www.pharmvar.org/gene/CYP21A2). Due to founder effects increased frequencies of some pseudogene-independent mutations are observed in some populations. Deletions, the splice site mutation in intron 2 (c.293-13A/C>G) and p.Ile172Asn (I172N) are the most common

mutations in most populations (Fig. 3.3.2.13). Novel or rare mutations in single families or small populations account for about 3%–5% of detected mutations in large cohorts.

Between 1% and 2% of *CYP21A2* disease causing mutations arise de novo.

The molecular genetic description of CYP21 deficiency allows genotype-phenotype correlations to be made. The gene locus is complex and high throughput techniques of massive gene sequencing are not applicable (Arriba and

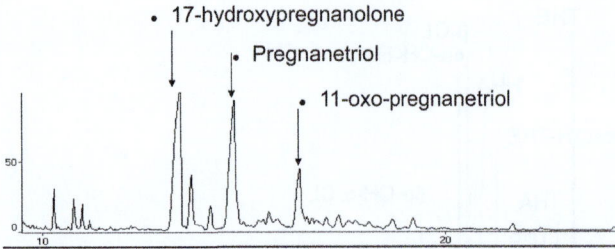

FIG. 3.3.2.12 GC-MS profile of urine steroids after deconjugation and formation of MO-TMS derivatives from patient with CAH due to 21-hydroxylase deficiency. Three steroids are in gross excess compared what is seen in a normal subject where most steroids would be at similar level to the internal standards. *(Author original.)*

Ezquieta, 2022). The disease has been classified clinically according to age of onset and symptoms, thus severe disease presenting in the newborn period with virilization of the external genitalia (females) and salt losing crises (both sexes) has been described as the salt wasting form of classical CAH. Presentation may be slighter earlier in the salt-wasting form than in the simple virilizing form, severe enough to lead to problems initially.

3.3.2.4.3.2 3β-Hydroxysteroid dehydrogenase-isomerase (3βHSD) deficiency

This membrane bound enzyme uses NAD^+ as a cofactor to catalyze the conversion of the Δ5-hydroxysteroids to the Δ4-3-oxo steroids in both the adrenals and gonads and is the only non-P450 enzyme involved in glucocorticoid and mineralocorticoid biosynthesis. The phenotypic expression of the disease is variable ranging from early death as a result of severe salt loss, to ambiguous genitalia with hypospadias in males and clitoral enlargement and labial fusion in females. Markedly elevated plasma DHEA, 17-hydroxypregnenolone and ACTH levels typify the disorder (Guran et al., 2020). The 17-hydroxypregnenolone to cortisol ratio by LC-MS/MS in 31 patients was 1000 or more (median 2.28 IQR 0.66–9.5 vs. 0.002, IQR 0.001–0.007). Some patients also have apparent raised 17-hydroxyprogesterone levels by immunoassay which had confused the diagnosis with that of CYP21 deficiency. This may be an artifact of the assay or due to reduction of 17-hydroxypregnenolone by an HSD3B isoform. A urine steroid profile will show metabolites of DHEA and

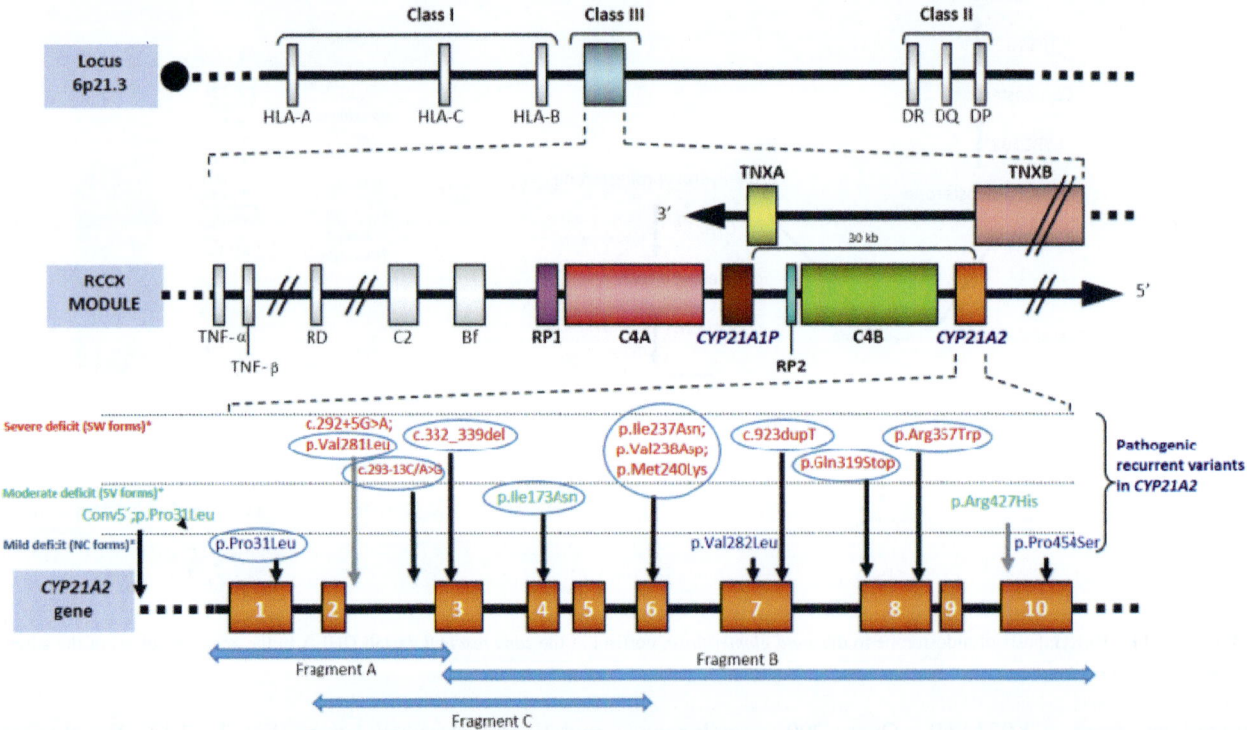

FIG. 3.3.2.13 Scheme of the RCCX module located on the short arm of chromosome 6 within the HLA class III region. Tandem duplication affects CYP21 and C4 genes. In humans, only CYP21A2 gives rise to the functional protein, whereas CYP21P is a homologous pseudogene that includes several inactivating point variants that can be transferred to the active gene by small gene conversion events. Both C4A and C4B are functional. Tenascin, also duplicated, is encoded in the complementary chain. The bottom of the image shows the recurrent variants grouped according to how they affect the enzymatic functionality: severely (*red*), moderately severe (*green*) or mildly (*blue*). Recurrent variants in all populations are *circled*. The complete nomenclature of each variant including the cDNA position (NM_000500.9) would be: c.92C>T [p.Pro31Leu], c.292+5C>A, c.293-13C>G, c.332-339del, c.518T>A [p.Ile173Asn], c.(710T>A; 713T>A; 719T>A) p.[Ile237Asn; Val238Glu; Met240Lys], c.844G>T [p.Val282Leu], c.923dupT, c.955C>T [p.Gln319*], c.1069C>T [p.Arg357Trp], c.1280G>A [p.Arg427His] and c.1360C>T [p.Pro454Ser]. The *arrows* "Fragment A," "Fragment B" and "Fragment C" represents specific amplicons for CYP21A2 amplification. *(From Arriba M, Ezquieta B. Molecular diagnosis of steroid 21-hydroxylase deficiency: a practical approach. Front Endocrinol (Lausanne) 2022;13:834549. Fig. 1 p. 3.)*

A

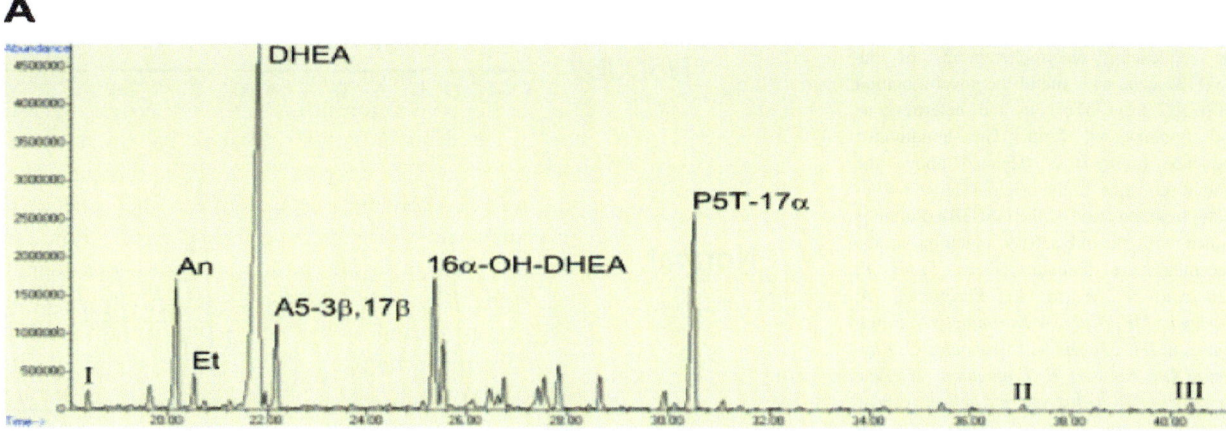

B

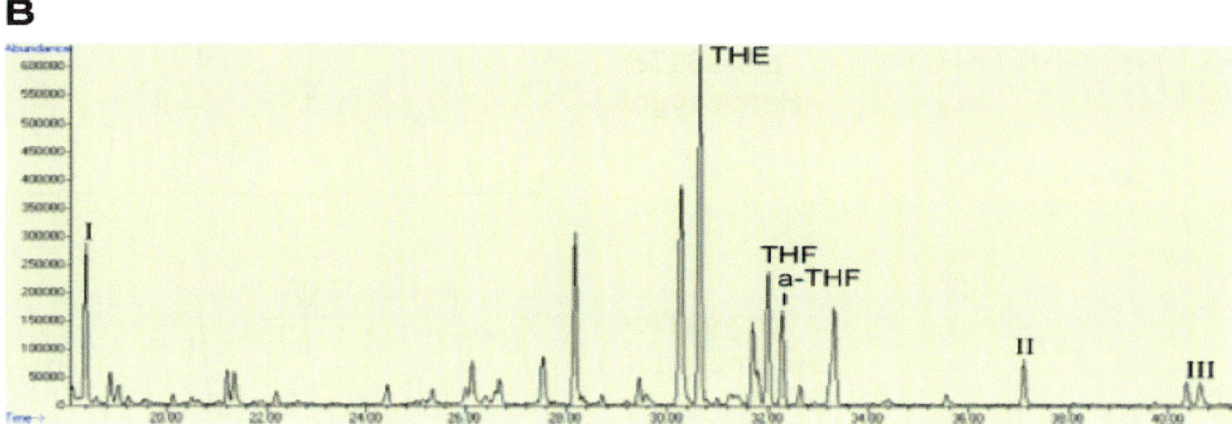

FIG. 3.3.2.14 Scans of the gas chromatography-mass spectrometry (GC-MS) steroid metabolome analysis of the patient with defect of HSD3B2 (A) as well as a healthy control (B). I, II, II: internal standards (5α-androstane-3α,17α-diol, stigmasterol, and cholesterylbutyrate); An (androsterone, 5α-androstane-3α-ol17-one), Et (etiocholanolone, 5β-androstane-3α-ol-17-one), DHEA (dehydroepiandrosterone, 5-androstene-3β-ol-17-one), A5-3β,17β (androstenediol, 5-androstene-3β,17β-diol), 16α-OH-DHEA (5-androstene-3β,16α-diol-17-one), P5T-17α (pregnenetriol, 5-pregnene3β,17α,20α-triol). Cortisol metabolites: THE (tetrahydrocortisone, 5β-pregnane-3α,17α,21-triol-11,20-dione), THF (tetrahydrocortisol, 5β-pregnane-3α,11β,17α,21-tetrol-20-one) and a-THF (5α-tetrahydrocortisol, 5α-pregnane-3α,11β,17α,21-tetrol-20-one). *(From Fanis P, Neocleous V, Kosta K, Karipiadou A, Hartmann MF, Wudy SA, Karantaglis N, Papadimitriou DT, Skordis N, Tsikopoulos G, Phylactou LA, Roilides E, Papagianni M. Late diagnosis of 3β-hydroxysteroid dehydrogenase deficiency: the pivotal role of gas chromatography-mass spectrometry urinary steroid metabolome analysis and a novel homozygous nonsense mutation in the HSD3B2 gene. J Pediatr Endocrinol Metab 2020;34(1):131–6. Fig. 1 P.)*

17-hydroxypregnenolone (Fanis et al., 2020) (Fig. 3.3.2.14).

The molecular basis of 3β-HSD deficiency has been defined in a number of patients, in all cases mutations occurred in HSD3B2. Three infants from Afghanistan and Pakistan were found to be homozygous for a frame shift mutation caused by a deletion of 2 adenosine residues at codon 273 and a resulting termination codon at 279 (Simard et al., 1994). Polymorphism analysis showed that they all shared a common haplotype suggesting a possible founder effect. Mutations throughout the HSD3B2 gene have been reported. To date, at least 50 mutations (including frameshift, nonsense, in-frame deletion, splicing, and missense mutations) have been identified in the HSD3B2 gene (Fig. 3.3.2.15). The findings in 82 individuals from 66 families suffering from classical 3βHSD deficiency were summarized in a 2018 review (Baquedano et al., 2018). Mutations in a further 31 patients have been reported (Guran et al., 2020). Missense mutations are associated with some residual HSD3b2 activity and non-SW phenotype (Al Alawi et al., 2019).

3.3.2.4.3.3 Apparent cholesterol side chain cleavage enzyme (CYP11A) deficiency

This is the most severe form of CAH and is called congenital lipoid adrenal hyperplasia because of the large, cholesterol ester laden appearance of the adrenals at autopsy. The disorder is often lethal and is distinguished by insignificant circulatory steroids and low urinary steroid excretion rates, high basal concentrations of ACTH and high plasma renin activity but low aldosterone concentrations. Enzymological

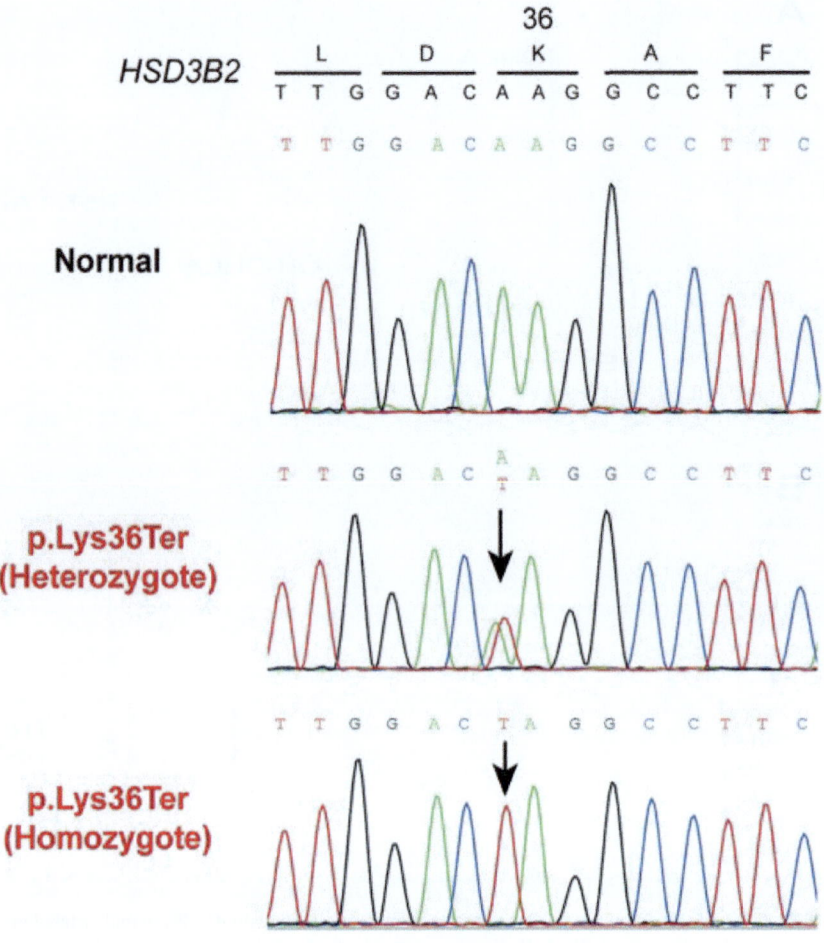

FIG. 3.3.2.15 Identification of a novel HSD3B2 mutation. The figure shows part of the sequencing electropherograms of the HSD3B2 gene with site of the novel mutation (HSD3B2:p.Lys36Ter) in the heterozygous and homozygous form. The nonmutated sequence (normal) is depicted above the mutant sequence. In the original figure a schematic representation of the HSD3B2 protein is shown with the p.lys36Ter mutation in the transmembrane domain. *(From Fanis P, Neocleous V, Kosta K, Karipiadou A, Hartmann MF, Wudy SA, Karantaglis N, Papadimitriou DT, Skordis N, Tsikopoulos G, Phylactou LA, Roilides E, Papagianni M. Late diagnosis of 3β-hydroxysteroid dehydrogenase deficiency: the pivotal role of gas chromatography-mass spectrometry urinary steroid metabolome analysis and a novel homozygous nonsense mutation in the HSD3B2 gene. J Pediatr Endocrinol Metab 2020;34 (1):131–6. Fig. 2 p. 135.)*

studies have shown a decreased conversion of cholesterol to pregnenolone in adrenal and testicular tissue in vitro. CYP11A deficiency is a rare genetic disease causing primary adrenal insufficiency with or without a 46XY difference in sexual development (Maharaj et al., 2018). Steroids will be absent in blood and urine so inevitably genetic testing is needed.

The P450scc enzyme has a key role in the initial steps of steroidogenesis by catalyzing the conversion of cholesterol to pregnenolone in steroidogenic tissues and is encoded by the CYP11A1 gene, localized on chromosome 15q23-q24 (Miller and Auchus, 2011). P450scc deficiency (OMIM 613743) leads to reduced production of gonadal and adrenal steroids. Complete or partial adrenal insufficiency leads to a wide range of clinical manifestations. Less than 40 patients with P450scc deficiency have been reported. The deficiency presents with severe early onset adrenal insufficiency in the neonatal period of an individual with female external genitalia in 46,XY individuals but partial defects result in late-onset adrenal insufficiency or glucocorticoid insufficiency alone, associated with either normal genitalia or variable degrees of poor androgenization. Compound heterozygous

variants in CYP11A1 involving rs6161 (c.940G.A; p. Glu314Lys) are surprisingly common and altered splicing should be considered when predicted benign or very rare synonymous changes are found (Maharaj et al., 2019) (Fig. 3.3.2.16).

3.3.2.4.3.4 StAR defect

Congenital lipoid adrenal hyperplasia (LCAH), because of a mutation of the STAR protein (*Steroidogenic Acute Regulatory protein*), often leads to a rapid death of female phenotypic XY subjects. This condition might be confused with effects of SF-1 mutation. Steroid production is low so ACTH and renin are elevated. Salt wasting is a problem in addition to the effects of cortisol deficiency and later of sex steroids. Patients present with low body weight, vomiting, dehydration and skin pigmentation. The disorder is common in Japan, Korea and Palestinian Arab populations. Young men and women with hypogonadism should be investigated for an Arg188Cys mutation that reduced their gonadal function and infertility might occur during puberty or in early adulthood. Patients with early onset

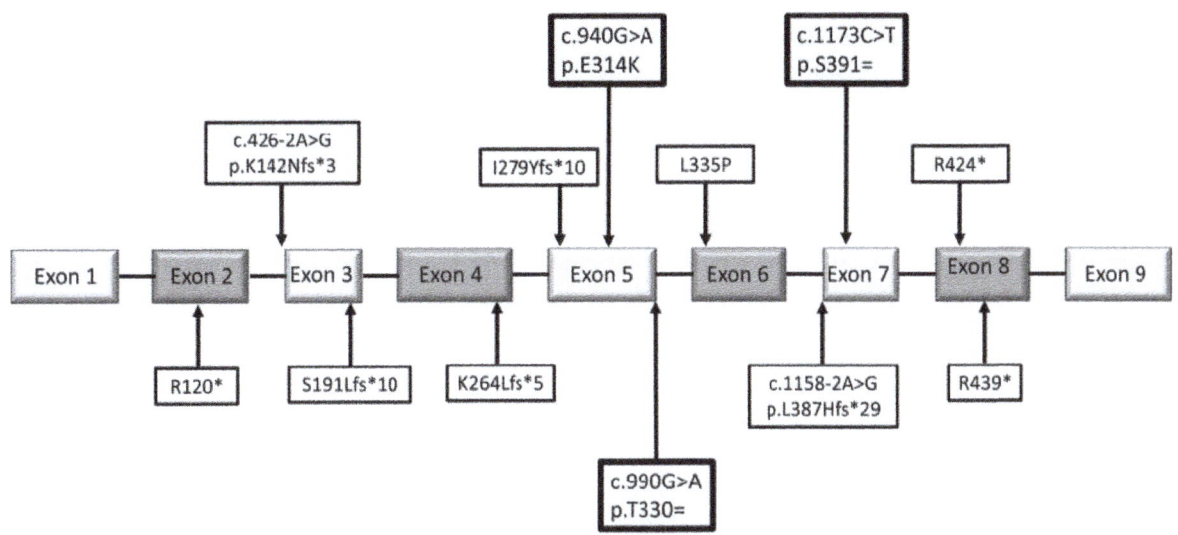

FIG. 3.3.2.16 Mutations CYP11A1 gene for side chain cleavage. Position of variants in CYP11A1 genomic/pre-mRNA sequence found a series of patients with PAI. Boxed in *bold*, the three predicted benign or synonymous variants assessed for their effect on splicing; SNP rs6161 (c.940G.A, p. Glu314Lys) is 110 bp from the start and 51 bp from the end of exon 5, c.990G.A (p.Thr330 =) occurs at the last base of exon 5, and c.1173C.T (p. Ser391 =) is 16 bp from the start of exon 7. *(From Maharaj A, Buonocore F, Meimaridou E, Ruiz-Babot G, Guasti L, Peng HM, Capper CP, Burgos-Tirado N, Prasad R, Hughes CR, Maudhoo A, Crowne E, Cheetham TD, Brain CE, Suntharalingham JP, Striglioni N, Yuksel B, Gurbuz F, Gupta S, Lindsay R, Couch R, Spoudeas HA, Guran T, Johnson S, Fowler DJ, Conwell LS, McInerney-Leo AM, Drui D, Cariou B, Lopez-Siguero JP, Harris M, Duncan EL, Hindmarsh PC, Auchus RJ, Donaldson MD, Achermann JC, Metherell LA. Predicted benign and synonymous variants in CYP11A1 cause primary adrenal insufficiency through missplicing. J Endocr Soc 2018;3(1):201–21. Fig. 1 p. 213.)*

are classified as classical (CLCAH). The nonclassic form may present as either (a) preserved masculinization of the external genitalia to be reared as a male with XY karyotype, (b) the preserved ability of mineralocorticoid secretion or (c) the onset of PAI at 1 year of age or older (Ishii et al.,

2020). Mutations in StAR gene are found in all seven exons (Fig. 3.3.2.17).

In a comprehensive study of 50 patients with LCAH, the genetic analysis of *STAR* was conducted in 20/27 XY patients with CLCAH (74.0%), all of the 16 XX patients

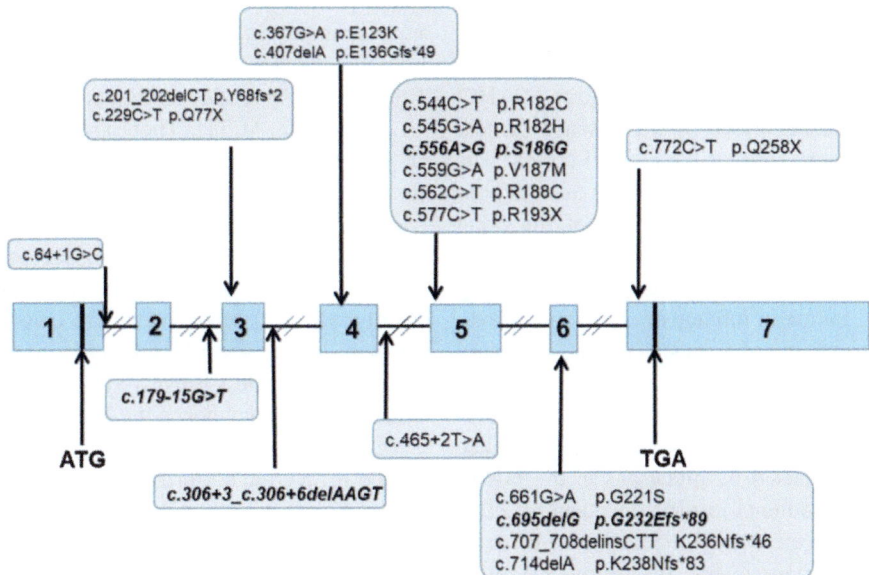

FIG. 3.3.2.17 Diagram of the STAR gene showing the location of mutations identified in patients with LCAH and novel variants are in *bold*. *(From Zhang T, Ma X, Wang J, Jia C, Wang W, Dong Z, Ye L, Sun S, Hu R, Ning G, Li C, Lu W. Clinical and molecular characterization of thirty Chinese patients with congenital lipoid adrenal hyperplasia. J Steroid Biochem Mol Biol 2021;206:105788. Fig. 1 p. 6.)*

with CLCAH, all of the 3 XY patients with NCLCAH, and the 8 XX patients with NCLCAH (Ishii et al., 2019, 2020). A complete loss-of-function variant with founder effect in East Asia, STAR-Gln258*, was identified in 31/36 genotype-known patients with CLCAH (86.1%) and 7/11 patients with NCLCAH (63.6%). Nonsense or frameshift variants in both alleles were identified only in patients with CLCAH (24/36, 66.7%) but not in those with NCLCAH (0/11, 0.0%). A partial loss-of-function variant, STAR-Arg272Cys or -Met225Thr, was not identified in patients with CLCAH but was observed in 8/11 patients with NCLCAH (72.7%). Four out of the 11 patients with NCLCAH were compound heterozygotes of STAR-Gln258* and -Arg272Cys, and 1 patient was compound heterozygote with STAR-Gln258* and -Met225Thr. A de novo pathogenic variant of STARGly22-Leu59del was already shown to have a dominant negative effect as previously reported in a patient with NCLCAH (Ishii et al., 2020). Other large studies include 30 LCAH (Zhang et al., 2021a,b), 42 patients (Kang et al., 2017), 25 cases (Kim, 2014) with quite similar findings.

3.3.2.4.3.5 Adrenal hypoplasia congenita (AHC)

When CAH due to 21-hydroxylase deficiency has been excluded all male patients should be screened for common **mutations in NR0B1 gene** that causes **DAX-1 deficiency** in **X-linked adrenal hypoplasia congenita (AHC)** (Buonocore et al., 2021; Wijaya et al., 2019).

Two forms can be identified according to the anatomy of the adrenal gland at autopsy and according to the genetic transmission mode. The primary forms of adrenal hypoplasia appear as X-linked and autosomal recessive disorders with different adrenal morphologies. The rarest form is autosomal recessive and this form is often linked with hypogonadotropic hypogonadism. In the X-linked form of AHC, the adrenal gland is characterized by the absence or near absence of the permanent zone of the adrenal cortex. The X-linked form is referred to as cytomegalic as there is poor development (aplasia) of the cortical zone of the adult adrenal gland and is filled with larger fetal adrenal-like cells (thus the cytomegalic designation). Affected boys usually present with adrenal failure in infancy or childhood. Meanwhile, hypogonadotropic hypogonadism (HHG), a primary defect in spermatogenesis, is usually associated with X-linked AHC in patients who have survived with appropriate steroid hormone replacement therapy. The X-linked form of AHC/HHG is caused by mutations in the DAX1 gene. This rare cause of adrenal insufficiency can be lethal without diagnosis then mineralocorticoid and glucocorticoid hormone replacement is for life. Adrenal insufficiency caused by the hypoplasia of the adrenal gland and hypogonadotropic hypogonadism was linked to an abnormality of DAX-1 (*Dosage sensitive sex reversal—AHC*).

Patients with this condition usually present with primary adrenal failure, salt-wasting, hyperpigmentation, failure to thrive, reduced serum cortisol and aldosterone with increased plasma adrenocorticotropic hormone (ACTH) and renin. The majority of patients present with salt loss in the first 2 months of their life, and the remainder present later in childhood when as well as adrenal insufficiency, hypogonadotropic hypogonadism (HHG) is frequent in X-linked AHC. In adolescence, normal pubertal development is absent. The condition is lethal if not untreated with steroid hormone replacement therapy (Suntharalingham et al., 2015).

X-linked AHC was mapped to Xp21 and **the nuclear receptor subfamily 0, group B, member 1 (NR0B1)** was found as the gene responsible for X-linked AHC and HHG. DAX1 is an orphan member of the nuclear receptor superfamily. The NR0B1 gene consists of two exons and encodes a 470-amino acid protein. The carboxy-terminal domain of DAX-1 (CTD) is homologous to the ligand-binding domain (LBD) of other nuclear receptors, whereas the amino-terminal domain (NTD), with no typical zinc finger DNA-binding motif, is composed of three short repeats, each containing an LXXLL motif. DAX1 is predominantly expressed in the adrenal cortex, gonads, hypothalamus and anterior pituitary. Functional studies suggested that DAX1 is a repressor of gene transcription, acting in part by **inhibiting the activity of another orphan nuclear receptor, steroidogenic factor 1 (SF-1)**, encoded by the NR5A1 gene. The disorder has been called **dosage-sensitive sex-reversal, adrenal hypoplasia congenital critical region on the X chromosome, protein 1 (DAX1)**.

In the majority of patients, first signs appear in the neonatal period, with repeated vomiting and dehydration, accompanied by hyperpigmentation and hypoglycemic convulsions. The diagnosis of adrenal insufficiency is confirmed by a mineralo- and gluco-corticoid deficiency with an increase in ACTH. There are associations of unilateral or bilateral cryptorchidism with X-linked AHC. The age distribution of the diagnosis of X-linked AHC is bimodal. Most of the patients are diagnosed during the neonatal period. Other patients are diagnosed between the age of 2 and 9 years old, which is not explained by a different genotype. Aldosterone needs are constant throughout life but with a peak at birth; subjects who survive the neonatal period could be less sensitive to a mineralocorticoid deficit throughout infancy hence the later diagnosis and treatment.

The diagnosis of **X-linked AHC** might also be suggested for a child suffering from unlabeled adrenal insufficiency whose puberty does not occur at the normal age. The significant variability of onset of illness, symptoms, serious clinical expression, hypogonadotropic hypogonadism has often been noticed with no correlation with the genetic abnormality. The diagnosis of AHC is suggested because of the association between congenital adrenal insufficiency

DAX-1/NR0B1

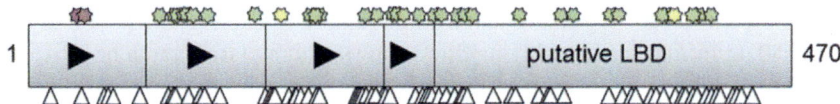

FIG. 3.3.2.18 Schematic of DAX-1 structure and a selection of the nonsense (*stars*), frameshift (*triangles*) and missense changes reported. *(From Suntharalingham JP, Buonocore F, Duncan AJ, Achermann JC. DAX-1 (NR0B1) and steroidogenic factor-1 (SF-1, NR5A1) in human disease. Best Pract Res Clin Endocrinol Metab 2015;29(4):607–19. Fig. 1 p. 609.)*

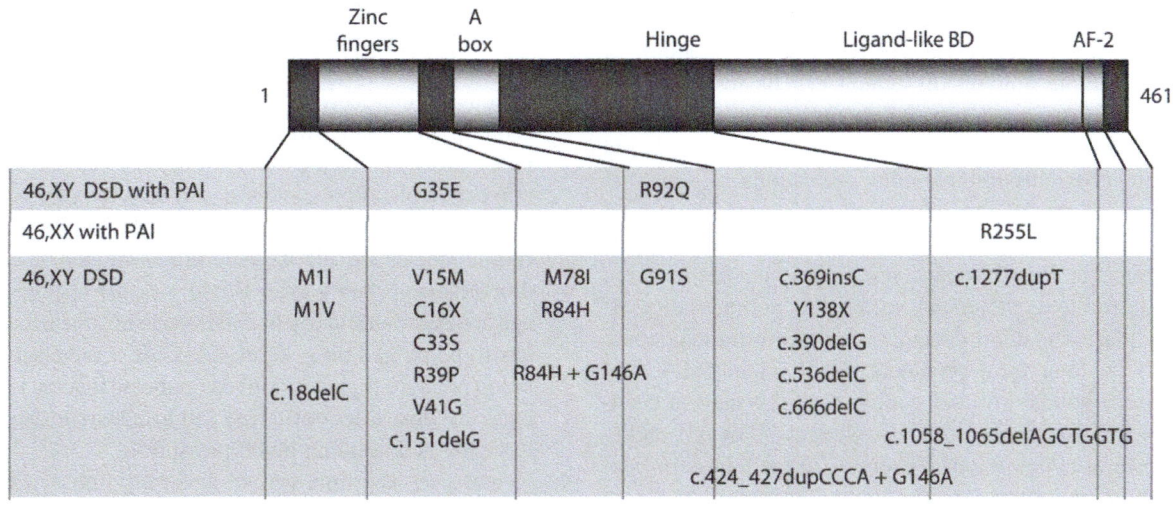

	Zinc fingers	A box	Hinge		Ligand-like BD	AF-2
46,XY DSD with PAI		G35E		R92Q		
46,XX with PAI					R255L	
46,XY DSD	M1I M1V c.18delC	V15M C16X C33S R39P V41G c.151delG	M78I R84H R84H + G146A	G91S	c.369insC Y138X c.390delG c.536delC c.666delC	c.1277dupT c.1058_1065delAGCTGGTG
				c.424_427dupCCCA + G146A		

FIG. 3.3.2.19 SF-1 structure and a selection of the changes reported in patients with adrenal insufficiency and/or reproductive phenotypes. Nonsense, frameshift and missense changes are shown. *BD*, binding domain; *DSD*, disorder of sex development; *PAI*, primary adrenal insufficiency. Data for primary ovarian insufficiency (POI) not shown here. *(From El-Khairi R, Martinez-Aguayo A, Ferraz-de-Souza B, Lin L, Achermann JC. Role of DAX-1 (NR0B1) and steroidogenic factor-1 (NR5A1) in human adrenal function. Endocr Dev 2011;20:38. Fig. 2 p. 42.)*

and delayed puberty with hypogonadotropic hypogonadism. The family history may reveal a relative who died of unspecified cause after birth probably (in retrospect) of acute adrenal insufficiency within the context of an X-linked congenital hypoplasia and transmission through the female line.

The gene causing this condition was first located in Xp21 hence the possible link with other pathologies (e.g., glycerol kinase deficiency, Duchenne muscular dystrophy, ornithine transcarbamylase deficiency and mental retardation). It is therefore necessary to rule out any of those diseases in every X-linked AHC affected boy through measurements in blood of CPK, glycerol, triglycerides (pseudo hypertriglyceridemia in glycerol kinase deficiency) and a mental assessment. One boy presented with a variant form of AHC associated with a W105C mutation in the amino terminus of DAX-1 and mineralocorticoid deficiency (Suntharalingham et al., 2015; Verrijn Stuart et al., 2007) (Fig. 3.3.2.18).

On expression, the protein exhibited mild loss of function for gene it represses and for genes it activates. The missense mutation may impair protein function. The mechanism for DAX-l function is not well known but there is an interaction with SF-1 (*Steroidogenic factor 1*) as the two molecules bind together. Treatment with

gonadotrophins does not restore spermatogenesis (Jadhav et al., 2011).

SF-1 is also an orphan nuclear receptor (NR5A1) that regulates the expression of steroidogenic enzymes and sexual determination proteins (especially SRY—*Sex determination Region of the Y chromosome*). The disruption of SF-1 gene in mice leads to an animal of female phenotype without gonads and adrenal glands. DAX-1 could therefore repress SF-1-mediated transactivation. Nonsense, frameshift and missense changes are found (Fig. 3.3.2.19). Girls present with ovarian insufficiency.

3.3.2.5 Treatment of CAH

Since cortisol production is compromised in CAH, treatment with hydrocortisone is essential but presents many problems as discussed in Chapter 3.1.2 on Hypocortisolism. Measurements of 17-hydroxyprogesterone (in the case of CYP21) are used to monitor the impact of glucocorticoid on steroids in excess due to the defective enzyme. Suppression of male sex steroids (androstenedione or DHEAS) may also be required and this is discussed in more detail in Chapter 3.2.1 on Hypergonadism. A 24h cortisol profile is needed to ensure long periods of hypocortisolism or hypercortisolism are minimized.

Replacement of mineralocorticoid is needed in cases with SWCAH and it is appropriate in this chapter on hypoaldosteronism to discuss treatment of salt loss. A typical dose of fludrocortisone is $100\,\mu g/m^2/day$ but $150\,\mu g/m^2/day$ because renal tubules are less mature than later in childhood. There are several ways to monitor replacement therapy:

1. Plasma electrolyte measurements. The concentration of sodium in blood can be easily measured and is a very useful measure after a salt-wasting crisis but it does not give a good overview of the total body sodium. A considerable amount of sodium can be lost before it shows in a blood test. Measurement of the plasma sodium coupled with the plasma renin activity or plasma renin concentration provides indications of the total body sodium. In situations where fludrocortisone is used because of an absence of aldosterone, if the dosing is correct then the plasma renin activity or plasma renin concentration should be within the normal range. If the plasma sodium concentration is not normal, there will be long term problems. Firstly, the total blood volume will be low and symptoms of this may be headaches or dizziness particularly when standing up quickly (postural hypotension). Secondly, in the long-term low sodium concentrations (less than 135 mmol/L) lead to osteoporosis. If the fludrocortisone dose is too high, the sodium will move toward the upper end of the normal range but will not go over 145 mmol/L, as the body will always compensate by retaining water which will have the effect of increasing blood pressure. However, high plasma sodium will be accompanied by low plasma potassium which is not good for heart function or blood pressure. In addition to measuring plasma sodium, the plasma potassium can be measured. These two electrolytes tend to go in opposite directions, if the fludrocortisone dose is too low, sodium will be lost in the urine and the plasma potassium will be high, as is seen in a salt-wasting crisis.

2. Plasma renin activity (PRA) or plasma renin concentration (PRC). Plasma renin and PRA need to be measured carefully. Blood samples are taken usually with the patient in the resting position after lying down for a period of 2–3 h and then after standing or being active for 2–3 h. These measurements give important information because plasma renin or PRA tends to increase slightly in the standing position, in order to cope with the change in blood volume resulting from the upright position. When patients are treated with fludrocortisone, the resting and the standing plasma renin or PRA within the normal range (less than 10 ng/mL/h or 7.7 nmol/L/h for PRA). A high value might indicate under replacement with fludrocortisone and the blood volume is low which often leads to postural hypotension. A drop in systolic blood pressure on standing of more than

20 mmHg, or in diastolic blood pressure of more than 10 mmHg, is considered abnormal. If taking fludrocortisone and a blood sample is taken after a period of rest and lying down shows a high plasma renin or PRA, the dose needs increasing. If the lying down value is normal but the standing up value is high, then an increased dose is needed, as this indicates the blood volume is lower than normal, as when standing up the blood pressure falls further. The low blood volume increases renin signaling to the RAAS.

3. Blood pressure. This is more of a long-term measure of effective fludrocortisone dosage. The drug retains sodium, and water will also be retained which leads to an increase in blood pressure. If renin is within the normal range then blood pressure will also be within the normal range for height. If over time the blood pressure starts to rise into the upper half of the normal range, too much fludrocortisone is being used. Blood pressure changes with age and size particularly in children, so any measurement has to be plotted on a set of standards. Conversely low blood pressure (hypotension) can indicate that the fludrocortisone dose is too low. This will be apparent also from the lying and standing blood pressures.

4. Hematocrit. The final way of assessing fludrocortisone replacement is to measure the hematocrit which is the ratio of total red blood cell volume and the total blood volume and is a guide to how the circulating blood volume is maintained. If the hematocrit is raised, this suggests that the circulating blood volume is reduced and often indicates that more fludrocortisone is needed. Hematocrit is really useful for fine tuning treatment as it can show changes before, they manifest in terms of blood pressure or renin/PRA.

Fludrocortisone is a potent glucocorticoid and if used in excess of mineralocorticoid requirement will lead to hyperglycemia, weight gain, hypertension and osteoporosis.

Abiraterone is an inhibitor of CYP17 and is used in CAH to reduce androgen production. Cyproterone acetate or flutamide blocks the action of androgens. Drugs are now available to block the action of CRF at the CRF1 receptor or ACTH with monoclonal antibodies and receptor antagonists and these are now in use in patients with CAH (Mallappa and Merke, 2022; Prete et al., 2021).

3.3.2.6 Secondary hypoaldosteronism

When renin production is suppressed then aldosterone synthesis will be low. Renin is suppressed when a mineralocorticoid such as 11-deoxycorticosterone is produced in excess. Two defects of cortisol synthesis causing congenital adrenal hyperplasia are associated with hyporeninemic hypoaldosteronism, this scenario characterizes secondary hypoaldosteronism (Fig. 3.3.2.20).

Growth chart.
CYP11B1
defect

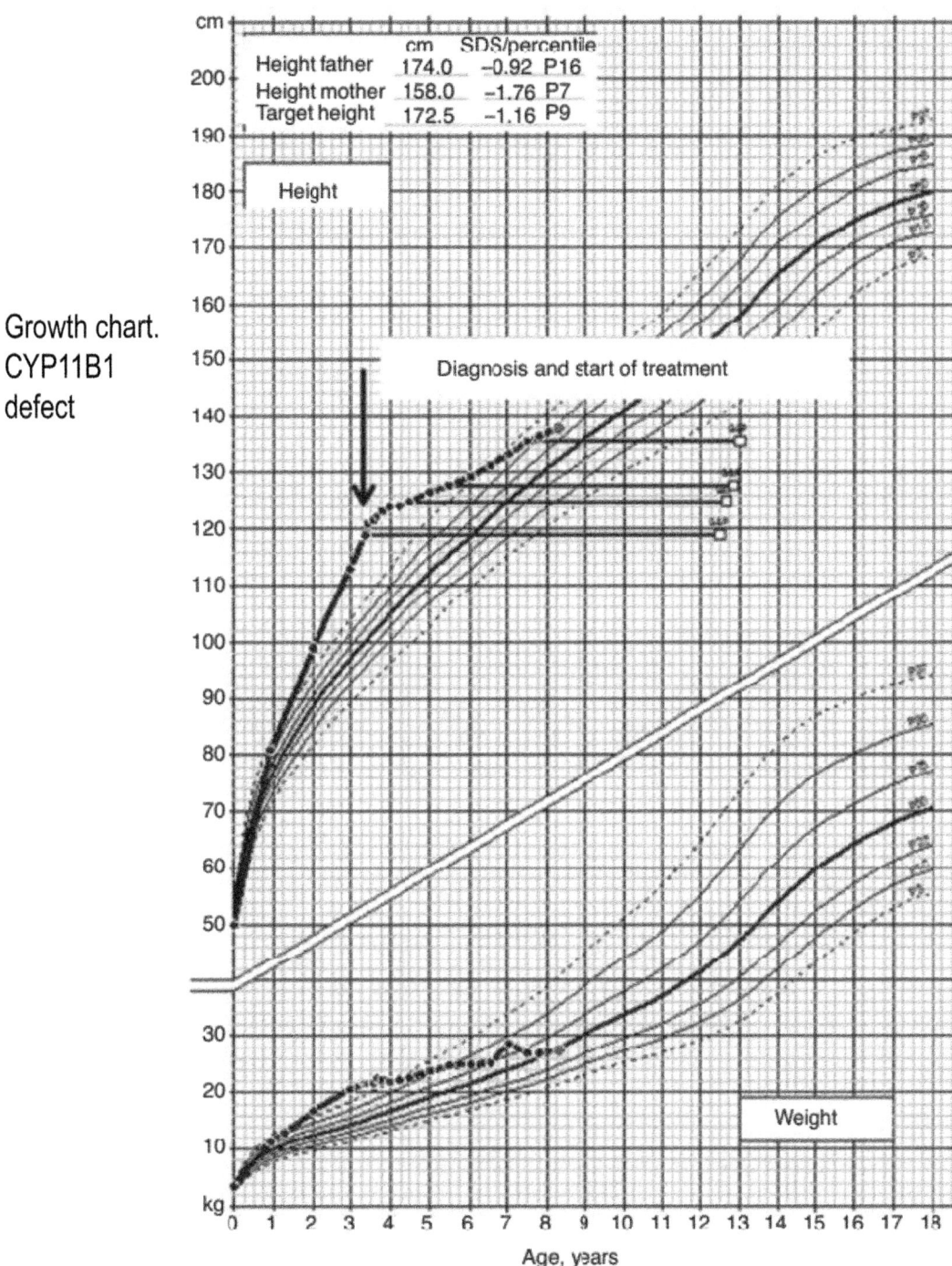

	cm	SDS/percentile
Height father	174.0	−0.92 P16
Height mother	158.0	−1.76 P7
Target height	172.5	−1.16 P9

Height

Diagnosis and start of treatment

Weight

Age, years

FIG. 3.3.2.20 Height and weight chart of boy with 11-hydroxylase deficiency before and after treatment plotted with German references. Repetitive bone ages (Greulich and Pyle) are depicted as arrows from distinct chronological ages to illustrate bone age acceleration by indicating height for bone ages. *SDS*, standard deviation score. Growth chart. *(From Breil T, Yakovenko V, Inta I, Choukair D, Klose D, Mittnacht J, Schulze E, Alrajab A, Grulich-Henn J, Bettendorf M. Typical characteristics of children with congenital adrenal hyperplasia due to 11β-hydroxylase deficiency: a single-centre experience and review of the literature. J Peditr Endocrinol Metab 2019;32:259. Fig. 1 p. 263.)*

3.3.2.6.1 Defect of 11-hydroxylase (CYP11B1)

When cortisol synthesis is blocked due to 11-hydroxylase deficiency there is increased production of sex steroids and 11-deoxycorticosterone (DOC) (Alzahrani et al., 2017; Khattab et al., 2017). Salt retention due to DOC leads to suppression of renin production in the kidney and reduced aldosterone synthesis. This is discussed in more detail in Chapter 3.1.1 on cortisol deficiency and Chapter 3.2.1 on

REGULATION OF ALDOSTERONE IN CAH

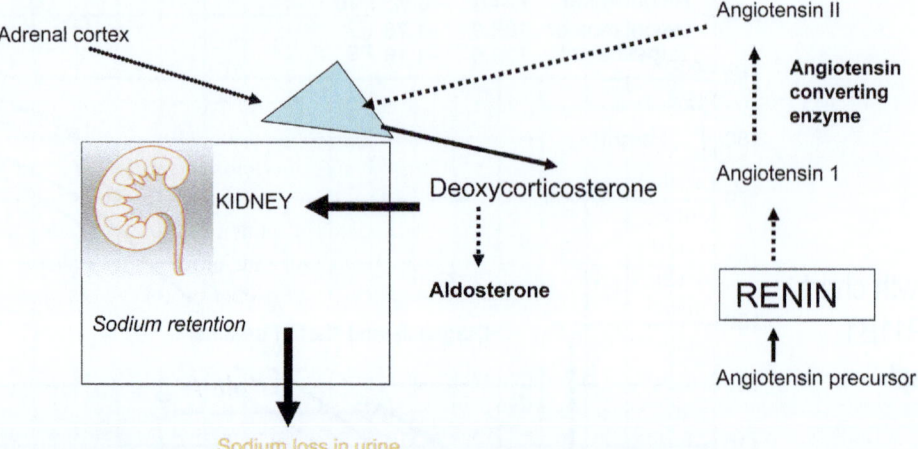

FIG. 3.3.2.21   If a defect in cortisol synthesis leads to high production of 11-deoxycorticosterone the renin angiotensin system is not activated because salt is retained in the kidney by the action of DOC.

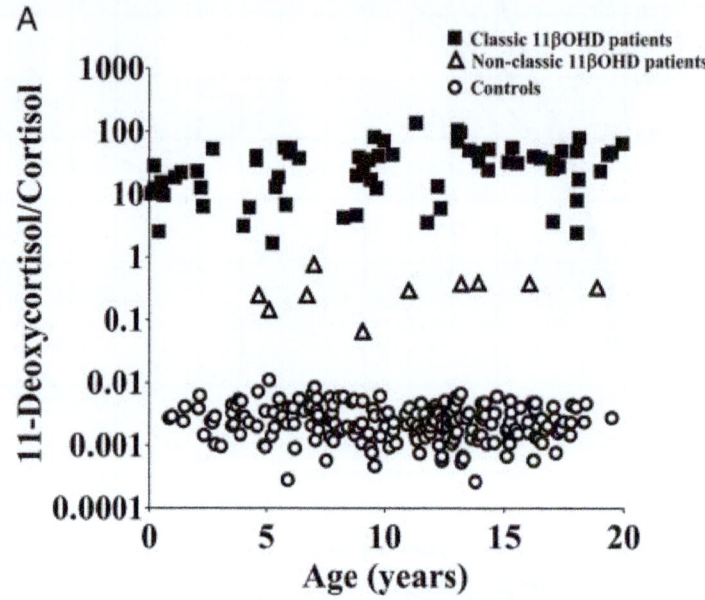

FIG. 3.3.2.22   Steroid ratio (11-deoxycortisol/cortisol) across ages in patients with classic and nonclassic 11β-hydroxylase deficiency. Data are shown as linear regressions of steroids normalized to cortisol concentrations as a function of age (in years). The ratios showed a significant difference between patients and controls. In original figure, 11-deoxycorticosterone/cortisol and androstenedione/cortisol ratios, but not 17OH-pregnenolone/cortisol were significantly different, 17OH-progesterone/cortisol, and DHEA/cortisol ratios could discriminate classic and nonclassic 11β-hydroxylase deficiency. *(From Yildiz M, Isik E, Abali ZY, Keskin M, Ozbek MN, Bas F, Ucakturk SA, Buyukinan M, Onal H, Kara C, Storbeck KH, Darendeliler F, Cayir A, Unal E, Anik A, Demirbilek H, Cetin T, Dursun F, Catli G, Turan S, Falhammar H, Baris T, Yaman A, Haklar G, Bereket A, Guran T. Clinical and hormonal profiles correlate with molecular characteristics in patients with 11β-hydroxylase deficiency. J Clin Endocrinol Metab 2021;106(9):e3714–24. Fig. 2 p. 3722.)*

Hyperandrogenism. There is acceleration of growth due to exposure to androgens (Breil et al., 2019) (Fig. 3.3.2.21).

11-Deoxycortisol is the substrate for 11-hydroxylase (CYP11B1) and this steroid in excess is a marker for the condition. A nonclassic form is distinguished from the classic form and normal by intermediate increases in concentrations of DOC and 11-deoxycortisol (Yildiz et al., 2021) (Fig. 3.3.2.22).

3.3.2.6.2   Defect of 17-hydroxylase

When cortisol synthesis is blocked due to 17-hydroxylase deficiency there is reduced production of sex steroids and an increase in synthesis of 11-deoxycorticosterone. Salt retention due to DOC leads to suppression of renin production in the kidney and reduced aldosterone synthesis (Lee et al., 2020). This is discussed in more detail in

Chapter 3.1.1 on cortisol deficiency and Chapter 3.2.2 on Hypoandrogenism.

3.3.2.6.3 Syndrome of apparent mineralocorticoid excess (AME)

A defect of HSD11B2 prevents the oxidation of cortisol to cortisone such that cortisol can act at the mineralocorticoid receptor to retain sodium at the kidney and cause hypertension. This is a feature of the condition with low PRA and low aldosterone concentration (Stewart et al., 1996) and is discussed in more detail in Chapter 3.3.1 on hyperaldosteronism due to the apparent aldosterone activity and Chapter 3.1.1 on hypercortisolism with lack of conversion to inactive cortisone. The defect is defined by increase in cortisol and reduction of cortisone in blood or the metabolites in urine (Fig. 3.3.2.23).

AME has been reported in a wide age range and fatal in 5 months old who died before results of a urine steroid analysis would have confirmed the diagnosis (Honour et al., 1983).

Patients with AME may be either homozygous or compound heterozygous carriers of HSD11B2 mutations (Fig. 3.3.2.24). Published mutations are mostly missense mutations clustering in exons 3, 4, and 5. Other genetic changes include nonsense, splicing, insertion, and deletion mutations. The identified compound heterozygous

mutations were situated in exons 2 and 5, and resulted in deletions of amino acid residues causing truncated 11βHSD2.

Classic and nonclassic forms are now distinguished (Carvajal et al., 2020) (Fig. 3.3.2.25). Nonclassic apparent mineralocorticoid excess differs from classic AME syndrome because it has a milder phenotype, is diagnosed later in life in adolescents and adults, and can be observed even in normotensive subjects. The NC-AME phenotype is mainly associated to genetic variants in HSD11B2 and epigenetic modifications that, along with a possible second hit, are able to impair the proper cortisol metabolism.

3.3.2.6.4 Triple A syndrome

The autosomal recessive form of AHC is also exceptional in clinical practice. There is a possibility of neonatal adrenal insufficiency with increased ACTH in cases of **hereditary ACTH resistance** (alacrima, Addison's and achalasia with neurological or pigmentation disorders and normal mineralocorticoid production). This genetic disease is transmitted in the autosomal recessive mode. ACTH resistance has also been seen in an Irish traveling community (O'Riordan et al., 2008) with **defects of mini chromosome maintenance deficient 4 homolog (MCM4) gene**. The MCM4 protein is important for DNA replication and genome integrity. Mutations of the gene and protein deficiency lead to

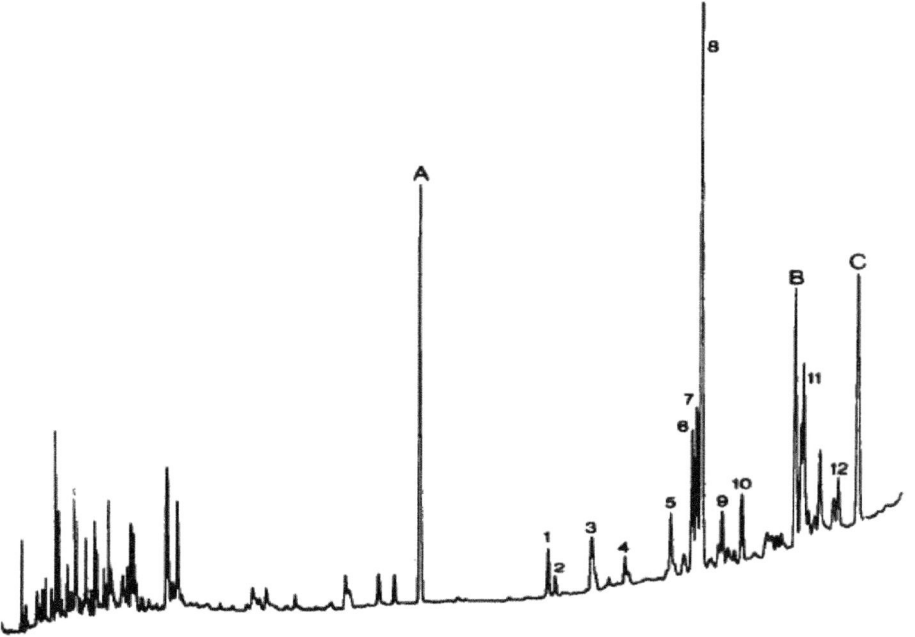

FIG. 3.3.2.23 Capillary gas chromatographic urinary steroid profile of a 4 year old patient with apparent mineralocorticoid excess. A, B, and C, Internal standards (5α-androstane-3α,17α-diol, stigmasterol, and cholesteryl butyrate, respectively). Peak 1, 11β-hydroxyandrosterone; 2, 11β-hydroxyetiocholanolone; 3, pregnanetriol; 4, tetrahydro-Substance S; 5, tetrahydrocortisone; 6, 5α-tetrahydrocorticosterone; 7, tetrahydrocortisol; 8, 5α-tetrahydrocortisol; 9, β-cortol; 10, α-cortol; 11, cortisol (doublet); 12, 20α-dihydrocortisol (doublet). The peak between 11 and 12 is 6β-hydroxycortisol. *(From Shackleton CH, Honour JW, Dillon MJ, Chantler C, Jones RW. Hypertension in a four-year-old child: gas chromatographic and mass spectrometric evidence for deficient hepatic metabolism of steroids. J Clin Endocrinol Metab 1980;50:786. Fig. 2 p. 788.)*

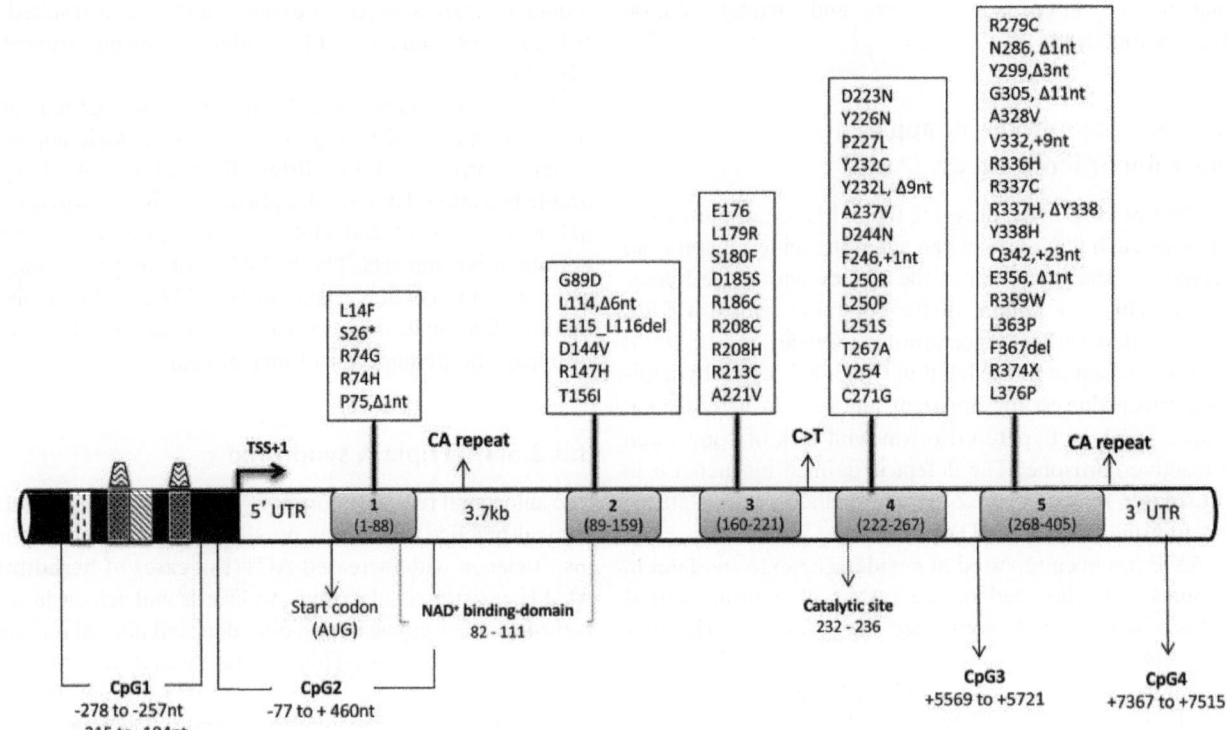

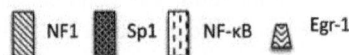

FIG. 3.3.2.24 Genetic variants in the human HSD11B2 gene. The diagram shows the human HSD11B2 gene and the location of identified pathogenic variants causing AME. The HSD11B2 gene is located on chromosome 16 and contains 5 exons, 4 CpG regions, 2 main (CA) repeat regions, and a promoter with SP1/SP3, NF1, NF-κB, and EGR-1 sites. Exons are represented by *gray boxes*. Mutations are listed relative to their position in the gene. *(From Carvajal CA, Tapia-Castillo A, Vecchiola A, Baudrand R, Fardella CE. Classic and nonclassic apparent mineralocorticoid excess syndrome. J Clin Endocrinol Metab. 2020;105(4):dgz315.)*

developmental defects including the adrenal cortex. Mineralocorticoid treatment may be needed, probably when cortisol replacement is not optimized (Flokas et al., 2019; Roucher-Boulez et al., 2018).

3.3.2.7 Defects in aldosterone action (pseudohypoaldosteronism) (PHA)

Patients with PHA present with the symptoms of hypoaldosteronism but aldosterone concentrations are not low. Renal salt wasting, hyperkalemia, hyponatremia, metabolic acidosis with markedly elevated renin and aldosterone was originally attributed to **aldosterone resistance** and confirmed with assays of electrolyte transfer by mononuclear leucocytes (Kuhnle et al., 1995; Armanini et al., 1995) but this is a research tool and not generally available for diagnosis. A normal serum cortisol excludes Addison's disease and congenital adrenal hyperplasia. Corticosterone, 18-hydroxycorticosterone and aldosterone are elevated through renin/angiotensin stimulation. The profile of steroids in urine of a patient with pseudohypoaldosteronism (PHA) shows an increase in aldosterone and 18-hydroxycorticosterone metabolites (tetrahydroaldosterone and 18-hydroxy-tetrahydro Compound A) (Fig. 3.3.2.26).

The higher excretion of both corticosterone and aldosterone metabolites particularly distinguishes PHA from a defect in aldosterone biosynthesis although PHA salt loss is due to renal tubules being refractory to aldosterone, hence the high excretion rates of tetrahydroaldosterone and 18-hydroxytetrahydro Compound A in urine.

The two distinct forms of **PHA1** have been recognized according to a **systemic** (autosomal recessive, and multiple target organs) or **renal** (autosomal dominant) basis (Riepe, 2013).

3.3.2.7.1 Autosomal dominant PHAI (defects in ENaC)

In target tissues, aldosterone binds to the MR and the complex is translocated to the nucleus where it binds to

Phenotype severity

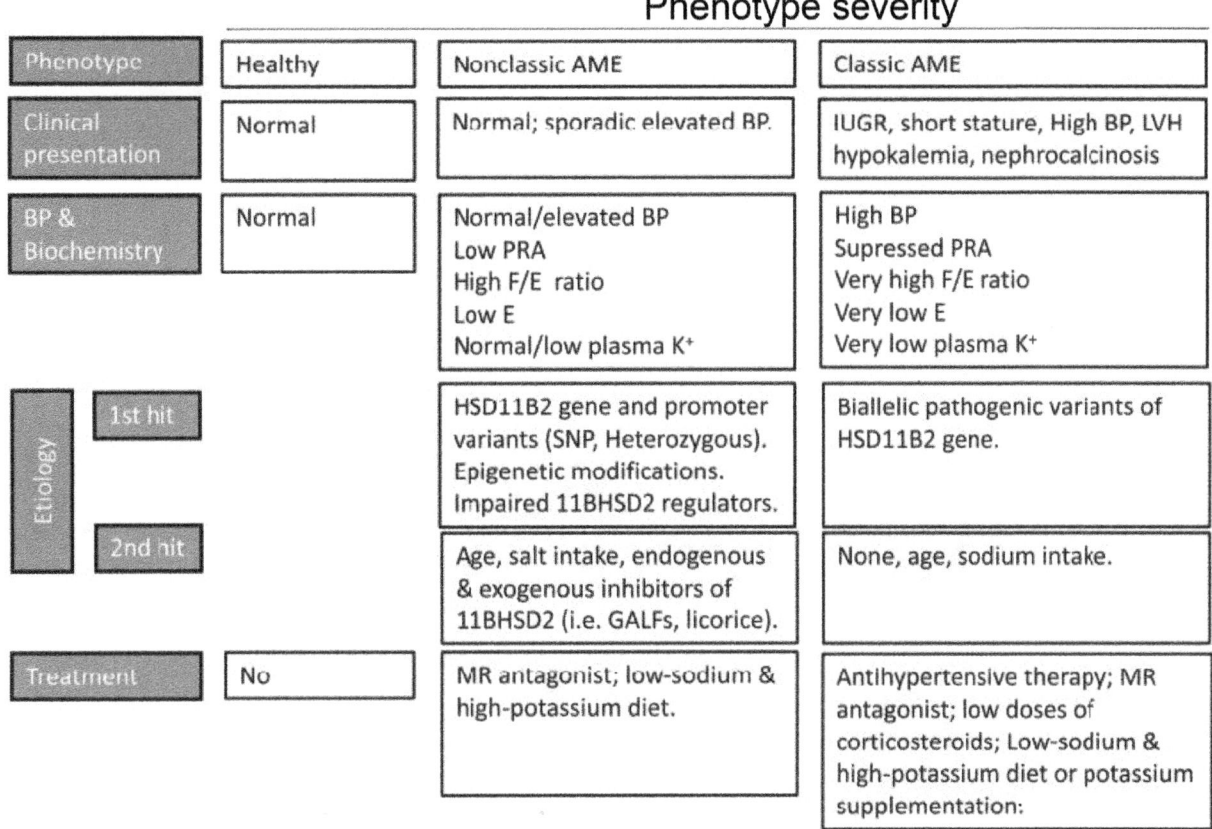

Phenotype	Healthy	Nonclassic AME	Classic AME
Clinical presentation	Normal	Normal; sporadic elevated BP.	IUGR, short stature, High BP, LVH hypokalemia, nephrocalcinosis
BP & Biochemistry	Normal	Normal/elevated BP Low PRA High F/E ratio Low E Normal/low plasma K⁺	High BP Supressed PRA Very high F/E ratio Very low E Very low plasma K⁺
Etiology — 1st hit		HSD11B2 gene and promoter variants (SNP, Heterozygous). Epigenetic modifications. Impaired 11BHSD2 regulators.	Biallelic pathogenic variants of HSD11B2 gene.
Etiology — 2nd hit		Age, salt intake, endogenous & exogenous inhibitors of 11BHSD2 (i.e. GALFs, licorice).	None, age, sodium intake.
Treatment	No	MR antagonist; low-sodium & high-potassium diet.	Antihypertensive therapy; MR antagonist; low doses of corticosteroids; Low-sodium & high-potassium diet or potassium supplementation:

FIG. 3.3.2.25 Phenotypical spectrum of AME and nonclassic AME. Classic AME and nonclassic AME are shown as a continuous spectrum. The scheme highlight the clinical and biochemical presentation of both conditions associated with first and second hits. Nonclassic apparent mineralocorticoid excess is a phenotype mainly related to an 11βHSD2 deficiency associated with genetic and epigenetic modifications affecting the HSD11B2 gene expression (first hit) and the potential additive action of endogenous or exogenous inhibitors (i.e., GALFS) (second hit). Treatment for NC-AME and AME is also shown. Abbreviations: *BP*, blood pressure; *GALFs*, glycyrrhetinic acid-like factors; *IUGR*, intrauterine growth restriction; *LVH*, left ventricular hypertrophy; *NR3C1*, human glucocorticoid receptor (GR) gene. *(From Carvajal CA, Tapia-Castillo A, Vecchiola A, Baudrand R, Fardella CE. Classic and nonclassic apparent mineralocorticoid excess syndrome. J Clin Endocrinol Metab. 2020;105(4):dgz315.)*

the regulator regions of several responsive gene with an increase in expression of the epithelial sodium transport channel (ENaC) and the Na⁺K⁺-ATPase. ENaC consists of three subunits; α, β and γ coded by SCNN1A on chromosome 12p13.13 and SCNN1B and SCNN1G on chromosome 16p12.1. Each subunit has an extracellular loop and intracellular N and C termini and two transmembrane sections.

The **epithelial sodium transporter channel (ENaC)** is comprised of three subunits alpha, beta and gamma. The channel has a critical role in sodium homeostasis in the distal nephron and late distal convoluted tubule triggered by aldosterone. ENaC allows the flow of sodium ions from the lumen across the apical cell membrane into the epithelial cell (Fig. 3.3.2.27).

Sodium ions are pumped out of the cell into the interstitial fluid by the action of sodium/potassium ATPase located in the basolateral membrane. Binding of a protein NEDD4-2 (for **neural precursor cell expressed developmentally downregulated protein 4**) to ENaC leads to

channel ubiquitylation, endocytosis and degradation (Fig. 3.3.2.28). NEDD4-2 is itself regulated by kinases of which SGK-1 is an aldosterone responsive kinase that enhances ENaC function.

3.3.2.7.1.1 Liddle's syndrome

The now more common systemic form of PHA is due to mutations of the **epithelial sodium transport channel** (ENaC-α, β-, or γ-) genes (Tetti et al., 2018; Belot et al., 2008; Hanukoglu et al., 2008). Salt loss occurs at the sweat glands, epithelium of lung and intestines in addition to the distal nephron. Potassium restriction is effective treatment (Adachi et al., 2020).

Patients with Liddle's syndrome present with early and typically moderate to severe hypertension. **Low PRA and PAC** suggest the effect of other mineralocorticoids but is a consequence of sodium retention. Transmission is in autosomal dominant manner. The pathogenesis of hypertension

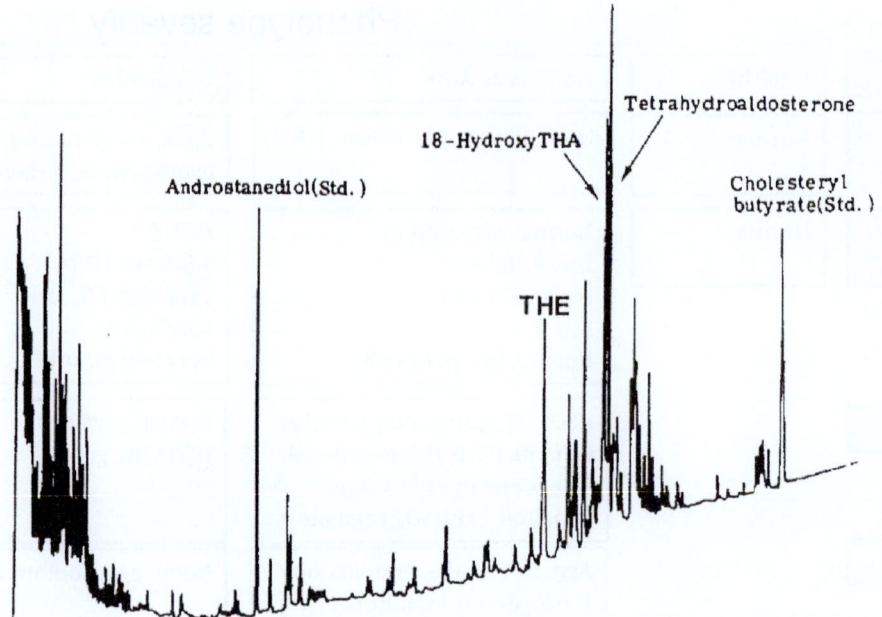

FIG. 3.3.2.26 Steroid excretion in the urine of an infant with pseudohypoaldosteronism with high excretion of metabolites of 18-hydroxycorticosterone and aldosterone. *(From Honour JW, Dillon MJ, Shackleton CH. Analysis of steroids in urine for differentiation of pseudohypoaldosteronism and aldosterone biosynthetic defect. J Clin Endocrinol Metab 1982;54(2):325–31. Fig. 2 p. 328.)*

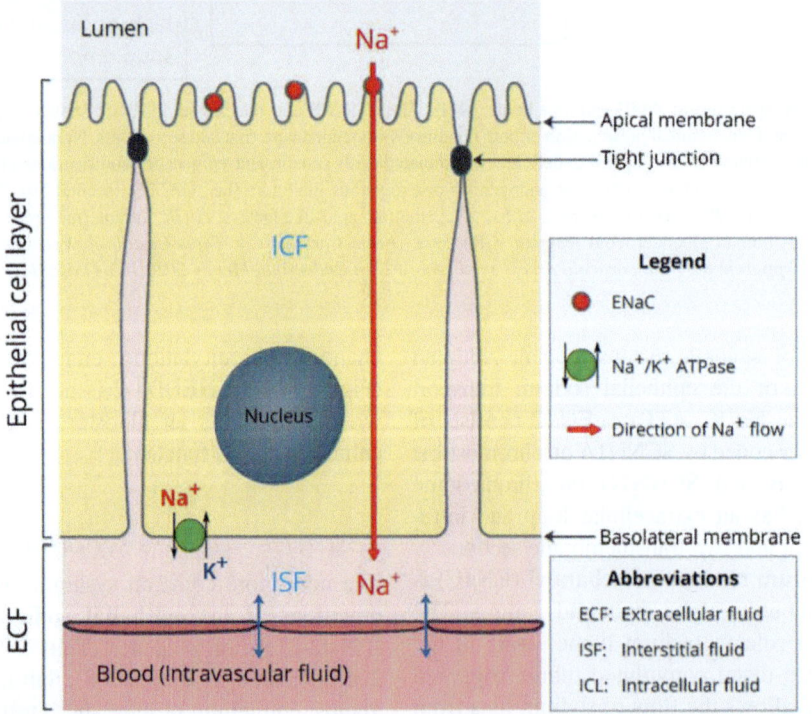

FIG. 3.3.2.27 Schematic illustration of the location and function of ENaC in epithelia. *(From Hanukoglu I, Hanukoglu A. Epithelial sodium channel (ENaC) family: phylogeny, structure-function, tissue distribution, and associated inherited diseases. Gene 2016;579(2):95–132. Fig. 1 p. 97.)*

in some cases entails increased sodium reabsorption of salt and water. Patients fail to gain weight in the first years of life (Welzel et al., 2013) (Fig. 3.3.2.29). These patients do not respond to mineralocorticoid receptor antagonists but low salt diet plus amiloride or triamterene treatments are effective.

Renal transplantation has corrected the defect in some patients supporting primary involvement of the kidneys.

The ENaC is a heteromeric protein complex made up of α-, β-, and γ-subunits. The α-subunit is mainly involved in sodium transport, while the β- and γ-subunits take part in the

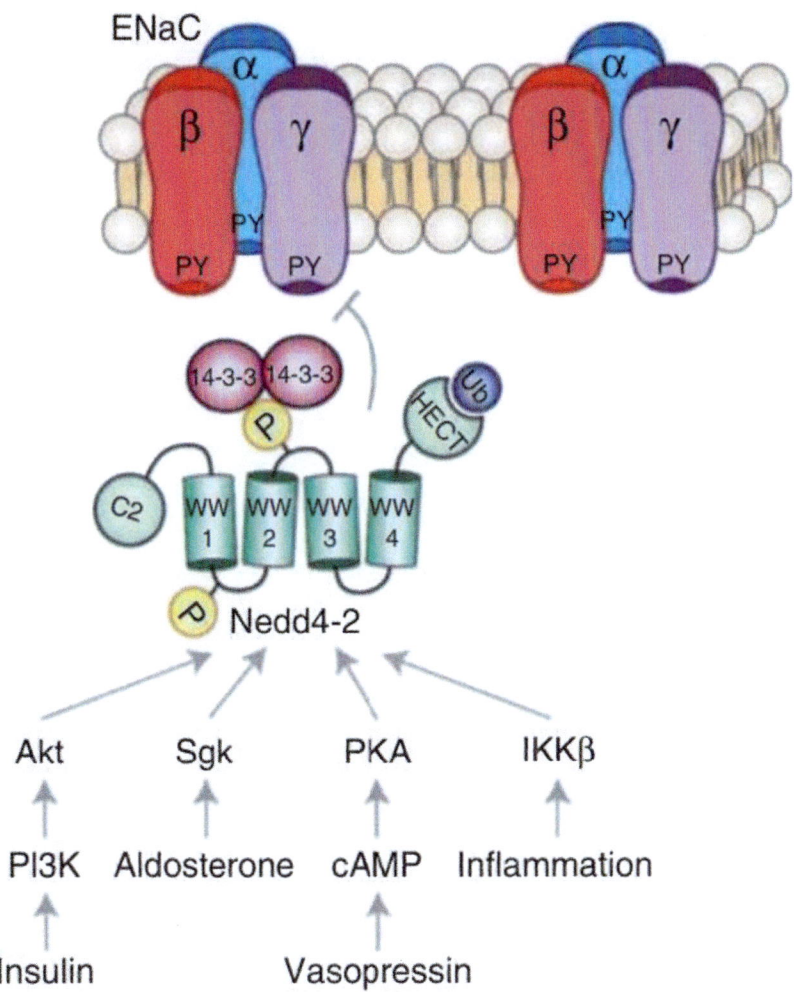

FIG. 3.3.2.28 Regulation of ENaC by NEDD4-2 and by ubiquitylation. The ubiquitin ligase NEDD4-2 (Nedd4-2) binds, via its WW domains, to the PY motifs of the ENaC subunits, leading to ENaC ubiquitylation and endocytosis, hence reduced Na^+ entry into cells. Several kinases (activated by specific hormones or inflammation) phosphorylate Nedd4-2, leading to binding of 14-3-3 proteins to the phosphorylated sites and thus impaired ability of Nedd4-2 to bind ENaC and ubiquitylate it, causing ENaC retention at the plasma membrane and elevated Na^+ influx. In Liddle's syndrome, where the PY motif of β (or γ) ENaC is mutated or deleted, the ability of Nedd4-2 to bind ENaC is impaired, resulting in channel retention at the cell surface and increased Na^+ entry into cells. *(From Rotin D, Staub O. Function and regulation of the epithelial Na^+ channel ENaC. Compr Physiol 2021;11(3):2017–45. Fig. 3 p. 2027.)*

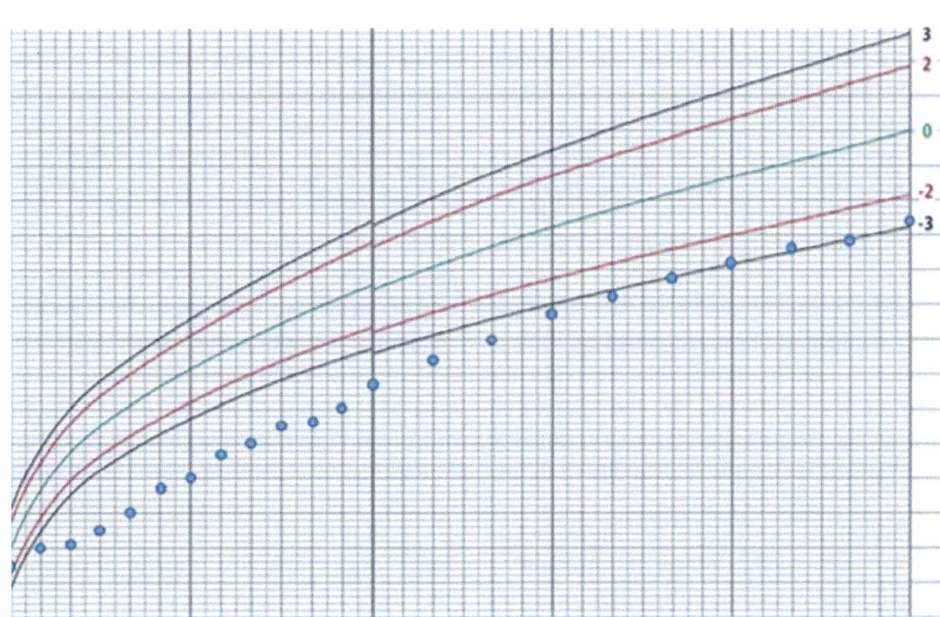

FIG. 3.3.2.29 Growth chart showing the 3rd, 50th, and 97th centile for height from birth to the age of 5 years old from WHO Child Growth Standards and results for infants with pseudohypoaldosteronism. *(From Coelho Almeida A, Bastos Gomes M, Martins SA, Marques OP, Gomes MM, Antunes AM. A case of severe systemic type 1 pseudohypoaldosteronism with 10 years of evolution. J Pediatr Endocrinol Metab 2022;35(11):1448–52.)*

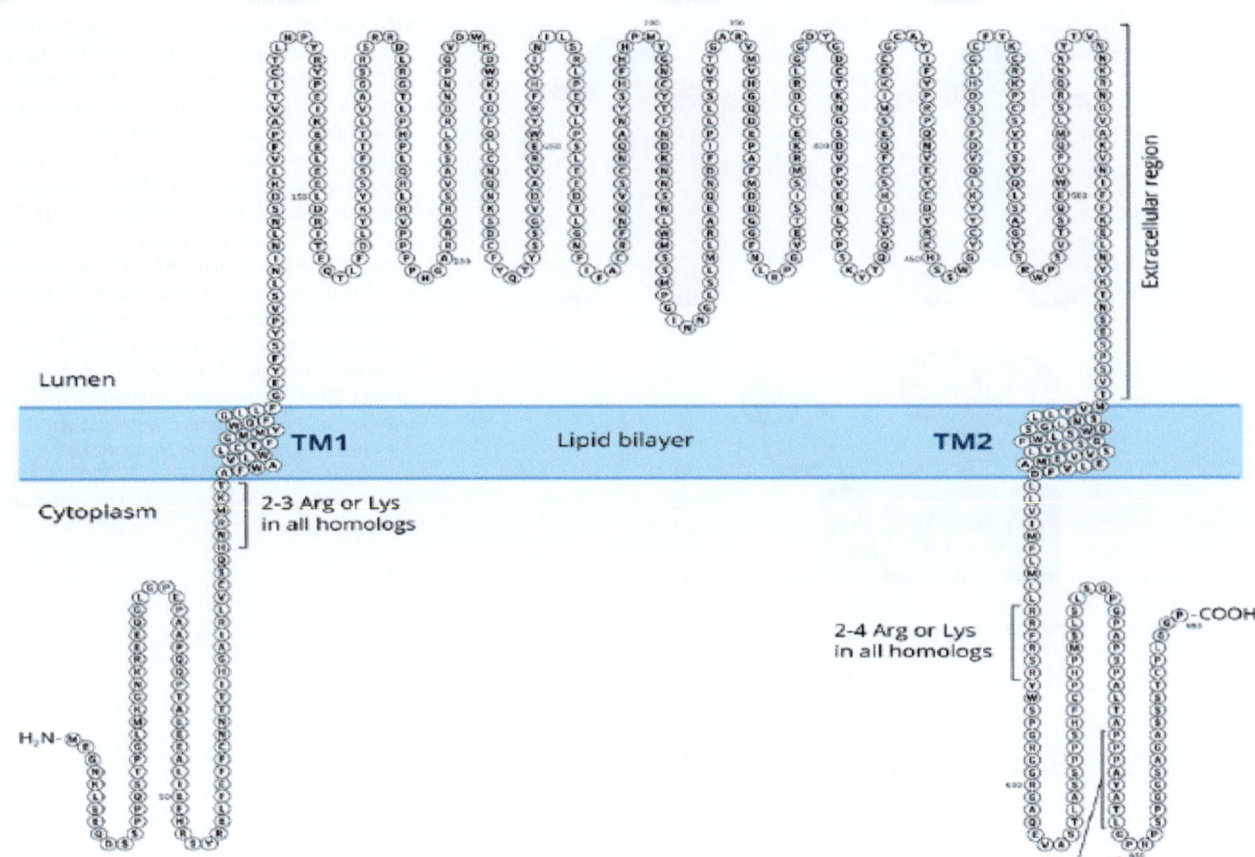

FIG. 3.3.2.30 Schematic illustration of the transmembrane localization of an ENaC subunit. The sequence shown is of human α subunit. All homologous ENaC subunits have two predicted transmembrane segments. The extracellular domain includes about 70% of the sequence of amino acids of an ENaC subunit. *(From Hanukoglu I, Hanukoglu A. Epithelial sodium channel (ENaC) family: phylogeny, structure-function, tissue distribution, and associated inherited diseases. Gene 2016;579(2):95–132. Fig. 3 p. 99.)*

regulation of transcellular fluid. Each ENaC subunit comprises two transmembrane segments, with one extracellular and one intracellular N and C-terminal domain (Fig. 3.3.2.30).

ENaCs are ubiquitous channels expressed in the kidneys and various extrarenal organs including the lungs, colon, salivary glands, and sweat glands. While ENaC regulates renal sodium reabsorption, thereby maintaining electrolyte homeostasis, in the lungs it primarily regulates the reabsorption of pulmonary interstitial fluid on the cell surface.

The gene causing Liddle's syndrome was initially localized to a small segment of chromosome 16 by linkage analysis and mutations in subunits of ENaC explain the syndrome (Fig. 3.3.2.31). Mutations in the β subunit (**SCNN1B**) cluster in the cytoplasmic carboxy termini of the subunits and result in constitutive activation of the ENaC channels.

Those mutations result in deletion of the subunits (Seyhanli et al., 2020; Voilley et al., 1997) or the introduction of amino acid substitutions into a proline-rich segment of the carboxy terminus (Cayir et al., 2019) (Fig. 3.3.2.32).

Similar mutations have been found in **SCNN1G** leading to amino acid sequences in the γ-subunits with an increase

in channel activity due to an inability to remove active channels from the apical cell surface possibly due to changes in interactions with other proteins (Kozina et al., 2019; Nur et al., 2017).

SCNN1B and SCCN1G (Yin et al., 2019) are expressed primarily in the kidneys but **SCNN1A** is expressed systemically including kidney, lungs, colon, sweat glands and salivary glands. SCNN1A is a homologous subunit encoded by the gene located at chromosome 12p.13.31, inherited in autosomal recessive manner. Racial disparities are seen. A T594T mutation which affects the last exon of the sodium channel β-subunit occurs more frequently in patients of African descent by preventing phosphorylation of protein kinase C and inhibiting channel activity. The reported mutations lead to truncated or abnormal nonfunctional proteins. Mutations in SCNN1A have been commonly found in patients with Turkish origin (Welzel et al., 2013). They are mainly deletions, insertions, or splicing mutations, which may cause abolition or severe malfunctioning of the encoded protein (subunit of EnaC) (Fig. 3.3.2.33). Correlations between missense mutations and milder forms of disease are hypothesized. By converse, nonmissense mutations (deletions, insertions and splicing) are observed

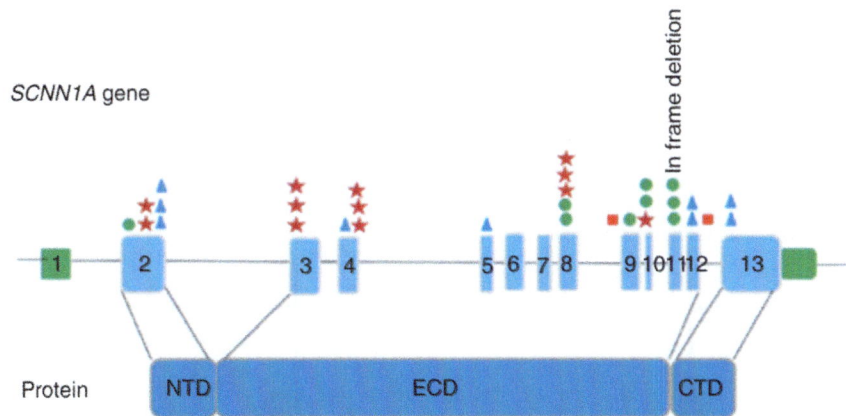

FIG. 3.3.2.31 Schematic representation of the SCNN1A gene (NM_001159575.1). The *blue boxes* represent the 13 exons, and the *green boxes* represent untranscribed regions. The homozygous mutations detected in this patient series are shown in approximate positions. Missense mutations are represented as blue triangles, nonsense mutations as green circles, frameshift mutations as red stars and intronic mutations as red squares. The protein has three domains: the N-terminal domain (NTD), the extracellular domain (ECD) and the C-terminal domain (CTD). *(From Gopal-Kothandapani JS, Doshi AB, Smith K, Christian M, Mushtaq T, Banerjee I, Padidela R, Ramakrishnan R, Owen C, Cheetham T, Dimitri P. Phenotypic diversity and correlation with the genotypes of pseudohypoaldosteronism type 1. J Pediatr Endocrinol Metab 2019;32(9):959–67. Fig. 2 p. 965.)*

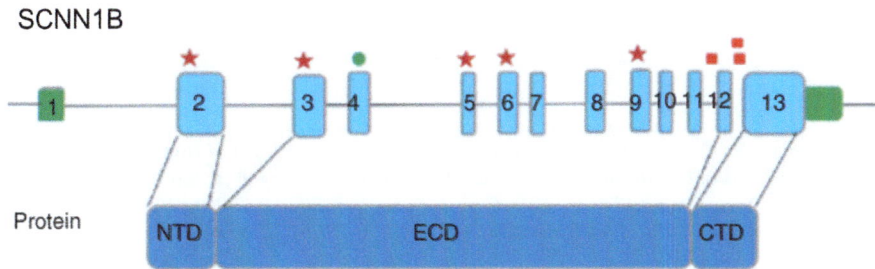

FIG. 3.3.2.32 Schematic representation of the SCNN1B gene (NM_000336.2). The *blue boxes* represent the 13 exons, and the *green boxes* represent untranscribed regions. The homozygous mutations detected in this patient series are shown in approximate positions. Missense mutations are represented as blue triangles, nonsense mutations as green circles, frameshift mutations as red stars and intronic mutations as red squares. The protein has three domains: the N-terminal domain (NTD), the extracellular domain (ECD) and the C-terminal domain (CTD). *(From Gopal-Kothandapani JS, Doshi AB, Smith K, Christian M, Mushtaq T, Banerjee I, Padidela R, Ramakrishnan R, Owen C, Cheetham T, Dimitri P. Phenotypic diversity and correlation with the genotypes of pseudohypoaldosteronism type 1. J Pediatr Endocrinol Metab 2019;32(9):959–67. Fig. 3 p. 965.)*

in subjects with a more severe clinical picture (Serra et al., 2021).

Survival with normal physical and neurological outcome is possible (Bandhakavi et al., 2021). The main treatment of PHA is a high sodium intake with ion exchange resins and dietary manipulation to reduce potassium levels. The most widely used potassium binding agents are sodium polystyrene sulfonate (kayexalate) and calcium polystyrene sulfonate although sodium containing resin is preferred to calcium because hyponatremia and hyperkalemia are corrected simultaneously. Potassium binding polymers are available (e.g., Patiromer) (Blair, 2018) but have a risk for hypokalemia.

3.3.2.7.2 Autosomal dominant PHAI (defects in MR)

In patients with **disorders of the mineralocorticoid receptor (MR)**, the secretion of corticosterone, 18-hydroxycorticosterone and aldosterone are elevated through renin/angiotensin stimulation due to mutations in the **gene NR3C2** encoding the mineralocorticoid receptor (Riepe et al., 2006; Pujo et al., 2007; Balsamo et al., 2007; Uchida et al., 2009). This is an **autosomal dominant** renal form of **PHA1a**. In children with the receptor defect, the salt loss seems to be partially correctable by increasing the dietary salt intake to 15–20 g/day to satisfy salt craving. Ion exchange resins and dietary manipulation may be needed to reduce potassium concentrations. The biochemical diagnosis is straightforward with a steroid profile if the child is sodium deplete (Honour et al., 1982) through obvious increase in the excretion rates of aldosterone and 18-hydroxycorticosterone metabolites.

Many young patients grow out of the condition with time although aldosterone concentrations are always raised (Hanukoglu et al., 2020). Patients with chromosome 4q deletions should be investigated for PHA1a (Barone Pritchard et al., 2020).

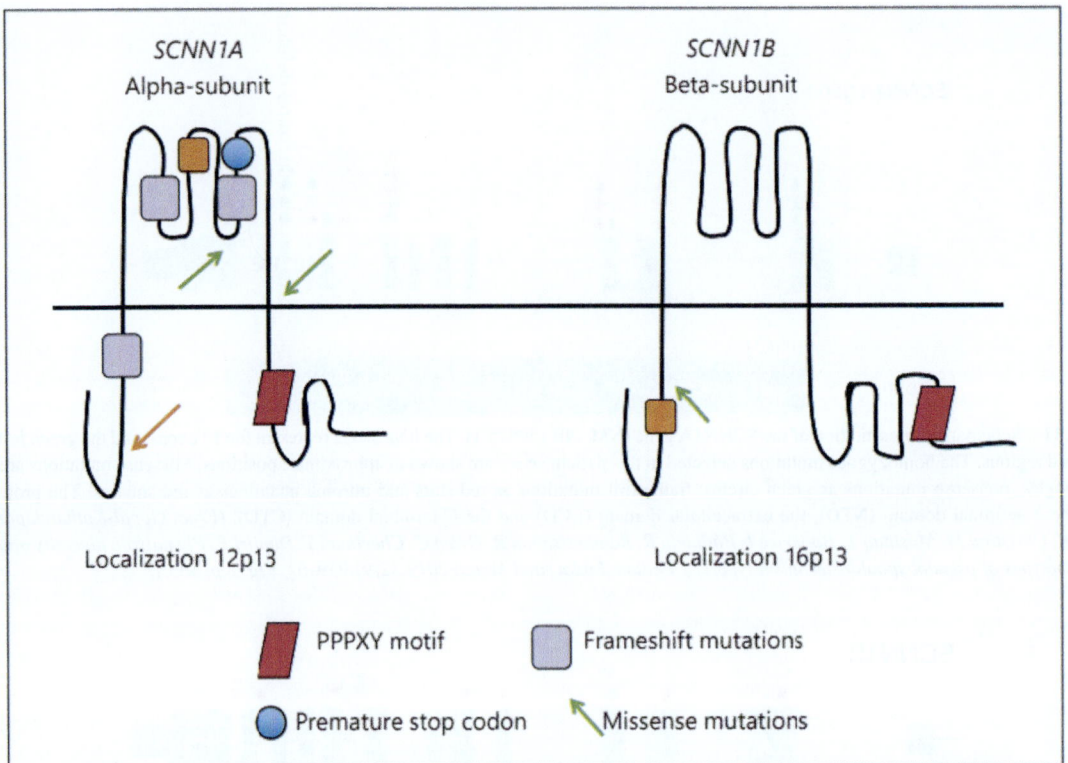

FIG. 3.3.2.33 Schematic illustration of the two ENAC subunits encoded by genes SCNN1A and SCNN1B. Premature stop codons, frameshift, and missense mutations may cause truncation of the proteins and change the PPPXY motif, resulting in the loss of interaction with other required proteins. *Orange indicates* the mutations identified in the present study. *(From Cayir A, Demirelli Y, Yildiz D, Kahveci H, Yarali O, Kurnaz E, Vuralli D, Demirbilek H. Systemic pseudohypoaldosteronism type 1 due to 3 novel mutations in SCNN1A and SCNN1B genes. Horm Res Paediatr 2019;91(3):175–85. Fig. 5 p. 183.)*

The MR is a member of the steroid hormone nuclear receptor family with 984 amino acids in three functional domains. The N terminal domain interacts with co-activators, co-repressors and the LBD. In the DBD, two zinc fingers are used in DNA binding and MR dimerization. The LBD is required for nuclear localization and MR dimerization. The MR is encoded by the NR3C2 gene of 10 exons, although only exons 2–9 are translated to the active protein. More than 100 mutations in the human NR3C2 gene causing rPHA1 have been described with familial or de novo mutations in a considerable part of patients, spread throughout the gene. Mutations are found in the heterozygous state. Identical NR3C2 gene mutations in rPHA1 lead to a very broad disease expression within an affected family. Some examples include termination in exon 2 p.G645* (Fig. 3.3.2.34) (Goda et al., 2020); missense mutation (L924P) (Tajima et al., 2017); frameshift (Kawashima Sonoyama et al., 2017); individual amino acids (Nishizaki et al., 2016; Uchida et al., 2009) and G633R, Q776R and L979P MR mutations (Sartorato et al., 2004). The mutations create truncated proteins or affect individual amino acids that affect receptor function.

3.3.2.8 Gordon's syndrome (GS)

A rare syndrome first described by Gordon is characterized by familial hyperkalemia despite normal glomerular filtration, hypertension and correction of abnormalities with thiazide diuretics. Short stature is a presenting feature in many cases. GS is also known as **pseudohypoaldosteronism type 2 (PHA2-II)**. Mild hyperchloremia, metabolic acidosis and suppressed PRA are variable associated findings that affect neurological and skeletal development. Gordon's syndrome is the only form of monogenic hypertension that manifests as hypoaldosteronism because of increased volume expansion, through increased sodium chloride reabsorption in the distal convoluted tubule (DCT), which decreases luminal sodium flow in the more distal nephron and makes ENaC less functional. Analysis of the pedigree in families has shown locus heterogeneity of this trait for linkage to chromosomes 1q31-q42 and 17p11-q21. (Of note the chromosome 17 locus in rat overlaps a syntenic interval that contains a blood pressure quantitative trait locus.) This was possibly the first step in the identification of a molecular basis for PHA-II (Mabillard and Sayer, 2019).

NR3C2

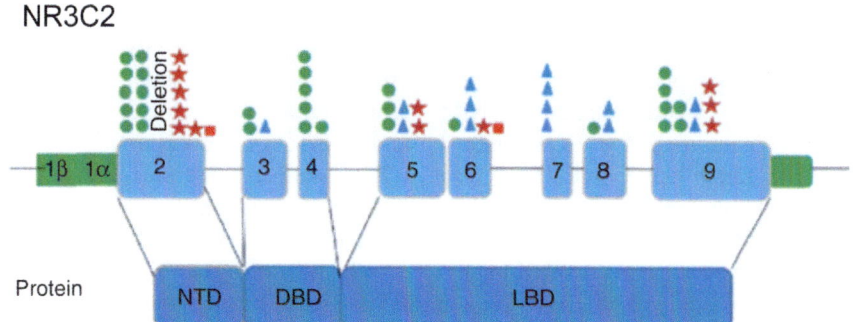

FIG. 3.3.2.34 Schematic representation of the NR3C2 gene (NM_000901.4). The *large blue boxes* represent the nine exons, and the *smaller green boxes* represent untranscribed regions. The heterozygous mutations detected in this patient series are shown in approximate positions. Missense mutations are represented as blue triangles, nonsense mutations as green circles, frameshift mutations as red stars and intronic mutations as red squares. The protein has three domains: the N-terminal domain (NTD), the DNA-binding domain (DBD) and the ligand-binding domain (LBD). *(From Gopal-Kothandapani JS, Doshi AB, Smith K, Christian M, Mushtaq T, Banerjee I, Padidela R, Ramakrishnan R, Owen C, Cheetham T, Dimitri P. Phenotypic diversity and correlation with the genotypes of pseudohypoaldosteronism type 1. J Pediatr Endocrinol Metab 2019;32(9):959–67. Fig. 1 p. 964.)*

In the distal nephron, kinases WNK1 and WNK4 regulate the expression of the thiazide sensitive sodium/chloride co-transporter (NCCT). The WNK (with no lysine) kinases are in a group of serine/threonine kinases that owe the name to the atypical position of a catalytic lysine in subdomain I whereas in most serine/threonine kinases the lysine residue is located in domain II. WNK4 is a positive regulator of the thiazide sensitive NaCl co-transporter (NCCT) (Mabillard and Sayer, 2019) (Fig. 3.3.2.35).

WNK4 can stimulate NCC trafficking to the plasma membrane of the DCT achieved by phosphorylation of NCC by SPAK/OSR1, stimulating NaCl transport. WNK1 and WNK4 can be ubiquitinated by Kelch-like 3 and Cullin 3. WNK4 is inhibited by WNK1. WNK1 also acts via SPAK to increase the phosphorylation of NCC. In Gordon's syndrome, the purple arrows represent an increased abundance of mutant WNK1 and wild-type WNK1 and WNK4 and the green arrows (marked +) indicate increased expression of NCC on the apical membrane of the distal convoluted tubule (DCT), NKCC1 on the basolateral membrane of the collecting duct (CD), and NKCC2 on the apical membrane of the thick ascending loop of Henle (TAL). The red crosses

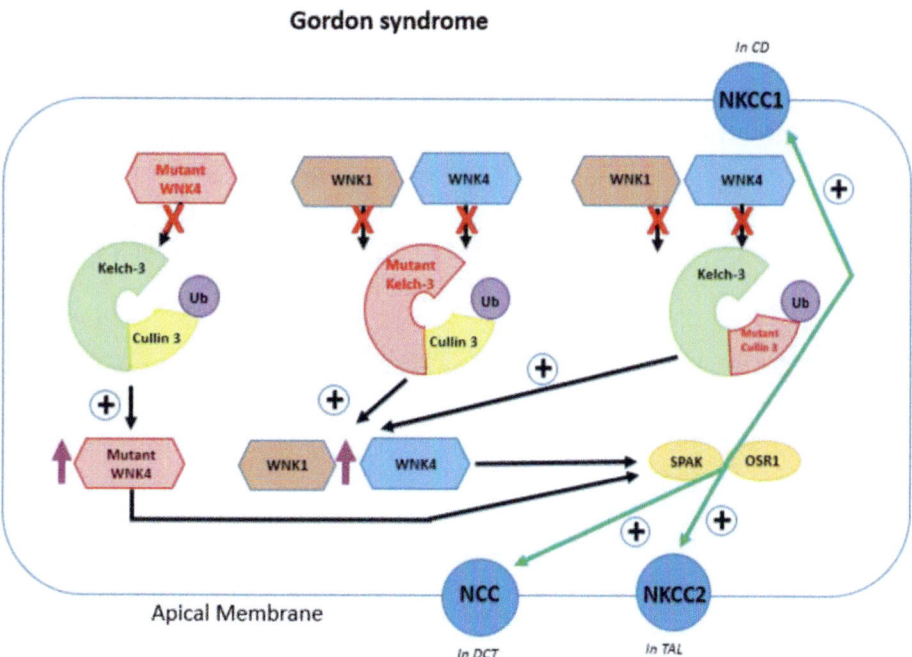

FIG. 3.3.2.35 Summary of pathophysiology of Gordon's syndrome in a tubular cell representative of different parts of the nephron. *(From Mabillard H, Sayer JA. The molecular genetics of Gordon syndrome. Genes (Basel) 2019;10(12):986. Fig. 4 p. 11 of 18.)*

signify the inability of WNK4 to bind with the Cullin 3-RING ubiquitin ligase complex.

Mutant WNK4 overstimulates plasma membrane proteins NCC, NKCC1, and NKCC2 because mutant WNK4 is unable to bind and be ubiquitinated by Kelch-like 3 and Cullin 3; mutations in Kelch-like-3 or Cullin 3 lead to WNK1 and WNK4 accumulation because of a failure of the Cullin 3-RING ubiquitin ligase complex to form and ubiquitinate WNK. Wild-type and mutant WNK4 phosphorylates SPAK and OSR1 to activate NCC, NKCC1, and NKCC2. Mutations in WNK1 are intronic deletions and do not affect the protein structure but lead to changes in the expression of a WNK1 isoform, which can phosphorylate and stimulate NCC.

Missense mutations as well as intronic deletions in WNK4 have been found in patients with Gordon's syndrome (Wilson et al., 2001). The disease increases salt reabsorption in the distal convoluted tubule and indirectly impairs potassium excretion (somewhat like the action of aldosterone). WNK 1 and 4 exert their effects on blood pressure through phosphorylation of serine/threonine kinase SPAK and oxidative stress response gene (OSR1). These in turn phosphorylate NCC (sodium chloride transporter) and NKCC (thiazide sensitive sodium chloride transporter). A gain of function in WNK4 causes familial hyperkalemic hypertension (see Chapter 3.3.1). Four types of Gordon's syndrome are recognized so far and the clinical features vary with age and severity (Table 3.3.2.2).

3.3.2.8.1 CUL3

The sodium/chloride transporters in the distal tubule of the kidney are activated by a phosphorylated cascade via activation of WNK to increase reabsorption of Na and Cl. In PHAII, the activity of ENaC is promoted but the volume of Na that should be reabsorbed decreases ad reabsorption of Na is lower and the secretion of potassium and hydrogen in the collecting ducts are reduced. Cullin 3 (CUL3, and scaffold protein) and Kelch3 (KLHL3 gene, and adaptor protein) have roles in ubiquitination and proteosomal processing of WNK kinases (Murillo-de-Ozores et al., 2021). The KLHL3 protein belongs to the BTB-BACK-kelch family of actin binding proteins that recruit substrates for Cullin-3 based ubiquitin ligase complexes. KLHL3 is co-expressed with NCC and downregulates NCC expression at the cell surface. Mutations in CUL3 or KLHL3 lead to abnormal activation of WNK/SPAK-NCC signaling cascade in the kidney (Boyden et al., 2012), disrupting electrolyte homeostasis and eliciting PHA phenotype (Nakano et al., 2020; Shao et al., 2018). Patients present with hyperkalemic metabolic acidosis, growth impairment and hypertension before age of 18 years.

3.3.2.8.2 Other causes of hyporeninemic hypoaldosteronism

3.3.2.8.2.1 Diabetes

Hyporeninemic hypoaldosteronism (HH) is found in patients with diabetic nephropathy (Sousa et al., 2016). The prevalence of HH in these patients can be difficult to ascertain because the patients are often on several medications due to varying degrees of impaired renal function. Patients with type 1 diabetes may have hyperkalemia due to concurrent adrenal insufficiency from autoimmune

TABLE 3.3.2.2 Phenotype-genotype correlation in Gordon's syndrome.

	WNK1	WNK4	KHL3	CUL3
Hypertension	Least severe phenotype and metabolic disorder often precedes hypertension	Metabolic disorder often precedes hypertension	Recessive mutations are more severe and diagnosed at an earlier age than dominant mutations	Most severe phenotype. Presents at age <18 years in majority of cases.
Hyperkalemia	Least severe	Yes	Dominant mutations have significantly higher serum K^+ than recessive mutations	Most severe. Presents at youngest age.
Metabolic acidosis	Least severe	Yes	Yes	Most severe
Other features		Hypercalciuria. Hypocalcemia. Decreased bone mineral density. Renal calcium stones.		Fertility likely affected in de novo mutations. Growth impairment most likely.

From Mabillard H, Sayer JA. The molecular genetics of Gordon syndrome. Genes (Basel) 2019;10(12):986. Table 1 p. 3 of 18.

polyglandular syndrome. Mild hyperkalemia is seen when prescribed NSAIDs or angiotensin inhibitors and potassium sparing diuretics. Poor diabetes control can lead to damage of juxtaglomerular cells or impaired beta-adrenergic stimulation. Fludrocortisone is effective for the treatment of the hyperkalemia.

3.3.2.8.2.2 *Human immunodeficiency virus (HIV)*

Patients with the acquired immunodeficiency syndromes (AIDS) due to HIV commonly present with fatigue, nausea, vomiting, postural hypotension, hyponatremia, pigmentation and anorexia, which are nonspecific signs similar to the presentation of AI (Mayo et al., 2002; Honour, 1998). Low aldosterone concentrations in patients infected with the HIV may be due to infection per se of the adrenal gland by the virus causing adrenal insufficiency and infection of the intestine causing chronic diarrhea (Kaile et al., 2008). The diagnosis, effects and management of the disease is complicated by malnutrition and cachexia. In addition, opportunistic infections and the use of drugs need to be considered. Ketoconazole inhibits steroidogenesis, Rifampin enhances cortisol metabolism, megestrol acetate has intrinsic glucocorticoid activity (Steer et al., 1995). Now that the condition is managed with antiretroviral drugs these findings are less common and only now seen in patients with advanced disease (Srinivasa et al., 2021). Emerging data suggests activation of the RAAS leading to increased risk of cardiovascular disease. These changes are associated with visceral adiposity and insulin resistance. Adipose tissue can synthesize renin, angiotensinogen, ACE and aldosterone which can promote inflammation and modulate cardiometabolic disease (Srinivasa et al., 2015).

3.3.2.9 Secondary PHA type 3

PHA is also seen with congenital anomalies of the kidney (uteropelvic obstruction, and renal hypoplasia) and urinary tract (bilateral megaureter) that may be complicated with urinary tract infections (acute pyelonephritis) (Delforge et al., 2019; Nandagopal et al., 2009). Oligohydramnios (lack of amniotic fluid) or hydronephrosis (swollen kidney) many be forerunners. Radiographic imaging may be needed. Symptoms are nonspecific such as failure to thrive, vomiting, dehydration and polyuria. Renal sodium loss and low potassium excretion is associated with hyponatremia and metabolic acidosis. Urinary tract obstruction increases intra-renal synthesis of TGF-β1 which inhibits aldosterone action. The electrolyte disorders can be life threatening through possible cardiac arrest so intravenous fluids is needed.

3.3.2.9.1 Urine tract infections

Secondary pseudohypoaldosteronism in children is associated with severe urinary tract infection sometimes with urinary tract malformation (Graziano et al., 2022; Kaninde et al., 2021; Krishnappa et al., 2016; Nandagopal et al., 2009). The presentation of hyponatremia during the first year of life is later than seen in children with CAH, and those with mutations in the MR or ENaC. Those genes have been shown to be normal in these patients. Bacterial endotoxin increases production of interleukin-1 (IL-1) which causes vasoconstriction and reduced GFR. The mechanism for apparent aldosterone resistance however is unknown.

Infections need antibiotic therapy. Endotoxins produced by bacteria are associated with release of prostaglandins and thromboxane A2 which contribute to renal vasoconstriction, decreased GFR, and increased sodium loss with hyperkalemia and metabolic acidosis. Urinary sodium, plasma renin and aldosterone should be monitored. The estimated incidence of PHA Type 3 in Ireland is 1 per 13,200 live births (Kaninde et al., 2021). Antimicrobial treatment and fluid resuscitation with an isotonic solution usually restores the acid-base and electrolyte imbalance.

3.3.2.9.2 Licorice and PHA

For many years, licorice was thought to act like aldosterone to cause hypertension. Glycyrrhetinic acid, the active component of licorice is now known to inhibit renal HSD11B2 that controls the inactivation of cortisol to cortisone (Stewart et al., 1988; Gallacher et al., 2017; McHugh et al., 2021) and like the syndrome of apparent mineralocorticoid excess is associated with low renin activity and aldosterone production (see Chapter 3.2.1). The defect can be recognized from analysis of blood or urine where the cortisol to cortisone ratio is high (Table 3.3.2.3) (Dellow et al., 1999).

3.3.2.9.3 Azoles

Ketoconazole was initially introduced for systemic applications (Heeres et al., 1979; Symoens et al., 1980). In several studies, low testosterone concentrations, gynecomastia, and glucocorticoid suppression were reported (DeFelice et al., 1981; Pont et al., 1982a,b; Santen et al., 1983). The inhibition of adrenal and gonadal activities of CYP11A1, CYP17A1 and CYP11B1 were found to cause these adverse effects (Engelhardt et al., 1991; Feldman, 1986; Loose et al., 1983). Ketoconazole was used firstly on an off-label basis for the treatment of CS (reviewed by Daniel and Newell-Price, 2015). Sustained hypertension was observed in a small subset of patients who had elevated concentrations of 11-deoxycortisol and DOC and decreased cortisol along

TABLE 3.3.2.3 Urinary steroid profile, blood pressure, plasma electrolytes, plasma aldosterone and renin activity of patient taking licorice and after 9 month without licorice.

	On licorice	On licorice	Off licorice	Off licorice	
Date	02/04/20	15/04/20	20/01/21	14/05/21	Normal ranges
Urine cortisol metabolites					
THE	1140	940	860	1350	980–3240
THF	1290	1120	1120	2090	440–1610
allo-THF	300	310	410	430	160–640
THF+allo-THF	1590	1450	1420	2520	610–2140
THF+allo-THF/THE	1.39	1.54	1.65	1.85	0.47–0.87
THE/THF	0.72	0.65	0.61		0.79–1.92
Allo-THF/THF	0.23	0.27	0.27		0.29–0.72
Total C21 cortisol metabolites	3330	3440	3260		2630–7300
Blood pressure	180/100	226/120	138/82	136/88	
Electrolytes					
Plasma potassium		2.4	4.7	4.1	3.3–4.8
Plasma sodium		145	132	142	137–145
Aldosterone (pmol/L)	170	89	295		600–1200
Plasma renin activity (nmol/L/h)	1.86	1.01	4.21		2.8–4.5

Data from Dellow EL, Unwin RJ, Honour JW. Pontefract cakes can be bad for you: refractory hypertension and liquorice excess. Nephrol Dial Transplant 1999; 14(1):218–20. Table 1 p. 219.

with normal aldosterone levels during treatment long-term or at high-dose (Aabo and De Coster, 1987; Leal-Cerro et al., 1989). In a retrospective multicenter study involving 200 patients, about 60% of the patients receiving ketoconazole as presurgical treatment or primary or secondary treatment of CS exhibited hypertension and hypokalemia (Castinetti et al., 2008). Due to concerns of severe hepatotoxicity, the EMA and the FDA restricted the systemic application of ketoconazole for the therapy of fungal infections; although, the EMA approved oral ketoconazole for the therapy of CS due to the limited alternative options. Recent advances in CS therapy such as the approval of osilodrostat now provide alternative treatment options (Pivonello et al., 2020).

The azole group of antifungals now comprises many more variants of the drug widely used in patients with infections such as with cystic fibrosis. Hypokalemia has been reported and attributed to inhibition of CYP11B1 and HSD11B2 on the basis of serum steroid measurements (Barton et al., 2018; Thompson III et al., 2017, 2019) and urinary steroid profiles (Boughton et al., 2018; Agarwal et al., 2020). Itraconazole also causes symptoms of psuedo-hypoaldosteronism with suppressed renin (Beck et al., 2017) and showed reduced HSD11B2 in a transfection assay

(Beck and Odermatt, 2021). Three other antifungals fluconazole, voriconazole and isavuconazole did not cause hypokalemia or hypertension (Ji et al., 2022).

3.3.2.10 Summary

Primary hypoaldosteronism with hyperkalemia, increased PRA and low aldosterone concentrations is due to **Addison's disease**, **congenital adrenal hypoplasia** or **defects of aldosterone synthesis** most commonly due to CAH. These disorders can be detected by analysis of multiple steroids in plasma (Li et al., 2021) or urine (Schiffer et al., 2019; Storbeck et al., 2019). Plasma renin activity (PRA) and aldosterone are rarely subnormal but this secondary hypoaldosteronism is found in association with diabetes, chronic renal disease and as an isolated occurrence. A salt losing state in infants was, on biochemical grounds, originally thought to be due to a defect of the aldosterone receptor and thus described as resistance to aldosterone or **pseudohypoaldosteronism**. Following genetic analysis a more common cause can now be defined due to mutations in the epithelial sodium transport channel (ENaC). There is much to be learnt about the actions of aldosterone.

TABLE 3.3.2.4 Affected genes, protein and transmittance of hypoaldosteronism due to isolated aldosterone deficiency, aldosterone resistance and other causes.

	Transmission	Defective protein	Gene
Isolated aldosterone deficiency			
Aldosterone synthase defect	AR	Aldosterone synthase	CYP11B2
Cortisol and aldosterone deficiency			
Congenital adrenal hypoplasia	X	DAX-1	NROB1
	AD	SF-1	SF-1
Congenital lipoid adrenal hyperplasia	AR	StAR	STAR
Congenital adrenal hyperplasia	AR	3β-Hydroxysteroid dehydrogenase	HSD3B2
	AR	21-Hydroxylase	CYP21A2
Aldosterone resistance syndromes			
Pseudohypoaldosteronism (PHA) type 1 (Liddle)	AR	α, β, γ-subunits epithelial sodium transport	SCNN1A, B, C
	AD	Mineralocorticoid receptor (MR)	NR3C2
Transient PHA	Acquired		
Secondary sodium loss (low renin)			
Congenital adrenal hyperplasia	AR	11-Hydroxylase	CYP11B1
	AR	17-Hydroxylase	CYP17A1
Diabetes, HIV infection, licorice, azoles	Acquired		
Pseudohypoaldosteronism type 3			
Congenital kidney development disorders			

Author original.

Studies of patients with pseudohypoaldosteronism have provided much new information about the roles of genes to the disease process. Mutations induced in mice have also created animals with human disease features. Genetic changes in WNK kinases have been found in patients with Gordon's syndrome. The Cullin ring E3 ligases mediate WNK ubiquitination. Electrolyte disorders are precipitated by mutations in genes such as WNK kinases and related Cullin 3 and KLHL3 that increases sodium reabsorption leading to the development of hypertension. These mutation studies are examples of emerging mechanisms of pathogenesis (Castel, 2022) due to dysregulation of protein degradation and accumulation of proteins. The causes of hypoaldosteronism, pseudohypoaldosteronism and salt loss are summarized in Table 3.3.2.4. Steroid analysis has an important place in the diagnosis of most of these conditions. A number of new channels, tubules, transporters and co transporters have been discovered around the handling of sodium, potassium, calcium, magnesium and chloride. The discovery of further mechanisms is anticipated through close interactions between basic science, laboratory diagnostics and clinical medicine.

Arising from the genetic studies there are limitations to relying on common in silico prediction tools for a defective protein and individual assessment of a polymorphism. Protein function should be assessed in a suitable cell line and assay system. The consequences of the change at the nucleic acid level must also be considered.

As with investigations for causes of hypertension, exosomes in urine are useful sources for protein and microRNA analysis (Maggio et al., 2021; Hoorn et al., 2005) to characterize potential biomarkers using mass spectrometry (Thongboonkerd, 2020; Barros and Carvajal, 2017) that may prove to be helpful in studies of patients with hypoaldosteronism or salt loss.

References

Aabo K, De Coster R. Hypertension during high-dose ketoconazole treatment: a probable mineralocorticosteroid effect. Lancet 1987;2 (8559):637–8.

Adachi M, Tajima T, Muroya K. Dietary potassium restriction attenuates urinary sodium wasting in the generalized form of pseudohypoaldosteronism type 1. CEN Case Rep 2020;9(2):133–7. Erratum in: CEN Case Rep. 2020;9(3):294.

Agarwal N, Apperley L, Taylor NF, Taylor DR, Ghataore L, Rumsby E, et al. Posaconazole-induced hypertension masquerading as congenital adrenal hyperplasia in a child with cystic fibrosis. Case Rep Med 2020;2020:8153012.

Al Alawi AM, Nordenström A, Falhammar H. Clinical perspectives in congenital adrenal hyperplasia due to 3 β-hydroxysteroid dehydrogenase type 2 deficiency. Endocrine 2019;63(3):407–21.

Alzahrani AS, Alswailem MM, Murugan AK, Alhomaidah DS, Capper CP, Auchus RJ, et al. A high rate of novel CYP11B1 mutations in Saudi Arabia. J Steroid Biochem Mol Biol 2017;174:217–24.

Arif S, Vallian S, Farzaneh F, Zanone MM, James SL, Pietropaolo M, et al. Identification of 3 beta-hydroxysteroid dehydrogenase as a novel target of steroid cell autoantibodies: association of autoantibodies with endocrine autoimmune disease. J Clin Endocrinol Metab 1996;81 (12):4439–45.

Armanini D, Karbowiak I, Zennaro CM, Zovato S, Pratesi C, De Lazzari P, et al. Pseudohypoaldosteronism: evaluation of type I receptors by radioreceptor assay and by antireceptor antibodies. Steroids 1995;60 (1):161–3.

Arriba M, Ezquieta B. Molecular diagnosis of steroid 21-hydroxylase deficiency: a practical approach. Front Endocrinol (Lausanne) 2022;13, 834549.

Balsamo A, Cicognani A, Gennari M, Sippell WG, Menabò S, Baronio F, et al. Functional characterization of naturally occurring NR3C2 gene mutations in Italian patients suffering from pseudohypoaldosteronism type 1. Eur J Endocrinol 2007;156(2):249–56.

Bandhakavi M, Wanaguru A, Ayuk L, Kirk JM, Barrett TG, Kershaw M, et al. Clinical characteristics and treatment requirements of children with autosomal recessive pseudohypoaldosteronism. Eur J Endocrinol 2021;184(5):K15–20.

Baquedano MS, Guercio G, Costanzo M, Marino R, Rivarola MA, Belgorosky A. Mutation of HSD3B2 gene and fate of dehydroepiandrosterone. Vitam Horm 2018;108:75–123.

Barone Pritchard A, Rittr A, Kearney HM, Izumi K. Interstitial 4q deletion syndrome including NR3C2 causing pseudohypoaldosteronism. Mol Syndromol 2020;10(6):327–31.

Barros ER, Carvajal CA. Urinary exosomes and their cargo: potential biomarkers for mineralocorticoid arterial hypertension? Front Endocrinol (Lausanne) 2017;8:230.

Barton K, Davis TK, Marshall B, Elward A, White NH. Posaconazole-induced hypertension and hypokalemia due to inhibition of the 11β-hydroxylase enzyme. Clin Kidney J 2018;11(5):691–3.

Beck KR, Odermatt A. Antifungal therapy with azoles and the syndrome of acquired mineralocorticoid excess. Mol Cell Endocrinol 2021;524: 111168.

Beck KR, Bächler M, Vuorinen A, Wagner S, Akram M, Griesser U, et al. Inhibition of 11β-hydroxysteroid dehydrogenase 2 by the fungicides itraconazole and posaconazole. Biochem Pharmacol 2017;130: 93–103.

Belot A, Ranchin B, Fichtner C, Pujo L, Rossier BC, Liutkus A, et al. Pseudohypoaldosteronisms, report on a 10-patient series. Nephrol Dial Transplant 2008;23(5):1636–41.

Blair HA. Patiromer: a review in hyperkalaemia. Clin Drug Investig 2018;38(8):785–94.

Boughton C, Taylor D, Ghataore L, Taylor N, Whitelaw BC. Mineralocorticoid hypertension and hypokalaemia induced by posaconazole. Endocrinol Diabetes Metab Case Rep 2018;2018:17–0157.

Boyd-Shiwarski CR, Shiwarski DJ, Roy A, Namboodiri HN, Nkashama LJ, Xie J, et al. Potassium-regulated distal tubule WNK bodies are kidney-specific WNK1 dependent. Mol Biol Cell 2018;29(4):499–509.

Boyden LM, Choi M, Choate KA, Nelson-Williams CJ, Farhi A, Toka HR, et al. Mutations in kelch-like 3 and cullin 3 cause hypertension and electrolyte abnormalities. Nature 2012;482(7383):98–102.

Breil T, Yakovenko V, Inta I, Choukair D, Klose D, Mittnacht J, et al. Typical characteristics of children with congenital adrenal hyperplasia due to 11β-hydroxylase deficiency: a single-centre experience and review of the literature. J Pediatr Endocrinol Metab 2019;32(3): 259–67.

Buonocore F, Maharaj A, Qamar Y, Koehler K, Suntharalingham JP, Chan LF, et al. Genetic analysis of pediatric primary adrenal insufficiency of unknown etiology: 25 years' experience in the UK. J Endocr Soc 2021;5(8):bvab086.

Bystrom CE, Salameh W, Reitz R, Clarke NJ. Plasma renin activity by LC-MS/MS: development of a prototypical clinical assay reveals a subpopulation of human plasma samples with substantial peptidase activity. Clin Chem 2010;56(10):1561–9.

Capalbo D, Moracas C, Cappa M, Balsamo A, Maghnie M, Wasniewska MG, et al. Primary adrenal insufficiency in childhood: data from a large nationwide cohort. J Clin Endocrinol Metab 2021;106(3): 762–73.

Carvajal CA, Tapia-Castillo A, Vecchiola A, Baudrand R, Fardella CE. Classic and nonclassic apparent mineralocorticoid excess syndrome. J Clin Endocrinol Metab 2020;105(4), dgz315. Erratum in: J Clin Endocrinol Metab. 2021;106(4):e1934.

Castel P. Defective protein degradation in genetic disorders. Biochim Biophys Acta Mol Basis Dis 2022;1868(5), 166366.

Castinetti F, Morange I, Jaquet P, Conte-Devolx B, Brue T. Ketoconazole revisited: a preoperative or postoperative treatment in Cushing's disease. Eur J Endocrinol 2008;158(1):91–9.

Cayir A, Demirelli Y, Yildiz D, Kahveci H, Yarali O, Kurnaz E, et al. Systemic pseudohypoaldosteronism type 1 due to 3 novel mutations in SCNN1A and SCNN1B genes. Horm Res Paediatr 2019;91(3):175–85.

Chappell DL, McAvoy T, Weiss B, Weiner R, Laterza OF. Development and validation of an ultra-sensitive method for the measurement of plasma renin activity in human plasma via LC-MS/MS. Bioanalysis 2012;4(23):2843–50.

Chen S, Sawicka J, Betterle C, Powell M, Prentice L, Volpato M, et al. Autoantibodies to steroidogenic enzymes in autoimmune polyglandular syndrome, Addison's disease, and premature ovarian failure. J Clin Endocrinol Metab 1996;81(5):1871–6.

Chen F, Cheng Z, Peng Y, Wang Z, Huang C, Liu D, et al. A liquid chromatography-tandem mass spectrometry (LC-MS/MS)-based assay for simultaneous quantification of aldosterone, renin activity, and angiotensin II in human plasma. J Chromatogr B Analyt Technol Biomed Life Sci 2021;1179, 122740.

Chen F, Cheng Z, Wang Z, Peng Y, Wang B, Guo W, et al. Liquid chromatography-tandem mass spectrometry (LC-MS/MS) based assay for the simultaneous quantification of 18-hydroxycorticosterone, 18-hydroxycortisol and 18-oxocortisol in human plasma. J Chromatogr B Analyt Technol Biomed Life Sci 2022;1188, 123030.

Cheng CJ, Yoon J, Baum M, Huang CL. STE20/SPS1-related proline/alanine-rich kinase (SPAK) is critical for sodium reabsorption in isolated, perfused thick ascending limb. Am J Physiol Renal Physiol 2015;308(5):F437–43.

Daniel E, Newell-Price JD. Therapy of endocrine disease: steroidogenesis enzyme inhibitors in Cushing's syndrome. Eur J Endocrinol 2015;172 (6):R263–80.

DeFelice R, Johnson DG, Galgiani JN. Gynecomastia with ketoconazole. Antimicrob Agents Chemother 1981;19(6):1073–4. https://doi.org/ 10.1128/AAC.19.6.1073. 6267997. PMC181611.

Delforge X, Kongolo G, Cauliez A, Braun K, Haraux E, Buisson P. Transient pseudohypoaldosteronism: a potentially severe condition affecting infants with urinary tract malformation. J Pediatr Urol 2019;15(3):265.e1–7.

Dellow EL, Unwin RJ, Honour JW. Pontefract cakes can be bad for you: refractory hypertension and liquorice excess. Nephrol Dial Transplant 1999;14(1):218–20.

Engelhardt D, Weber MM, Miksch T, Abedinpour F, Jaspers C. The influence of ketoconazole on human adrenal steroidogenesis: incubation studies with tissue slices. Clin Endocrinol (Oxf) 1991;35(2): 163–8.

Fanis P, Neocleous V, Kosta K, Karipiadou A, Hartmann MF, Wudy SA, et al. Late diagnosis of 3β-hydroxysteroid dehydrogenase deficiency: the pivotal role of gas chromatography-mass spectrometry urinary steroid metabolome analysis and a novel homozygous nonsense mutation in the HSD3B2 gene. J Pediatr Endocrinol Metab 2020;34 (1):131–6.

Fardella CE, Hum DW, Rodriguez H, Zhang G, Barry FL, Ilicki A, et al. Gene conversion in the CYP11B2 gene encoding P450c11AS is associated with, but does not cause, the syndrome of corticosterone methyloxidase II deficiency. J Clin Endocrinol Metab 1996;81(1):321–6.

Feldman LD. Ketoconazole for male metastatic breast cancer. Ann Intern Med 1986;104(1):123–4.

Fiselier T, Monnens L, van Munster P, Jansen M, Peer P, Lijnen P. The renin-angiotensin-aldosterone system in infancy and childhood in basal conditions and after stimulation. Eur J Pediatr 1984;143(1):18–24.

Flokas ME, Tomani M, Agdere L, Brown B. Triple A syndrome (Allgrove syndrome): improving outcomes with a multidisciplinary approach. Pediatric Health Med Ther 2019;10:99–106.

Gallacher SD, Tsokolas G, Dimitropoulos I. Liquorice-induced apparent mineralocorticoid excess presenting in the emergency department. Clin Med (Lond) 2017;17(1):43–5.

Gaudl A, Kratzsch J, Bae YJ, Kiess W, Thiery J, Ceglarek U. Liquid chromatography quadrupole linear ion trap mass spectrometry for quantitative steroid hormone analysis in plasma, urine, saliva and hair. J Chromatogr A 2016;1464:64–71.

Goda T, Komatsu H, Nozu K, Nakajima H. An infantile case of pseudohypoaldosteronism type 1 (PHA1) caused by a novel mutation of NR3C2. Clin Pediatr Endocrinol 2020;29(3):127–30.

Graziano N, Agostoni C, Chiaraviglio F, Betti C, Piffer A, Bianchetti MG, et al. Pseudo-hypoaldosteronism secondary to infantile urinary tract infections: role of ultrasound. Ital J Pediatr 2022;48(1):14.

Guran T, Kara C, Yildiz M, Bitkin EC, Haklar G, Lin JC, et al. Revisiting classical 3β-hydroxysteroid dehydrogenase 2 deficiency: lessons from 31 pediatric cases. J Clin Endocrinol Metab 2020;105(3), dgaa022.

Hanukoglu A, Edelheit O, Shriki Y, Gizewska M, Dascal N, Hanukoglu I. - Renin-aldosterone response, urinary Na/K ratio and growth in pseudohypoaldosteronism patients with mutations in epithelial sodium channel (ENaC) subunit genes. J Steroid Biochem Mol Biol 2008;111(3–5):268–74.

Hanukoglu A, Vargas-Poussou R, Landau Z, Yosovich K, Hureaux M, Zennaro MC. Renin-aldosterone system evaluation over four decades

in an extended family with autosomal dominant pseudohypoaldosteronism due to a deletion in the NR3C2 gene. J Steroid Biochem Mol Biol 2020;204, 105755.

Heeres J, Backx LJ, Mostmans JH, Van Cutsem J. Antimycotic imidazoles. Part 4. Synthesis and antifungal activity of ketoconazole, a new potent orally active broad-spectrum antifungal agent. J Med Chem 1979;22 (8):1003–5.

Honour JW. HIV and adrenal function. Curr Opin Endocrinol Diabetes 1998;5:162–7.

Honour JW, Shackleton CH. Mass spectrometric analysis for tetrahydroaldosterone. J Steroid Biochem 1977;8(4):299–305.

Honour JW, Valman HB, Shackleton HL. Aldosterone and sodium homeostasis in preterm infants. Acta Paediatr Scand 1977;66(1):103–9.

Honour JW, Dillon MJ, Shackleton CH. Analysis of steroids in urine for differentiation of pseudohypoaldosteronism and aldosterone biosynthetic defect. J Clin Endocrinol Metab 1982;54(2):325–31.

Honour JW, Dillon MJ, Levin M, Shah V. Fatal, low renin hypertension associated with a disturbance of cortisol metabolism. Arch Dis Child 1983;58(12):1018–20.

Hoorn EJ, Pisitkun T, Zietse R, Gross P, Frokiaer J, Wang NS, et al. Prospects for urinary proteomics: exosomes as a source of urinary biomarkers. Nephrology (Carlton) 2005;10(3):283–90.

Hoorn EJ, Nelson JH, McCormick JA, Ellison DH. The WNK kinase network regulating sodium, potassium, and blood pressure. J Am Soc Nephrol 2011;22(4):605–14.

Ishii T, Hori N, Amano N, Aya M, Shibata H, Katsumata N, et al. Pubertal and adult testicular functions in nonclassic lipoid congenital adrenal hyperplasia: a case series and review. J Endocr Soc 2019;3(7): 1367–74.

Ishii T, Tajima T, Kashimada K, Mukai T, Tanahashi Y, Katsumata N, et al. Clinical features of 57 patients with lipoid congenital adrenal hyperplasia: criteria for nonclassic form revisited. J Clin Endocrinol Metab 2020;105(11), dgaa557.

Jadhav U, Harris RM, Jameson JL. Hypogonadotropic hypogonadism in subjects with DAX1 mutations. Mol Cell Endocrinol 2011;346(1–2): 65–73.

Ji HH, Tang XW, Zhang N, Huo BN, Liu Y, Song L, et al. Antifungal therapy with azoles induced the syndrome of acquired apparent mineralocorticoid excess: a literature and database analysis. Antimicrob Agents Chemother 2022;66(1), e0166821.

Kaile T, Zulu I, Lumayi R, Ashman N, Kelly P. Inappropriately low aldosterone concentrations in adults with AIDS-related diarrhoea in Zambia: a study of response to fluid challenge. BMC Res Notes 2008;1:10.

Kang E, Kim YM, Kim GH, Lee BH, Yoo HW, Choi JH. Mutation spectrum of STAR and a founder effect of the p.Q258* in Korean patients with congenital lipoid adrenal hyperplasia. Mol Med 2017;23: 149–54.

Kaninde A, Grace ML, Joyce C, Taylor NF, Ghataore L, Riordan MF, et al. The incidence of transient infantile pseudohypoaldosteronism in Ireland: a prospective study. Acta Paediatr 2021;110(4):1257–63.

Kawashima Sonoyama Y, Tajima T, Fujimoto M, Hasegawa A, Miyahara N, Nishimura R, et al. A novel frameshift mutation in NR3C2 leads to decreased expression of mineralocorticoid receptor: a family with renal pseudohypoaldosteronism type 1. Endocr J 2017;64(1):83–90.

Khattab A, Haider S, Kumar A, Dhawan S, Alam D, Romero R, et al. Clinical, genetic, and structural basis of congenital adrenal hyperplasia due to 11β-hydroxylase deficiency. Proc Natl Acad Sci U S A 2017;114(10):E1933–40.

Kim CJ. Congenital lipoid adrenal hyperplasia. Ann Pediatr Endocrinol Metab 2014;19(4):179–83.

Ko B, Mistry AC, Hanson L, Mallick R, Wynne BM, Thai TL, et al. Aldosterone acutely stimulates NCC activity via a SPAK-mediated pathway. Am J Physiol Renal Physiol 2013;305(5):F645–52.

Kozina AA, Trofimova TA, Okuneva EG, Baryshnikova NV, Obuhova VA, Krasnenko AY, et al. Liddle syndrome due to a novel mutation in the γ subunit of the epithelial sodium channel (ENaC) in family from Russia: a case report. BMC Nephrol 2019;20(1):389.

Krishnappa V, Ross JH, Kenagy DN, Raina R. Secondary or transient pseudohypoaldosteronism associated with urinary tract anomaly and urinary infection: a case report. Urol Case Rep 2016;8:61–2.

Kuhnle U, Hinkel GK, Akkurt HI, Krozowski Z. Familial pseudohypoaldosteronism: a review on the heterogeneity of the syndrome. Steroids 1995;60(1):157–60.

Le Goff C, Farre-Segura J, Stojkovic V, Dufour P, Peeters S, Courtois J, et al. The pathway through LC-MS method development: in-house or ready-to-use kit-based methods? Clin Chem Lab Med 2020;58(6):1002–9.

Leal-Cerro A, García-Luna PP, Villar J, Miranda ML, Pereira JL, Gomez-Pan A, et al. Arterial hypertension as a complication of prolonged ketoconazole treatment. J Hypertens Suppl 1989;7(6):S212–3.

Lee C, Kim JH, Moon SJ, Shim J, Kim HI, Choi MH. Selective LC-MRM/SIM-MS based profiling of adrenal steroids reveals metabolic signatures of 17α-hydroxylase deficiency. J Steroid Biochem Mol Biol 2020;198, 105615.

Li Z, Liang Y, Du C, Yu X, Hou L, Wu W, et al. Clinical applications of genetic analysis and liquid chromatography tandem-mass spectrometry in rare types of congenital adrenal hyperplasia. BMC Endocr Disord 2021;21(1):237.

Lin W, Yao Z, Li Y, Liao Z, Xiao J, Chen Y, et al. Developing an ultra-performance liquid chromatography-tandem mass spectrometry for detecting aldosterone in human plasma. J Clin Lab Anal 2021;35 (11), e24029.

Loose DS, Kan PB, Hirst MA, Marcus RA, Feldman D. Ketoconazole blocks adrenal steroidogenesis by inhibiting cytochrome P450-dependent enzymes. J Clin Invest 1983;71(5):1495–9.

Mabillard H, Sayer JA. The molecular genetics of Gordon syndrome. Genes (Basel) 2019;10(12):986.

Maggio S, Polidori E, Ceccaroli P, Cioccoloni A, Stocchi V, Guescini M. Current methods for the isolation of urinary extracellular vesicles. Methods Mol Biol 2021;2292:153–72.

Maharaj A, Buonocore F, Meimaridou E, Ruiz-Babot G, Guasti L, Peng HM, et al. Predicted benign and synonymous variants in CYP11A1 cause primary adrenal insufficiency through missplicing. J Endocr Soc 2018;3(1):201–21.

Maharaj A, Maudhoo A, Chan LF, Novoselova T, Prasad R, Metherell LA, et al. Isolated glucocorticoid deficiency: genetic causes and animal models. J Steroid Biochem Mol Biol 2019;189:73–80. https://doi.org/10.1016/j.jsbmb.2019.02.012. Epub 2019 Feb 25 30817990.

Makowski N, Burckhardt BB. Enabling insights into the maturation of the renin-angiotensin-aldosterone system in children-development of a low-volume LC-MS assay for the simultaneous determination of aldosterone, its precursor, and main metabolite. Steroids 2019;148:73–81.

Mallappa A, Merke DP. Management challenges and therapeutic advances in congenital adrenal hyperplasia. Nat Rev Endocrinol 2022;18(6):337–52.

Martinerie L, Pussard E, Foix-L'Hélias L, Petit F, Cosson C, Boileau P, et al. Physiological partial aldosterone resistance in human newborns. Pediatr Res 2009a;66(3):323–8.

Martinerie L, Viengchareun S, Delezoide AL, Jaubert F, Sinico M, Prevot S, et al. Low renal mineralocorticoid receptor expression at birth contributes to partial aldosterone resistance in neonates. Endocrinology 2009b;150(9):4414–24.

Mayo J, Collazos J, Martínez E, Ibarra S. Adrenal function in the human immunodeficiency virus-infected patient. Arch Intern Med 2002;162 (10):1095–8.

McHugh J, Nagabathula R, Kyithar MP. A life-threatening case of pseudoaldosteronism secondary to excessive liquorice ingestion. BMC Endocr Disord 2021;21(1):158.

Merakou C, Fylaktou I, Sertedaki A, Dracopoulou M, Voutetakis A, Efthymiadou A, et al. Molecular analysis of the CYP11B2 gene in 62 patients with hypoaldosteronism due to aldosterone synthase deficiency. J Clin Endocrinol Metab 2021;106(1):e182–91.

Miao H, Yu Z, Lu L, Zhu H, Auchus RJ, Liu J, et al. Analysis of novel heterozygous mutations in the CYP11B2 gene causing congenital aldosterone synthase deficiency and literature review. Steroids 2019;150, 108448.

Miller WL, Auchus RJ. The molecular biology, biochemistry, and physiology of human steroidogenesis and its disorders. Endocr Rev 2011;32(1):81–151. Erratum in: Endocr Rev. 2011;32(4):579.

Mitsuuchi Y, Kawamoto T, Naiki Y, Miyahara K, Toda K, Kuribayashi I, et al. Congenitally defective aldosterone biosynthesis in humans: the involvement of point mutations of the P-450C18 gene (CYP11B2) in CMO II deficient patients. Biochem Biophys Res Commun 1992;182(2):974–9. Erratum in: Biochem Biophys Res Commun 1992;184(3):1529–1530.

Murillo-de-Ozores AR, Rodríguez-Gama A, Carbajal-Contreras H, Gamba G, Castañeda-Bueno M. WNK4 kinase: from structure to physiology. Am J Physiol Renal Physiol 2021;320(3):F378–403.

Nakano K, Kubota Y, Mori T, Chiga M, Mori T, Sonoda S, et al. Familial cases of pseudohypoaldosteronism type II harboring a novel mutation in the Cullin 3 gene. Nephrology (Carlton) 2020;25(11):818–21.

Nandagopal R, Vaidyanathan P, Kaplowitz P. Transient pseudohypoaldosteronism due to urinary tract infection in infancy: a report of 4 cases. Int J Pediatr Endocrinol 2009;2009, 195728.

Nguyen HH, Hannemann F, Hartmann MF, Wudy SA, Bernhardt R. Aldosterone synthase deficiency caused by a homozygous L451F mutation in the CYP11B2 gene. Mol Genet Metab 2008;93(4):458–67.

Nguyen HH, Hannemann F, Hartmann MF, Malunowicz EM, Wudy SA, Bernhardt R. Five novel mutations in CYP11B2 gene detected in patients with aldosterone synthase deficiency type I: functional characterization and structural analyses. Mol Genet Metab 2010;100 (4):357–64.

Nishizaki Y, Hiura M, Sato H, Ogawa Y, Saitoh A, Nagasaki K. A novel mutation in the human mineralocorticoid receptor gene in a Japanese family with autosomal-dominant pseudohypoaldosteronism type 1. Clin Pediatr Endocrinol 2016;25(4):135–8.

Nur N, Lang C, Hodax JK, Quintos JB. Systemic pseudohypoaldosteronism type I: a case report and review of the literature. Case Rep Pediatr 2017;2017:7939854.

O'Riordan SM, Lynch SA, Hindmarsh PC, Chan LF, Clark AJ, Costigan C. A novel variant of familial glucocorticoid deficiency prevalent among the Irish Traveler population. J Clin Endocrinol Metab 2008;93(7):2896–9.

Ozeki Y, Tanimura Y, Nagai S, Nomura T, Kinoshita M, Shibuta K, et al. Development of a new chemiluminescent enzyme immunoassay using a two-step Sandwich method for measuring aldosterone concentrations. Diagnostics (Basel) 2021;11(3):433.

Pascoe L, Curnow KM, Slutsker L, Rösler A, White PC. Mutations in the human CYP11B2 (aldosterone synthase) gene causing corticosterone methyloxidase II deficiency. Proc Natl Acad Sci U S A 1992;89(11): 4996–5000.

Pivonello R, Ferrigno R, De Martino MC, Simeoli C, Di Paola N, Pivonello C, et al. Medical treatment of Cushing's disease: an overview of the current and recent clinical trials. Front Endocrinol (Lausanne) 2020;11:648.

Pont A, Williams PL, Azhar S, Reitz RE, Bochra C, Smith ER, et al. Ketoconazole blocks testosterone synthesis. Arch Intern Med 1982a;142 (12):2137–40.

Pont A, Williams PL, Loose DS, Feldman D, Reitz RE, Bochra C, et al. Ketoconazole blocks adrenal steroid synthesis. Ann Intern Med 1982b;97(3):370–2.

Prete A, Auchus RJ, Ross RJ. Clinical advances in the pharmacotherapy of congenital adrenal hyperplasia. Eur J Endocrinol 2021;186(1): R1–R14.

Pujo L, Fagart J, Gary F, Papadimitriou DT, Claës A, Jeunemaître X, et al. Mineralocorticoid receptor mutations are the principal cause of renal type 1 pseudohypoaldosteronism. Hum Mutat 2007;28(1):33–40.

Ray JA, Kushnir MM, Palmer J, Sadjadi S, Rockwood AL, Meikle AW. Enhancement of specificity of aldosterone measurement in human serum and plasma using 2D-LC-MS/MS and comparison with commercial immunoassays. J Chromatogr B Analyt Technol Biomed Life Sci 2014;970:102–7.

Riepe FG. Pseudohypoaldosteronism. Endocr Dev 2013;24:86–95.

Riepe FG, Finkeldei J, de Sanctis L, Einaudi S, Testa A, Karges B, et al. Elucidating the underlying molecular pathogenesis of NR3C2 mutants causing autosomal dominant pseudohypoaldosteronism type 1. J Clin Endocrinol Metab 2006;91(11):4552–61.

Rotin D, Staub O. Nedd4-2 and the regulation of epithelial sodium transport. Front Physiol 2012;3:212.

Rotin D, Staub O. Function and regulation of the epithelial Na^+ channel ENaC. Compr Physiol 2021;11(3):2017–45.

Roucher-Boulez F, Brac de la Perriere A, Jacquez A, Chau D, Guignat L, Vial C, et al. Triple-A syndrome: a wide spectrum of adrenal dysfunction. Eur J Endocrinol 2018;178(3):199–207.

Santen RJ, Van den Bossche H, Symoens J, Brugmans J, DeCoster R. Site of action of low dose ketoconazole on androgen biosynthesis in men. J Clin Endocrinol Metab 1983;57(4):732–6. https://doi.org/10.1210/ jcem-57-4-732. 6309882.

Sartorato P, Khaldi Y, Lapeyraque AL, Armanini D, Kuhnle U, Salomon R, et al. Inactivating mutations of the mineralocorticoid receptor in type I pseudohypoaldosteronism. Mol Cell Endocrinol 2004;217(1–2): 119–25.

Schiffer L, Barnard L, Baranowski ES, Gilligan LC, Taylor AE, Arlt W, et al. Human steroid biosynthesis, metabolism and excretion are differentially reflected by serum and urine steroid metabolomes: a comprehensive review. J Steroid Biochem Mol Biol 2019;194, 105439.

Serra G, Antona V, D'Alessandro MM, Maggio MC, Verde V, Corsello G. Novel SCNN1A gene splicing-site mutation causing autosomal recessive pseudohypoaldosteronism type 1 (PHA1) in two Italian patients belonging to the same small town. Ital J Pediatr 2021;47 (1):138.

Seyhanli M, Ilhan O, Gumus E, Bor M, Karaca M. Pseudohypoaldosteronism type 1 newborn patient with a novel mutation in *SCNN1B*. J Pediatr Intensive Care 2020;9(2):145–8.

Shackleton CH, Honour JW. Identification and measurement of 18-hydroxycorticosterone metabolites by gas chromatography-mass spectrometry. J Steroid Biochem 1977;8(3):199–203.

Shackleton CH, Honour JW, Dillon M, Milla P. Multicomponent gas chromatographic analysis of urinary steroids excreted by an infant with a defect in aldosterone biosynthesis. Acta Endocrinol 1976;81(4): 762–73.

Shackleton CH, Swift PG, Savage DC, Honour JW. Deficient 3 beta-hydroxy-5-ene steroid secretion by newborn infants. J Clin Endocrinol Metab 1979;49(2):247–51.

Shao L, Cui L, Lu J, Lang Y, Bottillo I, Zhao X. A novel mutation in exon 9 of Cullin 3 gene contributes to aberrant splicing in pseudohypoaldosteronism type II. FEBS Open Bio 2018;8(3):461–9.

Simard J, Rhéaume E, Leblanc JF, Wallis SC, Joplin GF, Gilbey S, et al. Congenital adrenal hyperplasia caused by a novel homozygous frameshift mutation 273 delta AA in type II 3 beta-hydroxysteroid dehydrogenase gene (HSD3B2) in three male patients of Afghan/Pakistani origin. Hum Mol Genet 1994;3(2):327–30.

Soundararajan R, Zhang TT, Wang J, Vandewalle A, Pearce D. A novel role for glucocorticoid-induced leucine zipper protein in epithelial sodium channel-mediated sodium transport. J Biol Chem 2005;280 (48):39970–81.

Sousa AG, Cabral JV, El-Feghaly WB, de Sousa LS, Nunes AB. Hyporeninemic hypoaldosteronism and diabetes mellitus: pathophysiology assumptions, clinical aspects and implications for management. World J Diabetes 2016;7(5):101–11.

Srinivasa S, Fitch KV, Wong K, Torriani M, Mayhew C, Stanley T, et al. RAAS activation is associated with visceral adiposity and insulin resistance among HIV-infected patients. J Clin Endocrinol Metab 2015;100 (8):2873–82.

Srinivasa S, deFilippi C, Fitch KV, Iyengar S, Shen G, Burdo TH, et al. Evaluation of mineralocorticoid receptor antagonism on changes in NT-proBNP among persons with HIV. J Endocr Soc 2021;6(1), bvab175.

Steer KA, Kurtz AB, Honour JW. Megestrol-induced Cushing's syndrome. Clin Endocrinol (Oxf) 1995;42(1):91–3.

Stewart PM, Corrie JE, Shackleton CH, Edwards CR. Syndrome of apparent mineralocorticoid excess. A defect in the cortisol-cortisone shuttle. J Clin Invest 1988;82(1):340–9.

Stewart PM, Krozowski ZS, Gupta A, Milford DV, Howie AJ, Sheppard MC, et al. Hypertension in the syndrome of apparent mineralocorticoid excess due to mutation of the 11 beta-hydroxysteroid dehydrogenase type 2 gene. Lancet 1996;347(8994):88–91.

Storbeck KH, Schiffer L, Baranowski ES, Chortis V, Prete A, Barnard L, et al. Steroid metabolome analysis in disorders of adrenal steroid biosynthesis and metabolism. Endocr Rev 2019;40(6):1605–25.

Suessenbach FK, Tins J, Burckhardt BB, LENA Consortium. Customisation and validation of a low-volume plasma renin activity immunoassay: enabling of regulatory compliant determination in paediatric trials. Pract Lab Med 2019;17, e00144.

Suntharalingham JP, Buonocore F, Duncan AJ, Achermann JC. DAX-1 (NR0B1) and steroidogenic factor-1 (SF-1, NR5A1) in human disease. Best Pract Res Clin Endocrinol Metab 2015;29(4):607–19.

Symoens J, Moens M, Dom J, Scheijgrond H, Dony J, Schuermans V, et al. An evaluation of two years of clinical experience with ketoconazole. Rev Infect Dis 1980;2(4):674–87.

Tajima T, Morikawa S, Nakamura A. Clinical features and molecular basis of pseudohypoaldosteronism type 1. Clin Pediatr Endocrinol 2017;26 (3):109–17.

Tan X, Li F, Wang X, Wang Y. Quantitation and clinical evaluation of plasma aldosterone by ultra-performance liquid chromatography-mass spectrometry. J Chromatogr A 2020;1609:460456.

Tetti M, Monticone S, Burrello J, Matarazzo P, Veglio F, Pasini B, et al. Liddle syndrome: review of the literature and description of a new case. Int J Mol Sci 2018;19(3):812.

Thompson III GR, Chang D, Wittenberg RR, McHardy I, Semrad A. *In vivo* 11β-hydroxysteroid dehydrogenase inhibition in posaconazole-induced hypertension and hypokalemia. Antimicrob Agents Chemother 2017;61(8), e00760-17.

Thompson III GR, Beck KR, Patt M, Kratschmar DV, Odermatt A. -Posaconazole-induced hypertension due to inhibition of 11β-hydroxylase and 11β-hydroxysteroid dehydrogenase 2. J Endocr Soc 2019;3(7):1361–6.

Thongboonkerd V. Roles for exosome in various kidney diseases and disorders. Front Pharmacol 2020;10:1655.

Travers S, Martinerie L, Bouvattier C, Boileau P, Lombès M, Pussard E. Multiplexed steroid profiling of gluco- and mineralocorticoids pathways using a liquid chromatography tandem mass spectrometry method. J Steroid Biochem Mol Biol 2017;165(Pt B):202–11.

Tsilosani A, Gao C, Zhang W. Aldosterone-regulated sodium transport and blood pressure. Front Physiol 2022;13, 770375.

Turan I, Kotan LD, Tastan M, Gurbuz F, Topaloglu AK, Yuksel B. Molecular genetic studies in a case series of isolated hypoaldosteronism due to biosynthesis defects or aldosterone resistance. Clin Endocrinol (Oxf) 2018;88(6):799–805.

Turan H, Dağdeviren Çakır A, Özer Y, Tarçın G, Özcabi B, Ceylaner S, et al. Clinical and genetic characteristics of patients with corticosterone methyloxidase deficiency type 2: novel mutations in *CYP11B2*. J Clin Res Pediatr Endocrinol 2021;13(2):232–8.

Uchida N, Shiohara M, Miyagawa S, Yokota I, Mori T. A novel nonsense mutation of the mineralocorticoid receptor gene in the renal form of pseudohypoaldosteronism type 1. J Pediatr Endocrinol Metab 2009;22(1):91–5.

Uibo R, Aavik E, Peterson P, Perheentupa J, Aranko S, Pelkonen R, et al. Autoantibodies to cytochrome P450 enzymes P450scc, P450c17, and P450c21 in autoimmune polyglandular disease types I and II and in isolated Addison's disease. J Clin Endocrinol Metab 1994;78(2):323–8.

Üstyol A, Atabek ME, Taylor N, Yeung MC, Chan AO. Corticosterone methyl oxidase deficiency type 1 with normokalemia in an infant. J Clin Res Pediatr Endocrinol 2016;8(3):356–9.

Valinsky WC, Touyz RM, Shrier A. Aldosterone and ion channels. Vitam Horm 2019;109:105–31.

Verrijn Stuart AA, Ozisik G, de Vroede MA, Giltay JC, Sinke RJ, Peterson TJ, et al. An amino-terminal DAX1 (NROB1) missense mutation associated with isolated mineralocorticoid deficiency. J Clin Endocrinol Metab 2007;92(3):755–61.

Voilley N, Galibert A, Bassilana F, Renard S, Lingueglia E, Coscoy S, et al. The amiloride-sensitive Na+ channel: from primary structure to function. Comp Biochem Physiol A Physiol 1997;118(2):193–200.

Welzel M, Akin L, Büscher A, Güran T, Hauffa BP, Högler W, et al. Five novel mutations in the SCNN1A gene causing autosomal recessive pseudohypoaldosteronism type 1. Eur J Endocrinol 2013;168(5): 707–15.

Wijaya M, Huamei M, Jun Z, Du M, Li Y, Chen Q, et al. Etiology of primary adrenal insufficiency in children: a 29-year single-center experience. J Pediatr Endocrinol Metab 2019;32(6):615–22.

Wilson FH, Disse-Nicodème S, Choate KA, Ishikawa K, Nelson-Williams C, Desitter I, et al. Human hypertension caused by mutations in WNK kinases. Science 2001;293(5532):1107–12.

Yildiz M, Isik E, Abali ZY, Keskin M, Ozbek MN, Bas F, et al. Clinical and hormonal profiles correlate with molecular characteristics in patients with 11β-hydroxylase deficiency. J Clin Endocrinol Metab 2021;106 (9):e3714–24.

Yin LP, Zhu H, Zhu RY, Huang L. A novel SCNN1G mutation in a PHA I infant patient correlates with nephropathy. Biochem Biophys Res Commun 2019;519(2):415–21.

Zhang Q, Han L, Zheng S, Ouyang F, Wu X, Yan J, et al. An isotope dilution liquid chromatography-tandem mass spectrometry candidate reference measurement procedure for aldosterone measurement in human plasma. Anal Bioanal Chem 2021a;413(17):4471–81.

Zhang R, Zhang S, Luo Y, Li M, Wen X, Cai X, et al. A case report of pseudohypoaldosteronism type II with a homozygous KLHL3 variant accompanied by hyperthyroidism. BMC Endocr Disord 2021b;21 (1):103.

Further reading

Amin N, Alvi NS, Barth JH, Field HP, Finlay E, Tyerman K, et al. Pseudohypoaldosteronism type 1: clinical features and management in infancy. Endocrinol Diabetes Metab Case Rep 2013;2013, 130010.

Bandulik S, Schmidt K, Bockenhauer D, Zdebik AA, Humberg E, Kleta R, et al. The salt-wasting phenotype of EAST syndrome, a disease with multifaceted symptoms linked to the KCNJ10 K+ channel. Pflugers Arch 2011;461(4):423–35.

Barker PM, Nguyen MS, Gatzy JT, Grubb B, Norman H, Hummler E, et al. Role of gamma ENaC subunit in lung liquid clearance and electrolyte balance in newborn mice. Insights into perinatal adaptation and pseudohypoaldosteronism. J Clin Invest 1998;102(8):1634–40.

Bongers EMHF, Shelton LM, Milatz S, Verkaart S, Bech AP, Schoots J, et al. A novel hypokalemic-alkalotic salt-losing tubulopathy in patients with *CLDN10* mutations. J Am Soc Nephrol 2017;28(10): 3118–28.

Dillon MJ, Leonard JV, Buckler JM, Ogilvie D, Lillystone D, Honour JW, et al. Pseudohypoaldosteronism. Arch Dis Child 1980;55(6):427–34.

Downie ML, Lopez Garcia SC, Kleta R, Bockenhauer D. Inherited tubulopathies of the kidney: insights from genetics. Clin J Am Soc Nephrol 2021;16(4):620–30. Erratum in: Clin J Am Soc Nephrol. 2021;16 (7):1100.

Hanukoglu A, Hanukoglu I. Clinical improvement in patients with autosomal recessive pseudohypoaldosteronism and the necessity for salt supplementation. Clin Exp Nephrol 2010;14(5):518–9.

Morikawa S, Komatsu N, Sakata S, Nakamura-Utsunomiya A, Okada S, Tajima T. Two Japanese patients with the renal form of pseudohypoaldosteronism type 1 caused by mutations of NR3C2. Clin Pediatr Endocrinol 2015;24(3):135–8.

Murthy M, Kurz T, O'Shaughnessy KM. WNK signalling pathways in blood pressure regulation. Cell Mol Life Sci 2017;74(7):1261–80.

Reinalter SC, Jeck N, Peters M, Seyberth HW. Pharmacotyping of hypokalaemic salt-losing tubular disorders. Acta Physiol Scand 2004;181(4): 513–21.

Riepe FG, Krone N, Morlot M, Ludwig M, Sippell WG, Partsch CJ. Identification of a novel mutation in the human mineralocorticoid receptor gene in a German family with autosomal-dominant pseudohypoaldosteronism type 1: further evidence for marked interindividual clinical heterogeneity. J Clin Endocrinol Metab 2003;88(4):1683–6.

Schlingmann KP, Renigunta A, Hoorn EJ, Forst AL, Renigunta V, Atanasov V, et al. Defects in KCNJ16 cause a novel tubulopathy with hypokalemia, salt wasting, disturbed acid-base homeostasis, and sensorineural deafness. J Am Soc Nephrol 2021;32(6):1498–512.

Seyberth HW. Pathophysiology and clinical presentations of salt-losing tubulopathies. Pediatr Nephrol 2016;31(3):407–18.

Shabbir W, Topcagic N, Aufy M. Activation of autosomal recessive pseudohypoaldosteronism1 ENaC with aldosterone. Eur J Pharmacol 2021;901, 174090.

Sonino N. The endocrine effects of ketoconazole. J Endocrinol Invest 1986;9(4):341–7.

Speiser PW, Arlt W, Auchus RJ, Baskin LS, Conway GS, Merke DP, et al. Congenital adrenal hyperplasia due to steroid 21-hydroxylase deficiency: an Endocrine Society clinical practice guideline. J Clin Endocrinol Metab 2018;103(11):4043–88.

Strautnieks SS, Thompson RJ, Gardiner RM, Chung E. A novel splice-site mutation in the gamma subunit of the epithelial sodium channel gene in three pseudohypoaldosteronism type 1 families. Nat Genet 1996;13(2): 248–50.

Turpeinen U, Hämäläinen E, Stenman UH. Determination of aldosterone in serum by liquid chromatography-tandem mass spectrometry. J Chromatogr B Analyt Technol Biomed Life Sci 2008;862(1–2):113–8.

Viering DHHM, de Baaij JHF, Walsh SB, Kleta R, Bockenhauer D. Genetic causes of hypomagnesemia, a clinical overview. Pediatr Nephrol 2017;32(7):1123–35.

Voilley N, Bassilana F, Mignon C, Merscher S, Mattéi MG, Carle GF, et al. Cloning, chromosomal localization, and physical linkage of the beta and gamma subunits (SCNN1B and SCNN1G) of the human epithelial amiloride-sensitive sodium channel. Genomics 1995;28(3): 560–5.

Walker BR, Andrew R, Escoubet B, Zennaro MC. Activation of the hypothalamic-pituitary-adrenal axis in adults with mineralocorticoid receptor haploinsufficiency. J Clin Endocrinol Metab 2014;99(8): E1586–91.

Wijaya M, Ma H, Zhang J, Du M, Li Y, Chen Q, et al. Aldosterone signaling defect in young infants: single-center report and review. BMC Endocr Disord 2021;21(1):149.

Index

Note: Page numbers followed by *f* indicate figures, *t* indicate tables, and *b* indicate boxes.